Genetic Diseases of the Kidney

*During the development of this book, the editors were confronted with the
unexpected death of our colleague and friend, Steven Hebert.
This meant not only the loss of an individual of great innovative and
creative ability, but also of a wise counselor and warm personal friend.
It is to him that we dedicate this book.*

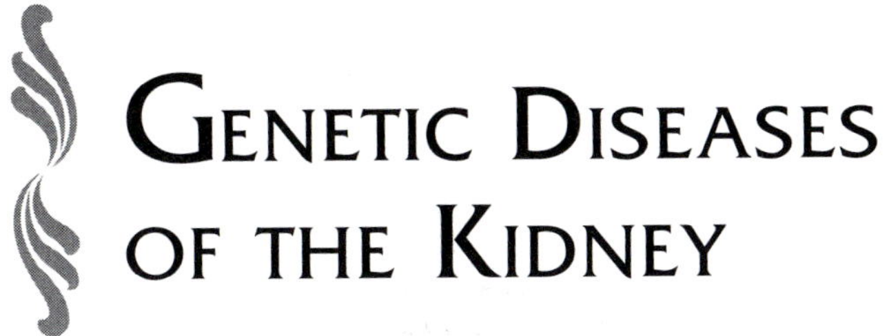

Genetic Diseases of the Kidney

Edited by

RICHARD P. LIFTON
Yale University School of Medicine

STEFAN SOMLO
Yale University School of Medicine

GERHARD H. GIEBISCH
Yale University School of Medicine

DONALD W. SELDIN
University of Texas

ELSEVIER

AMSTERDAM • BOSTON • HEIDELBERG • LONDON • NEW YORK • OXFORD
PARIS • SAN DIEGO • SAN FRANCISCO • SINGAPORE • SYDNEY • TOKYO

Academic Press is an imprint of Elsevier

Academic Press is an imprint of Elsevier
30 Corporate Drive, Suite 400, Burlington, MA 01803, USA
32 Jamestown Road, London NW1 7BY, UK
525 B Street, Suite 1900, San Diego, California 92101-4495, USA
360 Park Avenue South, New York, NY 10010-1710, USA

First edition 2009

Library of Congress Cataloging in Publication Data
A catalog record for this book is available from the Library of Congress

British Library Cataloguing in Publication Data
A catalogue record for this book is available from the British Library

ISBN: 978-0-12-449851-8

For all information on all Elsevier Academic Press publications
visit our Web site at www.elsevierdirect.com

Typeset by Charon Tec Ltd., A Macmillan Company.
(www.macmillansolutions.com)

Printed and bound in the United States of America

09 10 11 12 13   10 9 8 7 6 5 4 3 2 1

Working together to grow
libraries in developing countries

www.elsevier.com | www.bookaid.org | www.sabre.org

ELSEVIER   BOOK AID
             International   Sabre Foundation

Contents

Contents

PART VI

Systemic Hereditary Diseases with Renal Involvement: Multifactorial Diseases

Contributors

CORINNE ANTIGNAC, Institut National de la Santé, Hôpital Necker – Enfants Malades, Paris, France

PETER S. ARONSON, Department of Internal Medicine, Yale University School of Medicine, New Haven, CT, USA

WILLIAM S. ASCH, Yale University School of Medicine, Howard Hughes Medical Institute, New Haven, CT, USA

MICHEL BAUM, Departments of Internal Medicine, Pediatrics, University of Texas Southwestern Medical Center, Dallas, TX, USA

ISABEL BEERMAN, Howard Hughes Medical Institute, Departments of Genetics, Yale University School of Medicine, New Haven, CT, USA

ANNE BERGERON, Laboratory of Cell and Developmental Genetics, Department of Medicine, Université of Laval, Quebec, Canada

JÜRG BIBER, Physiologisches Institut, Universitat Zurich, Zurich, Switzerland

DANIEL G. BICHET, Centre de Recherche, Hopital de Sacré-Coeur de Montréal, Montréal, Quebec, Canada

ANTHONY J. BLEYER, Section on Nephrology, Wake Forest University School of Medicine, Winston-Salem, NC, USA

DONALD W. BOWDEN, Wake Forest University School of Medicine, Center for Human Genomics, Professor of Biochemistry and Internal Medicine, Section of Endocrinology and Metabolism, Winston-Salem, NC, USA

LYNN M. BOYDEN, Department of Genetics, Yale University School of Medicine, New Haven, CT, USA

EDWARD M. BROWN, Division of Endocrinology, Diabetes and Hypertension, Brigham and Women's Hospital, Boston, MA, USA

LLOYD G. CANTLEY, Department of Internal Medicine/Nephrology, Yale University School of Medicine, New Haven, CT, USA

JANICE Y. CHOU, Chief, Section on Cellular Differentiation, HDB, NICHD, NIH, Bethesda, MD, USA

STEVEN G. COCA, Department of Nephrology, Yale University School of Medicine, New Haven, CT, USA

TIMOTHY M. COX, Department of Medicine, University of Cambridge, Addenbrooke's NHS Foundation Hospitals Trust, Cambridge, United Kingdom

SCOTT D. CRAMER, Department of Urology, Wake Forest University School of Medicine, Medical Center Boulevard, Winston-Salem, NC, USA

COR WJR CREMERS, University Medical Center Nijmegen, Departments of Human Genetics and Otorhinolaryngology, Nijmegen, The Netherlands

PETER M.T. DEEN, Department of Physiology, RUNMC Nijmegen/NCMLS, Nijmegen, The Netherlands

ROBERT J. DESNICK, Human Genetics, Mt. Sinai School of Medicine, One Gustave L. Levy Place, New York, NY, USA

HITOSHI ENDOU, Department of Pharmacology and Toxicology, Kyourin University School of Medicine, Tokyo, Japan

ERNIE L. ESQUIVEL, Hôpital Necker – Enfants Malades, Paris, France

IAN C. FORSTER, Physiologisches Institut, Universitat Zurich, Zurich, Switzerland

BARRY I. FREEDMAN, Department of Internal Medicine/Nephrology, Wake Forest University School of Medicine, Medical Center Boulevard, Winston-Salem, NC, USA

DAVID S. GELLER, Department of Internal Medicine, Yale University School of Medicine, New Haven, CT, USA

ALI GHARAVI, Division of Nephrology, Columbia University, College of Physicians and Surgeons, New York, NY. USA

GERHARD H. GIEBISCH, Department of Cellular and Molecular Physiology, Yale University School of Medicine, New Haven, CT, USA

ROBERT L. GRUBB III, Urologic Oncology Branch, Center for Cancer Research, National Cancer Institute, Bethesda, MD, USA

LISA M. GUAY-WOODFORD, Inherited Renal Disorders Clinic, Children's Hospital of Alabama, Department of Medicine and Pediatrics, The University of Alabama at Birmingham, Cellular & Molecular Biology Program, Birmingham, AL, USA

ALI HARIRI, Department of Internal Medicine/Nephrology, Yale School of Medicine, New Haven, CT, USA

THOMAS C. HART, National Institute of Dental and Craniofacial Research, National Institutes of Health, Division of Intramural Research, Bethesda, MD, USA

SUNNY HARTWIG, Division of Nephrology, Department of Medicine, Children's Hospital, Boston, MA, USA

STEVEN C. HEBERT[†], Cellular & Molecular Physiology, Yale University School of Medicine, New Haven, CT, USA

NATI HERNANDO, Physiologisches Institut, Universitat Zurich, Zurich, Switzerland

FRIEDHELM HILDEBRANDT, University of Michigan Health Center, Department of Pediatrics and Communicable Diseases, Section of Pediatric Nephrology, Ann Arbor, MI, USA

TATSUO HOSOYA, Division of Kidney and Hypertension, Department of Internal Medicine, Jikei University School of Medicine, Tokyo, Japan

MAKOTO HOSOYAMADA, Department of Pharmacotherapeutics, Kyoritsu University of Pharmacy, Tokyo, Japan

KIMIYOSHI ICHIDA, Division of Kidney and Hypertension, Department of Pathophysiology, Tokyo University of Pharmacy and Life Sciences, Japan

PETER IGARASHI, Departments of Internal Medicine and Pediatrics, Division of Nephrology, The University of Texas Southwestern Medical Center, Dallas, TX, USA

ROSSANA JORQUERA[†], Laboratory of Cell and Developmental Genetics, Department of Medicine, Université of Laval, Quebec, Canada

KRISTOPHER T. KAHLE, Department of Neurosurgery, Massachusetts General Hospital, Harvard Medical School, Boston, MA, USA

VASILIKI KALATZIS, Institut Génétique Moléculaire de Montpellier, Montpellier, France

FIONA E. KARET, Department of Medical Genetics and Division of Renal Medicine, University of Cambridge, Cambridge Institute for Medical Research, Addenbrooke's Hospital, Cambridge, United Kingdom

PAUL E. KLOTMAN, Samuel Bronfman Department of Medicine, Mount Sinai Medical School, New York, NY, USA

NINE V.A.M. KNOERS, Departments of Human Genetics and Otorhinolaryngology, University Medical Center Nijmegen, Nijmegen, The Netherlands

JORDAN A. KREIDBERG, Division of Nephrology, Department of Medicine, Children's Hospital, Boston, MA, USA

DAVID JOSEPH KWIATKOWSKI, Harvard Medical School, Hematology Division, Boston, MA, USA

BRENDAN LEE, HHMI Molecular Genetics, Baylor College of Medicine, Houston, TX, USA

RICHARD P. LIFTON, Department of Genetics, Yale University School of Medicine, New Haven, CT, USA

FANGMING LIN, Departments of Pediatrics and Division of Basic Science, University of Texas Southwestern Medical Center at Dallas, Dallas, TX, USA

W. MARSTON LINEHAN, Urologic Oncology Branch, National Cancer Institute, Bethesda, MD, USA

BRIAN C. MANSFIELD, Section on Cellular Differentiation, Heritable Disorders Branch, National Institute of Child Health and Human Development, National Institutes of Health, Bethesda, MD, USA

ARNAUD MARLIER, Section of Nephrology, Department of Internal Medicine, Yale University School of Medicine, New Haven, CT, USA

PRAMOD KUMAR MISTRY, Department of Internal Med. (Digestive Diseases), Yale University School of Medicine, New Haven, CT, USA

ORSON W. MOE, Department of Internal Medicine, University of Texas Southwestern Medical Center, Dallas, TX, USA

CHANDRA MOHAN, Department of Immunology and, Division of Rheumatology and, Department of Internal Medicine, University of Texas Southwestern Medical Center, Dallas, TX, USA

ROY MORELLO, Department of Molecular Genetics, Baylor College of Medicine, Houston, TX, USA

HEINI MURER, Physiologisches Institut, Universitat Zurich, Zurich, Switzerland

MANUEL PALACIN, Department of Biochemistry and Physiology, University of Barcelona, Barcelona, Spain

VISHAL PATEL, Department of Internal Medicine, University of Texas Southwestern Medical Center at Dallas, Dallas, TX, USA

JAAKKO PATRAKKA, Division of Matrix Biology, Department of Medical Biochemistry & Biophysics, Karolinska Institutet, Stockholm, Sweden

MARTIN R. POLLAK, Renal Division, Brigham & Women's Hospital, Harvard Medical School, Boston, MA, USA

KRISHNA R. POLU, Clinical and Research Fellow, Renal Division, Brigham and Women's Hospital, Harvard Medical School, Boston, MA, USA

ROBERT F. REILLY, Department of Nephrology, Yale University School of Medicine, New Haven, CT, USA

STEPHEN S. RICH, Director, Center for Public Health Genomics; Harrison Professor of Public Health Sciences, Professor of Internal Medicine (Cardiovascular Medicine) University of Virginia, Charlottesville, VA, USA

BERNARD C. ROSSIER, Université de Lausanne, Institut de Pharmacologie et de Toxicologie, Lausanne, Switzerland

STEVEN J. SCHEINMAN, Department of Medicine, State University of New York, Health Science Center, Syracuse, NY, USA

LAURENT SCHILD, Institute of Pharmacology and Toxicology, Lausanne, Switzerland

MICHAEL L. SCHILSKY, Departments of Medicine and Surgery and Pediatrics, Yale New Haven Transplantation Center, New Haven, CT, USA

UTE I. SCHOLL, Departments of Genetics and Internal Medicine, Howard Hughes Medical Institute, Yale University School of Medicine, New Haven, CT, USA

FRANCESCO SCOLARI, Chair and Division of Nephrology, Divisione di Nefrologia, Spedali Civili di Brescia, Brescia, Italy

---

[†]deceased

DARYL SCOTT, Department of Molecular Genetics, Baylor College of Medicine, Houston, TX, USA

DONALD W. SELDIN, Department of Internal Medicine, The University of Texas, Southwestern Medical Center, Dallas, TX, USA

STEFAN SOMLO, Department of Internal Medicine/Nephrology, Yale University School of Medicine, New Haven, CT, USA

TATSUYA TAKAYAMA, Department of Urology, Wake Forest University School of Medicine, Medical Center Boulevard, Winston-Salem, NC, USA

ROBERT M. TANGUAY, Laboratory of Cell and Developmental Genetics, Department of Medicine, Université of Laval, Quebec, Canada

KARL TRYGGVASON, Division of Matrix Biology, Department of Medical Biochemistry & Biophysics, Karolinska Institutet, Stockholm, Sweden

CARSTEN A. WAGNER, Institute of Physiology, University of Zurich, Zurich, Switzerland

EDWARD K. WAKELAND, Department of Immunology and, Division of Rheumatology, Department of Internal Medicine, University of Texas Southwestern Medical Center, Dallas, TX, USA

McCLELLAN M. WALTHER, Urologic Oncology Branch, Center for Cancer Research, National Cancer Institute, Bethesda, MD, USA

ANDREW WANG, Departments of Immunology and, Division of Rheumatology and, Department of Internal Medicine, University of Texas Southwestern Medical Center, Dallas, TX, USA

DAVID A. WEINSTEIN, Division of Endocrinology, Children's Hospital and Harvard Medical School, Boston, MA, USA

PERRIN C. WHITE, Department of Pediatrics, University of Texas Southwestern Medical Center, Dallas, TX, USA

FREDERICK H. WILSON, Department of Medicine, Brigham and Women's Hospital, Harvard Medical School, Boston, MA, USA

ERNEST M. WRIGHT, Department of Physiology, University of California – Los Angeles School of Medicine, Los Angeles, CA, USA

CHRISTINA M. WYATT, Mount Sinai School of Medicine, One Gustave L. Levy Place, New York, NY, USA

CAREL H. VAN OS, Department of Cell Physiology, Faculty of Medical Science, Nijmegen, The Netherlands

Preface

In a broad sense, the kidney has as its principal function the maintenance of the volume and composition of the body fluids in the face of varied physiologic demands and pathologic disturbances. The mechanisms by which the kidney discharges these homeostatic responsibilities have been greatly elucidated by the reduction of gross macroscopic processes to the molecular level by the application of the conceptual and technical tools of molecular biology. Processes such as active and passive transport, diffusion, and bulk movement, can now be characterized by proteins which function as transporters, channels, receptors, and other biologic entities. At the same time, ensembles, such as the counter-current system, the peritubular regulation of proximal reabsorption, and the cellular adjustments to acidification, have disclosed how isolated cellular events can be integrated to perform complicated biologic processes. Finally, the characterization of feed-back systems between the kidney and internal environment has greatly clarified how regulation between internal disturbances and renal mechanisms can accomplish appropriate homeostatic adjustment.

These advances in renal physiology have been paralleled by a revolution in human genetics, highlighted by the sequencing of the human genome. This has not only provided many of the tools for the development of renal physiology but has facilitated the identification of gene mutations that underlie genetic diseases in human subjects. This interaction between physiology and genetics has greatly enhanced the understanding of normal kidney function as well as the molecular pathogenesis of renal genetic diseases.

The purpose of the present text is to identify and analyze genetic abnormalities causing renal diseases in human subjects. Although in a sense the genome contains all the instructions required for the formation of a phenotype, the information is encoded in an extremely complicated fashion. In *primary genetic diseases*, the genetic instruction specifies a phenotype clearly linked with a discrete lesion confined to the kidney. However, the genetic disturbance may be imbedded in a complicated physiologic ensemble, so that the nexus between the genetic disturbance and the phenotype may be obscured; in consequence, the causal sequence is extremely difficult to unravel. In many instances the renal disease is one component of a complicated *systemic hereditary disease*, either monogenic or polygenic. Indeed, renal disease may arise as the sum of minor inputs from many different, seemingly unrelated genes, so that the genetic contributions may be difficult to identify. Confounding the problem further are environmental influences, originating either in the chromosomal environment from modifier genes, or in the extra-chromosomal environmental from intrauterine or postnatal influences. These environmental influences confer upon the fetus a striking developmental plasticity, so that with varying environmental conditions a given genotype can eventuate in different phenotypes. Such considerations have determined both the organization of the text as well as the detailed description of the genetic disorders and the physiologic derangements that emerge.

Part I functions as an orientation to the genetic analysis of disease. After a detailed treatment of the principles underlying genetic disturbances, the clinical approach to genetic analysis is described. **Part II** contains a comprehensive analysis of genetic diseases of the nephron. A general description of renal function is first presented, emphasizing the homeostatic function of the kidney. The varied transport pathways and ensembles are described. Emphasis is placed on the manner in which this machinery responds to signals from the internal environment so as to maintain the volume and composition of the body. Twenty chapters then follow with an in-depth presentation of the genetic diseases in each segment of the nephron – glomerulus, proximal tubule, thick ascending limb of Henle, and distal convoluted tubule and collecting duct. In **Part III**, the genetic abnormalities of renal development are presented. There is a comprehensive presentation of normal renal development, followed by detailed treatment of the various developmental disorders that express themselves early or late in life. **Part IV** is devoted to inherited neoplastic diseases of the kidney – their genetic basis and clinical presentation. **Parts V and VI**

present a broad array of monogenic and polygenic systemic diseases with renal involvement. Here clinical issues of the most pervasive type – diabetes, hyperuricemia, autoimmunity, HIV – are examined in detail and their genetic components are explored.

It is our hope that the analysis presented here will lead to a better understanding of the complex genetic background of renal diseases and their often complicated and sometimes obscure clinical presentations. Inevitably, this will lead to more informed diagnosis and more enlightened prognosis and treatment.

Richard P. Lifton
Stefan Somlo
Gerhard H. Giebisch
Donald W. Seldin

PART I

General Background

CHAPTER 1

Genetic Approaches to Human Disease

RICHARD P. LIFTON AND LYNN M. BOYDEN

INTRODUCTION

The broad history of medicine reveals that fundamental understanding of the pathogenesis of human disease provides the framework for development of new and effective approaches to prevention and treatment. This principle has motivated biomedical research and the quest for insights into the underlying causes of human disease. Diverse approaches to achieve this end have been applied, with many stunning successes that range from public health efforts to personalized medicine. Prior to the genetic era that began 25 years ago, endeavors to understand human disease included diverse tools, each with substantial power. These can be categorized broadly as 'top down' approaches, searching for factors shared by disease subjects that distinguish them from disease-free subjects, and 'bottom up' approaches, which start with the identification of an interesting molecule and work logically toward the elucidation of its role in normal and/or disease physiology. Thus epidemiology and pathology have sought factors that are shared more often by patients concordant for disease than expected by chance when compared to control cohorts without disease. Conversely, the field of endocrinology made spectacular advances by characterizing circulating hormones with potent biological effects, whose measurement and replacement proved critical in the diagnosis and treatment of diverse diseases.

Despite stunning advances in the understanding of many diseases, the pathogenesis of many other disorders has remained elusive. Frequently, despite excellent descriptions of disease, the underlying pathophysiology has remained obscure. Moreover, in many cases specific factors have be associated with disease, but distinguishing whether these are causally related to disease processes, or alternatively are secondary consequences of disease, has proved extremely difficult.

In the current era, for virtually every disease whose causation is unknown, genetic approaches have the capacity to elucidate fundamental disease mechanisms. The importance of this approach is that no starting assumptions about disease pathogenesis are required except that genetic variation contributes to disease, and that starting from this recognition, the genes that are causally related to disease pathophysiology can be reliably identified. The last 20 years of biomedical investigation have demonstrated the enormous power of this concept with the identification of more than 2000 disease genes. These have revolutionized the understanding of diverse human diseases affecting every organ system, including the kidney.

BRIEF HISTORY OF GENETICS

Human genetics is inextricably linked to the history of genetics in biology. The concept that 'like begets like' reflects recognition of the general concept of inheritance and dates back more than 10 000 years. The cultivation of grains and the development of domesticated animals by selective breeding from wild species reflect an acute understanding and active use of genetics in the pre-historic era. Despite this gross recognition, the mechanisms underlying the transmission of distinct traits from one generation to another were entirely unknown.

Discovery of Mendelian Principles

The work of Gregor Mendel in the 1860s, breeding peas in his monastery, provided the first fundamental insight into

Genetic Diseases of the Kidney
ISBN: 978-0-12-449851-8

the mechanisms of genetic transmission (Mendel 1866). Heralding approaches taken in studies of humans decades later, Mendel selected pea plants that showed extreme variation in traits (or *phenotypes*) and bred them to one another, drawing conclusions about genetic inheritance from the resulting progeny. For example, he found that crossing plants that produced round peas with plants that produced wrinkled peas resulted in a first generation of progeny that all produced only round peas (Figure 1.1). However, if he intercrossed the plants arising from this first cross with each other, 75% of the offspring produced only round peas and 25% produced only wrinkled peas. He found similar behavior for a total of 16 traits. From these observations he concluded that these traits were determined by pairs of factors (which we now call *genes*) that occur in different forms (e.g. 'A' and 'a') in each plant, and that precisely one of these factors was transmitted from each parent to its offspring in unchanged form (we now call these different forms *alleles*). Importantly, the likelihood of transmission of either factor was equally likely, approximating 50%. The different combinations of these pairs of factors (now called *genotypes*) determined each trait – the combinations 'AA' or 'Aa' resulted in plants with round peas, whereas those with the combination 'aa' produced wrinkled peas. Simple statistics explained the results of the breeding experiments – with

a starting cross of 'AA' (round) × 'aa' (wrinkled), 100% of the offspring (known as the F_1 generation) will have genotype 'Aa' and produce round peas. Intercrossing of these 'Aa' F_1 generation plants will produce an F_2 generation in which 25% are 'AA', 50% are 'Aa' and 25% are 'aa'; hence the 25% of the progeny with the 'aa' genotype will produce wrinkled peas. In addition, Mendel found that the factors for the different traits that he studied segregated independently of one another, providing the principle of independent assortment.

These observations created the foundations of what is now referred to as Mendelian genetics, in which variation in single genes is necessary and sufficient to produce a phenotype. While Mendel provided groundbreaking insights into the principles of genetic transmission, his work did not elucidate the location in the cell nor the nature of the factors responsible for inheritance.

The Chromosome Theory of Inheritance

Mendel's work was largely unrecognized in biology until it was rediscovered in the early 20th century and built upon by Thomas Hunt Morgan, working with the fruit fly *Drosophila melanogaster*. Morgan, like Mendel, found variations in traits that showed simple patterns of transmission. His work, however, made two fundamental advances: first, he demonstrated that these factors, which Morgan called genes, are carried on chromosomes; second, he showed that genes are arranged in a linear order on chromosomes.

In one critical experiment Morgan showed that variation in eye color (white vs. red) segregated with gender, a trait which had been shown to be associated with differences in chromosome number (Figure 1.2). In a cross of white-eyed males and red-eyed females, all the offspring of both genders had red eyes. In the intercross of this F_1, however, 50% of the male progeny had white eyes, while all of the female progeny retained red eyes. This was the first demonstration of a sex-linked trait, and together with the observation that male and female fruit flies have different numbers of chromosomes, it provided evidence for the chromosomal basis of inheritance (Morgan 1910).

Using the fast breeding time of the fruit fly, Morgan and his students were able to study a large set of traits, and performed systematic crosses comparing the joint inheritance of these traits with one another. By this approach, they discovered that, in contrast to Mendel's law of independent assortment, in fact some alleles specifying different traits do not show independent assortment, but instead show either complete or incomplete co-segregation with one another. For example, alleles specifying white eyes and abnormal wings traveled together most of the time, but not all the time; this co-segregation was referred to as *linkage*, and the fraction of progeny that obeyed and violated linkage

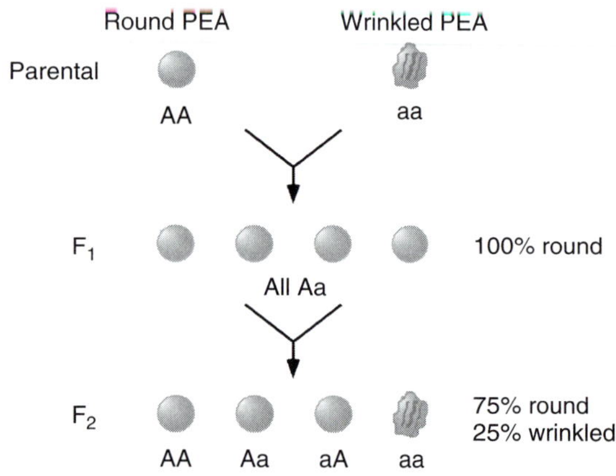

FIGURE 1.1 Mendelian segregation. An example of one of Mendel's pea plant crosses is shown, in which plants that produced round peas were crossed with plants that produced wrinkled peas, resulting in a first generation of progeny (F_1) that all produced round peas. Intercrossing of these F_1 plants resulted in an F_2 generation in which 75% of plants produced round peas and 25% produced wrinkled peas. From such experiments Mendel concluded that traits like pea shape were determined by pairs of factors (now called genes) that occur in different forms (or alleles – 'A' and 'a' in the figure), that one allele was transmitted from each parent to its offspring, and that the different combinations of alleles (now called genotypes) determined each trait. In this example, genotypes 'AA', 'Aa', and 'aA' result in round peas, and genotype 'aa' results in wrinkled peas. (see also Plate 1)

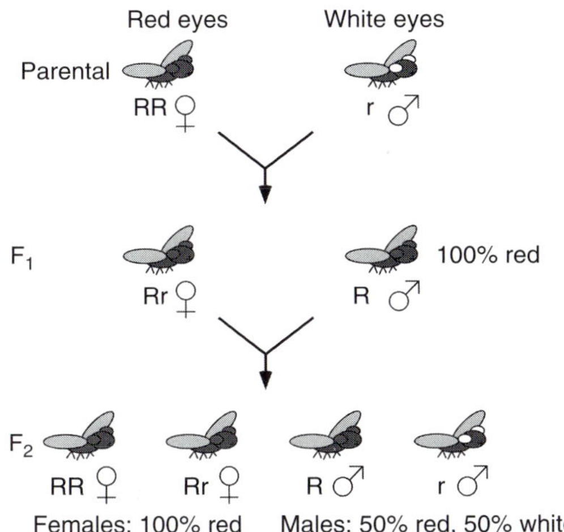

FIGURE 1.2 Morgan's demonstration of the chromosomal basis of inheritance. An example of one of Morgan's fruit fly crosses is shown, in which a cross of red-eyed females with white-eyed males resulted in F₁ offspring that all had red eyes. Intercrossing of the F₁ flies resulted in an F₂ generation in which all of the females and 50% of the males had red eyes, but the other 50% of males had white eyes. From these experiments, and the observation that male and female fruit flies have different numbers of chromosomes, Morgan concluded that genes are carried on chromosomes, and that genes for eye color in the fruit fly are on the sex-linked chromosome. In the figure, females inherit two alleles, one from each parent, and their genotypes 'RR' and 'Rr' result in red eyes. Males inherit only one allele, from their mother; genotype 'R' results in red eyes, while genotype 'r' results in white eyes. (see also Plate 2)

the gene. This work began with Oswald Avery's classic demonstration that heredity in bacteria is determined by DNA, a universal component of chromosomes (Avery et al 1944). At this point the abstract studies of classical genetics began to merge with chemistry, and subsequent discoveries describing the biological mechanisms for genetic inheritance are well-known to all students of biology. Seminal events included: (1) the determination of the structure of DNA as a double-stranded molecule comprised of antiparallel strings of four *bases* (or *nucleotides*), in which the bases on one strand form complementary pairs with those on the other; this provided a mechanism for the copying and transmission of alleles, and therefore the basis of heredity (Watson & Crick 1953); (2) elucidation of *central dogma*: DNA is self-replicating and RNA copies of DNA segments specify the synthesis of specific proteins (Crick 1958, 1970); (3) recognition that gene and protein sequences are co-linear and that the DNA bases are utilized in a triplet code (Crick et al 1961), by which each of 64 possible 3-base codons specify the insertion of a specific amino acid into a growing peptide chain, or alternatively specify chain termination (Nirenberg & Matthaei 1961, Nirenberg & Leder 1964); (4) description of the biochemical mechanisms of DNA replication, transcription to produce RNA, and translation of the triplet code to produce proteins (Bessman et al 1958, Lehman et al 1958, Furth et al 1962, Siekevitz & Zamecnik 1981). These events transpired between 1953 and 1965.

was highly reproducible (Morgan 1911). Other alleles that showed linkage to the white locus demonstrated various fractions of incomplete linkage. Morgan and his student Alfred Sturtevant deduced that linkage reflected the co-localization of alleles on the same chromosome and that violations of complete linkage were observed when meiotic *recombination* of homologous chromosomes occurred in the interval between two linked alleles. They correctly inferred that this recombination was the genetic representation of the physical event of chromosome crossing over, which had been previously described by cytologists (Figure 1.3A). From these observations, Morgan and Sturtevant recognized that by construction of appropriate crosses, the relative locations of genes could be deduced, and using this approach they produced the first genetic maps (Figure 1.3B). These maps revealed that genes lie in linear arrays and that the number of linkage groups equals the number of chromosomes (Sturtevant 1913).

DNA and Central Dogma

Morgan's work established the chromosomal theory of inheritance, but did not determine the chemical nature of

Recombinant DNA and Positional Cloning

The next major revolution in genetics was the advent of recombinant DNA, which utilized bacteria and extrachromosomal plasmids as tools to enable the isolation and propagation of specific chromosome segments from any species (Morrow et al 1974). With this broadly applicable new technological development, David Hogness recognized the ability to, for the first time, relate genetic inheritance of traits to exact physical positions on chromosomes. Hogness foresaw that by establishing the relationship between genetic and physical maps, it would be possible to identify genes, and mutations that underlie genetic traits, based solely on their chromosome position (Wensink et al 1974). His group ultimately achieved the first success of so-called *positional cloning* by 'chromosome walking,' using overlapping fragments of *Drosophila* DNA to tile along the chromosome from a known starting position to the position of a key developmental gene, *Ultrabithorax*, which determines body segment identity. This positional cloning enabled the characterization of the gene and its encoded product, a transcription factor (Bender et al 1983).

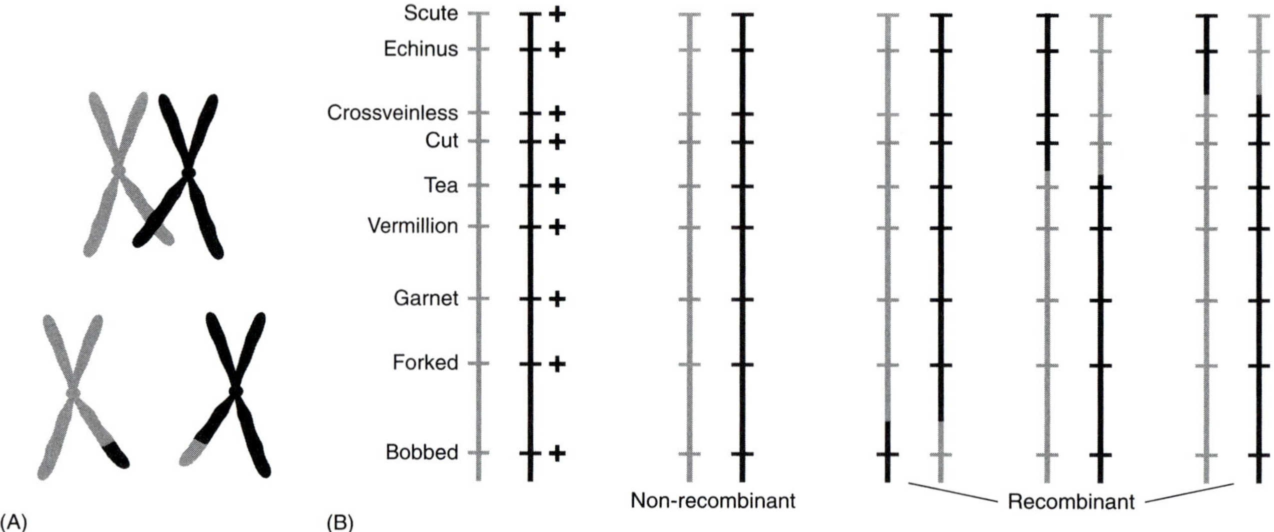

FIGURE 1.3   Genes are arranged in a linear order on chromosomes, and crossing over between homologous chromosomes results in recombination of alleles. (A) A schematic of crossing over between homologous chromosomes (shown in gray and black), in which the distal ends of one of each pair of chromosomes recombine (top), resulting in two recombinant chromosomes (bottom). (B) Schematic of the linear order of fruit fly genes on a chromosome. On the left, segments of two fruit fly chromosomes carrying nine genes are shown (gray and black), where '+' represents the wild-type allele and names represent mutant alleles. These nine genes are physically close to each other, and their alleles segregate together most of the time. This cosegregation is called linkage. When crossing over results in recombination between genes, recombinant chromosomes are produced, which have new combinations of wild-type and mutant alleles. The fractions of progeny with recombination between a given pair of genes (observed as new combinations of wild-type and mutant traits) is reproducible, and because the likelihood of recombination between the genes is a function of the physical distance between them, measuring recombination frequencies allows the relative locations of genes to be deduced

TRANSMISSION OF SINGLE-GENE DISEASES IN HUMANS

From the last century, it has been recognized that the genetic principles that apply to simple systems like peas and fruit flies also apply to humans. Archibald Garrod provided detailed evidence for Mendelian transmission of a number of metabolic diseases, including alkaptonuria in 1902, followed by description of a number of others including cystinuria, pentosuria, and albinism (Garrod 1902, 1909). Subsequent to this work, a very large number of human diseases and traits were recognized to be inherited and to have simple Mendelian transmission. These are catalogued in the Online Mendelian Inheritance in Man (http://www.ncbi.nlm.nih.gov/sites/entrez?db=omim), and over the last century nearly 4000 human Mendelian traits have been described.

Human traits attributable to single-gene variation typically show simple *autosomal dominant, autosomal recessive,* or *sex-linked* transmission. For traits that impair reproductive fitness, the underlying specific disease alleles are subject to *negative selection* (selection against alleles that reduce fitness) and consequently are maintained at very low frequencies in the population.

Autosomal Dominant Traits

For a typical autosomal dominant disease, affected subjects harbor one *wild-type* (normal) allele (+) and one disease-causing allele (D). When such individuals have offspring, by chance the wild-type allele will be transmitted to 50% of offspring, and these will be disease-free, while the disease-causing allele will be transmitted to the other 50% of offspring, and these will develop disease. The offspring who inherit the wild-type allele are not at increased risk of transmitting the disease to their offspring, whereas those inheriting the disease allele will pass it and its associated risk to 50% of their offspring. As a consequence, dominant traits show vertical transmission from generation to generation, with about 50% of the offspring of affected subjects developing the disease (Figure 1.4). Autosomal dominant traits can be caused by mutations that result in a genetic *gain of function* – i.e., a function not seen in the wild-type protein – or a genetic *loss of function*. Loss of function mutations can cause dominantly inherited disease due to *haploinsufficiency*, in which one dose of the normal gene is not sufficient to prevent disease, or by a so-called 'two-hit' mechanism, in which a somatic mutation in the remaining wild-type allele occurs, resulting in a cell-autonomous

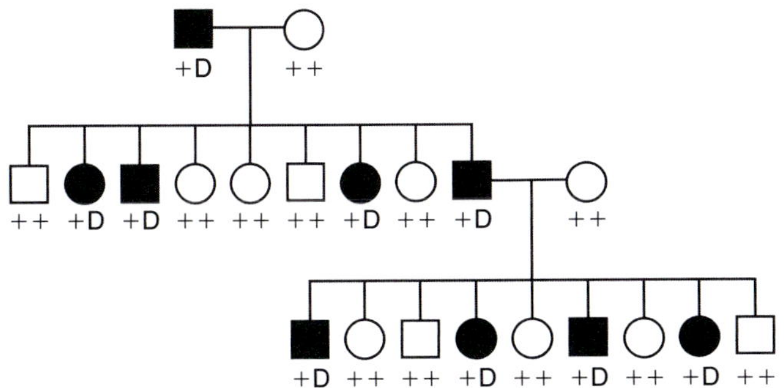

FIGURE 1.4 Autosomal dominant inheritance. An example of a kindred with an autosomal dominant disease is shown. Males are represented by squares, females by circles; those with disease are shown with filled black symbols. Genotypes of each individual for the disease gene are shown under each symbol. For autosomal dominant disease, affected subjects have one wild-type allele ('+') and one disease-causing allele ('D'). Offspring of affected individuals have a 50% chance of inheriting the '+' allele from the affected parent and being disease-free (genotype '++'), and a 50% chance of inheriting the 'D' allele and developing disease (genotype '+D'); autosomal dominant traits are commonly transmitted to multiple generations

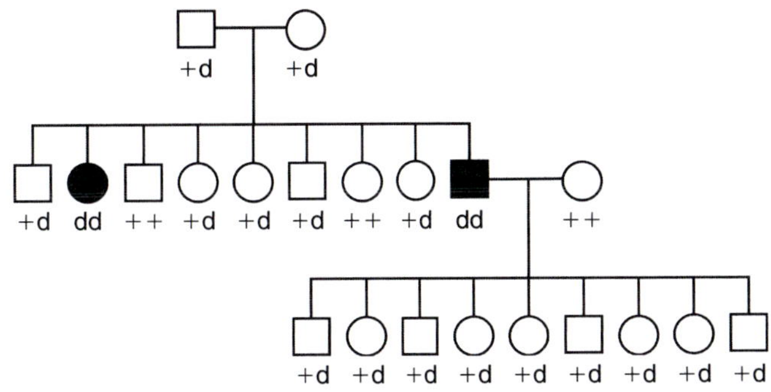

FIGURE 1.5 Autosomal recessive inheritance. An example of a kindred with an autosomal recessive disease is shown. For a typical autosomal recessive disease, in which the disease allele is rare, unaffected parents are both heterozygous for the disease allele ('d'). Offspring have a 75% chance of inheriting either zero or one disease allele and being disease-free (genotypes '++' and '+d'), and a 25% chance of inheriting two 'd' alleles and developing disease (genotype 'dd'). Offspring of affected individuals are unlikely to develop disease, as it would require that their unaffected parent also carry a disease allele. For diseases with autosomal recessive transmission, in small families it is common to observe only one affected individual, and it is uncommon for multiple generations to have affected members

phenotype. This mechanism was first described for inherited cancers caused by mutations in tumor suppressor genes, in which one defective allele is inherited, and tumors subsequently develop from cells in which the other allele incurs somatic mutation (Knudson 1971). However, the 'two-hit' or 'tumor suppressor' mechanism is also relevant to the pathogenesis of other genetic diseases, including polycystic kidney disease (Qian et al 1996, Wu et al 1998).

Autosomal Recessive Traits

Recessive diseases require that both alleles of a gene are mutant. These mutations are nearly always genetic loss of function, and, in such diseases, loss of function of both alleles must occur to produce disease. Thus subjects who are genotype 'd/d' (*homozygous* for the disease allele) are affected with disease, whereas those who are genotype '+/d' or '+/+' (*heterozygous* for the disease allele, or homozygous for the wild-type allele) are unaffected. For a typical rare recessive trait, unaffected parents are both heterozygous for the disease allele, and produce offspring in which 25% have the 'd/d' genotype and are affected, while 75% do not have the disease genotype and are disease-free (Figure 1.5).

The offspring of subjects with a recessive disease are typically at low risk for the disease; though they will invariably inherit one disease allele from the affected parent, they are only at risk of disease if the unaffected parent is a heterozygous carrier. For example, for a recessive trait with population frequency of 1/40 000, the trait allele frequency is 1/200 (the square root of the disease frequency) and therefore the frequency of heterozygous carriers is ~1/100.

Thus, the likelihood of a chance mating of an affected subject ('d/d') with a heterozygote ('+/d') is ~1/100, resulting in an overall disease risk to the affected subject's offspring of ~1/200. As a consequence of these considerations, in small nuclear families it is not common for there to be more than one affected child, and vertical transmission of disease is commonly not seen.

Sex-Linked Traits

Gender in humans is determined by the X and Y chromosomes, with females having the 'X,X' genotype and males 'X,Y'. As a consequence, the gender of offspring is determined by the male parent, who transmits an X chromosome to 50% of his offspring, who are genetically female, and a Y chromosome to 50%, who are genetically male. Aside from genes involved in sex determination and fertility, there are few genes on the Y chromosome. In contrast, there are many on the X. Because males are *hemizygous* for the X chromosome (i.e., they have only one copy), a loss of function mutation on the X chromosome often results in a disease phenotype in males, but not in females, who have a remaining wild-type copy of the X chromosome. The pattern of transmission of sex-linked (commonly referred to as *X-linked*) traits is consequently distinctive. Affected males do not transmit the disease phenotype to their offspring, and unaffected females who carry the mutation transmit the disease to 50% of their male offspring but to none of their female offspring. X-linked diseases are consequently typically manifest only in males, and skip generations

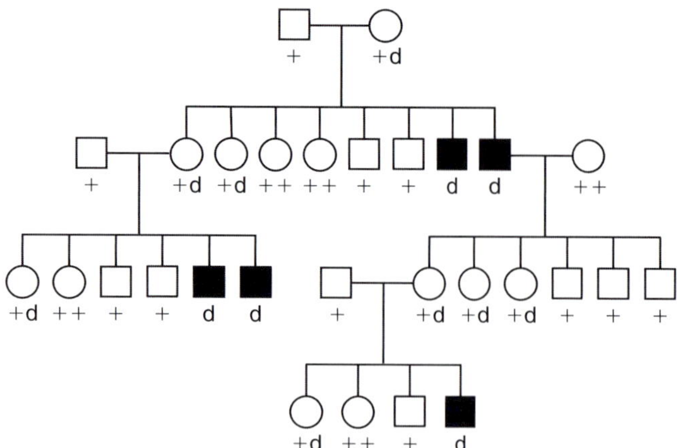

FIGURE 1.6   Sex-linked inheritance. An example of a kindred with a sex-linked, or X-linked, disease is shown. Loss of function alleles on the X chromosome often result in disease in males, who only have one X chromosome and consequently are hemizygous (genotype 'd'), but not in females, who have two X chromosomes and therefore retain a wild-type copy despite carrying the disease allele (genotype '+d'). The pattern of X-linked traits is distinctive. Unaffected female carriers transmit the disease allele to 50% of their male and female offspring, but only male offspring are affected. Affected males transmit the disease allele to all of their female offspring (who are not affected, but all carry the disease allele), and to none of their male offspring (who inherit a Y chromosome from their father). Consequently, X-linked disease is seen only in males, transmitted only from unaffected mothers, and skips generations as the disease allele passes through heterozygous female carriers

as disease mutations pass through heterozygous females (Figure 1.6).

Mitochondrial Transmission

In addition to these classical modes of transmission, there are several more that can be encountered. One of these is disease caused by mutation in the mitochondrial genome (Wallace 2005). Mitochondria contain their own genomes, which in humans are only 16000 base pairs and encode 13 proteins as well as ribosomal and transfer RNAs. All other gene products found in the mitochondria are encoded in the nuclear genome. Mitochondrial genetics has several unique features. First, in contrast to the fixed number of copies of each chromosome of the nuclear genome, there are hundreds to thousands of mitochondrial genomes per cell, and these are transmitted in bulk to daughter cells upon cell division. As a result, there is no precise segregation of alleles in mitosis or meiosis. Second, mitochondrial genomes can vary within single subjects; this variation is referred to as *heteroplasmy*. In particular, some mutations may be incompatible with cellular or organismal survival, and can only be found in the heteroplasmic state. Finally, virtually all mitochondrial genomes in the zygote are inherited from the mother's

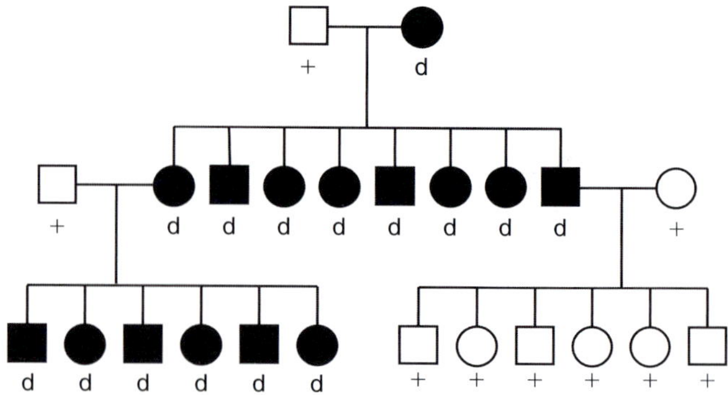

FIGURE 1.7   Mitochondrial inheritance. An example of a kindred with a mitochondrial disease is shown. Because virtually all mitochondria within a zygote are inherited from the mother's egg, diseases that result from mutations in mitochondrial DNA are transmitted only from mothers, almost never from fathers. If the mutation is homoplasmic (i.e. all of the mother's mitochondria have the mutation), all of her offspring will receive 100% mutant genomes (genotype 'd') and be affected with disease, as shown in the pedigree. If the mutation is heteroplasmic (i.e. there is variation in the mother's mitochondrial genomes at the mutation site), offspring may vary in the proportion of mutant mitochondria they receive, and thus may vary in phenotype

egg rather than the father's sperm. As a consequence of the exclusive maternal transmission of mitochondria, mitochondrial diseases are transmitted only from mothers to their offspring, virtually never from fathers. Because many copies of the mitochondrial genome are transmitted without segregation, virtually all offspring of a mother with a mitochondrial mutation will receive mutant genomes (Figure 1.7). If the mutation is homoplasmic, 100% of offspring will be affected with disease. If the mutation is heteroplasmic, offspring may vary in the proportion of mutant mitochondria they receive, and may consequently vary in their phenotype.

Genetic Anticipation

A second non-classical form of Mendelian inheritance is exemplified by *genetic anticipation*, in which a disease phenotype becomes increasingly severe with successive generations. Though this pattern of transmission was recognized over 90 years ago for myotonic dystrophy, the proof of this principle only came with the identification of mutations underlying the fragile X syndrome of X-linked mental retardation (Fleischer 1918, Fu et al 1991). For this disease and about a dozen others, it has been shown that trinucleotide repeat sequences are relatively stable over generations at a certain length, but if they increase in size to a particular length they are highly likely to increase still further in subsequent generations. If these longer length repeats alter the function of a gene, a disease phenotype may result. Those with intermediate repeat lengths may show mild phenotypic expression of the disease,

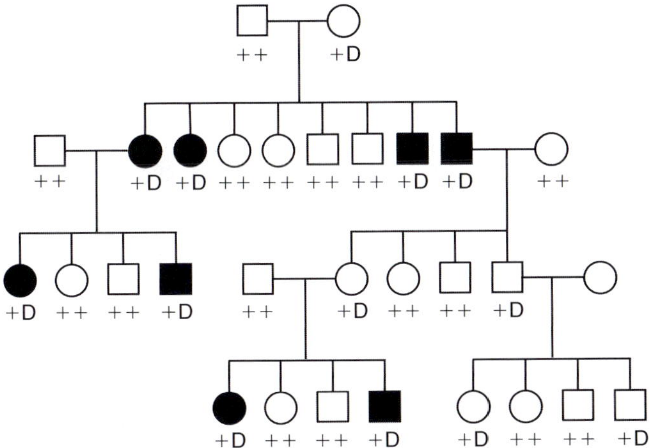

FIGURE 1.8 Genetic imprinting. An example of a kindred with a paternally imprinted disease is shown. The paternal allele of the disease gene is silenced; only the maternal allele is active. Those that inherit the disease allele 'D' from their mother have disease, whereas those that inherit 'D' from their father are disease-free, since the disease allele is silenced. Consequently 50% of offspring of mothers with a 'D' allele (whether the mother be affected or unaffected) are affected with disease (genotype '+D'), and all offspring of fathers with a 'D' allele (whether the father be affected or unaffected) are disease-free regardless of genotype, though 50% will be heterozygous carriers of the 'D' allele. This results in the disease skipping generations, as in X-linked disorders, but both genders are typically equally affected by disease

and as the repeat length increases the disease worsens in successive generations (McInnis 1996).

Genetic Imprinting

There are presently a small number of genetic diseases that show parent-of-origin effects, e.g. a specific mutation will result in disease if it passes through the female germline, but not the male germline. This arises as a consequence of *imprinting*, specific modification of a gene that silences the allele from one parent while leaving the other allele active. For example, in a gene with paternal imprinting (in which the paternal allele is silenced), a loss of function mutation will be phenotypically silent when passed through the father's germline, since the father's allele is already inactive. Conversely, because the zygote can only utilize the maternal copy of this gene, if the maternal allele is mutant, a disease phenotype will result. In this setting, males and females are at equal risk of disease, but only those who inherit the mutation from their mother will have disease. For example, an unaffected female mutation carrier will transmit the mutation and disease trait to half of her offspring. Her affected male offspring will transmit the mutation to 50% of their offspring, none of whom will be affected, whereas affected female offspring will transmit the mutation and disease to 50% of their offspring (Figure 1.8). While this pattern of transmission shares some similarities to sex-linked diseases, both genders are typically equally affected by the disease, a feature not commonly found for sex-linked disorders (Reik & Walter 2001).

Modifiers of Mendelian Transmission

Disease traits do not always demonstrate patterns of inheritance as simple as those described in the above discussion. For example, there may be *incomplete penetrance*, in which some carriers of mutant genotypes might nonetheless not display the disease trait. Reasons for incomplete penetrance include the requirement of a particular environmental exposure in order for mutation carriers to develop the disease. For example, an individual may be genetically susceptible to disease, but only develop it if exposed to a specific infectious agent, or if the diet is deficient in a particular nutrient. Similarly, there may be genetic modifiers of the trait, such that inheritance at other loci may prevent otherwise susceptible subjects from developing disease. Finally, a disease may show age-dependence, in which case younger mutation carriers might be disease-free, but become increasingly likely to develop disease with advancing age. Examples of Mendelian transmission with age-dependent penetrance include end-stage renal disease in patients with polycystic kidney disease and development of breast cancer in carriers of BRCA1 mutations (Hall et al 1990, Gabow 1993).

Patterns of inheritance may also violate Mendelian expectations owing to *phenocopies* in which some subjects in a pedigree have the disease despite not having the mutation. This can be particularly problematic for mapping genes contributing to common traits in the population, for which there may be multiple susceptibility factors, both genetic and environmental, present in affected kindreds.

Finally, mutations may also show *variable expressivity* in carriers, with different carriers of the identical mutation in families having different phenotypes. For example, while the pulmonary disease cystic fibrosis is caused by mutation in the gene encoding CFTR, the cystic fibrosis transmembrane regulator, it is now recognized that some males with mutations in CFTR have congenital absence of the vas deferens without pulmonary disease (Cutting 2005). Similarly, patients with mutations that cause polycystic kidney disease are at increased risk of developing intracranial aneurysm (Gabow 1993). These are both examples of variable expressivity.

NON-MENDELIAN OR COMPLEX TRAITS

Most common diseases are multifactorial or complex, with contributions from a multiplicity of genes and/or environmental factors in individual subjects. These include almost all diseases commonly encountered in clinical medicine, including end-stage renal disease, hypertension, diabetes, hypercholesterolemia, obesity, asthma, autism, and schizophrenia, among many others. Evidence that genetic factors contribute to disease susceptibility derives from several lines of evidence. For example, monozygotic twins, who share 100% of their genes, are concordant for these common disease phenotypes more often than dizygotic twins, who share only 50% of their genes. Similarly, siblings of affected subjects are more likely to have the same disease more often than the disease prevalence in the general population. Also, adoption studies often demonstrate greater concordance of disease in biological siblings than adoptive siblings raised in the same household, and in biological siblings raised apart, all suggesting a role of genes in disease pathogenesis. Analysis of the segregation of these traits within families, however, does not support Mendelian transmission among families ascertained from the general population, indicating that while genetic variation has substantial impact on disease risk, more than one gene and/or environmental factor influences the trait. Furthermore, while such studies of familial resemblance can provide estimates of the fraction of disease susceptibility that is genetic, they do not provide insight into how many genes influence the trait, the number of risk alleles in each gene, nor the magnitude of the effect imparted by any single gene.

The complexity of common disease traits has posed a vexing problem for genetic analysis. While single genes with very large effect can often be ruled out based on the patterns of transmission within families, the observed levels of familial recurrence could be accounted for by the combined effects of, for example, three or four genes, each with moderate effects in individual patients, or alternatively, the combined effects of 50–100 genes, each with very small effects. Similarly, at the outset it is unclear whether common alleles, identical by descent from a remote common ancestor, account for risk alleles, or alternatively whether many different rare mutations in the same gene account for risk alleles. For these reasons, the genetics of common diseases in the general population have been more difficult to dissect than their Mendelian counterparts.

THE HUMAN GENOME

The human genome has been sequenced to near completion and with high accuracy, revealing a haploid genome of about 3 billion base pairs (International Human Genome Sequencing Consortium 2001, 2004, Venter et al 2001). In addition, several other mammals have been sequenced to high levels of completion; as of March 2008 at least 35 mammalian species have either complete or draft sequence of their genomes finished or in progress (NHGRI 2008). The comparison of these sequences allows assessment of which sequences are under evolutionary constraint (i.e., those

sequences that are not permitted to diverge because variation is under negative selection). From such analysis, it is estimated that there are 20 000–25 000 protein-coding genes (International Human Genome Consortium 2004). Many additional genes do not encode proteins, but instead produce non-coding RNAs, including ribosomal RNAs, transfer RNAs, and an increasingly large set of micro RNAs that regulate the expression of other genes (Eddy 2001, Chen & Rajewsky 2007). There are many other non-coding transcripts produced which are of unknown function (The ENCODE Project Consortium 2007).

While protein-coding DNA accounts for only ~1.5% of the human genome, it is estimated that ~5% of the genome is under negative selection and therefore likely to be of functional importance (Mouse Genome Sequencing Consortium 2002). Some of this non-coding DNA under evolutionary constraint is involved in the regulation of gene expression. For example, a number of ultraconserved sequence elements that regulate key developmental genes have been identified (Bejerano et al 2004). Furthermore, it is possible that some fraction of the non-conserved genome might also be of functional importance, but not conserved across species because the gene has only recently emerged, has changed in function across species, or only has functional importance in humans.

NATURE OF MUTATIONS

There are many different types of mutation that can alter the function of a gene. All, of course, change the DNA sequence. New mutations occur infrequently. The rates of new inherited mutations are estimated to be $\sim 2 \times 10^{-8}$ at each base, suggesting that each individual carries ~120 new changes in the genome sequence (Nachman & Crowell 2000). The vast majority of these will lie outside gene coding regions or highly conserved genomic elements and are expected be of no phenotypic consequence.

The most common mutations are single base substitutions in which one base pair is altered to another (Figure 1.9). All base changes can occur, though some occur more commonly than others. For example, *transitions*, consisting of purine-purine (A to G or G to A) or pyrimidine-pyrimidine (C to T or T to C) substitutions, are more common than *transversions*, in which mutations alter a purine to a pyrimidine or vice versa. Because of the relatively frequent deamination of methyl-cytosine to thymidine in CpG dinucleotides, C to T transitions are the most common mutations in the human genome. Owing to the redundancy of the genetic code, single base substitutions within coding regions can be silent if they do not alter the sequence of the encoded protein, although such substitutions can sometimes have phenotypic effect owing to varying levels of transfer RNAs decoding each amino acid, as this may alter the level of protein expression

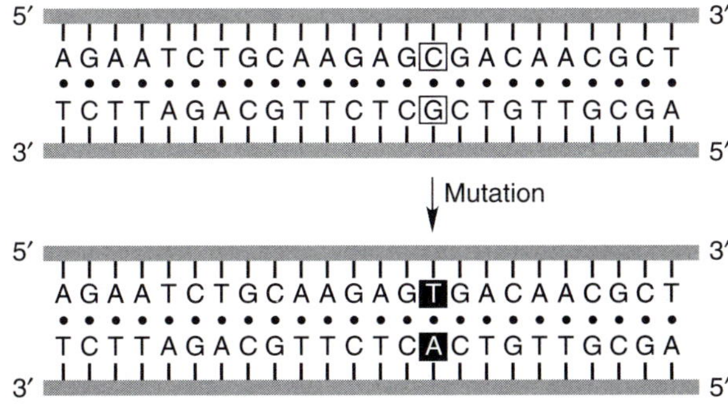

FIGURE 1.9 Single base substitution mutation in DNA. Example of a single base substitution, in which one DNA base pair is mutated to another. A 'C to T' transition mutation, the most common single base substitution owing to deamination of methyl-cytosine to thymidine in CpG dinucleotides, is shown. A C to T transition mutation results in transition of G to A on the opposite strand

or the timing of a protein complex assembly; similarly, substitutions can introduce new splice sites that might alter the encoded protein.

Other relatively mutable motifs exist. For example, simple sequences that are repeated in tandem many times (e.g. the dinucleotide GT repeated many times) are relatively frequently mutated, such that the same site will often be found to have a diverse number of alleles with different numbers of repeats in a large population of unrelated subjects (Weber & May 1989). These variations, which usually occur in non-coding regions, are called *microsatellites* or *simple tandem repeats* (*STRs*). Insertions and deletions also occur in the genome, and commonly insert one base or a small number of bases, though much larger insertions and deletions can occur. In addition, segmental duplications commonly occur in the genome, and these duplications are often in tandem, adjacent to one another in the genome (Bailey & Eichler 2006). If different copies of these repeats recombine with one another, one of the progeny chromosomes will be missing one of the repeat elements, while the other will have an additional copy of the repeated element. These so-called unequal crossing-over events can consequently give rise to either gain or loss of function alleles (Inoue & Lupski 2002). Finally, translocations between non-homologous chromosomes can also occur. Translocations that transect a gene can result in loss of function. Alternatively, translocations that fuse a novel regulatory element to a gene can produce a genetic gain of function – the translocations involving the *abl* kinase that cause chronic myelogenous leukemia represents one well-known example (Rowley 2001).

CONSEQUENCES OF MUTATIONS

Mutations with functional effect are usually classified as genetic gain of function or loss of function. In complex

cases, mutations can have gain of function effects on some traits and loss of function effects at others. Uncommonly, alleles can have *dominant negative* effects, in which the mutant allele impairs the function of the remaining wild-type allele.

In general, there are a far larger number of ways to cause loss of function than gain of function. For example, within protein coding regions, a single base insertion or deletion at any position will produce a frameshift mutation that is likely to cause loss of function. Similarly, in a typical protein 5% to 25% of its amino acids are completely conserved among all metazoan orthologs (proteins identical by descent from a common ancestor via speciation). Mutations that change amino acids at these positions will most frequently lead to loss of function, particularly if the substitutions are non-conservative (e.g., changing charge, hydrophobicity or bulkiness of the side group).

In contrast, there are far fewer sites in a protein that when mutated will lead to gain of function. Examples of gain of function mutations include mutations that prevent clearance of protein from the cell surface via endocytosis due to mutation of the required recognition sequence, mutations that result in constitutive activation of a receptor, and mutations in regulatory sites that lead to a protein's constitutive expression or expression in the wrong cell type. There are typically very few amino acid substitutions or DNA sequence changes that will result in genetic gain of function, and for this reason the effective target size for gain of function mutations is much smaller than the target size for loss of function mutations. As a result, a much larger number of mutations in a gene will result in loss of function rather than gain of function. Exceptions to this general rule are situations in which the mutant protein is part of a larger complex; in this context mutations at many positions may disrupt the assembly or normal function of the complex and thereby impart a gain of function. For example, mutations in many proteins of the cardiac sarcomere that result in hypertrophic cardiomyopathy are widely distributed through the proteins, yet appear to be dominant gain of function mutations – these are likely acting by disrupting assembly/function of larger protein complexes (Seidman & Seidman 2001).

Natural selection has differential effects on dominant and recessive alleles. As discussed earlier, gain of function mutations typically result in dominant traits, requiring only one mutant allele, while loss of function mutations are commonly recessive, requiring mutations in both alleles. As a consequence, gain of function mutations that impair reproductive fitness can be directly acted upon by natural selection, and they are typically rapidly eliminated from the population. For example, about 50% of patients with the dominant trait neurofibromatosis have de novo mutations (mutations that were not present in either parent and appeared anew in the affected subject) (Stephens 2003). In contrast, recessive loss of function alleles may be of little or no consequence in the heterozygous state, but when they are present in the homozygous state they cause severe disease and are subject to negative selection. In this case, selection will keep mutant allele frequencies at a low level, but will not eliminate them from the population. Thus recessive loss of function mutations are both more frequent and more tolerated in the population than dominant alleles. Moreover, recessive alleles are rarely attributable to de novo mutations, whereas dominant alleles that impair reproductive fitness are commonly de novo.

Because of the strong effects of natural selection, in any collection of independent disease subjects, it is likely that a wide spectrum of mutations will be observed for a recessive trait. Less diversity is likely to be seen for gain of function dominant traits owing to the restricted target for mutation that is often seen.

Exceptions to the rarity of specific mutations that cause Mendelian disease occur in several settings. Some are due to so-called *balancing selection*, in which the same allele may predispose to one disease but protect against another. The classic example of balancing selection is the sickle allele of beta-globin (Charlesworth 2006). The heterozygous state protects against malaria in endemic regions, whereas the homozygous state results in sickle cell anemia. The common lethality of sickle cell anemia at early ages would imply that the specific mutation should be eliminated from the population or maintained at low levels. The beneficial effects of the heterozygous state, however, have the opposite effect, permitting carriers to survive and leave more offspring than non-carriers in areas with endemic malaria. As a consequence of this balancing selection, the sickle allele is maintained at high frequency in malaria-endemic areas (Aidoo et al 2002).

IDENTIFYING DISEASE-CAUSING MUTATIONS

Pre-Genomic Era

As the catalog of Mendelian diseases in humans grew in the 20th century and the central dogma of biology was elucidated, there were many efforts to link disease phenotypes to specific mutations, first by discovering mutations revealed as alterations in protein sequence, and later as mutations in DNA sequence. The causes of a number of diseases were elucidated by insightful combinations of clinical investigation and protein biochemistry. For example, the molecular variation in hemoglobin that causes sickle cell anemia was identified in 1949, 4 years before the structure of DNA was elucidated (Pauling et al 1949). Similarly, a number of inborn errors of metabolism were solved by demonstration of enzymatic deficiencies that segregated as simple Mendelian traits. With the advent of recombinant DNA in the mid-1970s, genes encoding proteins important for

human physiology were identified by positional cloning, and some of these were found to be mutated in patients with specific Mendelian diseases. For example, the low-density lipoprotein receptor gene was recognized from careful clinical and biochemical studies to be deficient in patients with familial hypercholesterolemia; when the gene encoding the protein was cloned and sequenced, it was found to be mutated in patients with disease, proving the causal relationship between deficiency of this receptor, high LDL cholesterol levels and early coronary artery disease (Lehrman et al 1985, Brown & Goldstein 1986).

Through this period, however, there was no generalized, systematic approach to identifying disease-causing genes. Although it had long since been shown by the work of Morgan and Sturtevant in fruit flies that genetic maps of chromosomes could be constructed and that genes responsible for Mendelian traits could be reliably localized onto these maps, this required extensive breeding to get all the appropriate visible phenotypic markers into the same cross. Such schemes were of course impossible to execute in humans. As a consequence, while the mathematical basis for genetic mapping had been well-established by the 1940s, there was no practical application (Morton 1955).

This situation changed abruptly in 1980 with the publication of a seminal paper by Botstein, White, Skolnik, and Davis (1980). This paper recognized that variation in the human genome DNA sequence was likely to be relatively common and stable, and consequently these variations would constitute ideal *genetic markers*, with each variant deriving from a unique position in the genome. These variations in DNA sequence would nearly all be simple *polymorphisms* (variants without functional significance), but they would permit distinction of the two copies at a specific locus of a chromosome, allowing one to follow the inheritance of each allele, and the corresponding chromosome segment, through pedigrees. With a sufficiently dense map of genetic markers, one would be able to follow the inheritance of every segment of every chromosome through such pedigrees, and compare the inheritance of disease traits to the inheritance of every chromosome segment. With sufficient information, one would be able to unambiguously map the chromosomal location of disease loci. This map location would not itself be informative about the nature of the underlying disease gene, but would unambiguously provide evidence of its chromosomal position.

This recognition spurred an intense effort to develop human genetic maps. Initial efforts to identify polymorphic markers focused on discovery of restriction fragment length polymorphisms (RFLPs), in which a variation in DNA sequence results in allelic variants which differ by the presence or absence of DNA sites that are recognized and cleaved by sequence-specific restriction enzymes (Wyman & White 1980). These markers could be genotyped by cutting DNA with restriction enzymes and examining the size of the resulting DNA fragments to determine whether cleavage had occurred.

Subsequently, microsatellites (or STRs, described earlier) were recognized to be fairly frequent in the human genome. Microsatellite markers had the advantages of being more informative, with many allelic variants and a higher frequency of heterozygosity in the population, and of being simple to genotype. Using the *polymerase chain reaction* (*PCR*), many copies of a DNA fragment encompassing a microsatellite could be made, and the relative sizes of the resulting copies could be examined (Weber & May 1989).

As technology evolved, the laborious identification of RFLPs, which at a DNA sequence level often represent *single nucleotide polymorphisms* (*SNPs*), was replaced by direct sequencing of genomic DNA to capture SNP variation (The International HapMap Consortium 2003, 2005, 2007, Hinds et al 2005). In parallel, the genotyping of SNPs has evolved from the extremely labor-intensive method of RFLP analysis, where detection of a single SNP could take many hours, to massively parallel approaches (described later) which permit genotyping of up to a million SNPs distributed across the genome in a single experiment.

To localize genetic markers on the human genetic map, it was necessary to have a collection of families with a large number of informative meiotic events. A set of genomic DNAs from 60 reference families composed of four grandparents, two parents and 8–12 children in the third generation were collected and distributed to mapping laboratories. Genotypes of polymorphic markers were determined in these reference kindreds, and compared to one another. If two loci lie on different chromosomes, they should segregate independently in kindreds, just like alleles for different traits in Mendel's pea crosses. In contrast, if they lie extremely close to one another such that recombination rarely if ever occurs in the interval between them, they should show complete linkage, with alleles co-segregating with one another in pedigrees. Finally, if they are linked at some distance from one another, they should show significant linkage, but will show evidence of recombination with one another, and the frequency of recombination (or *recombination fraction*) will reflect the physical distance between the two loci on the chromosome, like traits in fly crosses performed by Morgan and Sturtevant. From these principles, complete genetic maps of the human genome were constructed, initially localizing hundreds of genetic markers to the map and permitting the mapping of disease loci (White et al 1985, Donnis-Keller et al 1987, NIH/CEPH Collaborative Mapping Group 1992, Weissenbach et al 1992, Cooperative Human Linkage Center 1994, Gyapay et al 1994).

MAPPING MENDELIAN TRAITS IN HUMAN PEDIGREES

With complete genetic maps, it is possible to determine the location genes for Mendelian traits by comparing the inheritance of the trait to the inheritance of each segment of each chromosome. The goal in mapping a disease locus is first to determine unequivocally a chromosomal interval in which it lies, and then to determine its location as precisely as possible as a prelude to identifying the disease gene itself.

For an autosomal dominant trait, at the underlying genetic locus affected subjects are heterozygous for the disease allele (genotype '+/D'), while their unaffected family members are homozygous for the wild-type allele (genotype '+/+') (Figure 1.4). The disease trait thus serves as a proxy for the underlying genotype at the disease gene locus. By genotyping polymorphic DNA markers distributed across the genome, one can then compare the inheritance of each segment of every chromosome to the inheritance of the disease in an extended kindred. If a genetic marker lies on a different chromosome than the disease-causing gene, alleles of the marker and the disease should show independent assortment (Figure 1.10A). Alternatively, if the DNA marker lies very close to the disease locus, alleles of the marker locus should show precise co-segregation with the trait such that all affected subjects will inherit the identical allele of the marker locus, and all unaffected subjects will not (Figure 1.10B). Finally, if the marker lies some distance from the disease gene, recombination in the interval between the disease and marker loci can occur in meiosis, and the frequency of these recombination events will be proportionate to the distance between them.

The statistical basis for mapping Mendelian traits is well established, and utilizes comparison of the likelihoods of the two alternative explanations of observed data: the likelihood that the data are explained by linkage between the trait and marker at a particular recombination fraction, versus the likelihood that the observed data are explained by random segregation of trait and marker loci (e.g., the likelihood that the trait and marker loci are unlinked). This statistical comparison is referred to as the *likelihood ratio* in favor of linkage. Consider, for example, the autosomal dominant pedigree in Figure 1.10. Panel 'A' shows the genotypes in each kindred member of a genetic marker that is unlinked to the disease trait. It is readily apparent that there is no fixed relationship between the inheritance of a particular allele of this marker and the presence or absence of disease. In contrast, in panel 'B', it is clear that every affected individual in the kindred has inherited allele '1' of the marker locus whereas every unaffected member has not. In this latter case we can calculate the evidence in favor of linkage. Under the most likely model of linkage, the marker and the trait would be linked at a recombination fraction of zero, and the likelihood under this model that we would see the experimentally observed results equals 1. We then calculate the likelihood that we would obtain these observed results entirely by chance if the trait and marker loci were unlinked. At the outset, knowing that the index case has disease and genotype '1, 2', we don't know if the disease

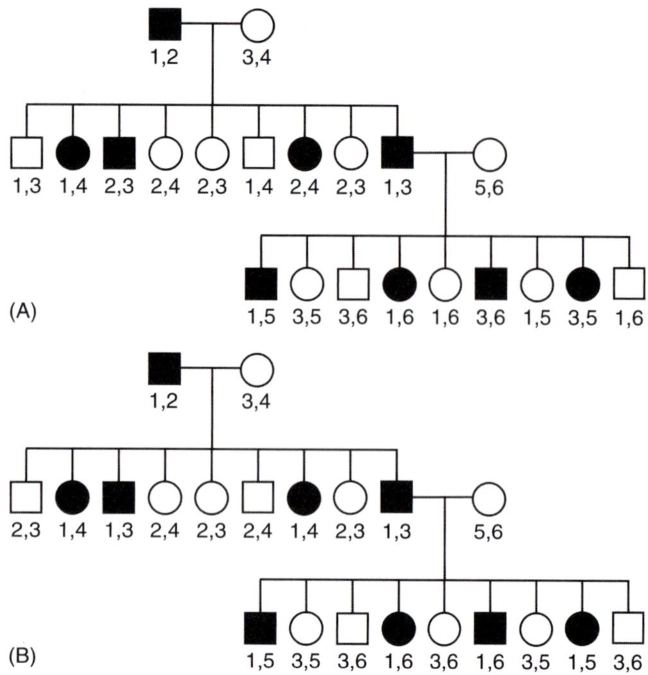

FIGURE 1.10 Linkage of a marker and trait with autosomal dominant inheritance. (A) Random segregation of an autosomal dominant trait and a marker with alleles numbered '1' to '6'. There is no relationship between inheritance of a particular allele of the marker and the presence or absence of disease; the trait and marker are unlinked, consistent with the disease and marker loci being on different chromosomes. (B) Linkage of the trait and marker. Every affected individual in the kindred has inherited marker allele '1' from their affected father, whereas every unaffected member has not. After determining the phase of the trait and marker alleles, one can use the remaining 17 meiotic events to determine the odds that the segregation of the marker and trait is explained by linkage, versus the alternative of random segregation. If the trait and marker are linked at a recombination frequency of zero, the likelihood of the observed segregation of allele '1' (in affecteds only) and '2' or '3' (in unaffecteds only) is 1. If the trait and marker were randomly segregating, the odds that all 17 individuals would have inherited their allele by chance is $(0.5)^{17} = 1/132\,000$. The ratio of these odds is $1/(0.5)^{17}$ or 132 000:1 in favor of linkage, and the $\log_{10}$ of the ratio (the lod score) is 5.12

allele is on the same chromosome as allele '1' or allele '2'. Once we look at any single transmission event, however, we can set the phase of these alleles – seeing that an affected offspring inherited allele 1 from the index case sets phase under the model of linkage with no recombination. We can then ask, 'What is the likelihood that another affected offspring would, by chance, also inherit allele "1"?' That likelihood is 0.5. Further, the likelihood that an unaffected offspring would, by chance, inherit allele '2' is also 0.5. The joint likelihood of both events occurring is $(0.5)^2$, or 0.25. Thus the overall likelihood of seeing precise co-segregation of allele 1 with disease in all 17 phase-known meiotic events, by chance, is $(0.5)^{17}$, or 1 in 132 000. The likelihood

ratio in favor of linkage is consequently $1/(0.5)^{17}$, or 132 000:1 in favor of linkage. For convenience, these likelihood ratios are commonly expressed as their decimal logarithm, which in this case would be 5.12. This $\log_{10}$ of the odds ratio is referred to as the *lod score*.

What thresholds should we apply to declare that a result provides significant evidence of linkage? It is important to recognize that with 22 autosomes and a 3000 cM genetic map in humans, absence of linkage is much more likely than linkage for any single DNA marker, but with so many loci being tested, there is likely to be nominal evidence of linkage at a few loci purely by chance. As a consequence, modest likelihood ratios of 20:1 or 100:1 would not constitute sufficient evidence of true linkage, and using such thresholds would yield many false-positive results. Taking this into consideration, it is widely accepted that a lod score of 3.0, which corresponds to a likelihood ratio in favor of linkage of 1000:1, is an appropriate threshold for declaration of significant linkage that yields few false-positive results for known Mendelian traits.

From this threshold, one can determine how many phase-known meiotic events one would need to observe to find significant evidence of linkage if a dominant trait and marker loci are completely linked. This number is 10 (for a lod score of 3.0, $\log[1/(0.5)^x] = 3$, and $x = 10$). These 10 informative phase-known meioses can be derived from a single large kindred, or alternatively from many smaller kindreds. The use of many smaller kindreds will be predictably successful if a disease shows *locus homogeneity*, i.e., the same gene is mutated in different disease families, but power will be diminished if in fact mutation in more than one gene can produce the same phenotype (*locus heterogeneity*). In this latter setting, though power is reduced, analytic techniques allow simultaneous detection of linkage to multiple loci among different families (so-called multilocus linkage analysis). Suffice to say, collection of all available at-risk kindred members in individual pedigrees maximizes the power to detect linkage, minimizes the risk and impact of locus heterogeneity, and provides the best opportunity to refine the location of the gene to the smallest interval.

How many markers need to be genotyped to detect linkage? This question is dependent principally upon two factors: the informativeness of the pedigree and the informativeness of the markers genotyped. To illustrate the problem, in human pedigrees there is often missing information. Specifically, at-risk kindred members may be deceased or unavailable, and it can consequently be difficult to be certain whether an allele seen in two different affected members of a pedigree represents alleles inherited *identically by descent* from a common ancestor, or inherited *identically by state* from two different branches of the kindred, in which case the allele in one affected subject might be tightly linked to the disease-causing mutation while the allele in another affected subject would not. In this setting, the genotyping of additional closely linked markers can allow the formation of *haplotypes* (strings of alleles at adjacent marker loci that are linked on a particular chromosome); the finding that two subjects in a pedigree share the same rare haplotype provides evidence that both have been inherited identically by descent from a common ancestor. Similarly, the *informativeness* of genetic markers varies and this critically affects their utility. For a linkage study, alleles transmitted from a heterozygous parent can be distinguished from one another and are informative; in contrast, if a marker locus in an affected parent is homozygous, the two copies of the chromosome at that position cannot be distinguished, and that marker makes no contribution to linkage information in resulting meiotic events. Consequently, markers that have high probabilities of being heterozygous are more informative than those that are less likely to be heterozygous. For example, SNPs with two alleles will have a maximum heterozygosity of 50% (as occurs when the two alleles each have frequency of 50% in the population). As the minor allele frequency is reduced, heterozygosity is diminished (e.g., 18% heterozygosity if the minor allele frequency is 10%). Microsatellite markers with many alleles, in contrast, have higher heterozygosities. For example, a marker with four equally prevalent alleles has a heterozygosity of 75% among unrelated subjects. Importantly, informativeness across a locus may be maximized by typing additional linked markers to produce haplotypes which are individually rare in the population. Taking these considerations into account, a general approach has been to type ~400 microsatellite markers distributed across the genome. More recently, with the advent of inexpensive SNP genotyping, linkage using 1500–10000 SNPs across the genome, which compensates for these markers' reduced informativeness, has also proved highly useful.

Recessive Traits

While large extended kindreds can often be characterized for dominant traits, this is rarely the case for recessive traits. One typically encounters disease restricted to a single sibship, and these often contain only one or two affected siblings with additional unaffected siblings. Mapping recessive traits in outbred kindreds is consequently challenging. In addition, unlike fully penetrant dominant traits, affected and unaffected subjects contribute differently to linkage information. Consider a recessive trait with two affected and four unaffected siblings (Figure 1.11). Each unaffected parent is likely to be heterozygous at the trait locus, with a wild-type and a mutant allele. For a marker tightly linked to the trait locus, the phases of the marker and trait alleles in each parent are unknown until the genotype of an affected subject is inspected. Once phase is set in this manner, additional affected subjects in a sibship are expected to have the identical marker genotype, which is expected to occur by chance with likelihood 0.25. Unaffected sibs are expected to have any of the three non-identical genotypes, which together would be expected to occur by chance with likelihood 0.75.

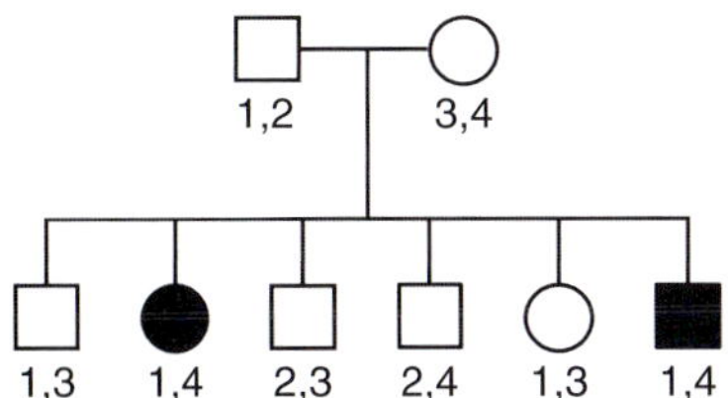

FIGURE 1.11 Linkage of a marker and trait with autosomal recessive inheritance. In this kindred, the unaffected parents are heterozygous at the trait locus, and offspring with the trait must have inherited two mutant alleles at the trait locus. Once phase is set by observing the genotype of one affected subject, all other affected siblings should have the identical genotype (in this example '1,4'), while all unaffected siblings should have a different genotype. In this kindred one can calculate that the odds ratio in favor of linkage of the trait to this marker at a recombination frequency of zero is $1/(0.25)(0.75)^4$ or only about 13:1 in favor of linkage; the lod score is only 1.1. Because the more prevalent unaffected sibs give much less information for linkage than affected sibs (since the likelihood of their having inherited their genotype by chance is greater), the ability to map recessive loci in outbred kindreds is impaired

Consequently, the more prevalent unaffected sibs give much less information for linkage than affected sibs. In this kindred, the maximum lod score with complete linkage information would consequently be log $(0.25)(0.75)^4 = 1.1$. Unfortunately, among sibships of 3, only 1 in 7 will have two or more affected sibs, seriously impairing the ability to map recessive loci in outbred kindreds.

Fortunately, however, there are situations in which a great deal of linkage information can be obtained from single affected subjects. This occurs in the setting of consanguineous union. In such settings, the occurrence of a rare recessive disease in the offspring of related subjects strongly implies that the affected subject has inherited not just mutations in both alleles of the same gene, but in fact has inherited the identical mutation on both chromosomes, as the mutation has traveled down both arms of the pedigree and become homozygous by inheritance from both parents (Figure 1.12). Importantly, not only will the patient be homozygous for the disease-causing mutation, he/she will also be homozygous for the flanking chromosome segment covering a long stretch, averaging ~10cM in the offspring of first cousins. The likelihood of such a homozygous segment occurring by chance in offspring of first cousins is 1 in 16. Thus the finding of such a segment in a single affected subject yields a lod score of 1.2, and it is apparent that if single affected sibs from even three such families were all homozygous for the same chromosome segment, the lod score would be 3.6, sufficient to establish linkage (Lander & Botstein 1987).

Identifying the Disease Gene within the Linked Interval

Linkage is a first step toward identifying the disease gene. Once the chromosomal position is defined and bounded,

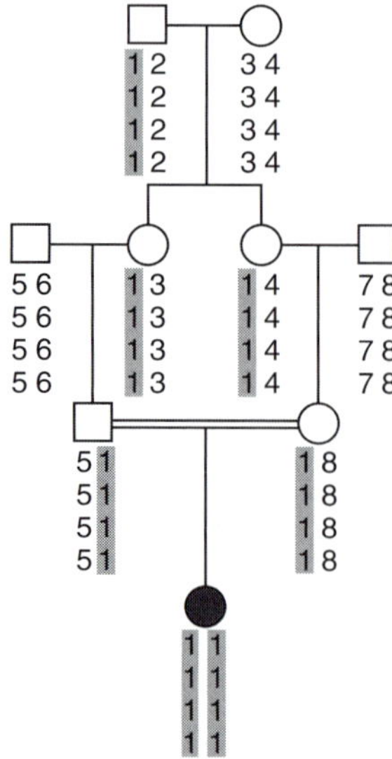

FIGURE 1.12   Homozygosity mapping in an inbred kindred with autsomal recessive disease. When a rare recessive disease occurs in the offspring of related parents, it strongly implies that the affected subject has inherited the identical disease allele from both parents, originating from a common ancestor. In this case, the affected subject will likely be homozygous for a long stretch of the chromosome flanking the disease gene, averaging ~10 cM in the offspring of first cousins. In the figure, the affected subject is homozygous for a chromosome segment containing four markers (genotype '1,1' at each marker). The identical chromosome segment (shaded gray) has been inherited from both first-cousin parents, originating from their common grandfather. The likelihood of a marker being identical by descent in a child of first cousins by chance is 1 in 16, resulting in an odds ratio of the likelihood of disease and marker linkage versus random segregation of 1/(1/16), or odds of 16:1 in favor of linkage (lod score 1.2). Finding single affected children of three such families to be homozygous for the same chromosome segment would result in a lod score of 3 × 1.2 = 3.6, sufficient to establish linkage of the chromome segment with disease

it can be searched for functional mutations in disease patients. To this end, it is critical to define the interval within which the disease gene is most likely to reside. With sufficient marker density, it is usually possible to find informative markers or haplotypes that are completely linked to a Mendelian disease. From this starting point, the first markers proximally and distally that show recombination with the disease locus define the boundaries within which the gene is likely to lie (Figure 1.13). It is intuitive that the more informative meioses analyzed, the more likely one is to have a recombination event that narrows the interval containing the disease locus. Consequently, while a lod score of 3.0 might be sufficient to establish linkage to a chromosome segment, if the resulting interval is very large, the search for the disease gene might be far less efficient than it would be if a larger set of families were analyzed and defined a much smaller interval. It is common to see reference to the

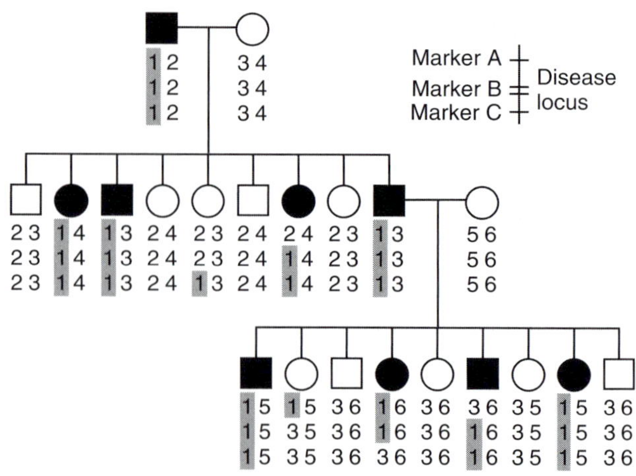

FIGURE 1.13   Recombination events refine the location of a disease gene. A kindred with an autosomal dominant disease is shown; genotypes at three markers surrounding the disease locus are shown under each symbol (markers A to C, each with alleles '1' to '4'; order of markers on the chromosome segment is shown). Chromosome segments that cosegregate with the disease are shaded gray. Recombination events between markers A and B localize the disease gene distal to marker A (an affected female in generation 2 and an affected male in generation 3 have the disease-associated segment only at markers B and C, and an unaffected female in generation 3 has the disease-associated segment at marker A). Recombination events between markers B and C localize the disease gene proximal to marker C (an affected female in generation 3 has the disease-associated segment only at markers A and B, and an unaffected female in generation 2 has the disease associated segment at marker C). The only marker for which complete cosegregation of marker and disease is observed is marker B, which is tightly linked to the disease locus. The more informative meioses analyzed, the greater the chance of observing recombination events that narrow the interval within which the disease locus may lie

'lod-1 interval' or the 'lod-3 interval' as representations of the most likely intervals of the disease gene. These represent the interval across which the lod score drops below and remains below 1 (or 3) lod units below its peak value. These roughly approximate the 95% confidence interval and the 1000:1 likely interval in which the disease gene is almost certainly located.

Once the linked interval has been well-defined, efforts can proceed toward identification of the disease gene. For the first 10–15 years of the positional cloning era, this was often an extraordinary challenge, as the identities of genes within the linked interval were frequently entirely unknown. Progress was made by construction of physical maps, and identification of transcribed elements or protein-coding *exons* from chromosome segments. Finding genes expressed in tissues appropriate to the disease prioritized genes for investigation. With the determination of the human genome sequence and the identification of nearly all highly conserved elements, this extremely laborious phase has been eliminated, replaced by rapid searches of online databases that provide annotation of all the genes in an interval, their precise genomic sequence, and their tissue expression patterns. From this information, it is straightforward to amplify desired segments of any gene by PCR and determine the DNA sequence. Detection of homozygous sequence variation is straightforward, and methods for detecting and confirming heterozygous sequence changes from diploid sequence are increasingly robust.

For the vast majority of Mendelian traits, different disease-causing mutations are found in different families. This provides a substantial advantage in identifying the disease gene. For example, for a typical recessive trait caused by loss of function mutations, the finding of diverse mutations that alter the encoded amino acid sequence at highly conserved positions, which segregate with the disease trait in pedigrees, and which are rare, showing high specificity for their association with disease alleles, constitutes extremely strong evidence that the disease gene has been identified. The finding of premature termination codons, frameshift mutations, large deletions, and mutations at canonical splice sites may constitute a minority of loss of function mutations but are very compelling. It is worth pointing out that a significant number of loss of function mutations lie outside the coding region of genes, disrupting regulatory elements or splicing by cryptic mechanisms. For many genes, these non-coding mutations constitute about 15% of disease-causing alleles (Cooper et al 1995). Consequently, even the most diligent searches of coding regions will often fail to identify both mutations in some kindreds with recessive disease. For dominant traits the same general rules apply, though the spectrum of mutations that cause gain of function alleles is likely to be much more restricted. For many dominant traits, de novo mutations can be identified, and these provide very compelling evidence implicating the mutation in disease causation because such events occur uncommonly by chance.

These considerations lead to the very practical questions of what evidence constitutes sufficient proof to be convincing that a Mendelian disease gene has been identified. The combination of significant linkage along with multiple independent mutations that meet the above criteria for significant effect and which are rare in the population are typically sufficient to make a convincing case from a pure analytic approach. Correlative evidence that the disease gene is expressed in the proper location to impart its effect, and biochemical evidence of altered function of the mutated gene product can be helpful. It is worth emphasizing, however, that in the presence of compelling genetic evidence, the absence of biochemical data or a detailed mechanism of how the mutated gene product imparts its effect does not detract from genetic evidence. In fact, for many of the most important genetic discoveries, the normal functions of the encoded genes and the mechanisms by which their mutations cause disease have required years of investigation to elucidate, and this work was only spurred by the genetic discovery convincingly linking the gene to disease.

It is worth considering that as technology has changed, the ability to identify disease-causing mutations in single patients/families is evolving. For example, within a linked interval, it is increasingly possible to sequence all the genes and/or all the DNA. If the linkage evidence is convincing, and one sequences the entire linkage interval and finds a single mutation of functional effect, the evidence that this is the cause of disease can be compelling. Similarly, dominant lethal diseases, which have heretofore been extremely difficult to approach because of the absence of extended kindreds, come onto the radar screen as the ability to sequence entire genomes becomes tractable and affordable. The ability to demonstrate de novo mutations in such settings is a powerful means of providing evidence that the disease gene has been identified – finding, for example, even two independent de novo mutations in unrelated patients with a dominant lethal trait would constitute convincing genetic evidence in a small cohort of patients.

New Insights from Mendelian Genetics

The success of the positional cloning paradigm has had profound impact on the understanding of human disease in every branch of medicine. More than 1600 Mendelian traits have been solved at the molecular level. These include well-known genetic disorders such as polycystic kidney disease, cystic fibrosis, hemochromatosis, and Huntington's disease (Rommens et al 1989, MacDonald et al 1993, The European Polycystic Kidney Disease Consortium 1994, Feder et al 1996). In each case, the gene product was novel, and its identification sparked intense investigation of the biochemical and cellular mechanisms that link genotype to disease phenotype; these studies have been enormously informative. In addition, knowledge of these disease genes has permitted development and implementation of new diagnostic tests to identify subjects at risk of transmitting disease or who are destined to develop disease. Although therapeutic advances may lie years in the future, defining disease biology is a first step toward understanding potential therapeutic interventions.

Equally interestingly, rare Mendelian traits have provided important insight into basic pathways that underlie common complex traits. Thus rare Mendelian forms of hypertension (Lifton et al 2001, Kahle et al 2008), hypercholesterolemia (Hobbs et al 1992), diabetes (Florez et al 2003), osteoporosis (Cohen 2006, Balemans & Van Hul 2007), obesity (Bell et al 2005), cardiac arrhythmias (Keating & Sanguinetti 2001), Alzheimer's disease (Hardy & Orr 2006, and colon and breast cancer (Narod & Foulkes 2004, Rustgi 2007) have identified key genes and pathways in which mutation can have very large effects on disease risk. These discoveries have defined key molecules and biological pathways that might be manipulated for health benefit in the general population; importantly, they have defined which gene products in complex pathways are at key nodes in system behavior such that their mutation can impart large biological effect. Such knowledge is very difficult to achieve by other approaches. Importantly, the effects of gain or loss of function mutations in a particular gene are often outstanding proxies for the effects that would be produced by agonism or antagonism of the respective gene product, thereby anticipating both therapeutic potential and possible adverse effects that are mechanism-based.

IDENTIFICATION OF COMMON ALLELES THAT CONTRIBUTE TO COMPLEX TRAITS

As discussed above, the genetic architecture of common disease has been largely unknown. One idea has been that common alleles might play a role in common disease. This could be the case if diseases were of sufficiently late onset to escape natural selection, if recent changes in the environment resulted in alleles that were neutral or adaptive recently becoming deleterious, or if there were balancing selection.

Testing of the common variant hypothesis has been tried repeatedly by candidate gene studies in which common polymorphisms in genes of interest are identified and their allele frequencies are compared in a cohort of cases and controls. There have been a modest number of spectacular successes with this approach. For example, common variants in the major histocompatibility complex confer large effect on the risk of type I diabetes mellitus (Florez et al 2003); the ApoE4 allele increases risk of development of Alzheimer's disease (Herz & Beffert 2000); factor V Leiden increases risk of deep vein thrombosis (Seligsohn & Lubetsky 2001). Despite these successes, the investigation of candidate gene variants in common diseases has for the most part been extraordinarily unproductive. Literally thousands of such studies have been published, and the overwhelming majority of those reporting positive results have proven not to be replicable (Lohmueller et al 2003). There are many reasons for this poor record of replication. First, unlike Mendelian traits, the lines of evidence supporting a positive association study are slender. For Mendelian traits, segregation studies within families provide convincing evidence that there is a single gene with large effect on disease risk within a family. For complex traits, there is usually no indication of how large the effect of variants at any locus is likely to be, and consequently the power of any study is unknown a priori. Second, linkage mapping in Mendelian kindreds provides convincing evidence that the disease locus lies within a specified interval in the genome; this linkage evidence is independent of the allelic effect of the underlying mutation. Candidate gene studies have no such positional information to accompany estimates of the effect of specific variants in the gene. Further, Mendelian studies have the great benefit of being able to find many independent functional mutations in the same gene in different pedigrees, providing convincing evidence that mutation and disease are strongly associated. Candidate gene studies of common alleles do not have this luxury. Finally, linkage studies are not influenced by *population stratification* (differences in allele frequencies related to ethnic background, which can exist even between very closely related ethnic groups). Linkage relies upon demonstration of inheritance of chromosome segments identically by descent within families, and hence upon explicit evidence of co-segregation of disease and marker loci. In contrast, candidate gene association studies critically rely upon the assumption that case and control cohorts derive from sufficiently similar genetic backgrounds that any observed differences in allele frequency are not simply the consequence of unrecognized population stratification. One can test for stratification by genotyping a battery of anonymous SNPs across the genome and looking for a greater-than-expected overall difference in allele frequencies between cases and controls, but such testing has almost never been performed in candidate gene studies. When the literature is evaluated, one finds many allelic associations that are not replicated in subsequent studies, or in which so-called replication studies use different definitions of the trait; this introduces new degrees of freedom into the analysis that are typically not accounted for. Similarly, some studies find association with different alleles or haplotypes than the original study; this again is not replication. For all of these reasons, reports of positive results from candidate gene association studies are usually false positive, with estimates of a false-positive rate of ~95% in the literature (Lohmueller et al 2003, Morgan et al 2007). For those few studies that have proved true positive, the hallmarks have been robust replication, consistent association with the same phenotype, and consistent association with the same allelic variants.

Genome-Wide Association Studies

It has been recognized that in principle one could perform *genome-wide association studies* (*GWAS*) of common variants with common disease. There were several obstacles to such studies. First, because a very large number of tests would be performed, correction for multiple testing would require the study of very large cohorts to be able to detect significant results. The power of such studies varies with the risk of developing disease in the presence or absence of the risk allele, and also the frequency of the risk allele in the population (Risch & Merikangas 1996). Second, it was necessary to identify all or a very substantial fraction of the common variants in human populations. Third, it was essential to determine the architecture of the human genome in order to determine how many SNPs needed to be genotyped in order to have coverage of a substantial fraction of the genome. Finally, the development of highly efficient and inexpensive methods for genotyping SNPs was required.

By resequencing segments of human genomes, more than 10 million common variants in the human population have been identified (The International HapMap Consortium 2007). It is worth noting that native Africans have significantly greater diversity than other populations, consistent with this population being older, and the site of the ancestral human population (Gabriel et al 2002). Most common human SNPs in any population can be found in African populations, indicating that contemporary SNPs in these populations are generally ancient and were carried out of Africa.

A key feature of the human genome that has emerged from recent studies is that there is substantial *linkage disequilibrium (LD)*; this has important implications for association studies (Gabriel et al 2002, The International HapMap Consortium 2003, 2005, 2007, de Bakker et al 2005, Hinds et al 2005). Linkage disequilibrium reflects the ability of knowledge of the specific allele at locus 'A' to correctly predict the specific allele at locus 'B'. Linkage disequilibrium will exist across chromosome segments if a population is young, if the frequency of recombination across the chromosome is non-random, or some combination of the two exists. Take for example, the situation in which a disease-susceptibility mutation is introduced into the population. At that time, the mutation exists on a specific haplotype in the population. As that chromosome is transmitted to subsequent generations, it will recombine in meiosis with other chromosomes; after a few generations, likely recombination events will introduce new alleles and haplotypes at chromosome segments distant from the mutation, but those closer to the mutation will retain the ancestral alleles and haplotype (Figure 1.14). If one genotyped any of the closely linked SNPs, its alleles would be predictive of the presence or absence of the disease allele (or of any other closely linked allele in the same haplotype). After many more generations, only those alleles most tightly linked to the mutation will still reflect the ancestral haplotype. The length of the segment that reflects the ancestral haplotype will depend upon the number of generations and the rate of recombination in the vicinity of the disease allele. Given a sufficient number of generations and random recombination, a state of linkage equilibrium will eventually be achieved in which alleles at one locus are not predictive of the alleles at even very nearby loci.

The study of haplotypes in the human genome has demonstrated that there is, in fact, substantial linkage disequilibrium. This can be quantitated in a number of ways that measure the non-random association of alleles at neighboring loci. These studies demonstrate that there are segments of the genome, commonly referred to as *haplotype blocks*, in which all markers are in strong linkage disequilibrium with one another (Gabriel et al 2002, The International HapMap Consortium 2003, 2005, 2007, de Bakker et al 2005, Hinds et al 2005). Within these blocks, there has been little or no meiotic recombination since the emergence of the human species. Some of these

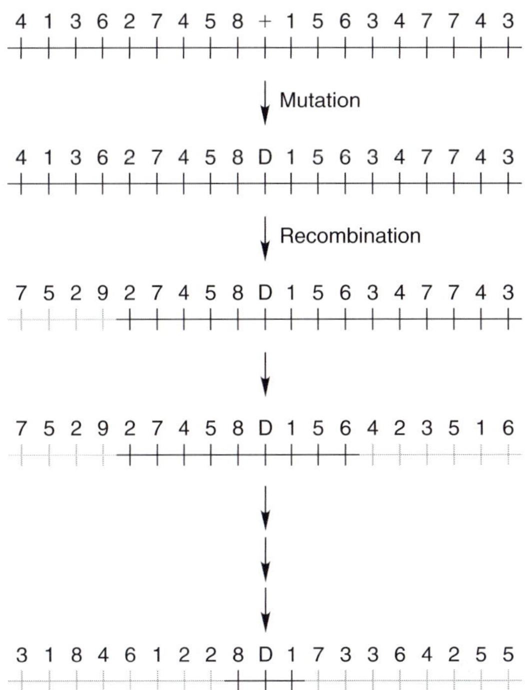

FIGURE 1.14 Linkage disequilibrium between markers and a disease locus. Shown is a chromosome segment containing many markers (represented with hatch marks; the allele at each marker is represented with a number) and a locus that was originally wild-type ('+') but was subsequently mutated to a disease allele ('D'). At the time of mutation and shortly thereafter, the disease allele exists on a specific haplotype (shown in black). As the chromosome is transmitted through many generations, it recombines in meiosis with other chromosomes (shown in gray). At chromosome segments distant from the mutation, new alleles and haplotypes are introduced (gray), but within the segment closely surrounding the disease allele, the ancestral marker alleles and haplotype are retained (black). If a marker closely linked to the disease allele were genotyped, the marker alleles would be predictive of the presence or absence of the disease allele, reflecting the linkage disequilibrium (LD) between the disease and marker. The length of the segment in LD depends on the number of generations since the mutation arose and on the rate of recombination in the vicinity of the disease allele

reflect significant variation in rates of recombination across a chromosome segment, while others may simply reflect chance events in a relatively young species. As a corollary of this effect of population age, it would be expected that Africans, the most ancient human population, should have less linkage disequilibrium than other populations that only left Africa ~100 000 years ago. This is in fact observed – Africans have less LD and shorter blocks of LD (Gabriel et al 2002).

The importance of this linkage disequilibrium in the human genome is that alleles of certain SNPs are highly predictive of alleles of many other SNPs; as a consequence, these other SNPs need not be genotyped directly. It is estimated that ~80% of the common variation in the human genome can be captured by genotyping as little as 5% of the variation (de Bakker et al 2005). As a result, GWAS comparing allele frequencies in disease cases and controls were recognized to have substantial power to detect loci that influence disease susceptibility.

High-Throughput SNP Genotyping

The ability to perform such studies required the development of large-scale and inexpensive SNP genotyping methods. These have been achieved on several platforms. One of these operates by differential hybridization of genomic DNA to microarrays containing oligonucleotides representing the sequences of alternative alleles at each SNP (NCBI (a)). Another approach uses a bead-based hybridization and single-base extension reaction at the queried SNP site (NCBI (b)). Both approaches in their current forms can genotype up to one million different SNPs per patient in a single reaction, at a cost of less that 0.1 cents per genotype (Hirschhorn & Daly 2005).

Analysis of Genome-Wide Association Studies

Analytic methods for GWAS are still evolving, but a number of critical methods have emerged that greatly strengthen analysis. At the outset, it is worth pointing out that careful study design, with meticulous matching of cases and controls, is key and greatly increases the power to detect a weak signal amid background noise. Nonetheless, it has been recognized that with hundreds of thousands of SNP genotypes in thousands of cases and controls, one can make very precise assessments of the degree to which cases and controls are genetically well matched. One approach to this is to assess the expected and observed chi-square statistic for differences in allele frequencies at each SNP across the genome in cases and controls (Pritchard & Donnelly 2001). For example, by chance one expects ~5% of SNPs to have chi-square P values <0.05, ~0.1% to have P values <0.001, etc. If there is unrecognized stratification in the genetic backgrounds of cases and controls, there will be many more SNPs exceeding these values than expected. This genome-wide inflation factor, λ, provides an estimate of how well matched cases and controls are, and corrections for the inflation resulting from subtle mismatches can be attempted. Other methods of assessing matching of cases and controls and correcting for differences have been developed. Nonetheless, it is always a concern that unrecognized stratification might account for seemingly significant P values in such studies, and as a consequence replication studies greatly strengthen evidence that identified alleles truly contribute to disease susceptibility.

As noted above, the power of GWAS to detect association at a single locus depends upon the number of cases and controls, the frequency of the risk allele in the population, the magnitude of the effect imparted by the risk allele, and whether the disease-related SNP is directly genotyped or in LD with another SNP that is typed. The effect size of the risk genotype is often specified by *genotype relative risk* (the ratio of the probability of disease with the risk genotype to the probability of disease without it – a relative risk of 1.5 would indicate that those with the risk genotype are 1.5 times, or 50%, more likely to develop disease than are those without the risk genotype). It is estimated that with 2000 cases and 3000 controls, using a P value threshold of 5×10^{-7}, there is ~80% power to detect alleles of average frequency that have a relative risk of ~1.5; power decreases to ~43% for alleles with a relative risk of ~1.3 (Wellcome Trust Case Control Consortium 2007). The ability to detect alleles of smaller effect size is reduced, but if there is a large number of such variants, some will be detected by chance.

Utility of Genome-Wide Association Studies

The first spectacular success with GWAS came with the study of age-related macular degeneration, a common cause of blindness in the elderly population. Using a small set of cases and controls, common variants in the coding sequence of complement factor H were found to be strongly associated with disease risk and impart very large effects on disease risk (Klein et al 2005). This association has been widely replicated (DeWan et al 2007), and underscores the point that common variants that contribute to disease risk well beyond reproductive age can escape negative selection. Similarly, GWAS of inflammatory bowel disease has implicated common variants in several genes, including those for NOD2 and the IL23 receptor (Ogura et al 2001, Duerr et al 2006). In these examples, it is possible that recent changes in the environment account for the presence of common disease-susceptibility alleles in the population.

Most recently, there has been an explosion of GWAS for common disease, including studies of type II diabetes mellitus (Diabetes Genetics Initiative of Broad Institute of Harvard and MIT, Lund University, and Novartis Institutes of Biomedical Research 2007, Wellcome Trust Case Control Consortium 2007), coronary artery disease (McPherson et al 2007, Wellcome Trust Case Control Consortium 2007, WTCCC and the Cardiogenics Consortium 2007), hypertension (Wellcome Trust Case Control Consortium 2007), asthma (Moffatt et al 2007), obesity (Frayling et al 2007), and many others. For some traits, such as hypertension, no significant associations have been discovered. For most, however, a number of significant associations have been found. Some of these have been replicated in very large cohorts, providing strong evidence that these loci contribute to disease susceptibility. For example, for type II diabetes mellitus, over 30000 cases and controls have been studied, identifying at least eight well-validated loci (Diabetes Genetics Initiative of Broad Institute of Harvard and MIT,

Lund University, and Novartis Institutes of Biomedical Research 2007).

It is interesting and instructive to examine the effects of these loci associated with type II diabetes risk. For most there are multiple alleles in linkage disequilibrium with *P* values that depart from the null distribution. While there is no doubt that a susceptibility allele lies within the block of LD, the identity of the functional SNP is for the most part unknown. This is made all the more challenging because most of these associated SNPs lie outside gene coding regions. As a consequence, it may be exceedingly difficult to determine which variant(s) is truly the functional allele conferring disease susceptibility. The susceptibility alleles are common, in some cases the more frequent SNP allele. The magnitude of effect of each risk allele is typically very small, with odds ratios of disease increased by 10–20% for all but one allele, which showed an ~40% increase. In aggregate, the eight genes identified thus far account for only ~2.5% of the interindividual risk of developing type II diabetes. Many of the variants identified lie within the transcription unit or adjacent to loci that are plausibly related to the regulation of insulin secretion, though for some the functional link to type II diabetes is presently unknown. Effects of different risk alleles appear to be additive. Similar patterns are observed for many other common diseases. These findings suggest that for many diseases a small fraction of disease risk will be accounted for by common variants. These may nonetheless provide important insight into disease biology and raise the question of whether rare mutations in these same genes might impart larger effects of greater importance to individual patients.

PERSPECTIVES ON THE FUTURE OF HUMAN DISEASE GENETICS

The progress in human genetics over the last 20 years has been staggering and revolutionary. As one contemplates the current state of the field and the prospects for the next decade, a number of themes emerge.

Discovery of New Mendelian Diseases and Genes

While the genes responsible for virtually all of the most common Mendelian traits have been identified, it continues to be apparent that more Mendelian disease genes remain to be discovered. Equally importantly, it is apparent that there are many more Mendelian traits that have yet to be described. These may represent traits for diseases of late onset such that the familial nature of disease may not be recognized, or represent genes with reduced penetrance, or traits that are beneficial, and which do not routinely come to clinical attention. The tools for identifying these genes are now superb. Mapping is routine, and recessive traits with incomplete penetrance can potentially be identified by mapping studies in affected subjects arising from consanguineous union. The ability to resequence all the genes or even the entire linked interval is at hand. While single mutant alleles can remain challenging to identify, this is happening with increasing frequency (Boyden et al 2002, Mani et al 2007). Finally, dominant lethal traits, for which one may rarely encounter even two affected members in the same family, have heretofore been a great challenge. New opportunities including all-exon or whole genome sequencing, as well as indirect methods for mutation detection, hold promise for identifying genes for such diseases.

Identification of Alleles Contributing to Common Disease

The genetic architecture of common disease is still being defined. It is clear that common alleles that contribute to common disease can be identified. It is equally clear that the insight deriving from these discoveries will be varied. In some cases, particular genes and pathways will be identified that will change the fundamental understanding of disease pathophysiology. For others, the new insight may be very modest. It is unclear to what extent common variants that impart very small effects will be useful as diagnostic tools or for assessing risk in individual patients. Nonetheless, as risk alleles are identified, the risk to individuals harboring different numbers of these alleles may stratify subjects into low- and high-risk groups. From the data emerging over the last year, it seems apparent that for many diseases there are few common variants that increase risk as much as 50%, and it seems likely that for many only a small fraction of inherited risk will prove to be attributable to common alleles.

These observations suggest that alternative models of complex genetic traits will frequently contribute to disease susceptibility. One alternative would be the presence of many individually rare alleles in disease genes. For example, a combination of individually rare alleles with relatively large effect size could contribute to, or be the major determinants of, common disease. In this case, each rare mutation would be on different haplotype backgrounds, and hence would be difficult or impossible to detect by linkage disequilibrium studies. This genetic architecture would be consistent with the expectation that disease-causing alleles are under negative selection.

It will be a significant challenge to identify these disease loci. One possibility is that some are the same loci that cause related Mendelian traits. For example, extreme phenotypes resulting from recessive traits are typically very rare; however, the heterozygous carrier state is much more common (allele frequency is the square root of the recessive disease frequency, so if a recessive disease has a frequency of 1 in 40 000, the overall disease allele frequency will be 1 in 200, and the heterozygous carrier state will be nearly 1% in the population). It is possible that the heterozygous state has significant, but much less extreme phenotypic effects. Similarly, genes implicated by common variants may also harbor rare mutations with larger effects on individual risk. Thus far, only a small number of studies have explored this possibility. Rare heterozygous mutations in *ABCA1*, the gene responsible for Tangier disease, a recessive trait featuring very low HDL levels, result in low HDL levels (Cohen et al 2004). Similarly, heterozygous variants in *NCCT*, *NKCC2* and *ROMK*, genes responsible for the recessive salt-wasting traits Gitelman and Bartter syndromes, reduce blood pressure and the incidence of hypertension (Ji et al 2008).

From the few deep resequencing studies done to date, a number of challenges are evident. First, the ability to identify functional mutations among the large set of rare and common variants in a gene is critical. Comparative genomic approaches show promise for identifying loss of function alleles. Another hurdle is the likely need for very large cohorts – even for well-validated candidates, thousands of subjects may be needed to detect a signal from rare alleles. Finally, improved technology for large-scale and cost-effective resequencing is required. In recent years, two new sequencing technologies have been commercialized that can reduce the cost of resequencing genomic DNA by one to two orders of magnitude (Bentley 2006), and several other technologies are in development that may further lower the cost of deep resequencing, approaching the goal of making resequencing of complete sets of genes and genomes first a research tool, and then a useful clinical application.

SUMMARY AND PERSPECTIVE

Application of genetic approaches to human disease has revolutionized the understanding of virtually every area of disease pathophysiology by establishing causal links between specific genes and disease. Success with these approaches has provided fundamental insight into disease biology, yielded new diagnostic tests for disease risk, and identified new targets for therapeutic intervention that are highly likely to be efficacious. The era of Mendelian genetics has been extraordinarily productive in identifying genes for a large number of human diseases. New approaches to defining the genetics of common complex genetic traits are emerging for identification of both common and rare disease alleles. These advances hold great promise for improving the diagnosis and treatment of diseases of the human species.

References

Aidoo M, Terlouw DJ, Kolczak MS, et al. Protective effects of the sickle cell gene against malaria morbidity and mortality. Lancet 2002; 359: 1311–12.

Avery OT, MacLeod CM, McCarty M. Studies on the chemical nature of the substance inducing transformation of pneumococcal types. J. Exp. Med. 1944; 79: 137–58.

Bailey JA, Eichler EE. Primate segmental duplications: crucibles of evolution, diversity and disease. Nat. Rev. Genet. 2006; 7: 552–64.

Balemans W, Van Hul W. Minireview: the genetics of low-density lipoprotein receptor-related protein 5 in bone: a story of extremes. Endocrinology 2007; 148: 2622–9.

Bejerano G, Pheasant M, Makunin I, et al. Ultraconserved elements in the human genome. Science 2004; 304: 1321–5.

Bell CG, Walley AJ, Froguel P. The genetics of human obesity. Nat. Rev. Genet. 2005; 6: 221–34.

Bender W, Akam M, Karch F, et al. Molecular genetics of the bithorax complex in *Drosophila melanogaster*. Science 1983; 221: 23–9.

Bentley DR. Whole-genome re-sequencing. Curr. Opin. Genet. Dev. 2006; 16: 545–52.

Bessman MJ, Lehman IR, Simms ES, Kornberg A. Enzymatic synthesis of deoxyribonucleic acid. II. General properties of the reaction. J. Biol. Chem. 1958; 233: 171–7.

Botstein D, White RL, Skolnick M, Davis RW. Construction of a genetic linkage map in man using restriction fragment length polymorphisms. Am. J. Hum. Genet. 1980; 32: 314–31.

Boyden LM, Mao J, Belsky J, et al. High bone density due to a mutation in LDL-receptor-related protein 5. N. Engl. J. Med. 2002; 346: 1513–21.

Brown MS, Goldstein JL. A receptor-mediated pathway for cholesterol homeostasis. Science 1986; 232: 34–47.

Charlesworth D. Balancing selection and its effects on sequences in nearby genome regions. PLoS Genet. 2006; 2: e64.

Chen K, Rajewsky N. The evolution of gene regulation by transcription factors and microRNAs. Nat. Rev. Genet. 2007; 8: 93–103.

Cohen JC, Kiss RS, Pertsemlidis A, Marcel YL, McPherson R, Hobbs HH. Multiple rare alleles contribute to low plasma levels of HDL cholesterol. Science 2004; 305: 869–72.

Cohen MM Jr. The new bone biology: pathologic, molecular, and clinical correlates. Am. J. Med. Genet. 2006; 140A: 2646–706.

Cooperative Human Linkage Center (CHLC). A comprehensive human linkage map with centimorgan density. Science 1994; 265: 2049–54.

Cooper DN, Krawczak M, Antonorakis SE. The nature and mechanisms of human gene mutation. In: Scriver C, Beaudet AL, Sly WS, Valle D, eds. The Metabolic and Molecular Bases of Inherited Disease. New York: McGraw-Hill, 1995: pp. 259–91.

Crick F. Central dogma of molecular biology. Nature 1970; 227: 561–3.

Crick FHC. On protein sythesis. Symp. Soc. Exp. Biol. 1958; 12: 139–63.

Crick FHC, Barnett L, Brenner S, Watts-Tobin RJ. General nature of the genetic code for proteins. Nature 1961; 192: 1227–32.

Cutting GR. Modifier genetics: cystic fibrosis. Annu. Rev. Genomics Hum. Genet. 2005; 6: 237–60.

de Bakker PI, Yelensky R, Pe'er I, Gabriel SB, Daly MJ, Altshuler D. Efficiency and power in genetic association studies. Nat. Genet. 2005; 37: 1217–23.

DeWan A, Bracken MB, Hoh J. Two genetic pathways for age-related macular degeneration. Curr. Opin. Genet. Dev. 2007; 17: 228–33.

Diabetes Genetics Initiative of Broad Institute of Harvard and MIT, Lund University, and Novartis Institutes of BioMedical Research. Genome-wide association analysis identifies loci for type 2 diabetes and triglyceride levels. Science 2007; 316: 1331–6.

Donnis-Keller D, Green P, Helms C, et al. A genetic linkage map of the human genome. Cell 1987; 51: 319–37.

Duerr RH, Taylor KD, Brant SR, et al. A genome-wide association study identifies IL23R as an inflammatory bowel disease gene. Science 2006; 314: 1461–3.

Eddy S. Non-coding RNA genes and the modern RNA world. Nat. Rev. Genet. 2001; 2: 919–29.

The ENCODE Project Consortium. Identification and analysis of functional elements in 1% of the human genome by the ENCODE pilot project. Nature 2007; 447: 799–816.

The European Polycystic Kidney Disease Consortium. The polycystic kidney disease 1 gene encodes a 14 kb transcript and lies within a duplicated region on chromosome 16. Cell 1994; 77: 881–94.

Feder JN, Gnirke A, Thomas W, et al. A novel MHC class I-like gene is mutated in patients with hereditary haemochromatosis. Nat. Genet. 1996; 13: 399–408.

Fleischer B. Ober myotonische dystrophie mit katarakt: eine heriditare, familiare degeneration. Archiv fur Ophthalmologie 1918; 96: 91–133.

Florez JC, Hirschhorn J, Altshuler D. The inherited basis of diabetes mellitus: implications for the genetic analysis of complex traits. Annu. Rev. Genomics Hum. Genet. 2003; 4: 257–91.

Frayling TM, Timpson NJ, Weedon MN, et al. A common variant in the FTO gene is associated with body mass index and predisposes to childhood and adult obesity. Science 2007; 316: 889–94.

Furth JJ, Hurwitz J, Anders M. The role of deoxyribonucleic acid in ribonucleic acid synthesis. I. The purification and properties of ribonucleic acid polymerase. J. Biol. Chem. 1962; 237: 2611–19.

Fu Y-H, Kuhl DPA, Pizzuti A, et al. Variation of the CGG repeat at the fragile X site results in genetic instability: resolution of the Sherman paradox. Cell 1991; 67: 1047–58.

Gabow PA. Autosomal dominant polycystic kidney disease. N. Engl. J. Med. 1993; 329: 332–42.

Gabriel SB, Schaffner SF, Nguyen H, et al. The structure of haplotype blocks in the human genome. Science 2002; 296: 2225–9.

Garrod AE. The incidence of alkaptonuria: a study in chemical individuality. Lancet 1902; 2: 1616–30.

Garrod AE. Inborn Errors of Metabolism. London: Henry Frowde, Hodder and Stoughton, 1909.

Gyapay G, Morisette J, Vignal A, et al. The 1993–94 Genethon human genetic linkage map. Nat. Genet. 1994; 7: 246–339.

Hall JM, Lee MK, Newman B, et al. Linkage of early-onset familial breast cancer to chromosome 17q21. Science 1990; 250: 1684–9.

Hardy J, Orr H. The genetics of neurodegenerative diseases. J. Neurochem. 2006; 97: 1690–9.

Herz J, Beffert U. Apolipoprotein E receptors: linking brain development and Alzheimer's disease. Nat. Rev. Neurosci. 2000; 1: 51–8.

Hinds DA, Stuve LL, Nilsen GB, et al. Whole-genome patterns of common DNA variation in three human populations. Science 2005; 307: 1072–9.

Hirschhorn JN, Daly MJ. Genome-wide association studies for common diseases and complex traits. Nat. Rev. Genet. 2005; 6: 95–108.

Hobbs HH, Brown MS, Goldstein JL. Molecular genetics of the LDL receptor gene in familial hypercholesterolemia. Hum. Mutat. 1992; 1: 445–66.

Inoue K, Lupski JR. Molecular mechanisms for genomic disorders. Annu. Rev. Genomics Hum. Genet. 2002; 3: 199–242.

The International HapMap Consortium. The International HapMap Project. Nature 2003; 426: 789–96.

The International HapMap Consortium. A haplotype map of the human genome. Nature 2005; 437: 1299–320.

The International HapMap Consortium. A second generation human haplotype map of over 3.1 million SNPs. Nature 2007; 449: 851–61.

International Human Genome Sequencing Consortium. Initial sequencing and analysis of the human genome. Nature 2001; 409: 860–921.

International Human Genome Sequencing Consortium. Finishing the euchromatic sequence of the human genome. Nature 2004; 431: 931–45.

Ji W, Foo JN, O'Roak B, et al. Rare independent mutations in renal salt handling genes contribute to blood pressure variation. Nat. Genet. 2008; in press.

Kahle KT, Ring AM, Lifton RP. Molecular physiology of the WNK kinases. Annu. Rev. Physiol. 2008; 70: 329–55.

Keating MT, Sanguinetti MC. Molecular and cellular mechanisms of cardiac arrhythmias. Cell 2001; 104: 569–80.

Klein RJ, Zeiss C, Chew EY, et al. Complement factor H polymorphism in age-related macular degeneration. Science 2005; 308: 385–9.

Knudson AG Jr. Mutation and cancer: statistical study of retinoblastoma. Proc. Natl Acad. Sci. USA, 1971; 68: 820–3.

Lander E, Botstein D. Homozygosity mapping: a way to map human recessive traits with the DNA of inbred children. Science 1987; 236: 1567–70.

Lehman IR, Bessman MJ, Simms ES, Kornberg A. Enzymatic synthesis of deoxyribonucleic acid. I. Preparation of substrates and partial purification of an enzyme from *Escherichia coli*. J. Biol. Chem. 1958; 233: 163–70.

Lehrman MA, Schneider WJ, Südhof TC, Brown MS, Goldstein JL, Russell DW. Mutation in LDL receptor: Alu-Alu recombination deletes exons encoding transmembrane and cytoplasmic domains. Science 1985; 227: 140–6.

Lifton RP, Gharavi AG, Geller DS. Molecular mechanisms of human hypertension. Cell 2001; 104: 545–56.

Lohmueller KE, Pearce CL, Pike M, Lander ES, Hirschhorn JN. Meta-analysis of genetic association studies supports a contribution of common variants to susceptibility to common disease. Nat. Genet. 2003; 33: 177–82.

MacDonald ME, Ambrose CM, Duyao MP, et al. A novel gene containing a trinucleotide repeat that is expanded and unstable on Huntington's disease chromosomes. Cell 1993; 72: 971–83.

Mani A, Radhakrishnan J, Wang H, et al. LRP6 mutation in a family with early coronary disease and metabolic risk factors. Science 2007; 315: 1278–82.

McInnis MG. Anticipation: an old idea in new genes. Am. J. Hum. Genet. 1996; 59: 973–9.

McPherson R, Pertsemlidis A, Kavaslar N, et al. A common allele on chromosome 9 associated with coronary heart disease. Science 2007; 316: 1488–91.

Mendel G. Versuche über pflanzen-hybriden. Verhandlungen des Naturforschenden Vereines. Brunn 1866; 4: 3–47.

Moffatt MF, Kabesch M, Liang L, et al. Genetic variants regulating ORMDL3 expression contribute to the risk of childhood asthma. Nature 2007; 448: 470–3.

Morgan TH. Sex limited inheritance in *Drosophila*. Science 1910; 32: 120–2.

Morgan TH. The origin of five mutations in eye color in *Drosophila* and their modes of inheritance. Science 1911; 33: 534–7.

Morgan TM, Krumholz HM, Lifton RP, Spertus JA. Nonvalidation of reported genetic risk factors for acute coronary syndrome in a large-scale replication study. JAMA 2007; 297: 1551–61.

Morrow JF, Cohen SN, Chang ACY, Boyer HW, Goodman HM, Helling RB. Replication and transcription of eukaryotic DNA in *Escherichia coli*. Proc. Natl Acad. Sci. USA 1974; 71: 1743–7.

Morton NE. Sequential tests for the detection of linkage. Am. J. Hum. Genet. 1955; 7: 277–318.

Mouse Genome Sequencing Consortium. Initial sequencing and comparative analysis of the mouse genome. Nature 2002; 420: 520–62.

Nachman MW, Crowell SL. Estimate of the mutation rate per nucleotide in humans. Genetics 2000; 156: 297–304.

Narod SA, Foulkes WD. BRCA1 and BRCA2: 1994 and beyond. Nat. Rev. Cancer 2004; 4: 665–76.

NCBI (a). Probe database of reagents for functional genomics: sequence specifioligonucleotide (SSO) probes. http://www.ncbi.nlm.nih.gov/projects/genome/probe/doc/TechSSO.shtml

NCBI (b). Probe database of reagents for functional genomics: Illumina http://www.ncbi.nlm.nih.gov/projects/genome/probe/doc/DistrIllumina.shtml

NHGRI. Approved sequencing targets, March 2008. http://www.genome.gov/10002154

NIH/CEPH Collaborative Mapping Group. A comprehensive genetic linkage map of the human genome. Science 1992; 258: 67–86.

Nirenberg M, Leder P. RNA codewords and protein synthesis: the effect of trinucleotides upon the binding of sRNA to ribosomes. Science 1964; 145: 1399–407.

Nirenberg MW, Matthaei JH. The dependence of cell-free protein synthesis in *E. coli* upon naturally occurring or synthetic polyribonucleotides. Proc. Natl Acad. Sci. USA 1961; 47: 1588–602.

Ogura Y, Bonen DK, Inohara N, et al. A frameshift mutation in NOD2 associated with susceptibility to Crohn's disease. Nature 2001; 411: 603–6.

Pauling L, Itano HA, Singer SJ, Wells IC. Sickle cell anemia, a molecular disease. Science 1949; 110: 543–8.

Pritchard JK, Donnelly P. Case-control studies of association in structured or admixed populations. Theor. Popul. Biol. 2001; 60: 227–37.

Qian F, Watnick TJ, Onuchic LF, Germino GG. The molecular basis of focal cyst formation in human autosomal dominant polycystic kidney disease type I. Cell 1996; 87: 979–87.

Reik W, Walter J. Genomic imprinting: parental influence on the genome. Nat. Rev. Genet. 2001; 2: 21–32.

Risch N, Merikangas K. The future of genetic studies of complex human diseases. Science 1996; 273: 1516–17.

Rommens JM, Iannuzzi MC, Kerem B, et al. Identification of the cystic fibrosis gene: chromosome walking and jumping. Science 1989; 245: 1059–65.

Rowley JD. Chromosome translocations: dangerous liaisons revisited. Nat. Rev. Cancer, 2001; 1: 245–50.

Rustgi AK. The genetics of hereditary colon cancer. Genes Dev. 2007; 21: 2525–38.

Seidman JG, Seidman C. The genetic basis for cardiomyopathy: from mutation identification to mechanistic paradigms. Cell 2001; 104: 557–67.

Seligsohn U, Lubetsky A. Genetic susceptibility to venous thrombosis. N. Engl. J. Med. 2001; 344: 1222–31.

Siekevitz P, Zamecnik PC. Ribosomes and protein synthesis. J. Cell Biol. 1981; 91: 53s–65s.

Stephens K. Genetics of neurofibromatosis 1-associated peripheral nerve sheath tumors. Cancer Invest. 2003; 21: 897–914.

Sturtevant AH. The linear arrangement of six sex-linked factors in *Drosophila*, as shown by their mode of association. J. Exp. Zool. 1913; 14: 43–59.

Venter JC, Adams MD, Myers EW, et al. The sequence of the human genome. Science 2001; 291: 1304–51.

Wallace DC. A mitochondrial paradigm of metabolic and degenerative diseases, aging, and cancer: a dawn for evolutionary medicine. Annu. Rev. Genet. 2005; 39: 359–407.

Watson JD, Crick FHC. Molecular structure of nucleic acids: a structure for deoxyribose nucleic acid. Nature 1953; 171: 737–8.

Weber JL, May PE. Abundant class of human DNA polymorphisms which can be typed using the polymerase chain reaction. Am. J. Hum. Genet. 1989; 44: 388–96.

Weissenbach J, Gyapay G, Dib C, et al. A second-generation linkage map of the human genome. Nature 1992; 359: 794–801.

Wellcome Trust Case Control Consortium. Genome-wide association study of 14,000 cases of seven common diseases and 3,000 shared controls. Nature 2007; 447: 661–78.

Wensink PC, Finnegan DJ, Donelson JE, Hogness DS. A system for mapping DNA sequences in the chromosomes of *Drosophila melanogaster*. Cell 1974; 3: 315–25.

White R, Leppert M, Bishop DT, et al. Construction of linkage maps with DNA markers for human chromosomes. Nature 1985; 313: 101–5.

WTCCC and the Cardiogenics Consortium. Genomewide association analysis of coronary artery disease. N. Engl. J. Med. 2007; 357: 443–53.

Wu G, D'Agati V, Cai Y, et al. Somatic inactivation of Pkd2 results in polycystic kidney disease. Cell 1998; 93: 177–88.

Wyman AR, White R. A highly polymorphic locus in human DNA. Proc. Natl Acad. Sci. USA 1980; 77: 6754–8.

CHAPTER 2

Clinical Applications of Genetics

LISA M. GUAY-WOODFORD AND NINE V.A.M. KNOERS

BASIC PRINCIPLES OF GENETIC TESTING

Genetic testing involves the analysis of chromosomes, deoxyribonucleic acid (DNA), ribonucleic acid (RNA), or specific metabolites in order to detect variants that are associated with human disease or a pharmacological response. Direct testing refers to gene-specific evaluation of a DNA and/or RNA sample. In comparison, linkage-based testing is an indirect method that examines the segregation of genetic markers located near or within the disease-causing gene with disease phenotypes within a family. Biochemical genetics involves assays for metabolites to identify defects in specific metabolic pathways. Cytogenetic testing examines chromosomes for large-scale disruptions, such as numerical abnormalities, insertions, deletions, duplications, and translocations.

In the broadest sense, genetic tests may be characterized as: (1) clinical tests, in which results are reported to the provider to aid with diagnosis, pre-symptomatic detection, or treatment of an individual patient; or (2) research tests, which are conducted to facilitate better understanding of disease pathogenesis or to develop and validate a specific test for clinical application.

There are four principal applications for clinical genetic testing:

- diagnosis;
- pre-symptomatic evaluation;
- determining disease predisposition (susceptibility); and
- pharmacogenetic analysis (Robin et al 2007).

Diagnostic testing is applied in individuals who manifest signs and/or symptoms of disease. In this context, genetic testing offers the advantages of precise diagnosis without the need for invasive tissue sampling or complicated physiological analysis. This diagnostic precision can in turn provide the basis for specific therapeutic interventions, anticipatory guidance, and genetic counseling (Korf 2001). *Pre-symptomatic or predictive testing* is used to identify disease-causing mutations in at-risk family members before symptoms appear. The primary utility of such testing is to ascertain disease risk, allowing disease-associated morbidity to be minimized through early therapeutic intervention. However, disease-causing mutations identified in this context cannot always predict whether the test individual will actually develop symptomatic disease and cannot portend specific disease expression. Moreover, therapeutic intervention is not always possible. In these cases, pre-symptomatic testing may be used to end uncertainty about gene-carrier status in the individual at risk. *Predispositional or susceptibility testing* pertains to polygenic disorders, e.g. diabetic nephropathy, in which the interaction of multiple genes with one another and with the environment contributes to disease expression. With methodological developments and technologies spurred by the International HapMap Project (The International HapMap Consortium 2003, Altshuler et al 2005), genome-wide association studies (GWAS) are beginning to identify genetic determinants that can be used to predict disease susceptibility in these disorders. *Pharmacogenetic testing* assesses genetic variants that are responsible for the variable response to drugs and other xenobiotics. In the last decade or so, pharmacogenetics has evolved to become a major driving force in clinical pharmacology (Brockmoller & Tzvetkov 2008), and will have increasing impact on therapeutic decision-making in clinical nephrology, improving drug efficacy and minimizing side effects.

Genetic Diseases of the Kidney
ISBN: 978-0-12-449851-8

Extensive information about genetic testing and testing laboratories is available through GeneTests (http://www.genetests.org) (Pagon 2006), a publicly funded medical genetics information resource developed for physicians, other healthcare providers, and researchers. The Laboratory Directory of the GeneTests site provides a listing of US and international laboratories offering molecular genetic testing, specialized cytogenetic testing, and biochemical testing for inherited disorders. The genetic tests offered by these laboratories may be used for medical management of a specific patient, reproductive decision-making, and assessment of future health risks.

While the armamentarium of genetic tests continues to rapidly increase, availability alone is not sufficient reason to include genetic testing in routine medical practice (Robin et al 2007). In particular, clinicians must evaluate the *clinical validity* of a specific test; that is the accuracy with which the test assesses disease risk. For example, in single-gene disorders, genetic tests may reveal sequence variants of unknown significance (potential false-positive result), and often do not identify the full spectrum of naturally occurring mutations (false-negative result).

In addition, clinicians must consider the *clinical utility* of a specific test; that is the power of a genetic test to guide medical management that prevents or minimizes disease-related morbidity and mortality. Potential applications include screening, diagnostic, and carrier testing for single-gene disorders, testing of multiple loci to develop profiles of disease risk, and pharmacogenomic testing to predict drug–genome interactions (Grosse & Khoury 2006). It is important to recognize that a screening or diagnostic test alone does not have inherent utility. Rather test-related utlity results from the implementation of preventative or therapeutic interventions that reduce the burden of disease or treatment. Therefore, the clinical utility of a test is inextricably linked to access to such interventions. This point is particularly important in the context of incorporating genetic testing into clinical nephrology practice, as the link between identifying disease risk and implementing targeted therapeutic interventions is currently limited for most single-gene disorders. In addition, the application of genetic tests to clinical management must take into account the ethical, legal, and social implications of such testing (McDonald & Avner 1996, EC Expert Group 2004, Lissemore 2005, Sanderson et al 2005).

SERVICES OFFERED BY GENETIC LABORATORIES

When considering a genetic testing laboratory, it is important for healthcare providers to distinguish those services offered by *clinical laboratories* versus *research laboratories*.

In a *clinical laboratory*, patient specimens are examined and the results are reported in writing to the healthcare professional to enable clinical decision-making and counseling for an individual patient. In the United States, clinical laboratories are licensed under the Clinical Laboratory Improvement Amendments (CLIA). All genetic tests that are intended to be used in clinical decision-making must be performed in a licensed clinical laboratory (Pagon 2006). Non-US laboratories, used when testing is not available in the USA, are not subject to CLIA regulation, but are accredited or certified by local authoritative bodies. The charge structure and the turnaround time for clinical genetic testing vary, depending, at least in part, on the complexity of the methodologies used.

In comparison, a *research laboratory* collects and examines patient specimens for the purpose of better understanding disease pathogenesis or to develop a clinical genetic test. Only a few research laboratories are CLIA-licensed. Test results from unlicensed laboratories should not be provided to patients or their providers. However, at the patient's request, a research laboratory may share potentially useful genetic data with a clinical laboratory so that the test results can be validated and a formal report issued to the patient and/or provider (Korf 2001). Research laboratories are not required to accept all requests for testing. Requests may be denied if sufficient samples have been obtained previously from other patients or if the clinical disease in the patient/family does not correspond to the focus of the research study. The costs associated with research testing are typically covered by the investigator.

COUNSELING AND INFORMED CONSENT

Before proceeding with a genetic test, clinicians need to be sure that the patient (or parent/guardian) understands the value and limitations of the test, its implications for medical management, the procedure(s) for sample collection and any associated risks, and the costs of the test. In addition, this pre-test counseling should include a discussion of alternative diagnostic methodologies, the relative accuracy of genetic testing versus the other available testing methods, a plan to discuss test results, and a provision for follow-up counseling. If a competent patient (or parent/guardian) agrees to the proposed genetic test after a full discussion of these issues, this constitutes informed consent. Many clinical laboratories provide documentation for this informed consent process and require patient/parent/guardian signatures.

Genetic testing results are returned to the requesting provider and the results should be discussed only with the individual tested, or his/her parent or guardian, unless explicit permission has been granted to share the results.

ETHICAL CONSIDERATIONS IN GENETIC TESTING

Genetic testing offers many potential benefits to individuals, families, and the greater society. However, it is important to recognize that genetic testing also raises significant ethical, legal, and social concerns, particularly in the context of pre-symptomatic screening (Hodges 2004).

Take for example the case of single-gene disorders. When an individual is found to carry a clinically important allele, related family members are at increased risk of carrying the same allele and of sharing the consequent disease risk. Therefore, if there are interventions that could modify the risk of disease progression, it could be asserted that the individual as well as their family should be informed. However, this assertion raises important concerns regarding privacy and confidentiality as well as patient autonomy (the right not to share the information with family members) (Kakuk 2008). Indeed, some individuals oppose sharing their genetic information even with family members, fearing discrimination and stigmatization (Minkoff & Ecker 2008). Moreover, if there is a gap between genetic information and possible therapeutic interventions, knowledge that one is carrying a clinically important allele may create a psychological burden.

Moreover, there are special issues to take into account when considering pre-symptomatic screening in children. Children are not able to make informed decisions on the risks and benefits of this information. Therefore, parents make the request on the child's behalf. However, this approach raises an important ethical principle. Do children have autonomy with regard to their genetic information? Recently, Borry et al. (2008) reported the responses of European clinical geneticists to a survey regarding presymptomatic/predictive genetic testing in minors. Testing that provides a clear medical benefit for childhood-onset disorders registered the strongest support. However, there was not clear consensus about when to recommend such testing, with some supporting the rule of earliest onset and others allowing broader parental discretion. For those childhood-onset disorders not associated with targeted therapeutic interventions, there was limited enthusiasm for presymptomatic/predictive genetic testing. With regard to adolescents and the risk of adult-onset disease, the majority favored presymptomatic and predictive genetic testing only when the adolescent and parents jointly made the request.

Clinicians who seek to integrate genetic information into their clinical practices must keep abreast of advances in genomics as well as the ethical, legal, and social issues associated with the expansion of genetic testing. In recognition that latter issues would burgeon in parallel with the rapid evolution of Human Genome Project (HGP), the National Human Genome Research Institute (NHGRI) devoted 3–5% of the annual HGP budget toward studying the ethical, legal, and social issues (ELSI) surrounding availability of genetic information and established the ELSI Research Program (http://www.genome.gov/10001618) as an integral part of the HGP (Collins et al 2003). The ELSI Research Program focuses on issues related to privacy, integration of genetic services into clinical health care, and educational preparation of the healthcare workforce.

CURRENT APPLICATIONS OF GENETIC TESTING

More than 2000 human genes involved in monogenic or Mendelian diseases have been identified to date (GeneTests). These disorders include those that disrupt the structure, function, and/or developmental patterning of the glomerulus, tubules, and urogenital tract, as well as those that predispose to renal cell tumors. Genetic testing increasingly holds the potential to inform clinical practice and impact the outcomes of patients with these disorders. Tatsioni et al. (2005) have proposed a framework for evaluating the potential benefit and the role of genetic testing in clinical practice, which includes four considerations: (1) diagnostic thinking, (2) therapeutic choice, (3) patient outcome, and (4) societal impact. Diagnostic thinking refers to the value of genetic test information in understanding disease-related diagnosis, cause, and prognosis. Therapeutic choice refers to the use of test results in the clinical management of a specific individual. Patient outcome refers to endpoints such as mortality or quality of life, and societal impact includes considerations of cost-effectiveness.

Within this framework, diagnostic thinking and therapeutic choice are the most relevant to current clinical nephrology practice. Therefore, the following discussion will highlight specific single-gene disorders and consider the clinical utility of genetic testing in the context of diagnostic evaluation and management guidance. An expanded list single-gene disorders and relevant genetic testing information is provided in Table 2.1.

Diagnostic Testing

PRENATAL DIAGNOSIS

Prenatal testing is performed during a pregnancy to assess the genetic status of a fetus at risk for a specific heritable condition, especially those associated with a high risk of perinatal mortality, e.g. autosomal recessive polycystic kidney disease (ARPKD), or morbidity, e.g. nephrogenic diabetes insipidus. Routine prenatal sampling procedures currently focus on chorionic villus sampling (CVS) and amniocentesis (Ball 2004). However, even in experienced centers, these prenatal testing procedures have an associated

TABLE 2.1 Single-gene disorders of the kidney

Disorder	Inheritance	MIM	Genes	Clinical genetic testing	Other testing
Glomerular disorders					
Congenital nephrotic syndrome	AR	256300	*NPHS1*	Gene-based dx; prenatal dx; PGD	
WT1-related disorders	AR	256370	*WT1*	Gene-based dx; prenatal dx;	FISH analysis
SRNS					
SRN1	AR	600995	*NPHS2*	Gene-based dx; prenatal dx	
FSGS1	AD	603278	*ACTN4*	Gene-based dx; prenatal dx	
FSGS2	AD	603965	*TRPC6*	Gene-based dx; prenatal dx	
FSGS3	AD	607832	*CD2AP*	Gene-based dx; prenatal dx	
Alport syndrome					
X-linked		301050	*COL4A5*	Linkage; gene-based dx; prenatal dx; PGD	Immunohistochemistry
AR		203780	*COL4A3; COL4A4*	Linkage; gene-based dx; prenatal dx; PGD	
AD		104200	*COL4A3; COL4A4*	Linkage; gene-based dx; prenatal dx; PGD	
Renal cystic diseases					
ADPKD	AD	173900			
PKD1			*PKD1*	Linkage; gene-based dx; prenatal dx	
PKD2			*PKD2*	Linkage; gene-based dx; prenatal dx	
ARPKD	AR	263200	*PKHD1*	Linkage; gene-based dx; prenatal dx; PGD	
Nephronophthisis (NPHP)	AR				
NPHP1		256100	*NPHP1*	Linkage; gene-based dx; prenatal dx	
NPHP2		602088	*NPHP2/INVS*	Linkage; gene-based dx; prenatal dx	
NPHP3		604387	*NPHP3*	Linkage; gene-based dx; prenatal dx	
NPHP4		606966	*NPHP4*	Linkage; gene-based dx; prenatal dx	
NPHP5-7			*NPHP5; NPHP6/CEP290; NPHP7/GLIS2; NPHP8/ RPGRIP1L; NPHP9/NEK8*	Research testing only	
MCKD	AD	603860	*UMOD (MCKD2)*	Linkage; gene-based dx; prenatal dx	Immunohistochemistry
Proximal tubular disorders					
Cystinosis	AR	219800	*CTNS*	Linkage; gene-based dx; prenatal dx; PGD	Analyte
Lowe syndrome	X-linked	309000	*OCRL*	Linkage; gene-based dx; prenatal dx; PGD	Enzyme assay
Dent disease	X-linked	300009	*CLCN5; OCRL*	Linkage; gene-based dx; prenatal dx	Enzyme assay
Cystinuria	AR	220100	*SLC3A1; SLC7A9*	Linkage; gene-based dx; prenatal dx	Analyte

Distal nephron disorders					
Bartter syndrome	AR				
Type I		601678	*SLC12A1*	Gene-based dx; prenatal dx	
Type II		241200	*KCNJ1*	Gene-based dx; prenatal dx	
Type III		607364	*CLCNKB*	Gene-based dx; prenatal dx	
Type IV		602522	*BSND*	Gene-based dx; prenatal dx	
Gitelman syndrome	AR	263800	*SLC12A3*	Gene-based dx; prenatal dx	
Low renin hypertension					
Liddle syndrome	AD	177200	*SCNN1B; SCNN1G*	Gene-based dx; prenatal dx	Urinary steroid profile
GRA	AD	103900	*CYP11B2; CYP11B1*	Gene-based dx; prenatal dx	Urinary steroid profile
AME	AR	218030	*HSD11B2*	Gene-based dx; prenatal dx	
Distal RTA					
AD RTA		179800	*SLC4A1*	Gene-based dx; prenatal dx	
AR RTA		602722	*ATP6V0A4; SLC4A1*	Gene-based dx; prenatal dx	
RTA with progressive deafness	AR	267300	*ATP6V1B1; ATP6V0A4*	Gene-based dx; prenatal dx	
Osteopetrosis with RTA	AR	259730	*CA2*	Gene-based dx; prenatal dx	
Nephrogenic diabetes insipidus					
X-linked		304800	*AVPR2*	Linkage; gene-based dx; prenatal dx	
AR, AD		125800	*AQP2*	Linkage; gene-based dx; prenatal dx	
Metabolic disorders					
Fabry disease	X-linked	301500	*GLA*	Gene-based dx	Analyte; enzyme assay
Primary oxaluria					
Type I	AR	259900	*AGXT*	Linkage; gene-based dx; prenatal dx	Analyte; enzyme assay
Type II	AR	260000	*GRHPR*	Linkage; gene-based dx; prenatal dx	Analyte; enzyme assay
Renal tumor predisposition disorders					
Tuberous sclerosis complex	AD	191100			
		605284	*TSC1*	Linkage; gene-based dx; prenatal dx; PGD	Deletion analysis
		191092	*TSC2*	Linkage; gene-based dx; prenatal dx; PGD	Deletion analysis
von Hippel Lindau syndrome	AD	193300	*VHL*	Gene-based dx; prenatal dx; PGD	Deletion analysis
Birt-Hogg-Dube syndrome	AD	135150	*FLCN*	Linkage; gene-based dx; prenatal dx	

MIM: Mendelian inheritance in Man (http://www.ncbi.nlm.nih.gov/sites/entrez?db=omim); AD: autosomal dominant; AR: autosomal recessive; PGD: Pre-implantation genetic diagnosis; SRNS: steroid resistant nephritic syndrome; FSGS: focal and segmental glomerulosclerosis; ADPKD: autosomal dominant polycystic kidney disease; ARPKD: autosomal recessive polycystic kidney disease; MCKD: medullary cystic kidney disease; GRA: glucocorticoid-remediable aldosteronism; AME: apparent mineralocorticoid excess risk of causing miscarriage of about 0.5–1% (Mujezinovic & Alfirevic 2007). Therefore, in recent years, there has been considerable effort toward developing non-invasive prenatal diagnostic sampling methods, including protocols to extract cell-free fetal DNA from maternal plasma or serum and isolating fetal cells from the maternal circulation (Sekizawa et al 2007).

Prenatal testing for adult-onset conditions is controversial (Steinbock 2007). Thus, the American College of Medical Genetics recommends that couples seeking prenatal diagnosis for adult-onset conditions should be referred for formal genetic counseling and complete discussion of the associated issues (Wilfond & ASHG Social Issues Committee 1995).

Preimplantation genetic diagnosis (PGD) is performed on early embryos resulting from in vitro fertilization before implantation in order to decrease the chance of a specific genetic disorder occurring in the fetus (Fragouli 2007). PGD has the advantage that, by selecting only those embryos identified as unaffected for implantation into the uterus, couples know from the start that any pregnancy should be unaffected. As such, PGD provides an alternative to prenatal diagnosis and the termination of affected pregnancies. The disadvantages of PGD are related to the in vitro fertilization procedure, with a low pregnancy rate (~35%), and the increased risk of twin or triplet pregnancies. Gigarel et al. (2008) have recently reported the successful development of a standardized single-cell diagnostic procedure that enabled PGD using linkage-based analysis to be offered to couples at risk of transmitting ARPKD.

Prior to pursuing either prenatal diagnosis or PGD, clinical laboratories require that the specific genetic mutation is first identified in an affected relative or a carrier parent or the disease-associated haplotype is established within the family. However, both prenatal diagnosis and PGD can be confounded by technical difficulties. Specifically, standard PCR-based strategies not infrequently fail to provide adequate DNA templates from the limited cells harvested for PGD and non-invasive prenatal diagnosis. In recent years, progress with whole-genome amplification (WGA) techniques has been successfully applied to PGD and prenatal diagnosis (Peng et al 2007). These WGA protocols allow ample amplification of genomic sequences from single cells for subsequent PCR analyses and/or comparative genomic hybridization (CGH), thus opening up a new area for prenatal diagnosis.

GUIDING CLINICAL MANAGEMENT
Autosomal dominant polycystic kidney disease (ADPKD)
The diagnosis of ADPKD is usually established by renal imaging studies (e.g., ultrasonography, computed tomography, or magnetic resonance imaging) (Ravine et al 1994, Nascimento et al 2001). However, there is a role for molecular diagnosis, especially in patients with equivocal imaging results, when a definite diagnosis is required in younger (<30 years old) at-risk individuals undergoing evaluation as potential living-related kidney donors, and in patients suspected of carrying de novo mutations who want to understand the transmission risks in future pregnancies. In addition, current clinical trials are evaluating the efficacy of several targeted interventions (details available at http://www.clinicaltrials.gov). If these trials are successful, gene-based diagnosis may become a critical factor in clinical decision-making.

Two large studies have recently examined the sensitivity of gene-based diagnosis in ADPKD (Garcia-Gonzalez et al 2007, Rossetti & Harris 2007). Protein-truncating mutations were ascertained in 41–63% of the patients, thus establishing the diagnosis in this subset. In an additional 26–37% of the cohort, bioinformatic analyses suggested that the sequence variants were likely to be pathogenic. Finally, no pathogenic mutations were identified in 11–22% of the cohorts. Therefore, one set of authors (Garcia-Gonzalez 2007) concluded that using current methodologies, gene-based testing should be employed with caution in clinical decision-making.

Steroid resistant nephrotic syndrome (SRNS)
SRNS results from a diverse set of pathogenic mechanisms and affected individuals can be categorized into at least three distinct subsets: (1) patients with a single-gene disorder involving recessively transmitted mutations, e.g. *NPHS2*, *LAMB2* (Karle et al 2002, Caridi et al 2004, Weber et al 2004), de novo mutations, e.g. *WT1* (Ruf et al 2004a), or dominantly transmitted mutations, e.g. *ACTN4*, *TRPC6* (Yao et al 2004, Winn et al 2005); (2) patients with a multifactorial defect that includes heterozygous mutations, e.g. the *NPHS2* R229Q substitution, or other polymorphisms in podocyte-related genes (Tsukaguchi et al 2002, Aucella et al 2005, Orloff et al 2005); and (3) patients with a T-cell disorder that causes production of circulating permeability factor(s) which alters the filtration barrier (reviewed in Eddy & Symons 2003).

There are significant therapeutic implications for this pathogenetic heterogeneity. Studies performed in European and North African cohorts demonstrate that children with two pathogenic *NPHS2* mutations have aggressive disease that essentially does not respond to second-line or third-line immunosuppressive therapies and yet, the risk of disease recurrence in renal allografts is very low (Bertelli et al 2003, Cattran 2003, Ruf et al 2004b). In comparison, when all SRNS patients are considered, 20–30% respond to second-line immunosuppressive therapy (Cattran 2003, Ponticelli & Passerini 2003) and the disease recurs in ~25% of renal allografts (Schnaper 2003). While further study is required, several groups have begun to recommend that patients with steroid-resistant nephrotic syndrome, particularly those diagnosed at less than 6 years of age, should undergo molecular analysis at diagnosis (Niaudet 2004, Ruf et al 2004, Weber et al 2004, Colquitt et al 2007, Hinkes et al 2008) in order to guide clinical decision-making regarding further immunosuppressive therapy and the timing of renal transplantation.

In contrast, in adult-onset SRNS, dominantly transmitted *ACTN4* and *TRPC6* mutations account for ~5% of cases (Daskalakis & Winn 2006) and *NPHS2* mutations have been detected in ~7% sporadic cases (He et al 2007). Therefore, routine use of mutational screening for clinical management is not advised (Deegens et al 2008).

X-linked Disorders

For disorders that are transmitted as X-linked recessive traits, carrier testing can be performed to identify females who carry the genetic mutation but who are often themselves asymptomatic. Such carrier testing can provide critical information for reproductive decision-making and neonatal management, as in the case of nephrogenic diabetes insipidus, or prompt diagnosis in female carriers who often have unrecognized disease, as in the case of Fabry disease.

Nephrogenic diabetes insipidus (NDI) is a hereditary disorder, which is characterized by insensitivity of the renal collecting duct cells to the antidiuretic effect of arginine vasopressin (AVP). As a result, there is a severe renal concentrating defect, which may lead to dehydration and electrolyte imbalance. In 90% of cases, NDI is an X-linked disorder due to mutations in *AVPR2*, the gene encoding the vasopressin type-2 receptor. In the remainder, the disease is caused by mutations in the autosomal gene, *AQP2*, which encodes the water channel aquaporin-2. Early diagnosis and treatment is essential to prevent periods of severe dehydration, especially in the infant. Since both *AVPR2* and *AQP2* are small genes, genetic testing in a child suspected for the disease is the most accurate procedure for diagnosis (Knoers & Deen 2001).

Bichet (2006) has proposed that all families with NDI should have their molecular defect identified so that women in X-linked families can consider carrier testing and prenatal testing. In this context, genetic testing can facilitate early diagnosis and treatment of affected male offspring, thus averting the physical and neurocognitive deficits that have been reported to be associated with repeated episodes of dehydration in this disorder.

Neurocognitive deficits are generally considered to be the principal extrarenal morbidity associated with X-linked NDI. However, given the paucity of psychometric studies in NDI patients, the evidence base for this assertion is limited. In one small study, Knoers and co-workers (Hoekstra et al 1996) evaluated 17 male NDI patients. The cohort was found to have a high prevalence of attention deficit hyperactivity disorder (8/17; 47.1%) and low short-term memory scores (7/10 tested). However, there were no correlations between test performances and age at diagnosis or hypernatremia, save for a negative correlation between age at therapy initiation and verbal IQ in one age group.

While this study suggests that there is limited association between dehydration episodes and cognitive impairment in NDI, it is important to note that the study was small and the participants were drawn from a center with longstanding expertise in early recognition and management of NDI patients. Moreover, the data from this study do not contradict the recommendation for molecular diagnosis in at-risk carrier females to facilitate early diagnosis in their male offspring.

Fabry disease (FD) is an X-linked, inborn error of metabolism due to mutations in the gene encoding the lysosomal enzyme α-galactosidase A. The enzyme deficiency causes progressive accumulation of globotriaosylceramide (GL-3) and related glycolipids in many cell types, including vascular endothelium, podocytes and various renal cell types, smooth-muscle cells of the cardiovascular system, cardiac myocytes, and neurons of autonomic nervous system. In affected males, the disease presents in childhood and culminates in cardiac, cerebrovascular, and end-stage renal disease. The clinical presentation can be highly variable and the diagnosis is frequently delayed or missed (Desnick et al 2003, Desnick & Brady 2004, Kotanko et al 2004). In affected males, demonstration of α-galactosidase A deficiency in leukocytes or plasma confirms the diagnosis.

While females were long considered to be asymptomatic carriers, recent studies demonstrate that asymptomatic female FD carriers are the exception, not the rule (Wilcox et al 2008). Female heterozygotes often suffer from significant multisystemic disease and reduced quality of life. The standard diagnostic approach based on enzymatic detection is often inconclusive in FD females. Gene-based analyses are more accurate.

Terryn et al (2008) screened a female-predominant, hemodialysis population using a two-tier approach. In the first tier, α-galactosidase A levels were determined and in the second tier, patients with the lowest α-galactosidase A levels were subjected to *GLA* mutation screening. These studies demonstrated that the prevalence of *GLA* carriers in this hemodialysis cohort was 0.3%. Additional family-based screening resulted in the identification of nine mutation carriers, who were not suspected to have FD. The authors concluded that FD screening is a cost-effective, technically feasible and clinically valuable objective, particularly since early intervention is crucial to minimize disease progression (Desnick & Brady 2004, Wilcox et al 2004, Germain et al 2007, Schiffmann et al 2007).

Renal Tumor Predisposition Disorders

One area where the use of genetic testing has expanded rapidly is in the area of hereditary cancer, particularly in relation to hereditary breast/ovarian and hereditary bowel cancer syndromes. But also for diseases characterized by an inherited predisposition to develop epithelial tumors of the kidney, such as tuberous sclerosis (TSC), von Hippel-Lindau (VHL) disease, and Birt-Hogg-Dubé (BHD) syndrome, genetic testing is becoming part of standard care. It is sensitive and specific, affordable, assists in the management of the individual patient, and allows early (presymptomatic) identification of gene carriers within the family, which is essential to improve prognosis and survival in affected family members.

Genetic screening may also be of help in determining the chance of developing renal tumors in these disorders. For instance, Ong et al (2007) have demonstrated in a very large cohort of patients with VHL disease, that patients carrying truncating mutations in the *VHL* gene have an earlier age of onset and higher age-related risk of renal cell carcinoma compared to those carrying misssense mutations. In patients with complete deletion of *VHL*, the frequency of renal cell carcinoma is much lower.

MOLECULAR ANALYSES: CURRENT APPROACHES, NEW INNOVATIONS, FUTURE DIRECTIONS

For many monogenic disorders, diagnostic evaluation increasingly incorporates molecular analyses. Molecular diagnostic laboratories have developed numerous methods to scan DNA and/or RNA for mutations in specific disease genes or, in certain disorders, e.g. cystic fibrosis, to screen for specific known mutations that cause disease in a high percentage of affected individuals. The current mainstay for molecular diagnosis involves sequence-based protocols that examine the exons and adjacent intronic regions of specific genes.

While this approach has proven to be robust in identifying disease-associated mutations in multiple disorders, current gene-based detection strategies have limitations. First, in contrast to routine laboratory studies, genetic testing typically does not detect all disease-causing alterations. There are several potential explanations for this 'false-negative result': (1) the pathogenic alteration cannot be detected by sequence analysis (e.g. a large deletion); (2) the sequence alteration lies in a region of the gene (e.g. a regulatory region or an intron) not assessed by the laboratory test; and (3) the genetic defect lies in a different gene than that tested (e.g. the patient represents a 'phenocopy'). Thus, a negative genetic test does not rule out the diagnosis. Furthermore, gene-based sequencing approaches can be confounded by false-positive results, since detection of a previously undescribed mutation may suggest, but does not establish, the diagnosis. For example, with rare mutations (e.g., so-called private mutations specific to a given family), there may be

insufficient data to judge the pathogenic potential of that specific sequence variant. In addition, for disorders, e.g., nephronophthisis, in which mutations in a number of different genes cause similar disease phenotypes (phenocopies), standard Sanger-based sequencing methods can be too costly to provide comprehensive molecular analyses of all known disease genes. Third, while a sequence change that causes premature termination of the growing peptide is likely to be pathogenic, many putative mutations involve a substitution of one amino acid for another. Numerous algorithms have been developed to assist with these analyses (Herrgard et al 2003, Ng & Henikoff 2003, Xi 2004, Balasubramanian et al 2005, Bromberg & Rost 2007, Barenboim et al 2008), but predicting the pathogenic potential of these amino acid substitutions remains an imprecise science.

In recent years, the issue of missing large intragenic deletions by sequencing has been partly overcome by the introduction of a method called multiplex-ligation-dependent probe amplification (MLPA) (Sellner & Taylor 2004). For renal disorders, this technique has been very successful in detecting large intragenic deletions in the genes for Bartter-Gitelman syndromes (*SLC12A3*, *CLCNKB*), nail-patella syndrome (*LMX1B*) and NDI (*AVPR2*) (NVAM Knoers, unpublished data).

In addition, over the past several years, there has been increasing recognition that copy number variation (CNV) is the most prevalent type of structural variation in the human genome and contributes significantly to genetic heterogeneity and human disease phenotypes, including monogenic disorders (Pinto et al 2007). CNVs result from insertions or deletions of genomic regions encompassing thousands to millions of base pairs of contiguous DNA sequences. Recent estimates suggest that these gene dosage imbalances comprise approximately 12% of the entire genome and involve 10% of all known genes (Kehrer-Sawatzki 2007, Shelling & Ferguson 2007). Standard Sanger sequencing-based strategies are not robust to detect or characterize CNVs.

To address these issues, newer strategies and technologies are being developed and applied in molecular diagnostic laboratories. For example, in cystic fibrosis, 70% of patients carry a deletion of codon 508 in the *CFTR* transcript. However, more than 1200 mutations have been identified in the *CFTR* gene (Eshaque & Dixon 2006). To facilitate comprehensive, high-throughput screening, Johnson et al (2007) examined a test cohort of 150 samples from CF carriers with known *CFTR* mutations on five technologically diverse platforms: (1) eSensor (Osmetech), an electronic detection assay system in which DNA fragments are attached to electrodes on the surface of a small circuit board; (2) InPlex (Third Wave Technologies Inc.), a signal amplification methodology that uses a microfluidics card; (3) oligonucleotide ligation assay, another method for detecting well-defined alleles that differ by a single base; (4) Signature, a direct hybridization assay using allele-specific capture probes; and (5)

Tag-It (Tm Bioscience Corp), an assay based on a universal bead array and allele-specific primer extension. All of the platforms performed comparably with respect to sensitivity, specificity, and missing data rates, suggesting that each of these technologies could serve as an appropriate platform for high-throughput *CFTR* mutation screening.

In disorders characterized by mutations in numerous genes, tiered screening strategies are becoming quite valuable. Stone (2007) has described a system that combines allele-specific detection (e.g., single-strand conformational polymorphism analysis (SSCP) and multiplex allele-specific ligation analysis (SNPlex) assays), with quantitative PCR and high-density SNP genotyping (to detect CNVs), and automated DNA sequencing. In Leber congenital amaurosis (LCA), this approach significantly increased the efficiency of mutation detection among the seven recessively transmitted LCA genes when compared with DNA sequencing alone. Thus, this algorithm may be adaptable to comprehensive mutation screening in disorders such as nephronophthisis, Bardet-Biedl syndrome, and Bartter-Gitelman syndrome.

In single-gene disorders where a high frequency of reported mutations is predicted to result in protein truncation, the protein truncation test (PTT) provides a fast method to screen for biologically relevant gene mutations (Hauss & Muller 2007). The method is based on size analysis of products resulting from in vitro transcription and translation; proteins of lower mass than the expected full-length protein represent translation products derived from truncating frame shift or stop mutations in the gene. Recent modifications, such as fluorescence labeling and the digital PTT, have improved the sensitivity and applicability of this method in molecular diagnosis.

Finally, recent innovations with array-based sequence analysis promise a technological breakthrough in mutational screening, by reducing cost, improving throughput, and allowing deep sequencing of the entire genomic sequence of specific genes (Bentley 2006, Beaudet & Belmont 2008). The National Human Genome Research Institute (Jung et al 1999) is currently championing an effort to sequence and analyze target regions in the genomes of individuals with mapped, uncloned, autosomal Mendelian disorders and X-linked disorders. The project combines genome resequencing of specific genes, characterization of CNVs, and comprehensive gene-expression profiling. Ideally, this broad-based approach should significantly advance the mechanistic understanding of a particular disease, with characterization of the genome- and transcriptome-level features that are associated with specific disease phenotypes.

Ultimately, these technological advances, as well as the continued development of bioinformatic tools and computational algorithms, will usher in the next era of genome sequencing technology, commonly referred to as 'the $1000 genome,' based on the projected price per genome in US dollars (Mardis 2006). In this paradigm, determining the full genomic sequence in an individual patient would become the initial step in a molecular diagnosis testing strategy that comprehensively catalogs genomic biomarkers, genetic mutations, copy-number changes, genomic rearrangements, and other alterations that are diagnostic for the disease in question. Between here and there, key issues related to economic, ethical, social, and legal principles must be carefully considered and developed into guidelines that inform clinical care, including personalized diagnosis, prognosis, and therapeutic management, while minimizing patients' genetic liability.

PHARMOCOGENOMICS

Pharmacogenomics is the fusion of pharmacology and genomics, designed to elucidate how genetic variations affect the ways in which people respond to drugs. In this context, genetic variation (e.g., single nucleotide polymorphisms (SNP), copy number variation (CNV)) can be assessed directly through genotyping or indirectly through protein, metabolite, or other biomarker assays (Swen et al 2007).

Numerous recent studies demonstrate that genetic variation can modulate individual response to specific medications, as well as the susceptibility to drug toxicity (Evans &McLeod 2003, Phillips & Van Bebber 2004, Toffoli & Cecchin 2004, Sconce et al 2005, van den Akker-van Marle et al 2006). Specifically, these studies have examined the association between SNPs and the clinical therapeutic response that involves the interplay between drug metabolism (pharmacokinetics) and drug response (pharmodynamics).

With the clinical application of pharmacogenomics, the era of 'one size fits all' for drugs and dosages promises to give way to personalized drug treatment, with improved efficacy and safety (Chung 2007). In the last several years, the US Food and Drug Administration (FDA) has approved several pharmacogenomic tests to guide individualized therapy. These tests detect variations in genes encoding hepatic enzymes involved in drug metabolism: cytochrome P450 CYP2C19 and CYP2D6 (Roche AmpliChip, http://www.roche.com/), and UDP-glucuronosyltransferase (Invader UGT1AI Molecular Assay; Third Wave Technologies, http://www.twt.com/) (Swen et al 2007). These enzymes direct the inactivation of many agents including tricyclic antidepressants, selective serotonin reuptake inhibitors, antipsychotics, β-blockers, benzodiazepines, and proton pump inhibitors (Chung 2007). Slow, rapid, and ultra-rapid metabolizers can be distinguished by different SNP profiles. In clinical practice, slow metabolizers are more likely to experience side effects at standard doses, whereas ultra-rapid metabolizers will need higher than standard doses to achieve therapeutic efficacy (Singh 2007).

Despite these FDA-approved assays and other promising data regarding anticoagulant therapies and antineoplastic agents, the incorporation of pharmacogenomic testing into patient care has been slow, largely because population-based studies have yet to demonstrate sufficient impact on outcome and cost-effectiveness (Chung 2007, Swen et al 2007). To be clinically useful, a pharmacogenomic test must be cost-effective; that is there must be (1) a significant prevalence of the relevant polymorphism in the target population; (2) good correlation between genotype (polymorphism) and phenotype (drug response); (3) significant morbidity or mortality if the disease state is left untreated; and (4) significant reduction in the adverse reaction rate by individualized pharmacogenomic testing (Flowers & Veenstra 2004).

To spur the introduction of pharmacogenomic testing into clinical nephrology, research must focus on the development of diagnostic tests for clinically important problems that do not currently have less-expensive alternatives for individualizing drug dosing (Swen et al 2007). For those tests shown to improve patient care, guidelines directing the clinical use of test results must then be developed. Analysis of cost-effectiveness and cost-consequences should then be performed and the data provided to insurance companies to facilitate reimbursement. If these challenges are met, the incorporation of pharmacogenomic testing into clinical practice may be achieved in the near future.

SUMMARY

The Human Genome Project has ushered in a new era of molecular medicine and in this new era, the paradigm for genetic testing is evolving. In the near future, genetic testing will increasingly move from its current focus on Mendelian disorders towards identifying susceptibility alleles that predispose individuals to developing complex traits, such as diabetic nephropathy, and incorporating pharmacogenomic testing to predict drug–genome interactions. In this context, the paradigm of genetic testing will shift from establishing diagnoses to predicting risk for individuals who may or may not actually develop disease or suffer an adverse drug reaction. The clinical relevance of genetic testing will expand beyond considerations of sensitivity and specificity, to increasingly emphasize the predictive value of the test.

This paradigm shift in genetic testing opens new opportunities for pre-emptive intervention in susceptible individuals. But this new role for genetic testing also has significant implications for privacy, family relationships, insurability, and the very meaning of health risks (Clayton 2003, Grosse & Khoury 2006). These issues must be carefully considered as nephrologists seek to incorporate the new tools of genomic medicine into their clinical practice.

References

Altshuler D, Brooks L, Chakravarti A, et al. A haplotype map of the human genome. Nature 2005; 437: 1299–320.

Aucella F, De Bonis P, Gatta G, et al. Molecular analysis of NPHS2 and ACTN4 genes in a series of 33 Italian patients affected by adult-onset nonfamilial focal segmental glomerulosclerosis. Nephron. Clin. Pract. 2005; 99: c31–6.

Balasubramanian S, Xia Y, Freinkman E, Gerstein M. Sequence variation in G-protein-coupled receptors: analysis of single nucleotide polymorphisms. Nucleic Acids Res. 2005; 33: 1710–21.

Ball RH. Invasive fetal testing. Curr. Opin. Obstet. Gynecol. 2004; 16: 159–62.

Barenboim M, Masso M, Vaisman II, Jamison DC. Statistical geometry based prediction of nonsynonymous SNP functional effects using random forest and neuro-fuzzy classifiers. Proteins 2008. (in press)

Beaudet AL, Belmont JW. Array-based DNA diagnostics: let the revolution begin. Annu. Rev. Med. 2008; 59: 113–29.

Bentley DR. Whole-genome re-sequencing. Curr. Opin. Genet. Dev. 2006; 16: 545–52.

Bertelli R, Ginevri F, Caridi G, et al. Recurrence of focal segmental glomerulosclerosis after renal transplantation in patients with mutations of podocin. Am. J. Kidney Dis. 2003; 41: 1314–21.

Bichet DG. Hereditary polyuric disorders: new concepts and differential diagnosis. Semin. Nephrol. 2006; 26: 224–33.

Borry P, Goffin T, Nys H, Dierickx K. Attitudes regarding predictive genetic testing in minors: a survey of European clinical geneticists. Am. J. Med. Genet. C. Semin. Med. Genet. 2008; 148: 78–83.

Brockmoller J, Tzvetkov MV. Pharmacogenetics: data, concepts and tools to improve drug discovery and drug treatment. Eur. J. Clin. Pharmacol. 2008; 64: 133–57.

Bromberg Y, Rost B. SNAP: predict effect of non-synonymous polymorphisms on function. Nucleic Acids Res. 2007; 35: 3823–35.

Caridi G, Berdeli A, Dagnino M, et al. Infantile steroid-resistant nephrotic syndrome associated with double homozygous mutations of podocin. Am. J. Kidney Dis. 2004; 43: 727–32.

Cattran D. Cyclosporine in the treatment of idiopathic focal segmental glomerulosclerosis. Semin. Nephrol. 2003; 23: 234–41.

Chung WK. Implementation of genetics to personalize medicine. Gend. Med. 2007; 4: 248–65.

Clayton E. Ethical, legal, and social implications of genomic medicine. N. Engl. J. Med. 2003; 349: 562–9.

Collins FS, Green ED, Guttmacher AE, Guyer MS. A vision for the future of genomics research. Nature 2003; 422: 835–47.

Colquitt JL, Kirby J, Green C, Cooper K, Trompeter RS. The clinical effectiveness and cost-effectiveness of treatments for children with idiopathic steroid-resistant nephrotic syndrome: a systematic review. Health Technol. Assess. 2007; 11: iii–iv, ix–xi, 1–93

Daskalakis N, Winn MP. Focal and segmental glomerulosclerosis. Cell. Mol. Life Sci. 2006; 63: 2506–11.

Deegens JK, Steenbergen EJ, Wetzels JF. Review on diagnosis and treatment of focal segmental glomerulosclerosis. Neth. J. Med. 2008; 66: 3–12.

Desnick RJ, Brady RO. Fabry disease in childhood. J. Pediatr. 2004; 144: S20–6.

Desnick RJ, Brady R, Barranger J, et al. Fabry disease, an under-recognized multisystemic disorder: expert recommendations for diagnosis, management, and enzyme replacement therapy. Ann. Intern. Med. 2003; 138: 338–46.

EC Expert Group. Ethical, legal and social implications of genetic testing. Bull. Med. Ethics. 2004; Jan(204): 9–11.

Eddy AA, Symons JM. Nephrotic syndrome in childhood. Lancet 2003; 362: 629–39.

Eshaque B, Dixon B. Technology platforms for molecular diagnosis of cystic fibrosis. Biotechnol. Adv. 2006; 24: 86–93.

Evans W, McLeod H. Pharmacogenomics – drug disposition, drug targets, and side effects. N. Engl. J. Med. 2003; 348: 538–49.

Flowers CR, Veenstra D. The role of cost-effectiveness analysis in the era of pharmacogenomics. Pharmacoeconomics 2004; 22: 481–93.

Fragouli E. Preimplantation genetic diagnosis: present and future. J. Assist. Reprod. Genet. 2007; 24: 201–7.

Garcia-Gonzalez MA, Jones JG, Allen SK, et al. Evaluating the clinical utility of a molecular genetic test for polycystic kidney disease. Mol. Genet. Metab. 2007; 92: 160.

Germain DP, Waldek S, Banikazemi M, et al. Sustained, long-term renal stabilization after 54 months of agalsidase beta therapy in patients with Fabry disease. J. Am. Soc. Nephrol. 2007; 18: 1547–57.

Gigarel N, Frydman N, Burlet P, et al. Preimplantation genetic diagnosis for autosomal recessive polycystic kidney disease. Reprod. Biomed. Online 2008; 16: 152–8.

Grosse SD, Khoury MJ. What is the clinical utility of genetic testing?. Genet. Med. 2006; 8: 448–50.

Hauss O, Muller O. The protein truncation test in mutation detection and molecular diagnosis. Method. Mol. Biol. 2007; 375: 151–64.

He N, Zahirieh A, Mei Y, et al. Recessive NPHS2 (Podocin) mutations are rare in adult-onset idiopathic focal segmental glomerulosclerosis. Clin. J. Am. Soc. Nephrol. 2007; 2: 31–7.

Herrgard S, Cammer SA, Hoffman BT, et al. Prediction of deleterious functional effects of amino acid mutations using a library of structure-based function descriptors. Proteins 2003; 53: 806–16.

Hinkes B, Vlangos C, Heeringa S, et al. Specific podocin mutations correlate with age of onset in steroid-resistant nephrotic syndrome. J. Am. Soc. Nephrol. 2008; 19: 365–71.

Hodges JG. Ethical issues concerning genetic testing and screening in public health. Am. J. Med. Genet. C. Semin. Med. Genet. 2004; 125C: 66–70.

Hoekstra JA, van Lieburg AF, Monnens LA, Hulstijn-Dirkmaat GM, Knoers VV. Cognitive and psychosocial functioning of patients with congenital nephrogenic diabetes insipidus. Am. J. Med. Genet. 1996; 61: 81–8.

The International HapMap Consortium. The International HapMap Project. Nature 2003; 426: 789–96.

Johnson MA, Yoshitomi MJ, Richards CS. A comparative study of five technologically diverse CFTR testing platforms. J. Mol. Diagn. 2007; 9: 401–7.

Jung G, Benz-Bohm G, Kugel H, Keller K-M, Querfeld U. MR cholangiography in children with autosomal recessive polycystic kidney disease. Pediatr. Radiol. 1999; 29: 463–6.

Kakuk P. Gene concepts and genethics: Beyond exceptionalism. Sci. Eng. Ethics 2008; (online)

Karle S, Uetz B, Ronner V, et al. Novel mutations in NPHS2 are detected in familial as well as sporadic steroid-resistant nephrotic syndrome. J. Am. Soc. Nephrol. 2002; 13: 388–93.

Kehrer-Sawatzki H. What a difference copy number variation makes. Bioessays 2007; 29: 311–13.

Knoers NV, Deen PM. Molecular and cellular defects in nephrogenic diabetes insipidus. Pediatr. Nephrol. 2001; 16: 1146–52.

Korf BR. Genetic testing and medical practice. Curr. Opin. Pediatr. 2001; 13: 547–9.

Kotanko P, Kramar R, Devrnja D, et al. Results of a nationwide screening for Anderson-Fabry disease among dialysis patients. J. Am. Soc. Nephrol. 2004; 15: 1323–9.

Lissemore JL. Linkage of genetics and ethics: more crossing over is needed. Biol. Cell 2005; 97: 599–604.

Mardis ER. Anticipating the 1,000 dollar genome. Genome Biol. 2006; 7: 112.

McDonald R, Avner E. Mouse models of polycystic kidney disease. In: Watson M, Torres V, eds. Polycystic Kidney Disease. Oxford, UK: Oxford University Press, 1996: pp. 63–87.

Minkoff H, Ecker J. Genetic testing and breach of patient confidentiality: Law, ethics, and pragmatics. Am. J. Obstet. Gynecol. 2008; (online)

Mujezinovic F, Alfirevic Z. Procedure-related complications of amniocentesis and chorionic villous sampling: a systematic review. Obstet. Gynecol. 2007; 110: 687–94.

Nascimento AB, Mitchell DG, Zhang XM, Kamishima T, Parker L, Holland GA. Rapid MR imaging detection of renal cysts: age-based standards. Radiology 2001; 221: 628–32.

Ng PC, Henikoff S. SIFT: Predicting amino acid changes that affect protein function. Nucleic Acids Res. 2003; 31: 3812–14.

Niaudet P. Podocin and nephrotic syndrome: implications for the clinician. J. Am. Soc. Nephrol. 2004; 15: 832–4.

Ong KR, Woodward ER, Killick P, Lim C, Macdonald F, Maher ER. Genotype-phenotype correlations in von Hippel-Lindau disease. Hum. Mutat. 2007; 28: 143–9.

Orloff M, Iyengar S, Winkler C, et al. Variants in the Wilms' tumor gene are associated with focal segmental glomerulosclerosis in the African American population. Physiol. Genomics 2005; 21: 212–21.

Pagon RA. GeneTests: an online genetic information resource for health care providers. J. Med. Libr. Assoc. 2006; 94: 343–8.

Peng W, Takabayashi H, Ikawa K. Whole genome amplification from single cells in preimplantation genetic diagnosis and prenatal diagnosis. Eur. J. Obstet. Gynecol. Reprod. Biol. 2007; 131: 13–20.

Phillips KA, Van Bebber SL. A systematic review of cost-effectiveness analyses of pharmacogenomic interventions. Pharmacogenomics 2004; 5: 1139–49.

Pinto D, Marshall C, Feuk L, Scherer SW. Copy-number variation in control population cohorts. Hum. Mol. Genet. 2007; 16(Spec No. 2): R168–73.

Ponticelli C, Passerini P. Other immunosuppressive agents for focal segmental glomerulosclerosis. Semin. Nephrol. 2003; 23: 242–8.

Ravine D, Gibson R, Walker R, Sheffield L, Kincaid-Smith P, Danks D. Evaluation of ultrasonographic diagnostic criteria for autosomal dominant polycystic kidney disease 1. Lancet 1994; 343: 824–7.

Robin NH, Tabereaux PB, Benza R, Korf BR. Genetic testing in cardiovascular disease. J. Am. Coll. Cardiol. 2007; 50: 727–37.

Rossetti S, Harris PC. Genotype-phenotype correlations in autosomal dominant and autosomal recessive polycystic kidney disease. J. Am. Soc. Nephrol. 2007; 18: 1374–80.

Ruf R, Schultheiss M, Lichtenberger A, et al. Prevalence of WT1 mutations in a large cohort of patients with steroid-resistant and steroid-sensitive nephrotic syndrome. Kidney Int. 2004a; 66: 564–70.

Ruf RG, Lichtenberger A, Karle SM, et al. Patients with mutations in NPHS2 (podocin) do not respond to standard steroid treatment of nephrotic syndrome. J. Am. Soc. Nephrol. 2004b; 15: 722–32.

Sanderson S, Zimmern R, Kroese M, Higgins J, Patch C, Emery J. How can the evaluation of genetic tests be enhanced? Lessons learned from the ACCE framework and evaluating genetic tests in the United Kingdom. Genet. Med. 2005; 7: 495–500.

Schiffmann R, Askari H, Timmons M, et al. Weekly enzyme replacement therapy may slow decline of renal function in patients with Fabry disease who are on long-term biweekly dosing. J. Am. Soc. Nephrol. 2007; 18: 1576–83.

Schnaper H. Idiopathic focal segmental glomerulosclerosis. Semin. Nephrol. 2003; 23: 183–93.

Sconce EA, Khan TI, Wynne HA, et al. The impact of CYP2C9 and VKORC1 genetic polymorphism and patient characteristics upon warfarin dose requirements: proposal for a new dosing regimen. Blood 2005; 106: 2329–33.

Sekizawa A, Purwosunu Y, Matsuoka R, et al. Recent advances in non-invasive prenatal DNA diagnosis through analysis of maternal blood. J. Obstet. Gynaecol. Res. 2007; 33: 747–64.

Sellner LN, Taylor GR. MLPA and MAPH: new techniques for detection of gene deletions. Hum. Mutat. 2004; 23: 413–19.

Shelling AN, Ferguson LR. Genetic variation in human disease and a new role for copy number variants. Mutat. Res. 2007; 622: 33–41.

Singh A. Pharmacogenomics–the potential of genetically guided prescribing. Aust. Fam. Physician 2007; 36: 820–4.

Steinbock B. Prenatal testing for adult-onset conditions: qui bono? Reprod. Biomed. Online 2007; 15(Suppl. 2): 38–42.

Stone EM. Leber congenital amaurosis – a model for efficient genetic testing of heterogeneous disorders: LXIV Edward Jackson Memorial Lecture. Am. J. Ophthalmol. 2007; 144: 791–811.

Swen JJ, Huizinga TW, Gelderblom H, et al. Translating pharmacogenomics: challenges on the road to the clinic. PLoS Med. 2007; 4: e209.

Tatsioni A, Zarin DA, Aronson N, et al. Challenges in systematic reviews of diagnostic technologies. Ann. Intern. Med. 2005; 142: 1048–55.

Terryn W, Poppe B, Wuyts B, et al. Two-tier approach for the detection of alpha-galactosidase A deficiency in a predominantly female haemodialysis population. Nephrol. Dial. Transplant 2008; 23: 294–300.

Toffoli G, Cecchin E. Uridine diphosphoglucuronosyl transferase and methylenetetrahydrofolate reductase polymorphisms as genomic predictors of toxicity and response to irinotecan-, antifolate- and fluoropyrimidine-based chemotherapy. J. Chemother. 2004; 16(Suppl 4): 31–5.

Tsukaguchi H, Sudhakar A, Le TC, et al. NPHS2 mutations in late-onset focal segmental glomerulosclerosis: R229Q is a common disease-associated allele. J. Clin. Invest. 2002; 110: 1659–66.

van den Akker-van Marle ME, Gurwitz D, Detmar SB, et al. Cost-effectiveness of pharmacogenomics in clinical practice: a case study of thiopurine methyltransferase genotyping in acute lymphoblastic leukemia in Europe. Pharmacogenomics 2006; 7: 783–92.

Weber S, Gribouval O, Esquivel E, et al. NPHS2 mutation analysis shows genetic heterogeneity of steroid-resistant nephrotic syndrome and low post-transplant recurrence. Kidney Int. 2004; 66: 571–9.

Wilcox WR, Banikazemi M, Guffon N, et al. Long-term safety and efficacy of enzyme replacement therapy for Fabry disease. Am. J. Hum. Genet. 2004; 75: 65–74.

Wilcox WR, Oliveira JP, Hopkin RJ, et al. Females with Fabry disease frequently have major organ involvement: lessons from the Fabry Registry. Mol. Genet. Metab. 2008; 93: 112–18.

Wilfond B for the ASHG Social Issues Committee. Ethical, legal, and psychosocial implications of genetic testing in children and adolescents (ACMG/ASHG Report). Am. J. Hum. Genet. 1995; 57: 1233–41.

Winn MP, Conlon PJ, Lynn KL, et al. A mutation in the TRPC6 cation channel causes familial focal segmental glomerulosclerosis. Science 2005; 308: 1801–4.

Xi T, Jones IM, Mohrenweiser HW. Many amino acid substitution variants identified in DNA repair genes during human population screenings are predicted to impact protein function. Genomics 2004; 83: 970–9.

Yao J, Le T, Kos C, et al. Alpha-actinin-4-mediated FSGS: an inherited kidney disease caused by an aggregated and rapidly degraded cytoskeletal protein. PLoS Biol. 2004; 2: e167.

A. Primary Genetic Diseases of Nephron Function

CHAPTER 3

Logic of the Kidney

ORSON W. MOE, GERHARD H. GIEBISCH AND DONALD W. SELDIN

KIDNEY FUNCTION: A SYSTEMS APPROACH

To discharge its responsibilities, the kidney must first be appraised of the state of the internal environment in a setting of varied intake, physiologic alterations, and even pathologic disturbances. This is accomplished in a feedback system where the information concerning the volume and composition of the blood is transmitted to the kidney via a variety of signaling pathways. The kidney has in place a host of transport systems which respond to the signals by activating or suppressing specific functions so as to maintain a fairly constant volume and composition in body fluid. Finally, when the normal set point is reached, the feedback signals restore the transport processes to a normal state.

More than 150 years ago, Claude Bernard in a seminal publication (Figure 3.1) described the stability of the internal environment as the necessary condition for independent existence. In fact, Bernard's *milieu intérieur* is one of the reasons for evolution of multicellularity. From the point of view of water and electrolyte metabolism, it is the kidney which provides the principal regulatory mechanisms maintaining constancy of volume and composition in the face of diverse physiologic demands and pathologic assaults.

In recent years, application of analytic and conceptual tools of molecular biology has served as a powerful tool to explore how the kidney discharges its homeostatic functions. By traversing a reductive cascade (Figure 3.2), many physiologic studies have risen to the level of the study of genes. Beginning with the most basic processes of the activity of genes and their products, many formidable novel normal and abnormal aspects of renal function have been elucidated, and the whole field of renal physiology was catapulted forward at an unprecedented pace. The return and impact of these efforts cannot be overemphasized. Bernard himself, who was a most vigorous experimentalist, certainly

THE CONSTANCY OF THE INTERNAL ENVIRONMENT

The living organism does not really exist in the *milieu extérieur* (the atmosphere, if it breathes air, salt or fresh water, if that is its element), but in the liquid *milieu intérieur* formed by the circulating organic liquid which surrounds and bathes all the tissue elements; this is the lymph or plasma, the liquid part of the blood, which in the higher animals is diffused through the tissues and forms the ensemble of the intracellular liquids and is the basis of all local nutrition and the common factor of all elementary exchanges. The stability of the *milieu intérieur* is the primary condition for freedom and independence of existence; the mechanism which allows of this is that which ensures in the *milieu intérieur* the maintenance of all the conditions necessary to the life of the elements.

Claude Bernard, *Leçons sur les phénomènes de la vie communs aux animaux et aux végétaux.* Baillière, Paris, 1878, transl. by J.F. Fulton.

FIGURE 3.1   Excerpt from Claude Bernard's monograph

Genetic Diseases of the Kidney
ISBN: 978-0-12-449851-8

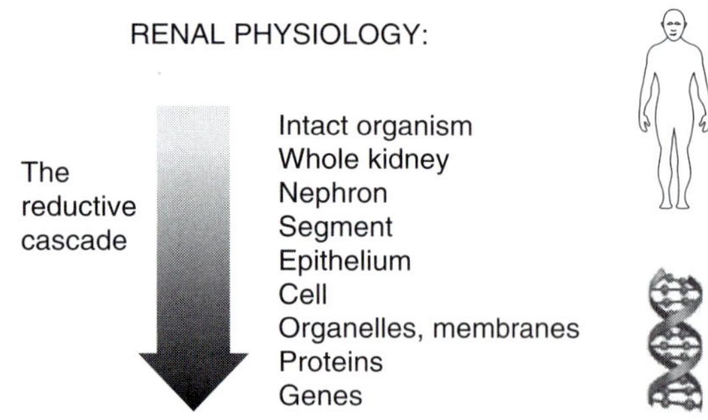

FIGURE 3.2   Reductive cascade in renal physiology

would have applauded these extraordinary achievements (Bernard 1865).

At the same time, it must be recognized that the molecular approach, while enormously fruitful, has not addressed the integrated function of ensembles so striking a feature of renal function. However powerful the reductive process, it will always be necessary to examine the integrated ensemble so as to understand how a net effect emerges from a discrete disturbance that has been identified in an isolated system. For example, even major discoveries of the properties of the isolated transporters, channels, and regulatory proteins from the renal medulla could never explain the integrated function of the countercurrent mechanism and urinary concentrating system. The countercurrent system must be analyzed as an intact integrated system, including transporters, circulation, pressure gradients, and complicated architecture, and many other system features. It is therefore necessary for an understanding of intact renal function to examine the properties of an ensemble so as to appreciate the effects of an isolated disturbance within the context of an integrated system.

Nor can intricate knowledge of every single gene and protein that participates in sodium handling, cardiac contractility, and vascular resistance reconstitute the knowledge of blood pressure homeostasis, which is a physiologic entity derived from the product of parameters such as effective arterial blood volume, cardiac output, and vascular resistance. While isolation is a mandatory prerequisite for a reductive investigative endeavor to succeed, one can only ascribe meaning to the discovery through backtracking up the reductive cascade to the integrated ensemble.

This concept is expounded in numerous instances in many different disciplines. In the entomologic study of social insects, such as ants and bees, populating colonies with highly developed complex hierarchy with coordinated division of labor, the study of the structure and function of a single insect in exquisite detail to complete understanding will not reveal how the colony works. In fact, that is an

impossible task unless the colony is examined in its entirety. At a more fundamental level, theoretical physicist and Nobel Laureate Robert B. Laughlin presented this notion in detail, clarity, and eloquence in his book *A Different Universe: Reinventing Physics: From the Bottom Down* (Laughlin 2007). Laughlin sees reductionism as not only a method but a philosophy and conviction 'that all things will necessarily be clarified when they are divided into smaller and smaller component parts.' After defending the importance of a reductionistic experimental approach, he explicitly states that his purpose is not to impugn technical reductionism. He emphasizes that laws of the universe can only come about with progressively higher levels of organization. Microscopic knowledge of elementary particles should spawn concepts and theories towards laws of higher organization. The last paragraph of his book reads: 'We live not at the end of discovery but at the end of Reductionism, a time in which the false ideology of human mastery of all things through microscopic is being swept away by events and reason. This is not to say that microscopic law is wrong or has no purpose, but only that it is rendered irrelevant in many circumstances by its children, and its children's children, the higher organizational laws of the world.'

There is an astounding parallel between Laughlin's view of physics and renal physiology. Denis Noble, who recently hailed Bernard as the first pioneer systems biologist in his monograph, cited Bernard: 'The control of the milieu intérieur meant not that the individual molecules did anything different from what they would do in non-living systems, but rather that the ensemble behaves in a controlled way, the controls being those that maintain the constancy of the internal environment' (Noble 2008). While it is entirely appropriate to promote reduction approaches as means to uncover knowledge, one must bear in mind the importance of putting the discovery back in the context of where the gene product normally executes its function and how that gene product interacts with the many other processes in the integrated ensemble that eventually defends the integrity of the volume and composition of the *milieu intérieur* as pointed out by Bernard.

To discharge its regulatory responsibilities, the kidney deploys several critical systems (Figure 3.3). The first component of this feedback loop is from the internal environment to the kidney consisting of volume of the circulation, mainly the effective arterial volume, systemic blood pressure, plasma composition of various electrolytes which can be monitored either directly by the kidney (e.g. ionized calcium) or by distant sensors (e.g. osmolarity), and various communications to the kidney via neuroendocrine pathways.

Extrarenal signals while important are not the sole input. The second component consists of important intrarenal processes that significantly adjust the feedback. Examples

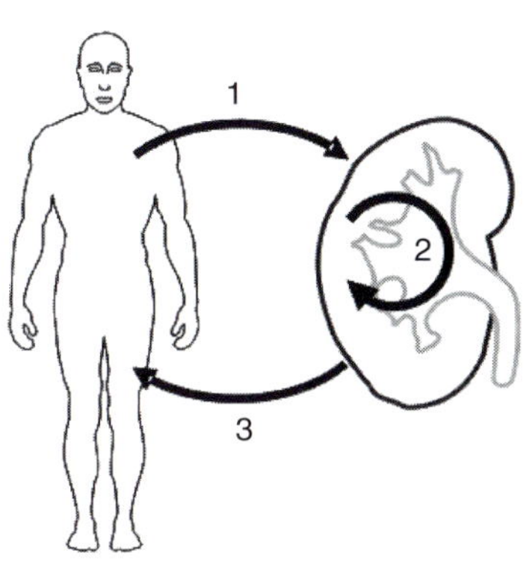

FIGURE 3.3   Systems approach to renal function. Input to the kidney consists of: (1) signals originating from the internal environment outside the kidney which are modified by (2) intrarenal processes, finally culminating in (3) the kidney's response to the internal environment

include hemodynamics of the renal circulation, autoregulation of glomerular filtration rate, peritubular Starling forces that alter transport, intrinsic intranephron control systems such as glomerulotubular balance and tubuloglomerular feedback, and intrarenal paracrine/autocrine systems such as renin-angiotensin, dopamine, and endothelin. Finally, the geographic relationship of the various loci of transport systems along the nephron exerts a critical role in the intrarenal mechanisms of regulation. This concept refers to the fact that upstream events can modify or be modified by downstream processes.

The third and last system to consider is the kidney's role as an effector to maintenance and alterations of the internal environment. One can view this as a mendicant role of the kidney where homeostatic and corrective operations are deployed as instructed by the incoming signals. These mechanisms involve the kidney's ability to adjust glomerular filtration rate, body fluid volume and solute concentration, and the ability to elaborate a number of endocrine hormones such as renin, 1,25-dihydroxyvitamin D_3, and erythropoietin.

In this chapter, our main aim is to explain how the structural and functional organization of the kidney is deployed to maintain homeostasis. First, the structural organization of the kidney is briefly described. Next, different transport systems along the nephron are characterized. Finally, the integrated adjustments whereby feedback regulation between the system circulation and internal environment and kidney functions to maintain constancy are analyzed. The most abundant extracellular cation, sodium, and the most abundant intracellular cation, potassium, are chosen as illustrative examples.

WALK THROUGH THE NEPHRON

Anatomy of the Nephron

KIDNEY ORGANIZATION

As much as the skin, lung, and intestinal tract, the kidney forms an interface between the organism and the outside but at this interface, one deals primarily with discharge of waste from within. As one 'walks' axially down the nephron, the structure of the various components will be reviewed from the level of the organ to the cell. One purpose is to provide a general background of renal anatomy but the main objective is to recognize that knowledge of the individual cellular structure and molecules that constitute the nephron must be appreciated in the light of the architecture and coordinated function of the entire kidney. The appreciation of the juxtaposition of various structures is the key to understanding the excretory and endocrine functions of the kidney. The fact cannot be overemphasized as one considers monogenetic disease of the kidney which typically starts with the malfunction of a single protein. Despite the highly focal nature of this type of lesion, the whole organ and whole organism phenotype which confronts the clinician can only be understood when the lesion is appraised in the light of the intact kidney.

To fulfill its homeostatic role, the kidney interfaces directly with the intravascular compartment where it modifies the composition and quantity of body fluid via its excretory and endocrine functions. Each human kidney is endowed with approximately 10^6 nephrons (Smith 1951). Each nephron starts with the glomerulus where an ultrafiltrate is produced from the plasma which subsequently journeys through the urinary lumen separated from the blood compartment by the tubular epithelium. This same epithelium mediates continual solute and water exchange between urinary lumen and capillaries until the desired quantity and composition of urine and venous blood is discharged into the ureter and the renal vein respectively.

The bisected surface of the kidney consists of the lighter-colored outer cortex and the darker-appearing inner medulla (Figure 3.4). The medulla is further divided radially into outer and inner regions with the outer medulla subdivided into outer and inner stripes. Grossly, the medulla assumes multiple conical contours called renal pyramids with their apices abutting on the renal pelvis to form a papilla. The contact points of the renal pelvis with the renal parenchyma are cup-like structures called calyces. Intercalated between pyramids are centripetal extensions of cortical tissue into the medulla called columns of Bertin. This is where the renal arterial system enters the renal parenchyma. In contrast to the human kidney, the rodent kidney which is used extensively as an experimental model has only one pyramid and papilla. Small mammals tend to be unipapillary. At the other end of the spectrum, as the size of the organism increases, kidney size does not simply increase accordingly, preserving the same anatomic features and relationships. As one example, the collecting duct cannot elongate endlessly as at some point, luminal pressure will fall to null and urine flow stops. In larger mammals, the large kidneys are actually composed of many small and relatively independent kidneys called renicules, separated from each other by interrenicular tissue (Abdelbaki et al 1984).

Renal Vasculature

One should note that the renal circulation is unique in several aspects (Table 3.1). First, the kidney receives approximately 20–25% of the cardiac output even though it is less than 1% of body weight (Thomson & Blantz 2007). Blood flows through most organs to provide oxygen and nutrients but blood traverses through the kidney primarily to be cleansed. Second, there are no anastomoses between the segmental branches of the renal artery (Figure 3.4) (Kriz & Kaissling 2007). Occlusion of these rather proximal yet terminal arteries will create a segmental infarction. Third, this high-magnitude blood flow courses through a countercurrent system where arterial blood is brought into close proximity

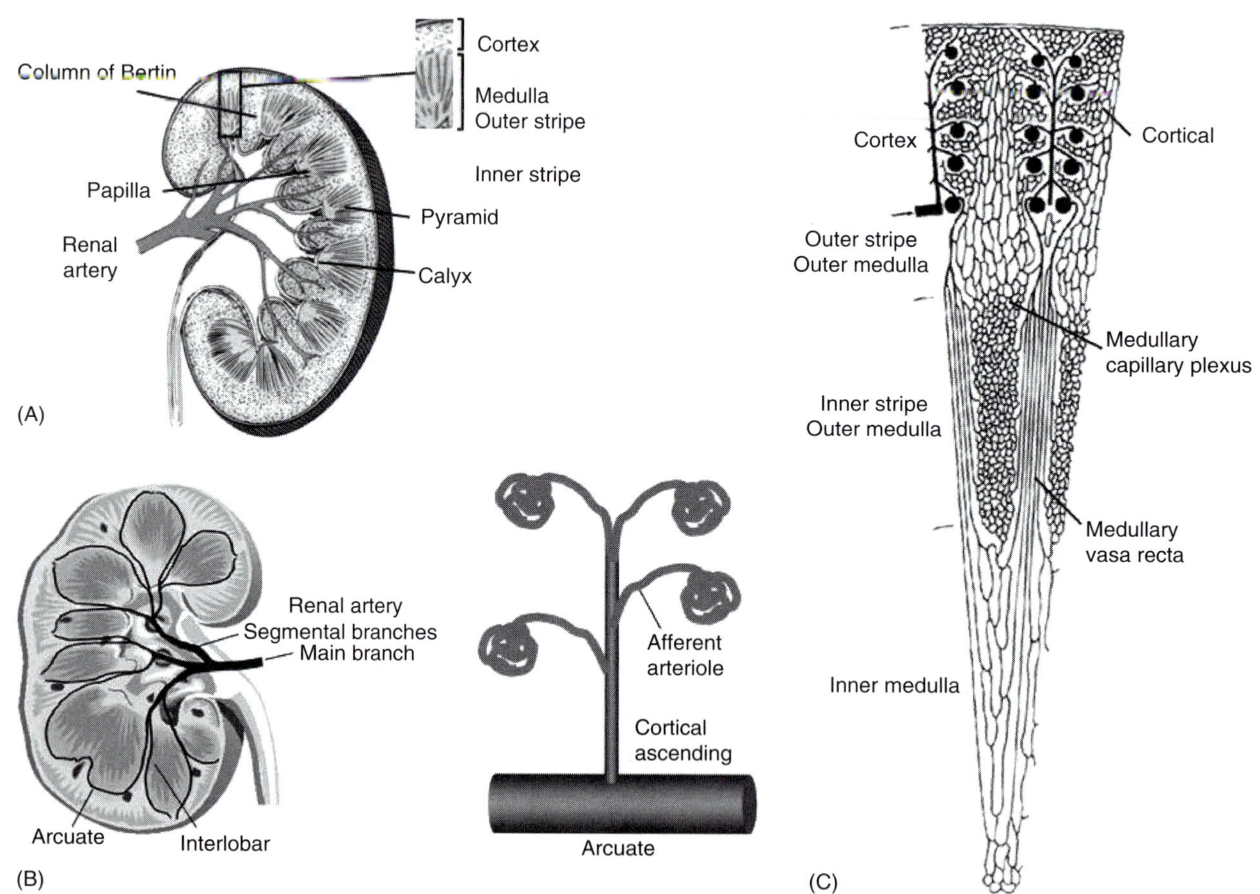

FIGURE 3.4   (A) Gross anatomy of the kidney. (B) Major renal arteries leading to afferent arteriole. (C) Unique dual capillary systems (glomerular and peritubular) in the kidney in tandem

with venous blood flowing in the opposite direction (Kriz & Kaissling 2007). This creates one of the largest functional arteriovenous shunts in any organ circulation. This is best exemplified by the fact that O_2 extraction by the kidney is only about 10–15% compared to skeletal muscle, which is 35% at rest, 75% at exercise, and up to 90% for a trained athlete at peak exercise (O'Connor 2006). However, despite this seemingly 'luxurious' O_2 supply, because of the functional arteriovenous shunt, O_2 tension can dip below 50 mmHg in deep cortex (Brezis et al 1989). This has been touted as one explanation for the kidney's propensity for ischemic damage. Fourth, the renal circulation is the only one that has two capillary circulations in tandem (Figure 3.4) (Kriz & Kaissling 2007). Both the pre- and post-capillary blood vessels in the glomerular circulation are arterioles. The efferent arteriole is structurally and functionally not a venule as it promptly ramifies into the peritubular capillaries. In addition, the peritubular capillaries that embrace the proximal tubule in cortex are quite distinct from the vasa recta system that travels down and back up the medulla.

The macrocirculation starts with the main renal artery branching into several segments in the human which are short ramifications that divide into the interlobar arteries running up the columns of Bertin between neighboring pyramids (Figure 3.4). In the corticomedullary junction, the interlobar arteries give rise to the arcuate arteries, which trace the corticomedullary junction radially and send off the cortical ascending arteries towards the kidney surface in a branching fashion. From the smaller branches of the cortical ascending arteries emerge the afferent arterioles that feed the glomerular ensemble. Renal venous drainage of the peritubular capillaries commences at all levels in the cortex and medulla and the drainage follows a path flowing counter direction to but anatomically exquisitely similar and closely apposed to the arterial system (arcuate, interlobar veins). This is the basis for the functional arteriovenous shunt described earlier. At the renal hilus, the interlobar veins coalesce to form the main renal vein. The glomerular ensemble and peritubular microcirculation are distinct and will be discussed separately.

GLOMERULAR ENSEMBLE

The glomerulus was named by Marcello Malpighi in the mid 1600s. The integral structural unit of the glomerular ensemble is the renal corpuscle which consists of vascular, epithelial, and mesangial components and is remarkably conserved in all species (Figure 3.5). In the human, the corpuscle is about 200 μm in diameter (Kriz & Kaissling 2007). Afferent arterioles emanate from the cortical ascending arteries, lose their smooth muscle layer upon entering the glomerular tuft, and immediately ramify extensively into a capillary network forming the glomerular tuft. The smooth muscle layers of the afferent and efferent arterioles however, are critical in determining arteriolar tone. The glomerular capillary has several salient characteristics. It is organized into lobules between the afferent and efferent vascular poles. The capillary tuft is suspended in close apposition to the mesangium on one side and separated from the foot processes of the visceral epithelium of Bowman's capsule

---

TABLE 3.1 Characteristics of the renal circulation

1. Highest blood flow per unit organ weight
2. Little or no anastomoses beyond the large segmental renal artery
3. Large functional arteriovenous shunt
4. Multiple capillary networks in tandem

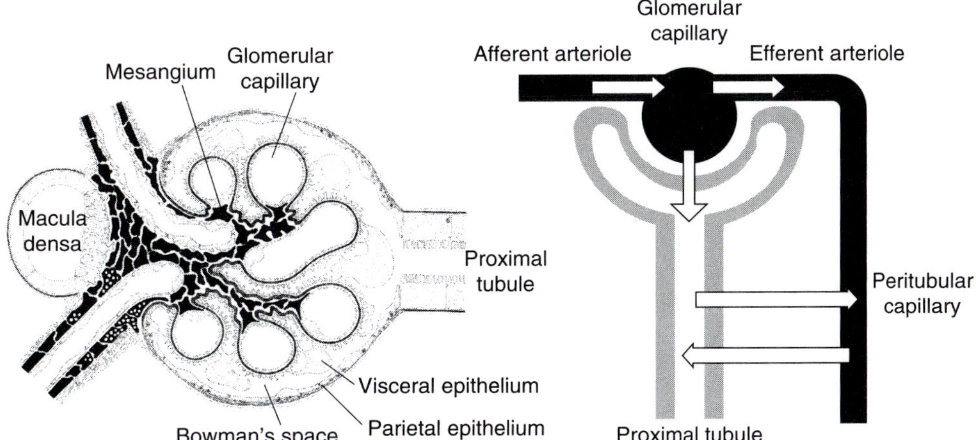

FIGURE 3.5 Glomerular ensemble. The left panel shows the structural components of the glomerulus. The right panel is a schematic to emphasize the structural proximity and relationship between the afferent arteriole, glomerular capillary, efferent arteriole, proximal tubule, and peritubular capillary. The specialized part of the distal nephron called the macula densa, the glomerular arterioles, and the extraglomerular mesangium form a tripartite structure termed the juxtaglomerular apparatus

(podocyte) on the opposite side by a characteristic basement membrane (Figure 3.5). The capillaries themselves have large lumina, are void of internal elastic layer, and have specialized endothelia with attenuated and fenestrated endothelium on the urinary side but not the mesangial side. These fenestrations (50–100 nm in diameter) are likely the reason for the highest hydraulic permeability of any known capillary. Smooth muscle cells of the arterioles are replaced by characteristic granular cells in the capillaries likely of smooth muscle origin.

Between the parietal and visceral layer of Bowman's capsule is Bowman's space where the nascent ultrafiltrate resides. The convergence of capillaries into the efferent arterioles is as abrupt as its divergence from the afferent arteriole. This coalescence can occur fairly deep in the tuft but the exit of the efferent arteriole at the vascular pole is in close proximity to the entry point of the afferent arteriole.

A specialized structure resides in the vascular pole of the glomerulus which consists of the afferent arteriole, efferent arteriole, extraglomerular mesangium and mesangial cells, and a specialized part of the thick ascending limb that physically contacts the glomerulus, called the macula densa (Figure 3.5). The granular cells in the vascular wall and mesangium are of particular importance. These cells have been classified as myo-epithelial cells due to their dual features of contractile and secretory epithelial cells. This collection of structures is the juxtaglomerular apparatus (JGA). The unique juxtaposition of the glomerular arterioles, mesangium, free access to systemic circulation, and sodium-transporting tubular epithelium allows the JGA to fulfill its critical role in systemic regulation of sodium balance and blood pressure via the endocrine renin-angiotensin system and the intrarenal regulation of glomerular filtration rate via tubular glomerular feedback. These homeostatic processes will be discussed in more detail below.

At the interface between the capillary blood and the beginning of the urinary space is the site of glomerular filtration. Glomerular filtration occurs at a tripartite structure of the capillary endothelium, glomerular basement membrane, and the podocyte with its quintessential slit diaphragm (Figure 3.6). The traditional view is that the endothelium contributes little to the permselectivity of the glomerulus although there has been some recent controversy as to whether that is entirely correct (Ballermann 2007). A previously unrecognized glycocalyx composed of negatively charged sulfated proteoglycans spans the fenestrations forming projections of brush-like structures called sieve plugs (Figure 3.6). Thus it is likely that the fenestrations are not really 'holes' at all but may in fact provide a charge and size barrier (Rostgaard & Qvortrup 1997, Haraldsson et al 2008).

Between the endothelium and the podocyte is the glomerular basement membrane (GBM) (Figure 3.6). GBM is a gel-like material (90–93% water by volume) and is negatively charged. It is a heteropolymeric network of type IV collagen, laminin, fibronectin, entactin, and abundant amounts of heparan sulfate proteoglycans ($\sim$1% of the dry weight of the GBM). Type IV (α3, α4, α5 in equimolar ratios) is a triple helical polypeptide and forms an interconnected network of fibers, to which other matrix components are attached. Laminin interacts with the cellular layers of the capillary wall and anchors both endothelia and podocytes to the GBM. The sulfated glycoprotein entactin/nidogen binds to collagen IV, heparan sulfate proteoglycan, and laminin – linking GBM constituents together. Mutations of type IV collagen lead to

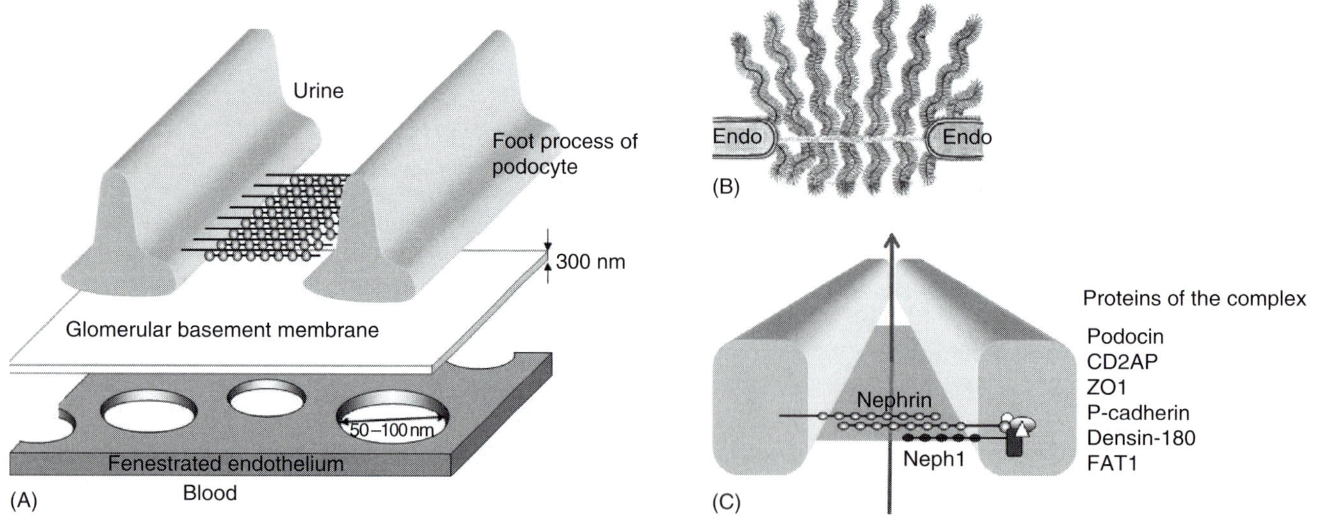

FIGURE 3.6 Podocyte and slit diaphragm. (A) Three structures separate the blood in the glomerular capillary from the urinary space. The fenestrated endothelium excludes cellular elements, the glomerular basement membrane (GBM) provides a coarse first filter, and the slit diaphragm between foot processes performs the stringent filtering. The slit diaphragm is like a zipper composed of homotypic meshwork of podocin. (B) Endothelial fenestration with extensive glycocalyx and filamentous molecular sieve. (C) Examples of some proteins of the complex at the podocyte

Alport syndrome where the architecture of GBM is altered but the permselectivity barrier is not significantly disturbed. The role of the GBM as a barrier is not established although there is clinical, experimental, and theoretical evidence for its role (Gubler 2008, Haraldsson et al 2008).

The epithelial part of the glomerulus consists of visceral and parietal cells of the Bowman's capsule. The parietal epithelium is flat and squamous, and very sparsely endowed with organelles or vesicles (Kriz & Kaissling 2007). Cilia have been noted in cells but their function is yet undefined. At the vascular pole, the parietal epithelium is immediately contiguous with a completely different epithelium – the proximal convoluted tubule (see below). On the opposite side of Bowman's space is the parietal epithelium, which is also known as the podocytes. These cells are much better characterized. In contradistinction to the visceral counterpart, the podocyte has singular and highly complex morphology. Long cytoplasmic extensions called foot processes come into direct contact with glomerular basement membrane. The cell bodies have well-developed organelles including Golgi, lysosomes, microtubules, and microfilaments. The most notable structure is the gap between foot processes, which is the filtration slit pore bridged by the slit diaphragm (Figure 3.6). As early as 1972, Karnovsky and Ainsworth proposed that the glomerular basement membrane may provide a coarse screening filter while the slit diaphragm acts as a stringent zipper-like molecular sieve that defines the composition of the ultrafiltrate (Karnovsky & Ainsworth 1972).

The traditional view and analysis of the glomerular filter has been largely biophysical in nature with emphasis on the kinetics and solute and solvent movement. A plethora of data and models have infused a much more biologic concept into this field. In the last decade, there have been momentous advances in the molecular identification of the components of the slit diaphragm complex and their function although the current model is far from complete.

Some of the proteins of the growing list are listed in Figure 3.6. There are numerous documented and even more proposed functions of the slit diaphragm multi-protein complex in addition to filtration barrier. These include establishment of contact and anchor for the foot processes and provision of a platform for a variety of outside-in signaling and actin filament organization. An unexplored but potential role will be inside-out signaling where the podocyte has the capability to regulate glomerular filtration. Several of the podocyte-specific proteins such as nephrin, podocin, neph1, and podoplannin have been shown to be candidates for diseases either as naturally occurring mutations in humans, targeted deletions in rodents, or via manipulation of the protein in experimental models (Garg et al 2007).

Interposed between the glomerular capillary loops is the mesangium (Figure 3.5). The mesangium is composed of mesangial cells embedded in the mesangial matrix. The mesangial matrix contains primarily sulfated glycosaminoglycans and various collagens. The mesangial cell has an impressive array of microfilaments including actin, α-actinin, and myosin. Mesangial cells harbor contractile elements and have contractile functions in response to agonists such as angiotensin II, norepinephrine, and vasopressin (Kreisberg et al 1985, Johnson et al 1992). These cells can alter intraglomerular renal blood flow as well as single nephron glomerular filtration rate. In addition to supporting the glomerular capillaries and regulating flow, the mesangial cells produce the mesangial matrix, have phagocytic capabilities, and perform paracrine functions such as synthesis of prostaglandins (Hao & Breyer 2007). Part of the mesangium extends beyond the glomerular corpuscle forming the extraglomerular mesangium, which is an integral part of the juxtaglomerular apparatus (Figure 3.5).

PERITUBULAR MICROCIRCULATION

The peritubular circulation is directly downstream from the glomerular capillary and of significant functional consequence because it provides the conduit for solutes and water to return to the systemic circulation as the tubular epithelium modifies the ultrafiltrate formed at Bowman's space. The cortical and medullary capillaries are quite distinct in structure and function (Figure 3.4).

There is great variation in the post-glomerular capillaries depending on the locale of the glomerulus. In the outermost cortical nephron, the efferent arteriole gives rise to a dense capillary network surrounding the proximal convoluted tubule and to a much lesser extent the distal convoluted tubule and cortical collecting duct. While there is some association between the initial portion of the peritubular capillary and the early proximal tubule of the same glomerulus (Kriz & Kaissling 2007), it is important to note that the peritubular capillary of a specific proximal tubule need not come exclusively from the efferent arteriole of the same parent glomerulus from the same nephron (Kriz & Kaissling 2007). Blood from several glomeruli may end up bathing several proximal tubules. The relationship of the efferent arteriole, peritubular capillary and the proximal tubule is of major significance in terms of proximal Na^+ and water reabsorption. The hydraulic permeability of the peritubular capillary is in the same order as that of a regular capillary rather than the extremely high conductivity found in the glomerular capillary (Wiederhielm & Weston 1973).

A total of 7–15% of postglomerular blood may be destined for the medullary capillaries. The efferent arterioles coming off the juxtamedullary glomeruli supply the entire renal medulla. In the outer medulla, the efferent arteriole ramifies into an intricate network consisting of coalescence of the vas recta and the interbundle capillary plexus. In the inner medulla, straight largely unbranching long descending and ascending capillaries (vasa recta) run down to and up from the medullary tip in bundles. Simple ramifications exist between the vasa rectae. The anatomic relationship between the vasa recta, the loop of Henle, and the collecting duct is critical for water absorption, and for urea, ammonium, and potassium recycling.

TUBULAR EPITHELIA

The glomerular ultrafiltrate is a voluminous but relatively non-selective product of plasma consisting of the non-cellular fluid phase minus macromolecules. Its modification to final urine of the target composition and volume requires multiple segments of renal tubules. In the glomerulus, the parietal epithelium although contiguous with the visceral epithelium, is quite distinct in structure consisting mostly of squamous cells of poorly understood function. The transporting renal tubule commences when the parietal epithelium transforms into the tubule performing transport, enzymatic, endocrine, paracrine, and autocrine roles. The structural features of the renal tubule are listed in Figure 3.7. The renal tubule is a prime example of polarized epithelia with distinct structural and functional properties on the opposite membranes bridging the urinary lumen to renal interstitium and eventually the peritubular capillary effecting vectoral transport. A polarized cell positions its plasma membrane transporters in a fashion such that net movement of specific solutes is ensured to proceed in a specific direction – whether it is from urine to blood (absorption) or blood to urine (secretion). Therefore a role served by specific transporter protein will completely depend on its apical vs. basolateral locale on the cell, where on the nephron segment is this cell located.

As one courses down the nephron, the renal epithelium is structurally distinct and highly specialized to perform specific functions. The renal tubular architecture coalesces

as one travels downstream (Figure 3.8). Approximately 11 glomeruli converge into one cortical collecting duct (Kriz & Kaissling 2007) and collecting ducts on the average merges about eight times so 2^8 cortical collecting ducts drain into one papillary duct. The single most important feature to note regarding the renal tubule is the juxtaposition of the different parts of the nephron to each other. One simply cannot overemphasize the importance of understanding the geographic relationship between tubular structures. The thick ascending limb loops back and physically communicates with its parent glomerulus as the macula densa, thereby constituting the juxtaglomerular apparatus. The positioning of the loop of Henle, the vas recta, and the collecting duct is critical for the countercurrent multiplication to create the high medullary tonicity. The surface area is amplified in the apical membrane by microvilli in the form of a fully developed brush border or simple protrusions and the basolateral counterpart is achieved by either inward infoldings into the cell or outward interdigitations with the neighboring cell. The cells are separated by specialized

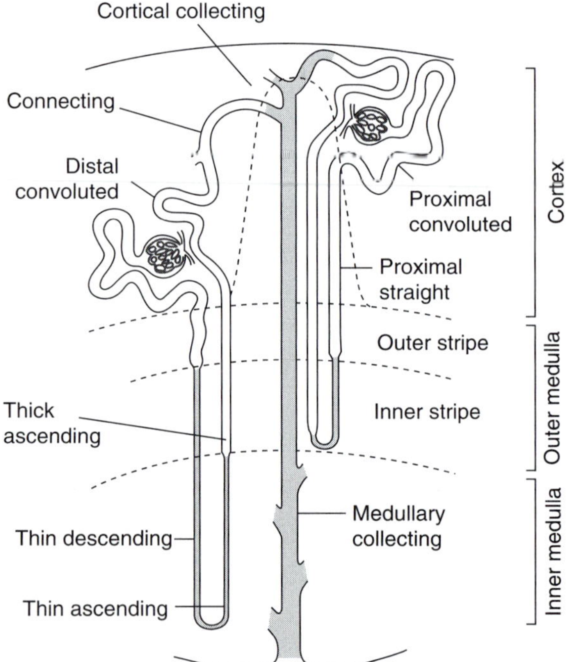

FIGURE 3.8 Overview of nephron segments. Two representative nephrons are depicted with a superficial or a deep glomerulus. The proximal tubule is in both cortex and medulla. The thin descending limb starts at the junction of the outer and inner stripe of the outer medulla. The tips of the loops of surface glomeruli never quite reach the inner medulla. The longer loops of the deeper glomeruli go all the way to the tip of the medulla. The distal nephron starts with the thick ascending limb within the outer medulla and ascends past the macula densa and becomes the distal convoluted tubule. Between the distal convoluted tubule and the cortical collecting duct is the connecting tubule. The collecting tubule spans from cortex to the renal pelvis

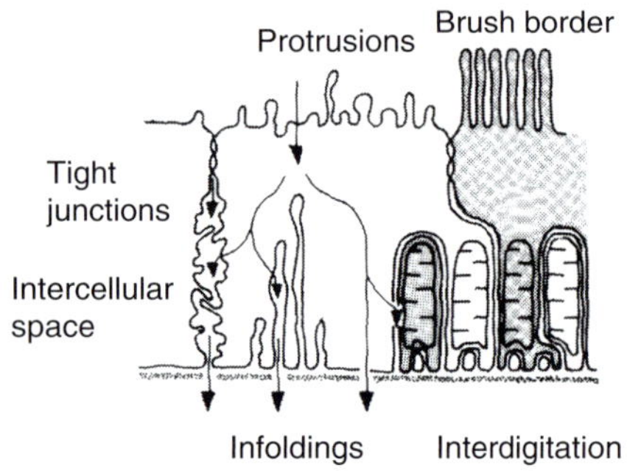

Tubule
Polarized epithelium
Axially differentiated for specialized function
Coalescence
Strategically positioned in relationship to other structures

Cell
Surface area amplification
  Brush border apical protrusions,
  Basolateral infoldings, interdigitations
Intercellular junctions
Intercellular space

FIGURE 3.7   Characteristics of a renal tubule and renal tubular cell

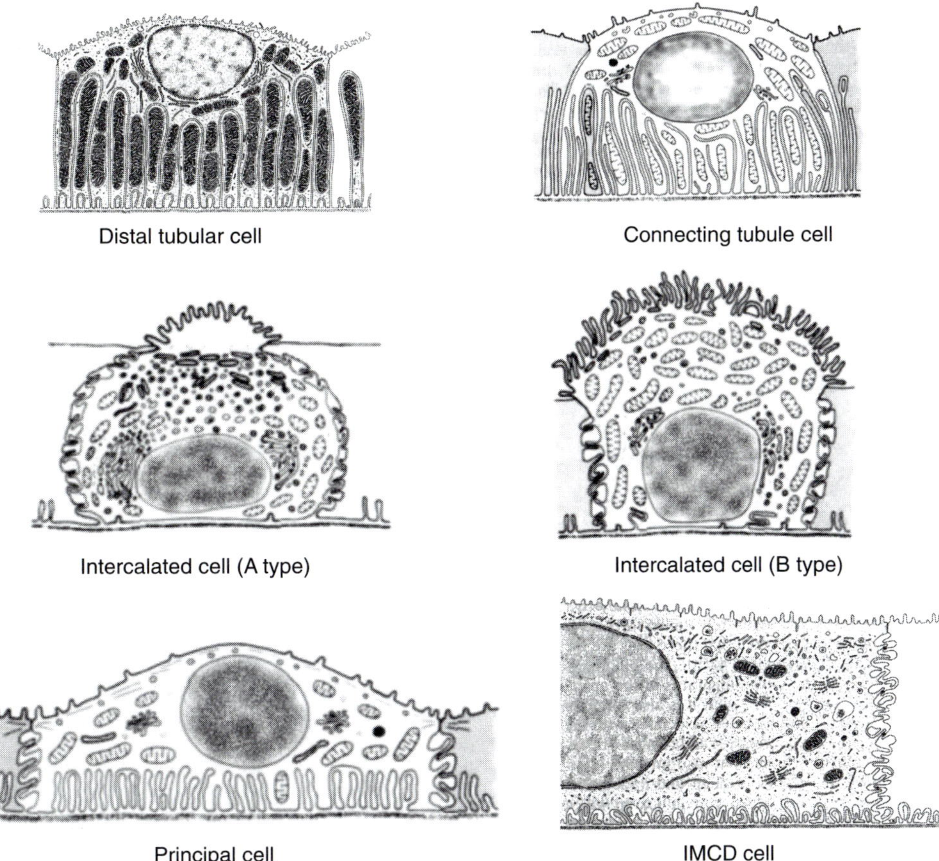

FIGURE 3.11   Cells of the distal nephron. The distal convoluted tubule showing extensive association is observed between mitochondria and the basolateral infoldings and interdigitations. The distinction between early DCT (DCT1) and late DCT (DCT2) is only possible by immunostaining for proteins with segment-specific expression. The basal surface area augmentation in the connecting tubule is achieved by true infoldings of the basal cell membrane inward to the cell instead of outward protrusions of interdigitating processes from adjacent cells. The mitochondria are fewer and smaller, and lack the close association with infoldings seen in the distal convoluted tubule. The principal cell contains characteristic basal infolding referred to as basal labyrinth in the bottom quarter above which lies a scatter of mitochondria. Extensive subapical microtubules and microfilaments create a network compatible with highly regulated protein trafficking. Tight junctional belt complexes with multiple strands are in direct communication with intercellular space, a finding congruent with the significant paracellular transport in these cells. Intercalated cells are extremely rich in mitochondria and ATP sustains largely the H^+-ATPase. The morphologically distinct type A and type B intercalated cells correspond to different functions in terms of acid–base regulation. There is an extensive array of tubular vesicle structures which are likely related to their regulation of transporter by protein trafficking. The IMCD cell resembles the principal cell with fewer mitochondria

RENAL INTERSTITIUM

The renal interstitium surrounds and accommodates the vascular and tubular elements. Because of its interposition between the tubules and the vessels, the renal interstitium provides a conduit for solute and water transport via multiple modes including urine-to-blood, blood-to-blood, blood-to-urine, and urine-to-urine. It is also the ground where electrochemical and hydrostatic gradients are installed in conjunction with the tubules and vasculature to facilitate homeostatic transport functions, such as hypertonicity to drive water reabsorption. The renal interstitium occupies about 10–15% of the renal cortex (Kriz & Kaissling 2007). The interstitial volume increases to 20–25% in the outer medulla and may be up to 30–40% in the inner medulla.

The renal interstitium is composed of extracellular matrix and various cell types.

The matrix consists of sulfated and non-sulfated glycosaminoglycans, type I and II collagen, and fibronectin (Couchman et al 1994). In addition to provision of support, this medium likely has minimal resistance to diffusive and convective flux although careful measurements have not been made. Although there is the potential, there is no evidence to date that changes in composition of the interstitium are involved with regulation of solute and water transport. Macromolecules can be hygroscopic and retain water. Polyanions can chelate ions such as calcium and magnesium.

The study of cellular components of the renal interstitium is still an evolving field. This unique compartment can harbor cells that are poised in a unique position in close apposition to the transport epithelial cells to influence urinary excretion, as well as in an organ with the largest blood flow per gram tissue to secrete endocrine hormones. Table 3.2 provides a summary of these interstitial cells and their possible functions. The role of renal interstitial cells can only be understood if the geographical location with the rest of the nephron is appreciated. Consider the cortical type I cell as an example. Erythropoietin transcript has been documented in these cells at baseline in the deep cortex (Eckardt et al 1993). Upon stimulation by either hypoxia or anemia, the number of erythropoietin-expressing cells increases and spreads geographically towards the superficial cortex. The stellate conformation of these cells enables a single cell to come into physical contact with both the proximal tubule and its peritubular capillary. The functional importance of this structural design will be discussed below.

In addition, a lipid-laden cell in the medulla has been shown to be the site of synthesis of prostaglandins and is likely related to water homeostasis (Yang 2003). There are also phagocytic and immune cells in the interstitium.

Physiologic Cell Models Along the Nephron

GENERAL PRINCIPLES OF EXCRETION

The kidney fulfills its homeostatic function in the capacity both as an excretory and endocrine organ (Figure 3.3).

Excretion is vital to all living things from single-cell prokaryotes to complex multicellular organisms. From the excretory point of view, the kidney maintains proper concentration of normal solutes, fluid volume of various compartments, and removes endogenous metabolic end products and exogenous substance acquired by the organism. The kidney has the task of recognizing what to excrete with fidelity and how much to excrete with precision.

In the simplest unicellular organisms, this can be achieved by simple diffusion into the ambient environment. Some rudimentary excretory mechanism can be found even in unicellular organisms where wastes are accumulated into vacuoles in the cytoplasm and the vacuoles are discharged into the exterior by contractile mechanisms. In multicellular organisms, excretion is achieved by specialized organs through primarily three processes: filtration, reabsorption, and secretion (Figure 3.12). The Malpighian tubule in insects, for example, is primarily secretory with some small components of filtration and reabsorption. Mammalian nephrons are functionally complex and perform a combination of filtration, reabsorption, and secretion. Filtration is often set by requirement of solute that requires high clearance such as creatinine. High filtration rate naturally mandates filtration–reabsorption design to reclaim and conserve valuable solutes that would have been lost due to the high filtration rate. Sodium and bicarbonate are such examples.

Table 3.3 provides some examples of the relative magnitudes of these three processes in the mammalian kidney. Note that the filtration and reabsorption of potassium virtually nullify each other and excretion is determined almost exclusively by secretion. Each of these components has different capacities for each solute and each is regulated

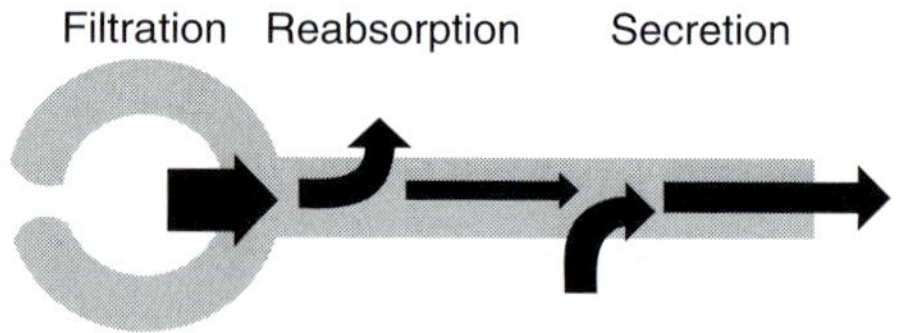

FIGURE 3.12   Three types of transport. A nephron can utilize a combination of the three processes. The relative contribution of each process may vary for individual solutes

TABLE 3.2   Renal interstitial cells

Cell type	Morphology	Function
Cortex		
Type I	Stellate Well developed ER Between tubule and capillary	Erythropoietin production
Type II	Mononuclear round	Antigen presentation
Medulla		
Type I	Lipid-laden	Water homeostasis, Prostaglandin synthesis
Type II	Mononuclear round	Phagocytic
Type III	Pericyte	Unknown

TABLE 3.3   Examples of relative magnitudes of filtered, reabsorbed, secreted, and excreted solutes and water

	Filtered	Reabsorbed	Secreted	Excreted
Na^+	24 000	23 900	0	100 mmoles
Ca^{2+}	260	255	0	5 mmoles
K^+	720	720	50	50 mmoles
HCO_3^-	4300	4300	0	0 mmoles
Water	170	169	0	1 L
Urea	600	80	0	520 mmoles
Creatinine	12	0	2	14 mmoles

All values varies with dietary intake and physiologic state.

to various extents ranging from largely constitutive operations to exquisitely regulated systems. For example, sodium and calcium is handled by a filtration–reabsorption design where the regulation lies mostly in the reabsorptive phase. Bicarbonate is handled by filtration–reabsorption but the kidney can also secrete bicarbonate in its distal segments. Potassium is filtered, completely reabsorbed, and largely regulated by secretion. The high filtration rate enables large turnover and clearance of the extracellular fluid volume, which is of clear benefit to an organism with high metabolic rate. High glomerular filtration rate also mandates recovery of this colossal filtrate, resulting in a fractional excretion of sodium of about 1%.

The excretory process starts with the incoming blood and ends with elaboration of urine in the collecting system. The tubule modifies urine composition and volume. Renal blood flow, glomerular filtration rate, and tubular transport will be discussed sequentially.

RENAL BLOOD FLOW AND GLOMERULAR FILTRATION

Only a few selected aspects of these topics will be covered in a limited fashion. These are excellent illustrations of intrarenal processes that significantly influence the feedback homeostatic loops (Figure 3.3). As indicated earlier, the human kidney constitutes less than 1% of body weight but receives more than 20% of cardiac output. The calibers of the vessels are not different from other organs but the high flow (low resistance) can be attributed to the extensive parallel conductances. Autoregulation refers to the ability of the kidney to maintain relatively constant blood flow despite changes in arterial pressure (Figure 3.13). This occurs in both the cortical and medullary circulations over the range of perfusion pressures from 80–160 mmHg. This phenomenon is unequivocally established for the whole kidney and cortex but has remained controversial for the medulla. The term autoregulation has been applied to describe the constancy of renal blood flow as well as glomerular filtration rate. Autoregulation stems from the ability of the kidney vasculature to decrease its resistance when perfusion pressure is reduced. Two mechanisms have been proposed. An intrinsic myogenic mechanism built into the preglomerular circulation relaxes the resistance vessels as perfusion pressure falls without change (or slight increase in resistance) in the post-glomerular vessels (Figure 3.13). The second mechanism is tubuloglomerular feedback, which is an ingeniously poised system where the distal delivery of sodium chloride to the specialized portion of the thick ascending limb, the macula densa, feeds back onto the glomerular ensemble of the same nephron and alters arteriolar tone, renal blood flow, and glomerular filtration rate (Figure 3.13). This is indeed a quintessential example of anatomic and functional intranephron feedback. In this set-up, proximal solute absorption dictates sodium chloride exit from the proximal nephron and eventually delivery to the macula densa. The physiologic response of the proximal tubule to extracellular fluid volume disturbance is recruited to provide a signal for the macula densa to adjust glomerular filtration rate so the organism can maintain high and constant glomerular filtration rate in the presence of low and fluctuating levels of sodium intake, an adaptation that is critical for terrestrial existence.

Ultrafiltration in the glomerular capillaries is unique. Whereas in most systemic capillaries, the 'filtration fraction' is about 0.25%, the same parameter is about 25%, a hundred-fold higher in the glomerular circulation. The hydrostatic pressure driving ultrafiltration is about 20–30 mmHg but the combined hydraulic permeability and surface area of the glomerular capillary tuft is 50–100 times higher than for a capillary in skeletal muscle. There are many determinants of glomerular filtration rate. Three of these factors are depicted in Figure 3.14. It is evident that the preglomerular and post-glomerular resistance vessels have drastically different effects on plasma flow and filtration rate (Figure 3.14). If afferent arteriolar tone is primarily increased, SNGFR falls sharply, even to a greater extent than does GPF, since preglomerular vasoconstriction will cause both a fall in glomerular capillary

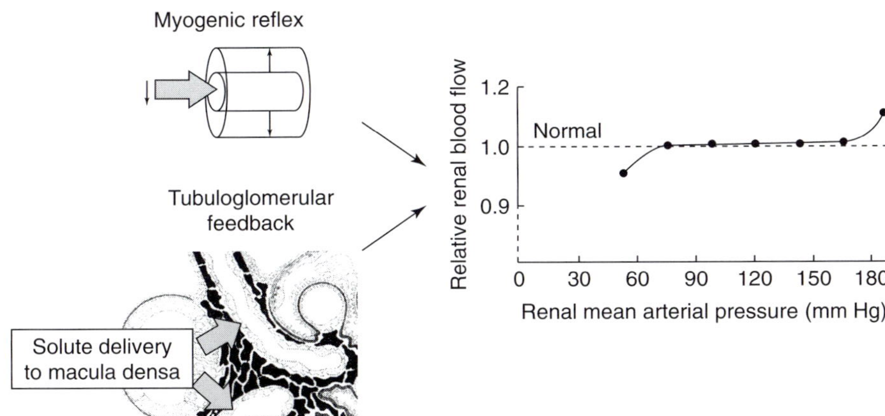

FIGURE 3.13 Autoregulation of renal blood flow. At least two processes are responsible for the autoregulation of renal blood flow – an intrinsic ability of the resistance vessels to dilate in the presence of diminishing perfusion pressure, and the ability of the macula densa to influence arterioles of its parent glomerulus based on sodium chloride delivery out of the proximal tubule

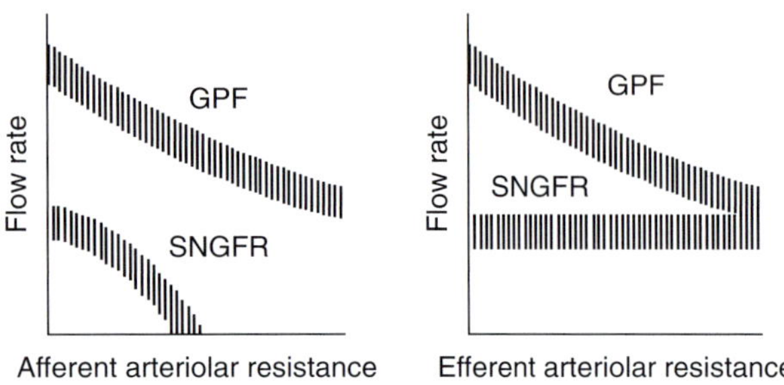

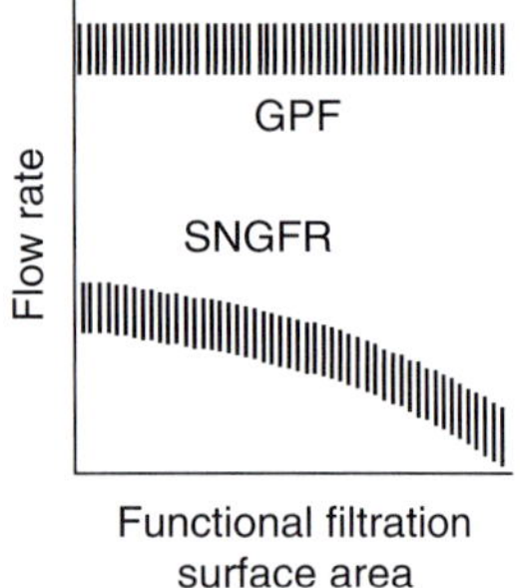

FIGURE 3.14 Three determinants of glomerular plasma flow (GPF) and single nephron glomerular filtration rate (SNGFR): afferent arteriolar resistance, efferent arteriolar resistance, and functional filtration surface area

pressure and a fall in GPF. The same reduction in GPF can be produced by selective efferent arteriolar constriction, but SNGFR is well preserved, since glomerular capillary pressure rises and thereby offsets the effect of reduced GPF. In addition, the filtration fraction rises, an effect that promotes proximal reabsorption. The effect of reduction of glomerular filtering surface reduces SNGFR but GPF is well preserved. The principal mechanisms regulating these arteriolar resistances include incoming systemic signals and intrinsic renal factors (Figure 3.3) including myogenic reflexes autoregulation (Haberle 1988), tubuloglomerular feedback (Schnermann & Briggs 1985), angiotensin (Myers et al 1975), renal nerves and catechols (Kon et al 1985), and prostaglandins (Haberle 1988).

TYPES OF TRANSPORT PROCESSES

Solutes can be moved in a number of ways. It is imperative to note that all solute or solvent movement can eventually be traced back to some type of energy requiring process that set up the physicochemical gradients. From this viewpoint, the division between active vs. passive transport can be viewed as somewhat arbitrary (Table 3.4). Passive transport involves movement of solutes either via diffusion (synonymous with conduction) down its chemical or electrical (for charged solutes only) gradient. Alternatively, as solvent (water) moves across an epithelium driven by either osmotic or hydraulic forces, solutes can move along with the bulk phase by convection (synonymous with solvent drag) provided the reflection coefficient of the solute across that membrane is sufficiently low to allow its transport as

TABLE 3.4 Types of solute transport

Passive transport (usually paracellular)

Diffusion –	Chemical
	Voltage
Convective –	Osmotic
	Hydraulic

Active transport (usually transcellular)

Primary active –	Direct coupling to ATP hydrolysis
Secondary and tertiary active –	Ion and voltage gradients set up by ATPase

TABLE 3.5 Metabolic substrates for different segments of the nephron

Proximal tubule

Glutamine	Aceto-acetate	Fatty acids
Glutamic acid	β-hydroxybutyrate	Glycerol
Alanine	Lactate	
	Pyruvate	
	Acetate	
	Citrate	

Thin limb

Glucose	Pyruvate
	Lactate
	β hydroxybutyrate

Thick ascending limb

Glucose	Lactate	Fatty acids
	Pyruvate	
	Acetate	
	Aceto-acetate	
	β-hydroxybutyrate	

Distal convoluted tubule

Glucose	Lactate
	β-hydroxybutyrate

Cortical collecting duct

Glucose	Lactate	Fatty acids
	Pyruvate	
	β-hydroxybutyrate	

Medullary collecting duct

Glucose	Lactate
Glutamine	β-hydroxybutyrate

The three columns are grouped into glucose and amino acids, organic anions, and fat.

a passenger. Glomerular ultrafiltration is a pure convective isotonic movement of solutes and water. There is however some diffusive movement of macromolecules such as albumin from the blood to Bowman's space as the intracapillary volume is reduced by ultrafiltration.

Active transport requires energy. Like many other organs, the kidney can utilize a broad range of metabolic substrates. The substrates utilized by the kidney are derived solely from the plasma and encompass a wide range of organic molecules. Some examples are shown in Table 3.5. There is some preference for specific substrates by different nephron segments although it is not absolute. This wide range of metabolic substrate is advantageous so the epithelial cells can subsist under a diverse range of metabolic states. For a given segment, the utilization of different substrates can vary under varying physiologic or pathophysiologic states. Most segments metabolize organic anions and some other alternatives such as glucose, amino acids, and fatty acids. Glucose is utilized by all segments except the proximal tubule where glucose is reabsorbed and actually generated as part of total body gluconeogenesis. Glutamine is used by the proximal tubule for ATP generation, gluconeogenesis, and ammoniagenesis. So when the demand for ammonia is increased such as during metabolic acidosis, glutamine will be preferred to generate ATP.

The currency for energy is converted once in the renal epithelium to the next level. The diverse metabolic substrates go through different metabolic pathways and the universal currency for cellular energy converges onto ATP, which sustains all cellular processes. Primary active transport directly couples the energetically uphill solute movement to the ATP hydrolysis. At the superficial level, one may intuitively think that each transported solute is served by its own private ATPase so dedicated regulation can be easily deployed. However, in reality, very few transporters and very few transport processes in the kidney are actually directly coupled to ATP hydrolysis. Examples include Na^+-K^+-ATPase, H^+-ATPase, Ca^{2+}-ATPase, and other ATP-binding proteins such as P-glycoprotein and cystic fibrosis transmembrane regulator (CFTR) that binds and hydrolyzes ATP but it is less clear whether the energy is directly coupled to move solutes. The majority of transport processes simply harness the electrochemical gradient set up by ATPases.

The energy stored in phosphate bonds in ATP is translated into a different currency in the form of electrochemical gradients. In the proximal tubule for example (Figure 3.15), the basolateral Na^+-K^+-ATPase converts the energy harbored in ATP into the most common currency of this cell, which is the chemical gradient for Na^+ and a negative interior cell voltage. Figure 3.16 depicts a large range of transporters in the kidney. Cotransporters and counter transporters usually couple the uphill movement of a target solute to the downhill movement of solute (e.g. Na^+) recruited to drive the process. Parallel coupled co- and/or countertransporters can expand the use of electrochemical ion gradients. Figure 3.16 also shows the coupling of two countertransporters to functionally constitute a 'new' transporter. Channels primarily move single species, charged (cations, anions) or uncharged (urea, water), across a lipid bilayer. Paracellular proteins also regulate passage solutes and water between cells.

It is clearly advantageous for energetic as well as genetic economy to have secondary transport running on the electrochemical currency rather than each solute with a dedicated ATPase. This design may be prudent from an energetic and genetic point of view but it places the cell in a position

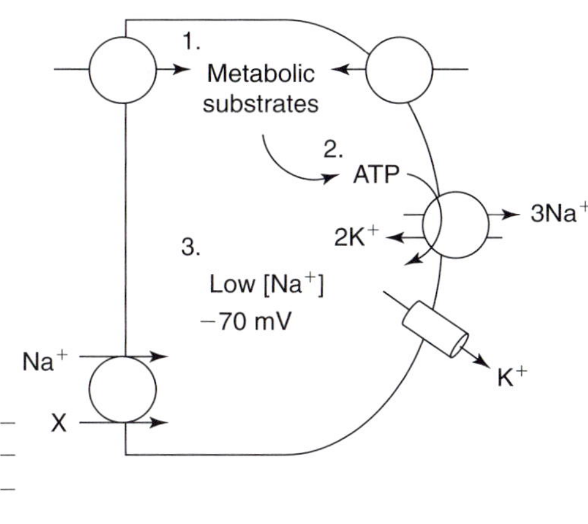

FIGURE 3.15 Substrate utilization and currency of energy expenditure in tubular epithelial cell. (1) The initial currency includes all metabolic substrates such as those catalogued in Table 3.5. (2) All of these various fuels converge into the high-energy ATP. (3) While a few transporters are directly coupled to ATP, much more commonly, energy from the phosphate bonds are translated to electrochemical gradients such as low cell sodium concentration and negative voltage from the $3Na^+$-$2K^+$-ATPase coupled to K^+ diffusion potential. The electrochemical gradient is then utilized as a generic currency to drive a whole array of transport processes. As an example, the luminal uptake of Na^+ and electroneutral X is energized by the electrochemical gradient. The negative charge in the lumen generated by this process then drives chloride around the cell

where a whole host of transport processes are coupled to the same electrochemical driving force, which renders control of individual solute transport rather challenging. One recurrent theme in this epithelium is the utilization of the apical membrane as the regulatory site where sodium-coupled transporters have restricted substrate-coupling. From this point of view, the apical transporters along with any specialized basolateral exit mechanism become the target of regulation by hormones akin to specific executive departments of a government and the Na^+-K^+-ATPase is relegated in the role of maintaining the energetic status of the cell, very similar to the financial health of a nation. For example, phosphate depletion is dealt with by up-regulation of the apical membrane sodium-coupled transporters (Forster et al 2006). The response to metabolic acidosis involves an increase in apical membrane sodium/proton exchange (bicarbonate absorption, ammonium excretion, titrate luminal phosphate and citrate to facilitate absorption), sodium–citrate cotransport (base conservation), sodium–phosphate cotransport (increase in urinary closed buffer), sodium–sulfate cotransport (increased distal delivery of non-absorbable anion to facilitate distal proton secretion), and basolateral sodium–bicarbonate cotransport (base exit to plasma). This regulation allows the proximal tubule to cater to a specific need and not disturb the integrity and equilibrium of other transport systems.

MECHANISMS OF SODIUM TRANSPORT

Many genes have been cloned that code for sodium transporters or proteins that regulate sodium transporters. These will not be covered in this chapter but instead, general concepts of regulation will be highlighted. The kidney is the major organ of external sodium chloride balance as there is limited regulated NaCl absorption in the intestine. Sodium handling follows primarily a filtration – reabsorption design. There is little or no documented sodium secretion in the mammalian nephron. Sodium absorption exemplifies a lot of the general properties of absorption along the nephron (Figure 3.17). Both transcellular and paracellular pathways are used extensively, driven largely by electrochemical forces. Minor contributions from convective flux are likely present in the proximal tubule. While the proximal nephron has high capacity and low gradient, the distal nephron

exhibits higher gradient and lower capacity modes of transport (Figure 3.18). This difference is reflected in the cellular array of transporters along the nephron (Figure 3.18). Relative magnitudes of NaCl flux are shown in Figure 3.18 axially along the nephron. The proximal nephron has high capacity for transport but low gradient. In contrast, the distal nephron has much lower capacity but can sustain a high sodium gradient between lumen and plasma and thus can render the urine literally sodium-free. This shift from capacity to gradient is reflected in the transporter design. The basolateral mechanism that converts ATP into electrochemical energy and mediates Na^+ efflux from the cell is the same for all segments – the Na^+/K^+-ATPase. In contrast, the apical mechanism varies widely. In the proximal nephron, an array of Na^+-coupled transporters in the proximal tubule harnesses the Na^+ gradient to reclaim various

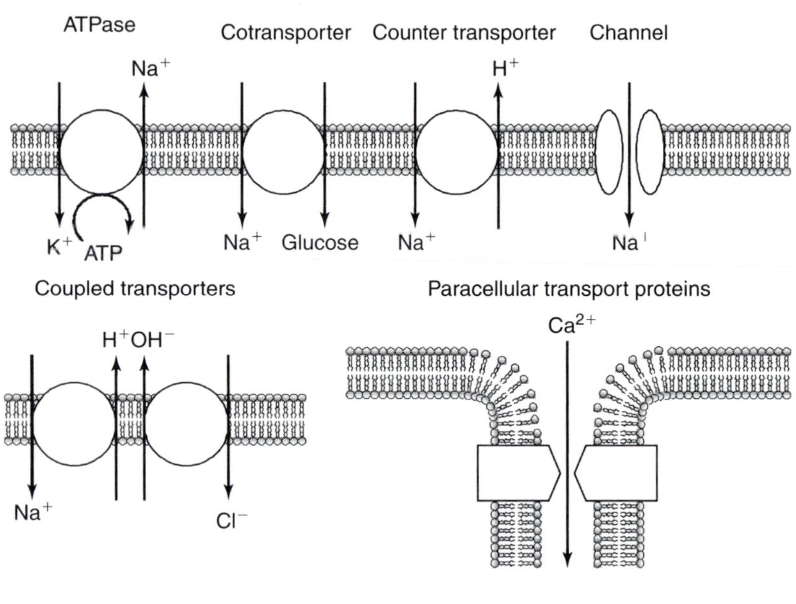

FIGURE 3.16 Transport proteins in renal epithelia. ATPases directly couple solute movement to the energy released from ATP hydrolysis while all others utilize electrochemical gradients ($\Delta\mu$) generated by ATPase. The low cell [Na^+] and negative interior voltage drive a number of cotransporters, countertransporters (exchangers, antiporters), and channels. Cotransporters and countertransporters can also be coupled to constitute new formats of solute movement, e.g. parallel coupling of Na^+/H^+ exchange and Cl^-/OH^- exchange with recycling of water creates $Na^+–Cl^-$ cotransport. Finally, paracellular proteins can control paracellular solute permeability and movement

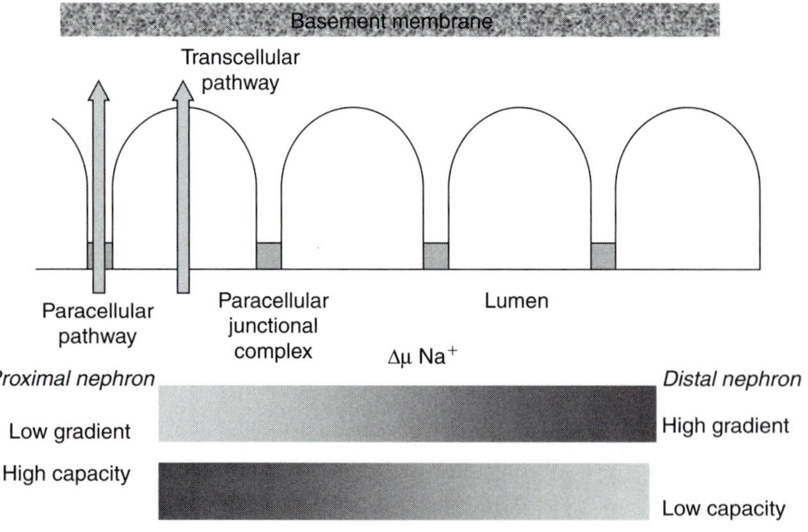

FIGURE 3.17 General features of absorption along the nephron. Transport can be transcellular or paracellular driven by the electrochemical gradient of Na^+ ($\Delta\mu$). Proximal nephron functions to reclaim the bulk of glomerular filtrate while the distal nephron performs more fine-tuning functions

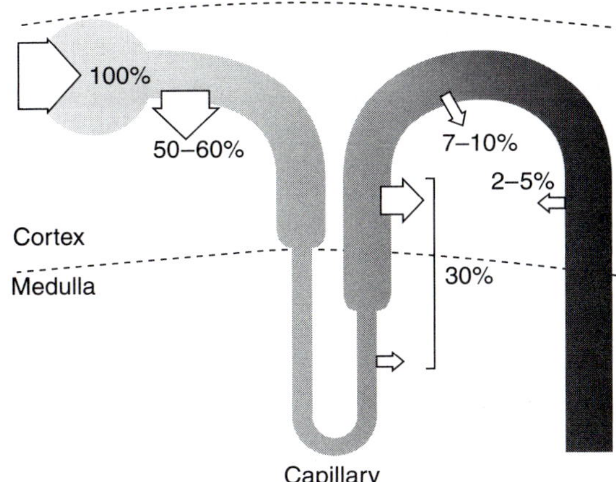

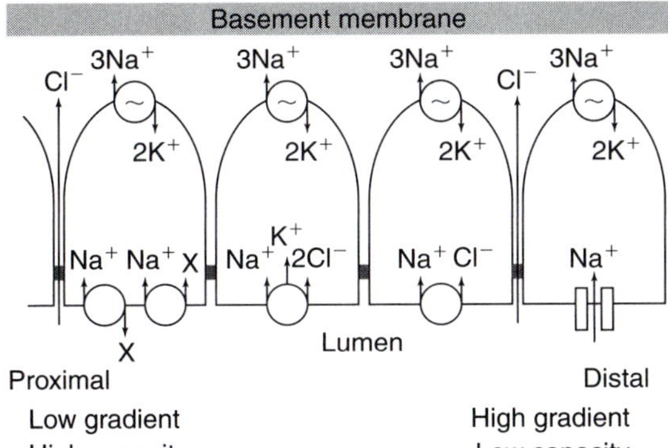

FIGURE 3.18 Sodium reabsorption in the nephron. Top panel shows the percentage of filtered sodium reabsorbed in the different segments. The bottom panel depicts a common basolateral active sodium exit mechanism with varying apical entry mechanisms for sodium with highly varied coupling to broad substrates to direct coupling with chloride and finally just a sodium channel

solutes. Apical mechanisms become more dedicated to Na$^+$ and Cl$^-$ transport in the distal nephron (Figure 3.18).

Proximal tubule

In the proximal tubule, large volume of isotonic fluid is absorbed driven by solute absorption and high water permeability (Figure 3.18). The mechanism of sodium absorption is the most complex of all the nephron segments. The reason for this is that the $\Delta\mu$ of sodium is utilized to reabsorb a variety of solutes in the proximal tubule, most of which are not related to extracellular fluid volume homeostasis. Sodium chloride can be absorbed through the transcellular or paracellular route (Figure 3.17). Both processes are driven by the 'master pump', the Na$^+$/K$^+$-ATPase on the basolateral membrane. Transcellular NaCl absorption is mediated by an array of parallel cotransporters in the apical membrane (Figure 3.19). Passive NaCl reabsorption was originally proposed by Rector and associates in 1966. While sodium concentration stays virtually invariant along the length of the proximal tubule, luminal chloride concentration rises considerably, thus providing a driving force for paracellular chloride diffusion (Figure 3.20). The small luminal positive potential difference in the initial proximal tubule generated by electrogenic (positive net charge) sodium-coupled transport of various organic and inorganic solutes also facilitates chloride diffusion. Due to the greater paracellular permeability to chloride than to bicarbonate, the potential difference in the late proximal tubule is actually lumen-positive, which provides the driving force for paracellular sodium movement since there is no chemical concentration gradient. This is one segment of the nephron

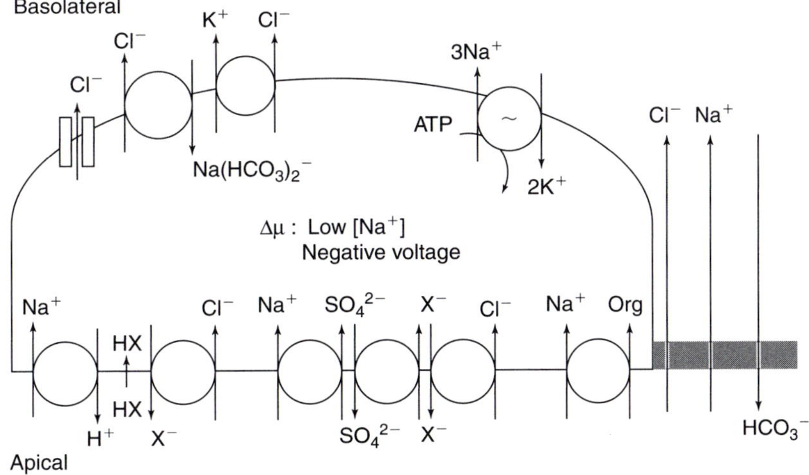

FIGURE 3.19 Cell model for proximal tubule sodium chloride absorption. Basolateral Na$^+$/K$^+$-ATPase translates ATP to electrochemical driving force ($\Delta\mu$). Apical sodium entry proceeds via different types of coupling of cotransporters and exchangers that results in net NaCl uptake. Electrogenic sodium-coupled transport uptake of organic solutes also provides the driving force for paracellular chloride movement

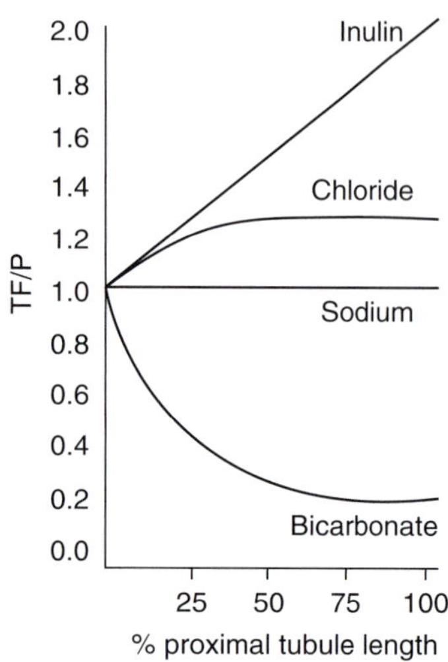

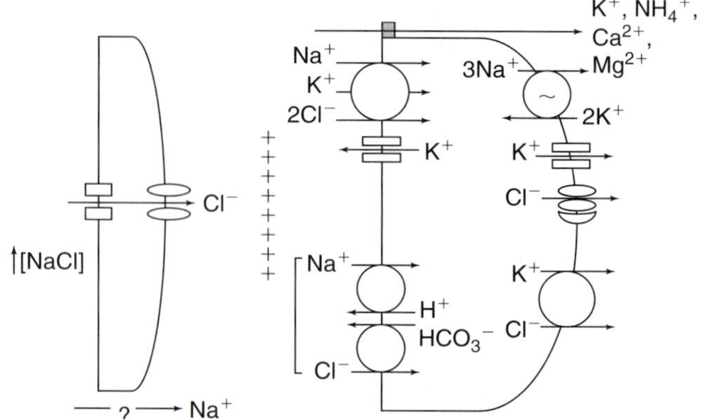

FIGURE 3.21 Sodium chloride transport at the loop of Henle. The thin ascending limb is show on the left, where there is transcellular chloride movement. The thick ascending limb is on the right showing apical and basolateral transport mechanisms

FIGURE 3.20 Axial change in composition of proximal tubule fluid. TF/P = tubular fluid to plasma ratio of the solute. Bicarbonate absorption concentrates chloride in the lumen and creates a reciprocal profile in luminal chloride concentration (adapted and redrawn from Rector FC Jr)

where hydrostatic pressure has been shown to affect sodium chloride and water transport. The intricate dynamics of per itubular physical forces will be discussed in a subsequent section.

Loop of Henle

Sodium chloride handling in the loop of Henle differs drastically from the proximal tubule. The descending limbs have extremely low Na^+/K^+-ATPase activity, and low ionic permeability with high water permeability. Sodium chloride absorption is insignificant in the descending limb but the high water transport serves to concentrate the luminal concentration of sodium and chloride. After the hairpin turn the thin ascending limb is similar in that it has a flat endothelium-like epithelium without significant mitochondria or Na^+/K^+-ATPase activity; therefore it does not transport solutes actively. However, in contrast to the descending limb, the ascending limb is water impermeable, moderately urea permeable, and highly sodium chloride permeable. In fact, the ascending loop has the highest chloride permeability reported in the nephron (Chou & Knepper 1993). The chloride channel ClC-K1 (ClC-KA in humans) is localized in the basolateral membrane of the thin ascending limb (Uchida et al 1993, 1995).

The thick ascending limb of the loop of Henle reabsorbs approximately 30% of the NaCl filtered at the glomerulus. Because of the relatively low hydraulic permeability, the solute flux greatly exceeds the water flux. The accepted model is depicted in Figure 3.21. The lumen-positive potential difference is due to the electrogenic transcellular transport. The energy for the apical membrane $Na^+/K^+/2Cl^-$ cotransporter is provided by the inward chemical concentration gradient

for Na^+ and Cl^-. As in the proximal nephron, the low cell Na^+ concentration is maintained by the basolateral Na^+/K^+-ATPase. At the apical membrane, a large potassium diffusion pathway exists in parallel with the $Na^+/K^+/2Cl^-$ cotransporter (Liu et al 2002). Almost all of the apical membrane conductance can be accounted for by potassium. As a result of this parallel arrangement, all the potassium that enters the cell recycles from the cell back into the tubule fluid across the apical membrane. Chloride exits the cell either via a channel or a K^+–Cl^- cotransporter (Figure 3.21). The lumen PD is positive at 3–10 mV, even under conditions of identical chemistry, due to apical membrane potassium diffusion potential (Sasaki & Imai 1980, Hebert et al 1981), which provides an electrical driving force via the low-resistance paracellular pathway for Na^+ transport across the TAL and for other cations, such as Ca^{2+}, Mg^{2+}, K^+, and NH_4^+ (Figure 3.21). Since the paracellular pathways are cation-selective and there is a high sodium concentration gradient from the interstitium to the lumen, the lumen-positive voltage can increase to as much as 30 mV due to diffusion potential of Na^+ from peritubular space to tubule fluid via the paracellular pathway. The thick ascending limb harbors the largest number of human mutations causing monogenic tubulopathies and will be covered in detail in this textbook.

The distal convoluted tubule is heterogeneous and can be divided into four segments as discussed before. Sodium reabsorption can proceed via electroneutral or electrogenic modes (Figure 3.22). The electroneutral mode predominates in the early and the electrogenic mode predominates in the late distal tubule. Electroneutral sodium absorption can be clearly dissociated from the electrogenic mode (Velazquez & Wright 1986, Ellison et al 1987). Although less important than direct sodium–chloride cotransport, the parallel coupling of sodium protein exchange and chloride bicarbonate exchange contributes to electroneutral sodium chloride

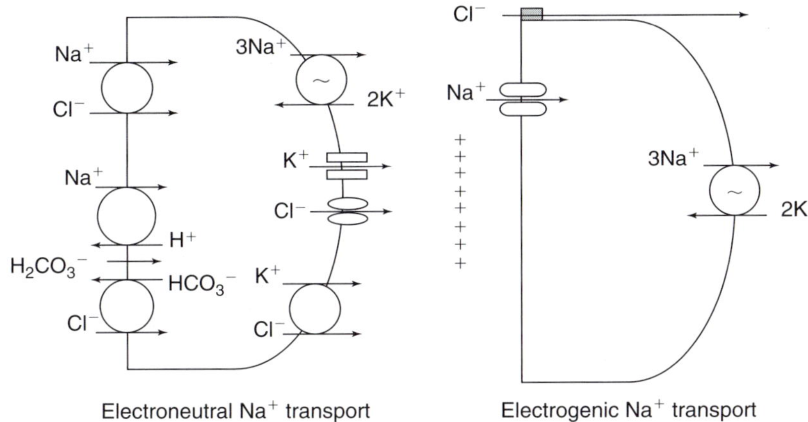

FIGURE 3.22 Sodium transport in the distal nephron. The electroneutral mode is used in the early while the electrogenic mode is employed in the late distal nephron. The electroneutral mode consists mainly of sodium chloride cotransport which can be mediated by the thiazide-sensitive cotransporter or to a lesser extent coupling of two parallel exchangers constituting net sodium chloride cotransport. Electrogenic sodium transport involves apical sodium channel with paracellular chloride movement driven by the negative luminal voltage

TABLE 3.6   Comparison of sodium and potassium homeostasis

	Sodium	Potassium
1. Systemic Factors	Effective arterial blood volume	Potassium concentration
	Blood pressure	Intracellular
		Extracellular
	Hormones	Hormones
	Sympathetic innervation	Acid–base status
2. Intrarenal Processes	GFR	Distal delivery of sodium
	Tubuloglomerular feedback	Luminal non-absorbable anions
	Peritubular physical factors	Potassium recycling
3. Renal		
Transport	Filtration – Regulated	Filtration – Freely passively filtered
	Reabsorption – Principal regulation	Reabsorption – Complete
	Secretion – None	Distal secretion – Principal regulation
		Distal reabsorption
Hormones	Renin release	Renin release

contransport (Stanton 1988). Sodium transport mechanisms in the collecting duct closely resemble the electrogenic cell in the distal tubule (Figure 3.22). Sodium reabsorption is under the control of aldosterone. Sodium enters the cell and exits at the basolateral membrane through the action of Na$^+$/K$^+$-ATPase, an enzyme which is activated directly by aldosterone. The reabsorption of sodium generates a negative voltage that secondarily influences three processes thusly: (a) chloride reabsorption through the paracellular pathway is enhanced; (b) potassium ions, copiously secreted into the tubular urine as a result of high intracellular potassium concentrations produced by the action of the basolateral ATPase, are trapped there; and (c) hydrogen ions, secreted by the intercalated A cells under the action of a sodium-independent aldosterone-sensitive proton ATPase are trapped in the tubular urine. This is also a segment of nephron that harbors a plethora of human monogenic defects that affect steroid hormone metabolizing proteins, their receptors, upstream regulators of transporters, and transporter proteins themselves.

Mechanisms of Potassium Transport

At the level of systems biology, the regulation of renal sodium excretion is governed by effective arterial blood volume and it is the same entity that is defended by renal sodium excretion or retention. In contrast to the quantity of a volume of fluid, renal potassium excretion is governed by both intracellular and extracellular potassium concentrations (Table 3.6).

The kidney is responsible for the bulk of potassium excretion but a modest fraction of ingested potassium may be excreted by secretion in the colon. The capacity of the latter to secrete potassium may increase when renal excretory mechanisms fail, but it is insufficient to maintain adequate external balance.

Clearance and balance studies demonstrated that potassium is reabsorbed and secreted by renal tubules and that urinary potassium excretion largely depends on secretion of potassium in distal nephron segments (Berliner 1961). Clearance studies also demonstrated the dependence of potassium secretion upon sodium reabsorption, potassium intake, adrenal hormones, and acid–base factors (Berliner

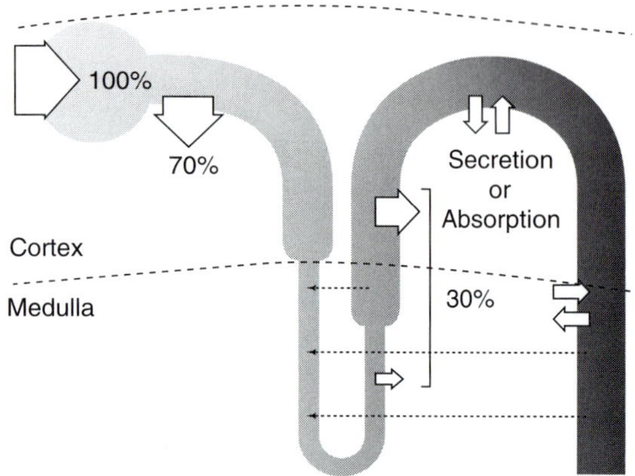

FIGURE 3.23   Potassium handling by the kidney. Of the amount filtered, the majority is reabsorbed by the proximal tubule and thick ascending limb. The distal nephron usually secretes potassium but can also engage in an absorptive mode. Dotted lines represent potassium recycling from one nephron segment to another through the interstitium

1961). The analysis of potassium transport by single nephron studies fully supports the view that potassium is extensively reabsorbed along the proximal tubule and the loop of Henle and that regulated potassium secretion along the connecting and collecting ducts is the source of potassium in the urine. Figure 3.23 summarizes the sites and magnitudes of potassium transport along the nephron and underscores the role of potassium secretion in potassium excretion (Berliner 1961, Giebisch 1998, 2004, Giebisch et al 2003). Note this is in sharp contrast to renal sodium handling which is mediated solely by filtration–reabsorption (Table 3.6). It is important to note that those nephron segments endowed with the ability to secrete potassium may also reabsorb potassium (Okusa et al 1992). Such reversal of net potassium transport can occur during states of prolonged potassium deficiency and provides an effective mechanism for potassium retention.

It is well established that distal nephron possesses at least two cell types and that potassium secretion and potassium reabsorption are carried out by different cells in connecting tubules and collecting ducts. Principal tubule cells secrete potassium, whereas a special type of intercalated cell reabsorbs potassium (Giebisch 1998, 2004, Giebisch et al 2003). Potassium handling by the kidney embraces all processes of filtration, reabsorption, and secretory mechanism but the principal regulatory step is secretion.

The proximal tubule

Potassium reabsorption along the proximal tubule is extensive, and is functionally linked to Na^+ and fluid absorption (Giebisch 1998, 2004, Giebisch et al 2003) (Figure 3.24). Basolateral potassium transport is similar in all tubule segments and mediated by active uptake of potassium via Na^+/K^+-ATPase. Potassium channels have been identified in the apical membrane of the proximal tubule cell (Hebert et al

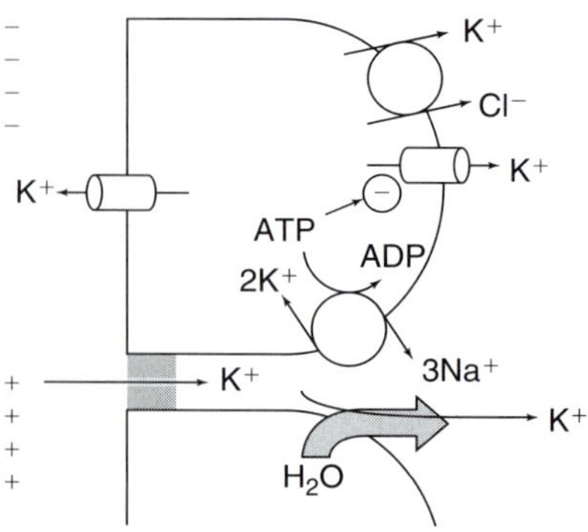

FIGURE 3.24   Potassium transport by the proximal tubule. Although an apical potassium channel is clearly present, it likely functions to contribute to the resting potential of the apical membrane rather than potassium secretion. The basolateral Na^+/K^+-ATPase lowers the potassium concentration of the lateral space, which facilitates diffusion from the lumen. Convective movement of water transport promotes potassium movement to the plasma. The positive luminal voltage in the late proximal tubule also contributes to the driving force. The basolateral potassium channel is gated by ATP/ADP ratio and functions as a coupler. When Na^+/K^+-ATPase activity is driven high by transcellular transport, the lower ATP/ADP ratio will activate basolateral potassium conductance to increasing recycling

2005). They do not mediate potassium secretion but play an important role in stabilizing the cell-negative electrical potential of the proximal tubule cell. Two distinct transport mechanisms have been identified to account for potassium reabsorption: diffusion and solvent drag (Weinstein 1988). Diffusion is driven by the lumen-positive potential in the second half of the proximal tubule. Solvent drag depends on active sodium and fluid absorption which generates small osmotic gradients in the epithelium that drive tubule fluid across cells and interspaces with potassium following because of its entrainment in the reabsorbate. The facts that the tubule wall has a high potassium permeability (low reflection coefficient) (Wareing et al 1995), and that net potassium reabsorption is dependent on and determined by fluid movement, support the view that potassium reabsorption depends on sodium and fluid transport (Weinstein 1988). The basolateral potassium channel in the proximal tubule is an excellent example of coupling to ATP consumption of the Na^+/K^+-ATPase (Beck et al 1991, Welling 1995) (Figure 3.24). Specific regulation of potassium transport appears to be absent in proximal tubules.

Loop of Henle

Different mechanisms of potassium transport are present in the thin and thick limbs of Henle's loop (Figure 3.25). Potassium transport in the thin limbs is limited to passive diffusion that mediates equilibration with the surrounding interstitial fluid as part of potassium recycling. Passive movement of potassium occurs especially in juxtamedullary nephrons with long loops of Henle, and depends on the presence of a

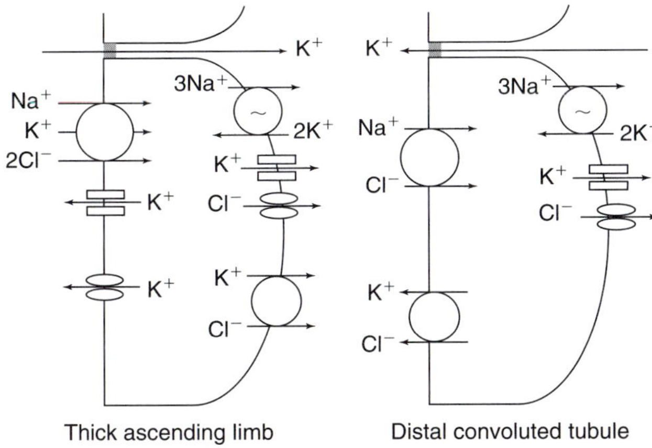

Thick ascending limb   Distal convoluted tubule

FIGURE 3.25 Potassium transport in the thick ascending limb and distal convoluted tubule. In the thick ascending limb apical membrane, sodium and potassium are transported in equimolar amounts despite much higher potassium delivery in the lumen. The apical potassium channel provides recycling as well as a diffusion potential to generate a luminal positive voltage that drives the paracellular diffusion of potassium. Potassium exits via several mechanisms on the basolateral membrane as shown. In the distal convoluted tubule, potassium is secreted via both transcellular and paracellular pathways

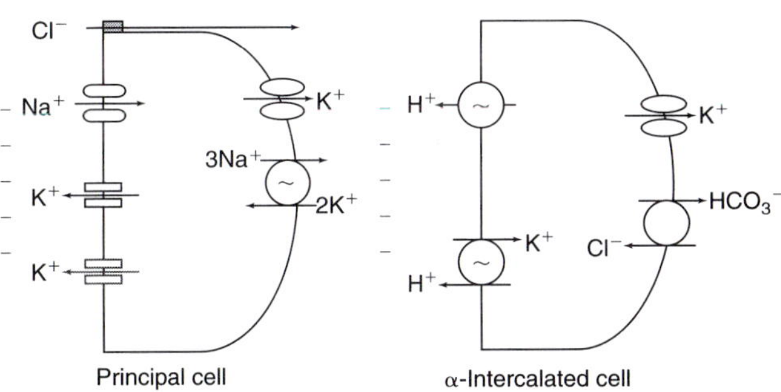

FIGURE 3.26 Potassium transport at the collecting duct. Principal cell on the left performs the bulk of potassium secretion. The simultaneous increase in apical membrane sodium and potassium conductances is critical to increase potassium secretion. The α-intercalated cell on the right depicts how the collecting duct can assume the role of a potassium reabsorbing cell which assumes particular importance in potassium deficiency

distinct corticomedullary potassium gradient with increasing interstitial potassium concentrations towards the papilla. Passive potassium transport plays a role in medullary potassium recycling (Jamison 1987). The latter allows some of the reabsorbed potassium from medullary collecting ducts to enter descending thin limbs of juxtamedullary nephrons by passive secretion. In contrast, potassium may leave thin ascending limbs by outward diffusion as a consequence of the diminishing potassium concentrations in the interstitial fluid as the corticomedullary junction is approached (Jamison 1987).

Two mechanisms are involved in the reabsorption of potassium along the thick ascending limbs of Henle's loop (Figure 3.25) (Hebert et al 1984, Greger 1985, Giebisch et al 2003, Giebisch 2004). Potassium is transported across the apical membrane by an electroneutral transporter that tightly binds one sodium and potassium ion to two chloride ions. A second component of potassium reabsorption involves paracellular transport mediated by the lumen positive transepithelial potential difference. The latter depends critically on apical potassium channels, as demonstrated by the sharp reduction of potassium reabsorption following inhibition of apical potassium channels. Potassium channels play an additional role by allowing recycling of potassium across the apical membrane. Such cell supply of potassium is important because the normal low concentrations of potassium in the tubule fluid ($\sim 2\,\mathrm{mM}$) are insufficient for maintaining $Na^+/2Cl^-/K^+$ transport in the absence of an additional supply of potassium. Indeed significant reduction of sodium chloride and potassium reabsorption follows

inhibition of apical potassium channels. Potassium transport across the basolateral membrane occurs either by potassium channels or possibly by KCl cotransport (Malnic et al 2004).

Collecting duct principal cells

Initial, cortical and connecting tubules are lined with principal cells that secrete potassium (Giebisch 1998, 2004, Giebisch et al 2003). As summarized in Figure 3.26, secretion of potassium depends on active, Na^+/K^+-ATPase-dependent uptake across the basolateral membrane, followed by passive apical potassium transport through ion-selective channels along a favorable electrochemical gradient. A second apical transport mechanism involves electroneutral K–Cl cotransport (Velazquez et al 1987). Its transport activity is normally low but increases when chloride concentrations in the lumen decline, for instance following the delivery of poorly reabsorbable anions such as bicarbonate or sulfate. Sodium channels play a key role in potassium secretion because their reabsorption not only supplies sodium for basolateral sodium–potassium exchange, but also drives apical potassium uptake. Diffusion of sodium from lumen to cell depolarizes the interior apical membrane potential, lowering it to below the potassium diffusion potential and thus stimulates potassium secretion (Giebisch 1998, 2004, Giebisch et al 2003). Effective enhancement of potassium transport into the lumen involves the coordinated increase in both potassium and sodium channel activity in the apical membrane.

Collecting duct intercalated cells

The sequential transition of potassium reabsorption in the proximal tubule and the loop of Henle, followed by variable and regulated secretion of potassium in the connecting and collecting tubules, may change to universal reabsorption of potassium along the whole nephron in states of potassium depletion. This is achieved by coordinated cessation of potassium secretion in principal cells and the emergence of active reabsorption of potassium in a distinct subfamily of intercalated cells (Doucet & Marsy 1987, Wang 2004). This process, shown in the right-hand panel of Figure 3.26, involves an active, ATP-coupled electroneutral exchange process in which potassium is reabsorbed and proton is secreted. Such K^+/H^+ exchange is activated by potassium depletion, but may also be stimulated by sodium depletion and metabolic acidosis (Silver et al 1996, 1998). When the entire nephron is engaged in potassium reabsorption, the urine can be rendered almost potassium-free. Indirect evidence suggests that under physiological conditions net potassium secretion may be the result of simultaneous secretory and reabsorptive fluxes of potassium.

Ca-sensitive (maxi) K channel has been identified in both principal and intercalated cells of the connecting and collecting tubule (Hunter et al 1984, Morita et al 1997, Palmer 1999). The channel is stimulated by membrane stretch, membrane depolarization and high cell calcium, and inhibited by tetraethylammonium (TEA) and iberiotoxin. The channel is activated by increasing tubule flow rate and high K intake (Taniguchi & Imai 1998, Woda et al 2001) and can be the basis for the flow-activated potassium secretion in this segment of the nephron.

INTEGRATION OF ANATOMY AND PHYSIOLOGY: REGULATION

In this section, we will use sodium and potassium homeostasis as examples to illustrate the systems design as illustrated in Figure 3.3 involving input from the internal environment to the kidney with modifications and adjustments by intrarenal processes, and finally a complex response is delivered by the kidney to rectify any disturbances that have been incurred. The differences between these two cations are contrasted in Table 3.6. Failure of full rectification constitutes and manifests as disease states. Monogenic disorders of steroid hormones, ion transporters, or signaling molecules all lead to disorders of sodium and potassium and understanding their pathophysiology requires knowledge of complex systems.

Regulation of Sodium Homeostasis

CONCEPT OF BALANCE AND STEADY STATE

Under normal circumstances, renal adjustments in sodium excretion serve to maintain extracellular volume within a relatively constant range despite enormous variations in salt intake. The kidney adjusts to low or high salt intake by increasing or reducing reabsorption in parallel fashion, so that intake and output come into balance and extracellular volume is stabilized, although not necessarily at one singular discrete level. Presumably, renal adjustments to different levels of salt intake reflect changes in extracellular volume. A low sodium diet leads to a transient negative sodium balance, thereby reducing extracellular volume, a process that activates the sodium-conserving mechanisms of the kidney. A high sodium diet, by transiently resulting in sodium retention, expands extracellular volume, thereby suppressing renal reabsorption. In both directions, a steady state is reached, where sodium intake and output are the same, but extracellular volume shrinks during salt restriction and is expanded during salt loads. Steady state refers to a status whereby a parameter of interest, such as extracellular fluid volume or its surrogate body weight, remains constant over time. In the normal individual, the kidney behaves as though extracellular volume was the regulatory stimulus articulating renal excretion with salt intake. This concept is central to understanding sodium and extracellular fluid volume homeostasis and is schematically illustrated in Figure 3.27.

The major edematous states such as congestive heart failure, cirrhosis with ascites, and the nephrotic syndrome, epitomize how the kidney deviates strikingly from these constraints. Tenacious renal retention of sodium does not reflect diminished intake, since it occurs when dietary salt is abundant. Moreover, total extracellular and plasma volumes are expanded not contracted, in the presence of renal salt retention. Despite massive overexpansion of extracellular volume, the kidney is behaving as though it were responding to a low-volume stimulus.

Primary edema, which reflects a state of overflow, results from an intrinsic abnormality in the kidney rather than a secondary response of the kidney to a shrunken effective circulatory volume. As a result of renal sodium retention, plasma volume expands and edema supervenes. Acute nephritic syndrome from post-streptococcal glomerulonephritis is a quintessential example of primary edema. In this book, numerous conditions of excessively active renal sodium reabsorption from monogenic lesions all represent primary renal sodium retention. *Secondary edema* is much more common, which reflects a state of underfill is the consequence of a sensed shrinkage of effective circulatory volume. Many of the features of underfill edema, serving to defend the circulation by increasing vascular tone and preserving volume by stimulating renal sodium retention, are shared by normal subjects on a low salt diet. However, key differences exist.

In underfill edema, the low fractional excretion of sodium is associated with ample salt intake, so that salt balance is positive, whereas in normal subjects on a low sodium diet, they are in balance. Replenishing salt in salt-depleted normal subjects results in only transient retention, after which salt excretion equals intake. In underfill edema, salt retention tends to be unrelenting and unresponsive in salt intake. Lastly, features of circulatory insufficiency persist in underfill edema despite overexpansion of extracellular volume, whereas the circulation is normalized after transient salt retention in normal subjects.

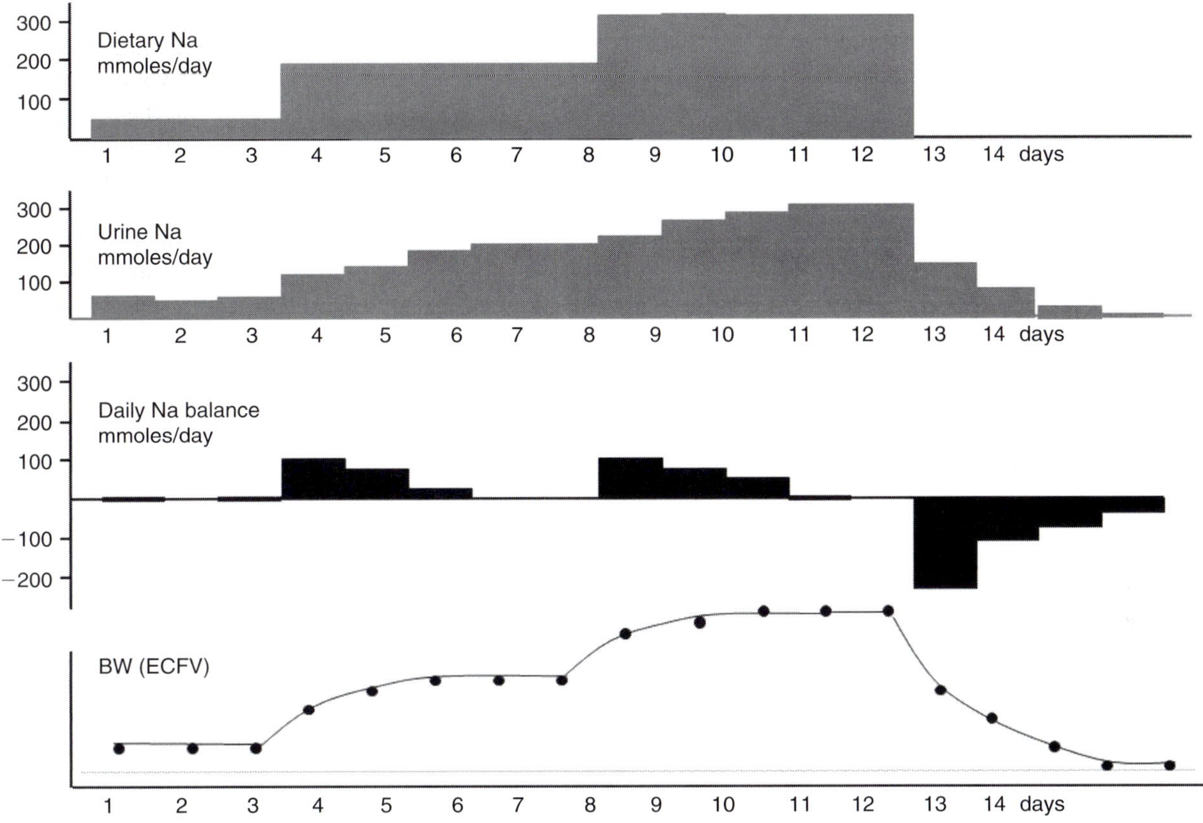

FIGURE 3.27   Sodium and extracellular fluid balance. Under normal circumstances, urinary sodium excretion always matches sodium intake thus achieving a state of balance. Extracellular fluid volume (ECFV) remains constant at a steady state. There is a time lag between changes in intake and output resulting in either ECFV expansion or contraction when dietary sodium is changed. Under normal conditions, the higher the sodium intake, the higher the ECFV

CONCEPT OF EFFECTIVE ARTERIAL BLOOD VOLUME (EABV)

Any theory of underfill sodium retention must identify and explain two central concepts: (1) an obligatory persistent low-volume stimulus that is responsible for unrelenting salt retention in the face of ample salt intake and overexpansion of the extracellular space; (2) a derangement in the extracellular compartment that prevents the retained extra sodium and fluid from terminating the low-volume stimulus (Seldin 1975).

Despite the fact that the kidney is responsible for the regulation of the total amount of sodium in the extracellular space, the principal sensing mechanisms that signal the kidney appear to reside outside the kidney but within the circulatory compartment. This afferent system does not represent the entire intravascular volume but only a critical subcompartment on the arterial side. Arterial blood volume is better correlated with renal sodium regulation than is total blood volume, which in turn is better than extracellular fluid volume. However, the sensing system seems to respond not only to the actual physical arterial volume but rather to the 'fullness of the blood stream.' The term effective arterial blood volume (EABV) was introduced to identify that theoretical volume that is sensed, and regulates sodium reabsorption by the kidney.

There are two major determinants of EABV (Seldin 1975). The first is the filling of the arterial tree, which is a function of venous return and left ventricular performance. The second is the magnitude of arteriolar runoff and the size and compliance of the vessels. A reduction in EABV, therefore, can mean an actual diminution of arterial volume or excessive peripheral runoff despite increased arterial filling as in the AV shunts and vasodilatation, or an increased size or compliance of the arterial vasculature out of proportion to the volume of blood, as in the slack circulation of pregnancy (in contrast to the tight circulation of primary aldosteronism where saline infusions cause amplified natriuresis). The relation of total extracellular volume to its subcompartments and to renal sodium excretion can be characterized schematically in normal and edematous states to provide a better insight into the determinants of EABV.

Normally, EABV is well correlated with extracellular fluid volume (Figure 3.28). The various subcompartments of the extracellular space expand or shrink in concert, as a high

	NaCl in	ECFV		Effective ECFV		Total IVV		ABV		Effective ABV		NaCl out
A. Normal	100	→ N	→	N	→	N	→	N	→	N	→	100
B. Low salt	10	→ ⬇	→	⬇	→	⬇	→	⬇	→	⬇	→	10
C. Peritonitis	100	→ ⬆	✖	⬇		⬇		⬇		⬇		10
D. Nephrosis	100	→ ⬆	→	⬆	✖	⬇		⬇		⬇		10
E. Heart failure	100	→ ⬆	→	⬆	→	⬆	✖	⬇		⬇		10
F. AV fistula	100	→ ⬆	→	⬆	→	⬆	→	⬆	✖	⬇		10

FIGURE 3.28   Sodium balance and theoretical consideration of pathologic blockade of transmission of signal under different conditions (C to F) where sodium intake is not matched by sodium output. Despite expansion of various compartments (C to F), effective arterial blood volume is sensed as low and the kidney responds by sodium retention

salt or low salt diet, respectively, is administered. On any constant salt intake, balance is reached, as evidenced by the fact that intake and output are the same. The institution of a low salt diet (Figure 3.28, condition B) results in a transient negative balance of sodium, causing a commensurate shrinkage of total extracellular volume and its subcompartments, diminution in venous return to the heart, fall in cardiac output, arterial filling, and in EABV. The reduced EABV stimulates renal sodium reabsorption, so that sodium excretion matches the low level of intake. Balance is now restored and extracellular volume is stabilized at a lower level. With a high salt load, the exact opposite sequence occurs. Note that renal excretion initially lags behind the increased intake so extracellular volume and its subcompartments are expanded resulting in an increase in venous return, cardiac output, arterial filling, and EABV (Figure 3.28). Increased sodium output comes into balance with increased intake and all the compartments of the extracellular compartment are now stabilized at a higher level but still within normalcy. In the normal individual, this system of volume regulation can be considered *open*. The compartments of the extracellular fluid expand and shrink in concert and in the steady state, the organism is in sodium balance; the set-point at which balance is attained is dictated by the salt intake.

In contrast in the major edematous states, urinary excretion may be far below intake, even though total extracellular volume is greatly expanded (Figure 3.28). In consequence, unrelenting salt retention prevails, and balance may not be achieved. The kidney behaves as if it were responding perpetually to a low-volume stimulus. EABV remains functionally contracted because of a disturbance in the forces governing fluid distribution within the extracellular space. EABV is not well correlated with either salt intake or total extracellular volume.

In Figure 3.28, the hypothetical behavior in the edematous states of the various subcompartments of the extracellular space is portrayed. Broad types of disturbed fluid distribution can be distinguished: (a) Starling disturbances within the interstitial space or between interstitial space and vascular tree; and (b) disturbances within the circulation. *Trapped fluid* is also referred to by some as 'third space' as seen in peritonitis, representing a sequestration of salt and water in inflamed tissue, vesicles, and bullae, so that it cannot contribute to effective extracellular volume. From a functional point of view, the trapped fluid behaves as if lost from the body so that EABV cannot be reexpanded and salt retention becomes unrelenting. The *nephrotic syndrome* represents dissociation between the vascular and interstitial compartments. The loss of circulating albumin causes a translocation of fluid from capillary bed to interstitial space. Blood volume and EABV shrink and renal sodium retention is stimulated, but the Starling block (hypoalbuminemia) across the capillary bed results in a leakage of the retained fluid into the interstitial space. Total blood volume, arterial volume, and EABV are reduced. Congestive heart failure represents a disturbance within the vascular tree, where impaired cardiac performance leads to a reduced cardiac output and high intraventricular pressures. In consequence, blood is segregated and distributed to the venous compartment, and arterial volume and EABV fall. Sodium retention is stimulated, but the retained fluid increases cardiac filling pressures and venous volume, causing a movement of fluid into the interstitial space. Arterial volume and EABV, therefore, remain contracted. *Arteriovenous shunts* as seen in burns, exfoliative dermatitis, Paget's disease, beri-beri, thyrotoxicosis, and cirrhosis result in an increased venous return to the heart, thereby augmenting cardiac output and arterial filling. However, shunting leaves 'critical areas' of circulation underperfused. This is sensed as a low EABV, and salt retention results. The exact nature of these 'critical areas' remains elusive.

The common element of all these edematous states appears to be a shrinkage of EABV in a setting in which the retention of salt and water cannot accomplish reexpansion owing to Starling or circulatory blocks within the extracellular space. In consequence, dietary salt intake becomes dissociated from urinary sodium excretion, and more or less unrelenting sodium retention ensues. In contrast to normal individuals, in whom a reduced EABV can be expanded by administration of dietary salt, the system of volume regulation in underfill edema can be regarded as *clamped*: salt administration cannot reexpand the contracted EABV.

AFFERENT PATHWAYS OF VOLUME SENSING

The mechanism by which EABV is sensed is only partially understood. Several sensing mechanisms have been identified. *Low-pressure baroreceptors* in the vena cava, heart, and pulmonary veins have been implicated as volume-sensors. Their activity is diminished by a decrease in pressure and stretch, reducing their tonic inhibition of the integrating centers in the central nervous system and thereby eliciting an increased sympathetic discharge. In normal subjects, low EABV might well be sensed by the low-pressure baroreceptor system. However, in heart failure, the reduced EABV is associated with a dilated heart and increased pressures in the central circulation. Therefore, the low-pressure system cannot be a major determinant in those edematous states characterized by increased pressure and volume in the heart and pulmonary vasculature. Analysis of effective arterial volume requires that the major sensing system be on the arterial side of the circulation. *High-pressure baroreceptors* in the carotid sinus and aortic arch reduce their tonic inhibition of central nervous system sympathetic outflow in response to both a fall in blood pressure on the arterial side of the circulation and a flattening of the contour of the arterial pressure profile. Such a system responds to diminished EABV in a setting of a normal vascular system or one with an underfill defect associated with a low cardiac output or a diminished arterial blood volume.

Highly controversial *intrahepatic baroreceptors* have been postulated and are thought to activate sympathetic afferent pathways in response to intrahepatic hypertension. *Chemosensitive receptors* have been identified in the heart and skeletal muscle as well as the kidney and seem to respond to lactate or other high-energy breakdown products as surrogates of underperfusion of key organs.

All the above sensors belong to number 1 in Figure 3.3. Processes residing within the kidney (Figure 3.3 number 2) also participate. *Renal sensing mechanisms* are of several kinds. The juxtaglomerular apparatus functions as a pressure-sensitive system, as evidenced by the reciprocal relationship between renin release and renal perfusion pressure.

A second intrarenal mechanism is the macula densa, which responds to increased sodium chloride concentration by transmitting a signal to adjust single nephron filtration rate (Schnermann & Briggs 1985). Renal mechanoreceptors have been suggested that respond to alterations in renal artery pressure, venous occlusion, and ureteral occlusion. Renal chemoreceptors that respond to ischemia can send afferent inputs from the kidney to converge with carotid sinus input in the central integrating centers of the medulla and hypothalamus (DiBona 1982).

THE EFFERENT LIMB OF VOLUME CONTROL

In secondary edema, where EABV is consistently contracted, the persistent stimulation of the afferent volume-sensing system described previously leads to the activation of at least four critical efferent pathways. *Increased sympathetic outflow* through nerves and circulating catechols acts both systemically and on the kidney directly. The systemic action promotes increased cardiac contractility and peripheral vasoconstriction. In the kidney, sympathetic activation directly increases renal vascular resistances (by increasing afferent and efferent arteriolar tone and diminishing functional glomerular surface area), augments the secretion of renin, and activates proximal tubule sodium absorption through α-adrenergic stimulation (Pellayo et al 1983). The net effect is to sustain systemic blood pressure and renal perfusion, to prevent a drastic fall in GFR by preferential efferent arteriolar constriction, to enhance proximal sodium reabsorption by increasing the filtration fraction, and proximal and distal reabsorption by directly augmenting epithelial transport, and, finally, to amplify these effects by contributing to the activation of the renin-angiotensin system.

Increased renin secretion results from stimulation of the juxtaglomerular apparatus by renal underperfusion, enhanced adrenergic nerve activity (Kon et al 1985), prostaglandin secretion (Yun et al 1977), and also by a tubuloglomerular feedback via the macula densa (Schnermann & Briggs 1985). The resulting angiotensin II production reinforces the catechol-induced vasoconstriction systemically, whereas in the kidney, it disproportionately increases efferent arteriolar resistance and proximal sodium transport. In addition, angiotensin stimulates aldosterone secretion which increases sodium reabsorption in the distal nephron. The concerted activation of the renin-angiotensin-aldosterone axis, like sympathetic nerve stimulation, supports systemic blood pressure, mitigates renal underperfusion, helps to prevent a drastic reduction in GFR, and enhances sodium reabsorption in the proximal and distal nephron.

Increased renal prostaglandin production is associated with shrinkage of EABV. Mesangial cells in the glomerulus, cortical vessels, and interstitial cells of the medulla synthesize prostaglandins in response to various hormones such as angiotensin II. Prostaglandins are potent renal vasodilators which antagonize adrenergic and angiotensin II-induced vasoconstriction at the same time that they enhance renin secretion (Yun et al 1977). The net effect is sustenance of renal blood flow in a setting of falling systemic blood pressure. By increasing medullary blood flow, a medullary interstitial washout is produced, resulting in lower concentrations of NaCl reaching the thick ascending limb and, therefore, reduced NaCl reabsorption (Reineck & Parma 1982). Prostaglandins may have a direct inhibitory effect on sodium reabsorption in the thick ascending limb and the cortical collecting duct. In concert, these influences moderate the intense vasoconstriction and enhanced sodium transport resulting from the action of adrenergic nerves and angiotensin II. *Nonosmotic release of ADH* (Seldin 1975) is promoted through reduced baroreceptor activity by contraction of EABV. The hormone promotes the reabsorption of water in the kidney and, in high concentrations, may exert a systemic vasoconstrictor effect. ADH generally is elevated in the same circumstances of volume contraction responsible for renin release and sympathetic nervous system activation. These three efferent responses are important signposts serving as evidence for volume contraction, even in the presence of edema. *Atrial natriuretic peptide* (Needleman & Greenwald 1986), from the cardiac atria is released in response to salt loads in euvolemic normal subjects. ANP produces systemic and renal vasodilatation, and increased GFR if hypotension is prevented, and directly inhibits renin release, aldosterone synthesis, and ADH release. It also may inhibit sodium reabsorption in the inner medullary collecting duct. If renal function is well preserved, ANP produces generous natriuresis. Part of the natriuresis can be attributed to renal vascular effects, i.e., on increase in GFR with a fall in filtration fraction which will reduce proximal reabsorption. Part of the natriuresis may be due to inhibition of NaCl transport in the inner medullary collecting duct. In states of underfill edema, such as nephrotic syndrome, and cirrhosis with ascites, ANP levels are often normal or depressed. In congestive heart failure, ANP levels can be high and administration of ANP usually elicits no significant diuretic response. One principal natriuretic effector is the intrarenal dopamine system. DA is synthesized in the mammalian kidney and maintains sodium homeostasis as an autocrine/paracrine natriuretic hormone. Dopamine increases renal plasma flow and decreases filtration fraction (Hollenberg et al 1973) thereby altering Starling driving forces that favor reduction of proximal tubular absorption. In addition, DA directly inhibits transepithelial sodium absorption by the renal tubule in the proximal tubule and thick ascending limb (Baum & Quigley 1998, Grider et al 1998).

RENAL MECHANISMS FOR SODIUM HOMEOSTASIS

The foregoing analysis assigns a key role to EABV in the elicitation of signals instructing the kidney to retain or excrete sodium. Shrunken EABV promotes the release of catecholamines, activation of the renin-angiotensin-aldosterone and sympathetic nervous systems, and secretion of ADH. These defenses raise the blood pressure but usually not to the original level, and renal underperfusion results. This state of renal underperfusion in the setting of humoral secretions and neurocirculatory reflexes alters the glomerular and postglomerular circulation, stimulates intrarenal paracrine secretions, and activates various transport systems throughout the nephron.

The glomerular and postglomerular capillary network regulates the entry of filtrate into the tubular lumen and influences net sodium and fluid reabsorption in the proximal tubule. Figure 3.14 illustrates three models where single nephron filtration rate (SNGFR) and glomerular plasma flow rate (GPF) are affected by preferential changes in vasomotor tone on several resistances. The principal mechanisms regulating these resistances appear to be intrinsic myogenic reflexes, tubuloglomerular feedback, angiotensin II, renal nerves, catecholamines, and prostaglandins. In states of contracted EABV, the efferent resistance is primarily increased perhaps because renal underperfusion elicits an autoregulatory vasodilatory response in the afferent arteriole and because prostaglandins are released in the afferent arterioles as a result of a low concentration of sodium chloride at the macula densa as part of tubuloglomerular feedback response. Both promote preferential afferent arteriolar dilatation.

Downstream from GFR lies the pivotal step by which the glomerular proto-urine is modified into the final excreted product. The tubular epithelium is one of the centerpieces of renal sodium homeostasis. The proximal tubule presents one of the most complex and also most elegant examples of regulation of epithelial sodium transport. With significant contraction of EABV, GFR tends to fall although the magnitude of the decrement is greatly buffered by efferent arteriolar constriction. The reduced entry of filtrate into the nephron is met with enhanced proximal reabsorption contributed by increased active outward transport, increased passive outward transport, and decreased passive backleak.

Figure 3.29 portrays the complex mechanisms involved in proximal sodium transport. A key role in both $NaHCO_3$ and NaCl reabsorption can be assigned to the apical Na^+/H^+ exchanger NHE_3. Transcellular $NaHCO_3$ absorption contributes to expansion and restoration of EABV. NaCl reabsorption also occurs via the coupled Na^+/H^+ exchange Cl^-/base exchange (Figure 3.19). As mentioned, HCO_3^- and isotonic water reabsorption result in luminal hyperchloruria which drives passive NaCl reabsorption (Figure 3.20). Increase in angiotensin II and α-adrenergic stimulation and decrease in dopamine can all increase Na^+/H^+ exchange activity and increase active and passive NaCl reabsorption. Net proximal reabsorption is also influenced by a backleak along the paracellular pathway through the tight junction into the tubular lumen (Rector 1983, Chantrelle et al 1985). The magnitude of the backleak is thought to be regulated by the Starling forces – effective hydrostatic pressure and effective oncotic pressure – acting across the peritubular capillary wall (Figure 3.29). When normal individuals

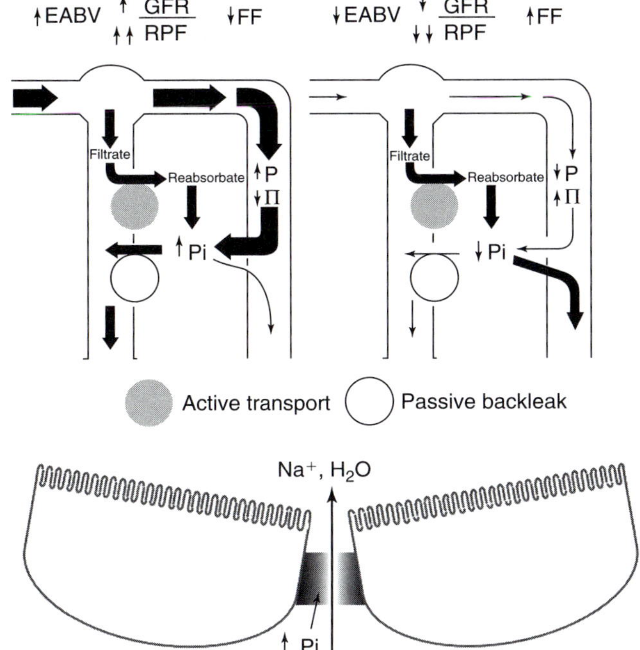

FIGURE 3.29 Peritubular factors controlling proximal tubule sodium chloride absorption and backleak. The two figures depict effective arterial blood volume (EABV) expansion and contraction. Pi = interstitial pressure; P = hydrostatic pressure; Π = oncotic pressure in the peritubular capillary. Indirect data suggest that signal from the peritubular space can affect proximal tubule paracellular permeability and backleak

ingest a diet with plentiful salt, both plasma volume and EABV expand. The plasma albumin concentration is only minimally reduced but the kidney responds by increasing renal plasma flow while GFR remains relatively constant; the filtration fraction (FF), therefore, falls. A proportionately greater volume of fluid will enter the efferent arteriole and peritubular capillary and peritubular oncotic pressure will fall as a result of reduced fraction of blood flow undergoing ultrafiltration at the glomerulus. Concomitantly, the hydrostatic pressure in the peritubular capillary will increase as more of the filtration pressure is transmitted to the postglomerular circulation. The uptake of reabsorbate into the peritubular capillary bed will be reduced because of a fall in peritubular capillary oncotic pressure and a rise in peritubular hydrostatic pressure. It has also been postulated that the accumulation of reabsorbate also raises interstitial pressure, thereby opening the tight junction between proximal tubular cells and enhancing a backleak of reabsorbate into the tubular lumen. The opposite sequence of events occurs when a low salt diet leads to contraction of plasma volume and EABV and an increase in FF. The avid uptake of reabsorbate as a result of high oncotic pressure and reduced hydrostatic pressure in the peritubular capillary bed leads to a fall in PI, a closure of the tight junction, and a reduction of the backleak. Although an increase in peritubular capillary flow

rate at low rates of glomerular capillary flow may influence the uptake of reabsorbate independent of the FF, the roughly linear correlation between FF and proximal reabsorption indicates the central role of FF as the major determinant of postglomerular Starling forces and PI (Brenner et al 1969, Brenner & Galla 1971).

Three types of experiment evidence are compatible with a backleak process. Volume expansion causes a fall in electrical resistance across the proximal tubule without any change in cell resistance. Permeability to nonreabsorbable markers is increased when isolated perfused tubules are exposed to a protein-free bath simulating low peritubular oncotic pressure in volume expansion. The most convincing evidence is the demonstration that volume expansion increases the permeability of the proximal tubule to $NaHCO_3$, the only naturally occurring substance that has been examined. Increased permeability of the tight junction would allow for a back-diffusion of HCO_3 from interstitium to tubular lumen along a favorable gradient. The situation for chloride is far more complicated. The diffusion gradient of chloride from interstitium to tubular lumen is not favorable. The peritubular control system could influence passive movement of chloride through the tight junction only by regulating solvent drag. It is not clear how significant quantitatively this effect could be, since the tight junctions constitute a very small part of the luminal surface of the proximal tubule. Indeed, in studies of isolated perfused rabbit tubules, variations in the concentration of bath protein could be shown to influence chloride movement only if the tubules were warmed. There was no influence on chloride movement if the tubules were cooled.

The factors regulating NaCl reabsorption in the thick ascending limb in response to changes in EABV are less well characterized; what is known is also an excellent example of why knowledge of the nephron intact geography is important in understanding function. Inhibition of endogenous prostaglandin production in volume contraction enhances net sodium reabsorption. Two mechanisms have been proposed to explain prostaglandin effects. Studies of the isolated perfused tubule to which PGE_2 has been added indicated inhibition of chloride transport out of the medullary thick ascending limb and a reduction in transepithelial voltage (Stokes & Kokko 1977). The increase in medullary blood flow and a reduction of interstitial chloride concentrations in the medulla may account for part of the observed effects. Saline dieresis is an effect that can be prevented if saline diuresis is induced while prostaglandin production is inhibited (Higashihara et al 1979). Medullary washout can also reduce medullary NaCl concentration, thereby diminishing water abstraction by the long-loop nephrons, concentration of sodium chloride in the lumen at the tip, and reduction of passive sodium chloride absorption in the ascending limb.

The fluid leaving the ascending limb then traverses the distal convolution, cortical collecting duct, and outer and inner medullary collecting ducts. These distal nephron segments, though of low capacity, are capable of reabsorbing NaCl against very steep gradients. They are, therefore, the final determinant in the excretion of the essentially NaCl-free urine that characterizes states of profound contraction of EABV. If sodium is reabsorbed more proximally as in volume contraction, so that little reaches the collecting duct, high levels of aldosterone do not augment potassium or acid loss (Stokes 1981). Similarly, high distal delivery rates of sodium do not appreciably accelerate potassium or acid excretion in the absence of aldosterone.

Regulation of Potassium Homeostasis

Challenges to potassium metabolism are related to several of its unique properties which are listed in Table 3.7. Most potassium resides within cells and a very small fraction in the

extracellular fluid (Figure 3.30). The unique distribution of potassium between intra- and extracellular fluid depends on the balance between regulated, active uptake of potassium by the ubiquitous Na^+/K^+-ATPase and passive backleak through potassium channels in cell membranes. Dietary potassium absorption takes place along the upper gastrointestinal tract, especially in the small intestine, and such potassium absorption lacks specific regulation leaving the burden of regulation mostly in the kidney. Most of the ingested potassium reaches the extracellular fluid from where it is temporarily shifted into cells followed by excretion into the urine by the kidney or into intestinal fluid by the colon.

DISTRIBUTION OF POTASSIUM

Given the large proportion of potassium in the cell, changes in distribution of potassium will obviously have great impact on potassium homeostasis. While clinicians primarily measure plasma potassium concentration, the body senses and regulates both intracellular and extracellular concentrations of potassium. Figure 3.30 provides an overview of the variables that maintain internal potassium balance, i.e., distribution between the intracellular and extracellular compartments. Such regulation depends on changes in extracellular potassium concentration, activation, or deactivation of hormone release brought about by changes in plasma potassium levels, by acid–base changes, exercise, and factors related to membrane integrity. Such rapid and effective shifts of potassium between body fluid compartments provide an important, fast,

---

TABLE 3.7 Features of potassium homeostasis

1. Majority of potassium is intracellular and not directly accessible to regulation by the kidney
2. Maintenance of steep concentration gradients of potassium across cell membranes is critical for generating and maintaining the membrane potential in excitable tissues
3. Narrow range of extracellular potassium concentrations in the presence of widely varying potassium intake

---

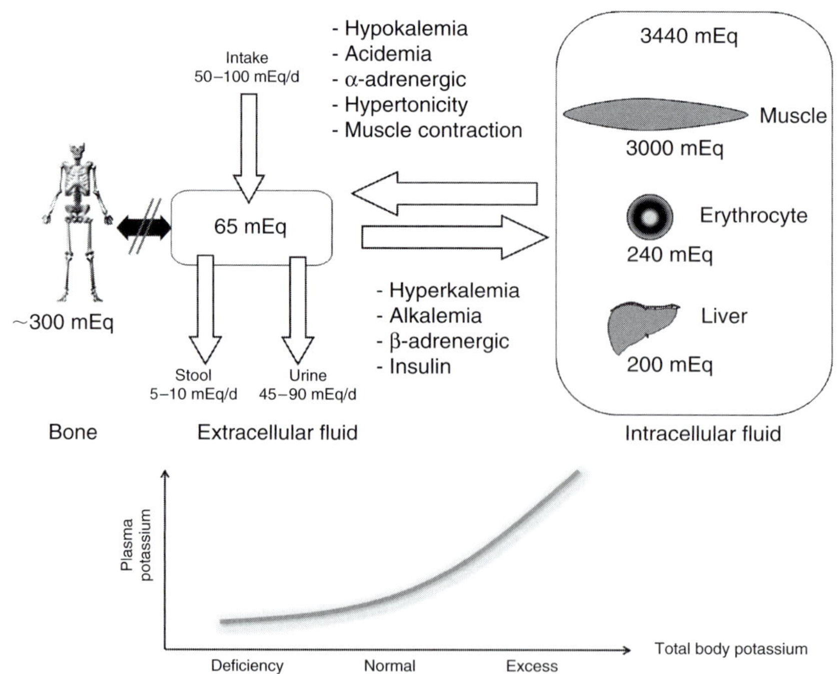

FIGURE 3.30 Distribution and flux of potassium and factors that control shift between the intracellular and extracellular space. There is a significant amount of potassium in bone that is not freely exchangeable so it is not usually considered

and protective response and constitute an effective first line of defense against disturbances of internal potassium balance. Significant and potentially dangerous changes in plasma potassium concentrations may occur whenever the control of internal potassium balance is inadequate. Physiologic potassium loads are buffered by the intracellular compartment long before the kidney has a chance to establish external balance. A simple hypothetical consideration is presented in Figure 3.31. If it were not for intracellular buffering, a simple meal of gathered berries for our terrestrial vertebrate ances-

tors could have provided enough of a potassium bolus to trigger a fatal cardiac event. Although this buffering mechanism is life-saving, external potassium balance is achieved largely by renal secretion.

SIGNALS THAT MODULATE RENAL POTASSIUM HANDLING

The systemic incoming signals that regulate renal potassium are diverse. For years, there have been suggestions that intestinal potassium is sensed directly either in the lumen of the gut or in the portal system (Thomas & Kumar 2008). Plasma potassium definitely is sensed and so is intracellular potassium. The pathways by which intestinal luminal or intracellular potassium communicate with the kidney are elusive to date. The relationship between plasma potassium concentration and total cellular potassium which is represented best by intracellular potassium concentration is not linear (Figure 3.30). Mild to moderate total body or intracellular potassium deficiency is frequently not detectable by clinical measures of plasma potassium. The fact that the body can respond to this mild degree of total body potassium depletion without frank hypokalemia indicates that more than just the plasma potassium is being sensed. Another excellent example that illustrates that plasma potassium is not the sole controller of distal potassium is provided in Figure 3.32 (Stanton & Giebisch 1982b). For any given plasma potassium concentration, animals that have previously ingested a high potassium diet secrete and excrete more potassium than those fed a control diet.

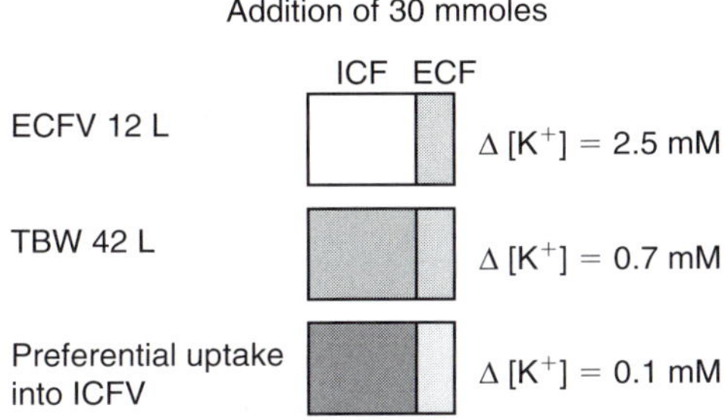

FIGURE 3.31 Cellular buffering of potassium. Theoretical consideration of a typical dietary load of potassium of 30 moles added to the body. Top two scenarios represent equilibration over extracellular fluid volume (ECFV) or total body water (TBW) and the predicted resultant increment in plasma potassium concentration. The bottom scenario represents preferential uptake of potassium into cells

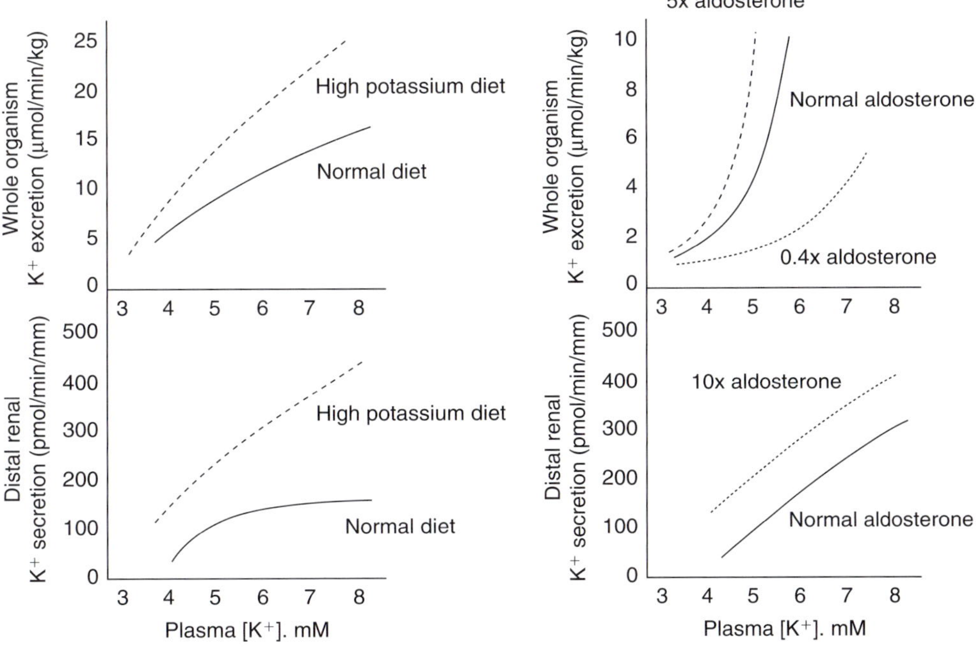

FIGURE 3.32 Renal potassium excretion and distal potassium secretion as a function of plasma potassium concentration under a normal diet vs. chronic feeding with a high potassium diet (left); and under different aldosterone states

TABLE 3.8   Factors modulating renal potassium excretion

1. Potassium intake and total body potassium status
2. Acid–base disturbances
3. Adrenal steroids and vasopressin
4. Delivery of sodium to distal tubule secretory sites
5. Delivery of poorly reabsorbable anions to distal tubule secretory sites

The systemic signals that control renal potassium transport include aldosterone, antidiuretic hormone, glucagon, and plasma pH (Tables 3.6 and 3.8) (Malnic et al 2007).

In addition to the circulating factors, intrarenal processes also modify the signal. The most important ones are distal sodium delivery, flow rate, the presence of non-absorbable anions, and luminal pH (Stanton & Giebisch 1982a, Malnic et al 2007). The cellular mechanisms by which sodium delivery stimulates distal potassium secretion is discussed below. This effect is best exemplified by natriuretics that work on the proximal tubule and thick ascending limb, which are not principal sites of potassium secretion. The increased sodium delivery in conjunction with high aldosterone greatly stimulates distal potassium secretion so a universal finding in these agents is renal potassium wasting. As long as this mismatched combination of high aldosterone and distal delivery exist, distal potassium secretion cannot be suppressed despite the ongoing potassium loss and deficiency.

MECHANISMS OF REGULATION OF RENAL POTASSIUM TRANSPORT

This section focuses on the distal nephron where the majority of regulation occurs. First, one should consider a few facts about the principal cell (Figure 3.26). This regulation is determined by input from circulating factors such as aldosterone (pathway 1 in Figure 3.3), intrarenal factors such as sodium delivery from more proximal sites and direct effects of plasma potassium on the tubule (pathway 2 in Figure 3.3). The final effectors (pathway 3 in Figure 3.3) involve multiple channels and transporters of which the potassium channels plays a major part.

Basolateral potassium channels are an important transport element because they modulate the extent of potassium recycling, thereby controlling the extent to which potassium, taken up by Na^+/K^+-ATPase, can be secreted into the lumen across the apical membrane. The basolateral electrical potential is normally quite close to the potassium equilibrium potential and basolateral potassium recycling is modest or absent. However, if the basolateral potential declines in magnitude, and falls below the potassium equilibrium potential (depolarize), a larger fraction of potassium taken up by the Na^+/K^+-ATPase will leak back into the peritubular fluid. This may occur whenever Na^+/K^+-ATPase activity declines, which is exactly what one expects when

potassium secretion is not necessary. In contrast, stimulation of Na^+/K^+-ATPase may drive the basolateral potential above the potassium equilibrium potential (hyperpolarize). As a result, potassium will be taken up synergistically into principal cells both by Na^+/K^+-ATPase and by passive diffusion through potassium channels (Giebisch 1998, 2004, Giebisch et al 2003). Therefore, these are intrinsic mechanisms where basolateral coupling between potassium channels and Na^+/K^+-ATPase regulates the direction of potassium movement in the basolateral membrane.

In the apical membrane, the electrical potential is equally important. In contrast to the basolateral membrane where sodium conductance is absent, abundant sodium channels reside on the apical membrane. This allows sodium ions to diffuse into principal cells from the lumen and depolarize the membrane potential significantly below the potassium equilibrium potential. This stimulates potassium secretion into the lumen and explains the critical dependence of potassium secretion on luminal sodium delivery. Increasing sodium delivery by enhancement of distal tubule flow rate stimulates potassium secretion. Flow-dependent activation of potassium secretion is also modulated by potassium intake. For a given flow rate, secretion rises with enhanced K intake (Giebisch 1998, 2004, Giebisch et al 2003).

Increased sodium delivery augments potassium secretion, not just by modulating the apical driving force for potassium transport, but also by providing additional sodium to the basolateral Na^+/K^+-ATPase because the latter is unsaturated at physiological levels of sodium delivery. Examples for the influence of apical sodium administration on potassium secretion include: apical sodium channel blockers lower potassium secretion whereas increased sodium delivery stimulates potassium secretion. Extracellular fluid volume expansion and administration of loop diuretics enhance distal delivery of sodium and stimulate potassium secretion. In addition to the urinary lumen, sodium may also enter principal cells by basolateral entry pathways that depend on Na^+/H^+ or $3Na^+/Ca^{2+}$ exchange. This explains the observation that potassium secretion, albeit decreased, continues when apical Na entry is curtailed by Na channel blockers or low sodium delivery (Yeyati et al 1990).

The main factors modulating potassium secretion are summarized in Table 3.8. Included are the lumen supply of sodium, the basolateral concentration of potassium, hormones such as aldosterone and vasopressin, acid–base balance, and potassium intake.

Regulation of potassium secretion in principal cells frequently involves the coordinated interaction of apical and basolateral changes in membrane transport. High potassium not only stimulates the release of aldosterone from the adrenal cortex, but also enhances Na^+/K^+-ATPase turnover, pump number, and basolateral membrane area. Moreover, high potassium intake also enhances apical sodium and potassium channel activity, both factors that favor potassium diffusion from K^+ cell to tubule lumen (Muto 1995, 2001).

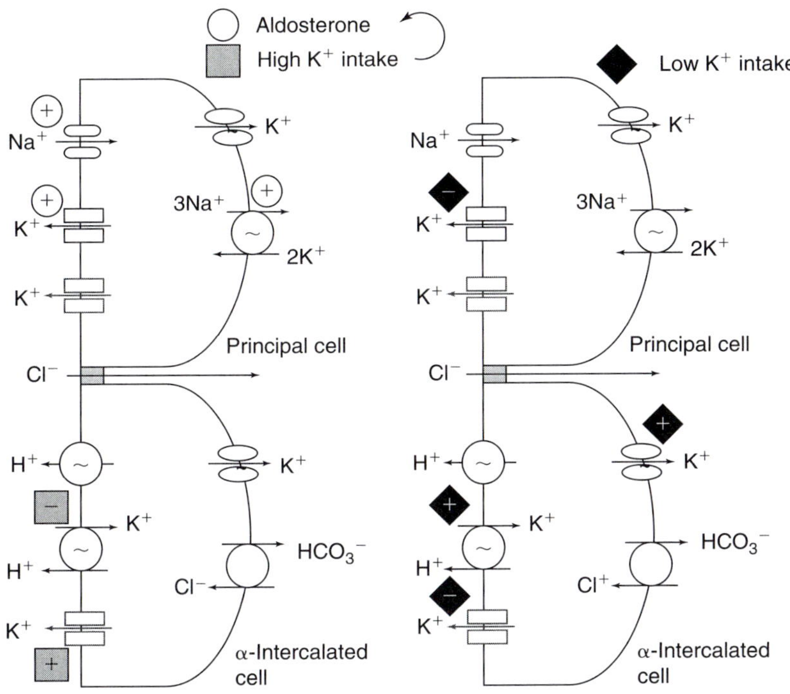

FIGURE 3.33 Collecting duct's response to dietary changes in potassium. High dietary potassium is shown in the left and low dietary potassium is on the right. High potassium also stimulates aldosterone release. The effects on specific transporters are shown as open circles (aldosterone), gray squares (high potassium), and black diamonds (low potassium)

The main effects of aldosterone are stimulation of ATPase and activation of apical sodium channels.

Potassium secretion rises with plasma potassium until plateau levels are reached at about 7 mM (Figure 3.32). The stimulating action of aldosterone on potassium secretion is well established, and this effect is maintained at different plasma potassium levels (Figure 3.32) suggesting that plasma potassium concentration and adosterone are independent factors. It is apparent that alkalemia stimulates potassium secretion, whereas acidemia has the opposite effect. Such pH-dependent modulation of potassium transport is consistent with the known effects on basolateral potassium uptake and apical potassium channel activity. Alkalemia enhances cell potassium uptake and stimulates apical potassium channels, whereas acidemia has the opposite effect. Moreover, cell pH also affects apical sodium channels, and acidosis reducing and alkalosis enhancing channel activity. These coordinated effects of apical potassium and sodium channel activity, together with altered basolateral Na^+/K^+-ATPase activity, can account for pH-induced changes in distal potassium secretion.

Vasopressin has been shown to raise potassium secretion in collecting ducts, an effect involving both basolateral and apical receptors and stimulation of apical sodium and potassium channels (Amorim & Malnic 2000). Vasopressin may prevent a fall in potassium secretion during volume contraction when distal flow rate and the supply of sodium diminish.

The complexity of physiological regulatory pathways of potassium transport is further demonstrated by studies on the modulation of potassium channel activity in tubules from animals on different potassium intake, which have provided insight into an aldosterone-independent direct regulation of potassium secretion. First, it has been demonstrated in tubule perfusion studies of cortical collecting ducts that potassium transport in adrenalectomized animals maintained on a small and constant infusion of aldosterone maintains the appropriate response concerning potassium secretion when their potassium intake is switched from a low to a high potassium diet (Muto 1995, 2001). Second, patch-clamp studies of apical potassium channels in principal cells of animals on a low potassium diet have shown increased endocytosis of apical secretory potassium channels, which critically depends on activation of a specific tyrosine phosphorylation site (Wang et al 2000, Wang 2004, Hebert et al 2005). Figure 3.33 shows cell models of the effects of sustained high and low potassium intake on apical and basolateral potassium channels and transporters in both principal and intercalated cells. Key elements include the stimulation of apical ATP-dependent K^+/H^+ exchange, diminished apical potassium recycling and activation of basolateral potassium channels (Wang 2004).

Figure 3.34 shows how plasma potassium concentrations can directly activate potassium secretion in the principal cells in the cortical collecting duct. Two points are noteworthy in this aspect. First is that distal potassium secretion cannot rely on single factors or factors that can be influenced by multiple signals such as aldosterone levels or distal sodium delivery. An excellent example can be found in the Yanomami Indians where a daily excretion of potassium of up to 300 mmoles per day can be achieved with 2–3 mmoles of sodium per day. Therefore with virtually no

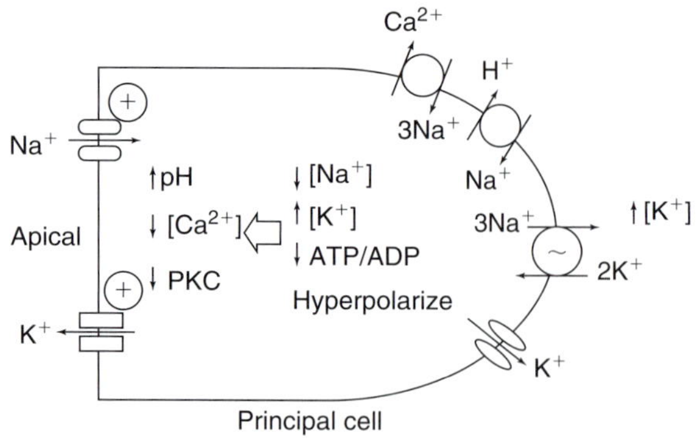

FIGURE 3.34 Hyperkalemia-stimulated apical potassium secretion as an illustration of cross of the basolateral and apical membranes. Increased plasma potassium stimulates the Na^+/K^+-ATPase, which lowers cell sodium concentration, increases cell potassium concentration and decreases ATP/ADP ratio. The high cellular potassium will hyperpolarize the cell. The low cell sodium will activate the basolateral sodium–calcium exchange and sodium–proton exchange. The net effect is increase cell pH, lower cell calcium, and suppression of protein kinase C (PKC) activity. All three of these signals stimulate apical sodium and potassium channel activities resulting in increase apical potassium secretion

distal sodium delivery, potassium excretion is not jeopardized. Second is that this seemingly simple direct effect on the cell involves multiple signaling mechanisms and crosstalk between the basolateral membrane and apical membrane. Figure 3.34 illustrates how a single primary event of increasing plasma potassium can trigger a series of secondary and tertiary changes that eventuate in activation of apical sodium and potassium channels which increase luminal potassium secretion. Even at the level of the single cell or along a cell membrane, there is a considerable build-up of hierarchy from single gene products.

Recent studies have also provided evidence for the involvement of WNK4 (with no lysine) kinase in the regulation of distal tubule potassium and sodium transport (Kahle et al 2003, 2004, 2005). Human mutations in WNK will be discussed in detail elsewhere. WNK4 belongs to a group of serine-threonine kinases, is present in renal tubules, and has been shown to have distinct effects on several transport processes. It affects both transcellular and paracellular transport pathways by inhibiting both NaCl cotransport in the distal convoluted tubule and secretory potassium channels in principal cells. An additional effect involves an increase in paracellular chloride conductance (Kahle et al 2005). Inhibiting sodium chloride reabsorption as well as distal potassium secretion provides a mechanism that uncouples and balances potassium secretion from sodium delivery in principal cells. Mutations in WNK4 have been responsible for altered sodium chloride and potassium transport in several syndromes of hypertension and hyperkalemia.

CONCLUSION

In this chapter, we presented a short description of the anatomic and functional design of the mammalian kidney and cited examples of regulation of sodium and potassium to illustrate the complexity of the system depicted in Figure 3.3. Within the construct portrayed in Figure 3.3 lie numerous diseased genes strewn over diverse pathways but converging onto the eventual phenotypes of disturbances in sodium or potassium homeostasis. While the primary defects of rare Mendelian monogenic diseases originate in highly restricted foci, the comprehension of the pathophysiology of the clinical manifestations requires knowledge of interactions of multiple systems at many levels within and outside the kidney in the context of the *ensemble harmonique*, as articulated by Claude Bernard (1865). A rather popular citation of Bernard's writing in translation reads:

'Physiologists are inclined to acknowledge in living organisms the existence of a harmonious and pre-established unity where all components are inter-dependent and influence each other. We must appreciate that when we break up an organism by taking the different components apart it is only for the sake of convenient experimentation and by no means because we consider them as separate entities. Indeed when we wish to ascribe to a physiologic property its significance we must always refer it to the whole organism and draw any conclusions only in relation to the effect of this property on the organism as a whole.'

Anderson assembled an interesting linear ascending hierarchy commencing with elementary particles to many bodies – physics, chemistry, biochemistry, molecular biology, cell biology, physiology, psychology, and social sciences (Anderson 1972). At each level, new complexities, patterns, and laws emerge, laws that cannot be comprehended or even fathomed when equipped solely with mastery of knowledge of even the highest precision and sophistication derived from a different level. Complexity is an inherent, pervasive, and perpetual impediment in all biologic sciences. Reduction is necessary and in fact indispensable for knowledge acquisition when confronted with the complexity of multi-system biology. No one will cast doubt on the fact that reduction is a sheer methodologic triumph. However, the ability to reduce everything down to fundamental molecules does not guarantee unhindered and automatic reconstruction up the hierarchy. Richard Dawkins once wrote that 'the biologist tries to explain the workings, and coming into existence of complex things in terms of simpler things. He can regard his task as done when he has arrived at entities so simple that they can be safely handed over to physicists.' Dawkins' observation, minted with a hint of sarcasm at the biologists' version of reductionism, is not entirely wrong. Of course not all biologists necessarily embrace that view.

As one embarks on an expedition through a textbook dedicated to, and owing its existence primarily to, elucidation of monogenic lesions in humans and experimental animals, one should be constantly cognizant of the conviction that with each reductive step, the new knowledge secured should be applied back to the whole cell, organ, and eventually organism level. The purpose of the gene and gene product is preservation of physiology and conversely, the disruption of physiology provides the clue and pathway by which these genes are discovered.

We can clone and study the gene in isolation but the wisdom that we beget from this exercise must bear the understanding that the gene does not exist in vacuo but rather under the sanctuary and harmony of the *milieu intérieur.*

References

Abdelbaki YZ, Henk WG, Haldiman JT, Albert TF, Henry RW, Duffield DW. Macroanatomy of the renicule of the bowhead whale (*Balaena mysticetus*). Anat. Rec. 1984; 208: 481–90.

Amorim JB, Malnic G. V(1) receptors in luminal action of vasopressin on distal K(+) secretion. Am. J. Physiol. Renal Physiol. 2000; 278: F809–16.

Anderson PW. More is different. Science 1972; 177: 393–6.

Ballermann BJ. Contribution of the endothelium to the glomerular permselectivity barrier in health and disease. Nephron. Physiol. 2007; 106: 19–25.

Baum M, Quigley R. Inhibition of proximal convoluted tubule transport by dopamine. Kidney Int. 1998; 54: 1593–600.

Beck JS, Breton S, Mairbaurl H, Laprade R, Giebisch G. Relationship between sodium transport and intracellular ATP in isolated perfused rabbit proximal convoluted tubule. Am. J. Physiol. 1991; 261: F634–9.

Berliner RW. Renal mechanisms for potassium excretion. Harvey Lectures 1961; 55: 141–71.

Bernard C. Introduction a L'etude de la Medecine Experimentale. Flammarion, Paris: 1865.

Brenner BM, Falchuk KH, Keimowitz RI, Berliner RW. The relationship between peritubular capillary protein concentration and fluid reabsorption by the renal proximal tubule. J. Clin. Invest. 1969; 48: 1519–31.

Brenner BM, Galla JH. Influence of postglomerular hematocrit and protein concentration on rat nephron fluid transfer. Am. J. Physiol. 1971; 220: 148–61.

Brezis M, Rosen SN, Epstein FH. The pathophysiological implications of medullary hypoxia. Am. J. Kidney Dis. 1989; 13: 253–8.

Chantrelle BM, Cogan MG, Rector FC, Jr. Active and passive components of NaCl absorption in the proximal convoluted tubule of the rat kidney. Miner. Electrolyte Metab. 1985; 11: 209–14.

Chou CL, Knepper MA. In vitro perfusion of chinchilla thin limb segments: urea and NaCl permeabilities. Am. J. Physiol. 1993; 264: F337–43.

Couchman JR, Beavan LA, McCarthy KJ. Glomerular matrix: synthesis, turnover and role in mesangial expansion. Kidney Int. 1994; 45: 328–35.

DiBona GF. Functions of renal nerves. Rev. Physiol. Biochem. Pharmacol. 1982; 94: 75–181.

Dieterich HJ, Barrett JM, Kriz W, Bulhoff JP. The ultrastructure of the thin loop limbs of the mouse kidney. Anat. Embryol. (Berl) 1975; 147: 1–18.

Doucet A, Marsy S. Characterization of K-ATPase activity in distal nephron: stimulation by potassium depletion. Am. J. Physiol. 1987; 253: F418–23.

Eckardt KU, Koury ST, Tan CC, et al. Distribution of erythropoietin producing cells in rat kidneys during hypoxic hypoxia. Kidney Int. 1993; 43: 815–23.

Ellison DH, Velazquez H, Wright FS. Thiazide-sensitive sodium chloride cotransport in early distal tubule. Am. J. Physiol. 1987; 253: F546–54.

Forster IC, Hernando N, Biber J, Murer H. Proximal tubular handling of phosphate: A molecular perspective. Kidney Int. 2006; 70: 1548–59.

Garg P, Verma R, Holzman LB. Slit diaphragm junctional complex and regulation of the cytoskeleton. Nephron Exp. Nephrol. 2007; 106: e67–72.

Giebisch G. Renal potassium transport: mechanisms and regulation. Am. J. Physiol. 1998; 274: F817–33.

Giebisch G. Challenges to potassium metabolism: internal distribution and external balance. Wien. Klin. Wochenschr. 2004; 116: 353–66.

Giebisch G, Hebert SC, Wang WH. New aspects of renal potassium transport. Pflugers Arch. 2003; 446: 289–97.

Greger R. Ion transport mechanisms in thick ascending limb of Henle's loop of mammalian nephron. Physiol. Rev. 1985; 65: 760–97.

Grider J, Kilpatrick E, Ott C, Jackson B. Effect of dopamine on NaCl transport in the medullary thick ascending limb of the rat. Eur. J. Pharmacol. 1998; 342: 281–4.

Gubler MC. Inherited diseases of the glomerular basement membrane. Nat. Clin. Pract. Nephrol. 2008; 4: 24–37.

Haberle DA. Hemodynamic interactions between intrinsic blood flow control mechanisms in the rat kidney. Ren. Physiol. Biochem. 1988; 11: 289–315.

Haraldsson B, Nystrom J, Deen WM. Properties of the glomerular barrier and mechanisms of proteinuria. Physiol. Rev. 2008; 88: 451–87.

Hao CM, Breyer MD. Physiologic and pathophysiologic roles of lipid mediators in the kidney. Kidney Int. 2007; 71: 1105–15.

Hebert SC, Andreoli TE. Control of NaCl transport in the thick ascending limb. Am. J. Physiol. 1984; 246: F745–56.

Hebert SC, Culpepper RM, Andreoli TE. NaCl transport in mouse medullary thick ascending limbs. I. Functional nephron heterogeneity and ADH-stimulated NaCl cotransport. Am. J. Physiol. 1981; 241: F412–31.

Hebert SC, Desir G, Giebisch G, Wang W. Molecular diversity and regulation of renal potassium channels. Physiol. Rev. 2005; 85: 319–71.

Higashihara E, Stokes JB, Kokko JP, Campbell WB, DuBose TD, Jr. Cortical and papillary micropuncture examination of chloride transport in segments of the rat kidney during inhibition of prostaglandin production. Possible role for prostaglandins in the chloruresis of acute volume expansion. J. Clin. Invest. 1979; 64: 1277–87.

Hollenberg NK, Adams DF, Mendell P, Abrams HL, Merrill JP. Renal vascular responses to dopamine: haemodynamic and angiographic observations in normal man. Clin. Sci. Mol. Med. 1973; 45: 733–42.

Hunter M, Lopes AG, Boulpaep EL, Giebisch GH. Single channel recordings of calcium-activated potassium channels in the apical membrane of rabbit cortical collecting tubules. Proc. Natl Acad. Sci. USA 1984; 81: 4237–9.

Jamison RL. Potassium recycling. Kidney Int. 1987; 31: 695–703.

Johnson RJ, Floege J, Yoshimura A, Iida H, Couser WG, Alpers CE. The activated mesangial cell: a glomerular "myofibroblast"? J. Am. Soc. Nephrol. 1992; 2: S190–7.

Kahle KT, Macgregor GG, Wilson FH, et al. Paracellular Cl-permeability is regulated by WNK4 kinase: insight into normal physiology and hypertension. Proc. Natl Acad. Sci. USA 2004; 101: 14877–82.

Kahle KT, Wilson FH, Leng Q, et al. WNK4 regulates the balance between renal NaCl reabsorption and K+ secretion. Nat. Genet. 2003; 35: 372–6.

Kahle KT, Wilson FH, Lifton RP. Regulation of diverse ion transport pathways by WNK4 kinase: a novel molecular switch. Trends Endocrinol. Metab. 2005; 16: 98–103.

Karnovsky MJ, Ainsworth SK. The structural basis of glomerular filtration. Adv. Nephrol. Necker. Hosp. 1972; 2: 35–60.

Kon V, Yared A, Ichikawa I. Role of renal sympathetic nerves in mediating hypoperfusion of renal cortical microcirculation in experimental congestive heart failure and acute extracellular fluid volume depletion. J. Clin. Invest. 1985; 76: 1913–20.

Kreisberg JI, Venkatachalam M, Troyer D. Contractile properties of cultured glomerular mesangial cells. Am. J. Physiol. 1985; 249: F457–63.

Kriz W, Kaissling B Structural Organization of the Mammalian Kidney, 4th ed. London: Academic Press, 2007.

Laughlin R. A Different Universe: Reinventing Physics from the Bottom Down. Basic Books, 2007.

Liu W, Morimoto T, Kondo Y, et al. Analysis of NaCl transport in thin ascending limb of Henle's loop in CLC-K1 null mice. Am. J. Physiol. Renal. Physiol. 2002; 282: F451–7.

Malnic G, Bailey MA, Giebisch G. Control of Renal Potassium Transport, 7th ed. London: W.B. Saunders, 2004.

Malnic G, Muto S, Giebisch G. Regulation of Potassium Excretion. Burlington, MA: Elsevier, 2007.

Morita T, Hanaoka K, Morales MM, Montrose-Rafizadeh C, Guggino WB. Cloning and characterization of maxi K+ channel alpha-subunit in rabbit kidney. Am. J. Physiol. 1997; 273: F615–24.

Muto S. Action of aldosterone on renal collecting tubule cells. Curr. Opin. Nephrol. Hypertens. 1995; 4: 31–40.

Muto S. Potassium transport in the mammalian collecting duct. Physiol. Rev. 2001; 81: 85–116.

Myers BD, Deen WM, Brenner BM. Effects of norepinephrine and angiotensin II on the determinants of glomerular ultrafiltration and proximal tubule fluid reabsorption in the rat. Circ. Res. 1975; 37: 101–10.

Needleman P, Greenwald JE. Atriopeptin: a cardiac hormone intimately involved in fluid, electrolyte, and blood-pressure homeostasis. N. Engl. J. Med. 1986; 314: 828–34.

Noble D. Claude Bernard, the first systems biologist, and the future of physiology. Exp. Physiol. 2008; 93: 16–26.

O'Connor PM. Renal oxygen delivery: matching delivery to metabolic demand. Clin. Exp. Pharmacol. Physiol. 2006; 33: 961–7.

Okusa MD, Unwin RJ, Velazquez H, Giebisch G, Wright FS. Active potassium absorption by the renal distal tubule. Am. J. Physiol. 1992; 262: F488–93.

Palmer LG. Potassium secretion and the regulation of distal nephron K channels. Am. J. Physiol. 1999; 277: F821–5.

Pellayo JC, Ziegler MG, Jose PA, Blantz RC. Renal denervation in the rat: analysis of glomerular and proximal tubular function. Am. J. Physiol. 1983; 244: F70–7.

Rector FC, Jr. Sodium, bicarbonate, and chloride absorption by the proximal tubule. Am. J. Physiol. 1983; 244: F461–71.

Rector FC, Jr., Martinez-Maldonado M, Brunner FP, Seldin DW. Evidence for passive reabsorption of NaCl in proximal tubule of rat kidney. J. Clin. Invest. 1966; 45: 1060.

Reineck HJ, Parma R. Effect of medullary tonicity on urinary sodium excretion in the rat. J. Clin. Invest. 1982; 69: 971–8.

Rostgaard J, Qvortrup K. Electron microscopic demonstrations of filamentous molecular sieve plugs in capillary fenestrae. Microvasc. Res. 1997; 53: 1–13.

Sasaki S, Imai M. Effects of vasopressin on water and NaCl transport across the in vitro perfused medullary thick ascending limb of Henle's loop of mouse, rat, and rabbit kidneys. Pflugers Arch. 1980; 383: 215–21.

Schnermann J, Briggs J. Function of the Juxtaglomerular Apparatus: Local Control of Glomerular Hemodynamics. New York: Raven Press, 1985.

Seldin DW. Sodium Balance and Fluid Volume. New York: Science and Medicine, 1975.

Silver RB, Choe H, Frindt G. Low-NaCl diet increases H-K-ATPase in intercalated cells from rat cortical collecting duct. Am. J. Physiol. 1998; 275: F94–102.

Silver RB, Mennitt PA, Satlin LM. Stimulation of apical H-K-ATPase in intercalated cells of cortical collecting duct with chronic metabolic acidosis. Am. J. Physiol. 1996; 270: F539–47.

Smith H. Structure and Function in Health and Disease. New York: Oxford University Press, 1951.

Stanton BA. Electroneutral NaCl transport by distal tubule: evidence for Na^+/H^+-Cl-/HCO_3- exchange. Am. J. Physiol. 1988; 254: F80–106.

Stanton BA, Giebisch G. Effects of pH on potassium transport by renal distal tubule. Am. J. Physiol. 1982a; 242: F544–51.

Stanton BA, Giebisch GH. Potassium transport by the renal distal tubule: effects of potassium loading. Am. J. Physiol. 1982b; 243: F487–93.

Stokes JB. Potassium secretion by cortical collecting tubule: relation to sodium absorption, luminal sodium concentration, and transepithelial voltage. Am. J. Physiol. 1981; 241: F395–402.

Stokes JB, Kokko JP. Inhibition of sodium transport by prostaglandin E2 across the isolated, perfused rabbit collecting tubule. J. Clin. Invest. 1977; 59: 1099–104.

Taniguchi J, Imai M. Flow-dependent activation of maxi K^+ channels in apical membrane of rabbit connecting tubule. J. Membr. Biol. 1998; 164: 35–45.

Thomson SC, Blantz RC. Biophysical Basis of Glomerular Filtration, 4th ed. Academic Press, 2007.

Thomas L, Kumar R. Control of renal solute excretion by enteric signals and mediators. J. Am. Soc. Nephrol. 2008; 19: 207–12.

Uchida S, Sasaki S, Furukawa T, et al. Molecular cloning of a chloride channel that is regulated by dehydration and expressed predominantly in kidney medulla. J. Biol. Chem, 1993; 268: 3821–4.

Uchida S, Sasaki S, Nitta K, et al. Localization and functional characterization of rat kidney-specific chloride channel, ClC-K1. J. Clin. Invest. 1995; 95: 104–13.

Velazquez H, Ellison DH, Wright FS. Chloride-dependent potassium secretion in early and late renal distal tubules. Am. J. Physiol. 1987; 253: F555–62.

Velazquez H, Wright FS. Effects of diuretic drugs on Na, Cl, and K transport by rat renal distal tubule. Am. J. Physiol. 1986; 250: F1013–23.

Wang W, Lerea KM, Chan M, Giebisch G. Protein tyrosine kinase regulates the number of renal secretory K channels. Am. J. Physiol. Renal. Physiol. 2000; 278: F165–71.

Wang WH. Regulation of renal K transport by dietary K intake. Annu. Rev. Physiol. 2004; 66: 547–69.

Wareing M, Wilson RW, Kibble JD, Green R. Estimated potassium reflection coefficient in perfused proximal convoluted tubules of the anaesthetized rat in vivo. J. Physiol. 1995; 488(Pt 1): 153–61.

Weinstein AM. Modeling the proximal tubule: complications of the paracellular pathway. Am. J. Physiol. 1988; 254: F297–305.

Welling PA. Cross-talk and the role of KATP channels in the proximal tubule. Kidney Int. 1995; 48: 1017–23.

Wiederhielm CA, Weston BV. Microvascular, lymphatic, and tissue pressures in the unanesthetized mammal. Am. J. Physiol. 1973; 225: 992–6.

Woda CB, Bragin A, Kleyman TR, Satlin LM. Flow-dependent K^+ secretion in the cortical collecting duct is mediated by a maxi-K channel. Am. J. Physiol. Renal Physiol. 2001; 280: F786–93.

Yang T. Regulation of cyclooxygenase-2 in renal medulla. Acta Physiol. Scand. 2003; 177: 417–21.

Yeyati NL, Etcheverry JC, Adrogue HJ. Kaliuretic response to potassium loading in amiloride-treated dogs. Ren. Physiol. Biochem. 1990; 13: 190–9.

Yun J, Kelly G, Bartter FC, Smith H, Jr. Role of prostaglandins in the control of renin secretion in the dog. Circ. Res. 1977; 40: 459–64.

B. Primary Genetic Diseases of the Glomerulus

CHAPTER 4

Alport's Disease and Thin Basement Membrane Nephropathy

KARL TRYGGVASON AND JAAKKO PATRAKKA

INTRODUCTION

Alport's disease (AD) is an inherited form of nephritis characterized by progressive nephropathy leading to end-stage renal disease (ESRD) usually at middle age at the latest (Hudson et al 2003). The syndrome was first described in 1927 by A.C. Alport as an inherited renal disease associated with sensorineural deafness (Alport 1927). During the 1990s, mutations in genes coding for type IV collagen chains α3-α5 were shown to cause the disease (Barker et al 1990, Mochizuki et al., 1994). More recently, monoallelic mutations in the same genes have been associated with thin basement membrane nephropathy (TBMN) (Lemmink et al 1996). TBMN manifests as persistent hematuria, but in contrast to AD, only rarely progresses to ESRD (Savige et al 2003). Type IV collagen molecules composed of α3-α5 chains form the core network of the postnatal glomerular basement membrane (GBM), while α1 and α2 are the main collagen IV components during embryogenesis (Hudson et al 2003). The mutated collagen proteins lead to impaired assembly of the collagen IV network in the GBM, which results in a dysfunctional filtration barrier, primarily associated with hematuria and glomerulopathy.

Identification of the causative genes and mutations has improved possibilities for diagnosis and elucidated the molecular pathobiology behind the development of nephropathy. It has become clear that the pathomechanisms of impaired renal function resulting from mutated type IV collagen chains are very complex, and satisfactory treatment options are still lacking. In the following, we describe various molecular and clinical aspects of these 'type IV collagen nephropathies.'

RENAL GLOMERULUS

The wall of the glomerular capillaries forms the sieving structure of the kidney. The filter allows for passage of water and low-molecular-weight molecules into the urinary space, but restricts the passage of larger molecules such as albumin (Tryggvason & Wartiovaara 2005). The capillary wall is composed of fenestrated endothelial cells and highly specialized epithelial cells termed podocytes, and the glomerular basement membrane (GBM) between the two cell layers (Figure 4.1A). The endothelial cells are extensively fenestrated, and in contrast to most fenestrated endothelia seen in other capillaries of the body, these fenestrations lack diaphragms (Abrahamson 1999). The endothelial cells adhere tightly to the GBM and are involved in the synthesis of GBM components (see below). Mesangial cells are pericytes or vascular smooth muscle cells of the glomerular capillary tuft, and they are located in the mesangial matrix between individual capillaries of the glomerular tuft.

The GBM is a 300–350 nm thick acellular sheet-like network structure approximately twice as thick as most basement membranes located elsewhere in the body. The double thickness is thought to result, at least in part, from fusion of the endothelial and podocyte epithelial basement membranes during glomerular development (Sariola et al 1984, Abrahamson 1985). Based on electron micrographs, the GBM has been divided into a dense central layer (lamina densa), and two electrolucent layers (lamina rara interna and externa). The molecular composition and developmental aspects of the GBM will be discussed in more detail later.

The outermost layer of the glomerular capillary wall is composed of highly specialized podocytes (Pavenstadt et al

Genetic Diseases of the Kidney
ISBN: 978-0-12-449851-8

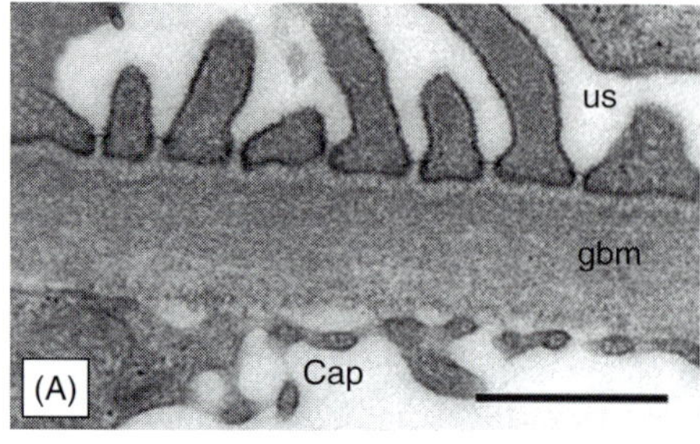

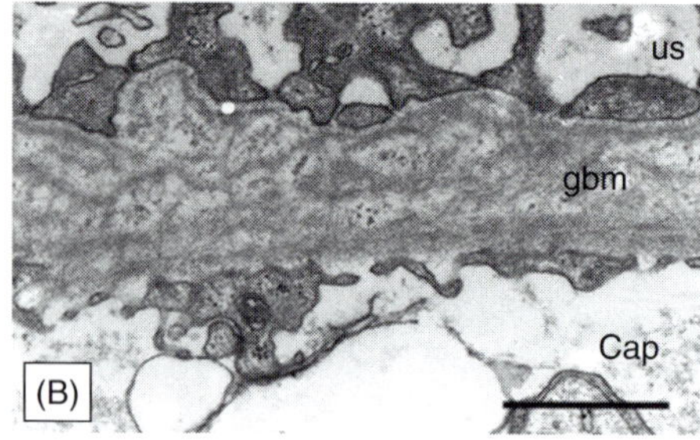

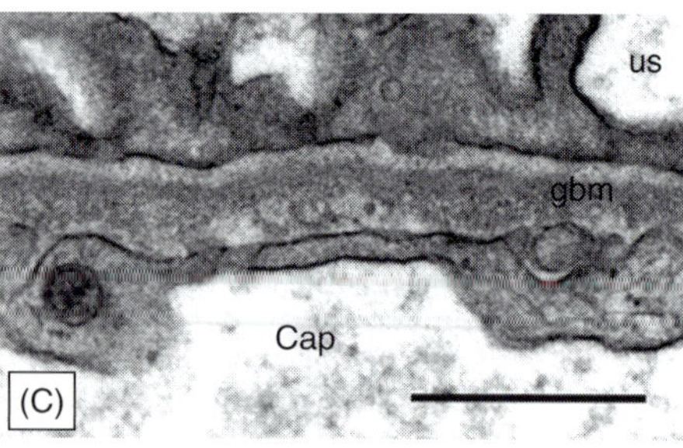

FIGURE 4.1 Morphological nature of the GBM in normal, AD, and TBMN. (A) The GBM is positioned in the middle of the glomerular capillary wall between podocytes and endothelial cells. In the normal glomerulus, the width of the GBM is uniformly between 300 and 350 nm. (B) In Alport nephropathy, electron microscopy reveals focal thickening and thinning of the GBM. Occasionally, lamination of the GBM is observed. (C) In TBMN, the GBM does not reveal any structural abnormalities. Three layers of the GBM, lamina rara interna, lamina densa, and lamina rara externa are obvious. However, the GBM is characteristically thinned, having only about half of the thickness present in a normal kidney. Bars = 500 nm. Cap, capillary lumen; us = urinary space. (A) courtesy of Dr Finn P. Reinholt and (B) of Dr Kjell Hultenby, Karolinska University Hospital, Huddinge, Sweden

2003, Tryggvason & Wartiovaara, 2005). The cell body and major processes of the podocytes 'float' in the urinary space of the Bowman's capsule, whereas the podocyte foot processes enclose capillaries in a comb-like fashion. Adjacent foot processes form an interrupted sheet around the GBM, and just above it, the processes are connected by a specialized cell–cell junctional complex termed slit diaphragm. Mutations in several podocyte genes, especially in those coding for proteins associated with the slit diaphragm, have been shown to cause proteinuric disorders both in man

and mouse (Tryggvason & Wartiovaara 2005), indicating the importance of this unique cell type for the functional glomerular filtration barrier (see Chapters 6 and 7).

Podocyte cells adhere tightly to the GBM. Integrins and dystroglycans, integral membrane proteins at the basal side of the foot process cell membrane, are thought to be key players in anchoring podocytes to the GBM (Korhonen et al 1990, Regele et al 2000). Via these two membrane receptors, the GBM is connected to the actin cytoskeleton of the foot processes. Integrins bind directly to collagen and laminin in the extracellular matrix, while dystroglycans have been shown to interact with laminin and agrin (Kretzler 2002). This molecular chain from the core GBM network to the foot process actin cytoskeleton, and further to the slit diaphragm complex, is essential for maintenance of the glomerular filtration barrier.

GLOMERULAR BASEMENT MEMBRANE

Basement membranes separate different cellular and tissue compartments, providing structural support and substrates for cellular signaling, migration, and differentiation. Basement membranes are composed of large collagenous and noncollagenous glycoproteins. The major collagenous component is type IV collagen, which forms the structural network of all basement membranes. Basement membrane is normally synthesized by a single cell type positioned on top of it. The GBM, however, is produced by two different cell types, podocytes and endothelial cells (Sariola et al 1984).

Type IV Collagen

Type IV collagens are the major structural component of all basement membranes in animal phyla. Individual type IV collagen molecules are composed of three long triple-helical collagen α-chains (Figure 4.2). Each α-chain contains a short 7S domain at the N-terminal, followed by an approximately 1400-residue collagenous domain with Gly-X-Y repeats, and a noncollagenous domain (NC1) positioned at the C-terminal (Timpl 1989, Hudson et al 1993, 2003). The presence of glycine residues in every third position in the collagenous domain is essential, since it is the only possible amino acid small enough to fit into the center of the tightly packed triple helix. The type IV collagen α-chains are highly glycosylated, with approximately 50 hydroxylysine-linked disaccharide units and an asparagine-linked oligosaccharide unit in the 7S-domain (Langeveld et al 1991, Nayak & Spiro 1991). Unlike rod-like fibrillar collagens, the type IV collagen collagenous domain contains several interruptions in the Gly-X-Y-repeat sequence which makes the molecule flexible.

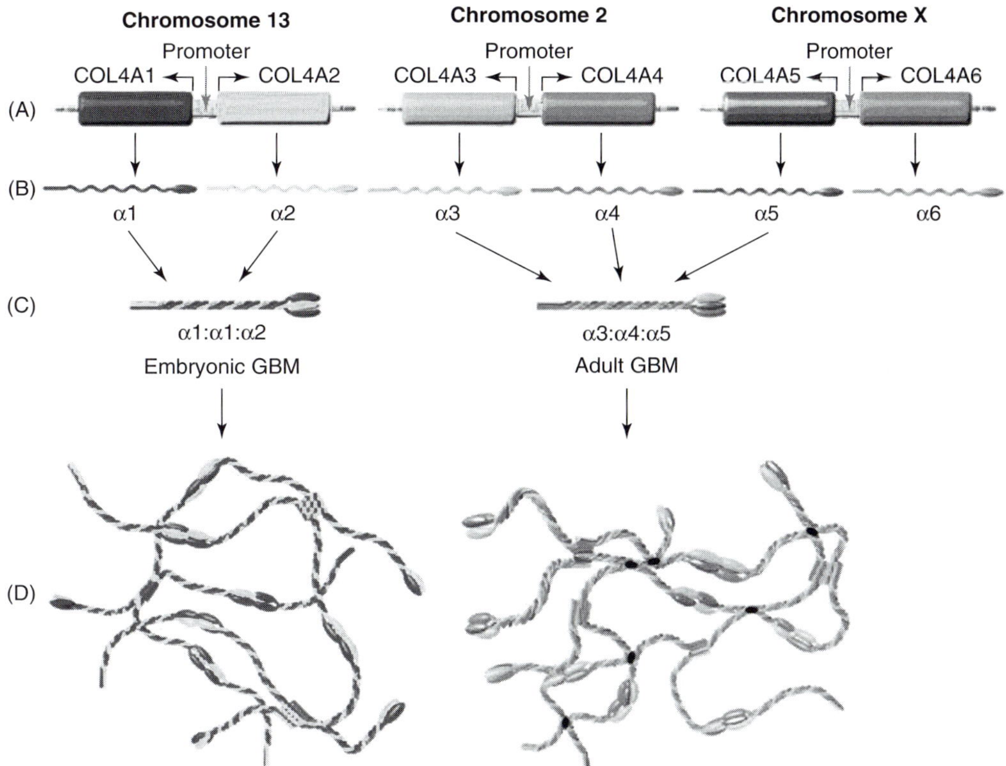

FIGURE 4.2   Type IV collagen genes, α chains and GBM specific isoforms. (A) The six collagen IV genes (*COL4A1–COL4A6*) located pairwise in a head-to-head fashion on three different chromosomes generate six different α chains that have a globular noncollagenous domain at their C-terminus (B). Three chains form triple-helical molecules that can have different combinations (C). (D) Extracellularly, the triple-helical type IV collagen molecules form a network by associating with each other at their ends so that two molecules are cross-linked through their C-terminal globular domain (NC1) and for trimers, associate with each other at the N-termini. In the embryonic GBM, the ubiquitous α1:α1:α2 trimer is the only isoform. After birth, this isoform is gradually replaced by an α3:α4:α5 isoform, which is more cross-linked through disulfide bonds within the collagenous regions (illustrated by black spots) and more resistant to extracellular proteolysis. Defects in the α3:α4:α5 trimers lead to AD or TBMN, depending on the extensiveness of alleles involved. Reproduced from Tryggvason & Patrakka (2006). (see also Plate 3)

The biosynthesis of type IV collagen involves similar posttranslational modifications as those present in all collagenous proteins in general (Kivirikko & Myllyla 1987). These include enzymatic hydroxylation of most prolyl and lysyl residues, followed by glycosylation of certain hydroxylysyl residues. Subsequently, the hydroxylated and glycosylated α chains can assemble intracellularly into triple-helical molecules, which are secreted into the extracellular space. This is followed by self-assembly of triple-helical monomers into a complex network structure (Figure 4.2). Individual trimers connect at their C-termini through two NC1 domains to form dimers, and four trimers assemble at the N-terminal 7S domains to form tetramers (Timpl et al 1981, Hudson et al 2003). The dimers and tetramers further assemble to form a complex three-dimensional network to which other basement membrane components can bind (Yurchenco & O'Rear 1994, Hudson et al 2003).

The basement membrane collagen network interacts with numerous molecules, including other basement membrane components and cell surface receptors (Yurchenco & O'Rear 1994, Timpl & Brown 1996). Among other matrix components, laminin and proteoglycans are probably most important, and among cell surface receptors, and it has been shown that collagen molecules also bind directly to integrins, perhaps the most crucial integral membrane receptors connecting the extracellular matrix to the cytosolic compartment.

There are six different type IV collagen α chains, α1–α6. Although these chains could theoretically form many combinations, only three sets of triple helical molecules have been confirmed to exist in vivo. These are α1:α1:α2, α3:α4:α5 and α5:α5:α6 (Figure 4.2) (Hudson et al 2003). Additionally, only three canonical sets of hexamers made from NC1 domains of two triple-helical molecules form networks: α1:α1:α2-α1:α1:α2,   α3:α4:α5-α3:α4:α5, α1:α1:α2-α5:α5:α6 (Hudson et al 2003). Assembly of these collagen IV networks is developmentally regulated (see below). The most abundant hexamer, α1:α1:α2-α1:α1:α2, is present in all mammalian basement membranes (Guo et al

1991, Zhang et al 1994, Netzer et al 1998). In contrast, α3:α4:α5-α3:α4:α5 and α1:α1:α2-α5:α5:α6 networks have a restricted distribution in mammalian tissues. In kidney, an α3:α4:α5 collagen scaffold is found in the GBM and some tubular basement membranes. Outside the kidney, it occurs in lung, testis, cochlea, and eye (Kalluri et al 1997, 1998, Cosgrove et al 1998). α1:α1:α2-α5:α5:α6 hexamers are components of basement membranes in skin, smooth muscle, esophagus, and kidney (Bowman's capsule) (Yoshioka et al 1994, Ninomiya et al 1995, Peissel et al 1995, Borza et al 2001, 2002).

Type IV Collagen Genes

The six mammalian type IV collagen genes have evolved from a common ancestor gene. They have an exceptional arrangement in the genome as they are located pairwise in a head-to-head fashion on three different chromosomes (Figure 4.2) (Hudson et al 2003). The human *COL4A1* and *COL4A2* genes are located on chromosome 13, *COL4A3* and *COL4A4* are on chromosome 2, and *COL4A5* and *COL4A6* are on the X chromosome. Each gene in a pair has its 5′ end opposing the partner gene, both genes being transcribed from opposite DNA strands. Thus, they share a common bidirectional promoter (Soininen et al 1988, Heikkila et al 1993).

The size of all type IV collagen genes is huge due to the large size of introns. Based on the current database predictions, the sizes of the *COL4A3*, *COL4A4*, and *COL4A5* are approximately 170 kb, 159 kb, and 278 kb, respectively. The number of exons coding for type IV collagen chains α3, α4, and α5 are 52, 48, and 51, respectively (Zhou et al 1994, Boye et al 1998, Heidet et al 2001). Because of the large size and high number of coding exons, analysis of these genes is rather difficult.

Other Glomerular Basement Membrane Components

Laminins

The laminins are complex extracellular glycoproteins that are core components of the basement membranes along with type IV collagens (Timpl & Brown 1994, Yurchenco & O'Rear 1994). Laminin molecules are heterotrimers composed of α, β, and γ chains. The sizes of the chains vary between 130 and 400 kDa. To date, five α, three β and two γ chains have been identified in man, these forming at least 14 different trimer combinations (Aumailley et al 2005). The assembly of trimers is initiated at the carboxyterminal ends of the chain which form a coiled coil structure referred to as the long arm.

Expression of different laminin isoforms is temporally and spatially regulated in the kidney (Hansen and Abrass 1999; Miner 1998). Laminin-11 (laminin-521) (Aumailley et al 2005), composed of α5, β2, and γ1 chains, is the most abundant isoform in the mature GBM (Miner et al 1997). This isoform is found also in the basement membrane of the neuromuscular junctions (Sanes & Lichtman 1999). The importance of the β2 chain for GBM function has been demonstrated in transgenic mice, where absence of the gene leads to severe protein leakage through the glomerular filter and renal failure (Noakes et al 1995). Also, mutations in the human *LAMB2* gene have been identified in patients with Pierson syndrome, a disease that manifests as congenital nephrotic syndrome, histological analysis revealing diffuse mesangial sclerosis (Zenker et al 2005).

Laminin binds to cells through various cell surface receptors. Integrins and dystroglycans, receptors found in podocyte foot processes, have been shown to bind to laminin (Sonnenberg et al 1990, Yamada et al 1994). In addition to binding membrane receptors, laminins can bind to other basement membrane components, including type IV collagen, and nidogen (Timpl & Brown 1996).

Nidogen/entactin

Nidogen is a 150-kDa single-chain glycoprotein found in most basement membranes (Timpl et al 1983). It has been shown to bind both laminins and type IV collagens (Mann et al 1988, Aumailley et al 1989, Fox et al 1991, Gerl et al 1991, Mayer et al 1993). In kidney, it is expressed by all basement membranes, and in the GBM, it is distributed along the whole width of the GBM (Katz et al 1991, Dziadek 1995). A homolog of nidogen/entactin, termed nidogen-2, has also been identified (Kimura et al 1998, Kohfeldt et al 1998). Nidogen-2 expression overlaps with that of nidogen-1 in the kidney and the GBM.

Proteoglycans

Proteoglycans are heterogeneous molecules composed of a core protein where glycosaminoglycan (GAG) chains are bound by 0-glycosidic linkage (Miner 1998). In the basement membranes, proteoglycans have typically heparan and chondroitin sulfate side chains. So far, three proteoglycans, perlecan, agrin, and bamacan, have been shown to be integral components of the GBM (Hassell et al 1980, Couchman et al 1996, Groffen et al 1997, 1998, Miner 1998). Due to their sulfate-rich side chains, proteoglycans have a highly negative charge, which has been speculated to form a charge barrier preventing passage of large plasma proteins (Tryggvason & Wartiovaara 2005).

Human perlecan has a core protein of about 470 kDa (Kallunki and Tryggvason 1992), to which three large heparan sulfate side chains are attached at the N-terminus (Noonan et al 1991). Previously, perlecan was thought to be the major proteoglycan of the GBM with a key role in glomerular permselectivity. However, it has been shown that perlecan expression is actually rather low (and restricted to subendothelial aspects) in the mature GBM (Groffen et al

1997) and, additionally, transgenic mice containing heparin-sulfate-deficient perlecan do not exhibit proteinuria (Rossi et al 2003). However, these mice seem to be more prone to develop proteinuria in protein overload challenge (Morita et al 2005), suggesting that perlecan may somehow contribute to the filtration barrier.

At the present, agrin is thought to be the major proteoglycan of the mature GBM. This large protein has a core protein sized 200 kDa, which is glycosylated and contains GAG side chains, leading to a total mass of 500 kDa. In contrast to perlecan, agrin is distributed along the whole width of the GBM (Groffen et al 1998), and it has been shown to interact with laminins and dystroglycans (Yamada et al 1996, Denzer et al 1997). Therefore, agrin has been proposed to have a role in the glomerular permselectivity and in the podocyte foot process adhesion to the GBM. However, in a very recent study, mice with ablation of the agrin gene in podocytes lack negative charges in the GBM, but they do not develop proteinuria (Harvey et al 2005). Together, the results obtained from genetically modified mice strongly question the role of the negative charges of GBM proteoglycans in the filtration process.

Maturation of the GBM

Developmentally, the GBM originates from the fusion of two basement membrane layers, one produced by developing podocytes, and the other by developing capillary endothelium (Sariola et al 1984, Abrahamson 1985). These extracellular matrix layers fuse in the developing vascular cleft region, and generate one heterogeneous but continuous sheet. The mechanisms of GBM fusion are not well understood, but they are likely to involve many molecular binding interactions between endothelial and podocytic basement membrane layers.

There is a developmental switch in the expression of different type IV collagen α chains in the maturing GBM (Miner & Sanes 1994, Miner 1998) (Figure 4.2). The expression of α1 and α2 is first detected at the vesicle stage glomerulus, and these two are dominant collagen chains in the early stages of glomerular development. In contrast, the expression of α3, α4, and α5 chains is detectable later, first at the capillary stage glomerulus, where they colocalize with the α1 and α2 chains. Further maturation of the glomerulus leads to a progressive loss of the α1 and α2 chains from the GBM, and instead, α3, α4, and α5 chains become the most prominent chains of the mature GBM. In humans (but not in mice), low levels of α1 and α2 chain expression can be detected still in the mature GBM.

Developmentally regulated expression has also been shown for the laminin α chains of the GBM (Miner et al 1997). At the early stages, α1 and α4 chains can be detected in the immature basement membrane. During development, α1, β1, and α4 laminin chains disappear from the GBM, and the α5 and β2 chains become expressed, forming the laminin-11 (laminin-521)

molecule (Miner et al 1997, Aumailley et al 2005). This laminin isoform is the main form in the mature GBM.

ALPORT'S DISEASE

Alport's disease (AD) is an inherited kidney disease manifesting with hematuria and sensorineural deafness (Alport 1927). After the initial description, ocular lesions were also shown to be associated with the syndrome. After introduction of electron microscopy, it was discovered that abnormalities of the GBM were typical for this disorder (Kinoshita et al 1969). The kidney lesion in AD is progressive, leading to renal failure latest at middle age. Prevalence of the disease in Utah is 1:5000, AD being the diagnosis in 2.3% of the patients undergoing renal transplantation (Atkin et al 1988). In Europe, AD is the cause of ESRD in 1–2% of cases (Wing & Brunner 1989).

Clinical Features

AD is clinically quite a heterogeneous disease (Hasstedt et al 1986, Colville & Savige 1997, Jais et al 2000, 2003, Gross et al 2002). Characteristically, recurrent microscopic/macroscopic hematuria is the first sign of Alport nephropathy, usually being detectable already in childhood. The disease develops generally earlier in males than in females. Alport nephropathy progresses and most patients end up with ESRD either in early or middle adulthood. Clinical features of nephropathy are usually more severe and progressive in male than female patients.

Extrarenal manifestations are frequently observed in patients with AD, and these most commonly include hearing impairment and ocular lesions. As in the case of Alport nephropathy, these extrarenal features are more common and severe in male than female patients. The hearing loss is sensorineural, and it mainly affects high tones. In the eye, a conical protrusion of the crystalline lens, termed lenticonus, is the pathognomonic feature of AD. Some other ocular lesions, including abnormal pigmentation of the macular region, have also been reported in patients with AD. Association of leiomyomatosis of the esophagus and tracheobronchial tree has also been observed in several families with AD (Antignac et al 1992), and in affected females, leiomyomas have been found in genital areas. Lastly, thrombocytopenia is sometimes associated with the syndrome.

Renal Pathology

Kidney biopsy specimens from patients with AD show only nonspecific changes by light microscopy. At early stages, mild glomerular changes, including mesangial proliferation and expansion, can be observed. At later stages, tubulointerstitial fibrosis is a more prominent finding. Electron microscopic evaluation of the kidney sample reveals the characteristic and diagnostic changes of AD (Kinoshita et al 1969, Meleg-Smith 2001) (Figure 4.1B). At early stages, the thinning and focal thickening or splitting of the GBM can be observed. At more advanced stages, the GBM pathology is more global, and the GBM is extensively thickened and multilamellinated. The thickening is the result of reduplication and splitting of the lamina densa, and it produces a pattern referred to as 'basket-weaving.' Less-specific changes seen in electron microscopy include expansion of the mesangial areas and podocyte foot process effacement.

The composition of type IV collagen chains in the GBM is altered in AD. In most cases, immunohistochemistry reveals absence of $\alpha3$, $\alpha4$, and $\alpha5$ chains, and instead, $\alpha1$ and $\alpha2$ chains are detected in the Alport GBM (Yoshioka et al 1994, Ninomiya et al 1995, Peissel et al 1995). This resembles the situation seen during glomerulogenesis, when $\alpha1$ and $\alpha2$ chains are expressed in immature GBM but no $\alpha3$, $\alpha4$, or $\alpha5$ chains are expressed. From a diagnostic standpoint it is, however, important to notice that not all Alport patients lack the $\alpha3$, $\alpha4$, and $\alpha5$ chains from the GBM.

Diagnosis

Due to the heterogeneous clinical features of AD (see above), diagnosis of this disease can be quite difficult in many cases. Earlier, it was proposed that AD can be diagnosed in a hematuric patient if three of the following criteria are fulfilled; (a) familial history, (b) hearing loss, (c) ocular lesions, and (d) GBM abnormalities (Atkin et al 1988). However, it has been shown that many patients with confirmed mutations in type IV collagen genes do not have family history or any extrarenal complications. Jais et al analyzed the findings in 195 European Alport families with confirmed COL4A5 mutations (Jais et al 2000), and these are summarized in Table 4.1. We have proposed previously that AD should be defined as a progressive hematuric hereditary nephritis caused by mutations in genes coding

for the $\alpha3$, $\alpha4$, or $\alpha5$ chains of type IV collagen (Hudson et al 2003). However, genetic tests sequencing all exon regions are, as yet, not commercially available for clinical diagnostics, and even if available, the mutation detection rate might still be expected to be only about 85% (Martin et al 1998). Therefore, it is currently practically impossible to set up completely satisfactory criteria for the clinical diagnosis of AD. Every patient with persistent hematuria of glomerular origin should be investigated carefully to diagnose or rule out the diagnosis of AD.

Genetics

As mentioned above, AD is caused by mutations in the genes coding for type IV collagen α-chains, i.e. the COL4A3, COL4A4, and COL4A5 genes (Figure 4.3). In a great majority (about 85%) of patients with AD, there is X-linked inheritance of mutations in the COL4A5 gene on chromosome X (Barker et al 1990, Hostikka et al 1990, Hudson et al 2003). The remaining cases are explained by being compound heterozygotes or homozygous mutations in COL4A3 or COL4A4 genes on chromosome 2q35-37. These patients follow autosomal-recessive inheritance. In addition, an autosomal-dominant form of AD associated with dominant COL4A3 or COL4A4 mutations has been reported in a few rare cases. These cases pose a major problem from the point of view of diagnosis of TBMN (see below).

MUTATIONS IN THE *COL4A5* GENE IN X-LINKED AD

In 1990, mutations in a novel type IV collagen $\alpha5$ chain gene (Hostikka et al 1990) were identified in X-linked AD (Barker et al 1990). Since then, approximately 300 mutations have been described by several different groups (Barker et al 1990, Martin et al 1998, Jais et al 2000, 2003). These mutations include gene rearrangements, such as deletions, insertions, inversions, and duplications, as well as single base changes leading to single amino acid changes. In some cases, the gene has been completely deleted. An interesting finding is that the mutations identified in COL4A5 are scattered in this huge gene, which contains 51 exons and exceeds about 250 kb in size. No 'hot spots' prone to mutagenic changes have been identified in the gene. Also, the same mutation has been detected only in a few cases in two or more unrelated kindreds.

AD has a heterogeneous clinical picture with onset of nephropathy at varying ages, and presence or absence of extrarenal manifestations, such as hearing loss and ocular lesions. Since a huge number of individual mutations have been described in the COL4A5 gene, it is of importance to elucidate the possible link between the COL4A5 mutation and clinical presentation of the disease. The European Community Alport Syndrome Concerted Action reported in 2000 a large study elucidating the genotype–phenotype

TABLE 4.1 Findings in 195 AD families with COL4A5 mutation. Adapted from reference (Jais et al 2000)

Variable	Frequency (%)
Hematuria	99
GBM changes detectable in EM	98
Familial history	89
IF reveals changes in collagens of the GBM	85
Hearing loss	83
ESRD	76
Ocular changes	44

EM, electron microscopy; IF, immunofluorescence for collagen type IV α-chains.

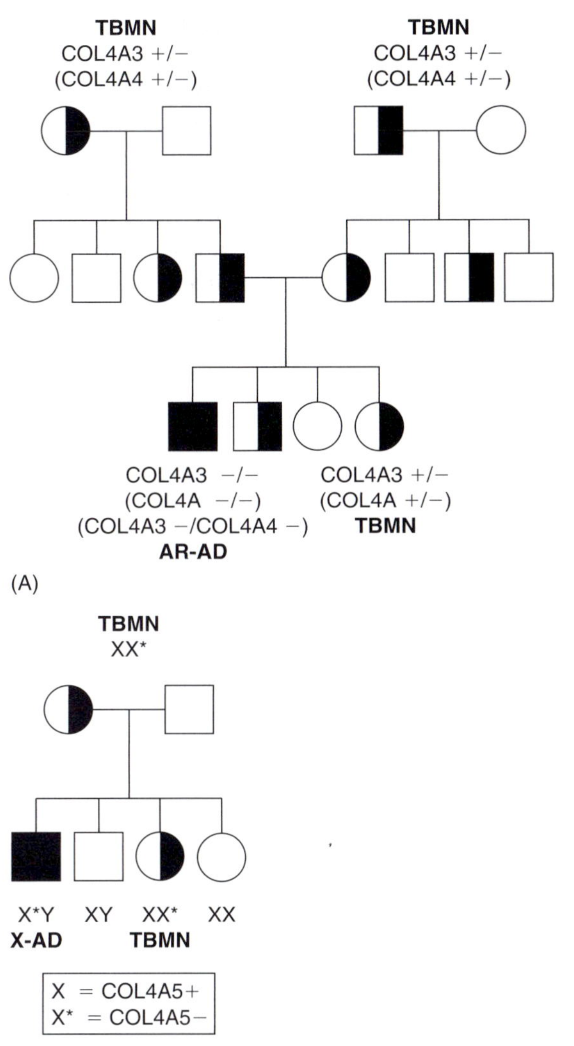

FIGURE 4.3 Interrelationship of inheritance for autosomal recessive and X-linked forms of TBMN and AD. (**A**) TBMN and autosomal recessive AD in two intermarried families with a mutation (−) in *COL4A3* (or *COL4A4*) located on chromosome 2. Half-shaded symbols represent hematuria and heterozygosity state for a mutation in *COL4A3* + /2 (or *COL4A4* + /2), and fully shaded symbols represent homozygosity for a mutation (−/−) or compound heterozygosity for mutations in *COL4A3* and *COL4A4* resulting in AD. First generation heterozygotes with TBMN can be considered carriers for recessive AD. Half of the second generation individuals can be expected to have TBMN. A third generation offspring of a couple where both parents have TBMN can be expected to get a child with AD, half of the children being expected to be TBMN patients. However, it has not yet been demonstrated that the offspring of TBMN affected parents are at risk to for developing AD. (**B**) TBMN simulating nephropathy and AD in a single family where the mother is heterozygous for a mutation in the X chromosomal *COL4A5* gene (XX*). Each child has a 25% risk of inheriting the mutant allele. Half-shaded symbols represent heterozygosity and TBMN mimicking disease in females, and fully shaded symbols represent AD (X-AD). Reproduced from Tryggvason & Patrakka (2006).

correlation in a large number of European AD families (Jais et al 2000). Importantly, they found that 95% of male patients with large deletions or nonsense mutations (changing the reading frame) in the *COL4A5* gene ended up with renal failure before the age of 30 years, whereas the same number in patients with missense and splice mutations was 50% and 70%, respectively. Also, patients with missense mutations had a 60% risk of developing hearing loss before the age of 30 years, whereas the same risk for patients with other mutations increased to 90%.

One clinically important group is the patients who develop anti-GBM antibodies after transplantation. So far, there has not been segregation of mutations in some parts of the *COL4A5* gene in patients with post-tranplant anti-GBM nephritis.

MUTATIONS IN THE *COL4A3* AND *COL4A4* GENES IN AUTOSOMAL RECESSIVE AD

Autosomal inheritance has been estimated to account for up 15% of affected AD families. In 1994, autosomal AD was shown to be caused by abnormalities in the *COL4A3* and *COL4A4* genes (Mochizuki et al 1994, Lemmink et al 1996), and since then over 50 different mutations have been characterized in these genes (Nagel et al 2005, Rana et al 2005, Longo et al 2006).

Similarly to X-linked AD, mutations in the *COL4A3* and *COL4A4* genes are also scattered along the coding exons without any obvious clustering into 'hot spots.' Mutations have been found to be both homozygous or combined heterozygous. Males and females are equally affected, and mutations differ usually between families. All *COL4A3* and *COL4A4* mutations cause nephropathy similar to that observed in X-linked AD, but extrarenal manifestations seem to be quite rare in these patients. The result of the mutations is also similar with loss of the α3, α4, and α5 chains from the GBM.

AUTOSOMAL-DOMINANT AD

An autosomal-dominant form of AD has been reported in a few families around the world. Analysis of the *COL4A3* and *COL4A4* genes have shown that various mutations can cause dominant form of AD (Jefferson et al 1997, van der Loop et al 2000, Ciccarese et al 2001, Longo et al 2002, Pescucci et al 2004). Many dominant mutations include substitution of glycine in the collagenous domain, leading to a disruption of collagen helix. Also, mutations in conserved cysteines in the C-terminal noncollagenous domain have been described, as well as a splice site mutation leading to skipping of exon 21.

The clinical picture of patients with autosomal-dominant AD is milder than that seen in recessive or X-linked forms. These patients usually manifest early with hematuria, but the progression towards renal insufficiency is generally slower. The disease may lead to renal failure and need for dialysis at adulthood, but in many cases, no development of ESRD is observed. Autosomal domainant AD represents a significant diagnostic problem as the disease may be different to separate from TBMN caused by monoallelic mutations in the *COL4A3* and *COL4A4* genes (see below).

Animal Models of AD

Several animal models of AD have been characterized. The models include spontaneous Alport-like diseases in dogs as well as transgenic models created in mice. Animal models for AD have provided us opportunities to study the pathogenesis of Alport nephropathy, and to reveal possible key pathogenic events which could be targeted by pharmacological intervention.

Both mouse and canine models are suitable for the studies. The advantage of mouse models is short gestation period and very fast progress of the nephropathy. This allows rather quick analysis of the development of the disease and follow-up of the outcome of the treatment trials. The main advantage of the canine models is, on the other hand, the rather large size of the animal, which facilitates the development of therapeutic interventions.

X-linked AD

There are two canine models and a transgenic mouse model for X-linked AD. All three X-linked AD models are caused by mutations in the gene coding for the type IV collagen α5 chain. The canine models are spontaneous, while the mouse model was recently generated by gene targeting in embryonic stem cells.

Samoyed dogs

The first and best described X-linked model is a spontaneous Alport-like nephropathy in Samoyed dogs (Zheng et al 1994, Thorner et al 1996). Nephropathy is caused by a mutation in exon 35 of the collagen α5 chain gene, which leads to a premature stop codon. The dogs develop proteinuria by the age of 2 months and hematuria is commonly observed. Nephropathy progresses to renal failure by 5 months of age, and this leads to death by 8–10 months of age. These dogs do not appear to have any extrarenal manifestations. Typical kidney pathology, and thickening and lamellination of the GBM, is first detected at the age of 1 month in some diseased dogs, and these changes progress so that all affected dogs have abnormal GBM structure as seen in electron microscopy by the age of 4–5 months.

Expression analysis of the type IV collagen chains (using Western blot) has shown that the α3, α4, and α5 chains are absent from the GBM of the affected dogs.

Navasoto dogs

Another canine model following the features of X-linked AD was reported by Lees and colleagues (Lees et al 1999). Analysis of the collagen α5 chain gene in these dogs resulted in identification of a 10-bp deletion in exon 9, which causes a frame-shift and premature stop codon. Kidney disease described in these dogs follows essentially the same course as described in Samoyed model.

Immunohistochemistry revealed absence of α3 and α4 chains in the GBM, but the expression of the α5 chain was still detectable in the GBM of the affected dogs. In contrast, the expression of α1 and α2 chains in the GBM was significantly increased. These immunohistochemical findings are in line with those of the X-linked Samoyed dog model (see above).

Knock-out mouse for *Col4a5* gene

In addition to spontaneous AD models found in dogs, a transgenic mouse model for X-linked AD has been developed recently (Rheault et al 2004). A mouse model was generated by targeting a human nonsense mutation, G5X, to the mouse *Col4a5* gene. This mutation leads to a null allele. Hemizygous mutant male and carrier female mice develop characteristic nephropathy with proteinuria and renal insufficiency within a few months of age, and die due to renal failure at 6–34 weeks of age (male), or carrier females at 8–45 weeks of age. Typical lamellation of the GBM can be detected in electron microscopic evaluation already at the age of 4 weeks.

Effects of the targeted *Col4a5* mutation on type IV collagen α-chain expression are consistent with those seen in humans, but somewhat different to those described in canine models. Immunohistochemically, the α5 chain is not detected, and immunoreactivity for the α3 chain is also lost from the GBM. Instead, expression of α1 and α2 chains in the GBM was comparable to that seen in wild-type mice. At the mRNA level, α5 expression was significantly reduced, while the levels for α1, α3, and α4 were maintained at normal values.

Autosomal-Recessive AD

Three mouse models and one canine model for autosomal-recessive AD have been characterized. Two of the mouse models were targeted to inactivate the *Col4a3* gene, and one affects both *Col4a3* and *Col4a4* genes. The mutation in the canine autosomal-recessive Alport-like nephropathy remains to be determined.

Col4a3 knockout mouse

Transgenic mouse absent of *Col4a3* gene is a murine model of the autosomal-recessive form of AD. Two different mouse models affecting the *Col4a3* gene have been generated (Cosgrove et al 1996, Miner & Sanes 1996). The first symptoms of nephropathy in the transgenes are observed already at 2 weeks of age, when microhematuria is observed, and proteinuria is detectable at around 5 weeks of age. Disease progresses to renal failure by 14 weeks of age. Ultrastructural pathology of the kidney includes focal multilamellinated thickening (and thinning) of the GBM at 4 weeks, and GBM pathology has progressed to more global changes at 8 weeks of age. Collapsed glomerular capillary loops and fibrotic changes are dominant features at the later stages when ESRD has developed.

Immunohistochemistry reveals absence of the $\alpha3$, $\alpha4$, and $\alpha5$ chains in the GBM of mutant mice, whereas increased immunoreactivity for $\alpha1$ and $\alpha2$ chains is detected in mature GBM. At the mRNA level, Northern analysis shows complete absence of $\alpha3$ mRNA (as expected), while $\alpha4$ and $\alpha5$ mRNAs are expressed at normal levels.

Mouse lacking both *Col4A3* and *Col4A4* chains

A transgenic mouse line lacking exons 1–12 of the *Col4a4* and exons 1–2 of the *Col4a3* gene, as well as the intergenic promoter region has been created (Lu et al 1999). The transgene was generated by random insertional mutagenesis. Homozygous mice for the deletion develop persistent proteinuria and hematuria as early as 2 weeks of age, and the renal disease progresses to azotemia and death by 10–14 weeks of age. Ultrastructural analysis of the kidney samples reveals occasional thinning and duplication of the GBM at 2 weeks of age, and in older animals, the GBM becomes thicker, and characteristic 'basketweaving' is observed.

Immunohistochemical analyses show that the $\alpha3$, $\alpha4$, and $\alpha5$ chains are not expressed in the GBM of the mutant mice. Instead, abnormally high levels of $\alpha1$ and $\alpha2$ chains are detected. At the mRNA level, *Col4a3* and *Col4a4* transcripts are also missing from the transgenes. In contrast, transcripts of *Col4aa1* and *Col4a5* were present normally in mutant kidneys.

English Cocker Spaniel

A naturally occurring hereditary nephritis in English Cocker Spaniels presents the features of the autosomal-recessive form of AD (Lees et al 1998). The affected dogs appear to be healthy at 3–4 months of age. Proteinuria associated with hematuria is the first abnormality detected at 5–8 months of age, and the disease in the affected dogs progresses to terminal renal failure at 8–27 months of age. Kidney pathology shows early thinning and thickening of the GBM (at 8 months of age), and later, GBM thickening and lamellation are dominant features. No signs of impaired hearing or ocular abnormalities have been found in these dogs despite thorough examination.

Immunohistochemical analysis of the type IV collagen α-chains showed that the affected dogs lack $\alpha3$- and $\alpha4$-chains in their GBM and the expression of the $\alpha5$ chain is also decreased. In contrast, immunoreactivity for the $\alpha1$

and $\alpha2$ chains is increased in the GBM of diseased dogs. The manifesting AD-like nephropathy combined with autosomal-recessive inheritance in these dogs suggests strongly that mutation either in *Col4a3* or *Col4a4* genes is behind the disease. However, the mutation(s) has not yet been identified in these dogs.

AUTOSOMAL-DOMINANT AD

Two spontaneous canine models for autosomal-dominant AD have been reported by Hood and colleagues (Hood et al 1995, 2002a, 2002b). In Bull Terriers and Dalmatian dogs, successive generations are affected; also males and females are affected equally, consistent with an autosomal-dominant inheritance. Inherited nephropathies in the two canine models highly resemble each other. Affected dogs develop hematuria and proteinuria within the first year of life, and nephropathy leads to renal failure. Kidney pathology includes lamellation and increased width of the GBM at early stages, and later, glomerular sclerosis and fibrotic changes become more dominant features.

Immunohistochemistry for type IV collagen α-chains show that the $\alpha1$-$\alpha5$ chains are expressed in the GBM of affected dogs at normal levels. Similar findings have been reported in human patients with an autosomal-dominant form of AS. The mutations behind these two autosomal-dominant Alport-like diseases have not yet been identified.

Pathogenesis

The present theory about the pathomechanism leading from mutated collagen chain into Alport-nephropathy is illustrated in Figure 4.2. Mutations in $\alpha3$, $\alpha4$, or $\alpha5$ chains may result in impaired folding and assembly of monomers, and these are rapidly degraded inside the cell. Therefore, these mutations arrest the normal developmental switch with persistence of $\alpha1$:$\alpha1$:$\alpha2$ networks in the mature GBM (Kalluri et al 1997). This is supported by the findings in AD patients and animal models, in which the $\alpha3$-$\alpha5$ chains are absent in the mature GBM, and instead, embryonic $\alpha1$ and $\alpha2$ chains are strongly expressed. Some $\alpha5$ mutations allow partial formation of $\alpha3$:$\alpha4$:$\alpha5$ networks, and result in less severe phenotypes (Jais et al 2000).

Why is the embryonic $\alpha1$:$\alpha1$:$\alpha2$ network not sufficient to maintain normal glomerular function postnatally? The replacement of the embryonic $\alpha1$:$\alpha1$:$\alpha2$ network with the adult type $\alpha3$:$\alpha4$:$\alpha5$ network could be important to withstand physical, proteolytic, and oxidative stress in the GBM (Hudson et al 2003). Thus, the adult type $\alpha3$:$\alpha4$:$\alpha5$ collagen IV has a higher amount of cysteines (Leinonen et al 1994, Mariyama et al 1994) and intermolecular disulfide bonds (Hudson et al 2003) making the molecules more resistant to increased intraglomerular pressure and proteolytic attacks. In the mature kidney, plasma traverses the glomerular capillaries, and leads to an increase in protein content, including proteases. Embryonic collagen network seems to be more susceptible to endoproteolysis than mature network, and, therefore, the GBM possibly needs the protection of a resistant collagen IV α3:α4:α5 network. This increased proteolysis would explain the pathology of the AD – unevenly thickened and split GBM that finally deteriorates.

Management

At the present, there is no curative or satisfactory treatment for AD, and the disease progresses gradually to renal failure (Adler et al 1996). The progression of kidney disease can, however, be slowed down with antihypertensive drugs and angiotensin-converting-enzyme inhibitors (Adler 2002). The positive effect of cyclosporine A treatment in slowing AD progression has also been reported (Callis et al 1999). Patients who develop renal failure require dialysis treatment and are candidates for kidney transplantation. Posttransplant survival and graft function of patients with AD seem to be comparable to other patient groups with a graft, although the special (and rare) problem of anti-GBM glomerulonephritis can occur in patients with AD and a kidney graft (Byrne et al 2002).

THIN BASEMENT MEMBRANE NEPHROPATHY

Thin basement membrane nephropathy (TBMN) is the most common cause of persistent hematuria in children and adults (Savige et al 2003, Tryggvason & Patrakka 2006), and it is a major diagnostic problem. In addition to hematuria, TBMN patients usually have minimal proteinuria, normal renal function and uniformly thinned GBM, as determined by electron microscopy. TBMN, which affects at least 1% of the population, is a lifelong nonprogressive disorder associated with family history.

TBMN was first described about 80 years ago as a curable form of hemorrhagic nephritis (Baehr 1926). Later, many cases of this microscopic, painless hematuria with good prognosis were shown to be inherited, and this fact is emphasized in the numerous names used for the disease in the literature, such as 'congenital hereditary hematuria' (Reyersbach & Butler 1954), 'hereditary hematuria' (Russell & Smith 1959), 'familial hematuric nephritis' (Russell & Smith 1959), 'familial benign hematuria' (McConville & McAdams 1966), 'benign familial hematuria' (Tina et al 1982, Lemmink et al 1996), 'familial benign essential hematuria' (Rogers et al 1973, Gubler et al 1980), 'benign hereditary nephritis' (Peterson & Schubert 1977), and 'benign essential hematuria' (Mihatsch & Zollinger 1980). TBMN has also been referred to as familial hematuria (Longo et al

2002), but that is a misnomer, as it does not distinguish it from progressive AD (Gregory 2005). Commonly used names are 'thin membrane nephropathy' (Aarons et al 1989, Perry et al 1989, Dische et al 1990), 'thin basement membrane disease' (Aarons et al 1989, Perry et al 1989, Basta-Jovanovic et al 1990, Dische et al 1990, Colville et al 1997, Buzza et al 2001, 2003, Ueda et al 2002, Sakai et al 2003), 'thin glomerular basement membrane (GBM) nephropathy' (Nieuwhof et al 1997), and 'thin glomerular basement membrane syndrome' (Abe et al 1987).

The term 'thin basement membrane nephropathy' is currently most widely used (Gauthier et al 1989, Savige et al 2003, Gregory 2004, Kashtan 2005, Tryggvason & Patrakka 2006, Wang et al 2004), and this name is to be preferred, as it refers to a renal disorder associated with observable structural changes in the basement membrane, and without necessarily being a 'true disease' (Gregory 2005).

The connection between recurrent benign hematuria and thin GBM was first noted in electron microscopic analysis of kidney specimens in 1973 (Rogers et al 1973). This typical histopathological feature of TBMN, i.e. uniform thinning of the GBM, could be found also at the early stages of AD, which suggested that the pathomechanisms of the two diseases might overlap. This connection was verified at the gene level during the early 1990s when the type IV collagen genes *COL4A3, COL4A4,* and *COL4A5* were discovered and shown to be mutated in X-linked and autosomal AD (Barker et al 1990, Lemmink et al 1994, Mochizuki et al 1994), with subsequent demonstration of mutations in *COL4A3* and *COL4A4* in TBMN (Lemmink et al 1996).

Clinical Features

The characteristic clinical manifestation of TBMN is persistent microscopic hematuria (Savige et al 2003). Most patients with TBMN present only with hematuria, without additional symptoms or progression to renal impairment, and the condition is usually incidentally detected during health control. The age at diagnosis varies considerably, from as early as 1 year of age (McConville & McAdams 1966, Yoshikawa et al 1988) up to an 86-year-old (Rogers et al 1973). TBMN, which does not progress, has been documented for up to 30 years (Perry et al 1989). Microscopic analysis of the urine samples reveals red blood cells in most patients. At least a single episode of macroscopic hematuria is observed in 5–22% of patients, typically manifesting after exercise or during infection. Occasionally, the hematuria has disappeared with time (Tiebosch et al 1989, Goel et al 1995, Gregory 2005, Packham et al 2005).

In spite of hematuria, the individuals usually do not have proteinuria or only minimally so, indicating that the podocyte slit diaphragm is not really affected. The proteinuria develops normally later than hematuria, and it is rarely seen in children (Gregory 2005). A significant proportion of adult patients, however, show mild to moderate proteinuria (Rogers et al 1973, Dische et al 1985, Blumenthal et al 1988, Aarons et al 1989, Perry et al 1989, Gregory 2005). On the other hand, nephrotic-range proteinuria is rare even in older patients.

Renal function in children with TBMN is normal (Yoshikawa et al 1988, Schroder et al 1990, Roth et al 2001), whereas adults have been reported to have low prevalence of renal insufficiency (Abe et al 1987, Perry et al 1989, Dische et al 1990, Goel et al 1995, Nieuwhof et al 1997, Auwardt et al 1999, van Paassen et al 2004). The incidence of renal insufficiency in these patients might partially reflect the complicated differential diagnosis from autosomal or X-linked AD, or concurrent additional renal disease. Similarly, rare occurrence of hearing loss in TBMN may also well reflect the difficulties in differential diagnosis (Aarons et al 1989). Also, hypertension has been diagnosed in 11–31% of adults with TBMN (Dische et al 1985, Goel et al 1995, Auwardt et al 1999, Badenas et al 2002, van Paassen et al 2004), but very infrequently in pediatric patients (Roth et al 2001). The interpretation of these results is difficult, but hypertension in TBMN patients could be coincidental. Generally, the prognosis for the nephropathy in true TBMN is excellent.

Epidemiology

TBMN has been reported in all races, although most of the cases have thus far been reported in developed countries. Hematuria has been diganosed at all ages (McConville & McAdams 1966, Rogers et al 1973, Yoshikawa et al 1988), and several studies have indicated that the disease is somewhat more common in females than males, both among children and adults (Marks et al 1970, Abe et al 1987, Yoshikawa et al 1988, Aarons et al 1989, Perry et al 1989, Goel et al 1995, Piqueras et al 1998, Auwardt et al 1999, Badenas et al 2002), but other studies have not revealed such findings (Blumenthal et al 1988, Tiebosch et al 1989, Nieuwhof et al 1997, Roth et al 2001). Exact prevalence of the disease is difficult to assess, as the diagnosis is mostly made based on persistent hematuria combined with minimal proteinuria, while the number of electron microscopic analyses of renal biopsies showing thinned basement membrane, have become less common (Wang & Savige 2005). Thus, most of the cases remain undiagnosed. However, the prevalence may be estimated from known frequencies of persistent hematuria in the population, and from the number of TBMN cases in archival series of renal biopsies, together with the knowledge of prevalence of autosomal AD. Several studies have addressed the prevalence of hematuria and persistent hematuria in children and adults (Dodge et al 1976, Vehaskari et al 1979, Ritchie et al 1986, Hogg et al 1998, Chadban et al 2003). Persistent hematuria is commonly defined as hematuria observed on at least two occasions, and in TBMN a useful additional criterion could be that these two incidences occurred at least 2 years

apart. Persistence of hematuria is important from the point of view of TBMN diagnosis, as it distinguishes from other more acute renal disorders, such as hematuria associated with streptococcal infections. The prevalence of hematuria in children has been estimated to be about 1–2% (Dodge et al 1976, Vehaskari et al 1979), but the prevalence in adults is not well known. According to Wang and Savige (Wang & Savige 2005), persistent hematuria occurs consistently in as much as 6% of both children and adults. Based on both direct and indirect approaches, the overall prevalence of TBMN in the population has been estimated to be about 1% (Gregory 2005). However, based on observations of frequencies of persistent hematuria, thin basement membrane in renal biopsies and autosomal recessive AD, another estimate indicates a higher prevalence of over 1% but less than 10% (Wang & Savige 2005). When making estimations about the prevalence of TBMN by analyzing incidence of hematuria, it should be kept in mind that not all TBMN patients have hematuria (or it is intermittent), some patients with persistent hematuria have other signs of renal impairment (excluding the diagnosis of TBMN), and lastly that hematuria is not always of glomerular origin. It can, however, be concluded that TBMN is the most common inherited renal disorder.

Renal Pathology

Rogers et al were the first to associate familial hematuria with thinning of the glomerular basement membrane (Rogers et al 1973). Light microscopy of renal samples in TBMN shows almost normal glomerular histology with only occasional mild mesangial cellular proliferation and matrix expansion (Kincaid-Smith 1995, Foster et al 2005). Slight attenuation of the GBM can sometimes be observed by Jones methenamine silver or periodic acid–Schiff stains, suggesting GBM thinning. Erythrocytes may be identified in the urinary space. In about 5–25% of the cases, focal glomerular sclerosis and tubular fibrosis may be found with aging (Nieuwhof et al 1997, Foster et al 2005) but all histopathological changes observed in light microscopic evaluation are nonspecific. Usually, direct immunofluorescence staining is negative for immunoglobulins and complement, but there are sometimes traces of segmental mesangial positivity for IgM or C3, and rarely IgG or IgA.

Electron microscopy reveals the typical feature of TBMN, i.e. thinning of the GBM (Figure 4.1C), but it does not distinguish between pure TBMN and thin GBM in early stages of AD. Normally, the GBM thickness varies with age and gender, and it is also influenced by the method and tissue preparations used. The GBM thinning in TBMN is uniform, which distinguishes it from patients having focal thinning process as seen sometimes in normal children (Bloom et al 1984), and in patients with minimal change nephrosis and some other forms of glomerulonephritides

(Hill et al 1974). In normally fixed kidney samples (such as with glutaraldehyde), the GBM appears as a trilaminar structure with a central lamina densa and an inner lamina rara interna and an outer lamina rara externa. However, in EM preparations made using quick-freezing, the GBM has a uniform appearance (Inoue 1994). According to general agreement, the GBM thickness is determined from fixed samples as the distance between the outer limits of the endothelial cell and the base of podocyte foot process cell membranes which are partially embedded in the GBM (Foster et al 2005). As an additional rule, the thickness should be measured only in the peripheral capillaries, and multiple measurements should be made over several capillaries of multiple glomeruli (Foster et al 2005).

Normal GBM thickness has been estimated in several reports. Osawa et al (1966) compared 587 measurements from normal adult biopsies and 254 from autopsy samples and showed the average thickness to be 315 nm with a range of means from 239 to 453 nm. Haynes (1981) reported the thickness in normal individuals aged between 11 and 26 years to be 394 nm (range 372–632 nm). Using the quite complicated orthogonal intercept technique (Gundersen et al 1978, Jensen et al 1979, Steffes et al 1983), the mean GBM thickness in adult females was found to be 326 nm and that in males 373 nm. According to those reports, the thickness increased until a plateau at age 40 years. According to Vogler et al (1997) the GBM thickness increases progressively in children until age 11. At ages 2 days to 1 year, the thickness is 132–208 nm, at ages 1–6 years 208–245 nm, and at ages 6–11 years 244–307 nm. As general guidelines concerning GBM thickness based on several published results, the mean thickness in adult men is 370 +/− 50 nm, and in adult females 320 +/− 50 nm. In children the GBM thickness at birth is 150 nm, 200 nm at age 1 year, and approaches thickness in adults at age 11 years.

A central question is then how one should determine what constitutes thin glomerular basement membrane in an individual. The World Health Organization has proposed a threshold of 250 nm for adults and 180 nm for children between 2 and 11 years of age (Churg et al 1995). According to Vogler et al (1987), the criteria for TBMN in children vary between <200 nm and <250 nm, and in adults from <200 nm (Cosio et al 1994) and <250 nm (Tiebosch et al 1989) or up to <264 nm. This variation in recommendations is due to difficulties in standardizing the technical methods. This means that the morphological criteria for thin GBM have to be carefully evaluated by the pathologist, and one has to take into account that the width of the GBM varies according to age and gender of the patient, and also between different laboratories, due to technical differences in sample preparation. Individual criteria should be established for each histological laboratory.

In TBMN, the cardinal findings are that the GBM is thinned in most of the glomerular capillaries, and that there is absence of other significant glomerular pathology. The GBM thickness varies in individuals with TBMN (Abe et al 1987, Shindo et al 1988, Saxena et al 1990, Dische 1992, Cosio et al 1994, Marquez et al 1999, Nogueira et al 2000), at least 50% of the glomeruli having abnormally thin GBM. Only rare regions are observed with lamellation or regional thickening, a feature typically seen in Alport kidneys. This may pose a problem in differentiating between TBMN and AD, as electron microscopic analysis at early stages of AD can reveal similar uniform thinning of the GBM (see below for differential diagnosis).

Immunohistochemical evaluation of the type IV collagen $\alpha3-\alpha5$ chains in renal biopsy has become of major importance as a method to use in differentiating between TBMN and early stages of AD with microscopic hematuria and thin GBM, as these chains usually are either absent or abnormally distributed in AD (Gubler et al 1995, Kashtan 2004). These findings are summarized in Table 4.2.

Genetics

TBMN mainly manifests as an inherited disorder with dominant transmission affecting approximately one-half of successive generations (Rogers et al 1973, Blumenthal et al 1988, Gauthier et al 1989). About two thirds of patients with TBMN have at least one other relative with hematuria (Buzza et al 2001). The remaining one-third of cases may have de novo mutations, or nonpenetrance in other members of the family (Rana et al 2005).

Although TBMN is currently viewed as a common inherited disorder of type IV collagen, the genetic basis of this GBM disease was not understood until the late 1990s. The abnormally thin GBM could be associated with defective structure of the GBM, but this finding could not be connected with the loss of any GBM proteins, Thus, all the GBM type IV collagen chains, $\alpha1-\alpha5$, are present, as are all other known GBM proteins (Pettersson et al 1990, Lajoie 2001).

The causes of TBMN started to unravel first through the identification of type IV collagen genes mutated in AD (see above). Males with progressive X chromosome-linked AD were shown to contain mutations in the *COL4A5* gene (Barker et al 1990), while their mothers carrying one normal allele in addition to the abnormal allele most often exhibited mild hematuria. Subsequently, the autosomal recessive forms of AD were shown to be due to mutations in the *COL4A3* and *COL4A4* genes that are located head-to-head on chromosome 2 (Figure 4.2) (Lemmink et al 1994, Mochizuki et al 1994). Autosomal AD can be caused by homozygous mutations in either *COL4A3* or *COL4A4*, or by combined heterozygosity for both.

Lemmink et al noted that some carriers of the autosomal forms of Alport had thin GBM, and this prompted them to analyze the *COL4A3* and *COL4A4* genes in patients with TBMN (Lemmink et al 1996). They first showed linkage of the disease with the *COL4A3/COL4A4* locus, and

TABLE 4.2 Expression of collagen IV α3, α4, and α5 chains in normal, TBMN, and AD kidneys*

| | | Autosomal X-linked AD | | Recessive AD**** |
	Normal	TBMN	Male**	Female***
GBM				
α3	+	+	−	mosaic	−
α4	+	+	−	mosaic	−
α5	+	+	−	mosaic	−
TBM (distal and collecting)				
α3	+	+	−	mosaic	−
α4	+	+	−	mosaic	−
α5	+	+	−	mosaic	+
Bowman's capsule				
α3	+	+	−	mosaic	−
α4	+	+	−	mosaic	−
α5	+	+	−	mosaic	+
Epidermal BM				
α5	+	+	−	mosaic	+

*Modified from reference 161.

**About 20% of males with X-linked AS show positive renal basement membrane staining for α3, α4, and α5 and for a5 in epidermal basement membranes.

***About 30% of female carriers for X-linked AS exhibit positive uninterrupted staining for renal basement membranes for the α3, α4, and α5 chains.

****Some patients with autosomal recessive AS show positive staining for α3, α4, and α5 in renal basement membranes.

subsequently identified a mutation leading to a substitution of glycine to glutamic acid in the collagenous domain of the *COL4A4* gene (Lemmink et al 1996). Thus, patients with TBMN linked to chromosome 2 are heterozygous for mutations in either *COL4A3* or *COL4A4*, and they represent a carrier status for autosomal recessive AD (Figure 4.3), in a similar fashion as females with mutations in one *COL4A5* allele are carriers for X-linked AD in males (Figure 4.3). Savige and colleagues have shown that up to 36% of TBMN cases associate with the *COL4A3/COL4A4* locus for autosomal-recessive AD (Buzza et al 2001). More recently, several studies aiming to identify mutations in the *COL4A3* and *COL4A4* genes, as well as their common promoter region, have revealed that heterozygous mutations in these two genes are found in many patients with TBMN (Ozen et al 2001, Badenas et al 2002, Longo et al 2002, Buzza et al 2003, Gross et al 2003, Tazon Vega et al 2003, Wang & Savige 2005). The current understanding of the mode of inheritance for autosomal TBMN and autosomal recessive AD is depicted in Figure 4.3.

Thus far, at least 21 *COL4A3* and *COL4A4* gene mutations have been identified in patients with TBMN (Rana et al 2005). Mutations in TBMN are scattered throughout the *COL4A3* and *COL4A4* genes without any 'hot spots,'

similarly to what has been observed in the same genes and *COL4A5* in autosomal recessive and X-linked AD patients, respectively. Most mutations result in single nucleotide substitutions and lead to missense or nonsense mutations. Many of the mutations lead to a replacement of a glycine in the collagenous domain, but also missense mutations in non-glycine residues of the collagenous domain. Single amino acid substitutions have been identified in the noncollagenous domain, and nonsense and splice site mutations have also been identified. In addition, six insertion or deletion mutations have been identified. To date, no mutations have been described in the promoter region of *COL4A3/COL4A4* genes in TBMN, in contrast to that observed in AD. Heterozygous mutations found in patients with TBMN are very similar, and in some cases, identical, to those identified in patients with autosomal-recessive AD.

So far, only few studies have attempted to correlate the type of mutation in type IV collagen to the clinical features of TBMN. Interestingly, Wang et al reported that the same mutations in different family members can result in different clinical features (Wang et al 2004). Autosomal dominant AD is also caused by heterozygous mutations in the *COL4A3* or *COL4A4* genes. As described above, the nephropathy in autosomal dominant AD follows a much more deleterious course, leading to renal failure and lamellated GBM, in contrast to the symptomless hematuria observed in TBMN. It would be crucial to understand the cause of this difference in the clinical outcome of these patients with heterozygous *COL4A3/COL4A4* mutations, but, so far, too few mutations have been described in autosomal dominant AD and TBMN to distinguish the two based on the genotype.

The *COL4A3* and *COL4A4* genes are huge in size and, therefore, the identification of mutations in the genes is difficult. In studies made by the Savige group, the mutation detection rate for *COL4A3* and *COL4A4* mutations in patients with TBMN was 11% and 14–17%, respectively, using SCCP (Buzza et al 2001, Wang et al 2004). However, the rate increased to as high as 67% when only individuals from families where TBMN was linked to the *COL4A3/COL4A4* locus were included. The low detection rate may be due to many other reasons. First, the use of SSCP instead of direct sequencing of exons lowers the detection rate (Martin et al 1998). Thus, sequencing of all exon regions has been shown to reveal over 80% of mutations in *COL4A5* in X-linked Alport syndrome. Secondly, the primers used in the studies were not designed to detect possible splice site mutations. Furthermore, due to numerous polymorphisms in the genes, some pathogenic alterations in the gene sequence may be missed and interpreted as non-pathogenic polymorphisms.

Practically all female carriers for X-linked AD with a mutation in one *COL4A5* allele exhibit hematuria. In a study on 288 heterozygous females from 329 families with X-linked AD, 96% had hematuria, and a relatively large proportion (30%) had progressive disease leading to chronic renal failure and ESRD (Jais et al 2003). This type of progression is not considered typical for TBMN. However, the female carriers for X-linked Alport syndrome do exhibit thinning and thickening or diffuse thinning of the GBM. Thus, such cases with thin GBM that mimics TBMN could be considered as TBMN cases. The relationship of thin GBM and X-linked AD is depicted in Figure 4.3.

TBMN is a disorder of the trimeric $\alpha3:\alpha4:\alpha5$ isoform of type IV collagen. Therefore, it could be anticipated that a single mutation in any of the alleles for these three genes would lead to a similar phenotype, as they all affect the same trimeric protein. However, it appears to be a common view in the literature that only mutations in *COL4A3* or *COL4A4* lead to TBMN as depicted in Figure 4.3A. It is still likely that X-linked TBMN exists with inheritance as depicted in Figure 4.3B. For example, clinical studies performed prior to identification of the *COL4A5* gene have indicated the existence of such cases (Rumpelt et al 1974, Habib et al 1982) and, similarly, a more recent study by Liapis et al (2002) suggests that a connection between some cases of TBMN and *COL4A5* could exist. It would be important to solve this controversy by more extensive sequence analyses of the *COL4A5* gene in individuals with TBMN not showing association with chromosome 2.

Many families with TBMN do not show linkage to the *COL4A3/COL4A4* or *COL4A5* loci (Yamazaki et al 1995, Piccini et al 1999). This may be explained by several factors. A high rate of de novo mutations has been shown in the *COL4A5* gene in X-linked AD, and this might be the case in *COL4A3* and *COL4A4* genes as well (Lemmink et al 1997). Incomplete penetrance of hematuria in heterozygous carriers has also been reported (Dagher et al 2002), and this can influence linkage analyses. Also, coincidental hematuria without proteinuria in family members makes it more difficult to establish linkage. Lastly, it is possible that some TBMN cases are due to mutations in another, as yet, unknown gene. In relation to this, Rana et al (2005) recently studied the linkage of TBMN in nine families which did not show segregation to *COL4A3/COL4A4*, and ruled out linkage to *MYH9* (mutations in this gene result in Alport-like nephropathy) and to genes coding for GBM components laminin-5 (laminin-332), perlecan, and fibronectin. So far, there are no data showing any genetic linkage to loci other than *COL4A3/COL4A4* and *COL4A5*, supporting the view that TBMN is a disease of the adult form of GBM type IV collagen ($\alpha3:\alpha4:\alpha5$).

Pathogenesis

In general, development of autosomal nonprogressive TBMN involves heterozygous mutations in either *COL4A3* or *COL4A4*, while homozygosity or combined heterozygosity of mutations in these genes results in autosomal recessive AD characterized by deterioration of the GBM (Tazon Vega et al 2003, Gregory 2005). Similarly, a TBMN-like phenotype may be caused in a female with heterozygosity

for a mutation in the X-chromosomal *COL4A5* gene. This could be explained by a 'dose effect' where absence of one normal allele leads to lower production of the $\alpha3:\alpha4:\alpha5$ trimer and thin GBM, whereas the loss of two alleles results in lack of the $\alpha3:\alpha4:\alpha5$ trimer and AD (Gregory 2004). However, there are reports that a single heterozygous mutation in *COL4A3* or *COL4A4* can cause AD in adult life (van der Loop et al 2000, Ciccarese et al 2001, Longo et al 2002). This could mean that some mutations are more severe than others, but it can still not yet be excluded that in these cases there is another unknown mutation somewhere else in the other allele or gene that could lead to absence of the chain. To date, too few mutations have been identified in TBMN and autosomal-dominant Alport for making conclusions about the differences in pathogenesis.

Differential Diagnosis of TBMN and AD

The characteristic manifestation of TBMN is persistent hematuria of glomerular origin, and therefore it is important to distinguish this rather benign disease from other superimposing causes of glomerular bleeding. These include IgA nephropathy, postinfectious glomerulonephritis, mesangiocapillary glomerulonephritis and lupus nephritis. While IgA nephropathy and TBMN may be difficult to distinguish on clinical grounds alone, other forms of glomerulonephritis are frequently associated with additional clinical features rarely seen in TBMN, such as proteinuria, hypertension, renal impairment and systemic symptoms.

For initial diagnosis of TBMN, it is critical to be able to distinguish the condition from early stages of AD that can yield very similar findings as TBMN. Differential diagnosis of these two is extremely important due to the dramatically different outcomes of the two disorders. X-linked AD that accounts for about 85% of AD cases generally occurs in 1 in 50 000 live births (Levy and Feingold 2000), but it can affect as many as 1 in 5000 males, such as in Utah (Barker et al 1990). Typical Alport syndrome findings, such as hearing loss, lenticonus, and severe structural changes in the GBM develop usually first during adolescence. For early diagnosis of AD, it very important to clarify family history. About 95% of female carriers of AD have hematuria (Dagher et al 2001), but they cannot be distinguished from individuals with TBMN (Savige et al 2003). However, knowledge about hearing loss, lenticonus, or retinopathy in other relatives can give a hint about X-linked AD.

If the existence of X-linked AD is not clear, renal biopsy should be performed. The characteristic uniform thinning of the GBM observed in electron microscopy often confirms the diagnosis of TBMN. However, the importance of this finding must not be overemphasized, as abnormally thin GBM can also be observed at the early stages of AD. Therefore, differential diagnosis from X-linked or autosomal recessive AD should include immunohistochemical analysis of the type IV collagen $\alpha3–\alpha5$ chains (Churg et al 1995,

Kashtan 2004, 2005). In both forms of AD, the expression of all three chains is usually significantly reduced or absent, whereas in TBMN the expression is comparable to normal levels (Table 4.2). Such analysis can in many cases be performed on skin biopsies, such as in X-linked AD, where the α5 chain may be absent from the epidermal basement membrane, while it is not absent in TBMN (Churg et al 1995, Kashtan 2004, 2005). However, immunohistochemical analysis can in some cases yield a false-negative result, since it has been shown that about 20% of autosomal AD cases and 30% of cases of female carriers for X-linked AD express low or even normal levels (Gubler et al 1995, van der Loop et al 2000).

Genetic analyses can verify the diagnosis, but DNA analysis is unfortunately not generally available. Segregation of the disease with the *COL4A5* locus and sequencing of all the exon regions can reveal over 80% of the mutations in X-linked AD (Martin et al 1998), while more easily performed methods such as SSCP only yield between 37 and 50% of the mutations (Kawai et al 1996, Knebelmann et al 1996, Renieri et al 1996). Unfortunately, an assay for screening *COL4A5* mutations is currently not commercially available. Similar sequencing of the *COL4A3* and *COL4A4* genes is also possible in specialized laboratories. It would be important to be able to provide DNA sequencing of the *COL4A3*, *COL4A4,* and *COL4A5* genes for making diagnosis of TBMN and AD, which together form a very large group of renal disorders. With today's technologies, such analysis methods should not have to be very costly.

Management of TBMN

Currently, there are no clear evidence-based treatment protocols available for TBMN. Although the condition generally has an excellent outcome, and in many cases cannot really be considered a disease, it must be noted that it may not always be benign, and the worsening of the clinical picture has been reported (Jais et al 2003, Frasca et al 2004, van Paassen et al 2004). It is still a dilemma for the physician to know whether such cases are in fact true TBMN cases or misdiagnosed AD. In any case, patients diagnosed with TBMN should be monitored for the appearance of hypertension, proteinuria or renal insufficiency, and patients developing these complications should be treated accordingly.

Conclusions

TBMN is one of the most common disorders of the kidney, affecting at least 1% of the total population. It appears to be a disease of the adult GBM type IV collagen trimer α3:α4:α5. Genetic evidence indicates that autosomal TBMN is caused by heterozygous mutations in either *COL4A3* or *COL4A4*, while homozygous or combined heterozygous mutations in the same genes lead to autosomal-recessive AD. Thus, individuals with autosomal TBMN are carriers

for autosomal AD. In some rare cases, monoallelic mutations can lead to autosomal-dominant AD. Heterozygosity for mutations in *COL4A5* in female carriers for X-linked AD can also mimic a TBMN condition, although some of those individuals will later develop progressive hematuria (Jais et al 2003). It is not clear if the individuals with TBMN and female carriers simulating TBMN that develop progressive disease could have a second mutation in some other of the six alleles that encode the α3:α4:α5 collagen form. The main clinical problem is to differentiate between TBMN and X-linked or autosomal recessive AD, as AD has a severe outcome. While family history and electron microscopy of renal biopsies are helpful for diagnosis, immunohistological examination of expression of the type IV collagen α3, α4, and α5 chains is currently the most informative method. It would be of utmost importance to make sequencing analyses of the *COL4A3, COL4A4,* and *COL4A5* genes available to the clinic as a routine diagnostic method of TBMN and AD.

References

Aarons I, Smith PS, Davies RA, Woodroffe AJ, Clarkson AR. Thin membrane nephropathy: a clinico-pathological study. Clin. Nephrol. 1989; 32: 151–8.

Abe S, Amagasaki Y, Iyori S, et al. Thin basement membrane syndrome in adults. J. Clin. Pathol. 1987; 40: 318–22.

Abrahamson DR. Origin of the glomerular basement membrane visualized after in vivo labeling of laminin in newborn rat kidneys. J. Cell. Biol. 1985; 100: 1988–2000.

Abrahamson DR. Glomerular endothelial cell development. Kidney Int. 1999; 56: 1597–8.

Adler L, et al. Angioconverting enzyme inhibitor therapy in children with Alport syndrome: effect on urinary albumin, TGF-beta, and nitrite excretion. BMC Nephrology 2002; 3: 2.

Adler S, Cohen A, Glassock R. Secondary glomerular disease. In: Brenner BM, ed. The Kidney, 5th edn. Boston, MA: WB Saunders, 1996.

Alport AC. Hereditary familial congenital haemorrhagic nephritis. BMJ 1927: 404–506.

Antignac C, Zhou J, Sanak M, et al. Alport syndrome and diffuse leiomyomatosis: deletions in the 5' end of the COL4A5 collagen gene. Kidney Int. 1992; 42: 1178–83.

Atkin CL, Gregory MC, Border WA. Alport Syndrome. Boston, MA: Little, Brown, 1988.

Aumailley M, Bruckner-Tuderman L, Carter WG, et al. A simplified laminin nomenclature. Matrix Biol. 2005; 24: 326–32.

Aumailley M, Wiedemann H, Mann K, Timpl R. Binding of nidogen and the laminin-nidogen complex to basement membrane collagen type IV. Eur. J. Biochem. 1989; 184: 241–8.

Auwardt R, Savige J, Wilson D. A comparison of the clinical and laboratory features of thin basement membrane disease (TBMD) and IgA glomerulonephritis (IgA GN). Clin. Nephrol. 1999; 52: 1–4.

Badenas C, Praga M, Tazon B. Mutations in the COL4A4 and COL4A3 genes cause familial benign hematuria. J. Am. Soc. Nephrol. 2002; 13: 1248–54.

Baehr G. Benign and curable form of hemorrhagic nephritis. JAMA 1926; 86: 1001–4.

Barker DF, Hostikka SL, Zhou J. Identification of mutations in the COL4A5 collagen gene in Alport syndrome. Science 1990; 248: 1224–7.

Basta-Jovanovic G, Venkataseshan VS, Gil J, Kim DU, Dikman SH, Churg J. Morphometric analysis of glomerular basement membranes (GBM) in thin basement membrane disease (TBMD). Clin. Nephrol. 1990; 33: 110–14.

Bloom P, Hartmann J, Vernier R. An electron microscopic evaluation of the width of normal basement membrane in children with hematuria. J. Pathol. 1984; 142: 263–7.

Blumenthal SS, Fritsche C, Lemann J Jr. Establishing the diagnosis of benign familial hematuria. The importance of examining the urine sediment of family members. JAMA 1988; 259: 2263–6.

Borza DB, Bondar O, Ninomiya Y, et al. The NC1 domain of collagen IV encodes a novel network composed of the alpha 1, alpha 2, alpha 5, and alpha 6 chains in smooth muscle basement membranes. J. Biol. Chem. 2001; 276: 28532–40.

Borza DB, Bondar O, Todd P, et al. Quaternary organization of the goodpasture autoantigen, the alpha 3(IV) collagen chain. Sequestration of two cryptic autoepitopes by intrapromoter interactions with the alpha4 and alpha5 NC1 domains. J. Biol. Chem. 2002; 277: 40075–83.

Boye E, Mollet G, Forestier L, et al. Determination of the genomic structure of the COL4A4 gene and of novel mutations causing autosomal recessive Alport syndrome. Am. J. Hum. Genet. 1998; 63: 1329–40.

Buzza M, Dagher H, Wang YY, et al. Mutations in the COL4A4 gene in thin basement membrane disease. Kidney Int. 2003; 63: 447–53.

Buzza M, Wilson D, Savige J. Segregation of hematuria in thin basement membrane disease with haplotypes at the loci for Alport syndrome. Kidney Int. 2001; 59: 1670–6.

Byrne MC, Budisavljevic MN, Fan Z, Self SE, Ploth DW. Renal transplant in patients with Alport's syndrome. Am. J. Kidney Dis. 2002; 39: 769–75.

Callis L, Vila A, Carrera M, Nieto J. Long-term effects of cyclosporine A in Alport's syndrome. Kidney Int. 1999; 55: 1051–6.

Chadban SJ, Briganti EM, Kerr PG. Prevalence of kidney damage in Australian adults: The AusDiab kidney study. J. Am. Soc. Nephrol. 2003; 14: S131–8.

Churg J, Bernstein J, Glassock R. Renal Disease: Classification and Atlas of Glomerular Diseases. New York: Igaku-Shoin, 1995.

Ciccarese M, Casu D, Ki Wong F. Identification of a new mutation in the alpha4(IV) collagen gene in a family with autosomal dominant Alport syndrome and hypercholesterolaemia. Nephrol. Dial. Transplant. 2001; 16: 2008–12.

Colville D, Savige J, Branley P, Wilson D. Ocular abnormalities in thin basement membrane disease. Br. J. Ophthalmol. 1997; 81: 373–7.

Colville DJ, Savige J. Alport syndrome. A review of the ocular manifestations. Ophthalmic Genet. 1997; 18: 161–73.

Cosgrove D, Meehan DT, Grunkemeyer JA, et al. Collagen COL4A3 knockout: a mouse model for autosomal Alport syndrome. Genes Dev. 1996; 10: 2981–92.

Cosgrove D, Samuelson G, Meehan DT, et al. Ultrastructural, physiological, and molecular defects in the inner ear of a gene-knockout mouse model for autosomal Alport syndrome. Hear. Res. 1998; 121: 84–98.

Cosio FG, Falkenhain ME, Sedmak DD. Association of thin glomerular basement membrane with other glomerulopathies. Kidney Int. 1994; 46: 471–4.

Couchman JR, Kapoor R, Sthanam M, Wu RR. Perlecan and basement membrane-chondroitin sulfate proteoglycan (bamacan) are two basement membrane chondroitin/dermatan sulfate proteoglycans in the Engelbreth-Holm-Swarm tumor matrix. J. Biol. Chem. 1996; 271: 9595–602.

Dagher H, Buzza M, Colville D, et al. A comparison of the clinical, histopathologic, and ultrastructural phenotypes in carriers of X-linked and autosomal recessive Alport's syndrome. Am. J. Kidney Dis. 2001; 38: 1217–28.

Dagher H, Yan Wang Y, Fassett R, Savige J. Three novel COL4A4 mutations resulting in stop codons and their clinical effects in autosomal recessive Alport syndrome. Hum. Mutat. 2002; 20: 321–2.

Denzer AJ, Brandenberger R, Gesemann M, Chiquet M, Ruegg MA. Agrin binds to the nerve-muscle basal lamina via laminin. J. Cell Biol. 1997; 137: 671–83.

Dische FE. Measurement of glomerular basement membrane thickness and its application to the diagnosis of thin-membrane nephropathy. Arch. Pathol. Lab. Med. 1992; 116: 43–9.

Dische FE, Anderson VE, Keane SJ, Taube D, Bewick M, Parsons V. Incidence of thin membrane nephropathy: morphometric investigation of a population sample. J. Clin. Pathol. 1990; 43: 457–60.

Dische FE, Weston MJ, Parsons V. Abnormally thin glomerular basement membranes associated with hematuria, proteinuria or renal failure in adults. Am. J. Nephrol. 1985; 5: 103–9.

Dodge WF, West EF, Smith EH, Bruce H, 3rd. Proteinuria and hematuria in schoolchildren: epidemiology and early natural history. J. Pediatr. 1976; 88: 327–47.

Dziadek M. Role of laminin-nidogen complexes in basement membrane formation during embryonic development. Experientia 1995; 51: 901–13.

Foster K, Markowitz GS, D'Agati VD. Pathology of thin basement membrane nephropathy. Semin. Nephrol. 2005; 25: 149–58.

Fox JW, Mayer U, Nischt R, et al. Recombinant nidogen consists of three globular domains and mediates binding of laminin to collagen type IV. Embo. J. 1991; 10: 3137–46.

Frasca GM, Soverini L, Gharavi AG, et al. Thin basement membrane disease in patients with familial IgA nephropathy. J. Nephrol. 2004; 17: 778–85.

Gauthier B, Trachtman H, Frank R, Valderrama E. Familial thin basement membrane nephropathy in children with asymptomatic microhematuria. Nephron. 1989; 51: 502–8.

Gerl M, Mann M, Aumailley M, Timpl R. Localization of a major nidogen-binding site to domain III of laminin B2 chain. Eur. J. Biochem. 1991; 202: 167–74.

Goel S, Davenport A, Goode NP. Clinical features and outcome of patients with thin and ultrathin glomerular membranes. QJM 1995; 88: 785–93.

Gregory MC. Alport syndrome and thin basement membrane nephropathy: unraveling the tangled strands of type IV collagen. Kidney Int. 2004; 65: 1109–10.

Gregory MC. The clinical features of thin basement membrane nephropathy. Semin. Nephrol. 2005; 25: 140–5.

Groffen AJ, Buskens CA, van Kuppevelt TH. Primary structure and high expression of human agrin in basement membranes of adult lung and kidney. Eur. J. Biochem. 1998; 254: 123–8.

Groffen AJ, Hop FW, Tryggvason K. Evidence for the existence of multiple heparan sulfate proteoglycans in the human glomerular basement membrane and mesangial matrix. Eur. J. Biochem. 1997; 247: 175–82.

Gross O, Netzer KO, Lambrecht R, Seibold S, Weber M. Meta-analysis of genotype-phenotype correlation in X-linked Alport syndrome: impact on clinical counselling. Nephrol. Dial. Transplant 2002; 17: 1218–7.

Gross O, Netzer KO, Lambrecht R, Seibold S, Weber M. Novel COL4A4 splice defect and in-frame deletion in a large consanguine family as a genetic link between benign familial haematuria and autosomal Alport syndrome. Nephrol. Dial. Transplant, 2003; 18: 1122–7.

Gubler MC, Knebelmann B, Beziau A. Autosomal recessive Alport syndrome: immunohistochemical study of type IV collagen chain distribution. Kidney Int. 1995; 47: 1142–7.

Gubler MC, Levy M, Naizot C, Habib R. Glomerular basement membrane changes in hereditary glomerular diseases. Ren. Physiol. 1980; 3: 405–13.

Gundersen HJ, Jensen TB, Osterby R. Distribution of membrane thickness determined by lineal analysis. J. Microsc. 1978; 113: 27–43.

Guo XD, Johnson JJ, Kramer JM. Embryonic lethality caused by mutations in basement membrane collagen of C. elegans. Nature 1991; 349: 707–9.

Habib R, Gubler MC, Hinglais N, et al. Alport's syndrome: experience at Hopital Necker. Kidney Int. Suppl. 1982; S20–8.

Hansen K, Abrass CK. Role of laminin isoforms in glomerular structure. Pathobiology 1999; 67: 84–91.

Harvey SJ, Burgess R, Miner JH. Podocyte-derived agrin is responsible for the glomerular basement membrane anionic charge. J. Am. Soc. Nephrol. 2005; 16: 1A.

Hassell JR, Robey PG, Barrach HJ, Wilczek J, Rennard SI, Martin GR. Isolation of a heparan sulfate-containing proteoglycan from basement membrane. Proc. Natl Acad. Sci. USA 1980; 77: 4494–8.

Hasstedt SJ, Atkin CL, San Juan AC Jr. Genetic heterogeneity among kindreds with Alport syndrome. Am. J. Hum. Genet. 1986; 38: 940–53.

Haynes WD. The normal human renal glomerulus. Virchows Arch. B Cell Pathol. Incl. Mol. Pathol. 1981; 35: 133–58.

Heidet L, Arrondel C, Forestier L, et al. Structure of the human type IV collagen gene COL4A3 and mutations in autosomal Alport syndrome. J. Am. Soc. Nephrol. 2001; 12: 97–106.

Heikkila P, Soininen R, Tryggvason K. Directional regulatory activity of cis-acting elements in the bidirectional alpha 1(IV) and alpha 2(IV) collagen gene promoter. J. Biol. Chem. 1993; 268: 24677–82.

Hill GS, Jenis EH, Goodloe S Jr. The nonspecificity of the ultrastructural alterations in hereditary nephritis with additional observations on benign familial hematuria. Lab. Invest. 1974; 31: 516–32.

Hogg RJ, Harris S, Lawrence DM, Henning PH, Wigg N, Jureidini KF. Renal tract abnormalities detected in Australian preschool children. J. Paediatr. Child Health, 1998; 34: 420–4.

Hood JC, Dowling J, Bertram JF, et al. Correlation of histopathological features and renal impairment in autosomal dominant Alport syndrome in Bull terriers. Nephrol. Dial Transplant 2002a; 17: 1897–908.

Hood JC, Huxtable C, Naito I, Smith C, Sinclair R, Savige J. A novel model of autosomal dominant Alport syndrome in Dalmatian dogs. Nephrol. Dial Transplant 2002b; 217: 2094–8.

Hood JC, Savige J, Hendtlass A, Kleppel MM, Huxtable CR, Robinson WF. Bull terrier hereditary nephritis: a model for autosomal dominant Alport syndrome. Kidney Int. 1995; 47: 758–65.

Hostikka SL, Eddy RL, Byers MG, Hoyhtya M, Shows TB, Tryggvason K. Identification of a distinct type IV collagen alpha chain with restricted kidney distribution and assignment of its gene to the locus of X chromosome-linked Alport syndrome. Proc. Natl Acad. Sci. USA 1990; 87: 1606–10.

Hudson BG, Reeders ST, Tryggvason K. Type IV collagen: structure, gene organization, and role in human diseases. Molecular basis of Goodpasture and Alport syndromes and diffuse leiomyomatosis. J. Biol. Chem. 1993; 268: 26033–6.

Hudson BG, Tryggvason K, Sundaramoorthy M, Neilson EG. Alport's syndrome, Goodpasture's syndrome, and type IV collagen. N. Engl. J. Med. 2003; 348: 2543–56.

Inoue S. Ultrastructural architecture of basement membranes. Contrib. Nephrol. 1994; 107: 21–8.

Jais JP, Knebelmann B, Giatras I, et al. X-linked Alport syndrome: natural history in 195 families and genotype- phenotype correlations in males. J. Am. Soc. Nephrol. 2000; 11: 649–57.

Jais JP, Knebelmann B, Giatras I, et al. X-linked Alport syndrome: natural history and genotype-phenotype correlations in girls and women belonging to 195 families: a "European Community Alport Syndrome Concerted Action" study. J. Am. Soc. Nephrol. 2003; 14: 2603–10.

Jefferson JA, Lemmink HH, Hughes AE, et al. Autosomal dominant Alport syndrome linked to the type IV collage alpha 3 and alpha 4 genes (COL4A3 and COL4A4). Nephrol. Dial. Transplant 1997; 12: 1595–9.

Jensen EB, Gundersen HJ, Osterby R. Determination of membrane thickness distribution from orthogonal intercepts. J. Microsc. 1979; 115: 19–33.

Kallunki P, Tryggvason K. Human basement membrane heparan sulfate proteoglycan core protein: a 467-kD protein containing multiple domains resembling elements of the low density lipoprotein receptor, laminin, neural cell adhesion molecules, and epidermal growth factor. J. Cell Biol. 1992; 116: 559–71.

Kalluri R, Gattone VH 2nd, Hudson BG. Identification and localization of type IV collagen chains in the inner ear cochlea. Connect. Tissue Res. 1998; 37: 143–50.

Kalluri R, Shield CF, Todd P, Hudson BG, Neilson EG. Isoform switching of type IV collagen is developmentally arrested in X-linked Alport syndrome leading to increased susceptibility of renal basement membranes to endoproteolysis. J. Clin. Invest. 1997; 99: 2470–8.

Kashtan CE. Familial hematuria due to type IV collagen mutations: Alport syndrome and thin basement membrane nephropathy. Curr. Opin. Pediatr. 2004; 16: 177–81.

Kashtan CE. The nongenetic diagnosis of thin basement membrane nephropathy. Semin. Nephrol. 2005; 25: 159–62.

Katz A, Fish AJ, Kleppel MM, Hagen SG, Michael AF, Butkowski RJ. Renal entactin (nidogen): isolation, characterization and tissue distribution. Kidney Int. 1991; 40: 643–52.

Kawai S, Nomura S, Harano T, Harano K, Fukushima T, Osawa G. The COL4A5 gene in Japanese Alport syndrome patients: spectrum of mutations of all exons. The Japanese Alport Network. Kidney Int. 1996; 49: 814–22.

Kimura N, Toyoshima T, Kojima T, Shimane M. Entactin-2: a new member of basement membrane protein with high homology to entactin/nidogen. Exp. Cell Res. 1998; 241: 36–45.

Kincaid-Smith P. Thin basement membrane disease, 1995.

Kinoshita Y, Osawa G, Morita T, Kobayashi N, Wada J. Hereditary chronic nephritis (alport) complicated by nephrotic syndrome. Light, Fluorescent and electronmicroscopic studies of renal biopsy specimens. Acta Med. Biol. (Niigata) 1969; 17: 101–17.

Kivirikko KI, Myllyla R. Recent developments in posttranslational modification: intracellular processing. Methods Enzymol. 1987; 144: 96–114.

Knebelmann B, Breillat C, Forestier L, et al. Spectrum of mutations in the COL4A5 collagen gene in X-linked Alport syndrome. Am. J. Hum. Genet. 1996; 59: 1221–32.

Kohfeldt E, Sasaki T, Gohring W, Timpl R. Nidogen-2: a new basement membrane protein with diverse binding properties. J. Mol. Biol. 1998; 282: 99–109.

Korhonen M, Ylanne J, Laitinen L, Virtanen I. The alpha 1-alpha 6 subunits of integrins are characteristically expressed in distinct segments of developing and adult human nephron. J. Cell Biol. 1990; 111: 1245–54.

Kretzler M. Regulation of adhesive interaction between podocytes and glomerular basement membrane. Microsc. Res. Tech. 2002; 57: 247–53.

Lajoie G. Approach to the diagnosis of thin basement membrane nephropathy in females with the use of antibodies to type IV collagen. Arch. Pathol. Lab. Med. 2001; 125: 631–6.

Langeveld JP, Noelken ME, Hard K, et al. Bovine glomerular basement membrane. Location and structure of the asparagine-linked oligosaccharide units and their potential role in the assembly of the 7 S collagen IV tetramer. J. Biol. Chem. 1991; 266: 2622–31.

Lees GE, Helman RG, Kashtan CE, et al. A model of autosomal recessive Alport syndrome in English cocker spaniel dogs. Kidney Int. 1998; 54: 706–19.

Lees GE, Helman RG, Kashtan CE, et al. New form of X-linked dominant hereditary nephritis in dogs. Am. J. Vet. Res. 1999; 60: 373–83.

Leinonen A, Mariyama M, Mochizuki T, Tryggvason K, Reeders ST. Complete primary structure of the human type IV collagen alpha 4(IV) chain. Comparison with structure and expression of the other alpha (IV) chains. J. Biol. Chem. 1994; 269: 26172–7.

Lemmink HH, Mochizuki T, van den Heuvel LP, et al. Mutations in the type IV collagen alpha 3 (COL4A3) gene in autosomal recessive Alport syndrome. Hum. Mol. Genet. 1994; 3: 1269–73.

Lemmink HH, Nillesen WN, Mochizuki T, et al. Benign familial hematuria due to mutation of the type IV collagen alpha4 gene. J. Clin. Invest. 1996; 98: 1114–18.

Lemmink HH, Schroder CH, Monnens LA, Smeets HJ. The clinical spectrum of type IV collagen mutations. Hum. Mutat. 1997; 9: 477–99.

Levy M, Feingold J. Estimating prevalence in single-gene kidney diseases progressing to renal failure. Kidney Int. 2000; 58: 925–43.

Liapis H, Gokden N, Hmiel P, Miner JH. Histopathology, ultrastructure, and clinical phenotypes in thin glomerular basement membrane disease variants. Hum. Pathol. 2002; 33: 836–45.

Longo I, Porcedda P, Mari F, et al. COL4A3/COL4A4 mutations: from familial hematuria to autosomal-dominant or recessive Alport syndrome. Kidney Int. 2002; 61: 1947–56.

Longo I, Scala E, Mari F, et al. Autosomal recessive Alport syndrome: an in-depth clinical and molecular analysis of five families. Nephrol. Dial. Transplant. 2006; 21: 665–71.

Lu W, Phillips CL, Killen PD, et al. Insertional mutation of the collagen genes Col4a3 and Col4a4 in a mouse model of Alport syndrome. Genomics 1999; 61: 113–24.

Mann K, Deutzmann R, Timpl R. Characterization of proteolytic fragments of the laminin-nidogen complex and their activity in ligand-binding assays. Eur. J. Biochem. 1988; 178: 71–80.

Mariyama M, Leinonen A, Mochizuki T, Tryggvason K, Reeders ST. Complete primary structure of the human alpha 3(IV) collagen chain. Coexpression of the alpha 3(IV) and alpha 4(IV) collagen chains in human tissues. J. Biol. Chem. 1994; 269: 23013–17.

Marks MI, McLaine PN, Drummond KN. Proteinuria in children with febrile illnesses. Arch. Dis. Child 1970; 45: 250–3.

Marquez B, Stavrou F, Zouvani I, et al. Thin glomerular basement membranes in patients with hematuria and minimal change disease. Ultrastruct. Pathol. 1999; 23: 149–56.

Martin P, Heiskari N, Zhou J, et al. High mutation detection rate in the COL4A5 collagen gene in suspected Alport syndrome using PCR and direct DNA sequencing. J. Am. Soc. Nephrol. 1998; 9: 2291–301.

Mayer U, Nischt R, Poschl E, et al. A single EGF-like motif of laminin is responsible for high affinity nidogen binding. Embo. J. 1993; 12: 1879–85.

McConville JM, McAdams AJ. Familial and nonfamilal benign hematuria. J. Pediatr. 1966; 69: 207–14.

Meleg-Smith S. Alport disease: a review of the diagnostic difficulties. Ultrastruct. Pathol. 2001; 25: 193–200.

Mihatsch MJ, Zollinger HU. Kidney disease. Pathol. Res. Pract. 1980; 167: 88–117.

Miner JH. Developmental biology of glomerular basement membrane components. Curr. Opin. Nephrol. Hypertens. 1998; 7: 13–19.

Miner JH, Patton BL, Lentz SI, et al. The laminin alpha chains: expression, developmental transitions, and chromosomal locations of alpha1-5, identification of heterotrimeric laminins 8–11, and cloning of a novel alpha3 isoform. J. Cell Biol. 1997; 137: 685–701.

Miner JH, Sanes JR. Collagen IV alpha 3, alpha 4, and alpha 5 chains in rodent basal laminae: sequence, distribution, association with laminins, and developmental switches. J. Cell Biol. 1994; 127: 879–91.

Miner JH, Sanes JR. Molecular and functional defects in kidneys of mice lacking collagen alpha 3(IV): implications for Alport syndrome. J. Cell Biol. 1996; 135: 1403–13.

Mochizuki T, Lemmink HH, Mariyama M, et al. Identification of mutations in the alpha 3(IV) and alpha 4(IV) collagen genes in autosomal recessive Alport syndrome. Nat. Genet. 1994; 8: 77–81.

Morita H, Yoshimura A, Inui K, et al. Heparan sulfate of perlecan is involved in glomerular filtration. J. Am. Soc. Nephrol. 2005; 16: 1703–10.

Nagel M, Nagorka S, Gross O. Novel COL4A5, COL4A4, and COL4A3 mutations in Alport syndrome. Hum. Mutat. 2005; 26: 60.

Nayak BR, Spiro RG. Localization and structure of the asparagine-linked oligosaccharides of type IV collagen from glomerular basement membrane and lens capsule. J. Biol. Chem. 1991; 266: 13978–87.

Netzer KO, Suzuki K, Itoh Y, Hudson BG, Khalifah RG. Comparative analysis of the noncollagenous NC1 domain of type IV collagen: identification of structural features important for assembly, function, and pathogenesis. Protein Sci. 1998; 7: 1340–51.

Nieuwhof CM, de Heer F, de Leeuw P, van Breda Vriesman PJ. Thin GBM nephropathy: premature glomerular obsolescence is associated with hypertension and late onset renal failure. Kidney Int. 1997; 51: 1596–601.

Ninomiya Y, Kagawa M, Iyama K, et al. Differential expression of two basement membrane collagen genes, COL4A6 and COL4A5, demonstrated by immunofluorescence staining using peptide-specific monoclonal antibodies. J. Cell Biol. 1995; 130: 1219–29.

Noakes PG, Miner JH, Gautam M, Cunningham JM, Sanes JR, Merlie JP. The renal glomerulus of mice lacking s-laminin/laminin beta 2: nephrosis despite molecular compensation by laminin beta 1. Nat. Genet. 1995; 10: 400–6.

Nogueira M, Cartwright J Jr., Horn K, et al. Thin basement membrane disease with heavy proteinuria or nephrotic syndrome at presentation. Am. J. Kidney Dis. 2000; 35: E15.

Noonan DM, Fulle A, Valente P, et al. The complete sequence of perlecan, a basement membrane heparan sulfate proteoglycan, reveals extensive similarity with laminin A chain, low density lipoprotein-receptor, and the neural cell adhesion molecule. J. Biol. Chem. 1991; 266: 22939–47.

Osawa G, Kimmelstiel P, Seiling V. Thickness of glomerular basement membranes. Am. J. Clin. Pathol. 1966; 45: 7–20.

Ozen S, Ertoy D, Heidet L, et al. Benign familial hematuria associated with a novel COL4A4 mutation. Pediatr. Nephrol. 2001; 16: 874–7.

Packham DK, Perkovic V, Savige J, Broome MR. Hematuria in thin basement membrane nephropathy. Semin. Nephrol. 2005; 25: 146–8.

Pavenstadt H, Kriz W, Kretzler M. Cell biology of the glomerular podocyte. Physiol. Rev. 2003; 83: 253–307.

Peissel B, Geng L, Kalluri R, et al. Comparative distribution of the alpha 1(IV), alpha 5(IV), and alpha 6(IV) collagen chains in normal human adult and fetal tissues and in kidneys from X-linked Alport syndrome patients. J. Clin. Invest. 1995; 96: 1948–57.

Perry GJ, George CR, Field MJ, et al. Thin-membrane nephropathy – a common cause of glomerular haematuria. Med. J. Aust. 1989; 151: 638–42.

Pescucci C, Mari F, Longo I, et al. Autosomal-dominant Alport syndrome: natural history of a disease due to COL4A3 or COL4A4 gene. Kidney Int. 2004; 65: 1598–603.

Peterson AS, Schubert JJ. Benign hereditary nephritis. J. Fam. Pract. 1977; 4: 437–41.

Pettersson E, Tornroth T, Wieslander J. Abnormally thin glomerular basement membrane and the Goodpasture epitope. Clin. Nephrol. 1990; 33: 105–9.

Piccini M, Casari G, Zhou J, et al. Evidence for genetic heterogeneity in benign familial hematuria. Am. J. Nephrol. 1999; 19: 464–7.

Piqueras AI, White RH, Raafat F, Moghal N, Milford DV. Renal biopsy diagnosis in children presenting with haematuria. Pediatr. Nephrol. 1998; 12: 386–91.

Rana K, Wang YY, Buzza M, et al. The genetics of thin basement membrane nephropathy. Semin. Nephrol. 2005; 25: 163–70.

Regele HM, Fillipovic E, Langer B, et al. Glomerular expression of dystroglycans is reduced in minimal change nephrosis but not in focal segmental glomerulosclerosis. J. Am. Soc. Nephrol. 2000; 11: 403–12.

Renieri A, Bruttini M, Galli L, et al. X-linked Alport syndrome: an SSCP-based mutation survey over all 51 exons of the COL4A5 gene. Am. J. Hum. Genet. 1996; 58: 1192–204.

Reyersbach GC, Butler AM. Congenital hereditary hematuria. N. Engl. J. Med. 1954; 251: 377–80.

Rheault MN, Kren SM, Thielen BK, et al. Mouse model of X-linked Alport syndrome. J. Am. Soc. Nephrol. 2004; 15: 1466–74.

Ritchie CD, Bevan EA, Collier SJ. Importance of occult haematuria found at screening. Br. Med. J. (Clin. Res. Ed.) 1986; 292: 681–3.

Rogers PW, Kurtzman NA, Bunn SM Jr., White MG. Familial benign essential hematuria. Arch. Intern. Med. 1973; 131: 257–62.

Rossi M, Morita H, Sormunen R, et al. Heparan sulfate chains of perlecan are indispensable in the lens capsule but not in the kidney. EMBO J. 2003; 22: 236–45.

Roth KS, Amaker BH, Chan JC. Pediatric hematuria and thin basement membrane nephropathy: what is it and what does it mean? Clin. Pediatr. (Phila) 2001; 40: 607–13.

Rumpelt HJ, Langer KH, Scharer K, Straub E, Thoenes W. Split and extremely thin glomerular basement membranes in hereditary nephropathy (Alport's syndrome). Virchows Arch. A. Pathol. Anat. Histol. 1974; 364: 225–33.

Russell EP, Smith NJ. Hereditary hematuria. Am. J. Dis. Child 1959; 98: 353–8.

Sakai K, Muramatsu M, Ogiwara H, et al. Living related kidney transplantation in a patient with autosomal-recessive Alport syndrome. Clin. Transplant. 2003; 17(Suppl. 10): 4–8.

Sanes JR, Lichtman JW. Development of the vertebrate neuromuscular junction. Annu. Rev. Neurosci. 1999; 22: 389–442.

Sariola H, Timpl R, von der Mark K, et al. Dual origin of glomerular basement membrane. Dev. Biol. 1984; 101: 86–96.

Savige J, Rana K, Tonna S, Buzza M, Dagher H, Wang YY. Thin basement membrane nephropathy. Kidney Int. 2003; 64: 1169–78.

Saxena S, Davies DJ, Kirsner RL. Thin basement membranes in minimally abnormal glomeruli. J. Clin. Pathol. 1990; 43: 32–8.

Schroder CH, Bontemps CM, Assmann KJ, et al. Renal biopsy and family studies in 65 children with isolated hematuria. Acta Paediatr. Scand. 1990; 79: 630–6.

Shindo S, Yoshimoto M, Kuriya N, Bernstein J. Glomerular basement membrane thickness in recurrent and persistent hematuria and nephrotic syndrome: correlation with sex and age. Pediatr. Nephrol. 1988; 2: 196–9.

Soininen R, Huotari M, Hostikka SL, Prockop DJ, Tryggvason K. The structural genes for alpha 1 and alpha 2 chains of human type IV collagen are divergently encoded on opposite DNA strands and have an overlapping promoter region. J. Biol. Chem. 1988; 263: 17217–20.

Sonnenberg A, Linders CJ, Modderman PW, Damsky CH, Aumailley M, Timpl R. Integrin recognition of different cell-binding fragments of laminin (P1, E3, E8) and evidence that alpha 6 beta 1 but not alpha 6 beta 4 functions as a major receptor for fragment E8. J. Cell Biol. 1990; 110: 2145–55.

Steffes MW, Barbosa J, Basgen JM, Sutherland DE, Najarian JS, Mauer SM. Quantitative glomerular morphology of the normal human kidney. Lab. Invest. 1983; 49: 82–6.

Tazon Vega B, Badenas C, Ars E, et al. Autosomal recessive Alport's syndrome and benign familial hematuria are collagen type IV diseases. Am. J. Kidney Dis. 2003; 42: 952–9.

Thorner PS, Zheng K, Kalluri R, Jacobs R, Hudson BG. Coordinate gene expression of the alpha3, alpha4, and alpha5 chains of collagen type IV. Evidence from a canine model of X-linked nephritis with a COL4A5 gene mutation. J. Biol. Chem. 1996; 271: 13821–8.

Tiebosch AT, Frederik PM, van Breda Vriesman PJ, et al. Thin-basement-membrane nephropathy in adults with persistent hematuria. N. Engl. J. Med. 1989; 320: 14–18.

Timpl R. Structure and biological activity of basement membrane proteins. Eur. J. Biochem. 1989; 180: 487–502.

Timpl R, Brown JC. The laminins. Matrix Biol. 1994; 14: 275–81.

Timpl R, Brown JC. Supramolecular assembly of basement membranes. Bioessays 1996; 18: 123–32.

Timpl R, Dziadek M, Fujiwara S, Nowack H, Wick G. Nidogen: a new, self-aggregating basement membrane protein. Eur. J. Biochem. 1983; 137: 455–65.

Timpl R, Wiedemann H, van Delden V, Furthmayr H, Kuhn K. A network model for the organization of type IV collagen molecules in basement membranes. Eur. J. Biochem. 1981; 120: 203–11.

Tina L, Jenis E, Jose P, Medani C, Papadopoulou Z, Calcagno P. The glomerular basement membrane in benign familial hematuria. Clin. Nephrol. 1982; 17: 1–4.

Tryggvason K, Patrakka J. Thin basement membrane nephropathy. J. Am. Soc. Nephrol. 2006; 17: 813–22.

Tryggvason K, Wartiovaara J. How does the kidney filter plasma? Physiology (Bethesda) 2005; 20: 96–101.

Ueda T, Nakajima M, Akazawa H, et al. Quantitative analysis of glomerular type IV collagen alpha3–5 chain expression in children with thin basement membrane disease. Nephron. 2002; 92: 271–8.

van der Loop FT, Heidet L, Timmer ED, et al. Autosomal dominant Alport syndrome caused by a COL4A3 splice site mutation. Kidney Int. 2000; 58: 1870–5.

van Paassen P, van Breda Vriesman PJ, van Rie H, Tervaert JW. Signs and symptoms of thin basement membrane nephropathy: a prospective regional study on primary glomerular disease– The Limburg Renal Registry. Kidney Int. 2004; 66: 909–13.

Vehaskari VM, Rapola J, Koskimies O, Savilahti E, Vilska J, Hallman N. Microscopic hematuria in school children: epidemiology and clinicopathologic evaluation. J. Pediatr. 1979; 95: 676–84.

Vogler C, McAdams AJ, Homan SM. Glomerular basement membrane and lamina densa in infants and children: an ultrastructural evaluation. Pediatr. Pathol. 1987; 7: 527–34.

Wang YY, Rana K, Tonna S, Lin T, Sin L, Savige J. COL4A3 mutations and their clinical consequences in thin basement membrane nephropathy (TBMN). Kidney Int. 2004; 65: 786–90.

Wang YY, Savige J. The epidemiology of thin basement membrane nephropathy. Semin. Nephrol. 2005; 25: 136–9.

Wing AJ, Brunner FP. Twenty-three years of dialysis and transplantation in Europe: experiences of the EDTA Registry. Am. J. Kidney Dis. 1989; 14: 341–6.

Yamada H, Denzer AJ, Hori H, et al. Dystroglycan is a dual receptor for agrin and laminin-2 in Schwann cell membrane. J. Biol. Chem. 1996; 271: 23418–23.

Yamada H, Shimizu T, Tanaka T, Campbell KP, Matsumura K. Dystroglycan is a binding protein of laminin and merosin in peripheral nerve. FEBS Lett. 1994; 352: 49–53.

Yamazaki H, Nakagawa Y, Saito A. No linkage to COL4A3 gene locus in Japanese thin basement membrane disease families. Nephrology 1995; 1: 315–21.

Yoshikawa N, Matsuyama S, Iijima K, Maehara K, Okada S, Matsuo T. Benign familial hematuria. Arch. Pathol. Lab. Med. 1988; 112: 794–7.

Yoshioka K, Hino S, Takemura T, et al. Type IV collagen alpha 5 chain. Normal distribution and abnormalities in X-linked Alport syndrome revealed by monoclonal antibody. Am. J. Pathol. 1994; 144: 986–96.

Yurchenco PD, O'Rear JJ. Basal lamina assembly. Curr. Opin. Cell Biol. 1994; 6: 674–81.

Zenker M, Pierson M, Jonveaux P, Reis A. Demonstration of two novel LAMB2 mutations in the original Pierson syndrome family reported 42 years ago. Am. J. Med. Genet. A 2005; 138: 73–4.

Zhang X, Hudson BG, Sarras MP Jr. Hydra cell aggregate development is blocked by selective fragments of fibronectin and type IV collagen. Dev. Biol. 1994; 164: 10–23.

Zheng K, Thorner PS, Marrano P, Baumal R, McInnes RR. Canine X chromosome-linked hereditary nephritis: a genetic model for human X-linked hereditary nephritis resulting from a single base mutation in the gene encoding the alpha 5 chain of collagen type IV. Proc. Natl Acad. Sci. USA 1994; 91: 3989–93.

Zhou J, Leinonen A, Tryggvason K. Structure of the human type IV collagen COL4A5 gene. J. Biol. Chem. 1994; 269: 6608–14.

CHAPTER 5

Idiopathic Nephrotic Syndrome

ERNIE L. ESQUIVEL AND CORINNE ANTIGNAC

INTRODUCTION

The nephrotic syndrome is characterized by the appearance of massive proteinuria, accompanied by hypoalbuminemia, hypercholesterolemia, and edema. Regardless of whether the etiology is a primary glomerular disease or a systemic disorder, the leakage of albumin in the urine reflects a breach of the permselectivity barrier in the kidney and dysregulation of plasma ultrafiltration. The glomerular filtration barrier consists of the interdigitating ramifications or foot processes of terminally differentiated epithelial cells called podocytes, fenestrated endothelial cells, and the intervening glomerular basement membrane (Figure 5.1A). As early as 1957, Farquhar and colleagues (1974) demonstrated that the characteristic pathologic feature of

nephrosis involves the effacement or morphologic distortion of these foot processes (Figure 5.1B). Later, Rodewald and Karnofsky (1974) proposed a zipper-like model of the glomerular filtration barrier, in which structures called slit diaphragms bridge neighboring foot processes. Effacement of foot processes involves the loss of these intervening slit diaphragms. Controversy exists, however, since animal models of renal disease have shown that albumin leakage may occur in the setting of preserved foot process architecture (Kalluri 2006).

Idiopathic nephrotic syndrome represents a heterogeneous group of glomerular disorders occurring mainly in children. Classically, nephrotic syndrome in children has been designated congenital when onset is at birth or within the first 3 months of life, infantile between 3 months and 1 year, and childhood-onset thereafter. Recent studies have revealed, however, that this classification scheme may not help distinguish the different monogenic causes of NS, since each may present with a broad spectrum of age of disease onset (Matejas et al 2006, Hinkes et al 2007, Philippe et al 2008). Additionally, the response to an often empiric therapy consisting of three intravenous doses of methylprednisolone, followed by 1 month of oral prednisone, differentiates between steroid-sensitive (SSNS) and steroid-resistant (SRNS) forms of nephrotic syndrome. The prevailing model is that a still-unidentified circulating permeability factor that compromises the glomerular permeability barrier accounts for steroid-sensitive and some steroid-resistant forms, whereas steroid resistance reflects primary defects in the filtration barrier in some patients (Figure 5.2) (Antignac 2002). Most patients show a favorable outcome to steroid therapy, although they may be subject to more or less frequent relapses of the disease. However, 10–20% of patients fail to respond and may develop end-stage renal failure.

The causes of nephrotic syndrome in the neonate and infant can be divided broadly into acquired or hereditary etiologies. Acquired cases may be due to a variety of infectious, toxic, and immunologic etiologies and tests to rule

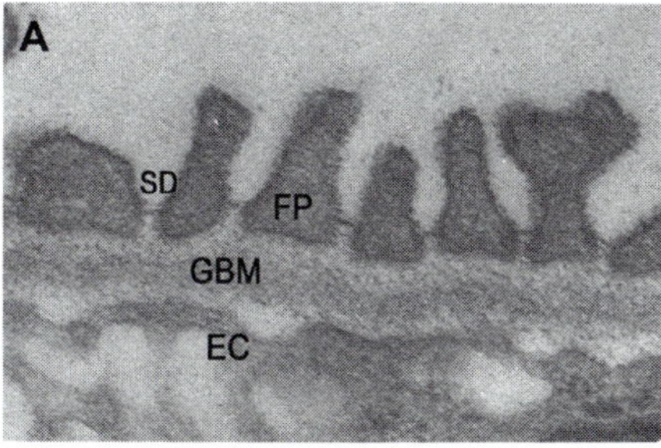
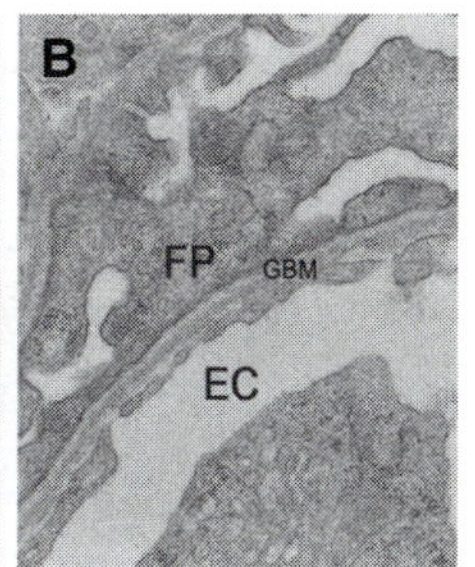

FIGURE 5.1 The glomerular filtration barrier in health and disease. (A) In a healthy individual, the glomerular filtration barrier is composed by the interdigitating foot processes (FP) of epithelial cells called podocytes, a fenestrated endothelial cell (EC), and an intervening glomerular basement membrane (GBM), as seen by electron microscopy. Foot processes are linked by structures called the slit diaphragm (SD). Magnification × 22 000 (B) In patients with nephrotic syndrome, foot process morphology is altered, here with diffuse effacement and loss of slit diaphragms. Magnification × 12 000

Genetic Diseases of the Kidney
ISBN: 978-0-12-449851-8

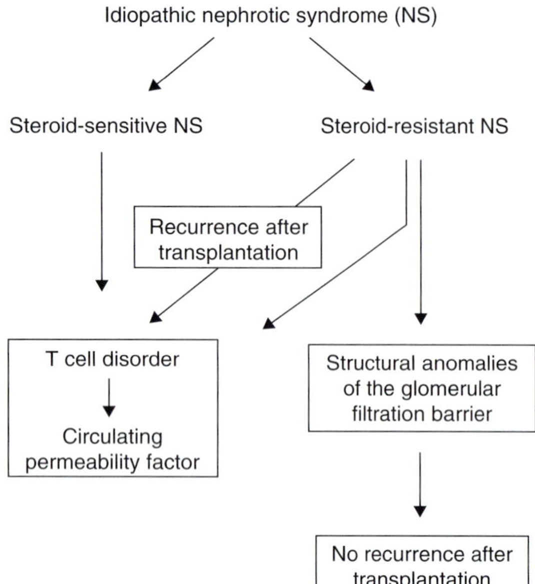

FIGURE 5.2 Mechanisms underlying idiopathic nephrotic syndrome (NS). A circulating permeability factor due to a T cell disorder may underlie steroid-sensitive and -resistant forms. By contrast, steroid-resistant forms have been shown to be due to primary defects in the glomerular filtration barrier, and frequently characterized by lack of recurrence after renal transplantation. Recurrence of disease after transplantation in some cases of steroid-resistant NS has been postulated to result from the development of a serum permeability factor

out syphilis, toxoplasmosis, hepatitis B, cytomegalovirus, or rubella infections constitute part of the initial work-up (Eddy & Symons 2003). Although the genetic underpinnings of nephrotic syndrome have long been appreciated, it has only been over the past decade that studies of monogenic forms of nephrotic syndrome have been fruitful and have since helped decipher the pathophysiologic mechanisms of the glomerular filtration process. These gene discovery efforts have resulted in the identification of mutations in novel genes which encode proteins crucial for the establishment and maintenance of the glomerular filtration barrier. Consequently, the structural protein constituents of the glomerular slit diaphragm, revealed by Rodewald and Karnovsky (1974), have been greatly elucidated. Subsequent functional studies have shown that these proteins are involved in dynamic interaction networks crucial for intracellular signaling events and for podocyte–endothelial cell–mesangial cell cross-talk (Benzing 2004). These studies have to-date identified genetic mutations responsible for SRNS, although one study recently demonstrated linkage of SSNS to chromosome 2p12-p13.2 (Ruf et al 2003). However, it is evident that genetic determinants likely impact upon the development and progression of nephrotic syndrome in complex inheritance mechanisms and in concert with a multitude of environmental factors (Ratelade et al 2008), particularly in those with later onset in childhood.

A recent study by Hinkes et al (2007) revealed that nephrotic syndrome presenting in children during the first year of life may largely be attributed to mutations in the *NPHS1, NPHS2, WT1,* or *LAMB2* genes, encoding nephrin, podocin, Wilms' tumor 1, and laminin β2, respectively. This chapter will focus on the mechanisms by which mutations in these four genes lead to proteinuric renal disease, as well as on recently discovered genes responsible for rarer cases of hereditary nephrotic syndrome, summarized in Table 5.1. The proteins encoded by these genes have been shown to serve a multitude of functions, encompassing structural, signaling, transcriptional regulatory roles, as well as the control of cellular metabolic pathways.

GENETIC DEFECTS OF STRUCTURAL PROTEINS

Nephrin and Congenital Nephrotic Syndrome of the Finnish Type

Congenital nephrotic syndrome of the Finnish type (CNF; MIM 256300) is an autosomal recessive disorder characterized by massive proteinuria detectable in utero, a large placenta and marked edema manifesting often at birth or within the first 3 months of life. There are no pathognomonic pathologic features of CNF, but the most constant light microscopic findings are mesangial hypercellularity and/or sclerosis and proximal tubular dilatation (Figure 5.3) (Huttunen et al 1980, Kuusniemi et al 2006). At the ultrastructural level, glomerular development is not impaired, but podocytes demonstrate flattening and shortening of epithelial cell processes and diffuse foot process effacement (Huttunen et al 1980, Autio-Harmainen et al 1981).

In 1998, Kestila et al. used a positional cloning approach to identify the *NPHS1* gene, localized on chromosome 19q13.1, and the mutations in a cohort of CNF patients (Kestila et al 1998). CNF is a progressive disease usually leading to end-stage renal failure during the first 2 years of life and the only life-saving treatment available is renal transplantation (Holmberg et al 1995). In a significant fraction of affected children, the development of less severe proteinuria is observed post-renal transplantation. As many as 20–25% of Finnish children developed NS after transplantation, with a high percentage displaying antiglomerular and antinephrin antibodies (Wang et al 2001, Patrakka et al 2002a). Loss of large amounts of α-fetoprotein (AFP) in the urine leads to elevated levels in the amniotic fluid, a useful diagnostic tool in the antenatal period (Seppala et al 1976, Morris et al 1995). However, a recent study showed prenatal proteinuria and elevated AFP in fetuses both heterozygous and homozygous for *NPHS1* mutations (Patrakka et al 2002b).

TABLE 5.1   Genetic causes of idiopathic nephrotic syndrome

Gene	Protein	Disease	Locus
NPHS1	Nephrin	Congenital nephrotic syndrome of the Finnish type	19q13.1
NPHS2	Podocin	Autosomal recessive steroid-resistant nephrotic syndrome	1q25-31
WT1	Wilms' tumor 1	Denys-Drash syndrome, Frasier syndrome, isolated FSGS and diffuse mesangial sclerosis, WAGR syndrome	11p13
LAMB2	Laminin-β2	Pierson syndrome	3p21
PLCE1	Phospholipase C epsilon 1	Diffuse mesangial sclerosis and FSGS	10q23
SMARCAL1	SW1/SNF2-related, matrix-associated, actin-dependent regulator of chromatin, subfamily a-like 1	Schimke immuno-osseous dysplasia	2q34-q36
SCARB2	Scavenger receptor 2	Action myoclonus renal failure	4q13-21
LMX1B	LIM-homeodomain transcription factor 1, beta	Nail-patella syndrome	9q34.1
COQ2	Parahydroxygenzoate-polyprenyltransferase enzyme	COQ2 deficiency	4q21-q22
PDSS1	Decaprenyl diphosphate synthase-1	COQ10 deficiency	10p12.1
PDSS2	Decaprenyl diphosphate synthase-2	COQ10 deficiency	6q21
MTTL1	Mitochondrial tRNA for leucine (UUR)	FSGS, with or without nephrotic syndrome	Mitochondrial

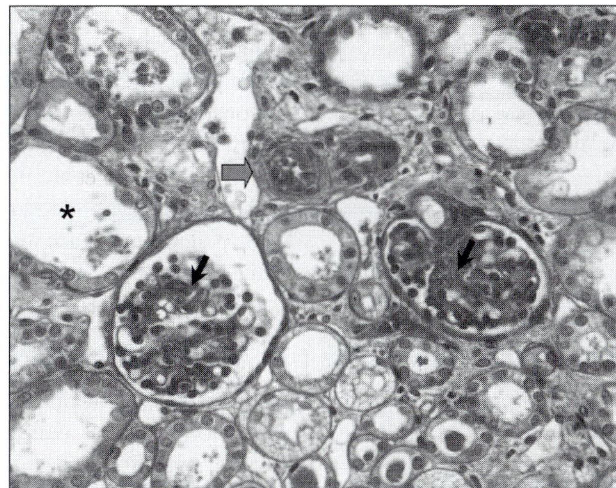

FIGURE 5.3   Histologic findings in congenital nephrotic syndrome of the Finnish type (CNF). A periodic acid–Schiff-stained renal biopsy of a 10-day-old male infant bearing mutations in the *NPHS1* gene, with proteinuria at birth, is representative of frequent findings in CNF, including mesangial proliferation (denoted by black arrows), diffuse tubular dilatation (*) and arteriolar lesions due to medial thickening (gray arrow). Magnification $\times$ 400

The *NPHS1* gene spans 26kb, consists of 29 exons, and encodes nephrin, a predicted 135-kDa protein with 1241 amino acid residues. The mouse and rat orthologs are encoded by 30 exons and bear an additional 14 amino acid residues but share 83% sequence identity with human nephrin (Putaala et al

2000). Nephrin belongs to the immunoglobulin (Ig) super-family and contains an N-terminal signal peptide, an extracellular domain containing eight Ig-like modules, a fibronectin type III-like module, a single transmembrane domain, and a cytosolic C-terminal end. Endogenous and exogenous nephrin have an apparent molecular mass of $\sim$185-kDa, suggesting posttranslational modification, including N-linked glycosylation, which is crucial for proper folding and plasma membrane targeting (Holzman et al 1999, Ruotsalainen et al 1999, Topham et al 1999).

Immunolocalization studies showed expression of nephrin exclusively in podocytes, at the glomerular slit diaphragm (Holthofer et al 1999, Holzman et al 1999, Ruotsalainen et al 1999). Endogenous nephrin mRNA has been detected as early as the S-shaped stage of glomerular development by in situ hybridization (Wong et al 2000), but immunolocalization of nephrin demonstrated expression beginning at the capillary stage (Holzman et al 1999, Ruotsalainen et al 2000). Additional experimental evidence suggests expression in discrete areas of the brain (Putaala et al 2000) and beta cells of the pancreas (Palmen et al 2001, Putaala et al 2001). Tissue-specific expression of nephrin may be driven by the use of alternative promoters (Beltcheva et al 2003). A PAX2 binding site and several E-box consensus sequences, which recognize basic helix-loop-helix proteins, have been identified in the nephrin promoter (Wong et al 2000). Additionally, nephrin has been demonstrated to be a target of the WT1 transcription factor (Wagner et al 2004).

Mutations in the nephrin gene leading to CNF were initially identified in the Finnish population, where the incidence of the disease is 1:10 000 newborns. However, *NPHS1* mutations have since been sequenced worldwide (Beltcheva et al 2001, Gigante et al 2002, Sako et al 2005). The spectrum of mutations described includes protein-truncating nonsense and frameshifting insertion/deletion mutations, splice-site changes and missense variants. In Finland, >90% of CNF patients are accounted for by the Fin-major (c.121delCT) mutation, which leads to a frameshifting deletion of two base pairs in exon 2, resulting in a premature stop codon in the same exon, and by the Fin-minor (p.R1109X) mutation, which leads to a premature truncation of the terminal 132 aa of the protein (Kestila et al 1998, Lenkkeri et al 1999). By contrast, most non-Finnish patients have 'private' mutations which include insertions, deletions, nonsense, missense and splice site mutations. Recently, we have identified a cohort of patients bearing at least one 'mild' (resulting in a protein with partial function) *NPHS1* mutation, mostly missense variants, who presented with late-onset SRNS and a protracted course of disease (Philippe et al 2008a).

Immunolocalization studies revealed the absence of nephrin expression in kidneys of patients bearing either the Fin-major or Fin-minor mutations (Holthofer et al 1999, Ruotsalainen et al 2000). This may be the case, as well, for other protein truncating mutations, although the p.R1160X mutant appears to be expressed and may account for a milder disease phenotype in some female patients (Koziell et al 2002). When studied in cell culture systems the majority of missense nephrin mutant proteins fail to address correctly to the plasma membrane, and are retained in the endoplasmic reticulum (Liu et al 2001). This defective intracellular transport of nephrin accounts for the severe loss-of-function phenotype in patients. Additionally, these missense mutations may result in cellular ER stress or imbalance due to an influx of unfolded proteins in the cell which overwhelms the folding/processing capacity of the endoplasmic reticulum. ER stress has been shown to perturb protein biogenesis, and in the case of nephrin, manifests as an alteration in N-glycosylation (Fujii et al 2006).

Interestingly, sodium 4-phenylbutyrate (4-PBA), a chemical chaperone, has been utilized in in vitro systems to rescue the cellular mislocalization of presumably misfolded missense nephrin mutants, thereby allowing for transport from the ER to the plasma membrane (Liu et al 2004). Successful targeting of these mutants to the plasma membrane resulted in tyrosine phosphorylation and binding to the NEPH1 protein, suggesting gain-of-function (Liu et al 2004). This study points to a potential therapeutic approach to delaying the onset and ameliorating the severity of renal disease in patients with missense mutations in the nephrin gene.

Two mouse models, in which the *Nphs1* gene has been inactivated, confirmed the crucial role of nephrin in the establishment of the glomerular filtration barrier as slit diaphragms failed to form in these mice, resulting in massive, nonselective proteinuria and edema at birth, leading to death within 24 hours (Putaala et al 2001, Rantanen et al 2002).

Histologic characterization revealed enlarged kidneys with mesangial cell hypercellularity and dilated proximal and distal tubules, reminiscent of lesions in CNF patients. No histologic evidence of brain abnormalities was seen.

Functional studies of the nephrin protein have revealed an important structural role in the formation of the glomerular slit diaphragm, and have also shown the slit diaphragm to be actively involved in signaling events. The slit diaphragm is a modified adherens junction (Reiser et al 2000), and nephrin co-localizes with the cell junction protein ZO-1 at the basal margin and in the cell–cell adhesion sites between developing podocytes, as early as the capillary loop stage, especially in junctions with ladder-like structures (Ruotsalainen et al 2000). Homophilic interactions between nephrin molecules are crucial to the formation of the slit diaphragm (Barletta et al 2003, Gerke et al 2003). Transfection of HEK293 cells in culture with full-length nephrin constructs leads to cell aggregation and the establishment of cell–cell contacts, which when studied using electron tomography revealed structural characteristics which recapitulate the electron microscopic characteristics of the glomerular slit diaphragm (Khoshnoodi et al 2003).

In addition, nephrin forms hetero-oligomers with the protein NEPH1, a transmembrane protein of the Ig superfamily, via interactions involving the extracellular and cytoplasmic domains (Barletta et al 2003, Gerke et al 2003, Liu et al 2003). The extracellular domain of NEPH1 contains five Ig domains and the integrin recognition motif RGD. Indeed, immunolocalization studies show both NEPH1 and nephrin at the slit diaphragm (Barletta et al 2003). Furthermore, the deletion of *Neph1* in mice leads to an identical phenotype found in nephrin null mice, with proteinuria, diffuse foot process effacement, and perinatal lethality (Donoviel et al 2001). The interaction of nephrin with ZO-1 may be mediated via direct interactions between NEPH1 and ZO-1 (Liu et al 2003). Nephrin exists in a complex with cell junction-associated proteins, including αII spectrin, βII spectrin, IQGAP1, CASK, MAGI-2, and α-actinin-4, which are known to play important roles in regulating cell adhesion, signaling, and actin cytoskeleton dynamics (Lehtonen et al 2005). Mutations in α-actinin-4 have been associated with nephrotic syndrome in hereditary focal segmental glomerulosclerosis (FSGS).

In cells, nephrin may be found in lipid rafts, the detergent-resistant membrane fraction involved in signaling events (Simons et al 2001). These signaling events are involved in actin cytoskeletal remodeling and in regulating podocyte apoptosis. Interaction of nephrin with β-arrestin 2 leads to internalization of nephrin in cells, and may be a mechanism for modulating nephrin signaling events (Quack et al 2006). Phosphorylation of nephrin, involving at least five tyrosine residues in the mouse ortholog (Verma et al 2006), is mediated by Fyn, a member of the Src family of kinases and occurs during foot process effacement, in response to podocyte injury (Verma et al 2003). Phosphorylated nephrin binds, via a functional SH2-binding domain, to Nck, an SH2-SH3 adapter protein. This interaction results in the recruitment of Nck to the plasma membrane and induces actin polymerization (Jones et al 2006, Verma et al 2006). The conditional inactivation of *Nck1* and *Nck2* in mouse podocytes results in nephrotic-range proteinuria and focal sclerosis, highlighting the importance of Nck–nephrin interaction in nephrin-dependent actin reorganization. Fyn, likewise, phosphorylates NEPH1 (at least on four tyrosine residues), allowing it to bind to Grb2, an event observed after podocyte injury. Activation of the NEPH1–nephrin complex leads to recruitment of Grb2 and Nck, inducing actin polymerization at the podocyte intercellular junction (Garg et al 2007). Finally, actin cystoskeleton remodeling by nephrin appears to involve the activation of the Rho GTPase Rac1 and leads to an increase in membrane ruffles or lamellipodia and a decrease in stress fibers (Zhu et al 2008).

An additional binding partner of nephrin is the 80-kDa adapter protein CD2AP (Shih et al 2001). The interaction between tyrosine phosphorylated nephrin and CD2AP results in recruitment of phosphatidylinositol-3-kinase to the membrane and the subsequent activation of AKT in podocytes (Huber et al 2003b). This pathway confers a survival advantage to podocytes mediated by the phosphorylation and subsequent inactivation of BAD, a pro-apoptotic molecule that interacts with pro-survival Bcl-2 family members (Huber et al 2003). Recently, a novel function has been attributed to dendrin, a proline-rich protein, which interacts directly with nephrin and CD2AP via its C-terminal region. Podocyte injury results in nuclear localization of dendrin, where it enhances podocyte apoptosis mediated by staurosporine and TGF-β (Asanuma et al 2007).

Podocin and Steroid-Resistant Nephrotic Syndrome

A genetic locus on chromosome 1q25-31 had previously been linked in 1995 to a familial form of steroid-resistant nephrotic syndrome characterized by autosomal recessive inheritance (SRN, MIM 600995) (Fuchshuber et al 1995). Using a positional cloning approach, Boute and colleagues identified mutations in the *NPHS2* gene, encoding a novel protein podocin (Boute et al 2000). These patients present with a disease characterized by onset between 3 months and 5 years, rapid progression to ESRD, absence of recurrence after renal transplantation, and absence of extrarenal disorders. Histologically, minimal glomerular changes are observed on early biopsy samples (Figure 5.4A) and FSGS at later stages (Figure 5.4B) (Boute et al 2000). Subsequent studies have shown that 30–42% of familial and 10–30% of sporadic cases of SRNS may be attributed to mutations in the podocin gene (Karle et al 2002, Weber et al 2004). In addition, rare patients presenting with onset during the congenital period have been reported, thus broadening the spectrum of clinical presentations (Karle et al 2002, Hinkes et al 2007). Recessive *NPHS2* mutations are, however, rare in adult-onset cases of FSGS (He et al 2007). Furthermore, several studies have demonstrated inter- and intrafamilial variability in disease phenotypes, therein suggesting potential roles for genetic modifiers and environmental factors (Fuchshuber et al 2001, Caridi et al 2003, Ruf et al 2004, Hinkes et al 2007).

The full-length *NPHS2* cDNA encodes podocin, a 383-amino-acid protein of approximately 42 kDa. Its strongest homologies were found with human stomatin and the *C. elegans* mechanosensory protein MEC-2 (Boute et al 2000). It encodes an integral membrane protein with

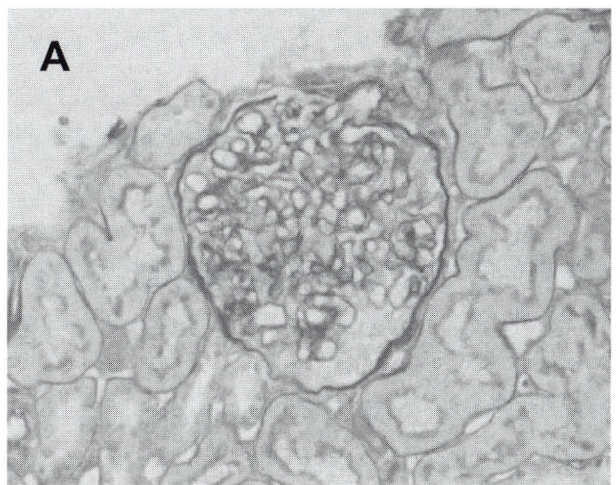
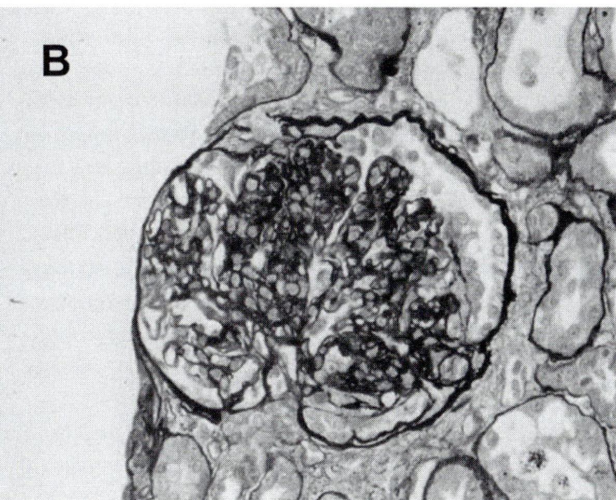

FIGURE 5.4 Spectrum of histologic findings in patients bearing podocin mutations. Renal biopsies in patients with mutations in the *NPHS2* gene may range from minimal change disease in early stages (A), shown in this periodic acid–Schiff stained section, to focal and segmental glomerulosclerosis (B) at more advanced stages, demonstrating segmental collagen deposition (black) in a periodic acid–silver-Azan-stained renal biopsy specimen. Magnification × 630

one transmembrane domain. Immunolocalization studies showed that podocin is localized exclusively in the slit diaphragm of podocytes (Schwarz et al 2001), with both N- and C-termini facing the cytosolic side (Roselli et al 2002). At the mRNA level, podocin is expressed as early as the S-shaped stage, concomitant with vascularization of the interior cleft of the developing nephron (Boute et al 2000); whereas protein expression has been documented beginning at the later capillary stage (Roselli et al 2002). Indeed, a 2.5-kb fragment of the *NPHS2* promoter has been shown to drive podocyte-specific gene expression in mice (Moeller et al 2002, Shigehara et al 2003), and has since been used to effect podocyte-specific gene targeting (Moeller et al 2003). The podocin promoter has been shown to be a downstream target of the LIM-homeodomain transcription factor LMX1B (Rohr et al 2002, Miner et al 2002).

Given its homology with the lipid raft-associated protein stomatin, podocin was demonstrated to be present in detergent-resistant membrane fractions in the cell, in complex with nephrin and CD2AP (Schwarz et al 2001). This suggests a potential role of podocin as a scaffolding protein, assembling the components of the slit diaphragm. Podocin directly interacts with the C-terminus of nephrin (Huber et al 2001, Schwarz et al 2001), and binding of podocin augments the activation of the AP-1 transcription factor by nephrin (Huber et al 2001). Phosphorylation of nephrin by the Src-family kinase Fyn increases its binding to podocin (Li et al 2004). In addition, podocin interacts with the transient receptor potential C channel protein TRPC6, mutated in some familial cases of autosomal dominant FSGS (Reiser et al 2005, Huber et al 2006). Regulation of the channel activity of TRPC6, in response to mechanostimulation, for instance, requires the binding of podocin to cholesterol in the plasma membrane (Huber et al 2007).

Mutations in the *NPHS2* gene have been described worldwide. A potential founder mutation, p.R138X, may be found in Israeli–Arab populations (Frishberg et al 2002), and has been associated with a high incidence of cardiac abnormalities in children (Frishberg et al 2006). Podocin mutations encompass a full spectrum of protein-truncating nonsense and frameshifting mutations, splice-site variants, and missense changes and involve all eight coding exons (Boute et al 2000, Ruf et al 2004, Weber et al 2004). Individuals bearing pathogenic mutations in the homozygous or compound heterozygous states manifested earlier than those without two pathogenic mutations, for both familial and sporadic cases (Weber et al 2004). Weber et al. (2004) also found that patients with frameshifting or nonsense mutations in the homozygous or compound heterozygous states led to an earlier mean age of onset of nephrotic syndrome. Furthermore, individuals homozygous for the most common missense variant p.R138Q tended to present with early-onset disease, as shown in the analyses of two patient cohorts (Weber et al 2004, Hinkes et al 2008). However, genotype did not correlate with the rate of progression to ESRD (Hinkes et al 2008).

Functional studies have revealed part of the mechanism by which missense podocin mutations lead to disease. In vitro studies have shown that podocin missense variants may either maintain proper intracellular targeting to the plasma membrane or be retained in the endoplasmic reticulum (ER) (Roselli et al 2004). Patients with missense mutations retained in the ER had an earlier onset of disease than patients with mutations that traffick to the membrane (20.8±4.0 months vs. 128.7±9.3 months) (Roselli et al 2004). Moreover, in cells expressing ER-retained mutants, nephrin is similarly retained in the endoplasmic reticulum (Nishibori et al 2004). Additional studies have also shown that although some mutants, such as p.R138X, are able to traffick to the plasma membrane, nephrin is not recruited to lipid rafts, from which downstream signaling events are generated (Huber et al 2003). Indeed, a mouse model in which the p.R138Q mutant is expressed developed early-onset, severe renal disease, owing to the presence of a mislocalized, functionally inactive protein (Philippe et al 2008). This model recapitulates the phenotype of an *Nphs2* null mouse, in which the gene has been constitutively inactivated by homologous recombination (Roselli et al 2004). However, as opposed to the human disease, podocin null mice present with lesions of mesangiolysis and mesangial sclerosis, which lead to death within the first 5 weeks of life, depending upon the underlying mouse genetic background (Roselli et al 2004, Ratelade et al 2008). Progression of renal disease in these mice appears to be subject not only to genetic modification, but also to the effects of the maternal environment in which mice are nourished prior to weaning (Ratelade et al 2008). In humans, phenotypic modification has been reported in rare patients in which triallellic inheritance of *NPHS1* and *NPHS2* mutations occurs (Koziell et al 2002), but additional studies are needed to better understand the complex genetics of renal disease progression in the setting of nephrotic syndrome.

Finally, the functional significance of the R229Q polymorphism in the *NPHS2* gene has been the subject of great interest. In one study, an allelic frequency of 3.6% was found in a control population, while the frequency was 6.0% in a population with primary FSGS (Tsukaguchi et al 2002). One subsequent study, however, found no increased frequency of the R229Q allele in a North American cohort of patients with adult-onset idiopathic FSGS (He et al 2007). The R229Q variant has also been associated with microalbuminuria in a cohort of Brazilian individuals of mixed European and African ancestries (Pereira et al 2004), whereas another study showed that R229Q may confer increased risk only in European-derived populations, but not in African-derived populations (Franceschini et al 2006). In vitro studies demonstrated that the R229Q podocin had a decreased interaction with nephrin (Tsukaguchi et al 2002). The R229Q polymorphism is more frequently found in the compound heterozygous state with a pathogenic *NPHS2* mutation in patients with the nephrotic syndrome than in control subjects, likely suggesting a deleterious effect

(Weber et al 2004). However, although R229Q mutations in the homozygous state have been reported, it has been difficult to be certain of their pathogenicity, and more likely they have a modulatory effect on the risk of developing renal disease.

The identification of *NPHS2* mutations in children presenting with nephrotic syndrome may have important clinical implications. First, patients with mutations in the *NPHS2* gene do not respond to standard steroid therapy and only partially to cyclosporine A or cyclophosphamide (Ruf et al 2004). This would suggest that these patients should be spared the potential risks and side-effects of immunosuppressive therapies. Secondly, several studies support the absence of recurrence of disease in patients with podocin mutations after renal transplantation. Among 32 patients with pathogenic *NPHS2* mutations who underwent kidney transplantation, only one developed a recurrence of proteinuria. There was neither evidence of rejection nor the presence of antipodocin antibodies and renal histology was consistent with the combined effects of FSGS and calcineurin toxicity (Weber et al 2004). In another series, three of five patients with sporadic SRNS developed recurrence after transplantation. The development of SRNS when it recurs early after transplantation is commonly regarded as arising from one or more suggested circulating permeability factor(s) that alter renal permeability, leading to proteinuria. Similarly, there was no evidence to support a role for antipodocin antibodies (Carraro et al 2002). Indeed, recurrence in one patient with a single pathogenic *NPHS2* mutation occurred early, with a quick response to increased immunosuppression, consistent with an immune-mediated pathomechanism (Billing et al 2004).

PLCE1 and Diffuse Mesangial Sclerosis

Most recently, a new gene locus, *NPHS3*, was identified on chromosome 10q23.32-q24.1 in a cohort of 22 consanguineous SRNS families (Hinkes et al 2006). In the 12 affected individuals, proteinuria and edema manifested at a median age of 0.8 years (range 0.2–4.0 years) and progressed to ESRF by 5 years of age (Hinkes et al 2006). A positional cloning approach coupled to gene expression profiling in rat glomeruli identified the *PLCE1* gene, encoding phospholipase C epsilon 1, as a good candidate. Mutational analysis subsequently revealed truncating and missense mutations in several of its 34 exons (Hinkes et al 2006). Furthermore, individuals bearing truncating mutations demonstrated lesions of diffuse mesangial sclerosis on renal biopsy (Figure 5.5), whereas FSGS was noted in the affected cases homozygous for missense mutations (Hinkes et al 2006). Diffuse mesangial sclerosis is characterized by a combination of thickening of the glomerular basement membranes and massive enlargement of mesangial areas, leading to reduction of the capillary lumens. The mesangial sclerosis eventually contracts the glomerular tuft into a sclerotic mass within a dilated urinary space. Subsequently these

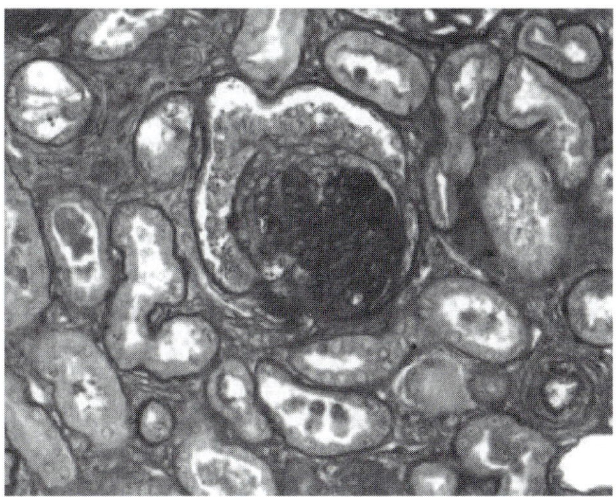

FIGURE 5.5 Histologic lesions in diffuse mesangial sclerosis. Diffuse mesangial sclerosis is a histologic entity characterized by thickening of the glomerular basement membranes and enlargement of the mesangial areas, leading to obliteration of the capillary lumens and subsequent contraction of the sclerotic glomerular tuft. The image shows a periodic acid–Schiff-stained renal biopsy. Magnification × 630

investigators have shown that mutations in the *PLCE1* gene account for 28.6% of cases of isolated DMS. Interestingly, two affected individuals, bearing truncating mutations, appeared to have responded to immunosuppressive therapy when treated early, hence potentially opening a window of opportunity for therapy of some forms of hereditary NS (Hinkes et al 2006, Quaggin 2006).

PLCε1 is a phospholipase enzyme that catalyzes the hydrolysis of phosphatidylinositol-4,5-bispohosphate and generates two second messengers: inositol 1,4,5-triphosphate (IP3) and diacylglycerol (DAG) (Wing et al 2003). IP3 releases Ca^{2+} from intracellular stores, and DAG stimulates protein kinase C. The PLCε1 protein contains the following putative proteins domains: RasGEF_CDC25 (guanine nucleotide exchange factor for Ras-like small GTPases domain), PH domain (pleckstrin homology domain), EF hand, phospholipase catalytic domains (PLC_X and PLC_Y), C2 motif (protein kinase C conserved region 2, subgroup 2) and RA1 and RA2 domains (RasGTP binding domain from guanine nucleotide exchange factors) (Wing et al 2003). Based on the observation that nephrin levels were diminished in the glomeruli of patients bearing mutations in *PLCE1*, it has been shown that PLCε1 interacts with the C-terminal half of IQGAP-1 (Hinkes et al 2006), a cell junction-associated protein and binding partner of nephrin involved in cell morphology and adhesion (Lehtonen et al 2005). An additional, although indirect, interacting partner of PLCε1 appears to be BRAF, a serine-threonine kinase protein, belonging to the *RAF* family of genes, as demonstrated by GST pull-down assays and coimmunoprecipitation experiments (Chaib et al 2008). Both proteins are expressed and co-localized in developing podocytes.

Finally, in the affected individuals presenting with DMS, immunofluorescence studies revealed that *PLCE1* mutations may lead to an arrest of glomerular development at the S-shaped stage, suggesting a potential role not only in cell junction and signaling events, but in development, as well (Hinkes et al 2006).

LAMB2 and Pierson Syndrome

Although several isolated case reports had previously appeared in the literature, Zenker and colleagues (2004) designated a syndromic disorder consisting of ocular abnormalities, most commonly microcoria, and congenital nephrotic syndrome as Pierson syndrome (MIM 609049). Children usually present with proteinuria at birth or during the first few days of life and succumb within the first year of life. Renal histologic findings on biopsy are characterized by diffuse mesangial sclerosis. In two consanguineous families, a locus was mapped to a region on chromosome 3p and positional cloning was greatly aided by the previous description of the development of nephrotic syndrome in *Lamb2* null mice (Noakes et al 1995). Affected cases in these families had truncating or missense mutations of the *LAMB2* gene, in either the homozygous or compound heterozygous states, leading to absent or reduced expression, respectively, of laminin β2 in the kidneys. Interestingly, mutational screening in additional families revealed patients bearing missense mutations in *LAMB2* who presented with minor or no ocular anomalies, but did develop congenital nephrotic syndrome characterized by focal and segmental glomerulosclerosis on biopsy (Hasselbacher et al 2006). Childhood-onset of Pierson syndrome has likewise been reported in a nonconsanguineous family with seven individuals presenting with nephrotic syndrome and ocular abnormalities (Matejas et al 2006).

Laminins provide the basic scaffold for assembly of the other components of the glomerular basement membrane, including type IV collagen, nidogen/entactin and sulfated proteoglycans. Laminins are heterotrimeric extracellular matrix proteins that are composed of α, β, and γ chains. At least 15 isoforms are assembled from 5α, 4β, and 3γ chains in mammals. The glomerular basement membrane is composed exclusively of laminin-521 (α5β2γ1) (Miner 2005). The α5 chain is required for GBM integrity and glomerular vascularization (Miner & Li 2000), whereas the β2 chain is dispensable for glomerulogenesis, since laminin-511 (α5β1γ1) can compensate structurally for the missing laminin-521 (Noakes et al 1995). However, *Lamb2* null mice develop progressive nephropathy, in which the appearance of proteinuria may be primarily attributable to ectopic deposition of several laminins and mislocalization of anionic sites in the GBM resulting in severe disorganization (Jarad et al 2006). Curiously, at the onset of proteinuria, protein components of the slit diaphragm, including podocin and nephrin, and integrins, do not appear to be altered, hence suggesting that structural integrity of the slit diaphragm alone is not sufficient to maintain the glomerular filtration barrier (Jarad et al 2006).

GENETIC DEFECTS OF TRANSCRIPTION FACTORS

WT1 and Glomerular Diseases

Mutations in the *WT1* gene, located on chromosome 11p13, encoding the Wilms' tumor 1 transcription factor, may be associated with several forms of renal disease. Denys-Drash syndrome (DDS, MIM 194080) is a rare urogenital disorder associated with male pseudohermaphroditism, a high risk for Wilms' tumors, and diffuse mesangial sclerosis (DMS), which presents before the age of 2 years and progresses rapidly to ESRD (Drash et al 1970, Habib et al 1985). On the other hand, Frasier syndrome (FS, MIM 136680) is characterized by male pseudohermaphroditism, increased susceptibility to gonadoblastomas, an association with childhood-onset proteinuria, usually between 2 and 6 years of age, slowly progressing to ESRD, and the histologic finding of FSGS on renal biopsy (Frasier et al 1964, Moorthy et al 1987). In both cases, no recurrence occurs after renal transplantation. De novo deletion of the 11p13 locus leads to WAGR syndrome (MIM 194072), characterized by Wilms' tumors, aniridia, genitourinary abnormalities, and mental retardation (Miller et al 1964). Isolated cases of diffuse mesangial sclerosis, similar to the lesions seen in Denys-Drash syndrome, have also been attributed to mutations in the *WT1* gene (Jeanpierre et al 1998, Ito et al 2001, Hahn et al 2006). Finally, mutations in WT1 have been associated with the development of isolated SRNS or FSGS in some phenotypic females (Demmer et al 1999, Denamur et al 1999, Ruf et al 2004).

The *WT1* gene contains ten exons; exons 1–6 encode a proline-/glutamine-rich transcriptional regulatory region, whereas exons 7–10 encode the four zinc fingers of the DNA-binding domain (Wang et al 1993, Moffett et al 1995). Up to 24 different isoforms of the protein may be synthesized as a result of alternative translation start sites, alternative splicing and RNA editing (Haber et al 1991). About 80% of transcripts include a sequence of three amino acids (lysine, threonine, and serine: +KTS) between zinc fingers 3 and 4 in exon 9. De novo mutations in the donor splice site in intron 9 of the WT1 gene are causative of Frasier syndrome, and lead to alternative splicing and loss of a +KTS isoform of the protein (Barbaux et al 1997). This results in an alteration of the normal ratio of +KTS/-KTS isoforms in the cell. WT1 isoforms which either contain or lack the KTS motif have different DNA-binding affinities and, therefore, possibly, different regulatory functions. Patients with Denys-Drash syndrome bear heterozygous mutations, mostly de novo, in exons 8 and 9 of the gene, encoding the second and third zinc finger domains (Pelletier et al 1991, Hastie et al 1992). In vitro studies have confirmed that missense mutations in the *WT1* gene lead to a change in the structural organization of the zinc finger domains, leading to loss or alteration of their DNA-binding abilities (Little & Wells 1997).

The crucial role of WT1 in kidney development has been highlighted by the development of animal and in vitro models showing the failure of, arrest in, or delayed development of nephrogenesis in the absence of WT1 expression (Kreidberg et al 1993, Moore et al 1998, Davies et al 2004). In the mature kidney, WT1 expression persists only in podocytes and epithelial cells of Bowman's capsule. The target genes of WT1 are numerous and include the transcription factors PAX2, PAX8, and NovH, as well as growth factors or their receptors: IGF2, IGFR, PDGF-A, TGF-β, EGFR (Rauscher et al 1993, Reddy & Licht 1996). WT1 acts in an inhibitory manner on the *PAX2* gene, and the development of glomerular disease in mice in which PAX2 is over-expressed is consistent with this regulatory model (Dressler et al 1993, Wagner et al 2006). Additional targets appear to be nephrin (Wagner et al 2004) and vascular endothelial growth factor (VEGF) (Gao et al 2005). A recent study demonstrated that the high level of expression of VEGF-165 and the lack of an inhibitory isoform of VEGF-165 (VEGF-165b) result in impaired glomerular maturation in kidneys of patients with DDS (Schumacher et al 2007).

LMX1B and Nail-Patella Syndrome

Nail-patella syndrome (NPS; MIM 161200) is an autosomal dominant disorder characterized by pleiotropic developmental defects of dorsal limb structures such as the nails, patellae, elbows, and bony outgrowths of the dorsal ilium, termed iliac horns. In addition, ocular abnormalities and hearing impairment appear to be part of the phenotypic spectrum of the disease (Bongers et al 2005). Approximately 30% of patients develop proteinuria and, in some cases, ESRD. Ultrastructural abnormalities include thickened and split glomerular basement membranes, disorganized and fused podocyte foot processes, and inappropriate deposition of fibrillar collagens. An *Lmx1b* mutant mouse model recapitulated the phenotype of NPS (Chen et al 1998), thus facilitating the identification of mutations in the *LMX1B* gene, located on chromosome 9q34, in patients with NPS (Dreyer et al 1998). The majority of mutations result in protein truncation, due either to nonsense or frameshifting variants, but also include splice-site and missense mutations (Dreyer et al 1998, McIntosh et al 1998, Vollrath et al 1998). Missense variants frequently involve changes in amino acid residues in a homeodomain region critical for DNA binding (Dreyer et al 1998, McIntosh et al 1998, Vollrath et al 1998). These putative loss-of-function mutations indicate that haploinsufficiency of the LMX1B transcription factor underlies this disorder.

The LMX1B protein contains two zinc-binding LIM domains at the N-terminus, thought to represent an interface for interaction with other proteins, and a homeodomain in the middle for DNA binding and a putative transcriptional activation domain at the C-terminus. Several potential targets of the LMX1B transcription factor in the kidney have been demonstrated by studies performed on the *Lmx1b* null mouse model. In *Lmx1b*-deficient mice, the expression levels of α(3)IV and α(4)IV chains of collagen were markedly diminished (Morello et al 2001), as were the levels of podocin (Miner et al 2002, Rohr et al 2002), CD2AP (Miner et al 2002), synaptopodin (Rohr et al 2002), and VEGF (Rohr et al 2002). LIM homeodomain-binding sites were identified in the shared 5′ regulatory regions of *Col4a3* and *Col4a4* genes (Morello et al 2001), in the promoter regions of the *Nphs2* (Rohr et al 2002) and *Cd2ap* (Miner et al 2002) genes, and LMX1B may be critical for glomerular development. Interestingly, immunohistochemical studies in two patients bearing heterozygous mutations in the *LMX1B* gene revealed no downregulation in the expression of the α3 and α4 chains of type IV collagen, and in podocin and CD2AP (Heidet et al 2003). Indeed, similar results were observed in mice in which *Lmx1b* had been inactivated specifically in podocytes using *Cre-loxP* technology (Suleiman et al 2007). Although the targets of LMX1B in the podocyte remain to be clarified, it appears to be essential for the development of precursor cells into podocytes and possibly also for the maintenance of the differentiated status of podocytes (Witzgall 2008).

SMARCAL1 AND SCHIMKE IMMUNO-OSSEOUS DYSPLASIA

Schimke immuno-osseous dysplasia (SIOD, OMIM 242900) is an autosomal recessive pleiotropic disorder with the diagnostic features of spondyloepiphyseal dysplasia, renal dysfunction due to focal glomerulosclerosis, and T-cell immunodeficiency. One study has shown that the characteristic disproportionate growth failure in these patients is not attributable to the renal failure which ensues (Lucke et al 2006). A genome-wide scan in four families identified significant linkage at chromosome 2q35, and mutations were identified in the *SMARCAL1* gene, encoding a member of an SNF2 subfamily of proteins which mediate DNA-nucleosome restructuring during gene regulation and DNA replication, recombination, methylation and repair. The gene consists of 18 exons and encodes a 106-kDa protein with 954 amino acid residues. The majority of mutations identified involve nonsense and frameshifting mutations, likely leading to loss-of-function (Boerkoel et al 2002). Several missense mutations were, likewise, detected involving crucial functional domains of the protein. However, patients with two missense mutations tended to have a milder course of disease, surviving beyond 15 years of age. To date, the functional targets of *SMARCAL1* are unknown.

GENETIC DEFECTS OF METABOLIC GENES

SCARB2/LIMP2 and Action Myoclonus-Renal Failure Syndrome

Action myoclonus–renal failure syndrome (AMRF, MIM 254900) is a lethal inherited form of progressive myoclonus epilepsy associated with renal failure. It typically presents at 15–25 years with proteinuria evolving into renal failure or with neurologic symptoms (tremor, action myoclonus, seizures, and later ataxia). The renal pathology is of focal glomerulosclerosis, sometimes with features of glomerular collapse. In three unrelated families, Berkovic and colleagues (2008) identified a region on chromosome 4q13-21 linked to the disease and using gene expression profiling to prioritize the genes within the region, they identified homozygous mutations in the *SCARB2* (Scavenger Reporter B2) gene leading to a downregulation in expression at the mRNA level. Its mouse ortholog, the *Limp2* gene, is expressed in a range of tissues including brain and kidney. Although its functional role is not well-established, it is thought to play a role in the biogenesis and maintenance of endosomal and lysosomal compartments. Interestingly, a *Limp2*-deficient mouse model presents with neurologic abnormalities and uretero–pelvic junction obstruction (Gamp et al 2003), but does not recapitulate the glomerular lesions seen in humans. Proteinuria is present and is associated with mesangial expansion, but occurs only with aged mice (Berkovic et al 2008). The mechanisms by which mutations in a lysosomal membrane protein results in proteinuria and FSGS will need further study.

Mitochondropathies and Nephrotic Syndrome

Renal dysfunction associated with mitochondropathies is well-known, but generally a rare event, and may result from mutations encoded for by the nuclear or mitochondrial genomes. In four patients presenting a heterogeneous pattern of glomerular lesions, ranging from FSGS to collapsing glomerulopathy, mutations in the *COQ2* gene were identified (Diomedi-Camassei et al 2007). The *COQ2* gene encodes the parahydroxybenzoate-polyprenyltransferase (EC 2.5.1.39) enzyme, which catalyzes the prenylation of parahydroxybenzoate with a polyprenyl group. It is part of the coenzyme Q_{10} pathway, a component of the mitochondrial respiratory chain vital for the transport of electrons from complexes I and II to complex III. The clinical presentation varied from severe oliguric renal failure due to crescentic GN on the fifth day of life to development of SRNS at 18 months in association with collapsing glomerulopathy

and anasarca (Diodemi-Camassei et al 2007). Neurologic manifestations were variable. In all renal biopsies, dysmorphic mitochondria were characteristic (Diodemi-Camassei et al 2007). The nephrotic syndrome has, likewise, been described in coenzyme Q_{10} (CoQ_{10}) deficiency (MIM 607426), in association with severe neurologic disorders (Rotig et al 2000, Salviati et al 2005, Quinzii et al 2006). In one affected individual, pathogenic mutations in the *PDSS2* gene, encoding a subunit of the decaprenyl diphosphate synthase, were identified (Lopez et al 2006). The *kd/kd* mouse, which is predisposed to developing a collapsing glomerulopathy, carries mutations in the murine ortholog of the human *PDSS2* gene (Barisoni et al 2005). Moreover, three infants have been described with congenital nephrotic syndrome in association with deficiencies of the mitochondrial respiratory chains II + V (Goldenberg et al 2005).

The mitochondrial genome encodes for 13 essential subunits of the mitochondrial respiratory chain, as well as the 22 transfer RNA (tRNA) and two ribosomal RNA (rRNA) genes enabling intraorganellar protein synthesis. The A3243G point mutation in the tRNA$^{Leu(UUR)}$ gene is the most frequently identified mitochondrial DNA defect. Although it has been associated with the MELAS (mitochondrial myopathy, encephalopathy, lactic acidosis, and stroke-like episodes) syndrome, it has been described in patients with isolated renal disease, frequently presenting as FSGS in adulthood, with (Lowik et al 2005) or without the nephrotic syndrome (Cheong et al 1999, Doleris et al 2000, Hotta et al 2001, Hirano et al 2002, Yamagata et al 2002, Guery et al 2003). It alters the A14 nucleotide that is highly conserved in tRNA, and may result in the disruption of the native tRNA conformation required for protein synthesis efficiency, aminoacylation, posttranscriptional modification and processing. In addition, a deletion in the mitochondrial DNA has been associated with FSGS in a cohort of Japanese patients (Yamagata et al 2002) and in a Turkish patient (Unal et al 2005).

GALLOWAY-MOWAT SYNDROME

The Galloway-Mowat syndrome was originally described in 1968 in two siblings who presented with the triad of nephrotic syndrome, congenital microcephaly, and hiatal hernia (Galloway et al 1968). Subsequently, additional reports have consistently associated the nephrotic syndrome with neurologic manifestations and craniofacial abnormalities, which may involve the ears, eyes, or the skull. The onset of renal disease varies widely from birth to as late as 2 years of age. Similarly, histologic findings on renal biopsy range from mesangial lesions, including sclerosis, hypercellularity, and increased mesangial matrix, to minimal change disease and focal segmental sclerosis. Although an autosomal dominant inheritance with variable penetrance cannot be excluded, most cases are consistent with an autosomal recessive pattern of inheritance. The underlying genetic abnormality has not yet been identified (Cooperstone et al 1993, Garty et al 1994).

CONCLUSIONS

Efforts over the past decade have successfully identified the genetic basis of several monogenic forms of hereditary nephrotic syndrome presenting in children. These studies have revealed that the glomerular filtration barrier consists of a dynamic network of proteins, involved in maintaining structural integrity and in signaling events. The identification of mutations in the genes encoding these proteins has revealed genotype–phenotype correlation bearing important implications in disease prognostication. Additionally, it has resulted in the development of animal models of disease which will be useful in unraveling the pathophysiologic mechanisms of renal disease development, as well as providing important tools on which to test novel therapeutic strategies. It has become evident, however, that the underlying genetic defects in several other forms of hereditary nephrotic syndrome remain to be identified, and that complex genetic inheritance patterns are similarly important in fully understanding the determinants of disease development and progression.

References

Antignac C. Genetic models: clues for understanding the pathogenesis of idiopathic nephrotic syndrome. J. Clin. Invest. 2002; 109(4): 447–9.

Asanuma K, Campbell KN, Kim K, Faul C, Mundel P. Nuclear relocation of the nephrin and CD2AP-binding protein dendrin promotes apoptosis of podocytes. Proc. Natl Acad. Sci. USA 2007; 104(24): 10134–9.

Autio-Harmainen H, Vaananen R, Rapola J. Scanning electron microscopic study of normal human glomerulogenesis and of fetal glomeruli in congenital nephrotic syndrome of the Finnish type. Kidney Int. 1981; 20(6): 747–52.

Barbaux S, Niaudet P, Gubler MC, et al. Donor splice-site mutations in WT1 are responsible for Frasier syndrome. Nat. Genet. 1997; 17(4): 467–70.

Barisoni L, Madaio MP, Eraso M, Gasser DL, Nelson PJ. The kd/kd mouse is a model of collapsing glomerulopathy. J. Am. Soc. Nephrol. 2005; 16(10): 2847–51.

Barletta GM, Kovari IA, Verma RK, Kerjaschki D, Holzman LB. Nephrin and Neph1 co-localize at the podocyte foot process intercellular junction and form cis hetero-oligomers. J. Biol. Chem. 2003; 278(21): 19266–71.

Beltcheva O, Martin P, Lenkkeri U, Tryggvason K. Mutation spectrum in the nephrin gene (NPHS1) in congenital nephrotic syndrome. Hum. Mutat. 2001; 17(5): 368–73.

Beltcheva O, Kontusaari S, Fetissov S, et al. Alternatively used promoters and distinct elements direct tissue-specific expression of nephrin. J. Am. Soc. Nephrol. 2003; 14(2): 352–8.

Benzing T. Signaling at the slit diaphragm. J. Am. Soc. Nephrol. 2004; 15(6): 1382–91.

Berkovic SF, Dibbens LM, Oshlack A, et al. Array-based gene discovery with three unrelated subjects shows SCARB2/LIMP-2 deficiency causes myoclonus epilepsy and glomerulosclerosis. Am. J. Hum. Genet. 2008; 82(3): 673–84.

Billing H, Muller D, Ruf R, et al. NPHS2 mutation associated with recurrence of proteinuria after transplantation. Pediatr. Nephrol. 2004; 19(5): 561–4.

Boerkoel CF, Takashima H, John J, et al. Mutant chromatin remodeling protein SMARCAL1 causes Schimke immuno-osseous dysplasia. Nat. Genet. 2002; 30(2): 215–20.

Bongers EM, Huysmans FT, Levtchenko E, et al. Genotype–phenotype studies in nail-patella syndrome show that LMX1B mutation location is involved in the risk of developing nephropathy. Eur. J. Hum. Genet. 2005; 13(8): 935–46.

Boute N, Gribouval O, Roselli S, et al. NPHS2, encoding the glomerular protein podocin, is mutated in autosomal recessive steroid-resistant nephrotic syndrome. Nat. Genet. 2000; 24(4): 349–54.

Caridi G, Bertelli R, Di Duca M, et al. Broadening the spectrum of diseases related to podocin mutations. J. Am. Soc. Nephrol. 2003; 14(5): 1278–86.

Carraro M, Caridi G, Bruschi M, et al. Serum glomerular permeability activity in patients with podocin mutations (NPHS2) and steroid-resistant nephrotic syndrome. J. Am. Soc. Nephrol. 2002; 13(7): 1946–52.

Chaib H, Hoskins BE, Ashraf S, Goyal M, Wiggins RC, Hildebrandt F. Identification of BRAF as a new interactor of PLCepsilon1, the protein mutated in nephrotic syndrome type 3. Am. J. Physiol. Renal Physiol. 2008; 294(1): F93–99.

Chen H, Lun Y, Ovchinnikov D, et al. Limb and kidney defects in Lmx1b mutant mice suggest an involvement of LMX1B in human nail patella syndrome. Nat. Genet. 1998; 19(1): 51–55.

Cheong HI, Chae JH, Kim JS, et al. Hereditary glomerulopathy associated with a mitochondrial tRNA(Leu) gene mutation. Pediatr. Nephrol. 1999; 13(6): 477–80.

Cooperstone BG, Friedman A, Kaplan BS. Galloway-Mowat syndrome of abnormal gyral patterns and glomerulopathy. Am. J. Med. Genet. 1993; 47(2): 250–4.

Davies JA, Ladomery M, Hohenstein P, et al. Development of an siRNA-based method for repressing specific genes in renal organ culture and its use to show that the Wt1 tumour suppressor is required for nephron differentiation. Hum. Mol. Genet. 2004; 13(2): 235–46.

Demmer L, Primack W, Loik V, Brown R, Therville N, McElreavey K. Frasier syndrome: a cause of focal segmental glomerulosclerosis in a 46,XX female. J. Am. Soc. Nephrol. 1999; 10(10): 2215–18.

Denamur E, Bocquet N, Mougenot B, et al. Mother-to-child transmitted WT1 splice-site mutation is responsible for distinct glomerular diseases. J. Am. Soc. Nephrol. 1999; 10(10): 2219–23.

Diomedi-Camassei F, Di Giandomenico S, Santorelli FM, et al. COQ2 nephropathy: a newly described inherited mitochondri-opathy with primary renal involvement. J. Am. Soc. Nephrol. 2007; 18(10): 2773–80.

Doleris LM, Hill GS, Chedin P, et al. Focal segmental glomerulosclerosis associated with mitochondrial cytopathy. Kidney Int. 2000; 58(5): 1851–8.

Donoviel DB, Freed DD, Vogel H, et al. Proteinuria and perinatal lethality in mice lacking NEPH1, a novel protein with homology to NEPHRIN. Mol. Cell. Biol. 2001; 21(14): 4829–36.

Drash A, Sherman F, Hartmann WH, Blizzard RM. A syndrome of pseudohermaphroditism, Wilms' tumor, hypertension, and degenerative renal disease. J. Pediatr. 1970; 76(4): 585–93.

Dressler GR, Wilkinson JE, Rothenpieler UW, Patterson LT, Williams-Simons L, Westphal H. Deregulation of Pax-2 expression in transgenic mice generates severe kidney abnormalities. Nature 1993; 362(6415): 65–7.

Dreyer SD, Zhou G, Baldini A, et al. Mutations in LMX1B cause abnormal skeletal patterning and renal dysplasia in nail patella syndrome. Nat. Genet. 1998; 19(1): 47–50.

Eddy AA, Symons JM. Nephrotic syndrome in childhood. Lancet 2003; 362(9384): 629–39.

Farquhar MG, Vernier RL, Good RA. An electron microscope study of the glomerulus in nephrosis, glomerulonephritis, and lupus erythematosus. J. Exp. Med. 1957; 106(5): 649–60.

Franceschini N, North KE, Kopp JB, McKenzie L, Winkler C. NPHS2 gene, nephrotic syndrome and focal segmental glomerulosclerosis: a HuGE review. Genet. Med. 2006; 8(2): 63–75.

Frasier SD, Bashore RA, Mosier HD. Gonadoblastoma associated with pure gonadal dysgenesis in monozygous twins. J. Pediatr. 1964; 64: 740–5.

Frishberg Y, Rinat C, Megged O, Shapira E, Feinstein S, Raas-Rothschild A. Mutations in NPHS2 encoding podocin are a prevalent cause of steroid-resistant nephrotic syndrome among Israeli-Arab children. J. Am. Soc. Nephrol. 2002; 13(2): 400–5.

Frishberg Y, Feinstein S, Rinat C, et al. The heart of children with steroid-resistant nephrotic syndrome: is it all podocin?. J. Am. Soc. Nephrol. 2006; 17(1): 227–31.

Fuchshuber A, Jean G, Gribouval O, et al. Mapping a gene (SRN1) to chromosome 1q25-q31 in idiopathic nephrotic syndrome confirms a distinct entity of autosomal recessive nephrosis. Hum. Mol. Genet. 1995; 4(11): 2155–58.

Fuchshuber A, Gribouval O, Ronner V, et al. Clinical and genetic evaluation of familial steroid-responsive nephrotic syndrome in childhood. J. Am. Soc. Nephrol. 2001; 12(2): 374–8.

Fujii Y, Khoshnoodi J, Takenaka H, et al. The effect of dexamethasone on defective nephrin transport caused by ER stress: a potential mechanism for the therapeutic action of glucocorticoids in the acquired glomerular diseases. Kidney Int. 2006; 69(8): 1350–9.

Gamp AC, Tanaka Y, Lullmann-Rauch R, et al. LIMP-2/LGP85 deficiency causes ureteric pelvic junction obstruction, deafness and peripheral neuropathy in mice. Hum. Mol. Genet. 2003; 12(6): 631–46.

Galloway WH, Mowat AP. Congenital microcephaly with hiatus hernia and nephrotic syndrome in two sibs. J. Med. Genet. 1968; 5(4): 319–21.

Gao X, Chen X, Taglienti M, Rumballe B, Little MH, Kreidberg JA. Angioblast-mesenchyme induction of early kidney development is mediated by Wt1 and Vegfa. Development 2005; 132(24): 5437–49.

Garg P, Verma R, Nihalani D, Johnstone DB, Holzman LB. Neph1 cooperates with nephrin to transduce a signal that induces actin polymerization. Mol. Cell. Biol. 2007; 27(24): 8698–712.

Garty BZ, Eisenstein B, Sandbank J, Kaffe S, Dagan R, Gadoth N. Microcephaly and congenital nephrotic syndrome owing to diffuse mesangial sclerosis: an autosomal recessive syndrome. J. Med. Genet. 1994; 31(2): 121–5.

Gerke P, Huber TB, Sellin L, Benzing T, Walz G. Homodimerization and heterodimerization of the glomerular podocyte proteins nephrin and NEPH1. J. Am. Soc. Nephrol. 2003; 14(4): 918–26.

Gigante M, Monno F, Roberto R, et al. Congenital nephrotic syndrome of the Finnish type in Italy: a molecular approach. J. Nephrol. 2002; 15(6): 696–702.

Goldenberg A, Ngoc LH, Thouret MC, et al. Respiratory chain deficiency presenting as congenital nephrotic syndrome. Pediatr. Nephrol. 2005; 20(4): 465–9.

Guery B, Choukroun G, Noel LH, et al. The spectrum of systemic involvement in adults presenting with renal lesion and mitochondrial tRNA(Leu) gene mutation. J. Am. Soc. Nephrol. 2003; 14(8): 2099–108.

Haber DA, Sohn RL, Buckler AJ, Pelletier J, Call KM, Housman DE. Alternative splicing and genomic structure of the Wilms tumor gene WT1. Proc. Natl Acad. Sci. USA 1991; 88(21): 9618–22.

Habib R, Loirat C, Gubler MC, et al. The nephropathy associated with male pseudohermaphroditism and Wilms' tumor (Drash syndrome): a distinctive glomerular lesion – report of 10 cases. Clin. Nephrol. 1985; 24(6): 269–78.

Hahn H, Cho YM, Park YS, You HW, Cheong HI. Two cases of isolated diffuse mesangial sclerosis with WT1 mutations. J. Korean Med. Sci. 2006; 21(1): 160–4.

Hasselbacher K, Wiggins RC, Matejas V, et al. Recessive missense mutations in LAMB2 expand the clinical spectrum of LAMB2-associated disorders. Kidney Int. 2006; 70(6): 1008–12.

Hastie ND. Dominant negative mutations in the Wilms tumour (WT1) gene cause Denys-Drash syndrome – proof that a tumour-suppressor gene plays a crucial role in normal genitourinary development. Hum. Mol. Genet. 1992; 1(5): 293–5.

He N, Zahirieh A, Mei Y, et al. Recessive NPHS2 (Podocin) mutations are rare in adult-onset idiopathic focal segmental glomerulosclerosis. Clin J. Am. Soc. Nephrol. 2007; 2(1): 31–7.

Heidet L, Bongers EM, Sich M, et al. *In vivo* expression of putative LMX1B targets in nail-patella syndrome kidneys. Am. J. Pathol. 2003; 163(1): 145–55.

Hinkes B, Wiggins RC, Gbadegesin R, et al. Positional cloning uncovers mutations in PLCE1 responsible for a nephrotic syndrome variant that may be reversible. Nat. Genet. 2006; 38(12): 1397–405.

Hinkes BG, Mucha B, Vlangos CN, et al. Nephrotic syndrome in the first year of life: two thirds of cases are caused by mutations in 4 genes (NPHS1, NPHS2, WT1, and LAMB2). Pediatrics 2007; 119(4): e907–19.

Hinkes B, Vlangos C, Heeringa S, et al. Specific podocin mutations correlate with age of onset in steroid-resistant nephrotic syndrome. J. Am. Soc. Nephrol. 2008; 19(2): 365–71.

Hirano M, Konishi K, Arata N, et al. Renal complications in a patient with A-to-G mutation of mitochondrial DNA at the 3243 position of leucine tRNA. Intern. Med. 2002; 41(2): 113–18.

Holmberg C, Antikainen M, Ronnholm K, Ala Houhala M, Jalanko H. Management of congenital nephrotic syndrome of the Finnish type. Pediatr. Nephrol. 1995; 9(1): 87–93.

Holthofer H, Ahola H, Solin ML, et al. Nephrin localizes at the podocyte filtration slit area and is characteristically spliced in the human kidney. Am. J. Pathol. 1999; 155(5): 1681–87.

Holzman LB, St John PL, Kovari IA, Verma R, Holthofer H, Abrahamson DR. Nephrin localizes to the slit pore of the glomerular epithelial cell Kidney Int. 1999; 56(4): 1481–91.

Hotta O, Inoue CN, Miyabayashi S, Furuta T, Takeuchi A, Taguma Y. Clinical and pathologic features of focal segmental glomerulosclerosis with mitochondrial tRNALeu(UUR) gene mutation. Kidney Int. 2001; 59(4): 1236–43.

Huber TB, Kottgen M, Schilling B, Walz G, Benzing T. Interaction with podocin facilitates nephrin signaling. J. Biol. Chem. 2001; 276(45): 41543–46.

Huber TB, Simons M, Hartleben B, et al. Molecular basis of the functional podocin-nephrin complex: mutations in the NPHS2 gene disrupt nephrin targeting to lipid raft microdomains. Hum. Mol. Genet. 2003a; 12(24): 3397–405.

Huber TB, Hartleben B, Kim J, Schmidts M, Schermer B, Keil A, et al. Nephrin and CD2AP associate with phosphoinositide 3-OH kinase and stimulate AKT-dependent signaling. Mol. Cell. Biol. 2003b; 23(14): 4917–28.

Huber TB, Schermer B, Muller RU, et al. Podocin and MEC-2 bind cholesterol to regulate the activity of associated ion channels. Proc. Natl Acad. Sci. USA 2006; 103(46): 17079–86.

Huttunen NP, Rapola J, Vilska J, Hallman N. Renal pathology in congenital nephrotic syndrome of Finnish type: a quantitative light microscopic study on 50 patients. Int. J. Pediatr. Nephrol. 1980; 1(1): 10–16.

Ito S, Takata A, Hataya H, et al. Isolated diffuse mesangial sclerosis and Wilms tumor suppressor gene. J. Pediatr. 2001; 138(3): 425–7.

Jarad G, Cunningham J, Shaw AS, Miner JH. Proteinuria precedes podocyte abnormalities inLamb2-/- mice, implicating the glomerular basement membrane as an albumin barrier. J. Clin. Invest. 2006; 116(8): 2272–9.

Jeanpierre C, Denamur E, Henry I, et al. Identification of constitutional WT1 mutations, in patients with isolated diffuse mesangial sclerosis, and analysis of genotype/phenotype correlations by use of a computerized mutation database. Am. J. Hum. Genet. 1998; 62(4): 824–33.

Jones N, Blasutig IM, Eremina V, et al. Nck adaptor proteins link nephrin to the actin cytoskeleton of kidney podocytes. Nature 2006; 440(7085): 818–23.

Kalluri R. Proteinuria with and without renal glomerular podocyte effacement. J. Am. Soc. Nephrol. 2006; 17(9): 2383–9.

Karle SM, Uetz B, Ronner V, Glaeser L, Hildebrandt F, Fuchshuber A. Novel mutations in NPHS2 detected in both familial and sporadic steroid-resistant nephrotic syndrome. J. Am. Soc. Nephrol. 2002; 13(2): 388–93.

Kestila M, Lenkkeri U, Mannikko M, et al. Positionally cloned gene for a novel glomerular protein – nephrin – is mutated in congenital nephrotic syndrome. Mol. Cell 1998; 1(4): 575–82.

Khoshnoodi J, Sigmundsson K, Ofverstedt LG, et al. Nephrin promotes cell–cell adhesion through homophilic interactions. Am. J. Pathol. 2003; 163(6): 2337–46.

Koziell A, Grech V, Hussain S, et al. Genotype/phenotype correlations of NPHS1 and NPHS2 mutations in nephrotic syndrome advocate a functional inter-relationship in glomerular filtration. Hum. Mol. Genet. 2002; 11(4): 379–88.

Kreidberg JA, Sariola H, Loring JM, et al. WT-1 is required for early kidney development. Cell 1993; 74(4): 679–91.

Kuusniemi AM, Merenmies J, Lahdenkari AT, et al. Glomerular sclerosis in kidneys with congenital nephrotic syndrome (NPHS1). Kidney Int. 2006; 70(8): 1423–31.

Lehtonen S, Ryan JJ, Kudlicka K, Iino N, Zhou H, Farquhar MG. Cell junction-associated proteins IQGAP1, MAGI-2, CASK, spectrins, and alpha-actinin are components of the nephrin multiprotein complex. Proc. Natl Acad. Sci. USA 2005; 102(28): 9814–19.

Lenkkeri U, Mannikko M, McCready P, et al. Structure of the gene for congenital nephrotic syndrome of the finnish type (NPHS1) and characterization of mutations. Am. J. Hum. Genet. 1999; 64(1): 51–61.

Li H, Lemay S, Aoudjit L, Kawachi H, Takano T. SRC-family kinase Fyn phosphorylates the cytoplasmic domain of nephrin and modulates its interaction with podocin. J. Am. Soc. Nephrol. 2004; 15(12): 3006–15.

Little M, Wells C. A clinical overview of WT1 gene mutations. Hum. Mutat. 1997; 9(3): 209–25.

Liu L, Done SC, Khoshnoodi J, Bertorello A, Wartiovaara J, Berggren PO, et al. Defective nephrin trafficking caused by missense mutations in the NPHS1 gene: insight into the mechanisms of congenital nephrotic syndrome. Hum. Mol. Genet. 2001; 10(23): 2637–44.

Liu G, Kaw B, Kurfis J, Rahmanuddin S, Kanwar YS, Chugh SS. Neph1 and nephrin interaction in the slit diaphragm is an important determinant of glomerular permeability. J. Clin. Invest. 2003; 112(2): 209–221.

Liu XL, Done SC, Yan K, Kilpelainen P, Pikkarainen T, Tryggvason K. Defective trafficking of nephrin missense mutants rescued by a chemical chaperone. J. Am. Soc. Nephrol. 2004; 15(7): 1731–8.

Lopez LC, Schuelke M, Quinzii CM, et al. Leigh syndrome with nephropathy and CoQ10 deficiency due to decaprenyl diphosphate synthase subunit 2 (PDSS2) mutations. Am. J. Hum. Genet. 2006; 79(6): 1125–9.

Lowik MM, Hol FA, Steenbergen EJ, Wetzels JF, van den Heuvel LP. Mitochondrial tRNALeu(UUR) mutation in a patient with steroid-resistant nephrotic syndrome and focal segmental glomerulosclerosis. Nephrol. Dial. Transplant. 2005; 20(2): 336–41.

Lucke T, Franke D, Clewing JM, et al. Schimke versus non-Schimke chronic kidney disease: an anthropometric approach. Pediatrics 2006; 118(2): e400–7.

Matejas V, Al-Gazali L, Amirlak I, Zenker M. A syndrome comprising childhood-onset glomerular kidney disease and ocular abnormalities with progressive loss of vision is caused by mutated LAMB2. Nephrol. Dial. Transplant. 2006; 21(11): 3283–6.

McIntosh I, Dreyer SD, Clough MV, et al. Mutation analysis of LMX1B gene in nail-patella syndrome patients. Am. J. Hum. Genet. 1998; 63(6): 1651–8.

Miller RW, Fraumeni JF Jr, Manning MD. Association of Wilms's tumor with aniridia, hemihypertrophy and other congenital malformations. N. Engl. J. Med. 1964; 270: 922–7.

Miner JH. Building the glomerulus: a matricentric view. J. Am. Soc. Nephrol. 2005; 16(4): 857–61.

Miner JH, Li C. Defective glomerulogenesis in the absence of laminin alpha5 demonstrates a developmental role for the kidney glomerular basement membrane. Dev. Biol. 2000; 217(2): 278–89.

Miner JH, Morello R, Andrews KL, et al. Transcriptional induction of slit diaphragm genes by Lmx1b is required in podocyte differentiation. J. Clin. Invest. 2002; 109(8): 1065–72.

Moeller MJ, Sanden SK, Soofi A, Wiggins RC, Holzman LB. Two gene fragments that direct podocyte-specific expression in transgenic mice. J. Am. Soc. Nephrol. 2002; 13(6): 1561–7.

Moeller MJ, Sanden SK, Soofi A, Wiggins RC, Holzman LB. Podocyte-specific expression of cre recombinase in transgenic mice. Genesis 2003; 35(1): 39–42.

Moffett P, Bruening W, Nakagama H, et al. Antagonism of WT1 activity by protein self-association. Proc. Natl Acad. Sci. USA 1995; 92(24): 11105–09.

Moore AW, Schedl A, McInnes L, Doyle M, Hecksher-Sorensen J, Hastie ND. YAC transgenic analysis reveals Wilms' tumour 1 gene activity in the proliferating coelomic epithelium, developing diaphragm and limb. Mech. Dev. 1998; 79(1–2): 169–84.

Moorthy AV, Chesney RW, Lubinsky M. Chronic renal failure and XY gonadal dysgenesis: "Frasier" syndrome – a commentary on reported cases. Am. J. Med. Genet. Suppl. 1987; 3: 297–302.

Morello R, Zhou G, Dreyer SD, et al. Regulation of glomerular basement membrane collagen expression by LMX1B contributes to renal disease in nail patella syndrome. Nat. Genet. 2001; 27(2): 205–8.

Morris J, Ellwood D, Kennedy D, Knight J. Amniotic alpha-fetoprotein in the prenatal diagnosis of congenital nephrotic syndrome of the Finnish type. Prenat. Diagn. 1995; 15(5): 482–5.

Nishibori Y, Liu L, Hosoyamada M, et al. Disease-causing missense mutations in NPHS2 gene alter normal nephrin trafficking to the plasma membrane. Kidney Int. 2004; 66(5): 1755–65.

Noakes PG, Miner JH, Gautam M, Cunningham JM, Sanes JR, Merlie JP. The renal glomerulus of mice lacking s-laminin/laminin beta 2: nephrosis despite molecular compensation by laminin beta 1. Nat. Genet. 1995; 10(4): 400–6.

Palmen T, Ahola H, Palgi J, et al. Nephrin is expressed in the pancreatic beta cells. Diabetologia 2001; 44(10): 1274–80.

Patrakka J, Ruotsalainen V, Reponen P, et al. Recurrence of nephrotic syndrome in kidney grafts of patients with congenital nephrotic syndrome of the Finnish type: role of nephrin. Transplantation 2002a; 73(3): 394–403.

Patrakka J, Martin P, Salonen R, et al. Proteinuria and prenatal diagnosis of congenital nephrosis in fetal carriers of nephrin gene mutations. Lancet 2002b; 359(9317): 1575–77.

Pelletier J, Bruening W, Kashtan CE, et al. Germline mutations in the Wilms' tumor suppressor gene are associated with abnormal urogenital development in Denys-Drash syndrome. Cell 1991; 67(2): 437–47.

Pereira AC, Pereira AB, Mota GF, et al. NPHS2 R229Q functional variant is associated with microalbuminuria in the general population. Kidney Int. 2004; 65(3): 1026–30.

Philippe A, Nevo F, Esquivel EL, et al. Spectrum of disease due to nephrin mutations includes childhood-onset steroid-resistant nephrotic syndrome. J. Am. Soc. Nephrol. 2008a, in press

Philippe A, Weber S, Esquivel EL, et al. A missense mutation in podocin leads to early and severe renal disease in mice. Kidney Int. 2008b, in press

Putaala H, Sainio K, Sariola H, Tryggvason K. Primary structure of mouse and rat nephrin cDNA and structure and expression of the mouse gene. J. Am. Soc. Nephrol. 2000; 11(6): 991–1001.

Putaala H, Soininen R, Kilpelainen P, Wartiovaara J, Tryggvason K. The murine nephrin gene is specifically expressed in kidney, brain and pancreas: inactivation of the gene leads to massive proteinuria and neonatal death. Hum. Mol. Genet. 2001; 10(1): 1–8.

Quack I, Rump LC, Gerke P, et al. beta-Arrestin2 mediates nephrin endocytosis and impairs slit diaphragm integrity. Proc. Natl Acad. Sci. USA 2006; 103(38): 14110–15.

Quaggin SE. A new piece in the nephrotic puzzle. Nat. Genet. 2006; 38(12): 1360–1.

Quinzii C, Naini A, Salviati L, et al. A mutation in para-hydroxy-benzoate-polyprenyl transferase (COQ2) causes primary coenzyme Q10 deficiency. Am. J. Hum. Genet. 2006; 78(2): 345–9.

Rantanen M, Palmen T, Patari A, et al. Nephrin TRAP mice lack slit diaphragms and show fibrotic glomeruli and cystic tubular lesions. J. Am. Soc. Nephrol. 2002; 13(6): 1586–94.

Ratelade J, Lavin TA, Muda AO, et al. Maternal environment interacts with modifier genes to influence progression of nephrotic syndrome. J. Am. Soc. Nephrol. 2008, in press

Rauscher FJ 3rd. The WT1 Wilms tumor gene product: a developmentally regulated transcription factor in the kidney that functions as a tumor suppressor. FASEB J. 1993; 7(10): 896–903.

Reddy JC, Licht JD. The WT1 Wilms' tumor suppressor gene: how much do we really know?. Biochim. Biophys. Acta. 1996; 1287(1): 1–28.

Reiser J, Kriz W, Kretzler M, Mundel P. The glomerular slit diaphragm is a modified adherens junction. J. Am. Soc. Nephrol. 2000; 11(1): 1–8.

Reiser J, Polu KR, Moller CC, et al. TRPC6 is a glomerular slit diaphragm-associated channel required for normal renal function. Nat. Genet. 2005; 37(7): 739–44.

Rodewald R, Karnovsky MJ. Porous substructure of the glomerular slit diaphragm in the rat and mouse. J. Cell Biol. 1974; 60(2): 423–33.

Rohr C, Prestel J, Heidet L, et al. The LIM-homeodomain transcription factor Lmx1b plays a crucial role in podocytes. J. Clin. Invest. 2002; 109(8): 1073–82.

Roselli S, Gribouval O, Boute N, et al. Podocin localizes in the kidney to the slit diaphragm area. Am. J. Pathol. 2002; 160(1): 131–9.

Roselli S, Heidet L, Sich M, et al. Early glomerular filtration defect and severe renal disease in podocin-deficient mice. Mol. Cell. Biol. 2004a; 24(2): 550–60.

Roselli S, Moutkine I, Gribouval O, Benmerah A, Antignac C. Plasma membrane targeting of podocin through the classical exocytic pathway: effect of NPHS2 mutations. Traffick 2004b; 5(1): 37–44.

Rotig A, Appelkvist EL, Geromel V, et al. Quinone-responsive multiple respiratory-chain dysfunction due to widespread coenzyme Q10 deficiency. Lancet 2000; 356(9227): 391–5.

Ruf RG, Fuchshuber A, Karle SM, et al. Identification of the first gene locus (SSNS1) for steroid-sensitive nephrotic syndrome on chromosome 2p. J. Am. Soc. Nephrol. 2003; 14(7): 1897–900.

Ruf RG, Lichtenberger A, Karle SM, et al. Patients with mutations in NPHS2 (podocin) do not respond to standard steroid treatment of nephrotic syndrome. J. Am. Soc. Nephrol. 2004a; 15(3): 722–32.

Ruf RG, Schultheiss M, Lichtenberger A, et al. Prevalence of WT1 mutations in a large cohort of patients with steroid-resistant and steroid-sensitive nephrotic syndrome. Kidney Int. 2004b; 66(2): 564–70.

Ruotsalainen V, Ljungberg P, Wartiovaara J, et al. Nephrin is specifically located at the slit diaphragm of glomerular podocytes. Proc. Natl Acad. Sci. USA 1999; 96(14): 7962–7.

Ruotsalainen V, Patrakka J, Tissari P, et al. Role of nephrin in cell junction formation in human nephrogenesis. Am. J. Pathol. 2000; 157(6): 1905–16.

Sako M, Nakanishi K, Obana M, et al. Analysis of NPHS1, NPHS2, ACTN4, and WT1 in Japanese patients with congenital nephrotic syndrome. Kidney Int. 2005; 67(4): 1248–55.

Salviati L, Sacconi S, Murer L, et al. Infantile encephalomyopathy and nephropathy with CoQ10 deficiency: a CoQ10-responsive condition. Neurology 2005; 65(4): 606–8.

Schumacher VA, Jeruschke S, Eitner F, et al. Impaired glomerular maturation and lack of VEGF165b in Denys-Drash syndrome. J. Am. Soc. Nephrol. 2007; 18(3): 719–29.

Schwarz K, Simons M, Reiser J, et al. Podocin, a raft-associated component of the glomerular slit diaphragm, interacts with CD2AP and nephrin. J. Clin. Invest. 2001; 108(11): 1621–9.

Seppala M, Rapola J, Huttunen NP, Aula P, Karjalainen O, Ruoslahti E. Congenital nephrotic syndrome: prenatal diagnosis and genetic counselling by estimation of aminotic-fluid and maternal serum alpha-fetoprotein. Lancet 1976; 2(7977): 123–5.

Shigehara T, Zaragoza C, Kitiyakara C, et al. Inducible podocyte-specific gene expression in transgenic mice. J. Am. Soc. Nephrol. 2003; 14(8): 1998–2003.

Shih NY, Li J, Cotran R, Mundel P, Miner JH, Shaw AS. CD2AP localizes to the slit diaphragm and binds to nephrin via a novel C-terminal domain. Am. J. Pathol. 2001; 159(6): 2303–8.

Simons M, Schwarz K, Kriz W, et al. Involvement of lipid rafts in nephrin phosphorylation and organization of the glomerular slit diaphragm. Am. J. Pathol. 2001; 159(3): 1069–77.

Suleiman H, Heudobler D, Raschta AS, et al. The podocyte-specific inactivation of Lmx1b, Ldb1 and E2a yields new insight into a transcriptional network in podocytes. Dev. Biol. 2007; 304(2): 701–12.

Topham PS, Kawachi H, Haydar SA, et al. Nephritogenic mAb 5-1-6 is directed at the extracellular domain of rat nephrin. J. Clin. Invest. 1999; 104(11): 1559–66.

Tsukaguchi H, Sudhakar A, Le TC, et al. NPHS2 mutations in late-onset focal segmental glomerulosclerosis: R229Q is a common disease-associated allele. J. Clin. Invest. 2002; 110(11): 1659–66.

Unal S, Kalkanoglu HS, Kocaefe C, et al. Four-month-old infant with focal segmental glomerulosclerosis and mitochondrial DNA deletion. J. Child. Neurol. 2005; 20(1): 83–4.

Verma R, Wharram B, Kovari I, et al. Fyn binds to and phosphorylates the kidney slit diaphragm component Nephrin. J. Biol. Chem. 2003; 278(23): 20716–23.

Verma R, Kovari I, Soofi A, Nihalani D, Patrie K, Holzman LB. Nephrin ectodomain engagement results in Src kinase activation, nephrin phosphorylation, Nck recruitment, and actin polymerization. J. Clin. Invest. 2006; 116(5): 1346–59.

Vollrath D, Jaramillo-Babb VL, Clough MV, et al. Loss-of-function mutations in the LIM-homeodomain gene, LMX1B, in nail-patella syndrome. Hum. Mol. Genet. 1998; 7(7): 1091–8.

Wagner N, Wagner KD, Xing Y, Scholz H, Schedl A. The major podocyte protein nephrin is transcriptionally activated by the Wilms' tumor suppressor WT1. J. Am. Soc. Nephrol. 2004; 15(12): 3044–51.

Wagner KD, Wagner N, Guo JK, et al. An inducible mouse model for PAX2-dependent glomerular disease: insights into a complex pathogenesis. Curr. Biol. 2006; 16(8): 793–800.

Wang ZY, Qiu QQ, Deuel TF. The Wilms' tumor gene product WT1 activates or suppresses transcription through separate functional domains. J. Biol. Chem. 1993; 268(13): 9172–5.

Wang SX, Ahola H, Palmen T, Solin ML, Luimula P, Holthofer H. Recurrence of nephrotic syndrome after transplantation in CNF is due to autoantibodies to nephrin. Exp. Nephrol. 2001; 9(5): 327–31.

Weber S, Gribouval O, Esquivel EL, et al. NPHS2 mutation analysis shows genetic heterogeneity of steroid-resistant nephrotic syndrome and low post-transplant recurrence. Kidney Int. 2004; 66(2): 571–9.

Wing MR, Bourdon DM, Harden TK. PLC-epsilon: a shared effector protein in Ras-, Rho-, and G alpha beta gamma-mediated signaling. Mol. Interv. 2003; 3(5): 273–80.

Witzgall R. How are podocytes affected in nail-patella syndrome?. Pediatr Nephrol 2008, in press

Wong MA, Cui S, Quaggin SE. Identification and characterization of a glomerular-specific promoter from the human nephrin gene. Am. J. Physiol. Renal Physiol. 2000; 279(6): F1027–32.

Yamagata K, Muro K, Usui J, et al. Mitochondrial DNA mutations in focal segmental glomerulosclerosis lesions. J. Am. Soc. Nephrol. 2002; 13(7): 1816–23.

Zenker M, Tralau T, Lennert T, et al. Congenital nephrosis, mesangial sclerosis, and distinct eye abnormalities with microcoria: an autosomal recessive syndrome. Am. J. Med. Genet. A. 2004; 130(2): 138–45.

Zhu J, Sun N, Aoudjit L, et al. Nephrin mediates actin reorganization via phosphoinositide 3-kinase in podocytes. Kidney Int. 2008; 73(5): 556–66.

CHAPTER 6

Focal Segmental Glomerulosclerosis

KRISHNA R. POLU AND MARTIN R. POLLAK

INTRODUCTION

The role of genetic factors in the development of focal segmental glomerulosclerosis in humans has become increasingly apparent in recent years. Genetic studies have also helped strengthen the notion that glomerular visceral epithelial cell (or podocyte) disorders lead to a spectrum of clinical presentations, from congenital nephrotic syndrome (CNF), to minimal change disease (MCD), and focal segmental glomerulosclerosis (FSGS). In addition, the heterogeneity of these disorders is underscored by the variability in phenotypes of disease. Age of presentation, severity of proteinuria, time to dialysis, response to steroids, and recurrence after transplantation may vary depending on the mechanism of injury. This complexity supports the notion that focal segmental glomerulosclerosis is likely a common downstream effect resulting from a number of mechanisms leading to glomerular injury, rather than a single disease entity. FSGS is seen in a variety of disease states and can exist in primary, secondary, and familial forms. Understanding the mechanisms of the development of FSGS has been advanced by our increasing understanding of podocyte biology through the study of familial patterns of disease. Mutations in both podocin gene (*NPHS2*) alleles lead to a range of human disease patterns, from child onset steroid-resistant FSGS and minimal change disease to adult onset FSGS. Dominant inheritance of mutations in ACTN4, the α-actinin-4 gene, can lead to a slowly progressive adult-onset form of FSGS. In addition, FSGS is observed as part of several rare multi-system inherited syndromes. Here we review recent progress in the understanding of the role of genetic factors in these disorders of podocytes related to the development of kidney disease, primarily FSGS.

THE PODOCYTE

Recent advances in the biology of inherited forms of proteinuric renal disease suggest that the podocyte, or glomerular visceral epithelial cell, is a central culprit in the development of FSGS. The podocyte is a non-replicating polarized epithelial cell that stems from precursor mesenchymal cells and is a key component of the filtration apparatus of the glomerulus. During nephrogenesis, these cells undergo modification from a classic epithelial cell phenotype to a highly specialized octopus-shaped cell essential in glomerular filtration (Gubler 2003). The podocyte covers the external surface of adjacent capillaries and interacts with the glomerular basement membrane (GBM) through finger-shaped extensions called pedicels or foot processes. These foot processes interdigitate in a complex manner forming a network of narrow gaps. The gaps span a space of 30–40 nm and are bridged by a mesh-like network that stretches between these processes called the slit diaphragm. Originally described by Rodewald and Karnofsky in mice, the slit diaphragm forms the glomerular ultrafiltration barrier that allows for the passage of water and solutes while inhibiting loss of larger plasma molecules such as albumin and immunoglobins (Rodewald & Karnovsky 1974). The slit diaphragm (SD) appears early on in nephrogenesis at the capillary stage of development. A modified adherens junction, the SD is a zipperlike structure composed of a complex interplay of a number of proteins. ZO-1, FAT, and nephrin are all transmembrane proteins shown to localize to the slit diaphragm. ZO-1 localizes to the cytoplasmic face of the SD and interacts with cell junction components as well as proteins in the cytoskeleton. FAT, a member of the cadherin superfamily, colocalizes with nephrin at the slit diaphragm. Within the kidney, the integral membrane protein nephrin is expressed exclusively in podocytes. Nephrin-deficient mice and humans develop severe nephrosis. Additional slit diaphragm-associated proteins include podocin and CD2AP which both interact with nephrin, and P-cadherin, which is involved in junction–cytoskeleton attachment (Pavenstadt et al 2003) (Figure 6.1).

The architecture of the podocyte and its unique shape is further characterized by a cytoskeleton composed of microtubules, intermediate filaments, and microfilaments. The cell

Genetic Diseases of the Kidney
ISBN: 978-0-12-449851-8

body cytoskeleton is composed mostly of microtubules and intermediate filaments, with prominent vimentin and desmin. Cytoplasmic extensions or foot process extensions arise from the podocyte cell body. These extensions contain both a microfilament-based contractile apparatus and an actin cytoskeletal network (Drenckhahn & Franke 1988). A group of intracellular cytoskeleton-associated proteins (talin, paxillin, and vinculin) link actin filaments to the cell membrane associated integrins at the glomerular basement membrane (Drenckhahn & Franke 1988, Kretzler 2002).

The role of the podocyte and the slit diaphragm as functional barriers to protein filtration is further defined by its polarity. The exposed surface of the podocyte is covered with a cell coat that contains the sialic acid protein podocalyxin that is thought to give rise to its anionic charge (Gubler 2003). A decrease in the content of sialic acid has been observed in rats with aminoglycoside nephrosis and in humans with proteinuria or glomerulonephritis (Ryan et al 1975). In addition, neutralization of the anionic charge of the sialoprotein coat results in loss of epithelial foot processes that resembles the loss of foot processes seen in humans with nephrotic syndrome (Charest & Roth 1985).

The complexity of interactions of multiple podocyte proteins provides several targets for cell injury and disruption of the filtration barrier. Loss of the integrity of this intricate structure can be seen in patients with nephrotic syndrome and FSGS who clinically present with hypertension, proteinuria, and progressive kidney disease. Electron microscopic examination of kidney biopsy specimens reveals foot process effacement, loss of normal podocyte structure, and podocyte detachment from the basement membrane.

Changes seen under light microscopy may show variable degrees of mesangial expansion and matrix deposition in a focal and segmental pattern (Figure 6.2) (D'Agati et al 2004). While initiation of this process is not fully understood, it can result from defects in any of a number of proteins that comprise the podocyte cytoskeleton and slit diaphragm.

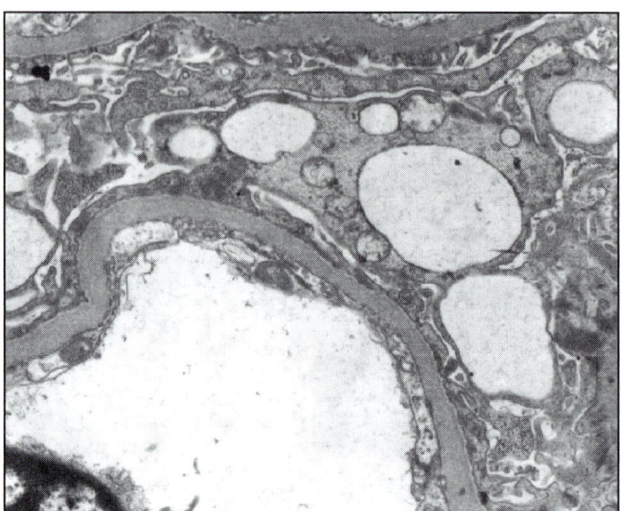

FIGURE 6.2 Light microscopy of a renal biopsy in a patient with ACTN4 mutation. A single glomerulus is depicted which shows segmental glomerulosclerosis. A prominent area of capillary loop hyalinosis is noted on the right lower quadrant. Courtesy of Joel Henderson. Department of Pathology, Brigham and Women's Hospital

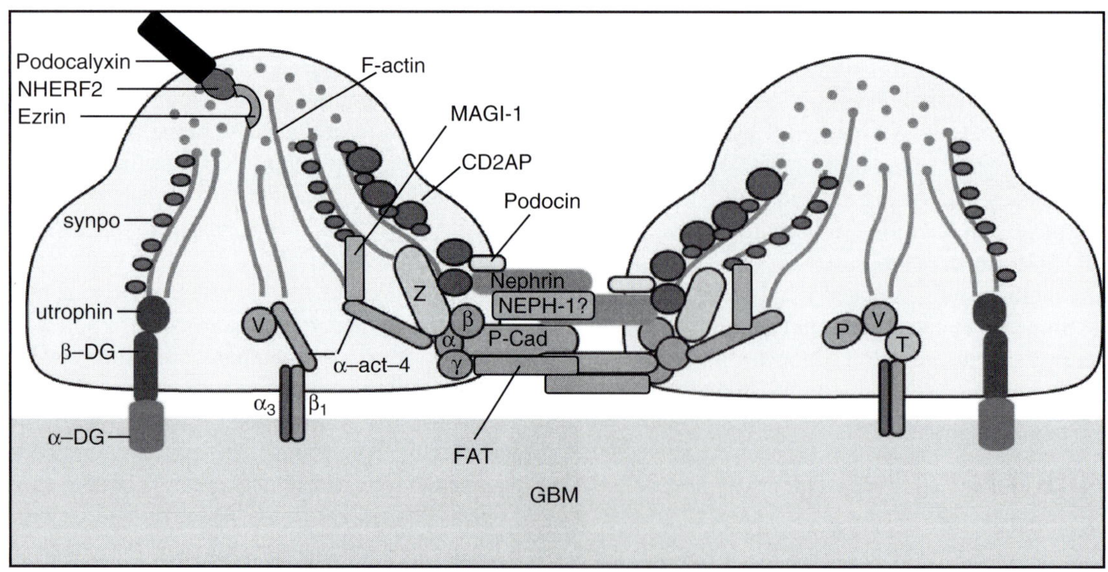

FIGURE 6.1 Electromicrograph of renal biopsy in a patient with ACTN4 mutation. A portion of the glomerular capillary loop is depicted. There is extensive foot process effacement with areas of preservation. Significant vacuolization is also present indicating podocyte injury. Courtesy of Joel Henderson, Department of Pathology, Brigham and Women's Hospital. (see also Plate 4)

FOCAL AND SEGMENTAL GLOMERULOSCLEROSIS

Focal and segmental glomerulosclerosis (FSGS) describes a pathologic lesion observed in a disparate group of clinical disorders, rather than a unique disease entity. Defined by a pathologic finding of foot process effacement and glomerulosclerosis in parts of the tufts of some glomeruli, FSGS occurs in primary (idiopathic) and secondary forms and appears to be increasing in incidence as a cause of proteinuric renal disease (Haas et al 1995). Secondary forms can occur in patients with HIV, sickle cell disease, diabetes, and reflux nephropathy. The complexity of the pathogenesis of FSGS is further illustrated by the variations of lesions seen histologically (D'Agati 2003). These pathologic variants include the classic FSGS subtype, perihilar and tip lesions, as well as collapsing or cellular forms. The collapsing variant carries a worse prognosis and is more commonly seen in patients with FSGS of African-American descent and individuals infected with HIV (Laurinavicius & Rennke 2002).

Primary FSGS accounts for approximately 15% of children and 30% of adult patients with idiopathic nephrotic syndrome (Ichikawa & Fogo 1996). A significant percentage of affected adults and children go on to develop ESRD. Individuals of African descent are disproportionately represented among adults with FSGS as well as with collapsing nephropathy. While the pathogenesis of primary FSGS is not yet clearly defined, a fraction of individuals with FSGS show evidence of a circulating 'glomerular permeability' factor (Sharma et al 2004). Steroids, an imperfect agent in this disease, have nevertheless been a mainstay of treatment, with 30% of patients responding to therapy and undergoing clinical remission.

Typically, clinicians differentiate primary and secondary forms of FSGS. An unclear percentage of cases of FSGS reflect an underlying genetic predisposition towards this condition. Much of what is termed primary FSGS may represent a secondary response to a primary defect in the podocyte. In addition, mutations in proteins that comprise the slit diaphragm and cytoskeleton have been identified. These defects have been shown to be causative in a substantial number of individuals with familial and nonfamilial patterns of FSGS.

MENDELIAN DISEASE

Familial aggregation of proteinuric disease has been noted for over 50 years, though its recognition has not been widespread. For several decades, there have been scattered reports of familial nephrosis in the medical literature (Werner 1942). A 1957 report described four siblings with nephrotic syndrome (Vernier 1957). Pathology showed minimal change disease in some children, FSGS in others. The absence of disease in the parents suggested recessive inheritance. Reports of both single and multigenerational disease have continued to appear in the case literature (McCurdy et al 1987, Mathis et al 1992). Of course, multiple members of a family exposed to the same environmental insults may develop similar diseases for environmental rather than genetic causes. However, recent studies of Mendelian disease have begun to clarify the clinical spectrum of the group of disorders that make up familial FSGS as well as familial nephrotic syndrome. Genetic manipulations in mice have identified other genes (Table 6.1) involved in maintaining the normal podocyte phenotype and the development of FSGS. By means of purely positional genetic approaches, novel proteins have been identified in childhood forms of nephritic. One of these autosomal recessive forms, congenital nephrotic syndrome of the Finnish type (CNF) caused by mutations in the slit diaphragm protein nephrin, is discussed in detail in Chapter 7.

RECESSIVE FSGS: NPHS2

Autosomal recessive forms of FSGS may be caused by defects in the podocyte slit-diaphragm protein podocin. Congenital nephrotic syndrome of the Finnish type (CNF) is caused by mutations in NPHS1, encoding the slit-diaphragm and podocin-associated protein nephrin. CNF, the most clinically severe of the inherited podocytopathies, presents in utero with severe NS and is resistant to therapy (Mannikko et al 1995, Kestila et al 1998). The podocin-associated form of disease appears to be the most common form of inherited FSGS, at least in children (Ruf et al 2004). Fuchshuber et al were the first to describe this form

TABLE 6.1   Identified non-syndromic FSGS/NS genes

Syndrome	Locus	Inheritance	Gene	Protein	MIM number	Reference
Congenital nephrotic syndrome	19q13	AR	NPHS1	Nephrin	602716	Kestila et al 1998
Steroid-resistant NS (FSGS)	1q25-32	AR	NPHS2	Podocin	604766	Boute et al 2000
FSGS	19q13	AD	ACTN4	α-actinin-4	604638	Kaplan et al 2000
FSGS	11q21-22	AD	TRPC6	Transient receptor potential channel, subfamily C	603965	Winn et al 1999b, 2004

*Mendelian Inheritance in Man Number.

of nephrosis characterized by recessive transmission, early onset, resistance to steroid therapy, and rapid progression to end-stage kidney failure (Fuchshuber et al 1995). The majority of the affected children showed an FSGS pattern on renal biopsy, though some showed minimal change disease (MCD). The gene for this recessive form of FSGS was identified by positional methods; the locus was mapped to chromosome 1q25-31 and subsequently cloned (Fuchshuber et al 1995, Boute et al 2000). The responsible gene (NPHS2) encodes podocin, a 383 amino acid integral membrane protein (Boute et al 2000). Podocin is a member of the stomatin family of lipid raft-associated proteins with significant homology to human stomatin family and to MEC-2, part of the *Caenorhabditis elegans* mechano-sensing apparatus (Huang et al 1995).

Podocin expression is localized to the podocyte foot process at the slit-diaphragm (SD) (Roselli et al 2002). Podocin plays a role in the structural organization of the SD through its direct interactions with other key podocyte proteins, nephrin and CD2AP (Huber et al 2001, 2003, Schwarz et al 2001, Salem et al 2002). NPHS2-deficient mice develop proteinuria shortly after birth and die within days (Roselli et al 2004). Kidneys from these mice reveal mesangial sclerosis and fusion of foot processes. Nephrin is downregulated and CD2AP and ZO1 are upregulated revealing how defects in NPHS2 may alter the expression of integral slit diaphragm proteins and lead to proteinuric renal disease (Roselli et al 2004).

The relatively small number of NPHS2 exons (eight) facilitates mutational analysis of human DNA and familial FSGS. Several recent papers have helped define the spectrum of NPHS2-associated disease (Fuchshuber et al 2001, Wu et al 2001a, 2001b, Frishberg et al 2002, Karle et al 2002, Koziell et al 2002, Tsukaguchi et al 2002, Maruyama et al 2003, Billing et al 2004, Caridi et al 2004, Pereira et al 2004, Ruf et al 2004, Yu et al 2004). Many of the disease-associated mutations create truncated proteins, suggesting that disease results from a loss of podocin function (Fuchshuber et al 2001, Wu et al 2001b). Most affected individuals in these reports presented with disease in early childhood, perhaps reflecting the ages of the population studied. R138Q appears to be a common disease-causing variant, and has been observed in several families without recent common ancestors. R138X seems to be particularly common in Arab-Israeli children with steroid-resistant nephrosis (Frishberg et al 2002). Data suggest that an R229Q variant, common in the general population, can cause late-onset FSGS when it occurs together with a second mutant (and probably more severely altered) allele (Tsukaguchi et al 2002). Podocin mutations underlie disease in a sizable fraction of both familial and nonfamilial instances of childhood-onset recessive FSGS. Fuchschuber et al found NPHS2 mutations in 46% of such families (Fuchshuber et al 2001). A larger, more recent study from the same group suggests that NPHS2 mutations cause disease in 20–30% of

children with what appears to be sporadic steroid-resistant nephrotic syndrome (Ruf et al 2004). A number of other studies have also identified mutations in the NPHS2 gene in sporadic cases of FSGS. Furthermore, the known podocin variant R229Q has been identified in greater frequency in sporadic cases of FSGS suggesting a possible association with disease susceptibility with this variant (Caridi et al 2003a). In an analysis of a large cohort of patients with steroid-resistant nephrotic syndrome, the allele frequency for R229Q was shown to be 5.13% compared to 3.75% in healthy control (Weber et al 2004).

Patients with mutations in NPHS2 do not appear to respond to steroids regardless of the type of mutation. In one large study, Ruf et al examined 190 cases from 165 families with steroid-resistant nephrotic syndrome and found that 26% had homozygous or compound heterozygous NPHS2 mutations (Ruf et al 2004). None of the control patients with steroid-sensitive nephrotic syndrome (124 patients from 120 families) had either homozygous or compound heterozygous mutations in NPHS2. This study in addition to another large series by Caridi et al demonstrates that individuals who are homozygotes or compound heterozygotes for NPHS2 mutations are steroid-resistant (Caridi et al 2003b, Ruf et al 2004).

The role of the R229Q polymorphism in NPHS2 may also confer risk in the development of subnephrotic proteinuria and the development of slowly progressive renal failure. The presence of a common polymorphism, R229Q, may be an example of a risk factor for the development of FSGS. Ruf et al showed that there was no statistically significant difference between the presence of this polymorphism in patients with steroid-resistant nephrotic syndrome (7%), steroid-sensitive nephrotic syndrome (6%), and healthy control subjects (11%) (Ruf et al 2004). However, in a cross-sectional study evaluating the R229Q variant in the general population in Brazil, there was a strong association found between the 229Q allele ($P = 0.008$) and microalbuminuria, a surrogate marker for the development of proteinuric renal disease and development of ESRD. The presence of this allele was associated with a 2.77-fold increased risk for microalbuminuria when adjusted for ethnicity, hypertension, obesity, and diabetes (Pereira et al 2004). The presence of this functional variant may have important implications in understanding proteinuric renal disease in the general population and be an important indicator for those at highest risk for developing ESRD.

The functional significance of the variant is supported in individuals who are homozygous for the R229Q allele. When present in the homozygous state, the R229Q variant may be disease-causing. In one study, the presence of two homozygous R229Q alleles was identified in families with steroid-resistant nephrotic syndrome and two individuals with sporadic disease with no additional NPHS2 mutations detected in these cases. Despite its relatively high frequency, this polymorphism was not identified in a homozygous state in any control individual, nor has it been reported by others in control groups (Weber et al 2004).

Direct interactions between podocin and other slit diaphragm proteins, including nephrin and CD2AP, have been demonstrated (Schwarz et al 2001, Palmen et al 2002, Huber et al 2003). Human data also suggest a genetic interaction between variations in different genes and different loci. Koziell et al have reported that the presence of a single NPHS2 may modify the course of NPHS1-associated congenital nephrosis (Koziell et al 2002). Recessive and steroid-resistant NS is genetically heterogeneous. Fuchshuber et al identified one large family unlinked to the chromosome 1q (NPHS1) locus in their report of locus identification. Genetic heterogeneity in human disease is not surprising, given the existence of several recessive loci for nephrotic syndrome in mice.

DOMINANT FSGS

ACTN4

Autosomal dominant forms of FSGS caused by mutations in the gene ACTN4 typically present later in life and with slower progression when compared to recessive forms (Mathis et al 1992, 1998, Winn et al 1999a, Vats et al 2000). Initially described in three families, mutations in ACTN4, encoding α-actinin-4, cause a slowly progressive form of disease with dominant inheritance, non-nephrotic proteinuria, and renal insufficiency (Mathis et al 1992). The phenotype is characterized by a mild increase in urine protein excretion starting in the teenage years or later with slowly progressive renal dysfunction and the development of ESRD in some (but not all) mutation-carrying individuals. The penetrance of ACTN4-associated disease is high but not 100%: in these families, several individuals carry disease-associated mutations but have no proteinuria or renal insufficiency. There is variable expressivity of disease; some family members have shown severe proteinuria or developed ESRD by the fourth decade while others only mild microalbuminuria in late adulthood (Mathis et al 1998).

ACTN4 is one of four α-actinin genes and located on chromosome 19q13. These genes encode biochemically similar, highly homologous actin crosslinking proteins. α-Actinin isoforms 2 and 3 are expressed almost exclusively in the sarcomere while α-actinins 1 and 4 are widely expressed. However, only α-actinin-4 is significantly expressed in the kidney and localizes to the podocyte foot processes (Kaplan et al 2000, Goode et al 2004). The α-actinins all form ~100 kDa head-to-tail homodimers that crosslink and bundle actin filaments and interact with a large number of cytoskeletal, cell-surface, and signaling molecules. The encoded protein contains an N-terminal actin-binding domain followed by four spectrin-like repeats. The ACTN4 mutations identified in FSGS families are all missense mutations and occur in close proximity to each other in an evolutionary conserved region of the encoded protein. These mutations increase the affinity of the encoded protein to actin filaments (Kaplan et al 2000). α-Actinin/actin affinity affects mechanical properties of actin gels, suggesting that these mutations may alter the mechanical properties of the podocyte (Wachsstock et al 1993). This form of disease appears to be rare compared to NPHS1- and NPHS2-associated nephrosis and may account for 4–5% of familial disease.

The role of α-actinin-4 in kidney function has been further demonstrated by mouse models. Mice homozygous for inactivated Actn4 show progressive proteinuria, glomerular disease, and typically death by several months of age. Electron microscopy shows focal areas of foot process effacement and duplication of glomerular basement membrane in young mice and diffuse effacement and globally disrupted podocyte morphology in older mice (Kos et al 2003). Absence of α-actinin-4 may alter the mechanical properties of the cytoskeleton and alter cell motility and cell adhesion via alterations in the podocyte–GBM and podocyte–podocyte interactions (Kos et al 2003). By contributing to the organization of the actin cytoskeleton and anchoring it to the plasma membrane, α-actinin-4 is thought to be important in the maintenance of cell shape, adhesion, and movement. Unlike mice with defects in the slit diaphragm proteins, alterations in the expression in α-actinin-4 may lead to podocyte injury and the development of FSGS through subtle cytoskeletal changes, rather than gross defects in the slit diaphragm.

Michaud et al developed a transgenic mouse model by overexpressing a human disease-associated ACTN4 missense mutation (Michaud et al 2003). Different transgenic strains developed showed different degrees of disease, with only three of eight proteinuric lines displaying reduced renal function and FSGS-like features. This suggests that fine regulation of α-actinin-4 expression is important for maintaining normal podocyte function. These mice showed evidence of reduced nephrin mRNA expression suggesting the close relationship between the regulation of the actin cytoskeleton and the maintenance of key components of the slit diaphragm complex.

Yao et al showed that mutant α-actinin-4 forms aggregates and suggested that these aggregates themselves may contribute to podocyte dysfuncton. One plausible model explains the development of podocyte injury as a direct effect of protein aggregation and toxic effects of such aggregation, as is found neurodegenerative disorders such as Alzheimer's and Parkinson's disease (Stefani 2004). The other model is one of loss-of-function disease with mutations leading to increased α-actinin degradation (Yao et al 2004). It is possible that these mechanisms in fact contribute to disease pathogenesis. Like CD2AP, α-actinin-4 is widely expressed. However, for unclear reasons, the human phenotype associated with ACTN4 mutations is apparent only in the kidney, which suggests either that a podocyte specific protein–protein interaction is altered by human disease-associated mutations, or that the unique structure of the podocyte makes it susceptible to subtle changes in cytoskeletal architecture (Kaplan et al 2000).

There are suggestions that altered α-actinin-4 expression may play a role in secondary forms of glomerular injury as well. At least in animal models, increased expression of α-actinin-4 may actually precede foot process effacement and proteinuria as seen in puromycin aminonucleoside induced nephrotic syndrome (Smoyer et al 1997). Aberrant expression of α-actinin-4 has been described in patients with membranous nephropathy (Goode et al 2004). Patients show increased expression of ACTN 4 as well as condensation of cytoskeletal actin bundles in effaced podocytes. The increase in α-actinin-4 expression may be due to changes in the distribution of actin filaments and associated molecules in membranous nephropathy. It is unclear if this is a primary or secondary response to immune deposits formation. By contrast, in those patients with minimal change nephropathy, labeling of α-actinin-4 was no different than those of normal glomeruli. In a study of 19 children with primary nephrotic syndrome and foot-process effacement on biopsy, the expression of podocyte proteins nephrin, podocin, α-actinin-4, and WT1 were evaluated. A dramatic decrease in podocin expression and abnormal distribution of nephrin, podocin, and α-actinin-4 was observed (Guan et al 2003). In particular, expression of α-actinin-4 in the disease differed greatly from what was observed in controls. This study (and others) raises the possibility that alterations of the expression of α-actinin-4 and other podocyte proteins (Figure 6.3) may also play a role in the pathobiology of a wider range of proteinuric renal disease independent of FSGS. In addition, α-actinin-4 has functions independent of its role in the actin cytoskeleton and interacts with transmembrane receptors (integrins, intercellular adhesion molecules) that

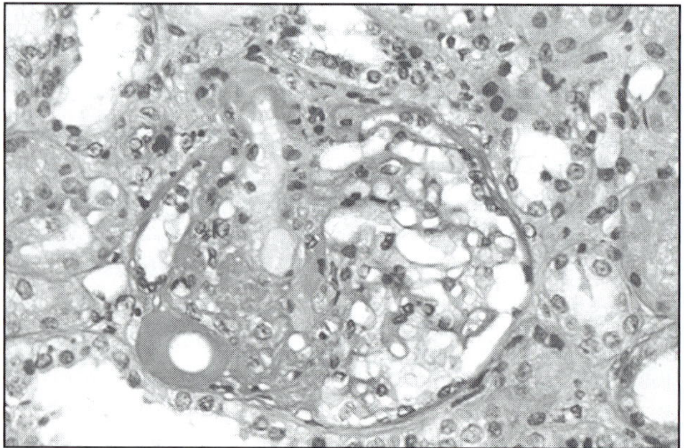

FIGURE 6.3 Podocyte proteins. Schematic diagram showing proteins of demonstrated importance in podocyte and slit-diaphragm function. Courtesy of Peter Mundel, Mt Sinai School of Medicine. (see also Plate 5)

are susceptible to influence by factors outside the cell. The biologic diversity of these proteins suggests that α-actinin-4 may modulate changes in the podocyte cytoskeleton either directly or through external factors (Goode et al 2004).

TRPC6

ACTN4 mutations appear to explain only a fraction of autosomal dominant forms of FSGS. A disease locus on chromosome 11q has been implicated in the pathogenesis of another autosomal dominant form of FSGS (Winn et al 1999b). This 11q locus segregates with disease in a large family with adult-onset disease from New Zealand (Winn et al 1999b). The responsible gene defect has recently been identified and encodes a member of the transient receptor potential (TRP) cation channel superfamily, TRPC6 (Winn et al 2004).

TRP channels are a superfamily of nonvoltage-gated nonselective cation channels that were first identified via studies of *Drosophila* phototransduction. A number of similar channels in mammals have been described bearing structural homology to the original TRP gene. These channels play an essential role in processes ranging from sensory physiology to male fertility. Disruption of several members of this channel family can lead to human disease as seen with mutations in the mammalian TRPP channel (also known as PKD2) that lead to a common form of polycystic kidney disease. TRP channels are expressed in a wide variety of cell types and vary greatly in their selectivity and mode of activation.

Among members of the TRP superfamily, the TRPC group carries the greatest sequence similarity to the *Drososphila* trp channel and consists of seven different members. Structurally, members in this family have six transmembrane segments with N- and C-terminal tails similar in topology to voltage-gated potassium, sodium, and calcium channels. The N-terminal domain of some TRP channels contains three to four ankyrin binding repeats that mediate cytoskeletal anchoring and protein–protein interactions.

TRPC6 shares 70–80% homology with TRPC3 and TRPC7 and primarily has been found in the central nervous system. Its role in the kidney has largely been unknown until recently. Winn and colleagues first described in a large Caucasian family from the south of New Zealand of British origin with an autosomal dominant form of FSGS (Walker et al 1982, Winn et al 1999a, 1999b). The mean age of disease presentation was 33 years (range 16–61), the mean amount of proteinuria on presentation was 3300 µg/24 hours (range 600–6500), and the mean serum creatinine was 1.6 mg/dl (range 0.6–4.1). The average time between the initial presentation and the development of ESRD was 10 years. Linkage analysis identified a disease locus at 11q21-22. Using a combination of haplotype and candidate gene analysis, a novel mutation in the ion channel protein, TRPC6, was identified as the likely disease-causing defect in this family. This mutation results in a substitution of glycine for a highly conserved proline at position 112 in the TRPC6 gene. Immunohistochemistry demonstrated expression of TRPC6 in the glomerulus. Enhanced calcium influx was seen in those cells expressing the mutant P112Q TRPC6 channel (compared with wild-type) in response to an analog of diacylglycerol (DAG), a known activator of TRPC6. Upon exposure to angiotensin II, cells expressing the mutant channels showed enhanced calcium signals (Winn et al 2004). Alterations in TRPC6-mediated calcium influx may alter signaling within glomerular cells leading to susceptibility to injury and development of FSGS.

SYNDROMIC FSGS

Genes encoding podocyte structural proteins may not be the only genes responsible for FSGS. Genes required for podocyte development can also lead to human disease. Mutations in the transcription factor WT1 cause a spectrum of urogenital disorders, including (in Frasier's syndrome) FSGS (Ruf et al 2004). Mutations in the transcription factor Lmx1b cause nail-patella syndrome with a renal phenotype characterized by basement-membrane and podocyte abnormalities (Fuchshuber et al 1995). Lmx1b appears to be important for correct expression of at least some slit diaphragm proteins. Binding sites for this transcription factor have been identified in the regulatory regions of CD2AP and NPHS2 (Huang et al 1995, Boute et al 2000). In the Buffalo/Mna rat, transplantation studies suggest that circulating proteinuric factors are responsible for the FSGS phenotype (Schwarz et al 2001, Roselli et al 2002). This observation strengthens the hypothesis that in humans, genetic alterations could cause FSGS by a similar extrarenal mechanism (e.g. genetic alteration creating a circulating proteinuric factor) (Rodewald & Karnovsky 1974, Pavenstadt et al 2003).

FSGS and related podocyte disorders are also seen as part of well-defined inherited syndromes. The spectrum of disease seen with WT1 mutations is the best studied of these disorders (Little & Wells 1997). The WT1 transcription factor was cloned on the basis of its role in the development of Wilms' tumor (Gessler et al 1990, Haber et al 1990). Frasier syndrome and Denys-Drash syndrome are related and overlapping syndromes caused by mutations in WT1 (Pelletier et al 1991, Barbaux et al 1997, Koziell et al 2000, McTaggart et al 2001). Both syndromes are characterized by glomerular disease and the development of male pseudohermaphroditism. Frasier syndrome is caused by donor splice mutations in intron 9 of WT1 (Barbaux et al 1997). An FSGS pattern is seen on renal biopsy. Frasier syndrome can present as FSGS in 46,XX females in association with gonadal malignancy (Demmer et al 1999, McTaggart et al 2001). WT1 mutations are not a significant cause of glomerular disease in the absence of other genitourinary features (Michaud et al 2003). Denys-Drash syndrome (DDS) is defined by diffuse mesangial sclerosis on renal biopsy, genitourinary tumors, and pseudohermaphroditism. A different spectrum of mutations is seen in DDS, most commonly within exon 9 of WT1 (Coppes et al 1993, Schmitt et al 1995, Denamur et al 2000, Kaltenis et al 2004, Orloff et al 2005). Recently, Orloff reported an association between specific noncoding WT1 SNPs and FSGS in the African-American population (Orloff et al 2005).

Nail-patella syndrome is generally regarded as a disease of the basement membrane rather than the podocyte, though both mechanisms are likely involved in the development of disease. Affected individuals typically demonstrate nephropathy, as well as dysplastic nails, and absent or hypoplastic patellae (Sabnis et al 1980). While an altered GBM typically predominates on histologic analysis, the glomerulopathy is variable and can present as nephrotic syndrome (Smeets et al 1996). Defects in the lmx1b transcription factor are responsible for disease (Dreyer et al 1998, Morello & Lee 2002). Lmx1b helps control the transcriptional regulation of matrix proteins by the podocyte as well as the podocyte genes *CD2AP* and *NPHS2* (Miner et al 2002, Rohr et al 2002). A variety of other inherited syndromes are associated with an increased frequency of FSGS. For example, Charcot-Marie-Tooth disease and Galloway-Mowat syndrome are both inherited neuropathies in which nephrosis and/or FSGS are seen with increased frequency (Chance & Fischbeck 1994, Cohen & Turner 1994).

MITOCHONDRIAL PATTERNS OF INHERITANCE

Mitochondrial disease is discussed in detail in Chapter 35. The mitochondrial genome is a set of extra-chromosomal genes present in a circular genome. Mutations in mitochondrial genes have been associated with maternally inherited diabetes and deafness (MIDD), myopathic syndromes, chloramphenicol resistance and toxicity, and aminoglycoside hearing loss. Expression of mitochondrial gene mutations follows a pattern of maternal inheritance; there is no transmission through affected males. Offspring of affected individuals who receive a copy of the mutant mitochondrial gene may show different levels of phenotypic expression of these mutations. This variability of expression of mitochondrial mutations in affected individuals is due to heteroplasmy, a reflection of different proportions of mutant/wild-type mitochondrial DNA in different individuals and different cell types. These factors make identification of familial forms of disease transmission difficult.

Several cases of mitochondrial patterns of FSGS inheritance have been described in the literature, associated with A to G point mutation at position 3243 in the mitochondrial genome found in the mitochondrial gene MTTL1 which encodes the tRNA[Leu(URR)] (Doleris et al 2000, Hotta et al 2001, Guery et al 2003, Ireland et al 2004, Lowik et al 2005). This mutation has also been associated with MELAS syndrome (myopathy, encephalopathy, lactic acidosis, and stroke-like episodes) as well as maternally inherited diabetes and deafness (MIDD) (Goto et al 1990, Jansen et al 1997). Diabetes mellitus, hearing loss, cardiomyopathy, and renal tubular dysfunction are uncommon manifestations of the A3243G mutation (Hotta et al 2001).

Patients with A3243G mtDNA mutations present with subnephrotic proteinuria, slowly progressive renal dysfunction, diabetes, and hearing loss. Proteinuria presents in young adulthood, estimated at a mean ages of 16 and 26 in two different studies of individuals with idiopathic FSGS (Jansen et al 1997, Hotta et al 2001). In both studies, patients were resistant to treatment with steroids with some developing diabetes after steroid therapy. Diabetes and hearing loss were the only features consistent with mitochondrial cytopathy and none of the patients showed evidence of MELAS syndrome. However, features of MELAS syndrome associated with the A3243G mutation, including stroke and lactic acidosis, have been described in patients with progressive renal disease, indicating that the true incidence of renal involvement of this syndrome is unknown (Ireland et al 2004).

Biopsies of patients with the A3243G mutation reveal glomerular podocytes abnormalities, with cell body attenuation, pseudocyst formation, and foot process effacement (Hotta et al 2001). The cytoplasm of podocytes shows an increase in the number of mitochondria with dysmorphic features including variations in size and shape, irregular outlines, mitochondrial swelling, and increases in cristae and lamellar structures. Abnormalities are not observed in mesangial cells, the capillary endothelial cells, or the glomerular basement membrane (Hotta et al 2001). The pathogenesis of this form of FSGS is not clearly understood. One hypothesis is that because podocytes are terminally differentiated cells they may be more susceptible to the effects of mitochondrial dysfunction, as seen in neural and muscle cells. With the accumulation of these abnormal mitochondria, the synthesis of mitochondrial proteins and appropriate oxidative phosphorylation to generate energy for proper podocyte function may be absent (Hotta et al 2001). This deficiency may lead to podocyte dysfunction and the development of FSGS.

ANIMAL MODELS

CD2AP

CD2-associated protein, another component of the slit diaphragm, may play a crucial role in podocyte function. CD2AP is an 80-kilodalton protein originally identified on the basis of its role in T cell activation. CD2AP stabilizes the interaction between T cells to antigen-presenting cells. Structural features of the protein indicate that it may also play a key role in cytoskeletal regulation (Welsch et al 2001, Lehtonen et al 2002). The N-terminal end contains three conserved SH3 domains (60–70 amino acids) that mediate protein–protein interactions and are thought to link signaling pathways with the cytoskeleton. The C-terminal end includes a coiled coil domain also believed to be important in mediating protein–protein interactions (Shaw & Miner 2001).

CD2AP is found in all tissues but is expressed primarily in epithelial cells. In the kidney, CD2AP is expressed in the epithelial cells of the collecting duct, proximal, and distal tubular epithelium and has been detected in cortical ureteric bud epithelium suggesting a potential role in mediating nephrogenesis (Li et al 2000). In the glomerulus, it is expressed exclusively in the podocyte foot processes and found to play a role in maintaining the integrity of the slit diaphragm through interactions with other key podocyte proteins (Li et al 2000, Shih et al 2001). CD2AP has also been shown to interact directly with both nephrin and podocin. CD2AP colocalizes with nephrin in cultured podocyte lines and is tightly associated with nephrin and podocin in lipid rafts (Schwarz et al 2001, Shih et al 2001, Simons et al 2001, Huber et al 2003). In addition to its structural role in the organization of the slit diaphragm complex, CD2AP may also participate in cell signaling pathways. Variants in CD2AP may lead to early podocyte cell death as well as deficiencies in endocytosis and vesicle trafficking leading to increased susceptibility to toxic injury (Wolf & Stahl 2003).

Mouse models demonstrate the significance of mutations in the gene encoding CD2AP and the development of proteinuric renal disease. Shih and colleagues showed that CD2AP-deficient mice are born proteinuric, show clinical features of nephrotic syndrome, and die within 6 weeks of age from kidney failure. During this time period mice developed cardiac hypertrophy, splenic/thymic atrophy, worsening azotemia, and evidence of growth retardation (Shih et al 1999). Biopsies reveal increased mesangial matrix deposition in almost all glomeruli with distended capillary loops, foot process effacement, and progressive glomerulosclerosis.

Haploinsufficiency of CD2AP may prove to be more relevant in the development of adult-onset FSGS. Kim and colleagues demonstrated that in heterozygous mice, CD2AP protein expression was decreased (Kim et al 2003). Predicting that this would lead to kidney defects, haploinsufficient mice were further evaluated at fixed time points for glomerular changes. Mice with CD2AP haploinsufficiency demonstrated evidence of glomerular abnormalities at 9 months of age, including mesangial expansion and hypercellularity, and variable degrees of injury. The kidneys of these mice also developed increased amounts of IgG deposition in the glomerulus. This accumulation may reflect an impaired ability of the podocyte to clear plasma proteins from the GBM rather than an abnormal immunologic process. In addition, haploinsufficient mice were more susceptible to persistent proteinuria when subjected to low doses of nephrotoxic antibody indicating its role in mediating secondary causes of injury (Kim et al 2003).

The relevance of mutations in CD2AP in humans has not been extensively documented in the literature. In a screen of 30 African-Americans with primary FSGS and 15 African-Americans with HIV nephropathy, two individuals with primary FSGS were found to have a single variant of CD2AP that was found to alter expression of CD2AP (Kim et al 2003). These heterozygous variants may confer increased susceptibility to podocyte injury. Familial patterns of FSGS arising from mutations in CD2AP have not been described. Identification of such families may help further characterize the significance of variants in CD2AP expression and function.

Additional Animal Models

Studies of a variety of podocyte proteins in model organisms such as mice have elucidated other potential biologic pathways important in the development of FSGS. Mice lacking NEPH-1, a nephrin homolog, develop severe nephrosis and die perinatally. Electron microscopy of kidneys in normal mice shows significant expression of NEPH1 in the podocyte. In NEPH-1-deficient mice, histology reveals diffuse foot process effacement and a phenotype similar to nephrin-deficient mice (Donoviel et al 2001). Studies of this and other nephrin family members in these model organisms may facilitate an increase of the understanding and biology of these apparently non-redundant proteins (Liu et al 2003, Sellin et al 2003).

The development of podocyte abnormalities in other mouse models has also been studied. RhoGDIα, a regulator of the Rho-GDP dissociation inhibitor family, is thought to mediate cytoskeletal rearrangements in the kidney. Mice deficient in RhoGDIα develop massive nephrosis emphasizing the importance of cytoskeletal regulation in the maintenance of podocyte function (Togawa et al 1999). Mice deficient in Fyn, a member of the Src family of tyrosine kinases, develop a lymphocyte-independent form of proteinuria (Yu et al 2001). Mice with an interruption of the *MPV17* gene, which encodes a preoxisomal protein that resulates MMP2 production, develop FSGS lesions (Weiher et al 1990, Weiher 1993, Zwacka et al 1994). Podocalyxin-deficient mice exhibit multiple renal and nonrenal abnormalities including the failure to form foot processes (Doyonnas et al 2001). Mice deficient in GLEPP1 (glomerular epithelial protein 1), a tyrosine phosphatase by the podocyte, display severely altered podocyte morphology. Interestingly, these mice, despite an absence of albuminuria, developed impaired GFR (Wharram et al 2000). This model in particular supports the notion that specific and separable functions can be assigned the various gene products that cause mouse and human podocytopathies.

Elevations in the circulating levels of TGF-β have been implicated in diabetic nephropathy and chronic allograft nephropathy. The role of TGF-β in FSGS has been studied in TGF-β transgenic mice that exhibit glomerulosclerosis (Kopp et al 1992). Although an intrinsic podocyte defect is not the primary cause of disease, podocyte depletion does occur in these mice as a direct effect of Smad-7 amplified TGF-β signaling (Schiffer et al 2001).

A variety of rat models develop proteinuria and progressive renal disease. Among the most interesting is the Buffalo/ Mna rat. These rats develop proteinuria and an FSGS-like histology at 2 months of age. Disease recurs in kidneys transplanted into these rats, but when Buf/Mna serve as kidney donors, the glomerulopathy regresses (Le Berre et al 2002). The phenotype thus supports the notion that a circulating factor can be responsible for the development of FSGS. One locus potentially responsible for the proteinuric phenotype has been mapped to a region of the rat chromosome 13 named *Pur1*. The region partially overlaps the rat region syntenic to the NPHS2 locus in humans (Murayama et al 1998). These rats also develop thymoma and antiryanodine receptor antibodies. Genetic differences that alter the activity of a circulating factor in rats increase the suspicion that variation in genes involved in the encoding or the metabolism of such factors may also be important in human disease (Savin et al 1996, Sharma et al 2004).

SECONDARY FSGS

The role of human FSGS and NS genes in acquired disease is a subject of ongoing investigation. Some studies have reported increased nephrin expression in some animal models of disease, others decreased expression in other models (Haltia et al 1999, Luimula et al 2000, 2002). Results from human studies have not yet provided a clear picture of the nature and role of nephrin expression in acquired glomerulopathies (Yuan et al 2002, Schmid et al 2003, Smeets et al 2003, Benigni et al 2004, Hingorani et al 2004).

Alterations in the expression and function of proteins comprising the slit diaphragm and cytoskeleton have been implicated in a number of diseases. In patients with diabetic nephropathy, nephrin mRNA and protein expression were markedly reduced without significant changes in CD2AP and podocin (Benigni et al 2004). In a rat model of type II diabetes, angiotensin-converting enzyme (ACE) inhibitors improved nephrin expression and corrected podocyte phenotypic changes (Blanco et al 2005). The alterations in nephrin expression and response to ACE inhibitors offer insight into one mechanism in the development of diabetic nephropathy.

HIV infection is also associated with an FSGS-like lesion (Ross & Klotman 2004). In patients with HIV-associated nephropathy there may be several podocyte-specific targets for HIV-mediated disease. In addition, variants in podocyte proteins may alter the response to these cells to an altered T cell response. In a study of patients with HIV-associated nephropathy and idiopathic collapsing nephropathy, abnormal distribution of the transcription factor WT1 and loss of podocyte markers were associated with podocyte proliferation without detectable apoptosis (Yang et al 2002). Sequence homology between a number of cytoskeletal proteins such as vimentin and α-actinin-4 has been identified as a target as the site of cleavage of human immunodeficiency virus type 1 protease. The cleavage of specific components of cytoskeletal proteins may contribute to development of HIV nephropathy (Shoeman et al 1991).

CLINICAL SPECTRUM OF DISEASE

Different defects in the podocyte lead to different clinical presentations. Further elucidation of the molecular mechanisms of these diseases will be required to fully understand these differences. Disease caused by defects in NPHS1, NPHS2, TRPC6, and ACTN4 forms a spectrum ranging from prenatal-onset to adult-onset disease. In addition, the disease severity caused by defects in any one of these genes is quite variable as well. It is unclear if the difference between FSGS and NS genes has to do simply with the severity of the resulting podocyte defect, or if FSGS genes perturb different biological pathways than NS genes. Interestingly, one recent paper observed that patients with two defective NPHS1 alleles and a third mutation in NPHS2 show a congenital FSGS phenotype (Koziell et al 2002). Some genes (such as *NPHS1*) may encode proteins whose major (or sole) function is to maintain the glomerular filtration barrier, whereas others encode proteins that function primarily to establish or maintain the normal podocyte architecture (such as ACTN4). Defects in genes that alter the filtration barrier may also alter the podocyte's production of GBM matrix proteins, leading to variations in glomerulosclerosis.

SPORADIC FSGS

What causes most cases of FSGS and minimal change disease? A significant fraction of sporadic FSGS in children is due to NPHS2 mutations. However, a greater fraction still remains unexplained by defects in known genes. Likely, complex combinations of genetic and environmental factors contribute to the development of much of this disease. African-Americans are a subset of the population where the interplay between these two elements may be critical. In a genome-wide scan of ESRD in black families with nondiabetic nephropathy, several loci were identified that were associated with early-onset ESRD, increased BMI, and early- and late-onset hypertension (Freedman et al 2004). Specific genetic variants at these loci have not been reported in the literature to date, nor do they appear to overlap with the known FSGS loci. The nature of the relationship between the increased incidence of FSGS in and the increased risk of ESRD in individuals of African descent is unclear.

APPROACH TO THERAPY

Familial forms of FSGS and nephrotic syndrome, which can be caused by mutations in the genes encoding nephrin, podocin, α-actinin-4, and Trpc6, and in the mitochondrial genome, typically manifest as steroid-resistant forms of a heterogeneous disease process. Research to date has led to the identification of podocyte proteins through elucidation of familial forms of FSGS, but these advances have not yet yielded new insights into therapy. At this time, management of hypertension and attention to modification of risk factors for kidney failure progression (such as hyperlipidemia) are the only forms of therapy that may alter the course of familial FSGS. Selective use of ACE inhibitors and angiotensin II receptor blockers may reduce proteinuria and further delay progression of kidney disease in individuals with familial forms of FSGS.

For those with familial forms of FSGS who progress to ESRD, renal transplantation is a viable option. Idiopathic FSGS recurs in approximately 30% of kidney transplants into FSGS-affected recipients (Cameron 1993). Some renal allograft recipients have experienced recurrence of proteinuria almost immediately after transplantation (Hoyer et al 1972). Patients with familial forms of FSGS, however, appear to have a very low risk for recurrence after transplantation. It is of course critical to be certain that a living-related kidney donor does not in fact carry a disease-associated allele. Because not all FSGS genes are known, and because genetic testing for the known genes is not straightforward, clinical testing of potential related donors must be especially rigorous.

Documentation of the potential for recurrence in patients with mutations in ACTN4 and TRPC6 has not been extensively described, though it appears rare. However, in patients with mutations in NPHS2, there does in fact appear to be some small risk of recurrence of FSGS after transplantation in patients, suggesting that the pathogenesis of this disease may be more complex that simply inherited defects in podocyte proteins. In a study by Ruf et al, of patients with SRNS, only two of 24 had evidence of recurrence (Ruf et al 2004). In contrast, Bertelli et al reported a 38% recurrence rate in 13 FSGS patients with NPHS2 mutations (Bertelli et al 2003). Five of these patients had recurrence of proteinuria and in two of the patients, biopsies revealed FSGS. Testing for antipodocin was performed in this group and found to be negative. Interestingly, in this study, the rate of recurrence was similar to that in non-NPHS2 FSGS children. Caridi et al showed that, in a cohort of nine children with NPHS2 mutations and FSGS, two had recurrence after transplantation which remitted after plasmapheresis and cyclophosphamide (Caridi et al 2001). The mechanism by which a disease that appears to be caused by an altered slit diaphragm protein recurs in an allograft is not obvious.

IMPLICATIONS

Prenatal and presymptomatic diagnosis is possible for inherited diseases with known genetic bases. The practical value of such testing depends on the specifics of the disease. Prenatal testing for specific disease-associated NPHS1 alleles has already been shown to be a useful clinical tool. The utility of NPHS2 testing to determine response to treatment now appears well established. The value of genetic testing for other forms of FSGS or NS will depend on the frequency of these forms of disease and their implications for response to specific treatments. At present, genetic testing for FSGS and NS genes remains primarily a research tool, rather than a clinical test. In the next several years, testing for at least some of these genes will likely evolve into useful clinical tools. As with other forms of inherited kidney disease, care must be taken to avoid using an affected relative as a renal transplant donor. Because the familial pattern of inheritance may not always be obvious in inherited FSGS, clinicians should take particular care. In those families where affected members are being evaluated for renal transplantation, caution should be taken when evaluating other family members as donors. As a practical matter, at the present time the best method for ensuring this is to be certain that the potential donor has absolutely no microalbuminuria in multiple repeated measurements. However, given the variability in expressivity and the later presentation of kidney disease, prospective donors may not reveal the phenotype upon evaluation. Later onset of renal disease as seen in mutations in NPHS2, ACTN4, and TRPC6 are examples. As our understanding of the biology and clinical manifestations become more apparent, screening for the mutations in donors where family members are affected with FSGS may be warranted. In the immediate future, this may be the most relevant application.

Does the human variation in the renal response to primary insults (such as diabetes, hypertension, reflux) involve common differences in genes that regulate podocyte structure and function? Variations in certain genes may be involved in the heritable response to podocyte injury, while other variation may cause altered podocyte function directly. Progress in the genetic and biologic understanding of inherited FSGS and NS will continue. In the future, we may come to regard much of the NS/FSGS group of diseases as a collection of inherited defects in the podocyte, the immune system, and genes involved in the response to injury.

References

Artero M, Biava C, Amend W, Tomlanovich S, Vincenti F. Recurrent focal glomerulosclerosis: natural history and response to therapy. Am. J. Mcd. 1992; 92(4): 375–83.

Barbaux S, Niaudet P, Gubler MC, et al. Donor splice-site mutations in WT1 are responsible for Frasier syndrome. Nat. Genet. 1997; 17(4): 467–70.

Benigni A, Gagliardini E, Tomasoni S, et al. Selective impairment of gene expression and assembly of nephrin in human diabetic nephropathy. Kidney Int. 2004; 65(6): 2193–200.

Bertelli R, Ginevri F, Caridi G, et al. Recurrence of focal segmental glomerulosclerosis after renal transplantation in patients with mutations of podocin. Am. J. Kidney Dis. 2003; 41(6): 1314–21.

Billing H, Muller D, Ruf R, et al. NPHS2 mutation associated with recurrence of proteinuria after transplantation. Pediatr. Nephrol. 2004; 19(5): 561–4.

Blanco S, Bonet J, Lopez D, Casas I, Romero R. ACE inhibitors improve nephrin expression in Zucker rats with glomerulosclerosis. Kidney Int. Suppl. 2005; 93: S10–14.

Boute N, Gribouval O, Roselli S, et al. NPHS2, encoding the glomerular protein podocin, is mutated in autosomal recessive steroid-resistant nephrotic syndrome. Nat. Genet. 2000; 24(4): 349–54.

Bucciantini M, Giannoni E, Chiti F, et al. Inherent toxicity of aggregates implies a common mechanism for protein misfolding diseases. Nature 2002; 416(6880): 507–11.

Cameron JS. Recurrent disease in renal allografts. Kidney Int. Suppl. 1993; 43: S91–4.

Caridi G, Bertelli R, Carrea A, et al. Prevalence, genetics, and clinical features of patients carrying podocin mutations in steroid-resistant nonfamilial focal segmental glomerulosclerosis. J. Am. Soc. Nephrol. 2001; 12(12): 2742–6.

Caridi G, Berdeli A, Dagnino M, et al. Infantile steroid-resistant nephrotic syndrome associated with double homozygous mutations of podocin. Am. J. Kidney Dis. 2004; 43(4): 727–32.

Caridi G, Bertelli R, Di Duca M, et al. Broadening the spectrum of diseases related to podocin mutations. J. Am. Soc. Nephrol. 2003a; 14(5): 1278–86.

Caridi G, Bertelli R, Scolari F, Sanna-Cherchi S, Di Duca M, Ghiggeri GM. Podocin mutations in sporadic focal-segmental glomerulosclerosis occurring in adulthood. Kidney Int. 2003b; 64(1): 365.

Chance PF, Fischbeck KH. Molecular genetics of Charcot-Marie-Tooth disease and related neuropathies. Hum. Mol. Genet. 1994(3 Spec No): 1503–7.

Charest PM, Roth J. Localization of sialic acid in kidney glomeruli: regionalization in the podocyte plasma membrane and loss in experimental nephrosis. Proc. Natl Acad. Sci. USA 1985; 82(24): 8508–12.

Cohen AH, Turner MC. Kidney in Galloway-Mowat syndrome: clinical spectrum with description of pathology. Kidney Int. 1994; 45(5): 1407–15.

Coppes MJ, Huff V, Pelletier J. Denys-Drash syndrome: relating a clinical disorder to genetic alterations in the tumor suppressor gene WT1. J. Pediatr. 1993; 123(5): 673–8.

D'Agati V. Pathologic classification of focal segmental glomerulosclerosis. Semin. Nephrol. 2003; 23(2): 117–34.

D'Agati VD, Fogo AB, Bruijn JA, Jennette JC. Pathologic classification of focal segmental glomerulosclerosis: a working proposal. Am. J. Kidney Dis. 2004; 43(2): 368–82.

Demmer L, Primack W, Loik V, Brown R, Therville N, McElreavey K. Frasier syndrome: a cause of focal segmental glomerulosclerosis in a 46,XX female. J. Am. Soc. Nephrol. 1999; 10(10): 2215–18.

Denamur E, Bocquet N, Baudouin V, et al. WT1 splice-site mutations are rarely associated with primary steroid-resistant focal and segmental glomerulosclerosis. Kidney Int. 2000; 57(5): 1868–72.

Doleris LM, Hill GS, Chedin P, et al. Focal segmental glomerulosclerosis associated with mitochondrial cytopathy. Kidney Int. 2000; 58(5): 1851–8.

Donoviel DB, Freed DD, Vogel H, et al. Proteinuria and perinatal lethality in mice lacking EPH1, a novel protein with homology to NEPHRIN. Mol. Cell. Biol. 2001; 21(14): 4829–36.

Doyonnas R, Kershaw DB, Duhme C, et al. Anuria, omphalocele, and perinatal lethality in mice lacking the CD34-related protein podocalyxin. J. Exp. Med. 2001; 194(1): 13–27.

Drenckhahn D, Franke RP. Ultrastructural organization of contractile and cytoskeletal proteins in glomerular podocytes of chicken, rat, and man. Lab. Invest. 1988; 59(5): 673–82.

Dreyer SD, Zhou G, Baldini A, et al. Mutations in LMX1B cause abnormal skeletal patterning and renal dysplasia in nail patella syndrome. Nat. Genet. 1998; 19(1): 47–50.

Dustin ML, Olszowy MW, Holdorf AD, et al. A novel adaptor protein orchestrates receptor patterning and cytoskeletal polarity in T-cell contacts. Cell 1998; 94(5): 667–77.

Freedman BI, Langefeld CD, Rich SS, et al. A genome scan for ESRD in black families enriched for nondiabetic nephropathy. J. Am. Soc. Nephrol. 2004; 15(10): 2719–27.

Frishberg Y, Rinat C, Megged O, Shapira E, Feinstein S, Raas-Rothschild A. Mutations in NPHS2 encoding podocin are a prevalent cause of steroid-resistant nephrotic syndrome among Israeli-Arab children. J. Am. Soc. Nephrol. 2002; 13(2): 400–5.

Fuchshuber A, Gribouval O, Ronner V, et al. Clinical and genetic evaluation of familial steroid-responsive nephrotic syndrome in childhood. J. Am. Soc. Nephrol. 2001; 12(2): 374–8.

Fuchshuber A, Jean G, Gribouval O, et al. Mapping a gene (SRN1) to chromosome 1q25-q31 in idiopathic nephrotic syndrome confirms a distinct entity of autosomal recessive nephrosis. Hum. Mol. Genet. 1995; 4(11): 2155–8.

Gessler M, Poustka A, Cavenee W, Neve RL, Orkin SH, Bruns GA. Homozygous deletion in Wilms tumours of a zinc-finger gene identified by chromosome jumping. Nature 1990; 343(6260): 774–8.

Goode NP, Shires M, Khan TN, Mooney AF. Expression of alpha-actinin-4 in acquired human nephrotic syndrome: a quantitative immunoelectron microscopy study. Nephrol. Dial. Transplant. 2004; 19(4): 844–51.

Goto Y, Nonaka I, Horai S. A mutation in the tRNA(Leu)(UUR) gene associated with the MELAS subgroup of mitochondrial encephalomyopathies. Nature 1990; 348(6302): 651–3.

Guan N, Ding J, Zhang J, Yang J. Expression of nephrin, podocin, alpha-actinin, and WT1 in children with nephrotic syndrome. Pediatr. Nephrology 2003; 18(11): 1122–7.

Gubler MC. Podocyte differentiation and hereditary proteinuria/nephrotic syndromes. J. Am. Soc. Nephrol. 2003; 14 Suppl 1: S22–6.

Guery B, Choukroun G, Noel LH, et al. The spectrum of systemic involvement in adults presenting with renal lesion and mitochondrial tRNA(Leu) gene mutation. J. Am. Soc. Nephrol. 2003; 14(8): 2099–108.

Haas M, Spargo BH, Coventry S. Increasing incidence of focal-segmental glomerulosclerosis among adult nephropathies: a 20-year renal biopsy study. Am. J. Kidney Dis. 1995; 26(5): 740–50.

Haber DA, Buckler AJ, Glaser T, et al. An internal deletion within an 11p13 zinc finger gene contributes to the development of Wilms' tumor. Cell 1990; 61(7): 1257–69.

Haltia A, Solin M, Luimula P, Kretzler M, Holthofer H. mRNA differential display analysis of nephrotic kidney glomeruli. Exp. Nephrol. 1999; 7(1): 52–8.

Hingorani SR, Finn LS, Kowalewska J, McDonald RA, Eddy AA. Expression of nephrin in acquired forms of nephrotic syndrome in childhood. Pediatr. Nephrol. 2004; 19(3): 300–5.

Hotta O, Inoue CN, Miyabayashi S, Furuta T, Takeuchi A, Taguma Y. Clinical and pathologic features of focal segmental glomerulosclerosis with mitochondrial tRNALeu(UUR) gene mutation. Kidney Int. 2001; 59(4): 1236–43.

Hoyer JR, Vernier RL, Najarian JS, Raij L, Simmons RL, Michael AF. Recurrence of idiopathic nephrotic syndrome after renal transplantation. Lancet 1972; 2(7773): 343–8.

Huang M, Gu G, Ferguson EL, Chalfie M. A stomatin-like protein necessary for mechanosensation in C. elegans. Nature 1995; 378(6554): 292–5.

Huber TB, Kottgen M, Schilling B, Walz G, Benzing T. Interaction with podocin facilitates nephrin signaling. J. Biol. Chem. 2001; 276(45): 41543–6.

Huber TB, Simons M, Hartleben B, et al. Molecular basis of the functional podocin-nephrin complex: mutations in the NPHS2 gene disrupt nephrin targeting to lipid raft microdomains. Hum. Mol. Genet. 2003; 12(24): 3397–405.

Ichikawa I, Fogo A. Focal segmental glomerulosclerosis. Pediatr. Nephrol. 1996; 10(3): 374–91.

Ireland J, Rossetti S, Haugen E, Michels V, Harris P. Mitochondrial causes of renal insufficiency and hearing loss. Kidney Int. 2004; 65(6): 2444–5.

Jansen JJ, Maassen JA, van der Woude FJ, et al. Mutation in mitochondrial tRNA(Leu(UUR)) gene associated with progressive kidney disease. J. Am. Soc. Nephrol. 1997; 8(7): 1118–24.

Kaltenis P, Schumacher V, Jankauskiene A, Laurinavicius A, Royer-Pokora B. Slow progressive FSGS associated with an F392L WT1 mutation. Pediatr. Nephrol. 2004; 19(3): 353–6.

Kaplan JM, Kim SH, North KN, et al. Mutations in ACTN4, encoding alpha-actinin-4, cause familial focal segmental glomerulosclerosis. Nat. Genet. 2000; 24(3): 251–6.

Karle SM, Uetz B, Ronner V, Glaeser L, Hildebrandt F, Fuchshuber A. Novel mutations in NPHS2 detected in both familial and sporadic steroid-resistant nephrotic syndrome. J. Am. Soc. Nephrol. 2002; 13(2): 388–93.

Kestila M, Lenkkeri U, Mannikko M, et al. Positionally cloned gene for a novel glomerular protein – nephrin – is mutated in congenital nephrotic syndrome. Mol. Cell 1998; 1(4): 575–82.

Kim JM, Wu H, Green G, et al. CD2-associated protein haploinsufficiency is linked to glomerular disease susceptibility. Science 2003; 300(5623): 1298–300.

Kopp JB, Klotman ME, Adler SH, et al. Progressive glomerulosclerosis and enhanced renal accumulation of basement membrane components in mice transgenic for human immunodeficiency virus type 1 genes. Proc. Natl Acad. Sci. USA 1992; 89(5): 1577–81.

Kos CH, Le TC, Sinha S, et al. Mice deficient in alpha-actinin-4 have severe glomerular disease. J. Clin. Invest. 2003; 111(11): 1683–90.

Koziell A, Charmandari E, Hindmarsh PC, Rees L, Scambler P, Brook CG. Frasier syndrome, part of the Denys Drash continuum or simply a WT1 gene associated disorder of intersex and nephropathy? Clin. Endocrinol. (Oxf) 2000; 52(4): 519–24.

Koziell A, Grech V, Hussain S, et al. Genotype/phenotype correlations of NPHS1 and NPHS2 mutations in nephrotic syndrome advocate a functional inter-relationship in glomerular filtration. Hum. Mol. Genet. 2002; 11(4): 379–88.

Kretzler M. Regulation of adhesive interaction between podocytes and glomerular basement membrane. Microsc. Res. Tech. 2002; 57(4): 247–53.

Laurinavicius A, Rennke HG. Collapsing glomerulopathy – a new pattern of renal injury. Semin. Diagn. Pathol. 2002; 19(3): 106–15.

Le Berre L, Godfrin Y, Gunther E, et al. Extrarenal effects on the pathogenesis and relapse of idiopathic nephrotic syndrome in Buffalo/Mna rats. J. Clin. Invest. 2002; 109(4): 491–8.

Lehtonen S, Zhao F, Lehtonen E. CD2-associated protein directly interacts with the actin cytoskeleton. Am. J. Physiol. Renal Physiol. 2002; 283(4): F734–43.

Li C, Ruotsalainen V, Tryggvason K, Shaw AS, Miner JH. CD2AP is expressed with nephrin in developing podocytes and is found widely in mature kidney and elsewhere. Am. J. Physiol. Renal Physiol. 2000; 279(4): F785–92.

Little M, Wells C. A clinical overview of WT1 gene mutations. Hum. Mutat. 1997; 9(3): 209–25.

Liu G, Kaw B, Kurfis J, Rahmanuddin S, Kanwar YS, Chugh SS. Neph1 and nephrin interaction in the slit diaphragm is an important determinant of glomerular permeability. J. Clin. Invest. 2003; 112(2): 209–21.

Lowik MM, Hol FA, Steenbergen EJ, Wetzels JF, van den Heuvel LP. Mitochondrial tRNALeu(UUR) mutation in a patient with steroid-resistant nephrotic syndrome and focal segmental glomerulosclerosis. Nephrol. Dial. Transplant. 2005; 20(2): 336–41.

Luimula P, Ahola H, Wang SX, et al. Nephrin in experimental glomerular disease. Kidney Int. 2000; 58(4): 1461–8.

Luimula P, Sandstrom N, Novikov D, Holthofer H. Podocyte-associated molecules in puromycin aminonucleoside nephrosis of the rat. Lab. Invest. 2002; 82(6): 713–18.

Mathis BJ, Calabrese KE, Slick GL. Familial glomerular disease with asymptomatic proteinuria and nephrotic syndrome: a new clinical entity. J. Am. Osteopath. Assoc. 1992; 92(7): 875–80, 883–4.

Mathis BJ, Kim SH, Calabrese K, et al. A locus for inherited focal segmental glomerulosclerosis maps to chromosome 19q13. Kidney Int. 1998; 53(2): 282–6.

Mannikko M, Kestaila M, Holmberg C, et al. Fine mapping and haplotype analysis of the locus for congenital nephrotic syndrome on chromosome 19q13.1. Am. J. Hum. Genet. 1995; 57(6): 1377–83.

Maruyama K, Iijima K, Ikeda M, et al. NPHS2 mutations in sporadic steroid-resistant nephrotic syndrome in Japanese children. Pediatr. Nephrol. 2003; 18(5): 412–16.

McCurdy FA, Butera PJ, Wilson R. The familial occurrence of focal segmental glomerular sclerosis. Am. J. Kidney Dis. 1987; 10(6): 467–9.

McTaggart SJ, Algar E, Chow CW, Powell HR, Jones CL. Clinical spectrum of Denys-Drash and Frasier syndrome. Pediatr. Nephrol. 2001; 16(4): 335–9.

Michaud JL, Lemieux LI, Dube M, Vanderhyden BC, Robertson SJ, Kennedy CR. Focal and segmental glomerulosclerosis in mice with podocyte-specific expression of mutant alpha-actinin-4. J. Am. Soc. Nephrol. 2003; 14(5): 1200–11.

Miner JH, Morello R, Andrews KL, et al. Transcriptional induction of slit diaphragm genes by Lmx1b is required in podocyte differentiation. J. Clin. Invest. 2002; 109(8): 1065–72.

Morello R, Lee B. Insight into podocyte differentiation from the study of human genetic disease: nail-patella syndrome and transcriptional regulation in podocytes. Pediatr. Res. 2002; 51(5): 551–8.

Murayama S, Yagyu S, Higo K, et al. A genetic locus susceptible to the overt proteinuria in BUF/Mna rat. Mammalian Genome 1998; 9: 886–8.

Orloff MS, Iyengar SK, Winkler CA, et al. Variants in the Wilms tumor gene are associated with focal segmental glomerulosclerosis in the African American population. Physiol. Genomics 2005.

Palmen T, Lehtonen S, Ora A, et al. Interaction of endogenous nephrin and CD2-associated protein in mouse epithelial M-1 cell line. J. Am. Soc. Nephrol. 2002; 13(7): 1766–72.

Pavenstadt H, Kriz W, Kretzler M. Cell biology of the glomerular podocyte. Physiol. Rev. 2003; 83(1): 253–307.

Pelletier J, Bruening W, Kashtan CE, et al. Germline mutations in the Wilms' tumor suppressor gene are associated with abnormal urogenital development in Denys-Drash syndrome. Cell 1991; 67(2): 437–47.

Pereira AC, Pereira AB, Mota GF, et al. NPHS2 R229Q functional variant is associated with microalbuminuria in the general population. Kidney Int. 2004; 65(3): 1026–30.

Rodewald R, Karnovsky MJ. Porous substructure of the glomerular slit diaphragm in the rat and mouse. J. Cell Biol. 1974; 60(2): 423–33.

Rohr C, Prestel J, Heidet L, et al. The LIM-homeodomain transcription factor Lmx1b plays a crucial role in podocytes. J. Clin. Invest. 2002; 109(8): 1073–82.

Roselli S, Gribouval O, Boute N, et al. Podocin localizes in the kidney to the slit diaphragm area. Am. J. Pathol. 2002; 160(1): 131–9.

Roselli S, Heidet L, Sich M, et al. Early glomerular filtration defect and severe renal disease in podocin-deficient mice. Mol. Cell. Biol. 2004a; 24(2): 550–60.

Roselli S, Moutkine I, Gribouval O, Benmerah A, Antignac C. Plasma membrane targeting of podocin through the classical exocytic pathway: effect of NPHS2 mutations. Traffic 2004b; 5(1): 37–44.

Ross MJ, Klotman PE. HIV-associated nephropathy. Aids 2004; 18(8): 1089–99.

Ruf RG, Lichtenberger A, Karle SM, et al. Patients with mutations in NPHS2 (podocin) do not respond to standard steroid treatment of nephrotic syndrome. J. Am. Soc. Nephrol. 2004; 15(3): 722–32.

Ryan GB, Rodewald R, Karnovsky MJ. An ultrastructural study of the glomerular slit diaphragm in aminonucleoside nephrosis. Lab. Invest. 1975; 33(5): 461–8.

Sabnis SG, Antonovych TT, Argy WP, Rakowski TA, Gandy DR, Salcedo JR. Nail-patella syndrome. Clin. Nephrol. 1980; 14(3): 148–53.

Saleem MA, Ni L, Witherden I, et al. Co-localization of nephrin, podocin, and the actin cytoskeleton: evidence for a role in podocyte foot process formation. Am. J. Pathol. 2002; 161(4): 1459–66.

Savin VJ, Sharma R, Sharma M, et al. Circulating factor associated with increased glomerular permeability to albumin in recurrent focal segmental glomerulosclerosis. N. Engl. J. Med. 1996; 334(14): 878–83.

Schiffer M, Bitzer M, Roberts IS, et al. Apoptosis in podocytes induced by TGF-beta and Smad7. J. Clin. Invest. 2001; 108(6): 807–16.

Schmid H, Henger A, Cohen CD, et al. Gene expression profiles of podocyte-associated molecules as diagnostic markers in acquired proteinuric diseases. J. Am. Soc. Nephrol. 2003; 14(11): 2958–66.

Schmitt K, Zabel B, Tulzer G, Eitelberger F, Pelletier J. Nephropathy with Wilms tumour or gonadal dysgenesis: incomplete Denys-Drash syndrome or separate diseases? Eur. J. Pediatr. 1995; 154(7): 577–81.

Schwarz K, Simons M, Reiser J, et al. Podocin, a raft-associated component of the glomerular slit diaphragm, interacts with CD2AP and nephrin. J. Clin. Invest. 2001; 108(11): 1621–9.

Sellin L, Huber TB, Gerke P, Quack I, Pavenstadt H, Walz G. NEPH1 defines a novel family of podocin interacting proteins. FASEB J 2003; 17(1): 115–17.

Sharma M, Sharma R, McCarthy ET, Savin VJ. The focal segmental glomerulosclerosis permeability factor: biochemical characteristics and biological effects. Exp. Biol. Med. (Maywood) 2004; 229(1): 85–98.

Shaw AS, Miner JH. CD2-associated protein and the kidney. Curr. Opin. Nephrol. Hypertens. 2001; 10(1): 19–22.

Shih NY, Li J, Karpitskii V, et al. Congenital nephrotic syndrome in mice lacking CD2-associated protein. Science 1999; 286(5438): 312–15.

Shih NY, Li J, Cotran R, Mundel P, Miner JH, Shaw AS. CD2AP localizes to the slit diaphragm and binds to nephrin via a novel C-terminal domain. Am. J. Pathol. 2001; 159(6): 2303–8.

Shoeman RL, Kesselmier C, Mothes E, Honer B, Traub P. Nonviral cellular substrates for human immunodeficiency virus type 1 protease. FEBS Lett. 1991; 278(2): 199–203.

Simons M, Schwarz K, Kriz W, et al. Involvement of lipid rafts in nephrin phosphorylation and organization of the glomerular slit diaphragm. Am. J. Pathol. 2001; 159(3): 1069–77.

Smeets B, Dijkman HB, te Loeke NA, et al. Podocyte changes upon induction of albuminuria in Thy-1.1 transgenic mice. Nephrol. Dial. Transplant. 2003; 18(12): 2524–33.

Smeets HJ, Knoers VV, van de Heuvel LP, Lemmink HH, Schroder CH, Monnens LA. Hereditary disorders of the glomerular basement membrane. Pediatr. Nephrol. 1996; 10(6): 779–88.

Smoyer WE, Mundel P, Gupta A, Welsh MJ. Podocyte alpha-actinin induction precedes foot process effacement in experimental nephrotic syndrome. Am. J. Physiol. 1997; 273(1 Pt 2): F150–7.

Stefani M. Protein misfolding and aggregation: new examples in medicine and biology of the dark side of the protein world. Biochim. Biophys. Acta 2004; 1739(1): 5–25.

Togawa A, Miyoshi J, Ishizaki H, et al. Progressive impairment of kidneys and reproductive organs in mice lacking Rho GDIalpha. Oncogene 1999; 18(39): 5373–80.

Tsukaguchi H, Sudhakar A, Le TC, et al. NPHS2 mutations in late-onset focal segmental glomerulosclerosis: R229Q is a common disease-associated allele. J. Clin. Invest. 2002; 110(11): 1659–66.

Vats A, Nayak A, Ellis D, et al. Familial nephrotic syndrome: clinical spectrum and linkage to chromosome 19q13. Kidney Int. 2000; 57(3): 875–81.

Vernier. Studies on familial nephrosis. Am. J. Dis. Child 1957; 93: 469.

Wachsstock DH, Schwartz WH, Pollard TD. Affinity of alpha-actinin for actin determines the structure and mechanical properties of actin filament gels. Biophys. J. 1993; 65(1): 205–14.

Walker R, Bailey RR, Lynn KL, Burry AF. Focal glomerulosclerosis – another familial renal disease? NZ Med. J. 1982; 95(717): 686–8.

Weber S, Gribouval O, Esquivel EL, et al. NPHS2 mutation analysis shows genetic heterogeneity of steroid-resistant nephrotic syndrome and low post-transplant recurrence. Kidney Int. 2004; 66(2): 571–9.

Weiher H. Glomerular sclerosis in transgenic mice: the Mpv-17 gene and its human homologue. Adv. Nephrol. Necker. Hosp. 1993; 22: 37–42.

Weiher H, Noda T, Gray DA, Sharpe AH, Jaenisch R. Transgenic mouse model of kidney disease: insertional inactivation of ubiquitously expressed gene leads to nephrotic syndrome. Cell 1990; 62(3): 425–34.

Welsch T, Endlich N, Kriz W, Endlich K. CD2AP and p130Cas localize to different F-actin structures in podocytes. Am. J. Physiol. Renal Physiol. 2001; 281(4): F769–77.

Werner M. Berlin: Springer; 1942.

Wharram BL, Goyal M, Gillespie PJ, et al. Altered podocyte structure in GLEPP1 (Ptpro)-deficient mice associated with hypertension and low glomerular filtration rate. J. Clin. Invest. 2000; 106(10): 1281–90.

Winn MP, Conlon PJ, Lynn KL, et al. Clinical and genetic heterogeneity in familial focal segmental glomerulosclerosis. International Collaborative Group for the Study of Familial Focal Segmental Glomerulosclerosis. Kidney Int. 1999a; 55(4): 1241–6.

Winn MP, Conlon PJ, Lynn KL, et al. Linkage of a gene causing familial focal segmental glomerulosclerosis to chromosome 11 and further evidence of genetic heterogeneity. Genomics 1999b; 58(2): 113–20.

Winn MP, Rosenberg P, Conlon PJ, et al. Mutation in TRPC6 causes familial focal segmental glomerulosclerosis. J. Am. Soc. Nephrol. 2004; 15: 33A.

Wolf G, Stahl RA. CD2-associated protein and glomerular disease. Lancet 2003; 362(9397): 1746–8.

Wu MC, Wu JY, Lee CC, Tsai CH, Tsai FJ. A novel polymorphism (c288C > T) of the NPHS2 gene identified in a Taiwan Chinese family. Hum. Mutat. 2001a; 17(1): 81–2.

Wu MC, Wu JY, Lee CC, Tsai CH, Tsai FJ. Two novel polymorphisms (c954T > C and c1038A > G) in exon8 of NPHS2 gene identified in Taiwan Chinese. Hum. Mutat. 2001b; 17(3): 237.

Yang Y, Gubler MC, Beaufils H. Dysregulation of podocyte phenotype in idiopathic collapsing glomerulopathy and HIV-associated nephropathy. Nephron 2002; 91(3): 416–23.

Yao J, Le TC, Kos CH, et al. Alpha-actinin-4-mediated FSGS: an inherited kidney disease caused by an aggregated and rapidly degraded cytoskeletal protein. PLoS Biol. 2004; 2(6): E167.

Yuan H, Takeuchi E, Taylor GA, McLaughlin M, Brown D, Salant DJ. Nephrin dissociates from actin, and its expression is reduced in early experimental membranous nephropathy. J. Am. Soc. Nephrol. 2002; 13(4): 946–56.

Yu CC, Yen TS, Lowell CA, DeFranco AL. Lupus-like kidney disease in mice deficient in the Src family tyrosine kinases Lyn and Fyn. Curr. Biol. 2001; 11(1): 34–8.

Yu ZH, Ding J, Guan N, et al. A novel mutation of NPHS2 identified in a Chinese family with steroid-resistant nephrotic syndrome. Zhonghua Er Ke Za Zhi 2004; 42(2): 108–12.

Zwacka RM, Reuter A, Pfaff E, et al. The glomerulosclerosis gene Mpv17 encodes a peroxisomal protein producing reactive oxygen species. EMBO J. 1994; 13(21): 5129–34.

C. Primary Genetic Diseases of the Proximal Renal Tubules

CHAPTER 7

Diseases of Renal Glucose Handling

ERNEST M. WRIGHT

INTRODUCTION

Blood glucose is normally maintained within narrow limits, 4–12 mM, in the face of variable inputs and outputs. On average 125 grams of glucose are absorbed each day from digested food in the gut, but the amount consumed can very from none to much greater than 300 grams/day. The body uses about 250 grams to fuel normal daily activity, but this amount varies widely depending upon the individual's work load. Irrespective of the amount of glucose in the diet and the level of physical activity the brain requires 125 grams to sustain neuronal function. The gap between the amount of glucose needed as fuel and the amount consumed is bridged by the use of glucose stores (glycogen) and gluconeogenesis. Surprisingly, glucose is not an essential nutrient and this is best exemplified by people who are intolerant of glucose (and galactose) in their diet. These individuals thrive on a high-fat and -protein diet (see below). The kidneys play two important roles in glucose homeostasis, i.e. salvaging glucose from the glomerular filtrate and by gluconeogenesis. Each day the kidney filters ~180 liters of plasma and this could potentially result in the loss of ~180 grams of glucose to the urine. Virtually all of the filtered glucose is reabsorbed in the proximal tubule and less than 0.5 grams are lost. Thus greater than 99% of the filtered load is reabsorbed.

This chapter summarizes the normal mechanisms of renal glucose reabsorption and their genetic disturbances. Glucose transporters were among the first to be cloned and genetic disorders of glucose transporters were among the first to be identified. The study of patients with specific defects in glucose transporters has clarified, but not completely resolved, how the kidney handles glucose. Furthermore, the benign nature of familial renal @mednet.ucla.eduinet mutations in the transporter gene but a through genetotyping and phenotyping rmal Tmal oral glycosuria and the advances in the molecular biology have stimulated the pharmaceutical industry into targeting renal glucose transporters in the fight against hyperglycemia in diabetes (Yasuda et al 2002).

PHYSIOLOGY OF RENAL GLUCOSE TRANSPORT

Glucose is freely filtered at the glomerulus and so the concentration in the glomerular filtrate is close to that in plasma. Given a GFR of ~180 liters/day/1.73 m^2 and plasma glucose concentration of 4 mM, the filtered glucose load is ~180 grams (1 mole)/day/1.73 m^2, but only (0.1–0.3 grams)/day/1.73 m^2 are excreted in the urine (Keller 1968, Maddox & Gennari 1987). Renal glucose titration studies provide an estimate of the maximal reabsorptive capacity (T$_{max}$~425 grams/day) and the minimum filtered load (F$_{min}$~325 grams/day) at which significant amounts of glucose (>1.4 grams/day) appear in the urine. That is, significant amounts of glucose are excreted when the plasma glucose concentration exceeds 9 mM and the maximum reabsorption capacity is reached when the plasma glucose concentration reaches 15 mM. Above 15 mM the rate of excretion is directly proportional to the filtered load.

Our present understanding of the physiological mechanisms involved in glucose reabsorption in the nephron comes largely from animal studies and analogy to glucose absorption from the small intestine. Micropuncture studies of amphibian and mammalian nephrons demonstrated that glucose is reabsorbed in the proximal tubule. In the rat the concentration of glucose in the glomerular filtrate falls by more than 50% within 1 mm and 90% within 2 mm of the glomerulus (see Figure 7.1). This leads to the conclusion that more than 98% of the filtered glucose is reabsorbed in the rat proximal tubule against a steep concentration gradient.

The kinetics of glucose absorption in the 'early' and 'late' segments of rabbit proximal tubule was examined in isolated, perfused nephron segments (Barfuss & Schafer 1981). The capacity of the 'active' component of glucose absorption decreases from 83 pmoles/min/mm in the proximal convoluted tubule to 8 pmoles/min/mm in the late proximal straight tubule. The apparent affinity for active glucose transport increased along the tubule: the K$_m$ values

Genetic Diseases of the Kidney
ISBN: 978-0-12-449851-8

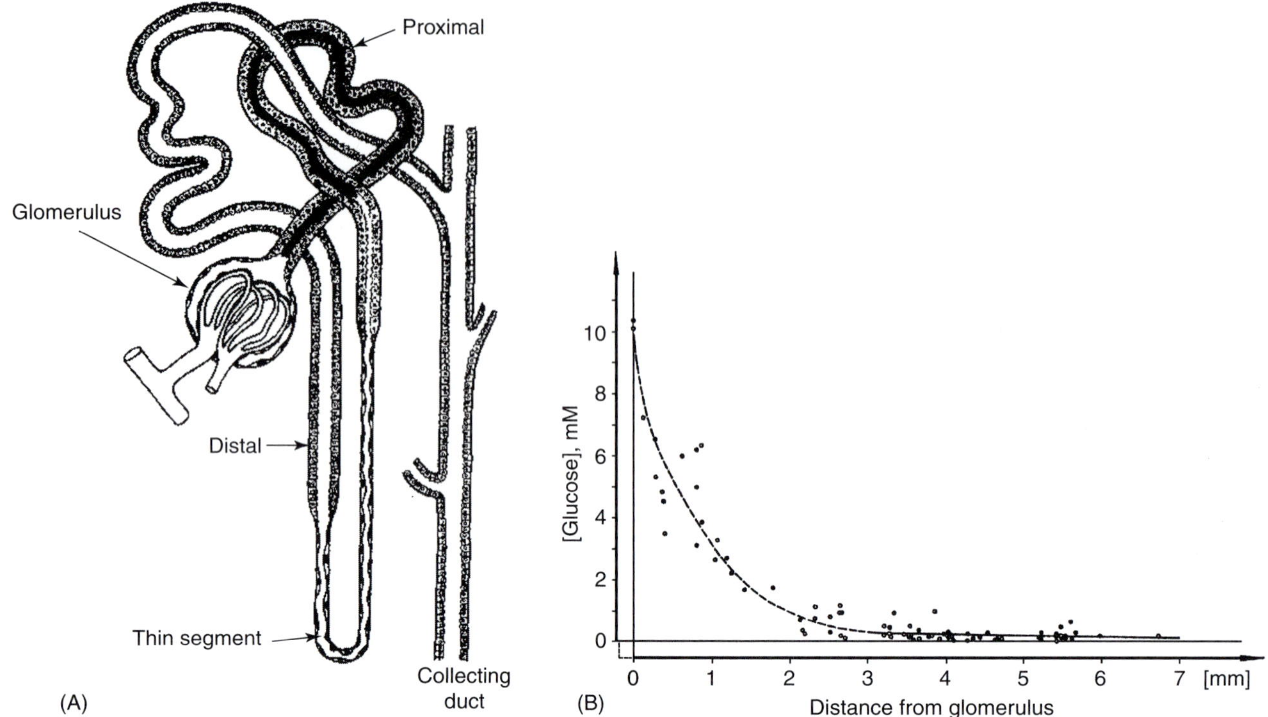

FIGURE 7.1   The glucose concentration profile ($TF_{glucose}$) along the rat proximal tubule from the glomerulus. Tubular fluid was collected from the proximal tubule (A) by micropuncture on normal male rats under tubular free flow conditions and assayed for glucose. The serum glucose concentration was 7.5 ± 1.5 (26) mM. (B) was modified from Frohnert et al (1970)

decreased from 1.6 to 0.35 mM. It should be noted that the perfused tubules were isolated from the superficial cortex and little is known about the heterogeneity of glucose transport among superficial and deep nephrons. Indirect support for differences in kinetics of glucose reabsorption along the proximal tubule was obtained by Turner & Moran (1982) in their study of brush border membrane vesicles from the rabbit outer cortex and outer medulla. They found that the K_m for glucose transport was 6 mM in the outer cortex and 0.3 mM in the inner medulla, and that the outer medulla glucose transporter was more sensitive to competition by galactose than the outer cortex transporter.

Studies of the molecular mechanisms of active glucose transport across the proximal tubule were advanced in the mid-1970s by the introduction of isolated brush border and basolateral membrane vesicle preparations (Kinne et al 1975). As in the small intestine, glucose uptake across the brush border membrane was concentrative, sodium-dependent, and phlorizin-sensitive, i.e. occurred by Na^+/glucose-cotransport. Glucose transport across the basolateral membrane was neither concentrative, nor sodium-dependent, nor phlorizin-sensitive, i.e. occurred by facilitated diffusion. A model for glucose transport across the proximal tubule is shown in Figure 7.2 where the basolateral Na/K-pump maintains the low intracellular sodium in the face of sodium entry across the apical membrane through the Na^+/glucose cotransporter.

The quest for the molecular identity of the brush border and basolateral membrane glucose transporters was advanced further in the mid-1980s by the cloning of membrane proteins, specifically the red cell facilitated glucose transporter (GLUT1, Mueckler et al 1985) and the intestinal brush border sodium/glucose cotransporter (SGLT1, Hediger et al 1987). With further advances in cloning, especially homology and expression cloning, and the completion of the human genome project, a clear picture has emerged about the two major families of human glucose transporters, the GLUT (SLC2) and SGLT (SLC5) families (Uldry & Thorens 2004, Wright & Turk 2004). The SLC2 family is comprised of 13 members in three classes: Class I GLUT1–4; Class II GLUT 5, 7, 9, and 11; and Class 3 GLUT 6, 8, 10, 12 and HMIT. All are postulated to have similar secondary structures with 12 transmembrane-spanning helices and a large number of conserved motifs. High-resolution structural models of the GLUTs are lacking but X-ray crystal structures of two members of the larger gene superfamily (MFS) have been solved, for *lac permease* (Abramson et al 2003) and GlpT (Huang et al 2003).

Initially, GLUTs 1–4 were functionally expressed and characterized as glucose transporters, and efforts are continuing to

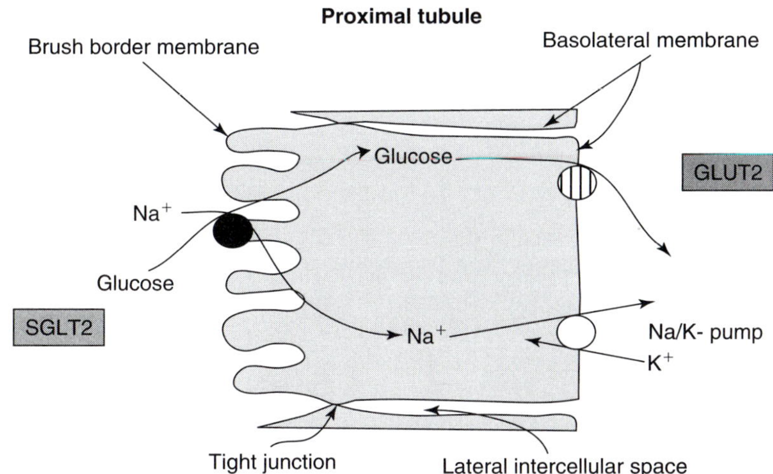

FIGURE 7.2 A model for glucose transport across the epithelial cells of the renal proximal tubule. Glucose is transported from the tubule lumen into the epithelial cell across the apical, brush border, membrane by an Na/glucose cotransporter (SGLT2). The sugar then exits the cell across the basolateral membrane by facilitated diffusion through GLUT2. Other transport proteins may play a part in glucose transport across specific segments of the nephron, e.g. SGLT1 may be the most important one in the S3 segment

document the functions of the other family members. Human GLUT1, 3, and 4 are high-affinity glucose transporters, K_m 1–5 mM. Of these three only GLUT1 appears to be expressed in the kidney, where it is found all along the nephron. GLUT2 is a low-affinity glucose transporter, K_m 17 mM, and this protein is found in the basolateral membranes of the proximal tubule and in the small intestine amongst other places. GLUT2 transports galactose and 2-deoxy-glucose, but not alpha-methyl-glucopyranoside.

The SLC5 family is comprised of 11 genes and two code for sodium/glucose cotransporters expressed in the kidney, SGLT2 and SGLT1 (Wright & Turk 2004). Other genes in the family with high sequence similarity to SGLT1 and two code for a glucose sensor (SGLT3), a mannose transporter (SGLT4, Tazawa et al 2005) and a myo-inositol transporter (SGLT6 or SMIT2). The function of SGLT5 is not yet known. All are expressed in the kidney, but the level of expression relative to that in the small intestine varies: SGLT1 K/I 0.05; SGLT2 Kid/Intest ~ infinite; SGLT3 Kid/Intest ~ 6; SGLT4 Kid/Intest ~ 1.25; SGLT5 Kid/Intest ~ infinite; and SGLT6 Kid/Intest ~ 9 (Bing, Martin, Turk & Wright, unpublished). There is no homology of the SLC5 genes with the SLC2 family members. The SGLTs have 14 transmembrane helices (see Figure 7.3) and there are no structural motifs in common with those in the GLUT family. The crystal structure of a bacterial SGLT has just been reported (Faham et al 2008).

Most functional studies have been carried out on SGLT1 expressed in oocytes, cultured mammalian and insect cells and in bacteria (Wright et al 2004). Transport of glucose is tightly coupled to sodium with a coupling coefficient of 2 and is driven by the sodium and electrical potential gradients across the brush border membrane. Under the maximal driving forces the K_m is 0.6 mM for glucose and 4 mM for sodium. Transport is competitively inhibited by phlorizin with a K_i of 0.6 mM. SGLT1 also transports galactose and alpha-methyl-glucopyranoside with identical kinetics to glucose, but does not transport fructose. Alpha-methyl-glucopyranoside is not transported by GLUTs and 2-deoxy-D-glucose is very

poorly transported by SGLT1 (K_m 110 mM). There is evidence that SGLT1 is located on the brush border membrane of the intestinal epithelium and the S3 segment of the proximal tubule.

Studies of SGLT2 have been severely limited owing to the poor expression of the clone in heterologous expression systems; e.g. *Xenopus laevis* oocytes and COS-7 cells. Na-dependent glucose uptake is barely above background and in oocytes we have been unable to detect SGLT2 in the plasma membrane, using electron microscopic methods. Nevertheless, it is reported that SGLT2 transports both glucose and alpha-methyl-glucopyranoside with a K_m of 2 mM, but galactose is not transported (Wright et al 2001, Pajor 2008). It is claimed the sodium to glucose coupling ratio is 1. These properties resemble those obtained on brush border membrane vesicles prepared from the rabbit outer cortex (see above). Direct proof for the expression of SGLT2 protein in the apical membrane of the human proximal tubule is lacking in the absence of SGLT2 antibodies.

What can we conclude about the identity of the glucose transporters in the brush border and basolateral membrane of the human proximal tubule other than that they are likely to be members of the SLC2 and SLC5 gene families? Some headway in answering this question comes from a comparison of the kinetic properties of the recombinant transporters and those in native membranes, and from studies of inherited disorders of glucose transport. Additional insights come from the study of rodent gene knockout models.

INHERITED DISORDERS OF RENAL GLUCOSE TRANSPORT

The urine from normal healthy subjects contains a small amount of glucose, ranging from 0.3 to 1.1 moles (or 100–300 mg)/day/1.73 m^2 (Keller 1968, Elsas & Rosenberg 1969). This is a small fraction of the normal filtered glucose

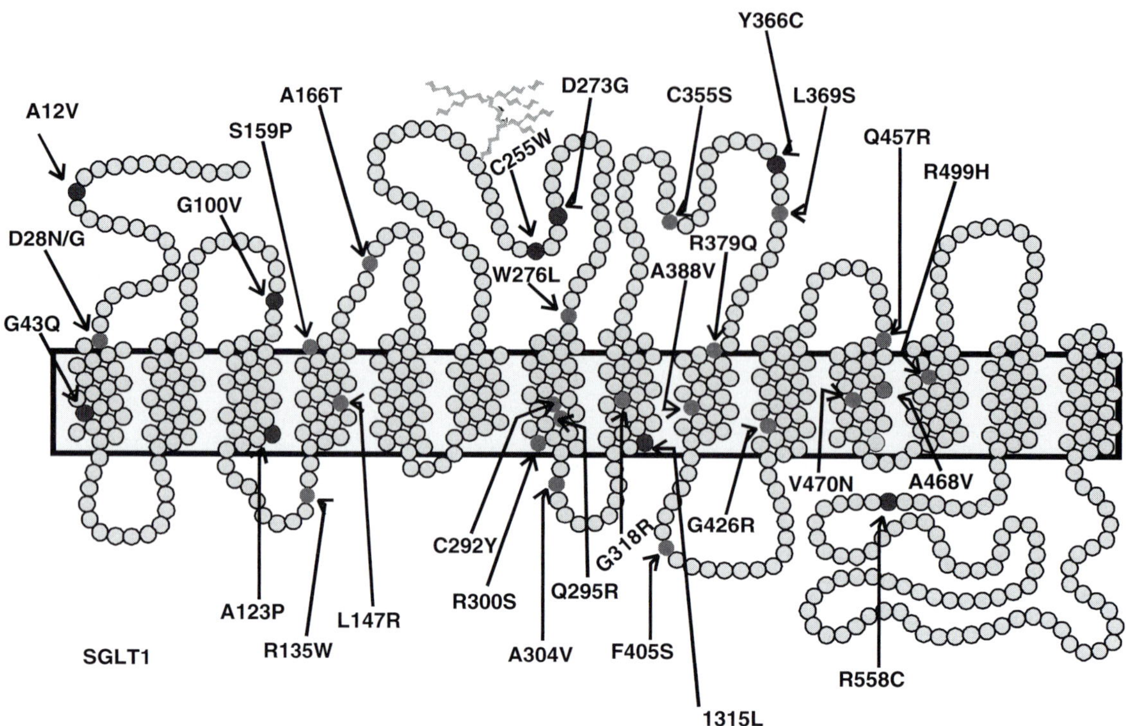

FIGURE 7.3   A secondary structure model of SGLT1 showing the location of the missense mutations identified in patients with glucose-galactose malabsorption (GGM). The 664 amino acid protein is predicted to contain 14 transmembrane helices and one external N-linked glycosylation site. Both the N- and C-termini are shown on the extracellular side of the membrane. (see also Plate 6)

load, ~180 grams/day and, given the volume of urine produced in 24 hours, ~950 ml, the glucose concentration in urine is generally only ~8% of the plasma concentration, i.e. barely sufficient to register on a dip-stick. There are two genetic disorders where glucose excretion ranges up to a high of 250 grams/day/1.73 m^2, i.e. in subjects with Fanconi-Bickel syndrome (MIM 227810) and with familial renal glucosuria (MIM 233100). A mild renal glucosuria may also occur in subjects with glucose-galactose malabsorption (MIM 182380). Familial renal glucosuria (FRG) is a rather benign condition whereas Fanconi-Bickel syndrome (FBS) subjects present with an array of other symptoms ranging from failure to grow to hepatomegaly. Glucose-galactose malabsorption (GGM) presents in newborn children as a massive life-threatening diarrhea and this is caused by variations in the SGLT1 gene (Turk et al 1991, Martin et al 1996, Wright et al 2001). This story will be presented first, followed by those for familial renal glucosuria (FRG) and the Fanconi-Bickel syndrome (FBS).

Glucose-Galactose Malabsorption (GGM) (MIM 182380)

The primary carbohydrate present in breast milk is lactose and this is hydrolyzed by lactase on the intestinal brush border to glucose and galactose. The liberated sugars are then normally transported across the brush border membrane by SGLT1 and across the basolateral membrane by GLUT2 (see also Figure 7.2). In patients with GGM, glucose and galactose are not absorbed and diarrhea results from the osmotic load in the gut. The diarrhea resolves on eliminating lactose, glucose, and galactose from the diet, but promptly resumes on adding one or more of the sugars back. This forms the basis for the rapid diagnosis of the disease. GGM patients are subject to a life-long diet free of glucose, galactose, and lactose.

Arguably the most complete evaluation of GGM was carried out during the first year of life of an American patient in the late 1960s (Schneider et al 1966). Clinical tests showed that there was malabsorption of glucose and galactose, while there was normal absorption of fructose. The histology of duodenal biopsies was normal, but biochemical and electron microscopic studies revealed that there was a defect in glucose and galactose transport across the brush border membrane due to the reduction in the density of transporters (phlorizin binding) in the brush border. This child thrived on a sugar-free diet and even as an adult she is intolerant of carbohydrate; ingestion of as little as 6 grams of table sugar produce malabsorption symptoms. Our oldest GGM subject, a 54-year-old male, lives an apparently normal, healthy life on a sugar-free diet but malabsorption symptoms also return rapidly if he consumes a modest amount of food containing the offending sugars (Phillips & McGill 1973). The only other symptom that we are aware of is that patients frequently have a mild renal glucosuria. This is consistent with a low level of SGLT1 expression in the proximal tubule.

In our genetic studies we have only screened patients with a clear diagnosis of GGM, i.e. patients with diarrhea who: (1) fully respond to removal of glucose and galactose (and lactose) from their diet; (2) show the diarrhea returning on the introduction of glucose; (3) exhibited flat oral glucose tolerance tests; (4) test positive with glucose hydrogen breath tests ($>100\,$ppm); and (5) have a normal intestinal biopsy. Although GGM is a rare disorder, about 10% of the normal adult population in the USA and the UK (first-year medical students) test positive on glucose hydrogen breath tests (Montes et al 1992, BH Hirst, personal communication). As yet we have not been unable to screen these subjects for SGLT1 mutations.

This rare autosomal recessive disorder is caused by variations in the gene coding for SGLT1 (Turk et al 1991, Martin et al 1996, Wright et al 2001). We are aware of some 300 patients worldwide and have carried out genetic testing on 83 patients in 75 unrelated families. About 65% of our patients are the products of a consanguineous union as revealed by the fact that they have the same mutation on each allele. The remainder are compound heterozygotes, i.e. they have inherited different mutations from their mother and their father. Only one mutation was found in multiple (three), unrelated families, and there are several instances where the same mutation was found in two unrelated families with similar racial and ethnic backgrounds. This may signal an increase in the frequency of these variants in these populations. No mutations have been found in one patient with a clear diagnosis of GGM. It is unclear if mutations lie outside of the DNA sequenced (the coding regions, the exon/intron boundaries and the 527 bp of the promoter) or if mutations in another gene cause the defect.

There are 34 missense mutations (Figure 7.3) and these are highly conserved residues in the 18 closely genes in the SGLT gene family including SGLT1 genes in eight different species. We have tested 24 of these in the *Xenopus laevis* oocyte expression system and all but three showed a serious defect in Na/glucose cotransport ($<10\%$ of wild-type activity). Using a combination of electrophysiological and electron microscopic methods we have established that the defect in transport is due to miss-sorting of the transporter in the cell, i.e. the protein is produced but is not delivered to the plasma membrane. In six patients we have been able to examine the distribution of the transporter in duodenal biopsies and have confirmed that the mutant proteins fail to reach the enterocyte brush border membrane. Five of the six patients with the three missense mutations with unimpaired transport in the oocyte assay have other mutations that cause the transport defect. In addition to the missense mutations, three nonsense, seven frame-shift and seven splice site mutations have been identified. These produce nonfunctional truncated transporters.

Overall, we have identified mutations on both alleles that account for the severe glucose (and galactose) malabsorption in 35 out of 39 patients. This leads to the conclusion that SGLT1 is the primary mechanism for glucose transport across the brush border of enterocytes. Other potential brush border glucose transporters include SGLT4, 5, and 6, but not SGLT2 which is poorly expressed in the human small intestine and not SGLT3 which is a glucosensor and not a transporter (Diez-Sampedro et al 2003).

The frequency of GGM is low due to the low frequency of SGLT1 mutations in the general population. As part of the Pharmacogenetics of Membrane Transporters project at UCSF (www.pharmacogentics.ucsf.edu) we have found that none of the GGM mutations has been found in the SGLT1 gene in the 552 alleles in the Coriell Institute genomic DNA collection. Only 12 nonsynonymous mutations are found, and only six have a frequency greater than 1%. The frequency of three variants reaches 9% (Asn51Ser, Ala411Thre, and His611Gln) in the 160 alleles in the European-American population. The first two of these were found in our pool of GGM patients but on the basis of the oocyte expression assay these were judged to be polymorphisms (Martin et al 1996). GGM patients with these polymorphisms were found to have other SGLT1 mutations that caused the defect in sugar transport.

Patients with GGM frequently have a mild renal glucosuria. For example in one patient with SGLT1 mutations G426R and S159P that abolish Na/glucose transport (kindred #6, Martin et al 1996) careful analysis of renal glucose titration demonstrates a low minimal threshold (F_{minG}) of 82 vs. a normal of 224 mg/min/$1.73\,m^2$ and excretion of more than 1 mg of glucose/min/$1.73\,m^2$ at all filtered loads above F_{minG} (Elsas et al 1970). This suggests that SGLT1 only plays a modest role in the reabsorption of glucose from the glomerular filtrate. Finally, patients with type 0 renal glucosuria have no intestinal defect in sugar absorption suggesting that another SGLT gene is responsible for the bulk of glucose reabsorption in the kidney (see below).

Familial Renal Glucosuria (FRG) (MIM 233100)

FFG is an autosomal recessive renal tubular disorder characterized by urinary glucose excretion in the presence of normal blood glucose levels and normal oral glucose tolerance. This is an isolated renal tubular disorder, where the glomerular filtration rate (GFR) and reabsorption of salt, water, and other nutrients is normal. It is considered as a non-disease as the majority of affected individuals are otherwise asymptomatic. The level of glucose excretion ranges from 1 up to 169 grams/day/$1.73\,m^2$ (Table 7.1). In the most severe cases there is no reabsorption of the filtered load of glucose. Three types of FFG have been defined, type A with a low threshold and low Tm, type B with a low threshold and

TABLE 7.1   Index patients with severe glucosuria and SGLT2 mutation on both chromosomes

Subject	Glucose excretion (g/1.73 m^2/d)	Allele 1	Allele 2
I	*	W440X	W440X
II	12	186X	N654S
III	>30	G304K	G304K
IV	9	T200K	N654S
01-1	126–162	347X	347X
02-1	74	G272R	G272R
03-1	51	IVS7+ 5 g > a	IVS7+ 5 g > a
04-1	21	IVS7+ 5 g > a	L307P
05-1	43	G449D	G449D
06-1	69	W440X	W440X
07-1	21	W487C, Δ448-506	?
09-1	39	R368W	R368W
10-1	2–5	186X	WT
11-1	0.75	IVS7+ 5 g > a	WT
12-1	15	IVS7+ 5 g > a	K311R
13-1	6	F72L	WT
15-1	202	R137H	Δ385-8
15-2	80	R137H	Δ385-8
16-1	2	T51P	WT
17-1	30–92	T543P	T543P
20-1	32	Y150H	R499C
21-1	1	P453L	WT
22-1	2	Δ385-8	WT
23-1	0.75	IVS7+ 5 g > a	WT
V	83	K321R	K321R
VI-I	101	K321R	K321R
VI-2	95	K321R	K321R
VI-3	114	K321R	K321R
VII-1	124	K321R	K321R
VII-2	169	K321R	K321R

* Glucosuria 62 g/l

I, van den Heuvel et al (2002); II, Caldo et al (2004); III, Francis et al (2004); Kleta et al (2004); 01-1 to 23-1, Santer et al (2003b); V–VII, Margen et al (2005)

normal T$_m$, and type 0 with a complete absence of reabsorption (Elsas & Rosenberg 1969, Oemar et al 1987). These may be explained by different mutations in the transporter gene but a thorough genotyping and phenotyping of FRG patients is lacking. What is the effect of renal glycosuria on the subject? In the case of one type 0 patient the long-term follow up indicates that at age 31 he is normal, in good clinical condition, with no nephrologic complications (Scholl-Burgi et al 2004).

Since genetic defects in SGLT1 are not associated with severe renal glucosuria, SGLT2 is expressed in the S1 and S2 segments of rodent tubules, and severe renal glycosuria is considered a nondisease, it has been suggested that mutations in the SGLT2 gene caused the defect in glucose transport across the apical membrane of the early proximal tubule. This together with cloning of the SGLT2 cDNA (Wells et al 1992) and mapping of the SGLT1 gene (Turk et al 1994) facilitated genetic screening of the SGLT2 by several groups (Calado et al 2004, van den Heuvel et al 2002, Santer et al 2003b, Francis et al 2004, Kleta et al 2004, Magen et al 2005). To date the SGLT2 gene has been screened for mutations in 31 families with cases of FRG.

The SGLT2 mutations identified in FRG index cases in 26 families are summarized in Table 7.1 along with the glucose excretion rates. There are 12 FRG cases with homozygous mutations and seven who are compound heterozygotes, where the glucosuria is greater then 9 grams/day/1.73 m^2. No mutations were found in three index cases where the glycosuria ranged from 3 to 17 grams/day/1.73 m^2. Mutations on only one allele were identified in eight cases with glucosuria ranging from 0.75 to 6 grams/day/1.73 m^2. The mutations include 17 missense, three nonsense, three frame-shift, one abnormal splice site variant, and four deletions distributed throughout the gene (Figure 7.4). So far, virtually every FRG family has its own private mutation as few mutations occur in unrelated families. Exceptions include the missense mutation W440X in two unrelated families and the missense mutation K321 in three unrelated Israeli-Arab families.

None of the SGLT2 mutations has been evaluated in a heterologous expression system and so it is difficult to relate the missense mutations to the severity of the renal glucosuria. However, it does appear that homozygous K321R mutations produced type 0 glucosuria in six patients (V–VII, Table 7.1). Excretion rates substantially less than the filtered loads do occur in homozygous and compound heterozygote subjects, e.g. G304K in subject III, G272R in 02-1, G449D in 05-1, R368W in 09-01, and T543P in 17-1. This could be due to retention of partial function by these mutant proteins. More perplexing are the excretion rates seen with homozygous stop codons. On one hand 347X is associated with the type 0 renal glucosuria (subject 01-1) but on the other hand much lower excretion rates are observed for two unrelated subjects with W440X (I and 06-1). In the absence of technical difficulties, this suggests that there are other transporters responsible for glucose reabsorption in the proximal tubule. Possibilities include SGLT4 and SGLT6 which appear to be low-affinity glucose transporters and SGLT5, that all expressed in the human renal cortex (Wright & Turk 2004). Yet another possibility is that a Na/glucose cotransporter in another gene family is involved, e.g. NaGLT1 (Horiba et al 2003).

Fanconi-Bickel Syndrome (FBS) (MIM 227810)

FBS is a rare autosomal recessive disorder characterized by failure to thrive, short stature, hepatomegaly, hepatorenal glycogen accumulation, and, most prominently, glucosuria ranging from 40 to 200 grams/day/1.73 m^2 (Fanconi & Bickel 1949, Manz et al 1987, Santer et al 1998). There may also be hyperaminoaciduria, moderate hyperphosphaturia,

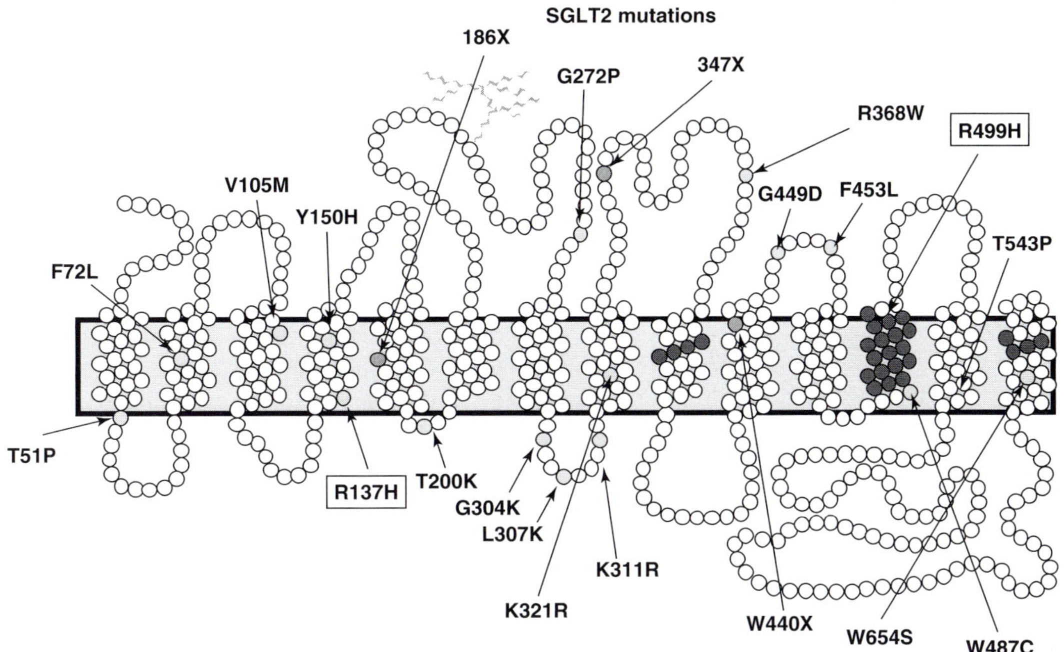

FIGURE 7.4 A secondary structure model of SGLT2 showing the location of the missense, nonsense, and deletion mutations identified in patients with familial renal glucosuria (FRG). The 672 membrane protein is assumed to have the same secondary structure as SGLT1 (Figure 7.3). It should be noted that there are no functional data available for the SGLT2 mutants (see Table 7.1) as it has been difficult to express in heterologous expression systems. The boxed mutation R137W is the only residue that is found mutated in GGM patients (R135H).. (see also Plate 7)

hypercalciuria, and fasting hypoglycemia. Oral glucose tolerance tests appear normal and there may be mild glucose malabsorption.

In their 1998 review of 82 cases from 70 families Santer et al (1998) highlight that a massive renal glucosuria, typically 40–200 grams/day/1.73 m^2, is the most prominent observation. Glucose excretion ranges from a low of 12 to a high of 323 grams/day/1.73 m^2, but two 'FBS' patients have no glucose in their urine. FBS patients have a low fasting blood glucose level and normal to low-normal GFR. Given the symptoms of FBS, Santer and his colleagues reasoned that GLUT2 gene was a prime candidate for molecular genetic studies. In their 2002 review (Santer et al 2002) 34 novel mutations in GLUT2 were identified in 55 FBS patients. Homozygous mutations were found in 47 cases and this is consistent with the known consanguinity in the FBS families.

Of the 34 mutations identified, 24 of these result in truncated GLUT2 proteins (seven nonsense, 10 frame-shift, and seven splice-site mutations) and, as might be expected, FBS patients with homozygous truncated mutations have a massive glucosuria (74X, 106X, 178X and 301X). Surprisingly, one sibling with the 74X mutation and another patient with the 365X mutation have a less severe renal glycosuria (80 and 12–43 grams/day/1.73 m^2). This raises questions about the presence of other glucose transporters in the basolateral membrane of these patients. There are eight missense mutations of highly conserved residues in the GLUT transporter

family. One is found on both alleles of five patients and for two, S326K and P417L, a massive glucosuria is reported. Functional assays on the eight mutant GLUT2 proteins have yet to be carried out and so it is premature to carry out a genotype–phenotype correlation. This and the fact that the GLUT2 mutations occur throughout the coding region of the gene makes it impractical to carry out a molecular diagnosis of FBS.

As in the case of GGM and FRG there are a number of FBS patients where no mutations have been identified in the coding regions, intron–exon boundaries, or in a limited stretch of the GLUT2 promoter. This and the imperfect genotype–phenotype correlation noted above suggest that other mechanisms of glucose transport across the basolateral membrane have to be considered. One potential transporter is GLUT1 as this is expressed in the basolateral membranes of the kidney and another is the exocytotic pathway proposed by Stumpel et al (2001) for the small intestine. It is unlikely to be GLUT1 as subjects with mutations in this gene (MIM 606777) do not appear to have renal glucosuria.

An interesting picture has emerged from studies of patients with the Fanconi-Bickel syndrome (FBS) and from SGLT2 gene knockout studies on rodents about differences between glucose transport in the intestine and the proximal tubule. It appears that GLUT2 is not essential for intestinal glucose absorption whereas it is much more important in the kidney. FBS patients may have a normal tolerance to glucose, i.e. normal oral glucose tolerance tests and no severe osmotic carbohydrate diarrhea, whereas these patients have renal glucosuria. Likewise mice lacking GLUT2 have normal oral glucose tolerance and severe renal glucosuria (Thorens et al 2000). In the intestine there is evidence for a transport pathway requiring glucose phosphorylation and transfer into the endoplasmic reticulum (Stumpel et al 2001, Santer et al 2003a). Essentially, glucose absorption may be fairly normal in FBS patients while absorption of the nonmetabolized sugar 3-O-methyl-glucoside is severely impaired. However, these same patients and GLUT2 knockout mice have severe renal glucosuria. Evidently, SGLT2 plays a much more important role in glucose transport across the proximal tubule than in the small intestine. The role of the exocytotic pathway could be tested in FBS subjects by comparing the clearances of glucose and a nonphosphorylated sugar that is a substrate for both SGLTs and GLUTs, 3-O-methyl-glucoside.

SUMMARY AND OUTLOOK

Studies of the rare genetic disorders of renal glucose transport have provided important insights into normal human renal physiology. It seems reasonable to conclude that SGLT2 and GLUT2 each play central roles in glucose transport across the apical and basolateral membranes of the proximal tubule as visualized in Figure 7.2. However, the less than perfect correlation between the genotyping and phenotyping of SGLT2 and GLUT2 in subjects with FRG and FBS suggest the presence of other glucose transporters. The clearest examples are those where subjects reabsorb substantial fractions of their filtered glucose loads despite having severely truncated SGLT2 or GLUT2 proteins. While it is unlikely that SGLT1 and SGLT3 fill the gap at the apical membrane, other potential candidates include SGLT4 and SGLT6, transporters that behave as low-affinity glucose transporters. SGLT5, which is expressed almost exclusively in the renal cortex, could be important but its function is not known. Potential candidates for glucose transport across the basolateral membrane include other members of the SLCA2 gene family. However, it is unlikely that GLUT1 is important as mutations in this gene (Brockmann et al 2001, MIM 606777) are not associated with renal glucosuria.

A number of technical advances could further our understanding of glucose transport in the human kidney: first, it is important to find methods to express SGLT2 and SGLT5 for functional studies; second, specific antibodies for SGLT1–6 are needed to determine the subcellular location of these proteins in the nephron; and third, noninvasive methods to visualize and quantitate the functional expression of SGLT and GLUT genes in the kidney would be an important step forward. One approach to the latter is to design specific positron emission tomography (PET) tracers for SGLTs and GLUTs: 2-^{18}F-2-deoxy-glucose (FDG) is already in use as a general tracer for GLUT activity (Phelps 2004); and there has been some movement towards the development of 11C and 18F tracers for SGLTs (Borman et al 2003, de Groot et al 2003).

Finally, a comment is warranted about the employment of molecular diagnostics for genetic disorders of renal glucose transport. This is impractical due to the very low frequency of FRG, FBS, and GGM, the fact that FRG is a nondisease, and the autosomal recessive character of all three disorders. Virtually every kindred harboring a case of FRG, FBS, or GGM has a private mutation. Furthermore, given the relatively poor correlation between genotype and phenotype for FBS and FRG it would be both costly and of limited value to carry out genetic screening, especially since the diagnosis of each disorder is fairly simple and straightforward. One exception may be the value of prenatal screening for FBS and GGM in families at risk. Quite recently 21 additional cases of familial renal glucosuria have been reported (Calado et al 2008.)

ACKNOWLEDGMENTS

The studies underlying this review were carried out by a very talented group of students, post-doctoral fellows, and collaborators who are the co-authors on many of the cited papers, and were made possible by grants from the National Institutes of Health (DK 19567, DK 44602, DK 44582 and GM 19567).

References

Abramson J, Smirnova I, Kasho V, Verner G, Kaback HR, Iwata S. Structure and mechanism of the lactose permease of *Escherichia coli*. Science 2003; 301: 610–61.

Barfuss DW, Schafer JA. Differences in active and passive glucose transport along the proximal nephron. Am. J. Physiol. 1981; 241: F322–32.

Bormans GM, Van Oosterwyck G, De Groot TJ, Veyhl M, Mortelmans L, Koepsell H. Synthesis and biologic evaluation of (11)C-methyl-d-glucoside, a tracer of the sodium-dependent glucose transporters. J. Nucl. Med. 2003; 44(7): 1075–81.

Brockmann K, Wang D, Korenke CG, et al. Autosomal dominant glut-1 deficiency syndrome and familial epilepsy. Ann. Neuro. 2001; 50: 476–85.

Calado J, Soto K, Clemente C. A novel compound heterozygous mutations in SLC5A2 are responsible for autosomal recessive renal glucosuria. Hum. Genet. 2004; 114: 314–6.

Calado J, Sznajer Y, Metzger D, Rita A, Hogan MC, Kattamis A, Scharf M, Tasic V, Greil J, Brinkert F, Kemper MJ, Santer R. Twenty-one additional cases of familial renal glucosuria: absence of genetic heterogeneity, high prevalence of private mutuations and further evidence of volume depletion. Nephrol. Dial. Transplant. 2008. Jul 12. [Epub ahead of print].

de Groot TJ, Veyhl M, Terwinghe C, et al. Synthesis of 18F-fluoroalkyl-beta-D-glucosides and their evaluation as tracers for sodium-dependent glucose transporters. J. Nucl. Med. 2003; 44(12): 1973–81.

Diez-Sampedro A, Hirayama BA, Oswald C, et al. A glucose sensor hiding in a family of transporters. PNAS 2003; 100: 11753–8.

Elsas LJ, Hillman RE, Patterson JH, Rosenberg LE. Renal and intestinal hexose transport in familial glucose-galactose malabsorption. J. Clin. Invest. 1970; 49: 576–85.

Elsas LJ, Rosenberg LE. Familial renal glycosuria: a genetic reappraisal of hexose transport by kidney and intestine. J. Clin. Invest. 1969; 48: 1845–54.

Faham S, Watanabe A, Mercado Besserer G, Cascio D, Specht A, Hirayama BA, Wright EM, Abramson, J. 2008; Science 321; 810–14.

Fanconi G, Bickel H. Die chronische aminoacidurie (aminosaeurediabetes oder nephrotisch-glukosurischer zwergwuchs) bei der glykogenose und der cystinkrankheit. Helv. Paediat. Acta. 1949; 4: 359–96.

Francis J, Zhang J, Farhi A, Carey H, Geller DS. A novel SGLT2 mutation in a patient with autosomal recessive renal glucosuria. Nephrol. Dial. Transplant. 2004; 19: 2893–5.

Frohnert PP, Hohmann B, Zwiebel R, Baumann K. Free flow micropuncture studies of glucose transport in the rat nephron. Pflugers Arch. 1970; 315(1): 66–85.

Hediger MA, Cody MJ, Ikeda TS, Wright EM. Expression cloning and cDNA sequencing of the Na^+/glucose cotransporter. Nature 1987; 330: 379–81.

Horiba N, Masuda S, Ohnishi C, Takeuchi D, Okuda M, Inui K. Na^+-dependent fructose transport via rNaGLT1 in rat kidney. FEBS Lett. 2003; 546: 276–80.

Huang Y, Lemieux MJ, Song J, Auer M, Wang DN. Structure and mechanism of the glycerol-3-phosphate transporter from *Escherichia coli*. Science 2003; 301: 616–20.

Keller DM. Glucose excretion in man and dog. Nephron 1968; 5(1): 43–66.

Kinne R, Murer H, Kinne-Saffran E, Thees M, Sachs G. Sugar transport by renal plasma membrane vesicles. Characterization of the systems in the brush-border microvilli and basal-lateral plasma membranes. J. Membr. Biol. 1975; 21: 375–95.

Kleta R, Stuart C, Gill FA, Gahl WA. Renal glucosuria due to SGLT2 mutations. Mol. Genet. Metab. 2004; 82: 56–8.

Maddox DA, Gennari FJ. The early proximal tubule: a high-capacity delivery-responsive reabsorptive site. Am. J. Physiol. 1987; 252: F573–84.

Magen D, Sprecher E, Zelikovic I, Skorecki K. A novel missense mutation in SLC5A2 encoding SGLT2 underlies autosomal-recessive renal glucosuria and aminoaciduria. Kidney Int. 2005; 67: 34–41.

Manz F, Bickel H, Brodehl J, et al. Fanconi-Bickel syndrome. Pediatr. Nephrol. 1987; 1: 509–18.

Martín GM, Turk E, Lostao MP, Kerner C, Wright EM. Defects in Na^+/glucose cotransporter (SGLT1) trafficking and function cause glucose-galactose malabsorption. Nat. Genet. 1996; 12: 216–20.

Montes RG, Gottal RF, Bayless TM, Hendrix TR, Perman JA. Breath hydrogen testing as a physiology laboratory exercise for medical students. Am. J. Physiol. 1992; 262: S25–8.

Mueckler M, Caruso C, Baldwin SA, et al. Sequence and structure of a human glucose transporter. Science 1985; 229(4717): 941–5.

Oemar BS, Byrd J, Brodehl J. Complete absence of tubular glucose reabsorption: a new type of renal glucosuria (type 0). Clin. Nephrol. 1987; 27(3): 156–60.

Pajor AM, Randolph KM, Kernere SA, Smith CD. Inhibitor binding in the human renal low- and high-affinity Na/glucose cotransporters. J. Pharmacology and Experimental Therapeutics 2008; 324: 985–91.

Phelps ME PET Molecular Imaging and its Biological Applications. New York: Springer, 2004.

Phillips S, McGill D. Glucose-galactose malabsorption in an adult: perfusion studies of sugar, electrolyte, and water transport. J. Digest. Dis. 1973; 18(12): 1017–23.

Santer R, Schneppenheim R, Suter D, Schaub J, Steinmann B. Fanconi-Bickel syndrome – the original patient and his natural history, historical steps leading to the primary defect, and a review of the literature. Eur. J. Pediatr. 1998; 157: 783–97.

Santer R, Groth S, Kinner M. The mutation spectrum of the facilitative glucose transporter gene SLC2A2 (GLUT2) in patients with Fanconi-Bickel syndrome. Hum. Genet. 2002; 110: 21–9.

Santer R, Hillebrand G, Steinmann B, Schaub J. Intestinal glucose transport: evidence for a membrane traffic-based pathway in humans. Gastroenterology 2003a; 124: 34–9.

Santer R, Kinner M, Lassen CL, et al. Molecular analysis of the SGLT2 gene in patients with renal glucosuria. J. Am. Soc. Nephrol. 2003b; 14: 2873–82.

Schneider AJ, Kinter WB, Stirling CE. Glucose-galactose-malabsorption: report of a case with autoradiographic studies of a mucosal biopsy. N. Engl. J. Med. 1966; 274: 305–12.

Scholl-Burgi S, Santer R, Ehrich JHH. Long-term outcome of renal glucosuria type 0: the original patient and his natural history. Nephrol. Dial. Transplant. 2004; 19: 2394–6.

Stumpel F, Burcelin R, Jungermann K, Thorens B. Normal kinetics of intestinal glucose absorption in the absence of GLUT2: Evidence for a transport pathway requiring glucose phosphorylation and transfer into the endoplasmic reticulum. PNAS 2001; 98: 11330–5.

Tazawa S, Yamato T, Fujikura H, et al. SLC5A9/SGLT4, a new Na^+-dependent glucose transporter, is an essential transporter for mannose, 1,5-anhydro-D-glucitol, and fructose. Life Science 2005; 76: 1039–50.

Thorens B, Gulliam M-T, Beermann F, Burcelin R. Transgenic expression of GLUT1 or GLUT2 in pancreatic ß cells rescues GLUT2-null mice from early death and restores normal glucose-stimulated insulin secretion. J. Biol. Chem. 2000; 275: 23751–8.

Turk E, Martín MG, Wright EM. Structure of the human Na^+/glucose cotransporter gene SGLTI. J. Biol. Chem. 1994; 269: 15204–9.

Turk E, Zabel B, Mundlos S, Dyer J, Wright EM. Glucose-galactose-malabsorption caused by a defect in the Na^+/glucose cotransporter. Nature 1991; 350: 354–6.

Turner RJ, Moran A. Heterogeneity of sodium dependent D-glucose transport sites along the proximal tubule: evidence from vesicle studies. Am. J. Physiol. Renal Electrolyte Physiol. 1982; 242: F406–11.

Uldry M, Thorens B. The SLC2 family or facilitated hexose and polyol transporters. Pflugers Arch. 2004; 447: 480–9.

Van den Heuvel LP, Assink K, Willemsen M, Monnens L. Autosomal recessive renal glucosuria attributable to a mutation in the sodium glucose cotransporter (SGLT2). Hum. Genet. 2002; 111: 544–7.

Wells RG, Kanai Y, Pajor AM, Turk E, Wright EM, Hediger MA. The cloning of a human kidney cDNA with similarity to the sodium/glucose cotransporter. Am. J. Physiol. Renal Electrolyte Physiol. 1992; 263: F459–65.

Wright EM, Martín GM, Turk E. Familial glucose-galactose malabsorption and hereditary renal glycosuria. In: Scriver CR, Beaudet AL, Sly WS, Valle D, eds. Metabolic Basis of Inherited Disease, 8th Edition, Volume III. 2001; 190: 4891–8.

Wright EM, Loo DDF, Hirayama BA, Turk E. Surprising versatility of Na^+/glucose cotransporters (SLC5). Physiology 2004; 9: 370–6.

Wright EM, Turk E. The sodium glucose cotransport family SLC5. Pflugers Arch. 2004; 447: 510–18.

Yasuda NK, Adachi T, Okamoto Y, et al. Beneficial effect of T-1095, a selective inhibitor of renal Na^+-glucose cotransporters, on metabolic index and insulin secretion in spontaneously diabetic GK rats. Clin. Exp. Pharmacol. Physiol. 2002; 29: 386–90.

CHAPTER 8

Primary Inherited Aminoacidurias: Genetic Defects in the Renal Handling of Amino Acids

MANUEL PALACÍN

PRIMARY INHERITED AMINOACIDURIAS

Amino acids are filtered by the glomerulus, and more than 98% are subsequently reabsorbed by the proximal tubule (mainly in the convoluted part; S1–S2 segments) (Silbernagl 1988). Primary inherited aminoacidurias (PIA) are caused by defective amino acid transport activities, which affect renal reabsorption of these compounds and may also affect their transport during intestinal absorption and in other organs as well. Several PIA have been described (Table 8.1). Inherited disorders of renal tubule like the renal Fanconi syndrome (MIM: 134600), which is a generalized dysfunction of the proximal tubule that results in wasting of phosphate, glucose, amino acid and bicarbonate, or cystinosis (MIM: 219800; 219900), affecting lysosomal efflux of cystine, are not discussed in this chapter. Neither are the inherited defects of amino acid metabolism resulting in aminoaciduria (e.g. homocystinuira, MIM 236200) described in this chapter.

Plasma membrane transport of dibasic amino acids (i.e. basic amino acids) is abnormal in four inherited diseases:

1. Cystinuria (MIM 220100; 600918), in which patients present hyperexcretion of cystine and dibasic amino acids (first described by Sir Archibald Garrod in 1908

TABLE 8.1   Primary inherited aminoacidurias

	Prevalence	Inheritance	Gene	Chromosome	Mutations	Transport system
Cystinuria*	1:7000	AR /ADIP	*SLC3A1*	2p16.3	112	
			SLC7A9	19q13.1	73	$b^{0,+}$
Isolated cystinuria	very rare	AR?	?	?	?	?
LPI	~200 cases	AR	*SLC7A7*	14q11	26	y^+L
Hyperdibasic aminoaciduria type I	very rare	AD	?	?	?	?
Isolated lysinuria	very rare	AR?	?	?	?	?
Hartnup disorder	1:26000	AR	*SLC6A19*	5p15	10	B^0
Renal familial iminoglycinuria	1:15000	AR	?	?	?	Imino(?) $
Dicarboxylic aminoaciduria	very rare	AR?	*SLC1A1* (?)	9p24	KO null #	$X_{AG}^{\,-}$

AR, autosomal recessive; ADIP, autosomal dominant with incomplete penetrance; AD, autosomal dominant; AR?, familial studies in the very few cases described for these diseases suggest an autosomal recessive mode of inheritance

*Three phenotypes of cystinuria, depending on the obligate heterozygotes, are considered: type I (with AR inheritance), type non-I (ADIP inheritance) and mixed type (combination of both). $, the amino acids hyperexcreted in patients with renal familial iminoglycinuria (glycine and proline) suggest defects in imino system. #, *Slc1a1*-null knockout mice present dicarboxylic aminoaciduria (Peghini et al 1997), pointing to this gene as a candidate for the human disease

Genetic Diseases of the Kidney
ISBN: 978-0-12-449851-8

(Garrod 1908)). There is phenotypic variability in obligate heterozygotes (i.e. silent or hyperexcretors of amino acids) (Palacín et al 2001).

2. Lysinuric protein intolerance (LPI) (also named hyperdibasic aminoaciduria type II, or familial protein intolerance; MIM 222700) (first described in Finland (Perheentupa & Visakorpi 1965)).

3. Autosomal dominant hyperdibasic aminoaciduria type I (MIM 222690) (Whelan & Scriver 1968).

4. Isolated lysinuria described in one Japanese patient (Omura et al 1976).

Cystinuria and LPI are caused by defective amino acid transporter systems $b^{0,+}$ and $y^{+}L$ respectively. These two transporters belong to the family of heteromeric amino acid transporters (HAT) (Palacín et al 1998, Chillarón et al 2001). Mutations in the two subunits of system $b^{0,+}$ (rBAT and $b^{0,+}$AT) causes cystinuria (Calonge et al 1994, Feliudabaló et al 1999) while mutations in one of the two subunits of system $y^{+}L$ (y^{+}LAT1), but not in the other subunit (4F2hc), produce LPI (Borsani et al 1999, Torrents et al 1999). At the molecular level, the relationship between LPI, and the very rare autosomal dominant hyperdibasic aminoaciduria type I, and isolated lysinuria is unknown.

Plasma membrane transport of zwitterionic amino acids (i.e. neutral amino acids at physiological pH) is defective in three inherited diseases:

1. Hartnup disorder (MIM 234500), in which patients present hyperexcretion of neutral amino acids (first described in two siblings of the Hartnup family (Baron et al 1956)).

2. Renal familial iminoglycinuria (MIM 242600) is an autosomal recessive benign disorder in which individuals present hyperexcretion of proline and glycine (first described in the sixties (Scriver et al 1964, Rosenberg et al 1968)). There is phenotypic complexity in this disorder (Scriver 1968, Chesney 2001): (i) renal iminoglycinuria with defective intestinal absorption and normal heterozygotes; (ii) renal iminoglycinuria without intestinal phenotype and normal heterozygotes; and (iii) renal iminoglycinuria without intestinal phenotype and isolated glycinuria in heterozygotes.

3. Isolated cystinuria (MIM 238200), in which patients present hyperexcretion of cystine but not dibasic amino acids (Brodehl et al 1967).

Hartnup disorder is due to a defective amino acid transport system B^{0} (also named neutral brush border) caused by mutations in B^{0}AT1 (*SLC6A19*) (Kleta et al 2004, Seow et al 2004). The relationship between isolated cystinuria and cystinuria at the molecular level is unknown. The molecular basis of iminoglycinuria is unknown but candidate genes are (Broër et al 2006): *SLC36A1* (coding for transporter PAT1) (Anderson et al 2004), *SLC6A20* (coding

for transporter SIT) (also called IMINO) (Kowalczuk et al 2005, Takanaga et al 2005) and *SLC6A18* (coding for the orphan transporter XT2).

Plasma membrane transport of dicarboxylic amino acids is defective in dicarboxylic aminoaciduria (MIM 222730) (Teijema et al 1974, Melançon et al 1977). The molecular basis of this disease is unknown but the glutamate transporter EAAC1 (*SLC1A1*; Kanai et al 1994) is an obvious candidate since the murine knockout of *Slc1a1* presents dicarboxylic aminoaciduria (Peghini et al 1997).

DEFECTS ASSOCIATED WITH HETEROMERIC AMINO ACID TRANSPORTERS

Heteromeric amino acid transporters (HATs) are composed of a heavy subunit and a light subunit (Table 8.2) (Chillarón et al 2001, Palacín & Kanai 2004, Verrey et al 2004). These are unique features among mammalian plasma membrane amino acid transporters. Two homologous heavy subunits from the SLC3 family have been cloned, rBAT (i.e. related to $b^{0,+}$ amino acid transport) and 4F2hc (i.e. heavy chain of the surface antigen 4F2hc, also named CD98 or fusion regulatory protein 1 (FRP1) (Ohgimoto et al 1995)). Ten light subunits (SLC7 family members from *SLC7A5* to *SLC7A14*) have been identified. Six of these are partners of 4F2hc (LAT1, LAT2, y^{+}LAT1, y^{+}LAT2, asc1, and xCT); one forms a heterodimer with rBAT ($b^{0,+}$AT); two (asc2 and AGT-1) appear to interact with as-yet-unknown heavy subunits (Chairoungdua et al 2001, Matsuo et al 2002); and the last one (arpAT) may interact with rBAT, 4F2hc or an unidentified heavy subunit (Fernández et al 2005). Two light subunits are not present in humans: asc2 is not found in the genome sequence and arpAT is heavily inactivated in this genome (Fernández et al 2005). Members *SLC7A1–4* of family SLC7 correspond to system y^{+} isoforms (i.e. cationic amino acid transporters; CATs) and related proteins, which on average show $<25\%$ amino acid identity with the light subunits of HATs.

The general features of HATs are as follows (Palacín & Kanai 2004, Verrey et al 2004):

- The heavy subunits (molecular mass of $\sim$90 and $\sim$80 kDa for rBAT and 4F2hc, respectively) are type II membrane N-glycoproteins with a single transmembrane domain, an intracellular N-terminus, and an extracellular C-terminus significantly homologous to insect and bacterial glucosidases (Figure 8.1). X-ray diffraction of the extracellular domain of human 4F2hc revealed a three-dimensional structure similar to that of bacterial glucosidases (a triose phosphate isomerase (TIM) barrel $[(\alpha\beta)8]$ and eight antiparallel β-strands; Figure 8.1) (unpublished results).

- The light subunits ($\sim$50 kDa) are highly hydrophobic and not glycosylated. This results in anomalously high

TABLE 8.2   Cystinuria type and genetic frequencies of the probands from the International Cystinuria Consortium

			Cystinuria phenotype				
Genotype	I	Non-I	Non-I carriers*	Mixed	Untyped	Total probands (%)	
AA	29			2	25	56 (34.1)	
AA(B)				1		1 (0.6)	
BB	1	34`		7	23	65 (39.6)	
B+			3			3 (1.8)	
BB(A)				1		1 (0.6)	126 (76.8)
A?	5			1	5	11 (6.7)	
B?	2	7		2	11	22 (13.3)	33 (20.1)
??		2			3	5 (3.0)	5 (3.0)
Total probands (%)	37 (22.6)	43 (26.2)	3 (1.8)	14 (8.5)	67 (40.9)	164 (100)	
Total alleles	74	89	3	28	134		
Explained alleles (%)	67 (90.5)	78 (87.6)	3 (100)	25 (89.3)	112 (83.6)		

A total of 164 probands have been studied. In 126 probands (76.8%) the alleles causing the disease have been identified. In 33 probands (20.1%) one of the two mutated alleles has been identified. In 5 probands (3.0%) no mutated alleles have been identified

* Heterozygote probands with cystine lithiasis. For patients AA(B) and BB(A), two alleles causing the disease and two explained alleles in each case have been taken into account in the calculations. A, allele *SLC3A1* mutated; B, allele *SLC7A9* mutated; +, normal allele; ?, unknown allele. Extracted from Font-Llitjós et al 2005

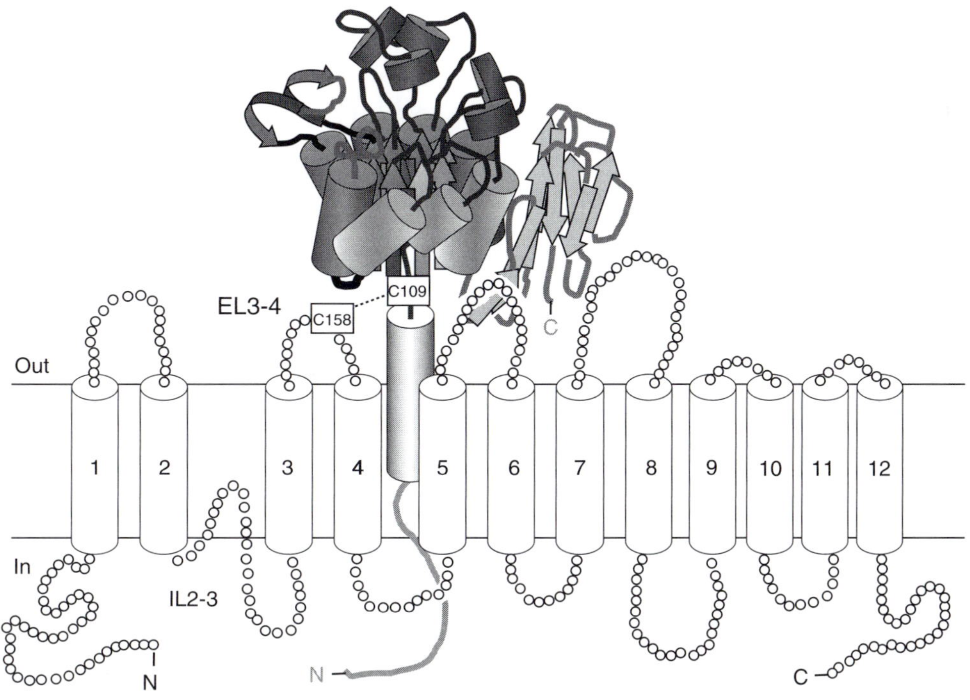

FIGURE 8.1   Schematic representation of a heteromeric amino acid transporter. The heavy subunit (gray) and the light subunit (white) are linked by a disulfide bridge with conserved cysteine residues (cysteine 158 for the human xCT and cysteine 109 for human 4F2hc). The heavy subunits (4F2hc or rBAT) are type II membrane glycoproteins with an intracellular NH_2 terminus, a single transmembrane domain, and a bulky COOH terminal domain (~50kDa without glycosylation; i.e. similar to the size of the light subunits). This part of the protein shows homology with bacterial glycosidases (a schematic representation of the TIM barrel and the all-β domain is shown). The light subunits are polytopic proteins with 12 transmembrane domains, with the NH_2 and COOH terminals located intracellularly and with a reentrant loop-like structure in the intracellular loop IL2-3 (on the basis of studies on xCT; Gasol et al 1998)

mobility in SDS-PAGE (35–40 kDa). Cysteine-scanning mutagenesis studies of xCT, as a model for the light subunits of HATs, support a 12-transmembrane-domain topology, with the N- and C-terminals located inside the cell and with a reentrant-like structure in the intracellular loop IL2–3 (Figure 8.1) (Gasol et al 2004).

- The light and the corresponding heavy subunit are linked by a disulfide bridge (Figure 8.1). For this reason, HATs are also named glycoprotein-associated amino acid transporters. The intervening cysteine residues are located in the putative extracellular loop EL3–4 of the light subunit and a few residues away from the transmembrane domain of the heavy subunit (Figure 8.1).
- The light subunit cannot reach the plasma membrane unless it interacts with the heavy subunit.
- The light subunit confers specific amino acid transport activity to the heteromeric complex (LAT1 and LAT2 for system L isoforms, y^+LAT1 and y^+LAT2 for system y^+L isoforms, asc1 and asc2 for system asc isoforms, xCT for system x_c^-, $b^{0,+}$AT for system $b^{0,+}$, AGT-1 for a system serving aspartate and glutamate transport, and arpAT for a transport system with aromatic amino acids as preferred substrates). Moreover, reconstitution in liposomes showed that the light subunit $b^{0,+}$AT is fully functional in the absence of the heavy subunit rBAT (Reig et al 2002).
- The light subunit $b^{0,+}$AT stabilizes the heavy subunit rBAT. No data are available as to whether this also holds for 4F2hc and associated light subunits.
- HATs are, with the exception of system asc isoforms, tightly coupled amino acid antiporters (Pineda et al 2004).

The transport characteristics of two HAT-associated transport systems are relevant to two PIA: cystinuria and LPI (Table 8.1): system $b^{0,+}$ [due to the rBAT (*SLC3A1*) and $b^{0,+}$AT (*SLC7A9*) heterodimer] is a tertiary active mechanism of renal reabsorption and intestinal absorption of diabasic amino acid and cystine in the apical plasma membrane. It mediates the electrogenic exchange of dibasic amino acids (influx) for neutral amino acids (efflux) (see Palacín et al 1998 for a review) (see Figure 8.3). System y^+L (4F2hc/y^+LAT1 heterodimer) mediates the electroneutral exchange of dibasic amino acids (efflux) for neutral amino acids plus sodium (influx) (Chillarón et al 1996, Broer et al 2000, Kanai et al 2000). This transport system allows the efflux of dibasic amino acids against the membrane potential at the basolateral domain of epithelial cells (see Figure 8.3). The 4F2hc/y^+LAT2 heterodimer also mediates system y^+L in many other cell types (see Verrey et al 2004 for a review).

Cystinuria

Cystinuria is an autosomal inherited disorder, characterized by impaired transport of cystine and dibasic amino acids in the proximal renal tubule and the gastrointestinal tract (see Palacín

et al 2001 for a review). The overall prevalence of the disease is 1 in 7000 neonates, ranging from 1 in 2500 neonates in Libyan Jews to 1 in 100 000 among Swedes. Patients present normal to low-normal levels in blood (see Table 8.4), hyperexcretion in urine, and intestinal malabsorption of these amino acids. High cystine concentration in the urinary tract most often causes the formation of recurring cystine stones (i.e. urolithiasis) as a result of the low solubility of this amino acid. This is the only symptom associated with the disease. Therefore, treatment attempts to increase cystine solubility in urine (increased hydration, urine alkalinization, and formation of soluble cystine adducts with thiol drugs). Cystinuria is not accompanied by malnutrition, suggesting that intestinal malabsorption is not severe. The transport defect occurs in the apical plasma membrane of renal and intestinal epithelial cells (Coicadan et al 1980). Absorption of di- and tripeptides via PEPT1 (*SLC15A1*) may prevent malnutrition in cystinuria (Daniel 2004) (Figure 8.3).

Traditionally, three types of cystinuria have been recognized in humans: type I, type II, and type III (Segal & Thier 1995). This classification correlates poorly with molecular findings, and it has recently been revised to type I (MIM 220100) and non-type I (MIM 600918) cystinuria (with the latter corresponding to old types II and III). These two are distinguished on the basis of the cystine and dibasic aminoaciduria of the obligate heterozygotes (Palacín et al 2001): type I heterozygotes are silent, whereas non-type I heterozygotes present a variable degree of urinary hyperexcretion of cystine and dibasic amino acids that is higher in type II than in type III. This indicates that type I cystinuria is transmitted as an autosomal recessive trait, whereas non-type I is transmitted dominantly, with incomplete penetrance. Not surprisingly, urolithiasis has been reported in a minority of non-type I heterozygotes. Thus, in the cohort of patients of the International Cystinuria Consortium (ICC) three type non-I heterozygotes with cystine urolithiasis have been identified from 164 cystinuria probands (Font-Llitjós et al 2005) (Table 8.2). Patients with a mixed type, inheriting type I and non-type I alleles from either parent, have also been described (Goodyer et al 1993). Data on the relative proportion of the two types in specific populations are scarce. In 97 well-characterized families of the ICC cohort of patients, mainly from Italy, Spain, and Israel, 38, 47, and 14% transmitted type I, non-type I, and mixed cystinuria, respectively (Table 8.2) (Font-Llitjós et al 2005). This cohort is not a registry, and therefore it may not be representative of the whole population within these countries (Dello Strologo et al 2002).

The characteristics of rBAT (expression in the plasma membrane of the epithelial cells of kidney proximal tubule and small intestine, and induction of system $b^{0,+}$ in oocytes (for a review see Palacín & Kanai 2004) pointed to this gene as a candidate for cystinuria. In 1994 it was demonstrated that mutations in *SLC3A1* cause type I cystinuria (Calonge et al 1994). Since then 112 distinct rBAT mutations have been described, including nonsense, missense, splice-site, and frame-shift mutations, as well as large deletions and chromosome rearrangements (mutations are listed in Font-Llitjós et al 2005; for more recently described mutations: Skopkova et al 2005, Jaeneken et al 2006, Yuen et al 2006). Cystinuria resembling type I caused by mutations in canine *Slc3a1* has been reported in Newfoundland dogs (Henthorn et al 2000). Similarly, *Pebbels* mice (homozygous for the rBAT mutation D140G) develop type I cystinuria with urolithiasis (Peters et al 2003).

Most of the cystinuria-specific rBAT missense mutations occur in the ectodomain (Font-Llitjós et al 2005). Unfortunately, amino acid sequence identity of the ectodomains of rBAT and 4F2hc is very low. This precludes the generation of a structural model of the ectodomain of rBAT using the crystal structure of that of 4F2hc. Therefore, at present, functional analysis is the easiest study available for missense rBAT mutations. The most common *SLC3A1* mutation, M467T, showed a trafficking defect, with the protein reaching the plasma membrane inefficiently (Chillarón et al 1997). A trafficking defect, also suggested for other *SLC3A1* mutations (Chillarón et al 1997, Saadi et al 1998, Pineda et al 2004), is consistent with the proposed role of rBAT as an ancillary subunit of $b^{0,+}$AT. *SLC3A1* mutations may also affect transport properties of the holotransporter $b^{0,+}$: the cystinuria-specific mutation R365W, in addition to temperature-sensitive protein stability and trafficking defect, shows a defect in the efflux of arginine but not in its influx (Pineda et al 2004). This observation indicates two pathways for the transport unit of system $b^{0,+}$, one for influx and the other for efflux. This scenario is consistent with two additional sets of results: (1) the intestinal system $b^{0,+}$ of the chicken has a sequential mechanism of exchange, compatible with the formation of a ternary complex (i.e. the transporter bound to its intracellular and extracellular amino acid substrates) (Torras-Llort et al 2001); and (2) the analog aminoisobutyrate (AIB) induces an unequal exchange with other substrates through the rBAT-induced system $b^{0,+}$ in oocytes (i.e. using the endogenous $b^{0,+}$AT subunit) (Coady et al 1996). The oligomeric structure of system $b^{0,+}$ (i.e. rBAT-$b^{0,+}$AT heteromer) is unknown. Functional coordination of two rBAT-$b^{0,+}$AT heterodimers in a heterotetrameric structure would explain these results. If this is not the case, the transport defect associated with mutation R365W would suggest that a single $b^{0,+}$AT subunit contains two translocation pathways.

The gene causing non-type I cystinuria was assigned to 19q12-13.1 by linkage analysis (Bisceglia et al 1997, Wartenfeld et al 1997, Stoller et al 1999). *SLC7A9* was a positional and functional candidate gene for non-type I cystinuria (i.e. appropriate chromosomal location, rBAT-associated amino acid transport activity (system $b^{0,+}$), and proper tissue expression in kidney and small intestine (reviewed in Palacín et al 2005)). In 1999 the non-type I cystinuria gene was identified as *SLC7A9* (Feliubadaló et al 1999). The protein product encoded by *SLC7A9* was termed $b^{0,+}$AT for $b^{0,+}$ amino acid transporter. Seventy-three *SLC7A9* mutations causing cystinuria have been described (mutations are listed in Font-Llitjós et al 2005; for more recently described mutations see: Skopkova et al 2005, Jaeken et al 2006, Yuan et al 2006). Mutation G105R is the most frequent *SLC7A9* mutation in the ICC cohort of patients (~27% of the *SLC7A9* alleles identified). Similarly to the human disease, the *Slc7a9*-knockout mouse *Stones* presents non-type I cystinuria with urolithiasis (Feliubadaló et al 2003). Several cystinuria-specific *SLC7A9* missense mutations have been reported to lead to a defect in transport function (Font et al 2001). Reconstitution in proteoliposomes showed that A182T-mutated $b^{0,+}$AT is active but with a trafficking defect to the plasma membrane, whereas mutation A354T renders the transporter inactive (Reig et al 2002).

Very recently, the ICC performed an exhaustive mutational analysis of 164 probands (Font-Llitjós et al 2005): ~87% of the independent alleles were identified. The coverage of identified alleles was similar in all cystinuria types (Table 8.2). The unidentified alleles (~13%) may be due to mutations in intronic or promoter regions (e.g. two cystinuria-specific sequence variants in the promoter region of *SLC3A1* have been reported in Boutros et al 1999), to *SLC3A1* or *SLC7A9* polymorphisms in combination with cystinuria-specific mutations in the other allele (Schmidt et al 2003a) or to unidentified genes. These three possibilities have not been confirmed or ruled out. Of particular interest is the possibility of a third cystinuria gene. In this regard, Goodyer's group proposed *SLC7A10* as a candidate (Leclerc et al 2001). This gene is located near the cystinuria gene *SLC7A9* on chromosome 19q13.1 and codes for asc1, a 4F2hc-associated renal light subunit with substrate specificity for cysteine and other small neutral amino acids. Moreover, these authors identified the missense mutation E112D associated with cystinuria. In contrast, recent studies have ruled out this hypothesis (Pineda et al 2004): (1) cystinuria-specific mutations are not found in patients with alleles not explained by mutations in the two cystinuria genes, (2) the conservative mutation E112D does not affect transport of 4F2hc-asc1, and (3) asc1 mRNA is expressed in the distal tubule where renal reabsorption of amino acids is not relevant but where asc1 may have a role in osmoregulation. The possibility of a third cystinuria gene cannot be discarded, but it would be relegated to a very small proportion of patients (only 3% of the probands of the ICC show no mutation in either of the two cystinuria genes (Table 8.2)).

GENOTYPE/PHENOTYPE CORRELATIONS

Initial data suggested a close correlation between the phenotype and the mutated gene (mutations in *SLC3A1* resulted in type I, and mutations in *SLC7A9* resulted in type non-I) (Calonge et al 1995, Gasparini et al 1995). In contrast to this simple view, recent data show a more complex scenario. On the one hand, all *SLC3A1* mutations in well-characterized families cause type I cystinuria, with the exception of mutation dupE5–E9, which shows the non-type I phenotype in four out of six heterozygotes studied (Font-Llitjós et al 2005). This mutation consists of a gene rearrangement c.(891^{+}1524_1618-1600)dup, which results in the duplication of exons 5–9 and the corresponding in-frame duplication of amino acid residues E298–D539 of rBAT, as shown by RNA studies (Schmidt et al 2003b, Font-Llitjós et al 2005). Functional studies are required to explain the dominant negative effect of dupE5–E9 mutation on the rBAT/b$^{0,+}$AT heteromeric complex. On the other hand, most of the heterozygotes carrying a *SLC7A9* mutation have a phenotype non-I (i.e. hyperexcretion of dibasic amino acids and cystine), but may also have a phenotype I (i.e. silent heterozygotes). Approximately 14% of the *SLC7A9* heterozygotes have phenotype I (Dello Strologo et al 2002). *SLC7A9* mutations associated with phenotype I in some families are I44T, G63R, G105R, T123M, A126T, V170M (the Libyan Jewish mutation), A182T, G195R, Y232C, P261L, W69X, and c.614dupA (Leclerc et al 2002, Font-Llitjós et al 2005). There is no clear explanation of why these mutations associate with phenotype I, since some proteins show residual transport activity when expressed in heterologous expression systems while others do not (Font et al 2001). A182T is the most frequent *SLC7A9* mutation associated with phenotype I (i.e. 6 out of 11 A182T heterozygotes in the ICC cohort), and this mutation leads to a protein with 50% residual transport activity at the plasma membrane. Moreover, in the ICC cohort the 11 mixed cystinuria patients and the single patient with cystinuria type I, which all carry two mutations in *SLC7A9*, presented aminoaciduria in the lower range of non-type I

patients with two mutations in this gene (Font-Llitjós et al 2005) (Table 8.3). This suggests that, in addition to individual and population variability, mild *SLC7A9* mutations may be more prone to associate with silent phenotype in heterozygotes (i.e. phenotype I).

The lack of a direct relationship between the mutated cystinuria gene and the type of cystinuria led the ICC to propose a parallel classification to describe cystinuria on the basis of the genotype of the patients (type A due to mutations in *SLC3A1*, type B due to mutations in *SLC7A9*, and type AB to define a possible digenic cystinuria) (Dello Strologo et al 2002). Table 8.2 summarizes the double classification for 78 cystinuria probands by the ICC as follows: (1) most type I patients have two mutations in *SLC3A1* (i.e. individuals AA); (2) all non-type I patients (including type non-type I heterozygotes with urolithiasis) have mutations in *SLC7A9* (i.e. individuals BB and B^{+}); (3) patients with mixed cystinuria carry mutations in *SLC3A1* (2 probands AA) or in *SLC7A9* (7 probands BB); and (4) two out of 126 fully genotyped probands carry mutations in both genes.

To my knowledge, only four patients with mutations in both cystinuria genes have been described (Harnevik et al 2003, Font-Llitjós et al 2005). There is no report of the urine phenotype of the Swedish patient AA(B). Two sisters AA(B) and one male BB(A) from two families out of 126 fully genotyped families in the ICC cohort have been identified and classified as mixed cystinuria patients (i.e. each of the two mutated alleles in the same gene is associated with phenotype I or non-phenotype I in the obligate heterozygotes). The aminoaciduria levels of these patients and their double-heterozygote (i.e. AB) relatives indicate that digenic inheritance in cystinuria has only a partial effect on the phenotype, restricted to a variable impact on the

TABLE 8.3   Urine amino acid excretion in patients classified by genotype and type of cystinuria

Genotype	Cystinuria type	n	Cystine	Lysine	Arginine	Ornithine
				Urine amino acid excreted (mmol/g creatinine)		
AA	I	34	1.66 [0.65–3.40]	6.58 [2.65–11.6]	3.14 [0.23–8.37]	1.74 [0.59–3.44]
AA	Mixed	3	0.78, 2.12, 5.56	3.31, 5.72, 11.4	1.23, 2.82, 7.03	0.72, 1.64, 1.92
AA(B)	Mixed	1	2.57	9.84	2.95	5.17
BB	I	1	2.69	2.28	111	0.30
BB	non-I	37	1.62 [0.50–3.30]	6.51 [1.72–14.7]	3.45 [0.50–6.15]	2.20 [0.30–4.77]
B+	non-I carriers	3	0.26*, 0.44*, 0.80	1.64*, 2.45, 3.88	0.02*, 0.12*, 0.15*	0.04*, 0.27*, 0.29*
BB	Mixed	11	1.82 [0.43–3.18]	4.58 [1.57–8.72]	1.54 [0.21–3.51]	1.33 [0.47–2.45]
BB(A)	Mixed	1	0.43	3.27	489	603

The mean of the amino acid levels for each group is indicated, with the exception of categories with less than 11 patients, where individual data are shown. When applicable, the 5th and 95th centile limits are shown in square brackets

* Excretion values below 5th centile of homozygotes of cystinuria type non-I (BB) in carriers of cystinuria type non-I. Genotypes are as described in the legend to Table 2. n, number of patients. Extracted from Font-Llitjós et al 2005

aminoaciduria. Indeed, none of the individuals AB presented urolithiasis. Given that the frequencies of type A and B alleles are similar in this cohort, if digenic inheritance was the rule in cystinuria, we would expect a quarter of patients to be AA, a quarter to be BB, and half to be AB. This indicates that digenic inheritance affecting phenotype is an exception in cystinuria. However, the possibility that some combinations of mutations A and B produce enough cystine hyperexcretion to cause urolithiasis cannot be ruled out.

A working hypothesis on the biogenesis of the rBAT/$b^{0,+}$AT heterodimer may explain the urine phenotypes and the apparent lack of full digenic inheritance in cystinuria: $b^{0,+}$AT controls the amount of active holotransporter at the plasma membrane. Thus, the rBAT protein would be produced in excess in kidney, and therefore an rBAT mutation in heterozygosis in humans and in mice does not lead to hyperexcretion of amino acids (phenotype I). The only exception to this rule that has been identified to date is the human rBAT mutation dupE5–E9, thereby indicating a dominant effect for this mutation. $b^{0,+}$AT controls the expression of the functional rBAT/$b^{0,+}$AT heterodimeric complex: interaction with $b^{0,+}$AT stabilizes rBAT, and the excess of rBAT is degraded, as shown in transfected cells (Bauch et al 2002, Reig et al 2002). As a result, a half dose of $b^{0,+}$AT (i.e. heterozygotes of severe human *SLC7A9* mutations or of the *Slc7a9*-knockout mice) causes hyperexcretion of cystine and dibasic amino acids (i.e. phenotype non-I) as a result of a significant decrease in the expression of functional rBAT/$b^{0,+}$AT (system $b^{0,+}$). In this scenario, the lack of a full cystinuria phenotype because of digenic inheritance indicates that in double heterozygotes (AB), the mutated rBAT does not compromise the heterodimerization and trafficking to the plasma membrane of the half dose of wild-type $b^{0,+}$AT with the half dose of wild-type rBAT. Thus individuals AB behave as heterozygotes B with a variable degree of aminoaciduria, which could be greater than that of single heterozygotes within the family, depending on the particular combination of mutations. Demonstration of this hypothesis requires an in-depth study of the effect of cystinuria-specific rBAT and $b^{0,+}$AT mutations on the biogenesis of the heteromeric complex rBAT/$b^{0,+}$AT both in cell culture studies and in vivo, using double heterozygote mice (*Slc3a1* D140G/$^+$, *Slc7a9* $-/-$).

Urolithiais shows a clear gender and individual variability among cystinuria patients (Dello Strologo et al 2002). In the ICC cohort, the age of onset of lithiasis ranges from 2 to 40 years with a median of 12 and 15 years for males and females, respectively. Similarly, the number of total stone events (i.e. spontaneously emitted stones plus those surgically removed) is higher in males than females (0.42 and 0.21 events per year in males and females, respectively). Of the 224 patients studied, ten with full genetic confirmation of the disease and presenting aminoaciduria did not develop renal stones, and two of these patients were over 40 years

of age. In contrast, clinical symptoms (i.e. urolithiais and its consequences) are almost identically represented in the two cystinuria types when either the clinical or the genetic classification is considered. The differences in severity between the genders and marked differences between siblings sharing the same mutations (Dello Strologo et al 2002) suggest that other lithogenic factors, genetic or environmental, contribute to the urolithiasis phenotype. Indeed, only about half of the *Slc7a9*-knockout mice, in a mixed genetic background, develop urolithiasis (Feliubadaló et al 2003). Moreover, lithiasic and nonlithiasic *Slc7a9*-knockout mice hyperexcrete similar levels of cystine. Studies in *Slc7a9*-knockout mice with distinct genetic backgrounds may unravel the genetic factors, in addition to mutations in *Slc7a9* and the cystine levels in urine that contribute to urolithiasis.

Lysinuric protein intolerance

LPI is a primary inherited aminoaciduria with an autosomal recessive mode of inheritance predominantly reported in Finland where the prevalence of the disorder is 1 in 60 000 (for a review see Simell 2001). Two other geographic locations with a relatively high prevalence are southern Italy and Japan (Palacín et al 2001), with the northern part of Iwate (Japan) registering a prevalence of 1 in 50 000 (Koizumi et al 2003). The diagnosis of LPI is often difficult because of an unspecific clinical presentation. Therefore it is not surprising that LPI is mainly known in Finland, Italy, and Japan ($\sim$200 patients described) where clinicians are accustomed to diagnose this disorder (Palacín et al 2001).

In LPI there is massive urinary excretion of dibasic amino acids, especially lysine, and the intestinal absorption of these amino acids is poor; therefore the concentration of dibasic amino acids in plasma is low (Kekomaki et al 1967, Oyanagi et al 1970, Simell et al 1975) (Table 8.4). Arginine and ornithine are intermediates of the urea cycle that provide the carbon skeleton to the cycle. Their reduced availability results in a functional deficiency of the urea cycle (for a review see Palacín et al 2004). Other characteristics of the LPI phenotype are as follows (for a review see Simell 2001). Protein malnutrition and deficiency of the essential amino acid lysine contribute to the patient's failure to thrive. Patients with LPI are usually asymptomatic while breastfeeding, and symptoms (e.g. vomiting, diarrhea, and hyperammonemic coma when force-fed high-protein food) appear only after weaning. After infancy, patients with LPI reject high-protein diets, and show a delay in bone growth and prominent osteoporosis, hepatosplenomegaly, muscle hypotonia, and sparse hair. Most patients have a normal mental development, but some may show moderate retardation. Low-protein diet and citrulline, a urea cycle intermediate, are used to correct the functional deficiency of intermediates of the urea cycle. The final height in treated patients is

TABLE 8.4   Plasma and urine amino acids in LPI and cystinuria

Plasma amino acids (μM)

Amino acid	Range in normal children	Patients with LPI		Controls cystinuria (mean$\pm$SD)	Patients with cystinuria (mean$\pm$SD)
		Mean	Range		
Lysine	71–151	70	32–179	171$\pm$26	121$\pm$30
Arginine	23–86	27	12–58 2–83	82$\pm$16	46$\pm$12
Ornithine	27–86	21	57–105	58$\pm$11	36$\pm$11
Cystine	48–140	80	3644–7161	79$\pm$12	43$\pm$12
Glutamine	57–467	5583	417–1017	n.d.	n.d.
Alanine	173–305	772		n.d.	n.d.

Amino acids in urine

Amino acid	Range in controls (mmol/g creatinine)		Patients with LPI (mmol/1.73 m2/24 h)		Patients with cystinuria type B (mmol/g creatinine)	
	Mean	Range#	Mean	Range	Mean	Range#
Lysine	0.18	0.04–0.50	4.13	1.02–7.00	6.51	1.72–14.7
Arginine	0.02	0.00–0.05	0.36	0.08–0.69	3.45	0.50–6.15
Ornithine	0.03	0.01–0.07	0.11	0.09–0.13	2.20	0.30–4.77
Cystine	0.05	0.02–0.11	0.12	0.06–0.21	1.62	0.50–3.30

Plasma amino acids are expressed in μM. Data for normal children, and patients with LPI (n = 20) are from reference (Simell 2001). Plasma glutamine data also include asparagine concentration. Plasma amino acids from controls (n = 12) and patients with cystinuria (n = 8) are from Morin et al 1971. In patients with LPI, urinary excretion is expressed in mmol/1.73 m^2/24 h (n = 4) (Simell 2001), whereas in controls (n = 83) (Dello Strologo et al 2002) and patients with cystinuria type B (i.e. due to mutations in SLC7A9) (n = 37) (Font-Llitjós et al 2005; extracted from Table 2) it is expressed in mmol/g creatinine

5th–95th centile range. SD, standard deviation. n.d., not determined

slightly subnormal or low-normal. This treatment does not correct all symptoms such as poor growth, hepatospleno-megaly, delayed bone age, and osteoporosis, which are all probably due to the lysine deficiency. Recently, Simell's group reported recovery of plasma lysine by oral supplementation with the amino acid (Lukkarinen et al 2003).

About two-third of patients with LPI have interstitial changes in chest radiographs, and some develop acute or chronic respiratory insufficiency (Parto et al 1993) that can lead to fatal pulmonary alveolar proteinosis and to multiple organ dysfunction syndrome. Further symptoms suggesting that the immune system is affected are glomerulonephritis and erythroblastophagia (Nagata et al 1987, DiRocco et al 1993).

System y$^+$L transports dibasic amino acids with high affinity (Km in the micromolar range) in a sodium-independent way, but requires sodium to transport neutral amino acids with high affinity (Devés & Boyd 1998): (i) in the absence of sodium, the transport of neutral amino acids through system y$^+$L is of very low affinity; and (ii) system y$^+$L catalyzes the electroneutral efflux of cationic amino acids in exchange for neutral amino acids plus sodium, using the driving force of the sodium concentration gradient. In the early 1990s, two groups described the expression of a system y$^+$-like transport activity in *Xenopus* oocytes after injection of 4F2hc cRNA (Bertran et al 1992, Wells et al 1992). Two closely related proteins (y$^+$LAT-1 and y$^+$LAT-2) that induce y$^+$L transport activity when expressed together with 4F2hc were identified by homology screening (Torrents et al 1998), using the light subunit of HAT LAT-1 (Kanai et al 1998, Mastroberardino et al 1998). The transport characteristics of 4F2hc/y$^+$LAT-1 have been studied in heterologous expression systems, where co-immunoprecipitation of these two proteins has been substantiated (Torrents et al 1998, Pfeiffer et al 1999, Kanai et al 2000). The transport activity elicited matches the characteristics of system y$^+$L; electroneutral exchange of dibasic amino acids for neutral amino acids plus sodium with a 1:1:1 stoichiometry. y$^+$LAT-1 is expressed in the basolateral plasma membrane in the epithelial cells of kidney tubules and polarized cellular models (Bauch et al 2003) (Figure 8.3).

The gene responsible for LPI was localized to 14q11.2 in Finnish and non-Finnish populations (Lauteala et al 1997, 1998). The cloning of y$^+$LAT-1, encoded by *SLC7A7*, revealed characteristics that made this gene an excellent candidate for LPI (i.e. appropriate chromosome location, co-expression of system y$^+$L with 4F2hc and proper expression in LPI affected tissues).

In 1999 two consortiums (Borsani et al 1999, Torrents et al 1999) independently reported the first mutational analysis of

SLC7A7 in patients with LPI. A single Finnish mutant allele (1181-2A > T) was found with an A > T transversion at position −2 of the acceptor splice site in intron 6 of *SLC7A7*. This inactivates the normal splice site acceptor and activates a cryptic acceptor 10 bp downstream with the result that 10 bp of the ORF are deleted and the reading frame is shifted. This mutation has been found in all Finnish LPI patients (i.e. 'the Finnish mutation') (Mykkanen et al 2000). These two seminal studies also identified LPI-specific *SLC7A7* mutations in Spanish and Italian patients, and established that mutations in *SLC7A7* cause LPI. The fact that system y$^+$L activity is present in LPI erythrocytes or fibroblasts (Smith et al 1988, Dall'Asta et al 2000) indicates the expression of a distinct y$^+$L transporter isoform in these cells, most probably y$^+$LAT-2. Additional studies showed the nonsense mutation W242X and the insertion 1625insATAC as the most prevalent mutations in the south of Italy (Sperandeo et al 2000), and the nonsense mutation R410X as the most prevalent in Japan (Noguchi et al 2000). A total of 26 *SLC7A7* mutations of any kind (large genomic rearrangements, missense and nonsense mutations, splicing mutations, insertions and deletions) has been described in 106 patients with LPI (>90% allele explained) (Sperandeo et al 2005). No LPI-associated mutations have been reported in *SLC3A2*, coding for the heavy subunit of y$^+$LAT-1 (4F2hc). This strongly suggests that *SLC7A7* is the only gene involved in the primary cause of LPI. It is believed that mutations in *SLC3A2* would be deleterious. 4F2hc serves as the heavy subunit of six other HAT (see above). Therefore, a defect in 4F2hc will result in six defective amino acid transport activities expressed in many cell types and tissues. Indeed, the murine *Slc3a2*-knockout is lethal (Tsumura et al 2003).

Functional studies in oocytes and transfected cells showed that frameshift mutations (e.g. 1291delCTTT, 1548delC, and the Finnish mutation) produce a severe trafficking defect (e.g. the mutated proteins do not localize to the plasma membrane when co-expressed with 4F2hc) (Mykkanen et al 2000, Toivonen et al 2002). In contrast, the missense mutations G54V and L334R inactivate the transporter (e.g. the mutated proteins reach the plasma membrane when co-expressed with 4F2hc but no transport activity is elicited) (Mykkanen et al 2000, Toivonen et al 2002). Mutation E36del showed a dominant negative effect when expressed in *Xenopus* oocytes (Sperandeo et al 2005b). The molecular basis for this effect is not yet fully understood.

PATHOPHYSIOLOGY OF LYSINURIC PROTEIN INTOLERANCE

LPI is a multisystemic disease. Some of the symptoms of this disease, like the renal and intestinal phenotypes, are easily explained by a defect in the basolateral amino acid transport system y$^+$L. Urea cycle malfunction is a characteristic of patients with LPI after weaning. Patients with LPI have a decreased tolerance for nitrogen and present with hyperammonemia after ingestion of even moderate amounts

of protein. The malfunction of the urea cycle in LPI is less severe than that caused by defects in the enzymes of the cycle. y$^+$LAT1 is not expressed in hepatocytes (Torrents et al 1998, Pfeiffer et al 1999). It is believed that urea cycle malfunction is due to diminished availability of the intermediates of this cycle because their low concentration in plasma 'intermediate functional deficiency hypothesis' (for a review see Palacín et al 2004). The mechanisms underlying the LPI-associated immune-related disorders (e.g. alveolar proteinosis, erythroblastophagia, and glomerulonephritis) are unknown. In addition, individual phenotypic variability precluded establishment of genotype/phenotype correlations (Mykkanen et al 2000, Sperandeo et al 2000). Thus, Finnish patients with LPI, all with the same Finnish mutation in homozygosis, show a wide range of phenotypic severity ranging from nearly normal growth with minimal protein intolerance to severe cases with hepatosplenomegaly, osteoporosis, alveolar proteinosis, and severe protein intolerance. In the following part of this section, the mechanisms that explain the renal and intestinal pathophysiology in LPI are discussed.

Table 8.4 compares plasma and urine levels for several amino acids in patients with LPI and cystinuria. Plasma concentrations of the dibasic amino acids (i.e. lysine, arginine, and ornithine) are usually subnormal (1/3 to 1/2 of the normal values), but occasionally may fall within the normal range (Simell 2001). Similarly, but to a lesser extent, plasma dibasic and cystine concentrations are lower in patients with cystinuria (Morin et al 1971). This observation indicates that the defects in renal reabsorption and intestinal absorption of dibasic amino acids may have a greater impact in LPI than in cystinuria, and therefore produce a larger depletion of these amino acids in plasma. In contrast to dibasic amino acids, the plasma concentrations of the neutral amino acids glutamine and alanine are increased in patients with LPI (Table 8.4), and to a lesser extent serine, glycine, citrulline and proline (Simell 2001). The considerable increase in plasma glutamine and alanine in LPI is believed to be the result of the large amount of waste nitrogen not incorporated into urea as a result of urea cycle malfunction.

In LPI, urinary excretion and renal clearance of lysine is massively increased, while that of arginine and ornithine is moderately augmented (Simell & Perheentupa 1974): lysine excretion is 10-fold and 30-fold that of arginine and ornithine in LPI patients respectively (Table 8.4). In contrast, lysine excretion is only 2- to 3-fold higher than that of arginine and ornithine in patients with cystinuria (Tables 8.3 and 8.4). Renal reabsorption of lysine is comparable in LPI and cystinuria, whereas hyperexcretion of arginine and ornithine are lower in LPI than in cystinuria (Table 8.4). These observations indicate that the LPI-defective transporter (y$^+$LAT-1/ 4F2hc) may have a more pronounced role in the reabsorption of lysine than of the other dibasic amino acids. In contrast to cystinuria, where cystine excretion in urine is four times lower than that of lysine, in LPI there is only a slight increase of renal cystine excretion (Table 8.4). This may be explained by the large tubular lysine load (i.e. caused by the reabsorption defect of lysine) that competes for absorption through the apical system $b^{0,+}$, and which shares uptake of cystine and dibasic amino acids in exchange with other neutral amino acids. The increased plasma concentration of serine, glycine, citrulline, proline, alanine, and glutamine in LPI explains hyperexcretion of these amino acids, and their renal clearance is within the normal range (Simell 2001).

The defect in kidney and intestine in LPI is located in the basolateral membrane and thus affects the basolateral efflux of dibasic amino acids (Desjeux et al 1980, Rajantie & Simell 1981). An oral load with the dipeptide lysyl-glycine increased glycine plasma concentrations, but plasma lysine remained almost unchanged in patients with LPI, while both amino acids increased in plasma of control subjects or in patients with cystinuria (Rajantie et al 1980a, b). Figure 8.3 shows the present knowledge on the molecular bases of the intestinal absorption of dibasic amino acids. At the luminal membrane of the enterocyte, the transport of oligopeptides (not shared with amino acids) is mediated by PEPT1 (Groneberg et al 2001). A major route for dibasic amino acids across the apical membrane is system $b^{0,+}$ (i.e. the transporter defective in cystinuria). The absorbed peptides are hydrolyzed to release amino acids in the cytoplasm of the enterocyte (Adibi et al 1971, Asatoor et al 1971, Mattheus & Adibi 1976), and are able to cross the basolateral membrane only as free amino acids. The lack of increased plasma lysine after the lysyl-glycine load, but normal increase in plasma glycine, shows that the basolateral efflux of the intracellularly delivered lysine is defective in LPI. In patients with cystinuria, the cleaved glycine and lysine cross the epithelial cell normally because the defect is apical (i.e. system $b^{0,+}$) (see Figure 8.3). The defect in the basolateral system $y^{+}L$ explains the renal and the intestinal phenotypes in LPI. The protein y^{+}LAT-1 has a basolateral location in epithelial cells. System $y^{+}L$ (i.e. the 4F2hc/y^{+}LAT-1 heteromeric complex) mediates the efflux of cationic amino acids by exchange with extracellular neutral amino acids and sodium (see Figure 8.3). Thus, the loss of transport function of the LPI-associated y^{+}LAT-1 mutations results in a defective basolateral efflux of dibasic amino acids in the intestinal absorptive and renal reabsorptive epithelial cells.

HARTNUP DISORDER

The original patients with Hartnup disorder presented cerebellar ataxia, tremor, nystagmus, pellagra-like photosensitive skin rash, and delayed intellectual development. Hartnup disorder affects the renal reabsorption and intestinal absorption of neutral amino acids with the exception of proline, hydroxyproline, glycine, and cystine (see Levy 2001 for a review). Pellagra-like symptoms (i.e. niacin deficiency) are frequent in patients with this disorder. Low tryptophan availability (i.e. defective renal and intestinal reabsorption of the amino acids) appears to be at the basis of the niacin deficiency: tryptophan and niacin deficiencies are thought to generate similar symptoms because this amino acid is a major source of NAD(P)H in humans. In this regard, pellagra-like symptoms respond to nicotinic acid supplementation.

The incidence of Hartnup disorder has been estimated at 1 in 26 000 in newborn screening programs (Levy 1973). The trait is transmitted in autosomal recessive fashion, but clinical manifestations are probably modulated by environmental and genetic factors (Scriver et al 1987).

System B^{0} neutral amino acid transporter has been considered the defective transporter in Hartnup disorder. Large neutral amino acids are mainly absorbed in the small intestine and reabsorbed in the proximal convoluted tubule (i.e. S1–S2 segments) by the apical system B^{0} (reviewed in Broër et al 2006). Functional studies in renal and intestinal brush-border membrane vesicles and derived cell models defined system B^{0} (B for broad and ^{0} for neutral charge; Mailliard et al 1995) as a transporter serving a broad spectrum of neutral amino acids. System B^{0} mediates co-transport of Na^{+} and neutral amino acids with 1:1 stoichiometry, where Na^{+} and amino acid affects each other's kinetic parameters (reviewed in Broer et al 2006).

Broer's group demonstrated that mouse B^{0}AT1 (previously the orphan XTR2-related transporter) when expressed in *Xenopus* oocytes induces Na^{+}-dependent and Cl-independent transport of neutral amino acids with broad specificity, matching the characteristic of system B^{0} (Broer et al 2004, Bohmer et al 2005). Apparent K_m for neutral amino acids ranges from 1 to 10 mM with the following substrate specificity (one letter code for amino acids): M = L = I = V > Q = N = C = F = A > S = G = Y = T = H = P > W (Bohmer et al 2005). The human ortholog showed similar transport characteristics (Kleta et al 2004, Seow et al 2004). Human B^{0}AT1 mRNA is expressed mainly in kidney and small intestine, and to a lesser extent in colon, pancreas, and prostate (Kleta et al 2004, Seow et al 2004). Mouse B^{0}AT1 was localized to the brush-border membrane of the epithelial cells of the renal proximal convoluted tubule (S1–S2 segments) and of the small intestine with a gradient of expression from the crypts toward the tip of the microvilli (Broer et al 2004, Kleta et al 2004). Human B^{0}AT1 gene (*SLC6A19*) localized to chromosome 5p15.33 (Seow et al 2004), and Hartnup disorder to chromosome 5p15 in Japanese families transmitting the disease (Nozaki et al 2001). Thus, *SLC6A19* was an obvious functional and positional candidate gene for Hartnup disorder.

In 2004, two independent studies demonstrated that mutations in *SLC6A19* are associated with Hartnup disorder and confirmed the recessive mode of inheritance (Kleta et al 2004,

Seow et al 2004). Patients from the Hartnup family were homozygotes for mutation IVS8 + 2T > G affecting the donor splice consensus sequence of exon 8 (Kleta et al 2004). In seven Australian pedigrees, six distinct mutations that cosegregated with the disorder were identified (three missense, one nonsense and two splice site mutations), including one Australian family transmitting the Hartnup family mutation in one allele. In the Australian population, D173N and R240X mutations occur at a frequency of 1 in 140 and 1 in 1000 people respectively. Four further mutations were identified in three Japanese families (one missense, one nonsense and two small deletions causing frameshift) (Kleta et al 2004). In total, ten Hartnup disorder-specific *SLC6A19* mutations have been identified in 13 independent pedigrees. This implies that ~73% of the independently studied alleles have been identified (19 out of 26 alleles).

Recently, the crystal structure of a prokaryotic homolog (LeuT$_{Aa}$ from *Aquifex aeolicus*) of the SLC6 family has been reported (Yamashita et al 2005). This structure will be very useful to ascertain the molecular events underlining the defects associated with Hartnup disorder mutations. The four Hartnup disorder-specific *SLC6A19* missense mutations (R57C, D173N, L242P, E501K) were checked for function in oocytes (Kleta et al 2004, Seow et al 2004). These mutations showed no transport function, with the exception of the most common mutation D173N, which has residual transport activity (~50%). Figure 8.2 shows the location of these mutations within the topology of SLC6

transporters. R57C destroys a saline bridge with residue Asp486. This bond helps to hold the position of TM1b, which interacts with the amino acid substrate and the two Na$^+$ ions. Leu242 involves the first of two residues, thereby constituting the extracellular β1 sheet, and mutation L242P, most probably, disrupt this structure. Glu501 in TM10 interacts with one of two water molecules that hold the structure of the unwound residues between TM6a and TM6b, which interact with the amino acid substrate and one of the Na$^+$ ions. Then, mutation E501K most probably affects the folding of this unwound region. Finally, mutation D173N is a conservative amino acid substitution affecting a residue not conserved among the SLC6 transporters in the extracellular α-helix EL2. Then, not surprisingly, this mutation retains significant transport activity (Seow et al 2004).

GENETIC HETEROGENEITY AND PHENOTYPE VARIABILITY
Taken together, these results demonstrated that mutations in *SLC6A19* cause Hartnup disorder. However, individuals that display Hartnup-like aminoaciduria without apparent mutations in *SLC6A19* (in two American pedigrees (Kleta et al 2004)) have been reported; similar results have been described by the Australian Hartnup Consortium (Broer et al 2006). Indeed, genetic linkage of Hartnup disorder with the 5p15 region has been excluded in an American family (Kleta et al 2004). This finding indicates that additional Hartnup disorder genes may be involved and remain to be

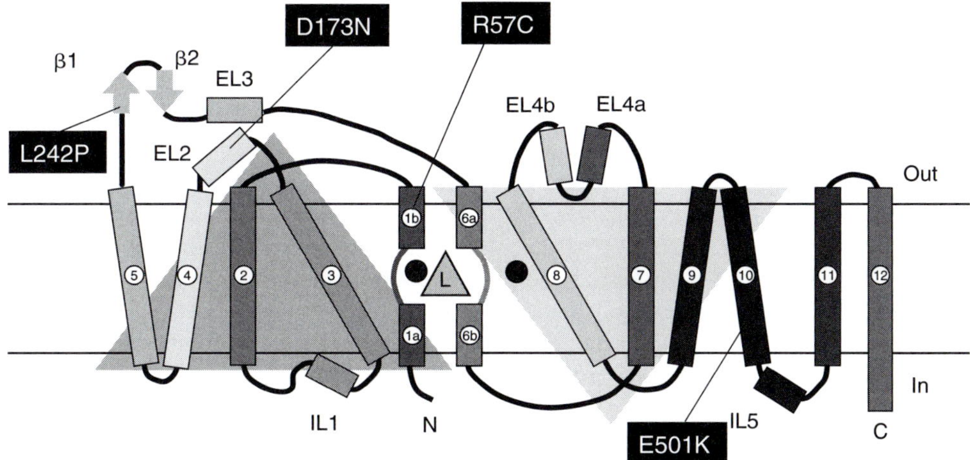

FIGURE 8.2 The topology of the SLC6 transporters. This topology is based on the crystal structure of LeuT$_{Aa}$ from bacterium *Aquifex aeolicus* (Yamashita et al 2005). There is a structural repeat, not based on amino acid sequence, in the first ten transmembrane (TM) helices of LeuT$_{Aa}$, relating TM1–TM5 (pink triangle) and TM6–TM10 (blue triangle) by a pseudo-two-fold axis located in the plane of the membrane. LeuT$_{Aa}$ is a bacterial homolog of Na$^+$/Cl$^-$-dependent neurotransmitter transporters (family SLC6), to which the defective Hartnup disorder transporter (B^{0}AT1) belongs. Amino acid sequence homology of B^{0}AT1 and LeuT$_{Aa}$ is ~20% and covers the whole sequences (CLUSTAL alignment; data not shown). Major amino acid sequence differences between LeuT$_{Aa}$ and the eukaryotic SLC6 transporters are located at the N- and C-termini, between TM3 and TM4, and between extracellular α-helices EL4a and EL4b (these segments are longer in B^{0}AT1 and in other SLC6 eukaryotic transporters). The positions of the substrate leucine and the two sodium ions are shown as a yellow triangle and two blue circles, respectively. Residues interacting with the substrate and ions are located within and surrounding the unwound regions between TM1a and TM1b, and TM6a and TM6b, as well as in TM3 and TM8. Hartnup disorder-specific missense B^{0}AT1 mutations (black boxes) are indicated within the LeuT$_{Aa}$ topology. Rectangle, α-helix. Arrow, β-sheet. Figure modified from Yamashita et al 2005

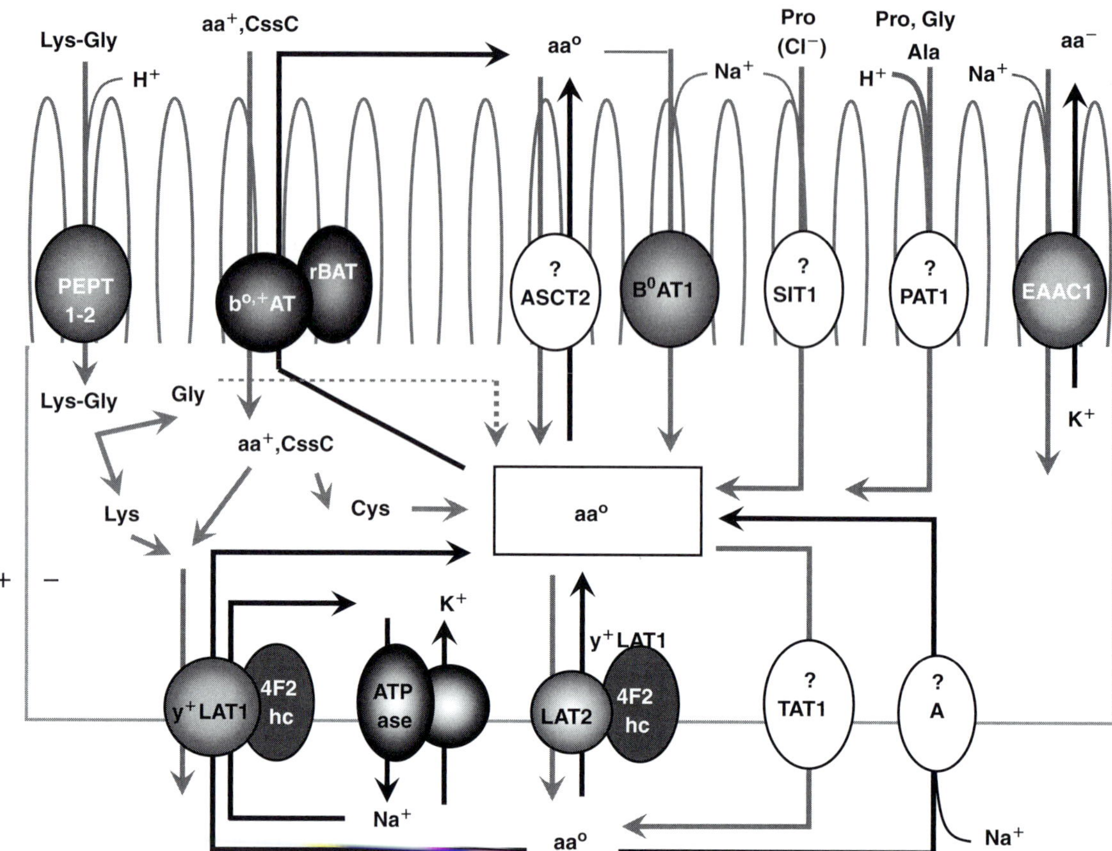

FIGURE 8.3 Transporters involved in the renal and intestinal reabsorption of amino acids. Transporters with a proved role in renal reabsorption or intestinal absorption of amino acids are colored. Transporters present in the apical or basolateral plasma membrane of the epithelial cells of the proximal convoluted tubule or of the small intestine, but with no direct experimental evidence supporting their role in reabsorption, are shown uncolored. Amino acid fluxes in the reabsorption direction are in red. Peptide transporters (PEPT1-2) mediate H^+-co-transport of di- and tripeptides (e.g. Rajantie et al 1980a, used lysyl-glycine as the substrate in the experiments in patients with LPI). PEPT1 and PEPT2 are expressed in the small intestine and kidney respectively. See text for details

identified. Five candidate neutral amino acid transporters have been excluded (genetic linkage exclussion and/or lack of co-segregating mutations) as causative genes of the disorder (Seow et al 2004): *SLC3A2* (4F2hc), *SLC7A8* (LAT2), *SLC1A5* (ASCT2 or ATB0), *SLC6A18* (orphan Xtrp2) and *SLCA20* (orphan XT3). A system B^0-like activity in the proximal straight tubule (S3 segment), which has not yet been identified, is an obvious candidate for Hartnup disorder (Broer et al 2006).

Patients with Hartnup disorder display a wide phenotype range. This was described in the original report of the Hartnup family: of the four siblings with clear aminoaciduria, two presented severe clinical symptoms, one had mild symptoms and one was asymptomatic. Symptoms most likely appear in individuals with subnormal plasma amino acid levels (reviewed in Scriver et al 1987). Intestinal absorption of peptides, via PEPT1, is thought to compensate for the lack of amino acid transport in Hartnup disorder (Daniel 2004) (Figure 8.3). This compensation has two consequences. On the one hand, in developed societies, characterized by high protein intake, most patients will remain asymptomatic. Only a limited number of patients will display symptoms (e.g. subnormal body weight, episodes of diarrhea, pellagra-like rash) (Wilcken et al 1977). On the other hand, genetic factors may predispose individuals to a more severe deficiency in amino acid uptake. The phenotype of Hartnup disorder could be influenced by the amino acid transporters which participate in the renal reabsorption and intestinal absorption of amino acids: other apical transporters for neutral amino acids (e.g. the B^0-like activity in the proximal straight tubule) and basolateral transporters (Table 8.3). Polymorphisms in these transporters may contribute to heterogeneity in the phenotype of Hartnup disorder.

THE MOLECULAR BASES OF INTESTINAL ABSORPTION AND RENAL REABSORPTION OF AMINO ACIDS

The renal reabsorption of amino acids occurs mainly in the proximal convoluted tubule (S1–S2 segments) (Silbernagl 1988), and the absorption of these compounds occurs in the small intestine (Mariotti et al 2000). The plasma membrane of epithelial cells in these two locations has a similar set of amino acid transporters (Figure 8.3). Transepithelial flux of amino acids from the intestinal or tubular lumen to the intercellular space requires transport through apical and basolateral domains of this plasma membrane. Several amino acid transporters have been identified in the apical domain: (i) for neutral amino acids B^0AT1 (system B^0), ASCT2 (system ASC), SIT (system Imino), and PAT1 (also representing system Imino; reviewed in Broer et al 2006); (ii) for dibasic amino acids, the heterodimer complex rBAT/$b^{0,+}$AT (system $b^{0,+}$); and (iii) for dicarboxylic amino acids, EAAC1 (system X_{AG}^-). Transporters localized in the basolateral domain of these cells are the heterodimers 4F2hc/y^+LAT1 (system y^+L) and 4F2hc/LAT2 (exchanger L for all neutral amino acids), and TAT1 (*SLC16A10*; aromatic amino acid (Trp) transporter). Several of these transporters present higher expression in the renal proximal convoluted (S1 and S2 segments) than in the straight tubule (S3 segment): rBAT/$b^{0,+}$AT (Fernández et al 2002), 4F2hc/y^+LAT1 (Bauch et al 2003) and 4F2hc/LAT2 (Pineda et al 1999, Bauch et al 2003) B^0AT1 (Broer et al 2004, Kleta et al 2004), ASCT2 (Avissar et al 2001), and SIT (Kowalczuk et al., 2005; Takanaga et al 2005). PAT1 has been identified in kidney, but its expression along the nephron has not been studied (Anderson et al 2004). Transporter ATB$^{0,+}$ (*SLC6A14*; system $B^{0,+}$: Na$^+$ and Cl- dependent co-transporter for neutral and dibasic amino acids), not shown in Figure 8.3, is expressed in distal ileum and colon but not in kidney, indicating a role for this transporter in the absorption of amino acids produced by bacterial metabolism (Nakanishi et al 2001).

The study of PIA is contributing to our understanding of the role of the above-mentioned transporters in the intestinal absorption and renal reabsorption of amino acids. Neutral amino acids are mainly absorbed in the small intestine and reabsorbed in the proximal convoluted tubule by system B^0. B^0AT1 accounts for system B^0 activity (electrogenic Na$^+$ co-transport of neutral amino acids) (Figure 8.3). Two additional B^0-like activities are expressed in the proximal straight tubule; the molecular identity of these transporters is unknown (reviewed in Broër et al 2006). Mutations in B^0AT1 cause Hartnup disorder, characterized by wastage of all neutral amino acids in urine, with the exception of proline, hidroxyproline, glycine and cystine (Levy 2001). This observation suggests that other transporters also mediate the reabsorption of proline. Indeed, renal iminoglycinuria, characterized by aminoaciduria of proline and glycine, also indicates that specific transporters contribute to the reabsorption of these amino acids. PAT1 and SIT are candidate transporters underlining the molecular bases of this disorder. PAT1 is a H$^+$ co-transporter of proline, glycine, and alanine (Boll et al 2002) while SIT is Na$^+$ co-transporter of proline and hydroxyproline (Kowalczuk et al 2005, Takanaga et al 2005) (Figure 8.3). Although

PAT1 is proton-dependent, sustained uptake in epithelial cells appears to be Na$^+$-dependent because removal of H$^+$ is coupled to the Na$^+$-gradient via the Na$^+$/H$^+$ exchanger (Anderson et al 2004). Recently, Broer and collaborators (Broer et al 2006) proposed a model for renal reabsorption of proline and glycine. PAT1, SIT, and B^0AT1 together will reabsorb proline at the convoluted tubule with a capacity exceeding normal kidney load. In contrast, reabsorption of glycine will approach the capacity of two transporters (glycine is not a substrate for SIT): PAT1 and B^0AT1. SIT would be the major player for intestinal reabsorption of proline in the small intestine. Following this model, mutations in PAT1 would result in iminoglycinuria: iminoglycinuria without intestinal phenotype may be caused by two mutated alleles in PAT1, whereas one mutated PAT1 allele would lead to isolated glycinuria. In contrast, iminoglycinuria with a defect in intestinal proline transport may be due to mutations in SIT. The possibility that a third gene is involved in renal iminoglycinuria can not be ruled out. Indeed, the murine *Slc6a18*-knockout model presents hyperglycinuria (Quan et al 2004). *SLC6A18* codes for the orphan transporter XT2, which is expressed in the proximal straight tubule (Obermuller et al 1997). Verification of the role of PAT1, SIT, and XT2 in the intestinal absorption and renal reabsorption of proline and glycine will require the clarification of the genotype of these genes in patients with iminoglycinuria.

Mutations in one of the two subunits of system $b^{0,+}$ (rBAT and $b^{0,+}$AT) causes cystinuria. System $b^{0,+}$ mediates the influx of cystine and dibasic amino acids in exchange with neutral amino acids (efflux) (Figure 8.3). The high intracellular concentration of neutral amino acids drives the direction of this exchange. The membrane potential (negative inside) favors the influx of dibasic amino acids (i.e. with a net positive charge at neutral pH) and the intracellular reduction of cystine to cysteine favors the influx of the former. As a result, patients with cystinuria present urinary hyperexcretion of cystine and dibasic amino acids but not other neutral amino acids. Interestingly, the mean and range (i.e., 5th–95th centile limits) of cystine, lysine, arginine, and ornithine in the urine of patients with mutations rBAT and in $b^{0,+}$AT are almost identical (see patients AA with phenotype I and patients BB with non-I phenotype in Table 8.3). This result is expected since all $b^{0,+}$AT heterodimerizes with rBAT in renal brush-border membranes, constituting the holotransporter $b^{0,+}$ (Fernandez et al 2002). Cystinuria patients may show almost null cystine reabsorption in kidney, whereas dibasic reabsorption in this organ remains significant (reviewed in Fernández et al 2002). This observation indicates that $b^{0,+}$ is the main reabsorption system for cystine, but other transporters also participate in the reabsorption of dibasic amino acids. The molecular identity of these transporters is unknown. The study of apical dibasic amino acid transport activities in the *Slc7a9*-knockout mouse should help to identify these transporters.

As mentioned above, the intracellular concentration of neutral amino acids is a major energy determinant of the active uptake of cystine and dibasic amino acids via system $b^{0,+}$. Apical (e.g. B^0AT1) and basolateral (e.g. system A) co-transporters of Na^+ and neutral amino acids should contribute to the high intracellular concentration of neutral amino acids (Figure 8.3). Moderate hyperexcretion of dibasic amino acids occurs in Hartnup disorder (Levy 1973), suggesting a coordinated function between systems B^0 and $b^{0,+}$: a defective system B^0 will reduce the intracellular concentration of neutral amino acids which drives the influx of dibasic amino acids via system $b^{0,+}$. In contrast, the impact of system A on renal reabsorption is unknown. The electrochemical gradient of Na^+ produced by the hydrolysis of ATP by basolateral Na^+/K^+ ATPase (i.e. primary active transporter) drives the active transport of the Na^+ co-transporters of neutral amino acids B^0 and A (i.e. secondary active transporters). Thus, system $b^{0,+}$ mediates active transport of cystine and dibasic amino acids with a tertiary active mechanism of transport.

Apical PEPT1 (*SLC15A1*) and PEPT2 (*SLC15A1*) are expressed in the small intestine and in kidney, respectively (Daniel 2004). These transporters co-transport H^+ with di- and tripeptides. The physiological role of PEPT2 in kidney is largely unknown (Rubio-Aliaga et al 2003). The contribution of PEPT1 to the assimilation of amino acids has not been properly evaluated in mammals or humans, but it is assumed that absorption of di- and tripeptides accounts for a significant proportion of the intestinal absorption of amino acids (Daniel 2004). A more extensive study of the phenotype of the *Slc15A2*-knockout mouse (Rubio-Aliaga et al 2003) and generation and study of the PEPT1 model may answer these questions. Meanwhile, the role of PEPT1 in amino acid nutrition is supported by observations of the lack of pathology associated with amino acid malabsorption in cystinuria and in many patients with Hartnup disorder. Patients with cystinuria do not show pathology, with the exception of cystine urolithiasis. It is believe that absorption of di- and tripeptides via PEPT1 compensate for the defective absorption of cystine and dibasic amino acids via system $b^{0,+}$. Similarly, phenotype severity in Hartnup disorder is reduced in well-nourished patients.

The study of LPI teaches us about the role of basolateral transporters in intestinal absorption and renal reabsorption of dibasic amino acids. The heterodimer 4F2hc/y^+LAT1 has a basolateral location and accounts for system y^+L activity (the electroneutral efflux of dibasic amino acids in exchange with neutral amino acids plus sodium) (Figure 8.3). Mutations in y^+LAT1 cause LPI, which is characterized by hyperdibasic aminoaciduria and malabsorption of dibasic amino acids. On the one hand, wastage of lysine in urine in LPI and cystinuria are similar, whereas that of arginine and ornithine are less severe in LPI than in cystinuria (Table 8.4). Regarding the renal reabsorption of dibasic amino acids, these findings indicate that: (1) lysine appears to be a preferred substrate for basolateral efflux via 4F2hc/y^+LAT1, and (2) other basolateral transporters mediate efflux of arginine and ornithine. The molecular identity of these transporters is unknown. On the other hand, LPI produces a larger depletion of the three dibasic amino acids in plasma than cystinuria (Table 8.4). All these observations suggest that malabsorption of dibasic amino acids is more severe in LPI than in cystinuria. Two reasons may account for this: (1) the contribution of the apical peptide transporter PEPT1 cannot compensate for the basolateral defect associated with LPI (Figure 8.3); and (2) 4F2hc/y^+LAT1 is probably the main basolateral system for intestinal absorption of dibasic amino acids.

The basolateral 4F2hc-LAT2 heterodimer is an exchanger with broad specificity for small and large neutral amino acids with characteristics of system L (Pineda et al 1999) (Figure 8.3). This transporter may be involved in intestinal absorption and renal reabsorption of neutral amino acids. Indeed, LAT2 knockdown experiments in the polarized opossum kidney cell line OK, derived from proximal tubule epithelial cells, demonstrated that LAT2 participates in the transepithelial flux of cystine, and the basolateral efflux of cysteine and influx of alanine, serine, and threonine (Fernández et al 2003). To my knowledge, no inherited human disease has yet been related to LAT2 mutations. Therefore, a final demonstration of the role of LAT2 in reabsorption requires the generation of LAT2-knockout mouse models.

The model proposed in Figure 8.3 for intestinal absorption and renal reabsorption of neutral amino acids requires a basolateral efflux system for neutral amino acids. A defective amino acid transport system for this efflux would increase the intracellular concentration of these compounds, resulting in their hyperexcretion in urine and intestinal malabsorption. Candidate transporters for this function may be found within transporter families SLC16 and SLC43. Amino acid transporters in these families mediate facilitated diffusion and may therefore mediate the efflux of neutral amino acids from the high intracellular concentration to the interstitial space. T-type amino acid transporter 1 (TAT1; *SLC16A10*) transports aromatic amino acids in a Na^+- and H^+-independent manner (Kim et al 2001, 2002). TAT1 is expressed in human kidney and small intestine with a basolateral location and can function as a net efflux pathway for aromatic amino acids (Ramadan et al 2006). Thus, TAT1 may supply parallel exchangers (systems y^+L and L) with recycling uptake substrates that could drive the efflux of other amino acids. The SLC16 family (also named MCT for monocarboxylate transporters) holds members that transport monocarboxylates and also thyroid hormones. Several members within this family are orphan transporters (Halestrap & Meredith 2004). Knockout mouse models for TAT1, and their related orphan transporters expressed in kidney cortex and small intestine, may help to identify the basolateral transporters involved in reabsorption of neutral amino acids. LAT3 (Babu et al 2003) and LAT4 (Bodoy et al 2005) within family SLC43 mediate the facilitated diffusion of neutral amino acids with characteristics of system L. Neither of these two transporters is expressed in epithelial cells of the renal proximal convoluted tubule or the small intestine. Interestingly, the SLC43 family has a third member with no identified transport function (EEG1; Stuart et al 2001). Functional and tissue-expression studies are required to ascertain the role of EEG1 in the reabsorption of amino acids.

The molecular basis of the renal and intestinal reabsorption of dicarboxylic amino acids is less known. The bulk ($>90\%$) of filtered acidic amino acids is reabsorbed within segment S1 (i.e. the first part of the proximal convoluted tubule) (Silbernagl & Wolkl 1983, Silbernagl 1983). Two apical acidic transport systems have been described in the proximal tubule: one of high capacity and low affinity and the other of low capacity and high affinity (Hediger et al 1999). The Na^+/K^+-dependent acidic amino acid transporter EAAC1 (also named EAAT3), which localized to chromosome 9p24 (Smith et al 1994) (system X_{AG}^-) is expressed mainly in the brush-border membranes of segments S2 and S3 of the nephron (Shayakul et al 1997) (Figure 8.3). The transport characteristics of *SLC1A1* correspond to the high-affinity system (Kanai & Hediger 1992). The *Slc1a1*-knockout mouse develops dicarboxylic aminoaciduria (Peghini et al 1997), demonstrating the role of this transporter in renal reabsorption of dicarboxylic amino acids. To my knowledge, mutational analysis of *SLC1A1* in patients with dicarboxylic aminoaciduria has not been performed. The apical low-affinity transport system for acidic amino acids in kidney has been characterized in brush-border membrane preparations (Weiss et al 1978), but its molecular entity remains elusive. At renal basolateral plasma membranes, a high-affinity Na^+/K^+-dependent transport system for acidic amino acids has been reported (Sacktor et al 1981), but its molecular structure has not been identified. GLT1 (i.e. the glial high-affinity glutamate transporter (Pines et al 1992, Shashidharan et al 1994) (also named EAAT2; *SLC1A2*) may be responsible for this activity. GLT1 mRNA is expressed in rat kidney cortex and porcine small intestine (Fan et al 2004, Welbourne & Matthews 1999), but the expression of GLT1 protein has not been studied in kidney or intestine. *Slc1a2*-knockout mice show lethal spontaneous epileptic seizures (Tanaka et al 1997). To my knowledge, the renal phenotype in these mice has not been examined.

The last decade has produced a wealth of knowledge about the intestinal and renal reabsorption of amino acids. A description of all the transporters participating in these processes requires: (i) the identification of the genes responsible for isolated cystinuria, autosomal dominant hyperdibasic aminoaciduria type I, isolated lysinuria, renal familial iminoglycinuria, dicarboxylic aminoaciduria, and Hartnup disorder in some patients, (ii) the study of new amino acid transporters expressed in the epithelial barrier of kidney and small intestine, and (iii) renal and intestinal phenotyping of their knockouts. Having addressed these points, the field will reach the stage of system biology: quantification of the simultaneous fluxes across the epithelial cell during intestinal absorption and renal reabsorption.

References

Adibi SA. Intestinal transport of dipeptides in man: relative importance of hydrolysis and intact absorption. J. Clin. Invest. 1971; 50: 2266–75.

Anderson CM, Grenade DS, Boll M, et al. H^+/amino acid transporter 1 (PAT1) is the imino acid carrier: an intestinal nutrient/drug transporter in human and rat. Gastroenterology 2004; 127: 1410–22.

Asatoor AM, Crouchman MR, Harrison AR, et al. Intestinal absorption of oligopeptides in cystinuria. Clin. Sci. 1971; 41: 23–33.

Avissar NE, Ryan CK, Ganapathy V, Sax HC. Na($^+$)-dependent neutral amino acid transporter ATB(0) is a rabbit epithelial cell brush-border protein. Am. J. Physiol. Cell Physiol. 2001; 281: C963–71.

Babu E, Kanai Y, Chairoungdua A, et al. Identification of a novel system L amino acid transporter structurally distinct from heterodimeric amino acid transporters. J. Biol. Chem. 2003; 278: 43838–45.

Baron DN, Dent CE, Harris H, Hart EW, Jepson JB. Hereditary pellagra-like skin rash with temporary cerebellar ataxia, constant renal amino-aciduria, and other bizarre biochemical features. Lancet 1956; 271: 421–8.

Bauch C, Forster N, Loffing-Cueni D, Summa V, Verrey F. Functional cooperation of epithelial heteromeric amino acid transporters expressed in madin-darby canine kidney cells. J. Biol. Chem. 2003; 278: 1316–22.

Bauch C, Verrey F. Apical heterodimeric cystine and cationic amino acid transporter expressed in MDCK cells. Am. J. Physiol. Renal Physiol. 2002; 283: F181–9.

Bertran J, Magagnin S, Werner A, et al. Stimulation of system y($^+$)-like amino acid transport by the heavy chain of human 4F2 surface antigen in *Xenopus* laevis oocytes. Proc. Natl Acad. Sci. USA 1992; 89: 5606–10.

Bisceglia L, Calonge MJ, Totaro A, et al. Localization, by linkage analysis, of the cystinuria type III gene to chromosome 19q13.1. Am. J. Hum. Genet. 1997; 60: 611–16.

Bodoy S, Martin L, Zorzano A, Palacín M, Estevez R, Bertran J. Identification of LAT4, a novel amino acid transporter with system L activity. J. Biol. Chem. 2005; 280: 12002–11.

Bohmer C, Broer A, Munzinger M, et al. Characterization of mouse amino acid transporter B0AT1 (slc6a19). Biochem. J. 2005; 389: 745–51.

Boll M, Foltz M, Rubio-Aliaga I, Kottra G, Daniel H. Functional characterization of two novel mammalian electrogenic proton-dependent amino acid cotransporters. J. Biol. Chem. 2002; 277: 22966–73.

Borsani G, Bassi MT, Sperandeo MP, et al. SLC7A7, encoding a putative permease-related protein, is mutated in patients with lysinuric protein intolerance. Nat. Genet. 1999; 21: 297–301.

Boutros M, Ong P, Saadi I, et al. The human rBAT promoter mutations in cystinuria (Abstract). Am. J. Hum. Genet. 1999; 65(Suppl): A94.

Brodehl J, Gellissen K, Kowalewski S. An isolated defect of the tubular cystine reabsorption in a family with idiopathic hypoparathyroidism. Klin. Wochenschr. 1967; 45: 38–40.

Broër A, Cavanaugh JA, Rasko JEJ, Broër S. The molecular basis of neutral aminoacidurias. Pflugers Arch. – Eur. J. Physiol. 2006; 451: 511–17.

Broer A, Klingel K, Kowalczuk S, Rasko JE, Cavanaugh J, Broer S. Molecular cloning of mouse amino acid transport system B^0, a neutral amino acid transporter related to Hartnup disorder. J. Biol. Chem. 2004; 279: 24467–76.

Broer A, Wagner CA, Lang F, Broer S. The heterodimeric amino acid transporter 4F2hc/y$^+$LAT2 mediates arginine efflux in exchange with glutamine. Biochem. J. 2000; 349: 787–95.

Calonge MJ, Gasparini P, Chillarón J, et al. Cystinuria caused by mutations in rBAT, a gene involved in the transport of cystine. Nat. Genet. 1994; 6: 420–5.

Calonge MJ, Volpini V, Bisceglia L, et al. Genetic heterogeneity in cystinuria: the SLC3A1 gene is linked to type I but not to type III cystinuria. Proc. Natl Acad. Sci. USA 1995; 92: 9667–71.

Chairoungdua A, Kanai Y, Matsuo H, Inatomi J, Kim DK, Endou H. Identification and characterization of a novel member of the heterodimeric amino acid transporter family presumed to be associated with an unknown heavy chain. J. Biol. Chem. 2001; 276: 49390–9.

Chesney RW. Iminoglycinuria. In: Scriver CR, Beaudet AL, Sly SW, Valle D, eds. The Metabolic and Molecular Bases of Inherited Disease. New York: McGraw Hill, 2001: pp. 4971–82.

Chillarón J, Estevez R, Mora C, et al. Obligatory amino acid exchange via systems $b^{0,+}$-like and y$^+$L-like. A tertiary active transport mechanism for renal reabsorption of cystine and dibasic amino acids. J. Biol. Chem. 1996; 271: 17761–70.

Chillarón J, Estevez R, Samarzija I, et al. An intracellular trafficking defect in type I cystinuria rBAT mutants M467T and M467K. J. Biol. Chem. 1997; 272: 9543–9.

Chillarón J, Roca R, Valencia A, Zorzano A, Palacín M. Heteromeric amino acid transporters: biochemistry, genetics, and physiology. Am. J. Physiol. Renal Physiol. 2001; 281: F995–F1018.

Coady MJ, Chen XZ, Lapointe JY. rBAT is an amino acid exchanger with variable stoichiometry. J. Membr. Biol. 1996; 149: 1–8.

Coicadan L, Heyman M, Grasset E, Desjeux JF. Cystinuria: reduced lysine permeability at the brush border of intestinal membrane cells. Pediatr. Res. 1980; 14: 109–12.

Dall'Asta V, Bussolati O, Sala R, et al. Arginine transport through system y($^+$)L in cultured human fibroblasts: normal phenotype of cells from LPI subjects. Am. J. Physiol. Cell Physiol. 2000; 279: C1829–37.

Daniel H. Molecular and integrative physiology of intestinal peptide transport. Annu. Rev. Physiol. 2004; 66: 361–84.

Dello Strologo L, Pras E, Pontesilli C, et al. Comparison between SLC3A1 and SLC7A9 cystinuria patients and carriers: a need for a new classification. J. Am. Soc. Nephrol. 2002; 13: 2547–53.

Desjeux JF, Rajantie J, Simell O, Dumontier AM, Perheentupa J. Lysine Fluxes across the jejunal epithelium in lysinuric protein intolerance. J. Clin. Invest. 1980; 65: 1382–87.

Deves R, Boyd CA. Transporters for cationic amino acids in animal cells: discovery, structure, and function. Physiol. Rev. 1998; 78: 487–545.

DiRocco M, Garibotto G, Rossi GA, et al. Role of haematological, pulmonary and renal complications in the long-term prognosis of patients with lysinuric protein intolerance. Eur. J. Pediatr. 1993; 152: 437–40.

Fan MZ, Matthews JC, Etienne NM, Stoll B, Lackeyram D, Burrin DG. Expression of apical membrane L-glutamate transporters in neonatal porcine epithelial cells along the small intestinal crypt-villus axis. Am. J. Physiol. Gastrointest. Liver Physiol. 2004; 287: G385–98.

Feliubadaló L, Arbones ML, Manas S, et al. Slc7a9-deficient mice develop cystinuria non-I and cystine urolithiasis. Hum. Mol. Genet. 2003; 12: 2097–108.

Feliubadaló L, Font M, Purroy J, et al. Non-type I cystinuria caused by mutations in SLC7A9, encoding a subunit (b$^{0,+}$AT) of rBAT. International Cystinuria Consortium. Nat. Genet. 1999; 23: 52–7.

Fernandez E, Carrascal M, Rousaud F, et al. rBAT-b$^{(0,+)}$AT heterodimer is the main apical reabsorption system for cystine in the kidney. Am. J. Physiol. Renal Physiol. 2002; 283: F540–8.

Fernandez E, Torrents D, Chillarón J, Martin Del Rio R, Zorzano A, Palacín M. Basolateral LAT-2 has a major role in the transepithelial flux of L-cystine in the renal proximal tubule cell line OK. J. Am. Soc. Nephrol. 2003; 14: 837–47.

Fernandez E, Torrents D, Zorzano A, Palacín M, Chillarón J. Identification and functional characterization of a novel low affinity aromatic-preferring amino acid transporter (arpAT). One of the few proteins silenced during primate evolution. J. Biol. Chem. 2005; 280: 19364–72.

Font MA, Feliubadaló L, Estivill X, et al. Functional analysis of mutations in SLC7A9, and genotype-phenotype correlation in non-Type I cystinuria. International Cystinuria Consortium. Hum. Mol. Genet. 2001; 10: 305–16.

Font-Llitjós M, Jimenez-Vidal M, Bisceglia L, et al. New insights into cystinuria: 40 new mutations, genotype-phenotype correlation, and digenic inheritance causing partial phenotype. J. Med. Genet. 2005; 42: 58–68.

Garrod AE. Inborn errors of metabolism (lectures I–IV). Lancet 1908; 2: 1–214.

Gasol E, Jimenez-Vidal M, Chillarón J, Zorzano A, Palacín M. Membrane topology of system xc-light subunit reveals a re-entrant loop with substrate-restricted accessibility. J. Biol. Chem. 2004; 279: 31228–36.

Gasparini P, Calonge MJ, Bisceglia L, et al. Molecular genetics of cystinuria: identification of four new mutations and seven polymorphisms, and evidence for genetic heterogeneity. Am. J. Hum. Genet. 1995; 57: 781–8.

Goodyer PR, Clow C, Reade T, Girardin C. Prospective analysis and classification of patients with cystinuria identified in a newborn screening program. J. Pediatr. 1993; 122: 568–72.

Groneberg DA, Doring F, Eynott PR, Fischer A, Daniel H. Intestinal peptide transport: ex vivo uptake studies and localization of peptide carrier PEPT1. Am. J. Physiol. Gastrointest. Liver Physiol. 2001; 281: G697–G704.

Halestrap AP, Meredith D. The SLC16 gene family–from monocarboxylate transporters (MCTs) to aromatic amino acid transporters and beyond. Pflügers Arch. 2004; 447: 619–28.

Harnevik L, Fjellstedt E, Molbaek A, Denneberg T, Soderkvist P. Mutation analysis of SLC7A9 in cystinuria patients in Sweden. Genet. Test. 2003; 7: 13–20.

Hediger MA. Glutamate transporters in kidney and brain. Am. J. Physiol. Renal Physiol. 1999; 277: F487–92.

Henthorn PS, Liu J, Gidalevich T, et al. Canine cystinuria: polymorphism in the canine SLC3A1 gene and identification of a nonsense mutation in cystinuric Newfoundland dogs. Hum. Genet. 2000; 107: 295–303.

Jaeken J, Martens K, Francois I, et al. Deletion of PREPL, a gene encoding a putative serine oligopeptidase, in patients with hypotonia-cystinuria syndrome. Am. J. Hum. Genet. 2006; 78: 38–51.

Kanai Y, Fukasawa Y, Cha SH, et al. Transport properties of a system y$^+$L neutral and basic amino acid transporter. Insights into the mechanisms of substrate recognition. J. Biol. Chem. 2000; 275: 20787–93.

Kanai Y, Hediger MA. Primary structure and functional characterization of a high-affinity glutamate transporter. Nature 1992; 360: 467–71.

Kanai Y, Segawa H, Miyamoto K, Uchino H, Takeda E, Endou H. Expression cloning and characterization of a transporter for large neutral amino acids activated by the heavy chain of 4F2 antigen (CD98). J. Biol. Chem. 1998; 273: 23629–32.

Kanai Y, Stelzner M, Nussberger S, et al. The neuronal and epithelial human high affinity glutamate transporter. Insights into structure and mechanism of transport. J. Biol. Chem. 1994; 269(32): 20599–606.

Kekomaki M, Visakorpi JK, Perheentupa J, Saxen L. Familial protein intolerance with deWcient transport of basic amino acids. An analysis of 10 patients. Acta Paediatr. Scand. 1967; 56: 617–30.

Kim DK, Kanai Y, Chairoungdua A, Matsuo H, Cha SH, Endou H. Expression cloning of a Na$^+$-independent aromatic amino acid transporter with structural similarity to H$^+$/monocarboxylate transporters. J. Biol. Chem. 2001; 276: 17221–8.

Kim DK, Kanai Y, Matsuo H, et al. The human T-type amino acid transporter-1: characterization, gene organization, and chromosomal location. Genomics 2002; 79: 95–103.

Kleta R, Romeo E, Ristic Z, et al. Mutations in SLC6A19, encoding B0AT1, cause Hartnup disorder. Nat. Genet. 2004; 36: 999–1002.

Koizumi A, Matsuura N, Inoue S, et al (Mass Screening Group). Evaluation of a mass screening program for lysinuric protein intolerance in the northern part of Japan. Genet. Test. 2003; 7: 29–35.

Kowalczuk S, Broer A, Munzinger M, Tietze N, Klingel K, Broer S. Molecular cloning of the mouse IMINO system: an Na$^+$- and Cl-dependent proline transporter. Biochem. J. 2005; 386: 417–22.

Lauteala T, Mykkanen J, Sperandeo MP, et al. Genetic homogeneity of lysinuric protein intolerance. Eur. J. Hum. Genet. 1998; 6: 612–15.

Lauteala T, Sistonen P, Savontaus ML, et al. Lysinuric protein intolerance (LPI) gene maps to the long arm of chromosome 14. Am. J. Hum.Genet. 1997; 60: 1479–86.

Leclerc D, Boutros M, Suh D, et al. SLC7A9 mutations in all three cystinuria subtypes. Kidney Int. 2002; 62: 1550–9.

Leclerc D, Wu Q, Ellis JR, Goodyer P, Rozen R. Is the SLC7A10 gene on chromosome 19 a candidate locus for cystinuria? Mol. Genet. Metab. 2001; 73: 333–9.

Levy HL. Genetic screening. Adv. Hum. Genet. 1973; 4: 1–104.

Levy LL. Hartnup disorder. In: Scriver CR, Beaudet AL, Sly SW, Valle D, eds. The Metabolic and Molecular Bases of Inherited Disease. New York: McGraw Hill, 2001: pp. 4957–69.

Lukkarinen M, Nänö-Salonen K, Pulkki K, Aalto M, Simell O. Oral supplementation corrects plasma lysine concentrations in lysinuric protein intolerance. Metabolism 2003; 52: 935–8.

Mailliard ME, Stevens BR, Mann GE. Amino acid transport by small intestinal, hepatic, and pancreatic epithelia. Gastroenterology 1995; 108: 888–910.

Mariotti F, Huneau JF, Mahe S, Tome D. Protein metabolism and the gut. Curr. Opin. Clin. Nutr. Metab. Care 2000; 3: 45–50.

Mastroberardino L, Spindler B, Pfeiffer R, et al. Amino-acid transport by heterodimers of 4F2hc/CD98 and members of a permease family. Nature 1998; 395: 288–91.

Mathews DM, Adibi SA. Peptide absorption. Gastroenterology 1976; 71: 151–61.

Matsuo H, Kanai Y, Kim JY, et al. Identification of a novel Na$^+$-independent acidic amino acid transporter with structural similarity to the member of a heterodimeric amino acid transporter family associated with unknown heavy chains. J. Biol. Chem. 2002; 277: 21017–26.

Melançon SB, Dallaire L, Lemieux B, Robitaille P, Potier M. Dicarboxylic aminoaciduria: an inborn error of amino acid conservation. J. Pediatr. 1977; 91: 422–7.

Morin CL, Thompson MW, Jackson SH, Sass-Kortsak A. Biochemical and genetic studies in cystinuria: observations on double heterozygotes of genotypes I-II. J. Clin. Invest. 1971; 50: 1961–76.

Mykkanen J, Torrents D, Pineda M, et al. Functional analysis of novel mutations in y($^+$)LAT-1 amino acid transporter gene causing lysinuric protein intolerance (LPI). Hum. Mol. Genet. 2000; 9: 431–8.

Nagata M, Suzuki M, Kawamura G, et al. Immunological abnormalities in a patient with lysinuric protein intolerance. Eur. J. Pediatr. 1987; 146: 427–8.

Nakanishi T, Hatanaka T, Huang W, et al. Na$^+$- and Cl-coupled active ransport of carnitine by the amino acid transporter ATB$^{(0,+)}$ from mouse colon expressed in HRPE cells and *Xenopus* oocytes. J. Physiol. 2001; 532: 297–304.

Noguchi A, Shoji Y, Koizumi A, et al. SLC7A7 genomic structure and novel variants in three Japanese lysinuric protein intolerance families. Hum. Mutat. 2000; 15: 367–72.

Nozaki J, Dakeishi M, Ohura T, et al. Homozygosity mapping to chromosome 5p15 of a gene responsible for Hartnup disorder. Biochem. Biophys. Res. Commun. 2001; 284: 255–60.

Obermuller N, Kranzlin B, Verma R, Gretz N, Kriz W, Witzgall R. Renal osmotic stress-induced cotransporter: expression in the newborn, adult and post-ischemic rat kidney. Kidney Int. 1997; 52: 1584–92.

Ohgimoto S, Tabata N, Suga S, et al. Molecular characterization of fusion regulatory protein-1 (FRP-1) that induces multinucleated giant cell formation of monocytes and HIV gp160 mediated cell fusion. FRP-1 and 4F2/CD98 are identical molecules. J. Immunol. 1995; 155: 3585–92.

Omura K, Yamanaka N, Higami S, et al. Lysine malabsorption syndrome: a new type of transport defect. Pediatrics 1976; 57: 102–5.

Oyanagi K, Miura R, Yamanouchi T. Congenital lysinuria: a new inherited transport disorder of dibasic amino acids. J. Pediatr. 1970; 77: 259–66.

Palacín M, Bertran J, Chillarón J, Estévez R, Zorzano A. Lysinuric protein intolerance: mechanisms of pathophysiology. Mol. Genet. Metab. 2004; 81: S27–37.

Palacín M, Borsani G, Sebastio G. The molecular bases of cystinuria and lysinuric protein intolerance. Curr. Opin. Genet. Dev. 2001a; 11: 328–35.

Palacín M, Estevez R, Bertran J, Zorzano A. Molecular biology of mammalian plasma membrane amino acid transporters. Physiol. Rev. 1998; 78: 969–1054.

Palacín M, Goodyer P, Nunes V, Gasparini P. Cystinuria. In: Scriver CR, Beaudet AL, Sly SW, Valle D, eds. The Metabolic and Molecular Bases of Inherited Disease. New York: McGraw Hill, 2001b: ch. 191, pp. 4957–69. (Updated chapter available at http://genetics.accessmedicine.com

Palacín M, Kanai Y. The ancillary proteins of HATs: SLC3 family of amino acid transporters. Pflügers Arch. 2004; 447: 490–4.

Palacín M, Nunes V, Font-Llitjós M, et al. The genetics of heteromeric amino acid transporters. Physiology (Bethesda) 2005; 20: 112–24.

Parto K, Svedstrom E, Majurin ML, Harkonen R, Simell O. Pulmonary manifestations in lysinuric protein intolerance. Chest 1993; 104: 1176–82.

Peghini P, Janzen J, Stoffel W. Glutamate transporter EAAC-1-deficient mice develop dicarboxylic aminoaciduria and behavioral abnormalities but no neurodegeneration. EMBO J. 1997; 16: 3822–32.

Perheentupa J, Visakorpi JK. Protein intolerance with deficient transport of basic amino acids. Lancet 1965; 2: 813–16.

Peters T, Thaete C, Wolf S, et al. A mouse model for cystinuria I. Hum. Mol. Genet. 2003; 12: 2109–20.

Pfeiffer R, Rossier G, Spindler B, Meier C, Kuhn L, Verrey F. Amino acid transport of y⁺L-type by heterodimers of 4F2hc/CD98 and members of the glycoprotein-associated amino acid transporter family. EMBO J. 1999; 18: 49–57.

Pineda M, Fernandez E, Torrents D, et al. Identification of a membrane protein, LAT-2, that co-expresses with 4F2 heavy chain, an L-type amino acid transport activity with broad specificity for small and large zwitterionic amino acids. J. Biol. Chem. 1999; 274: 19738–44.

Pineda M, Font M, Bassi MT, et al. The amino acid transporter asc-1 is not involved in cystinuria. Kidney Int. 2004a; 66: 1453–64.

Pineda M, Wagner CA, Broer A, et al. Cystinuria-specific rBAT(R365W) mutation reveals two translocation pathways in the amino acid transporter rBAT-b⁰,⁺AT. Biochem. J. 2004b; 377: 665–74.

Pines G, Danbolt NC, Bjoras M, et al. Cloning and expression of a rat brain L-glutamate transporter. Nature 1992; 360: 464–67.

Quan H, Athirakul K, Wetsel WC, et al. Hypertension and impaired glycine handling in mice lacking the orphan transporter XT2. Mol. Cell Biol. 2004; 24: 4166–73.

Rajantie J, Simell O, Perheentupa J. Basolateral membrane transport defect for lysine in lysinuric protein intolerance. Lancet 1980a; 1: 1219–21.

Rajantie J, Simell O, Perheentupa J. Intestinal absorption in lysinuric protein intolerance: impaired for diamino acids, normal for citrulline. Gut 1980b; 21: 519–24.

Rajantie J, Simell O. Lysinuric protein intolerance. Basolateral membrane transport defect in renal tubuli. J. Clin. Invest. 1981; 67: 1078–82.

Ramadan T, Camargo SM, Summa V, et al. Basolateral aromatic amino acid transporter TAT1 (Slc16a10) functions as an efflux pathway. J. Cell Physiol. 2006; 206: 771–9.

Reig N, Chillarón J, Bartoccioni P, et al. The light subunit of system b⁰,⁺ is fully functional in the absence of the heavy subunit. EMBO J. 2002; 21: 4906–14.

Rosenberg LE, Durant JL, Elsas LJ. Familial iminoglycinuria. An inborn error of renal tubular transport. N. Engl. J. Med. 1968; 278: 1407–13.

Rubio-Aliaga I, Frey I, Boll M, et al. Targeted disruption of the peptide transporter Pept2 gene in mice defines its physiological role in the kidney. Mol. Cell. Biol. 2003; 23: 3247–52.

Saadi I, Chen XZ, Hediger M, et al. Molecular genetics of cystinuria: mutation analysis of SLC3A1 and evidence for another gene in type I (silent) phenotype. Kidney Int. 1998; 54: 48–55.

Sacktor B, Rosenbloom IL, Liang CT, Cheng L. Sodium gradient- and sodium plus potassium gradient-dependent L-glutamate uptake in renal basolateral membrane vesicles. J. Membr. Biol. 1981; 60: 63–71.

Schmidt C, Tomiuk J, Botzenhart E, et al. Genetic variations of the SLC7A9 gene: allele distribution of 13 polymorphic sites in German cystinuria patients and controls. Arbeitsgemeinschaft für Padiatrische Nephrologie. Clin. Nephrol. 2003a; 59: 353–9.

Schmidt C, Vester U, Wagner CA, et al. Significant contribution of genomic rearrangements in SLC3A1 and SLC7A9 to the etiology of cystinuria. Arbeitsgemeinschaft für Padiatrische Nephrologie. Kidney Int. 2003b; 64: 1564–72.

Scriver CR. Renal tubular transport of proline, hydroxyproline, and glycine. Genetic basis for more than one mode of transport in human kidney. J. Clin. Invest. 1968; 47: 823–35.

Scriver CR, Efron ML, Schafer IA. Renal tubular transport of proline, hydroxyproline, and glycine in health and in familial hyperprolinemia. J. Clin. Invest. 1964; 43: 374–85.

Scriver CR, Mahon B, Levy HL, et al. The Hartnup phenotype: Mendelian transport disorder, multifactorial disease. Am. J. Hum. Genet. 1987; 40: 401–12.

Segal S and Their SO. Cystinuria. In: Scriver CH, Beaudet AL, Sly WS, Valle D, eds. The Metabolic and Molecular Bases of Inherited Diseases, 7th ed., vol. III. New York: McGraw-Hill, 1995: ch. 117. pp. 3581–601.

Seow HF, Broer S, Broer A, et al. Hartnup disorder is caused by mutations in the gene encoding the neutral amino acid transporter SLC6A19. Nat. Genet. 2004; 36: 1003–7.

Shashidharan P, Wittenberg I, Plaitakis A. Molecular cloning of human brain glutamate/aspartate transporter II. Biochim. Biophys. Acta 1994; 1191: 393–6.

Shayakul C, Kanai Y, Lee WS, Brown D, Rothstein JD, Hediger MA. Localization of the high affinity glutamate transporter EAAC1 in rat kidney. Am. J. Physiol. Renal Physiol. 1997; 273: F1023–9.

Silbernagl S, Volkl H. Molecular specificity of the tubular resorption of "acidic" amino acids. A continuous microperfusion study in rat kidney in vivo. Pflügers Arch. 1983; 396: 225–30.

Silbernagl S. Kinetics and localization of tubular resorption of "acidic" amino acids. A microperfusion and free flow micropuncture study in rat kidney. Pflügers Arch. 1983; 396: 218–24.

Silbernagl S. The renal handling of amino acids and oligopeptides. Physiol. Rev. 1988; 68: 911–1007.

Simell O. Lysinuric protein intolerance and other cationic aminoacidurias. In: Scriver CR, Beaudet AL, Sly SW, Valle D, eds. The Metabolic and Molecular Bases of Inherited Disease.

New York: McGraw Hill, 2001, ch. 192. (Available from http://genetics.accessmedicine.com)

Simell O, Perheentupa J. Renal handling of diamino acids in lysinuric protein intolerance. J. Clin. Invest. 1974; 54: 9–17.

Skopkova Z, Hrabincova E, Stastna S, Kozak L, Adam T. Molecular genetic analysis of SLC3A1 and SLC7A9 genes in Czech and Slovak cystinuric patients. Ann. Hum. Genet. 2005; 69: 501–7.

Smith CP, Weremowicz S, Kanai Y, Stelzner M, Morton CC, Hediger MA. Assignment of the gene coding for the human high-affinity glutamate transporter EAAC1 to 9p24: potential role in dicarboxylic aminoaciduria and neurodegenerative disorders. Genomics 1994; 20: 335–6.

Smith DW, Scriver CR, Simell O. Lysinuric protein intolerance mutation is not expressed in the plasma membrane of erythrocytes. Hum. Genet. 1988; 80: 395–6.

Sperandeo MP, Annunziata P, Ammendola V, et al. Lysinuric protein intolerance: identification and functional analysis of mutations of the SLC7A7 gene. Hum. Mutat. 2005a; 25: 410–11.

Sperandeo MP, Bassi MT, Riboni M, et al. Structure of the SLC7A7 gene and mutational analysis of patients affected by lysinuric protein intolerance. Am. J. Hum. Genet. 2000; 66: 92–9.

Sperandeo MP, Paladino S, Maiuri L, et al. A y($^{+}$)LAT-1 mutant protein interferes with y($^{+}$)LAT-2 activity: implications for the molecular pathogenesis of lysinuric protein intolerance. Eur. J. Hum. Genet. 2005b; 13: 628–34.

Stoller ML, Bruce JE, Bruce CA, Foroud T, Kirkwood SC, Stambrook PJ. Linkage of type II and type III cystinuria to 19q13.1: codominant inheritance of two cystinuric alleles at 19q13.1 produces an extreme stone-forming phenotype. Am. J. Med. Genet. 1999; 86: 134–9.

Stuart RO, Pavlova A, Beier D, Li Z, Krijanovski Y, Nigam SK. EEG1, a putative transporter expressed during epithelial organogenesis: comparison with embryonic transporter expression during nephrogenesis. Am. J. Physiol. Renal Physiol. 2001; 281: F1148–56.

Takanaga H, Mackenzie B, Suzuki Y, Hediger MA. Identification of Mammalian proline transporter SIT1 (SLC6A20) with characteristics of classical system imino. J. Biol. Chem. 2005; 280: 8974–84.

Tanaka K, Watase K, Manabe T, et al. Epilepsy and exacerbation of brain injury in mice lacking the glutamate transporter GLT-1. Science 1997; 276: 1699–702.

Teijema HL, van Gelderen HH, Giesberts MA, Laurent de Angulo MS. Dicarboxylic aminoaciduria: an inborn error of glutamate and aspartate transport with metabolic implications, in combination with a hyperprolinemia. Metabolism 1974; 23: 115–23.

Toivonen M, Mykkanen J, Aula P, Simell O, Savontaus ML, Huoponen K. Expression of normal and mutant GFP-tagged y($^{+}$)L amino acid transporter-1 in mammalian cells. Biochem. Biophys. Res. Commun. 2002; 291: 1173–9.

Torras-Llort M, Torrents D, Soriano-Garcia JF, et al. Sequential amino acid exchange across b$^{0,+}$-like system in chicken brush border jejunum. J. Membr. Biol. 2001; 180: 213–20.

Torrents D, Estévez R, Pineda M, et al. Identification and characterization of a membrane protein (y^{+}L amino acid transporter-1) that associates with 4F2hc to encode the amino acid transport activity y^{+}L. A candidate gene for lysinuric protein intolerance. J. Biol. Chem. 1998; 273: 32437–45.

Torrents D, Mykkanen J, Pineda M, et al. Identification of SLC7A7, encoding y^{+}LAT-1, as the lysinuric protein intolerance gene. Nat. Genet. 1999; 21: 293–6.

Tsumura H, Suzuki N, Saito H, et al. The targeted disruption of the CD98 gene results in embryonic lethality. Biochem. Biophys. Res. Commun. 2003; 308: 847–51.

Verrey F, Closs EI, Wagner CA, Palacín M, Endou H, Kanai Y. CATs and HATs: the SLC7 family of amino acid transporters. Pflügers Arch. 2004; 447: 532–42.

Wartenfeld R, Golomb E, Katz G, et al. Molecular analysis of cystinuria in Libyan Jews: exclusion of the SLC3A1 gene and mapping of a new locus on 19q. Am. J. Hum. Genet. 1997; 60: 617–24.

Weiss SD, McNamara PD, Pepe LM, Segal S. Glutamine and glutamic acid uptake by rat renal brushborder membrane vesicles. J. Membr. Biol. 1978; 43: 91–105.

Welbourne TC, Matthews JC. Glutamate transport and renal function. Am. J. Physiol. Renal Physiol. 1999; 277: F501–5.

Wells RG, Lee WS, Kanai Y, Leiden JM, Hediger MA. The 4F2 antigen heavy chain induces uptake of neutral and dibasic amino acids in *Xenopus* oocytes. J. Biol. Chem. 1992; 267: 15285–8.

Whelan DT, Scriver CR. Hyperdibasicaminoaciduria: an inherited disorder of amino acid transport. Pediatr. Res. 1968; 2: 525–34.

Wilcken B, Yu JS, Brown DA. Natural history of Hartnup disease. Arch. Dis. Child. 1977; 52: 38–40.

Yamashita A, Singh SK, Kawate T, Jin Y, Gouaux E. Crystal structure of a bacterial homologue of Na^{+}/Cl^{-}-dependent neurotransmitter transporters. Nature 2005; 437: 215–23.

Yuen YP, Lam CW, Lai CK, et al. Heterogeneous mutations in the SLC3A1 and SLC7A9 genes in Chinese patients with cystinuria. Kidney Int. 2006; 69: 123–8.

CHAPTER 9

Primary Renal Uricosuria

MAKOTO HOSOYAMADA, KIMIYOSHI ICHIDA, TATSUO HOSOYA AND HITOSHI ENDOU

INTRODUCTION

Uric acid is excreted as the end product of purine metabolism after reabsorption and secretion in human kidney. Renal uricosuria, caused by the attenuation of its reabsorption or by the enhancement of its secretion, is categorized as primary and secondary forms (listed in Table 9.1). In this chapter, primary form of renal uricosuria is mainly discussed. Primary renal uricosuria, or hereditary renal hypouricemia, a persistent marked hypouricemia due to high urinary urate excretion, is a common inherited and heterogeneous disorder (MIM 220150, 242050 and 307830). In 2002, urate transporter 1 (URAT1) was identified as a transporter for urate reabsorption at the brush-border membrane of the proximal tubule in human kidney. Moreover, *SLC22A12*, the URAT1 gene, is responsible for hereditary renal hypouricemia (Enomoto et al 2002); i.e. most renal hypouricemic Japanese patients were found to carry mutations in *SLC22A12* (Tanaka et al 2003, Ichida et al 2004, Iwai et al 2004, Komoda et al 2004).

BASICS OF URATE HANDLING IN HUMAN KIDNEY

Urate Excretion by Human Kidney

For the evaluation of urate handling in the kidney, fractional excretion of urate (FEua), the ratio of urate clearance of final

TABLE 9.1 Causes of hypouricemia due to increased renal excretion of uric acid

Condition	Genetic contribution	
1 Inherited renal hypouricemia (primary renal uricosuria)	URAT1 (*SLC22A12*)	Thermal burns
		Primary hyperparathyroidism
2 Secondary increased renal excretion of		Acute renal tubular necrosis
uric acid		Renal transplant rejection
(a) Inherited causes of the Fanconi		(c) Drugs
syndrome and its variants		Uricosuric agents (sulfinpyrazone,
Cystinosis (accumulation of	Cystinosin (*CTNS*)	probenecid, benzbromarone)
intralysosomal cystine)		Non-steroidal anti-inflammatory
Galactosemia (galactose-	Galactose-1-phosphate	drugs with uricosuric properties:
1-phosphate toxicity)	Uridyl transferase	phenylbutazone, azapropazone,
Hereditary fructose intolerance	Aldolase B	aspirin (> 4 g/day)
(fructose-1-phosphate toxicity)		Coumarin anticoagulants
Glycogen storage disease type 1	Glucose-6-phosphatase	(e.g. warfarin)
Wilson's disease (copper toxicity)	*ATP7B*	Outdated tetracycline (5-alpha-6-
Cytochrome C oxidase deficiency	Cytochrome C oxidase	anhydro-4-epitetracycline)
(b) Acquired causes of the Fanconi		(d) Others
syndrome and its variants		SIADH
Metal poisoning (Cd, Zn, Cu, Hg,		Carcinomatosis
uranium)		Diabetes mellitus
Multiple myelomatosis		Pregnancy
Nephrotic syndrome		
Malignant disease (paraneoplastic		
syndrome)		
Autoimmune disease (e.g. Sjogren's		
syndrome)		

Genetic Diseases of the Kidney
ISBN: 978-0-12-449851-8

urine to creatinine clearance, has been used in the clinical and experimental situation. Although urate is probably bound to plasma proteins, urate is freely filtered practically at the glomerulus (Holmes & Blondet 1979). Indeed, determinations of the concentration of urate in glomerular ultrafiltrate and plasma indicated that they are almost equal (Bordley & Richard 1933, Roch-Ramel et al 1980, Weinman et al 1981), and we can consider the urate concentration of glomerular filtrate can be substituted with plasma urate level. Therefore, if the tubules of the kidney are not reabsorbing or secreting urate, FEua of final urine must be 100%. When FEua of final urine is below 100%, the tubules must reabsorb filtered urate. Contrarily, when the FEua is more than 100%, the urate secretion must occur in the kidney. In normal humans, the FEua is about 8–10%, indicating very efficient urate reabsorption occurs in human kidneys (Coombs et al 1940).

Tubular Transport of Urate

From many physiological studies using experimental animals, urate is considered to be mainly reabsorbed at the proximal tubule in the kidney (Abramson & Levitt 1975, Kramp & Lenoir 1975, Roch-Ramel et al 1980). Although most filtered urate is reabsorbed in the proximal tubule, urate transport in distal nephron segments has been controversial. From micropuncture study with the kidney of the Cebus monkey,

which reabsorbs urate similar to the human kidney, 18% of filtered urate was reabsorbed between late proximal and early distal tubule, and 4% of filtered urate was reabsorbed between late distal tubule and pelvic urine (Roch-Ramel et al 1980). Based on the fact that the urate reabsorption was not different between superficial and deep nephrons of Munich-Wistar rat (Frommer et al 1982), only nephron heterogeneity is not favorable for interpretation of 4% of filtered urate loss, and urate reabsorption by the collecting duct cannot be ruled out. Although 18% of urate reabsorption between late proximal and early distal tubule can be explained by the reabsorption of the late proximal tubule segment after the puncture site, the possibility of urate transport at the Henle's loop segment has remained.

Before the identification of *SLC22A12*, renal handling of urate had been explained by the '4-component model,' which separated renal urate transport into four components: glomerular filtration, presecretory reabsorption, secretion, and postsecretory reabsorption (Diamond & Paolino 1973, Steele 1973, Steele & Boner 1973, Rieselbach & Steele 1974) (Figure 9.1A). This model hypothesized that the anti-uricosuric effect of pyrazinamide, which reduces urinary urate to very low levels without a change in the glomerular filtration rate (GFR), was the inhibition of urate secretion by pyrazinoate (PZA), the metabolite of pyrazinamide (Weiner & Tinker 1972) (Figure 9.1B). Moreover, based on the attenuation of the uricosuric effect of probenecid by

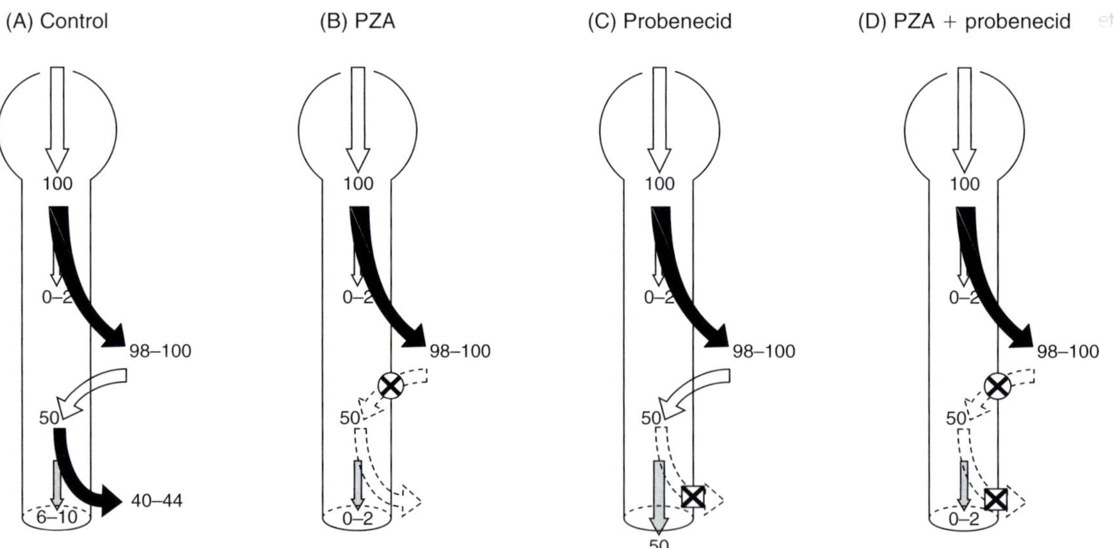

FIGURE 9.1   Obsolete '4-component model' for urate handling in human kidney. (A) Normal urate handling of single nephron. Numerical values indicate hypothetical order of magnitude of the transport processes. The upper open arrow represents glomerular filtration as the first component and urate is filtered freely at the glomerulus. The upper closed arrow represents pre-secretory reabsorption as the second component and 98–100% of the filtered urate is reabsorbed before the secretion site. Therefore the urate remaining in the lumen is only 0–2% of the filtered urate. The lower open arrow represents secretion as the third component and approximately 50% of the filtered urate is secreted. The lower closed arrow represents post-secretory reabsorption as the fourth component and 40–44% of the filtered urate is reabsorbed after secretion. The gray arrow represents urate excretion in final urine, 6–10% of the filtered urate. (B) The inhibition of urate secretion by PZA decreases urate excretion in the final urine. (C) The inhibition of post-secretory reabsorption by probenecid increases urate excretion in the final urine. (D) The '4-component model' can interpret the loss of uricosuric effect of probenecid by PZA pretreatment.

pretreatment of pyrazinamide (Diamond & Paolino 1973), probenecid was considered to inhibit the reabsorption of secreted urate (Figure 9.1D). This interpretation introduced the concept of presecretory and postsecretory reabsorption.

However, some reports using membrane vesicles have recently indicated that the antiuricosuria induced by pyrazinamide was due to enhanced urate reabsorption through exchange of PZA via the urate/anion exchanger at the brush-border membrane (Guggino et al 1983, Guggino & Aronson 1985, Roch-Ramel et al 1996, 1997). Furthermore, renal hypouricemic patients with homozygous *SLC22A12* defect revealed a deficit of both the antiuricosuric action of pyrazinamide and the uricosuric action of benzbromarone (Ichida et al 2004). This suggested that the target molecule for PZA is URAT1, the transporter for urate reabsorption. Thus, the hypothesis that PZA inhibited the transporter for urate secretion emerged as unfavorable, and the model of renal urate handling has been reconsidered. Based on the previous vesicle studies, we present an arranged model of renal urate handling: reabsorption by URAT1, reabsorption by urate transporters other than URAT1, and secretion, as shown in Figure 9.2.

Urate Reabsorption by URAT1

Using Northern blot analysis, human URAT1 mRNA is expressed in adult and fetal kidney. No expression was

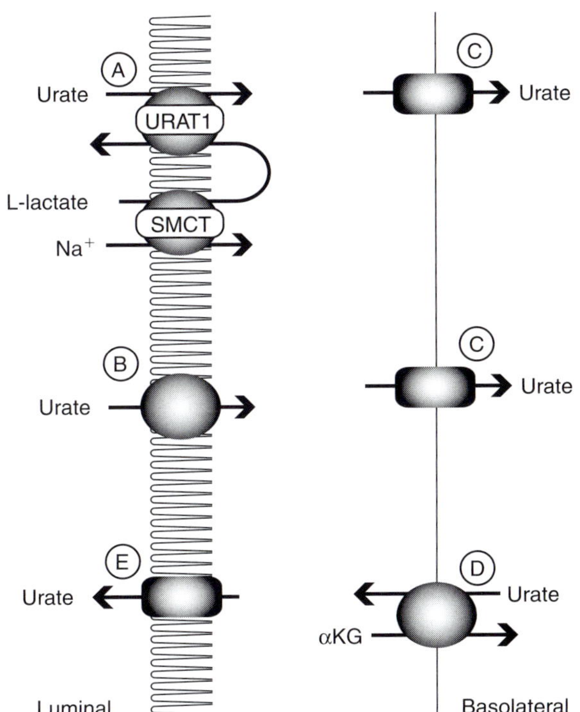

FIGURE 9.2 New concept for a model of urate handling in human kidney. (A) Urate reabsorption by URAT1. (B) Urate reabsorption by urate transporters other than URAT1. (C) Urate efflux transporter at the basolateral membrane for reabsorption. (D) Urate influx transporter at the basolateral membrane for secretion. (E) Urate efflux transporter at the brush border membrane for secretion.

detected in brain, heart, skeletal muscle, small and large intestine, thymus, spleen, leukocyte, liver, lung, or placenta. In human kidney, the expression of the human URAT1 molecule was specifically demonstrated at the brush-border membrane of proximal tubule epithelia by immunohistochemistry. The urate transport by human URAT1 was demonstrated with *Xenopus* oocytes which expressed human URAT1 protein by the injection of human URAT1 RNA. The urate transport by human URAT1 was characterized as its K_m value of $371 \pm 28\,\mu M$. From the cis-inhibition profile of human URAT1-dependent urate transport, clinically used uricosuric agents such as probenecid, benzbromarone, sulfinpyrazone, and salicylic acid inhibit urate transport by human URAT1 in vitro. Therefore, these agents are thought to express their uricosuric effect through the inhibition of urate reabsorption by URAT1 in human kidney. L-lactate, pyrazinoate, and nicotinate trans-stimulated urate transport by human URAT1 in vitro. Therefore, the clinical antiuricosuric effect of these drugs is reckoned to be due to enhancement of urate reabsorption by human URAT1 (Enomoto et al 2002).

On human brush-border membrane vesicles, the urate/Cl^- exchanger was expressed and cis-inhibited by pyrazinoate, probenecid, lactate, ketone bodies, succinate, and monovalent forms of alfa-ketoglutarate (Roch-Ramel et al 1994). Urate uptake by human URAT1 was also enhanced by an outward-directed Cl^- gradient and cis-inhibited by the same organic anions. The insensitivity of PAH or OH^- for the cis-inhibition of the urate/Cl^- exchanger of human brush-border membrane vesicles was also observed for urate uptake by human URAT1 (Enomoto et al 2002). Therefore, human URAT1 (Figure 9.3) is applicable to the urate/anion exchanger at the brush border of human kidney.

Since the influx of urate monoanion from luminal fluid to proximal tubule cells is uphill transport because of negative membrane potential, driving force is necessary for urate reabsorption by human URAT1. Two experiments support the idea that the concentration gradient of lactate produced by Na^+-lactate cotransporter at the brush-border membrane is the physiological driving force of urate reabsorption by human URAT1: trans-stimulation of human URAT1 dependent urate transport by lactate in vitro (Enomoto et al 2002) and suppression of FEua by lactate infusion in vivo (Yamamoto et al 1993). Recently, the gene product of *SLC5A8*, SMCT, was identified as one of the molecules of the Na^+-lactate co-transporter in mouse kidney (Coady et al 2004, Gopal et al 2004).

The larger difference of lactate concentration between the inside and the outside of the brush-border membrane induces the stronger driving force of urate reabsorption by human URAT1; therefore, lower lactate levels in luminal fluid produced by SMCT and higher intracellular lactate levels are thought to induce the larger urate reabsorption in human kidney. The intracellular lactate level is affected by blood lactate level, gluconeogenesis in proximal tubule cells, and the efflux of lactate from the basolateral membrane.

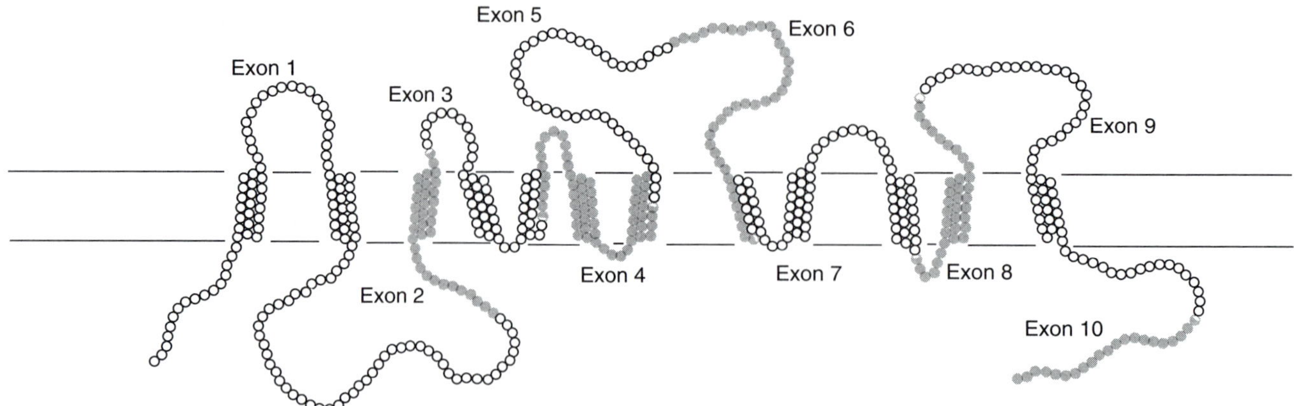

FIGURE 9.3 The predicted membrane-spanning secondary structure of URAT1 and positions of the mutations. Predicted secondary structure of URAT1 is shown. Each circle represents one amino acid. A chain of open or gray circles indicates that corresponding amino acids were derived from the same exon.

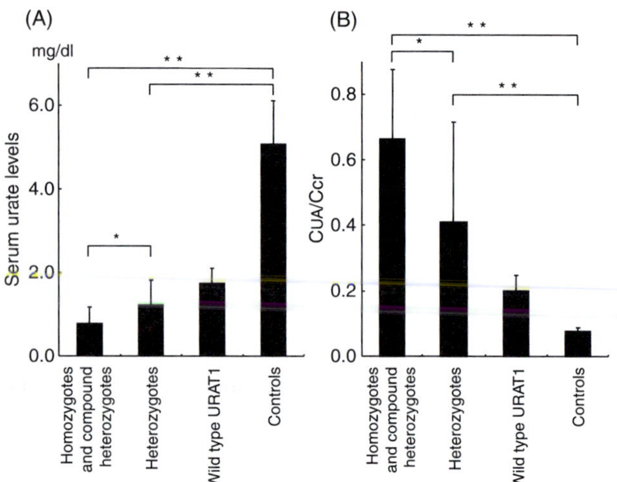

FIGURE 9.4 (A) Serum urate levels of renal hypouricemic patients with homozygous and compound heterozygous *SLC22A12* mutations, renal hypouricemic patients with heterozygous *SLC22A12* mutations, renal hypouricemic patients with wild type URAT1 and controls. (B) CUA/Ccr of the patients with homozygous and compound heterozygous *SLC22A12* mutations, the patients with heterozygous *SLC22A12* mutations, the patients with wild type URAT1 and controls. In this figure, serum urate levels and CUA/Ccr in the patients with wild type URAT1 were not statistically analyzed with those of other groups as only 2 patients were classified as patients with wild type URAT1. *: $p < 0.05$, **: $p < 0.001$.

FEua of renal hypouricemic patients with homozygous *SLC22A12* defect increased between 47% and 130% and the average was 66%. In the heterozygotes, the average FEua also increased to 41%. This relation of FEua among the homozygotes, the heterozygotes, and the controls demonstrated a gene dosage effect (Ichida et al 2004); therefore, the expression level of URAT1 is also thought to affect urate reabsorption in human kidney (Figure 9.4).

URATE REABSORPTION BY URATE TRANSPORTER OTHER THAN URAT1

Even in renal hypouricemic patients with homozygous *SLC22A12* defect, most patients demonstrated FEua below 100% (Ichida et al 2004); therefore, another urate transporter for reabsorption must be expressed on the brush-border membrane other than URAT1. This urate transporter molecule for reabsorption has not yet been identified.

Although the urate transport by URAT1 was not pH dependent, pH-dependent urate uptake was demonstrated with the brush-border membrane vesicle of rat or dog (Kahn & Aronson 1983, Kahn et al 1983). However, using brush-border membrane vesicle of human kidney, urate/OH$^-$ exchange was less effective than that of rat or dog. Although the small amount of urate uptake enhanced by pH difference was explained by the urate uptake through the voltage-driven urate channel at the human brush-border membrane (Roch-Ramel et al 1994), the possibility of a pH-dependent urate transporter at the human brush-border membrane has remained.

Moreover, intraluminal acidity in the late portion of the proximal tubule may increase the concentration of nonionized uric acid in luminal fluid higher than that in the interstitium; therefore, nonionized uric acid may also be able to be reabsorbed through the transcellular pathway.

Although the size of the urate molecule seems to be too large to pass through the paracellular pathway, the possibility of urate reabsoption through the paracellular pathway has not yet been dismissed. Since luminal potential of the early portion of the proximal tubule segment is negative to interstitium, the urate monoanion may be able to be reabsorbed through the paracellular pathway.

URATE EFFLUX TRANSPORTER AT THE BASOLATERAL MEMBRANE

The urate efflux from tubular cells to the interstitium has not been studied enough. Using rat basolateral membrane vesicles, urate/chloride anion exchanger was demonstrated (Kahn et al 1985). This urate/chloride anion exchanger at the basolateral membrane was not inhibited by PAH and lactate. Since urate transport by human OAT1 was inhibited by PAH and external chloride depletion, human OAT1 seems to be unfavorable as the molecule for urate/chloride anion exchange (Ichida et al 2003). Although human OAT3 was a candidate for urate/chloride anion exchanger at the basolateral membrane, human OAT3 was indicated to mediate secretory urate influx as the urate/alfa-ketoglutarate exchanger (Bakhiya et al 2003). These OAT transporters on the basolateral membrane seem to operate as urate transporters for secretory urate influx rather than for reabsorptive urate efflux; therefore, another transporter molecule for reabsorptive urate efflux is thought to be expressed at the basolateral membrane.

TUBULAR SECRETION OF URIC ACID

In some renal hypouricemic patients, FEua was reported to be above 100%. Although this fact could be explained by urate production in the tubule, urate loading experiments suggested that the urate secretory pathway exists in the human kidney (Gutman et al 1959). The report of the unusual antiuricosuric effect of probenecid in a renal hypouricemic patient also indicates the expression of a urate transporter molecule for secretion in human kidney (Matsuda et al 1982). However, it is unknown whether a urate secretory pathway always functions independently of urate reabsorptive pathways, or a urate reabsorptive pathway changes to secrete urate in a particular situation such as a high purine diet (Hisatome et al 1998). No supporting observation has been reported about the paracellular pathway for urate secretion.

URATE INFLUX TRANSPORTER AT THE BASOLATERAL MEMBRANE

Most urate-secreting animals demonstrate that PAH inhibits their urate secretion in vitro (Roch-Ramel et al 1980, Schali & Roch-Ramel 1980). Urate accumulation in kidney slices of urate-secreting animals was observed, which was inhibited by PAH (Kim et al 1992). Therefore, the PAH transporter can contribute uphill transport of urate at the basolateral membrane of the proximal tubule. Although rat and human OAT1, one of the PAH transporters localized at the basolateral membrane, demonstrated urate transport activity in vitro (Ichida et al 2003), urate accumulation in kidney slices of rat, a urate-reabsorptive animal, was not observed (Kim et al 1992). Moreover, the administration of PAH did not attenuate urate excretion in normal human kidney in vivo (Boner & Steele 1973). Therefore, urate efflux at the basolateral membrane mediated by urate efflux transporter (Figure 9.2C) may overcome urate influx at the basolateral membrane mediated by urate influx transporter (Figure 9.2D) in human kidney.

URATE EFFLUX TRANSPORTER AT THE BRUSH BORDER MEMBRANE

Since intracellular electrical potential is negative, facilitated transport can occur as urate monoanion efflux. This voltage-driven urate pathway was demonstrated with human brush-border membrane vesicles. As the channel molecule for this voltage-driven urate pathway at the brush border membrane, UAT, or galectin 9 was propounded (Lipkowitz et al 2001). Since PAH was also effluxed from brush-border membrane, the PAH channel at the brush-border membrane, OATv1 (Jutabha et al 2003) or NPT1 (Uchino et al 2000) is also one of candidates for the voltage-driven urate pathway. Moreover, the ATP-dependent efflux pump, ABCC4, was also propounded as a urate efflux pump at the brush-border membrane (Van Aubel et al 2005). The contribution of these candidate molecules toward the renal handling of urate in the human kidney will be evaluated by the discovery of a patient who lacks these molecules.

HEREDITARY RENAL HYPOURICEMIA

Clinical Entity

Serum urate concentration below 2.0 mg/dl has been most widely used as the criterion for hypouricemia classified as either the overexcretion type or underproduction type. Hereditary renal hypouricemia is a common overexcretion-type hypouricemia and heterogeneous disorder and the feature is a persistent marked hypouricemia due to high urinary urate excretion above about 25% of FEua characterized by impaired tubular urate transport. Secondary renal hypouricemia should be excluded (Table 9.1).

Incidence

Praetorius and Kirk first reported a hypouricemic man whose uric acid clearance was higher than inulin clearance (Praetorius & Kirk 1950). The majority of case reports were in patients of Japanese descent (Kawabe et al 1976, Akaoka et al 1977, Fujiwara et al 1980, Suzuki et al 1981, Matsuda et al 1982, Shichiri et al 1982, Tofuku et al 1982, Takeda et al 1985, Fujieda et al 1995, Ninomiya et al 1996, Hirasaki et al 1997, Oi et al 1998, Sato et al 1998, Hisanaga et al 1999, Kikuchi et al 2000, Watanabe et al 2000, Takeda et al 2001, Ishikawa 2002, Ohta et al 2002, Watanabe 2002, Ito et al 2003, Tanaka et al 2003, Ichida et al 2004, Komoda et al 2004), while most of the remainder were non-Ashkenazi Jews (Yeun & Hasbargen 1995). Although the frequency of this disorder, 0.12–0.40%, in Japanese was reported (Hisatome et al 1989, Igarashi 1993, Kotake et al 1993), the incidence in non-Ashkenazi Jews and others is unknown. The frequency of renal hypouricemia in other populations is thought to be lower because of the lower number of patients reported, the lower frequency of hypouricemia (Gresser et al 1990), and no renal hypouricemic patients in hospitalized patients with hypouricemia (Ramsdell & Kelley 1973, Weinberger et al 1977, Bairaktari et al 2003).

Classification of Renal Hypouricemia

Renal hypouricemia has been classified into the following five types by the response to the antiuricosuric drug pyrazinamide, and the uricosuric drug, probenecid: (1) a defect with an attenuated response to both pyrazinamide and probenecid (Greene et al 1972); (2) a defect with a response to pyrazinamide and no response to probenecid (Barrientos et al 1979); (3) a defect with a response to both drugs and probenecid inducing urate clearance over the GFR (Shichiri et al 1982, Munoz Sanz et al 1983); (4) a defect with an attenuated response to pyrazinamide and no or paradoxical antiuricosuric response to probenecid (Simkin et al 1974); and (5) a defect without any response to either drug (Shichiri et al 1990).

However, the identification of URAT1 as a urate transporter for reabsorption and as the target of PZA demonstrated that the '4-component model' was unfavorable as described above; therefore, new classification of renal hypouricemia will be proposed according to new paradigm.

Gene Responsible

URAT1, one of the organic anion transporter family, was discovered to be a transporter for urate reabsorption in exchange of lactate at apical membrane of a proximal tubular cell (Enomoto et al 2002). Enomoto demonstrated that URAT1 regulates serum urate levels by showing that three patients with renal hypouricemia had defects in *SLC22A12*, encoding URAT1 (Enomoto et al 2002). In a study of 32 patients with renal hypouricemia, 30 had *SLC22A12* defects (Ichida et al 2004). This fact demonstrates that most renal hypouricemia results from URAT1 defect, though several transporters are expected to participate in urate handling in the kidney. No transporter other than URAT1 has been identified as a transporter responsible for renal hypouricemia. Since Enomoto et al reported *SLC22A12* responsible for renal hypouricemia, *SLC22A12* mutation research for renal hypouricemia has been performed in several reports (Enomoto et al 2002, Tanaka et al 2003, Ichida et al 2004, Iwai et al 2004, Komoda et al 2004). Fifty-two patients with renal hypouricemia including two siblings have been analyzed (Table 9.2; Figure 9.4). Forty-seven of 52 patients had *SLC22A12* mutation (90.4%), whereas five patients had none in the *SLC22A12*-coding region. Sixty-six W258X mutations of 104 alleles (63.5%) were found; 22 homozygotes, 15 compound heterozygotes and seven heterozygotes of W258X. Six R90H mutations stood at 5.8%. All seven missense URAT1 mutants had no activity or significantly decreased in expressing URAT1 mutant in *Xenopus laevis* oocytes (Enomoto et al 2002, Ichida et al 2004).

Symptoms and Complications

No symptoms due to hypouricemia itself have been clarified. Exercise-induced acute renal failure (ARF) and urolithiasis, however, are significant complications. Most renal

TABLE 9.2

SLC22A12 mutations		Enomoto et al	Tanaka et al	Ichida et al	Komoda et al	Iwai et al	Total
Nucleotide	**Amino acid**						
G774A/G774A	W258X	1	2*	11	5	3	22
G774A/C650T	W258X/T217M			1	1		2
G774A/C889T	W258X/Q297X			1			1
G774A/ IVS2 + 1G → A	W258X/Frameshift			1			1
G774A/G269A	W258X/R90H			5		1	6
G774A/G412A	W258X/V138M			2			2
G774A/G490A	W258X/G164S			1			1
G774A/1639-1643delGTCCT	W258X/Frameshift			1			1
G774A/A1145T	W258X/Q382L			1			1
G774A/+	W258X			5		2	7
C650T/C650T	T217M/T217M	1					1
G894T/G894T	E298D/E298D	1					1
T1289C/+	M430T			1			1
No detectable mutation in *SLC22A12*				2	1	2	5
Total		3	2	32	7	8	52

*: Siblings.

hypouricemic patients have been identified from these complications. Urinary urate excretion in most renal hypouricemic patients is somewhat excessive at about 600 mg/day or over (Greene et al 1972, Sperling 2000). Incorporation of [^{15}N] glycine into urinary urate after oral administration of [^{15}N] glycine for a renal hypouricemic patient was increased (Akaoka et al 1975). These results mean acceleration of de novo purine biosynthesis pathway or attenuation of extrarenal disposal of urate.

Exercise-induced ARF

INCIDENCE OF EXERCISE-INDUCED ARF

The incidences of exercise-induced ARF in patients with renal hypouricemia have not been determined. Out of 32 patients with renal hypouricemia, three patients had a history of exercise-induced ARF (Ichida et al 2004). In this study, 31 patients were diagnosed from medical examination or other diseases. Accordingly, 6.5% of the 31 patients diagnosed by chance have a past history of exercise-induced ARF. An association study between genetic polymorphisms of *SLC22A12* and serum urate levels was conducted in an epidemiologic cohort representing the general Japanese population (Iwai et al 2004). In this study, eight of 1875 subjects were hypouricemic and six had mutations in *SLC22A12* (three homozygous, one compound heterozygous, two heterozygous). This means that the incidence of renal hypouricemia due to *SLC22A12* abnormality is 0.32%, similar to that of renal hypouricemia previously reported (Hisatome et al 1989, Igarashi 1993, Kotake et al 1993). The number of reported exercise-induced ARF in patients with renal hypouricemia in Japan (Ohta et al 2004) has been much smaller than the number estimated from the frequency of renal hypouricemia due to *SLC22A12* abnormality and the frequency of exercise-induced ARF in them. Thus, a number of renal hypouricemic patients with exercise-induced ARF might have been overlooked due to the fact that the serum urate levels in these patients increase to the normal to high-normal range, during an exercise-induced ARF episode. Furthermore, most events might not have been reported even if diagnosed as the prognosis of exercise-induced ARF is generally favorable.

SYMPTOMS OF EXERCISE-INDUCED ARF

After exercise, symptoms such as nausea, loin pain, abdominal pain, fatigue, and fever appear in 24 hours in most patients. The features of this form of ARF are mostly mild, nonoliguric, no or mild rise of serum creatine phosphokinase, and good prognosis. Recovery is achieved in a few weeks, at the least in one month, though hemodialysis is sometimes performed.

MECHANISM OF EXERCISE-INDUCED ARF

Exercise-induced ARF in renal hypouricemic patients was characterized by patchy renal vasoconstriction observable by enhanced abdominal computed tomography scan (Ishikawa et al 1990). This type of exercise-induced ARF without rhabdomyolysis was reported in the 1980s (Ishikawa et al 1981, 1982). As infusion of catecholamine and angiotensin induced patchy cortical vasoconstriction of the kidney in dogs (Carriere 1969, Carriere & Friborg 1969), exercise-inducing substances are suggested to act on renal blood vessels. As renal oxygen consumption is about 10% of whole body oxygen consumption and renal blood flow decreases with exercise, the kidney easily becomes ischemic and is damaged. The mechanism, however, is as yet unknown. Though this type of ARF is developed in healthy subjects and patients with renal hypouricemia, the frequency in patients with renal hypouricemia is obviously higher than that in healthy subjects in view of the proportion of renal hypouricemic patients to healthy subjects. Ishikawa reviewed 118 cases of exercise-induced ARF caused by patchy renal vasoconstriction. Forty-nine of 96 cases whose serum uric acid levels were described revealed renal hypouricemia (51.0%) (Ishikawa 2002). The reason for the high frequency in renal hypouricemic patients is unknown. One hypothesis proposed is that renal reperfusion injury due to vasoconstriction results from an exercise-induced increase in oxygen free radicals and a lack of urate free radical scavengers (Murakami et al 1993). This hypothesis, however, does not correspond to a fact that exercise-induced ARF has not been reported in xanthinuric patients who lack xanthine dehydrogenase and whose serum urate levels are almost 1 mg/dl or below, similar to renal hypouricemic patients. The mechanism which promotes exercise-induced ARF in renal hypouricemic patients needs to be reconsidered from the viewpoint of URAT1 functional loss.

Erley et al first reported ARF in a patient with renal hypouricemia and suggested that ARF resulted from uric acid nephropathy because of the observation of uric acid crystals in some renal tubular lumen (Erley et al 1989). In 1990, Ishikawa et al described exercise-induced ARF in renal hypouricemic patients having patchy renal vasoconstriction (Ishikawa et al 1990). Yeun et al advocated that urate nephropathy resulted from an increase in uric acid production by exercise-induced ATP breakdown (Yeun & Hasbargen 1995). Consequently, two mechanisms for exercise-induced ARF in patients with renal hypouricemia were proposed, though the first report did not refer to exercise prior to the episode. Tubular urate precipitation, however, has not been identified in renal biopsy specimens, with the exception of the first report (Erley et al 1989, Ishikawa et al 1990, Numabe et al 1992, Igarashi 1993, Ninomiya et al 1996, Tazawa et al 1996, Oi et al 1998, Kikuchi et al 2000, Watanabe et al 2000, Ito et al 2003). These facts demonstrate that patchy renal vasoconstriction is mainly responsible for exercise-induced ARF and uric acid nephropathy is exceptional.

PROGNOSIS OF EXERCISE-INDUCED ARF

The prognosis for a single episode of exercise-induced ARF is generally favorable, though the long-term prognosis is unknown. In the Japanese survey study, not one of 54 renal hypouricemic patients with exercise-induced ARF developed chronic renal failure (Ohta et al 2004). However, one patient who had at least four episodes of exercise-induced ARF had urine-concentrating disability even after recovery of renal function and presented chronic renal lesions such as thickening of the tubular basement membrane and interstitial fibrosis (Ohta et al 2004). Kikuchi et al classified seven renal hypouricemic patients into two groups by histories of exercise-induced ARF and calculus. The creatinine clearance of the group with complications was significantly lower than that of the group without complications, though creatinine clearances of all patients were within normal limits (Kikuchi et al 2000). Furthermore, two of 32 patients with renal hypouricemia had chronic renal failure and the origin of one patient's chronic renal failure was unknown in spite of no obvious past history of exercise-induced ARF (Ichida et al 2004). The patient might have had recurrent exercise-induced ARF without symptoms. The final long-term outcome of exercise-induced ARF might be underestimated and remains unknown.

Urolithiasis

The prevalence of urolithiasis in renal hypouricemic patients is markedly higher than the 2–3% prevalence in the general population. Seven of the 28 propositi with renal hypouricemia reported from 1972 to 1990 had urolithiasis (25%) (Sperling 2000). Urolithiasis in renal hypouricemic patients in Japan from 1977 to 1994 was summarized and 17 of 67 patients, 25.4%, had urinary calculi. Five patients had uric acid calculi, four patients calcium calculi, one patient magnesium ammonium phosphate calculus and the stone constituents in the other seven patients were not analyzed (Tabe 1996). A recent report described that four of 32 patients, 12.5%, had a past history of urolithiasis (Ichida et al 2004). Since the correlation of amount of urinary urate with the prevalence of urolithiasis is well known (Yu & Gutman 1967, Grover et al 1990), high prevalence in these reports is probably related to the increased amount of urinary urate in patients with renal hypouricemia compared with healthy subjects as mentioned above.

Treatment of Renal Hypouricemia

Specific treatment for exercise-induced ARF has not been established. Though Yeun et al described the possibility of allopurinol for prophylaxis, this has not been verified (Yeun & Hasbargen 1995). Restriction of exercise may be effective for prophylaxis; however, it is not entirely appropriate, because exercise-induced ARF does not always follow exercise and is believed to stem from other events or condition such as dehydration.

For preventing urolithiasis, high fluid intake and the neutralization of acidic urine are effective. Uric acid-decreasing agents such as allopurinol might be necessary when urinary urate excretion is excessive and urinary calculus develops recurrently.

References

Abramson RG, Levitt MF. Micropuncture study of uric acid transport in rat kidney. Am. J. Physiol. 1975; 228(5): 1597–605.

Akaoka I, Nishizawa T, et al. Renal urate excretion in five cases of hypouricemia with an isolated renal defect of urate transport. J. Rheumatol. 1977; 4(1): 86–94.

Akaoka I, Nishizawa T, et al. Familial hypouricaemia due to renal tubular defect of urate transport. Ann. Clin. Res. 1975; 7(5): 318–24.

Bairaktari ET, Kakafika AI, et al. Hypouricemia in individuals admitted to an inpatient hospital-based facility. Am. J. Kidney Dis. 2003; 41(6): 1225–32.

Bakhiya A, Bahn A, et al. Human organic anion transporter 3 (hOAT3) can operate as an exchanger and mediate secretory urate flux. Cell. Physiol. Biochem. 2003; 13(5): 249–56.

Barrientos A, Perez-Diaz V, et al. Hypouricemia by defect in the tubular reabsorption. Arch. Intern. Med. 1979; 139(7): 787–9.

Boner G, Steele TH. Relationship of urate and p-aminohippurate secretion in man. Am. J. Physiol. 1973; 225(1): 100–4.

Bordley JI, Richard A. Quantitative studies of the composition of glomerular urine. J. Biol. Chem. 1933; 101: 193.

Carriere S. Effect of norepinephrine, isoproterenol, and adrenergic blockers upon the intrarenal distribution of blood flow. Can. J. Physiol. Pharmacol. 1969; 47(2): 199–208.

Carriere S, Friborg J. Intrarenal blood flow and PAH extraction during angiotensin infusion. Am. J. Physiol. 1969; 217(6): 1708–15.

Coady MJ, Chang MH, et al. The human tumour suppressor gene SLC5A8 expresses a Na^+-monocarboxylate cotransporter. J. Physiol. 2004; 557(Pt 3): 719–31.

Coombs FS, Pecora LJ, et al. Renal function in patients with gout. J. Clin. Invest. 1940; 19(3): 525–35.

Diamond HS, Paolino JS. Evidence for a postsecretory reabsorptive site for uric acid in man. J. Clin. Invest. 1973; 52(6): 1491–9.

Enomoto A, Kimura H, et al. Molecular identification of a renal urate anion exchanger that regulates blood urate levels. Nature 2002; 417(6887): 447–52.

Erley CM, Hirschberg RR, et al. Acute renal failure due to uric acid nephropathy in a patient with renal hypouricemia. Klin. Wochenschr. 1989; 67(5): 308–12.

Frommer JP, Sheth AU, et al. Free-flow micropuncture study of renal urate transport in the Munich-Wistar rat. Miner. Electrolyte Metab. 1982; 7(6): 324–30.

Fujieda M, Yokoyama W, et al. Acute renal failure after exercise in a child with renal hypouricemia. Acta Paediatr. Jpn 1995; 37(5): 642–4.

Fujiwara Y, Takamitsu Y, et al. Hypouricemia due to an isolated defect in renal tubular urate reabsorption. Clin. Nephrol. 1980; 13(1): 44–8.

Gopal E, Fei YJ, et al. Expression of slc5a8 in kidney and its role in Na(+)-coupled transport of lactate. J. Biol. Chem. 2004; 279(43): 44522–32.

Greene ML, Marcus R, et al. Hypouricemia due to isolated renal tubular defect. Dalmatian dog mutation in man. Am. J. Med. 1972; 53(3): 361–7.

Gresser U, Gathof B, et al. Uric acid levels in southern Germany in 1989. A comparison with studies from 1962, 1971, and 1984. Klin Wochenschr 1990; 68(24): 1222–8.

Grover PK, Ryall RL, et al. Effect of urate on calcium oxalate crystallization in human urine: evidence for a promotory role of hyperuricosuria in urolithiasis. Clin. Sci. (Lond) 1990; 79(1): 9–15.

Guggino SE, Aronson PS. Paradoxical effects of pyrazinoate and nicotinate on urate transport in dog renal microvillus membranes. J. Clin. Invest. 1985; 76(2): 543–7.

Guggino SE, Martin GJ, et al. Specificity and modes of the anion exchanger in dog renal microvillus membranes. Am. J. Physiol. 1983; 244(6): F612–21.

Gutman AE, Yu TF, et al. Tubular secretion of urate in man. J. Clin. Invest. 1959; 38: 1778–81.

Hirasaki S, Koide N, et al. Two cases of renal hypouricemia with nephrolithiasis. Intern. Med. 1997; 36(3): 201–5.

Hisanaga S, Ueno N, et al. Exercise-induced acute renal failure associated with renal vasoconstriction. Nippon Jinzo Gakkai Shi 1999; 41(4): 406–12.

Hisatome I, Ogino K, et al. Cause of persistent hypouricemia in outpatients. Nephron 1989; 51(1): 13–16.

Hisatome I, Tanaka Y, et al. Excess urate excretion correlates with severely acidic urine in patients with renal hypouricemia. Intern. Med. 1998; 37(9): 726–31.

Holmes EW, Blondet P. Urate binding to serum albumin: lack of influence on renal clearance of uric acid. Arthritis Rheum. 1979; 22(7): 737–9.

Ichida K, Hosoyamada M, et al. Clinical and molecular analysis of patients with renal hypouricemia in Japan – influence of URAT1 gene on urinary urate excretion. J. Am. Soc. Nephrol. 2004; 15(1): 164–3.

Ichida K, Hosoyamada M, et al. Urate transport via human PAH transporter hOAT1 and its gene structure. Kidney Int. 2003; 63(1): 143–55.

Igarashi T. Normal serum uric acid concentrations for age and sex and incidence of renal hypouricaemia in Japanese school children. Pediatr. Nephrol. 1993; 7(2): 239–40.

Ishikawa I. Acute renal failure with severe loin pain and patchy renal ischemia after anaerobic exercise in patients with or without renal hypouricemia. Nephron 2002; 91(4): 559–70.

Ishikawa I, Onouchi Z, et al. Acute renal failure with severe loin pain and patchy renal vasoconstriction. Acute renal failure H. Elaihou. London: John Libbey, 1982: pp. 224–9.

Ishikawa I, Saito Y, et al. Evidence for patchy renal vasoconstriction in man: observation by CT scan. Nephron 1981; 27(1): 31–4.

Ishikawa I, Sakurai Y, et al. Exercise-induced acute renal failure in 3 patients with renal hypouricemia. Nippon Jinzo Gakkai Shi 1990; 32(8): 923–8.

Ito O, Hasegawa Y, et al. A case of exercise-induced acute renal failure in a patient with idiopathic renal hypouricemia developed during antihypertensive therapy with losartan and trichlormethiazide. Hypertens Res. 2003; 26(6): 509–13.

Iwai N, Mino Y, et al. A high prevalence of renal hypouricemia caused by inactive SLC22A12 in Japanese. Kidney Int. 2004; 66(3): 935–44.

Jutabha P, Kanai Y, et al. Identification of a novel voltage-driven organic anion transporter present at apical membrane of renal proximal tubule. J. Biol. Chem. 2003; 278(30): 27930–8.

Kahn AM, Aronson PS. Urate transport via anion exchange in dog renal microvillus membrane vesicles. Am. J. Physiol. 1983; 244(1): F56–63.

Kahn AM, Branham S, et al. Mechanism of urate and p-aminohippurate transport in rat renal microvillus membrane vesicles. Am. J. Physiol. 1983; 245(2): F151–8.

Kahn AM, Shelat H, et al. Urate and p-aminohippurate transport in rat renal basolateral vesicles. Am. J. Physiol. 1985; 249 (5 Pt 2): F654–61.

Kawabe K, Murayama T, et al. A case of uric acid renal stone with hypouricemia caused by tubular reabsorptive defect of uric acid. J. Urol. 1976; 116(6): 690–2.

Kikuchi Y, Koga H, et al. Patients with renal hypouricemia with exercise-induced acute renal failure and chronic renal dysfunction. Clin. Nephrol. 2000; 53(6): 467–72.

Kim YK, Jung JS, et al. Uptake of uric acid and p-aminohippurate (PAH) by renal cortical slices of various mammals. Comp. Biochem. Physiol. A. 1992; 101(1): 53–8.

Komoda F, Sekine T, et al. The W258X mutation in SLC22A12 is the predominant cause of Japanese renal hypouricemia. Pediatr. Nephrol. 2004; 19(7): 728–33.

Kotake T, Miura N, et al. Renal tubular hypouricemia and calcium urolithiasis. Scanning Microsc. 1993; 7(1): 417–21.

Kramp RA, Lenoir RH. Characteristics of urate influx in the rat nephron. Am. J. Physiol. 1975; 229(6): 1654–61.

Lipkowitz MS, Leal-Pinto E, et al. Functional reconstitution, membrane targeting, genomic structure, and chromosomal localization of a human urate transporter. J. Clin. Invest. 2001; 107(9): 1103–15.

Matsuda O, Shiigai T, et al. A case of familial renal hypouricemia associated with increased secretion of para-aminohippurate and idiopathic edema. Nephron 1982; 30(2): 178–86.

Murakami T, Kawakami H, et al. Recurrence of acute renal failure and renal hypouricaemia. Pediatr. Nephrol. 1993; 7(6): 772–3.

Ninomiya M, Ito Y, et al. Recurrent exercise-induced acute renal failure in renal hypouricemia. Acta Paediatr. 1996; 85(8): 1009–11.

Numabe A, Tsukada H, et al. A case of acute renal failure in a patient with idiopathic hypouricemia. Nippon Jinzo Gakkai Shi 1992; 34(7): 841–5.

Ohta T, Sakano T, et al. Exercise-induced acute renal failure associated with renal hypouricaemia: results of a questionnaire-based survey in Japan. Nephrol. Dial. Transplant. 2004; 19(6): 1447–53.

Ohta T, Sakano T, et al. Exercise-induced acute renal failure with renal hypouricemia: a case report and a review of the literature. Clin. Nephrol. 2002; 58(4): 313–16.

Oi K, Ichida H, et al. Case of renal hypouricemia causing exercise-induced acute renal failure. Nippon Naika Gakkai Zasshi 1998; 87(4): 732–4.

Praetorius E, Kirk JE. Hypouricemia: with evidence for tubular elimination of uric acid. J. Lab. Clin. Med. 1950; 35(6): 865–8.

Ramsdell CM, Kelley WN. The clinical significance of hypouricemia. Ann. Intern. Med. 1973; 78(2): 239–42.

Rieselbach RE, Steele TH. Influence of the kidney upon urate homeostasis in health and disease. Am. J. Med. 1974; 56(5): 665–75.

Roch-Ramel F, Diezi-Chomety F, et al. A micropuncture study of urate excretion by Cebus monkeys employing high performance liquid chromatography with amperometric detection of urate. Pflugers Arch. 1980; 383(3): 203–7.

Roch-Ramel F, Guisan B, et al. Effects of uricosuric and antiuricosuric agents on urate transport in human brush-border membrane vesicles. J. Pharmacol. Exp. Ther. 1997; 280(2): 839–45.

Roch-Ramel F, Guisan B, et al. Indirect coupling of urate and p-aminohippurate transport to sodium in human brush-border membrane vesicles. Am. J. Physiol. 1996; 270(1 Pt 2): F61–8.

Roch-Ramel F, Werner D, et al. Urate transport in brush-border membrane of human kidney. Am. J. Physiol. 1994; 266(5 Pt 2): F797–805.

Roch-Ramel F, White F, et al. Micropuncture study of tubular transport of urate and PAH in the pig kidney. Am. J. Physiol. 1980; 239(2): F107–12.

Sato T, Kuno T, et al. Exercise-induced acute renal failure in a girl with renal hypouricemia. Acta Paediatr. Jpn 1998; 40(1): 93–5.

Schali C, Roch-Ramel F. Accumulation of [14C]urate and [3H]PAH in isolated proximal tubular segments of the rabbit kidney. Am. J. Physiol. 1980; 239(3): F222–7.

Shichiri M, Itoh H, et al. Renal tubular hypouricemia: evidence for defect of both secretion and reabsorption. Nephron 1990; 56(4): 421–6.

Shichiri M, Matsuda O, et al. Hypouricemia due to an increment in renal tubular urate secretion. Arch. Intern. Med. 1982; 142(10): 1855–7.

Simkin PA, Skeith MD, et al. Suppression of uric acid secretion in a patient with renal hypouricemia. Adv. Exp. Med. Biol. 1974; 41: 723–8.

Sperling O. Hereditary Renal Hypouricemia. The Metabolic and Molecular Bases of Inherited Disease e. a. C. Scriver, New York, McGraw-Hill: 2000; pp. 5069–83.

Steele TH. Urate secretion in man: the pyrazinamide suppression test. Ann. Intern. Med. 1973; 79(5): 734–7.

Steele TH, Boner G. Origins of the uricosuric response. J. Clin. Invest. 1973; 52(6): 1368–75.

Suzuki T, Kidoguchi K, et al. Genetic heterogeneity of familial hypouricemia due to isolated renal tubular defect. Jinrui Idengaku Zasshi 1981; 26(3): 243–8.

Tabe A. Study on the physiopathology of hypouricemia. Tokyo Jikeikai Medical Journal 1996; 111: 821–39.

Takeda E, Kuroda Y, et al. Hereditary renal hypouricemia in children. J. Pediatr. 1985; 107(1): 71–4.

Takeda Y, Fujimoto T, et al. Two cases of exercise-induced acute renal failure with idiopathic renal hypouricemia. Nippon Jinzo Gakkai Shi 2001; 43(5): 384–8.

Tanaka M, Itoh K, et al. Two male siblings with hereditary renal hypouricemia and exercise-induced ARF. Am. J. Kidney Dis. 2003; 42(6): 1287–92.

Tazawa M, Morooka M, et al. Exercise-induced acute renal failure observed in a boy with idiopathic renal hypouricemia caused by postsecretory reabsorption defect of uric acid. Nippon Jinzo Gakkai Shi 1996; 38(9): 407–12.

Tofuku Y, Kuroda M, et al. Hypouricemia due to renal urate wasting. Two types of tubular transport defect. Nephron 1982; 30(1): 39–44.

Uchino H, Tamai I, et al. p-aminohippuric acid transport at renal apical membrane mediated by human inorganic phosphate transporter NPT1. Biochem. Biophys. Res. Commun. 2000; 270(1): 254–9.

Van Aubel RA, Smeets PH, et al. Human organic anion transporter MRP4 (ABCC4) is an efflux pump for the purine end metabolite urate with multiple allosteric substrate binding sites. Am. J. Physiol. Renal Physiol. 2005; 288(2): F327–33.

Watanabe T. Patchy renal vasoconstriction after exercise in a child without renal hypouricemia. Pediatr. Nephrol. 2002; 17(4): 284–6.

Watanabe T, Abe T, et al. Exercise-induced acute renal failure in a patient with renal hypouricemia. Pediatr. Nephrol. 2000; 14 (8–9): 851–2.

Weinberger A, Pinkhas J, et al. Frequency and causes of hypouricemia in hospital patients. Isr. J. Med. Sci. 1977; 13(5): 529–30.

Weiner IM, Tinker JP. Pharmacology of pyrazinamide: metabolic and renal function studies related to the mechanism of drug-induced urate retention. J. Pharmacol. Exp. Ther. 1972; 180(2): 411–34.

Weinman EJ, Steplock D, et al. Use of high-performance liquid chromatography for determination of urate concentrations in nanoliter quantities of fluid. Kidney Int. 1981; 19(1): 83–5.

Yamamoto T, Moriwaki Y, et al. Effect of lactate infusion on renal transport of purine bases and oxypurinol. Nephron 1993; 65(1): 73–6.

Yeun JY, Hasbargen JA. Renal hypouricemia: prevention of exercise-induced acute renal failure and a review of the literature. Am. J. Kidney Dis. 1995; 25(6): 937–46.

Yu T, Gutman AB. Uric acid nephrolithiasis in gout. Predisposing factors. Ann. Intern. Med. 1967; 67(6): 1133–48.

CHAPTER 10

The Fanconi Syndrome

ORSON W. MOE, DONALD W. SELDIN AND MICHEL BAUM

INTRODUCTION

The proximal tubule is an ancient part of the vertebrate nephron that is responsible for many of the homeostatic properties of the kidney. It possesses high-capacity reabsorptive function for a broad range of filtered solutes and water which is permissive and critical for ensuring that a high glomerular filtration rate does not eventuate in urinary losses of critical body constituents. It harbors secretory function for a multitude of endogenous and exogenous substances such as uric acid and various toxins. It serves important endocrine and paracrine functions such as synthesis of 1,25-dihydroxyvitamin D_3 and dopamine respectively. Lastly, it furnishes important systemic metabolic functions such as gluconeogenesis. Thus it is not surprising that disorders of the proximal tubule have profound impact on the homeostatic function of the kidney.

Monogenic defects of proximal tubule transporters that lead to discrete transporter dysfunctions are covered in detail in several other chapters. This chapter focuses on a group of congenital and acquired disorders that are not due to a single solute transporter defect but rather appear to result from more generalized and heterogeneous dysfunctions of the entire proximal tubule. The 'complete' or 'classical' Fanconi syndrome may be defined as an impairment of proximal tubule reabsorption of sodium, bicarbonate, potassium, phosphate, glucose, amino acids, uric acid and low-molecular-weight proteins, and peptides, as well as other organic solutes (Table 10.1). Diseases involving renal leak of some but not all of these solutes are termed 'partial' Fanconi syndromes. There is no single well-circumscribed lesion described to date, genetic or acquired, that leads to generalized proximal tubule dysfunction. The list of clinical and experimental etiologies is extremely diverse and the cellular mechanism by which these conditions impair tubular function is by and large elusive. An account of the current database and postulates of the pathophysiology will be presented.

HISTORY OF THE FANCONI SYNDROME

The cardinal features of renal solute wasting described above were gradually and cumulatively unraveled in the early accounts of this syndrome over several decades. The earliest reports consisted of cases that were compatible with cystinosis, which is one cause of the proximal tubule wasting disorder. Emil Abderhalden found cystine crystals in the liver and spleen of a 21-month-old infant in 1903 (Abderhalden 1903) and coined the term 'familial cystine diathesis' but no references were made to renal involvement although it would not be surprising at all if this patient had proximal dysfunction. In 1924, George Otto Emil Lignac described three children, with similar findings, but noted features of severe rickets and growth retardation (Lignac 1924). Swiss pediatrician and biochemist Guido Fanconi (1892–1979) described in 1931 a child with rickets, retarded growth, albuminuria, and glucosuria, thus implicating renal involvement for the very first time (Fanconi

TABLE 10.1   Solute wastage in Fanconi syndrome

I. Inorganic solutes	
Cation	
Sodium	
Potassium	
Calcium	
Magnesium	
Anion	
Chloride	
Bicarbonate	
Phosphate	
II. Organic solutes	
Glucose	
Amino acids	
Uric acid	
Organic acids	
Low molecular weight proteins and peptides	

Genetic Diseases of the Kidney
ISBN: 978-0-12-449851-8

1936). In 1933, Giovanni de Toni described a 5-year-old 'dwarf' with glucosuria, rickets, 'abundant urine,' and most importantly hypophosphatemia (De Toni 1933). Another case was described by Robert Debré et al a year later of an 11-year-old girl where the added feature of urinary organic acids leakage was presented (Debre 1934).

In 1936, Fanconi presented two additional patients, recognized the similarities between his cases and those of de Toni and Debré, and suggested that the organic acids found in the urine may be amino acids (Fanconi 1936). Since there are many eponyms associated with Fanconi's name, the syndrome referring to generalized renal tubular defect is sometimes called the Debré-de Toni-Fanconi syndrome acknowledging all three investigators. In a classic monograph in 1943, McCune et al presented an in-depth clinical and chemical analysis of one patient and a scholarly summary of all the historical cases of what he collectively referred to with the shorter eponym, the 'Fanconi syndrome' (McCune 1943). In the complete form of Fanconi syndrome, there is wastage of a wide variety of solutes (Table 10.1). Various limited forms of the syndrome occur where some but not all of the solutes listed in Table 10.1 are wasted in the urine. These disorders bear the name 'partial' Fanconi syndrome. This multisolute wastage sets the Fanconi syndrome distinct from a myriad of single transporter defects in the renal tubules. In common clinical parlance today, any syndrome characterized by generalized proximal tubule transport disfunction, whether complete or partial, is designated as the Fanconi syndrome.

ETIOLOGY

The etiology of the Fanconi syndrome must account for the impaired reabsorption of an extremely broad group of substances; it therefore seems likely that the proximal tubule can be injured by a variety of etiologies. The causes are listed in Table 10.2 and are divided into two broad categories (primary vs. secondary) depending on whether an underlying cause can be identified. Within each category, one can also subclassify the causes depending on whether a genetic disorder (identifiable or presumed) is operative.

Primary Fanconi Syndrome

This category embraces not only well-characterized genetic disturbances but also constitutes a default receptacle for all forms of clinical and biochemical Fanconi syndrome where no underlying causes of any kind can be identified. The classical inherited Fanconi syndrome has been described as autosomal recessive or in some instances autosomal dominant (Bergeron et al 1995, Bonnardeaux & Bichet 1999, McCune 1943). One note of caution is that not all but most

---

TABLE 10. 2   Causes of the Fanconi Syndrome

I. Primary	4. Nephrotic syndrome
A. *Genetic Fanconi Syndrome*	5. Renal vein thrombosis
1. Complete	6. Interstitial nephritis
2. Partial	7. Allograft rejection
B. *Sporadic Fanconi Syndrome*	8. Sjögren's syndrome
1. Complete	9. Balkan endemic nephropathy
2. Partial	10. Paroxysmal nocturnal
II. Secondary	hemoglobinuria
A. *Genetic diseases (details in Table 10.3)*	D. *Exogenous substances*
1. Cystinosis	1. **Heavy metals**
2. Hereditary fructose intolerance	Lead
3. Wilson's disease	Cadmiun
4. Tyrosinemia type I	Mercury
5. Glycogen storage disease type I	Uranium
6. Galactosemia	2. **Drugs**
7. Cytochrome C oxidase deficiency	Iphosphamide
8. Phosphenolpyruvate carboxykinase deficiency	Valproic acid
9. Metachromatic leukodustrophy	Cisplatinum
10. Acyl-CoA dehydrogenase deficiency	Tetracycline (outdated)
11. Dent's disease	Aminoglycosides
12. Fanconi-Bickel syndrome	6-mercaptopurine
13. Oculocerebrorenal syndrome of Lowe	Sulfanilamide
C. *Systemic and renal diseases*	3. **Toxins**
1. Amyloidosis	Paraquat
2. Malignant plasma cell dyscrasias	Lysol
3. Paraproteinemic states	Methyl-3-chrome

of these statements are based on observations from small kindreds. There are instances where the disease-causing genes have not or cannot be established. This group may not stem from a single inherited defect but rather from heterogeneous loci.

In addition to genetic causes, there are sporadic cases that present with the classic feature of Fanconi syndrome but after extensive evaluation, no causes or family history can be elicited. Although frequently classified as nonheritable, one cannot be absolutely certain that these cases are nongenetic in nature. The possibility exists for de novo mutations. In addition, the proband can be the sole homozygote offspring of two heterozygous carriers. Heterozygote carriers are likely phenotypically silent and genotyping is not feasible without known candidate genes. If the full-blown Fanconi syndrome is not present, mildly affected siblings may go completely undiagnosed. Finally, all the acquired causes listed in Table 10.2 can be misdiagnosed as primary if exogenous damage may have occurred in the past or without the individual's knowledge.

Secondary Fanconi Syndrome

In contrast to the group of disorders outlined above, a significant portion of patients with Fanconi syndrome have identifiable causes. Table 10.2 lists the known causes of Fanconi syndrome. The exogenous agents are arbitrarily classified into heavy metals, drugs, or environmental toxins.

GENETIC CAUSES

Cystinosis (#219800 nephropathic; #219750 adult non-nephropathic; #219900 late-onset juvenile)

Cystinosis is the most common cause of the Fanconi syndrome in childhood and serves as a prototype of this syndrome. Cystinosis is a lysosomal storage disorder where the defective cystine transporter, cystinosin, fails to extrude cystine from the lysosomes. Cystine crystallizes in tissues including kidney, liver, intestine, spleen, and cornea, which exhibit variable susceptibilities to injury. In the nephropathic or infantile form of cystinosis, renal tubular Fanconi syndrome occurs. Children with this disorder are normal for the first 6 months of life after which time they fail to thrive. They have polydipsia and polyuria due to the failure of proximal reabsorption and consequent solute diuresis. These children are prone to severe volume depletion and fever especially after relatively minor episodes of vomiting or diarrhea. The onset of the Fanconi syndrome in cystinosis likely occurs after 6 months, as it is at this point that enough cystine has accumulated to injure the proximal tubule. When left untreated, these children develop severe growth retardation and rickets develops (Gahl et al 2002).

Hereditary Fructose Intolerance (HFI) (OMIM 22960)

This is an autosomal recessive inborn error of carbohydrate metabolism caused by a catalytic deficiency of fructose 1-phosphate aldolase (aldolase B). Patients with HFI develop hypoglycemia and abdominal pain after ingestion of foods containing fructose, sucrose, or sorbitol. They also develop a transient Fanconi syndrome immediately after the ingestion of fructose (Morris 1968, Ali et al 1988, Wong 2005) that is reversible. Parenteral administration of fructose or sorbitol may be fatal. Continued ingestion of fructose and cognate sugars leads to permanent hepatic and renal injury.

Wilson's Disease (Hepatolenticular Degeneration) (OMIM #277900)

Fanconi syndrome is a well-known complication in Wilson's disease (Morgan et al 1962, Tabet 1969, Hodges & Kirkendall 1972), which is due to inactivating mutations of the copper-transporting ATPase (Das & Ray 2006). In addition to Fanconi's syndrome, hypercalciuric kidney stones and distal renal tubular acdiosis have been described (Wibers et al 1979, Hoppe et al 1993). Improvement or reversal of the tubulopathy has followed chelation therapy (Leu et al 1970, Elsas et al 1971, Schonheyder et al 1971, Asatoor et al 1983).

Tyrosinemia Type 1 (OMIM +276700)

Fumarylacetoacetate hydrolase (FAH) in liver and kidney is the last enzyme in the tyrosine catabolic pathway. Patients with mutations of FAH have tyrosinemia (Grompe et al 1994) and develop the Fanconi syndrome after ingestion of tyrosine or phenylalanine (Larochelle et al 1967, Cochat et al 1994). One of the accumulated metabolites, succinylacetone, is nephrotoxic but may not be the sole agent. Over the short term, the Fanconi syndrome is reversible upon removal of tyrosine or phenylalanine from the diet. Chronically, both human and murine forms of the disease show involution and apoptosis of renal tubular cells (Russo & O'Regan 1990, Sun et al 2000). The failure of reversal of the renal lesion after successful hepatic transplant (Laine et al 1995) suggests that either replacement of renal FAH is necessary, which is achievable by combined liver–kidney transplant, or the tubular lesion is extremely slow to recover or irreversible.

Glycogen Storage Disease Type 1 (Von Gierke's Disease) (OMIM +232200)

Although Cori and Cori (1952) suggested that glucose-6-phosphatase is responsible for type I glycogen storage disease, the genetic proof did not emerge for four decades

(Lei et al 1993). With deficiency of hepatic and renal glucose-6-phosphatase, glycogen storage occurs in the kidney most profoundly in the proximal tubule. It is presumed that the presence of glycogen leads to the Fanconi syndrome. In addition to Fanconi syndrome, these patients also have nephrocalcinosis, nephrolithiasis, hypertension, progressive loss of GFR and glomerular proteinuria due to a focal segmental glomerulosclerotic lesion (Chen 1991, 1998, Reitsma-Bierens 1993).

Galactosemia (OMIM #230400)

Galactosemia from galactose 1-phosphate uridyltransferase (Reichardt & Berg 1988, Reichardt et al 1992, Elsas & Lai 1998, Leslie 2003) is a rare condition. The cardinal features of classic galactosemia are hepatomegaly, cataracts, severe sepsis, and mental retardation, but the Fanconi syndrome has also been described. The reason for the proximal dysfunction is unknown.

Cytochrome C Oxidase Deficiency (#256000 Leigh Syndrome; #220110 Mitochondrial Complex IV Deficiency; #530000 Kearns-Sayre Syndrome)

Because of the rare nature of this disease, the entire literature comprises a handful of case reports but in combination they are quite compelling (Zeviani et al 1985, Kitano et al 1986, Ogier et al 1988, Mori et al 1991, Campos et al 1995, Kuwertz-Broking et al 2000, Berio & Piazzi 2001). Deficiency in mitochondrial complex IV, which is a cytochrome C oxidase, results from a heterogeneous group of diseases which can be caused by over 20 mitochondrial- and nuclear-encoded genes. Presentations vary widely but myopathy and neuropathy are quite prevalent along with the nephropthy. Renal biopsy has revealed nonspecific chronic tubulointerstitial with dysmorphic enlarged mitochondria with abnormal arborization and disorientation of the cristae in the proximal tubular cells (Mori et al 1991, Szabolcs et al 1994, Kuwertz-Broking et al 2000).

Phosphoenolpyruvate Carboxykinase (PEPCK) Deficiency (OMIM 261650)

This is a disorder of liver and kidney. Markedly impaired hepatic gluconeogenesis leads to profound hypoglycemia and lactic acidosis. Massive fat infiltration (Hommes et al 1976) has been described in the kidney along with clinical Fanconi syndrome (Clayton et al 1986). PEPCK also mediates ammoniagenesis and gluconeogenesis in the proximal tubule, the latter leading to net generation of ATP. PEPCK deficiency may or may not lead to ATP limitation as there are other substrates and pathways for ATP synthesis (Mandel 1985). It is more likely that the cellular steatosis is toxic to the proximal tubule.

Metachromic Leukodystrophy (OMIM #250100)

Metachromic leukodystrophy belongs in the category of diseases called sufatide lipidoses (Moser 1972). Metachromatic leukodystrophy itself is not a homogeneous disease. Several loci have been described including arylsulfatase A deficiency and the lack of an activator prosaposin (Kihara 1982, Kihara et al 1982, 1986, Kondo et al 1991). These are primarily neurologic diseases, but galactosphingosulfatide deposition in the kidney has been described which presumably leads to the Fanconi syndrome (Austin 1960).

Acyl-CoA Dehydrogenase Deficiency (ACAD) (OMIM #201450 Medium-chain acyl-CoA Dehydrogenase Deficiency; #201470 Short-chain Acyl-CoA Dehydrogenase Deficiency: #231680. Multiple Acyl-CoA Dehydrogenase deficiency; #201475 Very Long Chain Acyl-CoA Dehydrogenase)

This is another group of heterogeneous disorders (Matsubara et al 1990, Frerman & Goodman 2001, Madsen et al 2006). Mitochondrial fatty acid β-oxidation is an important energy source for many cell types, including those in the liver and kidney. The Acyl-CoA dehydrogenases (ACADs) are flavoproteins essential for β-oxidation of fatty acyl-CoA derivatives. Deficiency of these ACADs primarily causes organ lipidoses, in the kidney, this can be the basis for the proximal tubule dysfunction (Morris et al 1997). Cystic disease in the kidney has also been described (Chisholm et al 2001).

Oculocerebrorenal Syndrome of Lowe (OMIM #309000)

This X-linked syndrome is characterized by hydrophthalmia, cataract, mental retardation, and the Fanconi syndrome, and caused by mutation of the OCRL gene which codes for a phosphatidylinositol 4,5-bisphosphate 5-phosphatase (Lowe et al 1952, Auricchio et al 1961, Acker et al 1967, Charnas et al 1991). This syndrome is discussed in detail in a separate chapter.

Dent's Disease (OMIM #300009)

This is another X-linked disease that is caused by mutation of the ClC-5 chloride channel (Pook et al 1993, Lloyd et al 1996). There is considerable variability in the presentation, which may include hypercalciuria, low-molecular-weight proteinuria, and nephrocalcinosis (Dent & Friedman 1964, Scheinman 1998). Interestingly, phosphate and bicarbonate wasting are quite uncommon in this syndrome. Some have considered this spectrum of presentation as phenotypic variants of a single disease, and refer to these disorders as the 'Dent disease complex.' In addition to clinical heterogeneity, there may also be heterogeneity of genetic loci (Hoopes et al 2004). This genetic disease is covered in detail in a separate chapter.

Fanconi-Bickel Syndrome (OMIM 227810)

The original description was actually made by Guido Fanconi in 1949 (Fanconi & Bickel 1949). This disease is distinct from the Debré-de Toni-Fanconi syndrome that is under the spotlight in this chapter. Patients with Fanconi-Bickel syndrome present with hepatomegaly secondary to glycogen accumulation, glucose and galactose intolerance, hypoglycemia, generalized proximal tubulopathy, and severe growth retardation (Santer et al 2005). This is also the first genetic disorder known to be caused by a detectable genetic defect of one of the facilitative glucose transporters-GLUT2 (Manz et al 1987, Sakamoto et al 2000, Santer et al 1997, 1998b, 2005). Although this is a quintessential membrane transporter disease, the multiple solute transport defects in the proximal tubule is not directly related to the defective solute transporter. The proximal tubulopathy is more likely secondary to the glycogen accumulation caused by failure of basolateral glucose exit which is toxic to the proximal tubule. In addition to tubulopathy, patients also have a glomerulopathy somewhat resembling early diabetic glomerulopathy (Berry et al 1995).

SYSTEMIC AND RENAL DISEASES

There is a large group of systemic and renal conditions that are associated with the Fanconi syndrome where no exogenous substance can be identified. One makes the assumption that the renal proximal tubule is injured by an endogenous substance. These disorders will be mentioned but not discussed in detail in the chapter focused on genetics.

The Fanconi syndrome has been described in patients with amyloidosis and a variety of paraproteinemic states such as plasma cell dyscrasias and Waldenstrom's macroglobulinemia (Thorner et al 1983, Rruong et al 1989, Orfila et al 1991, Aucouturier et al 1993, Ronco et al 2000, Bridoux et al 2005). In addition to exposure to the plasma concentrations, the proximal tubule has the added task of having to reabsorb and metabolize excess paraproteins. The toxic effects of paraprotein metabolism on proximal tubule function have been documented by clinical and experimental data. The generalized proximal dysfunction described in the nephrotic syndrome may also result from an overload of filtered proteins (Yagame et al 1986, Bonsib & Horvath 1999) but it is unclear and intriguing why the majority of patients with nephrotic proteinuria do not develop the Fanconi syndrome. A number of conditions that involve immunologic tubular damage have been reported to cause the Fanconi syndrome including interstitial nephritis (Igarashi et al 1992), allograft rejection (Friedman & Chesney 1981), and Sjögren's syndrome (Bridoux et al 2005, Kobayashi et al 2006). The Fanconi syndrome has also been described in association with Balkan endemic nephropathy, and (Cvoriscec et al 1998) paroxysmal nocturnal hemoglobinuria (Saito et al 1975, Clark et al 1981).

EXOGENOUS SUBSTANCES

Heavy metals (cadmium, mercury, lead, chromium, and platinum) are transported by the proximal tubule and are also toxic to the same epithelia (Barbier et al 2005). Because of its ability to accumulate metals, the kidney is a prime target of heavy metal toxicity. An acute nephrotoxic syndrome has been described in experimental animals. In humans, chronic exposure can result in a range of disorders from the Fanconi syndrome to renal failure. Cadmium-induced Fanconi syndrome usually only occurs after years of industrial exposure, while lead-induced Fanconi syndrome can occur after a large acute exposure or after prolonged chronic low levels of exposure.

The most common acquired causes of the Fanconi syndrome are due to drugs such as iphosphamide (Rossi 1997, Rossi et al 1999, Woodland et al 2000), valproic acid (Lenoir et al 1981, Lande et al 1993, Watanabe et al 2005), cisplatinum (Daugaard et al 1988, Cachat et al 1998), tetracycline (outdated) (Gross 1963, Cleveland et al 1965, Montoliu et al 1981), and aminoglycosides (Gainza et al 1997), and toxins such as paraquat, lysol, methyl-3-chrome, and aristolochic acid (Finzer 1961, Tanaka et al 2000, Yang et al 2002, Lee et al 2004, Gil et al 2005).

BRIEF OVERVIEW OF PROXIMAL TUBULE TRANSPORT

The glomerulus produces 150–170 l/day of a plasma ultrafiltrate of near-identical ionic composition to that of plasma except that it is relatively devoid of protein. This is delivered to the proximal tubule, which reabsorbs all of the filtered glucose and amino acids, 80% of the filtered bicarbonate, 60% of the filtered chloride, and 90% of the filtered phosphate (Rector 1983, Liu & Cogan 1984). This nephron segment absorbs or secretes organic anions and cations. In addition, the proximal tubule reabsorbs filtered low-molecular-weight proteins. Most of the filtered bicarbonate, glucose and amino acid absorption occurs in the first 20% of the proximal tubule as depicted in Figure 10.1 (Rector 1983, Liu & Cogan 1984) (Table 10.3). The paracellular permeabilities of phosphate and various organic anions are very low so paracellular back-leak under physiologic conditions is negligible.

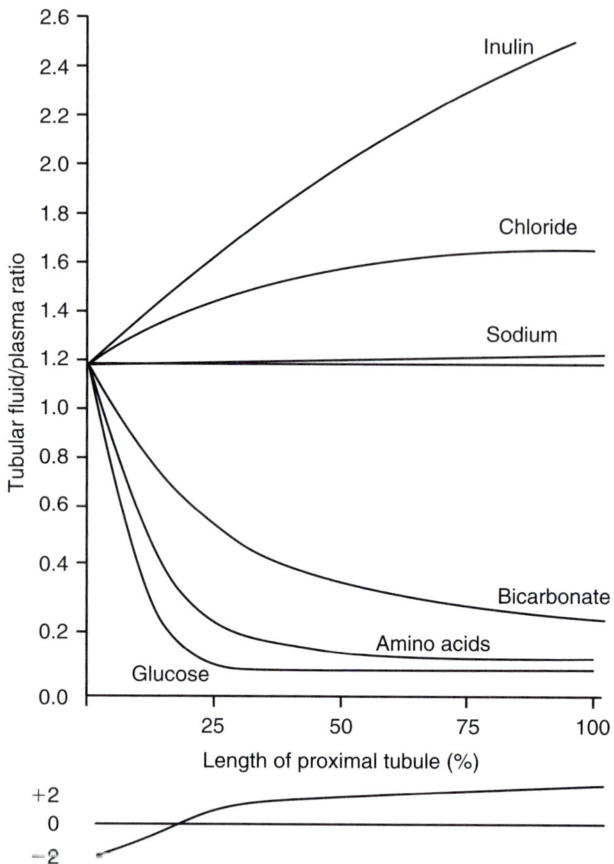

FIGURE 10.1   Axial changes in proximal tubule luminal composition. Depicted are the changes in luminal composition of major solutes along the length of the proximal tubule. There is preferential reabsorption of glucose, amino acids and bicarbonate in the early proximal tubule which results in an increase in the luminal chloride concentration above plasma. The luminal potential difference in the early proximal tubule is lumen negative due to the electrogenic reabsorption of sodium with glucose and amino acids. The transepithelial potential difference becomes lumen positive in the late proximal tubule as chloride diffuses from the lumen into the blood across the paracellular pathway down its concentration gradient

The basic mechanism for active sodium absorption in the proximal tubule is shown in Figure 10.2 (Moe et al 2004). The principal source of energy (save of a contribution from a proton ATPase) for the entire proximal transport system is the basolateral Na-K-ATPase. The currency of energy is converted from metabolic substrates to ATP. The basolateral Na^+-K^+-ATPase next translates the energy currency into electrochemical gradients by generating a low intracellular sodium concentration (~10 mEq/liter) and a high intracellular potassium concentration (~140 mEq/liter). The diffusion potential of the outward potassium gradient results in a negative potential difference of about −60 mV (Figure 10.2). The low intracellular sodium concentration and the large

negative potential difference provide a driving force for apical sodium entry. Most solute transport in the proximal tubule is either directly or indirectly linked to the downhill movement of sodium into the cell from the lumen. The transepithelial potential difference in the early proximal tubule is lumen negative due to the electrogenic reabsorption of sodium with glucose and amino acids (Figure 10.2). This lumen negative potential along with the chemical gradient contributes to the electrical diving force for the paracellular reabsorption of chloride (Rector 1983).

Two-thirds of the H^+ secretion mediating bicarbonate reabsorption is mediated by the apical membrane Na^+/H^+ exchanger utilizing the low cell $[Na^+]$ and one-third is directly coupled to ATP via a H^+-ATPase (Preisig et al 1987, Baum 1992). All the lumen-to-plasma transport of bicarbonate is transcellular. The Na^+/H^+ exchanger in parallel with a Cl^-/base exchanger mediates active transcellular NaCl transport (Baunm & Berry 1984, Aronson & Giebisch 1997, Shah et al 1999). As shown in Figure 10.2, the preferential reabsorption of organic solutes and sodium bicarbonate obligates water reabsorption, thereby elevating the concentration of luminal chloride. This leaves the luminal fluid devoid of these solutes resulting in a higher chloride and a lower bicarbonate concentration than in the peritubular fluid (Rector 1983, Liu & Cogan 1984). The hyperchloruia drives passive chloride transport across the paracellular pathway. Beyond the initial part of the proximal tubule, the lumen-to-plasma chloride and plasma-to-lumen bicarbonate chemical gradient lead to a lumen-positive transepithelial potential that provides the driving force for passive paracellular sodium movement. Half of NaCl transport is active and transcellular and half is passive paracellular (Rector 1983, Baum & Berry 1984, Preisig & Rector 1988).

CLINICAL PATHOPHYSIOLOGY

The clinical signs and symptoms can reflect the underlying conditions that cause Fanconi syndrome or the consequence of proximal dysfunction. Since the etiology of Fanconi syndrome is extremely diverse (Table 10.2), the manifestation of the underlying diseases can be quite varied and will not be discussed further here. This section will focus on the consequences of the proximal tubule defect.

Most of the clinically evident defects are related to the failure of reabsorption of filtered solutes. Interestingly, there are many fewer clinical features directly due to secretory, endocrine, or metabolic malfunctions of the proximal tubule. One can conceptualize the proximal tubule solute wasting defects in three categories: (1) Solutes that are solely handled by the proximal tubule with no means of distal compensation (glucose, amino acids). (2) The sheer magnitude of the increase of solutes exiting from the proximal tubule

TABLE 10.3   Monogenic Inherited Causes of the Fanconi Syndrome

Disease (ref)	Gene (chromosome) Protein (putative function)	Inheritance pattern	Possible pathophysiology
Cystinosis (Trevisan & Gardin 2005)	CTNS (17p13) Cystinosin (lysosomal cystine transporter)	Autosomal recessive	Lysosomal storage
Hereditary Fructose Intolerance (Cross, et al 1988) (Morris 1968) (Town et al 1998) (Morris et al 1971)	ALDOB (9q21.3-q22.2) Aldolase B	Autosomal recessive	Acute: functional chronic: steatosis
Wilson's Disease (Pfister et al 1998) (Thorner et al 1983)	ATP7B (13q14.3) Copper transporting P-type ATPase (β subunit)	Autosomal recessive	Deposition of non-ceruloplasmin copper
Tyrosinemia Type 1 (Grompe et al. 1994) (Piwon et al 2000) (Sabolic et al 1985) (Sussman et al 1980)	FAHD2A (15q23-q25) FAH (fumarylacetoacetate hydrolase)	Autosomal recessive	Apoptosis and involution of tubular cells
Glycogen Storage Disease-Type 1 (Garty et al 1974) (Lei et al 1993)	G6PC (17q21) Glucose-6-phosphatase	Autosomal recessive	Glycogen deposition
Galactosemia (Reichardt & Berg 1988) (Reitsma-Bierens 1993)	Galactose 1-phosphate uridyltransferase (9p13)	Autosomal recessive	Unknown
Cytochrome C Oxidase Deficiency (Campos et al 1995) (Kitano et al 1986) (Morris 1968)	Cytochrome C Oxidase Deficiency (Multiple loci)	Variable. Loci-dependent	Abnormal proximal tubule mitochondria. Possible defective energy generation
Phosphoenolpyruvate carboxykinase deficiency (Clayton et al 1986) (Hommes et al 1976) (Leonard et al 1991)	PEPCK2 Mitochondrial phosphoenolpyruvate carboxykinase (14q4)	Autosomal recessive	Fat deposition
Metachromatic leukodystrophy	Arylsulfatase A (22q13) Prosaposin (10q22.1)	Autosomal recessive	Sulfatide deposition
Acyl-CoA dehydrogenase deficiency	Multiple loci. Short chain, medium chain and multiple acyl-CoA dehydrogenase deficiency	Variable	Steatosis
Oculocerebrorenal Syndrome of (Lowe Lin, Orrison, & Leahey et al. 1997) (Nykjaer et al 2001)	OCRL1 (Xq26.1) Phosphatidylinositol 4,5-bisphosphate 5-phosphatase	X-linked recessive	Protein trafficking? Disturbed brush border organization?
Dent's Disease (Fisher et al 1994) (Fisher et al 1995) (Lloyd et al 1996, 1997)	CLCN5 ClC-5 (Chloride channel)	X-linked recessive	Protein trafficking?
Fanconi-Bickel Syndrome (Santer et al 1997, 1998b, 2005)	SLC2A2 (3q26.1-q26.3) GLUT2 (facilitative glucose transporter)	Autosomal recessive	Glycogen accumulation
Idiopathic Fanconi syndrome	No known gene(s) to date	Recessive, X-Linked Dominant	Unknown

which overwhelms the distal nephron's ability to compensate (sodium, chloride, bicarbonate, phosphate, water). (3) Inappropriate distal responses secondary to derangements of the proximal tubule (potassium). These defects account for the bulk of the clinical manifestations of the Fanconi syndrome.

Solutes Solely Handled by the Proximal Tubule

These transport defects are better understood if one thinks of solute reabsorption and excretion in terms of a renal threshold (Figure 10.3). The filtered load is the product of the glomerular filtration rate and the ultrafilterable concentration

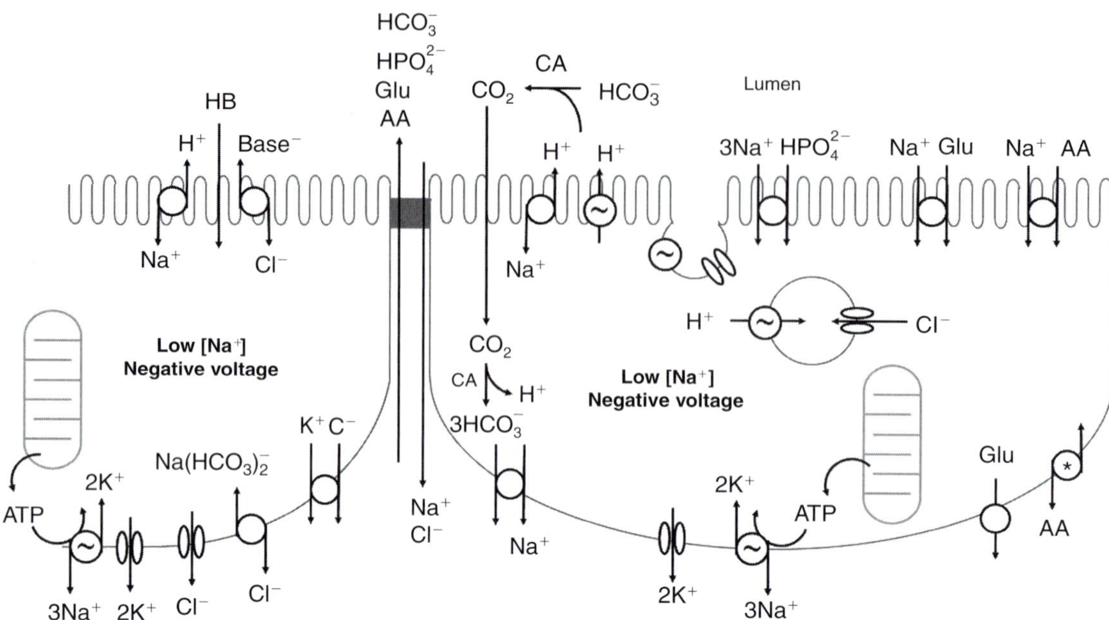

FIGURE 10.2 Normal proximal tubule cell. The reabsorption of luminal solutes in the proximal tubule cell is driven by electrochemical gradient. The basolateral Na$^+$-K$^+$-ATPase constitute the primary workhorse which produces a low cell [Na$^+$] (10 mM) and negative potential (-60 mV; K$^+$ diffusion potential via basolateral K$^+$ channel). The lumen to cell Na$^+$ gradient is 140:10 and a voltage of -60 mV provided the driving force for Na$^+$ entry. Apical entry of NaCl is medated by parallel Na$^+$/H$^+$ and Cl^{2-}/base (B) exchange and HB recycling. Additional apical NaCl entry mechanisms have been described and are not shown here. Basolateral Cl$^-$ exit is mediated by a Na(HCO$_3$)$_2^{2-}$/Cl$^-$ exchanger, Cl$^-$ channel, and a KCl cotransporter. Cl can also pass through the paracellular pathway. Phosphate (HPO$_4^{2-}$), glucose (Glu), and amino acids (AA) are directly coupled to Na$^+$ and are electrogenic. HCO$_3$ absorption is mediated by apical H$^+$ secretion via coupling to the Na$^+$ gradient (Na$^+$/H$^+$ exchange) or directly to ATP hydrolysis via the H$^+$-ATPase. Endocytic vesicles are acidified by the H$^+$-ATPase and Cl$^-$ channels. There is a finite amount of backleak of absorbed solutes into the lumen. Basolateral exit of amino acids occurs via a complex set of exchangers (*) and exit of phosphate is unknown and therefore not shown

of the solute in the serum. As delivery increases, more of the solute is reabsorbed. At some point as the filtered load is increased, the reabsorptive capacity of the kidney (tubular maximum; Tm) is surpassed and urinary solute excretion ensues. A splay may occur before Tm is reached, giving a lower threshold of excretion than the Tm would dictate. Hence both the Tm and the splay determine the threshold of excretion. The beginning of the splay point defines the threshold of excretion for that solute. In the presence of a normal glomerular filtration rate, this threshold sets the level of plasma ultrafilterable concentration of that solute above which urinary excretion will be observed. Figure 10.3 shows the two theoretical defects in the Fanconi syndrome. The first defect is decreased Tm leading to decreased threshold for the solute reabsorbed by the proximal tubule (Figure 10.3 upper panel). A second defect is decrease in affinity of the transport system for a solute (Figure 10.3 lower panel). In this model, Tm is preserved but not reached until the filtered load is very high. However, spillage and wastage in urine occurs at very low filtered loads. In either example, the excretion of a given solute is always higher in the diseased state (Figure 10.3, red lines of excretion) for a given filtered load compared to the normal tubule. These theoretical defects are not mutually exclusive.

Glucose is cited in Table 10.4 as an example. At a normal glomerular filtration rate and normal plasma glucose concentration, a normal individual does not have glucosuria until the plasma glucose concentration is increased beyond a certain point (plasma threshold) when reabsorptive capacity is exceeded and glucosuria results. In hyperfiltration, a lower plasma glucose concentration is required to exceed tubular reabsorptive capacity (Table 10.4). In the Fanconi syndrome, the primary defect is in the proximal tubule so even a normal serum concentration of solutes and near-normal glomerular filtration rate can exceed the threshold and solutes such as glucose, amino acids, and phosphate are excreted in the urine. The ramifications of the reduction in glucose and amino acid transport resulting in a generalized amino aciduria and glucosuria are relatively minor and serum levels of these solutes are minimally or not reduced. Asymptomatic hypoglycemia can present as a laboratory abnormality.

Low-molecular-weight proteins are normally filtered and reabsorbed by the proximal tubule. Low-molecular-weight proteinuria is almost always seen in patients with the Fanconi syndrome and was noted in Fanconi's manuscript in 1936 (Fanconi 1936). These include retinol-binding protein, vitamin D binding protein, transferrin, albumin,

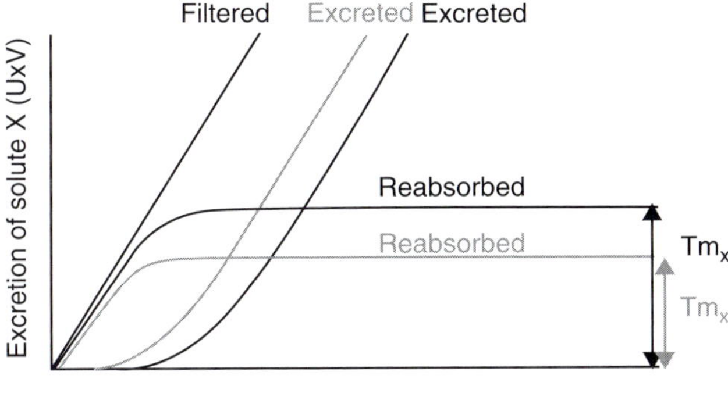

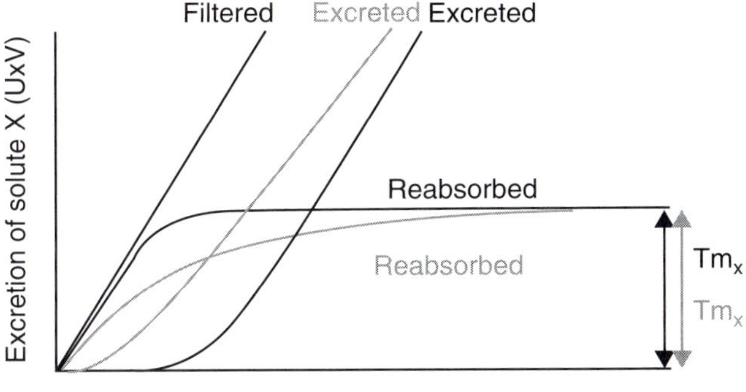

FIGURE 10.3 Threshold of renal reabsorption of solute X. The black lines depict normal filtration, reabsorption, and excretion of solutes as functions of the filtered load (product of the glomerular filtration rate and the solute concentration in the filtrate). Reabsorption of solutes increases as more solute is delivered to the nephron until the tubular reabsorptive capacity (Tm$_x$) is reached so further increases in filtered load results in excretion of X in the urine. The red lines represent the possible defects in the proximal tubule. The upper panel shows a reduction in transport capacity while the lower panel shows a reduction in affinity for substrate. In either case, increased excretion of X is observed for a given filtered load. The two types of theoretical defects are not mutually exclusive. (see also Plate 8)

apolipoprotein, hormones, and α1- and β2-microglobulins. The reabsorption of low-molecular-weight proteins requires endosomal trafficking, a process dependent on the proximal tubule multiligand receptors, megalin and cubilin, and endosomal acidification (Hilpert et al 1999, Leheste et al 1999, Christensen and Birn 2001, Kozyraki et al 2001, Nykjaer et al 2001). Endosomal acidification is mediated by a V type H^+-ATPase and a chloride channel (CLC-5) that shunts the electrical potential generated by the proton pump (Sabolic et al 1985, Brown & Stow 1996, Gunther et al 1998, 2003, Piwon et al 2000, Wang et al 2000b, 2005, Christensen et al 2003). The Na^+/H^+ exchanger NHE3 may also play a role in endosomal acidification of <1 pH unit (Gekle 2005).

Decreased Proximal Absorption Exceeding Distal Compensation

Patients with the Fanconi syndrome have renal salt wasting. This is not at all surprising in view of the fact that the proximal tubule reabsorbs two-thirds of the NaCl in the glomerular ultrafiltrate. The delivery to the distal nephron of the filtered NaCl normally reabsorbed by the proximal tubule surpasses the distal nephron's NaCl reabsorptive capacity (Houston et al 1968). Since 80% of the filtered HCO_3^- is reabsorbed by the proximal tubule, the distal nephron is overwhelmed by the large HCO_3^- load exiting the proximal tubule. The compensatory NaCl retention to restore volume in conjunction with the proximal tubule's inability to maintain a normal plasma HCO_3^- concentration and content leads to hyperchloremic acidosis. It is important to realize that in the steady state, there is no bicarbonaturia since the lower plasma HCO_3^- concentration (and to a minor extent lower GFR) reduces the filtered load to the point where the proximal and distal nephron can reabsorb the entire delivered bicarbonate. Since the distal nephron is intact, patients with the Fanconi syndrome can excrete urine with a pH less than 5.5 in the steady state. However, any attempts to

TABLE 10.4 Glycosuria due to mismatch between filtered load and tubular reabsorption of glucose

Condition	Glucose				
	Glomerular × filtration rate (ml/min)	Plsama = concentration (mg/100 ml)	Filtered load (mg/min)	Maximal tubular reabsorption (mg/min)	Urinary excretion
Normal renal function	100	100	100	220	No
	100	>220	>220	220	Yes
Glomerular problem, e.g. *hyperfiltration*	150	100	150	220	No
	150	>150	>220	220	Yes
Tubular problem, e.g. *Fanconi's*	90	90	81	81	No
	90	100	90	81	Yes

The shaded box represents the primary defect.

raise plasma HCO_3^- concentration will elevate the filtered load and result in massive bicarbonaturia (see section on treatment).

Although phosphate reabsorption has been described in the distal tubule (Lassiter & Colindres 1982), its magnitude is modest compared to the proximal tubule and is not sufficient to compensate for proximal phosphate absorption defect. Renal phosphate wasting leads to hypophosphatemia, vitamin D resistant rickets or osteomalacia, and muscle weakness.

In addition to the hyperphosphaturia, some patients with Fanconi syndrome also have hypercalciuria. While somewhat protected from renal stones and nephrocalcinosis because of the polyuria and increased citrate excretion manifested by these patients, a study using renal ultrasound has found a higher prevalence of nephrocalcinosis in patients with nephropathic cystinosis (Theodoropoulos et al 1995). Patients with Dent's disease have severe hypercalciuria and are prone to nephrolithiasis and nephrocalcinosis (Fisher et al 1994, Lloyd et al 1996, 1997, Scheinman et al 1999). Other causes of Fanconi syndrome are also associated with nephrocalcinosis and nephrolithiasis (Sliman et al 1995, Trevisan & Gardin 2005).

Solute Wasting by the Distal Nephron as a Result of Proximal Dysfunction

Hypokalemia can be severe in patients with Fanconi syndrome which is due to renal potassium wasting (Sebastian

et al 1971a, b, Tsai et al 2005). The pathophysiology of the hypokalemia is multifactorial. Extrarenal factors such as metabolic acidosis can result in intracellular potassium depletion (Poole-Wilson & Cameron 1975) and some decrease in renal potassium reabsorption (Tabei et al 1995). Most importantly, patients with Fanconi syndrome have secondary hyperreninemic hyperaldosteronism due to hypovolemia (Sebastian et al 1971). High mineralocorticoid activity, high distal sodium delivery, along with increased HCO_3^- and phosphate delivery, can result in significant distal potassium wasting. In addition, the hypokalemia can be exacerbated greatly if alkali is administered in the form of sodium salts. The excess bicarbonate exits the proximal tubule and provides a luminal negative anion to facilitate potassium excretion in the distal nephron. Hypokalemia contributes to the renal concentrating defect manifested by these patients and can cause muscle weakness and paralysis. Occasionally, potassium depletion can be the basis of fatality.

Relationship of Solute Wastage to the Salient Clinical Features

The solute wastage due to the proximal tubulopathy and its relationship to the clinical syndrome is shown in Figure 10.4. Not all the solutes wastage described above lead to clinical symptoms. The delivery of the nontransported organic solutes to the distal nephron does contribute

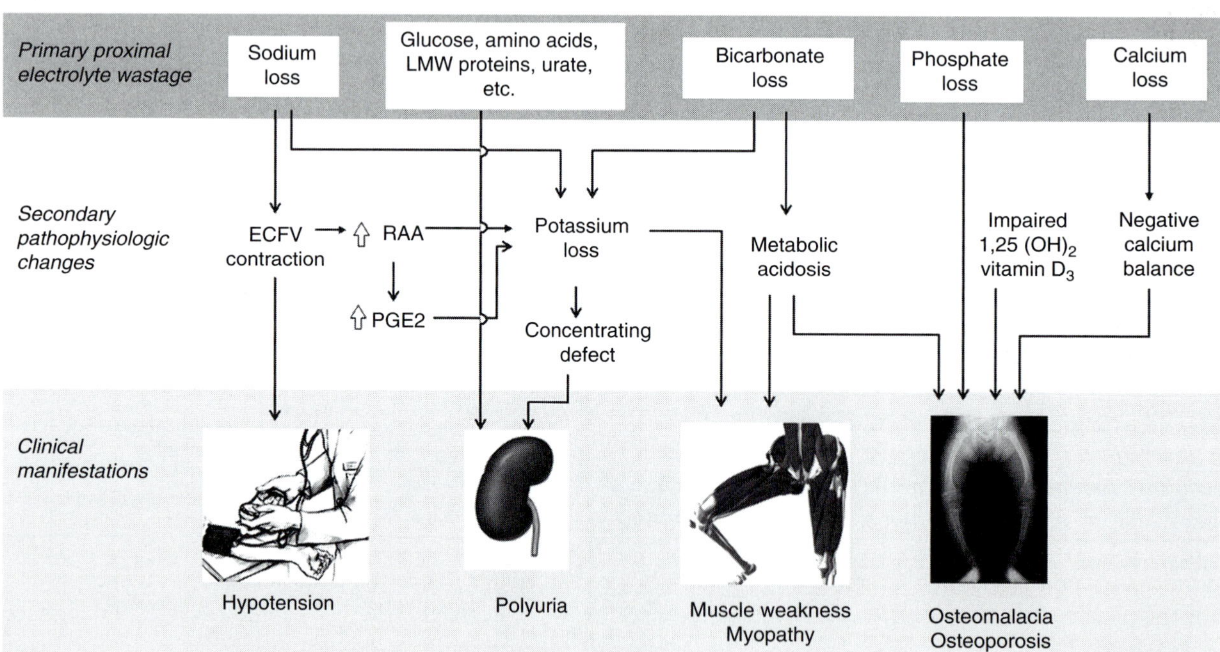

FIGURE 10.4 Relationship between primary disturbance, secondary pathophysiologic changes, and major clinical manifestations of the Fanconi syndrome. Wastage of the major groups of solutes is shown on the top panel (boxes). Their secondary pathophysiologic changes are shown in the middle panel. How these changes lead to four of the cardinal clinical features of the syndrome is depicted in the bottom panel. Not all the pathophysiologic and clinical defects are shown. Abbreviations: LMW = Low molecular weight, ECFV = Extracellular fluid volume, RAA = Renin angiotensin aldosterone system, PGE2 = Prostaglandin E2

to osmotic diuresis with a hypotonic (in regard to electrolytes) and hyperosmolar (in regard to total osmoles) urine. Glycosuria is primarily a self-correcting condition. As soon as the plasma levels falls, the filtered load will no longer exceed the tubular reabsorptive capacity and glycosuria ceases. One clinical manifestation of glucose wasting is reduction of the plasma glycosuric threshold, which is not a significant health problem. The same notion applies to the other organic solutes listed in Table 10.1. Organic solute wastage does cause solute diuresis to some degree since these osmoles are not reabsorbed in the distal nephron. The water loss is amplified by the fact that maximal urinary osmolarity is frequently not achieved owing to a concentrating defect possibly due to hypokalemia and the hyperprostaglandin state which is secondary to the activated renin-angiotensin system (Figure 10.4). This can manifest clinically as polydipsia, polyuria, dehydration, hypernatremia, and constipation.

The principal clinical consequences stem from the wastage of inorganic solutes (Figure 10.4). Renal excretion of NaCl results in volume contraction, a reduction in the glomerular filtration rate and filtered Na load. Patients with Fanconi syndrome are in Na^+ balance at the cost of chronic volume contraction. The chronic volume contraction predisposes patients with Fanconi syndrome to severe and potentially life-threatening hypotension when stressed, with vomiting or diarrhea. Other secondary symptoms include salt craving and constipation.

One of the most devastating clinical ordeals for patients with Fanconi syndrome is bone disease; particularly in children. The bone lesions in Fanconi syndrome are pleomorphic with osteomalacia being a major finding. Patients can also get osteoporosis and are at risk for nontraumatic fractures. The bone disease is not solely caused by the hypophosphatemia. Levels of 1, 25-dihydroxyvitamin D3 are either inappropriately low for the hypophosphatemia or frankly reduced (Brewer et al 1977, Baran & Marcy 1984, Colussi et al 1985, Clarke et al 1995). The etiology for the 1-α-hydroxylase deficiency is unclear. The same lesion can be reproduced in the maleic acid model (Brewer et al 1977). Experimentally induced acute metabolic acidosis in animals impairs 1-hydroxylation but such an effect has not been shown in chronic acidosis or in humans (Cunningham et al 1984). In addition to the mineralization defect of osteomalacia, the chronic metabolic acidosis also accelerates bone resorption independent of the other defects in Fanconi syndrome. The chronic acidosis, hypophosphatemia, and relative vitamin D deficiency all contribute to the short stature manifested by patients with Fanconi syndrome. Bone disease is not limited to children. Adults with acquired Fanconi syndrome can also develop debilitating skeletal abnormalities (Clarke et al 1995, Pham et al 2002, Earle et al 2004, Yang et al 2006). One of the largest series of cases is from the Mayo Clinic (Clarke et al 1995).

Although hypophosphatemia can present acutely with rhabdomyolysis (Knochel 1993), this complication has not been reported in Fanconi syndrome to date. The more common presentation is a chronic myopathy due mainly to phosphate and to some extent potassium depletion. Myopathy from phosphate depletion is seen in Fanconi syndome but can also result from various causes of renal phosphate wasting or intestinal malabsorption (Insogna et al 1980, Lian et al 1994, Rago et al 1994, Campos et al 1995, Wang et al 2000a, Parsonage et al 2005). Renal phosphate wasting has been discovered as a cause of chronic nonspecific fatigue syndrome (De Lorenzo et al 1998). Phosphate depletion is associated with neuromuscular dysfunction due to changes in mitochondrial respiration and defects of intracellular oxidative metabolism.

CELLULAR MECHANISMS OF PROXIMAL TUBULE TRANSPORT DEFECTS

Several points deserve emphasis before one embarks on a discussion of the cellular mechanisms responsible for proximal transport defects. First, the most prominent feature of Fanconi syndrome is the generalized defect in proximal tubule transport involving multiple solutes. The challenge is to understand how any given lesion can lead to such extensive pantubular dysfunction. Second, there is no single unifying model to date that explains all the transport defects in Fanconi syndrome. Many isolated mechanisms have been elucidated in animal and cell models that can potentially explain part of the picture but definitive relationship of these mechanisms to human Fanconi syndrome is still largely elusive. Third, there is no evidence that a single set of pathophysiologic mechanisms underlies the human syndrome. Fanconi syndrone is unlikely to have a unique underlying pathophysiologic mechanism. Certainly, considerable heterogeneity exists in the clinical picture and etiologies so there is no a priori reason to assume the cellular pathophysiology should be homogeneous in all patients. This section journeys through an exercise to explore the cellular mechanisms of proximal tubule transport defects in Fanconi syndrome backed by experimental data. Figures 10.5 and 10.6 classify the possible underlying defects empirically into two categories emphazing how a single discrete lesion can ramify into multiple transport defects.

Figure 10.5 summarizes the theoretical defects that abrogate the quintessential electrochemical gradient that drives most proximal tubule transport. With the exception of H^+-ATPase, most apical proximal tubule transport is Na^+-dependent with the driving force being the low intracellular Na^+, the negative cell interior voltage, or both. Thus, a generalized proximal tubular transport defect could be mediated by a primary failure to generate this electrochemical gradient (Figure 10.5). This can theoretically result from decreased ATP generation, decreased Na^+ pump activity, or dissipation of electrochemical gradient by processes other

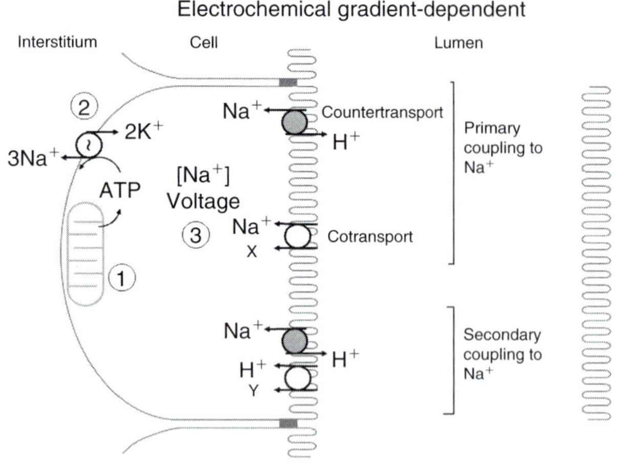

1. Decreased ATP generation
2. Decreased Na⁺ pump activity
3. Dissipation of electrochemical gradient

FIGURE 10.5   Potential cellular mechanism of multiple transport disorders in the Fanconi syndrome: Electrochemical gradient-dependent defects. Apical transporters are coupled by primary or secondary mechanisms to the electrochemical gradient generated by low cell [Na⁺] and negative voltage. There are three possible ways that the electrochemical driving force of Na⁺-coupled transport can be disrupted

than Na⁺ coupled transport (Figure 10.5), An increase in the permeability of the proximal tubule to sodium or potassium can dissipate the electrochemical gradient. In the defects illustrated in Figure 10.5, all apical membrane transporters are intrinsically normal but are working at a lower driving force.

In contrast, Figure 10.6 depicts the possible mechanisms whereby one lesion can cause multiple transporter defects without disrupting the electrochemical gradient. Defects in transcription factors that control multiple transporters can theoretically cause multiple transport defects. Disorders of protein trafficking, targeting, or scaffold mechanisms that maintain the integrity and function of apical membrane proteins, can result in faulty delivery and function of several apical transporters. Since transmembrane proteins can be affected by the lipid environment, alteration in the composition of the lipid bilayer can also affect multiple apical membrane transporter proteins. Finally, since there is a large lumen-to-plasma gradient for most solutes, an expanded paracellular backleak pathway can cause multiple solute transport defects.

Electrochemical Gradient-dependent Mechanisms

The electrochemical gradient-dependent mechanism is supported by a reasonable body of experimental data. In 1950, Robert Berliner was conducting experiments on the effect of organic acids on titratable acid excretion (Berliner et al 1950). Unexpectedly, these acidemic dogs developed paradoxical

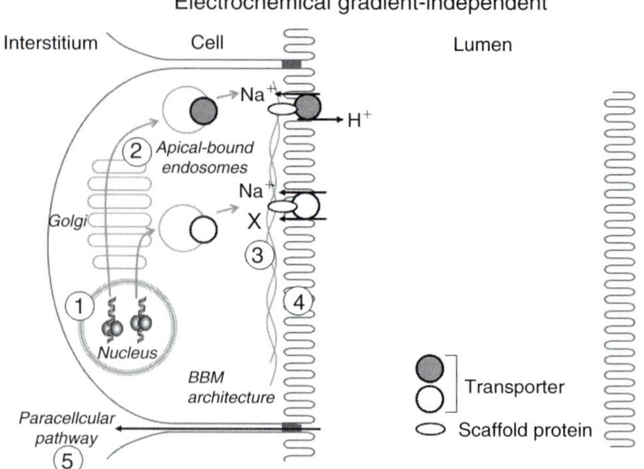

1. Defective transcription of several transporters
2. Defective protein trafficking for apical targeting
3. Defects in scaffold proteins or BBM architecture
4. Changes in membrane lipid and fluidity
5. Increased paracellular backleak

FIGURE 10.6   Potential cellular mechanisms of multiple transport disorders in the Fanconi syndrome: Electrochemical gradient-independent defects. The intracellular electrochemical driving forces are normal. A transcription factor that is required for gene expression of multiple transporters can lead to multiple transport defects. Protein trafficking abnormalities can affect apical targeting of multiple transporters. Disruption of the apical architecture or scaffolding proteins that organize the apical transporters can affect multiple transporters. Changes in membrane fluidity can affect multiple transport sytems. Theoretically, an abnormal paracellular backleak can affect multiple solutes

bicarbonaturia. Berliner also noted an increase in the urinary excretion of glucose and phosphate without a consistent change in glomerular filtration rate. Now we know the culprit was maleic acid-induced Fanconi syndrome. A lot of our knowledge of the pathogenesis of Fanconi syndrome comes from the study of the effect of maleic acid in rats and dogs (Berliner et al 1950, Kramer & Gonick 1970, 1973, Scharer et al 1972, Rosen et al 1973, Bergeron et al 1976, Brewer et al 1977, Reynolds et al 1978, Gunther et al 1979, Maesaka & McCaffery 1980, Al Bander et al 1985, Bank et al 1986, Guntupalli et al 1991, Castano et al 1997) and a second in vitro model of proximal tubule cystine accumulation as seen in cystinosis (Coor et al 1991, Foreman et al 1987, 2001, Foreman & Benson 1990a, 1990b, Salmon & Baum 1990, Sakarcan et al 1992, 1994, Bajaj & Baum 1996). Injection of maleic acid in rats results in Fanconi syndrome (Bergeron et al 1976, Maesaka & McCaffery 1980). While some micropuncture studies have found normal proximal tubule transport of amino acids and phosphate in the early proximal tubule, subsequent studies demonstrated a reduction in active proximal tubule transport (Kramer & Gonick 1970, 1973, Scharer et al1972, Reynolds et al 1978, Gunther et al 1979, Al Bander et al 1985, Bank et al 1986, Guntupalli et al 1991, Castano et al 1997). Inulin,

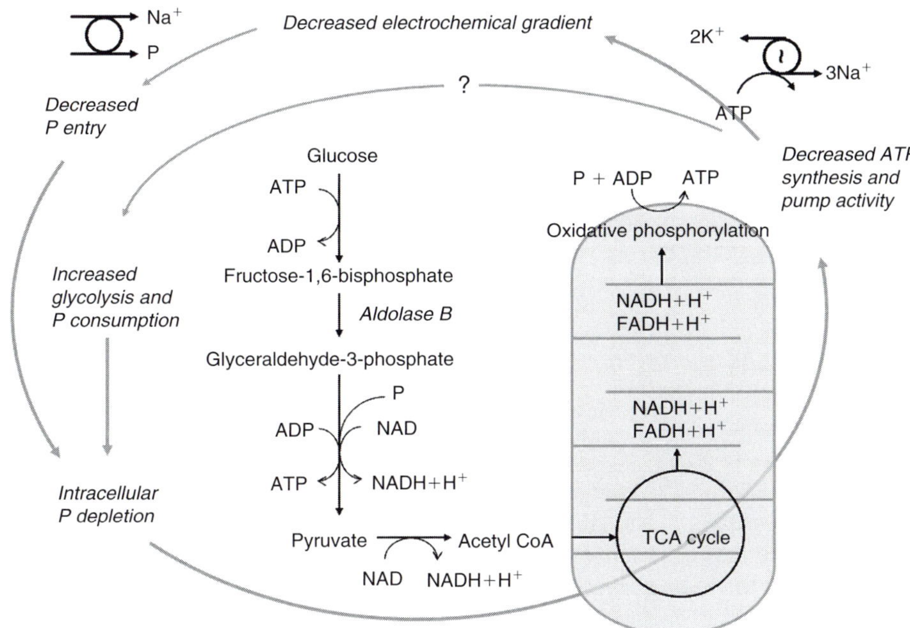

FIGURE 10.7 Potential vicious cycle of phosphate depletion, increased glycolysis and decreased ATP generation in the proximal tubule. Depicted is a 'circular model' where intracellular phosphate depletion can amplify the effects of impaired ATP generation and/or increased glycolysis. Regardless of where the primary defect is situated, this vicious cycle amplifies phosphate trapping and ATP depletion. Glycolysis is divided into two stages separated by the aldolase step. The initial part of glycolysis consumes ATP with no net exchange of phosphate. The second part consumes phosphate, and generates ATP and reduced equivalents. ATP generation by oxidative phosphorylation requires adequate free intracellular phosphate. (see also Plate 9)

bicarbonate and glycine permeabilities were unaffected by maleic acid (Gunther et al 1979, Bank et al 1986). In the cystine-loaded in vitro microperfused proximal tubule model, bicarbonate and mannitol permeabilities were also unaffected. Cystine-loaded proximal tubules had reduced active sodium, glucose, and bicarbonate transport (Salmon & Baum 1990). Neither direct incubation of maleic acid with brush border membrane vesicles nor brush border membranes harvested from animals that received maleic acid affected rate of transport, supporting the notion that normal brush transporters were running on lower electrochemical driving force in the intact tubule (Reynolds et al 1978, Silverman 1981). It is important to note that these findings are not universal and heterogeneity cannot be overemphasized. In constrast to maleic acid, in a model where Fanconi symdrome was induced by anhydro-4-epitetracycline in rabbits, glycosuria was accompanied by dramatic decrease in Na^+-dependent glucose transport (assayed under Vmax Na^+ gradients) in brush border vesicles, suggesting that the Na^+-gradient is not the cause (Yanase et al 1983, Orita et al 1984).

Intracellular ATP Depletion

The reduction in intracellular ATP will result in a decrease in Na^+-K^+-ATPase and H^+-ATPase activities and reduction in all active transport by the proximal tubule. There is evidence supporting depletion of intracellular ATP as a factor for inhibition of proximal tubule transport in Fanconi syndrome (Kramer & Gonick 1970, Coor et al 1991, Kellerman 1993). The reduction in ATP by infusion of maleic acid not only decreases the rate of proximal tubule transport but also disrupts the actin cytoskeleton; these effects can be ameliorated by ATP infusion (Kellerman 1993).

Intracellular cystine loading of proximal tubules resulted in a reduction in intracellular ATP which could be normalized by incubating the tubules with extracellular ATP (Coor et al 1991). Note that these rescue experiments may reflect the energy effect of ATP but they do not rule out purinergic receptor-mediated or adenosine-mediated effects.

There is evidence that the reduction in intracellular ATP can be secondary to limitation in intracellular phosphate. Crabtree described glucose-induced and fructose-induced inhibition of oxidative phosphorylation in malignant cells (Crabtree 1929). A decrease in proximal tubule transport can potentially be mediated by the Crabtree effect, an inhibition of oxidative phosphorylation by increased glycolysis (Crabtree 1929, Sussman et al 1980, Brazy et al 1982). Both glycolysis and oxidative phosphorylation are dependent on availability of intracellular phosphate. Addition of glucose to phosphate-depleted cells results in the accumulation of phosphorylated glycolytic intermediates which further decreases intracellular phosphate (Koobs 1972), resulting in inhibition of oxidative phosphorylation. A limitation in intracellular phosphate impairs proximal tubule oxidative phosphorylation of ADP to ATP by complex V. In addition, phosphate is also necessary for the entry of tricarboxylic acid cycle intermediates into the mitochondria (McGivan & Klingenberg 1971, Meijer & Van Dam 1974, LaNoue & Schoolwerth 1979). Figure 10.7 is a schematic depiction of the proposed model where intracellular phosphate depletion may trigger a vicious cycle of ATP depletion regardless of the inciting primary defect. One can start anything in the circle and end up with the same consequences. Impaired mitochondrial ATP synthesis can activate glycolysis, which increases phosphate trapping and worsens reduction of ATP synthesis. Decreased Na^+-K^+-ATPase can impair phosphate uptake and further augment phosphate depletion.

The bulk of the experimental data supporting the Crabtree effect are centered around the observation that transport is worsened by phosphate depletion and ameliorated by phosphate loading. Phosphate-depleted proximal tubules displays normal rates of transport in the absence of luminal glucose but when glucose was added to the luminal perfusate of phosphate-depleted proximal tubules, transport was inhibited and the tubule developed the equivalent of Fanconi syndrome (Brazy et al 1982). However, if the phosphate-deprived tubule was incubated with an inhibitor of glycolysis even in the presence of glucose, transport was preserved (Brazy et al 1982). There is evidence that the Crabtree effect may play a role in the pathogenesis of Fanconi syndrome due to maleic acid (Al Bander et al 1985, 1986, Shvil et al 1987, Eiam-Ong et al 1995), the cystine-loaded tubule (Bajaj & Baum 1996), the fructose loaded rat model of the Fanconi syndrome of hereditary fructose intolerance (Morris et al 1978), and hereditary fructose intolerance in humans (Morris et al 1971). In the maleic acid model, there is an increase in glucose uptake and rate of glycolysis which would utilize intracellular phosphate, and a reduction in oxidative phosphorylation compared to controls (Angielski & Rogulski 1962, Rogulski et al 1975, 1976). The cystine-loaded tubule has a reduction in intracellular phosphate and ATP (Bajaj & Baum 1996) and phosphate loading preserved intracellular ATP levels to control levels (Bajaj & Baum 1996). In rat models of the Fanconi syndrome produced by maleic acid, cystine, or fructose and in human hereditary fructose intolerance, there is an attenuation of the proximal tubule transport defect by prior phosphate loading (Baum 1998, Al Bander et al 1985, 1986, Eiam-Ong et al 1995, Bajaj & Baum 1996). It may also explain the transport defect in the Fanconi-Bickel syndrome where basolateral proximal tubule facilitated diffusion of glucose via GLUT2 is impaired (Santer et al 1997, 1998a, 2002). These data are consistent with but do not prove that the Crabtree effect is the principal mechanism in the reduction in oxidative phosphorylation and transport in some forms of Fanconi syndrome (Figure 10.8).

Na^+-K^+-ATPase

There is evidence that maleic acid can directly inhibit the Na^+-K^+-ATPase (Kramer & Gonick 1970, Mujais 1993, Eiam-Ong et al 1995, Castano et al 1997) in the proximal convoluted tubule but not in other nephron segments (Mujais 1993, Eiam-Ong et al 1995). In these studies, Na^+-K^+-ATPase was assayed with exogenous ATP added to the reaction mixture so that ATP was not a limiting factor. This is at variance to the cystine-loaded tubule where dissected proximal tubules had comparable Na^+-K^+-ATPase activity with and without cystine loading suggesting an ATP depletion rather than a defective pump mechanism (Coor et al 1991). Incubation of cultured proximal tubule cells with light chains from a patient with paraprotein-related Fanconi syndrome resulted in inhibition of Na^+-K^+-ATPase activity

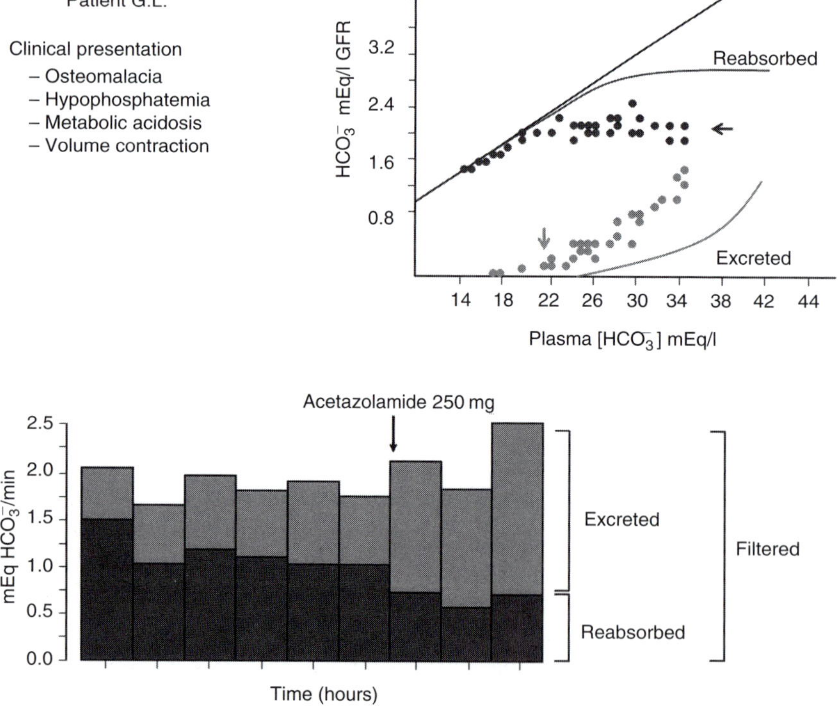

FIGURE 10.8 Clinical investigation of a patient with idiopathic Fanconi syndrome. Presenting features are shown in the upper left. The upper right panel depicts a titration study where excreted and reabsorbed HCO_3^- was measured and calculated over a range of varying filtered loads. The expected normal relationship is shown as solid lines. The actual data points are shown as dots. Both proximal capacity (blue arrow) and affinity defects (red arrow) were present. In the bottom panel, the patient's serum HCO_3^- was maintained between 22–24 mM by continuous HCO_3^- infusion and glomerular filtration was measured by inulin clearance. HCO_3^- excretion and reabsorption rates were measured and calculated respectively. A carbonic anhydrase inhibitor was given at the time indicated which resulted in further HCO_3^- wasting (unpublished studies: Seldin DW and Rector FC Jr.). (see also Plate 10)

and gene expression (Guan et al 1999). Loss of basolateral infolding and invaginations and Na^+-K^+-ATPase has also been described in cadmium-induced Fanconi's syndrome (Sabolic et al 2006).

Dissipation of the electrochemical gradient generated by the Na^+-K^+-ATPase as a cause of the Fanconi syndrome remains a theoretical consideration with no definitive experimental data to date describing increases in passive permeabilities to Na^+ or K^+.

Electrochemical Gradient-independent Mechanisms

It is possible for lesions to cause pantubular transport defects without affecting the electrochemical gradient (Figure 10.6).

TRANSCRIPTION FACTORS

A single genetic defect in a transcription factor can theoretically affect multiple transporter proteins. Hepatocyte nuclear factor 1 (HNF-1) controls the transcription of a number of genes in the liver and the kidney, and mice with deletion of HNF-1α have impaired insulin secretion and Fanconi syndrome (Pontoglio et al 1996, 1998). While HNF-1α null mice have clear proximal tubular dysfunction, patients with HNF-1α mutations have a more subtle phenotype manifested by lower threshold for glucose and amino acids (Pontoglio et al 2000, Bingham et al 2001). The cause of Fanconi syndrome in HNF-1α null mice is likely gradient independent as the proximal tubule ATP concentration and Na^+-K^+-ATPase activity are normal (Pontoglio et al 2000). These mice have a reduction in SGLT2 (the high-capacity/low-affinity glucose transporter on the proximal tubule brush border membrane transcription) (Pontoglio et al 2000). It has been hypothesized that glucose absorption in HNF-1α-deficient mice and patients with MODY3 may depolarize the proximal tubule cell resulting in a decrease in electrogenic amino acid transport (Bingham et al 2001). However, it is unclear how a reduction in SGLT2 abundance would depolarize the proximal tubular cell.

PROTEIN TRAFFICKING

Impaired endocytotic function can result in a generalized proximal tubular transport defect as seen in Dent's disease and will be discussed in detail in the chapter related to that disorder (Piwon et al 2000, Wang et al 2000b, 2005, Christensen et al 2003, Gunther et al 2003). A disruption in endosomal function can potentially exert a generalized effect on vesicular trafficking and lead to the Fanconi syndrome as seen in Dent's disease. Endosomes play an important role in the reabsorption of filtered low-molecular-weight proteins (Marshansky et al 2002, Gunther et al 2003). Megalin and cubilin are two such multiligand receptors for low-molecular-weight proteins (Hilpert et al 1999, Leheste et al 1999, Christensen & Birn 2001, Nykjaer

et al 2001). The endosome has a V-type H^+-ATPase that acidifies the interior (Sabolic et al 1985, Brown & Stow 1996); the potential is shunted by a chloride channel (ClC-5) (Gunther et al 1998, 2003, Piwon et al 2000, Wang et al 2000b, 2005, Christensen et al 2003). Disruption of either the H^+-ATPase, as has been demonstrated in the cadmium model of Fanconi syndrome (Herak-Kramberger et al 1998), or in ClC-5, the chloride channel mutated in Dent's disease (Lloyd et al 1996), leads to a generalized proximal tubule transport disorder (Gunther et al 1998, 2003, Piwon et al 2000, Wang et al 2000b, 2005, Christensen et al 2003). Endosomal vesicles from ClC-5 knockout mice have impaired rates of vesicular acidification (Piwon et al 2000, Gunther et al 2003).

Megalin and cubilin bind filtered low-molecular-weight proteins which are endocytosed and shuttled to the lysosomes where these filtered proteins are degraded (Marshansky et al 2002). Megalin is reduced in amount and its pattern of expression is perturbed with decreased brush border staining in the maleic acid model of Fanconi syndrome (Bergeron et al 1996). Disruption of endosomal trafficking can explain the low-molecular-weight proteinuria seen in Dent's disease and in Fanconi syndrome. Disruption of endocytic vesicles may also result in an inhibition in proximal tubule transport from a number of mechanisms, though at this point the mechanism is not entirely clear. Endocytic vesicles play an important role in the recycling of proximal tubular transporters and a defect in endocytic recycling may result in a decrease in apical transporter abundance. Urinary shedding of megalin is dramatically decreased in patients with Lowe's and Dent's disease but they appeared to be maintained in idiopathic congenital Fanconi's (Norden et al 2002).

Normal processing of endosomes requires acidification of its contents and subsequent degradation of the protein contents by lysosomes. Thus, it is possible that a defect in this system could result in accumulation of unprocessed material that could damage the proximal tubule cell (Christensen & Gburek 2004). In addition, some of the low-molecular-weight proteins include those involved in an inflammatory response that may result in proximal tubular injury (Christensen & Gburek 2004).

SCAFFOLD PROTEINS AND BRUSH BORDER INTEGRITY

Ezrin-radixin-moesin (ERM) proteins orchestrate the structure and function of specific cortical structures in polarized epithelial cells by provision of linkage between filamentous (F)-actin to transmembrane and plasma membrane associated proteins and are important for integrity of the brush border microvilli (Yonemura et al 1999, Van Furden et al 2004). One ligand for the ERM protein is EBP50 or NHERF1 (Reczek et al 1997). In EBP50 null mice, ERM proteins are decreased specifically in BBM from kidney and small intestine epithelial cells, but they remained unchanged in the cytoplasm (Morales et al 2004). Abnormalities of proximal Na^+-dependent phosphate transport, Na^+/H^+ exchanger, and urate transport have been described in the EBP50 null mice (Shenolikar et al 2002, Cunningham et al 2004). Fibroblasts from patients with mutations of phosphatidylinositol 4,5-bisphosphate 5-phosphatase (Lowe's syndrome) show a decrease in long actin stress fibers, heightened sensitivity to actin depolymerizing agents, and increase in punctate F-actin staining in a distinctly anomalous distribution in the central part of the cell. Distribution of gelsolin and alpha-actinin, which are actin-binding proteins and regulated by PIP2, are also altered (Suchy & Nussbaum 2002). Infusion of maleic acid that decreases the rate of proximal tubule transport also disrupts the actin cytoskeleton (Kellerman 1993). Collectrin is not a transporter itself, but when deleted can lead to generalized amino acid wasting affecting multiple amino acid transporters (Malakauskas et al 2006). Cadmium-induced Fanconi syndrome is associated with microtubule depolymerization and loss of clathrin in the subapical domain (Sabolic et al 2006).

MEMBRANE FLUIDITY

Hsu and coworkers found increased membrane fluidity and higher cholesterol to phospholipid molar ratio in renal cortical membrane from dogs with naturally occurring genetic Fanconi syndrome (Hsu et al 1991, 1992). In the model of Fanconi syndrome induced by succinylacetone which is a metabolite of tyrosine excreted in excess in hereditary tyrosinemia, increased membrane fluidity has also been found (Spencer et al 1988, Roth et al 1999). Alteration of apical membrane lipid and fluidity remains an intriguing explanation of multiple transport defects in the apical membrane.

PTH AND VITAMIN D HYDROXYLATION

These two calcitropic hormones are considered separately. Parathyroid hormone (PTH) is a small peptide which is filtered, reabsorbed and degraded by the proximal tubule. Megalin and endosomal trafficking are necessary for this process (Hilpert et al 1999, Gunther et al 2003) and megalin-deficient mice have increased levels of urinary parathyroid hormone excretion (Hilpert et al 1999). Failure to clear luminal parathyroid hormone could result in increased activation of luminal parathyroid receptors resulting in internalization of the sodium phosphate cotransporter (NaPi-2) and Na^+/H^+ exchanger responsible for phosphate transport (Pfister et al 1998, Fan et al 1999, Collazo et al 2000). This could be a factor mediating the phosphate, Na^+ and HCO_3^- wasting in Fanconi syndrome (Hilpert et al 1999, Gunther et al 2003).

Increased PTH action should stimulate 1-α-hydroxylase but inappropriately low $1,25(OH)_2$ vitamin D levels are seen in Fanconi syndrome. This may be explained by the limited substrate for 1-α-hydroxylation. Normal production of $1,25-(OH)_2$ vitamin D by the proximal tubule requires the binding of 25-(OH) vitamin D to its low-molecular-weight binding protein in the plasma which is filtered and reabsorbed by the proximal tubule. Internalization of the 25-(OH) vitamin D–vitamin D binding protein complex is mediated by cubilin/megalin-dependent endocytosis. Megalin-deficient mice and cubilin-deficient dogs have hypovitaminosis D (Nykjaer et al 2001). This could be a factor in the bone disease in Fanconi syndrome.

ANIMAL MODELS OF FANCONI SYNDROME

Naturally Occurring Fanconi Syndrome in Dogs

The most common genetic renal disorders in urban domesticated animals (mainly cats and dogs) are amyloidosis, renal dysplasia, polycystic kidney disease, basement membrane disorders, and Fanconi syndrome (Greco 2001). With inbreeding practices, the prevalence of Fanconi syndrome is estimated to be up to 10% in Basenji dogs (Noonan & Kay 1990, Yearley et al 2004). The transport defects in Basenjis are strikingly similar to those observed in humans with deranged renal tubular handling of glucose, phosphate, sodium, potassium, uric acid, and amino acids (Bovee et al 1978, Breitschwerdt et al 1983, Hsu et al 19992). In addition, some have reported extrarenal features such as hypothyroidism, hypergastrinemia, and a diffuse lymphocytic-plasmacytic enteritis suggesting the possibility of a systemic syndrome (Breitschwerdt et al 1983). The reason for this high prevalence in Basenjis is not known and neither is the genetic lesion. Similar findings have been reported in Labrador retrievers (Settles & Schmidt 1994, Jamieson & Chandler 2001, Hostutler et al 2004). These naturally occurring canine diseases can be utilized to identify loci for genetic generalized Fanconi syndrome.

Genetic Experimental Models

Hepatocyte nuclear factor 1 (HNF-1) is a transcription factor that increases the transcription of a number of genes including albumin, phenylalanine hydroxylase, and α1-antrypsin. Mice with deletion of HNF-1α have impaired insulin secretion and also have Fanconi syndrome. The most common inherited cause of diabetes is the autosomal dominant disorder maturity-onset diabetes of the young (MODY3). Patients with MODY3 have a mutation in HNF-1α (Yamagata et al 1996). While HNF-1α null mice have overt Fanconi syndrome, patients with MODY3 have been shown to have a lower threshold for glucose and amino acids (Bingham et al 2001, Pontoglio et al 2000). The cause for Fanconi syndrome in HNF-1α null mice is unclear. The proximal tubule ATP concentration and Na^+-K^+-ATPase activity are reportedly normal (Pontoglio et al 2000). These mice have a reduction in SGLT2 (the high-capacity/low-affinity glucose transporter on the proximal tubule brush border membrane transcription) (Pontoglio et al 2000).

It has been hypothesized that glucose absorption in HNF-1α-deficient mice and patients with MODY3 may depolarize the proximal tubule cell resulting in a decrease in electrogenic amino acid transport (Bingham et al 2001). However, it is unclear how a reduction in SGLT2 abundance would depolarize the proximal tubular cell.

ClC-5 chloride channel deletions in mice is an experimental model of Dent's disease (Piwon et al 2000, Wang et al 2000b, 2005, Christensen et al 2003, Gunther et al 2003). The proximal tubules from these mice have impaired endodyocytic function, which results in a generalized proximal tubular transport defect (Piwon et al 2000, Wang et al 2000b, 2005, Christensen et al 2003, Gunther et al 2003). The pathogenesis of the transport defect in Dent's disease will be discussed in the chapter on this disorder.

In the adult population, acquired Fanconi syndrome from paraproteins is far more common than all the other forms. One transgenic model of Fanconi syndrome from monoclonal gammopathies results in monoclonal light chain accumulation in the proximal tubule and a generalized dysfunction of the proximal tubule. These mice have their endogenous mouse Jκ cluster replaced by a human VκJκ rearranged gene cloned from a patient with myeloma-associated Fanconi syndrome. Association of the human VκI domain with a murine κ constant domain in the transgene resulted in nephrotoxicity similar to that observed in patients. The causal effect was demonstrated by a reduction of proximal tubule crystalline inclusions after conditional deletion of the human VκI transgene (Sirac et al 2006).

Mice with homozygous disruption of fumarylacetoacetate hydrolase die as neonates precluding this as a model of the study of Fanconi syndrome of tyrosinemia (Sun et al 2000, Endo & Sun 2002). The generation of fumarylacetoacetate from tyrosine requires 4-hydroxyphenylpyruvate dioxygenase. Mice with deletion of both fumarylacetoacetate hydrolase and 4-hydroxyphenylpyruvate dioxygenase (Fah$^{-/-}$Hpd$^{-/-}$) thrive and do not have liver or renal disease (Sun et al 2000, Endo & Sun 2002). Administration of homogentisate, a precursor of fumarylacetoacetate hydrolase, bypasses 4-hydroxyphenylpyruvate dioxygenase to Fah$^{-/-}$Hpd$^{-/-}$ in mice results in glucosuria and phosphaturia. These mice develop release of cytochrome C from mitochondria and have apoptosis of the proximal tubule (Sun et al 2000, Endo & Sun 2002). Fah$^{-/-}$-deficient mice also survive if 4-hydroxyphenylpyruvate dioxygenase is inhibited with 2-(2-nitro-4-trifluoromethyl-benzoyl)-1,3 cyclohexanedione and have proximal tubule toxicity after administration of homogentisate (Luijerink et al 2004, Jacobs et al 2006). Previous studies have demonstrated that administration of succinylacetone to rats, a metabolite of tyrosine that is excreted in the urine in high concentrations in patients with tyrosinosis, produces Fanconi syndrome (Wyss et al 1992).

Not all attempts to create a genetic model of Fanconi syndrome have been entirely successful. Cystinosis is due to a mutation in the gene CTNS which encodes cystinosin (Town et al 1998). Cystinosin is a lysosomal protein which is an H$^+$-cystine transporter that allows cystine to be transported out of lysosomes (Kalatzis et al 2001, Cherqui et al 2002, Haq et al 2002) Mice where CTNS is knocked out have high lysosomal cystine content but fail to develop the Fanconi syndrome (Cherqui et al 2002). Investigators have used cystine dimethyl ester to acutely cystine load proximal tubules to examine how intracellular cystine inhibits proximal tubular transport (Foreman et al 1987, 1995, Foreman & Benson 1990a, b, Salmon & Baum 1990, Coor et al 1991, Sakarcan et al 1992, 1994, Bajaj &Baum 1996).

Pharmacologic Animal Models

Two models were discussed in some detail in the previous section: Maleic acid administration in rats and dogs (Berliner et al 1950, Scharer et al 1972, Kramer & Gonick 1970, 1973, Rosen et al 1973, Bergeron et al 1976, Brewer et al 1977, Reynolds et al 1978, Gunther et al 1979, Maesaka & McCaffery 1980, Al Bander et al 1985, Bank et al 1986, Guntupalli et al 1991, Castano et al 1997) and in vitro cystine loading of proximal tubules (Foreman et al 1987, 1995, Foreman & Benson 1990a, b, Salmon & Baum 1990, Coor et al 1991, Sakarcan et al 1992, 1994, Bajaj & Baum 1996). In vivo injection of maleic acid in rats results in reduction of active proximal tubule transport (Berliner et al 1950, Scharer et al 1972, Kramer & Gonick 1970, 1973, Rosen et al 1973, Bergeron et al 1976, Brewer et al 1977, Reynolds et al 1978, Gunther et al 1979, Maesaka & McCaffery 1980, Al Bander et al 1985, Bank et al 1986, Guntupalli et al 1991, Castano et al 1997) without affecting permeabilities (Gunther et al 1979, Bank et al 1986) or transporter activites assayed in brush border membranes (Reynolds et al 1978, Silverman 1981). Cystine-loaded proximal tubules exhibit reduced active sodium, glucose, and bicarbonate transport (Salmon & Baum 1990). Other animal models of the Fanconi syndrome include agents that are known to produce proximal tubule transport defects in humans. For example, administration of heavy metals in vivo or in vitro (Gonick et al 1975, 1980, Kramer et al 1986, Blumenthal et al 1990, Herak-Kramberger et al 1996, 1998, Tabatabai et al 2001, 2003, 2005, Sabolic et al 2002) can induce alterations in proximal tubular transport. Iphosphamide is a cancer chemotherapeutic agent which is now a common cause for the Fanconi syndrome in humans (Cachat & Guignard 1996, Skinner 2003). Administration of iphosphamide to rodents produces a similar picture (Springate & Van Liew 1995, Nissin & Weinberg 1996). The most studied and best characterized model is still the maleic acid model of the Fanconi syndrome which was discussed in detail above in dogs (Berliner et al 1950, Scharer et al 1972, Kramer & Gonick 1970, 1973, Rosen et al 1973, Bergeron et al 1976, Brewer et al 1977, Reynolds et al 1978,

Gunther et al 1979, Maesaka & McCaffery 1980, Al Bander et al 1985, Bank et al 1986, Guntupalli et al 1991, Castano et al 1997).

TREATMENT

The management of the Fanconi syndrome can be divided into treatment of the specific underlying etiology and replacing the solute loss regardless of the cause.

Specific Treatment

Every effort should be made to delineate a specific diagnosis. For certain inherited disorders such as hereditary fructose intolerance, tyrosinemia and galactosemia, dietary changes aimed at removal of fructose, tyrosine and phenylalanine, and lactose, respectively, can effectively ameliorate the Fanconi syndrome. Patients with cystinosis have a progressive decline in glomerular filtration rate with age, which can be effectively treated with cysteamine which decreases intracellular cystine accumulation (Markello et al 1993). Unfortunately, cysteamine therapy is not very effective for the Fanconi syndrome in patients with impaired renal function. In acquired forms of Fanconi syndrome, removal of the toxic agent usually results in amelioration of the defect in proximal tubule transport.

Replacing Solute Loss

For patients with Fanconi syndrome for which the causative agent is unknown, or for which there is no therapy for the underlying disease, treatment can be extremely difficult and is often ineffective. At first blush oral replacement of excreted solutes would seem like a logical and feasible solution. However, the problem with this therapy becomes apparent when one peruses Figure 10.3 and thinks of the magnitude of solute reabsorption by the proximal tubule. Since serum bicarbonate level is set by the renal threshold, normal subjects who ingest bicarbonate or alkali, we will not substantively increase our serum bicarbonate level above baseline because of bicarbonaturia. The same principle holds true for patients with Fanconi syndrome except that their threshold is set at a lower level than normal. Attempts to substantively increase serum bicarbonate levels would require massive doses of bicarbonate administered frequently. This would not be without consequences for, if bicarbonate is given as the sodium salt, the distal delivery of sodium bicarbonate would result in severe potassium wasting. The distal excretion of potassium is profound because patients with Fanconi syndrome are volume-depleted and have high aldosterone levels. Hypokalemia can be obviated by administration of alkali as the potassium salt but nonetheless serum potassium levels need to be monitored closely and significant increases in serum bicarbonate levels are rarely achieved.

Glucosuria and aminoaciduria do not usually lead to symptoms and require no therapy. However, acidosis and hypophosphatemia result in severe growth retardation and bone disease. Rickets in patients with Fanconi syndrome can be profound. Rickets is due to hypophosphatemia, which is secondary to renal phosphate wasting, and low or inappropriately normal levels of vitamin D (Brewer et al 1977, Baran & Marcy 1984, Colussi et al 1985, Clarke et al 1995). Treatment of hypophosphatemia with oral phosphate with 1,25-dihydroxyvitamin D can improve the bone disease but complete normalization of the serum phosphate levels is difficult in patients with normal glomerular filtration rates and severe proximal tubule impairment. Supplemental vitamin D will increase gastrointestinal phosphate absorption even in patients with normal levels of 1,25-dihydroxyvitamin D.

In addition to pure replacement, an approach used by many clinicians is to treat patients with Fanconi syndrome by interrupting part of the pathophysiology (Figure 10.4). Two such examples are the cyclooxygenase inhibitor indomethacin (Betend et al 1979, Haycock et al 1982, Usberti et al 1985) and natriuretic hydrochlorthiazide (Rampini et al 1968, Baran & Marcy 1984). Indomethacin will reduce the glomerular filtration rate and thereby reduce the filtered solute load. This approach can make oral replacement therapy significantly easier. Indometacin can also alleviate the contribution of prostaglandins to potassium and water wasting (Figure 10.4). However, it must be appreciated that indometacin can cause gastritis, ulcers, and gastrointestinal bleeding as well as renal failure. Thiazide or loop natriuretics can contract effective volume to the point that renal bicarbonate threshold can be raised. A major drawback is that diuretics can result in severe potassium wasting. Indometacin and thiazide should be used judiciously and these patients require close and frequent monitoring.

The following patient with idiopathic Fanconi syndrome was attended by one of the authors and exemplifies the extreme difficulty in managing the most severely affected patients. GL was a Hispanic girl who presented with short stature, agonizing bone pain, and plasma and urine biochemistry indicative of complete Fanconi syndrome. A bicarbonate titration curve performed in the Clinical Research Center (Figure 10.8) showed two main features. The reduced Tm for bicarbonate is characteristic of the capacity defect shown in Figure 10.3. However, bicarbonaturia was also occurring at plasma bicarbonate concentrations considerably below the threshold, which is illustrative of the affinity (splay) defect shown in Figure 10.3. It appears that this patient had a combined capacity and affinity defect for bicarbonate absorption (Figure 10.8). When normal plasma bicarbonate concentration was kept normal by intravenous bicarbonate infusion, addition of acetozolamide resulted in further increment in bicarbonaturia, indicating that carbonic anhydrase deficiency was not responsible for the proximal renal tubule acidosis (Figure 10.8).

The treatment of GL posed a formidable challenge. A large dose of sodium bicarbonate resulted in prompt massive bicarbonaturia and no appreciable rise in the plasma bicarbonate concentration. Urinary postassium and phosphorus were also sharply augmented. To enhance bicarbonate absorption, bicarbonate administration was combined with measures to reduce extracellular fluid volume. A diuretic was given in a setting of NaCl restriction. This program resulted in a sharp rise in plasma bicarbonate to levels of a frank metabolic alkalosis. Hypophosphatemia was treated with sodium-neutral phosphate. This regimen to salvage metabolic acidosis and hypophosphatemia resulted in severe hypokalemia, which could not be adequately corrected with potassium chloride supplement. Eventually, the plasma potassium was rectified with a combination of potassium chloride and aldactone. While on this complex combination of drugs and under the strict dietary supervision in the in-patient setting, GL achieved normal plasma values.

Upon discharge from hospital and assuming a different diet, despite compliance with the prescribed medicines, the patient developed life-threatening hyperkalemia requiring an emergency admission to hospital. Not all cases of Fanconi's syndrome are of this level of difficulty in terms of management but this case highlights the importance of a solid command of physiology in designing treatment as well as rectifying complications from treatment.

SUMMARY

Fanconi syndrome remains one of the most challenging and intriguing groups of disorders of renal epithelial transport. Many known genetic causes of Fanconi syndrome remain to be understood and others await discovery still. The quest to understand the biology of Fanconi syndrome, however, is vastly different from the many monogenic epithelial transporter inactivations that result in disease phenotype. The remarkably diverse group of causes begs the notion that one is witnessing some form of final common pathway of convergence of pathophysiologic events in the proximal tubule. Interestingly, one does not encounter a syndrome composed of pantubular dysfunction of the thick ascending limb or collecting duct. A significant fraction of the known causes of Fanconi syndrome, whether genetic or acquired, probably end up involving infiltration of the proximal tubule cell by ectopic or excess metabolites, or exogenous toxins that exceed the defense and reserve of the proximal tubule. This chapter has discussed how single or few lesion(s) when strategically placed can eventuate in disruption of energy generation, protein processing, or brush border intergrity; all of which can theoretically culminate in the multiple transport failure seen in the full-blown Fanconi syndrome. The clinical syndrome and clinical pathophysiology of Fanconi syndrome can be directly and accurately traced back to specific proximal tubule solute transport defects not only in the form of overt proximal wastage physiology as in glycosuria, but also in sophisticated paradigms invoking profound endocrine changes and proximal–distal interactions as in kaliuresis. The serious systemic ramifications of proximal dysfunction can be appreciated from proximal phosphate wasting inducing severe osteopathy and myopathy. The treatment of the multiple solute-wasting state is as equally challenging as the understanding of the pathobiology. Even with regimens based on sound physiology, the outcome can be quite variable with high risk of iatrogenic complications. Under no circumstances can therapy be contemplated and delivered without clear knowledge of the pathiophysiology. As one studies the spectrum of clinical diseases resulting from perturbed epithelial transport physiology, Fanconi syndrome qualifies as the ultimate sine qua non. As a postscript, it seems appropriate to quote Dr Guido Fanconi himself: 'It is the doctor's privilege to pursue an occupation which is also his favourite pastime.'

References

Abderhalden E. Familiäre cystindiathese. Hoppe-Seylers Zeitschrift für physiologische Chemie. Strassburg 1903; 38: 557–61.

Acker KJ, Roels H, Beelaerts W, Pasternack A, Valcke R. The histologic lesions of the kidney in the oculo-cerebro-renal syndrome of Lowe. Nephron 1967; 4: 193–214.

Al Bander H, Etheredge SB, Paukert T, Humphreys MH, Morris RC Jr.. Phosphate loading attenuates renal tubular dysfunction induced by maleic acid in the dog. Am. J. Physiol. 1985; 248: F513–21.

Al Bander HA, Mock DM, Etheredge SB, Paukert TT, Humphreys MH, Morris RC Jr. Coordinately increased lysozymuria and lysosomal enzymuria induced by maleic acid. Kidney Int. 1986; 30: 804–12.

Ali M, Rellos P, Cox TM. Hereditary fructose intolerance. J. Med. Genet. 1988; 35: 353–65.

Angielski S, Rogulski J. Effect of maleic acid on the kidney. I. Oxidation of Krebs cycle intermediates by various tissues of maleate-intoxicated rats. Acta Biochim. Pol. 1962; 9: 357–65.

Aronson PS, Giebisch G. Mechanisms of chloride transport in the proximal tubule. Am. J. Physiol. 1997; 273: F179–92.

Asatoor AM, Milne MD, Walshe JM. The effect of chelation therapy on the amino aciduria and peptiduria of Wilson's disease. J. R. Coll. Physicians Lond. 1983; 17: 122–5.

Aucouturier P, Bauwens M, Khamlichi AA, et al. Monoclonal Ig L chain and L chain V domain fragment crystallization in myeloma-associated Fanconi's syndrome. J. Immunol. 1993; 150: 3561–8.

Auricchio S, Frischknecht W. Shmerlingd. Primary tubulopathies. III. A case of oculo-cerebro-renal syndrome (Lowe syndrome). Helv. Paediatr. Acta 1961; 16: 647–55.

Austin JH. Metachromatic form of diffuse cerebral sclerosis. III. Significance of sulfatide and other lipid abnormalities in white matter and kidney. Neurology 1960; 10: 470–83.

Bajaj G, Baum M. Proximal tubule dysfunction in cystine-loaded tubules: effect of phosphate and metabolic substrates. Am. J. Physiol. 1996; 271: F717–22.

Bank N, Aynedjian HS, Mutz BF. Microperfusion study of proximal tubule bicarbonate transport in maleic acid-induced renal tubular acidosis. Am. J. Physiol. 1986; 250: F476–82.

Baran DT, Marcy TW. Evidence for a defect in vitamin D metabolism in a patient with incomplete Fanconi syndrome. J. Clin. Endocrinol. Metab. 1984; 59: 998–1001.

Barbier O, Jacquillet G, Tauc M, Cougnon M, Poujeol P. Effect of heavy metals on, and handling by, the kidney. Nephron Physiol. 2005; 99: 105–10.

Baum M. Developmental changes in rabbit juxtamedullary proximal convoluted tubule acidification. Pediatr. Res. 1992; 31: 411–14.

Baum M. The Fanconi syndrome of cystinosis: insights into the pathophysiology. Pediatr. Nephrol. 1998; 12: 492–7.

Baum M, Berry CA. Evidence for neutral transcellular NaCl transport and neutral basolateral chloride exit in the rabbit convoluted tubule. J. Clin. Invest. 1984; 74: 205–11.

Bergeron M, Dubord L, Hausser C, Schwab C. Membrane permeability as a cause of transport defects in experimental Fanconi syndrome. A new hypothesis. J. Clin. Invest. 1976; 57: 1181–9.

Bergeron M, Gougoux A, Vinay P. In: The Metabolic and Molecular Basis of Inherited Disease. New York: McGraw-Hill, 1995: pp. 2691–3716.

Bergeron M, Mayers P, Brown D. Specific effect of maleate on an apical membrane glycoprotein (gp330) in proximal tubule of rat kidneys. Am. J. Physiol. 1996; 271: F908–16.

Berio A, Piazzi A. Kearns-Sayre syndrome associated with de Toni-Debre-Fanconi syndrome due to cytochrome-c-oxidase (COX) deficiency. Panminerva Med. 2001; 43: 211–14.

Berliner RW, Kennedy TJ, Hilton JG. Effect of maleic acid on renal function. Proc. Soc. Exp. Biol. Med. 1950; 75: 791–4.

Berry GT, Baker L, Kaplan FS, Witzleben CL. Diabetes-like renal glomerular disease in Fanconi-Bickel syndrome. Pediatr. Nephrol. 1995; 9: 287–91.

Betend B, David L, Vincent M, Hermier M, Francois R. Successful indomethacin treatment of two paediatric patients with severe tubulopathies. A boy with an unusual hypercalciuria and a girl with cystinosis. Helv. Paediatr. Acta 1979; 34: 339–44.

Bingham C, Ellard S, Nicholls AJ, et al. The generalized aminoaciduria seen in patients with hepatocyte nuclear factor-1alpha mutations is a feature of all patients with diabetes and is associated with glucosuria. Diabetes 2001; 50: 2047–52.

Blumenthal SS, Lewand DL, Buday MA, Kleinman JG, Krezoski SK, Petering DH. Cadmium inhibits glucose uptake in primary cultures of mouse cortical tubule cells. Am. J. Physiol. 1990; 258: F1625–33.

Bonnardeaux A, Bichet DG. Inherited disorders of the renal tubule. In: Brenner BM, ed. The Kidney. Philadelphia: W.B. Saunders, 1999: pp. 1656–98.

Bonsib SM, Horvath F Jr.. Multinucleated podocytes in a child with nephrotic syndrome and Fanconi's syndrome: A unique clue to the diagnosis. Am. J. Kidney Dis. 1999; 34: 966–71.

Bovee KC, Joyce T, Reynolds R, Segal S. The fanconi syndrome in Basenji dogs: a new model for renal transport defects. Science 1978; 201: 1129–31.

Brazy PC, Gullans SR, Mandel LJ, Dennis VW. Metabolic requirement for inorganic phosphate by the rabbit proximal tubule. J. Clin. Invest. 1982; 70: 53–62.

Breitschwerdt EB, Ochoa R, Waltman C. Multiple endocrine abnormalities in Basenji dogs with renal tubular dysfunction. J. Am. Vet. Med. Assoc. 1983; 182: 1348–53.

Brewer ED, Tsai HC, Szeto KS, Morris RC Jr.. Maleic acid-induced impaired conversion of 25(OH)D3 to 1,25(OH)2D3: implications for Fanconi's syndrome. Kidney Int. 1977; 12: 244–52.

Bridoux F, Sirac C, Hugue V, et al. Fanconi's syndrome induced by a monoclonal Vkappa3 light chain in Waldenstrom's macroglobulinemia. Am. J. Kidney Dis. 2005; 45: 749–57.

Brown D, Stow JL. Protein trafficking and polarity in kidney epithelium: from cell biology to physiology. Physiol. Rev. 1996; 76: 245–97.

Cachat F, Guignard JP. The kidney in children under chemotherapy. Rev. Med. Suisse Romande 1996; 116: 985–93.

Cachat F, Nenadov-Beck M, Guignard JP. Occurrence of an acute Fanconi syndrome following cisplatin chemotherapy. Med. Pediatr. Oncol. 1998; 31: 40–1.

Campos Y, Garcia-Silva T, Barrionuevo CR, Cabello A, Muley R, Arenas J. Mitochondrial DNA deletion in a patient with mitochondrial myopathy, lactic acidosis, and stroke-like episodes (MELAS) and Fanconi's syndrome. Pediatr. Neurol. 1995; 13: 69–72.

Castano E, Marzabal P, Casado FJ, Felipe A, Pastor-Anglada M. Na+,K(+)-ATPase expression in maleic-acid-induced Fanconi syndrome in rats. Clin. Sci. (Lond) 1997; 92: 247–53.

Charnas LR, Bernardini I, Rader D, Hoeg JM, Gahl WA. Clinical and laboratory findings in the oculocerebrorenal syndrome of Lowe, with special reference to growth and renal function. N. Engl. J. Med. 1991; 324: 1318–25.

Chen YT. Type I glycogen storage disease: kidney involvement, pathogenesis and its treatment. Pediatr. Nephrol. 1991; 5: 71–6.

Chen YT, Coleman RA, Scheinman JI, Kolbeck PC, Sidbury JB. Renal disease in type I glycogen storage disease. N. Engl. J. Med. 1998; 318: 7–11.

Cherqui S, Sevin C, Hamard G, et al. Intralysosomal cystine accumulation in mice lacking cystinosin, the protein defective in cystinosis. Mol. Cell Biol. 2002; 22: 7622–32.

Chesney RW, Rosen JF, Hamstra AJ, Deluca HF. Serum 1,25 dihydroxyvitamin D levels in normal children and in vitamin D disorders. Am. J. Dis. Child. 1980; 134: 135–9.

Chisholm CA, Vavelidis F, Lovell MA, et al. Prenatal diagnosis of multiple acyl-CoA dehydrogenase deficiency: association with elevated alpha-fetoprotein and cystic renal changes. Prenat. Diagn. 2001; 21: 856–9.

Christensen EI, Birn H. Megalin and cubilin: synergistic endocytic receptors in renal proximal tubule. Am. J. Physiol. Renal Physiol. 2001; 280: F562–73.

Christensen EI, Devuyst O, Dom G, et al. Loss of chloride channel ClC-5 impairs endocytosis by defective trafficking of megalin and cubilin in kidney proximal tubules. Proc. Natl Acad. Sci. USA 2003; 100: 8472–7.

Christensen EI, Gburek J. Protein reabsorption in renal proximal tubule-function and dysfunction in kidney pathophysiology. Pediatr. Nephrol. 2004; 19: 714–21.

Clark DA, Butler SA, Braren V, Hartmann RC, Jenkins DE Jr.. The kidneys in paroxysmal nocturnal hemoglobinuria. Blood 1981; 57: 83–9.

Clarke BL, Wynne AG, Wilson DM, Fitzpatrick LA. Osteomalacia associated with adult Fanconi's syndrome: clinical and diagnostic features. Clin. Endocrinol. (Oxf) 1995; 43: 479–90.

Clayton PT, Hyland K, Brand M, Leonard JV. Mitochondrial phosphoenolpyruvate carboxykinase deficiency. Eur. J. Pediatr. 1986; 145: 46–50.

Cleveland WW, Adams WC, Mann JB, Nyhan WL. Acquired Fanconi syndrome following degraded tetracyline. J. Pediatr. 1965; 66: 333–42.

Cochat P, Guibaud P, Baverel G. Renal involvement in type I tyrosinemia. Arch. Pediatr. 1994; 1: 417–18.

Collazo R, Fan L, Zhao H, Wiederkehr M, Moe OW. Acute regulation of Na+/H+ exchanger NHE3 by parathyroid hormone via NHE3 phosphorylation and dynamin-dependent endocytosis. J. Biol. Chem. 2000; 275: 31601–8.

Colussi G, De Ferrari ME, Surian M, et al. Vitamin D metabolites and osteomalacia in the human Fanconi syndrome. Proc. Eur. Dial. Transplant Assoc. Eur. Ren. Assoc. 1985; 21: 756–60.

Coor C, Salmon RF, Quigley R, Marver D, Baum M. Role of adenosine triphosphate (ATP) and NaK ATPase in the inhibition of proximal tubule transport with intracellular cystine loading. J. Clin. Invest. 1991; 87: 955–61.

Cori GT, Cori CF. Glucose-6-phosphatase of the liver in glycogen storage disease. J. Biol. Chem. 1952; 199: 661–7.

Crabtree HG. Observations on the carbohydrate metabolism of tumors. Biochem. J. 1929; 23: 536–45.

Cross NC, Tolan DR, Cox TM. Catalytic deficiency of human aldolase B in hereditary fructose intolerance caused by a common missense mutation. Cell 1988; 53: 881–5.

Cunningham J, Bikle DD, Avioli LV. Acute, but not chronic, metabolic acidosis disturbs 25-hydroxyvitamin D3 metabolism. Kidney Int. 1984; 25: 47–52.

Cunningham R, Steplock D, Wang F, et al. Defective parathyroid hormone regulation of NHE3 activity and phosphate adaptation in cultured NHERF-1-/- renal proximal tubule cells. J. Biol. Chem. 2004; 279: 37815–21.

Cvoriscec D, Ceovic S, Borso G, Rukavina AS. Endemic nephropathy in Croatia. Clin. Chem. Lab. Med. 1998; 36: 271–7.

Das SK, Ray K. Wilson's disease: an update. Nat. Clin. Pract. Neurol. 2006; 2: 482–93.

Daugaard G, Abildgaard U, Holstein-Rathlou NH, Bruunshuus I, Bucher D, Leyssac PP. Renal tubular function in patients treated with high-dose cisplatin. Clin. Pharmacol. Ther. 1988; 44: 164–72.

De Lorenzo F, Hargreaves J, Kakkar VV. Phosphate diabetes in patients with chronic fatigue syndrome. Postgrad. Med. J. 1998; 74: 229–32.

De Toni G. Remarks on the relations between renal rickets (renal dwarfism) and renal diabetes. Acta Paediatr. 1933; 16: 479.

Debre R. Rachitisme tardif coexistant avec une nephrite chronique et une glycosurie. Arch. Med. Enf. 1934; 37: 597.

Dent CE, Friedman M. Hypercaliuric rickets associated with renal tubular damage. Arch. Dis. Child. 1964; 39: 240–9.

Earle KE, Seneviratne T, Shaker J, Shoback D. Fanconi's syndrome in HIV+ adults: report of three cases and literature review. J. Bone Miner. Res. 2004; 19: 714–21.

Eiam-Ong S, Spohn M, Kurtzman NA, Sabatini S. Insights into the biochemical mechanism of maleic acid-induced Fanconi syndrome. Kidney Int. 1995; 48: 1542–8.

Elsas LJ, Hayslett JP, Spargo BH, Durant JL, Rosenberg LE. Wilson's disease with reversible renal tubular dysfunction. Correlation with proximal tubular ultrastructure. Ann. Intern. Med. 1971; 75: 427–33.

Elsas LJ, Lai K. The molecular biology of galactosemia. Genet. Med. 1998; 1: 40–8.

Endo F, Sun MS. Tyrosinaemia type I and apoptosis of hepatocytes and renal tubular cells. J. Inherit. Metab. Dis. 2002; 25: 227–34.

Fan L, Wiederkehr MR, Collazo R, et al. Dual mechanisms of regulation on Na/H exchanger NHE-3 by parathyroid hormone in rat kidney. J. Biol. Chem. 1999; 274: 11289–95.

Fanconi G. Die nicht diabetischen Glykosurien und Hyperglykamien des altern Kindes. Jahrb. Kinderheilk 1931; 133: 257.

Fanconi G. Der fruhinfantile nephrotsch-glykosurische Zwergwuchs mit hypophospahtamischer Rachitis. Jahrb. Kinderheilk 1936; 147: 299.

Fanconi G, Bickel H. Die chronische aminoacidurie (aminosaeurediabetes oder nephrotisch-glukosurisscher zwergwuchs) ber der glykogenose und cystinkrankheit. Helv. Paediatr. Acta 1949; 4: 359–96.

Finzer KH. Lower nephron nephrosis due to concentrated Lysol vaginal douches: a report of two cases. Can. Med. Assoc. J. 1961; 84: 549.

Fisher SE, Black GC, Lloyd SE, et al. Isolation and partial characterization of a chloride channel gene which is expressed in kidney and is a candidate for Dent's disease (an X-linked hereditary nephrolithiasis). Hum. Mol. Genet. 1994; 3: 2053–9.

Fisher SE, Lloyd SE, Pearce SH, Thakker RV, Craig IW. Cloning and characterization of CLCN5, the human kidney chloride channel gene implicated in Dent disease (an X-linked hereditary nephrolithiasis). Genomics 1995; 29: 598–606.

Foreman JW, Benson L. Effect of cystine loading and cystine dimethylester on renal brushborder membrane transport. Biosci. Rep. 1990a; 10: 455–9.

Foreman JW, Benson LL. Effect of cystine loading on substrate oxidation by rat renal tubules. Pediatr. Nephrol. 1990b; 4: 236–9.

Foreman JW, Benson LL, Wellons M, et al. Metabolic studies of rat renal tubule cells loaded with cystine: the cystine dimethylester model of cystinosis. J. Am. Soc. Nephrol. 1995; 6: 269–72.

Foreman JW, Bowring MA, Lee J, States B, Segal S. Effect of cystine dimethylester on renal solute handling and isolated renal tubule transport in the rat: a new model of the Fanconi syndrome. Metabolism 1987; 36: 1185–91.

Frerman FE, Goodman SI. Defects of electron transfer flavoprotein and electron transfer flavoprotein-ubiquinone oxidoreductase: gllutaric acidemia type II. In: Scriver CR, Beaudet AL, Sly WS, Valle D, eds. The Metabolic and Molecular Bases of Inherited Disease. New York: McGraw-Hill, 2001: pp. 2357–65.

Friedman A, Chesney R. Fanconi's syndrome in renal transplantation. Am. J. Nephrol. 1981; 1: 45–7.

Gahl WA, Thoene JG, Schneider JA. Cystinosis. N. Engl. J. Med. 2002; 347: 111–21.

Gainza FJ, Minguela JI, Lampreabe I. Aminoglycoside-associated Fanconi's syndrome: an underrecognized entity. Nephron 1997; 77: 205–11.

Garty R, Cooper M, Tabachnik E. The Fanconi syndrome associated with hepatic glycogenosis and abnormal metabolism of galactose. J. Pediatr. 1974; 85: 821–3.

Gekle M. Renal tubule albumin transport. Annu. Rev. Physiol. 2005; 67: 573–94.

Gieselmann V, Franken S, Klein D, et al. Metachromatic leukodystrophy: consequences of sulphatide accumulation. Acta Paediatr. Suppl. 2003; 92: 74–9.

Gil HW, Yang JO, Lee EY, Hong SY. Paraquat-induced Fanconi syndrome. Nephrology (Carlton) 2005; 10: 430–2.

Gonick H, Indraprasit S, Neustein H, Rosen V. Cadmium-induced experimental Fanconi syndrome. Curr. Probl. Clin. Biochem. 1975; 4: 111–18.

Gonick H, Indraprasit S, Rosen V, Neustein H, Van de Velde R, Raghavan SRV. Experimental Fanconi syndrome: III. Effect of cadmium on renal tubular function, the ATP-Na-K-ATPase transport system and renal tubular ultrastructure. Miner. Electrolyte Metab. 1980; 3: 21–35.

Greco DS. Congenital and inherited renal disease of small animals. Vet. Clin. North Am. Small Anim. Pract. 2001; 31: 393–9, viii.

Grompe M, St Louis M, Demers SI, al Dhalimy M, Leclerc B, Tanguay RM. A single mutation of the fumarylacetoacetate hydrolase gene in French Canadians with hereditary tyrosinemia type I. N. Engl. J. Med. 1994; 331: 353–7.

Gross JM. Fanconi syndrome (adult type) developing secondary to the ingestion of outdated tetracycline. Ann. Intern. Med. 1963; 58: 523–8.

Guan S, el Dahr S, Dipp S, Batuman V. Inhibition of Na-K-ATPase activity and gene expression by a myeloma light chain in proximal tubule cells. J. Investig. Med. 1999; 47: 496–501.

Gunther R, Silbernagl S, Deetjen P. Maleic acid induced aminoaciduria, studied by free flow micropuncture and continuous microperfusion. Pflugers Arch. 1979; 382: 109–14.

Gunther W, Luchow A, Cluzeaud F, Vandewalle A, Jentsch TJ. ClC-5, the chloride channel mutated in Dent's disease, colocalizes with the proton pump in endocytotically active kidney cells. Proc. Natl Acad. Sci. USA 1998; 95: 8075–80.

Gunther W, Piwon N, Jentsch TJ. The ClC-5 chloride channel knockout mouse – an animal model for Dent's disease. Pflugers Arch. 2003; 445: 456–62.

Guntupalli J, Delaney V, Bourke E. Studies on the maleic acid-induced Fanconi syndrome in the rat: mechanism of phosphaturia and its mitigation by dietary phosphate restriction. Contrib. Nephrol. 1991; 92: 83–92.

Haq MR, Kalatzis V, Gubler MC, et al. Immunolocalization of cystinosin, the protein defective in cystinosis. J. Am. Soc. Nephrol. 2002; 13: 2046–51.

Haycock GB, Al Dahhan J, Mak RH, Chantler C. Effect of indomethacin on clinical progress and renal function in cystinosis. Arch. Dis. Child. 1982; 57: 934–9.

Herak-Kramberger CM, Brown D, Sabolic I. Cadmium inhibits vacuolar H(+)-ATPase and endocytosis in rat kidney cortex. Kidney Int. 1998; 53: 1713–26.

Herak-Kramberger CM, Spindler B, Biber J, Murer H, Sabolic I. Renal type II Na/Pi-cotransporter is strongly impaired whereas the Na/sulphate-cotransporter and aquaporin 1 are unchanged in cadmium-treated rats. Pflugers Arch. 1996; 432: 336–44.

Hilpert J, Nykjaer A, Jacobsen C, et al. Megalin antagonizes activation of the parathyroid hormone receptor. J. Biol. Chem. 1999; 274: 5620–5.

Hodges RE, Kirkendall WM. Renal tubule and Wilson's disease. Ann. Intern. Med. 1972; 76: 835–6.

Hommes FA, Bendien K, Elema JD, Bremer HJ, Lombeck I. Two cases of phosphoenolpyruvate carboxykinase deficiency. Acta Paediatr. Scand. 1976; 65: 233–40.

Hoopes RR Jr., Raja KM, Koich A, et al. Evidence for genetic heterogeneity in Dent's disease. Kidney Int. 2004; 65: 1615–20.

Hoppe B, Neuhaus T, Superti-Furga A, Forster I, Leumann E. Hypercalciuria and nephrocalcinosis, a feature of Wilson's disease. Nephron 1993; 65: 460–2.

Hostutler RA, DiBartola SP, Eaton KA. Transient proximal renal tubular acidosis and Fanconi syndrome in a dog. J. Am. Vet. Med. Assoc. 2004; 224: 1611–14.

Houston IB, Boichis H, Edelmann CM Jr. Fanconi syndrome with renal sodium wasting and metabolic alkalosis. Am. J. Med. 1968; 44: 638–46.

Hsu BY, McNamara PD, Mahoney SG, et al. Membrane fluidity and sodium transport by renal membranes from dogs with spontaneous idiopathic Fanconi syndrome. Metabolism 1992; 41: 253–9.

Hsu BY, Wehrli SL, Yandrasitz JR, et al. Renal brush border membrane lipid composition in Basenji dogs with spontaneous idiopathic Fanconi syndrome. Identification of a mutation in the arylsulfatase A gene of a pation with adult-type metachromatic leukodystrophy. Am. J. Hum. Genet. 1991; 48: 971–8.

Igarashi T, Kawato H, Kamoshita S, Nosaka K, Seiya K, Hayakawa H. Acute tubulointerstitial nephritis with uveitis syndrome presenting as multiple tubular dysfunction including Fanconi's syndrome. Pediatr. Nephrol. 1992; 6: 547–9.

Insogna KL, Bordley DR, Caro JF, Lockwood DH. Osteomalacia and weakness from excessive antacid ingestion. JAMA 1980; 244: 2544–6.

Jacobs SM, van Beurden DH, Klomp LW, Berger R. Kidneys of mice with hereditary tyrosinemia type I are extremely sensitive to cytotoxicity. Pediatr Res 2006; 59: 365–70.

Jamieson PM, Chandler ML. Transient renal tubulopathy in a Labrador retriever. J. Small Anim. Pract. 2001; 42: 546–9.

Kalatzis V, Cherqui S, Antignac C, Gasnier B. Cystinosin, the protein defective in cystinosis, is a H(+)-driven lysosomal cystine transporter. EMBO J. 2001; 20: 5940–9.

Kellerman PS. Exogenous adenosine triphosphate (ATP) preserves proximal tubule microfilament structure and function in vivo in a maleic acid model of ATP depletion. J. Clin. Invest. 1993; 92: 1940–9.

Kihara H. Genetic heterogeneity in metachromatic leukodystrophy. Am. J. Hum. Genet. 1982; 34: 171–81.

Kihara H, Fluharty AL, O'Brien JS, Fish CH. Metachromatic leukodystrophy caused by a partial cerebroside sulfatase. Clin. Genet. 1982; 21: 253–61.

Kihara H, Meek WE, Fluharty AL. Attenuated activities and structural alterations of arylsulfatase A in tissues from subjects with pseudo arylsulfatase A deficiency. Hum. Genet. 1986; 74: 59–62.

Kitano A, Nishiyama S, Miike T, Hattori S, Ohtani Y, Matsuda I. Mitochondrial cytopathy with lactic acidosis, carnitine deficiency and DeToni-Fanconi-Debre syndrome. Brain Dev. 1986; 8: 289–95.

Knochel JP. Mechanisms of rhabdomyolysis. Curr. Opin. Rheumatol. 1993; 5: 725–31.

Kobayashi T, Muto S, Nemoto J, et al. Fanconi's syndrome and distal (type 1) renal tubular acidosis in a patient with primary Sjogren's syndrome with monoclonal gammopathy of undetermined significance. Clin. Nephrol. 2006; 65: 427–32.

Kondo R, Wakamatsu N, Yoshino H, Fukuhara N, Miyatake T, Tsuji S. Identification of a mutation in the arylsulfatase A gene of a patient with adult-type metachromatic leukodystrophy. Am. J. Hum. Genet. 1991; 48: 971–8.

Koobs DH. Phosphate mediation of the Crabtree and Pasteur effects. Science 1972; 178: 127–33.

Kozyraki R, Fyfe J, Verroust PJ, et al. Megalin-dependent cubilin-mediated endocytosis is a major pathway for the apical uptake of transferrin in polarized epithelia. Proc. Natl Acad. Sci. USA 2001; 98: 12491–6.

Kramer HJ, Gonick HC. Experimental Fanconi syndrome. I. Effect of maleic acid on renal cortical Na-K-ATPase activity and ATP levels. J. Lab. Clin. Med. 1970; 76: 799–808.

Kramer HJ, Gonick HC. Effect of maleic acid on sodium-linked tubular transport in experimental Fanconi syndrome. Nephron 1973; 10: 306–19.

Kramer HJ, Gonick HC, Lu E. In vitro inhibition of Na-K-ATPase by trace metals: relation to renal and cardiovascular damage. Nephron 1986; 44: 329–36.

Kuwertz-Broking E, Koch HG, Marquardt T, et al. Renal Fanconi syndrome: first sign of partial respiratory chain complex IV deficiency. Pediatr. Nephrol. 2000; 14: 495–8.

Laine J, Salo MK, Krogerus L, Karkkainen J, Wahlroos O, Holmberg C. The nephropathy of type I tyrosinemia after liver transplantation. Pediatr. Res. 1995; 37: 640–5.

Lande MB, Kim MS, Bartlett C, Guay-Woodford LM. Reversible Fanconi syndrome associated with valproate therapy. J. Pediatr. 1993; 23: 320–2.

LaNoue KF, Schoolwerth AC. Metabolite transport in mitochondria. Annu. Rev. Biochem. 1979; 48: 871–922.

Larochelle J, Mortezai A, Belanger M, Tremblay M, Claveau JC, Aubin G. Experience with 37 infants with tyrosinemia. Can. Med. Assoc. J. 1967; 97: 1051–4.

Lassiter WE, Colindres RE. Phosphate reabsorption in the distal convoluted tubule. Adv. Exp. Med. Biol. 1982; 151: 21–32.

Lee S, Lee T, Lee B, et al. Fanconi's syndrome and subsequent progressive renal failure caused by a Chinese herb containing aristolochic acid. Nephrology (Carlton) 2004; 9: 126–9.

Leheste JR, Rolinski B, Vorum H, et al. Megalin knockout mice as an animal model of low molecular weight proteinuria. Am. J. Pathol. 1999; 155: 1361–70.

Lei KJ, Shelly LL, Pan CJ, Sidbury JB, Chou JY. Mutations in the glucose-6-phosphatase gene that cause glycogen storage disease type 1a. Science 1993; 262: 580–3.

Lenoir GR, Perignon JL, Gubler MC, Broyer M. Valproic acid: a possible cause of proximal tubular renal syndrome. J. Pediatr. 1981; 98: 503–4.

Leonard JV, Hyland K, Furukawa N, Clayton PT. Mitochondrial phosphoenolpyruvate carboxykinase deficiency. Eur. J. Pediatr. 1991; 150: 198–9.

Leslie ND. Insights into the pathogenesis of galactosemia. Annu. Rev. Nutr. 2003; 23: 59–80.

Leu ML, Strickland GT, Gutman RA. Renal function in Wilson's disease: response to penicillamine therapy. Am. J. Med. Sci. 1970; 260: 381–98.

Lian LM, Chang YC, Yang CC, Yang JC, Kao KP, Chung MY. Adult Fanconi syndrome with proximal muscle weakness and hypophosphatemic osteomalacia: report of a case. J. Formos. Med. Assoc. 1994; 93: 709–14.

Lignac GO. Über störing des cystinstoffwechsels bei kindern. Deutsches Archiv für Klinische Medizin 1924; 145: 139–50.

Lin T, Orrison BM, Leahey AM, et al. Spectrum of mutations in the OCRL1 gene in the Lowe oculocerebrorenal syndrome. Am. J. Hum. Genet. 1997; 60: 1384–8.

Liu FY, Cogan MG. Axial heterogeneity in the rat proximal convoluted tubule. I. Bicarbonate, chloride, and water transport. Am. J. Physiol. 1984; 247: F816–21.

Lloyd SE, Gunther W, Pearce SH, et al. Characterisation of renal chloride channel, CLCN5, mutations in hypercalciuric nephrolithiasis (kidney stones) disorders. Hum. Mol. Genet. 1997; 6: 1233–9.

Lloyd SE, Pearce SH, Fisher SE, et al. A common molecular basis for three inherited kidney stone diseases. Nature 1996; 379: 445–9.

Lowe CU, Terrey M, MacLachlan EA. Organic-aciduria, decreased renal ammonia production, hydrophthalmos, and mental retardation; a clinical entity. AMA Am. J. Dis. Child. 1952; 83: 164–84.

Luijerink MC, van Beurden EA, Malingre HE, et al. Renal proximal tubular cells acquire resistance to cell death stimuli in mice with hereditary tyrosinemia type 1. Kidney Int. 2004; 66: 990–1000.

Madsen PP, Kibaek M, Roca X, et al. Short/branched-chain acyl-CoA dehydrogenase deficiency due to an IVS3 + 3A > G mutation that causes exon skipping. Hum. Genet. 2006; 118: 680–90.

Maesaka JK, McCaffery M. Evidence for renal tubular leakage in maleic acid-induced Fanconi syndrome. Am. J. Physiol. 1980; 239: F507–13.

Malakauskas SM, Quan H, Fields TA, et al. Aminoaciduria and altered renal expression of luminal amino acid transporters in mice lacking novel gene collectrin. Am. J. Physiol. Renal Physiol. 2006.

Mandel LJ. Metabolic substrates, cellular energy production, and the regulation of proximal tubular transport. Annu. Rev. Physiol. 1985; 47: 85–101.

Manz F, Bickel H, Brodehl J, et al. Fanconi-Bickel syndrome. Pediatr. Nephrol. 1987; 1: 509–18.

Markello TC, Bernardini IM, Gahl WA. Improved renal function in children with cystinosis treated with cysteamine. N. Engl. J. Med. 1993; 328: 1157–62.

Marshansky V, Ausiello DA, Brown D. Physiological importance of endosomal acidification: potential role in proximal tubulopathies. Curr. Opin. Nephrol. Hypertens. 2002; 11: 527–37.

Matsubara Y, Narisawa K, Miyabayashi S, Tada K, Coates PM. Molecular lesion in patients with medium-chain acyl-CoA dehydrogenase deficiency. Lancet 1990; 335: 1589.

McCune DJ. Intractable hypophosphatemic rickets with renal glycosuria and acidosis (the Fanconi syndrome). Am. J. Dis. Child. 1943; 65: 81.

McGivan JD, Klingenberg M. Correlation between H+ and anion movement in mitochondria and the key role of the phosphate carrier. Eur. J. Biochem. 1971; 20: 392–9.

Meijer AJ, Van Dam K. The metabolic significance of anion transport in mitochondria. Biochim. Biophys. Acta 1974; 346: 213–44.

Moe OW, Baum M, Berry CA, Rector FC. Renal Transport of Glucose, Amino Acids, Sodium, Chloride and Water. In: Brenner BM, ed. The Kidney. Philadelphia, PA: Saunders, 2004: pp. 413–52.

Montoliu J, Carrera M, Darnell A, Revert L. Lactic acidosis and Fanconi's syndrome due to degraded tetracycline. Br. Med. J. (Clin. Res. Ed.) 1981; 283: 1576–7.

Morales FC, Takahashi Y, Kreimann EL, Georgescu MM. Ezrinradixin-moesin (ERM)-binding phosphoprotein 50 organizes ERM proteins at the apical membrane of polarized epithelia. Proc. Natl Acad. Sci. USA 2004; 101: 17705–10.

Morgan HG, Stewart WK, Lowe KG, Stowers JM, Johnstone JH. Wilson's disease and the Fanconi syndrome. Q. J. Med. 1962; 31: 361–84.

Mori K, Narahara K, Ninomiya S, Goto Y, Nonaka I. Renal and skin involvement in a patient with complete Kearns-Sayre syndrome. Am. J. Med. Genet. 1991; 38: 583–7.

Morris AA, Olpin SE, Van't Hoff WG, Johnson AW, Leonard JV. Renal tubular dysfunction in multiple acyl-CoA dehydrogenase deficiency. J. Inherit. Metab. Dis. 1997; 20: 604–5.

Morris AA, Taylor RW, Birch-Machin MA, et al. Neonatal Fanconi syndrome due to deficiency of complex III of the respiratory chain. Pediatr. Nephrol. 1995; 9: 407–11.

Morris RC Jr. An experimental renal acidification defect in patients with hereditary fructose intolerance. I. Its resemblance to renal tubular acidosis. J. Clin. Invest. 1968a; 47: 1389–98.

Morris RC Jr. An experimental renal acidification defect in patients with hereditary fructose intolerance. II. Its distinction from classic renal tubular acidosis; its resemblance to the renal acidification defect associated with the Fanconi syndrome of children with cystinosis. J. Clin. Invest. 1968b; 47: 1648–63.

Morris RC Jr., McSherry E, Sebastian A. Modulation of experimental renal dysfunction of hereditary fructose intolerance by circulating parathyroid hormone. Proc. Natl Acad. Sci. USA 1971; 68: 132–5.

Morris RC, Nigon K, Reed EB. Evidence that the severity of depletion of inorganic phosphate determines the severity of the disturbance of adenine nucleotide metabolism in the liver and renal cortex of the fructose-loaded rat. J. Clin. Invest. 1978; 61: 209–20.

Moser HW. Sulfatide lipidosis: metachromatic leukodystrophy. In: Stanbury JB, Wyngaarden JB, Fredrickson DS, eds. The Metabolic Basis of Inherited Disease. New York: McGraw-Hill, 1972: pp. 688–729.

Mujais SK. Maleic acid-induced proximal tubulopathy: Na:K pump inhibition. J. Am. Soc. Nephrol. 1993; 4: 142–7.

Nissim I, Weinberg JM. Glycine attenuates Fanconi syndrome induced by maleate or ifosfamide in rats. Kidney Int. 1996; 49: 684–95.

Noonan CH, Kay JM. Prevalence and geographic distribution of Fanconi syndrome in Basenjis in the United States. J. Am. Vet. Med. Assoc. 1990; 197: 345–9.

Norden AG, Lapsley M, Igarashi T, et al. Urinary megalin deficiency implicates abnormal tubular endocytic function in Fanconi syndrome. J. Am. Soc. Nephrol. 2002; 13: 125–33.

Nussbaum RL, Orrison BM, Janne PA, Charnas L, Chinault AC. Physical mapping and genomic structure of the Lowe syndrome gene OCRL1. Hum. Genet. 1997; 99: 145–50.

Nykjaer A, Fyfe JC, Kozyraki R, et al. Cubilin dysfunction causes abnormal metabolism of the steroid hormone 25(OH) vitamin D(3). Proc. Natl Acad. Sci. USA 2001; 98: 13895–900.

Ogier H, Lombes A, Scholte HR, et al. de Toni-Fanconi-Debre syndrome with Leigh syndrome revealing severe muscle cytochrome c oxidase deficiency. J. Pediatr. 1988; 112: 734–9.

Orfila C, Lepert JC, Modesto A, Bernadet P, Suc JM. Fanconi's syndrome, kappa light-chain myeloma, non-amyloid fibrils and cytoplasmic crystals in renal tubular epithelium. Am. J. Nephrol. 1991; 11: 345–9.

Orita Y, Fukuhara Y, Yanase M, et al. The mechanism of decreased Na+ dependent D-glucose transport in brush-border membrane vesicles from rabbit kidneys with experimental Fanconi syndrome. Biochim. Biophys. Acta 1984; 771: 195–200.

Parsonage MJ, Wilkins EG, Snowden N, Issa BG, Savage MW. The development of hypophosphataemic osteomalacia with myopathy in two patients with HIV infection receiving tenofovir therapy. HIV Med. 2005; 6: 341–6.

Petrukhin K, Lutsenko S, Chernov I, Ross BM, Kaplan JH, Gilliam TC. Characterization of the Wilson disease gene encoding a P-type copper transporting ATPase: genomic organization, alternative splicing, and structure/function predictions. Hum. Mol. Genet. 1994; 3: 1647–56.

Pfister MF, Ruf I, Stange G, Ziegler U, Lederer E, Biber J, Murer H. Parathyroid hormone leads to the lysosomal degradation of the renal type II Na/Pi cotransporter. Proc. Natl Acad. Sci. USA 1998; 95: 1909–14.

Pham T, Furno-Steib S, Daumen-Legre V, Acquaviva PC, Lafforgue P. Bilateral symmetric polyarthralgia revealing Fanconi's syndrome. Joint Bone Spine 2002; 69: 209–13.

Phaneuf D, Labelle Y, Berube D, Arden K, Cavenee W, Gagne R, Tanguay RM. Cloning and expression of the cDNA encoding human fumarylacetoacetate hydrolase, the enzyme deficient in hereditary tyrosinemia: assignment of the gene to chromosome 15. Am. J. Hum. Genet. 1991; 48: 525–35.

Piwon N, Gunther W, Schwake M, Bosl MR, Jentsch TJ. ClC-5 Cl- channel disruption impairs endocytosis in a mouse model for Dent's disease. Nature 2000; 408: 369–73.

Pontoglio M, Barra J, Hadchouel M, et al. Hepatocyte nuclear factor 1 inactivation results in hepatic dysfunction, phenylketonuria, and renal Fanconi syndrome. Cell 1996; 84: 575–85.

Pontoglio M, Prie D, Cheret C, et al. HNF1alpha controls renal glucose reabsorption in mouse and man. EMBO Rep. 2000; 1: 359–65.

Pontoglio M, Sreenan S, Roe M, et al. Defective insulin secretion in hepatocyte nuclear factor 1alpha-deficient mice. J. Clin. Invest. 1998; 101: 2215–22.

Pook MA, Wrong O, Wooding C, Norden AG, Feest TG, Thakker RV. Dent's disease, a renal Fanconi syndrome with nephrocalcinosis and kidney stones, is associated with a microdeletion involving DXS255 and maps to Xp11.22. Hum. Mol. Genet. 1993; 2: 2129–34.

Poole-Wilson PA, Cameron IR. Intracellular pH and K+ of cardiac and skeletal muscle in acidosis and alkalosis. Am. J. Physiol. 1975; 229: 1305–10.

Preisig PA, Ives HE, Cragoe EJ Jr., Alpern RJ, Rector FC Jr. Role of the Na+/H+ antiporter in rat proximal tubule bicarbonate absorption. J. Clin. Invest. 1987; 80: 970–8.

Preisig PA, Rector FC Jr.. Role of Na+ -H+ antiport in rat proximal tubule NaCl absorption. Am. J. Physiol. 1988; 255: F461–5.

Rago RP, Miles JM, Sufit RL, Spriggs DR, Wilding G. Suramin-induced weakness from hypophosphatemia and mitochondrial myopathy. Association of suramin with mitochondrial toxicity in humans. Cancer 1994; 73: 1954–9.

Rampini S, Fanconi A, Illig R, Prader A. Effect of hydrochlorothiazide on proximal renal tubular acidosis in a patient with idiopathic "de tonidebre-fanconi syndrome". Helv. Paediatr. Acta 1968; 23: 13–21.

Rector FC Jr. Sodium, bicarbonate, and chloride absorption by the proximal tubule. Am. J. Physiol. 1983; 244: F461–71.

Reczek D, Berryman M, Bretscher A. Identification of EBP50: A PDZ-containing phosphoprotein that associates with members of the ezrinradixin-moesin family. J. Cell. Biol. 1997; 139: 169–79.

Reichardt JK, Belmont JW, Levy HL, Woo SL. Characterization of two missense mutations in human galactose-1-phosphate uridyltransferase: different molecular mechanisms for galactosemia. Genomics 1992; 12: 596–600.

Reichardt JK, Berg P. Cloning and characterization of a cDNA encoding human galactose-1-phosphate uridyl transferase. Mol. Biol. Med. 1988; 5: 107–22.

Reitsma-Bierens WC. Renal complications in glycogen storage disease type I. Eur. J. Pediatr. 1993; 152(Suppl 1): S60–2.

Reynolds R, McNamara PD, Segal S. On the maleic acid induced Fanconi syndrome: effects on transport by isolated rat kidney brushborder membrane vesicles. Life Sci. 1978; 22: 39–43.

Rogulski J, Pacanis A, Strzelecki T, Kaminska E, Angielski S. Effects of maleate on carbohydrate metabolism in rats. Am. J. Physiol. 1976; 230: 1163–7.

Rogulski J, Strzelecki T, Pacanis A, Kaminska E, Angielski S. Effects of maleate on renal carbohydrate metabolism in vivo and in vitro. Curr. Probl. Clin. Biochem. 1975; 4: 106–10.

Ronco P, Aucouturier P, Mougenot B. Plasma cell dyscrasia-related glomerulopathies and Fanconi's syndrome: a molecular approach. J. Nephrol. 2000; 13(Suppl 3): S34–44.

Rosen VJ, Kramer HJ, Gonick HC. Experimental Fanconi syndrome. II. Effect of maleic acid on renal tubular ultrastructure. Lab. Invest. 1973; 28: 446–55.

Rossi R. Nephrotoxicity of ifosfamide – moving towards understanding the molecular mechanisms. Nephrol. Dial. Transplant. 1997; 12: 1091–2.

Rossi R, Pleyer J, Schafers P, et al. Development of ifosfamide-induced nephrotoxicity: prospective follow-up in 75 patients. Med. Pediatr. Oncol. 1999; 32: 177–82.

Roth KS, Medow MS, Moses LC, Spencer PD, Schwarz SM. Renal Fanconi syndrome: developmental basis for a new animal model with relevance to human disease. Biochim. Biophys. Acta 1989; 987: 38–46.

Russo P, O'Regan S. Visceral pathology of hereditary tyrosinemia type I. Am. J. Hum. Genet. 1990; 47: 317–24.

Sabolic I, Haase W, Burckhardt G. ATP-dependent H+ pump in membrane vesicles from rat kidney cortex. Am. J. Physiol. 1985; 248: F835–44.

Sabolic I, Herak-Kramberger CM, Antolovic R, Breton S, Brown D. Loss of basolateral invaginations in proximal tubules of cadmium-intoxicated rats is independent of microtubules and clathrin. Toxicology 2006; 218: 149–63.

Sabolic I, Ljubojevic M, Herak-Kramberger CM, Brown D. Cd-MT causes endocytosis of brush-border transporters in rat renal proximal tubules Am. J. Physiol. Renal Physiol. 2002; 283: F1389–402.

Saito T, Furyama T, Onodera S, Saito H, Shioji R. Renal impairment in patients with paroxysmal nocturnal hemoglobinuria. Tohoku J. Exp. Med. 1975; 116: 267–75.

Sakamoto O, Ogawa E, Ohura T, et al. Mutation analysis of the GLUT2 gene in patients with Fanconi-Bickel syndrome. Pediatr. Res. 2000; 48: 586–9.

Sakarcan A, Aricheta R, Baum M. Intracellular cystine loading causes proximal tubule respiratory dysfunction: effect of glycine. Pediatr. Res. 1992; 32: 710–13.

Sakarcan A, Timmons C, Baum M. Intracellular distribution of cystine in cystine-loaded proximal tubules. Pediatr. Res. 1994; 35: 447–50.

Salmon RF, Baum M. Intracellular cystine loading inhibits transport in the rabbit proximal convoluted tubule. J. Clin. Invest. 1990; 85: 340–4.

Santer R, Groth S, Kinner M, et al. The mutation spectrum of the facilitative glucose transporter gene SLC2A2 (GLUT2) in patients with Fanconi-Bickel syndrome. Hum. Genet. 2002; 110: 21–9.

Santer R, Schneppenheim R, Dombrowski A, Gotze H, Steinmann B, Schaub J. Mutations in GLUT2, the gene for the liver-type glucose transporter, in patients with Fanconi-Bickel syndrome. Nat. Genet. 1997; 17: 324–6.

Santer R, Schneppenheim R, Dombrowski A, Gotze H, Steinmann B, Schaub J. Fanconi-Bickel syndrome – a congenital defect of the liver-type facilitative glucose transporter. SSIEM Award. Society for the Study of Inborn Errors of Metabolism. J. Inherit. Metab. Dis. 1998a; 21: 191–4.

Santer R, Schneppenheim R, Suter D, Schaub J, Steinmann B. Fanconi-Bickel syndrome – the original patient and his natural history, historical steps leading to the primary defect, and a review of the literature. Eur. J. Pediatr. 1998b; 157: 783–97.

Scharer K, Yoshida T, Voyer L, Berlow S, Pietra G, Metcoff J. Impaired renal gluconeogenesis and energy metabolism in maleic acid-induced nephropathy in rats. Res. Exp. Med. (Berl.) 1972; 157: 136–52.

Scheinman SJ. X-linked hypercalciuric nephrolithiasis: clinical syndromes and chloride channel mutations. Kidney Int. 1998; 53: 3–17.

Scheinman SJ, Guay-Woodford LM, Thakker RV, Warnock DG. Genetic disorders of renal electrolyte transport. N. Engl. J. Med. 1999; 340: 1177–87.

Schonheyder F, Gregersen G, Hansen HE, Skov PE. Renal clearances of different amino acids in Wilson's disease before and after treatment with penicillamine. Acta Med. Scand. 1971; 190: 395–9.

Sebastian A, McSherry E, Morris RC Jr. On the mechanism of renal potassium wasting in renal tubular acidosis associated with the Fanconi syndrome (type 2 RTA). J. Clin. Invest. 1971a; 50: 231–43.

Sebastian A, McSherry E, Morris RC Jr. Renal potassium wasting in renal tubular acidosis (RTA): its occurrence in types 1 and 2 RTA despite sustained correction of systemic acidosis. J. Clin. Invest. 1971b; 50: 667–78.

Settles EL, Schmidt D. Fanconi syndrome in a Labrador retriever. J. Vet. Intern. Med. 1994; 8: 390–3.

Shah M, Quigley R, Baum M. Neonatal rabbit proximal tubule basolateral membrane Na+/H+ antiporter and Cl-/base exchange. Am. J. Physiol. 1999; 276: R1792–7.

Shapiro LJ, Aleck KA, Kaback MM, et al. Metachromatic leukodystrophy without arylsulfatase A deficiency. Pediatr. Res. 1979; 13: 1179–81.

Shenolikar S, Voltz JW, Minkoff CM, Wade JB, Weinman EJ. Targeted disruption of the mouse NHERF-1 gene promotes internalization of proximal tubule sodium-phosphate cotransporter type IIa and renal phosphate wasting. Proc. Natl Acad. Sci. USA 2002; 99: 11470–5.

Shvil Y, Wald H, Popovtzer MM. Effect of bicarbonate and phosphate on renal phosphate leak in experimental Fanconi syndrome. Am. J. Physiol. 1987; 252: F310–16.

Silverman M. The mechanism of maleic acid nephropathy: investigations using brush border membrane vesicles. Membr. Biochem. 1981; 4: 63–9.

Sirac C, Bridoux F, Carrion C, et al. Role of the monoclonal kappa chain V domain and reversibility of renal damage in a transgenic model of acquired Fanconi syndrome. Blood 2006; 108: 536–43.

Skinner R. Chronic ifosfamide nephrotoxicity in children. Med. Pediatr. Oncol. 2003; 41: 190–7.

Sliman GA, Winters WD, Shaw DW, Avner ED. Hypercalciuria and nephrocalcinosis in the oculocerebrorenal syndrome. J. Urol. 1995; 153: 1244–6.

Spencer PD, Medow MS, Moses LC, Roth KS. Effects of succinylacetone on the uptake of sugars and amino acids by brush border vesicles. Kidney Int. 1988; 34: 671–7.

Springate JE, Van Liew JB. Nephrotoxicity of ifosfamide in rats. J. Appl. Toxicol. 1995; 15: 399–402.

Stevens RL, Fluharty AL, Kihara H, et al. Cerebroside sulfatase activator deficiency induced metachromatic leukodystrophy. Am. J. Hum. Genet. 1981; 33: 900–6.

Suchy SF, Nussbaum RL. The deficiency of PIP2 5-phosphatase in Lowe syndrome affects actin polymerization. Am. J. Hum. Genet. 2002; 71: 1420–7.

Sun MS, Hattori S, Kubo S, Awata H, Matsuda I, Endo F. A mouse model of renal tubular injury of tyrosinemia type 1: development of de Toni Fanconi syndrome and apoptosis of renal tubular cells in Fah/Hpd double mutant mice. J. Am. Soc. Nephrol. 2000; 11: 291–300.

Sussman I, Erecinska M, Wilson DF. Regulation of cellular energy metabolism: the Crabtree effect. Biochim. Biophys. Acta 1980; 591: 209–23.

Szabolcs MJ, Seigle R, Shanske S, Bonilla E, DiMauro S, D'Agati V. Mitochondrial DNA deletion: a cause of chronic tubulointerstitial nephropathy. Kidney Int. 1994; 45: 1388–96.

Tabatabai NM, Blumenthal SS, Lewand DL, Petering DH. Differential regulation of mouse kidney sodium-dependent transporters mRNA by cadmium. Toxicol. Appl. Pharmacol. 2001; 177: 163–73.

Tabatabai NM, Blumenthal SS, Lewand DL, Petering DH. Mouse kidney expresses mRNA of four highly related sodium-glucose cotransporters: regulation by cadmium. Kidney Int. 2003; 64: 1320–30.

Tabatabai NM, Blumenthal SS, Petering DH. Adverse effect of cadmium on binding of transcription factor Sp1 to the GC-rich regions of the mouse sodium-glucose cotransporter 1, SGLT1, promoter. Toxicology 2005; 207: 369–82.

Tabei K, Muto S, Furuya H, Sakairi Y, Ando Y, Asano Y. Potassium secretion is inhibited by metabolic acidosis in rabbit cortical collecting ducts in vitro. Am. J. Physiol. 1995; 268: F490–5.

Tabet M. De Toni-Debre-Fanconi syndrome in Wilson's disease. J. Med. Liban. 1969; 22: 55–65.

Tanaka A, Nishida R, Maeda K, Sugawara A, Kuwahara T. Chinese herb nephropathy in Japan presents adult-onset Fanconi syndrome: could different components of aristolochic acids cause a different type of Chinese herb nephropathy? Clin. Nephrol. 2000; 53: 301–6.

Theodoropoulos DS, Shawker TH, Heinrichs C, Gahl WA. Medullary nephrocalcinosis in nephropathic cystinosis. Pediatr. Nephrol. 1995; 9: 412–18.

Thomas GR, Forbes JR, Roberts EA, Walshe JM, Cox DW. The Wilson disease gene: spectrum of mutations and their consequences. Nat. Genet. 1995; 9: 210–17.

Thorner PS, Bedard YC, Fernandes BJ. Lambda-light-chain nephropathy with Fanconi's syndrome. Arch. Pathol. Lab. Med. 1983; 107: 654–7.

Tolan DR, Penhoet EE. Characterization of the human aldolase B gene. Mol. Biol. Med. 1986; 3: 245–64.

Town M, Jean G, Cherqui S, et al. A novel gene encoding an integral membrane protein is mutated in nephropathic cystinosis. Nat. Genet. 1998; 18: 319–24.

Trevisan A, Gardin C. Nephrolithiasis in a worker with cadmium exposure in the past. Int. Arch. Occup. Environ. Health 2005; 78: 670–2.

Truong LD, Mawad J, Cagle P, Mattioli C. Cytoplasmic crystals in multiple myeloma-associated Fanconi's syndrome. A morphological study including immunoelectron microscopy. Arch. Pathol. Lab. Med. 1989; 113: 781–5.

Tsai CS, Chen YC, Chen HH, Cheng CJ, Lin SH. An unusual cause of hypokalemic paralysis: aristolochic acid nephropathy with Fanconi syndrome. Am. J. Med. Sci. 2005; 330: 153–5.

Usberti M, Pecoraro C, Federico S, et al. Mechanism of action of indomethacin in tubular defects. Pediatrics 1985; 75: 501–7.

Van Furden D, Johnson K, Segbert C, Bossinger O. The C. elegans ezrin-radixin-moesin protein ERM-1 is necessary for apical junction remodelling and tubulogenesis in the intestine. Dev. Biol. 2004; 272: 262–76.

Wang LC, Lee WT, Tsai WY, Tsau YK, Shen YZ. Mitochondrial cytopathy combined with Fanconi's syndrome. Pediatr. Neurol. 2000a; 22: 403–6.

Wang SS, Devuyst O, Courtoy PJ, et al. Mice lacking renal chloride channel, CLC-5, are a model for Dent's disease, a nephrolithiasis disorder associated with defective receptor-mediated endocytosis. Hum. Mol. Genet. 2000b; 9: 2937–45.

Wang Y, Cai H, Cebotaru L, et al. ClC-5: role in endocytosis in the proximal tubule. Am. J. Physiol. Renal Physiol. 2005; 289: F850–62.

Watanabe T, Yoshikawa H, Yamazaki S, Abe Y, Abe T. Secondary renal Fanconi syndrome caused by valproate therapy. Pediatr. Nephrol. 2005; 20: 814–17.

Wiebers DO, Wilson DM, McLeod RA, Goldstein NP. Renal stones in Wilson's disease. Am. J. Med. 1979; 67: 249–54.

Wong D. Hereditary fructose intolerance. Mol. Genet. Metab. 2005; 85: 165–7.

Woodland C, Ito S, Granvil CP, Wainer IW, Klein J, Koren G. Evidence of renal metabolism of ifosfamide to nephrotoxic metabolites. Life Sci. 2000; 68: 109–17.

Wyss PA, Boynton SB, Chu J, Spencer RF, Roth KS. Physiological basis for an animal model of the renal Fanconi syndrome: use of succinylacetone in the rat. Clin. Sci. (Lond) 1992; 83: 81–7.

Yagame M, Tomino Y, Miura M, Suga T, Nomoto Y, Sakai H. An adult case of Fanconi's syndrome associated with membranous nephropathy. Tokai. J. Exp. Clin. Med. 1986; 11: 101–6.

Yamagata K, Oda N, Kaisaki PJ, et al. Mutations in the hepatocyte nuclear factor-1alpha gene in maturity-onset diabetes of the young (MODY3). Nature 1996; 384: 455–8.

Yanase M, Orita Y, Okada N, et al. Decreased Na+-gradient-dependent D-glucose transport in brush-border membrane vesicles from rabbits with experimental Fanconi syndrome. Biochim. Biophys. Acta 1983; 733: 95–101.

Yang SS, Chu P, Lin YF, Chen A, Lin SH. Aristolochic acid-induced Fanconi's syndrome and nephropathy presenting as hypokalemic paralysis. Am. J. Kidney Dis. 2002; 39: E14.

Yang YS, Peng CH, Sia SK, Huang CN. Acquired hypophosphatemia osteomalacia associated with Fanconi's syndrome in Sjogren's syndrome. Rheumatol Int. 2006.

Yearley JH, Hancock DD, Mealey KL. Survival time, lifespan, and quality of life in dogs with idiopathic Fanconi syndrome. J. Am. Vet. Med. Assoc. 2004; 225: 377–83.

Yonemura S, Tsukita S, Tsukita S. Direct involvement of ezrin/radixin/moesin (ERM)-binding membrane proteins in the organization of microvilli in collaboration with activated ERM proteins. J. Cell. Biol. 1999; 145: 1497–509.

Zeviani M, Nonaka I, Bonilla E, et al. Fatal infantile mitochondrial myopathy and renal dysfunction caused by cytochrome c oxidase deficiency: immunological studies in a new patient. Ann. Neurol. 1985; 17: 414–17.

CHAPTER 11

Proximal Renal Tubular Acidosis

PETER S. ARONSON AND GERHARD GIEBISCH

INTRODUCTION

The proximal tubule is the site where most of the filtered bicarbonate is retrieved, ammonium is produced and secreted, and critical organic anions like citrate are reabsorbed. Proximal renal tubular acidosis (pRTA) is a disease of defective proximal tubule function resulting in metabolic acidosis. This chapter will review the transport processes for acid–base equivalents in the proximal tubule, the clinical features of pRTA, and the specific molecular defects that can give rise to pRTA.

CELLULAR AND MOLECULAR MECHANISMS OF ACID–BASE TRANSPORT IN THE PROXIMAL TUBULE

Role of Proximal Tubule in Acid–base Homeostasis

Ingestion of a normal Western diet results in the generation of approximately 0.5–1.0 mEq/kg body weight of nonvolatile acid (Sebastian et al 2002, Lemann et al 2003). This would lead to the relentless titration of bicarbonate and thereby would cause severe metabolic acidosis if not for the process of renal net acid excretion. Net renal acid excretion is defined as follows (Relman et al 1961, Lennon et al 1966, Brown et al 1989, Lemann et al 2003): net acid excretion = urinary NH_4^+ + urinary titratable acid–urinary base (HCO_3^- and metabolizable anions like citrate). The process of glomerular filtration (160–180 l/day) removes huge quantities of bicarbonate from the plasma (4000–4500 mEq/day). Maintenance of acid–base balance requires the reabsorption of virtually all of the filtered bicarbonate so that net acid excretion can match endogenous acid production.

The proximal tubule is the site where most of the filtered bicarbonate is reabsorbed (Gottschalk et al 1960, Malnic et al 1972). Similarly, it is the site for reabsorption of filtered organic anions like citrate that when retained in the body and metabolized consume acid or generate base (Unwin et al 2004). Moreover, the proximal tubule is the principal site for production of ammonium, which is clearly critical to the overall process of net acid excretion (DuBose et al 1991). Finally, in lowering tubule fluid pH to approximately 6.5, the proximal tubule is also a site for formation of significant quantities of titratable acid as secreted H^+ binds to nonbicarbonate buffers as exemplified by the titration of filtered phosphate: $H^+ + HPO_4^- = H_2PO_4$.

Mechanisms of Apical Acid Secretion

The most important mechanisms for transmembrane transport of H^+ and HCO_3^- in the proximal tubule are summarized in Figure 11.1. Extrusion of H^+ from the cell into the tubule fluid occurs by at least two pathways: Na^+-H^+ exchange, and H^+-ATPase (Aronson 1983, Alpern 1990). Na^+-H^+ exchange is a secondary active process in which the driving force for uphill H^+ translocation is provided by the inward Na^+ gradient that results from its primary active transport across the basolateral membrane by the Na^+, K^+-ATPase. Since the inward Na^+ concentration gradient across the apical membrane in the proximal tubule is no more than 10-fold (Yoshimoto & Fromter 1985), the largest H^+ gradient that can be generated and maintained cannot exceed 10-fold, consistent with the observation that tubule fluid pH in the proximal tubule does not fall below 6.4 (Gottschalk et al 1960, Malnic et al 1972). Plasma membrane Na^+-H^+ exchange is mediated by isoforms of the NHE (SLC9) gene family (Orlowski & Grinstein 2004, Brett et al 2005). Immunolocalization studies have identified the expression of NHE3 on the apical membrane of proximal tubule cells (Biemesderfer et al 1993, 1997, Amemiya et al 1995). Several experimental approaches have been used to estimate the functional contribution of NHE3 to apical membrane acid extrusion and transtubular

Genetic Diseases of the Kidney
ISBN: 978-0-12-449851-8

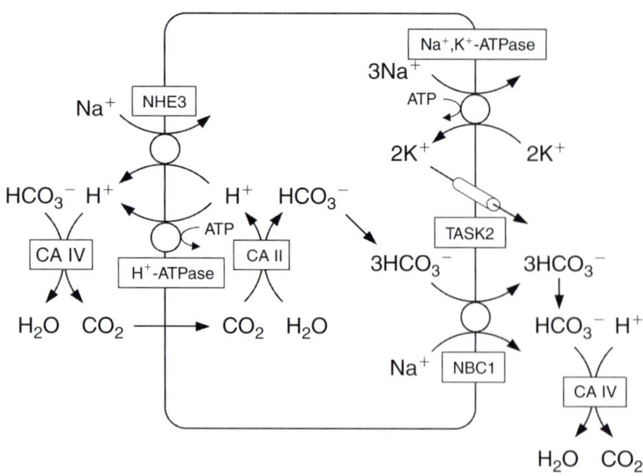

FIGURE 11.1 Major pathways for acid–base transport in the proximal tubule. Not shown is basolateral Cl^--HCO_3^- exchange, which plays a role in mediating base exit in the S3 segment

bicarbonate reabsorption in the proximal tubule. For example, use of NHE-selective inhibitors has indicated that virtually all measured Na^+-H^+ exchange activity in renal brush border vesicles can be accounted for by NHE3 (Wu et al 1996), and has demonstrated that approximately 60% of transtubular bicarbonate absorption is mediated by NHE3 (Vallon et al 2000, Wang et al 2001). This conclusion has been supported by microperfusion studies in NHE3 null mice showing a defect of similar magnitude in transtubular bicarbonate absorption (Schultheis et al 1998, Wang et al 1999).

Additional studies in NHE3 null mice have investigated the apical membrane pathways responsible for the remaining component of bicarbonate reabsorption. The vacuolar H^+-ATPase inhibitor bafilomycin reduced the bicarbonate reabsorption in NHE3 null mice by approximately 60%, indicating a substantial contribution of primary active H^+ secretion to the process of proximal acidification (Wang et al 1999). In fact, prior immunolocalization studies have demonstrated expression of vacuolar H^+-ATPase on the apical surface of proximal tubule cells, and a component of Na^+-independent acid extrusion from proximal tubule cells has been shown by intracellular pH measurements following cellular acid loading (Zimolo et al 1992).

It is presently uncertain what accounts for the remaining bicarbonate reabsorption in the proximal tubule not attributable to NHE3 or H^+-ATPase. A component of Na^+-dependent acid extrusion across the apical membrane of proximal tubule cells persists in NHE3 null mice, raising the possibility of yet another NHE isoform operating at this site (Choi et al 2000). Recent studies have identified expression of NHE8 on the brush border membrane of proximal tubule cells (Goyal et al 2003, 2005). It is therefore possible that NHE8 makes a modest contribution to proximal bicarbonate absorption.

Mechanisms of Basolateral Base Exit

At least two mechanisms contribute to bicarbonate exit across the basolateral membrane of proximal tubule cells, namely Na^+-HCO_3^- cotransport and Cl^--HCO_3^- exchange (Alpern 1990, Boron 2006). Na^+-HCO_3^- cotransport is the predominant mechanism through most of the length of the proximal tubule (S1 and S2) (Kondo & Fromter 1987, Abuladze et al 1998), whereas $Cl^-HCO_3^-$ is important in the S3 segment (Kondo & Fromter 1990). Since the Na^+ gradient is directed inward (Yoshimoto & Fromter 1985), and intracellular pH and HCO_3^- are not higher than in the plasma (Alpern 1990, Boron 2006), what then is the driving force to effect HCO_3^- exit across the basolateral membrane by Na^+-HCO_3^- cotransport? Because most but not all studies have indicated that the stoichiometry of Na^+-HCO_3^- cotransport in the proximal tubule is 1:3 (Yoshimoto et al 1985, Soleimani et al 1987, Seki et al 1993), the process is highly electrogenic and driven by the inside-negative membrane potential. The inside-negative membrane potential in turn arises from the outward K^+ gradient generated by the Na^+, K^+-ATPase and the presence of a high basolateral membrane K^+ conductance (Biagi et al 1981, Welling & O'Neil 1990, Beck et al 1993). It should be noted that there is no detectable basolateral HCO_3^- conductance in the absence of Na^+ (Yoshitomi et al 1985). In the case of Cl^--HCO_3^- exchange, the driving force to effect HCO_3^- exit from the cell is the inward Cl^- concentration gradient (Kondo & Fromter 1990).

Plasma membrane Na^+-HCO_3^- cotransport is mediated by NBC isoforms that are members of the SLC4 family of bicarbonate transporters (Boron 2006), as shown in Figure 11.2. It may be noted that there are two major subdivisions of the SLC4 bicarbonate transporter family, the Na^+-independent Cl^--HCO_3^- exchangers, and the Na^+-coupled HCO_3^- transporters. The latter group includes the electrogenic Na^+-HCO_3^- cotransporter, NBCe1 (NBC1, SLC4A4), which has been shown to be expressed on the basolateral membrane of cells of the S1 and S2 segments of the proximal tubule (Schmitt et al 1999). There are at least three splice isoforms of NBCe1 (NBCe-A, B and C), of which the NBCe1-A variant (also known as kNBC1) is the one expressed in the proximal tubule. This kidney variant of NBCe1 has been shown to have a variable Na^+:HCO_3^- stoichiometry of 1:2 or 1:3 (Gross et al 2001a, 2001b, Muller-Berger et al 2001), depending on such regulatory factors as intracellular calcium and phosphorylation by protein kinase A.

As discussed earlier, a Na^+:HCO_3^- stoichiometry of 1:3 is required for the transporter to mediate net HCO_3 efflux across the basolateral membrane of proximal tubule cells. Therefore, a regulatory change in stoichiometry from 1:3 to 1:2 would be expected to change the net direction of transport from outward to inward across the basolateral membrane and thereby impede net bicarbonate reabsorption. Phosphorylation of NBCe1 also affects its association with

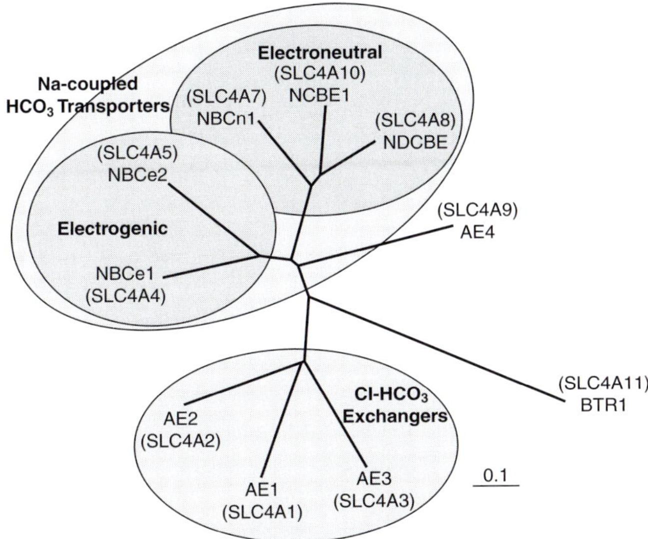

FIGURE 11.2 SLC4 family of bicarbonate transporters (from Boron 2006). (see also Plate 11)

carbonic anhydrase II (see below) (Gross et al 2002). As discussed above, the electrogenic exit of HCO_3 across the basolateral membrane by Na^+-HCO_3^- cotransport is strongly dependent on the inside-negative membrane potential as a driving force. In the proximal tubule, the predominant ion conductance responsible for the inside-negative membrane potential is that of K^+ (Biagi et al 1981, Welling & O'Neil 1990, Beck et al 1993). Although multiple types of basolateral membrane K^+ channels with different properties have been characterized electrophysiologically in the proximal tubule, to date only two have been identified at the molecular level: Kir6.1 and TASK2 (Brochiero et al 2002, Warth et al 2004, Hebert et al 2005). Interestingly, the K^+ conductance mediated by TASK2 is activated by basolateral alkalinization as results from Na^+-HCO_3^- cotransport activity (Warth et al 2004). Moreover, TASK2 null mice have evidence of mild metabolic acidosis and increased urinary HCO_3 excretion, consistent with a renal tubular acidosis (Warth et al 2004). In contrast to the identification of the Na^+-HCO_3^- cotransporter, the molecular basis for the basolateral membrane Cl^--HCO_3^- exchange activity expressed in the proximal tubule has not been established with certainty. Na^+-independent Cl^--HCO_3^- exchange can be mediated by specific members of the SLC4 family (e.g. AE1, AE2, AE3) as well as be certain members of the SLC26 family of anion transporters (e.g. SLC26A3 (DRA), SLC26A4 (pendrin), SLC26A6 (CFEX)) (Mount & Romero 2004, Boron 2006). To date, none of these transporters has been localized to the basolateral membrane of proximal tubule cells, except for SLC26A1 (SAT1), a sulfate/oxalate transporter that does not mediate Cl^--HCO_3^- exchange (Markovich & Aronson 2007).

Roles of Carbonic Anhydrase

It is well-established that carbonic anhydrase activity is essential for normal rates of proximal tubule bicarbonate reabsorption (Purkerson & Schwartz 2007). Inspection of Figure 11.1 indicates at least three distinct mechanisms by which this enzyme facilitates proximal acidification. First, carbonic anhydrase activity on the luminal face of the apical membrane accelerates the disposition of secreted H^+ by catalyzing its combination with filtered HCO_3^- to form water and CO_2, the latter leaving the lumen by passive transport across the apical membrane. Thus, inhibition of luminal carbonic anhydrase causes a decline in intratubular pH due to accumulation of secreted H^+, the 'disequilibrium pH,' which in turn inhibits the rate of H^+ secretion due to the change in driving force for Na^+-H^+ exchange and thereby reduces transtubular HCO_3^- absorption (Lucci et al 1983). Second, as shown in Figure 11.1, intracellular carbonic anhydrase has a dual effect to provide H^+ and HCO_3 from water and CO_2, thus facilitating apical H^+ extrusion and basolateral HCO_3^- exit. Carbonic anhydrase inhibition induces intracellular alkalinization as apical H^+ secretion then leads to depletion of intracellular H^+ (Sasaki & Marumo 1989). Whereas the transporters mediating apical acid secretion directly transport H^+ and are not HCO_3 dependent, the basolateral base exit mechanisms, particularly Na^+-HCO_3^- cotransport, require HCO_3 (Soleimani & Aronson 1989). Thus, carbonic anhydrase plays a pivotal role in coupling apical acid extrusion to basolateral base exit. Third, carbonic anhydrase at the external face of the basolateral membrane facilitates the exit of HCO_3 by catalyzing its disposition from the peritubular interstitial compartment, thereby reducing the peritubular alkalinization that would otherwise occur (Tsuruoka et al 2001). Thus, the driving force for HCO_3^- exit is optimized by avoiding its peritubular accumulation. To date, 15 known isoforms of carbonic anhydrase have been identified (Purkerson & Schwartz 2007). Carbonic anhydrase II (CAII) is a soluble cytosolic enzyme that accounts for over 95% of renal carbonic anhydrase activity (Purkerson & Schwartz 2007). CAII is abundantly expressed in the proximal tubule (Purkerson & Schwartz 2007). Interestingly, the regulated association of CAII with an internal domain of NBC1 has been described (Gross et al 2002, Pushkin et al 2004), facilitating production of HCO_3^- in the immediate vicinity of the transporter as part of a CO_2/HCO_3^- 'metabolon' (Pushkin et al 2004). However, the functional significance of association of CAII with NBC1 has been disputed in studies from another laboratory (Lu et al 2006, Piermarini et al 2007). In addition to CAII, CAIV is expressed in proximal tubule cells where it is located both on the apical and basolateral cell membranes (Purkerson & Schwartz 2007, Purkerson et al 2007). CAIV is attached to the membrane by a GPI anchor and its catalytic site faces the outside of the cell (Purkerson & Schwartz 2007, Purkerson et al 2007). CAIV has been reported to associate with an external domain of NBC1, thereby forming a 'metabolon' to dispose of HCO_3 from the immediate vicinity of the external face of the transporter (Alvarez et al 2003). It should be noted that proximal tubule HCO_3 absorption can be inhibited by application of impermeant carbonic anhydrase inhibitors to either the luminal or basolateral cell surfaces (Lucci et al 1983, Tsuruoka et al 2001). These findings underscore the importance of carbonic anhydrase activity at the membrane surface to dispose of secreted H^+ in the lumen and prevent accumulation of absorbed HCO_3 at the basolateral side of the proximal tubule cell as illustrated in Figure 11.1. Although regulated association of CAII with NHE1 has been described (Li et al 2002), to date there are no reports of association of CAII or CAIV with the apical Na^+-H^+ exchanger NHE3. CAXII is an additional isoform expressed on the basolateral surface of proximal tubules cells, although its expression at this site is more significant in rodent than in human kidney (Purkerson & Schwartz 2007). Similarly, in rodent kidney CAXIV is abundantly expressed on the apical membrane of proximal tubule cells (Purkerson & Schwartz 2007). Based on low or absent expression of these isoforms in the proximal tubule of human kidney, it is unlikely that they contribute significantly to HCO_3 reabsorption in this nephron segment in man.

Mechanisms of Ammonium Production and Secretion

As mentioned earlier, urinary excretion of ammonium is a key component of net acid excretion as required for maintaining acid–base balance while ingesting a Western acid-forming diet. Virtually all of the ammonium that ultimately appears in the urine is first produced and secreted by the proximal tubule (Karim et al 2005, Weiner & Hamm

2007). Ammonium is produced in the proximal tubule through the metabolism of glutamine to glutamate catalyzed by mitochondrial phosphate-dependent glutaminase, and then the metabolism of glutamate to α-keto-glutarate catalyzed by glutamate dehydrogenase (Karim et al 2005, Weiner & Hamm 2007). Metabolism of the divalent anion α-keto-glutarate subsequently generates two HCO_3 that exit across the basolateral membrane (Karim et al 2005, Weiner & Hamm 2007). Ammonium production in the proximal tubule is markedly enhanced by acidosis and hypokalemia, and inhibited by alkalosis and hyperkalemia (Nagami 2000). The most likely explanation is that ammonium production is regulated by intracellular pH, with stimulation when cell pH is acidified as in states of acidosis and hypokalemia (Nagami 2000). As shown in the inset of Figure 11.3, ammonium secretion across the apical membrane of proximal tubule cells can occur by two processes. First, ammonium can be exchanged for luminal Na^+ on the apical membrane Na^+-H^+ exchanger (Kinsella & Aronson 1981). Second, ammonium can cross the cell membrane by non-ionic diffusion of NH_3 and be trapped in the urine as NH_4^+ due to the low pH resulting from H^+ secretion. A major role for direct NH_4^+ secretion by the Na^+-H^+ exchanger is indicated by the findings that NH_4^+ secretion in the early proximal tubule takes place before appreciable luminal acidification has occurred, and that a major fraction of apical NH_4^+ secretion can be blocked by inhibition of the apical Na^+-H^+ exchanger with amiloride (Nagami 1988).

As indicated in Figure 11.3, ammonium secreted by the proximal tubule is extensively reabsorbed by the thick ascending limb, resulting in the generation of a cortico-medullary NH_3 gradient in the interstitium (Karim et al 2005, Weiner & Hamm 2007). Final secretion of ammonium from the interstitium to the urine then occurs in the collecting duct (Karim et al 2005, Weiner & Hamm 2007).

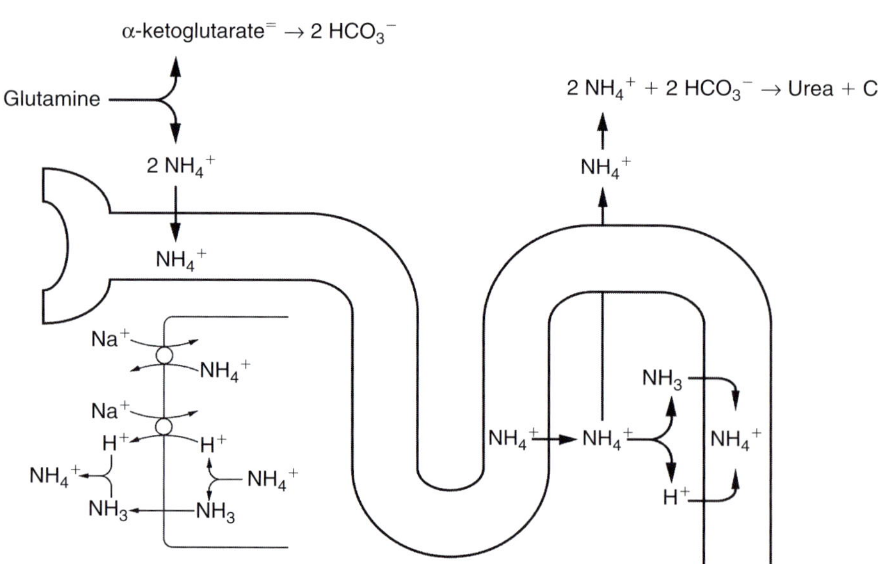

FIGURE 11.3   Itinerary of ammonium excretion, with production and secretion in the proximal tubule, reabsorption in the loop of Henle, and final secretion in the collecting duct. Mechanisms of apical membrane ammonium secretion in the proximal tubule are shown in the inset

It should be noted that the new HCO_3^- generated as the result of ammonium secretion is actually added by the proximal tubule to the peritubular blood. However, collecting duct ammonium secretion is essential to prevent NH_4^+ reabsorbed in the loop of Henle from accumulating in the medullary interstitium from where it would be carried to the liver and converted to urea. Such urea production would consume the HCO_3^- that had been generated by the proximal tubule (Weiner & Hamm 2007). Thus, proximal tubule ammonium production and collecting duct ammonium secretion are together essential for net acid excretion and generation of new HCO_3^- to neutralize endogenous acid production.

Reabsorption of Citrate

Citrate, which at physiological pH is a trivalent anion, is freely filtered and substantially (65–90%) reabsorbed in the proximal tubule. After entering the cell, reabsorbed citrate is completely metabolized by mitochondrial oxidation to yield HCO_3^-, which is returned to the systemic circulation. Thus, proximal tubule reabsorption of citrate accomplishes the retention of base, and citrate excretion represents loss of base from the body. Urinary citrate normally inhibits calcium nephrolithiasis by complexing calcium in a soluble form as well as by inhibiting crystal growth (De Yoreo et al 2006). The processes involved in reabsorption and metabolism of citrate are summarized in Figure 11.4. As shown in Figure 11.4, citrate is reabsorbed by secondary active Na^+-cotransport. Citrate reabsorption occurs as the divalent species by an electrogenic Na^+-cotransport pathway shared by such divalent Krebs cycle intermediates as succinate (Pajor 1999, Hamm & Hering-Smith 2002). Because citrate is reabsorbed as a divalent ion, its reabsorption is enhanced by luminal acidification (Brennan et al 1988) and is thus coupled to acid secretion which, as discussed above, principally occurs by Na^+-H^+ exchange. Also, as illustrated in Figure 11.4, metabolism of each citrate has the effect of generating 3 HCO_3^- that exit the basolateral membrane. In addition to the luminal membrane Na^+-cotransport of citrate, this anion is also accumulated across the basolateral membrane by

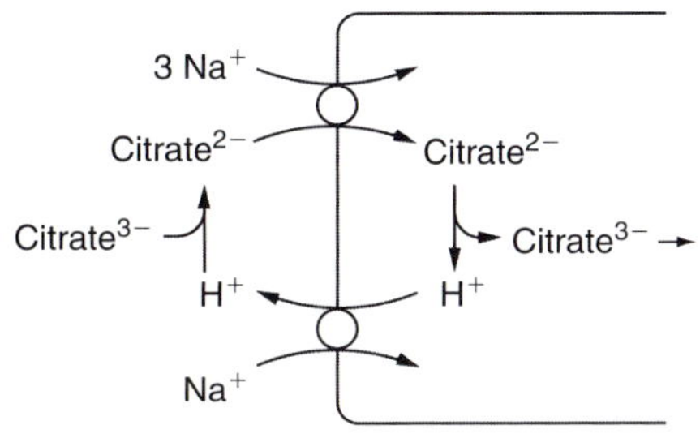

Na^+-cotransport (Pajor 1999, Hamm & Hering-Smith 2002). The apical membrane transporter responsible for Na^+-coupled citrate reabsorption is NaDC1 (Na-dicarboxylate-cotransporter 1) (Pajor 2000). NaDC1 (SLC13A2) is a member of the SLC13 transporter family that includes Na^+-coupled sulfate transporters as well as Na^+-coupled transporters for di- and tricarboxylates (Pajor 2006). NaDC1 cotransports divalent citrate and Na^+ with a stoichiometry of 3 Na^+:citrate^{2-} so that it is electrogenic. Basolateral membrane citrate uptake is mediated by NaDC3 (SLC13A3) (Pajor 2006). Proximal tubule citrate reabsorption is greatly enhanced by acidosis and reduced by alkalosis. Several mechanisms may underlie these effects. First, as discussed above, citrate reabsorption is sensitive to luminal pH (Brennan et al 1988), which will tend to be lower during acidosis and higher in alkalosis. Second, acidosis increases citrate lyase activity, which leads to a decrease in intracellular citrate that in turn augments citrate reabsorption (Melnick et al 1996). Third, expression of NaDC1 is up-regulated in states of acidosis. Hypokalemia resembles the effect of acidosis to up-regulate brush border citrate transport and reduce renal citrate excretion, presumably due to its effect to cause cellular acidification (Adler et al 1974, Levi et al 1991).

DEFINITION AND CLINICAL FEATURES OF PROXIMAL RTA

Reduced Apparent Threshold for Bicarbonate Reabsorption with Retained Ability for Acidification of Urinary pH

Proximal renal tubular acidosis (pRTA) is defined as a hyperchloremic metabolic acidosis due to a defect in bicarbonate transport capacity in this nephron segment (Morris 1969, Rodriguez Soriano 1967, 2002, Rodriguez Soriano & Edelmann 1969). This concept is based on interpretation of the relationship between bicarbonate reabsorption and the filtered load of bicarbonate presented to the proximal tubule. The relationships among plasma concentration, renal tubular reabsorption and urinary excretion of bicarbonate as a function of bicarbonate loading are illustrated in Figure 11.5 for normal persons and an individual with pRTA (Brenes et al 1977). In normal individuals, as indicated by the dotted lines, essentially all of the filtered bicarbonate is reabsorbed until plasma bicarbonate surpasses a threshold of about 26 mM, at which point bicarbonate reabsorption reaches a plateau, and bicarbonate appears in the urine. These results are consistent with the concept that bicarbonate reabsorption in normal individuals saturates at a filtered load corresponding to a plasma bicarbonate of approximately 26 mM. In contrast, in a patient with pRTA there is a greatly reduced threshold of plasma bicarbonate at which bicarbonate appears in the urine, and the maximal rate of reabsorption of bicarbonate is markedly depressed.

FIGURE 11.4 Mechanism of apical membrane citrate reabsorption in the proximal tubule and its dependence on luminal acidification

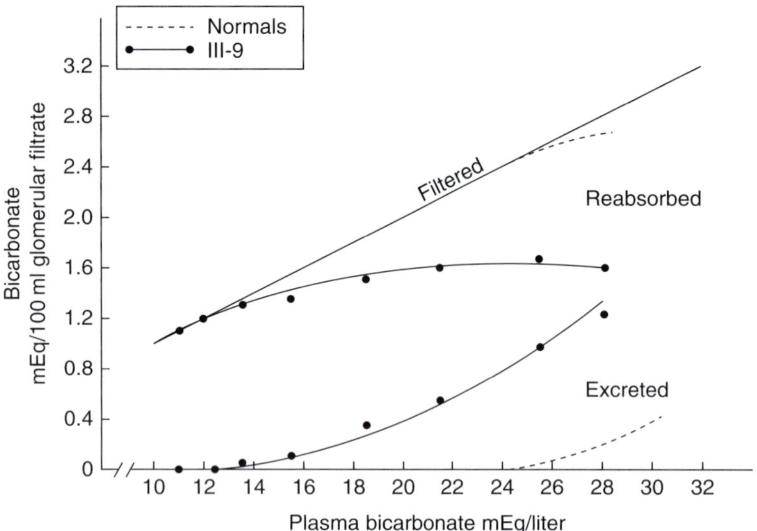

FIGURE 11.5   Relationship between plasma concentration, renal tubular reabsorption and urinary excretion of bicarbonate in normal subjects and a patient with proximal renal tubular acidosis (from Brenes et al 1977). In the patient with pRTA there is a greatly reduced threshold of plasma bicarbonate at which bicarbonate appears in the urine

However, such bicarbonate titration analyses are usually performed with sodium bicarbonate loading that would also cause a measure of volume expansion. Studies in humans and experimental animals have shown that if care is taken to minimize volume expansion, tubule bicarbonate reabsorption does not reach a plateau even at plasma bicarbonate levels exceeding 35 mM, and the threshold for bicarbonate appearance in the urine is shifted to higher plasma bicarbonate values (Purkerson et al 1969, Slatopolsky et al 1970). Nevertheless, it has been customary to compare renal reabsorption of bicarbonate as a function of bicarbonate loading in controls and pRTA patients without taking into account the possible effects of concurrent volume expansion. Since the volume expansion due to sodium bicarbonate loading would be, if anything, less in pRTA patients due to increased urinary excretion of sodium bicarbonate, the difference in titration curves between normal and pRTA individuals must be due to an intrinsic difference in proximal tubule bicarbonate transport capacity.

An important point is that distal acidification processes are normal in pRTA. Therefore, when plasma bicarbonate is at or below the threshold at which bicarbonate is excreted in the urine, all the filtered bicarbonate will be reabsorbed, and distal acidification can achieve appropriately low urinary pH. This is illustrated in Figure 11.6A, which depicts the relationship between urinary pH and plasma bicarbonate in patients with pRTA (Rodriguez Soriano & Edelmann 1969). The relationship for normal persons (dashed line) is shown for comparison. Due to the reduced threshold for bicarbonate reabsorption in pRTA patients, bicarbonate is excreted and the urine is alkaline at reduced levels of plasma bicarbonate compared to normals. However, at progressively lower plasma bicarbonate concentrations, urine pH in pRTA patients becomes appropriately acidified to pH values below 5.5. These findings stand in sharp contrast to the relationship between urinary pH and plasma

bicarbonate in patients with distal RTA (dRTA), as demonstrated in Figure 11.6B (Rodriguez Soriano & Edelmann 1969). Here we see that urinary pH fails to acidify to levels below 6.5 even in the presence of low plasma bicarbonate values. A schematic diagram integrating the data shown in Figures 11.5 and 11.6 is illustrated in Figure 11.7 (Morris et al 1972). As shown in Figure 11.7A, in a normal individual with normal plasma bicarbonate, the filtered load of bicarbonate is largely reabsorbed in the proximal tubule and the remainder in the distal nephron. Additional distal H^+ secretion then acidifies urinary pH. As seen in Figure 11.7B, untreated patients with pRTA will generally present with metabolic acidosis as reflected by low plasma bicarbonate, but no urinary bicarbonate wasting because the filtered load of bicarbonate is reduced to the level at which bicarbonate is no longer excreted and acidification of urinary pH can occur normally. However, if such a patient is treated with alkali leading to a rise in plasma bicarbonate (Figures 11.7C and 11.7D), the filtered load of bicarbonate now exceeds the capacity for its reabsorption in the proximal tubule as well as downstream nephron segments, resulting in bicarbonaturia and alkalinization of urinary pH. Importantly, it should be noted that in the untreated patient with pRTA and a low plasma HCO_3^- (e.g. Figure 11.7B), net acid excretion maintains acid–base balance since distal acidification is normal. Thus, balance studies have shown that patients with pRTA and metabolic acidosis actually maintain acid–base balance whereby urinary net acid excretion matches endogenous fixed acid production (Lemann et al 2000). This is in contrast to dRTA, in which acid–base balance is not achieved owing to defective ability to acidify the urine and thereby excrete appropriate quantities of titratable acid and ammonium (Morris 1969, Rodriguez Soriano & Edelmann 1969, Rodriguez Soriano 2002).

Although patients with pRTA maintain acid–base balance with 'normal' rates of ammonium excretion, in fact

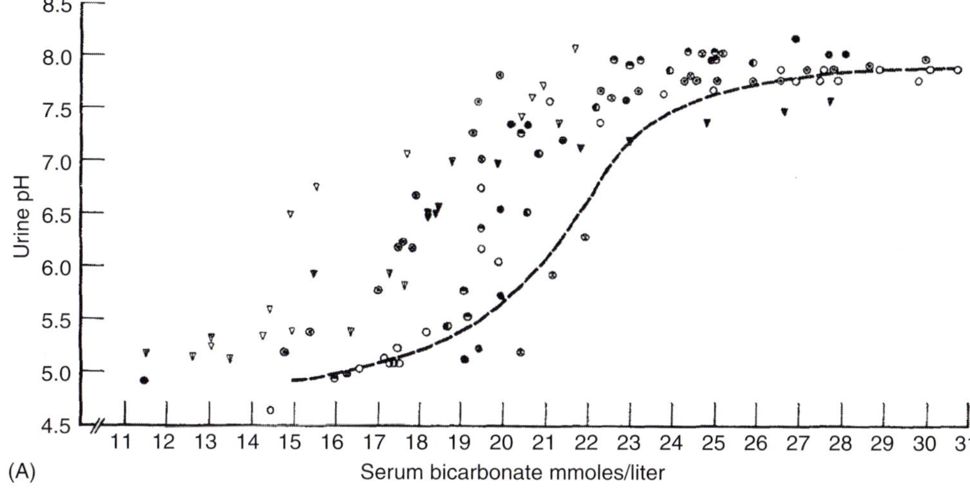

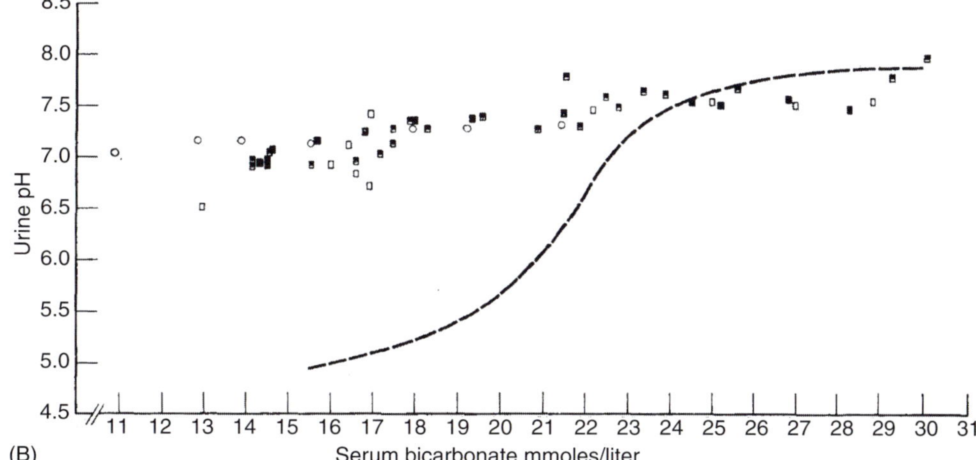

FIGURE 11.6 (A) Relationship between urinary pH and serum bicarbonate concentration in patients with proximal renal tubular acidosis (from Rodriguez Soriano & Edelmann 1969). (B) Relationship between urinary pH and serum bicarbonate concentration in patients with distal renal tubular acidosis (from Rodriguez Soriano & Edelmann 1969). The dashed line represents the response of normal subjects

ammonium excretion is subnormal for the state of chronic metabolic acidosis (Brenes et al 1977, Brenes & Sanchez 1993). As mentioned above, ammonium is produced and first secreted into the urine in the proximal tubule, and its production is strongly affected by cell pH. It was therefore proposed that alkalinization of proximal tubule cells may be a primary defect leading to reduced ammonium excretion (Halperin et al 1989). In fact, as will be discussed below, the principal identified cause of isolated pRTA in humans is a defect in basolateral base exit, which would be predicted to lead to relative alkalinization of intracellular pH. This in turn would prevent up-regulation of ammonium production as would normally occur in chronic metabolic acidosis.

Citrate and Calcium Excretion

In contrast to the hypocitraturia, hypercalciuria, and propensity to nephrolithiasis and nephrocalcinosis observed in patients with dRTA, such problems are much less severe or absent in patients with pRTA (Lemann et al 2000,

Rodriguez Soriano 2002). As described above, proximal tubule citrate reabsorption is markedly up-regulated in states of metabolic acidosis and reduced in metabolic alkalosis (Melnick et al 1996, Aruga et al 2000). These adaptive changes are likely mediated in part by corresponding changes in intracellular pH. As just discussed, the principal cause of isolated pRTA is a defect in basolateral base exit that would be predicted to lead to relative alkalinization of intracellular pH in proximal tubule cells. Such intracellular alkalinization would tend to mitigate the effects of systemic acidosis and thereby maintain citrate excretion at a higher level than observed in dRTA. In addition, as explained earlier, citrate reabsorption in the proximal tubule is strongly influenced by luminal pH (Brennan et al 1988). The defect in proximal acidification in pRTA would tend to cause a higher luminal pH, which would inhibit citrate reabsorption and promote its excretion. It should be noted that the importance of this mechanism may diminish as the filtered load of bicarbonate falls with attainment of steady state acidosis.

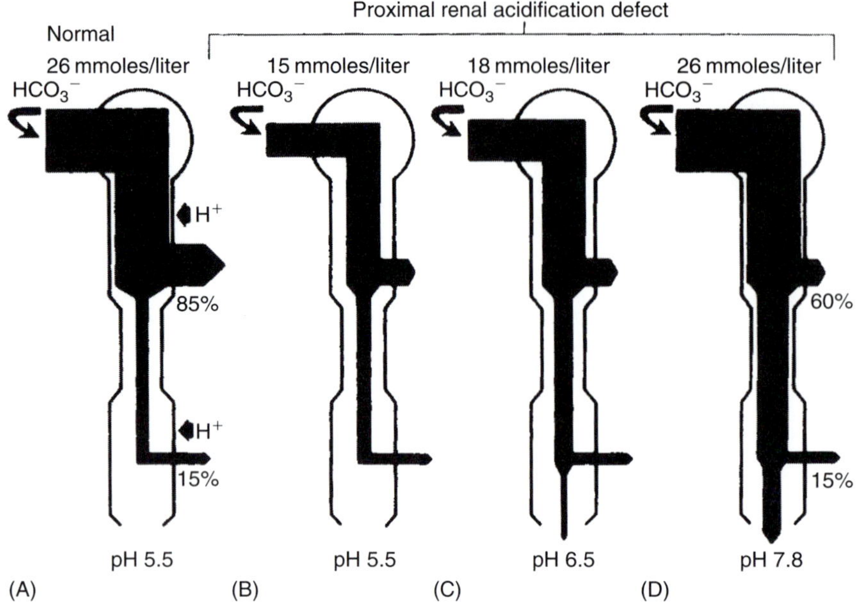

FIGURE 11.7 Schematic representation of bicarbonate reabsorption and excretion in proximal renal tubular acidosis (from Morris et al 1972). (A) Normal subject. (B) Patient with untreated pRTA and steady state reduction in serum bicarbonate. (C) Patient with pRTA with partial correction of serum bicarbonate by base administration. (D) Patient with pRTA with complete correction of serum bicarbonate by base administration. Not shown is the adaptive increase in distal nephron bicarbonate reabsorption that occurs when there is increased delivery of bicarbonate out of the proximal tubule (Nakamura et al 1999, Bailey et al 2004)

Patients with pRTA generally have minimal hypercalciuria and are not in negative calcium balance as is the case in dRTA (Lemann et al 2000, Rodriguez Soriano 2002). A possible explanation is that the luminal pH of fluid delivered to the distal tubule, the major site for active calcium reabsorption, would be higher than normal in pRTA due to the defect in proximal bicarbonate reabsorption. Calcium reabsorption in the distal tubule is stimulated by increased luminal pH (Mori et al 1992). Thus, it is inhibited during acidosis and stimulated by bicarbonate infusion (Sutton et al 1979). These effects are likely secondary to pH regulation of the apical Ca channel TRPV5 (Yeh et al 2003, Topala et al 2007). Regardless of the underlying mechanisms, the lower calcium excretion and higher citrate excretion in pRTA compared to dRTA presumably explains the far greater propensity to nephrocalcinosis and nephrolithiasis in dRTA. Similarly, abnormal bone metabolism reflected by growth retardation is more marked in untreated dRTA, although it is also a feature of pRTA (Lemann et al 2000, Rodriguez Soriano 2002).

Potassium Excretion

Hypokalemia due to urinary potassium wasting is a frequent though not invariable feature of pRTA (Sebastian et al 1971, Brenes et al 1977, Igarashi et al 2002). Multiple mechanisms may contribute to this disorder. First, due to the reduction in transport of Na^+ and HCO_3^- in the proximal tubule, there is diminished fluid reabsorption, which in turn leads to diminished passive paracellular potassium reabsorption in this nephron segment. Second, the reduction in Na^+ and fluid reabsorption in the proximal tubule will tend to cause volume depletion, leading to secondary hyperaldosteronism (Sebastian et al 1971a, b). Third, increased delivery

of Na^+ and fluid to the distal nephron in the presence of high aldosterone will stimulate distal potassium secretion (Giebisch 1998). Fourth, to the extent tubular fluid delivered to the distal nephron is more alkaline in pRTA, this will also result in stimulation of distal potassium secretion (Giebisch 1998, Hebert et al 2005). Finally, the delivery of tubule fluid enriched in non-chloride anion will independently augment distal potassium secretion by increasing apical K^+-Cl^- cotransport (Velazquez et al 1987, Amorim et al 2003).

The tendency to potassium wasting in pRTA is minimized in the steady state when there is metabolic acidosis and a low filtered load of bicarbonate. However, with attempted treatment by base administration, the Na^+ and bicarbonate load presented to the distal nephron is enhanced, and urinary potassium wasting is greatly aggravated (Sebastian et al 1971a).

SPECIFIC DISORDERS CAUSING PROXIMAL RENAL TUBULAR ACIDOSIS

Proximal renal tubular acidosis most commonly occurs as a manifestation of a generalized defect in proximal tubule function known as the Fanconi syndrome. The specific causes of the Fanconi syndrome are described in Chapter 10. We will focus on causes of isolated pRTA with preservation of other aspects of proximal tubule function.

Isolated pRTA in humans is quite rare and occurs in several patterns. First, pRTA may occur sporadically as a transient disorder in infants and young children (Rogriguez et al 1967, Nash et al 1972). This disorder typically presents as growth retardation that responds to alkali therapy, which can eventually be discontinued without recurrence of metabolic

TABLE 11.1   Molecular and clinical features of different SLC4A4 mutations

Mutation	Sex	Ocular pathology	MR	ED	NBCe1 variants	Effect on oocyte NBCe1 protein
Q29X	Female	Glc	+	−	kNBCe1	Loss of function due to protein truncation
R298S	Female	Glc, Bk, Cat	+	−	kNBCe1 pNBCe1	About 41% of wild-type activity
S427L	Female	Glc, Bk, Cat	−	+	kNBCe1 pNBCe1	About 10% of wild-type activity
T485S	Male	Glc(nd), Bk, Cat	nd	−	kNBCe1 pNBCe1	About 50% of wild-type activity in ECV304 Ce1ls; no activity in oocytes
R510H	Female	Glc, Bk, Cat	nd	−	kNBCe1 pNBCe1	About 57% of wild-type activity in ECV304 Ce1ls; 4% of wild-type activity in oocytes
L522P*	Male	Glc, Bk, Cat	−	+	kNBCe1 pNBCe1	Loss of function due to impaired translational processing and membrane trafficking
2311delA	Male	Glc, Bk, Cat	−	+	kNBCe1 pNBCe1	Loss of function due to protein truncation
A799V	Female	Glc, Bk, Cat	+	−	kNBCe1 pNBCe1	About 14% of wild-type activity
R881C	Female	Glc(nd), Bk, Cat	nd	−	kNBCe1 pNBCe1	About 39% of wild-type activity

All patients had proximal renal tubular acidosis and short stature. Listed are additional findings affecting at least 30% of patients, such as mental retardation (MR), enamel defects (ED), glaucoma (Glc), band keratopathy (Bk), and cataract (Cat). The asterisk indicates primary data from Demirci et al (2006). Some information from other studies was not determined (nd) From Demirci et al 2006

acidosis (Rodriguez Soriano 2002). The specific transport defect underlying this type of pRTA is not known. It is possible that there is an underlying defect in proximal acid–base transport that is actually persistent but that can be compensated for as the nephron matures and alternative mechanisms of bicarbonate reabsorption become more prominent. For example, there is recent evidence that the relative abundance of NHE8 compared to NHE3 is greater in the neonatal than in the adult kidney (Becker et al 2007). It is therefore conceivable that an inherited defect in NHE8 would only be clinically evident in the immature kidney.

NBC1 Mutations

Proximal renal tubular acidosis can present as an autosomal recessive inherited disorder associated with ocular abnormalities (Igarashi et al 2002). These patients have recently been found to have mutations in the electrogenic Na^+-bicarbonate cotransporter (NBCe1, SLC4A4) that mediates exit of bicarbonate across the basolateral membrane of proximal tubule cells (Igarashi et al 1999, 2002). Patients with autosomal recessive pRTA due to SLC4A4 mutations typically have severe metabolic acidosis with arterial pH 7.1–7.2 and

bicarbonate 5–10 mEq/l (Igarashi et al 2002). Urine pH during acidosis is <5.5, consistent with retained ability of the distal nephron to acidify urinary pH effectively (Igarashi et al 1999). Bicarbonate administration induces urinary bicarbonate wasting (Shiohara et al 2000, Igarashi et al 2001). Importantly, other aspects of proximal tubule function are normal, as there is no glycosuria, amino aciduria or proteinuria (Shiohara et al 2000). Associated features of this disorder include short stature, and such ocular abnormalities as glaucoma, cataracts and band keratopathy (Igarashi et al 2002). Ocular abnormalities are observed in all patients. Additional features may include enamel defects of the teeth, mental retardation, hypothyroidism and elevation of serum amylase without overt pancreatitis (Igarashi et al 2002). Basal ganglia calcification has also been reported. It should be noted that one or more splice isoforms of NBCe1 are expressed in multiple tissues including various compartments of the eye, brain and pancreas (Usui et al 2001, Romero et al 2004).

Mutations in SLC4A4 causing pRTA with ocular abnormalities are listed in Table 11.1 as tabulated by Demirci et al (2006). This table also lists the functional defects of mutant transporters noted using in vitro expression systems.

To date, there has not been a close correlation between the severity of transport defect and the clinical phenotype. For example, the Q29X mutation (Igarashi et al 2001), which results in truncation near the N-terminus of the NBCe1-A (kNBC1) splice variant expressed in the proximal tubule, would be expected to cause a much more severe pRTA phenotype than the R298S mutant that retains approximately 50% of normal transport activity (Igarashi et al 1999). Yet the latter also causes a similarly severe phenotype (Igarashi et al 1999). A potential explanation for the severity of clinical phenotype in patients whose mutations appear to cause only partial loss of function is that in vitro expression systems have been used that may not reflect in vivo conditions. Supporting this possibility are recent findings concerning mutant R881C (Toye et al 2006). When expressed in *Xenopus* oocytes, this mutant was noted to retain about 40% of wild-type transport activity, a defect entirely attributable to reduced surface expression. However, when expressed in polarized MDCK renal epithelial cells, this mutant was completely retained in the endoplasmic reticulum and was not at all expressed on the basolateral membrane (Toye et al 2006). Thus, the severe clinical phenotype of pRTA in patients with R881C mutation does indeed correlate with defective polarized expression when renal epithelial cells are used as the test expression system. Future studies using proximal tubule cells for expression of mutant transporters would be useful to establish genotype–phenotype relationships. Severe metabolic acidosis has also been observed in mice with targeted disruption of the NBC1 (Slc4a4) gene (Gawenis et al 2007). Because these mice died before weaning, detailed renal physiological studies were not performed. But the observation of severe metabolic acidosis in the absence of diarrhea is entirely consistent with pRTA, as observed in humans. Additional features of the human syndrome found in the null mice included growth retardation and abnormal dentition. Interestingly, although an abnormal gastrointestinal phenotype has not been reported in humans with pRTA due to NBC1 mutations, null mice suffered from intestinal obstruction attributed to a defect in NBC1-mediated anion and fluid secretion in the colon (Gawenis et al 2007).

CAII

An autosomal recessive syndrome of renal tubular acidosis, osteopetrosis and cerebral calcification results from inherited deficiency of carbonic anhydrase II (Sly et al 1983, Sly & Hu 1995). Osteopetrosis has been attributed to a defect in acid secretion and bone resorption by osteoclasts, which express high levels of CAII (Sly et al 1983). The renal tubular acidosis is most typically of the distal type, with inability to appropriately acidify urine pH in the face of severe acidemia (Nagai et al 1997). However, a proximal defect has also been identified reflected by bicarbonate wasting when plasma bicarbonate is increased by base administration

(Bourke et al 1981, Nagai et al 1997). A mixed picture of distal (type I) and proximal (type II) renal tubular acidosis is referred to as type III renal tubular acidosis (Rodriguez Soriano 2002).

The defect in proximal tubular bicarbonate reabsorption in CAII deficiency is a consequence of the role described earlier for intracellular carbonic anhydrase in providing H^+ and HCO_3^- from water and CO_2, thereby facilitating apical H^+ extrusion and basolateral HCO_3^- exit. Moreover, as also discussed above, the regulated association of CAII with an internal domain of NBC1 has been described, facilitating production of HCO_3^- in the immediate vicinity of the transporter as part of a CO_2/HCO_3^- 'metabolon' (Pushkin et al 2004). Metabolic acidosis due to a defect in renal acidification has also been observed in mice deficient for expression of CAII (Lewis et al 1988). Interestingly, other clinical features observed in human patients with CAII deficiency, such as osteopetrosis and cerebral calcification, were absent in these mice (Lewis et al 1988).

NHE3 Mutations

Mutations in apical membrane Na^+-H^+ exchanger NHE3 have so far not been identified as a cause of pRTA in humans. However, NHE3 null mice have a profound defect in proximal tubule HCO_3^- reabsorption associated with significant metabolic acidosis (Schultheis et al 1998). Proximal tubule HCO_3^- reabsorption capacity is reduced by approximately 60% in NHE3 null mice, with much of the remainder mediated by bafilomycin-sensitive vacuolar H^+-ATPase activity (Wang et al 1999). Interestingly, despite this profound defect in proximal tubule bicarbonate reabsorption, the metabolic acidosis in NHE3 null mice is surprisingly mild (plasma HCO_3^- 20 mEq/l in null vs. 26 mEq/l in wild-type).

Several mechanisms of compensation have been described that limit the impact of this defect on systemic acid–base balance. First, there is a large decrement in GFR in NHE3 null mice due to activation of tubuloglomerular feedback (Lorenz et al 1999), thereby greatly reducing the filtered load of HCO_3^-. Second, NHE2-mediated HCO_3^- reabsorption in the distal convoluted tubule plays an important role in preventing large urinary loss of bicarbonate when a high luminal bicarbonate load is presented to the distal tubule, as in NHE3 null mice (Bailey et al 2004). Third, in NHE3 null mice there is increased HCO_3^- reabsorption capacity in the cortical and outer medullary collecting duct segments (Nakamura et al 1999).

As mentioned earlier, NHE8 has recently been identified as an additional Na^+-H^+ exchanger isoform expressed on the apical membrane of proximal tubule cells (Goyal et al 2005). However, to date no studies have been reported characterizing the phenotype of NHE8 mutations in humans, or NHE8 null mice.

NHE1 is expressed on the basolateral membrane of proximal tubule cells (Biemesderfer et al 1992), where it would not be expected to directly facilitate proximal tubule HCO_3 reabsorption. But studies of HCO_3 reabsorption in medullary thick ascending limb segments have demonstrated an important functional interaction by which basolateral NHE1 regulates apical NHE3 activity via changes in actin cytoskeleton (Watts et al 2005). Consequently, NHE1 null mice have a defect in NHE3-mediated HCO_3^- absorption in medullary thick ascending limb segments (Good et al 2004). Whether a similar functional interaction between NHE1 and NHE3 occurs in the proximal tubule is not known. No significant abnormalities in plasma pH or HCO_3 were detected in NHE1 null mice (Bell et al 1999).

K^+ Channel Mutations

As discussed earlier, base exit across the basolateral membrane via NBC1 is strongly dependent on the inside-negative potential as a driving force, which itself is dependent on the basolateral K^+ conductance. TASK2 is a pH-sensitive expressed in the proximal tubule (Warth et al 2004) and is a member of a family of K^+ channels with two pore-forming domains. Studies on primary cultures of proximal tubule cells from wild-type and TASK2 null mice have shown that TASK2 is a basolateral K^+ channel activated by the alkalinization of the basolateral extracellular space pH resulting from NBC activity (Warth et al 2004). These findings are consistent with the concept that TASK2 mediates an outward K^+ current closely coupled to NBC activity across the basolateral membrane of proximal tubule cells.

Supporting a role for TASK2 in renal acid–base homeostasis is the finding that TASK2 null mice have a modest but significant metabolic acidosis (venous pH and HCO_3 7.33 and 24.7 in wild-type mice vs. 7.30 and 21.9 in TASK2 null mice, respectively) (Warth et al 2004). Given the evidence for the role of TASK2 in maintaining the basolateral membrane potential during HCO_3 reabsorption in cultured proximal tubule cells, it is reasonable to attribute the metabolic acidosis in TASK2 null mice to proximal renal tubular acidosis. However, since urine pH was significantly higher in TASK2 null mice than in wild-type mice (6.77 vs. 6.51, respectively), a distal defect may have also been present. To date, urinary acidification defects due to TASK2 mutations have not been identified in patients.

TREATMENT

In principle, the metabolic acidosis of proximal RTA could be treated by vigorous base administration. However, as discussed earlier and summarized in Figure 11.7, normalization of the plasma HCO_3^- in a patient with severe pRTA would result in excretion of large quantities of bicarbonate in the urine since the filtered load of HCO_3^- would greatly exceed the HCO_3^- reabsorption capacity of the proximal

tubule. Consequently, huge quantities of base administration would be required to maintain a normal plasma HCO_3^- and pH in such a patient with severe pRTA. This contrasts with the situation in dRTA, in which there is failure to excrete the normal endogenous acid load. These patients require an amount of base administration equal only to endogenous acid production because proximal HCO_3 reabsorption capacity is not defective and there is no urinary HCO_3^- wasting when plasma HCO_3^- is normalized. Whereas it may be possible to maintain normal plasma HCO_3^- by oral base administration in patients with mild pRTA, patients with severe pRTA may require quantities of base replacement that are not tolerated orally. As described earlier, the apparent threshold for HCO_3^- reabsorption in the proximal tubule is affected by the state of extracellular fluid volume (Purkerson et al 1969, Slatopolsky et al 1970).

Accordingly, inducing a state of volume contraction by administration of a thiazide diuretic can reduce HCO_3^- wasting in pRTA and reduce the amount of alkali required to maintain an increased plasma HCO_3^- concentration (Rampini et al 1968, Rodriguez Soriano 2002). It should be noted, however, that thiazides may aggravate K^+ depletion in pRTA so enhanced K^+ replacement may be required. Finally, it should be emphasized that although full normalization of plasma HCO_3^- is not usually possible, intensive alkali therapy has been found to improve growth in children even in the absence of full correction of acidemia (Shiohara et al 2000).

References

Abuladze N, Lee I, Newman D, Hwang J, Pushkin A, Kurtz I. Axial heterogeneity of sodium-bicarbonate cotransporter expression in the rabbit proximal tubule. Am. J. Physiol. 1998; 274: F628–33.

Adler S, Zett B, Anderson B. Renal citrate in the potassium-deficient rat: role of potassium and chloride ions. J. Lab. Clin. Med. 1974; 84: 307–16.

Alpern RJ. Cell mechanisms of proximal tubule acidification. Physiol. Rev. 1990; 70: 79–114.

Alvarez BV, Loiselle FB, Supuran CT, Schwartz GJ, Casey JR. Direct extracellular interaction between carbonic anhydrase IV and the human NBC1 sodium/bicarbonate co-transporter. Biochemistry 2003; 42: 12321–9.

Amemiya M, Loffing J, Lotscher M, Kaissling B, Alpern RJ, Moe OW. Expression of NHE-3 in the apical membrane of rat renal proximal tubule and thick ascending limb. Kidney Int. 1995; 48: 1206–15.

Amorim JB, Bailey MA, Musa-Aziz R, Giebisch G, Malnic G. Role of luminal anion and pH in distal tubule potassium secretion. Am. J. Physiol. Renal Physiol. 2003; 284: F381–8.

Aronson PS. Mechanisms of active H^+ secretion in the proximal tubule. Am. J. Physiol. 1983; 245: F647–59.

Aruga S, Wehrli S, Kaissling B, Moe OW, Preisig PA, Pajor AM, Alpern RJ. Chronic metabolic acidosis increases NaDC-1 mRNA and protein abundance in rat kidney. Kidney Int. 2000; 58: 206–15.

Bailey MA, Giebisch G, Abbiati T, et al. NHE2-mediated bicarbonate reabsorption in the distal tubule of NHE3 null mice. J. Physiol. 2004; 561: 765–75.

Beck JS, Hurst AM, Lapointe JY, Laprade R. Regulation of basolateral K channels in proximal tubule studied during continuous microperfusion. Am. J. Physiol. 1993; 264: F496–501.

Becker AM, Zhang J, Goyal S, et al. Ontogeny of NHE8 in the rat proximal tubule. Am. J. Physiol. Renal Physiol. 2007; 293: F255–61.

Bell SM, Schreiner CM, Schultheis PJ, et al. Targeted disruption of the murine Nhe1 locus induces ataxia, growth retardation, and seizures. Am. J. Physiol. 1999; 276: C788–95.

Biagi B, Kubota T, Sohtell M, Giebisch G. Intracellular potentials in rabbit proximal tubules perfused in vitro. Am. J. Physiol. 1981; 240: F200–10.

Biemesderfer D, Pizzonia J, Abu-Alfa A, et al. NHE3: a Na^+/H^+ exchanger isoform of renal brush border. Am. J. Physiol. 1993; 265: F736–42.

Biemesderfer D, Reilly RF, Exner M, Igarashi P, Aronson PS. Immunocytochemical characterization of Na^+-H^+ exchanger isoform NHE-1 in rabbit kidney. Am. J. Physiol. 1992; 263: F833–40.

Biemesderfer D, Rutherford PA, Nagy T, Pizzonia JH, Abu-Alfa AK, Aronson PS. Monoclonal antibodies for high-resolution localization of NHE3 in adult and neonatal rat kidney. Am. J. Physiol. 1997; 273: F289–99.

Boron WF. Acid–base transport by the renal proximal tubule. J. Am. Soc. Nephrol. 2006; 17: 2368–82.

Bourke E, Delaney VB, Mosawi M, Reavey P, Weston M. Renal tubular acidosis and osteopetrosis in siblings. Nephron 1981; 28: 268–72.

Brenes LG, Brenes JN, Hernandez MM. Familial proximal renal tubular acidosis. A distinct clinical entity. Am. J. Med. 1977; 63: 244–52.

Brenes LG, Sanchez MI. Impaired urinary ammonium excretion in patients with isolated proximal renal tubular acidosis. J. Am. Soc. Nephrol. 1993; 4: 1073–8.

Brennan S, Hering-Smith K, Hamm LL. Effect of pH on citrate reabsorption in the proximal convoluted tubule. Am. J. Physiol. 1988; 255: F301–6.

Brett CL, Donowitz M, Rao R. Evolutionary origins of eukaryotic sodium/proton exchangers. Am. J. Physiol. 2005; 288: C223–9.

Brochiero E, Wallendorf B, Gagnon D, Laprade R, Lapointe JY. Cloning of rabbit Kir6.1, SUR2A, and SUR2B: possible candidates for a renal K(ATP) channel. Am. J. Physiol. 2002; 282: F289–300.

Brown D, Hirsch S, Gluck S. Localization of a proton pumping ATPase in rat kidney. J. Clin. Invest. 1988; 82: 2114–26.

Brown JC, Packer RK, Knepper MA. Role of organic anions in renal response to dietary acid and base loads. Am. J. Physiol. 1989; 257: F170–6.

Choi JY, Shah M, Lee MG, et al. Novel amiloride-sensitive sodium-dependent proton secretion in the mouse proximal convoluted tubule. J. Clin. Invest. 2000; 105: 1141–6.

De Yoreo JJ, Qiu SR, Hoyer JR. Molecular modulation of calcium oxalate crystallization. Am. J. Physiol. Renal Physiol. 2006; 291: F1123–31.

Demirci FY, Chang MH, Mah TS, Romero MF, Gorin MB. Proximal renal tubular acidosis and ocular pathology: a novel missense mutation in the gene (SLC4A4) for sodium bicarbonate cotransporter protein (NBCe1). Mol. Vis. 2006; 12: 324–30.

DuBose TD Jr., Good DW, Hamm LL, Wall SM. Ammonium transport in the kidney: new physiological concepts and their clinical implications. J. Am. Soc. Nephrol. 1991; 1: 1193–203.

Gawenis LR, Bradford EM, Prasad V, et al. Colonic anion secretory defects and metabolic acidosis in mice lacking the NBC1 Na^+/HCO_3^- cotransporter. J. Biol. Chem. 2007; 282: 9042–52.

Giebisch G. Renal potassium transport: mechanisms and regulation. Am. J. Physiol. 1998; 274: F817–33.

Good DW, Watts BA 3rd, George T, Meyer JW, Shull GE. Transepithelial HCO_3^- absorption is defective in renal thick ascending limbs from Na^+/H^+ exchanger NHE1 null mutant mice. Am. J. Physiol. Renal Physiol. 2004; 287: F1244–9.

Gottschalk CW, Lassiter WE, Mylle M. Localization of urine acidification in the mammalian kidney. Am. J. Physiol. 1960; 198: 581–5.

Goyal S, Mentone S, Aronson PS. Immunolocalization of NHE8 in rat kidney. Am. J. Physiol. Renal Physiol. 2005; 288: F530–8.

Goyal S, Vanden Heuvel G, Aronson PS. Renal expression of novel Na^+/H^+ exchanger isoform NHE8. Am. J. Physiol. Renal Physiol. 2003; 284: F467–73.

Gross E, Hawkins K, Abuladze N, et al. The stoichiometry of the electrogenic sodium bicarbonate cotransporter NBC1 is cell-type dependent. J. Physiol. 2001a; 531: 597–603.

Gross E, Hawkins K, Pushkin A, et al. Phosphorylation of Ser(982) in the sodium bicarbonate cotransporter kNBC1 shifts the HCO_3^-:Na^+ stoichiometry from 3:1 to 2:1 in murine proximal tubule cells. J. Physiol. 2001b; 37: 659–65.

Gross E, Pushkin A, Abuladze N, Fedotoff O, Kurtz I. Regulation of the sodium bicarbonate cotransporter kNBC1 function: role of Asp(986), Asp(988) and kNBC1-carbonic anhydrase II binding. J. Physiol. 2002; 544: 679–85.

Halperin ML, Kamel KS, Ethier JH, Magner PO. What is the underlying defect in patients with isolated, proximal renal tubular acidosis? Am. J. Nephrol. 1989; 9: 265–8.

Hamm LL, Hering-Smith KS. Pathophysiology of hypocitraturic nephrolithiasis. Endocrinology and Metabolism Clinics of North America 2002; 31: 885–93, viii.

Hebert SC, Desir G, Giebisch G, Wang W. Molecular diversity and regulation of renal potassium channels. Physiol. Rev. 2005; 85: 319–71.

Igarashi T, Inatomi J, Sekine T, et al. Mutations in SLC4A4 cause permanent isolated proximal renal tubular acidosis with ocular abnormalities. Nat. Genet. 1999; 23: 264–6.

Igarashi T, Inatomi J, Sekine T, et al. Novel nonsense mutation in the Na^+/HCO_3^- cotransporter gene (SLC4A4) in a patient with permanent isolated proximal renal tubular acidosis and bilateral glaucoma. J. Am. Soc. Nephrol. 2001; 12: 713–18.

Igarashi T, Sekine T, Inatomi J, Seki G. Unraveling the molecular pathogenesis of isolated proximal renal tubular acidosis. J. Am. Soc. Nephrol. 2002; 13: 2171–7.

Karim Z, Szutkowska M, Vernimmen C, Bichara M. Renal handling of NH_3/NH_4^+: recent concepts. Nephron 2005; 101: 77–81.

Kinsella JL, Aronson PS. Interaction of NH_4^+ and Li^+ with the renal microvillus membrane Na^+-H^+ exchanger. Am. J. Physiol. 1981; 241: C220–6.

Kondo Y, Fromter E. Axial heterogeneity of sodium bicarbonate cotransport in proximal straight tubule of rabbit kidney. Pflugers Arch. 1987; 410: 481–6.

Kondo Y, Fromter E. Evidence of chloride/bicarbonate exchange mediating bicarbonate efflux from S3 segments of rabbit renal proximal tubule. Pflugers Arch. – Eur. J. Physiol. 1990; 415: 726–33.

Lemann J Jr., Adams ND, Wilz DR, Brenes LG. Acid and mineral balances and bone in familial proximal renal tubular acidosis. Kidney Int. 2000; 58: 1267–77.

Lemann J Jr., Bushinsky DA, Hamm LL. Bone buffering of acid and base in humans. Am. J. Physiol. 2003; 285: F811–32.

Lennon EJ, Lemann J Jr., Litzow JR. The effects of diet and stool composition on the net external acid balance of normal subjects. J. Clin. Invest. 1966; 45: 1601–7.

Lesage F, Lazdunski M. Molecular and functional properties of two-pore-domain potassium channels. Am. J. Physiol. Renal Physiol. 2000; 279: F793–801.

Levi M, McDonald LA, Preisig PA, Alpern RJ. Chronic K depletion stimulates rat renal brush-border membrane Na-citrate cotransporter. Am. J. Physiol. 1991; 261: F767–73.

Lewis SE, Erickson RP, Barnett LB, Venta PJ, Tashian RE. N-ethyl-N-nitrosourea-induced null mutation at the mouse Car-2 locus: an animal model for human carbonic anhydrase II deficiency syndrome. Proc. Natl Acad. Sci. USA 1988; 85: 1962–6.

Li X, Alvarez B, Casey JR, Reithmeier RA, Fliegel L. Carbonic anhydrase II binds to and enhances activity of the Na^+/H^+ exchanger. J. Biol. Chem. 2002; 277: 36085–91.

Lorenz JN, Schultheis PJ, Traynor T, Shull GE, Schnermann J. Micropuncture analysis of single-nephron function in NHE3-deficient mice. Am. J. Physiol. 1999; 277: F447–53.

Lu J, Daly CM, Parker MD, et al. Effect of human carbonic anhydrase II on the activity of the human electrogenic Na/HCO_3 cotransporter NBCe1-A in *Xenopus* oocytes. J. Biol. Chem. 2006; 281: 19241–50.

Lucci MS, Tinker JP, Weiner IM, DuBose TD Jr. Function of proximal tubule carbonic anhydrase defined by selective inhibition. Am. J. Physiol. 1983; 245: F443–9.

Malnic G, De Mello Aires M, Giebisch G. Micropuncture study of renal tubular hydrogen ion transport in the rat. Am. J. Physiol. 1972; 222: 147–58.

Markovich D, Aronson PS. Specificity and regulation of renal sulfate transporters. Annu. Rev. Physiol. 2007; 69: 361–75.

Melnick JZ, Srere PA, Elshourbagy NA, Moe OW, Preisig PA, Alpern RJ. Adenosine triphosphate citrate lyase mediates hypocitraturia in rats. J. Clin. Invest. 1996; 98: 2381–7.

Mori Y, Machida T, Miyakawa S, Bomsztyk K. Effects of amiloride on distal renal tubule sodium and calcium absorption: dependence on luminal pH. Pharmacol. Toxicol. 1992; 70: 201–4.

Morris RC Jr. Renal tubular acidosis. Mechanisms, classification and implications. N. Engl. J. Med, 1969; 281: 1405–13.

Morris RC Jr., Sebastian A, McSherry E. Renal acidosis. Kidney Int. 1972; 1: 322–40.

Mount DB, Romero MF. The SLC26 gene family of multifunctional anion exchangers. Pflugers Arch. 2004; 447: 710–21.

Muller-Berger S, Ducoudret O, Diakov A, Fromter E. The renal $Na-HCO_3^-$ cotransporter expressed in *Xenopus laevis* oocytes: change in stoichiometry in response to elevation of cytosolic Ca_2^+ concentration. Pflugers Arch. 2001; 442: 718–28.

Nagai R, Kooh SW, Balfe JW, Fenton T, Halperin ML. Renal tubular acidosis and osteopetrosis with carbonic anhydrase II deficiency: pathogenesis of impaired acidification. Pediatr. Nephrol. 1997; 11: 633–6.

Nagami GT. Luminal secretion of ammonia in the mouse proximal tubule perfused in vitro. J. Clin. Invest. 1988; 81: 159–64.

Nagami NT. Renal Ammonia Production and Excretion. In: Seldin DW, Giebisch G, eds. The Kidney. Philadelphia: Lippincott, 2000: pp. 1995–2013.

Nakamura S, Amlal H, Schultheis PJ, Galla JH, Shull GE, Soleimani M. HCO_3^- reabsorption in renal collecting duct of NHE-3-deficient mouse: a compensatory response. Am. J. Physiol 1999; 276: F914–21.

Nash MA, Torrado AD, Greifer I, Spitzer A, Edelmann CM Jr. Renal tubular acidosis in infants and children. Clinical course, response to treatment, and prognosis. J. Pediatr. 1972; 80: 738–48.

Orlowski J, Grinstein S. Diversity of the mammalian sodium/proton exchanger SLC9 gene family. Pflugers Arch. 2004; 447: 549–65.

Pajor AM. Sodium-coupled transporters for Krebs cycle intermediates. Annu. Rev. Physiol. 1999; 61: 663–82.

Pajor AM. Molecular properties of sodium/dicarboxylate cotransporters. J. Membrane Biol. 2000; 175: 1–8.

Pajor AM. Molecular properties of the SLC13 family of dicarboxylate and sulfate transporters. Pflugers Arch. 2006; 451: 597–605.

Piermarini PM, Kim EY, Boron WF. Evidence against a direct interaction between intracellular carbonic anhydrase II and pure C-terminal domains of SLC4 bicarbonate transporters. J. Biol. Chem. 2007; 282: 1409–21.

Purkerson JM, Kittelberger AM, Schwartz GJ. Basolateral carbonic anhydrase IV in the proximal tubule is a glycosylphosphatidylinositol-anchored protein. Kidney Int. 2007; 71: 407–16.

Purkerson JM, Schwartz GJ. The role of carbonic anhydrases in renal physiology. Kidney Int. 2007; 71: 103–15.

Purkerson ML, Lubowitz H, White RW, Bricker NS. On the influence of extracellular fluid volume expansion on bicarbonate reabsorption in the rat. J. Clin. Invest. 1969; 48: 1754–60.

Pushkin A, Abuladze N, Gross E, et al. Molecular mechanism of kNBC1-carbonic anhydrase II interaction in proximal tubule cells. J. Physiol. 2004; 559: 55–65.

Rampini S, Fanconi A, Illig R, Prader A. Effect of hydrochlorothiazide on proximal renal tubular acidosis in a patient with idiopathic "De Toni-Debre-Fanconi Syndrome". Helv. Paediatr. Acta 1968; 23: 13–21.

Relman AS, Lennon EJ, Lemann J Jr. Endogenous production of fixed acid and the measurement of the net balance of acid in normal subjects. J. Clin. Invest. 1961; 40: 1621–30.

Rodriguez Soriano J. Renal tubular acidosis: the clinical entity. J. Am. Soc. Nephrol. 2002; 13: 2160–70.

Rodriguez Soriano J, Boichis H, Stark H, Edelmann CM Jr. Proximal renal tubular acidosis. A defect in bicarbonate reabsorption with normal urinary acidification. Pediatr. Res. 1967; 1: 81–98.

Rodriguez Soriano J, Edelmann CM Jr. Renal tubular acidosis. Annu. Rev. Med. 1969; 20: 363–82.

Romero MF, Fulton CM, Boron WF. The SLC4 family of HCO_3^- transporters. Pflugers Arch. 2004; 447: 495–509.

Sasaki S, Marumo F. Effects of carbonic anhydrase inhibitors on basolateral base transport of rabbit proximal straight tubule. Am. J. Physiol. 1989; 257: F947–52.

Schmitt BM, Biemesderfer D, Romero MF, Boulpaep EL, Boron WF. Immunolocalization of the electrogenic Na^+-HCO_3^- cotransporter in mammalian and amphibian kidney. Am. J. Physiol. 1999; 276: F27–38.

Schultheis PJ, Clarke LL, Meneton P, et al. Renal and intestinal absorptive defects in mice lacking the NHE3 Na^+/H^+ exchanger. Nat. Genet. 1998; 19: 282–5.

Sebastian A, Frassetto LA, Sellmeyer DE, Merriam RL, Morris RC Jr. Estimation of the net acid load of the diet of ancestral preagricultural *Homo sapiens* and their hominid ancestors. Am. J. Clin. Nutr. 2002; 76: 1308–16.

Sebastian A, McSherry E, Morris RC Jr. On the mechanism of renal potassium wasting in renal tubular acidosis associated with the Fanconi syndrome (type 2 RTA). J. Clin. Invest. 1971a; 50: 231–43.

Sebastian A, McSherry E, Morris RC Jr. Renal potassium wasting in renal tubular acidosis (RTA): its occurrence in types 1 and 2 RTA despite sustained correction of systemic acidosis. J. Clin. Invest. 1971b; 50: 667–78.

Seki G, Coppola S, Fromter E. The Na^+-HCO_3^- cotransporter operates with a coupling ratio of 2 HCO_3^- to 1 Na^+ in isolated rabbit renal proximal tubule. Pflugers Arch. 1993; 425: 409–16.

Shiohara M, Igarashi T, Mori T, Komiyama A. Genetic and long-term data on a patient with permanent isolated proximal renal tubular acidosis. Eur. J. Pediatr. 2000; 159: 892–4.

Slatopolsky E, Hoffsten P, Purkerson M, Bricker NS. On the influence of extracellular fluid volume expansion and of uremia on bicarbonate reabsorption in man. J. Clin. Invest. 1970; 49: 988–98.

Sly WS, Hewett-Emmett D, Whyte MP, Yu YS, Tashian RE. Carbonic anhydrase II deficiency identified as the primary defect in the autosomal recessive syndrome of osteopetrosis with renal tubular acidosis and cerebral calcification. Proc. Natl Acad. Sci. USA 1983; 80: 2752–6.

Sly WS, Hu PY. Human carbonic anhydrases and carbonic anhydrase deficiencies. Annu. Rev. Biochem. 1995; 64: 375–401.

Soleimani M, Aronson PS. Ionic mechanism of Na^+-HCO_3^- cotransport in rabbit renal basolateral membrane vesicles. J. Biol. Chem. 1989; 264: 18302–8.

Soleimani M, Grassi SM, Aronson PS. Stoichiometry of Na^+-HCO_3^- cotransport in basolateral membrane vesicles isolated from rabbit renal cortex. J. Clin. Invest. 1987; 79: 1276–80.

Sutton RA, Wong NL, Dirks JH. Effects of metabolic acidosis and alkalosis on sodium and calcium transport in the dog kidney. Kidney Int. 1979; 15: 520–33.

Topala CN, Bindels RJ, Hoenderop JG. Regulation of the epithelial calcium channel TRPV5 by extracellular factors. Curr. Opin. Nephrol. Hypertens. 2007; 16: 319–24.

Toye AM, Parker MD, Daly CM, et al. The human NBCe1-A mutant R881C, associated with proximal renal tubular acidosis, retains function but is mistargeted in polarized renal epithelia. Am. J. Physiol. Cell. Physiol. 2006; 291: C788–801.

Tsuruoka S, Swenson ER, Petrovic S, Fujimura A, Schwartz GJ. Role of basolateral carbonic anhydrase in proximal tubular fluid and bicarbonate absorption. Am. J. Physiol. 2001; 280: F146–54.

Unwin RJ, Capasso G, Shirley DG. An overview of divalent cation and citrate handling by the kidney. Nephron 2004; 98: 15–20.

Usui T, Hara M, Satoh H, et al. Molecular basis of ocular abnormalities associated with proximal renal tubular acidosis. J. Clin. Invest. 2001; 108: 107–15.

Vallon V, Schwark JR, Richter K, Hropot M. Role of Na^+/H^+ exchanger NHE3 in nephron function: micropuncture studies with S3226, an inhibitor of NHE3. Am. J. Physiol. Renal Physiol. 2000; 278: F375–9.

Velazquez H, Ellison DH, Wright FS. Chloride-dependent potassium secretion in early and late renal distal tubules. Am. J. Physiol. 1987; 253: F555–62.

Wang T, Hropot M, Aronson PS, Giebisch G. Role of NHE isoforms in mediating bicarbonate reabsorption along the nephron. Am. J. Physiol. Renal Physiol. 2001; 281: F1117–22.

Wang T, Yang CL, Abbiati T, et al. Mechanism of proximal tubule bicarbonate absorption in NHE3 null mice. Am. J. Physiol. 1999; 277: F298–302.

Warth R, Barriere H, Meneton P, et al. Proximal renal tubular acidosis in TASK2 K^+ channel-deficient mice reveals a mechanism for stabilizing bicarbonate transport. Proc. Natl Acad. Sci. USA 2004; 101: 8215–20.

Watts BA 3rd, George T, Good DW. The basolateral NHE1 Na^+/H^+ exchanger regulates transepithelial HCO_3^- absorption through actin cytoskeleton remodeling in renal thick ascending limb. J. Biol. Chem. 2005; 280: 11439–47.

Weiner ID, Hamm LL. Molecular mechanisms of renal ammonia transport. Annu. Rev. Physiol. 2007; 69: 317–40.

Welling PA, O'Neil RG. Ionic conductive properties of rabbit proximal straight tubule basolateral membrane. Am. J. Physiol. 1990; 58: F940–50.

Wu MS, Biemesderfer D, Giebisch G, Aronson PS. Role of NHE3 in mediating renal brush border Na^+-H^+ exchange. Adaptation to metabolic acidosis. J. Biol. Chem. 1996; 271: 32749–52.

Yeh BI, Sun TJ, Lee JZ, Chen HH, Huang CL. Mechanism and molecular determinant for regulation of rabbit transient receptor potential type 5 (TRPV5) channel by extracellular pH. J. Biol. Chem. 2003; 278: 51044–52.

Yoshitomi K, Burckhardt BC, Fromter E. Rheogenic sodium bicarbonate cotransport in the peritubular cell membrane of rat renal proximal tubule. Pflugers Arch. 1985; 405: 360–6.

Yoshitomi K, Fromter E. How big is the electrochemical potential difference of Na^+ across rat renal proximal tubular cell membranes in vivo? Pflugers Arch. 1985; 405(Suppl 1): S121–6.

Zimolo Z, Montrose MH, Murer H. H^+ extrusion by an apical vacuolar-type H^+-ATPase in rat renal proximal tubules. J. Membrane Biol. 1992; 126: 19–26.

CHAPTER 12

Dent's Disease

STEVEN J. SCHEINMAN

INTRODUCTION

Dent's disease (OMIM 300009) is an inherited disorder whose clinical features result from renal tubular epithelial dysfunction. The full constellation of findings include proximal tubular solute loss, particularly LMW proteinuria; hypercalciuria; nephrocalcinosis and kidney stones; and progressive renal insufficiency. In addition, some patients manifest rickets. Inheritance is X-linked recessive, with a wide range of severity of phenotype both within and among families.

The first full descriptions of the syndrome were published by separate groups in 1991 and 1994 (Frymoyer et al 1991, Wrong et al 1994). Two boys with features consistent with this phenotype had been reported much earlier by Dent and Friedman (1964), and 30 years later mutation analysis confirmed that one of them had the same disease (Lloyd et al 1996). A genetic mapping approach led to identification of the gene *CLCN5* that is now known to be mutated in over 100 families with Dent's disease. The causal role of *CLCN5* in Dent's disease is supported by observations in knockout mice, in which targeted disruption of the mouse gene reproduces the clinical findings of the human disease. Discovery of this gene has allowed for extensive study of its protein product, CLC-5, a member of the CLC family of chloride channels, and its role in renal physiology.

More recently genetic heterogeneity has been established in Dent's disease, in that mutations in *CLCN5* do not account for all patients (Hoopes et al 2004). Identification of *OCRL1* as a second gene for Dent's disease has led to a reconsideration of the diagnostic criteria for Lowe's syndrome. Of patients with Dent's disease but lacking mutation in *CLCN5*, *OCRL1* is mutated in a proportion, though not all, of the remainder (Hoopes et al 2005).

PHYSIOLOGY OF CLC CHLORIDE CHANNELS

CLC-5 is a member of the CLC family that includes several proteins that are mutated in important human diseases

(Figure 12.1). CLC-1, the major chloride channel in mammalian muscle, is mutated in congenital myotonia. CLC-Kb, the basolateral chloride channel in the thick ascending limb of Henle, is mutated in the largest subset of patients with Bartter's syndrome (its associated protein barttin is mutated in another subset of Bartter's) (Chapter 13). CLC-7, which is expressed broadly across tissues, is mutated in congenital infantile osteopetrosis. The branch of this family that contains CLC-1 and CLC-Kb encodes proteins that are essential for chloride transport across plasma membranes. The branch that contains CLC-5 and CLC-7 encodes proteins that are crucial to the acidification by the vacuolar proton-ATPase of subcellular organelles, including endosomes (CLC-5) and the resorption lacunae at the ruffled surface of osteoclasts (CLC-7). Members of this branch appear to be chloride-proton antiporters rather than channels (Jentsch 2005, Jentsch et al 2005a).

In the human nephron, CLC-5 is expressed primarily in the proximal tubule, but also in the α-intercalated cells of the collecting duct and to a lesser extent in the thick ascending limb (Devuyst et al 1999). (Low levels of expression are found in some extrarenal tissues such as liver, brain, and colon (Steinmeyer et al 1995, Piwon et al 2000, Vandewalle et al 2001).) Much more is known about the expression and function of CLC-5 in proximal tubule than in other nephron segments.

This understanding has been particularly enriched by observations in CLC-5 knockout mice. The major manifestations of Dent's disease have been reproduced in three murine models in which expression of the ClC-5 channel was either reduced by antisense (Luyckx et al 1999) or eliminated by targeted disruption (Piwon et al 2000, Wang et al 2000, Silva et al 2003). In the aggregate, these mice exhibit LMW proteinuria, aminoaciduria, glycosuria, renal phosphate wasting, polyuria, renal failure, hypercalciuria and nephrocalcinosis.

In the proximal tubule, CLC-5 is expressed in subapical endosomes together with the vacuolar proton-ATPase and with markers of early endosomes such as Rab 5a, as well as with internalized proteins (Gunther et al 1998,

Genetic Diseases of the Kidney
ISBN: 978-0-12-449851-8

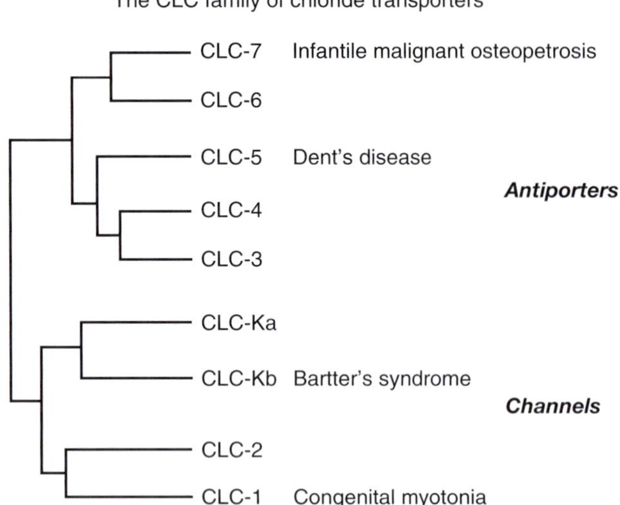

FIGURE 12.1 The CLC family of chloride transporters. CLC-1, -2, -Ka and -Kb are expressed on plasma membranes and function as voltage-gated chloride channels. CLC-3, -4, -5, -6 and -7 are expressed on subcellular organelles; CLC-4 and CLC-5 (and possibly others) function as Cl^-/H^+ exchangers

Luyckx et al 1998, Devuyst et al 1999). This localization is consistent with a role for the CLC-5 protein in endosomal acidification, and this acidification is impaired in endosomes of cells from CLC-5 knockout mice (Gunther et al 2003, Hara-Chikuma et al 2005). Both fluid-phase and receptor-mediated reabsorption of proteins are impaired in CLC-5 knockout mice (Piwon et al 2000, Wang et al 2000). Acidification of the lumen of these endosomes is an essential step in the processing of reabsorbed proteins. Activation of the chloride channel was presumed to facilitate this acidification by providing a mechanism to shunt the positive charge generated by the action of the proton pump (Figure 12.2).

This hypothesis was consistent with the view of CLC-5 as a chloride channel but has been difficult to reconcile with electrophysiological observations. Measurement of chloride channel function of CLC-5 expressed in oocytes established that the current is strongly outwardly-rectifying. No single-channel recordings of CLC-5 activity have been recorded. Furthermore, the chloride currents are triggered only at voltages more positive than +20 mV, a condition not likely to occur physiologically in vesicles, and are inhibited by 'extracellular' (i.e. intravesicular) acidic pH (Jentsch et al 2005b). Thus, if the properties of CLC-5 when expressed in oocytes resemble those in the endosomal membrane, a CLC-5 channel would be more likely to mediate chloride exit than entry.

Recent observations from two independent laboratories indicate that CLC-5, and the closely homologous CLC-4, function not as chloride channels but rather as Cl^-/H^+ exchangers. In the presence of a chloride concentration

gradient, CLC-5 is able to drive protons against an electrochemical gradient, a characteristic of active transport rather than channel function (Picollo & Pusch 2005, Scheel et al 2005). It would appear that the current previously associated with channel function arises from the electrogenic exchange of a cation (H^+) for an anion (Cl^-). While the branch of the CLC family that includes the plasma membrane CLCs (CLC-1, -2, -Ka and -Kb) clearly function as chloride channels (Jentsch et al 2005a, Picollo & Pusch, 2005), CLC-4 and CLC-5 now appear to function as antiporters rather than channels (as may other members of that branch). As a Cl^-/H^+ antiporter CLC-5 could effectively dissipate the lumen-positive charge generated by the proton-ATPase, since the proton exit coupled with chloride entry would represent a net charge transfer of +2 out of the endosome, though at some additional energy cost for acidification (Figure 12.2). As a Cl^-/H^+ exchanger, CLC-5 would introduce an outward leak of protons that could be a brake limiting acidification (Picollo & Pusch 2005).

Acidification of the nascent endosome as it first buds off from the apical surface does not involve the proton-ATPase (Gekle et al 1997). Thus, if the role of CLC-5 in endosomal function were as a chloride channel it would not be needed for this early acidification. As a Cl^-/H^+ exchanger, however, CLC-5 could acidify the nascent endosome directly, since the initially high luminal chloride concentration would drive proton entry across the exchanger (Scheel et al 2005) (Figure 12.2). However, this exchange would be highly electrogenic, and would generate an inside-positive charge that would need to be shunted to permit significant accumulation of protons. (An alternative mechanism for acidification of nascent endosomes lacking the H^+-ATPase could be through the NHE3 sodium-proton exchanger, which is expressed in these endosomes and which would acidify in an electroneutral manner. NHE3-knockout mice have tubular proteinuria, indicating a role for NHE3 in protein reabsorption (Gekle et al 2004).)

CLC-5 may be important in endosomal function beyond its role as a chloride transporter. There is substantial evidence for defective membrane trafficking in CLC-5 knockout mice and in patients with Dent's disease. Pharmacologic inhibition of endosomal acidification inhibits endocytosis (Wang et al 2005), but it is not clear that impaired acidification is a sufficient explanation for the trafficking dysfunction. There are substantial and specific decreases in the abundance of the apical protein receptors megalin and cubulin in the apical brush border in CLC-5 knockout mice, with these receptors redistributed to intracellular endocytic organelles (Christensen et al 2003). Reduction in apical expression of megalin has been reported in biopsy specimens in Dent's disease (Santo et al 2004, Yanagida et al 2004). Urinary excretion of megalin is reduced in CLC-5 knockout mice (Piwon et al 2000) and in patients with Dent's disease and Lowe's syndrome, though cubulin excretion in these patients was normal (Norden et al 2002). In proximal tubules of patients

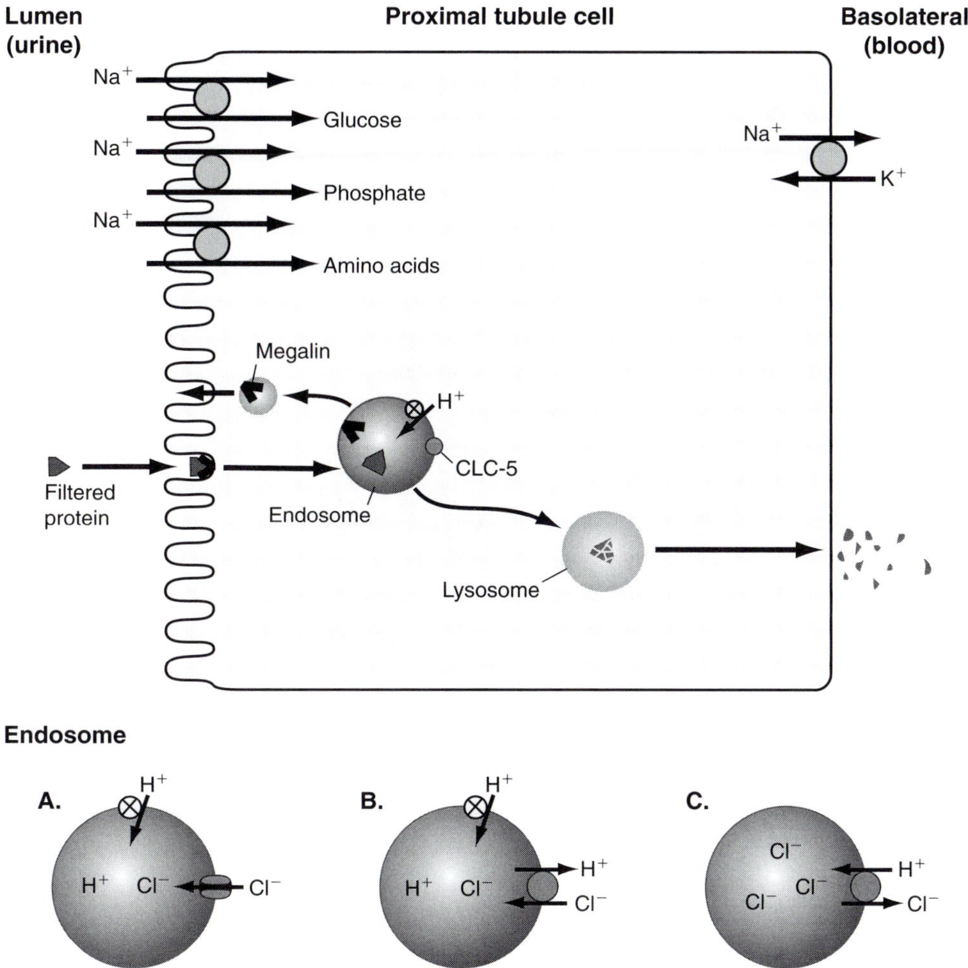

FIGURE 12.2   Possible roles of chloride–proton antiporter function in endosomal acidification. In proximal tubular epithelial cells, CLC-5 is expressed in subapical endosomes in which the V-type H^+-ATPase is responsible for electrogenic transport of protons into the endosomal lumen. CLC-5 facilitates this acidification. (A) Until recently CLC-5 was thought to function as a chloride channel, allowing entry of chloride which would shunt the positive charge. (B and C) CLC-5 is now known to act as a Cl^-/H^+ exchanger. (B) In the presence of an active H^+-ATPase, the proton gradient would drive exit of H^+ coupled with entry of Cl^-, and this would effectively dissipate the interior positive charge. (C) In the nascent endosome, an initially high luminal Cl^- concentration would drive exit of Cl^- coupled with entry of H^+ and thereby achieve acidification in the absence of a proton pump, if there were a mechanism to shunt the inside-positive charge generated by the antiporter (Brandt & Jentsch 1995, Picollo & Pusch 2005, Scheel et al 2005). (see also Plate 12)

with Dent's disease, the H^+-ATPase is mistargeted to the basolateral surface of the epithelial cells, in contrast with its normal localization at the apical pole. This defect is specific in that other membrane proteins such as the Na^+/K^+-ATPase and aminopeptidase are appropriately targeted. These latter observations suggest a specific interaction between C-terminus of CLC-5 and appropriate targeting of the H^+-ATPase protein (Moulin et al 2003).

CLC-5 may play a similar role in the collecting duct, where it is expressed in the acid-secreting α-intercalated cells. Transfection of cultured collecting duct cells with antisense CLC-5 causes arrest of endocytosis (Sayer et al 2003). In kidneys of patients with Dent's disease, the H^+-ATPase is not expressed in the α-intercalated cells of the collecting duct (Moulin et al 2003). While this is not associated with any defect in urinary acidification, Sayer and colleagues have proposed that it may be relevant to the mechanism of medullary nephrocalcinosis in these patients (Sayer et al 2003, 2004).

Guggino and colleagues have demonstrated **a** potentially important interaction between the C-terminal tail of CLC-5 and the actin-binding protein cofilin. Experimental interference with this interaction between cofilin and the CLC-5 C-terminus resulted in an inhibition of albumin uptake (Hryciw et al 2003). Cofilin modulates actin polymerization and participates in the regulation of albumin uptake by the proximal tubule. The interaction between CLC-5 and cofilin strongly suggests that independently of its function as a transporter, CLC-5 may be essential to the assembly and formation of the endocytic complex. Disruption of this interaction by mutation may be the explanation for the mistargeting of specific endosome-related membrane proteins such as the H^+-ATPase, megalin, and cubulin (Hryciw et al 2003).

CLC-5 is expressed early in nephrogenesis in both mouse and human kidney. In the mouse this involves a rapid induction at an early stage of nephrogenesis, together with the H^+-ATPase protein, followed by further maturation until birth. In human kidney, segmental distribution of CLC-5 together with the H^+-ATPase is detectable in the second trimester. These early events coincide with the onset of glomerular filtration, and are consistent with descriptions of the time course of expression of other brush-border membrane proteins (Jouret et al 2005). The concentration of LMW proteins in amniotic fluid falls during gestation, consistent with progressive maturation of the apparatus for protein reabsorption (Mussap et al 1996). That CLC-5 is essential for protein reabsorption by the time of birth is consistent with descriptions of LMW proteinuria in infancy in Dent's disease (Nakazato et al 1997, Bosio et al 1999).

The fact that the phenotype of Dent's disease occurs in patients with mutations in *OCRL1* raises the important question of overlap between the physiological functions of CLC-5 and the OCRL1 protein that would produce a common phenotypic end-point. *OCRL1* encodes a phosphatidyl-inositol 4,5-bisphosphate 5-phosphatase located in the Golgi apparatus. The OCRL1 phosphatase affects endosomal trafficking in several potential ways. Impairment of phosphatase function results in elevated cellular levels of PIP_2. PIP_2 levels affect vesicle trafficking at the Golgi (Dressman et al., 2000) and may also account for alterations in the actin cytoskeleton seen in fibroblasts from patients with Lowe's syndrome (Suchy & Nussbaum 2002). This would be predicted to alter endosomal membrane trafficking (Apodaca 2001). Beyond this, the OCRL1 protein is localized to early endosomes and interacts directly with clathrin to promote assembly of the clathrin-mediated endosomal apparatus. Over- and under-expression of *OCRL1* alters protein trafficking at the interface between endosomes and the trans-Golgi network (Choudhury et al 2005).

Insights from *CLCN5* Mutation Analysis

The gene *CLCN5* has 13 exons (11 coding), a transcript of 9.5 kb, and an open reading frame of 2238 bases. It encodes a protein product, designated CLC-5, of 746 amino acids (Fisher et al 1995). So far a total of 81 distinct mutations have been identified in 103 probands. These mutations consistently segregate with disease in families. They represent missense (27), nonsense (17), frameshift (22), in-frame codon insertion (1), splice-site (8), and deletion (6) mutations (Pook et al 1993; Lloyd et al 1996, 1997a,b, Akuta et al 1997, Nakazato et al 1997, 1999, Oudet et al 1997, Hoopes et al 1998, 2004, Igarashi et al 1998, 2000,

Kelleher et al 1998, Morimoto et al 1998, Bosio et al 1999, Cox et al 1999, Yamamoto et al 2000, Takemura et al 2001, Carballo-Trujillo et al 2003, Claverie-Martin et al 2003, Brakemeier et al 2004, Forino et al 2004, Matsuyama et al 2004, Besbas et al 2005, Cheong et al 2005, Ludwig et al 2005, Tosetto et al 2006). Heterologous expression of 28 of these mutations (including 15 of the missense mutations) demonstrates reduction or elimination of chloride currents by the expressed proteins (Lloyd et al 1996, 1997a, Igarashi et al 1998, Morimoto et al 1998, Yamamoto et al 2000, Ludwig et al 2005). Most mutations are unique; the most frequently reported mutation (S244L) has been described in no more than eight families. These eight families exhibit the full range of phenotype despite sharing an identical mutation (Lloyd et al 1996, Oudet et al 1997, Hoopes et al 1998, 2004, Kelleher et al 1998). The nature of the mutation also does not correlate with severity of phenotype, as evidenced by large families in which multiple males share the same mutation but with widely varying severity of features. Some patients with missense mutations have more severe clinical abnormalities such as renal failure or rickets than patients with complete deletion of the gene (Scheinman 1998).

Dutzler, MacKinnon and colleagues have described the X-ray crystallographic structure of homologous CLCs from salmonella and *E. coli* (Dutzler et al 2002). These observations confirmed previous speculation that these CLCs are homodimeric membrane proteins that form a channel with two pores. They also clarified the number of α-helices (18), and identified regions that form the interface between the two identical subunit proteins. Homology between these two bacterial CLCs and human CLC-5 is only 22% and 24%, respectively. In view of the recent observations that CLC-5 functions not as a channel but as an antiporter, extrapolating the structures of the bacterial proteins to the human must be considered tentative. With this caveat, Wu et al (2003) have mapped a number of the missense mutations in CLC-5 to a hypothetical 3-dimensional structure for the CLCs, and I have used that structure in the following discussion.

Missense mutations represent experiments of nature that reveal individual codons at which a single base change resulting in an amino acid substitution have significant functional consequences. Of the 22 codons at which 26 missense mutations have been described, over half occur in regions that form the interface between the two identical dimeric subunits (Figure 12.3). Two occur in the signal sorting sequence at helix A near the cytoplasmic amino-terminus; these reduce or abolish chloride transport (Lloyd et al 1997a, Morimoto et al 1998), possibly through altered trafficking of CLC-5 into the membrane (this sequence has no homology in the bacterial CLCs, and is not thought to be part of the interface). Four missense mutations occur at three codons in the selectivity filter in helices F and N. Two of these have been found not only to abolish transport but

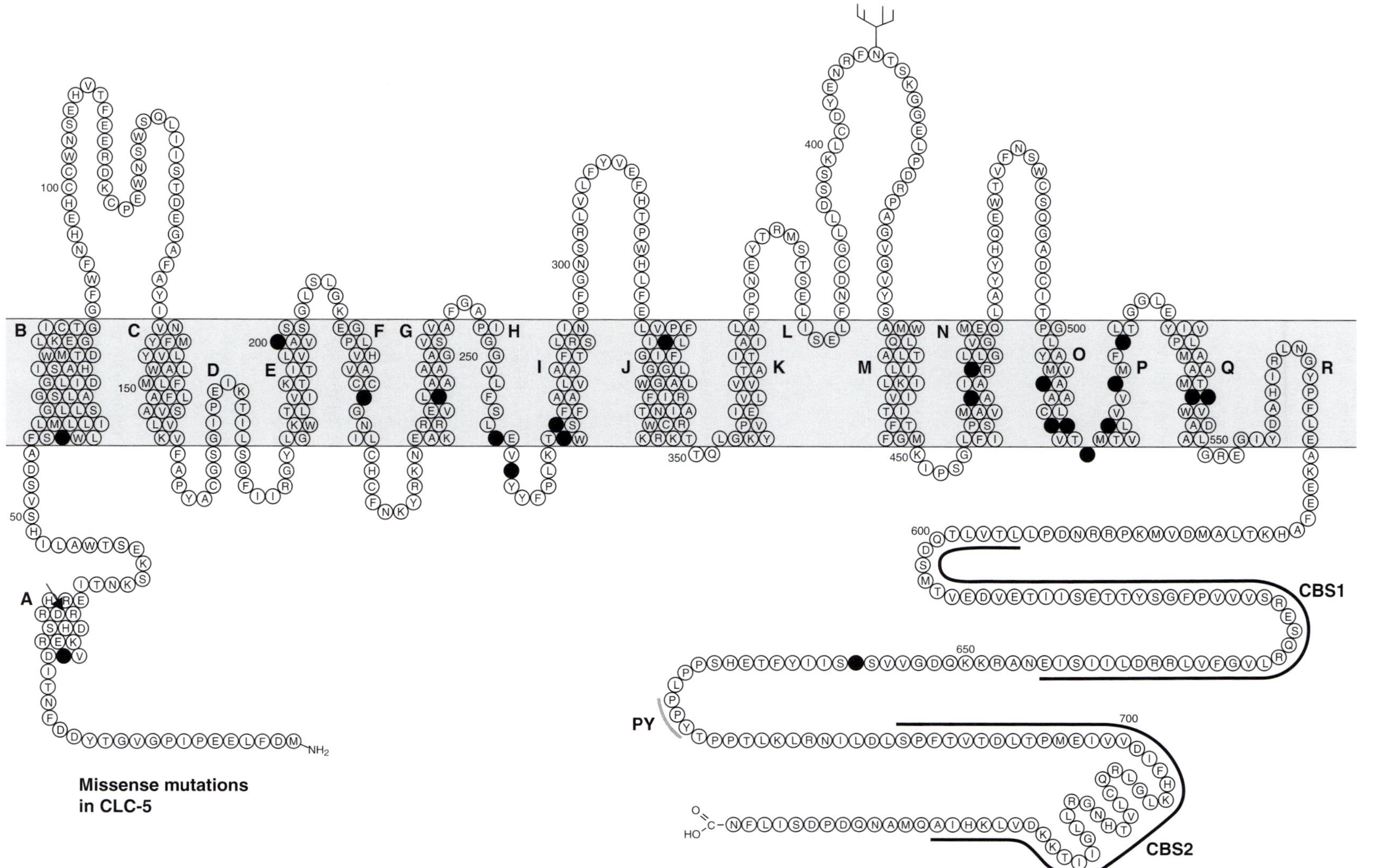

FIGURE 12.3 Missense mutations in *CLCN5*. The CLC-5 protein is shown schematically with 18 α-helices as predicted by Dutzler et al (2002) and Wu et al (2003). Helices are labeled A through R, and represented relative to the membrane (blue shading); intracellular domains are below and extracellular above the membrane. Helices B, G, H, I, P, and Q form the interface between subunits. Other functionally significant domains include a signal sorting sequence at helix A in the intracellular N-terminal region, two CBS domains in the C-terminal chain, and a PY motif between them. Twenty-six missense mutations, in which a single base change results in an amino acid substitution, have been reported at 22 codons, indicated in filled circles; an additional in-frame insertion of histidine at codon 30 is represented by a triangle. (see also Plate 13)

also to be associated with absence of CLC-5 protein at the membrane, suggesting that the mutations alter either trafficking or processing of the protein (Ludwig et al 2005).

Nonsense and frameshift mutations that truncate the protein, not shown on the figure, occur throughout the sequence, as late as the very end of the C-terminal tail. A frameshift mutation in a child with proteinuria that disrupts the amino acid sequence from codon 728 abolishes chloride transport when expressed (Yamamoto et al 2000); a nonsense mutation at codon 718 produces a full phenotype of Dent's disease (Carballo-Trujillo et al 2003, Hoopes et al 2004). The C-terminus contains two cystathione β-synthase (CBS) domains, as is the case in all eukaryotic CLCs, though bacterial CLCs lack CBS domains and are still functional. The role of CBS domains is poorly understood, but studies of mutations truncating the CLC-5 C-terminus, and co-expression studies with complementary fragments, have pointed a role for these C-terminal CBS domains in trafficking of CLC-5 to endosomes and regulation of channel activity (Carr et al 2003, Ludwig et al 2005). Between the CBS domains is a PY motif that resembles sequences in barttin and ENaC. PY motifs interact with ubiquitin ligases, and this has been confirmed for the CLC-5 PY motif (Schwake et al 2001). It has been speculated that ubiquitination of PY may regulate endocytic retrieval of CLC-5 from the plasma membrane, but its functional significance *in vivo* is yet unknown. A further function of the C-terminal tail is the interaction with cofilin, which may mediate a role for CLC-5 in the assembly of the endocytic complex (Hryciw et al 2003).

GENETICS

Genetic linkage was initially established to a region at Xp11.22 in a single very large pedigree (Scheinman et al 1993). This localization was subsequently refined by identification of a small (515 kb) deletion in this region in another family (Pook et al 1993) that encompassed a single candidate gene that was identified as a CLC chloride channel (Fisher et al 1994, 1995).

Mutations in *CLCN5* were initially identified in 11 kindreds (Lloyd et al 1996), and subsequently in many more representing the full range of phenotype of Dent's disease in over 100 probands. Although heterologous expression studies of these mutations were designed to measure function as a chloride channel, the observed differences between mutated and wild-type CLC-5 support their significance in this disease. Furthermore, the features of Dent's disease are reproduced in CLC-5 knockout mice (Luyckx et al 1999, Piwon et al 2000, Wang et al 2000, Silva et al 2003). Thus, the association between mutation in *CLCN5* and Dent's disease is firmly established.

However, it is now clear that there is genetic heterogeneity in Dent's disease. We studied a cohort of unrelated patients

in whom the diagnosis of Dent's disease was defined using strict clinical criteria. Only 60% of these patients had mutations in *CLCN5*. In the remainder, mutations in the *CLCN5* coding sequence and promoter were excluded by direct sequencing, and in one large pedigree with apparent X-linked inheritance, the *CLCN5* locus at Xp11.22 was further excluded by haplotype analysis (Hoopes et al 2004). In this pedigree, genetic mapping identified linkage between Dent's disease and another region at Xq25-Xq27. Sequencing of candidate genes revealed mutations in the gene *OCRL1* in this family and four others. The physiological significance of these mutations was confirmed by the finding of reduced or absent protein abundance, and marked reduction in the activity of the PIP(2)-5 phosphatase, in fibroblasts grown from these patients (Hoopes et al 2005).

OCRL1 is the gene responsible for the oculocerebrorenal syndrome of Lowe (see Chapter 10). The finding of mutations in *OCRL1* in these patients was surprising in that the clinical criteria for Dent's disease in this study excluded patients with cataracts or mental developmental delay. Formal slit-lamp examination was normal in all five probands, which is particularly remarkable since the presence of cataracts is considered a requirement for the diagnosis of Lowe's syndrome and its absence excludes the diagnosis on clinical grounds (Roschinger et al 2000, Charnas & Nussbaum 2001). In addition, formal mental developmental testing revealed only mild mental retardation in three of the probands; two were normal. This contrasts with the serious mental impairment that is typical of Lowe's syndrome. Furthermore, none of these probands had renal tubular acidosis, another feature that is a hallmark of Lowe's syndrome but typically absent in Dent's disease (Hoopes et al 2005).

Thus, Dent's disease is associated with mutations in *CLCN5* in about 60% of cases, and with mutations in *OCRL1* in another 15%. Presumably other gene(s) account(s) for the remainder. This is reminiscent of the genetic heterogeneity in Bartter's syndrome. Autosomal recessive inheritance of a syndrome consistent with Dent's disease has been reported (Magen et al 2004).

The discovery of patients with mutations in *OCRL1* with an isolated renal phenotype of Dent's disease, lacking cataracts and even renal tubular acidosis, raises questions regarding the pathophysiology of these features that need to be explained. *OCRL1* knockout mice have a normal phenotype, but studies involving concurrent inactivation of both OCRL1 and a second homologous phosphatase suggest strongly that the other phosphatase(s) can compensate for the loss of function of OCRL1 (Janne et al 1998). This implies that the phenotypic variability in patients with mutations in *OCRL1* may be the consequence of variations in activity of modifying gene(s) (Hoopes et al 2005).

The chromosomal segment containing *CLCN5* at Xp11.23 has been useful to population geneticists because it contains both a region with a very low female recombination rate (within one of the *CLCN5* introns) and a highly variable microsatellite, both within a space of only 2.8 kb, allowing for the use of combined haplotypes to study the population dynamics of humans using different time scales. The results (based on a group that included Pygmies, Basques, British, and Japanese) indicate that there is relatively little selection pressure on *CLCN5*. This is not surprising since symptomatic renal failure typically would not occur until well past the onset of reproductive age. Sequence diversity is much less in Eurasians than in Africans, consistent with a model of population bottleneck, with a relatively small number of ancestral Africans populating Eurasia, followed by exponential population growth (Alonso & Armour 2004).

CLINICAL PHENOTYPE AND PATHOPHYSIOLOGY OF DENT'S DISEASE

The major features of Dent's disease reflect disturbances in proximal tubular solute reabsorption. The most consistent of these is low-molecular-weight (LMW) proteinuria, which is essentially a universal feature in all affected males and, in less extreme degrees, in many carrier females. Beyond LMW proteinuria, a spectrum of clinical findings has been described in a group of reports that overlap broadly with those of Dent's disease. Other than LMW proteinuria, the severity of phenotype ranges widely. The ability to define the diagnosis by the presence of mutations in *CLCN5* has allowed the syndrome to be defined with clarity (Scheinman 1999, Scheinman & Thakker 2000). The major features of Dent's disease are listed in Table 12.1.

Proximal Tubular Solute Wasting

LMW proteinuria is the most consistent finding in patients with Dent's disease, and the degree of protein loss is extreme. For example, proteins such as ß2-microglobulin and retinol-binding protein appear in urine at levels that exceed normal exretion rates by 300- to 3000-fold (Wrong et al 1994, Reinhart et al 1995, Scheinman 1998). It occurs in infancy before any other evidence of renal dysfunction (Wrong et al 1994, Reinhart et al 1995), and even in adults can be an isolated finding (Lloyd et al 1997a). LMW proteinuria is dramatically milder in females, but it is sufficiently consistent as to have value as a screening tool, though some carriers have normal excretion of LMW proteins (Reinhart et al 1995, Scheinman 1998, Matsuyama et al 2004). This is consistent with X-linked recessive inheritance, with random inactivation of the X chromosome in somatic cells (lyonization) yielding a more variable and milder phenotype in females than in males.

The urinary proteome in Dent's disease reflects failure to reabsorb filtered proteins. It does not include proteins

TABLE 12.1 Major features of Dent's disease in affected male patients with documented mutations in *CLCN5*

Proximal solute wasting:	
LMW proteinuria	100%
Aminoaciduria	76%
Glycosuria	54%
Hypophosphatemia	50%
Dysregulation of calcium metabolism:	
Hypercalciuria*	95%
Nephrocalcinosis	74%
Nephrolithiasis	49%
Renal insufficiency	64%
Rickets	30%
Other:	
Hematuria	94%
Concentrating defect*	81%
Hypokalemia	35%
Hypomagnesemia	22%

Asterisks (*) indicate features for which patients were excluded if any degree of renal failure was present. (Adapted from Scheinman 1999, Scheinman and Thakker 2000.)

that are secreted into the urine in response to tubular injury (Wrong et al 1994). Among the LMW proteins excreted in excess are a number of hormones and cytokines that are potentially bioactive and might explain some of the clinical findings in this disease (Norden et al 2001). These include the monocyte chemoattractant protein MCP-1, speculated to contribute to interstitial inflammation and renal failure (Norden et al 2001) and parathyroid hormone, which has been proposed as an explanation for the hypercalciuria and phosphaturia (Piwon et al 2000). These are discussed below. Overall, the urine of patients with Dent's disease contain excessive amounts of vitamin and prosthetic group carrier proteins such as vitamin D-binding protein, transthyretin, retinol-binding protein, and IGF-binding protein, as well as complement components, apolipoproteins, and cytokines; in contrast, proteins of renal origin represent a much smaller proportion of the urinary proteome in the urine of patients with Dent's disease than in normal urine (Cutillas et al 2004). The effectiveness of the normal mechanism for protein reabsorption is evidenced by the fact that in health, filtered proteins are almost completely removed from the urine.

Since the proteinuria is a very early feature of Dent's disease, and renal failure is not evident until later, Norden and colleagues have used the urinary protein profile in young patients with Dent's disease and Lowe's syndrome to calculate glomerular sieving coefficients for a range of filtered proteins. These proteins include albumin, which represents about half of the proteinuria in these patients. The urinary protein profile in these patients is consistent with an intact glomerular filter and specific dysfunction in protein reabsorption. These in-vivo estimates of glomerular sieving coefficient match well with other measurements based on dextran clearances or direct measurements in rats (Norden et al 2001).

Other clinical signs of proximal reabsorptive dysfunction, such as glycosuria, aminoaciduria, and hypophosphatemia, are common but not universal. Their presence may vary among family members who share a common mutation (Wrong et al 1994, Igarashi et al 1995, Reinhart et al 1995, Scheinman 1998). When hypophosphatemia and hypokalemia occur, they can be severe and may require replacement (Frymoyer et al 1991, Bolino et al 1993, Wrong et al 1994, Oudet et al 1997).

Jentsch and colleagues have proposed a mechanism to explain the hypophosphatemia and hypercalciuria. They postulate that failure to reabsorb PTH in the early proximal tubule may lead to activation of apical PTH receptors later in the proximal tubule, and that this may explain both excessive activation of vitamin D (see below) and inhibition of phosphate transport, with recycling of the sodium-phosphate cotransporter NaPi2 to the endosomal compartment. Their hypothesis is based on elegant observations in their CLC-5 knockout mice, in whom NaPi2 was expressed apically in the S1 segment of proximal tubule but was internalized in the S2 and S3 segments, even though CLC-5 was absent in all cells. In the $clcn5^{+/-}$ female mice heterozygous for an intact *CLCN5* gene, NaPi2 localization resembled that in wild-type mice in all cells of the proximal tubule regardless of whether the cell expressed CLC-5 (Piwon et al 2000). Urinary excretion of PTH is excessive in CLC-5 knockout mice (Piwon et al 2000) and in patients with Dent's disease (Norden et al 2001). Recycling of the sodium-proton exchanger NHE3 is also impaired in CLC-5 knockout mice, also consistent with a PTH effect (Piwon et al 2000), although there is no clinical evidence of defective reabsorption of sodium or bicarbonate (Reinhart et al 1995, Anton-Gamero et al 2005). Excessive, if suppressible, PTH activity is consistent with the report of two boys with Dent's disease in whom urinary cAMP levels were elevated on fasting but suppressed with dietary calcium loading (Igarashi et al 1995). Jentsch's hypothesis would not, however, explain defective reabsorption of glucose and amino acids, which, like phosphate, are reabsorbed through the activity of secondary-active substrate-specific Na-dependent transporters. Neither phosphaturia nor aminoaciduria, nor glycosuria occurs as consistently as LMW proteinuria in patients with Dent's disease, and they can be intermittent (Wrong et al 1994, Reinhart et al 1995, Scheinman 1998). This suggests that the defect(s) may be a secondary phenomenon, perhaps a consequence of abnormal trafficking of these transporters.

Hypercalciuria, Nephrocalcinosis, and Stones

Although kidney stones occur in only half of patients, nephrolithiasis is one of the distinctive features of Dent's disease. Hypercalciuria is nearly universal before the onset

of renal failure, and the stones are composed of calcium, either oxalate or phosphate (Scheinman 1998). There are few published observations of excretion of citrate or oxalate in patients with Dent's disease, but these have been normal (Scheinman 1998); this is notable with regard to citrate, excretion of which is typically increased in other forms of the Fanconi syndrome (Wrong et al 1994). Thus, available evidence indicates that hypercalciuria is the principal risk factor for stones in these patients.

Hypercalciuria is moderate in adolescents and young adults with Dent's disease, in the range of 4–6 mg/kg body weight, but in children under age 10 years it can be much greater and can exceed 10 mg/kg. In either age group there is a tendency to an exaggerated calciuretic response to dietary calcium loading. Boys, but not adults, typically are hypercalciuric on fasting (Igarashi et al 1995, Reinhart et al 1995).

Serum levels of PTH are consistently below the mean of the normal range in patients with Dent's disease, and often frankly low. Conversely, levels of 1,25-dihydroxyvitamin D tend to exceed the mean of normal and are often elevated. These observations hold true in a number of reports (Murakami & Kawakami 1990, Bolino et al 1993, Wrong et al 1994, Igarashi et al 1995, Reinhart et al 1995, Hoopes et al 1998, Scheinman 1998, Schurman et al 1998). Serum phosphorus levels were normal in many of these descriptions, but they have not been studied under standardized conditions and therefore hypophosphatemia cannot be excluded as an explanation for the high levels of 1,25-dihydroxyvitamin D. As noted above, Jentsch has proposed that failure to reabsorb parathyroid hormone in the early proximal tubule may lead to activation of luminal PTH receptors. These, if coupled to stimulation of 1α-hydroxylation of vitamin D, could explain excessive serum levels of 1,25-dihydroxyvitamin D, which could drive excessive absorption of dietary calcium (Piwon et al 2000). At the same time, however, reduced expression of megalin at the apical surface would impair reabsorption of vitamin-D-binding protein; this is the finding in megalin knockout mice, and in that model it results in vitamin D-deficiency and bone disease (Christensen & Willnow 2003). Jentsh proposes that this balance between excessive PTH effect and vitamin D loss may vary among patients and could explain why hypercalciuria is not universally found in patients with Dent's disease, particularly in adults (Piwon et al 2000). One patient has been reported in whom serum levels of 25-hydroxyvitamin D and of 1,25-dihydroxyvitamin D were at the low end of normal, in contrast to the typical patient with Dent's disease, and in whom hypercalciuria was associated with hypocalcemia, which is very unusual in Dent's disease (Yanagida et al 2004). This case is consistent with Jentsch's hypothesis. In this patient, immunostaining demonstrated marked reduction in megalin expression at the apical surface of cells (Yanagida et al 2004).

In one CLC-5 knockout mouse model, a small degree of hypercalciuria persisted on dietary calcium restriction, suggesting that a component of the hypercalciuria resulted from abnormal bone mineral balance or renal reabsorption

(Silva et al 2003). Elevated serum levels of alkaline phosphatase, osteocalcin, and urinary deoxypyridinoline in these mice were consistent with increased bone turnover. In this respect these mice, which were studied while still growing, resemble children with Dent's disease, in whom the degree of hypercalciuria is greater than in adults, and whose hypercalciuria has a fasting component (Reinhart et al 1995). In these knockout mice (Silva et al 2003), another mouse model (Luyckx et al 1999), and patients with Dent's disease (Reinhart et al 1995), however, the largest component of hypercalciuria is diet-dependent.

The physiology of hypercalciuria in Dent's disease resembles that in a subset of stone-formers with idiopathic hypercalciuria. The identification of one man with isolated hypercalciuria, lacking LMW proteinuria, who had a disease-causing mutation in *CLCN5* in a family with multiple affected males, raised the question of whether some patients with idiopathic hypercalciuria might have mutations in this gene. However, in a series of 107 adults and children with idiopathic hypercalciuria and stones we found no evidence of Dent's disease in any (Scheinman et al 2000). Another study yielded similar results, but in a smaller sample (Rebelo et al 2005).

Nephrocalcinosis may reflect a disturbance beyond hypercalciuria, as the degree of hypercalciuria in Dent's disease (except in young boys, in whom it is marked) is no more severe than in a typical patient with idiopathic hypercalciuria. Sayer and colleagues have studied mouse IMCD collecting duct cells and found that inactivation of CLC-5 by transient transfection with antisense CLC-5 results in adherence of calcium phosphate and calcium oxalate crystals, leading to crystal agglomerates forming on the surface of these cells. They speculate that inhibition of endocytosis and of membrane trafficking alters apical membrane proteins and secreted molecules such as osteopontin, hyaluronan, and annexin, and that this could result in increased binding and attachment of crystals (Sayer et al 2003, 2004). They have confirmed an increase in surface expression of annexin A2 in CLC-5-deficient cells, and found that pretreatment with anti-annexin A2 antibodies attenuated surface crystal retention and agglomeration (Carr et al 2006).

Renal Failure

Renal failure develops in two-thirds of patients, but variably, and some patients reach an advanced age with little or only modest renal impairment. In the most aggressive cases, GFR declines measurably in late childhood and may reach end-stage in the early twenties, but more typically advanced renal failure occurs in the fourth or fifth decade of life. Affected males in the same family and sharing the same mutation may develop renal failure early, or not at all, exemplifying the phenotypic variability in this disease (Frymoyer et al 1991, Scheinman 1998).

From childhood, the degree of albuminuria ranges 1–2 grams per day, representing about half of the proteinuria. Nephrotic syndrome does not occur, and there is no clinical or histological evidence of a primary glomerular disturbance. The most prominent features on renal histopathology are tubular atrophy and interstitial fibrosis, with modest inflammation. Multiple small cysts have been described (Wrong et al 1994, Neild et al 2005). Glomeruli exhibit varying degrees of sclerosis, with some hypertrophic, suggesting that the glomerular changes are secondary to nephron damage. Electron microscopy reveals normal basement membrane ultrastructure, and immunofluorescence studies have been negative (Frymoyer et al 1991, Wrong et al 1994, Igarashi et al 1995, Scheinman 1998). These findings are nonspecific, and it is reasonable to speculate that unrecognized Dent's disease might constitute a proportion of patients with end-stage renal disease (ESRD) with the vague diagnosis of chronic tubulointerstitial nephritis. One study attempted to address this through mutation analysis in 25 male Italian patients with ESRD and stones, and found one mutation (G260V) in a patient not previously diagnosed as having Dent's disease (Tosetto et al 2006).

The mechanism of the renal failure is unknown. Nephrocalcinosis develops in a majority of patients with Dent's disease, but the correlation with renal failure is poor. Anecdotal reports of patients with advanced renal insufficiency but modest nephrocalcinosis, and of adult patients with marked nephrocalcinosis yet preserved renal function, make it difficult to ascribe the renal failure solely to tissue calcification as a consequence of hypercalciuria (Scheinman 1998). While it is unlikely that nephrocalcinosis alone is a sufficient explanation for the renal insufficiency, it may still be a contributing factor. In CLC-5 knockout mice, renal failure with interstitial fibrosis and nephrocalcinosis is accelerated if the mice are maintained on a citrate-free diet, and can be delayed substantially by feeding the mice a high-citrate diet (Cebotaru et al 2005). These observations lend strong support to the view that nephrocalcinosis does contribute to the renal failure.

Norden and colleagues have speculated that the presence of pro-inflammatory cytokines in the lumen of the nephron, as a consequence of the failure to reabsorb filtered proteins, may contribute to tubulointerstitial damage, but this speculation requires further study (Norden et al 2001).

The experience with transplantation in Dent's disease is anecdotal. Transplant-related tubulointerstitial damage can produce some features that resemble those of Dent's disease, and persistence of LMW proteinuria or aminoaciduria could reflect nonspecific allograft injury or residual native kidney function. However, it appears that stones and nephrocalcinosis do not recur and that transplant outcomes are good (Frymoyer et al 1991, Wrong et al 1994, Scheinman 1998).

Bone Disease

While rickets only occurs in a minority of patients with Dent's disease, in those patients it is often severe and disabling. In Wrong's report of Dent's original patients, only one of the affected males in each family had rickets (Wrong et al 1994). The most common of the *CLCN5* mutations in patients with Dent's disease, the missense mutation S244L, has been reported in eight families; in two of these families, all nine affected males had severe rickets (Bolino et al 1993, Oudet et al 1997), but in the other six families only one patient had rickets (Hoopes et al 1998, Kelleher et al 1998). While hypophosphatemia may well be a factor contributing to rickets in some patients, serum phosphate levels were normal and replacement was not necessary in a number of the reported rachitic patients (Wrong et al 1994, Oudet et al 1997). Hypocalcemia has not been reported in Dent's disease, nor is acidosis a feature of the disease (Scheinman 1998).

Patients with clinical bone disease have elevated serum levels of alkaline phosphatase, but those without clinical rickets have normal alkaline phosphatase levels and normal bone density by dual-energy X-ray absorptiometry (Wrong et al 1994, Reinhart et al 1995). Serum levels of 1,25-dihydroxyvitamin D are normal or high in patients with Dent's disease, but Wrong and colleagues have described healing of bone disease following treatment with large doses of vitamin D (Wrong et al 1994).

Other Disturbances in Renal Function

One remarkable feature of Dent's disease is the absence of systemic acidosis. Except in the presence of renal insufficiency or nephrocalcinosis, these patients are able to acidify the urine normally (Wrong et al 1994, Reinhart et al 1995). Growth failure is not a feature of Dent's disease except when explained by rickets. In this regard, Dent's disease differs strikingly from other causes of the Fanconi syndrome, and particularly Lowe's syndrome. It is all the more remarkable in view of the observation that CLC-5 is expressed in the acid-secreting intercalated cells of the collecting duct (Obermuller et al 1998). Furthermore, those patients defined clinically as having Dent's disease but found to have mutations in *OCRL1* also lacked acidosis, which is typically a significant problem for patients with Lowe's syndrome.

Nocturia is often the first symptom in affected boys, and moderate polyuria in the range of 1–3 liters per day is common. The urinary concentrating response to vasopressin is subnormal, and this does not correlate with the occasional occurrence of hypokalemia (Wrong et al 1994, Reinhart et al 1995). The concentrating defect precedes renal insufficiency or nephrocalcinosis, but could be a consequence of the hypercalciuria, with activation of calcium-sensing receptors on the apical surface of the collecting duct (Hebert et al 1997).

Patients with Dent's disease appear to be able to conserve sodium normally, and osmolar clearance is normal (Reinhart et al 1995, Anton-Gamero et al 2005). This suggests that the proximal tubular dysfunction in Dent's disease does not alter the function of the NHE3 sodium-proton exchanger, which accounts for the bulk of proximal sodium reabsorption. One patient was reported to have a dramatic natriuretic response to hydrochlorothiazide (Wrong et al 1994).

Hypokalemia occurs in fewer than half of patients with Dent's disease. This is another clinical feature that varies widely. Some patients require more than 100 mEq per day of potassium replacement, while the majority of patients do not require potassium replacement (Frymoyer et al 1991, Wrong et al 1994, Reinhart et al 1995).

TREATMENT

Treatment recommendations are empirical and based on anecdotal observations and only one controlled trial. That trial addressed the effects of diuretics on hypercalciuria, which is at least the major risk factor for stones in these patients, although we do not know to what extent if any it contributes to the progressive renal insufficiency. That study demonstrated that over a short (two-week) time frame, a thiazide diuretic (chlorthalidone) was as effective in correcting hypercalciuria in patients with Dent's disease as in an age-matched group of males with idiopathic hypercalciuria. Amiloride had no effect either on its own or to augment the effect of the thiazide. This study did not address any long-term effect of thiazide to reduce stone recurrence, nor any effect on renal function (Raja et al 2002).

Based on these findings, and general principles for management of hypercalciuria, the following recommendations are currently accepted. Patients with recurrent stones and hypercalciuria may benefit from treatment with a thiazide diuretic, which should promote a positive calcium balance through its effect to stimulate distal tubular calcium transport, and is known to reduce stone recurrence and improve bone mineralization in patients with idiopathic hypercalciuria. In any patient with hypercalciuria, dietary sodium restriction may reduce calcium excretion, and is especially important in those treated with thiazide in order to reduce the likelihood of hypokalemia and hypocitraturia which would increase stone risk. Although it does not reduce calcium excretion in patients with Dent's disease, amiloride may be helpful in the management of a hypokalemic patient. In view of one report of a dramatic natriuretic and kaliuretic response to a large dose of hydrochlorothiazide (Wrong et al 1994), patients should be followed closely when starting the diuretic.

Dietary calcium should not be restricted, even though the major component of the hypercalciuria is excessive intestinal absorption. Lower calcium intakes in general are associated with an increased risk of stone events (Curhan et al 1993, 1997) and a higher intestinal oxalate absorption (Lemann et al 1996) and may exacerbate a tendency to bone demineralization.

Wrong et al have described healing of rickets in Dent's disease following treatment with vitamin D, and recommend small doses of vitamin D (Wrong et al 1994). However, this can exacerbate hypercalciuria, and should only be considered in patients with clinical bone disease. If vitamin D is given, it is particularly important to monitor urinary calcium excretion as well as clinical markers of bone disease during this therapy.

As in any patient with nephrolithiasis or renal failure, careful attention to urinary infection and management of obstruction are important. Any possible value of dietary protein restriction or therapy with angiotensin-converting enzyme inhibitors is unknown. While it has been speculated that proteinuria contributes to the renal failure through excessive excretion of cytokines, or to hypercalciuria and phosphaturia through the effects of luminal PTH, it is not at all known whether these would be benefited by the hemodynamic effects of these therapies, in patients with an intact glomerular filter and without nephrotic syndrome.

References

Akuta N, Lloyd SE, Igarashi T, et al. Mutations of CLCN5 in Japanese children with idiopathic low molecular weight proteinuria, hypercalciuria and nephrocalcinosis. Kidney Int. 1997; 52: 911–16.

Alonso S, Armour JA. Compound haplotypes at Xp11.23 and human population growth in Eurasia. Ann. Hum. Genet. 2004; 68: 428–37.

Anton-Gamero M, Claverie-Martin F, Garcia-Nieto V, Vela-Enriquez F, Garcia-Martinez E, Perez-Navero JL. Chloride and sodium renal tubular handling in Dent's disease. Pediatr. Nephrol. 2005; 20: 1198–9.

Apodaca G. Endocytic traffic in polarized epithelial cells: role of the actin and microtubule cytoskeleton. Traffic 2001; 2: 149–59.

Besbas N, Ozaltin F, Jeck N, Seyberth H, Ludwig M. CLCN5 mutation (R347X) associated with hypokalaemic metabolic alkalosis in a Turkish child: an unusual presentation of Dent's disease. Nephrol. Dial. Transplant. 2005; 20: 1476–9.

Bolino A, Devoto M, Enia G, Zoccali CJW, Romeo G. A new form of X-linked hypophosphatemic rickets with hypercalciuria (HPDR II) maps in the Xp11 region. Eur. J. Hum. Genet. 1993; 1: 269–79.

Bosio M, Bianchi ML, Lloyd SE, Thakker RV. A familial syndrome due to Arg648Stop mutation in the X-linked renal chloride channel gene. Pediatr. Nephrol. 1999; 13: 278–83.

Brakemeier S, Si H, Gollasch M, Hoffler D, Buhl M, Kohler R, Hoyer J, Eichler I. Dent's disease: identification of a novel mutation in the renal chloride channel CLCN5. Clin. Nephrol. 2004; 62: 387–90.

Brandt S, Jentsch TJ. ClC-6 and ClC-7 are two novel broadly expressed members of the CLC chloride channel family. FEBS Lett. 1995; 377: 15–20.

Carballo-Trujillo I, Garcia-Nieto V, Moya-Angeler FJ, et al. Novel truncating mutations in the ClC-5 chloride channel gene in patients with Dent's disease. Nephrol. Dial. Transplant. 2003; 18: 717–23.

Carr G, Simmons N, Sayer J. A role for CBS domain 2 in trafficking of chloride channel CLC-5. Biochem. Biophys. Res. Commun. 2003; 310: 600–5.

Carr G, Simmons NL, Sayer JA. Disruption of clc-5 leads to a redistribution of annexin A2 and promotes calcium crystal

agglomeration in collecting duct epithelial cells. Cell. Mol. Life Sci. 2006; 63: 367–77.

Cebotaru V, Kaul S, Devuyst O, et al. High citrate diet delays progression of renal insufficiency in the ClC-5 knockout mouse model of Dent's disease. Kidney Int. 2005; 68: 642–52.

Charnas LR, Nussbaum RL. The oculocerebrorenal syndrome of Lowe (Lowe syndrome). In: Scriver CR, Beaudet AL, Sly WS, Valle D, eds. The Metabolic and Molecular Bases of Inherited Disease. New York: McGraw-Hill, 2001: pp. 6257–66.

Cheong HI, Lee JW, Zheng SH, et al. Phenotype and genotype of Dent's disease in three Korean boys. Pediatr. Nephrol. 2005; 20: 455–9.

Choudhury R, Diao A, Zhang F, et al. Lowe syndrome protein OCRL1 interacts with clathrin and regulates protein trafficking between endosomes and the trans-Golgi network. Mol. Biol. Cell. 2005; 16: 3467–79.

Christensen EI, Devuyst O, Dom G, et al. Loss of chloride channel ClC-5 impairs endocytosis by defective trafficking of megalin and cubilin in kidney proximal tubules. Proc. Natl Acad. Sci. USA 2003.

Christensen EI, Willnow TE. Essential role of megalin in renal proximal tubule for vitamin homeostasis. J. Am. Soc. Nephrol. 2003; 10: 2224–36.

Claverie-Martin F, Gonzalez-Acosta H, Flores C, Anton-Gamero M, Garcia-Nieto V. De novo insertion of an Alu sequence in the coding region of the CLCN5 gene results in Dent's disease. Hum. Genet. 2003; 113: 480–5.

Cox JP, Yamamoto K, Christie PT, et al. Renal chloride channel, CLCN5, mutations in Dent's disease. J. Bone Miner. Res. 1999; 14: 1536–42.

Curhan GC, Willett WC, Rimm EB, Stampfer MJ. A prospective study of dietary calcium and other nutrients and the risk of symptomatic kidney stones. N. Engl. J. Med. 1993; 328: 833–8.

Curhan GC, Willett WC, Speizer FE, Spiegelman D, Stampfer MJ. Comparison of dietary calcium with supplemental calcium and other nutrients as factors affecting the risk for kidney stones in women. Ann. Intern. Med. 1997; 126: 497–504.

Cutillas PR, Chalkley RJ, Hansen KC, et al. The urinary proteome in Fanconi syndrome implies specificity in the reabsorption of proteins by renal proximal tubule cells. Am. J. Physiol. Renal Physiol. 2004; 287: F353–64.

Dent CE, Friedman M. Hypercaliuric rickets associated with renal tubular damage. Arch. Dis. Child. 1964; 39: 240–9.

Devuyst O, Christie PT, Courtoy PJ, Beauwens R, Thakker RV. Intra-renal and subcellular distribution of the human chloride channel, CLC-5, reveals a pathophysiological basis for Dent's disease. Hum. Mol. Genet. 1999; 8: 247–57.

Dressman MA, Olivos-Glander IM, Nussbaum RL, Suchy SF. Ocrl1, a PtdIns(4,5)P(2) 5-phosphatase, is localized to the trans-Golgi network of fibroblasts and epithelial cells. J. Histochem. Cytochem. 2000; 48: 179–90.

Dutzler R, Campbell EB, Cadene M, Chait BT, MacKinnon R. X-ray structure of a ClC chloride channel at 3.0 A reveals the molecular basis of anion selectivity. Nature 2002; 415: 287–94.

Fisher SE, Black GCM, Lloyd SE, Wrong OM, Thakker RV, Craig IW. Isolation and partial characterization of a human chloride channel gene which is expressed in kidney and is a candidate for Dent's disease (an hereditary nephrolithiasis). Hum. Mol. Genet. 1994; 3: 2053–9.

Fisher SE, Van Bakel I, Lloyd SE, Pearce SHS, Thakker RV, Craig IW. Cloning and characterization of CLCN5, the human kidney chloride channel gene implicated in Dent disease (an X-linked hereditary nephrolithiasis). Genomics 1995; 29: 598–606.

Forino M, Graziotto R, Tosetto E, Gambaro G, D'Angelo A, Anglani F. Identification of a novel splice site mutation of CLCN5 gene and characterization of a new alternative 5' UTR end of ClC-5 mRNA in human renal tissue and leukocytes. J. Hum. Genet. 2004; 49: 53–60.

Frymoyer PA, Scheinman SJ, Dunham PB, Jones DB, Hueber P, Schroeder ET. X-linked recessive nephrolithiasis with renal failure. N. Engl. J. Med. 1991; 325: 681–6.

Gekle M, Mildenberger S, Freudinger R, Schwerdt G, Silbernagl S. Albumin endocytosis in OK cells: dependence on actin and microtubules and regulation by protein kinases. Am. J. Physiol. 1997; 272: F668–77.

Gekle M, Volker K, Mildenberger S, Freudinger R, Shull GE, Wiemann M. NHE3 Na+/H+ exchanger supports proximal tubular protein reabsorption in vivo. Am. J. Physiol. Renal Physiol. 2004; 287: F469–73.

Gunther W, Luchow A, Cluzeaud F, Vandewalle A, Jentsch TJ. ClC-5, the chloride channel mutated in Dent's disease, colocalizes with the proton pump in endocytotically active kidney cells. Proc. Natl Acad. Sci. USA 1998; 95: 8075–80.

Gunther W, Piwon N, Jentsch TJ. The ClC-5 chloride channel knock-out mouse – an animal model for Dent's disease. Pflugers Arch. 2003; 445: 456–62.

Hara-Chikuma M, Wang Y, Guggino SE, Guggino WB, Verkman AS. Impaired acidification in early endosomes of ClC-5 deficient proximal tubule. Biochem. Biophys. Res. Commun. 2005; 329: 941–6.

Hebert S, Brown E, Harris H. Role of the Ca^{2+}-sensing receptor in divalent mineral ion homeostasis. J. Exp. Biol. 1997; 200: 295–302.

Hoopes RR, Hueber PA, Reid RJ, et al. CLCN5 chloride-channel mutations in six new North American families with X-linked nephrolithiasis. Kidney Int. 1998; 54: 698–705.

Hoopes RR Jr., Raja KM, Koich A, et al. Evidence for genetic heterogeneity in Dent's disease. Kidney Int. 2004; 65: 1615–20.

Hoopes RR Jr., Shrimpton AE, Knohl SJ, et al. Dent disease with mutations in OCRL1. Am. J. Hum. Genet. 2005; 76: 260–7.

Hryciw DH, Wang Y, Devuyst O, Pollock CA, Poronnik P, Guggino WB. Cofilin interacts with ClC-5 and regulates albumin uptake in proximal tubule cell lines. J. Biol. Chem. 2003; 278: 40169–76.

Igarashi T, Gunther W, Sekine T, et al. Functional characterization of renal chloride channel, CLCN5, mutations associated with Dent's$_{Japan}$ disease. Kidney Int. 1998; 54: 1850–6.

Igarashi T, Hayakawa H, Shiraga H, et al. Hypercalciuria and nephrocalcinosis in patients with idiopathic low-molecular-weight proteinuria in Japan: is the disease identical to Dent's disease in United Kingdom? Nephron 1995; 69: 242–7.

Igarashi T, Inatomi J, Ohara T, Kuwahara T, Shimadzu M, Thakker RV. Clinical and genetic studies of CLCN5 mutations in Japanese families with Dent's disease. Kidney Int. 2000; 58: 520–7.

Janne PA, Suchy SF, Bernard D, et al. Functional overlap between murine Inpp5b and Ocrl1 may explain why deficiency of the murine ortholog for OCRL1 does not cause Lowe syndrome in mice. J. Clin. Invest. 1998; 101: 2042–53.

Jentsch TJ. Chloride transport in the kidney: lessons from human disease and knockout mice. J. Am. Soc. Nephrol. 2005; 16: 1549–61.

Jentsch TJ, Maritzen T, Zdebik AA. Chloride channel diseases resulting from impaired transepithelial transport or vesicular function. J. Clin. Invest. 2005a; 115: 2039–46.

Jentsch TJ, Poet M, Fuhrmann JC, Zdebik AA. Physiological functions of CLC Cl^- channels gleaned from human genetic disease and mouse models. Annu. of Rev. Physiol. 2005b; 67: 779–807.

Jouret F, Igarashi T, Gofflot F, et al. Comparative ontogeny, processing, and segmental distribution of the renal chloride channel, ClC-5. Kidney Int. 2004; 65: 198–208.

Kelleher CL, Buckalew VM, Frederickson ED, et al. CLCN5 mutation Ser244Leu is associated with X-linked renal failure without X-linked recessive hypophosphatemic rickets. Kidney Int. 1998; 53: 31–7.

Lemann J Jr., Pleuss JA, Worcester EM, Hornick L, Schrab D, Hoffmann RG. Urinary oxalate excretion increases with body size and decreases with increasing dietary calcium intake among healthy adults [published erratum appears in Kidney Int. 1996; 50(1): 341]. Kidney Int. 1996; 49: 200–8.

Lloyd SE, Gunther W, Pearce SHS, et al. Characterisation of renal chloride channel, CLCN5, mutations in hypercalciuric nephrolithiasis (kidney stones) disorders. Hum. Mol. Genet. 1997a; 6: 1233–9.

Lloyd SE, Pearce SH, Gunther W, et al. Idiopathic low molecular weight proteinuria associated with hypercalciuric nephrocalcinosis in Japanese children is due to mutations of the renal chloride channel (CLCN5). J. Clin. Invest. 1997b; 99: 967–74.

Lloyd SE, Pearce SHS, Fisher SE, et al. A common molecular basis for three inherited kidney stone diseases. Nature 1996; 379: 445–9.

Ludwig M, Doroszewicz J, Seyberth HW, et al. Functional evaluation of Dent's disease-causing mutations: implications for ClC-5 channel trafficking and internalization. Hum. Genet. 2005; 117: 228–37.

Luyckx VA, Goda FO, Mount DB, et al. Intrarenal and subcellular localization of rat CLC5. Am. J. Physiol. (Renal Fluid Electrol. Physiol.) 1998; 275: 761–9.

Luyckx VA, Leclercq B, Dowland LK, Yu ASL. Diet-dependent hypercalciuria in transgenic mice with reduced CLC5 chloride channel expression. Proc. Natl Acad. Sci. USA 1999; 21: 12174–9.

Magen D, Adler L, Mandel H, Efrati E, Zelikovic I. Autosomal recessive renal proximal tubulopathy and hypercalciuria: a new syndrome. Am. J. Kidney Dis. 2004; 43: 600–6.

Matsuyama T, Awazu M, Oikawa T, Inatomi J, Sekine T, Igarashi T. Molecular and clinical studies of Dent's disease in Japan: biochemical examination and renal ultrasonography do not predict carrier state. Clin. Nephrol. 2004; 61: 231–7.

Morimoto T, Uchida S, Sakamoto H, et al. Mutations in CLCN5 chloride channel in Japanese patients with low molecular weight proteinuria. J. Am. Soc. Nephrol. 1998; 9: 811–18.

Moulin P, Igarashi T, Van der Smissen P, et al. Altered polarity and expression of H^+-ATPase without ultrastructural changes in kidneys of Dent's disease patients. Kidney Int. 2003; 63: 1285–95.

Murakami T, Kawakami H. The clinical significance of asymptomatic low molecular weight proteinuria detected on routine screening of children in Japan: a survey of 53 patients. Clin. Nephrol. 1990; 33: 12–19.

Mussap M, Fanos V, Piccoli A, Zaninotto M, Padovani EM, Plebani M. Low molecular mass proteins and urinary enzymes in amniotic fluid of healthy pregnant women at progressive stages of gestation. Clin. Biochem. 1996; 29: 51–6.

Nakazato H, Hattori S, Furuse A, et al. Mutations in the CLCN5 gene in Japanese patients with familial idiopathic low-molecular-weight proteinuria. Kidney Int. 1997; 52: 895–900.

Nakazato H, Yoshimuta J, Karashima S, et al. Chloride channel CLCN5 mutations in Japanese children with familial idiopathic low molecular weight proteinuria. Kidney Int. 1999; 55: 63–70.

Neild GH, Thakker RV, Unwin RJ, Wrong OM. Dent's disease. Nephrol. Dial. Transplant. 2005; 20: 2284–5.

Norden AG, Lapsley M, Igarashi T, et al. Urinary megalin deficiency implicates abnormal tubular endocytic function in fanconi syndrome. J. Am. Soc. Nephrol. 2002; 13: 125–33.

Norden AG, Lapsley M, Lee PJ, et al. Glomerular protein sieving and implications for renal failure in Fanconi syndrome. Kidney Int. 2001; 60: 1885–92.

Obermuller N, Gretz N, Kriz W, Reilly RF, Witzgall R. The swelling-activated chloride channel ClC-2, the chloride channel ClC-3, and ClC-5, a chloride channel mutated in kidney stone disease, are expressed in distinct subpopulations of renal epithelial cells. J. Clin. Invest. 1998; 101: 635–42.

Oudet C, Martin-Coignard D, Pannetier S, Praud E, Champion G, Hanauer A. A second family with XLRH displays the mutation S244L in the CLCN5 gene. Hum. Genet. 1997; 99: 781–4.

Picollo A, Pusch M. Chloride/proton antiporter activity of mammalian CLC proteins ClC-4 and ClC-5. Nature 2005; 436: 420–3.

Piwon N, Gunther W, Schwake M, Bosl MR, Jentsch TJ. ClC-5 Cl⁻-channel disruption impairs endocytosis in a mouse model for Dent's disease. Nature 2000; 408: 369–73.

Pook MA, Wrong O, Wooding C, Norden AGW, Feest TG, Thakker RV. Dent's disease, a renal Fanconi syndrome with nephrocalcinosis and kidney stones, is associated with a microdeletion involving DXS255 and maps to Xp11.22. Hum. Mol. Genet. 1993; 2: 2129–34.

Raja KM, Schurman S, D'Mello RG, et al. Responsiveness of hypercalciuria to thiazide in Dent's disease. J. Am. Soc. Nephrol. 2002; 13: 2938–44.

Rebelo MA, Tostes V, Araujo NC, et al. Screening for CLCN5 mutation in renal calcium stone formers patients. An. Acad. Bras. Cienc. 2005; 77: 95–101.

Reinhart SC, Norden AGW, Lapsley M, et al. Characterization of carrier females and affected males with X-linked recessive nephrolithiasis. J. Am. Soc. Nephrol. 1995; 5: 1451–61.

Roschinger W, Muntau AC, Rudolph G, Roscher AA, Kammerer S. Carrier assessment in families with Lowe oculocerebrorenal syndrome: novel mutations in the OCRL1 gene and correlation of direct DNA diagnosis with ocular examination. Mol. Genet. Metab. 2000; 69: 213–22.

Santo Y, Hirai H, Shima M, et al. Examination of megalin in renal tubular epithelium from patients with Dent disease. Pediatr. Nephrol. 2004; 19: 612–15.

Sayer JA, Carr G, Pearce SH, Goodship TH, Simmons NL. Disordered calcium crystal handling in antisense CLC-5-treated collecting duct cells. Biochem. Biophys. Res. Commun. 2003; 300: 305–10.

Sayer JA, Carr G, Simmons NL. Calcium phosphate and calcium oxalate crystal handling is dependent upon CLC-5 expression in mouse collecting duct cells. Biochim. Biophys. Acta 2004; 1689: 83–90.

Scheel O, Zdebik AA, Lourdel S, Jentsch TJ. Voltage-dependent electrogenic chloride/proton exchange by endosomal CLC proteins. Nature 2005; 436: 424–7.

Scheinman SJ. X-linked hypercalciuric nephrolithiasis: Clinical syndromes and chloride channel mutations. Kidney Int. 1998; 53: 3–17.

Scheinman SJ. Nephrolithiasis. Sem. Nephrol. 1999; 19: 381–8.

Scheinman SJ, Cox JPD, Lloyd SE, et al. Isolated hypercalciuria with mutation in CLCN5: relevance to idiopathic hypercalciuria. Kidney Int. 2000; 57: 232–9.

Scheinman SJ, Pook MA, Wooding C, Pang JT, Frymoyer PA, Thakker RV. Mapping the gene causing X-linked recessive nephrolithiasis to Xp11.22 by linkage studies. J. Clin. Invest. 1993; 91: 2351–57.

Scheinman SJ, Thakker RV. X-linked nephrolithiasis/Dent's disease and mutations in the ClC-5 chloride channel. In: Econs MJ, ed. Genetic Aspects of Osteoporosis and Metabolic Bone Disease. Totowa, N.J.: Humana Press, 2000; pp. 133–52.

Schurman SJ, Norden AGW, Scheinman SJ. X-linked recessive nephrolithiasis: Presentation and diagnosis in children. J. Pediatr. 1998; 132: 859–62.

Schwake M, Friedrich T, Jentsch TJ. An internalization signal in ClC-5, an endosomal Cl-channel mutated in dent's disease. J. Biol. Chem. 2001; 276: 12049–54.

Silva IV, Cebotaru V, Wang H, et al. The CLC-5 knockout mouse model of Dent's disease has renal hypercalciuria and increased bone turnover. J. Bone Miner. Res. 2003; 18: 615–23.

Steinmeyer K, Schwappach B, Bens M, Vandewalle A, Jentsch TJ. Cloning and functional expression of rat CLC-5, a chloride channel related to kidney disease. J. Biol. Chem. 1995; 270: 31172–7.

Suchy SF, Nussbaum RL. The deficiency of PIP2 5-phosphatase in Lowe syndrome affects actin polymerization. Am. J. Hum. Genet. 2002; 71: 1420–7.

Takemura T, Hino S, Ikeda M, et al. Identification of two novel mutations in the CLCN5 gene in Japanese patients with familial idiopathic low molecular weight proteinuria (Japanese Dent's disease). Am. J. Kidney Dis. 2001; 37: 138–43.

Tosetto E, Graziotto R, Artifoni L, et al. Dent's disease and prevalence of renal stones in dialysis patients in Northeastern Italy. J. Hum. Genet. 2006; 51: 25–30.

Vandewalle A, Cluzeaud F, Peng KC, et al. Tissue distribution and subcellular localization of the ClC-5 chloride channel in rat intestinal cells. Am. J. Physiol. Cell. Physiol. 2001; 280: C373–81.

Wang SS, Devuyst O, Courtoy PJ, et al. Mice lacking renal chloride channel, CLC-5, are a model for Dent's disease, a nephrolithiasis disorder associated with defective receptor-mediated endocytosis. Hum. Mol. Genet. 2000; 9: 2937–45.

Wang Y, Cai H, Cebotaru L, et al. ClC-5: role in endocytosis in the proximal tubule. Am. J. Physiol. Renal Physiol. 2005; 289: F850–62.

Wrong O, Norden AGW, Feest TG. Dent's disease: a familial proximal renal tubular syndrome with low-molecular-weight proteinuria, hypercalciuria, nephrocalcinosis, metabolic bone disease, progressive renal failure, and a marked male predominance. Q. J. Med. 1994; 87: 473–93.

Wu F, Roche P, Christie PT, et al. Modeling study of human renal chloride channel (hCLC-5) mutations suggests a structural-functional relationship. Kidney Int. 2003; 63: 1426–32.

Yamamoto K, Cox JP, Friedrich T, et al. Characterization of renal chloride channel (CLCN5) mutations in Dent's disease. J. Am. Soc. Nephrol. 2000; 11: 1460–8.

Yanagida H, Ikeoka M, Kuwajima H, et al. A boy with Japanese Dent's disease exhibiting abnormal calcium metabolism and osseous disorder of the spine: defective megalin expression at the brush border of renal proximal tubules. Clin. Nephrol. 2004; 62: 306–12.

D. Primary Genetic Diseases of the Thick Ascending Limb of Henle

Molecular Genetics of Gitelman's and Bartter's Syndromes and their Implications for Blood Pressure Variation

UTE I. SCHOLL AND RICHARD P. LIFTON

INTRODUCTION

Gitelman's and Bartter's syndromes represent classic Mendelian disorders in humans. Their clinical descriptions presented a complex and bewildering array of physiologic abnormalities which thwarted efforts to define the underlying causes from physiologic analysis alone. Identification of the genes that cause these diseases revealed without question the underlying primary abnormalities and established that the diverse features seen must derive from these primary defects. As will be discussed, the mutations that cause these diseases have primary effects that impair renal electrolyte homeostasis. Importantly, they establish the dependence in vivo of renal salt and water homeostasis and blood pressure on specific channels and transporters (Figures 13.1 and 13.2; Table 13.1); moreover, the detailed understanding of disease causation defines the therapeutic opportunities for mutation carriers. Finally, they identify several new targets for therapeutic intervention in the general population that may have specific beneficial properties.

HISTORY OF BARTTER'S AND GITELMAN'S SYNDROMES

In 1962, Bartter (Barrter et al 1962) described two patients with severe hypokalemia, metabolic alkalosis, hyper-reninemic hyperaldosteronism and so-called normal blood

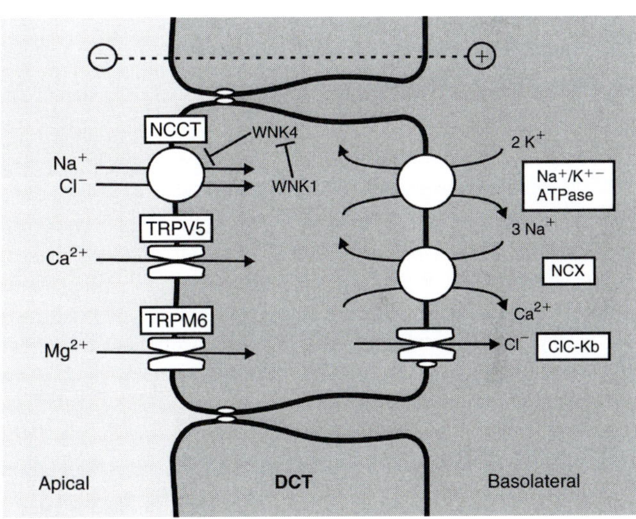

FIGURE 13.1 Salt reabsorption in the distal convoluted tubule. Na$^+$ and Cl$^-$ enter the cell via the apical cotransporter NCCT, which is regulated by the WNK kinases. Calcium reabsorption occurs via the apical TrpV5 Ca^{2+} channel and the basolateral NCX Na$^+$-Ca^{2+} exchanger. ClC-K channels serve basolateral chloride reabsorption; the apical TrpM6 channel is involved in magnesium reabsorption

pressure. Renal biopsy was normal except for hyperplasia of the juxtaglomerular apparatus and glomerular atrophy. Metabolic evaluation revealed marked urinary potassium and sodium chloride wasting. Dwarfism and mental retardation

Genetic Diseases of the Kidney
ISBN: 978-0-12-449851-8

229

were accompanying features and believed to be related to chronic potassium depletion. This is commonly referred to as the first description of what was later called Bartter's and Gitelman's syndromes. Rosenbaum and Hughes (1957) had reported a pediatric patient with persistent, probably congenital hypokalemic alkalosis as early as 1957; however, aldosterone levels were reported as normal, and the nature of this patient's syndrome remains unclear.

From the time of these original reports until 1995, over 600 cases of Bartter's syndrome were reported in the literature (Simon & Lifton 1996). The signs and symptoms of disease in these reports were quite variable, as is the age of presentation. Presentation includes intravascular volume depletion (Bettinelli et al 1992) with low blood pressure, which can vary from a medical emergency in the neonatal period, often with a history of polyhydramnios, to an incidental finding in adulthood; nephrocalcinosis (Matsumoto et al 1989) potentially leading to end-stage renal disease in the first decade of life; seizures (Hetzel et al 1991); tetany (Bettinelli et al 1992); muscular weakness (Marco-Franko et al 1994); paresthesias (Zarraga Larrondo et al 1992); and joint pain with chondrocalcinosis (Smilde et al 1994). Stunted growth and mental retardation are present in a minority of patients (Simopoulos 1979). In addition to hypokalemia and metabolic alkalosis, plasma renin activity and aldosterone levels are uniformly elevated. Diverse other features were identified in some patients, including: defective renal tubular sodium and chloride reabsorption in the loop of Henle (Chaimovitz et al 1973), hyperprostaglandinuria (Dunn 1981), impaired platelet function (Rodriguez Pereira & van Wersch 1983), insensitivity to the vasoconstrictive effects of angiotensin II and norepinephrine (Barrter et al 1962, Silverberg et al 1978) and elevated levels of atrial natriuretic peptide (Imai et al 1969, Graham et al 1986).

Gitelman et al (1966) recognized that Bartter's syndrome is in fact a heterogeneous entity, and described a distinct subset of patients having hypokalemic metabolic alkalosis with the additional features of hypomagnesemia and hypocalciuria; Gitelman's syndrome now refers to patients with these features.

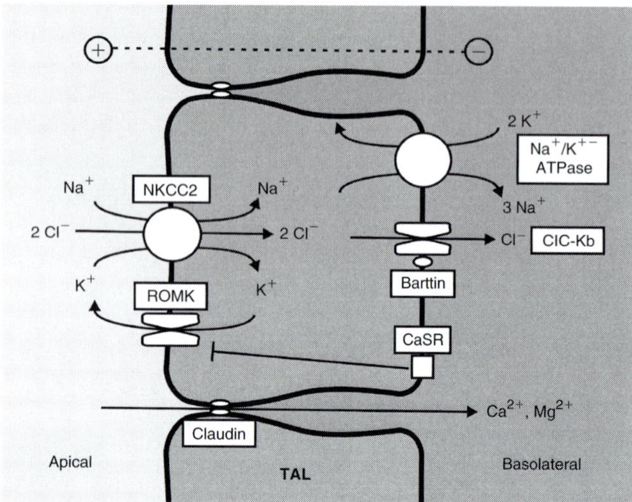

FIGURE 13.2 Salt reabsorption in the thick ascending limb of Henle's loop. Na^+, K^+ and Cl^- enter the cell via the apical cotransporter NKCC2 using the favorable Na^+ gradient established by the Na^+-K^+ ATPase. Chloride leaves via basolateral chloride channels, while potassium recirculates to the lumen via the K^+ channel ROMK. ROMK is inhibited by the calcium sensing receptor CaSR. Paracellular Ca^{2+} and Mg^{2+} reabsorption occurs in this nephron segment via paracellular flux across tight junctions, driven by the lumen-positive potential established by the asymmetric reabsorption of 1 Na^+ and 2 Cl^- via the net effects of NKCC2 and ROMK activities

TABLE 13.1   Distinguishing features of patients with Bartter's and Gitelman's syndromes due to mutations in different genes

Disease	Bartter's type I	Bartter's type II	Bartter's type III	Bartter's type IV	Bartter's type V	Gitelman's
Gene (encoded protein)	SLC12A1 (NKCC2)	KCNJ1 (ROMK)	CLCNKB (ClC-Kb)	BSND (barttin)	CASR (CaSR)	SLC12A3 (NCCT)
Age at diagnosis	Infancy	Infancy	Childhood	Infancy	Adolescence/ adulthood	Adolescence/ adulthood
K^+ as neonate	Low	High	Low	Low	NA	NA
Mg^{2+}	~nl	~nl	nl to low	nl to low	Low	Very low
Urinary Ca^{2+}/Cre	High	High	Variable	Variable	High	Very low
Nephrocalcinosis	+	+	Infrequent	Infrequent	+	−
Prematurity	+	+	−/(+)	+	−	−
Sensorineural deafness	−	−	−	+	−	−

+ and − denote the presence and the absence of the trait, respectively, −/(+) denotes variability with the trait being predominantly absent; nl = normal; NA = not applicable. Exceptions are discussed in the text

In 1992, Bettinelli et al (1992) provided the most definitive evidence that what had been referred to as Bartter's syndrome represents at least two distinct entities: Gitelman's syndrome and Bartter's syndrome. They studied 34 pediatric patients with hypokalemic metabolic alkalosis. The subgroup with Gitelman's syndrome had hypomagnesemia and profound hypocalciuria (urine calcium/creatinine $\leq$ 0.10 mmol/mmol) and a median age at diagnosis of 10.5 years. The Bartter's syndrome patients had normal to hypercalciuria (urine calcium/creatinine $\geq$ 0.40 mmol/mmol) and a far younger median age at diagnosis of 0.8 years. Despite this careful clinical classification, the overlapping features of these disorders resulted in continuing confusion and controversy, with a large number of patients with features of what is now recognized as Gitelman's syndrome being labeled as Bartter's syndrome in the literature.

Since the earliest reports of these disorders, both Gitelman's syndrome and Bartter's syndrome have been recognized to be familial. A variety of models of inheritance have been proposed, however the pattern of transmission in most reported families has been consistent with autosomal recessive inheritance (Rodriguez Pereira & van Wersch 1983). Once the genes responsible for these traits were identified, recessive transmission was proven.

MUTATIONS IN *NCC* CAUSE GITELMAN'S SYNDROME

Dissection of the physiology of renal electrolyte homeostasis suggested a number of potential candidate genes for Gitelman's syndrome and Bartter's syndrome; studies included investigation of genes encoding atrial natriuretic peptide and the angiotensin II receptor (AT1) (Graham et al 1986, Yoshida et al 1994), however these genes did not prove to be responsible for these traits.

Another attractive candidate gene was *SLC12A3* (aka *NCC*, *NCCT*, *TSC*), which encodes the thiazide-sensitive Na^{+}-Cl^{-} cotransporter. This cotransporter is localized to the distal convoluted tubule and believed to account for a significant fraction of net renal sodium reabsorption (Ellison 1991) (Figure 13.1). NCC is also the target of thiazide diuretics, one of the major classes of agents used in the treatment of high blood pressure. cDNAs encoding NCC were originally cloned from flounder bladder and rat kidney by Hebert and colleagues (Gamba et al 1993, 1994). The encoded protein from rat comprises 1002 amino acids, and contains 12 putative transmembrane domains, with long intracellular amino and carboxy termini. Similarities in features of patients with Gitelman's syndrome and patients receiving thiazide diuretics raised the possibility that mutations in *NCC* causing loss of function could result in Gitelman's syndrome. This consideration motivated examination of *NCC* as a candidate gene for Gitelman's syndrome.

At that time, the human *NCC* had not been cloned, nor were there informative genetic markers to enable its precise mapping in the human genome. Simon et al (1996) initially cloned the human *NCC* locus by hybridization of *NCC* cDNAs to human cosmid clone libraries. This led to identification of five overlapping cosmid clones that contained the entire genomic locus. They went on to demonstrate that the *NCC* protein was encoded in 26 exons. By hybridization of radiolabeled oligonucleotides to subclones of the *NCC* locus, they identified a polymorphic microsatellite marker at this locus, and used this to localize *NCC* on the human genetic map. This localized *NCC* to a 3 cM interval on chromosome 16q13 with very high likelihood, and identified additional linked polymorphisms that could be used to test for linkage of *NCC* to Gitelman's syndrome in kindreds with this disorder. Simon et al characterized eight multiplex families with Gitelman's syndrome. These included 26 affected members. In six kindreds all affected members were confined to one sibship; in one there was an affected parent and in one other there were affected second cousins with none of the other kindred members affected. All affected members had hypokalemic alkalosis with elevated renin and aldosterone, hypocalciuria, and hypomagnesemia; all had symptomatic presentation after age 8.

Markers across the interval containing *NCC* were genotyped in these eight families, and analysis of linkage between trait and marker was performed. Multipoint linkage analysis resulted in a maximum lod score of 9.5 (odds ratio > 3 billion:1) in favor of linkage of Gitelman's syndrome and the *NCC* locus at a recombination fraction of zero; the lod-1 interval spanned 7 cM encompassing the interval containing *NCC*. This finding provided extremely strong evidence that this interval contained the gene causing Gitelman's syndrome in these kindreds, and strongly motivated further investigation of *NCC* for mutations.

Affected members from 12 multiplex Gitelman's kindreds, who were inferred to harbor 26 mutant alleles, were comprehensively screened for mutations in *NCC*. Seventeen different mutations that alter the encoded protein were identified in affected subjects, and these accounted for 21 of the 26 mutant alleles; it was presumed that the remaining five alleles harbored mutations that escaped detection, as is common for virtually all inherited diseases. Thirteen of the identified mutations altered amino acids and one deleted an amino acid at positions that showed no variation between human and flounder, species that last shared a common ancestor ~400 million years ago. Two altered canonical splice donor or acceptor sites, and one introduced a premature termination codon. Importantly, these mutations co-segregated with Gitelman's syndrome in the pedigrees, none of these mutations was detected on 80 independent control chromosomes, and no such mutations were detected in a control sample of 26 nondisease subjects.

The finding of many independent mutations at highly conserved positions that cosegregate with and show specificity for the disease provided genetic proof that mutations in *NCC* cause Gitelman's syndrome. The finding of splice site and premature termination codons, along with the recessive transmission of disease, provides strong evidence that these mutations cause genetic loss of function. Biochemical evaluation of numerous disease-associated missense alleles has confirmed this expectation. Most of these mutant alleles have been assessed by expression in *Xenopus* oocytes (Kunchaparty et al 1999, Sabath et al 2004) and show variable activity in this system; whether this variation translates directly to the level of loss of function seen in vivo is not clear.

Schultheis et al (1998) generated and analyzed mice deficient for *NCC*. Interestingly, their phenotype is slightly different from human Gitelman's syndrome. *NCC* -/- mice develop hypocalciuria and hypomagnesemia; they do not, however, display an overt disease phenotype and lack hypokalemic alkalosis. Perturbations in fluid and electrolyte balance seem to be largely compensated, with increased renin, but normal aldosterone values. Importantly, the DCT of *NCC* -/- mice becomes hypoplastic, with a smaller number of DCT cells identified, reduced cell height and less elongated mitochondria. A similar phenomenon was observed with thiazide therapy in rats, wherein Loffing et al (1996) found DCT cell apoptosis. These findings indicate that DCT mass is determined at least substantially by activity of NCC and salt reabsorption, thus there may be effects in vivo that extend beyond the degree of the biochemical effect seen in vitro.

Since the identification of *NCC* as the gene for Gitelman's syndrome, many additional mutations have been identified in patients with this disease. The largest single study, only recently published, studied 148 subjects with biochemically defined Gitelman's syndrome and identified 91 different mutations in 246 alleles (Ji et al 2008). These included 29 missense mutations at positions conserved throughout all metazoa, including invertebrates, 24 at positions conserved only in vertebrates, eight splice site, five premature termination and ten frameshift mutations. As is well known for most Mendelian traits, a fraction of mutations will not be detected due to location outside coding regions. This does not constitute evidence for the effects of another gene. Convincing evidence against genetic homogeneity would require evidence excluding genetic linkage of Gitelman's syndrome to the *NCC* locus. No such kindreds have been identified. It thus appears that Gitelman's syndrome is rarely, if ever, caused by mutation in genes other than *NCC*.

UNIFORMITY OF ELECTROLYTE PHENOTYPES OF PATIENTS WITH GITELMAN'S SYNDROME

Since the identification of *NCC* as the gene for Gitelman's syndrome, it has become apparent that patients ascertained with the clinical syndrome of hypokalemic alkalosis with salt wasting, activation of the renin-angiotensin system, hypocalciuria, and hypomagnesemia essentially invariably have Gitelman's syndrome with mutation in *NCC*. The major exception is the group of patients who are taking thiazide diuretics; these patients share all of these hallmarks of Gitelman's syndrome.

These observations leave open the possibility of ascertainment bias; i.e. the patients who are evaluated for mutations are those with the full-blown phenotype. Subjects who are nonpenetrant or show incomplete phenotypes might not be investigated. Such possibilities are best assessed by investigation of extended families in which there are many at-risk family members. This is rare for recessive traits, but can occur in populations with high inbreeding coefficients. From the identification of an index case with typical Gitelman's syndrome who was found to be a compound heterozygote for an R642G mutation and a deletion that removes exons 1–7 of the gene, Cruz et al investigated 199 members of the extended kindred (Cruz et al 2001). Kindred members were systematically phenotyped and genotyped for these two mutations. Ultimately, 60 kindred members were found to be wild-type for *NCC*, 113 were heterozygotes with one mutant and one wild-type allele, and 26 were homozygous for *NCC* mutations. Most importantly, it was found that the biochemical features of hypokalemia, metabolic alkalosis, hypomagnesemia, and hypocalciuria were uniform among those with two mutant *NCC* alleles. All these subjects had $K^+ < 3.5\,\text{mM}$ (mean 2.8, range 1.9–3.4 mM) while none of the heterozygous or wild-type relatives had $K^+ < 3.5\,\text{mM}$ (mean of 4.3 mM in both groups). Mean serum Mg^{2+} was 1.1 mg/dl (0.45 mmol/l, range 0.6–1.8 mg/dl or 0.25–0.74 mmol/l) among homozygous mutant subjects vs. mean values of 1.9 mg/dl (0.78 mmol/l) in both heterozygous and wild-type subjects; mutation carriers showed a marked increase in 24-hour urinary Na^+ excretion, with homozygotes showing a 65% increase in Na^+ excretion compared with wild-type family members. Finally, 24-hour urinary Ca^{2+}:Cre ratio was < 0.25 (mean 0.15 mmol: mmol) in all subjects homozygous for *NCC* mutation but in no other family members (wild-type with mean ratio of 0.42 mmol:mmol). These findings document only modest variability in electrolyte values among patients homozygous for *NCC* mutations, and make clear that the classic electrolyte abnormalities apply to essentially all carriers of these mutations in the kindred studied. Nonetheless, there will undoubtedly be exceptional situations. For example, in one recent case report (Phillips et al 2006), a patient receiving K^+ supplementation developed acute renal failure and consequently presented with severe hyperkalemia. Ring et al (2002) reported a patient who was confirmed to be a compound heterozygote for *NCC* mutations G741R and F536L. However, urine samples failed to show hypocalciuria during magnesium supplementation or water diuresis, suggesting the use of samples with a high U_{osm} to establish the diagnosis. These findings underscore the importance of understanding the clinical situation in which patients present.

PATHOPHYSIOLOGY OF GITELMAN'S SYNDROME

The identification of mutations in *NCC* as the cause of Gitelman's syndrome provides an explanation for many of the clinical features of the disease, and suggests explanations for many others. The loss of *NCC* function results in reduced salt reabsorption in the DCT, eliminating ~6% of the normal salt reabsorbing capacity of the kidney in adults. This results in net loss of renal salt reabsorption and as a consequence an initial reduction in intravascular volume. This is perceived in the kidney via a mechanism proposed to be reduced delivery of chloride to the macula densa (Schnermann 2001), promoting increased secretion of the active aspartyl protease renin, which cleaves angiotensinogen to produce angiotensin I. Angiotensin I is further processed by angiotensin converting enzyme to angiotensin II (AII), which binds to its cognate receptors in the adrenal glomerulosa, increasing the secretion of aldosterone. Aldosterone in turn activates the mineralocorticoid receptor which acts in part by increasing activity of the epithelial sodium channel (ENaC), thereby increasing electrogenic Na^+ reabsorption. This increased Na^+ reabsorption produces the lumen-negative potential in the distal nephron that drives three processes: K^+ and H^+ secretion and Cl^- reabsorption. Thus, the loss of NCC function is compensated in part by activation of the renin-angiotensin-aldosterone system, which augments salt reabsorption at the expense of increased secretion of K^+ and H^+, thereby accounting for the hypokalemia and metabolic alkalosis observed in patients.

It is noteworthy that these patients consequently have chronic elevations in AII and aldosterone, yet remain free of hypertension, indicating that high AII and aldosterone alone are insufficient to sustain hypertension in the presence of even modest renal salt wasting. These observations underscore the primacy of renal salt handling, rather than AII/aldosterone levels, in blood pressure determination. One case study Schmitz et al (1994) noted cardiovascular remodeling without hypertension in a patient clinically diagnosed with Bartter's syndrome. However, more recent studies

(Calò et al 2007) involving Bartter's/Gitelman's patients observed altered signaling pathways of AII on a molecular level (increased expression of RGS-2 (see below), reduced expression of the α subunit of the Gq binding protein transducing the AII signal, reduced Ca^{2+} and IP3 release and reduced PKC activity); furthermore, the nitric oxide system seems to be upregulated while the RhoA/Rho-kinase pathway is downregulated.

The mechanisms of the accompanying electrolyte abnormalities in Gitelman's subjects (e.g. hypomagnesemia and hypocalciuria) are interesting and still not fully explained. Patients uniformly have hypocalciuria. Different models have been proposed to explain this observation on the basis of direct effects of NCC dysfunction on Ca^{2+} handling in the DCT (Kurtz 1998) (Figure 13.1). For example, reduced apical entry of Na^+ could result in reduced intracellular Na^+ level, leading to increased basolateral Na^+/Ca^{2+} exchange and reduction of intracellular Ca^{2+}; this in turn could promote increased Ca^{2+} influx via apical Ca^{2+} channels. Bindels and colleagues (Nijenhuis et al 2005) have recently proposed another explanation: they have shown that mice with genetic deficiency for TrpV5 (Hoenderop et al 2003), the apical, voltage-dependent calcium channel in the DCT that mediates Ca^{2+} entry, display severe hypercalciuria. Importantly, however, TrpV5 deficiency does not reverse the hypocalciuria induced by thiazide diuretics (Nijenhuis et al 2005), indicating that increased TrpV5 activity is not the cause of hypocalciuria in Gitelman's syndrome. Using WT and *TRPV5*-/- mice, they further demonstrated that the time course of development of hypocalciuria after initiation of thiazide treatment paralleled instead the increase in Na^+ reabsorption in the proximal tubule as a consequence of volume contraction. They thus propose an increase in proximal paracellular Ca^{2+} reabsorption that occurs as a result of compensatory enhanced Na^+ transport in the proximal tubule; the physiologic details underlying this effect remain uncertain.

Most interestingly, subjects with Gitelman's syndrome also have hypomagnesemia, a finding shared by patients taking thiazide diuretics. These findings unequivocally indicated an important role of the DCT in Mg^{2+} homeostasis, begging the question of mechanism. This link remained obscure until the elucidation of the genetic basis of another genetic disorder featuring magnesium deficiency (Schlingmann et al 2002). This disease, hypomagnesemia with secondary hypocalcemia (HSH) proved to be caused by homozygous loss-of-function mutations in TrpM6, a member of the Trp family of cationic channels. TrpM6 is expressed in the DCT and is believed to mediate Mg^{2+} flux (see Chapter 14 and Figure 13.1). To explain the hypomagnesemia that results from *NCC* deficiency, substantial efforts have been devoted to link the activities of NCC and TrpM6. Although Nijenhuis et al (2005) were able to show reduced TrpM6 expression in NCC knockout mice, a convincing link has not yet been made at the biochemical level. What has become increasingly apparent, however, is the remarkable plasticity of the mass of the DCT. Mice deficient for *NCC* show decreased DCT mass (Schultheis et al 1998), and rats treated with thiazide diuretics also show reduced DCT mass (Loffing et al 1996) (see discussion above). More recently, it has been shown that mutations in the protein kinase WNK4 that cause pseudo-hypoaldosteronism type II increase the mass of the DCT and increase NCC activity (Lalioti et al 2006). Intriguingly, combining the PHAII mutation with *NCC* deficiency converts the hyperplasia of the DCT to hypoplasia. These findings in concert indicate that DCT mass is directly regulated by NCC activity. Consequently, any other activity of the DCT may be affected by altered mass of this nephron segment. As a result, one might speculate that the hypomagnesemia of Gitelman's syndrome may reflect loss of TrpM6 via simple loss of DCT mass, and may not require a more complicated link between NCC and TrpM6 activities.

Patients with Gitelman's syndrome show reduced responsiveness to several vasoactive substances, including AII and epinephrine. This may occur for a number of reasons, including down-regulation of the AII receptor attributable to chronic exposure to ligand, as well as blunting of the response to ligand (Calò 2006). Recently, increased levels and activity of RGS-2, a regulator of G-protein signaling, have been demonstrated in Bartter's and Gitelman's syndromes (Calò et al 2008). RGS-2 acts as a GAP (GTPase activating protein), increasing the rate of conversion of G-protein-bound GTP to GDP, and appears to act on G-proteins coupled to the angiotensin II receptor. Consequently, increased RGS-2 activity may blunt AII signaling and response. The mechanism by which RGS-2 activity is increased in Bartter's and Gitelman's syndromes remains uncertain, but could arise as a direct consequence of chronically elevated AII levels.

CLINICAL SIGNS AND SYMPTOMS IN GITELMAN'S SYNDROME

Although Gitelman's syndrome is commonly referred to as featuring few or mild symptoms, patient-reported symptoms belie this notion and indicate a disease that adversely affects daily life. Part of the difficulty in assessing the complications of a rare disease is obtaining a sufficiently large cohort and providing proper control groups given that affected subjects are likely to be from diverse locations. Cruz et al (2001) have performed such a study in a cohort of 50 subjects with genotype-proven Gitelman's syndrome. Medical histories and qualities of life of these patients were assessed using standardized questionnaires; these values were compared to an age-, sex- and ethnically matched cohort and to cohorts with other chronic conditions such as hypertension,

diabetes, and congestive heart failure. The major symptom complexes that strongly distinguished Gitelman's syndrome subjects from controls included generalized weakness and dizziness, including fainting episodes in 34% of patients vs. none in controls. Muscle weakness of the extremities associated with cramping, carpal pedal spasm, and tetany, as well as a small number with paralysis also distinguished cases from controls. Paresthesias were present in more than 75% of cases. Salt and/or potassium craving was striking and present in 90% of cases, with descriptions of unusual dietary preferences including pickle brine, salted lemons, and salted cucumbers, as well as tomatoes and oranges. Affected subjects also reported a significantly increased prevalence of polyuria and nocturia, polydipsia, and thirst. Many patients had presented to emergency rooms for complaints of weakness, cramps, and paralysis, and had required intravenous treatment with potassium and/or magnesium. A number of subjects had required multiple visits for such problems. No significant differences were found between symptoms of males and females, although a number of women had encountered difficulty with pregnancy, predominantly volume depletion.

On standardized quality-of-life questionnaires, Gitelman's cases showed levels of overall health assessment that were worse than for subjects with hypertension and coronary artery disease, and comparable to those with congestive heart failure and diabetes. Most notably, perhaps, their reports of emotional wellbeing were lower than those of any of the above groups. These findings all underscore the point that patients with Gitelman's syndrome, while not as severely affected as those with Bartter's syndrome, nonetheless are far from asymptomatic.

GITELMAN'S SYNDROME PATIENTS HAVE REDUCED BLOOD PRESSURE AND INCREASED DIETARY SALT INTAKE

Patients with hyperaldosteronism have long been recognized to have hypertension with hypokalemia and metabolic alkalosis. Similarly, patients with Liddle syndrome have a primary increase in ENaC associated with hypertension with hypokalemia and metabolic alkalosis (Schild et al 1996). Patients with Gitelman's and Bartter's syndromes have been distinctive in that, despite having high levels of aldosterone with hypokalemia and metabolic alkalosis, they are typically free of hypertension. The finding that increased renal salt reabsorption led to an increase in blood pressure raised the question of whether reduced salt reabsorption would do the opposite – lowering blood pressure and protecting from the development of hypertension. Recognition that Gitelman's syndrome is caused by a primary defect in renal salt reabsorption posed the question of whether these patients might actually have reduced blood pressure. Formal examination of this point was performed by Cruz et al (2001), who systematically measured the blood pressure in subjects with Gitelman's syndrome (n = 26) and compared their blood pressures to those of relatives with no *NCC* mutations. Both Gitelman's males and females had low blood pressure, with means of 109/68 and 113/66 in adult males and females, respectively. These blood pressures were significantly lower than those of their unaffected relatives.

Cruz et al also measured 24-hour urinary sodium excretion in these subjects while they consumed their normal ad-lib diet; in steady-state conditions, output must closely matched intake. Gitelman's subjects had markedly increased urinary Na^+ excretion compared with their wild-type relatives, with an average 65% increase in dietary salt intake. Many Gitelman's subjects reported salt craving. These findings support the notion that the primary salt wasting caused by loss of NCC function induces an increased drive for salt consumption in order to compensate for the primary defect.

Cruz et al also studied the effect of the heterozygous state for *NCC* mutation on dietary salt intake and blood pressure. They noted that heterozygotes also had significantly increased salt intake compared to their wild-type relatives. They found a significant reduction in blood pressure among young (under age 18) heterozygotes, with an average 8.7 mmHg reduction in diastolic blood pressure. This within-family study was limited by the fact that patients were studied in the field, and that there was wide variation in ages of kindred members.

Recently, Fava et al (2008) compared systolic and diastolic blood pressure values of heterozygous carriers to those of matched controls in a Swedish population. They reported slightly higher fasting glucose and significantly lower blood pressure in heterozygotes and suggested that this effect might be more pronounced than in the report of Cruz et al (2001) because of less compensatory increase in salt intake.

These findings collectively indicate that *NCC* mutations result in an increase in dietary salt intake, and lower blood pressure in both the homozygous and heterozygous states; they raise the question of the prevalence and impact of heterozygous mutations in the general population (see below, 'Effects of the heterozygous state for Gitelman's and Bartter's syndromes on blood pressure in the population').

These observations make several other interesting points. First, they indicate that a primary defect in renal salt reabsorption influences the drive for dietary salt intake. The mechanisms underlying this effect are uncertain, but may well owe to activation of the renin-angiotensin system (Geerling & Loewy 2008). This has potential significance with regard to the compensatory response to thiazide diuretics – the diuretics are likely inducing increased drive for salt consumption, blunting their antihypertensive effects. It is also worth noting that patients with Gitelman's syndrome have high salt intake but lower blood pressure than unaffected family members. This is relevant to understanding the confounded relationship between salt and blood pressure in the general population.

GITELMAN'S SYNDROME PATIENTS DISPLAY INCREASED BONE DENSITY

Comparing 24 patients with confirmed Gitelman's syndrome to their non-affected family members, Cruz et al (2001) demonstrated that affected and, more importantly, even heterozygous family members, showed significantly higher bone density at the lumbar spine and hip compared with controls who carried no mutations. This points to a considerable protective effect on hip and spine fractures in these subjects as well as in patients treated with thiazide diuretics. The mechanism of this effect is not fully understood; additional processes apart from mere hypercalcemia due to hypocalciuria might be involved. In addition, affected patients can develop chondrocalcinosis (Volpe et al 2007), which is, however, not a common feature of Gitelman's syndrome.

Bartter's syndrome is caused by mutations that impair salt reabsorption in the thick ascending limb of Henle. The discovery that *NCC* mutations cause Gitelman's syndrome immediately suggested that Bartter's syndrome would result from related mechanisms. These patients also had evidence of salt wasting with activation of the renin-angiotensin system. They differed from Gitelman's subjects in having increased, rather than decreased, urinary calcium levels (Bettinelli et al 1992). This latter finding suggested the possibility of a primary defect in the thick ascending limb of Henle (TAL) (Figure 13.2), where disruption in any of the channels and transporters working in concert would lead to decreased paracellular Ca^{2+} reabsorption. In the TAL, the bumetanide-sensitive Na^+-K^+-$2Cl^-$ cotransporter NKCC2 (product of the *SLC12A1* gene) mediates electroneutral uptake of these ions from the lumen. It uses the electrochemical Na^+ gradient established by the basolateral Na^+/K^+ ATPase that transports Na^+ from the TAL to the interstitium. While Cl^- that enters the TAL via this cotransporter exits through a then-unknown mechanism in the basolateral membrane, K^+ is 'recycled' to the lumen via the renal outer medullary K^+ channel ROMK (the product of the *KCNJ1* gene). The net result of these activities is electrogenic, leaving a lumen-positive potential. This provides the electrical driving force for the reabsorption of Ca^{2+} and Mg^{2+}[47] which occurs via the paracellular pathway. The observance that furosemide, an inhibitor of NKCC2 activity, produced a phenotype with features similar to aspects of Bartter's syndrome, was consistent with this idea.

MUTATIONS IN *NKCC2*

To investigate this possibility, Simon et al (1996) collected samples from nine affected members of five kindreds with well-characterized Bartter's syndrome, along with 12 unaffected siblings. Patients had presented with hypokalemia and metabolic alkalosis in the neonatal period with high plasma renin activity and aldosterone levels, had marked hypercalciuria, with most affected subjects having evidence of nephrocalcinosis. All had been born prematurely with polyhydramnios. In four of these kindreds, there was known parental consanguinity, providing substantial power for analysis of linkage. Analysis of linkage in each of these families excluded linkage to *NCC* on chromosome 16. Simon et al. cloned and characterized the human *NKCC2* locus and identified informative microsatellite markers, permitting localization of *NKCC2* by analysis of linkage to chromosome 15q. Genotyping of informative markers across the *NKCC2* locus provided strong evidence of linkage – affected offspring of consanguineous union were homozygous across the *NKCC2* interval, affected sibs were all of identical genotype, and unaffected sibs all differed in genotype from their affected sibs. These findings provided strong evidence supporting linkage of *NKCC2* and Bartter's syndrome in these kindreds. Investigation of *NKCC2* for mutation in these kindreds identified six mutant alleles, accounting for all disease chromosomes in affected subjects. These included three frameshift mutations and three nonconservative amino acid substitutions at positions conserved among *NKCC2* orthologs and the related protein NKCC1 in species as distantly related as shark.

Further evidence supporting the pathogenetic role of *NKCC2* in Bartter's syndrome came from the study of a mouse model deficient for *NKCC2* (Takahashi et al 2000). *NKCC2 -/-* mice, in contrast to the human disease phenotype, did not develop obvious polyhydramnios. Salt loss with severe dehydration, however, was noticed 24 h postnatally, with subsequent failure to thrive and poor outcome of less than 2 weeks survival. Intriguingly, administration of indomethacin, which is the established treatment of Bartter's syndrome (see below), resulted in survival of 10% of *NKCC2 -/-* mice beyond weaning, with considerable long-term survival even after indomethacin had been stopped at 3 weeks of age. These mice, however, invariably developed hydronephrosis. Hypokalemia was less severe than in humans, and initial acidosis evolved to alkalosis. These mice also had hypercalciuria and nephrocalcinosis; proteinuria was also observed. In summary, mice deficient for *NKCC2* display a similar phenotype as patients with mutations in *NKCC2*.

MUTATIONS IN ROMK

Not all patients with the clinical diagnosis of Bartter's syndrome proved to have mutations in *NKCC2*. Simon et al (1996) studied another nine Bartter's syndrome kindreds and were able to formally exclude linkage to *NKCC2* in two of these, providing evidence of at least one additional gene. *ROMK* was considered an attractive candidate because of evidence that it is the 'recycling'

K^+ channel in the TAL (Giebisch 1995, Hebert 1995). In two of these kindreds, homozygous frameshift or premature termination codons were found in affected subjects from consanguineous union. In two others, compound heterozygotes were found – one with a premature termination codon and a mutation at a highly conserved protein kinase A phosphorylation site known to be essential for normal cotransporter activity, the other with a frameshift mutation and a missense mutation at a conserved position. These findings established mutations in *ROMK* as a second cause of Bartter's syndrome. As for *NKCC2*, *ROMK* knockout mice were generated (Lorenz et al 2002), showing a similar, but less severe phenotype with 5% survival to weaning. Interestingly, however, indomethacin treatment was not beneficial in this model.

MUTATIONS IN *CLCNKB*

Simon et al (1997) went on to investigate 66 kindreds with Bartter's syndrome, and found mutations in *NKCC2* and *ROMK* in only 22 of these, findings which suggested at least one additional gene that can cause this disease. Analysis of linkage in 11 consanguineous kindreds without mutation in either of these genes showed homozygosity at *NKCC2* or *ROMK* in only one instance, essentially excluding linkage in these kindreds and implicating other loci. One candidate for such a gene was the Cl^- channel believed to lie on the basolateral membrane that was used for Cl^- exit from the TAL epithelium. While the identity of this channel was unknown, two Cl^- channels expressed in the kidney had been identified – *CLCNKA* and *CLCNKB* (Kieferle et al 1994). Simon et al (1997) cloned and characterized these two genes, and demonstrated that they lie in tandem with *CLCNKA* 5' to *CLCNKB*. By identification of polymorphisms at this locus and linkage in CEPH kindreds, these genes were localized to human chromosome 1. Linkage analysis in the 11 consanguineous Bartter's kindreds that exclude *NKCC2* and ROMK demonstrated significant evidence of linkage with Bartter's syndrome, with five kindreds showing homozygosity across this locus.

Examination of these genes in Bartter's patients without mutation in *NKCC2* or *ROMK* revealed mutations affecting *CLCNKB* in 17 kindreds. In ten kindreds, these mutations were homozygous deletions of all or a segment of the coding region of *CLCNKB*. In at least one case, the mechanism of deletion was shown to be unequal, crossing over between *CLCNKA* and *CLCNKB*. In four others, there were homozygous missense mutations, and in three others compound heterozygous missense, splice site, or premature termination mutations. No significant missense mutations were observed in *CLCNKA*. These findings established loss-of-function mutations in *CLCNKB* as a cause of Bartter's syndrome, and suggested this Cl^- channel as the functional mediator of basolateral Cl^- exit from epithelia of the TAL.

MUTATIONS IN BARTTIN CAUSE BARTTER'S SYNDROME WITH SENSORINEURAL DEAFNESS

In 1995, Landau et al (1995) reported a Bedouin kindred with five members affected with Bartter's syndrome and sensorineural deafness. The phenotype of these patients appeared indistinguishable from other cases of Bartter's syndrome except for the occurrence of deafness, which was detected as early as 3 weeks of age. Brennan et al (1998) used homozygosity mapping to localize the gene responsible for disease in this family to a 3.4 cM segment of chromosome 1p31. They were able to exclude mutation in *CLCNKB* as the cause, establishing the presence of a new locus for Bartter's syndrome. Vollmer et al (2000) subsequently demonstrated that nine additional kindreds with Bartter's syndrome and sensorineural deafness also mapped to this locus, providing strong evidence that a single gene was responsible for both the renal and otic phenotypes. Birkenhäger et al (2001) refined the location of the disease gene and identified it as a novel gene, *Barttin* (*BSND*); the encoded protein is predicted to have two transmembrane domains. In 11 kindreds with this disease they identified eight different mutations. These included mutations that mutated the translation initiation codon, a splice site mutation, an intragenic deletion, and multiple missense mutations. This provided strong evidence that the mutations caused genetic loss of function. They demonstrated that *BSND* is expressed in both the thin and thick limbs of Henle's loop. *BSND* was also expressed in the cochlea, in marginal cells, and dark cells of the stria vascularis.

Estévez et al (2001) established that barttin is an accessory subunit for both ClC-Ka and ClC-Kb. Neither human Cl^- channel supported Cl^- flux when expressed alone in *Xenopus* oocytes, however co-expression with barttin resulted in Cl^- channel activity, resolving a discussion in the field, that was going back to the cloning of ClC-K channels (Adachi et al 1994, Kieferle et al 1994), in 1995. At least part of this effect appeared to relate to increased trafficking of the Cl^- channel to the cell surface in the presence of barttin. Most of the mutations found in disease patients resulted in loss of Cl^- channel function. Using specific antibodies, Estévez et al (2001) further demonstrated that barttin is expressed on the basolateral membrane of the thin and thick limb of Henle, as well as cells of the stria vascularis. They proposed a compelling model in which NKCC1 mediates basolateral entry of Na^+, K^+ and Cl^- into these cochlear cells; Na^+ and Cl^- exit across the basolateral membrane whereas apical K^+ channels mediate exit of K^+ into the endolymph. The further demonstration that both ClC-Ka and ClC-Kb are expressed in the cochlea suggested an explanation as to why mutations in either gene might not lead to deafness, whereas mutation in barttin would.

Scholl et al (2006) demonstrated that the amino terminal segment of barttin, which contains the two transmembrane domains, is sufficient for promoting trafficking of ClC-Kb from endoplasmic reticulum to the cell surface, which might explain the clustering of disease-causing mutations in these segments. Further evidence supported an additional effect of the cytoplasmic C-terminus to increase the unitary conductance and open probability of rat ClC-K1 (homolog of ClC-Ka) channels.

COMBINED MUTATIONS IN *CLCNKA* AND *CLCNKB* CAUSE BARTTER'S SYNDROME WITH SENSORINEURAL DEAFNESS

The findings with barttin suggested that this protein is a beta subunit for both ClC-Ka and ClC-Kb; since both of these Cl^- channels are expressed in the cochlea, it was logical to propose that their functions were redundant such that loss of either alone would not produce deafness while mutation in barttin would. This raised the question of whether Bartter's syndrome plus deafness could result from the mutation of both *CLCNKA* and *CLCNKB*. Schlingmann et al (2004) reported the case of a patient presenting with the characteristic features of Bartter's syndrome with sensorineural deafness. Remarkably, they found a homozygous *CLCNKB* deletion combined with a *CLCNKA* missense mutation (ClC-Ka W80C); expression of the mutated ClC-Ka protein in oocytes resulted in a reduction of whole-cell current amplitudes to about half of the wild-type value.

Whereas parental consanguinity of the patient described by Schlingmann et al still left open the possibility that deafness was resulting from recessive inheritance at an independent locus, Nozu et al (2008) reported a patient with neonatal Bartter's syndrome with sensorineural deafness and renal dysfunction (and, interestingly, hypocalciuria and hypomagnesemia) born from a nonconsanguineous family. They excluded mutations in *BSND* and found a *CLCNKA* nonsense mutation (Q260X) and a *CLCNKB* splicing site mutation (IVS17 + 1 g > a) in the paternal alleles, combined with a ~12 kbp deletion extending from *CLCNKA* to *CLCNKB* on the maternal allele. Their report confirmed digenic inheritance in Bartter's syndrome with sensorineural deafness, further supporting the in vivo role of both ClC-Ka and ClC-Kb in the process of hearing. Mutations in ClC-Ka without additional ClC-Kb mutations have not been described; however, the diabetes insipidus phenotype of a *CLCNK1* (the homolog of human *CLCNKA*) knockout mouse generated by Matsumura et al (1999) suggests a mild diabetes insipidus that might have evaded detection in humans thus far.

MUTATIONS IN THE CALCIUM-SENSING RECEPTOR (*CASR*)

The calcium-sensing receptor (Brown et al 1993) regulates both PTH secretion in the hypothalamus as well as renal Ca^{2+} reabsorption in the TAL. In the setting of high plasma calcium levels, activation of the CaSR directly inhibits Ca^{2+} reabsorption and also indirectly inhibits reabsorption by inhibition of ROMK, which reduces salt reabsorption, thereby diminishing the lumen-negative potential which helps drive Ca^{2+} reabsorption (see discussion above; Figure 13.2). As a consequence, it is theoretically possible that activation mutations in the CaSR, which cause autosomal dominant familial hypercalciuric hypocalcemia (FHH), could cause a salt-wasting syndrome similar to Bartter's syndrome. Five FHH patients with potent activating mutations in CASR have been reported to have such a phenotype (Vargas-Poussou et al 2002, Watanabe et al 2002, Vezzoli et al 2006). These patients all came to medical attention because of hypocalcemia, but at later ages were also found to have variable degrees of salt wasting, hypokalemia, hypercalciuria, and hypomagnesemia. Levels of renin and aldosterone were moderately elevated. Affected patients were diagnosed with their Bartter's-like phenotypes at ages 7, 19, 22 (twins) and 26. The marked and symptomatic hypocalcemia seen in these patients, the relatively modest consequences of salt wasting and the late presentation suggest that there would be little ambiguity in distinguishing these patients from those with other types of Bartter's syndrome; their hypercalciuria would also ensure distinction from patients with Gitelman's syndrome, who uniformly have hypocalciuria.

OTHER GENETIC DISEASES WITH FEATURES THAT OVERLAP WITH GITELMAN'S AND BARTTER'S SYNDROMES

A number of other diseases have features that overlap with aspects of Bartter's syndrome, but rarely cause diagnostic confusion. For example, subjects with Dent's disease have hypercalciuria and nephrocalcinosis, but rarely (Besbas et al 2005) show severe salt wasting with activation of the renin-angiotensin system.

Secondary causes of Bartter's syndrome have been reported as a consequence of several other diseases including Kearns-Sayre syndrome and cystinosis. Kearns-Sayre syndrome is a mitochondrial disease with ophthalmoplegia, pigmentary degeneration of the retina, and heart block as the leading findings. Tubular impairment is rare, however, Emma et al (2006) recently reported a patient who, at adolescent age, was additionally diagnosed with renal salt loss, hypercalciuria, hypokalemic metabolic alkalosis, and

activation of the RAS. They suggested an interesting mechanism – the development of renal heteroplasmy, with impaired mitochondrial function causing reduced salt reabsorption in the thick ascending limb and compensatory increased salt reabsorption in the distal nephron that might well account for similar cases reported previously. Similarly, hypokalemic metabolic alkalosis with low blood pressure despite hyperreninemic hyperaldosteronism has also been reported in patients with cystinosis, a metabolic disease leading to the deposition of cystine in various tissues (Whyte et al 1985). Impaired proximal tubular function (Fanconi syndrome) is the characteristic renal manifestation of cystinosis, and it seems possible that altered function of the TAL might lead to Bartter-like features in rare cases, possibly linked to distinct mutations in the *CTNS* gene (Pennesi et al 2005). Though these reports provide interesting genetic and pathophysiological observations, the characteristic features of the primary disease distinguish them from primary Bartter's syndrome.

Finally, the feature of hypokalemic metabolic alkalosis due to increased ENaC activity is also present in glucocorticoid remediable aldosteronism (Lifton et al 1992), Liddle's syndrome (Schild et al 1996), and the syndrome of apparent mineralocorticoid excess (Mune et al 1995). However, these diseases are readily distinguished from Bartter's syndrome by presenting with hypertension, suppressed rather than elevated renin activity and the absence of salt wasting.

PATHOPHYSIOLOGY OF BARTTER'S SYNDROME

All of the mutations that cause Bartter's syndrome impair renal salt reabsorption in the thick ascending limb of Henle, a nephron segment that mediates the reabsorption of $\sim30\%$ of the filtered load of salt. As a consequence, there is massive renal salt wasting accompanied by volume loss. This process begins in utero, accounting for the polyhydramnios that is typically encountered, which may in turn contribute to the preterm delivery that is typical of affected infants. While affected children are born in a euvolemic state due to placental function, they rapidly develop dramatic intravascular volume depletion after birth. This leads to activation of the renin-angiotensin system, with high levels of its active components, including plasma renin activity and aldosterone. Aldosterone promotes electrogenic Na^+ reabsorption via ENaC, which partially, but incompletely, compensates for reduced Na^+ reabsorption in the TAL. However, this compensation occurs at the expense of increased secretion of K^+ and H^+ as a consequence of the increased lumen-negative potential arising from ENaC activity. This accounts for the hypokalemia and metabolic alkalosis typically encountered in Bartter's patients.

Because mutations that impair Na-Cl reabsorption in the TAL prevent formation of the lumen-positive potential of this nephron segment, these mutations also inhibit reabsorption of Ca^{2+} and Mg^{2+}. This results in marked hypercalciuria in patients with mutations in *NKCC2* and *ROMK*, respectively. Normal serum Ca^{2+} levels are nonetheless maintained. Possible mechanisms might include increased PTH secretion and 1,25 dihydroxy vitamin D production (Proesmans 1997), intestinal calcium absorption as well as increased retrieval from bone (Rogriguez Soriano et al 2005).

Most patients maintain normal serum Mg^{2+} levels, and those with hypomagnesemia rarely reach the levels seen in patients with Gitelman's syndrome. This strongly suggests the ability to markedly augment Mg^{2+} reabsorption in other nephron segments; the DCT is a strong candidate via activity of the TrpM6 channel (Schlingmann et al 2002) (Figure 13.1). The recognition that NCCT activity regulates mass of the DCT (Schultheis et al 1998) strongly suggests that upregulation of NCCT activity via aldosterone may contribute to the increase in TrpM6 levels and be protective against hypomagnesemia in Bartter's patients. This could provide an explanation as to why these patients are protected from severe hypomagnesemia whereas patients with paracellin-1 (aka claudin-16) mutations (Simon et al 1999), who share marked Mg^{2+} wasting from the TAL, but have no salt wasting, develop dramatic hypomagnesemia.

Bartter's patients commonly develop nephrocalcinosis. This is presumed to be due to their hypercalciuria, however the mechanisms that account for development of nephrocalcinosis versus nephrolithiasis, which has only been reported in few cases (Colussi et al 2002, Zelikovic et al 2003), are poorly understood. Consistent with this notion, nephrocalcinosis is less prevalent among patients with *CLCNKB* and *BSND* mutations, who have less hypercalciuria. It has been suggested that the polyuria associated with the renal concentration defect would protect from the development of nephrolithiasis, but the mechanism by which nephrocalcinosis occurs (for example, low urinary citrate, tubulointerstitial damage (Naesens et al 2004)) remains unclear. It is possible that bone reabsorption from elevated PTH levels leads to increased urinary phosphate wasting, and that this feature promotes nephrocalcinosis rather than nephrolithiasis, though this is poorly documented in patients. Consistent with this possibility, Alon and collegues (Pattaragarn et al 2004) showed that in the young rat, furosemide-induced nephrocalcinosis seems to be at least partially PTH-dependent, because suppression of PTH secretion with an activator of the CaSR resulted in a significant decrease in furosemide-induced calcification.

Another potential mechanism contributing to the occurrence of nephrocalcinosis rather than nephrolithiasis might involve renal luminal pH; the increased acidification of the distal nephron due to increased ENaC activity might protect from calcium phosphate crystal formation in the distal parts of the nephron.

An additional distinctive feature of Bartter's patients is that they produce high levels of prostaglandin E2 (PGE2). This appears to occur as a consequence of increased renal elaboration of this prostaglandin, and is consistent with an increase in COX2 expression in the macula densa. The physiologic effects of increased PGE2 are diverse and include fever, diarrhea, and vomiting (Reinalter et al 2002). Its effects on the kidney can increase renal salt loss – renal blood flow and GFR are increased, enhancing the filtered salt load, and salt reabsorption along the nephron is further diminished, exacerbating renal salt wasting. Thus PGE2 can be maladaptive in Bartter's syndrome. It has been proposed that reduced Cl^- entry into the macula densa and/or cells of the TAL via NKCC2 is the signal for increased expression of COX2 and microsomal prostaglandin E2 synthase (Kömhoff et al 2004). If correct, one might anticipate that mutations causing Bartter's syndrome that impair apical Cl^- entry (i.e. *NKCC2* and *ROMK*) should have high PGE2, whereas mutations that impair Cl^- exit across the basolateral membrane (i.e. *CLCNKB*, *BSND*) might have high intracellular Cl^- levels. In contrast to this expectation, a recent report by Jeck et al (2005) found highest PGE2 levels in patients with *BSND* mutations, followed by patients with mutations in *NKCC2* and *ROMK*, while patients with *CLCNKB* had levels near normal. There was however, wide variation in urinary PGE2 level in each group, thwarting any simple genotype–phenotype correlation. It seems clear that the mechanism underlying high PGE2 levels in Bartter's syndrome remains an important but elusive problem.

GENOTYPE–PHENOTYPE CORRELATIONS (TABLE 13.1) AND DIAGNOSIS

Even before the identification of the corresponding genes, phenotypic variation had been noted in patients with Bartter's syndrome, distinguishing one group of patients presenting early, with severe salt loss and marked prostaglandinuria, from another group with a milder phenotype and later onset. This prompted Seyberth et al (1987) to introduce the term 'hyperprostaglandin E syndrome,' for the first group, with the second group commonly referred to as 'classic' Bartter's syndrome. However, after the identification of the responsible genes, it has become clear that patients with mutations in the same gene can present early or late and with mild or severe disease. The mutated genes and their biochemical consequences, however, provide a more useful context for understanding the varied clinical presentations; the genotype can most often be predicted from the phenotypic presentation, consistent with this notion (Table 13.1).

The first and most clear-cut distinction is between patients with Gitelman's and Bartter's syndromes. These patients differ in their ages of presentation, severity, presence of hypo- versus hypercalciuria and hypomagnesemia. Patients with Gitelman's syndrome virtually never present in the neonatal period with life-threatening hypovolemia, as often occurs with Bartter's syndrome. Presentation around the time of adolescence or in adulthood is typical of Gitelman's syndrome; hypokalemic alkalosis may be an incidental finding, or the syndrome might be discovered in the evaluation of muscular weakness, tetany, or seizures. Gitelman's patients almost always show the full spectrum of salt wasting with hypokalemic alkalosis, elevated renin and aldosterone, hypocalciuria and hypomagnesemia; recently, a small number of exceptions have been reported in Asian males, an interesting observation that should be pursued (Lin et al 2004).

In contrast, patients with Bartter's syndrome often present in the neonatal period with a history of prematurity and polyhydramnios; the finding of typically massive renal salt wasting with hypokalemia, metabolic alkalosis, high renin and aldosterone levels, and hypercalciuria strongly support the diagnosis. There is variation in the clinical presentation, however. Patients may present at later ages, and the degree of hypercalciuria can be variable. In addition, some patients can present with modest hypomagnesemia, though the severely low levels often seen in Gitelman's syndrome are uncommon.

Among Bartter's syndrome patients, there are a number of clinical features that are distinctive for the underlying mutant gene and which account for much of the phenotypic variability. Patients with *NKCC2* or *ROMK* mutations are usually diagnosed early, sometimes even prenatally, with polyhydramnios leading to prematurity and severe salt loss resulting in life-threatening volume depletion in the neonatal period. They have high renin and aldosterone levels, hypercalciuria, and almost always nephrocalcinosis. Patients with mutations in *ROMK* are unique in often presenting with hyper- rather than hypokalemia at birth, sometimes causing diagnostic confusion. A special feature with regard to their potassium homeostasis was noted by Simon et al (1996) and extensively studied by other groups (Peters et al 2002, Finer et al 2003): Patients with mutations in *ROMK* are unique in presenting with hyper- rather than hypokalemia at birth, sometimes making the diagnosis challenging. This hyperkalemia later evolves into hypokalemia, which, however, is modest, often not requiring potassium supplementation. This interesting feature can be explained by the expression of ROMK in the distal nephron, where, similar to its function in the TAL, the channel is involved in apical potassium secretion. As a consequence, patients with *ROMK* mutations have impaired distal K^+ secretion, limiting renal K^+ loss. It is likely that the neonatal hyperkalemia reflects the late development of K^+ secretion via the BK K^+ channel, which can protect the kidney from hyperkalemia in the setting of ROMK deficiency (Satlin 1994, Satlin & Palmer 1997, Gurkan et al 2007).

Patients with mutation in *CLCNKB* are distinctive for presenting at somewhat later ages, with less severe hypercalciuria and the less common occurrence of nephrocalcinosis (about 20%). These patients have a generally more variable presentation, which ranges from severely affected neonates to diagnosis as an adult. Urinary calcium levels can vary from very high with nephrocalcinosis, to even hypocalciuria that potentially could be confused with levels seen in Gitelman's syndrome. Magnesium levels are most often normal, but again can encroach upon levels seen in patients with Gitelman's syndrome. In our experience, there is nonetheless little confusion between patients with Gitelman's and Bartter's syndromes due to *CLCNKB* mutation – the latter most often present in early childhood, they rarely have hypocalciuria, and magnesium levels < 1.6 mg/dl are uncommon. The reason for the varied clinical expression is at present a matter of speculation. ClC-Kb appears to be expressed in both the TAL and the DCT, suggesting potential balancing effects of increased Ca^{2+} loss in the TAL offset by decreased loss due to effects in the DCT. Similarly, extension of ClC-Ka expression into the TAL could compensate for loss of ClC-Kb, providing variable severity of loss of salt reabsorption in this nephron segment – this may occur in rat (Uchida et al 1993).

A generally milder phenotype could also be accounted for by mutations that result in partial, rather than complete loss of function. For example, a founder mutation in *CLCNKB* with variable presentation accounts for many cases of Bartter's syndrome in Spain (Rodriguez Soriano et al 2005); one of these patients presented in adolescence, and another adult diagnosis was reported (Gorgojo et al 2006), consistent with an incomplete loss of function mutation. This mechanism – incomplete loss of function – might account as well for occasional patients with mutations in *NKCC2* (Pressler et al 2006) or *BSND* (Miyamura et al 2003, Brum et al 2007) who present in adolescence or as adults.

Patients with mutations in *BSND* (Birkenhäger et al 2001) usually present with a severe, antenatal phenotype and – unique among this group – sensorineural deafness. They also tend to have less hypercalciuria and nephrocalcinosis, consistent with a role in Cl^- reabsorption in both the TAL and DCT.

Finally, patients with activating mutations in the CaSR typically present early with symptomatic hypocalcemia, hypercalciuria and nephrocalcinosis (Watanabe et al 2002). Evidence of renal salt wasting is typically found later in life. The predominant hypocalcemia, as well as evidence of autosomal dominant transmission, rarely results in confusion with other types of Bartter's syndrome.

Although some groups have suggested an overlapping phenotype of Bartter's and Gitelman's syndrome (reviewed by Zelikovic (2003)) associated with mutations in the *CLCNKB* gene, in our experience with 148 genotypically proven cases with NCC mutations and 70 with *CLCNKB* mutation, there were zero patients prospectively diagnosed as Gitelman's syndrome who had mutation in *CLCNKB* and none referred with a diagnosis of Bartter's syndrome who had mutation in *NCC*.

Knowing the pathophysiology of these diseases, it is obvious that treatment with or abuse of loop or thiazide diuretics will mimic the Bartter's and Gitelman's phenotype; this may require exclusion by means of urinary assay. Other causes of hypokalemic alkalosis are surreptitious vomiting (which is readily distinguished from Gitelman's and Bartter's syndromes since patients with vomiting but normal renal function will have low urinary chloride excretion) and, less frequently, laxative abuse. An interesting observation is a Bartter's like phenotype as a side effect of gentamicin treatment where resolution of the symptoms on withdrawal is characteristic. The molecular basis of this aminoglycoside effect is still elusive, however, based on the clinical features (hypokalemic alkalosis, hypermagnesiuria, hypocalcemia, hypercalciuria, and, surprisingly, low PTH), Chou et al (2005) proposed an effect on the CaSR similar to the activating mutations in Bartter's syndrome type V.

Genetic testing for all known genes is available in theory. However, considering the cost and limited benefit to the patient, we suggest that patients with late-onset hypokalemic alkalosis, who have high renin and aldosterone levels, renal salt wasting, normal or low blood pressure, hypomagnesemia, and hypocalciuria, after exclusion of thiazide diuretic use, can be clinically diagnosed with Gitelman's syndrome; genetic testing should not be generally necessary. Because the heterozygous state for Gitelman's syndrome is present in ~1% of most outbred populations, the risk of Gitelman's syndrome to offspring of Gitelman's patients married to unrelated unaffected subjects is ~1 in 200. Because of the later presentation of Gitelman's syndrome, request for prenatal genetic testing for recurrence in kindreds with a known case is not common.

In the case of Bartter's syndrome, the specific gene mutated can often be deduced from the clinical presentation (see above). Cases of Bartter's syndrome more frequently occur in the setting of parental consanguinity; as a result, homozygosity across disease loci can be a useful additional screen to identify the responsible gene, focusing the mutation search. Because of the high frequency of early presentation, requests for prenatal testing in kindreds with a previous child with Bartter's syndrome may occur more frequently. If definitive mutations can be identified in the first case, prenatal diagnosis may be useful for parental decisions.

As the cost of mutation identification diminishes, it may become routine to perform mutation testing to definitively establish molecular diagnoses in individual patients.

The outcome for patients with neonatal Bartter's syndrome was extremely poor before the establishment of neonatal intensive care allowed for careful volume and electrolyte substitution, and indomethacin therapy was introduced to reduce salt loss (Verberckmoes et al 1976). As a consequence, the natural history of the disease has dramatically changed in the past 30 years. The long-term prognosis and clinical outcome of patients presenting with Bartter's syndrome in the neonatal period has not been well characterized. It is sobering that there are few patients who have been followed from the neonatal period through survival to adolescence, and fewer still who have been followed to adulthood; indeed, most Bartter's patients encountered as adults did not have recognized signs and symptoms as neonates, quite unlike the typically turbulent history in neonates. Death due to difficulty in fluid and electrolyte management, particularly with intercurrent illness, along with progressive renal failure are likely contributing factors to early death among severely affected subjects. End-stage renal disease has been reported in some cases clinically diagnosed as Bartter's syndrome (Chaudhuri et al 2006) and seems to be more common (Birkenhäger et al 2001), but not inevitable (Shalev et al 2003) in Bartter's syndrome with sensorineural deafness. Certainly, a considerable fraction of the patients reported with renal failure may be attributed to nephrocalcinosis, NSAIDs, or both (Schachter et al 1998, Shalev et al 2003). Proteinuria was a somewhat unexpected finding in patients with mutations in the *CLCNKB* gene followed-up for 5–24 years (Bettinelli et al 2007). Further studies are needed to determine and compare the development, outcome and, more importantly, therapeutic options for patients with Bartter's syndrome types I–V.

TREATMENT OF PATIENTS WITH GITELMAN'S AND BARTTER'S SYNDROMES

Though a milder disease than Bartter's syndrome, Gitelman's syndrome patients do have symptoms owing to their primary renal salt wasting. They partially compensate via activation of the renin-angiotensin system; however, they chronically maintain lower blood pressure than their unaffected relatives. These patients are often salt cravers and self-select a particularly high salt diet. This appears likely to be adaptive, and salt restriction of these patients is generally discouraged unless they have hypertension, which is uncommon. Similarly, well-intentioned efforts to ameliorate their hypokalemia with antagonists of the mineralocorticoid receptor or the epithelial sodium channel can sometimes produce symptomatic orthostasis, hypotension, or lethargy. Because blood pressure is generally low, with chronic activation of the renin-angiotensin system, intercurrent illness associated with volume depletion may be poorly tolerated – increased dehydration can be coupled with blunted response to vasoconstrictors, exacerbating the problem. Similarly, women may encounter difficulties in pregnancy (Cruz et al 2001) due to diminished capacity to expand plasma volume, sometimes leading to worsening of hypokalemia. In our experience, however, even high intravenous NaCl infusion rates fail to significantly alter net salt balance and do not succeed in suppressing the elevated renin and aldosterone levels. Because Gitelman's patients do not have increased PGE2 levels, NSAIDs are not generally beneficial (Lüthy et al 1995).

Aside from salt and water homeostasis, hypokalemia and hypomagnesemia are major clinical complications of Gitelman's syndrome. The mainstays of treatment are oral potassium and magnesium supplementation. In our experience, oral replacement is often sufficient to return K^+ levels to or near normal levels, whereas oral replacement often fails to restore normal Mg^{2+} levels. Because some patients report severe fatigue and malaise associated with persistent hypomagnesemia, physicians may resort to regular intravenous Mg^{2+} infusion; it is not well-documented that hypomagnesemia is the cause of these symptoms, nor that aggressive treatment has clinical benefit; careful evaluation of each patient is necessary.

The care of Bartter's syndrome patients can be very challenging, as these patients typically have much more severe salt wasting, owing to the normal reabsorption of 30% of the filtered salt load in this nephron segment. They typically require volume resuscitation and careful attention to electrolyte management owing to hypokalemia (Jeck et al 2005). Particularly problematic are patients with ROMK mutations who may be hyperkalemic at birth but then switch to a hypokalemic state with volume repletion and maturation of the nephron. The notion that Bartter's patients often have elevated PGE2 levels led to the recognition that indomethacin is an important element in their treatment, markedly reducing renal salt loss and polyuria in many patients and ameliorating systemic symptoms including fever, diarrhea and vomiting (Reinalter et al 2002). As discussed above, there appears to be marked variability in PGE2 production, and it is consequently difficult to predict which patients will be most likely to benefit. Although in principle more selective COX2 inhibitors should be ideal agents for these patients, clinical experience is limited and rapid progression of renal failure on change from indomethacin to rofecoxib (Fletcher et al 2006) has been reported. Nephrotoxicity, however, was also noted for ibuprofen in Bartter's syndrome (Schachter et al 1998), and there are hints that, apart from showing less response to indomethacin treatment (Jeck et al 2001, Shalev et al 2003), patients with mutations in BSND are more vulnerable to the development of renal failure after therapy with NSAIDs (Jeck et al 2001, Ozlu et al 2006). When administering indomethacin during pregnancy to reduce polyhydramnios (Konrad et al 1999), special attention has to be devoted to renal function, constriction of the ductus arteriosus and gastrointestinal side effects (which may also limit the generally beneficial therapy with indomethacin after birth). Vaisbich et al (2004) studied 12 patients with Bartter's syndrome under treatment with potassium and a cyclooxygenase inhibitor for up to 17 years. Recovery of growth velocity as well as improved electrolyte balance was attributed to treatment. However, many side effects were noted, including

gastric ulcer, gastritis, gastrocolic fistulae and reduction in creatinine clearance. Another interesting finding was transient hypophosphatemia in five patients.

Other therapeutic options include administration of aldosterone antagonists, potassium-sparing diuretics and ACE inhibitors (Jest et al 1991) for the treatment of hypokalemia; however, these agents counteract the compensatory mechanisms for maintenance of intravascular volume, and can be problematic. For similar reasons, thiazides are not generally a suitable treatment for the hypercalciuria seen in Bartter's syndrome (Kleta & Bockenhauer 2006). It is clear that more systematic evaluation of the effects and outcomes of therapies in Bartter's syndrome will be necessary to provide better treatment guidelines.

While correction of fluid and electrolyte abnormalities can help promote growth and development, treatment with recombinant growth hormone has been reported to be of benefit in both Bartter's and Gitelman's syndromes (Ko & Koo 1999).

The definitive demonstration that Bartter's syndrome is a primary renal disease has suggested that the disease could be cured by bilateral nephrectomy combined with renal transplantation, with the admitted risk of surgery and long-term immunosuppressive therapy. Several reports of this procedure have been published, with correction of electrolyte abnormalities and improved growth and development (Chaudhuri et al 2006). A major challenge remains to determine the appropriate age and indications for transplantation in these patients. Better determination of the natural history of patients with defined mutations is imperative to guide such critical decisions.

EFFECTS OF THE HETEROZYGOUS STATE FOR GITELMAN'S AND BARTTER'S SYNDROMES ON BLOOD PRESSURE IN THE POPULATION

Gitelman's and Bartter's syndromes have an estimated prevalence of 1 in 40 000 and 1 per million, respectively. The frequency of Gitelman's syndrome implies an allele frequency of 0.5% and a heterozygote frequency of ~1%; the prevalence of Bartter's syndrome suggests an allele frequency of 0.1% and a carrier frequency of 1 in 500. It is recognized, however, that prenatal and neonatal demise could result in an underestimate of the frequency of Bartter's alleles in the population. Ji et al (2008) recently sought to identify heterozygous carriers of functional mutations in *NCC*, *NKCC2* and *ROMK* in 3125 members of the Framingham Heart Study. These subjects were all members of the Framingham offspring cohort, who were ascertained in 1971 and have been followed with periodic examinations since that time. These patients have had careful and systematic measurement of blood pressure at regular intervals. They constitute 1985 unrelated subjects and 1140 kindred members. Members of the cohort were screened for sequence variation by temperature-gradient capillary electrophoresis, and identified variants were subjected to DNA sequence analysis. A total of 138 different variants in these were identified in 2492 subjects. Among these, ten different variants in 23 subjects were previously characterized mutations or likely null alleles that resulted in loss of function. In addition, by analysis of mutations in these genes in a large cohort of Gitelman's and Bartter's subjects, the investigators were able to define empiric criteria that yielded high sensitivity and specificity for functional mutations in these genes. This resulted in identification of 19 additional mutations present in 25 subjects – all were missense mutations at positions conserved in all vertebrates and invertebrate orthologs.

Analysis of mutation carriers demonstrated significantly reduced blood pressure at age 40, 50, and 60, and in analysis of longitudinal blood pressure. The mean longitudinal systolic blood pressure was reduced by 6.3 mmHg among mutation carriers compared with non-carriers, and the effect was maximal at age 60, with a 9 mmHg reduction in SBP. Mutations in each of the three genes showed similar blood pressure-lowering effects. Because there were kindred members discordant for genotype, the differences in blood pressure within-family were assessed as an independent test. The results demonstrated similar estimates – heterozygotes had 6.6 mmHg lower systolic blood pressure than their wild-type siblings. Finally, the prevalence of hypertension was compared in mutation carriers and non-carriers by Kaplan-Meyer analysis. The results demonstrated that mutation carriers were significantly protected from development of hypertension, with a hazard ratio of 0.41, corresponding to a 59% reduction in risk of hypertension among mutation carriers.

Among the unrelated subjects in FHS, the frequency of functional mutations in NCC is 0.48%, resulting in an estimated frequency of Gitelman's syndrome of ~1/40 000. This corresponds nicely to previous estimates of disease prevalence. For Bartter's syndrome, the frequency of mutations in *NKCC2* and *ROMK* combined was 0.43%, which would predict a disease frequency of ~1/100 000 from these two genes alone (assuming trans heterozygotes do not have clinical Bartter's syndrome). This is considerably higher than the estimated disease frequency from all Bartter's genes combined, and is consistent with the notion that many conceptions with a Bartter's syndrome genotype die before birth or early in life.

Collectively, these findings imply that there are ~100 million people worldwide who are heterozygous for mutations in one of these three genes, and that these subjects have reduced blood pressure and are protected from development of hypertension. In certain populations, this effect might be even larger. For example, Bouwer et al (2007) determined an overall carrier rate of 2% for a splice site mutation in *NCCT* (IVS9 + 1 G > T) in a Roma/Gypsy population

from Bulgaria. Moreover, polymorphisms in other genes involved in Bartter's syndrome (e.g. *CLCNKB* (Jeck et al 2004), *CLCNKA* (Barlassina et al 2007), and *BSND* (Sile et al 2007)) might also influence blood pressure regulation. It will be of interest to determine in long-term follow-up whether carriers of mutations in genes involved in renal salt handling achieve the expected protection from morbidity and mortality from cardiovascular disease and stroke.

Of particular interest in these findings, the demonstration that heterozygous mutations in ROMK lower blood pressure validates ROMK as a target for development of novel antihypertensive diuretic agents; because Bartter's patients with these mutations have reduced hypokalemia compared with those with mutations in NKCC2, ROMK inhibitors would be anticipated to have less K^+ loss than in patients treated with furosemide.

These findings demonstrate the relevance of inherited variation in renal salt handling in blood pressure variation in the general population. Moreover, the study of Gitelman's and Bartter's has illuminated aspects of integrated human physiology that can be used to advance the understanding and treatment of common diseases such as hypertension and congestive heart failure.

ACKNOWLEDGMENTS

We gratefully acknowledge the many patients, families, and physicians from around the world whose participation in research studies has directly advanced the understanding and treatment of these diseases. We are thankful to Weizhen Ji, Jia Nee Foo, Anita Farhi and William Asch for helpful discussions in the preparation of this manuscript.

References

Adachi S, Uchida S, Ito H, et al. Two isoforms of a chloride channel predominantly expressed in thick ascending limb of Henle's loop and collecting ducts of rat kidney. J. Biol. Chem. 1994; 269(26): 17677–83.

Barlassina C, Dal Fiume C, Lanzani C, et al. Common genetic variants and haplotypes in renal CLCNKA gene are associated to salt-sensitive hypertension. Hum. Mol. Genet. 2007; 16(13): 1630–8.

Bartter FC, Pronove P, Gill JR, MacCardle RC. Hyperplasia of the juxtaglomerular complex with hyperaldosteronism and hypokalemic alkalosis. A new syndrome. Am. J. Med. 1962; 33: 811–28.

Besbas N, Ozaltin F, Jeck N, Seyberth H, Ludwig M. CLCN5 mutation (R347X) associated with hypokalaemic metabolic alkalosis in a Turkish child: an unusual presentation of Dent's disease. Nephrol. Dial. Transplant. 2005; 20(7): 1476–9.

Bettinelli A, Bianchetti MG, Girardin E, et al. Use of calcium excretion values to distinguish two forms of primary renal tubular hypokalemic alkalosis: Bartter and Gitelman syndromes. J. Pediatr. 1992; 120(1): 38–43.

Bettinelli A, Borsa N, Bellantuono R, et al. Patients with biallelic mutations in the chloride channel gene CLCNKB: long-term management and outcome. Am. J. Kidney Dis. 2007; 49(1): 91–8.

Birkenhäger R, Otto E, Schurmann MJ, et al. Mutation of BSND causes Bartter syndrome with sensorineural deafness and kidney failure. Nat. Genet. 2001; 29(3): 310–14.

Bouwer ST, Coto E, Santos F, Angelicheva D, Chandler D, Kalaydjieva L. The Gitelman syndrome mutation, IVS9 + 1G > T, is common across Europe. Kidney Int. 2007; 72(7): 898.

Brennan TM, Landau D, Shalev H, et al. Linkage of Infantile Bartter Syndrome with Sensorineural Deafness to Chromosome 1p. Am. J. Hum. Genet. 1998; 62(2): 355–61.

Brown EM, Gamba G, Riccardi D, et al. Cloning and characterization of an extracellular Ca(2+)-sensing receptor from bovine parathyroid. Nature 1993; 366: 575–80.

Brum S, Rueff J, Santos JR, Calado J. Unusual adult-onset manifestation of an attenuated Bartter's syndrome type IV renal phenotype caused by a mutation in BSND. Nephrol. Dial. Transplant. 2007; 22(1): 288–9.

Calò LA, Davis PA, Palatini P, Semplicini A, Pessina AC. Urinary albumin excretion, endothelial dysfunction and cardiovascular risk: study in Bartter's/Gitelman's syndromes and relevance for hypertension. J. Hum. Hypertens. 2007; 21(11): 904–6.

Calò LA. Vascular tone control in humans: Insights from studies in Bartter's/Gitelman's syndromes. Kidney Int. 2006; 69(6): 963–6.

Calò LA, Pagnin E, Ceolotto G, et al. Silencing regulator of G protein signaling-2 (RGS-2) increases angiotensin II signaling: insights into hypertension from findings in Bartter's/Gitelman's syndromes. J. Hypertens. 2008; 26(5): 938–45.

Chaimovitz C, Levi J, Better OS, Oslander L, Benderli A. Studies on the site of renal salt loss in a patient with Bartter's syndrome. Pediatr. Res. 1973; 7(2): 89–94.

Chaudhuri A, Salvatierra O, Alexander SR, Sarwal MM. Option of pre-emptive nephrectomy and renal transplantation for Bartter's syndrome. Pediatr. Transplant. 2006; 10(2): 266–70.

Chou CL, Chen YH, Chau T, Lin SH. Acquired Bartter-Like Syndrome Associated with Gentamicin Administration. Am. J. Med. Sci. 2005; 329(3): 144–9.

Colussi G, De Ferrari ME, Tedeschi S, Prandoni S, Syrén ML, Civati G. Bartter syndrome type 3: an unusual cause of nephrolithiasis. Nephrol. Dial. Transplant. 2002; 17(3): 521–3.

Cruz DN. The renal tubular Na-Cl co-transporter (NCCT): a potential genetic link between blood pressure and bone density? Nephrol. Dial. Transplant. 2001; 16(4): 691–4.

Cruz DN, Shaer AJ, Bia MJ, Lifton RP, Simon DB. Gitelman's syndrome revisited: An evaluation of symptoms and health-related quality of life. Kidney Int. 2001a; 59(2): 710–17.

Cruz DN, Simon DB, Nelson-Williams C, et al. Mutations in the Na-Cl Cotransporter Reduce Blood Pressure in Humans. Hypertension 2001b; 37(6): 1458–64.

Dunn MJ. Prostaglandins and Bartter's syndrome. Kidney Int. 1981; 19(1): 86–102.

Ellison DH. The physiologic basis of diuretic synergism: its role in treating diuretic resistance. Ann. Intern. Med. 1991; 114(10): 886–94.

Emma F, Pizzini C, Tessa A, et al. "Bartter-like" phenotype in Kearns-Sayre syndrome. Pediatr. Nephrol. 2006; 21(3): 355–60.

Estévez R, Boettger T, Stein V, et al. Barttin is a Cl-channel beta-subunit crucial for renal Cl-reabsorption and inner ear K$^+$ secretion. Nature 2001; 414(6863): 558–61.

Fava C, Montagnana M, Rosberg L, et al. Subjects heterozygous for genetic loss of function of the thiazide-sensitive cotransporter have reduced blood pressure. Hum. Mol. Genet. 2008; 17(3): 413–18.

Finer G, Shalev H, Birk OS, et al. Transient neonatal hyperkalemia in the antenatal (ROMK defective) Bartter syndrome. J. Pediatr 2003; 142(3): 318–23.

Fletcher JT, Graf N, Scarman A, Saleh H, Alexander SI. Nephrotoxicity with cyclooxygenase 2 inhibitor use in children. Pediatr. Nephrol. 2006; 21(12): 1893–7.

Gamba G, Saltzberg SN, Lombardi M, et al. Primary structure and functional expression of a cDNA encoding the thiazide-sensitive, electroneutral sodium-chloride cotransporter. Proc. Natl Acad. Sci. USA 1993; 90(7): 2749–53.

Gamba G, Miyanoshita A, Lombardi M, et al. Molecular cloning, primary structure, and characterization of two members of the mammalian electroneutral sodium-(potassium)-chloride cotransporter family expressed in kidney. J. Biol. Chem. 1994; 269(26): 17713–22.

Geerling JC, Loewy AD. Central regulation of sodium appetite. Exp. Physiol. 2008; 93(2): 177–209.

Giebisch G. Renal potassium channels: an overview. Kidney Int. 1995; 48(4): 1004–9.

Gitelman HJ, Graham JB, Welt LG. A new familial disorder characterized by hypokalemia and hypomagnesemia. Trans. Assoc. Am. Physicians 1966; 79: 221–35.

Gorgojo JJ, Donnay S, Jeck N, Konrad M. A Spanish founder mutation in the chloride channel gene, CLCNKB, as a cause of atypical Bartter Syndrome in Adult Age. Horm. Res. 2006; 65(2): 62–8.

Graham RM, Bloch KD, Delaney VB, Bourke E, Seidman JG. Bartter's syndrome and the atrial natriuretic factor gene. Hypertension 1986; 8(6): 549–51.

Guay-Woodford LM. Bartter syndrome: unraveling the pathophysiologic enigma. Am. J. Med. 1998; 105(2): 151–61.

Gurkan S, Estilo GK, Wei Y, Satlin LM. Potassium transport in the maturing kidney. Pediatr. Nephrol. 2007; 22(7): 915–25.

Hebert SC. An ATP-regulated, inwardly rectifying potassium channel from rat kidney (ROMK). Kidney Int. 1995; 48(4): 1010–16.

Hetzel W, Molitor H. Paresen, Schmerz syndrome, Bewußtseinsstörungen: zum neurologischen Erscheinungsbild des Bartter-Syndroms. Nervenarzt 1991; 62(8): 500–5.

Hoenderop JG, van Leeuwen JP, van der Eerden BC, et al. Renal Ca^{2+} wasting, hyperabsorption, and reduced bone thickness in mice lacking TRPV5. J. Clin. Invest. 2003; 112(12): 1906–14.

Imai M, Yabuta K, Murata H, Takita S, Ohbe Y, Sokabe H. A case of Bartter's syndrome with abnormal renin response to salt load. J. Pediatr. 1969; 74(5): 738–49.

Jeck N, Reinalter SC, Henne T, et al. Hypokalemic salt-losing tubulopathy with chronic renal failure and sensorineural deafness. Pediatrics 2001; 108(1): E5.

Jeck N, Waldegger S, Lampert A, et al. Activating mutation of the renal epithelial chloride channel ClC-Kb predisposing to hypertension. Hypertension 2004; 43(6): 1175–81.

Jeck N, Schlingmann KP, Reinalter SC, et al. Salt handling in the distal nephron: lessons learned from inherited human disorders. Am. J. Physiol. Regul. Integr. Comp. Physiol. 2005; 288(4): R782–95.

Jest P, Pedersen KE, Klitgaard NA, Thomsen N, Kjaer K, Simonsen E. Angiotensin-converting enzyme inhibition as a therapeutic principle in Bartter's syndrome. Eur. J. Clin. Pharmacol. 1991; 41(4): 303–5.

Ji W, Foo JN, O'Roak BJ, et al. Rare independent mutations in renal salt handling genes contribute to blood pressure variation. Nat. Genet. 2008; 40(5): 592–9.

Kieferle S, Fong P, Bens M, Vandewalle A, Jentsch TJ. Two highly homologous members of the ClC chloride channel family in both rat and human kidney. Proc. Natl Acad. Sci. USA 1994; 91(15): 6943–7.

Kleta R, Bockenhauer D. Bartter syndromes and other salt-losing tubulopathies. Nephron Physiol. 2006; 104(2): 73–80.

Ko CW, Koo JH. Recombinant human growth hormone and Gitelman's syndrome. Am. J. Kidney Dis. 1999; 33(4): 778–81.

Kömhoff M, Reinalter SC, Gröne HJ, Seyberth HW. Induction of microsomal prostaglandin E2 synthase in the macula densa in children with hypokalemic salt-losing tubulopathies. Pediatr. Res. 2004; 55(2): 261–6.

Konrad M, Leonhardt A, Hensen P, Seyberth HW, Köckerling A. Prenatal and postnatal management of hyperprostaglandin E syndrome after genetic diagnosis from amniocytes. Pediatrics 1999; 103(3): 678–83.

Kunchaparty S, Palcso M, Berkman J, et al. Defective processing and expression of thiazide-sensitive Na-Cl cotransporter as a cause of Gitelman's syndrome. Am. J. Physiol. Renal Physiol. 1999; 277(4): F643–9.

Kurtz I. Molecular pathogenesis of Bartter's and Gitelman's syndromes. Kidney Int. 1998; 54(4): 1396–410.

Lalioti MD, Zhang J, Volkman HM, et al. Wnk4 controls blood pressure and potassium homeostasis via regulation of mass and activity of the distal convoluted tubule. Nat. Genet. 2006; 38(10): 1124–32.

Landau D, Shalev H, Ohaly M, Carmi R. Infantile variant of Bartter syndrome and sensorineural deafness: a new autosomal recessive disorder. Am. J. Med. Genet. 1995; 59: 454–9.

Lifton RP, Dluhy RG, Powers M, et al. A chimaeric 11 beta-hydroxylase/aldosterone synthase gene causes glucocorticoid-remediable aldosteronism and human hypertension. Nature 1992; 355(6357): 262–5.

Lin SH, Cheng NL, Hsu YJ, Halperin ML. Intrafamilial phenotype variability in patients with Gitelman syndrome having the same mutations in their thiazide-sensitive sodium/chloride cotransporter. Am. J. Kidney Dis. 2004; 43(2): 304–12.

Loffing J, Loffing-Cueni D, Hegyi I, et al. Thiazide treatment of rats provokes apoptosis in distal tubule cells. Kidney Int. 1996; 50(4): 1180–90.

Lorenz JN, Baird NR, Judd LM, et al. Impaired renal NaCl absorption in mice lacking the ROMK potassium channel, a model for type II Bartter's syndrome. J. Biol. Chem. 2002; 277(40): 37871–80.

Lüthy C, Bettinelli A, Iselin S, et al. Normal prostaglandinuria E2 in Gitelman's syndrome, the hypocalciuric variant of Bartter's syndrome. Am. J. Kidney Dis. 1995; 25(6): 824–8.

Marco-Franco JE, Morey A, Ventura C, Gascó JM, Alarcón A. Long-term evolution and growth patterns in a family with Bartter's syndrome. Clin. Nephrol. 1994; 42(1): 33–7.

Matsumoto J, Han BK, Restrepo de Rovetto C, Welch TR. Hypercalciuric Bartter syndrome: resolution of nephrocalcinosis with indomethacin. Am. J. Roentgenol. 1989; 152(6): 1251–3.

Matsumura Y, Uchida S, Kondo Y, et al. Overt nephrogenic diabetes insipidus in mice lacking the CLC-K1 chloride channel. Nat. Genet. 1999; 21: 95–8.

Miyamura N, Matsumoto K, Taguchi T, et al. Atypical Bartter syndrome with sensorineural deafness with G47R mutation of the ß-subunit for ClC-Ka and ClC-Kb chloride channels, barttin. J. Clin. Endocrinol. Metab. 2003; 88(2): 781–6.

Mune T, Rogerson FM, Nikkilä H, Agarwal AK, White PC. Human hypertension caused by mutations in the kidney isozyme of 11-beta-hydroxysteroid dehydrogenase. Nat. Genet. 1995; 10(4): 394–9.

Naesens M, Steels P, Verberckmoes R, Vanrenterghem Y, Kuypers D. Bartter's and Gitelman's syndromes: from gene to clinic. Nephron Physiol. 2004; 96(3): 65–78.

Nijenhuis T, Vallon V, van der Kemp AW, Loffing J, Hoenderop J, Bindels RJ. Enhanced passive Ca^{2+} reabsorption and reduced Mg^{2+} channel abundance explains thiazide-induced hypocalciuria and hypomagnesemia. J. Clin. Invest. 2005; 115: 1651–8.

Nozu K, Inagaki T, Fu XJ, et al. Molecular analysis of digenic inheritance in Bartter syndrome with sensorineural deafness. J. Med. Genet. 2008; 45(3): 182–6.

Ozlu F, Yapicioglu H, Satar M, et al. Barttin mutations in antenatal Bartter syndrome with sensorineural deafness. Pediatr. Nephrol. 2006; 21(7): 1056–7.

Pattaragarn A, Fox J, Alon US. Effect of the calcimimetic NPS R-467 on furosemide-induced nephrocalcinosis in the young rat. Kidney Int. 2004; 65(5): 1684–9.

Pennesi M, Marchetti F, Crovella S, et al. A new mutation in two siblings with cystinosis presenting with Bartter syndrome. Pediatr. Nephrol. 2005; 20(2): 217–19.

Peters M, Jeck N, Reinalter S, et al. Clinical presentation of genetically defined patients with hypokalemic salt-losing tubulopathies. Am. J. Med. 2002; 112(3): 183–90.

Phillips DR, Ahmad KI, Waller SJ, Meisner P, Karet FE. A serum potassium level above 10 mmol/l in a patient predisposed to hypokalemia. Nat. Clin. Pract. Nephrol. 2006; 2(6): 340–6.

Pressler CA, Heinzinger J, Jeck N, et al. Late-onset manifestation of antenatal bartter syndrome as a result of residual function of the mutated renal Na^+ -K^+-2Cl-co-transporter. J. Am. Soc. Nephrol. 2006; 17(8): 2136–42.

Proesmans W. Bartter syndrome and its neonatal variant. Eur. J. Pediatr. 1997; 156(9): 669–79.

Reinalter SC, Jeck N, Brochhausen C, et al. Role of cyclooxygenase-2 in hyperprostaglandin E syndrome/antenatal Bartter syndrome. Kidney Int. 2002; 62(1): 253–60.

Ring T, Knoers N, Oh MS, Halperin ML. Reevaluation of the criteria for the clinical diagnosis of Gitelman syndrome. Pediatr. Nephrol. 2002; 17(8): 612–16.

Rodrigues Pereira R, van Wersch J. Inheritance of Bartter syndrome. Am. J. Med. Genet. 1983; 15: 79–84.

Rodríguez-Soriano J, Vallo A, Aguirre M. Bone mineral density and bone turnover in patients with Bartter syndrome. Pediatr. Nephrol. 2005; 20(8): 1120–5.

Rodríguez-Soriano J, Vallo A, Pérez de Nanclares G, Bilbao JR, Castaño L. A founder mutation in the CLCNKB gene causes Bartter syndrome type III in Spain. Pediatr. Nephrol. 2005; 20(7): 891–6.

Rosenbaum P, Hughes M. Persistent, probably congenital hypokalemic metabolic alkalosis with hyaline degeneration of renal tubules and normal urinary aldosterone. Am. J. Dis. Child. 1957; 94: 560.

Sabath E, Meade P, Berkman J, et al. Pathophysiology of functional mutations of the thiazide-sensitive Na-Cl cotransporter in Gitelman disease. Am. J. Physio. Renal Physiol. 2004; 287(2): F195–203.

Satlin LM. Postnatal maturation of potassium transport in rabbit cortical collecting duct. Am. J. Physio. Renal Physiol. 1994; 266(1): F57–65.

Satlin LM, Palmer LG. Apical K^+ conductance in maturing rabbit principal cell. Am. J. Physiol. Renal Physiol. 1997; 272(3): F397–404.

Schachter AD, Arbus GS, Alexander RJ, Balfe JW. Non-steroidal anti-inflammatory drug-associated nephrotoxicity in Bartter syndrome. Pediatr. Nephrol. 1998; 12(9): 775–7.

Schild L, Lu Y, Gautschi I, Schneeberger E, Lifton RP, Rossier BC. Identification of a PY motif in the epithelial Na channel subunits as a target sequence for mutations causing channel activation found in Liddle syndrome. EMBO J. 1996; 15(10): 2381–7.

Schlingmann KP, Weber S, Peters M, et al. Hypomagnesemia with secondary hypocalcemia is caused by mutations in TRPM 6, a new member of the TRPM gene family. Nat. Genet. 2002; 31(2): 166–70.

Schlingmann KP, Konrad M, Jeck N, et al. Salt wasting and deafness resulting from mutations in two chloride channels. N. Engl. J. Med. 2004; 350(13): 1314–19.

Schmitz U, Ko Y, Becher H, Ludwig M, Vetter H, Düsing R. Evidence for cardiovascular remodeling in a patient with Bartter's syndrome. Clin. Invest. 1994; 72(11): 874–7.

Schnermann J. Cyclooxygenase-2 and macula densa control of renin secretion. Nephrol. Dial. Transplant. 2001; 16(9): 1735–8.

Scholl U, Hebeisen S, Janssen A, Müller-Newen G, Alekov A, Fahlke C. Barttin modulates trafficking and function of ClC-K channels. Proc. Natl Acad. Sci. USA 2006; 103(30): 11411–16.

Schultheis PJ, Lorenz JN, Meneton P, et al. Phenotype resembling Gitelman's syndrome in mice lacking the apical Na^+-Cl-cotransporter of the distal convoluted tubule. J. Biol. Chem. 1998; 273(44): 29150–5.

Seyberth HW, Königer SJ, Rascher W, Kühl PG, Schweer H. Role of prostaglandins in hyperprostaglandin E syndrome and in selected renal tubular disorders. Pediatr. Nephrol. 1987; 1(3): 491–7.

Shalev H, Ohali M, Kachko L, Landau D. The neonatal variant of Bartter syndrome and deafness: preservation of renal function. Pediatrics 2003; 112(3): 628–33.

Sile S, Gillani NB, Velez DR, et al. Functional BSND variants in essential hypertension. Am. J. Hypertens. 2007; 20(11): 1176–82.

Silverberg AB, Mennes PA, Cryer PE. Resistance to endogenous norepinephrine in Bartter's syndrome: reversion during indomethacin administration. Am. J. Med. 1978; 64(2): 231–5.

Simon DB, Lifton RP. The molecular basis of inherited hypokalemic alkalosis: Bartter's and Gitelman's syndromes. Am. J. Physiol. Renal Physiol. 1996; 271(5): F961–6.

Simon DB, Karet FE, Hamdan JM, Di Pietro A, Sanjad SA, Lifton RP. Bartter's syndrome, hypokalaemic alkalosis with hypercalciuria, is caused by mutations in the Na–K–2 Cl cotransporter NKCC2. Nat. Genet. 1996; 13(2): 183–8.

Simon DB, Karet FE, Rodriguez-Soriano J, et al. Genetic heterogeneity of Bartter's syndrome revealed by mutations in the K channel, ROMK. Nat. Genet. 1996; 14: 152–6.

Simon DB, Nelson-Williams C, Johnson Bia M, et al. Gitelman's variant of Bartter's syndrome, inherited hypokalaemic alkalosis, is caused by mutations in the thiazide-sensitive Na–Cl cotransporter. Nat. Genet. 1996; 12(1): 24–30.

Simon DB, Bindra RS, Mansfield TA, et al. Mutations in the chloride channel gene, CLCNKB, cause Bartter's syndrome type III. Nat. Genet. 1997; 17(2): 171–8.

Simon DB, Lu Y, Choate KA, et al. Paracellin-1, a renal tight junction protein required for paracellular Mg^{2+} resorption. Science 1999; 285(5424): 103–6.

Simopoulos AP. Growth characteristics in patients with Bartter's syndrome. Nephron 1979; 23(2–3): 130–5.

Smilde TJ, Haverman JF, Schipper P, et al. Familial hypokalemia/hypomagnesemia and chondrocalcinosis. J. Rheumatol. 1994; 21(8): 1515–19.

Takahashi N, Chernavvsky DR, Gomez RA, Igarashi P, Gitelman HJ, Smithies O. Uncompensated polyuria in a mouse model of Bartter's syndrome. Proc. Natl Acad. Sci. USA 2000; 97(10): 5434–9.

Uchida S, Sasaki S, Furukawa T, et al. Molecular cloning of a chloride channel that is regulated by dehydration and expressed predominantly in kidney medulla. J. Biol. Chem. 1993; 268(6): 3821–4.

Vaisbich MH, Fujimura MD, Koch VH. Bartter syndrome: benefits and side effects of long-term treatment. Pediatr. Nephrol. 2004; 19(8): 858–63.

Vargas-Poussou R, Huang C, Hulin P, et al. Functional characterization of a calcium-sensing receptor mutation in severe autosomal dominant hypocalcemia with a Bartter-like syndrome. J. Am. Soc. Nephrol. 2002; 13(9): 2259–66.

Verberckmoes R, van Damme B, Clement J, Amery A, Michielsen P. Bartter's syndrome with hyperplasia of renomedullary cells: successful treatment with indomethacin. Kidney Int. 1976; 9(3): 302–7.

Vezzoli G, Arcidiacono T, Paloschi V, et al. Autosomal dominant hypocalcemia with mild type 5 Bartter syndrome. J. Nephrol. 2006; 19(4): 525–8.

Vollmer M, Jeck N, Lemmink HH, et al. Antenatal Bartter syndrome with sensorineural deafness: refinement of the locus on chromosome 1p31. Nephrol. Dial. Transplant. 2000; 15(7): 970–4.

Volpe A, Caramaschi P, Thalheimer U, et al. Familiar association of Gitelman's syndrome and calcium pyrophosphate dihydrate crystal deposition disease – a case report. Rheumatology (Oxford) 2007; 46(9): 1506–8.

Watanabe S, Fukumoto S, Chang H, et al. Association between activating mutations of calcium-sensing receptor and Bartter's syndrome. Lancet 2002; 360(9334): 692–4.

Whyte MP, Shaheb S, Schnaper HW. Cystinosis presenting with features suggesting Bartter syndrome. Case report and literature review. Clin. Pediatr. (Phila) 1985; 24(8): 447–51.

Yoshida H, Kakuchi J, Yoshikawa N, et al. Angiotensin II type 1 receptor gene abnormality in a patient with Bartter's syndrome. Kidney Int. 1994; 46(6): 1505–9.

Zarraga Larrondo S, Vallo A, Gainza J, Muñiz R, Garcia Erauzkin G, Lampreabe I. Familial hypokalemia-hypomagnesemia or Gitelman's syndrome: a further case. Nephron 1992; 62(3): 340–4.

Zelikovic I. Hypokalaemic salt-losing tubulopathies: an evolving story. Nephrol. Dial. Transplant. 2003; 18(9): 1696–700.

Zelikovic I, Szargel R, Hawash A, et al. A novel mutation in the chloride channel gene, CLCNKB, as a cause of Gitelman and Bartter syndromes. Kidney Int. 2003; 63(1): 24–32.

CHAPTER 14

Molecular Genetics of Magnesium Homeostasis

WILLIAM S. ASCH AND RICHARD P. LIFTON

INTRODUCTION

Magnesium is the second most abundant intracellular cation and the predominant divalent cation. The body, primarily through controlled renal magnesium reabsorption, maintains magnesium levels in a narrow homeostatic range. Without this finely regulated reabsorptive mechanism, magnesium levels would fall below the homeostatic range, and possibly result in severe complications such as carpopedal spasm and seizures. These complications of hypomagnesemia clearly indicate a role for magnesium in maintaining normal neuromuscular excitability. In addition, magnesium is a cofactor for a number of enzymes and plays a role in diverse cellular functions including protein synthesis, nucleic acid stability, oxidative phosphorylation and maintenance of the structural integrity of bone. Our understanding of the molecules and pathways necessary to maintain magnesium homeostasis has largely derived from intensive efforts to identify and characterize families that have abnormalities in the magnesium reabsorption process. Linkage analysis and positional cloning has led to the discovery of genes and mutant alleles that cause hypomagnesemia. This chapter on the genetics of the hereditary magnesium wasting disorders will focus on the discovery of these mutant alleles, and will review the cell and molecular biology of magnesium as it is currently understood. As there remain families with magnesium wasting disorders that have not been shown to have mutations in the known genes, it is likely that our understanding of the molecular repertoire remains incomplete.

Unit Conversions and Equivalents

For magnesium: $1\,\text{mmol/l} = 2\,\text{mEq/l}$
$$= 24.3\,\text{mg/l} \left(2.43\,\text{mg/dl}\right).$$

Conversion equation: $\text{mmol}/l \times 2.433 = \text{mg of mg}/\text{dl}.$

PHYSIOLOGY OF MAGNESIUM HOMEOSTASIS

Magnesium uptake occurs across two absorptive epithelia in the body, the tubular epithelial cells of the kidney and the mucosa of the gastrointestinal tract. Balance studies have shown that on a typical Western diet, approximately 12 mmol of magnesium are ingested daily. Of this dietary magnesium, net intestinal transport is approximately 4 mmol per day (6 mmol is absorbed by the intestine, while 2 mmol are simultaneously secreted). To maintain magnesium levels in the normal range at steady state in the adult, a quantity equivalent to net GI absorption is excreted by the kidneys. In this manner, serum magnesium levels are tightly regulated in the normal range of 0.7–1.1 mmol/l (Quamme & de Rouffignac 2000). As a consequence, a reduction in dietary magnesium intake is balanced by a matched reduction in renal magnesium excretion. Conversely, on a high magnesium diet, renal excretion is increased.

Within the gastrointestinal tract, the small intestine is the major site of magnesium absorption. A substantially smaller quantity of magnesium is also absorbed by the colon. In 1991, Fine et al, by monitoring magnesium excretion following an oral load and using nonlinear regression analysis, hypothesized that two separate and distinct molecular pathways existed to absorb intestinal magnesium. A saturable transcellular magnesium absorptive pathway and an electrochemical gradient-dependant passive paracellular transport pathway were proposed (Fine et al 1991). When luminal magnesium concentrations are low, the transcellular pathway is predicted to be the predominant absorptive mechanism. However, as the concentration of magnesium rises, the paracellular transport pathway would play an increasingly greater role. Overall, this results in total magnesium absorption having a curvilinear relationship to increasing magnesium intake (Fine et al 1991).

Early studies to elucidate the renal magnesium reabsorptive pathways showed that magnesium reabsorption

Genetic Diseases of the Kidney
ISBN: 978-0-12-449851-8

occurred unequally across three physiologically distinct nephron segments. It has been estimated that the loop of Henle, and, more specifically, the cortical thick ascending limb, is responsible for 65–75% of renal magnesium reabsorption. The proximal tubule and distal convoluted tubule reabsorb the remainder, 15–20% and 5–10%, respectively (de Rouffignac & Quamme 1994). It is worth commenting that the relative contribution of these nephron segments in adults differs from that measured in the newborn. In neonates, it is estimated that the proximal tubule accounts for 70%, or more, of magnesium reabsorption, and it has been noted that this is a similar proportion to the sodium and calcium reabsorbed in this nephron segment (de Rouffignac & Quamme 1994). The fact that the relative contribution of each nephron segment is a dynamic value as a neonate matures likely results in the variable onset of presenting symptoms, and needs to be considered when comparing magnesium reabsorption across age groups (de Rouffignac & Quamme 1994).

While most of our current understanding of magnesium reabsorption stems from the elucidation of the inheritable hypomagnesemic disorders, several seminal observations were made prior to the era of molecular genetics. Notably, in vitro microperfusion and micropuncture studies using isolated nephron segments clearly showed that the bulk of magnesium transport occurred in the cTAL and that DCT, while only reabsorbing about 10% of the total filtered magnesium, was pivotal in determining the final urine composition. Transport in the cTAL was appreciated to be a passive process which likely occurred via a paracellular route (Figure 14.1). Furthermore, insights gained from Bartter's syndrome suggested that the driving force for magnesium movement in the cTAL was the positive luminal potential, and that this voltage was itself dependent upon potassium entering the cell from the tubular lumen via apical NKCC2 activity being returned to the lumen. Additional isolated nephron segment studies, along with studies focusing on immortalized cell lines, led investigators to speculate that magnesium transport in the DCT occurred via a unique apical channel and was driven by a favorable transmembrane voltage (Figure 14.2). The fact that magnesium exit out of the DCT into the blood occurred against its electrochemical gradient at the basolateral membrane led to the hypothesized existence of an active transport process occurring at this location. Subsequent work suggested that a sodium dependent exchange, driven by low intracellular sodium concentrations established by the Na^+/K^+-ATPase, resulted in magnesium transport across the basolateral membrane. Thus, like magnesium absorption in the gastrointestinal tract, magnesium reabsorption in the kidney appeared to occur via both paracellular and transcellular pathways.

While there are many similarities between intestinal and renal magnesium transport, the early classification of the inherited disorders showed that there were some key clinical differences requiring further explanation. In particular, physiologic studies utilizing radioactive magnesium isotopes showed that hypomagnesemia with secondary hypocalciuria, in contrast to all other inheritable forms of hypomagnesemia, was primarily a disorder of the gastrointestinal tract (Lombeck et al 1975). The disorder was fittingly dubbed 'primary intestinal hypomagnesemia' to reflect this finding. In support of this observation many, but not all, patients with this disorder had evidence of renal magnesium conservation. And, while still with some serum evidence of hypomagnesemia, the symptoms in these patients could be corrected with high oral magnesium intake (Shalev et al 1998). This was compatible with the

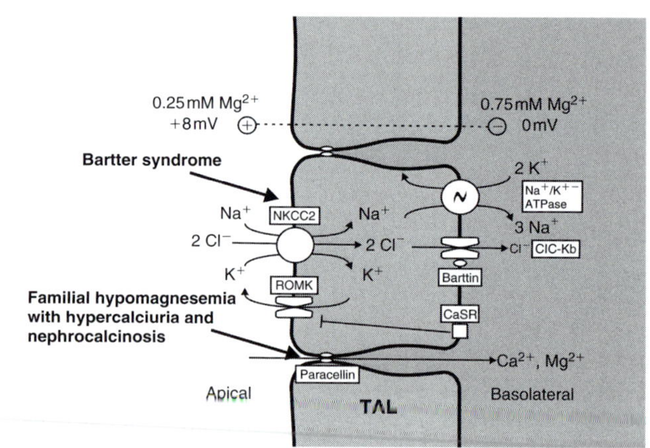

FIGURE 14.1 Magnesium reabsorption in the thick ascending limb of the loop of Henle. It is estimated that ~70% of the filtered magnesium load is reabsorbed in the TAL. This occurs via paracellular flux across the tight junction owing to the favorable electrochemical gradient and the selective permeability of the tight junctions of the TAL for divalent cations. A diagram of the epithelium of the TAL is shown, along with channels and transporters that are involved in magnesium flux. The electrical gradient is established by the asymmetric transcellular reabsorption of cations and anions: the entry of sodium, potassium and chloride across the apical membrane occurs via the electroneutral Na-K-2Cl cotransporter NKCC2; while sodium and chloride exit across the basolateral membrane via the action of the sodium pump and the chloride channel CLCNKB, potassium exits across the apical membrane via the potassium channel ROMK, resulting in a lumen-positive potential. Mutations in any of the genes required for establishing the electrical gradient in the TAL result in Bartter's syndrome, a disease dominated by severe salt wasting, but which also features renal magnesium and calcium wasting. Magnesium and calcium are reabsorbed in this nephron segment by their selective passage across the tight junction. This nephron segment expresses the tight junction proteins paracellin-1 (claudin 16) and claudin 19, which are each required for paracellular magnesium and calcium flux. Loss of function mutations in either claudin 16 or 19 causes severe renal magnesium and calcium wasting and the disease known as familial hypomagnesemia with hypercalciuria and nephrocalcinosis (FHHNC)

conclusion that the proposed unsaturable paracellular route of intestinal magnesium absorption, which was believed to account for transport when gut levels were very high, was intact in this disorder while transcellular transport was defective. In contrast, familial hypomagnesemia with hypercalciuria and nephrocalcinosis could not be abated by excessive administration of magnesium, but could be corrected with renal transplantation (Praga et al 1995).

As noted above, significant deviations in magnesium concentrations from the normal range can have severe consequences. Fortunately, the inherited disorders of magnesium wasting are rare. The prevalence of Gitelman syndrome is estimated to be ~1 per 50 000 in Sweden, and this is likely the most common of the magnesium disorders (Fava et al 2008). Several studies have attempted to determine the prevalence of hypomagnesemia in large populations (Iannello & Belfiore 2001). Indeed, latent

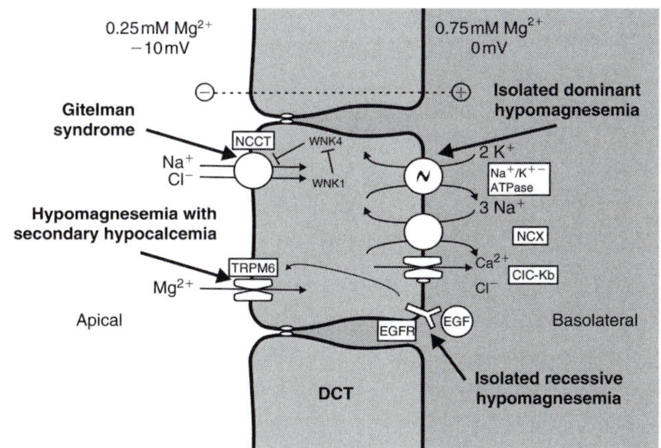

FIGURE 14.2   Magnesium reabsorption in the distal convoluted tubule (DCT). While only ~5–10% of the filtered magnesium load is reabsorbed in the DCT, this nephron segment appears to be the normal determinant of net magnesium reabsorption. Unlike in the TAL where magnesium reabsorption is a passive process occurring down an electrochemical gradient, in the DCT magnesium transport is a transcellular process that likely has an active component dependent upon ATP and the action of the Na⁺/K⁺-ATPase. An epithelial cell of the DCT is shown, along with components involved in magnesium reabsorption in this nephron segment. The entry step is the apical magnesium channel TRPM6, whose loss of function mutation results in the disease hypomagnesemia with secondary hypocalcemia. Mutation in the Na-Cl cotransporter NCCT results in hypomagnesemia and hypocalciuria in conjunction with salt wasting, the disease known as Gitelman's syndrome. Mutations in the gamma subunit of the sodium pump result in isolated dominant hypomagnesemia, and mutations in the mitochondrial-encoded isoleucine tRNA yield a similar phenotype. These mutations suggest an essential energy-consuming step in magnesium reabsorption that has not yet been defined. Finally, mutation in the gene encoding epidermal growth factor (EGF) causes isolated recessive hypomagnesemia, implicating EGF signaling in magnesium reabsorption in the DCT

hypomagnesemia is more common that might have been predicted. Schimatschek et al analyzed a German population of 16 000 individuals and found that 14.5% had evidence of hypomagnesemia (as defined by a serum magnesium concentration below 0.76 mmol/l) (Schimatschek & Rempis 2001). Interestingly, while the hypomagnesemia in this population may be latent, it is not necessarily innocuous. There are an increasing number of disease associations with hypomagnesemia (Sanders et al 1993, Innerarity 2000). Type II diabetes mellitus, hypertension, coronary heart disease, and asthma are examples of diseases believed to be either aggravated or precipitated by hypomagnesemia.

Chakraborti et al published a review focusing on the protective role of magnesium in cardiovascular disease in 2002 (Chakraborti et al 2002). They suggested that the available clinical, experimental, pathophysiological, and epidemiological data collectively warranted placing magnesium at the top of the cardiovascular risk factors. While this stance may be extreme, especially when the more traditional cardiac risk factors such as hypertension, diabetes, and obesity are considered, several possible protective effects, including antiplatelet aggregation, coronary vasodilatation, inhibition of calcium influx, and reduced peripheral resistance, have been proposed (Hampton et al 1994). It is nonetheless worth noting that the long-term clinical consequences of relatively mild hypomagnesemia are presently uncertain, as are the prospects for augmenting levels with oral replacement in the general population. These considerations, however, underscore the importance of understanding the pathways governing normal and abnormal magnesium reabsorption.

In the following, we discuss the clinical syndromes, genetic aspects, identified mutant proteins, and where possible, the molecular pathophysiological mechanisms of the hereditary magnesium wasting disorders. Note that hypomagnesemia associated with mitochondrial tRNA mutations and the calcium/magnesium sensing receptor each have a dedicated chapter elsewhere in this book. Consistent with prior reviews on this topic (Konrad & Weber 2003), the inherited disorders of magnesium wasting in this chapter will be grouped into five categories: familial hypomagnesemia with hypercalciuria and nephrocalcinosis (FHHNC), the Gitelman and Bartter syndromes, hypomagnesemia with secondary hypocalciuria (HSH), and isolated hypomagnesemia of either dominant or recessive transmission.

FAMILIAL HYPOMAGNESEMIA WITH HYPERCALCIURIA AND NEPHROCALCINOSIS (FHHNC)

Michelis et al, in 1972, described a sibship with a decreased bicarbonate threshold, renal magnesium wasting, and distal renal tubular acidosis as part of their studies evaluating the pathophysiological role of parathyroid hormone. This newly described disorder was associated with both excessive renal magnesium and calcium wasting. Affected individuals were also noted to have a progressive impairment in renal function, likely a consequence of severe nephrocalcinosis. Additional case series were required to fully characterize this disease entity, and allowed it to be more clearly separated from the other magnesium wasting disorders (Manz et al 1978, Milazzo et al 1981, Richard & Freycon 1992, Nicholson et al 1995, Praga et al 1995, Torralbo et al 1995). FHHNC is an autosomal recessive disorder, and characteristically presents as a syndrome with multi-organ system involvement. Presentation with clinical features of this disease generally occurs early in life, certainly before adolescence in most cases. While not the most distinctive clinical feature of this disease, recurrent urinary tract infection is the most common complaint prompting medical attention (Weber et al 2001). The ophthalmologic abnormalities associated with this disorder stand out when compared with the other inherited magnesium wasting disorders. Severe myopia, nystagmus, and chorioretinitis have all been described. Also reported, but less commonly, are tetany, seizures, hypertension, gouty arthritis, hearing impairments, chondrocalcinosis, and rickets. On a biochemical level, these patients invariably have hypomagnesemia (plasma magnesium 0.59 +/− 0.06 mmol/l), inappropriately high daily urine magnesium (2.07 +/− 0.73 mmol/24 h), high fractional excretion of magnesium despite hypomagnesemia (12.5 +/− 4.7%), and elevated urine calcium/creatinine ratio (1.88 +/− 0.67 mol/mol) (Kari et al 2003). As noted above, increased intestinal magnesium intake does typically not correct this order. Rather, kidney transplantation has been the only effective means of reducing the excessive magnesium and calcium wasting (Praga et al 1995), and clearly demonstrated that the molecular defect resided in the kidney.

Mutations in Paracellin-1 (Claudin-16)

In 1999, Simon et al reported the identification of twelve kindreds with renal hypomagnesemia (Simon et al 1999). In ten of these families, the affected individuals were the offspring of consanguineous unions. These families were genotyped for polymorphic markers covering the human genome. Initial, low resolution results suggested linkage to chromosome 3q (Simon et al 1999). They took advantage of the presence of 11 distantly related individuals with this disease in a village in Saudi Arabia to refine the location of the disease gene. They found that these affected subjects were all homozygous for the identical allele in a short segment within the linked interval. They then isolated bacterial artificial chromosomes across the linked interval, and used exon trapping to identify putative genes from these segments. By this approach, paracellin-1 (*PCLN-1*) was identified as a novel protein expressed virtually exclusively in the kidney. Structurally, paracellin-1 has four transmembrane spanning domains, and shows sequence homology

(albeit only 10–18% amino acid identity) with members of the claudin gene family (paracellin-1 is now also referred to as claudin-16). Diverse nonconservative missense mutations, premature termination and splice site mutations were identified and homozygous in consanguineous kindreds; these mutations showed specificity for disease and cosegregated with the disease phenotype. Microdissection of nephron segments showed that paracellin-1 expression was restricted to the thick ascending limb (TAL) of the loop of Henle, with lower expression in the distal convoluted tubule (DCT). Paracellin-1 is not expressed in the proximal tubule, an important consideration given that the majority of magnesium reabsorption occurs in this segment during the neonatal period (Simon et al 1999). Antibodies to PCLN-1 were made and used to localize the protein. The protein was confined to the TAL and showed co-localization with occludin, a well-characterized tight junction protein.

These findings established PCLN-1 as a novel tight junction protein of the TAL that when mutated results in renal Mg^{2+} and Ca^{2+} wasting (Figure 14.2). Given the known important role of paracellular Mg^{2+} and Ca^{2+} reabsorption in the TAL and the established role of the tight junction as the barrier to paracellular flux (Simon et al 1999), these findings suggested that paracellin-1 participates in the formation of a cation selective tight junction pore that permits passage of magnesium and calcium down their electrochemical gradients in the thick ascending limb of Henle (Claude 1978, Simon et al 1999). Furthermore, it was hypothesized that mutations in paracellin-1 would affect magnesium reabsorption more than calcium reabsorption if either (a) paracellin-1 also possessed an unidentified magnesium sensor, or (b) Ca^{2+} homeostasis was less dependent on this pathway because of an ability to either reabsorb more Ca^{2+} elsewhere in the nephron, or to increase intestinal absorption and mobilization of Ca^{2+} from bone via effects of parathyroid hormone (Claude 1978, Simon et al 1999). These findings were quickly confirmed by Weber et al, who identified additional mutations in this gene in affected patients (Weber et al 2000).

The identification of a physical association between paracellin-1 and ZO-1 bolstered the claim that paracellin-1 was part of the tight junction complex (Ikari et al 2004). Using an in vitro binding assay, Ikari et al showed that paracellin-1 physically interacted with ZO-1, a membrane associated guanylate kinase. More specifically, the authors provided convincing evidence that an intact C-terminal PDZ binding domain is necessary for the paracellin-1–ZO-1 interaction. Moreover, it appeared that the C-terminal threonine and valine were indispensable, while the arginine residue could be substituted with alanine and not disrupt the protein–protein interaction. Intriguingly, immunocytochemistry studies showed that the association of ZO-1 with paracellin-1 was necessary for proper paracellin-1 subcellular localization to the tight junction complex. This was unique in that while mutations of this domain in claudins 1–8 similarly fail to associate with ZO-1, the mutant claudins are still correctly localized at the tight junction. Paracellin-1 with a threonine to arginine mutation in the C-terminal PDZ binding domain misdirected the mutant protein toward lysosomal degradation (Ikari et al 2004). Thus, an association with ZO-1 is required for proper subcellular trafficking of paracellin-1 to the tight junction complex.

Physiological studies have demonstrated that expression of wild-type paracellin-1 increased peak and steady state transepithelial resistance (TER) across MDCK transwell monolayers (Ikari et al 2004). Perhaps the most convincing evidence for paracellin-1's role in paracellular magnesium and calcium transport is the finding of increased calcium transport from the apical to basolateral compartment in cells expressing paracellin-1 compared with those that do not (Ikari et al 2004). Furthermore, the addition of magnesium chloride to the apical compartment reduced calcium transport from the apical to basolateral space. Thus, it would appear that magnesium and calcium compete for paracellular transport through paracellin-1 (Ikari et al 2004).

The discovery that paracellin-1 was phosphorylated by protein kinase A (PKA) was the first molecular evidence to support the hypothesis that while paracellular cation transport was passive and unsaturable, it could nonetheless be a regulated process (Ikari et al 2006). The differential labeling of paracellin-1 with radioactive phosphate in the presence and absence of pharmacological inhibitors of PKA suggested a role for this kinase in the regulatory process. Indeed, claudin-3 was known to be phosphorylated by PKA in ovarian cancer cells (Agarwal et al 2005). Alanine substitution studies defined S217 as the target of PKA (Ikari et al 2006). The fact that the adenylate cyclase inhibitor DDA and PKA inhibitors lead to a reduced TER further supported a regulatory role for PKA in paracellin-1 function. Furthermore, in the presence of PKA inhibitors, magnesium transport from the apical to basolateral compartment of an MDCK monolayer did not differ when compared with mock cultures not expressing paracellin-1. Similar to studies using PKA inhibitors, an S217A paracellin-1 mutant showed no difference in TER and magnesium transport compared with mock transfected cells (Ikari et al 2006).

In agreement with the prior studies which demonstrated the need for an intact PDZ binding domain to target paracellin-1 to tight junction complexes, dephosphorylation of paracellin-1 resulted in localization to the lysosomal pathway. Similarly, the unphosphorylatable paracellin-1 S217A mutant protein was found in lysosomal structures rather than in tight junctions (Ikari et al 2006). Preliminary studies suggest that dephosphorylated paracellin-1 is mono-ubiquitinated (Ikari et al 2006). Taken together, the clinical and functional data show that paracellin-1, in the phosphorylated state, is localized to tight junctions and behaves as an ion barrier with magnesium-permeable pores. Dephosphorylation removes paracellin-1 from the tight

junction complex and likely provides a level of regulatory control over magnesium homeostasis.

Carriers with heterozygous mutations may show some clinical features, though hypomagnesemia may only be mild, if present at all. Hypercalciuria, nephrocalcinosis, and nephrolithiasis have all been described in this population (Praga et al 1995, Simon et al 1999, Weber et al 2001), raising the question of whether the heterozygous state for this disease predisposes to hypercalciuric stone formation. These, as well as further exploration of the consequences of specific mutations, remain as areas for further exploration (Muller et al 2002; Zimmerman et al 2006).

Mutations in Claudin-19

Recently, Konrad et al reported eight Spanish/Hispanic families and one Swiss family which appeared to have all the hallmark clinical symptoms of the FHHNC syndrome (Konrad et al 2006). However, these families were distinctive in that in addition to these symptoms, affected individuals showed multiple ophthalmologic abnormalities including significant myopia, macular colobomata, and horizontal nystagmus, features that do not appear to be present or severe among patients with claudin-16 mutations. No mutations in paracellin-1 were found in these affected individuals. However, a genome wide scan for shared homozygous regions led to linkage with chromosome 1p34.2. *CLDN19* was identified as the most promising positional candidate gene within the linked interval, and mutations in *CLDN19* were found (Konrad et al 2006).

While Claudin-19 is expressed in few organs, the highest levels of expression are seen in the kidney and eye, further supporting its role in this unique magnesium wasting disorder. Sequence analysis of the affected family members identified two different homozygous missense mutations. The CLDN19 G20D mutation, located in the first transmembrane domain, was identified, as was a mutation in the first extracellular loop, Q57E. Subsequently, a third mutation, L90P, was identified in another consanguineous family. These three amino acid residues have been highly conserved throughout evolution. Furthermore, the two initially discovered mutations result in a switch from neutral charge to negatively charged amino acids, and SIFT (Sorting Intolerant From Tolerant) computational analysis indicated that both mutations had the potential to be deleterious. The L90P mutation is located in the second transmembrane alpha-helical domain of claudin-19. As proline residues are rigid, they are not well tolerated in the context of alpha-helical structures. Thus, it is predicted that this mutation alters claudin-19 function by disrupting the alpha-helical structure of the second transcellular domain. Similar to paracellin-1, claudin-19 is normally trafficked to the cell membrane. However, expression of the claudin-19 G20D mutant in MDCK cells resulted in perinuclear retention, consistent with loss of the normal signal sequence of this protein.

GITELMAN AND BARTTER SYNDROMES

Our understanding of renal sodium reabsorption and potassium excretion is, in part, a result of intense efforts to identify families with blood pressure and serum potassium levels that markedly deviate from the population means. Several of these traits have shown patterns consistent with classic Mendelian transmission within families, enabling genetic mapping and gene identification. Gitelman and Bartter syndromes, discussed in detail elsewhere in this volume, are excellent examples which have proved instructive for understanding magnesium homeostasis (Bartter et al 1962, Gitelman et al 1966).

It is now recognized that patients with Bartter's syndrome have salt wasting due to defects in the thick ascending limb (TAL) of the loop of Henle (Figure 14.1). These genes include *SLC12A1* which encodes the Na-K-2Cl cotransporter NKCC2 through which the apical entry of these ions from the lumen of the TAL occurs (Quaggin et al 1995); *KCNJ1*, which encodes the potassium channel ROMK and which mediates the return of potassium across the apical membrane (Ho et al 1993); and genes encoding either of two subunits of the chloride channel that mediates basolateral exit of chloride in the TAL (*CLCNKB* and *BSND*) (Kierferle et al 1994, Birkenhager et al 2001). As discussed above, this mechanism results in the asymmetric reabsorption of cations and anions in the TAL, with development of a lumen-positive potential that provides the electrical driving force for paracellular magnesium and calcium reabsorption. Mutations that cause Bartter syndrome result in homozygous loss of one of these gene products, with consequent massive salt wasting. Despite the fact that a very large fraction of renal magnesium reabsorption occurs in the TAL, patients with Bartter syndrome do not have the marked hypomagnesemia seen in patients with mutations in claudin 16 and 19, and indeed frequently have normal magnesium levels (Simon et al 1997). Those who do develop hypomagnesemia are most often those with mutations in CLCNKB, which also appears to play a role in the DCT. This surprising finding suggests either: (1) that loss of the electrical driving force for magnesium reabsorption in the TAL in Bartter syndrome leaves a sufficiently strong concentration gradient that the intact paracellular pathway can still reabsorb considerable magnesium, which is not believed to be the case; or (2) that the defect in Bartter syndrome results in secondary compensation that promotes increased magnesium reabsorption in other nephron segments, while mutations in claudin 16 and 19 do not. The severe salt wasting of Bartter's syndrome strongly activates the renin-angiotensin system, markedly augmenting salt reabsorption in other nephron segments. In particular, there is expansion of the distal convoluted tubule, at least in mouse models of Bartter's syndrome; because the DCT is the site of the magnesium channel TRPM6, it seems possible that up-regulation of this channel could augment magnesium reabsorption in Bartter's syndrome in ways that would not occur in patients with claudin 16 and 19 mutations. Both explanations could of course contribute.

While Bartter's syndrome is a disease of the TAL, Gitelman's syndrome results from homozygous loss of function of *SLC12A3*, which encodes the thiazide-sensitive Na-Cl contransporter NCC in the distal convoluted tubule (DCT; Figure 14.2) (Simon et al 1996). Perhaps the most surprising feature resulting from these mutations in the renal salt reabsorption pathway is the uniform development of hypomagnesemia among affected patients (Gitelman et al 1966). This finding parallels the development of hypomagnesemia among patients taking thiazide diuretics, specific inhibitors of NCC, and the later observation of hypomagnesemia in mice with targeted mutations in SLC12A3. These observations indicate the important role of NCC activity in the normal maintenance of magnesium homeostasis. The molecular details of this mechanism are still a matter of uncertainty; however, it has become clear that while only a small fraction of magnesium reabsorption occurs in the DCT, this site plays a vital role in net magnesium balance. Two key observations provide insights to the likely explanations. First, the cation channel TRPM6, which localizes to the DCT, is the likely magnesium channel (see below) (Schlingmann et al 2002, Walder et al 2002). Second, for reasons that are as yet not understood, the mass of the DCT is highly malleable, and specifically varies with the activity of NCC. Thus, mutations in NCC that cause Gitelman's syndrome result in loss of DCT mass (Schultheis et al 1998), while mutations in WNK4 that cause increased activity of NCC (see chapter on WNK kinases in this volume) cause increased DCT mass (Lalioti et al 2006), as do Bartter's mutations that activate the renin-angiotensin system (Lu et al 2002). These observations afford the speculation that loss of NCC results in diminished DCT mass, and that this event has, as a secondary consequence, reduced net TRPM6 activity. Consistent with this possibility, the *NCC* deficient mouse has been shown to have markedly reduced TRPM6 expression (Nijenhuis et al 2005).

HYPOMAGNESEMIA WITH SECONDARY HYPOCALCEMIA (HSH)

HSH is an autosomal recessive disorder of intestinal and renal magnesium transport. This disorder, characterized by reduced serum magnesium and calcium levels, was first described by Paunier et al, in 1968 (Paunier et al 1968).

Similarly to the original report, patients usually present during infancy with the neurological manifestations of hypomagnesemic hypocalcemia, including seizures, tetany, and muscle spasms. Typical biochemical derangements include severely low serum magnesium levels (0.24 +/− 0.11 mmol/l) and reduced serum calcium concentrations (1.58 +/− 0.33 mmol/l). The severity of the disease is variable, but can result in permanent neurological injury, and in some cases even death.

The hypomagnesemia of this disorder is now recognized to be attributable to defects in both gastrointestinal absorption and renal reabsorption. Oral magnesium administration typically fails to raise magnesium levels to normal and does not raise urinary magnesium levels as much as would be expected from a simple renal deficiency, indicating an intestinal absorption defect. In addition, however, while intravenous administration of magnesium can increase serum levels, the fractional excretion of magnesium is elevated in the setting of hypomagnsemia, reflecting a renal defect as well (Shalev et al 1998).

In contrast to FHHNC, in which renal magnesium wasting is accompanied by severe hypercalciuria but maintenance of normal serum calcium levels, HSH features hypocalcemia and hypocalciuria. These features are a consequence of impaired parathyroid hormone (PTH) secretion (Agus 1999) with some patients having undetectable levels of parathyroid hormone despite hypocalcemia (Anast et al 1972, Suh et al 1973, Chase & Slatopolsky 1974). This appears to be a secondary consequence of the severe hypomagnesemia. Magnesium replacement restores PTH secretion and calcium levels toward normal. Furthermore, release of parathyroid hormone from parathyroid cells in vitro is significantly reduced when the concentration of magnesium in the culture media falls below 0.35 mmol/l (Targovnik et al 1971). Hypomagnesemia also attenuates the biological effect of PTH by interfering with its signal transduction (Agus 1999). These PTH-related effects are not seen with the other hypomagnesemic disorders, suggesting that this feature of HSH is a consequence of the extremely low serum magnesium levels encountered with this order. Finally, bone resistance to PTH may also be a contributor to the hypocalcemia associated with hypomagnesemia (Agus 1999). Given the central role of PTH in regulating bone turnover and renal calcium reabsorption, it is not surprising that the hypocalcemia associated with this disorder cannot be easily corrected with calcium or vitamin D therapy alone. Magnesium levels up to 20 times the normal intake are required to achieve normalization of calcium levels, but this form of therapy is not tolerated well by affected individuals (diarrhea is a common side effect of oral magnesium administration even in doses far less than what would be needed to correct this disorder) (Shalev et al 1998). In an attempt to reduce the side effects associated with bolus oral consumption of magnesium, continuous overnight nasogastric administration has been attempted with some success (Milla et al 1979).

Identification of Mutations in TRPM6

Walder et al identified three inbred kindreds from the Bedouin population in Israel with clinical findings consistent with HSH. Most of the affected individuals presented in infancy, with restlessness, tremor, neuromuscular hyperexcitability, and overt seizures (Walder et al 1997). Serum magnesium concentrations were between 0.2–0.8 mg/dl, and serum calcium levels were 3.7–7.6 mg/dl at the time of presentation. Using a DNA pooling method and searching for short tandem repeat polymorphic markers (STRPs) that were enriched among affected subjects, linkage was established to chromosome 9q (Walder et al 1997). These data contrasted with earlier predictions that HSH was an X-linked recessive disorder (these earlier conclusions were based on the predominance of male patients and the report of a single female patient with a balanced X:9 chromosome translocation) (Mattei et al 1978, Chery et al 1994). The linkage interval was further narrowed to an interval less than 1cM that contained a small number of genes. Subsequently, Schlingmann et al and Walder et al identified *TRPM6* as the causative gene in HSH (Schlingmann et al 2002, Walder et al 2002). The identified mutations in *TRPM6* are often homozygous as a consequence of consanguinity and include an in-frame stop codon at Ser590, which truncates the protein prior to the first transmembrane domain as well as an array of splice-site, frameshift and missense mutations (Schlingmann et al 2002). These mutations show specificity for disease families and segregate with the trait in families, unequivocally establishing loss of function mutations in *TRPM6* as the cause of this form of hypomagnesemia.

TRPM6 encodes an ion channel that by sequence homology is highly similar to the transient receptor potential (TRP) channel family. TRP ion channels possess a common ultrastructural organization characterized by six transmembrane segments, a highly conserved pore-forming region and a Pro-Pro-Pro motif following the last transmembrane segment (Harteneck et al 2000, Kraft & Harteneck 2005). *TRPM6* can be further classified as a member of the TRPM family of ion channels (where 'M' stands for 'melastatin') (Kraft & Harteneck 2005). TRPM channels are believed to function as tetramers (Jiang 2007). Unlike the TRPC and TRPV subfamilies, TRPM channels do not contain N-terminal ankyrin repeat motifs. Rather, these channels possess additional functional domains at their C-termini. *TRPM6*, and its closest relative, *TRPM7*, both contain functional serine/threonine protein kinase segments (Ryazanov et al 1997).

TRPM6 is strongly expressed in both the small intestine (duodenum, jejunum, and ileum) and the kidney (Schlingmann et al 2002). Furthermore, using rat microdissected nephrons it was demonstrated that *TRPM6* expression within the kidney is greatest in the distal convoluted tubule (DCT; Figure 14.2), with much weaker expression in other nephron segments (Schlingmann et al 2002). *TRPM6* is also highly expressed in both the small intestine (duodenum, jejenum, and ileum) and colon (Schlingmann et al 2002).

The finding that mutation in *TRPM6* causes hypomagnesemia strongly suggested that it would itself be a magnesium channel. This story is still unfolding, but is nonetheless highly constrained by the genetic evidence of the consequence of TRPM6 mutation. TRPM6 appears to be capable of mediating magnesium influx via channel activity, although, like other TRP channels, it is not highly selective (Voets et al 2004, Li et al 2006). Substantial evidence has suggested that TRPM6 may act in heteromultimers with TRPM7 or require interaction with TRPM7 to reach the apical membrane (Chubanov et al 2005, Schmitz et al 2005), however there is not consensus on this point. TRPM6 and TRPM7 are sensitive to magnesium and magnesium-nucleotide concentrations, with activation occurring when magnesium ions are depleted from cells. While it was tantalizing to speculate that the intrinsic kinase activity was necessary for proper channel activation, this does not appear to be the case (Matsushita et al 2005, Demeuse et al 2006).

Cao et al recently reported the identification of the receptor for activated C-kinase 1 (RACK1) as a TRPM6 regulatory protein that associates with the kinase domain (Cao et al 2008). RACK1 is an anchoring protein for protein kinase C, and is believed to pay a central role in several critical biological responses, such as cell growth (MaCahill et al 2002). RACK1, like TRPM6, is expressed in DCT (Cao et al 2008). RACK1 appears to inhibit activity of the channel in a kinase activity dependent manner. When RACK1 levels were reduced experimentally using small interfering RNA (siRNA), TRPM6 currents were increased (Cao et al 2008). Furthermore, the inhibitory effect of RACK1 is mediated by the phosphorylation state of Thr1851 of TRPM6, an identified autophosphorylation site. The authors also showed that Thr1851 is a crucial component of TRPM6 with regard to autophosphorylation and channel activity occurring in a magnesium sensitive fashion (Cao et al 2008).

It is clear that the identification of TRPM6 has provided fundamental new insight into magnesium homeostasis, playing an important role in both transcellular gastrointestinal absorption and renal reabsorption.

AUTOSOMAL DOMINANT HYPOMAGNESEMIA DUE TO MUTATION IN *FXYD2*

A novel syndrome featuring autosomal dominant hypomagnesemia due to renal magnesium wasting in conjunction with hypocalciuria was reported in two apparently unrelated Dutch families in 1987 (Geven et al 1987). Magnesium infusion studies demonstrated increased fractional excretion of magnesium despite hypomagnesemia, evidence of a renal reabsorption defect. Seizures were the presenting symptom in both of the two index cases. While the magnesium and calcium values of this disorder resembled those of Gitelman's syndrome, these patients had no evidence of renal salt wasting, hypokalemia or metabolic alkalosis, suggesting a distinct entity.

Genome-wide linkage studies provided evidence of linkage to a segment of Chr11q23 (Meij et al 2000a). While ROMK is located on Chr11q, its distance from the linked interval was considered too large to be a candidate gene causing this unique magnesium wasting disorder. Further screening of the mapped genes and ESTs in the linked interval led to the identification of a mutation in the γ-subunit of the Na^+-K^+-ATPase (Meij et al 2000b). As prior analysis had suggested that the haplotype linked to the disease in the two families was identical by descent from a common ancestor, it was satisfying to find that the same rare mutation was present on the disease-linked chromosome in both families. The mutation, Gly41Arg, was found in all affected individuals in both kindreds, segregated with the disease phenotype and was not seen in unaffected subjects (Meij et al 2000b). This mutation alters the charge of the amino acid, and occurs at a position that is highly conserved in *FXYD2* from many species. The Na^+-K^+-ATPase is a ubiquitous protein required for the maintenance of resting cell membrane potentials. Composed of two subunit isoforms, α and β, it is commonly the underlying driving force for the active transport of ions across cell membranes. The γ-subunit is not obligatory for function of the Na^+-K^+-ATPase, but is expressed predominantly in the kidney. While the α-subunit is expressed in the TAL, DCT, connecting tubule (CNT), cortical collecting duct (CCD), and proximal convoluted tubule (PCT), the γ-subunit appears restricted to the DCT, CNT, and PCT (Wetzel & Sweadner 2001). In vitro studies show that the affinity of the Na^+-K^+-ATPase for sodium and potassium is reduced in the presence of the γ-subunit (Arystarkhova et al 1999), and isolated nephron segments from rabbit and rat do show higher sodium affinity in the TAL and CCD (Figure 14.2). The rarity of the syndrome and its dominant transmission suggest a genetic gain-of-function mutation. Consistent with this notion, individuals who have genomic deletions that span *FXYD2* do not have evidence of hypomagnesemia, arguing against haploinsufficiency as the mechanism (Meij et al 2000b).

Cell culture studies demonstrate that the Gly41Arg point mutation, which placed a charged residue within the transmembrane domain of the protein, exerts a dominant effect by mistrafficking the γ-subunit to a perinuclear location (possibly retained in the Golgi complex) (Meij et al 2000b). Thus, one would predict from the above findings that individuals with the Gly41Arg mutation should have a Na^+-K^+-ATPase with increased affinity for sodium and potassium which indirectly reduces magnesium transport in the DCT.

Recent studies have shown that the γ-subunit oligomerizes, and that the Gly41Arg mutant maintains its ability to form complexes with wild-type subunits, resulting in the cellular mistrafficking of both, suggesting a dominant-negative mechanism (Cairo et al 2008). Other families with a similar phenotype have been described that do not have mutations in *FXYD2*, suggesting genetic heterogeneity of this trait (Kantorovich et al 2002). This leaves open the possibility that other mutant genes affecting this disease pathway will be discovered and provide more detailed functional insights.

Indeed, an additional family with a similar syndrome of renal hypomagnesemia and hypocalciuria has been characterized. As discussed elsewhere in this volume, in this kindred, hypomagnesemia with hypocalciuria recurs in multiple generations of a kindred; however, the trait is only passed from affected females, never affected males, to offspring and more that the expected 50% of offspring have the trait. These findings are the hallmarks of traits encoded in the mitochondrial genome, since we inherit all of our mitochondria from our mothers and none from our fathers. Sequencing of the mitochondrial genome in this family identified a novel mutation in the gene encoding the isoleucine tRNA at a position known to be key for the proper secondary structure of the molecule. This mutation strongly implicates altered mitochondrial function in the pathogenesis of hypomagnesemia.

These two findings together suggest an energy-dependent role, perhaps involving the sodium pump, in magnesium reabsorption in the DCT. The molecular nature of this step may be involved in magnesium exit across the basolateral membrane; however, this is not established.

ISOLATED RECESSIVE RENAL HYPOMAGNESEMIA

A unique kindred featuring severe isolated renal hypomagnesemia has led to identification of another gene required for renal magnesium homeostasis. Two sisters in a consanguineous family were identified with hypomagnesemia. Despite their hypomagnesemia, they maintained an elevated fractional excretion of magnesium, suggesting a renal defect. Interestingly, serum and urinary calcium levels were reported to be normal. Homozygosity mapping provided evidence of linkage to a segment of chromosome 4, though the evidence was inconclusive given the small size of the kindred (Lod score 2.66). Interestingly, however, a tantalizing candidate gene was in this chromosome segment, the epidermal growth factor (*EGF*) gene. *EGF* was of particular interest because pharmacologic antagonists have produced hypomagnesemia in humans (Schrag et al 2005, Fakih et al 2006). Moreover, EGF is expressed in DCT epithelium and the EGFR is expressed on the basolateral cell surface of this nephron segment. Sequencing of the gene identified a homozygous missense mutation that substitutes leucine for proline in the cytoplasmic tail of EGF at a position that is conserved in mammalian EGF. Expression of the gene product in MDCK cells suggested defective sorting to the basolateral membrane, the site of EGFR localization (Figure 14.2). Moreover, the studies suggested that EGF regulates TRPM6, the presumptive magnesium channel of the DCT.

These findings collectively implicate autocrine or paracrine EGF signaling in magnesium homeostasis. These findings provide important new insight but raise many new questions for further study. Is EGF expression modulated in the DCT, for example by magnesium level? What is the pathway that links activation of the EGFR to increased activity of TRPM6? Does altered EGF signaling play a role in other disorders of magnesium homeostasis? Are the effects of the identified EGF mutation confined to the DCT or are there more widespread consequences?

CONCLUSIONS AND FUTURE DIRECTIONS

Serum magnesium levels are tightly regulated, a feature critical to the normal function of diverse physiologic processes. Despite the importance of this regulation, decades of physiologic analysis alone were of only modest success in defining the pathways that govern this parameter. In the last 12 years, a host of genes that play critical roles in this process have been identified by human genetic studies. These advances have occurred because of clinical insight in the recognition of unique patients, careful clinical investigation to study extended kindreds with these traits to define their patterns of recurrence and penetrance of phenotypes, and molecular studies to map and identify the underlying disease genes. These studies have demonstrated the key role of specific claudin proteins (claudin 16 and 19) in the paracellular flux of magnesium and calcium in the TAL, and in fact provided the first demonstration of specific proteins in selective paracellular flux. They have also underscored the important role of specific mediators of salt reabsorption in the TAL in establishing the electrical potential that contributes to magnesium reabsorption in this nephron segment. In addition, however, these studies have revealed the key role of the DCT in magnesium homeostasis. At least 5 genes believed to impart their effects via action in the DCT are now recognized to affect magnesium homeostasis. These include the thiazide-sensitive Na-Cl cotransporter (NCC), the TRPM6 cation channel, the gamma subunit of the Na^+-K^+-ATPase, epidermal growth factor and the mitochondrial isoleucine tRNA.

From these studies, one can put together the speculative outlines of the magnesium handling pathway of the DCT. TRPM6 appears to mediate the apical entry of magnesium, and the activity of this channel appears to be regulated at least in part by EGF. NCC activity is required for magnesium reabsorption in this nephron segment, but whether this is due to a direct effect on TRPM6 or an indirect effect through its regulation of DCT mass is presently unclear. As its name implies, it is possible that EGF also exerts part of its effect through alteration of DCT mass. Mutations in the gamma subunit of the Na^+-K^+-ATPase and the mitochondrial isoleucine tRNA both produce similar phenotypes of increased renal magnesium wasting and hypocalciuria. The DCT has the highest density of mitochondria along the nephron, reflecting its very high metabolic demand. There is evidence that magnesium reabsorption in this nephron segment is an active process consuming ATP; it is consequently possible that the mitochondrial mutation impairs this process and that the mutation in the Na^+-K^+-ATPase similarly prevents establishment of an electrochemical gradient required for magnesium reabsorption.

It is clear that much remains to be done to understand magnesium homeostasis. How altered magnesium levels result in altered renal reabsorption is uncertain. What signaling pathways are employed? Is EGF part of this pathway, and if so, what is regulating the altered secretion of EGF and what is the downstream pathway? Is the paracellular pathway in the TAL regulated by PKA and if so, is this via trafficking to the tight junction or recruitment of these proteins to or out of the tight junction?

Perhaps most importantly, it seems clear that a key piece of the puzzle remains unsolved – the mechanism by which magnesium traverses the basolateral membrane of the DCT. Similarly, the specific mechanisms by which the gamma subunit and mitochondrial mutations alter magnesium homeostasis remain to be elucidated. These are important questions for the future of the field.

From a clinical perspective, several themes have emerged. Perhaps most importantly, each of these diseases results from a primary renal abnormality and consequently renal transplantation would be expected to correct the defects, likely in conjunction with removal of native kidneys. For some of these diseases (e.g. for many patients with Gitelman's syndrome), the risk of transplantation may not outweigh the benefits. However for others, such as patients with claudin 16 mutations, who appear to commonly develop end-stage renal disease at young ages, transplantation would likely confer marked benefit.

This research has also identified poorly understood links between the renal handling of salt, magnesium and calcium. For example, defects in salt handling in the TAL produce hypomagnesemia in conjunction with hypercalciuria, and defects in tight junction proteins in this segment produce parallel defects in reabsorption of magnesium and calcium. This is attributable to these divalent cations sharing the electrical driving force for reabsorption derived from asymmetric reabsorption of sodium and chloride in this nephron segment, and the fact that they also traverse the tight junction using the same paracellular pathway. In contrast,

hypomagnesemia resulting from a defect in the DCT is in most cases accompanied by hypocalciuria. The mechanism of this latter effect is incompletely understood at present.

Diagnostically, these observations are extremely helpful in the characterization of patients. When confronted with a patient with hypomagnesemia, the finding of an elevated fractional excretion of magnesium establishes a renal defect. The finding of concomitant hypercalciuria points to a primary defect in the TAL while hypocalciuria points to the DCT. The presence or absence of renal salt wasting allows distinction of Gitelman and Bartter syndromes from other magnesium-wasting diseases. As resequencing of genomic DNA becomes progressively less expensive, the ability to routinely resequence these genes to establish molecular diagnosis will become more routine.

Finally, the elucidation of these genes and pathways will undoubtedly provide continuing insight into the mechanisms of drug-induced hypomagnesemia. The hypomagnesemia resulting from inhibitors of the EGFR is one clear example that can now be understood, but others will likely be equally important.

It is clear that genetic studies have provided fundamental new insight into magnesium homeostasis; further studies starting from the framework provided by these human studies will complete a detailed picture of the homeostasis of this important biological ion.

ACKNOWLEDGMENTS

We thank the many patients and families from around the world whose participation in clinical research has enabled the new insights into magnesium homeostasis. We also thank our many colleagues whose work is represented in this chapter, as well as those whose work was not explicitly presented due to space constraints. WSA is a Research Fellow of the National Kidney Foundation. RPL is an Investigator of the Howard Hughes Medical Institute.

References

Agarwal R, D'Souza T, Morin PJ. Claudin-3 and claudin-4 expression in ovarian epithelial cells enhances invasion and is associated with increased matrix metalloproteinase-2 activity. Cancer Res. 2005; 65(16): 7378–85.

Agus ZS. Hypomagnesemia. J. Am. Soc. Nephrol. 1999; 10(7): 1616–22.

Anast CS, Mohs JM, Kaplan SL, Burns TW. Evidence for parathyroid failure in magnesium deficiency. Science 1972; 177(49): 606–8.

Arystarkhova E, Wetzel RK, Asinovski NK, Sweadner KJ. The gamma subunit modulates Na(+) and K(+) affinity of the renal Na,K-ATPase. J. Biol. Chem. 1999; 274(47): 33183–5.

Barlet-Bas C, Cheval L, Khadouri C, Marsy S, Doucet A. Difference in the Na affinity of Na(+)-K(+)-ATPase along the rabbit nephron: modulation by K. Am. J. Physiol. 1990; 259 (2 Pt 2): F246–50.

Bartter FC, Pronove P, Gill JR, Jr., Maccardle RC. Hyperplasia of the juxtaglomerular complex with hyperaldosteronism and hypokalemic alkalosis. A new syndrome. Am. J. Med. 1962; 33: 811–28.

Birkenhager R, Otto E, Schurmann MJ, et al. Mutation of BSND causes Bartter syndrome with sensorineural deafness and kidney failure. Nat. Genet. 2001; 29(3): 310–14.

Cairo ER, Friedrich T, Swarts HG, et al. Impaired routing of wild type FXYD2 after oligomerisation with FXYD2-G41R might explain the dominant nature of renal hypomagnesemia. Biochim. Biophys. Acta 2008; 1778(2): 398–404.

Cao G, Thebault S, van der Wijst J, et al. RACK1 inhibits TRPM6 activity via phosphorylation of the fused alpha-kinase domain. Curr. Biol. 2008; 18(3): 168–76.

Chakraborti S, Chakraborti T, Mandal M, Mandal A, Das S, Ghosh S. Protective role of magnesium in cardiovascular diseases: a review. Mol. Cell Biochem. 2002; 238(1–2): 163–79.

Chase LR, Slatopolsky E. Secretion and metabolic efficacy of parthyroid hormone in patients with severe hypomagnesemia. J. Clin. Endocrinol. Metab. 1974; 38(3): 363–71.

Chery M, Biancalana V, Philippe C, et al. Hypomagnesemia with secondary hypocalcemia in a female with balanced X;9 translocation: mapping of the Xp22 chromosome breakpoint. Hum. Genet. 1994; 93(5): 587–91.

Chubanov V, Gudermann T, Schlingmann KP. Essential role for TRPM6 in epithelial magnesium transport and body magnesium homeostasis. Pflugers Arch. 2005; 451(1): 228–34.

Claude P. Morphological factors influencing transepithelial permeability: a model for the resistance of the zonula occludens. J. Membr. Biol. 1978; 39(2–3): 219–32.

de Rouffignac C, Quamme G. Renal magnesium handling and its hormonal control. Physiol. Rev. 1994; 74(2): 305–22.

Demeuse P, Penner R, Fleig A. TRPM7 channel is regulated by magnesium nucleotides via its kinase domain. J. Gen. Physiol. 2006; 127(4): 421–34.

Fakih MG, Wilding G, Lombardo J. Cetuximab-induced hypomagnesemia in patients with colorectal cancer. Clin. Colorectal Cancer 2006; 6(2): 152–6.

Fava C, Montagnana M, Rosberg L, et al. Subjects heterozygous for genetic loss of function of the thiazide-sensitive cotransporter have reduced blood pressure. Hum. Mol. Genet. 2008; 17(3).

Fine KD, Santa Ana CA, Porter JL, Fordtran JS. Intestinal absorption of magnesium from food and supplements. J. Clin. Invest. 1991; 88(2): 396–402.

Geven WB, Monnens LA, Willems HL, Buijs WC, ter Haar BG. Renal magnesium wasting in two families with autosomal dominant inheritance. Kidney Int. 1987; 31(5): 1140–4.

Gitelman HJ, Graham JB, Welt LG. A new familial disorder characterized by hypokalemia and hypomagnesemia. Trans. Assoc. Am. Physicians 1966; 79: 221–35.

Hampton EM, Whang DD, Whang R. Intravenous magnesium therapy in acute myocardial infarction. Ann. Pharmacother. 1994; 28(2): 212–19.

Harteneck C, Plant TD, Schultz G. From worm to man: three subfamilies of TRP channels. Trends Neurosci. 2000; 23(4): 159–66.

Ho K, Nichols CG, Lederer WJ, et al. Cloning and expression of an inwardly rectifying ATP-regulated potassium channel. Nature 1993; 362(6415): 31–8.

Iannello S, Belfiore F. Hypomagnesemia. A review of pathophysiological, clinical and therapeutical aspects. Panminerva Med. 2001; 43(3): 177–209.

Ikari A, Hirai N, Shiroma M, et al. Association of paracellin-1 with ZO-1 augments the reabsorption of divalent cations in renal epithelial cells. J. Biol. Chem. 2004; 279(52): 54826–32.

Ikari A, Matsumoto S, Harada H, et al. Phosphorylation of paracellin-1 at Ser217 by protein kinase A is essential for localization in tight junctions. J. Cell Sci. 2006; 119(Pt 9): 1781–9.

Innerarity S. Hypomagnesemia in acute and chronic illness. Crit. Care Nurs. Q. 2000; 23(2): 1–19; quiz 87.

Jiang LH. Subunit interaction in channel assembly and functional regulation of transient receptor potential melastatin (TRPM) channels. Biochem. Soc. Trans. 2007; 35(Pt 1): 86–8.

Kantorovich V, Adams JS, Gaines JE, et al. Genetic heterogeneity in familial renal magnesium wasting. J Clin Endocrinol Metab 2002; 87(2): 612–17.

Kari JA, Farouq M, Alshaya HO. Familial hypomagnesemia with hypercalciuria and nephrocalcinosis. Pediatr. Nephrol. 2003; 18(6): 506–10.

Kieferle S, Fong P, Bens M, Vandewalle A, Jentsch TJ. Two highly homologous members of the ClC chloride channel family in both rat and human kidney. Proc. Natl Acad. Sci. USA 1994; 91(15): 6943–7.

Konrad M, Weber S. Recent advances in molecular genetics of hereditary magnesium-losing disorders. J. Am. Soc. Nephrol. 2003; 14(1): 249–60.

Konrad M, Schaller A, Seelow D, et al. Mutations in the tight-junction gene claudin 19 (CLDN19) are associated with renal magnesium wasting, renal failure, and severe ocular involvement. Am. J. Hum. Genet. 2006; 79(5): 949–57.

Kozak JA, Cahalan MD. MIC channels are inhibited by internal divalent cations but not ATP. Biophys. J. 2003; 84(2 Pt 1): 922–7.

Kraft R, Harteneck C. The mammalian melastatin-related transient receptor potential cation channels: an overview. Pflugers Arch. 2005; 451(1): 204–11.

Lalioti MD, Zhang J, Volkman HM, et al. Wnk4 controls blood pressure and potassium homeostasis via regulation of mass and activity of the distal convoluted tubule. Nat. Genet. 2006; 38(10): 1124–32.

Li M, Jiang J, Yue L. Functional characterization of homo- and heteromeric channel kinases TRPM6 and TRPM7. J. Gen. Physiol. 2006; 127(5): 525–37.

Lombeck I, Ritzl F, Schnippering HG, et al. Primary hypomagnesemia. I. Absorption Studies. Z. Kinderheilkd 1975; 118(4): 249–58.

Lu M, Wang T, Yan Q, et al. Absence of small conductance K+ channel (SK) activity in apical membranes of thick ascending limb and cortical collecting duct in ROMK (Bartter's) knockout mice. J. Biol. Chem. 2002; 277(40): 37881–7.

McCahill A, Warwicker J, Bolger GB, Houslay MD, Yarwood SJ. The RACK1 scaffold protein: a dynamic cog in cell response mechanisms. Mol. Pharmacol. 2002; 62(6): 1261–73.

Manz F, Scharer K, Janka P, Lombeck J. Renal magnesium wasting, incomplete tubular acidosis, hypercalciuria and nephrocalcinosis in siblings. Eur. J. Pediatr. 1978; 128(2): 67–79.

Matsushita M, Kozak JA, Shimizu Y, et al. Channel function is dissociated from the intrinsic kinase activity and autophosphorylation of TRPM7/ChaK1. J. Biol. Chem. 2005; 280(21): 20793–803.

Mattei MG, Mattei JF, Ayme S, Malpuech G, Giraud F. A dynamic study in two new cases of X chromosome translocations. Hum. Genet. 1978; 41(3): 251–7.

Meij I, Illy KE, Monnens L. Severe hypomagnesemia in a neonate with isolated renal magnesium loss. Nephron 2000a; 84(2): 198.

Meij IC, Koenderink JB, van Bokhoven H, et al. Dominant isolated renal magnesium loss is caused by misrouting of the Na(+),K(+)-ATPase gamma-subunit. Nat. Genet. 2000b; 26(3): 265–6.

Mercer RW, Biemesderfer D, Bliss DP, Jr., Collins JH, Forbush B, 3rd. Molecular cloning and immunological characterization of the gamma polypeptide, a small protein associated with the Na,K-ATPase. J. Cell Biol. 1993; 121(3): 579–86.

Michelis MF, Drash AL, Linarelli LG, De Rubertis FR, Davis BB. Decreased bicarbonate threshold and renal magnesium wasting in a sibship with distal renal tubular acidosis. (Evaluation of the pathophysiological role of parathyroid hormone). Metabolism 1972; 21(10): 905–20.

Milazzo SC, Ahern MJ, Cleland LG, Henderson DR. Calcium pyrophosphate dihydrate deposition disease and familial hypomagnesemia. J. Rheumatol. 1981; 8(5): 767–71.

Milla PJ, Aggett PJ, Wolff OH, Harries JT. Studies in primary hypomagnesaemia: evidence for defective carrier-mediated small intestinal transport of magnesium. Gut 1979; 20(11): 1028–33.

Muller D, Claverie-Martin F, Eggert P, Garcia-Nieto V. Mutationen im PDZ-Motif von Paracellin-1 als Ursache der Hyperkalziurie im Kindesalter [Abstract]. Nieren-und Hochdruck-krankheiten 2002; 31: 52.

Nadler MJ, Hermosura MC, Inabe K, et al. LTRPC7 is a Mg.ATP-regulated divalent cation channel required for cell viability. Nature 2001; 411(6837): 590–5.

Nicholson JC, Jones CL, Powell HR, Walker RG, McCredie DA. Familial hypomagnesaemia – hypercalciuria leading to end-stage renal failure. Pediatr. Nephrol. 1995; 9(1): 74–6.

Nijenhuis T, Vallon V, van der Kemp AW, Loffing J, Hoenderop JG, Bindels RJ. Enhanced passive Ca^{2+} reabsorption and reduced Mg^{2+} channel abundance explains thiazide-induced hypocalciuria and hypomagnesemia. J. Clin. Invest. 2005; 115(6): 1651–8.

Paunier L, Radde IC, Kooh SW, Conen PE, Fraser D. Primary hypomagnesemia with secondary hypocalcemia in an infant. Pediatrics 1968; 41(2): 385–402.

Praga M, Vara J, Gonzalez-Parra E, et al. Familial hypomagnesemia with hypercalciuria and nephrocalcinosis. Kidney Int. 1995; 47(5): 1419–25.

Quaggin SE, Payne JA, Forbush B, 3rd, Igarashi P. Localization of the renal Na-K-Cl cotransporter gene (Slc12a1) on mouse chromosome 2. Mamm. Genome 1995; 6(8): 557–8.

Quamme GA. Renal magnesium handling: new insights in understanding old problems. Kidney Int. 1997; 2(5): 1180–95.

Quamme GA, de Rouffignac C. Epithelial magnesium transport and regulation by the kidney. Front Biosci. 2000; 5: D694–711.

Richard O, Freycon MT. Congenital tubulopathy with magnesium loss. Pediatrie 1992; 47(7–8): 557–63.

Ryazanov AG, Ward MD, Mendola CE, et al. Identification of a new class of protein kinases represented by eukaryotic elongation factor-2 kinase. Proc. Natl Acad. Sci. USA 1997; 94(10): 4884–9.

Sanders GT, Huijgen HJ, Sanders R. Magnesium in disease: a review with special emphasis on the serum ionized magnesium. Clin. Chem. Lab. Med. 1999; 37(11–12): 1011–33.

Schimatschek HF, Rempis R. Prevalence of hypomagnesemia in an unselected German population of 16,000 individuals. Magnes. Res. 2001; 14(4): 283–90.

Schlingmann KP, Weber S, Peters M, et al. Hypomagnesemia with secondary hypocalcemia is caused by mutations in TRPM6, a new member of the TRPM gene family. Nat. Genet. 2002; 31(2): 166–70.

Schmitz C, Dorovkov MV, Zhao X, Davenport BJ, Ryazanov AG, Perraud AL. The channel kinases TRPM6 and TRPM7 are functionally nonredundant. J. Biol. Chem. 2005; 280(45): 37763–71.

Schrag D, Chung KY, Flombaum C, Saltz L. Cetuximab therapy and symptomatic hypomagnesemia. J. Natl Cancer Inst. 2005; 97(16): 1221–4.

Schultheis PJ, Lorenz JN, Meneton P, et al. Phenotype resembling Gitelman's syndrome in mice lacking the apical Na^+-Cl^- cotransporter of the distal convoluted tubule. J. Biol. Chem. 1998; 273(44): 29150–5.

Shalev H, Phillip M, Galil A, Carmi R, Landau D. Clinical presentation and outcome in primary familial hypomagnesaemia. Arch. Dis. Child. 1998; 78(2): 127–30.

Simon DB, Bindru RS, Mansfield TA, et al. Mutations in the chloride channel gene, CLCNKB, cause Bartter's syndrome type III. Nat. Genet. 1997; 17(2): 171–8.

Simon DB, Nelson-Williams C, Bia MJ, et al. Gitelman's variant of Bartter's syndrome, inherited hypokalaemic alkalosis, is caused by mutations in the thiazide-sensitive Na-Cl cotransporter. Nat. Genet. 1996; 12(1): 24–30.

Simon DB, Lu Y, Choate KA, et al. Paracellin-1, a renal tight junction protein required for paracellular Mg^{2+} resorption. Science 1999; 285(5424): 103–6.

Suh SM, Tashjian AH, Jr., Matsuo N, Parkinson DK, Fraser D. Pathogenesis of hypocalcemia in primary hypomagnesemia: normal end-organ responsiveness to parathyroid hormone, impaired parathyroid gland function. J. Clin. Invest. 1973; 52(1): 153–60.

Targovnik JH, Rodman JS, Sherwood LM. Regulation of parathyroid hormone secretion in vitro: quantitative aspects of calcium and magnesium ion control. Endocrinology 1971; 88(6): 1477–82.

Torralbo A, Pina E, Portoles J, Sanchez-Fructuoso A, Barrientos A. Renal magnesium wasting with hypercalciuria, nephrocalcinosis and ocular disorders. Nephron 1995; 69(4): 472–5.

Voets T, Nilius B, Hoefs S, et al. TRPM6 forms the Mg^{2+} influx channel involved in intestinal and renal Mg^{2+} absorption. J. Biol. Chem. 2004; 279(1): 19–25.

Walder RY, Shalev H, Brennan TM, et al. Familial hypomagnesemia maps to chromosome 9q, not to the X chromosome: genetic linkage mapping and analysis of a balanced translocation breakpoint. Hum. Mol. Genet. 1997; 6(9): 1491–7.

Walder RY, Landau D, Meyer P, et al. Mutation of TRPM6 causes familial hypomagnesemia with secondary hypocalcemia. Nat. Genet. 2002; 31(2): 171–4.

Weber S, Hoffmann K, Jeck N, et al. Familial hypomagnesaemia with hypercalciuria and nephrocalcinosis maps to chromosome 3q27 and is associated with mutations in the PCLN-1 gene. Eur. J. Hum. Genet. 2000; 8(6): 414–22.

Weber S, Schneider L, Peters M, et al. Novel paracellin-1 mutations in 25 families with familial hypomagnesemia with hypercalciuria and nephrocalcinosis. J. Am. Soc. Nephrol. 2001; 12(9): 1872–81.

Wetzel RK, Sweadner KJ. Immunocytochemical localization of Na-K-ATPase alpha- and gamma-subunits in rat kidney. Am. J. Physiol. Renal Physiol. 2001; 281(3): F531–45.

Zimmermann B, Plank C, Konrad M, et al. Hydrochlorothiazide in CLDN16 mutation. Nephrol. Dial. Transplant. 2006; 21(8): 2127–32.

CHAPTER 15

Inherited Diseases of the Calcium-Sensing Receptor: Impact on Parathyroid and Renal Function

EDWARD M. BROWN AND STEVEN C. HEBERT

INTRODUCTION

Calcium (Ca^{2+}) plays key roles in numerous bodily processes (Brown & MacLeod 2001). Intracellular Ca^{2+} is essential for hormonal secretion, cardiac contraction, neurotransmission and the formation of memory, to name a few (Berridge et al 2003). In excess of 99% of Ca^{2+} in the body resides as insoluble salts within the bones and teeth, where it enables locomotion, protects vital organs, permits mastication and serves as a reservoir for calcium and phosphate ions (Bringhurst et al 1998). Extracellular Ca^{2+} also plays key roles in blood clotting and cell adhesion.

Tetrapods (four-legged animals) possess a complicated homeostatic mechanism that maintains virtual constancy of the extracellular ionized Ca^{2+} concentration (Ca_e^{2+}), thereby ensuring the availability of Ca^{2+} for its diverse physiological roles (Bringhurst et al 1998). Since free-living terrestrial organisms ingest Ca^{2+} intermittently, abundant skeletal calcium stores provide a key reservoir of Ca^{2+} that can be drawn upon as needed. The Ca_e^{2+} homeostatic system finely regulates the movements of calcium ions into and out of the skeleton, as well as into and out of the body via the kidneys and gastrointestinal tract (Figure 15.1). The homeostatic mechanism has three important elements: (1) cells and tissues that translocate Ca^{2+} into or out of the ECF, (2) hormones {e.g. parathyroid hormone (PTH), 1,25-dihydroxyvitamin D_3 [1,25(OH)$_2$D$_3$], and calcitonin (CT)} that regulate these Ca^{2+} fluxes, and (3) cells that sense Ca_e^{2+} (i.e. parathyroid cells) and transduce this information into changes in hormonal secretion and, in some cases, Ca^{2+} translocation per se, thereby regulating the fluxes of Ca^{2+} into and out of the ECF.

Figure 15.1 illustrates the system's response to hypocalcemia, which evokes PTH secretion by the Ca_e^{2+}–sensing parathyroid chief cells. PTH's renal actions include (1) enhancing distal tubular Ca^{2+} reabsorption, (2) promoting phosphaturia, and (3) increasing the synthesis of 1,25(OH)$_2$D$_3$ from its largely inactive precursor, 25-hydroxyvitamin D$_3$ (Bringhurst et al 1998). The Ca_e^{2+}-retaining action of PTH 'resets' the kidney to sustain a higher level of serum Ca^{2+}, shifting the curve relating serum to urine Ca^{2+} rightward. The higher level of 1,25(OH)$_2$D$_3$ stimulates gastrointestinal Ca^{2+} absorption and, together with PTH, promotes skeletal Ca^{2+} release. Translocation of Ca^{2+} ions into the ECF from GI tract and bone, along with enhanced renal tubular reabsorption of Ca^{2+}, will, except when there is severe Ca^{2+} deficiency, normalize Ca_e^{2+}.

Ca_e^{2+}-sensing by the parathyroid is of sufficient sensitivity to permit physiologically appropriate responses to changes in Ca_e^{2+} of only a few percent (Brown 1991, 2001). Figure 15.1 shows that the capacity to sense Ca_e^{2+} resides not only in PTH-secreting cells but also in most, if not all, the tissues that translocate Ca^{2+} into or out of the ECF. For example, Ca^{2+} directly inhibits distal tubular Ca^{2+} reabsorption (Quamme 1982) and stimulates bone formation and inhibits its resorption, although the physiological importance of these actions on bone is uncertain (Brown & MacLeod 2001, Berndt et al 2005).

Recent studies have identified novel regulators of mineral ion homeostasis that are being actively investigated in order to fully characterize their homeostatic functions. A detailed discussion of these factors is beyond the scope of this chapter, but one such molecule is FGF-23, a

Genetic Diseases of the Kidney
ISBN: 978-0-12-449851-8

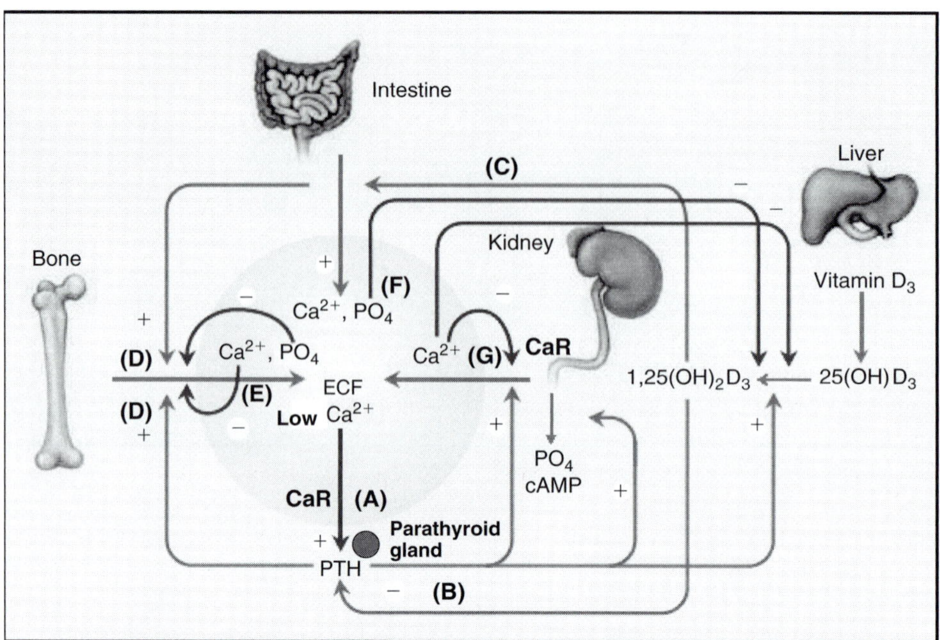

FIGURE 15.1   Schematic diagram showing the principal components involved in extracellular calcium homeostasis. Actions that elevate extracellular calcium levels are illustrated by purple arrows; actions that reduce those levels are indicated by red arrows. In this example, (A) a reduction in blood calcium concentration promotes parathyroid hormone secretion, which (B) stimulates the kidney to increase tubular calcium reabsorption in the distal tubule, enhance phosphaturia and promote synthesis of 1,25-dihydroxyvitamin D$_3$. The 1,25-dihydroxyvitamin D$_3$ can (C) stimulate intestinal calcium and phosphate absorption and (D) act with PTH to enhance skeletal calcium and phosphate release. In addition to being controlled by PTH and 1,25-dihydroxyvitamin D$_3$, extracellular calcium and phosphate themselves can act as extracellular or 'first' messengers, regulating many of the same functions of kidney and parathyroid that PTH and 1,25-dihydroxyvitamin D$_3$ act on. Increases in extracellular calcium and phosphate levels both (E) decrease bone resorption and enhance bone formation and also (F) inhibit the 1-hydroxylation of 25-hydroxyvitamin D$_3$. An additional effect of extracellular calcium on the kidney is (G) to inhibit its own reabsorption in the distal tubule. Among these first messenger functions of extracellular calcium, the regulation of PTH release and renal tubular calcium reabsorption are known to be mediated by the CaR in the figure. The mechanisms underlying the other effects of extracellular calcium and phosphate remain to be elucidated. Modified from Brown EM et al. (1995) Calcium-ion-sensing cell-surface receptors. N. Engl. J. Med. 333: 234–240. © 1995 Massachusetts Medical Society with permission. cAMP, cyclic AMP; ECF, extracellular fluid; PTH, parathyroid hormone. (see also Plate 14)

phosphaturic hormone originally identified as the cause of inherited hypophosphatemic disorders (i.e. autosomal dominant hypophosphatemic rickets) (Berndt et al 2005). Its production is stimulated by hyperphosphatemia, as is chronic renal insufficiency; it also inhibits 1-hydroxylation of 25-hydroxyvitamin D$_3$, enabling it to participate in both Ca^{2+} and phosphate homeostasis. Another recently identified regulator is alpha-klotho, which was originally characterized as an antiaging molecule, but also serves as a regulator of Ca^{2+} and phosphate metabolism. It serves as a co-receptor for FGF-23 and also modulates parathyroid and renal function (Nabeshima & Imura 2007).

THE CALCIUM-SENSING RECEPTOR

The calcium-sensing receptor (CaR or CaSR) represents the molecular mechanism for Ca$_e^{2+}$-sensing by parathyroid,

C-cell, kidney and, perhaps, some bone and intestinal cells. It was first cloned from bovine parathyroid (Brown et al 1993) and then from other mammalian species, the domestic chicken, and several fish species. The CaR shares structural similarity with the metabotropic glutamate (mGluR) and gamma aminobutyric acid (GABA$_B$) receptors, pheromone receptors, odorant (in fish) and taste receptors (Brauner-Osborne et al 2007). A structurally related receptor, GPRC6A, was first cloned as an orphan receptor but later found to sense basic amino acids, e.g. lysine and arginine. Recent studies have shown that it senses Ca$_e^{2+}$, although at supraphysiological levels (Pi et al 2005). Ongoing studies are addressing whether it is primarily a Ca$_e^{2+}$-modulated amino acid receptor or a physiologically relevant Ca$_e^{2+}$-sensing receptor in its own right.

These receptors comprise family C of the superfamily of G protein-coupled receptors. They have large extracellular domains (ECDs), 612 amino acids in the case of the human

CaR, which participate in sensing their respective ligands. They are evolutionarily related to the bacterial periplasmic binding proteins (PBPs) (Pin et al 2003, Brauner-Osborne et al 2007). Both the family C receptors and PBPs have extracellular domains that fold into the so-called venous fly trap motif, a bilobed structure that closes upon ligand binding, thereby initiating signal transduction (Hu & Spiegel 2003). The CaR, therefore, can be thought of as a fusion protein, with an extracellular domain that senses its ligands as well as transmembrane and intracellular domains that participate in signal transduction. Recent studies have proposed candidate Ca_e^{2+}-binding pockets in the CaR ECD, one of which is homologous to the site where the mGluRs bind glutamate in the cleft between the lobes of the VFT (Silve et al 2005). Ca_e^{2+}-sensing by the CaR, however, does not occur solely in the ECD, as a 'headless' receptor ECD still exhibits some responsiveness to Ca_e^{2+} (Hu et al 2005). The CaR can also sense certain amino acids, particularly aromatics, which may enable it to integrate Ca^{2+} and protein metabolism.

The active, cell surface form of the CaR is a glycosylated homodimer, linked by cysteines 129 and 131 within each CaR monomer (Hu & Spiegel 2003). In the parathyroid, most of the cell surface receptor resides within caveolae, flask-shaped invaginations of the plasma membrane that concentrate receptors, G proteins and other signaling molecules (Kifor et al 1998). The receptor links to several G-proteins, including G_i, $G_{q/11}$ and $G_{12/13}$, and, in turn, multiple intracellular signaling pathways – activating phospholipases A_2, C and D, and several mitogen-activated protein kinases (MAPKs) [i.e. extracellular signal-regulated kinase 1/2 (ERK1/2), p38, and c-jun N-terminal kinase (JNK)], and inhibiting adenylate cyclase (Brown & MacLeod 2001, Ward 2004). Protein kinase C phosphorylation sites within the CaR's intracellular domains may provide a mechanism by which activation of phospholipase C exerts negative feedback control of the receptor's function (Ward 2004). The CaR desensitizes relatively little, which may be important for its function as the body's 'thermostat for Ca_e^{2+}' or 'calciostat.' Its resistance to desensitization results, at least in part, from binding to the actin-binding, scaffold protein, filamin-A, presumably tethering the receptor to the actin-based cytoskeleton and making it physically less available for desensitization (Zhang & Breitwieser 2005). Various MAPK components bind to filamin-A, and the CaR's binding to filamin-A is important for its activation of ERK1/2 (Awata et al 2001). Thus the CaR's subcellular distribution (e.g. within caveolae, which bind filamin-A via the intrinsic caveolar protein, caveolin) (Williams & Lisanti 2004) and its presence within organized supramolecular structures (e.g. bound to the cytoskeleton via filamin-A) are likely important determinants of its functional properties. The precise mechanisms through which CaR signaling regulates its downstream biological responses in parathyroid, kidney and other tissues, however, remain to be fully elucidated in most cases.

ROLE OF THE CaR IN THE PARATHYROID

The CaR regulates three important parathyroid functions that are key for Ca_e^{2+} homeostasis – the secretion and synthesis of PTH and parathyroid cellular proliferation (Brown 2007). As described in more detail later, individuals homozygous for inactivating mutations of the CaR (Pollak et al 1994) and mice homozygous for knockout (KO) of the CaR gene (Ho et al 1995) exhibit dramatic increases in their circulating PTH levels despite severe hypercalcemia, indicating a marked increase in the set-point for Ca_e^{2+}-regulated PTH release or even frankly autonomous secretion (e.g. independent of Ca_e^{2+}). In both cases, there is dramatic parathyroid cellular hyperplasia. Thus the CaR normally tonically inhibits PTH secretion and restrains parathyroid cellular proliferation. Additional studies have documented that the receptor likewise controls expression of the PTH gene (Levi et al 2006). It is not currently known whether the CaR is the mediator of the increase in intracellular degradation of PTH that takes place during hypercalcemia, which further decreases the availability of intact, biologically active PTH for secretion. In contrast to the inhibitory action of the CaR on parathyroid function, the receptor stimulates calcitonin (CT) secretion, a homeostatically appropriate action given CT's hypocalcemic action, which is exerted principally via its direct inhibition of osteoclastic bone resorption (Fudge & Kovacs 2004).

ROLE OF THE CaR IN THE KIDNEY

Numerous segments of the kidney tubules express the CaR, including the apical plasma membrane of the proximal tubule, the basolateral surface of the medullary (MTAL) and cortical thick ascending limbs (CTAL) and distal convoluted tubule (DCT), and the apical surface of the inner medullary collecting duct (IMCD) (Riccardi et al 1995, 1996). The CaR is thought to exert the following actions along the nephron: it blunts the phosphaturic action of PTH (Ba et al 2003) and upregulates the vitamin D receptor (VDR) in the proximal tubule (Maiti & Beckman 2007). The latter action could account, in part, for the inhibition of the 1-hydroxylation of 25-hydroxyvitamin D_3 by high Ca^{2+} (Weisinger et al 1989), since activating the VDR downregulates the 1-hydroxylase gene. In some (De Jesus Ferreira & Bailly 1998), but not all (Motoyama & Friedman 2002) studies, the CaR inhibits the reabsorption of both mono- and divalent cations in the CTAL. CaR-mediated inhibition of the apical recycling K^+ channel and, perhaps, the

Na^+-K^+-$2Cl^-$ cotransporter in this nephron segment are thought to reduce the passive, paracellular reabsorption of Na^+, Ca^{2+} and Mg^{2+} (Hebert et al 1997, Wang et al 1997, De Jesus Ferreira & Bailey 1998), thereby promoting the excretion of all three ions.

The CaR may also inhibit Ca^{2+} reabsorption in the DCT, probably by inhibiting PTH-stimulated adenylate cyclase activity, but this has not been examined in detail. Finally, the CaR reduces vasopressin-stimulated water reabsorption in IMCD (Sands et al 1997, 1998) and decreases expression of aquaporin-2 protein but not mRNA (Sands et al 1998). This action may limit maximal urinary concentration in this nephron segment, thereby mitigating the risk of stone formation resulting from reabsorption of water without Ca^{2+} in IMCD. CaR-induced reduction in sodium chloride reabsorption in the MTAL, as it does in the CTAL, would also reduce urinary concentrating ability by 'washing out' the medullary concentration gradient (Hebert et al 1997). These actions of Ca^{2+} on renal water and salt transport in MTAL and IMCD probably account for the impaired urinary concentrating capacity observed in some hypercalcemic patients (Gill & Bartter 1961). Reduced renal concentrating ability can occur with hypercalciuria alone, even without hypercalcemia. Hypercalciuric children can have enuresis due to decreased urinary concentrating capacity (Valenti et al 2002). This symptom improves upon lowering urinary Ca^{2+} excretion by institution of a low Ca^{2+} diet, which is accompanied by increased excretion of aquaporin-2, presumably owing to a higher AQP-2 level in the IMCD apical membrane. The apical AQP-2 promotes greater water reabsorption, thereby decreasing nocturnal formation of urine.

ROLE OF THE CAR IN OTHER TISSUES INVOLVED IN CALCIUM HOMEOSTASIS

Elevated levels of Ca_e^{2+} inhibit bone resorption (Raisz 1965) and enhance bone formation (Quarles 1997) – effects that could participate in defending against hypercalcemia. Some have identified CaR protein and/or message in osteoblasts and stromal cells (Yamaguchi et al 1998, 2001), as well as in osteoclasts (Kameda et al 1998) and cells of the osteoclast lineage (e.g. monocytes) (Yamaguchi et al 1998). Others, however, have not and have instead postulated the existence of distinct calcium sensors in both osteoclasts and osteoblasts (Pi et al 1999, 2005, Zaidi et al 1999). The family C receptor, GPRC6A, in fact, is present in bone and senses Ca_e^{2+} at high levels (Pi et al 2005). Further characterizing GPRC6A and other putative Ca_e^{2+}-sensors in bone cells could provide key insights into bone biology and could provide new drug targets, as, for example, in the treatment of osteoporosis.

The CaR is expressed along the entire GI tract (Chattopadhyay et al 1998, Cheng et al 1999) and probably mediates the stimulation of gastrin (which enhances gastric action secretion) (Dinbar & Tulcinsky 1978), gastric acid and cholecystokinin secretion (which stimulates pancreatic enzyme secretion) (Layer et al 1985) by Ca^{2+} and/or amino acids. These effects could potentially coordinate the digestive functions of small intestine and stomach following the ingestion of these nutrients. While the receptor is present in intestinal segments that absorb Ca^{2+}, it is unclear whether the CaR modulates Ca^{2+} absorption. Recent studies, however, have shown that raising Ca^{2+} intake in VDR KO mice enhances TRPV6 expression, the vitamin-D-responsive, apical intestinal calcium uptake channel (Hoenderop et al 2004), a potentially CaR-mediated action. The CaR also markedly inhibits fluid secretion by the colonic crypts (Cheng et al 2004), an action that could contribute to the constipation that can be present in hypercalcemic individuals and serve as the basis for novel antidiarrheal therapies. The remainder of this chapter will focus on inherited diseases of the CaR and their impact on parathyroid and renal function. However, greater understanding of the CaR's roles in other tissues may inform the search for new phenotypes in these conditions involving other sites where the receptor is expressed, such as bone and intestine.

FAMILIAL HYPOCALCIURIC HYPERCALCEMIA (FHH) [OMIM 14598]

In 1972, Foley et al (1972) first described the characteristic clinical features of the hereditary condition now known as familial hypocalciuric hypercalcemia (FHH) (it was initially called familial benign hypercalcemia). FHH is, in the great majority of cases, a benign, autosomal dominant form of hypercalcemia with characteristic abnormalities in the regulation of parathyroid and renal function by extracellular Ca^{2+} (Marx et al 1981, Law & Heath 1985). Two decades later, linkage analysis showed that the disease gene for the most common form of FHH resides on the long arm of chromosome 3 (band q21-24) (Chou et al 1992). Shortly thereafter, heterozygous inactivating mutations of the CaR were identified in FHH families with linkage to chromosome 3 (Pollak et al 1993). (About 30% of FHH families do not have an identifiable mutation in the CaR gene. These presumably harbor mutations in regulatory regions of the CaR gene that control its expression.) However, FHH is not always linked to chromosome 3. Notably, two families with clinical features similar to FHH showed linkage to the short and long arms of chromosome 19 (Heath et al 1993, Lloyd et al 1991), respectively; the variant form of FHH linked to chromosome 19q13 was called the Oklahoma variant and exhibited a tendency for the biochemical abnormalities to worsen with time (Lloyd et al 1999).

The typical case of FHH presents with mild-to-moderate, PTH-dependent hypercalcemia averaging approximately 11 mg/dl (total calcium) and an inappropriately normal or even overtly low rate of urinary Ca^{2+} excretion given the patient's hypercalcemia (Marx et al 1978, 1981, Law & Heath 1985). In some cases, however, the disorder presents with more severe hypercalcemia, averaging 12–13 or even 14 mg/dl (Marx et al 1981, Bai et al 1996, 1997). Serum Mg^{2+} levels are often high-normal to mildly elevated, and there is a positive, linear relationship between serum Ca^{2+} and Mg^{2+}, unlike primary hyperparathyroidism (PHPT) in which there is an inverse relationship between these two parameters (Marx et al 1978, 1981). The increase in not only serum Ca^{2+} but also Mg^{2+} in FHH patients suggests that the CaR contributes to 'setting' Mg_0^{2+} as well as Ca_0^{2+} (Strewler 1994). Serum PTH is generally normal, although in about 15–20% of cases, PTH levels are frankly elevated (Heath 1989). Serum phosphate is usually normal or mildly decreased, while $1,25(OH)_2D_3$ levels are normal (Law et al 1984). Bone mineral density (BMD) is similar to that in age-matched controls, although markers of bone turnover may be mildly elevated (Law & Heath 1985). Because the disorder most commonly has an asymptomatic, benign clinical course, patients with FHH are often not diagnosed until a routine measurement of the blood Ca^{2+} concentration reveals hypercalcemia, or family screening is performed following the birth of a child with neonatal severe hyperparathyroidism (NSHPT), the homozygous form of FHH (Marx et al 1982) (see below). Because of the benign natural history of FHH, and because the disorder usually recurs rapidly following anything less than total parathyroidectomy, the standard of care is expectant follow up without medical or surgical intervention. Only in rare families do complications related to unusually severe hypercalcemia in the newborn period (Bai et al 1997), the occurrence of pancreatitis later in life (Pearce et al 1996), or the presence of hypercalciuria and overtly elevated serum PTH levels (Carling et al 2000) dictate a more aggressive surgical approach.

The inappropriately normal (i.e. nonsuppressed) PTH level in the hypercalcemic FHH patient reflects a right-shift in the set-point for Ca_0^{2+}-regulated PTH release (i.e. the level of Ca_0^{2+} producing half of the maximal inhibition of PTH release) (Auwer et al 1984, Khosla et al 1993). This has been demonstrated by comparing this relationship to that in normal subjects or patients with primary hyperparathyroidism during induced hypo- and/or hypercalcemia. These studies showed that the inverse sigmoidal relationship between serum Ca^{2+} and PTH levels is shifted to the right in FHH patients, and, consequently, a higher than normal Ca_0^{2+} is needed to suppress PTH to any given extent. This led to the suggestion that parathyroid function in FHH is characterized principally by an abnormality in only one of the four

parameters that can be used to describe Ca_0^{2+}-regulated PTH release, namely the set-point (Auwer et al 1984, Khosla et al 1993). This derangement in the control of PTH secretion in FHH can be thought of as 'resistance' of the parathyroid to Ca_0^{2+} and contributes importantly to hypercalcemia in this condition (Brown 2007). In PHPT (or in tertiary hyperparathyroidism in patients with end stage renal disease), in contrast, several or even all of the four parameters – the set-point, maximal and minimal secretory rates, and slope of the curve relating Ca_0^{2+} to PTH secretion – can be abnormal (Brown 1983). Not surprisingly, the parathyroid glands of patients with FHH have been found to be either essentially normal or to have subtle increases in the mass of parathyroid chief cells (Thogeirsson et al 1981, Law et al 1984). Based on these data, one might reasonably ask: How can a normal level of PTH and a normal or nearly normal mass of parathyroid cells in FHH sustain hypercalcemia of a degree similar to that of patients with PHPT, who nearly universally have an increased mass of parathyroid cells? The answer likely lies in the characteristic changes in renal handling of Ca^{2+} (and Mg^{2+}) in persons with FHH that are described below.

An additional key feature of FHH patients is their normal or even overtly decreased urinary Ca^{2+} excretion despite their coexistent hypercalcemia (Marx et al 1978, 1981, Attie et al 1983, Law & Heath 1985). Marx and coworkers have carried out detailed studies of the renal handling of divalent cations in FHH that have clarified these abnormalities in renal function (Marx et al 1978, Attie et al 1983) There is a substantial reduction in the clearance of Ca^{2+} in this condition, with the ratio of Ca^{2+} clearance to that of creatinine being less than 0.01 in about 80% of patients. This is less than the clearance ratio in patients with PHPT, about 80% of whom have values > 0.01 and commonly > 0.02 (Marx et al 1978). The clearance of Mg^{2+} is also reduced but to a lesser extent (about 30%) in patients with FHH (Marx et al 1978). Therefore, even with a degree of hypercalcemia comparable to that in many patients with PHPT, and a lower PTH level than that in PHPT, patients with FHH excrete less Ca^{2+} and Mg^{2+}, emphasizing the characteristically avid renal tubular reabsorption of these divalent cations in this condition. The Ca^{2+} to creatinine clearance ratio of hypercalcemic patients with a suppressed level of PTH, such as in multiple myeloma or immobilization, is greater still, reflecting loss of the Ca^{2+}-conserving action of PTH on the renal tubule, combined with a hypercalcemia-induced, CaR-mediated reduction in renal tubular reabsorption of Ca^{2+}.

Elegant studies by Attie et al. carried out more than 20 years ago identified a key nephron segment that contributed to the deranged handling of divalent cations in FHH (Attie et al 1983). Ca_0^{2+}-regulated PTH release in FHH patients with intact parathyroid glands confounds characterization of the intrinsic renal contribution to their relative or absolute hypocalciuria. That is, changes in PTH secretion accompanying experimentally induced alterations in serum Ca^{2+} concentration will secondarily modify renal tubular reabsorption of Ca^{2+}. For this reason, these investigators compared divalent cation handling in hypoparathyroid FHH patients and hypoparathyroid controls, in whom the impact of PTH on this process was absent. By determining renal Ca^{2+} excretion at various serum Ca^{2+} concentrations during infusion of Ca^{2+} or spontaneous episodes of hypercalcemia occurring during treatment with oral Ca^{2+} and vitamin D, it was possible to derive the relationship between serum and urine Ca^{2+} (e.g. the set-point for Ca_0^{2+}-regulated renal Ca^{2+} excretion) (Attie et al 1983). There was a marked rightward and downward shift in this relationship in the FHH patients, illustrating the characteristic PTH-independent reduction in renal Ca^{2+} excretion in FHH patients at any given level of serum Ca^{2+}. This most likely represents an intrinsic renal defect due to partial inactivation of the CaR in the kidney, a conclusion supported by the work of others (Davies et al 1984). Interestingly, renal handling of Mg^{2+} was similar in the FHH and control patients, suggesting that PTH is required to express the enhanced renal tubular Mg^{2+} reabsorption in FHH patients with intact parathyroid glands (Attie et al 1983).

Use of the loop diuretic, ethacrynic acid, suggested that the thick ascending limb (TAL) of Henle's loop was the likely site of this renal defect (Attie et al 1983). This diuretic produced an exaggerated calciuric and natriuretic response in the FHH patients, suggesting enhanced reabsorption of Ca^{2+} and Na^+ in the TAL, the site of the Na^+-K^+-$2Cl^-$ cotransporter that is inhibited by ethacrynic acid. This result can be interpreted as resulting from constitutive overactivity of the cotransporter in FHH owing to reduced activity of the CaR in this nephron segment, thereby increasing the paracellular reabsorption of Ca^{2+}. As noted earlier, the latter is driven in large part by the lumen positive potential generated by the concerted action of the cotransporter, the apical recycling (ROMK) K^+ channel and basolateral sodium and chloride transporters/channels (Hebert et al 1997). As will be discussed shortly, heterozygous loss of the CaR in FHH is the key contributor to altered divalent cation transport in the CTAL.

Thus, like the parathyroid cell, the kidney exhibits 'resistance' with respect to the usual calciuric action of hypercalcemia, which is manifested by a shift in the positive relationship between serum Ca^{2+} and urinary Ca^{2+} excretion to the right (e.g. an increase in the 'set-point' for Ca^{2+}-induced hypercalciuria). Of note in this regard, there was a marked decrease in not only the calciuric, but also the natriuretic response to Ca^{2+} infusion in the hypoparathyroid FHH patients (Attie et al 1983). This result likely reflects reduced activity of the CaR in the TAL in FHH and is consistent with the 'signature' of linked cation handling in this nephron segment, since the transport of Na^+ and Ca^{2+} (and Mg^{2+}) change in parallel in CTAL (Hebert et al 1997). It is possible

that other nephron segments contribute to the perturbation in renal Ca^{2+} transport in FHH, such as proximal tubule, where hypercalcemia modestly inhibits tubular reabsorption of Ca^{2+} (Quamme 1982), and the DCT, a key site of regulated Ca^{2+} reabsorption, but these have not been studied carefully in this regard. The avid renal tubular reabsorption of Ca^{2+} in FHH provides a plausible explanation for why these patients can maintain elevated serum Ca^{2+} concentrations despite their normal PTH levels. The argument is as follows: in FHH, the kidney 'collaborates' with the parathyroid glands, promoting normal or even low rates of renal Ca^{2+} excretion that obviate the need for a frankly elevated PTH level to sustain hypercalcemia. In PHPT, in contrast, the normal CaR in the kidney is activated by the concomitant hypercalcemia and promotes renal Ca^{2+} excretion, necessitating higher PTH and $1,25(OH)_2D_3$ levels to sustain hypercalcemia in the face of the associated renal Ca^{2+} losses.

The renal handling of not only monovalent and divalent cations but also water is altered in FHH. As noted above, raising the level of Ca_0^{2+} at the apical pole in the IMCD substantially inhibits vasopressin-stimulated water flow (Sands et al 1997, 1998). Given the CaR's apical localization in this nephron segment, this action of Ca_0^{2+} has been ascribed to activation of the receptor. The activated CaR has been suggested to alter the configuration of the cytoskeleton at the apical pole of the IMCD, thereby interfering with the capacity of vasopressin to promote fusion of aquaporin-2-containing endosomes with the apical plasma membrane (Valenti et al 2005). Marx et al compared the ability of patients with FHH and PHPT to concentrate their urine (Marx et al 1981). Patients with PHPT showed about a 20% reduction in their maximal urinary concentration during an 18–22-hour dehydration test relative to FHH patients with a similar degree of hypercalcemia. This study illustrates 'resistance' of the urinary concentrating mechanism, most likely in the IMCD, to Ca_0^{2+} in FHH. It is possible that a reduced inhibitory action of hypercalcemia on NaCl reabsorption in MTAL in FHH patients may also contribute to their normal or near normal urinary concentrating ability in the face of hypercalcemia (Hebert et al 1997).

Since the initial identification of three unique missense mutations in the CaR gene in three FHH families in 1993 (Pollak et al 1993), more than 100 additional mutations have been described (Figure 15.2). Most are unique to individual families, but a few apparently unrelated kindreds harbor identical mutations (e.g. gly552arg) (see web site http://www.casrdb.mcgill.ca). The majority are missense mutations and reside within the first half of the ECD or within the receptor's TMDs (Hauache 2001). However, splice site, insertion, truncation, and deletion mutations have also been described (Janicic et al 1995, D'Souza-Li et al 1998, Carling et al 2000, Hendy et al 2000). Individual mutations can be associated with distinct phenotypes, as noted below, although in general there are not strong-genotype phenotype relationships in FHH. Most mutant receptors, when

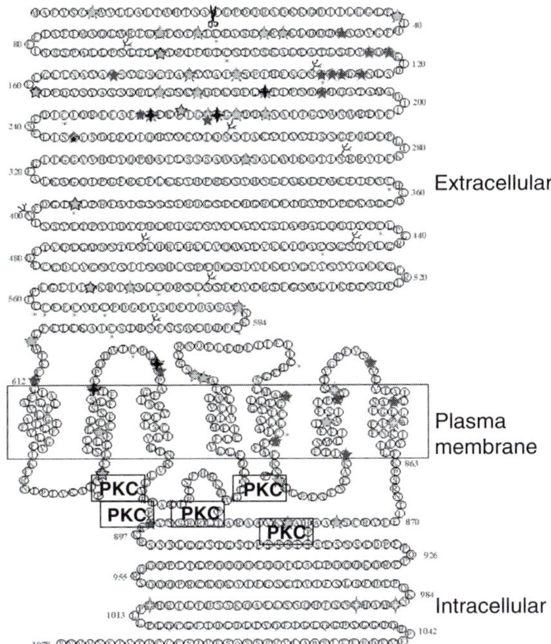

FIGURE 15.2 Schematic diagram of the CaR, delineating the locations of activating and inactivating mutations and polymorphisms. The circles indicate individual amino acids; blue symbols are polymorphisms; red symbols are activating mutations; black symbols over green are two inactivating mutations at the same codon; asterisks show conserved cysteines; the 'P' symbols shown in circles that are attached to the main amino-acid chain are protein kinase C phosphorylation sites and are highlighted by 'PKC' within boxes; branched symbols are glycosylation sites; scissors denote the cleavage site of the signal peptide. Reproduced with permission from Pidasheva S et al (2004) CASRdb: calcium-sensing receptor locus-specific database for mutations causing familial (benign) hypocalciuric hypercalcemia, neonatal severe hyperparathyroidism, and autosomal dominant hypocalcemia. Hum Mutat 24: 107–111. Copyright © 2004 Wiley-Liss, Inc. (see also Plate 15)

studied by expression in heterologous cell systems (i.e. HEK293 cells), range in their degree of inactivation from modest, ~2-fold increases in EC_{50} (the concentration of Ca_o^{2+} needed for half of the maximal biological response) to receptors totally lacking in biological activity (Bai et al 1996, Pearce et al 1996). In most cases, the simplest interpretation of the 'Ca_o^{2+}-resistance' in FHH patients is that the ambient levels of Ca_o^{2+} present in vivo are insufficient to activate the mutant receptors. Therefore, the only active receptors are the remaining normal ones, whose number appears to be about 50% of normal based on the complement of receptors found in mice heterozygous for knockout of the CaR gene (see below) (Ho et al 1995). A comparable example of this pathophysiology in FHH patients would be mutations producing a severely truncated receptor, which never reaches the cell surface. In this situation, the only biologically active receptors would arise from the normal

CaR allele. The ~50% reduction in the number of normal receptors in the heterozygous CaR knockout mice produces about a 10–15% increase in the set-point of the parathyroid and pari passu the serum Ca^{2+} concentration.

In some cases mutant CaRs found in FHH exert a dominant-negative action on the remaining wild type receptors within mutant-wild type heterodimers, since the CaR functions as a dimer, thereby producing a more severe phenotype (Bai et al 1996, 1997). This dominant-negative action presumably reflects the specific properties of the mutant receptor protein residing on the cell surface. Such properties may include receptors that (1) are severely impaired in their function, (2) reach the cell surface at robust levels comparable to those of the wild type receptor and (3) heterodimerize efficiently with the normal receptor. When a mutant receptor exerts a dominant negative action, the only fully functional receptors, the wild type CaR homodimers, likely represent about one-fourth of all cell surface receptors, based on the predicted 1:2:1 ratio of wild-type homodimers, wild-type-mutant heterodimers, and mutant homodimers, respectively. An interesting mutation producing a different type of functional abnormality was described in a Swedish family by Carling et al (2000). The affected members of this family were unusual for their overt hyperparathyroidism with frankly enlarged parathyroid glands and hyper- rather than hypocalciuria. This mutation (Carling et al 2000) resided within the receptor's C-terminal tail. One interpretation of this family's phenotype would be that the mutant receptor caused resistance to Ca_o^{2+} in the parathyroid, thereby producing hypercalcemia and hyper-parathyroidism, but coupled more efficiently to the CaR in the kidney, leading to a hypercalciuric response to hypercalcemia. It should be kept in mind while evaluating the literature describing in vitro characterization of mutant CaRs that the cells used for these experiments differ substantially from the chief cells of the parathyroid glands or CaR-expressing kidney cells in vivo. Thus the experimental data likely represent only an approximation of how the receptors would function in their native environment.

The development of mice with targeted inactivation or 'knockout' (KO) of the CaR has provided useful models of FHH and NSHPT (see next section) (Ho et al 1995). The heterozygous KO mouse provides a model of FHH. These mice exhibit mild increases in their serum Ca^{2+} and PTH concentrations as well as a reduction in their urinary calcium concentration. As noted above, they provide strong evidence that a reduced complement of normal CaRs can cause the Ca_o^{2+}-resistance of FHH, because, based on immunohistochemistry or Western blotting, the concentration of the CaR in parathyroid and kidney are both reduced by about 50% (Ho et al 1995). The further use of these mice or those with conditional KO of the CaR (Chang et al 2007) in parathyroid and/or kidney will provide useful models to study the impact of reduced CaR expression on parathyroid and renal function in vivo and in vitro.

NEONATAL SEVERE PRIMARY HYPERPARATHYROIDISM (NSHPT) [OMIM 239200]

NSHPT typically presents within the first six months of life (Rhone 1975, Marx et al 1985, Hauache 2001). Affected infants have severe, symptomatic, PTH-dependent hypercalcemia as well as the bony changes of severe hyperparathyroidism. Infants with NSHPT can also manifest polyuria, dehydration, hypotonia, and failure to thrive (Grantmyre 1973, Eftekhari & Yousefzadeh 1982, Marx et al 1985, Heath 1989, Brown et al 1997) A prominent component of the disorder is the hyperparathyroid bone disease, which can lead to multiple fractures of the long bones and other skeletal sites (Grantmyre 1973, Eftekhari & Yousefzadeh 1982). Rib fractures can sometimes produce a 'flail chest' syndrome in which multiple rib fractures cause respiratory distress as a result of inability of the affected infant to expand its chest wall and thereby generate sufficient negative intrathoracic pressure for normal respiration (Grantmyre 1973).

Laboratory studies in NSHPT reveal marked hypercalcemia and hyperparathyroidism, as well as relative or absolute hypocalciuria (Marx et al 1985, Cole et al 1997), although in some cases Ca^{2+} excretion is elevated, presumably owing to the marked increase in filtered load of Ca^{2+}. Total serum Ca^{2+} concentrations range from moderately elevated (e.g. $\sim$12–14 mM) to levels as high as 25-30 mg/dl in the most severely affected cases (Marx et al 1985, Heath 1989, Kobayashi et al 1997, Brown 2000). PTH levels are frequently 10-fold or more higher than the upper normal limit. The mass of the parathyroid glands in NSHPT is usually increased several-fold, and they exhibit striking chief cell hyperplasia. This severe form of hyperparathyroidism in the neonatal period is most commonly caused by the presence of homozygous (Pollak et al 1994a, b, Hauache 2001) or compound heterozygous mutations in the CaR (Kobayashi et al 1997) (in the latter, an infant inherits one inactivating CaR mutation from one parent and a second from the other parent). There is a similar severe phenotype in mice homozygous for KO of the Ca_o^{2+}-sensing receptor gene (Ho et al 1995). These mice exhibit severe hypercalcemia, hyperparathyroidism and bone disease, and they generally die within the first 1–2 weeks of life.

Early diagnosis is critical, as untreated NSHPT can have a fatal outcome without parathyroidectomy to alleviate the hyperparathyroidism and hypercalcemia (Heath 1994, Cole et al 1997). Total parathyroidectomy produces hypoparathyroidism, as in other forms of hyperparathyroidism, documenting that the absence of Ca^{2+} receptors in tissues other than the parathyroid (e.g. kidney, C-cell) is insufficient to produce hypercalcemia. A potentially useful temporizing measure in a severely ill neonate with NSHPT is the use of a bisphosphonate, such as pamidronate, which can lower serum Ca^{2+} concentration substantially and allow the patient to stabilize prior to surgery, if the latter is needed (Waller et al 2004, Fox et al 2007). Remission of hyperparathyroidism following parathyroidectomy leads to a rapid improvement in clinical status and healing of bony lesions within weeks to months; the prognosis thereafter is excellent. Failure to diagnose NSHPT in a timely manner in infants that nevertheless survived untreated has been associated in some cases with severe impairment of mental, skeletal and overall somatic growth (Cole et al 1997).

It is possible to 'rescue' mice homozygous for KO of the CaR gene by crossing heterozygotes with mice heterozygous for knockout of PTH gene or Gcm-2 (which is needed for development of the parathyroid glands) (Kos et al 2003, Tu et al 2003). As with human infants with NSHPT that have been rendered aparathyroid, mice lacking both the CaR and either PTH or parathyroid glands are hypoparathyroid but viable and can be maintained in good health by adding Ca^{2+} to their food and/or water. This mouse model could be useful in studying the roles of the CaR outside of the parathyroid gland in the absence of the severe hyperparathyroidism that is present in mice with KO of only the CaR.

More recent studies have emphasized a wider clinical spectrum for NSHPT than initially appreciated, largely because of the availability of genetic testing of the CaR gene. Consequently, several studies have now identified infants with less severe hyperparathyroidism and a substantially milder clinical presentation and course (Heath 1989, 1994, Pearce et al 1995, 1996). This latter form of the disease has been termed neonatal hyperparathyroidism (NHPT), in order to emphasize the milder phenotype that these infants manifest (Pearce 2002, Brown 2007). Several cases of NHPT have been found to harbor heterozygous inactivating CaR mutations, and in some cases the presence of a mutation exerting a dominant negative action may explain why an infant presents with NHPT rather than simply the benign FHH phenotype otherwise expected in the neonatal period with heterozygous inactivating CaR mutations (Pearce et al 1995). Another explanation for the more severe phenotype of NHPT compared to FHH is that an affected fetus in an unaffected mother would experience calcium concentrations that the fetal parathyroid glands would read as hypocalcemic relative to their elevated set-point, thereby stimulating 'secondary' hyperparathyroidism (Bai et al 1997). Over time, NHPT can revert to a phenotype resembling FHH with only medical management (Heath 1994, 1998). Thus currently parathyroidectomy should only be reserved for the most severely affected NHPT infants, in whom substantial hypercalcemia and hyperparathyroid bone disease persist despite intensive medical treatment (e.g. with vigorous hydration and, if needed, bisphosphonates). Such cases, however, are the exception rather than the rule in NHPT.

Recent reports have described two patients with homozygous mutations in the CaR gene who escaped detection until adulthood (Aida et al 1995, Chikatsu et al

1999). Both were products of consanguineous unions of parents who were heterozygous for inactivating CaR mutations. These patients were intermediate in their phenotype between FHH and NSHPT and are instructive in what they tell us about the normal functions of the CaR in vivo. These two homozygous patients did not have the usual symptoms and signs of hypercalcemia and were only identified serendipitously by routine biochemical screening. At presentation, both patients had what were presumably lifelong serum Ca^{2+} concentrations of 15–17 mg/dl, hypermagnesemia [in the case in which it was measured, it was 1.36 mmol/l (normal 0.74–0.99)] (Aida et al 1995), overt hypophosphatemia, and a PTH level in the upper normal range in one case and frankly elevated in the other. Urinary Ca^{2+} excretion was in the range observed in FHH despite the much more severe hypercalcemia (clearance ratios of <0.01), and renal function was apparently normal, although this was not addressed explicitly. The lack of the symptomatology and the usual complications of hypercalcemia in these two cases suggest that: (1) hypercalcemic symptoms (e.g. lethargy, anorexia, nausea) may be CaR-mediated, and (2) known effects of hypercalcemia on renal function, including hypercalciuria, impaired urinary concentrating capacity, reduction in GFR (in fact, FHH patients have increased GFR relative to patients with PHPT) (Marx et al 1978), and perhaps decreased renal blood flow, can apparently be ascribed to the CaR. The mild clinical presentation of these cases of 'NSHPT' that escaped detection in the neonatal period was likely the result of both mutant CaRs (Gln27Arg and Pro39Ala) exhibiting mildly decreased function in vivo, with EC_{50}s of 4.9 mM and 4.4 mM, respectively, vs. 3.7 mM for the wild-type receptor) (Chikatsu et al 1999). In fact, the modest increase in the EC_{50} of the two mutant receptors (i.e. ~30%) might have predicted even milder hypercalcemia in the two affected homozygotes. Perhaps an associated reduction in the level of expression of the mutant CaR in parathyroid and kidney, or other factors, produced a greater functional defect in vivo.

What can we say about the regulation of PTH secretion by Ca_0^{2+} in NSHPT and the homozygous CaR KO mouse based on available data? In vivo studies, similar to those performed in patients with FHH, utilizing dynamic testing, i.e. inducing hyper- and/or hypocalcemia, are not available for obvious reasons given the fragility of infants with NSHPT and homozygous CaR KO mice. Available in vivo data, showing marked increases in serum PTH in the face of severe hypercalcemia, demonstrate a severe defect in Ca_0^{2+}-regulated PTH secretion, with a potentially total or near total failure of suppression of secretion at high Ca_0^{2+} (Marx et al 1985, Hendy et al 2000). The only in vitro data that are available are in two cases of NSHPT that underwent parathyroidectomy, enabling performance of studies with dispersed parathyroid cells, similar to those performed earlier in various other types of hyperparathyroidism (Cooper et al 1986, Marx et al 1986). No comparable studies have

been carried out in FHH, since the vast majority of these patients should be, and generally are, followed without surgical intervention. In first of the two NSHPT cases studied in vitro (Marx et al 1986), the infant presented with severe hypercalcemia (~20 mg/dl). A total parathyroidectomy, with autotransplantation of a portion of one of the two glands, was carried out at the age of 6.5 months. The portion of the tissue studied in vitro as dispersed parathyroid cells revealed a set-point of 2.5 mM, more than twice the normal set-point of about 1 mM. In the second case, there was consanguinity, as the infant's father was also his grandfather, and a serum Ca^{2+} concentration of 26.4 mg/dl was documented in the neonatal period. As in the first case, hyperplastic tissue was obtained at the time of parathyroidectomy, dispersed and incubated with various levels of Ca^{2+}. There was minimal suppressibility at 2.0 mM Ca^{2+} (Cooper et al 1986); unfortunately the small amount of tissue available precluded studies at higher levels of Ca_0^{2+}. Since these studies were performed prior to the cloning of the CaR and the identification of mutations in the CaR gene in FHH and NSHPT, these mutant receptors were not studied to determine whether they were totally inactivated or harbored some residual activity. Perhaps, survival of neonates with NSHPT requires some residual activity of the CaR to avoid serum Ca^{2+} elevations incompatible with life – a point that has not been addressed.

There are even more limited data studying the relationship between serum Ca^{2+} concentration and urinary Ca^{2+} excretion in patients with NSHPT or in the homozygous CaR KO mice, again owing to the severity of the clinical presentation early in life. Accordingly, there are only scattered observations in vivo which do not clearly delineate the details of tubular handling of Ca^{2+} in these two settings other than to document inappropriately avid reabsorption of Ca^{2+} in the face of marked hypercalcemia. The Ca^{2+} to creatinine clearance ratios of the two patients discovered to be homozygous for inactivating mutations of the CaR as adults were in the same range as those of their heterozygous relatives with FHH (Aida et al 1995, Chikatsu et al 1999). Given the fact that the two homozygotes had serum calcium concentrations 1.5-fold or more higher than the heterozygous family members, this suggests that the curve relating serum to urine Ca^{2+} in the homozygotes is shifted farther to the right and downward relative to the typical FHH patient. Of interest in this regard, one of the three FHH patients in the study of Attie, et al. very likely had NSHPT, because he presented at the age of 14 days with a serum Ca^{2+} concentration of 21 mg/dl (Attie et al 1983) and was, therefore, subjected to removal of all parathyroid tissue with resultant hypoparathyroidism. The other two patients in this study had serum Ca^{2+} concentrations of ~14 mg/dl early in life, and it is possible that they had NHPT rather than NSHPT. Both, however, were rendered surgically hypoparathyroid and their diagnosis is somewhat uncertain. In the studies examining the relationship of serum to urine Ca^{2+} in these three patients, all had comparable shifts to the right and downward relative to the hypoparathyroid controls (Attie et al 1983). Therefore, it is possible that this study examined renal Ca^{2+} handling in the absence of PTH in unusually severe cases of FHH or, in fact, in NSHPT in one of the three patients. Similar studies in vivo and in vitro in CaR KO mice rescued by knockout of the PTH gene or ablation of the parathyroid glands could provide useful clues to the importance of the CaR on a quantitative basis in enabling upregulation of renal Ca^{2+} excretion in the face of hypercalcemia as well as the relative importance of various nephron segments in this regard. One caveat concerning this; however, is the not-yet-fully-documented assertion that KO of exon five in the mice used to date to study the pathophysiology of CaR KO mice, including in the two rescued CaR KO models, is an incomplete knockout (Rodriguez et al 2005). Alternative splicing can produce an exon 5-less CaR, of uncertain biological activity, that could confound results using this KO model. Studies are ongoing developing mice with conditional KO of exon 7 (Chang et al 2007, Tu et al 2007), which encodes the entire transmembrane domain and C-tail, to try to resolve this issue. It should be pointed out, however, that the phenotype of the non-rescued, exon 5 KO mouse has many of the features expected of complete, or nearly complete, KO of the CaR (Ho et al 1995).

AUTOSOMAL DOMINANT HYPOPARATHYROIDISM (ADH) (OMIM) [#601298]

Patients with this autosomal dominant form of hypocalcemia/hypoparathyroidism are commonly asymptomatic, similar to the majority of patients with FHH (Pearce et al 1996, Hendy et al 2000, Hauache 2001, Lienhardt et al 2001). Some, however, exhibit neuromuscular irritability, seizures and calcification of the basal ganglia, complications commonly seen in other forms of hypoparathyroidism in which the CaR gene is normal (Pearce et al 1996). During febrile episodes, patients with ADH, particularly children, can exhibit symptomatic hypocalcemia and, in some cases, develop seizures. Patients generally manifest mild to moderate hypocalcemia, with levels of serum PTH that are inappropriately low-normal given their hypocalcemia or are frankly subnormal (Pollak et al 1994b, Heath 1998). Affected individuals in the untreated state commonly have relative (i.e. inappropriately normal given their hypocalcemia) or absolute hypercalciuria (Baron et al 1996, Pearce et al 1996, Winer et al 2003). In several reports urinary Ca^{2+} excretion in untreated ADH patients is about twice that in patients with hypoparathyroidism of other causes (Baron et al 1996, Pearce et al 1996, Winer et al 2003). Conceptually, therefore, this syndrome is the inverse of FHH, i.e. it could be termed familial hypercalciuric hypocalcemia. However, not all studies have shown this difference in renal calcium excretion (Yamamoto et al 2000).

Patients with ADH appear particularly susceptible to complications during treatment with Ca^{2+} and vitamin D analogs aimed at increasing their serum Ca^{2+} concentrations toward normal (Pearce et al 1996, Lienhardt et al 2001). They are especially prone to developing renal complications, including nephrocalcinosis, nephrolithiasis, and renal impairment during treatment of ADH patients with Ca^{2+} and vitamin D (Pearce et al 1995, Lienhardt et al 2001). For example, in ADH patients described by Pearce et al (1995), four affected individuals developed long term reductions in renal function with creatinine clearances of 30 ml/min or less. A striking clinical observation seen in some ADH patients is that they develop the symptoms and complications of hypercalcemia while still frankly hypocalcemic, presumably because this represents 'hypercalcemia' in the ADH patient whose 'calciostat' is reset downward (Watanabe et al 1998). The renal complications observed during treatment of ADH generally occur in a setting in which the clinician has tried to correct the serum Ca^{2+} concentration to or close to the normal range. Treatment with Ca^{2+} supplements and vitamin D metabolites should be reserved for those patients with symptomatic ADH; the goal should be to increase the serum Ca^{2+} concentration only to a level sufficient to render the patient asymptomatic (Lienhardt et al 2001). Renal excretion of Ca^{2+} must be monitored carefully in treated patients in order to minimize the risk of renal complications. If a serum Ca^{2+} concentration high enough to ameliorate symptoms cannot be achieved with calcium and vitamin D supplementation without inducing frank hypercalciuria (generally 4 mg/kg/24 h), it may be necessary to co-administer a hypocalciuric agent, such as a thiazide diuretic or injectable PTH (Winer et al 2003).

ADH is a rare syndrome, although in index cases it may comprise a sizeable fraction of cases of idiopathic hypoparathyroidism, perhaps representing as many as a third of such cases (Lienhardt et al 2001). Within a year after the cloning of the CaR, Finegold et al (1994) showed linkage of ADH to a locus on chromosome 3q13 – the same locus containing the CaR gene. Shortly afterward, a heterozygous missense mutation, Glu127Ala, was shown to be the cause of ADH in an unrelated family (Pollak et al 1994b). Since these first reports, more than 30 mutations have been characterized causing ADH (Figure 15.2) (see also CaR Database at http://www.casrdb.mcgill.ca/). Most of these are missense mutations within the CaR's ECD and TMD. When expressed in heterologous systems, these mutations cause a left-shift in the activation of the CaR by Ca_o^{2+}; they only rarely induce constitutive activation of the receptor (Pollak et al 1994b, Baron et al 1996, Pearce et al 1996, Hauache 2001, De'Souza-Li). A recent report identified a family with a large deletion of 181 amino acids within the C-terminus of the CaR, which increased the sensitivity of the receptor for Ca_o^{2+} (Lienhardt et al 2000), at least in part due to an increased level of expression of the mutant receptor. This family contained the only individual to date known to be homozygous for an activating mutation, but this individual exhibited a phenotype very similar to that of the heterozygous family members. Thus one mutated allele may be enough to induce a maximal shift in the set-point of Ca_o^{2+}-regulated PTH secretion, and the presence of the second mutated allele does not alter the biochemical properties of the receptor dimers any further. Perhaps the mutant receptor exerts a 'dominant positive' effect on its wild type partner within wild type-mutant CaR heterodimers. Another activating mutation of the CaR changed a cysteine at amino acid 129 to a serine (Cys129Ser) (Hauache 2001, Hirai et al 2001). Because this cysteine participates in dimerization of the CaR, this result suggests that this cysteine constrains the receptor in its inactive state.

There are a paucity of data directly assessing the regulation of PTH secretion and urinary Ca^{2+} excretion by Ca_o^{2+} in patients with activating mutations of the CaR. The family in which an activating mutation of the CaR was first identified had been presented in abstract form in 1981 under the title, Familial Idiopathic Hypocalcemia (Estep et al 1981). In one affected family member, lowering serum Ca^{2+} concentration below its basal, hypocalcemic level elicited a substantial increase in serum PTH level. This syndrome was presciently postulated to result from a reduced set-point for Ca_o^{2+}-regulated PTH secretion. Further characterization of such families that has been carried out since the cloning of the CaR and the identification of activating mutations in the receptor point out an important point differentiating ADH from typical hypoparathyroidism. In the latter, there has been sufficient damage to the parathyroid glands from a variety of developmental defects or postnatal insults (viz. parathyroid agenesis, autoimmunity, surgical damage, radiation, overloading with heavy metals, such as copper, etc.) that the remaining parathyroid secretory capacity, if any, is insufficient to maintain normocalcemia. In ADH, in contrast, there is a true 'resetting' of the Ca_o^{2+} homeostatic system, including Ca_o^{2+}-regulated PTH secretion, so as to maintain and defend a lower than normal serum Ca^{2+} concentration. It seems highly likely that there is a leftward shift in the relationship between Ca_o^{2+} and urinary Ca^{2+} excretion in ADH, similar to the reduction in set-point for Ca_o^{2+}-regulated PTH secretion in this disorder. Evidence in support of this assertion, however, is largely anecdotal and there have not been systematic studies, such as those in FHH, defining quantitatively the 'set-point' of the kidney for Ca_o^{2+}-regulated renal Ca^{2+} handling. These data include the following: (1) Some studies have shown a greater rate of urinary Ca^{2+} excretion in untreated ADH cases compared to that in other forms of hypoparathyroidism, as noted above (Pearce et al 1996, Winer et al 2003), and (2) ADH patients seem more prone to develop nephrolithiasis and nephrocalcinosis during therapy with oral calcium and vitamin D supplementation than other hypoparathyroid subjects (Pearce et al 1996, Lienhardt et al 2001). Surprisingly, however, while Yamamoto et al. found substantially higher rates of urinary Ca^{2+} excretion in untreated ADH patients compared to those with idiopathic hypoparathyroidism, a plot of serum Ca^{2+} vs. urinary Ca^{2+}/creatinine during treatment showed no difference between the two groups (Yamamoto et al 2000). It is possible that if individual patients with ADH or idiopathic hypoparathyroidism were studied under more carefully controlled condition, such as during Ca^{2+} infusion, such a difference might emerge.

BARTTER'S SYNDROME WITH ACTIVATING CAR MUTATIONS

Recent reports have described three patients with activating mutations of the CaR whose clinical presentation included features resembling Bartter's syndrome (Vargas-Poussou et al 2002, Watanabe et al 2002). These patients presented with the typical features of ADH (i.e. hypocalcemia, hypomagnesemia, low serum PTH, and increased renal excretion of Ca^{2+} and Mg^{2+}), but also exhibited hypokalemia with renal K^+ wasting, hyperreninemia, and hyperaldosteronemia indicating volume depletion. The three mutant CaRs showed markedly left-shifted Ca^{2+} concentration–response curves. It was postulated that inhibition of paracellular reabsorption of Na^+, Ca^{2+} and Mg^{2+} in the CTAL by these unusually activated CaRs causes, in addition to hypercalciuria and hypermagnesuria, Na^+ wasting, with resultant hyperreninemia, hyperaldosteronism and K^+ wasting.

CaR-BASED THERAPEUTICS

The development of allosteric activators ('calcimimetics') and antagonists ('calcilytics') (Gowen et al 2000) of the CaR has enabled CaR-based therapy of disorders of calcium homeostasis. ®Cinacalcet hydrochloride (also known as ®Sensipar) was approved in 2004 by the FDA for treating severe secondary hyperparathyroidism in patients receiving dialysis therapy for end stage renal disease (Block et al 2004) as well as in parathyroid cancer (Silverberg et al 2007). The drug is also effective in mild primary hyperparathyroidism but has not received FDA approval for this indication (Peacock et al 2005). It is possible that the occasional family with FHH and a more severe than usual elevation in serum Ca^{2+} concentration, or complication such as pancreatitis (Pearce et al 1996), might benefit from administration of cinacalcet, either to reduce serum Ca^{2+} concentration or as a therapeutic trial to assess whether associated symptoms could be ascribed to the patient's biochemical abnormalities. Since most mutant CaRs with inactivating mutations show a left shift in their sensitivity to Ca_o^{2+} in response to a calcimimetic (Zhang et al 2002), it is also possible that this class of drug might be useful as an initial trial in patients with NSHPT, to assess to what extent the drug lowered serum Ca_o^{2+} concentration and could be of therapeutic utility in stabilizing an infant's condition.

Ca_o^{2+} receptor antagonists, so-called calcilytics, are also available and are currently in clinical trials. In the presence of a calcilytic, a higher than normal level of Ca_o^{2+} is needed to suppress PTH levels to any given extent (Nemeth et al 1998, Gowen et al 2000). In other words, the CaR reads normocalcemia as hypocalcemia and secretes a pulse of PTH. When PTH is injected as a once daily regimen, it produces an anabolic effect on the skeleton, and is the only truly anabolic agent available for the treatment of osteoporosis (Neer et al 2001). Current clinical trials are assessing whether once or twice daily oral dosing with a calcilytic exerts a similar action owing to release of a pulse of endogenous PTH. It is possible that a calcilytic might also be helpful in the treatment of ADH, or ADH with the features of Bartter's syndrome. By right-shifting the activation of the mutant CaR by extracellular Ca^{2+}, the patient's calciostat would be shifted to the right, leading to a more normal set-point for Ca_o^{2+}-regulated PTH release and urinary Ca^{2+} excretion. Normal or near normal levels of serum Ca^{2+} concentration and urinary Ca^{2+} excretion levels would be the desired result of this approach.

SUMMARY AND FUTURE ISSUES

The CaR is a membrane-bound, seven-membrane-spanning receptor that is expressed in all tissues regulating Ca_o^{2+} homeostasis. It 'senses' minute changes in the level of Ca_o^{2+} in the blood, thereby acting as the body's 'calciostat.' The CaR, in turn, controls the functions of its target tissues so as to maintain the level of blood Ca^{2+} concentration within a very narrow range. Patients with loss-of-function mutations in the CaR gene manifest a generally benign form of hypercalcemia that is accompanied by absolute or hypocalciuria in the heterozygous state. The presence of homozygous mutations, in contrast, causes NSHPT, in which the resultant severe hypercalcemia and hyperparathyroidism can be lethal if it is not treated surgically. Gain-of-function mutations cause a generally benign form of hypocalcemia with relative or absolute hypercalciuria, ADH. The alterations in the control of parathyroid and renal function by Ca_o^{2+} present in patients with these inherited abnormalities in Ca_o^{2+}-sensing have taught us much about the CaR's roles in Ca_o^{2+} homeostasis. Finally, symptomatic patients with these conditions may benefit from the new calcimimetic CaR activators and, perhaps in the future, calcilytics.

DEDICATION

I (EMB) would like to dedicate this chapter to the late Steven C. Hebert, MD, who died just after it was completed. My scientific career was greatly influenced by working with Steve. The cloning of the calcium-sensing receptor by the two of us was very much a joint undertaking, with each of us bringing crucial elements to the project that ensured its success. Steve became a close friend of mine, and we shared many scientific and nonscientific interests that established a strong bond between us. He was a superb scientist and a wonderful human being – warm, positive, and supportive. I will miss him greatly as will his family and his many friends, former students, and colleagues. We were all privileged to know him.

References

Aida K, Koishi S, Inoue M, Nakazato M, Tawata M, Onaya T. Familial hypocalciuric hypercalcemia associated with mutation in the human Ca(2+)-sensing receptor gene. J. Clin. Endocrinol. Metab. 1995; 80: 2594–8.

Attie MF, Gill J Jr., Stock JL, et al. Urinary calcium excretion in familial hypocalciuric hypercalcemia. Persistence of relative hypocalciuria after induction of hypoparathyroidism. J. Clin. Invest. 1983; 72: 667–76.

Auwerx J, Demedts M, Bouillon R. Altered parathyroid set point to calcium in familial hypocalciuric hypercalcaemia. Acta Endocrinologica (Copenh.) 1984; 106: 215–18.

Awata H, Huang C, Handlogten ME, Miller RT. Interaction of the calcium-sensing receptor and filamin, a potential scaffolding protein. J. Biol. Chem. 2001; 276: 34871–9.

Ba J, Brown D, Friedman PA. Calcium-sensing receptor regulation of PTH-inhibitable proximal tubule phosphate transport. Am. J. Physiol. Renal Physiol. 2003; 285: F1233–43.

Bai M, Pearce SH, Kifor O, et al. In vivo and in vitro characterization of neonatal hyperparathyroidism resulting from a de novo, heterozygous mutation in the Ca^{2+}-sensing receptor gene: normal maternal calcium homeostasis as a cause of secondary hyperparathyroidism in familial benign hypocalciuric hypercalcemia. J. Clin. Invest. 1997; 99: 88–96.

Bai M, Quinn S, Trivedi S, et al. Expression and characterization of inactivating and activating mutations in the human Ca^{2+} o-sensing receptor. J. Biol. Chem. 1996; 271: 19537–45.

Baron J, Winer KK, Yanovski JA, et al. Mutations in the Ca(2+)-sensing receptor gene cause autosomal dominant and sporadic hypoparathyroidism. Hum. Mol. Genet. 1996; 5: 601–6.

Berndt TJ, Schiavi S, Kumar R. 'Phosphatonins' and the regulation of phosphorus homeostasis. Am. J. Physiol. Renal Physiol. 2005; 289: F1170–82.

Berridge MJ, Bootman MD, Roderick HL. Calcium signalling: dynamics, homeostasis and remodelling. Nat. Rev. Mol. Cell. Biol. 2003; 4: 517–29.

Block GA, Martin KJ, de Francisco AL, et al. Cinacalcet for secondary hyperparathyroidism in patients receiving hemodialysis. N. Engl. J. Med. 2004; 350: 1516–25.

Brauner-Osborne H, Wellendorph P, Jensen AA. Structure, pharmacology and therapeutic prospects of family C G-protein coupled receptors. Curr. Drug Targets 2007; 8: 169–84.

Bringhurst FR, Demay MB, Kronenberg HM. Hormones and disorders of mineral metabolism. In: Wilson JD, Foster DW, Kronenberg HM, Larsen PR, eds. Williams Textbook of Endocrinology, 9th edn. Philadelphia: W.B. Saunders, 1998: pp. 1155–209.

Brown EM. Four parameter model of the sigmoidal relationship between parathyroid hormone release and extracellular calcium concentration in normal and abnormal parathyroid tissue. J. Clin. Endocrinol. Metab. 1983; 56: 572–81.

Brown EM. Extracellular Ca^{2+} sensing, regulation of parathyroid cell function, and role of Ca^{2+} and other ions as extracellular (first) messengers. Physiol. Rev. 1991; 71: 371–411.

Brown EM. Familial hypocalciuric hypercalcemia and other disorders with resistance to extracellular calcium. Endocrinol. Metab. Clin. Am. 2000; 29: 503–22.

Brown EM. Physiology of Calcium homeostasis. In: Biliezikian JP, Raisz LG, Rodan G, eds. The Parathyroids, 2nd edn. San Diego: Academic Press, 2001: pp. 167–81.

Brown EM. Clinical lessons from the calcium-sensing receptor. Nat. Clin. Pract. Endocrinol. Metab. 2007; 3: 122–33.

Brown EM, MacLeod RJ. Extracellular calcium sensing and extracellular calcium signaling. Physiol. Rev. 2001; 81: 239–97.

Brown EM, Bai M, Pollak MR. Familial benign hypocalciuric hypercalcemia and other syndromes of altered responsiveness to extracellular calcium. In: Krane S, Avioli LV, eds. Metabolic Bone Diseases and Clinically Related Disorders, third edn. San Diego, CA: Academic Press, 1997: pp. 479–99.

Brown EM, Gamba G, Riccardi D, et al. Cloning and characterization of an extracellular Ca(2+)-sensing receptor from bovine parathyroid. Nature 1993; 366: 575–80.

Carling T, Szabo E, Bai M, et al. Familial hypercalcemia and hypercalciuria caused by a novel mutation in the cytoplasmic tail of the calcium receptor [see comments]. J. Clin. Endocrinol. Metab. 2000; 85: 2042–7.

Chang W, Tu C, Chen T, et al. Conditional knockouts in early and mature osteoblasts reveals a critical role for Ca^{2+} receptors in bone development. J. Bone Mineral Res. 2007; 22(S79): Abst. 1284.

Chattopadhyay N, Cheng I, Rogers K, et al. Identification and localization of extracellular Ca(2+)-sensing receptor in rat intestine. Am. J. Physiol. 1998; 274: G122–30.

Cheng I, Qureshi I, Chattopadhyay N, et al. Expression of an extracellular calcium-sensing receptor in rat stomach. Gastroenterology 1999; 116: 118–26.

Cheng SX, Geibel JP, Hebert SC. Extracellular polyamines regulate fluid secretion in rat colonic crypts via the extracellular calcium-sensing receptor. Gastroenterology 2004; 126: 148–58.

Chikatsu N, Fukumoto S, Suzawa M, et al. An adult patient with severe hypercalcaemia and hypocalciuria due to a novel homozygous inactivating mutation of calcium-sensing receptor [In Process Citation]. Clin. Endocrinol. (Oxf.) 1999; 50: 537–43.

Chou YH, Brown EM, Levi T, et al. The gene responsible for familial hypocalciuric hypercalcemia maps to chromosome 3q in four unrelated families. Nat. Genet. 1992; 1: 295–300.

Cole DE, Janicic N, Salisbury SR, Hendy GN. Neonatal severe hyperparathyroidism, secondary hyperparathyroidism, and familial hypocalciuric hypercalcemia: multiple different phenotypes associated with an inactivating Alu insertion mutation of the calcium-sensing receptor gene. Am. J. Med. Genet. 1997; 71: 202–10 [published erratum appears in Am. J. Med. Genet. 1997 Oct 17; 72(2): 251–2].

Cole DE, Quamme GA. Inherited disorders of renal magnesium handling. J. Am. Soc. Nephrol. 1937; 11: 1937–47.

Cooper L, Wertheimer J, Levey R, et al. Severe primary hyperparathyroidism in a neonate with two hypercalcemic parents: management with parathyroidectomy and heterotopic autotransplantation. Pediatrics 1986; 78: 263–8.

Davies M, Adams PH, Lumb GA, Berry JL, Loveridge N. Familial hypocalciuric hypercalcaemia: evidence for continued enhanced renal tubular reabsorption of calcium following total parathyroidectomy. Acta Endocrinol. (Copenh.) 1984; 106: 499–504.

De Jesus Ferreira MC, Bailly C. Extracellular Ca^{2+} decreases chloride reabsorption in rat CTAL by inhibiting cAMP pathway. Am. J. Physiol. 1998; 275: F198–203.

DeLuca HF. Overview of general physiologic features and functions of vitamin D. Am. J. Clin. Nutr. 2004; 80: 1689S–96S.

Dinbar A, Tulcinsky DB. Effect of induced hypercalcemia on gastric acid secretion and serum gastrin levels in duodenal ulcer patients. Isr. J. Med. Sci. 1978; 14: 992–4.

D'Souza-Li L, Canaff L, Janicic N, Cole DEC, Hendy GN. An acceptor splice site mutation in the calcium-sensing receptor gene in familial hypocalciuric hypercalcemia and neonatal severe hyperparathyroidism. Submitted, 1998.

D'Souza-Li L, Yang B, Canaff L, et al. Identification and functional characterization of novel calcium-sensing receptor mutations in familial hypocalciuric hypercalcemia and autosomal dominant hypocalcemia. J. Clin. Endocrinol. Metab. 2002; 87: 1309–18.

Eftekhari F, Yousefzadeh D. Primary infantile hyperparathyroidism: clinical, laboratory, and radiographic features in 21 cases. Skeletal Radiol. 1982; 8: 201–8.

Estep H, Mistry Z, Burke P. Familial idiopathic hypocalcemia. In: Proceedings and abstracts of the 63rd Annual Meeting of the Endocrine Society, Cincinnati, Endocrine Society, 1981, p. 275 (abstract).

Finegold DN, Armitage MM, Galiani M, et al. Preliminary localization of a gene for autosomal dominant hypoparathyroidism to chromosome 3q13. Pediatr. Res. 1994; 36: 414–17.

Foley T Jr., Harrison H, Arnaud C, Harrison H. Familial benign hypercalcemia. J. Pediatr. 1972; 81: 1060–7.

Fox L, Sadowsky J, Pringle KP, et al. Neonatal hyperparathyroidism and pamidronate therapy in an extremely premature infant. Pediatrics 2007; 120: e1350–4.

Fudge NJ, Kovacs CS. Physiological studies in heterozygous calcium sensing receptor (CaSR) gene-ablated mice confirm that the CaSR regulates calcitonin release in vivo. BMC Physiol. 2004; 4: 5.

Gill JJ, Bartter F. On the impairment of renal concentrating ability in prolonged hypercalcemia and hypercalciuria in man. J. Clin. Invest. 1961; 40: 716–22.

Gowen M, Stroup GB, Dodds RA, et al. Antagonizing the parathyroid calcium receptor stimulates parathyroid hormone secretion and bone formation in osteopenic rats. J. Clin. Invest. 2000; 105: 1595–604.

Grantmyre E. Roentgenographic features of 'primary' hyperparathyroidism in infancy. J. Can. Assoc. Radiol. 1973; 24: 257–60.

Harris SS, D'Ercole AJ. Neonatal hyperparathyroidism: the natural course in the absence of surgical intervention. Pediatrics 1989; 83: 53–6.

Hauache OM. Extracellular calcium-sensing receptor: structural and functional features and association with diseases. Braz. J. Med. Biol. Res. 2001; 34: 577–84.

Heath D. Familial benign hypercalcemia. Trends Endocrinol. Metab. 1989a; 1: 6–9.

Heath Hd. Familial benign (hypocalciuric) hypercalcemia. A troublesome mimic of mild primary hyperparathyroidism. Endocrinol. Metab. Clin. North Am. 1989b; 18: 723–40.

Heath Hd, Jackson CE, Otterud B, Leppert MF. Genetic linkage analysis in familial benign (hypocalciuric) hypercalcemia: evidence for locus heterogeneity. Am. J. Hum. Genet. 1993; 53: 193–200.

Heath DA. Familial hypocalciuric hypercalcemia. In: Bilezikian JP, Marcus R, Levine MA, eds. The Parathyroids. New York, NY: Raven Press, 1994: pp. 699–710.

Heath DA. Clinical manifestations of abnormalities of the calcium sensing receptor. Clin. Endocrinol. (Oxf.) 1998; 48: 257–8.

Hebert SC, Brown EM, Harris HW. Role of the Ca(2+)-sensing receptor in divalent mineral ion homeostasis. J. Exp. Biol. 1997; 200: 295–302.

Hendy GN, D'Souza-Li L, Yang B, Canaff L, Cole DE. Mutations of the calcium-sensing receptor (CASR) in familial hypocalciuric hypercalcemia, neonatal severe hyperparathyroidism, and autosomal dominant hypocalcemia [In Process Citation]. Hum. Mutat. 2000; 16: 281–96.

Hirai H, Nakajima S, Miyauchi A, et al. A novel activating mutation (C129S) in the calcium-sensing receptor gene in a Japanese family with autosomal dominant hypocalcemia. J. Hum. Genet. 2001; 46: 41–4.

Ho C, Conner DA, Pollak MR, et al. A mouse model of human familial hypocalciuric hypercalcemia and neonatal severe hyperparathyroidism [see comments]. Nat. Genet. 1995; 11: 389–94.

Hoenderop JG, Chon H, Gkika D, et al. Regulation of gene expression by dietary Ca^{2+} in kidneys of 25-hydroxyvitamin D3-1 alpha-hydroxylase knockout mice. Kidney Int. 2004; 65: 531–9.

Hu J, Spiegel AM. Naturally occurring mutations in the extracellular Ca^{2+}-sensing receptor: implications for its structure and function. Trends endocrinol. metabol. 2003; 14: 282–8.

Hu J, McLarnon SJ, Mora S, et al. A region in the seven-transmembrane domain of the human Ca^{2+} receptor critical for response to Ca^{2+}. J. Biol. Chem. 2005; 280: 5113–20.

Janicic N, Pausova Z, Cole DE, Hendy GN. Insertion of an Alu sequence in the Ca(2+)-sensing receptor gene in familial hypocalciuric hypercalcemia and neonatal severe hyperparathyroidism. Am. J. Hum. Genet. 1995; 56: 880–6.

Kameda T, Mano H, Yamada Y, et al. Calcium-sensing receptor in mature osteoclasts, which are bone resorbing cells. Biochem. Biophys. Res. Commun. 1998; 245: 419–22.

Khosla S, Ebeling PR, Firek AF, Burritt MM, Kao PC, Heath Hd. Calcium infusion suggests a 'set-point' abnormality of parathyroid gland function in familial benign hypercalcemia and more complex disturbances in primary hyperparathyroidism. J. Clin. Endocrinol. Metab. 1993; 76: 715–20.

Kifor O, Diaz R, Butters R, Kifor I, Brown EM. The calcium-sensing receptor is localized in caveolin-rich plasma membrane domains of bovine parathyroid cells. J. Biol. Chem. 1998; 273: 21708–13.

Kobayashi M, Tanaka H, Tsuzuki K, et al. Two novel missense mutations in calcium-sensing receptor gene associated with neonatal severe hyperparathyroidism. J. Clin. Endocrinol. Metab. 1997; 82: 2716–19.

Kos CH, Karaplis AC, Peng JB, et al. The calcium-sensing receptor is required for normal calcium homeostasis independent of parathyroid hormone. J. Clin. Invest. 2003; 111: 1021–8.

Law WM Jr., Bollman S, Kumar R, Heath H III. Vitamin D metabolism in familial benign hypercalcemia (hypocalciuric hypercalcemia) differs from that in primary hyperparathyroidism. J. Clin. Endocrinol. Metab. 1984a; 58: 744–7.

Law WM Jr., Carney JA, Heath H III. Parathyroid glands in familial benign hypercalcemia (familial hypocalciuric hypercalcemia). Am. J. Med. 1984b; 76: 1021–6.

Law WM Jr., Heath H III. Familial benign hypercalcemia (hypocalciuric hypercalcemia). Clinical and pathogenetic studies in 21 families. Ann. Intern. Med. 1985; 105: 511–19.

Layer P, Hotz J, Eysselein VE, et al. Effects of acute hypercalcemia on exocrine pancreatic secretion in the cat. Gastroenterology 1985; 88: 1168–74.

Levi R, Ben-Dov IZ, Lavi-Moshayoff V, Dinur M, Martin D, Naveh-Many T, Silver J. Increased parathyroid hormone gene expression in secondary hyperparathyroidism of experimental uremia is reversed by calcimimetics: correlation with post-translational modification of the trans acting factor AUF1. J. Am. Soc. Nephrol. 2006; 17: 107–12.

Lienhardt A, Bai M, Lagarde JP, et al. Activating mutations of the calcium-sensing receptor: management of hypocalcemia. J. Clin. Endocrinol. Metab. 2001; 86: 5313–23.

Lienhardt A, Garabedian M, Bai M, et al. A large homozygous or heterozygous in-frame deletion within the calcium-sensing receptor's carboxylterminal cytoplasmic tail that causes autosomal dominant hypocalcemia. J. Clin. Endocrinol. Metab. 2000; 85: 1695–702.

Lloyd SE, Pannett AA, Dixon PH, Whyte MP, Thakker RV. Localization of familial benign hypercalcemia, Oklahoma variant (FBHOk), to chromosome 19q13. Am. J. Hum. Genet. 1999; 64: 189–95.

Maiti A, Beckman MJ. Extracellular calcium is a direct effecter of VDR levels in proximal tubule epithelial cells that counter-balances effects of PTH on renal Vitamin D metabolism. J. Steroid Biochem. Mol. Biol. 2007; 103: 504–8.

Marx S, Spiegel A, Brown E, et al. Divalent cation metabolism. Familial hypocalciuric hypercalcemia versus typical primary hyperparathyroidism. Am. J. Med. 1978; 65: 235–42.

Marx SJ, Attie MF, Levine MA, Spiegel AM, Downs RW Jr., Lasker RD. The hypocalciuric or benign variant of familial hypercalcemia: clinical and biochemical features in fifteen kindreds. Medicine (Baltimore) 1981a; 60: 397–412.

Marx SJ, Attie MF, Stock JL, Spiegel AM, Levine MA. Maximal urine-concentrating ability: familial hypocalciuric hypercalcemia versus typical primary hyperparathyroidism. J. Clin. Endocrinol. Metab. 1981b; 52: 736–40.

Marx S, Attie M, Spiegel A, Levine M, Lasker R, Fox M. An association between neonatal severe primary hyperparathyroidism and familial hypocalciuric hypercalcemia in three kindreds. N. Engl. J. Med. 1982; 306: 257–84.

Marx SJ, Fraser D, Rapoport A. Familial hypocalciuric hypercalcemia. Mild expression of the gene in heterozygotes and severe expression in homozygotes. Am. J. Med. 1985; 78: 15–22.

Marx S, Lasker R, Brown E, et al. Secretory dysfunction in parathyroid cells from a neonate with severe primary hyperparathyroidism. J. Clin. Endocrinol. Metab. 1986; 62: 445–9.

Motoyama HI, Friedman PA. Calcium-sensing receptor regulation of PTH-dependent calcium absorption by mouse cortical ascending limbs. Am. J. Physiol. Renal Physiol. 2002; 283: F399–406.

Nabeshima YI, Imura H. alpha-klotho: a regulator that integrates calcium homeostasis. Am. J. Nephrol. 2007; 28: 455–64.

Neer RM, Arnaud CD, Zanchetta JR. Effect of parathyroid hormone (1-34) on fractures and bone mineral density in postmenopausal women with osteoporosis. N. Engl. J. Med. 2001; 344: 1434–41.

Nemeth EF, Steffey ME, Hammerland LG, et al. Calcimimetics with potent and selective activity on the parathyroid calcium receptor. Proc. Natl Acad. Sci. USA 1998a; 95: 4040–5.

Nemeth EF, Fox J, Delmar EG, et al. Stimulation of parathyroid hormone secretion by a small molecule antagonist of the calcium receptor. J. Bone Miner. Res. 1998b; 23: S156 (abstract #1030).

Peacock M, Robertson WG, Nordin BE. Relation between serum and urinary calcium with particular reference to parathyroid activity. Lancet 1969; 1: 384–6.

Peacock M, Bilezikian JP, Klassen PS, Guo MD, Turner SA, Shoback D. Cinacalcet hydrochloride maintains long-term normocalcemia in patients with primary hyperparathyroidism. J. Clin. Endocrinol. Metab. 2005; 90: 135–41.

Pearce SH. Clinical disorders of extracellular calcium-sensing and the molecular biology of the calcium-sensing receptor. Ann. Med. 2002; 34: 201–6.

Pearce SH, Bai M, Quinn SJ, Kifor O, Brown EM, Thakker RV. Functional characterization of calcium-sensing receptor mutations expressed in human embryonic kidney cells. J. Clin. Invest. 1996a; 98: 1860–6.

Pearce S, Coulthard M, Kendall-Taylor P, Thakker R. Autosomal dominant hypocalcemia associated with a mutation in the calcium-sensing receptor. J. Bone Miner. Res. 1995a; 10: S176 (abstract 149).

Pearce SH, Trump D, Wooding C, et al. Calcium-sensing receptor mutations in familial benign hypercalcemia and neonatal hyperparathyroidism. J. Clin. Invest. 1995b; 96: 2683–92.

Pearce SH, Wooding C, Davies M, Tollefsen SE, Whyte MP, Thakker RV. Calcium-sensing receptor mutations in familial hypocalciuric hypercalcaemia with recurrent pancreatitis. Clin. Endocrinol. (Oxf.) 1996b; 45: 675–80.

Pearce SH, Williamson C, Kifor O, et al. A familial syndrome of hypocalcemia with hypercalciuria due to mutations in the calcium-sensing receptor [see comments]. N. Engl. J. Med. 1996c; 335: 1115–22.

Pi M, Hinson TK, Quarles L. Failure to detect the extracellular calcium-sensing receptor (CasR) in human osteoblast cell lines. J. Bone Miner. Res. 1999; 14: 1310–19.

Pi M, Faber P, Ekema G, et al. Identification of a novel extracellular cation-sensing G-protein-coupled receptor. J. Bio.l Chem. 2005; 280: 40201–9.

Pin JP, Galvez T, Prezeau L. Evolution, structure, and activation mechanism of family 3/C G-protein-coupled receptors. Pharmacol. Ther. 2003; 98: 325–54.

Pollak MR, Brown EM, Chou YH, et al. Mutations in the human Ca(2+)-sensing receptor gene cause familial hypocalciuric hypercalcemia and neonatal severe hyperparathyroidism. Cell 1993; 75: 1297–1303.

Pollak MR, Brown EM, Estep HL, et al. Autosomal dominant hypocalcaemia caused by a Ca(2+)-sensing receptor gene mutation. Nat. Genet. 1994a; 8: 303–7.

Pollak MR, Chou YH, Marx SJ, et al. Familial hypocalciuric hypercalcemia and neonatal severe hyperparathyroidism. Effects of mutant gene dosage on phenotype. J. Clin. Invest. 1994b; 93: 1108–12.

Quamme GA. Effect of hypercalcemia on renal tubular handling of calcium and magnesium. Can. J. Physiol. Pharmacol. 1982; 60: 1275–80.

Quarles LD. Cation-sensing receptors in bone: A novel paradigm for regulating bone remodeling? J. Bone Miner. Res. 1997; 12: 1971–4.

Raisz LG. Bone resorption in tissue culture. Factors influencing the response to parathyroid hormone. J. Clin. Invest. 1965; 44: 103–16.

Rhone DP. Primary neonatal hyperparathyroidism. Report of a case and review of the literature. Am. J. Clin. Path. 1975; 64: 488–99.

Riccardi D, Park J, Lee WS, Gamba G, Brown EM, Hebert SC. Cloning and functional expression of a rat kidney extracellular calcium/polyvalent cation-sensing receptor. Proc. Natl Acad. Sci. USA 1995; 92: 131–5.

Riccardi D, Lee WS, Lee K, Segre GV, Brown EM, Hebert SC. Localization of the extracellular Ca(2+)-sensing receptor and PTH/PTHrP receptor in rat kidney. Am. J. Physiol. 1996; 271: F951–6.

Rodriguez L, Tu C, Cheng Z, et al. Expression and functional assessment of an alternatively spliced extracellular Ca^{2+}-sensing receptor in growth plate chondrocytes. Endocrinology 2005; 146: 5294–303.

Sands JM, Naruse M, Baum M, et al. Apical extracellular calcium/polyvalent cation-sensing receptor regulates vasopressin-elicited water permeability in rat kidney inner medullary collecting duct. J. Clin. Invest. 1997; 99: 1399–405.

Sands JM, Flores FX, Kato A, et al. Vasopressin-elicited water and urea permeabilities are altered in IMCD in hypercalcemic rats. Am. J. Physiol. 1998; 274: F978–85.

Silve C, Petrel C, Leroy C, et al. Delineating a Ca^{2+} binding pocket within the venus flytrap module of the human calcium-sensing receptor. J. Biol. Chem. 2005; 280: 37917–23.

Silverberg SJ, Rubin MR, Faiman C, et al. Cinacalcet hydrochloride reduces the serum calcium concentration in inoperable parathyroid carcinoma. J. Clin. Endocrinol. Metab. 2007; 92: 3803–8.

Spiegel AM, Harrison HE, Marx SJ, Brown EM, Aurbach GD. Neonatal primary hyperparathyroidism with autosomal dominant inheritance. J. Pediatr. 1977; 90: 269–72.

Strewler GJ. Familial benign hypocalciuric hypercalcemia – from the clinic to the calcium sensor [editorial; comment]. West J. Med. 1994; 160: 579–80.

Thogeirsson U, Costa J, Marx SJ. The parathyroid glands in familial hypocalciuric hypercalcemia. Hum. Pathol. 1981; 12: 229–37.

Tu C, Elalieh H, Chen T, et al. Expression of Ca^{2+} receptors in cartilage is essential for embryonic skeletal develoment in vivo. J. Bone Miner. Res. 2007; 22(Suppl 1): S50 (abstract 1176).

Tu Q, Pi M, Karsenty G, Simpson L, Liu S, Quarles LD. Rescue of the skeletal phenotype in CasR-deficient mice by transfer onto the Gcm2 null background. J. Clin. Invest. 2003; 111: 1029–37.

Valenti G, Laera A, Gouraud S, et al. Low-calcium diet in hypercalciuric enuretic children restores AQP2 excretion and improves clinical symptoms. Am. J. Physiol. Renal Physiol. 2002; 283: F895–903.

Valenti G, Procino G, Tamma G, Carmosino M, Svelto M. Minireview: aquaporin 2 trafficking. Endocrinology 2005; 146: 5063–70.

Vargas-Poussou R, Huang C, Hulin P, et al. Functional characterization of a calcium-sensing receptor mutation in severe autosomal dominant hypocalcemia with a Bartter-like syndrome. J. Am. Soc. Nephrol. 2002; 13: 2259–66.

Waller S, Kurzawinski T, Spitz L, et al. Neonatal severe hyperparathyroidism: genotype/phenotype correlation and the use of pamidronate as rescue therapy. Eur. J. Pediatr. 2004; 163: 589–94.

Wang W, Lu M, Balazy M, Hebert SC. Phospholipase A2 is involved in mediating the effect of extracellular Ca^{2+} on apical K^+ channels in rat TAL. Am. J. Physiol. 1997; 273: F421–9.

Ward DT. Calcium receptor-mediated intracellular signalling. Cell Calcium 2004; 35: 217–28.

Watanabe T, Bai M, Lane CR, et al. Familial hypoparathyroidism: identification of a novel gain of function mutation in transmembrane domain 5 of the calcium-sensing receptor. J. Clin. Endocrinol. Metab. 1998; 83: 2497–502.

Watanabe S, Fukumoto S, Chang H, et al. Association between activating mutations of calcium-sensing receptor and Bartter's syndrome. Lancet 2002; 360: 692–4.

Weisinger JR, Favus MJ, Langman CB, Bushinsky D. Regulation of 1,25-dihydroxyvitamin D3 by calcium in the parathyroidectomized, parathyroid hormone-replete rat. J. Bone. Miner. Res. 1989; 4: 929–35.

Williams TM, Lisanti MP. The Caveolin genes: from cell biology to medicine. Ann. Med. 2004; 36: 584–95.

Winer KK, Ko CW, Reynolds JC, et al. Long-term treatment of hypoparathyroidism: a randomized controlled study comparing parathyroid hormone-(1-34) versus calcitriol and calcium. J. Clin. Endocrinol. Metab. 2003; 88: 4214–20.

Yamaguchi T, Chattopadhyay N, Kifor O, Brown EM. Extracellular calcium (Ca^{2+}(o))-sensing receptor in a murine bone marrow-derived stromal cell line (ST2): potential mediator of the actions of Ca^{2+} (o) on the function of ST2 cells. Endocrinology 1998a; 139: 3561–8.

Yamaguchi T, Chattopadhyay N, Kifor O, et al. Expression of extracellular calcium-sensing receptor in human osteoblastic MG-63 cell line. Am. J. Physiol. Cell Physiol. 2001; 280: C382–93.

Yamaguchi T, Olozak I, Chattopadhyay N, et al. Expression of extracellular calcium ($Ca2+_o$)-sensing receptor in human peripheral blood monocytes. Biochem. Biophys. Res. Commun. 1998b; 246: 501–6.

Yamamoto M, Akatsu T, Nagase T, Ogata E. Comparison of hypocalcemic hypercalciuria between patients with idiopathic hypoparathyroidism and those with gain-of-function mutations in the calcium-sensing receptor: is it possible to differentiate the two disorders? J. Clin. Endocrinol. Metab. 2000; 85: 4583–91.

Zaidi M, Adebanjo OA, Moonga BS, Sun L, Huang CL. Emerging insights into the role of calcium ions in osteoclast regulation. J. Bone Miner. Res. 1999; 14: 669–74.

Zhang Z, Jiang Y, Quinn SJ, Krapcho K, Nemeth EF, Bai M. L-phenylalanine and NPS R-467 synergistically potentiate the function of the extracellular calcium-sensing receptor through distinct sites. J. Biol. Chem. 2002; 277: 33736–41.

Zhang M, Breitwieser G. High affinity interaction with filamin A protects against calcium-sensing receptor degradation. J. Biol. Chem. 2005; 280: 11140–6.

E. Primary Genetic Diseases of the Distal Convoluted Tubule and Collecting Duct

CHAPTER 16

Liddle's Syndrome (Pseudoaldosteronism)

LAURENT SCHILD AND BERNARD C. ROSSIER

INTRODUCTION

Since the pioneer work of AC Guyton, the kidney is considered to play the major causal role in the pathophysiology of essential hypertension. As proposed by Guyton, within hours or days, a kidney pressure control system is induced that increases body fluid volume when the pressure falls (or decreases the volume when the pressure rises). This kidney-fluid system is the dominant physiological mechanism establishing long-term control of arterial blood pressure (Guyton 1987).

In this respect, the kidney plays the key role in controlling the sodium balance, and thereby the extracellular volume, blood volume and blood pressure. Blood pressure is a quantitative trait, distributed normally in the population, which greatly varies according to many factors including age, gender and physical activity. Hypertension has been defined as systolic blood pressure above 140 mmHg and/or diastolic pressure above 90 mmHg. Hypertension is a risk factor for cardiovascular diseases. Its pathogenesis is multifactorial with genetic determinants accounting for about 30% and environmental factors (i.e. smoking, drugs or salt intake) for about 70% of blood pressure variations. The understanding of the complex interplay between genetic and environmental factors has greatly benefited from the study of mendelian forms of human hypertension characterized by their extreme phenotype (Lifton 1996).

Liddle's syndrome (pseudoaldosteronism) is a good example of a rare mendelian form of human hypertension that responds to treatment by the K^+-sparing diuretic amiloride, associated with a low sodium diet. Genetic, physiological and biochemical studies of Liddle's syndrome have provided a new understanding of the cellular and molecular basis of the regulation of blood pressure and volume by the kidney.

PSEUDOALDOSTERONISM

In 1963, G.W. Liddle reported the case of a 16-year-old girl with elevated blood pressure, hypokalemia, metabolic alkalosis (Liddle et al 1963). She was provisionally diagnosed with primary aldosteronism. Her mother and grandmother both were severely hypertensive and died before their fourth decade. Her younger brother exhibited the same clinical picture with severe hypertension, hypokalemia and alkalosis.

The clinical investigations by G.W. Liddle of both this young girl and her brother, revealed that the urinary excretion of aldosterone under low salt diet was minimal, 15–50 µg/day (normal 200–1300 µg/day). Since these patients did not respond to a low Na^+ diet by increasing the urinary aldosterone excretion, their ability to respond to exogenous aldosterone was investigated by G.W. Liddle. When given aldosterone i.m., the urinary sodium excretion fell to less than 1 mEq/day, indicating that the ability of the kidney to retain Na^+ in response to aldosterone was maintained. Furthermore, when compared to patients with Addison disease kept under low dose of aldosterone to match the aldosterone secretory rates found in patients with pseudoaldosteronism, the latter excreted much less sodium than the Addison patients. G.W. Liddle concluded from these investigations that his patients had an abnormal aptitude to retain sodium and excrete potassium in the kidney, even in the presence of minimal aldosterone secretion.

Pharmacological investigations revealed that triamterene, a K^+-sparing diuretic, causes a significant increase in Na^+ excretion and a decrease in K^+ excretion, and, when associated with a low Na^+ diet, the drug normalized blood pressure. The mineralocorticoid antagonist, spironolactone, was without effects on the renal handling of Na^+ and K^+, and also on blood pressure. G.W. Liddle concluded that

Genetic Diseases of the Kidney
ISBN: 978-0-12-449851-8

the primary defect in these patients lies in the renal tubular transporters for Na^+ and K^+ ions.

The extended pedigree of the probands investigated originally by G.W. Liddle was revisited 30 years later by Botero-Velez et al (1994). This study basically confirmed the original phenotype described by Liddle, consisting of hypertension, a low rate of urinary aldosterone excretion, and a low ratio of aldosterone to potassium in the urine. The severity and the onset of hypertension were variable in this kindred, although the onset of hypertension usually occurs in adolescence. The chronic suppression of aldosterone secretion, secondary to the chronic expansion of the extracellular fluid volume, was a constant observation. However, hypokalemia was not a constant finding among the affected members of this family, some of them having serum K^+ concentration above 4 mmol/l. In addition, the extensive study of this pedigree clearly demonstrated that pseudoaldosteronism was transmitted in an autosomal dominant manner.

One proband originally investigated by G.W. Liddle developed end-stage renal failure and received a kidney transplant. She was studied after renal transplantation, and, at that time, blood pressure and serum potassium concentration were normal, and plasma renin activity and plasma aldosterone concentration increased normally, in response to a low salt diet. This further supported the conclusion of G.W. Liddle that pseudoaldosteronism was primarily a Na^+ and K^+ transport defect of the kidneys. Together with the amelioration of the syndrome with blockers of the epithelial sodium channel, it became quite likely that the specific transport defect resides in the distal nephron, where sodium and potassium excretion respond to aldosterone, amiloride, or triamterene.

ENaC AND Na^+ ABSORPTION IN THE ASDN

Morphological and functional studies on rodent and human kidneys indicate that at least three successive portions of the distal nephron, i.e. the late portion of the distal convoluted tubule (DCT), the connecting tubule (CNT), and the collecting duct (CD), contribute to the ASDN (Bachmann et al 1996, Reilly & Ellison 2000). Although these segments have distinct structural and functional features, they have in common the expression of the epithelial sodium channel (ENaC), the mineralocorticoid receptor (MR) and the 11-β hydroxysteroid dehydrogenase type II (11-βHSD2) proteins.

Sodium absorption in the ASDN occurs via ENaC and is electrogenic: Na^+ ions enter the epithelial cell from the tubule lumen through ENaC and are extruded from the cell across the basolateral membrane by the Na^+/K^+ATPase. Increasing electrogenic Na^+ absorption in the ASDN depolarizes the apical membrane of principal cells, favoring K^+ secretion through apical K^+ channels (ROMK) along a favorable electrochemical gradient. This electrical coupling between Na^+ absorption and K^+ secretion accounts for the increased urinary K^+ loss in patients with high Na^+ reabsorption in the ASDN.

The elevation of plasma aldosterone levels in response to changes in the sodium diet increases electrogenic Na^+ absorption in the ASDN that correlates with changes in the subcellular localization of ENaC in principal cells and along the axis of the ASDN (Masilamani et al 1999, Loffing et al 2000). In rodents, kept under a high dietary sodium intake with low plasma aldosterone levels, ENaC subunits are barely detectable at the luminal membrane and are found almost exclusively at intracellular sites. On a standard dietary sodium intake with moderate plasma aldosterone levels, ENaC subunits are traceable at the apical pole (luminal membrane and subapical vesicles) of late DCT and early CNT. However, in farther downstream segments (i.e. the collecting duct), β and γ ENaC subunits remain almost exclusively localized at intracellular sites. Under a severe dietary sodium restriction with high plasma aldosterone levels, ENaC subunits become detectable in the apical pole all along the ASDN from the late DCT, CNT to the CD. Nevertheless, the axial gradient along the entire ASDN for apical ENaC still prevails and the apical localization of ENaC subunits is more prominent in early ASDN than in late ASDN. This parallels the magnitude of the amiloride-sensitive currents recorded along the ASDN, which predominates in the CNT, then decreases along the collecting duct (Frindt & Palmer 2004). The aldosterone-dependent adaptation of renal sodium excretion to dietary sodium intake occurs predominantly in the early ASDN (day/night, short-term regulation), while the late ASDN gets recruited only under chronically (long-term regulation) high plasma aldosterone levels.

It should be mentioned that in the DCT, aldosterone also increases the protein expression of the NaCl cotransporter sensitive to thiazides (NCC or TSC) (Kim et al 1998, Masilamani et al 2002). Consistently, aldosterone has been shown to stimulate electroneutral Na^+ transport in the DCT (Velazquez et al 1996). It needs to be determined whether these stimulatory effects occur along the entire DCT or only in the late DCT.

The epithelial sodium channel ENaC is highly selective for Na^+ over K^+ ions ($P_{Na}/P_K > 100$), and has a single channel conductance (4–5 pS), with Na^+ as the charge carrier. From patch clamp recordings, ENaC usually shows long open and closed times. The open probability (P_o) of ENaC is, however, quite variable under similar physiological conditions and ranges from ≤0.05 to >0.95; this variation in P_o could reflect different gating modes of the channel (Palmer & Frindt 1996). A number of variables have been proposed to affect ENaC gating that include intracellular pH, Na^+ and Ca^{2+} (Palmer & Frindt 1987, Frindt et al 2001). High extracellular concentrations of Na^+ ions tend to inhibit ENaC activity, a phenomenon called self-inhibition (Chraibi & Horisberger 2002). This fast inhibition of ENaC upon increasing extracellular Na^+ concentration may be important to avoid massive Na^+ absorption in the early part of the ASDN, when Na^+ delivery rises in this nephron segment.

The identification of the primary structure of ENaC by functional cloning revealed a channel made of homologous α, β, and γ subunits (Canessa et al 1994). These homologous subunits are arranged pseudosymmetrically around a channel in a likely 2α, 1β and 1γ configuration, each subunit contributing to the ionic selectivity and the sensitivity to block by amiloride (Firsov et al 1998, Kellenberger et al 2001). The homologous ENaC subunits have two transmembrane segments connected by a large extracellular loop and intracellular NH and COOH termini (Figure 16.1). The coexpression of the three α, β, and γ ENaC subunits is necessary for maximal expression of the amiloride-sensitive current (Canessa et al 1994). Pharmacological blockers of the ENaC used as K^+-sparring diuretics include amiloride and triamterene. They bind at the same site in the extracellular entry of the channel pore, blocking the flux of Na^+ ions through the channel (Schild et al 1997).

Because of the favorable response of the elevated blood pressure to treatment by amiloride, but not by spironolactone, it became quite obvious that the primary defect in pseudoaldosteronism lies in the channel itself or in the regulatory pathway between the MR and ENaC.

GENOTYPE–PHENOTYPE RELATIONSHIP

The human α ENaC gene (SCNN1A) covers 17 kb on chromosome 12p13. The human β and γ ENaC genes (SCNN1B and SCNN1G) are physically and genetically tightly linked, both being located in close proximity on chromosome 16p12. The use of genetic markers in this segment of chromosome 16 allowed the demonstration of a complete linkage of the gene encoding the β ENaC or γ ENaC subunits to Liddle's syndrome (Shimkets et al 1994, Hansson et al 1995a, 1995b, Tamura et al 1996).

Analysis of the β ENaC gene of subject with Liddle's syndrome showed different types of mutations, all located in the cytosolic C-terminus end of the β or γ ENaC subunit (exon XIII of SCNN1B and SCNN1G). These mutations include premature stop codons, frameshift or missense mutations that delete or change a conserved proline-rich sequence (see Table 16.1 and Figure 16.1). Liddle's syndrome also arises from mutations in the γ ENaC subunit at corresponding positions in the cytosolic C-terminus of the subunit (exon XIII). In one case Liddle's syndrome

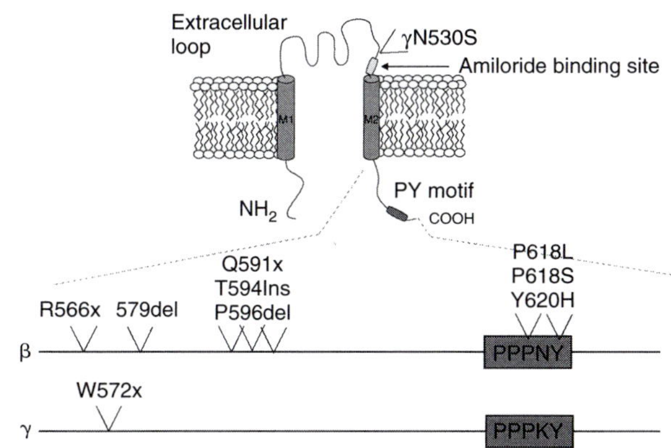

FIGURE 16.1 Membrane topology of an ENaC subunit and locations of the mutations causing Liddle syndrome. Mutations are found in the intracellular carboxy terminus of the β or γ ENaC subunit, deleting or mutating a conserved proline-rich PY motif. One mutation (γN530S) has been described in the extracellular loop of γENaC, close to the amiloride binding site. (see also Plate 16)

TABLE 16.1 Mutations on the SCNN1B and SCNN1G genes encoding the β and γ ENaC subunits and causing Liddle syndrome

Mutations	Type	Ref.
βR566 X	Premature stop	Shimkets et al 1994 Kyuma et al 2001, Melander et al 1998
βQ591X	Premature stop	Shimkets et al 1994
βT594InsC	Frameshift	Shimkets et al 1994, Jackson et al 1998, Nakano et al 2002
βP596delC	Frameshift	Shimkets et al 1994
βP618L	Missense	Hansson 1995b, Uehara et al 1998
βP618S	Missense	Inoue et al 1998, Uehara et al 1998
βY620H	Missense	Tamura et al 1996
β579del32	Deletion – premature stop	Jeunemaître et al 1997
γW573X	Premature stop	Hansson et al 1995
γW575X	Premature stop	Yamashita et al 2001
γN530S	Missense	Hiltunen et al 2002

X, stop codon; Ins, insertion; del, deletion.

was associated with a missense mutation in the exon XIII encoding the γ ENaC subunit (Hiltunen et al 2002).

Except for one case, all the mutations associated with Liddle syndrome delete or modify the sequence of a conserved proline-rich motif xPPxY (x being any amino acid, P proline and Y tyrosine) in the C-terminus of the β and γ ENaC subunits (Figure 16.1). The exception is a case of Liddle's syndrome without mutations in the PY motif, but with a mutation in a conserved region of the extracellular loop preceding the second transmembrane segment of the γ ENaC subunit (γN530S).

The identification of ENaC mutations causing Liddle's syndrome allowed the analysis of the genotype–phenotype relationships in kindreds of affected patients. Liddle's syndrome is genetically heterogeneous, showing a variable penetration of hypertension and hypokalemia among the different kindreds studied so far. In particular, among the members of a large extended kindred with Liddle's syndrome, blood pressure was highly variable, ranging from 118/72 to 160/108 (Findling et al 1997). In this kindred with mild hypertension, elevated blood pressure and hypokalemia were not obligatory among the subjects carrying a mutation in the gene coding for βENaC. Some affected subjects, including one with a normal blood pressure, showed normal renin levels. The features that clearly discriminate among the subjects with and without the mutation causing the Liddle's syndrome, were the low 24 h urinary aldosterone levels and the blunted response of the plasma aldosterone to stimulation with ACTH (Findling et al 1997).

Complete normalization of blood pressure and hypokalemia could be observed after treatment with amiloride (Jeunemaitre et al 1997). In these cases, the treatment had almost no effects on plasma aldosterone levels, and resulted only in a partial correction of the plasma renin activity. This may represent what G.W. Liddle called a 'disuse atrophy' of the aldosterone secretory mechanism in adrenal glands, due to the chronic suppression of renin secretion, resulting in a progressive hypoplasia of the adrenal cortex, a down-regulation of angiotensin II receptors, and a resetting of the aldosterone secretion levels (Liddle et al 1963).

Besides the renal phenotype leading to alterations in the extracellular fluid and electrolyte balance, no clinical signs of altered functions in the lungs or exocrine glands have been reported in Liddle's patients. Besides the kidney, the ENaC is also expressed in the pulmonary epithelium, where it serves to maintain the composition and the level of the air-surface liquid that covers the bronchial epithelium. Gain of function of ENaC targeted in the lung results in mice in a cystic fibrosis-like syndrome, but a similar clinical feature has not been reported in patients with Liddle's syndrome.

DIFFERENTIAL DIAGNOSIS

Essential hypertension represents the most common form of hypertension (98% of the cases). Primary aldosteronism (Conn's syndrome) is common in hypertensive subjects: its prevalence increases with increasing the severity of hypertension and can reach approximately 20% among patients with resistant hypertension (Nishizaka & Calhoun 2005). By contrast to Liddle's syndrome, patients with primary aldosteronism exhibit an elevated serum aldosterone to plasma renin activity ratio. Genetic disorders with excess of mineralocorticoids due to disorders in the biosynthesis of steroid hormones, include glucocorticoid-remediable aldosteronism

(GRA), inactivating mutations of hydroxylase enzymes involved in cortisol biosynthesis (P450C11β deficiency), and congenital adrenal hyperplasia (Dluhy 2002).

In addition to Liddle's syndrome, hereditary renal disorders associated with elevated blood pressure, expansion of the extracellular fluid volume due to excessive distal Na^+ absorption include the 'apparent mineralocorticoid excess' (AME), 'the hypertension exacerbated in pregnancy', or the 'pseudohypoaldosteronism type II' (see for review Lifton et al (2001), Dluhy (2002), New et al (2005)).

AME occurs as an autosomal recessive disorder and results from mutations in 11β-hydroxysteroid dehydrogenase type 2 (11β-HSD2), an enzyme specifically expressed in the ASDN and essential for excluding cortisol from occupying the mineralocorticoid receptor. The 11β HSD2 enzyme converts cortisol, found at plasma concentrations more than 100-fold higher than those of aldosterone, into biologically inactive cortisone on the mineralocorticoid receptor. Loss of function mutations of 11β-HSD2, as found in AME, results from an overstimulation of the MR by cortisol, leading to an increase in Na^+ absorption, an elevated blood pressure with hypokalemia, low plasma levels of aldosterone and a suppressed plasma renin activity (Mune et al 1995). By contrast with patients with Liddle's syndrome, patients with AME have an increased urinary ratio of cortisol to cortisone and respond to treatment with spironolactone.

Hypertension exacerbated by pregnancy is an autosomal dominant disorder with hypertension, hypokalemia and suppressed plasma renin activity (Geller et al 2000). The causative mutation is found in the gene encoding the MR, leading to a constitutive activity of the MR receptor and to the activation of the MR by non-physiological ligands, such as progesterone or spironolactone.

Finally, pseudohypoaldosteronism type 2 (PHA-type 2), or Gordon's syndrome, is an autosomal dominant disorder with hypertension, hyperkalemia, hyperchloremia and suppressed plasma renin activity, and normal or elevated aldosterone levels. Hypertension and the electrolyte abnormalities are reversed by thiazide diuretics (Disse-Nicodeme et al 2001). Mutations in serine-threonine kinases WNK1 and WNK4 that control the activity of different renal transporters Na-Cl and ROMK as well as the ionic permeability of tight junctions in the ASDN, are responsible for PHA-type 2 (Wilson et al 2001).

PATHOPHYSIOLOGY

The functional expression of ENaC carrying the mutations causing the Liddle's syndrome in *Xenopus* oocytes and in renal epithelial cell lines, reproduces the abnormally high activity of ENaC, postulated by G.W. Liddle in his seminal paper on pseudoaldosteronism (Schild et al 1995, Snyder et al

1995, Auberson et al 2003). Indeed, deletion or missense mutations targeting the xPPPxY sequence (PY motif) in β or γ ENaC, increase up to 4-fold the activity of ENaC, as measured by the amiloride-sensitive Na$^+$ current. This higher activity expressed by ENaC channels with mutations in the PY motif at the plasma membrane is associated with a higher number of ENaC channels. However, the magnitude of the amiloride-sensitive current through these ENaC mutants increases more than the actual number of the mutant channels found at the cell surface. Since the single channel conductance was not altered in these ENaC mutants, it was concluded that the mutations in the PY motif of ENaC also increase the open probability of the channel. These observations indicate that the PY motif in the cytoplasmic carboxy-terminus of ENaC critically modulates both channel density and channel gating at the cell surface (Snyder et al 1995, Firsov et al 1996).

One reported case with the clinical features characteristic of the Liddle's syndrome was shown to harbor a gain of function mutation (γN530S) in the extracellular domain of γ ENaC, and no mutations in the PY motif (Hiltunen et al 2002). The Asn 530 to Ser substitution is located in a stretch of highly conserved amino acids preceding the second transmembrane segment, and was found to enhance ENaC activity in the *Xenopus* expression system, essentially by increasing the open probability of the channel A corresponding amino acids' substitution in the *C. elegans* ortholog of ENaC (MEC4, DEG-1 genes) and causes cell swelling and neuronal degeneration that was interpreted as a mutation causing channel gain of function (Driscoll & Chalfie 1991).

MUTATIONS OF THE PY MOTIF: A DUAL EFFECT ON ENaC FUNCTION

The tyrosine residue of the PY motif sequence (xPPxY) conforms a consensus internalization signal (NPxY) found in receptors removed from the plasma membrane via clathrin-coated pits. Evidence supporting the role of the clathrin-coated pit pathway for ENaC endocytosis comes from the ENaC retention at the cell surface obtained with the co-expression of a dominant-negative dynamin mutant (Shimkets et al 1997). Mutations in the PY motif decreasing ENaC internalization result in channel retention at the cell surface.

Furthermore, proline-rich motifs such as the PY motif are involved in specific protein–protein interactions. A binding partner for the PY motif of β ENaC (Staub et al 1996) was isolated from a rat cDNA library, using a two-hybrid screen and identified as a ubiquitin protein ligase called Nedd4. This first suggested that ubiquitylation was a possible mechanism for regulation of ENaC activity at the cell surface (Staub et al 1996). The function of ubiquitin ligase

proteins is to catalyze the final steps in the ubiquitylation cascade, in particular the attachment of ubiquitin moieties onto lysine residues. Ubiquitylation of ENaC could be demonstrated, as well as the endocytosis, and degradation of the channel by the lysosomal and proteasomal pathway (Staub et al 1997), further supporting a role of Nedd4 in regulating ENaC activity at the cell surface.

In a cortical collecting duct cell line (mpkCCD cells), it was later demonstrated that the likely candidate for interaction with ENaC is a homolog of Nedd4, called Nedd4-2 (Kamynina et al 2001). Nedd4-2 mRNA is found in brain, heart, liver, and also in the kidney, where it is expressed along the entire nephron. At the protein level, Nedd4-2 shows a high expression in the ASDN and an axial gradient of expression that is inversely correlated with the apical localization of ENaC (Loffing-Cueni et al 2006). Although Nedd4-2 is also found in cells that do not express ENaC, these observations suggest that Nedd4-2 contributes to the regulation of ENaC activity at the apical cell surface, where high abundance of Nedd4-2 decreases ENaC localization at the cell surface.

Coexpression of Nedd4-2 with ENaC in the *Xenopus* oocytes almost suppressed ENaC activity at the cell surface. Although direct evidence for ubiquitylation of ENaC in cells expressing robust ENaC-mediated Na$^+$ current (i.e. active ENaC at the plasma membrane) is still lacking, the contribution of Nedd4-2 in the pathogenesis of the Liddle's syndrome was suggested by the observation that a catalytically inactive Nedd4-2 obtained by site-directed mutagenesis, reproduced in *Xenopus* oocytes a similar stimulatory effect of ENaC activity as the one obtained by mutations in the ENaC PY motif (Abriel et al 1999). The prevailing idea underlying the molecular mechanisms causing the Liddle's syndrome is that mutations in the PY motif of ENaC, inhibit ENaC endocytosis and degradation by impairing the binding interaction with the ubiquitin ligase Nedd4-2, leading to an increased stability and therefore a longer half life of ENaC activity at the cell surface.

Nedd4-2 is composed of four WW domains which are the protein-protein interaction modules that bind the PY motif of ENaC β and γ subunits. WW domains, such as those found in Nedd4-2, recognize a large variety of proteins carrying proline-rich motifs. The specificity of the binding interaction between the WW domains of Nedd4-2 and the PY motif of ENaC has been addressed in functional experiments and in vitro binding assays. They show that the third WW (WW3) and the fourth WW (WW4) domains are required for the suppressor effect of Nedd4-2 on ENaC activity and that WW3 determines the high affinity of interaction with the ENaC PY motif (Kamynina et al 2001, Henry et al 2003). The structure of the WW3 domain of Nedd4 with the βENaC PY motif has been solved by NMR and reveals a three-stranded anti-parallel β-sheet with a hydrophobic binding surface (Kanelis et al 2001). The two prolines preceding the tyrosine of the PY motif form a polyproline type II helix and are accommodated in a groove of the WW3 domain. Substitution of the tyrosine residue 618 in β ENaC (Tyr-618 to Leu), as found in the Liddle's syndrome, disrupts the hydrophobic interaction essential for recognition between the PY motif and Nedd4. Thus, the solved structure of the Nedd4 WW3 domain can fairly well explain the impact of ENaC mutations in patients with the Liddle's syndrome on the binding interaction between Nedd4 and the PY motif of ENaC. The final demonstration that Nedd4-2 is the physiological partner of ENaC that regulates ENaC activity via interactions with the PY motif, awaits future genetic evidence in humans or in mouse models.

Recent evidence further support that mutations in the PY motif have a distinct effect on ENaC number and P_o, explaining the observation that the Liddle's mutations increased I_{Na} significantly more than the number of channels measured at the cell surface (Firsov et al 1996, Michlig et al 2004). By surface labeling of ENaC subunits, it was observed that not only the surface expression of α, β, and γ ENaC subunits were increased as expected, but, in addition, α and γ ENaC were cleaved, consistent with a proteolytic processing proposed by Hughey and colleagues (Hughey et al 2004, Knight et al 2006). Thus, the Liddle's syndrome mutations had two distinct effects on ENaC function at the cell surface: first, they increase the number of channel proteins and, second, they enhance the fraction of ENaC at the cell surface that are cleaved and active with a high P_o gating mode.

PY MOTIFS AND ALDOSTERONE RESPONSE

In mice, a low sodium diet associated with high plasma aldosterone levels reduces expression of Nedd4-2 in the ASDN and increases cell surface abundance of ENaC, suggesting that aldosterone regulates ENaC ubiquitylation and internalization via Nedd4-2 (Loffing-Cueni et al 2006). Such regulation of Nedd4-2 likely takes places during the late response to aldosterone (>3 hours). It has been reported that at the early response, aldosterone increases phosphorylation of Nedd4-2, impairing the binding interaction between Nedd4-2 and ENaC ((Loffing-Cueni et al 2006). These observations suggest that Nedd4-2, via its interaction with the PY motif of β and γ ENaC, contributes to the regulation of ENaC activity by aldosterone.

This hypothesis was further investigated since the serum and glucocorticoid-regulated protein kinase (SGK) induced during the early response to aldosterone contains in its amino-acid sequence a PY motif like ENaC that could potentially be involved in binding interactions with the WW domains of Nedd4-2. It was shown in the *Xenopus* oocyte expression system that Nedd4-2 interacts with SGK and that

this interaction requires the PY motif of SGK (Debonneville et al 2001, Snyder et al 2002). The Nedd4-2–SGK interaction leads to the phosphorylation of Nedd4-2 by the kinase, which then reduces the Nedd4-2 ability to interact with ENaC. The consequence is an increase in activity of ENaC at the cell surface. These experiments point to a possible contribution of the PY motif in increasing the number of active ENaC at the cell surface in response to aldosterone. According to this hypothesis, one would expect that mutations in the ENaC PY motif increase the basal activity of ENaC at the surface of principal cells, but make the channel less responsive to further stimulation by aldosterone. Thus, the principal cells in the ASDN of Liddle's patients are expected to behave as Na^+ absorbing cells constitutively activated, despite the absence of aldosterone, but with a reduced ability to further stimulate the Na^+ transport in the presence of aldosterone.

The contribution of the PY motif in β and γ ENaC in the ability of the channel to respond to aldosterone was investigated in a cortical collecting duct cell line (mpk-CCD cells) that responds to aldosterone by increasing an ENaC mediated transepithelial Na^+ transport (Bens et al 1999, Auberson et al 2003). These cells were engineered to express ENaC Liddle's mutants lacking the β and/or γ ENaC C-termini or substitution of Thyr in the β ENaC PY motif cells: as expected these cells used as a model of Liddle's syndrome showed a 3–4 fold increase in baseline amiloride-sensitive Na^+ current. In addition, the rate of the increase of Na^+ transport mediated by the ENaC mutants during the first 3 hours of aldosterone action, was identical to that mediated by ENaC wild type, indicating that mutations in the PY motif do not alter the ENaC response to aldosterone stimulation. Thus, mutations in the PY motif of ENaC do not impair the ability of the channel to respond to aldosterone, supporting the early observation by G.W. Liddle that affected patients with pseudoaldosteronism maintain their ability to respond to an acute aldosterone injection by decreasing the urinary sodium excretion to almost nil.

MOUSE MODELS OF LIDDLE'S SYNDROME

A mouse model of Liddle's syndrome was generated by homologous recombination (Pradervand et al 1999). Under a standard sodium diet, these mice have a normal blood pressure, as well as normal plasma electrolyte concentrations and acid–base status. In these mice, ENaC activity, as measured by the amiloride-sensitive rectal potential difference, was increased by approximately 60%. The lower plasma level of aldosterone was consistent with a significant degree of extracellular volume expansion. However, the hypokalemic hypertension, as in Liddle's syndrome, could be reproduced in these mice only under a high Na^+ diet.

Isolated cortical collecting duct principal cells from wild-type mice and homozygote Liddle's mutant mice (L/L mice), have ENaC sensitivity to aldosterone during the late response phase (Dahlmann et al 2003). Under a high salt diet and a low plasma aldosterone concentration, amiloride-sensitive whole cell current could not be recorded in principal cells by patch clamp technique either in wild-type or in Liddle's homozygote mice; this is probably due to the density of active ENaC wt as well as mutants in the apical membrane of single principal cells that is too low for ENaC mediated Na^+ currents to be detectable by the patch-clamp technique. When mice were placed under low Na^+ diet for 6–8 days or were chronically infused with aldosterone, ENaC-mediated Na^+ currents could be recorded in principal cells. Interestingly, the L/L mice exhibited considerably higher ENaC currents (up to five times), compared to wild-type mice. These observations show not only that ENaC mutants causing Liddle's syndrome retain their sensitivity to respond to aldosterone but also demonstrate that the ENaC response to an elevation of plasma aldosterone is exacerbated in Liddle's mice, in particular during the late aldosterone response.

From studies in aldosterone-responding cells and in animal models, it turns out that sensitivity to aldosterone of ENaC with mutations in the PY motif is conserved during the early response, and during the late phase of the aldosterone response, the activity at the cell surface of these ENaC mutants is considerably increased compared to ENaC wild type. This is consistent with the notion that mutations in PY motif of β and γ ENaC impair essentially the retrieval of active ENaC from the apical surface of principal cells, whereas the aldosterone-controlled targeting of the channels to the cell surface is only minimally affected by these mutations. During the late aldosterone response, ENaC with mutations in the PY motif are efficiently addressed at the cell surface and tend to accumulate at the apical membrane, potentiating the response to aldosterone in terms of the maximal effect (E_{max}) and duration (see Figure 16.2).

PATHOGENESIS

Experiments in heterologous expression systems, in aldosterone-responding epithelial cells or in mouse models, analyzing the function and regulation of ENaC with mutations causing Liddle's syndrome, have generated a considerable amount of knowledge: how is this relevant for our understanding of the pathogenesis of this syndrome in humans? Clearly the mutations in the PY motif of β or γ ENaC subunits associated with Liddle's syndrome result in ENaC gain of function at the apical surface of principal cells (Figure 16.2). The channel defect due to mutations of the PY motifs of β and γ ENaC subunits essentially involves a retention of hyperactive ENaC at the cell surface, most likely due to

an impaired internalization of the channels rather than an increase in the ENaC targeting at the cell surface. Mutations in the PY motif decrease the ability of ENaC to bind the ubiquitin ligase Nedd4-2. This interaction has two major consequences on channel function: first, it closes ENaC, second, it promotes ENaC internalization presumably mediated by clathrin-coated pits. The ubiquitylation likely targets ENaC for degradation in the proteasome and/or the lysosome, but it is not excluded that a significant fraction of the internalized channels are recycled to the plasma membrane.

A consequence of the retention of hyperactive ENaC mutants with a longer half-life at the cell surface, is that the biological effect of aldosterone is amplified and prolonged (Figure 16.2C, D). The clinical history of Liddle's syndrome patients shows that hypertension typically appears in adolescence, suggesting that several years are necessary for extracellular volume expansion and hypertension to develop. At an early stage of the disease, before the complete suppression of aldosterone secretion, it is possible that the abnormally high and prolonged Na^+ absorption in response to aldosterone contributes significantly in the development of volume expansion.

The abnormally high Na^+ absorption in the ASDN observed in Liddle's syndrome, despite low plasma levels of aldosterone, suggests that aldosterone is not absolutely needed for ENaC synthesis and expression at the cell surface to allow Na^+ absorption in the ASDN. In this respect, the knockout of the mineralocorticoid receptor in mice does not completely abolish ENaC-mediated amiloride-sensitive Na^+ absorption in the colon and the kidney, further supporting that ENaC is not exclusively controlled by the mineralocorticoid receptor (Berger et al 1998). It is also possible that in Liddle's syndrome, some degree of constitutive activity of MR receptors may also contribute to the renewal and maintenance constitutive pool of long-lasting hyperactive ENaC at the cell surface. These mechanisms likely result in a high and aldosterone independent Na^+ absorption in the ASDN observed in Liddle's syndrome, and certainly contribute to maintain a state of chronic volume expansion. To recapitulate Guyton's theory, in Liddle's syndrome with an abnormally high ENaC-mediated Na^+ absorption in the ASDN, a higher arterial blood pressure is needed to balance sodium intake and urinary sodium excretion, compared with normal persons. This difference is exacerbated when Na^+ intake is high, justifying the association of amiloride and a sodium diet in the treatment of Liddle's syndrome.

Like other hereditary forms of hypertension, such as hypertension exacerbated in pregnancy, AME or GRA, Liddle's syndrome beautifully demonstrates that the aldosterone-sensitive distal nephron is the most important site for the control of sodium, potassium and fluid balance by the kidney, in order to maintain the extracellular fluid at a constant level, and blood pressure within physiological ranges.

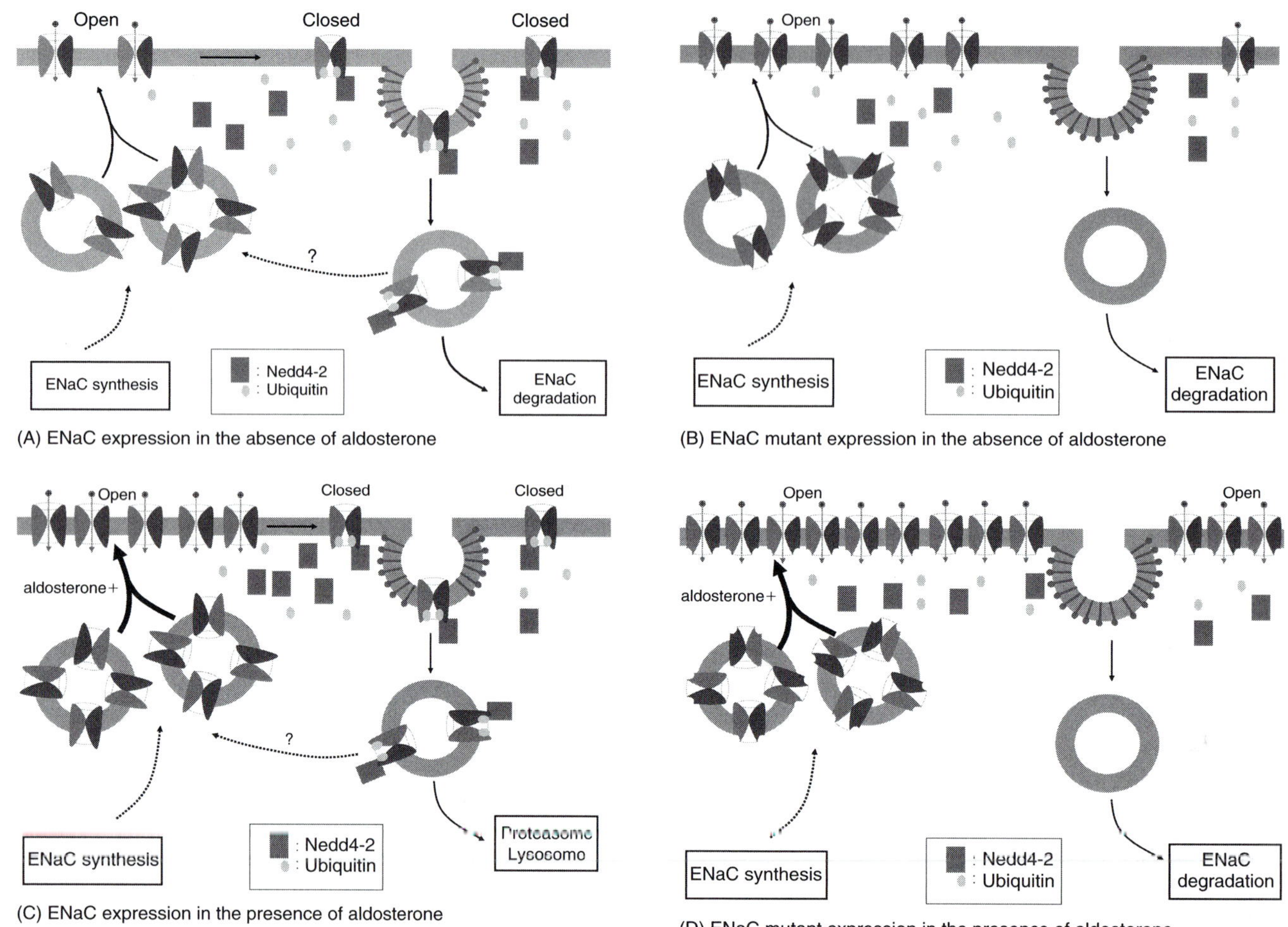

FIGURE 16.2 This figure illustrates, in principal cells of the ASDN, the apical and subapical pool of ENaC wt and ENaC with mutations causing Liddle's syndrome. (A, B): principal cells are not stimulated by aldosterone and a few ENaC wt (A) are active at the apical membrane with a 1/2 life of approximately 0.5 to 1 h; upon binding Nedd4-2 the ubiquitylated channels close, are internalized in clathrin-coated pits, and may either be degraded or recycled back to the plasma membrane. The ENaC mutants (B) lacking the PY motif have lost their ability to interact with Nedd4-2, remain open, and are less efficiently internalized in clathrin-coated pits, leading to the retention of active ENaC at the apical membrane. (C, D) In the presence of aldosterone the subapical pool of ENaC is translocated into the apical membrane, resulting in a higher number of active channels at the cell surface. Aldosterone may potentially also control ENaC removal from the cell surface by decreasing ENaC retrieval (C). ENaC lacking the PY motif are less efficiently removed from the apical membrane and tend to accumulate, leading to an increased and prolonged (>1/2 life of ENaC wt) Na^+ entry into the cell. (see also Plate 17)

References

Abriel H, Loffing J, Rebhun JF, et al. J. Clin. Invest. 1999; 103: 667–73.

Auberson M, Hoffmann-Pochon N, Vandewalle A, Kellenberger S, Schild L. Am. J. Physiol. 2003; 285: F459–71.

Bachmann S, Bostanjoglo M, Schmitt R, Ellison DH. Anat. Embryol. (Berl) 1996; 200: 447–68.

Bens M, Vallet V, Cluzeaud F, et al. J. Am. Soc. Nephrol. 1999; 10: 923–34.

Berger S, Bleich M, Schmid W, et al. Proc. Natl. Acad. Sci. USA 1998; 95: 9424–9.

Botero-Velez M, Curtis J, Warnock DG. N. Engl. J. Med. 1994; 330: 178–81.

Canessa CM, Schild L, Buell G, et al. Nature 1994; 367: 463–7.

Chraibi A, Horisberger AD. J. Gen. Physiol. 2002; 120: 133–45.

Dahlmann A, Pradervand S, Hummler E, Rossier BC, Frindt G, Palmer LG. Am J. Physiol. 2003; 285: F310–18.

Debonneville C, Flores SY, Kamynina E, et al. EMBO J. 2001; 20: 7052–9.

Disse-Nicodeme S, Desitter I, Fiquet-Kempf B, et al. J. Hypertens. 2001; 19: 1957–64.

Dluhy RG. Curr. Hypertens. Rep. 2002; 4: 439–44.

Driscoll M, Chalfie M. Nature 1991; 349: 588–93.

Findling JW, Raff H, Hansson JH, Lifton RP. J. Clin. Endocrinol. Metabol. 1997; 82: 1071–4.

Firsov D, Schild L, Gautschi I, Mérillat A-M, Schneeberger E, Rossier BC. Proc. Natl. Acad. Sci. USA 1996; 93: 15370–5.

Firsov D, Gautschi I, Merillat AM, Rossier BC, Schild L. . EMBO J. 1998; 17: 344–52.

Frindt G, Masilamani S, Knepper MA, Palmer LG. Am. J. Physiol. 2001; 280: F112–18.

Frindt G, Palmer LG. Am. J. Physiol. 2004; 286: F669–74.

Geller DS, Farhi A, Pinkerton N, et al. Science 2000; 289: 119–23.

Guyton AC. Hypertension 1987; 10: 1–6.

Hansson J, Nelson-Williams C, Suzuki H, et al. Nature Genet. 1995a; 11: 76–82.

Hansson JH, Schild L, Lu Y, et al. Proc. Natl. Acad. Sci. USA 1995b; 92: 11495–9.

Henry PC, Kanelis V, O'Brien MC, et al. J. Biol. Chem. 2003; 278: 20019–28.

Hiltunen TP, Hannila-Handelberg T, Petajaniemi N, Kantola I, Tikkanen I, Virtamo J, Gautschi I, Schild L, Kontula K. J. Hypert. 2002; 20: 2383–90.

Hughey R, Bruns JB, Kinlough CL, et al. J. Biol. Chem. 2004; 279: 18111–14.

Inoue J, Iwaoka T, Tokunaga H, et al. J. Clin. Endocrinol. Metabol. 1998; 83: 2210–13.

Jackson SJ, Williams B, Houtman P, Trembath RC. J. Med. Genet. 1998; 35: 510–12.

Jeunemaitre X, Bassilana F, Persu A, et al. J. Hypertens. 1997; 15: 1091–100.

Kamynina E, Debonneville C, Bens M, Vandewalle A, Staub O. FASEB J. 2001a; 15: 204–14.

Kamynina E, Tauxe C, Staub O. Am. J. Physiol. 2001b; 281: F469–77.

Kanelis V, Rotin D, Forman-Kay JD. Nat. Struct. Biol. 2001; 8: 407–12.

Kellenberger S, Auberson M, Gautschi I, Schneeberger E, Schild L. J. Gen. Physiol. 2001; 118: 679–92.

Kim GH, Masilamani S, Turner R, Mitchell C, Wade JB, Knepper MA. Proc. Natl. Acad. Sci. USA 1998; 95: 14552–7.

Knight KK, Olson DR, Zhou R, Snyder PM. Proc. Natl. Acad. Sci. USA 2006; 103: 2805–8.

Kyuma M, Ura N, Torii T, et al. Clin. Exper. Hypertens. 2001; 23: 471–8.

Liddle GW, Bledsoe T, Coppage WS. Trans. Assoc. Am. Physicians 1963; 76: 199–213.

Lifton RP. Science 1996; 272: 676–80.

Lifton RP, Gharavi AG, Geller DS. Cell 2001; 104: 545–56.

Loffing J, Pietri L, Aregger F, et al. Am. J. Physiol. 2000; 279: F252–8.

Loffing-Cueni D, Flores SY, Sauter D, et al. J. Am. Soc. Nephrol. 2006; 17: 1264–74.

Masilamani S, Kim GH, Mitchell C, Wade JB, Knepper MA. J. Clin. Invest. 1999; 104: R19–23.

Masilamani S, Wang XY, Kim GH, et al. Am. J. Physiol. 2002; 283: F648–57.

Melander O, Orho M, Fagerudd J, et al. Hypertension 1998; 31: 1118–24.

Michlig S, Mercier A, Doucet A, et al. J. Biol. Chem. 2004; 279: 51002–12.

Mune T, Rogerson FM, Nikkila H, Agarwal AK, White PC. Nat. Genet. 1995; 10: 394–9.

Nakano Y, Ishida T, Ozono R, et al. J. Hypertens. 2002; 20: 2379–82.

New MI, Geller DS, Fallo F, Wilson RC. Trends Endocrinol. Metab. 2005; 16: 92–7.

Nishizaka MK, Calhoun DA. Curr. Cardiol. Rep. 2005; 7: 412–17.

Palmer LG, Frindt G. Am. J. Physiol. 1987; 253: F333–9.

Palmer LG, Frindt G. J. Gen. Physiol. 1996; 107: 35–45.

Pradervand S, Wang Q, Burnier M, et al. J. Am. Soc. Nephrol. 1999; 10: 2527–33.

Reilly RF, Ellison DH. Phys. Rev. 2000; 80: 277–313.

Schild L, Canessa CM, Shimkets RA, Gautschi I, Lifton RP, Rossier BC. Proc. Natl. Acad. Sci. USA 1995; 92: 5699–703.

Schild L, Schneeberger E, Gautschi I, Firsov D. J. Gen. Physiol. 1997; 109: 15–26.

Shimkets RA, Warnock D, Bositis CM, et al. Cell 1994; 79: 407–14.

Shimkets RA, Lifton RP, Canessa CM. J. Biol. Chem. 1997; 272: 25537–41.

Snyder PM, Price MP, McDonald FJ, et al. Cell 1995; 83: 969–78.

Snyder PM, Olson DR, Thomas BC. J. Biol. Chem. 2002; 277: 5–8.

Staub O, Dho S, Henry PC, et al. EMBO J. 1996; 15: 2371–80.

Staub O, Gautschi I, Ishikawa T, et al. . EMBO J. 1997; 16: 6325–36.

Tamura H, Schild L, Enomoto N, et al. J. Clin. Invest. 1996; 97: 1780–4.

Uehara Y, Sasaguri M, Kinoshita A, et al. J. Hypertens. 1998; 16: 1131–5.

Velazquez H, Bartiss A, Bernstein P, Ellison DH. Am. J. Physiol. 1996; 270: F211–19.

Wilson FH, Disse-Nicodeme S, Choate KA, et al. Science 2001; 293: 1107–12.

Yamashita Y, Koga M, Takeda Y, et al. Am. J. Kidney Dis. 2001; 37: 499–504.

CHAPTER 17

The Syndrome of Apparent Mineralocorticoid Excess

PERRIN C. WHITE

CLINICAL FEATURES OF THE SYNDROME OF APPARENT MINERALOCORTICOID EXCESS (AME)

Clinical Presentation

A hypertensive syndrome was first described 30 years ago in which children presented with hypertension, hypokalemia, and reduced plasma renin activity, all signs of mineralocorticoid excess. A low salt diet or blockade of mineralocorticoid receptors with spironolactone ameliorated the hypertension, whereas ACTH exacerbated it. Therefore, it was initially suspected that these patients had elevated levels of an unknown ACTH-inducible mineralocorticoid, but levels of all known mineralocorticoids were low (Werder et al 1974, New et al 1977, Ulick et al 1977). For this reason, this was termed a syndrome of 'apparent mineralocorticoid excess' (AME) ((Ulick et al 1979), reviewed in White et al (1997)). This disorder could be distinguished clinically from the similar disorder, Liddle's syndrome, by the therapeutic response to spironolactone. Patients with Liddle's syndrome do not respond to blockade of the mineralocorticoid receptor but can be treated by direct blockade of the renal tubular sodium channel with triamterine or amiloride. It is now known that Liddle's syndrome is caused by mutations in the β or γ subunits of the sodium channel, leading to constitutive activation of this channel (see Chapter 16).

Other clinical features seen in patients with AME include moderate intrauterine growth retardation and postnatal failure to thrive. Hypokalemia is often more severe than that seen in primary aldosteronism, with consequences including nephrocalcinosis and nephrogenic diabetes insipidus; the latter condition causes polyuria and polydipsia. Hypokalemia also has deleterious effects on muscles ranging from elevations in serum creatine phosphokinase to frank rhabdomyolysis.

In some patients, the hypertension tends to be labile or paroxysmal with severe emotional stress as a precipitating factor. Complications of hypertension have included cerebrovascular accidents. Several patients have died during infancy or adolescence. The reasons for their demise are not known, but might include either electrolyte imbalances leading to cardiac arrhythmias or vascular sequelae of hypertension.

Several affected sibling pairs have been reported but parents have usually been asymptomatic, consistent with a genetic disorder with an autosomal recessive mode of inheritance. However, occasional abnormalities have been observed in parents including hypertension and hypokalemic alkalosis (Stewart et al 1988). Approximately 50 kindreds have been reported worldwide (Werder et al 1974, New et al 1977, Winter & McKenzie 1977, Igarashi et al 1979, Ulick et al 1979, Shackleton et al 1980, Fiselier et al 1982, Honour et al 1983, Harinck et al 1984, Shackleton et al 1985, Monder et al 1986, Batista et al 1986, Dimartino-Nardi et al 1987, Stewart et al 1988, Milford et al 1995, Mune et al 1995, Wilson et al 1995a, 1995b, 1998, Kitanaka et al 1996, Stewart et al 1996, Mune & White 1996, Li et al 1997, 1998, Nunez et al 1999, Odermatt et al 2001, Carvajal et al 2003, Lavery et al 2003, Lin-Su et al 2004, Quinkler et al 2004). A clinically milder form of the syndrome has been referred to as 'type 2' AME (Mantero et al 1994) but it is merely an allelic variant and this term is not widely used.

Biochemical Findings

Patients with AME have abnormal cortisol metabolism (Figure 17.1). Plasma cortisol half life is prolonged from the normal of approximately 80 minutes to 120–190 minutes (Ulick et al 1979). This is due to a marked deficiency in 11β-dehydrogenation. In research studies, very low levels of cortisone metabolites are excreted when labeled cortisol is administered. However, 11-reduction is unimpaired; labeled cortisone is excreted entirely as cortisol and other 11β-reduced metabolites (Harinck et al 1984, Batista et al 1986, Shackleton et al 1985). Total urinary excretion of cortisol metabolites is markedly decreased but serum cortisol levels are normal due to its prolonged serum half-life.

Genetic Diseases of the Kidney
ISBN: 978-0-12-449851-8

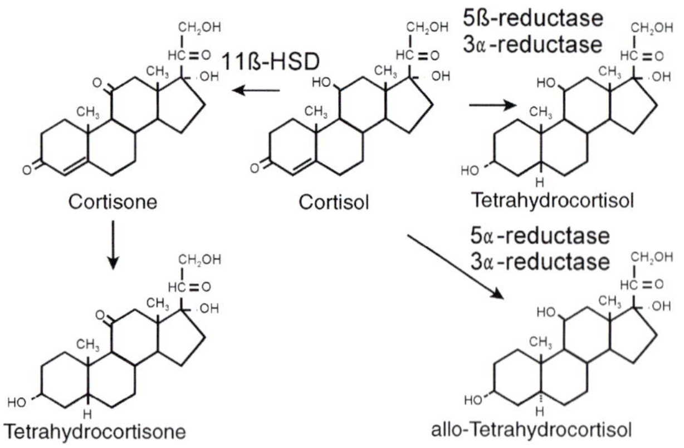

FIGURE 17.1   Metabolic pathway of cortisol

These findings suggest that cortisol itself acts as a strong mineralocorticoid in these patients. Indeed, administering hydrocortisone (i.e. cortisol) aggravates the hypertension and hypokalemia in a dose-dependent manner (Oberfield et al 1983). In contrast, administering aldosterone has no effect on blood pressure, presumably because mineralocorticoid receptors are fully occupied by cortisol.

In practice, the biochemical parameter most frequently used to diagnose this disorder has been a precursor:product ratio reflecting 11β-dehydrogenase activity — the sum of the urinary concentrations of tetrahydrocortisol and allo-tetrahydrocortisol, divided by the concentration of tetrahydrocortisone, abbreviated (THF + aTHF)/THE. This is invariably elevated in patients with AME. The ratio of free urinary cortisol to free urinary cortisone (F/E) is also decreased (Palermo et al 1996). The relative activity of 5β-reductase, measured by the ratio of the 5α reduced to 5β reduced metabolites (aTHF/THF) is usually abnormal as well (Bournot et al 1982, Monder et al 1986). However, aTHF/THF and (THF + aTHF)/THE are linearly related and the latter ratio is usually much higher, implying that the primary defect in this disorder is indeed one of 11β-dehydrogenation.

Treatment

A low salt diet, potassium supplementation, and spironolactone have been the most often used therapies for AME. Triamterine and amiloride, other potassium-sparing diuretics, have also been successfully used, particularly because it is difficult to distinguish AME from Liddle's syndrome without specialized steroid assays. These two agents block the epithelial sodium channel. The hypertension of AME has also been treated with nifedipine. Although ACTH exacerbates the signs of this disorder, suppressing ACTH secretion with dexamethasone has been effective in ameliorating hypokalemia and hypertension only in mild cases.

The lack of efficacy of this agent may be related to incomplete suppression of cortisol secretion (Igarashi et al 1979). Angiotensin-converting enzyme inhibitors such as captopril enhance renal 11-HSD activity and might theoretically be useful in patients with partially active enzymes (Riddle & McDaniel 1994).

Licorice Intoxication

Clinical and biochemical abnormalities that are similar to, but milder than, those seen in AME occur with licorice intoxication (Stewart et al 1987). The active component of licorice, glycyrrhetinic acid, inhibits 11-HSD in isolated rat kidney microsomes (Monder et al 1989). Comparable effects are seen with carbenoxolone, the hemisuccinate derivative of glycyrrhetinic acid, which has been used in Europe and Japan for treatment of ulcers. Thus, it appears that licorice intoxication is a reversible pharmacological counterpart to the inherited syndrome of apparent mineralocorticoid excess.

Why Does 11-HSD Deficiency Cause Hypertension?

Aldosterone acts through transcriptional effects mediated by a specific nuclear receptor referred to as the mineralocorticoid or 'type 1 steroid' receptor. These receptors are expressed at high levels in renal distal tubules and cortical collecting ducts but also in other mineralocorticoid target tissues, including salivary glands and the colon. The mineralocorticoid receptor has a high degree of sequence identity with the glucocorticoid or 'type 2' receptor (Arriza et al 1987), and it has very similar in vitro binding affinities for aldosterone and for glucocorticoids such as corticosterone and cortisol (Krozowski & Funder 1983, Arriza et al 1987).

It has been proposed (Stewart et al 1987, Edwards et al 1988, Funder et al 1988) that oxidation by 11-HSD of cortisol or corticosterone to cortisone or 11-dehydrocorticosterone, respectively, represents the most important physiological mechanism conferring specificity for aldosterone upon the mineralocorticoid receptor (Figure 17.2). Although cortisol and corticosterone bind the receptor well in vitro, cortisone and 11-dehydrocorticosterone are poor agonists for this receptor. Aldosterone is a poor substrate for 11-HSD because, in solution, its 11-hydroxyl group is normally in a hemiacetal conformation with the 18-aldehyde group. Thus, in AME or licorice intoxication, 11-HSD deficiency permits cortisol to occupy the mineralocorticoid receptor. Because cortisol normally circulates at levels 100–1000 times those of aldosterone, this leads to signs of mineralocorticoid excess even though aldosterone secretion is suppressed.

This hypothesis has been confirmed by molecular genetic studies (see later).

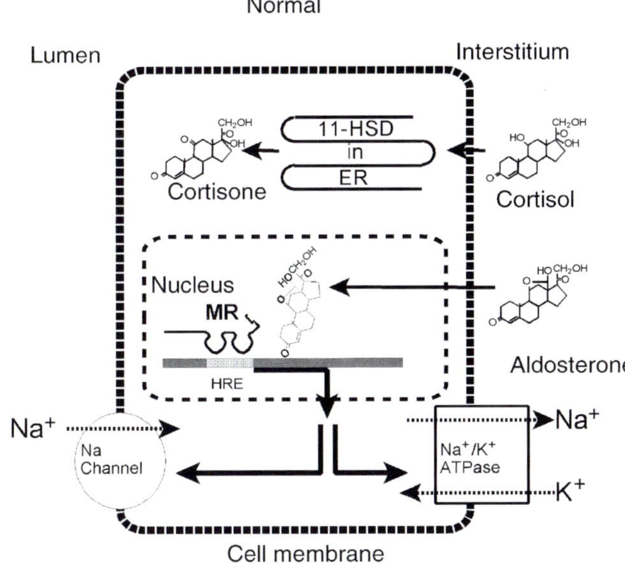

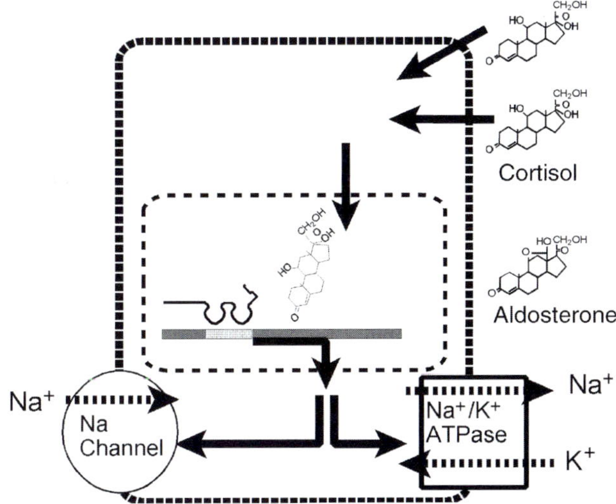

FIGURE 17.2 Schematic of mineralocorticoid action. Top, a normal mineralocorticoid target cell in a renal cortical collecting duct. Aldosterone occupies nuclear receptors (MR) that bind to hormone response elements (HRE), increasing transcription of genes and directly or indirectly increasing activities of apical sodium (Na) channels and the basolateral sodium–potassium (Na/K)-ATPase. This increases resorption of sodium from and excretion of potassium into the tubular lumen. Cortisol, which circulates at higher levels than aldosterone, cannot occupy the receptor because it is oxidized to cortisone by 11β-hydroxysteroid dehydrogenase(11-HSD) in the endoplasmic reticulum (ER). Bottom, a cell from a patient with the syndrome of apparent mineralocorticoid excess. Because 11-HSD is absent, cortisol inappropriately occupies mineralocorticoid receptors, leading to increased gene transcription, increased activity of sodium channels and the Na/K ATPase, increased resorption of sodium and excretion of potassium, and hypertension

TABLE 17.1 Characteristics of 11β-hydroxysteroid dehydrogenase isozymes and genes

	Liver (11-HSD1)	Kidney (11-HSD2)
Highest expression	Liver, adipose	Kidney, placenta
Cofactor dependence	NADPH	NAD$^+$
Catalyzes reversible reactions?	Yes	No
K_m for steroids	1 μM	1–50 nM
# of amino acids	291	405
MW, kDa	34	41
Chromosomal location	1q32.2	16q22.1
# of exons in gene	6	5
Involvement in AME	No	Yes

MOLECULAR STUDIES OF 11β-HYDROXYSTEROID DEHYDROGENASE (11-HSD)

Cloning of cDNA Encoding Isozymes of 11-HSD

There are two distinct isozymes of 11-HSD (Table 17.1). The first, termed the liver or 11-HSD1 isozyme, was originally isolated from rat liver microsomes (Lakshmi & Monder 1988). It is a glycoprotein with a molecular weight of 34 kDa that requires NADP$^+$ as a cofactor. The full length cDNA predicts a protein of 292 amino acids (Agarwal et al 1989). The recombinant enzyme expressed from cloned cDNA exhibits both 11β-dehydrogenase and the reverse oxoreductase activity (conversion of 11-dehydrocorticosterone to corticosterone) when expressed in mammalian cells (Agarwal et al 1989); it seems that this enzyme functions primarily as a reductase in vivo. The corresponding human gene for this isozyme, *HSD11B1*, is located on chromosome 1, and contains six exons with a total length of over 9 kb (Tannin et al 1991). No mutations were identified in this gene in patients with AME (Nikkila et al 1993). Instead, 11-HSD1 isozyme serves to modulate active glucocorticoid levels in target tissues such as the liver and adipose tissue, and it is now considered a drug target for treatment of the metabolic syndrome (reviewed in Tomlinson et al (2004)).

Accordingly, a second isozyme was sought in mineralocorticoid target tissues. Biochemical studies of isolated rabbit kidney cortical collecting duct cells demonstrated 11-HSD activity in the microsomal fraction that was almost exclusively NAD$^+$ dependent and had a high affinity for corticosterone (K_m of 26 nM) (Rusvai & Naray-Fejes-Toth 1993). There was almost no reduction of 11-dehydrocorticosterone to corticosterone, suggesting that, unlike 11-HSD1, the kidney or 11-HSD2 isozyme only catalyzed dehydrogenation. The enzyme in the human placenta had similar characteristics (Brown et al 1993).

The cDNA encoding this isoform was isolated by expression cloning techniques (Agarwal et al 1994,

Albiston et al 1994). Recombinant 11-HSD2 has properties that are virtually identical to the activity found in mineralocorticoid target tissues: it functions exclusively as a dehydrogenase, has an almost exclusive preference for NAD^+ as a cofactor and has a high affinity for glucocorticoids. Corticosterone is the preferred substrate, with first-order rate constants ten times higher than those for cortisol, even in mammalian species in which cortisol is the predominant glucocorticoid. Reported K_m values are ~10 and ~50 nM for corticosterone and cortisol, respectively.

The protein is predicted to have a M_r ~41 000. It is a member of the short-chain alcohol dehydrogenase superfamily, but is only 21% identical to 11-HSD1.

Regions of sequence similarity between the two isozymes include part of the putative binding site for the nucleotide cofactor (residues 85–95 in 11-HSD2) and absolutely conserved tyrosine and lysine residues (Y232 and K236) that function in catalysis (Persson et al 1991, Obeid & White 1992). The region immediately to the N-terminal side of the catalytic residues forms part of a putative steroid-binding pocket in a related short chain dehydrogenase that has been analyzed by X-ray crystallography, 3α,20β-HSD (Ghosh et al 1991). This region is notably well conserved (10/18 identical residues) between the two isozymes of 11-HSD, consistent with a role in binding the substrate.

Hydropathicity plots suggest that 11-HSD2 has three successive hydrophobic segments of approximately 20 amino acids each in the N-terminal region prior to the cofactor binding domain (Agarwal et al 1994). These function as transmembrane segments anchoring the enzyme to the membrane of the endoplasmic reticulum and control its orientation; the catalytic domain of 11-HSD2 is cytoplasmic, whereas the corresponding domain of 11-HSD1 is luminal. Orientation is critical for function; a chimeric 11-HSD2 enzyme with the N-terminus of 11-HSD1 has an inverted orientation but is catalytically inactive (Odermatt et al 2001). This may reflect availability of reduced and oxidized cofactors in different cellular compartments. 11-HSD1 functions as a reductase due to high levels of NADPH within the lumen of the endoplasmic reticulum which are generated by activity of a specific lumenal enzyme, hexose 6-phosphate dehydrogenase (Draper et al 2003). Orientation of 11-HSD2 into the cytoplasm may allow it to interact physically with the mineralocorticoid receptor and more efficiently regulate access of glucocorticoids to this receptor (Odermatt et al 2001).

THE *HSD11B2* GENE

The *HSD11B2* gene encoding 11-HSD2 contains five exons spaced over approximately 6.2 kb (Agarwal et al 1995) (Figure 17.3). The putative binding site for the NAD^+ cofactor (including the core sequence, GxxxGxG) is split

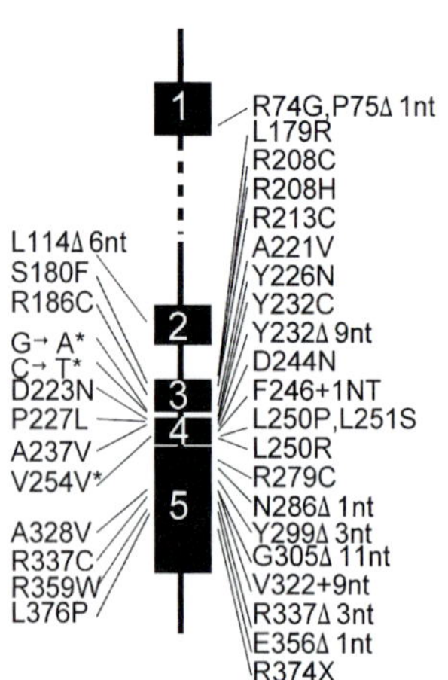

FIGURE 17.3 Locations of mutations in *HSD11B2* causing apparent mineralocorticoid excess. Intron 1 is not drawn to scale. Mutants with >5% of wild-type enzymatic activity when expressed in vitro are on the left side of the figure, as are mutations that affect pre-mRNA splicing (denoted by asterisks)

between exons 1 and 2, whereas the putative catalytic residues, Y232 and K236, are encoded by exon 4. This structure is different from that of the *HSD11B1* gene encoding 11-HSD1 (Tannin et al 1991), indicating that the two isozymes belong to different gene families.

The 5′ flanking region and the first exon are rich in CpG dinucleotides (Agarwal et al 1995). Using luciferase reporter constructs, the region from −2 to −330 nt relative to the initial ATG codon has been identified as an essential region for basal transcription of *HSD11B2* in JEG-3 human choriocarcinoma cells (Agarwal & White 1996) and renal cell lines (Agarwal 2000). Although the sites identified vary depending on cell line studied and use of in vitro (Agarwal & White 1996) or in vivo (Nawrocki et al 2002) footprinting methods, several consensus binding sites for the Sp1 family of transcription factors are apparently important for gene expression, including sites in the 5′ flanking region and in the 5′ untranslated region of exon 1. Both Sp1 and Sp3 bind to these sites. Because Sp1 and Sp3 are ubiquitous proteins, it is not obvious how interactions with these proteins could permit gene regulation in a tissue-specific manner. However, it appears that HSD11B2 is subject to epigenetic regulation. Tissue-specific methylation of CpG dinucleotides is inversely correlated with gene expression, and inhibitors of DNA methylases increase HSD11B2 expression both in cultured cells and in vivo. This raises the intriguing possibility that individual variations in methylation of HSD11B2 could represent a risk factor for development of hypertension (Alikhani-Koopaei et al 2004).

Expression of 11-HSD2

The tissue distribution of expression of 11-HSD2 has been examined by RNA blot hybridization in human adults (Albiston et al 1994) and fetuses (Casey et al 1994, Agarwal et al 1995) and several other mammals (Agarwal et al 1994, Zhou et al 1995, Cole 1995). In all species, this isozyme is expressed in placenta and mineralocorticoid target tissues, particularly the kidney, whereas it is not detected in the liver, heart, or adult testis. It is expressed at high levels in sheep and rat adrenal glands, but is not detected in the mouse or in human fetal adrenals.

Expression within mineralocorticoid target tissues has been further localized by immunohistochemistry and in situ hybridization. In the kidney, 11-HSD2 is expressed predominantly in distal convoluted tubules and cortical collecting ducts and is thus colocalized with the mineralocorticoid receptor (Naray-Fejes-Toth & Fejes-Toth 1995, Cole 1995, Roland et al 1995a, Whorwood et al 1995, Krozowski et al 1995, Kyossev et al 1996), although expression in glomeruli has also been reported (Kataoka et al 2002). In salivary glands, it is expressed in tubular elements with no or minimal expression in acini (Roland & Funder 1996). In the colon, it is found in the mucosa but not in the submucosa or the muscularis (Roland & Funder 1996, Kyossev et al 1996). Among tissues that are not classic targets of mineralocorticoids, 11-HSD2 is strongly expressed in the syncytiotrophoblast of the placenta (Krozowski et al 1995, Roland & Funder 1996). In the rat (Roland & Funder 1996) and sheep (Yang & Matthews 1995) adrenal cortex, it is expressed in the zona fasciculata and to a lesser extent in the zona reticularis, but not in the zona glomerulosa. In the rat ovary, it is expressed in the corpus luteum but not in follicles, and it is expressed in stromal cells in the rat oviduct and the lamina propria of the uterus.

Within the brain, this isozyme is expressed in the commissural portion of the nucleus tractus solitarius, subcommissural organ, and ventrolateral and ventromedial hypothalamus (Roland et al 1995b). The role(s) of this isozyme in these locations is/are uncertain. Lesions of the commissural portion of the nucleus tractus solitarius rapidly cause hypertension, suggesting that this structure regulates the cardiovascular system. The function of the subcommissural organ is not known with certainty. It also contains angiotensin receptors (but not mineralocorticoid receptors) and it may help regulate salt and water balance.

MOLECULAR GENETIC ANALYSIS OF APPARENT MINERALOCORTICOID EXCESS

Mutations in *HSD11B2* are Detected in All AME Patients

Mutations in HSD11B2 have been detected on both chromosomes in all AME patients examined thus far, thus confirming in its entirety the hypothesis that 11-HSD protects the mineralocorticoid receptor from high concentrations of cortisol (Mune et al 1995, Wilson et al 1995a, 1995b, 1998, Stewart

et al 1996, Mune & White 1996, Li et al 1997, 1998, Nunez et al 1999, Odermatt et al 2001, Carvajal et al 2003, Lavery et al 2003, Lin-Su et al 2004, Quinkler et al 2004) (Figure 17.3).

All but two reported mutations cluster in exons 3–5, a region representing 60% of the coding sequence (Agarwal et al 1995, Nunez et al 1999). This is unlikely to occur by chance, but an explanation for it is not immediately obvious. Although approximately half of AME mutations are CpG to TpG changes, exon 1 has a higher proportion of CpG dinucleotides (20% vs. 4%) than the remainder of the gene. However, CpG to TpG mutations occur by deamination of methylcytosine, so hypomethylation of the 5′ end of the gene in germ cells may partially explain the observed distribution of mutations. Additionally, several codons including R208, L250, and R337 have undergone several independent mutations as evidenced both by occurrence in unrelated ethnic groups and by the presence of different mutations in each codon, suggesting that these codons are mutational hotspots.

Most patients have carried homozygous mutations; few patients have been compound heterozygotes for different mutations. This suggests that the prevalence of AME mutations in the general population is low, so that the disease is found mostly in limited populations in which inbreeding is relatively high. For example, three Zoroastrian kindreds from India and Iran are all homozygous for the same mutation (R337Δ3nt).

It is noteworthy that six kindreds are of Native American origin. Three from either Minnesota or Canada carry the same mutation (L250S, L251P), consistent with a founder effect, but the others are each homozygous for a different mutation. The reason for the relatively high prevalence of this very rare disease among Native Americans is not immediately apparent. It might represent a bias in access to physicians with an interest in, and ability to make, this diagnosis as compared with similarly inbred populations elsewhere in the world. Alternatively, heterozygosity for mutations in *HSD11B2* might confer a selective advantage. Few obligate heterozygotes have been studied in detail, and it is possible that such individuals have an increased ability to conserve sodium under conditions of extreme sodium deprivation. Indeed, there have been anecdotal reports of heterozygotes with mild-to-moderate hypertension (Li et al 1997, Lavery et al 2003) but the numbers are so small that this might be due to chance.

Mouse Model of Apparent Mineralocorticoid Excess

To further confirm the role of HSD11B2 mutations in causing AME, this gene was knocked out in mice. Mutant mice recapitulate AME with hypokalemia that is fatal in 50% of newborns, hypertension and nephrogenic diabetes insipidus. The epithelium of the distal tubule is hyperplastic and hypertrophic, and these changes are not reversed upon treatment with mineralocorticoid receptor blockers (Kotelevtsev et al 1999).

Functional Effects of Mutations

Of the human mutations identified thus far, five shift the reading frame of translation, one is a nonsense mutation, and another deletes three amino acids including a crucial catalytic residue (Y232). These mutations are all presumed to completely destroy enzymatic activity. One mutation in the third intron leads to skipping of the fourth exon during processing of pre-mRNA (Mune et al 1995), another leads to utilization of a cryptic splice site in exon 3 with a shift of the reading frame, and a mutation in exon 4 (V254V) does not change the coding sequence but instead activates a cryptic splice site and also shifts the reading frame (Lavery et al 2003). As the fourth exon encodes the catalytic site, the resulting enzymes are all presumably inactive. The remaining mutations are missense mutations or small deletions that maintain the reading frame of translation. Several of these affect residues known to be involved in binding the nucleotide cofactor or in catalysis, but the reasons why others affect enzymatic activity is not immediately apparent. Several mutations are partially or fully active in whole cells (Wilson et al 1998, Nunez et al 1999) but have decreased enzymatic activity in cell lysates, suggesting that they affect stability of the enzyme. Consistent with this, decreased amounts of these mutant proteins are detected on Western blots (Mune & White 1996, Nunez et al 1999). It is not surprising that altered stability of certain mutant enzymes could adversely affect activity in vivo but not in cultured cells. Presumably 11-HSD2 is synthesized at lower levels in renal tubular epithelial cells than in cultured cells transfected with a high-level expression vector, and steady-state levels of the enzyme may be more sensitive to its rate of degradation when levels of synthesis are relatively low.

Correlations Between Genotype and Biochemical Phenotype

In many inherited diseases, the nature of the particular mutations causing the disease is the best predictor of disease severity, especially when the latter is assessed with biochemical measurements of protein function (i.e. intermediate phenotypes). Product:precursor ratios (e.g. for 11-HSD2, THE/[THF + aTHF], a ratio of cortisone to cortisol metabolites in the urine) are expected to be directly correlated with enzymatic activity. In contrast, precursor:product ratios should be inversely correlated, making it more difficult to fit the data to a linear relationship.

Indeed, when activity of mutant 11-HSD2 isozymes is measured by transfecting cDNA constructs into mammalian cells, THE/(THF + aTHF) is highly correlated with percent of wild-type enzymatic activity, particularly when cortisol

is used as the substrate (Figure 17.4) (Nunez et al 1999). It should be noted that the intercept of the regression line in this analysis differs significantly from 0; an average of 5% of the normal amount of cortisone metabolites is produced in patients who are predicted to completely lack 11-HSD2 activity. In such patients, an enzyme other than 11-HSD2 must produce cortisone. The most likely candidate is the liver isozyme of 11-HSD, 11-HSD1, which functions predominantly as a reductase in vivo but catalyzes both oxidation and reduction when expressed in cultured cells (Agarwal et al 1990).

The ratio of free urinary cortisone to free urinary cortisol (E/F) (Palermo et al 1996) is also highly correlated with enzymatic activity using cortisol as substrate (Figure 17.4). This ratio seems to have equivalent validity to the THE/(THF + aTHF) ratio for evaluation of AME.

Correlations Between Genotype and Clinical Phenotype

Two parameters of clinical severity are strongly correlated with whole-cell enzymatic activity (Figure 17.4), namely birth weight and the age at which each patient was diagnosed (Nunez et al 1999). The correlation with birth weight is presumably a consequence of the normal expression of 11-HSD2 in the placenta, where it can metabolize maternal glucocorticoids and thus regulate fetal glucocorticoid levels.

Serum potassium levels are weakly correlated with genotype.

Because blood pressure varies with age, this parameter is best expressed as a Z-score (number of standard deviations above the mean for age) using published normative data (Task Force on Blood Pressure Control in Children 1987). However, no correlation between genotype and either systolic or diastolic blood pressure can be established; all patients are markedly hypertensive. This may reflect the inherent 'noisiness' of random blood pressure measurements compared with many other clinical parameters, or the pathophysiology of AME may result in maximal stimulation of mineralocorticoid receptors with even moderate compromise of 11-HSD2 enzymatic activity.

Reduced 11-HSD2 Activity and Essential Hypertension

Because the hypertension in AME is salt-sensitive, it is reasonable to speculate that reduced activity of 11-HSD2 from whatever cause could increase the sensitivity of blood pressure to salt loading in otherwise normal individuals and/or be a risk factor for developing hypertension. Altered ratios of cortisol to cortisone metabolites (THF + aTHF/THE) have indeed been associated with sensitivity in some (Lovati et al 1999, Ferrari et al 2001) although not other (Kerstens et al 2004, Melander et al 2003) studies. Moreover, 11-HSD2

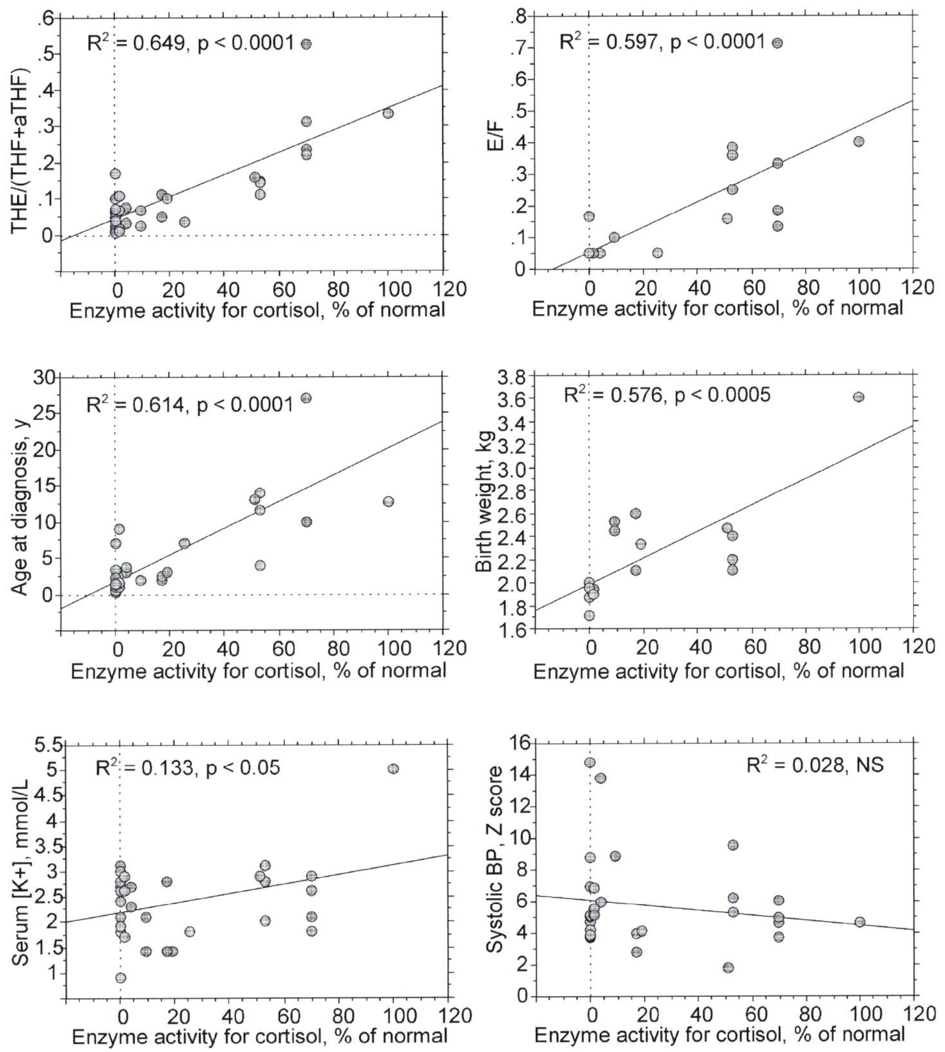

FIGURE 17.4 Biochemical and clinical parameters in AME patients correlated with enzymatic activity in intact cells using cortisol as the substrate (Nunez et al 1999). Each circle represents a single patient. THE/(THF + aTHF), ratio of urinary tetrahydrocortisone to the sum of tetrahydrocortisol plus allo-tetrahydrocortisol. E/F, ratio of urinary free cortisone to free cortisol. The systolic BP Z-score represents the number of standard deviations by which the blood pressure exceeds the mean for age. The diastolic BP (not shown) is also not correlated with enzymatic activity

activity is reduced in sweat glands in skin biopsies from hypertensive patients (Bocchi et al 2004).

Coding sequence mutations that mildly affect enzymatic activity (i.e. a forme fruste of AME) are the most obvious way in which 11-HSD2 activity could be reduced (Wilson et al 1998, Li et al 1998). However, AME clearly behaves as an autosomal recessive disease in the great majority of kindreds. Thus, it will be very rare unless there are frequent unsuspected mutations in the population. This is also unlikely; most patients including those who have mild disease are homozygous for a single mutation and come from relatively inbred populations, suggesting that matings between two carriers are rare in the general population. Moreover, there are few coding sequence polymorphisms in normal individuals (Smolenicka et al 1998, Zaehner et al 2000, Poch et al 2001). Thus routine genetic characterization of this

locus in patients with low renin essential hypertension cannot be recommended, although it may be worthwhile in individuals with a suggestive history and altered cortisol: cortisone metabolite ratios.

Polymorphisms outside of the coding sequence that affect gene expression have also been sought. The most extensively studied is a CA repeat polymorphism in the first intron. A minigene construct containing a 'short' allele (few repeats) is expressed at higher levels than a construct with a 'long' allele (large number of repeats) when transfected into cultured cells. However, the polymorphism has little effect on urinary cortisone:cortisol metabolite ratios, and shorter alleles are associated with increased salt-sensitivity in normal or hypertensive individuals, which is opposite to what would be expected if short alleles increased gene expression in vivo as they do in vitro (Agarwal et al 2000).

Other studies have likewise failed to demonstrate any association between polymorphisms in HSD11B2 and essential hypertension (Smolenicka et al 1998, Sugiyama et al 2001).

References

Agarwal AK. Expression of HSD11K (NAD+ dependent 11beta-hydroxysteroid dehydrogenase) promoter constructs in renal cell lines. Endocr. Res. 2000; 26: 289–302.

Agarwal AK, Giacchetti G, Lavery G, et al. CA-repeat polymorphism in intron 1 of HSD11B2: Effects on gene expression and salt sensitivity. Hypertension 2000; 36: 187–94.

Agarwal AK, Monder C, Eckstein B, White PC. Cloning and expression of rat cDNA encoding corticosteroid 11 beta-dehydrogenase. J. Biol. Chem. 1989; 264: 18939–43.

Agarwal AK, Mune T, Monder C, White PC. NAD$^+$-dependent isoform of 11 beta hydroxysteroid dehydrogenase: cloning and characterization of cDNA from sheep kidney. J. Biol. Chem. 1994; 269: 25959–62.

Agarwal AK, Rogerson FM, Mune T, White PC. Gene structure and chromosomal localization of the human HSD11K gene encoding the kidney (type 2) isozyme of 11β-hydroxysteroid dehydrogenase. Genomics 1995; 29: 195–99.

Agarwal AK, Tusie-Luna MT, Monder C, White PC. Expression of 11 beta-hydroxysteroid dehydrogenase using recombinant vaccinia virus. Mol. Endocrinol. 1990; 4: 1827–32.

Agarwal AK, White PC. Analysis of the promoter of the NAD$^+$ dependent 11β-hydroxysteroid dehydrogenase (HSD11K) gene in JEG-3 human choriocarcinoma cells. Mol. Cell. Endocrinol. 1996; 121: 93–99.

Albiston AL, Obeyesekere VR, Smith RE, Krozowski ZS. Cloning and tissue distribution of the human 11-HSD type 2 enzyme. Mol. Cell. Endocrinol. 1994; 105: R11–17.

Alikhani-Koopaei R, Fouladkou F, Frey FJ, Frey BM. Epigenetic regulation of 11 beta-hydroxysteroid dehydrogenase type 2 expression. J. Clin. Invest. 2004; 114: 1146–57.

Arriza JL, Weinberger C, Cerelli G, et al. Cloning of human mineralocorticoid receptor complementary DNA: structural and functional kinship with the glucocorticoid receptor. Science 1987; 237: 268–75.

Batista MC, Mendonca BB, Kater CE, et al. Spironolactone-reversible rickets associated with 11 beta- hydroxysteroid dehydrogenase deficiency syndrome. J. Pediatr. 1986; 109: 989–93.

Bocchi B, Kenouch S, Lamarre-Cliche M, et al. Impaired 11-beta hydroxysteroid dehydrogenase type 2 activity in sweat gland ducts in human essential hypertension. Hypertension 2004; 43: 803–8.

Bournot P, Pitoizet N, Zachmann M, Maume BF. Partial characterization of unusual polar steroids in the urine of a child with low renin hypertension. J. Steroid Biochem. 1982; 16: 467–77.

Brown RW, Chapman KE, Edwards CR, Seckl JR. Human placental 11 beta-hydroxysteroid dehydrogenase: evidence for and partial purification of a distinct NAD-dependent isoform. Endocrinology 1993; 132: 2614–21.

Carvajal CA, Gonzalez AA, Romero DG, et al. Two homozygous mutations in the 11 beta-hydroxysteroid dehydrogenase type 2 gene in a case of apparent mineralocorticoid excess. J. Clin. Endocrinol. Metab. 2003; 88: 2501–7.

Casey ML, MacDonald PC, Andersson S. 17-HSD type 2: chromosomal assignment and progestin regulation of gene expression in human endometrium. J. Clin. Invest. 1994; 94: 2135–41.

Cole TJ. Cloning of the mouse 11 beta-hydroxysteroid dehydrogenase type 2 gene: tissue specific expression and localization in distal convoluted tubules and collecting ducts of the kidney. Endocrinology 1995; 136: 4693–6.

Dimartino-Nardi J, Stoner E, Martin K, Balfe JW, Jose PA, New MI. New findings in apparent mineralocorticoid excess. Clin. Endocrinol. (Oxf) 1987; 27: 49–62.

Draper N, Walker EA, Bujalska IJ, et al. Mutations in the genes encoding 11beta-hydroxysteroid dehydrogenase type 1 and hexose-6-phosphate dehydrogenase interact to cause cortisone reductase deficiency. Nat. Genet. 2003; 34: 434–9.

Edwards CR, Stewart PM, Burt D, et al. Localisation of 11 beta-hydroxysteroid dehydrogenase – tissue specific protector of the mineralocorticoid receptor. Lancet 1988; 2: 986–9.

Ferrari P, Sansonnens A, Dick B, Frey FJ. In vivo 11beta-HSD-2 activity: variability, salt-sensitivity, and effect of licorice. Hypertension 2001; 38: 1330–6.

Fiselier TJ, Otten BJ, Monnens LA, Honour J, van Munster PJ. Low-renin, low-aldosterone hypertension and abnormal cortisol metabolism in a 19-month-old child. Hormone Res. 1982; 16: 107–14.

Funder JW, Pearce PT, Smith R, Smith AI. Mineralocorticoid action: target tissue specificity is enzyme, not receptor, mediated. Science 1988; 242: 583–5.

Ghosh D, Weeks CM, Grochulski P, et al. Three-dimensional structure of holo 3 alpha,20 beta-hydroxysteroid dehydrogenase: a member of a short-chain dehydrogenase family. Proc. Natl Acad. Sci. USA, 1991; 88: 10064–8.

Harinck HI, van Brummelen P, van Seters AP, Moolenaar AJ. Apparent mineralocorticoid excess and deficient 11 beta-oxidation of cortisol in a young female. Clin. Endocrinol. (Oxf) 1984; 21: 505–14.

Honour JW, Dillon MJ, Levin M, Shah V. Fatal, low renin hypertension associated with a disturbance of cortisol metabolism. Arch. Dis. Child. 1983; 58: 1018–20.

Igarashi Y, Egi S, Takehiro A, Ohzeki T, Kawaguchi H. Studies on the metabolic abnormality of cortisol and corticosterone in a case of dexamethasone responsive mineralocorticoid excess. Folia Endocrinol. Jap. 1979; 55: 1341–57.

Kataoka S, Kudo A, Hirano H, et al. 11beta-hydroxysteroid dehydrogenase type 2 is expressed in the human kidney glomerulus. J. Clin. Endocrinol. Metab. 2002; 87: 877–82.

Kerstens MN, Navis G, Dullaart RP. Salt sensitivity and 11beta-hydroxysteroid dehydrogenase type 2 activity. Am. J. Hypertens. 2004; 17: 283–4.

Kitanaka S, Tanae A, Hibi I. Apparent mineralocorticoid excess due to 11β-hydroxysteroid dehydrogenase deficiency: a possible cause of intrauterine growth retardation. Clin. Endocrinol. (Oxf) 1996; 44: 353–9.

Kotelevtsev Y, Brown RW, Fleming S, et al. Hypertension in mice lacking 11beta-hydroxysteroid dehydrogenase type 2. J. Clin. Invest. 1999; 103: 683–9.

Krozowski Z, MaGuire JA, Stein-Oakley AN, Dowling J, Smith RE, Andrews RK. Immunohistochemical localization of the 11 beta-hydroxysteroid dehydrogenase type II enzyme in human kidney and placenta. J. Clin. Endocrinol. Metab. 1995; 80: 2203–9.

Krozowski ZS, Funder JW. Renal mineralocorticoid receptors and hippocampal corticosterone binding species have identical intrinsic steroid specificity. Proc. Natl Acad. Sci. USA 1983; 80: 6056–60.

Kyossev Z, Walker PD, Reeves WB. Immunolocalization of NAD-dependent 11β-hydroxysteroid dehydrogenase in human kidney and colon. Kidney Int. 1996; 49: 271–81.

Lakshmi V, Monder C. Purification and characterization of the corticosteroid 11 beta-dehydrogenase component of the rat liver 11 beta-hydroxysteroid dehydrogenase complex. Endocrinology 1988; 123: 2390–8.

Lavery GG, Ronconi V, Draper N, et al. Late-onset apparent mineralocorticoid excess caused by novel compound heterozygous mutations in the HSD11B2 gene. Hypertension 2003; 42: 123–9.

Li A, Li KX, Marui S, et al. Apparent mineralocorticoid excess in a Brazilian kindred: hypertension in the heterozygote state. J. Hypertens. 1997; 15: 1397–402.

Li A, Tedde R, Krozowski ZS, et al. Molecular basis for hypertension in the "type II variant" of apparent mineralocorticoid excess. Am. J. Hum. Genet. 1998; 63: 370–9.

Lin-Su K, Zhou P, Arora N, Betensky BP, New MI, Wilson RC. In vitro expression studies of a novel mutation delta299 in a patient affected with apparent mineralocorticoid excess. J. Clin. Endocrinol. Metab. 2004; 89: 2024–7.

Lovati E, Ferrari P, Dick B, et al. Molecular basis of human salt-sensitivity: the role of the 11β-hydroxysteroid dehydrogenase type 2. J. Clin. Endocrinol. Metab. 1999; 84: 3745–9.

Mantero F, Tedde R, Opocher G, Dessi Fulgheri P, Arnaldi G. Apparent mineralocorticoid excess type II. Steroids 1994; 59: 80–3.

Melander O, Frandsen E, Groop L, Hulthen UL. No evidence of a relation between 11beta-hydroxysteroid dehydrogenase type 2 activity and salt sensitivity. Am. J. Hypertens. 2003; 16: 33.

Milford DV, Shackleton CH, Stewart PM. Mineralocorticoid hypertension and congenital deficiency of 11 beta-hydroxysteroid dehydrogenase in a family with the syndrome of 'apparent' mineralocorticoid excess. Clin. Endocrinol. (Oxf) 1995; 43: 241–6.

Monder C, Shackleton CH, Bradlow HL, et al. The syndrome of apparent mineralocorticoid excess: its association with 11 beta-dehydrogenase and 5 beta-reductase deficiency and some consequences for corticosteroid metabolism. J. Clin. Endocrinol. Metab. 1986; 63: 550–7.

Monder C, Stewart PM, Lakshmi V, Valentino R, Burt D, Edwards CR. Licorice inhibits corticosteroid 11 beta-dehydrogenase of rat kidney and liver: in vivo and in vitro studies. Endocrinology 1989; 125: 1046–53.

Mune T, Rogerson FM, Nikkila H, Agarwal AK, White PC. Human hypertension caused by mutations in the kidney isozyme of 11 beta-hydroxysteroid dehydrogenase. Nat. Genet. 1995; 10: 394–9.

Mune T, White PC. Apparent mineralocorticoid excess: genotype is correlated with biochemical phenotype. Hypertension 1996; 27: 1193–9.

Naray-Fejes-Toth A, Fejes-Toth G. Expression cloning of the aldosterone target cell-specific 11 beta-hydroxysteroid dehydrogenase from rabbit collecting duct cells. Endocrinology 1995; 136: 2579–86.

Nawrocki AR, Goldring CE, Kostadinova RM, Frey FJ, Frey BM. In vivo footprinting of the human 11beta-hydroxysteroid dehydrogenase type 2 promoter: evidence for cell-specific regulation by Sp1 and Sp3. J. Biol. Chem. 2002; 277: 14647–56.

New MI, Levine LS, Biglieri EG, Pareira J, Ulick S. Evidence for an unidentified steroid in a child with apparent mineralocorticoid hypertension. J. Clin. Endocrinol. Metab. 1977; 44: 924–33.

Nikkila H, Tannin GM, New MI, et al. Defects in the HSD11 gene encoding 11β-hydroxysteroid dehydrogenase are not found in patients with apparent mineralocorticoid excess or 11-oxoreductase deficiency. J. Clin. Endocrinol. Metab. 1993; 77: 687–91.

Nunez BS, Rogerson FM, Mune T, et al. Mutants of 11beta-hydroxysteroid dehydrogenase (HSD11B2) with partial activity: improved correlations between genotype and biochemical phenotype in apparent mineralocorticoid excess. Hypertension 1999; 34: 638–42.

Obeid J, White PC. Tyr-179 and Lys-183 are essential for enzymatic activity of 11 beta-hydroxysteroid dehydrogenase. Biochem. Biophys. Res. Commun. 1992; 188: 222–7.

Oberfield SE, Levine LS, Carey RM, Greig F, Ulick S, New MI. Metabolic and blood pressure responses to hydrocortisone in the syndrome of apparent mineralocorticoid excess. J. Clin. Endocrinol. Metab. 1983; 56: 332–9.

Odermatt A, Arnold P, Frey FJ. The intracellular localization of the mineralocorticoid receptor is regulated by 11beta-hydroxysteroid dehydrogenase type 2. J. Biol. Chem. 2001a; 276: 28484–92.

Odermatt A, Arnold P, Stauffer A, Frey BM, Frey FJ. The N-terminal anchor sequences of 11beta-hydroxysteroid dehydrogenases determine their orientation in the endoplasmic reticulum membrane. J. Biol. Chem. 1999; 274: 28762–70.

Odermatt A, Dick B, Arnold P, et al. A mutation in the cofactor-binding domain of 11beta-hydroxysteroid dehydrogenase type 2 associated with mineralocorticoid hypertension. J. Clin. Endocrinol. Metab. 2001b; 86: 1247–52.

Palermo M, Shackleton CH, Mantero F, Stewart PM. Urinary free cortisone and the assessment of 11 beta-hydroxysteroid dehydrogenase activity in man. Clin. Endocrinol. (Oxf) 1996; 45: 605–11.

Persson B, Krook M, Jornvall H. Characteristics of short-chain alcohol dehydrogenases and related enzymes. Eur. J. Biochem. 1991; 200: 537–43.

Poch E, Gonzalez D, Giner V, Bragulat E, Coca A, de La SA. Molecular basis of salt sensitivity in human hypertension. Evaluation of renin-angiotensin-aldosterone system gene polymorphisms. Hypertension 2001; 38: 1204–9.

Quinkler M, Bappal B, Draper N, et al. Molecular basis for the apparent mineralocorticoid excess syndrome in the Oman population. Mol. Cell. Endocrinol. 2004; 217: 143–9.

Riddle MC, McDaniel PA. Renal 11 beta-hydroxysteroid dehydrogenase activity is enhanced by ramipril and captopril. J. Clin. Endocrinol. Metab. 1994; 78: 830–4.

Roland BL, Funder JW. Localization of 11β-hydroxysteroid dehydrogenase type 2 in rat tissues: in situ studies. Endocrinology 1996; 137: 1123–8.

Roland BL, Krozowski ZS, Funder JW. Glucocorticoid receptor, mineralocorticoid receptors, 11 beta-hydroxysteroid dehydrogenase-1 and -2 expression in rat brain and kidney: in situ studies. Mol. Cell. Endocrinol. 1995a; 111: R1–7.

Roland BL, Li KX, Funder JW. Hybridization histochemical localization of 11 beta-hydroxysteroid dehydrogenase type 2 in rat brain. Endocrinology 1995b; 136: 4697–700.

Rusvai E, Naray-Fejes-Toth A. A new isoform of 11 beta-hydroxysteroid dehydrogenase in aldosterone target cells. J. Biol. Chem. 1993; 268: 10717–20.

Shackleton CH, Honour J, Dillon MJ, Chantler C, Jones RW. Hypertension in a four-year-old child: gas chromatographic and mass spectrometric evidence for deficient hepatic metabolism of steroids. J. Clin. Endocrinol. Metab. 1980; 50: 786–92.

Shackleton CH, Rodriguez J, Arteaga E, Lopez JM, Winter JS. Congenital 11 beta-hydroxysteroid dehydrogenase deficiency associated with juvenile hypertension: corticosteroid metabolite profiles of four patients and their families. Clin. Endocrinol. (Oxf) 1985; 22: 701–12.

Smolenicka Z, Bach E, Schaer A, et al. A new polymorphic restriction site in the human 11 beta-hydroxysteroid dehydrogenase type 2 gene. J. Clin. Endocrinol. Metab. 1998; 83: 1814–17.

Stewart PM, Corrie JE, Shackleton CH, Edwards CR. Syndrome of apparent mineralocorticoid excess. A defect in the cortisol–cortisone shuttle. J. Clin. Invest. 1988; 82: 340–9.

Stewart PM, Krozowski ZS, Gupta A, et al. Hypertension in the syndrome of apparent mineralocorticoid excess due to mutation of the 11β-hydroxysteroid dehydrogenase type 2 gene. Lancet 1996; 347: 88–91.

Stewart PM, Wallace AM, Valentino R, Burt D, Shackleton CH, Edwards CR. Mineralocorticoid activity of liquorice: 11-beta-hydroxysteroid dehydrogenase deficiency comes of age. Lancet 1987; 2: 821–4.

Sugiyama T, Kato N, Ishinaga Y, Yamori Y, Yazaki Y. Evaluation of selected polymorphisms of the Mendelian hypertensive disease genes in the Japanese population. Hypertens. Res. 2001; 24: 515–21.

Tannin GM, Agarwal AK, Monder C, New MI, White PC. The human gene for 11 beta-hydroxysteroid dehydrogenase. Structure, tissue distribution, and chromosomal localization. J. Biol. Chem. 1991; 266: 16653–8.

Task Force on Blood Pressure Control in Children. Report of the Second Task Force on blood pressure control in children. Pediatrics 1987; 79: 1–25.

Tomlinson JW, Walker EA, Bujalska IJ, et al. 11beta-hydroxysteroid dehydrogenase type 1: a tissue-specific regulator of glucocorticoid response. Endocr. Rev. 2004; 25: 831–66.

Ulick S, Levine LS, Gunczler P, et al. A syndrome of apparent mineralocorticoid excess associated with defects in the peripheral metabolism of cortisol. J. Clin. Endocrinol. Metab. 1979; 49: 757–64.

Ulick S, Ramirez LC, New MI. An abnormality in steroid reductive metabolism in a hypertensive syndrome. J. Clin. Endocrinol. Metab. 1977; 44: 799–802.

Werder EA, Zachmann M, Vollmin JA, Veyrat R, Prader A. Unusual steroid excretion in a child with low-renin hypertension. Res. Steroids 1974; 6: 385–9.

White PC, Mune T, Agarwal AK. 11β-hydroxysteroid dehydrogenase and the syndrome of apparent mineralocorticoid excess. Endocr. Rev. 1997; 18: 135–56.

Whorwood CB, Mason JI, Ricketts ML, Howie AJ. Detection of human 11 beta-hydroxysteroid dehydrogenase isoforms using reverse-transcriptase-polymerase chain reaction and localization of the type 2 isoform to renal collecting ducts. Mol. Cell. Endocrinol. 1995; 110: R7–12.

Wilson RC, Dave-Sharma S, Wei JQ, et al. A genetic defect resulting in mild low-renin hypertension. Proc. Natl. Acad. Sci. USA 1998; 95: 10200–205. [Review].

Wilson RC, Krozowski ZS, Li K, et al. A mutation in the HSD11B2 gene in a family with apparent mineralocorticoid excess. J. Clin. Endocrinol. Metab. 1995a; 80: 2263–6.

Wilson RC, Harbison MD, Krozowski ZS, et al. Several homozygous mutations in the gene for 11β-hydroxysteroid dehydrogenase type 2 in patients with apparent mineralocorticoid excess. J. Clin. Endocrinol. Metab. 1995b; 80: 3145–50.

Winter JS, McKenzie JK. A syndrome of low renin hypertension in children. In: New MI, Levine LS, eds. Juvenile Hypertension. New York: Raven Press, 1977: pp. 123–31.

Yang K, Matthews SG. Cellular localization of 11 beta-hydroxysteroid dehydrogenase 2 gene expression in the ovine adrenal gland. Mol. Cell. Endocrinol. 1995; 111: R19–23.

Zaehner T, Plueshke V, Frey BM, Frey FJ, Ferrari P. Structural analysis of the 11beta-hydroxysteroid dehydrogenase type 2 gene in end-stage renal disease. Kidney Int. 2000; 58: 1413–19.

Zhou MY, Gomez-Sanchez EP, Cox DL, Cosby D, Gomez-Sanchez CE. Cloning, expression, and tissue distribution of the rat nicotinamide adenine dinucleotide-dependent 11 beta-hydroxysteroid dehydrogenase. Endocrinology 1995; 136: 3729–34.

CHAPTER 18

Pseudohypaldosteronism Type 1 and Hypertension Exacerbated in Pregnancy

DAVID S. GELLER

MINERALOCORTICOID RECEPTOR MUTATIONS IN HUMAN DISEASE

Since its original isolation by Simpson and Tait (Grundy et al 1952), aldosterone has occupied a prominent place in our understanding of physiologic mechanisms by which the kidney regulates salt and electrolyte balance. As the final effector molecule in the renin-angiotensin-aldosterone pathway, aldosterone plays a critical role in the regulation of blood pressure and electrolyte homeostasis. Secreted in response to hypotension or hyperkalemia, aldosterone binds to the mineralocorticoid receptor (MR) in the distal nephron, triggering increased sodium reabsorption via the epithelial sodium channel (ENaC) to restore intravascular volume. The electrical gradient created by sodium reabsorption provides a driving force for potassium and proton secretion, giving aldosterone an important role in electrolyte homeostasis as well. Recent years have witnessed an upsurge in interest in aldosterone biology, triggered by clinical studies demonstrating substantial benefits of mineralocorticoid blockade in a variety of important clinical conditions, including heart failure (Pitt et al 1999, 2003) and renal disease (Epstein 2003, Sato et al 2003, Rachmani et al 2004). Furthermore, the finding that primary aldosteronism, characterized by renin independent secretion of aldosterone from the adrenal gland, can cause hypertension in the absence of hypokalemia has led to increased screening and diagnosis of this treatable form of hypertension, leading to the suggestion that aldosteronism may underlie a significant proportion of what has previously been considered to be essential hypertension (Stowasser & Gordon 2004); specific measures to decrease aldosterone effect in these patients may be highly effective in reducing blood pressure. Due to these and other findings, there has been a marked upsurge in the use of antimineralocorticoid agents, and consequently, in the incidence of their side effects as well (Juurlink et al 2004).

Although the benefits of mineralocorticoid antagonism in these and other clinical conditions are clear, the mechanism by which antimineralocorticoid agents exert their beneficial effects remains in question. It has been proposed that mineralocorticoid antagonism exerts beneficial effects via blockade of mineralocorticoid receptor specific effects in the heart and vascular tissue itself (Brilla et al 1990, Brilla & Weber 1992, Brilla 2000), but it has been difficult to prove that the principal benefit of mineralocorticoid blockade is not simply achieved via the alteration of aldosterone-dependent renal sodium reabsorption. Clarifying this issue is of critical importance, as it will suggest mechanisms by which outcomes may be further improved in these patients. In all likelihood, further insights will require improved understanding of aldosterone biology and physiology. The study of patients bearing mutations in the mineralocorticoid receptor holds promise to elucidate underlying mechanisms of aldosterone-induced disease. Here, we review two such syndromes, autosomal dominant pseudohypoaldosteronism type 1 and hypertension exacerbated by pregnancy.

ALDOSTERONE BIOLOGY

Aldosterone represents the final effector molecule of the renin-angiotensin-aldosterone system. The kidney, in response to low blood volume or reduced blood pressure, secretes the aspartyl protease renin. Renin cleaves angiotensinogen to angiotensin I, which is subsequently converted to angiotensin II (AngII) by the angiotensin-converting enzyme. Angiotensin II acts in the adrenal zona glomerulosa to upregulate expression of aldosterone. Alternatively, hyperkalemia is a potent inducer of adrenal aldosterone production in the absence of AngII.

Aldosterone exerts its effects in the distal nephron via its actions on MR, a member of the steroid hormone receptor family. Steroid hormone receptors, a subgroup of the larger family of nuclear receptors (NRs), are ligand-activated transcription factors which play a critical regulatory role in a wide variety of physiologic processes, including development,

Genetic Diseases of the Kidney
ISBN: 978-0-12-449851-8

cellular differentiation, and maintenance of cellular homeostasis and blood pressure. They possess a conserved modular structure, with a long N-terminal domain thought to be important for transactivation, a central DNA binding domain, and a C-terminal hormone-binding domain.

MR is encoded by a gene lying on chromosome 4 and spanning approximately 400 kb. Exon 2 is a large exon encoding the translational start site as well as an N-terminal transactivational domain. Exons 3 and 4 encode the DNA-binding domain, while exons 5–9 encode the ligand-binding domain (LBD). Alternate promoters direct transcription of two distinct isoforms in aldosterone target tissues (Zennaro et al 1997). Alternative transcription of two untranslated exons generates two mRNA isoforms which are subjected to distinct tissue-specific and hormonal regulation (Zennaro et al 1997, Le Menuet et al 2000).

The LBDs of NRs share a common architecture, with 12 alpha helices and one beta turn arranged around a central hydrophobic core. In the absence of a steroid ligand, the receptor lies in the cytoplasm in association with a multiprotein complex composed of heat shock proteins, such as hsp90 and hsp70, as well as immunophilins such as FKBP52, FKBP54, cyclophilin CyP40 (Zennaro & Lombes 2004). Ligand-binding induces a conformational change in the receptor, resulting in dissociation of the receptor from this complex and translocation into the nucleus, where it binds to specific DNA sequences regulating hormone-responsive genes. The ligand-bound receptor is believed to assist in the recruitment of transcriptional co-regulators, which assist in chromatin remodeling, histone acetylation, and recruitment of the transcriptional machinery (Zennaro & Lombes 2004), inducing transcription of a number of genes, including the serum and glucocorticoid regulated kinase (sgk), c-Fos, Fra-2, channel-inducing factor (CHIF), and the N-myc down-regulated gene (NDRG2) (Verrey et al 2000, Boulkroun et al 2002). The exact mechanisms by which mineralocorticoid receptor activation leads to increased distal nephron sodium reabsorption has not been elucidated to date, however.

It has been proposed that some effects of aldosterone may be mediated via rapid nongenomic mechanisms, acting perhaps via an as yet unidentified membrane receptor (Losel et al 2004). For example, aldosterone triggers an increase in intracellular calcium in vascular smooth muscle cells, as well as in mesangial and distal tubule epithelial cells, via a nongenomic mechanism (Koppel et al 2003); these effects may be mediated via influences of aldosterone on cellular kinases (Gekle et al 2001). Clinically, aldosterone can be demonstrated to trigger rapid alterations in systemic vascular resistance (Gunaruwan et al 2002) and cardiac output, as well as to modulate intracardiac monophosphate action potentials in healthy volunteers (Losel et al 2004); these actions are not inhibitable by spironolactone. While the concept of nongenomic actions of aldosterone is attractive, for aldosterone levels rise and fall much

more rapidly in response to changes in body position than would seem necessary to trigger slow genomic responses, the true physiologic significance of these pathways is not clear, as we shall see below.

Further insights into the biological role(s) of aldosterone have come from the study of human diseases caused by mutation in MR. Mutations of the mineralocorticoid receptor have been linked with two human genetic diseases, autosomal dominant pseudohypoaldosteronism type 1 and hypertension exacerbated by pregnancy. Here, we review these diseases with an eye towards insights these diseases give towards an improved understanding of aldosterone biology.

PSEUDOHYPOALDOSTERONISM TYPE 1

First described by Cheek and Perry, pseudohypoaldosteronism type 1 (PHA1) is a rare metabolic disorder characterized by renal salt wasting, hyperkalemia, hyponatremia, and metabolic acidosis, all of these occurring despite markedly elevated renin and aldosterone levels. The syndrome is thus one of renal resistance to the antinatriuretic and kaliuretic actions of aldosterone. There is great phenotypic variability among patients with this disorder. Some patients are quite ill, and exhibit salt-wasting from diverse tissues including the colon and the sweat and salivary glands; these patients require massive sodium supplementation throughout life. In contrast, other patients are only mildly affected, with salt-wasting primarily from the kidney; these patients generally discontinue sodium supplementation after their childhood years. The reasons underlying this phenotypic diversity were clarified with the recognition that there are two distinct forms of PHA1, an autosomal recessive form (arPHA1) and an autosomal dominant form (adPHA1). ArPHA1 is caused by homozygous loss-of-function mutations in any one of the three subunits of the epithelial sodium channel (Chang et al 1996). This finding confirmed ENaC as the principal mediator of aldosterone-sensitive sodium transport in the distal nephron. AdPHA1 is caused by heterozygous mutations in the mineralocorticoid receptor. A sporadic form of PHA1 has also been described; whether these cases represent individuals with a non-genetic form of PHA1, an unidentified de novo mutation in a gene associated with PHA1, or a mutation in some other gene is not known.

Autosomal recessive PHA1 (arPHA1) is characterized by severe salt-wasting and hyperkalemia. Infants typically present in the first week of life, frequently with vomiting, dehydration, and failure to thrive. Laboratory evaluation reveals potentially life-threatening hyperkalemia accompanied by hyponatremia due to volume depletion. After volume resuscitation, renal function normalizes, but renal salt wasting persists. The clinical picture of hyponatremia, hyperkalemia, and failure to thrive, resembles that seen in adrenal insufficiency, but serum cortisol levels are normal. The diagnosis of PHA1 is confirmed by the finding of markedly elevated renin and aldosterone levels in the face of normal renal function and continued renal salt wasting.

Patients with arPHA1 have salt-wasting from the kidney, colon, and sweat and salivary glands. They have high rates of neonatal death from volume depletion and hyperkalemia, and they require massive lifelong sodium supplementation and potassium binding resins to survive. They also suffer from a distinct pulmonary disorder characterized by excess airway liquid and an asthmatic-like appearance. Survival to adulthood is possible, but this requires aggressive therapy, with lifelong provision of potassium binding resins and salt, coupled with aggressive saline infusion for even mild febrile illness; even with this regimen, hyperkalemia and ventricular arrest is an everpresent threat (U. Kuhnle, personal communication). Autosomal dominant PHA1 shares many of the same clinical features at the recessive form of the disease, including salt-wasting, hyperkalemia, and acidosis despite elevated aldosterone levels, but it is typically much milder in its course. AdPHA1 patients may either be asymptomatic or may have evidence of significant salt wasting during the neonatal period, but symptoms generally subside after early childhood. Unlike patients with arPHA1, patients with adPHA1 do not have elevated sweat or salivary sodium levels, and there is no described pulmonary component either. As with loss-of-function mutations in ENaC, loss-of-function mutations in the mineralocorticoid receptor lead to salt-wasting and volume depletion, leading to elevated serum renin and aldosterone levels in affected individuals. Elevated aldosterone levels are not sufficient to normalize sodium and potassium balance early in life, and patients usually require salt supplementation and perhaps potassium binding resins. After childhood years, however, adPHA1 patients maintain electrolyte homeostasis without salt supplementation. This is in marked contrast to arPHA1 patients, who require lifelong treatment with potassium binding resins and sodium. Children with adPHA1 may be small for their age, but they experience rapid catchup growth after salt therapy is initiated (Hanukoglu 1991).

Genetics

To determine the molecular basis of adPHA1, we collected a series of PHA1 patients with sporadic PHA1 or autosomal dominant PHA1. We first screened subunits of the epithelial sodium channel (ENaC) for mutations in these patients, and we identified no disease-causing mutations, suggesting that mutations in ENaC rarely, if ever, cause adPHA1. We next screened the mineralocorticoid receptor (MR). MR had previously been screened in patients with PHA1 via sequencing of MR cDNA, but this approach can miss certain types of genomic mutations (Zennaro et al 1994, Arai et al 1995, Komesaroff 1995). We identified multiple independent mutations in the gene coding for MR which cosegregated with disease (Geller et al 1998). In addition, we identified a de novo mutation which introduces a premature stop codon in the N-terminal region of the receptor in the affected child of unaffected parents. These findings confirmed that heterozygous mutations in *MR* cause adPHA1.

To date, twenty-one different PHA1 causing mutations in MR have been identified in twenty-three kindreds (Figure 18.1), including nonsense, frameshift, missense, and splice-site mutations distributed throughout the gene (Geller et al 1998, Tajima et al 2000, Viemann et al 2001, Riepe et al 2003, Sartorato et al 2003, Nystrom et al 2004). It has been proposed that adPHA1 may be a genetically heterogeneous disorder based on the failure to identify disease-causing mutations in MR after sequencing of all exonic sequences in adPHA1 patients (Viemann et al 2001, Huey et al 2004). However, certain types of mutations, such as non-exonic mutations, (e.g. promoter, intronic, and 3′ untranslated region mutations) or large deletions, would not be detected by these means. Furthermore, linkage analysis, which could have definitively excluded the gene, was not performed in these studies. As such, disease-causing mutations in MR cannot be ruled out in these kindreds. Indeed, Sartorato et al used similar sequencing methods to rule out an exonic mutation in an adPHA1 kindred, but linkage analysis ultimately led to the identification of a large disease-causing deletion in *MR* spanning exons 2 and 3 (Sartorato et al 2004). In our own experience, we have identified MR

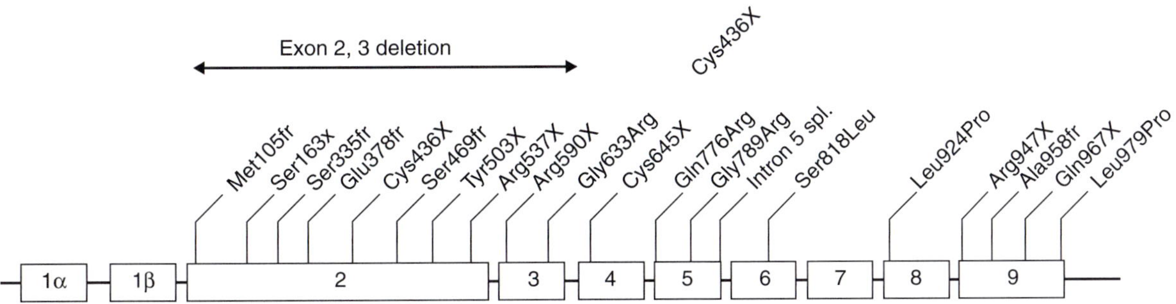

FIGURE 18.1 AdPHA1 is caused by mutations in *MR*: The identity and location of all published mutations in the mineralocorticoid receptor causing adPHA1 are depicted. fr: frameshift; spl: splice-site mutation; X: stop codon

mutations in seven of nine kindreds with evidence of dominant disease transmission, and we could not exclude linkage to MR in the remaining kindreds. As such, we believe that mutation in *MR* represents the principal, if not the only, cause of adPHA1.

Intriguingly, of the 21 MR mutations described in adPHA1, five have been confirmed in our laboratory as being true de novo mutations. Furthermore, two children described in the literature have mutations in MR not carried by their parents, although formal genotyping to confirm parental identity was not confirmed. As such, one quarter to one-third of all index cases represent true de novo mutations in MR.

The mutations shown to cause adPHA1 have shed light on functional roles of distinct residues to mineralocorticoid receptor function. Disease-causing mutations in the ligand-binding domain of the receptor have identified specific residues necessary for specific binding of aldosterone (Tajima et al 2000, Viemann et al 2001, Sartorato et al 2004; Geller, unpublished observations). G663R, an adPHA1-causing mutation in the DNA binding domain of the receptor, has the interesting phenotype of altered intracellular trafficking, the mutant receptor residing in the nucleus in the absence of hormone (Sartorato et al 2004); this mutation does not change the affinity of the DNA binding domain for the glucocorticoid response element (GRE), but instead, decreases the dissociation rate of the receptor from the DNA.

It is perhaps surprising that heterozygous loss-of-function mutations in MR result in a clinical phenotype, as one functional MR copy is still present and would presumably make up for the missing allele. Since MR is known to function as a dimer, we therefore wondered whether mutant MR peptides might be produced by PHA1 patients which could interfere with the function of the wild-type allele, either by forming an inactive heterodimer or perhaps by binding to and inactivating necessary transcription factors. We answered this question via the study of kindred, PHA30, a large dominant French kindred which has been well described in the literature (Zennaro et al 1994). We identified a disease-causing mutation in *MR* in an individual in this kindred, a C- > T substitution which converts Arg590 to a stop codon, resulting in termination of translation prior to the DNA binding and hormone binding domains of the receptor. We screened *MR* cDNA prepared from peripheral blood lymphocytes from a PHA30 kindred member bearing this mutation. Interestingly, we demonstrated that while genomic DNA shows evidence of the heterozygous mutation, the mutation is absent in cDNA prepared from RNA derived from peripheral blood lymphocytes (Geller et al, manuscript in preparation)). This suggests that the mutant RNA has been degraded, most likely via nonsense-mediated decay (Hentze & Kulozik 1999), and that it is therefore not expressed. It is thus clear that haploinsufficiency of MR is sufficient to cause the adPHA1 phenotype.

Genotype–Phenotype Correlations in adPHA1

The ability to assign affection status on the basis of genotype rather than phenotype allowed us recently to perform genotype–phenotype correlation studies. We extended two large kindreds from a small village in the northwest of Spain by recruiting all first degree relatives of genotypically affected individuals. While not known to be related, these two kindreds share the same R537X mutation (Geller et al 1998). In all, we studied 14 affected and 22 unaffected adult kindred members. We found no difference among all indices of aldosterone function measured – the groups were indistinguishable from each other in terms of systolic blood pressure, diastolic blood pressure, serum sodium, serum potassium, fractional excretion of sodium, and trans-tubular potassium gradient. The only significant difference between the two groups was in the serum aldosterone level. Adult family members with PHA1 had serum aldosterone levels approximately 15-fold higher than in their unaffected brethren, indicating that they were able to maintain salt homeostasis by markedly upregulating aldosterone synthesis (Geller et al, manuscript in preparation).

Pregnancy and infancy are two periods of intrinsic aldosterone resistance, and so we were curious whether patients with adPHA1 would be at risk during these two phases of life. While we noted a number of spontaneous miscarriages in pregnant women with adPHA1, the incidence of miscarriage did not differ from what is seen in the general population, and so we cannot assert that PHA1 played a role in antepartum difficulties. On the other hand, we identified deaths in four neonates at risk for adPHA1; others have noted neonatal deaths in infants at risk for adPHA1 as well (Tajima et al 2000). While genotypic data is not available on the deceased infants, the high infant mortality rate coupled with the known importance of aldosterone in the neonatal period make it reasonable to presume that adPHA1 may have played a role in these infants' deaths. As such, we recommend prophylactic salt supplementation and early definitive diagnosis for infants known to be at risk for adPHA1 (Geller et al, manuscript in preparation).

The demonstration of neonatal death in infants at-risk for adPHA1 may help explain one of the mysteries of the disease: why it is so rare. Ordinarily, one would expect that a relatively benign genetic disease which is transmitted in autosomal dominant fashion would spread quickly through the population, and would become relatively common. However, only 21 disease-causing mutations have been identified, suggesting this remains an extraordinarily rare disorder. One possibility could be that MR mutations interfere with the fertility process and are not transmitted from parent to child; we see no evidence for this in kindreds identified, in which the diseased allele is transmitted at least as often as the wild-type allele, regardless of which parent is affected. Alternatively, one could imagine that there is ascertainment bias – that many patients with adPHA1 are asymptomatic and never come to clinical attention. While this no doubt occurs, we believe it is not especially common. The high percentage of proven de novo mutations identified in affected individuals argues that a substantial fraction of affected individuals do come to clinical attention. Thus, unless the *MR* locus is somehow an infrequent site for mutations, we believe that the primary reason adPHA1 is such a rare disorder is that the disease was commonly a fatal neonatal disorder in previous generations, when perinatal care was not as aggressive as it currently is. We believe that as more of these children come to clinical attention and survive to procreate themselves, the incidence of the disease may well increase, forcing physicians to be aware of the potential perinatal dangers.

The clinical severity of disease in adPHA1 patients early in life highlights the essential role of aldosterone-sensitive sodium transport in the neonate and raises the question as to the reasons underlying the apparently diminished requirement for this system in later years. One possibility relates to the low sodium content of human breast milk (Neville et al 1988, Morton 1994), which may render neonates particularly sensitive to renal salt-wasting; this sensitivity lessens as the infant transitions to the high salt intake characteristic in the industrialized world. This suggests a gene-by-environment interaction, in that, on a low-sodium diet, humans are dependent on maximal activation of the renin-angiotensin-aldosterone system, and MR haploinsufficiency results in volume depletion and hyperkalemia, but on a high salt diet, adPHA1 is clinically silent. An alternative explanation for the improvement in the adPHA1 phenotype after the neonatal years involves the development of the renal tubule. Aldosterone-mediated sodium transport through ENaC in the cortical collecting duct (CCD) is coupled to K^+ secretion via a potassium secretory channel ROMK. ROMK is expressed postnatally (Benchimol et al 2000), potentially allowing improved efficiency of the renin-angiotensin-aldosterone system, and possibly providing a physiologic mechanism for enhanced mineralocorticoid sensitive sodium and potassium transport after the perinatal period.

ANIMAL MODELS OF PHA1

Further insights into the role of the mineralocorticoid receptor in electrolyte homeostasis have come from the study of mice deficient in MR. Mice homozygous for MR deletion are born essentially normal, but they grow more slowly than their wild-type and heterozygous littermates and die approximately ten days after birth. Biochemical data demonstrate marked hyponatremia, hyperkalemia, hyperreninemia, and hyperaldosteronemia. Mice heterozygous for the MR deletion, similar to humans with adPHA1, have a mild increase in the fractional excretion of sodium coupled with 3–4-fold elevations of renin and aldosterone, but they develop normally (Berger et al 1998). Interestingly, MR knockout mice

can be rescued via provision of extra sodium via subcutaneous injection prior to weaning (Bleich et al 1999). After weaning, they stay alive and grow somewhat normally via upregulation of dietary sodium intake, but they remain hyperkalemic and hyperaldosteronemic. These data point to an important role of aldosterone in mediating perinatal sodium homeostasis, perhaps due to the low sodium content of maternal milk, and indicate the decreased necessity for aldosterone in an adult with a high sodium intake. Interestingly, mice thus rescued exhibit a significant antinatriuretic response to dexamethasone, clarifying that glucocorticoids have important renal sodium retentive properties not mediated via the mineralocorticoid receptor (Schulz-Baldes et al 2001).

HYPERTENSION EXACERBATED BY PREGNANCY

Autosomal recessive PHA1, caused by homozygous loss of function mutations in ENaC, results in salt-wasting and hypotension, whereas Liddle's syndrome, characterized by gain-of-function mutations in ENaC, leads to Mendelian hypertension. Given that loss-of-function mutations in *MR* cause renal salt-wasting and hypotension, we wondered whether there might be individuals in whom gain-of-function mutations cause renal salt retention and hypertension. Exhaustive screening of *MR* coding regions in hundreds of kindreds referred for evaluation of familial hypertension revealed a single kindred with a mutation of interest, a missense mutation which cosegregates with severe, early onset hypertension in the family. The alteration of thymidine for cytosine at base 2645 leads to a substitution of leucine for serine at codon 810 in the hormone-binding domain of the receptor. The child carrying the allele was a 15-year-old boy with severe hypertension (210/120 mmHg) in whom an extensive workup for a secondary cause of hypertension had been unrevealing. To determine the effect of this mutation on clinical phenotypes, we clinically characterized the family. In all, we identified 21 members of the kindred, 11 of whom had evidence of severe, early onset hypertension. Genotypic evaluation of family members indicated that all individuals with severe, early-onset hypertension carried the MR_{L810} allele, while no unaffected family members did. Furthermore, the mutation was not detected in 150 control chromosomes from an ethnically matched population. There was thus complete correlation of early-onset HTN with the MR_{L810} allele, conclusively implicating this allele as the cause of hypertension in this kindred (Geller et al 2000).

To better understand the effect of the MR_{L810} allele on clinical outcomes in this kindred, we clinically characterized the family members (Table 18.1). We found that systolic blood pressure was 41 mmHg higher, and diastolic blood pressure was 32 mmHg higher in MR_{L810}, despite these individuals taking, on average, 1.5 antihypertensive agents more per person. This suggests a dramatic effect of this mutation on blood pressure regulation. Interestingly, while there was a trend towards lower serum potassium concentrations in MR_{L810} carriers, this difference was not statistically significant. Serum aldosterone and 24 h urinary aldosterone levels were uniformly suppressed in MR_{L810} carriers (Geller et al 2000).

The clinical picture of MR_{L810} carriers suggested either that the mutant MR was either constitutively active, or that it was being activated by a novel mineralocorticoid. To determine how this mutation might predispose individuals to early-onset hypertension, we performed cotransfection reporter assays, assaying receptor activity in the presence

of related steroid hormones. We first determined the activity of the mutant receptor in the presence of aldosterone, the normal receptor agonist. Both the wild-type and mutant receptors are similarly activated by aldosterone. However, in the absence of aldosterone, the mutant receptor retains approximately 25% of maximal activity, indicating significant constitutive activity (Figure 18.2A)(Geller et al 2000).

Further characterization of receptor activity demonstrated that normal mineralocorticoid agonists, such as desoxycorticosterone (DOC), corticosterone, and cortisol, activate both the wild-type and mutant receptors, while steroids which normally do not bind to MR, such as estrogen and testosterone, had no activity on the mutant receptor. Intriguingly, however, progesterone and related progestagens, which normally function in vivo and in vitro as antagonists of the mineralocorticoid receptor, are potent agonists of the mutant receptor (Figure 18.2B). The activity of MR_{L810} in the presence of progesterone is indistinguishable from its activity in the presence of aldosterone. Progesterone is distinguished from the mineralocorticoid agonist DOC by the presence of a 21-OH group, a group common to all mineralocorticoid agonists, indicating that the S810L mutation confers 21-OH independent activity on MR.

Progesterone levels rise 100–1000-fold during pregnancy. As such, the demonstration that progesterone functions as an agonist of MR_{L810} led to the prediction that pregnancy would have important consequences in an MR_{L810} carrier. Consistent with this, two family members had carried three pregnancies to term. One woman had two pregnancies; during each, she experienced severe hypertension which necessitated premature delivery of the child by the seventh month. Unfortunately, medical records are not available for these pregnancies. The sister of the index case, on the other hand, had five pregnancies, each of which was characterized by severe hypertension and hypokalemia. Four of these pregnancies were electively terminated,

TABLE 18.1   Clinical features of MR_{L810} carriers (+) and non-carriers (−) in kindred 503

Clinical parameter	MR_{L810} $^+$ ($n = 8$)	MR_{L810} $^-$ ($n = 11$)	P
Age	29.1±6.3	32.9±8.1	0.88
HTN < Age 20	100%	0%	<0.0001
Anti-HTN meds	1.5±0.27	0.2±0.12	0.001
SBP (mmHg)	167±11	126±10	0.014
DBP (mmHg)	110±6	78±6	0.002
Serum K$^+$ (mM)	3.91±0.18	4.36±0.11	0.08
Serum HCO$_3$ (mM)	27.1±0.87	26.4±0.83	0.59
Serum aldosterone (ng/dl)	2.48±0.68	12.1±2.96	0.008
Urinary aldosterone (μg/24 h)	<2	7.75±1.55	0.03

All values are mean ± SEM. Statistical comparisons were assessed by the Mann-Whitney U test. Normal values: SBP < 140, DBP < 90, serum K$^+$: 3.8–5.2, HCO$_3^-$: 23–30, serum aldosterone 4–15, urinary aldosterone 6–31. Urinary aldosterone was measured in three $MR_{L810}$$^+$ and four $MR_{L810}$$^-$ individuals. HTN, hypertension; SBP, systolic blood pressure; DBP, diastolic blood pressure; HCO$_3^-$, bicarbonate.

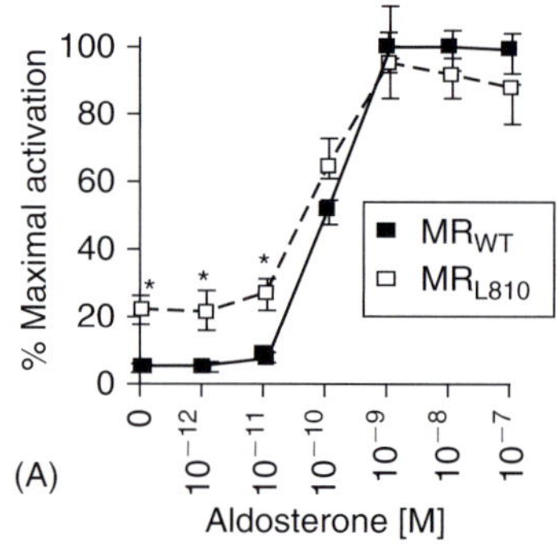

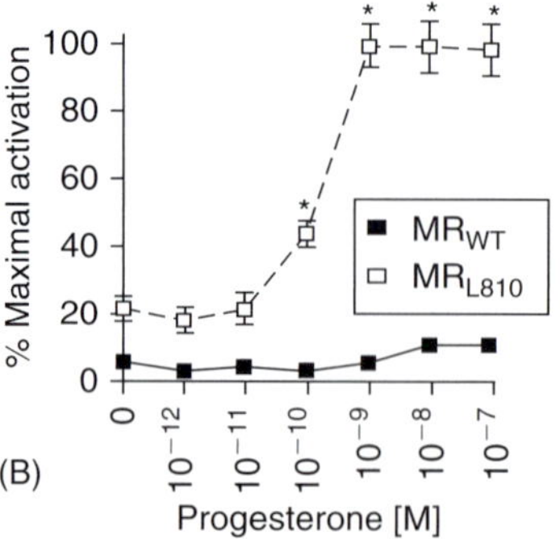

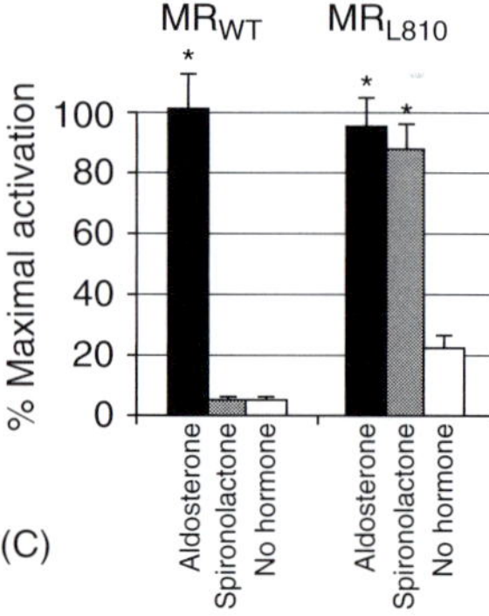

FIGURE 18.2   Transcriptional activation by MR_{WT} and MR_{L810}. The ability of MR_{WT} and MR_{L810} to induce luciferase expressed under control of the MMTV promoter was assessed in Cos7 cells in the absence or presence of indicated steroids. Luciferase activity is expressed as percent of maximal induction of MR_{WT} by aldosterone. (A) Dose–response curve for induction of luciferase by MR_{WT} and MR_{L810} in response to aldosterone (*: $P < 0.001$ MR_{L810} vs. MR_{WT}). (B) Dose–response curve for induction of luciferase by MR_{WT} and MR_{L810} in response to progesterone (*: $P < 0.001$ vs. vehicle control). (C) Luciferase induction in response to spironolactone (*: $P < 0.001$ vs. vehicle)

but one was carried to 34 weeks. During this pregnancy, she experienced extraordinarily high blood pressure, coupled with hypokalemia, and hyperkaliuria (Table 18.2). Serum and urinary aldosterone levels, which normally rise 10–20-fold during pregnancy to counteract the higher concentrations of the antimineralocorticoid effects of progesterone, were undetectable (Table 18.2). This clinical picture is highly atypical for pregnancy, but is the exact picture one would expect in the setting of a mutant MR which is activated by progesterone. These data thus provide dramatic clinical confirmation of the in vitro data. It should be noted that neither woman experienced neurologic symptoms, proteinuria, or edema, indicating that their problem had no connection with pre-eclampsia (Geller et al 2000).

It is intriguing to note that while hypokalemia developed during the first trimester, severe hypertension did not develop until midway through the second trimester. The hypokalemia in the first trimester suggests that the moderate elevations of progesterone during the first trimester were sufficient to increase mineralocorticoid-sensitive sodium reabsorption and potassium secretion, but that the magnitude of increased sodium reabsorption was not enough to induce hypertension in the setting of the marked decrease in systemic vascular resistance during pregnancy.

While progesterone is no doubt the functional mineralocorticoid during pregnancy, the presence of hypertension in male and non-pregnant female MR_{L810} carriers suggests another agent is responsible for hypertension in these individuals. A prime candidate for this is cortisone, which has been shown to be a potent agonist of MR_{L810} (Rafestin-Oblin et al 2003). When MR was first isolated, it was noted that cortisol is a potent agonist of MR in vitro (Arriza et al 1987). This presented a conundrum, as serum cortisol levels are normally 100–1000-fold higher than aldosterone levels in the body, leaving it unclear how aldosterone could affect MR in

the face of much higher cortisol levels. The answer proved to be that the kidney expresses an enzyme, 11β-hydroxysteroid dehydrogenase type 2 (11β-HSD2), which metabolizes cortisol to the inactive cortisone. Proof of the significance of this system is that individuals lacking this enzyme have the syndrome of apparent mineralocorticoid excess, with severe hypertension and hypokalemia (Mune et al 1995). The finding that cortisone is a potent agonist of MR_{L810} coupled with the high levels of cortisone present in the principal cell makes cortisone a likely functional mineralocorticoid responsible for hypertension in MR_{L810} carriers.

If cortisone is, in fact, the principal functional mineralocorticoid in MR_{L810} carriers, it is reasonable to wonder why hypertension in MR_{L810} carriers is not even worse. Cortisone levels in the principal cell in normal patients presumably rival those of cortisol in patients with the syndrome of apparent mineralocorticoid excess, and yet, 11β-HSD2 deficient patients have a much more severe clinical phenotype, with growth deficits, severe HTN with evidence of end-organ failure as children, and risk of premature death; these findings have not been observed in MR_{L810} carriers. There appear to be a number of reasons for this dichotomy. First of all, in contrast to cortisol in AME patients, cortisone in MR_{L810} carriers has only one MR to bind to, as the wild-type MR allele is resistant to cortisone; thus, functional haploinsufficiency at MR leads to a decrease in mineralocorticoid-mediated sodium retention, as in adPHA1 (see above). Secondly, cortisone appears to be a slightly less potent agonist of MR_{L810} than cortisol or progesterone (Rafestin-Oblin et al 2003); thus, the rise in progesterone levels during pregnancy can increase blood pressure and potassium excretion although at this concentration, levels would be similar to standard cortisone levels because it is a better MR_{L810} agonist. Finally, it appears that the S810L receptor is expressed at lower levels than the

TABLE 18.2   Clinical course of pregnancy in an MR_{L810} heterozygote

Week	BP	Medications prescribed	Serum K⁺ (mM)	Urine K⁺ (mmol/24 h)	Serum aldosterone (ng/dl)	Urine aldosterone (µg/24 h)
11	160/90		3.2			
16	140/90		NA			
21	160/120	α-methyldopa 500 mg qid KCl 40 mmol/d	2.1		<2.5	
28	170/130	α-methyldopa 500 mg qid Hydralazine 25 mg qid KCl 40 mmol/d	3.0	102	<2.5	<2.5
34	210/120	α-methyldopa 500 mg qid Hydralazine 25 mg qid KCl 40 mmol/d	3.5			

The clinical course of the first pregnancy of a member of K503 heterozygous for MR_{L810} is shown. Blood pressure rose dramatically despite treatment, and serum potassium levels dropped, with continuing urinary potassium wasting. This occurred despite undetectable levels of aldosterone in serum and urine. Caesarean section was performed at 34 weeks. BP, blood pressure wild-type receptor, further decreasing mineralocorticoid effect via this receptor (see below).

Having determined that excess activity of the mineralocorticoid receptor was responsible for hypertension in this kindred, we sought to determine whether an antimineralocorticoid agent might provide a specific form of therapy. However, like progesterone, spironolactone proved to be a potent agonist of the mutant receptor, suggesting that its use is contraindicated in MR_{L810} carriers (Figure 18.2C). Similarly, a wide variety of other mineralocorticoid antagonists, including eplerenone, drospirenone and others, have all proven to be MR_{L810} agonists (data not shown).

THE S810L MUTATION ALTERS RECEPTOR SPECIFICITY VIA A NEW INTRAMOLECULAR CONTACT

The altered activity of the MR_{L810} receptor led us to wonder how a single amino acid substitution could so drastically alter receptor specificity. The LBDs of NRs share a common architecture, with 12 alpha helices and one beta turn arranged around a central hydrophobic core (Moras & Gronemeyer 1998). Comparisons of the crystal structures of ligand-free and ligand bound NRs have shown that, upon ligand binding, NRs undergo a conformational change, including movement of helix 12 (H12) and a bending of H3 towards H5. The repositioned H12 combines with H3 and H5 to form a pocket where transcriptional coactivators can bind (Moras & Gronemeyer 1998). To better understand how a single base pair change could so dramatically alter receptor structure, we created a structural model of the mutant receptor ligand-binding domain based on the previously reported crystal structure of the closely related progesterone receptor.

In our structural model, serine 810 on helix 5 lay in close proximity to alanine 773 on helix 3 and the C19 methyl group of aldosterone, but the distances were too great to allow for productive van der Waals (vdW) interaction, which normally require a 3–4 Å separation. When we substituted leucine for serine in this model, however, the longer hydrophobic sidechain of leucine was well positioned to form vdW contact with both of these moieties, suggesting that one of these contacts might be responsible for the observed activity of the mutant receptor.

This model was tested via biochemical study of altered steroids and altered receptors. 19-Norprogesterone, a progesterone analog lacking the C19 methyl group, was a potent agonist of MR_{L810}, but not MR_{WT}, indicating that vdW contact between L810 and C19 is not necessary for progesterone-mediated activation of the mutant receptor (Figure 18.3A, B). Substitution of methionine at residue 810, which, like leucine, has a long side chain, resulted in no loss of progesterone-mediated activation. However, progressive

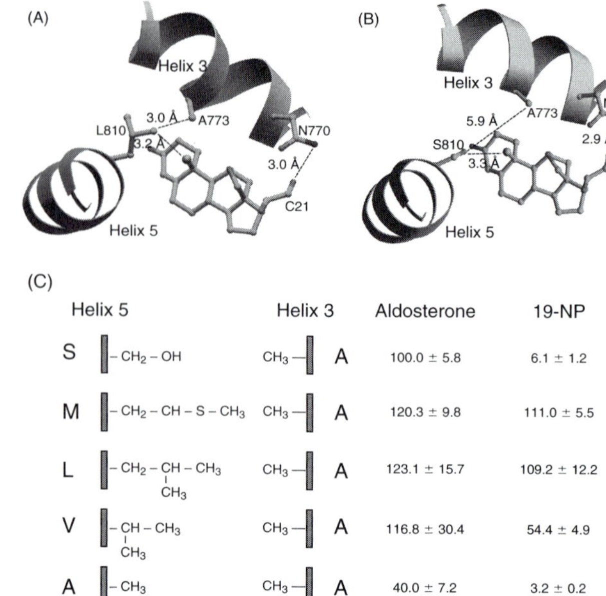

Helix 5		Helix 3		Aldosterone	19-NP
S	–CH₂–OH	CH₃–	A	100.0 ± 5.8	6.1 ± 1.2
M	–CH₂–CH–S–CH₃	CH₃–	A	120.3 ± 9.8	111.0 ± 5.5
L	–CH₂–CH–CH₃ (CH₃)	CH₃–	A	123.1 ± 15.7	109.2 ± 12.2
V	–CH–CH₃ (CH₃)	CH₃–	A	116.8 ± 30.4	54.4 ± 4.9
A	–CH₃	CH₃–	A	40.0 ± 7.2	3.2 ± 0.2
V	–CH–CH₃ (CH₃)	H–	G	138.2 ± 38.9	5.7 ± 2.0
L	–CH₂–CH–CH₃ (CH₃)	H–	G	108.1 ± 8.4	96.8 ± 6.9

FIGURE 18.3 Helix 3–Helix 5 interaction in progesterone-mediated activation of MR_{L810}. (A) Structural model of a portion of the hormone-binding domain (HBD) of MR_{L810} bound to aldosterone. Using the crystal structure of the progesterone receptor (PR), a model of the MR LBD was created by substituting MR-specific residues in the ligand-binding cavity for their corresponding residues in PR. The side chain of L810 lies in sufficiently close proximity to A773 and the C-19 methyl group of the steroid to form van der Waals interactions. (B) Model of MR_{WT}. The side chain of S810 does not interact with either A773 or the steroid. (C) Activity of MRs with various amino acid substitutions at residues 810 and 773. Mutant receptors containing the indicated substitutions at positions 810 in helix 5 and 773 in helix 3 were tested for their ability to induce luciferase activity in the presence of 1 nM aldosterone or 19-norprogesterone. The length of each side chain is approximated. Each data point represents the mean of 9 independent transfections and is expressed as mean ± SEM of the percentage of luciferase induction by MR_{WT} in response to aldosterone. (see also Plate 18)

shortening of helix 5 residue length, via substitution of isoleucine, valine and alanine for serine at position 810, led to a progressive decrease in 19-NP, but not aldosterone, mediated transcriptional activity (Figure 18.3C). This effect was likely due to interaction of the helix 5 residue with the helix 3 residue at residue 773. MR with a V810-G773 pair showed no activation by 19-NP, but activation was restored with a V810-A773 pair (Figure 18.3C). The second-site complementation supports the importance of helix 3–helix 5 interaction in the activation of MR_{L810} by steroids lacking 21-hydroxyl groups (Geller et al 2000).

Recent work has confirmed the importance of this intramolecular interaction. The L810 mutation renders MR relatively resistant to chymotrypsin digestion, suggesting a conformational change in the receptor, and also stabilizes the interaction of progesterone with the receptor. Rafestin-Oblin and colleagues created a structural model of the receptor based on homology with PR, and found evidence for bending of helix 3 which was triggered apparently by steric hindrance between Leu810 and Gln776 on helix 3, resulting in H3 shift of 1.3Å for Gln776 and 0.8Å for Ala773. They surmised that steroid ligands with short side chains that could be anchored by MR_{L810} without allowing stabilizing contacts with MR_{L810} might function as receptor antagonists, and screened such molecules. In this way, they identified 5-pregnan-20-one, a progesterone analog lacking the 3-ketone group, and 4,9-androstadiene-3,17-dione, and the progesterone antagonist RU486 as inhibitors of MR_{L810}. While this finding has not yet translated into a therapeutic benefit for patients harboring MR_{L810}, it does highlight our growing ability to use insights from receptor structural models to direct the design of receptor agonists and antagonists.

Recently, however, Pinon et al. used structural analysis of the receptor-ligand interaction to design compounds capable of inhibiting MR_{L810}. Analysis of an MR_{L810} structural model suggested that the S810L mutation causes a bending of helix 3 because of steric hindrance of L810 with Gln776, thus altering the positioning of MR_{WT} agonists within the ligand-binding cavity. They used this model to identify ligands with short side chains which could be anchored by MR_{L810}, but not by MR_{WT}. They identified two steroids in this fashion, 5α-pregnan-20-one (P1), lacking the 3-ketone group but harboring the 17-methyl ketone group of progesterone, and 4,9-androstadiene-3,17-dione (A1), characterized by a 3- and 17-ketone. Consistent with their model, these agents proved capable of binding to and inhibiting aldosterone-mediated MR_{L810} activity, and furthermore, A1 had no activity, and P1 had decreased inhibitory activity on MR_{WT} (Pinon et al 2004).

In a related approach, Pinon et al questioned whether RU486, a potent antiglucocorticoid and antiprogestin which lacks antimineralocorticoid properties, could affect MR_{L810}. RU486 is a progesterone derivative carrying a bulky side chain at the 11-hydroxyl position. Previous work had demonstrated that alteration of helix 3 residues in PR brings on steric hindrance with this bulky side chain and prevents antiprogestin activity in PR. They surmised that the bending of helix 3 induced by the S810L mutation might allow RU486 to bind to the receptor and inhibit its activity, and they were able to show inhibition of MR_{L810}, but not MR_{WT}, by RU486 (Pinon et al 2004). Unfortunately, while the 1 μM dose which inhibited MR_{L810} in vitro is likely achievable in humans, the antiglucocorticoid and antiprogestin effects of RU486 would likely prevent its chronic use as a specific antihypertensive agent in these individuals (Sarkar 2002).

GENETICS

Hypertension exacerbated by pregnancy is an extraordinarily rare disease, having been identified only in the one kindred described here. This is despite the screening of hundreds of other kindreds with clinical histories suggestive of Mendelian hypertension as well as significant numbers of females with hypertension during pregnancy (D. Geller, unpublished observations; Sarkar 2002). While it is possible that there are other families with this clinical picture, we suspect this is unlikely. Biochemical and genetic arguments suggest why the disease is so rare.

In diseases stemming from loss-of-function mutations, such as pseudohypoaldosteronism type 1, Bartter's syndrome, and polycystic kidney disease to name a few, a wide variety of missense, frameshift, nonsense, and splice-site mutations can result in a functionless allele. Even in other genetic kidney diseases bearing a gain-of-function of the relevant protein, such as Liddle's syndrome and pseudohypoaldosteronism type 2, a number of permissive mutations have been identified. Our analysis of MRs bearing mutations on helices 3 and 5 suggests that there are only two permissive residues for progesterone-mediated activation of MR, leucine and methionine, at residue 810. Leucine is coded for by TTA, TTG, or CTN, while methionine is exclusively coded for by ATG. The nucleotide sequence of *MR* codon 810 is TCA, so the only realistic mutation capable of creating a permissive residue is the C2645T mutation seen in this kindred. Thus, hypertension exacerbated by pregnancy seems to be that rare syndrome in which only a single, highly specific nucleotide substitution exists which is capable of causing the genetic syndrome. Furthermore, in addition to the incredible genotypic precision required for the hypertension exacerbated by pregnancy phenotype, there is an inherent negative selection bias which may have limited the incidence of this disease. Female carriers of the MR_{L810} allele were likely incapable of transmitting the allele to their children until recent times, as successful pregnancy requires a high degree of medical sophistication. Thus, even if the MR_{L810} mutation arose previously in history, the likelihood of the mutation's persistence into the modern era is substantially lower than for other diseases. These factors likely account for the rarity of the disease.

A SPLICING DEFECT IN MR_{L810} CARRIERS

Efforts to create a mouse model of hypertension exacerbated by pregnancy have been frustrated by a surprising finding. While we have successfully knocked the MR_{L810} allele into the mouse genome, the mice are not hypertensive, nor are their aldosterone levels suppressed (D. Geller, unpublished observations). Evaluation of RNA expression in the mutant mice revealed the reason why – there is clear evidence of exon skipping of the mutant allele, suggesting that the C2645T mutation disrupts an exonic splicing enhancer necessary for proper splicing of the exon. Our estimate is that the mutant mRNA is present at 20% of the level of the wild-type allele.

Due to this finding, we were curious as to whether exon skipping might play a role in humans as well. Analysis of RNA prepared from lymphocytes from a human MR_{L810} carrier again demonstrated exon skipping of the mutant allele. In contrast to the situation in mice, however, our estimate is that the mutant allele is expressed at approximately 50% of the level of expression of the wild-type allele (data not shown), perhaps offering an explanation why the MR_{L810} allele has much larger effects on humans than on mice. On the other hand, the significantly decreased quantity of mutant receptor suggests that hypertension in MR_{L810} carriers might be much worse if there were no such deficit in splicing. To our knowledge, this is the first example of a detrimental disease phenotype being attenuated by an RNA splicing defect. Unfortunately, at present, mechanisms underlying RNA splicing in this context are poorly understood, and so it is unclear whether this process is regulated, or whether, for example, splicing efficiency differs among individuals. As such, it is not clear whether maneuvers to alter MR mRNA splicing could affect blood pressure in MR_{L810} carriers, or conceivably, in the general population.

PERSPECTIVES

The demonstration that loss-of-function mutations in MR lead to salt-wasting and hypotension, while gain-of-function mutations lead to hypertension, is consistent with insights gained from all other Mendelian disorders primarily affecting blood pressure. Mutations which increase renal sodium reabsorption increase blood pressure, while those that decrease renal sodium reabsorption decrease blood pressure (Lifton et al 2001). Among these monogenic forms of hypertension, there is no clear relation between potassium handling, calcium handling, or vascular mechanisms to blood pressure, although these have all been proposed to be relevant to the development of hypertension.

The findings described above, coupled with the findings derived from patients with Liddle's syndrome and arPHA1, add insight to questions concerning extrarenal and non-genomic actions of aldosterone. The finding that aldosterone blockade connotes significant morbidity and mortality benefits on patients with congestive heart failure and/or renal disease in the absence of a blood pressure altering effect (Pitt et al 1999, Sato et al 2003) has raised the question of whether aldosterone may have deleterious effects on cardiac physiology above and beyond its effects on renal sodium handling. The data from these Mendelian forms of HTN suggest that the primary effect of aldosterone in human disease is mediated via its effects on renal sodium handling. Patients with Liddle's syndrome or hypertension exacerbated by pregnancy have significant cardiovascular and renal morbidity and mortality despite a virtual absence of circulating angiotensin II and aldosterone, while patients with pseudohypoaldosteronism type 1 have no evidence of structural heart disease despite high circulating levels of these hormones. Stated simply, excess renal sodium reabsorption is both necessary and sufficient for the induction of cardiovascular disease while aldosterone (or angiotensin II) is neither necessary nor sufficient. Whether aldosterone induces excess cardiovascular morbidity in the salt-overloaded condition above and beyond its effects on renal sodium reabsorption remains an open question.

Similarly, these data shed light on the role of non-genomic activities of aldosterone. Patients with PHA1 have some of the highest aldosterone levels ever observed, and yet, they are at risk of death from salt-wasting and hypotension. In contrast, patients with Liddle's syndrome and hypertension exacerbated by pregnancy have minimal aldosterone production, and yet, they suffer from severe hypertension and cardiovascular disease (Warnock 2001). These findings suggest that the non-genomic mechanisms mediated by aldosterone via an as yet unidentified membrane receptor (Losel et al 2004), while they may have relevant short-term effects on cardiovascular indices such as vascular tone or heart rate, do not have major effects on the ultimate development of cardiovascular disease in these patients.

References

Arai K, Tsigos C, Suzuki Y, et al. No apparent mineralocorticoid receptor defect in a series of sporadic cases of pseudohypoaldosteronism. J. Clin. Endocrinol. Metab. 1995; 814–17.

Arriza JL, Weinberger C, Cerelli G, et al. Cloning of human mineralocorticoid receptor complementary DNA: structural and functional kinship with the glucocorticoid receptor. Science 1987; 268–75.

Benchimol C, Zavilowitz B, Satlin LM. Developmental expression of ROMK mRNA in rabbit cortical collecting duct. Pediatr. Res. 2000; 46–52.

Berger S, Bleich M, Schmid W, et al. Mineralocorticoid receptor knockout mice: pathophysiology of Na^+ metabolism. Proc. Natl Acad. Sci. USA, 1998; 9424–9.

Bleich M, Warth R, Schmidt-Hieber M, Schulz-Baldes A, et al. Rescue of the mineralocorticoid receptor knock-out mouse. Pflugers Arch. 1999; 245–54.

Boulkroun S, Fay M, Zennaro MC, et al. Characterization of rat NDRG2 (N-Myc downstream regulated gene 2), a novel early mineralocorticoid-specific induced gene. J. Biol. Chem. 2002; 31506–15.

Brilla CG. Renin-angiotensin-aldosterone system and myocardial fibrosis. Cardiovasc. Res. 2000; 1–3.

Brilla CG, Pick R, Tan LB, Janicki JS, Weber KT. Remodeling of the rat right and left ventricles in experimental hypertension. Circ. Res. 1990; 1355–64.

Brilla CG, Weber KT. Reactive and reparative myocardial fibrosis in arterial hypertension in the rat. Cardiovasc. Res. 1992; 671–7.

Chang SS, Grunder S, Hanukoglu A, et al. Mutations in subunits of the epithelial sodium channel cause salt wasting with hyperkalaemic acidosis, pseudohypoaldosteronism type 1. Nat. Genet. 1996; 248–53.

Epstein M. Aldosterone receptor blockade and the role of eplerenone: evolving perspectives. Nephrol. Dial. Transplant. 2003; 1984–92.

Gekle M, Freudinger R, Mildenberger S, Schenk K, Marschitz I, Schramek H. Rapid activation of Na^+/H^+-exchange in MDCK cells by aldosterone involves MAP-kinase ERK1/2. Pflugers Arch. 2001; 781–6.

Geller DS, Farhi A, Pinkerton N, et al. Activating mineralocorticoid receptor mutation in hypertension exacerbated by pregnancy. Science 2000; 119–23.

Geller DS, Rodriguez-Soriano J, Vallo Boado A, et al. Mutations in the mineralocorticoid receptor gene cause autosomal dominant pseudohypoaldosteronism type I. Nature Genetics 1998; 279–81.

Geller DS, Zhang J, Zennaro CM, et al. Genotype–phenotype correlations in autosomal dominant pseudohypoaldosteronism type 1. (Manuscript in preparation.)

Grundy HM, Simpson SA, Tait JF. Isolation of a highly active mineralocorticoid from beef adrenal extract. Nature 1952; 795–6.

Gunaruwan P, Schmitt M, Taylor J, Lee L, Struthers A, Frenneaux M. Lack of rapid aldosterone effects on forearm resistance vasculature in health. J. Renin Angiotensin Aldosterone Syst. 2002; 123–5.

Hanukoglu A. Type I pseudohypoaldosteronism includes two clinically and genetically distinct entities with either renal or multiple target organ defects. J. Clin. Endocrinol. Metab. 1991; 936–44.

Hentze MW, Kulozik AE. A perfect message: RNA surveillance and nonsense-mediated decay. Cell 1999; 307–10.

Huey CL, Riepe FG, Sippell WG, Yu AS. Genetic heterogeneity in autosomal dominant pseudohypoaldosteronism type I: Exclusion of claudin-8 as a candidate gene. Am. J. Nephrol. 2004; 483–7.

Juurlink DN, Mamdani MM, Lee DS, et al. Rates of hyperkalemia after publication of the Randomized Aldactone Evaluation Study. N. Engl. J. Med. 2004; 543–51.

Komesaroff PA. Pseudohypoaldosteronism: options for consideration. Steroids 1995; 168–72.

Koppel H, Christ M, Yard BA, Bar PC, van der Woude FJ, Wehling M. Nongenomic effects of aldosterone on human renal cells. J. Clin. Endocrinol. Metab. 2003; 1297–302.

Le Menuet D, Viengchareun S, Penfornis P, Walker F, Zennaro MC, Lombes M. Targeted oncogenesis reveals a distinct tissue-specific utilization of alternative promoters of the human mineralocorticoid receptor gene in transgenic mice. J. Biol. Chem. 2000; 7878–86.

Lifton RP, Gharavi AG, Geller DS. Molecular mechanisms of human hypertension. Cell 2001; 545–56.

Losel R, Schultz A, Boldyreff B, Wehling M. Rapid effects of aldosterone on vascular cells: clinical implications. Steroids 2004; 575–8.

Moras D, Gronemeyer H. The nuclear receptor ligand-binding domain: structure and function. Curr. Opin. Cell Biol. 1998; 384–91.

Morton JA. The clinical usefulness of breast milk sodium in the assessment of lactogenesis. Pediatrics 1994; 802–6.

Mune T, Rogerson FM, Nikkila H, Agarwal AK, White PC. Human hypertension caused by mutations in the kidney isozyme of 11 beta-hydroxysteroid dehydrogenase. Nat. Genet. 1995; 394–9.

Neville MC, Keller R, Seacat J, et al. Studies in human lactation: milk volumes in lactating women during the onset of lactation and full lactation. Am. J. Clin. Nutr. 1988; 1375–86.

Nystrom AM, Bondeson ML, Skanke N, et al. A novel nonsense mutation of the mineralocorticoid receptor gene in a Swedish family with pseudohypoaldosteronism type I (PHA1). J. Clin. Endocrinol. Metab. 2004; 227–31.

Pinon GM, Fagart J, Souque A, Auzou G, Vandewalle A, Rafestin-Oblin ME. Identification of steroid ligands able to inactivate the mineralocorticoid receptor harboring the S810L mutation responsible for a severe form of hypertension. Mol. Cell Endocrinol. 2004; 181–8.

Pitt B, Remme W, Zannad F, et al. Eplerenone, a selective aldosterone blocker, in patients with left ventricular dysfunction after myocardial infarction. N. Engl. J. Med. 2003; 1309–21.

Pitt B, Zannad F, Remme WJ, et al. The effect of spironolactone on morbidity and mortality in patients with severe heart failure. Randomized Aldactone Evaluation Study Investigators. N. Engl. J. Med. 1999; 709–17.

Rachmani R, Slavachevsky I, Amit M, et al. The effect of spironolactone, cilazapril and their combination on albuminuria in patients with hypertension and diabetic nephropathy is independent of blood pressure reduction: a randomized controlled study. Diabet. Med. 2004; 471–5.

Rafestin-Oblin ME, Souque A, Bocchi B, Pinon G, Fagart J, Vandewalle A. The severe form of hypertension caused by the activating S810L mutation in the mineralocorticoid receptor is cortisone related. Endocrinology 2003; 528–33.

Riepe FG, Krone N, Morlot M, Ludwig M, Sippell WG, Partsch CJ. Identification of a novel mutation in the human mineralocorticoid receptor gene in a German family with autosomal-dominant pseudohypoaldosteronism type 1: further evidence for marked interindividual clinical heterogeneity. J. Clin. Endocrinol. Metab. 2003; 1683–6.

Sarkar NN. Mifepristone: bioavailability, pharmacokinetics and use-effectiveness. Eur. J. Obstet. Gynecol. Reprod. Biol. 2002; 113–20.

Sartorato P, Khaldi Y, Lapeyraque AL, et al. Inactivating mutations of the mineralocorticoid receptor in Type I pseudohypoaldosteronism. Mol. Cell Endocrinol. 2004; 119–25.

Sartorato P, Lapeyraque AL, Armanini D, et al. Different inactivating mutations of the mineralocorticoid receptor in fourteen families affected by type I pseudohypoaldosteronism. J. Clin. Endocrinol. Metab. 2003; 2508–17.

Sato A, Hayashi K, Naruse M, Saruta T. Effectiveness of aldosterone blockade in patients with diabetic nephropathy. Hypertension 2003; 64–8.

Schulz-Baldes A, Berger S, Grahammer F, et al. Induction of the epithelial Na^+ channel via glucocorticoids in mineralocorticoid receptor knockout mice. Pflugers Arch. 2001: 297–305.

Stowasser M, Gordon RD. Primary aldosteronism – careful investigation is essential and rewarding. Mol. Cell Endocrinol. 2004; 33–9.

Tajima T, Kitagawa H, Yokoya S, et al. A novel missense mutation of mineralocorticoid receptor gene in one Japanese family with a renal form of pseudohypoaldosteronism type 1. J. Clin. Endocrinol. Metab. 2000; 4690–4.

Verrey F, Pearce D, Pfeiffer R, et al. Pleiotropic action of aldosterone in epithelia mediated by transcription and post-transcription mechanisms. Kidney Int. 2000; 1277–82.

Viemann M, Peter M, Lopez-Siguero JP, Simic-Schleicher G, Sippell WG. Evidence for genetic heterogeneity of pseudo-hypoaldosteronism type 1: identification of a novel mutation in the human mineralocorticoid receptor in one sporadic case and no mutations in two autosomal dominant kindreds. J. Clin. Endocrinol. Metab. 2001; 2056–9.

Warnock DG. Liddle syndrome: genetics and mechanisms of Na$^+$ channel defects. Am. J. Med. Sci. 2001; 302–7.

Zennaro MC, Borensztein P, Jeunemaitre X, Armanini D, Soubrier F. No alteration in the primary structure of the mineralocorticoid receptor in a family with pseudohypoaldosteronism. J. Clin. Endocrinol. Metab. 1994; 32–8.

Zennaro MC, Farman N, Bonvalet JP, Lombes M. Tissue-specific expression of alpha and beta messenger ribonucleic acid isoforms of the human mineralocorticoid receptor in normal and pathological states. J. Clin. Endocrinol. Metab. 1997; 1345–52.

Zennaro MC, Lombes M. Mineralocorticoid resistance. Trends Endocrinol. Metab. 2004; 264–70.

CHAPTER 19

The Syndrome of Hypertension and Hyperkalemia (Pseudohypoaldosteronism Type II): WNK Kinases Regulate the Balance Between Renal Salt Reabsorption and Potassium Secretion

KRISTOPHER T. KAHLE, FREDERICK H. WILSON AND RICHARD P. LIFTON

THE ROLE OF THE KIDNEY IN THE REGULATION OF BLOOD PRESSURE AND ELECTROLYTE HOMEOSTASIS

The mechanisms governing the homeostasis of water, sodium, chloride, and potassium in the extracellular fluid compartment are interdependent processes that are essential for the survival of all land vertebrates. Normal sodium and chloride homeostasis is critical for maintaining an effective circulating volume (and hence blood pressure), which provides tissues with adequate oxygen and nutrients. The precise regulation of serum potassium is essential for the electrical activity of excitable cells such as neurons in the central and peripheral nervous systems, and myocytes in smooth, skeletal, and cardiac muscle. The nephron determines the level of sodium, chloride, and potassium in the extracellular space, in addition to the volume of fluid in the intravascular compartment, by regulating the reasborption of water and the transport of ions. This complex homeostatic process is achieved by a diverse cohort of proteins that function as ion channels, transporters, pumps, and paracellular pores. Over the last half century, the genetic identity and electrophysiological properties of these individual proteins have been well elucidated; however, the mechanisms that allow their individual activities to be coordinated

to maintain homeostasis during periods of physiologic perturbation are largely unknown.

The fine-tuning of electrolyte and fluid homeostasis occurs in the distal nephron, encompassing the distal convoluted tubule (DCT), the connecting tubule (CNT), and the collecting duct (CD). In these nephron segments, sodium reabsorption occurs via both electroneutral and electrogenic pathways, mediated by the Na-Cl cotransporter NCC and the epithelial sodium channel ENaC, respectively. Electrogenic sodium reabsorption via ENaC creates a lumen-negative electrical potential, driving both chloride reabsorption via a paracellular pathway and the secretion of potassium and hydrogen ions. Potassium secretion normally occurs via the inwardly rectifying renal outer medullary potassium channel ROMK, and proton secretion occurs via the multisubunit H^+-ATPase.

The steroid hormone aldosterone, secreted from the glomerulosa of the adrenal cortex, is the major regulator of salt reabsorption, and potassium and hydrogen ion secretion, in the distal nephron. Aldosterone increases the activity of ENaC, which creates an electronegative intraluminal potential that drives the reabsorption of chloride along the paracellular pathway, and secretion of potassium and hydrogen ions across the apical membrane of CCD cells. The secretion of aldosterone is coupled to two

Genetic Diseases of the Kidney
ISBN: 978-0-12-449851-8

different physiologic stimuli. In the setting of decreased intravascular volume, activation of the renin-angiotensin system leads to increased angiotensin II levels, which stimulate aldosterone secretion via activation of the angiotensin II (AT) receptor. Alternatively, elevated levels of serum potassium can also trigger the secretion of aldosterone; in this case, potassium ions act as a direct secretagog of the hormone. Aldosterone exerts its genomic effects by binding the mineralocorticoid receptor in cells of the distal nephron; the activated mineralocorticoid receptor, in turn, binds specific response elements encoded in the DNA of promoter regions of various aldosterone-responsive genes, and alters mRNA transcription. Aldosterone also has nongenomic effects, such as stimulating the insertion or retrieval of protein ion channels and transporters into and out of the plasma membrane. How the kidney distinguishes aldosterone signaling in the distinct physiologic states of intravascular volume depletion versus elevated serum potassium to provide an appropriate homeostatic response to each perturbation, is unclear. Prevailing views have been that this is achieved secondarily as a result of alterations in fluid flow and/or salt delivery to the distal nephron; however, an indirect mechanism for such a critical process seems unlikely.

PSEUDOHYPOALDOSTERONISM TYPE II: A HUMAN MODEL TO EXPLORE COORDINATED REGULATION OF BLOOD PRESSURE AND ELECTROLYTE HOMEOSTASIS

The term 'pseudohypoaldosteronism' has been used to describe the phenotype of chronic hyperkalemia despite normal glomerular filtration rate and normal or elevated levels of aldosterone (Dillon et al 1980, Popow et al 1988). The persistent hyperkalemia in these patients suggests that aldosterone levels are below normal, but they are in fact within the normal range or elevated. The classic form of pseudohypoaldosteronism (PHA), PHA type I (PHAI), is a Mendelian form of hypotension due to renal salt wasting, and is also characterized by hyperkalemia, metabolic acidosis, and elevated aldosterone levels. PHAI can be transmitted as either an autosomal dominant or recessive trait. Loss-of-function mutations in the mineralocorticoid receptor cause autosomal dominant PHAI (Geller et al 1998). These mutations prevent the activation of ENaC via the renin-angiotensin system, thus impairing sodium reabsorption. Because ENaC activity establishes the lumen-negative potential that supports potassium and hydrogen ion secretion in the CCD, decreased ENaC activity reduces the driving force for the secretion of these ions, resulting in hyperkalemia and metabolic acidosis in affected patients. Autosomal recessive PHAI is caused by loss-of-function mutations in any of the three different subunits (alpha, beta, and gamma) of ENaC (Chang et al 1996).

These mutations render ENaC functionally inactive and unresponsive to the stimulatory action of aldosterone.

Though in fact a misnomer, a second clinical entity of 'pseudohypoaldosteronism' has been described in which patients present with hyperkalemia despite normal glomerular filtration rate and hypertension, instead of the hypotension, observed in PHAI. This disease is termed pseudohypoaldosteronism type II (PHAII; Online Mendelian Inheritance in Man#145260; also known as familial hyperkalemia and hypertension or Gordon's syndrome). The disorder has been of particular interest because the unusual association of hypertension with hyperkalemia is not readily explained by known physiological mechanisms. The pathophysiological basis of PHAI is the distal nephron's inability to respond to aldosterone, leading to salt wasting and decreased potassium secretion. In contrast, in PHAII, only the kaliuretic effect of aldosterone is impaired, and patients actually present with hypertension due to increased renal salt reabsorption. As such, the association of hypertension with hyperkalemia in PHAII appears to represent a dissociation of the two primary functions of aldosterone: salt reabsorption and potassium secretion. This rare pathophysiologic event raises the possibility that the genetic lesion underlying PHAII, unlike other Mendelian forms of hypertension whose genetic bases have been identified, might reside within a novel regulatory pathway other than the renin angiotensin system, since the PHAII phenotype is inconsistent with a simple up- or downregulation of the renin-angiotensin pathway. Alternatively, the genetic defect underlying PHAII could provide insight into the ways in which existing systems, like the renin-angiotensin system, might be 'switched' into previously unrecognized regulatory states.

In 1964, the first case report documenting the association of hyperkalemia and hypertension in the absence of renal insufficiency in a 15-year-old Australian male was published (Paver & Pauline 1964). The patient had hypertension with a blood pressure measurement of 180/120 mmHg, and measurement of serum electrolytes revealed elevated serum potassium levels, ranging from 7.0 to 8.2 mM (3.5–5.0 mM, normal) in the setting of normal glomerular filtration. Based on these findings, the authors proposed a novel congenital disorder of the renal tubule. Subsequent laboratory examinations of this patient revealed suppressed plasma renin levels and normal aldosterone levels (Stokes et al 1968).

A second patient, also Australian, was identified by Gordon and colleagues (Gordon et al 1970). This individual was a 10-year-old female unrelated to the initial patient reported by Paver and Pauline. The child presented with an elevated blood pressure of 160/110 mmHg, a serum potassium of 8.5 mM, and metabolic acidosis. The child also had short stature and suffered from muscular weakness and periodic paralysis. Aldosterone levels were within the normal range despite severely elevated serum potassium levels.

The authors noted that the patient's hypertension, hyperkalemia, and metabolic acidosis were all corrected by restricting dietary sodium intake.

Since the description of these two case reports in Australia, over 70 patients with hypertension and persistent hyperkalemia despite normal renal function have been reported worldwide. Greater than 70% of these patients have at least one family member also affected with the disorder, indicating a genetic basis for the condition. With the identification and thorough characterization of more affected patients, the heterogeneity of the phenotype has become apparent: while hypertension is present in most but not all adults, chronic hyperkalemia in the absence of renal insufficiency is the most consistent feature of the disorder. Age, gender, and dietary salt intake appear to influence the presence of hypertension in patients with PHAII, as is often the case for essential hypertension in the general population. Also similar to patients with essential hypertension, many PHAII patients under the age of 20 have normal blood pressure, but then develop hypertension in adulthood. Other associated features of PHAII include suppressed plasma renin levels, hyperchloremia, and metabolic acidosis. Serum aldosterone levels vary widely, but most would consider aldosterone levels low relative to the degree of hyperkalemia. The variability of serum aldosterone levels among patients with PHAII has led some to question the utility (and accuracy) of the term 'pseudohypoaldosteronism' to describe the disease.

An intriguing feature of PHAII that has proven to be consistent in affected individuals is the observation that all clinical phenotypes can be corrected with thiazide diuretics, drugs that specifically inhibit the activity of the sodium-chloride cotransporter (NCC) in the DCT (Gordon et al 1995). Thiazide diuretics are also an effective therapeutic tool in the treatment of hypertension in the general population (Unwin et al 1995). This similarity, together with the influences of age and sex on the development of hypertension in both PHAII patients and the general population, has led to speculation that the mechanistic pathway(s) altered in PHAII might prove directly relevant to hypertension in the general population. The sensitivity of PHAII patients to the action of thiazide diuretics also suggests that a primary abnormality underlying the pathophysiology of the disease involves an increase in the activity of NCC.

HYPOTHESES OF PHAII PATHOPHYSIOLOGY

In light of the conflicting clinical descriptions of PHAII, it is perhaps not surprising to find a number of different proposals in regard to the primary pathophysiological abnormality underlying the disease. An early proposal suggested a defect in the aldosterone-sensitive distal nephron's

secretory pathway for potassium (Arnold & Healy 1969, Spitzer et al 1973, Weinstein et al 1974). This would explain the apparent 'uncoupling' of the effects of aldosterone, since the expected increase in sodium reabsorption would occur in response to aldosterone, but the kaliuretic response to aldosterone would be impaired. Clinical studies, however, have since demonstrated that PHAII patients are in fact capable of normal kaliuresis under certain conditions (Farfel et al 1978a, 1978b, Schambelan et al 1981).

Another theory by Farfel and colleagues proposed a generalized cell membrane defect that impedes the influx of potassium into cells (Farfel et al 1978a). In support of this hypothesis, the authors reported that glucose and insulin infusions failed to reduce serum potassium levels in two related PHAII patients. However, this proposal failed to address the hypertension or any of the other observed electrolyte abnormalities in PHAII. Moreover, normal responses to these clinical tests were reported in other PHAII patients (Bravo et al 1980).

Two other prominent theories concerning PHAII pathogenesis are based on conflicting data sets derived from clinical studies that were designed to evaluate the response of PHAII patients to mineralocorticoids. While several studies reported a normal response to the administration of mineralocorticoids in PHAII patients (i.e. serum potassium levels were reduced to normal upon treatment) (Gordon et al 1970, Licht et al 1985, Travis & Cushner 1986), studies in other PHAII patients have indicated an impaired response to mineralocorticoids (since serum potassium levels are not reduced to normal) (Weinstein et al 1974, Schambelan et al 1981, Nahum et al 1986). Gordon and colleagues reported that all abnormalities in their PHAII patients were corrected with dietary salt restriction (Gordon et al 1970). A significant reduction in dietary sodium intake is expected to increase aldosterone secretion via the renin-angiotensin system. A correction of all abnormalities suggested a normal response to mineralocorticoid, leading the authors to conclude that the primary pathophysiological defect in PHAII does not reside at the site of aldosterone action (the distal nephron), because patients exhibited the expected kaliuretic response to aldosterone. Instead, Gordon and colleagues proposed that the primary defect in PHAII is excessive sodium reabsorption at a site in the nephron *proximal* to the aldosterone-sensitive nephron, thus causing plasma volume expansion (leading to hypertension), and subsequent down-regulation of renin and aldosterone (leading to hyperkalemia from aldosterone suppression). In support of this hypothesis, Gordon reported increased sodium reabsorption in the proximal tubules of three different PHAII patients based on measurements of lithium clearance relative to nine controls (Klemm et al 1991); however, these findings have not been confirmed by other investigators.

An alternative theory was proposed by Schambelan and colleagues after their characterization of a PHAII patient who showed resistance to the effects of exogenously administered mineralocorticoids (Schambelan et at 1981). In this patient, the hyperkalemia was not corrected with exogenous mineralocorticoid administration, suggesting that the primary defect lies in the aldosterone-sensitive distal nephron. Because mineralocorticoid-resistant hyperkalemia is also present in PHAI (though PHAI patients also have hypotension due to renal wasting), Schambelan et al proposed the name 'pseudohypoaldosteronism type II' to distinguish the two syndromes. Importantly, Schambelan's group also made the observation that renal potassium clearance increased normally when the distal delivery of sodium was increased with sulfate or bicarbonate as counter-ions, demonstrating that the potassium secretory mechanism of the distal nephron is operational in PHAII patients. However, an abnormally low kaliuretic response was observed when the distal delivery of sodium was increased with chloride as the counter-ion, suggesting that the presence of chloride in the distal nephron affects potassium secretion in patients with PHAII.

These findings led Schambelan and colleagues to postulate the existence of a pathophysiologic 'chloride shunt' in the distal nephron that is responsible for the clinical features of PHAII. This shunt is an abnormal chloride conductance caused by increased chloride reabsorption in the distal nephron. Mechanistically, increased chloride reabsorption would reduce the lumen-negative electrical potential difference that normally opposes sodium reabsorption and promotes potassium and hydrogen ion secretion in the distal nephron. Dissipation of this electrochemical potential difference would increase sodium reabsorption, leading to plasma volume expansion and hypertension, and would also contribute to hyperkalemia and metabolic acidosis. The hyperchloremia in PHAII patients could be directly attributed to increased chloride reabsorption along the 'shunt.' Such a shunt could explain why a normal kaliuretic response was observed after administering an increased sodium delivery to the distal nephron with sulfate or bicarbonate, but not with chloride: sulfate and bicarbonate are poorly reabsorbed anions, and replacement of chloride with either of these anions in a patient with PHAII would restore the lumen-negative electrochemical potential, thus correcting all associated physiological abnormalities due to elimination of the chloride shunt. The clinical findings reported by Schambelan and colleagues have been confirmed in three affected individuals from a different PHAII family by other investigators (Take et al 1991).

Perhaps the simplest model to account for the pathogenesis of PHAII is one that has invoked over-activity of the electroneutral Na–Cl cotransporter NCC. PHAII phenotypes are corrected by the administration of thiazide diuretics, inhibitors of NCC, and the phenotypes of PHAII are almost the exact mirror image of Gitelman's syndrome, an autosomal recessive disease featuring hypotension, hypokalemia, metabolic alkalosis, and hypocalciuria due to loss-of-function mutations in NCC (Simon et al 1996, Mayan et al 2002). However,

no significant linkage was observed between PHAII and the *SLC123* locus (encoding NCC) on human chromosome 16 (Mansfield et al 1997). An alternative possibility to account for these phenotypes would be if a gene defect causing PHAII resulted in dysregulation of NCC activity, resulting in an increase in NCC-mediated salt reabsorption.

MOLECULAR GENETICS OF PSEUDOHYPOALDOSTERONISM TYPE II: EVIDENCE FOR GENETIC HETEROGENEITY AND DISCOVERY OF THE WNK PROTEIN KINASES

In the setting of uncertain pathophysiology of PHAII, a strategy of mapping anonymous disease loci using a genome-wide linkage analysis in large PHAII pedigrees was employed. Positional cloning techniques could then be used to identify disease-causing genes within loci linked to the PHAII phenotype.

Mansfield et al performed a genome-wide linkage analysis in eight families with PHAII. Multilocus analysis of linkage provided significant evidence for linkage of PHAII to either of two loci- one on chromosome 17, and another on chromosome 1. The chromosome 17 linkage was supported by significant evidence for linkage in a single family, providing fairly strong support. In contrast, the chromosome 1 linkage was mainly derived from small families, leaving somewhat greater ambiguity about the significance of this locus (Mansfield et al 1997). Disse-Nicodeme et al subsequently found significant linkage of PHAII to a third locus on chromosome 12 in a single large kindred (Disse-Nicodème et al 2000). Additional families have suggested the presence of at least one additional locus. Together, these studies revealed that PHAII is a genetically heterogeneous disorder, with at least four genes responsible for a similar phenotype.

In 2001, Wilson et al identified two genes for PHAII (Figure 19.1) (Wilson et al 2001). The first of these was identified in the fine mapping of the chromosome 12p interval. A newly identified PHAII kindred showed linkage to distal 12p; the genotyping of additional markers in the interval unexpectedly demonstrated 'pseudomisinheritance' – in multiple branches of the kindred, three consecutive markers in affected subjects appeared to be homozygous for alleles inherited from unaffected parents, with no allele transmitted from the affected parent. This led to the recognition that the corresponding chromosome segment was deleted on the disease-linked chromosome. Wilson et al. went on to demonstrate that this deletion was a 41 000 base pair deletion in the first intron of human WNK1, a novel serine-threonine kinase that had just been identified by Xu and colleagues, in a search for MAP (mitogen activated protein)-kinase relatives (Xu et al 2000). Examination of the 12p-linked kindred of Disse-Nicodeme et al identified an independent but

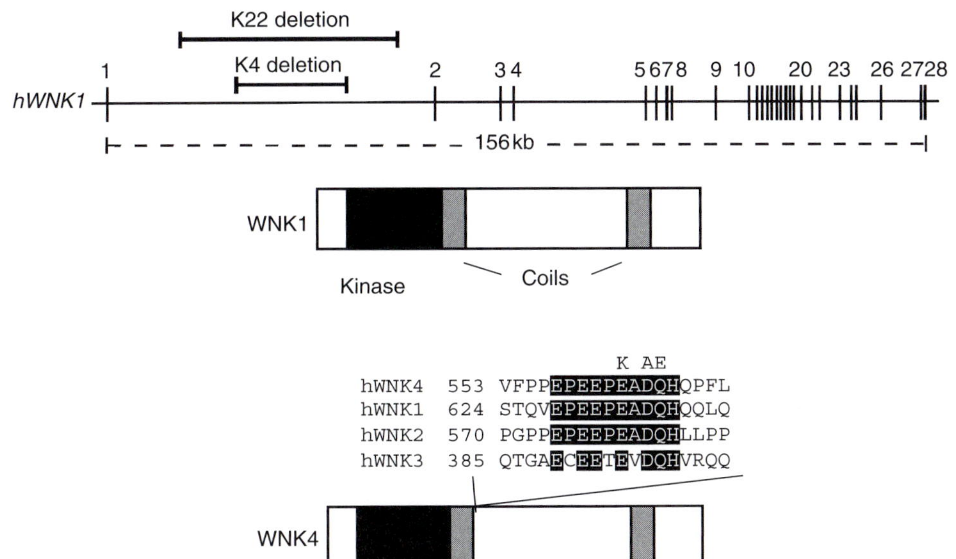

FIGURE 19.1   Mutations in WNK1 and WNK4 cause pseudohypoaldosteronism type II. The general protein structures of WNK1 (encoded on human chromosome 12) and WNK4 (encoded on human chromosome 17) are shown; they have highly homologous N-terminal kinase domains (shown in black), and two coil-coil domains (shown in gray). At the top of the figure, the genomic organization of WNK1 is shown, with its 28 exons spanning 156 kb in genomic DNA. The positions of deletions in the first intron of WNK1 found in two families (K22 and K4) that show genetic linkage to 12p are shown. Above the diagram of WNK4, the amino acid sequence of a short acidic segment of human WNK4 (hWNK4) is shown and compared to the other human WNKs. Missense mutations found in three different PHAII kindreds linked to chromsome 17 are indicated above the sequence – all occur at completely conserved positions and change the charge of the encoded amino acid. (see also Plate 19)

overlapping deletion of 22 000 base pairs in this kindred as well (Wilson et al 2001). These deletions precisely coseg-regated with the disease phenotype in individuals in these families, and were not found in a large series of control subjects, providing strong evidence that these mutations are the cause of PHAII. Expression studies in peripheral blood leukocytes of affected and unaffected relatives of one of the large PHAII families demonstrated that the intronic dele-tion resulted in a five-fold increase in *WNK1* transcript lev-els, implicating increased WNK1 activity as a mechanism of disease (Wilson et al 2001).

In silico database searches by Wilson et al identified three WNK paralogs in the human genome – WNK3 on the X chromosome, WNK2 on chromosome 9, and WNK4 on chromosome 17. The encoded proteins all showed high conservation in the kinase domain and have the distinctive substitution of cysteine for lysine in the active site. *WNK4* mapped to the maximum lod score interval for PHAII on chromosome 17. Characterization of *WNK4* showed it to be encoded in 19 exons contained within 16 kb of genomic DNA; the encoded amino acid sequence of *WNK4* was found to be 76% identical to *WNK1* across a 370 amino acid segment spanning the kinase domain and the first puta-tive coil domain, 51% identical across an 83 amino acid segment encompassing the COOH-terminal putative coil domain, and 52% identical across a 102 amino acid segment from residues 640 to 741 of WNK4 (Wilson et al 2001).

Examination of the WNK4 sequence in PHAII kin-dreds that showed evidence of linkage to chromosome 17 revealed four different missense mutations, all of which cosegregated with the disease (Wilson et al 2001). Three of these were charge-changing substitutions that clustered in a span of four amino acids just distal to the first putative coil domain within a negatively charged 10 amino acid segment that is highly conserved among all members of the WNK family in human, as well as their known orthologs in mouse and rat. A fourth mutation, in PHAII kindred K21, was found to lie just distal to the second putative coil domain, and also changed charge, mutating arginine at position 1185 to cysteine at a residue conserved in three of four WNK kinases. None of these mutations was identified among unrelated unaffected control subjects. The clustering of these mutations and the autosomal dominant nature of the trait argued that these are likely to have a genetic gain-of-function component. Subsequent studies have reported addi-tional missense mutations within the acidic motif of WNK4 in other PHAII families (Golbang et al 2005). Interestingly, the WNK4 gene lies only 1 Mb from locus *D17S1299*, the site showing the strongest linkage to blood pressure varia-tion in the Framingham Heart Study population.

The fact that WNK mutations: (1) segregated with the PHAII phenotype and were not present in unaffected family members; (2) altered gene expression level (as in WNK1) or protein structure (as in WNK4); and (3) were not found in the general population, constituted proof that mutations in WNK1 and WNK4 caused PHAII (Wilson et al 2001). These findings established that the WNK kinases were important novel regulators of renal salt and potassium homeostasis in humans; however, they did not provide ready insight into the mechanisms by which these effects are imparted.

THE WNKs ARE A NOVEL FAMILY OF SALT-SENSITIVE SERINE-THREONINE KINASES WITH A UNIQUE CATALYTIC STRUCTURE

Melanie Cobb and colleagues identified WNK1 as the prototype of a distinct family of serine-threonine kinases in rat. Cloned using PCR with degenerate primers homologous to segments of MAP kinases, WNK1 was identified as a novel kinase with the unique substitution of cysteine for lysine at a nearly invariant position within subdomain II of its catalytic core (hence the name *with no K* = lysine) (Xu et al 2000). This same substitution is found in the four other members of the WNK family. Despite its unique amino acid substitution, WNK1 is an active kinase in vitro. Min et al, by solving the crystal structure of the low-activity conformation of the WNK1 kinase domain at 1.8 angstroms, demonstrated the unique location of the catalytic lysine (K233): in WNK1 this is in the active site arising from beta strand 2 rather than beta strand 3 as in other protein kinases (Min et al 2004). WNK1 and its paralogs all contain an N-terminal kinase domain, two putative coiled-coil domains, multiple SH3 domain binding motifs, and proline-rich segments (Verissimo & Jordan 2001). A conserved autoinhibitory domain distal to the kinase domain has also been identified in WNK paralogs, and expression of this isolated domain is sufficient to inhibit WNK catalytic activity in *trans* (Xu et al 2002, Lenertz et al 2005). In addition to these features, the human WNK family members and their mammalian counterparts all have a very short negatively charged segment distal to their kinase domains; PHAII-causing mutations in WNK4 cluster within this domain (Wilson et al 2001).

Treatment of HEK-293 cells with salt increases WNK1 autophosphorylation, and changes in WNK1 autophosphorylation correlate with the phosphorylation of the substrate myelin basic protein. Subsequent studies have shown that multiple hypertonic stresses, including glucose, sucrose, mannitol, sorbitol, potassium chloride, sodium chloride, and urea, all increase WNK1 activity in mammalian DCT cells (Lenertz et al 2005). In these cells, addition of 0.5 M salt to the culture medium produces a 10-fold increase in WNK1 activity, while 0.5 M mannitol, sorbitol, glucose, sucrose, or urea produce a 3.0–5.0-fold increase in WNK1

activity. Hypotonic stress (reductions of osmolarity below 40 mOsM) also increase WNK1 kinase activity, but to a lesser extent than hypertonic stress. The effect of altered tonicity on WNK1 kinase activity appears to be specific, because forskolin, dexamethasone, fetal bovine serum, aldosterone, H_2O_2, parathyroid hormone, epidermal growth factor, insulin, vasopressin, and transforming growth factor have no effect on WNK1 kinase activity (Lenertz et al 2005). However, there are as yet few data about the regulation of the kinase activity of other WNKs.

WNK1 AND WNK4 LOCALIZE TO THE ALDOSTERONE-SENSITIVE DISTAL NEPHRON AND EXTRARENAL CHLORIDE-TRANSPORTING EXTRARENAL EPITHELIA

WNK4 transcripts are most highly expressed in the kidney. In the kidney, WNK4 localizes to the aldosterone-sensitive distal nephron (the DCT, CNT, and CCD). In these tubule segments, WNK4 localizes to the tight junction and extends along the lateral membrane; WNK4 is also expressed at lower levels in the cytoplasm (Wilson et al 2001, Kahle et al 2004). O'Reilly and co-workers found expression of *WNK4* transcripts in both cortical and medullary loops of Henle via in situ hybridization and rt-PCR (O'Reilly et al 2006). In addition to its renal expression, WNK4 is expressed in chloride-transporting epithelia including bile ducts, pancreatic ducts, the epididymis, colonic crypts, and sweat glands. In these epithelia, WNK4 localizes to tight junctions, with extension along the lateral membrane (Kahle et al 2004).

WNK1's expression pattern is more complicated, in part attributable to the existence of multiple alternatively spliced isoforms. A *WNK1* transcript containing the kinase domain (*L-WNK1*) is expressed in many tissues, with highest expression in the kidney, heart, skeletal muscle, and testis (Wilson et al 2001). In contrast, a shorter kidney-specific isoform (*KS-WNK1*) contains a unique amino-terminal exon and lacks the kinase domain (Delaloy et al 2003). Two different promoters upstream of exon 1 initiate transcription of the full-length isoform, whereas the kidney-specific transcript is generated by a downstream promoter that resides in intron 4.

L-WNK1 transcripts are expressed along the entire nephron at low levels, while *KS-WNK1* transcripts are highly expressed in the distal nephron. In the distal nephron, *KS-WNK1* is expressed most highly in DCT_2, with lower expression in more distal segments (O'Reilly et al 2006). WNK1 localizes predominantly to the cytoplasm in these tubule segments (Wilson et al 2001). WNK1 also localizes to polarized epithelia in extrarenal tissues, with variable expression in the cytoplasm, intercellular junctions, and along the lateral membrane (Choate et al 2003).

WNK1 AND WNK4 REGULATE DIVERSE ALDOSTERONE-SENSITIVE MEDIATORS OF ION TRANSPORT VIA DISTINCT MECHANISMS

The phenotype of PHAII, coupled with the localization of WNK1 and WNK4 to distal nephron segments known to be involved in salt and potassium homeostasis, suggested that these proteins might regulate the activities of one or more mediators of sodium, chloride, and potassium flux. Logical targets include the sodium and chloride reabsorption pathways mediated by NCC and ENaC; the mediators and regulators of distal nephron potassium secretion, including ROMK, the BK potassium channel, and ENaC; the paracellular chloride flux pathway; and the apical H$^+$-ATPase involved in distal proton secretion (Figure 19.2).

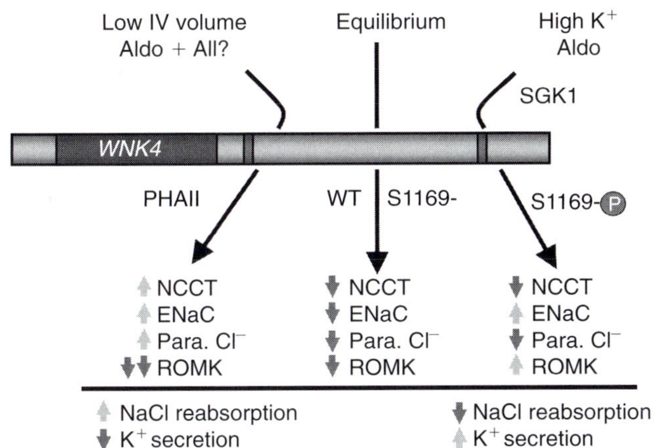

FIGURE 19.2 WNK4 is a switch that regulates the balance between renal salt reabsorption and K$^+$ secretion. WNK4 is shown to exist in three distinct states that enable it to direct the kidney to modulate the balance between renal Na–Cl reabsorption and K$^+$ secretion. In a ground or equilibrium state, WNK4 inhibits mediators of both Na–Cl reabsorption and K$^+$ secretion. In a second state, exemplified by WNK4 harboring PHAII mutations, WNK4 loses inhibition of Na–Cl reabsorbing pathways, including NCCT, ENaC and paracellular Cl$^-$ flux, but augments inhibition of K$^+$ secretion via ROMK, resulting in increased salt reabsorption and decreased K$^+$ secretion. This state is the appropriate response to intravascular volume depletion, a condition accompanied by high levels of aldosterone and angiotensin II. A third state, characterized by WNK4 with phosphorylation at S1169 via the kinase SGK1, results in inhibition of the Na–Cl cotransporter NCCT, but increased activity of the epithelial sodium channel (ENaC) and increased activity of the K$^+$ channel ROMK, resulting in increased K$^+$ secretion without increased Na$^+$ reabsorption. (Reprinted with permission from Annual Reviews of Physiology, 2007 15:48). (see also Plate 20)

NCC stood out as a potential target because PHAII phenotypes constitute a mirror image of Gitelman syndrome, the phenotype resulting from homozygous loss of NCC function (Simon et al 1996). Subjects with Gitelman syndrome uniformly have reduced blood pressure, hypokalemia and metabolic alkalosis (Cruz et al 2001), the opposite of PHAII. This observation raised the question of whether increased NCC function might account for the physiologic features of PHAII. Moreover, the physiologic abnormalities in PHAII are all corrected by thiazide diuretics, which are specific pharmacologic antagonists of NCC (Hadchouel et al 2006).

The marked impairment in renal potassium secretion is the most consistent feature of PHAII and suggests several potential targets. The potassium channel ROMK is the major mediator of renal potassium secretion in the CNT/CCD under normal circumstances (Wang et al 1997), and is one possible target of WNK4. ENaC, the major pathway for electrogenic sodium reabsorption in the CNT and CCD (Reilly & Ellison 2000), is another potential target of the WNKs. ENaC activity produces the lumen-negative potential and is required for normal potassium secretion, as revealed by inherited and acquired alteration in ENaC activity. States featuring increased ENaC activity often cause hypokalemia owing to the increased lumen-negative potential. Conversely, mutations that result in loss of ENaC function lead to impaired production of a lumen-negative potential and often profound hyperkalemia (Lifton et al 2001). Thus decreased ENaC activity could explain the hyperkalemia of PHAII, but this effect alone could not explain the hypertension seen in this disorder. The paracellular chloride flux pathway is another possible target of WNK activity. Electrogenic sodium reabsorption via ENaC can be used either for cation secretion or for anion reabsorption. The major electrogenic anionic flux pathway in the distal nephron is selective paracellular chloride flux across the tight junction (Reilly & Ellison 2000); increased permeability of this pathway could have the dual effects of maximizing reabsorption and also dissipating the electronegative potential of the lumen, thereby inhibiting potassium secretion.

WNK4 and NCC

The first target studied in detail was NCC, using expression studies in *Xenopus* oocytes, the system from which this transporter was initially identified by expression cloning. Expression of NCC in these oocytes induces robust thiazide-sensitive influx of sodium as measured via ^{22}Na flux (Gamba et al 1993). Similar work was performed by Wilson et al and Yang et al. Coexpression of wild-type WNK4 with NCC produced a marked inhibition of thiazide-sensitive ^{22}Na flux (Wilson et al 2003, Yang et al 2003). Addition of a green fluorescent protein tag on NCC permitted examination of the mechanism of this inhibition, and demonstrated that WNK4's inhibitory effect is mediated via loss of NCC at the cell surface (Wilson et al 2003). Introduction of a single amino acid substitution essential for kinase activity eliminated WNK4's inhibition of NCC, demonstrating a required role for catalytic activity for inhibition of this target (Wilson et al 2003). Critically, missense mutations in WNK4 that recapitulate mutations seen in patients with PHAII lose their inhibitory effect on NCC (Wilson et al 2003, Yang et al 2003). As a consequence, these in vitro experiments predicted that PHAII-mutant WNK4 would result in increased NCC activity in vivo. Subsequent work in mammalian cell lines has led to similar conclusions, showing that WNK4 reduces NCC activity by reducing its cell surface expression via a clathrin-independent pathway (Cai et al 2006, Golbang et al 2006). Inhibition of the lysosomal pathway with baflomycin also reduces WNK4's inhibition of NCC, implicating the lysosomal pathway in WNK4's inhibition of NCC (Cai et al 2006).

WNK4 and ROMK

The second target tested was ROMK (Kahle et al 2003), which was also expression cloned in the *Xenopus* oocyte system (Wang et al 1997). Co-expression of wild-type WNK4 with ROMK in oocytes demonstrated robust inhibition of ROMK (Wang et al 1997). As for NCC, reduced expression of ROMK at the cell surface was shown to be the mechanism. This reduction in surface expression was dependent on endocytosis via clathrin coated pits, as elimination of the NPXY element in the C-terminus of ROMK (which is necessary for ROMK1's endocytic internilization) or inhibition of the endocytic pathway with a dominant-negative dynamin mutant prevented inhibition of ROMK by WNK4. In contrast to the kinase-dependence of NCC inhibition, however, kinase-dead WNK4 continued to show inhibition of ROMK1 (Wang et al 1997). Most strikingly, however, PHAII-mutant WNK4 showed markedly increased inhibition of ROMK relative to wild type WNK4, the opposite of PHAII-mutant WNK4's effect on NCC (Kahle et al 2003). These effects of wild type and PHAII-mutant WNK4 on ROMK have been subsequently confirmed in mammalian cells by other groups. Recent studies have demonstrated that WNK4 can form a complex with intersectin, a protein involved in clathrin-mediated endocytosis, and that this interaction is enhanced by PHAII mutations (He et al 2007). Finally, the observation that RNA interference directed against endogenous WNK4 increases basal ROMK activity argues that these observed effects are not an artifact of over-expression of these proteins in vitro (Lazrak et al 2006).

These findings together suggest a mechanism that could explain the hyperkalemia in patients with PHAII mutations (Figure 19.2); moreover, they indicate that WNK4 can have distinct effects on different, structurally unrelated targets. Moreover, these data suggest that WNK4 is a molecular switch capable of altering nephron function from one in which both salt reabsorption and potassium secretion are inhibited to one in which salt reabsorption is augmented while potassium secretion is further inhibited. These effects resulting from PHAII-mutant WNK4 would likely mimic the body's normal response to volume depletion (Kahle et al 2003).

WNK4 and Paracellular Chloride Flux

Increased permeability to chloride flux via the paracellular pathway could augment the reabsorption of salt and reduce the lumen-negative potential in the CCD, inhibiting potassium secretion. Yamauchi et al and Kahle et al independently produced Madin Darby Canine Kidney (MDCK) cell lines with inducible expression of WNK4 and studied its effect on paracellular flux in monolayers using either tracer methods (^{22}Na and ^{36}Cl flux studies) or electrophysiology with measurement of transepithelial resistance and dilution potential (Kahle et al 2004, Yamauchi et al 2004). Both groups came to similar conclusions. Expression of wild-type WNK4 induced a modest and selective increase in paracellular chloride permeability (Kahle et al 2004). PHAII-mutant WNK4, in contrast, produced a much greater increase in chloride permeability, sufficient to switch the permeability of the paracellular pathway from one that favors sodium flux, to one that favors chloride flux (Kahle et al 2004, Yamauchi et al 2004). This effect was seen when WNK4 was expressed at the time of monolayer formation or if WNK4 was induced after mature monolayers had formed, indicating a *dynamic* effect of WNK4 of paracellular chloride permeability (Kahle et al 2004). In contrast, induction of WNK4 did not alter the flux of uncharged solutes (Kahle et al 2004, Yamauchi et al 2004).

WNK4 did not alter the expression or localization of selected tight-junction proteins, and did not alter tight-junction depth or fibril number (Kahle et al 2004). However, expression of WNK4 increased phosphorylation of claudin-4 and was shown to physically associate with its C-terminal YV motif (Yamauchi et al 2004). WNK4's effects on the paracellular pathway were not seen with a catalytically inactive WNK4 (Kahle et al 2004, Yamauchi et al 2004). Because aldosterone increases claudin-4 phosphorylation and paracellular chloride conductance in CCD cells, it has been postulated that WNK4 might be positioned downstream of aldosterone in this pathway (Le Moellic et al 2005).

WNK4 and ENaC

Like NCC, wild-type WNK4 inhibited ENaC activity in oocytes and this inhibition was dependent upon an active kinase domain. Inhibition required the PPXY domain in ENaC subunits, suggesting a requirement for clathrin-mediated endocytosis for inhibition. PHAII mutations relieved inhibition of ENaC activity (Ring et al 2007).

The findings with ENaC, ROMK and paracellular chloride flux collectively suggest that PHAII mutations increase reabsorption of sodium with chloride, while inhibiting ROMK activity (Figure 19.2). Like the effects on NCC, these effects would maximize salt reabsorption while minimizing potassium loss. These findings also identified two distinct states of WNK4 – one of basal inhibition of diverse flux pathways, and another, mimicked by PHAII mutations, which would promote an appropriate response to volume depletion. These findings, in conjunction with the human genetic data, implicated WNK4 in the integrated regulation of salt and potassium balance, and further suggested that WNK4 is a molecular switch that determines the balance between salt reabsorption and potassium secretion in the distal nephron (Kahle et al 2003). The 'PHAII state' of WNK4 appeared to be the appropriate response to aldosterone in the setting of volume depletion, and raised the question of whether WNK4 might have a third state that could drive the appropriate response to the other setting in which aldosterone secretion is elevated – hyperkalemia. Recent findings support such a 'third state' of the WNK4 switch.

Phosphorylation of WNK4 by SGK1 Promotes Potassium Secretion

The C-terminus of WNK4 contains a highly predicted site for phosphorylation by SGK1 at serine 1169. This site is far removed from the kinase domain and the domain of clustered PHAII mutations and is highly conserved in WNK4 orthologs extending from human to chicken. In vitro studies have shown that serine 1169 is specifically phosphorylated by SGK1. Coimmunoprecipitation studies have demonstrated that SGK1 and WNK4 associate in a complex when expressed in mammalian cells (Ring et al 2007). The effects of phosphorylation at serine 1169 on WNK4 function were tested in vitro by expressing WNK4 harboring the substitution of aspartate for serine at position 1169 (WNK4$_{S1169D}$), mimicking the SGK1 phosphorylation event. In *Xenopus* oocytes, WNK4$_{S1169D}$ loses its ability to inhibit both ENaC and ROMK (Ring et al 2007). The expected consequence of this effect would be increased electrogenic sodium reabsorption with increased ROMK-mediated potassium secretion; both effects would promote a net increase in potassium secretion, and would constitute a desired physiologic response to hyperkalemia. Consistent with this idea, SGK1 is an immediate transcriptional target of aldosterone, providing a potential mechanism whereby WNK4 could be phosphorylated in response to high serum potassium level downstream of aldosterone (Pearce 2001).

As shown in Figure 19.2, WNK4 thus appears to have 3 distinct states that might allow it to switch distal nephron function from (1) a basal state (wild type WNK4), to (2) one that promotes maximal salt reabsorption with minimal potassium secretion, or alternatively (PHAII-mutant WNK4), to (3) one that promotes maximal potassium secretion (WNK4$_{S1169D}$) (Ring et al 2007). This would allow WNK4 to coordinate distal nephron function to regulate the balance of salt reabsorption and potassium secretion, possibly downstream of aldosterone signaling, in the distal nephron.

WNK1 and NCC

The phenotypes of PHAII patients with mutations in either WNK1 or WNK4 are similar, suggesting that these kinases act on similar targets in series or in parallel pathways. Current evidence suggests that WNK1 and WNK4 interact in a complex signaling pathway to regulate multiple targets of distal nephron ion transport. Coexpression studies of L-WNK1 with NCC in *Xenopus* oocytes indicated that WNK1 does not inhibit NCC activity by itself (Yang et al 2003). In the presence of L-WNK1, however, WNK4 no longer inhibits NCC, suggesting that L-WNK1 is an upstream inhibitor of WNK4 action at NCC (Yang et al 2005). The kinase-deficient isoform of WNK1, KS-WNK1, is a dominant-negative regulator of L-WNK1. As a consequence, KS-WNK1 inhibits NCC activity *indirectly* by inhibiting L-WNK1 and thereby increasing WNK4 activity, which increases NCC inhibition (Yang et al 2005). These data suggest that WNK4, L-WNK1, and KS-WNK1 interact in a signaling cascade that regulates electroneutral salt reabsorption in the DCT/CNT. L-WNK1 has been shown to phosphorylate WNK4 (Lenertz et al 2005), suggesting that this phosphorylation diminishes WNK4 inhibition of NCC. These findings are consistent with the human genetic data indicating that increased WNK1 expression results in PHAII (Wilson et al 2001).

WNK1 and ROMK

Multiple groups have documented differential effects of the WNK1 isoforms on ROMK. L-WNK1 inhibits ROMK, while KS-WNK1 reverses the inhibition of ROMK caused by L-WNK1 (Lazrak et al 2006). Inhibition of ROMK by L-WNK1 is synergistic with, but not dependent on, WNK4. Like WNK4, WNK1 is necessary for the basal endocytosis of ROMK1 (Lazrak et al 2006), and stimulates ROMK1 internalization through its association with the scaffolding protein Intersectin (He et al 2007). Dietary potassium experiments in rats corroborate these data: rats placed on high potassium diet have increased expression of *KS-WNK1*, which would be expected to alleviate WNK1 inhibition of ROMK surface expression and facilitate potassium secretion (O'Reilly et al 2006). These data suggest that the antagonistic modulation of L-WNK1 by KS-WNK1 is not only relevant to the regulation of NCC co-transport, but is also important for the regulation of ROMK-mediated potassium secretion.

WNK1 and ENaC

Xu et al have shown that L-WNK1, like WNK4, regulates ENaC activity (Xu et al 2005). PI3-kinase-dependent phosphorylation of L-WNK1 at threonine-58 results in the L-WNK1-mediated activation of SGK-1 (Xu et al 2005). In turn, activated SGK-1 phosphorylates the E3 ubiquitin–protein ligase Nedd4-2, thereby inactivating it and preventing its induction of clearance of ENaC from the plasma membrane (Xu et al 2005).

WNK1 and Paracellular Chloride Flux

Analogous to WNK4, WNK1 has been shown to increase paracellular chloride flux about threefold and induces the phosphorylation of the paracellular tight junction protein claudin-4. Co-expression of wild-type WNK4 did not further increase WNK1-enhanced chloride permeability (Ohta et al 2006).

Together, these findings generally place WNK1 upstream of WNK4 and/or indicate parallel pathways, depending upon the specific target; moreover, the biochemical findings provide a logical explanation by which increased expression of L-WNK1 could produce the phentoype seen in PHAII patients.

IN VIVO MOUSE MODELS OF WNK1 AND WNK4 FUNCTION

The in vitro studies mentioned above make a number of distinct predictions. First, they predict that expression of PHAII-mutant WNK4 should recapitulate features of PHAII and that wild-type WNK4 should lead to increased inhibition of NCC with a consequent lowering of blood pressure. Lalioti et al constructed mice carrying one additional genomic copy of either wild-type or PHAII-mutant WNK4 as bacterial artificial chromosome (BAC) transgenes (Lalioti et al 2006). As predicted from the in vitro data, the effects of transgenic wild-type and PHAII-mutant WNK4 were opposite on blood pressure, potassium homeostasis, and urinary calcium. PHAII mice showed significantly elevated blood pressure compared with wild-type littermates, as measured by radiotelemetry, (1) at baseline, (2) with salt loading, and (3) with salt depletion. They showed significant hyperkalemia at baseline (5.3 mM vs. 4.5 mM in wild type) and dramatic hyperkalemia on high potassium diet (8.4 mM), as well as hypercalciuria that was dramatically exacerbated by high salt diet. Conversely, mice expressing an additional copy of wild-type WNK4 showed lower blood pressure and modestly reduced potassium levels on low potassium diet.

Examination of renal structure in these mice unexpectedly revealed that PHAII mice had markedly increased mass of the distal convoluted tubule, with expansion of both DCT_1 and DCT_2 tubule segments, along with increased NCC expression. Immunostaining for ENaC and ROMK revealed no discernible difference from wild-type mice, and the morphology of the CCD and other nephron segments appeared normal. Conversely, mice expressing the wild-type WNK4 transgene showed markedly diminished mass of the DCT and reduced NCC expression (Lalioti et al 2006). In addition to these effects, ENaC-mediated sodium

currents were shown to be increased in the aldosterone-sensitive distal colons of PHAII mice (Ring et al 2007). This effect was interpreted as a potential means by which the body could maximize salt and water absorption. The effects of PHAII-mutant WNK4 might mimic a physiological state in which a *global* defense of intravascular volume is needed – say, for example, in severe volume depletion.

The finding of dramatic alteration of NCC expression, with variation in expression between wild-type and PHAII-mutant WNK4 mice, suggested that inhibition of NCC might reverse PHAII phenotypes. This was tested by either genetic deletion of NCC using targeted disruption of the structural gene, or administration of thiazide diuretics as a direct pharmacologic inhibitor. Either of these maneuvers completely corrected the hypertension, hyperkalemia, hypercalciuria, and hyperplasia/hypertrophy of the DCT (Lalioti et al 2006). These findings indicate that the essential features of PHAII require NCC activity and can be reversed by a deficiency of NCC activity.

A second model of PHAII, a knock-in of a PHAII mutation at the endogenous *WNK4* locus, has recently been reported (Yang et al 2007). These mice also recapitulate features of the human disease. These mice also exhibit increased apical NCC expression by immunogold staining, and reversal of phenotypes by hydrochlorothiazide, although the required doses were 25 mg/kg, ~100-fold higher than those effective in humans. In this model, increased phosphorylation of OSR1/SPAK and NCC was reported, providing insight into the elements of a pathway from mutation to downstream effector. An increased amiloride-sensitive lumen-negative potential in the CCD was shown in the PHAII knock-in mouse, along with normal levels of ROMK and increased levels of BK potassium channels. The authors of the paper speculated that the increased ENaC activity was a secondary effect, despite aldosterone levels being no different from those in wild-type littermates (Yang et al 2007).

The function of WNK1 has been explored in vivo through the creation of a WNK1 knockout mouse. Although the homozygous *WNK1* knockout is embryonic lethal at ~day E13, the heterozygote is viable and exhibits a reduction of blood pressure of ~12 mmHg (mean arterial pressure) (Zambrowicz et al 2003). The finding of decreased blood pressure in *WNK1* +/− mice is consistent with the finding that gain-of-WNK1 function causes *increased* blood pressure in humans and supports the findings of biochemical studies that L-WNK1 promotes the activities of NCC and ENaC, either directly or by antagonizing WNK4's inhibition.

PATHOPHYSIOLOGICAL MECHANISMS OF PHAII

In vitro experiments provide compelling evidence that WNK4 is a molecular switch that regulates distal nephron function to achieve specific homeostatic effects in response to different physiologic perturbations. At a basal, equilibrium state, WNK4 inhibits ion transport pathways involved in salt reabsorption and potassium secretion (Kahle et al 2004). The single amino acid substitutions in WNK4 that cause PHAII alleviate inhibition of NCC and ENaC, and increase paracellular chloride permeability, while simultaneously increasing inhibition of ROMK (Kahle et al 2003, 2004, Wilson et al 2003, Yang et al 2003, Yamauchi et al 2004, Ring et al 2007). These effects would be expected to translate into maximal salt reabsorption with minimal potassium secretion in the kidney; comparable effects on ENaC in the colon would favor increased gastrointestinal absorption of salt. All of these effects would correspond to the desired response to intravascular volume depletion.

These findings have been recapitulated in in vivo models in human and mouse. In PHAII mouse models, increased NCC levels are seen, and either pharmacologic or genetic deficiency of NCC reverses the PHAII phenotypes (Lalioti et al 2006). Increased ENaC activity is seen in the colon and CCD (Ring et al 2007, Yang et al 2007), and a reported increase of an amiloride-sensitive nasal potential difference in humans with PHAII (Farfel et al 2005) suggests a possible increase in ENaC activity in the respiratory epithelia. Both humans and mice with PHAII mutations exhibit profound impairment of renal potassium secretion, and develop severe hyperkalemia.

While increased NCC activity could increase net renal salt reabsorption and thereby contribute to the hypertension seen in PHAII patients, it is less obvious how this could lead to hyperkalemia. One possibility is that hyperkalemia results from a reduction in sodium delivery to more distal nephron segments, thereby resulting in decreased electrogenic sodium reabsorption via ENaC, and the consequent impairment of potassium secretion (as is known to occur with genetic deficiency of ENaC) (Chang et al 1996). This interpretation seems plausible as an acute response, but does not seem plausible at steady state, in which distal nephron sodium delivery and flow are restored to normal. Normal distal sodium delivery is presumed to be the case in humans with PHAII, and is directly shown to be the case in PHAII mice, which are in salt balance with fractional excretions of sodium identical to those of their wild-type littermates (Lalioti et al 2006, Yang et al 2007). In this setting, sodium delivery to the CCD is at normal levels, yet potassium secretion is nonetheless dramatically impaired.

In light of these findings, it is important to consider that the key site of ENaC and ROMK activity may lie more *proximally* in the nephron, in the DCT$_2$/CNT than in more distal segments. Data from selective ENaC deficiency in the CCD indicate that loss of ENaC activity at this site has little effect on net renal salt and potassium homesostasis, consistent with a more proximal site of essential activity (Rubera et al 2003). It is known that a lumen-negative potential develops in the late DCT$_2$/CNT, and that substantial potassium

secretion occurs in this nephron segment (Reilly & Ellison 2000). Owing to the fact that many DCT/CNT segments feed into single CCD segments, modeling studies have suggested that the former accounts for quantitatively much more sodium reabsorption and potassium secretion than the latter.

Thus, a more accurate model for nephron function may be one in which NCC and ENaC effectively 'compete' for sodium in the same nephron segment. An increase in NCC activity would limit electrogenic sodium reabsorption, and thereby inhibit potassium secretion. This would be expected to result in the observed hypertension and hyperkalemia of PHAII. In this case, genetic or pharmacologic elimination of NCC function would uncover underlying ENaC activity, permitting restoration of the lumen-negative potential and restoration of potassium secretion. This model may also be consistent with WNK4 having an inhibitory effect on ROMK, as occurs in vitro. Both animal models of PHAII appear to show grossly normal levels of ROMK1; however, in the setting of the profound hyperkalemia that these animals exhibit, such 'normal' levels are far from what should be achieved (Wang et al 1997), suggesting an effect of PHAII mutation on ROMK. The observation, however, that BK levels are elevated (Yang et al 2007) provides a potential explanation for why the loss of NCC reverses hyperkalemia but does not result in hypokalemia: BK channels are believed to provide a 'brake' to hyperkalemia, becoming activated with high potassium levels (Wang et al 1997). Like ROMK, BK channels can only operate with a lumen-negative potential, which is restored with inhibition of NCC.

POSSIBLE ROLE OF WNK1 AND WNK4 IN ESSENTIAL HYPERTENSION AND WNKs AS POTENTIAL TARGETS OF NOVEL THERAPEUTICS

The findings that altered levels or function of WNK1 and WNK4 alter blood pressure in humans, along with the observation that very modest quantitative variation in WNK1 and WNK4 expression in mouse can have a significant impact on blood pressure (Lalioti et al 2006), raise the question of whether inherited or acquired variation in *WNK1* or *WNK4* might commonly contribute to blood pressure variation in the general population. Since many individuals with essential hypertension have low plasma renin levels and exhibit a robust treatment response to thiazide diuretics – characteristics similar to those of patients with PHAII – this speculation is not remote. *WNK4* lies only 1 Mb from locus *D17S1299*, the site showing the strongest linkage to blood pressure variation (Mayan et al 2002) in the Framingham Heart Study population (Levy et al 2000); this same chromosomal segment has shown evidence of genetic linkage to hypertension in two other studies (Julier et al 1997, Baima et al 1999). It will be of interest to determine whether variants that alter WNK4 activity account for these findings. A common single nucleotide polymorphism (SNP) in *WNK4* has been reported to show association with essential hypertension Japan (Kokubo et al 2004), and common variants in *WNK1* show association with altered blood pressure in Caucasians (Newhouse et al 2005, Tobin et al 2005). Nonetheless, until such studies have shown robust replication, these findings must be viewed with caution owing to the extremely high false-positive rate of such association studies (Lohmueller et al 2003).

Because mice heterozygous for *WNK1* deficiency appear to have significant decreases in blood pressure without other overt phenotypes (Zambrowicz et al 2003), inhibition of WNK1 might have therapeutically useful blood pressure-lowering effects in humans. Current data would suggest that inhibition of the kinase activity of WNK1 would produce this effect. The recently solved crystal structure of the WNK1 kinase domain could prove useful for the design of highly specific antagonists (Min et al 2004). Such agents might be of substantial benefit, as the effects of commonly used diuretic agents such as thiazides and loop diuretics are limited by their activation of the renin-angiotensin-aldosterone system, which causes a secondary increase in sodium reabsorption by ENaC. The demonstration that WNK1 regulates both NCC and ENaC, suggests that modulation of WNK1 activity might have substantial blood pressure lowering effects. Whether these effects would have a favorable or unfavorable effect on potassium homeostasis compared with existing agents, and whether there might be limiting toxicities due to effects in other organs, is uncertain from the current data available. The embryonic lethality of the *WNK1* homozygous null mouse indicates a requirement in normal embryonic development, but this does not mean such agents would necessarily have adverse effects in the adult. Further investigation of a conditional knockout allele of WNK1 could help clarify this situation.

The observation that the increased NCC activity of PHAII results in marked hypercalciuria that is exacerbated by increased dietary salt intake (Lalioti et al 2006) is reminiscent of a subset of hypertensive subjects who display marked hypercalciuria that is not explained simply by their dietary salt intake (Yamakawa et al 1992). This observation raises the question of whether hypercalciuria in hypertensive humans might be a useful biomarker of increased NCC activity. A simple urine test measuring the level of calcium might be able to identify a subset of patients with hypertension in which the inexpensive thiazide diuretics might be *particularly* beneficial.

Elucidation of the WNK signaling pathway has shed insight into an apparent obligatory reciprocal relationship between salt reabsorption and potassium secretion. When WNK4 promotes increased salt reabsorption, this occurs at the expense of reduced potassium secretion, and vice versa (Take et al 1991). Recognition that renal excretion of a dietary potassium load carries an obligatory reduction in renal salt reabsorption provides an explanation for the well-described, but poorly understood, blood-pressure-lowering effect of supplemental dietary potassium, as occurs with the DASH diet (Sacks et al 2001). Perhaps the measurement of salt balance in the early stages of this diet would reveal an early loss of salt, with eventual return to salt balance at a lower set point with reduced blood pressure.

THE ROLE OF THE WNK KINASES IN THE REGULATION OF CELL VOLUME AND INTRACELLULAR CHLORIDE HOMEOSTASIS

WNK1 and WNK4 are expressed in epithelia other than the renal tubule, raising the question of what their function is in these extrarenal sites. Moreover, WNK2 and WNK3, recently cloned WNK family members, are highly expressed in nonrenal tissues. Recent data have demonstrated that the WNK kinases, in association with the oxidative-stress-responsive kinase 1 (OSR1) and the Ste20/SPS1-related proline/alanine-rich kinase (SPAK), play key roles in an evolutionarily-conserved chloride-sensitive signaling cascade that regulates the activity of the Na–K–2Cl and K–Cl cotransporters that modulate the level of intracellular chloride to defend the volume of cells during osmotic stress, and provide the proper electrical response to GABA neurotransmission (because the $GABA_A$ receptor is a ligand-gated chloride channel).

The Na–K–2Cl and K–Cl cotransporters are cation/chloride cotransporters of the *SLC12* gene family that play critical roles in salt and water transport across renal and numerous other chloride-transporting epithelia, and intracellular chloride homeostasis in neurons, red blood cells, and most other cells throughout the body. By dynamically altering the level of intercellular chloride in cells, NKCCs and the KCCs play essential roles in cell-volume regulation. In hypotonic extracellular conditions, cell swelling results in a homeostatic counter-response termed regulatory volume decrease (RVD), which results in the activation of KCCs (a major cellular chloride efflux pathway) and coordinate inhibition of NKCCs (the chloride influx pathway) – thus promoting the net efflux of KCl and water. Cell shrinkage activates regulatory volume increase (RVI), whereby activation of NKCCs, and inhibition of KCCs, promotes the net uptake of NaCl, KCl, and water into cells.

Activation of the NKCCs and KCCs is achieved by serine/threonine phosphorylation and dephosphorylation, respectively. Current evidence suggests that a common chloride-sensitive kinase mediates the cell volume-dependent phosphorylation of both cotransporters. In this model, cell shrinkage and/or decreases in intracellular chloride activate the kinase(s), which in turn phosphorylates the NKCCs and KCCs, promoting their activation and inactivation, respectively. Cell swelling and/or increases in intracellular chloride inhibits kinase activity, allowing protein phosphatases to dephosphorylate both cotransporters, thereby promoting the inhibition of NKCC and activation of KCC. The identity of this kinase has been elusive for over 20 years.

VOLUME SENSITIVITY OF CATION/CHLORIDE COTRANSPORTERS IS REGULATED BY AN INTERACTION BETWEEN THE WNK AND STE20-TYPE KINASES SPAK/OSR1

The mammalian Ste(sterile)20-type kinases, named after their yeast homolog originally discovered in a genetic analysis of *Saccharomyces cerevisiae* mating, are divided into two groups: the p21-activated kinases (PAKs) and the germinal center kinases (GCKs). The Ste20-type kinases oxidative-stress-responsive kinase 1 (OSR1) and the Ste20/Sps1-related proline/alanine-rich kinase (SPAK), genes orthologous to the *Drosophila melanogaster* Fray the *Caenorhabditis elegans* GCK-3, are GCKs (subfamily VI) highly expressed in neurons and chloride-transporting epithelia, including the distal nephron. Recent work has shown that both SPAK and OSR1 are key substrates downstream of the WNK kinases in a signaling cascade that coordinately regulates the activity of the NKCCs and KCCs to defend the volume of cells against osmotic stress.

Delpire and colleagues first demonstrated that a dominant-negative PASK mutant and OSR1 interact with NKCC1 and KCCs via a specific conserved carboxy-terminal (CCT) domain (Piechotta et al 2002, 2003). This motif of OSR1 and SPAK recognizes an Arg-Phe-Xaa-Val (RFXV) domain harbored in the cytoplasmic amino terminal tails of NKCCs and the KCCs. While the activity of NKCC1, measured in both isotonic (basal) and hypertonic (stimulated) conditions, was not shown to be significantly affected by co-expression of SPAK in oocytes or mammalian cells, the finding that PASK dramatically reduced the regulatory phosphorylation and activity of NKCC1 in low intracellular chloride (activating) conditions suggested that SPAK might be a key element downstream of the chloride-sensing mechanism that regulates NKCC1 (Dowd et al 2003).

An important link occurred when yeast two-hybrid experiments revealed that the SPAK/OSR1 CCT domain also bound RFXV domains embedded within the carboxy termini of WNK1 and WNK4 (Piechotta et al 2002, 2003). This interaction was found to be of important functional significance, as co-expression of PASK and WNK4 together, but not alone, robustly increased NKCC1 activity in both oocytes and mammalian cells in a manner insensitive to conditions of external osmolarity (Gagnon et al 2006a, 2006b). This activation was dependent on the catalytic activities of PASK and WNK4, and PASK's physical interaction with both WNK4 and NKCC1. OSR1 exhibited similar functional activation of NKCC1 when co-expressed with WNK1. In a reciprocal manner, co-expression of WNK4 and PASK has been shown to inhibit KCC activity in isotonic (basal) and hypotonic (stimulating) conditions, while co-expression of dominant-negative PASK and WNK4 activates KCC. These studies revealed that a novel interaction between the WNK and Ste20-type kinases coordinately regulates the NKCCs and KCCs.

THE WNK KINASE/STE20-TYPE KINASE/NKCC PHOSPHORYLATION CASCADE

Insight into the mechanism by which the WNKs interact with SPAK/OSR1 to regulate the cotransporters has been achieved through detailed biochemical and structural studies (Moriguchi et al 2005, Vitari et al 2005). WNKs have been shown to activate SPAK/OSR1 via regulatory phosphorylation. Phosphopeptide mapping studies revealed that WNK1 phosphorylates SPAK and OSR1 at equivalent residues located within the T-loop of their catalytic domains (threonine 233 in SPAK, threonine 185 in OSR1) and a serine residue located within their non-catalytic carboxy terminal region (serine 373 in SPAK, serine 325 in OSR1). Mutation of threonine 185 to alanine in OSR1 prevented the activation of this kinase by WNK1, whereas a mutation mimicking the phosphorylation of threonine 185 (threonine to glutamic acid) increased the basal activity of OSR1 greater than 20-fold. Activated SPAK/OSR1, in turn, has been shown to phosphorylate NKCC1 at three key residues that are conserved in NCC and NKCC2 (threonine 175, 179, and 184 in shark or threonine 203, 207, and 212 in human). This phosphorylation event robustly increases NKCC1 activity. The residues phosphorylated by SPAK/OSR1 in response to WNK activation are the same ones that have been shown to be essential for NKCC1 activation during conditions of extracellular hypertonicity (cell shrinkage) and decreased intracellular chloride. Critically, WNK1 initiates this phosphorylation cascade after exposure to hyperosmotic (cell-shrinking) conditions, which triggers the autophosphorylation of serine 382 in WNK1's own T-loop.

WNK3: A BRAIN-ENRICHED KINASE WITH RECIPROCAL ACTIONS ON THE NKCCs AND KCCs

Investigators have focused on studying WNK1 and WNK4 because of their established roles in human disease. However, data suggest other WNK family members may have equally important functions. WNK3 is a recently cloned gene that exists in several alternatively spliced isoforms, including a brain-specific variant that contains a unique exon 22. Similarly to WNK1 and WNK4, WNK3 localizes to the intercellular junctions of chloride-transporting epithelia throughout the body (including the crypts of Lieberkühn in the small intestine, parietal cells of gastric glands, and the kidney's thick ascending limb) (Kahle et al 2005). However, in contrast to WNK1 and WNK4, WNK3 is most highly expressed in brain, including the hippocampus, several cortical layers, thalamic and hypothalamic nuclei (including the supraoptic and suprachiasmatic nuclei), dorsal raphe nucleus, locus ceruleus, reticular formation, and cerebellum. WNK3 localizes to neurons expressing ionotropic GABA$_A$ receptors, an intriguing finding given the importance of intracellular chloride concentration for the subsequent neuronal response (excitatory versus inhibitory) to GABA binding its receptor and associated chloride channel. Also compelling is the dynamic developmental shift in the expression of WNK3 in the brain, which parallels that of the neuronal-specific KCC2 in the hippocampus and cerebellum.

Investigators have shown that WNK3 modulates the balance between cellular chloride influx and efflux via reciprocal actions on the NKCCs and KCCs, and is able to override the normal effects of variation in extracellular osmolarity on cotransporter activity (Kahle et al 2005). Coexpression of wild-type WNK3 with NKCC1 in oocytes resulted in a potent increase in NKCC1 activity not only in hypertonic media (the normal condition for assaying NKCC1 activity), but also in hypotonic media (a condition in which NKCC1 is normally silenced). Conversely, wild-type WNK3 inhibited each of the KCCs, even in hypotonic media (a condition in which KCCs are normally maximally active). A catalytically silent WNK3 mutant produced the mirror-image effects of wild-type WNK3 on the cotransporters, inhibiting NKCC1 but activating the KCCs. Interestingly, kinase-inactive WNK3's stimulation of the KCCs in hypertonic conditions is abolished by calyculin A and cyclosporine A, implicating protein phosphatases PP1 and 2B in this effect (de Los Heros et al 2006). Similar to the effect of the WNKs in association with SPAK/OSR1, wild type WNK3 increases the phosphorylation of threonine 212 and threonine 217 in NKCC1, and kinase-inactive WNK3 prevented the phosphorylation of these essential residues, even in hypertonic (activating) conditions.

These data suggest that WNK3, along with WNK1 and WNK4, are likely the long-sought volume/chloride sensitive kinases, which in combination with the downstream SPAK and OSR1 kinases, are important for maintaining cell volume during osmotic stress. The fact that WNK3 is highly expressed in neurons expressing ionotropic GABA$_A$ receptors and is able to modulate intracellular chloride concentration via its actions on NKCCs and KCCs suggests neurons may use this same mechanism to modulate their response (excitation versus inhibition) to GABA.

CONCLUSIONS AND FUTURE DIRECTIONS

The molecular genetic discovery that mutations in WNK1 and WNK4 cause PHAII provided a key entry point into unraveling the complex mechanism by which the kidney regulates the balance between salt reabsorption and potassium secretion, and provides an excellent example of how the identification of rare human mutations can point to new aspects of complex physiological pathways. It seems apparent that the diverse functions of these proteins would have been difficult to identify without the direct clues resulting from their mutation in humans. To date, driven by the consequences of human mutations, studies have revealed a role of WNKs in orchestrating the functions of different ion transport pathways to promote appropriate responses to physiologic perturbations. Given that the kidney is exposed to aldosterone in two distinct physiologic conditions – hypovolemia and hyperkalemia – WNK1 and WNK4 appear poised to help determine whether aldosterone action should be used to drive maximal salt reabsorption or maximal potassium secretion. The evidence supporting these propositions is solid, with consistent findings from human genetics, in vitro physiology and biochemistry, and in vivo genetically engineered mouse models.

Despite the significant experimental progress, important elements are still poorly understood. For example, the clustered missense mutations in WNK4 define a key component of a molecular switch that drives nephron function to promote increased salt reabsorption and decreased potassium secretion. It is likely that PHAII mutations mimic an in vivo effect that is achieved by the binding of an endogenous ligand or the addition of some sort of modification near the site of WNK4 that harbors these mutations. The identity of the modifcation/ligand at this site, and the regulation of this event, is at present unknown. While signaling pathways downstream of angiotensin II would seem a logical possibility, the nature of the downstream effector has not been identified. The observations that other signaling pathways including the Akt and Erk5 proteins have been implicated in WNK regulation, and fact that WNKs have been shown to be involved in the regulation of TGF-beta and Wnt signaling, hints there is considerable complexity that is yet to be defined. Similarly, the mechanisms by which PHAII mutations alter WNK activity and function are very poorly understood – do these changes directly alter or allow modification of the activity of the kinase domain and/or allow recruitment of phosphatases or interactions with other proteins?

The unexpected finding in animal models that altered WNK4 activity can be directed to either increase or decrease the mass of the distal convoluted tubule through its effects on NCC activity has revealed striking plasticity in this nephron segment that is dependent upon apical salt entry. The mechanism underlying this effect is unknown but suggests a number of possible approaches.

The functions of the WNK kinases have a number of implications for blood pressure homeostasis in humans. Further work examining the effects of increased dietary potassium on salt balance in humans, for example, seem important given the new appreciation of the reciprocal nature of renal salt reabsorption and potassium secretion. Similarly, investigation of hypertensive subjects with hypercalciuria as a potential subset with increased activity of NCC and a particular susceptibility to the beneficial effects of thiazide diuretics is another implication of these new findings. Finally, the potential of WNK1 kinase inhibitors as potent antihypertensive agents can be further explored. To this end, use of conditional knockout mice as a means to avoid the embryonic lethality would be a useful start to assess whether there would be adverse effects of marked reduction of WNK1 in adult animals.

ACKNOWLEDGMENTS

We thank our many colleagues whose work and discussions contributed to this review. In particular, we thank members of the Lifton laboratory (particularly Jesse Rinehart, Aaron Ring, and Carol Nelson-Williams), and Steven Hebert, Gerardo Gamba, Xavier Jeunemaitre, and their colleagues for very productive collaborations that contributed immensely to our own work in this field. We apologize in advance to the authors of many studies on WNK kinases whose work was not directly discussed owing to space limitations. Supported in part by an NIH Specialized Center of Research in Hypertension grant to R.P.L.

DEDICATION

This chapter is dedicated to the memory Stephen Hebert, a brilliant, dedicated and generous friend and colleague who contributed greatly to the elucidation of WNK kinase function. Without his towering contributions to basic renal physiology, this research would not have been possible.

References

Arnold JE, Healy JK. Hyperkalemia, hypertension and systemic acidosis without renal failure associated with a tubular defect in potassium excretion. Am. J. Med. 1969; 47: 461–72.

Baima J, Nicolaou M, Schwartz F, et al. Evidence for linkage between essential hypertension and a putative locus on human chromosome 17. Hypertension 1999; 34: 4–7.

Brautbar N, Levi J, Rosler A, Leitesdorf, et al. Familial hyperkalemia, hypertension, and hyporeninemia with normal aldosterone levels. A tubular defect in potassium handling. Arch. Intern. Med. 1978; 138: 607–10.

Bravo E, Textor S, Mujais S, Cotton D. Chronic hyperkalemia in siblings associated with enhanced renal chloride absorption [abstract]. Clin. Res. 1980; 28: 782A.

Cai H, Cebotaru V, Wang YH, et al. WNK4 kinase regulates surface expression of the human sodium chloride cotransporter in mammalian cells. Kidney Int. 2006; 69: 2162–70.

Chang SS, Grunder S, Hanukoglu A, et al. Mutations in subunits of the epithelial sodium channel cause wasting with hyperkalaemic acidosis, pseudohypoaldosteronism type 1. Nat. Genet. 1996; 12: 248–53.

Choate KA, Kahle KT, Wilson FH, Nelson-Williams C, Lifton RP. WNK1, a kinase mutated in inherited hypertension with hyperkalemia, localizes to diverse Cl-transporting epithelia. Proc. Natl Acad. Sci. USA 2003; 100: 663–8.

Cruz DN, Simon DB, Nelson-Williams C, et al. Mutations in the Na-Cl cotransporter reduce blood pressure in humans. Hypertension 2001; 37: 1458–64.

de Los Heros P, Kahle KT, Rinehart J, et al. WNK3 bypasses the tonicity requirement for K-Cl cotransporter activation via a phosphatase-dependent pathway. Proc. Natl Acad. Sci. USA 2006; 103: 1976–81.

Delaloy C, Lu J, Houot AM, et al. Multiple promoters in the WNK1 gene: one controls expression of a kidney-specific kinase-defective isoform. Mol. Cell. Biol. 2003; 23: 9208–21.

Dillon MJ, Leonard JV, Buckler JM, et al. Pseudohypoaldosteronism. Arch. Dis. Child. 1980; 55: 427–34.

Disse-Nicodème S, Achard JM, Desitter I, et al. A new locus on chromosome 12p13.3 for pseudohypoaldosteronism type II, an autosomal dominant form of hypertension. Am. J. Hum. Genet. 2000; 67: 302–10.

Dowd BF, Forbush B. PASK (prolinealanine-rich STE20-related kinase), a regulatory kinase of the Na-K-Cl cotransporter (NKCC1). J. Biol. Chem. 2003; 278: 27347–53.

Farfel Z, Iaina A, Rosenthal T, Waks U, Shibolet S, Gafni J. Familial hyperpotassemia and hypertension accompanied by normal plasma aldosterone levels: possible hereditary cell membrane defect. Arch. Intern. Med. 1978a; 138: 1828–32.

Farfel Z, Iaina A, Levi J, Gafni J. Proximal renal tubular acidosis: association with familial normaldosteronemic hyperpotassemia and hypertension. Arch. Intern. Med. 1978b; 138: 1837–40.

Farfel Z, Mayan H, Yaacov Y, et al. WNK4 regulates airway Na$^+$ transport: study of familial hyperkalaemia and hypertension. Eur. J. Clin. Invest. 2005; 35: 410–15.

Gagnon KB, England R, Delpire E. Volume sensitivity of cation-chloride cotransporters is modulated by the interaction of two kinases: SPAK and WNK4. Am. J. Physiol. Cell Physiol. 2006a; 290: C134–C42.

Gagnon KB, England R, Delpire E. Characterization of SPAK and OSR1, regulatory kinases of the Na-K-2Cl cotransporter. Mol. Cell. Biol. 2006b; 26: 689–98.

Gamba G, Saltzberg SN, Lombardi M, et al. Primary structure and functional expression of a cDNA encoding the thiazide-sensitive, electroneutral sodium-chloride cotransporter. Proc. Natl Acad. Sci. USA 1993; 90: 2749–53.

Geller DS, Rodriguez-Soriano J, Vallo Boado A, et al. Mutations in the mineralocorticoid receptor gene cause autosomal dominant pseudohypoaldosteronism type I. Nat. Genet. 1998; 19: 279–81.

Golbang AP, Murthy M, Hamad A, et al. A new kindred with pseudohypoaldosteronism type II and a novel mutation (564D > H) in the acidic motif of the WNK4 gene. Hypertension 2005; 46: 295–300.

Golbang AP, Cope G, Hamad A, et al. Regulation of the expression of the Na/Cl cotransporter by WNK4 and WNK1: evidence that accelerated dynamin-dependent endocytosis is not involved. Am. J. Physiol. Renal Physiol. 2006; 291: F1369–F76.

Gordon RD, Geddes RA, Pawsey CG, O'Halloran MW. Hypertension and severe hyperkalaemia associated with suppression of renin and aldosterone and completely reversed by dietary sodium restriction. Australas. Ann. Med. 1970; 19: 287–94.

Gordon R, Klemm S, Tunny T, Stowasser M. Gordon's syndrome: a sodium-volume dependent form of hypertension with a genetic basis. In: Laragh JH, Brenner BM, eds. Hypertension: Pathophysiology, Diagnosis and Management, 2nd edn. New York: Raven Press, 1995: pp. 2111–23.

Hadchouel J, Delaloy C, Faure S, Achard JM, Jeunemaitre X. Familial hyperkalemic hypertension. J. Am. Soc. Nephrol. 2006; 17: 208–17.

He G, Wang HR, Huang SK, Huang CL. Intersectin links WNK kinases to endocytosis of ROMK1. J. Clin. Invest. 2007; 117: 1078–87.

Julier C, Delépine M, Keavney B, et al. Genetic susceptibility for human familial essential hypertension in a region of homology with blood pressure linkage on rat chromosome 10. Hum. Mol. Genet. 1997; 6: 2077–85.

Kahle KT, Wilson FH, Leng Q, et al. WNK4 regulates the balance between renal NaCl reabsorption and K⁺ secretion. Nat. Genet. 2003; 35: 372–6.

Kahle KT, Gimenez I, Hassan H, et al. WNK4 regulates apical and basolateral Cl⁻ flux in extrarenal epithelia. Proc. Natl Acad. Sci. USA 2004a; 101: 2064–9.

Kahle KT, Wilson FH, Lalioti M, Toka H, Qin H, Lifton RP. WNK kinases: molecular regulators of integrated epithelial ion transport. Curr. Opin. Nephrol. Hypertens. 2004b; 13: 557–62.

Kahle KT, Macgregor GG, Wilson FH, et al. Paracellular Cl⁻ permeability is regulated by WNK4 kinase: insight into normal physiology and hypertension. Proc. Natl Acad. Sci. USA 2004c; 101: 14877–82.

Kahle KT, Rinehart J, de Los Heros P, et al. WNK3 modulates transport of Cl⁻ in and out of cells: implications for control of cell volume and neuronal excitability. Proc. Natl Acad. Sci. USA, 2005; 102: 16783–8.

Klemm SA, Gordon RD, Tunny TJ, Thompson RE. The syndrome of hypertension and hyperkalemia with normal GFR (Gordon's syndrome): is there increased proximal sodium reabsorption? Clin. Invest. Med. 1991; 14: 551–8.

Kokubo Y, Kamide K, Inamoto N et al. Identification of 108 SNPs in TSC, WNK1, and WNK4 and their association with hypertension in a Japanese general population. J. Hum. Genet. 2004; 49: 507–15.

Lalioti MD, Zhang J, Volkman HM, et al. Wnk4 controls blood pressure and potassium homeostasis via regulation of mass and activity of the distal convoluted tubule. Nat. Genet. 2006; 38: 1124–32.

Lazrak A, Liu Z, Huang CL. Antagonistic regulation of ROMK by long and kidney-specific WNK1 isoforms. Proc. Natl Acad. Sci. USA 2006; 103: 1615–20.

Le Moellic C, Boulkroun S, González-Nunez D, et al. Aldosterone and tight junctions: modulation of claudin-4 phosphorylation in renal collecting duct cells. Am. J. Physiol. Cell Physiol. 2005; 289: C1513–C21.

Lenertz LY, Lee BH, Min X, et al. Properties of WNK1 and implications for other family members. J. Biol. Chem. 2005; 280: 26653–8.

Levy D, DeStefano AL, Larson MG, et al. Evidence for a gene influencing blood pressure on chromosome 17. Genome scan linkage results for longitudinal blood pressure phenotypes in subjects from the Framingham Heart Study. Hypertension 2000; 36: 477–83.

Licht JH, Amundson D, Hsueh WA, Lombardo JV. Familiar hyperkalaemic acidosis. Q. J. Med. 1985; 54: 161–76.

Lifton RP, Gharavi AG, Geller DS. Molecular mechanisms of human hypertension. Cell 2001; 104: 545–56.

Lohmueller KE, Pearce CL, Pike M, Lander ES, Hirschhorn JN. Meta-analysis of genetic association studies supports a contribution of common variants to susceptibility to common disease. Nat. Genet. 2003; 33: 177–82.

Mansfield TA, Simon DB, Farfel Z, et al. Multilocus linkage of familial hyperkalaemia and hypertension, pseudohypoaldosteronism type II, to chromosomes 1q31-42 and 17p11-q21. Nat. Genet. 1997; 16: 202–5.

Mayan H, Vered I, Mouallem M, Tzadok-Witkon M, Pauzner R, Farfel Z. Pseudohypoaldosteronism type II: marked sensitivity to thiazides, hypercalciuria, normomagnesemia, and low bone mineral density. J. Clin. Endocrinol. Metab. 2002; 87: 3248–54.

Min X, Lee BH, Cobb MH, Goldsmith EJ. Crystal structure of the kinase domain of WNK1, a kinase that causes a hereditary form of hypertension. Structure 2004; 12: 1303–11.

Moriguchi T, Urushiyama S, Hisamoto N, et al. WNK1 regulates phosphorylation of cation-chloride-coupled cotransporters via the STE20-related kinases, SPAK and OSR1. J. Biol. Chem. 2005; 280: 42685–93.

Nahum H, Paillard M, Prigent A, et al. Pseudohypoaldosteronism type II: proximal renal tubular acidosis and dDAVP-sensitive renal hyperkalemia. Am. J. Nephrol. 1986; 6: 253–62.

Newhouse SJ, Wallace C, Dobson R, et al. Haplotypes of the WNK1 gene associate with blood pressure variation in a severely hypertensive population from the British Genetics of Hypertension study. Hum. Mol. Genet. 2005; 14: 1805–14.

Ohta A, Yang SS, Rai T, Chiga M, Sasaki S, Uchida S. Overexpression of human WNK1 increases paracellular chloride permeability and phosphorylation of claudin-4 in MDCKII cells. Biochem. Biophys. Res. Commun. 2006; 349: 804–8.

O'Reilly M, Marshall E, Macgillivray T, et al. Dietary electrolyte-driven responses in the renal WNK kinase pathway in vivo. J. Am. Soc. Nephrol. 2006; 17: 2402–13.

Paver WK, Pauline GJ. Hypertension and hyperpotassaemia without renal disease in a young male. Med. J. Aust. 1964; 2: 305–6.

Pearce D. The role of SGK1 in hormone-regulated sodium transport. Trends Endocrinol. Metab. 2001; 12: 341–7.

Piechotta K, Lu J, Delpire E. Cation chloride cotransporters interact with the stress-related kinases Ste20-related proline-alanine-rich kinase (SPAK) and oxidative stress response 1 (OSR1). J. Biol. Chem. 2002; 277: 50812–19.

Piechotta K, Garbarini N, England R, Delpire E. Characterization of the interaction of the stress kinase SPAK with the Na⁺-K⁺-2Cl⁻ cotransporter in the nervous system: evidence for a scaffolding role of the kinase. J. Biol. Chem. 2003; 278: 52848–56.

Popow C, Pollak A, Herkner K, Scheibenreiter S, Swoboda W. Familial pseudohypoaldosteronism. Acta Paediatr. Scand. 1988; 77: 136–41.

Reilly RF, Ellison DH. Mammalian distal tubule: physiology, pathophysiology, and molecular anatomy. Physiol. Rev. 2000; 80: 277–313.

Ring AM, Cheng SX, Leng Q, et al. WNK4 regulates activity of the epithelial Na^+ channel in vitro and in vivo. Proc. Natl Acad. Sci. USA 2007a; 104: 4020–4.

Ring AM, Cheng SX, Leng Q, et al. An SGK1 site in WNK4 regulates Na^+ channel and K^+ channel activity and has implications for aldosterone signaling and K^+ homeostasis. Proc. Natl Acad. Sci. USA 2007b; 104: 4025–9.

Rubera I, Loffing J, Palmer LG, et al. Collecting duct-specific gene inactivation of alphaENaC in the mouse kidney does not impair sodium and potassium balance. J. Clin. Invest. 2003; 112: 554–65.

Sacks FM, Svetkey LP, Vollmer WM, et al. DASH-Sodium Collaborative Research Group. Effects on blood pressure of reduced dietary sodium and the Dietary Approaches to Stop Hypertension (DASH) diet. DASH-Sodium Collaborative Research Group. N. Engl J. Med. 2001; 344: 3–10.

Schambelan M, Sebastian A, Rector FC Jr. Mineralocorticoid-resistant renal hyperkalemia without salt wasting (type II pseudohypoaldosteronism): role of increased renal chloride reabsorption. Kidney Int. 1981; 19: 716–27.

Simon DB, Nelson-Williams C, Bia MJ, et al. Gitelman's variant of Bartter's syndrome, inherited hypokalaemic alkalosis, is caused by mutations in the thiazide-sensitive Na-Cl cotransporter. Nat. Genet. 1996; 12: 24–30.

Spitzer A, Edelmann CM, Goldberg LD, Henneman PH. Short stature, hyperkalemia and acidosis: A defect in renal transport of potassium. Kidney Int. 1973; 3: 251–7.

Stokes GS, Gentle JL, Edwards KD, Stewart JH. Syndrome of idiopathic hyperkalaemia and hypertension with decreased plasma renin activity: effects on plasma renin and aldosterone of reducing the serum potassium level. Med. J. Aust. 1968; 2: 1050–4.

Take C, Ikeda K, Kurasawa T, Kurokawa K. Increased chloride reabsorption as an inherited renal tubular defect in familial type II pseudohypoaldosteronism. N. Engl. J. Med. 1991; 324: 472–6.

Tobin MD, Raleigh SM, Newhouse S, et al. Association of WNK1 gene polymorphisms and haplotypes with ambulatory blood pressure in the general population. Circulation 2005; 112: 3423–9.

Travis PS, Cushner HM. Mineralocorticoid-induced kaliuresis in type-II pseudohypoaldosteronism. Am. J. Med. Sci. 1986; 292: 235–40.

Unwin RJ, Ligueros M, Shakelton C, Wilcox CS. Diuretics in the management of hypertension. In: Laragh JH, Brenner BM, eds. Hypertension: Pathophysiology, Diagnosis and Management, 2nd edn. New York: Raven Press, 1995: pp. 2785–99.

Verissimo F, Jordan P. WNK kinases, a novel protein kinase subfamily in multi-cellular organisms. Oncogene 2001; 20: 5562–9.

Vitari AC, Deak M, Morrice NA, Alessi DR. The WNK1 and WNK4 protein kinases that are mutated in Gordon's hypertension syndrome phosphorylate and activate SPAK and OSR1 protein kinases. Biochem. J. 2005; 391: 17–24.

Wang W, Hebert SC, Giebisch G. Renal K^+ channels: structure and function. Annu. Rev. Physiol. 1997; 59: 413–36.

Weinstein SF, Allan DM, Mendoza SA. Hyperkalemia, acidosis, and short stature associated with a defect in renal potassium excretion. J. Pediatr. 1974; 85: 355–8.

Wilson FH, Disse-Nicodème S, Choate KA, et al. Human hypertension caused by mutations in WNK kinases. Science 2001; 293: 1107–12.

Wilson FH, Kahle KT, Sabath E, et al. Molecular pathogenesis of inherited hypertension with hyperkalemia: the Na-Cl cotransporter is inhibited by wild-type but not mutant WNK4. Proc. Natl Acad. Sci. USA 2003; 100: 680–4.

Xu B, English JM, Wilsbacher JL, Stippec S, Goldsmith EJ, Cobb MH. WNK1, a novel mammalian serine/threonine protein kinase lacking the catalytic lysine in subdomain II. J. Biol. Chem. 2000; 275: 16795–801.

Xu BE, Min X, Stippec S, Lee BH, Goldsmith EJ, Cobb MH. Regulation of WNK1 by an autoinhibitory domain and autophosphorylation. J. Biol. Chem. 2002; 277: 48456–62.

Xu BE, Stippec S, Chu PY, et al. WNK1 activates SGK1 to regulate the epithelial sodium channel. Proc. Natl Acad. Sci. USA 2005a; 102: 10315–20.

Xu BE, Stippec S, Lazrak A, Huang CL, Cobb MH. WNK1 activates SGK1 by a phosphatidylinositol 3-kinase-dependent and non-catalytic mechanism. J. Biol. Chem. 2005b; 280: 34218–23.

Yamakawa H, Suzuki H, Nakamura M, Ohno Y, Saruta T. Disturbed calcium metabolism in offspring of hypertensive parents. Hypertension 1992; 19: 528–34.

Yamauchi K, Rai T, Kobayashi K, et al. Disease-causing mutant WNK4 increases paracellular chloride permeability and phosphorylates claudins. Proc. Natl Acad. Sci. USA 2004; 101: 4690–4.

Yang CL, Angell J, Mitchell R, Ellison DH. WNK kinases regulate thiazide-sensitive Na-Cl cotransport. J. Clin. Invest. 2003; 111: 1039–45.

Yang CL, Zhu X, Wang Z, Subramanya AR, Ellison DH. Mechanisms of WNK1 and WNK4 interaction in the regulation of thiazide-sensitive NaCl cotransport. J. Clin. Invest. 2005; 115: 1379–87.

Yang SS, Morimoto T, Rai T, et al. Molecular pathogenesis of pseudohypoaldosteronism type II: generation and analysis of a Wnk4(D561A/+) knockin mouse model. Cell Metab. 2007; 5: 331–44.

Zambrowicz BP, Abuin A, Ramirez-Solis R, et al. Wnk1 kinase deficiency lowers blood pressure in mice: a gene-trap screen to identify potential targets for therapeutic intervention. Proc. Natl Acad. Sci. USA 2003; 100: 14109–14.

CHAPTER 20

Distal Renal Tubular Acidosis

FIONA E. KARET

INTRODUCTION

Distal renal tubular acidosis is a disease of defective urinary acidification, which is caused by insufficient net acid excretion by the kidney. Unfortunately, the classification of the RTAs can be confusing, because within the overall historical scheme of nomenclature, types 1 and 4 RTA are both caused by distal nephron dysfunction. In this chapter we are mainly concerned with type 1 RTA, which is directly due to dysfunction of the acid-handling α-intercalated cells (α-IC) in the collecting system, and this distal form is associated with hypokalemia. In contrast, type 4 RTA is associated with hypoaldosteronism (true and pseudo-) and defective distal nephron sodium handling by principal cells in this same nephron segment, and therefore with hyperkalemia, and only secondarily with inadequate acid excretion. Both types 1 and 4 dRTA actually cover large groups of different conditions, both inherited and acquired. Thus it is type 1 RTA that is most commonly referred to as distal RTA (dRTA).

DEFINITION AND CLINICAL FEATURES

Distal RTA arises when the collecting duct fails to remove the excess acid load of a normal diet (about 70–100 mmol H^+ daily on an omnivorous diet) into the urine.

Thus RTA is characterized by failure of the kidney to produce an appropriately acid urine either in the presence of systemic metabolic acidosis, or following acid loading. Distal RTA can be diagnosed biochemically if urine pH is greater than 5.3 when systemic pH is less than 7.34, with a normal anion gap and renal function; or following the administration of ammonium chloride as an oral acid challenge. One hundred mg/kg body weight of NH_4Cl should serve to reduce urine pH to less than 5.3 over the following 6 hours (Wrong and Davies 1959).

The unifying features of this form of RTA are hypokalemic metabolic acidosis, metabolic bone disease, and nephrocalcinosis and/or nephrolithiasis. It was first reported in the mid 1930s (Lightwood 1935, Butler et al 1936). As with nephrogenic diabetes insipidus dRTA is an acquired condition in the majority of cases. It is most often seen in association with autoimmune diseases such as Sjögren syndrome (Wrong et al 1993), and in adult clinical practice this is much more likely than primary disease. It is sometimes associated with conditions such as myeloma or chronic liver disease.

Of the inherited forms, autosomal dominant dRTA (ddRTA) in general (but not always) presents later, sometimes not until adulthood, and with milder phenotype, than does the recessive form (Karet 2002). Sensorineural hearing loss is a common accompaniment, but is only ever a feature of recessive disease. Primary dRTA is relatively rare in Western populations, but occurs more commonly in areas of the world where rates of parental consanguinity are high.

In its severest form, affected infants with recessive dRTA are very acidotic and volume-depleted, with marked hypokalemia but otherwise normal renal excretory function. Calcium and phosphate are usually normal. Growth is poor, and rickets is common in untreated cases, both because of leaching of mineral from bone in an attempt to buffer the acidosis, and because bone resorption by osteoclasts is increased in acidosis. Nephrocalcinosis may be observed from a very early age, and may lead to accelerated renal impairment. Urinary citrate is low in dRTA, because citrate reabsorption is up-regulated in the proximal tubule to provide new bicarbonate (each citrate molecule yielding three bicarbonate). Abnormal calcium deposition in RTA is attributed in large part to this hypocitraturia and to urine alkalinity, but the exact mechanisms for, and precise sites of, this deposition remain unclear.

Genetic Diseases of the Kidney
ISBN: 978-0-12-449851-8

Oral alkali replacement (by administration of citrate or bicarbonate at a dose of 1–3 mg/kg/day) is sufficient to reverse most of the biochemical abnormalities and the associated bone disease in both dominant and recessive disease, leading to the resumption of growth (Santos & Chan 1986, Peces 2000). Potassium salts are preferable to sodium as the latter can exacerbate hypokalemia. In untreated cases, elements of proximal tubular dysfunction may be present, such as low-molecular-weight proteinuria, but these resolve with correction of acid–base status.

The majority of affected children with recessive dRTA also have progressive sensorineural hearing loss that, unlike the growth failure and renal calcium deposition, is not ameliorated by alkali therapy (Zakzouk et al 1995). One report (Berrettini et al 2001) describes the radiological observation of vestibular aqueductal widening in association with recessive dRTA, but this abnormality is not pathognomonic, also being seen in branchio-oto-renal syndrome (Chapter 37), Pendred syndrome, and in isolation.

At the opposite end of the clinical spectrum, affected individuals with dominant or sporadic dRTA may actually be asymptomatic, with renal calcium deposition discovered by chance, or may present with nephrolithiasis. It may require an oral acid challenge to reveal an underlying, but compensated, metabolic acidosis (sometimes termed 'incomplete' RTA). Genotype–phenotype correlation is variable even within families, but to date, the potential role of modifier genes has not been examined.

In hyperkalemia RTA, the unifying features are hyporeninemic hypoaldosteronism, with secondary acidosis. The acidosis here is dependent on sodium homeostasis, and acquired forms are again commoner than inherited disorders. It is therefore most commonly found in adults with diabetic nephropathy, or other causes of chronic renal impairment. The clinical parameters are those of the associated underlying conditions. Drugs such as trimethoprim or amiloride may also mimic hyperkalemia RTA by inhibiting principal cell function.

Examples of inherited disorders resulting in hyperkalemia RTA are PHA1 (caused by loss of function of the epithelial sodium channel ENaC in the neighboring principal cells), and Gordon syndrome.

α-INTERCALATED CELL FUNCTION

In the human distal nephron, most of the acid–base regulatory cells are α-ICs, because of the net acid load generated by our diet. However, in some circumstances, for example in the rodent kidney, a significant proportion of the cells is of the opposite morphology, termed β-IC, and are responsible for bicarbonate extrusion into the urine. Their resident membrane proteins are similar but not always identical to

those of the α-IC, but to date no human reral disorders have been associated with non-α-ICs.

All of the epithelial cells that form the renal tubule are necessarily highly polarized, one surface being in contact with the urine and the other with interstitial fluid. There is thus a strict requirement for vectorial transport of ions and other molecules across both apical and basolateral plasma membranes, and the cell surface resident proteins via which this transport occurs are different on the opposite sides of the cell, and their repertoires also diverge in the different cell types found in different nephron segments. Moreover, highly specialized machinery is required to bring these transport proteins to their appropriate site of action. This is a dynamic process, with regulated recycling between the cell surface and intracellular vesicular structures.

Thus in the α-IC (Figure 20.1A), protons are secreted actively across the apical surface of the cell into the collecting duct lumen by a vacuolar-type H^+-ATPase that is present at high density in the plasma membrane. This pump is composed of at least 13 different subunits organized into a membrane-anchored V_0 domain (containing subunits a, c, c″, d, and e) across which protons are pumped, and the V_1 domain, comprised of subunits A–H. This protrudes into the cytosol and is responsible for the ATP hydrolysis that fuels the pump and for coupling the two domains (Figure 20.1C) (Stevens & Forgac 1997). H^+-ATPases are inhibitable by the macrolide bafilomycins. There is also a P-type ATPase in the apical membrane that exchanges H^+ for K^+, together with an ill-defined K^+ channel, and basolateral chloride and KCl transporters.

The secretion of H^+ ions into the urine is coupled to the reclamation of bicarbonate ions across the basolateral plasma membrane, in exchange for chloride, via the chloride–bicarbonate exchanger, AE1. In the α-IC, AE1 is a multiple transmembrane-spanning protein composed of 856 amino acids. It exchanges chloride and bicarbonate with 1:1 stoichiometry that is pharmacologically sensitive to DIDS (4,4′-di-isothiocyanato-stilbene-2,2-disulfonate). The other major site of AE1 expression is at the red cell membrane, where it is 911 residues long, and participates not only in volume regulation by means of its ion exchange capacity, but also in stabilization of red cell biconcavity through interaction with various cytoskeletal proteins such as ankyrin and spectrin. This latter function is made possible by a 65-amino-acid N-terminal extension of the protein, thanks to an alternative promoter. Heterozygous mutations in AE1 are associated with red cell dysmorphologic disorders such as hereditary spherocytosis and ovalocytosis (Tanner 1997).

The group of disorders collectively known as 'distal RTA' could therefore in theory result either from loss of function in the α-IC of any of the essential molecules involved in acid–base homeostasis, or from abnormalities of the

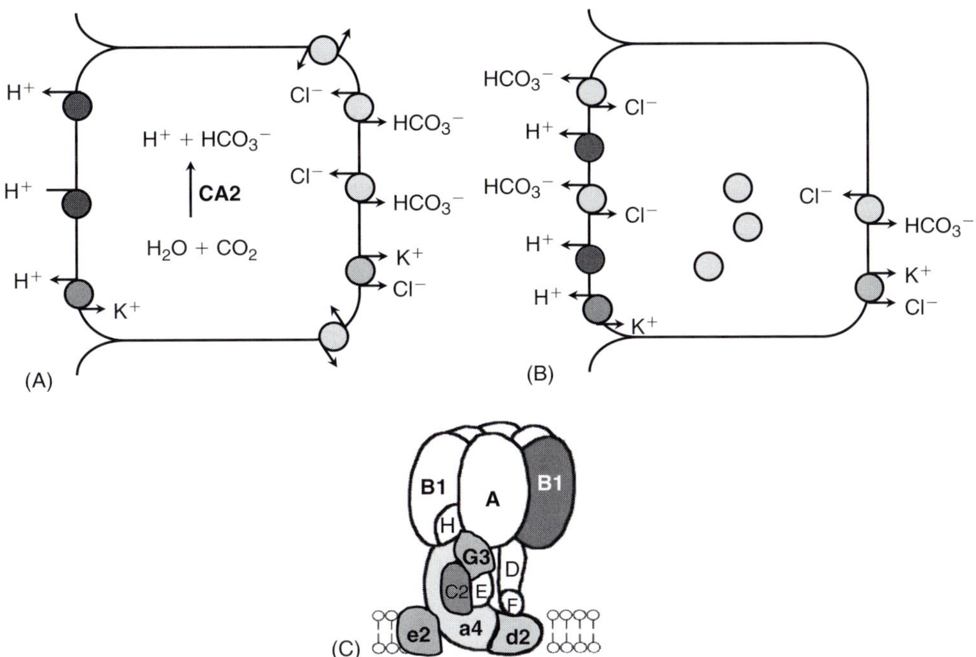

FIGURE 20.1 Polarized function of α-IC and molecular contributors to dRTA. (A) In cortical collecting duct α-IC, acid–base balance is tightly regulated by proton pumps on the apical surface of the cell that actively secrete acid into the duct. This process is coupled to the reclamation of bicarbonate ions into the interstitium, via the chloride–bicarbonate exchanger AE1, whose expression in the kidney is confined to the basolateral surface of these cells. Other molecular residents are also illustrated, with the exception of the ubiquitously expressed Na/K-ATPase. The KCC4 potassium-chloride co-transporter has to date only been localized in rodents. (B) In at least two forms of autosomal dominant distal renal tubular acidosis, loss of a C-terminal basolateral targeting sequence results in a non-polarized distribution of AE1, such that a the protein reaches the apical surface, thus probably disrupting the electrochemical balance of the cell in other cases. (C) The apical proton pump is a multimeric complex of at least 13 subunits. At least six of these are the product of alternate genes (shown in color) that are either solely or mainly expressed in the kidney. (see also Plate 21)

machinery that gets them to their proper site of action. In fact, at the time of early descriptions, dRTA was thought to be due to 'back-leak' of normally secreted protons across a leaky tubular epithelium, as in the case of amphotericin (see below). Over the following five decades it became evident that acid secretion itself is abnormal, due to overall failure of hydrogen ion secretion in the cortical collecting duct. Whether this was a primary or secondary event at the molecular level remained an enigma until recently.

AUTOSOMAL DOMINANT DISTAL RENAL TUBULAR ACIDOSIS

Genetics

To date, the only known genetic cause of dominantly inherited distal RTA (ddRTA) is heterozygous mutation in *SLC4A1*, encoding the basolateral anion exchanger AE1 (Bruce et al 1997, Jarolim et al 1998, Karet et al 1998, Weber et al 2000, Sritippayawan et al 2003, Cheidde et al 2003, Rungroj et al 2004). There is no current evidence for

genetic heterogeneity in ddRTA. Dominant disease is sometimes also referred to as type 1a.

Interestingly, of the 16 reported dominant or sporadic kindreds, the majority harbor sequence alterations affecting the arginine residue at position 589 (Table 20.1). Arg589 lies at the intracellular border of the sixth transmembrane domain of the protein, between Leu588 and Lys590; this latter residue is the target for the specific AE1 inhibitor phenyl isothiocyanate. These basic residues are conserved in all the known vertebrate anion exchanger isoforms, and are thought to form part of the site of intracellular anion binding. The kindreds harboring R589 mutations are not geographically or ethnically linked, and two are de novo findings, suggesting a mutational hotspot. This is probably because arginine at this point is encoded as CGC and L588 by CTG, giving CpG dinucleotides on both strands. In all the substitutions, C is replaced by T, which can arise as a result of endogenous methylation and spontaneous deamination (Duncan & Miller 1980).

The other residues affected by missense alterations that are associated with dominant disease are G609, S613, and A858. These are all either within, or at the boundary of, predicted transmembrane domains of the protein. In addition,

TABLE 20.1   Mutations in *SLC4A1* causing type 1 dRTA

Amino acid	Substitution	Number of kindreds	Dominant (D) or recessive (R) allele	Reference
R589	H	5	D	(Bruce et al 1997, Jarolim et al 1998, Karet et al 1998)
	C	4	D	(Bruce et al 1997, Weber et al 2000, Sritippayawan et al 2003)
	S	1	D	(Karet et al 1998)
V488	M	1	R	(Ribeiro et al 2000)
R602	H	1**	R	(Sritippayawan et al 2004)
G609	R	1	D	(Rungroj et al 2004)
S613	F	1	D	(Bruce et al 1997)
G701	D	9*	R	(Tanphaichitr et al 1998, Vasuvattakul et al 1999, Bruce et al 2000, Yenchitsomanus et al 2002, Sritippayawan et al 2004)
S773	P	1	R	(Sritippayawan et al 2004)
V850	deleted	6**	R	(Bruce et al 2000)
A858	D	2	D	(Bruce et al 2000)
D889	stop	1	D	(Cheidde et al 2003)
R901	stop	1	D	(Karet et al 1998)

*One as compound heterozygote with S773P, two with South East Asian ovalocytosis (SAO) mutation

**Compound heterozygotes with SAO mutation except one homozygous kindred and one heterozygous with A858D

two complex mutations share the similar end result of truncating the cytoplasmic C-terminal tail of the protein, at D889 or R901.

Mechanisms of Disease

With respect to the mechanism of dominant disease, it had been realized for several years that ddRTA is unlikely to result from simple haploinsufficiency. The evidence for this was first, that numerous heterozygous AE1 mutations have also been described that cause the autosomal dominant erythrocyte diseases hereditary spherocytosis and ovalocytosis (the red cell being the other major site of expression of AE1) including mutations that result in very early termination (codon 81) or frameshift (codon 170) (Jarolim et al 1996). These would likely result in severe disruption or absence of the encoded protein, but are not usually associated with a urine acidification defect.

Secondly, it is notable that when expressed in *Xenopus* oocytes, all of the dominant dRTA-associated AE1 mutations thus far studied have demonstrated normal or at worst only mildly diminished anion exchange function (Bruce et al 1997, Jarolim et al 1998, Toye et al 2002, Rungroj et al 2004), including those within the sixth and seventh transmembrane domains. The majority of those studied are the missense mutations of R589 (R589H, R589S and R589C) as well as S613F, G609R and a complex mutation resulting in an 11 amino acid truncation at the C-terminus, R901X.

The hypothesis of an AE1 trafficking abnormality as a mechanism of disease therefore arose (Tanphaichitr et al 1998). Possible explanations included the protein being retained intracellularly, or it reaching the surface but losing its specific basolateral distribution, such that the electrochemical balance of the cell might be disturbed (Figure 20.1B).

Given that 50% of functional AE1 appears to be sufficient for normal urinary acidification, and that AE1 functions as a dimer, intracellular retention of the mutant AE1 would also have to be associated with retention of a significant proportion of the wild-type AE1 encoded by the normal allele. Indeed, expression of AE1 mutants in (nonpolarized) HEK293 cell line has suggested that they are retained intracellularly, and that they exert a dominant negative influence by preventing co-expressed wild-type AE1 from reaching the cell surface (Quilty et al 2002). Similarly, Toye and colleagues have suggested intracellular retention of the R901X mutant, this time in Madin Darby canine kidney (MDCK) cells grown in non-polarizing conditions (Toye et al 2002). Since then, this group has gone on to examine various mutants in MDCK cells grown to polarity, and demonstrated that AE1 containing R589H or S613F do indeed continue to be intracellularly retained, even after adequate polarization is achieved (Toye et al 2004).

In pursuit of the alternative hypothesis, we demonstrated that in transiently transfected and adequately polarized MDCK cells, two mutants, R901X and G609R, appear inappropriately at the apical surface. We went on to confirm these findings in the rat renal epithelial cell line, IMCD (Devonald et al 2003, Rungroj et al 2004). In addition, when transplanted onto CD8, an apical reporter protein, in place of CD8's own C-terminus, the isolated R901X tail caused basolateral redistribution of a proportion of the CD8 chimera (Devonald et al 2003). Notably, nephrocalcinosis was particularly severe in affected individuals with these particular mutations, in two cases leading to premature renal failure and dialysis dependency followed by transplantation (Rungroj et al 2004).

The R901X findings have been confirmed in a stably transfected MDCK system (Toye et al 2004) and here apical expression alone was observed for mutant but not wild-type AE1 in polarized monolayers. Interestingly, transfected wild-type AE1 resulted in an alteration of transepithelial permeability. This was not exhibited by mutant protein, which suggests some interaction under normal circumstances between AE1 and molecules involved in cell–cell adhesion. In addition, R901X exhibited a more complex N-glycosylation pattern, which these authors suggest may act as a neo-apical targeting signal.

Erroneous targeting of mutant proteins to the wrong cell surface, as opposed to intracellular retention, is an unusual mechanism of disease, having been observed only in one form of hypercholesterolemia, and in the kidney, in one form of nephrogenic diabetes insipidus (Koivisto et al 2001, Asai et al 2003, Kamsteeg et al 2003). Within the C-terminal tail of AE1 there lies a motif, YDEV, which resembles the YXXØ motif (where X is any amino acid, and Ø is one possessing a hydrophobic side chain) associated with basolateral targeting. We found that the tyrosine component of the motif, Y904, was essential for basolateral targeting of wild-type AE1. When changed to alanine, a nonpolarized distribution was observed.

Thus, the C-terminal tail of AE1 contains at least one basolateral targeting signal. However, several different basolateral plasma membrane targeting motifs have been described. In addition to the YXXØ motif, these include the NPXY and di-leucine motifs (Brown & Breton 2000, Mostov et al 2003, Rodriguez-Boulan et al 2004).

Of all these, YXXØ is the best characterized. YXXØ-containing proteins associate with one of the adaptor protein (AP) complexes, which are hetero-tetrameric structures. AP-1 is found in the TGN and endosomes (Folsch et al 2001). In polarized epithelial cells, the AP-1B type and AP-4 have been implicated in sorting from the TGN to the basolateral cell surface (Folsch et al 1999, Mostov et al 1999, Simmen et al 2002), while AP-2 functions in the opposite direction to internalize proteins from the surface (Owen & Luzio 2000). However, the details of AP-complex mediated targeting in individual cells along the nephron are not evident, apart from the observation that LLC-PK$_1$ cells, which are derived from porcine proximal tubular cells, lack AP-1B (Folsch et al 1999). We used this cell-type difference to show that the missing 11 amino acid tail of AE1 cannot be interacting with AP-1B, as mutant AE1 behaves identically in LLC-PK1 transfection studies. Whether or not AP4 is involved remains to be determined, as no studies have yet identified binding partners for the C-terminus of AE1, apart from carbonic anhydrase 2 (Vince & Reithmeier 2000).

AUTOSOMAL RECESSIVE TYPE 1 RTA

Mutations in three genes are known to be associated with the recessive form of distal RTA: *SLC4A1*, *ATP6V1B1*, and *ATP6V0A4*.

SLC4A1

Because of its involvement in dominant disease, AE1 has been examined as a candidate in recessive disease both by linkage analysis and by direct screening. No R589 substitutions have been implicated in recessive disease. However, several recessive kindreds from Thailand and Papua New Guinea with homozygous mutations in *SLC4A1* (G701D, V488M or ΔV850) have now been described (Tanphaichitr et al 1998, Bruce et al 2000, Ribeiro et al 2000, Yenchitsomanus et al 2002). In addition, the coexistence of hemolytic anemia and RTA is described as a result of compound heterozygosity, involving residues R602H, G701D, S773P and A858D (Tanphaichitr et al 1998, Vasuvattakul et al 1999, Bruce et al 2000, Sritippayawan et al 2004). In some cases there is documented loss-of-function change in *SLC4A1* that manifests in in vitro *Xenopus* but not red cell anion transport. These alleles are acting recessively, as heterozygote parents for a single allele are not affected with dRTA. To date, however, the involvement of AE1 in recessive disease seems to be confined to South-East Asia, as in a cohort of kindreds of Middle Eastern and European Caucasian origin, linkage to *SLC4A1* was conclusively excluded (Karet et al 1998).

H$^+$-ATPase Subunit Genes

A genome-wide search to localize a gene or genes for recessive type 1 RTA in this same population revealed evidence for linkage to two loci, on 2p and 7q, in a cohort of mainly consanguineous kindreds. The genes responsible have both been identified and encode different kidney-specific subunits of the apical H$^+$-ATPase, accounting for RTA types 1b (early deafness) and 1c (milder hearing loss) respectively.

RECESSIVE dRTA WITH DEAFNESS (TYPE 1B)
ATP6V1B1, the gene encoding the B1-subunit of H$^+$-ATPase, resides on chromosome 2. This gene was of particular interest because in man, expression of the B1 isoform of the secretory proton pump had been demonstrated at significant levels only in kidney and male genital tract epithelia (van Hille et al 1994). Radiation hybrid mapping definitively placed *ATP6V1B1* within the linked dRTA interval. Screening for mutations in this gene revealed 15 different mutations in kindreds where almost all the affected

individuals had documented bilateral SNHL occurring at a young age, and in all but one kindred homozygous mutations were found (Karet et al 1999). The majority of these mutations are likely to disrupt the structure, or abrogate the production, of the normal B1-subunit protein, leading to loss of function of this specialized form of the proton pump (Figure 20.1C).

RT-PCR of material from the human inner ear, and immunolocalization in mouse tissue, showed that B1 mRNA and protein are both expressed within the cochlea and endolymphatic sac (Karet et al 1999). This is of interest in view of the unique ionic composition of endolymph, the fluid that fills these compartments, which has a high K^+ concentration of some 150 mM. This is thought necessary to maximize the sensitivity of the hair cells that transduce the endolymphatic fluid impulse into auditory nerve signals. Such a high K^+ concentration coexisting as it does with an endolymph pH of 7.4 (rather than the predicted alkalinity) implies a requirement for proton pumping into this fluid. Although it remains speculative, the idea is that defective H^+-ATPase function leads to raised endolymph pH and eventual irremediable functional loss of hair cells. This would explain why alkali therapy is of no help to the deafness of recessive dRTA.

B1 subunit expression is also reported in the eye, localizing to the ciliary epithelium in rat and rabbit at least (Wax et al 1997), though no human studies exist. Topical application of bafilomycin A1 (a proton pump inhibitor) reduced intraocular pressure in rabbits, indicating an essential role of the H^+-ATPase in ciliary epithelial ion transport.

RECESSIVE TYPE 1 RTA WITH LATER-ONSET HEARING LOSS (TYPE 1C)

By a similar linkage approach in a cohort of dRTA kindreds where hearing was essentially normal at the time of referral, the defective gene in this subset of families was found, and named *ATP6V0A4* because it proved to encode a newly identified kidney-specific isoform of the proton pump's 116 kDa 'accessory' a-subunit designated a4 (Smith et al 2000). It had previously been debated whether proton pumps in the kidney even contained an a-subunit (Gluck & Caldwell 1998, Gillespie et al 1991). Although this genetic finding showed that in the kidney at least, the a-subunit must be essential for proper proton pump function, its role within the multisubunit pump structure remains somewhat unclear. Yeast studies have suggested that it may assist not only in proton translocation, but also in assembling and stabilizing the pump (Leng et al 1998).

In humans, both the a4 and ubiquitously expressed a1 subunit have been shown to interact via their C-termini with the rate-limiting glycolytic enzyme PFK-1 (Su et al 2003). Interestingly, another H^+-ATPase component, the E subunit, interacts with aldolase, the next enzyme down the glycolytic cascade (Lu et al 2001, 2004). These findings suggest a direct link between proton pump function and glycolysis, which is important given the necessary presence of glucose for pump assembly (Kane 1995).

Additional Heterogeneity

A few recessive families do not appear to link to either *ATP6V1B1* or *ATP6V0A4*. There are numerous other theoretical candidate genes for recessive disease, including: (a) genes encoding the other subunits of the proton transporters; (b) genes whose products are required for trafficking of proton pumps to the apical membrane of the α-intercalated cell; and (c) genes encoding molecules necessary for the generation of protons, absorption of bicarbonate, recycling of chloride, or maintenance of the electrochemical gradient across both apical and basolateral membranes. Unfortunately, many of these potential candidates remain unidentified or uncharacterized, in humans. Alternatively, known genes may have as yet unrevealed novel isoforms in the kidney. Three proton pump subunit isoforms that are either solely or mainly expressed in the kidney, C2, G3, and d2, have been cloned and screened in a cohort of nine unlinked families, but no disease-attributable mutations have been identified (Smith et al 2002).

One candidate that arose from animal studies is that encoding the potassium chloride co-transporter KCC4 (Boettger et al 2002). Homozygous null *Kcc4* mice are both deaf and acidotic, and immunolocalization studies of wild-type animals suggest that this transporter is expressed on the basolateral surface of the intercalated cell. However, genetic analysis of the same cohort of recessive dRTA patients unlinked to either *ATP6V1B1* or *ATP6V0A4* did not reveal any potential disease-causing sequence alterations (T. Jentsch, personal communication).

GENOTYPE–PHENOTYPE CORRELATIONS IN PRIMARY DISTAL RTA

In dominant disease, it is notable that the degree of nephrocalcinosis in those affected by both R901X and G609R mutations in AE1 was especially severe, in the latter case leading to premature end-stage renal failure and transplantation (Rungroj et al 2004). Whether this is a result of futile H_2CO_3 recycling across the apical surface is not known and would be difficult to elucidate, partly because measurement of urine pCO_2 would be unreliable in the face of this level of calcification.

Apart from the presence or absence of hearing loss at diagnosis in recessive disease, which was used in the original linkage studies, there did not initially appear to be major phenotypic differences between recessive patients with

ATP6V1B1 and *ATP6V0A4* mutations, and treatment is identical. However, screening of a larger cohort later revealed that the latter mutations are indeed associated with hearing impairment. In general this has a later age of onset and is milder, though in some cases gives rise to profound deafness (Stover et al 2002). As predicted from these observations, the a4 subunit has also been revealed to be expressed within the cochlea (Stover et al 2002, Dou et al 2004).

OTHER FORMS OF dRTA

Carbonic Anhydrase II Deficiency

The protons and bicarbonate ions to be transported by the H^+-ATPase and AE1 in the α-IC are produced by the catalytic activity of carbonic anhydrase II (CAII) within the cytosol, and normal function of this enzyme is necessary both in the kidney and in osteoclasts for normal bone resorption. Because of its dual organ expression, loss of function of this enzyme causes the combination of renal tubular acidosis with osteopetrosis (also known as Guibaud-Vainsel syndrome or marble brain disease) (McMahon et al 2001, Shah et al 2004). As CAII is expressed both proximally and distally in the nephron, the RTA here is often of mixed proximal and distal type, though either alone or in combination with bone thickening is reported. Bone marrow transplantation may be curative of the bone problems (McMahon et al 2001).

Other Medical Conditions

In clinical practice, RTA may be found in autoimmune disorders, especially Sjorgren's syndrome. It often remits with successful immunosuppression. The relevant antigens are unknown. It is also associated (usually in its 'incomplete' form) with a proportion of calcium-stone-formers.

Amphotericin

Therapy with the antifungal agent amphotericin may result in hypokalemic dRTA. In this case, the mechanism is slightly different in that α-IC function proceeds normally, but the drug causes a back leak of protons from the urine to the interstitium via drug-induced pores in the intercellular barrier.

Vanadate

Ingestion of compounds containing the heavy metal vanadium can cause distal RTA, because of blockade of the P-type ATPase that exchanges H^+ for K^+ at the apical surface of α-IC (Figure 20.1A). This form of disease is usually at the milder end of the clinical spectrum. Vanadate is present at higher than usual concentration in the soil of parts of South East Asia, and consequently dRTA is endemic in these areas.

ACKNOWLEDGMENT

FEK is a Senior Clinical Research Fellow of the Wellcome Trust.

References

Asai T, Kuwahara M, Kurihara H, et al. Pathogenesis of nephrogenic diabetes insipidus by aquaporin-2 C-terminus mutations. Kidney Int. 2003; 64: 2–10.

Berrettini S, Neri E, Forli F, et al. Large vestibular aqueduct in distal renal tubular acidosis. High-resolution MR in three cases. Acta Radiol. 2001; 42: 320–2.

Boettger T, Hubner CA, Maler H, Rust MB, Beck FX, Jentsch TJ. Deafness and renal tubular acidosis in mice lacking the KCl cotransporter Kcc4. Nature 2002; 416: 874–8.

Brown D, Breton S. Sorting proteins to their target membranes. Kidney Int. 2000; 57: 816–24.

Bruce LJ, Cope DL, Jones GK, et al. Familial distal renal tubular acidosis is associated with mutations in the red cell anion exchanger (Band 3, AE1) gene. J. Clin. Invest. 1997; 100: 1693–707.

Bruce LJ, Wrong O, Toye AM, et al. Band 3 mutations, renal tubular acidosis and South-East Asian ovalocytosis in Malaysia and Papua New Guinea: loss of up to 95% band 3 transport in red cells. Biochem. J. 2000; 350: 41–51.

Butler AM, Wilson JL, Farber S. Dehydration and acidosis with calcification at renal tubules. J. Pediatr. 1936; 8: 489.

Cheidde L, Vieira TC, Lima PR, Saad ST, Heilberg IP. A novel mutation in the anion exchanger 1 gene is associated with familial distal renal tubular acidosis and nephrocalcinosis. Pediatrics 2003; 112: 1361–7.

Devonald MA, Smith AN, Poon JP, Ihrke G, Karet FE. Non-polarized targeting of AE1 causes autosomal dominant distal renal tubular acidosis. Nat. Genet. 2003; 33: 125–7.

Dou H, Xu J, Wang Z, et al. Co-expression of pendrin, vacuolar H^+-ATPase alpha4-subunit and carbonic anhydrase II in epithelial cells of the murine endolymphatic sac. J. Histochem. Cytochem. 2004; 52: 1377–84.

Duncan BK, Miller JH. Mutagenic deamination of cytosine residues in DNA. Nature 1980; 287: 560–1.

Folsch H, Ohno H, Bonifacino JS, Mellman I. A novel clathrin adaptor complex mediates basolateral targeting in polarized epithelial cells. Cell 1999; 99: 189–98.

Folsch H, Pypaert M, Schu P, Mellman I. Distribution and function of AP-1 clathrin adaptor complexes in polarized epithelial cells. J. Cell Biol. 2001; 152: 595–606.

Gillespie J, Ozanne S, Tugal B, et al. The vacuolar $H(+)$-translocating ATPase of renal tubules contains a 115-kDa glycosylated subunit. FEBS Lett. 1991; 282: 69–72.

Gluck S, Caldwell J. Proton-translocating ATPase from bovine kidney medulla: partial purification and reconstitution. Am. J. Physiol. 1998; 254: F71–9.

Jarolim P, Murray JL, Rubin HL, et al. Characterization of 13 novel band 3 gene defects in hereditary spherocytosis with band 3 deficiency. Blood 1996; 88: 4366–74.

Jarolim P, Shayakul C, Prabakaran D, et al. Autosomal dominant distal renal tubular acidosis is associated in three families with heterozygosity for the R589H mutation in the AE1 (band 3) Cl^-/HCO_3- exchanger. J. Biol. Chem. 1998; 273: 6380–8.

Kamsteeg EJ, Bichet DG, Konings IB, et al. Reversed polarized delivery of an aquaporin-2 mutant causes dominant nephrogenic diabetes insipidus. J. Cell Biol. 2003; 163: 1099–109.

Kane PM. Disassembly and reassembly of the yeast vacuolar H(+)-ATPase in vivo. J. Biol. Chem. 1995; 270: 17025–32.

Karet FE. Inherited distal renal tubular acidosis. J. Am. Soc. Nephrol. 2002; 13: 2178–84.

Karet FE, Gainza FJ, Gyory AZ, et al. Mutations in the chloride-bicarbonate exchanger gene AE1 cause autosomal dominant but not autosomal recessive distal renal tubular acidosis. Proc. Natl Acad. Sci. USA 1998; 95: 6337–42.

Karet FE, Finberg KE, Nelson RD, et al. Mutations in the gene encoding B1 subunit of H^+-ATPase cause renal tubular acidosis with sensorineural deafness. Nat. Genet. 1999; 21: 84–90.

Koivisto UM, Hubbard AL, Mellman I. A novel cellular phenotype for familial hypercholesterolemia due to a defect in polarized targeting of LDL receptor. Cell 2001; 105: 575–85.

Leng XH, Manolson MF, Forgac M. Function of the COOH-terminal domain of Vph1p in activity and assembly of the yeast V-ATPase. J. Biol. Chem. 1998; 273: 6717–23.

Lightwood R. Communication no. 1. Arch. Dis. Child. 1935; 10: 205.

Lu M, Holliday LS, Zhang L, Dunn WA Jr, Gluck SL. Interaction between aldolase and vacuolar H^+-ATPase: evidence for direct coupling of glycolysis to the ATP-hydrolyzing proton pump. J. Biol. Chem. 2001; 276: 30407–313.

Lu M, Sautin YY, Holliday LS, Gluck SL. The glycolytic enzyme aldolase mediates assembly, expression, and activity of vacuolar H^+-ATPase. J. Biol. Chem. 2004; 279: 8732–9.

McMahon C, Will A, Hu P, Shah GN, Sly WS, Smith OP. Bone marrow transplantation corrects osteopetrosis in the carbonic anhydrase II deficiency syndrome. Blood 2001; 97: 1947–50.

Mostov K, ter Beest MB, Chapin SJ. Catch the mu1B train to the basolateral surface. Cell 1999; 99: 121–2.

Mostov K, Su T, ter Beest M. Polarized epithelial membrane traffic: conservation and plasticity. Nat. Cell. Biol. 2003; 5: 287–93.

Owen DJ, Luzio JP. Structural insights into clathrin-mediated endocytosis. Curr. Opin. Cell Biol. 2000; 12: 467–74.

Peces R. Long-term follow-up in distal renal tubular acidosis with sensorineural deafness. Pediatr. Nephrol. 2000; 15: 63–5.

Quilty JA, Li J, Reithmeier RA. Impaired trafficking of distal renal tubular acidosis mutants of the human kidney anion exchanger kAE1. Am. J. Physiol. Renal Physiol. 2002; 282: F810–20.

Ribeiro ML, Alloisio N, Almeida H, et al. Severe hereditary spherocytosis and distal renal tubular acidosis associated with the total absence of band 3. Blood 2000; 96: 1602–4.

Rodriguez-Boulan E, Musch A, Le Bivic A. Epithelial trafficking: new routes to familiar places. Curr. Opin. Cell Biol. 2004; 16: 436–42.

Rungroj N, Devonald MA, Cuthbert AW, et al. A novel missense mutation in AE1 causing autosomal dominant distal renal tubular acidosis retains normal transport function but is mistargeted in polarized epithelial cells. J. Biol. Chem. 2004; 279: 13833–8.

Santos F, Chan JC. Renal tubular acidosis in children. Diagnosis, treatment and prognosis. Am. J. Nephrol. 1986; 6: 289–95.

Shah GN, Bonapace G, Hu PY, Strisciuglio P, Sly WS. Carbonic anhydrase II deficiency syndrome (osteopetrosis with renal tubular acidosis and brain calcification): novel mutations in CA2 identified by direct sequencing expand the opportunity for genotype-phenotype correlation. Hum. Mutat. 2004; 24: 272.

Simmen T, Honing S, Icking A, Tikkanen R, Hunziker W. AP-4 binds basolateral signals and participates in basolateral sorting in epithelial MDCK cells. Nat. Cell Biol. 2002; 4: 154–9.

Smith AN, Borthwick KJ, Karet FE. Molecular cloning and characterization of novel tissue-specific isoforms of the human vacuolar H(+)-ATPase C, G and d subunits, and their evaluation in autosomal recessive distal renal tubular acidosis. Gene 2002; 297: 169–77.

Smith AN, Skaug J, Choate KA, et al. Mutations in ATP6N1B, encoding a new kidney vacuolar proton pump 116-kD subunit, cause recessive distal renal tubular acidosis with preserved hearing. Nat. Genet. 2000; 26: 71–5.

Sritippayawan S, Kirdpon S, Vasuvattakul S, et al. A de novo R589C mutation of anion exchanger 1 causing distal renal tubular acidosis. Pediatr. Nephrol. 2003; 18: 644–8.

Sritippayawan S, Sumboonnanonda A, Vasuvattakul S, et al. Novel compound heterozygous SLC4A1 mutations in Thai patients with autosomal recessive distal renal tubular acidosis. Am. J. Kidney Dis. 2004; 44: 64–70.

Stevens TH, Forgac M. Structure, function and regulation of the vacuolar (H+)-ATPase. Annu. Rev. Cell Dev. Biol. 1997, 13. 779–808.

Stover EH, Borthwick KJ, Bavalia C, et al. Novel ATP6V1B1 and ATP6V0A4 mutations in autosomal recessive distal renal tubular acidosis with new evidence for hearing loss. J. Med. Genet. 2002; 39: 796–803.

Su Y, Zhou A, Al-Lamki RS, Karet FE. The a-subunit of the V-type H^+-ATPase interacts with phosphofructokinase-1 in humans. J. Biol. Chem. 2003; 278: 20013–18.

Tanner MJ. The structure and function of band 3 (AE1): recent developments (review). Mol. Membr. Biol. 1997; 14: 155–65.

Tanphaichitr VS, Sumboonnanonda A, Ideguchi H, et al. Novel AE1 mutations in recessive distal renal tubular acidosis. Loss-of-function is rescued by glycophorin A. J. Clin. Invest. 1998; 102: 2173–9.

Toye AM, Banting G, Tanner MJ. Regions of human kidney anion exchanger 1 (kAE1) required for basolateral targeting of kAE1 in polarized kidney cells: mis-targeting expalins dominant renal tubular acidosis. J. Cell Sci. 2004; 117: 1399–410.

Toye AM, Bruce LJ, Unwin RJ, Wrong O, Tanner MJ. Band 3 Walton, a C-terminal deletion associated with distal renal tubular acidosis, is expressed in the red cell membrane but retained internally in kidney cells. Blood 2002; 99: 342–7.

van Hille B, Richener H, Schmid P, Puettner I, Green JR, Bilbe G. Heterogeneity of vacuolar H(+)-ATPase: differential expression of two human subunit B isoforms. Biochem. J. 1994; 303 (Pt 1): 191–8.

Vasuvattakul S, Yenchitsomanus PT, Vachuanichsanong P, et al. Autosomal recessive distal renal tubular acidosis associated with Southeast Asian ovalocytosis. Kidney Int. 1999; 56: 1674–82.

Vince JW, Reithmeier RA. Identification of the carbonic anhydrase II binding site in the Cl(−)/HCO(3)(−) anion exchanger AE1. Biochemistry 2000; 39: 5527–33.

Wax MB, Saito I, Tenkova T, et al. Vacuolar H^+-ATPase in ocular ciliary epithelium. Proc. Natl Acad. Sci. USA 1997; 94: 6752–7.

Weber S, Soergel M, Jeck N, Konrad M. Atypical distal renal tubular acidosis confirmed by mutation analysis. Pediatr. Nephrol. 2000; 15: 201–4.

Wrong O, Davies HE. The excretion of acid in renal disease. Q. J. Med. 1959; 28: 259–313.

Wrong OM, Feest TG, MacIver AG. Immune-related potassium-losing interstitial nephritis: a comparison with distal renal tubular acidosis. Q. J. Med. 1993; 86: 513–34.

Yenchitsomanus PT, Vasuvattakul S, Kirdpon S, et al. Autosomal recessive distal renal tubular acidosis caused by G701D mutation of anion exchanger 1 gene. Am. J. Kidney Dis. 2002; 40: 21–9.

Zakzouk SM, Sobki SH, Mansour F, al Anazy FH. Hearing impairment in association with distal renal tubular acidosis among Saudi children. J. Laryngol. Otol. 1995; 109: 930–4.

CHAPTER 21

Nephrogenic Diabetes Insipidus: Vasopressin Receptor Defect

DANIEL G. BICHET

CELLULAR ACTIONS OF VASOPRESSIN

The neurohypophyseal hormone AVP has multiple actions, including the inhibition of diuresis, contraction of smooth muscle, aggregation of platelets, stimulation of liver glycogenolysis, modulation of adrenocorticotropic hormone release from the pituitary, and central regulation of somatic functions (thermoregulation, blood pressure). These multiple actions of AVP could be explained by the interaction of AVP with at least three types of G protein-coupled receptors; the V_{1a} (vascular hepatic) and V_{1b} (anterior pituitary) receptors act through phosphatidylinositol hydrolysis to mobilize calcium, and the V_2 (kidney) receptor is coupled to adenylate cyclase (Thibonnier et al 2001, Serradeil-Le Gal et al 2002).

The transfer of water across the principal cells of the collecting ducts is now known at such a detailed level that billions of molecules of water traversing the membrane could be represented and are useful teaching tools (http://www.mpibpc.gwdg.de/abteilungen/073/gallery.html; http://www.ks.uiuc.edu/research/aquaporins). The 2003 Nobel Prize in chemistry was awarded to Peter Agre and Roderick MacKinnon, who solved two complementary problems presented by the cell membrane: How does a cell let one type of ion through the lipid membrane to the exclusion of other ions? And how does it permeate water without ions? This contributed to a momentum and renewed interest in basic discoveries related to the transport of water and indirectly to diabetes insipidus. The first step in the action of AVP on water excretion is its binding to arginine vasopressin type 2 receptors (hereafter referred to as V_2 receptors) on the basolateral membrane of the collecting duct cells (Figure 21.1). The human gene that codes for the V_2 receptor (*AVPR2*) is located in chromosome region Xq28 and has three exons and two small introns (Birnbaumer et al 1992, Seibold et al 1992). The sequence of the cDNA predicts a polypeptide of 371 amino acids with seven transmembrane, four extracellular, and four cytoplasmic domains. The V_2 receptor is one of 701 members of the rhodopsin family within the superfamily of guanine-nucleotide (G) protein-coupled receptors ((Fredriksson et al 2003); see also perspective by Perez (2003)). The activation of the V_2 receptor on renal collecting tubules stimulates adenylyl cyclase via the stimulatory G protein (G_s) and promotes the cyclic adenosine monophosphate (cAMP)-mediated incorporation of water pores into the luminal surface of these cells. This process is the molecular basis of the vasopressin-induced increase in the osmotic water permeability of the apical membrane of the collecting tubule (Nielsen et al. 2002).

AVP also increases the water reabsorptive capacity of the kidney by regulating the urea transporter UT-A1 that is present in the inner medullary collecting duct, predominantly in its terminal part (Yang et al 2002). AVP also increases the permeability of principal collecting duct cells to sodium (Bankir et al 2005).

In summary, in the absence of AVP stimulation, collecting duct epithelia exhibit very low permeabilities to sodium urea and water. These specialized permeability properties permit the excretion of large volumes of hypotonic urine formed during intervals of water diuresis. By contrast, AVP stimulation of the principal cells of the collecting ducts leads to selective increases in the permeability of the apical membrane to water (P_f), urea (P_{urea}), and Na (P_{Na}).

These actions of vasopressin in the distal nephron are possibly modulated by prostaglandins E_2 (PGE_2s), nitric oxide (Morishita et al 2005), and by the luminal calcium concentration. High levels of E-prostanoid (EP_3) receptors are expressed in the kidney (Fleming et al 1998). However, mice lacking EP_3 receptors for PGE_2 were found to have quasi-normal regulation of urine volume and osmolality

Genetic Diseases of the Kidney
ISBN: 978-0-12-449851-8

in response to various physiological stimuli (Fleming et al 1998). An apical calcium/polycation receptor protein expressed in the terminal portion of the inner medullary collecting duct of the rat has been shown to reduce AVP-elicited osmotic water permeability when luminal calcium concentration rises (Sands et al 1997). This possible link between calcium and water metabolism may play a role in the pathogenesis of renal stone formation (Sands et al 1997).

RARENESS AND DIVERSITY OF *AVPR2* MUTATIONS

X-linked NDI is generally a rare disease in which the affected male patients do not concentrate their urine after administration of AVP (Bichet & Fujiwara 2001). Because this form is a rare, recessive X-linked disease, females are unlikely to be affected, but heterozygous females can exhibit variable degrees of polyuria and polydipsia because of skewed X chromosome inactivation. In Quebec, the incidence of this disease among males was estimated to be approximately 8.8 in 1 000 000 male live births (Arthus et al 2000). A founder effect of two particular *AVPR2* mutations (Bichet et al 1993), one in Ulster Scot immigrants (the 'Hopewell' mutation, W71X) and one in a large Utah kindred (the 'Cannon' pedigree) results in an elevated prevalence of X-linked NDI in their descendants in certain communities in Nova Scotia, Canada, and in Utah, USA (Bichet et al 1993). These founder mutations have now spread all over the North-American continent. To date, we have identified the W71X mutation in 42 affected males who reside

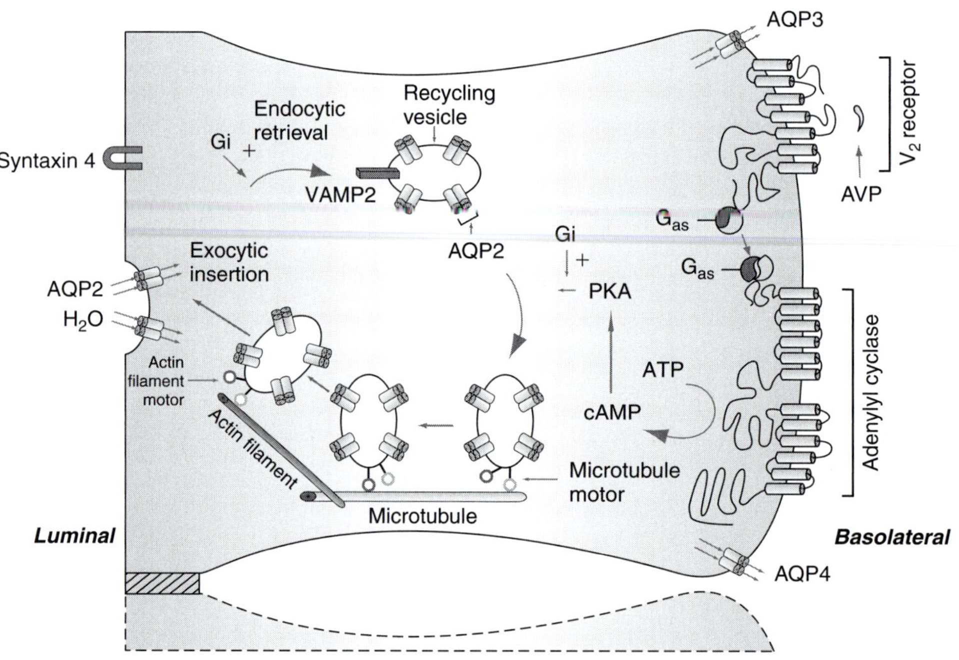

FIGURE 21.1   Schematic representation of the effect of arginine vasopressin (AVP) to increase water permeability in the principal cells of the collecting duct. AVP is bound to the V_2 receptor (a G-protein-linked receptor) on the basolateral membrane. The basic process of G-protein-coupled receptor signaling consists of three steps: a hepta-helical receptor that detects a ligand (in this case, AVP) in the extracellular milieu, a G-protein that dissociates into α-subunits bound to GTP and βγ-subunits after interaction with the ligand-bound receptor, and an effector (in this case, adenylyl cyclase) that interacts with dissociated G-protein subunits to generate small-molecule second messengers. AVP activates adenylyl cyclase increasing the intracellular concentration of cyclic adenosine monophosphate (cAMP). The topology of adenylyl cyclase is characterized by two tandem repeats of six hydrophobic transmembrane domains separated by a large cytoplasmic loop and terminates in a large intracellular tail. Generation of cAMP follows receptor-linked activation of the heteromeric G-protein (G_s) and interaction of the free $G_{\alpha s}$-chain with the adenylyl cyclase catalyst. Protein kinase A (PKA) is the target of the generated cAMP. Cytoplasmic vesicles carrying the water channel proteins (represented as homotetrameric complexes) are fused to the luminal membrane in response to AVP, thereby increasing the water permeability of this membrane. Microtubules and actin filaments are necessary for vesicle movement toward the membrane. The mechanisms underlying docking and fusion of aquaporin-2 (AQP2)-bearing vesicles are not known. The detection of the small GTP binding protein Rab3a, synaptobrevin 2, and syntaxin 4 in principal cells suggests that these proteins are involved in AQP2 trafficking (Valenti et al 1998). When AVP is not available, water channels are retrieved by an endocytic process, and water permeability returns to its original low rate. AQP3 and AQP4 water channels are expressed on the basolateral membrane. (see also Plate 22)

predominantly in the Maritime Provinces of Nova Scotia and New Brunswick, and the L312X mutation in eight affected males who reside in the central USA. We know of 98 living affected males of the Hopewell kindred and 18 living affected males of the Cannon pedigree. We also determined that the historical case report (Perry et al 1967) was related to the Hopewell pedigree and had the W71X mutation (Figure 21.2). To date, 183 putative disease-causing *AVPR2* mutations have been published in 287 NDI families (Figure 21.3) (Arthus et al 2004).

Approximately half of the mutations are missense mutations. Frameshift mutations owing to nucleotide deletions or insertions (25%), nonsense mutations (10%), large deletions (10%), in-frame deletions or insertions (4%), splice-site mutations, and one complex mutation account for the remainder of the mutations. Mutations have been identified in every domain, but on a per-nucleotide basis, about twice as many mutations occur in transmembrane domains compared with the extracellular or intracellular domains. We previously identified private mutations, recurrent mutations, and mechanisms of mutagenesis (Bichet et al 1994, Arthus et al 2000). The ten recurrent mutations (D85N, V88M, R113W, Y128S, R137H,

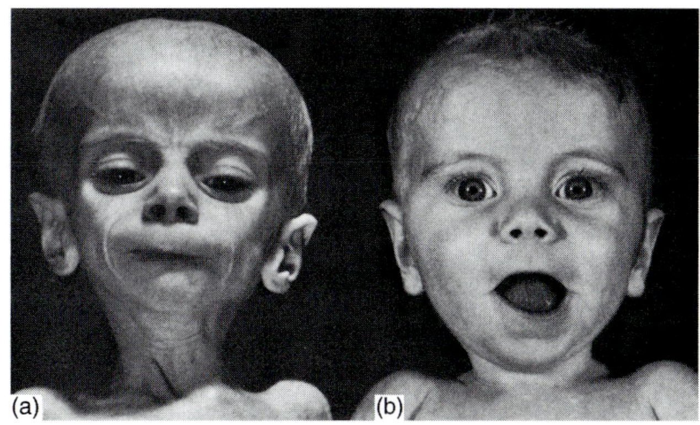

FIGURE 21.2 A typical 'historical' picture of a dehydrated and malnourished infant (a) with nephrogenic diabetes insipidus, looking healthy after rehydration and improved nutrition (b). This infant died a few years later due to repeated episodes of dehydration. This report was published years before the identification of the *AVPR2* gene. We were contacted by the mother and sister of this patient and we were able to reconstruct and link this family to the large Hopewell kindred (Bichet et al 1993; Bichet and Arthus, unpublished data). The mother and sister both had the W71X mutation. Photograph is Figure 2 of Perry et al reproduced with permission from the New England Journal of Medicine, 1967; 276

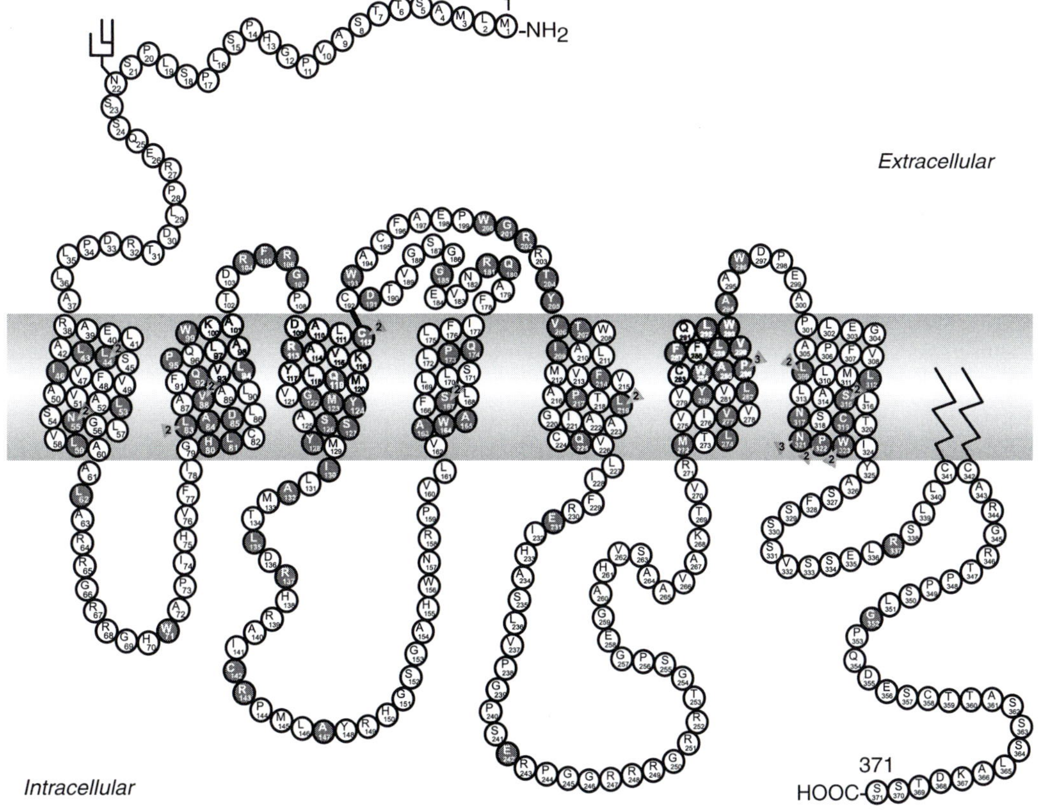

FIGURE 21.3 Schematic representation of the V$_2$ receptor and identification of 183 putative disease-causing *AVPR2* mutations. Predicted amino acids are given as the one-letter amino acid code. Solid symbols indicate missense or nonsense mutations; a number indicates more than one mutation in the same codon; other types of mutations are not indicated on the figure. The extracellular, transmembrane, and cytoplasmic domains are defined according to Mouillac et al (1995). There are 89 missense, 18 nonsense, 45 frameshift deletion or insertion, seven inframe deletion or insertion, four splice-site, and 19 large deletion mutations, and one complex mutation. See http://www.medicine.mcgill.ca/nephros for a list of mutations. (see also Plate 23)

S167L, R181C, R202C, A294P, and S315R) were found in 35 ancestrally independent families. The occurrence of the same mutation on different haplotypes was considered evidence for recurrent mutation. In addition, the most frequent mutations – D85N, V88N, R113W, R137H, S167L, R181C, and R202C – occurred at potential mutational hot spots (a C-to-T or G-to-A nucleotide substitution occurred at a CpG dinucleotide).

BENEFITS OF GENETIC TESTING

The natural history of untreated X-linked NDI includes hypernatremia, hyperthermia, mental retardation, and repeated episodes of dehydration in early infancy (Bichet & Fujiwara 2001). Mental retardation, a consequence of repeated episodes of dehydration (Figure 21.2), was prevalent in the Crawford and Bode study, in which only nine of 82 patients (11%) had normal intelligence (Crawford & Bode 1975), however, data from the Nijmegen group suggest that this complication was overestimated in their group of NDI patients (Hoekstra et al 1996). Early recognition and treatment of X-linked NDI with an abundant intake of water allows a normal lifespan with normal physical and mental development. Familial occurrence of males and mental retardation in untreated patients are two characteristics suggestive of X-linked NDI. Skewed X-inactivation is the most likely explanation for clinical symptoms of NDI in female carriers (Arthus et al 2000).

Identification of the molecular defect underlying X-linked NDI is of immediate clinical significance because early diagnosis and treatment of affected infants can avert the physical and mental retardation resulting from repeated episodes of dehydration. Affected males are immediately treated with abundant water intake, a low sodium diet, and hydrochlorothiazide. They do not experience severe episodes of dehydration and their physical and mental development remains normal; however, their urinary output is only decreased by 30% and a normal growth curve is still difficult to reach during the first 2–3 years of their lives despite the aforementioned treatments and intensive attention. Water should be offered every 2 hours day and night, and temperature, appetite, and growth should be monitored. Admission to hospital may be necessary for continuous gastric feeding. The voluminous amounts of water kept in patients' stomachs will exacerbate physiological gastrointestinal reflux as an infant and toddler, and many affected boys frequently vomit and have a strong positive 'Tuttle test' (esophageal pH testing). These young patients often improve with the absorption of an H-2 blocker and with metoclopramide (which could induce extrapyramidal symptoms) or with domperidone, which seems to be better tolerated and efficacious.

All polyuric states (whether neurogenic, nephrogenic, or psychogenic) can induce large dilatations of the urinary tract and bladder (Crawford & Bode 1975, Boyd et al 1980, Gautier et al 1981) and bladder function impairment has been well documented in patients bearing *AVPR2* or *AQP2* mutations (Shalev et al 2004, Ulinski et al 2004). Chronic renal failure secondary to bilateral hydronephrosis has been observed as a long-term complication in some of these patients. Renal and abdominal ultrasound should be done annually and simple recommendations including frequent urination and 'double voiding' could be important to prevent these consequences.

We propose that all families with hereditary diabetes insipidus should have their molecular defect identified. The molecular identification underlying X-linked NDI is of immediate clinical significance because early diagnosis and treatment of affected infants can avert the physical and mental retardation resulting from repeated episodes of dehydration. Diagnosis of X-linked NDI was accomplished by mutation testing of chorionic villous samples (n = 7), cultured amniotic cells (n = 6), or cord blood (n = 31). Three infants who had mutation testing done on amniotic cells (n = 1) or chorionic villous samples (n = 2) also had their diagnosis confirmed by cord blood testing. Of the 44 offspring tested, 21 were found to be affected males, 16 were unaffected males, and five were non-carriers (M.-F. Arthus, M. Lonergan, and D.G. Bichet, unpublished data). The affected males were immediately treated with abundant water intake, a low sodium diet, and hydrochlorothiazide. They have not experienced severe episodes of dehydration and their physical and mental development remain normal. Affected premature males may experience less severe polyuric symptoms and may only need increased hydration during their first week without a need for hydrochlorothiazide treatment.

MOST MUTANT V$_2$ RECEPTORS ARE NOT TRANSPORTED TO THE CELL MEMBRANE AND ARE RETAINED IN THE INTRACELLULAR COMPARTMENTS

Classification of the defects of naturally occurring mutant human V$_2$ receptors can be based on a similar scheme to that used for the low-density lipoprotein receptor. Mutations have been grouped according to the function and subcellular localization of the mutant protein whose cDNA has been transiently transfected in a heterologous expression system (Hobbs et al 1990). Using this classification, type 1 mutant V$_2$ receptors reach the cell surface but display impaired ligand binding and are consequently unable to induce normal cAMP production. The presence of mutant V$_2$ receptors on the surface of transfected cells can be determined pharmacologically. By carrying out saturation binding experiments using tritiated AVP, the number of cell surface mutant V_2 receptors and their apparent binding affinity can be compared with that of the wild-type receptor. In addition, the presence of cell surface receptors can be assessed directly by using immunodetection strategies to visualize epitope-tagged receptors in whole-cell immunofluorescence assays.

Type 2 mutant receptors have defective intracellular transport. This phenotype is confirmed by carrying out, in parallel, immunofluorescence experiments on cells that are intact (to demonstrate the absence of cell surface receptors) or permeabilized (to confirm the presence of intracellular receptor pools). In addition, protein expression is confirmed by western blot analysis of membrane preparations from transfected cells. It is likely that these mutant type 2 receptors accumulate in a pre-Golgi compartment, because they are initially glycosylated but fail to undergo glycosyl-trimming maturation.

Type 3 mutant receptors are ineffectively transcribed and lead to unstable mRNAs which are rapidly degraded. This subgroup seems to be rare, since Northern blot analysis of cells expressing mutant AVPR2 receptors showed mRNAs of normal quantity and molecular size.

Most of the *AVPR2* mutants that we and other investigators have tested are type 2 mutant receptors. They did not reach the cell membrane and were trapped in the interior of the cell (Morello et al 2000, Bernier et al 2004, Hermosilla et al 2004, Wuller et al 2004). Other mutant G-protein-coupled receptors (Schoneberg et al 2004) and gene products causing genetic disorders are also characterized by protein misfolding. Mutations that affect the folding of secretory proteins, integral plasma membrane proteins, or enzymes destined to the endoplasmic reticulum, Golgi complex, and lysosomes result in loss-of-function phenotypes irrespective of their direct impact on protein function because these mutant proteins are prevented from reaching their final destination (Romisch 2004). Folding in the endoplasmic reticulum is the limiting step: mutant proteins which fail to correctly fold are initially retained in the endoplasmic reticulum and subsequently often degraded. Key proteins involved in the urine countercurrent mechanisms are good examples of this basic mechanism of misfolding. *AQP2* mutations responsible for autosomal recessive NDI are characterized by misrouting of the misfolded mutant proteins and are trapped in the endoplasmic reticulum (Tamarappoo & Verkman 1998). Mutants encoding other renal membrane proteins that are responsible for Gitelman's syndrome (Kunchaparty et al 1999), Bartter's syndrome (Hayama et al 2003, Peters et al 2003), and cystinuria (Chillaron et al 1997) are also retained in the endoplasmic reticulum.

The *AVPR2* missense mutations are likely to impair folding and to lead to rapid degradation of the misfolded polypeptide and not to the accumulation of toxic aggregates (as is the case for AVP mutants), because the other important functions of the principal cells of the collecting duct (where

AVPR2 is expressed) are entirely normal. These cells express the epithelial sodium channel (ENaC). Decreased function of this channel results in a sodium-losing state (Bonnardeaux & Bichet 2004). This has not been observed in patients with *AVPR2* mutations. However, recent data showed that dDAVP could not stimulate sodium reabsorption in male patients with NDI bearing *AVPR2* mutations (Bankir et al 2005). By contrast, another type of conformational disease is characterized by the toxic retention of the misfolded protein. The relatively common Z mutation in α1-antitrypsin deficiency not only causes retention of the mutant protein in the endoplasmic reticulum but also affects the secondary structure by insertion of the reactive center loop of one molecule into a destabilized β sheet of a second molecule (Lomas et al 1992). These polymers clog up the endoplasmic reticulum of hepatocytes and lead to cell death and juvenile hepatitis, cirrhosis, and hepatocarcinomas in these patients (Lawless et al 2004).

NONPEPTIDE VASORESSIN RECEPTOR ANTAGONISTS ACT AS PHARMACOLOGICAL CHAPERONES TO FUNCTIONALLY RESCUE MISFOLDED MUTANT V_2 RECEPTORS RESPONSIBLE FOR X-LINKED NDI

If the misfolded protein/traffic problem responsible for so many human genetic diseases can be overcome and the mutant protein transported out of the endoplasmic reticulum to its final destination, these mutant proteins could be sufficiently functional (Cohen & Kelly 2003). Therefore, using pharmacological chaperones or pharmacoperones to promote escape from the endoplasmic reticulum is a possible therapeutic approach (Bernier et al 2004, Romisch 2004, Ulloa-Aguirre et al 2004). We used selective nonpeptide V_2 and V_1 receptor antagonists to rescue the cell surface expression and function of naturally occurring misfolded human V_2 receptors (Morello et al 2000). Since the beneficial effect of nonpeptide V_2 antagonists could be secondary to prevention and interference with endocytosis, we studied the R137H mutant previously reported to lead to constitutive endocytosis (Barak et al 2001). We found that the antagonist did not prevent the constitutive β-arrestin-promoted endocytosis (Bernier et al 2004). These results indicate that as for other *AVPR2* mutants, the beneficial effects of the treatment result from the action of the pharmacological chaperones. In clinical studies, we administered a nonpeptide vasopressin antagonist SR49059 to five adult NDI patients bearing the del62-64, R137H, and W164S mutations. SR49059 significantly decreased urine volume and water intake and increased urine osmolality while sodium, potassium, and creatinine excretions and plasma sodium were constant throughout the study (Bernier et al 2006) (Figure 21.4). This new therapeutic approach could be applied to the treatment of several hereditary diseases resulting from errors

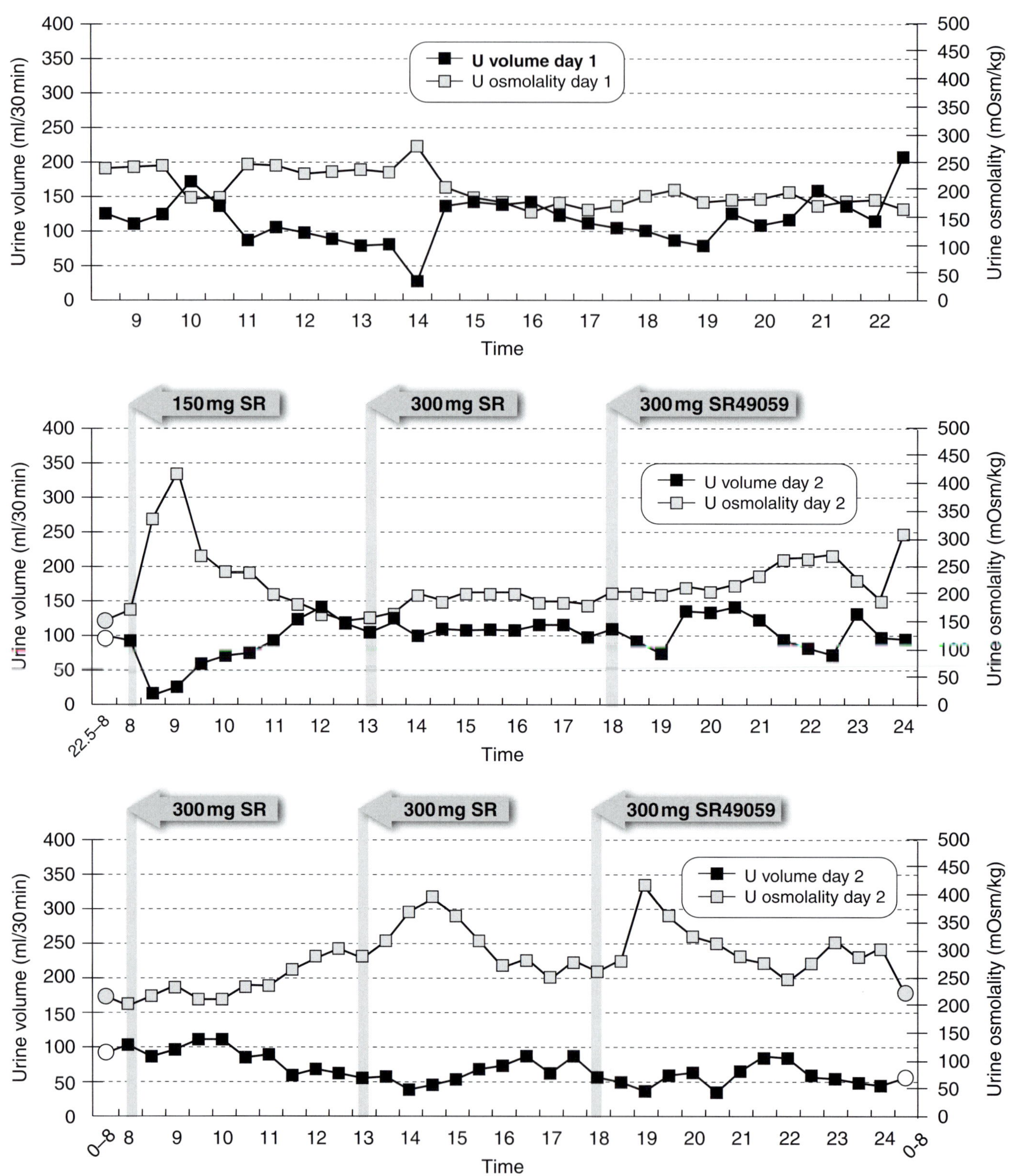

FIGURE 21.4   Urine volume and osmolality before (day 1) and after (days 2 and 3) SR49059 administration to a patient bearing the R137H mutation. Note that the distances observed between the two lines on days 2 and 3 represent the mirror images of urine volume and osmolality. Urine volume and osmolalities that were obtained during the control, second, and third nights are indicated by round circles. These data were obtained from 9:30 p.m. to 8:00 a.m. for the patient described here. Data from Bernier (2006) with permission from J Am Soc Nephrol

in protein folding and kinesis (Cohen & Kelly 2003, Ulloa-Aguirre et al 2004).

Since most human gene-therapy experiments using viruses to deliver and integrate DNA into host cells are potentially dangerous (Glover et al 2005), other treatments are being actively pursued. Torsten Schöneberg and colleagues (Sangkuhl et al 2004) used aminoglycoside antibiotics because of their ability to suppress premature termination codons (Mankin & Liebman 1999). They demonstrated that geneticin, a potent aminoglycoside antibiotic, increased AVP-stimulated cAMP in cultured collecting duct cells prepared from E242X mutant mice. The urine-concentrating ability of heterozygous mutant mice was also improved.

TESTING PATIENTS WITH NDI; PLEASE AVOID DEHYDRATION

Over the past few years it has become clear that 'pure' (i.e. loss of water only) congenital NDI is caused by an inactivating mutation(s) in the genes that code for the V_2 receptor or the AQP2 water channel (Fujiwara & Bichet 2005). The time of onset of the disease (shortly after birth) and the clinical symptoms are similar regardless of the molecular defect. The defective gene can be deduced by clinical testing: dDAVP elicits extrarenal (coagulation and vasodilatory) responses in male patients with NDI due to *AQP2* mutations, whereas patients with *AVPR2* mutations lack extrarenal responses (Bichet et al 1988, 1989, Lin et al 2002). Because this test is difficult to do in young infants, it has been replaced by *AVPR2* and *AQP2* mutation analysis. The small sizes of the genomic and coding regions of the genes involved (*AVPR2* and *AQP2*) allows for relatively easy mutational analysis, thereby allowing for carrier, prenatal and perinatal testing. If a dehydration test or an infusion of dDAVP is done these should be performed only during the day and under strict medical and nursing supervision.

In our clinical research unit, plasma sodium and plasma and urine osmolalities are measured at the beginning of each dehydration procedure and at regular intervals (usually hourly) thereafter, depending on the severity of the polyuric syndrome explored. In one case, an 8-year-old patient (31 kg body weight) with a clinical diagnosis of congenital NDI (later found to bear the de novo *AVPR2* mutant 274insG (Bichet et al 1994)) continued to excrete large volumes of urine (300 ml/h) during a short 4-hour dehydration test. During this time, the patient was suffering from severe thirst, his plasma sodium was 155 mEq/l, his plasma osmolality was 310 mmol/kg, and his urine osmolality was 85 mmol/kg. The patient received 1 μg of desmopressin IV and was allowed to drink water. Repeated urine osmolality measurements demonstrated a complete urinary resistance to desmopressin. It would have been dangerous and unnecessary to prolong the dehydration further in this young patient. Thus, the usual prescription of overnight dehydration should not be used in patients, and especially children, with severe polyuria and polydipsia. Great care should be taken to avoid any severe hypertonic state, arbitrarily defined as a plasma sodium >155 mEq/l.

ACKNOWLEDGMENTS

The author's work cited in this manuscript is supported by the Canadian Institutes of Health Research (MOP-8126), and by the Kidney Foundation of Canada. Dr Daniel G. Bichet holds a Canada Research Chair in Genetics of Renal Diseases. We thank Danielle Binette for computer graphics expertise and secretarial assistance.

References

Arthus M-F, Lonergan M, Crumley MJ, et al. Report of 33 novel AVPR2 mutations and analysis of 117 families with X-linked nephrogenic diabetes insipidus. J. Am. Soc. Nephrol. 2000; 11: 1044–54.

Arthus MF, Lonergan M, Fujiwara TM, Bichet DG. Clinical and genetic approaches to the diagnosis of congenital polyuro-polydipsic syndromes. The NDI Foundation 2004 Global Conference, Phoenix, Arizona; 2004.

Bankir L, Fernandes S, Bardoux P, Bouby N, Bichet DG. Vasopressin-V2 receptor stimulation reduces sodium excretion in healthy humans. J. Am. Soc. Nephrol. 2005; 16: 1920–8.

Barak LS, Oakley RH, Laporte SA, Caron MG. Constitutive arrestin-mediated desensitization of a human vasopressin receptor mutant associated with nephrogenic diabetes insipidus. Proc. Natl Acad. Sci. USA 2001; 98: 93–8.

Bernier V, Lagace M, Bichet DG, Bouvier M. Pharmacological chaperones: potential treatment for conformational diseases. Trends Endocrinol. Metab. 2004a; 15: 222–8.

Bernier V, Lagace M, Lonergan M, Arthus MF, Bichet DG, Bouvier M. Functional rescue of the constitutively internalized V2 vasopressin receptor mutant R137H by the pharmacological chaperone action of SR49059. Mol. Endocrinol. 2004b; 18: 2074–84.

Bernier V, Morello J-P, Zarruk A, et al. Pharmacological chaperones as a potential treatment for X-linked nephrogenic diabetes insipidus. J. Am. Soc. Nephrol. 2006; 17: 232–243.

Bichet DG, Arthus M-F, Lonergan M, et al. X-linked nephrogenic diabetes insipidus mutations in North America and the Hopewell hypothesis. J. Clin. Invest. 1993; 92: 1262–8.

Bichet DG, Birnbaumer M, Lonergan M, et al. Nature and recurrence of AVPR2 mutations in X-linked nephrogenic diabetes insipidus. Am. J. Hum. Genet. 1994; 55: 278–86.

Bichet DG, Fujiwara TM. Nephrogenic diabetes insipidus. In: Scriver CR, Beaudet AL, Sly WS, Vallee D, Childs B, Kinzler KW, Vogelstein B, eds. The Metabolic and Molecular Bases of Inherited Disease, 3rd edn. New York: McGraw-Hill, 2001: pp. 4181–204.

Bichet DG, Razi M, Arthus MF, et al. Epinephrine and dDAVP administration in patients with congenital nephrogenic diabetes insipidus. Evidence for a pre-cyclic AMP V2 receptor defective mechanism. Kidney Int. 1989; 36: 859–66.

Bichet DG, Razi M, Lonergan M, et al. Hemodynamic and coagulation responses to 1-desamino[8-D-arginine]vasopressin (dDAVP) infusion in patients with congenital nephrogenic diabetes insipidus. N. Engl. J. Med. 1988; 318: 881–7.

Birnbaumer M, Seibold A, Gilbert S, et al. Molecular cloning of the receptor for human antidiuretic hormone. Nature 1992; 357: 333–5.

Bonnardeaux A, Bichet DG. Inherited disorders of the renal tubule. In: Brenner BM, ed. Brenner & Rector's The Kidney, 2nd edn. Philadelphia: Saunders, 2004: pp. 1697–741.

Boyd SD, Raz S, Ehrlich RM. Diabetes insipidus and nonobstructive dilatation of urinary tract. Urology 1980; 16: 266–9.

Chillaron J, Estevez R, Samarzija I, et al. An intracellular trafficking defect in type I cystinuria rBAT mutants M467T and M467K. J. Biol. Chem. 1997; 272: 9543–9.

Cohen FE, Kelly JW. Therapeutic approaches to protein-misfolding diseases. Nature 2003; 426: 905–9.

Crawford JD, Bode HH. Disorders of the posterior pituitary in children. In: Gardner LI, ed. Endocrine and Genetic Diseases of Childhood and Adolescence. Philadelphia: W.B. Saunders, 1975: pp. 126–58.

Fleming EF, Athirakul K, Oliverio MI, et al. Urinary concentrating function in mice lacking EP3 receptors for prostaglandin E2. Am. J. Physiol. 1998; 275: F955–61.

Fredriksson R, Lagerstrom MC, Lundin LG, Schioth HB. The G-protein-coupled receptors in the human genome form five main families. Phylogenetic analysis, paralogon groups, and fingerprints. Mol. Pharmacol. 2003; 63: 1256–72.

Fujiwara TM, Bichet DG. Molecular biology of hereditary diabetes insipidus. J. Am. Soc. Nephrol. 2005; 16: 2836–46.

Gautier B, Thieblot P, Steg A. Mégauretère, mégavessie et diabète insipide familial. Sem. Hop. 1981; 57: 60–1.

Glover DJ, Lipps HJ, Jans DA. Towards safe, non-viral therapeutic gene expression in humans. Nat. Rev. Genet. 2005; 6: 299–310.

Hayama A, Rai T, Sasaki S, Uchida S. Molecular mechanisms of Bartter syndrome caused by mutations in the BSND gene. Histochem. Cell Biol. 2003; 119: 485–93.

Hermosilla R, Oueslati M, Donalies U, et al. Disease-causing V(2) vasopressin receptors are retained in different compartments of the early secretory pathway. Traffic 2004; 5: 993–1005.

Hobbs HH, Russell DW, Brown MS, Goldstein JL. The LDL receptor locus in familial hypercholesterolemia: mutational analysis of a membrane protein. Annu. Rev. Genet. 1990; 24: 133–70.

Hoekstra JA, van Lieburg AF, Monnens LA, Hulstijn-Dirkmaat GM, Knoers VV. Cognitive and psychosocial functioning of patients with congenital nephrogenic diabetes insipidus. Am. J. Med. Genet. 1996; 61: 81–8.

Kunchaparty S, Palcso M, Berkman J, et al. Defective processing and expression of thiazide-sensitive Na-Cl cotransporter as a cause of Gitelman's syndrome. Am. J. Physiol. 1999; 277: F643–9.

Lawless MW, Greene CM, Mulgrew A, et al. Activation of endoplasmic reticulum-specific stress responses associated with the conformational disease Z alpha 1-antitrypsin deficiency. J. Immunol. 2004; 172: 5722–6.

Lin SH, Bichet DG, Sasaki S, et al. Two novel aquaporin-2 mutations responsible for congenital nephrogenic diabetes insipidus in Chinese families. J. Clin. Endocrinol. Metab. 2002; 87: 2694–700.

Lomas DA, Evans DL, Finch JT, Carrell RW. The mechanism of Z alpha 1-antitrypsin accumulation in the liver. Nature 1992; 357: 605–7.

Mankin AS, Liebman SW. Baby, don't stop! Nat. Genet. 1999; 23: 8–10.

Morello JP, Salahpour A, Laperrière A, et al. Pharmacological chaperones rescue cell-surface expression and function of misfolded V2 vasopressin receptor mutants. J. Clin. Invest. 2000; 105: 887–95.

Morishita T, Tsutsui M, Shimokawa H, et al. Nephrogenic diabetes insipidus in mice lacking all nitric oxide synthase isoforms. Proc. Natl Acad. Sci. USA 2005; 102: 10616–21.

Mouillac B, Chini B, Balestre MN, et al. The binding site of neuropeptide vasopressin V1a receptor. Evidence for a major localization within transmembrane regions. J. Biol. Chem. 1995; 270: 25771–7.

Nielsen S, Frokiaer J, Marples D, Kwon TH, Agre P, Knepper MA. Aquaporins in the kidney: from molecules to medicine. Physiol. Rev. 2002; 82: 205–44.

Online Mendelian Inheritance in Man, OMIM (TM) (2000). McKusick-Nathans Institute for Genetic Medicine, Johns Hopkins University (Baltimore, MD) and National Center for Biotechnology Information, National Library of Medicine (Bethesda, MD). World Wide Web URL: http://www.ncbi. nlm.nih.gov/omim/

Perez DM. The evolutionarily triumphant G-protein-coupled receptor. Mol. Pharmacol. 2003; 63: 1202–5.

Perry TL, Robinson GC, Teasdale JM, Hansen S. Concurrence of cystathioninuria, nephrogenic diabetes insipidus and severe anemia. N. Engl. J. Med. 1967; 276: 721–5.

Peters M, Ermert S, Jeck N, et al. Classification and rescue of ROMK mutations underlying hyperprostaglandin E syndrome/antenatal Bartter syndrome. Kidney Int. 2003; 64: 923–32.

Romisch K. A cure for traffic jams: small molecule chaperones in the endoplasmic reticulum. Traffic 2004; 5: 815–20.

Sands JM, Naruse M, Baum M, et al. Apical extracellular calcium/polyvalent cation-sensing receptor regulates vasopressin-elicited water permeability in rat kidney inner medullary collecting duct. J. Clin. Invest. 1997; 99: 1399–405.

Sangkuhl K, Schulz A, Rompler H, Yun J, Wess J, Schoneberg T. Aminoglycoside-mediated rescue of a disease-causing nonsense mutation in the V2 vasopressin receptor gene in vitro and in vivo. Hum. Mol. Genet. 2004; 13: 893–903.

Schoneberg T, Schulz A, Biebermann H, Hermsdorf T, Rompler H, Sangkuhl K. Mutant G-protein-coupled receptors as a cause of human diseases. Pharmacol. Ther. 2004; 104: 173–206.

Seibold A, Brabet P, Rosenthal W, Birnbaumer M. Structure and chromosomal localization of the human antidiuretic hormone receptor gene. Am. J. Hum. Genet. 1992; 51: 1078–83.

Serradeil-Le Gal C, Wagnon J, Valette G, et al. Nonpeptide vasopressin receptor antagonists: development of selective and orally active V1a, V2 and V1b receptor ligands. Prog. Brain. Res. 2002; 139: 197–210.

Shalev H, Romanovsky I, Knoers NV, Lupa S, Landau D. Bladder function impairment in aquaporin-2 defective nephrogenic diabetes insipidus. Nephrol. Dial. Transplant. 2004; 19: 608–13.

Tamarappoo BK, Verkman AS. Defective aquaporin-2 trafficking in nephrogenic diabetes insipidus and correction by chemical chaperones. J. Clin. Invest. 1998; 101: 2257–67.

Thibonnier M, Coles P, Thibonnier A, Shoham M. The basic and clinical pharmacology of nonpeptide vasopressin receptor antagonists. Annu. Rev. Pharmacol. Toxicol. 2001; 41: 175–202.

Ulinski T, Grapin C, Forin V, et al. Severe bladder dysfunction in a family with ADH receptor gene mutation responsible for X-linked nephrogenic diabetes insipidus. Nephrol. Dial. Transplant. 2004; 19: 2928–9.

Ulloa-Aguirre A, Janovick JA, Brothers SP, Conn PM. Pharmacologic rescue of conformationally-defective proteins: implications for the treatment of human disease. Traffic 2004; 5: 821–37.

Valenti G, Procino G, Liebenhoff U, et al. A heterotrimeric G protein of the Gi family is required for cAMP- triggered trafficking of aquaporin 2 in kidney epithelial cells. J. Biol. Chem. 1998; 273: 22627–34.

Wuller S, Wiesner B, Loffler A, et al. Pharmacochaperones posttranslationally enhance cell surface expression by increasing conformational stability of wild-type and mutant vasopressin V2 receptors. J. Biol. Chem. 2004; 279: 47254–63.

Yang B, Bankir L, Gillespie A, Epstein CJ, Verkman AS. Urea-selective concentrating defect in transgenic mice lacking urea transporter UT-B. J. Biol. Chem. 2002; 277: 10633–7.

CHAPTER 22

Nephrogenic Diabetes Insipidus: Aquaporin-2 Defect

PETER M.T. DEEN, CAREL H. VAN OS AND NINE V.A.M. KNOERS

INTRODUCTION

An essential function of the mammalian kidney is its ability to conserve or excrete free water independent of changes in solute excretion. Early physiological studies have provided evidence for constitutive absorption of about 90% of the glomerular filtration rate (GFR) volume in proximal tubules and descending limbs of Henle's loop, whereas facultative water reabsorption occurs in collecting ducts under control of the antidiuretic hormone arginine vasopressin (AVP). Although all cell membranes have a surprisingly high diffusive water permeability, there is now extensive evidence that water transport in the nephron is facilitated by water channels (King & Agre 1996).

Since water movement across cell membranes is a fundamental property of life, it is of no surprise that the initial cloning of Aquaporin (AQP) 1, the archetypal water channel, has reanimated research in biology of water transport and in osmoregulation of mammals. There are ten well-characterized mammalian aquaporins and many more in plants. AQPs are members of a ubiquitous and ancient family of channel forming proteins. Based on sequence homology data, phylogenetic comparisons and permeability properties, aquaporins are subdivided into aquaporins and aquaglyceroporins. AQP 0, 1, 2, 4, 5, and 8 are water-selective channels, the orthodox aquaporins, and AQP 3, 7, and 9 are the aquaglyceroporins, which are channels with slightly less restrictive pores permeated by water, glycerol and other small non-electrolytes (Koyama et al 1998, Borgnia et al 1999).

MOLECULAR STRUCTURE OF AQUAPORINS

It is firmly established that aquaporins have six transmembrane domains, TM 1 through 6, and intracellular NH_2- and COOH-termini. The six TMs are connected by five loops (A through E; Figure 22.1). Loops B and E contain the highly conserved motifs asparagine-proline-alanine, NPA. The initial postulate that loops B and E fold back into the membrane and form the water pore, the so-called 'hourglass' hypothesis (Jung et al 1994), has been proven to be correct. Several utrastructural studies, using cryo-electron microscopy, have confirmed a barrel-like structure formed by the TMs enclosing loops B and E. For detailed information on the 3-dimensional structure of AQPs, the reader is referred to excellent papers and reviews written by the specialists in this field (Borgnia et al 1999, Heymann & Engel 2000, Verkman and Mitra 2000). Given the highly homologous primary structures of AQP1 and the other aquaporins, it is likely that all AQPs have a more or less similar three-dimensional structure. The less conserved region among the aquaporins is the COOH-terminus and this property has been exploited to generate highly specific polyclonal antibodies which have been very useful for localization studies.

Another well established feature of AQP structure is its occurrence in tetramers (Walz et al 1997, Kamsteeg et al 1999, Ringler et al 1999). While AQP2, and likely the other AQPs assemble to tetramers in the endoplasmic reticulum (Hendriks et al 2004), the structural features involved in tetramer assembly are still unresolved. It is, however, important to remember that all subunits in an AQP-tetramer form individual water pores (Van Hoek et al 1991, 1992, Jung et al 1994, Walz et al 1997, Borgnia et al 1999).

AQUAPORIN 2

AQP2 is the AVP-dependent water channel of the collecting duct (for review see Knepper & Inoue (1997)). It is well established that the collecting duct water permeability can be regulated in two ways: short-term regulation, which occurs in minutes, and long-term regulation, which takes hours.

Genetic Diseases of the Kidney
ISBN: 978-0-12-449851-8

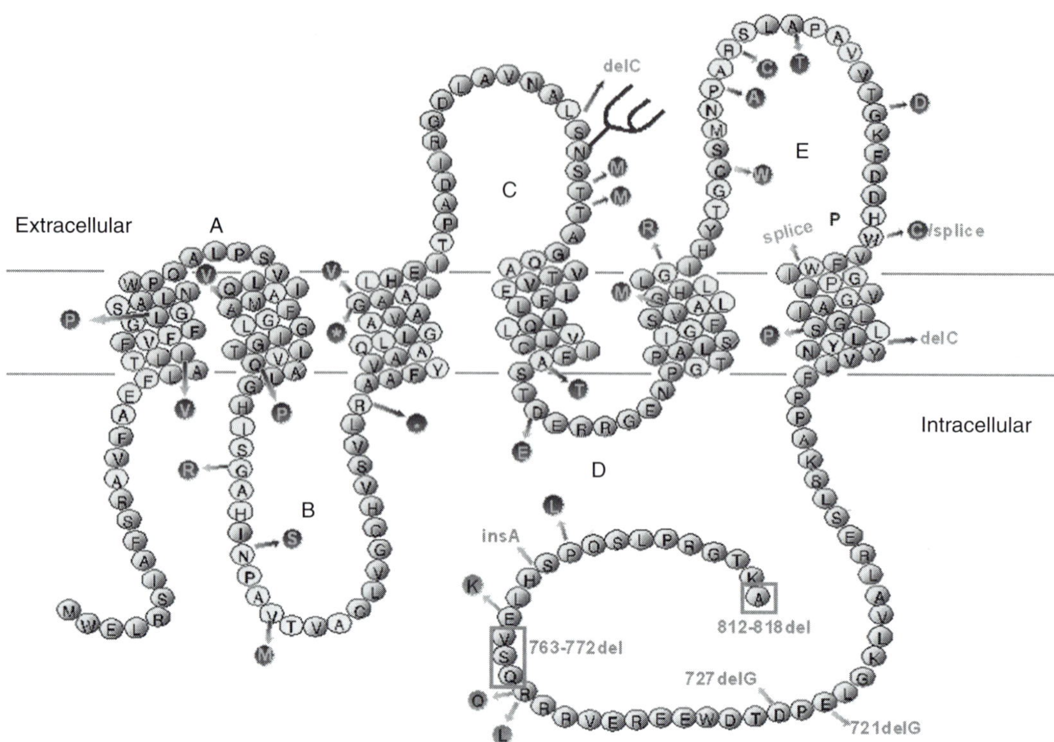

FIGURE 22.1   Mutations in Aquaporin-2 in nephrogenic diabetes insipidus. An AQP2 monomer consists of six transmembrane domains connected by five loops (A through E) and with its N- and C-termini located intracellularly. Loops B and E meet each other with their characteristic NPA boxes in the membrane and form the water selective pore. The N-glycosylation site (tree at N123) is indicated. Bright amino acids are conserved between different AQPs (Heymann & Engel 2000). Red amino acids and blue words or green amino acids and words outside the topology model represent mutations found in patients with recessive NDI or dominant NDI, respectively. Missense (amino acid), stop codon (*), nucleotide deletions (del) or insertions (ins), or splice site (splice) mutations are indicated. (see also Plate 24)

In short-term regulation, AVP, released from the pituitary in states of hypernatremia or hypovolemia, binds its V2R in the basolateral membrane of the renal principal cells (Figure 22.2). This hormone binding results in the switch of a V2R-bound GDP-coupled trimeric G-protein to its active GTP-bound form, which is subsequently cleaved in its α- and βγ-subunits. While the role of the βγ-subunit remains elusive, the α-subunit stimulates membrane-bound adenylate cyclase to convert ATP into cAMP, which in its turn activates protein kinase A. This kinase phosphorylates, besides other proteins, the AQP2 protein at its Ser256 residue, resulting in its steady state redistribution from intracellular vesicles to the apical membrane. Driven by the osmotic gradient of sodium, water will then be transcellularly reabsorbed by entering the principal cells through AQP2 in the apical membrane and leaving the cells for the interstitium through AQP3 and AQP4, which reside in the basolateral membrane (Figures 22.2 & 22.3) (Nielsen et al 1995). Although phosphorylation of Ser256 does not alter the water permeability of AQP2 (Lande et al 1996), it is essential for its translocation to the apical membrane, as AQP2-S256A, which mimics a constitutively non-phosphorylated AQP2 protein, is retained in intracellular vesicles, independent of the presence of the adenylate

cyclase activator, forskolin (Fushimi et al 1997, Katsura et al 1997, Kamsteeg et al 1999, Van Balkom et al 2002). Interesting in this respect is that in an unstimulated steady state condition, phosphorylated AQP2 could be detected in intracellular vesicles in the kidney (Christensen et al 2000). However, AQP2 is expressed as a homotetramer (Kamsteeg et al 1999, Werten et al 2001) and recent studies using oocytes as a model system indicated that for a plasma membrane localization three out of four monomers in an AQP2 tetramer need to be phosphorylated (Kamsteeg et al 2000).

This stimulatory process is rapidly reversible upon dissociation of AVP from its receptor, which might, but does not necessarily, coincide with de-phosphorylation of AQP2. Many hormones that counteract the antidiuretic action of AVP, such as endothelin, prostaglandin E2, TPA, epidermal growth factor and dopamine, are thought to reduce the apical expression of AQP2 by activation of protein kinase C (PKC) proteins, as in most cases, the inhibitory effects were absent upon co-treatment of the collecting ducts with protein kinase C (PKC) inhibitors (Maeda et al 1992, Nadler et al 1992, Deen et al 2000). In addition, the phorbol ester PMA, which activates several PKCs, also inhibits AVP-induced water permeability (Han et al 1994).

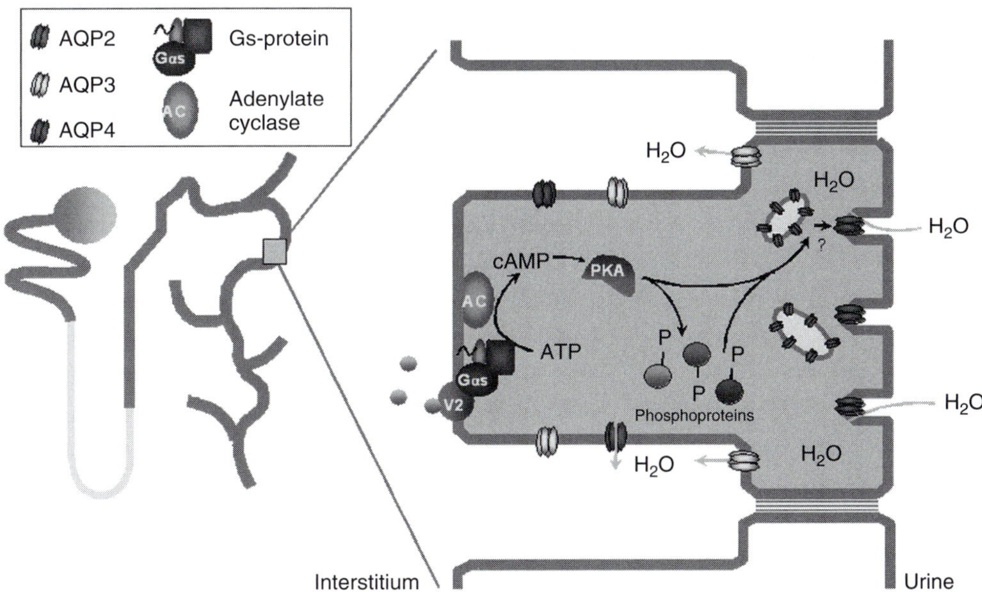

FIGURE 22.2   Regulation of the Aquaporin-2 mediated water transport by vasopressin. Vasopressin, vasopressin V2 receptor (V2), stimulatory GTP binding protein (Gs-protein), adenylate cyclase (AC), adenosine triphosphate (ATP), cyclic adenosine monophosphate (cAMP), protein kinase A (PKA) and phosphorylated proteins (O-P) are indicated. For details see text. (see also Plate 25)

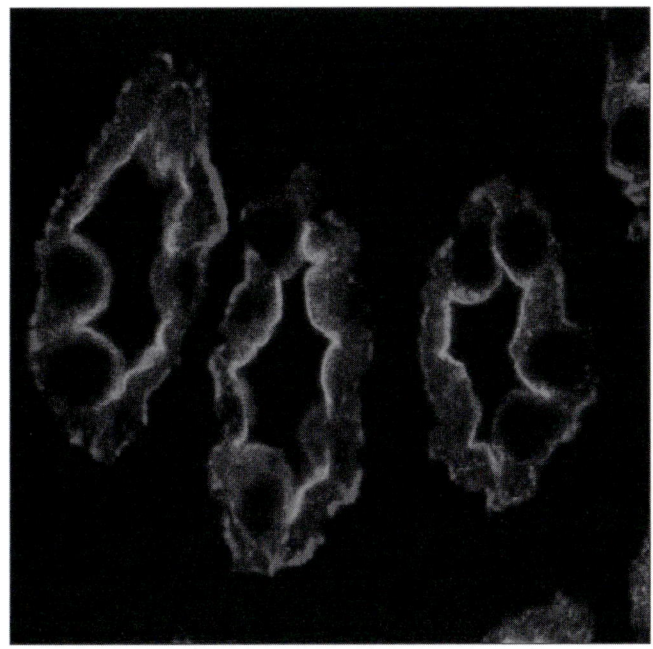

FIGURE 22.3   Expression of AQP2 and AQP3 in renal collecting ducts. Rabbit anti-AQP3 (red) and guinea pig anti-AQP2 (green) antibodies were used to specifically stain the kidney of rats dehydrated for 24 hours for AQP2 and AQP3. Note the apical membrane localization of AQP2 and the basolateral membrane localization of AQP3 in the collecting duct principal cells. (see also Plate 26)

Indeed, in later studies Zelenina and co-workers found that PGE2 induced the internalization of AQP2 in collecting ducts without changing the level of AQP2 phosphorylation (Zelenina et al 2001).

In fact, recent data from studies in polarized Madin-Darby canine kidney (MDCK) cells indicate that the action of these hormones occurs independently from AQP2 phosphorylation or de-phosphorylation. As in collecting ducts, human AQP2 expressed in these cells is translocated from intracellular vesicles to the apical membrane upon incubation of these cells with AVP or forskolin (Deen et al 1997). To study the role of phosphorylation in AQP2 trafficking and regulation, van Balkom et al changed the three putative casein kinase II (S148, S229, T244), one PKC (S231) and one PKA (S256) sites into alanines or aspartates, to mimic their constitutively nonphosphorylated or phosphorylated states, respectively (Van Balkom et al 2002). Upon expression in polarized MDCK cells, it appeared that, except for the S256 mutations, none of the other mutations did alter the subcellular localization, nor the translocation to the apical membrane with forskolin as compared to wild-type AQP2 (Van Balkom et al 2002). Indeed, only phosphorylation of the PKA site appeared to be relevant, as AQP2-S256A resided in intracellular vesicles and AQP2-S256D was located in the apical membrane, independent of forskolin treatment. Activation of PKC enzymes by PMA in the presence of forskolin, however, resulted in the re-distribution to intracellular vesicles of AQP2-S256D and all AQP2 kinase mutants, while they remained phosphorylated at S256. These data clearly indicated that, although phosphorylation of S256 is necessary for expression of AQP2 in the apical membrane, hormones that activate PKC enzymes can overcome the AVP-mediated phosphorylation and translocation of AQP2 to the apical membrane.

Likely, this internalization of even phosphorylated AQP2 with these hormones is mediated by the polymerization state of the actin cytoskeleton, as vasopressin is known to induce a de- and repolymerization of the actin cytoskeleton in collecting ducts (Simon et al 1993), and cytochalasins, which disrupt actin filaments, markedly inhibit the vasopressin response in target epithelia (Pearl & Taylor 1985). In addition, Valenti, Rosenthal and co-workers have provided compelling evidence that in cultured cells activation of RhoA kinase by PGE2 or other stimuli induces F-actin polymerization and inhibits AQP2 translocation to the plasma membrane, whereas inactivation of RhoA kinase by phosphorylation with forskolin or with a specific RhoA kinase inhibitor allowed translocation of AQP2 to the plasma membrane (Klussmann et al 2001, Tamma et al 2001, 2003a, 2003b).

In long-term regulation, adaptation to circulating AVP levels increases collecting duct water permeability by increasing the expression level of AQP2 (Digiovanni et al 1994, van Os et al 1994). In the promoter region of the AQP2 gene, different cis-acting elements have been reported that were shown to be able to confer AVP-induced expression of a reporter gene (Hozawa et al 1996, Yasui et al 1997). The main action of AVP seems to be a transcriptional regulation of the AQP2 gene, which is mediated by phosphorylation of a cAMP response element (CRE)-binding protein (CREB) and binding of phosphorylated CREB to the CRE in the promoter region of the AQP2 gene (Matsumura et al 1997).

The Role of AQP2 in Diseases

Several pathophysiological conditions exist which lead to a disturbed water homeostasis. Since AVP is the hormone that controls serum osmolality by decreasing free water clearance, any condition that interferes with AVP production, secretion and binding to V_2 receptors or with AQP2 synthesis and trafficking will result in loss of the ability to concentrate urine. In patients suffering from familial central diabetes insipidus, CDI, numerous mutations within the AVP gene have been identified (for review see: Robertson and Berl (1995)). This disorder is transmitted in an autosomal dominant manner, as mostly, the vasopressin-secreting magnocellular neurons become apoptotic and degenerate because of the difficulties of transporting the mutant AVP-neurophysin-glycopeptide pro-hormone form the endoplasmic reticulum (ER). Central DI typically presents in early childhood, but can be treated by DDAVP administration.

Besides central DI, a disturbed water balance can also be of renal origin, which is consequently called nephrogenic diabetes insipidus (NDI). Acquired NDI is the most common form. Lithium therapy, which is mostly prescribed to patients with bipolar disorders, but also conditions of hypokalemia (as is common in hypertensive patients treated with thiazide diuretics), hypercalcemia and bilateral urethral obstruction (BUO) often lead acquired forms of NDI, a disease in which the kidney does not respond to AVP. Paradoxically, diseased

states like congestive heart failure, liver cirrhosis, preeclampsia, and the syndrome of inappropriate release of the antidiuretic hormone (SIADH) induce excessive renal water reabsorption, which often can lead to life-threatening hypernatremia and edema formation. From numerous rat model studies, of which most were elegantly performed by Nielsen and Schrier and colleagues, it has become clear that all these forms of acquired NDI coincide with a strong reduction in expression of AQP2, whereas the states of excessive renal water uptake coincide with increased AQP2 expression (for review see: Marples et al 1999, Nielsen et al, 1999, Schrier et al 2001). So far, all data collected on AQP2 confirm its unique role, which is AVP-regulated control of the osmotic water permeability of the principal cell by insertion of AQP2 water channels in an otherwise extremely water impermeable apical plasma membrane.

Besides acquired NDI, a congenital form of NDI also exists, which is quite rare and has an occurrence of one per 100 000 humans. The recent discovery of different genetic causes of NDI has important implications for genetic counseling, especially in those families in which only one patient is affected.

HISTORY OF NDI

Although the X-linked type of NDI is by far the most frequent form of the disease encountered, there is ample evidence, also in the older literature, for the existence of other genetic forms of the disease. Thus, several authors have described the complete clinical picture to occur in females (Wiggelinkhuizen et al 1973, Zimmerman & Green 1975, Schreiner et al 1978, Aggarwal et al 1986). At present, we know that many, but certainly not all, of these female patients are carriers of V2R mutations (van Lieburg et al 1995, Moses et al 1995, Chan Seem et al 1999, Sato et al 1999, Arthus et al 2000). The full expression of the V2R defect in these females, in general rather unlikely to occur in an X-linked recessive disease, can be explained by skewed X-inactivation, resulting in predominant expression of the mutant V2R allele (Nomura et al 1997). In several female cases, however, X-linked inheritance could be excluded on the basis of the pedigree and/or DNA-analysis of the V2R region or mutation analysis of the V2R gene. One of the best examples in this respect is the case reported by Langley et al who described two sisters with NDI, born to consanguineous Pakistani parents. DNA analysis with one of the DNA markers known to be very closely linked to the NDI locus on the X chromosome showed that each girl inherited different Xq28 regions, ruling out a diagnosis of classical X-linked NDI (Langley et al 1991). On the basis of parental consanguinity along with the observation that both parents concentrated urine well and the results of DNA analysis, Langley et al suggested that NDI in these girls must be the result of an autosomal recessive mutation. In addition to the families in which NDI showed autosomal recessive transmission, pedigrees have been described in which NDI must be an autosomal dominant trait, since in these families father-to-son transmission of the disorder was observed (Cannon 1955).

Identification of the AQP2 Gene as the Gene Causing Autosomal Recessive NDI

The final proof for the existence of a non-X-linked type of NDI came from our studies in an apparently sporadic Dutch male NDI patient. Careful clinical testing of this NDI case unraveled the clue towards the elucidation of a new gene defect for NDI. The patient involved could be differentiated from patients with X-linked NDI on the basis of a 1-deamino-8-D-arginine vasopressin (DDAVP) infusion test. DDAVP is a selective agonist for the V2R. In addition to the renal V2R, which mediates the antidiuretic effect of AVP and DDAVP, there is evidence for an extrarenal V2R, which mediates extrarenal effects of these ligands, such as a rise in Factor VIII, von Willebrand factor (vWF) and tissue-type plasminogen activator (t-PA) and a fall in diastolic blood pressure (Mannucci et al 1977, Kaufmann et al 2000). In contrast to the blunted coagulation, fibrinolytic, and vasodilatory responses seen in most NDI patients after DDAVP infusion and reflecting a general V2R defect (Kobrinsky et al 1985, Bichet et al 1988, Knoers et al 1990), the Dutch male NDI patient showed completely normal extrarenal responses to DDAVP (Knoers & Monnens 1991). This finding indicated that the unresponsiveness to vasopressin in this boy was restricted to the kidney. The results of the DDAVP test, together with the finding that the coding region of the V2R gene did not harbor any potentially harmful mutation in this patient, pointed towards a defect in a kidney-specific protein in the cellular vasopressin-signaling cascade beyond the V2R. A candidate gene became available when Fushimi et al cloned a water channel of the rat renal collecting duct, WCH-CD, which was later renamed Aquaporin-2 (AQP2) (Fushimi et al 1993). Expression in *Xenopus* oocytes of its mRNA was found to increase osmotic swelling in response to hypotonic bathing, suggesting that it encoded the vasopressin-regulated water channel. Subsequent cloning and localization of the human equivalent AQP2 gene favored its potential involvement in autosomal NDI, as it localized on chromosome 12q13 (Deen et al 1994b, Saito et al 1995). Indeed, upon sequencing of the AQP2 gene in the Dutch variant NDI patient, he appeared to carry two point mutations in the AQP2 gene, one resulting in substitution of a cysteine for arginine 187 (R187C) in the third extracellular loop of the water channel and the other resulting in substitution of a proline for serine 216 (S216P) in the sixth transmembrane domain (Deen et al 1994a). The parents of the boy, who had completely normal concentrating abilities, each carried one of these two missense mutations. Functional expression studies in *Xenopus* oocytes revealed that each mutation

resulted in nonfunctional water channels (Deen et al 1994a). All together, the findings indicated that aberrant AQP2 water channels underlie autosomal recessive NDI, an assumption which was further strengthened by the subsequent detection of mutations in other families, in which patients were from consanguineous matings (van Lieburg et al 1994).

Differential Diagnosis Between Autosomal NDI Caused by AQP2 Mutations and X-linked NDI Caused by V2R Mutations

The X-linked and autosomal recessive forms of NDI do not differ with respect to the clinical symptoms of the disorder and, with the exception of a few cases, also not with respect to the time of onset of the disease. Only in a far minority of patients with X-linked NDI, namely in those patients that are carrying V2R mutations resulting in partial insensitivity to AVP, the disease-onset was reported to be not directly after birth but later in childhood. In general, the initial symptoms in most autosomal dominant cases also appear later, in some cases even not before early adulthood. After the determination of normal extra-renal responses to DDAVP in the first patient shown to have AQP2 mutations, it was suggested that intravenous DDAVP administration with measurement of von Willebrand factor, factor VIII, and tissue-type plasminogen activator levels, could be of use in the differentiation between the X-linked and autosomal recessive forms of NDI. Indeed, it was shown that male patients with autosomal recessive NDI show normal extrarenal responses to DDAVP, whereas in all studied patients with X-linked NDI these extrarenal responses are absent as a result of an extrarenal mutant vasopressin type 2 receptor (van Lieburg et al 1996). In female patients, however, the interpretation of this intravenous DDAVP test is more complicated. Although absence of the extrarenal responses to intravenous administration of DDAVP in females clearly points to the presence of a V2 receptor defect, a normal response cannot be interpreted as indicative of a defect beyond the V2 receptor, thus of an AQP2 defect (van Lieburg et al 1995). For instance, a symptomatic female patient described by Moses et al, who was shown to be heterozygous for a V2R mutation, showed a twofold increase in FVIII activity after administration of DDAVP (Moses et al 1995). This discrepancy between the extrarenal and renal response to DDAVP in these female V2 receptor mutation carriers may be due to variability of the pattern of X-chromosome inactivation between different tissues (Brown et al 1990).

THE AQP2 GENE

The role of the V2R gene in NDI is covered in Chapter 23. The human AQP2 gene (genbank accession number z29491) is located at chromosome 12q13 (Deen et al 1994b,

Sasaki et al 1994), as part of an aquaporin gene cluster to which also AQP0, AQP5, and AQP6 belong. This single copy gene comprises four exons distributed over approximately 5 kb genomic DNA. The 1.5 kb mRNA encodes a protein of 271 amino-acids, which has a predicted molecular weight of 29 kDa.

Autosomal Recessive NDI

Both the autosomal recessive and the autosomal dominant types of NDI are caused by mutations in the aquaporin-2 water channel gene. To date 27 mutations in AQP2 have been reported (Figure 22.1) in families with autosomal recessive NDI, which include 21 missense mutations, two nonsense mutations, two 1 bp deletions and two splice site mutations (Moses et al 1984, Deen et al 1994a, van Lieburg et al 1994, Deen et al 1995, Oksche et al 1996, Canfield et al 1997, Hochberg et al 1997, Mulders et al 1997, Goji et al 1998, Kuwahara 1998, Tamarappoo & Verkman 1998, Vargas-Poussou et al 1998, Marr et al 2001, 2002a, Lin et al 2002, Boccalandro et al 2004, De Mattia et al 2004a, Guyon et al 2004). Interestingly, nearly all these mutations in autosomal recessive NDI are found in the region encoding the AQP2 segment in between the first and the last transmembrane domain. To establish the involvement of an identified AQP2 gene mutation in NDI and to study the underlying mechanism why the mutation would lead to recessive NDI, the *Xenopus* oocyte expression system was mostly used, as these cells are easy to inject, because they have a diameter of 5 mm, have a translational capacity similar to 50 000 somatic cells, can be easily 'cultured' in a simple buffer for several days, and show an extensive swelling in hypotonic buffer when expressing functional AQP water channels. Two to three days after injection of AQP2 cRNAs, these cells show a proper trafficking and a marked expression of wt-AQP2 in the plasma membrane (Figure 22.4). In contrast, AQP2 mutants in recessive NDI show a dispersed 'reticular' staining (Figure 22.4), which is typical for ER-retained proteins. Although Goji et al reported that AQP2-T125M and AQP2-G175R are non-functional water channels that are unaffected in their routing to the plasma membrane (Goji et al 1998), others found that these and all other AQP2 missense mutants in recessive NDI are impaired in their export from the ER upon expression in *Xenopus laevis* oocytes. As found for most missense mutants in other disease, this ER retention is likely due to misfolding, which is usually followed by proteasomal degradation through the endoplasmic reticulum associated degradation pathway (ERAD; Tamarappoo & Verkman 1998; Hirano et al 2003, Buck et al 2004). A similar distribution for AQP2 mutants in recessive NDI was observed in mammalian cells (Figure 22.4), which underscores the conservation of the fundamental folding process during evolution and the stringency of the ER quality control mechanism in cells. The low stability

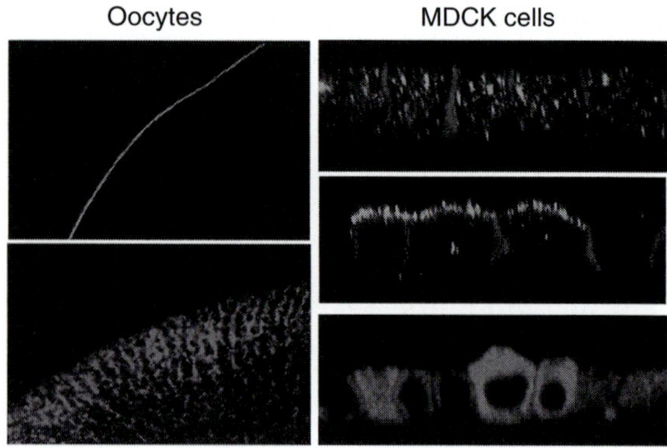

FIGURE 22.4 Localization of wild-type AQP2 and an AQP2 mutant in recessive NDI. Following injection of their cRNAs, wild-type AQP2 (upper left) or an AQP2 mutant in recessive NDI (lower left) were expressed in oocytes. Also, MDCK cells were stably transfected with an AQP2 mutant in recessive NDI or with wt-AQP2, and grown to confluence. All cells were subjected to immunocytochemistry and analyzed by confocal laser scanning microscopy. In oocytes, wt-AQP2 is located in the plasma membrane, whereas the mutant in recessive NDI shows the reticular pattern of the ER. In MDCK cells a similar reticular pattern (only the nucleus and plasma membranes are unstained) is observed for the AQP2 mutant in NDI (lower right). In MDCK cells, wild-type AQP2 resides in intracellular vesicles without stimulation (upper right), but is redistributed to the apical membrane with forskolin (middle right). Wild-type-AQP2 (green), the mutant in recessive NDI (red) and the basolateral marker E-cadherin (blue) are indicated. (see also Plate 27)

of these AQP2 mutant proteins is in line with in vivo data, as mutant AQP2 proteins could not be detected in urine of patient suffering from autosomal recessive NDI, while wt-AQP2 is well detectable in urine from healthy individuals (Kanno et al 1995, Deen et al 1996). Although retained in the ER, at high expression levels, at which some of the AQP2 mutant protein escapes from the ER and is routed to the plasma membrane (Kamsteeg & Deen 2000), seven mutants (L22V, A47V, G64R, T125M, T126M, A147T and V168M) appeared to be able to confer water permeability (Canfield et al 1997, Mulders et al 1997, Marr et al 2001, 2002a, Boccalandro et al 2004).

Autosomal Dominant NDI

At present, eight families have been described with autosomal dominant NDI, which were based on a father-to-son transmission of the disease or the absence of mutations in the V2R gene. Subsequent sequencing of the patient's genes revealed putative disease-causing mutations in the AQP2 gene of one allele, which were deletions of one (721delG, 727delG), seven (812-818del) or ten (763-772del) nucleotides resulting in a +1 frame shift, a one base insertion (779-780insA) resulting in a +2 frame shift and missense mutations (R254L, R254Q and E258K). Interestingly, all

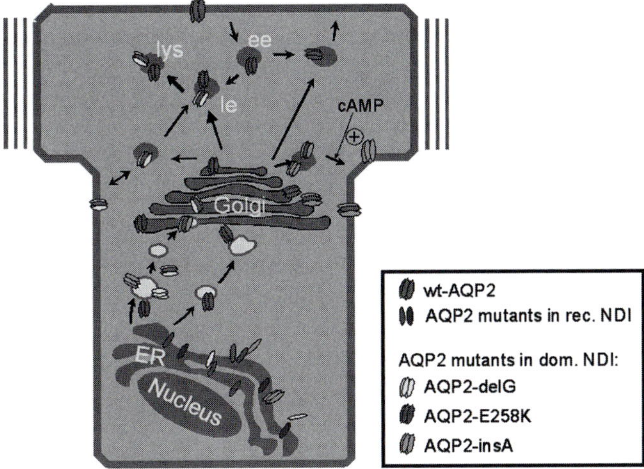

FIGURE 22.5 Cellular mechanism underlying recessive and dominant NDI. In healthy individuals, wild-type AQP2 is assembled to homotetramers in the ER, and traverses through the Golgi complex to intracellular vesicles, from where it will be re-distributed to the apical membrane upon activation of the cAMP signaling cascade by AVP. Upon removal of AVP, AQP2 recycles via early endosomes (ee) to its storage compartment. AQP2 mutants in recessive NDI are misfolded and retained in the endoplasmic reticulum (ER), from where they are targeted for degradation by proteasomes in the cytosol, and are not able to heteroligomerize with wt-AQP2. In contrast, AQP2 mutants in dominant NDI (here represented by AQP2-delG, AQP2-E258K and AQP2-insA), heteroligomerize with wt-AQP2 in the ER and, following trafficking through the Golgi complex, missort the formed wt-AQP2/mutant complex to the late endosomes (LE)/lysosomes (Lys) (AQP2-delG and AQP2-E258K) or to vesicles that translocate the AQP2 complexes to the basolateral membrane upon stimulation with AVP (AQP2-insA). (see also Plate 28)

these mutations were found in the coding region of the C-terminal tail of AQP2 (Figure 22.1), which is not a part of the water pore forming segment, but, as shown by the role of phosphorylation of S256 in AQP2, has an important role in AQP2 trafficking. Indeed, all AQP2 proteins in dominant NDI appeared to be functional water channels, while expression in oocytes and MDCK cells revealed that all these AQP2 mutants in dominant NDI were indeed sorted to other destinations in the cell than wt-AQP2. As wt-AQP2 in polarized MDCK cells is translocated from intracellular vesicles to the apical membrane by forskolin (Figure 22.4; Deen et al 1997), AQP2-R254L/Q mutants were sorted to forskolin-insensitive vesicles, AQP2-E258K was routed to the Golgi complex (oocytes) or late endosomes/lysosomes (mammalian cells), AQP2-727delG trafficked to the basolateral membrane and late endosomes/lysosomes, and AQP2-721delG, 812-818del, 763-772del and AQP2-insA were reported to target for the basolateral membrane (Mulders et al 1998, Kuwahara et al 2001, Hirano et al 2002, Marr et al 2002b, Kamsteeg et al 2003, De Mattia et al 2004b, Savelkoul et al 2004). Importantly, as none of these mutants was misfolded, they were able to interact with and

form heterotetramers with wt-AQP2, which is in contrast to AQP2 mutants in recessive NDI (Kamsteeg et al 1999). In addition, in elegant microscopic studies on cells co-expressing wt and mutant AQP2 proteins, it appeared that the wt/mutant complexes were also missorted, which was due to the wt-mutant interaction and dominancy of the missorting signals in the mutant proteins. Extrapolated to principal cells of the patients, this would likely lead to the lack of sufficient amounts of wt-AQP2 in the apical membrane, thereby explaining the dominant trait of NDI in these families (Figure 22.5).

Interestingly, six AQP2 mutants in dominant NDI are thought to be missorted due to a signal *introduced* with the mutation. In oocytes, deletion of the segment surrounding E258 greatly restored the plasma membrane expression of the AQP2-E258K mutant, which indicated that the dominant feature was caused by the introduced Lys (Mulders et al 1998). With AQP2-insA, the changed C-terminal tail targeted it to the basolateral membrane, which harbored two basolateral sorting signals (Kamsteeg et al 2003). Although starting at different positions, the AQP2 mutants with the 721delG, 727delG, del763-772 or del812-818 mutations have similar extended C-terminal tails. In polarized MDCK cells, AQP2-727delG appeared to accumulate in late endosomes/lysosomes and, to some extent, in the basolateral plasma membrane (Marr et al 2002b), whereas the other mutants were localized in the basolateral membrane (Asai et al 2003). Since the extended tail in AQP2-del812-818 starts only at the stopcodon of wt-AQP2 and the extended tails contain the known basolateral membrane targeting di-leucine motif (Heilker et al 1999), the missorting of these mutants was suggested to be due to this introduced basolateral sorting signal. Also, although AQP2-R254L could not be phosphorylated at S256, as the mutation destroys the PKA phosphorylation consensus site, the introduced leucine residue also appeared to introduce a basolateral sorting signal, because AQP2-R254L-S256D, which mimics constitutively phosphorylated AQP2-R254L, was sorted to the apical and basolateral membrane in polarized cells (De Mattia et al 2004b).

AQP2-R254Q, however, causes dominant NDI due to the *lack* of a key feature of wt-AQP2, because this mutant was impaired in its phosphorylation by PKA (Savelkoul et al 2004). Also, this is the sole cause for its missorting, because the introduction of the S256D mutation completely restored its apical membrane expression. Consistent with this, AQP2-R254Q seemed to accumulate in vesicles in which unstimulated wt-AQP2 resided, and therefore, dominant NDI in this family is likely not due to missorting, but to misregulation of the wt-AQP2/mutant complexes (reduced level of phosphorylated AQP2 monomers). Interestingly, the patient encoding the AQP2-R254Q mutant showed an impressive reduction in urine volume (277 to 68 ml/kg/day) and increased urine osmolality (<100 to >300 mOsm/kg) with a combined treatment of thiazide and indomethacin (Savelkoul et al 2004).

The observed cellular phenotype might provide an explanation for the unusual clinical phenotype of this patient. Many, if not all, membrane proteins are continuously transported between several organelles along microtubular and actin cytoskeletal systems (Ghosh et al 1998, Rohn et al 2000, Sonnichsen et al 2000). The predominant subcellular localization of the protein in such a dynamic equilibrium is then determined by several factors, including the activity of protein kinases and phosphatases (Zehavi Feferman et al 1995, Molloy et al 1999, Bao et al 2000). In parallel, it has been shown that in renal collecting duct and/or transfected cells, AQP2 is continuously shuttled between intracellular vesicles and the apical plasma membrane (Brown et al 1988, Strange et al 1988) and that the steady state localization of AQP2 (intracellular vesicles or plasma membrane) is determined by whether AQP2 is phosphorylated or not (without or with AVP stimulation, respectively) (Wade et al 1981). Indeed, AQP2-S256A, which mainly resided in intracellular vesicles of LLC-PK1 cells, appeared to continuously cycle between these vesicles and the plasma membrane, as an endocytosis block resulted in its accumulation at the plasma membrane (Sun et al 2002, Lu et al 2003). AQP2-R254Q, which seemed to accumulate in storage vesicles that also harbor nonphosphorylated wt-AQP2 or AQP2-S256A, is therefore also likely to cycle to the plasma membrane.

Recently, we reported that for a steady state membrane localization of AQP2, three out of four AQP2 monomers in a tetramer need to be phosphorylated (Kamsteeg et al 2000). If this also holds for renal cells, if wt-AQP2 and AQP2-R254Q are produced at equal levels, and if they randomly assemble into tetramers, then around 30% of the wt-AQP2/AQP2-R254Q complexes compared to 100% of wt-AQP2/wt-AQP2 complexes in a healthy individual are expected to be present in the apical membrane of the patient. Since in normal individuals a peak antidiuretic effect is obtained with 2–3 pg/ml and the patient had a basal blood AVP level of 11.1 pg/ml, wt-AQP2 in our patient is likely maximally phosphorylated. Therefore, the low osmolality of the patient's urine indicates that about 30% of apical membrane expression of AQP2 is not enough to confer urine concentration.

Recently, it has been reported that thiazide increases AQP2 expression in lithium-induced NDI (Kim et al 2004). Similarly, indomethacin likely also increases AQP2 expression as it prevents the inhibitory action of prostaglandins on AVP-mediated cAMP generation (Breyer & Breyer 2000). Although this will presumably not result in increased phosphorylation of AQP2, and thus not in a changed distribution of AQP2-R254Q/wt-AQP2 complexes in the renal cells, the increased wt-AQP2 and AQP2-R254Q expression levels will lead to higher total amounts of wt-AQP2/AQP2-R254Q complexes in the apical membrane. This might explain the observed reduction in urine volume and increased urine osmolality found in the patient.

To provide evidence that the observed clinical phenotype is indeed only due to the lack of phosphorylation of AQP2-R254Q, it would be important to compare this phenotype to that of other patients with dominant NDI treated similarly. Unfortunately, however, these clinical aspects have not been reported for any of the other patients with dominant NDI.

Exception to the Rule

Recently, we encountered two families with a recessive inheritance of NDI in which the two patients appeared heterozygous for an R187C or A190T mutation in one allele, combined with a P262L mutation in the other allele. This was rather surprising, as the P262l mutation was located in the AQP2 C-terminal tail, and was, based on the data above, anticipated to lead to dominant NDI. The answer to this intriguing situation came from our cell biological studies (De Mattia et al 2004a). AQP2-P262L appeared a functional water channel, which was retained in intracellular compartments different from the ER and which formed heteroligomers with wt-AQP2. These were all similar to AQP2 mutants in dominant NDI. Immunocytochemical analysis of cells co-expressing AQP2-P262L with wt-AQP2, however, revealed that these wt-mutant complexes were located in the apical membrane and thus that here, the apical sorting of wt-AQP2 was dominant over the missorting signals of AQP2-P262L. This was different from dominant NDI, as in that form the mutants retain wt-AQP2 in intracellular locations. Based on these data, the recessive inheritance of NDI in these families can be explained as follows: In the two patients, AQP2-R187C and AQP2-A190T mutants are retained in the ER, and do not interact with AQP2-P262L. AQP2-P262L folds properly and assembles into homotetramers, but will be mainly retained in intracellular vesicles. The consequent lack of sufficient AQP2 proteins in the apical membrane of the patient's collecting duct cells then explains their NDI phenotype. In the parents coding for wt-AQP2 and AQP2-R187C or A190T, wt-AQP2 will not interact with either mutant, but will form homotetrameric complexes, of which the insertion into the apical membrane in collecting duct cells will be properly regulated by vasopressin and will give a healthy phenotype. In the proband's healthy relatives encoding wt-AQP2 and AQP2-P262L, both proteins likely assemble into heterotetramers. However, the dominancy of wt-AQP2 sorting on the localization of AQP2-P262L will then result in a proper AVP-regulated trafficking of the heterotetrameric complexes to the apical membrane of their collecting duct cells.

TREATMENT FOR NDI

As described in Chapter 23, the most important component of treatment of NDI, whether acquired or congenital, is replacement of urinary water losses by adequate supply of fluid. Most infants with NDI, however, cannot drink the required amounts of fluid. One approach to reduce urine output is provision of a low-solute diet to reduce the renal osmolar load and decrease obligatory water excretion. Initially, a diet low in sodium (1 mmol/kg/day) as well as protein (2 g/kg/day) was recommended, but severe limitations of dietary protein may introduce serious nutritional deficiencies and therefore a dietary restriction of sodium only is preferable. Nowadays, the treatment usually consists of the thiazide diuretic, which blocks the Na/Cl cotransporter in the distal convoluted tubule, which is sometimes combined with amiloride, which blocks the epithelial sodium channel ENaC in the collecting duct, or with cyclo-oxygenase (COX) inhibitors, such as indomethacin.

During the last 10 years, different approaches to develop alternative therapies for NDI have been tested in vitro. One of the most promising approaches at the moment is the use of cell-permeable V2R or V1R inverse agonists in the rescue of ER-retained, but functional, V2R mutants in NDI (see Chapter 23; Morello et al 2000, Bernier et al 2004). Although the inverse agonists have to be competed for by endogenous AVP when the rescued receptor appears on the plasma membrane, these inverse agonists appear to be able to rescue several different V2R mutants. Also, the use of V1R or V2R inverse agonists might guarantee a high level of specificity and thus a low risk of side effects. In the coming years, it will be exciting to see how well these compounds work out in vivo. Since most AQP2 mutants in NDI are ER-retained, while many of them are functional, the rescue approach as described for V2R mutants above would also be very useful for AQP2 mutants. For AQP water channels, however, no specific interacting compounds are known. While the number of patients with recessive NDI due to AQP2 mutations is low and the conventional therapy relieves NDI to a great extent, efforts to identify compounds that can rescue AQP2 mutants in NDI is often considered not worth the effort. However, due to its role in excessive water reabsorption in congestive heart failure, liver cirrhosis, SIADH and pre-eclampsia, which are encountered quite frequently, efforts are on their way to identify AQP2-specific inhibitors. Possibly, cell permeable forms of these compounds might become the drugs to rescue AQP2 mutants in NDI and to provide an additional therapy for many of the NDI patients with AQP2 gene mutations.

References

Aggarwal R, Janakiramen N, Luken J, Kumar S. Nephrogenic diabetes insipidus in a female infant with hydrocephalus. Am. J. Dis. Child. 1986; 140: 1095–6.

Arthus MF, Lonergan M, Crumley MJ, et al. Report of 33 novel AVPR2 mutations and analysis of 117 families with X-linked nephrogenic diabetes insipidus. J. Am. Soc. Nephrol. 2000; 11: 1044–54.

Asai T, Kuwahara M, Kurihara H, et al. Pathogenesis of nephrogenic diabetes insipidus by aquaporin-2 C-terminus mutations. Kidney Int. 2003; 64: 2–10.

Bao J, Alroy I, Waterman H, Schejter ED, Brodie C, Yarden Y. Threonine phosphorylation diverts internalized EGF-receptors from a degradative pathway to the recycling endosome. J. Biol. Chem. 2000.

Bernier V, Lagace M, Lonergan M, Arthus MF, Bichet DG, Bouvier M. Functional rescue of the constitutively internalized V2 Vasopressin receptor mutant R137H by the pharmacological chaperone action of SR49059. Mol. Endocrinol. 2004; 18: 2074–84.

Bichet DG, Razi M, Lonergan M, et al. Hemodynamic and coagulation responses to 1-desamino [8-D-arginine] vasopressin in patients with congenital nephrogenic diabetes insipidus. N. Engl. J. Med. 1988; 318: 881–7.

Boccalandro C, De Mattia F, Guo DC, et al. Characterization of an aquaporin-2 water channel gene mutation causing partial nephrogenic diabetes insipidus in a Mexican family: evidence of increased frequency of the mutation in the town of origin. J. Am. Soc. Nephrol. 2004; 15: 1223–31.

Borgnia M, Nielsen S, Engel A, Agre P. Cellular and molecular biology of the aquaporin water channels. Annu. Rev. Biochem. 1999; 68: 425–58.

Breyer MD, Breyer RM. Prostaglandin E receptors and the kidney. Am. J. Physiol. Renal Physiol. 2000; 279: F12–23.

Brown D, Weyer P, Orci L. Vasopressin stimulates endocytosis in kidney collecting duct principal cells. Eur. J. Cell Biol. 1988; 46: 336–41.

Brown RM, Fraser NJ, Brown GK. Differential methylation of the hypervariable locus DXS255 on active and inactive X chromosomes correlates with the expression of a human X-linked gene. Genomics 1990; 7: 215–21.

Buck TM, Eledge J, Skach WR. Evidence for stabilization of aquaporin-2 folding mutants by N-linked glycosylation in endoplasmic reticulum. Am. J. Physiol. Cell Physiol. 2004; 287: C1292–9.

Canfield MC, Tamarappoo BK, Moses AM, Verkman AS, Holtzman EJ. Identification and characterization of aquaporin-2 water channel mutations causing nephrogenic diabetes insipidus with partial vasopressin response. Hum. Mol. Genet. 1997; 6: 1865–71.

Cannon JF. Diabetes insipidus; clinical and experimental studies with consideration of genetic relationships. AMA. Arch. Intern. Med. 1955; 96: 215–72.

Chan Seem CP, Dossetor JF, Penney MD. Nephrogenic diabetes insipidus due to a new mutation of the arginine vasopressin V2 receptor gene in a girl presenting with non-accidental injury. Ann. Clin. Biochem. 1999; 36(Pt 6): 779–82.

Christensen BM, Zelenina M, Aperia A, Nielsen S. Localization and regulation of PKA-phosphorylated AQP2 in response to V(2)-receptor agonist/antagonist treatment. Am. J. Physiol. Renal Physiol. 2000; 278: F29–42.

De Mattia F, Savelkoul PJ, Bichet DG, et al. A novel mechanism in recessive Nephrogenic Diabetes Insipidus: wild-type Aquaporin-2 rescues the apical membrane expression of intra-cellularly retained AQP2-P262L. Hum. Mol. Genet. 2004a.

De Mattia F, Savelkoul PJM, Kamsteeg EJ, et al. Lack of AVP-induced phosphorylation of the Aquaporin-2 mutant

AQP2-R254L explains dominant Nephrogenic Diabetes Insipidus. 2004b.

Deen PMT, Verdijk MAJ, Knoers NVAM, Wieringa B, Monnens LAH, van Os CH, van Oost BA. Requirement of human renal water channel aquaporin-2 for vasopressin-dependent concentration of urine. Science 1994a; 264: 92–5.

Deen PMT, Weghuis DO, Sinke RJ, Geurts van Kessel A, Wieringa B, van Os CH. Assignment of the human gene for the water channel of renal collecting duct Aquaporin 2 (AQP2) to chromosome 12 region q12-->q13. Cytogenet. Cell Genet. 1994b; 66: 260–2.

Deen PMT, Croes H, van Aubel RA, Ginsel LA, van Os CH. Water channels encoded by mutant aquaporin-2 genes in nephrogenic diabetes insipidus are impaired in their cellular routing. J. Clin. Invest. 1995; 95: 2291–6.

Deen PMT, van Aubel RA, van Lieburg AF, van Os CH. Urinary content of aquaporin 1 and 2 in nephrogenic diabetes insipidus. J. Am. Soc. Nephrol. 1996; 7: 836–41.

Deen PMT, Rijss JPL, Mulders SM, Errington RJ, van Baal J, van Os CH. Aquaporin-2 transfection of Madin-Darby canine kidney cells reconstitutes vasopressin-regulated transcellular osmotic water transport. J. Am. Soc. Nephrol. 1997; 8: 1493–501.

Deen PMT, Marr N, Kamsteeg EJ, Van Balkom BWM. Nephrogenic diabetes insipidus. Curr. Opin. Nephrol. Hypertens. 2000; 9: 591–5.

Digiovanni SR, Nielsen S, Christensen EI, Knepper MA. Regulation of collecting duct water channel expression by vasopressin in Brattleboro rat. Proc. Natl Acad. Sci. USA 1994; 91: 8984–8.

Fushimi K, Uchida S, Hara Y, Hirata Y, Marumo F, Sasaki S. Cloning and expression of apical membrane water channel of rat kidney collecting tubule. Nature 1993; 361: 549–52.

Fushimi K, Sasaki S, Marumo F. Phosphorylation of serine 256 is required for cAMP-dependent regulatory exocytosis of the aquaporin-2 water channel. J. Biol. Chem. 1997; 272: 14800–4.

Ghosh RN, Mallet WG, Soe TT, McGraw TE, Maxfield FR. An endocytosed TGN38 chimeric protein is delivered to the TGN after trafficking through the endocytic recycling compartment in CHO cells. J. Cell Biol. 1998; 142: 923–36.

Goji K, Kuwahara M, Gu Y, Matsuo M, Marumo F, Sasaki S. Novel mutations in aquaporin-2 gene in female siblings with nephrogenic diabetes insipidus: evidence of disrupted water channel function. J. Clin. Endocrinol. Metab. 1998; 83: 3205–9.

Guyon C, Bissonnette P, Lussier Y, et al. Novel Aquaporin-2 (AQP2) mutations responsible for autosomal recessive nephrogenic diabetes insipidus. J. Am. Soc. Nephrol. 2004.

Han JS, Maeda Y, Ecelbarger C, Knepper MA. Vasopressin-independent regulation of collecting duct water permeability. Am. J. Physiol. 1994; 266: F139–46.

Heilker R, Spiess M, Crottet P. Recognition of sorting signals by clathrin adaptors. Bioessays 1999; 21: 558–67.

Hendriks G, Koudijs M, van Balkom BW, et al. Glycosylation is important for cell surface expression of the water channel aquaporin-2 but is not essential for tetramerization in the endoplasmic reticulum. J. Biol. Chem. 2004; 279: 2975–83.

Heymann JB, Engel A. Structural clues in the sequences of the aquaporins. J. Mol. Biol. 2000; 295: 1039–53.

Hirano K, Roth J, Zuber C, Ziak M. Expression of a mutant ER-retained polytope membrane protein in cultured rat hepatocytes results in Mallory body formation. Histochem. Cell Biol. 2002; 117: 41–3.

Hirano K, Zuber C, Roth J, Ziak M. The proteasome is involved in the degradation of different aquaporin-2 mutants causing nephrogenic diabetes insipidus. Am. J. Pathol. 2003; 163: 111–20.

Hochberg Z, van Lieburg AF, Even L, et al. Autosomal recessive nephrogenic diabetes insipidus caused by an aquaporin-2 mutation. J. Clin. Endocrinol. Metab. 1997; 82: 686–9.

Hozawa S, Holtzman EJ, Ausiello DA. cAMP motifs regulating transcription in the aquaporin 2 gene. Am. J. Physiol. 1996; 39: C1695–702.

Jung JS, Preston GM, Smith BL, Guggino WB, Agre P. Molecular structure of the water channel through aquaporin CHIP. The hourglass model. J. Biol. Chem. 1994; 269: 14648–54.

Kamsteeg EJ, Bichet DG, Konings IB, et al. Reversed polarized delivery of an aquaporin-2 mutant causes dominant nephrogenic diabetes insipidus. J. Cell Biol. 2003; 163: 1099–109.

Kamsteeg EJ, Deen PMT. Importance of aquaporin-2 expression levels in genotype-phenotype studies in nephrogenic diabetes insipidus. Am. J. Physiol. Renal Physiol. 2000; 279: F778–84.

Kamsteeg EJ, Heijnen I, van Os CH, Deen PMT. The subcellular localization of an Aquaporin-2 tetramer depends on the stoichiometry of phosphorylated and nonphosphorylated monomers. J. Cell Biol. 2000; 151: 919–30.

Kamsteeg EJ, Wormhoudt TA, Rijss JPL, van Os CH, Deen PMT. An impaired routing of wild-type aquaporin-2 after tetramerization with an aquaporin 2 mutant explains dominant nephrogenic diabetes insipidus. EMBO J. 1999; 18: 2394–400.

Kanno K, Sasaki S, Hirata Y, et al. Urinary excretion of aquaporin-2 in patients with diabetes insipidus. N. Engl. J. Med. 1995; 332: 1540–5.

Katsura T, Gustafson CE, Ausiello DA, Brown D. Protein kinase A phosphorylation is involved in regulated exocytosis of aquaporin-2 in transfected LLC-PK1 cells. Am. J. Physiol. 1997; 41: F816–22.

Kaufmann JE, Oksche A, Wollheim CB, Gunther G, Rosenthal W, Vischer UM. Vasopressin-induced von Willebrand factor secretion from endothelial cells involves V2 receptors and cAMP. J. Clin. Invest. 2000; 106: 107–16.

Kim GH, Lee JW, Oh YK, et al. Antidiuretic effect of hydrochlorothiazide in lithium-induced nephrogenic diabetes insipidus is associated with upregulation of aquaporin-2, Na-Cl cotransporter, and epithelial sodium channel. J. Am. Soc. Nephrol. 2004; 15: 2836–43.

King LS, Agre P. Pathophysiology of the aquaporin water channels. Annu. Rev. Physiol. 1996; 58: 619–48.

Klussmann E, Tamma G, Lorenz D, et al. An inhibitory role of Rho in the vasopressin-mediated translocation of aquaporin-2 into cell membranes of renal principal cells. J. Biol. Chem. 2001; 276: 20451–7.

Knepper MA, Inoue T. Regulation of aquaporin-2 water channel trafficking by vasopressin. Curr. Opin. Cell Biol. 1997; 9: 560–4.

Knoers NVAM, Brommer EJ, Willems H, van Oost BA, Monnens LAH. Fibrinolytic responses to 1-desamino-8-D-arginine-vasopressin in patients with congenital nephrogenic diabetes insipidus [see comments]. Nephron 1990; 54: 322–6.

Knoers NVAM, Monnens LAH. A variant of nephrogenic diabetes insipidus: V2 receptor abnormality restricted to the kidney. Eur. J. Pediatr. 1991; 150: 370–3.

Kobrinsky NL, Doyle JJ, Israels ED, et al. Absent factor VIII response to synthetic vasopressin analogue (DDAVP) in nephrogenic diabetes insipidus. Lancet 1985; 1: 1293–4.

Koyama N, Ishibashi K, Kuwahara M, et al. Cloning and functional expression of human aquaporin8 cDNA and analysis of its gene. Genomics 1998; 54: 169–72.

Kuwahara M. Aquaporin-2, a vasopressin-sensitive water channel, and nephrogenic diabetes insipidus. Intern. Med. 1998; 37: 215–17.

Kuwahara M, Iwai K, Ooeda T, et al. Three families with autosomal dominant nephrogenic diabetes insipidus caused by aquaporin-2 mutations in the C-terminus. Am. J. Hum. Genet. 2001; 69: 738–48.

Lande MB, Jo I, Zeidel ML, Somers M, Harris HW. Phosphorylation of aquaporin-2 does not alter the membrane water permeability of rat papillary water channel-containing vesicles. J. Biol. Chem. 1996; 271: 5552–7.

Langley JM, Balfe JW, Selander T, Ray PN, Clarke JT. Autosomal recessive inheritance of vasopressin-resistant diabetes insipidus. Am. J. Med. Genet. 1991; 38: 90–4.

Lin SH, Bichet DG, Sasaki S, et al. Two novel aquaporin-2 mutations responsible for congenital nephrogenic diabetes insipidus in Chinese families. J. Clin. Endocrinol. Metab. 2002; 87: 2694–700.

Lu H, Sun TX, Bouley R, Blackburn K, McLaughlin M, Brown D. Inhibition of endocytosis causes phosphorylation (S256)-independent plasma membrane accumulation of AQP-2. Am. J. Physiol. Renal Physiol. 2003.

Maeda Y, Terada Y, Nonoguchi H, Knepper MA. Hormone and autacoid regulation of cAMP production in rat IMCD subsegments. Am. J. Physiol. 1992; 263: F319–27.

Mannucci PM, Ruggeri ZM, Pareti FI, Capitanio A. 1-Deamino-8-d-arginine vasopressin: a new pharmacological approach to the management of haemophilia and von Willebrands' diseases. Lancet 1977; 1: 869–72.

Marples D, Frokiaer J, Nielsen S. Long-term regulation of aquaporins in the kidney. Am. J. Physiol. 1999; 276: F331–9.

Marr N, Bichet DG, Hoefs S, et al. Cell-biologic and functional analyses of five new Aquaporin-2 missense mutations that cause recessive nephrogenic diabetes insipidus. J. Am. Soc. Nephrol. 2002a; 13: 2267–77.

Marr N, Bichet DG, Lonergan M, et al. Heteroligomerization of an Aquaporin-2 mutant with wild-type Aquaporin-2 and their misrouting to late endosomes/lysosomes explains dominant nephrogenic diabetes insipidus. Hum. Mol. Genet. 2002b; 11: 779–89.

Marr N, Kamsteeg EJ, Van Raak M, van Os CH, Deen PMT. Functionality of aquaporin-2 missense mutants in recessive nephrogenic diabetes insipidus. Pflugers Arch. 2001; 442: 73–7.

Matsumura Y, Uchida S, Rai T, Sasaki S, Marumo F. Transcriptional regulation of aquaporin-2 water channel gene by cAMP. J. Am. Soc. Nephrol. 1997; 8: 861–7.

Molloy SS, Anderson ED, Jean F, Thomas G. Bi-cycling the furin pathway: from TGN localization to pathogen activation and embryogenesis. Trends Cell Biol. 1999; 9: 28–35.

Morello JP, Salahpour A, Laperriere A, et al. Pharmacological chaperones rescue cell-surface expression and function of misfolded V2 vasopressin receptor mutants [see comments]. J. Clin. Invest. 2000; 105: 887–95.

Moses AM, Sangani G, Miller JL. Proposed cause of marked vasopressin resistance in a female with an X-linked recessive V2 receptor abnormality. J. Clin. Endocrinol. Metab. 1995; 80: 1184–6.

Moses AM, Scheinman SJ, Oppenheim A. Marked hypotonic polyuria resulting from nephrogenic diabetes insipidus with partial sensitivity to vasopressin. J. Clin. Endocrinol. Metab. 1984; 59: 1044–9.

Mulders SM, Bichet DG, Rijss JPL, et al. An aquaporin-2 water channel mutant which causes autosomal dominant nephrogenic diabetes insipidus is retained in the Golgi complex. J. Clin. Invest. 1998; 102: 57–66.

Mulders SM, Knoers NVAM, van Lieburg AF, et al. New mutations in the AQP2 gene in nephrogenic diabetes insipidus resulting in functional but misrouted water channels. J. Am. Soc. Nephrol. 1997; 8: 242–8.

Nadler SP, Zimpelmann JA, Hebert RL. PGE2 inhibits water permeability at a post-cAMP site in rat terminal inner medullary collecting duct. Am. J. Physiol. 1992; 262: F229–35.

Nielsen S, Chou CL, Marples D, Christensen EI, Kishore BK, Knepper MA. Vasopressin increases water permeability of kidney collecting duct by inducing translocation of aquaporin-CD water channels to plasma membrane. Proc. Natl Acad. Sci. USA 1995; 92: 1013–17.

Nielsen S, Kwon TH, Christensen BM, Promeneur D, Frokiaer J, Marples D. Physiology and pathophysiology of renal aquaporins. J. Am. Soc. Nephrol. 1999; 10: 647–63.

Nomura Y, Onigata K, Nagashima T, et al. Detection of skewed X-inactivation in two female carriers of vasopressin type 2 receptor gene mutation. J. Clin. Endocrinol. Metab. 1997; 82: 3434–7.

Oksche A, Moller A, Dickson J, et al. Two novel mutations in the aquaporin-2 and the vasopressin V2 receptor genes in patients with congenital nephrogenic diabetes insipidus. Hum. Genet. 1996; 98: 587–9.

Pearl M, Taylor A. Role of the cytoskeleton in the control of transcellular water flow by vasopressin in amphibian urinary bladder. Biol. Cell 1985; 55: 163–72.

Ringler P, Borgnia MJ, Stahlberg H, Maloney PC, Agre P, Engel A. Structure of the water channel AqpZ from *Escherichia coli* revealed by electron crystallography. J. Mol. Biol. 1999; 291: 1181–90.

Robertson GL, Berl T. Pathophysiology of water metabolism. In: Brenner BM, Rector RC, eds. Disturbances in Control of Body Fluid and Composition. Philadelphia: Saunders, 1995: pp. 873–928.

Rohn WM, Rouille Y, Waguri S, Hoflack B. Bi-directional trafficking between the trans-golgi network and the endosomal/lysosomal system [In Process Citation]. J. Cell Sci. 2000; 113(Pt 12): 2093–101.

Saito F, Sasaki S, Chepelinsky AB, Fushimi K, Marumo F, Ikeuchi T. Human AQP2 and MIP genes, two members of the MIP family, map within chromosome band 12q13 on the basis of two-color FISH. Cytogenet. Cell Genet. 1995; 68: 45–8.

Sasaki S, Fushimi K, Saito H, et al. Cloning, characterization, and chromosomal mapping of human aquaporin of collecting duct. J. Clin. Invest. 1994; 93: 1250–6.

Sato K, Fukuno H, Taniguchi T, Sawada S, Fukui T, Kinoshita M. A novel mutation in the vasopressin V2 receptor gene in a woman with congenital nephrogenic diabetes insipidus [In Process Citation]. Intern. Med. 1999; 38: 808–12.

Savelkoul PJ, De Mattia F, Kamsteeg EJ, et al. The Aquaporin-2 mutant AQP2-R254Q, which causes dominant nephrogenic diabetes insipidus reveals the essental role of AQP2 phosphorylation in vivo. 2004.

Schreiner RL, Skafish PR, Anand SK, Northway JD. Congenital nephrogenic diabetes insipidus in a baby girl. Arch. Dis. Child. 1978; 53: 906–8.

Schrier RW, Cadnapaphornchai MA, Ohara M. Water retention and aquaporins in heart failure, liver disease and pregnancy. J. R. Soc. Med. 2001; 94: 265–9.

Simon H, Gao Y, Franki N, Hays RM. Vasopressin depolymerizes apical F-actin in rat inner medullary. Am. J. Physiol. 1993; 265: C757–62.

Sonnichsen B, De Renzis S, Nielsen E, Rietdorf J, Zerial M. Distinct membrane domains on endosomes in the recycling pathway visualized by multicolor imaging of Rab4, Rab5, and Rab11. J. Cell Biol. 2000; 149: 901–14.

Strange K, Willingham MC, Handler JS, Harris HW Jr. Apical membrane endocytosis via coated pits is stimulated by removal of antidiuretic hormone from isolated, perfused rabbit cortical collecting tubule. J. Membr. Biol. 1988; 103: 17–28.

Sun TX, van Hoek A, Huang Y, Bouley R, McLaughlin M, Brown D. Aquaporin-2 localization in clathrin-coated pits: inhibition of endocytosis by dominant-negative dynamin. Am. J. Physiol. Renal Physiol. 2002; 282: F998–1011.

Tamarappoo BK, Verkman AS. Defective aquaporin-2 trafficking in nephrogenic diabetes insipidus and correction by chemical chaperones. J. Clin. Invest. 1998; 101: 2257–67.

Tamma G, Klussmann E, Maric K, et al. Rho inhibits cAMP-induced translocation of aquaporin-2 into the apical membrane of renal cells. Am. J. Physiol. Renal Physiol. 2001; 281: F1092–101.

Tamma G, Klussmann E, Procino G, et al. cAMP-induced AQP2 translocation is associated with RhoA inhibition through RhoA phosphorylation and interaction with RhoGDI. J. Cell Sci. 2003a; 116: 1519–25.

Tamma G, Wiesner B, Furkert J, et al. The prostaglandin E2 analogue sulprostone antagonizes vasopressin-induced antidiuresis through activation of Rho. J. Cell Sci. 2003b; 116: 3285–94.

Van Balkom BWM, Savelkoul PJ, Markovich D, et al. The role of putative phosphorylation sites in the targeting and shuttling of the aquaporin-2 water channel. J. Biol. Chem. 2002; 277: 41473–49.

Van Hoek AN, Hom ML, Luthjens LH, de Jong MD, Dempster JA, van Os CH. Functional unit of 30 kDa for proximal tubule water channels as revealed by radiation inactivation. J. Biol. Chem. 1991; 266: 16633–5.

Van Hoek AN, Luthjens LH, Hom ML, van Os CH, Dempster JA. A 30 kDa functional size for the erythrocyte water channel determined in situ by radiation inactivation. Biochem. Biophys. Res. Commun. 1992; 184: 1331–8.

van Lieburg AF, Knoers NVAM, Mallmann R, et al. Normal fibrinolytic responses to 1-desamino-8-D-arginine vasopressin in patients with nephrogenic diabetes insipidus caused by mutations in the aquaporin 2 gene. Nephron 1996; 72: 544–6.

van Lieburg AF, Verdijk MAJ, Knoers NVAM, et al. Patients with autosomal nephrogenic diabetes insipidus homozygous for mutations in the aquaporin 2 water-channel gene. Am. J. Hum. Genet. 1994; 55: 648–52.

van Lieburg AF, Verdijk MAJ, Schoute F, et al. Clinical phenotype of nephrogenic diabetes insipidus in females heterozygous for a vasopressin type 2 receptor mutation. Hum. Genet. 1995; 96: 70–8.

van Os CH, Deen PMT, Dempster JA. Aquaporins: water selective channels in biological membranes. Molecular structure and tissue distribution. Biochim. Biophys. Acta 1994; 1197: 291–309.

Vargas-Poussou R, Forestier L, Dautzenberg MD, Niaudet P, Dechaux M, Antignac C. Mutations in the vasopressin V2 receptor and aquaporin-2 genes in twelve families with congenital nephrogenic diabetes insipidus. Adv. Exp. Med. Biol. 1998; 449: 387–90.

Verkman AS, Mitra AK. Structure and function of aquaporin water channels. Am. J. Physiol. Renal Physiol. 2000; 278: F13–28.

Wade JB, Stetson DL, Lewis SA. ADH action: evidence for a membrane shuttle mechanism. Ann. NY Acad. Sci. 1981; 372: 106–17.

Walz T, Hirai T, Murata K, et al. The three-dimensional structure of aquaporin-1. Nature 1997; 387: 624–7.

Werten PJ, Hasler L, Koenderink JB, et al. Large-scale purification of functional recombinant human aquaporin-2. FEBS Lett. 2001; 504: 200–5.

Wiggelinkhuizen J, Retief PJ, Wolff B, Fisher RM, Cremin BJ. Nephrogenic diabetes insipidus and obstructive uropathy. Am. J. Dis. Child. 1973; 126: 398–401.

Yasui M, Zelenin SM, Celsi G, Aperia A. Adenylate cyclase-coupled vasopressin receptor activates AQP2 promoter via a dual effect on CRE and AP1 elements. Am. J. Physiol. 1997; 41: F443–50.

Zehavi Feferman R, Burgess JW, Stanley KK. Control of p62 binding to TGN38/41 by phosphorylation. FEBS Lett. 1995; 368: 122–4.

Zelenina M, Christensen BM, Palmer J, Nairn AC, Nielsen S, Aperia A. Prostaglandin E(2) interaction with AVP: effects on AQP2 phosphorylation and distribution. Am. J. Physiol. Renal Physiol. 2001; 278: F388–94.

Zimmerman D, Green CO. Nephrogenic diabetes insipidus type II. Defect distal to adenylate cyclase step. Pediatr. Res. 1975; 9: 381.

PART III

Genetic Abnormalities of Renal Development and Morphogenesis

CHAPTER 23

An Overview of Renal Development

ARNAUD MARLIER AND LLOYD G. CANTLEY

INTRODUCTION

The metanephros, the permanent mammalian kidney, develops in a manner that is unique among all of the organs of the body. Tracing this process affords us a window on the evolution of excretory organs from teleosts to amphibians to land dwellers. In the embryo, a single tissue, the intermediate mesoderm, contains the precursors for three separate kidneys, the pronephros, the mesonephros, and the metanephros. A linear group of cells in the intermediate mesoderm first undergoes mesenchymal-to-epithelial transformation to form an epithelial tube, the nephric, or Wolffian, duct. Along this tube, lateral outpouchings form the vestigial pronephros (the kidney of primitive fish) and the more developed, but still rudimentary, mesonephros (the typical amphibian kidney). However, the mammalian metanephros must be able to excrete nitrogenous waste in a concentrated urine to conserve water and salt, and thus has evolved a substantially more complex three-dimensional structure as well as an array of cells specialized to handle the reabsorption and/or secretion of select electrolytes, sugars, lipids, and peptides. This structure develops as the result of a highly regulated set of interactions between an outgrowth of the nephric duct, the ureteric bud, and a cluster of mesenchymal cells from the caudal region of the intermediate mesoderm, the metanephric mesenchyme. Even though both tissues are of intermediate mesoderm origin, the fact that one is epithelial and the other mesenchymal creates an environment in which distinct signals from each tissue can regulate the development of the other, with the metanephric mesenchyme exerting control over ureteric bud branching and thus the architecture of the collecting system, while signals from the ureteric bud regulate the discrete sites of initiation of mesenchymal–epithelial transformation within the metanephric mesenchyme that ultimately leads to proximal nephron formation.

The morphologic aspects of renal development have been extensively described elsewhere and will only be briefly reviewed here. (For a detailed anatomic description of mammalian kidney development, the reader is referred to works by Potter and by Saxen (Potter 1972, Saxen 1987).) These morphologic changes occur due to the control of four basic cellular processes; proliferation, programmed cell death (apoptosis), morphogenesis/migration (change in cell shape and position), and transformation (the initiation of a genetic program that changes the functional characteristics of the cell). Two major advances during the past 50 years have provided us with a host of new insights into the molecular control of these 4 events during kidney development. First, in the 1950s, Grobstein and collaborators developed the methodology to culture the embryonic kidney as an organ explant, allowing the identification and in vitro examination of specific factors that influence ureteric bud branching and nephrogenesis. Second, the introduction of methodologies to interrupt gene expression in vivo in the 1980s has led to the targeted disruption of multiple genes in the mouse embryo, allowing us to identify many of the genes that are critical for kidney development and to determine at what stage they exert their action. This chapter will provide an overview of the results of some, but by no means all, of these studies, and a description of what they tell us about how development of the kidney is controlled. Since most of these studies were performed in mice, references within this chapter to the day of embryonic development will refer to mouse development unless otherwise noted, with mouse gestation typically lasting 20–21 days.

While a large body of information has accumulated from these approaches, it should be noted that the two technologies frequently give conflicting results, making it difficult to judge the true importance of some of the findings. In many instances these conflicting results are ascribed, possibly correctly, to the redundant nature of many of the developmental regulatory pathways. For this reason, genetic disruption of a factor that stimulates nephrogenesis in vitro may be found to have no effect on nephrogenesis in the developing mouse due to the continued presence of a second factor with similar capabilities. A second aspect of these studies that somewhat limits their applicability to the

Genetic Diseases of the Kidney
ISBN: 978-0-12-449851-8

study of human disease is that development of the mouse kidney does not exactly recapitulate the human kidney, with some clear differences in regulatory events between the two. However, with these caveats in mind, the information gleaned from recent studies can provide the framework for a basic understanding of how the interactions between the ureteric bud and the metanephric mesenchyme result in the formation of a functional kidney.

ANATOMIC DESCRIPTION OF KIDNEY DEVELOPMENT

The formation of the kidney begins when a region of the embryonic mesoderm lying between the lateral plates and the somites, termed the intermediate mesoderm, is induced to form an elongated epithelial tube known as the nephric duct. As the embryo elongates, the nephric duct extends caudally where it ultimately connects with the digestive track forming the cloaca. Along the length of the nephric duct, the cells of the adjacent intermediate mesoderm (the nephrogenic cord) give rise to a succession of nephrons from anterior to posterior that are cumulatively known as the pronephros (the most cranial), mesonephros, and metanephros (the most caudal) (Figure 23.1). The pronephros appears

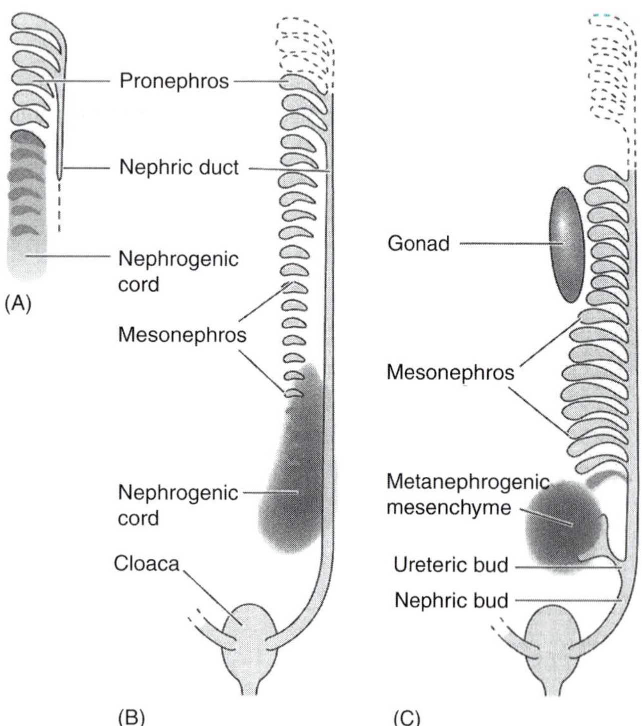

FIGURE 23.1 Development of the mammalian excretory system progresses in a cranial–caudal fashion with the formation of the nephric duct and associated pronephros (A), mesonephros (B), and finally the metanephros (C). Adapted from Saxen (1987). (see also Plate 29)

on embryonic day 8 in the mouse, but is a rudimentary structure that is of unclear significance. In most teleosts the pronephros is the sole excretory organ throughout life, although in most lower vertebrates it is functional only until formation of the mesonephros.

Between E9 and E9.5, cells from the nephrogenic cord just caudal to the pronephros aggregate medial to the nephric duct forming distinct vesicles (Croisille et al 1976). These vesicles develop into primitive nephrons with a glomerulus and a short tubule that fuses with the nephric duct. Capillaries appear to invade the glomerulus from the ventro-lateral regions of the aorta and form a venous plexus around the associated tubule prior to emptying into the mesonephric vein. Overall, forty to forty-two of these vesicles are formed in the human mesonephros, although cranial vesicles degenerate as the more caudal ones form, resulting in a maximum of thirty to thirty-two present at any given time (Potter 1972). These primitive nephrons may function briefly during the early stages of metanephric development, but ultimately the glomeruli degenerate and the remaining tubules become incorporated into the developing genital system.

Around E10.5–11 in the mouse, a protrusion arises from the caudal portion of the nephric duct, just anterior to the site of entry into the cloaca. This epithelial structure, termed the ureteric bud (UB), grows out into the adjacent nephrogenic cord accompanied by the condensation of a group of nephrogenic cord-derived cells around the tip of the UB. These condensing cells are collectively called the metanephric mesenchyme (MM), and are comprised of tubule precursors, endothelial precursors and stromal cells. The ureteric bud undergoes a series of regulated divisions to form the entire collecting system of the kidney. In the human, the first 3–5 generations of branches will dilate to form the renal pelvis and associated major calyces, the next 3–5 generations form the minor calyces and renal papillae, and the final 6–9 generations of branches form the collecting system (with the papillary duct comprising the first generation of these final branches) (al-Awqati & Goldberg 1998, Osathanondh & Potter 1963a).

As the ureteric bud branches, the nearby cells of the metanephric mesenchyme either proliferate and migrate with the UB tip, or remain associated with the stalk of the UB and differentiate into connective tissue. Prior to ureteric bud branching, the tip region swells to form an ampulla, and at the base of this ampulla a discrete group of cells of the metanephric mesenchyme further condense to form a vesicle similar to that seen in the mesonephros (Figure 23.2). The cells within this teardrop-shaped vesicle rapidly undergo mesenchymal–epithelial transformation to form a structure known as the comma-shaped body, followed by a series of carefully preserved morphogenic changes in the architecture of the vesicle itself leading to the formation of a more complex structure known as the S-shaped body. The lower limb of the S (the most distal from the UB) contains the cells that will eventually develop into the podocytes (the upper layer) and Bowman's capsule (the lower layer) and is separated from the middle portion of the S by the vascular cleft that will be the site of invasion by the endothelial and mesangial cells (Dorup & Maunsbach 1982). The upper limb of the S forms the distal convoluted tubule and fuses with the ureteric bud to form the connecting segment, while the proximal convoluted tubule and the loop of Henle develop from the mid-portion of the S-shaped body (Osathanondh & Potter 1966). Interestingly, in both humans and rodents, new nephron induction continues even after ureteric bud branching is complete, resulting in a single terminal ureteric bud branch having 9–11 attached

nephrons (Neiss 1982, Osathanondh & Potter 1963b). This process of new nephron induction is generally complete by 32 weeks gestation in the human, but continues for several weeks after birth in the mouse. Ultimately, the number of nephrons that form is roughly dependent on the mass of the animal, with rodents having approximately 35 000 nephrons/kidney whereas human kidneys contain approximately 600 000 to 1 million nephrons per kidney (Merlet-Benichou et al 1999).

SPECIFICATION OF THE NEPHRIC (WOLFFIAN) DUCT AND NEPHROGENIC CORD

Delineation of a primitive tissue that will undergo transition to become the adult kidney is an indispensable requirement for normal kidney formation. This initial specification of the nephrogenic primordia is believed to be mediated by specific patterns of gene expression in a defined group of cells within the intermediate mesoderm. While many of the genes that must be upregulated in order to begin the process of kidney development have not yet been identified, several of the transcription factors that regulate this process have now been determined, including Lim1 and Pax2. The localized expression of the homeodomain transcription factor Lim1 appears to mark the earliest precursor cells of the developing excretory system. Lim1 is initially expressed in several areas of the primitive streak (embryonic day 6.5 in the mouse), and then is subsequently expressed in the intermediate mesoderm and ultimately the nephrogenic cord and portions of the central nervous system by E9–10 (Barnes et al 1994, Taira et al 1994). On E9.5, Lim1 is found in both the nephric duct and the surrounding nephric blastema (Tsang et al 2000), and as kidney development progresses Lim1 expression is detectable in the ureteric bud as well as the adjacent condensed metanephric mesenchyme as it forms the renal vesicle and subsequent tubular structures (Karavanov et al 1998). Interruption of Lim1 expression results in loss of anterior head structures and embryonic death at approximately E10, with the few animals that survive longer demonstrating complete loss of kidney development and gonadal structures (Shawlot & Behringer 1995). Examination of *Lim1* null animals at E9.5 reveals that the precursor cells of the nephrogenic cord fail to aggregate normally and undergo degeneration (Tsang et al 2000). Thus Lim1 appears to be a critical inducer of the earliest stages of kidney specification in the embryo and is required for nephric duct formation and elongation. Its expression in the UB tips and the renal vesicle suggests that Lim1 may play an important role in the later stages of tubule formation as well.

A second transcription factor, *Pax2*, a member of the paired box family of transcription factors, is expressed in the intermediate mesoderm beginning at E8.5 just prior

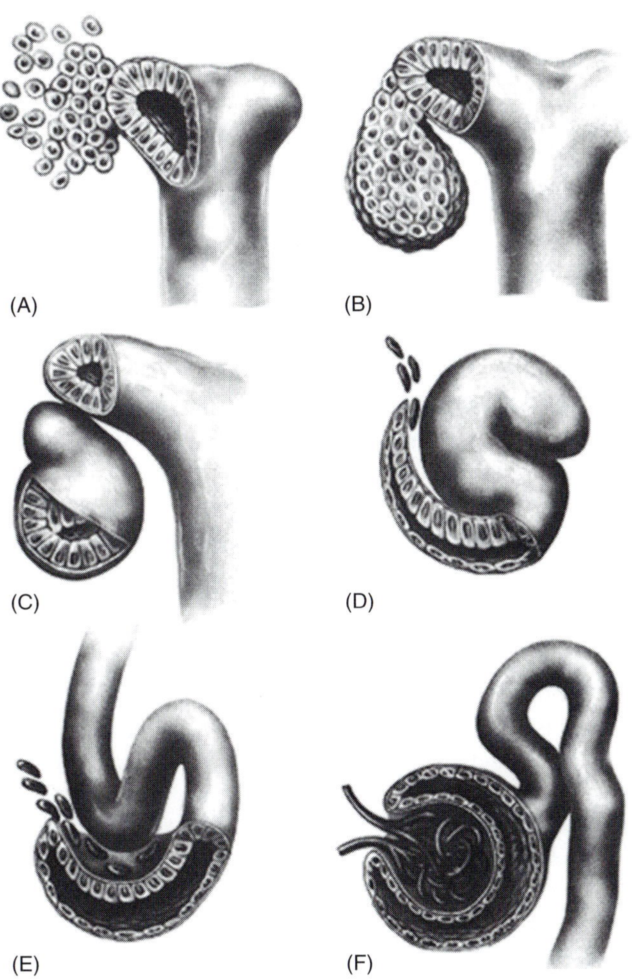

FIGURE 23.2 A depiction of the ureteric bud as it is undergoing the first round of branching (A) and the nearby mesenchyme is condensing near the tip to form the renal vesicle (B). The mesenchymal cells of the renal vesicle then undergo mesenchymal–epithelial transformation to form the comma-shaped body (C), followed by the development of the vascular cleft at the site furthest from the ureteric bud to form the S-shaped body (D). Endothelial precursors migrate into the vascular cleft in close approximation to the immature podocytes lining the upper surface of the lower limb (E), resulting in formation of the glomerulus (F). Modified with permission from Saxen 1987

to formation of the nephric duct (Dressler et al 1990). As the mesonephros and metanephros develop, Pax2 expression remains high in the nephric duct as well as in both the mesonephric tubules and the metanephric mesenchyme, especially in the condensing mesenchyme at the tips of the ureteric bud (Dressler et al 1993). In the *Lim1* null embryos, the expression of Pax2 is lost in the caudal intermediate mesoderm, coincident with loss of the formation of the nephric duct (Tsang et al 2000). In embryos lacking expression of Pax2 itself, the cranial portion of the nephric duct begins to form but then degenerates by E11.5 without the formation of either normal mesonephric tubules or the ureteric bud and associated metanephric mesenchyme (Torres et al 1995). This results in complete absence of kidney and genital development (Figure 23.3). A closely related member of the *Pax* gene family, *Pax8*, has been shown to be critical for pronephric development in *Xenopus* (Carroll & Vize 1999), but there is no apparent abnormality of kidney development in the *Pax8* null mouse (Mansouri et al 1998). A potentially cooperative role of *Pax2* and *Pax8* in the earliest stages of kidney development has been recently suggested by the finding that mice lacking both *Pax2* and *Pax8* have an even more severe phenotype than *Pax2* null animals, with failure to form even the earliest stages of the anterior nephric duct (Bouchard et al 2002).

The complete loss of mesonephric tubule development in *Pax2* null mice, along with the strong detection of Pax2 in the condensed mesenchyme around the tips of the ureteric bud, has led to the hypothesis that Pax2 expression is

required for mesenchymal–epithelial transformation during the early stages of tubule formation. In agreement with this idea, the use of antisense RNA to *Pax2* in E11.5 kidneys grown in culture revealed that condensation of the metanephric mesenchyme was inhibited and there was general failure of epithelialization (Rothenpieler & Dressler 1993). Later in development, when the renal vesicle has undergone epithelial transition to form the S-shaped body, Pax2 expression is significantly downregulated (Dressler et al 1990, Eccles et al 1992).

Examination of mice heterozygous for the loss of *Pax2* reveals that there is significant renal hypoplasia associated with both a decrease in ureteric bud branching and a diminution of nephron formation (Torres et al 1995). These findings have been attributed to increased rates of epithelial apoptosis in the setting of diminished levels of the Pax2 protein (Ostrom et al 2000, Porteous et al 2000). In support of a similar role for PAX2 in human kidney development, heterozygous mutations in the human *PAX2* gene have now been identified in several families, and can lead to optic nerve colobomas, high frequency hearing loss, and renal hypoplasia, collectively termed the renal–coloboma syndrome (Eccles & Schimmenti 1999, Porteous et al 2000).

The factors that upregulate Pax2 expression during development of the mammalian kidney are not yet well understood. The Six-Dach-Eya transcriptional module has been shown to regulate Pax2 expression under certain circumstances, but does not appear to directly influence Pax2 expression in the kidney. Six1 is a homeobox transcription factor that is expressed in the uninduced metanephric mesenchyme by E10.5, and its expression is maintained in the induced mesenchyme (E11–11.5) and during formation of the collecting tubules (Xu et al 2003). Mice with a genetic deletion of *Six1* demonstrate a reduction in the size of the metanephric mesenchyme associated with reduced levels of total Pax2 expression around the ureteric bud (Xu et al 2003), although the mesenchymal cells that are present appear to express relatively normal levels of Pax2/cell (Li et al 2003). These animals often display renal agenesis due to lack of invasion of the metanephric mesenchyme by the ureteric bud (although the bud does initially form and expresses normal levels of Pax2) (Xu et al 2003). Six1 associates with another transcription factor Dach, with the heterodimeric complex normally acting as a transcriptional repressor. However, the protein phosphatase Eya1, the mouse ortholog of the *Drosophilia eyes absent* gene, can interact with the Six1-Dach complex and convert it to a transcriptional activator, an effect that requires the Eya1 phosphatase activity (Li et al 2003). It is therefore hypothesized that it is this co-activator effect of the Eya1-Six1-Dach complex that is required for normal Pax2 expression, possibly by promoting survival and/or proliferation of the metanephric mesenchyme rather than via a direct effect on *Pax2* gene transcription per se.

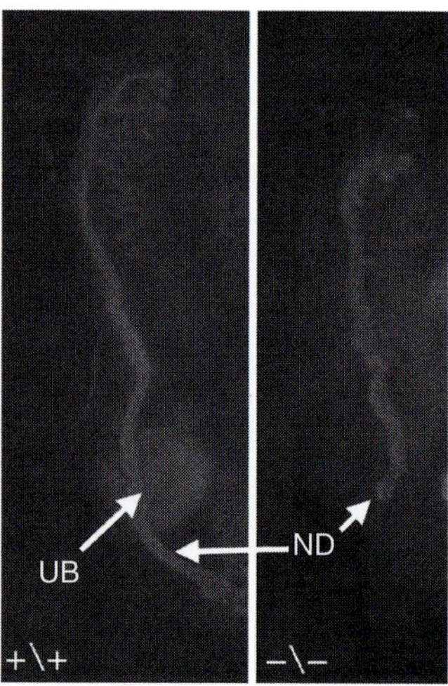

FIGURE 23.3 Examination of mesonephric and metanephric kidney development in wild-type (+/+) and *Pax2* deficient (-/-) mice. Longitudinal sections from E11 mice were immunostained with anti-pancytokeratin (green) and anti-Pax2 (red). Mice lacking Pax2 expression fail to develop both mesonephric tubules and the metanephric kidney. UB, ureteric bud; ND, nephric duct. Image courtesy of Dr Greg Dressler. (see also Plate 30)

METANEPHRIC KIDNEY DEVELOPMENT

As previously noted, the formation of the metanephric kidney begins with the protrusion of the ureteric bud from the caudal region of the nephric duct and its outgrowth into a loose group of nearby cells termed the metanephric mesenchyme. Landmark studies by Grobstein in the 1950s demonstrated that the formation of the ureteric bud and the condensation of the adjacent metanephric mesenchyme were the result of reciprocal interactions between the two tissues, with factors derived from the metanephric mesenchyme (or embryonic spinal cord in experimental models) acting to induce UB outgrowth and branching while factors derived from the UB induce MM condensation and renal vesicle formation (Grobstein 1953, 1956). Using co-culture of the explanted tissues as well as targeted disruption of gene expression in the mouse, major advances in the identification of these factors has been made in the past 10 years.

Formation of the Collecting System of the Kidney

Development of the collecting system of the kidney can be artificially divided into three stages: the initial outgrowth of the ureteric bud at a defined site on the nephric duct, the invasion of the bud into the metanephric mesenchyme, and the branching of the UB within the metanephric mesenchyme to form the individual collecting ducts. While some factors such as glial-derived neurotrophic factor (Gdnf) appear to influence all three of these steps, others seem to be selectively involved in the mesenchymal invasion and/or branching steps.

INITIATION OF THE URETERIC BUD

Approximately 10 years ago, a membrane associated receptor tyrosine kinase, Ret, was found to be expressed along the length of the nephric duct from E8.5–10.5, as well as in the newly forming ureteric bud on E11–11.5 and at the ureteric bud tips until E17.5 (Pachnis et al 1993). This pattern of expression was suggestive of a role in bud outgrowth and/or branching, and disruption of the *Ret* gene was found to result in complete renal agenesis in a high percentage of mice (Schuchardt et al 1994). Examination of these mice demonstrated that the ureteric bud had failed to form or had formed but then failed to branch normally. In addition to the loss of kidney development, Ret was also found to be expressed in the neural crest cells and the myenteric ganglia of the gut, and the null mouse developed a gastrointestinal motility disorder typical of that described in Hirschsprung's disease.

Tissue explant studies using the *Ret* null mouse revealed that the nephric duct from these animals failed to produce a ureteric bud in response to inductive signals from normal metanephric mesenchyme, although the metanephric mesenchyme from the *Ret* null mice responded normally to wild-type nephric duct induction (Schuchardt et al 1996). These studies have established Ret as a critical receptor for the ureteric bud inductive signal derived from the metanephric mesenchyme; however, the lack of a classic extracellular ligand-binding domain in Ret initially frustrated attempts to identify the exact MM-derived protein responsible for Ret receptor activation. The identity of this factor was determined when Gdnf, a potent survival factor for dopaminergic neurons, was found to activate Ret when associated with the GPI-linked transmembrane receptor Gfrα, (also called Gdnfrα) (Jing et al 1996, Trupp et al 1996). Mice lacking either Gdnf or Gfrα were subsequently found to have failure of ureteric bud outgrowth as well as defective enteric innervation, replicating the findings in *Ret* null mice (Cacalano et al 1998, Pichel et al 1996, Sanchez et al 1996). More recently, the *Ret* gene has been found to encode at least two isoforms of the receptor, Ret9 and Ret51, that differ in their carboxy terminus (Lorenzo et al 1995). Generation of mice that express only Ret9 or Ret51 has demonstrated that Ret9 is the isoform that is required for normal ureteric bud development (de Graaff et al 2001).

The addition of an exogenous point source of Gdnf can induce ectopic ureteric bud formation by the nephric duct, suggesting that Gdnf-mediated ureteric bud formation may occur as a result of a chemoattractant effect of Gdnf for epithelial cells from the nephric duct (Sainio et al 1997). Studies in cell culture models support this hypothesis, demonstrating that Gdnf can stimulate cell migration as well as three-dimensional organization into tubular structures (O'Rourke et al 1999, Tang et al 1998). Recent studies have demonstrated that in addition to direct signaling via Gfrα and Ret, Gdnf can activate a second morphogenic tyrosine kinase receptor, Met. This trans-activation does not appear to be mediated by direct interaction of Gdnf with Met, but rather may require the activation of a cytosolic tyrosine kinase such as Src which subsequently phosphorylates and activates Met independent of ligand binding (Popsueva et al 2003). (For a recent review of Gdnf-Ret signaling, the reader is referred to Sariola & Saarma (2003)). Met is the receptor tyrosine kinase for hepatocyte growth factor (Hgf), a potent morphogen that can induce epithelial cell tubulogenesis in vitro (Montesano et al 1991). Both Hgf and Met are expressed in the developing metanephric kidney (Defrances et al 1992, Kolatsi-Joannou et al 1997), and several studies have supported a role for Hgf/Met signaling in the in vitro stimulation of ureteric bud branching and mesenchymal-epithelial transformation of the metanephric mesenchyme (Karp et al 1994, Santos et al 1994, van Adelsberg et al 2001). However, mice lacking either the ligand or receptor appear to have normal initial metanephric kidney development and then die at E13-E15 due to defects in the placenta and liver (Bladt et al 1995, Schmidt et al 1995, Uehara et al 1995). It is therefore unclear what role, if any, Hgf and Met play in the normal process of ureteric bud branching.

REGULATION OF GDNF AND RET EXPRESSION

The finding that overexpression of Gdnf results in the formation of ectopic ureteric buds along the nephric duct suggests that the level and location of Gdnf expression must be carefully regulated in the embryo (Sainio et al 1997). Recent studies of *Pax2* null mice have revealed that Gdnf is not expressed in the MM from these animals, and cell culture models have confirmed that Pax2 can bind a *Gdnf* promoter element and initiate transcription of the Gdnf message (Brophy et al 2001). However, the addition of Gdnf to nephric ducts explanted from *Pax2* null mice does not rescue UB formation, suggesting that Pax2 is likely to regulate other genes critical for UB formation in addition to *Gdnf*.

As has been previously noted, the Six1-Dach-Eya1 transcription module can upregulate Pax2 levels in the metanephric mesenchyme, probably via an indirect effect involving increased mesenchymal cell survival and/or proliferation. In addition to this indirect effect on Pax2 expression, Six1 can also upregulate the expression of another Six transcription factor that is expressed in the metanephric mesenchyme, Six2. Recently it has been shown that Six2 can bind to the promoter region of *Gdnf* and directly activate *Gdnf* gene expression (Brodbeck et al 2004) (Figure 23.4) although *Six2* null embryos have normal *Gdnf* expression (Self et al 2006). It is not surprising then that mice lacking either *Six1* or *Eya1* have been found to have diminished levels of Pax2 in the metanephric mesenchyme, associated with decreased levels of Gdnf and failure of normal ureteric bud outgrowth and invasion of the metanephric mesenchyme (Li et al 2003, Xu et al 1999, 2003).

Furthermore, heterozygous mutations of human *EYA1* have been shown to be associated with brachio-oto-renal (BOR) syndrome, an autosomal dominant disorder with brachial arch abnormalities, renal dysplasia and collecting system anomalies (Abdelhak et al 1997, Melnick et al 1976, Rodriguez-Soriano et al 2001; see also Chapter 37). In contrast, mutations in the gene for the *Foxc1* transcription factor result in upregulation of Gdnf expression leading to multiple defects including duplication of the ureter and hydroureter (Kume et al 2000). In these mice, both Eya1 and Gdnf expression were found to extend cranial to that seen in normal mice, resulting in the formation of ectopic ureteric buds.

In addition to the effects of the Pax and Six transcription factors, a secreted protein, growth/differentiation factor 11 (Gdf11) has now been found to play a major role in Gdnf production. Gdf11 is a member of the TGFβ family of growth factors, and is expressed in the developing brain, eye and limb buds. In the E10.5 kidney, Gdf11 is expressed in the nephric duct and to a lesser extent in the uninduced metanephric mesenchyme. As the ureteric bud forms and invades the mesenchyme, Gdf11 expression is maintained in both structures. *Gdf11-/-* mice were found to die at birth with many of the newborns exhibiting bilateral renal agenesis (Esquela & Lee 2003). Examination of embryonic *Gdf11-/-* mice revealed failure of the ureteric bud to form in many cases, correlating with an absence of Gdnf expression by the metanephric mesenchyme in those kidneys. This failure of UB formation could be rescued in vitro by the addition of exogenous Gdnf. Interestingly, in embryos where the UB did form, Gdnf expression by the MM was found to be near normal, suggesting that factors other than

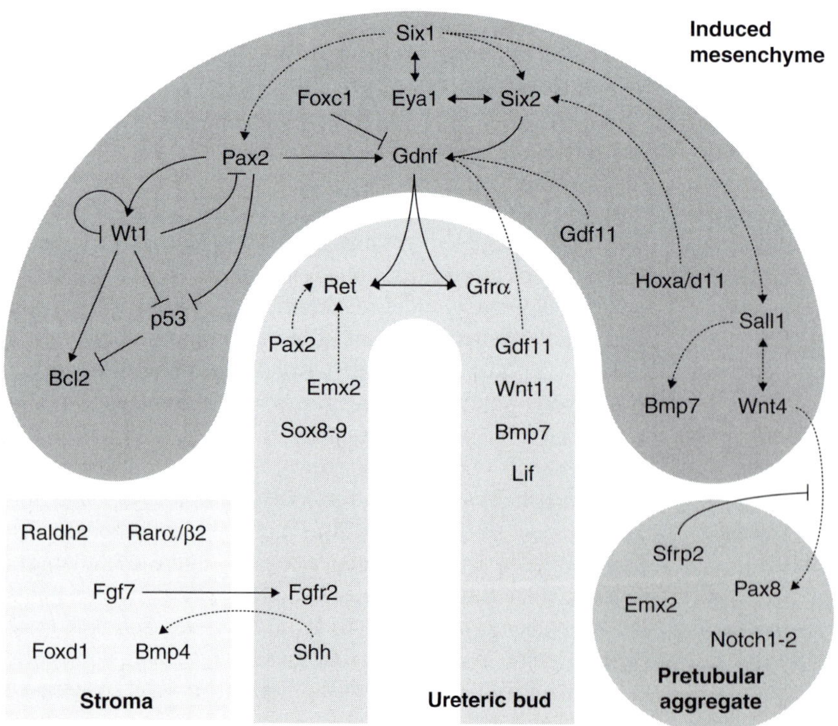

FIGURE 23.4 A schematic depiction of representative transcription factors and signaling pathways involved in the molecular cross-talk between the four cellular compartments in the early metanephros. (see also Plate 31)

Gdf11 are capable of stimulating Gdnf expression by the MM. Overall, these findings support the idea that a complex network of factors, acting both directly on *Gdnf* gene transcription and on each other, serve to tightly control the amount of Gdnf produced and the exact site of that production (Brodbeck & Englert 2004). This level of control normally allows only a single ureteric bud to be generated.

The regulation of Ret expression in the ureteric bud is less well understood. The observation that Ret expression in neural crest cells is regulated by Pax3 and Sox 10 (Lang and Epstein, 2003), transcription factors that are not expressed in the ureteric bud, has led to the search for homologs of these transcription factors. The Sox 10 family members Sox8 and Sox9 are expressed in the kidney, with Sox8 expression specifically detected at the tips of the ureteric bud. However, mice genetically null for either *Sox8* or *Sox9* apparently demonstrate normal early kidney development (Akiyama et al., 2004; Sock et al., 2001). Pax2 is expressed in the ureteric bud as well as the metanephric mesenchyme, and it has now been shown that $Pax2^{+/-};Ret^{+/-}$ compound heterozygotes have a greater decrease in Ret expression than that seen in simple $Ret^{+/-}$ heterozygotes with an increased incidence of renal agenesis (Clarke et al 2006), suggesting that Pax2 may regulate Ret expression in the UB.

Despite our present lack of understanding of the factors that directly regulate Ret expression, a significant amount of data has accumulated supporting an important role for vitamin A in the indirect regulation of Ret expression. Vitamin A deficiency has long been known to result in developmental defects in the kidney, with mild vitamin A deficiency causing a decrease in nephron number and severe deficiency resulting in marked growth retardation and even fetal death (Lelievre-Pegorier et al 1998, Wilson et al 1953). Rat embryos exposed to severe vitamin A deficiency in utero demonstrate marked downregulation of Ret expression in the early ureteric bud, followed shortly thereafter by intrauterine death (Batourina et al 2001). Furthermore, culture of explanted mouse kidneys in the absence of vitamin A resulted in the downregulation of Ret expression and a decrease in ureteric bud branching (Batourina et al 2001), while the addition of exogenous vitamin A to explanted rat kidneys stimulated ureteric bud branching in a dose dependant manner and resulted in an increase in nephron number (Vilar et al 1996).

Vitamin A is normally metabolized by retinoic acid alcohol dehydrogenases (abbreviated as Raldh or Aldh1a) to form its primary active metabolite, retinoic acid. Retinoic acid then binds to a family of nuclear receptors called the retinoic acid receptors (Rar). In the kidney, Rarα is expressed ubiquitously, but is co-expressed with Rarβ2 exclusively in the stromal cells of the metanephric mesenchyme (Mendelsohn et al 1999). These cells also express Aldh1a2 (Raldh2), resulting in the generation of retinoic acid in the immediate vicinity of the Rarα/Rarβ2 receptors. The simultaneous genetic inactivation of both the α and

β receptors results in a decrease in ureteric bud branching associated with greatly impaired development of the collecting system of the kidney (Mendelsohn et al 1999). Similar to the findings in vitamin-A-deficient rats, examination of the nephric duct in *Rarα/Rarβ2* deficient mice reveals downregulation of Ret protein expression. Furthermore, transgenic overexpression of Ret in the ureteric bud rescues the *Rarα/Rarβ2* null phenotype (Batourina et al 2001), suggesting that retinoic acid-mediated activation of the Rar in stromal cells is critical for expression of Ret in the nearby nephric duct, although the factor that mediates this cross talk between the stromal cell and the UB is presently unknown. Interestingly, nephron formation by the metanephric mesenchyme in *Rarα/Rarβ2* deficient mice appears to be normal in regions where UB branching has occurred, suggesting that vitamin A signaling is not directly involved in the induction of the mesenchyme itself.

Maintenance of Ret expression in the growing UB appears to require the expression of the homeobox gene, *Emx2*. Emx2 is normally expressed in the embryonic kidney in both the nephric duct and ureteric bud, as well in the renal vesicle during formation of the proximal tubular segments (Pellegrini et al 1997). In mice lacking *Emx2*, both the kidneys and the genital tracks fail to form. Interestingly, Gdnf expression in the metanephric mesenchyme and Ret expression in the ureteric bud are initially near normal, and the UB does begin to invade the metanephric mesenchyme. However, Ret, Gdnf, and Pax2 expression subsequently diminish, correlating with failure of the bud to branch and failure of formation of renal vesicles (Miyamoto et al 1997). The observation that metanephric mesenchyme derived from the *Emx2* null mouse can still be induced normally in response to co-culture with a wild-type ureteric bud, whereas the *Emx2* null ureteric buds failed to respond to or induce wild-type metanephric mesenchyme, suggests that the primary defect in these animals is in the bud where Ret expression is lost.

As the ureteric bud forms and begins to branch, Ret expression is maintained at the tips of the ampulla where subsequent branching will occur, but lost along the stalk of the UB. This spatial restriction appears to be important for normal patterning of UB branching and has been shown to be regulated by two transcription factors, Foxd1 (also called Bf-2) and Pod1. Similar to the retinoic acid receptors, both Foxd1 and Pod1 are expressed in the stromal cells surrounding the UB rather than in the nephric duct or UB itself. Inactivation of either transcription factor results in ectopic expression of Ret along the stalk of the UB, associated with hypoplastic kidneys with abnormal patterns of ureteric bud branching and decreased numbers of branches (Hatini et al 1996, Quaggin et al 1999). It is not fully understood why ectopic Ret expression would result in decreased UB branching, although the observation that decreased ureteric bud branching can be reproduced by transgenic overexpression of Ret in mice, and that additional Gdnf can partially rescue the phenotype, has resulted in the hypothesis that ectopic expression of Ret along the stalk of the ureteric bud results in sequestration of Gdnf in these regions, thereby decreasing the availability of Gdnf at the tips of the UB and subsequently diminishing the stimulus for UB branching (Srinivas et al 1999). In addition to ectopic Ret expression, *Pod1-/-* animals also demonstrate an abnormal accumulation of the condensed mesenchyme layer around the UB. Normally this condensed mesenchyme (referred to as the cap) is approximately two to three cell layers thick, but in *Pod1-/-* animals as many as 15 cell layers may accumulate around the UB (Quaggin et al 1999). These cells have normal expression of Pax2 and Wnt4, arguing that initial specification of the MM has occurred. However, the formation of comma and S-shaped bodies is significantly delayed, suggesting that the signal for mesenchymal–epithelial transformation has been diminished or interrupted (see below).

REGULATION OF UB BRANCHING

The finding that nephron induction occurs exclusively at the ampullae of the ureteric bud has led to the conclusion that the number of branches formed by the UB will ultimately determine the number of nephrons found in the adult kidney. Recent studies suggesting a correlation between low nephron number and the occurrence of nephrosclerosis and hypertension later in life (Hoy et al 2003, Keller et al 2003) have therefore led to extensive investigations into the identification of factors that can both positively and negatively regulate UB branching. (For a more complete discussion of the regulation of UB branching, the reader is referred to recent reviews of the subject (Piscione & Rosenblum 2002, Shah et al 2004).)

The evidence presented heretofore supports a primary role of Gdnf and its receptor Ret in UB initiation, as well as an ongoing role for normal Ret signaling in the subsequent patterning of UB branching. However, in ureteric bud explants Gdnf alone is insufficient to stimulate normal UB branching (Sainio et al 1997), whereas Gdnf in combination with conditioned media derived from metanephric mesenchyme-derived cells will facilitate this process even in the absence of cell contact between the MM and the UB (Qiao et al 1999a). These experiments suggest that soluble factors produced by the metanephric mesenchyme in addition to Gdnf must participate in the stimulation of UB branching morphogenesis. In vitro studies using either isolated UB explants or cultured cells derived from the UB have resulted in the identification of a significant number of growth factors that appear to be capable of stimulating ureteric bud cell proliferation and/or branching morphogenesis, including Fgf1, Fgf2, Fgf7, Fgf10, Hgf, Egf, and Tgfα (Cantley et al 1994, Qiao et al 1999b, 2001, Sakurai et al 1997, 2001). Of these, gene inactivation studies in the mouse suggest that the two ligands for the IIIb isoform of the fibroblast growth factor receptor 2 {Fgfr2(IIIb)}, Fgf7 and Fgf10, are important for this process in vivo, since the

loss of expression of either ligand results in smaller kidneys with an abnormal collecting system architecture (Ohuchi et al 2000, Qiao et al 1999b).

In contrast, inactivation of Hgf, Egf, or their respective receptors in the mouse results in no change in early renal development, suggesting that these factors are either not critical for organization of UB branching, or are redundant with other factors that are present in the embryonic kidney (Bladt et al 1995, Threadgill et al 1995). Alternatively, it may be that these growth factors play a more significant role in regulating the later stages of remodeling of the collecting system. For example, it has been proposed that programmed cell death (apoptosis) plays a major role in the dilation of the first 3–5 generations of UB branches to form the ureter and renal pelvis. Examination of rat kidneys has demonstrated that there is indeed a significant level of apoptosis in the central parts of the collecting system during development, and that treatment with Egf can substantially diminish this (Coles et al 1993). Interestingly, in certain genetic backgrounds, deletion of the *Egfr* in the mouse results in abnormally increased dilitation of the medullary collecting system (Threadgill et al 1995), suggesting that Egf signaling may normally prevent collecting system dilatation and remodeling, at least in part, by inhibiting tubular cell apoptosis.

The observation that isolated ureteric buds cultured with various combinations of Gdnf and the above growth factors fail to branch as extensively as those co-cultured with complete supernatant from metanephric mesenchyme has suggested the presence of other MM-derived soluble factors that participate in the stimulation of UB branching. Examination of the developing rat kidney has shown that the message for another growth factor, pleiotrophin (also called Hb-Gam), is expressed in the early metanephric mesenchyme (Vanderwinden et al 1992), and the protein has been purified from supernatants of metanephric mesenchyme-derived cells in culture (Sakurai et al 2001). Pleiotrophin, a member of the midkine family of heparin-binding growth factors, was initially described as a regulatory factor for neurite outgrowth, but its expression in multiple organs outside of the nervous system, including lung, pancreas and kidney, suggests a broader role (Mitsiadis et al 1995, Rauvala 1989). In agreement with this, the addition of purified pleiotrophin to cultured ureteric buds in the presence of Gdnf and Fgf1 has been found to stimulate marked UB branching, similar to that seen with co-culture of the UB with MM (Figure 23.5). Interestingly, pleiotrophin expression has been recently found to be induced by vitamin A treatment in neural-derived P19 cells (Brunet-de Carvalho et al 2003), suggesting that this might be a target for induction by vitamin A in the kidney as well. However, like Hgf and Egf, targeted deletion of *pleiotrophin* in the mouse fails to yield a major alteration in kidney development (Amet et al 2001), again supporting the idea that a cocktail of growth factors expressed by the metanephric mesenchyme

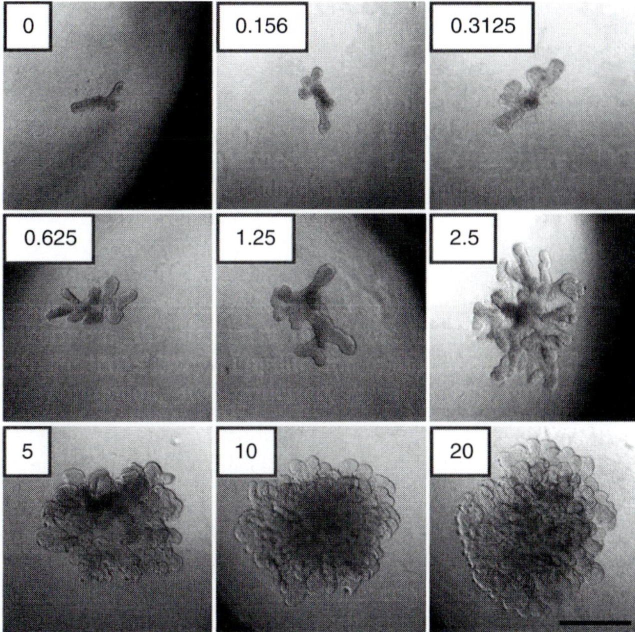

FIGURE 23.5 Culture of the isolated ureteric bud in vitro in the presence of Gdnf, Fgf1, and increasing doses of pleiotrophin (in μg/ml) results in extensive branching and bud elongation. Used with permission from H. Sakurai (Sakurai et al 2001). (see also Plate 32)

may have redundant abilities to support and stimulate ureteric bud branching.

While it has long been obvious that factors which stimulate ureteric bud branching are critical for normal kidney development, it has become increasingly clear that the formation of the mature collecting system of the kidney also requires regulated inhibition of branching in defined regions and at certain time points to allow elongation of specific segments of the collecting tubules before the next round of branching. For example, the papillary collecting ducts traversing the renal medulla are elongated with few branch points whereas those in the cortex exhibit shorter stalks with multiple branches. In vitro studies using UB cell lines and culture of kidney explants have demonstrated that several members of the transforming growth factor β (Tgfβ) family of secreted growth factors can exert inhibitory effects on UB outgrowth and branching, including Tgfβ itself and the bone morphogenetic proteins, Bmp2 and Bmp4 (Piscione et al 1997, Raatikainen-Ahokas et al 2000, Rogers et al 1993). Again, the in vivo role of these growth factors during renal development has been less easy to ascertain. Deletion of *Tgfβ* expression in the mouse has no detectable effect on collecting system development (Kulkarni et al 1993), while mice homozygous for deletions of the *Bmp2* or *Bmp4* genes exhibit embryonic lethality prior to kidney formation (Lawson et al 1999; Zhang & Bradley 1996). However, *Bmp4+/-* heterozygous mice exhibit a combination of defects in the collecting system of the kidney, including doubling of the collecting system, hydroureter, and dysplastic

kidneys (Miyazaki et al 2000). These defects are highly similar to those seen in the human CAKUT syndrome (congenital abnormalities of the kidney and urinary tract), although to date no mutations in *BMP4* have been reported in these patients (Nakano et al 2003).

Examination of Bmp mRNA expression in the mouse embryo reveals that Bmp4 is expressed in the mesenchyme surrounding the nephric duct prior to UB formation (E10.5), and primarily along the stalk of the UB rather than the tip once UB branching is underway (E12.5) (Miyazaki et al 2000). The Bmp4 receptors Alk3 and Alk6 are expressed to the greatest degree in the adjacent nephric duct and the UB itself. This pattern of expression is similar to Gdnf and its receptor Ret, suggesting that Bmps might normally act to negatively regulate Gdnf responses in the nephric duct. In agreement with this, the addition of Bmp4 to kidney explants was shown to inhibit Gdnf-stimulated ureteric bud branching, although levels of Gdnf itself were not affected (Miyazaki et al 2000). Furthermore, expression of a constitutively active form of the Bmp2/4 receptor *Alk3* in mice resulted in a decrease in ureteric bud branching and the formation of medullary duct cysts (Hu et al 2003). In contrast to its inhibitory effects on UB branching, Bmp4 was found to stimulate bud elongation. This finding, combined with the observation that the main UB trunk and the first few generations of branches were significantly shorter in Bmp4+/- animals than in controls, suggests that Bmp4 plays a regulatory role in promoting UB segment elongation without branching.

Gremlin and noggin are members of another subgroup of proteins in the Tgfβ superfamily that have been found to bind directly to Bmps and inhibit their actions during development. Expression of gremlin1 (Grem1), which is believed to antagonize both Bmp2 and Bmp4 (Avsian-Kretchmer & Hsueh 2004), appears to be critical for kidney development since mice lacking this protein have been found to die at birth due to complete renal agenesis (Michos et al 2004). Examination of these mice demonstrates that the ureteric bud begins to develop, but fails to invade the metanephric mesenchyme or undergo branching, despite normal expression of Ret. Interestingly, Gdnf expression is initially normal, but then is lost as the metanephric mesenchyme undergoes extensive apoptosis. Thus, it is believed that Grem1 is a survival factor for the metanephric mesenchyme, and that failure of ureteric bud growth and branching is due to the loss of sustained Gdnf expression.

This pattern of initial ureteric bud sprouting followed by failure of the bud to invade the metanephric mesenchyme and the coincident apoptosis of the mesenchyme itself was also observed in mice lacking the putative transcriptional repressor Sall1. Sall1 is a member of a family of zinc finger containing proteins that were first identified in *Drosophilia* as being critical for normal body segmentation during development. In the mammalian kidney, it is expressed by E10.5 in the uninduced mesenchyme and in the nephric duct (Nishinakamura et al 2001). Its expression is subsequently lost in the caudal nephric duct and ureteric bud, but maintained in the condensed mesenchyme. Mice that lack *Sall1* display either renal agenesis or severe dysgenesis and eventually die at birth. Examination of these mice reveals that, similar to *Grem1* deficient mice, Pax2 is initially expressed normally and the ureteric bud begins to sprout, but that invasion of the bud into the mesenchyme is absent or incomplete (Nishinakamura et al 2001). Under these circumstances the bud fails to branch normally and the metanephric mesenchyme undergoes apoptosis rather than its normal expansion. Again, similar to the *Grem1* null animals, Gdnf, Wnt4, and Bmp7 expression are reduced in the mesenchyme, while Ret expression in the ureteric bud was unaffected. Interestingly, co-culture of the explanted metanephric mesenchyme with embryonic spinal cord rescued the mesenchyme from apoptosis and induced apparently normal tubule formation in this compartment, implying that a survival signal derived from the ureteric bud is likely to be absent in these animals, although specific transcriptional targets of Sall1 have yet to be identified in the kidney. Defects in the human *SALL1* gene have been found to cause Townes-Brocks syndrome, a collection of developmental abnormalities including dysplastic ears, polydactily, imperforate anus and occasionally renal hypoplasia and heart abnormalities (Salerno et al 2000).

It is of interest that many of the growth factors that either positively or negatively regulate UB branching contain heparin-binding domains, allowing them to be purified on heparin affinity columns. These growth factors, including Hgf, Gdnf, pleiotrophin, Fgfs, Wnts, Bmps and Tgfβ, can interact with heparan sulfate proteoglycans (HSPGs) on the cell surface (such as syndecans and glypicans) or with heparan sulfate moieties expressed in the extracellular matrix (on perlecan or collagen XVIII for example). For a review of this subject, see Shah et al (2004). The importance of HSPGs for renal development was proven when inactivation of the heparan sulfate 2-O-sulfotransferase enzyme (which is required for formation of the heparan sulfate side chains) was found to result in perinatal death due to renal agenesis (Bullock et al 1998) (Figure 23.6). These animals exhibited initial UB outgrowth and Pax2 expression in the MM, but then failed to exhibit normal ureteric bud branching or condensation of the metanephric mesenchyme. Similarly, treatment with chlorate to deplete heparan sulfate side chains in kidney organ culture studies resulted in the inhibition of UB branching (Davies et al 1995). Cumulative data demonstrating that the location, type, and duration of HSPG expression is highly regulated in the developing kidney has led to the hypothesis that the expression of specific HSPGs can mediate the site-specific accumulation and/or sequestration of heparin-binding growth factors, thus providing an additional level of control regarding exactly which cell will respond to a given stimulatory or inhibitory signal (Bernfield et al 1993,

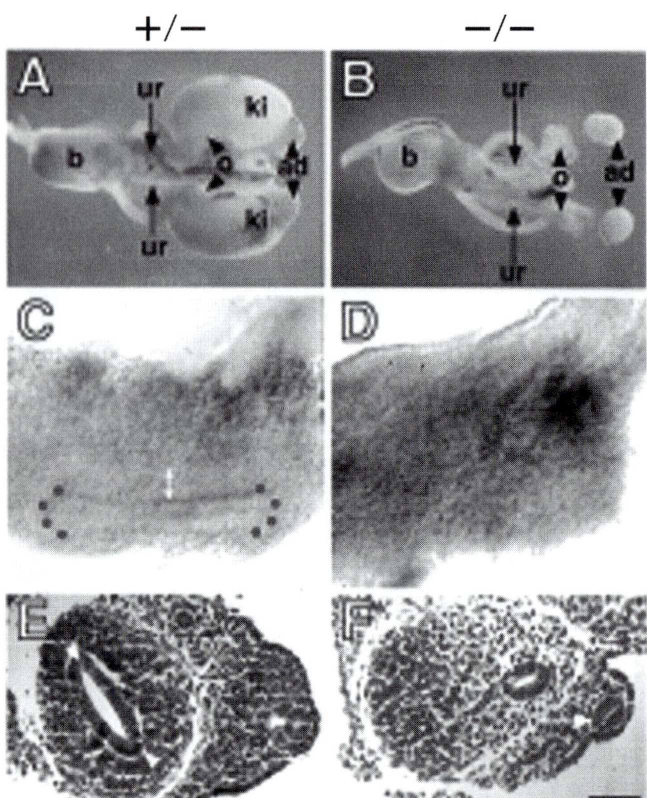

FIGURE 23.6   Dissected urogenital systems of normal (heterozygous) (A) and heparan sulfate 2-O-sulfotransferase homozygous mutant (B) newborn mice reveals bilateral renal agenesis in the absence of *Hs2st* expression. Homozygotes display bilateral renal agenesis and possess a blind-ended ureter (ur). The remainder of the mutant urogenital system is normal. (ad) Adrenal gland; (b) bladder; (ki) kidney; (o) ovary. (C–F) Morphology of representative E11.5 heterozygous (C, E) and mutant (D, F) embryonic kidney rudiments viewed in whole-mount (lateral view) (C, D) or as transverse sections stained with hematoxylin and eosin (E, F). (C) In heterozygotes, the ureteric bud has invaded the mesenchyme and bifurcated once to form a T-shape surrounded by a mantle of condensed mesenchyme (double-headed arrow). (D) In mutants, the ureteric bud has emerged from the Wolffian duct and invaded the mesenchymal blastema but failed to bifurcate and there are no overt signs of mesenchymal condensation. In C and D the position of the ureteric bud tip(s) is highlighted by red dots. (E) A dilated ureteric bud (arrowheads) is present in the heterozygotes surrounded by condensed mesenchyme. (F) In mutants, the ureteric bud (arrowhead) has not dilated, and there are no overt signs of condensation in the mesenchyme. (Arrow in E and F) Wolffian duct. Scale bar:1300 μm (A and B); 30 μm (C–F). Used with permission from Bullock et al 1998. (see also Plate 33)

Watanabe et al 1995, Lin et al 2001b, Shah et al 2004, Steer et al 2004). This idea is further supported by the observation that there appears to be significant specificity in binding partners and signaling responses for different HSPGs. For example, Hgf-stimulated epithelial tubule formation in vitro appears to be dependent on the expression of glypican-4 (Gpc4), and cannot be supported by other glypicans such as Gpc1 or Gpc3, whereas cell migration is supported by the expression of either Gpc3 or Gpc4 (Karihaloo et al 2001, in press). Furthermore, selective loss of expression of Gpc3 in the mouse results in a complex renal phenotype with an early increase in ureteric bud branching followed later by medullary cysts and kidney overgrowth (Cano-Gauci et al 1999), findings that have been attributed to the loss of normal Bmp2 and endostatin inhibitory effects on tubule branching but enhanced Fgf7 and Bmp7 stimulatory effects on tubule growth (Grisaru et al 2001, Karihaloo et al 2001). Mutations in the *GPC3* gene have been found to result in the Simpson-Golabi-Behmel syndrome (SGBS), an X-linked disorder characterized by developmental overgrowth and renal dysplasia (Pilia et al 1996).

Another group of proteins that bind to heparin and/or heparan sulfate moieties are the members of the Wnt family of secreted glycoproteins that are related to *Drosophilia wingless*, a gene that is required for heart and nervous system formation in the fly. Due to their binding to heparan sulfate proteoglycans on the cell surface and in the surrounding matrix, members of the Wnt family do not appear to circulate as free proteins, but rather are presented to their receptors (members of the Frizzled family of transmembrane proteins) in the context of cell–cell or cell–matrix interactions. Of the 18 known mammalian *Wnt* genes, Wnt-2b, Wnt-4, Wnt-6, Wnt-7b, and Wnt-11 have been found to be expressed in the developing mouse kidney. (Reviewed in Vainio (2003).) Of these, Wnt-2b is expressed in the cells surrounding the metanephric mesenchyme, Wnt-4 is expressed in the metanephric mesenchyme itself, and Wnt-6 and Wnt-11 are found in the ureteric bud with Wnt11 highly expressed at the bud tips where branching is most prominent (Kispert et al 1996). Targeted deletion of the *Wnt-11* gene results in small kidneys with reduced numbers of UB branches, even though initial UB invasion of the MM is normal (Majumdar et al 2003). Examination of the Ret/Gdnf pathway in these animals revealed the Gdnf levels were initially normal, but fell by E12.5, coincident with the decrease in later rounds of UB branching. These results suggest that Wnt-11 expression by the UB is required for normal Gdnf expression by the MM, and that the decrease in UB branching seen in the *Wnt-11-/-* animals may be due to the reduction of Gdnf rather than a direct effect of loss of Wnt-11 per se. Interestingly, Wnt-11 expression at the UB tips is also significantly reduced in the *Ret-/-* mice, supporting the idea that Wnt-11 and Gdnf/Ret participate in a cooperative loop in which Wnt-11 expression by the UB augments Gdnf expression by the MM which in turn activates Ret and induces further Wnt-11 expression (Majumdar et al 2003). This type of positive feedback loop could serve to regulate the timing of ureteric bud branching. For example, as the tip of the UB enlarges to form the ampulla, a Wnt-11 mediated increase in the local concentration of Gdnf could induce one round of branching, with the subsequent separation of the two new

tips resulting in lower local levels of Wnt-11, and therefore decreased induction of Gdnf adjacent to each tip. This would allow branch elongation and ampulla formation to occur again prior to the next round of branching.

CELL–MATRIX INTERACTIONS IN URETERIC BUD BRANCHING
Surrounding the nascent ureteric bud is a basement membrane composed of multiple matrix components including collagen IV, laminin-10, perlecan, nidogen, nephronectin, and TIN-Ag. These matrix components interact with specific cell surface receptors, including the heterodimeric α/β integrins and dystroglycan. These interactions provide the cells with information regarding orientation and basement membrane composition that is critical for regulating cell polarity as well as in the determination of cell identity. Unfortunately, genetic deletion of many of these matrix components and their receptors results in embryonic lethality prior to kidney development, meaning that their role in tubule maturation will remain unclear until deletions targeted selectively to the renal tubule are performed. However, studies using antibodies that neutralize the association of matrix proteins and their integrin receptors in the cultured kidney explant, as well as studies in which the loss of expression of some of these proteins did not prevent embryonic development, have confirmed that distinct matrix–receptor interactions are crucial for ureteric bud branching, mesenchymal-epithelial transition of the renal vesicle, and correct organization of the mature tubule and glomerulus.

Laminin is a heterotrimeric protein comprised of α, β, and γ chains, with laminin-10 containing the $\alpha5$, $\beta1$, and $\gamma1$ subunits. Mice with targeted disruption of the *Lama5* gene that encodes the laminin $\alpha5$ protein (and therefore lacking expression of laminin-10) display a significant incidence of renal agenesis, as well as severe abnormalities in glomerular formation (see below) (Miner & Li 2000). Examination of these mice revealed a decrease in ureteric bud invasion of the metanephric mesenchyme, but normal renal vesicle formation and mesenchymal–epithelial transition in regions where the UB was present and branching, although glomerular development was abnormal. Explant culture of the metanephric kidney confirmed a marked defect in ureteric bud branching in the absence of laminin $\alpha5$ expression.

The predominant cell surface receptors in renal epithelial cells for both laminin-10 and collagen IV are $\alpha2\beta1$ and $\alpha3\beta1$ integrins. Genetic disruption of $\alpha3$ integrin expression, like disruption of *Lama5*, results in severe abnormalities in glomerular development (Kreidberg et al 1996). In addition, these mice had smaller kidneys than heterozygous littermates with a decrease in the numbers of proximal ureteric bud branches, although the final numbers of nephrons were not appreciably reduced. A role for $\alpha3\beta1$ integrins in regulating ureteric bud branching is further supported by studies in which the branching of the ureteric bud in the explanted kidney was found to be markedly reduced in the presence of neutralizing antibodies to $\alpha3$ integrin

(Zent et al 2001). In this study, two other integrin heterodimers that act as laminin receptors, α6β1 and α6β4, were shown to be expressed in the ureteric bud and antibodies to α6 were also found to inhibit ureteric bud branching. Expression of another integrin subunit, α8 integrin, is induced in the cells of the metanephric mesenchyme following invasion of the ureteric bud (Muller et al 1997). Interruption of the α8 integrin gene, resulting in loss of expression of the nephronectin and osteopontin receptor α8β1 integrin, was found to inhibit ureteric bud enlargement and branching, as well as to decrease the epithelialization of the metanephric mesenchyme (Muller et al 1997, Brandenberger et al 2001).

The formation and turnover of a normal basement membrane also involves the regulated cleavage of matrix components and other extracellular proteins (including growth factors) by matrix metalloproteinases (MMPs) and their endogenous inhibitors (tissue inhibitors of metalloproteinases – TIMPs). There is a wealth of data in vitro suggesting that the expression of specific MMPs and TIMPs, including MMP2, MMP9, MT1-MMP, and TIMP-2, play a regulatory role in ureteric bud branching (Lelongt et al 1997, Kanwar et al 1999, Pohl et al 2000). However, genetic interruption of individual among these factors has so far failed to result in any obvious defect in renal development, suggesting complementary functions of these proteases in vivo (Itoh et al 1997, Andrews et al 2000, Zhou et al 2000).

DEVELOPMENT OF THE URETER

As the tips of the UB undergo branching to form the collecting system of the kidney, the stalk of the UB dilates to form the ureter and acquires a layer of smooth muscle that is necessary for the peristaltic contraction that propels urine into the bladder. This process, similar to that seen with UB branching and nephron induction, appears to involve reciprocal signals between the stalk of the UB and the adjacent mesenchyme. Sonic hedgehog (Shh), a secreted protein that is involved in multiple aspects of embryonic patterning and cell survival during development, is expressed by the ureteric bud stalk and the first generation of UB branches, while one of the Shh receptors, patched 1 (Ptch1), is expressed by the adjacent mesenchymal cells (Yu et al 2002). Deletion of Shh targeted specifically to the ureteric bud has been performed, allowing normal extra-renal development of the embryos with viable newborns. The kidneys of these mice were noted to be small with a moderate reduction in the numbers of glomeruli, but the most striking finding was the presence of a marked dilation of the ureter (hydroureter) and the eventual development of hydronephrosis after birth (Yu et al 2002). Examination of the embryonic kidneys from these mice revealed that Ptch (a transcriptional target of Shh as well as an Shh receptor) and Bmp4 were significantly downregulated in the mesenchyme immediately surrounding the stalk of the UB, correlating with a marked reduction in mesenchymal cell proliferation in this area. In addition to the decrease in mesenchymal cell proliferation, the differentiation of mesenchyme to smooth muscle

was also abnormal in the *Shh* null animals. Whereas wild-type mice demonstrated the investment of the entire ureter with smooth muscle cells by birth, the *Shh* null kidneys demonstrated both a reduction in the numbers of smooth muscle progenitors and in their caudal distribution. This loss of smooth muscle differentiation in the absence of Shh expression appears to be mediated by a complex series of events since Shh itself was found to inhibit mesenchyme-to-smooth muscle differentiation in vitro.

Development of the Nephron

MESENCHYMAL–EPITHELIAL TRANSFORMATION

As the ureteric bud elongates and branches to become the collecting system of the kidney, the loose collection of the cells surrounding the caudal nephric duct condenses around the tip of the invading UB to form the metanephric cap. Some of these cells proliferate and migrate with the invading UB tips, while others remain in place and differentiate into connective tissue elements surrounding the stalk of the UB. Recent studies have shown that the cap mesenchyme is comprised of a group of self-renewing nephron progenitor cells that can be identified by their expression of Six2 (Kobayashi et al 2008). One function of Six2 expression in these cells appears to be suppression of the epithelial differentiation signal provided by the UB. Loss of Six2 expression results in severe hypoplasia due to ectopic nephrogenesis of the cap mesenchyme with rapid depletion of the progenitor pool (Self et al 2006).

As the UB tips invade, they swell to form an ampulla prior to undergoing branching, and at the angle between the ampulla and the UB stalk, a collection of cells from the metanephric cap aggregate to form the renal vesicle. As has been previously noted, Pax2 expression is prominent in these cells that are fated to become the nephron, and the use of antisense to *Pax2* in the explanted kidney can prevent normal aggregation and epithelialization of the MM (Rothenpieler and Dressler, 1993). Shortly after Pax2 expression increases, another transcription factor, Wt1, is detected in the mesonephric tubules and glomeruli and in the caudal portions of the nephrogenic cord (between E9-E11). As the ureteric bud is formed (E11.5), Wt1 expression significantly increases in the surrounding condensed mesenchyme and it becomes highly expressed as the renal vesicle is generated adjacent to the ureteric bud tips (Armstrong et al 1993). A progressive increase in expression of Wt1 during mesenchymal-epithelial transformation in the renal vesicle coincides with downregulation of Pax2 expression, likely due to a direct repressive effect of high levels of Wt1 on *Pax2* gene expression (Ryan et al 1995). During formation of the S shaped body, Wt1 expression is lost in the more distal regions of the early tubule but is maintained in the cells that will eventually become the glomerular podocytes, where it is maintained into the adult kidney.

Genetic deletion of *Wt1* results in an embryonic lethal phenotype in the mouse with loss of formation of the caudal mesonephric tubules (although cranial mesonephric tubules form normally) and failure of UB outgrowth and absence of metanephric induction (Kreidberg et al 1993). Explant studies have shown that even though the earliest markers of metanephric development are expressed in the prospective metanephric mesenchyme, there is failure of normal mesenchymal condensation and subsequent mesenchymal–epithelial transformation, and that the cells eventually undergo apoptosis (Donovan et al 1999). The possibility that Pax2 and Wt1 act cooperatively in normal renal development is supported by studies showing that Pax2 can upregulate Wt1 expression and that animals carrying heterozygous mutations in both *Wt1* and *Pax2* exhibit a smaller kidney size than either parental strain due to a marked decrease in renal medullary development (Dehbi et al 1996, Discenza et al 2003).

Apoptosis is a tightly regulated process during development, with expression of anti-apoptotic proteins such as Bcl-2 and Bcl-x(L) inhibiting cell death while proteins such as p53 and Bax promote cell death (Gonzalez-Garcia et al 1994). Bcl-2 has been found to be highly expressed in the metanephric mesenchyme in the regions undergoing condensation to form the renal vesicle, with lower expression in the uninduced mesenchyme and collecting system where apoptosis is more prominent (Chandler et al 1994). Mice lacking Bcl-2 expression demonstrate increased rates of apoptosis in the metanephric mesenchyme, reduced numbers of forming nephrons, and development of multi-cystic kidneys with eventual death due to renal failure (Sorenson et al 1995). Expression of Wt1 has been found to coincide with Bcl-2 in the developing kidney, and Wt1 can directly inhibit apoptosis in vitro by upregulating the transcription of Bcl-2 (Mayo et al 1999). In addition, Wt1 can directly interact with the pro-apoptotic protein p53. This interaction results in the stabilization of p53 protein levels, but prevents p53-induced apoptosis (Maheswaran et al 1995). Thus expression of Wt1 appears to be a fundamentally important in the determination of cell survival in both the nephric duct and metanephric mesenchyme during kidney development.

Analysis of the effects of Wt1 on gene transcription demonstrate that it acts primarily as a transcriptional repressor (downregulating Pax2 and Egf receptor expression, for example (Scharnhorst et al 2001)), although, as noted above, it can also activate transcription in some genes (such as Bcl-2 and the Egf family member amphiregulin (Lee et al 1999)). Its role as a tumor suppressor in the adult kidney is supported by the finding that heterozygous mutations of *WT1* in humans result in the development of nephroblastomas derived from regions of undifferentiated mesenchyme, commonly called Wilms tumors (Discenza & Pelletier 2004, Haber et al 1990). These tumors can occur either sporadically or as a part of several syndromes including WAGR and the Denys Drash syndrome (DDS). For a more complete review of the genetic syndromes associated with mutations in *WT1*, the reader is referred to recent reviews by Bates (2000) and Discenza and Pelletier (2004) and to Chapter 31. WAGR (Wilms' tumor, aniridia, genitourinary

abnormalities, and mental retardation) is caused by genetic defects in both *WT1* (resulting in the development of Wilms' tumors and genitourinary abnormalities) and *PAX6* (resulting in aniridia) (Baird et al 1992, Brown et al 1992). The Denys Drash syndrome has been shown to be due to mutations in the zinc finger DNA binding region of *WT1* on one allele of the *WT1* gene that results in the loss of binding of the mutant protein (Pelletier et al 1991) as well as a dominant negative effect on DNA binding by the wild-type protein (Moffett et al 1995). This can lead to a more severe phenotype with sexual ambiguity, a higher incidence of Wilms' tumors, and diffuse mesangial sclerosis resulting in the development of end-stage renal disease.

A second population of cells can also be identified within the metanephric mesenchyme, the stromal cells. These cells have been characterized by their expression of the Foxd1 (also called Bf-2) transcription factor and are primarily seen at the periphery of the metanephric kidney and in the interstitium between the developing renal vesicles (Hatini et al 1996). In support of an important role of the stromal cells in normal kidney development, disruption of *Foxd1* expression results in perinatal death due to renal failure. These animals display significant abnormalities in both the collecting system and nephron formation, including a decrease in UB branching as well as a specific defect in mesenchymal–epithelial transformation resulting in the retention of large collections of condensed mesenchyme (Hatini et al 1996). Experimental approaches to identify the factors that regulate condensation and mesenchymal–epithelial transformation of the metanephric mesenchyme have suggested that both secreted factors as well as cell–cell and cell–matrix interactions play important roles in this process.

Cell–matrix interactions

As the pretubular aggregates begin to form in the condensing metanephric mesenchyme, the composition of the surrounding extracellular matrix undergoes a shift from expression of predominantly collagen I, collagen III, and fibronectin (typical components of the interstitial matrix) to expression of collagen IV, laminin-1, and laminin-10, matrix components present in the adult tubular and/or glomerular basement membranes (Figure 23.7 (Ekblom et al 1980, 1981), reviewed in Kanwar et al (2004) and Miner (1999)). As noted previously, loss of expression of the $\alpha5$ chain of laminin-10 resulted in defects in UB branching and glomerular formation, but no abnormalities in mesenchymal–epithelial transition (Miner & Li 2000). However, the appearance of expression of the laminin $\alpha1$ chain in the renal vesicle at the time of mesenchymal–epithelial transition led to the proposal that laminin-1 (comprised of the $\alpha1\beta1\gamma1$ chains) expression might be important for nephron formation, and indeed neutralizing antibodies directed against the laminin $\alpha1$ subunit have been found to prevent mesenchymal–epithelial transition in kidney explant culture (Klein et al 1988). Nidogen (also known as entactin) is another matrix component that is expressed in the

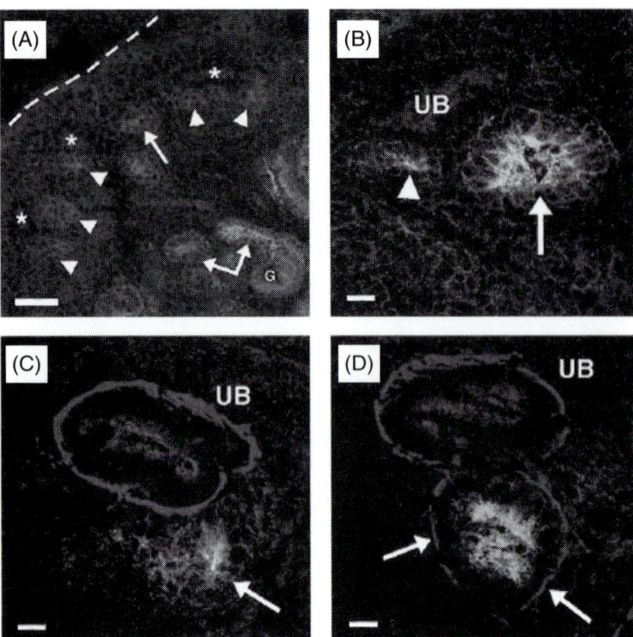

FIGURE 23.7 Mesenchymal–epithelial transformation of the renal vesicle is accompanied by upregulation of cadherin-6 and expression of laminin in the basement membrane. (A, B) Immunostaining for R-cadherin (green) and cadherin-6 (red) demonstrates primarily R-cadherin expression in the early mesenchymal aggregates (arrowheads) adjacent to the ureteric bud tips (asterisks in A, UB in B), whereas regions that have undergone MET to form tubules express increasing amounts of cadherin-6 (double-headed arrow in A, arrow in B; yellow indicates dual expression, red indicates primarily cadherin-6). G, glomerulus; dashed line, edge of the kidney. (C, D) Immunostaining the region of the ureteric bud tips for R-cadherin (green), Zo-1 (red), and laminin (blue) demonstrates that the ureteric bud (UB) is fully epithelialized with a laminin-containing basement membrane (blue ring in C and D) and Zo-1 expression at the apical cell junctions, while a newly formed mesenchymal aggregate (arrow in C) has no distinct basement membrane, and expresses primarily R-cadherin (green cells in C) with some areas of co-expression of R-cadherin and Zo-1 (yellow cells in C). In a vesicle undergoing mesenchymal–epithelial transformation (arrows in D), the laminin-containing basement membrane is detectable and Zo-1 expression is upregulated resulting in most of the cells appearing yellow. Used with permission from Mah et al 2000. (see also Plate 34)

metanephric mesenchyme as well as in the comma and S-shaped bodies. Nidogen binds to the $\gamma1$ chain of laminin-1 and can thus act to cross-link laminin-1 with other matrix proteins such as collagen IV and perlecan, a function that is believed to be required for normal basement membrane assembly. The possible importance of this cross-linking effect for normal tubule formation has been supported by evidence that neutralizing antibodies directed against nidogen prevent mesenchymal–epithelial transition of the metanephric mesenchyme in organ culture (Ekblom et al 1994), while the transgenic expression of a laminin $\gamma1$ chain that lacks the nidogen binding site results in renal

agenesis (Willem et al 2002). Despite these observations, the selective loss of nidogen-1 does not interrupt normal kidney formation, possibly due to redundant capacities of the other nidogen isoforms (Murshed et al 2000).

Secreted molecules that regulate mesenchymal–epithelial transition

As previously noted, Wnt proteins are secreted molecules that bind to heparan sulfate moieties in the extracellular matrix and/or on the cell surface. Wnt9b is secreted by the UB stalk cells and appears to act in a paracrine fashion on the mesenchymal cells near the tips of the UB (Carroll et al 2005). In contrast Wnt-4 expression is first seen in the mesenchymal cells adjacent to the UB as pre-tubular aggregates are beginning to form (Stark et al 1994). The expression of Wnt4 and Pax8 in these nascent structures is absent in the Wnt9b null embryos, correlating with loss of mesenchymal-epithelial transformation and complete failure of nephrogenesis (Carroll et al 2005). During subsequent rounds of UB branching, Wnt-4 is consistently detected in the mesenchyme near the tips, and in the comma shaped bodies, but is lost once fusion of the S-shaped body with the UB occurs. In mice expressing a targeted disruption of the *Wnt-4* gene, initial ureteric bud formation and branching appears to occur normally, but there is failure of formation of the pre-tubular aggregates, leading to the complete absence of comma or S-shaped bodies and therefore loss of nephron formation (Stark et al 1994). In these animals, the expression of both Wt-1 and Pax-2 in the metanephric mesenchyme was detected until E14.5, suggesting that induction of the metanephric mesenchyme was intact and that the failure to form tubules was due to a loss of mesenchymal-epithelial transformation. Wnt-6 is also expressed in the ureteric bud, and studies using co-culture of cells transgenically expressing Wnt-6 have demonstrated that this Wnt family member can also induce mesenchymal-epithelial transition (Itaranta et al 2002). In contrast, in vitro studies suggest that Wnt-2b, although expressed in the stromal cells surrounding the metanephric mesenchyme, does not appear to be critical for mesenchymal-epithelial transition and tubule formation, but rather stimulates ureteric bud branching (Lin et al 2001a).

The mechanism whereby Wnt-4 signaling regulates mesenchymal–epithelial transition is presently unknown. The classic Wnt signaling pathway involves binding of Wnt to a member of the Frizzled family of transmembrane receptors, resulting in activation of the cytosolic protein Dishevelled. Dishevelled, by inhibiting a second protein called glycogen synthase kinase-3β (GSK-3β), causes an increase in the cytosolic levels of β-catenin. This results in the translocation of β-catenin to the nucleus where it regulates the activity of the TCF/Lef family of transcription factors. This could result in proliferation of the cells of the pre-tubular aggregates, prevention of their apoptosis, or direct upregulation and/or stabilization of proteins involved in epithelial transformation such as cadherins. In vitro studies using cell lines transgenically expressing Wnt family members have argued for roles of both the classic Wnt signaling pathways and non β-catenin dependent Wnt signaling in the regulation of mesenchymal–epithelial transition and tubule formation (Kispert et al 1998, Lyons et al 2004).

Similar to the dual role of Wnt proteins in both UB expansion and nephron induction in the metanephric mesenchyme, several other growth factor families including the Fgfs and the Bmps have been found to participate in both processes. Expression of a soluble form of the Fgf receptor that can bind multiple Fgf family members and prevent normal cell surface receptor activation resulted in renal agenesis due to failure of both UB branching and mesenchymal development (Celli et al 1998). It appears likely that these growth factors act primarily to promote growth and prevent apoptosis of the metanephric mesenchyme, as neither Bmps nor Fgfs have been shown to directly regulate mesenchymal–epithelial transition (Dudley et al 1995, 1999).

Studies utilizing the approach of purifying and identifying factors secreted by the ureteric bud that can stimulate mesenchymal–epithelial transition and tubule formation in explanted metanephric mesenchyme have also identified lipocalin-2, leukemia inhibitory factor (Lif), and Timp-2 as potentially playing significant roles in this process (Barasch et al 1999a, 1999b, Yang et al 2002). Lipocalin-2 (also called Ngal or Sip-24) has been identified as a siderophore-binding protein that can mediate iron uptake and thereby regulate iron-dependent gene transcription (Yang et al 2002). It is proposed that lipocalin-2 plays a role in the early process of cell determination prior to actual mesenchymal–epithelial transition. Lif is a member of the interleukin-6 (Il-6) family of cytokines that acts via the gp130 cell surface receptor. Activation of this receptor by any of the Il-6 family members can result in activation of the signal transduction and activation of transcription (Stat) family member Stat-3. The role of this pathway in mesenchymal–epithelial transition appears to be complex since addition of Lif to the intact explanted kidney can result in inhibition of mesenchymal–epithelial transition (Bard & Ross 1991), whereas addition to isolated metanephric mesenchyme that has been pre-treated with other growth factors such as Fgf2 can stimulate mesenchymal–epithelial transition (Barasch et al 1999b).

Cell–cell interactions

The role of cell–cell interactions in defining mesenchymal–epithelial transition and tubule formation is presently less well understood. One of the major components of the cell–cell adherens junction in epithelial cells are the cadherins. There are multiple members of the cadherin family, and in the developing kidney it appears that the expression of specific cadherins is temporally regulated and is predictive of both mesenchymal–epithelial transformation and

tubule patterning. During the early stages of mesenchymal–epithelial transformation, cadherin-11 expression is lost, followed by the sequential upregulation of R-cadherin and cadherin-6 expression (Cho et al 1998). Genetic interruption of *Cadherin-6* results in a moderate defect in mesenchymal–epithelial transformation and in failure of some S-shaped bodies to fuse with the ureteric bud tip, leading to a reduced number of total nephrons (Mah et al 2000). P-cadherin is highly expressed in the bottom of the S-shaped body where glomerular differentiation will occur, while cadherin-6 is expressed more in the region of the S-shaped body that is fated to become the proximal tubule and E-cadherin is initially expressed to the highest degree in the distal parts of the nephron (Cho et al 1998). Kidney specific cadherin (Ksp-cadherin), a cadherin family member that is expressed almost exclusively in the kidney, appears very late compared to the other cadherins, possibly arguing for a role of this protein in terminal differentiation of tubular cells (Thomson & Aronson 1999).

Nephron Segmentation (Tubule Patterning)

The factors that regulate the patterning and morphology of the developing tubule are presently the least well understood, although recent studies have implicated several transcription regulators in this process, including Brn1, Notch, and Hnf1α. Brn1, also known as Pou3f3, is a member of the Pou family of transcription factors that have been shown to have important roles in the development of the brain. In addition to brain, Brn1 is expressed in the developing kidney, with expression appearing at the time of renal vesicle formation and becoming restricted to the cells of the loop of Henle, macula densa, and distal convoluted tubule (Figure 23.8; Nakai et al 2003). Mice lacking expression of Brn1 die at 1–2 days post-partum due to renal dysfunction. Examination of the kidneys from these mice revealed that although the primitive Henle's loop is detected, its elongation to form the immature loop is severely retarded, correlating with the loss of expression of TAL specific markers such as Umod and Nkcc2 (the furosemide-sensitive sodium, potassium, chloride cotransporter). Similarly, differentiation of the S-shaped body to form the macula densa and distal convoluted tubule was also defective. The cells of the primitive Henle's loop from *Brn1-/-* mice were found to have decreased rates of proliferation and increased rates of apoptosis as compared to heterozygous littermates. Interestingly, the decrease in expression of TAL specific markers preceded the induction of apoptosis, suggesting that Brn1 may directly regulate expression of distal tubule specific markers, and that in the absence of expression of these markers the cells are induced to undergo programmed cell death.

A second family of transcription factors that appear to play a role in regulating cell survival in the developing tubule are Ap-2α and Ap-2β. Both factors are expressed in an overlapping pattern during the earliest stages of tubule

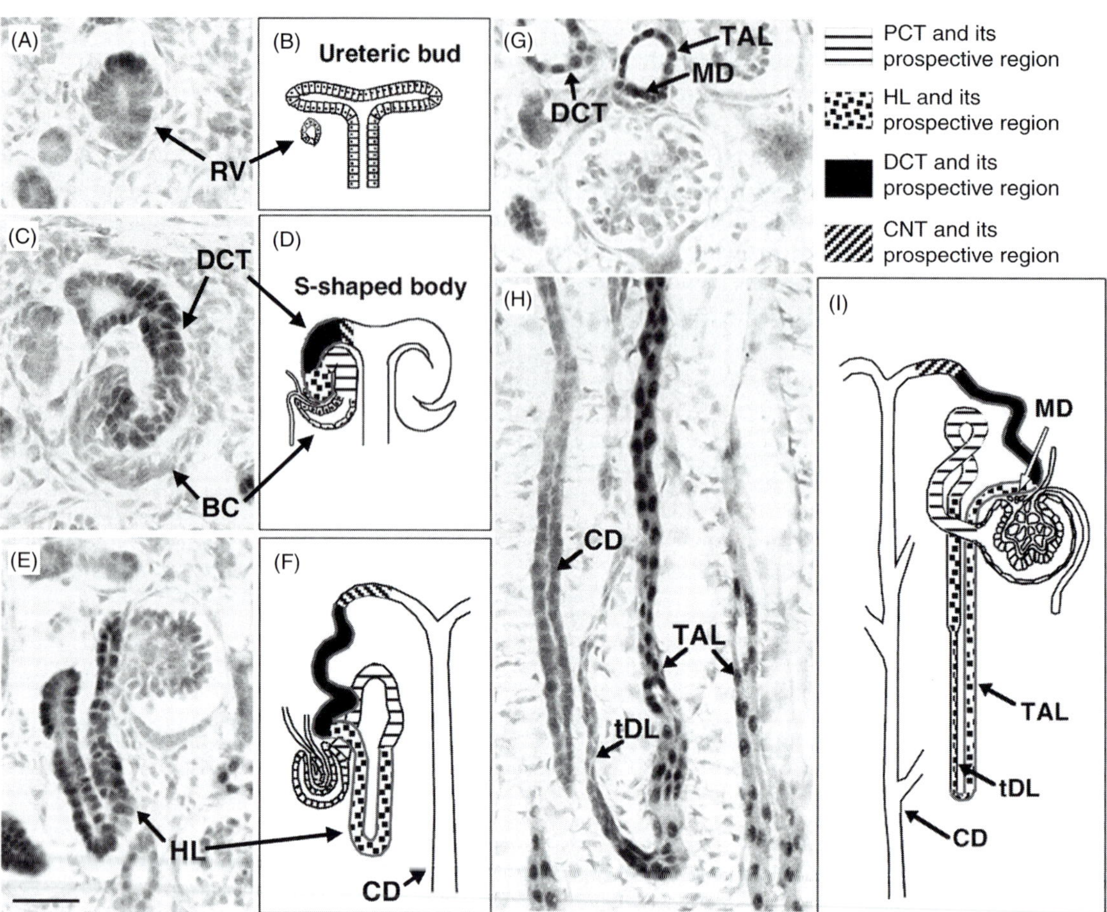

FIGURE 23.8 Expression of Brn1 marks the loop of Henle and distal convoluted tubule during nephrogenesis. Brn1 immunostaining is first detected in a restricted cell population at the time of mesenchymal–epithelial transition in the early renal vesicle (RV; A, B), and then in the connecting segment, as well as the progenitors of the loop of Henle and the distal convoluted tubule in the S-shaped body (DCT; C, D). Brn1 expression is maintained as the loop of Henle elongates (HL; E, F) and is detected in the macula densa (MD), thick ascending limb (TAL), and distal convoluted tubule (DCT) in the newborn (G–I). Brn1-positive regions of the nephron are indicated by a red outline in the schematic drawings (B, D, F, I). BC, Bowman's capsule; PCT, proximal convoluted tubule; CNT, connecting tubule; tDL, thin descending limb. Scale bar: 20 μm. Modified, with permission, from Nakai et al (2003)

development, but by E15 Ap-2α is primarily seen in the cells of the primitive proximal tubule whereas Ap-2β is confined to the distal tubule segments and collecting ducts (Moser et al 1997b). Mice in which *Ap-2β* has been genetically inactivated show normal initial mesenchymal–epithelial transition and tubule formation, but die soon after birth due to massive apoptosis in the distal tubules and collecting ducts, correlating with a downregulation of the expression of the anti-apoptotic factors Bcl-2 and Bcl-x(L) (Moser et al 1997a).

As noted previously, cadherin expression patterns suggest that these proteins are likely to participate in patterning of the tubule. In addition, another class of cell–cell interacting molecules, the Notch receptors and their intercellular ligands Delta-like (Dll) and Jagged (Jag), have been found to be expressed in a manner that suggests a role in tubule patterning. Notch2 is expressed on epithelial cells throughout the developing nephron (including the primitive

podocyte), whereas Notch1 is seen on proximal and distal tubule precursor cells but not on the cells fated to become the podocytes (Leimeister et al 2003). In this same study, the expression of the Notch ligand Dll1 was found to be restricted to the proximal tubule progenitors, whereas a second ligand, Jag1, was expressed on both proximal tubule and glomerular precursors. In contrast, Dll3, along with the Notch4 receptor, were expressed exclusively in the vascular endothelial cells. The binding of the Notch receptor to Jag or Dll on adjacent cells results in the cleavage of the intracellular carboxy terminus of Notch by a membrane associated protease complex called γ-secretase. This cleavage allows the carboxy terminal fragment of the receptor to translocate to the nucleus where it regulates gene transcription events, and is critical for the normal signaling of the receptor. Inhibition of γ-secretase activity in the explanted mouse kidney does not appear to markedly alter UB branching or renal vesicle formation, but did result in a significant reduction in both proximal tubule and glomerular formation (Cheng et al 2003). Interestingly, there is significant cross-talk between the Wnt and Notch signaling pathways that is likely to add a further layer of complexity to the developmental regulation of nephrogenesis. Activation of the Wnt-dependent transcription factor Lef1 has been shown to induce the expression of the Notch ligand Dll1 (Galceran et al 2004), while Wnt-dependent inhibition of Gsk3β can result in increased Notch mediated gene transcription (Espinosa et al 2003). Furthermore, presenilin, a critical component of the γ-secretase complex, has been found to act as a scaffold for β-catenin phosphorylation and thereby downregulate β-catenin nuclear signaling independent of the canonical Wnt pathway (Kang et al 2002).

The final aspects of proximal tubule differentiation appear to be regulated by a separate group of transcription factors, including hepatocyte nuclear factor 1α and 1β (Hnf1α and Hnf1β). Hnf1α and 1β are homeoproteins involved in the transcriptional regulation of several liver-specific genes, and are expressed in both liver and kidney. Hnf1β is expressed during the earliest stages of renal vesicle formation as mesenchymal–epithelial transition occurs, and persists in the developing proximal and distal tubules as well as the collecting duct (Lazzaro et al 1992). In contrast, Hnf1α is expressed later as nephron segmentation is occurring, and is initially restricted to the proximal and distal nephron, and later to the proximal nephron (Pontoglio et al 1996). Mice lacking *Hnf1α* gene expression died after a progressive wasting syndrome with marked liver enlargement (Pontoglio et al 1996). Interestingly, although renal tubular development was judged to be normal morphologically, examination of the urine from these mice demonstrated the presence of glycosuria, aminoaciduria, and phosphaturia, characteristics of the Fanconi Syndrome of proximal tubule transport dysfunction. Thus it appears that Hnf1α expression may be required for the expression of transporters critical for the fully differentiated phenotype of proximal tubule cells.

Glomerular Formation

The glomerulus forms the filtering unit of the nephron, with the glomerular basement membrane (GBM) as the core structure that separates the blood from the urinary space. The GBM is deposited and supported by two principal cell types, the epithelial podocyte and the glomerular capillary endothelial cell. A third cell type, the mesangial cell, acts both in a contractile manner to regulate glomerular capillary surface area and as a support cell to provide clearance for proteins deposited on or within the GBM. The podocyte is first detected during the earliest stages of formation of the S-shaped body, and is located along the inner layer of the lower (distal) limb of the S (Figure 23.9, reviewed in Kreidberg 2003, Quaggin 2002). The apical aspect of these cells is facing down into the future urinary space while the basal surface (at the top of the cells in the figure) is attached

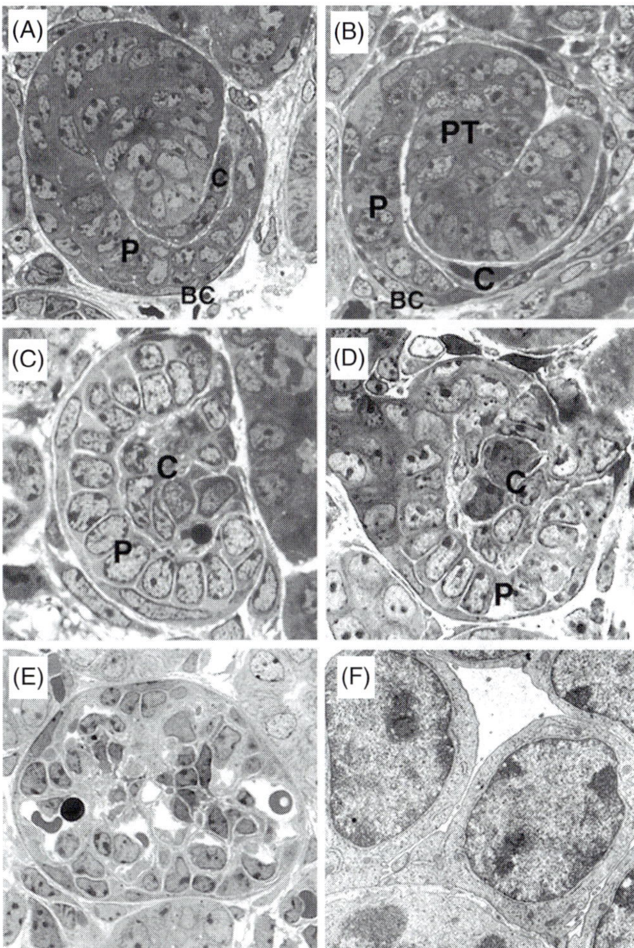

FIGURE 23.9 The developing glomerulus is first detected as a distinct structure when the lower limb of the S-shaped body elongates with phenotypically distinct inner and outer layers (A, B). The cells of the outer layer develop a flattened morphology and will comprise Bowman's capsule (BC), while the cells of the inner layer assume a cuboidal shape and represent the future podocytes (P). At this time, the angioblasts that will form the glomerular capillaries (C) migrate into the cleft between the podocytes and the middle limb of the S-shaped body. Capillary proliferation results in enlargement of the capillary tuft and outward displacement of the podocyte layer to form a cup (C, D). During the capillary loop stage, the podocytes lose their cuboidal shape, migrate around the expanding capillary loops, and begin to assume the unique cell–cell attachments that will eventually mature into the slit diaphragms (E, F). Used with permission from Kreidberg (2003)

to a rudimentary basement membrane that will ultimately become part of the GBM. At this early stage in development, the capillary endothelial cells are just beginning to invade the cleft between the lower and middle limb of the S-shaped body and the mesangial cells are not yet identifiable. As glomerulogenesis progresses, the invading endothelial cells and mesangial cells organize into an early capillary that elongates to form several vascular loops as the podocytes lose their classic columnar shape and normal epithelial cell–cell attachments and migrate to cover the expanded capillary surface (the capillary loop stage). The final stages of glomerular development involve the extension of processes from the podocyte cell body that interdigitate with processes from adjacent podocytes to form unique cell–cell associations that are critical for the ultimate formation of the filtration barrier. These extensions mature into the podocyte foot processes with the cell–cell junction ultimately made up of the proteins of the slit diaphragm. The development of each of these cell types is interdependent, although experimental models that disrupt podocyte differentiation suggest that this cell may play a key role in orchestrating the architectural development of the mesangial and endothelial cells.

PODOCYTE DIFFERENTIATION

As previously noted, Wt1 expression, which is initially detected in all of the cells of the renal vesicle, becomes progressively restricted to the visceral epithelial layer of the distal limb of the S-shaped body and thus serves as one of the earliest markers of the immature podocyte. Complete disruption of *Wt1* expression results in loss of all elements of nephron formation, including the podocyte. However, studies involving transgenic over-expression of Wt1 or loss of expression of only selected Wt1 isoforms also suggest an important role for Wt1 in normal glomerular maturation (Hammes et al 2001, Palmer et al 2001, Natoli et al 2002), including regulation of the expression of the slit diaphragm protein nephrin (Guo et al 2004). Recently, several other transcription factors have been identified as having important roles in the regulation of podocyte differentiation as well. Pod1 is a member of the basic-Helix-Loop-Helix (bHLH) family of transcription factors that is initially expressed in the condensing metanephric mesenchyme, and is then lost until it is again upregulated in the podocyte precursors of the S-shaped body. As previously noted, genetic inactivation of *Pod1* results in multiple defects in kidney development, including decreased ureteric bud branching and delayed mesenchyma–epithelial transformation (Quaggin et al 1999). Examination of glomerulogenesis in these animals reveals that the podocytes initially form, but do not lose their early columnar shape or extend normal numbers of foot processes around the capillary loop. This correlates with failure of the capillary loop to elongate and therefore premature arrest of glomerular formation.

A second transcription factor, MafB (also called Kreisler after the mouse bearing a mutation in this gene), is expressed in the podocyte beginning in the capillary loop stage (around E14.5) and extending into adulthood (Sadl et al 2002). Mice homozygous for a genetic mutation in *MafB* exhibit developmental defects in the hindbrain and kidney. These mice were found to have marked proteinuria and die within 24 hours of birth due to dehydration. Examination of the kidneys revealed normal formation of the capillary loops and a normal glomerular basement membrane, but failure of the podocytes to terminally differentiate. Although they had progressed beyond the columnar stage seen in the *Pod1* mutants, the foot processes had failed to mature as distinct structures, leaving a continuous sheet of fused podocytes without distinct slit diaphragms.

The podocyte foot processes attach to the GBM via classic integrin heterodimers, most notably $\alpha3\beta1$. This transmembrane complex serves as the receptor for laminin in the GBM, and mice lacking $\alpha3$ integrin (and therefore lacking $\alpha3\beta1$) displayed decreased capillary loop formation and failure of foot process formation (Kreidberg et al 1996). In addition, there were fewer podocytes detected, suggesting a role for $\alpha3\beta1$ integrin in mediating podocyte proliferation and/or in preventing apoptotic death of the podocytes. In support of a major role for the podocyte in the formation and organization of the GBM, these mice also displayed a widened and severely disorganized basement membrane.

The formation of the slit diaphragm between the interdigitated foot processes of adjacent podocytes marks the terminally differentiated state of the podocyte. Slit diaphragms provide the final barrier for glomerular filtration and are a critical component of the charge barrier that prevents massive proteinuria. In the last few years, several proteins have been identified as functional components of the slit diaphragm, primarily based on the occurrence of foot process fusion and proteinuria in their absence (reviewed in (Asanuma & Mundel 2003, Ly et al 2004)). These include nephrin, podocin, Cd2-ap, and Neph1-3. Mutations in *NPHS1* (the human gene encoding nephrin) and *NPHS2* (the human gene encoding podocin) are responsible for many of the cases of congenital nephrotic syndrome (see Chapter 6). It appears likely that these proteins exist as part of a large transmembrane complex that both supports the stability of the slit diaphragm and that can signal into the cell. This complex also includes more traditional cell–cell junction proteins including P-cadherin and Zo-1, and appears to require the presence of α-actinin-4, an actin cross-linking protein, since mutations in *ACTN4* can cause foot process fusion and nephrotic syndrome in humans (Kaplan et al 2000).

GLOMERULAR CAPILLARY FORMATION

Within the metanephric mesenchyme are a loose collection of cells, primarily at the margins of tissue and adjacent to the ureteric bud, that are the precursors for the renal vasculature. These cells, called angioblasts, coalesce to form the endothelium of primitive vessels within the metanephros in a process known as vasculogenesis. These primitive capillary networks form around the ureteric bud as well as in the cleft of the lower limb of the S-shaped body at the site of the future glomerulus. Following this de novo formation of capillary networks within the MM, vessels from outside of the kidney appear to invade the mesenchyme (a process known as angiogenesis), resulting in the formation of the permanent connections between the renal microcirculation and the aorta ((Robert et al 1996, 1998), reviewed in Abrahamson et al 1998).

Angioblasts in the E11 metanephric mesenchyme can be identified in the absence of any attached vessel by their expression of the trans-membrane tyrosine kinase receptors Flt1 and Flk1 (Robert et al 2000). These receptors are also known as Vegfr1 and Vegfr2 because they have been identified as the major receptors for vascular endothelial growth factor (Vegf-A), a key regulator of both vasculogenesis and angiogenesis. The immature podocytes that lie adjacent to the cleft in the lower limb of the S-shaped body have been found to secrete Vegf-A, which has been shown to act as a chemoattractant for the migration of endothelial cells. While the major stimulus for Vegf production is believed to be hypoxia, in cultured podocytes it has been found that engagement of integrin $\alpha 3\beta 1$ with laminin can result in transcriptional activation of Vegf-A independent of hypoxia (Datta et al 2004).

Mice lacking Vegf-A or Vegfr2 expression have multiple defects in vasculogenesis and die by E9.5 prior to kidney development (Shalaby et al 1995). However, the importance of Vegf expression for normal endothelial and mesangial cell investment in the glomerular tuft has now been demonstrated using a podocyte-specific knockout of Vegf-A (Eremina et al 2003). Mice that lacked podocyte-mediated Vegf-A expression died within 24 hours of birth due to kidney failure. Examination of glomerular development in these mice revealed that there were markedly fewer endothelial cells present in the capillary loop stage than normal, and no endothelial cells or mesangial cells present in the mature glomerulus (Figure 23.10). The fact that some endothelial cells were present initially, but none was detected in the mature glomerulus, suggests that Vegf is likely to be required for both endothelial cell migration into the cleft as well as endothelial cell proliferation and/or survival. The paucity of endothelial cells resulted in defects in the formation of the GBM (fusion of the endothelial component and epithelial component of the GBM apparently failed to occur), associated with poorly formed podocyte slit diaphragms as well as foot process fusion in many areas.

In addition to Vegfr1 and Vegfr2, another cell surface protein, neuropilin-1 (Np1), can bind Vegf-A and thereby potentiate Vegf-induced activation of Vegfr2. However, neuropilins are also the receptors for a second group of proteins, the semaphorins. The binding of semaphorins secreted by one cell to neuropilins on the surface of nearby cells typically acts in a chemorepulsive manner to prevent cell–cell association, and has been shown to be critical in normal axon guidance. It is likely that semaphorins play a similar role in the organization of renal capillary networks since they are present in the developing kidney and have been shown to inhibit endothelial cell migration (Miao et al 1999). The recent observation that semaphorin-3A and semaphorin-3F are expressed by the immature podocyte (Figure 23.11), while Np-1 and Np-2 are expressed by the invading endothelial cells, suggests that this receptor–ligand complex is likely to specifically be involved in determining the pattern of vascular and podocyte organization in the developing glomerular tuft (Villegas & Tufro 2002).

Another ligand-receptor pair, the Ephs and ephrins, are also likely to play a role in regulating vascular assembly in the developing kidney. The Ephs are transmembrane tyrosine kinase receptors that associate with specific ephrins expressed on the surface of adjacent cells. The Eph receptors can be divided into two classes, EphA1–A8 and EphB1–B6, that interact with ephrin-A1–A5 and ephrin-B1–B3, respectively. The regulated expression of specific Ephs and ephrins on the surface of endothelial cells has been shown to determine which cells assemble into vascular networks, and which are excluded. During development of the kidney, ephrin-B2 has been found to be expressed on the surface of the immature podocytes in the S-shaped body, and later on the invading endothelial and mesangial cells (Takahashi et al 2001), while the ephrin-B2 receptor, EphB4, was found exclusively on the endothelial cells. Thus, it appears that chemoattractive signals from Vegf and chemorepulsive signals from semaphorins serve to guide the invasion of endothelial cells into the vascular cleft, while specific cell–cell interactions such as those mediated by Ephs and ephrins act to regulate the organization of the invading cells into a capillary network.

Angiopoietin, another recently described endothelial growth factor, is expressed in the developing kidney

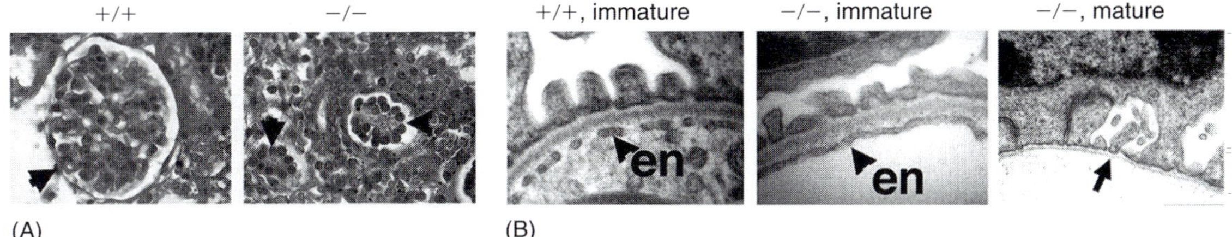

FIGURE 23.10 Endothelial cell development within the glomerulus requires Vegf-A expression by the podocyte. (A) Glomeruli from wild-type mice (+/+) demonstrate normal capillary loops (arrow) while glomeruli from mice lacking Vegf-A expression by the podocyte (−/−) reveal failure of endothelial cell expansion, with a simple layer of immature podocytes surrounding a rudimentary mesangium (arrowhead). (B) Transmission EM reveals a normal basement membrane and capillary endothelial fenestrations in the wild-type glomerulus (+/+, en), whereas endothelial fenestrations are absent in the immature Vegf-A deficient glomerulus (−/−, arrowhead) and the entire endothelial layer is lost in the mature glomerulus (−/−, arrow). Used with permission from Eremina et al (2003). (see also Plate 35)

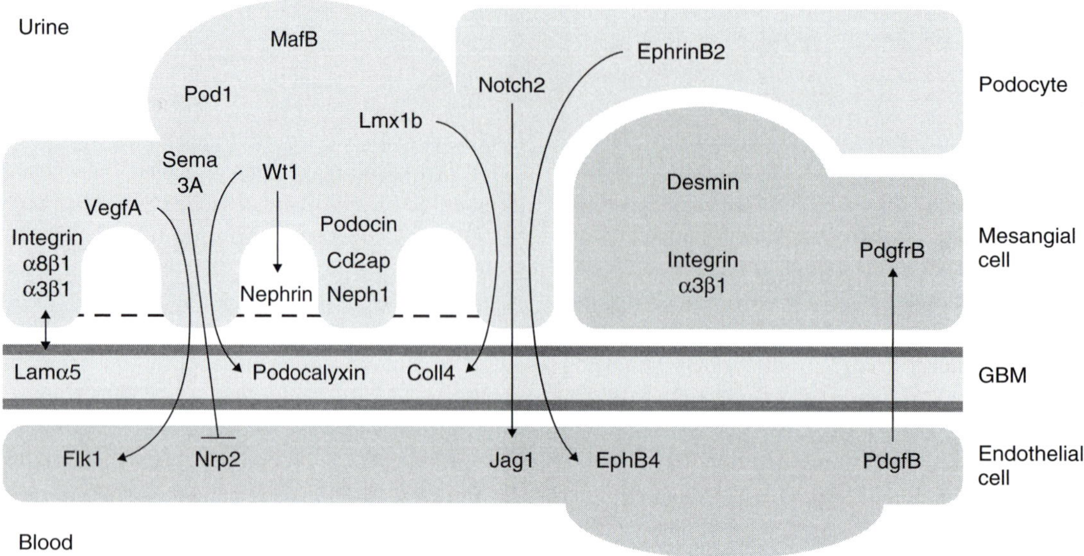

FIGURE 23.11   A schematic depiction of representative pathways involved in the molecular control of podocyte, endothelial and mesangial development in the glomerulus. (see also Plate 36)

as well (Kolatsi-Joannou et al 2001). Both angiopoietin-1 (Ang-1) and angiopoietin-2 (Ang-2) are detected by E12.5 and increase in expression throughout development. The angiopoietin receptors, Tie-1 and Tie-2, are expressed on the metanephric endothelial cells surrounding the ureteric bud, but are either absent or are expressed at very low levels in the endothelial cells that invade the vascular cleft of the S-shaped body (Kolatsi-Joannou et al 2001, Loughna et al 1997). However, Tie-2 is later upregulated in the capillary loop stage, suggesting that activation of angiopoietin signaling is not required for initial endothelial migration and assembly but may be more important in endothelial cell proliferation during capillary maturation. In agreement with this, cultures of explanted kidneys treated with angiopoietin demonstrate increased endothelial cell numbers in the developing glomeruli.

MESANGIAL CELL PARTICIPATION IN GLOMERULOGENESIS

Like endothelial cells, mesangial cells are recruited into the vascular cleft of the developing glomerulus where they play an important role in the organization of the glomerular capillary network. Similar to the capillary pericyte, mesangial cells can promote endothelial cell organization to form a capillary network, as well as having contractile properties that can serve to regulate the capillary surface area that is available for filtration in the mature glomerulus. The origin of glomerular mesangial cells is not fully understood. Most data support the idea that mesangial cells differentiate in situ from precursors present within the metanephric mesenchyme (Hyink et al 1996, Ricono et al 2003), while other studies suggest that bone marrow stem cells can be recruited to differentiate into mesangial cells in the glomerulus (although

whether this takes place during development or only in the adult is unclear) (Ito et al 2001, Masuya et al 2003). Immunostaining for Thy1.1 as a mesangial cell marker and PECAM as an endothelial cell marker has revealed that many of the endothelial cell precursors in the cleft of the S-shaped body co-express Thy1.1, suggesting that these cells may be precursors for either cell type (Ricono et al 2003).

In addition to Thy1.1, mesangial cells have been found to express Vegfr2 (Flk-1) and the receptor for platelet derived growth factor (Pdgf) (Ricono et al 2003). As noted previously, glomeruli from mice that lack podocyte-mediated expression of Vegf-A demonstrate a lack of both endothelial cells and mesangial cells (Eremina et al 2003), although it is unclear whether the absence of mesangial cells is due directly to the loss of Vegf-A signaling to the mesangial cell precursor, or whether the loss of the endothelial cells results in a secondary failure of mesangial cell differentiation. However, mesangial cell recruitment and/or survival does appear to specifically require Pdgf signaling. Pdgf is a well known chemoattractant for smooth muscle cells and fibroblasts, and Pdgf is secreted by the glomerular endothelial cells as well as the mesangial cells themselves (Alpers et al 1992, Lindahl et al 1998). Mice that lack either the B isoform of Pdgf (Pdgf-B) or the Pdgfβ receptor demonstrate normal S-shaped body formation, and develop immature glomeruli containing both podocytes and endothelial cells, but lacking mesangial cells (Leveen et al 1994, Soriano 1994). The capillary bed in these glomeruli fails to mature normally, and is typically comprised of a single widely dilated, blood engorged, capillary loop, albeit with morphologically normal podocytes and endothelial cells.

As previously noted, inhibition of γ-secretase, the enzyme that cleaves Notch receptors and activates their nuclear signaling, inhibits both proximal tubule and glomerular formation (Cheng et al 2003). Examination of the developing kidney reveals that Notch2 is expressed on the immature podocytes in the S-shaped body, whereas the adjacent endothelial and mesenchymal precursors express the notch ligand Jag1, suggesting that Notch2 and Jag1 might interact to regulate endothelial cell and/or mesangial cell maturation in the vascular cleft (McCright et al 2001). Animals that fail to express Notch2 die prior to E11, preventing examination of glomerular development. However, genetic deletion of a region of the extracellular domain of Notch2 (*Notch2^{del1}*) results in the expression of a mutant protein that partially supports development, with *Notch2^{del1}*/*Notch2^{del1}* homozygous embryos surviving until the time of birth (McCright et al 2001). Examination of the developing kidneys in these mice revealed that ureteric bud branching and renal vesicle formation were relatively normal (although there were fewer renal vesicles present than in wild-type littermates), but that glomerular maturation was severely impaired. The early glomerulus expressed both podocytes and endothelial cells, but no mesangial cells were detected. This resulted in a developmental arrest in most glomeruli just prior to the capillary loop stage, with the podocytes and endothelial cells present in a disorganized clump. Some glomeruli progressed further and resembled those seen in the *Pdgf-B* null mouse with one or two dilated, blood engorged, capillary loops that entirely lacked mesangial cells. Analysis of BrdU uptake and TUNEL staining revealed that glomeruli from the *Notch2^{del1}*/*Notch2^{del1}* homozygous mice exhibited both a decrease in cell proliferation and an increase in apoptotic rates, arguing that Notch signaling in the podocyte may be critical for both mesangial cell survival and proliferation.

In addition to Notch2 expression by the podocyte, high levels of Notch3 have also been found in cells at the center of the immature glomerulus, presumably either endothelial cell or mesangial cell precursors (Leimeister et al 2003). Notch3 is known to be expressed on the surface of pericytes and vascular smooth muscle cells where it is believed to play a role in cell survival and therefore in the development and maintenance of the vascular integrity (Claxton & Fruttiger 2004, Shawber & Kitajewski 2004). In contrast, Notch4 has been found to be expressed on the endothelial cells of the developing glomerulus (Leimeister et al 2003). The activation of Notch signaling is well described as a mechanism for the determination of cell fate, making it interesting to speculate that expression of either Notch3 or Notch4 might provide the signal for divergence of the vascular precursor cells into a mesangial or endothelial phenotype, respectively.

One of the putative Notch targets is the transcription factor Cux-1, a mammalian ortholog of the *Drosophilia* cut proteins. Examination of Cux-1 expression in the kidney reveals that it indeed overlaps with notch expression in the immature podocyte, and that renal epithelial cells expressing a constitutively active form of Notch1 have upregulated Cux-1 expression (Sharma et al 2004). Interestingly, Cux-1 expression results in transcriptional repression of several genes, including the cyclin kinase inhibitors p21 and p27^{kip}. Transgenic over-expression of Cux-1 has been shown to result in mesangial cell hyperplasia (Brantley et al 2003), suggesting that the dependence of mesangial cell proliferation on normal Notch signaling may be due, at least in part, to Notch-mediated activation of Cux-1 expression.

FORMATION OF THE GLOMERULAR BASEMENT MEMBRANE
During nephrogenesis, the GBM is believed to form as the result of fusion of the immature podocyte basement membrane and the closely apposed endothelial cell basement membrane. The composition of the nascent GBM undergoes a significant series of changes that roughly correlate with maturation of the glomerulus (reviewed in Kanwar et al 2004, Miner 1999). The initial basement membrane of the comma and S-shaped bodies is composed primarily of laminin-1 ($\alpha1\beta1\gamma1$) and laminin-8 ($\alpha4\beta1\gamma1$), as well as the $\alpha1$ and $\alpha2$ chains of type IV collagen and non-collagenous proteins such as perlecan and nidogen. During transition to the capillary loop stage, the $\alpha1$, $\alpha4$, and $\beta1$ chains of laminin are downregulated, while expression of the laminin $\alpha5$ and $\beta2$ chains is upregulated. Similarly, the $\alpha1$ and $\alpha2$ chains of type IV collagen are replaced by the $\alpha3$, $\alpha4$, and $\alpha5$ chains. This results in a mature GBM that contains laminin-11 ($\alpha5\beta2\gamma1$) along with type IV collagen composed primarily of the $\alpha3$–$\alpha5$ chains. The importance of the expression of the $\alpha5$ subunit of laminin-11 for both the integrity of the GBM, as well as for normal podocyte and endothelial cell arrangement, has been demonstrated in mice in which the *Lama5* gene (which encodes the $\alpha5$ chain) has been genetically inactivated (Miner & Li 2000). These mice exhibit multiple developmental defects resulting in embryonic lethality between E14 and E17, and have a reduction in ureteric bud branching as previously noted. Examination of glomerular development in the *Lama5-/-* mouse revealed that the podocytes formed normally, but failed to maintain a single monolayer along the GBM, while the endothelial cells and mesangial cells failed to maintain contact with the GBM and were lost from the glomerular tuft as the vascular pole underwent constriction in the final stages of glomerulogenesis. The GBM itself was noted to be difficult to detect once the $\alpha1$ chain of laminin-1 was downregulated in the early capillary loop stage. Interestingly, transgenic expression of a chimeric laminin α–chain that is composed of the laminin $\alpha5$ domains I-IV, but the laminin $\alpha1$ G domain, was able to rescue podocyte and GBM maturation, but not normal vascularization of the glomerulus (Kikkawa et al 2003). This was found to be due, at least in part, to the failure of mesangial cells to remain adherent to the hybrid GBM, with specific failure of the mesangial cell $\alpha3\beta1$ integrin to associate with the laminin $\alpha1$ G domain.

The correct collagen expression in the GBM also appears to be required for normal glomerular maturation. Lmx1b is a member of the LIM-homeodomain family of transcription factors, and is expressed in the podocyte beginning in the capillary loop stage. Mice with genetic disruption of *Lmx1b* develop skeletal and renal abnormalities typical of humans with nail-patella syndrome (NPS) (Chen et al 1998), and several mutations of the *LMX1B* gene have now been defined in NPS patients (Dreyer et al 1998). Examination of *Lmx1b-/-* mice revealed glomerular proteinuria with foot process fusion and thickening of the GBM adjacent to the podocytes. A search for potential transcriptional targets of Lmx1b has revealed that the promoter regions of the *Col4a3* and *Col4a4* genes (encoding the α3 and α4 chains of type IV collagen) contain binding sites for Lmx1b, and mice lacking *Lmx1b* expression demonstrate a marked reduction in the expression of both the α3 and α4 chains of type IV collagen in the GBM (Morello et al 2001).

CONCLUSION

Cumulatively, these studies (along with many others that have not been included due to space limitations) confirm our early suspicions that the development of an organ as complex as the kidney requires the precisely regulated expression of a whole host of transcription factors, growth factors, cytokines, cell adhesion molecules, and matrix components. The expression of these proteins must be regulated in both time and space to achieve the goal of inducing nephron formation in sufficient numbers and with the correct architecture and cell types. General themes that have emerged include the observation that the loss of expression of factors that determine cell survival (such as Pax2 and Wt1) typically results in complete failure of kidney development, whereas those that mediate cell morphogenesis and proliferation (such as Bmps) are primarily required for the proper organization of the renal architecture and control of nephron number. Precise determination of the developmental role of many of these factors is complicated by the fact that there appear to be redundant pathways capable of initiating the same general responses, while increasing evidence for multiple levels of cross-talk during the downstream signaling events initiated by these factors suggests an even greater level of complexity than originally thought.

While we have made substantial progress in understanding the molecular control of metanephrogenesis, much remains to be learned. With the emergence of the Cre-Lox system of cell type specific gene disruption, we are now beginning to address the mechanisms regulating the development of specific tubule segments as well as the guidance of the architectural arrangement of the tubule and glomerulus. These studies are of particular interest since understanding how these processes are regulated developmentally is likely to provide us with new clues as to how to therapeutically intervene in cases of tubular or glomerular dysfunction. Furthermore, investigation of the pluripotency of mesenchymal cells for differentiation into the varied cell types that populate the tubule segments will provide a better understanding of whether a renal stem cell exists, and what role it might play in the adult organism.

References

Abdelhak S, Kalatzis V, Heilig R, et al. A human homologue of the *Drosophila* eyes absent gene underlies branchio-oto-renal (BOR) syndrome and identifies a novel gene family. Nat. Genet. 1997; 15: 157–64.

Abrahamson DR, Robert B, Hyink DP, St John PL, Daniel TO. Origins and formation of microvasculature in the developing kidney. Kidney Int. 1998 (Suppl 67): S7–11.

Akiyama H, Chaboissier MC, Behringer RR, et al. Essential role of Sox9 in the pathway that controls formation of cardiac valves and septa. Proc. Natl Acad. Sci. USA 2004; 101: 6502–7.

al-Awqati Q, Goldberg MR. Architectural patterns in branching morphogenesis in the kidney. Kidney Int. 1998; 54: 1832–42.

Alpers CE, Seifert RA, Hudkins KL, Johnson RJ, Bowen-Pope DF. Developmental patterns of PDGF B-chain, PDGF-receptor, and alpha-actin expression in human glomerulogenesis. Kidney Int. 1992; 42: 390–9.

Amet LE, Lauri SE, Hienola A, et al. Enhanced hippocampal long-term potentiation in mice lacking heparin-binding growth-associated molecule. Mol. Cell Neurosci. 2001; 17: 1014–24.

Andrews KL, Betsuyaku T, Rogers S, Shipley JM, Senior RM, Miner JH. Gelatinase B. (MMP-9) is not essential in the normal kidney and does not influence progression of renal disease in a mouse model of Alport syndrome. Am. J. Pathol. 2000; 157: 303–11.

Armstrong JF, Pritchard-Jones K, Bickmore WA, Hastie ND, Bard JB. The expression of the Wilms' tumour gene, WT1, in the developing mammalian embryo. Mech. Dev. 1993; 40: 85–97.

Asanuma K, Mundel P. The role of podocytes in glomerular pathobiology. Clin. Exp. Nephrol. 2003; 7: 255–9.

Avsian-Kretchmer O, Hsueh AJ. Comparative genomic analysis of the eight-membered ring cystine knot-containing bone morphogenetic protein antagonists. Mol. Endocrinol. 2004; 18: 1–12.

Baird PN, Groves N, Haber DA, Housman DE, Cowell JK. Identification of mutations in the WT1 gene in tumours from patients with the WAGR syndrome. Oncogene 1992; 7: 2141–9.

Barasch J, Yang J, Qiao J, et al. Tissue inhibitor of metalloproteinase-2 stimulates mesenchymal growth and regulates epithelial branching during morphogenesis of the rat metanephros. J. Clin. Invest. 1999a; 103: 1299–307.

Barasch J, Yang J, Ware CB, et al. Mesenchymal to epithelial conversion in rat metanephros is induced by LIF. Cell 1999b; 99: 377–86.

Bard JB, Ross AS. LIF, the ES-cell inhibition factor, reversibly blocks nephrogenesis in cultured mouse kidney rudiments. Development 1991; 113: 193–8.

Barnes JD, Crosby JL, Jones CM, Wright CV, Hogan BL. Embryonic expression of Lim-1, the mouse homolog of Xenopus Xlim-1, suggests a role in lateral mesoderm differentiation and neurogenesis. Dev. Biol. 1994; 161: 168–78.

Bates CM. Kidney development: regulatory molecules crucial to both mice and men. Mol. Genet. Metab. 2000; 71: 391–6.

Batourina E, Gim S, Bello N, et al. Vitamin A controls epithelial/mesenchymal interactions through Ret expression. Nat. Genet. 2001; 27: 74–8.

Bernfield M, Hinkes MT, Gallo RL. Developmental expression of the syndecans: possible function and regulation. Dev. Suppl. 1993: 205–12.

Bladt F, Riethmacher D, Isenmann S, Aguzzi A, Birchmeier C. Essential role for the c-met receptor in the migration of myogenic precursor cells into the limb bud. Nature 1995; 376: 768–71.

Bouchard M, Souabni A, Mandler M, Neubuser A, Busslinger M. Nephric lineage specification by Pax2 and Pax8. Genes Dev. 2002; 16: 2958–70.

Brandenberger R, Schmidt A, Linton J, et al. Identification and characterization of a novel extracellular matrix protein nephronectin that is associated with integrin alpha8beta1 in the embryonic kidney. J. Cell Biol. 2001; 154: 447–58.

Brantley JG, Sharma M, Alcalay NI, Heuvel GB. Cux-1 transgenic mice develop glomerulosclerosis and interstitial fibrosis. Kidney Int. 2003; 63: 1240–8.

Brodbeck S, Besenbeck B, Englert C. The transcription factor Six2 activates expression of the Gdnf gene as well as its own promoter. Mech. Dev. 2004; 121: 1211–22.

Brodbeck S, Englert C. Genetic determination of nephrogenesis: the Pax/Eya/Six gene network. Pediatr. Nephrol. 2004; 19: 249–55.

Brophy PD, Ostrom L, Lang KM, Dressler GR. Regulation of ureteric bud outgrowth by Pax2-dependent activation of the glial derived neurotrophic factor gene. Development 2001; 128: 4747–56.

Brown KW, Watson JE, Poirier V, Mott MG, Berry PJ, Maitland NJ. Inactivation of the remaining allele of the WT1 gene in a Wilms' tumour from a WAGR patient. Oncogene 1992; 7: 763–8.

Brunet-de Carvalho N, Raulais D, Rauvala H, Souttou B, Vigny M. HB-GAM/Pleiotrophin and Midkine are differently expressed and distributed during retinoic acid-induced neural differentiation of P19 cells. Growth Factors 2003; 21: 139–49.

Bullock S, Fletcher JM, Beddington RS, Wilson VA. Renal agenesis in mice homozygous for a gene trap mutation in the gene encoding heparan sulfate 2-sulfotransferase. Genes Dev. 1998; 12: 1894–906.

Cacalano G, Farinas I, Wang LC, et al. GFRalpha1 is an essential receptor component for GDNF in the developing nervous system and kidney. Neuron 1998; 21: 53–62.

Cano-Gauci DF, Song HH, Yang H, et al. Glypican-3-deficient mice exhibit developmental overgrowth and some of the abnormalities typical of Simpson-Golabi-Behmel syndrome. J. Cell Biol. 1999; 146: 255–64.

Cantley LG, Barros EJG, Gandhi M, Rauchman M, Nigam SK. Regulation of mitogenesis, motogenesis and tubulogenesis by hepatocyte growth factor in renal collecting duct cells. Am. J. Physiol. 1994; 267: F271–80.

Carroll TJ, Park JS, Hayashi S, Majumdar A, McMahon AP. Wnt9b plays a central role in the regulation of mesenchymal to epithelial transitions underlying organogenesis of the mammalian urogenital system. Dev. Cell 2005; 9: 283–292.

Carroll TJ, Vize PD. Synergism between Pax-8 and lim-1 in embryonic kidney development. Dev. Biol. 1999; 214: 46–59.

Celli G, LaRochelle WJ, Mackem S, Sharp R, Merlino G. Soluble dominant-negative receptor uncovers essential roles for fibroblast growth factors in multi-organ induction and patterning. EMBO J 1998; 17: 1642–55.

Chandler D, el-Naggar AK, Brisbay S, Redline RW, McDonnell TJ. Apoptosis and expression of the bcl-2 proto-oncogene in the fetal and adult human kidney: evidence for the contribution of bcl-2 expression to renal carcinogenesis. Hum. Pathol. 1994; 25: 789–96.

Chen H, Lun Y, Ovchinnikov D, et al. Limb and kidney defects in Lmx1b mutant mice suggest an involvement of LMX1B in human nail patella syndrome. Nat. Genet. 1998; 19: 51–5.

Cheng HT, Miner JH, Lin M, Tansey MG, Roth K, Kopan R. Gamma-secretase activity is dispensable for mesenchyme-to-epithelium transition but required for podocyte and proximal tubule formation in developing mouse kidney. Development 2003; 130: 5031–42.

Cho EA, Patterson LT, Brookhiser WT, Mah S, Kintner C, Dressler GR. Differential expression and function of cadherin-6 during renal epithelium development. Development 1998; 125: 803–12.

Clarke JC, Patel SR, Raymond RM Jr, et al. Regulation of c-Ret in the developing kidney is response to Pax2 gene dosage. Hum. Mol. Genet. 2006; 15: 3420–8.

Claxton S, Fruttiger M. Periodic Delta-like 4 expression in developing retinal arteries. Gene Expr. Patterns 2004; 5: 123–7.

Coles HS, Burne JF, Raff MC. Large-scale normal cell death in the developing rat kidney and its reduction by epidermal growth factor. Development 1993; 118: 777–84.

Croisille Y, Gumpel-Pinot M, Martin C. Differentiation of secretory tubes of avian kidney: effect of heterogenous inducers. C. R. Acad. Sci. Hebd. Seances Acad. Sci. D. 1976; 282: 1987–90.

Datta K, Li J, Karumanchi SA, Wang E, Rondeau E, Mukhopadhyay D. Regulation of vascular permeability factor/vascular endothelial growth factor (VPF/VEGF-A) expression in podocytes. Kidney Int. 2004; 66: 1471–8.

Davies J, Lyon M, Gallagher J, Garrod D. Sulphated proteoglycan is required for collecting duct growth and branching but not nephron formation during kidney development. Development 1995; 121: 1507–17.

de Graaff E, Srinivas S, Kilkenny C, et al. Differential activities of the RET tyrosine kinase receptor isoforms during mammalian embryogenesis. Genes Dev. 2001; 15: 2433–44.

Defrances MC, Wolf HK, Michalopoulos GK, Zarnegar R. The presence of hepatocyte growth factor in the developing rat. Development 1992; 116: 387–95.

Dehbi M, Ghahremani M, Lechner M, Dressler G, Pelletier J. The paired-box transcription factor, PAX2, positively modulates expression of the Wilms' tumor suppressor gene (WT1). Oncogene 1996; 13: 447–53.

Discenza MT, He S, Lee TH, et al. WT1 is a modifier of the Pax2 mutant phenotype: cooperation and interaction between WT1 and Pax2. Oncogene 2003; 22: 8145–55.

Discenza MT, Pelletier J. Insights into the physiological role of WT1 from studies of genetically modified mice. Physiol. Genomics 2004; 16: 287–300.

Donovan MJ, Natoli TA, Sainio K, et al. Initial differentiation of the metanephric mesenchyme is independent of WT1 and the ureteric bud. Dev. Genet. 1999; 24: 252–62.

Dorup J, Maunsbach AB. The ultrastructural development of distal nephron segments in the human fetal kidney. Anat. Embryol. (Berl) 1982; 164: 19–41.

Dressler GR, Deutsch U, Chowdhury K, Nornes HO, Gruss P. Pax2, a new murine paired-box-containing gene and its expression in the developing excretory system. Development 1990; 109: 787–95.

Dressler GR, Wilkinson JE, Rothenpieler UW, Patterson LT, Williams-Simons L, Westphal H. Deregulation of Pax-2 expression in transgenic mice generates severe kidney abnormalities. Nature 1993; 362: 65–7.

Dreyer SD, Zhou G, Baldini A, et al. Mutations in LMX1B cause abnormal skeletal patterning and renal dysplasia in nail patella syndrome. Nat. Genet. 1998; 19: 47–50.

Dudley AT, Godin RE, Robertson EJ. Interaction between FGF and BMP signaling pathways regulates development of metanephric mesenchyme. Genes Dev. 1999; 13: 1601–13.

Dudley AT, Lyons KM, Robertson EJ. A requirement for bone morphogenetic protein-7 during development of the mammalian kidney and eye. Genes Dev. 1995; 9: 2795–807.

Eccles MR, Schimmenti LA. Renal-coloboma syndrome: a multisystem developmental disorder caused by PAX2 mutations. Clin. Genet. 1999; 56: 1–9.

Eccles MR, Wallis LJ, Fidler AE, Spurr NK, Goodfellow PJ, Reeve AE. Expression of the PAX2 gene in human fetal kidney and Wilms' tumor. Cell Growth Differ. 1992; 3: 279–89.

Ekblom P, Alitalo K, Vaheri A, Timpl R, Saxen L. Induction of a basement membrane glycoprotein in embryonic kidney: possible role of laminin in morphogenesis. Proc. Natl Acad. Sci. USA 1980; 77: 485–9.

Ekblom P, Ekblom M, Fecker L, et al. Role of mesenchymal nidogen for epithelial morphogenesis in vitro. Development 1994; 120: 2003–14.

Ekblom P, Lehtonen E, Saxen L, Timpl R. Shift in collagen type as an early response to induction of the metanephric mesenchyme. J. Cell Biol. 1981; 89: 276–83.

Eremina V, Sood M, Haigh J, et al. Glomerular-specific alterations of VEGF-A expression lead to distinct congenital and acquired renal diseases. J. Clin. Invest. 2003; 111: 707–16.

Espinosa L, Ingles-Esteve J, Aguilera C, Bigas A. Phosphorylation by glycogen synthase kinase-3 beta down-regulates Notch activity, a link for Notch and Wnt pathways. J. Biol. Chem. 2003; 278: 32227–35.

Esquela AF, Lee SJ. Regulation of metanephric kidney development by growth/differentiation factor 11. Dev. Biol. 2003; 257: 356–70.

Galceran J, Sustmann C, Hsu SC, Folberth S, Grosschedl R. LEF1-mediated regulation of Delta-like1 links Wnt and Notch signaling in somatogenesis. Genes Dev. 2004; 18: 2718–23.

Gonzalez-Garcia M, Perez-Ballestero R, Ding L, et al. bcl-XL is the major bcl-x mRNA form expressed during murine development and its product localizes to mitochondria. Development 1994; 120: 3033–42.

Grisaru S, Cano-Gauci D, Tee J, Filmus J, Rosenblum ND. Glypican-3 modulates BMP- and FGF-mediated effects during renal branching morphogenesis. Dev. Biol. 2001; 231: 31–46.

Grobstein C. Inductive epitheliomesenchymal interaction in cultured organ rudiments of the mouse. Science 1953; 118: 52–55.

Grobstein C. Trans-filter induction of tubules in mouse metanephrogenic mesenchyme. Exp. Cell Res. 1956; 10: 424–40.

Guo G, Morrison DJ, Licht JD, Quaggin SE. WT1 activates a glomerular-specific enhancer identified from the human nephrin gene. J. Am. Soc. Nephrol 2004; 15: 2851–56.

Haber DA, Buckler AJ, Glaser T, et al. An internal deletion within an 11p13 zinc finger gene contributes to the development of Wilms' tumor. Cell 1990; 61: 1257–69.

Hammes A, Guo JK, Lutsch G, et al. Two splice variants of the Wilms' tumor 1 gene have distinct functions during sex determination and nephron formation. Cell 2001; 106: 319–29.

Hatini V, Huh SO, Herzlinger D, Soares VC, Lai E. Essential role of stromal mesenchyme in kidney morphogenesis revealed by targeted disruption of Winged Helix transcription factor BF-2. Genes Dev. 1996; 10: 1467–78.

Hoy WE, Douglas-Denton RN, Hughson MD, Cass A, Johnson K, Bertram JF. A stereological study of glomerular number and volume: preliminary findings in a multiracial study of kidneys at autopsy. Kidney Int. 2003(Suppl): S31–7.

Hu MC, Piscione TD, Rosenblum ND. Elevated SMAD1/beta-catenin molecular complexes and renal medullary cystic dysplasia in ALK3 transgenic mice. Development 2003; 130: 2753–66.

Hyink DP, Tucker DC, St John PL, et al. Endogenous origin of glomerular endothelial and mesangial cells in grafts of embryonic kidneys. Am. J. Physiol. 1996; 270: F886–99.

Itaranta P, Lin Y, Perasaari J, Roel G, Destree O, Vainio S. Wnt-6 is expressed in the ureter bud and induces kidney tubule development in vitro. Genesis 2002; 32: 259–68.

Ito T, Suzuki A, Imai E, Okabe M, Hori M. Bone marrow is a reservoir of repopulating mesangial cells during glomerular remodeling. J. Am. Soc. Nephrol. 2001; 12: 2625–35.

Itoh T, Ikeda T, Gomi H, Nakao S, Suzuki T, Itohara S. Unaltered secretion of beta-amyloid precursor protein in gelatinase A (matrix metalloproteinase 2)-deficient mice. J. Biol. Chem. 1997; 272: 22389–92.

Jing S, Wen D, Yu Y, et al. GDNF-induced activation of the ret protein tyrosine kinase is mediated by GDNFR-alpha, a novel receptor for GDNF. Cell 1996; 85: 1113–24.

Kang DE, Soriano S, Xia X, et al. Presenilin couples the paired phosphorylation of beta-catenin independent of axin: implications for beta-catenin activation in tumorigenesis. Cell 2002; 110: 751–762.

Kanwar YS, Ota K, Yang Q, et al. Role of membrane-type matrix metalloproteinase 1 (MT-1-MMP), MMP-2, and its inhibitor in nephrogenesis. Am. J. Physiol. 1999; 277: F934–47.

Kanwar YS, Wada J, Lin S, et al. Update of extracellular matrix, its receptors, and cell adhesion molecules in mammalian nephrogenesis. Am. J. Physiol. Renal Physiol. 2004; 286: F202–15.

Kaplan JM, Kim SH, North KN, et al. Mutations in ACTN4, encoding alpha-actinin-4, cause familial focal segmental glomerulosclerosis. Nat. Genet. 2000; 24: 251–6.

Karavanov AA, Karavanova I, Perantoni A, Dawid IB. Expression pattern of the rat Lim-1 homeobox gene suggests a dual role during kidney development. Int. J. Dev. Biol. 1998; 42: 61–6.

Karihaloo A, Kale S, Rosenblum ND, Cantley LG. HGF mediated renal epithelial branching morphogenesis is regulated by Glypican-4 expression. Mol. Cell. Biol. (in press).

Karihaloo A, Karumanchi SA, Barasch J, et al. Endostatin regulates branching morphogenesis of renal epithelial cells and ureteric bud. Proc. Natl Acad. Sci. USA 2001; 98: 12509–14.

Karp SL, Ortiz-Arduan A, Li S, Neilson EG. Epithelial differentiation of metanephric mesenchymal cells after stimulation with hepatocyte growth factor or embryonic spinal cord. Proc. Natl Acad. Sci. USA 1994; 91: 5286–90.

Keller G, Zimmer G, Mall G, Ritz E, Amann K. Nephron number in patients with primary hypertension. N. Engl. J. Med. 2003; 348: 101–8.

Kikkawa Y, Virtanen I, Miner JH. Mesangial cells organize the glomerular capillaries by adhering to the G domain of laminin alpha5 in the glomerular basement membrane. J. Cell Biol. 2003; 161: 187–96.

Kispert A, Vainio S, McMahon AP. Wnt-4 is a mesenchymal signal for epithelial transformation of metanephric mesenchyme in the developing kidney. Development 1998; 125: 4225–34.

Kispert A, Vainio S, Shen L, Rowitch DH, McMahon AP. Proteoglycans are required for maintenance of Wnt-11 expression in the ureter tips. Development 1996; 122: 3627–37.

Klein G, Langegger M, Timpl R, Ekblom P. Role of laminin A chain in the development of epithelial cell polarity. Cell 1988; 55: 331–41.

Kobayashi A, Valerius MT, Mugford JW, Carroll TJ, Self M, Oliver G, McMahon AP. Six2 defines and regulates a multipotent self-renewing nephron progenitor population throughout mammalian kidney development. Cell Stem Cell 2008; 3: 169–181.

Kolatsi-Joannou M, Li XZ, Suda T, Yuan HT, Woolf AS. Expression and potential role of angiopoietins and Tie-2 in early development of the mouse metanephros. Dev. Dyn. 2001; 222: 120–6.

Kolatsi-Joannou M, Moore R, Winyard PJ, Woolf AS. Expression of hepatocyte growth factor/scatter factor and its receptor, MET, suggests roles in human embryonic organogenesis. Pediatr. Res. 1997; 41: 657–65.

Kreidberg JA. Podocyte differentiation and glomerulogenesis. J. Am. Soc. Nephrol. 2003; 14: 806–14.

Kreidberg JA, Sariola H, Loring JM, et al. WT-1 is required for early kidney development. Cell 1993; 74: 679–91.

Kreidberg JA, Donovan MJ, Goldstein SL, et al. Alpha 3 beta 1 integrin has a crucial role in kidney and lung organogenesis. Development 1996; 122: 3537–47.

Kulkarni AB, Huh CG, Becker D, et al. Transforming growth factor beta 1 null mutation in mice causes excessive inflammatory response and early death. Proc. Natl Acad. Sci. USA 1993; 90: 770–4.

Kume T, Deng K, Hogan BL. Murine forkhead/winged helix genes Foxc1 (Mf1) and Foxc2 (Mfh1) are required for the early organogenesis of the kidney and urinary tract. Development 2000; 127: 1387–95.

Lang D, Epstein JA. Sox10 and Pax3 physically interact to mediate activation of a conserved c-RET enhancer. Hum. Mol. Genet. 2003; 12: 937–45.

Lawson KA, Dunn NR, Roelen BA, et al. Bmp4 is required for the generation of primordial germ cells in the mouse embryo. Genes Dev. 1999; 13: 424–36.

Lazzaro D, De Simone V, De Magistris L, Lehtonen E, Cortese R. LFB1 and LFB3 homeoproteins are sequentially expressed during kidney development. Development 1992; 114: 469–79.

Lee SB, Huang K, Palmer R, et al. The Wilms tumor suppressor WT1 encodes a transcriptional activator of amphiregulin. Cell 1999; 98: 663–73.

Leimeister C, Schumacher N, Gessler M. Expression of Notch pathway genes in the embryonic mouse metanephros suggests a role in proximal tubule development. Gene Expr. Patterns 2003; 3: 595–8.

Lelievre-Pegorier M, Vilar J, Ferrier ML, et al. Mild vitamin A deficiency leads to inborn nephron deficit in the rat. Kidney Int. 1998; 54: 1455–62.

Lelongt B, Trugnan G, Murphy G, Ronco PM. Matrix metalloproteinases MMP2 and MMP9 are produced in early stages of kidney morphogenesis but only MMP9 is required for renal organogenesis in vitro. J. Cell Biol. 1997; 136: 1363–73.

Leveen P, Pekny M, Gebre-Medhin S, Swolin B, Larsson E, Betsholtz C. Mice deficient for PDGF B show renal, cardiovascular, and hematological abnormalities. Genes Dev. 1994; 8: 1875–87.

Li X, Oghi KA, Zhang J, et al. Eya protein phosphatase activity regulates Six1-Dach-Eya transcriptional effects in mammalian organogenesis. Nature 2003; 426: 247–54.

Lin Y, Liu A, Zhang S, et al. Induction of ureter branching as a response to Wnt-2b signaling during early kidney organogenesis. Dev. Dyn. 2001a; 222: 26–39.

Lin Y, Zhang S, Rehn M, et al. Induced repatterning of type XVIII collagen expression in ureter bud from kidney to lung type: association with sonic hedgehog and ectopic surfactant protein C. Development 2001b; 128: 1573–85.

Lindahl P, Hellstrom M, Kalen M, et al. Paracrine PDGF-B/PDGF-Rbeta signaling controls mesangial cell development in kidney glomeruli. Development 1998; 125: 3313–22.

Lorenzo MJ, Eng C, Mulligan LM, et al. Multiple mRNA isoforms of the human RET proto-oncogene generated by alternate splicing. Oncogene 1995; 10: 1377–83.

Loughna S, Hardman P, Landels E, Jussila L, Alitalo K, Woolf AS. A molecular and genetic analysis of renal glomerular capillary development. Angiogenesis 1997; 1: 84–101.

Ly J, Alexander M, Quaggin SE. A podocentric view of nephrology. Curr Opin Nephrol Hypertension 2004; 13: 299–305.

Lyons JP, Mueller UW, Ji H, et al. Wnt-4 activates the canonical beta-catenin-mediated Wnt pathway and binds Frizzled-6 CRD: functional implications of Wnt/beta-catenin activity in kidney epithelial cells. Exp. Cell Res. 2004; 298: 369–87.

Mah SP, Saueressig H, Goulding M, Kintner C, Dressler GR. Kidney development in cadherin-6 mutants: delayed mesenchyme-to-epithelial conversion and loss of nephrons. Dev. Biol. 2000; 223: 38–53.

Maheswaran S, Englert C, Bennett P, Heinrich G, Haber DA. The WT1 gene product stabilizes p53 and inhibits p53-mediated apoptosis. Genes Dev. 1995; 9: 2143–56.

Majumdar A, Vainio S, Kispert A, McMahon J, McMahon AP. Wnt11 and Ret/Gdnf pathways cooperate in regulating ureteric branching during metanephric kidney development. Development 2003; 130: 3175–85.

Mansouri A, Chowdhury K, Gruss P. Follicular cells of the thyroid gland require Pax8 gene function. Nat. Genet. 1998; 19: 87–90.

Masuya M, Drake CJ, Fleming PA, et al. Hematopoietic origin of glomerular mesangial cells. Blood 2003; 101: 2215–18.

Mayo MW, Wang CY, Drouin SS, et al. WT1 modulates apoptosis by transcriptionally upregulating the bcl-2 proto-oncogene. EMBO J 1999; 18: 3990–4003.

McCright B, Gao X, Shen L, et al. Defects in development of the kidney, heart and eye vasculature in mice homozygous for a hypomorphic Notch2 mutation. Development 2001; 128: 491–502.

Melnick M, Bixler D, Nance WE, Silk K, Yune H. Familial branchio-oto-renal dysplasia: a new addition to the branchial arch syndromes. Clin. Genet. 1976; 9: 25–34.

Mendelsohn C, Batourina E, Fung S, Gilbert T, Dodd J. Stromal cells mediate retinoid-dependent functions essential for renal development. Development 1999; 126: 1139–48.

Merlet-Benichou C, Gilbert T, Vilar J, Moreau E, Freund N, Lelievre-Pegorier M. Nephron number: variability is the rule. Causes and consequences. Lab. Invest. 1999; 79: 515–27.

Miao HQ, Soker S, Feiner L, Alonso JL, Raper JA, Klagsbrun M. Neuropilin-1 mediates collapsin-1/semaphorin III inhibition of endothelial cell motility: functional competition of collapsin-1 and vascular endothelial growth factor-165. J. Cell Biol. 1999; 146: 233–42.

Michos O, Panman L, Vintersten K, Beier K, Zeller R, Zuniga A. Gremlin-mediated BMP antagonism induces the epithelial-mesenchymal feedback signaling controlling metanephric kidney and limb organogenesis. Development 2004; 131: 3401–10.

Miner JH. Renal basement membrane components. Kidney Int. 1999; 56: 2016–24.

Miner JH, Li C. Defective glomerulogenesis in the absence of laminin alpha5 demonstrates a developmental role for the kidney glomerular basement membrane. Dev. Biol. 2000; 217: 278–89.

Mitsiadis TA, Salmivirta M, Muramatsu T, et al. Expression of the heparin-binding cytokines, midkine (MK) and HB-GAM (pleiotrophin) is associated with epithelial-mesenchymal interactions during fetal development and organogenesis. Development 1995; 121: 37–51.

Miyamoto N, Yoshida M, Kuratani S, Matsuo I, Aizawa S. Defects of urogenital development in mice lacking Emx2. Development 1997; 124: 1653–64.

Miyazaki Y, Oshima K, Fogo A, Hogan BL, Ichikawa I. Bone morphogenetic protein 4 regulates the budding site and elongation of the mouse ureter. J. Clin. Invest. 2000; 105: 863–73.

Moffett P, Bruening W, Nakagama H, et al. Antagonism of WT1 activity by protein self-association. Proc. Natl Acad. Sci. USA 1995; 92: 11105–9.

Montesano R, Matsumoto K, Nakamura T, Orci L. Identification of a fibroblast-derived epithelial morphogen as hepatocyte growth factor. Cell 1991; 67: 901–8.

Morello R, Zhou G, Dreyer SD, et al. Regulation of glomerular basement membrane collagen expression by LMX1B contributes to renal disease in nail patella syndrome. Nat. Genet. 2001; 27: 205–8.

Moser M, Pscherer A, Roth C, et al. Enhanced apoptotic cell death of renal epithelial cells in mice lacking transcription factor AP-2beta. Genes Dev. 1997a; 11: 1938–48.

Moser M, Ruschoff J, Buettner R. Comparative analysis of AP-2 alpha and AP-2 beta gene expression during murine embryogenesis. Dev. Dyn. 1997b; 208: 115–24.

Muller U, Wang D, Denda S, Meneses JJ, Pedersen RA, Reichardt LF. Integrin alpha8beta1 is critically important for epithelial-mesenchymal interactions during kidney morphogenesis. Cell 1997; 88: 603–13.

Murshed M, Smyth N, Miosge N, et al. The absence of nidogen 1 does not affect murine basement membrane formation. Mol. Cell Biol. 2000; 20: 7007–12.

Nakai S, Sugitani Y, Sato H, et al. Crucial roles of Brn1 in distal tubule formation and function in mouse kidney. Development 2003; 130: 4751–9.

Nakano T, Niimura F, Hohenfellner K, Miyakita E, Ichikawa I. Screening for mutations in BMP4 and FOXC1 genes in congenital anomalies of the kidney and urinary tract in humans. Tokai. J. Exp. Clin. Med. 2003; 28: 121–6.

Natoli TA, Liu J, Eremina V, et al. A mutant form of the Wilms' tumor suppressor gene WT1 observed in Denys-Drash syndrome interferes with glomerular capillary development. J. Am. Soc. Nephrol. 2002; 13: 2058–67.

Neiss WF. Morphogenesis and histogenesis of the connecting tubule in the rat kidney. Anat. Embryol. (Berl.) 1982; 165: 81–95.

Nishinakamura R, Matsumoto Y, Nakao K, et al. Murine homolog of SALL1 is essential for ureteric bud invasion in kidney development. Development 2001; 128: 3105–15.

O'Rourke DA, Sakurai H, Spokes K, et al. Expression of c-ret promotes morphogenesis and cell survival in mIMCD-3 cells. Am. J. Physiol. 1999; 276: F581–8.

Ohuchi H, Hori Y, Yamasaki M, et al. FGF10 acts as a major ligand for FGF receptor 2 IIIb in mouse multi-organ development. Biochem. Biophys. Res. Commun. 2000; 277: 643–9.

Osathanondh V, Potter EL. Development of human kidney as shown by microdissection. II. Renal pelvis, calyces, and papillae. Arch. Pathol. 1963a; 76: 277–89.

Osathanondh V, Potter EL. Development of human kidney as shown by microdissection. III. Formation and interrelationship of collecting tubules and nephrons. Arch. Pathol. 1963b; 76: 290–302.

Osathanondh V, Potter EL. Development of human kidney as shown by microdissection. IV. Development of tubular portions of nephrons. Arch. Pathol. 1966; 82: 391–402.

Ostrom L, Tang MJ, Gruss P, Dressler GR. Reduced Pax2 gene dosage increases apoptosis and slows the progression of renal cystic disease. Dev. Biol. 2000; 219: 250–8.

Pachnis V, Mankoo B, Costantini F. Expression of the c-ret proto-oncogene during mouse embryogenesis. Development 1993; 119: 1005–17.

Palmer RE, Kotsianti A, Cadman B, Boyd T, Gerald W, Haber DA. WT1 regulates the expression of the major glomerular podocyte membrane protein Podocalyxin. Curr. Biol. 2001; 11: 1805–9.

Pellegrini M, Pantano S, Lucchini F, Fumi M, Forabosco A. Emx2 developmental expression in the primordia of the reproductive and excretory systems. Anat. Embryol. (Berl.) 1997; 196: 427–33.

Pelletier J, Bruening W, Kashtan CE, et al. Germline mutations in the Wilms' tumor suppressor gene are associated with abnormal urogenital development in Denys-Drash syndrome. Cell 1991; 67: 437–47.

Pichel JG, Shen L, Sheng HZ, et al. Defects in enteric innervation and kidney development in mice lacking GDNF. Nature 1996; 382: 73–6.

Pilia G, Hughes-Benzie RM, MacKenzie A, et al. Mutations in GPC3, a glypican gene, cause the Simpson-Golabi-Behmel overgrowth syndrome. Nat. Genet. 1996; 12: 241–7.

Piscione TD, Rosenblum ND. The molecular control of renal branching morphogenesis: current knowledge and emerging insights. Differentiation 2002; 70: 227–46.

Piscione TD, Yager TD, Gupta IR, et al. BMP-2 and OP-1 exert direct and opposite effects on renal branching morphogenesis. Am. J. Physiol. 1997; 273: F961–75.

Pohl M, Sakurai H, Bush KT, Nigam SK. Matrix metalloproteinases and their inhibitors regulate in vitro ureteric bud branching morphogenesis. Am. J. Physiol. Renal Physiol. 2000; 279: F891–900.

Pontoglio M, Barra J, Hadchouel M, et al. Hepatocyte nuclear factor 1 inactivation results in hepatic dysfunction, phenylketonuria, and renal Fanconi syndrome. Cell 1996; 84: 575–85.

Popsueva A, Poteryaev D, Arighi E, et al. GDNF promotes tubulogenesis of GFRalpha1-expressing MDCK cells by Src-mediated phosphorylation of Met receptor tyrosine kinase. J. Cell Biol. 2003; 161: 119–29.

Porteous S, Torban E, Cho NP, et al. Primary renal hypoplasia in humans and mice with PAX2 mutations: evidence of increased apoptosis in fetal kidneys of Pax2(1Neu) +/− mutant mice. Hum. Mol. Genet. 2000; 9: 1–11.

Potter EL Normal and Abnormal Development of the Kidney. Chicago, IL: Year Book Medical Publishers, 1972.

Qiao J, Bush KT, Steer DL, et al. Multiple fibroblast growth factors support growth of the ureteric bud but have different effects on branching morphogenesis. Mech. Dev. 2001; 109: 123–35.

Qiao J, Sakurai H, Nigam SK. Branching morphogenesis independent of mesenchymal-epithelial contact in the developing kidney. Proc. Natl Acad. Sci. USA 1999a; 96: 7330–5.

Qiao J, Uzzo R, Obara-Ishihara T, Degenstein L, Fuchs E, Herzlinger D. FGF-7 modulates ureteric bud growth and nephron number in the developing kidney. Development 1999b; 126: 547–54.

Quaggin SE. Transcriptional regulation of podocyte specification and differentiation. Microsc. Res. Tech. 2002; 57: 208–11.

Quaggin SE, Schwartz L, Cui S, et al. The basic-helix-loop-helix protein pod1 is critically important for kidney and lung organogenesis. Development 1999; 126: 5771–83.

Raatikainen-Ahokas A, Hytonen M, Tenhunen A, Sainio K, Sariola H. BMP-4 affects the differentiation of metanephric mesenchyme and reveals an early anterior-posterior axis of the embryonic kidney. Dev. Dyn. 2000; 217: 146–58.

Rauvala H. An 18-kd heparin-binding protein of developing brain that is distinct from fibroblast growth factors. EMBO J 1989; 8: 2933–41.

Ricono JM, Xu YC, Arar M, Jin DC, Barnes JL, Abboud HE. Morphological insights into the origin of glomerular endothelial and mesangial cells and their precursors. J. Histochem. Cytochem. 2003; 51: 141–50.

Robert B, St John PL, Abrahamson DR. Direct visualization of renal vascular morphogenesis in Flk1 heterozygous mutant mice. Am. J. Physiol. 1998; 275: F164–72.

Robert B, St John PL, Hyink DP, Abrahamson DR. Evidence that embryonic kidney cells expressing flk-1 are intrinsic, vasculogenic angioblasts. Am. J. Physiol. 1996; 271: F744–53.

Robert B, Zhao X, Abrahamson DR. Coexpression of neuropilin-1, Flk1, and VEGF(164) in developing and mature mouse kidney glomeruli. Am. J. Physiol. Renal. Physiol. 2000; 279: F275–82.

Rodriguez-Soriano J, Vallo A, Bilbao JR, Castano L. Branchio-oto-renal syndrome: identification of a novel mutation in the EYA1 gene. Pediatr. Nephrol. 2001; 16: 550–3.

Rogers SA, Ryan G, Purchio AF, Hammerman MR. Metanephric transforming growth factor-beta 1 regulates nephrogenesis in vitro. Am. J. Physiol. 1993; 264: F996–1002.

Rothenpieler UW, Dressler GR. Pax-2 is required for mesenchyme-to-epithelium conversion during kidney development. Development 1993; 119: 711–20.

Ryan G, Steele-Perkins V, Morris JF, Rauscher FJ 3rd, Dressler GR. Repression of Pax-2 by WT1 during normal kidney development. Development 1995; 121: 867–75.

Sadl V, Jin F, Yu J, et al. The mouse Kreisler (Krml1/MafB) segmentation gene is required for differentiation of glomerular visceral epithelial cells. Dev. Biol. 2002; 249: 16–29.

Sainio K, Suvanto P, Davies J, et al. Glial-cell-line-derived neurotrophic factor is required for bud initiation from ureteric epithelium. Development 1997; 124: 4077–87.

Sakurai H, Bush KT, Nigam SK. Identification of pleiotrophin as a mesenchymal factor involved in ureteric bud branching morphogenesis. Development 2001; 128: 3283–93.

Sakurai H, Tsukamoto T, Kjelsberg CA, Cantley LG, Nigam SK. EGF receptor ligands are a large fraction of in vitro branching morphogens secreted by embryonic kidney. Am. J. Physiol. 1997; 273: F463–72.

Salerno A, Kohlhase J, Kaplan BS. Townes-Brocks syndrome and renal dysplasia: a novel mutation in the SALL1 gene. Pediatr. Nephrol. 2000; 14: 25–8.

Sanchez MP, Silos-Santiago I, Frisen J, He B, Lira SA, Barbacid M. Renal agenesis and the absence of enteric neurons in mice lacking GDNF. Nature 1996; 382: 70–3.

Santos OF, Barros EJ, Yang XM, et al. Involvement of hepatocyte growth factor in kidney development. Dev. Biol. 1994; 163: 525–9.

Sariola H, Saarma M. Novel functions and signalling pathways for GDNF. J. Cell Sci. 2003; 116: 3855–62.

Saxen L Organogenesis of the Kidney. Cambridge, UK: Cambridge University Press, 1987.

Scharnhorst V, van der Eb AJ, Jochemsen AG. WT1 proteins: functions in growth and differentiation. Gene 2001; 273: 141–61.

Schmidt C, Bladt F, Goedecke S, et al. Scatter factor/hepatocyte growth factor is essential for liver development. Nature 1995; 373: 699–702.

Schuchardt A, D'Agati V, Larsson-Blomberg L, Costantini F, Pachnis V. Defects in the kidney and enteric nervous system of mice lacking the tyrosine kinase receptor Ret. Nature 1994; 367: 380–3.

Schuchardt A, D'Agati V, Pachnis V, Costantini F. Renal agenesis and hypodysplasia in ret-k-mutant mice result from defects in ureteric bud development. Development 1996; 122: 1919–29.

Self M, Lagutin OV, Bowling B, Hendrix J, Cai Y, Dressler GR, Oliver G. Six2 is required for suppression of nephrogenesis and progenitor renewal in the developing kidney. Embo. J. 2006; 25: 5214–28.

Shah MM, Sampogna RV, Sakurai H, Bush KT, Nigam SK. Branching morphogenesis and kidney disease. Development 2004; 131: 1449–62.

Shalaby F, Rossant J, Yamaguchi TP, et al. Failure of blood-island formation and vasculogenesis in Flk-1-deficient mice. Nature 1995; 376: 62–6.

Sharma M, Fopma A, Brantley JG, Vanden Heuvel GB. Coexpression of Cux-1 and notch signaling pathway components during kidney development. Dev. Dyn. 2004; 231: 828–38.

Shawber CJ, Kitajewski J. Notch function in the vasculature: insights from zebrafish, mouse and man. Bioessays 2004; 26: 225–34.

Shawlot W, Behringer RR. Requirement for Lim1 in head-organizer function. Nature 1995; 374: 425–30.

Sock E, Schmidt K, Hermanns-Borgmeyer I, Bosl MR, Wegner M. Idiopathic weight reduction in mice deficient in the high-mobility-group transcription factor Sox8. Mol. Cell. Biol. 2001; 21: 6951–9.

Sorenson CM, Rogers SA, Korsmeyer SJ, Hammerman MR. Fulminant metanephric apoptosis and abnormal kidney development in bcl-2-deficient mice. Am. J. Physiol. 1995; 268: F73–81.

Soriano P. Abnormal kidney development and hematological disorders in PDGF beta-receptor mutant mice. Genes Dev. 1994; 8: 1888–96.

Srinivas S, Wu Z, Chen CM, D'Agati V, Costantini F. Dominant effects of RET receptor misexpression and ligand-independent RET signaling on ureteric bud development. Development 1999; 126: 1375–86.

Stark K, Vainio S, Vassileva G, McMahon AP. Epithelial transformation of metanephric mesenchyme in the developing kidney regulated by Wnt-4. Nature 1994; 372: 679–83.

Steer DL, Shah MM, Bush KT, et al. Regulation of ureteric bud branching morphogenesis by sulfated proteoglycans in the developing kidney. Dev. Biol. 2004; 272: 310–27.

Taira M, Otani H, Jamrich M, Dawid IB. Expression of the LIM class homeobox gene Xlim-1 in pronephros and CNS cell lineages of Xenopus embryos is affected by retinoic acid and exogastrulation. Development 1994; 120: 1525–36.

Takahashi T, Takahashi K, Gerety S, Wang H, Anderson DJ, Daniel TO. Temporally compartmentalized expression of ephrin-B2 during renal glomerular development. J. Am. Soc. Nephrol. 2001; 12: 2673–82.

Tang MJ, Worley D, Sanicola M, Dressler GR. The RET-glial cell-derived neurotrophic factor (GDNF) pathway stimulates migration and chemoattraction of epithelial cells. J. Cell Biol. 1998; 142: 1337–45.

Thomson RB, Aronson PS. Immunolocalization of Ksp-cadherin in the adult and developing rabbit kidney. Am. J. Physiol. 1999; 277: F146–56.

Threadgill DW, Dlugosz AA, Hansen LA, et al. Targeted disruption of mouse EGF receptor: effect of genetic background on mutant phenotype. Science 1995; 269: 230–4.

Torres M, Gomez-Pardo E, Dressler GR, Gruss P. Pax-2 controls multiple steps of urogenital development. Development 1995; 121: 4057–65.

Trupp M, Arenas E, Fainzilber M, et al. Functional receptor for GDNF encoded by the c-ret proto-oncogene. Nature 1996; 381: 785–9.

Tsang TE, Shawlot W, Kinder SJ, et al. Lim1 activity is required for intermediate mesoderm differentiation in the mouse embryo. Dev. Biol. 2000; 223: 77–90.

Uehara Y, Minowa O, Mori C, et al. Placental defect and embryonic lethality in mice lacking hepatocyte growth factor/scatter factor. Nature 1995; 373: 702–5.

Vainio SJ. Nephrogenesis regulated by Wnt signaling. J. Nephrol. 2003; 16: 279–85.

van Adelsberg J, Sehgal S, Kukes A, et al. Activation of hepatocyte growth factor (HGF) by endogenous HGF activator is required for metanephric kidney morphogenesis in vitro. J. Biol. Chem. 2001; 276: 15099–106.

Vanderwinden JM, Mailleux P, Schiffmann SN, Vanderhaeghen JJ. Cellular distribution of the new growth factor pleiotrophin (HB-GAM) mRNA in developing and adult rat tissues. Anat. Embryol. (Berl) 1992; 186: 387–406.

Vilar J, Gilbert T, Moreau E, Merlet-Benichou C. Metanephros organogenesis is highly stimulated by vitamin A derivatives in organ culture. Kidney Int. 1996; 49: 1478–87.

Villegas G, Tufro A. Ontogeny of semaphorins 3A and 3F and their receptors neuropilins 1 and 2 in the kidney. Mech. Dev. 2002; 119(Suppl 1): S149–53.

Watanabe K, Yamada H, Yamaguchi Y. K-glypican: a novel GPI-anchored heparan sulfate proteoglycan that is highly expressed in developing brain and kidney. J. Cell Biol. 1995; 130: 1207–18.

Willem M, Miosge N, Halfter W, et al. Specific ablation of the nidogen-binding site in the laminin gamma1 chain interferes with kidney and lung development. Development 2002; 129: 2711–22.

Wilson JG, Roth CB, Warkany J. An analysis of the syndrome of malformations induced by maternal vitamin A deficiency. Effects of restoration of vitamin A at various times during gestation. Am. J. Anat. 1953; 92: 189–217.

Xu PX, Adams J, Peters H, Brown MC, Heaney S, Maas R. Eya1-deficient mice lack ears and kidneys and show abnormal apoptosis of organ primordia. Nat. Genet. 1999; 23: 113–17.

Xu PX, Zheng W, Huang L, Maire P, Laclef C, Silvius D. Six1 is required for the early organogenesis of mammalian kidney. Development 2003; 130: 3085–94.

Yang J, Goetz D, Li JY, et al. An iron delivery pathway mediated by a lipocalin. Mol. Cell 2002; 10: 1045–56.

Yu J, Carroll TJ, McMahon AP. Sonic hedgehog regulates proliferation and differentiation of mesenchymal cells in the mouse metanephric kidney. Development 2002; 129: 5301–12.

Zent R, Bush KT, Pohl ML, et al. Involvement of laminin binding integrins and laminin-5 in branching morphogenesis of the ureteric bud during kidney development. Dev. Biol. 2001; 238: 289–302.

Zhang H, Bradley A. Mice deficient for BMP2 are nonviable and have defects in amnion/chorion and cardiac development. Development 1996; 122: 2977–86.

Zhou Z, Apte SS, Soininen R, et al. Impaired endochondral ossification and angiogenesis in mice deficient in membrane-type matrix metalloproteinase I. Proc. Natl Acad. Sci. USA 2000; 97: 4052–7.

CHAPTER 24

Polycystic Kidney Disease

STEFAN SOMLO AND LISA M. GUAY-WOODFORD

INTRODUCTION

Polycystic kidney disease (PKD) describes a heterogeneous set of disorders that result from single gene defects transmitted as either autosomal dominant or autosomal recessive traits. While the development of fluid-filled cysts and progressive impairment of renal function is a common feature, these disorders are distinguished by different ages of onset, variable rates of renal disease progression, and a diverse array of extrarenal manifestations.

Autosomal dominant polycystic kidney disease (ADPKD; MIM 173900) is characterized by age-dependent occurrence of bilateral, multiple renal cysts as well as a variety of extrarenal manifestations. The latter include cysts in the liver bile ducts, pancreatic ducts, seminal vesicles, and arachnoid membrane, as well as noncystic manifestations, such as intracranial aneurysms and dolichoectasias, aortic root dilatation and aneurysms, mitral valve prolapse, and abdominal wall hernias (Torres et al 2007). Recently, isolated autosomal dominant polycystic liver disease without kidney cysts (ADPLD; MIM 174050) has been identified as a genetically distinct but pathophysiologically related disease (Qian et al 2003).

Autosomal recessive polycystic kidney disease (ARPKD; MIM 263200) is a severe, typically early-onset form of cystic disease that primarily involves the kidneys and biliary tract. Affected patients have a spectrum of clinical phenotypes that depend in part on the age at presentation (Guay-Woodford & Desmond 2003). The basic defects observed in ARPKD suggest that the terminal differentiation of the renal collecting duct and intrahepatic biliary ducts is disordered.

The study of these diseases through human genetic approaches, animal models, and classical cellular biology and biochemistry has yielded remarkable progress and insights although the goal of finding effective treatment remains a work in progress. Major incremental advances in understanding the pathogenesis of human polycystic diseases included the discovery of the mutated genes and their novel protein products by positional cloning (The European Polycystic Kidney Disease Consortium 1994, The International Polycystic Kidney Disease Consortium 1995, Mochizuki et al 1996), the understanding that multiple somatic mutations are implicated in the molecular pathogenesis (Qian et al 1996), and the emergence of a neglected organelle, the primary cilium, as the focus of investigation not just in this disease but in a broad spectrum of biological processes (Yoder 2007). These specific advances coupled with broader scientific interest and investigations in the field have led to improved understanding of the clinical disease and the variation it exhibits and have culminated in early directed therapeutic clinical trials in patients.

DIAGNOSIS AND CLINICAL FEATURES

Autosomal Dominant Polycystic Kidney Disease

ADPKD affects between 1 in 400 and 1 in 1000 live births in all ethnic populations worldwide (Iglesias et al 1983, Torres et al 1985). ADPKD exhibits genetic heterogeneity in that mutations in either of two genes, *PKD1* or *PKD2*, result in the ADPKD clinical phenotype and are not readily distinguishable on clinical grounds alone. It is the most common single gene disorder that can lead to premature death in man. The yearly incidence rates for end stage renal disease (ESRD) caused by ADPKD in men and women, respectively, are 8.7 and 6.9 per million in United States with more than 2100 new ADPKD patients entering ESRD annually (Renal Data System 1999). Incidences are similar worldwide as is the implication that ADPKD is a more progressive disease in men than in women.

Diagnosis

The diagnosis of ADPKD relies on imaging testing (Kielstein & Sass 2002) (Figure 24.1). Renal ultrasound is commonly used because of cost and safety. Counseling should be done prior to testing an individual with a positive family history

Genetic Diseases of the Kidney
ISBN: 978-0-12-449851-8

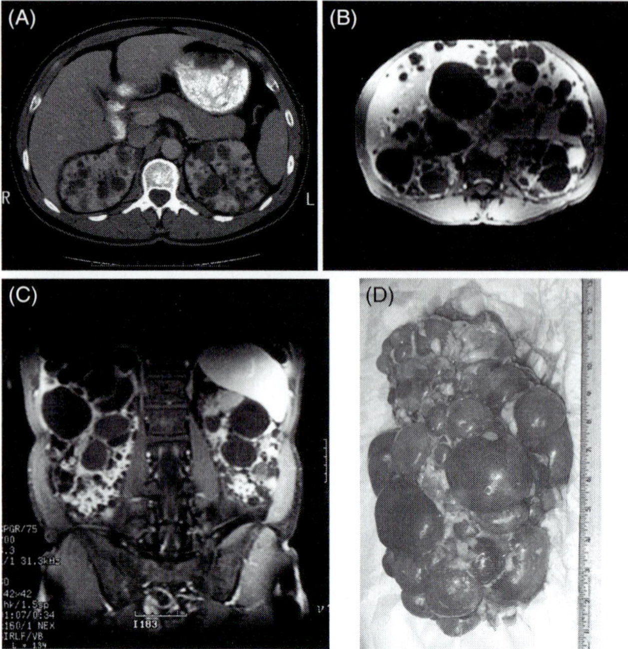

FIGURE 24.1 Autosomal dominant polycystic kidney disease. Axial CT (A) and MRI (B) and coronal MRI (C) images demonstrating moderate polycystic kidney disease and absence of hepatic cysts (A) and more advanced polycystic kidney and liver disease with marked enlargement of both organs (B, C). In both the CT and MR images, cysts appear as hypodense areas within the organ parenchyma. (D) Nephrectomy specimen from a patient with ADPKD and end stage renal disease showing profound cystic deformation of the kidney. (Modified from Somlo S, Torres VE and Caplan MJ. (2007). In: Seldin and Giebisch's The Kidney: Physiology and Pathophysiology, Alpern, RJ and Hebert SC. Academic Press, pp. 2283–2314). (see also Plate 37)

for the presence of ADPKD. The potential discrimination in terms of insurability and employment associated with a positive diagnosis has been reduced in the USA with the recent passage of the Genetic Information Nondiscrimination Act, although the psychological burden of being aware of being affected with a chronic disease should be considered in the decision of when to test.

The occurrence and severity of the cystic lesion is highly variable. Cysts have been observed in utero and have been detected as incidental findings in 80-year-old patients with otherwise normal blood pressure and kidney function. In genetic terms, the expressivity, or severity of the phenotype in this disease is highly variable. By contrast the penetrance is virtually complete if it is expressed as a function of age. Individuals over the age of 30 carrying a causative gene mutation will invariably manifest cysts sufficient for diagnosis when assessed with a sufficiently sensitive imaging study. The ultrasound criteria, referred to as the Ravine criteria (Ravine et al 1994), take into account the age dependent penetrance of the disease by requiring increasing

numbers of cysts with increasing age to make the diagnosis. For individuals with a known family history of ADPKD, the modified Ravine criteria for diagnosis by ultrasound include at least two unilateral or bilateral cysts in individuals younger than 30 years; two cysts in each kidney in individuals aged 30–59 years; four cysts in each kidney in individuals 60 years or older (Ravine et al 1994). In the absence of a family history of ADPKD, bilateral renal enlargement and cysts or the presence of multiple bilateral cysts with hepatic cysts together suggest diagnosis of ADPKD if other diseases associated with kidney cysts are excluded.

Genetic testing can be used when the imaging results are equivocal and/or when a definite diagnosis is required in a younger individual, such as a potential living related kidney donor. Prenatal and preimplantation genetic testing are rarely considered for ADPKD (Sujansky et al 1990, De Rycke et al 2005). Genetic testing is performed by direct sequence analysis. Direct sequencing yields mutation detection rates of approximately 85% (Harris et al 2006). However, as most mutations are unique and up to one-third of *PKD1* changes are predicted to be single amino acid substitution changes, the causative nature of some sequence changes is difficult to prove. Genetic testing can be helpful with de novo mutations in the absence of a family history. A unique variant in either *PKD1* or *PKD2* present in the affected child but absent in both parents in the setting of documented maternity and paternity can identify the new mutation.

Natural History of ADPKD

The Consortium for Radiologic Imaging Studies of Polycystic Kidney Disease (CRISP) has provided critical information on the relationship of cyst growth with ADPKD. Two hundred and forty-one nonazotemic patients were followed prospectively with yearly MRI examinations measuring progression kidney and cyst volume (Chapman et al 2003, Grantham et al 2006). Consistent with the tendency toward more rapid progression noted above, men had higher rates of kidney and cyst growth than women. The mean rate of increase in total kidney volume over 3 years was 5.3% per year. Baseline total kidney volume correlated with subsequent rate of increase in renal volume. Initial total kidney volume above 1500 ml was associated with declining glomerular filtration rate (GFR) in patients. Introduction of this method of volume measurement may offer added prognostic information in clinical practice.

Renal Manifestations of ADPKD

Hypertension is present in approximately 50% of ADPKD patients with normal renal function and increases to nearly 100% of patients with ESRD (Kelleher et al 2004). Early detection and treatment of hypertension is important because cardiovascular disease is the main cause of death and uncontrolled blood pressure increases the risk for proteinuria, hematuria, and faster decline of renal function (Fick et al 1995, Ecder et al 2000). The association between renal size and prevalence of hypertension has supported the hypothesis that deformation of the vascular tree by cyst expansion causes ischemia and activation of the local intrarenal renin-angiotensin system (Gabow et al 1990a, Torres et al 1992). In addition, defective polycystin protein function in the vasculature may also play a role in the pathogenesis of hypertension in ADPKD. Enhanced vascular smooth muscle contractility (Qian et al 2007) and impaired nitric oxide endothelium-dependent vasorelaxation (Wang et al 2003, Kocaman et al 2004, Clausen et al 2006) are among the proposed mechanisms.

Pain is the most commonly reported symptom (~60%) by adult patients (Bajwa et al 2004). Acute pain may occur as a consequence of renal hemorrhage, passage of stones and urinary tract infections; chronic flank pain can occur without identifiable etiology other than the cysts. Cyst hemorrhage as evidenced by hyperdense (CT) or high signal (MRI) cysts occurs in most patients although symptomatic episodes are less commonly reported. Approximately 20% of ADPKD patients have kidney stones (Torres et al 1988, 1993, Grampsas et al 2000). Their composition is usually uric acid and/or Ca^{2+} oxalate. Stones may be difficult to differentiate from cyst wall and parenchymal calcification, which also occur. As in the general population urinary tract infections are more frequent in females than males and cyst infections do occur (Elzinga & Bennett 1996). CT and MRI are sensitive to detect complicated cysts and provide anatomic definition, but the findings are not specific for infection. Cyst aspiration is sometimes required when the clinical setting and imaging are suggestive of infection and blood and urine cultures are negative.

The development of renal failure is highly variable and GFR is typically maintained until the fourth to sixth decades of life. Once GFR begins to decline, the progression is inexorable with an average rate of decrease of approximately 4.4–5.9 ml/min/year (Klahr et al 1995). Risk factors associated with a worse prognosis include the aforementioned kidney size and cyst burden and male gender, as well as black race, first episode of hematuria before the age of 30, onset of hypertension before the age of 35, hyperlipidemia, low HDL, and sickle cell trait (Gabow et al 1992, Yium et al 1994, Johnson & Gabow 1997). The CRISP study has shown that kidney and cyst volumes are the strongest predictors of renal functional decline (Grantham et al 2006) and it is anticipated that this analysis will find its way into more common practice.

Extrarenal Manifestations of ADPKD

Polycystic liver disease (PLD) is the most common extrarenal manifestation. It is associated with both *PKD1* and *PKD2* genotypes as well as with isolated ADPLD. The cysts arise by excessive proliferation and dilatation of biliary ductules and peribiliary glands. Estrogen receptors are expressed in the epithelium lining the hepatic cysts and estrogens stimulate hepatic cyst-derived cell proliferation (Alvaro et al 2006). Bile duct cyst growth is also promoted by growth factors and cytokines secreted into the cyst fluid (Nichols et al 2004, Fabris et al 2006). PLD in ADPKD is more common than previously appreciated. The prevalence by MRI in the CRISP study is 58%, 85% and 94% in 15–24-, 25–34-, and 35–46-year-old participants (Bae et al 2006). The severity of liver cysts is greater in women than in men and amongst women, is greater in those who have had multiple pregnancies or have used oral contraceptive agents or estrogen replacement therapy (Gabow et al 1990b, Sherstha et al 1997).

PLD is usually asymptomatic, but symptoms including dyspnea, early satiety, gastroesophageal reflux, and mechanical low back pain may result from mass effects due to massive enlargement of the liver or from a single or a limited number of dominant cysts. More localized mass effects can result in hepatic venous outflow obstruction, inferior vena cava compression, portal vein compression or bile duct compression presenting as obstructive jaundice (Torres et al 1994). Finally, liver cysts are subject to local complications including cyst hemorrhage, infection, and rupture. Rare association of PLD with congenital hepatic fibrosis (Grunfeld et al 1985) has been cited as evidence that the ADPKD and ARPKD are part of a pathologic continuum.

While rare, the vascular complications associated with ADPKD can be clinically very significant. These include intracranial aneurysms and dolichoectasias, thoracic aortic and cervicocephalic artery dissections, and coronary artery aneurysms. Intracranial aneurysms (ICA) occur in approximately 6% of patients with a negative, and 16% of those with a positive, family history of aneurysms (Pirson et al 2002). The familial clustering suggests that familial factors, either the germline mutation in the respective PKD gene or associated modifier genes, may play a role in this complication of ADPKD. ICA are most often asymptomatic in ADPKD. The risk of rupture depends on factors that apply to aneurysms in general and include the size of the aneurysm. Rupture carries a 35–55% risk of combined severe morbidity and mortality (Inagawa 2001). The mean age at rupture is lower than in the general population (39 years versus 51 years). Most patients with ICA have normal renal function and up to 29% will have normal blood pressure at the time of rupture.

Autosomal Recessive Polycystic Kidney Disease

ARPKD occurs in approximately 1 in 20000 live births (Zerres et al 1998). While older reports indicate that the disease is expressed primarily in Caucasians, more recent data indicate that ARPKD occurs in all ethnic groups (Guay-Woodford & Desmond 2003). The majority of patients are identified in utero or as infants with enlarged and echogenic kidneys. In fetal life, the associated decreased urine output leads to oligohydramnios, which in turn contributes to the development of pulmonary hypoplasia. Despite significant advances in neonatal care over the past two decades, approximately 30% of affected neonates die shortly after birth, due to respiratory insufficiency. Among neonatal survivors, morbidity results from severe systemic hypertension, renal insufficiency, and portal hypertension due to portal-tract hyperplasia and associated fibrosis.

Pathological analysis has been the mainstay for ARPKD diagnosis (Bernstein & Slovis 1992). In affected neonates, ARPKD kidneys can reach up to tenfold normal size, while retaining their reniform configuration. Dilated, fusiform collecting ducts extend radially through the cortex, while medullary collecting duct cysts are more spherical in configuration. Cysts are lined with a single layer of nondescript cuboidal epithelium. The glomeruli and more proximal nephron segments are generally normal in structure, but are often crowded between ectatic collecting ducts or displaced into subcapsular wedges. Cartilage or other dysplastic elements are not evident. In contrast to ADPKD, ARPKD cysts retain their afferent and efferent connections (Osathanondh & Potter 1964). With age, cysts, particularly those in the medullary collecting ducts, can expand up to 5 cm in diameter. Progressive interstitial fibrosis is likely responsible for secondary tubular obstruction. In older children, medullary ductal ectasia is the predominant finding.

The liver lesion in ARPKD is characterized by the ductal plate malformation (Desmet 1998). The liver can be either normal in size or somewhat enlarged. Bile ducts are dilated (biliary ectasia) and marked cystic dilatation of the entire intrahepatic biliary system (Caroli's disease) has been described. In neonatal ARPKD, the bile ducts are increased in number, tortuous in configuration, and often located around the periphery of the portal tract. In older children, the biliary ectasia is accompanied by increasing portal tract fibrosis and hypoplasia of the small portal vein branches. The hepatic parenchyma may be intersected by delicate fibrous septa that link the portal tracts, but the hepatocytes themselves are seldom affected.

Diagnosis

In recent years, clinical diagnosis has increasingly relied on imaging modalities (Figure 24.2) instead of histopathological analyses. ARPKD belongs to a group of disorders that are collectively described as hepatorenal fibrocystic diseases (Kerkar et al 2006). While most of these disorders are characterized by large, echogenic kidneys in the fetus and neonate, recent studies indicate that to a large extent they can be distinguished by sonography (Brun et al 2004, Chaumoitre et al 2006). ARPKD kidneys in utero are hyperechogenic but tend to be larger than ADPKD kidneys and display *decreased* corticomedullary differentiation due to the hyperechogenic medulla (Reuss et al 1991). Beyond 1–2

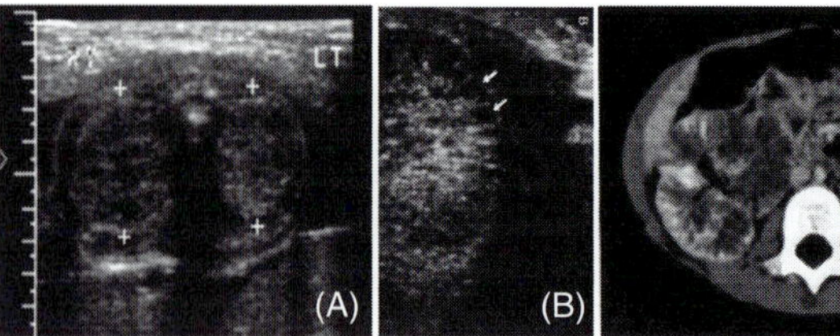

FIGURE 24.2 Autosomal recessive polycystic kidney disease. In utero sonogram (A) of an ARPKD fetus at 26 weeks gestational age. Transverse view reveals enlarged, diffusely echogenic kidneys. ARPKD in a neonate (B). High resolution sonography reveals radially arrayed dilated collecting ducts (arrowheads). ARPKD in a symptomatic four year old girl (C). Contrast-enhanced CT shows a striated nephrogram and prolonged corticomedullary differentiation. (Modified from Guay-Woodford LM, Jafri Z and Bernstein J. Other cystic diseases. In: Comprehensive Clinical Nephrology. Johnson R and Feehally J (eds), 3rd edn.). (see also Plate 38)

years of age, renal size decreases in ARPKD patients, and nephromegaly becomes a more specific feature of ADPKD. In comparison, ADPKD kidneys in utero tend to be moderately enlarged with a hyperechogenic cortex and relatively hypoechogenic medulla causing *increased* corticomedullary differentiation. While similar findings have been described in Bardet–Biedl syndrome, Meckel–Gruber syndrome (and Meckel-like syndrome), Ivemark II syndrome (reno-hepatico-pancreatic dysplasia), Jeune syndrome, short-rib polydactyly syndromes (Saldino-Noonan-I, Majewski-II, Beemer-IV), glutaric aciduria II, and trisomy 9, these disorders are distinguished from ARPKD and ADPKD by characteristic sets of extrarenal malformations (Table 24.1). The presence of hepatic cysts in young adults is pathognomonic for ADPKD, while the characteristic sonographic features of Caroli's disease or hepatic fibrosis are virtually diagnostic of ARPKD.

With the identification of *PKHD1* as the principal disease gene in ARPKD, genetic testing is available as a clinical diagnostic tool. Recent studies report mutation detection rates of 80–87% (Bergmann et al 2005, Losekoot et al 2005, Sharp et al 2005, Adeva et al 2006). Gene-based testing is primarily applied in the context of prenatal testing (Zerres et al 2004) and preimplantation genetic diagnosis (Gigarel et al 2008). To date, there is limited evidence for genotype–phenotype correlations. Patients with two truncating mutations have a high risk of perinatal demise (Bergmann et al 2003, Furu et al 2003, Sharp et al 2005). However, as in ADPKD, there is a high percentage of unique, missense changes in *PKHD1*, which can complicate the unequivocal interpretation of gene-based testing. Moreover, about 20% of ARPKD siblings have discordant clinical phenotypes (Kaplan et al 1988, Deget et al 1995, Bergmann et al 2005). These data potentially complicate genetic counseling and caution must be exercised in predicting the clinical outcome of future affected children (Bergmann et al 2005).

Table 24.1   Hepatorenal fibrocystic diseases

Disease	MIM	Inheritance	Gene	Renal disease	Hepatic disease	Associated features
ARPKD	263200	AR	*PKHD1*	Collecting duct dilatation	CHF; Caroli disease	Growth retardation
ADPKD	173900	AD	*PKD1;* *PKD2*	Cysts along entire nephron	Biliary cysts; CHF	Minimal in children
Nephronophthisis (NPHP2 and NPHP3)	602088; 608002	AR	*NPHP2;* *NPHP3*	Cysts at C-M junction	CHF	Tapetoretinal degeneration; situs inversus
Joubert syndrome (JBTS1)	213300	AR	–	Cystic dysplasia; NPHP	CHF; Caroli disease	Cerebellar vermis hypo/aplasia with episodic hyperpnea; abnormal eye movements; mental retardation
Bardet-Biedel syndrome	209900	AR; triallelic	*BBS1–12*	Cystic dysplasia; NPHP	CHF	Retinal degeneration; obesity; post-axial polydactyly; hypogonadism in males; mental retardation
McKusick-Kaufman syndrome	236700	AR	*BBS6*	Cystic dysplasia; NPHP	CHF	Hydrometrocolpos; post-axial polydactyly; congenital heart disease
Meckel syndrome (Types 1,3-5)	249000	AR	*MSK1;* *MSK3*	Cystic dysplasia	CHF	Occipital encephalocele; polydactyly
Jeune syndrome	208500	AR	*IFT80**	Cystic dysplasia	CHF; Caroli disease	Short stature; skeletal dysplasia; small thorax; limb shortness; polydactyly; hypoplastic pelvis
Short-rib polydactyly syndromes (Types I, II, IV)	263530; 263520; 269860	AR	–	Cystic dysplasia	Intrahepatic biliary dysgenesis	Median cleft lip; pre- and post-axial polysyndactyly; short ribs and limbs; pancreatic cysts; genital abnormalities
Zellweger syndrome	214100	AR	*PEX1-3;* *5,6;12;14;26*	Renal cortical microcysts, including glomerular cysts	Intrahepatic biliary dysgenesis	Hypotonia; seizures; agenesis/hypoplasia of corpus callosum; characteristic facies; skeletal abnormalies; neonatal death
Glutaric aciduria IIA	231680	AR	*ETFA*	Renal cortical microcysts	Hepatic periportal necrosis	Neonatal acidosis and hypoglycemia; muscular hypotonia; sweaty feet odor; hepatomegaly; neonatal death
Ivemark II	208540	AR	–	Cystic dysplasia	CHF; Caroli disease	Pancreatic fibrosis; situs inversus; polysplenia; cardiac and CNS anomalies

MIM: Mendelian Inheritance in Man (accessible at http://www.ncbi.nlm.nih.gov/omim/); AR: autosomal recessive; AD: autosomal dominant; CHF: congenital hepatic fibrosis

*Beales P, Bland E, Tobin J, Bacchelli C, Tuysuz B, Hill J, Rix S, Pearson C, Kai M, Hartley J, Johnson C, Irving M, Elcioglu N, Winey M, Tada, M, Scambler P. IFT80, which encodes a conserved intraflagellar transport protein, is mutated in Jeune asphyxiating thoracic dystrophy. Nat. Genet. 39:727–9, 2007

Natural History of ARPKD

While ARPKD has a variable spectrum of clinical manifestations, most patients are identified either in utero or at birth (Guay-Woodford & Desmond 2003). The most severely affected fetuses have enlarged, echogenic kidneys and oligohydramnios, due to poor fetal renal output. Interestingly, these sonographic features are often not evident until after 20 weeks gestation (Zerres et al 1998). While advances in neonatal critical care management have improved the survival of ARPKD neonates, approximately 30% of affected newborns die within the first week of life due to respiratory insufficiency (Guay-Woodford & Desmond 2003). Several small studies (n = 1–3) report some efficacy of unilateral or bilateral nephrectomy in ARPKD neonates with severe respiratory compromise (Sumfest et al 1993, Bean 1995, Munding 1999, Spechtenhauser 1999, Shukla 2004). However, this treatment modality has not been rigorously evaluated in controlled, adequately powered clinical trials and thus lacks the evidence base for recommended practice.

For neonatal survivors, the prognosis is encouraging with two large studies reporting overall patient survival of 85.8% at 1 month, 78.6% at 1 year, and 74.6% at 5 years in a North American cohort (Guay-Woodford & Desmond 2003) and 85% at 1 year and 82% at 10 years in a European cohort (Bergmann et al 2005). In the subset of patients who survive the perinatal period, morbidity is primarily associated with systemic hypertension, progressive renal insufficiency, and portal hypertension secondary to portal-tract fibrosis.

Renal Manifestations of ARPKD

Hypertension is a common feature of ARPKD, occurring in approximately 65–75% of affected children (Guay-Woodford & Desmond 2003, Bergmann et al 2005). Due to the paucity of studies and conflicting data, the pathogenic mechanisms remain to be established. Early clinical studies and follow-up cell culture experiments implicate dysregulated sodium transport in ARPKD collecting ducts as the primary mechanism, causing increased intravascular volume (Kaplan et al 1989, Rohatgi et al 2003). In contrast, immunohistochemical studies in ARPKD kidneys have demonstrated ectopic expression of renin-angiotensin system (RAS) components in cystic tubules, suggesting that overactivity of RAS could result in increased intrarenal angiotensin II production (Loghman-Adham et al 2005). Angiotensin converting enzyme inhibitors or angiotensin II receptor inhibitors are generally effective therapies (Guay-Woodford & Desmond 2003).

Chronic renal insufficiency occurs frequently in ARPKD. In one large study, actuarial renal survival rates (end point defined as start of dialysis/renal transplantation or by death due to end-stage renal disease) were 86% at 5 years, 71% at 10 years, 66% at 15 years, and 42% at 20 years (Bergmann et al 2005). Adeva et al (2006) expanded the analysis to examine three subsets of ARPKD patients defined by age at diagnosis (<1 year; 1–20 years; >20 years). In those patients diagnosed in the first year of life, the actuarial renal survival was 75% at 10 years and 50% at 15 years; whereas, among patients diagnosed between 1 and 20 years of age, 75% had functioning kidneys at 28 years. Overall, actuarial data differed significantly between the three diagnostic groups with 36%, 80%, and 88% renal survival respectively, 20 years after the diagnosis. Considerable evidence of intrafamilial phenotype variability was observed.

Extrarenal Manifestations of ARPKD

In a relative minority of patients, ARPKD is diagnosed in late childhood, adolescence, or early adulthood, usually with hepatosplenomegaly as the presenting feature (Fonck et al 2001, Adeva et al 2006). Later presentation is usually associated with less renal enlargement and more variability in renal cyst size (Blyth & Ockenden 1971) or even with isolated congenital hepatic fibrosis (CHF) (Adeva et al 2006). These patients are more prone to develop complications related to CHF and Caroli's disease, e.g. portal hypertension, variceal bleeding, hypersplenism, and ascending cholangitis (Guay-Woodford & Desmond 2003, Adeva et al 2006, Goilav et al 2006). Medical management is largely supportive and ursodeoxycholic acid can be effective in the management of Caroli's disease (Ros et al 1993, Shneider & Magid 2005). Patients with clinically significant portal hypertension may require porto-systemic shunting or even liver transplantation (De Kerckhove et al 2006). In rare cases, sequential or concurrent liver–kidney transplants can be an effective therapeutic option (Davis et al 2003).

In addition to liver-related complications, other ARPKD comorbidities include neonatal hyponatremia, chronic lung disease (primarily due to prolonged mechanical ventilation), growth retardation, and increased frequency of urinary tract infections (Guay-Woodford & Desmond 2003).

GENETICS OF POLYCYSTIC KIDNEY DISEASE

ADPKD

The two causative genes for ADPKD, *PKD1* located on chromosome 16p13.3 and *PKD2* located on chromosome 4q21, were isolated by positional cloning (The European Polycystic Kidney Disease Consortium 1994, The American PKD1 Consortium 1995, The International Polycystic Kidney Disease Consortium 1995, Mochizuki et al 1996). The respective protein products for *PKD1* and *PKD2* are polycystin-1 (PC1) and polycystin-2 (PC2 or TRPP2). Despite some early data to suggest that there is a third genetic locus for ADPKD, the existence of the third gene now appears unlikely (Paterson & Pei 1998). Compound heterozygous individuals with *PKD1* and *PKD2* mutations have been described (Pei et al 2001), but no individuals homozygous for mutation in either gene have been found. This fact, coupled with the embryonic lethality of homozygous knockout mice for either gene (Lu et al 1997, Wu et al 2000), suggests that individuals with mutations to both alleles of either gene would not be viable (Paterson et al 2002).

The *PKD1* and *PKD2* genes are of similar size (~45–50 kb), but *PKD1* contains 46 exons that encode ~12 kb of open reading frame, compared to 15 exons and ~3 kb of open reading frame for *PKD2* (The American PKD1 Consortium 1995, The International Polycystic Kidney Disease Consortium 1995). A feature of *PKD1* that complicates sequence based genetic testing is that the 5′ two-thirds of the genomic segment is duplicated multiple times with very high sequence fidelity (>95%) in more proximal regions of chromosome 16 (Peral et al 1997, Watnick et al 1997, Harris et al 1994, Martin et al 2004). *PKD1* accounts for approximately 85% of affected families, with the remaining 15% resulting from *PKD2* mutations (Peters & Sandkuijl 1992, Torra et al 1996).

Analysis of the gene locus dependent clinical phenotypes have highlighted the increased severity of *PKD1*. The mean age of the composite endpoint of death or ESRD is 53 years for *PKD1* and 69.1 years for *PKD2*; both differed significantly from the control population (78 years) (Torra et al 1996, Hateboer et al 1999). Given this difference, it would be expected that the relative prevalence of *PKD2* will increase in older patient subgroups. In one study, 39.1% of patients reaching ESRD after age 63 had *PKD2* mutations (Torra et al 2000). Men have a significantly shorter median survival than women in *PKD2* (67.3 vs. 71.0), but no such sex difference is found in *PKD1* (Hateboer et al 1999). The prevalence of hypertension is fourfold lower in *PKD2* compared to *PKD1* and the occurrence of urinary tract infections and hematuria is also reduced in the former compared to *PKD1* (Hateboer et al 1999, Torra et al 1996). CRISP demonstrated that the differences in the rate of progression between *PKD1* and *PKD2* were due to differences in the number of cysts, but not in the rate at which individual cysts grow (Harris et al 2006). This suggests that cyst initiation, but not cyst enlargement, is modulated by the disease gene. The value of these gene locus-based clinical differences in providing specific prognostic information to patients is limited due to the marked intrafamilial and interfamilial variation in clinical manifestations of ADPKD (Milutinovic et al 1983, Hateboer et al 1999, Persu et al 2004).

ARPKD

All typical forms of ARPKD are caused by mutations in a single gene, *PKHD1* (polycystic kidney and hepatic disease) (Guay-Woodford et al 1995). The protein product has been varyingly named either polyductin or fibrocystin. The gene was identified by three groups working independently. The first group identified *PKHD1* by characterizing its ortholog in the polycystic kidney (*pck*) rat model that phenotypically resembles human ARPKD (Ward et al 2002). Two other groups used positional cloning strategies to identify the *PKHD1* gene on chromosome 6p21.1-p12 (Onuchic et al 2002, Xiong et al 2002).

PKHD1 is among the largest and most transcriptionally complex genes identified to date in the human genome. The genomic sequence spans approximately 470 kb and contains a minimum of 86 exons, which are predicted to be assembled as multiple alternatively spliced transcripts that encode both membrane-bound and soluble protein products (Onuchic et al 2002). The longest transcript contains 67 exons with the open reading frame (ORF) encoded in exons 2–67. *PKHD1* is most abundantly expressed in fetal and adult kidneys, with lower levels detected in pancreas and even lower levels in fetal and adult liver (Onuchic et al 2002, Ward et al 2002).

Mouse *Pkhd1* shares key features with its human ortholog. It is a large gene extending over a 500 kb genomic interval and the longest ORF is encoded by a 67-exon transcript (Nagasawa et al 2002). The mouse gene is expressed not only in fetal and adult kidney, liver, and pancreas, but also at low levels in lung, trachea, sympathetic ganglia, vascular smooth muscle, and testis. Like its human counterpart, the gene is transcriptionally complex. Interestingly, while all gene-targeted models recapitulate the congenital hepatic fibrosis phenotype (Moser et al 2005, Woollard et al 2007, Gallagher et al 2008, Kim et al 2008), only two show late changes in the distal nephron (Garcia-Gonzalez et al 2007, Williams et al 2008) and none recapitulates the full spectrum of early-onset kidney enlargement and failure associated with truncating mutations in humans. The basis for this phenotypic variation may be some combination of hypomorphic alleles and possible species differences in gene product function.

Mechanisms of Cyst Formation in PKD

The recessive pattern of inheritance in ARPKD suggests homozygous complete or partial loss of *PKHD1* function as the basis for the disease. However, understanding the mechanism of cyst formation in ADPKD was more problematic given the observation that focal cystic dilatations occur in kidney tubules that otherwise appear normal along most of their length (Baert 1978). A molecular explanation for the focal origins of cysts despite the germline nature of heterozygous loss-of-function mutations came with the discovery that cyst lining cells from human ADPKD cysts have loss of heterozygosity (LOH) leading to homozygous inactivation of the respective *PKD* genes in both kidney tubular cysts (Qian et al 1996, Brasier & Henske 1997, Pei et al 1999, Torra et al 1999) and bile ductular cysts (Watnick et al 1998).

The conclusions from these studies were challenged (Ong & Harris 1997) on the basis of: immunolocalization studies showing evidence of PC1 expression in human cysts (Geng et al 1996, Ward et al 1996, Ong et al 1999a, 1999b); the low rate of detection of LOH in *PKD1* cysts; the high rate of somatic mutations required to account for all cysts; and the absence of a causal link in LOH analyses. The initial immunolocalization studies focused on cellular staining patterns and did not examine cilia expression of the protein; the latter is now thought to be the primary site of PC1 function in ADPKD (Davenport & Yoder 2005). The low detection rate of LOH in cysts reflected limitations in comprehensive screening of *PKD1* for pathogenic mutations. When applied to cysts from patients with *PKD2* mutations, the rate of detection of second hits was the same as the sensitivity of mutation detection suggesting that effectively all cysts had second hits (Watnick et al 2000). Furthermore, the rate of somatic mutations required to account for cysts that develop in ADPKD is in keeping with the rate of measured mutation frequencies in kidney epithelial cells (Martin et al 1996). Finally, a novel mouse model of *Pkd2* was the first to established a causal relationship between somatic inactivation of the normal allele and cyst formation (Wu et al 1998a). In this model, a modified allele, *Pkd2*^{WS25}, resulted from insertion of a disrupted exon 1 in tandem with the wild type exon1. *Pkd2*^{WS25} expresses functional PC2 but was prone to genomic rearrangement converting it to either a null or a true wild-type allele (Wu et al 1998a). Somatic conversion to a allele correlated with cyst formation in the kidney, liver and pancreas and cyst-lining cells did not express PC2 showing that somatic loss of *Pkd2* is sufficient for cyst formation (Wu et al 1998a). More recently the conditional Cre-lox system has been used to bypass the embryonic lethality of null mice (Lu et al 1997, Wu et al 2000) to inactivate genes of interest in a tissue-selective or temporally controlled manner and produce a number of orthologous gene-based postnatal polycystic kidney disease models (Lantinga-van et al 2004, Piontek et al 2004, Jiang et al 2006, Shibazaki et al 2008).

It is now clear that LOH is sufficient to initiate cyst formation. However, alternative mechanisms and determinants of cyst formation may also exist and the relative importance of these in human clinical disease remains to be determined. A unique chimeric animal model produced by aggregation of *Pkd1*$^{-/-}$ embryonic stem cells with wild type morulas expressing LacZ as the wild-type lineage marker (Nishio et al 2005) found that mice formed kidney cysts commensurate with the degree of *Pkd1*$^{-/-}$ chimerism. The cyst-lining epithelia were mosaic with both *Pkd1*$^{-/-}$ and *Pkd1*$^{+/+}$ epithelial cells present in the early stages of cystogenesis. Over time, *Pkd1*$^{-/-}$ cells replaced the *Pkd1*$^{+/+}$ cells by inducing apoptosis via the c-Jun N-terminal kinase (JNK)-mediated pathway (Nishio et al 2005). These findings suggest the possibility that cells that lose PC1 expression induce cystic degeneration and programmed cell death in surrounding tubular cells. Temporally controlled Cre-lox-based inactivation of *Pkd1* has shown that the timing of gene inactivation impacts the rate of cyst growth (Davenport & Yoder 2005, Piontek et al 2007). Inactivation of *Pkd1* in the developing kidney results in rapid cyst growth while inactivation in the adult kidney results

in markedly slower cyst growth (Davenport & Yoder 2005, Piontek et al 2007). A developmentally regulated change in the gene expression profile of the kidney coincides with the differential effect of *Pkd1* loss on cyst formation (Piontek et al 2007). The underlying proliferative potential of the cells as a function of developmental stage may account for part of the difference in the rate of cyst formation (Shibazaki et al 2008). This is supported by the finding that regenerating tubules after ischemia reperfusion injury have an increased rate of cyst formation (Patel et al 2008).

Cyst formation resulting from transheterozygous somatic mutations involving *PKD1* and *PKD2* have also been proposed (Watnick et al 2000, Koptides et al 2000). However, transheterozygous mutations are unlikely to be sufficient for cyst formation. Individuals with bilineal inheritance of *PKD1* and *PKD2* mutations (Pei et al 1999) and transheterozygous mice (Wu et al 2002) show somewhat more severe polycystic kidney disease, but the overall severity of disease in both man and mouse is within the range observed for single gene mutations. For LOH mechanisms, complete absence of the gene product may not be required and, at least for PC1, markedly reduced expression without complete absence of protein is sufficient for cyst formation (Lantinga-van et al 2004, Jiang et al 2006). Paradoxically, overexpression of PC1 as large insert genomic clone transgenes in mice results in cyst formation, the severity of which appears related to the degree of overexpression (Pritchard et al 2000, Thivierge et al 2006). The mechanism of this phenomenon is not understood.

Genotype–Phenotype Correlations

Most of the several hundred *PKD1* mutations described to date are unique to individual families. This suggests that each mutation essentially arose as a de novo event that was fixed in the population since it confers little or no significant reproductive disadvantage (Milutinovic et al 1983). The de novo mutation rate for *PKD1* is estimated to range from 1.8×10^{-5} to 6.9×10^{-5} per generation (Dobin et al 1993, Rossetti et al 2001). Genotype–phenotype correlations refer to the association between specific germline mutations (genotype) and the resulting spectrum of disease expression (phenotype). Genotype–phenotype correlations are largely absent in ADPKD. The one notable exception to this is the occurrence of contiguous gene deletion syndromes with loss of *PKD1* and *TSC2*, the gene mutated in tuberous sclerosis complex 2. This contiguous gene deletion results in early-onset severe APDKD and clinical features of tuberous sclerosis (Sampson et al 1997). The marked intrafamilial variation seen in ADPKD suggests that factors other than the germline mutation determine the course of the disease. Intrafamilial variability is illustrated by dizygotic twin pairs, one of whom developed severe polycystic kidneys in utero while the sibling showed a more typical course of adult onset disease (Peral et al 1996). In an analysis of age of renal death in 74 parent–offspring pairs, the median difference within the pairs was 0 years, but the range was −26.3 to +27.2 years in a patttern that followed a normal Gaussian distribution (Geberth et al 1995). A systematic analysis of genotype–renal function correlation in 461 affected individuals from 71 ADPKD families with known *PKD2* mutations revealed that the location of *PKD2* mutations did not influence the age of onset of ESRD (Magistroni et al 2003). This study also found that marked intrafamilial variability in disease severity occurs in *PKD2* as well, with several individuals showing presentation of ESRD before age 50 while affected family members had the typical late-onset disease.

In *PKD1*, about one-third of putative pathogenic variants are predicted to be amino acid substitution or other in-frame variants. No correlation between the type of mutation and disease severity has been found in *PKD1* (Rossetti et al 2002), suggesting that the nontruncating mutant alleles are nonetheless loss-of-function. Two studies have correlated 5′ location of mutations in *PKD1* with more severe phenotypes. In one study examining 324 patients from 80 families with known *PKD1* mutations, location of the mutations 5′ to the median position correlated significantly with about a 3 year earlier onset of ESRD (Rossetti et al 2002). In a second study, 51 patients with vascular complications and known *PKD1* mutations had the median position of the *PKD1* mutation significantly further 5′ than *PKD1* families without vascular complications (Rossetti et al 2003). These data correspond to a previous report of a smaller number of cases (Watnick et al 1999) and raise the possibility that the large 5′ extracellular domain of polycystin-1 may act as a dominant factor in aneurysmal complications in ADPKD.

There is limited evidence for genotype–phenotype correlations in ARPKD. In general, *PKHD1* mutations are distributed along the genomic sequence and most patients are compound heterozygotes for private mutations. Nonconservative amino acid substitutions account for single largest mutational group (42–77%), whereas truncating mutations have been observed in 23–39% of patients (Furu et al 2003; Bergmann et al 2005, Loosekoot et al 2005, Sharp et al 2005). All patients carrying two truncating mutations died shortly after birth, whereas the presence of at least one missense change was compatible with neonatal survival (Bergmann et al 2003, 2005, Furu et al 2003, Sharp et al 2005, Rossetti & Harris 2007). The converse was not observed, suggesting that some missense changes are as detrimental as truncating mutations. However, in general, patients with milder clinical phenotypes were more likely to have missense mutations and protein-truncating mutations were more frequently identified in patients with more severe phenotypes (Rossetti & Harris 2007).

Modifier Effects in PKD

Modifier effects, either genetic or environmental, may contribute to the phenotypic variation seen in ADPKD. Support for modifying effects comes from a study comparing intrafamilial variation in age to ESRD among siblings to that in monozygotic twins that found significant excess variability among the siblings relative to the twin pairs (Persu et al 2004). Two studies have examined the proportion of the variance explained by modifier genes (i.e. heritability). The first examined 406 patients from 66 *PKD1* families and estimated the heritability for creatinine clearance (C_{cr}) among those without ESRD to be 42% and the heritability for ESRD to be 78% (Paterson et al 2005). A second study examined variance of phenotypic traits in 315 affected relatives from 83 *PKD1* families. In this study, modifier genes accounted for 32% of the variance for C_{cr} and 43% to 50% of the variance for age of onset of ESRD (Fain et al 2005). Although conclusions from these data are limited by power, Paterson et al (2005) raise the possibility that genetic modifier genes contribute to differing extents to progression in early-stage disease and late-stage disease.

A commonly used alternative to genome-wide association studies has been candidate gene association studies. In general, these studies have had limited success and impact due to failure to replicate, and their exploration in ADPKD has resulted in similarly inconclusive outcomes. A meta-analysis of a large number of studies looking at the ACE I/D polymorphism in ADPKD concluded that evidence does not support the hypothesis that ACE activity modifies ADPKD outcome (Pereira et al 2006). The largest study to undertake the candidate gene association approach examined the effect of seven candidate gene polymorphisms (endothelial nitric oxide synthase (*NOS3*), angiotensin-converting enzyme (*ACE*), tumor growth factor-ß1 (*TGFB1*), bradikinin receptor B1 (*BDKRB1*) and B2 (*BDKRB2*), epidermal growth factor receptor (*EGFR*) and *PKD2*) on age of onset of ESRD in 355 patients from 131 families with known *PKD1* mutations (Tazon-Vega et al 2007). No significant association with disease progression for any of these functional candidate modifiers was found. Successful identification of modifier genes in ADPKD will benefit from use of a quantitative trait like rate of progression of kidney volume as determined by the CRISP protocol, larger studies powered for genome wide association, and improved understanding of the molecular pathways involved in disease progression to help identify candidate loci.

In ARPKD, the majority of affected siblings have relatively concordant clinical phenotypes. However, intrafamilial phenotypic variability is well-described (Kaplan et al 1988, Deget et al 1995, Bergmann et al 2005). On average, about 20% of ARPKD siblings have discordant clinical features, suggesting that genetic modifiers may modulate clinical disease expression. While genome-wide association studies have not been performed in human ARPKD largely due to small cohort sizes, genetic modifiers have been mapped in several mouse models of recessive PKD (Guay-Woodford 2003). These studies have identified specific loci to test for modifier gene effects in human ARPKD patients. For example, *Kif12*, the gene encoding a kinesin-12 family member protein, has been identified as a candidate genetic modifier in the *cpk* mouse model of recessive PKD (Mrug et al 2005). Breeding studies between *Pkhd1^{del3−4/del3−4}* and *Pkd^{+/−}* mice demonstrates that *Pkd1* may modulate ARPKD disease severity (Garcia-Gonzalez et al 2007). This suggests the possibility that the major PKD genes interact in a common genetic pathway and that *PKD1* variants may modulate disease expression in human ARPKD.

POLYCYSTIN-1

Polycystin-1 (PC1) was a completely novel protein when it was initially discovered. It is comprised of 4302 amino acids with a ~3000 amino acid extracellular NH$_2$-terminus and a ~220 amino acid cytosolic COOH-terminus (The International Polycystic Kidney Disease Consortium 1994, Hughes et al 1995) and 11 transmembrane domains (Nims et al 2007) (Figure 24.3). The extracellular NH$_2$-terminal domain contains a distinct combination protein motifs including leucine-rich repeats, a WSC homology domain,

C-type lectin domain, an LDL-A related domain, and 16 iterations of immunoglobin-like PKD repeats (The International Polycystic Kidney Disease Consortium 1994, Hughes et al 1995, Sandford et al 1997), a receptor egg jelly (REJ) domain (Moy et al 1996) and a GPCR proteolytic site (GPS domain) (Ponting et al 1999). Additional domains identified either by homology or computational prediction include a PLAT/lipoxygenase homology 2 (LH2) domain between the first and second transmembrane spans (Bateman & Sandford 1999) and a polycystin motif region in the extracellular loop between the sixth and seventh transmembrane spans (Li et al 2003b). The cytoplasmic COOH-terminus of PC1 contains a coiled-coil domain (Qian et al 1997) as well as a potential PEST sequence (Tsiokas et al 1997) and several putative phosphorylation sites (Li et al 1999, Parnell et al 1999, Wilson et al 1999, Roitbak et al 2004). The region of the last five transmembrane spans of PC1 share sequence homology with polycystin-2 (PC2) although there are critical differences in the sequence that suggest that PC1 does not function directly as a channel protein the way PC2 does (Li et al 2003b).

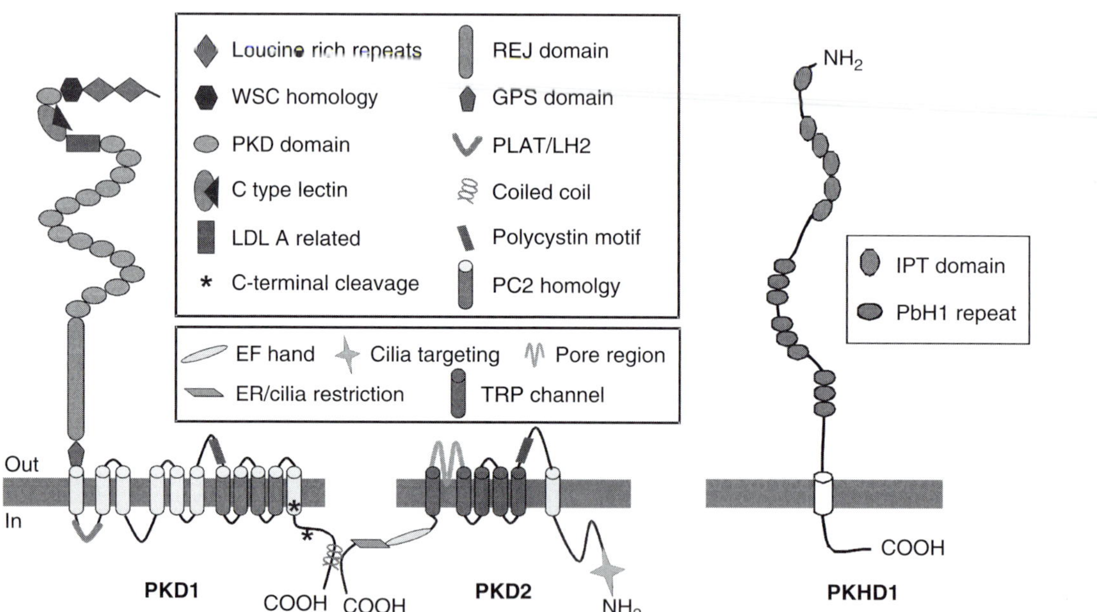

FIGURE 24.3   Schematic representation of the predicted structural domains of *PKD1, PKD2* and *PKHD1* gene products. There are a combination of structurally predicted domains and experimentally identified or verified functional domains shown. In PC1, the immunoglobulin-like structure of the PKD domains have been determined by structural studies. Cleavage at the GPS site, as well as at two C-terminal cleavage locations, have been demonstrated. The REJ has been experimentally implicated in the autoproteolytic GPS cleavage process. The interaction of the COOH-terminal coiled coil domain of PC1 with the COOH-terminus of PC2 is well established. The TRP channel function of PC2 has been shown in several systems to be a non-selective cation channel permeable to Ca^{2+}. The EF-hand can bind Ca^{2+} in vitro (S. Somlo, unpublished observations). The functions of the cilia targeting and the ER/cilia restriction domains in PC2 have been defined. Polyductin/fibrocystin has a short cytoplasmic carboxyl terminus, single transmembrane (TM) domain, and a long extracellular domain with tandemly repeated immunoglobulin-like-plexin transcription factor (IPT) domains and a minimum of 9 parallel β-helix 1 (PbHl) repeats clustered within three groups between the last IPT domain and the TM domain. The carboxy terminus includes putative cAMP/cGMP-dependent protein kinase phosphorylation sites. (Modified from Somlo S, Torres VE and Caplan MJ. (2007). In: Seldin and Giebisch's The Kidney: Physiology & Pathophysiology, Alpern RJ and Hebert SC. Academic Press, pp. 2283–2314). (see also Plate 39)

The large extracellular domain of PC1 has generated much interest from both a structural and functional perspective. The NMR solution structure of one PKD domain showed the domain to have a β-sandwich fold that belonged to a distinct family conserved among vertebrate polycystin homlogs (Bycroft et al 1999). Atomic force spectroscopy studies have shown that PKD domains have mechanical stability similar to that of immunoglobulin domains in the muscle protein titin (Forman et al 2005). Stretching the PC1 extracellular region results in a sawtooth pattern of equally spaced force peaks indicative of the sequential unfolding of individual PKD domains; the domains refold after mechanical force unloading (Qian et al 2005). PC1 signaling events might be regulated through force driven unfolding/refolding events with or without associated ligand interactions.

PC1 undergoes autoproteolytic cleavage at the GPS site to yield the extracellular NH_2-terminal fragment (NTF) and the intramembranous COOH-terminal fragment (CTF) (Qian et al 2002). This cleavage process requires an intact REJ and disease mutations in the REJ can result in failure of GPS cleavage. The NTF and CTF remain non-covalently associated with each other and functional GPS cleavage is required for PC1 to be able to stimulate spontaneous tubulogenesis of MDCK cells in three dimensional matrix culture (Qian et al 2002). GPS cleavage is conserved in proteins evolutionarily related to PC1 (Mengerink et al 2002) and has been reported as an essential step for folding, trafficking, and bioactivity of nonpolycystin proteins (Krasnoperov et al 2002).

PC1 is expressed in kidney, brain, heart, bone, and muscle (Geng et al 1997, Peters et al 1999) and its expression in the kidney appears to be downregulated after development (Geng et al 1997). The subcellular localization of PC1 has been difficult to establish and a broad range of sometimes conflicting expression patterns have been reported. Several studies identified PC1 in the plasma membrane of kidney tubule cells, particularly in the distal nephron segments including collecting ducts (Geng et al 1996, Palsson et al 1996, Ibraghimov-Beskrovnaya 1997). Within these segments, PC1 appeared at the lateral membrane at sites of intercellular adhesion (Foggensteiner et al 2000) and at desmosomes (Scheffers et al 2000). Overexpressed, epitope-tagged full length PC1 has been localized in the lateral membrane (Boletta et al 2000). Similar lateral location of a PC1–PC2 complex was reported in kidney tissue overexpressing PC1 as a transgene (Newby et al 2002). With the recognition of the importance of primary cilia in ADPKD, native PC1 has been localized to the primary cilia of epithelial cells in culture (Yoder et al 2002, Nauli et al 2003) and there is general agreement on this localization.

Proteins homologous to PC1 have provided additional insights into function. The discovery of the PC1 homolog in C. elegans, the location of vulva-1 (LOV-1), was one of the first to implicate cilia in the pathogenesis of cystic disease (Barr & Sternberg 1999). With the advent of the sequencing of the human and mouse genomes, four genes encoding proteins structurally related to PC1 have been identified in mammals (Hughes et al 1999, Yuasa et al 2002, Li et al 2003). Recently, a complex of the PC1 homolog PKD1L3 and a PC2 homolog, PKD2L1 have been found form the sour taste receptor (Huang et al 2006, Ishimaru et al 2006, Lopez-Jimenez et al 2006) as well as the cerebrospinal fluid chemosensory system that sense extracellular pH (Huang et al 2006). This is the first functional demonstration in a paralogous system of the prevailing hypothesis that PC1 and PC2 function as a receptor and channel signaling complex.

POLYCYSTIN-2

Polycystin-2 (PC2) is a non-selective cation ion channel permeable to Ca^{2+} that belongs to the TRPP subfamily of TRP cation channels (TRPP2, reviewed in Kiselyov et al 2007, Nilius et al 2007). The last five transmembrane spans in PC2 bear a strong TRP channel signature and the region between S5 and S6 (transmembrane segments 5 and 6) contains the putative pore region (Figure 24.3) (Li et al 2003b). The first extracellular loop between S1 and S2 contains a polycystin motif of unknown function that is highly conserved in all PC1- and PC2-related proteins. The cytoplasmic tail of PC2 contains a Ca^{2+} binding EF-hand (Mochizuki et al 1996). In addition to the domains identified using informatics, motifs involved in trafficking of PC2 have been experimentally defined. The RVxP motif in the NH_2-terminus of PC2 is necessary and sufficient for the cilia location of PC2 (Geng et al 2006). A domain in the COOH-terminus restricts the membrane location of PC2 to the ER and cilia (Aguiari et al 1999, Geng et al 2006). Mammalian PC2 has at least one phosphorylation site in its COOH-tail that modulates the Ca^{2+} dependence of channel activity (Cai et al 2004) and has a role in trafficking of PC2 between ER, Golgi and plasma membrane compartments (Kottgen et al 2005).

PC2 is expressed in the kidney, heart, ovary, testis, vascular smooth muscle, and small intestine (Markowitz et al 1999, Obermuller et al 1999). In the kidney, PC2 is expressed in all nephron segments, with the exception of the thin limbs of the loops of Henle and the glomerulus. Expression in the distal nephron segments may be more robust than in the proximal segments (Wu et al 1998). There is general agreement that PC2 is strongly expressed in the membrane of the ER (Cai et al 1999, Hanaoka et al 2000, Koulen et al 2002, Newby et al 2002, Scheffers et al 2002, Luo et al 2003, Hidaka et al 2004). There is also consensus and strong evidence to show that PC2 is localized to the primary cilia in kidney tissues and cultured epithelial cells (Pazour et al 2002, Yoder et al 2002). Native and overexpressed protein can be detected in the primary cilium. The NH_2-terminus domain of PC2 containing an RVxP amino acid motif has been identified as both necessary and sufficient for cilia location of type 2 membrane proteins (Geng et al 2006, Jenkins et al 2006). Trafficking of PC2 to cilia is independent of PC1. Native PC2, but not heterologously overexpressed protein, has been found on the plasma membrane both by anti-native protein antibodies (Scheffers et al 2000, Luo et al 2003) and by channel physiology (Ma et al 2005). This has not been the case for heterologously expressed epitope-tagged PC2; this has led to debate as to whether or not PC2 is present in the generalized plasma membrane (reviewed in Witzgall 2005). It has been proposed that PC2 requires co-assembly with PC1 to traffic to the cell membrane and coexpression of both polycystins results in appearance of a novel plasma membrane cation channel activity (Hanaoka et al 2000, Nauli et al 2003). Studies using the C-terminal fragment of PC1 suggested a possible reciprocal role for PC2 in the cellular distribution of PC1 (Grimm et al 2003).

PC2 is highly conserved in metazoan evolution and a homolog has even been identified in the unicellular green algae, *Chlamydomonas reinhardtii* (Huang et al 2007). Two paralogs of PC2, PKD2L1 (or PCL) and PKD2L2, have been identified in the mammalian genome (Nomura et al 1998, Wu et al 1998, Veldhuisen et al 1999, Geng et al 2006). As noted above, PKD2L1 functions as a sensor for sour taste and cerebrospinal fluid pH in association with PKD1L3 (Huang et al 2006, Ishimaru et al 2006, Lopez-Jimenez et al 2006). Most of the sequence conservation between mammalian species and lower organisms occurs in the intramembranous TRP domain regions. There is little sequence conservation of the cytosolic NH_2- and COOH-termini from mammals to invertebrates and there appear to be differences in the phosphorylation patterns and trafficking signals across species (Geng et al 2006, Hu et al 2006). In sea urchin, the PC2 homolog associates with suREJ3, the PC1 homolog, and both localize to the acrosomal region of spermatozoa suggesting a possible role in the ionic channel events that accompany the acrosome reaction during fertilization (Neill et al 2004). In *C. elegans*, mutations in the PC2 homolog result in defects in location-of-vulva and response behavior in the male worms that is indistinguishable from the phenotype seen in LOV-1 (PC1 homolog) mutants (Barr et al 2001). The PC2 homolog in *Drosophila* is expressed in sperm, and knockout of PC2 in *Drosophila* results in sperm that retain normal motility and ability to fertilize but lose the guidance mechanism required for reaching the sperm storage organs in the female (Gao et al 2003, Watnick et al 2003). Finally, zebrafish have proven to be a valuable vertebrate genetic model of polycystic kidney disease. Zebrafish have a pronephric kidney as well as left–right body axis asymmetry and knockdown or mutation of the PC2 homolog in fish results in cystic dilation of the pronephric tubule as well as abnormalities of left–right body axis formation (Sun et al 2004, Obara et al 2006, Schottenfeld et al 2007).

POLYDUCTIN/FIBROCYSTIN

When expressed from the longest ORF transcript, polyductin is a novel integral membrane protein of 4074 amino acids (4059 amino acids in the mouse) with a signal peptide, an extensive, highly glycosylated, 3858-amino acid extracellular NH_2-terminus, a single transmembrane domain, and a short (192 amino acid) cytoplasmic tail (Nagasawa et al 2002, Onuchic et al 2002, Ward et al 2002). Structural analysis indicates that polyductin contains a series of immunoglobulin-like-plexin-transcription factor (IPT) domains in its amino terminus and multiple parallel β-helix 1 (PbH1) repeats between the last IPT and the transmembrane domain (Figure 24.3) (Onuchic et al 2002). Multiple tandemly arranged IPT domains are also found in single-pass cell surface receptors that belong to the Sema superfamily of proteins, including the hepatocyte growth factor receptor and the plexins. The homology with the Sema protein family suggests that polyductin may similarly function in the regulation of cellular adhesion, repulsion, and proliferation. In addition, the multiple PbH1 repeats suggest additional polyductin functions, including recognition and/or modification of carbohydrate moieties. The short carboxyl tail includes putative cAMP/cGMP-dependent protein kinase phosphorylation sites. The same general domain structure is conserved in mouse polyductin and the human and mouse proteins share 73% identity over their complete length, although there are segments with considerably higher (87%) and lower (40%) identities (Nagasawa et al 2002).

Northern analysis indicates that *PKHD1* is transcriptionally complex (Nagasawa et al 2002, Onuchic et al 2002). Alternatively spliced transcripts are predicted to encode two broad categories of proteins, one membrane-bound and the other secreted. This complex splicing pattern is highly conserved in the mouse *Pkhd1* ortholog (Nagasawa et al 2002). Current speculation holds that these multiple splice variants may provide a mechanism for regulating the temporal and spatial functions of polyductin in a tissue-specific manner (Menezes et al 2004, Nagasawa et al 2002, Onuchic et al 2002). Immunohistochemical analyses of polyductin expression demonstrate staining of cortical and medullary collecting ducts and thick ascending limbs of Henle in the kidney, as well as of biliary, pancreatic, and salivary ductal epithelia (Gallagher et al 2000, Ward et al 2003, Menezes et al 2004). In the developing fetus, staining was observed in ureteric bud branches, but not in the metanephric mesenchyme, S-shaped bodies or glomeruli (Menezes et al 2004). Immunoreactivity was also observed in intra- and extrahepatic biliary ducts, pancreatic ducts and salivary gland ducts. This pattern of temporal and spatial expression largely mirrors the renal tubular and biliary pathology observed in ARPKD.

At the subcellular level, polyductin is expressed along the apical membrane in cortical and medullary collecting duct epithelia and the thick ascending limb of Henle

(Menezes et al 2004, Wang et al 2004, Zhang et al 2004, Nagano et al 2005, Gallagher et al 2008), as well as in biliary and pancreatic ductal epithelia. Furthermore, like PC1 and PC2, polyductin localizes to the primary apical cilia and basal bodies in renal epithelial cells (Menezes et al 2004, Wang et al 2004, Zhang et al 2004, Nagano et al 2005, Gallagher et al 2008) and in cholangiocytes of intrahepatic bile ducts (Masyuk et al 2003).

Informatic analyses have identified a mammalian *PKHD1* homolog, *PKHDL1*, on chromosome 8q23 (Hogan et al 2003). *PKHDL1* contains 78 exons and spans a genomic interval of approximately 168 kb. Like *PKHD1*, *PKHDL1* is transcriptionally complex and predicted to encode both membrane-spanning and soluble peptides. The longest 13.1 kb transcript encodes fibrocystin-L, which is predicted to be a large receptor protein (4243 amino acids; 466 kDa) with a signal peptide, large extracellular NH_2-terminus, single transmembrane domain, and short cytoplasmic COOH-tail. *PKHDL1* is widely expressed at a low level in most tissues, particularly immune cell subtypes. Mutational screening indicates that *PKHDL1* is unlikely to be associated with ARPKD.

CILIA

A major insight into understanding ADPKD and ARPKD was the discovery of the central role of cilia in the pathogenesis of these and other hepatorenal fibrocystic diseases. The 'cilia hypothesis' for human diseases has been reviewed extensively (e.g. Pazour 2004, Bisgrove & Yost 2006, Marshall & Nonaka 2006, Singla & Reiter 2006). Defects in proteins found in the cilial membrane, cilial axoneme, the basal body or pericentriolar region are varyingly associated with a clinical spectrum of disease that can include cysts in kidney tubules, bile ducts and pancreatic ducts, as well as retinal degeneration and retinitis pigmentosa, *situs inversus* (incorrect left–right body axis), anosmia, infertility, and hydrocephalus. This list is one that is likely to grow over time and will include metabolic disorders such as obesity (Davenport et al 2007) in addition to diseases of organ structure.

The apical surfaces of kidney tubular epithelial cells, with the exception of mature intercalated cells, have a single non-motile 9 + 0 primary cilium with nine peripheral microtubule doublets and no central pair (Figure 24.4). The cilium emanates from the centrosome, which is comprised of mother and daughter centrioles and pericentriolar material that together form the basal body complex. The microtubule-based cilial axoneme is covered by a membrane that is physically continuous with the apical plasma membrane but is functionally a discrete compartment (Meder et al 2005, Geng et al 2006). Cilia are devoid of protein synthetic machinery so all cilia components are produced in the cell

body and assembled into the cilium by intraflagellar transport (IFT) (Figure 24.4). Transition fibers at the base of cilia form a functional barrier that separates the cilial compartment from the rest of the cell. Kinesin-2 and cytoplasmic dynein motor proteins respectively mediate anterograde and retrograde traffic along the axoneme. Kinesin-2 is a heterotrimeric complex composed of two motor subunits, KIF3A and KIF3B or KIF3C, and a non-motor accessory subunit, KAP3. Cilia are present on terminally differentiated, non-mitotic cells – when the cell prepares to divide, the primary cilium is reabsorbed and each centriole divides into new mother and daughter centrioles that migrate to the poles of the mitotic spindle.

The central role of cilia in the pathogenesis of ADPKD evolved from studies of model organisms. The gene mutated in *C. elegans* defective for a stereotyped male mating behavior was identified as the PC1 homolog and was in turn localized to cilia in the nematode (Barr & Sternberg 1999). The *inv* mutant mouse (Mochizuki et al 1998) and the *Tg737* mutant mouse (Murcia et al 2000) were the first noted to share the phenotypic features of cystic kidneys and left–right axis abnormalities. This is significant since left–right axis determination had previously been established to be a cilia-dependent phenotype (Hirokawa et al 2006). The connection was further strengthened when a recessive mouse polycystic kidney disease gene, *Tg737* (also called polaris and IFT88), was identified as a component of the intraflagellar transport machinery necessary for ciliary biogenesis (Pazour et al 2000). These studies led to more direct evidence of the central role of cilia in ADPKD beginning with the direct demonstration of the localization of PC2 (Pazour et al 2002) and PC1 (Yoder et al 2002) in the cilia of kidney epithelia. Further direct evidence of the cilia link with ADPKD came with the discovery that $Pkd2^{-/-}$ mice have defects in left–right axis formation (Pennekamp et al 2002). PC2 is required for increased cellular Ca^{2+} in cells at the left margin of the node (McGrath et al 2003). This lateralized signal was absent in $Pkd2^{-/-}$ embryonic nodes. It was proposed that PC2 was part of a mechanosensory complex in the node that sensed the flow generated by cilial movement (McGrath et al 2003), although a more complex mechanism for the lateralized nodal Ca^{2+} signal may be operational (Tanaka et al 2005).

Subsequently, a number of genes have been identified in recessive mouse models that manifest cystic kidney disease as part of their phenotype. Although not all of these have homologs associated with human disease, the majority have protein products that can be linked to cilia (reviewed in Guay-Woodford 2003). Similarly, a forward genetic screen using random insertional mutageneis in zebrafish followed by cloning of disrupted genes that resulted in a cystic kidney phenotype in the fish identified a number of intraflagellar transport related proteins, as well as *Pkd2*, among the target genes (Sun et al 2004). This consistent association of the polycystic kidney phenotype with proteins involved in various aspects of cilia function provides

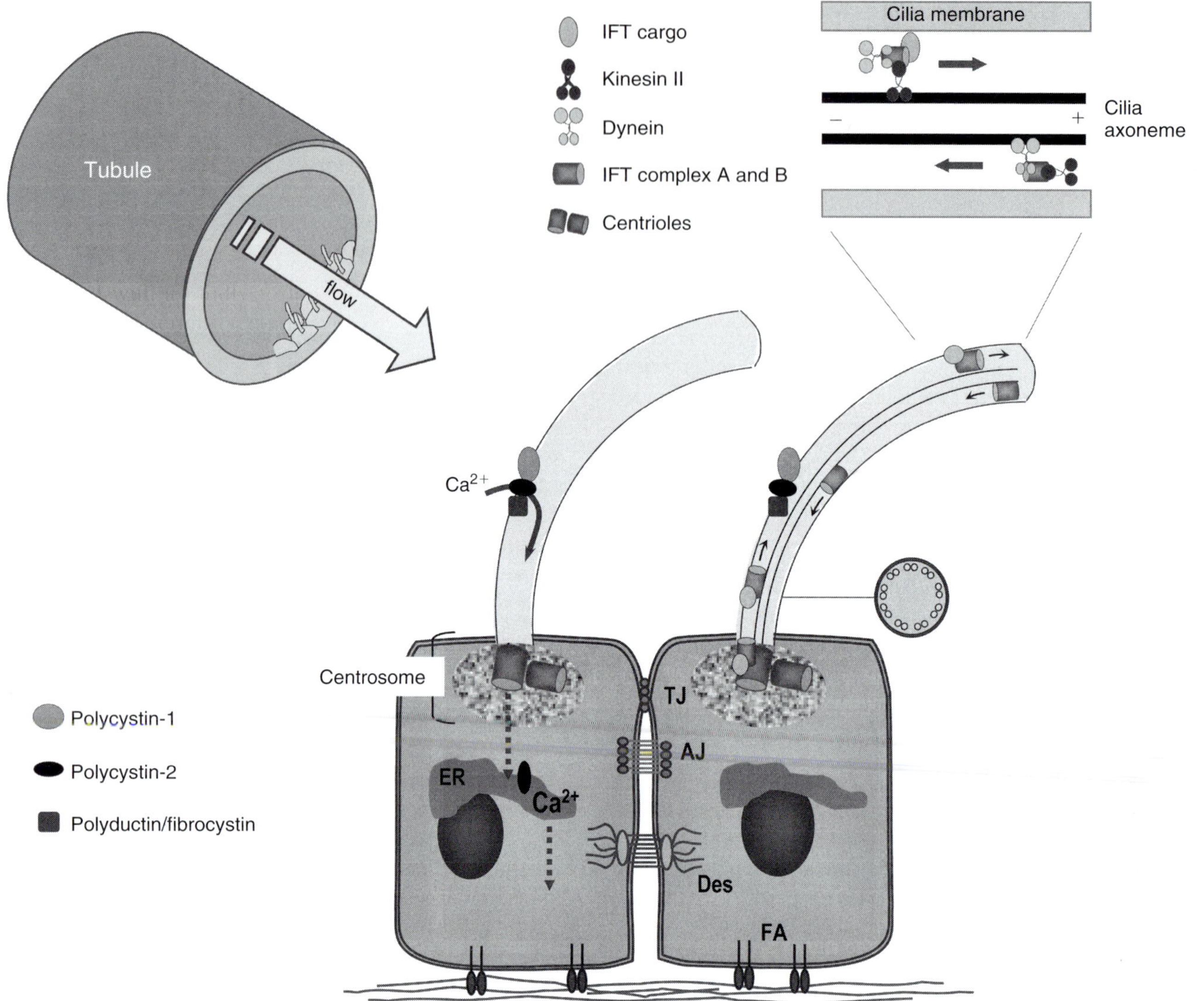

FIGURE 24.4   Structure and function of cilia. Protein synthetic machinery is not present in cilia so proteins must enter cilia by intraflagellar transport (IFT) from the cytosol. IFT involves the movement of large multiprotein complexes along the outer doublet microtubules beneath the ciliary membrane. Proteins destined for the ciliary membrane (e.g. the polycystins and polyductin) are synthesized in the endoplasmic reticulum, processed through the Golgi stack and trafficked into vesicles. These vesicles are transported to the base of the cilium comprised of the centrosome/basal body complex, where they interact with the IFT-particle proteins concentrated in a cytoplasmic pool. The vesicles fuse with the plasma membrane and the particles, and associated membrane are transported to the flagellar tip by the plus strand hetertrimeric kinesin-II motor. Once at the tip, the kinesin-II is replaced by cytoplasmic dynein lb/2 which acts as the minus strand motor to carry the particles back to the cell body. (see also Plate 40)

strong support for the central role of cilia in ADPKD and ARPKD. The hypothesis was formally put to the test by conditionally inactivating Kif3a, a component of the heterotrimeric kinesin-2 anterograde motor, in the distal nephron segments of the kidney (Lin et al 2003). Inactivation of the motor protein required for normal cilia formation results in loss of cilia only in the nephron segments where the gene was inactivated. Loss of cilia by this mechanism in turn resulted in cyst formation in the kidney showing prospectively that impairing cilia formation and function in kidney tubules results in cyst formation.

Primary cilia are increasingly implicated in a wide variety of important morphogenic signaling pathways (e.g. Corbit et al 2005, Haycraft et al 2005) which in turn account for the wide spectrum of clinical features associated with syndromes when their function is disturbed (Badano et al 2006). In addition to ADPKD and ARPKD, the cilia, basal body and pericentriolar complex has been implicated in renal fibrocystic disorders such as nephronophthisis (nine genes, *NPHP1–NPHP9*), Bardet-Biedl syndrome, (12 genes, *BBS1–BBS12*), Meckel-Gruber syndrome (two genes, *MKS1, MKS3*) and oro-facial-digital syndrome

(*OFD1*) (Badano et al 2005, Davenport & Yoder 2005, Praetorius &Spring 2005, Davis et al 2006, Marshall & Nonaka 2006, Singla & Reiter 2006) (Table 24.1). The roles of cilia–basal body–pericentriolar complex in maintaining epithelial cell differentiation and tubular morphology are now the focus of efforts to further understand polycystic kidney disease.

A notable exception to the direct cilia link among polycystic diseases is ADPLD (Qian et al 2003b). Loss of function mutations in two genes, *PRKCSH* and *SEC63*, result in this disease (Drenth et al 2003, Li et al 2003a, Davila et al 2004) which also requires a somatic second step mutation to initiate cyst formation (S. Somlo, unpublished observations). *PRKCSH* encodes the non-catalytic β-subunit of glucosidase II (GIIβ), an *N*-linked glycan processing (glucose trimming) enzyme in the ER (Trombetta et al 1996, D'Alessio et al 1999, Ellgaard et al 1999, Treml et al 2000, Helenius & Aebi 2001). The SEC63 protein functions as part of the protein translocation machinery (translocon) to deliver proteins into and through the ER membrane. The activities of the translocon are tightly coordinated with downstream events of protein folding, modification (e.g. *N*-glycosylation) and assembly (Schnell & Hebert 2003), placing both SEC63 and GIIβ in the same functional pathway. SEC63 and GIIβ function upstream of the delivery of integral membrane proteins to their respective destinations, including the cilial membrane. It has been propose that the cystic phenotypes associated with loss of either ADPLD gene may result indirectly from loss of effective delivery of proteins such as PC1 and PC2 to their functional destinations (e.g. cilia) (Davila et al 2004).

POLYCYSTIC KIDNEY DISEASES AND CELLULAR PATHWAYS

There are two general classes of outcome to disease gene positional cloning efforts. The first is to discover that the underlying disease genes have known function and the association with disease elucidates a novel role for this known protein and the pathways in which it functions (e.g. Rodriguez-Viciana et al 2006). The other, which characterizes the experience in ADPKD and ARPKD, is that the genes and their protein products are completely novel and do not easily fit into any known pathways or cellular processes. In the case of the latter, there are three general approaches used to define the functional connection between disease gene and disease phenotype. The first is to look for genetic or physical interacting partners of disease genes and proteins in the hopes of defining function through knowledge of the pathways in which associated genes or proteins participate. The second is to prospectively test the role of the new proteins in known pathways. The outcome of these efforts may in fact lead to the third approach, the

realization that the novel genes in fact define novel pathways that do not readily fit into any known paradigm. The latter discovery is most often attained after numerous false starts involving either or both of the former.

In the case of the polycystic kidney diseases, a good deal of progress in defining the importance of cilia in the disease was achieved through genetic approaches – overlapping disease phenotypes led to proteins associated with cilia and disruption of cilia proteins lead to the polycystic phenotype. In addition, extensive efforts at defining physical interacting partners and at tying the polycystins and polyductin into known cellular pathways have been enlightening although it may still be that there are novel, as yet incompletely understood pathways that will eventually be defined by these disease genes.

Interacting Proteins

The prediction that PC1 is a receptor protein (Hughes et al 1995) coupled with the discovery that PC2 is a channel protein suggested the hypothesis that the two proteins form a receptor–channel complex with PC1 regulating the activity of PC2 (Mochizuki et al 1996). In directed interaction analyses, PC1 and PC2 were shown to associate through their respective COOH-termini (Qian et al 1997, Tsiokas et al 1997) (Figure 24.3). This interaction depends on the integrity of the coiled coil domain in the COOH-terminus of PC1 and pathogenic truncating mutations in either gene result in loss of this interaction (Hanaoka et al 2000). The colocalization of both proteins in cilia (Yoder et al 2002) supports the existence of a PC1–PC2 complex.

Several other associated proteins have been identified for PC1. Not all of these have been validated by interaction with full length PC1 and several have only been examined in invertebrates, so their roles in the pathogenesis of human ADPKD remain to be elucidated. In *C. elegans*, the β-subunit of ATP synthase (ATP-2) co-localizes in cilia and interacts with the PLAT/LH2 domain of the nematode PC1 homolog, LOV-1 (Hu & Barr 2005). Casein kinase II also interacts with LOV-1 and this interaction may regulate trafficking of the *C. elegans* PC2 homolog into cilia in a phosphorylation-dependent manner (Hu et al 2006). Interaction with the COOH-terminal domain of PC1 inhibits degradation and results in re-localization of regulators of G protein signaling (RGS7) (Kim et al 1999), perhaps fitting with a postulated role for PC1 as an atypical G-protein coupled receptor (Delmas et al 2002, Parnell et al 2002). The cytoplasmic tail of PC1 also interacts with tuberin, the tuberous sclerosis complex 2 (*TSC2*) gene product (Shillingford et al 2006). PC1 interacts with intermediate filament components including vimentin, cytokeratin 8 and 18 and desmin (Xu et al 2001). PC1 has also been shown to colocalize and complex with the intercellular adhesion molecules E-cadherin and β-catenin suggesting a possible role in stabilization of adherens junctions and the maintenance of the differentiated polarized state of epithelia in kidney tubules (Huan et al 1999, Geng et al 2000, Roitbak et al 2004). PC1 may mediate communication with the extracellular matrix as suggested by colocalization with $\alpha2,\beta1$ integrin (Wilson et al 1999) and interaction via the COOH-terminal region of the $\beta3$ integrin with $\alpha v,\beta3$ integrin (Woods et al 2004). Finally, a finding that bears further investigation is the observation that extracellular domains of PC1 are capable of in vitro homotypic interactions via their PKD domain repeats (Ibraghimov-Beskrovnaya et al 2000).

In addition to interacting with PC1, PC2 homomultimerizes with itself via its COOH terminus but the stoichiometry of this interaction is unknown (Qian et al 1997, Tsiokas et al 1997, Somlo 1999). PC2 also associates with a number of Ca^{2+} channel proteins although the role of these interactions in ADPKD remains to be fully elucidated. PC2 interacts with at least one other known TRP channel, TRPC1 (Tsiokas et al 1999). In keeping with its putative role in regulating cellular Ca^{2+} homeostasis via intracellular Ca^{2+} pools, PC2 interacts directly with the inositol 1,4,5-trisphosphate receptor (IP_3R) (Li et al 2005) and also regulates the activity of the ryanodine receptor (RyR) through direct interaction (Anyatonwu et al 2007).

Yeast two hybrid screens have yielded several putative PC2 binding partners potentially involved in cytoskeletal interactions. PC2 binds Hax-1 which in turn associates with the F-actin-binding protein cortactin providing a potential link between PC2 and the actin cytoskeleton (Gallagher et al 2000). The COOH-terminus of PC2 interacts with CD2AP (Lehtonen et al 2000) which also potentially provides a link with the actin cytoskeleton (Lehtonen et al 2002). PC2 interacts with tropomyosin-1, troponin-1 and α-actinin, components of the actin microfilament complex (Li et al 2003a, 2003b, 2005a). Yeast two-hybrid screening also identified mDia1/Drf1 (mammalian Diaphanous or Diaphanous-related formin 1 protein) as a PC2-interacting protein (Rundle et al 2004). Interaction with mDia1 has been proposed as a regulator of the voltage dependent gating of PC2 (Bai et al 2008).

PIGEA-14 has been identified as an interacting partner for PC2 (Hidaka et al 2004). Coexpression of PIGEA-14 with PC2 resulted in translocation of both proteins to the trans-Golgi network, suggesting a role for PIGEA-14 in regulating the intracellular location of PC2. Association between PC2 and fibrocystin/polyductin has been investigated in an attempt to unify the mechanisms ADPKD and ARPKD (Guay-Woodford 2003, Zerres & Mucher 1998). Polyductin and PC2 colocalize and coimmunoprecipitate suggesting that they may exist in the same complex (Wang et al 2007). This complex may also include the kinesin-2 anterograde cilia transport motor and polyductin may regulate PC2 channel function (Wu et al 2006). It has even been proposed that polyductin regulates renal tubule formation by regulating PC2 expression and function (Kim et al 2008) although this is not universally accepted (Kaimori & Germino 2008).

Interacting Pathways

Beyond direct interaction, much effort has been devoted to understanding how the polycystins and polyductin may interact with known pathways mediating cellular processes such as proliferation, differentiation, polarity or apoptosis (Figure 24.5). Discovery of these pathways would suggest targets for therapy and several of these are already finding their way into clinical trials (Figure 24.5). One class of pathways that have received attention with good reason is cellular Ca^{2+} signaling. As a TRP channel, PC2 is a nonselective Ca^{2+}-permeable cation channel (Hanaoka et al 2000, Gonzalez-Perrett et al 2001, Koulen et al 2002, Luo et al 2003). While the channel is nonselective for cations, it does pass Ca^{2+} and most investigators believe that it functions in vivo in Ca^{2+}-based second messenger signaling (Delmas 2005). PC2 channel activity is stimulated by physiological concentrations of Ca^{2+} and is inhibited by higher concentrations of Ca^{2+} that may be achieved locally as a result of PC2 channel activity (Gonzalez-Perrett et al 2001, Koulen et al 2002). The importance of its channel function to the pathogenesis of ADPKD was demonstrated by an amino acid substitution mutation, D511V, that is pathogenic in patients (Reynolds et al 1999) and results in loss of channel function without apparently affecting other properties of the protein (Koulen et al 2002, Cai et al 2004). PC2 is a channel at the ER membrane. PC2 expression enhances the release of Ca^{2+} from ER stores (Koulen et al 2002). PC2 can physically interact directly with the IP_3R and, perhaps by virtue of this association, can alter the kinetics of cytoplasmic Ca^{2+} transients induced by IP_3 and by agonists of receptors that signal through IP_3 (Li et al 2005). PC2 also associates with the other major intracellular Ca^{2+} release channel, the RyR (Anyatonwu et al 2007). Further evidence of a role for PC2 in modulating intracellular Ca^{2+} stores comes from the finding that vascular smooth muscle cells isolated from $Pkd2^{+/-}$ mice have a reduction in releasable intracellular Ca^{2+} and reduced capacitative Ca^{2+} entry (Qian et al 2003). In keeping with this data and the RyR interaction data (Anyatonwu et al 2007), *Drosophila melanogaster* heterozygous for a loss-of-function PC2 allele have reduced contractility of the visceral smooth muscle cells via a RyR-dependent mechanism (Gao et al 2004). A role for PC1 in modulating intracellular Ca^{2+} stores is at least suggested by the observation that overexpression of PC1 in MDCK cells may inhibit capacitative Ca^{2+} entry and speed Ca^{2+} re-uptake by the ER (Hooper et al 2005). Despite mounting evidence of a role in cellular Ca^{2+} homeostasis, the role of the polycystins and particularly the PC2 ER channel in ADPKD has yet to be addressed directly.

There is general acceptance of the expression of both PC1 and PC2 in the unique sub-domain of the plasma membrane overlying the cilial axoneme. In light of the 'cilia hypothesis' of fibrocystic disease discussed above, it is generally believed that the role of this pool of polycystins

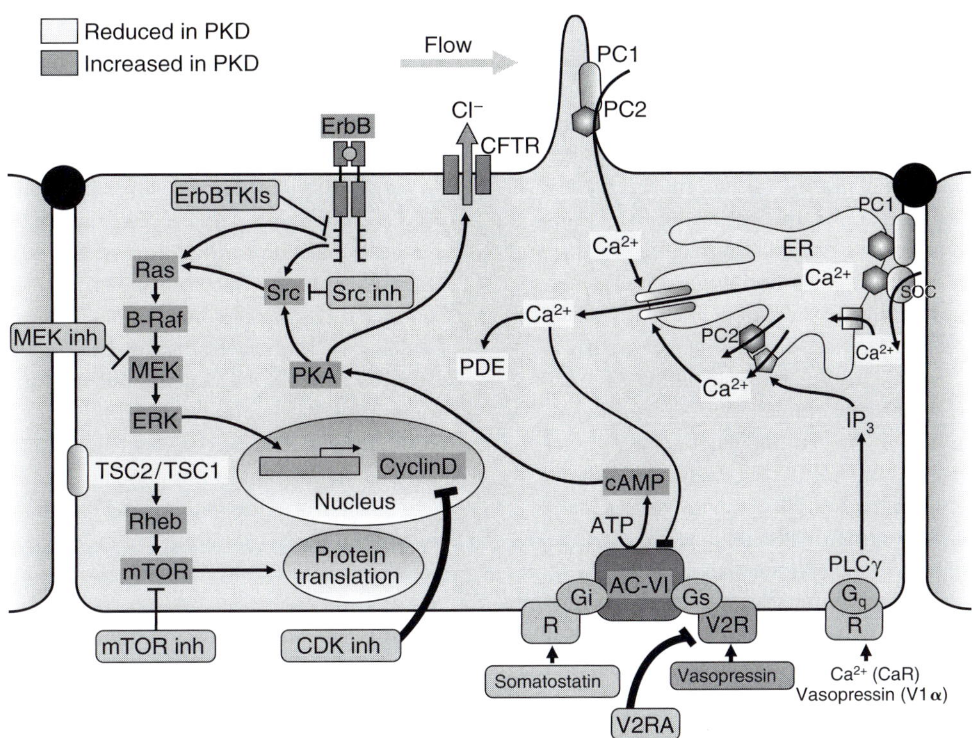

FIGURE 24.5 Polycystin signaling pathways and PKD. Schematic representation of cellular signaling pathways that involve polycystins and related potential targets for therapy. The polycystin complex appears in the cilium where it may participate in flow stimulated Ca^{2+} entry; PC2 is abundant in the ER and intracellular endoplasmic reticulum (ER) Ca^{2+} stores are likely subject to regulation by the polycystins as well. There are direct or indirect interactions between the polycystin proteins and components of the MAPK/ERK, cAMP and mTOR pathways. These response pathways have come to be seen as potentially valuable targets in the search for PKD therapeutics. AC–VI = adenylyl cyclase 6. CDK = cyclin-dependent kinase. ER = endoplasmic reticulum. MAPK = mitogen-activated protein kinase. mTOR = mammalian target of rapamycin. PC1 = polycystin-1. PC2 = polycystin-2. PDE = phosphodiesterase. PKA = protein kinase A. R = receptor. TSC = tuberous sclerosis proteins tuberin (TSC2) and hamartin (TSCl). V2R = vasopressin V2 receptor. V2RA = vasopressin V2 receptor antagonists. (Modified from Torres VE, Harris PC, Pirson Y. Autosomal dominant polycystic kidney disease. Lancet 2007; 369: 1287–1301). (see also Plate 41)

is essential to the pathogenesis of ADPKD. A clue to the a possible function of polycystins in cilia came from the work of Praetorius and Spring. They observed elevations of cellular Ca^{2+} concentrations in response to mechanical displacement of cilia or bending of cilia under laminar flow shear stress (Praetorius et al 2003, Praetorius & Spring 2001, 2003a, 2003b) (Figures. 24.4 and 24.5). This work gave rise to the hypothesis that cilia may serve as mechanosensors responsive to flow stimuli in renal tubules and, although Praetorius and Spring never tested it directly, they proposed the PC1/PC2 channel complex was necessary for the responses they observed. It is noteworthy that the cellular Ca^{2+} response they observed in MDCK cells required both outside Ca^{2+} as well as IP₃R activity. Subsequent studies examining flow responses in mouse kidney cell lines expressing mutant PC1 showed absence of Ca^{2+} transients in response to flow (Nauli et al 2003). It is interesting to note that the kinetics of the increase in cellular Ca^{2+} observed in the latter experiments (Nauli et al 2003) differ substantially from those observed in the original MDCK cell studies (Praetorius & Spring 2001). The PC1

deficient mouse cell line responses required RyR, but not IP₃R (Nauli et al 2003). Nauli et al also suggested that PC2 is absent from cilia of cells lacking functional PC1, although subsequent studies have shown that PC2 traffics to cilia independently of PC1 (Geng et al 2006). The role of cilia in regulating tubule cell Ca^{2+} in response to flow was also examined in isolated perfused collecting ducts from the *Tg737* mouse line with a mutation in an IFT protein (IFT88) that leads to stunted cilia. Loss of normal cilia in this mutant resulted in blunting of the flow-induced rise in cellular Ca^{2+} in perfused tubules (Liu et al 2005). The role of flow and mechanosensation in polycystin signaling remains a hypothesis that bears further testing and verification.

A role for polyductin in Ca^{2+} signaling has also been proposed. PC2 and polyductin appear to physically interact and antibodies directed against extracellular epitopes of polyductin block cellular Ca^{2+} responses to flow stimulation. This suggests that the polycystins and polyductin participate, at least in part, in a common mechanotransduction pathway (Wang et al 2007). As further support for this thesis,

Pkhd1-silencing in renal cell culture systems has been shown to lower cellular Ca^{2+} and prompt epidermal growth factor (EGF)-induced activation of the mitogen-activated protein kinase/extracellular regulated kinase (MAPK/ERK) cascade, causing these cells to become hyperproliferative in response to EGF (Yang et al 2007). Pharmacological elevation of cellular Ca^{2+} blocks this hyperproliferative response. Therefore, in ARPKD, loss of polyductin may cause abnormal proliferation of renal epithelial cells and cystogenesis, perhaps via disruption of PC2-regulated mechanisms that modulate cellular Ca^{2+}. A more direct link between EGF signaling has been proposed by a study showing that EGF binding to its receptor initiates elevations in cytosolic Ca^{2+} concentrations that appear to require participation of the channel activity of PC2 (Ma et al 2005).

The planar cell polarity (PCP) pathway has recently become a focus in fibrocystic diseases related to cilia dysfunction. PCP refers to polarization that occurs within the plane of an epithelial monolayer (Jones & Chen 2007, Seifert & Mlodzik 2007, Wang & Nathans 2007). The initial recognition of PCP and the subsequent genetic dissection of the PCP pathway has largely been achieved in two tissues in *Drosophila*, the wing and the eye (Seifert & Mlodzik 2007, Wang & Nathans 2007). The studies have identified the core PCP genes in flies that include *frizzled* (*fz3, fz6*), *dishevelled* (*dvl2*), *prickle* (*pk1/2*), *strabismus/Van Gogh* (*Vangl2*), *flamingo* (*Celsr1*), and *diego* (diversin, inversin) (Jones & Chen 2007, Seifert & Mlodzik 2007). The protein products of these genes are components of the non-canonical Wnt signaling pathway. As with apical basal polarity, the planar asymmetry at the tissue and cellular level is based on the asymmetric distribution of PCP and downstream proteins within the cell. PCP in vertebrates is less well understood. A process during embryonic development termed 'convergent extension' whereby cells move toward the midline and divide along the long axis of the body plan is a PCP process. Neural tube closure, eyelid closure, the orientation of the actin based hair bundles in the inner ear and the orientation of hair follicles are also PCP processes (Jones & Chen 2007, Wang & Nathans 2007). Most recently, it has been suggested that elongation of kidney tubules during nephrogenesis and their maintenance in the adult kidney may also be PCP processes, suggesting a potential link to PKD (Fischer et al 2006). The latter hypothesis has grown out of two general observations – the emerging role of cilia in PCP and the discovery that oriented cell division plays a role in polycystic kidney disease.

The role of primary cilia in PCP is likely to be unique to vertebrates since non-motile cilia are nearly ubiquitous in cells of vertebrate organisms but have only been found in a minority of specialized neurons in *Drosophila* and *C. elegans*. One line of evidence that cilia, and by extension cilia-related diseases, affect PCP, comes from studies of inversin (Simons et al 2005). Inversin, a cause of human nephronophthisis (*NPHP2*) (Otto et al 2003), is localized to cilia and basal bodies (Eley et al 2004) and causes kidney cysts and left–right axis abnormalities in mice (Mochizuki et al 1998). Inversin's role in PCP is suggested by its homology to core PCP protein *diego*. Inversin binds to the PCP core protein dishevelled and marks it for degradation, and inversin is required for convergent extension (Simons et al 2005). Similar overlapping roles in cilia function and PCP have also been described for several of the Bardet-Biedl syndrome gene products (Mykytyn & Sheffield 2004). For example, mutations in *Bbs1*, *Bbs4*, and *Bbs6* lead to misorientation of inner ear sensory hair bundles, a mammalian PCP phenotype. *Bbs4* knockdown in zebrafish results in convergent extension defects, and *Bbs1* and *Bbs6* interact genetically with the core PCP protein *Vangl2* (Ross et al 2005).

Tubule elongation or regeneration in the kidney undergoing proliferation as part of development or repair can be achieved either by oriented cell division, i.e. the nonrandom alignment of the mitotic spindle axis and the subsequent cytokinetic event, or by random division followed by directed local migration and integration. Oriented cell division is controlled by the PCP pathway through noncanonical Wnt signaling (Gong et al 2004). Several lines of evidence support the role of oriented cell division in shaping tissues (Ahringer 2003, Baena-Lopez et al 2005) and roles in ARPKD and ADPKD have been proposed (Germino 2005, Fischer et al 2006). The elongating nephron of the developing kidney shows significant cellular proliferation yet the tubules grow in a longitudinal direction without an appreciable increase in cross section (Fischer et al 2006). In this case, the mitotic spindles orient along the long axis of the tubule and cell division takes place in this orientation. However, in two cystic animal models, *Hnf1β* deficient mice and the *pck* rat, the alignment of the mitotic spindle orientation with the long axis of the tubule is lost before the nephron segments become frankly cystic (Fischer et al 2006). *Hnf1β* is a transcription factor that regulates expression of several PKD genes, including *Pkhd1* (Hiesberger et al 2004, Gresh et al 2004); the *pck* rat is a an orthologous model for human ARPKD (Ward et al 2007). Loss of oriented cell division has also been observed with kidney-specific loss of cilia due to mutation in *Kif3A* (Patel et al 2008). The data on loss of oriented cell division as in the *Hnf1β* and *pck* models are consistent with a role for PCP in polycystic kidney disease. The data so far are most compelling for ARPKD, and it remains to be seen if ADPKD is also associated with PCP in this manner.

While PCP uses components of the noncanonical Wnt signaling pathway, the canonical Wnt pathway may also play a role in PKD. The canonical Wnt pathway functions through β-catenin and controls cell proliferation and differentiation during development (Moon 2005). β-catenin is a modulator of intercellular adhesion in epithelial cells via direct interaction with E-cadherin and of cellular proliferation and differentiation via transcriptional activation (Gallagher et al 2000, Brembeck et al 2006). The pool of β-catenin that is not bound to E-cadherin is able to enter the nucleus in a regulated manner, where it binds to, and activates, the TCF transcription factor (Moon 2005). The canonical Wnt pathway actively regulates the availability of β-catenin for nuclear translocation. Evidence for a role of canonical Wnt signaling in PKD comes from studies showing that transgenic expression of constitutively active β-catenin or kidney-specific inactivation of the APC gene (which increases nuclear β-catenin) results in cyst formation (Saadi-Kheddouci et al 2001, Qian et al 2005a). A more direct role for PC1 in regulating the pool of membrane bound β-catenin is suggested by the finding that PC1 can be coimmunoprecipitated in a complex that includes E-cadherin and β-catenin (Huan & van Adelsberg 1999, Geng et al 2000, Roitbak et al 2004). Still, the connection between PKD and canonical Wnt signaling remains speculative. This is highlighted by conflicting reports on whether polycystic disease is associated with activation of canonical Wnt signaling (Lin et al 2003, Le et al 2004), with inactivation of Wnt signaling (Kim et al 1999, Muto et al 2002) or with no effect on Wnt signaling through β-catenin (Nishio et al 2005). Taken together, it appears that Wnt signaling constitutes a plausible downstream effector mechanism for some aspects of PKD, but it is unlikely that this mechanism uniformly underlies all manifestations of these diseases.

Reports from numerous laboratories have demonstrated that the epidermal growth factor (EGF)/transforming growth factor-alpha (TGF-alpha)/EGF receptor (EGFR) axis plays a key role in pathogenesis of cyst formation and progressive enlargement in human PKD, as well as several mouse PKD models (Sweeney & Avner 2006). Cystic epithelia have both quantitative (overexpression) and qualitative (apical mislocalization) defects in the EGFR-family of receptors. These receptors bind EGF/TGF-alpha with high affinity, autophosphorylate, and elicit a hyperproliferative response in both renal and biliary epithelia. In some mouse PKD models, EGFR tyrosine kinase inhibitors attenuate cystic disease progression in both the kidney and the liver (Sweeney et al 2003). Surprisingly, these agents were not reno-protective in the *pck* rat (Torres et al 2004a). However, the EGFR tyrosine kinase inhibitor, gefitinib, acting through the MEK5-ERK5 cascade, significantly inhibited the cystic dilatation of the intrahepatic bile ducts in this rat model of ARPKD and improved the associated biliary fibrosis (Sato et al 2005, 2006). These data suggest that signaling pathways mediated by the EGFR axis differ between the liver and kidney and the MEK5-ERK5 pathway represents a potential therapeutic target for ARPKD-associated liver disease.

Aberrant EGFR signaling also may be associated with the dysregulated sodium reabsorption observed in ARPKD. At present, however, the data are conflicting. One group has shown that in a nonorthologous mouse PKD model, the mislocalized EGFR signals through ERK1/2 to inhibit amiloride-sensitive sodium channels in the distal nephron, causing decreased sodium reabsorption (Veizis & Cotton 2005). A second group examined transepithelial sodium transport

in cultured human ARPKD cyst-lining epithelial cells and demonstrated that these cells not only absorb sodium but do so at a rate ~50% greater than that of normal control epithelia (Rohatgi et al 2003). This absorptive flux was only modestly inhibited by high doses of amiloride, an inhibitor of the epithelial Na$^+$ channel (ENaC). Avid sodium absorption could contribute to the genesis and maintenance of hypertension in ARPKD, but this mechanistic link remains to be established.

The mitogen activated protein kinase/extracellular regulated kinase (MAPK/ERK) cascade couples input from a variety of signaling pathways, including receptor tyrosine kinases such as EGF receptor, Ca^{2+}, cAMP, PKA, PKC, and integrins (Kolch 2005) to modulation of protein translation and activities of transcription factors and the cell cycle (Figure 24.5). The MAPK/ERK pathway has been implicated in the proliferation associated with ADPKD in nonorthologous (Nagao et al 2003, Yamaguchi et al 2003, Omori et al 2006) and orthologous (Shibazaki et al 2008) animal models. In the latter, the activated cascade begins upstream with Ras and proceeds through Raf-1 to ERK1/2 and includes at least P90^{RSK} as an effector. MAPK/ERK has been implicated in renal epithelial morphogenetic processes (Ishibe et al 2003) that are dependent on PC1 (Joly et al 2006). PC2 has been implicated in providing tonic suppression of MAPK/ERK signaling (Grimm et al 2006). ERK activation observed in cultured ADPKD patient cyst-derived cells has been attributed to cAMP-dependent activation of B-Raf (Yamaguchi et al 2003, 2004). Inhibition of the MAPK/ERK cascade has slowed cyst formation in a murine cystic model based on a gene for human nephronophthisis (Omori et al 2006), but the MEK1/2 inhibitor U0126 failed to alter the course of ADPKD in a conditional *Pkd1* gene inactivation model (Shibazaki et al 2008).

We investigated the role of ERK1/2 activation in cystic disease progression more directly using the MEK1/2 inhibitor U0126. U0126 is an inhibitor of MEK1/2 that blocks phosphorylation-dependent activation of ERK1/2 (Favata et al 1998). U0126 has recently been shown to improve a perinatal craniosynostosis phenotype associated with ERK1/2 activation in the setting of a mutation in fibroblast growth factor receptor 2 (Shukla et al 2007). We established a dosage regimen that returned active phosphorylated ERK1/2 levels to baseline in cystic kidneys and investigated whether this would have an effect on cyst progression. No effect on *Pkd1*-dependent ADPKD cyst progression was observed.

Secretory properties of cyst-lining cells have been implicated in fostering cyst growth (Riordan et al 1994, Calvet & Grantham 2001). Net secretion of NaCl across the cyst epithelia sets up an osmotic gradient that draws water into the cyst lumen and contributes to expansion of cyst volume. CFTR has been identified as the apical chloride channel involved in cyst fluid secretion (Brill et al 1996, Sullivan et al 1998). The finding that family members affected with both cystic fibrosis and ADPKD had a somewhat milder course of ADPKD than family members with only ADPKD supports a role for functional CFTR in cyst expansion (Xu et al 2001). The CFTR channel opens in response to protein kinase A-dependent phosphorylation which in turn is activated by cytoplasmic cAMP (Sheppard & Welsh 1999) (Figure 24.5). cAMP levels are known to be elevated in cystic kidneys and these elevated levels are thought to stimulate cyst expansion through fluid and electrolyte secretion (Yamaguchi et al 2000, Belibi et al 2004). In this formulation, CFTR is the conduit, but increased cAMP is the underlying cause for cyst growth. Cytosolic Ca^{2+} levels may be the connection between cAMP levels and polycystin function. The activities of certain isoforms of adenylate cyclase and phosphodiesterase are regulated by Ca^{2+}. Polycystins have a role in cellular Ca^{2+} homeostasis and therefore may modulate cAMP levels by impacting the relative rate of formation and degradation of cAMP, respectively. Independently, the polycystin proteins may directly or indirectly alter the activities of G-protein coupled receptors (GPCR) that signal through cAMP. For example, expression and activity of one such GPCR, the V2 vasopressin receptor, is elevated in animal models of PKD (Gattone et al 2003, Belibi et al 2004). V2 receptor antagonists are now being investigated as potential therapeutic agents that may reduce to cyst expansion by reducing cytosolic cAMP levels (Gattone et al 2003, Torres et al 2004b).

PC1 itself has been proposed to function as a non-traditional G-protein coupled receptor, although most of the data are based on use of constructs containing only the isolated COOH-terminal tail of the protein. This domain contains a domain capable of activating purified heterotrimeric G-proteins (Parnell et al 1998) and it can activate JNK and the AP-1 pathway in a heterotrimeric G-protein-dependent manner (Parnell et al 2002). The membrane tethered COOH-terminal tail of PC1 can also activate Gαq leading to increased activity of phospholipase C (PLC) activity and generation of IP_3. IP_3 biding to its receptor raises cytosolic Ca^{2+} levels resulting in activation of the Ca^{2+}-dependent calcineurin/NFAT-dependent transcription pathway (Puri et al 2004). One study using full-length PC1 expressed by microinjection found PC1 expression initiates activation of Gα subunits and dissociation of the Gβ and γ polypeptides (Delmas et al 2002). This system utilized measurement of the activities of Ca^{2+} and K^+ channels whose open probabilities are modulated by βγ interactions and it was also found that the presence of functional PC2 prevented PC1 from exerting this effect on G-proteins. It may be that PC1 can modulate the activities of heterotrimeric G-proteins and their downstream effectors in a manner similar to seven transmembrane domain G-protein coupled receptors, although in vivo validation will be required to understand the role of this function in PKD.

The mTOR (mammalian target of rapamycin) protein is a kinase whose activation leads to increased protein translation and cell growth (Sarbassov et al 2005) (Figure 24.5). Signals that upregulate mTOR activity increase the protein synthetic capacity and the size of cells. Cell surface receptors signal through phosphatidylinositol 3-kinase (PI3-K) and AKT kinase to phosphorylate the tuberous sclerosis complex (TSC) gene products, TSC1 (hamartin) and TSC2 (tuberin). TSC2 has a GTPase activating protein (GAP) domain whose target is small G protein Rheb. Phosphorylation of the TSC proteins inhibits the GAP activity, increasing the GTP bound form of Rheb and resulting in elevated levels of active mTOR kinase (Mamane et al 2006) (Figure 24.5). Several lines of evidence suggest that mTOR pathway dysregulation may have a role in the pathogenesis of PKD. The mTOR pathway may regulate with the MAPK/ERK pathway (see above) in that GTP-bound active Rheb is associated with decreased B-Raf and C-Raf phosphorylation and inhibited ERK signaling (Karbowniczek et al 2006). PC1 interacts with tuberin through its COOH-terminus (Kleymenova et al 2001, Shillingford et al 2006) and it has been suggested that tuberin may play a role in controlling the trafficking of newly synthesized PC1 to the cell surface (Kleymenova et al 2001). The downstream effectors of the mTOR pathway are inappropriately activated in cyst lining cells in PKD mouse models and human tissues (Moser et al 2005). Most interestingly, administration of mTOR pathway inhibitors has shown some promise in preclinical studies of PKD. Rapamycin markedly slows cyst development the Han:SPRD rat and *bpk* mouse models of PKD (Tao et al 2005, Wahl et al 2006). Retrospective studies from PKD patients who have received renal transplants and were treated with mTOR antagonists suggests that native cystic kidney size (Shillingford et al 2006) and liver size (Qian et al 2008) is smaller in those patients treated with mTOR inhibitors that in those patients treated with immunosuppressive regimens not including mTOR inhibitors. The mTOR pathway is a proven target for small molecule pharmacotherapy, and clinical studies in ADPKD are being pursued.

By virtue of the increased cystic cell mass and the massive enlargement of the kidneys associated with PKD, loss of appropriate regulation of cellular proliferation is thought to be central to the pathogenesis of the disease (Grantham 2000). The signaling pathways implicated in the pathogenesis of PKD, including Wnt, MAPK/ERK, mTOR, Ca^{2+} and cAMP, are all implicated in control of cellular proliferation at different stages. In addition to interacting with the above pathways, the polycystin proteins have been shown to play a role in regulation of cell cycle function. MDCK cells over-expressing full-length PC1 exhibit reduced growth rates, resistance to apoptosis and increased spontaneous tubulogenesis in three dimensional culture (Boletta et al 2000). The reduced cell growth in this model is mediated by PC1 through activation of JAK/STAT signaling and upregulation of p21[waf], a cell cycle inhibitor (Bhunia et al 2002). This effect is dependent on the presence of PC2, thus fulfilling the expectation that any function of PC1 important to PKD must also require PC2. Expression of p21[waf] is reduced by Id2 (Matsumura et al 2002), a helix-loop-helix protein that binds to the COOH-terminal tail of PC2 in a phosphorylation- and PC1-dependent process (Li et al 2005b). Loss of PC2 or PC1 can thus result in the translocation of Id2 to the nucleus, where it can suppress p21^{waf} and activate the cell cycle. PC1 over-expression in MDCK cells also results in activation of PI3-Kβ, Akt, and effector molecules such as FKHR (Boca et al 2006). This activation of Akt is necessary for PC1-induced resistance to apoptosis (Boka et al 2006). Cells with overexpression of PC1 have reduced proliferation. *Pkd1*$^{-/-}$ cells have showed increased rates of proliferation and decreased rates of apoptosis when compared to control heterozygous cell lines (Wei et al 2008). Furthermore, *Pkd1*$^{-/-}$ cells spontaneously formed cysts rather than tubules in an in vitro tubulogenesis assay (Wei et al 2008). Similarly, *Pkd2*$^{-/-}$ kidney epithelial cells lacking PC2 proliferate significantly more rapidly than null cells re-expressing PC2 (Grimm et al 2006). It is noteworthy that the effects on cell cycle active components in the all of the above cited studies occurred in proliferating cells which do not have cilia, suggesting that these effects of polycystin function do not depend on the their location in cilia. Nonetheless, these cell cycle functions may have clinical significance as illustrated by the effect of roscovitine, a cell cycle dependent kinase inhibitor, to inhibit cyst growth in the *jck* and *cpk* models of polycystic kidney disease (Bukanov et al 2006).

Studies of the *Drosophila* notch protein have led to a new signaling paradigm of communication between cell surface receptors and the nucleus known as regulated intramembranous proteolysis (RIP). Notch undergoes sequential proteolytic cleavages to ultimately release a cytoplasmic fragment that travels to the nucleus, where it influences transcriptional activities critical to cell fate determination (Schroeter et al 1998, Mumm et al 2000). In addition to cleavage at the GPS site, it appears that PC1 undergoes cleavage and release of COOH-terminal tail fragments capable of entering the nucleus (Chauvet et al 2004, Low et al 2006). The larger of these fragments contains the putative G-protein activation sequence, and corresponds to the fragment that leads to activation of JNK and the AP-1 pathway when transfected into cells (Chauvet et al 2004). The smaller COOH-terminal cleavage fragment can interact with STAT6 and the co-activator P100 and is translocated into nuclei by virtue of these associations (Low et al 2006). Nuclear accumulation of the PC1 COOH-terminal tail is detected in the renal epithelial cells of transgenic mice that overexpress PC1, and of mice that manifest kidney specific agenesis of cilia, as well as of wild-type mice that have been subjected to ureteral ligation (Chauvet et al 2004). Absence of cilia or loss of flow appears to initiate both of the reported PC1 COOH-terminal tail cleavage events (Chauvet et al 2004, Low et al 2006). Recent data indicate that the protein encoded by the longest open reading frame of *Pkhd1* also undergoes a complicated pattern of Notch-like proteolytic processing (Hiesberger et al 2006, Kaimori et al 2007). Cleavage releases the large extracellular domain which remains tethered to the COOH-terminal stalk by disulfide bridges. Further proteolytic action by a member of the ADAM metalloproteinase disintegrins family results in the concomitant release of the extracellular domain fragment and the intracellular COOH-terminal fragment. This process, regulated by intracellular Ca^{2+}, appears to represent a novel Ca^{2+}-dependent pathway that transduces extracellular signals from the primary cilium/apical membrane to the nucleus. The physiological functions of the PC1 and polyductin cleavage events remain an area of active investigation.

THERAPY IN PKD

Understanding the mechanisms of PKD will yield broad ranging benefits, elucidating interesting and novel biological insights in areas such as cilia function, cell signaling, organ development and maintenance of three-dimensional structures. Still, a major goal of these investigations remains the gaining of sufficient understanding to develop therapies to treat the millions of patients affected by these disorders worldwide.

Over a decade after the discovery of the first gene for ADPKD, the studies are beginning to bear fruit. The role of cAMP in cystogenesis provided the rationale for preclinical trials of vasopressin V2 receptor (VPV2R) antagonists (Figure 24.5). These preclinical trials demonstrated marked effects in reducing levels of cAMP and inhibiting cyst expansion in models of ARPKD, ADPKD and nephronophthisis (Gattone et al 1999, 2003, Torres et al 2004b). High water intake by itself also exerts a protective effect on the development of PKD in *pck* rats likely due to suppression of vasopressin (Nagao et al 2006). Phase II clinical trials with tolvaptan, an oral VPV2R antagonist, have been completed and a phase III clinical trial is in process. An alternative approach inhibiting cAMP levels is through somatostatin acting on SST2 receptors. Octreotide, a somatostatin analog, halts the expansion of kidney cysts in the *pck* rat (Masyuk et al 2007); VPV2R antagonists do not have effect on liver cysts due to the lack of VPV2R in bile duct epithelia. These observations are consistent with the inhibition of renal growth in a pilot study of long-acting octreotide for human ADPKD (Ruggenenti et al 2005) and provide support for further clinical trials for PKD and PLD.

The discovery of mTOR activation in polycystic kidneys (Shillingford et al 2006) coupled with the availability of well validated mTOR pathway inhibitors in clinical use and the success that one such inhibitor, rapamycin, has had in retarding cyst expansion in rodent models (Tao et al 2005, Shillingford et al 2006, Wahl et al 2006), has prompted interest in clinical use of these drugs in PKD. Phase 2 clinical trials of rapamycin and everolimus, two mTOR inhibitors, are being implemented (Walz 2006) (Figure 24.5). Antiproliferative agents, primarily developed to treat neoplastic diseases may also have a role in treatment of PKD, although the therapeutic window for use of these drugs in longer-term therapy of PKD patients may pose some practical problems. Among this class of agents that have been effective in preclinical trials and may be of potential value for the treatment of human PKD are Erb-B1 (epidermal growth factor receptor) (Sweeney et al 2000) and Erb-B2 (Wilson et al 2006) tyrosine kinase inhibitors, Src kinase inhibitors and MEK inhibitors (Omori et al 2006) (Figure 24.5). Roscovitine, a cell cycle dependent kinase inhibitor, has been shown to inhibit cyst growth in the *jck* model of polycystic kidney disease in which *Nek8* is mutated (Bukanov et al 2006). Another potential target for therapy is modulation of the cellular Ca^{2+} homeostasis. Recently, it has been reported that triptolide, the active diterpene in the traditional Chinese medicine *Lei Gong Teng*, induces Ca^{2+} release by a PC2-dependent mechanism, arrests cellular proliferation and attenuates overall cyst formation in an embryonic *Pkd1* murine animal model of polycystic kidney disease (Leuenroth et al 2007). It can be reasonably expected that additional targets and agents will be discovered in the near future as the drive to develop effective therapy for PKD continues.

The main challenge beyond identifying appropriate candidate therapies lies in the study design to validate any of these in humans. The use of renal function as the primary outcome is problematic as the natural history of ADPKD involves decades of normal renal function in the setting of progressive kidney enlargement and cystic transformation. By the time the GFR starts declining, the kidneys are markedly enlarged, distorted, and damaged by secondary processes of inflammation and fibrosis, making it unlikely that treatments aimed at slowing cyst growth will be effective in these late stages of the disease. On the other hand, early interventional trials are not feasible if renal function is to be used as the primary outcome. The results of CRISP (Grantham et al 2006b) have shown that the rate of renal growth is a good predictor of functional decline and justify the utilization of kidney volume as a surrogate marker of disease progression in clinical trials for ADPKD. The challenge is to convince regulatory agencies that the endpoints defined by CRISP are appropriate for defining the benefit of treatments under investigation.

The substantial scientific and clinical progress that has occurred in the study of PKD during the past decade and a half gives great promise that the future will bring deeper mechanistic understanding and effective treatment. For PKD, as for other inherited kidney diseases, this will be the legacy of the great success of the international human genome program and the deciphering of the genomes of a broad range of species.

References

Adeva M, El-Youssef M, Rossetti S, et al. Clinical and molecular characterization defines a broadened spectrum of autosomal recessive polycystic kidney disease (ARPKD). Medicine (Baltimore) 2006; 85: 1–21.

Aguiari G, Manzati E, Penolazzi L, et al. Mutations in autosomal dominant polycystic kidney disease 2 gene: Reduced expression of PKD2 protein in lymphoblastoid cells. Am. J. Kidney Dis. 1999; 33: 880–5.

Ahringer J. Control of cell polarity and mitotic spindle positioning in animal cells. Curr. Opin. Cell Biol. 2003; 15: 73–81.

Alvaro D, Mancino MG, Onori P, et al. Estrogens and the pathophysiology of the biliary tree. World J. Gastroenterol. 2006; 12: 3537–45.

The American PKD1 Consortium. Analysis of the genomic sequence for the autosomal dominant polycystic kidney disease gene (PKD1) predicts the presence of a leucine-rich repeat. Hum. Mol. Genet. 1995; 4: 575–82.

Anyatonwu GI, Estrada M, Tian X, Somlo S, Ehrlich BE. Regulation of ryanodine receptor-dependent calcium signaling by polycystin-2. Proc. Natl Acad. Sci. USA 2007; 104: 6454–9.

Badano JL, Mitsuma N, Beales PL, Katsanis N. The ciliopathies: An emerging class of human genetic disorders. Annu. Rev. Genomics Hum. Genet. 2006; 7: 125–48.

Badano JL, Teslovich TM, Katsanis N. The centrosome in human genetic disease. Nat. Rev. Genet. 2005; 6: 194–205.

Bae KT, Zhu F, Guay-Woodford LM, et al. Magnetic resonance imaging evaluation of hepatic cysts in early autosomal dominant polycystic kidney disease. Clin. J. Am. Soc. Nephrol. 2006; 1: 64–9.

Baena-Lopez LA, Baonza A, Garcia-Bellido A. The orientation of cell divisions determines the shape of *Drosophila* organs. Curr. Biol. 2005; 15: 1640–4.

Baert L. Hereditary polycystic kidney disease (adult form): a microdissection study of two cases at an early stage of the disease. Kidney Int. 1978; 13: 519–25.

Bai CX, Kim S, Li WP, Streets AJ, Ong AC, Tsiokas L. Activation of TRPP2 through mDia1-dependent voltage gating. EMBO J. 2008; 27: 1345–56.

Bajwa ZH, Sial KA, Malik AB, Steinman TI. Pain patterns in patients with polycystic kidney disease. Kidney Int. 2004; 66: 1561–9.

Barr MM, DeModena J, Braun D, Nguyen CQ, Hall DH, Sternberg PW. The *Caenorhabditis elegans* autosomal dominant polycystic kidney disease gene homologs lov-1 and pkd-2 act in the same pathway. Curr. Biol. 2001; 11: 1341–6.

Barr MM, Sternberg PW. A polycystic kidney disease gene homolog required for male mating behavior in *Caenorhabiditis elegans*. Nature 1999; 401: 386–9.

Bateman A, Sandford R. The PLAT domain: a new piece in the PKD1 puzzle. Curr. Biol. 1999; 9: R588–90.

Belibi FA, Reif G, Wallace DP, et al. Cyclic AMP promotes growth and secretion in human polycystic kidney epithelial cells. Kidney Int. 2004; 66: 964–73.

Bergmann C, Kupper F, Dornia C, et al. Algorithm for efficient PKHD1 mutation screening in autosomal recessive polycystic kidney disease (ARPKD). Hum. Mutat. 2005a; 25: 225–31.

Bergmann C, Senderek J, Sedlacek B, et al. Spectrum of mutations in the gene for autosomal recessive polycystic kidney disease (ARPKD/PKHD1). J. Am. Soc. Nephrol. 2003; 14: 76–89.

Bergmann C, Senderek J, Windelen E, et al. members of the APN (Arbeitsgemeinschaft fur Padiatrische Nephrologie). Clinical consequences of PKHD1 mutations in 164 patients with autosomal recessive polycystic kidney disease (ARPKD). Kidney Int. 2005b; 67: 829–48.

Bernstein J, Slovis T. Polycystic diseases of the kidney. In: Edelmann CM, ed. Pediatric Kidney Disease. Boston, MA: Little, Brown and Company, 1992: pp. 1139–58.

Bhunia AK, Piontek K, Boletta A, et al. PKD1 induces p21(waf1) and regulation of the cell cycle via direct activation of the JAK-STAT signaling pathway in a process requiring PKD2. Cell 2002; 109: 157–68.

Bisgrove BW, Yost HJ. The roles of cilia in developmental disorders and disease. Development 2006; 133: 4131–43.

Blyth H, Ockenden BG. Polycystic disease of kidneys and liver presenting in childhood. J. Med. Genet. 1971; 8: 257–84.

Boca M, Distefano G, Qian F, Bhunia AK, Germino GG, Boletta A. Polycystin-1 induces resistance to apoptosis through the phosphatidylinositol 3-kinase/Akt signaling pathway. J. Am. Soc. Nephrol. 2006; 17: 637–47.

Boletta A, Qian F, Onuchic LF, et al. Polycystin-1, the gene product of PKD1, induces resistance to apoptosis and spontaneous tubulogenesis in MDCK cells. Mol. Cell 2000; 6: 1267–73.

Boletta A, Qian F, Onuchic LF, et al. Biochemical characterization of bona fide polycystin-1 in vitro and in vivo. Am. J. Kidney Dis. 2001; 38: 1421–9.

Brasier JL, Henske EP. Loss of the polycystic kidney disease (PKD1) region of chromosome 16p13 in renal cyst cells supports a loss-of-function model for cyst pathogenesis. J. Clin. Invest. 1997; 99: 194–9.

Brembeck FH, Rosario M, Birchmeier W. Balancing cell adhesion and Wnt signaling, the key role of beta-catenin. Curr. Opin. Genet. Dev. 2006; 16: 51–9.

Brill SR, Ross KE, Davidow CJ, Ye M, Grantham JJ, Caplan MJ. Immunolocalization of ion transport proteins in human autosomal dominant polycystic kidney epithelial cells. Proc. Natl Acad. Sci. USA 1996; 93: 10206–11.

Brun M, Maugey-Laulom B, Eurin D, Didier F, Avni E. Prenatal sonographic patterns in autosomal dominant polycystic kidney disease: a multicenter study. Ultrasound Obstet. Gynecol. 2004; 24: 55–61.

Bukanov N, Smith LA, Klinger K, Ledbetter SR, Ibraghimov-Beskrovnaya O. Long lasting arrest of murine polycystic kidney disease with CDK inhibitor R-Roscovitine. Nature 2006; 444: 949–52.

Bycroft M, Bateman A, Clarke J, et al. The structure of a PKD domain from polycystin-1: implications for polycystic kidney disease. EMBO J. 1999; 18: 297–305.

Cai Y, Anyatonwu G, Okuhara D, et al. Calcium dependence of polycystin-2 channel activity is modulated by phosphorylation at Ser812. J. Biol. Chem. 2004; 279: 19987–95.

Cai Y, Maeda Y, Cedzich A, et al. Identification and characterization of polycystin-2, the PKD2 gene product. J. Biol. Chem. 1999; 274: 28557–65.

Calvet JP, Grantham JJ. The genetics and physiology of polycystic kidney disease. Semin. Nephrol. 2001; 21: 107–23.

Chapman A, Guay-Woodford L, Grantham JJ, et al. Renal structure in early autosomal dominant polycystic kidney disease (ADPKD); the Consortium for Radiologic Imaging Studies of Polycystic Kidney Disease (CRISP) Cohort. Kidney Int. 2003; 64: 1035–45.

Chaumoitre K, Brun M, Cassart M, et al. Differential diagnosis of fetal hyperechogenic cystic kidneys unrelated to renal tract anomalies: A multicenter study. Ultrasound Obstet. Gynecol. 2006; 28: 911–17.

Chauvet V, Tian X, Husson H, et al. Mechanical stimuli induce cleavage and nuclear translocation of the polycystin-1 C terminus. J. Clin. Invest. 2004; 114: 1433–43.

Clausen P, Feldt-Rasmussen B, Iversen J, Lange M, Eidemak I, Strandgaard S. Flow-associated dilatory capacity of the brachial artery is intact in early autosomal dominant polycystic kidney disease. Am. J. Nephrol. 2006; 26: 335–9.

Corbit KC, Aanstad P, Singla V, Norman AR, Stainier DY, Reiter JF. Vertebrate smoothened functions at the primary cilium. Nature 2005; 437: 1018–21.

D'Alessio C, Fernandez F, Trombetta ES, Parodi AJ. Genetic evidence for the heterodimeric structure of glucosidase II. The effect of disrupting the subunit-encoding genes on glycoprotein folding. J. Biol. Chem. 1999; 274: 25899–905.

Davenport JR, Yoder BK. An incredible decade for the primary cilium: a look at a once-forgotten organelle. Am. J. Physiol. Renal. Physiol. 2005; 289: 1159–69.

Davenport JR, Watts AJ, Roper VC, et al. Disruption of intraflagellar transport in adult mice leads to obesity and slow-onset cystic kidney disease. Curr. Biol. 2007; 17: 1586–94.

Davila S, Furu L, Gharavi AG, et al. Mutations in SEC63 cause autosomal dominant polycystic liver disease. Nat. Genet. 2004; 36: 575–7.

Davis EE, Brueckner M, Katsanis N. The emerging complexity of the vertebrate cilium: new functional roles for an ancient organelle. Dev. Cell 2006; 11: 9–19.

Davis ID, Ho M, Hupertz V, Avner ED. Survival of childhood polycystic kidney disease following renal transplantation: the impact of advanced hepatobiliary disease. Pediatr. Transplant. 2003; 7: 364–9.

De Kerckhove L, De Meyer M, Verbaandert C, et al. The place of liver transplantation in Caroli's disease and syndrome. Transplant. Int. 2006; 19: 381–8.

De Rycke M, Georgiou I, Sermon K, et al. PGD for autosomal dominant polycystic kidney disease type 1. Mol. Hum. Reprod. 2005; 11: 65–71.

Deget F, Rudnik-Schoneborn S, Zerres K. Course of autosomal recessive polycystic kidney disease (ARPKD) in siblings: a clinical comparison of 20 sibships. Clin. Genet. 1995; 47: 248–53.

Delmas P. Polycystins: polymodal receptor/ion-channel cellular sensors. Pflugers Arch. 2005; 451: 264–76.

Delmas P, Nomura H, Li X, et al. Constitutive activation of G-proteins by polycystin-1 is antagonized by polycystin-2. J. Biol. Chem. 2002; 277: 11276–83.

Desmet V. Pathogenesis of ductal plate abnormalities. Mayo Clin. Proc. 1998; 73: 80–9.

Dobin A, Kimberling WJ, Pettinger W, Bailey-Wilson JE, Shugart YY, Gabow P. Segregation analysis of autosomal dominant polycystic kidney disease. Genet. Epidemiol. 1993; 10: 189–200.

Drenth JP, Te Morsche RH, Smink R, Bonifacino JS, Jansen JB. Germline mutations in PRKCSH are associated with autosomal dominant polycystic liver disease. Nat. Genet. 2003; 33: 345–7.

Ecder T, Chapman A, Brosnahan G, Edelstein C, Johnson A, Schrier R. Effect of antihypertensive therapy on renal function and urinary albumin excretion in hypertensive patients with autosomal dominant polycystic kidney disease. Am. J. Kid. Dis. 2000; 35: 427–32.

Eley L, Turnpenny L, Yates LM, et al. A perspective on inversin. Cell Biol. Int. 2004; 28: 119–24.

Ellgaard L, Molinari M, Helenius A. Setting the standards: quality control in the secretory pathway. Science 1999; 286: 1882–8.

Elzinga LW, Bennett WM. Miscellaneous renal and systemic complications of autosomal dominant polycystic kidney disease including infection. In: Watson ML, Torres VE, eds. Polycystic Kidney Disease. Oxford, UK: Oxford Medical Publications, 1996: pp. 483–99.

The European Polycystic Kidney Disease Consortium. The polycystic kidney disease 1 gene encodes a 14 kb transcript and lies within a duplicated region on chromosme 16. Cell 1994; 77: 881–94.

Fabris L, Cadamuro M, Fiorotto R, et al. Effects of angiogenic factor overexpression by human and rodent cholangiocytes in polycystic liver diseases. Hepatology 2006; 43: 1001–12.

Fain PR, McFann KK, Taylor MR, et al. Modifier genes play a significant role in the phenotypic expression of PKD1. Kidney Int. 2005; 67: 1256–67.

Favata MF, Horiuchi KY, Manos EJ, et al. Identification of a novel inhibitor of mitogen-activated protein kinase kinase. J. Biol. Chem. 1998; 273: 18623–32.

Fick GM, Johnson AM, Hammond WS, Gabow PA. Causes of death in autosomal dominant polycystic kidney disease. J. Am. Soc. Nephrol. 1995; 5: 2048–56.

Fischer E, Legue E, Doyen A, et al. Defective planar cell polarity in polycystic kidney disease. Nat. Genet. 2006; 38: 21–3.

Foggensteiner L, Bevan AP, Thomas R, et al. Cellular and subcellular distribution of polycystin-2, the protein product of the PKD2 gene. J. Am. Soc. Nephrol. 2000; 11: 814–27.

Fonck C, Chauveau D, Gagnadoux M, Pirson Y, Grunfeld J. Autosomal recessive polycystic kidney disease in adulthood. Nephrol. Dial. Transplant. 2001; 16: 1648–52.

Forman JR, Qamar S, Paci E, Sandford RN, Clarke J. The remarkable mechanical strength of polycystin-1 supports a direct role in mechanotransduction. J. Mol. Biol. 2005; 349: 861–71.

Furu L, Onuchic LF, Gharavi A, et al. Milder presentation of recessive polycystic kidney disease requires presence of amino acid substitution mutations. J. Am. Soc. Nephrol. 2003; 14: 2004–14.

Gabow PA, Chapman AB, Johnson AM, et al. Renal structure and hypertension in autosomal dominant polycystic kidney disease. Kidney Int. 1990a; 38: 1177–80.

Gabow P, Johnson A, Kaehny W, Manco-Johnson M, Duley I, Everson G. Risk factors for the development of hepatic cysts in autosomal dominant polycystic kidney disease. Hepatology 1990b; 11: 1033–7.

Gabow P, Johnson A, Kaehny W, et al. Factors affecting the progression of renal disease in autosomal-dominant polycystic kidney disease. Kidney Int. 1992; 41: 1311–19.

Gallagher AR, Cedzich A, Gretz N, Somlo S, Witzgall R. The polycystic kidney disease protein PKD2 interacts with Hax-1, a protein associated with the actin cytoskeleton. Proc. Natl Acad. Sci. USA 2000; 97: 4017–22.

Gallagher AR, Esquivel EL, Briere TS, et al. Biliary and pancreatic dysgenesis in mice harboring a mutation in pkhd1. Am. J. Pathol. 2008; 172: 417–29.

Gao Z, Joseph E, Ruden DM, Lu X. *Drosophila* Pkd2 is haploid-insufficient for mediating optimal smooth muscle contractility. J. Biol. Chem. 2004; 279: 14225–31.

Gao Z, Ruden DM, Lu X. PKD2 cation channel is required for directional sperm movement and male fertility. Curr. Biol. 2003; 13: 2175–8.

Garcia-Gonzalez MA, Menezes LF, Piontek KB, et al. Genetic interaction studies link autosomal dominant and recessive polycystic kidney disease in a common pathway. Hum. Mol. Genet. 2007; 16: 1940–50.

Gattone VH, Maser RL, Tian C, Rosenberg JM, Branden MG. Developmental expression of urine concentration-associated genes and their altered expression in murine infantile-type polycystic kidney disease. Develop. Genet. 1999; 24: 309–18.

Gattone VH, Wang X, Harris PC, Torres VE. Inhibition of renal cystic disease development and progression by a vasopressin V2 receptor antagonist. Nat. Med. 2003; 9: 1323–6.

Geberth St, Ritz E, Zeier M, Stier E. Anticipation of age at renal death in autosomal dominant polycystic kidney disease (ADPKD)? Nephrol. Dial. Transplant. 1995; 10: 1603–6.

Geng L, Burrow CR, Li HP, Wilson PD. Modification of the composition of polycystin-1 multiprotein complexes by calcium and tyrosine phosphorylation. Biochim. Biophys. Acta 2000; 1535: 21–35.

Geng L, Okuhara D, Yu Z, et al. Polycystin-2 traffics to cilia independently of polycystin-1 by using an N-terminal RVxP motif. J. Cell Sci. 2006; 119: 1383–95.

Geng L, Segal Y, Pavlova A, et al. Distribution and developmentally regulated expression of murine polycystin. Am. J. Physiol. 1997; 272: F451–9.

Geng L, Segal Y, Peissel B, et al. Identification and localization of polycystin, the PKD1 gene product. J. Clin. Invest. 1996; 98: 2674–82.

Germino GG. Linking cilia to Wnts. Nat. Genet. 2005; 37: 455–7.

Gigarel N, Frydman N, Burlet P, et al. Preimplantation genetic diagnosis for autosomal recessive polycystic kidney disease. Reprod. Biomed. Online 2008; 16: 152–8.

Goilav B, Norton K, Satlin L, et al. Predominant extrahepatic biliary disease in autosomal recessive polycystic kidney disease: a new association. Pediatr Transplant 2006; 10: 294–8.

Gong Y, Mo C, Fraser SE. Planar cell polarity signalling controls cell division orientation during zebrafish gastrulation. Nature 2004; 430: 689–93.

Gonzalez-Perrett S, Kim K, Ibarra C, et al. Polycystin-2, the protein mutated in autosomal dominant polycystic kidney disease (ADPKD), is a Ca^{2+}-permeable nonselective cation channel. PNAS 2001; 98: 1182–7.

Grampsas SA, Chandhoke PS, Fan J, et al. Anatomic and metabolic risk factors for nephrolithiasis in patients with autosomal dominant polycystic kidney disease. Am. J. Kidney Dis. 2000; 36: 53–7.

Grantham JJ. Time to treat polycystic kidney diseases like the neoplastic disorders that they are. Kidney Int. 2000; 57: 339–40.

Grantham JJ, Chapman AB, Torres VE. Volume progression in autosomal dominant polycystic kidney disease: The major factor determining clinical outcomes. Clin. J. Am. Soc. Nephrol. 2006a; 1: 148–57.

Grantham JJ, Torres VE, Chapman AB, et al. Volume progression in polycystic kidney disease. N. Engl. J. Med. 2006b; 354: 2122–30.

Gresh L, Fischer E, Reimann A, et al. A transcriptional network in polycystic kidney disease. EMBO J. 2004; 23: 1657–68.

Grimm DH, Cai Y, Chauvet V, et al. Polycystin-1 distribution is modulated by polycystin-2 expression in mammalian cells. J. Biol. Chem. 2003; 278: 36786–93.

Grimm DH, Karihaloo A, Cai Y, Somlo S, Cantley LG, Caplan MJ. Polycystin-2 regulates proliferation and branching morphogenesis in kidney epithelial cells. J. Biol. Chem. 2006; 281: 137–44.

Grunfeld JP, Albouze G, Jungers P, et al. Liver changes and complications in adult polycystic kidney disease. Adv. Nephrol. Necker. Hosp. 1985; 14: 1–20.

Guay-Woodford L. Autosomal recessive polycystic kidney disease. In: Flinter F, Saggar-Malik A, eds. The Genetics of Renal Disease. Oxford, UK: Oxford University Press, 2003a: pp. 239–51.

Guay-Woodford LM. Murine models of polycystic kidney disease: molecular and therapeutic insights. Am. J. Physiol. Renal Physiol. 2003b; 285: F1034–49.

Guay-Woodford L, Desmond R. Autosomal recessive polycystic kidney disease (ARPKD): the clinical experience in North America. Pediatrics 2003; 111: 1072–80.

Guay-Woodford L, Muecher G, Hopkins S, et al. The severe perinatal form of autosomal recessive polycystic kidney disease (ARPKD) maps to chromosome 6p21.1-p12: Implications for genetic counseling. Am. J. Hum. Genet. 1995; 56: 1101–7.

Hanaoka K, Qian F, Boletta A, et al. Co-assembly of polycystin-1 and -2 produces unique cation-permeable currents. Nature 2000; 408: 990–4.

Harris PC, Bae KT, Rossetti S, et al. Cyst number but not the rate of cystic growth is associated with the mutated gene in autosomal dominant polycystic kidney disease. J. Am. Soc. Nephrol. 2006; 17: 3013–19.

Harris PC, Thomas S, MacCarthy AB, et al. A large duplicated area in the polycystic kidney disease 1 (PKD1) region of chromosome 16 is prone to rearrangement. Genomics 1994; 23: 321–30.

Hateboer N, Dijk MA, Bogdanova N, et al. Comparison of phenotypes of polycystic kidney disease types 1 and 2. European PKD1-PKD2 Study Group. Lancet 1999a; 353: 103–7.

Hateboer N, Lazarou LP, Williams AJ, Holmans P, Ravine D. Familial phenotype differences in PKD1. Kidney Int. 1999b; 56: 34–40.

Haycraft CJ, Banizs B, Ydin-Son Y, Zhang Q, Michaud EJ, Yoder BK. Gli2 and Gli3 localize to cilia and require the intraflagellar transport protein polaris for processing and function. PLoS Genet. 2005; 1: e53.

Helenius A, Aebi M. Intracellular functions of N-linked glycans. Science 2001; 291: 2364–9.

Hidaka S, Konecke V, Osten L, Witzgall R. PIGEA-14, a novel coiled-coil protein affecting the intracellular distribution of polycystin-2. J. Biol. Chem. 2004; 279: 35009–16.

Hiesberger T, Bai Y, Shao X, et al. Mutation of hepatocyte nuclear factor-1β inhibits Pkhd1 gene expression and produces renal cysts in mice. J. Clin. Invest. 2004; 113: 814–25.

Hiesberger T, Gourley E, Erickson A, et al. Proteolytic cleavage and nuclear translocation of fibrocystin is regulated by intracellular Ca^{2+} and activation of protein kinase C. J. Biol. Chem. 2006; 281: 34357–64.

Hirokawa N, Tanaka Y, Okada Y, Takeda S. Nodal flow and the generation of left-right asymmetry. Cell 2006; 125: 33–45.

Hogan M, Griffin M, Rossetti S, Torres V, Ward C, Harris P. PKHDL1, a homolog of the autosomal recessive polycystic kidney disease gene, encodes a receptor with inducible T lymphocyte expression. Hum. Mol. Genet. 2003; 12: 685–98.

Hooper KM, Boletta A, Germino GG, Hu Q, Ziegelstein RC, Sutters M. Expression of polycystin-1 enhances endoplasmic reticulum calcium uptake and decreases capacitative calcium entry in ATP-stimulated MDCK cells. Am. J. Physiol. Renal Physiol. 2005; 289: 521–30.

Hu J, Bae YK, Knobel KM, Barr MM. Casein kinase II and calcineurin modulate TRPP function and ciliary localization. Mol. Biol. Cell 2006; 17: 2200–11.

Hu J, Barr MM. ATP-2 interacts with the PLAT domain of LOV-1 and is involved in Caenorhabditis elegans polycystin signaling. Mol. Biol. Cell 2005; 16: 458–69.

Huan Y, van Adelsberg J. Polycystin-1, the PKD1 gene product, is in a complex containing E-cadherin and the catenins. J. Clin. Invest. 1999; 104: 1459–68.

Huang AL, Chen X, Hoon MA, et al. The cells and logic for mammalian sour taste detection. Nature 2006; 442: 934–8.

Huang K, Diener DR, Mitchell A, Pazour GJ, Witman GB, Rosenbaum JL. Function and dynamics of PKD2 in Chlamydomonas reinhardtii flagella. J. Cell Biol. 2007; 179: 501–14.

Hughes J, Ward CJ, Aspinwall R, Butler R, Harris PC. Identification of a human homologue of the sea urchin receptor for egg jelly: a polycystic kidney disease-like protein. Hum. Mol. Genet. 1999; 8: 543–9.

Hughes J, Ward CJ, Peral B, et al. The polycystic kidney disease 1 (PKD1) gene encodes a novel protein with multiple cell recognition domains. Nat. Genet. 1995; 10: 151–60.

Ibraghimov-Beskrovnaya O, Bukanov NO, Donohue LC, Dackowski WR, Klinger KW, Landes GM. Strong homophilic interactions of the Ig-like domains of polycystin-1, the protein product of an autosomal dominant polycystic kidney disease gene, PKD1 [In Process Citation]. Hum. Mol. Genet. 2000; 9: 1641–9.

Ibraghimov-Beskrovnaya O, Dackowski WR, Foggensteiner L, et al. Polycystin: In vitro synthesis, in vivo tissue expression, and subcellular localization identifies a large membrane-associated protein. Proc. Natl Acad. Sci. USA 1997; 94: 6397–402.

Iglesias CG, Torres VE, Offord KP, Holley KE, Beard CM, Kurland LT. Epidemiology of adult polycystic kidney disease, Olmsted County, Minnesota. Am. J. Kidney Dis. 1983; 2: 630–9.

Inagawa T. Trends in incidence and case fatality rates of aneurysmal subarachnoid hemorrhage in Izumo City, Japan, between 1980–1989 and 1990–1998. Stroke 2001; 32: 1499–507.

The International Polycystic Kidney Disease Consortium. Polycystic kidney disease: The complete structure of the PKD1 Gene and its Protein. Cell 1995; 81: 289–98.

Ishibe S, Joly D, Zhu X, Cantley LG. Phosphorylation-dependent paxillin-ERK association mediates hepatocyte growth factor-stimulated epithelial morphogenesis. Mol. Cell 2003; 12: 1275–85.

Ishimaru Y, Inada H, Kubota M, Zhuang H, Tominaga M, Matsunami H. Transient receptor potential family members PKD1L3 and PKD2L1 form a candidate sour taste receptor. Proc. Natl Acad. Sci. USA 2006; 103: 12569–74.

Jenkins PM, Hurd TW, Zhang L, et al. Ciliary targeting of olfactory CNG channels requires the CNGB1b subunit and the kinesin-2 motor protein, KIF17. Curr. Biol. 2006; 16: 1211–16.

Jiang ST, Chiou YY, Wang E, et al. Defining a link with autosomal-dominant polycystic kidney disease in mice with congenitally low expression of Pkd1. Am. J. Pathol. 2006; 8: 205–20.

Johnson A, Gabow P. Identification of patients with autosomal dominant polycystic kidney disease at highest risk for end-stage renal disease. J. Am. Soc. Nephrol. 1997; 8: 1560–7.

Joly D, Ishibe S, Nickel C, Yu Z, Somlo S, Cantley LG. The polycystin 1-C-terminal fragment stimulates ERK-dependent spreading of renal epithelial cells. J. Biol. Chem. 2006; 281: 26329–39.

Jones C, Chen P. Planar cell polarity signaling in vertebrates. Bioessays 2007; 29: 120–32.

Kaimori J, Nagasawa Y, Menezes L, et al. Polyductin undergoes notch-like processing and regulated release from primary cilia. Hum. Mol. Genet. 2007; 16: 942–56.

Kaimori JY, Germino GG. ARPKD and ADPKD: first cousins or more distant relatives? J. Am. Soc. Nephrol. 2008; 19: 416–18.

Kaplan B, Fay J, Shah V. Autosomal recessive polycystic kidney disease. Pediatr. Nephrol. 1989; 3: 43–9.

Kaplan BS, Kaplan P, de Chadarevian JP, Jequier S, O'Regan S, Russo P. Variable expression of autosomal recessive polycystic kidney disease and congenital hepatic fibrosis within a family. Am. J. Med. Genet. 1988; 29: 639–47.

Karbowniczek M, Robertson GP, Henske EP. Rheb inhibits C-raf activity and B-raf/C-raf heterodimerization. J. Biol. Chem. 2006; 281: 25447–56.

Kelleher CL, McFann KK, Johnson AM, Schrier RW. Characteristics of hypertension in young adults with autosomal dominant polycystic kidney disease compared with the general U.S. population. Am. J. Hypertens. 2004; 17: 1029–34.

Kerkar N, Norton K, Suchy F. The hepatic fibrocystic diseases. Clin. Liver Dis. 2006; 10: 55–71.

Kielstein R, Sass HM. Genetics in kidney disease: how much do we want to know? Am. J. Kidney Dis. 2002; 39: 637–52.

Kim E, Arnould T, Sellin L, et al. Interaction between RGS7 and polycystin. Proc. Natl Acad. Sci. USA 1999a; 96: 6371–6.

Kim E, Arnould T, Sellin LK, et al. The polycystic kidney disease 1 gene product modulates Wnt signaling. J. Biol. Chem. 1999b; 274: 4947–53.

Kim I, Fu Y, Hui K, et al. Fibrocystin/polyductin modulates renal tubular formation by regulating polycystin-2 expression and function. J. Am. Soc. Nephrol. 2008; 19: 455–6.

Kiselyov K, Soyombo A, Muallem S. TRPpathies. J. Physiol. 2007; 578: 641–53.

Klahr S, Breyer J, Beck G, et al. Dietary protein restriction, blood pressure control, and the progression of polycystic kidney disease modification of diet in renal disease study group. J. Am. Soc. Nephrol. 1995; 5: 2037–47.

Kleymenova E, Ibraghimov-Beskrovnaya O, Kugoh H, et al. Tuberin-dependent membrane localization of polycystin-1: a functional link between polycystic kidney disease and the TSC2 tumor suppressor gene. Mol. Cell 2001; 7: 823–32.

Kocaman O, Oflaz H, Yekeler E, et al. Endothelial dysfunction and increased carotid intima-media thickness in patients with autosomal dominant polycystic kidney disease. Am. J. Kidney Dis. 2004; 43: 854–60.

Kolch W. Coordinating ERK/MAPK signalling through scaffolds and inhibitors. Nat. Rev. Mol. Cell Biol. 2005; 6: 827–37.

Koptides M, Mean R, Demetriou K, Pierides A, DeltaS CC. Genetic evidence for a trans-heterozygous model for cystogenesis in autosomal dominant polycystic kidney disease [In Process Citation]. Hum. Mol. Genet. 2000; 9: 447–52.

Kottgen M, Benzing T, Simmen T, et al. Trafficking of TRPP2 by PACS proteins represents a novel mechanism of ion channel regulation. EMBO J. 2005; 24: 705–16.

Koulen P, Cai Y, Geng L, et al. Polycystin-2 is an intracellular calcium release channel. Nat. Cell Biol. 2002; 4: 191–7.

Krasnoperov V, Lu Y, Buryanovsky L, Neubert TA, Ichtchenko K, Petrenko AG. Post-translational proteolytic processing of CIRL, a natural chimera of cell adhesion protein and G protein-coupled receptor. The role of the GPS motif. J. Biol. Chem. 2002; [volume number: page span].

Lantinga-van L, Dauwerse JG, Baelde HJ, et al. Lowering of Pkd1 expression is sufficient to cause polycystic kidney disease. Hum. Mol. Genet. 2004; 13: 3069–77.

Le NH, van der Bent P, Huls G, et al. Aberrant polycystin-1 expression results in modification of activator protein-1 activity, whereas Wnt signaling remains unaffected. J. Biol. Chem. 2004; 279: 27472–81.

Lehtonen S, Ora A, Olkkonen VM, et al. In vivo interaction of the adapter protein CD2-associated protein with the type 2 polycystic kidney disease protein, polycystin-2. J. Biol. Chem. 2000; 275: 32888–93.

Lehtonen S, Zhao F, Lehtonen E. CD2-associated protein directly interacts with the actin cytoskeleton. Am. J. Physiol. Renal Physiol. 2002; 283: F734–43.

Leuenroth SJ, Okuhara D, Shotwell JD, et al. Triptolide is a traditional Chinese medicine-derived inhibitor of polycystic kidney disease. Proc. Natl Acad. Sci. USA 2007; 104: 4389–94.

Li HP, Geng L, Burrow CR, Wilson PD. Identification of phosphorylation sites in the PKD1-encoded protein C-terminal domain. Biochem. Biophys. Res. Commun. 1999; 259: 356–63.

Li A, Davila S, Furu L, et al. Mutations in PRKCSH cause isolated autosomal dominant polycystic liver disease. Am. J. Hum. Genet. 2003a; 72: 691–703.

Li A, Tian X, Sung SW, Somlo S. Identification of two novel polycystic kidney disease 1-like genes in human and mouse genomes. Genomics 2003b; 82: 498–500.

Li Q, Dai Y, Guo L, et al. Polycystin-2 associates with tropomyosin-1, an actin microfilament component. J. Mol. Biol. 2003c; 325: 949–62.

Li Q, Shen PY, Wu G, Chen XZ. Polycystin-2 interacts with troponin I, an angiogenesis inhibitor. Biochemistry 2003d; 42: 450–7.

Li Q, Montalbetti N, Shen PY, et al. Alpha-actinin associates with polycystin-2 and regulates its channel activity. Hum. Mol. Genet. 2005a; 14: 1587–603.

Li X, Luo Y, Starremans PG, McNamara CA, Pei Y, Zhou J. Polycystin-1 and polycystin-2 regulate the cell cycle through the helix-loop-helix inhibitor Id2. Nat. Cell Biol. 2005b; 7: 1202–12.

Li Y, Wright JM, Qian F, Germino GG, Guggino WB. Polycystin 2 interacts with type I inositol 1,4,5-trisphosphate receptor to modulate intracellular Ca^{2+} signaling. J. Biol. Chem. 2005c; 280: 41298–306.

Lin F, Hiesberger T, Cordes K, et al. Kidney-specific inactivation of the KIF3A subunit of kinesin-II inhibits renal ciliogenesis and produces polycystic kidney disease. Proc. Natl Acad. Sci. USA 2003; 100: 5286–91.

Liu W, Murcia NS, Duan Y, et al. Mechanoregulation of intracellular Ca^{2+} concentration is attenuated in collecting duct of monocilium-impaired orpk mice. Am. J. Physiol. Renal Physiol. 2005; 289: F978–88.

Loghman-Adham M, Soto C, Inagami T, Sotelo-Avila C. Expression of components of the renin-angiotensin system in autosomal recessive polycystic kidney disease. J. Histochem. Cytochem. 2005; 53: 979–88.

LopezJimenez ND, Cavenagh MM, Sainz E, Cruz-Ithier MA, Battey JF, Sullivan SL. Two members of the TRPP family of ion channels, Pkd1l3 and Pkd2l1, are co-expressed in a subset of taste receptor cells. J. Neurochem. 2006; 98: 68–77.

Losekoot M, Haarloo C, Ruivenkamp C, White S, Breuning M, Peters D. Analysis of missense variants in the PKHD1-gene in patients with autosomal recessive polycystic kidney disease (ARPKD). Hum. Genet. 2005; 118: 185–206.

Low SH, Vasanth S, Larson CH, et al. Polycystin-1, STAT6, and P100 function in a pathway that transduces ciliary mechanosensation and is activated in polycystic kidney disease. Dev. Cell 2006; 10: 57–69.

Lu W, Peissel B, Babakhanlou H, et al. Perinatal lethality with kidney and pancreas defects in mice with a targetted Pkd1 mutation. Nat. Genet. 1997; 17: 179–81.

Luo Y, Vassilev PM, Li X, Kawanabe Y, Zhou J. Native polycystin 2 functions as a plasma membrane Ca^{2+}-permeable cation channel in renal epithelia. Mol. Cell. Biol. 2003; 23: 2600–7.

Ma R, Li WP, Rundle D, Kong J, Akbarali HI, Tsiokas L. PKD2 functions as an epidermal growth factor-activated plasma membrane channel. Mol. Cell. Biol. 2005; 25: 8285–98.

Magistroni R, He N, Wang K, et al. Genotype-renal function correlation in type 2 autosomal dominant polycystic kidney disease. J. Am. Soc. Nephrol. 2003; 14: 1164–74.

Mamane Y, Petroulakis E, LeBacquer O, Sonenberg N. mTOR, translation initiation and cancer. Oncogene 2006; 25: 6416–22.

Markowitz GS, Cai Y, Li L, et al. Polycystin-2 expression is developmentally regulated. Am. J. Physiol. 1999; 277: F17–25.

Marshall WF, Nonaka S. Cilia: tuning in to the cell's antenna. Curr. Biol. 2006; 16: R604–14.

Martin GM, Ogburn CE, Colgin LM, Gown AM, Edland SD, Monnat RJ Jr. Somatic mutations are frequent and increase with age in human kidney epithelial cells. Hum. Mol. Genet. 1996; 5: 215–21.

Martin J, Han C, Gordon LA, et al. The sequence and analysis of duplication-rich human chromosome 16. Nature 2004; 432: 988–94.

Masyuk TV, Huang BQ, Ward CJ, et al. Defects in cholangiocyte fibrocystin expression and ciliary structure in the PCK rat. Gastroenterology 2003; 125: 1303–10.

Masyuk TV, Masyuk AI, Torres VE, Harris PC, LaRusso NF. Octreotide inhibits hepatic cystogenesis in a rodent model of polycystic liver disease by reducing cholangiocyte adenosine 3′,5′-cyclic monophosphate. Gastroenterology 2007; 132: 1104–16.

Matsumura ME, Lobe DR, McNamara CA. Contribution of the helix-loop-helix factor Id2 to regulation of vascular smooth muscle cell proliferation. J. Biol. Chem. 2002; 277: 7293–7.

McGrath J, Somlo S, Makova S, Tian X, Brueckner M. Two populations of node monocilia initiate left-right asymmetry in the mouse. Cell 2003; 114: 61–73.

Meder D, Shevchenko A, Simons K, Fullekrug J. Gp135/podocalyxin and NHERF-2 participate in the formation of a preapical domain during polarization of MDCK cells. J. Cell Biol. 2005; 168: 303–13.

Menezes LF, Cai Y, Nagasawa Y, et al. Polyductin, the PKHD1 gene product, comprises isoforms expressed in plasma membrane, primary cilium, and cytoplasm. Kidney Int. 2004; 66: 1345–55.

Mengerink KJ, Moy GW, Vacquier VD. suREJ3, a polycystin-1 protein, is cleaved at the GPS domain and localizes to the acrosomal region of sea urchin sperm. J. Biol. Chem. 2002; 277: 943–8.

Milutinovic J, Fialkow PJ, Agodoa LY, Phillips LA, Bryant JI. Fertility and pregnancy complications in women with autosomal dominant polycystic kidney disease. Obstet. Gynecol. 1983; 61: 566–9.

Milutinovic J, Rust PF, Fialkow PJ, et al. Intrafamilial phenotypic expression of autosomal dominant polycystic kidney disease. Am. J. Kidney Dis. 1992; 19: 465–72.

Mochizuki T, Saijoh Y, Tsuchiya K, et al. Cloning of inv, a gene that controls left/right asymmetry and kidney development. Nature 1998; 395: 177–81.

Mochizuki T, Wu G, Hayashi T, et al. PKD2, a gene for polycystic kidney disease that encodes an integral membrane protein. Science 1996; 272: 1339–42.

Moon RT. Wnt/beta-catenin pathway. Science 2005; 1: 2005.

Moser M, Matthiesen S, Kirfel J, et al. A mouse model for cystic biliary dysgenesis in autosomal recessive polycystic kidney disease (ARPKD). Hepatology 2005; 41: 1113–21.

Moy GW, Mendoza LM, Schulz JR, Swanson WJ, Glabe CG, Vacquier VD. The sea urchin sperm receptor for egg jelly is a modular protein with extensive homology to the human polycystic kidney disease protein, PKD1. J. Cell Biol. 1996; 133: 809–17.

Mrug M, Li R, Cui X, Schoeb T, Churchill G, Guay-Woodford L. Kinesin family member 12 Is a candidate polycystic kidney disease modifier in the cpk mouse. J. Am. Soc. Nephrol. 2005; 16: 905–16.

Mumm JS, Schroeter EH, Saxena MT, et al. A ligand-induced extracellular cleavage regulates gamma-secretase-like proteolytic activation of Notch1. Mol. Cell 2000; 5: 197–206.

Murcia NS, Richards WG, Yoder BK, Mucenski ML, Dunlap JR, Woychik RP. The Oak Ridge Polycystic Kidney (ORPKD) disease gene is required for left-right axis determination. Development 2000; 127: 2347–55.

Muto S, Aiba A, Saito Y, et al. Pioglitazone improves the phenotype and molecular defects of a targeted Pkd1 mutant. Hum. Mol. Genet. 2002; 11: 1731–42.

Mykytyn K, Sheffield VC. Establishing a connection between cilia and Bardet-Biedl Syndrome. Trends Mol. Med. 2004; 10: 106–9.

Nagano J, Kitamura K, Hujer K, et al. Fibrocystin interacts with CAML, a protein involved in Ca^{2+} signaling. Biophys. Res. Commun. 2005; 338: 880–9.

Nagao S, Yamaguchi T, Kusaka M, et al. Renal activation of extracellular signal-regulated kinase in rats with autosomal-dominant polycystic kidney disease. Kidney Int. 2003; 63: 427–37.

Nagao S, Kazuhiro N, Katsuyama M, et al. Increased water intake decreases progression of polycystic kidney disease in the PCK rat. J. Am. Soc. Nephrol. 2006; 17: 228–35.

Nagasawa Y, Matthiesen S, Onuchic LF, et al. Identification and characterization of Pkhd1, the mouse orthologue of the human ARPKD gene. J. Am. Soc. Nephrol. 2002; 13: 2246–58.

Nauli SM, Alenghat FJ, Luo Y, et al. Polycystins 1 and 2 mediate mechanosensation in the primary cilium of kidney cells. Nat. Genet. 2003; 33: 129–37.

Neill AT, Moy GW, Vacquier VD. Polycystin-2 associates with the polycystin-1 homolog, suREJ3, and localizes to the acrosomal region of sea urchin spermatozoa. Mol. Reprod. Dev. 2004; 67: 472–7.

Newby LJ, Streets AJ, Zhao Y, Harris PC, Ward CJ, Ong AC. Identification, characterization, and localization of a novel kidney polycystin-1-polycystin-2 complex. J. Biol. Chem. 2002; 277: 20763–73.

Nichols MT, Gidey E, Matzakos T, Dahl R, Stiegmann G, Shah RJ, Grantham JJ, Fitz JG, Doctor RB. Secretion of cytokines and growth factors into autosomal dominant polycystic kidney disease liver cyst fluid. Hepatology 2004; 40: 836–46.

Nilius B, Owsianik G, Voets T, Peters JA. Transient receptor potential cation channels in disease. Physiol. Rev. 2007; 87: 165–217.

Nims N, Vassmer D, Maser RL. Transmembrane domain analysis of polycystin-1, the product of the polycystic kidney disease-1 (PKD1) gene: evidence for 11 membrane-spanning domains. Biochemistry 2003; 42: 13035–48.

Nishio S, Hatano M, Nagata M, Horie S, Koike T, Tokuhisa T, Mochizuki T. Pkd1 regulates immortalized proliferation of rean tubular epithelial cells through p53 induction and JNK activation. J. Clin. Invest. 2005; 115: 910–18.

Nomura H, Turco AE, Pei Y, et al. Identification of PKDL, a novel polycystic kidney disease 2-like gene whose murine homologue is deleted in mice with kidney and retinal defects. J. Biol. Chem. 1998; 273: 25967–73.

Obara T, Mangos S, Liu Y, et al. Polycystin-2 immunolocalization and function in zebrafish. J. Am. Soc. Nephrol. 2006; 17: 2706–18.

Obermuller N, Gallagher AR, Cai Y, et al. The rat Pkd2 protein assumes distinct subcellular distributions in different organs. Am. J. Physiol. 1999; 277: F914–25.

Omori S, Hida M, Fujita H, Takahashi H, Tanimura S, Kohno M, Awazu M. Extracellular signal-regulated kinase inhibition slows disease progression in mice with polycystic kidney disease. J. Am. Soc. Nephrol. 2006; 17: 1604–14.

Ong AC, Harris PC. Molecular basis of renal cyst formation – one hit or two? Lancet 1997; 349: 1039–40.

Ong AC, Harris PC, Davies DR, et al. Polycystin-1 expression in PKD1, early-onset PKD1, and TSC2/PKD1 cystic tissue. Kidney Int. 1999a; 56: 1324–33.

Ong AC, Ward CJ, Butler RJ, et al. Coordinate expression of the autosomal dominant polycystic kidney disease proteins, polycystin-2 and polycystin-1, in normal and cystic tissue. Am. J. Pathol. 1999b; 154: 1721–9.

Onuchic LF, Furu L, Nagasawa Y, et al. PKHD1, the polycystic kidney and hepatic disease 1 gene, encodes a novel large protein containing multiple immunoglobulin-like plexin-transcription-factor domains and parallel beta-helix 1 repeats. Am. J. Hum. Genet. 2002; 70: 1305–17.

Osathanondh V, Potter E. Pathogenesis of polycystic kidneys. Type 1 due to hyperplasia of interstitial portions of the collecting tubules. Arch. Pathol. 1964; 77: 466–73.

Otto EA, Schermer B, Obara T, et al. Mutations in INVS encoding inversin cause nephronophthisis type 2, linking renal cystic disease to the function of primary cilia and left-right axis determination. Nat. Genet. 2003; 34: 413–20.

Palsson R, Sharma CP, Kim K, McLaughlin M, Brown D, Arnaout MA. Characterization and cell distribution of polycystin, the product of autosomal dominant polycystic kidney disease gene 1. Mol. Med. 1996; 2: 702–11.

Parnell SC, Magenheimer BS, Maser RL, et al. The polycystic kidney disease-1 protein, polycystin-1, binds and activates heterotrimeric G-proteins in vitro. Biochem. Biophys. Res. Commun. 1998; 251: 625–31.

Parnell SC, Magenheimer BS, Maser RL, Calvet JP. Identification of the major site of in vitro PKA phosphorylation in the polycystin-1 C-terminal cytosolic domain. Biochem. Biophys. Res. Commun. 1999; 259: 539–43.

Parnell SC, Magenheimer BS, Maser RL, Zien CA, Frischauf AM, Calvet JP. Polycystin-1 activation of c-Jun N-terminal kinase and AP-1 is mediated by heterotrimeric G proteins. J. Biol. Chem. 2002; 277: 19566–72.

Patel V, Li L, Cobo-Stark P, Shao X, Somlo S, Lin F, Igarashi P. Acute kidney injury and aberrant planar cell polarity induce cyst formation in mice lacking renal cilia. Hum. Mol. Genet. 2008; 17: 1578–90.

Paterson AD, Magistroni R, He N, et al. Progressive loss of renal function is an age-dependent heritable trait in type 1 autosomal dominant polycystic kidney disease. J. Am. Soc. Nephrol. 2005; 16: 755–62.

Paterson AD, Pei Y. Is there a third gene for autosomal dominant polycystic kidney disease? Kidney Int. 1998; 54: 1759–61.

Paterson AD, Wang KR, Lupea D, St George-Hyslop P, Pei Y. Recurrent fetal loss associated with bilineal inheritance of type 1 autosomal dominant polycystic kidney disease. Am. J. Kidney Dis. 2002; 40: 16–20.

Pazour GJ. Intraflagellar transport and cilia-dependent renal disease: the ciliary hypothesis of polycystic kidney disease. J. Am. Soc. Nephrol. 2004; 15: 2528–36.

Pazour GJ, Dickert BL, Vucica Y, Seeley ES, Rosenbaum JL, Witman GB, Cole DG. Chlamydomonas IFT88 and its mouse homologue, polycystic kidney disease gene tg737, are required for assembly of cilia and flagella. J. Cell Biol. 2000; 151: 709–18.

Pazour GJ, San Agustin JT, Follit JA, Rosenbaum JL, Witman GB. Polycystin-2 localizes to kidney cilia and the ciliary level is elevated in orpk mice with polycystic kidney disease. Curr. Biol. 2002; 12: R378–80.

Pei Y, Paterson AD, Wang KR, et al. Bilineal disease and trans-heterozygotes in autosomal dominant polycystic kidney disease. Am. J. Hum. Genet. 2001; 68: 355–63.

Pei Y, Watnick T, He N, et al. Somatic PKD2 mutations in individual kidney and liver cysts support a "two-hit" model of cystogenesis in type 2 autosomal dominant polycystic kidney disease. J. Am. Soc. Nephrol. 1999; 10: 1524–9.

Pennekamp P, Karcher C, Fischer A, et al. The ion channel polycystin-2 is required for left-right axis determination in mice. Curr. Biol. 2002; 12: 938–43.

Peral B, Gamble V, Strong C, et al. Identification of mutations in the duplicated region of the polycystic kidney disease 1 gene

(PKD1) by a novel approach. Am. J. Hum. Genet. 1997; 60: 1399–410.

Peral B, Ong AC, San Millan JL, Gamble V, Rees L, Harris PC. A stable, nonsense mutation associated with a case of infantile onset polycystic kidney disease 1 (PKD1). Hum. Mol. Genet. 1996; 5: 539–42.

Pereira TV, Nunes AC, Rudnicki M, et al. Influence of ACE I/D gene polymorphism in the progression of renal failure in autosomal dominant polycystic kidney disease: a meta-analysis. Nephrol. Dial. Transplant. 2006; 21: 3155–63.

Persu A, Duyme M, Pirson Y, et al. Comparison between siblings and twins supports a role for modifier genes in ADPKD. Kidney Int. 2004; 66: 2132–6.

Peters DJM, Sandkuijl LA. Genetic heterogeneity of polycystic kidney disease in Europe. In: Breuining MH, Devoto M, Romeo G, eds. Contributions to Nephrology: Polycystic Kidney Disease. Basel: Karger, 1992: pp. 128–39.

Peters DJM, van de Wal A, Spruit L, et al. Cellular localization and tissue distribution of polycystin-1. J. Pathol. 1999; 188: 439–46.

Piontek K, Menezes LF, Garcia-Gonzalez MA, Huso DL, Germino GG. A critical developmental switch defines the kinetics of kidney cyst formation after loss of Pkd1. Nat. Med. 2007; 13: 1490–5.

Piontek KB, Huso DL, Grinberg A, et al. A functional floxed allele of Pkd1 that can be conditionally inactivated in vivo. J. Am. Soc. Nephrol. 2004; 15: 3035–43.

Pirson Y, Chauveau D, Torres VE. Management of cerebral aneurysms in autosomal dominant polycystic kidney disease: unruptured asymptomatic intracranial aneurysms. J. Am. Soc. Nephrol. 2002; 13: 269–76.

Ponting CP, Hofmann K, Bork P. A latrophilin/CL-1-like GPS domain in polycystin-1. Curr. Biol. 1999; 9: R585–8.

Praetorius HA, Frokiaer J, Nielsen S, Spring KR. Bending the primary cilium opens Ca^{2+}-sensitive intermediate-conductance K^+ channels in MDCK cells. J. Membr. Biol. 2003; 191: 193–200.

Praetorius HA, Spring KR. Bending the MDCK cell primary cilium increases intracellular calcium. J. Membr. Biol. 2001; 184: 71–9.

Praetorius HA, Spring KR. Removal of the MDCK cell primary cilium abolishes flow sensing. J. Membr. Biol. 2003a; 191: 69–76.

Praetorius HA, Spring KR. The renal cell primary cilium functions as a flow sensor. Curr. Opin. Nephrol. Hypertens. 2003b; 12: 517–20.

Praetorius HA, Spring KR. A physiological view of the primary cilium. Annu. Rev. Physiol. 2005; 67: 515–29.

Pritchard L, Sloane-Stanley JA, Sharpe J, et al. A human *PKD1* transgene generates functional polycystin-1 in mice and is associated with a cystic phenotype. Hum. Mol. Genet. 2000; 9: 2617–27.

Puri S, Magenheimer BS, Maser RL, et al. Polycystin-1 activates the calcineurin/NFAT (nuclear factor of activated T-cells) signaling pathway. J. Biol. Chem. 2004; 279: 55455–64.

Qian CN, Knol J, Igarashi P, et al. Cystic renal neoplasia following conditional inactivation of apc in mouse renal tubular epithelium. J. Biol. Chem. 2005a; 280: 3938–45.

Qian F, Boletta A, Bhunia AK, et al. Cleavage of polycystin-1 requires the receptor for egg jelly domain and is disrupted by human autosomal-dominant polycystic kidney disease 1-associated mutations. Proc. Natl Acad. Sci. USA 2002; 99: 16981–6.

Qian F, Germino FJ, Cai Y, Zhang X, Somlo S, Germino GG. PKD1 interacts with PKD2 through a probable coiled-coil domain. Nat. Genet. 1997; 16: 179–83.

Qian F, Watnick TJ, Onuchic LF, Germino GG. The molecular basis of focal cyst formation in human autosomal dominant polycystic kidney disease type I. Cell 1996; 87: 979–87.

Qian F, Wei W, Germino G, Oberhauser A. The nanomechanics of polycystin-1 extracellular region. J. Biol. Chem. 2005b; 280: 40723–30.

Qian Q, Du H, King BF, Kumar S, Dean PG, Cosio FG, Torres VE. Sirolimus reduces polycystic liver volume in ADPKD patients. J. Am. Soc. Nephrol. 2008; 19: 631–8.

Qian Q, Hunter LW, Han YS, et al. Pkd2+/− vascular smooth muscles develop exaggerated vasocontraction in response to phenylephrine stimulation. JASN 2007; 18: 485–93.

Qian Q, Hunter LW, Li M, et al. Pkd2 haploinsufficiency alters intracellular calcium regulation in vascular smooth muscle cells. Hum. Mol. Genet. 2003a; 12: 1875–80.

Qian Q, Li A, King BF, et al. Clinical profile of autosomal dominant polycystic liver disease. Hepatology 2003b; 37: 164–71.

Ravine D, Gibson RN, Walker RG, Sheffield LJ, Kincaid-Smith P, Danks DM. Evaluation of ultrasonographic diagnostic criteria for autosomal dominant polycystic kidney disease 1. Lancet 1994; 343: 824–7.

Renal Data System US. USRDS 1999 Annual Data Report. Bethesda: National Institutes of Health, 1999.

Reuss A, Wladimiroff JW, Niermeyer MF. Sonographic, clinical and genetic aspects of prenatal diagnosis of cystic kidney disease. Ultrasound Med. Biol. 1991; 17: 687–94.

Reynolds DM, Hayashi T, Cai Y, et al. Aberrant splicing in the PKD2 gene as a cause of polycystic kidney disease. J. Am. Soc. Nephrol. 1990; 10: 2342–51.

Riordan JR, Forbush B3, Hanrahan JW. The molecular basis of chloride transport in shark rectal gland. J. Exp. Biol. 1994; 196: 405–18.

Rodriguez-Viciana P, Tetsu O, Tidyman WE, et al. Germline mutations in genes within the MAPK pathway cause cardio-facio-cutaneous syndrome. Science 2006; 311: 1287–90.

Rohatgi R, Greenberg A, Burrow C, Wilson P, Satlin L. Na transport in autosomal recessive polycystic kidney disease (ARPKD) cyst lining epithelial cells. J. Am. Soc. Nephrol. 2003; 14: 827–36.

Roitbak T, Ward CJ, Harris PC, Bacallao R, Ness SA, Wandinger-Ness A. A polycystin-1 multiprotein complex is disrupted in polycystic kidney disease cells. Mol. Biol. Cell 2004; 15: 1334–46.

Ros E, Navarro S, Bru C, Gilabert R, Bianchi L, Bruguera M. Ursodeoxycholic acid treatment of primary hepatolithiasis in Caroli's syndrome. Lancet 1993; 342: 404–6.

Ross AJ, May-Simera H, Eichers ER, et al. Disruption of Bardet-Biedl syndrome ciliary proteins perturbs planar cell polarity in vertebrates. Nat. Genet. 2005; 37: 1135–40.

Rossetti S, Burton S, Strmecki L, et al. The position of the polycystic kidney disease 1 (PKD1) gene mutation correlates with the severity of renal disease. J. Am. Soc. Nephrol. 2002; 13: 1230–7.

Rossetti S, Chauveau D, Kubly V, et al. Association of mutation position in polycystic kidney disease 1 (PKD1) gene and development of a vascular phenotype. Lancet 2003; 361: 2196–201.

Rossetti S. Genotype–phenotype correlations in autosomal dominant and autosomal recessive polycystic kidney disease. J. Am. Soc. Nephrol. 2007; 18: 1374–80.

Rossetti S, Strmecki L, Gamble V, et al. Mutation analysis of the entire PKD1 gene: genetic and diagnostic implications. Am. J. Hum. Genet. 2001; 68: 46–63.

Ruggenenti P, Remuzzi A, Ondei P, et al. Safety and efficacy of long-acting somatostatin treatment in autosomal dominant polcysytic kidney disease. Kidney Int. 2005; 68: 206–16.

Rundle DR, Gorbsky G, Tsiokas L. PKD2 interacts and co-localizes with mDia1 to mitotic spindles of dividing cells: role of mDia1 IN PKD2 localization to mitotic spindles. J. Biol. Chem. 2004; 279: 29728–39.

Saadi-Kheddouci S, Berrebi D, Romagnolo B, et al. Early development of polycystic kidney disease in transgenic mice expressing an activated mutant of the beta-catenin gene. Oncogene 2001; 20: 5972–81.

Sampson JR, Maheshwar MM, Aspinwall R, et al. Renal cystic disease in tuberous sclerosis: The role of the polycystic kidney disease 1 gene. Am. J. Hum. Genet. 1997; 61: 843–51.

Sandford R, Sgotto B, Aparicio S, et al. Comparative analysis of the polycystic kidney disease 1 (PKD!) gene reveals an integral membrane glycoprotein with multiple evolutionary conserved domains. Hum. Mol. Genet. 1997; 6: 1483–9.

Sarbassov DD, Ali SM, Sabatini DM. Growing roles for the mTOR pathway. Curr. Opin. Cell Biol. 2005; 17: 596–603.

Sato Y, Harada K, Furubo S, et al. Inhibition of intrahepatic bile duct dilation of the polycystic kidney rat with a novel tyrosine kinase inhibitor gefitinib. Am. J. Pathol. 2006; 169: 1238–50.

Sato Y, Harada K, Kizawa K, et al. Activation of the MEK5/ERK5 cascade is responsible for biliary dysgenesis in a rat model of Caroli's disease. Am. J. Pathol. 2005; 166: 49–60.

Scheffers MS, Le H, van der BP, et al. Distinct subcellular expression of endogenous polycystin-2 in the plasma membrane and Golgi apparatus of MDCK cells. Hum. Mol. Genet. 2002; 11: 59–67.

Scheffers MS, van der BP, Prins F, et al. Polycystin-1, the product of the polycystic kidney disease 1 gene, co-localizes with desmosomes in MDCK cells. Hum. Mol. Genet. 2000; 9: 2743–50.

Schnell DJ, Hebert DN. Protein translocons: multifunctional mediators of protein translocation across membranes. Cell 2003; 112: 491–5.

Schottenfeld J, Sullivan-Brown J, Burdine RD. Zebrafish curly up encodes a Pkd2 ortholog that restricts left-side-specific expression of southpaw. Development 2007; 134: 1605–15.

Schroeter EH, Kisslinger JA, Kopan R. Notch-1 signalling requires ligand-induced proteolytic release of intracellular domain. Nature 1998; 393: 382–6.

Seifert JR, Mlodzik M. Frizzled/PCP signalling: a conserved mechanism regulating cell polarity and directed motility. Nat. Rev. Genet. 2007; 8: 126–38.

Sharp A, Messiaen L, Page G, et al. Comprehensive genomic analysis for PKHD1 mutations in ARPKD cohorts. J. Med. Genet. 2005; 42: 336–49.

Sheppard DN, Welsh MJ. Structure and function of the CFTR chloride channel. Physiol. Rev. 1999; 79: 23–45.

Sherstha R, McKinley C, Russ P, et al. Postmenopausal estrogen therapy selectively stimulates hepatic enlargement in women with autosomal dominant polycystic kidney disease. Hepatology 1997; 26: 1282–6.

Shibazaki S, Yu Z, Nishio S, et al. Cyst formation and activation of the extracellular regulated kinase pathway after kidney specific inactivation of Pkd1. Hum. Mol. Genet. 2008; 17: 1505–16.

Shillingford JM, Murcia NS, Larson CH, et al. The mTOR pathway is regulated by polycystin-1, and its inhibition reverses renal cystogenesis in polycystic kidney disease. Proc. Natl Acad. Sci. USA 2006; 103: 5466–71.

Shneider B, Magid M. Liver disease in autosomal recessive polycystic kidney disease. Pediatr. Transplant. 2005; 9: 634–49.

Shukla V, Coumoul X, Wang RH, Kim HS, Deng CX. RNA interference and inhibition of MEK-ERK signaling prevent abnormal skeletal phenotypes in a mouse model of craniosynostosis. Nat. Genet. 2007; 39: 1145–50.

Simons M, Gloy J, Ganner A, et al. Inversin, the gene product mutated in nephronophthisis type II, functions as a molecular switch between Wnt signaling pathways. Nat. Genet. 2005; 37: 537–43.

Singla V, Reiter JF. The primary cilium as the cell's antenna: signaling at a sensory organelle. Science 2006; 313: 629–33.

Somlo S. The PKD2 gene: structure, interactions, mutations, and inactivation. Adv. Nephrol. Necker Hosp. 1999; 29: 257–5.

Sujansky E, Kreutzer SB, Johnson AM, Lezotte DC, Schrier RW, Gabow PA. Attitudes of at-risk and affected individuals regarding presymptomatic testing for autosomal dominant polycystic kidney disease. Am. J. Med. Genet. 1990; 35: 510–15.

Sullivan LP, Wallace DP, Grantham JJ. Epithelial transport in polycystic kidney disease. Physiol. Rev. 1998; 78: 1165–91.

Sumfest JM, Burns MW, Mitchell ME. Aggressive surgical and medical management of autosomal recessive polycystic kidney disease. Urology 1993; 42: 309–12.

Sun Z, Amsterdam A, Pazour GJ, Cole DG, Miller MS, Hopkins N. A genetic screen in zebrafish identifies cilia genes as a principal cause of cystic kidney. Development 2004; 131: 4085–93.

Sweeney WJ, Avner E. Molecular and cellular pathophysiology of autosomal recessive polycystic kidney disease (ARPKD). Cell Tissue Res. 2006; 326: 671–85.

Sweeney WE Jr, Chen Y, Nakanish K, Frost P, Avner ED. Treatment of polycystic kidney disease with a novel tyrosine kinase inhibitor. Kidney Int. 2000; 57: 33–40.

Sweeney WE Jr, Hamahira K, Sweeney J, Garcia-Gatrell M, Frost P, Avner ED. Combination treatment of PKD utilizing dual inhibition of EGF-receptor activity and ligand bioavailability. Kidney Int. 2003; 64: 1310–19.

Tanaka Y, Okada Y, Hirokawa N. FGF-induced vesicular release of Sonic hedgehog and retinoic acid in leftward nodal flow is critical for left-right determination. Nature 2005; 435: 172–7.

Tao Y, Kim J, Schrier RW, Edelstein CL. Rapamycin markedly slows disease progression in a rat model of polycystic kidney disease. J. Am. Soc. Nephrol. 2005; 16: 46–51.

Tazon-Vega B, Vilardell M, Perez-Oller L, et al. Study of candidate genes affecting the progression of renal disease in autosomal dominant polycystic kidney disease type 1. Nephrol. Dial. Transplant. 2007; 22: 1567–77.

Thivierge C, Kurbegovic A, Couillard M, Guillaume R, Cote O, Trudel M. Overexpression of PKD1 causes polycystic kidney disease. Mol. Cell. Biol. 2006; 26: 1538–48.

Torra R, Badenas C, Darnell A, et al. Linkage, clinical features, and prognosis of autosomal dominant polycystic kidney disease types 1 and 2. J. Am. Soc. Nephrol. 1996; 7: 2142–51.

Torra R, Badenas C, Perez-Oller L, et al. Increased prevalence of polycystic kidney disease type 2 among elderly polycystic patients. Am. J. Kidney Dis. 2000; 36: 728–34.

Torra R, Badenas C, San Millan JL, Perez-Oller L, Estivill X, Darnell A. A loss-of-function model for cystogenesis in human autosomal dominant polycystic kidney disease type 2. Am. J. Hum. Genet. 1999; 65: 345–52.

Torres V, Sweeney W, Wang X, et al. Epidermal growth factor receptor tyrosine kinase inhibition is not protective in PCK rats. Kidney Int. 2004a; 66: 1766–73.

Torres VE, Donovan KA, Scicli G, et al. Synthesis of renin by tubulocystic epithelium in autosomal-dominant polycystic kidney disease. Kidney Int. 1992; 42: 364–73.

Torres VE, Erickson SB, Smith LH, Wilson DM, Hattery RR, Segura JW. The association of nephrolithiasis and autosomal dominant polycystic kidney disease. Am. J. Kidney Dis. 1988; 11: 318–25.

Torres VE, Harris PC, Pirson Y. Autosomal dominant polycystic kidney disease. Lancet 2007; 369: 1287–301.

Torres VE, Holley KE, Offord KP. General features of autosomal dominant polycystic kidney disease. In: Grantham J, Gardner K, eds. Problems in Diagnosis and Management of Polycystic Kidney Disease. Kansas City, MO: PKD Foundation, 1985: pp. 49–69.

Torres VE, Wang X, Qian Q, Somlo S, Harris PC, Gattone VH. Effective treatment of an orthologous model of autosomal dominant polycystic kidney disease. Nat. Med. 2004b; 10: 363–4.

Torres VE, Wilson DM, Hattery RR, Segura JW. Renal stone disease in autosomal dominant polycystic kidney disease. Am. J. Kidney Dis. 1993; 22: 513–19.

Torres V, Rastogi S, King B, Stanson A, Gross J Jr., Nagorney D. Hepatic venous outflow obstruction in autosomal dominant polycystic kidney disease. J. Am. Soc. Nephrol. 1994; 5: 1186–92.

Treml K, Meimaroglou D, Hentges A, Bause E. The alpha- and beta-subunits are required for expression of catalytic activity in the hetero-dimeric glucosidase II complex from human liver. Glycobiology 2000; 10: 493–502.

Trombetta ES, Simons JF, Helenius A. Endoplasmic reticulum glucosidase II is composed of a catalytic subunit, conserved from yeast to mammals, and a tightly bound noncatalytic HDEL-containing subunit. J. Biol. Chem. 1996; 271: 27509–16.

Tsiokas L, Arnould T, Zhu C, Kim E, Walz G, Sukhatme VP. Specific association of the gene product of PKD2 with the TRPC1 channel. Proc. Natl Acad. Sci. USA 1999; 96: 3934–9.

Tsiokas L, Kim E, Arnould T, Sukhatme VP, Walz G. Homo- and heterodimeric interactions between the gene products of PKD1 and PKD2. Proc. Natl Acad. Sci. USA 1997; 94: 6965–70.

Veizis I, Cotton C. Abnormal EGF-dependent regulation of sodium absorption in ARPKD collecting duct cells. Am. J. Physiol. Renal Physiol. 2005; 288: F474–82.

Veldhuisen B, Spruit L, Dauwerse HG, Breuning MH, Peters DJM. Genes homologous to the autosomal dominant polycystic kidney disease genes (PKD1 and PKD2). Euro. J. Hum. Genet. 1999; 7: 860–72.

Wahl PR, Serra AL, Le Hir M, Molle KD, Hall MN, Wuthrich RP. Inhibition of mTOR with sirolimus slows disease progression in Han: SPRD rats with autosomal dominant polycystic kidney disease (ADPKD). Nephrol. Dial. Transplant. 2006; 21: 598–4.

Walz G. Therapeutic approaches in autosomal dominant polycystic kidney disease (ADPKD): is there light at the end of the tunnel?. Nephrol. Dial. Transplant. 2006; 21: 1752–7.

Wang D, Iversen J, Wilcox CS, Strandgaard S. Endothelial dysfunction and reduced nitric oxide in resistance arteries in autosomal-dominant polycystic kidney disease. Kidney Int. 2003; 64: 1381–8.

Wang S, Luo Y, Wilson P, Witman G, Zhou J. The autosomal recessive polycystic kidney disease protein is localized to primary cilia, with concentration in the basal body area. J. Am. Soc. Nephrol. 2004; 15: 592–602.

Wang S, Zhang J, Nauli SM, et al. Fibrocystin/polyductin, found in the same protein complex with polycystin-2, regulates calcium responses in kidney epithelia. Mol. Cell. Biol. 2007; 27: 3241–52.

Wang Y, Nathans J. Tissue/planar cell polarity in vertebrates: new insights and new questions. Development 2007; 134: 647–58.

Ward CJ, Hogan MC, Rossetti S, et al. The gene mutated in autosomal recessive polycystic kidney disease encodes a large, receptor-like protein. Nat. Genet. 2002; 30: 259–69.

Ward CJ, Turley H, Ong AC, et al. Polycystin, the polycystic kidney disease 1 protein, is expressed by epithelial cells in fetal, adult, and polycystic kidney. Proc. Natl Acad. Sci. USA 1996; 93: 1524–8.

Ward CJ, Yuan D, Masyuk TV, et al. Cellular and subcellular localization of the ARPKD protein fibrocystin is expressed on primary cilia. Hum. Mol. Genet. 2003; 12: 2703–10.

Watnick T, He N, Wang K, et al. Mutations of PKD1 in ADPKD2 cysts suggest a pathogenic effect of trans-heterozygous mutations. Nat. Genet. 2000; 25: 143–4.

Watnick T, Phakdeekitcharoen B, Johnson A, et al. Mutation detection of PKD1 identifies a novel mutation common to three families with aneurysms and/or very-early-onset disease. Am. J. Hum. Genet. 1999; 65: 1561–71.

Watnick T, Piontek K, Cordal T, et al. An unusual pattern of mutation in the duplicated portion of PKD1 is revealed by use of a novel strategy for mutation detection. Hum. Mol. Genet. 1997; 6: 1473–81.

Watnick TJ, Jin Y, Matunis E, Kernan MJ, Montell C. A flagellar polycystin-2 homolog required for male fertility in *Drosophila*. Curr. Biol. 2003; 13: 2179–84.

Watnick TJ, Torres VE, Gandolph MA, et al. Somatic mutation in individual liver cysts supports a two-hit model of cystogenesis in autosomal dominant polycystic kidney disease. Mol. Cell 1998; 2: 247–51.

Wei F, Karihaloo A, Yu Z, et al. Ngal suppresses cyst growth by *Pkd1* null cells *in vitro* and *in vivo*. Kidney Int. 2008 (in press).

Williams SS, Cobo-Stark P, James LR, Somlo S, Igarashi P. Kidney cysts, pancreatic cysts, and biliary disease in a mouse model of autosomal recessive polycystic kidney disease. Pediatr. Nephrol. 2008; 23: 733–41.

Wilson PD, Geng L, Li X, Burrow CR. The PKD1 gene product, "polycystin-1," is a tyrosine-phosphorylated protein that colocalizes with alpha2beta1-integrin in focal clusters in adherent renal epithelia. Lab. Invest. 1999; 79: 1311–23.

Wilson SJ, Amsler K, Hyink DP, et al. Inhibition of HER-2(neu/ErbB2) restores normal function and structure to polycystic kidney disease (PKD) epithelia. Biochim. Biophys. Acta 2006.

Witzgall R. Polycystin-2-an intracellular or plasma membrane channel? Naunyn Schmiedebergs Arch. Pharmacol. 2005; 371: 342–7.

Woods AJ, White DP, Caswell PT, Norman JC. PKD1/PKCmu promotes alphavbeta3 integrin recycling and delivery to nascent focal adhesions. EMBO J. 2004; 23: 2531–43.

Woollard J, Punyashtiti R, Richardson S, et al. A mouse model of autosomal recessive polycystic kidney disease with biliary duct and proximal tubule dilatation. Kidney Int. 2007; 72: 328–36.

Wu G, D'Agati V, Cai Y, et al. Somatic inactivation of Pkd2 results in polycystic kidney disease. Cell 1998a; 93: 177–88.

Wu G, Markowitz GS, Li L, et al. Cardiac defects and renal failure in mice with targeted mutations in Pkd2. Nat. Genet. 2000; 24: 75–8.

Wu G, Tian X, Nishimura S, et al. Trans-heterozygous Pkd1 and Pkd2 mutations modify expression of polycystic kidney disease. Hum. Mol. Genet. 2002; 11: 1845–54.

Wu GQ, Hayashi T, Park JH, et al. Identification of the human PKD2-related cDNA, *PKD2L*: tissue-specific expression and linkage mapping on chromosome 10q25. Genomics 1998b; 54: 564–8.

Wu Y, Dai XQ, Li Q, et al. Kinesin-2 mediates physical and functional interactions between polycystin-2 and fibrocystin. Hum. Mol. Genet. 2006; 15: 3280–92.

Xiong H, Chen Y, Yi Y, et al. A novel gene encoding a TIG multiple domain protein is a positional candidate for autosomal recessive polycystic kidney disease. Genomics 2002; 80: 96–104.

Xu GM, Sikaneta T, Sullivan BM, et al. Polycystin-1 interacts with intermediate filaments. J. Biol. Chem. 2001; 276: 46544–52.

Xu N, Glockner JF, Rossetti S, Babovich-Vuksanovic D, Harris PC, Torres VE. Autosomal dominant polycystic kidney disease coexisting with cystic fibrosis. J. Nephrol. 2006; 19: 529–34.

Yamaguchi T, Nagao S, Wallace DP, et al. Cyclic AMP activates B-Raf and ERK in cyst epithelial cells from autosomal-dominant polycystic kidneys. Kidney Int. 2003; 63: 1983–94.

Yamaguchi T, Pelling JC, Ramaswamy NT, et al. cAMP stimulates the in vitro proliferation of renal cyst epithelial cells by activating the extracellular signal-regulated kinase pathway. Kidney Int. 2000; 57: 1460–71.

Yamaguchi T, Wallace DP, Magenheimer BS, Hempson SJ, Grantham JJ, Calvet JP. Calcium restriction allows cAMP activation of the B-Raf/ERK pathway, switching cells to a cAMP-dependent growth-stimulated phenotype. J. Biol. Chem. 2004; 279: 40419–30.

Yang J, Zhang S, Zhou Q, et al. PKHD1 gene silencing may cause cell abnormal proliferation through modulation of intracellular calcium in autosomal recessive polycystic kidney disease. J. Biochem. Mol. Biol. 2007; 40: 467–74.

Yium J, Gabow P, Johnson A, Kimberling W, Martinez-Maldonado M. Autosomal dominant polycystic kidney disease in Blacks: clinical course and effects of sickle-cell hemoglobin. J. Am. Soc. Nephrol. 1994; 4: 1670–4.

Yoder BK. Role of primary cilia in the pathogenesis of polycystic kidney disease. J. Am. Soc. Nephrol. 2007; 18: 1381–8.

Yoder BK, Hou X, Guay-Woodford LM. The polycystic kidney disease proteins, polycystin-1, polycystin-2, polaris, and cystin, are co-localized in renal cilia. J. Am. Soc. Nephrol. 2002; 13: 2508–16.

Yuasa T, Venugopal B, Weremowicz S, Morton CC, Guo L, Zhou J. The sequence, expression, and chromosomal localization of a novel polycystic kidney disease 1-like gene, PKD1L1, in human. Genomics 2002; 79: 376–86.

Zerres K, Mucher G. Autosomal recessive polycystic kidney disease. J. Mol. Med. 1998; 76: 303–9.

Zerres K, Mucher G, Becker J, et al. Prenatal diagnosis of autosomal recessive polycystic kidney disease (ARPKD): molecular genetics, clinical experience, and fetal morphology. Am. J. Med. Genet. 1998; 76: 137–44.

Zerres K, Senderek J, Rudnik-Schoneborn S, et al. New options for prenatal diagnosis in autosomal recessive polycystic kidney disease by mutation analysis of the PKHD1 gene. Clin. Genet. 2004; 66: 53–7.

Zhang M, Mai W, Li C, et al. PKHD1 protein encoded by the gene for autosomal recessive polycystic kidney disease associates with basal bodies and primary cilia in renal epithelial cells. Proc. Natl Acad. Sci. USA 2004; 101: 2311–16.

CHAPTER 25

Nephronophthisis

FRIEDHELM HILDEBRANDT

OVERVIEW ON NEPHRONOPHTHISIS AND RELATED DISORDERS

Nephronophthisis (NPHP) is an autosomal recessive cystic kidney disease that represents the most frequent monogenic cause of end-stage renal disease (ESRD) in the first three decades of life. Three clinical forms of NPHP have been distinguished by onset of ESRD: infantile (Gagnadoux et al 1989), juvenile (Hildebrandt et al 1992), and adolescent NPHP (Omran et al 2000), which manifest with ESRD at the median ages of 1 year, 13 years, and 15 years, respectively. Symptoms are relatively mild, and consist of polyuria, polydipsia with regular fluid intake at night time, secondary enuresis, anemia, and growth retardation. In about 10% of patients there is an association of NPHP with retinitis pigmentosa, termed Senior-Loken syndrome (SLSN). There can also be an association with ocular motor apraxia type Cogan, cerebellar vermis aplasia (Joubert syndrome, JBTS), with liver fibrosis, or cone-shaped epiphyses of the phalanges. At an average age of 9 years a slightly raised serum creatinine is noted. Renal ultrasound reveals increased echogenicity, and occasionally cysts appear at the corticomedullary junction within kidneys of normal size (Blowey et al 1996). Renal histology reveals a characteristic triad of tubular basement membrane disruption, tubulointerstitial nephropathy, and corticomedullary cysts (Waldherr et al 1982, Zollinger et al 1980). In nephronophthisis cysts arise from the corticomedullary junction of the kidneys, in contrast to polycystic kidney disease, where cysts are evenly spread out over the entire organ.

The pathogenesis of nephronophthisis had been elusive since its first description in 1945 (Fanconi et al 1951, Smith & Graham 1945). Recently, the paradigm of positional cloning has been successful in the identification of genes that cause nephronophthisis if mutated. This approach has led to new insights into disease mechanisms of NPHP, which are related to sensory cilia, centrosomes, and planar cell polarity (Germino 2005, Hildebrandt & Otto 2005, Watnick & Germino 2003).

MODE OF INHERITANCE

NPHP, regardless of the presence or absence of extrarenal organ involvement, is inherited in an autosomal recessive pattern (Gardner 1971, Hildebrandt 1999). For Bardet-Biedl syndrome, in which NPHP-like kidney involvement can be seen, an oligogenic inheritance pattern has been described. This refers to the finding that mutations in more than one *BBS* gene may be required for full penetrance of some aspects of organ involvement (Badano & Katsanis 2002). NPHP has previously been grouped together with medullary cystic kidney disease (MCKD) under the term 'NPHP-MCKD complex' (Waldherr et al 1982, Hildebrandt et al 1992), since the different disease entities share several features regarding macroscopic pathology, microscopic pathology, and clinical symptoms (Gardner 1976). Both NPHP and MCKD primarily involve the corticomedullary region of the kidney. However, NPHP and MCKD are treated here in separate chapters, since there are three features which distinguish them among the disease entities of the NPHP–MCKD complex: (i) the mode of inheritance, (ii) the age at onset of ESRD, and (iii) the type of extrarenal organ involvement. Inheritance of NPHP is autosomal recessive, whereas MCKD is inherited in an autosomal dominant mode. In NPHP ESRD ensues in the first three decades of life, whereas in MCKD ESRD develops in the 3rd to 7th decade of life. Extrarenal manifestations such as retinitis pigmentosa, ocular motor apraxia, cerebellar vermis aplasia, hepatic fibrosis, and skeletal defects, have exclusively been described in association with nephronophthisis. The only extrarenal associations reported in MCKD are hyperuricemia and gout.

EPIDEMIOLOGY

NPHP has no gender predilection. It has been reported from virtually all regions of the world (Kleinknecht 1989).

Genetic Diseases of the Kidney
ISBN: 978-0-12-449851-8

The incidence of the disease has been given as 9 patients/8.3 million (Potter et al 1980) in the United States or 1 in 50 000 live births in Canada (Waldherr et al 1982, Pistor et al 1985). Although a rare disorder, NPHP constitutes the most frequent monogenic cause for ESRD in the first three decades of life. It represents a major cause of ESRD in children, accounting for 10–25% of these patients in Europe (Betts & Forest-Hay 1973, Cantani et al 1986, Kleinknecht 1989, Green et al 1990). In the North American pediatric end-stage renal disease population pooled data indicate a prevalence of approximately less than 5% of all children with ESRD (Avner 1994, Warady et al 1997).

CLINICAL FEATURES OF NPHP

Renal Involvement

The first case of juvenile NPHP was described by Smith and Graham in 1945 (Smith & Graham 1945). This report of a sporadic case was followed by publication of two large kindreds with familial disease by Fanconi et al (1951), who introduced the term 'familial juvenile nephronophthisis.' Since the first description over 300 cases have been published in the literature (Waldherr et al 1982). Five different genes have recently been identified by positional cloning as mutated in NPHP: *NPHP1, NPHP2/inversin, NPHP3, NPHP4,* and *NPHP5,* defining NPHP types 1, 2, 3, 4, and 5, respectively. This has made definite molecular genetic diagnostics possible (see below). Mutations in the *NPHP1* gene (see below) constitute the most frequent form of NPHP accounting for approximately 25% of all NPHP cases (Hoefele, unpublished data). Penetrance of the renal phenotype in affected individuals seems to be 100%. In juvenile NPHP the symptoms of polyuria, polydipsia, decreased urinary concentrating ability, and secondary enuresis are the earliest presenting symptoms in over 80% of cases (Kleinknecht 1989) and occur at around 4–6 years of age. Anemia and growth retardation occur later in the course of the disease and are usually pronounced (Ala-Mello et al 1996). Once renal failure has progressed further, pallor, weakness and generalized pruritus are also common. Regular fluid intake at nighttime is a characteristic feature of the patients' history, and starts around age 6 years. This characteristic symptom should actively be sought when taking the patients' history. The mild nature of symptoms together with lack of edema, hypertension and urinary tract infections may lead to a delay in diagnosis and therapy. Due to late detection of symptoms there is a small but definite risk of sudden death from fluid and electrolyte imbalance. Importantly, disease recurrence has never been reported in kidneys transplanted to NPHP patients (Steel et al 1980).

Onset of End-Stage Renal Disease

In NPHP, chronic renal failure develops within the first three decades of life (Hildebrandt et al 1997, Haider et al 1998, Omran et al 2000). Infantile NPHP, which is characterized by mutations in *NPHP2/inversin* leads to ESRD between birth and age 3 years (Haider et al 1998, Otto et al 2003a). In juvenile NPHP initial studies had shown a median age at ESRD of 11.5 years (Waldherr et al 1982). The rate of deterioration of renal function was homogeneous in a study of 29 patients with juvenile nephronophthisis (Gretz 1989). The median time lapse between a serum creatinine of 2 and 4 mg/dl was 32 months, between 4 and 6 mg/dl 10 months, and between 6 and 8 mg/dl 5 months (Gretz 1989). In a study conducted in 46 children with juvenile NPHP type 1 caused by mutations of the *NPHP1* gene, a serum creatinine of 6 mg/dl was reached at a median age of 13 years of age (range 4–20 years) (Hildebrandt et al 1992, 1997). Similarly, the median age of ESRD in patients with mutations in the *NPHP5* gene was 13 years (Otto et al 2005). In patients with adolescent NPHP due to mutations in the *NPHP3* gene ESRD developed by 19 years of age (Omran et al 2000), which was similar to patients with mutations in *NPHP4* (20 years) (Otto et al 2002). If renal failure has not developed by the age of 25 years, the diagnosis of recessive NPHP should be questioned and autosomal dominant MCKD considered as a differential diagnosis. In MCKD, which follows autosomal dominant inheritance, ESRD occurs later in life. MCKD types 1 and 2 show a median onset of ESRD at 62 years (Christodoulou et al 1998) and 32 years (Scolari et al 1998), respectively. MCKD type 2 can be positively diagnosed by detection of mutations in the *UMOD* gene encoding uromodulin/Tamm-Horsfall protein. In monozygotic twins a high concordance rate was noted in the development of renal failure from NPHP (Mongeau & Worthen 1967, Makker et al 1973).

NPHP with Extrarenal Associations

NPHP can occur in combination with extrarenal associations involving retina, cerebellum, liver, bones, and other organs. Specifically, it can be associated with ocular motor apraxia type Cogan (Saunier et al 1997b), with retinitis pigmentosa in about 10% of cases (Senior-Løken syndrome; SLSN) (Loken et al 1961, Senior et al 1961), with liver fibrosis (Boichis et al 1973), with cone-shaped epiphyses in Mainzer-Saldino syndrome (Mainzer et al 1970), and with cerebellar vermis aplasia and coloboma of the optic nerve in Joubert syndrome type B (Saraiva & Baraitser 1992, Valente et al 2005). Infantile NPHP type 2 can be associated with *situs inversus* (Otto et al 2003a), retinitis pigmentosa (O'Toole et al 2004), or cardiac ventricular septal defect (Otto et al 2003b). A detailed description of extrarenal symptoms is given in the following paragraphs.

RETINAL INVOLVEMENT (SENIOR-LOKEN SYNDROME)
Senior-Løken syndrome (SLSN), represented by concomitant occurrence of NPHP with retinitis pigmentosa, was first described by Contreras (Contreras & Espinoza 1960),

Senior (Senior et al 1961) and Løken (Loken et al 1961). The fact that three different terms, retinitis pigmentosa, tapetoretinal degeneration, and retinal-renal dysplasia have been used in the literature to describe the retinal findings of SLSN most likely reflects a spectrum within the pathogenesis that ranges from dysplasia to degeneration (Saraux et al 1970). The association of retinitis pigmentosa is found exclusively in recessive NPHP and is absent from dominant MCKD (Antoine et al 1963, Kleinknecht 1989). In NPHP types 1–4 retinitis pigmentosa occurs in about 10% of all affected families, without any overt genotype phenotype correlation. Whether patients with retinal involvement carry an unknown mutation in an additional modifier gene is an open question. Early-onset and late-onset types of SLSN have been distinguished. The early-onset type seems to represent a form of Leber's congenital amaurosis, since children exhibit coarse nystagmus and/or blindness at birth or develop these symptoms within the first 2 years of life (Medhioub et al 1994). Patients with mutations in the *NPHP5* gene exhibit retinits pigmentosa in all known cases. And in NPHP type 5, retinitis is early-onset with blindness occurring by the end of the 3rd year of life (Otto et al 2005).

Retinitis pigmentosa is diagnosed by its specific findings on ophthalmoscopy, which include increased pigment, attenuation of retinal vessels, and pallor of the optic disc, coupled with the results of electroretinography and electrooculography. Retinal degeneration is characterized by a constant and complete extinction of the electroretinogram, which precedes the development of visual and fundoscopic signs of retinitis pigmentosa (Biersdorf 1989). Whether the heterozygous state in the Senior-Løken syndrome can be determined by electroretinography and electrooculography as reported by Polak et al (1983) or whether it may cause adult-onset macular degeneration (AMD) awaits further studies. Fundoscopic alterations are present in all late-onset SLSN patients by the age of 10 years. The late-onset form is characterized by development of blindness during school age following the development of night blindness. Fundoscopy in these patients usually reveals sector retinitis pigmentosa or retinitis punctata albescens. Other eye symptoms besides tapeto-retinal degeneration comprise myopia, coloboma of the chorioidea, optical nerve atrophy, nystagmus, strabismus, hyperopia, and amblyopia (Fanconi et al 1951). Age of onset, symptoms, and histology of renal disease is identical to what is known from patients with juvenile nephronophthisis without ocular involvement (see above).

OCULOMOTOR APRAXIA TYPE COGAN

Ocular motor apraxia type Cogan is the transient inability of horizontal eye movements in the first few years of life. This symptom has been described in patients with mutations in the *NPHP1* (Saunier et al 1997b, Betz et al 2000) and *NPHP4* (Mollet al 2002) genes. It may be due to defects in the nuclei of the abducens nerve, which contain both ipsilaterally projecting motor neurons and contralaterally projecting interneurons, or supranuclear control regions such as the pontine paramedian reticular formation (PPRF) that projects to the abducens and oculomotor nuclei, as has been postulated for other forms of horizontal gaze palsy (Jen et al 2004).

CEREBELLAR VERMIS APLASIA (JOUBERT SYNDROME)

In Joubert syndrome type B (JBTS), a developmental disorder with multiple organ involvement, NPHP occurs in association with coloboma of the eye (or retinal degeneration), aplasia/hypoplasia of the cerebellar vermis with ataxia, and the facultative symptoms of psychomotor retardation, polydactyly, occipital encephalocele, and episodic neonatal tachy/dyspnea (Joubert et al 1969, Saraiva & Baraitser 1992). A pathognomonic diagnostic feature of JBTS on axial magnetic resonance imaging (MRI) of the brain is prominent superior cerebellar peduncles, termed the 'molar tooth sign' (MTS) of the midbrain–hindbrain junction (Gleeson et al 2004, Castori et al 2005). Two genes have now been shown to be mutated in JBTS: (i) A homozygous deletion of *NPHP1* was found in two siblings and a third patient with mild features of JBTS (Parisi et al 2004); (ii) The second gene defect (JBTS3) was found in the *Abelson Helper Integration Site* (*AHI1*) gene on chromosome 6q23.3 [OMIM*608894] (Dixon-Salazar et al 2004, Ferland et al 2004). Its protein product has been termed jouberin. The domain structure of the jouberin protein, which includes a coiled coil domain, six WD40 domains, and an SH3 domain, makes it a likely member of the nephrocystin complex of proteins. Mutations in *AHI1* occurred in JBTS patients with abnormal axonal decussation (crossing in the brain) affecting the corticospinal tract and superior cerebellar peduncles, thereby explaining the motor and behavioral abnormalities (Ferland et al 2004). Some patients also had abnormal cerebral structure with cortical polymicrogyria (Dixon-Salazar et al 2004). Whereas in the initial reports of *AHI1* mutations no renal phenotype was reported, NPHP was recently described in patients with *AHI1* mutations (Parisi et al 2005 in press, Utsch et al 2005). At least three other gene loci have been mapped for Joubert syndrome. These include the *JBTS1* locus on chromosome 9q34.3 [OMIM#213300] (Saar et al 1999), the *CORS2* locus on chromosome 11p12-q13.3 [OMIM%608091] (Keeler et al 2003), and the Smith-Magenis locus on 17p11.2 [OMIM#182290] (Natacci et al 2000).

LIVER FIBROSIS

The association of NPHP with *liver fibrosis* was first noted by Boichis (Boichis et al 1973) and was later also reported by others (Rayfield & McDonald 1972, Proesmans et al 1975, Delaney et al 1978). All patients had hepatomegaly and moderate portal fibrosis with mild bile duct proliferation. This pattern differs from that of classical congenital hepatic fibrosis, where biliary dysgenesis is prominent. A recessive mutation in the *NPHP3* gene was recently described in a patient with NPHP and liver fibrosis (Olbrich et al 2003). Hepatic involvement in NPHP type 2 (infantile NPHP) seems to only involve transient elevation of transaminases (Haider et al 1998).

SITUS INVERSUS

The presence of *situs inversus* was shown in a patient with infantile NPHP and mutations in the *NPHP2/inversin* gene (Otto et al 2003b). This would be expected from the mouse phenotype of *inversin* mutations, the *inv/inv* mouse, in which *situs inversus* was present in 90% of mice (Mochizuki et al 1998, Morgan et al 1998). Thus, the role of inversin for left–right axis specification was confirmed in humans. The patients with *situs inversus* also had a cardiac ventricular septal defect as a heterotaxy phenotype. This finding was analogous to the result that randomization of heart looping occurred in *nphp2/inversin* knockdown experiments in zebrafish. Recently, it has become apparent that products of other genes associated with renal cystic disease (in addition to inversin) are important in left–right axis determination of the body plan (Hirokawa et al 1998, Igarashi & Somlo 2002). The gene *PKD2*, mutations in which cause autosomal dominant PKD, and which encodes the calcium release channel polycystin-2, had been shown in a *Pkd2−/−* mouse model to represent a master gene regulating left–right axis determination, acting upstream of *Nodal*, *Ebaf*, *Leftb* and *Pitx2* (Pennekamp et al 2002). In addition to the malformations described previously (Wu et al 1998, 2000) homozygous mutant embryos showed right pulmonary isomerism, randomization of embryonic turning, heart looping, and abdominal *situs inversus*. In *Pkd2−/−* mice, *Leftb* and *Nodal* were not expressed in the left lateral plate mesoderm, and *Ebaf* was absent from the floorplate (Pennekamp et al 2002).

SKELETAL CHANGES

The association of NPHP with cone shaped epiphyses of the phalanges (types 28 and 28A) is known as *Mainzer-Saldino syndrome*, and was first published by Mainzer (Mainzer et al 1970) in patients that also had retinal degeneration and cerebellar ataxia. In Bardet-Biedl syndrome, a disorder related to NPHP, polydactyly may be present.

OTHER SYNDROMES WITH NPHP

Additional phenotypes have been described in association with NPHP (Table 25.1). These include Jeune syndrome (asphyxiating thoracic dysplasia) [OMIM %208500] (Jeune et al 1955, Donaldson et al 1985, Amirou et al 1998, Sarimurat et al 1998), Ellis van Creveld syndrome [OMIM #225500, *604831] (Moudgil et al 1998), RHYNS syndrome (retinitis pigmentosa, hypopituitarism, NPHP, skeletal dysplasia) [OMIM 602152] (Di Rocco et al 1997),

Alstrom syndrome (retinitis pigmentosa, deafness, obesity, and diabetes mellitus without mental defect, polydactyly, or hypogonadism) [OMIM #203800, *606844] (Alstrom et al 1959, Goldstein & Fialkow 1973), Sensenbrenner syndrome (cranioectodermal dysplasia) [OMIM%218330] (Tsimaratos et al 1998, Costet et al 2000), and Arima syndrome (cerebro-oculo-hepato-renal syndrome) [OMIM %243910] (Chance et al 1999, Satran et al 1999, Kumada et al 2004). In both Ellis-van Creveld syndrome and Alstrom syndrome, one causative gene has been identified so far (Table 25.1). NPHP has also been described in association with ulcerative colitis (Ala-Mello et al 2001). Bardet-Biedl syndrome (Alton & McDonald 1973, Green et al 1989) has been reported to exhibit renal histology findings reminiscent of NPHP. Recently, positional cloning of the genes responsible has revealed that the molecular relation between these syndromes may lie in coexpression of the respective gene products in primary cilia, basal bodies, or centrosomes of renal epithelial cells (Watnick & Germino 2003) (Table 25.1).

PATHOLOGY

Macroscopic Pathology

Macroscopic pathology of NPHP has been described in a study on 27 patients with juvenile nephronophthisis (Waldherr et al 1982). Kidney size is normal or moderately reduced. Cysts appear primarily at the cortico-medullary border of the kidneys (Figure 25.1). This is quite distinct from autosomal dominant and autosomal recessive polycystic kidney disease, where kidneys become grossly enlarged as a result of cystic dilatation. In NPHP there is always bilateral renal involvement. The external kidney surface has a finely granular appearance, most likely due to the protrusion of dilated cortical collecting ducts. Calices and pelvis appear completely normal. There are from 5 to over 50 cysts of 1–15 mm diameter, located preferentially at the corticomedullary border (Figure 25.1). The cysts seem to arise primarily from the distal convoluted and medullary collecting tubules as shown by microdissection (Sherman et al 1971), but may also appear in the papilla. Cysts are not always present but occur in about 70% of autopsy cases. They apparently develop late in the course of the disease (Sworn & Eisinger 1972). Cysts do not seem to be important for disease progression to renal failure (Bernstein & Gardner 1992), but rather the concomitant interstitial fibrosis seems to drive the progression of renal failure. Therefore, the presence of cysts is not a prerequisite for diagnosis. No cysts are present in organs other than the kidney.

Microscopic Pathology

Renal histological changes are characteristic but not pathognomonic for NPHP (Zollinger et al 1980, Waldherr et al 1982).

TABLE 25.1   Genetics and pathophysiology of nephronophthisis and other cystic diseases of the kidney (CDK): Disease groups are color coded as they relate to Figure 25.2. Columns represent disease names, genes, encoded proteins, cell biological data, clinical phenotypes, and evolutionary conservation of genes in mouse, zebrafish, and *Cenorrhabditis elegans*. Human genes mutated in CDK are marked in different colors for PKD (green), MCKD and GCKD (purple), NPHP, JBTS, ALMS (blue), BBS, and OFD (orange). Colors are repeated in animal model columns if mutations in the respective gene also lead to CDK in mouse or zebrafish, or if the gene is expressed in head and tail of *C. elegans* ciliated neurons. Genes known so far to be mutated in CDK of animal models only (and not in human disease), are marked for mouse models (yellow) and zebrafish models (dark blue). Citations of the literature are shown in italic. (see also Plate 42)

[a]Human cystic kidney disease	Human *gene* (inheritance)	Protein (domains)	Protein interaction partner	Subcellular localization	Human Renal disease (median ESRF)	Human extrarenal disease	Mouse CDK model	Mouse gene	Zebrafish CDK *gene* (phenotype)	*C. elegans* orthologous gene[b]
ADPKD1	*PKD1* (*Hughes et al 1995*) (AD)	Polycystin-1 (CC, multiple others)	[b]PKD2, [b]ATP2 (*Hu and Barr, 2005*), PACS1 and 2 (Kottgen et al 2005), [b]Siah-1 (*Kim et al 2004*)	Cilia, Golgi, FA, desmosomes	Cysts ($\approx$55 yr)	Liver cysts, pancreatic cysts, cerebral aneurysms	*Pkd1$-/-$ (Lu et al 1997)*	Pkd1	*pkd1 (Kim et al 1999)*	*lov-1 (Barr et al 2001)*
ADPKD2	*PKD2* (*Mochizuki et al 1996*) (AD)	Polycystin-2 (CC, multiple others)	[b]PKD1, [b]Hax-1 (*Gallagher et al 2000*), [b]PIGEA-14 (*Hidaka et al 2004*)	Cilia, Golgi, FA	Cysts ($\approx$70 yr)	Liver cysts, pancreatic cysts	*Pkd2$-/-$ (Wu et al 1998)*	Pkd2	*pkd2 (Sun et al 2004)* (body curvature only)	*pkd-2 (Barr et al 2001)*
							$Pkd1^{+/-}$ $Pkd2^{+/-}$ *(Wu et al 2002)*			
ARPKD	*PKHD1 (Ward et al 2002)* (AR)	Fibrocystin/ polyductin (multiple)	?	Cilia, cytoplasm (*Menezes et al 2004*)	Cysts, (42% at 20 yrs) (*Bergmann et al 2005*)	Liver fibrosis (>44%)	(rat model: *pck*)	Pkhd1	?	?
GCDK	*HNF1β/ TCF2* (AD)	HNF1β	?			MODY diabetes type 5	*Hnf-1β (Hiesberger et al 2004)*	Hnf-1β	*vhnf1 (Sun et al 2004)*	?
MCDK1	*UMOD (Hart et al 2002)* (AD/AR)	Uromodulin/ THP	?	Apical membrane	CM Cysts	Gout	$Umod^{-/-}$ *(Bachmann et al 2005)*	Umod	? (*crumbs* in fruit fly)	?
NPHP1	*NPHP1* (*Hildebrandt et al 1997a Saunier et al 1997a*) (AR)	NPHP1 (CC, SH3, E-rich, NPHP homology domain)	NPHP2/inversin, NPHP3, NPHP4, β-tubulin, [b]p130Cas, Pyk2, tensin, filamin A, filamin B	Cilia, BB, FA, AJ	NPHP (13 yr)	RP (10%), OMA ($\approx$2%), CVA (1/157) (*Parisi et al 2004*)	*Nphp1$-/-$ (Otto, unpubl.)*	Nphp1	*nphp1 mating phenotype)*	*nph-1*, M28.7
NPHP2	*INVS/NPHP2* (*Otto et al 2003b*) (AR)	Inversin/ NPHP2 (ANK, CC, NLS, SP)	NPHP1, [b]Apc2 [b]CALM, N-cadherin, catenins	Cilia, BB, CEN	CM cysts (<1 yr)	RP (10%), LF	*inv/inv (Mochizuki et al 1998, Morgan et al 1998)*	*Invs (Mochizuki et al 1998, Morgan et al 1998)*	*invs (Otto et al 2003b)* (CDK, laterality defect)	ZC15.7
NPHP3	*NPHP3* (*Olbrich et al 2003*) (AR)	NPHP3 (CC, SP, TPR, STAND, TL)	NPHP1	Cilia, retinal cilia	CM NPHP (19 yr)	RP (10%), LF?	*pcy (Olbrich et al 2003)*	Nphp3 (*Olbrich et al 2003*)	?	–

(*Continued*)

Table 25.1 (Continued)

[a]Human cystic kidney disease	Human gene (inheritance)	Protein (domains)	Protein interaction partner	Subcellular localization	Human Renal disease (median ESRF)	Human extrarenal disease	Mouse CDK model	Mouse gene	Zebrafish CDK gene (phenotype)	C. elegans orthologous gene[b]
NPHP4	*NPHP4* (AR)	NPHP4	NPHP1	Cilia, BB, AJ	CM NPHP (21 yr)	RP (10%), OMA, LF	–	*Nphp4*	?	*nph-4,* R13H4.1
NPHP5	*NPHP5* (AR)	NPHP5 (IQ, CC)	[b]CALM, RPGR	Cilia, retinal cilia	CM NPHP (13 yr)	RP with early onset (100%)	–	*Nphp5 (Otto, 2005)*	*zgc:101792*	?
JBTS3	*AHI1 (Dixon-Salazar et al 2004 Ferland et al 2004)* (AR)	AHI protein (SH3, WD40, SH3-binding, CC)	?	?	CM NPHP	CA	–	*Ahi1*	?	*C14B1.4?*
ALMS1	*ALMS1 (Collin et al 2002 Hearn et al 2002)* (AR)	ALMS1 (CC)	?	CEN	NPHP, UTM	MR, RP, HG/IF, OB, PDM				
BBS1	*BBS1 (Mykytyn et al 2002)* (AR)	?	?	Olfactory cilia	NPHP, UTM	MR, RP, PD, HG/IF, OB, DM, CHD		*Bbs1*	?	*bbs-1 (Ansley et al 2003)*
BBS2	*BBS2 (Nishimura et al 2004)* (AR)	BBS2 (CC)	?	Olfactory and retinal cilia *(Nishimura et al 2004)*	NPHP, UTM	MR, RP, PD, HG/IF, OB, DM, CHD	*Bbs2–/– (Nishimura et al 2004)*	*Bbs2*	*bbs2*	*bbs-2 (Ansley et al 2003)*
BBS3	*BBS3 (Fan et al 2004, ARL6* (AR)	ADP-ribosylation factor-like 6, ARF	?	Ciliated neurons	NPHP, UTM	MR, RP, PD, HG/IF, OB, DM, CHD	–	*Bbs3*	*zgc:101762*	C38D4.8
BBS4	*BBS4 (Mykytyn et al 2001)* (AR)	BBS4 (TPR)	?	Olfactory cilia, BB, CEN	NPHP, UTM	MR, RP, PD, HG/IF, OB, DM, CHD	*Bbs4–/– (Mykytyn et al 2004)*	*Bbs4*	*ttc8*	K04G7.3
BBS5	*BBS5 (Li et al 2004)* (AR)	BBS5	?	BB	NPHP, UTM	MR, RP, PD, HG/IF, OB, DM, CHD	–	*Bbs5*	*zgc:56578*	*bbs-5,* R01H10.6
BBS6, MKKS	*BBS6 (Katsanis et al 2000)* (AR)	BBS6, MKKS (chaperonin)	?	? BB	NPHP, UTM	MR, RP, PD, HG/IF, OB, DM, CHD	–	*Bbs6*	*zgc:55608*	?
BBS7	*BBS7 (Badano et al 2003)* (AR)	BBS7 (CC)	?	Ciliated neurons *(Blacque et al 2004)*	NPHP, UTM	MR, RP, PD, HG/IF, OB, DM, CHD	–	*Bbs7*	?	*bbs-7/osm-12 (Ansley et al 2003)*

BBS8	*BBS8 (Ansley et al 2003)* (AR)	BBS8, TTC8 (TPR)	?	Ciliated neurons *(Blacque et al 2004)*	NPHP, UTM	MR, RP, PD, HG/IF, OB, DM, CHD	–	*Bbs8*	?	*bbs-8 (Ansley et al 2003)*
OFD1	*OFD1 (Romio et al 2003)* (XLD)	OFD1 (CC)	?	Cilia, CEN *(Romio et al 2004)*	PKD	MR, PD, deafness, dental, hydroceph.	–	*Ofd1*	*ofd1*	?
?	*PKDR1*	Sam-Cystin					*cy/+* (rat)	*Pkdr1*		
?	?	?					*wpk* (rat)	*Wpk*		
?	*NEK8*	NEK8	?	?	?	?	*Jck (Atala et al 1993)*	*Nek8*	*jck (Atala et al 1993)* (CDK)	*ZC581.1R*
?	*BICC1*	bicaudal C homolog 1					*Bpk, Jcpk (Cogswell et al 2003)*	*Bicc1*	*LOC402785*	?
?	*NEK1*	NEK1	?	?	?	?	*Kat (Upadhya et al 2000 Upadhya et al 1999)*	*Nek1*	?	*Y39G10AR.3*
?		polaris	?	?	?	?	*Orpk (Moyer et al 1994), fxo (Huangfu et al 2003)*	*Tg737*	*ttc10*	*osm-5*
?	*KIF3A*	kinesin family member 3A	?	?	?	?	*Kif3A−/− (Lin et al 2003)*	*Kif3A*	*ENSDARG 00000021329*	*Klp-20, Y50D7A.6*
?	*KIF3B*	kinesin family member 3B	?	?	?	?	*Kif3B⁻/⁻*	*Kif3B*	*ENSDARG 00000012081*	*Klp-11, F20C5.2*
?	*TNS*	tensin	?	?	?	?	*Tns−/− (Lo et al 1997)*	*Tns*	*ENSDARG 00000024585*	*F46F11.3*
?	*ACE*	ACE	?	?	?	?	*Ace−/− (Carpenter et al 1996)*	*ACE*	*ENSDARG 00000002836*	*C42D8.5*
?	*EGFR*	EGFR	?	?	?	?	*Egfr −/−*	*Egfr*	?	?
?	*CYS*	cystin	?	?	?	?	*cpk*	*Cys1*	?	?
?	*CTNNB1*	β-catenin	?	?	?	?	*β-catenin over-expression*	*β-catenin*	*β-catenin*	?
?	*BCL2*	BCL-2	?	?	?	?	*Bcl-2 −/−*	*Bcl-2*	*bcl2l*	?
?	*PLMP*	preny 1 transferase-like mitochondrial protein	?	?	?	?	*kd*	*Plmp*	?	?

(Continued)

Table 25.1 (Continued)

[a]Human cystic kidney disease	Human *gene* (inheritance)	Protein (domains)	Protein interaction partner	Subcellular localization	Human Renal disease (median ESRF)	Human extrarenal disease	Mouse CDK model	Mouse gene	Zebrafish CDK *gene* (phenotype)	*C. elegans* orthologous gene[b]
?	CDX1	homoebox protein CDX-1	?	?	?	?	–	Cdx1	Cad1/caudal (Sun et al 2004)	?
?	ESRRBL1/ HIPPI	estrogen receptor β-like 1	?	?	?	?	–	Hippi	hi3417/IFT57 (Sun et al 2004)	IFT57 (Sun et al., 2004)
?	CDV1	CDV-1	?	?	?	?	–	Cdv1	hi409/IFT81 (Sun et al 2004)	IFT81 (Sun et al., 2004)
?	SLB	selective LIM binding factor	?	?	?	?	wim (Huangfu et al 2003)	Slb	Hi2211/IFT172 (Sun et al 2004)	IFT172 (Sun et al., 2004)
?	ARL2L1	ADP-ribosylation factor-like 2-like 1	?	?	?	?	–	Arl2l1	hi459/scorpion (Sun et al 2004)	?
?	RUVBL1	RuvB-like 1	?	?	?	?	–	?	hi1055b/pontin (Sun et al 2004)	?
?	RUVBL2	RuvB-like 2	?	?	?	?	–	?	reptin (Sun et al 2004)	?
?	LRRC6L		?	?	?	?	–	Lrrc6l	hi3308/seahorse (Sun et al 2004)	?
?	CLUAP1	Clusterin associated protein 1	BBS-7 (zebrafish)	?	?	?	–	Cluap1	hi3959/qilin (Sun et al 2004)	C04C3.5
?	?		?	?	?	?	–	?	hi1392/twister (Sun et al 2004)	?
?	PDLIM7	enigma protein	?	?	?	?	–	Pdlim7	hi2005/enigma (Sun et al 2004)	?

[a]ADPKD, autosomal dominant polycystic kidney disease; ARPKD, autosomal recessive polycystic kidney disease; NPHP, nephronophthisis; JBTS, Joubert syndrome; ALMS, Alstrom syndrome; BBS, Bardet-Biedl syndrome; MKKS, McKusick-Kaufman syndrome; OFD, oral-facial-digital syndrome
[b]Direct interaction.
AD, autosomal dominant; AJ, adherens junction; AR, autosomal recessive; BB, basal bodies; CC, coiled-coil; CEN, centrosomes; CHD (congenital heart defects); CVA, cerebellar vermis aplasia/hypoplasia; DM, diabetes mellitus; ESRF, end-stage renal failure; FA, focal adhesion; HG hypogenitalism; IF, infertility; MKKS, McKusick-Kaufman (hydrometrocolpos, postaxial polydactyly and congenital heart defects); MR, mental retardation; MTOC, microtubule organization center; NLS, nuclear localization signal; NPHP, nephronophthisis (tubular basement disruption, interstitial nephritis and fibrosis, cortico-medullary cysts); OMA, oculomotor apraxia type Cogan; PKD, polycystic kidneys; PD, polydactyly; RM, renal malformation; RP retinitis pigmentosa; SI, situs inversus; SP, signal peptide; TPR, tetratricopeptide repeat; TL, tyrosine ligase; UTM, malformations of the kidney and urinary tract; XLD, X-linked dominant

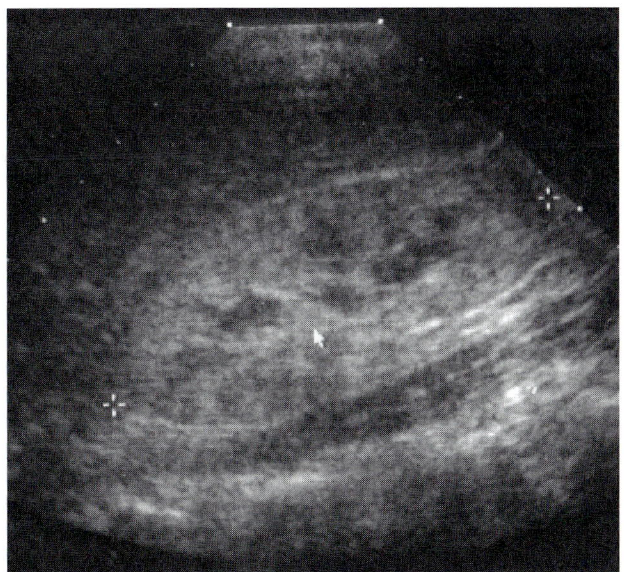

FIGURE 25.1   Renal ultrasound of juvenile nephronophthisis in a 14-year-old girl with Senior-Loken syndrome. Note normal kidney size, presence of increased echogenicity, lack of corticomedullary differentiation, and numerous cysts of varying size at the corticomedullary junction. (Courtesy Dr U. Vester, Essen, Germany)

The characteristic histological triad of NPHP consists of: (i) tubular basement membrane (TBM) disintegration with irregular thickening and attenuation of the TBM, (ii) interstitial round cell infiltration with marked fibrosis, and later on, (iii) tubular atrophy with cyst development occurring predominantly at the corticomedullary junction (Zollinger et al 1980, Waldherr et al 1982). Cysts seem to be the result rather than the cause of the atrophic process, although this time course could not be corroborated by statistical analysis. Occasionally, a communication between a cyst and a tubule was seen, indicating that some of the dilated tubules will have to be addressed as diverticula rather than cysts. TBM changes and diverticulum formation were most prominent in the distal tubules, where cysts were lined with a single layer of cuboidal or flattened epithelium. In the advanced stage the histologic picture merged into a diffuse sclerosing tubulo-interstitial nephropathy, the characteristic image of end-stage NPHP. The only significant glomerular change in early stages involved periglomerular fibrosis with splitting and thickening of Bowman's capsule. Glomerular obsolescence was only seen in nephrons that had been destroyed by the tubular alterations. An escape of Tamm-Horsfall (uromodulin) protein from damaged collecting tubules into the interstitium had been demonstrated in about 50% of patients with an NPHP/MCKD phenotype by specific immunofluorescence staining with an anti-THP antibody (Resnick et al 1978). Interestingly, mutations in uromodulin have recently been identified as causing MCKD type 2 (Hart et al 2002). Characteristic changes demonstrated by transmission electron microscopy include thickening, splitting, attenuation, and granular disintegration of the TBM. Transition between these alterations was abrupt (Zollinger et al 1980). Fibroblasts were seen in direct contact with the TBM. At the base of the tubular epithelial cells a marked increase of microfilaments was seen. The thickening was either homogeneous or had a lamellated, annular, ring-like appearance. The glomerular basement membrane was normal.

MOLECULAR GENETICS OF NPHP: AN ETIOLOGICAL CLASSIFICATION OF NPHP

An etiological classification of NPHP disease variants has recently become possible through positional cloning of distinct genes that are mutated in NPHP and related disorders (Table 25.1). This now allows for uniequivocal molecular genetic diagnosis of NPHP in about 34% of patients with NPHP/SLSN. Of the five genes identified by positional cloning four were novel. Studies into the function of the related gene products, termed 'nephrocystins,' led to new insights into the pathogenesis of NPHP and other renal cystic diseases, as discussed below. Briefly, nephrocystins are expressed in primary cilia, basal bodies and centrosomes of renal epithelial cells, and play a functional role in focal adhesion signaling and most likely in the regulation of planar cell polarity of renal epithelial cells. To date, five distinct genes (*NPHP1–5*) have been identified by positional cloning as mutated in NPHP.

Nephronophthisis Type 1 (NPHP1)

By total genome search for linkage a gene locus (*NPHP1*) for juvenile NPHP was localized to human chromosome 2q12-q13 (Antignac et al 1993, Hildebrandt et al 1993). The critical genetic region was subsequently cloned in YAC and PAC contigs (Gardner 1971, Haider et al 1998, Hildebrandt 1999, Hildebrandt et al 1997b, Omran et al 2000), which led to the identification of a novel gene termed *NPHP1*, defects in which are responsible for NPHP1 (Hildebrandt et al 1997a). The identity of this gene as mutated in juvenile nephronophthisis was since confirmed (Saunier et al 1997a). About 85% of children with juvenile nephronophthisis type 1 harbor large (250kb) homozygous deletions of the *NPHP1* gene, while some carry point mutations in combination with heterozygous deletions (Hildebrandt et al 2001, Konrad et al 1996, Saunier et al 2000). Through gene identification, molecular genetic diagnostics in NPHP1 has become possible (see below) (Caridi et al 2000, Hildebrandt et al 2001).

The *NPHP1* gene spans 83kb, consists of 20 exons and encodes an mRNA of 4.5kb. It is flanked by two large (330kb) inverted duplications. In addition, a second sequence of 45kb, which is located between the centromeric inverted duplication and the *NPHP1* gene, is repeated directly within the telomeric inverted duplication. In several *NPHP1* families, the deletion breakpoints have been localized to the 45 kb direct repeats by pulsed field gel electrophoresis (Saunier et al 2000). Chromosomal misalignment followed by unequal crossover or the formation of a loop structure on a single chromosome has been suggested as a potential cause for these deletions. In addition, there is a high degree of further rearrangements known to occur in this region of chromosome 2 (Saunier et al 2000). Thus, these duplications convey a 'genomic etiology' for the origination of *NPHP1* deletions, which give rise to NPHP type 1 in patients that inherit 2 copies of such deletions. Consequently, the *NPHP1* locus has been identified as one of 24 regions in the human genome, in which duplications lead to hot spots of genomic instability that are associated with genetic disease (Bailey et al 2002). In addition, an unusual maternal deletion in a child with *NPHP1* has recently been molecularly characterized. By direct sequencing it was shown that the centromeric breakpoint was localized within a long interspersed nuclear element-1 (LINE1), which belongs to an abundant group of transposable elements within the human genome (Otto et al 2000a). About 25% of all cases with NPHP are caused by mutations in *NPHP1*. Eighty-five percent of these represent *NPHP1* deletions on both alleles. The remaining 15% are caused by the combination of a hemizygous deletion with an *NPHP1* point mutation (Hildebrandt et al 2001). In about 10% of all patients with *NPHP1* mutations there is an association with retinitis pigmentosa (SLSN) (Caridi et al 2000), without any recognizable genotype/phenotype correlation as regards the presence or absence of retinitis pigmentosa. In NPHP type 1, a few cases have been described with associated oculomotor apraxia type Cogan (Saunier et al 1997a, Betz et al 2000, Hoefele, unpublished), and rarely also with the association of cerebellar vermis aplasia thereby representing the JBTS phenotype (Parisi et al 2004).

NPHP1 was identified as a novel gene, which was not related to any known gene families. Expression studies in man and mouse revealed a broad tissue expression pattern. In addition, in situ hybridization studies of whole mount mouse embryos showed ubiquitous but weak *Nphp1* expression at all embryonic stages between days 7.5 and 11.5 p.c. (Otto et al 2000b). In the adult mouse there was also strong expression in testis. The protein encoded by *NPHP1*, nephrocystin-1, is a novel protein that contains multiple putative protein–protein interaction domains. It was shown to play a role in signaling processes at focal adhesions and adherens junctions of renal epithelial cells (see below). *NPHP1* has a highly conserved ortholog, *nph-1*, in the nematode *C. elegans* (Otto et al 2000c). The tertiary structure of the nephrocystin-1 SH3 domain has in the meantime been resolved by NMR spectrometry (le Maire et al 2005).

Nephronophthisis Type 2 (NPHP2)

A gene locus (*NPHP2*) for infantile NPHP has been localized to chromosome 9q22-q31 by homozygosity mapping in a large Bedouin kindred (Haider et al 1998). The clinical course and histology in this disease are quite different from other forms of NPHP (Gagnadoux et al 1989). Terminal renal failure ensues within the first 2 years of life. The *inv* mouse model, in which disruption of the protein inversin led to consistent reversal of the left–right axis (Morgan et al 1998, 2002b) was noted to have cystic kidney disease (Mochizuki et al 1998). These observations led to the identification of mutations in the *inversin* (*INVS*) gene mutated in patients with NPHP2 with and without *situs inversus* (Otto et al 2003a). *INVS* encodes a 1062 amino acid protein containing 16 ankyrin repeats, a bipartite nuclear localization signal and an IQ calmodulin binding domain (Morgan et al 2002a). By yeast two-hybrid assay and co-immunoprecipitation direct interaction of inversin and calmodulin was confirmed. In addition, inversin was shown to coimmunoprecipitate with nephrocystin-1, and β-tubulin, and to colocalize with nephrocystin-1 to primary cilia of renal epithelial cells, thereby revealing a potential role of primary cilia in the pathogenesis of NPHP (Otto et al 2003a). The inversin protein had been demonstrated to localize to primary cilia, mitotic spindles and centrosomes (Morgan et al 2002a) in a cell-cycle-dependent manner. Inversin is intimately associated with the microtubule cytoskeleton (Nurnberger et al 2004). Using a knock-down of inversin in zebrafish a renal cystic phenotype was observed in addition to randomization of heart looping (Otto et al 2003a). Very recently, it was demonstrated that in renal tubules inversin may be involved in switching from the canonical to noncanonical Wnt signaling pathway in response to flow sensing by primary cilia of renal tubular cells, thereby helping to maintain renal epithelial cell polarity (Simons et al 2005). *INVS/NPHP2* mutations represent a rare cause of NPHP, accounting for less than 1% of cases (Otto et al 2003a).

Nephronophthisis Type 3 (NPHP3)

A third locus (*NPHP3*) for NPHP was mapped by total genome search for linkage to chromosome 3q21-q22 in a large Venezuelan kindred, applying the strategy of homozygosity mapping (Omran et al 1999, 2000, 2002). This disease variant was termed 'adolescent nephronophthisis' (NPHP3), since onset of end-stage renal disease occurs at median age 19 years. By positional cloning, a novel gene, *NPHP3*, was identified (Olbrich et al 2003). It encodes a 1330 amino acid protein, nephrocystin-3, which interacts with nephrocystin-1. *NPHP3* mutations were found in patients with isolated NPHP, but also in NPHP with hepatic fibrosis or tapeto-retinal degeneration. Mutations in *NPHP3* accounted for only 1.5% of NPHP cases in a worldwide cohort (Hoefele, unpublished). Murine *Nphp3* was shown to be expressed in the embryonic node, kidney tubules, retina, respiratory epithelium, liver, biliary tract, and neural tissues. This expression occurs most likely in primary cilia of these tissues. Ciliary expression reflects the common denominator of the pleiotropy in NPHP, since it seems to be essential for the integrity of renal tubular cell, bile duct, and photoreceptor structure and function, and possibly for decussating neuron integrity in the cerebellum. In parallel, a homozygous missense mutation in *Nphp3* was identified as the underlying defect of the polycystic kidney disease (*pcy*) mouse phenotype, which shows strong resemblance to human NPHP3 at the macroscopic and microscopic level (Olbrich et al 2003). Recently, treatment of *pcy* mice with an antagonists of the V2 receptor was shown to inhibit renal cystogenesis in the *pcy* mouse model (Gattone et al 2003).

Nephronophthisis Type 4 (NPHP4)

By total genome search for linkage a fourth gene locus (*NPHP4*) was localized to chromosome 1p36 (Schuermann et al 2002). The respective gene (*NPHP4*) was subsequently identified as causing NPHP type 4 and SLSN type 4 (Mollet et al 2002, Otto et al 2002). The gene and its gene product nephroretinin/nephrocystin-4 were novel and highly conserved in evolution, including a *C. elegans* ortholog (*nph-4*). Nephrocystin-4 protein expression has been demonstrated in primary cilia, centrosomes and in the vicinity of the cortical actin cytoskteton, with partial colocalization with β-catenin in polarized MDCK cells (Mollet et al 2005). Co-immunoprecipitation experiments showed that nephrocystin-1, p130Cas, and Pyk2 are in a complex with nephrocystin-4 (Mollet et al 2005).

Nephronophthisis Type 5 (NPHP5)

By positional cloning a novel gene *NPHP5* (*ICQB1*) was identified as mutated in NPHP type 5 (Otto et al 2005). Interestingly, in all patients bearing *NPHP5* mutations retinitis pigmentosa was present. In addition, retinitis was of the early-onset type, with blindness occurring before the third year of life (Otto et al 2005). All of the mutations found in *NPHP5* lead to a truncation of the encoded protein product 'nephrocystin-5.' This protein directly interacts with calmodulin, and is in a protein complex with the retinitis pigmentosa GTPase regulator (RPGR), mutations in which are the most frequent cause of X-linked retinitis pigmentosa. Both nephrocystin-5 and calmodulin, were coexpressed in primary cilia of renal epithelial cells. Coexpression of nephrocystin-5 and RPGR in connecting cilia of photoreceptors was demonstrated and may explain the retinal phenotype of the disease.

FUNCTION OF NEPHROCYSTINS AND PATHOGENESIS OF NPHP

Nephronophthisis: Animal Models

Genetic animal models that resemble nephronophthisis (Table 25.1) have been fruitful in the identification of underlying gene defects, and more recently in the experimental treatment of cystic kidney disease. In the *inv/inv* mouse model of renal cystic disease the *inversin* gene is mutated. This gives rise to renal cysts as well as left–right asymmetry, cardiovascular defects, hepatobiliary defects, and premature death (Mochizuki et al 1998, Morgan et al 1998, Morishima et al 1998). Collecting ducts of newborn *inv/inv* mice demonstrate diffuse cystic dilatation (Phillips et al 2004). Mutations in *INVS* cause human NPHP type 2, with and without *situs inversus* (Otto et al 2003a). When the human *NPHP3* gene was detected by positional cloning, it was shown that the *pcy* mouse model (Takahashi et al 1991) exhibiting cystic kidney disease and interstitial fibrosis, was created by a missense mutation in the murine ortholog *Nphp3* (Olbrich et al 2003). The *pcy* phenotype was shown to be improved by treatment with a vasopressin type 2 receptor antagonist (Gattone et al 2003) (see below). Additional transgenic mouse models for NPHP, such as the *tensin* knockout mouse (Lo et al 1997), and the Rho GDIα deficient mouse, contribute to the focal adhesion/adherens junction hypothesis of renal cystic diseases (Figure 25.2A, F, H). The *blc-2* knockout mouse (Veis et al 1993, Sorenson et al 1995) as well as the *bbs2* (Nishimura et al 2004) and *bbs4* (Fath et al 2005) knockout mouse models, suggests involvement of apoptosis in the renal cystic process. There may also be a relation to mechanisms of apoptosis in the *kd* ('kidney disease') model of NPHP (Lyon & Hulse 1971, Sibalic et al 1998), which shares several clinical and histologic (Sibalic et al 1998) features with human NPHP. The mice are born apparently healthy. Starting at 8 weeks of age they develop severe interstitial nephritis, which progresses to ESRD by age 4–8 months. The defect was recently found to be due to mutations in a novel gene, which encodes a mitochondrial protein, namely prenyltransferase-like mitochondrial protein (PLMP; GenBank AK017329) (Peng et al 2004). *Kd/kd* mice have dysmorphic mitochondria within renal tubular epithelia (Peng et al 2004). How renal cysts and interstitial fibrosis arises in the *Ace* knockout mouse (Carpenter et al 1996) is currently unclear (Hildebrandt & Otto 2000). A canine model of NPHP has also been reported (Finco 1976, Finco et al 1970, 1977).

Evolutionary Conservation of Nephrocystins

There is a high conservation of cystoproteins in evolution, since orthologs of these proteins are expressed in specific head and tail ciliated neurons of the nematode *C. elegans*. This applies for the orthologs of *PKD1* and *PKD2*, *lov-1* and *pkd-2*, respectively (Hildebrandt et al 1997a, Otto et al 2000b, 2002, Barr et al 2001) as well as *polaris* (*osm-5*) (Wolf et al 2005). Since human *NPHP1* and *NPHP4* have highly conserved orthologs, *nph-1* and *nph-4*, in the nematode *C. elegans* (Otto et al 2000b), expression of green fluorescent protein fusion constructs that contained the *C. elegans* promoter regions for *nph-1* and *nph-4* were examined

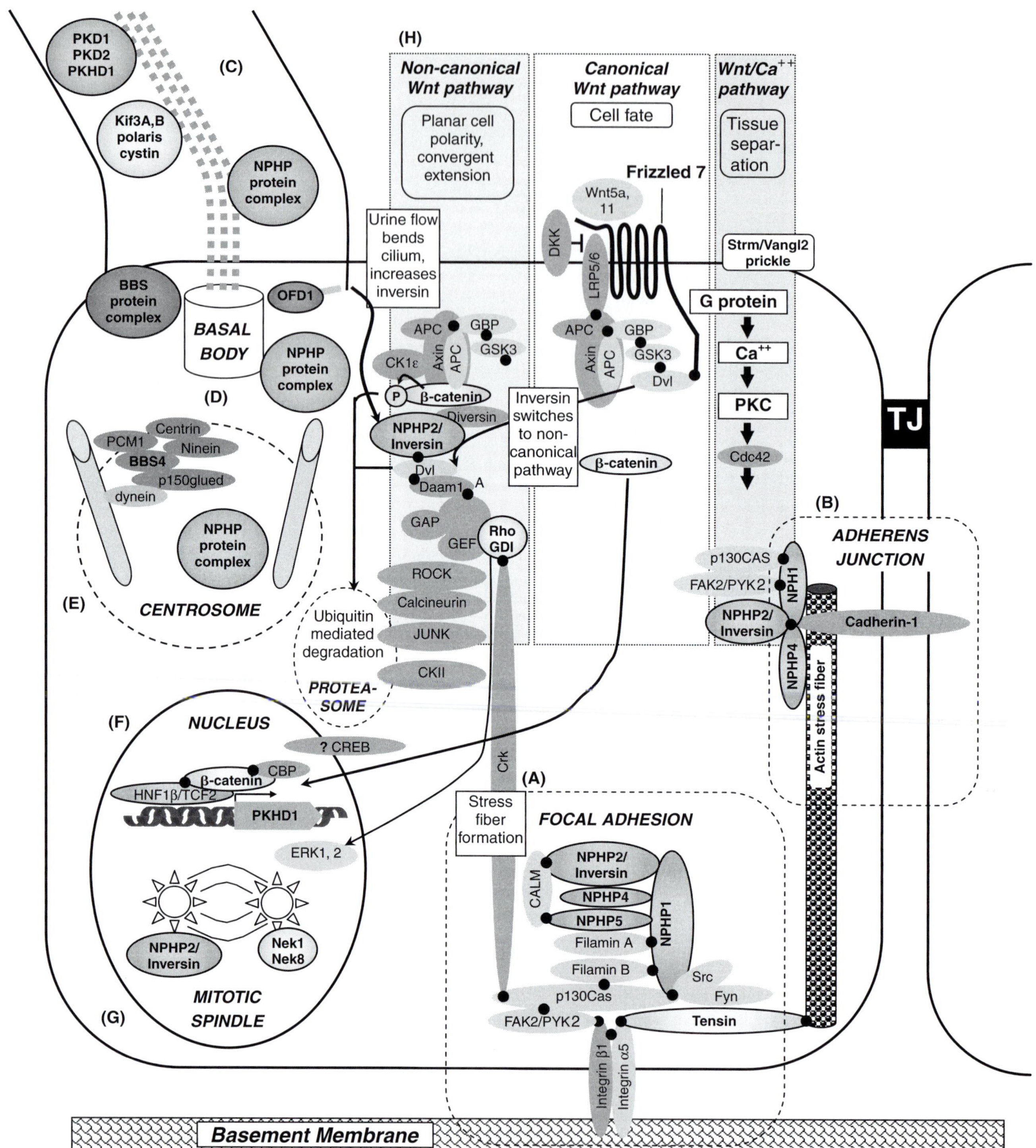

FIGURE 25.2 Localization of cystoproteins to subcellular locations that play a role in planar cell polarity (PCP) pathways (focal adhesion, adherens junction, cilia, basal body, centrosome, nucleus, mitotic spindle, and the Wnt pathways). Cystoproteins (proteins that are mutated in cystic kidney diseases) are shown on colored background and in bold type. Colors for groups of diseases or mouse models (yellow) match colors from Table 25.1. Proteins appear in non-bold if they have been described in the context of CDK pathogenesis (e.g. as a binding partner to a *bona fide* cystoprotein), but they have not been shown to be mutated in CDK. Associated proteins with no known role in CDK are shown on gray background. For references on protein function in the literature see Table 25.1. Black dots connect proteins that

are direct interaction partners. (A) *Focal adhesions.* (i) Nephrocystin-1 directly interacts with the focal adhesion adapter protein p130Cas ('crk-associated substrate') (Donaldson et al 2000, 2002, Hildebrandt & Otto 2000), which is a major mediator of focal adhesion assembly, binds to focal adhesion kinase, and mediates stress fiber formation (Brugge 1998). Nephrocystin-1 competes for binding to p130Cas with the proto-oncogene products Src and Fyn (Donaldson et al 2000). (ii) Nephrocystin-1 is in a protein complex with the focal adhesion proteins Pyk2/Fak2 (focal adhesion kinase 2), tensin (Benzing et al 2001), and filamin A and B. Its overexpression leads to activation of extracellular regulated kinases 1 and 2 (ERK1 and ERK2) (Benzing et al 2001). (iii) In children with NPHP overexpression of $\alpha5\beta1$ integrin was described in proximal tubules, which most likely results from defective $\alpha6$ integrin expression (Rahilly & Fleming 1995). (iv) The knockout mouse models for *tensin* (Lo et al 1997) and for the *Rho GDIα* gene (Togawa et al 1999) both exhibit an NPHP-like phenotype, thereby implicating further proteins of the focal adhesion signaling cascade in the pathogenesis of NPHP-like diseases. (B) *Adherens junctions.* (i) Nephrocystin-1 co-localizes with E-cadherin and p130Cas to adherens junctions. (ii) The C-terminal half of nephrocystin, the 'nephrocystin homology domain,' is able to promote nephrocystin self-association and epithelial adherens junctional targeting (Donaldson et al 2002). (iii) Disruption of this targeting leads to reduced transepithelial resistance (Donaldson et al 2002). (iv) Nephrocystin-4 is in a protein complex with nephrocystin-1, NPHP2/inversin, p130Cas, and Pyk2/Fak2, and has been localized to adherens junctions in confluent MDCK cells (Mollet et al 2005). (C) *Primary cilia.* Recently, the development of a unifying hypothesis of renal cystogenesis has been established (Watnick & Germino 2003). This hypothesis states that proteins that, if mutated, cause renal cystic disease in humans, in mice or in zebrafish, are part of a functional module, as defined by their subcellular localization to primary cilia, basal bodies or centrioles. This applies to polycystin-1 and -2, fibrocystin/polyductin, nephrocystin-1, -2 (inversin), -3, -4, -5, Bardet-Biedl syndrome (BBS)-associated proteins, cystin, polaris, ALMS1, OFD1, and many others (Table 25.1). Since nephrocystins interact, they are represented here as the 'nephrocystin complex.' Based on positional cloning, mutations in the *inversin* gene were identified as causing infantile NPHP (type 2) (Otto et al 2003b). This established a link between the pathogenesis of NPHP and disease mechanisms of polycystic kidney disease (Otto et al 2003a). Within these studies it was found that nephrocystin-1 interacts with both inversin and with β-tubulin. Since β-tubulin is a major component of primary cilia, this led to demonstration of colocalization of all three proteins in the primary renal cilia of epithelial cells (Otto et al 2003a). The complex also contains *NPHP3*, which is mutated in adolescent NPHP (type 3) and in the renal cystic mouse model *pcy*. The ciliary hypothesis of CDL was confirmed, by revealing that *Nphp3* mRNA was expressed in kidney, retinal connecting cilia, ciliated bile ducts, and the node, which regulates left–right body axis in mice (Olbrich et al 2003), and by identifying mutations in nephrocystin-5 (IQCB1), which colocalizes with calmodulin to primary cilia of renal epithelial cells and retinal connecting cilia (Otto et al 2005). (D) *Basal bodies.* Basal bodies are short cylindrical arrays of microtubules and other proteins, found at the base of a cilia, that organize the assembly of the ciliary axoneme. They are analogous to centrosomes. Proteins mutated in Bardet-Biedl syndrome (BBS) are components of the basal body transitional zone and are highly conserved in evolution. (E) *Centrosomes.* For a protein mutated in the related disease Bardet-Biedl syndrome type 4 (BBS4) it was shown that BBS4 is instrumental in recruiting proteins to the pericentrosomal matrix, confirming the role of centrosomal function in the pathogenesis of BBS (Ansley et al 2003). (F) *Transcriptional programs.* The transcription factor HNF1β regulates transcription of multiple CDK genes (Hiesberger et al 2005). (G) *Cell cycle regulation.* The hypotheses of functional involvement of the nephrocystin complex at focal adhesions and adherens junction on one hand and the ciliary/centrosomal hypotheses on the other hand may be integrated by demonstration that different locations of the complex may predominate in a cell-cycle-dependent manner. This is evidenced by the findings that inversin/nephrocystin-2 and nephrocystin-4 expression occurs in a cell-cycle-dependent manner. Inversin exhibits a dynamic expression pattern in MDCK cells showing expression at centrosomes in early prophase, at spindle poles in metaphase and anaphase, and at the midbody in cells undergoing cytokinesis (Morgan et al 2002a, Nurnberger et al 2002). Nephrocystin-4 was detected in MDCK cells at centrosomes of dividing cells, and in polarized cells close to the cytoskeleton and in the vicinity of the cortical actin cytoskteton, with colocalization of p130Cas, Pyk2, and β-catenin (Mollet et al 2005). (H) *Wnt pathways.* Recent data demonstrated that in renal tubules inversin may induce switching from the canonical to the non-canonical Wnt signaling pathway in response to flow sensing by primary cilia of renal tubular cells (Simons et al 2005). It is thought that this function of inversin is important to maintain renal epithelial cell polarity (Germino 2005). The hypothesis that the renal cystic disease phenotype is due to defects in the maintenance of planar cell polarity seems plausible at present for multiple reasons: (i) it would reconcile previous functional hypotheses, since focal adhesions, adherens junctions, cilia, centrosomes, and basal bodies, as well as regulation of the cell cycle, all play a pivotal role in the regulation and maintenance of planar cell polarity (Keller 2002); (ii) since planar cell polarity plays an important role in developmental morphogenesis and also in the regeneration of differentiated tissue, a defect in planar cell polarity may explain both occurrence of cysts during organogenesis and degenerative cystogenesis as it occurs in nephronophthisis; (iii) the mechanism of convergent extension, which may be central to renal tubular morphology, was shown to be disturbed in many renal cystic diseases (Ross et al 2005). APC, adenomatous polyposis coli protein; APC2, anaphase promoting complex subunit 2; CBP, CREB binding protein; CREB, cAMP response element binding protein; CK1ε, casein kinase epsilon; DKK, Dickkopf-related protein; CK2, casein kinase II; GAP, RhoGTPase activating protein; GEF, guanine nucleotide exchange factor; GSK3, glycogen synthase kinase 3; JUNK, JUN kinase; PCM1, pericentriolar material 1; p150, p150^{glued}/dynactin-1; PKC, protein kinase C; ROCK, Rho-associated protein kinase. (see also Plate 43) and expression was detected in ciliated sensory neurons of *C. elegans* in head and the tail of hermaphrodites (Jauregui & Barr 2005, Wolf et al 2005). RNAi knockdown of *lov-1* and *pkd-2* was shown to lead to impaired male mating behavior (Barr et al 2001). When *C. elegans* males were analyzed for this phenotype a similar phenotype was found in the *nph-1* and *nph-4* RNAi knockdown animals (Barr et al 2001, Wolf et al 2005). These data reflect a very high evolutionary conservation of functional modules related to roles of cystoproteins.

Focal Adhesion and Adherens Junction Signaling

When nephrocystin-1 was first identified as a novel protein mutated in NPHP1, one of its striking features was the presence of a src-homology 3 (SH3) domain (Hildebrandt et al 1997a, Saunier et al 1997a). Since SH3 domains are found in adapter proteins at focal adhesions of cell-basement membrane contacts, and since the basement membrane is disrupted in nephronophthisis, we hypothesized that nephrocystin might play a role in focal adhesions, which are involved in interacellular signaling from the basement membrane, and which regulate stress fiber formation for cell locomotion (Hildebrandt 1998, Hildebrandt & Otto 2000). This hypothesis gained initial support by the finding that nephrocystin directly interacts with the protein p130Cas ('crk-associated substrate') (Donaldson et al 2000, Benzing et al 2001), which is a major component of focal adhesions. By demonstration of colocalization of nephrocystin-1 and E-cadherin at adherens junctions of renal epithelial cells a functional hypothesis for nephrocystin was then extended. This 'focal adhesion/adherens junction hypothesis' is now supported by multiple findings (Figure 25.2A, B).

Focal Adhesions (Figure 25.2A)

(i) Nephrocystin-1 directly interacts with the focal adhesion adapter protein p130Cas ('crk-associated substrate') (Donaldson et al 2000, 2002, Hildebrandt & Otto 2000), which is a major mediator of focal adhesion assembly, binds to focal adhesion kinase, and mediates stress fiber formation (Brugge 1998). Nephrocystin-1 competes for binding to p130Cas with the proto-oncogene products Src and Fyn (Donaldson et al 2000).

(ii) Nephrocystin-1 is in a protein complex with the focal adhesion proteins Fak2/Pyk2 (focal adhesion kinase 2), tensin (Benzing et al 2001), and filamin A and B (Donaldson et al 2002). Its overexpression leads to activation of extracellular regulated kinases 1 and 2 (ERK1 and ERK2) (Benzing et al 2001).

(iii) In children with NPHP overexpression of $\alpha5\beta1$ integrin was described in proximal tubules, from which $\alpha5$ integrin is normally absent. This most likely results

from defective $\alpha6$ integrin expression (Rahilly & Fleming 1995).

(iv) The knockout mouse models for *tensin* (Lo et al 1997) and for the *Rho GDIα* gene (Togawa et al 1999) both exhibit an NPHP-like phenotype, thereby implicating further proteins of the focal adhesion signaling cascade in the pathogenesis of NPHP-like diseases (Figure 25.2A, B).

Adherens junctions (Figure 25.2B)

(i) Nephrocystin-1 co-localizes with E-cadherin and p130Cas to adherens junctions.

(ii) The C-terminal half of nephrocystin, the 'nephrocystin homology domain,' is able to promote nephrocystin self-association and epithelial adherens junctional targeting (Donaldson et al 2002).

(iii) Disruption of this targeting leads to reduced transepithelial resistance (Donaldson et al 2002).

(iv) Nephrocystin-4 is in a protein complex with nephrocystin-1, p130Cas, and Fak2/Pyk2, and has been localized to adherens junctions in confluent MDCK cells (Figure 25.2B) (Mollet et al 2005).

Primary Cilia, Basal Bodies, and Centrosomes

In addition to the focal adhesion/adherens junction hypothesis, recently another pathophysiologic hypothesis for NPHP and cystic diseases of the kidney has emerged. It focuses on the function of primary cilia, basal bodies, and centrosomes of renal epithelial cells (Figure 25.2C–E,): Based on positional cloning, mutations in the *inversin* gene were identified as causing infantile NPHP (type 2). Within these studies it was established that nephrocystin-1 interacts with both inversin and β-tubulin. Since β-tubulin is a major component of primary cilia, this leads to demonstration of colocalization of all three proteins in the primary renal cilia of epithelial cells (Otto et al 2003a) (Figure 25.2C). This established a link between the pathogenesis of NPHP and disease mechanisms of polycystic kidney disease (Otto et al 2003a). The knockdown of *invs* in the zebrafish embryo likewise caused a renal cystic phenotype. Inversin was also shown to be localized to basal bodies, centrosomes, and mitotic spindles (Figure 25.2D, E, G,) (Morgan et al 2002a, Nurnberger et al 2002). In addition, positional cloning of the novel gene *NPHP3*, mutated in adolescent NPHP (type 3) and the renal cystic mouse model *pcy* confirmed the ciliary hypothesis, by revealing that *Nphp3* mRNA was expressed in kidney, retinal connecting cilia, ciliated bile ducts, and the node, which regulates left–right body axis in mice (Olbrich et al 2003). Furthermore, nephrocystin-5 (IQCB1) colocalized with calmodulin and RPGR to primary cilia of renal epithelial cells and retinal connecting cilia and centrosomes (Otto et al 2005). For a protein mutated in the NPHP-related disease Bardet-Biedl syndrome type 4 (BBS4) it was shown that BBS4 is instrumental in recruiting proteins to the pericentrosomal matrix, pointing to a role of centrosomal function in the pathogenesis of BBS (Figure 25.2E). These data demonstrated that nephrocystins colocalize to primary cilia, basal bodies or centrosomes together with other proteins that, if defective, cause renal cystic diseases (Figure 25.2C–E,). Therefore, recently, the development of a unifying hypothesis of renal cystogenesis has been suggested (Watnick & Germino 2003). This hypothesis states that proteins that, if mutated, cause renal cystic disease in humans, in mice or in zebrafish, are part of a shared functional module, as defined by their subcellular localization to primary cilia, basal bodies or centrosomes. This applies to polycystin-1 and polycystin-2, fibrocystin/polyductin, nephrocystin-1, nephrocystin-2/inversin, nephrocystin-3, nephrocystin-4, nephrocystin-5, Bardet-Biedl syndrome (BBS)-associated proteins, cystin, polaris, ALMS1, OFD1, and many others (Table 25.1, Figure 25.2).

Relations to Cell Cycle

The hypotheses of functional involvement of the nephrocystin complex at focal adhesions and adherens junction on one hand and the ciliary/centrosomal hypotheses on the other hand may be integrated by demonstration that different locations of the complex may depend on different stages of the cell cycle. This is evidenced by the findings that inversin/nephrocystin-2 and nephrocystin-4 expression occurs in a cell cycle dependent manner. Inversin exhibits a dynamic expression pattern in MDCK cells showing expression at centrosomes in early prophase, at spindle poles in metaphase and anaphase (Figure 25.2G), and at the midbody in cells undergoing cytokinesis. Nephrocystin-4 was detected in MDCK cells at centrosomes of dividing cells, and in polarized cells in the vicinity of the cortical actin cytosketeton, with partial colocalization with p130Cas, Pyk2, and β-catenin (Figure 25.2B) (Mollet et al 2005).

Planar Cell Polarity: A Unifying Theory of NPHP and Other Cystic Disease of the Kidney?

Recent data demonstrated that in renal tubules inversin may induce switching from the canonical to the noncanonical Wnt signaling pathway in response to flow sensing by primary cilia of renal tubular cells (Figure 25.2F, H) (Simons et al 2005). It is thought that this switching of inversin toward the non-canonical Wnt pathway is important to maintain renal epithelial cell polarity (Germino 2005). The hypothesis that the renal cystic disease phenotype is due to defects in the maintenance of planar cell polarity seems plausible at present for multiple reasons: (i) it would reconcile previous functional hypotheses, since focal adhesions, adherens junctions, cilia, centrosomes, and basal bodies, as well as regulation of the cell cycle, all play a pivotal role in the regulation and maintenance of planar cell polarity

(Keller 2002); (ii) since planar cell polarity plays an important role in developmental morphogenesis and well as in the regeneration of differentiated tissue, a defect in planar cell polarity may explain both occurrence of cysts during organogenesis and degenerative cystogenesis as it occurs in nephronophthisis; (iii) the mechanism of convergent extension, which may be central to renal tubular morphology, was shown to be disturbed in many renal cystic diseases (Ross et al 2005) (Table 25.1, Figure 25.2).

DIAGNOSTICS: MOLECULAR GENETICS, IMAGING, AND LABORATORY STUDIES

Molecular Genetic Diagnosis in NPHP

Through identification of the responsible genes, nephronophthisis types 1–5 can now be diagnosed by molecular genetic analysis (www.renalgenes.org). This represents the only diagnostic procedure by which the diagnosis of NPHP can be made with certainty. If mutations are found this obviates the necessity for renal biopsy. However, since homozygous deletions in *NPHP1* only represent 25% of patients with NPHP1, and mutations in *NPHP2, 3, 4,* or *5* are only found in 1–3% of patients, the lack of detection of mutations in those genes does *not* exclude the diagnosis of NPHP, due to the existence of unknown genes that are mutated in NPHP. Molecular genetic testing should be performed only in the context of genetic counseling and within the guidelines of the National and International Societies for Human Genetics (http://www.ethics.ubc.ca). Prior to genetic counseling a thorough family history to distinguish recessive (early-onset) from dominant (late-onset) disease is mandatory, and extrarenal organ involvement should be sought. Healthy siblings of children with NPHP can be evaluated once a year by testing urinary concentrating ability in a morning urine specimen, by renal ultrasound, and serum creatinine measurement, to allow early detection and early prevention of complications. If pathologic values are found, and a mutation in an NPHP gene is known from the affected sib, mutational analysis can then be offered to the sibling. Symptom free siblings, however, should not be screened for mutations to avoid unnecessary stigma in the absence of symptoms (http://www.cps.ca; www.ama-assn.org).

Imaging Techniques

Renal ultrasound is a very useful imaging technique in NPHP. Kidney size is normal or moderately reduced. There is loss of corticomedullary differentiation and increased echogenicity. Later in the course of the disease cysts can be detected at the cortico-medullary junction (Aguilera et al 1997, Ala-Mello et al 1998, Blowey et al 1996). Medullary cysts were noted in renal ultrasound in 13 of 15 children studied at the time of renal failure (mean age 9.7 years)

(Garel et al 1984). Magnetic resonance tomography seems to be evolving as a very useful procedure in the diagnosis of NPHP. Medullary cysts can sometimes also be demonstrated on computed tomography (Fyhrquist et al 1980, McGregor & Bailey 1989, Elzouki et al 1996, Neumann et al 1997). Caution must be exercised when performing contrast studies in patients with renal failure.

Laboratory Studies

There are no chemical laboratory tests in NPHP that specifically establish the diagnosis other than molecular genetic diagnostics. Hematuria, proteinuria, and bacteriuria are typically absent in NPHP. In rare cases where proteinuria is present, it is usually mild and of the tubular type. Prolinuria was reported in 2 cases (Bennett et al 1975, Fernandez-Rodriguez et al 1990). Laboratory studies are needed to assess the severity of renal failure and generally demonstrate elevated serum creatinine, BUN, phosphorus, metabolic acidosis, and anemia at the characteristic ages of onset. Ophthalmoscopy should be performed in any patient to exclude SLSN. Liver function and hepatic ultrasonography should also be performed to facilitate detection of patients with hepatic fibrosis. Functional impairment of tubular function, with the constant finding of a renal concentration defect usually precedes any documentable reduction in glomerular filtration rate (Gardner 1971) and may be present with minimal histologic abnormalities (Kleinknecht & Habib 1992). A characteristic early finding in NPHP is the decreased ability to concentrate the urine following a water deprivation test (Kleinknecht 1989). An intermediate defect of urinary concentration ability has been inconsistently demonstrated in the parents and some siblings of children with NPHP, and has been suggested to reflect the heterozygous state of the disease (Mangos et al 1964, Brouhard et al 1977). NPHP has also become known as 'salt-losing nephritis.' However, the question of whether sodium loss is a typical finding in the disease complex, has not been clearly answered, upon critical evaluation of the literature (Gardner 1992). Poor renal uptake of 99ᵐ-Technetium-DMSA has been proposed as diagnostic of NPHP (Hecht et al 1996).

Differential Diagnosis of NPHP

Histopathologically NPHP needs to be differentiated from other forms of interstitial nephropathies such as chronic pyelonephritis or drug injury. In oligomeganephronic dysplasia (Schimke 1975) kidney size is reduced and histology is distinct from NPHP. The paucity of urinary abnormalities, the frequent lack of hypertension, and the localization of renal cysts readily differentiate NPHP from recessive or dominant polycystic kidney disease. Medullary sponge kidney (Cacchi & Ricci 1948) does usually not lead to chronic renal failure and shows calcifications and calculi on renal ultrasound.

THERAPY, PROGNOSIS, AND GENETIC COUNSELING

Therapy of NPHP is symptomatic and is directed to the correction of disturbances of electrolyte, acid–base, and water balance as well as the treatment of hypertension, if present. Hypokalemia may contribute to the polyuria. Oral potassium supplementation may alleviate this symptom. Metabolic acidosis should be corrected, and osteodystrophy and secondary hyperparathyroidism should be treated with adequate calcium supplementation, phosphorus restriction or binders, and vitamin D therapy. Anemia can be treated with erythropoietin. Growth retardation may require administration of growth hormone, if the diagnosis is made early enough for an intervention. Adequate nutrition (caloric and amino-acid supplementation) should be maintained with the help of a dietitian. Especially in the phase just preceding the development of end-stage renal disease patients are at risk for sudden water and electrolyte disturbances, due to the high urinary output and salt loss. Sufficient salt and water supplementation is important at this stage, but may have to be restricted as hypertension develops late in the course of renal failure. In some cases an event of severe dehydration with acute renal failure can abruptly precipitate chronic renal failure. Psychosocial counseling of the patients is an integral part of therapy, because of the poor self-image associated with growth retardation and to alleviate pressures resulting from the need to comply with complicated medications and dietary prescriptions.

Once ESRD has developed during childhood, adolescence or early adulthood, all patients will require renal replacement therapy. At present, chronic dialysis treatment and/or renal transplantation is the only available treatment for NPHP. If a transplant recipient's renal histology suggests NPHP and a living related donor is considered, an extensive search should be made to exclude or detect renal disease within the family. Future therapeutic strategies targeted at mitigation of interstitial fibrosis may lead to the successful delay of ESRD. Vasopressin, a major adenyl cyclase agonist, acts via V2 receptors in the collecting duct. Recently, an antagonist of the V2 receptor (OPC31260) has been shown to inhibit renal cystogenesis in the *pcy* mouse, the murine equivalent of human NPHP3 (Torres 2004). Similarly, the anti-proliferative agent rapamycin has been used in the Han:SPRD rat model of polycystic kidney disease, where a reduction in cyst volume density and a preservation of renal function was observed (Tao et al 2005).

References

Aguilera A, Rivera M, Gallego N, Nogueira J, Ortuno J. Sonographic appearance of the juvenile nephronophthisis-cystic renal medulla complex. Nephrol. Dial. Transplant. 1997; 12: 625–6.

Ala-Mello S, Jaaskelainen J, Koskimies O. Familial juvenile neph-ronophthisis. An ultrasonographic follow-up of seven patients. Acta Radiol. 1998; 39: 84–9.

Ala-Mello S, Kaariainen H, Koskimies O. Nephronophthisis and ulcerative colitis in siblings: a new association. Pediatr. Nephrol. 2001; 16: 507–9.

Ala-Mello S, Kivivuori SM, Ronnholm KA, Koskimies O, Siimes MA. Mechanism underlying early anaemia in children with familial juvenile nephronophthisis. Pediatr. Nephrol. 1996; 10: 578–81.

Alstrom CH, Hallgren B, Nilsson LB, Asander H. Retinal degen-eration combined with obesity, diabetes mellitus and neurog-enous deafness: a specific syndrome (not hitherto described) distinct from the Laurence-Moon-Biedl syndrome. A clinical endocrinological and genetic examination based on a large pedigree. Acta Psychiat. Neurol. Scand. 1959; 34(Suppl. 129): 1–35.

Alton DJ, McDonald P. Urographic findings in the Bardet-Biedl syndrome, formerly the Laurence-Moon-Biedl syndrome. Radiology 1973; 109: 659–63.

Amirou M, Bourdat-Michel G, Pinel N, Huet G, Gaultier J, Cochat P. Successful renal transplantation in Jeune syndrome type 2. Pediatr. Nephrol. 1998; 12: 293–4.

Ansley SJ, Badano JL, Blacque OE, et al. Basal body dysfunction is a likely cause of pleiotropic Bardet-Biedl syndrome. Nature 2003; 425: 628–33.

Antignac C, Arduy CH, Beckmann JS, et al. A gene for familial juvenile nephronophthisis (recessive medullary cystic kidney disease) maps to chromosome 2p. Nat. Genet. 1993; 3: 342–5.

Antoine B, Braun-Vallon S, Perrin D, Dunod JP, Ryckewaert A. Néphropathie familiale avec atteintes osseuse et chorio-retinienne. [Familial nephropathy with bone and chorioretinal involvement] J. Urol. Nephrol. (Paris) 1963; 69: 81–9.

Atala A, Freeman MR, Mandell J, Beier DR. Juvenile cystic kid-neys (jck): a new mouse mutation which causes polycystic kidneys. Kidney Int. 1993; 43: 1081–5.

Avner ED. Medullary cystic disease and medullary sponge kid-ney. In: Greenberg A, ed. Primer on Kidney Diseases. Boston: Academic Press, 1994.

Bachmann S, Mutig K, Bates J, et al. Renal effects of Tamm-Horsfall protein (uromodulin) deficiency in mice. Am. J. Physiol. Renal Physiol. 2005; 288: F559–67.

Badano JL, Ansley SJ, Leitch CC, Lewis RA, Lupski JR, Katsanis N. Identification of a novel Bardet-Biedl syndrome protein, BBS7, that shares structural features with BBS1 and BBS2. Am. J. Hum. Genet. 2003; 72: 650–8.

Badano JL, Katsanis N. Beyond Mendel: an evolving view of human genetic disease transmission. Nat. Rev. Genet. 2002; 3: 779–89.

Bailey JA, Gu Z, Clark RA, et al. Recent segmental duplications in the human genome. Science 2002; 297: 1003–7.

Barr MM, DeModena J, Braun D, Nguyen CQ, Hall DH, Sternberg PW. The *Caenorhabditis elegans* autosomal domi-nant polycystic kidney disease gene homologs lov-1 and pkd-2 act in the same pathway. Curr. Biol. 2001; 11: 1341–6.

Bennett WM, Simon NM, Krill AE, Weinstein RF, Carone FA. Cystic disease of the renal medulla associated with retinitis pigmentosa and imino acid abnormalities. Clin. Nephrol. 1975; 4: 25–31.

Benzing T, Gerke P, Hopker K, Hildebrandt F, Kim E, Walz G. Nephrocystin interacts with Pyk2, p130 (Cas), and tensin and triggers phosphorylation of Pyk2. Proc. Natl Acad. Sci. USA 2001; 98: 9784–9.

Bergmann C, Senderek J, Windelen E, et al. Clinical consequences of PKHD1 mutations in 164 patients with autosomal-recessive polycystic kidney disease (ARPKD). Kidney Int. 2005; 67: 829–48.

Bernstein J, Gardner KD. Hereditary Tubulo Interstitial Nephritis. In: Stein JH, ed. Oxford textbook of clinical nephrology. Oxford: Oxford University Press, 1992.

Betts PR, Forest-Hay I. Juvenile nephronophthisis. Lancet 1973; 2: 475–8.

Betz R, Rensing C, Otto E, et al. Children with ocular motor apraxia type Cogan carry deletions in the gene (NPHP1) for juvenile nephronophthisis. J. Pediatr. 2000; 136: 828–31.

Biersdorf WR. The clinical utility of the foveal electroretinogram: a review. Doc. Ophthalmol. 1989; 73: 313–25.

Blacque OE, Reardon MJ, Li C, et al. Loss of *C. elegans* BBS-7 and BBS-8 protein function results in cilia defects and compromised intraflagellar transport. Genes Dev. 2004; 18: 1630–42.

Blowey DL, Querfeld U, Geary D, Warady BA, Alon U. Ultrasound findings in juvenile nephronophthisis. Pediatr. Nephrol. 1996; 10: 22–4.

Boichis H, Passwell J, David R, Miller H. Congenital hepatic fibrosis and nephronophthisis. A family study. Q. J. Med. 1973; 42: 221–33.

Brouhard BH, Srivastava RN, Travis LB, et al. Nephronophthisis. Renal function and histologic studies in a family. Nephron 1977; 19: 99–112.

Brugge JS. Casting light on focal adhesions. Nat. Genet. 1998; 19: 309–11.

Cacchi R, Ricci V. Sopra una rara e forse ancora non descritta effezione cistica della piramidi renali ("rene a spugna"). Atti. Soc. Ital. Urol. 1948; 5: 59.

Cantani A, Bamonte G, Ceccoli D. Familial juvenile nephronoph-thisis. Clin. Pediatr. 1986; 25: 90–5.

Caridi G, Dagnino M, Gusmano R, et al. Clinical and molecular heterogeneity of juvenile nephronophthisis in Italy: insights from molecular screening. Am. J. Kidney Dis. 2000; 35: 44–51.

Carpenter C, Honkanen AA, Mashimo H, et al. Renal abnormali-ties in mutant mice. Nature 1996; 380: 292.

Castori M, Valente EM, Donati MA, et al. NPHP1 gene deletion is a rare cause of Joubert syndrome related disorders. J. Med. Genet. 2005; 42: e9.

Chance PF, Cavalier L, Satran D, et al. Clinical nosologic and genetic aspects of Joubert and related syndromes. J. Child. Neurol. 1999; 14: 660–6, discussion 669–72.

Christodoulou K, Tsingis M, Stavrou C, et al. Chromosome 1 localization of a gene for autosomal dominant medullary cystic kidney disease. Hum. Mol. Genet. 1998; 7: 905–11.

Cogswell C, Price SJ, Hou X, et al. Positional cloning of jcpk/bpk locus of the mouse. Mamm. Genome 2003; 14: 242–9.

Collin GB, Marshall JD, Ikeda A, et al. Mutations in ALMS1 cause obesity, type 2 diabetes and neurosensory degeneration in Alstrom syndrome. Nat. Genet. 2002; 31: 74–8.

Contreras DB, Espinoza JS. Discussion clinica y anatomopato-logica de enfermos que presentaron un problema diagnostico. Pediatrica (Santiago) 1960; 3: 271–82.

Costet C, Betis F, Berard E, et al. Pigmentosum retinis and tubulo-interstitial nephronophthisis in Sensenbrenner syndrome: a case report. J. Fr. Ophthalmol. 2000; 23: 158–60.

Delaney V, Mullaney J, Bourke E. Juvenile nephronophthisis, congenital hepatic fibrosis and retinal hypoplasia in twins. Q. J. Med. 1978; 47: 281–90.

Di Rocco M, Picco P, Arslanian A, et al. Retinitis pigmentosa, hypopituitarism, nephronophthisis, and mild skeletal dysplasia (RHYNS): a new syndrome? Am. J. Med. Genet. 1997; 73: 1–4.

Dixon-Salazar T, Silhavy JL, Marsh SE, et al. Mutations in the AHI1 gene, encoding jouberin, cause Joubert syndrome with cortical polymicrogyria. Am. J. Hum. Genet. 2004; 75: 979–87.

Donaldson JC, Dempsey PJ, Reddy S, et al. Crk-associated substrate p130 (Cas) interacts with nephrocystin and both proteins localize to cell–cell contacts of polarized epithelial cells. Exp. Cell Res. 2000; 256: 168–78.

Donaldson JC, Dise RS, Ritchie MD, Hanks SK. Nephrocystin-conserved domains involved in targeting to epithelial cell-cell junctions, interaction with filamins, and establishing cell polarity. J. Biol. Chem. 2002; 277: 29028–35.

Donaldson MD, Warner AA, Trompeter RS, Haycock GB, Chantler C. Familial juvenile nephronophthisis, Jeune's syndrome, and associated disorders. Arch. Dis. Child. 1985; 60: 426–34.

Elzouki AY, al-Suhaibani H, Mirza K, al-Sowailem AM. Thin-section computed tomography scans detect medullary cysts in patients believed to have juvenile nephronophthisis. Am. J. Kidney Dis. 1996; 27: 216–19.

Fan Y, Esmail MA, Ansley SJ, et al. Mutations in a member of the Ras superfamily of small GTP-binding proteins causes Bardet-Biedl syndrome. Nat. Genet. 2004; 36: 989–93.

Fanconi G, Hanhart E, Albertini A, Uhlinger E, Dolivo G, Prader A. Die familiäre juvenile Nephronophthise. Helv. Paediatr. Acta, 1951; 6: 1–49.

Fath MA, Mullins RF, Searby C, et al. Mkks-null mice have a phenotype resembling Bardet-Biedl Syndrome. Hum. Mol. Genet. 2005.

Ferland RJ, Eyaid W, Collura RV, et al. Abnormal cerebellar development and axonal decussation due to mutations in AHI1 in Joubert syndrome. Nat. Genet. 2004; 36: 1008–13.

Fernandez-Rodriguez R, Morales JM, et al. Senior-Loken syndrome (nephronophthisis and pigmentary retinopathy) associated to liver fibrosis: a family study. Nephron 1990; 55: 74–7.

Finco DR. Familial renal disease in Norwegian Elkhound dogs: physiologic and biochemical examinations. Am. J. Vet. Res. 1976; 37: 87–91.

Finco DR, Duncan JD, Crowell WA, Hulsey ML. Familial renal disease in Norwegian Elkhound dogs: morphologic examinations. Am. J. Vet. Res. 1977; 38: 941–7.

Finco DR, Kurtz HJ, Low DG, Perman V. Familial renal disease in Norwegian Elkhound dogs. J. Am. Vet. Med. Assoc. 1970; 156: 747–60.

Fyhrquist FY, Klockars M, Gordin A, Tornroth T, Kock B. Hyperreninemia, lysozymuria, and erythrocytosis in Fanconi syndrome with medullary cystic kidney. Acta Med. Scand. 1980; 207: 359–65.

Gagnadoux MF, Bacri JL, Broyer M, Habib R. Infantile chronic tubulo-interstitial nephritis with cortical microcysts: variant of nephronophthisis or new disease entity? Pediatr. Nephrol. 1989; 3: 50–5.

Gallagher AR, Cedzich A, Gretz N, Somlo S, Witzgall R. The polycystic kidney disease protein PKD2 interacts with Hax-1, a protein associated with the actin cytoskeleton. Proc. Natl Acad. Sci. USA 2000; 97: 4017–22.

Gardner KD. Evolution of clinical signs in adult-onset cystic disease of the renal medulla. Ann. Intern. Med. 1971; 74: 47–54.

Gardner KD. Medullary and miscellaneous renal cystic disorders. In: Gottschalk CW, ed. Diseases of the Kidney. Boston: Little Brown, 1992.

Gardner KD Jr. Juvenile nephronophthisis and renal medullary cystic disease. Perspect. Nephrol. Hypertens. 1976; 4: 173–85.

Garel LA, Habib R, Pariente D, Broyer M, Sauvegrain J. Juvenile nephronophthisis: sonographic appearance in children with severe uremia. Radiology 1984; 151: 93–5.

Gattone VH II, Wang X, Harris PC, Torres VE. Inhibition of renal cystic disease development and progression by a vasopressin V2 receptor antagonist. Nat. Med. 2003; 9: 1323–6.

Germino GG. Linking cilia to Wnts. Nat. Genet. 2005; 37: 455–7.

Gleeson JG, Keeler LC, Parisi MA, et al. Molar tooth sign of the midbrain-hindbrain junction: occurrence in multiple distinct syndromes. Am. J. Med. Genet. 2004; 125A: 125–34; discussion 117

Goldstein JL, Fialkow PJ. The Alstrom syndrome. Report of three cases with further delineation of the clinical, pathophysiological, and genetic aspects of the disorder. Medicine (Baltimore) 1973; 52: 53–71.

Green A, Allos M, Donohoe J, Carmody M, Walshe J. Prevalence of hereditary renal disease. Ir. Med. J. 1990; 83: 11–13.

Green JS, Parfrey PS, Harnett JD, et al. The cardinal manifestations of Bardet-Biedl syndrome, a form of Laurence-Moon-Biedl syndrome. N. Engl. J. Med. 1989; 321: 1002–9.

Gretz N. Rate of deterioration of renal function in juvenile nephronophthisis. Pediatr. Nephrol. 1989; 3: 56–60.

Haider NB, Carmi R, Shalev H, Sheffield VC, Landau D. A Bedouin kindred with infantile nephronophthisis demonstrates linkage to chromosome 9 by homozygosity mapping. Am. J. Hum. Genet. 1998; 63: 1404–10.

Hart TC, Gorry MC, Hart PS, et al. Mutations of the UMOD gene are responsible for medullary cystic kidney disease 2 and familial juvenile hyperuricaemic nephropathy. J. Med. Genet. 2002; 39: 882–92.

Hearn T, Renforth GL, Spalluto C, et al. Mutation of ALMS1, a large gene with a tandem repeat encoding 47 amino acids, causes Alstrom syndrome. Nat. Genet. 2002; 31: 79–83.

Hecht H, Ohlsson J, Starck SA. Poor renal uptake of 99^{m} technetium-dimercaptosuccinic acid and near-normal 99^{m} technetium-mercaptoacetyltriglycine renogram in nephronophthisis. Pediatr. Nephrol. 1996; 10: 167–70.

Hidaka S, Konecke V, Osten L, Witzgall R. PIGEA-14, a novel coiled-coil protein affecting the intracellular distribution of polycystin-2. J. Biol. Chem. 2004; 279: 35009–16.

Hiesberger T, Bai Y, Shao X, et al. Mutation of hepatocyte nuclear factor-1beta inhibits Pkhd1 gene expression and produces renal cysts in mice. J. Clin. Invest. 2004; 113: 814–25.

Hiesberger T, Shao X, Gourley E, Reimann A, Pontoglio M, Igarashi P. Role of the hepatocyte nuclear factor-1beta (HNF-1beta) C-terminal domain in Pkhd1 (ARPKD) gene transcription and renal cystogenesis. J. Biol. Chem. 2005; 280: 10578–86.

Hildebrandt F. Identification of a gene for nephronophthisis. Nephrol. Dial. Transplant. 1998; 13: 1334–6.

Hildebrandt F. Juvenile nephronophthisis. In: Harmon WE, ed. Pediatric Nephrology. Baltimore: Williams & Wilkins, 1999.

Hildebrandt F, Otto E. Molecular genetics of nephronophthisis and medullary cystic kidney disease. J. Am. Soc. Nephrol. 2000; 11: 1753–61.

Hildebrandt F, Otto EA. Primary cilia: a unifying pathogenic concept for cystic kidney disease? Nat. Rev. Genet. 2005.

Hildebrandt F, Otto E, Rensing C, et al. A novel gene encoding an SH3 domain protein is mutated in nephronophthisis type 1. Nat. Genet. 1973a; 17: 149–53.

Hildebrandt F, Rensing C, Betz R, et al. Establishing an algorithm for molecular genetic diagnostics in 127 families with juvenile nephronophthisis. Kidney Int. 2001; 59: 434–45.

Hildebrandt F, Singh-Sawhney I, Schnieders B, et al. Mapping of a gene for familial juvenile nephronophthisis: refining the map and defining flanking markers on chromosome 2. APN Study Group. Am. J. Hum. Genet. 1993; 53: 1256–61.

Hildebrandt F, Strahm B, Nothwang HG, et al. Molecular genetic identification of families with juvenile nephronophthisis type 1: rate of progression to renal failure. APN Study Group. Arbeitsgemeinschaft für Padiatrische Nephrologie. Kidney Int. 1997b; 51: 261–9.

Hildebrandt F, Waldherr R, Kutt R, Brandis M. The nephronophthisis complex: clinical and genetic aspects. Clin. Investig. 1992; 70: 802–8.

Hirokawa N, Noda Y, Okada Y. Kinesin and dynein superfamily proteins in organelle transport and cell division. Curr. Opin. Cell. Biol. 1998; 10: 60–73.

Hu J, Barr MM. ATP-2 Interacts with the PLAT domain of LOV-1 and is involved in *Caenorhabditis elegans* polycystin signaling. Mol. Biol. Cell 2005; 16: 458–69.

Huangfu D, Liu A, Rakeman AS, et al. Hedgehog signalling in the mouse requires intraflagellar transport proteins. Nature 2003; 426: 83–7.

Hughes J, Ward CJ, Peral B, et al. The polycystic kidney disease 1 (PKD1) gene encodes a novel protein with multiple cell recognition domains. Nat. Genet. 1995; 10: 151–60.

Igarashi P, Somlo S. Genetics and pathogenesis of polycystic kidney disease. J. Am. Soc. Nephrol. 2002; 13: 2384–98.

Jauregui AR, Barr MM. Functional characterization of the *C. elegans* nephrocystins NPHP-1 and NPHP-4 and their role in cilia and male sensory behaviors. Exp. Cell Res. 2005; 305: 333–42.

Jen JC, Chan WM, Bosley TM, et al. Mutations in a human ROBO gene disrupt hindbrain axon pathway crossing and morphogenesis. Science 2004; 304: 1509–13.

Jeune M, Beraud C, Carron R. Dystrophie thoracique asphyxiante de caractere familial. [Asphyxiating thoracic dystrophy with familial characteristics], Arch. Fr. Pediatr. 1955; 12: 886–91.

Joubert M, Eisenring JJ, Robb JP, Andermann F. Familial agenesis of the cerebellar vermis. A syndrome of episodic hyperpnea, abnormal eye movements, ataxia, and retardation. Neurology 1969; 19: 813–25.

Katsanis N, Beales PL, Woods MO, et al. Mutations in MKKS cause obesity, retinal dystrophy and renal malformations associated with Bardet-Biedl syndrome. Nat. Genet. 2000; 26: 67–70.

Keeler LC, Marsh SE, Leeflang EP, et al. Linkage analysis in families with Joubert syndrome plus oculo-renal involvement identifies the CORS2 locus on chromosome 11p12-q13.3. Am. J. Hum. Genet. 2003; 73: 656–62.

Keller R. Shaping the vertebrate body plan by polarized embryonic cell movements. Science 2002; 298: 1950–4.

Kim E, Arnould T, Sellin LK, et al. The polycystic kidney disease 1 gene product modulates Wnt signaling. J. Biol. Chem. 1999; 274: 4947–53.

Kim H, Jeong W, Ahn K, Ahn C, Kang S. Siah-1 interacts with the intracellular region of polycystin-1 and affects its stability via the ubiquitin-proteasome pathway. J. Am. Soc. Nephrol. 2004; 15: 2042–9.

Kleinknecht C. The inheritance of nephronophthisis. In: Avner ED, ed. Inheritance of Kidney and Urinary Tract Diseases, Vol. 9. Boston: Kluwer Academic Publishers, 1989: p. 464.

Kleinknecht C, Habib R. Nephronophthisis. In: Ritz E, ed. Oxford Textbook of Clinical Nephrology. Oxford: Oxford University Press, 1992.

Konrad M, Saunier S, Heidet L, et al. Large homozygous deletions of the 2q13 region are a major cause of juvenile nephronophthisis. Hum. Mol. Genet. 1996; 5: 367–71.

Kottgen M, Benzing T, Simmen T, et al. Trafficking of TRPP2 by PACS proteins represents a novel mechanism of ion channel regulation. EMBO J. 2005.

Kumada S, Hayashi M, Arima K, et al. Renal disease in Arima syndrome is nephronophthisis as in other Joubert-related cerebello-oculo-renal syndromes. Am. J. Med. Genet. A 2004; 131: 71–6.

le Maire A, Weber T, Saunier S, et al. Solution NMR structure of the SH3 domain of human nephrocystin and analysis of a mutation-causing juvenile nephronophthisis. Proteins 2005; 59: 347–55.

Li JB, Gerdes JM, Haycraft CJ, et al. Comparative genomics identifies a flagellar and basal body proteome that includes the BBS5 human disease gene. Cell 2004; 117: 541–52.

Lin F, Hiesberger T, Cordes K, et al. Kidney-specific inactivation of the KIF3A subunit of kinesin-II inhibits renal ciliogenesis and produces polycystic kidney disease. Proc. Natl Acad. Sci. USA 2003; 100: 5286–91.

Lo SH, Yu QC, Degenstein L, Chen LB, Fuchs E. Progressive kidney degeneration in mice lacking tensin. J. Cell Biol. 1997; 136: 1349–61.

Loken AC, Hanssen O, Halvorsen S, Jolster NJ. Hereditary renal dysplasia and blindness. Acta Paediatr. 1961; 50: 177–84.

Lu W, Peissel B, Babakhanlou H, et al. Perinatal lethality with kidney and pancreas defects in mice with a targetted Pkd1 mutation. Nat. Genet. 1997; 17: 179–81.

Lyon MF, Hulse EV. An inherited kidney disease of mice resembling human nephronophthisis. J. Med. Genet. 1971; 8: 41–8.

Mainzer F, Saldino RM, Ozonoff MB, Minagi H. Familial nephropathy associatdd with retinitis pigmentosa, cerebellar ataxia and skeletal abnormalities. Am. J. Med. 1970; 49: 556–62.

Makker SP, Grupe WE, Perrin E. Identical progression of juvenile hereditary nephronophthisis in monozygotic twins. J. Pediatr. 1973; 82: 773–9.

Mangos JA, Opitz JM, Lobeck CC, Cookson DU. Familial juvenile nephronophthisis. An unrecognized renal disease in the United States. Pediatrics 1964; 34: 337–45.

McGregor AR, Bailey RR. Nephronophthisis-cystic renal medulla complex: diagnosis by computerized tomography. Nephron 1989; 53: 70–2.

Medhioub M, Cherif D, Benessy F, et al. Refined mapping of a gene (NPH1) causing familial juvenile nephronophthisis and evidence for genetic heterogeneity. Genomics 1994; 22: 296–301.

Menezes LF, Cai Y, Nagasawa Y, et al. Polyductin, the PKHD1 gene product, comprises isoforms expressed in plasma membrane, primary cilium, and cytoplasm. Kidney Int. 2004; 66: 1345–55.

Mochizuki T, Saijoh Y, Tsuchiya K, et al. Cloning of inv, a gene that controls left/right asymmetry and kidney development. Nature 1998; 395: 177–81.

Mochizuki T, Wu G, Hayashi T, et al. PKD2, a gene for polycystic kidney disease that encodes an integral membrane protein. Science 1996; 272: 1339–42.

Mollet G, Salomon R, Gribouval O, et al. The gene mutated in juvenile nephronophthisis type 4 encodes a novel protein that interacts with nephrocystin. Nat. Genet. 2002; 32: 300–5.

Mollet G, Silbermann F, Delous M, Salomon R, Antignac C, Saunier S. Characterization of the nephrocystin/nephrocystin-4 complex and subcellular localization of nephrocystin-4 to primary cilia and centrosomes. Hum. Mol. Genet. 2005; 14: 645–56.

Mongeau JG, Worthen HG. Nephronophthisis and medullary cystic disease. Am. J. Med. 1967; 43: 345–55.

Morgan D, Turnpenny L, Goodship J, et al. Inversin, a novel gene in the vertebrate left-right axis pathway, is partially deleted in the inv mouse. Nat. Genet. 1998; 20: 149–56.

Morgan D, Eley L, Sayer J, et al. Expression analyses and interaction with the anaphase promoting complex protein Apc2 suggest a role for inversin in primary cilia and involvement in the cell cycle. Hum. Mol. Genet. 2002a; 11: 3345–50.

Morgan D, Goodship J, Essner JJ, et al. The left-right determinant inversin has highly conserved ankyrin repeat and IQ domains and interacts with calmodulin. Hum. Genet. 2002b; 110: 377–84.

Morishima M, Yasui H, Nakazawa M, Ando M, Ishibashi M, Takao A. Situs variation and cardiovascular anomalies in the transgenic mouse insertional mutation, inv. Teratology 1998; 57: 302–9.

Moudgil A, Bagga A, Kamil ES, et al. Nephronophthisis associated with Ellis-van Creveld syndrome. Pediatr. Nephrol. 1998; 12: 20–2.

Moyer JH, Lee-Tischler MJ, Kwon HY, et al. Candidate gene associated with a mutation causing recessive polycystic kidney disease in mice. Science 1994; 264: 1329–33.

Mykytyn K, Braun T, Carmi R, et al. Identification of the gene that, when mutated, causes the human obesity syndrome BBS4. Nat. Genet. 2001; 28: 188–91.

Mykytyn K, Mullins RF, et al. Bardet-Biedl syndrome type 4 (BBS4)-null mice implicate Bbs4 in flagella formation but not global cilia assembly. Proc. Natl Acad. Sci. USA 2004; 101: 8664–9.

Mykytyn K, Nishimura DY, Searby CC, et al. Identification of the gene (BBS1) most commonly involved in Bardet-Biedl syndrome, a complex human obesity syndrome. Nat. Genet. 2002; 31: 435–8.

Natacci F, Corrado L, Pierri M, et al. Patient with large 17p11.2 deletion presenting with Smith-Magenis syndrome and Joubert syndrome phenotype. Am. J. Med. Genet. 2000; 95: 467–72.

Neumann HP, Zauner I, Strahm B, et al. Late occurrence of cysts in autosomal dominant medullary cystic kidney disease. Nephrol. Dial. Transplant. 1997; 12: 1242–6.

Nishimura DY, Fath M, Mullins RF, et al. Bbs2-null mice have neurosensory deficits, a defect in social dominance, and retinopathy associated with mislocalization of rhodopsin. Proc. Natl Acad. Sci. USA 2004; 101: 16588–93.

Nurnberger J, Bacallao RL, Phillips CL. Inversin forms a complex with catenins and N-cadherin in polarized epithelial cells. Mol. Biol. Cell 2002; 13: 3096–106.

Nurnberger J, Kribben A, Opazo Saez A, et al. The Invs gene encodes a microtubule-associated protein. J. Am. Soc. Nephrol. 2004; 15: 1700–10.

Olbrich H, Fliegauf M, Hoefele J, et al. Mutations in a novel gene, NPHP3, cause adolescent nephronophthisis, tapetoretinal degeneration and hepatic fibrosis. Nat. Genet. 2003; 34: 455–9.

Omran H, Fernandez C, Jung M, et al. Identification of a new gene locus for adolescent nephronophthisis, on chromosome 3q22 in a large Venezuelan pedigree. Am. J. Hum. Genet. 2000; 66: 118–27.

Omran H, Haffner K, Vollmer M, et al. Exclusion of the candidate genes ACE and Bcl-2 for six families with nephronophthisis not linked to the NPH1 locus. Nephrol. Dial. Transplant. 1999; 14: 2328–31.

Omran H, Sasmaz G, Haffner K, et al. Identification of a gene locus for Senior-Loken syndrome in the region of the nephronophthisis type 3 gene. J. Am. Soc. Nephrol. 2002; 13: 75–9.

O'Toole JF, Otto E, Frishberg Y. Retinitis pigmentosa and renal failure in a patient with mutations in inversin. J. Am. Soc. Nephrol. 2004; 15: 215A.

Otto E, Betz R, Rensing C, et al. A deletion distinct from the classical homologous recombination of juvenile nephronophthisis type 1 (NPH1) allows exact molecular definition of deletion breakpoints. Hum. Mutat. 2000a; 16: 211–23.

Otto E, Kispert A, Schatzle LB, Rensing C, Hildebrandt F. Nephrocystin: gene expression and sequence conservation between human, mouse, and *Caenorhabditis elegans*. J. Am. Soc. Nephrol. 2000b; 11: 270–82.

Otto E, Kispert A, Schätzle LB, Rensing C, Hildebrandt F. Nephrocystin: gene expression and sequence conservation between human, mouse, and *Caenorhabditis elegans*. J. Am. Soc. Nephrol. 2000c; 11: 270–82.

Otto E, Hoefele J, Ruf R, et al. A gene mutated in nephronophthisis and retinitis pigmentosa encodes a novel protein, nephroretinin, conserved in evolution. Am. J. Hum. Genet. 2002; 71: 1167–71.

Otto EA, Schermer B, Obara T, et al. Mutations in INVS encoding inversin cause nephronophthisis type 2, linking renal cystic disease to the function of primary cilia and left-right axis determination. Nat. Genet. 2003a; 34: 413–20.

Otto EA, Schermer B, Obara T, et al. Mutations in INVS encoding inversin cause nephronophthisis type 2, linking renal cystic disease to the function of primary cilia and left-right axis determination. Nat. Genet. 2003b; 34: 413–20.

Otto EA, Loeys B, Khanna H, et al. Nephrocystin-5, a ciliary IQ domain protein, is mutated in Senior-Loken syndrome and interacts with RPGR and calmodulin. Nat. Genet. 2005; 37: 282–8.

Parisi MA, Bennett CL, Eckert ML, et al. The NPHP1 gene deletion associated with juvenile nephronophthisis is present in a subset of individuals with Joubert syndrome. Am. J. Hum. Genet. 2004; 75: 82–91.

Parisi MA, Doherty D, Eckert ML, et al. AHI1 mutations cause both retinal dystrophy and renal cystic disease in Joubert syndrome. J. Med. Genet. 2005.

Peng M, Jarett L, Meade R, et al. Mutant prenyltransferase-like mitochondrial protein (PLMP) and mitochondrial abnormalities in kd/kd mice. Kidney Int. 2004; 66: 20–8.

Pennekamp P, Karcher C, Fischer A, et al. The ion channel polycystin-2 is required for left-right axis determination in mice. Curr. Biol. 2002; 12: 938–43.

Phillips CL, Miller KJ, Filson AJ, et al. Renal cysts of inv/inv mice resemble early infantile nephronophthisis. J. Am. Soc. Nephrol. 2004; 15: 1744–55.

Pistor K, et al. Children with chronic renal failure in the Federal Republic of Germany: II. Primary renal diseases, age and intervals from early renal failure to renal death. Arbeitsgemeinschaft fur Padiatrische Nephrologie. Clin. Nephrol. 1985; 23: 278–84.

Polak BC, van Lith FH, Delleman JW, van Balen AT. Carrier detection in tapetoretinal degeneration in association with medullary cystic disease. Am. J. Ophthalmol. 1983; 95: 487–94.

Potter DE, Holliday MA, Piel CF, Feduska NJ, Belzer FO, Salvatierra O, Jr. Treatment of end-stage renal disease in children: a 15-year experience. Kidney Int. 1980; 18: 103–9.

Proesmans W, Van Damme B, Macken J. Nephronophthisis and tapetoretinal degeneration associated with liver fibrosis. Clin. Nephrol. 1975; 3: 160–4.

Rahilly MA, Fleming S. Abnormal integrin receptor expression in two cases of familial nephronophthisis. Histopathology 1995; 26: 345–9.

Rayfield EJ, McDonald FD. Red and blonde hair in renal medullary cystic disease. Arch. Intern. Med. 1972; 130: 72–5.

Resnick J, Sisson S. Tamm-Horsfall protein. Abnormal localization in renal disease. Lab. Invest. 1978; 38: 550.

Romio L, Fry AM, Winyard PJ, Malcolm S, Woolf AS, Feather SA. OFD1 is a centrosomal/basal body protein expressed during mesenchymal-epithelial transition in human nephrogenesis. J. Am. Soc. Nephrol. 2004; 15: 2556–68.

Romio L, Wright V, Price K, et al. OFD1, the gene mutated in oral-facial-digital syndrome type 1, is expressed in the metanephros and in human embryonic renal mesenchymal cells. J. Am. Soc. Nephrol. 2003; 14: 680–9.

Ross AJ, May-Simera H, Eichers ER, et al. Disruption of Bardet-Biedl syndrome ciliary proteins perturbs planar cell polarity in vertebrates. Nat. Genet. 2005; 37: 1135–40.

Saar K, Al-Gazali L, Sztriha L, et al. Homozygosity mapping in families with Joubert syndrome identifies a locus on chromosome 9q34.3 and evidence for genetic heterogeneity. Am. J. Hum. Genet. 1999; 65: 1666–71.

Saraiva JM, Baraitser M. Joubert syndrome: a review. Am. J. Med. Genet. 1992; 43: 726–31.

Saraux H, Dhermy P, Fontaine JL, et al. La dégénéréscence rétino-tubulaire de Senior et Loken [Senior-Loken retino-tubular degeneration]. Arch. Ophthalmol. Rev. Gen. Ophthalmol. 1970; 30: 683–96.

Sarimurat N, Elcioglu N, Tekant GT, Elicevik M, Yeker D. Jeune's asphyxiating thoracic dystrophy of the newborn. Eur. J. Pediatr. Surg. 1998; 8: 100–1.

Satran D, Pierpont ME, Dobyns WB. Cerebello-oculo-renal syndromes including Arima, Senior-Loken and COACH syndromes: more than just variants of Joubert syndrome. Am. J. Med. Genet. 1999; 86: 459–69.

Saunier S, Calado J, Heilig R, et al. A novel gene that encodes a protein with a putative src homology 3 domain is a candidate gene for familial juvenile nephronophthisis. Hum. Mol. Genet. 1997a; 6: 2317–23.

Saunier S, Morin G, Calado J, et al. Large deletions of the NPH1 region in Cogan syndrome (CS) associated with familial juvenile nephronophthisis (NPH). Am. J. Hum. Genet. 1997b; 61: A346.

Saunier S, Calado J, Benessy F, et al. Characterization of the NPHP1 locus: mutational mechanism involved in deletions in familial juvenile nephronophthisis. Am. J. Hum. Genet. 2000; 66: 778–89.

Schimke RN. Juvenile nephronophthisis-medullary cystic disease. In: Barrat TM, ed. Pediatric Nephrology. Baltimore: Williams and Wilkins, 1975.

Schuermann MJ, Otto E, Becker A, et al. Mapping of gene loci for nephronophthisis type 4 and Senior-Løken syndrome, to chromosome 1p36. Am. J. Hum. Genet. 2002; 70: 1240–6.

Scolari F, Ghiggeri GM, Amoroso A, Caridi GL, Aridon P. Genetic heterogeneity for autosomal dominant medullary cystic kidney disease (ADMCKD). J. Am. Soc. Nephrol. 1998; 9: 393A.

Senior B, Friedmann AI, Braudo JL. Juvenile familial nephropathy with tapetoretinal degeneration: a new oculorenal dystrophy. Am. J. Ophthalmol. 1961; 52: 625–33.

Sherman FE, Studnicki FM, Fetterman GH. Renal lesions of familial juvenile nephronophthisis examined by microdissection. Am. J. Clin. Pathol. 1971; 55: 391.

Sibalic V, Sun L, Sibalic A, et al. Characteristic matrix and tubular basement membrane abnormalities in the CBA/Ca-kdkd mouse model of hereditary tubulointerstitial disease. Nephron 1998; 80: 305–13.

Simons M, Gloy J, Ganner A, et al. Inversin, the gene product mutated in nephronophthisis type II, functions as a molecular switch between Wnt signaling pathways. Nat. Genet. 2005; 37: 537–43.

Smith CH, Graham JB. Congenital medullary cysts of the kidneys with severe refractory anemia. Am. J. Dis. Child 1945; 69: 369–77.

Sorenson CM, Rogers SA, Korsmeyer SJ, Hammerman MR. Fulminant metanephric apoptosis and abnormal kidney development in bcl-2-deficient mice. Am. J. Physiol. 1995; 268: F73–81.

Steel BT, Lirenman DS, Battie CW. Nephronophthisis. Am. J. Med. 1980; 68: 531–8.

Sun Z, Amsterdam A, Pazour GJ, et al. A genetic screen in zebrafish identifies cilia genes as a principal cause of cystic kidney. Development 2004; 131: 4085–93.

Sworn MJ, Eisinger AJ. Medullary cystic disease and juvenile nephronophthisis in separate members of the same family. Arch. Dis. Child. 1972; 47: 278.

Takahashi H, Calvet JP, Dittemore-Hoover D, et al. A hereditary model of slowly progressive polycystic kidney disease in the mouse. J. Am. Soc. Nephrol. 1991; 1: 980–9.

Tao Y, Kim J, Schrier RW, Edelstein CL. Rapamycin markedly slows disease progression in a rat model of polycystic kidney disease. J. Am. Soc. Nephrol. 2005; 16: 46–51.

Togawa A, Miyoshi J, Ishizaki H, et al. Progressive impairment of kidneys and reproductive organs in mice lacking Rho GDIalpha. Oncogene 1999; 18: 5373–80.

Torres VE. Therapies to slow polycystic kidney disease. Nephron. Exp. Nephrol. 2004; 98: 1–7.

Tsimaratos M, Sarles J, Sigaudy S, Philip N. Renal and retinal involvement in the Sensenbrenner syndrome. Am. J. Med. Genet. 1998; 77: 337.

Upadhya P, Birkenmeier EH, Birkenmeier CS, Barker JE. Mutations in a NIMA-related kinase gene, Nek1, cause pleiotropic effects including a progressive polycystic kidney disease in mice. Proc. Natl Acad. Sci. USA 2000; 97: 217–21.

Upadhya P, Churchill G, Birkenmeier EH, Barker JE, Frankel WN. Genetic modifiers of polycystic kidney disease in intersubspecific KAT2J mutants. Genomics 1999; 58: 129–37.

Utsch B, Sayer J, Attanasio M, et al. Identification of the first AHI1 gene mutations in families with joubert syndrome and nephronophthisis. Ped. Nephrol. 2006; 21: 32–5.

Valente EM, Marsh SE, Castori M, et al. Distinguishing the four genetic causes of jouberts syndrome-related disorders. Ann. Neurol. 2005; 57: 513–19.

Veis DJ, Sorenson CM, Shutter JR, Korsmeyer SJ. Bcl-2-deficient mice demonstrate fulminant lymphoid apoptosis, polycystic kidneys, and hypopigmented hair. Cell 1993; 75: 229–40.

Waldherr R, Lennert T, Weber HP, Fodisch HJ, Scharer K. The nephronophthisis complex. A clinicopathologic study in children. Virch. Arch. A. Pathol. Anat. Histol. 1982; 394: 235–54.

Warady BA, Hebert D, Sullivan EK, Alexander SR, Tejani A. Renal transplantation, chronic dialysis, and chronic renal insufficiency in children and adolescents. The 1995 Annual Report of the North American Pediatric Renal Transplant Cooperative Study. Pediatr. Nephrol. 1997; 11: 49–64.

Ward CJ, Hogan MC, Rossetti S, et al. The gene mutated in autosomal recessive polycystic kidney disease encodes a large, receptor-like protein. Nat. Genet. 2002; 30: 259–69.

Watnick T, Germino G. From cilia to cyst. Nat. Genet. 2003; 34: 355–6.

Wolf MT, Lee J, Panther F, Otto EA, Guan KL, Hildebrandt F. Expression and phenotype analysis of the nephrocystin-1 and nephrocystin-4 homologs in caenorhabditis elegans. J. Am. Soc. Nephrol. 2005; 16: 676–87.

Wu G, D'Agati V, Cai Y, et al. Somatic inactivation of Pkd2 results in polycystic kidney disease. Cell 1998; 93: 177–88.

Wu G, Markowitz GS, Li L, et al. Cardiac defects and renal failure in mice with targeted mutations in Pkd2. Nat. Genet. 2000; 24: 75–8.

Wu G, Tian X, Nishimura S, et al. Trans-heterozygous Pkd1 and Pkd2 mutations modify expression of polycystic kidney disease. Hum. Mol. Genet. 2002; 11: 1845–54.

Zollinger HU, Mihatsch MJ, Edefonti A, Gaboardi F, Imbasciati E, Lennert T. Nephronophthisis (medullary cystic disease of the kidney). A study using electron microscopy, immunofluorescence, and a review of the morphological findings. Helv. Paediatr. Acta 1980; 35: 509–30.

CHAPTER 26

Medullary Cystic Disease

ANTHONY J. BLEYER AND THOMAS C. HART

INTRODUCTION

Medullary cystic kidney disease (MCKD) is a collective term for a group of Mendelian genetic disorders with a similar autosomal dominant mode of inheritance and shared clinical findings. The primary manifestations are hyperuricemia and chronic progressive renal failure. Hyperuricemia, due to a decreased urinary excretion of uric acid, may present in adolescence and lead to tophaceous gout in some cases (Figure 26.1). Chronic kidney failure is associated with a bland urinary sediment. Renal failure has an insidious onset, is slowly progressive, with renal replacement therapy required between the 4th and 8th decades of life. Medullary cysts are not present in many individuals with the disorder, appearing sporadically on ultrasound (hence, the name MCKD is a misnomer). The disease is limited to the kidneys with no extrarenal manifestations.

The past clinical characterization and classification of these conditions has led to marked confusion for the practicing nephrologist. Various names have been ascribed

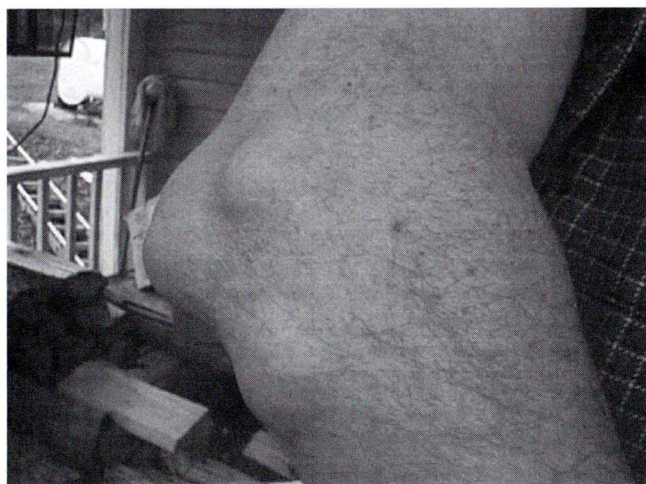

FIGURE 26.1   (see also Plate 44)

to MCKD, with controversy regarding terminology dating back to the early 1960s. Unfortunately, the relatively rare occurrence of MCKD, together with the confusion in nomenclature, has contributed to a poor understanding of this condition and has often resulted in clinical misdiagnosis and under-diagnosis. Careful characterization of cardinal clinical features, mode of inheritance and identification of the specific genetic etiology of kidney diseases is permitting development of a more refined and clinically robust nosology of MCKD. MCKD was previously grouped together with other conditions under the disease complex nephronophthisis/MCKD. Nephronophthisis and MCKD are now recognized as distinct disease entities. Nephronophthisis currently refers to a group of disorders of autosomal recessive inheritance which result in salt-wasting and end-stage kidney disease in the teenage years or early twenties (Hildebrandt & Omram 2001). MCKD refers to several autosomal dominant conditions associated with hyperuricemia and kidney failure, with end-stage kidney disease occurring quite variably from the 4th through the 8th decade (Bleyer et al 2003b).

There are at least two types of MCKD (Bichet & Fujiwara 2004), and they have a similar clinical presentation (see Figure 26.2 and Table 26.1). The genetic cause of MCKD type 1 (MCKD1) has not been determined, although a gene for the condition has been localized to chromosome 1q21 (Christodoulou et al 1998, Fuchshuber et al 2001, Kiser et al 2004). The genetic cause for MCKD type 2 (MCKD2) has been found to be mutations in the *UMOD* (uromodulin) gene (Hart et al 2002, Turner et al 2003, Bleyer et al 2003a, 2003b, Dahan et al 2003). Mutations in the *UMOD* gene have also been found to be responsible for a number of cases of familial juvenile hyperuricemic nephropathy (FJHN), a condition with a clinically indistinguishable phenotype from MCKD2. (Hart et al 2002, Dahan et al 2001). Glomerulocystic kidney disease is a rare disorder, and one family with this condition has been found to have mutations in the *UMOD* gene (Gusmano et al 2002).

Genetic Diseases of the Kidney
ISBN: 978-0-12-449851-8

447

HISTORY

In 1944, Thorn, Koepf, and Clinton described a case of salt-wasting, hypotension, and kidney failure occurring in a young man (Thorn et al 1944). The patient died in his early twenties of end-stage kidney disease with a blood urea nitrogen level of 304 mg/dl, and autopsy revealed many small cysts, mostly located at the corticomedullary junction. In 1951 Fanconi described two families with kidney failure occurring in childhood with a genetic transmission that appeared autosomal recessive. He termed this condition nephronophthisis (phthisis = Greek 'wasting'). A small number of cysts are present on the autopsy kidney specimen shown in Fanconi's original report (Fanconi et al 1951).

Subsequently, Strauss published a series of 18 cases of 'Cystic Disease of the Renal Medulla' (1962). While two

cases each had a sibling with renal disease, there was no family history of kidney disease in 16 of the 18 cases. The absence of proteinuria, the presence of anemia, salt-wasting, and kidney failure, and the presence of medullary cysts at the time of autopsy were identified as the cardinal manifestations of these cases. In 1966, Goldman and Gardner described a five-generation family with 60 members who suffered from kidney failure inherited in an autosomal dominant manner (Goldman et al 1966). The patient ages at the time of death from kidney failure ranged from 19 to 33, and medullary cysts were found at autopsy in six deceased family members (see Figure 26.3). This family likely represents the first report of a family suffering from the condition that we presently call medullary cystic kidney disease. Based on shared clinical findings, primarily related to decreased urine concentrating ability and progressive kidney failure, these cases were felt to be similar to the prior reports of MCKD by Strauss (1962). The existence of common clinical findings in these Mendelizing kidney disorders led some to conclude that juvenile nephronophthisis and medullary cystic disease were morphologically indistinguishable hereditary renal disorders (Chamberlin et al 1977), and what we now recognize as MCKD was grouped together with other conditions under the disease complex nephronophthisis/ MCKD. Over the next several decades, features of these diseases were better identified and delineated. Refinement of the clinical characterization of these conditions facilitated the use of the molecular genetic linkage studies that would ultimately help clarify the classification of these conditions. By the early 1990s, clinical features of the conditions were well-enough delineated that genetic linkage studies could localize disease loci to specific chromosomes. Based on differences in age of onset of renal failure, mode of inheritance, and gene mapping studies, it has become clear that the prior nephronophthisis/MCKD disease grouping consisted of a genetically and clinically heterogeneous group of conditions.

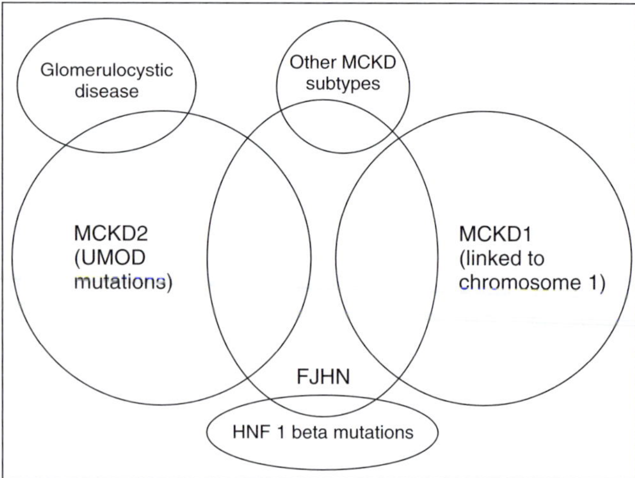

FIGURE 26.2   Relationship between familial juvenile hyperuricemic nephropathy (FJHN), glomerulocystic kidney disease, medullary cystic kidney disease (MCKD) subtypes, and hepatocyte nuclear factor 1beta (HNF 1beta) mutations

TABLE 26.1   Nomenclature concerning hereditary diseases of the tubulo-interstitium

Nephronophthisis	Refers to a group of autosomal recessive conditions associated with salt wasting, medullary cysts, and the development of renal failure in childhood or the late teenage years/early 3rd decade
Medullary cystic	Refers to several autosomal dominant conditions associated with kidney disease chronic kidney failure progressing to end stage kidney disease in the 3rd through 8th decade; associated with hyperuricemia and gout and the occasional presence of renal cysts
Medullary cystic kidney disease type 1	Autosomal dominant condition localized to chromosome 1q21
Medullary cystic kidney disease type 2	Autosomal dominant condition caused by a mutation in the gene encoding uromodulin (Tamm Horsfall protein)
Familial juvenile hyperuricemic nephropathy	Genetically heterogeneous group of disorders with autosomal dominant inheritance of chronic renal failure and hypouricosuric hyperuricemia. Clinically indistinguishable from MCKD. A number of cases have been found due to mutations in the gene encoding uromodulin
Uromodulin associated kidney disease	An alternative term used to describe families suffering from mutations in the uromodulin gene

Nephronophthisis has an earlier disease onset. It is transmitted as an autosomal recessive trait and, as a result of genetic studies, mutations identified in at least four different genes established nephronophthisis as a distinct condition from MCKD (OMIM 256100; 602088; 604387; 606966). Consequently the term nephronophthisis was used to describe a group of autosomal recessive illnesses characterized by salt-wasting in childhood, enuresis, polyuria, and medullary cysts (Hildebrant & Omram 2001). In contrast, the term MCKD was used to describe an autosomal dominant

condition characterized by kidney failure that occurred at a later age, most commonly in adults (Neumann et al 1997, Bleyer et al 2003a).

Similar application of genetic linkage studies facilitated a more refined nosology for MCKD, and currently two broad types, MCKD1 and MCKD2 are recognized, distinguished primarily by genetic findings. Two nonallelic loci for MCKD were identified: the first on chromosome 1q21 (Christodoulou et al 1998) was termed MCKD1, and subsequently linkage for MCKD on chromosome 16p12 (Christodoulou et al 1998) led to the MCKD2 designation. The locus found on chromosome 16 was found to be quite near to the locus identified for another kidney disease, familial juvenile hyperuricemic nephropathy (FJHN) (Dahan et al 2001). FJHN was first identified by Duncan and Dixon (1960). This condition is also an autosomal dominant disease characterized by hyperuricemia and gout with eventual development of kidney failure in the third through 6th decades of life (McBride et al 1997). It was soon realized that the presence of medullary cysts in MCKD2 had been overemphasized (Neumann et al 1997), and the occurrence of hyperuricemia in MCKD was under-emphasized, leading Dahan and co-workers to propose that MCKD and FJHN were similar, if not the same, conditions (Dahan et al 2001).

In 2001, mutations in the *UMOD* gene were identified in families suffering from MCKD and also families suffering from FJHN (Hart et al 2002). Subsequently, these findings were confirmed in other families with MCKD and FJHN (Bleyer et al 2003b, Wolf et al 2003b, Turner et al 2003, Dahan et al 2003, Rampoldi et al 2003, Kudo et al 2004, Rezende-Lima et al 2004) (see Table 26.2). One family with an extremely uncommon condition (Gusmano et al 2002) – glomerulocystic kidney disease – has also been found to have mutations in the *UMOD* gene (Rampoldi et al 2003). It should be noted that genetic linkage studies also support the existence of at least one additional gene locus for MCKD that does not localize to the MCKD1 region of chromosome 1q21 or to the region of chromosome 16q spanning the UMOD locus that is etiologic for MCKD2, and the genetic cause of this form of disease clinically diagnosed as MCKD has not yet been determined (Kudo et al 2004).

Many, but not all, cases of FJHN have been found to have mutations in the *UMOD* gene. For example, one family with a prior diagnosis of FJHN and diabetes was found to have a hepatocyte nuclear factor-1ß gene mutation (Bingham et al 2003). FJHN may be a heterogeneous group of conditions inherited in an autosomal dominant manner that shares with MCKD2 the common clinical findings of chronic kidney failure and hyperuricemia. A number of these cases likely will be the result of mutations in the gene responsible for MCKD1 (see Figure 26.2). The spectrum of diseases included under FJHN and their relationship to MCKD will likely become clearer as the genetic etiology of these conditions is identified and a nosology based on clinical and molecular genetic information is developed.

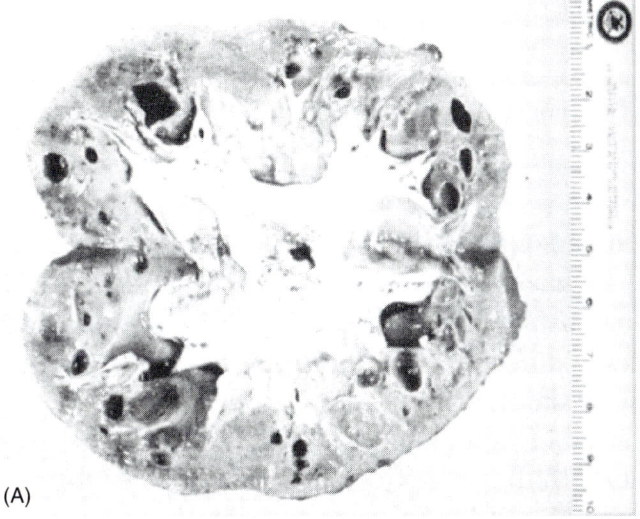

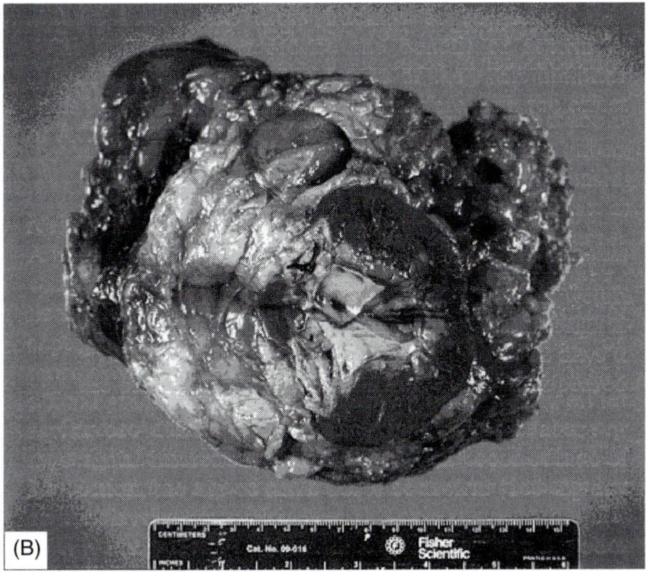

FIGURE 26.3   (A) Gross findings in a patient with medullary cystic kidney disease. Note cysts at cortico-medullary junction. Cysts are actually an inconsistent finding in medullary cystic kidney disease. Adapted from Goldman SH, Walker SR, Merigan TC, Gardner KD, Bull JMC. Hereditary occurrence of cystic disease of the renal medulla. N. Engl. J. Med. 1966; 274: 984–92. (B) Gross kidney specimen from patient with MCKD2. (see also Plate 45)

TABLE 26.2   Amino acid changes reported for uromodulin

Amino acid change	Country	Hyperuricemia	Gout	Onset dialysis	Presence of medullary cysts	Reference
Cys52Trp	Japan	2 of 2	2 of 2			Kudo 2004
Asp59Ala	Belgium		0 of 1	59	1 of 1	Dahan 2003
Cys77Tyr	Austria	2 of 2	1 of 2	35 to >47	0 of 2	Turner 2003
valcysproglugly 93-97alaalasercys		0 of 6	0 of 6	37 to 60	1 of 1	Wolf 2003b
G103Cys	USA	2 of 3	1 of 4			Hart 2002
Cys112Arg	Belgium		1 of 1	64		Dahan 2003
Cys126Arg	Austria	7 of 7	7 of 7	34 to 55	4 of 7	Turner 2003
Cys126Arg	Italy		2 of 2	52		Dahan 2003
Asn128Ser	Spain	6 of 6	6 of 8			Turner 2003
Cys135Ser	Japan	7 of 7	7 of 7			Kudo 2004
C148Y	USA	9 of 9	6 of 9	52, 62		Hart 2002
C148W	Italy	3 of 3	1 of 3	>39	0 of 3	Rampoldi 2003
C150S	Italy	13 of 17	11 of 17	50 to >65	4 of 16	Rampoldi 2003
Cys170Tyr	France		0 of 3	>72	1 of 2	Dahan 2003
H177_R185del	USA	35 of 39	17 of 39	44 to >60	1 of 5	Hart 2002
Arg185Ser	Belgium		6 of 11	28 to 63	3 of 3	Dahan 2003
Trp202Ser	Japan	4 of 4	4 of 4			Kudo 2004
AA188-221del	France		1 of 3	38	2 of 3	Dahan 2003
Cys195Phe	Japan	2 of 3	2 of 3			Kudo 2004
Arg204Gly	Morocco		1 of 1	>57		Dahan 2003
Cys217Arg	USA		0 of 1			Hart 2002
Cys217Gly	Belgium		1 of 1	36	0	Dahan 2003
Arg222Pro	Corsica		4 of 7	47	2 of 2	Dahan 2003
Cys223Tyr	USA	2 of 2	1 of 2	49	0 of 2	Bleyer 2003b
Thr225Meth	France		1 of 8	38 to 60	1 of 1	Dahan 2003
Thr225Lys		3 of 3	1 of 3	29,34	1 of 3	Wolf 2003b
Pro236Leu	Japan		15 of 20	45 to 73		Kudo 2004
cys248Trp		3 of 3	0 of 3	45	1 of 3	Wolf 2003b
Cys255Tyr	Spain	4 of 4	3 of 5			Turner 2003
Cys255Tyr	Spain	8 of 9	2 of 9	45 to >71	0 of 9	Rezende-Lima 2004
Cys282Arg	France		1 of 1	25		Dahan 2003
Cys300Gly	Spain					Turner 2003
K307T						Calado 2005
C315R	Italy	3 of 3	0 of 3		3 of 3	Rampoldi 2003
Cys317Y	Italy	6 of 8	2 of 8	>54	1 of 8	Rampoldi 2003
C347G						Tinschert et al 2004

MCKD1 is clinically similar to MCKD2, and identification of the gene for this disorder is still to be determined.

The ensuing chapter will focus on the findings of MCKD1 and MCKD2, with the latter being designated specifically as a disorder caused by *UMOD* gene mutations. When the term MCKD is used, it will refer to both MCKD1 and MCKD2.

EPIDEMIOLOGY

MCKD is a clinical diagnosis for a group of at least three uncommon Mendelian type genetic disorders. In the United States Renal Disease System database, it accounts for less than 1% of kidney diseases. Approximately 70 patients each year starting dialysis in the United States carry a diagnosis of gouty nephropathy or MCKD (US Renal Data System 2003). Unfortunately, this number is likely to be quite inaccurate due to incorrect and under-diagnosis. As of 2004, MCKD2 has been diagnosed in 35 families with 33 different mutations with families from Europe, the United States, and Japan having been identified (Table 26.2). As of this writing, there have been no cases described in African or African-American families. Since the disease is inherited in an autosomal dominant manner and results in death or kidney failure after the most productive years of child-bearing, there are a number of larger families suffering from MCKD1, and it is not uncommon to find that families with 'new mutations' are actually

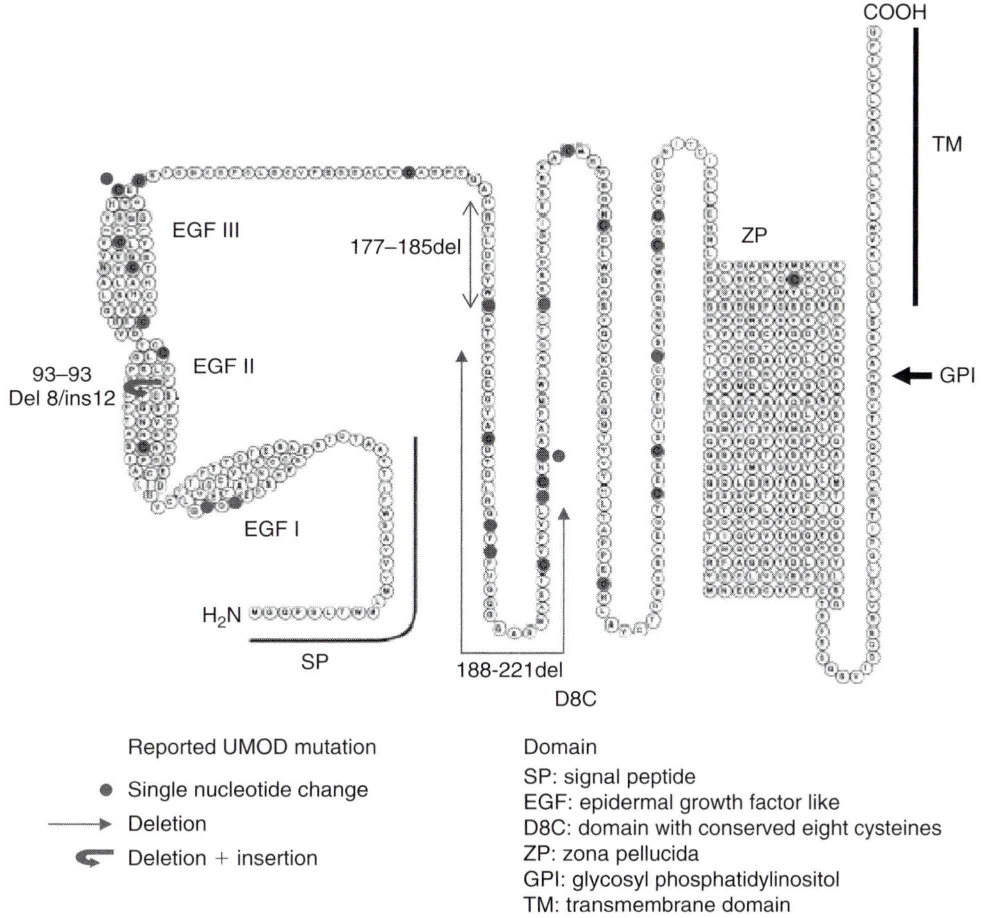

FIGURE 26.4   Reported amino acid changes caused by UMOD mutations showing affected domains. (see also Plate 46)

undiscovered branches of large, previously identified families (Hart et al 2002, Wolf et al 2003a). The frequency of occurrence of new mutations is unknown; however, screening of several hundred unselected dialysis patients did not reveal any patients with mutations in the *UMOD* gene, suggesting it is not a common cause of end stage renal disease.

GENETICS

Improved clinical delineation of MCKD has facilitated genetic linkage studies to localize etiologic genes. These studies have clearly indicated MCKD segregates in families as a genetically heterogeneous autosomal dominant condition. Two nonallelic loci for MCKD were identified: the first, MCKD1 on chromosome 1q21, and the second, MCKD2 on chromosome 16p12 (Christodoulou et al 1998). It should be noted that genetic linkage studies of families segregating MCKD also support the existence of at least one additional etiologic gene that does not localize to the MCKD1 region of chromosome 1q21, or to the region of chromosome 16q spanning the *UMOD* locus that is etiologic for MCKD2 (Kudo et al 2004). Because this condition,

clinically diagnosed as MCKD, appears due to a third gene, its identification may lead to delineation of MCKD3.

MCKD1

Although MCKD1 was the first to be mapped to a chromosomal location, the etiologic gene has not yet been identified. Similarly, while the existence of a third gene locus for MCKD has been established, the gene responsible has neither been localized nor identified.

MCKD2

Mutations of the uromodulin gene which encodes uromodulin (Tamm-Horsfall protein) are responsible for MCKD2 (Hart et al 2002). The uromodulin gene consists of 16 663 base pairs located on chromosome 16p12.3. The gene encodes 12 exons, although the translation initiation site (ATG) is located in exon 3. The two untranslated exons may play a role in the restricted tissue expression of the gene (Zbikowska et al 2002). Expression studies in human tissues indicate the uromodulin gene is only expressed in the kidney. The uromodulin gene encodes several functional domains (Figure 26.4). Multiple different disease-associated mutations have been identified in all three EGF-like domains and in the D8C domain. To date, only one mutation has been reported in the ZP domain.

At least 33 different uromodulin gene mutations have been reported in individuals with MCKD2 (Table 26.2), and the majority of these mutations (>80%) either abolish existing or introduce novel cysteine codons. These mutations occur in exons 4, 5, and 6. Uromodulin mutations usually act to cause disease in a highly penetrant dominant fashion. However, there are documented cases of very mild clinical expression associated with uromodulin mutations in females, indicating that there are likely other genetic or environmental factors that can modify the clinical expression of *UMOD* mutations, although these modifiers have not been identified (Bleyer et al 2004). Individuals who have inherited two mutant *UMOD* genes have been reported in a consanguineous Spanish family. These individuals have a more severe disease presentation including an earlier onset of hyperuricemia and end stage renal disease than heterozygote family members who have inherited only one mutated *UMOD* allele (Rezende-Lima et al 2004).

Uromodulin: Normal Structure and Function

Uromodulin is the most common urinary protein found in healthy individuals. It is highly conserved and found in all mammalian species studied to date. Despite this, the function of uromodulin remains unclear. Uromodulin was initially isolated by salt precipitation of urine by Tamm and Horsfall (1950), and the protein called Tamm-Horsfall protein. In 1985 Muchmore and Decker used lectin-affinity chromatography to isolate uromodulin from the urine of pregnant women (Brown et al 1986). Uromodulin and Tamm-Horsfall glycoprotein are two large glycoproteins with an identical amino acid sequence, and both are a product of the uromodulin gene (Pennica et al 1987). Uromodulin was isolated from the urine of pregnant women and has been found to have a small increase in Man7GlcNac2 molar percentage compared to Tamm-Horsfall protein, which is isolated by salt precipitation in non-pregnant individuals (Hildebrandt & Omram 2001). In this chapter we use the term uromodulin to refer to the protein product of the uromodulin gene. Uromodulin is a large glycoprotein with an extremely high cysteine content (Serafini-Cessi et al 2003). The protein initially is produced with 640 amino acids, the first 24 N-terminal amino acids constituting a signal peptide which is cleaved after synthesis. Of the remaining 616 amino acids, 48 are cysteine residues (Pennica et al 1987). These cysteine residues are believed to be extremely important in cross-linking and helping to make uromodulin an extremely insoluble, adherent protein. The molecule is composed of a number of identifiable functional domains. There is a glycosyl phosphatidylinositol (GPI) anchor that attaches the protein to the luminal membrane of the tubular cells. A zona pellucida-like domain is similar to that found in other adherent glycoproteins, including the ZP-2 and ZP-3

glycoproteins which form the outer coating (zona pellucida) of the ova of primate species. Most disease-associated *UMOD* mutations occur in the recently identified D8C domain and in the three epidermal growth factor domains (Figure 26.4). Uromodulin synthesis in the rough endoplasmic reticulum (ER) is highly regulated and requires the formation of a large number of disulfide bonds in order for the protein to achieve its final conformation and exit the ER (Malagolini et al 1997). Formation of a correct set of intrachain disulfide bonds appears to be a rate limiting step for precursor export out of the ER (Malagolini et al 1997, Serafini-Cessi et al 2003). After glycosylation, uromodulin is then directed to the luminal surface of the cell membrane.

Uromodulin is produced exclusively in the thick ascending limb of the renal tubule and the early distal convoluted tubule (Hoyer et al 1979). The molecule is adherent to the surface of the thick ascending limb throughout except for the macula densa. After proteolytic cleavage of the GPI anchor, uromodulin is excreted in the urine.

A number of functions have been proposed for uromodulin. Uromodulin is known to bind to the luminal surface of tubular cells in the thick ascending limb of Henle (Hoyer et al 1979). Its function there is likely related to salt and water transport, though the mechanism for this has not been determined. Experiments in uromodulin-deficient mice have identified significant sodium-wasting which may be related to this function (Bachmann et al 2005). Uromodulin has also been found to bind to fimbriated *E. coli* (Orskov et al 1980) and is thought to be of potential importance in preventing urinary tract infections. Experiments in uromodulin-deficient mice have also been performed to evaluate this potential function (Bates et al 2004). Strains of *E. coli* were introduced transurethrally into the bladder of the uromodulin-deficient and wild type mice. Deficient mice had a longer duration of colonization and higher colony counts than wild type mice. Uromodulin has also been shown to bind interleukin-1 and tumor necrosis factor (Brown et al 1986, Sherblom et al 1988), leading to speculation that the protein may have a role in modulating inflammation in the kidney. Finally, uromodulin may have a role in the development of nephrolithiasis. Clinical studies have demonstrated potential abnormalities in uromodulin aggregation in some recurrent stone formers. In vitro experiments demonstrate inhibition of calcium oxalate crystal aggregation at high pH and low calcium concentrations, while uromodulin acts as a promoter of calcium oxalate crystallization with high calcium concentrations (Hess 2004). Recent experiments in uromodulin knockout mice have demonstrated spontaneous precipitation of calcium crystals after treatment with ethylene glycol and vitamin D3 (Mo et al 2004). The above experiments in mice may not be clinically applicable to humans because of obvious species differences, as well as the use of laboratory conditions (water deprivation, bacterial installation into the bladder, and treatment with ethylene glycol) which may not be approximated in the human environment.

PATHOPHYSIOLOGY

MCKD2

In most cases of MCKD2, mutations in the *UMOD* gene are substitutions resulting in an exchange of a cysteine residue (Bleyer et al 2004). The cysteine residues are important in intra-chain disulfide bonding to form the tertiary structure of uromodulin. Experiments by Serafini-Cessi in HeLa cells transfected with uromodulin have shown that appropriate disulfide bonding is critical for uromodulin to exit the ER (Malagolini et al 1997). In experiments by Rampoldi et al (2003) mutant *UMOD* genes with mutations resulting in cysteine substitution were transfected into HeLa cells. Transfection resulted in a marked decrease in surface expression of uromodulin which was not due to a decrease in gene transcription or translation. Colocalization experiments using Golgi and ER markers showed that the mutant uromodulin remained mostly in the ER. Immunopathologic studies with antibody to uromodulin revealed cytoplasmic deposits in thick ascending limb and distal tubular cells (see Figure 26.5). Dahan also noted deposition of uromodulin in tubular cells of affected individuals, together with a decreased urinary excretion of uromodulin (Dahan et al 2003). Bleyer quantitatively demonstrated that uromodulin excretion was much less than half the normal expected uromodulin in individuals with *UMOD* mutations and preserved renal function (Bleyer et al 2004).

It thus appears that mutations in the *UMOD* gene result in amino acid changes that lead to defective folding of the uromodulin protein. Mis-folded and incompletely assembled proteins, particularly due to crosslinking by non-native inter-chain disulfide bonds, are prone to aggregation and retention in the endoplasmic reticulum. This results in deposition of uromodulin in the ER and interferes with the processing of normal uromodulin protein produced by the normal *UMOD* allele. Histological evidence from kidney biopsies of affected patients demonstrates that intracellular accumulation of uromodulin occurs in the ER (Dahan et al 2003, Scolari et al 2004, Rampoldi et al 2003, Tinschert et al 2004). MCKD2 is an example of a dominant negative condition, in which the inheritance of one abnormal gene interferes with the function of the normal gene. While a half complement of uromodulin could be produced from the normal gene allele, its production is interfered with by the mutated gene product. All mutations reported to date have resulted in an alteration in the uromodulin protein rather than a change in the genetic sequence resulting in truncation or loss of transcription. This adds further support to the hypothesis that the primary clinical manifestations of renal damage are the result of production of an abnormal uromodulin protein rather than a quantitative decrease in the amount of uromodulin protein.

These findings indicate the disease can be classified as an endoplasmic reticulum storage disease (Rutishauser & Spiess 2002), similar to cystic fibrosis, autosomal dominant neurohypophyseal diabetes insipidus, and protein C deficiency. Over time, accumulation of abnormal uromodulin in renal tubular cells likely leads to more rapid cell turnover and cell death and de-differentiation, resulting in loss of thick ascending limb tubular cells and subsequent destruction of the nephron. This in turn may lead to progressive renal failure.

While abnormal uromodulin protein may cause progressive renal damage, the loss of function of the normal protein may result in other physiologic consequences. The absence of uromodulin could be important in the development of hyperuricemia that affects patients with MCKD2 (Hart et al 2002). Uromodulin-deficient mice develop sodium-wasting together with an up-regulation of distal tubular transporters and a decrease in juxtaglomerular cyclooxygenase-2 and renin mRNA expression (Bachmann et al 2005). The tubular effects of absent uromodulin expression likely result in increased proximal tubular reabsorption and secondary hyperuricemia. As mice synthesize the enzyme uricase, serum uric acid levels are very low, and it is not possible to state with certainty whether these changes are responsible.

At present, it is unclear how the absence of uromodulin may result in the occasional appearance of medullary cysts. The death and de-differentiation of renal tubular cells may result in tubular dilatation that leads to cyst formation.

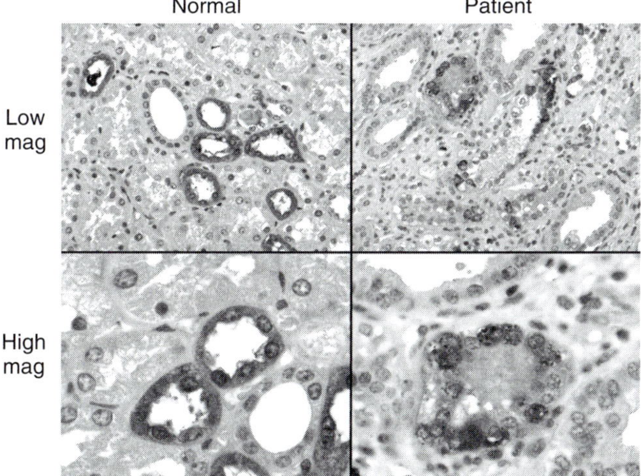

FIGURE 26.5 Uromodulin deposits seen in the thick ascending limb. Immunohistochemical staining with antibody to Tamm–Horsfall protein (uromodulin). In the normal section, antibody staining is diffuse throughout all tubular cells. In contrast, in the patient with a uromodulin mutation, tubular atrophy is present (see low magnification). Under high magnification, antibody staining reveals irregular, dense deposits in tubular cells. (Top panels magnified ×100, bottom panels magnified × 400). (see also Plate 47)

MCKD1

The gene causing MCKD1 has not yet been identified, and pathophysiologic studies of MCKD1 have not been performed.

CLINICAL MANIFESTATIONS

Clinical Manifestations of MCKD2

RENAL FAILURE

The cardinal manifestation of MCKD2 is the development of progressive kidney failure over time (Bleyer et al 2003a). The renal disease in these patients is very slowly progressive and is variable both within families and between families (see Figure 26.6). Renal insufficiency may manifest itself in childhood with a mildly to moderately increased blood urea nitrogen or serum creatinine. Most children we have studied have had a mild decrease in GFR, though for some individuals the estimated glomerular filtration rate (GFR) remains normal through adolescence. As patients will typically require dialysis between 40 and 70 years of age, the rate of decline in GFR is slow. Assuming that renal deterioration begins at age 10, GFR decline would be approximately 3 ml/min/year to 1.5 ml/min/year. This very slow rate of progression will result in the serum creatinine remaining very stable for a number of years during follow-up, with a more rapid rise in the serum creatinine as terminal kidney failure approaches. Renal disease progression appears to be slower in women than in men. For example, in one young female patient we have followed, the serum creatinine has remained stable from age 23 to 30 at 1.4 to 1.5 mg/dl without treatment. It is unclear what effects dietary protein or uric acid production have on the progression of kidney disease.

Patients with MCKD2 usually have a bland urinary sediment, as the disease is tubulo-interstitial in origin and only affects the glomerulus secondarily. Mild proteinuria may develop later in the course of the disease, perhaps reflecting hyperfiltration changes.

HYPERURICEMIA

Hyperuricemia is present in virtually every family studied, but may not be present in individual patients. The hyperuricemia in this condition is due to an increased proximal tubular reabsorption of uric acid (Bleyer et al 2003b). Individuals with MCKD2 and normal renal function typically have a reduced fractional excretion of uric acid, usually less than 5% (Moro et al 1991).

While hyperuricemia is a result of MCKD, there are a number of other factors that determine serum uric acid levels, as is the case in the general population. Meat consumption, age, genetic differences in the rate of uric acid production, the use of diuretics and aspirin, and gender differences all influence serum uric acid levels. Therefore, even when the genetic defect predisposing to hyperuricemia is present, individuals with low uric acid production (frequently females, vegetarians) may have normal serum uric acid levels. While these individuals may have a normal serum uric acid level, their fractional excretion of uric acid will still be low. Large individuals who suffer from the disease will tend to have very high serum uric acid levels, since they produce more uric acid endogenously and are more likely to consume larger amounts of meat, and this amplifies their inability to excrete uric acid.

In the normal child, the fractional excretion of uric acid slowly decreases (Stapleton et al 1978), and with

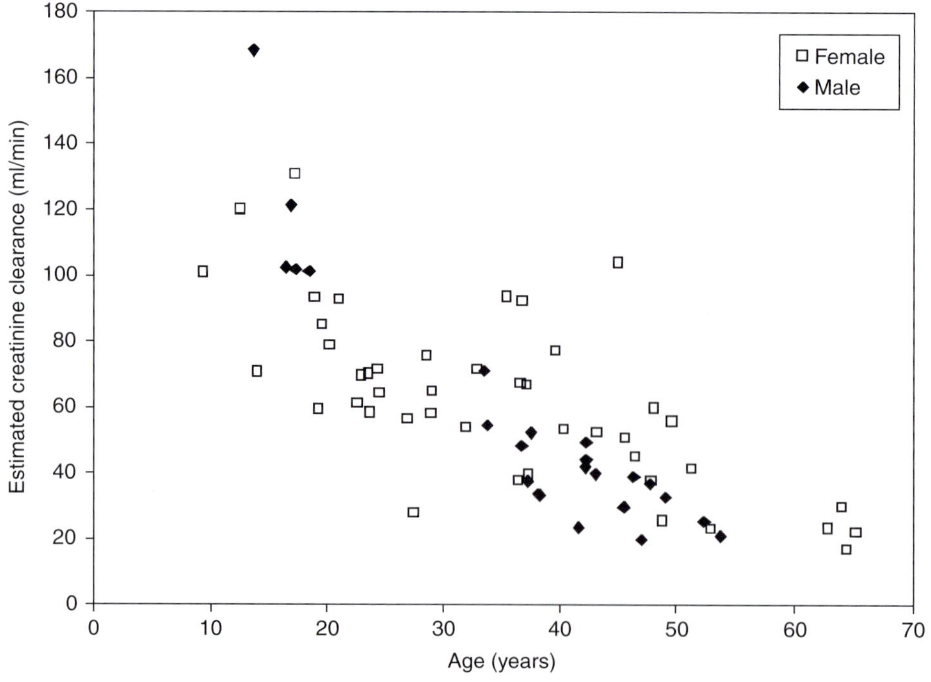

FIGURE 26.6 Estimated creatinine clearance (ml/min) using Cockroft-Gault formula vs. time. This figure shows creatinine clearance estimations from all available laboratory determinations according to age in a family with uromodulin mutations

adolescence in the male, the fractional excretion of uric acid decreases significantly. For this reason, hyperuricemia will tend to worsen in adolescence in young men (Wilcox 1996), and it is therefore the young male teenager who is more likely to develop gout. As the disease progresses, renal failure worsens and the retention of uric acid increases. Patients may develop repetitive attacks of gout, which untreated may result in large tophi (see Figure 26.1).

URINARY CONCENTRATION

Patients with MCKD2 may have mild deficits in urinary concentration. This may present as enuresis in childhood. We encountered a 25% enuresis rate (vs. 10% in controls) in a large MCKD2 family we have studied (Bleyer et al 2003a). However, enuresis is a variable finding, and is usually not identified as a problem in most families. Maximal urinary concentration after an overnight fast ranged from 523 to 736 mOsm/liter in seven individuals in one study, and in another study (Scolari et al 2004) urinary osmolalities were less than 600 mOsm/l in 21 of 25 affected individuals after an overnight fast. Moreover, these authors demonstrated a correlation between the level of serum uric acid and the ability to concentrate the urine.

While defects in urinary concentration can be determined in a clinical setting, polydipsia and polyuria are infrequent complaints in individuals with MCKD2. The large distended bladders characteristic of nephronophthisis are not seen.

HYPERTENSION

Hypertension usually develops after renal insufficiency in patients with MCKD. Problems with urinary concentration and possible salt-wasting may help to keep the blood pressure from rising. As renal insufficiency worsens, hypertension becomes more prevalent.

URINARY TRACT INFECTION AND UROLITHIASIS

While uromodulin-deficient mice have an increased susceptibility to urinary tract infections, families with MCKD2 have not been reported to have an increased incidence of infection. Likewise, an increased prevalence of urolithiasis has not been noted. The differences between uromodulin-deficient mice and humans lacking uromodulin may be secondary to physiologic differences between the species, or due to the fact that the artificial stimuli used to induce urinary tract infection and calcium oxalate crystallization in the laboratory are much greater in magnitude than the stressors found in the human environment.

MEDULLARY CYST FORMATION

Medullary cysts are an inconsistent finding in MCKD2, and this misnomer has helped to perpetuate a poor understanding and rampant misdiagnosis of this condition.

In all the initial investigations of this condition (Strauss 1962, Goldman et al 1966), medullary cysts were identified at autopsy in end-stage kidneys. In addition, the frequent finding of renal cysts in individuals with chronic renal failure of any cause (Sarasin et al 1995) leads to confusion as to whether cysts are present due to renal failure or due to the underlying disease. Cyst identification is variable in different case studies. In a study of 30 patients, Rampoldi noted medullary cysts in five patients. Dahan noted cysts in nine of twelve patients from eight families. The cysts were small and present in small kidneys (Dahan et al 2003). Bleyer noted no cysts on ultrasound and CT scan in five patients, but did note medullary cysts in one affected individual undergoing MRI at age 45 years (Bleyer et al 2003a).

Cysts are in general not present early in the disease and should not be used as a diagnostic criterion.

HOMOZYGOUS UROMODULIN MUTATIONS

One family has been reported with consanguineous mating and three progeny carrying two identical mutant *UMOD* genes. The homozygous male developed renal insufficiency at age 10, gout at age 14, and end-stage kidney disease at 22, compared to a mean age starting dialysis of 58 years in heterozygous male family members. In the two homozygous females, renal insufficiency was diagnosed at age 25 years, and end-stage kidney disease has developed in one at age 55 years, while the other is 43 years old and is not on dialysis. Heterozygous women in this family have a mean age of starting dialysis of 66 (range 61–74 years). All three homozygous family members had renal cysts on ultrasound which were not seen in ultrasounds on heterozygous family members.

Clinical Manifestations of MCKD1

In MCKD identified in six Cypriot families and linked to chromosome 1 (Stavrou et al 2002), hyperuricemia was found in 36 of 72 affected individuals, but in only two of 17 affected individuals with normal kidney function. Clinical gout occurred in five of 75 patients. Ultrasonography revealed cysts in 41% of the patients tested. The mean age of onset of kidney failure was 54 years.

In a large Native American population with MCKD linked to chromosome 1 (Kiser et al 2004), the median age of onset of renal disease was 47 years (range 35–66 years). No patients presented with salt-wasting or polyuria. Gout was present in 61%, anemia in 39%, and hypertension in 55%, though the age of the patients at the time of testing was not given. Urinary protein was negligible except for two patients with proteinuria of 0.57 g/d and 2.05 g/d. Of nine ultrasounds, one patient had corticomedullary cysts.

Linkage to chromosome 1 was also seen in five Finnish families (Auranen et al 2001). In one family end-stage kidney disease developed between the ages of 25 and 33 years.

None of the patients had gout, and in three patients the serum uric acid levels were normal. In the other families, age of onset of kidney failure was between 34 and 55 years of age. Gout was infrequent in all five families, and mild proteinuria was occasionally present.

In a Belgian cohort (Wolf et al 2004) with six affected individuals, the age of ESRD varied from 34 to 49 years (median 41 years). Cysts were seen on ultrasound in several patients.

Taken together, these cases suggest a similar age of development of end-stage kidney disease, with the exception of two families who developed kidney failure in predominantly the third decade (Goldman et al 1966, Auranen et al 2001). Gout and hyperuricemia appear to be less frequent than in individuals with MCKD2, though a more complete phenotyping will require gene identification.

PATHOLOGY

Gross examination of kidneys from individuals with MCKD are obtained late in the course of the disease and usually are small and contracted, consistent with chronic kidney disease. The presence of small cysts at the corticomedullary junction is frequently seen (see Figure 26.3), though it is not an invariable finding. Urate deposits are not seen either grossly or microscopically.

Glomerular changes in MCKD are secondary to the underlying tubular involvement. However, since kidney biopsies routinely obtain primarily cortical tissue, glomeruli are frequently the area of interest to the interpreting pathologist. Globally sclerotic glomeruli are often seen as a result of primary tubular cell death. Arteriolonephrosclerosis may also be noted. In one family, two cases of glomerulocystic changes have been described (Gusmano et al 2002), though these changes are not characteristic.

Tubular atrophy and interstitial fibrosis is the most common finding, and secondary tubular dilatation may be seen as well (see Figure 26.5). Thickening of the tubular basement membrane may also occur (Massari et al 1980, Dahan et al 2001), and a sheath of hyaline connective tissue may be present around tubules and collecting ducts. Medullary cysts are lined by a single layer of cuboidal or columnar epithelium (Thompson et al 1978). Occasionally, a mild inflammatory infiltrate will be seen in the interstitium. Immunofluorescence microscopy is negative, unless antibody staining to uromodulin is performed. When done, this will result in the appearance of dense deposits in tubular cells (see Figure 26.5).

Overall, pathologic changes are nonspecific in MCKD, and kidney biopsy is not indicated in the diagnosis or treatment of this disorder. If kidney tissue is available, one should consider antibody staining with uromodulin. This would demonstrate abnormal deposits in renal tubular cells.

DIAGNOSIS

MCKD is frequently overlooked as a potential diagnosis, mostly because of ignorance regarding the condition. MCKD is usually confused with childhood nephronophthisis, and nephrologists do not consider the diagnosis in kidney failure presenting in adulthood. In addition, if the renal ultrasound does not show medullary cysts, the diagnosis is wrongly excluded. If a kidney biopsy is obtained – normally the gold standard for the diagnosis of renal diseases – cortex is obtained, and secondary glomerulosclerosis is noted, leading to a diagnosis of focal glomerulosclerosis or hypertensive nephrosclerosis (Bleyer et al 2003a). At other times, the diagnosis of 'familial glomerulonephritis' is used, despite the fact that proteinuria is absent.

Another significant barrier in the diagnosis of the disease is the mundane nature of this disorder. The cardinal manifestation is chronic renal insufficiency with a bland urinary sediment, a presentation seen frequently in many patients visiting a nephrologist. The absence of a singular, consistent clinical finding (such as medullary cysts!) results in MCKD not coming to the forefront in the differential diagnosis.

A family history of chronic kidney failure of unknown cause is the most important clue leading to the correct diagnosis.

The most common presentation of this condition will be chronic renal insufficiency or gout. Renal insufficiency will usually be asymptomatic and identified on laboratory studies obtained for other reasons. Patients usually are not hypertensive early in the course of the disease, in contrast to most types of kidney failure. A history of enuresis may or may not be present. Urinalysis will typically reveal a bland urinary sediment, though mild hematuria and proteinuria may be present. A renal ultrasound will usually be normal. If the patient does not suffer from gout, a serum uric acid will likely not be obtained, but would be useful in diagnosis of the disorder. The most important tool in pursuing the diagnosis will be a careful family history. Since this disease is autosomal dominant in inheritance, a large proportion of family members may be affected. While gout may not be present in the affected individual, gout will be seen in some family members, frequently severe in nature. In the setting of a positive family history, the next appropriate test is to perform genetic analysis. Since most mutations to date have occurred in exon 4 or 5, sequencing of these exons will lead to a diagnosis of MCKD2 in most cases. This test is currently performed by Athena Diagnostics in Worcester, MA (www.athenadiagnostics.com). If the testing for MCKD2 is negative, this could mean the family members suffer from MCKD1. Referral of the family to geneticists and nephrologists studying MCKD1 may lead to a diagnosis through linkage and help to further the scientific advancement in this area.

There are several other diagnostic tests which may be of some benefit. First, a 24-hour urine collection for uric acid with determination of the fractional excretion of uric acid may be beneficial (McBride et al 1991). This collection should be performed on a low purine diet if possible, and the patient should avoid aspirin and nonsteroidal antiinflammatory agents in the week prior to testing. An elevated serum uric acid with a low fractional excretion of uric acid is suggestive of MCKD. However, there are several important caveats. First, hyperuricemia is frequent in the general population (Roubenoff 1990), and a low fractional excretion of uric acid is the most common reason for hyperuricemia (Roubenoff 1990). There may therefore be overlap between individuals with hyperuricemia for other reasons. Secondly, as renal failure progresses, the fractional excretion of uric acid rises (Danovitch 1972), even in individuals with MCKD2. For this reason, individuals with MCKD2 who have some renal insufficiency may have normal fractional excretions, and the fractional excretion of uric acid is helpful but not diagnostic in this condition.

Another diagnostic approach is to measure urinary uromodulin, since the excretion of urinary uromodulin is low in individuals with the disease (Bleyer et al 2004). There are several drawbacks to this approach. First, urinary uromodulin determination is not a commercially available test. Secondly, urinary uromodulin concentrations decrease as renal failure progresses in all types of renal disease (Thornley et al 1985). For this reason, urinary uromodulin will be low in many patients with renal insufficiency. The measurement of urinary uromodulin may be of benefit in the small child in whom blood sampling is difficult, and whose family has already been genotyped.

Is the pursuit of diagnostic testing worthwhile? At present, there is no cure for MCKD. Treatment with allopurinol may potentially prevent progression of the disease and is likely to prevent development of tophaceous gout (McBride et al 1998). For this reason, testing may be of benefit. In addition, if living related kidney donation is being considered, it would be important to make the diagnosis in the family and test the potential donor.

For many families, there is a strong desire to know the cause of the disease that has been causing early death and severe morbidity. However, there are many complex psychological factors at work which may affect the decision as to when genetic testing should be performed. These issues are similar to those faced by families with polycystic kidney disease (Sujansky et al 1990). We do not recommend prenatal diagnosis in this condition, as we believe this condition is treatable and it is likely that less invasive treatments will be available by the time current newborns reach an age where kidney failure will develop.

Diagnosis should also be considered in the patient without a family history, as a new mutation may be present. However, how should a nephrologist decide who to screen for mutations when he sees so many patients with chronic renal failure, gout, and a bland urinary sediment? The most appropriate patients for testing would include individuals with a history of precocious gout. Also, individuals with a history of chronic renal failure who have no other potential causes should be considered for testing. The absence of hypertension in a patient with renal insufficiency also suggests MCKD as a possible diagnosis. Genetic screening for a uromodulin mutation may rule out MCKD2 as a cause of disease. While there is no specific therapy for MCKD at present, as treatment options become available, even patients with a low suspicion of MCKD will require screening in order to take advantage of potential benefits of therapy.

TREATMENT

Treatment of Renal Failure

Evaluation of treatment efficacy for MCKD2 is hampered by a number of factors. Disease progression is extremely slow, with an annual decline in GFR for most patients of 2–5 ml/min. This is amplified by the crude measurements of glomerular filtration that we currently employ. In patients with renal reserve, changes in serum creatinine may not occur for some time. For instance, one of our patients with MCKD2 who has received no specific treatment has had a stable serum creatinine for 7 years. It is therefore clear that evaluation of treatment in a placebo-controlled prospective trial could take many years before achieving a significant difference, even if one exists.

Another barrier to evaluation of treatment efficacy is the small number of patients suffering from MCKD. Coupled with the slow rate of progression, this makes the performance of prospective randomized clinical trials following serum creatinine over time problematic.

In addition, patients with a clinical presentation of MCKD may have a number of heterogeneous genetic conditions, for which particular treatments may have different benefits. At this point, only MCKD2 can be determined genetically and therefore studied in a rigorous manner.

Identification of surrogate markers of kidney disease progression will be of great importance in determining potential therapies. In addition, the development of a mouse model with a *UMOD* mutation will be an important step in exploring therapeutic possibilities.

ALLOPURINOL

The most common treatment advocated for delaying progression of renal disease in MCKD has been the use of allopurinol. In patients with MCKD2 and known mutations in the *UMOD* gene, there are at present no laboratory or animal studies providing a theoretical basis for initiating treatment with this agent.

Allopurinol has been advocated by some as an agent which will actually stop progression of FJHN (a condition in which many families have been found to have *UMOD* mutations – see above) if started at an early age (McBride et al 1998). The English group, led by Drs Simmonds, McBride, and Cameron, has studied family members with FJHN for many years (Sebesta et al 1995). These authors have reported stable renal function over a number of years in individuals with the diagnosis of FJHN. They have begun allopurinol prior to the development of renal failure in a number of children and have documented compliance by the measurement of serum oxypurinol levels. Unfortunately, it is uncertain at this point if these families suffer from MCKD1, MCKD2 or perhaps another genetic cause of disease. Other authors have not found a stabilization of renal function in individuals with similar clinical manifestations (Miranda 1995, Bleyer & Hart 2003), with one group documenting an increase in serum creatinine in six patients receiving allopurinol compared to six patients who did not (Miranda 1995). It is unclear if allopurinol slows progression, but it does not appear to stop progression of renal damage in all studies.

Allopurinol clearly will prevent gout in individuals with MCKD2. Moreover, gout can progress in severity as renal function worsens, with the development of tophi and crippling arthritis. In our experience, allopurinol is of the most significant benefit in young men who are hyperuricemic. These individuals are at risk for recurring attacks and also at increased risk of progression of disease. We therefore recommend allopurinol in this setting. In young women, especially those with normal serum uric acid levels, the role of allopurinol is less clear. Patients with a slower rate of disease progression may not be willing to take a medication daily that is uncertain to slow the rate of progression of renal disease.

Benzbromarone

A recent report described the usage of benzbromarone together with allopurinol in two individuals with MCKD2 (Lhotta 2003). For these individuals, with baseline creatinine clearances of 17 ml/min and 58 ml/min, benzbromarone was given at a dosage of 100 mg/d together with allopurinol. Renal function stabilized in both individuals for a period of 68 months.

Angiotensin Converting Enzyme (ACE) Inhibitors

ACE inhibitors are a mainstay of treatment for chronic kidney disease (Levey 2002) but have not been studied in MCKD. Many patients with MCKD do not manifest hypertension, and if present, it may not be severe. Therefore, these individuals may be less likely to be started on ACE inhibitor therapy than other patients with chronic kidney failure.

ACE inhibitors may theoretically be beneficial in MCKD2 due to their ability to decrease production of uromodulin (Guidi et al 1998). A decrease in uromodulin production could theoretically decrease the build-up of abnormal uromodulin in tubular cells and prevent tubular cell death. However, the decreased production of normal uromodulin in individuals who already produce low levels of uromodulin could result in worsening hyperuricemia or other potential effects.

As with polycystic kidney disease, there is currently no specific treatment for MCKD. A primary goal of the nephrologist is to prevent the clinical sequelae of chronic renal failure and keep the patient as healthy as possible, so that they will be ideal transplant candidates when end-stage kidney failure develops.

Treatment of Hyperuricemia

Treatment of hyperuricemia and prevention of crippling or tophaceous gout is paramount. Most patients will tolerate the use of allopurinol and this is the agent of choice by most physicians. Unfortunately, some individuals may have gastric intolerance or allergic reactions. If these patients have significant gout, alternatives should be considered. If allergic reactions to allopurinol are problematic, referral to an allergist for desensitization may be useful (Fam et al 2001). Other possibilities include entering clinical trials with allopurinol metabolites such as oxypurinol (Freudenberger et al 2004). Probenecid or benzbromarone may also be used to treat hyperuricemia, and they have been shown to increase the fractional excretion of uric acid in MCKD2 patients (Lhotta et al 1998). As stated above, they may be potentially useful in the slowing of progression of kidney disease. Increased production of uric acid is not a mechanism causing hyperuricemia in MCKD, so the patients should not be at an increased risk of developing nephrolithiasis, though this should be monitored. Colchicine may be useful to prevent the development of gout attacks, but it will not prevent the further increase in the body burden of uric acid.

Treatment of Hypertension

When hypertension develops, aggressive treatment is indicated to prevent sequelae of hypertension and to hopefully slow progression of kidney disease. Aggressive treatments of lipids and other general health care will also be helpful.

Preparation for End-Stage Kidney Disease

Most patients with MCKD will be good kidney transplant candidates, as the disease is limited to the kidney. Transplantation is an effective cure for this condition, and MCKD will obviously not recur in the transplanted kidney. Since kidney disease progression is slow, patients should undergo transplant evaluation prior to the need for dialysis. Family members with MCKD2 can undergo *UMOD* mutational analysis to determine if they are appropriate kidney donors.

Hemodialysis and peritoneal dialysis are equally effective therapeutic options. Gout may persist in patients who have had severe gout prior to starting renal replacement therapy.

FUTURE HORIZONS

Significant progress has been made with the identification of the genetic cause of MCKD2, but there are still a significant number of other important areas of research. First, identification of the gene for MCKD1 is being pursued by several groups, with continued narrowing of the genetic interval (Wolf et al 2004). However, even with the identification of the genes for both major forms of MCKD, there are a number of obstacles to a definitive treatment.

First, development of an appropriate mouse model will be needed to better study pathophysiology and determine potential therapeutic treatments. Important issues include determination of factors which result in ER retention of abnormal uromodulin, identification of factors that alleviate ER retention, and determination of the effects of endoplasmic uromodulin retention. A mouse model that replicates clinical aspects of the human disease will be useful in identifying potential therapeutic interventions.

Secondly, identification of useful biomarkers will be important to develop surrogate endpoints for changes in glomerular filtration rate. Appropriate biomarkers can be used to evaluate the effect of interventions on affected individuals.

Finally, increased education of physicians and of patients with the disease is important. At present, there is no patient support group for MCKD, and this is vitally needed to improve patient understanding of the condition. The understanding of nephrologists regarding these diseases is also poor. Clarity in nomenclature will help to resolve this problem. Also, publications will increase physician and patient knowledge in this regard.

References

Auranen M, Ala-Mello S, Turunen JA, Jarvela I. Further evidence for linkage of autosomal-dominant medullary cystic kidney disease on chromosome 1q21. Kidney Int. 2001; 60: 1225–32.

Bachmann S, Mutig K, Bates J, et al. Renal effects of Tamm-Horsfall protein (uromodulin) deficiency in mice. Am. J. Physiol. Renal Physiol. 2005; 288: F559–67.

Bates JM, Raffi HM, Prasadan K, et al. Tamm-Horsfall protein knockout mice are more prone to urinary tract infection: rapid communication. Kidney Int. 2004; 65: 791–7.

Bichet DG, Fujiwara TM. The quest for the gene responsible for medullary cystic kidney disease type 1. Kidney Int. 2004; 66: 864–5.

Bingham C, Ellard S, van't Hoff WG, et al. Atypical familial juvenile hyperuricemic nephropathy associated with a hepatocyte nuclear factor-1beta gene mutation. Kidney Int. 2003; 63: 1645–51.

Bleyer AJ, Hart TC. Familial juvenile hyperuricaemic nephropathy. QJM 2003; 96: 867–8.

Bleyer AJ, Woodard AS, Shihabi Z, et al. Clinical characterization of a family with a mutation in the uromodulin (Tamm-Horsfall glycoprotein) gene. Kidney Int. 2003a; 64: 36–42.

Bleyer AJ, Trachtman H, Sandhu J, Gorry MC, Hart TC. Renal manifestations of a mutation in the uromodulin (Tamm Horsfall protein) gene. Am. J. Kidney Dis. 2003b; 42: 1–7.

Bleyer AJ, Hart TC, Shihabi Z, Robins V, Hoyer JR. Mutations in the uromodulin gene decrease urinary excretion of Tamm-Horsfall protein. Kidney Int. 2004; 66: 974–7.

Brown KM, Muchmore AV, Rosenstreich DL. Uromodulin, an immunosuppressive protein derived from pregnancy urine, is an inhibitor of interleukin 1. Proc. Natl Acad. Sci. 1986; 83: 9119–23.

Calado J, Gaspar A, Clemente C, Rueff J. A novel heterozygous missense mutation in the UMOD gene responsible for Familial Juvenile Hyperuricemic Nephropathy. BMC Med. Genet. 2005; 27(6): 5.

Chamberlin BC, Hagge WW, Stickler GB. Juvenile nephronophthisis and medullary cystic disease. Mayo Clin. Proc. 1977; 52: 485–91.

Christodoulou K, Tsingis M, Stavrou C, et al. Chromosome 1 localization of a gene for autosomal dominant medullary cystic kidney disease (ADMCKD). Hum. Mol. Med. Genet. 1998; 7: 905–11.

Dahan K, Fuchshuber A, Adamis S, et al. Familial juvenile hyperuricemic nephropathy and autosomal dominant medullary cystic kidney disease Type 2: Two facets of the same disease? J. Am. Soc. Nephrol. 2001; 12: 2348–57.

Dahan K, Devuyst O, Smaers M, et al. A cluster of mutations in the UMOD gene causes familial juvenile hyperuricemic nephropathy with abnormal expression of uromodulin. J. Am. Soc. Nephrol. 2003; 14: 2883–93.

Danovitch G. Uric acid transport in renal failure: a review. Nephron 1972; 9: 291–9.

Duncan H, Dixon SJ. Gout, familial hyperuricaemia and renal disease. Q. J. Med. 1960; 29: 127–35.

Fam A, Dunne S, Iazzetta J, Paton T. Efficacy and safety of desensitization to allopurinol following cutaneous reactions. Arthritis Rheum. 2001; 44: 231–8.

Fanconi VG, Hanhart E, Albertini AV, Uhlinger E, Dolivo G, Prader A. Die familiare juvenile nephronophthise. Helv. Paediatr. Acta, 1951; 6: 1–49.

Freudenberger RS, Schwarz RP Jr, Brown J, et al. Rationale, design and organisation of an efficacy and safety study of oxypurinol added to standard therapy in patients with NYHA class III–IV congestive heart failure. Expert Opin. Investig. Drugs 2004; 13: 1509–16.

Fuchshuber A, Kroiss S, Karle S, et al. Refinement of the gene locus for autosomal dominant medullary cystic kidney disease type 1 (MCKD1) and construction of a physical and partial transcriptional map of the region. Genomics 2001; 72: 278–84.

Goldman SH, Walker SR, Merigan TC, Gardner KD, Bull JM. Hereditary occurrence of cystic disease of the renal medulla. N. Engl. J. Med. 1966; 274: 984–92.

Guidi E, Giglioni A, Cozzi M, Minetti E. Which urinary proteins are decreased after angiotensin converting-enzyme inhibition? Renal Fail. 1998; 20: 243–8.

Gusmano R, Caridi G, Marini M, et al. Glomerulocystic kidney disease in a family. Nephrol. Dial. Transplant. 2002; 17: 813–18.

Hart TC, Gorry MC, Hart PS, et al. Mutations of the UMOD gene are responsible for medullary cystic kidney disease 2 and familial juvenile hyperuricaemic nephropathy. J. Med. Genet. 2002; 39: 882–92.

Hess B. Tamm-Horsfall glycoprotein and calcium nephrolithiasis. Miner. Electrolyte Metab. 2004; 20: 393–8.

Hildebrandt F, Omram H. New insights: Nephronophthisis-medullary cystic kidney disease. Pediatr. Nephrol. 2001; 16: 168–76.

Hoyer JR, Sisson SR, Vernier RL. Tamm-Horsfall glycoprotein ultrastructural immunoperoxidase localization in rat kidney. Lab. Invest. 1979; 41: 168–73.

Kiser RL, Wolf MT, Martin JL, et al. Medullary cystic kidney disease type 1 in a large Native-American kindred. Am. J. Kidney Dis. 2004; 44: 611–17.

Kroiss S, Huck K, Berthold S, et al. Evidence of further genetic heterogeneity in autosomal dominant medullary cystic kidney disease. Nephrol. Dial. Transplant. 2000; 15: 818–21.

Kudo E, Kamatani N, Tezuka O, et al. Familial juvenile hyperuricemic nephropathy: Detection of mutations in the uromodulin gene in five Japanese families. Kidney Int. 2004; 65: 1589–97.

Levey A. Clinical practice. Nondiabetic kidney disease. N. Engl. J. Med. 2002; 347: 1505–11.

Lhotta K, Gruber J, Sgonc R, Fend F, Konig P. Apoptosis of tubular epithelial cells in familial juvenile gouty nephropathy. Nephron 1998; 79: 340–4.

Lhotta K. Stopping progression in familial juvenile hyperuricemic nephropathy with benzbromarone?. Kidney Int. 2003; 64: 1920–1.

Malagolini N, Cavallone D, Serafini-Cessi F. Intracellular transport, cell-surface exposure and release of recombinant Tamm-Horsfall glycoprotein. Kidney Int. 1997; 52: 1340–50.

Massari PU, Hsu CH, Barnes RV, Fox IH, Gikas PW, Weller JM. Familial hyperuricemia and renal disease. Arch. Intern. Med. 1980; 140: 680–4.

McBride MB, Raman GV, Ogg CS, Chantler C, Cameron JS, Duley JA, Renal urate hypoexcretion preceding renal disease in a new kindred with familial juvenile gouty nephropathy (FJGN). In: Harkness. RA, et al, eds. Purine and Pyrimidine Metabolism in Man, Vol. VII. New York: Plenum Press, 1991: pp. 191–4.

McBride MB, Simmonds HA, Moro F. Familial renal disease or familial juvenile hyperuricaemic nephropathy. J. Inherit. Metab. Dis. 1997; 20: 351–3.

McBride MB, Simmonds HA, Ogg CS, et al. Efficacy of allopurinol in ameliorating the progressive renal disease in familial juvenile hyperuricaemic nephropathy (FJHN): A Six-Year Update. In: Griesmacher EA, ed. Purine and Pyrimidine Metabolism in Man,Vol. IX. New York: Plenum Press, 1998: pp. 7–11.

Miranda M. The influence of allopurinol on renal deterioration in familial nephropathy associated with hyperuricemia. Adv. Exp. Med. Biol. 1995; 370: 61–4.

Mo L, Huang H, Zhu X, Shapiro E, Hasty D, Xue-Ru W. Tamm-Horsfall protein is a critical renal defense factor protecting against calcium oxalate crystal formation. Kidney Int. 2004; 66: 1159–66.

Moro F, Ogg CS, Simmonds HA, et al. Familial juvenile gouty nephropathy with renal urate hypoexcretion preceding renal disease. Clin. Nephrol. 1991; 9: 263–9.

Neumann HP, Zauner I, Strahm B, et al. Late occurrence of cysts in autosomal dominant medullary cystic kidney disease. Nephrol. Dial. Transplant. 1997; 12: 1242–6.

Online Mendelian Inheritance in Man, OMIM (TM). Johns Hopkins University, Baltimore, MD. MIM Number: {256100}: {10/14/2004}: {602088}: {9/20/2004}:{604387}: {11/10/2003}:{606966}: {5/28/2003}. World Wide Web URL: http://www.ncbi.nlm.nih.gov/omim/

Orskov I, Ferencz A, Orskov F. Tamm-Horsfall protein or uromucoid is the normal urinary slime that traps type I fimbriated *Escherichia coli.* Lancet 1980; 1: 887.

Pennica D, Kohr WJ, Kuang WJ, et al. Identification of human uromodulin as the Tamm-Horsfall urinary glycoprotein. Science 1987; 236: 83–8.

Rampoldi L, Caridi G, Santon D, et al. Allelism of MCKD, FJHN and GCKD caused by impairment of uromodulin export dynamics. Hum. Mol. Genet. 2003; 12: 3369–84.

Rezende-Lima W, Parreira KS, Garcia-Gonzalez M, Riveira E, Banet JF, Lens XM. Homozygosity for uromodulin disorders: FJHN and MCKD-type 2. Kidney Int. 2004; 66: 558–63.

Roubenoff R. Gout and hyperuricemia. Rheum. Dis. Clin. North Am. 1990; 16: 539–50.

Rutishauser J, Spiess M. Endoplasmic reticulum storage diseases. Swiss Med. Wkly. 2002; 132: 211–22.

Sarasin F, Wong J, Levey A, Meyer K. Screening for acquired cystic kidney disease: a decision analytic perspective. Kidney Int. 1995; 48: 207–19.

Scolari F, Caridi G, Rampoldi L, et al. Uromodulin storage diseases: Clinical aspects and mechanisms. Am. J. Kidney Dis. 2004; 44: 987–99.

Sebesta I, Krijt J, Pavelka K, Maly J, Simmonds HA, McBride MB. Familial juvenile hyperuricaemic nephropathy in adolescents. In: Sahota A, Taylor M, eds. Purine and Pyrimidine Metabolism in Man, Vol. VIII. New York: Plenum Press, 1995: pp. 73–6.

Serafini-Cessi F, Malagolini N, Cavallone D. Tamm-Horsfall glycoprotein: Biology and clinical relevance. Am. J. Kidney Dis. 2003; 42: 658–76.

Sherblom AP, Decker JM, Muchmore AV. The lectin-like interaction between recombinant tumor necrosis factor and uromodulin. J. Biol. Chem. 1988; 263: 5418–24.

Stapleton FB, Linshaw MA, Hassanein K, Gruskin AB. Uric acid excretion in normal children. J. Pediatr. 1978; 92: 911–14.

Stavrou C, Koptides M, Tombazos C, et al. Autosomal-dominant medullary cystic kidney disease type 1: Clinical and molecular findings in six large Cypriot families. Kidney Int. 2002; 62: 1385–94.

Strauss MB. Clinical and pathological aspects of cystic disease of the renal medulla. Ann. Intern. Med. 1962; 57: 373–81.

Sujansky E, Kreutzer S, Johnson A, Lezotte D, Schrier R, Gabow P. Attitudes of at-risk and affected individuals regarding presymptomatic testing for autosomal dominant polycystic kidney disease. Am. J. Med. Genet. 1990; 35: 510–15.

Tamm I, Horsfall F. Characterization and separation of an inhibitor of viral hemagglutination present in urine. Proc. Soc. Exp. Biol. Med. 1950; 74: 108–14.

Thompson GR, Weiss JJ, Goldman RT, Rigg GA. Familial occurrence of hyperuricemia, gout, and medullary cystic disease. Arch. Intern. Med. 1978; 138: 1614–17.

Thorn G, Koepf G, Clinton M Jr. Renal failure simulating adrenocortical insufficiency. N. Engl. J. Med. 1944; 231: 76–85.

Thornley C, Dawnay A, Cattell WR. Human Tamm-Horsfall glycoprotein: urinary and plasma levels in normal subjects and patients with renal disease determined by a fully validated radioimmunoassay. Clin. Sci. 1985; 68: 529–35.

Tinschert S, Ruf N, Bernascone I, et al. Functional consequences of a novel uromodulin mutation in a family with familial juvenile hyperuricaemic nephropathy. Nephrol. Dial. Transplant. 2004; 19: 3150–4.

Turner JJ, Stacey JM, Harding B, et al. Uromodulin mutations cause familial juvenile hyperuricemic nephropathy. J. Clin. Endocrinol. Metab. 2003; 88: 1398–401.

US Renal Data System. Table H.16. Death Rates by Primary Cause of Death: Dialysis Patients, Unadjusted. In: U.S. Renal Data System, ed. USRDS 2003 Annual Data Report: Atlas of End-Stage Renal Disease in the United States. Bethesda, MD, 2003: p. 484.

Wilcox WD. Abnormal serum uric acid levels in children. J. Pediatr. 1996; 128: 731–41.

Wolf MT, Karle SM, Schwarz S, et al. Refinement of the critical region for MCKD 1 by detection of transcontinental haplotype sharing. Kidney Int. 2003a; 64: 788–92.

Wolf MT, Mucha BE, Attanasio M, et al. Mutations of the Uromodulin gene in MCKD type 2 patients cluster in exon 4, which encodes three EGF-like domains. Kidney Int. 2003b; 64: 1580–7.

Wolf MT, van Vlem B, Hennies HC, et al. Telomeric refinement of the MCKD1 locus on chromosome 1q21. Kidney Int. 2004; 66: 580–5.

Zbikowska HM, Soukhareva N, Behnam R, et al. The use of the uromodulin promoter to target production of recombinant proteins into urine of transgenic animals. Transgenic Res. 2002; 11: 425–35.

CHAPTER 27

Renal Dysgenesis

FANGMING LIN, VISHAL PATEL AND PETER IGARASHI

INTRODUCTION

Renal dysgenesis comprises a diverse group of disorders that are characterized by defects in the embryonic development of the kidney. Renal dysgenesis includes abnormalities in the size or position of the kidney (e.g. renal hypoplasia or ectopic kidney) and abnormalities in renal differentiation (renal dysplasia). Renal dysgenesis frequently occurs in concert with congenital abnormalities of the urinary tract (duplex ureters, ureteral obstruction at the ureteropelvic junction or ureterovesical junction, vesicoureteral reflux, ectopic ureter, bladder outflow obstruction and posterior urethral valves). Collectively, these disorders are referred to as congenital anomalies of the kidney and urinary track (CAKUT). The diseases that cause CAKUT carry a high morbidity and mortality and have significant financial impact. CAKUT is the most common cause of end-stage renal disease (ESRD) in children and is responsible for ~40% of cases of pediatric ESRD in the US and UK (McEnery et al 1992, Lewis & Shaw 2002, Smith et al 2007). This chapter will review genetic causes of renal dysgenesis in humans, focusing on monogenic disorders in which the disease genes have been identified and characterized.

RENAL DEVELOPMENTAL DEFECTS

Abnormalities in the Position or Size of the Kidney

Renal dysgenesis may present as abnormalities in the number, position, or size of the kidney. The complete absence of the kidney is referred to as renal agenesis and can be either unilateral or bilateral. Bilateral renal agenesis is incompatible with postnatal life, but unilateral renal agenesis is not uncommon. Renal agenesis is caused by the failure of nephrogenesis in utero. Since normal kidney development depends on the reciprocal interaction between the metanephric mesenchyme

and the ureteric bud (see below), defects in either of these embryonic anlage or their interaction can produce renal agenesis. Renal hypoplasia describes a condition in which the kidneys contain significantly fewer nephrons than normal and are small in size (below two standard deviations from the mean). In renal hypoplasia, the kidneys are small but the renal parenchyma is normal in appearance and does not show evidence of abnormal differentiation (dysplasia). When the number of nephrons is markedly reduced and the existing nephrons are hypertrophic, the condition is called oligomeganephronic renal hypoplasia. In addition to abnormalities in the size of the kidneys, abnormalities in the position of the kidneys may exist. The metanephric kidneys initially form in the pelvis and are oriented ventrally. During fetal development, the kidneys normally ascend in the abdomen and rotate medially until they reach their final destination in the retroperitoneum. Failure of normal migration produces ectopic kidneys. Occasionally, the kidneys may fuse forming a horseshoe kidney (Figure 27.1). These renal

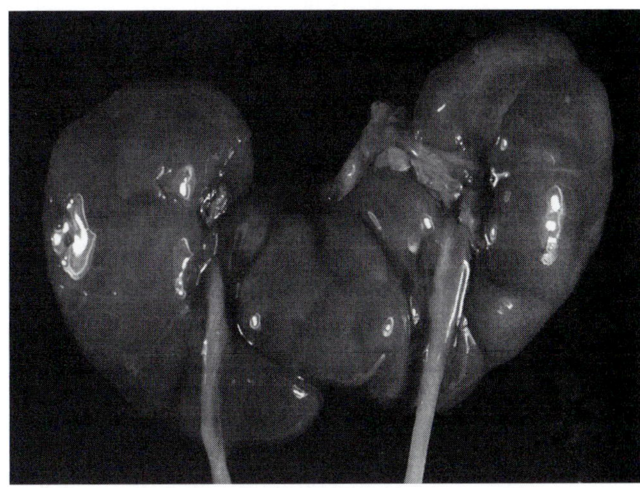

FIGURE 27.1 Horseshoe kidney. Horseshoe kidney formed by fusion of two kidney moieties, each drained by a separate ureter. Photo courtesy of Arthur Weinberg. (see also Plate 48)

Genetic Diseases of the Kidney
ISBN: 978-0-12-449851-8

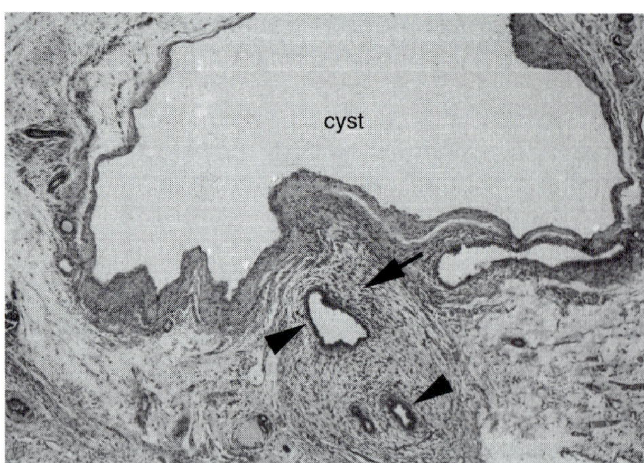

FIGURE 27.2   Renal dysplasia. Histology of a dysplastic kidney showing typical features including large cysts lined by flattened epithelium and primitive ducts (arrowheads) surrounded by fibromuscular collars (arrow). Photo courtesy of Arthur Weinberg. (see also Plate 49)

malformations generally do not cause renal failure and are often found incidentally during imaging studies. Most cases of renal malformations are sporadic. However, renal hypoplasia, renal agenesis, and horseshoe kidney are also seen in certain genetic disorders, e.g. renal-coloboma syndrome and MODY5 (Eccles & Schimmenti 1999, Bellanne-Chantelot et al 2004).

Renal Dysplasia

Renal dysplasia is defined as an abnormality in metanephric differentiation. Kidney size is variable, and normal renal tissue is often minimally present or absent. Classically, the histology of the dysplastic kidney shows primitive ducts surrounded by fibromuscular collars (Figure 27.2). Metaplastic cartilage, dysmorphic nerves and vessels, and erythropoietic cells may also be present (Potter 1972, Woolf et al 2004). Multiple kidney cysts often co-exist with dysplastic tissue, and this condition is called multicystic dysplastic kidney (MCDK). MCDK is the most frequent cause of an abdominal mass in newborns and is the most common cystic kidney disease during infancy (Figure 27.3). Most cases of MCDK are sporadic and unilateral, but bilateral MCDK can be seen in some genetic disorders (see below). MCDK is frequently associated with other urinary tract abnormalities, especially ipsilateral ureteral atresia. Malformations of the contralateral kidney and ureter, and especially vesicoureteral reflux, are also frequently seen. The pathogenesis of this abnormality may be related to intrauterine urinary tract obstruction, which in turn causes abnormal proliferation and differentiation of the metanephric mesenchyme. Multicystic dysplastic kidneys can enlarge over time and later involute producing an 'aplastic'

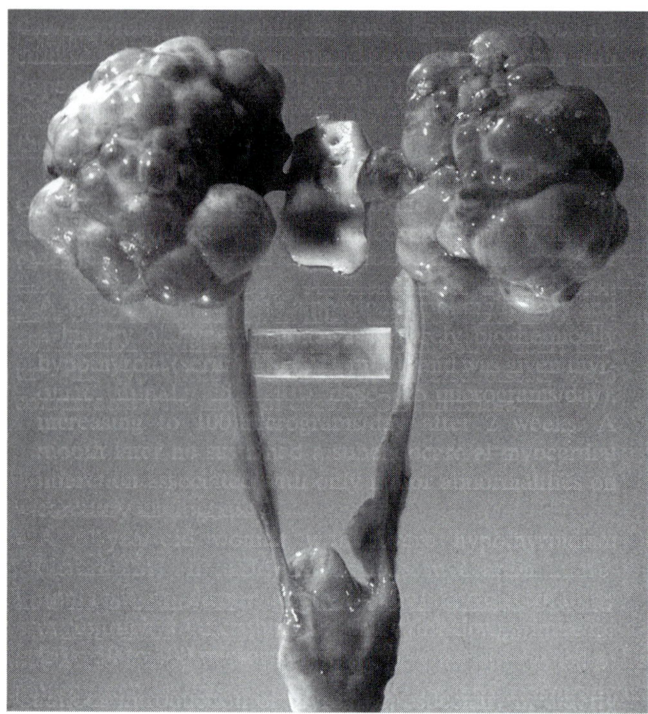

FIGURE 27.3   Bilateral multicystic dysplastic kidneys. Gross appearance of multicystic dysplastic kidneys showing numerous fluid-filled cysts. Photo courtesy of Arthur Weinberg. (see also Plate 50)

phenotype, suggesting that MCDK is a dynamic disorder (Winyard et al 1996, Woolf et al 2004).

Malformations of the Lower Urinary Tract

The development of the metanephric kidney depends on reciprocal inductive interactions between the metanephric mesenchyme, which gives rise to the nephrons proper, and the ureteric bud, which gives rise to the collecting ducts, renal pelvis, and ureter. Therefore, lower urinary tract malformations frequently co-exist with kidney developmental defects. The most common lower urinary tract malformation is primary vesicoureteral reflux (VUR). In this condition, the normal valve mechanism at the junction of and between ureter and bladder is impaired, which permits urine to flow retrograde from the bladder into the ureter, renal pelvis, and even the renal parenchyma. The prevalence of VUR may be as high as 1% in some groups (Dillon & Goonasekera 1998). VUR shows familial clustering and can be detected in 45% of the asymptomatic siblings of an affected child (Van den Abbeele et al 1987). Primary VUR is associated with an increased risk of urinary tract infection, renal scarring, and renal insufficiency. The decline in renal function associated with VUR is called reflux nephropathy.

Double or duplex ureters arise when two (or more) ureteric buds emerge from the Wolffian duct instead of the normal single bud. Each ureteric bud can induce metanephric development, producing either duplex kidneys or a single kidney with a double collecting system. Duplex ureters often produce urinary tract obstruction and reflux. Posterior urethral valves are exclusively seen in males. In this condition, obstruction of the lower urinary tract may lead to bilaterally dysplastic kidneys. Other common sites of congenital urinary tract obstruction are at the junction between the ureter and the renal pelvis (UPJ) and the junction between the ureter and bladder (UVJ).

Irrespective of the cause, intrauterine kidney failure may present clinically as Potter's syndrome (Potter 1972). Potter's syndrome (or Potter's sequence) may be seen in any condition where the fetal kidneys fail to produce sufficient urine. Since two-thirds of the amniotic fluid is normally produced by fetal urination in the second and third trimesters, intrauterine kidney failure results in oligohydramnios, a deficiency of amniotic fluid. Amniotic fluid is required for normal development of the lungs as well as to prevent compression of the fetus by the mother's uterus. The impairment in lung development results in pulmonary hypoplasia. Respiratory failure remains a significant cause of mortality in newborn infants with oligohydramnios. Azotemia does not occur until after birth, since the excretory function of the kidneys is performed by the placenta until parturition. Oligohydramnios produces a characteristic facies with low-set abnormal ears, prominent epicanthal folds, flattened nose, and micrognathia. Lower limb deformities include talipes equinovarus. The most common causes of Potter's syndrome are bilateral renal agenesis, bilateral severe renal dysplasia, polycystic kidney disease, and posterior urethral valves.

Gene Mutations Associated with CAKUT

As shown in Table 27.1, many known gene mutations are associated with renal dysgenesis. These disorders often have distinct extrarenal manifestations, which are also listed in the table. The actual frequency of renal dysgenesis caused by specific gene mutations is not known, since the renal malformations are often clinically silent. However, a recent study (ESCAPE Study) showed a high prevalence of HNF-1β (*TCF2*) mutations in European children with renal hypoplasia or dysplasia (Weber et al 2006).

Alterations in the renin-angiotensin system have also been associated with the development of CAKUT. Several studies have identified an association between polymorphisms of the angiotensin II type 2 receptor gene (*AT2R*) and an increased incidence of CAKUT. One study of Caucasian males found a significant association between CAKUT and a nucleotide transition (A1332G) within the first intron of the human *AT2R* gene (Nishimura et al 1999). This polymorphism perturbs the splicing of *AT2R* mRNA and results in decreased expression of *AT2R* (Nishimura et al 1999). Activation of the angiotensin II type 2 receptor stimulates apoptosis, which is important for the development of the metanephric mesenchyme and the branching of the ureteric

bud. The role of the angiotensin II type 2 receptor is further supported by studies of *At2r* mutant mice, which exhibit kidney and urogenital defects that are similar to those seen in humans with CAKUT (Nishimura et al 1999). An increased occurrence of the G allele of *AT2R* was also seen in 102 Italian children with CAKUT (Rigoli et al 2004). However, a study in a Japanese population failed to find an association between CAKUT and *AT2R* polymorphisms (Hiraoka et al 2001). Studies in Irish families with primary VUR also showed no clear association with the A to G transition of the *AT2R* gene (Yoneda et al 2002). Further studies will be needed to establish the roles of angiotensin and its receptors in human CAKUT.

KIDNEY DEVELOPMENT

Human Kidney Development

The normal development of the human kidney is discussed in greater detail elsewhere in this textbook (Chapter 23). Here, we provide a brief synopsis to assist with the discussion of abnormal kidney development. The development of the human kidney and urogenital tract are interrelated. The kidney and urogenital systems originate from three embryonic anlagen: the metanephric mesenchyme, the Wolffian duct (mesonephric duct), and the cloaca. The metanephric mesenchyme and Wolffian duct are derived from the intermediate mesenchyme (Saxen 1987), and the cloaca is derived from endoderm. The vertebrate kidney develops in three stages: pronephros, mesonephros, and metanephros. The pronephros and mesonephros are absent or transient structures in humans and have not been shown to have clear excretory functions, whereas the metanephros is the permanent kidney. At 4–5 weeks of gestation, the ureteric bud emerges as an evagination from the dorsal aspect of the Wolffian duct and invades the metanephric mesenchyme. Signals originating from the ureteric bud induce cells in the metanephric mesenchyme to proliferate and differentiate to form the nephrons. In a reciprocal manner, signals from the metanephric mesenchyme induce growth and branching of the ureteric bud (Saxen 1987). The ureteric bud gives rise to the collecting ducts, renal pelvis, ureter, and trigone of the urinary bladder. The metanephric mesenchyme undergoes a mesenchymal-to-epithelial transition to form the nephron proper, which includes the glomeruli, proximal tubules, loops of Henle, and distal convoluted tubules. Stereologic procedures have shown that 60% of the nephrons are formed in the third trimester, and the number of nephrons reaches a plateau at 36 weeks of gestation in normal pregnancy (Hinchliffe et al 1991). The number of glomeruli ranges from 227 327 to 1 825 380 per kidney, reflecting an 8-fold difference in the human population. Moreover, there is a direct correlation between the number of glomeruli and birth weight (Hughson et al 2003). Reduction of nephron

TABLE 27.1 Syndromes associated with renal dysgenesis

Syndrome	Mututed gene	Renal and urinary tract anomalies	Extra renal anomalies
Alagille syndrome*	*JAG1*	Renal dysplasia, and mesangiolipidosis	Paucity of intrahepatic bile duct, cardiac defect, facial and skeletal anomaly
Bardet-Biedl syndrome	*BBS1-14*	Renal dysplasia and calyceal malformations	Retinopathy, digit anomalies, obesity, diabetes, and male hypogonadism
Beckwith-Wiedemann syndrome	*p57KIP2* (cell cycle gene) mutation in a minority of patients	Large kidneys, cystic kidneys, and renal dysplasia	Widespread somatic overgrowth
Branchio-oto-renal syndrome	*EYA1*	Renal agenesis and dysplasia	Deafness and branchial arch defects
Campomelic dysplasia	*SOX9*	Multicystic dysplastic kidney	Skeletal malformations, and genital defects
Carnitine palmitoyltransferase II deficiency	*Carnitine* palmitoyltransferase II	Renal dysplasia and cyst formation	Lipid accumulation in multiple organs and metabolic defect
CHARGE syndrome*	*CHD7* and *SEMA3E* in most cases	Solitary kidney, hydronephrosis, renal hypoplasia, duplex kidneys, nephrolithiasis, ureteropelvic junction obstruction, and vesicoureteral reflux	Coloboma, heart malformation, choanal atresia, retardation, genital and ear anomalies
Denys-Drash syndrome*	*WT1*	Diffuse mesangial sclerosis and calyceal defects	Wilms' tumor, and pseudohermaphroditism, gonadoblastoma
Di George syndrome	Microdeletion at 22q11, probably several genes involved	Renal agenesis, dysplasia, and vesicoureteric reflux	Heart and branchial arch defects
Fraser syndrome	*FRAS1* mutation, putative cell adhesion molecule	Renal agenesis and dysplasia	Digital and ocular malformations
Frasier syndrome*	Altered ratio of splice isoforms of WT1 proteins	Focal segmental glomerulosclerosis	Intersex, gonadoblastoma (common), Wilms' tumor (rare)
Kallmann's syndrome 1	*KAL1* mutation (X-linked form)	Renal agenesis and dysplasia	Hypogonadotropic hypogonadism and anosmia
Meckel syndrome	*MKS1-6*	Cystic renal dysplasia	Central nervous system and digital malformations
Oral-Facial-Digital-Syndrome type 1*	*OFD1*	Glomerular cysts	Facial and digital anomalies
Renal-coloboma syndrome*	*PAX2*	Renal hypoplasia and vesicoureteric reflux	Coloboma of the optic nerve
Renal cyst and diabetes (RCAD)*	*HNF1β* (TCF2)	Renal dysplasia, cysts, and hypoplasia	Diabetes and genital tract anomalies
Simpson-Golabi-Behmel syndrome*	*GPC3*	Renal cysts, and dysplasia	Somatic overgrowth, craniofacial, limb and vertebral anomalies, cardiac defect and cryptorchidism
Smith-Lemli-Opitz syndrome	8 (7)-dehydrocholesterol reductase	Renal cysts, dysplasia and urinary tract obstruction and reflux	Failure to thrive, craniofacial and limb malformations
Townes-Brocks syndrome*	*SALL1*	Renal hypoplasia, dysplasia, agenesis, vesicoureteric reflux, and posterior urethral valve	Limb malformation, hearing loss and imperforate anus
VACTERL association	Genetic basis unknown apart from one report of mitochondrial	Renal dysplasia	Vertebral defects, anal atresia, tracheoesophageal fistula with esophageal atresia, and limb anomalies
WAGR syndrome	*WT1* and *PAX6* contiguous gene defect	Wilms tumor and renal cysts	Aniridia, genital malformations and mental retardation
Zellweger syndrome	Mutations of genes involved in peroxisome biogenesis	Cystic dysplastic kidneys	Malformations of the central nervous systems, skeleton, liver, and heart

Syndromes marked with * are discussed in this chapter

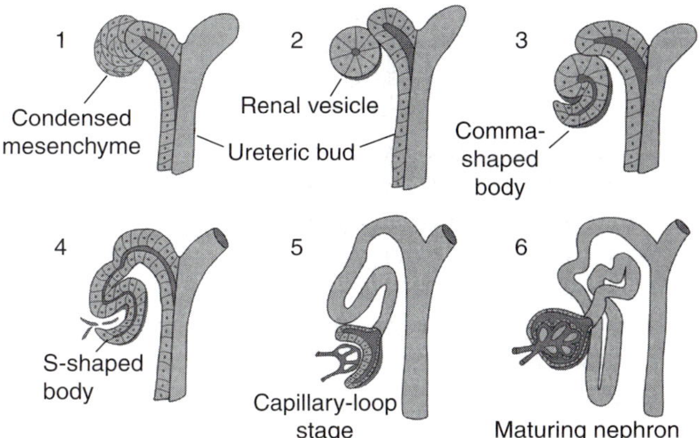

FIGURE 27.4 Stages of nephrogenesis. 1) Condensed mesenchyme. 2) Renal vesicle. 3) Comma-shaped body. 4) S-shaped body. 5) Capillary-loop stage. 6) Maturing nephron. (see also Plate 51)

number has been associated with increased risk of hypertension and renal functional insufficiency later in life (Keller et al 2003).

At about 4 weeks of gestation, the endoderm-derived cloaca is divided into the urogenital sinus and rectum by the urorectal fold. The caudal portion of the mesonephric duct opens into the bladder and forms the vesicoureteric orifice of the trigone. Between 5 and 6 weeks of gestation, the Müllerian duct appears. In male, the Müllerian duct subsequently regresses, whereas the Wolffian duct develops into the epididymis, vas deferens, seminal vesicles, and ejaculatory ducts. Conversely, in females the Wolffian duct regresses. The Müllerian duct develops into the ureterovaginal primordium, which merges with the urogenital sinus, and eventually gives rise to uterus, oviducts, and the proximal vagina.

The Mouse As a Useful Tool to Study Human Kidney Development

Animal models have proven to be indispensable tools for our understanding of the development of the kidney and urogenital tract in humans. Studies in rodent embryos have provided a detailed picture of the processes involved in kidney development. The process of nephrogenesis is similar in mice and humans, although the overall time frame of kidney development is much shorter in mice because gestation is 19 days. In mice, metanephric kidney development begins at embryonic day 11.5 (E11.5) when the ureteric bud emerges from the Wolffian duct and invades the metanephric mesenchyme. As depicted in Figure 27.4, mesenchymal cells in the metanephric mesenchyme condense around the tips (ampullae) of the ureteric bud. The condensed mesenchyme undergoes a mesenchymal-to-epithelial transition forming an epithelial structure called the renal vesicle. Invagination of the renal vesicle produces a comma-shaped body which further develops into an S-shaped body. The

S-shaped body consists of the glomerular anlage adjacent to the vascular cleft, a proximal limb that will give rise to the proximal tubule, and a distal limb that will give rise to the distal nephron. The distal end of the S-shaped body communicates with the tips of the branching ureteric bud producing a uriniferous tubule with a continuous lumen. Subsequent stages of nephrogenesis, termed the capillary-loop stage and maturing nephron, are characterized by further elongation and differentiation of the tubules as well as glomerular differentiation (Reeves et al 1978, Abrahamson 1991). The first functional nephrons form at E14.5 in mice and 9 weeks of gestation in humans (Saxen 1987).

Studies in mice have also revealed details about the development of the renal collecting system. After emerging from the Wolffian duct, the ureteric bud undergoes branching morphogenesis and differentiates into the collecting ducts, renal pelvis, ureter and trigone of the bladder. The ureteric bud initially branches several times until the most distal ampullae connect with the first induced nephrons. Ampullae that are not yet connected to nephrons continue to divide by lateral branching to induce the formation of more nephrons. Each nephron is transiently attached to an ampulla but then shifts its linkage to the connecting piece of the next-formed nephron, creating 'arcades' of nephrons. The first 6 to 10 generations of ureteric bud branches remodel to form the pelvis and calyces, whereas the final 6 to 9 generations form collecting ducts (Oliver 1968). The kidneys can be removed from mouse embryos and cultured in vitro for several days, which permits developmental processes to be monitored under controlled conditions. Studies using kidney organ culture have revealed that the ureteric bud undergoes a complex pattern of branching morphogenesis, including lateral branching, bifurcations, and trifurcations (Watanabe & Costantini 2004). Abnormalities in the location and number of the initial ureteric buds or their subsequent branching morphogenesis can result in renal agenesis, dysplasia, MCDK, and vesicoureteric reflux (al-Awqati & Goldberg 1998, Pohl et al 2002). Failure of canalization of the ureter that normally occurs at 8 weeks of gestation (Ruano-Gil et al 1975) causes obstruction of urine flow, a finding that is frequently associated with MCDK.

In addition to providing information about normal kidney development, mouse models have also provided valuable insights into the mechanisms of abnormal kidney development. Like humans, mice can also acquire spontaneous gene mutations that produce kidney dysgenesis or CAKUT. Identification of the genes by positional cloning has led to the discovery of molecules and pathways that are important for normal kidney development. A particularly powerful advantage of studies in mice compared to humans is that mice can be genetically manipulated. Transgenic mice can be produced in which wild-type or mutant genes of interest are expressed in specific tissues to determine their effects on embryonic development. Gene targeting by homologous recombination can be used to 'knock out' or inactivate a particular gene of interest and determine whether it plays an essential role in kidney development. Several examples of the use of these technologies will be provided later in this chapter.

Molecular Regulation of Kidney Development

The development of the metanephros involves reciprocal interactions between the cells in the metanephric mesenchyme and the ureteric bud (Birchmeier & Birchmeier 1993, Vainio and Muller 1997). Mesenchymal-to-epithelial transition of the metanephric mesenchyme requires the presence of the ureteric bud (Grobstein 1953, Grobstein 1953, Grobstein 1956, Saxen 1987). Conversely, branching morphogenesis of the ureteric bud is dependent on signals from the surrounding mesenchyme (Stuart et al 1995, Vainio & Muller 1997). Identification of the molecular signals that are required for kidney development has been the subject of intense investigation (Barasch et al 1999, Challen et al 2005, Schmidt-Ott et al 2005). These studies have shown that the induction of nephrons can be induced by factors, such as leukemia inhibitory factor and Wnt9b, which are produced by the ureteric bud and signal to the surrounding mesenchyme (Figure 27.5) (Barasch et al 1999; Carroll et al 2005). Conversely, glial cell line-derived neurotrophic factor (GDNF) produced in the metanephric mesenchyme is indispensable for ureteric bud branching morphogenesis (Moore et al 1996; Pichel et al 1996). GDNF interacts with a receptor complex consisting of the c-Ret receptor tyrosine kinase and GDNF family receptor α1 (GFRα1) expressed in the ureteric bud and the Wolffian duct.

Other molecules that are involved in renal organogenesis include extracellular matrix proteins, cell adhesion molecules, receptors, and growth factors whose expression is controlled by transcription factors. These molecules play a critical role in every step of kidney development: specification of nephrogenic mesenchyme from intermediate mesoderm, branching of the ureteric bud from the Wolffian duct, condensation of metanephric mesenchyme, mesenchymal–epithelial transition, proximal-distal patterning, glomerulogenesis, and vascular formation. An excellent review of the molecular basis of kidney development has been provided by Carroll and McMahon (2003). Mutations of several genes that are involved in kidney development have been identified in human syndromes associated with renal dysgenesis. In the following sections, we will discuss these syndromes focusing on the clinical features, molecular genetics, and insights into disease pathogenesis that have been gleaned from mouse models. Although many of the syndromes are relatively rare, together they have provided significant insights into the molecular mechanisms governing

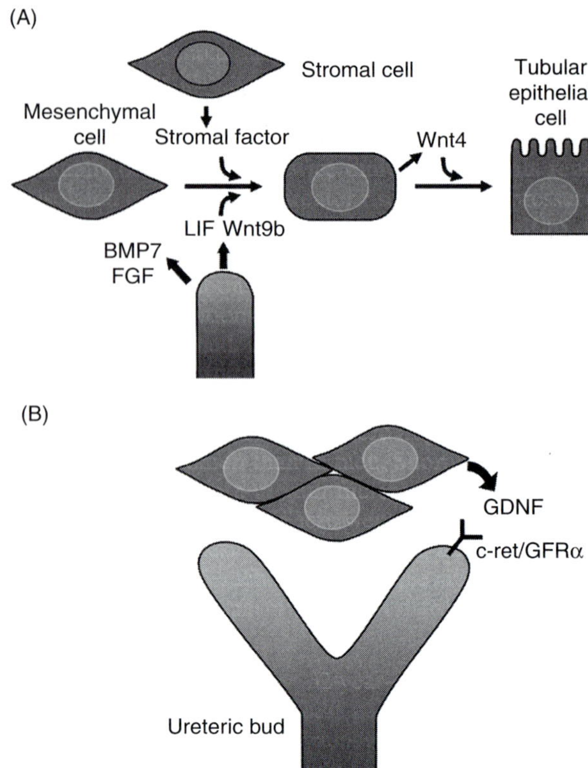

FIGURE 27.5 Molecular regulation of kidney development. (A) Nephron induction. The ureteric bud secretes leukemia inhibitory factor (LIF) and Wnt9b, which induce metanephric mesenchymal cells to differentiate into renal epithelial cells. Wnt4 and stromal factors are also required. BMP7 and FGF are required for cell survival and proliferation. (B) Branching morphogenesis of the ureteric bud. Metanephric mesenchymal cells secrete glial cell-derived neurotrophic factor (GDNF), which binds to its receptor, c-ret/GFRα, located on the ureteric bud. Deletion of either GDNF or c-ret prevents ureteric branching. (see also Plate 52)

kidney and urogenital tract development. Moreover, mutations of the same genes are sometimes also seen in sporadic cases of renal dysgenesis.

HUMAN SYNDROMES ASSOCIATED WITH RENAL DYSGENESIS

Renal-Coloboma Syndrome

CLINICAL FEATURES OF RENAL-COLOBOMA SYNDROME
Renal-coloboma syndrome (OMIM 120330) is an autosomal dominant disorder that affects the development of the kidneys and eyes. Affected individuals present with optic disc dysplasia and renal anomalies, especially renal hypoplasia. The ocular lesion is sometimes called the 'morning glory' anomaly because funduscopic examination shows

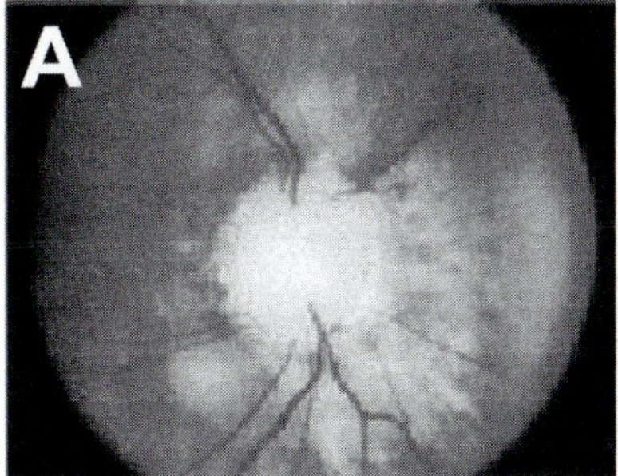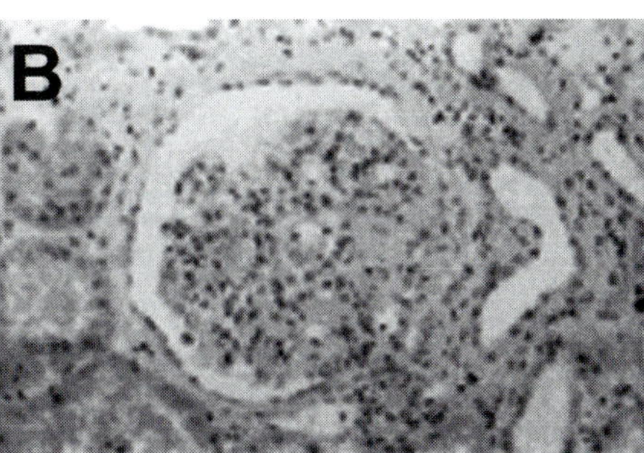

FIGURE 27.6 Optic nerve coloboma (A) and renal histology (B) of a patient with renal-coloboma syndrome. The optic nerve head contains a large excavation, and retinal vessels emerge from the periphery. The glomerulus shows mesangial sclerosis. Reproduced with permission from Schimmenti et al (1995). (see also Plate 53)

an unusually broad and cup-shaped appearance of the optic disc resembling the eponymous flower (Figure 27.6) (Karcher 1979). This anomaly was originally thought to be due to a colobomatous defect or a failure in the closure of the optic nerve, which led to the naming of the syndrome as renal-coloboma syndrome. Further studies have shown that the ocular defect may be due to optic disc dysplasia rather than coloboma, and therefore papillorenal syndrome has been suggested as an alternative designation (Bron et al 1989, Parsa et al 2001). It is also known as coloboma-ureteral-renal syndrome and optic nerve coloboma with renal disease.

In the typical renal-coloboma syndrome, patients exhibit visual field defects due to bilateral optic disc dysplasia and renal insufficiency due to renal hypoplasia and/or sclerosis. The ocular defect can range from a subtle optic abnormality to microphthalmia. Clinical variation is observed between family members who share the same mutation (Schimmenti et al 1997). Other clinical manifestations include proteinuria, bilateral vesicoureteral reflux, structural defects in the urogenital tract, high frequency hearing loss, central nervous system anomalies, and skin and joint laxity (Eccles & Schimmenti 1999).

The most frequent kidney abnormality is renal hypoplasia. Sixty-eight percent of affected individuals have reduced kidney size (Eccles & Schimmenti 1999). Cortical thinning with low numbers of glomeruli and collecting ducts in the cortex and hypoplastic papillae indicate renal hypoplasia (Devriendt et al 1998). Glomerulosclerosis, interstitial fibrosis, and tubular atrophy are not uncommon in renal biopsies (Figure 27.6) (Schimmenti et al 1997). The renal disease frequently progresses to end-stage renal disease, although minimal renal involvement has also been reported (Schimmenti et al 1995). Vesicoureteral reflux can be seen in 26% of patients and is a significant contributing factor to the development of renal failure because of reflux nephropathy. Reflux can occur early in utero, and the associated recurrent urinary tract infection leads to renal scarring and a further decline in renal function. Glomerulosclerosis, interstitial fibrosis, and tubular atrophy are also pathological features.

MUTATIONS OF *PAX2* ARE RESPONSIBLE FOR RENAL-COLOBOMA SYNDROME

The link between renal-coloboma syndrome and mutations of *PAX2* was discovered in 1995 (Schimmenti et al 1995). A deletion of nucleotide 1104 in the *PAX2* gene was found in a father and three of his five sons with optic nerve colobomas, high frequency hearing loss, vesicoureteral reflux, small dysplastic kidneys, and renal failure. The deletion causes a frameshift that results in a premature stop codon and truncation of the protein. The pattern of abnormalities in renal-coloboma syndrome parallels the expression pattern of *PAX2*, which is expressed in the developing eye, ear, kidney, ureteric bud, and central nervous system.

The Human PAX2 Allelic Variant Database (http://pax2.hgu.mrc.ac.uk/) lists the 21 different *PAX2* mutations that have been identified to date in individuals with renal-coloboma syndrome. Mutations in exon 2 account for 50% of the cases. The mutations include deletions, insertions with duplications, translocations, and frameshift and missense mutations. There are no clear genotype-phenotype correlations (Schimmenti et al 1997). Wide inter- and intra-familial phenotypic variability has been reported, even between individuals with the same mutation (Schimmenti et al 1997, Amiel et al 2000).

PAX2 AND ITS ROLE IN DEVELOPMENT

The *PAX* gene family has nine members in humans (*PAX1–PAX9*) (Lang et al 2007). The *PAX* genes encode transcription factors that contain a highly conserved 128-amino acid domain called the 'paired box.' The paired box functions as a DNA-binding domain that mediates the binding of PAX proteins to a specific DNA sequence located in the promoters of genes that are important in embryonic development. Human *PAX2* has been localized to chromosome 10q25 (Stapleton et al 1993). The complete genomic sequence of *PAX2* was obtained in 1996 and comprises 12 exons spanning ~70 kb (Sanyanusin et al 1996).

The *PAX* genes are highly evolutionarily conserved, and orthologs have been identified in the mouse genome. Mouse *Pax2* is expressed in the intermediate mesoderm at the 4–6 somite stage and can be detected in the nephric duct at the

12 somite stage (Bouchard et al 2002). *Pax2* continues to be expressed in the pronephric duct, mesonephric tubules, ureteric bud, and metanephric mesenchyme. High levels of expression are observed in the condensing mesenchyme surrounding the tip of the ureteric bud (Dressler et al 1993). As induced mesenchyme undergoes epithelialization, the expression of *Pax2* is down-regulated. *Pax2* is not detectable in mature proximal and distal tubules. In contrast, the renal collecting ducts, which are derived from the ureteric bud, continue to express *Pax2* throughout adulthood (Dressler & Woolf 1999). Consistent with the sites that are affected in renal-coloboma syndrome, *PAX2* is expressed in the retina and its anlage, the optic vesicle. *PAX2* is also expressed in the inner ear, adrenal glands, spinal cord, and hindbrain (Tellier et al 2000).

Studies in mice indicate that both loss-of-function and gain-of-function mutations of Pax2 produce kidney deformities. *Pax2* homozygous null mice completely lack kidneys, ureters, and genital tracts (Torres et al 1995). Renal agenesis arises from the failure of the ureteric bud to grow and contact the metanephric mesenchyme, which precludes any metanephric induction. In addition, the mutant metanephric mesenchyme contains an intrinsic defect in nephrogenesis, since it is unresponsive to induction by the wild-type ureteric bud. The Wolffian and Müllerian ducts develop only partially and degenerate during embryogenesis. Heterozygous *Pax2*$^{+/-}$ mice develop small kidneys, which is similar to the phenotype of humans with renal-coloboma syndrome, although the magnitude of renal hypoplasia varies (Torres et al 1995). Heterozygous mutant mice also develop optic nerve colobomas, which demonstrates that the renal-coloboma syndrome arises from haploinsufficiency of *PAX2* and explains the autosomal dominant pattern of inheritance. The renal hypoplasia appears to be due to defects in the growth and branching of the ureteric bud as well as defective epithelial transformation of the metanephric mesenchyme. The nephric duct forms normally in *Pax2* null mice, most likely as a result of the redundant expression of *Pax8*, a closely related gene that is also expressed in the intermediate mesoderm. *Pax8* mutant mice have normal kidneys, but *Pax2*$^{+/-}$*Pax8*$^{+/-}$ compound heterozygotes contain kidneys that are considerably smaller than individual *Pax2*$^{+/-}$ heterozygotes (Figure 27.7) (Bouchard et al 2002). These findings indicate that *Pax8* interacts genetically with *Pax2*.

Overexpression of *Pax2* in transgenic mice also produces kidney abnormalities, which manifest as renal insufficiency and severe proteinuria (Dressler et al 1993). Histological examination of the transgenic kidneys shows multifocal cystic dilatation of the renal tubules and atrophic glomeruli. These results suggest that repression of *Pax2* is required for normal kidney development, and persistent expression of *Pax2* may interfere with differentiation of tubular epithelial cells. These studies further underscore the critical importance of *Pax2* gene dosage for normal kidney development.

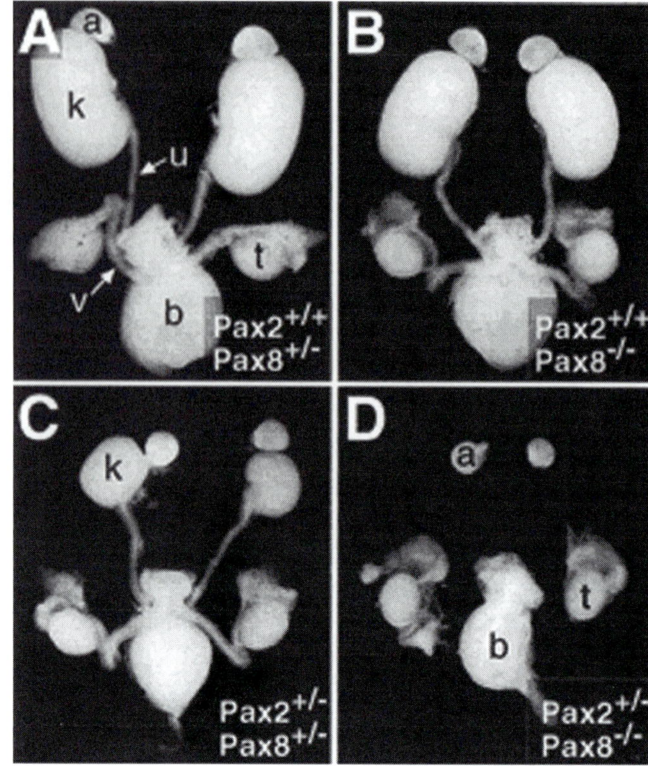

FIGURE 27.7 Renal hypoplasia in Pax2/Pax8 mutant mice. The kidneys in *Pax2*$^{+/-}$*Pax8*$^{+/-}$ embryos (C) are smaller than control kidneys (A, B). *Pax2*$^{+/-}$*Pax8*$^{-/-}$ embryos (D) fail to develop a kidney, ureter, and genital tract, whereas the adrenal gland, testis, and bladder form normally. a, adrenal gland; b, bladder; k, kidney; t, testis; u, ureter; v, vas deferens. Reproduced with permission from Bouchard et al (2002). (see also Plate 54)

MOLECULAR PATHOGENESIS OF RENAL-COLOBOMA SYNDROME

Pax2 is a sequence-specific DNA-binding protein that stimulates or inhibits the transcription of downstream target genes. Therefore, it occupies a key position in a regulatory network controlling kidney development. Antisense oligonucleotides that specifically decrease the abundance of Pax2 in kidney organ culture block nephron formation in the metanephric mesenchyme. This experiment suggests that Pax2 is required for mesenchyme-to-epithelium conversion (Rothenpieler & Dressler 1993). Pax2 has been shown to regulate the expression of several genes that are essential for normal kidney development. Because increased expression of Pax2 is followed by increased levels of *Wt1* expression, Dehbi et al. investigated the trans-activating activity of Pax2 on the *Wt1* promoter. They identified a Pax2 responsive element in the *Wt1* promoter and showed that PAX2 could stimulate expression of the endogenous *Wt1* gene, suggesting that PAX2 induces *Wt1* expression during the mesenchyme-to-epithelium transition in renal development (Dehbi et al 1996). In contrast, Pax2 appears to be repressed by *Wt1* during kidney development (Ryan et al 1995). Pax2 also regulates the expression of Wnt4, a secreted glycoprotein that is required for the formation of renal vesicles (Figure 27.5) (Torban et al 2006). As discussed above, branching morphogenesis of the ureteric bud is dependent on GDNF, which is secreted by the metanephric mesenchyme and binds to the c-Ret/GFRα1 receptor that is expressed on the ureteric bud (Figure 27.5). Pax2 regulates the transcription of the genes encoding both GDNF and c-ret, which explains the severe branching defects that are seen in *Pax2* mutants (Brophy et al 2001, Clarke et al 2006).

Another downstream target of *Pax2* is the tumor suppressor *P53*. The first exon of the *P53* gene contains a *Pax2* binding site, and *Pax2* inhibits *P53* promoter activity in transiently transfected rat embryo fibroblasts (Stuart et al 1995). These results suggest that *Pax2* may be involved in the regulation of p53-dependent apoptosis. Excessive apoptosis has been observed in the hypoplastic kidneys of *Pax2* heterozygous mutant mice (Porteous et al 2000). Moreover, inhibition of apoptosis, either by transgenic expression of *Bcl-2* or administration of a caspase inhibitor, corrects the abnormalities in kidney development (Clark et al 2004, Dziarmaga et al 2006). These results provide compelling evidence for a role of dysregulated apoptosis in the pathogenesis of renal-coloboma syndrome and suggest a possible therapeutic approach.

Denys-Drash Syndrome and Frasier Syndrome

CLINICAL FEATURES OF DENYS-DRASH SYNDROME

Denys-Drash syndrome (OMIM 194080) is an autosomal dominant disorder that is characterized by the triad of diffuse mesangial sclerosis, male pseudohermaphroditism, and Wilms' tumor. This syndrome was described in 1967 by Denys and in 1970 by Drash (Denys et al 1967, Drash et al 1970). The true incidence of Denys-Drash syndrome is not known, but it is a rare disorder with fewer than 200 cases reported since 1967 (Mueller 1994).

Denys-Drash syndrome usually manifests in early infancy with proteinuria leading to nephrotic syndrome. Edema and proteinuria can present between 2 weeks and 10 months of age. Renal function declines rapidly, and most patients develop end-stage renal disease before 3 years of age (Eddy & Mauer 1985, Jadresic et al 1990, Schumacher et al 1998). Complications related to nephrotic syndrome, such as recurrent infections, often contribute to morbidity before patients develop renal failure. Hypertension is common and can become severe as renal function declines. Kidney histology typically shows varying degrees of diffuse mesangial sclerosis. Early in the course, deposition of fibrillary material in the cytoplasm leads to mesangial cell expansion, obliteration of the capillary lumens, and thickening of the glomerular basement membrane. Hypertrophy of the podocytes can be prominent. As the disease progresses, histopathology shows mesangial sclerosis, glomerular tuft contraction, tubular atrophy and interstitial fibrosis reflecting end-stage renal disease (Mueller 1994).

In males with Denys-Drash syndrome, the external genitalia are usually ambiguous, although some individuals with dysgenic gonads have a female phenotype. Other anomalies of the external genitalia include hypospadia with cryptorchidism, enlarged clitoris with labial fusion, bifid scrotum and micropenis. The dysgenic gonads are always intra-abdominal and have a 20–30% risk of developing gonadoblastoma. The risk of developing Wilms' tumor is 55% in patients with male pseudohermaphrodism and glomerulopathy (Eddy & Mauer 1985). Unilateral or bilateral Wilms' tumor often presents as an abdominal mass. Tumors can develop in any residual renal tissue.

CLINICAL FEATURES OF FRASIER SYNDROME

Frasier syndrome is an autosomal dominant disorder that is closely related to Denys-Drash syndrome and consists of intersex disorders and nephropathy (OMIM 136680). However, unlike the diffuse mesangial sclerosis seen in Denys-Drash syndrome, the renal histology in Frasier syndrome shows focal segmental glomerulosclerosis (Koziell et al 2000). Proteinuria can occur early in life, and renal function typically deteriorates leading to renal failure by late childhood. The development of Wilms' tumor is very unusual in Frasier syndrome. However, gonadoblastomas are more common than in Denys-Drash syndrome.

WT1 gene mutations are associated with male pseudohermaphrodism in Denys-Drash syndrome and intersex in Frasier syndrome. Affected XY males with Frasier syndrome may appear to have a normal female phenotype including normal development of pubic hair and breast tissues. Therefore, karyo-type analysis should be included in the diagnostic evaluation. Due to the high incidence of gonadoblastoma, especially in Frasier syndrome, dysgenic gonads should be removed. The nephrotic syndrome associated with Denys-Drash and Frasier syndromes does not respond to conventional treatments, such as corticosteroids. Once patients reach ESRD, dialysis and transplantation are the only treatment options. No recurrence of nephrotic syndrome has been reported in either Denys-Drash or Frasier syndrome after kidney transplantation.

MUTATIONS OF WT1 CAUSE WILMS' TUMOR, DENYS-DRASH SYNDROME, AND FRASIER SYNDROME

Both Denys-Drash syndrome and Frasier syndrome are caused by heterozygous mutations of *WT1*. *WT1* is located on human chromosome 11p13 and contains 10 exons spanning ~50kb of genomic DNA. Exons 1–6 encode an amino-terminal proline/glutamine-rich transcriptional regulatory region, and exons 7–10 encode four C2H2 zinc

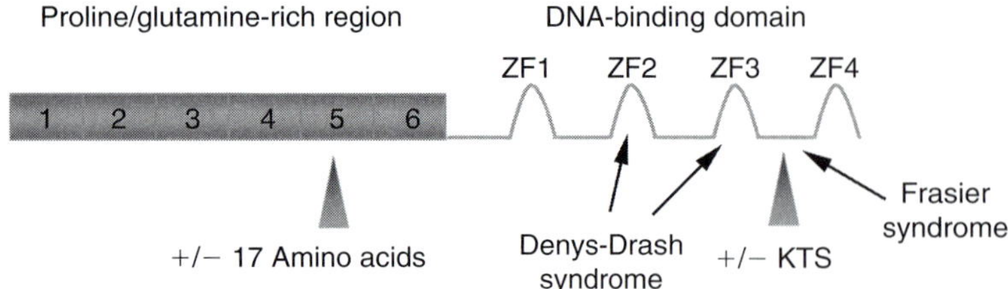

FIGURE 27.8 Mutations of WT1 in Denys-Drash and Frasier syndromes. Structure of WT1 is depicted showing the amino-terminal proline-glutamine-rich region and four zinc finger (ZF) DNA-binding domains. Locations of alternative splicing are indicated by the arrowheads. Arrows indicate the locations of mutations causing Denys-Drash syndrome and Frasier syndrome. (see also Plate 55)

fingers that mediate binding to DNA, RNA, and proteins (Figure 27.8) (Caricasole et al 1996, Kennedy et al 1996, Rae et al 2004). *WT1* is a classic tumor suppressor, and loss of its function is associated with the development of Wilms' tumor. Wilms' tumor is the most common solid malignancy in childhood and originates from metanephric mesenchymal cells that fail to differentiate. Evidence for the role of WT1 as a tumor suppressor came from the observation of germline deletions of chromosome 11p, which includes the *WT1* locus, in familial Wilms' tumors (Gessler et al 1990, Varanasi et al 1994). Analysis of the tumors reveals loss of heterozygosity of *WT1* due to inactivation of the remaining wild-type allele, a genetic signature of a tumor suppressor (Haber et al 1990). Mutations of *WT1* can also be detected in 10–15% of sporadic Wilms' tumors (Gessler et al 1994, Varanasi et al 1994) and are also seen in acute myeloid leukemia (Pritchard-Jones et al 1994, King-Underwood et al 1996) and mesothelioma (Park et al 1993). WT1 has been shown to inhibit the transcription of growth factor genes, such as insulin-like growth factor-2 (IGF2), further supporting its role in growth suppression (Drummond et al 1992).

WT1 undergoes complex post-transcriptional processing, including alternative splicing and RNA editing to produce multiple isoforms from a single gene. The isoforms have distinct biological functions. Alternative splicing generates four protein isoforms of molecular weight 52–54 kDa that contain the four zinc finger motifs (Morris et al 1991). Alternative splicing of exon 5 results in the inclusion or exclusion of 17 amino acids. Of particular interest is alternative splicing at exon 9, which results in the inclusion or exclusion of 3 amino acids: lysine, threonine, and serine (KTS) located between the third and fourth zinc fingers. Under normal conditions, the ratio of +KTS/−KTS isoforms is about 2:1, and the proper function of *WT1* gene products is dependent on the conserved ratio of the isoforms (Pelletier et al 1991). The alternative splice sites are highly conserved in vertebrates, and the ratio of the +KTS/−KTS isoforms does not change throughout kidney development (Kent et al 1995, Renshaw et al 1997, Hammes et al 2001). In addition to alternative splicing, RNA editing (Sharma et al 1994) and alternative initiation codons increase the potential number of WT1 proteins to at least 24 isoforms (Bruening & Pelletier 1996, Scharnhorst et al 1999, Hammes et al 2001).

Both Denys-Drash syndrome and Frasier syndrome are caused by mutations of *WT1*, but the mutations are distinct (Figure 27.8). More than 90% of the cases of Denys-Drash syndrome are caused by mutations located in the zinc finger region of WT1 (Little & Wells 1997). The mutations include missense mutations and deletions or insertions that result in a frameshift and truncation of the encoded protein. Mutations in the zinc finger region of WT1 prevent DNA binding. The most common mutation, observed in 40% of cases, is a missense mutation that results in the substitution of a tryptophan for arginine at codon 394 (R394W) located in the third zinc finger. Mutations that cause the deletion of the third and fourth zinc fingers are also found, and mouse studies (see below) suggest that these mutations may have a dominant-negative effect. In contrast, Frasier syndrome is caused by mutations of the second splice donor site in intron 9, which eliminates the production of the +KTS splice variant of *WT1* (Hammes et al 2001).

WT1 AND ITS ROLES IN DEVELOPMENT

During embryonic development, the expression of *WT1* is confined to the urogenital tract, mesothelial organs, and a subset of hematopoietic progenitor cells (Armstrong et al 1993, Baird & Simmons, 1997, Maurer et al 1997, Moore et al 1998, 1999). *WT1* transcripts are first detected in the intermediate mesoderm at E9.0 in mice. At E11.0, *WT1* is detected in uninduced metanephric mesenchyme, and its expression persists in induced mesenchyme at E12.5. The level of expression increases greatly in the comma- and S-shaped bodies and in the developing glomeruli (Armstrong et al 1993). WT1 expression is then down-regulated in mature tubular epithelial cells. In the adult kidney, WT1 is only expressed in the podocytes of the glomerulus (Pritchard-Jones et al 1990, Mundlos et al 1993). In humans, WT1 protein can be detected in the presumptive podocytes in the mesonephros and metanephros at 7–10 weeks of gestation. Expression continues in mature podocytes in postnatal kidneys. In addition, a low level of WT1 mRNA is also present in the condensed mesenchyme. No expression of WT1 protein or mRNA has been detected in the ureteric bud or its derivatives (Mundlos et al 1993). During gonadal development, WT1 is expressed in the sex cords of the genital ridge and in the mesenchymal compartments of the developing male and female gonads (Pritchard-Jones et al 1990,

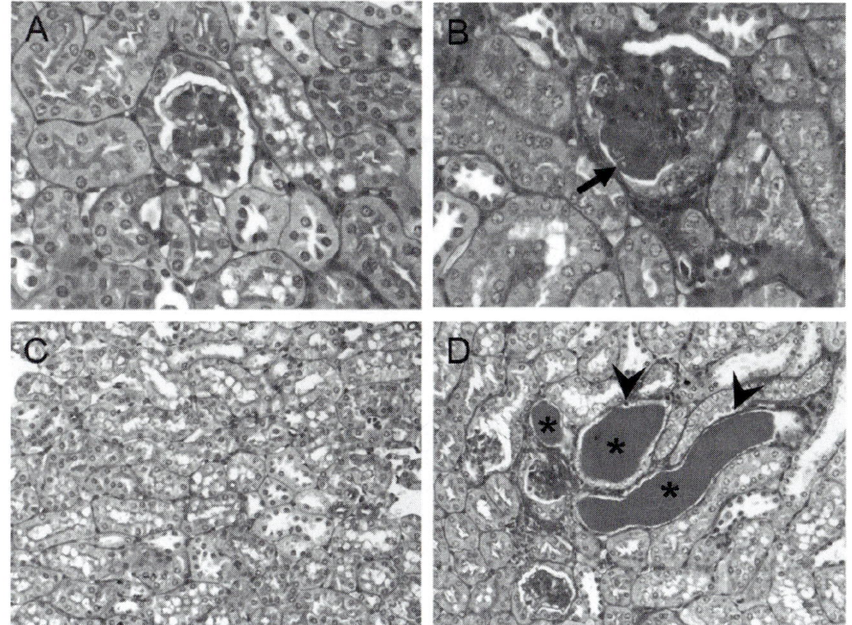

FIGURE 27.9 Mouse model of Denys-Drash syndrome. Normal glomerulus in wild-type mice (A) and glomerulosclerosis (arrow) in $Wt1^{+/R394W}$ mice (B). Renal tubules are dilated (arrowheads) and contain proteinaceous casts (asterisks) in $Wt1^{+/R394W}$ mice (D) compared to wild-type mice (C). Kidneys are stained with periodic acid and Schiff's (PAS). Photo courtesy of Vicki Huff and Hao Zhang. (see also Plate 56)

Pelletier et al 1991, Mundlos et al 1993). In adult males, expression of WT1 is confined to Sertoli cells of the testis. In females, WR1 is expressed in granulosa cells of the ovary as well as the oviduct and endometrium (Pelletier et al 1991).

WT1 gene products are essential for the normal development of the kidney and urogenital systems in humans and mice (Bruening et al 1992, Kreidberg et al 1993, Discenza & Pelletier 2004). Mice in which *Wt1* has been inactivated lack kidneys and gonads and die in utero. The ureteric bud fails to grow from the Wolffian duct at E11.0, and consequently cells within the metanephric blastema undergo apoptosis. In addition, *Wt1* mutant embryos have abnormal development of the mesothelium, heart, and lungs. (Kreidberg et al 1993). $Wt1^{+/-}$ mice develop late onset proteinuria and elevated serum BUN and creatinine. Proteinuria correlates with the thickening of glomerular basement membrane and fusion of the podocytes' foot processes. $Wt1^{+/-}$ mice subsequently develop diffuse glomerulosclerosis and loss of podocyte markers, nephrin, CD2AP, and α-actinin-4 (Menke et al 2003). These results suggest that continuous expression of Wt1 in podocytes is required for the maintenance of glomerular function in postnatal kidneys.

MOUSE MODELS OF DENYS-DRASH AND FRASIER SYNDROMES

Several mouse models of Denys-Drash syndrome have been produced. In one model, a truncation mutant of Wt1 (R362X) was expressed in transgenic mice (Natoli et al 2002). This mutant lacks the last two zinc fingers of Wt1

and affects the expression of growth factors known to regulate glomerular capillary development. The transgenic mice show abnormal capillary development leading to glomerular disease. Because the R362X mutant was expressed in mice carrying two wild-type *Wt1* alleles, the occurrence of disease suggests that the mutant functioned as a dominant-negative by interfering with the normal function of *Wt1*. Other mouse models of Denys-Drash syndrome have been produced by 'knocking-in' disease-causing mutations into the *Wt1* locus (Patek et al 1999, Gao et al 2004). The mice develop mesangial sclerosis, proteinuria and early-onset of renal failure characteristic of human Denys-Drash syndrome (Figure 27.9).

A mouse model of Frasier syndrome has been produced by altering the ratio of +KTS/−KTS isoforms. A mutation of the second splice donor site in intron 9 was introduced into the mouse genome using gene targeting methods (Hammes et al 2001). The mutation is identical to one found in humans with Frasier syndrome and prevents the expression of the +KTS isoform. Heterozygous mutant mice are born normally but develop proteinuria within 2 months and die from renal failure. Kidney histology shows severe mesangial sclerosis and focal segmental glomerular sclerosis with dilated tubules containing proteinaceous casts, which is similar to the histology of humans with Frasier syndrome. Although heterozygous mutant mice do not exhibit genital tract abnormalities, homozygous mutants that completely lack the +KTS isoform have complete XY sex reversal. Homozygous mutant mice also have more severe kidney disease and die within 24h after birth. Kidney size is reduced, and the bladder is empty indicating absent urine formation. Histological examination reveals increased stromal compartment, decreased nephrogenic zone, and decreased size of glomerular tufts. These findings highlight the importance of tight regulation of the ratio of the $+KTS/-KTS$ isoforms of WT1.

FUNCTIONS OF WT1

The proteins that are encoded by *WT1* are multifunctional. Different isoforms have distinct subcellular localizations and functions (Wagner et al 2003). WT1 proteins may act as transcriptional repressors or activators depending on the promoter, cell type, and cell cycle stage. There is also evidence showing that WT1 can bind to RNA and alter the expression of genes at the post-transcriptional level (Bardeesy & Pelletier 1998). In vitro transfection assays, microarray analysis, and subtractive hybridization have been used to identify downstream genes that are regulated by WT1. WT1 target genes include growth factors, growth factor receptors, transcription factors, extracellular matrix/secreted proteins, and others (Little et al 1999, Scharnhorst et al 2001, Rae et al 2004). Some WT1 target genes that are pertinent to kidney development are epidermal growth factor receptor (EGFR) (Englert et al 1995), the EGF family member *amphiregulin* (Lee et al 1999), Pax2 (Ryan et al 1995), E-cadherin (Hosono et al 1999) and Bcl-2 (Mayo et al 1999).

WT1 directly regulates the promoters of the *Podx1* and *Nphs1* genes, which encode the podocyte-specific proteins podocalyxin and nephrin, respectively (Palmer et al 2001, Guo et al 2004, Wagner et al 2004). Decreased expression of these genes and other podocyte-specific genes is found in Wt1 mutant mice and may form the basis for proteinuria in Denys-Drash and Frasier syndromes. Wt1 also regulates the expression of Dax-1 and Sry, which are essential for normal sexual differentiation and proper development of the genital tract (Kim et al 1999, Hossain & Saunders 2001). In the mouse model of Frasier syndrome described above, the male sex reversal appears to be due to reduced expression of Sry. The interaction between Wt1 and Pax2 is particularly intriguing, since mutations of Pax2 also produce renal dysgenesis (see preceding section). Wt1 inhibits Pax2 transcription, and overexpression of Pax2 has been observed in podocytes from Denys-Drash patients (Ryan et al 1995).

Townes-Brocks syndrome

CLINICAL FEATURES OF TOWNES-BROCKS SYNDROME

Townes-Brocks syndrome (TBS, OMIM 107480) is characterized by renal anomalies (mainly hypoplastic kidneys), deformity of the external ear, deafness, imperforate anus, and limb malformations. TBS is a rare autosomal dominant disorder with an estimated frequency of 1 in 250000 live births (Martinez-Frias et al 1999). In 1972, Townes and Brocks reported a father and five of his seven offspring who

had imperforate anus, triphalangeal thumbs, fusion of metatarsals, mild sensorineural deafness, and lop ears (Townes & Brocks 1972). The most common limb malformations are triphalangeal thumb and preaxial polydactyly with either a well-formed or vestigial digit. Bifurcation, ulnar deviation, or broad appearances of the distal phalanx of the thumb are also common. Toe anomalies, such as short or absent third toe, syndactyly of the third and fourth toes, and overlapping toes occur less frequently. Hearing loss is common, ranges from mild to profound, and is primarily sensorineural, although a small conductive component is often present. Hearing loss can be progressive and is worse in the high frequencies. The most common anal anomaly is imperforate anus. Anal stenosis without imperforate anus has also been reported (Powell & Michaelis 1999).

Subsequent reports have added cardiac, renal, and urogenital abnormalities to the syndrome (Kurnit et al 1978, Newman et al 1997, Salerno et al 2000, Surka et al 2001). Cardiac anomalies include ventricular septal defect, pulmonary atresia, truncus arteriosus, and tetra-logy of Fallot and have been reported in sporadic as well as familial cases (O'Callaghan & Young, 1990, Surka et al 2001). Most patients with TBS have normal intelligence, although learning disabilities or mental retardation have been described in a few cases (Salerno et al 2000). High intrafamilial variability has been observed (de Vries-Van der Weerd et al 1988, Surka et al 2001).

Approximately 60% of patients with TBS have anomalies in the kidney and urogenital tract (Salerno et al 2000, Ma et al 2001). The renal malformations include unilateral or bilateral renal hypoplasia, dysplasia, and renal agenesis. Multicystic dysplastic kidneys and vesicoureteral reflux are not uncommon. In males, posterior urethral valves, meatal stenosis, bifid scrotum, and hypospadia have been reported (Powell & Michaelis 1999). Affected individuals may develop renal failure in the neonatal period, childhood, or adulthood. Kidney transplant has been performed successfully in TBS patients (Newman et al 1997).

TBS has many overlapping features with other syndromes and associations, especially the VACTERL association (acronym for *v*ertebral anomalies, *a*nal atresia, *c*ongenital cardiac disease, *t*racheoesophageal fistula, *r*enal anomalies, and *l*imb defects). However, tracheoesophageal fistula and vertebral anomalies are not seen in TBS. In contrast, ear anomalies and hearing defects are not typical for VACTERL. Polydactyly is postaxial in VACTERL and preaxial in TBS (Powell & Michaelis 1999). These differences can be useful in differential diagnosis.

TBS IS CAUSED BY MUTATIONS OF *SALL1*

TBS is caused by heterozygous mutations of *SALL1*, a gene that is related to the Drosophila *sal* gene. Kohlhase et al. discovered that *SALL1* gene mutations were responsible for TBS in a family with three cases in two generations and in an unrelated family with a sporadic case. Both mutations are predicted to produce a truncated SALL1 protein lacking the C2H2 zinc finger-containing DNA binding domain (Kohlhase et al 1998). As of 2007, a total of 56 *SALL1* mutations have been described in TBS (Botzenhart et al 2007). Most mutations are private, and 75% are found in exon 2, 5′ to the region encoding the zinc finger domain. The mutations include frameshift mutations, nonsense mutations, and one mutation within intron 2 that creates an aberrant splice site (Blanck et al 2000). All mutations lead to premature stop codons and truncated proteins lacking DNA binding domains.

TBS is thought to be due to haploinsufficiency of *SALL1*. However, heterozygous *Sall1* knockout mice do not mimic human TBS (Nishinakamura et al 2001). Most truncated *SALL1* mutants in humans with TBS lack an intact DNA binding domain but retain a glutamine-rich domain that is necessary for dimerization. Transgenic mice expressing a C-terminally truncated Sall1 mutant recapitulate the abnormalities seen in human TBS, including high-frequency sensorineural hearing loss, renal hypoplasia, and limb abnormalities (Kiefer et al 2003). This finding suggests that mutations of human *SALL1* gene might have a dominant negative-effect by forming inactive heterodimers with wild-type *SALL1*. In addition, a recent study has shown that *SALL1* interacts with another Sal-related gene, *SALL4* (Sakaki-Yumoto et al 2006). Mutations of human *SALL4* cause an autosomal dominant disorder called Okihiro syndrome, which rarely includes kidney anomalies. *Sal4*$^{+/-}$ heterozygotes do not have kidney abnormalities. However, *Sall1*$^{+/-}$*Sall4*$^{+/-}$ double heterozygotes develop renal agenesis with an incidence more than 10-fold higher than *Sall1*$^{+/-}$heterozygotes, indicating a genetic interaction. C-terminally truncated mutant *SALL1* can form heterodimers with wild-type SALL4, which results in mislocalization of SALL4 in heterochromatin. These results suggest that TBS may arise from dominant-negative mutations of *SALL1* that inhibit the activity of wild-type SALL4.

SALL1 AND ITS ROLE IN DEVELOPMENT

Sal (spalt) of *Drosophila melanogaster* encodes an evolutionally conserved zinc finger protein. *Sal* is a homeotic gene that promotes the specification of posterior head and anterior tail structures. Sal-like genes (*SALL*) were first identified in humans in 1996 (Kohlhase et al 1996). At least four Sal-related genes exist in humans (*SALL1-4*) and mice (*Sall1-4*) (Kohlhase et al 1996, 1999, 2002, Ott et al 1996, Buck et al 2000). Human *SALL1* is located on Chromosome 16q12.1. The *SALL1* gene encodes a protein that is composed of 1325 amino acids and contains four characteristically arranged C2H2 double zinc finger domains (Kohlhase et al 1996, 1998, 1999, Netzer et al 2002).

In humans, *SALL1* is expressed in the kidney, brain, and liver with the highest levels of expression in the kidney (Kohlhase et al 1996). Immunohistochemical analysis of 12–14 week-old human fetal kidneys shows weak expression of SALL1 in the induced metanephric mesenchyme around the ureteric bud. SALL1 is detected in the ureteric bud and immature tubules in the nephrogenic zone of the developing kidney. In contrast, there is no expression of SALL1 in uninduced mesenchyme or glomeruli. Expression of SALL1 persists in tubules in the cortex and the loop of Henle in the medulla in the adult kidney. Consistent with the expression patterns in the fetal kidney, stromal cells and glomeruli in the adult kidney show no expression of SALL1 (Ma et al 2001).

Sall1-lacZ knock-in mice and *Sall1* knockout mice have provided a useful tool to study the role of *Sall1* in mouse kidney development (Nishinakamura et al 2001, Nishinakamura 2003). X-gal staining of heterozygous *Sall1-lacZ* knock-in mice shows *Sall1* expression in the nephrogenic primordium at E10.5. At E11.5, *Sall1* is expressed in mesonephric tubules and the Wolffian ducts and in the metanephric mesenchyme surrounding the ureteric bud. Expression of *Sall1* in the ureteric bud is not detectable in mice. At E14.5, *Sall1* is weakly expressed in comma-shaped bodies but not in glomeruli. Homozygous *Sall1*-deficient mice die at 24 hours after birth. One third of the mice have bilateral renal agenesis and absent ureters. The remaining mice have either bilateral renal hypoplasia or unilateral renal agenesis and contralateral renal hypoplasia. No urine is observed in the bladder, although the bladder develops normally.

The metanephric mesenchyme and ureteric bud form in *Sall1* null mice. However, in most mutant mice the ureteric bud fails to invade the metanephric mesenchyme (Nishinakamura et al 2001). In the absence of the ureteric bud, the metanephric mesenchyme fails to differentiate into nephrons and undergoes apoptosis. In some *Sall1* null mice, the ureteric bud reaches the metanephric mesenchyme, but mesenchymal condensation and ureteric branching are significantly impaired. The remnant kidneys contain poorly differentiated tubules, shrunken glomeruli, and multiple cysts. Molecular studies in E11.5 embryos reveal significant down-regulation of GDNF, BMP7, and Wnt4, signaling proteins that are essential for kidney development. The expression of Pax2 and Wt1 in the mesenchyme is reduced, but c-Ret expression in the ureteric bud is unchanged. When co-cultured with embryonic spinal cord, which can serve as a nephron-inducing tissue, the mutant mesenchyme is able to form tubules, indicating that it is competent in epithelial transformation. Thus, the failure of ureteric bud invasion into the metanephric mesenchyme is the direct cause of renal agenesis in *Sall1* null mice.

Sall1 is a transcriptional repressor that is localized in heterochromatin and is associated with histone deacetylases (Netzer et al 2001, Kiefer et al 2002). The direct DNA binding sites and targets of repression by Sall1 have not yet been elucidated. Sall1 contains 10 zinc-finger motifs, most of which are clustered in pairs or triplets. Sall1 interacts with β-catenin and partially colocalizes with β-catenin in pericentromeric heterochromatin. When *Sall1* is transfected into NIH-3T3 cells, it activates canonical Wnt signaling. Disease-causing truncation mutations of *Sall1* disturb the localization of native Sall1 to heterochromatin and down-regulate the activation of Wnt signaling (Sato et al 2004). Since Wnt signaling plays a crucial role in kidney development, down-regulation of this pathway may explain the abnormal kidney phenotype seen in humans with *SALL1* mutations and in Sall1-deficient mice.

Examination of extrarenal tissues in mice shows that Sall1 is expressed in the otic vesicles and endocardium between E10.5 and E11.5. At E14.5, Sall1 is expressed in limb buds, anorectal region, developing olfactory bulb, nose, and eye. In newborn mice, Sall1 is expressed in the kidney, heart, liver, ventricular regions of the brain and the olfactory bulbs (Ott et al 1996, Nishinakamura et al 2001). In humans, SALL1 protein is detected in several endocrine organs, including the cells that produce growth hormone and gonadotropins in the pituitary, Leydig and Sertoli cells in the testis, granulosa cells in the ovary, and the adrenal cortex (Ma et al 2002). The expression pattern correlates with the extrarenal malformations caused by mutations of *SALL1* in Townes-Brocks syndrome.

Maturity-Onset Diabetes of the Young type 5

CLINICAL FEATURES OF MODY5

Maturity-Onset Diabetes of the Young (MODY) was first described in 1974 by Fajans and Tattersall (Fajans et al 2001). MODY is a monogenic form of diabetes mellitus that is inherited as an autosomal dominant trait. The diabetes typically presents before the age of 25 and is not associated with obesity or ketosis. MODY is genetically heterogeneous, and eight genes that cause this disorder have been identified. In 1993, using a candidate gene approach based on its pivotal role in glycolysis, mutations of glucokinase (*GCK*) were identified as a cause of MODY in pedigrees of French and English decent (Froguel et al 1992). This subgroup of patients is now categorized as having MODY2. In 1996, the transcription factor HNF-1α (*TCF1*) was identified as the mutated gene in patients with MODY3 (Yamagata et al 1996). This discovery prompted a closer look at other transcription factors that regulate pancreatic beta cell function. Mutations of HNF-4α (*HNF4A*) and HNF-1β (*TCF2*) were subsequently identified as causes of MODY1 and MODY5, respectively (Yamagata et al 1996, Horikawa et al 1997). Based on the finding of pancreatic agenesis in *Ipf1* knockout mice and a patient with neonatal diabetes, *IPF1* was investigated and found to be mutated in patients with MODY4 (Stoffers et al 1997). Subsequently, *NEUROD1* (also called *BETA2*), Kruppel-like factor 11 (*KLF11*), and carboxyl-ester lipase (*CEL*)

have been identified as the genes responsible for MODY6-8 (Malecki et al 1999, Neve et al 2005, Raeder et al 2006). The different subtypes of MODY vary in their incidence and severity of diabetes. MODY2 and MODY3 are the most common subtypes, whereas the other forms are rare. Diabetes arises from a primary defect in pancreatic beta cell function due to abnormal transcriptional regulation and/or defects in glucose sensitivity (Bell & Polonsky 2001, Fajans et al 2001). Some of the subtypes of MODY have distinct extrapancreatic manifestations. For example, MODY3 is associated with renal glucosuria due to defects in the transcription of genes encoding Na/glucose cotransporters in the renal proximal tubule (Menzel et al 1998, Pontoglio et al 2000).

MODY5 (OMIM 137920) is characterized by the combination of non-ketotic diabetes mellitus and non-diabetic kidney disease. The diabetes arises from a defect in insulin secretion, although insulin resistance and hyperinsulinemia have also been described (Pearson et al 2004). The kidney disease in MODY5 is highly penetrant and may be present in up to 100% of affected individuals. Kidney disease usually precedes the onset of diabetes, which indicates that it is not secondary to diabetic nephropathy. The renal manifestations of MODY5 are diverse with cystic kidney disease being the most common (Figure 27.10). The coexistence of diabetes mellitus and cystic kidney disease is referred to as *R*enal *C*ysts and *D*iabetes (RCAD). The cystic kidney abnormalities include simple cysts, multicystic dysplastic kidney (MCDK), and the hypoplastic form of glomerulocystic kidney disease. In addition to the formation of kidney cysts, other congenital renal anomalies have been described including oligomeganephronia, solitary kidney, renal hypoplasia/dysplasia, and horseshoe kidney (Bellanne-Chantelot et al 2004). Renal involvement in MODY5 is often severe and frequently leads to renal failure. Abnormalities of the male and female genital tracts may also be present, including bicornuate uterus, atresia of the vas deferens, and hemi-uterus (Bellanne-Chantelot et al 2004, Edghill et al 2006). Findings outside the kidney and urogenital tract include pancreatic atrophy and abnormalities in liver enzymes. MODY5 is one of the causes of familial juvenile hyperuricemic nephropathy (Mache et al 2002, Bellanne-Chantelot et al 2004, Edghill et al 2006).

MODY5 is a rare disorder. A prevalence of 0.8% was found in Japanese subjects with type 2 diabetes and a strong family history of diabetes (Furuta et al 2002). However, MODY5 may be under-diagnosed, since some affected individuals present primarily with renal disease without clinically significant diabetes. Moreover, mutations of the gene that causes MODY5 may be a relatively common cause of sporadic renal dysgenesis (Weber et al 2006).

MODY5 IS CAUSED BY MUTATIONS OF HNF-1β

MODY5 is caused by heterozygous mutations of *TCF2*, which encodes the transcription factor HNF-1β (Figure 27.11).

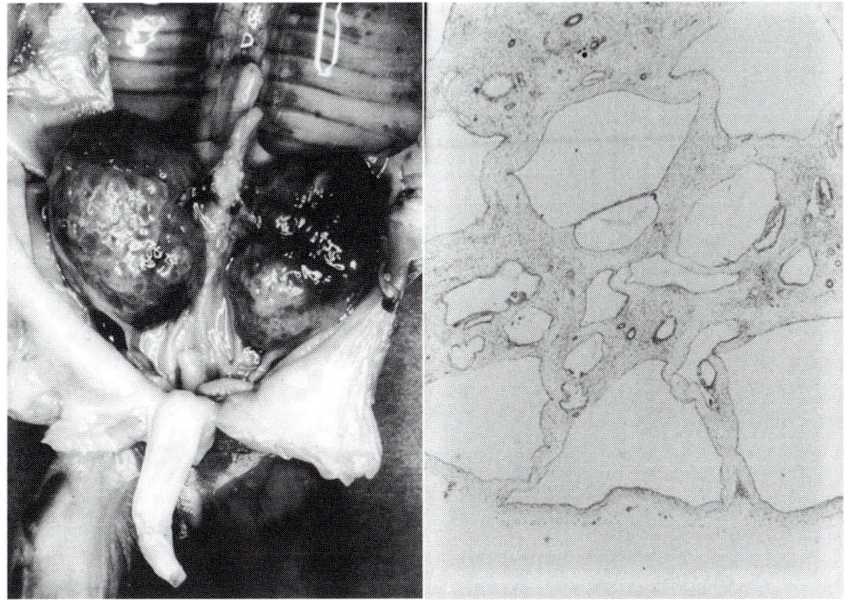

FIGURE 27.10 Renal dysplasia in MODY5. Left panel shows gross appearance of dysplastic kidneys in a human fetus with MODY5 due to mutation of *TCF2*. Kidney histology (right panel) shows kidney cysts, interstitial fibrosis, and absence of normal kidney tissue. Reproduced with permission from Bingham et al (2000).

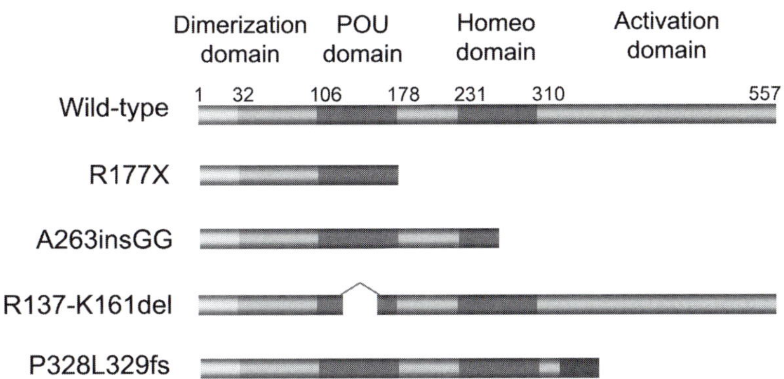

FIGURE 27.11 Mutations of TCF2 in MODY5. Upper panel shows wild-type HNF-1β containing an amino-terminal dimerization domain, DNA-binding POU/homeodomain, and carboxy-terminal activation domain. Four representative mutations of HNF-1β in families with MODY5 are shown. Most mutations affect the POU-homeodomain and prevent DNA binding. The P328L329fsdelCCTCT mutant (bottom) contains an intact DNA-binding domain but lacks the activation domain. Pink box indicates ectopic amino acids introduced by the frameshift mutation. (see also Plate 57)

TCF2 maps to the short arm of human chromosome 17 (17cen-q12.3). Mutations in *TCF2* were first identified in two Japanese families with MODY and non-diabetic kidney disease (Horikawa et al 1997, Nishigori et al 1998). One mutation was a nonsense mutation (R177X), and the second was a frameshift mutation that produced a truncated protein (A263fsinsGG). Both mutations disrupted the DNA-binding domain of HNF-1β, and functional studies revealed that the mutant proteins were unable to activate an HNF-1β-responsive reporter gene. When co-expressed with wild-type HNF-1β, the mutants decreased the activity of the wild-type protein, suggesting that they have dominant-negative effects (Tomura et al 1999).

Subsequently, a Norwegian family with diabetes, end-stage renal disease, and Müllerian aplasia was noted to have a 75 bp deletion in exon 2 of *TCF2*, which also disrupts the DNA-binding domain of HNF-1β (Lindner et al 1999). Analysis of this mutant revealed that it was a loss-of-function mutation. In 2000, a 5 bp deletion (P328L329fsdelCCTCT)

was identified in a woman with diabetes and renal insufficiency as well as two offspring with renal dysplasia (Bingham et al 2000). This mutation produces a protein with an intact DNA-binding domain and addition of 29 amino acids from the shifted reading frame. Expression of the 5 bp deletion mutant in *Xenopus* embryos produced defects in the development of the pronephros (Wild et al 2000). In addition, germline mosaicism and large genomic rearrangements of TCF2 have been described as the cause of MODY5 (Yorifuji et al 2004). To date, more than 30 different mutations of *TCF2* have been identified in families with MODY5. Most mutations affect the DNA-binding domain of HNF-1β and are predicted to cause loss-of-function or dominant-negative effects.

Although MODY5 is a rare disorder, mutations of *TCF2* have been increasingly identified in the sporadic cases of renal dysgenesis that occur not uncommonly in the general population. One genetic screen of 100 children with renal hypolasia/dysplasia identified mutations of *TCF2* in 8%

(Weber et al 2006). Another study found mutation rates of 29% in a pediatric cohort with renal cysts, hyperechogenicity, renal hypoplasia, or solitary kidney (Ulinski et al 2006). The most frequent mutation was a deletion of the entire *TCF2* gene. Mutations of *TCF2* were the most frequent genetic abnormality detected in fetuses with bilateral hyperechogenic kidneys (Decramer et al 2007). Taken together, these studies suggest that the prevalence of *TCF2* mutations among individuals with renal dysgenesis may be higher than had been appreciated originally. Mutations of *TCF2* may also be involved in the pathogenesis of renal cancer. Biallelic inactivation of *TCF2* has been found in the chromophobe variant of renal cell carcinoma (Rebouissou et al 2005).

HNF-1β AND ITS ROLES IN DEVELOPMENT

Hepatocyte nuclear factor-1β (HNF-1β, vHNF1, LF-B2) is a member of the HNF1 family of transcription factors and has a similar structure to the other member of the family, HNF-1α. The HNF-1β protein contains an amino-terminal dimerization domain, a Pit-1/Oct-1/Unc-86 (POU) domain and a homeodomain that mediate DNA binding, and a C-terminal transcriptional activation domain (Figure 27.11) (Mendel et al 1991). HNF-1β binds to the promoter regions of target genes as a homodimer or a heterodimer with HNF-1α. The sequence of the POU/homeodomain of HNF-1β is highly similar to that of HNF-1α, and both proteins recognize the identical consensus sequence 5′-GTTAATNATTAAC-3′ (Rey-Campos et al 1991). HNF-1β was originally identified as a liver-enriched transcription factor, but subsequent studies have revealed that it is also expressed in epithelia of the kidney, small intestine, liver, pancreas, and lung (Ott et al 1991, Coffinier et al 1999). In the developing kidney, HNF-1β is expressed in the ureteric bud that will give rise to the renal collecting system as well as the comma-shaped and S-shaped bodies which will give rise to the nephron proper (Ott et al 1991, Coffinier et al 1999). HNF-1β is also expressed in the mesonephric (Wolffian) duct, which is the anlage of the vas deferens, and epididymis in males and the Müllerian duct, which is the anlage of the oviduct, uterus, and cervix in females. In the mature kidney, HNF-1β is expressed in all the segments of the nephron including the collecting duct but no expression is observed in the glomerular tufts, blood vessels and the interstitium. In addition, HNF-1β is expressed in the liver, pancreatic islets, and small intestine (Coffinier et al 1999).

Insights into the roles of HNF-1β in embryonic development have been provided by animal models. *Danio rerio* (zebrafish) is an excellent model to study organogenesis because the embryo is transparent and develops ex utero. A mutation of zebrafish HNF-1β (*vhnf1*) has been identified in an insertional mutagenesis screen (Sun and Hopkins, 2001). The phenotype of the mutant animals includes cysts

in the pronephros, underdevelopment of the liver and pancreas, and reduction in size of the otic vesicle. Analysis of wild-type animals has revealed that HNF-1β is involved in the regional specification of organ primordia by directing the expression of tissue-specific genes. The severe phenotype observed in HNF-1β mutant animals underscores the important role of HNF-1β in the development of the kidney, liver, pancreas, and brain. In mice, HNF-1β-deficient embryos die soon after implantation due to a poorly organized ectoderm and absence of visceral and parietal endoderm (Barbacci et al 1999). HNF-1β-null embryoid bodies lack expression of HNF-4α, HNF-1α, and HNF-3γ, suggesting that HNF-1β acts upstream of these transcription factors. The lethal phenotype can be rescued by aggregation of HNF-1β-deficient embryonic stem cells with wild-type embryos (Barbacci et al 1999). These experiments indicate that similar to the zebrafish, HNF-1β is required for normal embryogenesis in mice.

MOLECULAR PATHOGENESIS OF MODY5

Several mouse models of MODY5 have been generated using techniques that circumvent embryonic lethality and permit studies of the effects of HNF-1β inactivation in the pancreas, liver, and kidney. Using a tetraploid rescue strategy, the role of HNF-1β in pancreatic development has been elucidated. These embryos exhibit a dramatic phenotype of pancreatic agenesis due to perturbation of the transcriptional network within pancreatic precursor cells (Haumaitre et al 2005). The expression of pancreatic and duodenal homeobox 1 (*Pdx1*), the earliest marker of pancreatic precursor cells, is regulated by HNF-6. The expression of HNF-6 is in turn regulated by the binding of HNF-1β to an intronic enhancer within the *Hnf-6* gene (Poll et al 2006). These studies indicate that HNF-1β is upstream of a transcriptional hierarchy that is required for the formation of pancreatic precursor cells and the development of the pancreas. Pancreatic beta cell-specific knockout of *Hnf-1β* in mice causes glucose intolerance but not overt diabetes (Wang et al 2004). These mice are viable and have normal development, but the expression of *Hnf-4* is decreased in the knockout cells, suggesting a role of HNF-1β in the transcriptional control required for the normal endocrine function of pancreatic beta cells.

The role of HNF-1β during liver and kidney development has been studied using conditional gene targeting by Cre/loxP recombination. Liver-specific inactivation of *Hnf-1β* in mice causes severe jaundice due to developmental abnormalities of the gallbladder and a paucity of intrahepatic bile ducts (IHBD) (Coffinier et al 2002). The defect in IHBD is linked to a failure in organization of ductal structures during liver organogenesis, suggesting an essential role of HNF-1β in bile duct morphogenesis. Kidney-specific inactivation of *Hnf-1β* results in cyst formation in the renal tubules (Gresh et al

2004). Analysis of these mutant animals revealed that several known cystic disease genes are down-regulated in the cyst epithelia, including *PKHD1*, which is mutated in autosomal recessive polycystic kidney disease (ARPKD). Another model of MODY5 was produced by transgenic expression of a dominant-negative HNF-1β mutant in the kidneys of mice (Hiesberger et al 2004). The transgenic mice develop cysts in the renal tubules and glomeruli, which is similar to the phenotype of humans with MODY5. Expression of *Pkhd1* is decreased in the kidney cysts, and HNF-1β was shown to directly regulate the *Pkhd1* promoter. A similar phenotype was observed in mice expressing an HNF-1β truncation mutant lacking the C-terminal activation domain (Hiesberger et al 2005). Since the mutant protein is expressed in animals with two wild-type alleles of HNF-1β, these results suggest that deletion of the C-terminal domain produces a dominant-negative mutation. Taken together, the studies suggest that kidney cyst formation in humans with mutations in *TCF2* may be due to down-regulation of *PKHD1*, thereby linking MODY5 and ARPKD in a common genetic pathway. Since mutations of *Pkhd1* also produce ductal plate malformations in the liver, down-regulation of *Pkhd1* may also underlie the liver abnormalities that are seen in HNF-1β mutant mice.

The mechanism of kidney cyst formation in rodents harboring mutations in *Hnf-1β* and *Pkhd1* may involve abnormalities in planar cell polarity (PCP). PCP is defined as polarity along a tissue plane that is perpendicular to the apical-basal axis (Zallen 2007). PCP signaling (also referred to as non-canonical Wnt signaling) is involved in regulating epithelial cell polarization, convergent extension, and oriented cell division. In mice, oriented cell division is required for normal postnatal development of the kidney. By orienting the angle of cell division parallel to the longitudinal axis of the kidney tubule, elongation of the tubule occurs without changing the tubule diameter. Aberrant PCP, evidenced by randomization of oriented cell division, precedes kidney cyst formation in mice and rats harboring mutations in *Hnf-1β* and *Pkhd1*, respectively (Fischer et al 2006). Randomization of the orientation of cell division is predicted to increase tubular diameter causing cyst formation. Whether aberrant PCP is common to all cystic kidney diseases or how HNF-1β and *Pkhd1* regulate PCP is not known at this time.

Relatively few genes that are regulated by HNF-1β in the kidney have been identified. Recently, chromatin immunoprecipitation and DNA microarray (ChIP-chip) have been used for genome-wide identification of HNF-1β target genes in the kidney (Ma et al 2007). Among the target genes that were identified is suppressor of cytokine signaling-3 (*SOCS3*). SOCS3 inhibits JAK/STAT and ERK signaling pathways that are important for kidney development. Normally, HNF-1β inhibits *SOCS3* expression, but in HNF-1β-deficient mice or cells expressing dominant-negative mutant HNF-1β, *SOCS3* expression is inappropriately elevated. The increased expression of *SOCS3* prevents renal tubulogenesis by blocking JAK/STAT and ERK signaling pathways. These studies identify *SOCS3* as an important target of HNF-1β in renal tubulogenesis. Since SOCS3 also inhibits insulin receptor signaling, dysregulation of *SOCS3* may underlie the insulin resistance seen in MODY5.

Oral-Facial-Digital Syndrome

CLINICAL FEATURES OF ORAL-FACIAL-DIGITAL SYNDROME Oral-facial-digital syndromes (OFDs) encompass malformations of the oral cavity, face, and digits (Figure 27.12) (Toriello 1993). Nine different types of OFD have been described to date. Oral-facial-digital type 1 (OFD1, OMIM 311200) is the most common type and is distinguished by the frequent association with cystic kidney disease and X-linked dominant inheritance, although some cases are sporadic. Almost all affected males die before birth, usually in the first and second trimesters (Doege et al 1964, Wettke-Schafer & Kantner 1983). The only male who survived the postnatal period in a kindred of 15 females with OFD1 had XXY Klinefelter syndrome (Doege et al 1964). OFD1 is rare with an estimated incidence of 1 in 250 000 live births, and the penetrance is highly variable (Wettke-Schafer & Kantner 1983).

Oral and facial abnormalities in OFD1 include facial asymmetry, hypertelorism, micrognathia, broadened nasal ridge, facial milia, median pseudoclefting of the upper lip, clefts of the palate and tongue, oral frenula, thickened alveolar ridges, and abnormal dentition. The digital abnormalities include asymmetric shortening of the digits, syndactyly,

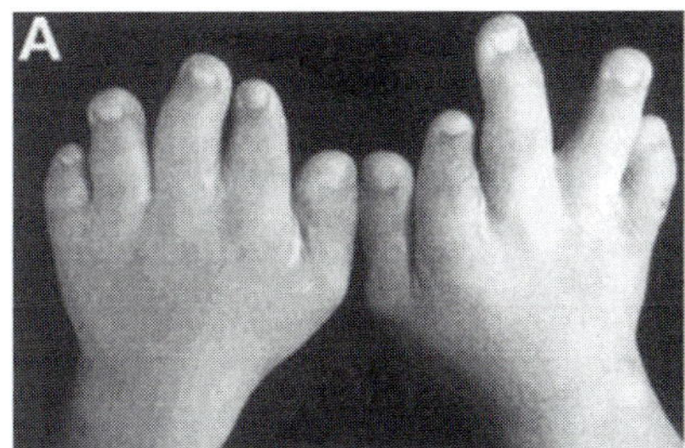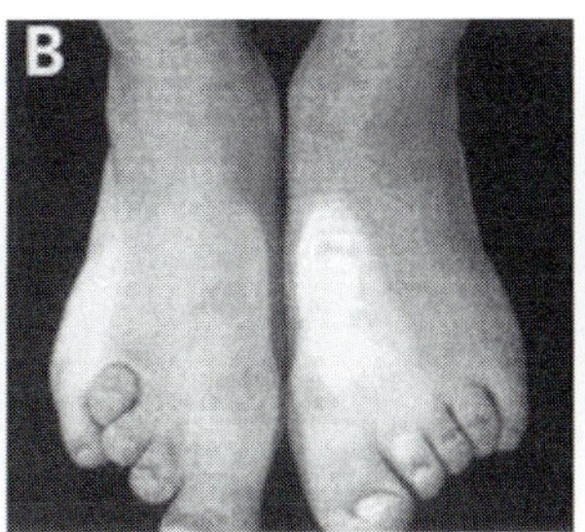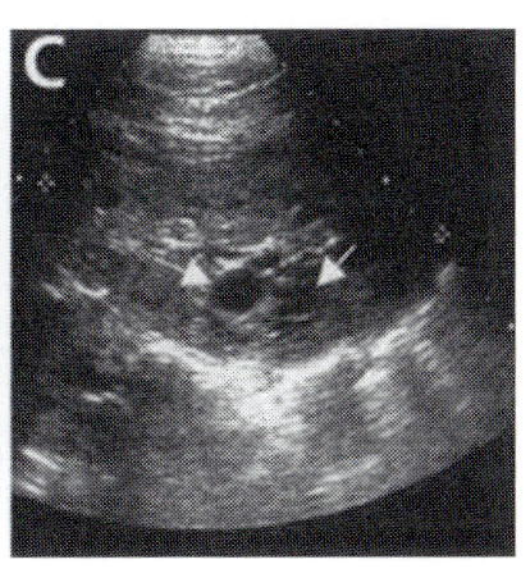

FIGURE 27.12 Oral-facial-digital syndrome. Panels A and B show abnormal digits including brachydactyly and syndactyly in a child with mutations of OFD1. Ultrasound image of the kidney (panel C) shows kidney cysts. Reproduced with permission from Ferrante et al (2001)

brachydactyly, clinodactyly, and preaxial or postaxial poly-dactyly. Involvement of the hands is usually more frequent and more severe than of the feet. Central nervous system involvement seen in 40% of cases includes hydrocephalus, cerebellar anomalies, porencephaly, and agenesis of the corpus callosum. Variable degrees of mental retardation are seen in about 57% of affected individuals (Towfighi et al 1985, Toriello 1993).

Cystic kidney disease is commonly seen in OFD1 (Figure 27.2) (Donnai et al 1987, Scolari et al 1997, Woolf et al 2002). Feather et al studied the clinical, radiological, and histopathological features of OFD1 in a family with five affected females in three generations. All five individuals had renal cysts. Kidney cysts were usually less than 1 cm in size as measured by renal ultrasound. Most individuals had bilateral kidney cysts. The size of the kidney was normal, but the contour was irregular by sonogram. The majority of the cysts were formed in the glomeruli, and a small percent-age were derived from the distal tubules. The cyst epithe-lial cells were flattened and had high rates of proliferation. Three of the five patients had normal plasma creatinine, and two had end-stage renal disease requiring renal replacement therapy. All affected individuals had malformations of the face and oral cavity and learning disabilities (Feather et al 1997). OFD1 should be included in the differential diagnosis of ADPKD in females who have concomitant malformations of the oral cavity, face, and digits.

Mutations of *OFD1* in Oral-Facial-Digital Syndrome Type1

The *OFD1* locus was mapped to the short arm of the X chromosome by linkage analysis in 1997 (Feather et al 1997). The *OFD1*-containing region was further narrowed down to 12 Mb flanked by markers DXS85 and DXS7105 in the Xp22 region (Gedeon et al 1999). Sequence analysis of three familial and four sporadic cases of OFD1 revealed that mutations of a novel gene, called *OFD1*, were responsible for the disease (Ferrante et al 2001). A missense mutation, a 19 bp deletion, and a single base pair deletion leading to a frameshift were discovered in the familial cases, whereas in the sporadic cases, de novo missense, nonsense, splicing, and frameshift mutations were identified.

OFD1 contains 23 exons and encodes an open read-ing frame of 3033 bp. Two alternative splice sites are located in intron 9, which result in two encoded proteins, OFD1a and OFD1b (de Conciliis et al 1998, Romio et al 2003). The longer form (OFD1a) is predicted to contain 1011 amino acids, and the shorter form (OFD1b) is pre-dicted to contain 367 amino acids. Both proteins contain a large number of coiled-coil domains, which are protein interaction domains that are found in a variety of different proteins, including structural proteins, transmembrane pro-teins, and transcription factors. However, a DNA binding domain has not been identified in OFD1. The function of the OFD1 protein is not clear. It is interesting to note that coiled-coil domains are also present in polycystin-1 and polycystin-2, the proteins that are mutated in auto-somal dominant polycystic kidney disease (ADPKD). The disease-causing mutations that have been identified in *OFD1* are predicted to truncate the protein and delete the coiled-coil domains that may be involved in protein inter-actions and gene regulation.

OFD1 and its Role in Development

The expression pattern of *OFD1* in humans and mice coincides well with the tissues that are affected in humans with OFD1. *OFD1* transcripts are expressed in all the tis-sues examined in humans (kidney, pancreas, skeletal mus-cle, liver, lung, brain, and heart) (de Conciliis et al 1998). A more recent study by Romio et al shows that *OFD1* mRNA is expressed in all developing tissues that are known to be affected in OFD1 (Romio et al 2003). RT-PCR and north-ern blot analyses reveal the presence of both OFD1a and OFD1b in the metanephros, brain, tongue, and limbs in nor-mal human embryos at 9–14 weeks of gestation. RT-PCR also shows the presence of both transcripts in the heart, lungs and gut, which are not known to be affected clinically. Using antiserum to OFD1a, the protein can be detected in human metanephros. Immunohistochemical analysis of human kidneys at 7–14 weeks of gestation shows expres-sion of OFD1a in the mesenchyme in the nephrogenic zone. Faint expression is seen in nephron precursors, including renal vesicles and comma- and S-shaped bodies. Consistent staining has not been observed in the developing glomer-uli. OFD1a protein is also expressed in a broad range of craniofacial tissues, including nasal and skull cartilage, oral mucosa, skin, brain, and tongue myocytes.

Mouse *Ofd1* contains 23 exons spanning 50 kb on the X chromosome. *Ofd1* encodes a protein containing 1017 amino acids. The sequence identity between human and mouse is 71% at the nucleotide level and 66% at the protein level. In situ hybridization shows that *Ofd1* is expressed at high levels in the heart, brain, liver, kidney, and testis of adult mice (Ferrante et al 2003). During development, the first site of expression is in the genital ridges at 12.5 days of gestation, and high levels of expression continue at E14.5 and E16.5. In addition, high-to-moderate levels of expres-sion are detected in various craniofacial structures and in the nervous system, especially in the epithelium lining the oral and nasal cavities. The lung and thymus show low lev-els of expression (Ferrante et al 2001).

OFD1 is a Ciliary Protein

OFD1a is localized to the apical region in cells of the renal vesicles during their acquisition of epithelial cell polarity. In cultured human proximal tubular cells, OFD1 is detected in the basal body of the primary cilium. When centrosomes are isolated from human embryonic kidney cells (HEK 293 cells) and stained with antibodies to OFD1 and α-tubulin, OFD1 colocalizes with α-tubulin in centrosomes (Romio et al 2004).

OFD1 is closely associated with the centrosome throughout the cell cycle as demonstrated by immunofluorescence in HEK293 cells. Transfection of HEK293 cells with mutant proteins containing various deletions of the carboxy terminus of OFD1 indicates that the coiled-coil domains are required for centrosomal targeting (Romio et al 2004). It is particularly interesting to note that most mutations of human *OFD1* are predicted to produce truncations leading to the loss of the coiled-coil domains, which may result in mistargeting of the protein. Polycystin-1 and polycystin-2 are also ciliary proteins, which suggests that mutations of OFD1 may share a common mechanism of cilium-dependent cystogenesis.

The role of the primary cilium in the pathogenesis of OFD1 is further supported by recent studies of *Ofd1* knockout mice (Ferrante et al 2006). Heterozygous Ofd1 mutant males are embryonic lethal, and female heterozygotes die at birth. Similar to humans with OFD1 mutations, female heterozygotes exhibit craniofacial and limb abnormalities including cleft palate, polydactyly, and skeletal defects. The kidneys from mutant females contain glomerular cysts that lack primary cilia. Cilia are present in the surrounding noncystic tubules. Examination of the mutant males shows that they die at E12.5 from cardiac defects due to situs inversus. The establishment of left–right asymmetry is dependent on primary cilia in the embryonic node. The embryonic nodes of Ofd1 mutant males lack cilia, which explains the laterality defects. These findings indicate that the cystic kidney disease in OFD1 arises from defective ciliogenesis. Interestingly, recent studies have shown that the primary cilium is also required for the establishment of planar cell polarity in the kidney (Patel et al 2008), raising the possibility that the cystogenic pathways in MODY5 and OFD1 may converge.

Alagille Syndrome

CLINICAL FEATURES OF ALAGILLE SYNDROME

Alagille syndrome (AGS, OMIM 118450) is an autosomal dominant disorder that is characterized by developmental abnormalities of the liver, heart, eye, skeleton, and several other organs including the kidney and urogenital system (Krantz et al 1997). About 95% of affected individuals have a typical facies consisting of deep-set eyes, broad forehead, prominent chin, and small and low-set or malformed ears. Most patients present with neonatal jaundice and cholestasis resulting from a paucity of intrahepatic bile ducts. Eight-five percent of patients have cardiac defects including pulmonic stenosis, atrial or ventricular septal defects, patent ductus arteriosus, and coarctation of the aorta. Complications

from liver and cardiac disease are responsible for 19–25% and 15–29% of deaths, respectively. Intracranial bleeding or stroke occur in 14% of patients and is responsible for 25% of the mortality. Skeletal defects, mainly butterfly vertebrae, occur in 87% of patients, and eye abnormalities (primarily anterior chamber defects such as posterior embryotoxon) occur in 78–88% of patients (Alagille et al 1987, Mueller 1987, Emerick et al 1999).

Renal abnormalities are observed in 70% of patients in the case series reviewed by Alagille and 40% of the patients reviewed by Emerick (Alagille et al 1987, Emerick et al 1999). Renal ultrasound shows small, hyperechoic kidneys consistent with renal dysplasia. The main histological finding is mesangiolipidosis (Habib et al 1987). In less severely affected patients, light microscopy shows a fibrillar appearance of the mesangium, and electron microscopy shows widespread lipid vacuoles in the mesangial matrix. Mesangial foam cells are seen in more severely affected individuals. The extent of mesangiolipidosis is related to the degree of cholestasis and may be the consequence of abnormal lipid metabolism. Other histological findings include cystic tubular dilatation and glomerulosclerosis. Unilateral cystic kidney disease with or without renal dysplasia may be associated with Alagille syndrome. Martin et al reported three children with Alagille syndrome and cystic kidney disease. Two of them had unilateral multicystic dysplastic kidneys detected by prenatal ultrasound, and the third had a solitary cortical cyst detected in childhood (Martin et al 1996). These authors suggest that Alagille syndrome should be included in the differential diagnosis of cystic kidney disorders associated with cholestatic liver disease, and patients with Alagille syndrome should be evaluated by renal ultrasound. Renal insufficiency is observed in 7% of patients in the case series reviewed by Emerick. However, renal failure can occur in neonates, and death due to chronic renal failure has been reported (Habib et al 1987). Up to 36% of Alagille patients manifest renal tubular acidosis (Emerick et al 1999). Other urogenital anomalies such as hypogonadism have been reported.

MUTATIONS OF *JAG1* AND *NOTCH2* IN ALAGILLE SYNDROME

Positional cloning first showed that Alagille syndrome could be caused by mutations in *JAG1*, which encodes a ligand in the Notch signaling pathway (Li et al 1997, Oda et al 1997). In a study of more than 300 patients with AGS, mutations of *JAG1* were identified in 60–75% (Spinner et al 2001). Most mutations (72%) are frameshift mutations that produce truncated proteins. Missense mutations occur less frequently (23%), and at least some of the mutated proteins are unable to reach the cell surface due to abnormal trafficking (Morrissette et al 2001). The mutations of *JAG1* are clustered in conserved regions at the 5′ end of the gene or in the

EGF repeats (see below). In 15% of the patients, mutations in consensus splice sites are identified, and complete gene deletions have been reported in 3–7% of patients. A high frequency of de novo mutations (60–70%) has also been reported. The phenotypes of patients with *JAG1* deletions are indistinguishable from the phenotypes of patients with intragenic *JAG1* mutations, suggesting that *JAG1* haploinsufficiency as at least one mechanism of Alagille syndrome.

Alagille syndrome is inherited as an autosomal dominant trait with high penetrance (94%) and highly variable expressivity. Phenotypic discordance has been observed in monozygotic twins with one twin developing severe pulmonary atresia and mild liver disease, while the other twin has tetralogy of Fallot and severe cholestatic liver disease (Kamath et al 2002). The variable expressivity may be explained by genetic or non-genetic factors. Heterozygous *Jag1* mutant mice exhibit only eye defects (Xue et al 1999). However, double heterozygous mice carrying a null allele of *Jag1* and a hypomorphic allele of *Notch2* exhibit developmental abnormalities characteristic of Alagille syndrome including kidney malformations, growth retardation, impaired differentiation of intrahepatic bile ducts, and defects in the heart and eye (McCright et al 2002). The mouse studies suggest that *NOTCH2* alleles might influence the severity of the AGS phenotype in humans. In addition, mutations of *JAG1* have been identified in only 70–94% of individuals with Alagille syndrome. Prompted by these findings, McDaniell et al analyzed 11 AGS patients without *JAG1* mutations and identified two independent mutations of *NOTCH2* (McDaniell et al 2006). The mutations consisted of a splice site mutation and a missense mutation affecting a conserved cysteine residue. The renal manifestations in these families included renal hypoplasia, cystic dysplasia, proteinuria, renal tubular acidosis, and renal insufficiency. The renal anomalies were accompanied by typical facies, cardiac abnormalities, and cholestatic liver disease. Taken together, these results indicate that Alagille syndrome is genetically heterogeneous and can arise from mutations in two genes in the Notch signaling pathway.

Notch Signaling Pathway

Notch gained its name from the discovery in 1919 of a *Drosophila* mutation that caused the formation of a notch at the wing margin (Mohr 1919). Later, it was discovered that all germ layers are affected in *Notch* mutants (Poulson 1937). The most dramatic change is the transformation of surface ectoderm to neurons and expansion of neuronal tissue due to the inability to suppress neurogenic differentiation (Artavanis-Tsakonas et al 1991, 1999). The *Notch* gene later became the founding member of the 'neurogenic genes,' in which mutations cause increased numbers of neurons. The *Drosophila Notch* gene was cloned and found to encode a transmembrane receptor containing multiple EGF-like repeats and ankyrin repeats (Wharton et al 1985, Kidd et al 1986). *Notch* is evolutionarily conserved. In mammals,

there are four NOTCH proteins (NOTCH 1, 2, 3 and 4) that are closely related to *Drosophila* Notch.

The ligands for Notch receptors are also integral membrane proteins that are expressed on the surface of neighboring cells. Two Notch ligands, Delta and Serrate, have been identified in *Drosophila* (Fleming 1998). Delta is composed of an extracellular domain containing a single cysteine-rich EGF-like repeat, named DSL (based on the ligands Delta, Serrate, and Lag-2) followed by nine EGF-like repeats and a transmembrane domain. There are three *Delta*-like genes (*Dll1, Dll3,* and *Dll4*) and a gene similar to Delta (*Dlk* or *pref-1*) in mammals. The proteins vary in the number of EGF-like repeats (Callahan & Egan 2004). Like Delta, Serrate is also a single transmembrane protein containing a DSL domain and EFG-like repeats, although the N- and C-terminal sequence of Serrate is less conserved with Delta. The mammalian homologs of *Serrate* are *Jagged1* and *Jagged2* (*Jag1* and *Jag2*) (Fleming 1998, Lissemore & Starmer 1999).

Notch signaling is initiated by the interaction of the Notch receptor with its ligands. Figure 27.13 depicts a simplified scheme of the Notch signaling pathway (Gridley

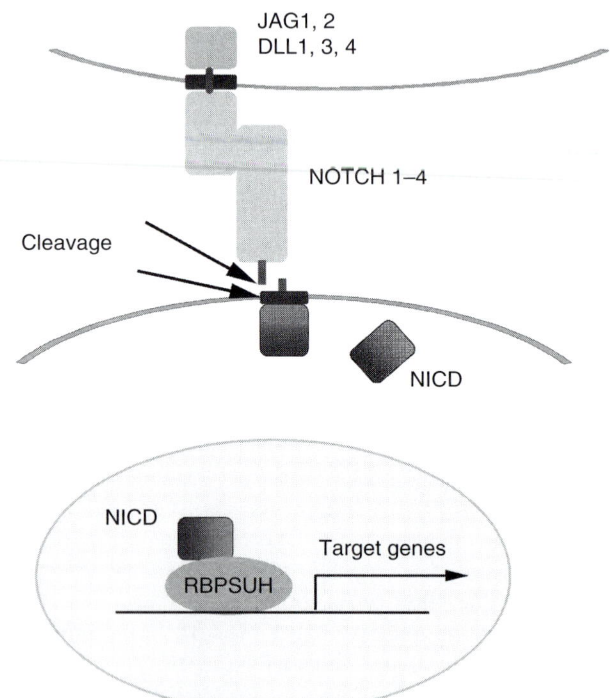

FIGURE 27.13   Notch signaling pathway. Notch family receptors (NOTCH1–NOTCH4) are expressed on the cell surface and exist as a proteolytically cleaved extracellular domain tethered to a membrane domain. Binding of ligands of the Jagged (JAG1 and JAG2) and Delta-like (DLL1, DLL3, DLL4) families induces two additional proteolytic cleavages that release the Notch intracellular domain (NICD). The Notch intracellular domain translocates to the nucleus where it forms a complex with RBPSUH, displacing a histone deacetylase/corepressor complex (not shown), leading to the transcriptional activation of Notch target genes. (see also Plate 58)

2003). These interactions induce proteolytic cleavage of the Notch receptors, which first results in membrane-bound Notch fragments. These fragments are then proteolytically cleaved within the membrane by γ-secretase. The released intracellular fragments of Notch (NICD) translocate to the nucleus and form a complex with the recombining binding protein suppressor of hairless (RBPSUH). The NICD displaces a histone deacetylase (HDAC)/corepressor (CoR) complex from RBPSUH to form transcriptional activator complexes. These complexes activate Notch target genes, such as *Hairy/Enhancer of split* (*Hes*), which encode basic-helix-loop-helix (bHLH) transcriptional repressors that down-regulate genes involved in cell differentiation (Artavanis-Tsakonas et al 1999). The Notch signaling pathway is an evolutionarily conserved intercellular signaling mechanism that is essential for cell fate specification, proliferation, differentiation, and apoptosis.

Notch Signaling and Kidney Development

Notch, its ligands, and several down-stream targets of the Notch signaling pathway are expressed in the kidney. Analysis of E14.5–E17.5 mouse kidney shows expression of *Notch1* and *Notch2* transcripts in renal vesicles, comma- and S-shaped bodies, and in the developing glomeruli. *Notch3* is highly expressed in the vascular smooth cells of renal blood vessels and glomeruli, whereas *Notch4* is only detected in vascular endothelial cells. Weak expression of *Notch3* and *Notch4* is detected in the collecting duct in E17.5 mouse kidneys (Leimeister et al 2003). The Notch ligand, *Jag1*, is expressed in the nephrogenic mesenchyme immediately after the differentiation of the intermediate mesoderm. At E10, *Jag1* is highly expressed in the mesonephric tubules, weakly expressed in the cranial Wolffian duct and periductal cells, but not expressed in the Wolffian duct at the site of the future metanephros. At E10.5, *Jag1* is expressed in both the Wolffian duct and the tip of the ureteric bud but is absent from the ureteric stalk. In the metanephric mesenchyme the expression of *Jag1* is gradually upregulated, and it is highly expressed in the peritubular mesenchyme (Kuure et al 2005). Both *Jag1* and *Dll1* are expressed in the renal vesicle, common- and S-shaped bodies, and the vasculature. McCright et al reported expression of *Jag1* in the collecting duct and ubiquitous expression of *Jag2* in the kidney (McCright et al 2001), but Leimeister et al found no expression of *Jag1* in the collecting duct and expression of *Jag2* only in a subset of vascular cells (Leimeister et al 2003). *Dll3* is not expressed in developing kidney, and *Dll4* expression is restricted to endothelia (Leimeister et al 2003). Similarly, the Notch target genes, *Hes1* and *Hes5*, are expressed in pre-tubular aggregates and comma-shaped bodies of E13.5 kidneys and in newborn kidneys (Piscione et al 2004).

Studies in *Drosophila*, *Xenopus*, and mice support the conserved role of Notch signaling in cell fate specification during the formation of kidney anlagen (McLaughlin et al 2000, Wan et al 2000, Cheng et al 2003, Wang et al 2003). Deletion of *Notch2* or *Jag1* is embryonic lethal prior to renal organogenesis (Hamada et al 1999, Xue et al 1999). Mice that are homozygous for a hypomorphic allele of *Notch2* (*Notch2del1/Notch2del1*) exhibit hypoplastic kidneys and defects in glomerular tuft formation (McCright et al 2001). About 50% of the mice die perinatally due to glomerular malformations. The kidneys of *Notch2/Jag1* double heterozygous mice (*Jag1dDSL/Notch2del1*) are only half the size of control littermates, lack glomerular tufts, and form capillary aneurysms (McCright et al 2002). Inhibition of proteolytic cleavage of Notch results in reduced ureteric bud branching and decreased formation of proximal tubules and podocytes (Cheng et al 2003, Wang et al 2003). More recent studies have confirmed a novel role of Notch signaling in the patterning of the nephron. Deletion of *Notch2* specifically in the metanephric mesenchyme results in hypoplastic kidneys that lack glomeruli and proximal tubules (Cheng et al 2007). These results indicate that *Notch2* is essential for the acquisition of podocyte and proximal tubule cell fates during kidney development.

Crosstalk between Jag1 and GDNF/C-ret/GFRα1 regulates ureteric bud growth and branching in the developing kidney (Kuure et al 2005). Transgenic mice with persistent expression of human JAG1 in the Wolffian duct and its derivatives display a spectrum of renal malformations including renal agenesis, dysplasia, hypoplasia, hydronephrosis, hydroureter, double ureter, duplication of collecting systems, and tubular cyst formation. At E11, RT-PCR analysis shows a 70% reduction in GDNF expression in the metanephric mesenchyme. The normal down-regulation of c-Ret expression in the Wolffian duct does not occur. Instead, there is a persistent expression of c-Ret in the Wolffian duct and in the ureter. Similarly, GFRα1 expression persists in the entire Wolffian duct and does not become restricted to the tips of the ureteric bud. A low level of GFRβ1 is also detected in the metanephric mesenchyme where GFRβ1 is not normally expressed. Ureteric bud budding and branching are significantly impaired in the transgenic mice. However, application of GDNF-soaked beads restores ureteric bud branching and results in an up-regulation of endogenous *Jag1* expression. Up-regulation of Jag1 expression by GDNF suggests that Ret/GFRβ1 signaling may be upstream to Jag1 during early ureteric bud morphogenesis.

Simpson-Golabi-Behmel Syndrome

Clinical Features of Simpson-Golabi-Behmel Syndrome

Simpson-Golabi-Behmel syndrome (SGBS, OMIM 312870) describes an X-liked congenital overgrowth syndrome that includes macrosomia, renal and skeletal abnormalities, and increased risk of developing embryonal cancers. In 1975, Simpson et al reported a family of previously unrecognized 'bulldog'-like dysmorphic features with broad stocky appearance, distinctive facies (large protruding jaw, widened nasal bridge, and upturned nasal tip), enlarged tongue, and broad short hands and fingers (Simpson et al 1975). In 1984, Behmel described the identical syndrome in a family with the new findings of increased birth weight and length (Behmel et al 1984). In the same year, Golabi and Rosen added a midline groove of the tongue, submucosal cleft palate, rib abnormalities, and polydactyly and syndactyly to this syndrome (Golabi & Rosen 1984). The incidence of this disorder is not clear, but it is believed to be rare. There is significant overlap with other overgrowth syndromes, such as Beckwith-Wiedemann syndrome.

Infants born with this syndrome usually have birth weight and length greater than 95% for their gestational ages. Skeletal abnormalities are prominent, and bone age can be advanced as the child grows. Two types of SGBS are recognized, a less severe form (SGBS type 1) and a more severe form (SGBS type 2). Mutations in glypican-3 gene (*GPC3*), which is located on chromosome Xq26, are associated with a relatively milder form of the disorder and account for most of the cases (Pilia et al 1996). A second locus, SGBS2, located on chromosome Xp22, has been shown to be responsible for the more severe from of SGBS (Brzustowicz et al 1999). In the more severe form, affected individuals die in utero or early infancy secondary to cor pulmonale, heart failure from congenital structural defects of the heart or defects in the conducting system, diaphragmatic hernia, overwhelming sepsis, or hypoglycemia due to overproduction of insulin (Terespolsky et al 1995). In the less severe form, affected children may live to adulthood although their height is usually above 195 cm (Hughes-Benzie et al 1992). There are only a few living adults with SGBS reported in the literature, making estimation of life expectancy and disease progression difficult.

SGBS includes a variety of renal and urogenital tract abnormalities. The anomalies include nephromegaly, multicystic dysplastic kidney, hydronephrosis, hydroureter, cryptorchidism, small penis, and adrenorenal fusion (Hughes-Benzie et al 1992). Multicystic dysplastic kidneys have been detected in a human fetus as early as 19 weeks of gestation (Terespolsky et al 1995). Renal function is variable. No kidney transplantation has been reported in the literature, presumably due to the small number of patients and early death of more severely affected infants. Similar to Beckwith-Wiedemann syndrome, individuals with SGBS have an increased risk of cancers, such as Wilms' tumor and neuroblastoma. However, unlike Beckwith-Wiedemann syndrome, SGBS appears to produce an increased risk of hepato-cellular carcinoma and testicular gonadoblastoma (Terespolsky et al 1995, DeBaun et al 2001).

GPC3 AND ITS ROLE IN DEVELOPMENT

Although a more severe form of SGBS has been mapped to the locus on Xp22 (Brzustowicz et al 1999), most

clinically observed cases of SGBS are linked to mutations of glypican-3 (*GPC3*) located on chromosome Xq26 (Pilia et al 1996). Glypicans belong to a family of glycosyl-phosphatidylinositol-linked cell surface heparan sulfate proteoglycans (HSPGs) that have been shown to play a critical role in cell growth during development. Six mammalian glypicans (GPC1–6) and a single *Drosophila* glypican, Dally (division abnormally delayed), have been identified (Nakato et al 1995, Veugelers et al 1998, Paine-Saunders et al 1999). *GPC3* and *GPC4* are contiguous on the X chromosome, which may bear significance to the variability of SGBS (Veugelers et al 1998). All *GPC* genes encode a core protein of molecular weight ~60 kDa, although the sequence identity at the amino acid level is only 20–50%. Two heparan sulfate chains are attached to the core protein.

Among all mammalian glypicans, only mutation of *GPC3* is associated with a phenotype in humans (Pilia et al 1996). *GPC3* spans more than 500 kb of genomic DNA and contains 8 exons. The mRNA for *Gpc3* is detected as early as E7.5 in mouse embryos. In the human fetus, *GPC3* mRNA is detected in mesenchyme-derived tissues in the lung, liver, and kidney (Pilia et al 1996). A detailed analysis of expression of glypicans in the developing mouse kidney shows weak expression of *Gpc3* in comma- and S-shaped bodies at E13.5. Strong expression of *Gpc4* is found in the differentiating tubular epithelial cells derived from metanephric mesenchyme at E15–E16 (Watanabe et al 1995).

Gpc3-deficient mice generated by the removal of the promoter and the first exon are valuable tools to study human SGBS (Cano-Gauci et al 1999). Male $Gpc3^{-/-}$ mice die soon after birth due to respiratory tract infection and mucus accumulation in the airway. The mutant mice exhibit several of the clinical features observed in SGBS patients, including somatic overgrowth, cystic dysplastic kidneys, and abnormal lung development. However, a proportion of the mutant mice also display mandibular hypoplasia and an imperforate vagina, which have not been described in humans with SGBS. Abnormal kidney development is detected at E12 when branching of the ureteric bud is markedly enhanced. At E13.5, the kidney is much larger compared to wild-type littermates and is associated with an increase in cell proliferation. The mesenchyme-derived glomeruli and tubules are disorganized in the cortex at E16.5. At this stage there is enhanced apoptosis and a paucity of tubules in the medulla. which has been replaced by highly proliferative cystic structures. All male $Gpc3^{-/-}$ and some female $Gpc3^{+/-}$ mice have cystic dysplastic kidneys at birth. Somatic overgrowth also begins in utero and the degree of overgrowth is similar to that in mice that are deficient in IGF receptor type 2 (IGF2R), a negative regulator of IGF2. Unlike IGF2R-deficient mice, the levels of IGF2 in Gpc3 knockout mice are similar to those in wild-type mice.

The function of GPC3 has not been clearly elucidated. It has been suggested that dysregulation of IGF2 may be responsible for the clinical manifestations of SGBS. GPC3 interacts with IGF2 and may regulate IGF2 function (Pilia et al 1996). Beckwith-Wiedemann syndrome, another overgrowth syndrome, appears to be caused by relative over-expression of IGF2. Thus overgrowth syndromes may share similar mechanisms. Studies in mouse kidney organ culture have shown that GPC3 deficiency abrogates the inhibitory activity of BMP2 on ureteric bud branching, converts BMP7-dependent inhibition to stimulation, and enhances the stimulatory effects of FGF7 on ureteric bud branching morphogenesis (Grisaru et al 2001). Since tight control of cell proliferation and apoptosis by transcription factors and growth factors is critical for kidney development, perturbation of these pro-cesses may result in renal developmental abnormalities.

CHARGE syndrome

CLINICAL FEATURES OF CHARGE SYNDROME

CHARGE syndrome (OMIM 124800) is an acronym for *c*oloboma of the retina or less commonly the iris, *h*eart anomaly, choanal *a*tresia, *r*etardation (mental), *g*enital and *e*ar anomalies. Facial palsy, cleft palate, and dysphagia are also commonly associated with this syndrome. The constellation of findings has been classified as a true syndrome rather than an association (Verloes 2005). The first descriptions of this syndrome were provided by Hall (Hall 1979) and Hittner (Hittner et al 1979), and therefore this syndrome is also referred to as Hall-Hittner syndrome (HHS). Choanal atresia can be unilateral or bilateral. Because newborn infants are obligate nose breathers, bilateral choanal atresia can cause significant respiratory distress and should be recognized promptly. Choanal atresia should be suspected in a newborn whose respiratory distress disappears when he or she opens the mouth to cry. The diagnosis can be confirmed by the failure of passage of a nasogastric tube. Immediate treatment consists of keeping the infant's mouth open with a large open nipple. A combined report from the California Birth Defects Monitoring Program in the Unites States, Institut Européen des Génomutations in France, and Tornblad Institute at University of Lund in Sweden, indicates that the average rate of choanal atresia is 0.82 per 10 000 births. Diagnosis of CHARGE should be restricted to infants with multiple malformations and choanal atresia and/or coloboma combined with other cardinal malformations (heart, ear, and genital) and with a total of at least three cardinal malformations (Harris et al 1997). Updated diagnostic criteria divide the anomalies into three major criteria (the three 'C:' Coloboma, Choanal atresia, semicircular Canals anomalies) and five minors (rhombencephalic anomalies, hypothalamo-hypophyseal dysfunction, external/middle ear malformations, malformation of mediastinal viscera, and mental retardation). The typical CHARGE syndrome diagnosis should include 3 major criteria, or 2/3 major criteria and 2/5 minors (Verloes 2005). The frequency of the anomalies has been found to be coloboma (79%), heart malformation (85%), choanal atresia (57%), growth and/or mental retardation (100%), genital anomalies (34%), ear anomalies (91%), and deafness (62%) (Tellier et al 1998).

Genitourinary tract abnormalities are diverse and occur in 69% of patients with CHARGE syndrome (Ragan et al 1999). Genital anomalies include micropenis; penile agenesis; hypospadias; chordee; cryptorchidism; bifid scrotum; atresia of the uterus, cervix and vagina; and hypoplastic labia majora, labia minora and clitoris. Renal and other urological malformations include solitary kidney, hydronephrosis, renal hypoplasia, duplex kidneys, nephrolithiasis, ureteropelvic junction obstruction, vesicoureteral reflux, and neurogenic bladder secondary to spinal dysraphism. Gonadotropin levels are at or below minimal detectable levels for age, suggesting that central hypogonadism is responsible not only for the genital hypoplasia in males but also for the lack of secondary sexual development in affected individuals of both sexes (Wheeler et al 2000).

MUTATIONS OF *CHD7* ARE RESPONSIBLE FOR MOST CASES OF CHARGE SYNDROME

Recently, it has been shown that mutations of the chromodomain helicase DNA-binding protein-7 (*CHD7*) are responsible for most cases of CHARGE (Vissers et al 2004). *CHD7* was first cloned in the adult human brain and was found to be highly expressed in the brain, kidney, and skeletal muscle (Nagase et al 2000). *CHD7* consists of 38 exons spanning 188 kb. *CHD7* is expressed throughout the CNS and neural crest-containing mesenchyme of the pharyngeal arches during early human embryonic development (day 26). At day 34, the expression in the CNS is more restricted to the otic vesicle. By 9 weeks, *CHD7* expression is localized to the nasal epithelial, neural retina, optic nerve sheath, and anterior and median lobes of the pituitary gland (Sanlaville et al 2006). Vissers et al found ubiquitous expression of *CHD7* in human fetuses and adults using semi-quantitative RT-PCR (Vissers et al 2004). At present, the function of *CHD7* in development is not clear.

In 2004, Vissers et al first identified mutations of *CHD7* in 12/17 patients with CHARGE syndrome (Vissers et al 2004). Ten heterozygous gene mutations were identified, including seven nonsense mutations, two missense mutations, and one mutation at an intron-exon boundary. Further analysis of 107 index patients with clinical features of CHARGE showed that heterozygous *CHD7* mutations occurred in 69 patients, indicating that haploinsufficiency of *CHD7* could account for most cases of CHARGE syndrome. However, a broad clinical spectrum was noted, and no genotype-phenotype correlation has been observed (Jongmans et al 2006). Most affected individuals have de novo mutations, although rarely, familial cases have been observed (Lalani et al 2006).

ENU mutagenesis has been used to generate multiple mutations of mouse *Chd7* (Bosman et al 2005). During early mouse development, *Chd7* is expressed in the eye, olfactory epithelium, inner ear, and vascular system, which correlates with the sites of anomalies in CHARGE syndrome. Heterozygous mutant mice have cleft palate, choanal atresia, septal defects of the heart and vulva, and clitoral defects and keratoconjunctivitis sicca, and have high incidence of prenatal death. Many of the defects mimic CHARGE syndrome. However, renal abnormalities are not observed. These mouse models may provide usefulness to study the function of Chd7 and the role of CHD7 in human CHARGE syndrome.

ACKNOWLEDGMENTS

We thank Michel Baum for critically reviewing the manuscript. Work from the authors' laboratories was supported by NIH grants K08 DK062839, R01DK42921, R01DK067565, and R01DK066535 and the UT Southwestern O'Brien Kidney Research Core Center (P30DK079328). VP was supported by an NIH training grant (T32DK007257).

References

Abrahamson DR. Glomerulogenesis in the developing kidney. Semin. Nephrol. 1991; 11(4): 375–89.

al-Awqati Q, Goldberg MR. Architectural patterns in branching morphogenesis in the kidney. Kidney Int. 1998; 54(6): 1832–42.

Alagille D, Estrada A, Hadchouel M, Gautier M, Odievre M, Dommergues JP. Syndromic paucity of interlobular bile ducts (Alagille syndrome or arteriohepatic dysplasia): review of 80 cases. J. Pediatr. 1987; 110(2): 195–200.

Amiel J, Audollent S, Joly D, Dureau P, Salomon R, Tellier AL, et al. PAX2 mutations in renal-coloboma syndrome: mutational hotspot and germline mosaicism. Eur. J. Hum. Genet. 2000; 8(11): 820–6.

Armstrong JF, Pritchard-Jones K, Bickmore WA, Hastie ND, Bard JB. The expression of the Wilms' tumour gene, WT1, in the developing mammalian embryo. Mech. Dev. 1993; 40(1–2): 85–97.

Artavanis-Tsakonas S, Delidakis C, Fehon RG. The Notch locus and the cell biology of neuroblast segregation. Annu. Rev. Cell Biol. 1991; 7: 427–52.

Artavanis-Tsakonas S, Rand MD, Lake RJ. Notch signaling: cell fate control and signal integration in development. Science 1999; 284(5415): 770–6.

Baird PN, Simmons PJ. Expression of the Wilms' tumor gene (WT1) in normal hemopoiesis. Exp. Hematol. 1997; 25(4): 312–20.

Barasch J, Yang J, Ware CB, Taga T, Yoshida K, Erdjument-Bromage H, et al. Mesenchymal to epithelial conversion in rat metanephros is induced by LIF. Cell 1999; 99(4): 377–86.

Barbacci E, Reber M, Ott MO, Breillat C, Huetz F, Cereghini S. Variant hepatocyte nuclear factor 1 is required for visceral endoderm specification. Development 1999; 126(21): 4795–805.

Bardeesy N, Pelletier J. Overlapping RNA and DNA binding domains of the wt1 tumor suppressor gene product. Nucleic Acids Res. 1998; 26(7): 1784–92.

Behmel A, Plochl E, Rosenkranz W. A new X-linked dysplasia gigantism syndrome: identical with the Simpson dysplasia syndrome? Hum. Genet. 1984; 67(4): 409–13.

Bell GI, Polonsky KS. Diabetes mellitus and genetically programmed defects in beta-cell function. Nature 2001; 414(6865): 788–91.

Bellanne-Chantelot C, Chauveau D, Gautier JF, Dubois-Laforgue D, Clauin S, Beaufils S, et al. Clinical spectrum associated with hepatocyte nuclear factor-1beta mutations. Ann. Intern. Med. 2004; 140(7): 510–17.

Bingham C, Ellard S, Allen L, Bulman M, Shepherd M, Frayling T, et al. Abnormal nephron development associated with a frameshift mutation in the transcription factor hepatocyte nuclear factor-1 beta. Kidney Int. 2000; 57(3): 898–907.

Birchmeier C, Birchmeier W. Molecular aspects of mesenchymal-epithelial interactions. Annu. Rev. Cell Biol. 1993; 9: 511–40.

Blanck C, Kohlhase J, Engels S, Burfeind P, Engel W, Bottani A, et al. Three novel SALL1 mutations extend the mutational spectrum in Townes-Brocks syndrome. J. Med. Genet. 2000; 37(4): 303–7.

Bosman EA, Penn AC, Ambrose JC, Kettleborough R, Stemple DL, Steel KP. Multiple mutations in mouse Chd7 provide models for CHARGE syndrome. Hum. Mol. Genet. 2005; 14(22): 3463–76.

Botzenhart EM, Bartalini G, Blair E, Brady AF, Elmslie F, Chong KL, et al. Townes-Brocks syndrome: twenty novel SALL1 mutations in sporadic and familial cases and refinement of the SALL1 hot spot region. Hum. Mutat. 2007; 28(2): 204–5.

Bouchard M, Souabni A, Mandler M, Neubuser A, Busslinger M. Nephric lineage specification by Pax2 and Pax8. Genes Dev. 2002; 16(22): 2958–70.

Bron AJ, Burgess SE, Awdry PN, Oliver D, Arden G. Papillo-renal syndrome. An inherited association of optic disc dysplasia and renal disease. Report and review of the literature. Ophth. Paediatr. Genet. 1989; 10(3): 185–98.

Brophy PD, Ostrom L, Lang KM, Dressler GR. Regulation of ureteric bud outgrowth by Pax2-dependent activation of the glial derived neurotrophic factor gene. Development 2001; 128(23): 4747–56.

Bruening W, Bardeesy N, Silverman BL, Cohn RA, Machin GA, Aronson AJ, et al. Germline intronic and exonic mutations in the Wilms' tumour gene (WT1) affecting urogenital development. Nat. Gen. 1992; 1(2): 144–8.

Bruening W, Pelletier J. A non-AUG translational initiation event generates novel WT1 isoforms. J. Bio. Chem. 1996; 271(15): 8646–54.

Brzustowicz LM, Farrell S, Khan MB, Weksberg R. Mapping of a new SGBS locus to chromosome Xp22 in a family with a severe form of Simpson-Golabi-Behmel syndrome. Am. J. Hum. Genet. 1999; 65(3): 779–83.

Buck A, Archangelo L, Dixkens C, Kohlhase J. Molecular cloning, chromosomal localization, and expression of the murine SALL1 ortholog Sall1. Cytogenet. Cell. Genet. 2000; 89(3–4): 150–3.

Callahan R, Egan SE. Notch signaling in mammary development and oncogenesis. J. Mammary Gland Biol. Neoplasia 2004; 9(2): 145–63.

Cano-Gauci DF, Song HH, Yang H, McKerlie C, Choo B, Shi W, et al. Glypican-3-deficient mice exhibit developmental overgrowth and some of the abnormalities typical of Simpson-Golabi-Behmel syndrome. J. Cell Biol. 1999; 146(1): 255–64.

Caricasole A, Duarte A, Larsson SH, Hastie ND, Little M, Holmes G, et al. RNA binding by the Wilms tumor suppressor zinc finger proteins. Proc. Nat. Acad. Sci. USA 1996; 93(15): 7562–6.

Carroll TJ, McMahon AP. The Kidney: From Normal Development to Congenital Disease. San Diego: Academic Press, 2003.

Carroll TJ, Park JS, Hayashi S, Majumdar A, McMahon AP. Wnt9b plays a central role in the regulation of mesenchymal to epithelial transitions underlying organogenesis of the mammalian urogenital system. Dev. Cell 2005; 9(2): 283–92.

Challen G, Gardiner B, Caruana G, Kostoulias X, Martinez G, Crowe M, et al. Temporal and spatial transcriptional programs in murine kidney development. Physiol. Genom. 2005.

Cheng HT, Kim M, Valerius MT, Surendran K, Schuster-Gossler K, Gossler A, et al. Notch2, but not Notch1, is required for proximal fate acquisition in the mammalian nephron. Development 2007; 134(4): 801–11.

Cheng HT, Miner JH, Lin M, Tansey MG, Roth K, Kopan R. Gamma-secretase activity is dispensable for mesenchyme-to-epithelium transition but required for podocyte and proximal tubule formation in developing mouse kidney. Development 2003; 130(20): 5031–42.

Clark P, Dziarmaga A, Eccles M, Goodyer P. Rescue of defective branching nephrogenesis in renal-coloboma syndrome by the caspase inhibitor, Z-VAD-fmk. J. Am. Soc. Nephrol. 2004; 15(2): 299–305.

Clarke JC, Patel SR, Raymond RM, Jr., Andrew S, Robinson BG, Dressler GR, et al. Regulation of c-Ret in the developing kidney is responsive to Pax2 gene dosage. Hum. Mol. Gen. 2006; 15(23): 3420–8.

Coffinier C, Barra J, Babinet C, Yaniv M. Expression of the vHNF1/HNF1beta homeoprotein gene during mouse organogenesis. Mech. Dev. 1999; 89(1–2): 211–13.

Coffinier C, Gresh L, Fiette L, Tronche F, Schutz G, Babinet C, et al. Bile system morphogenesis defects and liver dysfunction upon targeted deletion of HNF1beta. Development 2002; 129(8): 1829–38.

de Conciliis L, Marchitiello A, Wapenaar MC, Borsani G, Giglio S, Mariani M, et al. Characterization of Cxorf5 (71-7A), a novel human cDNA mapping to Xp22 and encoding a protein containing coiled-coil alpha-helical domains. Genomics 1998; 51(2): 243–50.

de Vries-Van der Weerd MA, Willems PJ, Mandema HM, ten Kate LP. A new family with the Townes-Brocks syndrome. Clin. Genet. 1988; 34(3): 195–200.

DeBaun MR, Ess J, Saunders S. Simpson Golabi Behmel syndrome: progress toward understanding the molecular basis for overgrowth, malformation, and cancer predisposition. Mol. Genet. Metab. 2001; 72(4): 279–86.

Decramer S, Parant O, Beaufils S, Clauin S, Guillou C, Kessler S, et al. Anomalies of the TCF2 gene are the main cause of fetal bilateral hyperechogenic kidneys. J. Am. Soc. Nephrol. 2007; 18(3): 923–33.

Dehbi M, Ghahremani M, Lechner M, Dressler G, Pelletier J. The paired-box transcription factor, PAX2, positively modulates expression of the Wilms' tumor suppressor gene (WT1). Oncogene 1996 Aug 1; 13(3): 447–53.

Denys P, Malvaux P, Van Den Berghe H, Tanghe W, Proesmans W. Association of an anatomo-pathological syndrome of male pseudohermaphroditism, Wilms' tumor, parenchymatous nephropathy and XX/XY mosaicism. Arch. Fr. Pediatr. 1967; 24(7): 729–39.

Devriendt K, Matthijs G, Van Damme B, Van Caesbroeck D, Eccles M, Vanrenterghem Y, et al. Missense mutation and hexanucleotide duplication in the PAX2 gene in two unrelated families with renal-coloboma syndrome (MIM 120330). Hum. Genet. 1998; 103(2): 149–53.

Dillon MJ, Goonasekera CD. Reflux nephropathy. J. Am. Soc. Nephrol. 1998; 9(12): 2377–83.

Discenza MT, Pelletier J. Insights into the physiological role of WT1 from studies of genetically modified mice. Physiol. Genomics 2004; 16(3): 287–300.

Doege TC, Thuline HC, Priest JH, Norby DE, Bryant JS. Studies of a family with the Oral-Facial-Digital syndrome. N. Engl. J. Med. 1964; 271: 1073–8.

Donnai D, Kerzin-Storrar L, Harris R. Familial orofaciodigital syndrome type I presenting as adult polycystic kidney disease. J. Med. Genet. 1987; 24(2): 84–7.

Drash A, Sherman F, Hartmann WH, Blizzard RM. A syndrome of pseudohermaphroditism, Wilms' tumor, hypertension, and degenerative renal disease. J. Pediatr. 1970; 76(4): 585–93.

Dressler GR, Wilkinson JE, Rothenpieler UW, Patterson LT, Williams-Simons L, Westphal H. Deregulation of Pax-2 expression in transgenic mice generates severe kidney abnormalities. Nature 1993; 362(6415): 65–7.

Dressler GR, Woolf AS. Pax2 in development and renal disease. Int. J. Dev. Biol. 1999; 43(5): 463–8.

Drummond IA, Madden SL, Rohwer-Nutter P, Bell GI, Sukhatme VP, Rauscher FJ, 3rd. Repression of the insulin-like growth factor II gene by the Wilms tumor suppressor WT1. Science 1992; 257(5070): 674–8.

Dziarmaga A, Eccles M, Goodyer P. Suppression of ureteric bud apoptosis rescues nephron endowment and adult renal function in Pax2 mutant mice. J. Am. Soc. Nephrol. 2006; 17(6): 1568–75.

Eccles MR, Schimmenti LA. Renal-coloboma syndrome: a multisystem developmental disorder caused by PAX2 mutations. Clin. Genet. 1999; 56(1): 1–9.

Eddy AA, Mauer SM. Pseudohermaphroditism, glomerulopathy, and Wilms tumor (Drash syndrome): frequency in end-stage renal failure. J. Pediatr. 1985; 106(4): 584–7.

Edghill EL, Bingham C, Ellard S, Hattersley AT. Mutations in hepatocyte nuclear factor-1beta and their related phenotypes. J. Med. Genet. 2006; 43(1): 84–90.

Emerick KM, Rand EB, Goldmuntz E, Krantz ID, Spinner NB, Piccoli DA. Features of Alagille syndrome in 92 patients: frequency and relation to prognosis. Hepatology 1999; 29(3): 822–9.

Englert C, Hou X, Maheswaran S, Bennett P, Ngwu C, Re GG, et al. WT1 suppresses synthesis of the epidermal growth factor receptor and induces apoptosis. EMBO J. 1995; 14(19): 4662–75.

Fajans SS, Bell GI, Polonsky KS. Molecular mechanisms and clinical pathophysiology of maturity-onset diabetes of the young. N. Engl. J. Med. 2001; 345(13): 971–80.

Feather SA, Winyard PJ, Dodd S, Woolf AS. Oral-facial-digital syndrome type 1 is another dominant polycystic kidney disease: clinical, radiological and histopathological features of a new kindred. Nephrol. Dial. Transplant. 1997a; 12(7): 1354–61.

Feather SA, Woolf AS, Donnai D, Malcolm S, Winter RM. The oral-facial-digital syndrome type 1 (OFD1), a cause of polycystic kidney disease and associated malformations, maps to Xp22.2–Xp22.3. Hum. Mol. Genet. 1997b; 6(7): 1163–7.

Ferrante MI, Barra A, Truong JP, Banfi S, Disteche CM, Franco B. Characterization of the OFD1/Ofd1 genes on the human and mouse sex chromosomes and exclusion of Ofd1 for the Xpl mouse mutant. Genomics 2003; 81(6): 560–9.

Ferrante MI, Giorgio G, Feather SA, Bulfone A, Wright V, Ghiani M, et al. Identification of the gene for oral-facial-digital type I syndrome. Am. J. Hum. Genet. 2001; 68(3): 569–76.

Ferrante MI, Zullo A, Barra A, Bimonte S, Messaddeq N, Studer M, et al. Oral-facial-digital type I protein is required for primary cilia formation and left-right axis specification. Nat. Genet. 2006; 38(1): 112–17.

Fischer E, Legue E, Doyen A, Nato F, Nicolas JF, Torres V, et al. Defective planar cell polarity in polycystic kidney disease. Nat. Genet. 2006; 38(1): 21–3.

Fleming RJ. Structural conservation of Notch receptors and ligands. Semin. Cell. Dev. Biol. 1998; 9(6): 599–607.

Froguel P, Vaxillaire M, Sun F, Velho G, Zouali H, Butel MO, et al. Close linkage of glucokinase locus on chromosome 7p to early-onset non-insulin-dependent diabetes mellitus. Nature 1992; 356(6365): 162–4.

Furuta H, Furuta M, Sanke T, Ekawa K, Hanabusa T, Nishi M, et al. Nonsense and missense mutations in the human hepatocyte nuclear factor-1 beta gene (TCF2) and their relation to type 2 diabetes in Japanese. J. Clin. Endocrinol. Metab. 2002; 87(8): 3859–63.

Gao F, Maiti S, Sun G, Ordonez NG, Udtha M, Deng JM, et al. The Wt1+/R394W mouse displays glomerulosclerosis and early-onset renal failure characteristic of human Denys-Drash syndrome. Mol. Cell. Biol. 2004; 24(22): 9899–910.

Gedeon AK, Oley C, Nelson J, Turner G, Mulley JC. Gene localization for oral-facial-digital syndrome type 1 (OFD1:MIM 311200) proximal to DXS85. Am. J. Med. Genet. 1999; 82(4): 352–4.

Gessler M, Konig A, Arden K, Grundy P, Orkin S, Sallan S, et al. Infrequent mutation of the WT1 gene in 77 Wilms' Tumors. Hum. Mutat. 1994; 3(3): 212–22.

Gessler M, Poustka A, Cavenee W, Neve RL, Orkin SH, Bruns GA. Homozygous deletion in Wilms tumours of a zinc-finger gene identified by chromosome jumping. Nature 1990; 343(6260): 774–8.

Golabi M, Rosen L. A new X-linked mental retardation-overgrowth syndrome. Am. J. Med. Genet. 1984; 17(1): 345–58.

Gresh L, Fischer E, Reimann A, Tanguy M, Garbay S, Shao X, et al. A transcriptional network in polycystic kidney disease. EMBO J. 2004; 23(7): 1657–68.

Gridley T. Notch signaling and inherited disease syndromes. Hum. Mol. Genet. 2003; Spec No 1:R9-13

Grisaru S, Cano-Gauci D, Tee J, Filmus J, Rosenblum ND. Glypican-3 modulates BMP- and FGF-mediated effects during renal branching morphogenesis. Dev. Biol. 2001; 231(1): 31–46.

Grobstein C. Inductive epitheliomesenchymal interaction in cultured organ rudiments of the mouse. Science 1953a; 118(3054): 52–5.

Grobstein C. Morphogenetic interaction between embryonic mouse tissues separated by a membrane filter. Nature 1953b; 172(4384): 869–70.

Grobstein C. Trans-filter induction of tubules in mouse metanephrogenic mesenchyme. Exp. Cell Res. 1956; 10(2): 424–40.

Guo G, Morrison DJ, Licht JD, Quaggin SE. WT1 activates a glomerular-specific enhancer identified from the human nephrin gene. J. Am. Soc. Nephrol. 2004; 15(11): 2851–6.

Haber DA, Buckler AJ, Glaser T, Call KM, Pelletier J, Sohn RL, et al. An internal deletion within an 11p13 zinc finger gene contributes to the development of Wilms' tumor. Cell 1990; 61(7): 1257–69.

Habib R, Dommergues JP, Gubler MC, Hadchouel M, Gautier M, Odievre M, et al. Glomerular mesangiolipidosis in Alagille syndrome (arteriohepatic dysplasia). Pediatr. Nephrol. 1987; 1(3): 455–64.

Hall BD. Choanal atresia and associated multiple anomalies. J. Pediatr. 1979; 95(3): 395–8.

Hamada Y, Kadokawa Y, Okabe M, Ikawa M, Coleman JR, Tsujimoto Y. Mutation in ankyrin repeats of the mouse Notch2 gene induces early embryonic lethality. Development 1999; 126(15): 3415–24.

Hammes A, Guo JK, Lutsch G, Leheste JR, Landrock D, Ziegler U, et al. Two splice variants of the Wilms' tumor 1 gene have distinct functions during sex determination and nephron formation. Cell 2001; 106(3): 319–29.

Harris J, Robert E, Kallen B. Epidemiology of choanal atresia with special reference to the CHARGE association. Pediatrics 1997; 99(3): 363–7.

Haumaitre C, Barbacci E, Jenny M, Ott MO, Gradwohl G, Cereghini S. Lack of TCF2/vHNF1 in mice leads to pancreas agenesis. Proc. Nat. Acad. Sci. USA 2005; 102(5): 1490–5.

Hiesberger T, Bai Y, Shao X, McNally BT, Sinclair AM, Tian X, et al. Mutation of hepatocyte nuclear factor-1beta inhibits Pkhd1 gene expression and produces renal cysts in mice. J. Clin. Invest. 2004; 113(6): 814–25.

Hiesberger T, Shao X, Gourley E, Reimann A, Pontoglio M, Igarashi P. Role of the hepatocyte nuclear factor-1beta (HNF-1beta) C-terminal domain in Pkhd1 (ARPKD) gene transcription and renal cystogenesis. J. Biol. Chem. 2005 Mar 18; 280(11): 10578–86.

Hinchliffe SA, Sargent PH, Howard CV, Chan YF, van Velzen D. Human intrauterine renal growth expressed in absolute number of glomeruli assessed by the disector method and Cavalieri principle. Lab. Invest. 1991; 64(6): 777–84.

Hiraoka M, Taniguchi T, Nakai H, Kino M, Okada Y, Tanizawa A, et al. No evidence for AT2R gene derangement in human urinary tract anomalies. Kidney Int. 2001; 59(4): 1244–9.

Hittner HM, Hirsch NJ, Kreh GM, Rudolph AJ. Colobomatous microphthalmia, heart disease, hearing loss, and mental retardation–a syndrome. J. Pediatr. Ophthalmol. Strabismus 1979; 16(2): 122–8.

Horikawa Y, Iwasaki N, Hara M, Furuta H, Hinokio Y, Cockburn BN, et al. Mutation in hepatocyte nuclear factor-1 beta gene (TCF2) associated with MODY. Nat. Genet. 1997; 17(4): 384–5.

Hosono S, Luo X, Hyink DP, Schnapp LM, Wilson PD, Burrow CR, et al. WT1 expression induces features of renal epithelial differentiation in mesenchymal fibroblasts. Oncogene 1999 Jan 14; 18(2): 417–27.

Hossain A, Saunders GF. The human sex-determining gene SRY is a direct target of WT1. J. Bio. Chem. 2001; 276(20): 16817–23.

Hughes-Benzie RM, Hunter AG, Allanson JE, Mackenzie AE. Simpson-Golabi-Behmel syndrome associated with renal dysplasia and embryonal tumor: localization of the gene to Xqcen-q21. Am. J. Med. Genet. 1992; 43(1–2): 428–35.

Hughson M, Farris AB, 3rd, Douglas-Denton R, Hoy WE, Bertram JF. Glomerular number and size in autopsy kidneys: the relationship to birth weight. Kidney Int. 2003; 63(6): 2113–22.

Jadresic L, Leake J, Gordon I, Dillon MJ, Grant DB, Pritchard J, et al. Clinicopathologic review of twelve children with nephropathy, Wilms tumor, and genital abnormalities (Drash syndrome). J. Pediatr. 1990; 117(5): 717–25.

Jongmans MC, Admiraal RJ, van der Donk KP, Vissers LE, Baas AF, Kapusta L, et al. CHARGE syndrome: the phenotypic spectrum of mutations in the CHD7 gene. J. Med. Genet. 2006; 43(4): 306–14.

Kamath BM, Krantz ID, Spinner NB, Heubi JE, Piccoli DA. Monozygotic twins with a severe form of Alagille syndrome and phenotypic discordance. Am. J. Med. Genet. 2002; 112(2): 194–7.

Karcher H. [The morning glory syndrome]. Klin. Monatsbl. Augenheilkd. 1979; 175(6): 835–40.

Keller G, Zimmer G, Mall G, Ritz E, Amann K. Nephron number in patients with primary hypertension. N. Engl. J. Med. 2003; 348: 101–8.

Kennedy D, Ramsdale T, Mattick J, Little M. An RNA recognition motif in Wilms' tumour protein (WT1) revealed by structural modelling. Nat. Genet. 1996; 12(3): 329–31.

Kent J, Coriat AM, Sharpe PT, Hastie ND, van Heyningen V. The evolution of WT1 sequence and expression pattern in the vertebrates. Oncogene 1995; 11(9): 1781–92.

Kidd S, Kelley MR, Young MW. Sequence of the notch locus of Drosophila melanogaster: relationship of the encoded protein to mammalian clotting and growth factors. Mol. Cell. Biol. 1986; 6(9): 3094–108.

Kiefer SM, McDill BW, Yang J, Rauchman M. Murine Sall1 represses transcription by recruiting a histone deacetylase complex. J. Bio. Chem. 2002; 277(17): 14869–76.

Kiefer SM, Ohlemiller KK, Yang J, McDill BW, Kohlhase J, Rauchman M. Expression of a truncated Sall1 transcriptional repressor is responsible for Townes-Brocks syndrome birth defects. Hum. Mol. Gen. 2003; 12(17): 2221–7.

Kim J, Prawitt D, Bardeesy N, Torban E, Vicaner C, Goodyer P, et al. The Wilms' tumor suppressor gene (wt1) product regulates Dax-1 gene expression during gonadal differentiation. Mol. Cell. Biol. 1999; 19(3): 2289–99.

King-Underwood L, Renshaw J, Pritchard-Jones K. Mutations in the Wilms' tumor gene WT1 in leukemias. Blood. 1996; 87(6): 2171–9.

Kohlhase J, Hausmann S, Stojmenovic G, Dixkens C, Bink K, Schulz-Schaeffer W, et al. SALL3, a new member of the human spalt-like gene family, maps to 18q23. Genomics 1999a; 62(2): 216–22.

Kohlhase J, Heinrich M, Schubert L, Liebers M, Kispert A, Laccone F, et al. Okihiro syndrome is caused by SALL4 mutations. Hum. Mol. Genet. 2002; 11(23): 2979–87.

Kohlhase J, Schuh R, Dowe G, Kuhnlein RP, Jackle H, Schroeder B, et al. Isolation, characterization, and organ-specific expression of two novel human zinc finger genes related to the Drosophila gene spalt. Genomics 1996; 38(3): 291–8.

Kohlhase J, Taschner PE, Burfeind P, Pasche B, Newman B, Blanck C, et al. Molecular analysis of SALL1 mutations in Townes-Brocks syndrome. Am. J. Hum. Genet. 1999b; 64(2): 435–45.

Kohlhase J, Wischermann A, Reichenbach H, Froster U, Engel W. Mutations in the SALL1 putative transcription factor gene cause Townes-Brocks syndrome. Nat. Genet. 1998; 18(1): 81–3.

Koziell A, Charmandari E, Hindmarsh PC, Rees L, Scambler P, Brook CG. Frasier syndrome, part of the Denys Drash continuum or simply a WT1 gene associated disorder of intersex and nephropathy? Clin. Endocrinol. (Oxf) 2000; 52(4): 519–24.

Krantz ID, Piccoli DA, Spinner NB. Alagille syndrome. J. Med. Genet. 1997; 34(2): 152–7.

Kreidberg JA, Sariola H, Loring JM, Maeda M, Pelletier J, Housman D, et al. WT-1 is required for early kidney development. Cell 1993; 74(4): 679–91.

Kurnit DM, Steele MW, Pinsky L, Dibbins A. Autosomal dominant transmission of a syndrome of anal, ear, renal, and radial congenital malformations. J. Pediatr. 1978; 93(2): 270–3.

Kuure S, Sainio K, Vuolteenaho R, Ilves M, Wartiovaara K, Immonen T, et al. Crosstalk between Jagged1 and GDNF/Ret/GFRalpha1 signalling regulates ureteric budding and branching. Mech. Dev. 2005; 122(6): 765–80.

Lalani SR, Safiullah AM, Fernbach SD, Harutyunyan KG, Thaller C, Peterson LE, et al. Spectrum of CHD7 mutations in 110 individuals with CHARGE syndrome and genotype–phenotype correlation. Am. J. Hum. Genet. 2006; 78(2): 303–14.

Lang D, Powell SK, Plummer RS, Young KP, Ruggeri BA. PAX genes: roles in development, pathophysiology, and cancer. Biochem. Pharmacol. 2007; 73(1): 1–14.

Lee SB, Huang K, Palmer R, Truong VB, Herzlinger D, Kolquist KA, et al. The Wilms tumor suppressor WT1 encodes a transcriptional activator of amphiregulin. Cell 1999; 98(5): 663–73.

Leimeister C, Schumacher N, Gessler M. Expression of Notch pathway genes in the embryonic mouse metanephros suggests a role in proximal tubule development. Gene. Expr. Patterns 2003; 3(5): 595–8.

Lewis M, Shaw J. Report from the Paediatirc Renal Registry 2002.

Li L, Krantz ID, Deng Y, Genin A, Banta AB, Collins CC, et al. Alagille syndrome is caused by mutations in human Jagged1, which encodes a ligand for Notch1. Nat. Genet. 1997; 16(3): 243–51.

Lindner TH, Njolstad PR, Horikawa Y, Bostad L, Bell GI, Sovik O. A novel syndrome of diabetes mellitus, renal dysfunction and genital malformation associated with a partial deletion of the pseudo-POU domain of hepatocyte nuclear factor-1beta. Hum. Mol. Genet. 1999; 8(11): 2001–8.

Lissemore JL, Starmer WT. Phylogenetic analysis of vertebrate and invertebrate Delta/Serrate/LAG-2 (DSL) proteins. Mol. Phylogenet. Evol. 1999; 11(2): 308–19.

Little M, Holmes G, Walsh P. WT1: what has the last decade told us? Bioessays 1999; 21(3): 191–202.

Little M, Wells C. A clinical overview of WT1 gene mutations. Hum. Mutat. 1997; 9(3): 209–25.

Ma Y, Chai L, Cortez SC, Stopa EG, Steinhoff MM, Ford D, et al. SALL1 expression in the human pituitary-adrenal/gonadal axis. J. Endocrinol. 2002; 173(3): 437–48.

Ma Y, Singer DB, Gozman A, Ford D, Chai L, Steinhoff MM, et al. Hsal 1 is related to kidney and gonad development and is expressed in Wilms tumor. Pediatr. Nephrol. 2001; 16(9): 701–9.

Ma Z, Gong Y, Patel V, Karner CM, Fischer E, Hiesberger T, et al. Mutations of HNF-1beta inhibit epithelial morphogenesis through dysregulation of SOCS-3. Proc. Nat. Acad. Sci. USA 2007; 104(51): 20386–91.

Mache CJ, Preisegger KH, Kopp S, Ratschek M, Ring E. De novo HNF-1 beta gene mutation in familial hypoplastic glomerulocystic kidney disease. Pediatr. Nephrol. 2002; 17(12): 1021–6.

Malecki MT, Jhala US, Antonellis A, Fields L, Doria A, Orban T, et al. Mutations in NEUROD1 are associated with the development of type 2 diabetes mellitus. Nat. Genet. 1999; 23(3): 323–8.

Martin SR, Garel L, Alvarez F. Alagille's syndrome associated with cystic renal disease. Arch. Dis. Child. 1996; 74(3): 232–5.

Martinez-Frias ML, Bermejo Sanchez E, Arroyo Carrera I, Perez Fernandez JL, Pardo Romero M, Buron Martinez E, et al. The Townes-Brocks syndrome in Spain: the epidemiological aspects in a consecutive series of cases. An. Esp. Pediatr. 1999; 50(1): 57–60.

Maurer U, Brieger J, Weidmann E, Mitrou PS, Hoelzer D, Bergmann L. The Wilms' tumor gene is expressed in a subset of CD34+ progenitors and downregulated early in the course of differentiation in vitro. Exp. Hematol. 1997; 25(9): 945–50.

Mayo MW, Wang CY, Drouin SS, Madrid LV, Marshall AF, Reed JC, et al. WT1 modulates apoptosis by transcriptionally upregulating the bcl-2 proto-oncogene. EMBO J. 1999; 18(14): 3990–4003.

McCright B, Gao X, Shen L, Lozier J, Lan Y, Maguire M, et al. Defects in development of the kidney, heart and eye vasculature in mice homozygous for a hypomorphic Notch2 mutation. Development 2001; 128(4): 491–502.

McCright B, Lozier J, Gridley T. A mouse model of Alagille syndrome: Notch2 as a genetic modifier of Jag1 haploinsufficiency. Development 2002; 129(4): 1075–82.

McDaniell R, Warthen DM, Sanchez-Lara PA, Pai A, Krantz ID, Piccoli DA, et al. NOTCH2 mutations cause Alagille syndrome, a heterogeneous disorder of the notch signaling pathway. Am. J. Hum. Genet. 2006; 79(1): 169–73.

McEnery PT, Stablein DM, Arbus G, Tejani A. Renal transplantation in children. A report of the North American Pediatric Renal Transplant Cooperative Study. N. Engl. J. Med. 1992; 326(26): 1727–32.

McLaughlin KA, Rones MS, Mercola M. Notch regulates cell fate in the developing pronephros. Dev. Biol. 2000; 227(2): 567–80.

Mendel DB, Hansen LP, Graves MK, Conley PB, Crabtree GR. HNF-1 alpha and HNF-1 beta (vHNF-1) share dimerization and homeo domains, but not activation domains, and form heterodimers in vitro. Genes Dev. 1991; 5(6): 1042–56.

Menke AL, Fleming S, Ross A, Medine CN, Patek CE, et al. The wt1-heterozygous mouse; a model to study the development of glomerular sclerosis. J. Pathol. 2003; 200(5): 667–74.

Menzel R, Kaisaki PJ, Rjasanowski I, Heinke P, Kerner W, Menzel S. A low renal threshold for glucose in diabetic patients with a mutation in the hepatocyte nuclear factor-1alpha (HNF-1alpha) gene. Diabet. Med. 1998; 15(10): 816–20.

Mohr O. Character changes caused by mutation of an entire region of a chromosome in *Drosophila*. Genetics 1919; 4: 274–82.

Moore AW, McInnes L, Kreidberg J, Hastie ND, Schedl A. YAC complementation shows a requirement for Wt1 in the development of epicardium, adrenal gland and throughout nephrogenesis. Development 1999; 126(9): 1845–57.

Moore AW, Schedl A, McInnes L, Doyle M, Hecksher-Sorensen J, Hastie ND. YAC transgenic analysis reveals Wilms' tumour 1 gene activity in the proliferating coelomic epithelium, developing diaphragm and limb. Mech. Dev. 1998; 79(1–2): 169–84.

Moore MW, Klein RD, Farinas I, Sauer H, Armanini M, Phillips H, et al. Renal and neuronal abnormalities in mice lacking GDNF. Nature 1996; 382(6586): 76–9.

Morris JF, Madden SL, Tournay OE, Cook DM, Sukhatme VP, Rauscher FJ 3rd. Characterization of the zinc finger protein encoded by the WT1 Wilms' tumor locus. Oncogene 1991; 6(12): 2339–48.

Morrissette JD, Colliton RP, Spinner NB. Defective intracellular transport and processing of JAG1 missense mutations in Alagille syndrome. Hum. Mol. Genet. 2001; 10(4): 405–13.

Mueller RF. The Alagille syndrome (arteriohepatic dysplasia). J. Med. Genet. 1987; 24(10): 621–6.

Mueller RF. The Denys-Drash syndrome. J. Med. Genet. 1994; 31(6): 471–7.

Mundlos S, Pelletier J, Darveau A, Bachmann M, Winterpacht A, Zabel B. Nuclear localization of the protein encoded by the Wilms' tumor gene WT1 in embryonic and adult tissues. Development 1993; 119(4): 1329–41.

Nagase T, Kikuno R, Ishikawa KI, Hirosawa M, Ohara O. Prediction of the coding sequences of unidentified human genes. XVI. The complete sequences of 150 new cDNA clones from brain which code for large proteins in vitro. DNA Res. 2000; 7(1): 65–73.

Nakato H, Futch TA, Selleck SB. The division abnormally delayed (dally) gene: a putative integral membrane proteoglycan required for cell division patterning during postembryonic development of the nervous system in Drosophila. Development 1995; 121(11): 3687–702.

Natoli TA, Liu J, Eremina V, Hodgens K, Li C, Hamano Y, et al. A mutant form of the Wilms' tumor suppressor gene WT1 observed in Denys-Drash syndrome interferes with glomerular capillary development. J. Am. Soc. Nephrol. 2002; 13(8): 2058–67.

Netzer C, Bohlander SK, Rieger L, Muller S, Kohlhase J. Interaction of the developmental regulator SALL1 with UBE2I and SUMO-1. Biochem. Biophys. Res. Commun. 2002; 296(4): 870–6.

Netzer C, Rieger L, Brero A, Zhang CD, Hinzke M, Kohlhase J, et al. SALL1, the gene mutated in Townes-Brocks syndrome, encodes a transcriptional repressor which interacts with TRF1/PIN2 and localizes to pericentromeric heterochromatin. Hum. Mol. Genet. 2001; 10(26): 3017–24.

Neve B, Fernandez-Zapico ME, Ashkenazi-Katalan V, Dina C, Hamid YH, Joly E, et al. Role of transcription factor KLF11 and its diabetes-associated gene variants in pancreatic beta cell function. Proc. Nat. Acad. Sci. USA 2005; 102(13): 4807–12.

Newman WG, Brunet MD, Donnai D. Townes-Brocks syndrome presenting as end stage renal failure. Clin. Dysmorphol. 1997; 6(1): 57–60.

Nishigori H, Yamada S, Kohama T, Tomura H, Sho K, Horikawa Y, et al. Frameshift mutation, A263fsinsGG, in the hepatocyte nuclear factor-1beta gene associated with diabetes and renal dysfunction. Diabetes 1998; 47(8): 1354–55.

Nishimura H, Yerkes E, Hohenfellner K, Miyazaki Y, Ma J, Hunley TE, et al. Role of the angiotensin type 2 receptor gene in congenital anomalies of the kidney and urinary tract, CAKUT, of mice and men. Mol. Cell 1999; 3(1): 1–10.

Nishinakamura R. Kidney development conserved over species: essential roles of Sall1. Semin. Cell. Dev. Biol. 2003; 14(4): 241–7.

Nishinakamura R, Matsumoto Y, Nakao K, Nakamura K, Sato A, Copeland NG, et al. Murine homolog of SALL1 is essential for ureteric bud invasion in kidney development. Development 2001; 128(16): 3105–15.

Oda T, Elkahloun AG, Meltzer PS, Chandrasekharappa SC. Identification and cloning of the human homolog (JAG1) of the rat Jagged1 gene from the Alagille syndrome critical region at 20p12. Genomics 1997; 43(3): 376–9.

Oliver J. Nephrons and Kidney. New York: Hoeber Medical Division, Harper & Row, 1968.

Ott MO, Rey-Campos J, Cereghini S, Yaniv M. vHNF1 is expressed in epithelial cells of distinct embryonic origin during development and precedes HNF1 expression. Mech. Dev. 1991; 36(1–2): 47–58.

Ott T, Kaestner KH, Monaghan AP, Schutz G. The mouse homolog of the region specific homeotic gene spalt of Drosophila is expressed in the developing nervous system and in mesoderm-derived structures. Mech. Dev. 1996; 56(1–2): 117–28.

O'Callaghan M, Young ID. The Townes-Brocks syndrome. J. Med. Genet. 1990; 27(7): 457–61.

Paine-Saunders S, Viviano BL, Saunders S. GPC6, a novel member of the glypican gene family, encodes a product structurally related to GPC4 and is colocalized with GPC5 on human chromosome 13. Genomics 1999; 57(3): 455–8.

Palmer RE, Kotsianti A, Cadman B, Boyd T, Gerald W, Haber DA. WT1 regulates the expression of the major glomerular podocyte membrane protein Podocalyxin. Curr. Biol. 2001; 11(22): 1805–9.

Park S, Schalling M, Bernard A, Maheswaran S, Shipley GC, Roberts D, et al. The Wilms tumour gene WT1 is expressed in murine mesoderm-derived tissues and mutated in a human mesothelioma. Nat. Genet. 1993; 4(4): 415–20.

Parsa CF, Silva ED, Sundin OH, Goldberg MF, De Jong MR, Sunness JS, et al. Redefining papillorenal syndrome: an underdiagnosed cause of ocular and renal morbidity. Ophthalmology 2001; 108(4): 738–49.

Patek CE, Little MH, Fleming S, Miles C, Charlieu JP, Clarke AR, et al. A zinc finger truncation of murine WT1 results in the characteristic urogenital abnormalities of Denys-Drash syndrome. Proc. Nat. Acad. Sci. USA 1999; 96(6): 2931–6.

Patel V, Li L, Cobo-Stark P, Shao X, Somlo S, Lin F, et al. Acute kidney injury and aberrant planar cell polarity induce cyst formation in mice lacking renal cilia. Hum. Mol. Genet. 2008; 17(11): 1578–90.

Pearson ER, Badman MK, Lockwood CR, Clark PM, Ellard S, Bingham C, et al. Contrasting diabetes phenotypes associated with hepatocyte nuclear factor-1alpha and -1beta mutations. Diabetes Care 2004; 27(5): 1102–7.

Pelletier J, Schalling M, Buckler AJ, Rogers A, Haber DA, Housman D. Expression of the Wilms' tumor gene WT1 in the murine urogenital system. Genes Dev. 1991; 5(8): 1345–56.

Pichel JG, Shen L, Sheng HZ, Granholm AC, Drago J, Grinberg A, et al. Defects in enteric innervation and kidney development in mice lacking GDNF. Nature 1996; 382(6586): 73–6.

Pilia G, Hughes-Benzie RM, MacKenzie A, Baybayan P, Chen EY, Huber R, et al. Mutations in GPC3, a glypican gene, cause the Simpson-Golabi-Behmel overgrowth syndrome. Nat. Genet. 1996; 12(3): 241–7.

Piscione TD, Wu MY, Quaggin SE. Expression of Hairy/Enhancer of Split genes, Hes1 and Hes5, during murine nephron morphogenesis. Gene Expr. Patterns 2004; 4(6): 707–11.

Pohl M, Bhatnagar V, Mendoza SA, Nigam SK. Toward an etiological classification of developmental disorders of the kidney and upper urinary tract. Kidney Int. 2002; 61(1): 10–19.

Poll AV, Pierreux CE, Lokmane L, Haumaitre C, Achouri Y, Jacquemin P, et al. A vHNF1/TCF2-HNF6 cascade regulates the transcription factor network that controls generation of pancreatic precursor cells. Diabetes 2006; 55(1): 61–9.

Pontoglio M, Prie D, Cheret C, Doyen A, Leroy C, Froguel P, et al. HNF1alpha controls renal glucose reabsorption in mouse and man. EMBO Reports 2000; 1(4): 359–65.

Porteous S, Torban E, Cho NP, Cunliffe H, Chua L, McNoe L, et al. Primary renal hypoplasia in humans and mice with PAX2 mutations: evidence of increased apoptosis in fetal kidneys of Pax2(1Neu)+/− mutant mice. Hum. Mol. Genet. 2000; 9(1): 1–11.

Potter E Normal and Abnormal Development of the Kidney. Chicago: Year Book Medical Publishers, 1972.

Poulson D. Chromosomal deficiencies and embryonic development of Drosophila melanogaster. Proc. Nat. Acad. Sci. USA 1937; 23: 133–7.

Powell CM, Michaelis RC. Townes-Brocks syndrome. J. Med. Genet. 1999; 36(2): 89–93.

Pritchard-Jones K, Fleming S, Davidson D, Bickmore W, Porteous D, Gosden C, et al. The candidate Wilms' tumour gene is involved in genitourinary development. Nature 1990; 346(6280): 194–7.

Pritchard-Jones K, Renshaw J, King-Underwood L. The Wilms tumour (WT1) gene is mutated in a secondary leukaemia in a WAGR patient. Hum. Mol. Genet. 1994; 3(9): 1633–7.

Rae FK, Martinez G, Gillinder KR, Smith A, Shooter G, Forrest AR, et al. Anlaysis of complementary expression profiles following WT1 induction versus repression reveals the cholesterol/fatty acid synthetic pathways as a possible major target of WT1. Oncogene 2004; 23(17): 3067–79.

Raeder H, Johansson S, Holm PI, Haldorsen IS, Mas E, Sbarra V, et al. Mutations in the CEL VNTR cause a syndrome of diabetes and pancreatic exocrine dysfunction. Nat. Genet. 2006; 38(1): 54–62.

Ragan DC, Casale AJ, Rink RC, Cain MP, Weaver DD. Genitourinary anomalies in the CHARGE association. J. Urol. 1999; 161(2): 622–5.

Rebouissou S, Vasiliu V, Thomas C, Bellanne-Chantelot C, Bui H, Chretien Y, et al. Germline hepatocyte nuclear factor 1alpha and 1beta mutations in renal cell carcinomas. Hum. Mol. Genet. 2005; 14(5): 603–14.

Reeves W, Caulfield JP, Farquhar MG. Differentiation of epithelial foot processes and filtration slits: sequential appearance of occluding junctions, epithelial polyanion, and slit membranes in developing glomeruli. Lab. Invest. 1978; 39(2): 90–100.

Renshaw J, King-Underwood L, Pritchard-Jones K. Differential splicing of exon 5 of the Wilms tumour (WTI) gene. Genes Chromosomes Cancer 1997; 19(4): 256–66.

Rey-Campos J, Chouard T, Yaniv M, Cereghini S. vHNF1 is a homeoprotein that activates transcription and forms heterodimers with HNF1. EMBO J. 1991; 10(6): 1445–57.

Rigoli L, Chimenz R, di Bella C, Cavallaro E, Caruso R, Briuglia S, et al. Angiotensin-converting enzyme and angiotensin type 2 receptor gene genotype distributions in Italian children with congenital uropathies. Pediatr. Res. 2004; 56(6): 988–93.

Romio L, Fry AM, Winyard PJ, Malcolm S, Woolf AS, Feather SA. OFD1 is a centrosomal/basal body protein expressed during mesenchymal-epithelial transition in human nephrogenesis. J. Am. Soc. Nephrol. 2004; 15(10): 2556–68.

Romio L, Wright V, Price K, Winyard PJ, Donnai D, Porteous ME, et al. OFD1, the gene mutated in oral-facial-digital syndrome type 1, is expressed in the metanephros and in human embryonic renal mesenchymal cells. J. Am. Soc. Nephrol. 2003; 14(3): 680–9.

Rothenpieler UW, Dressler GR. Pax-2 is required for mesenchyme-to-epithelium conversion during kidney development. Development 1993; 119(3): 711–20.

Ruano-Gil D, Coca-Payeras A, Tejedo-Mateu A. Obstruction and normal recanalization of the ureter in the human embryo. Its relation to congenital ureteric obstruction. Eur. Urol. 1975; 1(6): 287–93.

Ryan G, Steele-Perkins V, Morris JF, Rauscher FJ 3rd, Dressler GR. Repression of Pax-2 by WT1 during normal kidney development. Development 1995; 121(3): 867–75.

Sakaki-Yumoto M, Kobayashi C, Sato A, Fujimura S, Matsumoto Y, Takasato M, et al. The murine homolog of SALL4, a causative gene in Okihiro syndrome, is essential for embryonic stem cell proliferation, and cooperates with Sall1 in anorectal, heart, brain and kidney development. Development 2006; 133(15): 3005–13.

Salerno A, Kohlhase J, Kaplan BS. Townes-Brocks syndrome and renal dysplasia: a novel mutation in the SALL1 gene. Pediatr. Nephrol. 2000; 14(1): 25–8.

Sanlaville D, Etchevers HC, Gonzales M, Martinovic J, Clement-Ziza M, Delezoide AL, et al. Phenotypic spectrum of CHARGE syndrome in fetuses with CHD7 truncating mutations correlates with expression during human development. J. Med. Genet. 2006; 43(3): 211–317.

Sanyanusin P, Norrish JH, Ward TA, Nebel A, McNoe LA, Eccles MR. Genomic structure of the human PAX2 gene. Genomics 1996 Jul 1; 35(1): 258–61.

Sato A, Kishida S, Tanaka T, Kikuchi A, Kodama T, Asashima M, et al. Sall1, a causative gene for Townes-Brocks syndrome, enhances the canonical Wnt signaling by localizing to heterochromatin. Biochem. Biophys. Res. Commun. 2004 ; 319(1): 103–13.

Saxen L. Organogenesis of the Kidney. Cambridge: Cambridge University Press, 1987.

Scharnhorst V, Dekker P, van der Eb AJ, Jochemsen AG. Internal translation initiation generates novel WT1 protein isoforms with distinct biological properties. J. Biol. Chem. 1999 Aug 13; 274(33): 23456–62.

Scharnhorst V, van der Eb AJ, Jochemsen AG. WT1 proteins: functions in growth and differentiation. Gene. 2001 Aug 8; 273(2): 141–61.

Schimmenti LA, Cunliffe HE, McNoe LA, Ward TA, French MC, Shim HH, et al. Further delineation of renal-coloboma syndrome in patients with extreme variability of phenotype and identical PAX2 mutations. Am. J. Hum. Genet. 1997; 60(4): 869–78.

Schimmenti LA, Pierpont ME, Carpenter BL, Kashtan CE, Johnson MR, Dobyns WB. Autosomal dominant optic nerve colobomas, vesicoureteral reflux, and renal anomalies. Am. J. Med. Genet. 1995 Nov 6; 59(2): 204–8.

Schmidt-Ott KM, Yang J, Chen X, Wang H, Paragas N, Mori K, et al. Novel regulators of kidney development from the tips of the ureteric bud. J. Am. Soc. Nephrol. 2005; 16(7): 1993–2002.

Schumacher V, Scharer K, Wuhl E, Altrogge H, Bonzel KE, Guschmann M, et al. Spectrum of early onset nephrotic syndrome associated with WT1 missense mutations. Kidney Int. 1998; 53(6): 1594–600.

Scolari F, Valzorio B, Carli O, Vizzardi V, Costantino E, Grazioli L, et al. Oral-facial-digital syndrome type I: an unusual cause of hereditary cystic kidney disease. Nephrol. Dial. Transplant. 1997; 12(6): 1247–50.

Sharma PM, Bowman M, Madden SL, Rauscher FJ 3rd., Sukumar S. RNA editing in the Wilms' tumor susceptibility gene, WT1. Genes Dev. 1994; 8(6): 720–31.

Simpson JL, Landey S, New M, German J. A previously unrecognized X-linked syndrome of dysmorphia. Birth Defects Orig. Artic. Ser. 1975; 11(2): 18–24.

Smith JM, Stablein DM, Munoz R, Hebert D, McDonald RA. Contributions of the Transplant Registry: The 2006 Annual Report of the North American Pediatric Renal Trials and Collaborative Studies (NAPRTCS). Pediatr. Transplant. 2007; 11(4): 366–73.

Spinner NB, Colliton RP, Crosnier C, Krantz ID, Hadchouel M, Meunier-Rotival M. Jagged1 mutations in alagille syndrome. Hum. Mutat. 2001; 17(1): 18–33.

Stapleton P, Weith A, Urbanek P, Kozmik Z, Busslinger M. Chromosomal localization of seven PAX genes and cloning of a novel family member, PAX-9. Nat. Genet. 1993; 3(4): 292–8.

Stoffers DA, Ferrer J, Clarke WL, Habener JF. Early-onset type-II diabetes mellitus (MODY4) linked to IPF1. Nat. Genet. 1997; 17(2): 138–9.

Stuart ET, Haffner R, Oren M, Gruss P. Loss of p53 function through PAX-mediated transcriptional repression. EMBO J. 1995a; 14(22): 5638–45.

Stuart RO, Barros EJ, Ribeiro E, Nigam SK. Epithelial tubulogenesis through branching morphogenesis: relevance to collecting system development. J. Am. Soc. Nephrol. 1995b; 6(4): 1151–9.

Sun Z, Hopkins N. vhnf1, the MODY5 and familial GCKD-associated gene, regulates regional specification of the zebrafish gut, pronephros, and hindbrain. Genes Dev. 2001 Dec 1; 15(23): 3217–29.

Surka WS, Kohlhase J, Neunert CE, Schneider DS, Proud VK. Unique family with Townes-Brocks syndrome, SALL1 mutation, and cardiac defects. Am. J. Med. Genet. 2001 Aug 15; 102(3): 250–7.

Tellier AL, Amiel J, Delezoide AL, Audollent S, Auge J, Esnault D, et al. Expression of the PAX2 gene in human embryos and exclusion in the CHARGE syndrome. Am. J. Med. Genet. 2000; 93(2): 85–8.

Tellier AL, Cormier-Daire V, Abadie V, Amiel J, Sigaudy S, Bonnet D, et al. CHARGE syndrome: report of 47 cases and review. Am. J. Med. Genet. 1998 Apr 13; 76(5): 402–9.

Terespolsky D, Farrell SA, Siegel-Bartelt J, Weksberg R. Infantile lethal variant of Simpson-Golabi-Behmel syndrome associated with hydrops fetalis. Am. J. Med. Genet. 1995 ; 59(3): 329–33.

Tomura H, Nishigori H, Sho K, Yamagata K, Inoue I, Takeda J. Loss-of-function and dominant-negative mechanisms associated with hepatocyte nuclear factor-1beta mutations in familial type 2 diabetes mellitus. J. Bio. Chem. 1999; 274(19): 12975–8.

Torban E, Dziarmaga A, Iglesias D, Chu LL, Vassilieva T, Little M, et al. PAX2 activates WNT4 expression during mammalian kidney development. J. Bio. Chem. 2006; 281(18): 12705–12.

Toriello HV. Oral-facial-digital syndromes. Clin. Dysmorphol. 1993; 2(2): 95–105.

Torres M, Gomez-Pardo E, Dressler GR, Gruss P. Pax-2 controls multiple steps of urogenital development. Development 1995; 121(12): 4057–65.

Towfighi J, Berlin CM Jr., Ladda RL, Frauenhoffer EE, Lehman RA. Neuropathology of oral-facial-digital syndromes. Arch. Pathol. Lab. Med. 1985; 109(7): 642–6.

Townes PL, Brocks ER. Hereditary syndrome of imperforate anus with hand, foot, and ear anomalies. J. Pediatr. 1972; 81(2): 321–6.

Ulinski T, Lescure S, Beaufils S, Guigonis V, Decramer S, Morin D, et al. Renal phenotypes related to hepatocyte nuclear factor-1beta (TCF2) mutations in a pediatric cohort. J. Am. Soc. Nephrol. 2006; 17(2): 497–503.

Vainio S, Muller U. Inductive tissue interactions, cell signaling, and the control of kidney organogenesis. Cell 1997; 90(6): 975–8.

Van den Abbeele AD, Treves ST, Lebowitz RL, Bauer S, Davis RT, Retik A, et al. Vesicoureteral reflux in asymptomatic siblings of patients with known reflux: radionuclide cystography. Pediatrics 1987; 79(1): 147–53.

Varanasi R, Bardeesy N, Ghahremani M, Petruzzi MJ, Nowak N, Adam MA, et al. Fine structure analysis of the WT1 gene in sporadic Wilms tumors. Proc. Nat. Acad. Sci. USA 1994; 91(9): 3554–8.

Verloes A. Updated diagnostic criteria for CHARGE syndrome: a proposal. Am. J. Med. Genet., A. 2005; 133(3): 306–8.

Veugelers M, Vermeesch J, Watanabe K, Yamaguchi Y, Marynen P, David G. GPC4, the gene for human K-glypican, flanks GPC3 on xq26: deletion of the GPC3-GPC4 gene cluster in one family with Simpson-Golabi-Behmel syndrome. Genomics 1998; 53(1): 1–11.

Vissers LE, van Ravenswaaij CM, Admiraal R, Hurst JA, de Vries BB, Janssen IM, et al. Mutations in a new member of the chromodomain gene family cause CHARGE syndrome. Nat. Genet. 2004; 36(9): 955–57.

Wagner KD, Wagner N, Schedl A. The complex life of WT1. J. Cell Sci. 2003 May 1; 116(Pt 9): 1653–8.

Wagner N, Wagner KD, Xing Y, Scholz H, Schedl A. The major podocyte protein nephrin is transcriptionally activated by the Wilms' tumor suppressor WT1. J. Am. Soc. Nephrol. 2004; 15(12): 3044–51.

Wan S, Cato AM, Skaer H. Multiple signalling pathways establish cell fate and cell number in *Drosophila* malpighian tubules. Dev. Biol. 2000; 217(1): 153–65.

Wang L, Coffinier C, Thomas MK, Gresh L, Eddu G, Manor T, et al. Selective deletion of the Hnf1beta (MODY5) gene in beta-cells leads to altered gene expression and defective insulin release. Endocrinology 2004; 145(8): 3941–9.

Wang P, Pereira FA, Beasley D, Zheng H. Presenilins are required for the formation of comma- and S-shaped bodies during nephrogenesis. Development 2003; 130(20): 5019–29.

Watanabe K, Yamada H, Yamaguchi Y. K-glypican: a novel GPI-anchored heparan sulfate proteoglycan that is highly expressed in developing brain and kidney. J. Cell. Biol. 1995; 130(5): 1207–18.

Watanabe T, Costantini F. Real-time analysis of ureteric bud branching morphogenesis in vitro. Dev. Biol. 2004 Jul 1; 271(1): 98–108.

Weber S, Moriniere V, Knuppel T, Charbit M, Dusek J, Ghiggeri GM, et al. Prevalence of mutations in renal developmental genes in children with renal hypodysplasia: results of the ESCAPE study. J. Am. Soc. Nephrol. 2006; 17(10): 2864–70.

Wettke-Schafer R, Kantner G. X-linked dominant inherited diseases with lethality in hemizygous males. Hum. Genet. 1983; 64(1): 1–23.

Wharton KA, Johansen KM, Xu T, Artavanis-Tsakonas S. Nucleotide sequence from the neurogenic locus notch implies a gene product that shares homology with proteins containing EGF-like repeats. Cell 1985; 43(3 Pt 2): 567–81.

Wheeler PG, Quigley CA, Sadeghi-Nejad A, Weaver DD. Hypogonadism and CHARGE association. Am. J. Med. Genet. 2000 Sep 18; 94(3): 228–31.

Wild W, Pogge von Strandmann E, Nastos A, Senkel S, Lingott-Frieg A, Bulman M, et al. The mutated human gene encoding hepatocyte nuclear factor 1beta inhibits kidney formation in developing *Xenopus* embryos. Proc. Nat. Acad. Sci. USA 2000; 97(9): 4695–700.

Winyard PJ, Risdon RA, Sams VR, Dressler GR, Woolf AS. The PAX2 tanscription factor is expressed in cystic and hyperproliferative dysplastic epithelia in human kidney malformations. J. Clin. Invest. 1996 Jul 15; 98(2): 451–9.

Woolf AS, Feather SA, Bingham C. Recent insights into kidney diseases associated with glomerular cysts. Pediatr. Nephrol. 2002; 17(4): 229–35.

Woolf AS, Price KL, Scambler PJ, Winyard PJ. Evolving concepts in human renal dysplasia. J. Am. Soc. Nephrol. 2004; 15(4): 998–1007.

Xue Y, Gao X, Lindsell CE, Norton CR, Chang B, Hicks C, et al. Embryonic lethality and vascular defects in mice lacking the Notch ligand Jagged1. Hum. Mol. Genet. 1999; 8(5): 723–30.

Yamagata K, Oda N, Kaisaki PJ, Menzel S, Furuta H, Vaxillaire M, et al. Mutations in the hepatocyte nuclear factor-1alpha gene in maturity-onset diabetes of the young (MODY3). Nature 1996; 384(6608): 455–8.

Yoneda A, Cascio S, Green A, Barton D, Puri P. Angiotensin II type 2 receptor gene is not responsible for familial vesicoureteral reflux. J. Urol. 2002; 168(3): 1138–41.

Yorifuji T, Kurokawa K, Mamada M, Imai T, Kawai M, Nishi Y, et al. Neonatal diabetes mellitus and neonatal polycystic, dysplastic kidneys: Phenotypically discordant recurrence of a mutation in the hepatocyte nuclear factor-1beta gene due to germline mosaicism. J. Clin. Endocrinol. Metab. 2004; 89(6): 2905–8.

Zallen JA. Planar polarity and tissue morphogenesis. Cell 2007; 129(6): 1051–63.

PART IV

Inherited Neoplastic Diseases Affecting the Kidney

CHAPTER 28

The Genetic Basis of Cancer of the Kidney

ROBERT L. GRUBB III, McCLELLAN M. WALTHER AND W. MARSTON LINEHAN

INTRODUCTION

Renal Cell Cancer Epidemiology

There are an estimated 36 000 new cases of renal cancer each year, with nearly 12 000 deaths (Jemal et al 2003). Renal cell cancer incidence has increased during the last two decades (Chow et al 1999). Although the greatest increase has been in localized tumors, increases in advanced tumors and mortality were also seen suggesting that increased incidental detection by greater use of CT and MRI is not the sole explanation for the increase. With renal cancer becoming an increasing problem, knowledge of the molecular biology of renal cancers has likewise increased. At the root of this increased knotwledge is a greater understanding of the importance of heredity in the development of renal cancer.

Known hereditary renal cancer syndromes include von Hippel-Lindau (VHL), hereditary papillary renal cancer (HPRC), Birt-Hogg-Dubé (BHD), and hereditary leiomyomatosis and renal cell cancer (HLRCC). Each of these syndromes carries a varying risk of renal cancer and most are associated with characteristic tumor histology (Figure 28.1) and associated clinical abnormalities (Table 28.1).

Hereditary Renal Cancer

Four percent of renal cancer cases are estimated to be due to autosomal dominant hereditary causes (Choyke et al 2003). Having a first or second-degree relative with renal cell carcinoma confers an odds ratio of 2.9 times the risk of having renal cell carcinoma, compared to a matched group of controls (Gago-Dominguez et al 2001). Hereditary renal cancer is characterized by several important differences from sporadic renal cancer (Zbar 2000, Phillips et al 2001), earlier age of onset, frequent multifocality and bilaterality, and more equal sex distribution (Choyke et al 2003). Having numerous relatives, especially over multiple generations

with renal cell cancer should arouse suspicion of a hereditary cause of renal cell carcinoma (Figure 28.2).

VON HIPPEL-LINDAU DISEASE

Clinical Manifestations of VHL

Von Hippel-Lindau was initially described by Eugen von Hippel, and further characterized by Arvid Lindau (von Hippel 1904, Lindau 1926). Melmon and Rosen were the first to report renal cancer as part of the syndrome (Melmon & Rosen 1964). Von Hippel-Lindau is an autosomal dominantly inherited disorder that leads to development of hemangioblastomas of the central nervous system (CNS), endolymphatic sac tumors, pancreatic cysts and neoplasms, pheochromocytomas, renal cysts and cancers and cystadenomas of the epididymis and broad ligament (Table 28.2). VHL is estimated to occur in 1 in 36 000 births and has a greater than 90% penetrance by age 65 (Maher et al 1990, 1991, Neumann & Wiestler 1991). Manifestations can occur as early as the first decade of life.

Renal Cysts and Tumors

Renal cancers and cysts occur in 25–60% of patients and can be detected in patients as early as the second decade, with an average age of onset of 39 years (range, 16–67) (Maher et al 1990, Choyke et al 1995). Renal cancer was historically a major cause of mortality for VHL patients, whose life expectancy prior to widespread screening was less than 50 years (Lonser et al 2003). Before wide availability of CT 13–42% of patients with VHL died of metastatic renal cell carcinoma (Herring et al 2001). Renal cancers in VHL disease are uniformly of clear cell histology, usually of a low Fuhrman grade. Walther et al estimated that a kidney involved with VHL could harbor as

Genetic Diseases of the Kidney
ISBN: 978-0-12-449851-8

497

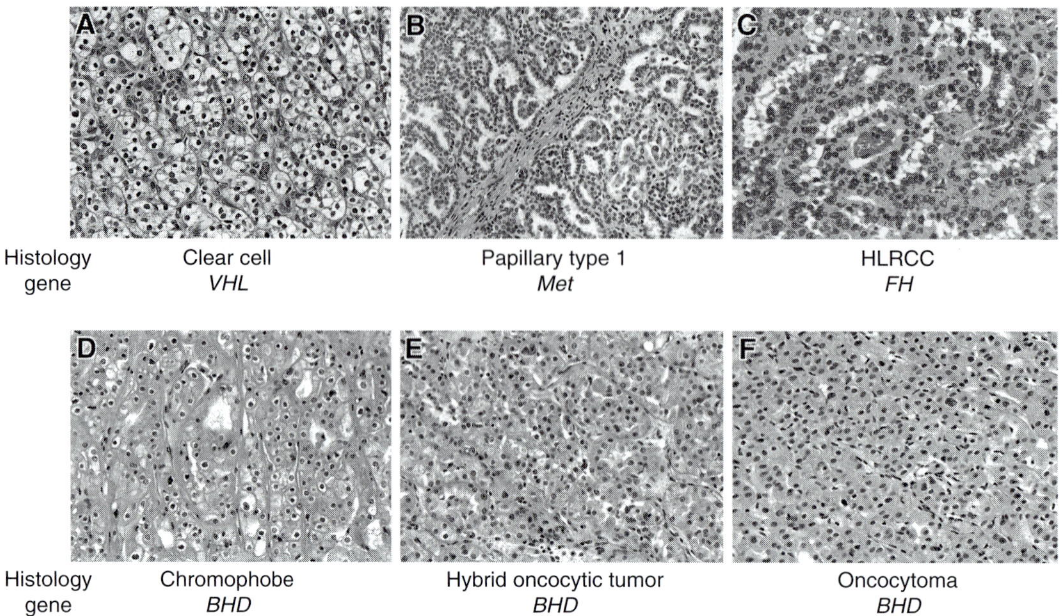

FIGURE 28.1   Hereditary cancer syndromes display distinct histology. Von Hippel-Lindau (VHL) is associated with clear cell RCC (A). Hereditary papillary renal cancer (HPRC) patients develop papillary type I RCC (B), while patients with hereditary leiomyomatosis and renal cell cancer (HLRCC) develop characteristic tubulopapillary tumors suggestive of type II papillary RCC (C). Birt-Hogg-Dubé (BHD)-associated tumors may have multiple histologies, including chromophobe (D), hybrid oncocytic tumors (E), and oncocytoma (F). (see also Plate 59)

TABLE 28.1   Hereditary renal cancer syndromes

Syndrome	Causative gene, location	Renal tumor histology	Extrarenal manifestations
Von Hippel-Lindau (VHL)	*VHL*, 3p25	Clear cell	CNS and retinal hemangioblastomas, endolymphatic sac tumors, pheochromocytomas, pancreatic cysts/tumors, epididymal and broad ligament cystadenomas
Hereditary papillary renal cancer (HPRC)	*Met*, 7q31	Papillary, type I	None known
Birt-Hogg-Dubé (BHD)	*BHD*, 17p11.2	Hybrid oncocytic, chromophobe, clear cell, oncocytoma	Cutaneous papules (fibrofolliculomas), lung cysts, spontaneous pneumothoraces
Hereditary leiomyomatosis and renal cell cancer (HLRCC)	*Fumarate hydratase*, 1q25-32	Characteristic HLRCC tumors (often suggestive of papillary type II)	Uterine leiomyomas, cutaneous nodules (leiomyomas)

many as 600 microscopic foci of renal cell cancer and up to 1100 cysts (Walther et al 1995a).

Other Visceral Manifestations: Pheochromocytoma and Pancreatic Cysts and Tumors

Other visceral manifestations of VHL include pheochromocytomas and pancreatic cysts and tumors. Pheochromocytomas can be multiple, bilateral and may also arise in extra-adrenal locations (sympathetic chain, carotid body). While VHL pheochromocytomas can be functional, it has been noted that nearly one-third of patients with VHL pheochromocytomas found in a screening protocol for affected families were without symptoms at presentation (Walther et al 1999b). Pancreatic manifestations include neuroendocrine tumors (NET), pancreatic cysts and serous cystadenomas. Pancreatic NETs can be malignant. Libutti et al found metastases present in four of 14 patients with pancreatic NET (Libutti et al 1998).

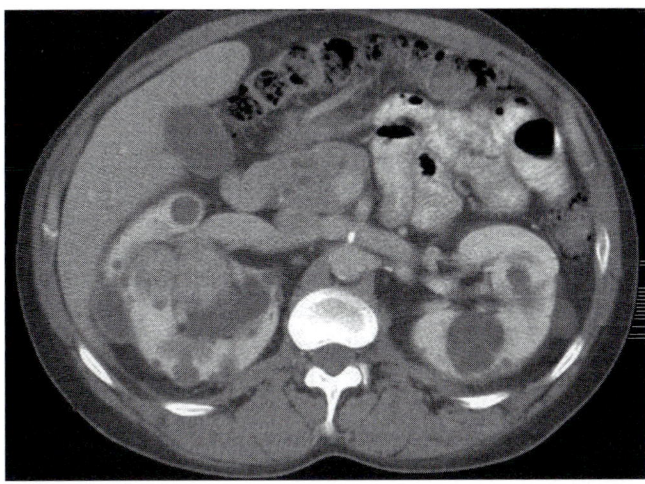

FIGURE 28.2 Multifocal bilateral renal tumors in a patient with von-Hippel-Lindau. Patients with VHL and other hereditary cancer syndromes frequently have multifocal tumors involving both kidneys

TABLE 28.2 Clinical features of VHL

Feature	Incidence (%)	Mean age of onset (range)
Kidney cysts and tumors	25–70	39 (16–67)
(clear cell RCC)	(RCC 40)	
Pheochromocytoma	10–20	30 (5–58)
Pancreatic tumors	35–70	36 (5–70)
CNS hemangioblastoma	60–80	33 (9–78)
Endolymphatic sac tumors	10	22 (12–50)
Retinal hemangioblastoma	2560	25 (1–67)
Epidymal cystadenoma	25–60	?
Broad ligament cystadenoma	?	?

RCC – Renal cell carcinoma. ? – unknown. Adapted from Lonser et al (2003)

Central Nervous System Manifestations, Endolymphatic Sac Tumors, and Retinal Hemangioblastomas

Hemangioblastomas of the CNS are the most common manifestation of VHL (Table 28.2) (Lonser et al 2003). These are most commonly found in the spinal cord and cerebellum. These are benign tumors, but can cause much morbidity. Lesions are resected when they become symptomatic.

Retinal angiomas occur in 25–60% of VHL patients (Lonser et al 2003). Lesions can be asymptomatic, but still risk visual impairment if not treated. Endolymphatic sac tumors occur in 10% of cases and can result in hearing loss, tinnitus, and balance problems (Lonser et al 2003).

TABLE 28.3 Subtypes of VHL

	Manifestations	Mutations
Type 1	No pheochromocytoma	Missense, complete deletion, loss of HIF ubiquitination
Type 2A	Low likelihood RCC	Missense, retain HIF ubiquitination
Type 2B	High likelihood RCC	Missense, loss of HIF ubiquitination
Type 2C	ONLY pheochromocytoma	Missense, single reported mutation, retain HIF ubiquitination

OTHER MANIFESTATIONS

Epididymal and broad ligament cystadenomas can be found in the epididymis of affected males and in the broad ligament of affected females. Treatment of these benign tumors is generally conservative, except in rare, symptomatic cases (Lonser et al 2003).

VHL Subtypes

Extensive study of VHL kindred and increasing knowledge of the variety of VHL mutations has resulted in identification of subtypes of the syndrome related to the type of mutation. VHL type 1 patients do not form pheochromocytomas but may exhibit all other typical VHL manifestations. Mutations are generally partial deletions leading to complete loss of VHL function. VHL type 2 patients include patients with at least one family member with pheochromocytoma. VHL type 2A patients have all tumor types, but have a low likelihood of forming renal tumors, while VHL type 2B patients form renal tumors. Missense mutations comprise the majority of mutations in VHL type 2A and type 2B. It is thought that mutations leading to type 2A retain some HIF regulatory activity (see below) (Clifford et al 2001). VHL type 2C is a rare subcategory where patients are predisposed to form only pheochromocytomas (Neumann et al 1995). Maranchie et al demonstrated a higher prevalence of RCC in patients with partial deletion *VHL* mutations than in patients with complete *VHL* deletions (Maranchie et al 2004) (Table 28.3).

VHL and Hypoxia Inducible Factor

Study of patients affected with VHL led to the discovery of the gene responsible for VHL in 1993 (Latif et al 1993). The VHL protein was found to form a complex with Elongin B and C (Kibel et al 1995), and Cul-2 (Pause et al 1997) to form a ubiquitin ligase complex. Under normoxic conditions this complex, through VHL, binds HIF 1-α via prolyl hydroxylation (Ivan et al 2001, Jaakkola et al 2001). This ubiquitinated complex is targeted for degradation by the proteasome (Ohh et al 2000). However, under hypoxic conditions prolyl hydroxylation and ubiquitination do not occur and HIF 1-α accumulates, is translocated into the nucleus and combines with HIF-1β. This complex can induce expression of a variety of HIF-responsive genes including VEGF (vascular endothelial growth factor), GLUT-1, PDGF (platelet derived growth factor), TGF-α (transforming growth factor-alpha) and erythropoietin. Many of these genes are thought to be involved in VHL-related tumorigenesis.

In vivo experiments have shown that downregulation of HIF 1-α is necessary for tumor formation (Iliopoulos et al 1995). $VHL^{-/-}$ cell lines produce tumors in nude mice while re-expression of VHL suppresses tumorigenesis (Kondo et al 2002). Thus, it appears that absence of VHL, as may occur in renal cells of VHL patients who have lost one copy of *VHL* via germline mutation and have inactivation of the second copy of *VHL*, lose the inability to degrade HIF, resulting in a situation identical to the cellular response to hypoxia and translation of the hypoxia-inducible genes (see above; Figure 28.3).

MANAGEMENT OF PATIENTS WITH VHL

Because of the diverse manifestations of VHL and the presence of neoplasms in multiple organ systems, multispecialty involvement can help to optimize management of patients with VHL.

Management of Renal Cysts and Masses in Von Hippel-Lindau Disease

The goals of management of renal cysts and masses in patients with von Hippel-Lindau disease should be to prevent morbidity and mortality from renal cancer. Physicians treating these patients need to remember important differences between patients with VHL and renal masses and patients with nonhereditary renal masses. Renal masses in VHL, as in other hereditary renal cancer syndromes, tend to be bilateral and multiple and can be recurrent. Cancer cure and control needs to be carefully balanced with quality of life issues and the effects of treatment. Goals of treatment should be prevention of metastatic disease, preservation of renal function and minimization of the number of operative procedures which patients must undergo (Walther et al 1995a, 1999a).

Despite the fact that VHL tumors tend to be low grade, the high historical rate of metastasis in poorly screened patients makes watchful waiting an unappealing management option. Some researchers have advocated bilateral nephrectomy as a result of the high rate of local recurrence

after nephron-sparing approaches managing patients with VHL (Novick & Streem 1992). This strategy, however, necessitates renal replacement therapy. Goldfarb et al reviewed a multicenter experience with renal transplant in patients with VHL; the 5-year survival in this cohort was 65% (Goldfarb et al 1997). Three of 32 patients in the study cohort died of metastatic disease and two had noncancer-related deaths. Another multicenter study of patients 65 patients with VHL found that 23% eventually developed end-stage renal disease (ESRD) requiring renal replacement therapy. Forty percent of the ESRD subset underwent successful transplantation, while the rest required chronic hemodialysis (Steinbach et al 1995).

Because of the high morbidity and mortality associated with renal replacement therapy, some researchers have developed a strategy of following patients closely with serial imaging until the largest tumor is three centimeters in diameter before recommending intervention regardless of the number or pattern of tumors (Walther et al 1995, 1999a, Duffey et al 2004). At-risk patients are screened with serial abdominal CT and are recommended to undergo nephron-sparing surgery (NSS) when the largest tumor reaches three centimeters regardless of tumor number or pattern (Walther et al 1999a).

Outcomes of 108 patients with VHL whose renal tumors were less than 3 cm at the time of presentation managed using this strategy were compared with outcomes of a group of 73 patients with VHL and tumors larger than 3 cm at the time of diagnosis (Duffey et al 2004). Median follow-up for both groups was approximately 5 years or greater. In the group of patients with tumors less than 3 cm at presentation no patients developed metastatic disease and no patients developed ESRD. Thirty-four percent of these patients (37/108) were managed with observation alone, as they did not develop tumors larger than 3 cm during the study period. Metastatic renal cancer developed in 27% (20/73) of patients with renal tumors larger than 3 cm at the time of presentation. The authors concluded that the strategy of observing tumors until they reached a 3 cm threshold provided good cancer control (the patients were not necessarily cured of cancer, but metastatic disease was prevented), while preserving renal function and minimizing the number of operative procedures for each patient. NSS for patients with VHL has also been advocated by multiple other researchers (Levine et al 1983, Loughlin & Gittes 1986, Spencer et al 1988).

Minimally Invasive Approaches

Ablative technologies such as cryotherapy and radiofrequency ablation (RFA) have been explored as alternatives to extirpative surgery, to minimize the morbidity of managing patients with hereditary renal cancer syndromes. Percutaneous MRI-guided cryoablation has been used to treat renal tumors in patients with VHL (Shingleton &

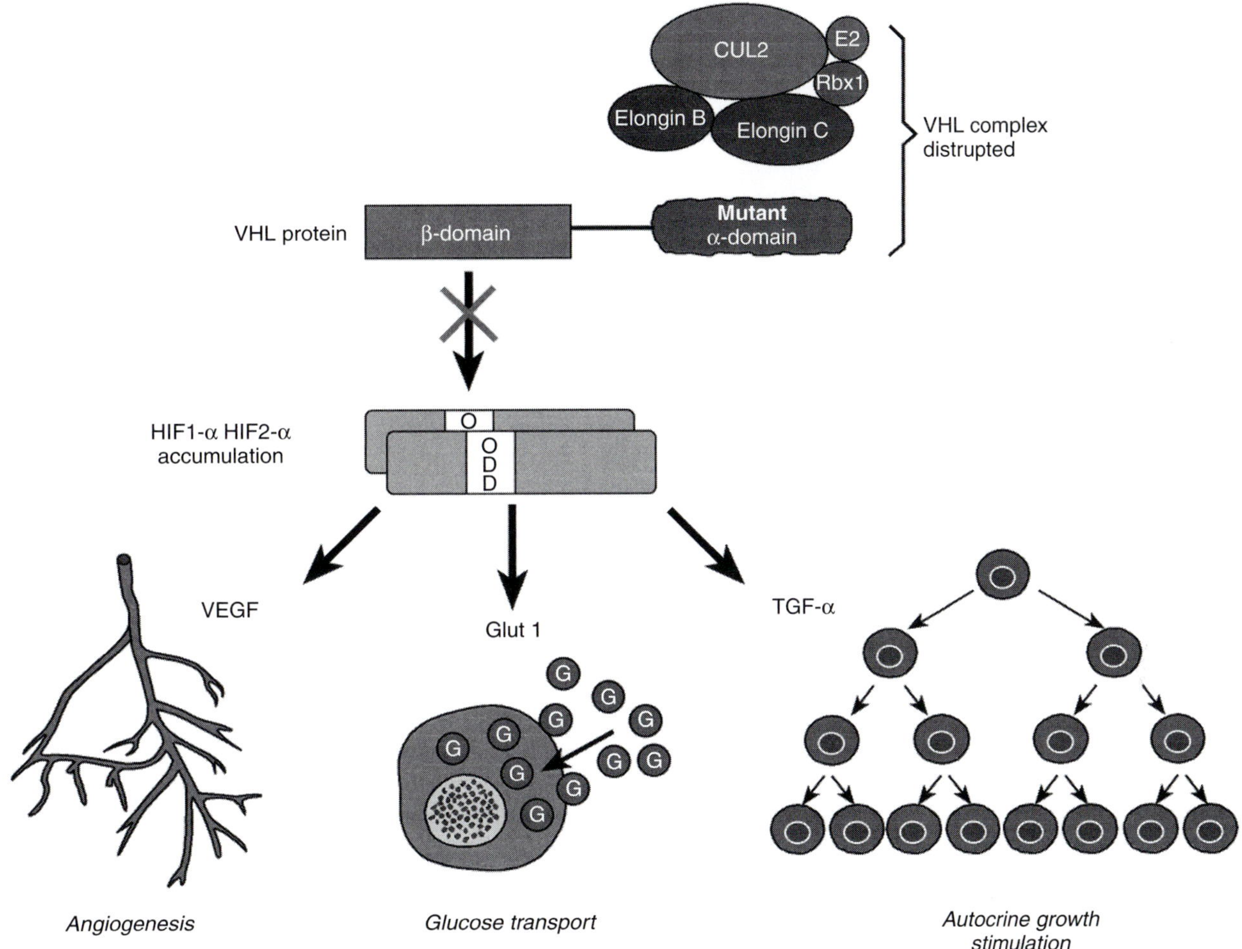

FIGURE 28.3 Molecular abnormalities of VHL. Absence of functional *VHL* gene product leads to decreased degradation of HIF 1-α and subsequent transcription of downstream hypoxia-inducible genes such as VEGF, EGFR, Glut-1 and TGF-α. Adapted from Linehan et al (2004). (see also Plate 60)

Sewell 2002). Five tumors were successfully treated in four patients with VHL; tumor sizes ranged from 2.8 to 5.0 cm in diameter. Percutaneous or laparoscopic RFA was used to treat 24 small (median size 2.26 cm) lesions in patients with VHL and HPRC with a 96% success rate (Hwang et al 2004). Other institutions have reported similar success rates in treating patients with VHL and renal masses (Farrell et al 2003, Su et al 2003).

TABLE 28.4 Clinical features of hereditary papillary renal carcinoma

	Incidence (%)	Mean age of onset (range)
Papillary renal cell carcinoma (often multifocal, bilateral	19–50	47 (23–83)

No known extrarenal manifestations.

HEREDITARY PAPILLARY RENAL CELL CANCER

Clinical Manifestations of HPRC

Hereditary papillary renal cancer (HPRC) was first described in 1994 (Zbar et al 1994). HPRC is an autosomal dominant syndrome with high penetrance marked by the predisposition of patients to form multifocal, bilateral renal tumors of papillary type I pathology (Figure 28.1B) (Schmidt et al 1998). The tumors are often appear later than in other hereditary renal cancer syndromes, with onset in the 4th, 5th, and 6th decades of life (Walther et al 1999a, Linehan et al 2003). Unlike other hereditary renal cancer syndromes, there do not appear to be associated extrarenal manifestations of the disease (Table 28.4).

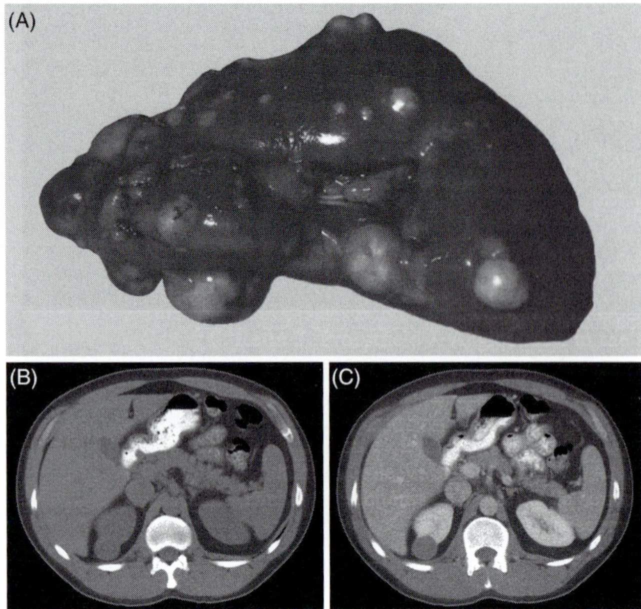

FIGURE 28.4 Hereditary papillary renal cancer. Patients with HPRC may form multiple bilateral tumors (A) (adapted from Linehan et al 2003). Papillary renal cancers may appear as hypovascular lesions on CT imaging (B). Adapted from Grubb et al (2005). (see also Plate 61)

Genetics of HPRC

Patients with HPRC were found to have mutations in the tyrosine kinase domain of the *MET* gene, located on chromosome 7q31, which is postulated to be an activating mutation, unlike *VHL*, a tumor suppressor gene (Schmidt et al 1997). *MET* encodes a tyrosine kinase activated by hepatocyte growth factor (HGF) (Bottaro et al 1991). The HGF-MET signaling pathway is thought to cause tumors by influencing proliferation, differentiation and morphogenesis (Pavlovich et al 2004b). Like mutations in other tyrosine kinases, these *MET* mutations are associated with a higher level of basal MET receptor activation (Jeffers et al 1997). Additionally, the threshold for HGF activation of the MET receptor is lowered (Michieli et al 1999). Duplication of the mutated *MET* allele is observed in tumors with very high frequency, suggesting that it may be necessary for clinical tumor development (Fischer et al 1998, Zhuang et al 1998). *MET* mutations similar to those found in HPRC are found in some sporadic papillary RCCs, but less frequently than VHL mutations are found in sporadic clear cell RCC (Schmidt et al 1998) (Figures 28.4, 28.5).

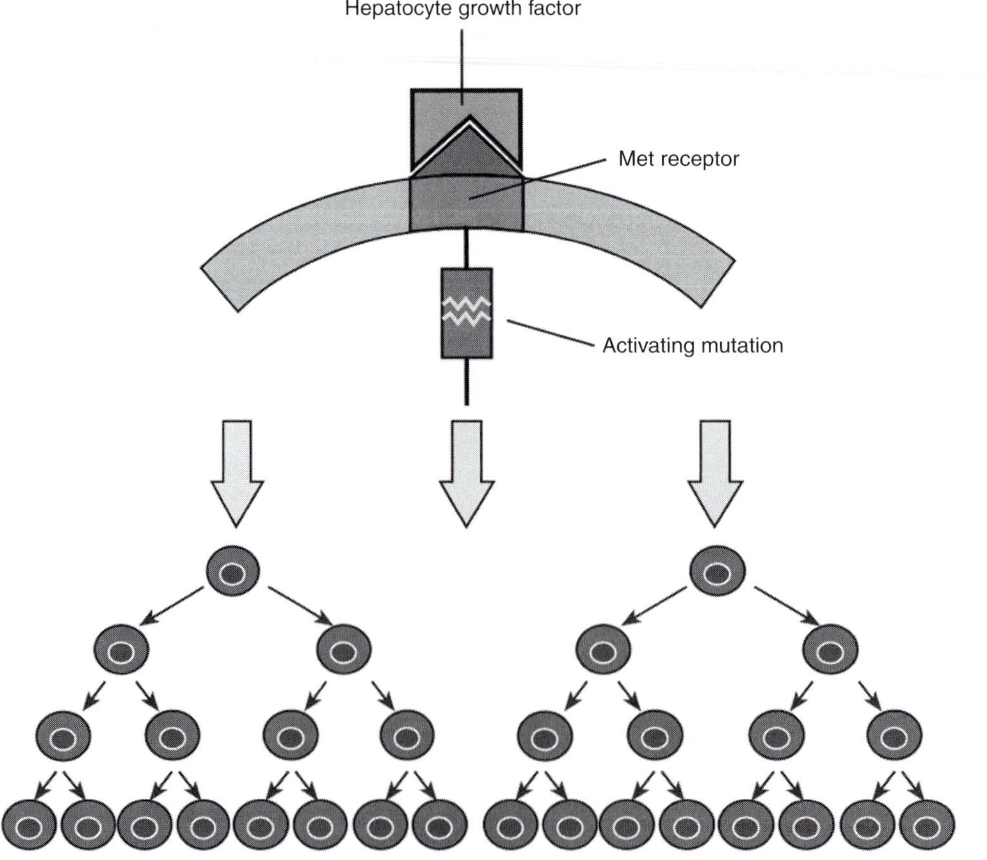

FIGURE 28.5 HPRC tumorigenesis results from activating mutations in *MET*. Adapted from Linehan et al (2003). (see also Plate 62)

Managing Renal Tumors in Hereditary Papillary Renal Cancer

Because renal tumors in HPRC are multifocal and bilateral like renal tumors in VHL, the same management strategies can be applied. Analysis of kidneys from HPRC patients has revealed the presence of up to 3400 microscopic lesions per kidney (Ornstein et al 2000). An observational strategy similar to that used for VHL has been applied to patients with HPRC. A series of 23 patients with HPRC was reported (Walther et al 1999a). Ten of the 23 patients had tumors less than 3 cm at initial presentation, including two who had undergone previous renal surgery. These 10 patients were followed for a mean of 61 months and did not require surgery. Two patients (15%) of the 13 who presented with tumors larger than 3 cm presented with metastatic disease (tumor sizes 7 and 15 cm), while no patients with tumors less than 3 cm at presentation developed metastatic disease (Walther et al 1999a). It appears that the 3 cm strategy is safe for HPRC; however, the metastatic potential of papillary renal cell carcinoma should not be ignored.

BIRT-HOGG-DUBÉ SYNDROME

Clinical Manifestations of BHD

The Birt-Hogg-Dubé syndrome, first described in 1977 (Birt et al 1977), is an autosomal dominant disorder in which affected individuals develop benign cutaneous tumors (fibrofolliculomas, Figure 28.6(A), pulmonary air-filled cysts (Figure 28.6C, D) and spontaneous pneumothoraces (Figure 28.6D) and renal neoplasms (Figure 28.6B). The hallmark lesion of BHD is the fibrofolliculoma, a benign tumor of the hair follicle, which appears in the third to fourth decade of life and manifests as multiple white or flesh-colored papules on the face, neck, back, and trunk. Reports of internal manifestations of the syndrome have included spontaneous pneumothorax (Binet et al 1986), renal tumors (Roth et al 1993, Toro et al 1999), colonic polyps (Hornstein 1976, Binet et al 1986, Khoo et al 2002), and parotid oncocytomas (Liu et al 2000). Zbar reported on the findings of families identified as having BHD by dermatologic criteria. Lung cysts, spontaneous pneumothorax, and solid renal tumors were associated with BHD (Zbar et al 2002). The odds ratios for developing spontaneous pneumothorax and renal tumors were 50.3 and 6.9 respectively, for affected family members compared to unaffected relatives. Renal tumors were strongly associated with the syndrome in this study, but colon polyps were not (Zbar et al 2002). Other groups have found BHD-affected patients to be at increased risk for colon polyps (Khoo et al 2002).

Patients with the clinical features of BHD (characteristic skin lesions, history of spontaneous pneumothorax, multiple bilateral renal tumors) and at-risk patients can now

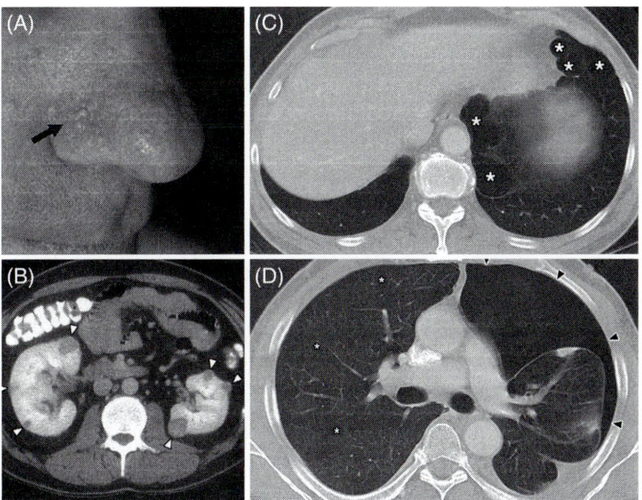

FIGURE 28.6 BHD manifestations. The characteristic lesion of Birt-Hogg-Dubé (BHD) is the cutaneous fibrofolliculoma, a benign hair follicle tumor of the face and upper trunk (A, arrow). Patients are also at risk for forming multifocal and bilateral renal tumors (B, arrowheads). Patients are also susceptible to form pulmonary cysts (C, D, asterisks) and spontaneous pneumothoraces (D, arrowheads). (see also Plate 63)

TABLE 28.5   Clinical characteristics of BHD syndrome

	Incidence (%)	Mean age of onset (range)
Fibrofolliculomas	?–100	?
Pulmonary cysts	83	?
Spontaneous pneumothorax	22	?
Renal tumors (chromophobe, HOT, clear cell, oncocytoma)	15–25	50 (31–74)
Colon polyps	18	?

HOT – hybrid oncocytic tumor; ? – unknown.

be definitively diagnosed by germline genetic testing for *BHD* mutations. There are reports of clinical phenocopies (patients in affected families with characteristics of the disease, without carrying the mutation) (Khoo et al 2002) as well as families affected by renal cancer without apparent skin lesions (Pavlovich et al 2005) (Table 28.5).

BHD Renal Tumors

BHD, in contrast to VHL and HPRC predisposes to a variety of renal tumor histologic types (Pavlovich et al 2002); 15–30% of BHD patients develop renal tumors (Choyke 2003). The predominant lesion found in a review of 130 tumors resected from 30 patients with BHD at the National Institutes of Health was the hybrid oncocytic tumor (50%), which contains a mixture of oncocytes and chromophobe cells arranged in patterns reminiscent of oncocytoma in some parts and chromophobe RCC in others (Figure 28.1E) (Tickoo et al 1999, Pavlovich et al 2002). Renal oncocytosis (small nodules of cells similar to those found in the larger hybrid tumors, peppered throughout the renal parenchyma) was seen in a majority of patients with BHD who had renal surgery. This may indicate that, as in VHL and HRPC, the whole parenchyma of these patients is at increased risk for neoplasia (Pavlovich et al 2002). Other histologies found in patients with BHD include chromophobe (34%) (Figure 28.1D), clear cell (9%) oncocytoma (5%) (Figure 28.1F), and papillary (2%).

Renal tumors can be seen as early as the fourth decade; annual screening of affected or at-risk patients should begin at age thirty with ultrasound or contrast-enhanced CT (Toro et al 1999). Imaging cannot reliably distinguish the differing histologic types of RCC in the BHD spectrum, therefore all tumors must be considered potentially aggressive, malignant types.

Molecular Studies of BHD

The *BHD* gene, located at 17p11.2, has recently been cloned and found to encode a novel protein, named folliculin (Khoo et al 2001, Schmidt et al 2001, Nickerson et al 2002). Somatic mutations in the gene in affected individuals generally cause protein truncations, but the exact role of *BHD* in the resultant phenotypic abnormalities remains to be elucidated. Mutations in *BHD* have been shown to occur in only a small percentage of sporadic renal tumors (Khoo et al 2003). Animal models of BHD, including Nihon rat RCC and canine hereditary multifocal cystadenoma and nodular dermatofibrosis, a naturally occurring inherited RCC syndrome of German Shepherd dogs, are expected to help advance the understanding of BHD renal tumor formation (Pavlovich & Schmidt 2004).

Management of Renal Tumors in Birt-Hogg-Dubé Syndrome

As BHD is newly described, little has been written about the management of affected patients. Several case reports have described the association of BHD with renal tumors (Roth et al 1993, Durrani et al 2002). Toro et al screened 152 patients at risk for BHD (Toro et al 1999). Thirteen patients were found to have the syndrome and seven were found to have renal tumors.

Initial reports indicate that observational strategies and nephron-sparing surgery which have been successfully used in VHL and HPRC patients may be efficacious in BHD (Pavlovich et al 2005). Fourteen BHD patients were prospectively managed using this strategy. Four patients had renal tumors <3 cm and these were observed. The remaining ten patients underwent 12 renal procedures; four underwent radical nephrectomy and eight underwent

TABLE 28.6 Clinical features of hereditary leiomyomatosis and renal cell cancer (HLRCC)

	Incidence (%)	Mean age at onset (range)
Cutaneous nodules (leiomyomas)	36–91	25 (10–47)
Uterine leiomyoma (fibroids)	75–98	30 (10–50)
Uterine leiomyosarcomas	12.5	38
Renal tumors (characteristic HLRCC tumors, often suggestive of papillary type II)	14–32	36–44 (23–?)

? – Unknown.

nephron-sparing surgery. At 38 months median follow-up five patients (50%) are without evidence of disease, three had developed small renal tumors and two had died of metastatic renal cancer (Pavlovich et al 2005).

HEREDITARY LEIOMYOMATOSIS AND RENAL CELL CARCINOMA

Discovery of the HRLCC Syndrome and Genetics

Studies of familial predisposition to multiple cutaneous leiomyomatosis and uterine leiomyomas noted family members with papillary type II renal cancer with an early age of onset (Kiuru et al 2001). The syndrome has been more fully characterized in subsequent reports and termed hereditary leiomyomatosis and renal cell cancer (HLRCC) (Launonen et al 2001, Toro et al 2003). The causative gene is autosomal dominant and was mapped to chromosome 1q42-44 (Launonen et al 2001) and was subsequently found to encode fumarate hydratase (FH), a mitochondrial Krebs' cycle enzyme (Tomlinson et al 2002). The result of these mutations and their possible effects on cellular functions and their role in the development of the clinical phenotype is not known at this time. As in HPRC and BHD, *FH* mutations have not been found in a majority of sporadic papillary type II renal cancers or uterine leiomyomas (Barker et al 2002, Kiuru et al 2002). It has been postulated that mitochondrial *FH* mutations might affect oxidative metabolism leading to increased cellular oxygen free radicals. The resultant state of hypoxia could lead to HIF-mediated activation of hypoxia-inducible genes (Eng et al 2003) (Table 28.6).

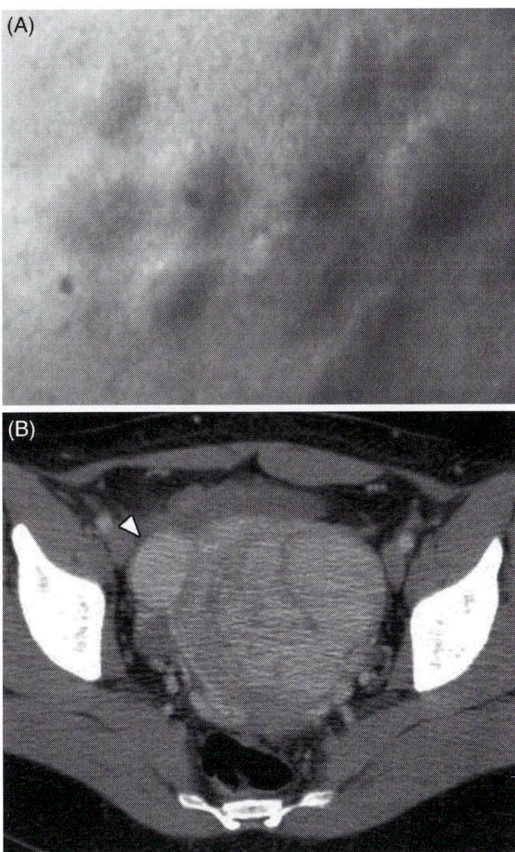

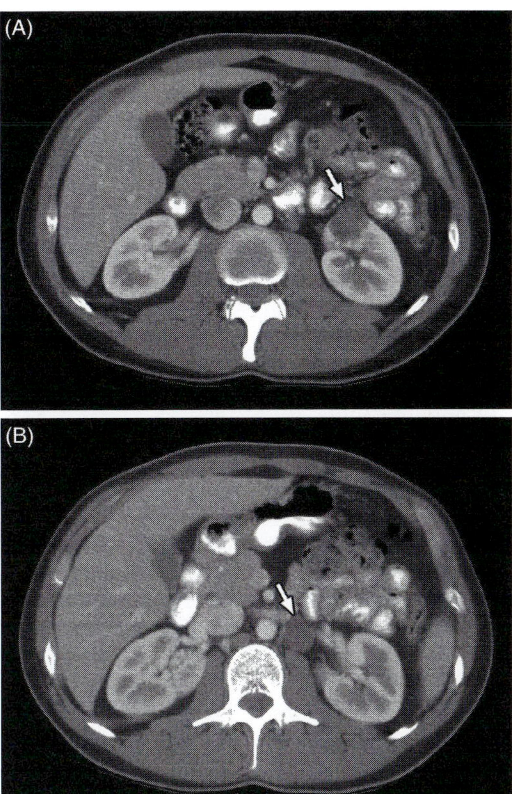

FIGURE 28.7   Cutaneous and uterine leiomyomas in HLRCC. The characteristic cutaneous lesion of HLRCC is the leiomyoma (A) (from Linehan et al 2003). Uterine leiomyomas are found in up to 98% of female patients affected with HLRCC (B, arrowhead). (see also Plate 64)

FIGURE 28.8   Renal tumors in HLRCC. CT demonstrates small left renal tumor (A, arrow) with regional nodal disease (B, arrow) in a patient affected by HLRCC. Adapted from Grubb et al (2005)

Clinical Manifestations of HLRCC

The extrarenal manifestations of HLRCC have been described in two series of patients (Launonen et al 2001, Toro et al 2003). The most frequent (75–98%) manifestation was uterine leiomyoma (Figure 28.7B). Nearly 90% of the women underwent myomectomy or hysterectomy and more than half had undergone hysterectomy by 30 years of age (Toro et al 2003). In the European cohort, 2/16 (12.5%) female patients had uterine leiomyosarcomas (Launonen et al 2001); this was not observed in the North American series (Toro et al 2003). Cutaneous leiomyomas (Figure 28.7(A) are firm skin-colored to light brown papules and can be segmental and multifocal and are mainly found on the trunk and extremities; they can be painful, occasionally requiring excision. Mean age of onset is 25 years (10–47) (Toro et al 2003). The incidence ranged from 36–91% (Launonen et al 2001, Toro et al 2003).

Renal Tumors in HLRCC

Kiuru first reported the link between cutaneous and uterine leiomyomas and renal cancer (Kiuru et al 2001). In the initial report of the syndrome, renal tumors were found in 6/19 (32%) patients in two families with HLRCC (Launonen et al 2001). In a North American cohort, 13 patients with renal tumors from the 95 mutational carriers evaluated (14%) (Toro et al 2003). Median age of onset in the two series ranged from 36 to 44 years, with tumors presenting as early as 23 years of age (Launonen et al 2001). Histologic evaluation of these tumors has revealed a variety of morphologic patterns including cysts, tubulopapillary formations, and pure papillary tumors suggestive of papillary type II renal cancer (Merino et al 2003). The hallmark of these tumors, however, is the presence of large prominent nuclei and nucleoli. Collecting duct tumors have been seen in a minority of patients (Toro et al 2003). These tumors have been high grade (Furman nuclear grade 3 or 4), in contrast to VHL and HPRC, where lower nuclear grades are the norm. Reported sizes have ranged from 3.8 to 22 cm

In the European series, all six patients with renal tumors had metastatic disease at presentation and all patients were dead of disease at the time of the study (Launonen et al 2001). Similarly aggressive behavior was seen in the North American study (Figure 28.8) (Toro et al 2003). Four of 13 patients were diagnosed with renal tumors at ≤30 years of age. Seventy percent (9/13) of patients were dead of metastatic disease within 5 years of diagnosis. It seems that

HLRCC renal tumors may represent a potentially more aggressive type of hereditary renal cancer than VHL, HPRC, or BHD, which is similar to the behavior of sporadic type II papillary RCC (Jiang et al 1998, Delahunt et al 2001). Observational strategies which have been used successfully for other forms of hereditary RCC may not be appropriate for HLRCC given the potentially more aggressive nature of these tumors and the ability to metastasize at small sizes.

CONCLUSIONS

Increased understanding of the role of genetics in renal cancer in the last two decades has led to the identification of at least four distinct hereditary syndromes associated with increased risk of RCC: VHL, HPRC, BHD, and HLRCC. The characteristic histopathology of each syndrome has aided in phenotypic differentiation and gene discovery with resultant elucidation of underlying molecular mechanisms. VHL is known to be mediated through decreased degradation of HIF and resultant increases in its downstream substrates. HPRC is caused by an activating mutation of *MET*. Efforts are underway to elucidate the important molecular mechanism associated with BHD and HLRCC. It is hoped that insights into the underlying biology will aid in our understanding of the molecular pathophysiology and potential treatments of different types of both sporadic and hereditary renal cancers.

Implementation of observational strategies for VHL, HPRC and BHD using a 3 cm threshold for operative intervention and employing nephron-sparing approaches where feasible has resulted in maintenance of native renal function and good oncologic control in a majority of patients with hereditary renal tumors. The success of this type of strategy is dependent on effective diagnosis and diligent screening. Due to its potentially more aggressive biologic behavior, early surgical intervention is recommended for patients with suspected HLRCC-associated renal tumors.

References

Barker KT, Bevan S, Wang R, et al. Low frequency of somatic mutations in the FH/multiple cutaneous leiomyomatosis gene in sporadic leiomyosarcomas and uterine leiomyomas. Br. J. Cancer 2002; 87: 446–8.

Binet O, Robin J, Vicart M, et al. Fibromes perifolliculaires, polypose coliqe familiale, pneumothorax spontanes familiaux. Ann. Dermatol. Venereol. 1986; 113: 928.

Birt AR, Hogg GR, Dube WJ. Hereditary multiple fibrofolliculomas with trichodiscomas and acrochordons. Arch. Dermatol. 1977; 113: 1674–7.

Bottaro DP, Rubin JS, Faletto DL, et al. Identification of the hepatocyte growth factor receptor as the c-met proto-oncogene product. Science 1991; 251: 802–4.

Chow WH, Devesa SS, Warren JL, et al. Rising incidence of renal cell cancer in the United States. JAMA 1999; 281: 1628–31.

Choyke PL. Imaging of hereditary renal cancer. Radiol. Clin. North Am. 2003; 41: 1037–51.

Choyke PL, Glenn GM, Walther MM, et al. von Hippel-Lindau disease: genetic, clinical, and imaging features. Radiology 1995; 194: 629–42.

Choyke PL, Glenn GM, Walther MM, et al. Hereditary renal cancers. Radiology 2003; 226: 33–46.

Clifford SC, Cockman ME, Smallwood AC, et al. Contrasting effects on HIF-1alpha regulation by disease-causing pVHL mutations correlate with patterns of tumourigenesis in von Hippel-Lindau disease. Hum. Mol. Genet. 2001; 10: 1029–38.

Delahunt B, Eble JN, McCredie MR, et al. Morphologic typing of papillary renal cell carcinoma: comparison of growth kinetics and patient survival in 66 cases. Hum. Pathol. 2001; 32: 590–5.

Duffey BG, Choyke PL, Glenn G, et al. The relationship between renal tumor size and metastases in patients with von Hippel-Lindau disease. J. Urol. 2004; 172: 63–5.

Durrani OH, Ng L, Bihrle W III. Chromophobe renal cell carcinoma in a patient with the Birt-Hogg-Dube syndrome. J. Urol. 2002; 168: 1484–5.

Eng C, Kiuru M, Fernandez MJ, et al. A role for mitochondrial enzymes in inherited neoplasia and beyond. Nat. Rev. Cancer 2003; 3: 193–202.

Farrell MA, Charboneau WJ, DiMarco DS, et al. Imaging-guided radiofrequency ablation of solid renal tumors. AJR Am. J. Roentgenol. 2003; 180: 1509–13.

Fischer J, Palmedo G, von Knobloch R, et al. Duplication and overexpression of the mutant allele of the MET proto-oncogene in multiple hereditary papillary renal cell tumours. Oncogene 1998; 17: 733–9.

Gago-Dominguez M, Yuan JM, Castelao JE, et al. Family history and risk of renal cell carcinoma. Cancer Epidemiol. Biomarkers Prev. 2001; 10: 1001–4.

Goldfarb DA, Neumann HP, Penn I, et al. Results of renal transplantation in patients with renal cell carcinoma and von Hippel-Lindau disease. Transplantation 1997; 64: 1726–9.

Grubb RL, Linehan WM, Walther MM. Management of inherited forms of renal cancer. In: Vogelzang NJ (ed.) Comprehensive Textbook of Genitourinary Oncology. Lippincott, Williams and Wilkins, Baltimore; in press.

Herring JC, Enquist EG, Chernoff A, et al. Parenchymal sparing surgery in patients with hereditary renal cell carcinoma: 10-year experience. J. Urol. 2001; 165: 777–81.

Hornstein OP. Generalized dermal perifollicular fibromas with polyps of the colon. Hum. Genet. 1976; 33: 193–7.

Hwang JJ, Walther MM, Pautler SE, et al. Radio frequency ablation of small renal tumors: intermediate results. J. Urol. 2004; 171: 1814–18.

Iliopoulos O, Kibel A, Gray S, et al. Tumour suppression by the human von Hippel-Lindau gene product. Nat. Med. 1995; 1: 822–6.

Ivan M, Kondo K, Yang H, et al. HIFalpha targeted for VHL-mediated destruction by proline hydroxylation: implications for O_2 sensing. Science 2001; 292: 464–8.

Jaakkola P, Mole DR, Tian YM, et al. Targeting of HIF-alpha to the von Hippel-Lindau ubiquitylation complex by O_2-regulated prolyl hydroxylation. Science 2001; 292: 468–72.

Jeffers M, Schmidt L, Nakaigawa N, et al. Activating mutations for the met tyrosine kinase receptor in human cancer. Proc. Natl Acad. Sci. USA 1997; 94: 11445–50.

Jemal A, Murray T, Samuels A, et al. Cancer statistics, 2003. CA Cancer J. Clin. 2003; 53: 5–26.

Jiang F, Richter J, Schraml P, et al. Chromosomal imbalances in papillary renal cell carcinoma: genetic differences between histological subtypes. Am. J. Pathol. 1998; 153: 1467–73.

Khoo SK, Bradley M, Wong FK, et al. Birt-Hogg-Dube syndrome: mapping of a novel hereditary neoplasia gene to chromosome 17p12-q11.2. Oncogene 2001; 20: 5239–42.

Khoo SK, Giraud S, Kahnoski K, et al. Clinical and genetic studies of Birt-Hogg-Dube syndrome. J. Med. Genet. 2002; 39: 906–12.

Khoo SK, Kahnoski K, Sugimura J, et al. Inactivation of BHD in sporadic renal tumors. Cancer Res. 2003; 63: 4583–7.

Kibel A, Iliopoulos O, DeCaprio JA, et al. Binding of the von Hippel-Lindau tumor suppressor protein to Elongin B and C. Science 1995; 269: 1444–6.

Kiuru M, Launonen V, Hietala M, et al. Familial cutaneous leiomyomatosis is a two-hit condition associated with renal cell cancer of characteristic histopathology. Am. J. Pathol. 2001; 159: 825–9.

Kiuru M, Lehtonen R, Arola J, et al. Few FH mutations in sporadic counterparts of tumor types observed in hereditary leiomyomatosis and renal cell cancer families. Cancer Res. 2002; 62: 4554–7.

Kondo K, Klco J, Nakamura E, et al. Inhibition of HIF is necessary for tumor suppression by the von Hippel-Lindau protein. Cancer Cell 2002; 1: 237–46.

Latif F, Tory K, Gnarra J, et al. Identification of the von Hippel-Lindau disease tumor suppressor gene. Science 1993; 260: 1317–20.

Launonen V, Vierimaa O, Kiuru M, et al. Inherited susceptibility to uterine leiomyomas and renal cell cancer. Proc. Natl Acad. Sci. USA 2001; 98: 3387–92.

Levine E, Weigel JW, Collins DL. Diagnosis and management of asymptomatic renal cell carcinomas in von Hippel-Lindau syndrome. Urology 1983; 21: 146–50.

Libutti SK, Choyke PL, Bartlett DL, et al. Pancreatic neuroendocrine tumors associated with von Hippel Lindau disease: diagnostic and management recommendations. Surgery 1998; 124: 1153–9.

Lindau A. Studien ber kleinbirncysten bau: pathogenese und beziehungen zur angiomatosis rentinae. Acta Radiol. Microbiol. Scandinavica 1926; 1: 1–128.

Linehan WM, Walther MM, Zbar B. The genetic basis of cancer of the kidney. J. Urol. 2003; 170: 2163–72.

Linehan WM, Zbar B. Focus on kidney cancer. Cancer Cell 2004; 6: 223–8.

Liu V, Kwan T, Page EH. Parotid oncocytoma in the Birt-Hogg-Dube syndrome. J. Am. Acad. Dermatol. 2000; 43: 1120–2.

Lonser RR, Glenn GM, Walther M, et al. von Hippel-Lindau disease. Lancet 2003; 361: 2059–67.

Loughlin KR, Gittes RF. Urological management of patients with von Hippel-Lindau's disease. J. Urol. 1986; 136: 789–91.

Maher ER, Yates JR, Harries R, et al. Clinical features and natural history of von Hippel-Lindau disease. Q. J. Med. 1990; 77: 1151–63.

Maher ER, Iselius L, Yates JR, et al. Von Hippel-Lindau disease: a genetic study. J. Med. Genet. 1991; 28: 443–7.

Maranchie JK, Afonso A, Albert PS, et al. Solid renal tumor severity in von Hippel Lindau disease is related to germline deletion length and location. Hum. Mutat. 2004; 23: 40–6.

Melmon KL, Rosen SW. Lindau's disease. Am. J. Med. 1964; 36: 595–617.

Merino MJ, Torres-Cabala C, Toro J, et al. The morphologic spectrum of renal tumors in hereditary leiomyomatosis and renal cell carcinoma (HLRCC) syndrome; Correlation with molecular findings. Mod. Pathol. 2003; 16(1): 739.

Michieli P, Basilico C, Pennacchietti S, et al. Mutant Met-mediated transformation is ligand-dependent and can be inhibited by HGF antagonists. Oncogene 1999; 18: 5221–31.

Neumann HP, Eng C, Mulligan LM, et al. Consequences of direct genetic testing for germline mutations in the clinical management of families with multiple endocrine neoplasia, type II. JAMA 1995; 274: 1149–51.

Neumann HP, Wiestler OD. Clustering of features and genetics of von Hippel-Lindau syndrome. Lancet 1991; 338: 258.

Nickerson ML, Warren MB, Toro JR, et al. Mutations in a novel gene lead to kidney tumors, lung wall defects, and benign tumors of the hair follicle in patients with the Birt-Hogg-Dube syndrome. Cancer Cell 2002; 2: 157–64.

Novick AC, Streem SB. Long-term followup after nephron sparing surgery for renal cell carcinoma in von Hippel-Lindau disease. J. Urol. 1992; 147: 1488–90.

Ohh M, Park CW, Ivan M, et al. Ubiquitination of hypoxia-inducible factor requires direct binding to the beta-domain of the von Hippel-Lindau protein. Nat. Cell Biol. 2000; 2: 423–7.

Ornstein DK, Lubensky IA, Venzon D, et al. Prevalence of microscopic tumors in normal appearing renal parenchyma of patients with hereditary papillary renal cancer. J. Urol. 2000; 163: 431–3.

Pause A, Lee S, Worrell RA, et al. The von Hippel-Lindau tumor-suppressor gene product forms a stable complex with human CUL-2, a member of the Cdc53 family of proteins. Proc. Natl Acad. Sci. USA 1997; 94: 2156–61.

Pavlovich CP, Walther MM, Eyler RA, et al. Renal tumors in the Birt-Hogg-Dube syndrome. Am. J. Surg. Pathol. 2002; 26: 1542–52.

Pavlovich CP, Schmidt LS. Searching for the hereditary causes of renal-cell carcinoma. Nat. Rev. Cancer 2004; 4: 381–93.

Pavlovich CP, Grubb RL, Hurley K, et al. Evaluation and management of renal tumors in the Birt-Hogg-Dubé syndrome. J. Urol. in press.

Phillips JL, Pavlovich CP, Walther M, et al. The genetic basis of renal epithelial tumors: advances in research and its impact on prognosis and therapy. Curr. Opin. Urol. 2001; 11: 463–9.

Roth JS, Rabinowitz AD, Benson M, et al. Bilateral renal cell carcinoma in the Birt-Hogg-Dube syndrome. J. Am. Acad. Dermatol. 1993; 29: 1055–6.

Schmidt L, Duh FM, Chen F, et al. Germline and somatic mutations in the tyrosine kinase domain of the MET proto-oncogene in papillary renal carcinomas. Nat. Genet. 1997; 16: 68–73.

Schmidt L, Junker K, Weirich G, et al. Two North American families with hereditary papillary renal carcinoma and identical novel mutations in the MET proto-oncogene. Cancer Res. 1998; 58: 1719–22.

Schmidt LS, Warren MB, Nickerson ML, et al. Birt-Hogg-Dube syndrome, a genodermatosis associated with spontaneous pneumothorax and kidney neoplasia, maps to chromosome 17p11.2. Am. J. Hum. Genet. 2001; 69: 876–82.

Shingleton WB, Sewell PE Jr. Percutaneous renal cryoablation of renal tumors in patients with von Hippel-Lindau disease. J. Urol. 2002; 167: 1268–70.

Spencer WF, Novick AC, Montie JE, et al. Surgical treatment of localized renal cell carcinoma in von Hippel-Lindau's disease. J. Urol. 1988; 139: 507–9.

Steinbach F, Novick AC, Zincke H, et al. Treatment of renal cell carcinoma in von Hippel-Lindau disease: a multicenter study. J. Urol. 1995; 153: 1812–16.

Su LM, Jarrett TW, Chan DY, et al. Percutaneous computed tomography-guided radiofrequency ablation of renal masses in high surgical risk patients: preliminary results. Urology 2003; 61: 26–33.

Tickoo SK, Reuter VE, Amin MB, et al. Renal oncocytosis: a morphologic study of fourteen cases. Am. J. Surg. Pathol. 1999; 23: 1094–101.

Tomlinson IP, Alam NA, Rowan AJ, et al. Germline mutations in FH predispose to dominantly inherited uterine fibroids, skin leiomyomata and papillary renal cell cancer. Nat. Genet. 2002; 30: 406–10.

Toro JR, Glenn G, Duray P, et al. Birt-Hogg-Dube syndrome: a novel marker of kidney neoplasia. Arch. Dermatol. 1999; 135: 1195–202.

Toro JR, Nickerson ML, Wei MH, et al. Mutations in the fumarate hydratase gene cause hereditary leiomyomatosis and renal cell cancer in families in North America. Am. J. Hum. Genet. 2003; 73: 95–106.

von Hippel E. Uber eine sehr self seltene erkrankung der netzhaut. Klinische Beobachtungen Archive fur Opthalmologie 1904; 59: 83–106.

Walther MM, Choyke PL, Weiss G, et al. Parenchymal sparing surgery in patients with hereditary renal cell carcinoma. J. Urol. 1995b; 153: 913–16.

Walther MM, Lubensky IA, Venzon D, et al. Prevalence of microscopic lesions in grossly normal renal parenchyma from patients with von Hippel-Lindau disease, sporadic renal cell carcinoma and no renal disease: clinical implications. J. Urol. 1995a; 154: 2010–14.

Walther MM, Reiter R, Keiser HR, et al. Clinical and genetic characterization of pheochromocytoma in von Hippel-Lindau families: comparison with sporadic pheochromocytoma gives insight into natural history of pheochromocytoma. J. Urol. 1999b; 162: 659–64.

Walther MM, Choyke PL, Glenn G, et al. Renal cancer in families with hereditary renal cancer: prospective analysis of a tumor size threshold for renal parenchymal sparing surgery. J. Urol. 1999a; 161: 1475–9.

Zbar B, Tory K, Merino M, et al. Hereditary papillary renal cell carcinoma. J. Urol. 1994; 151: 561–6.

Zbar B. Inherited epithelial tumors of the kidney: old and new diseases. Semin. Cancer Biol. 2000; 10: 313–18.

Zbar B, Alvord WG, Glenn G, et al. Risk of renal and colonic neoplasms and spontaneous pneumothorax in the Birt-Hogg-Dube syndrome. Cancer Epidemiol. Biomarkers Prev. 2002; 11: 393–400.

Zhuang Z, Park WS, Pack S, et al. Trisomy 7-harbouring nonrandom duplication of the mutant MET allele in hereditary papillary renal carcinomas. Nat. Genet. 1998; 20: 66–9.

CHAPTER 29

Wilms' Tumor

SUNNY HARTWIG AND JORDAN A. KREIDBERG

INTRODUCTION

The pediatric kidney malignancy Wilms' tumor (nephroblastoma) was first described in a 3-year-old patient by the German surgeon Max Wilms in 1899 (Wilms 1899). Accounting for 8% of all childhood cancers, Wilms' tumor affects 1 in 10 000 children in North America (Matsunaga 1989), making it the most common solid tumor in childhood and the fourth most common childhood malignancy (Bennington & Beckwith 1975, Haber 1998, Rivera & Haber 2005). Treatment protocols have improved tremendously over the past two decades, and what used to be a uniformly lethal disease now has a cure rate of over 90% for localized disease and 70% for metastatic disease (Rivera & Haber 2005). Most cases are treated with a combination of surgery and chemotherapy, with radiation therapy reserved for advanced disease (Haber 1998).

Inactivating mutations in the tumor suppressor gene, *Wilms' tumor suppressor-1* (*WT1*) are responsible for 10–15% of Wilms' tumor cases (Hastie 1994). Although Wilms' tumor may also develop in response to mutations at other chromosomal locations (*WTX* at Xq11.1 and *CTNNB1* at 3p22 encoding beta-catenin), thus far *WT1* on human chromosome 11p3 is the only gene shown to inhibit tumor growth (Haber et al 1993). Unlike other tumor suppressors, WT1 also plays a critical role during embryogenesis, particularly during development of the urogenital system (Pritchard-Jones et al 1990, Pelletier et al 1991a, Freidberg et al 1993). Much of our understanding of Wilms' tumor has been derived from the study of *WT1* and its connection to kidney development, and this chapter will discuss the *WT1* gene, its involvement in Wilms' tumorigenesis, and its role in normal kidney development.

WILMS' TUMOR: HISTOLOGY, GENETICS, AND ASSOCIATED SYNDROMES

Histology of Wilms' Tumor

Wilms' tumors are characterized by the presence of undifferentiated blastemal (mesenchymal) elements together with stromal and epithelial components (Rivera & Haber 2005, Scholz & Kirschner 2005). This so-called 'triphasic' morphology recapitulates sequential phases of normal nephrogenesis, insofar as the blastemal components in Wilms' tumor resemble condensing nephrogenic mesenchyme, while the epithelial components of the tumor resemble comma-shaped bodies, S-shaped bodies, and glomeruli (Beckwith 1998). In some tumors, further differentiation along myogenic or neural cell lineages is also observed (Miyagawa et al 1998).

The ontogeny of Wilms' tumor was first described more than 100 years ago, when Max Wilms recognized that 'similar to the growth of an embryo, all these tissues develop from a common and macroscopically undifferentiated germ cell' (Wilms 1981). More precisely, Wilms' tumor is thought to derive from a single transformed, pluripotent mesenchymal stem cell whose progeny fail to undergo normal differentiation. Instead, mutant blastemal cells in the tumor form 'residual islands of immature kidney tissue locked in a state of continual proliferation' known as nephrogenic rests (Reddy & Licht 1996), while a minority of cells able to overcome the differentiation blockade proceed to form primitive renal, neural or myogenic epithelial structures (Zhuang et al 1997, Li et al 2005).

Consistent with the model of impaired differentiation in Wilms' tumorigenesis, microarray gene-expression studies demonstrate that blastemal components of Wilms' tumor overexpress markers of the earliest (uninduced) stage of kidney development, including genes encoding developmentally essential transcription factors PAX2, SIX1, EYA1, HBF2, and HOXA11 (Li et al 2002, 2005). Conversely, genes corresponding to later stages of metanephric development, such as cadherin 6 and E-cadherin, tend to be under-represented in these cells (Li et al 2002, Rivera & Haber 2005).

The Genetics of Sporadic Wilms' Tumor

The vast majority of Wilms' tumors are unilateral (90%) and sporadic, presenting before age 5 in otherwise healthy children, with equal incidence between the sexes (Wiener et al 1998, Menke & Hastie 2001). In rare instances, genetic predisposition to Wilms' tumor conferred by either inherited (1–2% of cases) or de novo germline mutations in a tumor suppressor gene (Breslow & Beckwith 1982, Haber 1998), results in bilateral tumors which present earlier, at a median age of 32 months compared with 44 months for unilateral disease (Menke & Hastie 2001). In addition, several genetic syndromes are associated with increased risk of developing Wilms' tumor and account for 5–10% of all Wilms' tumor cases (Table 29.1).

Because of the strong association previously demonstrated between familial inheritance and bilateral, early-onset retinoblastoma, Knudson and Strong proposed a similar 'two-hit' model for the development of bilateral Wilms' tumor (Knudson & Strong 1972). According to the two-hit model, genetic predisposition to a pediatric tumor is conferred by an inherited or de novo germline mutation in one allele of a tumor suppressor gene, requiring only one additional 'hit' during somatic development to cause somatic homozygosity by inactivating the second allele. In contrast, individuals with no such predisposition would require two independent genetic events to delete both copies of the tumor suppressor gene, and would thus present at a later age (Reddy & Licht 1996). Consistent with this model, 35–40% of children with retinoblastoma have an inherited mutation in the retinoblastoma

TABLE 29.1   Human syndromes associated with predisposition to Wilms' tumor

	Clinical manifestations	Locus/gene	References
Overgrowth syndrome			
Beckwith-Wiedemann	EMG triad (exomphalos-macroglossia-gigantism), organomegaly, hemihypertrophy, hypoglycemia, earlobe pits or creases, facial nevus flammeus, and predisposition to several tumors including Wilms' tumor, adrenal cortex carcinoma, hepatoblastoma, rhabdomyosarcoma and Wilms' tumor; 5–10% risk of developing Wilms' tumor	**11p15** paternal duplications of 11p15, maternal translocations involving 11p15.3–p15.5. *Igf2, H19*, imprinting defects *p57, LIT1*	2, 6, 59, 61, 72
Simpson Golabi-Behmel	Organomegaly, macroglossia, coarse facies with mandibular overgrowth, cleft palate, heart defects, hernias, supernumerary nipples, renal dysplasia, polydactyly and other skeletal defects; 7% risk of developing Wilms' tumor	**X-linked** with heterogeneity: main locus is Xq26; one large pedigree maps to Xp22 (lethal form) *Gpc3* point mutations, microdeletions	64–67
Perlman	Fetal gigantism (macrosomia), renal dysplasia, visceromegaly, cryptorchidism, polyhydramnios, islet cell hypertrophy, distinct facies, mental retardation; 30% risk of developing Wilms' tumor	**Unknown** Cytogenetic studies show an extra band on the tip of the short arm of chromosome 11 (11p)	68–71
Sotos	Craniofacial dysmorphism, cerebral gigantism, mental retardation and predisposition to several tumors including Wilms' tumor, neuroblastoma and acute lymphocytic leukemia; 5% risk of developing Wilms' tumor	**5q35** *NSD1* point mutations, microdeletions	72, 73
Non-overgrowth syndrome			
WAGR	Aniridia (100%), genitourinary defects including hypospadias and cryptorchidism, mental retardation; 30% risk of developing Wilms' tumor	**11p13** band deletions *WT1, Pax6*, deletions	18, 52, 54, 63–67
Denys-Drash	Triad of congenital nephropathy (mesangial sclerosis) leading to renal failure by age 3 yrs, Wilms' tumor (90%), and intersex disorders (complete gonadal dysgenesis with male pseudohermaphroditism)	**11p13** *WT1*, missense changes in exons 8 or 9, which encode for the zinc fingers 2 and 3. Arginine-394, critical for DNA binding, is a mutational 'hot spot' for DDS	3, 29, 68, 69
Bloom's	Telangiectases and photosensitivity, growth deficiency, immunodeficiency, and predisposition to many neoplasms; 30% risk of developing Wilms' tumor	**15q26.1** *BLM*, point mutations	74, 75
Frasier	Nephrotic syndrome (focal segmental glomerulosclerosis) leading to renal failure in late childhood, complete XY gonadal dysgenesis, streak gonads, high risk of gonadoblastoma; <5% risk of developing Wilms' tumor	**11p13** *WT1*, mutations disrupt alternative splice site in intron 9intron alter the ratio of the +KTS: −KTS isoforms from 2:1 to 1:2	3, 44, 58

predisposition gene, *Rb1*, on chromosome 13 (Knudson et al 1976); bilateral tumor development in these children occurs due to the second 'hit' on the normal allele during retinal stem cell development. In contrast to the relatively simple genetics of familial retinoblastoma, the genetic basis of bilateral Wilms' tumor is poorly understood, as the tumor suppressor genes accounting for the majority of Wilms' tumors are not yet identified.

Loss of heterozygosity (LOH) at chromosome 11 occurs in 30–40% of Wilms' tumors, encompassing allelic loss of both the *WT1* locus at chromosome 11p13 and the locus telomeric to it at 11p15 (Koufos et al 1984, Reeve et al 1989, Lee & Haber 2001). LOH at chromosome 16 is observed in 20% of sporadic Wilms' tumors (Maw et al 1992), and at chromosome 1p at a frequency of 11% (Hing et al 2001, Lu et al 2002). Notably, patients with tumor-specific loss of these loci have significantly worse relapse-free and overall survival rates (Grundy et al 1994), indicating that potential tumor suppressor genes may exist at 1p and 16q. A further 5% of cases exhibit LOH at 4p, 7p, 8p, and 17, 17q and 18q (Sheng et al 1990, Grundy et al 1994).

The most common molecular abnormality associated with sporadic Wilms' tumor is genomic loss of imprinting (LOI), a feature common to several embryonal cancers and many other tumor types (Grundy et al 1991). LOI at chromosome 11p15 is observed in 30% of sporadic Wilms' tumors (Giannoukakis et al 1993, Ogawa et al 1993, Ravenel et al 2001), and involves activation of the normally silent maternal allele of *Insulin-like growth factor 2* (*Igf2*) resulting in biallelic *Igf2* expression (Giannoukakis et al 1993), silencing of the normally active maternal allele of *H19* (Bartolemei et al 1991, Hao et al 1993), and hypermethylation of the differentially methylated region (DMR) upstream of the maternal copy of *H19* (Cui et al 2001). Both *Igf2* and *H19* have been suggested to be Wilms' tumor suppressor genes, but direct evidence of such a role for either gene is lacking. Alternatively, 11p15 may harbor an as yet unidentified Wilms' tumor suppressor gene (*WT2*).

To Loss-of-function mutations in *WT1* account for 10–15% cases of sporadic Wilms' tumor (Haber et al 1990). *WT1* mutations include both nonsense and missense changes, as well as the generation of aberrant splice forms, and are distributed throughout the coding sequence (Brenner et al 1992, Bruening et al 1992, Brown et al 1993). However, most sporadic tumors express wild-type *WT1*, often at high levels[45], suggesting that downstream *WT1* signaling is affected in these tumors, or that genetic disruption of other cellular pathways is responsible for tumorigenesis. In fact, approximately 70% of Wilms' tumors exhibiting loss-of-function mutations in *WT1* are also associated with constitutive activation and nuclear translocation of β-catenin (Koesters et al 1999, Maiti et al 2000), both in the presence and notably, in the absence of accompanying activating mutations in the gene encoding β-catenin, *CTNBB1*. These data suggest WT1 may normally function as a transcriptional regulator of the *Wnt* signaling axis during renal development, and that pathologic, β-catenin-dependent activation of *Wnt* target genes may represent a key oncogenic step in *WT1*-dependent Wilms' tumorigenesis. Recently, a novel, X-linked Wilms' tumor gene *Wilms' Tumor Gene on the X Chromosome* (*WTX*) was shown to be deleted or mutated in 15–30% of Wilms' tumors, identified through a genome-wide scan for DNA copy-number changes in primary Wilms' tumors (Rivera et al, Science, 2007). While *WTX* does not share significant homology with other genes of known function, tandem-affinity purification assays have demonstrated direct binding of WTX to β-catenin and its E3 ubiquitin ligase adaptor β-TrCP1, a component of the β-catenin destruction complex, and WTX appears to negatively regulate β-catenin by promoting its ubiquitination both *in vitro* and *in vivo* (Major et al, Science 2007). Collectively, these data suggest potential genetic cross-talk between the *WT1*, *WTX* and *Wnt* signaling axes.

While the presence or absence of *WT1* mutations does not appear to influence prognosis, *WT1*-mutant tumors are more frequently associated with intralobular nephrogenic rests (located inside the renal lobule) than with perilobular rests (located at the periphery) (Park et al 1993, Guertl et al 2003). In addition, tumors with *WT1* mutations tend to exhibit stromal predominant histology accompanied by myogenic differentiation (Schumacher et al 1997, Miyagawa et al 1998). Mutations in *WT1* have also been detected in other malignancies, including acute myeloid leukemia (Miwa et al 1992, Inoue et al 1994, Bergmann et al 1997), rare cases of mesothelioma (Park et al 1993), adrenocortical carcinoma (Henry et al 1989) and gonadoblastoma associated with Frasier Syndrome (Barbaux et al 1997). Notably, high levels of *WT1* expression are markers of poor prognosis in leukemia (Bergmann et al 1997). High levels of *WT1* have also been detected in other malignancies including renal clear cell carcinoma (Campbell et al 1998), and may represent a tumor-associated expression abnormality, a cellular defense response against tumor growth, or a physiological property of immature (de-differentiated) cells.

Wilms' Tumor-Associated Syndromes

Wilms' tumor normally occurs in otherwise healthy children. However, 10–15% of cases occur in children with recognized malformations, including hemihypertrophy, cryptorchidism, and hypospadias, or in association with a recognizable genetic syndrome (Table 29.1). Genetic linkage between predisposition to Wilms' tumor and the two major chromosomal loci on chromosome 11, has been demonstrated for these Wilms' tumor-associated syndromes. Three of these syndromes, *Wilms' tumor, Aniridia, Genitourinary malformation, mental Retardation* (WAGR) syndrome (Riccardi et al 1978, Francke et al 1979, Fantes et al 1992), *Denys-Drash Syndrome* (DDS) (Pelletier et al 1991, Baird et al 1992, Coppes et al 1992), and Frasier Syndrome (FS) (Barbaux et al 1997, Klamt et al

1998) are associated with *WT1* at chromosome locus 11p13 (Pelletier et al 1991). *Beckwith-Wiedemann Syndrome* (BWS) is the most common Wilms' tumor-associated condition, affecting 1 in 13 000 children, and is linked with 11p15 (Henry et al 1989a, Ohlsson et al 1993, Weksberg et al 1993). Several other genetic syndromes unlinked to chromosome 11 are also associated with increased incidence of Wilms' tumor (Lapunzina 2005, Scott et al 2006), including Simpson-Golabi-Behmel Syndrome (linked to Xp26) (Hughes-Benzie et al 1996, Pilia et al 1996, Neri et al 1998, Pellegrini et al 1998), Perlman Syndrome (Neri et al 1984, Perlman 1986, Chitty et al 1998, Henneveld et al 1999), Sotos Syndrome (linked to 5q35) (Hersh et al 1992, Kurotaki et al 2002), and Bloom's Syndrome (linked to 15q26) (Ellis & German 1996, Pujana et al 2001).

WAGR syndrome is caused by a microdeletion at 11p13 that deletes both *WT1* and *Pax6* (Francke et al 1979, Fantes et al 1992). WAGR is associated with aniridia in all cases resulting from hemizygosity for *Pax6* (Hill et al 1991, Ton et al 1991, Jordan et al 1992), but is variably associated with Wilms' tumor (50% risk of Wilms' Tumor) (Narahara et al 1984) and genitourinary malformations due to hemizygosity for *WT1*. WAGR is also variably associated with mental retardation and congenital heart disease; however the gene(s) on 11p13 responsible for these defects have not been identified (Crolla et al 1997).

A second *WT1*-associated syndrome, DDS, includes the triad of Wilms' tumor (90% risk of Wilms' tumor development) (Menke & Hastie 2001), genitourinary malformations and nephropathy (mesangial sclerosis), but various combinations of theses features have been reported (Little & Wells 1997, Schumacher et al 1998, Royer-Pokora et al 2004). DDS is caused by intragenic *WT1* point mutations, most commonly an arginine-to-tryptophan transition in exon 3 (Arg 394), or missense alterations in the zinc-finger domains of exons 8 and 9 (Lee & Haber 2001, Rivera & Haber 2005). The severity of kidney disease associated with DDS compared with WAGR, has raised the possibility of a dominant-negative effect that is mediated by dimerization of mutant and wild-type proteins through their N-terminal domains (Little et al 1993, Bardeesy et al 1994).

FS is characterized by gonadal dysgenesis, often resulting in XY sex reversal in males (Rivera & Haber 2005), progressive glomerular nephropathy (focal segmental glomerulosclerosis) (Denamur et al 2000), gonadoblastoma, and a low risk (less than 5%) of developing Wilms' tumor. FS is caused by *WT1* splicing mutations in intron 9 that alter the ratio of WT1 isoforms (discussed below) (Barbaux et al 1997, Klamt et al 1998).

BWS is characterized by prenatal overgrowth and increased incidence of embryonal tumors of liver (hepatoblastoma), muscle (rhabdomyosarcoma), and kidney (10% risk of Wilms' tumor) (Hastie 1994). The genetics of Wilms' tumor formation in BWS are largely unknown. However, at least 20% of BWS patients exhibit paternal uniparental disomy (UPD) for 11p15, the *Igf2* and *H19* imprinting locus strongly associated with sporadic Wilms' tumor (Henry et al 1989b), and these children have a high risk (64%) of developing embryonal tumors (Weksberg et al 1993). In addition, familial BWS is linked to chromosome 11p15 (Koufos et al 1989). A third, unidentified tumor suppressor gene on 11p15 has been linked to rhabodomyosarcoma (Hu et al 1997), suggesting that at least three genes on 11p15 may predispose to growth abnormalities and Wilms' tumor as well as other embryonal tumors.

Familial Wilms' Tumor

True familial Wilms' tumor is extremely rare, accounting for only 1–2% of all cases (Knudson & Strong 1972), suggesting that de novo germline mutations, rather than familial transmission of a mutant allele underlie the genetic predisposition (Rivera & Haber 2005). The low number of familial cases may reflect lethality of this cancer before the advent of effective chemotherapy, and, in the future, familial cases may become more common with the curative treatment of mutation carriers. In addition, reduced fertility may be associated with germline mutations in genes that regulate urogenital development (Hastie 1994). Two familial Wilms' tumor genes have been mapped – *FWT1* (*Familial Wilms' tumor 1 locus*; also known as *WT4*) at 17q12-21 (Rahman et al 1996) and *FWT2* at 19q13.4 (McDonald et al 1998), but neither gene has been identified. In addition, familial cases unlinked to any of the known loci have been reported, indicating that other familial Wilms' tumor genes may exist (Rapley et al 2000).

THE *WT1* GENE, mRNAS, AND PROTEINS

The Discovery of a Wilms' Tumor Suppressor Gene at 11p13

Long before *WT1* was mapped and cloned, the existence of a Wilms' tumor suppressor gene was hypothesized (Knudson & Strong 1972). A major step towards the discovery of the *WT1* locus was the identification of cytogenetically visible, germline interstitial deletions in 11p13 as the locus for WAGR Syndrome (Riccardi et al 1978). The absence of this chromosomal band in WAGR patients suggested that loss of both alleles of a gene at 11p13 might cause Wilms' tumor (Reeve et al 1989). Thus, constitutional hemizygosity for a tumor suppressor gene at 11p13 would predispose to Wilms' tumor, while somatic LOH at 11p13 during kidney development would result in Wilms' tumor formation, consistent with Knudson's two-hit hypothesis. Together with the discovery that some individuals with sporadic, nonsyndromic Wilms' tumor also showed LOH in 11p13 (Lewis et al 1988, Guertl et al 2003), a minimal 30 kb region on 11p13 was identified as an critical locus in the development of Wilms' tumor, and was given the designation of the '*WT1*' locus (Compton et al 1990, Rose et al 1990). Molecular analysis of a large number of WAGR-associated deletions allowed the *WT1* locus to be mapped to a 350 kb region within 11p13 (Compton et al 1988). Subsequently, two independent groups identified the *WT1* gene by mapping the smallest region of overlap among chromosomal deletions in germline and sporadic tumor specimens (Call et al 1990, Gessler et al 1990). Shortly after it was cloned, *WT1* was shown to be mutated in several patients with sporadic Wilms' tumor (Little et al 1992). Notably, one patient exhibited a heterozygous mutation in *WT1*, suggesting that this mutant allele was behaving in a dominant negative fashion (Haber et al 1990). Other cases of Wilms' tumor exhibited absent or grossly reduced levels of *WT1* transcripts (Huang et al 1990), and one patient with bilateral tumor exhibited homozygous deletion of *WT1* (Huff et al 1991). Subsequently, it was shown that chemically inducible Wilms' tumorigenesis in rats (using the chemical mutagen N-nitroso-N'methylurea) was linked to *WT1* mutations (Sharma et al 1994b). Together with the demonstration that transfection of wild-type WT1 suppressed the growth of tumorigenic Wilms' tumor cells in vitro (Haber et al 1993), these data collectively identified *WT1* as a tumor suppressor gene whose loss results in Wilms' tumorigenesis in vivo (Little et al 1992).

Structure of WT1

In both humans and mice, the *WT1* gene comprises 10 exons, and encodes a 55 kDa zinc-finger transcription factor with both DNA- and RNA-binding activity (Figure 29.1) (Kreidberg 2002). Consistent with its role as a transcriptional regulator, the carboxy (C) terminus of WT1 contains four Krüppel-type (Cys_2His_2) zinc-finger domains and two nuclear localization domains that mediate DNA binding. The zinc-finger region and its alternative KTS splice site are critical for WT1 function, as mutations in the zinc-finger domain are associated with Wilms' tumorigenesis as well as the severe urogenital abnormalities observed in DDS (Bruening et al 1992). The amino (N) terminus of *WT1* contains a glutamine/proline-rich transactivation domain (Lewis et al 1999), a dimerization domain (Englert et al 1995, Reddy et al 1995) and putative RNA-binding domain (Call et al 1990, Gessler et al 1992).

Alternative Splicing of WT1

Alternative CUG or CTG translation initiation both upstream and downstream of the initial AUG (Bruening & Pelletier 1996, Scharnhorst et al 1999), alternative pre-mRNA splicing (Haber et al 1991) and RNA editing (Sharma et al 1994a) can theoretically result in the generation of twenty-four *WT1* isoforms. However, in mammals, two alternative splice inserts result in the production of four major *WT1* transcripts (Kent et al 1995), which are expressed at relatively constant ratios in all tissues throughout development and across species (Haber et al 1991, Brenner et al 1992).

The first alternative splice inserts exon 5, which encodes 17 amino acids unique to placental mammals (Kent et al 1995). However, the biological function of this isoform remains obscure as mice lacking this exon develop normally and are fertile (Natoli et al 2002a). However, *WT1(+exon5)* is frequently over-expressed in Wilms' tumor and other malignancies (Iben & Royer-Pokora 1999), and has been shown to regulate cell division and survival in vitro (Kudoh et al 1995), suggesting that it may play a subtle tumor suppressor role.

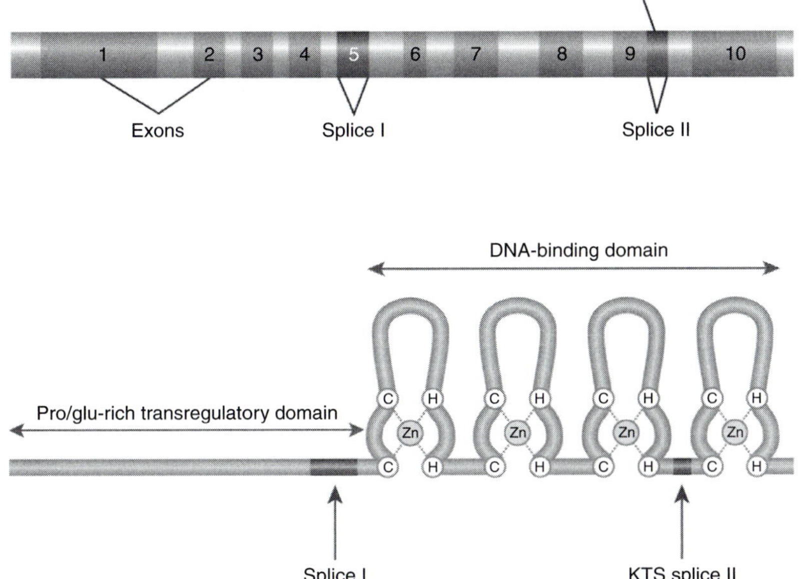

FIGURE 29.1 Structure of *WT1* mRNA and protein. Red boxes represent nucleotides or amino acids included as the result of alternative mRNA splicing. (see also Plate 65)

A second alternative splice donor site at exon 9 inserts the amino acids lysine, threonine and serine (KTS) in the H-C link region between the third and fourth zinc fingers, serving to alter the DNA-binding affinity of WT1 (Haber et al 1991, Gessler et al 1992). The +KTS/−KTS ratio is tightly regulated and highly conserved from zebrafish to humans (Kent et al 1995), and maintenance of the correct ratio between +KTS vs. −KTS isoforms is critical for normal urogenital (Hammes et al 2001) and olfactory (Wagner et al 2005) development.

THE ROLE OF WT1 (−KTS) *VS.* WT1(+KTS) ISOFORMS

WT1(−KTS) as a Transcriptional Repressor

An emerging body of evidence suggests that the +KTS and −KTS serve distinct functions in the nucleus, with (−KTS) isoforms playing a role in transcriptional regulation, and (+KTS) proteins implicated in posttranscriptional modification of target genes via RNA editing (Hammes et al 2001, Bor et al 2006). WT1(−KTS) exhibits characteristic properties of a DNA-binding transcription factor (Rauscher et al 1990), and several EGR1-like GC-rich DNA sequences (Rauscher et al 1990, Drummond et al 1994) and (TCC)n consensus sequences have been identified as cognate WT1 binding sites in vitro (Call et al 1990, Gessler et al 1990). The best predictor of endogenous gene transactivation is a high-affinity 10 bp EGR1-like (WTE) motif (5′GCGTGGGAGT3′) identified from genomic DNA (Nakagama et al 1995, Rivera & Haber 2005).

Early studies demonstrating over-expression of *Igf2* and *Platelet-Derived Growth Factor* (*Pdgf*) in Wilms' tumors, led to the hypothesis that WT1 functions as a transcriptional repressor of growth-promoting genes (Reeve et al 1985, Scott et al 1985, Madden et al 1991). Thus, loss of *WT1* function would result in *Igf2* and *Pdgf* over-expression and ultimately in Wilms' tumorigenesis.

Consistent with this hypothesis, transient transfection assays demonstrated that WT1 represses *Igf2* (Nakagama et al 1995) and *Pdgf* (Madden et al 1991) promoter activity in vitro, and antibodies directed against the IGF2 receptor were shown to inhibit tumor growth in Wilms' tumor explants (Gansler et al 1989). Subsequently, WT1 was shown to bind and repress promoters of numerous other growth-promoting genes expressed in the kidney including *Pax2* (Ryan et al 1995), *c-myc* (Hewitt et al 1995), *N-myc* (Madden et al 1991, Zhang et al 1999), *B-Cell CLL/lymphoma 2* (*Bcl-2*) (Lichnovsky et al 1999, Mayo et al 1999), *EGR* (Rauscher et al 1990), *epidermal growth factor receptor* (*EGFR*) (Englert et al 1995) and *colony stimulating factor-1* (*CSF-1*) (Harrington et al 1993, Thate et al 1998), as well as *WT1* itself (Rupprecht et al 1994) (Table 29.2). However the biological relevance of many of these regulatory events is unknown, as GC-rich sequences are found in the promoter regions of thousands of genes, including those that have CpG islands. Moreover, major discrepancies were found when evaluating WT1 function using promoter-reporter constructs vs. studies on the regulation of endogenous genes. In several cases, WT1 transfection failed to repress expression of the native gene, despite potently repressing the activity of its corresponding promoter-reporter constructs in transfection assays (Lee & Haber 2001, Scharnhorst et al 2001). In the case of EGFR, both whose endogenous gene expression and promoter activity were downregulated following WT1 transfection in Saos2 osteosarcoma cells (Englert et al 1995), results could not be confirmed in HEK293 cells (Thate et al 1998). Thus, WT1 may behave as either a transcriptional activator or repressor in vitro, depending on promoter architecture, experimental conditions, cell line, and even type of expression vector (Reddy et al 1995). Clearly, the ascription of bonafide WT1 gene targets awaits identification of endogenous kidney genes whose down-regulation by WT1 is linked to biologically relevant outcomes in vivo.

WT1(-KTS) as a Transcriptional Activator

While 'unbiased' microarray analyses have yet to identify biologically relevant genes that are repressed by WT1 (Wagner & Roberts 2004), several genes implicated in the control of kidney development appear to be upregulated either directly or indirectly by WT1 in vivo, including *Sprouty1* (*Spry1*) (Gross et al 2003, Basson et al 2005), *Amphiregulin* (*AREG*) (Sakurai et al 1997, Lee et al 1999) and the podocyte-specific genes *Nephrin* (*NRHS1*) (Liu et al 2003, Guo et al 2004, Wagner et al 2004, Wartiovaara et al 2004, Tryggvason et al 2006) and *Podocalyxin* (*PODXL*) (Orlando et al 2001, Palmer et al 2001, Koop et al 2003). Consistent with its role as a tumor suppressor, WT1 activates *p21* (Englert et al 1997), leading to G1 cell cycle arrest. WT1 has also been shown to activate the anti-apoptotic gene *Bcl2* (Lichnovsky et al 1999, Lin et al 1999, Mayo et al 1999), previously shown to be repressed by WT1 (Hewitt et al 1995). However, homozygous deletions of these genes fail to recapitulate the *WT1*−/− renal phenotype or induce Wilms' tumorigenesis, indicating that critical WT1 targets in the developing kidney have yet to be identified.

Microarray data also demonstrate that WT1 is a transcriptional activator of genes involved in gonadal or retinal development including *Nuclear Receptor Subfamily 0, group B, Member 1* (*NR0B1* or *DAX1*) (Kim et al 1999), *Sex Determining Region Y* (*Sry*) (Berta et al 1992, Shimamura et al 1997, Hossain & Saunders 2001), *Steroidogenic Factor-1* (*SF-1*) (Wilhelm & Englert 2002, Schepers et al 2003) and *POU domain, class4, transcription factor 2* (*POU4F2*) (Wagner et al 2002, 2003, 2005). Notably, WT1-dependent activation of *SF-1* induces *Müllerian Inhibiting Substance* (*MIS*) (Nachtigal et al 1998), a key determinant of male sex determination that acts downstream of *Sry* (Kim et al 1999, Schepers et al 2003). These data

TABLE 29.2 Candidate WT1 target genes

Gene	Promoter-reporter Assays	Target sequence	References
Growth factors			
Igf2	repression	EGR1 consensus	(Scott et al 1985, Drummond et al 1992)
Tgf-β	repression	EGR1 consensus	(Dey et al 1994)
CTGF	repression	Not defined	(Stanhope-Baker & Williams 2000)
PDGF-A	repression/activation	EGR1 consensus	(Gashler et al 1992, Wang et al 1992)
Csf-1	repression/activation	EGR1 consensus	(Harrington et al 1993, Thate et al 1998)
Amphiregulin	activation	WTE	(Sakurai et al 1997, Lee et al 1999)
MIS	activation	SF-1 site (indirect)	(Nachtigal et al 1998)
Growth factor receptors			
Igf-1R	repression	EGR1 consensus	(Werner et al 1993, 1994, 1995, Tajinda et al 1999)
RAR-α	repression	EGR1 consensus	(Goodyer et al 1995)
EGFR	repression/activation	(TCC)n repeats	(Wang et al 1993, Englert et al 1995)
Transcription factors			
WT1	repression	EGR1 consensus	(Rupprecht et al 1994)
c-myc, N-myc	repression	EGR1 consensus	(Hewitt et al 1995, Zhang et al 1999)
Pax2	repression	EGR1 consensus	(Ryan et al 1995)
VDR	repression	WTE	(Maurer et al 2001)
EGR-1	activation	EGR1 consensus	(Madden et al 1991, Reddy et al 1995)
Sry	activation	EGR1 consensus	(Hossain & Saunders 2001)
NR0B1/DAX-1	activation	EGR1 consensus	(Kim et al 1999)
POU4F2	activation	EGR1 consensus	(Wagner et al 2003)
Others			
$G\alpha_{i-2}$	repression	EGR1 consensus	(Kinane et al 1995)
ODC	repression	EGR1 consensus	(Moshier et al 1996)
MDR-1	repression	EGR1 consensus	(McCoy et al 1999)
HTERT	repression	EGR1 consensus	(Oh et al 1999)
Bcl-2	repression/activation	WTE	(Hewitt et al 1995, Mayo et al 1999)
Syndecan-1	activation	EGR1 consensus	(Cook et al 1996)
E-cadherin	activation	EGR1 consensus	(Hosono et al 2000)
Hsp70	activation	(Indirect via HSF)	(Maheswaran et al 1998)
p21	activation	Not defined	(Englert et al 1997)
RbAp46	activation	Not defined	(Guan et al 1998)
Sprouty1	activation	EGR1 consensus	(Gross et al 2003, Basson et al 2005)
Podocalyxin	activation	WTE	(Palmer et al 2001, Guo et al 2002)
Nephrin	activation	EGR1 consensus	(Guo et al 2004)

suggest that the XY sex reversal observed in *WT1*-associated syndromes DDS and FS may be due, in part, to both disruptions in *WT1(+KTS)*-dependent activation of *Sry* and *WT1 (−KTS)-SF-1-MIS* signaling.

WT1(+KTS) May Regulate RNA Processing

The KTS insertion disrupts the linker region between zinc fingers 3 and 4, thereby decreasing the DNA binding affinity of WT1 to known target sequences (Rauhscher et al 1990). The diminished DNA-binding affinity of (+KTS) isoforms relative to (−KTS) isoforms, suggests that WT1(+KTS) performs a unique function independent of direct DNA-binding and target gene regulation. While no clear function has been ascribed to this most abundant of WT1 isoforms (Haber et al 1991), several lines of evidence suggest that WT1(+KTS) functions posttranslationally to regulate RNA processing. Whereas WT1(−KTS) isoforms exhibit diffuse nuclear expression typical of transcription factors, punctate or 'speckled' sub-nuclear *WT1(+KTS)* mRNA expression is detected within RNA spliceosome domains; and this subnuclear localization of WT1(+KTS) appears to be RNase- and not DNase-sensitive (Caricasole et al 1996). *WT1(+KTS)* expression colocalizes with that of RNA splicing factors such as U2AF65 (Ladomery et al 1999) and WTAP (Little et al 2000), as well as small nuclear ribonucleoproteins (snRNPs) (Englert et al 1995, Larsson et al 1995). Direct WT1(+KTS)-mRNA binding has been demonstrated for *Igf2* mRNA in vitro (Caricasole et al 1996), and WT1(+KTS) directly interacts with RNA-export machinery to promote translation of unspliced RNA (Bor et al 2006). Remarkably, substitution of AAA for the KTS sequence does not diminish WT1(+KTS)-RNA interactions (Bor et al 2006). These data suggest that the sole function of the KTS insertion is to disrupt the linker between the third and fourth zinc fingers, thus altering protein structure to disrupt DNA-binding and preferentially promote RNA-binding (Bor et al 2006).

WT1 Protein Partners

Coimmunoprecipitation assays have demonstrated that WT1 binds several transcription factors, suggesting that WT1 may function as a co-factor in target gene regulation in vivo (Table 29.3). For example, p53-dependent activation of the *Muscle Creatine Kinase* (*MCK*) promoter is strongly augmented by co-expression with WT1 (Maheswaran et al 1993). However, WT1 expression alone fails to activate the promoter, which lacks WT1 binding sites, suggesting that WT1 may function as a p53 co-activator (Maheswaran et al 1995, Zhan et al 1998). Several other potential WT1 protein partners have been identified, including CREB-binding protein (CBP) (Maheswaran et al 1995), tumor suppressor proteins p73 and p63 (Scharnhorst et al 2000), prostate apoptosis response-4 protein (PAR4) (Johnstone

et al 1996), CIAO1 (Johnstone et al 1998, 1999), and others (Lee & Haber 2001, Scharnhorst et al 2001, Rivera & Haber 2005). However, the biological relevance of these protein–protein interactions remains to be established in vivo.

WT1 FUNCTION DURING KIDNEY DEVELOPMENT

Overview of Mammalian Kidney Development

During vertebrate embryonic development, both the kidneys and gonads arise from the urogenital ridge, an elongated structure along the posterior aspect of the abdominal cavity, that is itself derived from intermediate mesoderm. Kidney development proceeds in three sequential stages along the urogenital ridge in a rostro-caudal direction, forming first the pronephros, followed by the mesonephros, and finally the metanephric kidney at the caudal end of the urogenital ridge. While the pronephros and mesonephros are functional excretory organs in lower vertebrates, in mammals, these kidneys are transient, vestigial structures that degenerate prior to birth; and it is the metanephros that gives rise to the permanent kidney (Saxen & Sariola 1987).

TABLE 29.3 Candidate WT1 protein partners

Protein	WT1 interaction domain	Outcome of interactions	References
WT1	First 180 aa	Homodimerization is required for WT1-dependent transcriptional regulation	(Englert et al 1995, Reddy et al 1995)
Hsp70	First 180 aa	Promotes WT1-dependent growth arrest; promotes Hsp70 transcription by displacing HSF from its complex with Hsp70	(Maheswaran et al 1998)
Par4	Zinc fingers	Inhibits WT1-dependent transactivation and growth suppression; promotes WT1-dependent transcriptional repression	(Johnstone et al 1996)
Ciao-1	Zinc fingers	Inhibits WT1-dependent transactivation	(Johnstone et al 1998)
UBC9	N-terminus	S-phase and M-phase cyclin ubiquitination?	(Wang et al 1996)
SF-1	N-terminus	DAX-1 and WT1 compete for SF-1 binding; WT1-SF-1 interactions lead to MIS transactivation	(Nachtigal et al 1998)
CBP	First two zinc fingers	Mediates WT1-dependent transactivation (e.g. *amphiregulin* promoter); alters chromatin structure?	(Lee & Haber 2001)
p53	Zinc fingers	Stabilizes p53 leading to increased DNA-binding by p53; modulates transcriptional regulation by both WT1 and p53; inhibits p53-dependent apoptosis	(Maheswaran et al 1993)
p73	Zinc fingers	Inhibits DNA-binding of WT1 isoforms lacking exon5 and KTS motifs; modulates transcriptional regulation by WT1 and p73	(Scharnhorst et al 2000)
U2AF65	Zinc fingers	Promotes binding of (+KTS) isoforms	(Davies et al 1998, Ladmery et al 1999)
mRNA splicing factors	Not defined	Colocalization of splicing factors primarily with (+KTS) isoforms in nuclear speckles and coiled bodies; RNA processing/editing	(Larsson et al 1995, Bor et al 2006)

In humans, metanephric development commences at approximately days E35–37 of gestation (E11.5 in the mouse), through a complex series of reciprocal inductive events between the metanephric mesenchyme and the epithelial ureteric bud (Potter 1972). Initially, the ureteric bud grows out from the Wolffian duct into the mesenchymal portion of the urogenital ridge. Coincident with ureteric bud outgrowth, the metanephric mesenchyme becomes histologically distinct from the surrounding mesenchyme. The initially unbranched ureteric bud begins to bifurcate as it invades the metanephric mesenchyme, and simultaneously, the mesenchyme is induced to condense along the surface of the bud (Figure 29.2A). A portion of these mesenchymal condensates form caps of closely associated cells called pretubular aggregates, adjacent and inferior to the tips of the branching ureteric bud (Figure 29.2B). Pretubular aggregates then undergo a mesenchymal-to-epithelial transition to form early renal vesicles, which in turn give rise to the comma-shaped and S-shaped bodies (Figure 29.2C), and ultimately, to the segmented nephron whose composite parts include the glomerular podocytes, proximal tubules, loops of Henle and distal tubules of the kidney (Figure 29.2D, E) (Kreidberg 2006). Podocytes are cells within the glomerulus that extend interdigitating foot processes to form a scaffolding around the glomerular capillary loops. Between these foot processes exists a

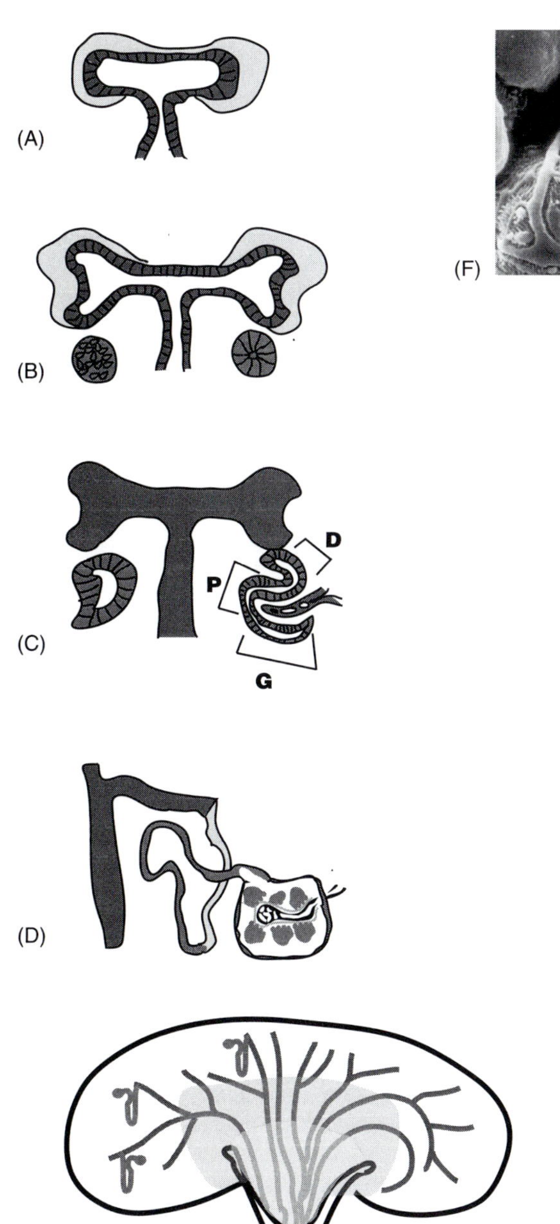

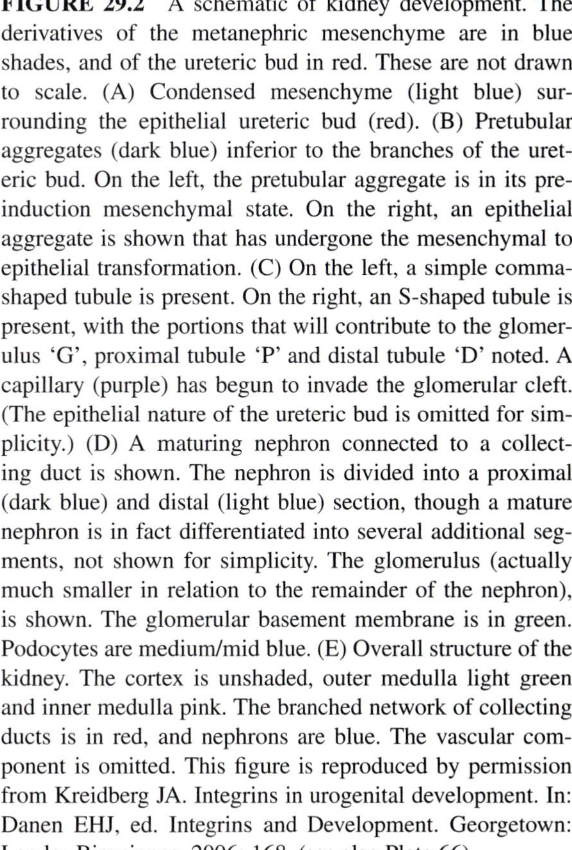

FIGURE 29.2 A schematic of kidney development. The derivatives of the metanephric mesenchyme are in blue shades, and of the ureteric bud in red. These are not drawn to scale. (A) Condensed mesenchyme (light blue) surrounding the epithelial ureteric bud (red). (B) Pretubular aggregates (dark blue) inferior to the branches of the ureteric bud. On the left, the pretubular aggregate is in its preinduction mesenchymal state. On the right, an epithelial aggregate is shown that has undergone the mesenchymal to epithelial transformation. (C) On the left, a simple comma-shaped tubule is present. On the right, an S-shaped tubule is present, with the portions that will contribute to the glomerulus 'G', proximal tubule 'P' and distal tubule 'D' noted. A capillary (purple) has begun to invade the glomerular cleft. (The epithelial nature of the ureteric bud is omitted for simplicity.) (D) A maturing nephron connected to a collecting duct is shown. The nephron is divided into a proximal (dark blue) and distal (light blue) section, though a mature nephron is in fact differentiated into several additional segments, not shown for simplicity. The glomerulus (actually much smaller in relation to the remainder of the nephron), is shown. The glomerular basement membrane is in green. Podocytes are medium/mid blue. (E) Overall structure of the kidney. The cortex is unshaded, outer medulla light green and inner medulla pink. The branched network of collecting ducts is in red, and nephrons are blue. The vascular component is omitted. This figure is reproduced by permission from Kreidberg JA. Integrins in urogenital development. In: Danen EHJ, ed. Integrins and Development. Georgetown: Landes Bioscience, 2006: 168. (see also Plate 66)

continuous junctional structure known as the slit-diaphragm, that functions as part of the glomerular filtration barrier that restricts passage of large molecules from the blood into Bowman's space.

In parallel, the nephrogenic mesenchyme induces iterative branching of the ureteric bud to yield the collecting duct system of the kidney (Figure 29.2F) (Saxen & Sariola 1987), which connects to the distal tubules of the nephron. Stromal elements develop in the periphery of the nephrogenic mesenchyme and between the branches of the ureteric bud (Rivera & Haber 2005). In humans, renal development is completed by week 34 of gestation (postnatal day 7 in mice).

Developmental Expression Pattern of WT1

WT1 mRNA and protein expression is first detected in the intermediate mesoderm lateral to the coelomic cavity, and is crucial for development of both the kidneys and gonads (Armstrong et al 1993). As Wilms' tumors derive from the condensing nephrogenic mesenchyme, it was expected that a Wilms' tumor suppressor gene would be expressed in the condensed mesenchyme and not in the reciprocally induced ureteric bud (Hastie 1994). Indeed, *WT1* expression in the developing kidney is limited to the metanephric mesenchyme and its derivatives. *WT1* is expressed at low levels in uninduced metanephric mesenchyme and expression increases dramatically in the condensed nephrogenic mesenchyme as induction proceeds. During epithelial reorganization, *WT1* expression becomes restricted to the posterior portion of induced nephrogenic structures (Pritchard-Jones et al 1990), the region that will form the epithelial podocyte population of the glomerulus by the S-shape stage of nephron development (Ryan et al 1995). Once kidney development is completed, *WT1* expression declines, except in podocytes, which express *WT1* throughout adulthood (Mundlos et al 1993).

Developmental expression of *WT1* is detected in several non-renal tissues. Not unexpectedly, given the common embryologic origin of the kidneys and gonads from intermediate mesoderm, *WT1* is strongly expressed in the developing gonads from the earliest stages of gonadal development within the thickening coelomic epithelium of the urogenital ridge (Pritchard-Jones et al 1990, Park et al 1993). As development proceeds, *WT1* expression is maintained in the bipotential gonads as well as in the differentiated female and male gonads, specifically within the Sertoli cells of the testis as well as the granulosa and epithelial cells of the fetal ovaries (Armstrong et al 1993). *WT1* expression declines once gonadal development is complete, but, in the female, persists in the granulosa cells of primary and secondary (but not antral) follicles throughout adulthood (Hsu et al 1995). The function of *WT1* in these immature follicles is unknown, however some strains of *WT1+/−* mice have fewer ova, suggesting that *WT1* gene dosage may influence survival of the egg (Kreidberg et al 1999). Other embryonic tissues that express

WT1 include fetal spinal cord, brain, heart, diaphragm, spleen, adrenals, eye and tongue (Reddy & Licht 1996).

Analysis of WT1 Function During Urogenital Development in WT1 Mutant

Mouse models

The developmental expression pattern of *WT1* in the kidneys and gonads suggests an essential role of *WT1* in these tissues. Indeed, targeted mutation of *WT1* in mice results in bilateral gonadal and renal agenesis, the latter characterized by a failure of the ureteric bud to invade the surrounding metanephric mesenchyme (Kreidberg et al, 1993). Kidney rudiments formed in a small subset of *WT1−/−* mice exhibit high levels of apoptosis. Isolated *WT1−/−* mesenchyme does not differentiate when co-cultured with wild-type ureteric bud cells, indicating that mesenchymal incompetence is the primary defect in *WT1−/−* mice (Donovan et al 1999). In the complementary gain-of-function experiment, microinjection of *WT1*-expressing plasmids into isolated embryonic kidneys stimulates kidney development (Gao et al 2005). These data suggest that *WT1* activates a permissive signaling cascade, enabling mesenchymal cells to respond to the ureteric bud and undergo differentiation. At least one of these downstream WT1 signals appears to be vascular endothelial growth factor-A (VegfA) expressed in the mesenchyme and its type II receptor Flk1 in renal angioblasts (Gao et al 2005), leading to activation of critical early mesenchymal transcription factors Pax2 and Gdnf, which in turn elicit the initial inductive signal from the ureteric bud.

WT1−/− mice on the mixed 129Sv/J-C57B/6 background do not survive beyond E13.5, presumably due to a failure of myocardial development leading to massive edema. However, outbred mouse strains (Herzer et al 1999), as well as strains developed by partial yeast artificial chromosome (YAC) rescue (YAC-*WT1* mice) (Guo et al 2002), exhibit a milder *WT1* knock-down phenotype, and survive until later stages of development. Kidneys from YAC-*WT1* mice express ~60% of wild-type *WT1* levels, and proceed beyond mesenchymal condensation to form comma-shaped bodies. However, S-shaped bodies are only rarely observed, and podocytes are not formed. An alternative strategy of bypassing the renal agenic *WT1−/−* phenotype using siRNA-mediated suppression of *WT1* at E11, also results in a marked decrease in mature nephrogenic structures in embryonic kidney explants (Davies et al 2004). Consistent with its continued focal expression pattern within mature nephrogenic structures, these data indicate that WT1 plays a critical role in the differentiation of mesenchyme into specialized tubular structures during later stages of renal development.

Persistent *WT1* expression is detected in glomerular podocytes into adulthood, indicating a continuing role for WT1 in maintaining podocyte integrity to scaffold glomerular capillaries throughout adulthood (Natoli et al 2002).

Indeed, despite normal kidney histology at birth and in the first weeks of life, heterozygous *WT1* +/− mice exhibit increased mortality due to adult-onset nephrotic syndrome, characterized by severe albuminuria and diffuse mesangial sclerosis (accumulation of fibrous tissue in the mesangial matrix between adjacent glomerular capillaries) (Menke et al 2003). Severe crescentic glomerulonephritis or mesangial sclerosis are observed even earlier in YAC-*WT1* mice (within three weeks of birth), and results in renal failure (Guo et al 2002).

WT1 IN HUMAN UROGENITAL SYNDROMES

In humans, sporadic heterozygous mutations of *WT1* may lead to Wilms' tumor, accompanied by mild urogenital defects including cryptorchidism and hypospadias in boys, or horse-shoe kidney in children of both sexes (Hastie 1994). In contrast, inherited dominant negative mutations of *WT1* observed in DDS and FS in humans are associated with severe urogenital anomalies and progressive glomerulosclerosis.

Frasier Syndrome

The essential role of WT1(+KTS) in gonadal and kidney development is evidenced by the phenotype of patients afflicted with FS, including gonadal dysgenesis, XY sex reversal (incomplete to complete) and progressive focal segmental glomerulosclerosis (Reddy & Licht 1996, Barbaux et al 1997, Klamt et al 1998). This disease is caused by specific dominant mutations in intron 9 of WT1 that result in an inability to splice in the KTS sequence, thus reducing the dosage of the (+KTS) isoforms (Barbaux et al 1997). Curiously, the incidence of Wilms' tumor is low in FS, suggesting that Wilms' tumorigenesis is not influenced by disruptions in the (+KTS) vs. (−KTS) ratio.

Denys-Drash Syndrome

The phenotype of FS is similar to, but distinct from that of DDS, suggesting that the different KTS isoforms perform both unique and overlapping functions during urogenital development. In DDS, mutations within the *WT1* zinc finger domain (often a tryptophan exchange at Arg394) disrupt DNA binding, resulting in a more severe renal phenotype than observed in FS (Pelletier et al 1991). For example, individuals afflicted with DDS also display severe urogenital defects, including gonadal dysgenesis, incomplete-to-complete XY sex reversal and progressive glomerular sclerosis. However, the nephropathy in DDS patients is more widespread and severe than in FS, and DDS is also associated with Wilms' tumor. Notably, podocytes are often under-developed in DDS patients (Jadresic et al 1990), consistent with a role for *WT1* in directing podocyte development.

(−KTS) AND (+KTS) EXPRESSING MOUSE MODELS OF KIDNEY DISEASE

Recently, analysis of mice expressing either *WT1(+KTS)* or *WT1(-KTS)* (but not both), has further elucidated the function of these two isoforms during renal development (Hammes et al 2001, Kreidberg 2002). Unlike *WT1−/−* mice which die early in embryonic development, homozygous mice for either *(+KTS)* or *(−KTS)* isoform survive to birth, with almost completely formed nephrons, as well as normal hearts, adrenals and spleens. However, *(+KTS)*-only mice (which do not express the −*KTS* isoform but *do* express the +*KTS* isoform) are characterized by small kidneys with a reduced nephrogenic zone. During late stages of renal development, these kidneys exhibit severe reductions in glomerular number and a near absence of podocytes, highlighting the importance of the (−KTS) isoform in podocyte development. In contrast, (−*KTS*)-only, FS-like kidneys (which do not express the +*KTS* isoform), exhibit a milder hypoplastic phenotype: glomeruli are small but present in normal numbers, and abnormalities in podocyte structure permit leakage of red blood cells. These results demonstrate distinct as well as overlapping functions of the (−KTS) and (+KTS) isoforms, during urogenital development (Hammes et al 2001).

Similarly, the two isoforms exhibit distinct and essential functions during gonadal development. Most female and male (+*KTS*)-only mice exhibit streak (unformed) gonads characterized by high levels of apoptosis. However, primitive gonad structures formed in a subset of XY males display male markers including SRY and SOX9, key transcription factors in the male sex determination signaling cascade (Hammes et al 2001). In contrast, 100% of XY male (−*KTS*)-only mice undergo complete XY sex reversal, recapitulating the male gonad phenotype in FS, and these gonads do not express *Sry* or *Sox9*. Female (−*KTS*)-only gonads develop normally. These data collectively suggest that the (+KTS) isoform may initiate the *Sry*-dependent male sex determination pathway, while the (−KTS) isoform controls survival of the gonadal primordium, and development of the undifferentiated gonad (Hammes et al 2001).

CONCLUSIONS

Cytogenetic and molecular studies have implicated several chromosomal loci in the development of Wilms' tumor; yet to date, only a single Wilms' tumor suppressor gene has been identified, and the mechanisms underlying tumorigenesis remain largely undefined. Recent advances in field of *WT1* research including the development of novel *WT1*-mutant mouse models, together with siRNA-based and microinjection experiments in embryonic kidney explants and cell lines, have increased our understanding of the role played by *WT1* in urogenital development. The task now ahead is to use these resources, in conjunction with ChIP-chip and proteomics approaches to identify biologically relevant gene targets of WT1 and modulators of WT1 activity during renal development in vivo; results from these future studies will help define how disruption of these interactions leads to Wilms' tumorigenesis and WT1-associated renal disease.

ACKNOWLEDGMENTS

The authors thank Dr Valerie Schumacher for reviewing the chapter. Work in Dr Kreidberg's laboratory is supported by the NIDDK and the National Kidney Foundation. Dr Hartwig is the recipient of a KRESCENT fellowship.

References

Armstrong JF, Pritchard-Jones K, Bickmore WA, Hastie ND, Bard JB. The expression of the Wilms' tumour gene, WT1, in the developing mammalian embryo. Mech. Dev. 1993; 40: 85–97.

Baird PN, Groves N, Haber DA, Housman DE, Cowell JK. Identification of mutations in the WT1 gene in tumours from patients with the WAGR syndrome. Oncogene 1992; 7: 2141–9.

Barbaux S, et al. Donor splice-site mutations in WT1 are responsible for Frasier syndrome. Nat. Genet. 1997; 17: 467–70.

Bardeesy N, Zabel B, Schmitt K, Pelletier J. WT1 mutations associated with incomplete Denys-Drash syndrome define a domain predicted to behave in a dominant-negative fashion. Genomics 1994; 21: 663–4.

Bartolomei MS, Zemel S, Tilghman SM. Parental imprinting of the mouse H19 gene. Nature 1991; 351: 153–5.

Basson MA, et al. Sprouty1 is a critical regulator of GDNF/RET-mediated kidney induction. Dev. Cell 2005; 8: 229–39.

Beckwith JB. Nephrogenic rests and the pathogenesis of Wilms tumor: developmental and clinical considerations. Am. J. Med. Genet. 1998; 79: 268–73.

Bennington J, Beckwith J. Tumors of the kidney, renal pelvis and ureter. Washington, DC: Armed Forces Institute of Pathology, 1975.

Bergmann L, et al. High levels of Wilms' tumor gene (wt1) mRNA in acute myeloid leukemias are associated with a worse long-term outcome. Blood 1997; 90: 1217–25.

Berta P, et al. Molecular analysis of the sex-determining region from the Y chromosome in two patients with Frasier syndrome. Horm. Res. 1992; 37: 103–6.

Bor Y, et al. The Wilms' tumor 1 (WT1) gene (+KTS isoform) functions with a CTE to enhance translation from an unspliced RNA with a retained intron. Genes Dev. 2006; 20: 1597–608.

Brenner B, Wildhardt G, Schneider S, Royer-Pokora B. RNA polymerase chain reaction detects different levels of four alternatively spliced WT1 transcripts in Wilms' tumors. Oncogene 1992; 7: 1431–3.

Breslow NE, Beckwith JB. Epidemiological features of Wilms' tumor: results of the National Wilms' Tumor Study. J. Natl Cancer Inst. 1982; 68: 429–36.

Brown KW, et al. Low frequency of mutations in the WT1 coding region in Wilms' tumor. Genes Chromosomes Cancer 1993; 8: 74–9.

Bruening W, et al. Germline intronic and exonic mutations in the Wilms' tumour gene (WT1) affecting urogenital development. Nat. Genet. 1992; 1: 144–8.

Bruening W, Pelletier J. A non-AUG translational initiation event generates novel WT1 isoforms. J. Biol. Chem. 1996; 271: 8646–54.

Call KM, et al. Isolation and characterization of a zinc finger polypeptide gene at the human chromosome 11 Wilms' tumor locus. Cell 1990; 60: 509–20.

Campbell CE, et al. Constitutive expression of the Wilms tumor suppressor gene (WT1) in renal cell carcinoma. Int. J. Cancer 1998; 78: 182–8.

Caricasole A, et al. RNA binding by the Wilms tumor suppressor zinc finger proteins. Proc. Natl Acad. Sci. USA 1996; 93: 7562–6.

Chitty LS, Clark T, Maxwell D. Perlman syndrome – a cause of enlarged, hyperechogenic kidneys. Prenat. Diagn. 1998; 18: 1163–8.

Compton DA, et al. Long range physical map of the Wilms' tumor-aniridia region on human chromosome 11. Cell 1988; 55: 827–36.

Compton DA, et al. Definition of the limits of the Wilms tumor locus on human chromosome 11p13. Genomics 1990; 6: 309–15.

Cook DM, Hinkes MT, Bernfield M, Rauscher FJ 3rd. Transcriptional activation of the syndecan-1 promoter by the Wilms' tumor protein WT1. Oncogene 1996; 13: 1789–99.

Coppes MJ, et al. Inherited WT1 mutation in Denys-Drash syndrome. Cancer Res. 1992; 52: 6125–8.

Crolla JA, et al. A FISH approach to defining the extent and possible clinical significance of deletions at the WAGR locus. J. Med. Genet. 1997; 34: 207–12.

Cui H, et al. Loss of imprinting of insulin-like growth factor-II in Wilms' tumor commonly involves altered methylation but not mutations of CTCF or its binding site. Cancer Res. 2001; 61: 4947–50.

Davies JA, et al. Development of an siRNA-based method for repressing specific genes in renal organ culture and its use to show that the Wt1 tumour suppressor is required for nephron differentiation. Hum. Mol. Genet. 2004; 13: 235–46.

Davies RC, et al. WT1 interacts with the splicing factor U2AF65 in an isoform-dependent manner and can be incorporated into spliceosomes. Genes Dev. 1998; 12: 3217–25.

Denamur E, et al. WT1 splice-site mutations are rarely associated with primary steroid-resistant focal and segmental glomerulosclerosis. Kidney Int. 2000; 57: 1868–72.

Dey BR, et al. Repression of the transforming growth factor-beta 1 gene by the Wilms' tumor suppressor WT1 gene product. Mol. Endocrinol. 1994; 8: 595–602.

Donovan MJ, et al. Initial differentiation of the metanephric mesenchyme is independent of WT1 and the ureteric bud. Dev. Genet. 1999; 24: 252–62.

Dowdy SF, et al. Suppression of tumorigenicity in Wilms tumor by the p15.5-p14 region of chromosome 11. Science 1991; 254: 293–5.

Drummond IA, et al. Repression of the insulin-like growth factor II gene by the Wilms tumor suppressor WT1. Science 1992; 257: 674–8.

Drummond IA, et al. DNA recognition by splicing variants of the Wilms' tumor suppressor, WT1. Mol. Cell. Biol. 1994; 14: 3800–9.

Ellis NA, German J. Molecular genetics of Bloom's syndrome. Hum. Mol. Genet. Spec No 1996; 5: 1457–63.

Englert C, et al. Truncated WT1 mutants alter the subnuclear localization of the wild-type protein. Proc. Natl Acad. Sci. USA 1995a; 92: 11960–4.

Englert C, et al. WT1 suppresses synthesis of the epidermal growth factor receptor and induces apoptosis. EMBO J. 1995b; 14: 4662–75.

Englert C, Maheswaran S, Garvin AJ, Kreidberg J, Haber DA. Induction of p21 by the Wilms' tumor suppressor gene WT1. Cancer Res. 1997; 57: 1429–34.

Fantes JA, et al. Submicroscopic deletions at the WAGR locus, revealed by nonradioactive in situ hybridization. Am. J. Hum. Genet. 1992; 51: 1286–94.

Francke U, Holmes LB, Atkins L, Riccardi VM. Aniridia-Wilms' tumor association: evidence for specific deletion of 11p13. Cytogenet. Cell Genet. 1979; 24: 185–92.

Gansler T, et al. Antibody to type I insulinlike growth factor receptor inhibits growth of Wilms' tumor in culture and in athymic mice. Am. J. Pathol. 1989; 135: 961–6.

Gao X, et al. Angioblast-mesenchyme induction of early kidney development is mediated by Wt1 and Vegfa. Development 2005; 132: 5437–49.

Gashler AL, et al. Human platelet-derived growth factor A chain is transcriptionally repressed by the Wilms tumor suppressor WT1. Proc. Natl Acad. Sci. USA 1992; 89: 10984–8.

Gessler M, et al. Homozygous deletion in Wilms tumours of a zinc-finger gene identified by chromosome jumping. Nature 1990; 343: 774–8.

Gessler M, Konig A, Bruns GA. The genomic organization and expression of the WT1 gene. Genomics 1992; 12: 807–13.

Gessler M, et al. Infrequent mutation of the WT1 gene in 77 Wilms' tumors. Hum. Mutat. 1994; 3: 212–22.

Giannoukakis N, Deal C, Paquette J, Goodyer CG, Polychronakos C. Parental genomic imprinting of the human IGF2 gene. Nat. Genet. 1993; 4: 98–101.

Goodyer P, Dehbi M, Torban E, Bruening W, Pelletier J. Repression of the retinoic acid receptor-alpha gene by the Wilms' tumor suppressor gene product, wt1. Oncogene 1995; 10: 1125–9.

Gross I, et al. The receptor tyrosine kinase regulator Sprouty1 is a target of the tumor suppressor WT1 and important for kidney development. J. Biol. Chem. 2003; 278: 41420–30.

Grundy P, et al. Chromosome 11 uniparental isodisomy predisposing to embryonal neoplasms. Lancet 1991; 338: 1079–80.

Grundy PE, et al. Loss of heterozygosity for chromosomes 16q and 1p in Wilms' tumors predicts an adverse outcome. Cancer Res. 1994; 54: 2331–3.

Guan LS, Rauchman M, Wang ZY. Induction of Rb-associated protein (RbAp46) by Wilms' tumor suppressor WT1 mediates growth inhibition. J. Biol. Chem. 1998; 273: 27047–50.

Guertl B, et al. Clonality and loss of heterozygosity of WT genes are early events in the pathogenesis of nephroblastomas. Hum. Pathol. 2003; 34: 278–81.

Guo JK, et al. WT1 is a key regulator of podocyte function: reduced expression levels cause crescentic glomerulonephritis and mesangial sclerosis. Hum. Mol. Genet. 2002; 11: 651–9.

Guo G, Morrison DJ, Licht JD, Quaggin SE. WT1 activates a glomerular-specific enhancer identified from the human nephrin gene. J. Am. Soc. Nephrol. 2004; 15: 2851–6.

Haber DA, et al. An internal deletion within an 11p13 zinc finger gene contributes to the development of Wilms' tumor. Cell 1990; 61: 1257–69.

Haber DA, et al. Alternative splicing and genomic structure of the Wilms tumor gene WT1. Proc. Natl Acad. Sci. USA 1991; 88: 9618–22.

Haber DA, et al. WT1-mediated growth suppression of Wilms tumor cells expressing a WT1 splicing variant. Science 1993; 262: 2057–9.

Haber DA. In: Vogelstein B, ed. The Genetic Basis of Human Cancer. New York: McGraw Hill, 1998: pp. 403–15.

Hammes A, et al. Two splice variants of the Wilms' tumor 1 gene have distinct functions during sex determination and nephron formation. Cell 2001; 106: 319–29.

Hao Y, Crenshaw T, Moulton T, Newcomb E, Tycko B. Tumour-suppressor activity of H19 RNA. Nature 1993; 365: 764–7.

Harrington MA, et al. Inhibition of colony-stimulating factor-1 promoter activity by the product of the Wilms' tumor locus. J. Biol. Chem. 1993; 268: 21271–5.

Hastie ND. The genetics of Wilms' tumor – a case of disrupted development. Annu. Rev. Genet. 1994; 28: 523–58.

Henneveld HT, van Lingen RA, Hamel BC, Stolte-Dijkstra I, van Essen AJ. Perlman syndrome: four additional cases and review. Am. J. Med. Genet. 1999; 86: 439–46.

Henry I, et al. Molecular definition of the 11p15.5 region involved in Beckwith-Wiedemann syndrome and probably in predisposition to adrenocortical carcinoma. Hum. Genet. 1989a; 81: 273–7.

Henry I, et al. Tumor-specific loss of 11p15.5 alleles in del11p13 Wilms tumor and in familial adrenocortical carcinoma. Proc. Natl Acad. Sci. USA 1989b; 86: 3247–51.

Hersh JH, Cole TR, Bloom AS, Bertolone SJ, Hughes HE. Risk of malignancy in Sotos syndrome. J. Pediatr. 1992; 120: 572–4.

Herzer U, Crocoll A, Barton D, Howells N, Englert C. The Wilms tumor suppressor gene wt1 is required for development of the spleen. Curr. Biol. 1999; 9: 837–40.

Hewitt SM, Hamada S, McDonnell TJ, Rauscher FJ 3rd, Saunders GF. Regulation of the proto-oncogenes bcl-2 and c-myc by the Wilms' tumor suppressor gene WT1. Cancer Res. 1995; 55: 5386–9.

Hill RE, et al. Mouse small eye results from mutations in a paired-like homeobox-containing gene. Nature 1991; 354: 522–5.

Hing S, et al. Gain of 1q is associated with adverse outcome in favorable histology Wilms' tumors. Am. J. Pathol. 2001; 158: 393–8.

Hosono S, et al. E-cadherin is a WT1 target gene. J. Biol. Chem. 2000; 275: 10943–53.

Hossain A, Saunders GF. The human sex-determining gene SRY is a direct target of WT1. J. Biol. Chem. 2001; 276: 16817–23.

Hsu SY, et al. Wilms' tumor protein WT1 as an ovarian transcription factor: decreases in expression during follicle development and repression of inhibin-alpha gene promoter. Mol. Endocrinol. 1995; 9: 1356–66.

Hu RJ, et al. A 2.5-Mb transcript map of a tumor-suppressing subchromosomal transferable fragment from 11p15.5, and isolation and sequence analysis of three novel genes. Genomics 1997; 46: 9–17.

Huang A, et al. Tissue, developmental, and tumor-specific expression of divergent transcripts in Wilms tumor. Science 1990; 250: 991–4.

Huff V, et al. Evidence for WT1 as a Wilms tumor (WT) gene: intragenic germinal deletion in bilateral WT. Am. J. Hum. Genet. 1991; 48: 997–1003.

Hughes-Benzi RM, et al. Simpson-Golabi-Behmel syndrome: genotype/phenotype analysis of 18 affected males from 7 unrelated families. Am. J. Med. Genet. 1996; 66: 227–34.

Iben S, Royer-Pokora B. Analysis of native WT1 protein from frozen human kidney and Wilms' tumors. Oncogene 1999; 18: 2533–6.

Inoue K, et al. WT1 as a new prognostic factor and a new marker for the detection of minimal residual disease in acute leukemia. Blood 1994; 84: 3071–9.

Jadresic L, et al. Clinicopathologic review of twelve children with nephropathy, Wilms tumor, and genital abnormalities (Drash syndrome). J. Pediatr. 1990; 117: 717–25.

Johnstone RW, et al. A novel repressor, par-4, modulates transcription and growth suppression functions of the Wilms' tumor suppressor WT1. Mol. Cell. Biol. 1996; 16: 6945–56.

Johnstone RW, et al. Ciao 1 is a novel WD40 protein that interacts with the tumor suppressor protein WT1. J. Biol. Chem. 1998; 273: 10880–7.

Johnstone RW, Tommerup N, Hansen C, Vissing H, Shi Y. Structural organization, tissue expression, and chromosomal localization of Ciao 1, a functional modulator of the Wilms' tumor suppressor, WT1. Immunogenetics 1999; 49: 900–5.

Jordan T, et al. The human PAX6 gene is mutated in two patients with aniridia. Nat. Genet. 1992; 1: 328–32.

Kent J, Coriat AM, Sharpe PT, Hastie ND, van Heyningen V. The evolution of WT1 sequence and expression pattern in the vertebrates. Oncogene 1995; 11: 1781–92.

Kim J, et al. The Wilms' tumor suppressor gene (wt1) product regulates Dax-1 gene expression during gonadal differentiation. Mol. Cell. Biol. 1999; 19: 2289–99.

Kinane TB, et al. LLC-PK1 cell growth is repressed by WT1 inhibition of G-protein alpha i-2 protooncogene transcription. J. Biol. Chem. 1995; 270: 30760–4.

Klamt B, et al. Frasier syndrome is caused by defective alternative splicing of WT1 leading to an altered ratio of WT1 +/− KTS splice isoforms. Hum. Mol. Genet. 1998; 7: 709–14.

Knudson AG Jr., Strong LC. Mutation and cancer: a model for Wilms' tumor of the kidney. J. Natl Cancer Inst. 1972; 48: 313–24.

Knudson A Jr., Meadows AT, Nichols WW, Hill R. Chromosomal deletion and retinoblastoma. N. Engl. J. Med. 1976; 295: 1120–3.

Koesters R, Kopp-Schneider A, Betts D, Adams V, Niggli F, Briner J, von Knebel Doeberitz M. Mutational activation of the beta-catenin proto-oncogene is a common event in the development of Wilms' tumors. Cancer Res. 1999; 59: 3880–2.

Koop K, Expression of podocyte-associated molecules in acquired human kidney diseases. J. Am. Soc. Nephrol. 2003; 14: 2063–71.

Koufos A, et al. Loss of alleles at loci on human chromosome 11 during genesis of Wilms' tumour. Nature 1984; 309: 170–2.

Koufos A, et al. Familial Wiedemann-Beckwith syndrome and a second Wilms tumor locus both map to 11p15.5. Am. J. Hum. Genet. 1989; 44: 711–19.

Kreidberg JA, et al. WT-1 is required for early kidney development. Cell 1993; 74: 679–91.

Kreidberg JA, et al. Coordinate action of Wt1 and a modifier gene supports embryonic survival in the oviduct. Mol. Reprod. Dev. 1999; 52: 366–75.

Kreidberg J. Kidneys and sex, the Wilms' tumor connection. Pediatr. Res. 2002; 51: 128.

Kreidberg J. In: Danen EHJ, ed. Integrins and Development. Georgetown: Landes Bioscience, 2006.

Kudoh T, Ishidate T, Moriyama M, Toyoshima K, Akiyama T. G1 phase arrest induced by Wilms tumor protein WT1 is abrogated by cyclin/CDK complexes. Proc. Natl Acad. Sci. USA 1995; 92: 4517–21.

Kurotaki N, et al. Haploinsufficiency of NSD1 causes Sotos syndrome. Nat. Genet. 2002; 30: 365–6.

Ladomery MR, Slight J, McGhee S, Hastie ND. Presence of WT1, the Wilm's tumor suppressor gene product, in nuclear poly(A)(+) ribonucleoprotein. J. Biol. Chem. 1999; 274: 36520–6.

Lapunzina P. Risk of tumorigenesis in overgrowth syndromes: a comprehensive review. Am. J. Med. Genet. C. Semin. Med. Genet. 2005; 137: 53–71.

Larsson SH, et al. Subnuclear localization of WT1 in splicing or transcription factor domains is regulated by alternative splicing. Cell 1995; 81: 391–401.

Lee SB, et al. The Wilms tumor suppressor WT1 encodes a transcriptional activator of amphiregulin. Cell 1999; 98: 663–73.

Lee SB, Haber DA. Wilms tumor and the WT1 gene. Exp. Cell Res. 2001; 264: 74–99.

Lewis WH, et al. Homozygous deletion of a DNA marker from chromosome 11p13 in sporadic Wilms tumor. Genomics 1988; 3: 25–31.

Li CM, et al. Gene expression in Wilms' tumor mimics the earliest committed stage in the metanephric mesenchymal-epithelial transition. Am. J. Pathol. 2002; 160: 2181–90.

Li W, Kessler P, Williams BR. Transcript profiling of Wilms tumors reveals connections to kidney morphogenesis and expression patterns associated with anaplasia. Oncogene 2005; 24: 457–68.

Lichnovsky V, Erdosova B, Punkt K, Zapletal M. Expression of BCL-2 in the developing kidney of human embryos and fetuses: qualitative and quantitative study. Acta Univ. Palacki. Olomuc Fac. Med. 1999; 142: 61–4.

Lin HH, Yang TP, Jiang ST, Yang HY, Tang MJ. Bcl-2 overexpression prevents apoptosis-induced Madin-Darby canine kidney simple epithelial cyst formation. Kidney Int. 1999; 55: 168–78.

Little MH, et al. Equivalent expression of paternally and maternally inherited WT1 alleles in normal fetal tissue and Wilms' tumours. Oncogene 1992; 7: 635–41.

Little MH, et al. Evidence that WT1 mutations in Denys-Drash syndrome patients may act in a dominant-negative fashion. Hum. Mol. Genet. 1993; 2: 259–64.

Little M, Wells CA. Clinical overview of WT1 gene mutations. Hum. Mutat. 1997; 9: 209–25.

Little NA, Hastie ND, Davies RC. Identification of WTAP, a novel Wilms' tumour 1-associating protein. Hum. Mol. Genet. 2000; 9: 2231–9.

Liu G, et al. Neph1 and nephrin interaction in the slit diaphragm is an important determinant of glomerular permeability. J. Clin. Invest. 2003; 112: 209–21.

Lu YJ, et al. Chromosome 1q expression profiling and relapse in Wilms' tumour. Lancet 2002; 360: 385–6.

Madden SL, et al. Transcriptional repression mediated by the WT1 Wilms tumor gene product. Science 1991; 253: 1550–3.

Maheswaran S, et al. Physical and functional interaction between WT1 and p53 proteins. Proc. Natl Acad. Sci. USA 1993; 90: 5100–4.

Maheswaran S, Englert C, Bennett P, Heinrich G, Haber DA. The WT1 gene product stabilizes p53 and inhibits p53-mediated apoptosis. Genes Dev. 1995; 9: 2143–56.

Maheswaran S, et al. Inhibition of cellular proliferation by the Wilms tumor suppressor WT1 requires association with the inducible chaperone Hsp70. Genes Dev. 1998; 12: 1108–20.

Maiti S, Alam R, Amos CI, Huff V. Frequent association of beta-catenin and WT1 mutations in Wilms tumors. Cancer Res. 2000; 60: 6288–92.

Major MB, Camp ND, Berndt JD, Yi X, Goldenberg SJ, Hubbert C, Biechele TL, Gingras AC, Zherg N, Maccoss MJ, Angers S, Moon RT. Wilms tumor supressor WTX negatively regulates WNT beta-catenin signaling. Science. 2007; 316: 1043–6.

Matsunaga E. Genetics of Wilms' tumor. Hum. Genet. 1989; 57: 231–46.

Maurer U, et al. The Wilms' tumor gene product (WT1) modulates the response to 1,25-dihydroxyvitamin D3 by induction of the vitamin D receptor. J. Biol. Chem. 2001; 276: 3727–32.

Maw MA, et al. A third Wilms' tumor locus on chromosome 16q. Cancer Res. 1992; 52: 3094–8.

Mayo MW, et al. WT1 modulates apoptosis by transcriptionally upregulating the bcl-2 proto-oncogene. EMBO J. 1999; 18: 3990–4003.

McCoy C, McGee SB, Cornwell MM. The Wilms' tumor suppressor, WT1, inhibits 12-O-tetradecanoylphorbol-13-acetate activation of the multidrug resistance-1 promoter. Cell Growth Differ. 1999; 10: 377–86.

McDonald JM, et al. Linkage of familial Wilms' tumor predisposition to chromosome 19 and a two-locus model for the etiology of familial tumors. Cancer Res. 1998; 58: 1387–90.

Menke AL, Hastie ND. In: Fisher DE, ed. Tumor Suppressor Genes in Human Cancer. New Jersey: Humana Press, 2001: pp. 307–50.

Menke AL, et al. The wt1-heterozygous mouse; a model to study the development of glomerular sclerosis. J. Pathol. 2003; 200: 667–74.

Miwa H, Beran M, Saunders GF. Expression of the Wilms' tumor gene (WT1) in human leukemias. Leukemia 1992; 6: 405–9.

Miyagawa K, et al. Loss of WT1 function leads to ectopic myogenesis in Wilms' tumour. Nat. Genet. 1998; 18: 15–17.

Moshier JA, et al. Regulation of ornithine decarboxylase gene expression by the Wilms' tumor suppressor WT1. Nucleic Acids Res. 1996; 24: 1149–57.

Mundlos S, et al. Nuclear localization of the protein encoded by the Wilms' tumor gene WT1 in embryonic and adult tissues. Development 1993; 119: 1329–41.

Nachtigal MW, et al. Wilms' tumor 1 and Dax-1 modulate the orphan nuclear receptor SF-1 in sex-specific gene expression. Cell 1998; 93: 445–54.

Nakagama H, Heinrich G, Pelletier J, Housman DE. Sequence and structural requirements for high-affinity DNA binding by the WT1 gene product. Mol. Cell. Biol. 1995; 15: 1489–98.

Narahara K, et al. Regional mapping of catalase and Wilms tumor-aniridia, genitourinary abnormalities, and mental retardation triad loci to the chromosome segment 11p1305–p1306. Hum. Genet. 1984; 66: 181–5.

Natoli TA, et al. A mammal-specific exon of WT1 is not required for development or fertility. Mol. Cell. Biol. 2002a; 22: 4433–8.

Natoli TA, et al. A mutant form of the Wilms' tumor suppressor gene WT1 observed in Denys-Drash syndrome interferes with glomerular capillary development. J. Am. Soc. Nephrol. 2002b; 13: 2058–67.

Neri G, Martini-Neri ME, Katz BE, Opitz JM. The Perlman syndrome: familial renal dysplasia with Wilms tumor, fetal gigantism and multiple congenital anomalies. Am. J. Med. Genet. 1984; 19: 195–207.

Neri G, Gurrieri F, Zanni G, Lin A. Clinical and molecular aspects of the Simpson-Golabi-Behmel syndrome. Am. J. Med. Genet. 1998; 79: 279–83.

Ogawa O, et al. Constitutional relaxation of insulin-like growth factor II gene imprinting associated with Wilms' tumour and gigantism. Nat. Genet. 1993; 5: 408–12.

Oh S, Song Y, Yim J, Kim TK. The Wilms' tumor 1 tumor suppressor gene represses transcription of the human telomerase reverse transcriptase gene. J. Biol. Chem. 1999; 274: 37473–8.

Ohlsson R, et al. IGF2 is parentally imprinted during human embryogenesis and in the Beckwith-Wiedemann syndrome. Nat. Genet. 1993; 4: 94–7.

Orlando RA, et al. The glomerular epithelial cell anti-adhesin podocalyxin associates with the actin cytoskeleton through interactions with ezrin. J. Am. Soc. Nephrol. 2001; 12: 1589–98.

Palmer RE, et al. WT1 regulates the expression of the major glomerular podocyte membrane protein Podocalyxin. Curr. Biol. 2001; 11: 1805–9.

Park S, et al. Inactivation of WT1 in nephrogenic rests, genetic precursors to Wilms' tumour. Nat. Genet. 1993a; 5: 363–7.

Park S, et al. The Wilms tumour gene WT1 is expressed in murine mesoderm-derived tissues and mutated in a human mesothelioma. Nat. Genet. 1993b; 4: 415–20.

Pellegrini M, et al. Gpc3 expression correlates with the phenotype of the Simpson-Golabi-Behmel syndrome. Dev. Dyn. 1998; 213: 431–9.

Pelletier J, et al. Expression of the Wilms' tumor gene WT1 in the murine urogenital system. Genes. Dev. 1991a; 5: 1345–56.

Pelletier J, et al. Germline mutations in the Wilms' tumor suppressor gene are associated with abnormal urogenital development in Denys-Drash syndrome. Cell 1991b; 67: 437–47.

Perlman M. Perlman syndrome: familial renal dysplasia with Wilms tumor, fetal gigantism, and multiple congenital anomalies. Am. J. Med. Genet. 1986; 25: 793–5.

Pilia G, et al. Mutations in GPC3, a glypican gene, cause the Simpson-Golabi-Behmel overgrowth syndrome. Nat. Genet. 1996; 12: 241–7.

Potter EL Normal and Abnormal Development of the Kidney. Chicago: Year Book Med, Publ., 1972.

Pritchard-Jones K, et al. The candidate Wilms' tumour gene is involved in genitourinary development. Nature 1990; 346: 194–7.

Pujana MA, et al. Additional complexity on human chromosome 15q: identification of a set of newly recognized duplicons (LCR15) on 15q11-q13, 15q24, and 15q26. Genome Res. 2001; 11: 98–111.

Rahman N, et al. Evidence for a familial Wilms' tumour gene (FWT1) on chromosome 17q12-q21. Nat. Genet. 1996; 13: 461–3.

Rapley EA, et al. Evidence for susceptibility genes to familial Wilms tumour in addition to WT1, FWT1 and FWT2. Br. J. Cancer 2000; 83: 177–83.

Rauscher FJ, Morris JF, Tournay OE, Cook DM, Curran T. Binding of the Wilms' tumor locus zinc finger protein to the EGR-1 consensus sequence. Science 1990; 250: 1259–62.

Ravenel JD, et al. Loss of imprinting of insulin-like growth factor-II (IGF2) gene in distinguishing specific biologic subtypes of Wilms tumor. J. Natl Cancer Inst. 2001; 93: 1698–703.

Reddy JC, et al. WT1-mediated transcriptional activation is inhibited by dominant negative mutant proteins. J. Biol. Chem. 1995a; 270: 10878–84.

Reddy JC, Hosono S, Licht JD. The transcriptional effect of WT1 is modulated by choice of expression vector. J. Biol. Chem. 1995b; 270: 29976–82.

Reddy JC, Licht JD. The WT1 Wilms' tumor suppressor gene: how much do we really know? Biochim. Biophys. Acta 1996; 1287: 1–28.

Reeve AE, Eccles MR, Wilkins RJ, Bell GI, Millow LJ. Expression of insulin-like growth factor-II transcripts in Wilms' tumour. Nature 1985; 317: 258–60.

Reeve AE, Sih SH, Raizis AM, Feinberg AP. Loss of allelic heterozygosity at a second locus on chromosome 11 in sporadic Wilms' tumor cells. Mol. Cell. Biol. 1989; 9: 1799–803.

Riccardi VM, Sujansky E, Smith AC, Francke U. Chromosomal imbalance in the Aniridia-Wilms' tumor association: 11p interstitial deletion. Pediatrics 1978; 61: 604–10.

Rivera MN, Haber DA. Wilms' tumour: connecting tumorigenesis and organ development in the kidney. Nat. Rev. Cancer 2005; 5: 699–712.

Rivera MN, Kim WJ, Wells J, Driscoll DR, Brannigan BW, Han M, Kim JC, Feinberg AP, Gerald WL, Vargas SO, Chin L, Iafrate AJ, Bell DW, Haber DA. An X chromosome gene, WTX, is commonly inactivated in Wilms tumor. Science. 2007; 315: 642–5.

Rose EA, et al. Complete physical map of the WAGR region of 11p13 localizes a candidate Wilms' tumor gene. Cell 1990; 60: 495–508.

Royer-Pokora B, et al. Twenty-four new cases of WT1 germline mutations and review of the literature: genotype/phenotype correlations for Wilms tumor development. Am. J. Med. Genet. A 2004; 127: 249–57.

Rupprecht HD, Drummond IA, Madden SL, Rauscher FJ 3rd, Sukhatme VP. The Wilms' tumor suppressor gene WT1 is negatively autoregulated. J. Biol. Chem. 1994; 269: 6198–206.

Ryan G, Steele-Perkins V, Morris JF, Rauscher FJ 3rd, Dressler GR. Repression of Pax-2 by WT1 during normal kidney development. Development 1995; 121: 867–75.

Sakurai H, Tsukamoto T, Kjelsberg CA, Cantley LG, Nigam SK. EGF receptor ligands are a large fraction of in vitro branching morphogens secreted by embryonic kidney. Am. J. Physiol. 1997; 273: F463–72.

Saxen L, Sariola H. Early organogenesis of the kidney. Pediatr. Nephrol. 1987; 1: 385–92.

Scharnhorst V, Dekker P, van der Eb AJ, Jochemsen AG. Internal translation initiation generates novel WT1 protein isoforms with distinct biological properties. J. Biol. Chem. 1999; 274: 23456–62.

Scharnhorst V, Dekker P, van der Eb AJ, Jochemsen AG. Physical interaction between Wilms tumor 1 and p73 proteins modulates their functions. J. Biol. Chem. 2000; 275: 10202–11.

Scharnhorst V, van der Eb AJ, Jochemsen AG. WT1 proteins: functions in growth and differentiation. Gene 2001; 273: 141–61.

Schepers G, Wilson M, Wilhelm D, Koopman P. SOX8 is expressed during testis differentiation in mice and synergizes with SF1 to activate the Amh promoter in vitro. J. Biol. Chem. 2003; 278: 28101–8.

Scholz H, Kirschner KM. A role for the Wilms' tumor protein WT1 in organ development. Physiology (Bethesda) 2005; 20: 54–9.

Schumacher V, et al. Correlation of germ-line mutations and two-hit inactivation of the WT1 gene with Wilms tumors of stromal-predominant histology. Proc. Natl Acad. Sci. USA 1997; 94: 3972–7.

Schumacher V, et al. Spectrum of early onset nephrotic syndrome associated with WT1 missense mutations. Kidney Int. 1998; 53: 1594–600.

Scott J, et al. Insulin-like growth factor-II gene expression in Wilms' tumour and embryonic tissues. Nature 1985; 317: 260–2.

Scott RH, Stiller CA, Walker L, Rahman N. Syndromes and constitutional chromosomal abnormalities associated with Wilms tumour. J. Med. Genet. 2006.

Sharma PM, Bowman M, Madden SL, Rauscher FJ 3rd, Sukumar S. RNA editing in the Wilms' tumor susceptibility gene, WT1. Genes Dev. 1994a; 8: 720–31.

Sharma PM, Bowman M, Yu BF, Sukumar S. A rodent model for Wilms tumors: embryonal kidney neoplasms induced by N-nitroso-N'-methylurea. Proc. Natl Acad. Sci. USA 1994b; 91: 9931–5.

Sheng WW, Soukup S, Bove K, Gotwals B, Lampkin B. Chromosome analysis of 31 Wilms' tumors. Cancer Res. 1990; 50: 2786–93.

Shimamura R, Fraizer GC, Trapman J, Lau YfC, Saunders GF. The Wilms' tumor gene WT1 can regulate genes involved in sex determination and differentiation: SRY, Mullerian-inhibiting substance, and the androgen receptor. Clin. Cancer Res. 1997; 3: 2571–80.

Stanhope-Baker P, Williams BR. Identification of connective tissue growth factor as a target of WT1 transcriptional regulation. J. Biol. Chem. 2000; 275: 38139–50.

Tajinda K, Carroll J, Roberts CT Jr. Regulation of insulin-like growth factor I receptor promoter activity by wild-type and mutant versions of the WT1 tumor suppressor. Endocrinology 1999; 140: 4713–24.

Thate C, Englert C, Gessler M. Analysis of WT1 target gene expression in stably transfected cell lines. Oncogene 1998; 17: 1287–94.

Ton CC, et al. Positional cloning and characterization of a paired box- and homeobox-containing gene from the aniridia region. Cell 1991; 67: 1059–74.

Tryggvason K, Pikkarainen T, Patrakka J. Nck links nephrin to actin in kidney podocytes. Cell 2006; 125: 221–4.

Wagner KD, Wagner N, Schley G, Theres H, Scholz H. The Wilms' tumor suppressor Wt1 encodes a transcriptional activator of the class IV POU-domain factor Pou4f2 (Brn-3b). Gene 2003; 305: 217–23.

Wagner KD, et al. The Wilms' tumor gene Wt1 is required for normal development of the retina. EMBO J. 2002; 21: 1398–405.

Wagner KJ, Roberts SG. Transcriptional regulation by the Wilms' tumour suppressor protein WT1. Biochem. Soc. Trans. 2004; 32: 932–5.

Wagner N, Wagner KD, Xing Y, Scholz H, Schedl A. The major podocyte protein nephrin is transcriptionally activated by the Wilms' tumor suppressor WT1. J. Am. Soc. Nephrol. 2004; 15: 3044–51.

Wagner N, et al. A splice variant of the Wilms' tumour suppressor Wt1 is required for normal development of the olfactory system. Development 2005; 132: 1327–36.

Wang ZY, Madden SL, Deuel TF, Rauscher FJ 3rd. The Wilms' tumor gene product, WT1, represses transcription of the platelet-derived growth factor A-chain gene. J. Biol. Chem. 1992; 267: 21999–2002.

Wang ZY, Qiu QQ, Deuel TF. The Wilms' tumor gene product WT1 activates or suppresses transcription through separate functional domains. J. Biol. Chem. 1993; 268: 9172–5.

Wang ZY, et al. Molecular cloning of the cDNA and chromosome localization of the gene for human ubiquitin-conjugating enzyme 9. J. Biol. Chem. 1996; 271: 24811–16.

Wartiovaara J, et al. Nephrin strands contribute to a porous slit diaphragm scaffold as revealed by electron tomography. J. Clin. Invest. 2004; 114: 1475–83.

Weksberg R, Shen DR, Fei YL, Song QL, Squire J. Disruption of insulin-like growth factor 2 imprinting in Beckwith-Wiedemann syndrome. Nat. Genet. 1993; 5: 143–50.

Werner H, et al. Increased expression of the insulin-like growth factor I receptor gene, IGF1R, in Wilms tumor is correlated with modulation of IGF1R promoter activity by the WT1 Wilms tumor gene product. Proc. Natl Acad. Sci. USA 1993; 90: 5828–32.

Werner H, et al. Transcriptional repression of the insulin-like growth factor I receptor (IGF-I-R) gene by the tumor suppressor WT1 involves binding to sequences both upstream and downstream of the IGF-I-R gene transcription start site. J. Biol. Chem. 1994; 269: 12577–82.

Werner H, et al. Inhibition of cellular proliferation by the Wilms' tumor suppressor WT1 is associated with suppression of insulin-like growth factor I receptor gene expression. Mol. Cell. Biol. 1995; 15: 3516–22.

Wiener JS, Coppes MJ, Ritchey ML. Current concepts in the biology and management of Wilms tumor. J. Urol. 1998; 159: 1316–25.

Wilhelm D, Englert C. The Wilms tumor suppressor WT1 regulates early gonad development by activation of Sf1. Genes Dev. 2002; 16: 1839–51.

Wilms M Die Mischgeschwulste der Niere. Leipzig: Verlag von Arthur Georgi, 1899.

Zhan Q, Chen IT, Antinore MJ, Fornace AJ Jr. Tumor suppressor p53 can participate in transcriptional induction of the GADD45 promoter in the absence of direct DNA binding. Mol. Cell. Biol. 1998; 18: 2768–78.

Zhang X, Xing G, Saunders GF. Proto-oncogene N-myc promoter is down regulated by the Wilms' tumor suppressor gene WT1. Anticancer Res. 1999; 19: 1641–8.

Zhuang Z, et al. Identical genetic changes in different histologic components of Wilms' tumors. J. Natl Cancer Inst. 1997; 89: 1148–52.

CHAPTER 30

Tuberous Sclerosis

DAVID J. KWIATKOWSKI

INTRODUCTION

Tuberous sclerosis complex (TSC) is an autosomal dominant disease characterized by the development of hamartomas and hamartias in several organ systems. It is due to inactivating mutations in either the TSC1 gene, which encodes the protein hamartin, or in the TSC2 gene, which encodes the protein tuberin. There are two recent books and a new pamphlet that are excellent sources of information on the clinical manifestations of tuberous sclerosis (Gomez et al 1999, Curatolo 2003, Alliance 2004).

CLINICAL FEATURES OF TSC

Overview and Diagnostic Criteria

TSC is a relatively variable clinical syndrome, in which several pathologic manifestations are highly specific for the disease (Gomez et al 1999). The cardinal feature is the cortical tuber, which is a distinctive form of brain hamartia. Hamartias and hamartomas usually involve several different organ systems in TSC patients, in addition to the brain. Involvement of the skin, heart, and kidney are particularly frequent, occurring at some point during the lifetime of most patients. We define hamartia as the occurrence of a group of dysplastic, malorganized cells within an organ, but without unusual growth. A hamartoma is the occurrence of cells in a similar aberrant pattern, but with some observed or inferred growth tendency. Generally hamartomas are made up of multiple types of cells, and this is the case for most but not all TSC hamartomas. Although hamartomas are classically considered 'tumor-like,' some grow so that they can be classified as tumors.

The diversity of hamartias/hamartomas and variability of clinical presentation in TSC has led to the development of formal diagnostic criteria (Table 30.1) (Roach et al 1998). Many clinical features of TSC appear as the patient ages (Table 30.2) (Jozwiak et al 2000).

Neurologic Manifestations

Cortical tubers, the hallmark finding in TSC, are found grossly as firm, smooth, somewhat raised, pale lesions of the cerebral cortex (Gomez et al 1999). They are seen in about 90% of TSC patients, and up to 40 such lesions have been seen in one brain. They range in size from a few millimeters to several centimeters, and can occupy any lobe as well as the cerebellum. The second most common brain lesion in TSC is the subependymal nodule (SEN), found in the wall of the lateral ventricles typically at the caudothalamic groove near the foramen of Monro. They occur grossly as smooth, round-to-ovoid elevations projecting into the ventricle. They also are very common in TSC, and typically calcify in patients after 10 years of age. Both types of lesion can be visualized by MRI and CT, although tubers are best seen by MRI using the FLAIR technique.

Histologically, cortical tubers are characterized by architectural disarray with particular disruption of cortical lamination (Crino & Henske 1999, Gomez et al 1999). Tubers contain a mixture of cells, including normal and dysplastic neurons, astrocytes, and the hallmark giant cell. Giant cells extend short aberrant processes and staining studies suggest they are of mixed glioneuronal lineage. Dysplastic neurons in tubers show disrupted radial orientation and abnormal dendritic arbors, and express both typical neuronal markers, and markers indicative of an undifferentiated state,

Genetic Diseases of the Kidney
ISBN: 978-0-12-449851-8

TABLE 30.1   Diagnostic criteria for TSC (Roach et al 1998)

Major features

1. Facial angiofibromas or forehead plaque
2. Nontraumatic ungual or periungual fibroma
3. Hypomelanotic macules (three or more)
4. Shagreen patch (connective tissue nevus)
5. Multiple retinal nodular hamartomas
6. Cortical tuber*
7. Subependymal nodule
8. Subependymal giant cell astrocytoma
9. Cardiac rhabdomyoma, single or multiple
10. Lymphangiomyomatosis+
11. Renal angiomyolipoma+

Minor features

1. Multiple, randomly distributed pits in dental enamel
2. Hamartomatous rectal polyps
3. Bone cysts
4. Cerebral white matter radial migration lines*
5. Gingival fibromas
6. Nonrenal hamartoma
7. Retinal achromic patch
8. 'Confetti' skin lesions
9. Multiple renal cysts

Definite tuberous sclerosis complex: either two major features or one major feature plus two minor features.

Probable tuberous sclerosis complex: one major plus one minor feature.

Possible tuberous sclerosis complex: either one major feature or two or more minor features

*When cerebral cortical dysplasia and cerebral white matter migration tracts occur together, they should be counted as one rather than two features of tuberous sclerosis.

+When both lymphangiomyomatosis and renal angiomyolipomas are present, other features of tuberous sclerosis should be present before a definite diagnosis is assigned.

TABLE 30.2   Frequency of diagnostic features of TSC in children, according to age (Jozwiak et al. 2000) (N = 13–40 for each age/diagnostic feature comparison)

Diagnostic feature	Age (y)				
	0–2	2–5	5–9	9–14	14–18
Hypomelanotic macules	90%	95%	97%	97%	97%
Cardiac rhabdomyoma	83%	21%	21%	41%	75%
Epilepsy, usually infantile spasms	83%	91%	94%	96%	96%
Subependymal nodules on CT/MRI	83%	97%	97%	100%	100%
Renal angiomyolipomas	17%	42%	64%	65%	92%
Facial angiofibromas	10%	63%	74%	77%	77%
Retinal hamartoma	8%	17%	21%	17%	31%
Shagreen patch	0	21%	35%	47%	48%
Forehead plaque	0	9%	16%	19%	19%
Liver AMLs (angiomyolipomas)	0	0	11%	24%	46%
Periungual fibromas	0	0	0	11%	15%

Note that these patients are from a pediatric neurology referral practice in Warsaw, Poland, and are likely relatively severely affected overall

e.g. nestin. SENs consist of large cells in a vascular stroma, with occasional cells that attain giant cell proportions with aberrant and/or multiple nuclei.

Subependymal giant cell astrocytomas (SEGAs) develop in 5–10% of TSC patients. These lesions represent a form of SEN with progressive growth and enlargement to a size >1 cm. Histologically there is no difference between a SEN and a SEGA, and both types of lesion express astrocytic and neuronal markers.

The major presenting symptoms and signs of cerebral TSC are a variety of epileptic seizures (Gomez et al 1999, Curatolo 2003, Crino 2004). Although many referral-based series cite an epilepsy rate of >90% in TSC patients, more recent experience suggests that this frequency is closer to 70% overall. Seizures have their origin in the cortical tubers or in nearby transitional cortex. The majority of TSC patients with epilepsy develop infantile spasms during the first year of life, a type of seizure characterized by flexion movements of the arms and head that tend to occur in clusters. Other types of seizures seen in TSC include partial, complex partial, myoclonic, and generalized tonic–clonic seizures. There is a general correlation between the number of cortical tubers in a TSC patient and both a younger age of seizure onset and seizure severity.

Mental retardation and developmental disabilities are major issues for many TSC patients. About half of TSC patients have normal intelligence, while the other half have mental disability ranging from mild to profound. Cognitive deficits are very common in TSC patients with normal overall intellectual abilities, and include attention deficit hyperactivity disorder (ADHD), executive control problems, language delay, learning and memory problems, and visuospatial difficulties. In those with mental retardation, ADHD, autism, and autism spectrum disorders are commonly seen. The association between TSC and autism is quite striking in that autism is rarely seen in other neurogenetic disorders. In fact, TSC is the most common genetic cause of autism as a whole, accounting for about 3% of all cases. Based upon MRI and PET scans, generalized epilepsy in early life and functional imbalance in subcortical circuits appear to be associated with autistic features in children with TSC (Asano et al 2001).

Growing SEGAs will cause significant intracranial hypertension due to obstruction of ventricular CSF flow, with symptoms of headache and vomiting. Surgical resection is required for symptomatic SEGAs, and is typically quite effective, although occasional cases require repeat procedures.

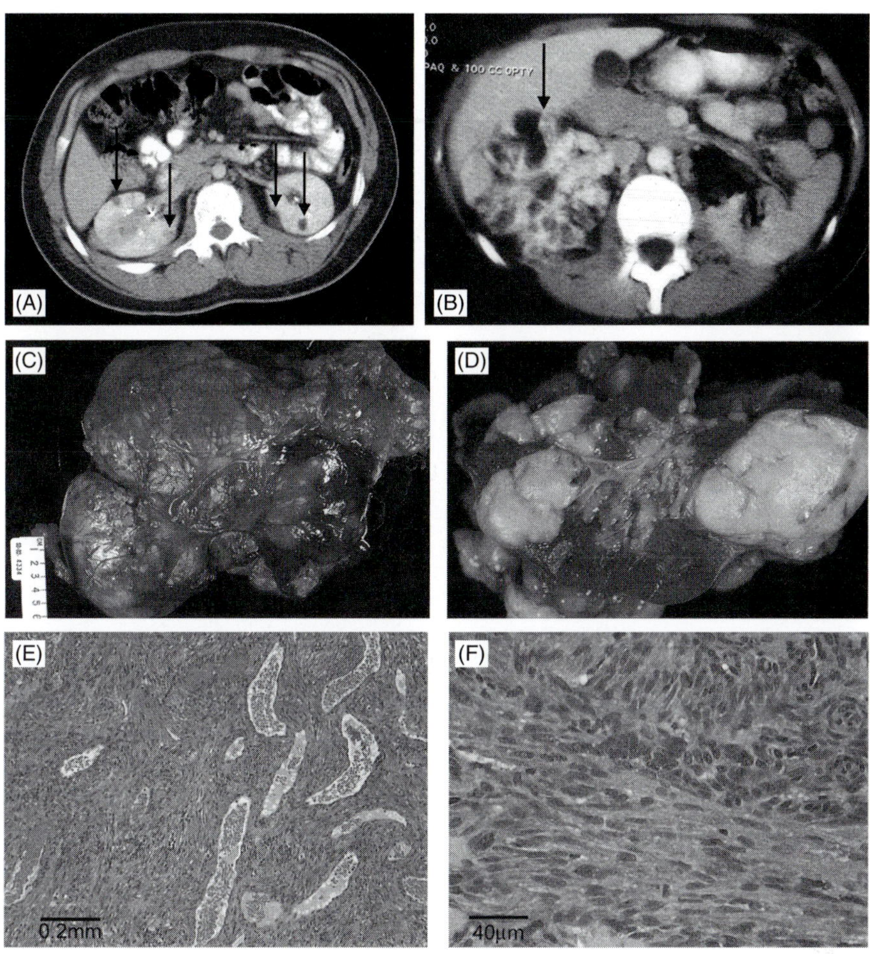

FIGURE 30.1   Renal angiomyolipoma. A, B. CT scans following contrast administration. A. Four small AML with variable fat and vascular content (arrows). B. Extensive AML involving the right kidney. Note areas of both high enhancement reflecting high vascularity and regions of low enhancement reflecting fat. Arrow points to kidney. C, D. Gross pictures of a massively enlarged and distorted kidney due to extensive involvement by AML. Yellowish material is fatty AML. C, external view; D, after sectioning. E, F. Histologic sections of AML. Note numerous vascular channels in E; and proliferating spindle-shaped SMCs in F. Images courtesy of John Bissler, Frank McCormack, Elizabeth P. Henske, Nisreen El-Hashemite. (see also Plate 67)

Renal Involvement

Angiomyolipomas (AMLs) are the most common TSC kidney lesion, seen in about 75% of children by adolescence using ultrasound imaging (Ewalt et al 1998; Bissler & Kingswood 2004, O'Callaghan et al 2004) (Figure 30.1). As suggested by the name, these lesions consist of a variable mixture of aberrant vessels, smooth muscle, and fat. Some smooth muscle cells and other more epithelioid cells seen in AMLs stain with the HMB-45 antibody, which recognizes an antigen normally expressed exclusively in the melanosome. AMLs are rare in TSC patients at birth, typically appear between the ages of 5–10, are usually bilateral, and often grow in the adolescent interval (Ewalt et al 1998).

AMLs often occur as multiple lesions in both kidneys, which can grow to become nearly confluent, making it difficult to distinguish by imaging studies between normal renal parenchyma and AML (Figure 30.1). In other patients, single lesions can predominate. Very large AMLs (>6 cm diameter) are likely to grow further, and are at significant risk to cause life-threatening hemorrhage. Large AML also often cause pain.

Renal ultrasound is an effective and simple method of monitoring the growth of AMLs when they are not extensive. For extensive tumors, CT or MRI of the kidney is preferred. Because of the infiltrating nature of many AMLs and the risk of bleeding, surgical resection is best avoided. Selective embolization is the preferred management approach for large lesions for which the risk of bleeding is high.

Renal cysts are seen in about 10–20% of young TSC adults (Ewalt et al 1998). The cysts have a distinctive lining epithelium (Gomez et al 1999), but are not a significant medical problem. About 3% of TSC patients have severe progressive polycystic kidney disease, due to combined mutation of both the TSC2 and PKD1 genes (see below).

Renal malignancy appears to occur at an elevated rate and at an unusually young age in TSC patients compared to the general population (Gomez et al 1999), with a lifetime risk of about 2–3%. A proportion of these tumors are typical renal cell carcinoma, while others are malignant AMLs, based upon spindle cell architecture and other histologic features, as well as positive staining with HMB-45.

Cutaneous Manifestations

Cutaneous findings are present in over 80% of TSC patients, and are the most easily recognized sign of the disease. Most lesions are of minor significance, but facial angiofibromas can be a significant cosmetic issue (Gomez et al 1999).

Hypomelanotic macules or white spots typically have a lance-ovate shape (ash-leaf) and are most common over the trunk and buttocks. Three or more of these lesions are very unusual in the general population and thus are considered a major diagnostic criterion (Table 30.1). Histopathological exam shows reduced number, pigment content, and size of melanosomes within the melanocytes in a hypomelanotic macule. The lesions are present at birth and do not change, though they are more easily observed following suntanning, or after using a UV light.

Facial angiofibromas (formerly known as adenoma sebaceum) are red to pink papules or nodules with a smooth surface that are typically seen in a malar distribution, and when more extensive extend down to the chin (Figure 30.1B). They typically first appear between the ages of 2 and 6 years, and progress to a variable extent during puberty. Histologic findings are dermal fibrosis and vasodilatation, with occasional large glial-appearing cells. They are found in about 85% of all TSC patients. Forehead fibrous plaque and shagreen patch are related lesions with more connective tissue than angiofibroma, found on the forehead and lumbar regions, respectively.

Periungual and gingival fibromas are also common cutaneous manifestations of TSC. Periungual fibromas often develop later in life, seen in up to 90% of TSC adults over the age of 30 years (Gomez et al 1999). These lesions are red- or flesh-colored papules or nodules occurring in the finger or toe nail beds, and along the proximal or distal nail fold or under the nail plate. Their histopathology is similar to that of facial angiofibromas.

Other Sites of Involvement

Cardiac rhabdomyomas are common in TSC infants and children (Gomez et al 1999). These lesions can be up to several centimeters in diameter and consist of glycogen-filled myocytes. With echocardiographic screening about 60% of TSC infants are found to have rhabdomyomas, and they are often detected prenatally. In most cases, there are no symptoms and the lesions spontaneously decrease in size or disappear over time. Occasionally, rhabdomyomas cause heart failure due to obstruction, and surgical resection is necessary. They can also affect myocardial function by replacement, or cause rhythm disturbances.

Lymphangioleiomyomatosis (LAM) is a disorder seen almost exclusively in females, which is characterized pathologically by smooth muscle and other cellular proliferation and cystic changes in the lung parenchyma. Some LAM cells stain positive with HMB-45 like AML cells, and these tend to be cells of an epithelioid nature (Matsumoto et al 1999). Most LAM patients also have AMLs in the kidney or elsewhere in the abdomen. LAM presents clinically as progressive dyspnea, or acute dyspnea due to pneumothorax. The natural history of LAM is poorly defined. There are women in whom it is progressive and fatal, and many are treated with lung transplant. Recent studies indicate that about 40% of adult TSC women have CT evidence of LAM. It is likely that most of those in whom it is asymptomatic and detected only by CT will not have significant symptoms from this condition. Multifocal micronodular pneumocyte hyperplasic (MMPH) is a hamartoma consisting of type II pneumocytes. It is common in adults with TSC, and does not produce clinical symptoms. Pulmonary cysts are also common in female TSC patients, and may be stable for many years or be a first sign of LAM.

Small retinal astrocytomas are seen in about 70% of TSC patients. Dental pits are very common in TSC patients. Hamartomas and polyposis of the intestinal tract (stomach through colon) also are common in TSC, but are not associated with a higher risk of malignancy in those organs.

CLINICAL AND MOLECULAR GENETICS OF TSC

Introduction and Clinical Genetics of TSC

TSC occurs as an autosomal dominant disorder. Large families with the condition are rare due to reduced reproduction among those affected. About two-thirds of TSC cases are sporadic, due to new mutations. TSC occurs in up to 1 in 6000–8000 live births without apparent ethnic clustering. Linkage of TSC to chromosome 9q34 was first reported in 1987, and the corresponding TSC1 gene was identified in 1997 (van Slegtenhorst et al 1997). Linkage of TSC to chromosome 16p13 was discovered in 1992, and the corresponding TSC2 gene identified in 1993 (European CTS Consortium 1993). Among families large enough to permit linkage analysis,

TABLE 30.3 Types of TSC1 and TSC2 mutations

	TSC1		TSC2	
Large deletions/ rearrangements	3	2%	69	16%
Insertions	29	17%	34	8%
Deletions	58	34%	97	23%
Nonsense	67	39%	86	20%
Point splicing	13	7%	52	12%
Missense	2	1%	84	20%
Total	172		422	

Note: derived from a literature summary and database maintained up through 2001

approximately half show linkage to 9q34 and half to 16p13 (Cheadle et al 2000). There is no evidence for a third locus.

Gene Structure

The TSC1 gene consists of 23 exons, of which 21 encode the 130 kDa hamartin, with an 8.6 kb mRNA including a 4.5 kb 3′ untranslated region. There is variable splicing of the second, untranslated exon. The gene is found in a genomic region of 55 kb on 9q34. The TSC2 gene consists of 42 exons, of which 41 encode the 198 kDa tuberin, with a 5.4 kb mRNA and relatively short 5′ and 3′ untranslated regions. There is alternative splicing of exon 25, the first 3 bp of exon 26, and exon 31, although the significance of these alternative forms of tuberin with respect to function is not known (Cheadle et al 2000).

Mutations in TSC1 and TSC2

Well over 600 mutations in the TSC1 and TSC2 genes have been identified (chromium.liacs.NL/LOUDZ/TSC/home.php). Here we discuss the first 594 TSC patients with defined mutations (172 in TSC1 and 422 in TSC2; Table 30.3), which were published prior to 2001 (Cheadle et al 2000, Dabora et al 2001). Among these 594 mutations, 427 are unique, 115 in TSC1 and 312 in TSC2. Fifty-four mutations have been reported as recurrent (21 in TSC1 and 33 in TSC2) accounting for 176 of the 594 (30%) mutations. Mutations have been identified in all but six of the 62 coding exons of TSC1 and TSC2 (Figure 30.2).

In TSC1, 48% of the mutations are single base substitutions, 82% of which are nonsense mutations (Table 30.3, Figure 30.2). Just over half the nonsense mutations are recurrent C to T transitions at 6 of the 7 CGA codons within the TSC1 coding region. Splicing mutations account for 16% of all single base changes in TSC1. Two missense mutations have been seen in TSC1.

In TSC2, 53% of the mutations are single base substitutions (Table 30.3, Figure 30.2). Nonsense mutations in TSC2 make up only 39% of this class, but again commonly (36%) occur at 5 of the 7 CpG sites which can transition to nonsense codons in TSC2. In contrast to TSC1, missense mutations account for 38% of the single base substitutions in TSC2. TSC2 has two CpG sites that are relatively common sites of mutation, leading to missense mutations – 1831-2 (611R > W and 611R > Q) and 5024-5 (1675P > L). These two CpG sites account for 35 of the 84 (42%) missense mutations in TSC2. Splice mutations make up the remaining 23% of point mutations in TSC2.

Thirty-four percent of TSC1 mutations are small deletions (<29 bp), while 17% are small insertions (<34 bp), and nearly all of these cause frameshifts in the TSC1 mRNA. Nearly all small insertions arise as duplication of an adjacent single or multiple bases. Similarly, most deletions (62%) are the deletion of an element of a tandem repeat. Similar observations apply to small deletions and insertions found in TSC2.

Large deletions and rearrangements are strikingly more common in TSC2 than in TSC1 (69 vs. 3). They show no consistency in either the size or junction fragments involved, ranging in size from 1 kb to over 100 kb (Sampson et al 1997, Dabora et al 2000).

Large sets of TSC patients (150 and 224) have been put through comprehensive screens for mutations in TSC1 and TSC2, with an overall rate of mutation detection of 80–90% (Jones et al 1999, Dabora et al 2001). TSC1 mutations have been identified in 10–15% of sporadic TSC patients, while TSC2 mutations have been found in about 70%. Several differences between the genes and their types of mutation contribute to the different mutation rates. First, both missense mutations and large rearrangements and deletions rarely occur in TSC1, in contrast to TSC2. Second, the TSC2 coding region is about 1.5 times longer than that of TSC1, and has approximately twice the number of splice sites. Nonetheless, even with these considerations, the intrinsic mutation rate is much higher in TSC2 than in TSC1, probably reflecting differences in chromosomal location, genomic structural features, or the specific sequences present at the two loci.

TSC Patients without Mutations in TSC1 or TSC2

Even with comprehensive effort, there remains 10–20% of TSC patients for whom no mutation can be identified (Jones et al 1999, Dabora et al 2001). There are several possible explanations for this, including incomplete sensitivity of mutation detection, and unidentified intronic and promoter region mutations. Through protein truncation test screening, four intronic mutations in TSC2 (at a distance of ≥15 bp from an exon–intron border) have been identified (Mayer

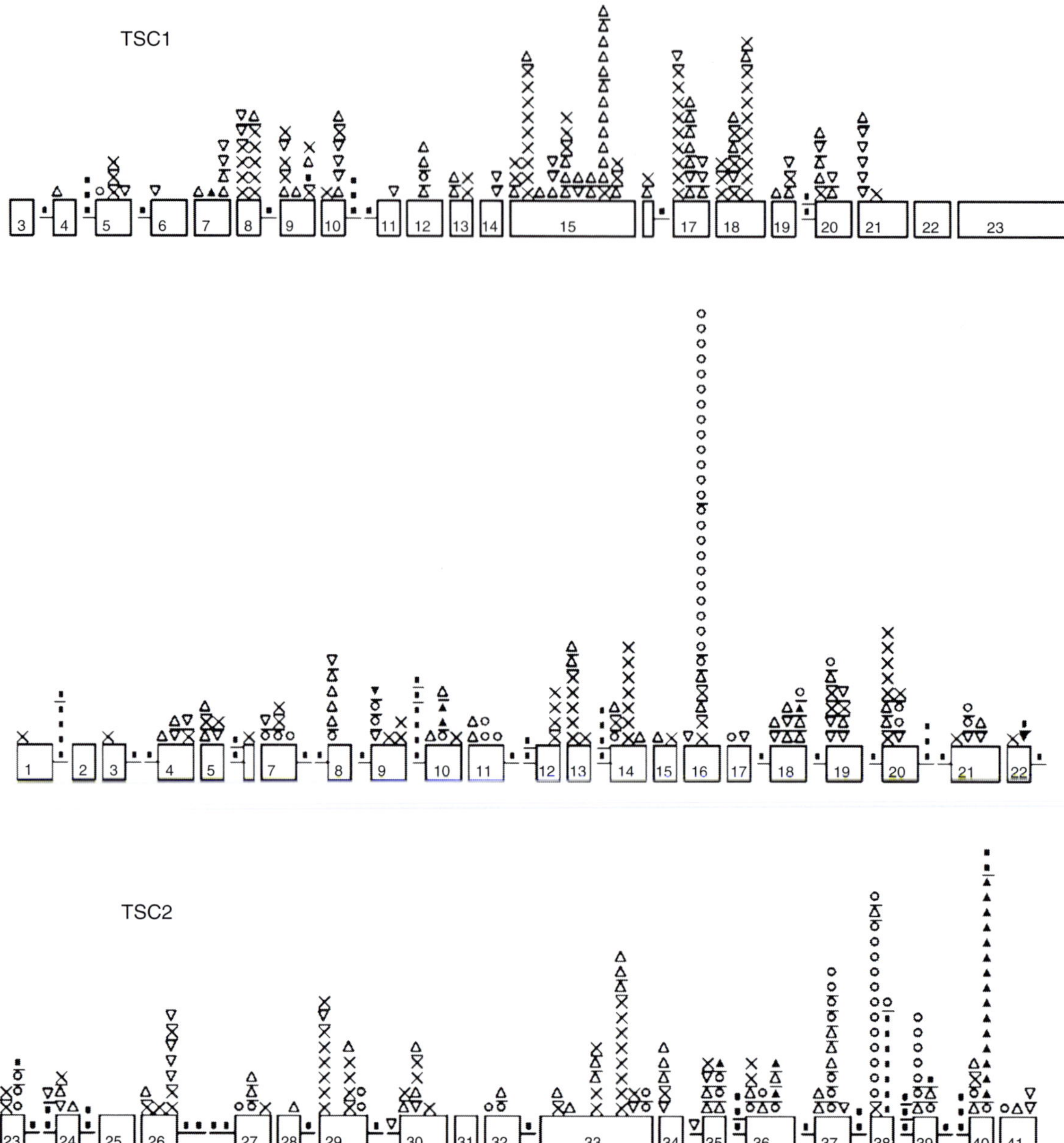

FIGURE 30.2   Map of the sites of several hundred small mutations in TSC1 and TSC2. Proportional drawings (boxes) of all of the exons of TSC1 and TSC2 are shown. Intron regions are not drawn to scale and are expanded only when a mutation is present. Mutation symbols are: ✕, nonsense; ■, splice site; △, deletion; ▽, insertion; ○, missense; ▲, in-frame deletion. A line separates mutations occurring at nearby but not identical nucleotide positions

et al 2002). Thus it seems likely that additional sequence variants in intronic or exonic regions are unrecognized pathogenic mutations causing aberrant splicing. Another type of 'missed' mutation is single or multiexonic deletions in TSC1/TSC2. These deletions can occur in a variety of sizes (100 bp to >100 kb), and are difficult to detect comprehensively. Low-level mosaicism of a mutation in a TSC gene may cause a full TSC phenotype and may be very difficult to detect. Last, it is possible that there are additional genes, mutations in which can cause the TSC syndrome. However, all reported large TSC families show linkage to either TSC1 or TSC2, and there is no other evidence to suggest there is a third gene.

We have observed that patients without identified mutations in TSC1 or TSC2 in our set of 224 patients carefully screened for mutations, had clinical features that were milder than those of patients with defined TSC2 mutations but similar to those of patients with defined TSC1 mutations (Dabora et al 2001). This observation seems most consistent with the occurrence of mosaicism for a TSC2 mutation in these patients, to explain both the lack of mutation identification and the milder clinical features (see below).

Pathogenic missense mutations in TSC1 are very rare. As there is no reason why such mutations would not occur, this observation suggests that missense variation in TSC1 might either have no effect, or cause a unique clinical phenotype distinct from that of TSC. It seems possible that some TSC1 missense mutations may have a dominant negative effect, leading to an embryonic lethal or other severe phenotype.

Mosaicism in TSC

Mosaicism for a TSC gene mutation occurs when the mutation occurs early during embryogenesis but after the fertilized egg undergoes one or more divisions. Both somatic (generalized) and germline (confined gonadal) mosaicism for TSC1 and TSC2 mutations have been described. The highest level of mosaicism reported was 26% of TSC families with large deletions in TSC2 (Sampson et al 1997). However, in most series much lower levels of mosaicism have been reported, likely reflecting the low sensitivity of detection techniques. Although mosaicism has been identified in some patients with relatively mild features of TSC, it appears in general that an individual who is mosaic for a TSC gene mutation will meet formal diagnostic criteria. For example, in a comprehensive assessment, all parents of a set of 30 TSC sporadic patients had no evidence of mosaicism at a level above 1% in their blood DNA (Roberts et al 2004).

Ten sets of siblings affected by TSC but with apparently normal parents have been reported (Rose et al 1999). In several cases in which mutation identification was performed, identical TSC2 mutations were found in DNA of affected siblings but not in blood-derived DNA of their parents. These observations strongly implicate gonadal mosaicism in a parent. However, the frequency of gonadal mosaicism is low in TSC, as the observed rate of recurrent sporadic TSC children born to normal parents is very low, about 1–2%.

Frequency of New Mutations in TSC1, TSC2

With a birth incidence of 1 in 6000–8000, and two-thirds of cases being sporadic in nature due to new mutations, we can calculate the frequency of new mutations in either TSC1 or TSC2 as about 1 in 10 000 per generation. Through analysis of SNPs and haplotypes in the 3' region of TSC2, it was shown that new mutations in that region of TSC2 occurred at approximately equally frequency in the father and mother of sporadic TSC patients (Roberts et al 2002).

GENOTYPE–PHENOTYPE CORRELATIONS

TSC1 vs. TSC2 Disease

The discordance between the frequency of TSC1 and TSC2 mutations in sporadic TSC patients (10–15% TSC1, 70% TSC2) and their frequencies in large TSC families (50% TSC1, 50% TSC2) has suggested that TSC due to TSC1 mutations might be less severe than TSC due to TSC2 mutations. Two series have verified this hypothesis (Jones et al 1999, Dabora et al 2001).

Using a comprehensive TSC clinical data instrument, we compared 22 sporadic TSC patients with TSC1 mutations and 129 sporadic cases with TSC2 mutations (Dabora et al 2001). Eight of 16 clinical features occurred at a significantly higher frequency and/or with greater severity in TSC2 than TSC1 sporadic cases (Table 30.4). These included: seizures, moderate to severe mental retardation, mean number of subependymal nodules, tuber count, kidney angiomyolipomas, mean grade of facial angiofibromas, forehead plaque, and retinal hamartomas. Thus, this study provides strong evidence that TSC1 disease is less severe than TSC2 disease in multiple respects.

The observation that there is a lower frequency of sporadic TSC patients with TSC1 mutations compared to TSC2, and that there is a smaller number of cortical tubers and subependymal nodules in TSC1 patients (4.9 and 1.7) compared to TSC2 patients (12.9 and 6.7) is consistent with the model that both germline and somatic mutations (see section VIIA for further discussion) occur less frequently in TSC1 than TSC2. However, brain MRI does not detect all cortical tubers or SENs. Thus, an alternative explanation that could contribute to the difference in brain phenotype is that TSC1 gene loss (thought to occur in the cortical tubers and SENs of a TSC1 patient) leads to development of smaller tubers, which are less often detected by brain MRI, in comparison to TSC2 patients.

A Contiguous Gene Deletion Syndrome Involving TSC2 and PKD1

Renal cysts are common in TSC patients, but in most patients are small in number and pose no significant health risk. About 3% of TSC patients develop renal cystic disease similar to but more severe than that seen in typical polycystic kidney disease (PKD). Large genomic deletions involving both TSC2 and PKD1, which are directly adjacent in the genome, are present in most of these patients (Sampson et al 1997). Interestingly, the polycystic kidney disease that develops in these patients with combined disruption of both TSC2 and PKD1 is more severe than the usual polycystic kidney disease associated with mutation in PKD1 alone, with cystic changes in the kidney in early childhood, and progression to renal failure by age 20 (Sampson et al 1997). It appears likely that this accelerated

TABLE 30.4   Comparison of clinical features in sporadic TSC patients with TSC1 vs. TSC2 mutations (Dabora et al. 2001)

	TSC1 sporadic cases	**TSC2** sporadic cases	**P value** TSC1 vs. TSC2
Number of patients	22	129	
Age range (Avg. age)	2–51y (13.4y)	1–44y (11.2y)	NS
Median age	9	10	
NEUROLOGIC			
Seizures	19/22 (86%)	127/128 (99%)	0.02
Mental handicap (age >6)			
Moderate + Severe	2/14 (14%)	41/90 (46%)	0.04
Mean MR gr (scale 0–3)	0.67	1.4	0.007
Subependymal nodules	15/20 (75%)	127/136 (93%)	0.02
Mean SEN number	1.7	6.7	0.0002
SEGA	2/21 (9%)	13/118 (12%)	NS
Tubers (any)	13/15 (87%)	55/60 (92%)	NS
>10 tubers	1/9 (11%)	29/42 (69%)	0.002
Mean tuber number	4.4	12.9	0.002
RENAL			
Kidney cysts (grade 1–4)	3/19 (16%)	30/122 (25%)	NS
grade 2–4	0/19 (0%)	19/122 (16%)	NS (0.08)
grade 4	0/19 (0%)	5/122 (4%)	NS
Mean grade (scale 0–4)	0.16	0.52	NS (0.14)
Kidney AMLs	6/19 (31%)	72/121 (60%)	0.03
Mean grade (scale 0–3)	0.32	0.97	0.006
SKIN			
Hypomelanotic macules	20/21 (95%)	124/128 (97%)	NS
Facial angiofibromas (age >2)	13/22 (59%)	95/121 (78%)	NS
Mean grade (scale 0–3)	0.9	1.5	0.02
Shagreen patch	7/20 (35%)	68/130 (52%)	NS
Ungual fibroma	5/20 (25%)	26/128 (20%)	NS
Forehead plaque	2/20 (10%)	51/128 (40%)	0.01
OTHER			
Liver AMLs	0/15	9/117 (8%)	NS
Retinal hamartomas	0/16	32/117 (27%)	0.01
Cardiac rhabdomyoma	8/17 (47%)	58/117 (50%)	NS
LAM (females, age >16)	0/5	3/17 (18%)	NS

pathogenesis is due to the genetic mechanism of genomic deletion of the second allele of both genes occurring as a single somatic event (see section below). In such cells there would be loss of expression of both tuberin and polycystin (the product of the PKD1 gene) to produce a synergistic effect in cystogenesis.

A Mild TSC Phenotype

Several TSC families have been identified in which clinical manifestations are unusually mild. Two extended families in which the TSC2 mutation 1503Q > P segregates had relatively mild TSC features, including some carriers who did not appear to have any clinical signs of TSC, but were not thoroughly clinically evaluated (Khare et al 2001). In another family, seven members carried a TSC2: 1036S > P mutation (O'Connor et al 2003). Only two of the seven met formal TSC diagnostic criteria, while three others had a seizure disorder, and two had no major findings of TSC at all. All had undergone complete evaluation including brain imaging, skin and eye exams, and cardiac and renal ultrasound. Such mild TSC families appear to be quite rare at this time, but under-recognition is possible. The finding of missense mutations in these two families suggests that there may be incomplete loss of TSC2 function, leading to a mild phenotype.

Apart from these two reports, there is no evidence of incomplete penetrance of TSC when complete clinical evaluation is performed.

GENETIC COUNSELING IN TSC

Genetic counseling is appropriate for every TSC patient and family (Gomez et al 1999). TSC patients should have a thorough investigation of their family history. Manifestations of TSC can go unrecognized. Specific inquiry should address all of the clinical features described in Table 30.1, and related symptoms and signs. If additional children are contemplated, the parents of sporadic cases of TSC should have a thorough evaluation, including general physical exam with UV light evaluation of the skin, ophthalmologic exam, renal ultrasound or CT, and brain MRI or CT. These studies should be interpreted by physicians familiar with TSC.

A parent who has one or two findings consistent with TSC but who does not meet formal diagnostic criteria should be considered cautiously. Such individuals may be full carriers of a TSC gene mutation with low expression, may be mosaic for a mutation, or may not carry a mutation at all. This parent's risk of having a second child with TSC should be considered elevated over the general population but the specific risk is uncertain.

Even after a completely normal clinical and radiographic evaluation, there is still the possibility that one of the parents has confined gonadal mosaicism for a TSC mutation. However, recurrent sporadic cases born to clinically normal parents are quite uncommon, so the recurrence risk in this situation should be estimated as about 2%, and definitely no greater than 5%.

Mutational testing with identification of a specific TSC1 or TSC2 mutation in a sporadic patient enables direct testing of parental DNA samples for the mutation, and this testing is available through Athena Diagnostics, Inc. However, low-level somatic mosaicism as well as gonadal mosaicism in a parent may be missed by current methods. An alternative approach that parents may wish to consider is prenatal diagnosis, by CVS or amniocentesis, to directly determine whether the fetus is carrying the mutation present in the sporadic child.

When one parent is affected with TSC, the risk of transmission of the mutant allele is 50%. However, prospective parents in this situation should be aware that the severity of tuberous sclerosis is not predictable, so an affected child may have relatively few or many medical problems. As above, about half of TSC patients have normal cognitive function, although they may have specific learning or developmental problems. If the specific mutation is known, then somewhat more accurate information can be given using the information in Table 30.4. Although many mothers with TSC have had uneventful pregnancies, there appears to be some additional risk due to AML hemorrhage and worsening of pulmonary LAM.

Prenatal diagnosis of TSC can be made on the basis of identification of cardiac rhabdomyomas by ultrasonography and/or cortical tubers by ultrasonography or rapid sequence MRI. However, the frequency of positive findings in fetuses with TSC using these modalities is uncertain. In addition, the earliest time at which definite findings have been identified is 22 weeks of gestation, so the potential for intervention is limited. Nonetheless such findings may be helpful in the management of delivery and in the care of the neonate.

MOLECULAR PATHOGENESIS OF TUBEROUS SCLEROSIS

Tumor Suppressor Gene Mechanism

The focal nature of the lesions and tumors which occur in tuberous sclerosis has suggested for many years that a fundamental pathogenetic mechanism in this disease was the classic Knudson model of a first germline hit and a second somatic hit, ablating both alleles of either TSC1 or TSC2 in tumors. Large somatic deletions of the wild-type TSC1 or TSC2 allele have been inferred by loss of heterozygosity (LOH) analysis using nearby microsatellite and SNP markers in numerous TSC tumors (Green et al 1994, Carbonara et al 1996, Henske et al 1996, Au et al 1999, Niida et al 2001). However, the rate of LOH varies considerably according to the site of the lesion, and is highest (about 75%) in renal AML.

Older studies had found LOH in brain lesions from TSC patients at relatively low rates, four of 28 (14%) cortical tubers and 13 of 36 (36%) SEGAs (Henske et al 1996, Sepp et al 1996, Niida et al 2001, Tucker & Friedman 2002). Since cortical tubers always contain a high proportion of normal or reactive cells, this could mask LOH. A more recent study of six SEGAs in which the germline mutation in TSC1 or TSC2 was known, however, demonstrated that five of six SEGAs showed either evidence of LOH (four cases) or a second TSC2 mutation confined to the SEGA (one case) (Chan et al 2004). Thus, SEGAs arise by the classic two hit mechanism in most cases. For tubers this pathogenic mechanism remains unproven (see further discussion below).

Functions of TSC1 and TSC2
Drosophila Insights

Although the TSC1 and TSC2 genes were identified in 1997 and 1993, respectively, little progress was made in understanding the function of their protein products until 2001. Earlier work had identified potential roles in the cell cycle, endocytosis, and cell adhesion, but the importance of the findings to the pathogenesis of the lesions and tumors developing in TSC patients was not clear. TSC1 and TSC2 had also been recognized to form a high-affinity complex (Plank et al 1998, van Slegtenhorst et al 1998). In addition, a parallel level of expression of the two proteins was observed in many tissues.

Then in 2001, using fly mutagenesis to perform a genome screen, three different groups reported the identification of a function for the *Drosophila* homologs of these genes in the regulation of cell and organ size (Gao & Pan 2001, Potter et al 2001, Tapon et al 2001). These seminal studies opened up the field of investigation into the biochemical and signaling properties of the protein products of these two genes. Although studies in *Drosophila* continue to be productive in highlighting novel genes and proteins that participate in the same signaling pathway as TSC1 and TSC2, they will not be discussed further here, as the mammalian work is extensive, and due to limitations of space. However, recent studies continue to demonstrate the value of the fly system (e.g. Kapahi et al 2004, Pan et al 2004, Reiling and Hafen 2004).

The PI3K-Akt-TSC1/TSC2-mTOR Signaling Pathway

The TSC1 and TSC2 protein products have an important role in a signal transduction pathway that regulates cell growth and metabolism, and is conserved to varying degrees back to *Schizosaccharomyces pombe*. We now describe this signaling pathway (Figure 30.3) with particular emphasis on the role of the TSC1 and TSC2 proteins (for recent detailed reviews, see Kwiatkowski 2003, Fingar & Blenis 2004, Li et al 2004, Mak & Yeung 2004, Pan et al 2004).

In normal cells, several growth factors can act on their cognate receptor tyrosine kinases (RTK) to induce intracellular phosphorylation events that lead to recruitment of PI3Kinase (PI3K) to the plasma membrane (Figure 30.3).

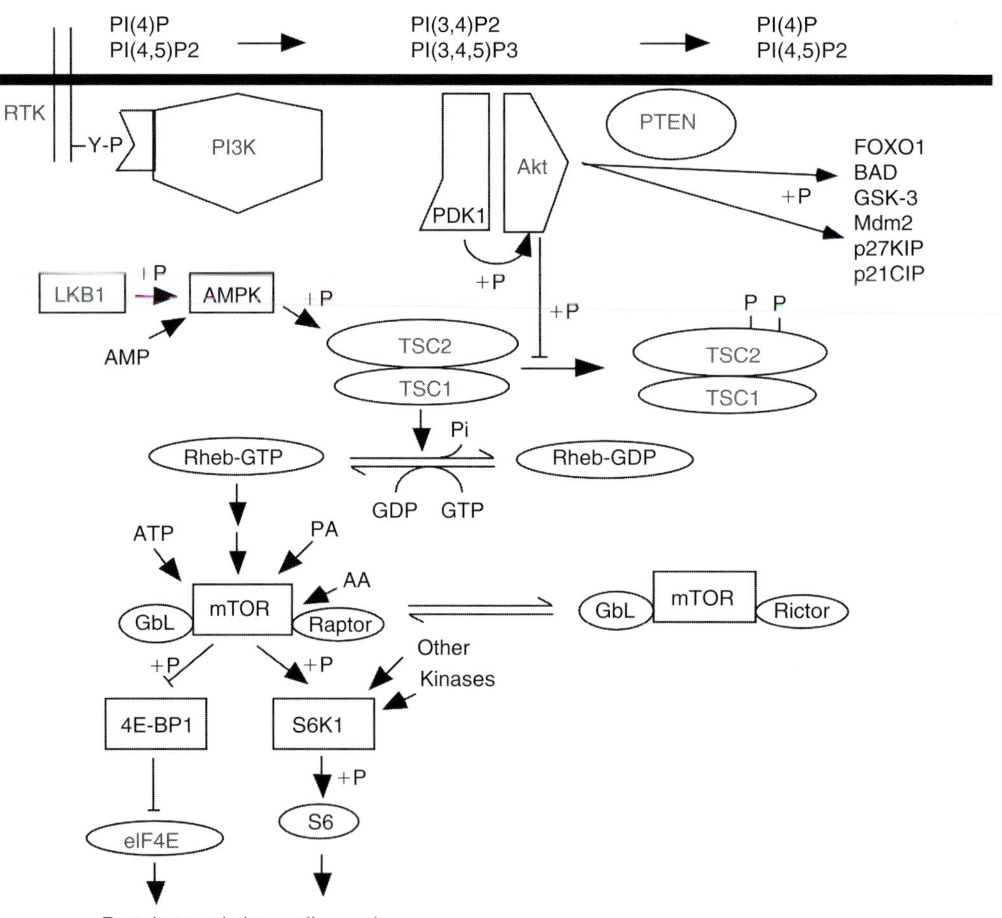

FIGURE 30.3   The PI3K-Akt-TSC1/TSC2-mTOR signaling pathway. This figure portrays a portion of a more complex pathway. Receptors illustrated by the parallel lines at left undergo tyrosine phosphorylation (Y-P), which leads to recruitment of PI3kinase, which phosphorylates phosphoinositides at the 3' position to generate PI(3,4)P$_2$ and PI(3,4,5)P$_3$. These molecules lead to recruitment of Akt to the membrane which is then activated by phosphorylation by PDK1, and phosphorylates many downstream targets including TSC2. Phosphorylation of TSC2 reduces its GAP activity, leading to high levels of Rheb-GTP, which through unknown mechanisms positively activates mTOR. mTOR then phosphorylates the S6Ks and 4E-BP1, leading to enhanced protein translation and cell growth. LKB1 connects with this pathway by phosphorylating AMPK in the presence of high AMP/ATP levels, and pAMPK then phosphorylates TSC2 and activates its GAP activity. See text for additional details. (see also Plate 68)

Although RTK activation also leads to activation of the ras-raf-MEK-MAPK kinase cascade, we focus on the PI3K pathway. PI3K recruitment to the cell membrane occurs either through direct binding of the regulatory subunit of PI3K to phosphorylated receptor or through intermediate adapter molecules such as IRS-1 and IRS-2. At the membrane, PI3K generates 3'-phosphoinositides, including phosphoinositide-(3,4,5)-phosphate ($PI(3,4,5)P_3$), that lead to recruitment of the Akt kinase to the membrane through its PH domain. At the membrane, Akt is activated by phosphorylation by the PDK1 kinase at T308, and a second phosphorylation event at S473, possibly mediated by mTOR. Activated Akt then phosphorylates a number of downstream targets, including p21CIP1, MDM2, GSK3β, BAD, AFX, FOXO1, and eNOS. Most of these targets are important in survival signaling in cells, relating to regulation of expression and/or activation of proteins involved in apoptosis. Activated Akt also phosphorylates TSC2 at S939 and T1462, and to a lesser extent other sites. In some studies, Akt-phosphorylated TSC2 appears to have a reduced half-life in cells with reduced binding to TSC1, and to undergo ubiquitination followed by proteasome-dependent degradation. Native, unphosphorylated TSC2 in complex with TSC1 functions as a GTPase activating protein (GAP) for the ras family member rheb. Akt-phosphorylated TSC2 is inactivated with respect to its GAP activity towards the rheb GTPase. Thus, Akt phosphorylation of TSC2 leads to increased rheb-GTP levels. Rheb-GTP is a major regulator of the kinase activity of mTOR, through a mechanism that is currently unknown. mTOR is a critical sensor and driver of cellular metabolism in that its activity is also influenced by levels of amino acids, ATP, phosphatidic acid, and hypoxia. It functions as a driver of cellular metabolism by phosphorylating its downstream targets: the p70S6 kinases (S6K1 and S6K2) at T389 and 4E-BP1 at multiple sites, and, likely, other targets. Activation of the S6Ks is actually a multi-site, intricate process that involves several other kinases. However, mTOR phosphorylation occurs at the last stage of S6K activation, and thus has a critical effect. Activated S6K1 and S6K2 phosphorylate the ribosomal protein S6 and other unknown targets, leading to enhanced cell growth and protein synthesis. Phosphorylated 4E-BP1 dissociates from eIF4E to permit eIF4E to participate in the formation of translation initiation complexes, also leading to accelerated cell growth and protein synthesis. Activation of mTOR also appears to reduce phosphatase activity, though the mechanism is unknown.

It is apparent that TSC1 and TSC2 have a critical role in the above signaling pathway, regulating the state of activation of Rheb, and thus mTOR. In the absence of either TSC1 or TSC2, the other is nonfunctional, and the following sequence ensues. There are constitutive high levels of rheb-GTP and constitutive activation of the mTOR kinase. This has the effect of increasing cell size, enhancing cell growth, and driving cellular metabolism in the absence of external signals and appropriate growth conditions. Participation in this growth signaling cascade appears to be the fundamental cellular function of the TSC1 and TSC2 proteins, and to explain the growth potential of cells in which there is complete absence of either protein. The role of these proteins in regulating cell size also connects nicely with the classic giant cell morphology seen in cortical tubers.

Signaling Pathways and Events Connecting with the PI3K-Akt-TSC1/TSC2-mTOR Pathway

Recently another tumor suppressor gene syndrome, Peutz-Jeghers syndrome (PJS), has been linked to this same pathway. PJS is an autosomal dominant disorder in which GI hamartomatous polyps develop along with a high predisposition to the development of GI cancer, and is due to mutations in the LKB1 gene. The LKB1 kinase has the normal function of phosphorylating the AMP-dependent protein kinase AMPK. In response to energy stress (high AMP/ATP levels) AMPK undergoes a conformational change, permitting phosphorylation by LKB1. Phosphorylated, activated AMPK phosphorylates many downstream targets. Recent work has identified TSC2 as a prominent target of AMPK phosphorylation, at S1227 and S1345, and the effect of this phosphorylation is to enhance the GAP activity of TSC2 toward rheb (Inoki et al 2003, Shaw et al 2004). Thus, in the absence of LKB1, there is a failure of normal upregulation of TSC1/TSC2 GAP activity in response to nutrient deprivation, resulting in inappropriate cell growth.

Recent observations indicate that there is significant feedback circuitry operating in this pathway. In mouse embryo fibroblasts lacking either TSC1 or TSC2, there is a relative lack of Akt activation in response to several different growth factors. In the case of the insulin or IGF1 stimulation, this is due to decreased amounts of IRS-1 and IRS-2, adaptor proteins that are critical for insulin receptor signaling (Harrington et al 2004, Shah et al 2004). The mechanism of this reduced expression is both reduced mRNA levels and phosphorylation of the IRS proteins by activated S6Ks. In the case of PDGF, reduced Akt activation is related to reduced expression of the PDGF receptor (PDGFR) due to reduced transcription (Zhang et al 2003). Both of these counter-regulatory circuit observations appear to contribute to the relative lack of malignancy that is seen in TSC patients despite their high burden of hamartomas.

Another aspect of this circuit with relevance to the vascular nature of TSC AML is a connection with the response to hypoxia and production of VEGF (El-Hashemite et al 2003). Normally, hypoxia inhibits mTOR in a mechanism that requires both TSC1 and TSC2 and the hypoxia-inducible gene REDD1. Loss of TSC1 or TSC2 blocks this inhibition so that mTOR is highly active, and results in high levels of hypoxia-inducible factor (HIF) (Brugarolas et al 2004). This results in cell growth in the presence of low oxygen levels and high level production of VEGF by TSC tumors.

There are also connections or crosstalk between this PI3K-Akt-TSC1/TSC2-mTOR pathway and the classic ras-raf-MEK-MAPK kinase pathway. Following MAPK activation by TPA, there is activation of the RSK1 kinase which, among other targets, phosphorylates TSC2 at T1462 (Roux et al 2004). Phosphorylation at this site is synergistic with phosphorylation at the S939/T1462 sites by Akt, to inhibit TSC2 and enhance mTOR activation during growth signaling. There may also be a direct connection from TSC1/TSC2 to MAPK. In ELT3 cells (uterine leiomyoma cell line from the Eker rat) lacking Tsc2, there is lower level activation of MAPK than in control cells expressing Tsc2 (Karbowniczek et al 2004). This appears to be due to rheb binding to and inactivation of the B-raf kinase, inhibiting activation of MAPK. However, in Tsc2 null primary renal epithelial cells from Eker rats, there is high ERK activity which is reduced by Tsc2 expression (Yoon et al 2004).

The STAT pathway of signal transduction is also affected in TSC cells and tumors. There is over-expression of STAT1, phospho-S727-STAT1, and phospho-Y705 STAT3, which is rapamycin sensitive and leads to apoptotic cell death in response to interferon-γ treatment (El-Hashemite et al 2004).

Insights from Rodent Models

Null mutations in each of Tsc1 and Tsc2 have been engineered in mice (Kobayashi et al 1999, 2001, Onda et al 1999, Kwiatkowski et al 2002), and there is a spontaneously occurring null allele of Tsc2 in the Eker rat (Eker et al 1981, Yeung et al 1994, Kobayashi et al 1995). The pathology in all of these animal models is similar.

Tsc1 null or Tsc2 null embryos do not survive past embryonic day 13, due to liver hypoplasia resulting in cardiac failure. Exencephaly has been seen in some null embryos, but is not consistent, and appears to correlate with strain in the rat. Mouse embryo fibroblast lines null for either Tsc1 or Tsc2 have been developed which have been widely used for signaling pathway studies.

Mice heterozygous for mutations in either Tsc1 or Tsc2, as well as the Eker rat, consistently develop two types of tumor. Renal cystadenomas develop by 6 months of age and grow progressively and slowly throughout the life of the animal. Histologically these lesions appear as a continuous spectrum from pure cysts, cysts with papillary projections, to solid adenomas. In both these rodents, progression of these lesions to carcinoma is a rare event given the number of cystadenomas per animal, and has been estimated at less than 1 in 1000 cystadenomas (Onda et al 1999). Such carcinomas are characterized by cellular and nuclear pleomorphism (common), and massive (>1 cm) growth (rare at age <1 year, more common in older mice), but never metastasis. The cell of origin of these tumors is tubular epithelial cells (Onda et al 1999).

Liver hemangiomas, characterized by endothelial and smooth muscle cell proliferation with large vascular spaces, are also commonly seen in both Tsc1 and Tsc2 heterozygote mice. These lesions have some similarity to the kidney angiomyolipomas and pulmonary lesions seen in TSC patients, and also express HMB-45. Interestingly, they are also more common, more extensive, and cause greater mortality in female compared with male Tsc1 heterozygote mice (Kwiatkowski et al 2002).

Brain hamartomas resembling TSC subependymal nodules have been identified in Eker rats at a rate of 63% at age 1.5–2 years (Yeung et al 1997). The lesions consist of a mixture of large and elongated cells that express glial markers, are often calcified, and are found in both subcortical and subependymal regions. No brain lesions have been seen in the Tsc1 and Tsc2 heterozygote mice (Onda et al 1999).

Variation in the severity of expression of the heterozygote phenotype according to genetic background has been shown for both the mouse and rat models of TSC, and influences kidney, liver, and brain lesions (Rennebeck et al 1998, Onda et al 1999, Kwiatkowski et al 2002). A locus Mot1 has been mapped genetically that accounts for 35% of the variation in the severity of renal tumor development in the Fischer344 and Brown Norway rat strains (Yeung et al 2001).

Other Potential Functions of Tuberin and Hamartin

Increased fluid phase endocytosis has been noted in a single Eker rat cell line, potentially consistent with hyperactivation of rab5 signalling, as rab5 is thought to regulate endosome fusion. In addition, in an Eker rat kidney-derived cell line in which there is homozygous loss of tuberin, there is aberrant retention of polycystin-1 (product of the PKD1 gene) in the Golgi, consistent with an endosomal pathway defect (Kleymenova et al 2001).

Hamartin binds to ezrin in endothelial cells (Lamb et al 2000). Transient loss of hamartin causes endothelial cell retraction with loss of focal adhesions and substrate detachment. Overexpression of hamartin or certain fragments induce the formation of actin stress fibers in Swiss 3T3 cells through activation of the rho GTPase (Lamb et al 2000). Changes in the actin cytoskeleton are seen in Tsc1 null NIH3T3 cells, but are relatively mild (Kwiatkowski et al 2002). In the Eker rat-derived ELT3 cell line, there is inhibition of cell motility, inhibition of rac1 activation, and reduced actin dynamics, which appears to be due to an inhibition of rac1 activation by Tsc1. This inhibition is relieved by reduced expression of Tsc1 or by expression of the Tsc1 binding domain of Tsc2 (Goncharova et al 2004).

Several other binding partners have been described for both Tsc1 and Tsc2. Pam, a protein identified as an interactor of Myc, binds to Tsc1-Tsc2 in the brain through an interaction between a RING zinc finger motif in PAM with tuberin. Pam is expressed in embryonic and adult brain as well as in cultured neurons, and this finding may be important in the brain-specific functions of TSC1/TSC2 (Murthy et al 2004). Tuberin has also been reported to bind to SMAD2 and SMAD3 proteins, such that Tsc2 augments TGF-1 signaling through SMADs in myeloid cells (Birchenall-Roberts et al 2004). In the absence of Tsc2 there is enhanced proliferation and impaired differentiation. Tuberin also binds to CDK1 (Catania et al 2001).

Involvement of the TSC Genes in Other Disorders

The wide variety of tumors and lesions associated with TSC has suggested the possibility that some forms of cancer might be associated with inactivation of either TSC1 or TSC2. There is evidence that both sporadic AMLs and sporadic LAM are due to two hit inactivation of either TSC1 or TSC2 (Henske et al 1995, Carsillo et al 2000). TSC1 and TSC2 may also be involved in atypical adenomatous hyperplasia of the lung and some lung adenocarcinomas (Takamochi et al 2001), and TSC1 may be involved in bladder carcinoma (Hornigold et al 1999). However, the TSC genes do not appear to be involved in the pathogenesis of renal cell carcinoma in patients without TSC, glial and glioneuronal cancers, or medulloblastoma.

Molecular Pathogenesis of Cortical Tubers

The pathogenesis of cortical tubers has great importance, given their major clinical consequences. As above, a two-hit pathogenic mechanism has not yet been clearly demonstrated by LOH studies, which may be due to the low proportion of giant cells in these lesions. Immunohistochemical analyses of tuberin and hamartin expression in tubers has yielded conflicting results in different laboratories, with some indicating robust expression of both proteins in giant cells, and others indicating reduced expression of both (Henske et al 1997, Johnson et al 1999, Mizuguchi et al 2000, Han et al 2004). However, many proteins are expressed similarly in TSC tuber giant cells and SEGA cells, including several proteins downstream of mTOR, suggesting that tuber and SEGA pathogenesis are similar (Baybis et al 2004, Chan et al 2004, Miyata et al 2004). However, the possibility that haploinsufficiency contributes to some aspects of the TSC phenotype, including in the brain, should not be overlooked.

The molecular basis of epileptogenesis in and nearby cortical tubers is uncertain. Single cell studies have shown that within tuber giant cells there is altered transcription of several genes important in determining neuronal excitability, including glutamate receptors, γ-aminobutyric acid receptor subunits, neuronal glutamate transporters and the vesicular GABA transporter. These expression changes likely contribute to epileptogenesis in TSC (White et al 2001, Crino 2004).

Gene expression differences have been observed in a mouse brain model of TSC in which there is loss of Tsc1 in astrocytes (Ess et al 2004). Expression changes were consistent among cultured Tsc1 null astrocytes, mouse brain lysates in which astrocytes lacked Tsc1, and cortical tubers, suggesting

that complete loss of Tsc1/Tsc2 in astrocytes contributes to the pathogenesis of cortical tuber development.

Molecular Pathogenesis of LAM

The development of LAM nearly exclusively in females has long suggested a significant role of sex hormones in its pathogenesis. AML and LAM cells in patients with both conditions but without TSC show the same biallelic mutation in the TSC2 gene, suggesting that cells can migrate from one lesion to the other (Carsillo et al 2000). In addition, LAM cells with TSC2 mutations have been isolated from the blood, lymphatic fluids, and urine from patients (Crooks et al 2004), confirming a benign metastasis mechanism of disease spread. Growth responses to estrogens have been shown to differ according to TSC2 genotype in both Eker rat-derived cells and an AML-derived cell line, and are due to PDGFR activation in the Eker cell line (Finlay et al 2004).

THERAPIES FOR TSC

Eventually gene therapy may be a potential form of treatment for TSC. However, neurologic morbidity in the form of seizures, mental retardation, and developmental disorders is the most serious aspect of TSC. Cortical tubers are the main cause of these neurologic morbidities, and the development of both the brain and cortical tubers is largely complete by birth. Therefore any approach directed at elimination or modification of tuber development is particularly challenging as therapy in utero would be required.

Although such a fundamental approach to the care of the TSC patient is not practical at present, effective management of TSC patients can improve the quality of life of the affected individual to a dramatic extent. Regular clinical evaluation is recommended to identify potential clinical problems at an early stage and permit effective intervention (Roach et al 1999). Neurodevelopmental testing is recommended at the time of diagnosis and at school entry. Renal ultrasonography is recommended every 1 to 3 years from the age of diagnosis until age 20, to permit monitoring of renal AMLs. In 10–20% of children these lesions grow to a significant extent and early intervention as described below may lead to preservation of renal function and avoidance of morbidity. Repeated brain MRI imaging or brain CT scans are recommended every 1–3 years from age of diagnosis until age 20, to permit serial follow-up of subependymal nodules, and early intervention for SEGAs.

Aggressive seizure control is mandatory for all TSC patients with seizures, due to the connection between poor seizure control and poor intellectual outcome. An expanding armamentarium of drugs is available for seizure control, with two relatively new drugs that are effective in TSC: vigabatrin and lamotrogine (Curatolo et al 2001, Franz et al 2001). Surgical resection of seizure foci is another approach that can be effective in TSC patients who have proven refractory to antiepileptic drugs.

Drugs designed to intervene in the mTOR pathway that is constitutively activated in TSC lesions and tumors have considerable promise as therapeutics for TSC. Rapamycin selectively inhibits mTOR function by binding to FKBP12 and then to mTOR. Rapamycin has been shown to be effective in rodent TSC models (Kenerson et al 2004, Lee et al 2004), and single and multiple institutional trials of the drug in TSC patients are underway. The requirement of rheb for farnesylation for its activity in stimulating mTOR makes farnesyl transferase inhibitors another promising therapeutic, for which trials in rodent models are underway. In addition, interferons may have benefit, as suggested by studies in rodent models (Lee et al 2004), and compounds that activate AMPK may also have unique benefit in causing apoptosis of TSC tumor cells.

Since this chapter was written, there have been many important additional studies on TSC and the TSC1/TSC2 genes and proteins. There is a better understanding of the function of these proteins in the regulation of the mTORC1 complex (see Huang and Manning 2008 for a recent review). The first clinical trials of rapamycin in TSC patients have been reported (Bissler et al 2008). Rapamycin caused a significant reduction in the size of kidney AMLs in the majority of patients on one trial. However, there were side effects, and lung function changes in the patients with LAM were modest. Nonetheless, there are multiple additional trials in progress assessing the potential benefits of rapamycin and related drugs in TSC/LAM.

ACKNOWLEDGMENTS

The author wishes to apologize for not citing all relevant references due to space limitations; to thank his colleagues, fellows, students, technicians, and TSC family members who have stimulated his thinking on this subject for many years; and to acknowledge the support of the NIH, the Rothberg Courage Fund, the LAM Foundation, and Novartis.

References

Alliance TS. A Review of Tuberous Sclerosis Complex with Differential Diagnosis; 2004.

Asano E, Chugani DC, Muzik O, et al. Autism in tuberous sclerosis complex is related to both cortical and subcortical dysfunction. Neurology 2001; 57: 1269–77.

Au KS, Hebert AA, Roach ES, Northrup H. Complete inactivation of the TSC2 gene leads to formation of hamartomas. Am. J. Hum. Genet. 1999; 65: 1790–5.

Baybis M, Yu J, Lee A, et al. mTOR cascade activation distinguishes tubers from focal cortical dysplasia. Ann. Neurol. 2004; 56: 478–87.

Birchenall-Roberts MC, Fu T, Bang OS, et al. Tuberous sclerosis complex 2 gene product interacts with human SMAD proteins. A molecular link of two tumor suppressor pathways. J. Biol. Chem. 2004; 279: 25605–13.

Bissler JJ, Kingswood JC. Renal angiomyolipomata. Kidney Int. 2004; 66: 924–34.

Bissler JJ, McCormack FX, Young LR, Elwing JM, Chuck G, Leonard JM, Schmithorst VJ, Laor T, Brody AS, Bean J, Salisbury S, Franz DN. Sirolimus for angiomyolipoma in tuberous sclerosis complex or lymphangioleiomyomatosis. N. Engl. J. Med. 2008; Jan 10; 358(2): 140–51.

Brugarolas J, Lei K, Hurley RL, et al. Regulation of mTOR function in response to hypoxia by REDD1 and the TSC1/TSC2 tumor suppressor complex. Genes Dev. 2004; 18: 2893–904.

Carbonara C, Longa L, Grosso E, et al. Apparent preferential loss of heterozygosity at TSC2 over TSC1 chromosomal region in tuberous sclerosis hamartomas. Genes Chromosomes Cancer 1996; 15: 18–25.

Carsillo T, Astrinidis A, Henske EP. Mutations in the tuberous sclerosis complex gene TSC2 are a cause of sporadic pulmonary lymphangioleiomyomatosis. Proc. Natl Acad. Sci. USA 2000; 97: 6085–90.

Catania MG, Mischel PS, Vinters HV. Hamartin and tuberin interaction with the G2/M cyclin-dependent kinase CDK1 and its regulatory cyclins A and B. J. Neuropathol. Exp. Neurol. 2001; 60: 711–23.

Chan JA, Zhang H, Roberts PS, et al. Pathogenesis of tuberous sclerosis subependymal giant cell astrocytomas: Biallelic inactivation of TSC1 or TSC2 leads to mTOR activation. J. Neuropath. Exp. Neurol. 2004.

Cheadle JP, Reeve MP, Sampson JR, Kwiatkowski DJ. Molecular genetic advances in tuberous sclerosis. Hum. Genet. 2000; 107: 97–114.

Crino PB. Molecular pathogenesis of tuber formation in tuberous sclerosis complex. J. Child. Neurol. 2004; 19: 716–25.

Crooks DM, Pacheco-Rodriguez G, Decastro RM, et al. Molecular and genetic analysis of disseminated neoplastic cells in lymphangioleiomyomatosis. Proc. Natl Acad. Sci. USA 2004; 101: 17462–7.

Curatolo P. Tuberous Sclerosis Complex: from basic science to clinical phenotypes. Cambridge: Mac Keith Press, 2003.

Curatolo P, Verdecchia M, Bombardieri R. Vigabatrin for tuberous sclerosis complex. Brain Dev. 2001; 23: 649–53.

Dabora SL, Jozwiak S, Franz DN, et al. Mutational analysis in a cohort of 224 tuberous sclerosis patients indicates increased severity of TSC2, compared with TSC1, disease in multiple organs. Am. J. Hum. Genet. 2001; 68: 64–80.

Dabora SL, Nieto AA, Franz D, Jozwiak S, Van Den Ouweland A, Kwiatkowski DJ. Characterisation of six large deletions in TSC2 identified using long range PCR suggests diverse mechanisms including Alu mediated recombination. J. Med. Genet. 2000; 37: 877–83.

Eker R, Mossige J, Johannessen JV, Aars H. Hereditary renal adenomas and adenocarcinomas in rats. Diagn. Histopath. 1981; 4: 99–110.

El-Hashemite N, Walker V, Zhang H, Kwiatkowski DJ. Loss of Tsc1 or Tsc2 induces vascular endothelial growth factor production through mammalian target of rapamycin. Cancer Res. 2003; 63: 5173–7.

El-Hashemite N, Zhang H, Walker V, Hoffmeister KM, Kwiatkowski DJ. Perturbed IFN-gamma-Jak-signal transducers and activators of transcription signaling in tuberous sclerosis mouse models: synergistic effects of rapamycin-IFN-gamma treatment. Cancer Res. 2004; 64: 3436–43.

Ess KC, Uhlmann EJ, Li W, Li H, Declue JE, Crino PB, Gutmann DH. Expression profiling in tuberous sclerosis complex (TSC) knockout mouse astrocytes to characterize human TSC brain pathology. Glia 2004; 46: 28–40.

The European Chromosome 16 Tuberous Sclerosis Consortium. Identification and characterization of the tuberous sclerosis gene on chromosome 16. Cell 1993; 75: 1305–15.

Ewalt DH, Sheffield E, Sparagana SP, Delgado MR, Roach ES. Renal lesion growth in children with tuberous sclerosis complex. J. Urol. 1998; 160: 141–5.

Fingar DC, Blenis J. Target of rapamycin (TOR): an integrator of nutrient and growth factor signals and coordinator of cell growth and cell cycle progression. Oncogene 2004; 23: 3151–71.

Finlay GA, York B, Karas RH, Fanburg BL, Zhang H, Kwiatkowski DJ, Noonan DJ. Estrogen-induced smooth muscle cell growth is regulated by tuberin and associated with altered activation of platelet-derived growth factor receptor-beta and ERK-1/2. J. Biol. Chem. 2004; 279: 23114–22.

Franz DN, Tudor C, Leonard J, et al. Lamotrigine therapy of epilepsy in tuberous sclerosis. Epilepsia 2001; 42: 935–940.

Gao X, Pan D. TSC1 and TSC2 tumor suppressors antagonize insulin signaling in cell growth. Genes Dev. 2001; 15: 1383–92.

Gomez M, Sampson J, Whittemore V (eds). The Tuberous Sclerosis Complex. Oxford University Press, Oxford, UK; 1999.

Goncharova E, Goncharov D, Noonan D, Krymskaya VP. TSC2 modulates actin cytoskeleton and focal adhesion through TSC1-binding domain and the Rac1 GTPase. J. Cell Biol. 2004; 167: 1171–82.

Green AJ, Smith M, Yates JR. Loss of heterozygosity on chromosome 16p13.3 in hamartomas from tuberous sclerosis patients. Nat. Genet. 1994; 6: 193–6.

Han S, Santos TM, Puga A, et al. Phosphorylation of tuberin as a novel mechanism for somatic inactivation of the tuberous sclerosis complex proteins in brain lesions. Cancer Res. 2004; 64: 812–16.

Harrington LS, Findlay GM, Gray A, et al. The TSC1-2 tumor suppressor controls insulin-PI3K signaling via regulation of IRS proteins. J. Cell Biol. 2004; 166: 213–23.

Henske EP, Neumann HP, Scheithauer BW, Herbst EW, Short MP, Kwiatkowski DJ. Loss of heterozygosity in the tuberous sclerosis (TSC2) region of chromosome band 16p13 occurs in sporadic as well as TSC-associated renal angiomyolipomas. Genes Chromosomes Cancer 1995; 13: 295–8.

Henske EP, Scheithauer BW, Short MP, et al. Allelic loss is frequent in tuberous sclerosis kidney lesions but rare in brain lesions. Am. J. Hum. Genet. 1996; 59: 400–6.

Henske EP, Wessner LL, Golden J, et al. Loss of tuberin in both subependymal giant cell astrocytomas and angiomyolipomas supports a two-hit model for the pathogenesis of tuberous sclerosis tumors. Am. J. Path. 1997; 151: 1639–47.

Hornigold N, Devlin J, Davies AM, Aveyard JS, Habuchi T, Knowles MA. Mutation of the 9q34 gene TSC1 in sporadic bladder cancer. Oncogene 1999; 18: 2657–61.

Huang J, Manning BD. The TSC1-TSC2 complex: a molecular switchboard controlling cell growth. Biochem. J. 2008; Jun 1; 412(2): 179–90.

Inoki K, Zhu T, Guan KL. TSC2 mediates cellular energy response to control cell growth and survival. Cell 2003; 115: 577–90.

Johnson MW, Emelin JK, Park SH, Vinters HV. Co-localization of TSC1 and TSC2 gene products in tubers of patients with tuberous sclerosis. Brain Pathol. 1999; 9: 45–54.

Jones AC, Shyamsundar MM, Thomas MW, et al. Comprehensive mutation analysis of TSC1 and TSC2 and phenotypic correlations in 150 families with tuberous sclerosis. Am. J. Hum. Genet. 1999; 64: 1305–15.

Jozwiak S, Schwartz RA, Janniger CK, Bielicka-Cymerman J. Usefulness of diagnostic criteria of tuberous sclerosis complex in pediatric patients. J. Child. Neurol. 2000; 15: 652–9.

Kapahi P, Zid BM, Harper T, Koslover D, Sapin V, Benzer S. Regulation of lifespan in Drosophila by modulation of genes in the TOR signaling pathway. Curr. Biol. 2004; 14: 885–90.

Karbowniczek M, Cash T, Cheung M, Robertson GP, Astrinidis A, Henske EP. Regulation of B-Raf kinase activity by tuberin and Rheb is mammalian target of rapamycin (mTOR)-independent. J. Biol. Chem. 2004; 279: 29930–7.

Kenerson H, Dundon TA, Yeung RS. Effects of rapamycin in the eker rat model of tuberous sclerosis complex. Pediatr. Res. 2004.

Khare L, Strizheva GD, Bailey JN, et al. A novel missense mutation in the GTPase activating protein homology region of TSC2 in two large families with tuberous sclerosis complex. J. Med. Genet. 2001; 38: 347–9.

Kleymenova E, Ibraghimov-Beskrovnaya O, Kugoh H, et al. Tuberin-dependent membrane localization of polycystin-1: a functional link between polycystic kidney disease and the TSC2 tumor suppressor gene. Mol. Cell 2001; 7: 823–32.

Kobayashi T, Hirayama Y, Kobayashi E, Kubo Y, Hino O. A germline insertion in the tuberous sclerosis (Tsc2) gene gives rise to the Eker rat model of dominantly inherited cancer. Nat. Genet. 1995; 9: 70–4.

Kobayashi T, Minowa O, Kuno J, Mitani H, Hino O, Noda T. Renal carcinogenesis, hepatic hemangiomatosis, and embryonic lethality caused by a germ-line Tsc2 mutation in mice. Cancer Res. 1999; 59: 1206–11.

Kobayashi T, Minowa O, Sugitani Y, et al. A germ-line Tsc1 mutation causes tumor development and embryonic lethality that are similar, but not identical to, those caused by Tsc2 mutation in mice. Proc. Natl Acad. Sci. USA 2001; 98: 8762–7.

Kwiatkowski DJ. Rhebbing up mTOR: new insights on TSC1 and TSC2 and the pathogenesis of tuberous sclerosis. Cancer Biol. Ther. 2003; 2: 471–6.

Kwiatkowski DJ, Zhang H, Bandura JL, et al. A mouse model of TSC1 reveals sex-dependent lethality from liver hemangiomas, and up-regulation of p70S6 kinase activity in Tsc1 null cells. Hum. Mol. Genet. 2002; 11: 525–34.

Lamb RF, Roy C, Diefenbach TJ, et al. The TSC1 tumour suppressor hamartin regulates cell adhesion through ERM proteins and the GTPase Rho. Nat. Cell Biol. 2000; 2: 281–7.

Lee L, Sudentas P, Donohue B, et al. Efficacy of a rapamycin analog (CCI-779) and IFN-gamma in tuberous sclerosis mouse models. Genes Chromosomes Cancer 2004.

Li Y, Corradetti MN, Inoki K, Guan KL. TSC2: filling the GAP in the mTOR signaling pathway. Trends Biochem. Sci. 2004; 29: 32–8.

Mak BC, Yeung RS. The tuberous sclerosis complex genes in tumor development. Cancer Invest. 2004; 22: 588–603.

Matsumoto Y, Horiba K, Usuki J, Chu SC, Ferrans VJ, Moss J. Markers of cell proliferation and expression of melanosomal antigen in lymphangioleiomyomatosis. Am. J. Respir. Cell. Mol. Biol. 1999; 21: 327–36.

Mayer K, Ballhausen W, Leistner W, Rott H. Three novel types of splicing aberrations in the tuberous sclerosis TSC2 gene caused by mutations apart from splice consensus sequences. Biochim. Biophys. Acta 2002; 1502: 495–507.

Miyata H, Chiang AC, Vinters HV. Insulin signaling pathways in cortical dysplasia and TSC-tubers: tissue microarray analysis. Ann. Neurol. 2004; 56: 510–19.

Mizuguchi M, Ikeda K, Takashima S. Simultaneous loss of hamartin and tuberin from the cerebrum, kidney and heart with tuberous sclerosis. Acta Neuropathol. (Berl) 2000; 99: 503–10.

Murthy V, Han S, Beauchamp RL, et al. Pam and its ortholog highwire interact with and may negatively regulate the TSC1. TSC2 complex. J. Biol. Chem. 2004; 279: 1351–8.

Niida Y, Stemmer-Rachamimov AO, Logrip M, et al. Survey of somatic mutations in tuberous sclerosis complex (TSC) hamartomas suggests different genetic mechanisms for pathogenesis of TSC lesions. Am. J. Hum. Genet. 2001; 69: 493–503.

O'Callaghan FJ, Noakes MJ, Martyn CN, Osborne JP. An epidemiological study of renal pathology in tuberous sclerosis complex. BJU Int. 2004; 94: 853–7.

O'Connor SE, Kwiatkowski DJ, Roberts PS, Wollmann RL, Huttenlocher PR. A family with seizures and minor features of tuberous sclerosis and a novel TSC2 mutation. Neurology 2003; 61: 409–12.

Onda H, Lueck A, Marks PW, Warren HB, Kwiatkowski DJ. Tsc2(+/−) mice develop tumors in multiple sites that express gelsolin and are influenced by genetic background. J. Clin. Invest. 1999; 104: 687–95.

Pan D, Dong J, Zhang Y, Gao X. Tuberous sclerosis complex: from *Drosophila* to human disease. Trends Cell. Biol. 2004; 14: 78–85.

Plank TL, Yeung RS, Henske EP. Hamartin, the product of the tuberous sclerosis 1 (TSC1) gene, interacts with tuberin and appears to be localized to cytoplasmic vesicles. Cancer Res. 1998; 58: 4766–70.

Potter CJ, Huang H, Xu T. *Drosophila* Tsc1 functions with Tsc2 to antagonize insulin signaling in regulating cell growth, cell proliferation, and organ size. Cell 2001; 105: 357–68.

Reiling JH, Hafen E. The hypoxia-induced paralogs Scylla and Charybdis inhibit growth by down-regulating S6K activity upstream of TSC in *Drosophila*. Genes Dev. 2004; 18: 2879–92.

Rennebeck G, Kleymenova EV, Anderson R, Yeung RS, Artzt K, Walker CL. Loss of function of the tuberous sclerosis 2 tumor suppressor gene results in embryonic lethality characterized by disrupted neuroepithelial growth and development. Proc. Natl Acad. Sci. USA 1998; 95: 15629–34.

Roach ES, DiMario FJ, Kandt RS, Northrup H. Tuberous Sclerosis Consensus Conference: recommendations for diagnostic evaluation. National Tuberous Sclerosis Association. J. Child. Neurol. 1999; 14: 401–7.

Roach ES, Gomez MR, Northrup H. Tuberous sclerosis complex consensus conference: revised clinical diagnostic criteria. J. Child. Neurol. 1998; 13: 624–8.

Roberts PS, Chung J, Jozwiak S, et al. SNP identification, haplotype analysis, and parental origin of mutations in TSC2. Hum. Genet. 2002; 111: 96–101.

Roberts PS, Dabora S, Thiele EA, Franz DN, Jozwiak S, Kwiatkowski DJ. Somatic mosaicism is rare in unaffected parents of patients with sporadic tuberous sclerosis. J. Med. Genet. 2004; 41: e69.

Rose VM, Au KS, Pollom G, Roach ES, Prashner HR, Northrup H. Germ-line mosaicism in tuberous sclerosis: How common?. Am. J. Hum. Genet. 1999; 64: 986–92.

Roux PP, Ballif BA, Anjum R, Gygi SP, Blenis J. Tumor-promoting phorbol esters and activated Ras inactivate the tuberous sclerosis tumor suppressor complex via p90 ribosomal S6 kinase. Proc. Natl Acad. Sci. USA 2004; 101: 13489–94.

Sampson JR, Maheshwar MM, Aspinwall R, et al. Renal cystic disease in tuberous sclerosis: role of the polycystic kidney disease 1 gene. Am. J. Hum. Genet. 1997; 61: 843–51.

Sepp T, Yates JRW, Green AJ. Loss of heterozygosity in tuberous sclerosis hamartomas. J. Med. Genet. 1996; 33: 962–4.

Shah OJ, Wang Z, Hunter T. Inappropriate activation of the TSC/Rheb/mTOR/S6K cassette induces IRS1/2 depletion, insulin resistance, and cell survival deficiencies. Curr. Biol. 2004; 14: 1650–6.

Shaw RJ, Bardeesy N, Manning BD, et al. The LKB1 tumor suppressor negatively regulates mTOR signaling. Cancer Cell 2004; 6: 91–9.

Takamochi K, Ogura T, Suzuki K, et al. Loss of heterozygosity on chromosomes 9q and 16p in atypical adenomatous hyperplasia concomitant with adenocarcinoma of the lung. Am. J. Pathol. 2001; 159: 1941–8.

Tapon N, Ito N, Dickson BJ, Treisman JE, Hariharan IK. The drosophila tuberous sclerosis complex gene homologs restrict cell growth and cell proliferation. Cell 2001; 105: 345–55.

Tucker T, Friedman JM. Pathogenesis of hereditary tumors: beyond the "two hit" hypothesis. Clin. Genet. 2002; 62: 345–57.

van Slegtenhorst M, de Hoogt R, Hermans C, et al. Identification of the tuberous sclerosis gene TSC1 on chromosome 9q34. Science 1997; 277: 805–8.

van Slegtenhorst M, Nellist M, Nagelkerken B, et al. Interaction between hamartin and tuberin, the TSC1 and TSC2 gene products. Hum. Mol. Genet. 1998; 7: 1053–8.

White R, Hua Y, Scheithauer B, Lynch DR, Henske EP, Crino PB. Selective alterations in glutamate and GABA receptor subunit mRNA expression in dysplastic neurons and giant cells of cortical tubers. Ann. Neurol. 2001; 49: 67–78.

Yeung RS, Gu H, Lee M, Dundon TA. Genetic identification of a locus, Mot1, that affects renal tumor size in the rat. Genomics 2001; 78: 108–12.

Yeung RS, Katsetos CD, Klein-Szanto A. Subependymal astrocytic hamartomas in the Eker rat model of tuberous sclerosis. Am. J. Pathol. 1997; 151: 1477–86.

Yeung RS, Xiao GH, Jin F, Lee WC, Testa JR, Knudson AG. Predisposition to renal carcinoma in the Eker rat is determined by germ-line mutation of the tuberous sclerosis 2 (TSC2) gene. Proc. Natl Acad. Sci. USA 1994; 91: 11413–16.

Yoon HS, Ramachandiran S, Chacko MA, Monks TJ, Lau SS. Tuberous sclerosis-2 tumor suppressor modulates ERK and B-Raf activity in transformed renal epithelial cells. Am. J. Physiol. Renal Physiol. 2004; 286: F417–24.

Zhang H, Cicchetti G, Onda H, et al. Loss of Tsc1/Tsc2 activates mTOR and disrupts PI3K-Akt signaling through downregulation of PDGFR. J. Clin. Invest. 2003; 112: 1223–33.

PART V

Systemic Diseases with Renal Involvement: Monogenic Disorders

CHAPTER 31

Nail-Patella Syndrome

ROY MORELLO, DARYL SCOTT AND BRENDAN LEE

CLINICAL FEATURES AND NATURAL HISTORY

Nail-patella syndrome (NPS) is a dominantly inherited malformation syndrome that affects multiple organs including the skeleton, nails, eyes, and kidneys (see Table 31.1). The first report of an individual with a combination of congenital abnormalities affecting the elbows, knees, and nails was made in 1820 and familial associations of these abnormalities were documented in the latter part of the 1800s (Lee & Morello 2004). Over the years the association of skeletal and nail abnormalities typical for NPS has been given several names including, Österreicher syndrome, Turner syndrome, Turner-Kieser syndrome, arthrodysplasia, Touraine syndrome, hereditary osteo-onychodysplasia, and Fong disease (Lee & Morello 2004). Renal involvement in NPS was described in 1950 (Brixey & Burke 1950, Hawkings & Smith 1950) and has since been recognized as an important cause of long-term morbidity and mortality among individuals with NPS.

NPS affects approximately 1 in every 50 000 newborns and is a highly penetrant, autosomal dominant disorder with the majority of individuals who inherit the affected gene having symptoms and signs of the disease (Bongers et al 2002, Lee & Morello 2004). However, the degree to which each organ system is affected can vary significantly between individuals, even within the same family. This variable expressivity makes presymptomatic counseling especially difficult.

Skeletal Dysplasia

The skeletal dysplasia of NPS includes abnormalities of the knees, elbows, pelvis, and feet (Guidera et al 1991). Knee and elbow dysplasia are the most common skeletal manifestation of the disease and were the first to be described

in the medical literature. Iliac horns – pyramidal processes projecting dorsolaterally from the ilia – are pathognomonic for NPS but are present in only about 70–80% of patients (Fong 1946, Hawkings & Smith 1950, Sartoris & Resnick 1987, Sweeney et al 2003). Unlike the knee and elbow dysplasia of NPS that can be associated with significant morbidity, iliac horns have not been shown to affect function or cause gait abnormalities (Bongers et al 2002). Foot abnormalities, including equinovarus and equinovalgus deformities, are found less frequently in patients with NPS but can be the symptom that first brings a patient to medical attention (Beals & Eckhardt 1969, Guidera et al 1991).

Nail Dysplasia

Nail dysplasia is often present at birth and may present as nail hypoplasia, anonychia, longitudinal ridging, splitting, and spooning of the nail (Dunston et al 2005, Stratigos & Baden 2001). Triangular-appearing lunulae – the white areas at the base of the nail – are characteristic of NPS nail dysplasia (Mellotte & Eastwood 1995). Often the severity of the nail dysplasia increases as one progresses from the 5th to the 1st finger of the upper extremities. The nails of the lower extremity are less likely to be affected. Although nail dysplasia may present a cosmetic problem, it does not cause significant long-term morbidity.

Ocular Findings

Primary open-angle glaucoma (POAG) has been shown to cosegregate with NPS in some families and is now considered a variable features of NPS (Lichter et al 1997). One study of 123 NPS patients revealed ocular hypertension in 7% and glaucoma in 10% (Sweeney et al 2003). A zone of darker pigmentation around the central part of the iris, known as Lester's sign, is present in 42–54% of patients with NPS but

Genetic Diseases of the Kidney
ISBN: 978-0-12-449851-8

can also be seen in the general population at a much lower frequency (Carbonara & Alpert 1964, Sweeney et al 2003).

Renal Disease

The renal disease that accompanies NPS has the greatest impact on long-term morbidity and mortality (Mihatsch & Zollinger 1980). Renal symptoms can start at any age and most often begin as microalbuminuria with or without hematuria. Proteinuria can be intermittent and may present during or be exacerbated by pregnancy (Sweeney et al 2003). In some cases, the proteinuria spontaneously

TABLE 31.1   Physical and clinical findings in nail-patella syndrome

Organ system	Manifestations of NPS	Comments
Skeletal	Knee dysplasia Patellar aplasia/hypoplasia (93%) Hypoplasia of the lateral femoral condyle	May cause recurrent subluxation or dislocation of the patellae, knee disability, knee pain, patello-femoral arthrosis
	Elbow dysplasia (93%) Hypoplasia of radial heads Hypoplasia of the capitellum	May cause recurrent posterolateral subluxation/dislocation of the radial heads (61%)
	Pelvic dysplasia Iliac horns (70–80%)	Iliac horns are pathognomonic for NPS
	Foot dysplasia Equinovarus Equinovalgus	Less frequent that other skeletal manifestations of NPS
Nails	Nail dysplasia (93%) Triangular nail lunulae Anonychia, hemianonychia Nail hypoplasia Longitudinal ridging Splitting Spoon-shaped nails Flaky nails	Cosmetic problem but not a source of long-term morbidity
Eye	Primary open-angle glaucoma (10%) and ocular hypertension (7%) Lester's sign of the iris (42–54%)	Lester's sign is not pathognomonic for NPS but is seen more frequently in individuals with NPS than in the general population
Kidney	Proteinuria (12–55%) with or without hematuria Renal failure (5–14%)	Renal disease is the main cause of long-term morbidity and mortality in NPS

Data from Bongers et al (2002), Sweeney et al (2003)

resolves or remains asymptomatic while in other cases there may be gradual or rapid progression to nephrotic syndrome, nephritis, or frank renal failure.

Estimates of the percent of NPS patients with signs of renal involvement range from 12% to 55% and estimates of the percent that progress to renal failure range from 5% to 14% (Knoers et al 2000, Sweeney et al 2003, Bongers et al 2005). It is important, however, to recognize that the cross-sectional nature of these studies and variations in clinical definition and methodology make it difficult to arrive at accurate estimates of life-time risk based on available data. It is clear, however, that the percentage of individuals with NPS who have signs or symptoms of renal involvement increases with age and there is considerable variation in the extent of renal involvement between and within NPS families.

Light microscopic evaluation of renal biopsies from patients with NPS often demonstrates focal thickening of the glomerular basement membrane, which is a consistent but nonspecific glomerular feature (Hoyer et al 1972). Other findings, such as focal and segmental sclerosis, proliferative glomerulonephritis with crescent formation, and hyalinization of the glomeruli are nonspecific and most likely are related to the degree of renal failure present at the time of biopsy (Hoyer et al 1972, Bennett et al 1973, Taguchi et al 1988).

Although light-microscopic and clinical signs of renal involvement may be absent, it is likely that all NPS patients have alterations at the ultrastructural level (Gubler et al 1980, Taguchi et al 1988, Drut et al 1992, Gubler & Levy 1993). Such changes have been well documented in both fetuses and adults and include fusion of podocyte foot processes, diffuse and irregular thickening and splitting of the glomerular basement membrane (GBM), and lucencies within the GBM that give it a 'moth-eaten' appearance by electron microscopy (Del Pozo & Lapp 1970). These electron-lucent areas contain a dark fibrillar material that have the periodicity of interstitial collagen (Del Pozo & Lapp 1970, Ben-Bassat et al 1971). This material can be labeled by antibodies to type III collagen but not by antibodies directed at other fibrillar collagens (Heidet et al 2003). The presence of type III collagen within the GBM was also demonstrated in asymptomatic patients whose glomeruli were normal by light microscopy.

At the present time, there is no means of accurately predicting whether a particular individual will develop clinical nephropathy or progression to end-stage renal disease, although some new data suggest there might be emerging molecular correlates (Bongers et al 2005). Death from renal failure has been documented in patients as young as 8 years old, but typically occurs after the fourth decade (Leahy 1966). Reports of renal transplantation in individuals with NPS have been favorable (Uranga et al 1973, Verdich 1980, Chan et al 1988, Bodziak et al 1994).

ESTABLISHING THE DIAGNOSIS

The diagnosis of NPS is usually made on a clinical basis. The diagnosis is straightforward when the classic tetrad of abnormal nails, elbows, knees, and iliac horns is found on physical exam. Nail dysplasia often present at birth can be easily evaluated. A skeletal survey can be used to identify skeletal abnormalities of the elbows and knees and other osseous structures that may not be evident on physical exam (Table 31.2; Figure 31.1). It should be noted, however, that the time at which the patellae ossify in childhood (starting approximately 42 and 32 months in males and females, respectively) varies between individuals and radiographic studies may miss patellar hypoplasia early in childhood (Walker et al 1998). In these cases, direct physical examination in conjunction with ultrasonography may prove helpful in documenting patellar abnormalities (Miller et al 1998). Magnetic resonance imaging can also be used to detect patellar hypoplasia. Iliac horns can sometimes be felt on physical exam, and they can be easily identified in infants and young children on standard radiographs.

Since NPS is an autosomal dominant disorder that is highly penetrant, a carefully obtained family history may help to identify other affected individuals within a family. When possible, parents and siblings should be carefully examined to determine if they have findings suggestive of NPS. It is important, however, to remember that individuals who do not have a positive family history may have NPS due to a de novo mutation.

When NPS is suspected, a formal ophthalmologic evaluation should be requested. This evaluation may reveal ocular abnormalities that can help to confirm the clinical suspicion of NPS. A formal ophthalmologic evaluation is also clinically important in identifying individuals with NPS who have ocular hypertension and glaucoma. A urine

TABLE 31.2 Radiographic findings in nail-patella syndrome

Common findings

Iliac horns
Hypoplasia/aplasia of the patella
Elongated radius
Hypoplasia of the radial head
Hypoplasia of the capitellum
Flaring of the iliac wing
Small iliac angle
Prominence of the medial femoral condyle

Less common findings

Joint asymmetry
Clinodactyly
Camptodactyly
Synphalangism
Calcaneovalgus deformity
Subluxation of the tarsal-metatarsal joint
Small acromion
Hypoplastic scapula

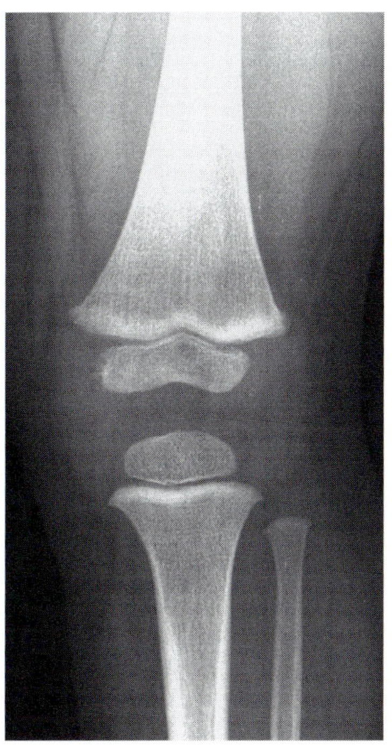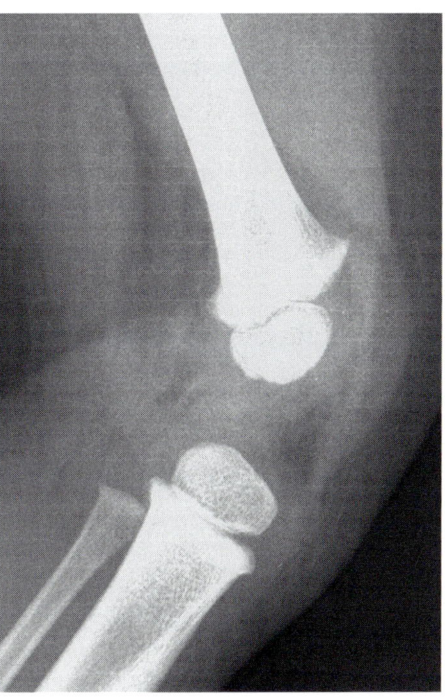

FIGURE 31.1 Radiographic features of NPS. Left-hand panel shows antero-posterior (A/P) radiographs of the pelvis and dorsal iliac horns (unmarked on the left and marked on the right). Right-hand panel shows A/P and lateral radiographs of the knee showing absence of the patella.

analysis and/or a 24-hour urine collection for protein can be used to detect the presence of proteinuria.

In individuals in whom the diagnosis remains in question, sequencing of the *LMX1B* gene may be a reasonable means of confirming the diagnosis. *LMX1B* DNA testing is available on both a clinical and a research basis (www.genestests. org). Direct sequencing of exons 2 through 6 of the *LMX1B* gene reveals a causative mutation in about 85% of individuals with NPS. Prenatal DNA testing is also available on a clinical basis. In cases in which an *LMX1B* mutation is not identified, prenatal diagnosis can sometimes be made by using linkage analysis or through the use of prenatal ultrasonography to detect skeletal abnormalities including iliac horns, absent patellae, and lower limb anomalies (Feingold et al 1998, McIntosh et al 1999, Pinette et al 1999).

There are a number of disorders that have physical exam and clinical findings that are similar to those seen in patients with NPS. These disorders should be included in the differential diagnosis and are summarized in Table 31.3.

MANAGEMENT, TREATMENT, AND COUNSELING

At diagnosis, it is important to provide patients with a full nephrologic, orthopedic, ophthalmologic evaluation. Although nephropathy usually does not become clinically significant until later in life, the onset of renal disease is variable and some individuals have been described with renal failure in childhood. Prospective monitoring should be instituted to identify microalbuminuria and proteinuria. In childhood, routine urinalysis is usually sufficient for this purpose. For follow-up of clinical nephropathy, a 24-hour urine collection to calculate proteinuria and creatinine clearance is indicated along with a full nephrology evaluation. The nephropathy seen in NPS is not immunologically related and is usually steroid-resistant. Kidney transplantation is curative due to the cell-autonomous nature of the underlying renal defect, and it should be considered for individuals with end-stage renal disease (Uranga et al 1973, Verdich 1980, Chan et al 1988, Bodziak et al 1994).

NPS is associated with a variety of orthopedic problems including chronic problems with elbows and knees and foot abnormalities (Hogh & Macnicol 1985, Fiedler et al 1987, Guidera et al 1991). In general, a conservative approach should be taken with surgical intervention being considered only when functional limitations outweigh risks for a particular individual. Prospective monitoring for the development of ocular hypertension and glaucoma is indicated in all patients with NPS. Few studies have addressed the efficacy of different treatment options for NPS patients with glaucoma; hence, a traditional ophthalmologic approach should be taken.

NPS is an autosomal dominant disease. It is important that families be counseled on the inheritance pattern and recurrence risk associated with NPS. Parents of newly diagnosed patients with NPS should be examined for subtle signs of the disease since this would affect recurrence risk and subtly affected parents may be at risk for the ophthalmologic and nephrologic complications of NPS. Although gonadal mosaicism has not been proven in association with NPS, this remains a formal possibility. Counseling should stress the lack of adequate genotype–phenotype correlations or clinical indices to predict the severity of orthopedic complications and the development of either glaucoma or nephropathy.

LMX1B: GENE STRUCTURE, MUTATIONS IN NPS AND THEIR PREDICTED EFFECT

Mutations in the *LMX1B* gene cause NPS. NPS was one of the first Mendelian conditions mapped by linkage using ABO blood groups (Renwick & Lawler 1955). Subsequently, fine mapping placed it to chromosome 9q34.1 where LMX1b was also mapped (Campeau et al 1995, Iannotti et al 1997, McIntosh et al 1997). *LMX1B* extends over 81 kilobases (kb) of which more than 70 kb constitutes the large second intron. It contains eight exons which encode a messenger RNA of about 7 kb. The open reading frame is 1119 base pairs in length and encodes a protein of 372 amino acids. However an alternative splice site has been described at the 3′end of exon 7, which yields an alternative protein product that is seven amino acids longer and whose function has not yet been explored (Seri et al 1999). More recently, Dunston et al have described an upstream, in-frame start codon that would generate a polypeptide with 23 additional amino acids; this stretch of nucleotides appears to be conserved in the genomic sequences of other species (mouse, rat, chicken, *Xenopus*) (Dunston et al 2004). Upstream of this start codon, and within the 5′ UTR, are two additional very short open reading frames of unknown function (Dunston et al 2004). While the *LMX1B* transcription start site has not been associated with a consensus TATA box, several putative binding sites for different transcription factors, including *SP1*, *OCT1*, *TCF/LEF1*, have been identified in the promoter region (Dunston et al 2004). The identification of an antisense transcript near the 5′ UTR of *LMX1B* is of particular interest. This transcript was first identified in the same location in the chick and then in the mouse *Lmx1b* locus and is known as *ALC* (Holmes et al 2003). Although both the avian and the murine transcripts are expressed in a similar pattern as *Lmx1b*, they do not share sequence similarity. However, the predicted human transcript does not contain an open reading frame (Figures 31.2–31.4).

TABLE 31.3   Syndromes with features similar to those in NPS

Syndrome	Similarities to NPS	Distinguishing features and other findings
Patella aplasia–hypoplasia syndrome (OMIM 168860)	Patellar aplasia or hypoplasia Dominant inheritance	Absence of nail, renal, and ocular finding and other skeletal changes
Ear patella short stature syndrome (Meier-Gorlin syndrome; OMIM 224690)	Patellar hypoplasia Elbow dislocation	Recessive inheritance Failure to thrive/short stature Microtia Dysmorphic features
Small patella syndrome (ischiopatellar dysplasia; OMIM 147891)	Patellar hypoplasia Dominant inheritance	Anomalies of the upper femur Ischial hypoplasia Delayed ossification of the ischiopubic junction
Familial posterior dislocation of the radial heads (OMIM 179200)	Radial head anomalies Dominant inheritance	Absence of nail, renal, and ocular finding and other skeletal changes
Genitopatellar syndrome (OMIM 606170)	Patellar aplasia Flexion deformities of the knees and hips Club foot	Genital anomalies Dysmorphic features Mental retardation Microcephaly Structural kidney anomalies such as multicystic kidneys and hydronephrosis Hypoplasia of the ischia and iliac bones
Deafness and onychodystrophy (DOOR syndrome, OMIM 220500; dominant form OMIM 124480)	Nail aplasia or malformation Some cases have dominant inheritance	Digital anomalies including triphalangeal thumbs Mental retardation Seizures Sensorineural deafness Some cases have recessive inheritance pattern
Trisomy 8 mosaicism	Patellar aplasia or hypoplasia Limited elbow supination Abnormal nails	Mental retardation/learning defects Dysmorphic features Camptodactyly Joint restriction usually affecting fingers and toes
Fifth digit syndrome (Coffin-Siris syndrome, OMIM 135900)	Patellar aplasia or hypoplasia Radial head dislocation Nail aplasia or hypoplasia especially affecting 5th fingers and toes Elbow dislocation	Dysmorphic features Mental retardation Short stature Recessive inheritance
RAPADILINO syndrome (OMIM 266280)	Patellar aplasia or hypoplasia Radial defects Joint dislocation	Cleft palate Dysmorphic features Short stature Absent or hypoplastic thumbs and radii Autosomal recessive inheritance
Brachymorphism-onychodysplasia-dysphalangism syndrome (OMIM 113477)	Nail aplasia or hypoplasia	Hypoplasia of the 5th digit Dysmorphic features Mild intellectual impairment Short stature

Data from Sweeney & McIntosh (2005).

The LMX1B protein contains, from the N-terminus to the C-terminus, four distinct functional domains: two LIM domains, LIM-A and LIM-B, that are important for protein–protein interactions; a homeodomain important for DNA binding; and a glutamine-rich and serine-rich domain with putative transcriptional activation functions. The vast majority of *LMX1B* mutations causing NPS are clustered in the LIM domains or in the homeodomain, and almost always affect conserved residues. In contrast, no NPS causing mutations have been found in the C-terminal putative

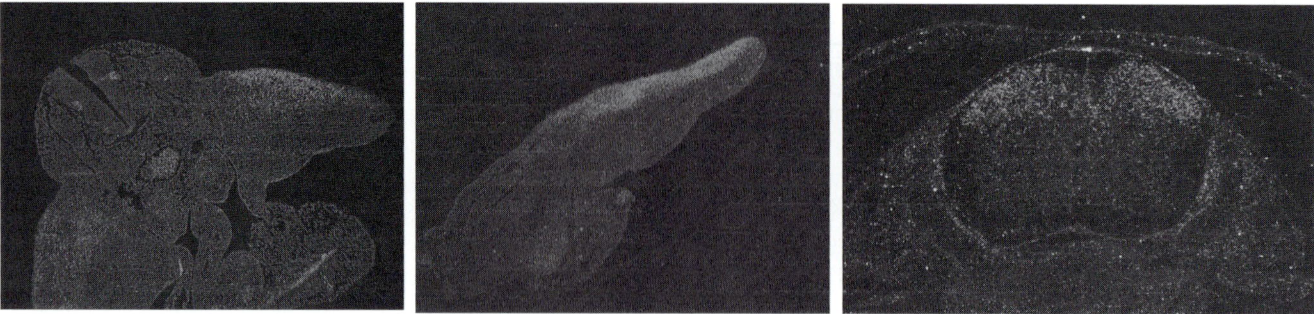

FIGURE 31.2   Temporal-spatial text of *Lmx1b* expression during mouse development. RNA in situ hybridization of *Lmx1b* in transverse sections at embryonic day 11.5 (left panel) and dorso-ventral sections of the forelimb at embryonic day 12.5 (middle panel), both showing strong expression in dorsal mesenchyme. *Lmx1b* expression in transverse section of the spinal cord at embryonic day 15.5 showing dorsal expression (right panel). (see also Plate 69)

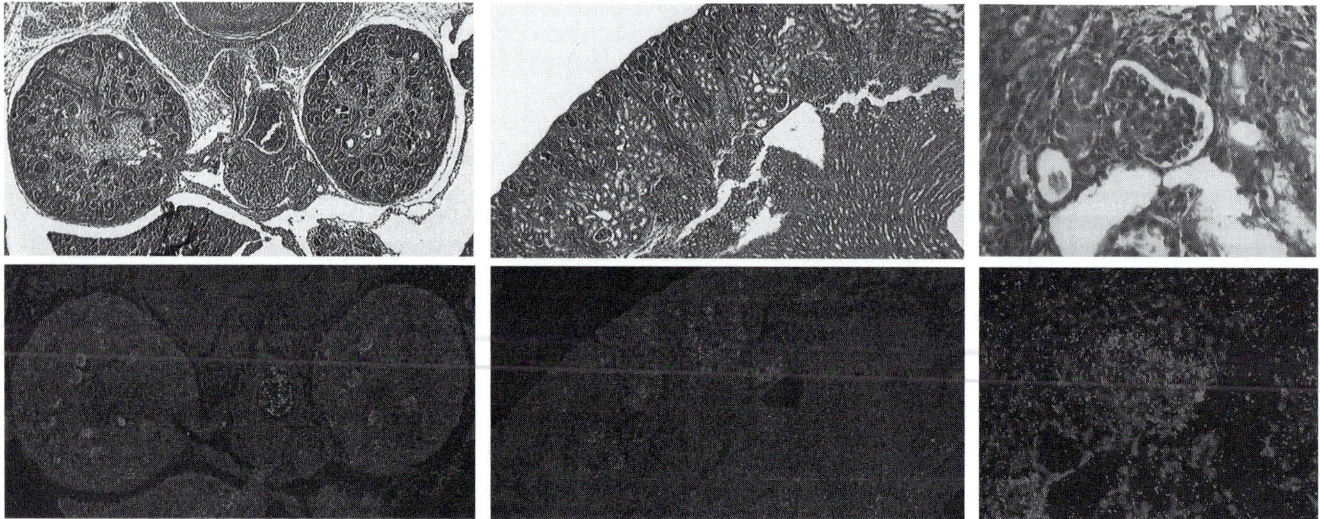

FIGURE 31.3   Expression of *Lmx1b* during mouse kidney development. Top panels are hematoxylin & eosin sections. Bottom panels show sequential sections of RNA in situ hybridization to *Lmx1b* in metanephric kidney at embryonic day 14.5 (left panel), postnatal day 3 (middle panel), and higher magnification of the postnatal glomerulus (right panel). (see also Plate 70)

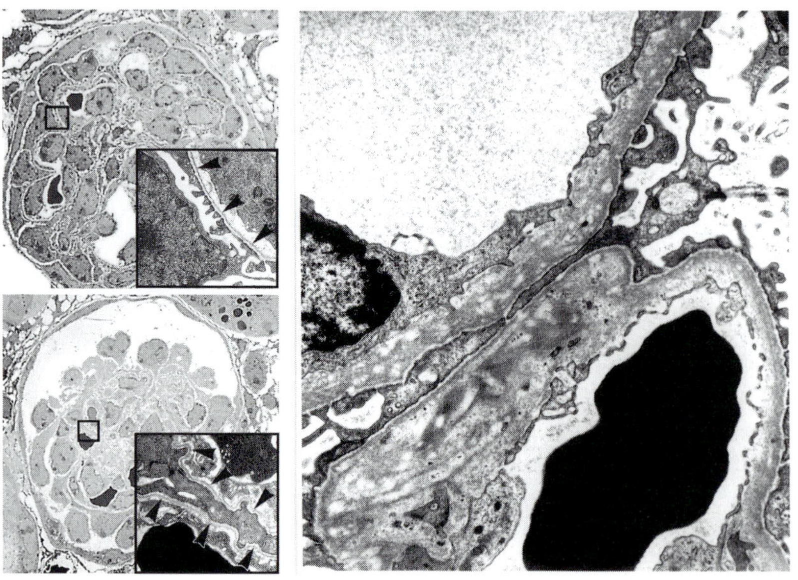

FIGURE 31.4   Ultrastructural changes in mouse and human NPS. Left panels show electron micrographs of wild type *Lmx1b* newborn mouse glomerulus with magnified inset of boxed area showing glomerular basement membrane and podocyte foot processes (top) and *Lmx1b* null newborn mouse glomerulus with magnified inset of boxed area showing thickened basement membrane and absence of normal foot processes. The right panel shows EM of human patient with NPS and heterozygous mutation in *LMX1B*. Thickened and split GBM is present and fibrillar deposits are apparent

transactivation domain and Dunston et al unsuccessfully searched for mutations in the 5′ UTR and in the conserved putative regulatory sequences in intron 2 of patients where no *LMX1B* coding region mutations or deletions could be found (Dunston et al 2004).

To date almost 150 different *LMX1B* mutations that cause NPS have been described (Dreyer et al 1998, 2000, McIntosh et al 1998, Vollrath et al 1998, Clough et al 1999, Seri et al 1999, Hamlington et al 2000, 2001, Knoers et al 2000, Sweeney et al 2001, Dunston et al 2004, 2005, Midro et al 2004, Bongers et al 2005, Sato et al 2005). They include nonsense, missense, and splice mutations as well as deletions and insertions. While point mutations in the LIM domains most likely alter their zinc finger structure, missense mutations in the homeodomain tend to interfere with the DNA binding ability of the protein. Nonsense mutations as well as frameshifts, caused by splice mutations, deletions, or insertions, generate premature termination codons. While such mutations could theoretically generate shorter aberrant transcripts, these transcripts are probably eliminated by the mechanism of nonsense mediated RNA decay. However, patient RNA has never been examined due to unavailability of suitable RNA sources from NPS patients. In vitro studies suggest that these transcripts are unlikely to interfere with the activity of the normal allele product in dominant negative fashion (Dreyer et al 2000, Sato et al 2005). Therefore, the pathogenetic mechanism underlying NPS is thought to be haploinsufficiency of the *LMX1B* gene. This is supported by reports of NPS caused by large deletions of the *LMX1B* coding sequence, including a whole gene deletion (Dunston et al 2004), and by a balanced translocation with one of the breakpoint involving the *LMX1B* locus (Duba et al 1998).

Despite several efforts, it has been impossible to establish a clear genotype–phenotype correlation in patients affected with NPS. In fact, the disease is characterized by high penetrance but also great intra- and interfamilial variability. An allele that, for instance, causes proteinuria in one patient may not cause it in another family member or in affected individuals in an unrelated family (Knoers et al 2000). However, a recent study suggests that at least in some populations, mutations in the homeodomain may be associated with a higher incidence of nephropathy and a greater magnitude of proteinuria than mutations in the LIM domains (Bongers et al 2005).

HISTORY OF THE *LMX1B* GENE AND ANIMAL MODELS OF NPS AND *LMX1B* FUNCTION

LMX1B belongs to the subfamily of LHX transcription factor genes. This subfamily, which is further divided into six subgroups based on a phylogenetic analysis of homeodomain conservation, has been well preserved through evolution and contains several genes isolated from both vertebrates and invertebrates (Hobert & Westphal 2000). It is characterized by the presence of a DNA binding domain (homeodomain) and two LIM domains, LIM-A and LIM-B, at the N-terminus of the protein (Hobert & Westphal 2000). The LIM domain is a cysteine-rich sequence organized as a tandem zinc-finger structure and is considered to act as the interface for protein–protein interactions (Retaux & Bachy 2002). Proteins containing LIM domains are usually involved in the control of gene expression and cytoskeletal function (Kadrmas & Beckerle 2004). A few proteins have been described that interact with LIM-homeodomain (HD) transcription factors and exert either a positive or negative regulatory effect on them. They include: E47, a ubiquitously expressed basic helix-loop-helix protein (German et al 1992), CHIP (van Meyel et al 1999, 2000), LIM-only or LMO proteins (Milan et al 1998, Milan & Cohen 1999), RLIM (Bach et al 1999, Ostendorff et al 2000) – which was also described to have ubiquitin ligase activity and to negatively regulate LIM HD activity (Ostendorff et al 2002) – CLIM1 and CLIM2/Ldb1/ NLI (Agulnick et al 1996, Jurata & Gill 1997, Semina et al 1998). LHX transcription factors likely function by interacting with these and probably other proteins in a transcriptional activation complex that may include additional transcription factors (Asbreuk et al 2002, Lee & Pfaff 2003, Marini et al 2005). CLIM2/Ldb1/NLI in particular represents a multifunctional, multiadaptor protein that is able to coordinate the formation of multimeric, transcriptionally active complexes (Matthews & Visvader 2003). The different protein partners that participate in these complexes could explain the context-dependent and tissue specific action of LHX factors.

There are at least two *Lmx1* genes in mammals, named *Lmx1a* (German et al 1992) and *Lmx1b* (Iannotti et al 1997). While their homeodomains are 100% identical, their function is quite different. They are both involved in CNS development (Kania et al 2000, Millonig et al 2000, Smidt et al 2000, Ding et al 2003, 2004, Chizhikov & Millen 2004, Chinta & Andersen 2005). *Lmx1b* is involved in specification of multiple neuronal subtypes including serotonergic and dopaminergic neurons. However, *Lmx1a*, which is less well characterized, seems to also regulate the expression of the insulin gene (German et al 1992). In contrast, *Lmx1b* has been implicated in multiple developmental functions including dorso-ventral (D/V) patterning of the limb bud and cellular differentiation in the kidney (Chen et al 1998, Morello et al 2001, Chen & Johnson 2002, Miner et al 2002, Morello & Lee 2002, Rohr et al 2002, Schweizer et al 2004).

The first insights into *Lmx1b* function came from studies in the chicken where mis-expression experiments of the avian ortholog *C-Lmx1* suggested its role for dorsal-ventral cell fate determination in the developing limb bud (Riddle et al 1995, Vogel et al 1996). Then in 1998, Chen et al created a targeted deletion of exons 2–7 of the *Lmx1b* gene, and subsequently, generated homozygous null mice (Chen et al 1998). While heterozygous mice are normal, *Lmx1b−/−* mice die a few hours after birth with multiple defects including, most importantly, a complete dorsal to ventral conversion of all mesenchymal-derived soft-tissue structures of the limb (Chen et al 1998). The limb defects seen in *Lmx1b−/−* mice include: absence of patellae (structures derived from the dorsal patellar tendon), absence of nails, hypoplasia of distal ulna, small scapulae, misoriented clavicles, and complete distal duplication of the ventral pattern of tendons, muscles, and sesamoid bones. The ventralization is particularly evident in the zeugopod and autopod elements of the limb (Chen et al 1998).

In situ hybridization studies have demonstrated that *Lmx1b* is expressed in the developing limb bud starting at embryonic day E8.5 through E16.5 (Cygan et al 1997, Dreyer et al 2000). Its expression shows a proximal–distal and anterior–posterior gradient and is restricted to the dorsal mesenchyme (Dreyer et al 2000, 2004) where it is induced by *Wnt7a* from the dorsal ectoderm (Riddle et al 1995, Vogel et al 1996). The striking similarity between the *Lmx1b* null mouse defects and the NPS phenotype, and the fact that *Lmx1b* was mapped to a syntenic region of NPS critical interval, led to the identification of heterozygous mutations in the *LMX1B* gene as the cause of NPS (Dreyer et al 1998, Vollrath et al 1998).

The *Lmx1b* null mouse remains the only animal model available for the study of *Lmx1b* function. Although *Lmx1b−/−* mice die shortly after birth, this model proved to be a useful developmental tool for correlating the *Lmx1b* expression pattern and the NPS spectrum of defects. One of these correlations is the expression of *Lmx1b* in the anterior chamber during eye formation and the finding of primary open angle glaucoma in some NPS patients (Vollrath et al 1998). *Lmx1b* expression has been detected in the periocular mesenchyme by E10.5 and then in the presumptive cornea until E14.5. *Lmx1b−/−* mice develop iris and ciliary body hypoplasia and corneal stromal defects with a reduction in depth of the anterior chamber (Pressman et al 2000). This suggests that *Lmx1b* plays an important role in the morphogenesis of the anterior segment of the murine eye (Gould et al 2004). Although these data also point to a developmental origin of the glaucoma seen in NPS patients, a role for *Lmx1b* in the adult eye cannot be ruled out due to the inability to study adult *Lmx1b−/−* mice.

Another correlation is the expression of *Lmx1b* starting in the developing kidney and extending into adulthood and the presence of glomerular disease in NPS patients. For details see the section 'Role of *LMX1B* during kidney development.'

The function of *Lmx1b* during CNS development has quickly emerged during the past few years. Initial studies in the chicken revealed an important role in the isthmic organizer, a morphological constriction of the neural tube that eventually defines the mesencephalic/metencephalic boundary (Adams et al 2000). In this structure, *Lmx1b* contributes to the formation and maintenance of the isthmic organizing activity together with *Fgf8* and *Wnt1* (Adams et al 2000, Matsunaga et al 2002). At the same time, *Lmx1b*

was shown to be implicated in the proper differentiation of mesencephalic dopaminergic neurons. This function is probably carried on in conjunction with the Ptx family of homeodomain transcription factors (Asbreuk et al 2002, Smidt et al 2000). Later in development, *Lmx1b* was found to be required for the development of all the serotonergic (5-HT) neurons in the central nervous system as indicated by loss of genes necessary for serotonin synthesis and transport in *Lmx1b* null mice (Cheng et al 2003, Ding et al 2003). *Lmx1b* is believed to be downstream of *Nkx2.2*, which mediates the early specification of serotonergic neurons and upstream of *Pet-1*, which mediates their terminal differentiation (Ding et al 2003). Finally, while it was previously known that *Lmx1b* is expressed in interneurons of the dorsal spinal cord, it was not until very recently that its role in the differentiation and migration of the superficial dorsal horn neurons was elucidated and several downstream targets important for this function were identified (Ding et al 2004). *Lmx1b* appears to be necessary in the dorsal horn cells to provide molecular cues that allow the migration of sensory axons into the dorsal spinal cord (Ding et al 2004). This expression pattern and function would correlate well with the diminished pain response that was observed in NPS patients (Dunston et al 2005).

In the limb bud, a role for *Lmx1b* has also been demonstrated in establishing the correct dorsal/ventral migration pattern of motor neuron projections originating from the lateral motor column of the spinal cord (Kania et al 2000). In the absence of *Lmx1b*, this pattern becomes random. Whether this function can be correlated with the muscle hypoplasia observed in some NPS patients is not known. However, *Lmx1b* does not appear to be expressed in muscle progenitor cells (Dreyer et al 2004) and skeletal muscle abnormalities in NPS are probably due to secondary effects of skeletal patterning defects.

ROLE OF *LMX1B* DURING KIDNEY DEVELOPMENT

The discovery of *LMX1B* as the gene responsible for causing NPS steered researchers' attention toward investigating its importance during kidney development. Starting at E13.5 during mouse development, *Lmx1b* is expressed in early stages of glomerular differentiation (S-shaped body) and continues to be expressed by glomerular visceral epithelial cells, the podocytes, of fully mature glomeruli, until after birth (Chen et al 1998, Morello et al 2001, Morello & Lee 2002). The *Lmx1b* mutant mouse is the only animal model established for the study of *Lmx1b* function and proved useful for evaluating *Lmx1b*'s role during metanephric nephron formation. Despite the fact that NPS is an autosomal dominant condition, *Lmx1b* heterozygous mice are healthy and do not develop nephropathy. The *Lmx1b−/−* kidney abnormalities, albeit more severe, recapitulate what has been observed in human biopsies of NPS patients: thickening and splitting of the glomerular basement membrane (GBM), fusion of podocyte secondary processes and proteinuria, visualized by PAS-positive stained material (glycoproteins) accumulated in the distal convoluted tubules (Chen et al 1998). Moreover, the mutant kidneys are smaller compared to those of wild-type mice and immunohistochemical assays showed decreased fumarylacetoacetase (FAH) staining suggesting poor proximal tubular development (Morello et al 2001).

The defects observed in the GBM prompted investigators to hypothesize that *Lmx1b* could control the expression levels of different extracellular matrix components typically expressed in this structure. Although the transcription level of several laminin subunits was unchanged, a significant downregulation, both at the transcriptional and protein level, of the α3 and α4 chains of type IV collagen, was observed in the GBM of *Lmx1b−/−* mice compared to wild-type littermates (Morello et al 2001). Furthermore, *Lmx1b* was shown to bind to an A-T rich putative enhancer element (FLAT element) at the 3′ end of *Col4a4* intron1 thus demonstrating that the tandemly arranged *Col4a4* and *Col4a3* genes are directly coregulated by *Lmx1b* transcriptional activity (Morello et al 2001). Interestingly, during normal embryogenesis of GBM, a coordinated temporal and spatial switch in gene expression from ubiquitously expressed α1(IV) and α2(IV) collagen to the α3(IV), α4(IV) and α5(IV) chains occurs (Miner & Sanes 1994). A similar switch among different laminin subunits also occurs. These molecular switches gradually take place beginning at the capillary loop stage (E14.5 in the mouse) and proceed after birth until the functional adult GBM composition is reached. Likely, *Lmx1b* alone is not sufficient to specify the developmentally coordinated switch in matrix composition in the GBM, but it is required for maintenance of *Col4a4* and *Col4a3* expression. This activity is, therefore, critical for the proper assembly of the GBM and its deficiency contributes to the observed proteinuria.

Because a simple downregulation of type IV collagen genes was not sufficient to explain the severity of the kidney phenotype in *Lmx1b−/−* mice, Miner et al focused their attention on the structure and developmental status of glomerular visceral epithelial cells. Podocytes from *Lmx1b−/−* mice were found to be developmentally arrested with immature foot processes and abnormal slit diaphragm formation (Miner et al 2002). Moreover, it was shown that two differentiation markers important for slit diaphragm assembly, *Nphs2* (podocin) and *Cd2ap*, were also downregulated in *Lmx1b* null mice. Much like the type IV collagen situation, putative *Lmx1b* binding sites were also identified in the genomic regions 5′ to these genes (Miner et al 2002, Rohr et al 2002). Interestingly, mutations in the protein podocin cause steroid-resistant nephrotic syndrome (Boute et al 2000), while mutations in CD2AP have been implicated in a genetic form of nephrotic syndrome in mice (Shih et al 1999) and in focal segmental glomelurosclerosis in humans (Kim et al 2003). Both of these proteins represent additional direct targets of

the *Lmx1b* transcriptional activation cascade where alteration of their function at least partially phenocopy *Lmx1b* dysregulation. These data underscore the importance of *Lmx1b* for podocyte terminal differentiation, and ultimately, glomerular filtration function. Conflicting evidence is present in the literature whether *Lmx1b* may directly regulate the expression of the *Nphs1* gene (nephrin), another slit diaphragm protein that is associated with congenital nephrotic syndrome of the Finnish type (Kestila et al 1998, Morello et al 2001, Hamano et al 2002, Miner et al 2002). No *Lmx1b* binding sites have been identified in the putative *Nphs1* promoter sequence, and recently, it has been shown that nephrin is transcriptionally activated by the Wilm's tumor suppressor gene *WT1* (Wagner et al 2004).

All the studies performed on the *Lmx1b−/−* mice seem to suggest that the pathogenetic mechanism of renal disease in NPS is derived from the underexpression of several podocyte-specific genes. Thus, *LMX1B* haploinsufficiency may cause subtle changes in the expression pattern of critical genes for glomerular filtration function and may lead to human nephrotic syndrome in the presence of specific genetic or environmental cofactors (Miner et al 2002). In contrast to data from *Lmx1b−/−* mice, immunohistochemical analyses of kidney biopsies from seven NPS patients presumably heterozygous for *LMX1B* mutations did not reveal a decrease in the expression of podocin, cd2ap or α3(IV) and α4(IV) collagen chains. Instead a GBM deposition of fibrillar material stained positive with an antitype III collagen antibody (Heidet et al 2003). These discrepancies may be due to differences of dosage sensitivity to loss of *Lmx1b* function during podocyte formation or to genetic modifiers present in the outbred human genetic background. Also, the *Lmx1b* null mouse is a developmental model of glomerular pathogenesis while the NPS patient biopsies represent diseased tissue where secondary changes may have occurred as a product of longstanding gene–environment interactions.

PATHOGENESIS OF NPS AND IMPORTANT QUESTIONS

The skeletal features of NPS likely reflect a relative loss of *Lmx1b* specification of dorsal patterning of the limb mesenchyme. This is best exemplified by the hypoplasia of the patella, an embryologic dorsal element of hindlimb morphogenesis. At the same time, dorsal–ventral (D/V) patterning is intimately linked to establishment of anterior–posterior (A/P) pattern. Hence, it is not surprising that subtle defects in human and mouse limb development in NPS also reflect disproportionate growth of anterior vs. posterior segments, e.g. ulnar vs. radial and medial vs. later femoral condyles. Similarly, the persistence of *Lmx1b* expression in distal limb mesenchymal derivatives in a graded pattern across the A/P axis is reflected in the nail dysplasia that is often more severe in the anterior digits (first and second digits). It is still unclear which transcriptional targets of *Lmx1b* can impart dorsal positional information when transiently expressed in the limb bud. Perhaps the progress made on the kidney front may lend insight into this. Surprisingly, *Lmx1b* remain the only mesenchymal determinant described to date of dorsal limb pattern.

Because of the sustained expression of *Lmx1b* during podocyte differentiation and glomerulogenesis, significantly more progress has been made elucidating its targets during kidney development as opposed to in other developmental programs. However, similarly to other transcription factors that are expressed both early and late in kidney development, e.g. *WT1*, *Lmx1b* may serve different functions early during nephrogenesis vs. later in the postnatal kidney. If this is the case, what are the components of the *Lmx1b* transcriptional complex that specify its context-dependent functions at different time points? For example, how does *Lmx1b* participate in the regulation of the developmental switch in collagen and laminin expression during GBM maturation? What are the transcriptional coregulators involved in *Lmx1b* regulation of slit diaphragm structural integrity? Ultimately, the generation of conditional mouse mutants will aid in answering some of these questions.

ACKNOWLEDGMENTS

The authors would like to thank M. Land for administrative support and S. Carter and A. Tran for clinical research support. The authors are supported by NIH grants ES11253, HD22657, and the Baylor College of Medicine MRDDRC.

References

Adams KA, Maida JM, Golden JA, Riddle RD. The transcription factor Lmx1b maintains Wnt1 expression within the isthmic organizer. Development 2000; 127: 1857–67.

Agulnick AD, Taira M, Breen JJ, Tanaka T, Dawid IB, Westphal H. Interactions of the LIM-domain-binding factor Ldb1 with LIM homeodomain proteins. Nature 1996; 384: 270–2.

Asbreuk CH, Vogelaar CF, Hellemons A, Smidt MP, Burbach JP. CNS expression pattern of Lmx1b and coexpression with ptx genes suggest functional cooperativity in the development of forebrain motor control systems. Mol. Cell Neurosci. 2002; 21: 410–20.

Bach I, Rodriguez-Esteban C, Carriere C, et al. RLIM inhibits functional activity of LIM homeodomain transcription factors via recruitment of the histone deacetylase complex. Nat. Genet. 1999; 22: 394–9.

Beals RK, Eckhardt AL. Hereditary onycho-osteodysplasia (Nail-Patella syndrome). A report of nine kindreds. J. Bone Joint Surg. Am. 1969; 51: 505–16.

Ben-Bassat M, Cohen L, Rosenfeld J. The glomerular basement membrane in the nail-patella syndrome. Arch. Pathol. 1971; 92: 350–5.

Bennett WM, Musgrave JE, Campbell RA, et al. The nephropathy of the nail-patella syndrome. Clinicopathologic analysis of 11 kindred. Am. J. Med. 1973; 54: 304–19.

Bodziak KA, Hammond WS, Molitoris BA. Inherited diseases of the glomerular basement membrane. Am. J. Kidney Dis. 1994; 23: 605–18.

Bongers EM, Gubler MC, Knoers NV. Nail-patella syndrome. Overview on clinical and molecular findings. Pediatr. Nephrol. 2002; 17: 703–12.

Bongers EM, Huysmans FT, Levtchenko E, et al. Genotype-phenotype studies in nail-patella syndrome show that LMX1B mutation location is involved in the risk of developing nephropathy. Eur. J. Hum. Genet. 2005; 13: 935–46.

Boute N, Gribouval O, Roselli S, et al. NPHS2, encoding the glomerular protein podocin, is mutated in autosomal recessive steroid-resistant nephrotic syndrome. Nat. Genet. 2000; 24: 349–54.

Brixey AM Jr., Burke RM. Arthro-onychodysplasia: hereditary syndrome involving deformity of head of radius, absence of patellas, posterior iliac spurs, dystrophy of finger nails. Am. J. Med. 1950; 8: 738–44.

Campeau E, Watkins D, Rouleau GA, et al. Linkage analysis of the nail-patella syndrome. Am. J. Hum. Genet. 1995; 56: 243–7.

Carbonara P, Alpert M. Hereditary Osteo-Onycho-Dysplasia (Hood). Am. J. Med. Sci. 1964; 248: 139–51.

Chan PC, Chan KW, Cheng IK, Chan MK. Living-related renal transplantation in a patient with nail-patella syndrome. Nephron 1988; 50: 164–6.

Chen H, Johnson RL. Interactions between dorsal-ventral patterning genes lmx1b, engrailed-1 and wnt-7a in the vertebrate limb. Int. J. Dev. Biol. 2002; 46: 937–41.

Chen H, Lun Y, Ovchinnikov D, et al. Limb and kidney defects in Lmx1b mutant mice suggest an involvement of LMX1B in human nail patella syndrome. Nat. Genet. 1998; 19: 51–5.

Cheng L, Chen CL, Luo P, et al. Lmx1b, Pet-1, and Nkx2.2 coordinately specify serotonergic neurotransmitter phenotype. J. Neurosci. 2003; 23: 9961–7.

Chinta SJ, Andersen JK. Dopaminergic neurons. Int. J. Biochem. Cell Biol. 2005; 37: 942–6.

Chizhikov VV, Millen KJ. Control of roof plate development and signaling by Lmx1b in the caudal vertebrate CNS. J. Neurosci. 2004; 24: 5694–703.

Clough MV, Hamlington JD, McIntosh I. Restricted distribution of loss-of-function mutations within the LMX1B genes of nail-patella syndrome patients. Hum. Mutat. 1999; 14: 459–65.

Cygan JA, Johnson RL, McMahon AP. Novel regulatory interactions revealed by studies of murine limb pattern in Wnt-7a and En-1 mutants. Development 1997; 124: 5021–32.

Del Pozo E, Lapp H. Ultrastructure of the kidney in the nephropathy of the nail–patella syndrome. Am. J. Clin. Pathol. 1970; 54: 845–51.

Ding YQ, Marklund U, Yuan W, et al. Lmx1b is essential for the development of serotonergic neurons. Nat. Neurosci. 2003; 6: 933–8.

Ding YQ, Yin J, Kania A, Zhao ZQ, Johnson RL, Chen ZF. Lmx1b controls the differentiation and migration of the superficial dorsal horn neurons of the spinal cord. Development 2004; 131: 3693–703.

Dreyer SD, Morello R, German MS, et al. LMX1B transactivation and expression in nail-patella syndrome. Hum. Mol. Genet. 2000; 9: 1067–74.

Dreyer SD, Naruse T, Morello R, et al. Lmx1b expression during joint and tendon formation: localization and evaluation of potential downstream targets. Gene Expr. Patterns 2004; 4: 397–405.

Dreyer SD, Zhou G, Baldini A, et al. Mutations in LMX1B cause abnormal skeletal patterning and renal dysplasia in nail patella syndrome. Nat. Genet. 1998; 19: 47–50.

Drut RM, Chandra S, Latorraca R, Gilbert-Barness E. Nail-patella syndrome in a spontaneously aborted 18-week fetus: ultrastructural and immunofluorescent study of the kidneys. Am. J. Med. Genet. 1992; 43: 693–6.

Duba HC, Erdel M, Loffler J, Wirth J, Utermann B, Utermann G. Nail patella syndrome in a cytogenetically balanced t(9;17)(q34.1;q25) carrier. Eur. J. Hum. Genet. 1998; 6: 75–9.

Dunston JA, Hamlington JD, Zaveri J, et al. The human LMX1B gene: transcription unit, promoter, and pathogenic mutations. Genomics 2004; 84: 565–76.

Dunston JA, Lin S, Park JW, Malbroux M, McIntosh I. Phenotype severity and genetic variation at the disease locus: an investigation of nail dysplasia in the nail patella syndrome. Ann. Hum. Genet. 2005; 69: 1–8.

Feingold M, Itzchak Y, Goodman RM. Ultrasound prenatal diagnosis of the Nail-Patella syndrome. Prenat. Diagn. 1998; 18: 854–6.

Fiedler BS, De Smet AA, Kling TF Jr., Fisher DR. Foot deformity in hereditary onycho-osteodysplasia. Can. Assoc. Radiol. J. 1987; 38: 305–8.

Fong EE. Iliac horns (symmetrical bilateral central posterior iliac processes) a case report. Radiology 1946; 47: 517–18.

German MS, Wang J, Chadwick RB, Rutter WJ. Synergistic activation of the insulin gene by a LIM–homeo domain protein and a basic helix-loop-helix protein: building a functional insulin minienhancer complex. Genes Dev. 1992; 6: 2165–76.

Gould DB, Smith RS, John SW. Anterior segment development relevant to glaucoma. Int. J. Dev. Biol. 2004; 48: 1015–29.

Gubler MC, Levy M. Prenatal diagnosis of nail-patella syndrome by intrauterine kidney biopsy [letter; comment]. Am. J. Med. Genet. 1993; 47: 122–4.

Gubler MC, Levy M, Naizot C, Habib R. Glomerular basement membrane changes in hereditary glomerular diseases. Renal Physiol. 1980; 3: 405–13.

Guidera KJ, Satterwhite Y, Ogden JA, Pugh L, Ganey T. Nail patella syndrome: a review of 44 orthopaedic patients. J. Pediatr. Orthop. 1991; 11: 737–42.

Hamano Y, Grunkemeyer JA, Sudhakar A, et al. Determinants of vascular permeability in the kidney glomerulus. J. Biol. Chem. 2002; 277: 31154–62.

Hamlington JD, Clough MV, Dunston JA, McIntosh I. Deletion of a branch-point consensus sequence in the LMX1B gene causes exon skipping in a family with nail patella syndrome. Eur. J. Hum. Genet. 2000; 8: 311–14.

Hamlington JD, Jones C, McIntosh I. Twenty-two novel LMX1B mutations identified in nail patella syndrome (NPS) patients. Hum. Mutat. 2001; 18: 458.

Hawkings CF, Smith OE. Renal dysplasia in a family with multiple hereditary abnormalities including iliac horns. Lancet 1950; I: 803–8.

Heidet L, Bongers EM, Sich M, et al. In vivo expression of putative LMX1B targets in nail-patella syndrome kidneys. Am. J. Pathol. 2003; 163: 145–55.

Hobert O, Westphal H. Functions of LIM-homeobox genes. Trends Genet. 2000; 16: 75–83.

Hogh J, Macnicol MF. Foot deformities associated with onycho-osteodysplasia. A familial study and a review of associated features. Int. Orthop. 1985; 9: 135–8.

Holmes G, Crooijmans R, Groenen M, Niswander L. ALC (adjacent to LMX1 in chick) is a novel dorsal limb mesenchyme marker. Gene Expr. Patterns 2003; 3: 735–41.

Hoyer JR, Michael AF, Vernier RL. Renal disease in nail-patella syndrome: clinical and morphologic studies. Kidney Int. 1972; 2: 231–8.

Iannotti CA, Inoue H, Bernal E, et al. Identification of a human LMX1 (LMX1.1)-related gene, LMX1.2: tissue-specific expression and linkage mapping on chromosome 9. Genomics 1997; 46: 520–4.

Jurata LW, Gill GN. Functional analysis of the nuclear LIM domain interactor NLI. Mol. Cell. Biol. 1997; 17: 5688–98.

Kadrmas JL, Beckerle MC. The LIM domain: from the cytoskeleton to the nucleus. Nat. Rev. Mol. Cell. Biol. 2004; 5: 920–31.

Kania A, Johnson RL, Jessell TM. Coordinate roles for LIM homeobox genes in directing the dorsoventral trajectory of motor axons in the vertebrate limb. Cell 2000; 102: 161–73.

Kestila M, Lenkkeri U, Mannikko M, et al. Positionally cloned gene for a novel glomerular protein – nephrin – is mutated in congenital nephrotic syndrome. Mol. Cell 1998; 1: 575–82.

Kim JM, Wu H, Green G, et al. CD2-associated protein haploinsufficiency is linked to glomerular disease susceptibility. Science 2003; 300: 1298–300.

Knoers NV, Bongers EM, van Beersum SE, Lommen EJ, van Bokhoven H, Hol FA. Nail-patella syndrome: identification of mutations in the LMX1B gene in Dutch families. J. Am. Soc. Nephrol. 2000; 11: 1762–6.

Leahy MS. The hereditary nephropathy of osteo-onychodysplasia. Nail-patella syndrome. Am. J. Dis. Child. 1966; 112: 237–41.

Lee B, Morello R. LMX1B and nail patella syndrome. In: Epstein CJ, Erickson RP, Wynshaw-Boris A, eds. Inborn Errors of Development. New York: Oxford University Press, 2004: pp. 615–22.

Lee SK, Pfaff SL. Synchronization of neurogenesis and motor neuron specification by direct coupling of bHLH and homeodomain transcription factors. Neuron 2003; 38: 731–45.

Lichter PR, Richards JE, Downs CA, Stringham HM, Boehnke M, Farley FA. Cosegregation of open-angle glaucoma and the nail-patella syndrome. Am. J. Ophthalmol. 1997; 124: 506–15.

Marini M, Giacopelli F, Seri M, Ravazzolo R. Interaction of the LMX1B and PAX2 gene products suggests possible molecular basis of differential phenotypes in Nail-Patella syndrome. Eur. J. Hum. Genet. 2005; 13: 789–92.

Matsunaga E, Katahira T, Nakamura H. Role of Lmx1b and Wnt1 in mesencephalon and metencephalon development. Development 2002; 129: 5269–77.

Matthews JM, Visvader JE. LIM-domain-binding protein 1: a multifunctional cofactor that interacts with diverse proteins. EMBO Rep. 2003; 4: 1132–7.

McIntosh I, Clough MV, Gak E, Frydman M. Prenatal diagnosis of nail-patella syndrome. Prenat. Diagn. 1999; 19: 287–8.

McIntosh I, Clough MV, Schaffer AA, et al. Fine mapping of the nail-patella syndrome locus at 9q34. Am. J. Hum. Genet. 1997; 60: 133–42.

McIntosh I, Dreyer SD, Clough MV, et al. Mutation analysis of LMX1B gene in nail-patella syndrome patients. Am. J. Hum. Genet. 1998; 63: 1651–8.

Mellotte GJ, Eastwood JB. Pathognomonic sign of triangular lunulae in the nail-patella syndrome. Nephrol. Dial. Transplant. 1995; 10: 300–1.

Midro AT, Panasiuk B, Tumer Z, et al. Interstitial deletion 9q22.32-q33.2 associated with additional familial translocation t(9;17)(q34.11;p11.2) in a patient with Gorlin-Goltz syndrome and features of Nail-Patella syndrome. Am. J. Med. Genet. A 2004; 124: 179–91.

Mihatsch MJ, Zollinger HU. Kidney disease. Pathol. Res. Prac. 1980; 167: 88–117.

Milan M, Cohen SM. Regulation of LIM homeodomain activity in vivo: a tetramer of dLDB and apterous confers activity and capacity for regulation by dLMO. Mol. Cell 1999; 4: 267–73.

Milan M, Diaz-Benjumea FJ, Cohen SM. Beadex encodes an LMO protein that regulates Apterous LIM-homeodomain activity in *Drosophila* wing development: a model for LMO oncogene function. Genes Dev. 1998; 12: 2912–20.

Miller TT, Shapiro MA, Schultz E, Crider R, Paley D. Sonography of patellar abnormalities in children. AJR Am. J. Roentgenol. 1998; 171: 739–42.

Millonig JH, Millen KJ, Hatten ME. The mouse Dreher gene Lmx1a controls formation of the roof plate in the vertebrate CNS. Nature 2000; 403: 764–9.

Miner JH, Morello R, Andrews KL, et al. Transcriptional induction of slit diaphragm genes by Lmx1b is required in podocyte differentiation. J. Clin. Invest. 2002; 109: 1065–72.

Miner JH, Sanes JR. Collagen IV alpha 3, alpha 4, and alpha 5 chains in rodent basal laminae: sequence, distribution, association with laminins, and developmental switches. J. Cell Biol. 1994; 127: 879–91.

Morello R, Lee B. Insight into podocyte differentiation from the study of human genetic disease: nail-patella syndrome and transcriptional regulation in podocytes. Pediatr. Res. 2002; 51: 551–8.

Morello R, Zhou G, Dreyer SD, et al. Regulation of glomerular basement membrane collagen expression by LMX1B contributes to renal disease in nail patella syndrome. Nat. Genet. 2001; 27: 205–8.

Ostendorff HP, Bossenz M, Mincheva A, et al. Functional characterization of the gene encoding RLIM, the corepressor of LIM homeodomain factors. Genomics 2000; 69: 120–30.

Ostendorff HP, Peirano RI, Peters MA, et al. Ubiquitination-dependent cofactor exchange on LIM homeodomain transcription factors. Nature 2002; 416: 99–103.

Pinette MG, Ukleja M, Blackstone J. Early prenatal diagnosis of nail-patella syndrome by ultrasonography. J. Ultrasound Med. 1999; 18: 387–9.

Pressman CL, Chen H, Johnson RL. LMX1B, a LIM homeodomain class transcription factor, is necessary for normal development of multiple tissues in the anterior segment of the murine eye. Genesis 2000; 26: 15–25.

Renwick JH, Lawler SD. Genetical linkage between the ABO and nail-patella loci. Ann. Hum. Genet. 1955; 19: 312–31.

Retaux S, Bachy I. A short history of LIM domains (1993-2002): from protein interaction to degradation. Mol. Neurobiol. 2002; 26: 269–81.

Riddle RD, Ensini M, Nelson C, et al. Induction of the LIM homeobox gene Lmx1 by WNT7a establishes dorsoventral pattern in the vertebrate limb. Cell 1995; 83: 631–40.

Rohr C, Prestel J, Heidet L, et al. The LIM-homeodomain transcription factor Lmx1b plays a crucial role in podocytes. J. Clin. Invest. 2002; 109: 1073–82.

Sartoris DJ, Resnick D. The horn: a pathognomonic feature of paediatric bone dysplasias. Aust. Paediatr. J. 1987; 23: 347–9.

Sato U, Kitanaka S, Sekine T, Takahashi S, Ashida A, Igarashi T. Functional characterization of LMX1B mutations associated with nail-patella syndrome. Pediatr. Res. 2005; 57: 783–8.

Schweizer H, Johnson RL, Brand-Saberi B. Characterization of migration behavior of myogenic precursor cells in the limb bud with respect to Lmx1b expression. Anat. Embryol. (Berl.) 2004; 208: 7–18.

Semina EV, Altherr MR, Murray JC. Cloning and chromosomal localization of two novel human genes encoding LIM-domain binding factors CLIM1 and CLIM2/LDB1/NLI. Mamm. Genome 1998; 9: 921–4.

Seri M, Melchionda S, Dreyer S, et al. Identification of LMX1B gene point mutations in Italian patients affected with nail-patella syndrome. Int. J. Mol. Med. 1999; 4: 285–90.

Shih NY, Li J, Karpitskii V, et al. Congenital nephrotic syndrome in mice lacking CD2-associated protein [see comments]. Science 1999; 286: 312–15.

Smidt MP, Asbreuk CH, Cox JJ, Chen H, Johnson RL, Burbach JP. A second independent pathway for development of mesencephalic dopaminergic neurons requires Lmx1b. Nat. Neurosci. 2000; 3: 337–41.

Stratigos AJ, Baden HP. Unraveling the molecular mechanisms of hair and nail genodermatoses. Arch. Dermatol. 2001; 137: 1465–71.

Sweeney E, Fryer A, Mountford R, Green A, McIntosh I. Nail patella syndrome: a review of the phenotype aided by developmental biology. J. Med. Genet. 2003; 40: 153–62.

Sweeney E, Fryer AE, Mountford RC, Green AJ, McIntosh I. Nail patella syndrome: a study of 123 patients from 43 British families and detection of 16 novel mutations of LMX1B. Am. J. Hum. Genet. 2001; 69: A72.

Taguchi T, Takebayashi S, Nishimura M, Tsuru N. Nephropathy of nail-patella syndrome. Ultrastruct. Pathol. 1988; 12: 175–83.

Uranga VM, Simmons RL, Hoyer JR, Kjellstrand CM, Buselmeier TJ, Najarian JS. Renal transplantation for the nail patella syndrome. Am. J. Surg. 1973; 125: 777–9.

van Meyel DJ, O'Keefe DD, Jurata LW, Thor S, Gill GN, Thomas JB. Chip and apterous physically interact to form a functional complex during *Drosophila* development. Mol. Cell 1999; 4: 259–65.

van Meyel DJ, O'Keefe DD, Thor S, Jurata LW, Gill GN, Thomas JB. Chip is an essential cofactor for apterous in the regulation of axon guidance in *Drosophila*. Development 2000; 127: 1823–31.

Verdich J. Nail-patella syndrome associated with renal failure requiring transplantation. Acta Dermato-Venereol. 1980; 60: 440–3.

Vogel A, Rodriguez C, Warnken W, Izpisua Belmonte JC. Dorsal cell fate specified by chick Lmx1 during vertebrate limb development [published erratum appears in Nature 1996 Feb 29; 379(6568): 848]. Nature 1996; 378: 716–20.

Vollrath D, Jaramillo-Babb VL, Clough MV, et al. Loss-of-function mutations in the LIM-homeodomain gene, LMX1B, in nail-patella syndrome. Hum. Mol. Genet. 1998; 7: 1091–8.

Wagner N, Wagner KD, Xing Y, Scholz H, Schedl A. The major podocyte protein nephrin is transcriptionally activated by the Wilms' tumor suppressor WT1. J. Am. Soc. Nephrol. 2004; 15: 3044–51.

Walker P, Harris I, Leicester A. Patellar tendon-to-patella ratio in children. J. Pediatr. Orthop. 1998; 18: 129–31.

CHAPTER 32

Mitochondrial Diseases of the Kidney

ALI HARIRI

INTRODUCTION

Despite advances made in the past 25 years in our understanding of the biological processes in mitochondria and the sequencing of the mitochondrial genome (Anderson et al 1981), the precise etiologies of mitochondrial diseases are still uncharacterized. This is due to the rarity and complexity of these disorders. Since many of these disorders were initially recognized in the nervous system, neurologists and neuroscientists have had a longer tradition in studying such diseases than have scientists in other fields. Recent advances in technology have allowed us to analyze the mitochondrial genome easily. The prevalence of mitochondrial disease in recent studies is believed to be 1 in 3500 in some populations (Schaefer et al 2004). While mitochondrial disorders could be a result of mutations in the nuclear genome (Mandel et al 2001, Saada et al 2001, Coenen et al 2004), this chapter focuses on diseases encoded by mutations in the mitochondrial DNA (mtDNA).

MITOCHONDRIAL GENOME AND GENETICS

The mitochondrial genome and genetics are different from the nuclear genome and Mendelian genetics in many aspects. These differences result in interesting biological and clinical consequences.

Mitochondrial Genome

The mitochondrial genome consists of a circular, double-stranded DNA molecule (16.6 kb). It lacks intron and encodes 37 genes: two rRNAs and 22 tRNAs, and 13 polypeptides. The tRNAs show minimal redundancy, the exceptions being tRNA^{Leu} and tRNA^{Ser}. These 13 polypeptides are responsible for the oxidative phosphorylation processes (OXPHOS) in the five complexes: complex I (seven subunits), complex III (one subunit), complex IV (three subunits), and complex V (two subunits). complexes I–IV are responsible for the respiratory chain and complex V is involved in ATP generation by ATP synthase. The rest of the proteins of the respiratory chain subunits and other mitochondrial functions are encoded by the nuclear genome and are specifically targeted and sorted to the mitochondria. For example, complex II is entirely encoded by the nuclear genome. There is also evidence that nuclear tRNAs are transported into mitochondria (Kolesnikova et al 2000).

Transcription initiation can occur on both strands. To initiate transcription, the mitochondrial RNA polymerase requires mitochondrial transcription factor A (TFAM) and either mitochondrial transcription factor B1 or B2 (Falkenberg et al 2002). A polycistronic precursor RNA is generated under the control of the mitochondrial promoter (controlling region). Replication of the mitochondrial genome requires mtDNA polymerase γ. This enzyme consists of two subunits, one for proofreading (PolgA) and the other for processing (PolgB) (Carrodeguas et al 2001). It is not surprising that mutation or lack of PolgA leads to accumulation of mutations in mtDNA (Trifunovic et al 2004).

Mitochondrial Genetics

Thousands of copies of the mitochondrial genome can exist in one cell; as a result mutated and wild-type DNA can coexist in the same cell, a concept referred to as heteroplasmy. In the presence of heteroplasmy, the threshold level of mutation is important for the clinical manifestation of the disease. Mitochondria are almost exclusively inherited maternally (Schwartz & Vissing 2002) As a result, fathers with mtDNA mutations are at no risk of transmitting the defect to their offspring. In addition to the fact that mtDNA is transmitted maternally, the risk of disease depends on the type of mutation and segregation of mutation in maternal tissues. In the case of homoplasmic mutations, all the maternal offspring inherit the DNA. However, variable penetrance is also observed. For example, homoplasmy is noted

Genetic Diseases of the Kidney
ISBN: 978-0-12-449851-8

in Leber hereditary optic neuropathy (LHON). Although all offspring inherit the mutation, only some will clinically manifest the disease (Carelli et al 2003). This is consistent with the role of nuclear genetic modifiers. These mutations could be a predisposing factor. Environmental factors could contribute to the manifestation of the phenotype as well. For example, a mutation in rRNA predisposes the patient to deafness when aminoglycosides are administered (Li et al 2005). Common mutations also predispose the patients to Parkinson disease (Ross et al 2003, van der Walt et al 2003) or Alzheimer's (Shoffner et al 1993). mtDNA mutations in colorectal cancer and prostate cancer have been reported (Polyak et al 1998, Jeronimo et al 2001). While a direct link between these mutations and the tumorogenesis has not been established, mitochondrial mutations could become useful in tumor detection (Fliss et al 2000).

As ATP is required for the function of the cells, it is not surprising that mitochondrial mutations could affect all the tissues. It is believed that tissues with high ATP demand are more susceptible to manifestations of disease due to mitochondrial disorders. Typically, the cells in the central nervous system, heart, muscle, endocrine organs, and kidney are considered to be at high risk. Neurological disorders could range from migraine, stroke, and epilepsy, to a common finding of sensorineural hearing loss. Common cardiac manifestations are heart block or cardiomyopathy. Endocrine disorders include thyroid disease, parathyroid disease, and ovarian failure, and the most common is diabetes. In this chapter, we will discuss renal manifestations of mitochondrial disorders. Many documented mitochondrial disorders are syndromic and have pleiotropic effects on many organs. For example, MERRF (myoclonic epilepsy ragged red fiber), MELAS (myopathy encephalopathy lactic acidosis stroke-like episodes), Pearson syndrome (pancytopenia, lactic acidodsis), NARP (neurogenic muscle weakness, ataxia, and retinitis pigmentosa) and Kearns-Sayre syndrome (myopathy, ophthalmoplegia, and cardiomyopathy) are among them. In general, the presence of maternal inheritance, the involvement of multiple organs, and the combination of known symptoms, should raise the prospect of a mitochondrial disorder. Although tRNA genes represent 10% of the mitochondrial genome, mutations in these genes account for 75% of known mitochondrial diseases (Schon et al 1997).

The mitochondrial theory of aging is strongly connected to the reactive oxidative species (ROS) theory of aging (Harman 1992). ROS cause spontaneous mutations in the mitochondrial genome; mitochondrial genome damage caused by ROS is more extensive and persists longer than damage to the nuclear damage (Mandavilli et al 2002). These mutations lead to mitochondrial dysfunction, which leads to a further increase in ROS production. This progressive damage in the mitochondrial genome is thought to contribute to aging (Michikawa et al 1999, Lee & Ruvkun 2002, Lenaz et al 2002, Hekimi & Guarente 2003, Lee et al 2003). As a result, patients with inherited mitochondrial mutation have a decreased reserve and are more prone to many common diseases (AD, PD). While the exact mechanism of effect of pathogenesis of ROS on the tissues is not clear, lack of energy, apoptosis, and replicative senescence are the proposed mechanisms. In an animal model lacking PolgA (proofreading enzyme for mitochondrial replication), there is an accumulation of mitochondrial mutations and severe aging (Trifunovic et al 2004). However, further analysis revealed that in a similar animal, the rate of ROS production is not increased and these mice also show an increase in apoptosis (Kujoth et al 2005).

FOCAL SEGMENTAL GLOMERULOSCLEROSIS (FSGS)

Many chronic renal progressive diseases result from nephron degeneration. Focal segmental glomerosclerosis could represent the final common pathway for this degenerative disorder. The initial changes are believed to occur in glomerular epithelial cells (Kriz & Lemley 1999). It is expected that one will observe renal epithelial damage due to mitochondrial dysfunction. Similar to neurons and muscular cells, which are considered the major sites of mtDNA mutations, the epithclial tissues of the glomeruli are terminally differentiated and do not proliferate (Fries et al 1989, Nagata et al 1992). Although mitochondrial-caused FSGS constitutes a minor portion of the individuals afflicted with this disease, determining its pathogenesis will aid us in understanding the mechanism of disease and possible treatment.

tRNA^Leu (A3243G)

The A3243G point mutation in the tRNA^Leu is considered to be the most common mtDNA defect in the population with a 1/7000 prevalence in Finnish adults (Majamaa et al 1998). The mutation was primarily observed in patients with MELAS syndrome (mitochondrial encephalomyopathy, lactic acidosis, and stroke-like episodes) (Johns & Hurko 1991, Jean-Francois et al 1994). Interestingly, MELAS can be caused by many other mitochondrial mutations (T3271C and A3253G) (Ireland et al 2004). The A3243G mutation is also shown to cause type II diabetes with deafness (van den Ouweland et al 1992), cardiomyopathy (Silvestri et al 1994) or progressive external ophthalmoplegia (Jean-Francois et al 1994). FSGS is also noted in patients with this mutation. FSGS could precede MELAS (Mochizuki et al 1996) or in some other patients with MELAS could be noted in autopsy follow-up. In addition, patients with this mutation could have clustering of FSGS, DM, and deafness (Jansen et al 1997, Dinour et al 2004). Furthermore, two more patients are reported who suffer from FSGS, deafness, DM, and cardiomyopathy (Kurogouchi et al 1998, Iwasaki et al 2001, Guery et al 2003).

The A3243 mutation is observed in clustering of deafness and FSGS. Deafness typically predates renal involvement at an early age. This is characterized by bilateral high-frequency neurosensory hearing loss and progressive worsening. As deafness is associated with renal diseases, Alport syndrome is overdiagnosed (Jansen et al 1997). It is worth differentiating Alport syndrome and FSGS based on microscopic hematuria. The DM in these patients is typically a late manifestation and is noted after the kidney disease. While maternally inherited DM and deafness are estimated to be present in 0.5–2.8% of diabetic patients, only 5.5% had renal disease (Guillausseau et al 2001). The prevalence of this mutation in Japanese patients with DM and ESRD is less than 1 to 5.9% (Iwasaki et al 2001). It is noteworthy that the prevalence of retinopathy in patients with this mutation is 8% in a 12-year follow-up while the prevalence is 30–75% in regular DM patients.

Renal biopsies have systematically ruled out the presence of diabetic nephropathy in affected patients. Moulonguet-deleris reported severe hyaline changes in cytoplasm of smooth muscle of afferent arterioles and small arteries. Mochizuki (Mochizuki et al 1996) and Takeda (Segawa-Takaeda et al 2002) also reported mitochondrial abnormalities in vascular smooth muscle cell in two cases. The authors attributed the changes to the hemodynamic variations due to vascular involvements (Doleris et al 2000). In most biopsies consistent with FSGS, electron microscopy revealed an accumulation of abnormal mitochondria in the foot processes of podocytes (Hotta et al 2001). The mitochondria were found to be normal in the endothelium of glomeruli. The mean glomeruli volume (MGV) was evaluated in one study (Hotta et al 2001) which found no changes in MGV in contrast to other patients with primary and secondary FSGS. This result is suggestive that glomerular hypertrophy may not be involved in the pathogenesis of mitochondrial FSGS due to tRNA^{Leu} mutation. Overall, the pathogenesis of FSGS in these patients is not clear.

This mutation is heteroplasmic, and there is a higher abundance of the mutation in DNA isolated from the urine than from leukocytes. As a result, the detection of mutated mtDNA from the sediment of urine is extremely helpful. However, there is no correlation between the severity, or presence of the disease, or the age of onset of renal problems and the ratio of mutant to normal mtDNA in the urine. Surprisingly, the patient with the highest ratio of mutant to normal mtDNA was completely asymptomatic.

tRNA^{Tyr} (A5843G)

In a single case report, a 34-month-old girl presented with steroid-resistant nephritic syndrome (Scaglia et al 2003). The renal biopsy was consistent with FSGS. Her family history was significant for migraine headaches and early onset of stroke on the maternal side, and at the age of 7 years, she developed ESRD.

In contrast to tRNA^{Leu}, no hyalinized lesions, blood vessel abnormalities, or abnormalities in mitochondria in podocytes or tubules were noted. Complexes I and III showed significant reduced activity in muscle. A homoplasmic mutation of A5843G was reported in this patient at a highly conserved position throughout the evolution in mitochondrial tRNA^{Tyr}. This mutation was also present in the asymptomatic mother of the patient. In addition to LHON, this is another example of variable penetrance in a homoplasmic mutation.

INTERSTITIAL NEPHRITIS

Mitochondrial interstitial nephritis constitutes a minority of clinical cases. However, the mechanism of the disease remains a mystery. Mutations in prenyltransferase-like protein (a mitochondrial protein encoded by the nuclear genome) in animal models lead to interstitial nephritis (Peng et al 2004). Dysregulation of apoptosis has been proposed as a mechanism of pathogenesis of inflammatory processes. It is interesting to note that a mitochondrial mutation in the renal epithelial tissue may lead to inflammation while the liver, heart, and brain are unaffected.

tRNA^{Phe} A608G

The case was reported in two brothers of one family (Tzen et al 2001). The two brothers were noted to have chronic renal insufficiency at 1 year old and 27 months old. They also suffered from multiple extrarenal disorders, such as atrial septal defect, muscle weakness, seizures, and ataxia.

Renal biopsies from both brothers revealed tubulointerstitial nephritis. Many tubules contained amorphous material positive for periodic acid–schiff (PAS). On electron microscopy, renal tubular epithelial cells revealed mitochondria of variable sizes and shapes with some of the mitochondria showing disintegration. mtDNA extracted from renal biopsy and leukocytes showed a homoplasmic mutation in A608G in the tRNA^{Phe} in a conserved position. This position led to shorting of the anticodon stem and lengthening of the antocodon loop. The mother of these children harbors the same mutation; however, she is unaffected. The presence of nuclear modifiers could explain the presence of incomplete penetrance. For example, a nuclear tRNA^{Phe} could be a potential source of this clinical presentation.

A5656G

Two siblings were noted to have chronic kidney disease with hyponatremia (Zsurka et al 1997). They suffered from mitral prolapse and cardiac conduction defects. One living cousin and two other relatives suffered from renal dysfunction.

The renal biopsy of the brothers and their cousin was consistent with tubulointerstitial nephritis and electron microscopy revealed dysmorphic mitochondrial changes in a large number of cells of distal tubules. DNA samples collected from leukocytes showed a homoplasmic mutation in A5656G. This is a single noncoding nucleotide separating tRNA^{Ala} and tRNA^{Asn}, and the exact function of this mutation is very unclear. It is possible to imagine that the processing of one or both of these tRNAs is affected by this mutation.

Deletion 10598–13206

A single 11-year-old boy presented with chronic kidney disease and moderate proteinuria (Rotig et al 1995). He was eventually started on hemodialysis and underwent cadaveric renal transplant. However, he developed leukodystrophy 8 years later.

His renal biopsy showed interstitial fibrosis with tubular obstruction by hyaline casts, and the glomeruli were sclerosed. A muscle biopsy revealed ragged-red fibers, but enzymatic studies in the muscle, skin fibroblast, and lymphocytes were normal.

Southern blot analysis revealed a heteroplasmic deletion of 2608 base pairs. This deletion spans positions 10598–13206, which includes tRNA^{His}, tRNA^{Ser}, tRNA^{Leu} CUN, ND4, ND5, and part of ND4L. This heteroplasmic deletion was present in cerebral white matter, muscle, and fibroblast at 86%, 62%, and 36% respectively.

Fragment Deletion

A 9-year-old girl with history of megaloblastic anemia was evaluated for 2+ proteinuria and progressive loss of renal function (Szabolcs et al 1994). Over the years, she developed decreased extraocular movement and bilateral ptosis.

Renal biopsy showed stripe-like rays associated with interstitial fibrosis. The tubules contained hyaline casts. Electron microscopy showed abnormalities in size, number, and morphology of mitochondria. Renal sections were negative for cytochrome C oxidase staining.

Genetic analysis revealed a 2.8 kb mtDNA heteroplasmic deletion (90% and 50% in renal cells and leukocytes respectively). While the location of the deletion was not fine mapped, it is between bases 9765 and 13692.

tRNA^{Leu}A3243G

While this mutation is associated predominantly with FSGS, there are three reports of association with interstitial nephritis (Guery et al 2003). Two sisters 24 and 39 years old with neurosensory deafness, DMII, hypertrophic cardiomyopathy, and multiple neurological disorders presented with chronic renal failure. The third patient also suffered from deafness, DMII, and hypertrophic cardiomyopathy. Renal biopsies of these patients only revealed severe tubular interstitial

nephritis. Electron microscopy revealed an increased number of enlarged mitochondria in the proximal tubules.

FANCONI'S SYNDROME

Although we will focus on the mitochondrial origin of Fanconi's syndrome, this disorder can result from many inherited metabolic diseases, toxic agents, and immune disorders. As tubular reabsorption is heavily energy-dependent in the proximal tubules, it is not surprising that mitochondrial disorders lead to Fanconi's syndrome. Affected patients develop hypokalemia, hypophosphotemic rickets, glycosuria, hypercalciuria, aminoaciduria, proteinuria and hypouricemia, and proximal renal tubular acidosis. Renal manifestation could be limited to only one aspect of these abnormalities. For example, there is a report of proximal renal acidosis with hypercalciuria (Matsutani et al 1992). The mitochondrial abnormalities alone or in combination with electrolyte abnormalities can lead to severe muscle weakness (Van Biervliet et al 1977, DiMauro et al 1980, Heiman-Patterson et al 1982, Muller-Hocker et al 1983, Zeviani et al 1985, Sperl et al 1988, Morris et al 1995, Ning et al 1996, Wang et al 2000). In addition to myopathy (Sperl et al 1988, Niaudet et al 1994), neurological symptoms (Ogier et al 1988), deafness (Egger et al 1981, Eviatar et al 1990), diabetes mellitus (Eviatar et al 1990, Majander et al 1991) or cardiac problems could be present. Renal abnormalities could be the first clinical sign of mitochondrial disorders without extrarenal manifestations (Eviatar et al 1990, Mochizuki et al 1996, Kuwertz-Broking et al 2000) since renal tubular cells are very ATP sensitive. Long-lasting mitochondrial disorders may lead to renal failure. The majority of patients presents with renal disorders before the age of two. In addition, this disorder is fatal in 40% of patients, primarily due to asphyxia secondary to muscular insufficiency (Van Biervliet et al 1977). Familial representation of this disease is only present in approximately 50% of cases.

Many of these patients also have Pearson's syndrome (Majander et al 1991, Niaudet et al 1994), Kearns-Sayre syndrome (Eviatar et al 1990, Mori et al 1991), or Leigh's syndrome (Ogier et al 1988). The exact etiology of disease is unclear. In several studies, large mtDNA deletions are reported (Eviatar et al 1990, Majander et al 1991, Mori et al 1991, Niaudet et al 1994). However, in most cases, these studies are not performed. Studies of the muscle mainly reveal decreased or deficient cytochrome C oxidase (Van Biervliet et al 1977, DiMauro et al 1980, Mori et al 1991, Mochizuki et al 1996, Kuwertz-Broking et al 2000). Enzymatic studies also show decreased activity of complex III or IV predominantly (Sperl et al 1988, Mori et al 1991, Morris et al 1995, Kuwertz-Broking et al 2000).

Renal biopsies show nonspecific abnormalities of tubular epithelium with obstruction with casts, and atrophy. Giant dysmorphic mitochondria are always present (Thorner et al 1985, Mori et al 1991, Mochizuki et al 1996).

RHABDOMYOLYSIS

Rare disorders of carbohydrate metabolism, and fatty acid oxidation, may present as muscle break-down. Recurrent rhabdomyolysis in a healthy patient or a patient with neurological disorders suggestive of mitochondrial dysfunction should raise the suspicion for further evaluation of mitochondrial disorders.

tRNA[Phe] A606G

A 30-year-old male developed muscle pain and weakness after exercise (Chinnery et al 1997). He subsequently had dark urine on one occasion and developed malaise and generalized muscle weakness. His creatine kinase (CK) was 21290 U/L. His laboratory exams were otherwise normal.

A muscle biopsy revealed that 1–2% of muscle fibers had subsarcolemmal accumulation of mitochondria and that 39% of the fibers lacked cytochrome C oxidase staining. Enzymatic studies were consistent with decreased complex IV activities.

A heteroplasmic mutation in the anticodon stem of tRNA[Phe] was noted (A606G). This mutation was present at 96% and 62% in skeletal muscle and leukocytes respectively. His mother was asymptomatic with 54% mutated mtDNA in leukocytes.

Cytochrome C Oxidase Deficiency

A 15-year-old female had four episodes of myoglobinuria after prolonged exercise, a viral illness, and two unidentified episodes (Keightley et al 1996). On two of these occasions, the patient had CK of 17 000–44 000. Her exercise stress test was normal.

A muscle biopsy demonstrated RRF with increased SDH staining. The biopsy showed 64% cytochrome C oxidase negative fibers. Electron microscopy showed an increased number of mitochondria, enlarged mitochondria, and irregular cristae. Enzymatic studies revealed a decrease in complex IV activity.

Further analysis showed a 15 bp in-frame deletion (9488–9503) in the third hydrophobic domain of COX III, which includes two phenylalanine conserved residues. This might explain why a mutation in tRNA[Phe] also causes rhabdomyolysis. This mutation was present at a 92% rate in skeletal muscle while in leukocytes and fibroblasts the rates were 0.7% and 0% respectively. Her mother lacked this mutation in leukocytes. This is consistent with a de novo deletion in this patient. In general, deleted mtDNA are rarely transmitted from affected women to their children

(Chinnery et al 1999). In this case, the causative mutation probably originated in the germ line.

A 3-week-old female became lethargic and had recurrent attacks of limb hypertonia (Saunier et al 1995). Upon evaluation, she had a CK of 96950. She had increased urinary succinate, and α-ketoglutarate. She had additional episodes at the age of 3 months and 6 months. In the last episode, she also had anemia, thrombocytopenia, hemorrhagic diarrhea, and schizocytosis. Enzymatic studies showed decreased cytochrome C oxidase activity and decreased oxygen consumption. While the exact mutation was not identified, the mitochondrial origin of the disorder is strongly speculated.

mtDNA Deletion

Two brothers had multiple episodes of rhabdomyolysis since age 18. These episodes were provoked by strenuous exercise, alcohol intake, or fasting (Ohno et al 1991). RRF was noted on muscle biopsy in both patients at a rate of 4.7% and 1.1%. The COX-negative fibers were present at a proportion of 8.7% and 7.9%. Electron microscopy revealed an increase in the number of dysmorphic mitochondria and a moderate increase of glycogen particles. Enzymatic studies on the muscle of one of the patients were normal. Further analysis demonstrated a deletion on the mitochondrial genome of these two patients. The boundaries of this heteroplasmic deletion are not clear; however, it at least includes ND4, ND4L, and part of ND5. These are all components of complex I.

ONCOCYTOMA

Oncocytomas are benign tumors predominantly occurring in the kidney. Histologically, they reveal epithelial cells with granular eosinophilic cytoplasm. These cells have an increased number of mitochondria and the mitochondria are dysmorphic. Several studies have linked mitochondria with carcinogenesis. Due to the changes of mitochondria in oncocytomas, six oncocytomas were evaluated (Welter et al 1989). This study revealed a 40 bp deletion in the mtDNA which was localized to the cytochrome C oxidase subunit I gene. Interestingly, the adjacent healthy tissue of the patients did not show this deletion.

ELECTROLYTE ABNORMALITIES

While patients with Fanconi's syndrome show many electrolyte abnormalities such as hypophosphatemia, hypokalemia, and hypomagnesemia, there is a large number of patients with electrolyte abnormalities and mitochondrial mutations who lack aminoaciduria, acidosis, and glucosuria.

The exact mechanism of these disorders is not clear. There are 26 cases of patients with known or suspected mitochondrial disorders and electrolyte abnormalities whose profiles do not fit the Fanconi's syndrome. While most of these patients have Kearns-Sayre syndrome (Simopoulos et al 1971, Toppet et al 1977, Horwitz & Roessmann 1978, Rheuban et al 1983, Dewhurst et al 1986, Goto et al 1990, Harvey & Barnett 1992, Wilichowski et al 1997, Francesco et al 2003, Katsanos et al 2001) there are patients with sensorineural deafness (Kendall 1965, Gomez et al 1972), myopathy (Spiro et al 1970, Simopoulos et al 1971, Gomez et al 1972, Wilichowski et al 1997), cardiomyopathy (Riggs et al 1992), and Leigh's syndrome (Simopoulos et al 1971). The disorder is equally distributed among females and males, and the average age of onset is 11 years old. None of the reported cases showed any evidence of familial electrolyte abnormality. The majority of these patients were evaluated for seizure, paresthesia, spasm, and tetany. Typically, ptosis predated the manifestation of electrolyte abnormalities. Further evaluation led to the finding of hypocalcemia, hyperphosphotamia, and hypoparathyroidism (Kendall 1965, Gomez et al 1972, Jonnard 1977, Toppet et al 1977, Horwitz & Roessmann 1978, Pellock et al 1978, Allen et al 1983, Dewhurst et al 1986, Wilichowski et al 1997, Katsanos et al 2001, Lee et al 2001). While magnesium was not checked in all the patients with hypoparathyroid disorder, serum magnesium was significantly decreased in the patients with available values (Toppet et al 1977, Dewhurst et al 1986, Goto et al 1990, Wilichowski et al 1997, Lee et al 2001, Francesco et al 2003). Hypomagnesemia was noted in patients without hypoparathyroidism. One patient also had spasms without hypocalcemia due to hypomagnesemia alone (Carboni et al 1981, Rheuban et al 1983, Goto et al 1990, Harvey & Barnett 1992). In addition, all patients with hypomagnesemia showed renal wasting hypokalemia (Simopoulos et al 1971, Toppet et al 1977, Goto et al 1990, Katsanos et al 2001, Lee et al 2001, Menegon et al 2004). It is quite possible that these patients also suffer from potassium malabsorption. Hyperaldosteronism and hypomagnesemia were associated in seven of these cases (Simopoulos et al 1971). In one study, a patient with myopathy and blindness showed potassium, magnesium and salt-wasting (Menegon et al 2004). Interestingly, a measurement of lithium clearance illuminated a distinction between the fractional excretion of the proximal and distal regions of the kidney. While the proximal excretion of sodium was decreased, the distal tubules were predominantly responsible for salt-wasting. In most of these cases, the mechanism of mitochondrial abnormalities was not described at all. Only seven of these cases demonstrate a clear mitochondrial abnormality (Goto et al 1990, Wilichowski et al 1997, Lee et al 2001, Francesco et al 2003). They all show mitochondrial heteroplasmic genomic rearrangements in the region of 7883–14288.

There is only one report of familial mitochondrial electrolyte abnormality without evidence of Fanconi's syndrome.

A kindred with 142 blood relatives was reported to have a novel syndrome of hypomagnesemia, hypertension, and hypercholesterolemia (Wilson et al 2004). When analyzed as a quantitative or qualitative trait, each is strikingly more prevalent in the maternal lineage. The analysis of the mitochondrial genome revealed a mutation in tRNA[Ile] immediately 5′ to the anticodon (T4291C). This position is conserved in all known tRNAs. In addition to the renal magnesium wasting, they showed hypokalemia due to renal potassium wasting. In contrast to other reported cases, they had normal calcium. However, similar to other familial causes of hypomagnesemia due to distal tubular abnormalities they had hypocalciuria. The high number of mitochondria and the ATP sensitivity of the distal tubules (Reilly & Ellison 2000) could explain how mitochondrial dysfunction could lead to the above abnormalities. It is not surprising that all the known genes of familial hypomagnesemia in the distal tubules are ATP-dependent (Simon et al 1996, Meij et al 2000, Schlingmann et al 2002). The maternal members of this kindred showed an increased prevalence of sensorineural hearing loss and migraine.

HYPERTENSION

While the etiology of hypertension is unclear, the interaction of the environment and genetic factors significantly contributes to this disorder (Lifton et al 2001). The role of genetic factors in the pathogenesis of hypertension is clear, but the mode of inheritance of essential hypertension is unclear and it is considered to be a complex trait. The analysis of Mendelian diseases of hypertension and hypotension has been fruitful in dissecting the genes involved in blood pressure regulation (Lifton et al 2001). Most of these genetic analyses have focused on the nuclear genome; as a result, the mitochondrial genome has not been investigated. This is surprising as the mitochondrial genome, which is significantly smaller than the nuclear genome, had been characterized almost two decades before the nuclear genome. The mitochondrial genome is significantly smaller than nuclear genome. Several early studies have supported the predominant effect of mother–offspring correlation (Annest et al 1979, Bengtsson et al 1979, Longini et al 1984, Brandao et al 1992, DeStefano et al 2001, Sun et al 2003). The original studies contributed this effect to the role of the placenta in the pathogenesis of hypertension in the offspring. Several studies in experimental animals provided evidence that increased reactive oxygen species (ROS) were involved in progression of hypertension (Hamilton et al 2004). Mitochondria have been recognized as major source of ROS. Finally, mitochondrial dysfunction is thought to contribute to the process of aging (Petersen et al 2003, Trifunovic et al 2004), and it is well recognized that the prevalence of hypertension increases with age (Vasan et al 2002).

While mitochondrial studies have been the focus of analysis in studying essential hypertension, a mitochondrial origin was suspected among patients with preeclampsia and eclampsia. The pattern of inheritance clearly fits a maternal pattern of inheritance with variable penetrance. A heteroplasmic mutation in tRNA^{Leu} (A3243G) was identified in a family with three generations of preeclampsia and eclampsia. In a second family, a mutation in another tRNA^{Leu} (A12308G) was identified (Folgero et al 1996). While the cause and effect were not established, this was very suggestive of contributing to the preeclampsia. The A3243G seems to be not only a common polymorphism but it also could be associated with many different renal involvements.

Further phenotypic analysis of a kindred with a tRNA^{Lys} mutation (A8344G) which causes myoclonic epilepsy and ragged red fibers syndrome (MERRF), showed lipomas, diabetes, ataxia, and peripheral neuropathy (Austin et al 1998). Six living affected family members in five generations were also noted to have hypertension. This was the first direct evidence linking any mitochondrial mutation to hypertension.

The mitochondrial genome was assessed as a contributing site for progression of end-stage renal disease due to hypertension among black Americans. Fifty-eight patients with end-stage renal disease due to hypertension were compared with 58 normotensive individuals without any kidney disease (Watson et al 2001). The investigators used a high-resolution restrict map analysis. A mutation in ND3 in complex I (A10086G, Asn→Asp) was found to occur at a significantly higher frequency in the patients compared to controls. This study suggests that mitochondrial mutations contribute to the development of essential hypertension and also to the progression of hypertension and ESRD with aging.

The analysis of the mitochondrial genome of 20 probands with essential hypertension from African-American, Caucasian, and Greek Caucasian backgrounds revealed several base changes (Schwartz et al 2004). These changes were noted in 9 tRNAs (Ile, Gln, Met, Ala, Cys, Arg, Ser, Leu2, Thr). There were four variations in tRNA^{Thr} and tRNA^{Leu2}. Eight of the probands had two or more variations in the tRNAs. The analysis identified 17 novel amino acid changes in all the complexes. Two previously reported amino acid substitutions were also identified. These substitutions include mutations in Leber's hereditary optic neuropathy, MELAS, and SIDS. While three of these probands lacked any amino acid changes, the rest had at least two changes and some, up to 15 changes. It is not clear whether these variations in the tRNAs and changes in the amino acids were heteroplasmic or homoplasmic. Schwartz et al also reported a variation in the control region. Many of these patients had multiple variations in the mitochondrial genome. This could be consistent with the known hypothesis of mitochondrial

reserve, where an original mitochondrial dysfunction due to a mutation may lead to additional changes in the mitochondrial genome and progressive deterioration. However, in this study, the age and genotype of other affected members were not compared with those of the probands.

In one kindred of 142 patients with hypertension, hypomagnesemia, and hypercholesterolemia, a homoplasmic mutation in tRNA^{Ile} was described (Wilson et al 2004). In contrast to all the known genetic causes of hypertension, the onset of hypertension was in adulthood. In addition, the prevalence of hypertension increased with age in a pattern that resembled essential hypertension. However, the prevalence of hypertension in the maternal members of this kindred was significantly higher compared to the Caucasians in the NHANES studies. This mutation contributed to an increase in systolic and diastolic blood pressure. There is no evidence of salt-wasting in this kindred. The common association of hypertension with hypomagnesemia in several epidemiological studies could be explained based on this study. Interestingly, despite the presence of a homoplasmic mutation, the penetrance was approximately 50%. The variable penetrance of the effect of mitochondrial mutation, late age of onset, and the high phenocopy rate of hypertension could explain why the effect of the mitochondrial genome has not been explored.

While these studies show strong genetic evidence for the role of mitochondria in the pathogenesis of hypertension, the exact mechanisms need to be explored further. Many have suggested a role based on the generation of ROS and their damage to the mitochondrial genome. Recent animal models have not supported a role of ROS in aging of the mice with mitochondrial mutations (Kujoth et al 2005).

TREATMENT

There is no effective treatment for diseases of the mitochondria. Treatment is primarily symptomatic. In addition, patients should avoid respiratory chain toxic medications (valproic, barbiturates, tetracycline, chloramphenicol). Patients with complex III deficiency may benefit from vitamin K-3 (40–160 mg/day) or coenzyme Q10 (80–300 mg/day). Patients with complex I deficiency may be treated with riboflavin (100 mg/day) (Niaudet 1998). Vitamin C has been prescribed to prevent ROS damage. Dichloroacetate, an activator of pyruvate dehydrogenase complex, accelerates the oxidation of glucose, lactate, and pyruvate to acetylcoenzyme A. This product has been used at 25–200 mg/kg and improvement has been observed in patients with lactic acidosis (Neiberger et al 2002, Duncan et al 2004). Mitochondrial diseases are rare and many are not limited to one specific complex. As a result, performing controlled trials is not possible. Moreover, the phenotype of the disease in each family alone is quite variable, and so are the responses to medications. Nuclear genetic modifiers could influence the response drastically.

GENETIC COUNSELING

Many centers do not offer any prenatal diagnostic testing through amniocentesis or chorionic villous biopsy for mitochondrial disorders. There are several problems with such analyses. The level of heteroplasmy measured in the fetus may not correspond to the other tissues and may change during the course of pregnancy. Similarly, the establishment of the threshold level for occurrence of the disease is not clear and is only based on past experiences (Jacobs et al 2005). Finally, there may not be any correlation between the level of heteroplasmy and the severity of disease. It has been suggested that there is little variation of mutation in the tissues in utero in a cause of NARP and Leigh syndrome: T8993G mutation (White et al 1999). However, these results cannot be extended to other mutations. Preimplantation genetic diagnosis could be used for mitochondrial disorders. Couples with strong reservation about embryo testing could use the polar body from an unfertilized oocyte. This could also be performed in 8-cell embryos, followed by implantation of healthy embryos. The high number of mitochondria in the oocyte and in the 8-cell embryo allows this technique to be possible. However, determining the threshold level is a problem.

The most reliable method to avoid the disease is using a donor oocyte. This approach is combined with in vitro fertilization using the partner's sperm. Relatives along the maternal lineage will be prevented from donating. The major limitation is lack of a child who is biologically related to both parents. Cytoplasmic transfer allows the dilution of the mutant mitochondria with the donor mitochondria (Kagawa & Hayashi 1997). In this approach, large volumes of cytoplasm need to be transferred. Nuclear transfer has been suggested as another possible way of preventing disease. The transfer of the nucleus of the oocyte of the mother to the enucleated donor oocyte will allow this (Roberts 1999). However, the transferred nucleus will carry a small portion of the original mitochondria and cytoplasm (Jansen 2000).

We still do not know the answer to many important questions about mitochondrial function and biology. Patients with mtDNA mutations do not have a classic phenotype. Many common disorders such as hypertension and chronic kidney disease may be affected by mitochondrial dysfunction. With respect to aging, decreased renal function with age could be further pursued. The mitochondrial mutations in common diseases are probably undiagnosed. Further awareness will provide the opportunity for referral and diagnosis of these disorders in the kidney. While current treatments of mitochondrial disorders have not proven to be effective, further research will allow us to explore new options.

ACKNOWLEDGMENTS

I am grateful to Dr Weizhen Ji, Hakan R. Toka, Carol Nelson-Williams and Aaron Ring for helpful discussions. This work was supported by American Heart Association Fellow-to- Faculty Transition Award (0475003N).

References

Allen RJ, DiMauro S, Coulter DL, Papadimitriou A, Rothenberg SP. Kearns-Sayre syndrome with reduced plasma and cerebrospinal fluid folate. Ann. Neurol. 1983; 13: 679–82.

Anderson S, Bankier AT, Barrell BG, et al. Sequence and organization of the human mitochondrial genome. Nature 1981; 290: 457–65.

Annest JL, Sing CF, Biron P, Mongeau JG. Familial aggregation of blood pressure and weight in adoptive families. I. Comparisons of blood pressure and weight statistics among families with adopted, natural, or both natural and adopted children. Am. J. Epidemiol. 1979; 110: 479–91.

Austin SA, Vriesendorp FJ, Thandroyen FT, Hecht JT, Jones OT, Johns DR. Expanding the phenotype of the 8344 transfer RNA lysine mitochondrial DNA mutation. Neurology 1998; 51: 1447–50.

Bengtsson B, Thulin T, Schersten B. Familial resemblance in casual blood pressure–a maternal effect? Clin. Sci. (London) 1979; 57(Suppl 5): 279s–81s.

Brandao AP, Brandao AA, Araujo EM, Oliveira RC. Familial aggregation of arterial blood pressure and possible genetic influence. Hypertension 1992; 19: II214–17.

Carboni P, Giacanelli M, Porro G, Sideri G, Paolella A. Kearns-Sayre syndrome. A case of the complete syndrome with encephalic leukodystrophy and calcification of basal ganglia. Ital. J. Neurol. Sci. 1981; 2: 263–8.

Carelli V, Giordano C, d'Amati G. Pathogenic expression of homoplasmic mtDNA mutations needs a complex nuclear-mitochondrial interaction. Trends Genet. 2003; 19: 257–62.

Carrodeguas JA, Theis K, Bogenhagen DF, Kisker C. Crystal structure and deletion analysis show that the accessory subunit of mammalian DNA polymerase gamma, Pol gamma B, functions as a homodimer. Mol. Cell 2001; 7: 43–54.

Chinnery PF, Howell N, Andrews RM, Turnbull DM. Clinical mitochondrial genetics. J. Med. Genet. 1999; 36: 425–36.

Chinnery PF, Johnson MA, Taylor RW, Lightowlers RN, Turnbull DM. A novel mitochondrial tRNA phenylalanine mutation presenting with acute rhabdomyolysis. Ann. Neurol. 1997; 41: 408–10.

Coenen MJ, Antonicka H, Ugalde C, et al. Mutant mitochondrial elongation factor G1 and combined oxidative phosphorylation deficiency. N. Engl. J. Med. 2004; 351: 2080–6.

DeStefano AL, Gavras H, Heard-Costa N, et al. Maternal component in the familial aggregation of hypertension. Clin. Genet. 2001; 60: 13–21.

Dewhurst AG, Hall D, Schwartz MS, McKeran RO. Kearns-Sayre syndrome, hypoparathyroidism, and basal ganglia calcification. J. Neurol. Neurosurg. Psychiatry 1986; 49: 1323–4.

DiMauro S, Mendell JR, Sahenk Z, et al. Fatal infantile mitochondrial myopathy and renal dysfunction due to cytochrome-c-oxidase deficiency. Neurology 1980; 30: 795–804.

Dinour D, Mini S, Polak-Charcon S, Lotan D, Holtzman EJ. Progressive nephropathy associated with mitochondrial tRNA gene mutation. Clin. Nephrol. 2004; 62: 149–54.

Doleris LM, Hill GS, Chedin P, et al. Focal segmental glomerulosclerosis associated with mitochondrial cytopathy. Kidney Int. 2000; 58: 1851–8.

Duncan GE, Perkins LA, Theriaque DW, Neiberger RE, Stacpoole PW. Dichloroacetate therapy attenuates the blood lactate response to submaximal exercise in patients with defects in mitochondrial energy metabolism. J. Clin. Endocrinol. Metab. 2004; 89: 1733–8.

Egger J, Lake BD, Wilson J. Mitochondrial cytopathy. A multisystem disorder with ragged red fibres on muscle biopsy. Arch. Dis. Child. 1981; 56: 741–52.

Eviatar L, Shanske S, Gauthier B, et al. Kearns-Sayre syndrome presenting as renal tubular acidosis. Neurology 1990; 40: 1761–3.

Falkenberg M, Gaspari M, Rantanen A, Trifunovic A, Larsson NG, Gustafsson CM. Mitochondrial transcription factors B1 and B2 activate transcription of human mtDNA. Nat. Genet. 2002; 31: 28–94.

Fliss MS, Usadel H, Caballero OL, et al. Facile detection of mitochondrial DNA mutations in tumors and bodily fluids. Science 2000; 287: 2017–19.

Folgero T, Storbakk N, Torbergsen T, Oian P. Mutations in mitochondrial transfer ribonucleic acid genes in preeclampsia. Am. J. Obstet. Gynecol. 1996; 174: 1626–30.

Francesco Emma CP, Tezza A, Santorelli F, Bertini E, Rizzoni G. Bartter-like phenotype and mitochondrial DNA heteroplasmy in Kearns-Sayre syndrome. In: J. Am. Soc. Nephrol. (Annual ASN meeting), 2003: pp. 748–9.

Fries JW, Sandstrom DJ, Meyer TW, Rennke HG. Glomerular hypertrophy and epithelial cell injury modulate progressive glomerulosclerosis in the rat. Lab. Invest. 1989; 60: 205–18.

Gomez MR, Engel AG, Dyck PJ. Progressive ataxia, retinal degeneration, neuromyopathy, and mental subnormality in a patient with true hypoparathyroidism, dwarfism, malabsorption, and cholelithiasis. Neurology 1972; 22: 849–55.

Goto Y, Itami N, Kajii N, Tochimaru H, Endo M, Horai S. Renal tubular involvement mimicking Bartter syndrome in a patient with Kearns-Sayre syndrome. J. Pediatr. 1990; 116: 904–10.

Guery B, Choukroun G, Noel LH, et al. The spectrum of systemic involvement in adults presenting with renal lesion and mitochondrial tRNA(Leu) gene mutation. J. Am. Soc. Nephrol. 2003; 14: 2099–108.

Guillausseau PJ, Massin P, Dubois-LaForgue D, et al. Maternally inherited diabetes and deafness: a multicenter study. Ann. Intern. Med. 2001; 134: 721–8.

Hamilton CA, Miller WH, Al-Benna S, et al. Strategies to reduce oxidative stress in cardiovascular disease. Clin. Sci. (London) 2004; 106: 219–34.

Harman D. Role of free radicals in aging and disease. Ann. NY Acad. Sci. 1992; 673: 126–41.

Harvey JN, Barnett D. Endocrine dysfunction in Kearns-Sayre syndrome. Clin. Endocrinol. (Oxford), 1992; 37: 97–103.

Heiman-Patterson TD, Bonilla E, DiMauro S, Foreman J, Schotland DL. Cytochrome-c-oxidase deficiency in a floppy infant. Neurology 1982; 32: 898–901.

Hekimi S, Guarente L. Genetics and the specificity of the aging process. Science 2003; 299: 1351–4.

Horwitz SJ, Roessmann U. Kearns-Sayre syndrome with hypoparathyroidism. Ann. Neurol. 1978; 3: 513–18.

Hotta O, Inoue CN, Miyabayashi S, Furuta T, Takeuchi A, Taguma Y. Clinical and pathologic features of focal segmental glomerulosclerosis with mitochondrial tRNALeu(UUR) gene mutation. Kidney Int. 2001; 59: 1236–43.

Ireland J, Rossetti S, Haugen E, Michels V, Harris P. Mitochondrial causes of renal insufficiency and hearing loss. Kidney Int. 2004; 65: 2444–5.

Iwasaki N, Babazono T, Tsuchiya K, et al. Prevalence of A-to-G mutation at nucleotide 3243 of the mitochondrial tRNA(Leu(UUR)) gene in Japanese patients with diabetes mellitus and end stage renal disease. J. Hum. Genet. 2001; 46: 330–4.

Jacobs LJ, de Coo IF, Nijland JG, et al. Transmission and prenatal diagnosis of the T9176C mitochondrial DNA mutation. Mol. Hum. Reprod. 2005; 11: 223–8.

Jansen JJ, Maassen JA, van der Woude FJ, et al. Mutation in mitochondrial tRNA(Leu(UUR)) gene associated with progressive kidney disease. J. Am. Soc. Nephrol. 1997; 8: 1118–24.

Jansen RP. Germline passage of mitochondria: quantitative considerations and possible embryological sequelae. Hum. Reprod. 2000; 15(Suppl 2): 112–28.

Jean-Francois MJ, Lertrit P, Berkovic SF, et al. Heterogeneity in the phenotypic expression of the mutation in the mitochondrial tRNA(Leu) (UUR) gene generally associated with the MELAS subset of mitochondrial encephalomyopathies. Aust. NZ J. Med. 1994; 24: 188–93.

Jeronimo C, Nomoto S, Caballero OL, et al. Mitochondrial mutations in early stage prostate cancer and bodily fluids. Oncogene 2001; 20: 5195–8.

Johns DR, Hurko O. Mitochondrial leucine tRNA mutation in neurological diseases. Lancet 1991; 337: 927–8.

Jonnard A, Machecourt D, Mouillon M, Ghislofi J, Bost M, Beaudoing A. Syndrome de Kearns avec hypocalcemie transitorie. Pediatrie. 1977; 32: 797–806.

Kagawa Y, Hayashi JI. Gene therapy of mitochondrial diseases using human cytoplasts. Gene. Ther. 1997; 4: 6–10.

Katsanos KH, Elisaf M, Bairaktari E, Tsianos EV. Severe hypomagnesemia and hypoparathyroidism in Kearns-Sayre syndrome. Am. J. Nephrol. 2001; 21: 150–3.

Keightley JA, Hoffbuhr KC, Burton MD, et al. A microdeletion in cytochrome c oxidase (COX) subunit III associated with COX deficiency and recurrent myoglobinuria. Nat. Genet. 1996; 12: 410–16.

Kendall D. Hypoparathyroidism. Proc. R. Soc. Med. 1965; 58: 1087–8.

Kolesnikova OA, Entelis NS, Mireau H, et al. Suppression of mutations in mitochondrial DNA by tRNAs imported from the cytoplasm. Science 2000; 289: 1931–3.

Kriz W, Lemley KV. The role of the podocyte in glomerulosclerosis. Curr. Opin. Nephrol. Hypertens. 1999; 8: 489–97.

Kujoth GC, Hiona A, Pugh TD, et al. Mitochondrial DNA mutations, oxidative stress, and apoptosis in mammalian aging. Science 2005; 309: 481–4.

Kurogouchi F, Oguchi T, Mawatari E, et al. A case of mitochondrial cytopathy with a typical point mutation for MELAS, presenting with severe focal-segmental glomerulosclerosis as main clinical manifestation. Am. J. Nephrol. 1998; 18: 551–6.

Kuwertz-Broking E, Koch HG, Marquardt T, et al. Renal Fanconi syndrome: first sign of partial respiratory chain complex IV deficiency. Pediatr. Nephrol. 2000; 14: 495–8.

Lee SS, Lee RY, Fraser AG, Kamath RS, Ahringer J, Ruvkun G. A systematic RNAi screen identifies a critical role for mitochondria in *C. elegans* longevity. Nat. Genet. 2003; 33: 40–8.

Lee SS, Ruvkun G. Longevity: don't hold your breath. Nature 2002; 418: 287–8.

Lee YS, Yap HK, Barshop BA, Rajalingam S, Loke KY. Mitochondrial tubulopathy: the many faces of mitochondrial disorders. Pediatr. Nephrol. 2001; 16: 710–12.

Lenaz G, Bovina C, D'Aurelio M, et al. Role of mitochondria in oxidative stress and aging. Ann. NY Acad. Sci. 2002; 959: 199–213.

Li Z, Li R, Chen J, et al. Mutational analysis of the mitochondrial 12S rRNA gene in Chinese pediatric subjects with aminoglycoside-induced and non-syndromic hearing loss. Hum. Genet. 2005; 117: 9–15.

Lifton RP, Gharavi AG, Geller DS. Molecular mechanisms of human hypertension. Cell 2001; 104: 545–56.

Longini IM Jr., Higgins MW, Hinton PC, Moll PP, Keller JB. Environmental and genetic sources of familial aggregation of blood pressure in Tecumseh, Michigan. Am. J. Epidemiol. 1984; 120: 131–44.

Majamaa K, Moilanen JS, Uimonen S, et al. Epidemiology of A3243G, the mutation for mitochondrial encephalomyopathy, lactic acidosis, and strokelike episodes: prevalence of the mutation in an adult population. Am. J. Hum. Genet. 1998; 63: 447–54.

Majander A, Suomalainen A, Vettenranta K, et al. Congenital hypoplastic anemia, diabetes, and severe renal tubular dysfunction associated with a mitochondrial DNA deletion. Pediatr. Res. 1991; 30: 327–30.

Mandavilli BS, Santos JH, Van Houten B. Mitochondrial DNA repair and aging. Mutat. Res. 2002; 509: 127–51.

Mandel H, Szargel R, Labay V, et al. The deoxyguanosine kinase gene is mutated in individuals with depleted hepatocerebral mitochondrial DNA. Nat. Genet. 2001; 29: 337–41.

Matsutani H, Mizusawa Y, Shimoda M, et al. Partial deficiency of cytochrome c oxidase with isolated proximal renal tubular acidosis and hypercalciuria. Child Nephrol. Urol. 1992; 12: 221–4.

Meij IC, Koenderink JB, van Bokhoven H, et al. Dominant isolated renal magnesium loss is caused by misrouting of the Na(+),K(+)-ATPase gamma-subunit. Nat. Genet. 2000; 26: 265–6.

Menegon LF, Amaral TN, Gontijo JA. Renal sodium handling study in an atypical case of Bartter's syndrome associated with mitochondriopathy and sensorineural blindness. Renal Fail. 2004; 26: 195–7.

Michikawa Y, Mazzucchelli F, Bresolin N, Scarlato G, Attardi G. Aging-dependent large accumulation of point mutations in the human mtDNA control region for replication. Science 1999; 286: 774–9.

Mochizuki H, Joh K, Kawame H, et al. Mitochondrial encephalomyopathies preceded by de-Toni-Debre-Fanconi syndrome or focal segmental glomerulosclerosis. Clin. Nephrol. 1996; 46: 347–52.

Mori K, Narahara K, Ninomiya S, Goto Y, Nonaka I. Renal and skin involvement in a patient with complete Kearns-Sayre syndrome. Am. J. Med. Genet. 1991; 38: 583–7.

Morris AA, Taylor RW, Birch-Machin MA, et al. Neonatal Fanconi syndrome due to deficiency of complex III of the respiratory chain. Pediatr. Nephrol. 1995; 9: 407–11.

Muller-Hocker J, Pongratz D, Deufel T, Trijbels JM, Endres W, Hubner G. Fatal lipid storage myopathy with deficiency of cytochrome-c-oxidase and carnitine. A contribution to the combined cytochemical-finestructural identification of cytochrome-c-oxidase in longterm frozen muscle. Virchows Arch. A Pathol. Anat. Histopathol. 1983; 399: 11–23.

Nagata M, Scharer K, Kriz W. Glomerular damage after uninephrectomy in young rats. I. Hypertrophy and distortion of capillary architecture. Kidney Int. 1992; 42: 136–47.

Neiberger RE, George JC, Perkins LA, Theriaque DW, Hutson AD, Stacpoole PW. Renal manifestations of congenital lactic acidosis. Am. J. Kidney Dis. 2002; 39: 12–23.

Niaudet P. Mitochondrial disorders and the kidney. Arch. Dis. Child. 1998; 78: 387–90.

Niaudet P, Heidet L, Munnich A, Schmitz J, Bouissou F, Gubler MC, Rotig A. Deletion of the mitochondrial DNA in a case of de Toni-Debre-Fanconi syndrome and Pearson syndrome. Pediatr. Nephrol. 1994; 8: 164–8.

Ning C, Kuhara T, Inoue Y, et al. Gas chromatographic-mass spectrometric metabolic profiling of patients with fatal infantile mitochondrial myopathy with de Toni-Fanconi-Debre syndrome. Acta Paediatr. Jpn 1996; 38: 661–6.

Ogier H, Lombes A, Scholte HR, et al. de Toni-Fanconi-Debre syndrome with Leigh syndrome revealing severe muscle cytochrome c oxidase deficiency. J. Pediatr. 1988; 112: 734–9.

Ohno K, Tanaka M, Sahashi K, et al. Mitochondrial DNA deletions in inherited recurrent myoglobinuria. Ann. Neurol. 1991; 29: 364–9.

Pellock JM, Behrens M, Lewis L, Holub D, Carter S, Rowland LP. Kearns-Sayre syndrome and hypoparathyroidism. Ann. Neurol. 1978; 3: 455–8.

Peng M, Jarett L, Meade R, et al. Mutant prenyltransferase-like mitochondrial protein (PLMP) and mitochondrial abnormalities in kd/kd mice. Kidney Int. 2004; 66: 20–8.

Petersen KF, Befroy D, Dufour S, et al. Mitochondrial dysfunction in the elderly: possible role in insulin resistance. Science 2003; 300: 1140–2.

Polyak K, Li Y, Zhu H, et al. Somatic mutations of the mitochondrial genome in human colorectal tumours. Nat. Genet. 1998; 20: 291–3.

Reilly RF, Ellison DH. Mammalian distal tubule: physiology, pathophysiology, and molecular anatomy. Physiol. Rev. 2000; 80: 277–313.

Rheuban KS, Ayres NA, Sellers TD, DiMarco JP. Near-fatal Kearns-Sayre syndrome. A case report and review of clinical manifestations. Clin. Pediatr. (Philadelphia) 1983; 22: 822–5.

Riggs JE, Klingberg WG, Flink EB, Schochet SS Jr., Balian AA, Jenkins JJ 3rd. Cardioskeletal mitochondrial myopathy associated with chronic magnesium deficiency. Neurology 1992; 42: 128–30.

Roberts RM. Prevention of human mitochondrial (mtDNA) disease by nucleus transplantation into an enucleated donor oocyte. Am. J. Med. Genet. 1999; 87: 265–6.

Ross OA, McCormack R, Maxwell LD, et al. mt4216C variant in linkage with the mtDNA TJ cluster may confer a susceptibility to mitochondrial dysfunction resulting in an increased risk of Parkinson's disease in the Irish. Exp. Gerontol. 2003; 38: 397–405.

Rotig A, Goutieres F, Niaudet P, et al. Deletion of mitochondrial DNA in patient with chronic tubulointerstitial nephritis. J. Pediatr. 1995; 126: 597–601.

Saada A, Shaag A, Mandel H, Nevo Y, Eriksson S, Elpeleg O. Mutant mitochondrial thymidine kinase in mitochondrial DNA depletion myopathy. Nat. Genet. 2001; 29: 342–4.

Saunier P, Chretien D, Wood C, et al. Cytochrome c oxidase deficiency presenting as recurrent neonatal myoglobinuria. Neuromuscul. Disord. 1995; 5: 285–9.

Scaglia F, Vogel H, Hawkins EP, Vladutiu GD, Liu LL, Wong LJ. Novel homoplasmic mutation in the mitochondrial tRNATyr gene associated with atypical mitochondrial cytopathy presenting with focal segmental glomerulosclerosis. Am. J. Med. Genet. A 2003; 123: 172–8.

Schaefer AM, Taylor RW, Turnbull DM, Chinnery PF. The epidemiology of mitochondrial disorders–past, present and future. Biochim. Biophys. Acta 2004; 1659: 115–20.

Schlingmann KP, Weber S, Peters M, et al. Hypomagnesemia with secondary hypocalcemia is caused by mutations in TRPM6, a new member of the TRPM gene family. Nat. Genet. 2002; 31: 166–70.

Schon EA, Bonilla E, DiMauro S. Mitochondrial DNA mutations and pathogenesis. J. Bioenerg. Biomembr. 1997; 29: 131–49.

Schwartz F, Duka A, Sun F, Cui J, Manolis A, Gavras H. Mitochondrial genome mutations in hypertensive individuals. Am. J. Hypertens. 2004; 17: 629–35.

Schwartz M, Vissing J. Paternal inheritance of mitochondrial DNA. N. Engl. J. Med. 2002; 347: 576–80.

Segawa-Takaeda C, Takeda S, Ieki Y, et al. Focal glomerulosclerosis expanding from the glomerular vascular pole in a Japanese male with mitochondrial-DNA mutation. Nephrol. Dial. Transplant. 2002; 17: 172–4.

Shoffner JM, Brown MD, Torroni A, et al. Mitochondrial DNA variants observed in Alzheimer disease and Parkinson disease patients. Genomics 1993; 17: 171–84.

Silvestri G, Santorelli FM, Shanske S, et al. A new mtDNA mutation in the tRNA(Leu(UUR)) gene associated with maternally inherited cardiomyopathy. Hum. Mutat. 1994; 3: 37–43.

Simon DB, Nelson-Williams C, Bia MJ, et al. Gitelman's variant of Bartter's syndrome, inherited hypokalaemic alkalosis, is caused by mutations in the thiazide-sensitive Na-Cl cotransporter. Nat. Genet. 1996; 12: 24–30.

Simopoulos AP, Delea CS, Bartter FC. Neurodegenerative disorders and hyperaldosteronism. J. Pediatr. 1971; 79: 633–41.

Sperl W, Ruitenbeek W, Trijbels JM, Sengers RC, Stadhouders AM, Guggenbichler JP. Mitochondrial myopathy with lactic acidaemia, Fanconi-De Toni-Debre syndrome and a disturbed succinate: cytochrome c oxidoreductase activity. Eur. J. Pediatr. 1988; 147: 418–21.

Spiro AJ, Prineas JW, Moore CL. A new mitochondrial myopathy in a patient with salt craving. Arch. Neurol. 1970; 22: 259–69.

Sun F, Cui J, Gavras H, Schwartz F. A novel class of tests for the detection of mitochondrial DNA-mutation involvement in diseases. Am. J. Hum. Genet. 2003; 72: 1515–26.

Szabolcs MJ, Seigle R, Shanske S, Bonilla E, DiMauro S, D'Agati V. Mitochondrial DNA deletion: a cause of chronic tubulointerstitial nephropathy. Kidney Int. 1994; 45: 1388–96.

Thorner PS, Balfe JW, Becker LE, Baumal R. Abnormal mitochondria on a renal biopsy from a case of mitochondrial myopathy. Pediatr. Pathol. 1985; 4: 25–35.

Toppet M, Telerman-Toppet N, Szliwowski HB, Vainsel M, Coers C. Oculocraniosomatic neuromuscular disease with hypoparathyroidism. Am. J. Dis. Child. 1977; 131: 437–41.

Trifunovic A, Wredenberg A, Falkenberg M, et al. Premature ageing in mice expressing defective mitochondrial DNA polymerase. Nature 2004; 429: 417–23.

Tzen CY, Tsai JD, Wu TY, et al. Tubulointerstitial nephritis associated with a novel mitochondrial point mutation. Kidney Int. 2001; 59: 846–54.

Van Biervliet JP, Bruinvis L, Ketting D, De Bree PK, Van der Heiden C, Wadman SK. Hereditary mitochondrial myopathy with lactic acidemia, a De Toni-Fanconi-Debre syndrome, and a defective respiratory chain in voluntary striated muscles. Pediatr. Res. 1977; 11: 1088–93.

van den Ouweland JM, Lemkes HH, Ruitenbeek W, et al. Mutation in mitochondrial tRNA(Leu)(UUR) gene in a large pedigree with maternally transmitted type II diabetes mellitus and deafness. Nat. Genet. 1992; 1: 368–71.

van der Walt JM, Nicodemus KK, Martin ER, et al. Mitochondrial polymorphisms significantly reduce the risk of Parkinson disease. Am. J. Hum. Genet. 2003; 72: 804–11.

Vasan RS, Beiser A, Seshadri S, et al. Residual lifetime risk for developing hypertension in middle-aged women and men: The Framingham Heart Study. JAMA 2002; 287: 1003–10.

Wang LC, Lee WT, Tsai WY, et al. Mitochondrial cytopathy combined with Fanconi's syndrome. Pediatr. Neurol. 2000; 22: 403–6.

Watson B Jr., Khan MA, Desmond RA, Bergman S. Mitochondrial DNA mutations in black Americans with hypertension-associated end-stage renal disease. Am. J. Kidney Dis. 2001; 38: 529–36.

Welter C, Kovacs G, Seitz G, Blin N. Alteration of mitochondrial DNA in human oncocytomas. Genes Chromosomes Cancer 1989; 1: 79–82.

White SL, Shanske S, Biros I, et al. Two cases of prenatal analysis for the pathogenic T to G substitution at nucleotide 8993 in mitochondrial DNA. Prenat. Diagn. 1999; 19: 1165–8.

Wilichowski E, Gruters A, Kruse K, et al. Hypoparathyroidism and deafness associated with pleioplasmic large scale rearrangements of the mitochondrial DNA: a clinical and molecular genetic study of four children with Kearns-Sayre syndrome. Pediatr. Res. 1997; 41: 193–200.

Wilson FH, Hariri A, Farhi A, et al. A cluster of metabolic defects caused by mutation in a mitochondrial tRNA. Science 2004; 306: 1190–4.

Zeviani M, Nonaka I, Bonilla E, et al. Fatal infantile mitochondrial myopathy and renal dysfunction caused by cytochrome c oxidase deficiency: immunological studies in a new patient. Ann. Neurol. 1985; 17: 414–17.

Zsurka G, Ormos J, Ivanyi B, et al. Mitochondrial mutation as a probable causative factor in familial progressive tubulointerstitial nephritis. Hum. Genet. 1997; 99: 484–7.

CHAPTER 33

Primary Hyperoxaluria

SCOTT D. CRAMER AND TATSUYA TAKAYAMA

INTRODUCTION

Oxalate is a two-carbon dihydroxyacid compound that is found in plants, animals, fungi, and bacteria. In the plant kingdom, oxalate can have both structural defensive roles, and it can maintain calcium homeostasis (Francheschi & Nakata 2005). It is found in many plant products that are consumed by humans such as legumes (e.g. peanuts) and leafy green vegetables (e.g. spinach). In addition, oxalate-degrading bacteria utilize oxalate as a carbon source (Allison et al 1985, Ito et al 1996, Hokama et al 2000, Kodama et al 2002, Hokama et al 2005), which may be useful in therapeutic intervention (see below). However, for mammals oxalate is an end product of metabolism of glyoxylate and possibly other simple carbon compounds. Because of the poor solubility characteristics of calcium oxalate, excess oxalate has potentially lethal consequences if crystallization occurs in sensitive organs such as the kidney.

Dietary oxalate overconsumption can lead to hyperoxaluria; however, this condition is easily managed by oxalate restriction. Secondary hyperoxaluria due to gastrointestinal diseases such as Crohn's disease can lead to defects in oxalate flux with subsequent hyperoxaluria. In contrast the primary hyperoxalurias (1 and 2) are genetic disorders in glyoxylate metabolism that lead to systemic overproduction of oxalate. This endogenous production of oxalate cannot be managed by restriction of dietary oxalate, although it may be possible to restrict dietary glyoxylate precursors, such as glycolate and hydroxyproline to limit endogenous oxalate production. To our knowledge this has not been extensively investigated.

PH1 was first described in the literature by the French physician Lepoutre in 1925 (Lepoutre 1925). However, significant research on the disease did not begin until the 1950s and the molecular genetic basis was not determined until the cloning of the *AGXT* gene in 1990 by Danpure and colleagues (Takada et al 1990). Patients with PH1 have increased urinary glycolate and oxalate that results from a loss of liver peroxisomal alanine:glyoxylate aminotransferase (AGT) enzyme. The intracellular buildup of glyoxylate due to loss of AGT activity leads to its conversion to oxalate by lactate dehydrogenase (LDH) or glycolate oxidase (GO), and its conversion to glycolate by glyoxylate reductase (GR). See Figure 33.1 and discussion below for more details on the biochemical pathways. PH1 is the most prevalent, and in general the most severe form of primary hyperoxaluria. Estimates of the prevalence of the disease are unreliable due to significant problems with uniformity in methods of diagnosis (see the section of diagnosis later in this chapter). Rough estimates suggest that the prevalence is approximately 1 in 10^6 individuals worldwide. However, this is likely an underestimate and the actual value certainly varies within different population groups throughout the world. Diagnosis can occur as early as the first few months of life due to nephrolithiasis. As mentioned previously the genetic basis of PH1 is due to mutations in the gene encoding AGT (*AGXT* gene). *AGXT* contains 11 exons spanning 10 kb on chromosome 2q37.3. There are in excess of 80 different mutations described in the *AGXT* gene from PH1 patients. These mutations result in frameshifts (deletion and insertion mutations), nonsense codons, missense codons, and splice site mutations. However, many of the mutations (particularly missense mutations) lack corresponding functional studies to demonstrate a change in enzymatic activity. AGT is a liver-specific peroxisomal enzyme that requires pyridoxal phosphate (PLP) as a cofactor. There are many interesting and unique aspects of the cell biology of AGT in the context of PH1, not the least of which is the presence of mutations that alter the intracellular targeting of AGT from peroxisomes to mitochondria. Even more intriguing is that the most common mutation in PH1, a 508G→A point mutation that results in a Gly170Arg amino acid substitution, by itself has no functional effect on the AGT enzyme. The mutation is only significant in the context of a specific allelic variant (polymorphism) of the *AGXT* gene (the so-called minor allele), which is present in approximately 20% of the normal population. The minor allele is a specific haplotype of *AGXT* composed of three polymorphisms, only

Genetic Diseases of the Kidney
ISBN: 978-0-12-449851-8

one of which (C154T) results in an amino acid substitution (Pro11Leu) with functional consequences on the enzyme. The minor allele in the absence of other mutations has near normal AGT activity. However, the G630A mutation in the context of the minor allele results in mistargeting of AGT to the mitochondria and leads to primary hyperoxaluria. A detailed discussion of this will be presented below.

PH2 was first reported in the literature in 1968 by Williams and Smith (Williams & Smith 1968). PH2 is the result of the absence of the protein that has D-glycerate dehydrogenase (DGDH) glyoxylate reductase (GR) and hydroxypyruvate reductase (HPR) enzymatic activities (Williams & Smith 1968, Chalmers et al 1984, Cramer et al 1999, Webster et al 2000). It was initially thought to be milder than PH1, but with the aging of the original patients identified with the disease, it is clear that a significant number develop end-stage renal disease and may die at a premature age (Mansell 1995). The characteristic features of the disease are elevated urinary excretions of oxalate and L-glycerate. The increased oxalate excretion can cause nephrolithiasis and nephrocalcinosis and can result in renal failure and systemic oxalosis. A loss of DGDH activity in leukocytes and liver tissue was originally reported as the etiological basis of the disease (Williams & Smith 1968, Chalmers et al 1984), suggesting that mutations in the gene coding for DGDH cause the disease. The protein that contains DGDH enzymatic activity also contains glyoxylate reductase (GR) and hydroxypyruvate reductase (HPR) activities. The reduction reactions are thought to be the predominant reactions in humans (Van Schaftingen et al 1989) and the human protein is now referred to as the GRHPR enzyme. The *GRHPR* gene was characterized in 1999 (Cramer et al 1999). Thirteen mutations have now been reported in the literature from various populations (Cramer et al 1999, Webster et al 2000, Lam et al 2001, Johnson et al 2002, Cregeen et al 2003). All mutations reported result in a loss of enzyme expression or function.

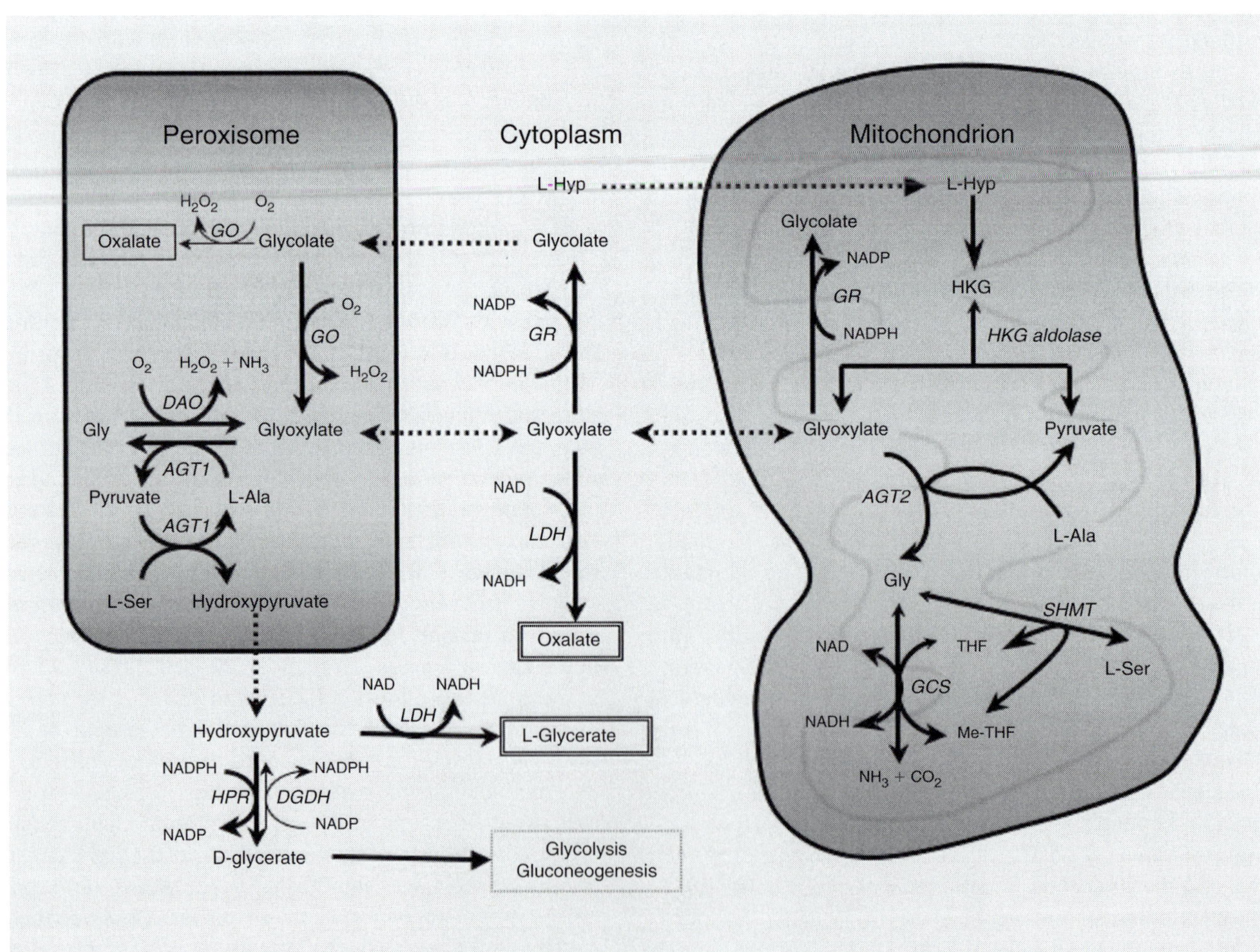

FIGURE 33.1 Pathways associated with oxalate synthesis. Transport processes are shown as the thin, dark arrows and enzymatic reactions as the thicker, lighter arrows. DAO, D-amino acid oxidase; GCS, glycine cleavage system; GO, glycolate oxidase; GOA, glutamate: oxaloacetate aminotransferase; GR, glyoxylate reductase; HKGA, 4-hydroxy-2-ketoglutarate lyase; LDH, lactate dehydrogenase; SHMT, serine hydroxymethyltransferase; THF, tetrahydrofolate. Modified from Baker et al (2004). (see also Plate 71)

BIOCHEMISTRY

The production of oxalate due to defects in glyoxylate metabolism is central to the etiology of both PH1 and PH2. An understanding of these pathways is essential for comprehension of the clinical features of the disease. Therefore, it is worth exploring the metabolism of glyoxylate in more detail. Figure 33.1 shows the intracellular relationships between glyoxylate, its precursors, and its catabolites. Figure 33.2 shows the structural relationships between the key molecules in these pathways. For a discussion of the pathways involved also see Baker et al (2004). Glyoxylate can be produced by deamidation of glycine in peroxisomes by D-amino acid oxidase, by oxidation of glycolate by peroxisomal glycolate oxidase (GO) or through catabolism of hydroxyproline by a pathway present in liver mitochondria. There may also be unknown sources of glyoxylate that have yet to be elucidated. In general there is a lack of definitive information on the roles of these biochemical pathways in glyoxylate production. Many factors may contribute to this relative scarcity of biochemical information including the technical hurdles with measurement of metabolites, the lack of suitable models, and up until recently a lack of enthusiasm by funding agencies to support this line of research. Another major hurdle conceptually has been the narrow focus on hepatic peroxisomal production of glyoxylate due to the historic focus on PH1, which is clearly a hepatic peroxisomal disease. However, the emerging molecular and biochemical data from work on the characterization of GRHPR and the genetic basis of PH2 suggest that both nonperoxisomal and nonhepatic production of glyoxylate should be an area of future focus (see below).

Glyoxylate produced in peroxisomes by GO can be converted to glycine by AGT. However, in the absence of AGT

(PH1 patients) the peroxisomal glyoxylate is converted to oxalate, either by peroxisomal GO or by LDH (presumably cytosolic). The relative contributions of these enzymes to the production of oxalate from peroxisomal glyoxylate in PH1 and PH2 are unresolved. GR is also able to convert some of the cytoplasmic glyoxylate to glycolate. In PH1 patients the high urinary glycolate in combination with hyperoxaluria is a key biochemical marker for the disease.

Glyxoylate produced in the mitochondria or present in the cytoplasm can be converted to glycolate by the GR activity of the GRHPR enzyme. This enzyme is present both in the cytoplasm and in the mitochondria. In the absence of GR activity (PH2 patients) the excess glyoxylate is converted to oxalate, presumably in the cytoplasm by LDH. Because of the specific intracellular localizations of these enzymes it is unlikely that the peroxisomal AGT is able to handle the glyoxylate produced at other sites within the cell, such as the mitochondrion. GRHPR also possesses HPR activity. A detailed diagram of the enzymatic activities of the GRHPR enzyme is depicted in Figure 33.1. The HPR activity reduces hydroxypyruvate to D-glycerate, presumably in a liver-specific manner. In PH2 patients, who lack both GR and HPR enzymatic activities, the lack of HPR results in an accumulation of hydroxypyruvate, which is converted to L-glycerate by LDH. Thus, the diagnostic biochemical features of PH2 are high urinary L-glycerate in conjunction with hyperoxaluria.

PH1

AGT was first identified as serine:pyruvate aminotransferase (SPT) in the 1950s (Sallack 1956). The protein was later shown to be a predominant enzyme of L-serine metabolism in human, rabbit, and dog livers (Xue et al 1999), However, Danpure and colleagues showed that the alanine:glyoxylate aminotransferase activity of this enzyme is indispensable for proper hepatic peroxisomal metabolism of glyoxylate (Danpure & Jennings 1986). Because of the prominent role of the AGT activity in PH1, the gene product is most often referred to as AGT. However, it is also variously referred to as SPT or SPT/AGT. In this review we use the most widely used convention of AGT.

The K_m of AGT for glyoxylate is as low as 0.01 mM in rats and humans (Yanagisawa et al 1983). AGT uses pyridoxal-phosphate (PLP) as a cofactor to catalyze the transamination of glyoxylate to glycine. PLP is an essential cofactor for AGT, as it is for all aminotransferases. Because pyridoxine (B6) can be taken orally and is converted to PLP, the vitamin is used therapeutically with success in some PH1 patients who have low levels of AGT (see further discussion of this below).

There is considerable species variability in the intracellular localization of AGT that appears to be related to adaptations to dietary intake of glyoxylate precursors from either plant or animal sources. AGT is located in liver peroxisomes

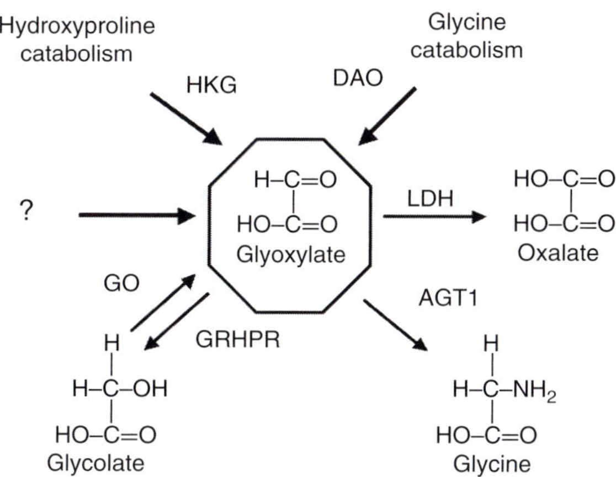

FIGURE 33.2 Glyoxylate flux represents the key pathway for oxalate production in primary hyperoxaluria. AGT1, alanine:glyoxylate amino transferase; DAO, D-amino acid oxidase; GO, glycolate oxidase; GRHPR, glyoxylate reductase/hydroxypruvate reductase, LDH, lactate dehydrogenase; HKG, 4-hydroxy-2-ketoglutarate lyase

in herbivores and humans, largely in mitochondria in carnivores (Noguchi 1978, Takada & Noguchi 1982, Noguchi 1987, Danpure et al 1990, Danpure et al 1994) and in both peroxisomes and mitochondria in omnivores, such as rats and mice. In herbivores, and humans (who probably evolved on a mostly herbivorous diet), a major source of glyoxylate is believed to be oxidation of glycolate by glycolate oxidase in liver peroxisomes. Glycolate is an intermediate of photorespiration and thus is much higher in content in plants than in animal tissues (Harris & Richardson 1980). Therefore, it has been proposed that the peroxisomal localization of AGT evolved as a protective mechanism to catabolize the glyoxylate produced from glycolate oxidation, thus preventing conversion of glyoxylate to oxalate by LDH.

Glyoxylate is also formed in liver and kidney mitochondria from 4-hydroxy-2-ketoglutarate (HKG), an intermediate of L-hydroxyproline metabolism (Maitra & Dekker 1964, Lowry et al 1985). Mitochondrial production of glyoxylate from hydroxyproline may be significant in carnivores (Takayama et al 2003) because the hydroxyproline content of collagen is about 10–13% and collagen accounts for about 30% of total animal protein. It has been hypothesized that mitochondrial localization of AGT is an adaptation in carnivores to the high mitochondrial glyoxylate content in these species (Birdsey et al 2005). The available evidence suggests that this hypothesis is a plausible explanation of AGT intracellular localization among different species, although it should be noted that mitochondrial GR may contribute to significant detoxification of glyoxylate produced by L-hydroxyproline catabolism (Knight & Holmes 2005). The model proposed by Danpure does not take into account the contribution of mitochondrial GR to the detoxification of glyoxlate produced in the mitochondria. The relationship between glyoxylate metabolism and the species-specific and food habit-dependent subcellular distribution of AGT has been extensively reviewed elsewhere (Birdsey et al 2005).

The nucleotide sequence of the AGT cDNA was first identified in 1990 (Nishiyama et al 1990, Purdue et al 1990) and consists of a continuous open reading frame of 1179 nucleotides encoding a 392-amino-acid protein with a calculated molecular weight of 43 kDa (Nishiyama et al 1990, Purdue et al 1990, 1991, Takada et al 1990). AGT crystallizes as a homodimer, the monomer consisting of a large N-terminal domain and a smaller C-terminal domain (Zhang et al 2003).

In all animal studies including human, AGT is encoded by a single gene that has the potential to encode N-terminal mitochondrial and C-terminal peroxisomal targeting sequences (PTS). The different organelle distribution of AGT among species arises from transcription from two potential initiation sites. Species that generate the longer transcript produce AGT with an additional 22 amino acids that serve as an N-terminal mitochondrial targeting sequence (MTS). This signal appears to have been deleted during evolution of humans by a point mutation of the first

initiation start codon ATG to ATA. Instead AGT is imported into peroxisomes, probably using the Pep5p-dependent pathway as a result of an atypical C-terminal type 1 PTS (PTS1, Lys-Lys-Leu) (Huber et al 2005). AGT mistargeting from peroxisomes to mitochondria, which is the most common phenotype in PH1, is produced by a combination of a specific mutation (Gly170Arg) and a common polymorphism (Pro11Leu) that together result in AGT mitochondrial targeting. This occurs by the Gly170Arg blocking AGT dimerization, and the Pro11Leu acting as an N-terminal MTS of monomer AGT (see below for further discussion).

PH2

Nucleotide sequencing of the cDNA for the GRHPR gene indicates that the gene encodes a 328-amino-acid protein with a calculated molecular weight of 35 563 Da (Cramer et al 1999, Rumsby & Cregeen 1999). The GRHPR gene is conserved throughout evolution (Cramer et al 1999). The human gene product has a high degree of identity with the homologous enzyme found in archeabacteria, eubacteria, fungi, and plants. It is worth noting that the enzyme has assumed divergent roles in different organisms. For instance in deep-sea vent methylotrophs the DGDH activity is essential for energy production in the form of NADH. These deep-sea vent organisms would not survive without this enzyme. In plants the enzyme is important for photosynthesis in C4 photosynthetic plants.

The enzyme utilizes pyridine nucleotides as cofactors and the mammalian enzyme has about a 100-fold higher affinity for NADPH over NADH (Van Schaftingen et al 1989). In studying its interactions with pyridine nucleotides Van Schaftingen et al (1989) concluded that the enzyme could only function in vivo (in mammals) as an NADPH-dependent hydroxypyruvate reductase when considering D-glycerate and hydroxypyruvate as substrates. Further evidence that this enzyme has difficulty in working in the reverse direction in vivo comes from individuals with an inborn error of metabolism and lacking glycerate kinase activity. These individuals excrete excessive amounts of D-glycerate in their urine (Van Schaftingen 1989). If GRHPR catalyzed the conversion of D-glycerate to hydroxypyruvate in vivo this excretion pattern would not be anticipated. One would expect excretion of L-glycerate as hydroxypyruvate accumulated. It should be noted that in vitro one can promote the DGDH reaction given appropriate reaction conditions (Cramer et al 1999). This forced reaction of the human enzyme can be highly sensitive and is extremely valuable for clinically evaluating potential PH2 patients (Knight et al 2006). GRHPR also has glyoxylate reductase activity (Dawkins & Dickens 1965). The enzyme from rat liver appears to have similar affinities for hydroxypyruvate and glyoxylate, with half maximal activity at 25 μM substrate (Van Schaftingen et al 1989). Its maximum velocity with hydroxypyruvate was about two times higher than for glyoxylate. Similar results were also obtained with beef liver GRHPR (Van Schaftingen et al 1989). These catalytic determinations were only preliminary and because of this and because intracellular concentrations of hydroxypyruvate are not known, which reaction predominates in mammalian liver is not known; however, it is presumed that both are functional and important.

Despite the wide acceptance of the role of GRHPR deficiency in PH2, the exact mechanism of oxalate overproduction that leads to the clinical manifestations of the disease is not known (Danpure & Purdue 1995, Holmes et al 1999). Figure 33.1 gives a diagrammatical representation of the pathways involved, showing the importance of GR activities in glyoxylate metabolism. It should be pointed out that many of the enzymatic activities depicted in Figure 33.1 are tissue-specific and organelle-specific. For instance AGT, the enzyme mutated in PH1 patients, is restricted primarily to the liver with some expression found in the kidney. Additionally, both AGT and glycolate oxidase (GO) are peroxisomal enzymes. Conversely, GRHPR is both cytoplasmic and mitochondrial (Knight & Holmes 2005) and is expressed in a wide variety of tissues in the body (Cramer et al 1999, Webster & Cramer 2000). The widespread expression of this gene in many cell types in humans suggests that it plays an important role in metabolism. This critical role is as yet undefined. The HPR reaction, the reduction of hydroxypyruvate to D-glycerate, may be important in the shuttling of serine to 2-phosphoglycerate which can enter either the gluconeogenic or glycolytic pathways (Kitagawa & Sugimoto 1979, Van Schaftingen et al 1989). However, this is probably only important in AGT-expressing tissues (liver and kidney) where there is a source of hydroxypyruvate. The role (if any) of HPR enzyme activity in nonAGT-expressing cells is not known. Absence of hepatic HPR activity in PH2 patients is thought to result in the conversion of hydroxypyruvate to L-glycerate by cytosolic lactate dehydrogenase (LDH), thereby elevating urinary excretion of this metabolite. The cytosolic GR reaction, the reduction of glyoxylate to glycolate, has an unknown role in metabolism but may contribute to an intracellular glycolate pool that may be important for other biosynthetic reactions. A major effect of this enzyme may be the maintenance of low cytosolic glyoxylate levels, particularly in tissues that do not express AGT (i.e. most cells in the body). Absence of GR activity (as occurs in PH2 patients) in nonAGT-expressing tissues could lead to an accumulation of glyoxylate that can be converted to oxalate by cytoplasmic LDH.

GENETICS

PH1

The *AGXT* gene spans 10kb on chromosome 2q37.3. The primary transcript is encoded by 11 exons and 10 introns

(Purdue et al 1991). Fourteen polymorphisms and 68 mutations have been described in the AGT gene from PH1 patients (Coulter-Mackie & Rumsby 2004, Yuen et al 2004, Coulter-Mackie et al 2005, Monico et al 2005). There are two common polymorphic variants of AGXT that are referred to as the major and minor alleles (Purdue et al 1990) based on their frequencies in Europeans and North Americans. The minor allele is defined by three polymorphisms; 32C→T(Pro11Leu), 1020A→G(Ile340Met) and a 74-bp duplication in intron 1. An African-specific variant of the minor allele has also been reported that differs from the minor allele in European populations by not having the Ile340Met variation (Coulter-Mackie et al 2003). The frequency of the minor allele varies among different populations. The frequency in Caucasians, in black South Africans and in Japanese is approximately 20%, 3%, and 2%, respectively (Danpure et al 1994, Coulter-Mackie et al 2003). In individuals with no *AGXT* mutations there are only minor differences in the functioning of the proteins produced from the major and minor alleles. The protein encoded by the minor allele has a slightly lower catalytic activity and a slower dimerization rate than the protein encoded by the major allele. These differences are attributable to the Pro11Leu substitution (Purdue et al 1990). However, the major and minor alleles have significant effects on the mistargeting of AGT when in the presence of some of the mutations that promote PH1 (see below).

Four of the 60-plus mutations identified in the gene for AGT account for more than 50% of PH1 alleles (Coulter-Mackie et al 2005). These four mutations are Gly170Arg (30–39.5%), 33_34insC (9.5–13%), Phe152Ile (1–5.8%) and Ile244Thr (3.5–9%). Two common mutations, Gly170Arg and 33_34insC, show no obvious ethnic association and have been found worldwide. On the other hand, although Phe152Ile and Ile244Thr are referred as common mutations, there are significant differences in the frequencies among different ethnic populations. Ile244Thr has a strong association with individuals of Spanish and North African descent. In PH1 patients from the Canary Islands the frequency represented 90% (Santana et al 2003).

The described mutations in *AGXT* are of all types, including small deletions, insertions, nonsense, missense, and splicing mutations. Two large partial gene deletions have also been reported (Nogueira et al 2000, Coulter-Mackie et al 2001). Interestingly, the majority of mutations are missense mutations. In addition, several missense mutations cosegregate with either the major or minor allele. Figure 33.3 shows cosegregation patterns of some of the more common missense mutations with the major and minor alleles. Those mutations that occur commonly on the minor allele include 508G→A (Gly170Arg) (Purdue et al 1990), 454T→A (Phe152Ile) (Danpure et al 1993), and 731T→C (Ile244Thr) (von Schnakenburg et al 1997). In contrast the 33_34insC (Asp167Stop) cosegregates with the major allele (Pirulli et al 1999, Milosevic et al 2002, Coulter-Mackie & Rumsby 2004).

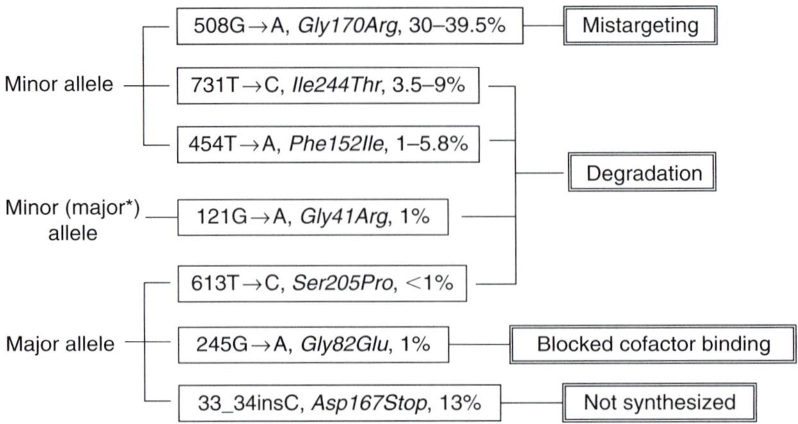

FIGURE 33.3   The relationship between polymorphisms and mutations on the properties of AGT. See text for discussion

The cosegregation of missense mutations with the major and minor alleles suggests functional interactions between the polymorphisms present on the major and minor alleles and the missense mutations that cause PH1. Elegant work by a number of groups has demonstrated these functional interactions and has enabled formulation of plausible biological explanations for these phenomena. The Gly170Arg, Ile244Thr and Phe152Ile are all predicted to have no distinct effects on enzymatic activity or intracellular localization in the absence of the minor allele. However, when these mutations are located on the minor allele variant they result in either intracellular targeting defects (Gly170Arg) or loss of enzyme stability (Ile244Thr, Phe152Ile).

The defect associated with the Gly170Arg mutation in combination with the minor allele (Pro11Leu) is the most well-characterized example. The result of this mutation in combination with the minor allele is an alteration in intracellular targeting from the normal human AGT peroxisomal localization to a predominantly mitochondrial localization pattern (Purdue et al 1990). The minor allele Pro11Leu substitution generates a weak mitochondrial targeting sequence at the N-terminus and attenuates dimerization of the protein. Efficient mitochondrial import requires unfolded monomer. In the absence of mutations, folding of AGT occurs too rapidly for efficient mitochondrial targeting of the protein encoded by the minor allele. However, in the context of the Gly170Arg mutation the efficiency of this mitochondrial targeting sequence is enhanced due to further attenuation of dimerization. Figure 33.4 shows a schematic of human AGT subcellular distribution as a function of genotype. The normal maturation of AGT wild-type (major allele) leads through unfolded monomer, folded monomer, dimerization, and peroxisomal targeting. The efficiency of this targeting is in the order of 100%. The minor allele has a slightly attenuated folding that in homozygous individuals results in about 5% of AGT targeted to the mitochondria (Purdue et al 1990). The Gly170Arg mutation in combination with Pro11Leu leads to greatly attenuated dimerization

(Leiper et al 1996, Lumb & Danpure 2000) and efficient mitochondrial targeting (90%) at the expense of peroxisomal targeting (Danpure et al 1989, Purdue et al 1990, 1991). The net result of this is loss of peroxisomal glyoxylate metabolism and hyperoxaluria, but only in the context of the minor allele.

Other mutations in the AGXT gene have less interesting biological effects on the enzyme. The 33_34insC insertion, which is the second most common mutation in PH1 patients, results in a frame shift and a premature stop codon (Asp167Stop) (Pirulli et al 1999, Milosevic et al 2002, Coulter-Mackie & Rumsby 2004). Patients homozygous for this mutation have no AGT immunoreactivity or catalytic activity. The Ile244Thr and Phe152Ile mutations both cause aggregation and degradation of the enzyme. The remaining described mutations each occur at or below 1% of PH1 patients. Many of the missense mutations have yet to be tested for their functional consequences on enzyme activity, intracellular localization, or protein stability. It is not the purpose of this review to systematically describe each of these mutations. Other reviews have described the spectrum of AGXT mutations thoroughly (Danpure 2001, Coulter-Mackie & Rumsby 2004, Danpure & Rumsby 2004). However, some mutations are worth noting. The Gly41Arg is associated with both the major and minor alleles and results in enhanced protein degradation. Ser205Pro is a rare mutation that occurs only in the major allele; homozygosity leads to accelerated protein degradation (Nishiyama et al 1990). Gly82Glu is major allele specific. This mutation results in a complete loss of catalytic activity due to the inhibition of pyridoxal phosphate binding (Purdue & Lumb 1992, Coulter-Mackie et al 2001).

PH2

Since the first reports on PH2, the mode of disease transmission has suggested an autosomal recessive mode of inheritance (Williams & Smith 1968, Chalmers et al 1984).

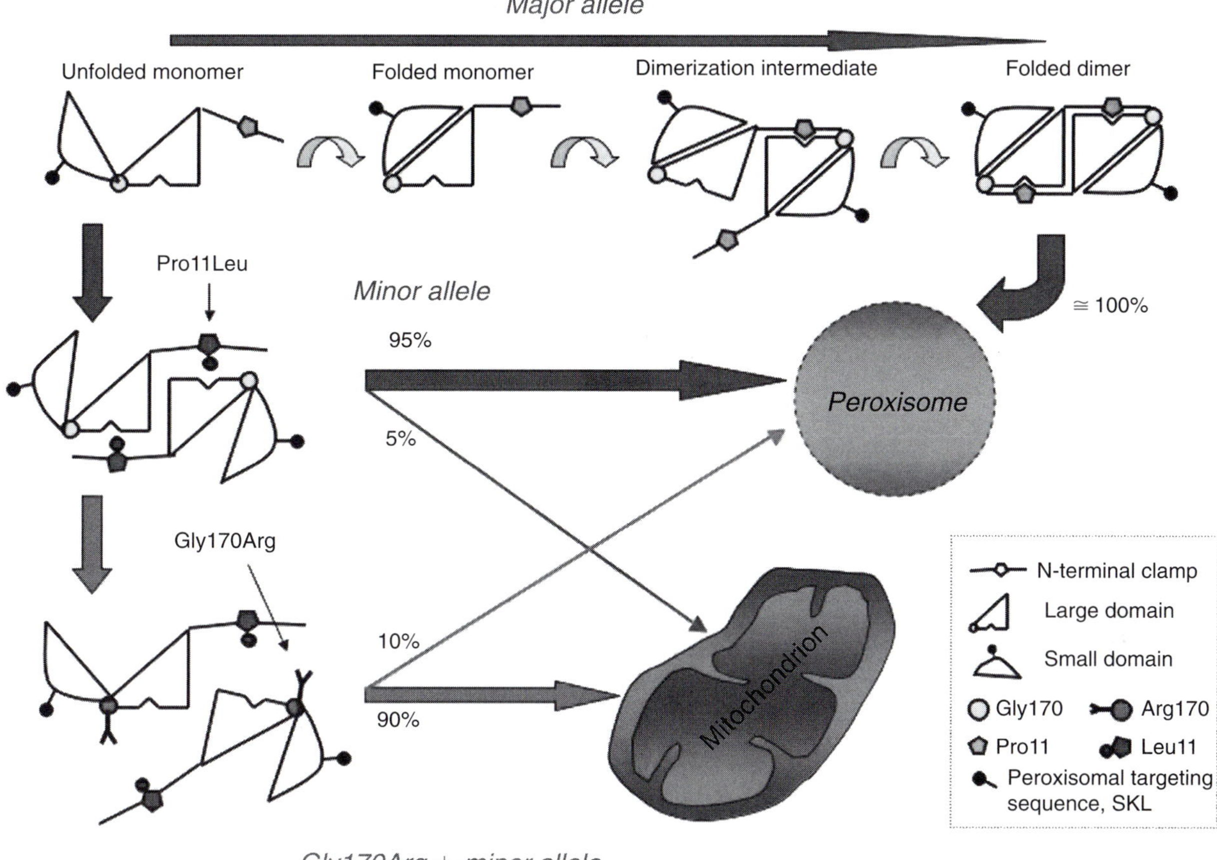

FIGURE 33.4   Schematic of human AGT subcellular distribution as a function of phenotype. The minor allele facilitates mistargeting of AGT with the Gly170Arg mutation to the mitochondrion. See text for discussion. (see also Plate 72)

However, unlike PH1, where the cDNA and genomic DNA for AGT were characterized in the early 1990s (Takada et al 1990, Purdue et al 1991), elucidation of the molecular basis for PH2 was considerably slower. We initiated work in 1997 that ultimately resulted in elucidating the genetic basis of PH2 (Cramer et al 1999, Webster & Cramer 2000, Webster et al 2000). We identified a cDNA through the expressed sequence tag (EST) database that encoded DGDH, GR, and HPR enzymatic activities (Cramer et al 1998, 1999). Using this cDNA sequence we mapped the genomic structure of the human GRHPR gene and developed PCR-single strand conformation polymorphism (SSCP) assays for each of the nine exons (Cramer et al 1999). Using these assays we have identified several mutations in the GRHPR gene from PH2 patients (Cramer et al 1999, Webster & Cramer 2000, Webster et al 2000). Subsequent to our initial reports other groups have reported additional mutations in the GRHPR gene from PH2 patients (Lam et al 2001, Johnson et al 2002, Cregeen et al 2003). In all PH2 patients that have been definitively genotyped, both alleles of the GRHPR contain inactivating mutations (complete loss of function). These mutations include missense, nonsense, and deletion mutations. Figure 33.5 shows a summary of these mutations. The most common mutation found in PH2 patients from northern Europe and North America is a single nucleotide deletion of a G at position 103 relative to the translation initiation codon (Cramer et al 1999). This mutation results in a frame shift and a premature stop codon at codon 44. This mutation is found in approximately 50% of all PH2 patients that have been genetically screened. The data with microsatellite analysis suggest a founder of northern European descent (Webster et al 2000). There are numerous additional mutations described (14 including unpublished data from our laboratory). Many of these have been described in one patient, or are confined to specific populations. Remarkably, all of the described mutations result in loss of enzyme function, either by lack of complete protein or by mutation of critical residues essential for enzyme function. There are no partial function mutants described for GRHPR. This is in contrast to many of the mutations found in AGXT, which may retain partial activity or near full activity but are mislocalized within the cell. Thus, the cell biology associated with PH2 genetics is unremarkable when compared to that of PH1.

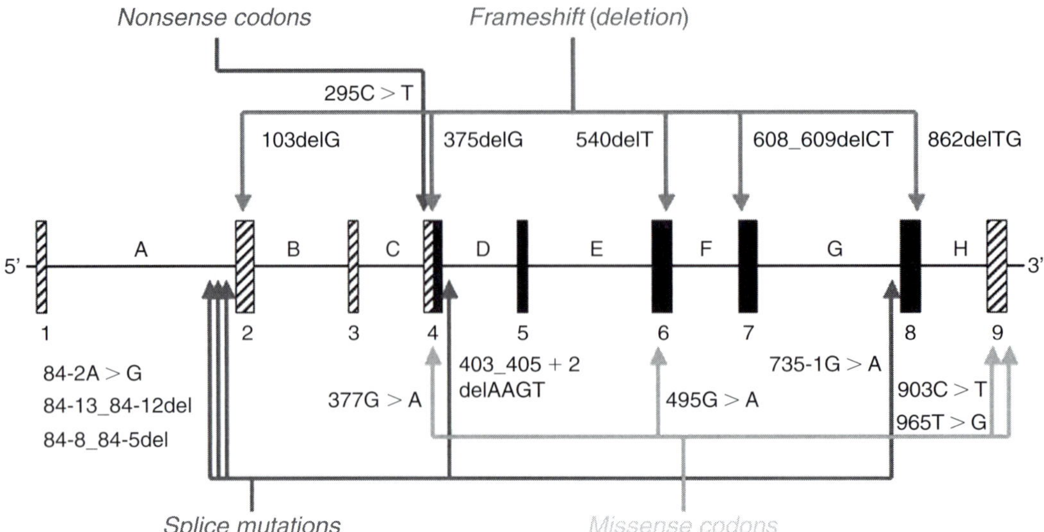

FIGURE 33.5   Summary of described GRHPR mutations. To date, 14 mutations have been identified in PH2 patients. A novel 377G→A missense mutation in exon 4 of a Japanese patient is also included (Takayama & Cramer, unpublished data). This patient was a compound heterozygote with the 495G→A missense mutation. All mutations identified thus far lead to complete loss of GRHPR enzyme activity. (see also Plate 73)

DIAGNOSIS/SCREENING

The diagnosis and screening of PH, and the further classification of PH into PH1 and PH2 utilizes biochemical analyte measurements of urine, genetic screening for mutations, and measurements of tissue enzyme functions. We will present each type of assay individually and at the end of this section present an integrated approach to diagnosis based on an algorithm originally proposed by Milliner.

Biochemical

The pathology of PH is that of progressive renal deposition of calcium oxalate in the form of urolithiasis and/or nephrocalcinosis that leads to end-stage renal failure (ESRF) and systemic oxalosis, finally resulting in death usually before the third decade of age. Therefore, the diagnostic approach depends on the observed renal function of the patients.

Urinary oxalate excretion is usually greatly elevated among PH patients with recurrent urolithiasis and/or nephrocalcinosis, exceeding 2 mmol/24 h per 1.73 m^2 or 100 mg/24 h (normal, <0.5 mmol/24 h per 1.73 m^2 or <45 mg/24 h per 1.73 m^2) (Cochat 1999, Danpure 2001). However, family studies have demonstrated that some untreated patients with PH1 may exhibit only slightly elevated (0.5 to 1 mmol/24 h per 1.73 m^2) or even normal urinary oxalate excretion (Koppe et al 1997). Frequent errors are made in the quantification of urinary oxalate by ignoring the age and body surface area of the patients. In ESRF cases, plasma oxalate determinations may be helpful. In patients with PH1, plasma oxalate levels are significantly higher (>80 to 100 μM) in ESRF, compared with those for patients without PH1 (40 to 60 μM) (Marangella et al 1993, Hoppe et al 1998, 1999). Interestingly, they are already elevated (>6.3 μM) with normal renal function (Leumann & Hoppe 2001). Measurement of urinary glycolate and L-glyceric acid excretion, necessary to distinguish PH1 from PH2, is performed in a small minority of patients. Glycolate levels are elevated in only approximately two-thirds of patients with PH1 (Cochat 1999, Danpure 2001). It is recommended that more widespread use of these distinguishing assays be used to help direct clinical management of the disease.

Enzymatic

A definitive diagnosis of PH1 and PH2 can be obtained by assessments of AGT and GRHPR enzymatic activities and immunoreactivity in hepatic tissue (minimum of 2 mg) respectively. Alternatively, the diagnosis can be established at the molecular level in the majority of patients with minimal invasiveness and associated morbidity (Rumsby et al 2004, Milliner 2005, Rumsby 2005). Enzymatic testing for both PH1 and PH2 can be performed with the same needle-biopsy specimen of the liver (Giafi & Rumsby 1998). However, assuming the AGT activity of glutamate:glyoxlate aminotransferase (GGT) to be 66% of its GGT activity (Noguchi et al 1977), the GGT activity should be concomitantly assayed, followed by appropriate correction of the apparent AGT activity with use of the activity catalyzed by GGT. In contrast, the SPT activity in human liver extract is almost entirely catalyzed by SPT/AGT; therefore the SPT has higher specificity than the AGT assay.

GRHPR has GR, HPR, and DGDH activity. Measurement of GR and DGDH activity in liver has been used for definitive diagnosis of PH2 (Giafi & Rumsby 1998). However, liver biopsy is now unnecessary for enzymatic diagnosis of PH2. The GRHPR enzyme is expressed in virtually every tissue in the body. This has led to the exploration of the use of more available cells, such as leukocytes and lymphocytes in the assay of GRHPR as a less invasive procedure than liver biopsy (Williams & Smith 1968, Knight et al 2006). Recently, we reported that the assay of GR or DGDH activity in blood mononuclear cells could be used as a minimally invasive diagnostic test for PH2 (Knight et al 2006). In unpublished work we have also used buccal swabs to measure DGDH activity, suggesting that virtually any tissue sample may be useful for PH2 diagnosis. This is a particularly important point when a definitive diagnosis might be warranted. Traditionally, the use of a liver biopsy has been the definitive assay for both PH1 and PH2 diagnosis. Given the minimal invasiveness of a blood draw it can easily be argued that a diagnosis of PH2 must be ruled out before proceeding to a liver biopsy to diagnose PH1. We also would strongly encourage genetic testing prior to liver biopsy. Given our current technologies and understanding of GRHPR tissue distribution there is little value to be obtained from liver biopsy for a PH2 diagnosis.

Genetic

The elucidation of the genetic structures of the *AGXT* and *GRHPR* genes and the identification of many common mutations for both PH1 and PH2 has advanced the field closer to routine mutation analysis for definitive diagnosis of PH1 and PH2 (Rumsby et al 2004, Milliner 2005, Rumsby 2005). These tests could be further enhanced by high throughput sequencing and genetic screening strategies currently available. A major advantage of this approach over enzymatic assay of liver biopsy is the lack of associated morbitity of the procedure. A disadvantage is the limited database on functional mutations for both PH1 and PH2. Currently, there remains resistance to wholly embrace genetic screening as a primary test for PH. This resistance comes from reluctance to relinquish the historic focus on enzymatic assay of liver biopsies and the current costs and labor involved in mutation analysis on a limited patient population.

Theoretically, one could sequence all of the exons and flanking intron boundaries for both *GRHPR* and *AGXT* from a suspected PH1 or PH2 patient. This is likely to yield a high sensitivity of mutation detection. We have performed these kinds of mutation screening specifically for suspected PH2 patients with a high degree of success (Cramer & Takayama, unpublished observations). However, the costs associated with this prohibit this approach on a widespread basis outside of that specifically for research purposes. There are, however, three common *AGXT* mutations in PH1 patients (c.33_34insC, c.508G > A, and c.731T > C) and one common GRHPR mutation in PH2 patients (c.103delG)

that together account for greater than 50% of the reported PH mutations in the literature. Rumsby et al. tested the sensitivity of diagnosis using these four mutations in DNA samples from 365 unrelated individuals referred for diagnosis of PH1 and/or PH2 (Rumsby et al 2004). In PH1, the sensitivity of genetic screening was 62%, compared with 33% in PH2 (Rumsby et al 2004). Based on these results one could expect to diagnose approximately 50% of PH patients using these four mutations. However, depending on the ethnicity of the referring population, the sensitivity of the assay could vary significantly. As our knowledge of the mutation spectrum in specific populations increases it might be better to change the selection of mutations in a population specific manner. If the first pass screen using the four mutations described above or another population-specific set fails to enable a molecular diagnosis, then further genetic screening of the *AGXT* and *GRHPR* genes would be required.

Prenatal diagnosis and testing can be performed by mutational or linkage analysis using chorionic villous (9–12 weeks' gestation) or amniocentesis (16 weeks' gestation). More than 60 prenatal diagnoses for PH1 have been carried out by the London group (Danpure & Rumsby 2004). Although prenatal diagnosis of PH2 has not yet been carried out, there are several candidate microsatellite markers for linkage analysis (Webster et al 2000, Johnson et al 2002, Cregeen et al 2003).

Algorithm for Diagnosis of Hyperoxaluria

The incidence and prevalence of PH are not accurately known. Rough estimates show the frequency of PH1 and PH2 at approximately 1 in 10^6 and 10^5, respectively. Due to the rarity of PH, a majority of nephrologists, urologists, and pediatricians may never encounter PH patients. Moreover, accurate diagnosis requires highly specialized procedures that are only available at a few locations worldwide. Due to these barriers many patients are misdiagnosed based either on incomplete clinical evaluation or a lack of awareness of the disease. Milliner recently proposed an algorithm for the differential diagnosis of primary hyperoxalurias (Milliner 2005). Figure 33.6 presents a modification of the algorithm proposed by Milliner. Our modification incorporates our recent findings that the assay of DGDH and GR activity in BMC could be used as a minimally invasive test for PH2 (Knight et al 2006). We also encourage a greater emphasis on genetic screening. Our emphasis is directed towards eliminating the need for invasive test, particularly liver biopsies.

Candidate patients are those who have nephrocalcinosis and/or CaOX stones, or a family history of primary hyperoxaluria. Pediatric patients in these categories should be especially targeted for evaluation. For those patients with normal renal function, the first diagnostic tool is 24 h urinary oxalate excretion. In candidates with ESRF, plasma oxalate concentration is the first diagnostic test. Secondary hyperoxaluria due to various intestinal diseases such as Crohn's

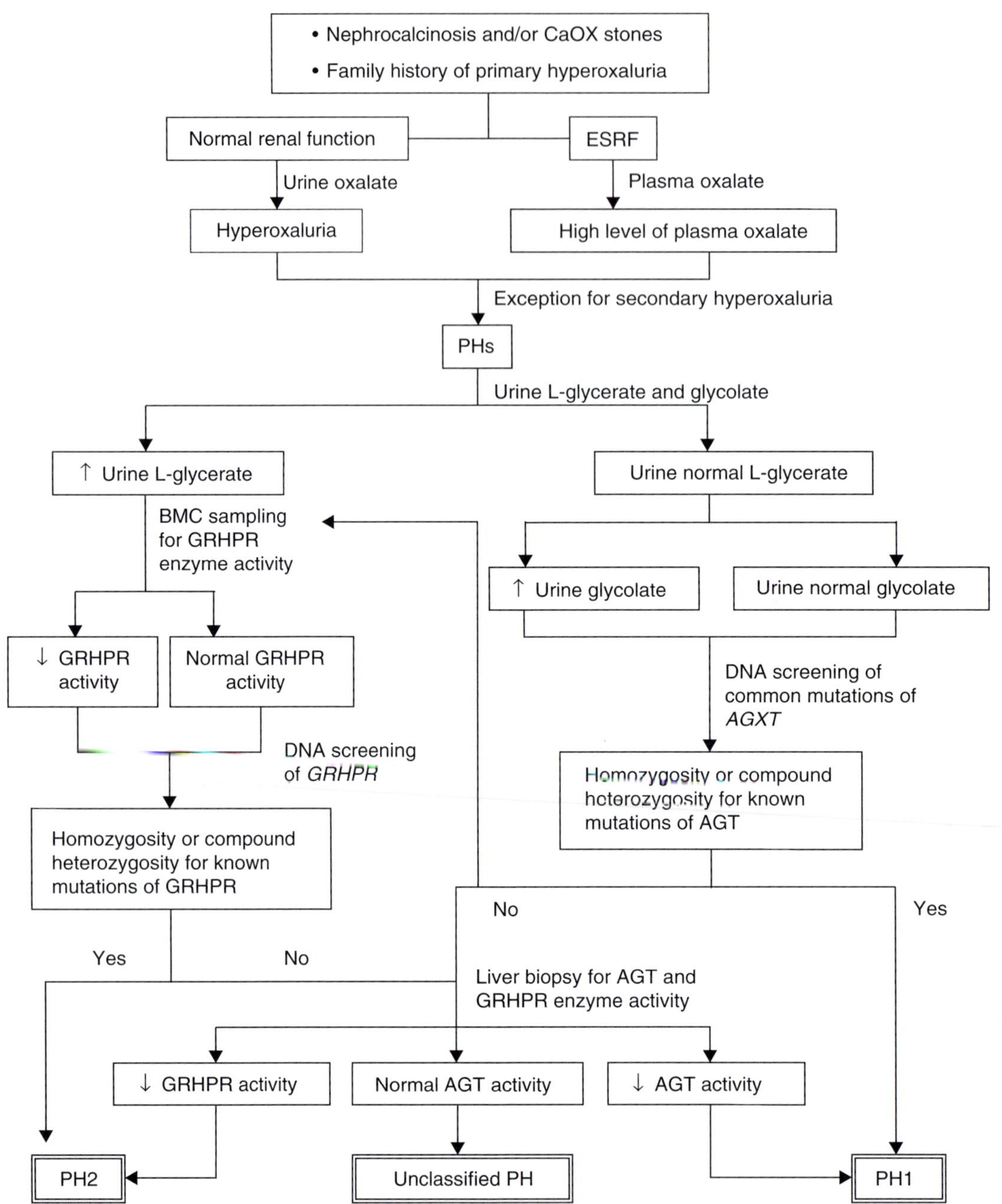

FIGURE 33.6   Diagnostic algorithm of primary hyperoxaluria. Modified from Milliner (2005). See text for discussion

disease excludes a diagnosis of primary hyperoxaluria. Otherwise, the remaining patients fall in the category of PH.

Further distinction between PH1 and PH2 is based on the measurement of urinary L-glycerate and glycolate excretion, GRHPR activity in BMC, and mutation analysis. Increased urinary excretion of L-glycerate should be followed by measurement of GRHPR (DGDH) activity in BMC. Given appropriate controls, a lack of DGDH activity in BMC with associated elevated urinary oxalate and L-glycerate should be definitive for diagnosis of PH2 (Williams & Smith 1968, Knight et al 2006). The defect of GRHPR

activity leads to PH2. This diagnosis can be verified by genetic analysis of the GRHPR gene.

Normal urinary excretion of L-glycerate should be followed by DNA screening of common mutations of *AGXT*, regardless of high urine glycolate or urine normal glycolate, although patients with increased urinary glycolate excretion are candidates for PH1. Genetic screening of those patients provides the definitive diagnosis in 34% of PH1 (Rumsby et al 2004). If the patient has homozygosity or compound heterozygosity for known mutations of *AGT*, the diagnosis should be PH1. Liver biopsy may be recommended if all other assays fail to generate a definitive diagnosis and a definitive diagnosis is necessary for management of the patient.

TREATMENT/MANAGEMENT

Treatment of PH is primarily management of renal handling of oxalate load. A subset of patients with PH1 will respond to pyridoxine therapy, however, even these patients need other management strategies. Therapeutic 'cure' is not achievable with current technology. The treatment of renal stones is not reviewed here but is covered elsewhere (Pais & Assimos 2005). Table 33.1 summarizes the following discussion.

Pyridoxine Phosphate

Pyridoxine (vitamin B6) is metabolized to PLP in the body. As discussed above, PLP is an essential co-factor for AGT. In patients with some forms of missense mutations abundant intracellular PLP appears to stabilize AGT and 'correct' the enzymatic defect. Oral administration of pyridoxine is used as a therapeutic intervention in PH1 to reduce hepatic oxalate production. The success of this treatment is remarkable in a subset of patients. However, in others the therapy is completely without therapeutic benefit. Doses of pyridoxine have varied from 2 to 30 mg/kg/day, which have been followed by significantly reduced urinary oxalate excretion in 25–30% of PH1 patients (Leumann & Hoppe 1999, Hoppe & Langman 2003). Up until recently the reasons behind the differential effects of pyridoxine therapy among PH1 patients remained elusive. Monico and colleagues studied genotype–phenotype relationships between *AGXT* mutations in PH1 patients and their responsiveness to pyridoxine. This study showed that pyridoxine responsiveness was clearly related to the presence of Gly170Arg mutation. In general, patients homozygous for Gly170Arg were more responsive than those expressing only one Gly170Arg allele (van Woerden et al 2004, Monico et al 2005). These findings support work by van Woerden et al who also showed that patients with the Phe153Ile mutation to be responsive to pyridoxine (van Woerden et al 2004). The mechanism of responsiveness for pyridoxine remains uncertain. However, in the PH1 patients who have Gly170Arg, AGT is mistargeted to mitochondria, but retains residual AGT activity. One possibility is that the pyridoxine overcomes the misfolding and dimerization defects associated with the Gly170Arg mutation and allows for correct peroxisomal targeting. This hypothesis has yet to be tested. Pyridoxine is not effective in PH2 and therefore is not utilized.

Enhancement of Solubility

Supersaturation of CaOX in the kidney results in precipitation and crystallization. One strategy for stone management in PH patients is hydration and pH management to reduce supersaturation levels. In order to enhance solubility of CaOX, it is beneficial to maintain hydration by constant generous fluid intake ($>2.5 l/m^2/day$). Additionally, drugs that increase the urinary solubility product, such as orthophosphate, citrate, and magnesium, may provide benefit in some patients (Leumann et al 1993, Cochat 1999, Cochat & Basmaison 2000).

Diet

The contribution of dietary oxalate to urinary oxalate excretion varies from 10 to 72% in healthy individuals (Holmes et al 2001). However, for patients with PH, oxalate of dietary origin contributes little to hyperoxaluria, because of their extremely high levels of endogenous oxalate production. Despite this, most physicians suggest avoidance of high-oxalate-content foods.

It is presumed that oxalate is generated from glyoxylate produced from endogenous sources in the body. Two major sources of glyoxylate are from catabolism of glycolate by peroxisomal GO and catabolism of hydroxyproline by mitochondrial enzymes in the liver and kidney (see above for discussion of sources of glyoxylate). It remains to be

TABLE 33.1   Primary hyperoxaluria management/treatment

Methods	Specifics	Target patient
High fluid intake	2.2–2.5 l/m²/day	Except for ESRF
Diet	Restriction of oxalate-content foods	All
	High calcium	Except for ESRF
	Hydroxyproline restriction	Especially in PH2 patients
Pyridoxine	2–30 mg/kg/day	Gly170Arg homozygous
Crystallization inhibitors	Citrate, 150 mg/kg/day	Except for ESRF
	Orthophosphate, 30 mg/kg/day	Except for ESRF
Oxalate-degrading bacteria	Lactic acid	All
	Oxalobacter formigenes	All
Gene therapy	Not available	All
Chemical chaperones	Not available	Missense mutations
Transplantation	Kidney transplantation	Pyridoxine responder
	Hepatocyte transplantation	PH1; but still on bench side
	Liver (and kidney) transplantation	PH1 patients

tested if dietary restriction of glycolate or hydroxyproline is feasible or effective in the management of PH.

It is known that loss of enteric oxalate degrading bacteria, especially *Oxalobacter formigenes*, is associated with secondary hyperoxaluria in patients with cystic fibrosis and inflammatory bowel disease (Sidhu et al 1998, Kumar et al 2004). Hoppe et al demonstrated that successful colonization of the intestinal tract by oxalate degrading bacteria could be a useful therapeutic intervention for patients with primary hyperoxaluria (Hoppe et al 2005). It has been suggested that the enteric oxalate degradation promotes a reverse flux of oxalate from the blood that is then excreted with the feces. Evidence for this flux in response to *Oxalobacter formigenes* has recently been demonstrated in isolated rate colon preparations (Hatch et al 2006). Colonization of patients with oral administration of *Oxalobacter formigenes* appears to be transient and requires follow-up administration to maintain sufficient bacterial titers (Hoppe et al 2005). This strategy is still in its infancy and the clinical usefulness of it will require larger trials.

Transplantation

Several transplantation strategies have been utilized for the treatment and management of PH including isolated kidney transplantation (KTx), preemptive liver transplantation (pLTx) or combined liver-kidney transplantation (LKTx). PH1 is a liver-specific loss of AGT activity. Therefore, pLTx is essentially an enzyme replacement strategy for PH1. Given the wide tissue distribution of GRHPR expression there is currently no justification for LTx, pLTx or in the combined setting, in PH2 patients.

In a recent survey of the success of transplantation strategies isolated KTx failed in more than 50% of patients (Hoppe & Langman 2003). However, in the subset of PH1 patients that are pyridoxine responsive the success is improved (Marangella 1999).

LKTx has two forms, concurrent (simultaneous) or sequential (first liver, then kidney). The advantages of concurrent LKTx are the rapid correction of both the metabolic defect and the renal failure leading to a better quality of life for the patients. Sequential LKTx is generally limited to infants and small children who are not considered candidates for concurrent LKTx because they are too unstable, or for whom concurrent LKTx is not possible due to anatomical reasons. The concept behind sequential LKTx is to decrease oxalate synthesis by LTx, improving morbidity and mortality until KTx is feasible.

Liver transplantation, before serious renal damage occurs, has been performed in PH1 children resulting in stable renal function or even improvement of residual renal function for the long term (Schurmann et al 1990, Gruessner 1998). Successful LTx reverses the systemic massive oxalate production rapidly. However, oxalate excretion can take much longer to normalize (often years). This lag is due to the time required to resolve the accumulated oxalate pools in organs such as bone. Therefore, after LTx it is essential to prevent oxalate crystallization in the kidney through suspersaturation management strategies. According to the European PH1 Transplant Registry, 127 LTx were performed in Europe between 1984 and 2004. The 1-, 5-, and 10-year patient survival rates were 86, 86, and 69%, respectively with liver graft survival rates of 80, 72, and 60% during the same intervals. It is notable that renal function can be maintained after 5 years, with creatinine clearance rates of 50–60 ml/min/1.73 m^2 (Jamieson & Group 2005). In Germany, more than a dozen cases of PH1 have been treated by pLTx with good results (Nolkemper et al 2000, Kemper 2005). However, there is still considerable controversy concerning the balance between the benefits to pLTX and the risks which are associated with surgery, long-term immunosuppression and oxalosis-related complications (Leumann & Hoppe 2000, Kemper 2005). Considerations also include the ideal timing of LTX intervention and ethical concerns.

Gene Therapy

Given that primary hyperoxaluria is a monogenic disease, the prospect of gene therapy to correct the defects has been proposed in the past (Danpurc 2001). PH1 is a liver-specific loss of correct AGT function. Liver-specific replacement of AGT is theoretically possibility either through in vivo transduction of the liver, or ex vivo transduction of hepatocytes and then reimplantation. Given the current limitations on both of these approaches due to technical and ethical issues, to our knowledge no serious work is being focused in this area. However, given appropriate breakthroughs and technological advances, gene therapy could form an attractive alternative approach to the treatment of PH1. Our current understanding of the tissue distribution of GRHPR (ubiquitous) and the uncertainty about the source(s) of oxalate in PH2 patients precludes any work directed on gene therapy for PH2.

UNCLASSIFIED HYPEROXALURIA

Monico and colleagues have described several patients with hereditary hyperoxaluria with an unusual pattern of presentation (Monico et al 2002). For instance, classically PH1 and PH2 have been distinguishable biochemically by analysis of urinary oxalate, glycolate, and L-glycerate. As mentioned previously, increased L-glycerate in PH2 patients is presumed to be the result of a loss of hepatic HPR activity and the resulting conversion of hydroxypyruvate to L-glycerate by LDH. Interestingly, the patients described by Monico et al have presumed hereditary oxalosis with normal urinary glycolate, L-glycerate, hepatic AGT, and hepatic GRHPR. The mechanism of hyperoxaluria is unknown and there are currently no candidate genes. Glycolate oxidase has recently been ruled out as a viable candidate (Monico et al 2002). We hypothesized that a specific class of GRHPR mutations might be responsible for uncPH. Specifically, if there was a selective loss of mitochondrial GRHPR such as by a missense mutation in a critical residue for targeting, but retention of cytoplasmic GRHPR, then one might expect to find elevated oxalate in the absence of elevated L-glycerate with hepatic GRHPR activity within the normal range. This hypothesis was tested in seven well-characterized UncPH patients from the Mayo Clinic. This was accomplished by completely sequencing all exons with intron boundaries in the GRHPR gene from this group. No mutations were found eliminating GRHPR as a candidate (Cramer et al, unpublished).

Recently, Jiang et al reported that Slc26a6-null mice showed a dramatic calcium oxalate stone phenotype (Jiang et al 2006). Slc26a6 is localized in kidney and intestine, and can function as a multifunctional anion exchanger, primarily exchanging chloride and oxalate. Slc26a6-null mice have a defect in intestinal oxalate secretion resulting in enhanced net absorption of oxalate. These mice have elevated plasma oxalate with significant hyperoxaluria. The hyperoxaluria is greatly attenuated by dietary oxalate restriction (Jiang et al 2006). This study clearly demonstrated that Slc26a6 plays a critical role in limiting net intestinal absorption of oxalate, preventing hyperoxaluria and reducing the risk of urolithiasis. Humans with mutations in Slc26a6 would be expected to have the urinary phenotype of unclassified hyperoxaluria and this gene could be considered a candidate for UncPH. However, in the well-characterized patients that have been described oxalate absorption was normal (Monico et al 2002), suggesting that mutations in Slc26a6 are unlikely to be causal for UncPH.

CONCLUSIONS

The primary hyperoxalurias are monogenic disorders in glyoxylate metabolism that lead to life-threatening destruction of the kidney. Considerable advances have occurred in the past several years both in the identification of the molecular nature of the disease and in the differential diagnosis between PH1 and PH2. Unfortunately, these advances have not been followed by parallel advances in the treatment and/or management of PH. Future research should be focused on further elucidation of genotype–phenotype relationships and the development of effective therapeutic strategies.

ACKNOWLEDGMENTS

This work was supported in part by a grant from the National Institutes of Health to Scott Cramer (DK069331).

References

Allison M, Dawson K, Mayberry W, Foss J. *Oxalobacter formigenes* gen. nov., sp. nov.: oxalate-degrading anaerobes that inhabit the gastrointestinal tract. Arch. Microbiol. 1985; 141: 1–7.

Baker P, Cramer S, Kennedy M, Assimos D, Holmes R. Glycolate and glyoxylate metabolism in HepG2 cells. Am. J. Physiol. Cell Physiol. 2004; 287: 1359–65.

Birdsey G, Lewin J, Holbrook J, Simpson V, Cunningham A, Danpure C. A comparative analysis of the evolutionary relationship between diet and enzyme targeting in bats, marsupials and other mammals. Proc. Biol. Sci. 2005; 22: 833–40.

Chalmers R, Tracey B, Mistry J, Griffiths K, Green A, Winterborn M. L-glyceric aciduria (primary hyperoxaluria type 2) in siblings in two unrelated families. J. Inherit. Metab. Dis. 1984; 7(Suppl. 2): 133.

Cochat P. Primary hyperoxaluria type I. Kidney Int. 1999; 55: 2533–47.

Cochat P, Basmaison O. Current approaches to the management of primary hyperoxaluria. Arch. Dis. Child. 2000; 82: 470–3.

Coulter-Mackie M, Lian Q, Applegarth D, Toone J. The major allele of the alanine:glyoxylate aminotransferase gene: nine novel mutations and polymorphisms associated with primary hyperoxaluria type 1. Mol. Genet. Metab. 2005; 86: 172–8.

Coulter-Mackie M, Rumsby G. Genetic heterogeneity in primary heperoxaluria type 1: impact on diagnosis. Mol. Genet. Metab. 2004; 83: 38–46.

Coulter-Mackie M, Rumsby G, Applegarth D, Toone J. Three novel deletions in the alanine: glyoxylate aminotransferase gene of three patients with type 1 hyperoxaluria. Mol. Genet. Metab. 2001; 74: 314–21.

Coulter-Mackie M, Tung A, Hernderson H, Toone J, Applegarth D. The AGT gene in Africa: a distinctive minor allele haplotype, a polymorphism (V326I), and a novel PH1 mutation (A112D) in Black Africans. Mol. Genet. Metab. 2003; 78: 44–50.

Cramer S, Ferree P, Lin K, Milliner D, Holmes R. The gene encoding hydroxypyruvate reductase is mutated in patients with primary hyperoxaluria type II. Hum. Mol. Genet. 1999; 8: 2063–9.

Cramer S, Lin K, Holmes R. Towards identification of the gene responsible for primary hyperoxaluria Type II: a cDNA encoding human D-glycerate dehydrogenase. J. Urol. 1998; 159(Suppl.): 173.

Cregeen D, Williams E, Hulton S, Rumsby G. Molecular analysis of the glyoxylate reductase (GRHPR) gene and description of mutations underlying primary hyperoxaluria type 2. Hum. Mutat. 2003; 22: 497.

Danpure C. Primary Hyperoxaluria. In: Scriver C, Beaudet A, Sly W, et al eds. The Metabolic and Molecular Bases of Inherited Disease. New York: McGraw-Hill, 2001: pp. 3323–67.

Danpure C, Birdsey G, Rumsby G, Lumb M, Purdue P. Allsop J. Molecular characterization and clinical use of a polymorphic tandem repeat in an intron of the human alanine:glyoxylate aminotransferase gene. Hum. Genet. 1994a; 94: 55–64.

Danpure C, Fryer P, Jennings P, Allsop J, Griffiths S, Cunningham A. Evolution of alanine: glyoxylate aminotransferase 1 peroxisomal and mitochondrial targeting. A survey of its subcellular distribution in the livers of various representatives of the classes Mammalia, Aves and Amphibia. Eur. J. Cell Biol. 1994b; 64: 295–313.

Danpure C, Guttridge K, Fryer P, Jennings P, Allsop T, Purdue P. Subcellular distribution of hepatic alanine:glyoxylate aminotransferase in various mammalian species. Cell Sci. 1990; 97: 669–78.

Danpure C, Jennings P. Peroxisomal alanine: glyoxylate aminotransferase deficiency in primary hyperoxaluria type I. FEBS 1986; 201: 20–4.

Danpure C, Jennings P, Fryer P, Purdue P, Allsop J. Primary hyperoxaluria type 1: genotypic and phenotypic heterogeneity. J. Inherit. Metab. Dis. 1994c; 17: 487–99.

Danpure C, Jennings P, Mistry J, et al. Enzymological characterization of a feline analoque of primary hyperoxaluria type 2: a model for the human disease. J. Inher. Metab. Dis. 1989; 12: 403–14.

Danpure C, Purdue P. Primary hyperoxaluria. In: Scriver C, Beaudet A, Sly W, Valle D, eds. The Metabolic Basis of Inherited Diseases. New York: McGraw-Hill, 1995: pp. 2385–424.

Danpure C, Purdue P, Fryer P, et al. Enzymological and mutational analysis of a complex primary hyperoxaluria type 1 phenotype involving alanine:glyoxylate aminotransferase peroxisome-to-mitochondrion mis-targeting and intraperoxisomal aggregation. Am. J. Hum. Genet. 1993; 53: 417–32.

Danpure C, Rumsby G. Molecular aetiology of primary hyperoxaluria and its implications for clinical management. Expert Rev. Mol. Med. 2004; 6: 1–16.

Dawkins P, Dickens F. The oxidation of D- and L-glycerate by rat liver. Biochem. J. 1965; 94: 353–67.

Francheschi V, Nakata PA. Calcium oxalate in plants: formation and function. Annu. Rev. Plant Biol. 2005; 56: 41–71.

Giafi CF, Rumsby G. Kinetic analysis and tissue distribution of human D-glycerate dehydrogenase/glyoxylate reductase and its relevance to the diagnosis of primary hyperoxaluria type 2. Ann. Clin. Biochem. 1998; 35: 104–9.

Gruessner R. Preemptive liver transplantation from a living related donor for primary hyperoxaluria type I. N. Engl. J. Med. 1998; 338: 1924.

Harris K, Richardson K. Glycolate in the diet and its conversion to urinary oxalate in the rat. Invest. Urol. 1980; 18: 106–9.

Hatch M, Cornelius J, Allison M, Sidhu H, Peck A, Freel R. *Oxalobacter* sp. reduces urinary oxalate excretion by promoting enteric oxalate secretion. Kidney Int. 2006; 69: 691–8.

Hokama S, Honma Y, Toma C, Ogawa Y. Oxalate-degrading *Enterococcus faecalis*. Microbiol. Immunol. 2000; 44: 235–40.

Hokama S, Toma C, Iwanaga M, Morozumi M, Sugaya K, Ogawa Y. Oxalate-degrading *Providencia rettgeri* isolated from human stools. Int. J. Urol. 2005; 12: 533–8.

Holmes R, Goodman H, Assimos D. Contribution of dietary oxalate to urinary oxalate excretion. Kidney Int. 2001; 59: 270–6.

Holmes RP, Sexton WJ, Applewhite JC, Kennedy M, Assimos DG. Glycolate metabolism by Hep G2 cells. J. Am. Soc. Nephrol. 1999; 10: S345–7.

Hoppe B, Kemper M, Bokenkamp A, Langman C. Plasma calcium-oxalate saturation in children with renal insufficiency and in children with primary hyperoxaluria. Kidney Int. 1998; 54: 921–5.

Hoppe B, Kemper M, Bokenkamp A, Portale A, Cohn R, Langman C. Plasma calcium-oxalate supersaturation in children with primary hyperoxaluria and end-stage renal failure. Kidney Int. 1999; 56: 268–74.

Hoppe B, Langman C. A United States survey on diagnosis, treament, and outcome of primary hyperoxaluria. Pediatr. Nephrol. 2003; 18: 986–91.

Hoppe B, von Unruh G, Laube N, Hesse A, Sidhu H. Oxalate degrading bacteria: new treatment option for patients with primary and secondary hyperoxaluria. Urol. Res. 2005; 33: 372–5.

Huber P, Birdsey G, Lumb M, Prowse D, Perkins T, Knight D. Peroxisomal import of human alanine: glyoxylate aminotransferase requires ancillary targeting information remote from its C terminus. J. Biol. Chem. 2005; 22: 27111–20.

Ito H, Miura M, Yamamoto K, Hara T. Reduction of oxalate content of foods by the oxalate degrading bacterium, *Eubacterium lentum* WYH-1. Int. J. Urol. 1996; 3: 31–4.

Jamieson N, Group EPTS. A 20-year experience of combined liver/kidney transplantation for primary hyperoxaluria (PH1): the European transplant registry experience 1984–2004. Am. J. Nephrol. 2005; 25: 282–9.

Jiang Z, Asplin J, Evan A, et al. Calcium oxalate urolithiasis in mice lacking anion transporter Slc26a6. Nat. Genet. 2006; 38: 474–8.

Johnson S, Rumsby G, Cregeen D, Hulton S. Primary hyperoxaluria type 2 in children. Pediatr. Nephrol. 2002; 17: 597–601.

Kemper M. The role of preemptive liver transplantation in primary hyperoxaluria type 1. Urol. Res. 2005; 33: 376–9.

Kitagawa Y, Sugimoto E. Possibility of mitochondrial-cytoplasmic cooperation in gluconeogenesis from serine via hydroxypyruvate. Biochim. Biophys. Acta 1979; 582: 276–82.

Knight D, Holmes R, Milliner D, Monico C, Cramer S. Glyoxylate reductase activity in blood mononuclear cells and the diagnosis of primary hyperoxaluria type 2. Nephrol. Dial. Transplant. 2006; [Epub ahead of print].

Knight J, Holmes R. Mitochondrial hydroxyproline metabolism: implications for primary hyperoxaluria. Am. J. Nephrol. 2005; 25: 171–5.

Kodama T, Akakura K, Mikami K, Ito H. Detection and identification of oxalate-degrading bacteria in human feces. Int. J. Urol. 2002; 9: 392–7.

Koppe B, Danpure C, Rumsby G, et al. A vertical (pseudodominant) pattern of inheritance in the autosomal recessive disease primary hyperoxaluria type I: Lack of relationship between genotype, enzymic phenotype and disease severity. Am. J. Kidney Dis. 1997; 29: 36–44.

Kumar R, Ghoshal U, Singh G, Mittal R. Infrequency of colonization with *Oxalobacter formigenes* in inflammatory bowel disease: possible role in renal stone formation. J. Gastroenterol. Hepatol. 2004; 19: 1403–9.

Lam C, Yeun Y, Lai C, et al. Novel mutation in the GRHPR gene in a Chinese patient with primary hyperoxaluria type 2 requiring renal transplantation from a living related donor. Am. J. Kidney Dis. 2001; 38: 1307–10.

Leiper J, Oatey P, Danpure C. Inhibition of alanine: glyoxylate amintransferase 1 dimerization is a prerequisite for its peroxisome-to-mitochondrion mistargeting in primary hyperoxaluria type 1. J. Cell Biol. 1996; 135: 939–51.

Lepoutre C. Calculs multiples chez un enfant: Infiltration du parenchyme renal par des depots cristallins. J. Urol. 1925; 20: 424.

Leumann E, Hoppe B. What is new in primary hyperoxaluria? Nephrol. Dial. Transplant. 1999; 14: 2556–8.

Leumann E, Hoppe B. Pre-emptive liver transplantation in primary hyperoxaluria type 1: a controversial issue. Pediatr. Transplant. 2000; 4: 161–4.

Leumann E, Hoppe B. The primary hyperoxalurias. J. Am. Soc. Nephrol. 2001; 12: 1986–93.

Leumann E, Hoppe B, Neuhaus T. Management of primary hyperoxaluria: efficacy of oral citrate administration. Pediatr. Nephrol. 1993; 7: 207–11.

Lowry M, Hall D, Brosnan J. Hydroxyproline metabolism by the rat kidney: Distribution of renal enzymes of hydroxyproline catabolism and renal conversion of hydroxyproline to glycine and serine. Metabolism 1985; 34: 955–61.

Lumb M, Danpure C. Functional synergism between the most common polymorphism in human alanine: glyoxylate aminotransferase and four of the most common disease-causing mutations. J. Biol. Chem. 2000; 17: 36415–22.

Maitra U, Dekker E. Purification and properties of rat liver 2-keto-4-hydroxyglutarate aldolase. J. Biol. Chem. 1964; 239: 1485–91.

Mansell M. Primary hyperoxaluria type 2. Nephrol. Dial. Transplant. 1995; 10(Suppl 8): 58–60.

Marangella M. Transplantation strategies in type 1 primary hyperoxaluria: the issue of pyridoxine responsiveness. Nephrol. Dial. Transplant. 1999; 14: 301–3.

Marangella M, Cosseddu D, Petrarulo M, Vitale C, Linari F. Thresholds of serum calcium oxalate supersaturation in relation to renal function in patients with or without primary hyperoxaluria. Nephrol. Dial. Transplant. 1993; 8: 1333–7.

Milliner D. The primary hyperoxalurias: an algorithm for diagnosis. Am. J. Nephrol. 2005; 25: 154–60.

Milosevic D, Rinat C, Batinic D, Frishberg Y. Genetic analysis – a diagnostic tool for primary hyperoxaluria. Pediatr. Nephrol. 2002; 17: 896–8.

Monico C, Olson J, Milliner D. Implications of genotype and enzyme phenotype in pyridoxine response of patients with type I primary hyperoxaluria. Am. J. Nephrol. 2005a; 25: 183–8.

Monico C, Persson M, Ford G, Rumsby G, Milliner D. Potential mechanisms of marked hyperoxaluria not due to primary hyperoxaluria I or II. Kidney Int. 2002; 62: 392–400.

Monico C, Rossetti S, Olson J, Milliner D. Pyridoxine effect in type I primary hyperoxaluria is associated with the most common mutant allele. Kidney Int. 2005b; 67: 1704–9.

Nishiyama K, Berstein G, Oda T, Ichiyama A. Cloning and nucleotide sequence of cDNA encoding human liver serine-pyruvate aminotransferase. Eur. J. Biochem. 1990; 194: 9–18.

Noguchi T. Peroxisomal localization of serine:pyruvate aminotransferase in human liver. J. Biol. Chem. 1978; 253: 7598–600.

Noguchi T. Amino acid metabolism in animal peroxisomes. In: Fahimi H, Sies H, eds. Peroxisomes in Biology and Medicine. Berlin: Springer-Verlag, 1987: pp. 234–43.

Noguchi T, Takada Y, Kido R. Glutamate-glyoxylate aminotransferase in rat liver cytosol. Purification, properties and identity with the alanine-2-oxoglutarate aminotransferase. Hoppe Seylers Z. Physiol. Chem. 1977; 358: 1533–42.

Nogueira P, Vuong T, Bouton O, et al. Partial deletion of the AGXT gene (EX1_EX7del): A new genotype in hyperoxaluria type 1. Hum. Mutat. 2000; 15: 384–5.

Nolkemper D, Kemper M, Burdelski M, et al. Long-term results of pre-emptive liver transplantation in primary hyperoxaluria type 1. Pediatr. Transplant. 2000; 4: 177–81.

Pais VJ, Assimos D. Pitfalls in the management of patients with primary hyperoxaluria: a urologist's perspective. Urol. Res. 2005; 33: 390–3.

Pirulli D, Puzzer D, Ferri L, et al. Molecular analysis of hyperoxaluria type 1 in Italian patients reveals eight new mutations in the alanine: glyoxylate aminotransferase gene. Hum. Genet. 1999; 104: 523–5.

Purdue P, Takada Y, Danpure C. Identification of mutations associated with peroxisome-to-mitochonrion mistargeting of alanine/ glyoxylate aminotransferase in primary hyperoxaluria type 1. J. Cell Biol. 1990; 111: 2341–51.

Purdue PE, Lumb MJ, Fox M, et al. Characterization and chromosomal mapping of a genomic clone encoding human alanine: glyoxylate aminotransferase. Genomics 1991; 10: 34–42.

Rumsby G. An overview of the role of genotyping in the diagnosis of the primary hyperoxalurias. Urol. Res. 2005; 33: 318–20.

Rumsby G, Cregeen DP. Identification and expression of a cDNA for human hydroxypyruvate/glyoxylate reductase. Biochim. Biophys. Acta – Gene Structure & Expression 1999; 1446: 383–8.

Rumsby G, Williams E, Coulter-Mackie M. Evaluation of muation screening as a first line test for the diagnosis of the primary hyperoxalurias. Kidney Int. 2004; 66: 959–63.

Sallack H. Formation of serine from hydroxypyruvate and L-alanine. J. Biol. Chem. 1956; 223.

Santana A, Salido E, Torres A, Shapiro L. Primary hyperoxaluria in the Canary Islands: a conformational disease due to I244T mutation in the P11L-containing alanine:glyoxylate aminotransferase. Proc. Natl Acad. Sci. USA 2003; 100: 7277–82.

Schurmann G, Scharer K, Wingen A, Otto G, Herfarth C. Early liver transplantation for primary hyperoxaluria type 1 in an infant with chronic renal failure. Nephrol. Dial. Transplant. 1990; 5: 825–7.

Sidhu H, Hoppe B, Hesse A, et al. Absence of *Oxalobacter formigenes* in cystic fibrosis patients: a risk factor for hyperoxaluria. Lancet 1998; 352: 1026–9.

Takada Y, Kaneko N, Esumi H, Purdue PE, Danpure CJ. Human peroxisomal L-alanine: glyoxylate aminotransferase. Evolutionary loss of a mitochondrial targeting signal by point mutation of the initiation codon. Biochem. J. 1990; 268: 517–20.

Takada Y, Noguchi T. Subcellular distribution, and physical and immunological properties of hepatic alanine: glyoxylate aminotransferase isozymes in different mammalian species. Comp. Biochem. Physiol. 1982; 72B: 597–604.

Takayama T, Fujita K, Suzuki K, et al. Control of oxalate formation from L-hydroxyproline in liver mitochondria. J. Am. Soc. Nephrol. 2003; 14: 939–46.

Van Schaftingen E. D-glycerate kinase deficiency as a cause of D-glyceric aciduria. FEBS Lett. 1989; 243: 127–31.

Van Schaftingen E, Draye J-P, van Hoof F. Coenzyme specificity of mammalian liver D-glycerate dehydrogenase. Eur. J. Biochem. 1989; 186: 355–9.

van Woerden C, Groothoff J, Wijburg F, Annink C, Wanders R, Waterham H. Clinical implications of mutation analysis in primary hyperoxaluria type 1. Kidney Int. 2004; 66: 746–52.

von Schnakenburg C, Weir T, Rumsby G. Linkage of microsatellites to the AGXT gene on chromosome 2q37.3 and their role in prenatal diagnosis of primary hyperoxaluria type 1. Ann. Hum. Genet. 1997; 61: 365–8.

Webster K, Cramer S. Genetic basis of primary hyperoxaluria type II. Mol. Urol. 2000; 4: 355–63.

Webster K, Ferree P, Holmes R, Cramer S. Identification of missense, nonsense, and deletion mutations in the GRHPR gene in patients with primary hyperoxaluria type II (PH2). Hum. Genet. 2000; 107: 176–85.

Williams H, Smith L. L-glyceric aciduria; a new genetic variant of primary hyperoxaluria. N. Engl. J. Med. 1968; 278: 233–9.

Xue H, Sakaguchi T, Fuije M, Ogawa H, Ichiyama A. Flux of the L-serine metabolism in rabbit, human, and dog livers. Substantial contributions of both mitochondrial and peroxisomal serine:pyruvate/alanine:glyoxylate aminotransferase. J. Biol. Chem. 1999; 274: 16028–33.

Yanagisawa M, Higashi S, Oda T, Ichiyama A. Properties and possible physiological role of rat liver serine:pyruvate aminotransferase. In: Lennon D, Stratman F, Zahlten R, eds. Biochemistry of Metabolic Processes. New York: Elsevier Biomedical, 1983.

Yuen Y, Lai C, Tong G, et al. Novel mutations of the AGXT gene causing primary hyperoxaluria type 1. J. Nephrol. 2004; 17: 436–40.

Zhang X, Roe S, Hou Y, et al. Crystal structure of alanine:glyoxylate aminotransferase and the relationship between genotype and enzymatic phenotype in primary hyperoxaluria type 1. J. Mol. Biol. 2003; 15: 643–52.

CHAPTER 34

The Oculocerebrorenal Syndrome of Lowe

STEVEN G. COCA AND ROBERT F. REILLY

INTRODUCTION

The oculocerebrorenal syndrome of Lowe (OCRL) is a rare human X-linked disorder manifested clinically by congenital cataracts, severe mental retardation, and renal tubular dysfunction (Fanconi's syndrome). Growth retardation, areflexia, hypotonia, and glaucoma are also prominent features. The syndrome was first described by Lowe, Terrey, and MacLachlan in 1952 (Lowe et al 1952). Although the syndrome can occur in all races, the majority of reported cases are in Whites (approximately 87%) (McSpadden 1991). The majority of individuals are diagnosed with OCRL by the age of one, with the average age at diagnosis 1.5 years (McSpadden 1991).

The prevalence of OCRL is low, only a few cases per 100 000 males, and occasional cases have been reported in females (Scholten 1960, Svorc 1967, Sagel et al 1970). Mutations of the gene named OCRL1 on chromosome Xq24-26 result in the characteristic syndrome. One case report of deletion of the entire gene was reported (Peverall et al 2000). These mutations uniformly lead to a deficiency of the 105 kDa enzyme phosphatidylinositol (4,5) biphosphate 5-phosphatase (belonging to the type II 5-phosphatase family). How deficiency of the enzyme leads to the phenotypic syndrome is not clear, but various hypotheses are outlined below.

GENETICS

The OCRL1 locus was mapped to Xq24-25 by molecular linkage analysis (Silver et al 1987). Attree and colleagues cloned the OCRL1 gene in 1992 (Attree et al 1992). The transcript contained an open reading frame of 2904 nucleotides and is predicted to encode a polypeptide of 968 amino acids with a calculated M_r of 112 000. A single transcript of 5.8 kb was expressed in brain, skeletal muscle, heart, kidney, lung, placenta, and fibroblasts, but not in lymphoblastoid cells. Northern blot analysis of fibroblasts from 15 affected male patients with cytogenetic abnormalities revealed no detectable transcript. These patients also showed a normal Southern blotting pattern suggesting that the genetic abnormality was due to either a point mutation, or a small insertion or deletion. Structural analysis, hydropathy calculations, and comparison to the ProSite protein motif database did not identify features of structural importance. Searches of the GeneBank database revealed, however, that the novel protein encoded by the gene was highly homologous to a partial sequence for a soluble 75-kDa inositol polyphosphate-5-phosphatase from human platelets (53% amino acid identity and 71% similarity throughout the entire sequence) (Ross et al 1991).

Physical mapping of the OCRL1 gene revealed that it spans 58 kb of the genome, and comprises 24 exons, numbered 1 through 23 plus exon 18a, a 24-bp exon that is differentially retained or spliced out of OCRL1 transcripts. It is transcribed in a centromeric to telomeric direction (Nussbaum et al 1997).

Several mutations, including truncation mutations (nonsense, splicesite, frame-shift), missense and occasionally, somatic mutations have been identified as causative for OCRL. Nucleotide mutations are concentrated in the 3′ part of the gene, whereas most of the large genomic deletions occur in the 5′ part of the gene. Ninety-three percent of mutations are concentrated in only half of the 24 exons (10–19, 21–23) (Satre et al 1999). Fifty-two percent of mutations are in five exons, and 20% have mutations in exon 15. There are almost no mutations in exons 1 through 9. Moreover, not all mutations are inherited. Genetic analysis of 36 families of Lowe's syndrome individuals revealed that 30% of the

Genetic Diseases of the Kidney
ISBN: 978-0-12-449851-8

mothers were not carriers for a mutation, implying that a spontaneous mutation occurred in the affected child (Monnier et al 2000).

The type of mutation results in some minor differences in its gene product. For example, the ocrl1 protein was diminished but detectable by immunoblotting in some patients with either missense mutations or a codon deletion but was not detectable in those with premature termination mutations (nonsense or frame-shift mutations) (Lin et al 1998). In general, with any of the mutations, the levels of ocrl1 protein are far lower (at least tenfold) than controls, and these abnormalities of the ocrl1 protein appear to be severe enough to result in the characteristic Lowe syndrome (Lin et al 1997) (see below). At the present time, there does not appear to be a correlation between the type of OCRL1 mutation and disease severity.

The OCRL-1 gene is expressed in almost all tissues except for hematopoietic cells (Olivos-Glander et al 1995). OCRL1 is also expressed throughout the nephron: the glomerulus, proximal tubule, medullary and cortical thick ascending limb, distal convoluted tubule, connecting tubule, cortical collecting duct, and outer medullary collecting duct (Erb et al 1997).

FUNCTION OF OCRL1 GENE PRODUCT (ocrl1)

The gene product of OCRL1, referred to as ocrl1, a 105 kD protein, deficient in Lowe syndrome, was identified in 1995 to be the enzyme phosphatidylinositol-4,5-biphosphate 5-phosphatase (Suchy et al 1995, Zhang et al 1995). This protein was localized to the Golgi complex in the same year (Olivos-Glander et al 1995). Antibodies to ocrl1 stained a juxtanuclear region by immunofluorescence in normal fibroblast cells, while immunofluroescence staining was absent from fibroblasts of a patient with OCRL. This juxtanuclear area was determined to be the Golgi complex by using a known monoclonal antibody against a Golgi-specific coat protein. Further delineation of the site of ocrl1 was obtained utilizing three techniques: immunofluorescence, subcellular fractionation, and a dynamic perturbation assay with brefeldin A. In this study, ocrl1 was localized to the transGolgi network within fibroblasts and kidney epithelial cell lines (Dressman et al 2000). The function of the trans-Golgi network is to sort proteins into vesicles destined for secretory vesicles or lysosomes. In polarized cells, the trans-Golgi network also serves as the key sorting site for proteins destined for either the apical or basolateral membrane.

Deficiency of the ocrl1 protein is the first inherited defect of phosphatidylinositol metabolism described. Although there are multiple 5-phosphatases in kidney proximal tubule cells and fibroblasts, phosphatidylinositol (4,5) biphosphate 5-phosphatase is the major 5-phosphatase in these cells. In kidney tubule cell lines and fibroblasts of a Lowe syndrome patient, there was tenfold less phosphatidylinositol (4,5) biphosphate 5-phosphatase activity than in kidney and fibroblast cell lines of a normal individual. The enzymatic deficiency resulted in abnormal intracellular accumulation the enzyme's substrate, phosphatidylinositol (4,5) biphosphate (Zhang et al 1998). Phosphatidylinositol (4,5) biphosphate regulates proteins involved in Golgi vesicular transport, ADP ribosylation factor GTPase-activating protein (ARF-GAP) and phospholipase D (Liscovitch et al 1994, DeCamilli et al 1996, Randazzo 1997). This may be one of the mechanisms by which cell function is disrupted in OCRL.

Phosphatidylinositol (4,5) biphosphate appears, however, to have other functions. It regulates the actin-binding proteins, actinin and gelsolin, which are involved in cytoskeleton assembly (Stossel 1989, Yu et al 1992, Janmey 1994). A recent study confirmed abnormal actin polymerization in Lowe syndrome (Suchy & Nussbaum 2002). Lowe fibroblasts from seven different cell lines of Lowe syndrome patients were studied. The Lowe fibroblasts had a decrease in number and length of actin stress fibers, enhanced sensitivity to actin depolymerizing agents, and an increase in punctate F-actin staining in 64% of cells, compared with only 1% of normal fibroblasts from controls. In addition, gelsolin, a major actin-severing protein (regulated by phosphatidylinositol (4,5) biphosphate and calcium) was found to be redistributed in an abnormal punctate pattern in the center of the Lowe fibroblasts. Punctate staining of gelsolin was absent from normal cells. Alpha-actinin, another actin-binding protein regulated by phosphatidylinositol (4,5) biphosphate and calcium, was also stained and had an abnormal accumulation of punctuate staining at the center of the cell in 61% of Lowe fibroblasts compared with only 4% of normal fibroblasts (Suchy & Nussbaum 2002). Finally, to prove that abnormal ocrl1 expression was responsible for these cellular derangements, these Lowe fibroblasts, which had no detectable OCRL1 mRNA, were transfected with OCRL1 cDNA. While the untransfected cells had the aforementioned punctuate staining, no transfected cells had punctate staining.

These alterations in actin-binding proteins may lead to the phenotypic abnormalities seen in Lowe syndrome. Disruption of actin polymerization leads to failure of epithelial cell adhesion. In cultured kidney epithelium, actin polymerization is necessary for the formation, maintenance, and functioning of tight junctions that are required for normal barrier function of renal proximal tubule cells (Stevenson & Begg 1994, Wittchen et al 1999).

Another potential pathogenetic role of abnormal metabolism of phosphatidylinositol (4,5) biphosphate may exist. Vesicular trafficking from lysosomes may be regulated by phosphatidylinositol (4,5) biphosphate. Double-immunostaining revealed that phosphatidylinositol (4,5) biphosphate co-localized with the lysosomal marker Lamp1 (Zhang et al 1998) (which is in contrast to previous studies that have localized phosphatidylinositol (4,5) biphosphate to the Golgi). Further support for the potential role of phosphatidylinositol (4,5) biphosphate negatively impacting lysosomal trafficking in OCRL was provided by a clinical study involving 15 patients with Lowe syndrome and 15 age-matched controls. A 1.6–2.0-fold increase in seven lysosomal enzymes was found in plasma of the Lowe patients. Ninety-five percent of values in the lowest quintile came from normal subjects while 85% of the values in the highest quintile were from patients with Lowe syndrome (Ungewickell & Majerus 1999).

Finally, ocrl1 may have functions in addition to its phosphatidylinositol (4,5) biphosphate 5-phosphatase activity. The c-terminal portion of orcl1 contains a RhoGAP (GTPase activating protein) domain (Peck et al 2002). Rho GTPases control multiple cellular functions including actin cytoskeleton organization, vesicular trafficking, transcriptional regulation, and cell cycle progression. Furthermore, various forms of mental retardation are attributed to mutations in genes of Rho pathways, including nonspecific forms (Billuart et al 1998, Kutsche et al 2000) and also genetic diseases such as Aarskog-Scott syndrome (Pasteris et al 1994), tuberous sclerosis (Astrinidis et al 1998), and Williams syndrome (Frangiskakis et al 1996). Several mutations of OCRL1 were mapped to the RhoGAP-like region (Lin et al 1997, 1998) although the functional role of the domain is not entirely clear. Preliminary work indicates that the ocrl1 c-terminus RhoGAP complexes with Rac GTPase in the transGolgi network and that the RhoGAP domain is capable of inhibiting RacGTP-dependent actin reorganization (Faucherre et al 2003).

In conclusion, a variety of mutations of the OCRL1 gene leads to a loss of function of its product, the enzyme phosphatidylinositol (4,5) biphosphate 5-phosphatase (orcl1), which leads to accumulation of phosphatidylinositol (4,5) biphosphate. The buildup of phosphatidylinositol (4,5) biphosphate may then impair cell function by: (a) interfering with proteins involved in Golgi vesicular transport; (b) interfering with actin-binding proteins, thereby disrupting cytoskeleton formation; (c) interfering with vesicular trafficking from lysosomes. Finally, the c-terminal domain of ocrl1, RhoGAP, binds to Rac GTPase, which is likely involved in localizing ocrl1 to the transGolgi network. Mutations resulting in deficiencies of the ocrl1 RhoGAP may disrupt actin and cytoskeleton formation.

Several questions remain unanswered. First, how do these cellular disturbances lead to the clinical phenotype of OCRL? Second, why does a mutation in a gene that is found in nearly all cells in the body result in phenotypic abnormalities of significant severity in only certain organ systems (predominantly the kidney, eyes, and nervous system)?

PHYSICAL FEATURES

Children with OCRL typically have coarse facial features with a low-bridged, broad nose, prominent ears, epicanthal folds, and occasional frontal bossing. The hands are short and stubby. High palate and undescended testicles are common in newborns (McSpadden 1991). Approximately 13% require treatment for undescended testicles. The rate of growth begins to fall at 3 months of age, and by 1 year, most children with OCRL are in the 10th percentile for height. The average height of adults with OCRL is 5'1" (McSpadden 1991). Weight similarly slows after birth. In one series, weights remained close to normal percentiles for the first 3 years of life, then fell below the third percentile (Charnas et al 1991), but another report of a larger population of individuals with OCRL indicated that by 9 months of age, most children are in the 5th percentile for weight (McSpadden 1991). Psychomotor retardation is often present, and speech is often delayed. When speech finally develops usually between the ages of 3–7 years, most are able to combine two words in a phrase (Schwartz et al 1964). Most children over the age of 3 years can follow two-step commands. Puberty is often delayed, as 31% do not reach puberty until the age of 18 (McSpadden 1991).

OPHTHALMOLOGIC ABNORMALITIES

Congenital cataracts are a cardinal finding of OCRL. They are bilateral in 90% of cases (Francois 1972). Fifty percent have cataracts detected at birth, and 99% have cataracts by the age of 9 months (Abbassi et al 1968, McSpadden 1991). The cataracts develop as a result of disordered migration of embryonic lens epithelium, which occurs as a primary defect rather than secondary to a metabolic defect (Ginsberg et al 1981). Glaucoma occurs in up to 47–60% of patients, and is often diagnosed in the first year of life, but can appear up until 20 years of age (Abbassi et al 1968, Francois 1972, McSpadden 1991). Several patients (68%) are treated surgically for glaucoma, and many require serial surgical procedures (McSpadden 1991). Corneal degeneration (keloids), hypothesized to occur secondary to trauma or inflammation, is seen with moderate frequency (28%) and is frequently bilateral when present (44%) (McSpadden 1991). Other ocular defects that can be seen in OCRL include miosis, nystagmus, enophthalmos, buphthalmos, iris atrophy, and strabismus (Falls 1959, Terslev 1960, Wilson et al 1963, Francois 1972, McSpadden 1991). Visual impairment, both secondary to cataracts, primary retinal dysfunction, and corneal degeneration, is universal in OCRL (Abbassi et al 1968).

NEUROLOGIC MANIFESTATIONS

Most patients with OCRL have some degree of mental retardation, most with moderate to severe impairment (McSpadden 1991, Kenworthy et al 1993). A small percentage of individuals (10%) have IQs in the normal or borderline range (McSpadden 1991, Kenworthy et al 1993). It is, however, important to note that IQ testing may be biased against patients with OCRL because of their underlying visual impairment.

Ability to walk is usually delayed (Richards et al 1965). While rare cases are reported when walking occurred at a normal age (Schwartz et al 1964), most children with OCRL have delayed development of walking by months to several years (Witzleben et al 1968). But, by the age of 6.8 years, most individuals have developed walking ability (McSpadden 1991). Behavioral abnormalities such as nonvisual hand flapping, love of and response to music, emotional rigidity or stubbornness, fearfulness, fixation with specific objects, and rejection of different textures are frequently present, and behaviors such as aggressiveness towards others, self hand biting, hand banging and head banging occur less commonly.

Histologic examination of the CNS can be normal (Chutorian & Rowland 1966) or may reveal changes including fibrosis of the leptomeninges, diffuse loss of cortical neurons (Garzuly et al 1973, 1974), neurofibrillary tangles (Matin & Sylvester 1980), vacuolization of subpial parenchyma, astrocytes with swollen nuclei, and ventriculomegaly (Richards et al 1965). CT scanning can reveal moderate enlargement of ventricles, and both CT and MRI can reveal characteristic changes of asymmetric demyclination or gliosis (high intensity T2 lesions) of cortical white matter (Charnas et al 1988, Demmer et al 1992, Carroll et al 1993) or low-intensity T1 lesions suggestive of cystic lesions (Demmer et al 1992, Ono et al 1996, Schneider et al 2001). Peripheral neuropathy, while rare, can occur. Nerve conduction studies can be slightly abnormal (Kornfeld et al 1975). Nerve biopsies reveal axonal degeneration (Wisniewski et al 1984), or decreased number of myelinated fibers, with either normal myelinization of the remaining nerve fibers (Charnas et al 1988) or rare demyelinization (Kornfeld et al 1975). A significant proportion (37%) of individuals with OCRL have a seizure disorder, with the majority of those (72%) having grand mal seizures (McSpadden 1991).

Clinically, hypotonia is almost universal (86%) (McSpadden 1991). Serum concentrations of creatine kinase, aspartate aminotransferase, and lactate dehydrogenase are elevated suggesting muscle involvement of the disorder (Charnas et al 1991, Fivush et al 1992). Muscle biopsy findings are varied however, and nonspecific. Biopsy reports range from normal (Richards et al 1960, Chutorian & Rowland 1966), or can reveal type 1 fiber predominance (similar to that observed in congenital fiber type disproportion myopathy), a decrease in number of muscle fibers (Abbassi et al 1968), proliferation of sarcolemmal nuclei (Abbassi et al 1968), patchy muscle atrophy (Rutsaert et al 1972), or replacement of muscle by fat (Richards et al 1965). Muscle necrosis is usually not present (Kohyama et al 1989). Electromyographic evidence of denervation has been reported in 40% of cases (Kornfeld et al 1975).

MUSCULOSKELETAL MANIFESTATIONS

Case series report several musculoskeletal abnormalities in OCRL. Rickets is very common occurring in approximately 50% (Abbassi et al 1968, McSpadden 1991). Several children suffer multiple fractures (Holtgrewe & Kalen 1986, McSpadden 1991). Scoliosis, kyphosis, platyspondyly, cervical spine abnormalities, and subluxed/dislocated hips were reported (Holtgrewe & Kalen 1986, McSpadden 1991). Arthritis is present in some individuals (8%), and occasionally results in clinically significant joint effusions (Athreya et al 1983, McSpadden 1991). It is not certain why these musculoskeletal abnormalities occur, but increased urinary excretion of mucopolysaccharides (Holtgrewe & Kalen 1986) and glycosaminoglycans (Akasaki et al 1978, Kieras et al 1984, Wisniewski et al 1984) are hypothesized.

RENAL MANIFESTATIONS

OCRL results in proximal tubule dysfunction (Fanconi syndrome) and progressive renal failure. The onset of renal dysfunction usually is manifest a few weeks to months after birth, presenting with renal tubular acidosis, proteinuria, phosphaturia, and aminoaciduria and carnitine wasting (Charnas et al 1991). Most patients with OCRL have proteinuria, with almost two-thirds having nephrotic-range proteinuria, but do not have other characteristics of the nephrotic syndrome. Urine electrophoresis in patients with OCRL reveals a moderate albumin band, α- and β-globulin zones that are variable in intensity, and a unique finding of discrete bands in the gamma-globulin region that is distinct from the urine electophoresis pattern seen in idiopathic Fanconi's syndrome (Papadopoulos et al 1989).

Mild potassium and phosphorus wasting occurs in nearly all patients with Lowe syndrome, although only approximately one-third of patients require supplementation (Charnas et al 1991). Interestingly, unlike Fanconi's syndrome, none to little glucosuria occurs in Lowe syndrome (Terslev 1960, Richards et al 1965, Charnas et al 1991).

Significant aminoaciduria occurs in Lowe syndrome. It is present very early in life and may worsen or remain stable over time (Abbassi et al 1968). Early studies reported various findings regarding aminoaciduria, ranging from a generalized aminoaciduria (Terslev 1960, Schwartz et al 1964, Richards et al 1965, Abbassi et al 1968) to aminoaciduria with sparing of the branched-chain amino acids (valine, leucine, isoleucine) (Schoen 1959), to an aminoaciduria with sparing of several amino acids that becomes exaggerated with ornithine loading (Schwartz et al 1964), and can be markedly reduced with a protein-free diet (Chutorian & Rowland 1966). More recent studies with modern technology may be more accurate. In the largest case series of 23 patients with in-depth laboratory values measured, urine amino acid indices ranged from two to 20 times normal values (mean values: 686 μmol/kg/day vs. 94 μmol/kg/day in controls) (Charnas et al 1991). Indeed, branched-chain amino acids were spared, as mean fractional excretions of valine, isoleucine, and leucine were the lowest of all the amino acids, and were up to 26 times normal for lysine and arginine. It should be noted that the aminoaciduria seen in OCRL is in general less than that seen in cystinosis. Furthermore, amino acid transport defects may be present in the small intestine. Jejunal tissue taken from two patients with OCRL, 2 and 23 years of age, had partial defects in lysine and arginine accumulation (Bartsocas et al 1969).

The acidosis in OCRL is multifactorial. Lowe et al. initially suggested that the acidosis resulted from increased organic acid production (Lowe et al 1952), as have others (Svorc et al 1967). Further studies, however, have not supported this hypothesis. Instead, the acidosis is likely multifactorial, resulting from abnormalities in bicarbonate reabsorption (Richards et al 1965, Abbassi et al 1968, Matsuda et al 1969, Sagel et al 1970), defective ammonia secretion (Schoen 1959, Richards et al 1965, Svorc et al 1967, Abbassi et al 1968, Sagel et al 1970, Holmes & Tucker 1972) or decreased titratable acid formation (Schoen 1959, Richards et al 1965, Holmes & Tucker 1972).

The majority of patients with OCRL have hypercalciuria (mean 1.98 ± 0.13-fold the upper limit of normal). Furthermore, 2/3rds of patients have nephrocalcinosis and/or calculi, as diagnosed by ultrasound. In contrast to other types of Fanconi syndromes, glycosuria is not a common manifestation of OCRL (Bockenhauer et al 2008).

Renal function during the first 10 years of life is typically normal, however, renal function declines gradually, with end-stage renal disease developing in the fourth decade (Charnas et al 1991).

RENAL PATHOLOGY

Pathologic changes in the kidneys of individuals with OCRL appear to be age dependent and can be divided into three groups. Patients from birth to 3 months of age usually show no evidence of renal pathology. Between 3 months and 5 years there is evidence of diffuse tubular changes with little or no glomerular pathology. Tubular abnormalities range from mild, i.e. dilation of the mitochondria and disruption of the cristae seen only on electron microscopic observation, to severe, i.e. diffusely dilated tubules with atrophy of the tubular epithelium, most of which are filled with proteinaceous casts. Distal nephron calcium deposition may also occur. In more advanced stages of the disease, usually in patients over the age of 5–7 years, the entire renal parenchyma shows focal involvement. The glomeruli may be fibrosed, hyalinized or diffusely hypercellular with thickening of the basement membrane and foot process fusion. There is multifocal tubular atrophy, basement membrane thickening and interstitial fibrosis (see Table 34.1).

DIAGNOSIS

The diagnosis of OCRL is largely clinical. Children with OCRL are born with congenital cataracts, hypotonia, and areflexia. Later, mental retardation, glaucoma, and renal tubular dysfunction manifest. The renal dysfunction may initially be subtle, and quantitative analysis of amino acid excretion may be performed to detect excessive aminoaciduria. Other characteristics outlined in the above text may be supportive of the diagnosis.

A definitive diagnosis of OCRL is made by performing an assay for phosphatidylinositol (4,5) biphosphate 5-phosphatase activity in cultured fibroblasts. Serum testing is not reliable as leukocytes do not possess phosphatidylinositol (4,5) biphosphate 5-phosphatase. Phosphatidylinositol (4,5) biphosphate 5-phosphatase activity is always less than 10% of normal in affected individuals (Lin et al 1997). In the United States, only one laboratory performs the enzyme assay on fibroblasts, which are cultured for 2–4 weeks. Skin biopsies from suspected cases of OCRL can be sent to the Biochemical Genetics Laboratory at Baylor College of Medicine in Houston, Texas for the test.

Molecular genetic testing for the Lowe syndrome OCRL1 gene is also available. Symptomatic males can be tested by automated fluorescent DNA sequence analyses of the OCRL1 gene at the Baylor College of Medicine. This test is 99% sensitive for mutations in OCRL1. A 20% up-front payment is required for the index case before the blood specimen can be processed. The turnaround time is 4 weeks.

CARRIER DETECTION

Carrier detection may be an important consideration for family members of OCRL patients. Carriers are usually asymptomatic. While assay of the enzymatic activity of phosphatidylinositol (4,5) biphosphate 5-phosphatase is the gold standard for diagnosis of OCRL in an affected individual, it is not an acceptable test for carrier detection. OCRL is x-linked, therefore the OCRL1 mutation is not expressed uniformly because of its random inactivation (lyonization), resulting in heterogeneous enzyme levels.

In family members of an affected individual whose mutation is known, direct mutation detection via targeted DNA sequencing can be performed (Lin et al 1997, 1999, Satre et al 1999, Monnier et al 2000). This can be accomplished by prescreening the probands with single-strand conformation analysis (SSCA) and subsequent haplotyping by two flanking microsatellite-repeat markers, followed by sequencing of the genomic DNA (Satre et al 1999, Monnier et al 2000) or by SSCA followed by PCR amplification of the abnormal coding exons identified in the proband (Lin et al 1999, Roschinger et al 2000). One alternative method for screening described in one report in the literature is to perform

TABLE 34.1   Summary of renal biopsy findings in Lowe's syndrome patients

| Patient | Age (mo) | Glomeruli | | | | | | Tubules | | | | | | |
| | | Light microscopy | | | EM | | | Light microscopy | | | | | EM | |
		Hypercellularity	BM thickening	Fibrosis	Swelling	Fusion	BM thickening	Casts	Dilatation	Calcium	Atrophy	Cell necrosis	Swelling of mitochondria	Rupture of cristae
Schoen (1959)	22	–	–	–	+	–	–	–	–	–	–	–	+	+
Dundas (1964)	54	–	+	–				–	–	–	+/–	–		
Richards et al (1965)	30	–	–	–				+	+	–	–	–		
Richards et al (1965)	62	–	–	+				+	+/–	–	–	+		
Richards et al (1965)	70	–	–	+				+	+/–	–	–	+		
Van Acker et al (1967)	3.5	–	–	–	+	+	+	–	–	–	–	–	–	–
Van Acker et al (1967)	9	–	–	–				–	–	–	+	–		
Van Acker et al (1967)	28.5	+	+	+				+	+	–	+	–		
Sagel et al (1970)	13	–	–	–	–	–	–	–	–	–	–	–	+	+
Abassi et al (1968)	1	–	–	–				–	–	–	–	–		
Abassi et al (1968)	36	–	–	–				+	+	–	–	–		
Abassi et al (1968)	78	+	–	+				–	–	+	+	+		
Witzleben et al (1968)	22	–	–	–	–	+/–	–	–	–	–	–	–	+/–	–
Witzelben et al (1968)	108	+	+	+	–	+	+	+	–	–	+	–	–	–
Ores (1970)	?					–	+	+					+	+
Rutsaert et al (1972)	4 (case 2)	–	–	–	–	+	–	–	–	–	–	–	–	–
Rutsaert et al (1972)	4 (case 3)	–	–	–	–	+	–	–	–	–	–	–	–	–
Rutsaert et al (1972)	6	–	–	–	–	–	–	+	+	–	–	–	–	–
Rutsaert et al (1972)	15	–	–	–				+	+	+	–	–		
Matin & Sylvester (1980)	396	+	+	+				+	–	–	+	+		
Garzuly et al (1973)	66	–	–	–				+	+	–	+	–		
Fivush et al (1992)	132	+	–	–	–	+	+	+	+	–	–	+	–	–
Fivush et al (1992)	144	+	–	–	–	–	+	–	+	–	–	+	–	–

fluorescence in situ hybridization (FISH), which has confirmed a deletion of the entire OCRL1 gene (the first case and only case published of this type of deletion) in both the proband and mother (Peverall et al 2000).

When considering genetic counseling, two factors are important to consider. First, approximately one-third of cases are a result of sporadic mutations (Monnier et al 2000, Addis et al 2004). Second, two cases of germline mosaicism were reported (Satre et al 1999, Monnier et al 2000).

Screening for heterozygotes can also be done clinically. Ophthalmologic slit-lamp exam was shown to be sensitive and specific for carrier detection (Reilly et al 1988, Wadelius et al 1989, Lin et al 1999, Roschinger et al 2000). Examination of carriers reveals several characteristic punctate white or gray lenticular opacities. The opacities are found peripherally, not centrally, and cluster in radial bands, and are also predominantly found anteriorly and in the equatorial region of the lens (Reilly et al 1988, Wadelius et al 1989, Lin et al 1999, Roschinger et al 2000).

Two early studies attempted to ascertain the sensitivity of the slit-lamp examination for determination of carrier status, utilizing family members of known affected individuals (Reilly et al 1988, Wadelius et al 1989). These studies were performed before the OCRL1 gene was cloned, and therefore used linkage analysis with markers in the Xq24-q26 region. Loci DXS42 and DXS10 were the most closely linked to the presumed location of the OCRL1 gene. The results of both studies indicated that the characteristic findings on slit-lamp examination were quite sensitive and specific for carrier status, utilizing the linkage analysis techniques. Using an arbitrary threshold of greater than 100 opacities, all eight of the subjects with greater than 100 opacities had inherited markers linked to the mutated OCRL1 gene (Wadelius et al 1989). Two of ten subjects with fewer than 100 opacities were found to have markers linked to the mutated gene, yielding a sensitivity of 80% and specificity of 100% (Wadelius et al 1989). The other series had slightly more subjects and yielded similar results, finding that slit-lamp examination had a sensitivity of 94% and specificity of 100% when confirmed with linkage analysis (Reilly et al 1988).

A more recent study, performed after the completion of positional cloning of OCRL1, reexamined the sensitivity and specificity of slit-lamp examination, utilizing allelic-specific detection of mutations in the DNA of 31 females from three Lowe syndrome families (Lin et al 1999). Slit-lamp examination provided a sensitivity of 93% for carrier detection. There was only one false negative, as the youngest individual in the study, 5 years of age, did not have an abnormal ophthalmologic exam, despite her DNA diagnosis as a carrier. There were no false positives, yielding a specificity of 100% for the ophthalmologic exam.

In summary, in family members of affected individuals, carrier detection can be performed via dilated slit-lamp examination and DNA testing for OCRL1 mutations. Because of the small potential for false negatives, slit-lamp examination should not be performed alone in high-risk individuals. Finally, the large incidence of sporadic mutations and small incidence of germline mosaicism should be kept cognizant when performing genetic counseling.

PRENATAL DIAGNOSIS

Some attempts have been made at diagnosing OCRL prenatally in families of individuals with known OCRL. One case series reported that maternal serum alpha-fetoprotein was detected at 2.5 times the upper limit of normal in five of twenty pregnancies with a fetus affected with OCRL (Miller et al 1994). These findings are very nonspecific, and provide a very low sensitivity and specificity. Detection of fetal cataracts was reported with prenatal ultrasound (Gaary et al 1993), but it is neither sensitive nor marginally successful in detecting the lens abnormalities in OCRL (Miller et al 1994).

Suchy and colleagues were the first to report a successful prenatal diagnosis of Lowe syndrome via amniocentesis (Suchy et al 1998). First, they confirmed by Western analysis that phosphatidylinositol-4,5-biphosphate 5-phosphatase is expressed in amniocytes and chorionic villi. Enzyme activity in amniocytes and chorionic villi was similar to the activity in control fibroblasts. The enzyme activity in 27 Lowe patient fibroblasts was greater than tenfold lower than in all three cell lines of the controls (fibroblasts, amniocytes, and chorionic villi). There was no evidence of correlation between enzyme activity and gestational age. Subsequently, amniocytes from three pregnancies at risk for Lowe syndrome were studied for enzyme activity. In two of the pregnancies, the enzyme activity in amniocytes was in the normal range, which ultimately resulted in the delivery of two unaffected, healthy males. The third pregnancy studied, however, had very low enzyme activity detected in amniocytes. This pregnancy was terminated and post-mortem autopsy was not performed. Molecular studies confirmed that this fetus was indeed affected with OCRL, as the fetus was found to be carrying the same mutation in exon 12 of OCRL1 as the affected proband in the family.

In conclusion, although there are few data on the subject, prenatal diagnosis may be accomplished through assessment of phosphatidylinositol-4,5-biphosphate 5-phosphatase activity in amniocytes obtained through amniocentesis in pregnant women with family members affected with Lowe syndrome. Alpha-fetoprotein levels and prenatal ultrasound are not useful or routinely performed for screening of Lowe syndrome.

TREATMENT

Treatment of OCRL is limited and primarily deals with symptoms produced by specific organ dysfunction. Over the past 50 years, little has been written nor have strategies changed in the management of the sequelae of OCRL.

Vision

Several ophthalmologic abnormalities are frequent and occur at an early age, therefore requiring early intervention. Cataracts should be removed when diagnosed, and the majority (90%) are removed in the first year of life (McSpadden 1991). Because of the rapid growth rate in childhood, it is difficult to implant artificial lenses. Therefore, the resultant aphakia should be treated with glasses (Kruger et al 2003). Approximately 90% of children require glasses. Even with glasses, however, only 25% of children have fair vision, with the majority having poor vision (63%), and 15% remain nearly blind (McSpadden 1991). Glaucoma is managed both medically and surgically. Patients usually require two or more medications to control intraocular pressure (Kruger et al 2003). Medical treatment fails in approximately two-thirds and surgical procedures such as goniotomy, cyclocryotherapy, and trabeculectomy must be performed (McSpadden 1991, Kruger et al 2003). The average number of surgeries for glaucoma is 4.6 per child (McSpadden 1991). Corneal keloids are frequently full thickness, and, therefore, lamellar keratoplasty is ineffective. Because of the potential stimulatory role of minor surface trauma in the development of corneal keloids, experts believe contact lenses should be avoided and glasses used instead. Strabismus can be treated with patching of the good eye to stimulate the weak eye. If necessary, surgical correction of the extraocular muscles is performed. It is important to recognize that even with ophthalmologic interventions, as discussed above, visual acuity is usually not normal and because of the inadequate visual development, several patients develop sensory nystagmus. In one series, after cataract extraction at a mean age of 1.25 years, visual acuity ranged from 20/40 to 20/80 and five of seven patients developed nystagmus (Kruger et al 2003). Finally, visual acuity is often difficult to assess because of mental retardation.

Neurologic Symptoms

Treatment for the mental retardation and hypotonia in OCRL is limited to conservative therapies, such as physical therapy, occupational therapy, and speech therapy. Seizures can be successfully treated with standard antiepileptics, such as phenobarbital, phenytoin, and valproic acid (Charnas 1989).

Musculoskeletal Disease

Scoliosis, kyphosis, platyspondyly, cervical spine abnormalities, and subluxed/dislocated hips should be treated with appropriate braces and, if needed, orthopedic procedures. Rickets and osteomalacia can occur for several reasons in OCRL. Two of the main underlying etiologies early in the disease are renal phosphate wasting and metabolic acidosis, which are nearly uniform in OCRL (due to the Fanconi-like syndrome) (McSpadden 1991). When progressive renal insufficiency develops, decreased formation of calcitriol and secondary hyperparathyroidism also contribute. Treatment may, therefore, require a multifaceted approach. In the early stages (before there is renal insufficiency) phosphorus supplementation (1–4 grams per day in divided doses) may be used. Modest doses of vitamin D (1000–4000 IU/day) may be added, which can aid in intestinal absorption of phosphorus (Chutorian & Rowland 1966) Laboratory values of calcium, phosphorus, alkaline phosphatase, and creatinine should be monitored every three months. The doses of phosphorus and vitamin D are titrated to normalization of alkaline phosphatase, not serum phosphorus. Improvement of bone disease is correlated with normalization of the alkaline phosphatase. In addition, if the serum bicarbonate is less than 20, alkali should also be administered at a dose of 2–3 mEq/kg/day given every 6–8 hours, until the serum bicarbonate approaches normal values. The addition of bicarbonate treatment may not only facilitate normalization of alkaline phosphatase, but also assist in accelerating growth (Schoen 1959).

Renal Disease

Renal function inexorably declines over years in OCRL. Dietary protein restriction remains controversial. In chronic kidney diseases, the role of protein restriction is inconclusive, with debate whether restriction to 0.6–0.8 g/kg/day of protein slows deterioration of renal function over time (Hebert et al 2001). In addition, it was shown that adherence

to a protein free diet can markedly reduced the aminoaciduria that occurs in OCRL (Chutorian & Rowland 1966). Given the difficulties with adherence to a protein-free diet, however, complexed with the reasoning that because of aminoaciduria and proteinuria, children with OCRL are in negative protein balance, it is not prudent to restrict dietary protein as this may cause further growth inhibition. The management of acidosis and hypophosphatemia were discussed in the above section. Some patients may require potassium repletion, which can be accomplished with potassium citrate formulations, as this aids in correction of the metabolic acidosis as well.

Gene Therapy

There have been no reports of failed or successful attempts at gene therapy in OCRL. One could hypothesize that as gene therapy techniques advance over the ensuing years, OCRL may be a prime candidate, as mutations within one single gene (OCRL1) result in the extensive phenotypic outcomes in various organ systems.

References

Abbassi V, Lowe CU, Calcagno PL. Oculo-cerebro-renal syndrome. A review. Am. J. Dis. Child. 1968; 115: 145–68.

Addis M, Loi M, Lepiani C, Cau M, Melis MA. OCRL mutation analysis in Italian patients with Lowe syndrome. Hum. Mutat. 2004; 23: 524–5.

Akasaki M, Fukui S, Sakano T, Tanaka T, Usui T, Yamashina I. Urinary excretion of a large amount of bound sialic acid and of under sulfated chondroitin sulfate A by patients with the Lowe syndrome. Clin. Chim. Acta 1978; 89: 119–25.

Astrinidis A, Kouvatsi A, Nahmias J, et al. Novel intragenic polymorphisms in the tuberous sclerosis 2 (TSC2) gene. Mutations in brief no. 184. Online. Hum. Mutat. 1998; 12: 217.

Athreya BH, Schumacher HR, Getz HD, Norman ME, Borden ST, Witzleben CL. Arthropathy of Lowe's (oculocerebrorenal) syndrome. Arthritis Rheum. 1983; 26: 728–35.

Attree O, Olivos IM, Okabe I, et al. The Lowe's oculocerebrorenal syndrome gene encodes a protein highly homologous to inositol polyphosphate-5-phosphatase. Nature 1992; 358: 239–42.

Bartsocas CS, Levy HL, Crawford JD, Thier SO. A defect in intestinal amino acid transport in Lowe's syndrome. Am. J. Dis. Child. 1969; 117: 93–5.

Billuart P, Bienvenu T, Ronce N, et al. Oligophrenin 1 encodes a rho-GAP protein involved in X-linked mental retardation. Pathol. Biol. (Paris) 1998; 46: 678.

Bockenhauer D, Bokenkamp A, van't Hoff W. et al. Clin. J. Am. Soc. Nephrol. 2008; 1430–6.

Carroll WJ, Woodruff WW, Cadman TE. MR findings in oculocerebrorenal syndrome. Am. J. Neuroradiol. 1993; 14: 449–51.

Charnas L. Seizures in the oculocerebrorenal syndrome of Lowe. Neurology 1989; 39(Suppl. 1): 362.

Charnas L, Bernar J, Pezeshkpour GH, Dalakas M, Harper GS, Gahl WA. MRI findings and peripheral neuropathy in Lowe's syndrome. Neuropediatrics 1988; 19: 7–9.

Charnas LR, Bernardini I, Rader D, Hoeg JM, Gahl WA. Clinical and laboratory findings in the oculocerebrorenal syndrome of Lowe, with special reference to growth and renal function. N. Engl. J. Med. 1991; 324: 1318–25.

Chutorian A, Rowland L. Lowe's syndrome. Neurology 1966; 16: 115–22.

DeCamilli P, Emr S, McPherson P, Novick P. Phosphoinositides as regulators in membrane traffic. Science 1996; 271: 1533–9.

Demmer LA, Wippold FJ 2nd, Dowton SB. Periventricular white matter cystic lesions in Lowe (oculocerebrorenal) syndrome. A new MR finding. Pediatr. Radiol. 1992; 22: 76–7.

Dressman MA, Olivos-Glander IM, Nussbaum RL, Suchy SF. Ocrl1, a PtdIns(4,5)P(2) 5-phosphatase, is localized to the trans-Golgi network of fibroblasts and epithelial cells. J. Histochem. Cytochem. 2000; 48: 179–90.

Dundas JB. Lowe's Syndrome. Proc. R. Soc. Med. 1964; 57: 837.

Erb BC, Velazquez H, Gisser M, Shugrue CA, Reilly RF. cDNA cloning and localization of OCRL-1 in rabbit kidney. Am. J. Physiol. 1997; 273: F790–5.

Falls HF. Ocular manifestations of the chronic renal tubular insufficiency syndromes. Am. Arch. Ophthalmol. 1959; 62: 188–95.

Faucherre A, Desbois P, Satre V, Lunardi J, Dorseuil O, Gacon G. Lowe syndrome protein OCRL1 interacts with Rac GTPase in the trans-Golgi network. Hum. Mol. Genet. 2003; 12: 2449–56.

Fivush BA, Racusen LC, Christenson MJ, Olson JL. Acute tubular necrosis associated with Lowe's syndrome: possible role of rhabdomyolysis. Am. J. Kidney Dis. 1992; 20: 396–9.

Focus on Lowe's Syndrome: Proceedings of the 1st International Conference on Lowe's Syndrome. Indianapolis, IN: 1986.

Francois J. Ocular manifestations in aminoacidopathies. Adv. Ophthalmol. 1972; 25: 28–103.

Frangiskakis JM, Ewart AK, Morris CA, et al. LIM-kinase1 hemizygosity implicated in impaired visuospatial constructive cognition. Cell 1996; 86: 59–69.

Gaary EA, Rawnsley E, Marin-Padilla JM, Morse CL, Crow HC. In utero detection of fetal cataracts. J. Ultrasound Med. 1993; 12: 234–6.

Garzuly F, Jellinger K, Szabo L, Toth K. Morbid changes in Lowe's oculo-cerebro-renal syndrome. Neuropadiatrie 1973; 4: 304–13.

Garzuly F, Szabo L, Toth K, Kurt J. [Neuropathological changes in the oculocerebrorenal (Lowe) syndrome]. Morphol Igazsagugyi Orv Sz 1974; 14: 252–8.

Ginsberg J, Bove KE, Fogelson MH. Pathological features of the eye in the oculocerebrorenal (Lowe) syndrome. J. Pediatr. Ophthalmol. Strabismus 1981; 18: 16–24.

Hebert LA, Wilmer WA, Falkenhain ME, et al. Renoprotection: one or many therapies?. Kidney Int. 2001; 59: 1211–26.

Holmes G, Tucker V. Oculo-cerbro-renal syndrome, A four generation family study and case reports of two living children. Clin. Pediatr. (Philadelphia) 1972; 11: 119–24.

Holtgrewe JL, Kalen V. Orthopedic manifestations of the Lowe (oculocerebrorenal) syndrome. J. Pediatr. Orthop. 1986; 6: 165–71.

Janmey PA. Phosphoinositides and calcium as regulators of cellular actin assembly and disassembly. Annu. Rev. Physiol. 1994; 56: 169–91.

Kenworthy L, Park T, Charnas LR. Cognitive and behavioral profile of the oculocerebrorenal syndrome of Lowe. Am. J. Med. Genet. 1993; 46: 297–303.

Kieras FJ, Houck GE Jr., French JH, Wisniewski K. Low sulfated glycosaminoglycans are excreted in patients with the Lowe syndrome. Biochem. Med. 1984; 31: 201–10.

Kohyama J, Niimura F, Kawashima K, Iwakawa Y, Nonaka I. Congenital fiber type disproportion myopathy in Lowe syndrome. Pediatr. Neurol. 1989; 5: 373–6.

Kornfeld M, Synder RD, MacGee J, Appenzeller O. The oculocerebral-renal syndrome of Lowe. Arch. Neurol. 1975; 32: 103–7.

Kruger SJ, Wilson ME Jr., Hutchinson AK, et al. Cataracts and glaucoma in patients with oculocerebrorenal syndrome. Arch. Ophthalmol. 2003; 121: 1234–7.

Kutsche K, Yntema H, Brandt A, et al. Mutations in ARHGEF6, encoding a guanine nucleotide exchange factor for Rho GTPases, in patients with X-linked mental retardation. Nat. Genet. 2000; 26: 247–50.

Lin T, Lewis RA, Nussbaum RL. Molecular confirmation of carriers for Lowe syndrome. Ophthalmology 1999; 106: 119–22.

Lin T, Orrison BM, Leahey AM, et al. Spectrum of mutations in the OCRL1 gene in the Lowe oculocerebrorenal syndrome. Am. J. Hum. Genet. 1997; 60: 1384–8.

Lin T, Orrison BM, Suchy SF, Lewis RA, Nussbaum RL. Mutations are not uniformly distributed throughout the OCRL1 gene in Lowe syndrome patients. Mol. Genet. Metab. 1998; 64: 58–61.

Liscovitch M, Chalifa V, Pertile P, Chen CS, Cantley LC. Novel function of phosphatidylinositol 4,5-bisphosphate as a cofactor for brain membrane phospholipase D. J. Biol. Chem. 1994; 269: 21403–6.

Lowe C, Terrey M, Maclachan E. Organic aciduria, decreased renal ammonia production, hydrophthalmos, and mental retardation: a clinical entitiy. Am. J. Dis. Child. 1952; 83: 164.

Matin MA, Sylvester PE. Clinicopathological studies of oculo cerebrorenal syndrome of Lowe, Terrey and MacLachlan. J. Ment. Defic. Res. 1980; 24: 1–16.

Matsuda I, Takeda T, Sugai M, Matsuura N. Oculocerebrorenal syndrome in a child with a normal urinary acidification and a defect in bicarbonate reabsorption. Am. J. Dis. Child. 1969; 117: 205–12.

McSpadden K. Report on the Lowe's Syndrome Comprehensive Survey. West Lafayette, IN: Lowe's Syndrome Association, 1991.

Miller RC, Wolf EJ, Gould M, Macri CJ, Charnas LR. Fetal oculocerebrorenal syndrome of Lowe associated with elevated maternal serum and amniotic fluid alpha-fetoprotein levels. Obstet. Gynecol. 1994; 84: 77–80.

Monnier N, Satre V, Lerouge E, Berthoin F, Lunardi J. OCRL1 mutation analysis in French Lowe syndrome patients: implications for molecular diagnosis strategy and genetic counseling. Hum. Mutat. 2000; 16: 157–65.

Nussbaum RL, Orrison BM, Janne PA, Charnas L, Chinault AC. Physical mapping and genomic structure of the Lowe syndrome gene OCRL1. Hum. Genet. 1997; 99: 145–50.

Olivos-Glander IM, Janne PA, Nussbaum RL. The oculocerebrorenal syndrome gene product is a 105-kD protein localized to the Golgi complex. Am. J. Hum. Genet. 1995; 57: 817–23.

Ono J, Harada K, Mano T, Yamamoto T, Okada S. MR findings and neurologic manifestations in Lowe oculocerebrorenal syndrome. Pediatr. Neurol. 1996; 14: 162–4.

Ores RO. Renal changes in oculo-cerebro-renal syndrome of Lowe. Electron microscopic study. Arch. Pathol. 1970; 89: 221–5.

Papadopoulos NM, Costello R, Charnas L, Adamson MD, Gahl WA. Electrophoretic examination of proteinuria in Lowe's syndrome and other causes of renal tubular Fanconi syndrome. Clin. Chem. 1989; 35: 2231–3.

Pasteris NG, Cadle A, Logie LJ, et al. Isolation and characterization of the faciogenital dysplasia (Aarskog-Scott syndrome) gene: a putative Rho/Rac guanine nucleotide exchange factor. Cell 1994; 79: 669–78.

Peck J, Douglas GT, Wu CH, Burbelo PD. Human RhoGAP domain-containing proteins: structure, function and evolutionary relationships. FEBS Lett. 2002; 528: 27–34.

Peverall J, Edkins E, Goldblatt J, Murch A. Identification of a novel deletion of the entire OCRL1 gene detected by FISH analysis in a family with Lowe syndrome. Clin. Genet. 2000; 58: 479–82.

Randazzo PA. Functional interaction of ADP-ribosylation factor 1 with phosphatidylinositol 4,5-bisphosphate. J. Biol. Chem. 1997; 272: 7688–92.

Reilly DS, Lewis RA, Ledbetter DH, Nussbaum RL. Tightly linked flanking markers for the Lowe oculocerebrorenal syndrome, with application to carrier assessment. Am. J. Hum. Genet. 1988; 42: 748–55.

Richards W, Donnell GA, Wilson WA, et al. The oculocerebro-renal syndrome of Lowe. Am. J. Dis. Child. 1960; 100: 707–9.

Richards W, Donnell GN, Wilson WA, Stowens D, Perry T. The oculo-cerebro-renal syndrome of Lowe. Am. J. Dis. Child. 1965; 109: 185–203.

Roschinger W, Muntau AC, Rudolph G, Roscher AA, Kammerer S. Carrier assessment in families with lowe oculocerebrorenal syndrome: novel mutations in the OCRL1 gene and correlation of direct DNA diagnosis with ocular examination. Mol. Genet. Metab. 2000; 69: 213–22.

Ross T, Jefferson A, Mitchell C, Majerus P. Cloning and expression of human 75-kDa inositol polyphosphate-5-phosphatase. J. Biol. Chem. 1991; 266: 20283–9.

Rutsaert J, S-C. A, Potvliege P. Lowe's syndrome: pathologic studies of four cases. Pathol Eur 1972; 7: 249–62.

Sagel I, Ores RO, Yuceoglu AM. Renal function and morphology in a girl with oculocerebrorenal syndrome. J. Pediatr. 1970; 77: 124–7.

Satre V, Monnier N, Berthoin F, et al. Characterization of a germline mosaicism in families with Lowe syndrome, and identification of seven novel mutations in the OCRL1 gene. Am. J. Hum. Genet. 1999; 65: 68–76.

Schneider JF, Boltshauser E, Neuhaus TJ, Rauscher C, Martin E. MRI and proton spectroscopy in Lowe syndrome. Neuropediatrics 2001; 32: 45–8.

Schoen E. Lowe's Syndrome. Abnormalities in renal tubular function in combination with other congenital defect. Am. J. Med. 1959; 27: 781–92.

Scholten H. Eeen meisje met Lowe-syndrome, maandschr. Kindergeneesk 1960; 28: 251.

Schwartz R, Hall PW 3rd, Gabuzda GJ Jr. Metabolism of ornithine and other amino acids in the cerebro-oculo-renal syndrome. Am. J. Med. 1964; 36: 778–86.

Silver DN, Lewis RA, Nussbaum RL. Mapping the Lowe oculocerebrorenal syndrome to Xq24-q26 by use of restriction fragment length polymorphisms. J. Clin. Invest. 1987; 79: 282–5.

Stevenson BR, Begg DA. Concentration-dependent effects of cytochalasin D on tight junctions and actin filaments in MDCK epithelial cells. J. Cell Sci. 1994; 107(Pt 3): 367–75.

Stossel TP. From signal to pseudopod. How cells control cytoplasmic actin assembly. J. Biol. Chem. 1989; 264: 18261–4.

Suchy SF, Lin T, Horwitz JA, O'Brien WE, Nussbaum RL. First report of prenatal biochemical diagnosis of Lowe syndrome. Prenat. Diagn. 1998; 18: 1117–21.

Suchy SF, Nussbaum RL. The deficiency of PIP2 5-phosphatase in Lowe syndrome affects actin polymerization. Am. J. Hum. Genet. 2002; 71: 1420–7.

Suchy SF, Olivos-Glander IM, Nussabaum RL. Lowe syndrome, a deficiency of phosphatidylinositol 4,5-bisphosphate 5-phosphatase in the Golgi apparatus. Hum. Mol. Genet. 1995; 4: 2245–50.

Svorc J, Masopust J, Komarkova A, Macek M, Hyanek J. Oculocerebrorenal syndrome in a female child. Am. J. Dis. Child. 1967a; 114: 186–90.

Terslev E. Two cases of aminoaciduria, ocular changes and retarded mental and somatic development (Lowe's syndrome). Acta Paediatr. 1960; 49: 635–44.

Ungewickell AJ, Majerus PW. Increased levels of plasma lysosomal enzymes in patients with Lowe syndrome. Proc. Natl Acad. Sci. USA 1999; 96: 13342–4.

Van Acker K, Roels H, Beelaerts W, Pasternack A, Valcke R. The histologic lesions of the kidney in the oculo-cerebro-renal syndrome of lowe. Nephron 1967; 4: 193–214.

Wadelius C, Fagerholm P, Pettersson U, Anneren G. Lowe oculocerebrorenal syndrome: DNA-based linkage of the gene to Xq24-q26, using tightly linked flanking markers and the correlation to lens examination in carrier diagnosis. Am. J. Hum. Genet. 1989; 44: 241–7.

Wilson W, Richards W, Donnell G. Oculo-cebebral-renal syndrome of lowe. Arch. Ophthalmol. 1963; 70: 49–55.

Wisniewski KE, Kieras FJ, French JH, Houck GE Jr., Ramos PL. Ultrastructural, neurological, and glycosaminoglycan abnormalities in Lowe's syndrome. Ann. Neurol. 1984; 16: 40–9.

Wittchen ES, Haskins J, Stevenson BR. Protein interactions at the tight junction. Actin has multiple binding partners, and ZO-1 forms independent complexes with ZO-2 and ZO-3. J. Biol. Chem. 1999; 274: 35179–85.

Witzleben CL, Schoen EJ, Tu WH, McDonald LW. Progressive morphologic renal changes in the oculo-cerebro-renal syndrome of Lowe. Am. J. Med. 1968; 44: 319–24.

Yu FX, Sun HQ, Janmey PA, Yin HL. Identification of a polyphosphoinositide-binding sequence in an actin monomer-binding domain of gelsolin. J. Biol. Chem. 1992; 267: 14616–21.

Zhang X, Hartz PA, Philip E, Racusen LC, Majerus PW. Cell lines from kidney proximal tubules of a patient with Lowe syndrome lack OCRL inositol polyphosphate 5-phosphatase and accumulate phosphatidylinositol 4,5-bisphosphate. J. Biol. Chem. 1998; 273: 1574–82.

Zhang X, Jefferson AB, Auethavekiat V, Majerus PW. The protein deficient in Lowe syndrome is a phosphatidylinositol-4, 5-bisphosphate 5-phosphatase. Proc. Natl Acad. Sci. USA 1995; 92: 4853–6.

CHAPTER 35

Fabry Disease (α-Galactosidase A Deficiency): An X-linked Nephropathy

R.J. DESNICK

INTRODUCTION

Fabry disease is an X-linked recessive inborn error of glycosphingolipid catabolism caused by the deficient activity of the lysosomal enzyme, α-galactosidase A (α-Gal A; Figure 35.1) (Brady et al 1967, Desnick et al 2001). This enzymatic defect results in the progressive accumulation of globotriaosylceramide (GL-3) and related glycosphingolipids with terminal α-galactosyl moieties in the lysosomes of endothelial, epithelial, perithelial and smooth muscle cells throughout the body. In classically affected males who have little, if any, α-Gal A activity, the GL-3 deposition in the vascular endothelium is primarily responsible for the major early clinical manifestations of the classic disease phenotype which is characterized by the onset in childhood of angiokeratomas, acroparesthesias, hypohidrosis, and corneal and lenticular changes. With advancing age, the progressive vascular GL-3 accumulation leads to renal failure, cardiac and cerebrovascular disease, and early demise (Desnick et al 2001). Most males with the classic phenotype require renal replacement therapy by 35–45 years of age (Branton et al 2002, Thadhani et al 2002). Prior to renal transplantation and dialysis, the average age of death for classically affected males was about 40 years (Colombi et al 1967, MacDermot et al 2001b, Branton et al 2002). With renal replacement therapy the median lifespan in classically affected males increased to 50 years (MacDermot et al 2001b).

The clinical spectrum of the disease has been expanded with the recent recognition of later-onset cardiac and renal variants who have residual α-Gal A activity, usually due to missense or splice site mutations (Elleder et al 1990a, von Scheidt et al 1991, Nakao et al 1995, 2003). These later-onset variants, recognized during the past decade, lack the early manifestations (e.g. angiokeratoma, acroparesthesias, hypohydrosis, and the corneal/lenticular abnormalities) that signal the diagnosis of classically affected patients. The cardiac variants typically present in the fifth to eighth decade of life with left ventricular hypertrophy (LVH), arrhythmias, mitral insufficiency and/or hypertrophic cardiomyopathy and may have mild to moderate proteinuria with normal renal function for age (von Scheidt et al 1991, Nakao et al 1995, Sachdev et al 2002). They have very low, but sufficient, residual α-Gal A activity (>1%) to preclude the vascular endothelial GL-3 deposits that are the pathologic hallmark of the classical disease phenotype (von Scheidt et al 1991, Desnick et al 2001). The renal pathology in the cardiac variant is limited GL-3 deposition in podocytes which is presumably responsible for their proteinuria (Meehan et al 2004). The recently recognized renal variants also lack the early manifestations of the classic phenotype, but develop proteinuria and end-stage renal disease (ESRD) usually after 50 years of age (Spada & Pagliardini 2002, Nakao et al 2003, Kotanko et al 2004, Rosenthal et al 2004).

Previously, the treatment of Fabry disease had been non-specific and directed to the palliative management of disease complications. Recently, however, enzyme replacement therapy (ERT) was shown to be safe and effective (Eng et al 2001a, 2001b, Schiffmann et al 2001, Wilcox et al 2004) and was approved by the EMEA in 2001 in the European Union and by the FDA in 2003 in the United States. For the later-onset cardiac and renal variants, a combination of ERT and 'chaperone-mediated' enzyme stabilization may prove effective (Frustaci et al 2001, Desnick & Schuchman 2002, Fan 2003). In this review, the clinical, pathologic and genetic features of the classic phenotype and later-onset renal and cardiac variants are described with particular emphasis on the renal manifestations, the recent evidence-based development of ERT, and future therapeutic prospects.

THE CLASSIC PHENOTYPE

In classically affected males, who have essentially no α-Gal A activity, clinical manifestations result predominantly from the progressive lysosomal deposition of GL-3

Genetic Diseases of the Kidney
ISBN: 978-0-12-449851-8

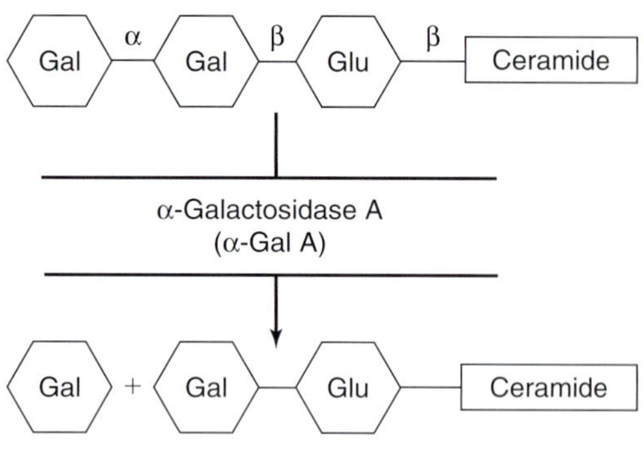

FIGURE 35.1 The metabolic defect in Fabry disease. The deficient α-Gal A activity leads to the accumulation of glycosphingolipids with terminal α-linked galactosyl moieties, including globotriaosylceramide (GL-3), galabiosylceramide, and blood group B substances

in the vascular endothelium (Figure 35.2A, B). Onset usually occurs in childhood or adolescence with periodic crises of severe pain in the extremities (acroparesthesias), the appearance of vascular cutaneous lesions (angiokeratomas), hypohidrosis, and the characteristic corneal and lenticular opacities. Gastrointestinal symptoms including postprandial abdominal cramping, diarrhea, vomiting and nausea, as well as temperature and exercise intolerance observed in children (Desnick & Brady 2004).

Progressive glycosphingolipid deposition in the kidney interferes with renal function, resulting in the early onset of proteinuria and the development of azotemia and renal insufficiency, typically occurring in the third to fifth decades of life. With maturity, most classically affected males experience cardiovascular and/or cerebrovascular disease including myocardial ischemia and infarctions, transient ischemic attacks and/or strokes. The progressive vascular involvement is a major cause of morbidity and mortality, particularly after treatment of the renal insufficiency by chronic dialysis or transplantation. Each of the major manifestations is presented in Table 35.1 and discussed below.

Pain (Acroparesthesias)

Episodic crises of agonizing, burning pain in the distal extremities most often begin in childhood or early adolescence and signal clinical onset of the disease. These crises typically last from minutes to several days and usually are triggered by exercise, fatigue, emotional stress, or rapid changes in temperature and humidity. Often the pain will radiate to the proximal extremities and other parts of the body. With increasing age, the crises usually decrease in frequency and severity; however, in some patients, they may occur more often, and the pain can be so excruciating and incapacitating that the patient may contemplate suicide.

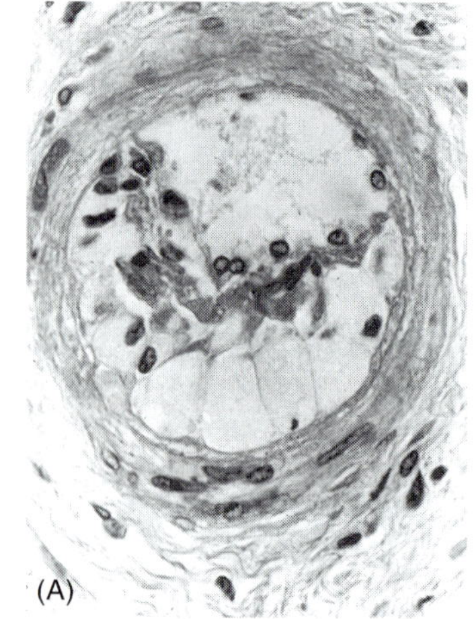

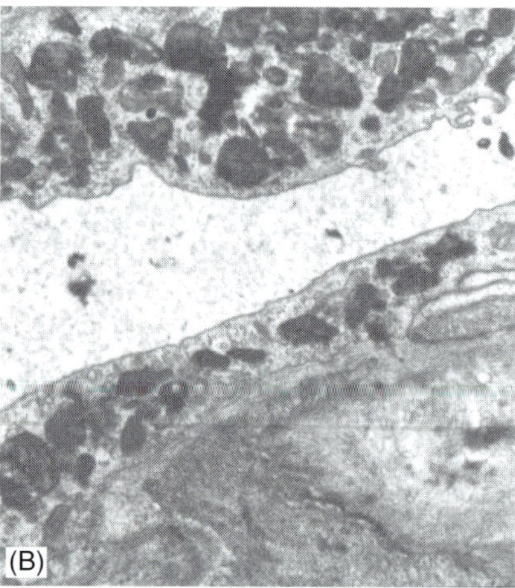

FIGURE 35.2 (A) Photomicrograph of a prostatic arteriole from a classically affected male, showing the hypertrophied, lipid-laden endothelial cells encroaching on the vascular lumen. (B) Electromicrograph showing an occluded vessesl due to the progressive lysosomal deposition of the glycosphingolipid substrate which leads to the narrowing and eventual occlusion of the vascular lumen. (see also Plate 74)

Because the pain usually is associated with a low-grade fever and an elevated erythrocyte sedimentation rate, these symptoms frequently have led to the misdiagnosis of rheumatic fever, neurosis, or erythromelalgia.

The early acroparesthesias that occur in children presumably result from glycosphingolipid deposition in the microvasculature that supplies blood to the peripheral nerve cells. The endothelial glycosphingolipid accumulation narrows the vascular lumen and vessel spasms or frank infarction cause the excruciating pain. It also has been suggested that the neuropathic pain results from GL-3 accumulation in peripheral nerves. In sural nerve biopsies, GL-3 deposits are present in perineural cells, and in myelinated and unmyelinated axons, and to a lesser extent in Schwann cells. Small fiber dysfunction has been described with Aδ fibers being more severely affected than C fibers (Dutsch

TABLE 35.1   Fabry disease: major manifestations in affected males with the classic and later-onset phenotypes

Manifestation	Classic	Renal variant	Cardiac variant
Age at onset	4–8 yr	?	>40 yr
Average age of death	~40 yr[¶]	?	>60 yr
Angiokeratoma	++	−	−
Acroparathesias	++	−	−
Hypohidrosis/anhidrosis	++	−	−
Corneal/lenticular opacity	+	−	−
Heart	Ischemia/MI*/myopathy	LVH[‡]	LVH/myopathy
Brain	TIA**/strokes	?	−
Kidney	Renal failure	Renal failure	Mild proteinuria
α-Gal A activity	<1%	>1%	>1%

[¶]pre-renal replacement therapy; *MI = myocardial infraction; [†]LVH = left ventricular hypertrophy
**TIA = transient ischemic attack; + = present; − = absent.

et al 2002). However, recent studies of the neuropathy revealed a reduced depolarized threshold electrotonus and superexcitability, but normal late subexcitability, indicating that ischemia and infarction occur first and that the poor nerve perfusion contributes to the axonal damage (Tan et al 2005). Thus, the ischemia is primary and the nerve pathology is secondary in the etiology of the Fabry neuropathy.

Angiokeratoma and Hypohidrosis

Angiokeratomas appear as clusters of individual punctate, dark red to blue-black angiectases in the superficial layers of the skin, and are often one of the earliest manifestations of the disease (Figure 35.3A,B). The number and size of these cutaneous vascular lesions progressively increase with age. The lesions may be flat or slightly raised and do not blanch with pressure. There is a slight hyperkeratosis notable in larger lesions. The clusters of lesions are most dense between the umbilicus and the knees, most commonly involve the hips, back, thighs, buttocks, penis, and scrotum, and have a tendency toward bilateral symmetry. However, there may be a wide variation in the distribution pattern and density of the lesions. The oral mucosa, conjunctiva, and other mucosal areas are commonly involved. Classically affected patients without any or with only a few isolated skin lesions have been reported (Clarke et al 1971a). Although the angiectases may not be readily apparent in some patients, careful examination of the skin, especially the scrotum and umbilicus, may reveal the presence of isolated lesions. In addition to these vascular lesions, anhidrosis, or more commonly hypohidrosis, is an early and almost constant finding.

Renal Involvement

The progressive glycosphingolipid accumulation in the kidney interferes with renal function, resulting in azotemia and renal insufficiency. During childhood and adolescence, protein, casts, red cells, and birefringent lipid globules with characteristic 'Maltese crosses' can be observed in the urinary sediment. Proteinuria, isosthenuria, and a gradual deterioration of tubular reabsorption, secretion, and excretion occur with advancing age (Pabico et al 1973). In a study of 105 Fabry males, 50% developed proteinuria by age 35 and 100% of surviving patients developed proteinuria by 52 years of age (Branton et al 2002). Polyuria and a syndrome similar to vasopressin-resistant diabetes insipidus occasionally develop. Gradual deterioration of renal function and the development of azotemia typically occur in the third to fifth decades of life, although renal failure has been reported in the second decade (Sheth et al 1983). Death most often results from uremia unless chronic hemodialysis or renal transplantation is undertaken. The mean age of death of affected males prior to dialysis and transplantation was about 40 years in two series (Colombi et al 1967, Branton et al 2002) and 48.5 years when renal replacement therapy was available (MacDermot et al 2001b). Occasionally, a classically affected male has survived into his sixties. Recent reviews of the renal manifestations are available (Branton et al 2002, Sessa et al 2003, Warnock, 2005).

Cardiac and Cerebrovascular Manifestations

Most classically affected males develop cardiovascular and/or cerebrovascular disease in middle age. The progressive vascular involvement is a major cause of morbidity and mortality, particularly after treatment of the renal insufficiency by chronic dialysis or transplantation (Desnick et al 2001). Left ventricular enlargement, valvular involvement, and conduction abnormalities are early findings. Two European surveys of classically affected males reported that 46% of 98 males and 88% of 200 males had left ventricular hypertrophy (LVH) with the mean age of onset of 42 and 38 years, respectively (MacDermot et al 2001b, Mehta et al

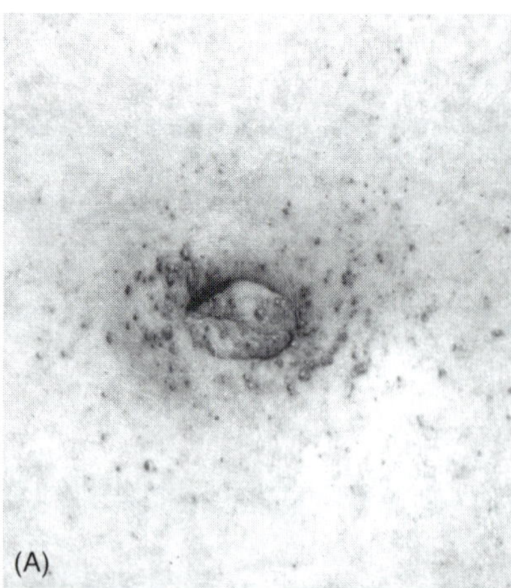

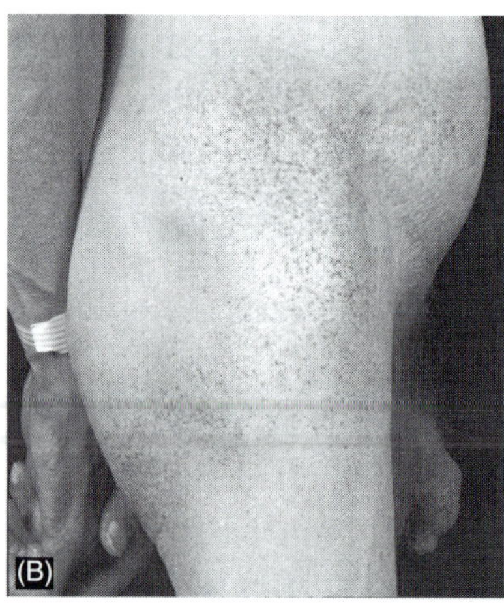

FIGURE 35.3   (A) Clusters of dark-red to blue angiokeratomas (telangiectases) in the umbilical area. (B) A lateral view showing the typical distribution of the skin lesions in the 'bathing trunk' area of a classically affected male with Fabry disease. (see also Plate 75)

2004). Mitral insufficiency may be present in childhood or adolescence. Dysrhythmias such as ST segment changes, T wave inversion, intermittent supraventricular tachycardias, and a short PR interval may be due to glycosphingolipid infiltration of the conduction system. Echocardiographic studies demonstrate an increased incidence of mitral valve prolapse and an increased thickness of the interventricular septum and the left ventricular posterior wall (Bass et al 1980, Goldman et al 1986, Linhart et al 2000, Kampmann et al 2002, Pieroni et al 2003, Weidemann et al 2005). In addition, hypertrophic obstructive cardiomyopathy secondary to glycosphingolipid infiltration in the interventricular septum is a common finding (Colucci et al 1982, Linhart et al 2000, Kampmann et al 2002, Pieroni et al 2003, Weidemann et al 2005). Hypertension, angina pectoris, myocardial ischemia and infarction, congestive heart failure, and severe mitral regurgitation typically occur late in the disease course.

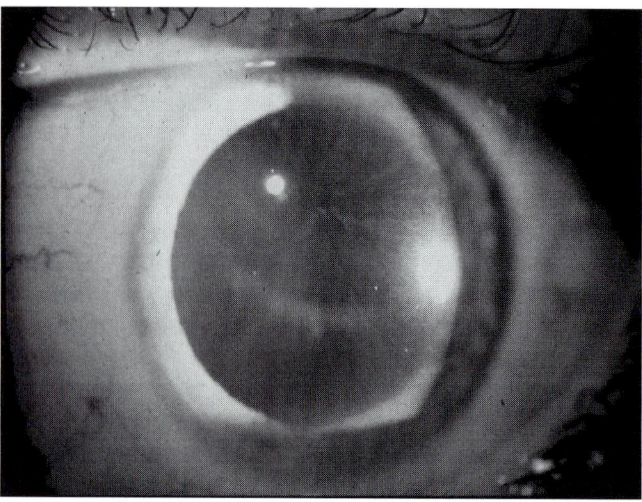

FIGURE 35.4   Corneal opacity in a heterozygous female for the classic phenotype observed by slit-lamp microscopy. (see also Plate 76)

Cerebrovascular manifestations result primarily from multifocal small vessel involvement and may include thromboses, transient ischemic attacks, basilar artery ischemia and aneurysm, seizures, hemiplegia, hemianesthesia, aphasia, labyrinthine disorders, or frank cerebral hemorrhage (Desnick et al 2001). Of patients with lesions on brain MRI, there was a significant association with the polymorphic genotypes of interleukein 6 (G174C), endothelial nitric oxide synthase (G894T), Factor V Leiden (G1691A), and protein Z (A-13G and G79A) (Altarescu et al 2005).

Ocular Features

The cornea, lens, conjunctiva, and retina all may be involved (Sher et al 1976, Orssaud et al 2003, Nguyen et al 2005). A characteristic corneal opacity, observed only by slit-lamp microscopy (Figure 35.4), is found in over 90% of classically affected males (Nguyen et al 2005) and is also present in most heterozygous females, but is typically absent in later-onset cardiac and renal variants. The earliest corneal lesion is a diffuse haziness in the subepithelial layer. With time, the opacities appear as whorled streaks extending from a central vortex to the periphery of the cornea. Typically, the whorl-like opacities are inferior and cream-colored; however, they range from white to golden-brown and may be very faint. Lenticular changes include a granular anterior capsular or subcapsular deposit seen in about 30% of affected males, and a unique, possibly pathognomonic, lenticular opacity, the 'Fabry cataract' (Sher et al 1976). The cataracts, which are best observed by retroillumination, are whitish, spoke-like deposits of fine granular material on or near the posterior lens capsule. These lines usually radiate from the central part of the posterior cortex. These lesions do not interfere with visual acuity. Aneurysmal dilatation and tortuosity of conjuctival and retinal vessels also occur.

Gastrointestinal Manifestations

Beginning in childhood, affected males may have mild to severe gastrointestinal disturbances (Desnick & Brady 2004). Glycosphingolipid deposition in intestinal small vessels and in the autonomic ganglia of the bowel may cause episodic diarrhea, flank pain and/or intestinal malabsorption. Gastrointestinal symptoms, which occur in 70–90% of classically affected males, may worsen with age (MacDermot et al 2001b, Galanos et al 2002). The most common symptoms are episodes of postprandial abdominal pain and bloating followed by multiple bowel movements and diarrhea. Attacks of abdominal or flank pain may simulate appendicitis or renal colic. Other gastrointestinal symptoms include nausea, vomiting, and early satiety. Achalasia and jejunal diverticulosis, which may lead to perforation of the small bowel, have been described (Friedman et al 1984, Roberts & Gilmore 1984, Field et al 2001). Radiographic studies may reveal thickened, edematous colonic folds, mild dilatation of the small bowel, a granular-appearing ileum, and the loss of haustral markings throughout the colon.

Other Clinical Features

Classically affected males may have auditory, pulmonary and other manifestations.. Hearing difficulties and vestibular abnormalities have been described. Bilateral high-frequency neurosensory hearing loss is common in affected males, but deafness is rare (MacDermot et al 2001b, Germain et al 2002). In addition, patients may have pulmonary involvement, which manifested clinically as chronic bronchitis, wheezing, or dyspnea (Brown et al 1997). Pulmonary function studies may show a mild obstructive component, and primary pulmonary involvement has been reported in the absence of cardiac or renal disease. Pitting edema of the lower extremities may be present in adulthood in the absence of hypoproteinemia, varices, or other clinically significant vascular disease. The edema is caused by the progressive glycosphingolipid deposition in the lymphatic vessels and lymph nodes. Varicosities, hemorrhoids, and priapism also have been reported.

HETEROZYGOTES FOR THE CLASSIC PHENOTYPE

Heterozygous females from families with the classic phenotype have a significantly different clinical course and prognosis than affected males (Desnick et al 2001). Their clinical manifestations are highly variable due to random X-chromosomal inactivation (Willard 2001), and range from asymptomatic throughout a normal lifespan to severe as affected males. To date, two studies have suggested that skewing of X-inactivation in heterozygotes may account for the variability of the presence/severity of the disease manifestations (Morrone et al 2003, Dobrovolny et al 2005).

Many heterozygous females experience some symptoms in their 30s to 50s when classically affected males already have severe renal, cerebrovascular, and/or cardiac involvement. However, with increasing age, many heterozygotes manifest disease symptoms. These symptoms can be classified as nonlife-threatening or more severe, incapacitating, and/or life-threatening. Using Kaplan-Meier survival methods, MacDermot determined the median age of death of classically affected males and heterozygous females to be 50 and 70 years, respectively (MacDermot et al 2001a, b).

Nonlife-threatening manifestations in heterozygotes include the characteristic corneal and lenticular opacities, angiokeratomas, acroparesthesias, and hypohidrosis. As shown in Table 35.2, recent surveys of heterozygotes indicate that 70–75% of heterozygous females have the whorl-like corneal opacities (cornea verticillata) which does not affect vision (Figure 35.4), approximately 10–50% have the skin lesions which typically are few and sparse, and 50–90% report acroparesthesias, mainly tingling in the fingers and/or toes, and 11–58% have chronic abdominal pain and diarrhea (Desnick et al 2001, MacDermot et al 2001a, Whybra et al 2001, Galanos et al 2002). With advancing age, heterozygous females may develop mild to moderate left ventricular enlargement and mitral valve prolapse (Kampmann et al 2002). However, a few heterozygotes have been reported in which the symptoms were comparable to those seen in classically affected males (Desnick et al 1972b).

As shown in Table 35.2, more serious, life-threatening manifestations in heterozygous females include significant LVH (19–42%), cardiomegaly (10–55%), myocardial ischemia and infarction, arrhythmias, transient ischemia attacks and strokes (5–18%), and renal failure (0–5%) (MacDermot et al 2001a, Whybra et al 2001, Galanos et al 2002). Conduction defects including a prolonged PR interval, bradycardia and various arrythmias are not uncommon. The occurrences of cardiac and cerebrovascular disease are consistent with the microvascular pathology of the disease.

Renal findings in heterozygotes include isothenuria, the presence of erythrocytes, leukocytes, and granular and hyaline casts in the urinary sediment, and proteinuria. Although about 10% of heterozygous females develop renal failure requiring dialysis or transplantation according to the US and European dialysis and transplant registries (Thadhani et al 2002), recent surveys estimate that 5% or less of heterozygotes develop renal failure (MacDermot et al 2001a, Galanos et al 2002), while classically affected males uniformly develop renal failure by their late 40s to early 50s (Branton et al 2002, Thadhani et al 2002) unless the cardiac or cerebrovascular complications cause demise before the kidneys fail.

Only limited information is available on the heterozygous relatives of males with the later-onset renal or cardiac phenotypes.

TABLE 35.2   Clinical manifestations in heterozygous females for the classic phenotype

	MacDermot et al (2001a)	Galanos et al (2002)	Whybra et al (2001)
Number of heterozygotes	60	38	20
Mean age and range (yr)	46 (17–60)	42 (16–78)	38 (12–65)
Corneal/lenticular opacity	–	76	70
Acroparesthesias	70	53	90
Hypohidrosis	33	–	–
Angiokeratomas	35	13	55
Abdominal pain/diarrhea	58	11	50
LVH*/cardiomyopathy	19	40	55
TIAs**/strokes	28	5	–
Renal failure	3	0	5

*LVH = left ventricular hypertrophy
**TIA = transient ischemic attack.

THE LATER-ONSET VARIANTS

Later-onset variants have been identified who have residual (>1%) α–Gal A activity and lack the characteristic early manifestations of the classic phenotype including the angiokeratomas, acroparesthesias, hypohidrosis, and ocular changes (Elleder et al 1990a, Nagao et al 1991, Nakao et al 1995, 2003, von Scheidt et al 1991). Since they primarily have cardiac manifestations or have been identified in hemodialysis clinics, they have been referred as the 'cardiac' and 'renal' variants (Table 35.1). However, the cardiac variants may have proteinurina or other manifestations, while the renal variants may have cardiac disease. The cardiac and renal variant nomenclature has been promulgated to increase the awareness of these often unrecognized Fabry variants by cardiologists and nephrologists.

The Cardiac Variant

Cardiac variants are essentially asymptomatic during most of their lives and lack the characteristic early manifestations of the classical phenotype including angiokeratoma, acroparesthesias, hypohidrosis, and corneal and lenticular opacities (Elleder et al 1990a, Nagao et al 1991, von Scheidt et al 1991, Nakao et al 1995) (for review, see Desnick et al 2001). Instead, they present in the fifth to seventh decades of life with cardiac disease including LVH, cardiomegaly, typically involving the left ventricular wall and interventricular septum and/or electrocardiographic abnormalities consistent with a cardiomyopathy. They have been misdiagnosed as having hypertrophic cardiomyopathy (Sachdev et al 2002). Cardiac variants may have proteinuria, detected as early as the third decade of life (Desnick et al 2001), but their renal function is typically normal for age or may reflect other age-related complications such

as hypertension. Of note, histological and ultrastructural examination of kidney, endomyocardium, skin, and other nonrenal tissues has consistently shown that cardiac variants lack the vascular endothelial glycosphingolipid deposits that are the pathologic hallmark of the classical disease phenotype (Elleder et al 1990a, von Scheidt et al 1991, Nakao et al 1995, Desnick et al 2001).

The frequency of the cardiac variant is unknown, but two studies suggest that it may be relatively prevalent (Nakao et al 1995, Sachdev et al 2002). In 1995, 3% (7 of 230) of Japanese males who had LVH, were found to have markedly decreased α-Gal A activities and α-Gal A missense mutations. More recently, ~3% (6 of 153) of English males diagnosed as having hypertrophic cardiomyopathy were found to be enzyme- and mutation-confirmed cardiac variants (Sachdev et al 2002).

The oldest cardiac variant described to date was a 75-year-old male who had the N215S α-Gal A mutation (Meehan et al 2004). In addition to cardiac disease, he had proteinuria which led to a renal biopsy, and the subsequent enzymatic diagnosis of Fabry disease. His renal biopsy lacked the renal endothelial, mesangial, epithelial, interstitial, and smooth muscle glycosphingolipid deposition seen in classically affected males, indicating that the residual α-Gal A activity protected these cells from GL-3 accumulation, thereby precluding the early development of renal insufficiency (Meehan et al 2004).

The Renal Variant

Later-onset renal variants were first identified among Japanese male chronic hemodialysis patients whose proteinuria and end-stage renal disease had been diagnosed as chronic glomerulonephritis (Nakao et al 2003). This survey of 516 consecutive unselected hemodialysis male patients found six patients (1.2%) who had absent or low plasma α-Gal A activity, and unrecognized Fabry disease. DNA analysis identified a mutation in each patient's α-Gal A gene. One of the six patients had the classic phenotype, while the others did not have angiokeratomas, acroparesthesias, hypohidrosis, or corneal opacities, the hallmarks of the classic phenotype (Table 35.1). These patients had begun hemodialysis at ages 25 to 56 years. Subsequent screening of male hemodialysis patients in the United States, Japan, and Europe revealed that 0.2–0.5% of patients actually had unrecognized Fabry disease (Utsumi et al 1999, Desnick 2002, Spada & Pagliardini 2002, Kotanko et al 2004, Rosenthal et al 2004). Thus, identification of hemodialysis patients with Fabry disease is important since these patients are likely to develop the Fabry cardiac and cerebrovascular complications in addition to those associated with dialysis. These patients should be considered for ERT to improve their survival on dialysis (Thadhani et al 2002, Pisani et al 2005). In addition, detection of renal variants will permit family studies to identify other affected relatives for therapeutic intervention.

PATHOLOGY

In classically affected patients, birefringent crystalline glycosphingolipid deposits are found in the lysosomes of endothelial, perithelial, and smooth-muscle cells of blood vessels (Figure 35.2A,B) and, to a lesser degree, in histiocytic and reticular cells of connective tissue. Lipid deposits are also prominent in epithelial cells of the cornea, in glomeruli and tubules of the kidney, in muscle fibers of the heart, and in ganglion cells of the autonomic system.

In the kidney, the earliest lesions are due to the accumulation of GL-3 in endothelial and epithelial cells of the glomerulus and of Bowman's space and in the epithelium of the loops of Henle and of distal tubules (Figures 35.5A,B and 35.6). In later stages, and to a lesser extent, proximal tubules, interstitial histiocytes, and fibrocytes may show lipid accumulation. Podocytes and distal tubular epithelial cells contain the most GL-3 deposits, while proximal tubular epithelial cells were relatively unaffected (Thurberg et al 2002). Renal blood vessels, especially the endothelial cells, are involved progressively and often extensively. An early finding is arterial fibrinoid deposits, which may result from the necrosis of severely involved muscular cells (Pompen et al 1947, Gubler et al 1978). Vascular smooth muscle accumulated moderate amounts of granular cytoplasmic GL-3, while in mesangial cells, interstitial fibroblasts and phagocytic cells of the renal cortex, GL-3 appeared as dense granules clustered around the nucleus in the cytoplasm (Thurberg et al 2002). Lipid-laden distal tubular epithelial cells desquamate and may be detected in the urinary sediment (Desnick et al 1971). These cells have been shown to

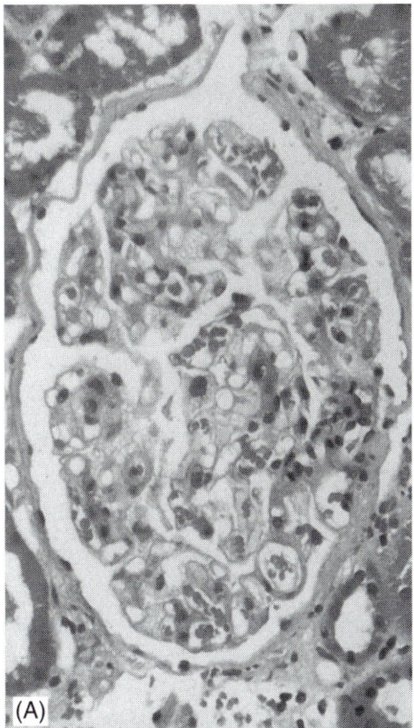

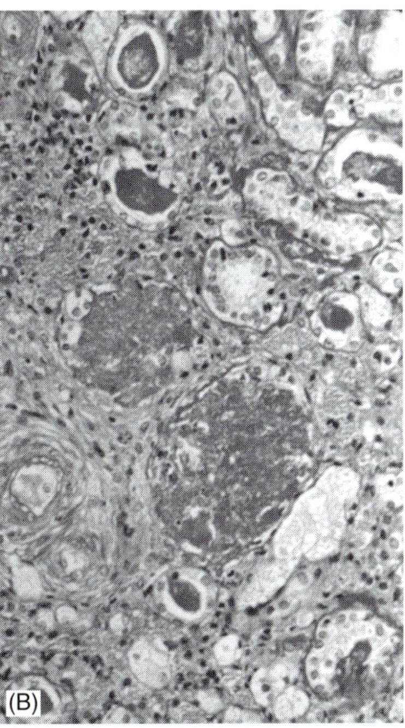

FIGURE 35.5 (A) Photomicrograph of a glomerulus from a 35-year-old classically affected male. (B) Photomicrograph of the end-stage renal disease in a 42-year-old classically affected male. Zenker's fixation, paraffin embedding, and H & E staining. (see also Plate 77)

account for about 75% of the urinary cells shed by a classically affected male (Chatterjee et al 1984).

Other histologic changes in the kidney are the sequelae of nonspecific, end-stage renal disease with evidence of severe arteriolar sclerosis, glomerular atrophy and fibrosis, pseudotubular proliferation of residual glomerular epithelium, tubular atrophy, and diffuse interstitial fibrosis

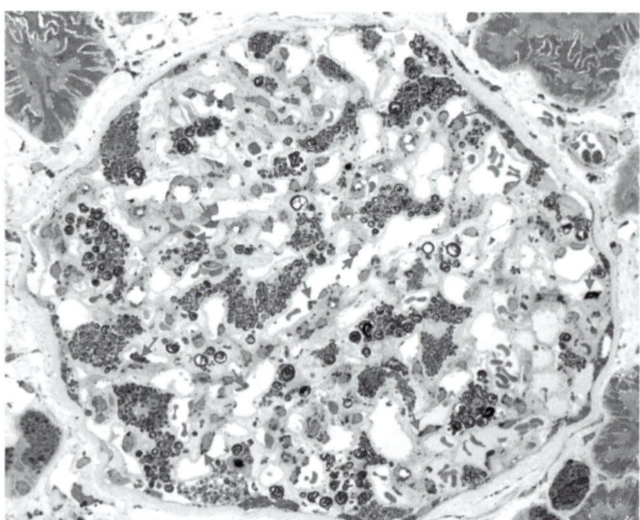

FIGURE 35.6   Photomicrograph of a glomeulus from a classically affected Fabry male showing GL-3 accumulation in multiple cell types including endothelial, interstitial, and mesangial cells as well as podoctyes. From Thurberg et al (2002) with permission

(Figure 35.5B). Kidney size increases during the third decade of life, followed by a decrease in the fourth and fifth decades. The renal involvement has been the subject of comprehensive reviews (McNary & Lowenstein 1965, Morel-Maroger et al 1966, Pabico et al 1973, Burkholder et al 1980, Desnick et al 2001, Branton et al 2002, Thurberg et al 2002).

In males with the cardiac variant phenotype, glycosphingolipid deposition has been documented only in the lysosomes of myocytes and podocytes (Elleder et al 1990b, Desnick et al 2001). Renal biopsies have revealed little, if any, glycosphingolipid deposition in the vascular endothelium, mesangial cells or interstitial cells, as noted above (Meehan et al 2004). To date, little information is available on the renal manifestations or pathology of the renal variant.

THE METABOLIC AND MOLECULAR DEFECTS IN FABRY DISEASE

The Metabolic Defect

The deficient activity of α-Gal A, a lysosomal exogalactosidase, leads to the progressive accumulation of GL-3 and related glycosphingolipids with terminal α-galactosyl moieties (i.e. galabiosylceramide and the blood group B glycolipids) (Sweeley & Klionsky 1963). Brady and coworkers were first to identify a deficient galactosidase activity in affected males with Fabry disease (Brady et al 1967). In 1970, Kint demonstrated that the deficient galactosidase was an α-galactosidase, using synthetic substrates (Kint

1970). Subsequent spectroscopic and enzymatic studies revealed that GL-3 from Fabry kidney had an $\alpha 1 \rightarrow 4$ linkage in the terminal galactosyl moiety (Clarke et al 1971b, Hakomori et al 1971), thereby confirming the enzyme's anomeric specificity. Human α-Gal A was recently crystallized to a resolution of 325 Å and shown to be a homodimer with each monomer containing a (beta/alpha 8) domain with the active site and an antiparallel beta domain. N-linked carbohydrate chains are located at six sites in the glycoprotein dimer, revealing the basis for lysosomal transport via the mannose-6-phosphate receptor (Bishop et al 1986, Ioannou et al 1998, Garman & Garboczi 2004). Recombinant α-Gal A expressed and secreted by CHO cells had oligosaccharide chains which were heterogeneous with high mannose (63%), complex (30%) and hybrid (5%) structures (Matsuura et al 1998). Characterization of the individual *N*-glycosylation sites (N139, N192, and N215) indicated the importance of the oligosaccharide chain at N215 for proper folding and stability (Ioannou et al 1998).

In contrast to classically affected males, who have essentially no detectable $\tilde{\alpha}$ Gal A activity, the later-onset cardiac and renal variants have residual α-Gal A activity ($>$1% of normal), consistent with the attenuation or absence of the characteristic clinical manifestations (von Scheidt et al 1991, Desnick et al 2001, Nakao et al 2003).

The Molecular Defect

The α-Gal A gene has been isolated, sequenced (Bishop et al 1986, 1988, Kornreich et al 1989), and localized to the X-chromosomal region, q22.1 (Srivastava et al 1999). Studies of the α-Gal A mutations in unrelated Fabry families have identified a variety of lesions indicating the molecular genetic heterogeneity of the disease. Over 480 α-Gal A mutations are listed in the Human Gene Mutation Database, as of September, 2005, including duplications, deletions, missense, nonsense, splice site, and complex mutations (Stenson et al 2003) (Figure 35.7). Most mutations are private, occurring only in single pedigrees. However, mutations at CpG dinucleotides have been found in unrelated families of different ethnic or geographic backgrounds. Haplotype analysis of mutant alleles that occurred in two or more families revealed that those with rare novel alleles were probably related, whereas those with mutations involving CpG dinucleotide 'hot spots' were not (Ashton-Prolla et al 2000).

Efforts to establish genotype/phenotype correlations have been limited since most Fabry patients had private mutations, and attempts to predict the phenotype requires more extensive clinical information from unrelated patients with the same genotype. Of the reported molecular lesions, classically affected males had a wide variety of α-Gal A mutations, including large and small gene rearrangements, splicing defects, and missense or nonsense mutations. In contrast, most cardiac and renal variants had missense

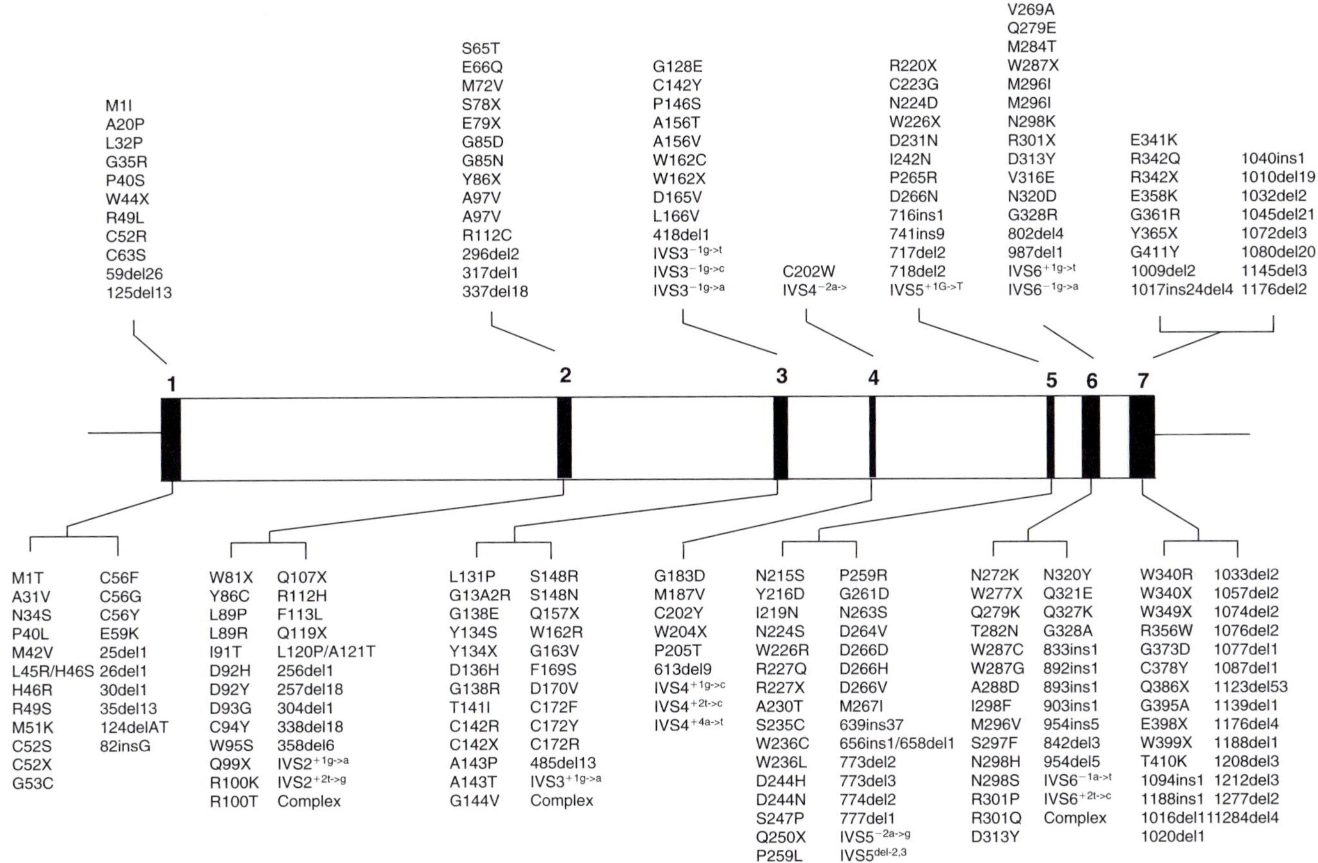

FIGURE 35.7 Schematic of the α-Gal A gene showing representative mutations identified in each exon. Numbers for deletions (del) or insertions (ins) refer to the nucleotide position in the α-Gal A cDNA sequence. IVS = intervening (intronic) sequence

mutations which expressed residual α-Gal A activity. Of note, the N215S mutation was found in several unrelated cardiac variants. Other cardiac variant mutations have included I91T, R112H, F113L, M296I, R301Q, and G328R (Eng et al 1997). Among the renal variants described to date, all had missense mutations (Desnick 2002, Spada & Pagliardini 2002, Nakao et al 2003, Spada et al 2003, Rosenthal et al 2004). The A143T missense mutation has been identified in several very-late-onset renal variants (Spada & Pagliardini 2002, Spada et al 2003).

Recently, molecular modeling studies of α-Gal A mutations suggested a basis for genotype/phenotype correlations (Matsuzawa et al 2005). Of note, three mutations (R112H, R301Q, and G328R) have been identified in patients with the classical disease and the cardiac variant phenotypes, suggesting that other modifying factors are involved in disease expression (Ashton-Prolla et al 2000). In vitro expression of missense mutations causing the classic phenotype had <1% of mean expressed wild-type activity, while various mutations causing the later-onset phenotypes had expressed from 3–50% of expressed wild-type activity (Yasuda & Desnick, unpublished results).

To date, only one coding region sequence variant, D313Y, has been identified which has reduced activity at neutral pH, and low plasma α-Gal A activity (Froissart et al 2003, Yasuda et al 2003). D313Y has been identified in 0.45% of normal Caucasian individuals (n = 800 alleles). Expression of D313Y in COS-7 cells resulted in approximately 60% of expressed wild-type enzymatic activity and showed normal lysosomal localization (Yasuda et al 2003). Thus, D313Y is a variant which does not cause Fabry disease, but would increase disease severity if it occurred in the same allele with another missense or splicing mutation.

DIAGNOSIS

Fabry disease should be considered in males and females with periodic crises of severe pain in the extremities (acroparesthesias), the appearance of vascular cutaneous lesions (angiokeratomas), hypohidrosis, characteristic corneal and lenticular opacities, strokes, cardiac involvement (especially LVH and hypertrophic cardiomyopathy), or renal insufficiency of unknown etiology. Affected males for all phenotypes can be reliably diagnosed by the demonstration of markedly deficient α-Gal A activity in plasma, isolated leukocytes, and/or cultured cells (Desnick et al

1973a). Classically affected males usually have no detectable α–Gal A activity when the assay is performed with synthetic substrates for α-Gal A with the addition of α-*N*–acetylgalactosamine in the reaction mixture to inhibit α-galactosidase B activity (Mayes et al 1981). Cardiac and renal variants may be detected by the presence of residual α-Gal A activity ranging from >1 to 10% of normal.

Heterozygous females for either the classic or later-onset phenotypes have markedly variable α-Gal A activities because of random X-chromosomal inactivation and, therefore, measurement of plasma and/or leukocyte α̃ Gal A activity may be misleading. For example, obligate heterozygotes (daughters of classically affected males) may have α̃ Gal A levels ranging from normal to very low activities similar to those of affected males. Most heterozygous females from classic families have the characteristic corneal dystrophy of Fabry disease (Desnick et al 2001). Accurate diagnosis of heterozygous females requires demonstration of the specific family mutation in the α-Gal A gene. Such testing is recommended for all at-risk females. Prenatal testing can be performed in chorionic villi or amniocytes using either enzymatic or molecular techniques.

TREATMENT

Medical Management

In Fabry disease, the chronicity of the clinical events causes severe debilitation and incapacity that extends over years. In young classically affected males, the single most debilitating aspect of the disease is the excruciating pain. Numerous drugs have been tried to relieve these agonizing pains (Wise et al 1962). With the exception of centrally acting narcotic analgesics such as morphine (Gorden et al 1995, Philippart 1995), which have been only partially effective, conventional analgesic agents have not been helpful. However, prophylactic administration of low maintenance dosages of diphenylhydantoin has been found to provide relief by reducing the frequency and severity of the periodic crises of excruciating pain and constant discomfort in affected males and heterozygotes (Lockman et al 1973). Carbamazepine has similar effects (Lenoir et al 1977). The combination of these two drugs also may significantly reduce the frequency and severity of the pain. The potential side effects of gingival hypertrophy with diphenylhydantoin and dose-related autonomic complications with carbamazepine, including urinary retention, nausea, vomiting, and ileus, have been recorded (Filling-Katz et al 1989). Gabapentin (neurontin) has also been shown to decrease the frequency and severity of neuropathic pain in Fabry disease, but long-term experience will determine its effectiveness (Ries et al 2003). Narcotic analgestics and non-steroidal anti-inflammatory drugs should be avoided.

Care of patients with regard to cardiac, pulmonary, and cerebrovascular manifestations remains nonspecific and symptomatic. Control of hypertension is essential to minimize renal, cardiovascular and cerebrovascular disease. Patients should be monitored for microalbuminuria and angiotensin converting enzyme (ACE) inhibitors and angiotension receptor blockers (ARBs) should be prescribed aggressively, even prophylactically. Successful heart transplantation of a heterozygous female with end-stage cardiomyopathy has been reported (Cantor et al 1998). Obstructive lung disease has been documented in older classically affected males and heterozygous females, with more severe impairment in smokers; therefore, patients should be discouraged from smoking. Patients with reversible obstructive airway disease may benefit from bronchodilation therapy. Prophylactic oral anticoagulants are recommended for affected males and stroke-prone heterozygotes. Pancrelipase or metoclopramide may improve gastrointestinal symptoms (Argoff et al 1998).

Dialysis and Renal Transplantation

Since renal insufficiency is the most frequent and serious late complication in classically affected patients and the renal variants, chronic hemodialysis and/or renal transplantation have become life-extending procedures. In the United States Renal Data System (USRDS) there were 42 patients with Fabry disease on dialysis between 1995 and 1998 (Thadhani et al 2002), while the European Renal Association-European Dialysis and Transplantation Association (ERA-EDTA) identified 83 Fabry patients who began dialysis between 1987 and 1993 (Tsakiris et al 1996). The frequencies of patients with Fabry disease in the USRDS and ERA-EDTA were 0.0167% and 0.0188%, respectively. Of note, heterozygous females who developed renal insufficiency accounted for 12% of the Fabry patients in both the American and European studies. The mean age at which Fabry patients started renal replacement therapy was 42 and 38 years in the USRDS and ERA-EDTA studies, respectively (Tsakiris et al 1996, Thadhani et al 2002). Similar 3-year survival rates for Fabry patients on dialysis, 63% and 60%, were reported in the USRDS (n = 95, 1985–1993) and ERA-EDTA (n = 83, 1987–1993) studies. In the US study, the survival rate for Fabry patients was higher (63%) than that (53%) for diabetic controls, but lower than that (74%) for non-diabetic controls. Cardiovascular complications (48%) and cachexia (17%) were the main causes of death. The 3-year survival data for the 42 USRDS Fabry patients who started dialysis from 1995 to 1998 improved: 70% for Fabry patients, 76% for nondiabetic controls, and 40% for diabetic controls (Tsakiris et al 1996). It is anticipated that the survival statistics will improve in patients on ERT (Kosch et al 2004, Pisani et al 2005).

The first renal transplant in a patient with Fabry disease was performed in 1967 (Buhler et al 1973), and experience has indicated the clear benefit of transplantation in this disease (Maizel et al 1981a, Friedlaender et al 1987, Ojo et al

2000). Recent reviews of the registries of the ERA-EDTA (Tsakiris et al 1996) and the United States Renal Data System (Ojo et al 2000) support excellent outcomes for renal transplantation in Fabry disease. For example, during the 10-year period from 1988 to 1998, the 93 transplanted Fabry patients reported to the US registry had an equivalent, if not better, 5-year patient (82% vs. 83%) and graft (67% vs. 75%) survival than a matched control group with the usual nephropathies (Ojo et al 2000). Also, immune function in Fabry affected males has been shown to be similar to that in other uremic patients, obviating any immunologic contraindication to transplantation in this disease (Donati et al 1984).

Successful transplantation corrects renal function, and the engrafted kidney will remain histologically free of glycosphingolipid deposition (e.g. Bannwart 1982, Mosnier et al 1991), since the normal α–Gal A activity in the allograft will catabolize endogenous renal glycosphingolipid substrates. Rare, isolated glycosphingolipid deposits have been detected in endothelial cells of transplanted cadaveric kidneys by electron microscopy as small myelin figures (Faraggiana et al 1981, Friedlaender et al 1987, Mosnier et al 1991). These cells were derived from the recipient and were α–Gal A deficient (Sinclair 1972). Also, the presence of glycosphingolipid was demonstrated in some cells of the allograft in a patient who expired following acute rejection. This recipient was cross-reacting immunologic material (CRIM)-negative for the enzyme protein and developed anti–α-Gal A antibodies, which apparently inhibited the endogenous α-Gal A activity and resulted in GL-3 accumulation (Faraggiana et al 1981).

Importantly, transplantation of kidneys from Fabry heterozygotes should be avoided as they already may contain significant substrate deposition. In one reported case, transplantation of an allograft from a heterozygous female to an affected male relative resulted in decreased renal function five years later (serum creatinine, 3.0 mg/dl) (Popli et al 1987). Therefore, all potential related donors must be carefully evaluated so that affected males and heterozygous females for the Fabry gene are excluded. Since heterozygous females may have low normal to normal α-Gal A activities due to random X-inactivation, it is necessary to demonstrate the absence of the recipient's α-Gal A gene mutation or haplotype in potential related female donors.

In addition to treatment of the renal failure, kidney transplantation has been undertaken to determine if the allograft could provide normal α-Gal A for systemic substrate metabolism (Desnick et al 1972b). Hypothetically, the normal kidney might metabolize the accumulated substrate by uptake and catabolism within the allograft and/or by the release of the active enzyme into the circulation for uptake and metabolism in other tissues, such as the vascular endothelium. Although biochemical and/or clinical improvement initially was reported (e.g. Desnick et al 1972a, 1973b, Maizel et al 1981b) no clear biochemical

effect could be demonstrated (Clarke et al 1972, Grunfeld et al 1975, Desnick et al 2001).

Of note, patients with successful engraftment who survived for 10 or more years have expired from complications of vascular disease (Kramer et al 1985) or other unrelated causes. Thus, renal allografts are not effective in altering the rate of progressive systemic substrate accumulation, and patients with allografts should benefit from ERT to prevent the late cardiac and cerebrovascular complications of the disease.

ENZYME REPLACEMENT THERAPY

Early Clinical Trials

Early pilot trials of enzyme replacement in Fabry disease indicated the feasibility of this therapeutic approach (Mapes et al 1970, Brady et al 1973, Desnick et al 1979). However, the rate-limiting obstacles were the inability to produce sufficient amounts of purified normal enzyme and the lack of an animal model to evaluate the pharmacokinetics and pharmacodynamics required for the design of clinical trials.

Preclinical Studies in Fabry Mice: Proof of Concept

These obstacles were overcome by the isolation of the human α-Gal A cDNA (Bishop et al 1986) and its high-level expression in Chinese hamster ovary cells (CHO) (Ioannou et al 1992), which provided a source of selectively secreted recombinant human enzyme that was highly sialylated and had mannose-6-phosphate residues for lysosomal targeting. In addition, the generation of an α-Gal A knockout mouse permitted the evaluation of enzyme replacement in an animal model of Fabry disease (Wang et al 1996, Ioannou et al 2001). With these advances, preclinical studies of four recombinant human α-Gal A glycoforms were performed in the enzyme deficient mice. The pharmacokinetics and biodistributions were determined for each glycoform, which differed in sialic acid and mannose-6-phosphate content (Ioannou et al 2001). The plasma half-lives of the glycoforms ranged from ~2 to 5 min, with the more sialylated glycoforms circulating longer. The tissue distributions of the glycoforms were dose-dependent. Following intravenous doses of 0.3–10 mg/kg, each glycoform was primarily recovered in the liver (~30% of dose), with low, but increasing levels of each detected in heart, lung, spleen, and kidney, but not brain. Importantly, GL-3 was cleared from liver, heart, spleen, kidney and plasma in a dose-dependent manner. Not only did these studies show that enzyme delivery and GL-3 clearance were dose-dependent, but that they also determined the rate of GL-3 reaccumulation which indicated an every 2-week dosing schedule. These preclinical studies provided 'proof of concept' and the pharmacokinetic and pharmacodynamic data critical for the design of clinical trials.

Clinical Trials of ERT

Trials of ERT in patients with classic Fabry disease have been been carried out using recombinant human α-Gal A produced by two different companies (Schiffmann et al 2000, 2001, Eng et al 2001a, 2001b, Desnick 2004, Wilcox et al 2004, Banikazemi, in review). A recent study comparing the structural and kinetic properties as well as the cell uptake and in vivo pharmacokinetics and biodistribution of the two preparations (agalsidase alfa [Replagal™] Transkaryotic Therapies, Inc. and agalsidase beta [Fabrazyme®], Genzyme Corp.), found that the enzyme glycoproteins were similar in specific activities and their in vivo pharmacokinetics and biodistributions in the mouse model. The only notable differences were the findings that Fabrazyme had three times more mannose-6-phosphate and higher sialylation, and that its uptake in the kidney and heart was greater (Hopkin et al 2003, Lee et al 2003, Elleder personal communication). Thus, the trials can be taken together as a body of evidence supporting the efficacy of ERT in Fabry disease.

PHASE 1 AND PHASE 1/2 CLINICAL TRIALS

A phase 1 trial involving ten 0 affected males demonstrated that a single dose of recombinant human α-Gal A (agalsidase alpha) could reduce the accumulated GL-3 in the liver and urinary sediment (Schiffmann et al 2000). A phase 1/2 open-label, dose-escalation trial involving 15 classically affected males evaluated the safety and effectiveness of five doses of agalsidase beta in dose regimens of 0.3, 1.0, or 3.0 mg/kg every 14 days or 1.0 or 3.0 mg/kg every 48 h (Eng et al 2001a). The enzyme was well-tolerated, and rapid, marked reductions in plasma and tissue GL3 were observed biochemically, histologically, and ultrastructurally. Notably, the clearance of plasma GL-3 was dose-dependent (Eng et al 2001a). Mean GL3 content decreased 84% in liver (n = 13) and was markedly reduced in the kidney in four of five patients who had pre- and posttreatment renal biopsies. Importantly, GL3 deposits also were reduced in the vascular endothelium of the kidney, heart, skin and liver by light and electron microscopic evaluation. The trial demonstrated that the dose-dependent GL-3 clearance by agalsidase beta seen in the preclinical studies was also seen in patients with Fabry disease. In addition, patients reported decreased acroparesthesias, decreased gastrointestinal problems, and increased sweating (Eng et al 2001a).

PHASE 2 AND PHASE 3 CLINICAL TRIAL

A single-center, double-blind, placebo-controlled phase 2 trial of agalsidase alpha involved 26 male patients with neuropathic pain who received 0.2 mg/kg every 2 weeks for 22 weeks (12 doses) (Schiffmann et al 2001). The primary efficacy endpoint was pain at its worst and pain medication was withdrawn before evaluations. The mean Brief Pain Inventory neuropathic pain severity score declined from 6.2 to 4 in patients treated with agalsidase alpha (n = 14) versus no significant change in the placebo group (n = 12) (*P* = 0.02). Mean creatinine clearance increased by 2.1 ml/min for patients receiving the enzyme versus a decrease of 16.1 ml/min for the patients receiving the placebo (*P* = 0.02). One patient in the placebo group advanced to renal failure. Plasma GL-3 levels decreased approximately 50% in patients treated with agalsidase alpha.

A phase 3 multinational, multicenter, randomized placebo-controlled, double-blind study evaluated the safety and effectiveness of agalsidase beta in 58 patients who received 1 mg/kg of the enzyme or placebo every 2 weeks for 20 weeks (11 doses) (Eng et al 2001b). The primary efficacy endpoint was the percentage of patients whose renal capillary endothelial GL-3 deposits cleared to normal or near normal (Figure 35.8). Also evaluated were the histological clearance of microvascular endothelial GL-3 deposits in the heart and skin, and changes in pain (Short Form McGill Pain Questionnaire) and in quality of life (Short Form-36 Health Status Survey). In this study, 20 of 29 enzyme-treated patients (69%) cleared the accumulated GL-3 from the renal capillary endothelium versus 0 of 29 placebo-treated patients (*P* < 0.001). Compared to the placebo group, enzyme-treated patients also had markedly decreased microvascular endothelial GL-3 in the skin (*P* < 0.001) and heart (*P* < 0.001). Patients receiving the enzyme cleared the accumulated GL-3 in plasma to nondetectable levels.

All 58 patients completed the phase 3 trial and received agalsidase beta in an extension study (Wilcox et al 2004). After 6 months of the open-label therapy, 98% of both former placebo and enzyme-treated patients who had biopsies achieved or maintained normal or near normal renal capillary endothelial histology. Importantly, the GL-3 was cleared to normal or near normal levels ('0' scores by three independent renal pathologists) in the renal interstitial, glomerular, and nonglomerular endothelial cells, mesangial cells and interstitial cells after 6 or 12 months of therapy (Thurberg et al 2002) (Table 35.3). Reduced GL-3 deposits also were histologically documented in the podocytes and tubular epithelial cells (Thurberg et al 2002).

Similar results (i.e. '0' scores) were found for GL-3 clearance in the skin (96%), and heart (75%) in the extension study. Thus, the study confirmed the results of the double-blind phase 3 trial and demonstrated that the patients treated with enzyme for 12 months had continued GL-3 clearance from the vascular endothelium in the kidney, heart, and skin, the key sites of pathology in this disease. Also, the accumulated plasma GL-3 in the former placebo group decreased to non-detectable levels after 6 months of enzyme treatment. Of note, the mean serum creatinine concentration remained normal (0.8–0.9 mg/dl) without changing from baseline after 12 months of treatment. However, three of the 58 patients who were older (42–48 years), had

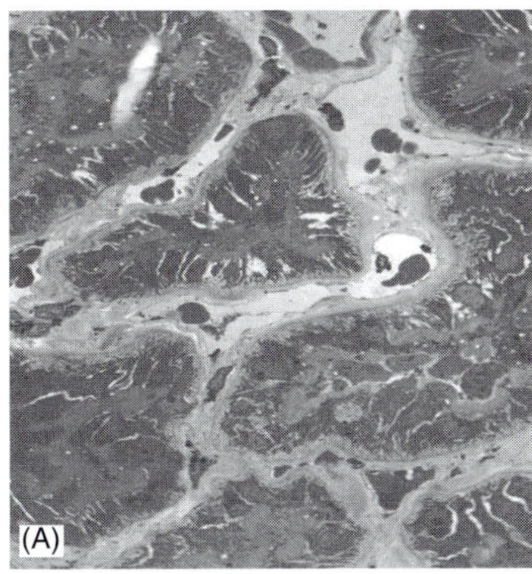

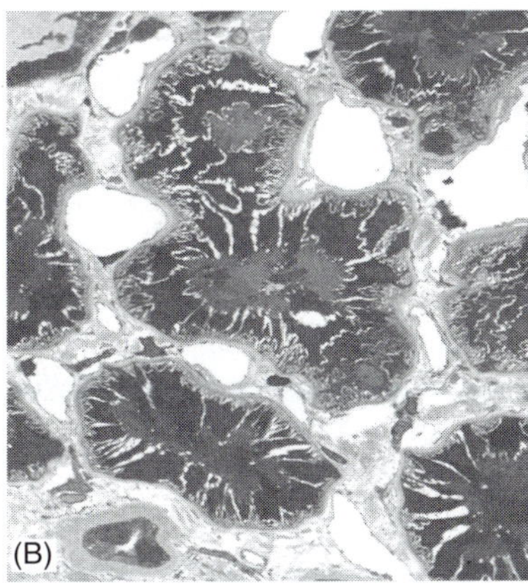

FIGURE 35.8   Paired photomicrographs of the renal cortex pre-treatment (A) and post-treatment (B) from a patient who received 11 treatments of agalsidase beta in the double-blind study (Eng et al 2001b). Pre-treatment, the intertubular capillaries demonstrate numerous GL-3 inclusions (arrows). Post-treatment the capillary endothelium has been cleared of GL-3 inclusions (arrows), resulting in a '0' score. (see also Plate 78)

more than 50% glomerulosclerosis (57–100%), and had urinary protein/urinary creatinine ratios >2.0 at baseline, increased their serum creatinine levels. These findings were instructive, as they indicated those patients who already had significant renal disease and whose renal disease would progress, presumably more slowly when treated with adequate doses of enzyme (Wilcox et al 2004).

Treatment with agalsidase alpha and agalsidase beta were well tolerated, except for mild to moderate infusion reactions (i.e. chills and fever) to the recombinant enzyme, which were managed conservatively. While IgG seroconversion occurred in 64–88% of enzyme-treated patients (Eng et al 2001b, Schiffmann et al 2001), GL-3 clearance was not impaired, based on plasma GL-3 levels, and antibody titers decreased with continued treatment. For example, in the agalsidase beta phase 3 extension study, after 3 years of treatment at 1 mg/kg, over 50% of the 58 patients had decreased their IgG titers more than four-fold and seven had tolerized. The relevance of the recent in vitro findings of neutralizing antibodies in some patients treated with agalsidase alpha or agalsidase beta is not clear (Linthorst et al 2004) since decreased clinical effectiveness of the treatments has not been reported. Based on these preclinical and clinical findings, both agalsidase alpha and agalsidase beta were approved by the European Agency for Evaluation of Medical Products (EMEA) in 2001, but only agalsidase beta was approved by the Food and Drug Administration (FDA) in 2003 in the United States (Desnick 2004).

CLINICAL STUDIES OF ERT

Since the approval of ERT in Europe and the United States, the clinical benefits of ERT in Fabry patients have been reported. These include results showing the effects of agalsidase alpha on pain, cardiac, hearing, gastrointestinal, and renal manifestations (Hajioff et al 2003, Beck et al 2004, Hoffmann et al 2004, 2005). Studies with agalsidase beta have demonstrated its effectiveness in stabilizing renal function, and improving cardiac function, gastrointestinal manifestations, the Fabry neuropathy, and quality of life (De Schoenmakere et al 2003, Waldek 2003, Weidemann

TABLE 35.3   GL-3 clearance from renal cells of Fabry patients on agalsidase beta

Renal cell type	6 months' treatment		12 months' treatment		
	Patients biopsied	% '0' scores*	Patients biopsied	% '0' scores*	
Endothelial cells					
Glomerular capillary	18	100	17	100	
Interstitial capillary	22	100	21	95	(100)**
Noncapillary	22	87 (100)**	22	95	(100)**
Interstitial cells	24	79 (92)**	24	100	
Mesangial cells	19	94 (100)**	17	100	

*'0' score indicates GL-3 clearance to normal or near normal levels as judged by three independent renal pathologists
**% of biopsies in which GL-3 scores were reduced by 1 to 2 points from baseline scores

et al 2003, Guffon & Fouilhoux 2004, Hilz et al 2004, Spinelli et al 2004, Banikazemi et al 2005).

Of note, agalsidase beta has been administered during hemodialysis without lost of the recombinant enzyme in the dialysate (Kosch et al 2004). Moreover, agalsidase beta treatment has been shown to improve the quality of life of Fabry patients on dialysis by decreasing pain, improving gastrointestinal symptoms, and slowing or reversing the progressive Fabry cardiomyopathy (Pisani et al 2005).

PHASE 4 CLINICAL TRIAL

Since the FDA approved agalsidase beta on the basis of a biologic marker, GL-3 clearance from the kidney and other sites, a phase 4 clinical trial was required by their Accelerated Approval Program to further confirm the clinical benefit of the treatment. This study was a multinational, multicenter, double-blind, placebo-controlled trial involving 82 patients with classical Fabry disease who had mild to moderate renal disease. The patients were randomized 2:1 (agalsidase beta:placebo) at each study site and the median study time was 18.5 months (35 months total). The primary endpoint compared the time to the first clinical event (renal, cardiac, cerebrovascular or death) between the two treatment groups. After a clinical event, a patient could be switched to open-label enzyme.

This trial showed that compared to placebo, agalsidase beta at 1 mg/kg every 2 weeks slowed the rate of progression of Fabry disease and substantially reduced the risk of renal, cardiac, and cerebrovascular events together and individually. After the prescribed adjustment for the baseline imbalance in proteinuria between the two treatment groups, patients randomized to agalsidase beta were 53% less likely than the placebo-treated patients to experience a clinically significant renal, cardiac, or cerebrovascular event ('Intent to Treat' population). Similarly to other renal diseases, baseline proteinuria was the most important determinant of outcome. Among the 74 patients who were compliant with the study protocol (the 'Per-Protocol' population), after the prescribed adjustment for the baseline proteinuria imbalance between the two treatment groups, the patients who received agalsidase beta were 61% less likely to experience a clinically significant event ($P = 0.034$). The most pronounced benefits of agalsidase beta were seen when therapy was started earlier in the course of the disease (i.e. with less renal dysfunction). These findings emphasize the importance of early treatment with 1 mg/kg of agalsidase beta.

Recommendations for Treatment

A group of physicians expert in Fabry disease established consensus recommendations for the diagnosis and treatment of the disease (Desnick et al 2003). These experts recommended that all males with Fabry disease (including those with end-stage renal disease) and heterozygous females with substantial disease manifestations should be treated with ERT and that the treatment should be initiated as early as possible.

FUTURE THERAPIES

Pharmacologic Chaperones

In addition to ERT, current efforts are focused on the development of other strategies to treat Fabry disease. An attractive approach for treating lysosomal and other genetic diseases resulting from protein misfolding and/or trafficking is 'pharmacologic chaperone therapy' involving the use of low-molecular-weight pharmacologic chaperones to rescue misfolded or unstable proteins, thereby increasing protein function and clinical benefit. Pharmacologic chaperones include substrate analogs, receptor agonists and antagonists, or other modulators, as well as epitope-directed ligands. Pharmacologic chaperones are designed to specifically stabilize their respective proteins. These compounds diffuse into the cell and bind site-specifically to folding intermediates during the biosynthesis of the mutant protein, thereby stabilizing the intermediate that is rate-limiting for the folding and/or trafficking of the mutant protein (Figure 35.9).

The clinical efficacy of pharmacologic chaperones for lysosomal disorders has been investigated in the cardiac variant of Fabry disease. These patients have residual α-Gal A activity and a later-onset phenotype (von Scheidt et al 1991, Nakao et al 1995, Desnick et al 2001). As proof of the concept, a cardiac variant for Fabry disease who had severe heart disease and was a candidate for a cardiac transplant was treated with intravenous galactose (1 g/kg) three times weekly. The galactose served as a competitive, reversible inhibitor that could bind to the active site and rescue the mutant α-Gal A enzyme, thereby promoting the proper folding, dimerization, and processing of the mutant enzyme, and preventing the proteasomal degradation of misfolded, mutant enzyme glycopeptides. The infusions were well tolerated, and the biochemical, histologic and clinical effects of the infusions were monitored at 3 months and 2 years (Frustaci et al 2001).

After 3 months of treatment, there was evidence of improvement, and after 2 years of continuous treatment, there was marked improvement in cardiac contractility (seen as an increase in the ventricular ejection fraction from 33% to 55%), a moderate reduction in ventricular-wall thickness, and a 20% reduction in cardiac mass. These improvements, which persisted for more than 3 years, were confirmed by the findings of independent observers, by two-dimensional echocardiography, and by cardiac MRI studies. Cardiac transplantation was no longer required in this patient, because of the clinical improvement (from NYHA functional class IV to class I). Thus, for patients with the later-onset variants of Fabry disease whose residual α-Gal A activity can be enhanced in vitro (Fan et al 1999), active site-specific

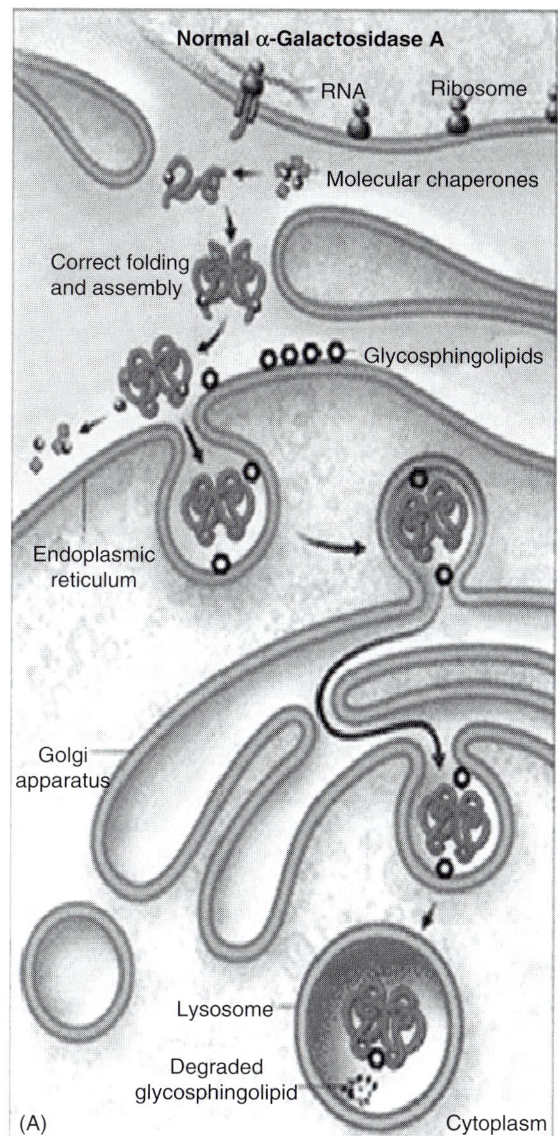

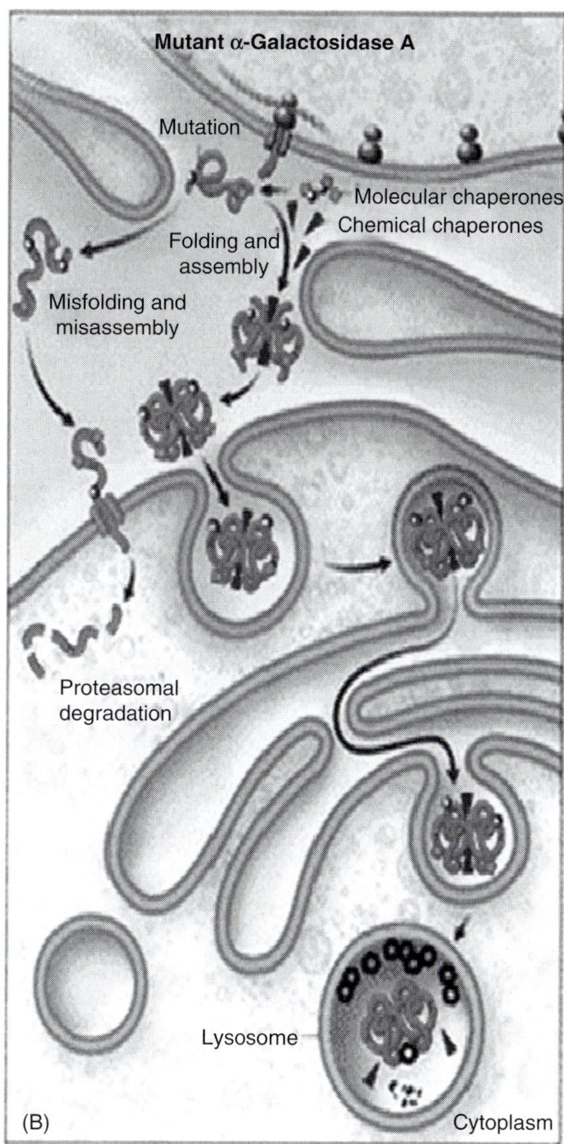

FIGURE 35.9   Proposed mechanism of enzyme stabilization by pharmacologic chaperones. (A) shows the processing of normal α-Gal A. Newly synthesized α-Gal A is translocated into the endoplasmic reticulum, where molecular chaperones facilitate its proper folding and dimerization by specialized processing enzymes. The molecular chaperones then dissociate from the folded, dimerized enzyme, which moves to the Golgi apparatus and then to lysosomes, where the enzyme is stable and active in the acidic environment of these organelles. (B) shows the processing of mutant α-Gal A. Most mutations in the α-Gal A gene encode enzyme molecules that are misfolded, misassembled, or aggregated in the endoplasmic reticulum, where they are degraded, presumably by the ubiquitin–proteasome pathway. However, certain missense mutations decrease the stability of the enzyme, but the conformation of the active site is retained. Although most of these mutant enzymes are degraded in the endoplasmic reticulum, they may be stabilized by site-specific pharmacologic chaperones, such as galactose or deoxygalactonorjirimycin (Fan et al 1999), that bind to the active site of the enzyme, promote folding, and stabilize the mutant enzyme. Some of the enzyme then reaches the lysosomes, where it retains low levels of activity. In the lysosomes, the accumulated glycosphingolipid substrates displace the pharmacologic chaperones and are hydrolyzed by the enzyme. From Frustaci et al 2001. (see also Plate 79)

pharmacologic chaperone-mediated therapy may prove safe and therapeutically effective.

An obvious advantage of this approach is the fact that pharmacologic chaperones are small molecules that can be administered orally, and may have markedly different biodistributions than enzymes whose tissue delivery is primarily dependent on their respective receptor-mediated delivery.

Experimental Gene Therapy

Experimental studies of gene therapy have been stimulated by the availability of the Fabry knockout mouse which has no α-Gal A activity and progressively accumulates GL-3. Both ex vivo and in vivo gene transfer studies have been reported using a variety of recombinant viral vectors (i.e. adeno, adeno-associated, retroviral, and lentiviral) and plasmid vectors (Takenaka et al 1999a, 1999b, Ziegler et al 1999, D'Costa et al 2003, Park et al 2003, Przybylska et al 2004, Yoshimitsu et al 2004, Ziegler et al, 2004). In general, these studies have demonstrated increased levels of α-Gal A activity and decreased concentrations of GL-3 in various murine organs. For example, recent reports described the use of a liver-restricted enhancer/promoter to drive an Adeno-associated viral vector which resulted in sustained hepatic $\tilde{\alpha}$ Gal A activity and reduced tissue GL-3 levels for 12 months, and a reduced immune response to the enzyme which was foreign to the Fabry mice (Ziegler et al 2004). Another study employed a Lentiviral vector which can integrate into nondividing cells including neuronal and bone marrow stem cells. A single injection into the temporal veins of newborn Fabry mice of a recombinant lentiviral vector containing the α-Gal A gene resulted in increased plasma and tissue α-Gal A enzymatic activities and decreased tissue GL-3 levels for at least 28 weeks (Yoshimitsu et al 2004). In addition, gene transfer using a nonviral vector has been reported (Przybylska et al 2004). The CpG-depleted α-Gal A plasmid DNA was complexed with cationic lipids and injected intravenously into Fabry mice. Increased α-Gal A enzymatic activity and GL-3 clearance were detected in various tissues for 3 months. Treatment with dexamethasone before and after injection of the complexed α-Gal A DNA increased $\tilde{\alpha}$ Gal A expression and GL-3 clearance.

FUTURE PROSPECTS

Enzyme replacement therapy using recombinant human α-Gal A has proven safe and effective in clinical trials and subsequent years of clinical experience. Other modes of therapy, such as biochemical stabilization and enhancement of enzymatic activity using α-Gal A inhibitors as 'pharmacologic chaperones' and gene therapy are being investigated. It is likely that the future medical outlook for Fabry patients will continue to improve. Emphasis should be on early recognition of patients affected by the disease, and identification of at-risk family members so that treatment can be instituted as early as possible.

ACKNOWLEDGMENTS

This work was supported in part by a grant from the National Institutes of Health including a research grant (5 R37 DK34045 Merit Award), a grant (5 MO1 RR00071) for the Mount Sinai General Clinical Research Center from the National Center of Research Resources, and a research grant from the Genzyme Corporation.

References

Altarescu G, Moore DF, Schiffmann R. Effect of genetic modifiers on cerebral lesions in Fabry disease. Neurology 2005; 64: 2148–50.

Argoff CE, Barton NW, Brady RO, Ziessman HA. Gastrointestinal symptoms and delayed gastric emptying in Fabry's disease: response to metoclopramide. Nucl. Med. Commun. 1998; 19: 887–91.

Ashton-Prolla P, Tong B, Astrin KH, Shabbeer J, Eng CM, Desnick RJ. 22 novel mutations in the α-galactosidase A gene and genotype/phenotype correlations including mild hemizygotes and severely affected heterozygotes. J. Invest. Med. 2000; 48: 227–35.

Banikazemi M, Ullman T, Desnick RJ. Gastrointestinal manifestations of Fabry disease: Clinical response to enzyme replacement therapy. Mol. Genet. Metab. 2005; 85: 255–9.

Bannwart F. Fabry's disease. Light and electron microscopic cardiac findings 12 years after successful kidney transplantation. Schweiz. Med. Wochenschr. 1982; 112: 1742–7.

Bass JL, Shrivastava S, Grabowski GA, Desnick RJ, Moller JH. The M-mode echocardiogram in Fabry's disease. Am. Heart J. 1980; 100: 807–12.

Beck M, Ricci R, Widmer U, et al. Fabry disease: overall effects of agalsidase alfa treatment. Eur. J. Clin. Invest. 2004; 34: 838–44.

Bishop DF, Calhoun DH, Bernstein HS, Hantzopoulos P, Quinn M, Desnick RJ. Human α-galactosidase A: Nucleotide sequence of a cDNA clone encoding the mature enzyme. Proc. Natl. Acad. Sci. USA 1986; 83: 4859–63.

Bishop DF, Kornreich R, Eng CM, Ioannou YA, Fitzmaurice TF, Desnick RJ. Human α-galactosidase: Characterization and eukaryotic expression of the full-length cDNA and structural organization of the gene. In: Salvayre R, Douste-Blazy L, Gatt S, eds. Lipid Storage Disorders. New York: Plenum, 1988: pp. 809–22.

Brady RO, Gal AE, Bradley RM, Martensson E, Warshaw AL, Laster L. Enzymatic defect in Fabry's disease. ceramidetrihexosidase deficiency. N. Engl. J. Med. 1967; 276: 1163–7.

Brady RO, Tallman JF, Johnson WG, et al. Replacement therapy for inherited enzyme deficiency. Use of purified ceramidetrihexosidase in Fabry's disease. N. Engl. J. Med. 1973; 289: 9–14.

Branton MH, Schiffmann R, Sabnis SG, et al. Natural history of Fabry renal disease: influence of alpha-galactosidase A activity and genetic mutations on clinical course. Medicine (Baltimore) 2002; 81: 122–38.

Brown LK, Miller A, Bhuptani A, et al. Pulmonary involvement in Fabry disease. Am. J. Respir. Crit. Care Med. 1997; 155: 1004–10.

Buhler FR, Thiel G, Dubach UC, Enderlin F, Gloor F, Tholen H. Kidney transplantation in Fabry's disease. Br. Med. J. 1973; 3: 28–9.

Burkholder PM, Updike SJ, Ware RA, Reese OG. Clinicopathologic, enzymatic, and genetic features in a case of Fabry's disease. Arch. Pathol. Lab. Med. 1980; 104: 17–25.

Cantor WJ, Daly P, Iwanochko M, Clarke JT, Cusimano RJ, Butany J. Cardiac transplantation for Fabry's disease. Can. J. Cardiol. 1998; 14: 81–4.

Chatterjee S, Gupta P, Pyeritz RE, Kwiterovich POJ. Immunohistochemical localization of glycosphingolipid in urinary renal tubular cells in Fabry's disease. Am. J. Clin. Pathol. 1984; 82: 24–8.

Clarke JT, Guttmann RD, Wolfe LS, Beaudoin JG, Morehouse DD. Enzyme replacement therapy by renal allotransplantation in Fabry's disease. N. Engl. J. Med. 1972; 287: 1215–18.

Clarke JT, Knaack J, Crawhall JC, Wolfe LS. Ceramide trihexosidosis (Fabry's disease) without skin lesions. N. Engl. J. Med. 1971a; 284: 233–5.

Clarke JTR, Wolfe LS, Perlin AS. Evidence for an a terminal α-D-galactopyranosyl residue in galactosyl-galactosyl-glucosyl ceramide from human kidney. J. Biol. Chem. 1971b; 246: 5563–9.

Colombi A, Kostyal A, Bracher R, Gloor F, Mazzi R, Tholen H. Angiokeratoma corporis diffusum – Fabry's disease. Helv. Med. Acta 1967; 34: 67–83.

Colucci WS, Lorell BH, Schoen FJ, Warhol MJ, Grossman W. Hypertrophic obstructive cardiomyopathy due to Fabry's disease. N. Engl. J. Med. 1982; 307: 926–8.

D'Costa J, Harvey-White J, Qasba P, et al. HIV-2 derived lentiviral vectors: gene transfer in Parkinson's and Fabry disease models in vitro. J. Med. Virol. 2003; 71: 173–82.

De Schoenmakere G, Chauveau D, Grunfeld JP. Enzyme replacement therapy in Anderson-Fabry's disease: beneficial clinical effect on vital organ function. Nephrol. Dial. Transplant. 2003; 18: 33–5.

Desnick RJ. Fabry disease: Unrecognized ESRD patients and effectiveness of enzyme replacement on renal pathology and function. J. Inherit. Metab. Dis. 2002; 25: 116.

Desnick RJ. Enzyme replacement therapy for Fabry disease: lessons from two alpha-galactosidase A orphan products and one FDA approval. Expert Opin. Biol. Ther. 2004; 4: 1167–76.

Desnick RJ, Allen KY, Desnick SJ, Raman MK, Bernlohr RW, Krivit W. Fabry's disease: Enzymatic diagnosis of hemizygotes and heterozygotes. α-Galactosidase activities in plasma, serum, urine, and leukocytes. J. Lab. Clin. Med. 1973a; 81: 157–71.

Desnick RJ, Allen KY, Simmons RL, et al. Fabry disease: correction of the enzymatic deficiency by renal transplantation. Birth Defects Orig. Artic. Ser. 1973b; 9: 88–96.

Desnick RJ, Brady R, Barranger J, et al. Fabry disease, an under-recognized multisystemic disorder: expert recommendations for diagnosis, management, and enzyme replacement therapy. Ann. Intern. Med. 2003; 138: 338–46.

Desnick RJ, Brady RO. Fabry disease in childhood. J. Pediatr. 2004; 144: S20–6.

Desnick RJ, Dawson G, Desnick SJ, Sweeley CC, Krivit W. Diagnosis of glycosphingolipidoses by urinary-sediment analysis. N. Engl. J. Med. 1971; 284: 739–44.

Desnick RJ, Dean KJ, Grabowski G, Bishop DF, Sweeley CC. Enzyme therapy in Fabry disease: Differential in vivo plasma clearance and metabolic effectiveness of plasma and splenic α-galactosidase A isozymes. Proc. Natl Acad. Sci. USA 1979; 76: 5326–30.

Desnick RJ, Ioannou YA, Eng CM. α-Galactosidase A deficiency: Fabry disease. In: Scriver CR, Beaudet AL, Sly WS, Valle D, Kinzler KE, Vogelstein B, eds. The Metabolic and Molecular Bases of Inherited Disease. New York: McGraw-Hill, 2001: pp. 3733–74.

Desnick RJ, Raman M, Allen KY, et al. Enzyme therapy in Fabry's disease by renal transplantation. Proc. Clin. Dial. Transplant Forum 1972a; 2: 27–35.

Desnick RJ, Schuchman EH. Enzyme replacement and enhancement therapies: lessons from lysosomal disorders. Nat. Rev. Genet. 2002; 3: 954–66.

Desnick RJ, Simmons RL, Allen KY, et al. Correction of enzymatic deficiencies by renal transplantation: Fabry's disease. Surgery 1972b; 72: 203–11.

Dobrovolny R, Dvorakova L, Ledvinova J, et al. Relationship between X-inactivation and clinical involvement in Fabry heterozygotes. Eleven novel mutations in the alpha-galactosidase A gene in the Czech and Slovak population. J. Mol. Med. 2005; 83: 647–54.

Donati D, Sabbadini M, Capsoni F, et al. Immune function and renal transplantation in Fabry's disease. Proc. Eur. Dialysis Transplant Assoc. 1984; 21: 686.

Dutsch M, Marthol H, Stemper B, Brys M, Haendl T, Hilz MJ. Small fiber dysfunction predominates in Fabry neuropathy. J. Clin. Neurophysiol. 2002; 19: 575–86.

Elleder M, Bradova V, Smid F, et al. Cardiocyte storage and hypertrophy as a sole manifestation of Fabry's disease. Report on a case simulating hypertrophic non-obstructive cardiomyopathy. Virchows Arch. A Pathol. Anat. Histopathol. 1990a; 417: 449–55.

Elleder M, Ledvinova J, Vosmik F, Zeman J, Stejskal D, Lageron A. An atypical ultrastructural pattern in Fabry's disease: a study on its nature and incidence in 7 cases. Ultrastruct. Pathol. 1990b; 14: 467–74.

Eng CM, Ashley GA, Burgert TS, Enriquez AL, D'Souza M, Desnick RJ. Fabry disease: Thirty-five mutations in the α-galactosidase A gene in patients with classic and variant phenotypes. Mol. Med. 1997; 3: 174–82.

Eng CM, Banikazemi M, Gordon R. A phase 1/2 clinical trial of enzyme replacement in Fabry disease: Pharmacokinetic, substrate clearance, and safety studies. Am. J. Hum. Genet. 2001a; 68: 711–22.

Eng CM, Guffon N, Wilcox WR, et al. Safety and efficacy of recombinant human α-galactosidase A replacement therapy in Fabry's disease. N. Eng. J. Med. 2001b; 345: 9–16.

Fan JQ. A contradictory treatment for lysosomal storage disorders: inhibitors enhance mutant enzyme activity. Trends Pharmacol. Sci. 2003; 24: 355–60.

Fan JQ, Ishii S, Asano N, Suzuki Y. Accelerated transport and maturation of lysosomal α-galactosidase A in Fabry lymphoblasts by an enzyme inhibitor. Nat. Med. 1999; 5: 112–15.

Faraggiana T, Churg J, Grishman E, et al. Light- and electron-microscopic histochemistry of Fabry's disease. Am. J. Pathol. 1981; 103: 247–62.

Field DG, Ostrov BE, Devenyi AG, Hoban TF. Achalasia in an adolescent with Fabry disease. J. Pediatr. Gastroenterol. Nutr. 2001; 32: 201–3.

Filling-Katz MR, Merrick HF, Fink JK, Miles RB, Sokol J, Barton NW. Carbamazepine in Fabry's disease: effective analgesia with dose-dependent exacerbation of autonomic dysfunction. Neurology 1989; 39: 598–600.

Friedlaender MM, Kopolovic J, Rubinger D, et al. Renal biopsy in Fabry's disease eight years after successful renal transplantation. Clin. Nephrol. 1987; 27: 206–11.

Friedman LS, Kirkham SE, Thistlethwaite JR, Platika D, Kolodny EH, Schuffler MD. Jejunal diverticulosis with perforation as a complication of Fabry's disease. Gastroenterology 1984; 86: 558–63.

Froissart R, Guffon N, Vanier MT, Desnick RJ, Maire I. Fabry disease: D313Y is an alpha-galactosidase A sequence variant that causes pseudodeficient activity in plasma. Mol. Genet. Metab. 2003; 80: 307–14.

Frustaci A, Chimenti C, Ricci R, et al. Improvement in cardiac function in the cardiac variant of Fabry's disease with galactose-infusion therapy. N. Eng. J. Med. 2001; 345: 25–32.

Galanos J, Nicholls K, Grigg L, Kiers L, Crawford A, Becker G. Clinical features of Fabry's disease in Australian patients. Intern. Med. J. 2002; 32: 575–84.

Garman SC, Garboczi DN. The molecular defect leading to Fabry disease: structure of human alpha-galactosidase. J. Mol. Biol. 2004; 337: 319–35.

Germain DP, Avan P, Chassaing A, Bonfils P. Patients affected with Fabry disease have an increased incidence of progressive hearing loss and sudden deafness: an investigation of twenty-two consecutive hemizygous male patients. BMC Med. Genet. 2002; 3: 10–20.

Goldman ME, Cantor R, Schwartz MF, Baker M, Desnick RJ. Echocardiographic abnormalities and disease severity in Fabry's disease. J. Am. Coll. Cardiol. 1986; 7: 1157–61.

Gorden KE, Ludman MD, Finley GA. Successful treatment of painful crises of Fabry disease with low dose morphine. Pediatr. Neurol. 1995; 12: 250–1.

Grunfeld JP, Le Porrier M, Droz D, Bensaude I, Hinglais N, Crosnier J. Renal transplantation in patients suffering from Fabry's disease. Kidney transplantation from an heterozygote subject to a subject without Fabry's disease. Nouv. Presse Med. 1975; 4: 2081–5.

Gubler MC, Lenoir G, Grunfeld JP, Ulmann A, Droz D, Habib R. Early renal changes in hemizygous and heterozygous patients with Fabry's disease. Kidney Int. 1978; 13: 223–35.

Guffon N, Fouilhoux A. Clinical benefit in Fabry patients given enzyme replacement therapy – a case series. J. Inherit. Metab. Dis. 2004; 27: 221–7.

Hajioff D, Enever Y, Quiney R, Zuckerman J, Mackermot K, Mehta A. Hearing loss in Fabry disease: the effect of agalsidase alfa replacement therapy. J. Inherit. Metab. Dis. 2003; 26: 787–94.

Hakomori SI, Siddiqui B, Li YT, Li SC, Hellerqvist CB. Anomeric structures of globoside and ceramide trihexoside of human erythrocytes and hamster fibroblasts. J. Biol. Chem. 1971; 246: 2271–7.

Hilz MJ, Brys M, Marthol H, Stemper B, Dutsch M. Enzyme replacement therapy improves function of C-, Adelta-, and Abeta-nerve fibers in Fabry neuropathy. Neurology 2004; 62: 1066–72.

Hoffmann B, Reinhardt D, Koletzko B. Effect of enzyme-replacement therapy on gastrointestinal symptoms in Fabry disease. Eur. J. Gastroenterol. Hepatol. 2004; 16: 1067–9.

Hoffmann B, Garcia de Lorenzo A, Mehta A, Beck M, Widmer U, Ricci R. Effects of enzyme replacement therapy on pain and health related quality of life in patients with Fabry disease: data from FOS (Fabry Outcome Survey). J. Med. Genet. 2005; 42: 247–52.

Hopkin RJ, Bissler J, Grabowski GA. Comparative evaluation of alpha-galactosidase A infusions for treatment of Fabry disease. Genet. Med. 2003; 5: 144–53.

Ioannou YA, Bishop DF, Desnick RJ. Overexpression of human α-galactosidase A results in its intracellular aggregation, crystallization in lysosomes and selective secretion. J. Cell Biol. 1992; 119: 1137–50.

Ioannou YA, Zeidner KM, Gordon RE, Desnick RJ. Fabry disease: Preclinical studies demonstrate the effectiveness of α-galactosidase A replacement in enzyme-deficient mice. Am. J. Hum. Genet. 2001; 68: 14–25.

Ioannou YA, Zeidner KM, Grace ME, Desnick RJ. Human α-galactosidase A: Glycosylation site 3 is essential for enzyme solubility. Biochem. J. 1998; 332: 789–97.

Kampmann C, Baehner F, Whybra C, et al. Cardiac manifestations of Anderson-Fabry disease in heterozygous females. J. Am. Coll. Cardiol. 2002; 40: 1668–74.

Kint JA. Fabry disease: alpha-galactosidase deficiency. Science 1970; 167: 1268–9.

Kornreich R, Desnick RJ, Bishop DF. Nucleotide sequence of the human α-galactosidase A gene. Nucl. Acids. Res. 1989; 17: 3301–2.

Kosch M, Koch HG, Oliveira JP, et al. Enzyme replacement therapy administered during hemodialysis in patients with Fabry disease. Kidney Int. 2004; 66: 1279–82.

Kotanko P, Kramar R, Devrnja D, et al. Results of a nationwide screening for Anderson-Fabry disease among dialysis patients. J. Am. Soc. Nephrol. 2004; 15: 1323–9.

Kramer W, Thormann J, Mueller K, Frenzel H. Progressive cardiac involvement by Fabry's disease despite successful renal allotransplantation. Int. J. Cardiol. 1985; 7: 72–5.

Lee K, Jin X, Zhang K, et al. A biochemical and pharmacological comparison of enzyme replacement therapies for the glycolipid storage disorder Fabry disease. Glycobiology 2003; 13: 305–13.

Lenoir G, Rivron M, Gubler MC, Dufier JL, Tome FS, Guivarch M. Fabry's disease. Carbamazepine therapy in acrodyniform syndrome. Arch. Fr. Pediatr. 1977; 34: 704–16.

Linhart A, Palecek T, Bultas J, et al. New insights in cardiac structural changes in patients with Fabry's disease. Am. Heart J. 2000; 139: 1101–8.

Linthorst GE, Hollak CE, Donker-Koopman WE, Strijland A, Aerts JM. Enzyme therapy for Fabry disease: neutralizing antibodies toward agalsidase alpha and beta. Kidney Int. 2004; 66: 1589–95.

Lockman LA, Hunninghake DB, Krivit W, Desnick RJ. Relief of pain of Fabry's disease by diphenylhydantoin. Neurology 1973; 23: 871–5.

MacDermot KD, Holmes A, Miners AH. Anderson-Fabry disease: clinical manifestations and impact of disease in a cohort of 60 obligate carrier females. J. Med. Genet. 2001a; 38: 769–75.

MacDermot KD, Holmes A, Miners AH. Anderson-Fabry disease: clinical manifestations and impact of disease in a cohort of 98 hemizygous males. J. Med. Genet. 2001b; 38: 750–60.

Maizel SE, Simmons RL, Kjellstrand C, Fryd DS. Advances in the treatment of inherited metabolic diseases. Adv. Hum. Genet. 1981a; 11: 281–369.

Maizel SE, Simmons RL, Kjellstrand C, Fryd DS. Ten-year experience in renal transplantation for Fabry's disease. Transplant Proc. 1981b; 13: 57–9.

Mapes CA, Anderson RL, Sweeley CC, Desnick RJ, Krivit W. Enzyme replacement in Fabry's disease, an inborn error of metabolism. Science 1970; 169: 987–9.

Matsuura F, Ohta M, Ioannou YA, Desnick RJ. Human α-galactosidase A: Characterization of the N-linked oligosaccharides on the intracellular and secreted glycoforms overexpressed by Chinese hamster ovary cells. Glycobiology 1998; 8: 329–39.

Matsuzawa F, Aikawa S, Doi H, Okumiya T, Sakuraba H. Fabry disease: correlation between structural changes in alpha-galactosidase, and clinical and biochemical phenotypes. Hum. Genet. 2005; 117: 317–28.

Mayes JS, Scheerer JB, Sifers RN, Donaldson ML. Differential assay for lysosomal alpha-galactosidases in human tissues and its application to Fabry's disease. Clin. Chim. Acta 1981; 112: 247–51.

McNary W, Lowenstein L. A morphological study of the renal lesion in angiokeratoma corporis diffusum universale (Fabry's disease). J. Urol. 1965; 93: 641.

Meehan SM, Junsanto T, Rydel JJ, Desnick RJ. Fabry disease: renal involvement limited to podocyte pathology and proteinuria in a septuagenarian cardiac variant. Pathologic and therapeutic implications. Am. J. Kidney Dis. 2004; 43: 164–71.

Mehta A, Ricci R, Widmer U, et al. Fabry disease defined: baseline clinical manifestations of 366 patients in the Fabry Outcome Survey. Eur. J. Clin. Invest. 2004; 34: 236–42.

Morel-Maroger L, Ganter P, Ardaillou R, Cathelineau G, Richet G. Histochemical study of a lipid thesaurismosis with renal, cutaneous and neurologic involvement. Its relation to Fabry's angiokeratosis and familial renal cytodystrophy. Bull. Mem. Soc. Med. Hop. Paris 1966; 117: 49–57.

Morrone A, Cavicchi C, Bardelli T, et al. Fabry disease: molecular studies in Italian patients and X inactivation analysis in manifesting carriers. J. Med. Genet. 2003; 40: e103.

Mosnier JF, Degott C, Bedrossian J, et al. Recurrence of Fabry's disease in a renal allograft eleven years after successful renal transplantation. Transplantation 1991; 51: 759–62.

Nagao Y, Nakashima H, Fukuhara Y, et al. Hypertrophic cardiomyopathy in late-onset variant of Fabry disease with high residual activity of α-galactosidase A. Clin. Genet. 1991; 39: 233–7.

Nakao S, Kodama C, Takenaka T, et al. Fabry disease: Detection of undiagnosed hemodialysis patients and identification of a "renal variant" phenotype. Kidney Int. 2003; 64: 801–7.

Nakao S, Takenaka T, Maeda M, et al. An atypical variant of Fabry's disease in men with left ventricular hypertrophy. N. Engl. J. Med. 1995; 333: 288–93.

Nguyen TT, Gin T, Nicholls K, Low M, Galanos J, Crawford A. Ophthalmological manifestations of Fabry disease: a survey of patients at the Royal Melbourne Fabry Disease Treatment Centre. Clin. Experiment. Ophthalmol. 2005; 33: 164–8.

Ojo A, Meier-Kriesche HU, Friedman G, et al. Excellent outcome of renal transplantation in patients with Fabry's disease. Transplantation 2000; 69: 2337–9.

Orssaud C, Dufier J, Germain D. Ocular manifestations in Fabry disease: a survey of 32 hemizygous male patients. Ophthalmic Genet. 2003; 24: 129–39.

Pabico RC, Atancio BC, McKenna BA, Pamukcoglu T, Yodaiken R. Renal pathologic lesions and functional alterations in a man with Fabry's disease. Am. J. Med. 1973; 55: 415–25.

Park J, Murray GJ, Limaye A, et al. Long-term correction of globotriaosylceramide storage in Fabry mice by recombinant adeno-associated virus-mediated gene transfer. Proc. Natl Acad. Sci. USA 2003; 100: 3450–4.

Philippart M. Treatment of painful crises of Fabry disease with morphine [letter; comment]. Pediatr. Neurol. 1995; 13: 268.

Pieroni M, Chimenti C, Ricci R, Sale P, Russo MA, Frustaci A. Early detection of Fabry cardiomyopathy by tissue Doppler imaging. Circulation 2003; 107: 1978–84.

Pisani A, Spinelli L, Sabbatini M, et al. Enzyme replacement therapy in Fabry disease patients undergoing dialysis: effects on quality of life and organ involvement. Am. J. Kidney Dis. 2005; 46: 120–7.

Pompen AWM, Ruiter M, Wyers JJG. Angiokertoma corporis diffusum (universale) Fabry, as a sign of internal disease: Two autopsy reports. Acta Med. Scand. 1947; 128: 234–55.

Popli S, Molnar ZV, Leehey DJ, et al. Involvement of renal allograft by Fabry's disease. Am. J. Nephrol. 1987; 7: 316–18.

Przybylska M, Wu IH, Zhao H, et al. Partial correction of the alpha-galactosidase A deficiency and reduction of glycolipid storage in Fabry mice using synthetic vectors. J. Gene Med. 2004; 6: 85–92.

Ries M, Mengel E, Kutschke G, et al. Use of gabapentin to reduce chronic neuropathic pain in Fabry disease. J. Inherit. Metab. Dis. 2003; 26: 413–14.

Roberts DH, Gilmore IT. Achalasia in Anderson-Fabry's disease. J. R. Soc. Med. 1984; 77: 430–1.

Rosenthal D, Lien YH, Lager D, et al. A novel alpha-galactosidase a mutant (M42L) identified in a renal variant of Fabry disease. Am. J. Kidney Dis. 2004; 85–9.

Sachdev B, Takenaka T, Teraguchi H, et al. Prevalence of Anderson-Fabry disease in male patients with late onset hypertrophic cardiomyopathy. Circulation 2002; 105: 1407–11.

Schiffmann R, Kopp JB, Austin HA 3rd, et al. Enzyme replacement therapy in Fabry disease: a randomized controlled trial. JAMA 2001; 285: 2743–9.

Schiffmann R, Murray GJ, Treco D, et al. Infusion of alpha-galactosidase A reduces tissue globotriaosylceramide storage in patients with Fabry disease. Proc. Natl Acad. Sci. USA 2000; 97: 365–70.

Sessa A, Meroni M, Battini G, et al. Renal involvement in Anderson-Fabry disease. J. Nephrol. 2003; 16: 310–13.

Sher NA, Letson RD, Desnick RJ. The ocular manifestations in Fabry's disease. Arch. Ophthalmol. 1976; 97: 671–6.

Sheth KJ, Roth DA, Adams MB. Early renal failure in Fabry's disease. Am. J. Kidney Dis. 1983; 2: 651–4.

Sinclair R. Origin of endothelium in human renal allograft. Br. Med. J. 1972; 11: 15.

Spada M, Marongiu A, Voglino G, et al. Molecular study in 20 unrelated male patients with Fabry disease: the A143T genotype correlates with the late-onset end-stage nephropathy. J. Inherit. Metab. Dis. 2003; 26: 171.

Spada M, Pagliardini S. Screening for Fabry disease in end-stage nephropathies. J. Inherit. Metab. Dis. 2002; 25: 113.

Spinelli L, Pisani A, Sabbatini M, et al. Enzyme replacement therapy with agalsidase beta improves cardiac involvement in Fabry's disease. Clin. Genet. 2004; 66: 158–65.

Srivastava AK, McMillan S, Jermak C, et al. Integrated STS/YAC physical, genetic, and transcript map of human Xq21.3 to q23/q24 (DXS1203-DXS1059). Genomics 1999; 58: 188–201.

Stenson PD, Ball EV, Mort M, et al. Human Gene Mutation Database (HGMD): 2003 update. Hum Mutat 2003; 21: 577–81.

Sweeley CC, Klionsky B. Fabry's disease: Classification as a sphingolipidosis and partial characterization of a novel glycolipid. J. Biol. Chem. 1963; 238: 3148–50.

Takenaka T, Hendrickson CS, Tworek DM, et al. Enzymatic and functional correction along with long-term enzyme secretion from transduced bone marrow hematopoietic stem/progenitor and stromal cells derived from patients with Fabry disease. Exp. Hematol. 1999a; 27: 1149–59.

Takenaka T, Qin G, Brady RO, Medin JA. Circulating alpha-galactosidase A derived from transduced bone marrow cells: relevance for corrective gene transfer for Fabry disease. Hum. Gene Ther. 1999b; 10: 1931–9.

Tan SV, Lee PJ, Walters RJ, Mehta A, Bostock H. Evidence for motor axon depolarization in Fabry disease. Muscle Nerve 2005.

Thadhani R, Wolf M, West ML, et al. Patients with Fabry disease on dialysis in the United States. Kidney Int. 2002; 61: 249–55.

Thurberg BL, Rennke H, Colvin RB, et al. Globotriaosylceramide accumulation in the Fabry kidney is cleared from multiple cell types after enzyme replacement therapy. Kidney Int. 2002; 62: 1933–46.

Tsakiris D, Simpson HK, Jones EH, et al. Report on management of renale failure in Europe, XXVI, 1995. Rare diseases in renal replacement therapy in the ERA-EDTA Registry. Nephrol. Dial. Transplant. 1996; 11(Suppl. 7): 4–20.

Utsumi K, Itoh K, Kase R, et al. Urinary excretion of the vitronectin receptor (integrin alpha V beta 3) in patients with Fabry disease. Clin. Chim. Acta 1999; 279: 55–68.

von Scheidt W, Eng CM, Fitzmaurice TF, et al. An atypical variant of Fabry's disease with manifestations confined to the myocardium. N. Engl. J. Med. 1991; 324: 395–9.

Waldek S. PR interval and the response to enzyme-replacement therapy for Fabry's disease. N. Engl. J. Med. 2003; 348: 1186–7.

Wang AM, Ioannou YA, Zeidner KM, et al. Generation of a mouse model with α-galactosidase A deficiency. Am J. Hum. Genet. 1996; 59: A208.

Warnock DG. Fabry disease: diagnosis and management, with emphasis on the renal manifestations. Curr. Opin. Nephrol. Hypertens. 2005; 14: 87–95.

Weidemann F, Breunig F, Beer M, et al. The variation of morphological and functional cardiac manifestation in Fabry disease: potential implications for the time course of the disease. Eur. Heart J. 2005; 26: 1221–7.

Weidemann F, Breunig F, Beer M, et al. Improvement of cardiac function during enzyme replacement therapy in patients with Fabry disease: a prospective strain rate imaging study. Circulation 2003; 108: 1299–301.

Whybra C, Kampmann C, Willers I, et al. Anderson-Fabry disease: clinical manifestations of disease in female heterozygotes. J. Inherit. Metab. Dis. 2001; 24: 715–24.

Wilcox WR, Banikazemi M, Guffon N, et al. Long-term safety and efficacy of enzyme replacement therapy for Fabry disease. Am. J. Hum. Genet. 2004; 75: 65–74.

Willard HF. The sex chromosome and X chromosome inactivation. In: Scriver CR, Beaudet AL, Sly WS, Valle D, Kinzler KE, Vogelstein B, eds. The Metabolic and Molecular Bases of Inherited Disease. New York: McGraw-Hill, 2001: pp. 1191–212.

Wise D, Wallace H, Jellinck E. Angiokeratoma corporis diffusum: A clinical study of eight affected families. Q. J. Med. 1962; 31: 177.

Yasuda M, Shabbeer J, Benson SD, Maire I, Burnett RM, Desnick RJ. Fabry disease: characterization of alpha-galactosidase A double mutations and the D313Y plasma enzyme pseudodeficiency allele. Hum. Mutat. 2003; 22: 486–92.

Yoshimitsu M, Sato T, Tao K, et al. Bioluminescent imaging of a marking transgene and correction of Fabry mice by neonatal injection of recombinant lentiviral vectors. Proc. Natl Acad. Sci. USA 2004; 101: 16909–14.

Ziegler RJ, Yew NS, Li C, et al. Correction of enzymatic and lysosomal storage defects in Fabry mice by adenovirus-mediated gene transfer. Hum. Gene Ther. 1999; 10: 1667–82.

Ziegler RJ, Lonning SM, Armentano D, et al. AAV2 vector harboring a liver-restricted promoter facilitates sustained expression of therapeutic levels of alpha-galactosidase A and the induction of immune tolerance in Fabry mice. Mol. Ther. 2004; 9: 231–40.

CHAPTER 36

Hereditary Fructose Intolerance

TIMOTHY M. COX

INTRODUCTION

Hereditary fructose intolerance (HFI) provides a vivid example of gene–environment interactions: newborns with hereditary deficiency of aldolase B experience no symptoms and are free of metabolic disturbance unless they are exposed to fructose or its congeners (Chambers & Pratt 1956, Froesch et al 1957, Wolf et al 1959, Nikkilä et al 1962, Levin et al 1963, Cornblath et al 1963, Lindén & Nisell 1964, Dubois et al 1965). Induction of the disorder by ingestion, or rarely by inappropriate medicinal administration, of fructose and the related sugars sucrose and sorbitol, injures those organs that assimilate fructose by the specialized pathway involving aldolase B (Hers & Kusaka 1953, Hers 1957), the enzyme that is deficient in this disease (Hers & Joassin 1961, Nikkilä et al 1962, Froesch et al 1963, Levin et al 1963, Cornblath et al 1963, Lindén & Nisell 1964, Morris et al 1967).

Thus hereditary fructose intolerance affects the liver, small intestine, and the kidney: although the toxicity is associated with symptomatic hypoglycemia, abdominal pain, and vomiting and is dominated by effects on fructose metabolism in the liver (Froesch et al 1957, 1963), it is clear that disturbed function of the kidney contributes importantly to the manifestations of this nutritional disease (Perheentupa et al 1962, Lelong et al 1962, Cornblath et al 1963, Mass et al 1966, Levin et al 1963, 1968, Lamière et al 1978, Mock et al 1983).

In the kidney, the principal target for the noxious effects of fructose is the renal tubule: administration of fructose disrupts the function of the proximal nephron located in the kidney cortex (Morris 1968a) and may cause the full-blown Lignat-De Toni-Debré-Fanconi syndrome (Morris 1968b). There is defective acidification of the urine with impaired re-absorption of bicarbonate, amino acids, glucose, and phosphate and other filtered solutes; characteristically, glomerular filtration is unaffected. Generalized malfunction of the proximal tubule in hereditary fructose intolerance has a brisk onset and is potentially reversible, as demonstrated by the experimental administration of intravenous fructose.

Thus from the renal aspect, hereditary fructose intolerance is associated with deranged proximal renal tubular function which differs from that encountered in other inborn errors of metabolism, such as galactosemia and cystinosis since, in the early stages at least, the functional abnormalities recover rapidly on withdrawal of the offending nutrient (Morris 1968a).

As with other causes of malfunction of the proximal renal tubule, the most severely affected patients develop metabolic bone disease associated with vitamin D deficiency due partly to a failure in the tubular reabsorption of this essential secosteroid. In growing children, rickets develops and in adults osteomalacia may occur. Proximal renal tubular acidosis with the Fanconi syndrome, develops in patients with hereditary fructose intolerance ingesting moderate to copious amounts of sugar (Mock et al 1984). When exposure to fructose is sustained for long periods, marked electrolyte disturbances with hypokalemia, dehydration, nephrocalcinosis, polydipsia, polyuria, and impaired growth requiring parenteral fluid and electrolyte replacement, develop (Mass et al 1966). Hereditary fructose intolerance should be excluded as a priority in the specific diagnosis of any patient with uncharacterized renal tubular acidosis or manifest disease of the proximal nephron (Quigley 2006).

Given the severity of the untreated disease, its potentially lethal consequences and rapid improvement induced by excluding noxious sugars, prompt recognition of hereditary fructose intolerance and institution of dietary treatment is mandatory.

HISTORY

Early Discoveries about the Disease

Hereditary fructose intolerance has been recognized for more than 50 years and was first described as 'Idiosyncrasy to fructose' by Chambers and Pratt in 1956. These investigators

Genetic Diseases of the Kidney
ISBN: 978-0-12-449851-8

617

reported a 24-year-old woman who, after taking sugar and fructose, complained of phobic symptoms, faintness, abdominal pain, and nausea. When she took glucose, these symptoms were absent – although it is noteworthy that she did not enjoy the sweet taste. Chambers and Pratt systematically tested a range of sugars whose identity was unknown to the patient and concluded that the unwanted effects were specific to fructose and sucrose but not lactose, maltose, and galactose; the effects of the cognate hexose alcohol, sorbitol, which is converted rapidly to fructose, were not explored.

Chambers and Pratt suspected, but never formally proved, that the symptoms experienced by their patient could in part be explained by fructose-induced hypoglycemia. The diagnosis of HFI in the original patient has latterly been confirmed in the author's laboratory by molecular analysis of the aldolase B gene from a sample of her genomic DNA; she is homozygous for the most prevalent mutant allele, a missense mutation termed A149P (Ali et al 1998).

Shortly after the original description, there were numerous reports by European investigators, led by Froesch and colleagues in 1957, of infants and children with HFI, who suffered a more alarming and severe course of disease (Wolf et al 1959, Dubois et al 1961, 1965, Nikkilå et al 1962, Perheentupa et al 1962, Levin et al 1963, Cornblath et al 1963, Lindén & Nisell 1964). Those patients with HFI who were exposed to dietary fructose immediately at weaning and thus less able to defend themselves against its toxic effects appeared to suffer most (Froesch et al 1957, 1963, Swales & Smith 1966).

These early studies quickly differentiated hereditary fructose intolerance as an inborn error of metabolism in children and adults associated with hypoglycemia: the inheritance pattern of disease suggested a recessive trait (Froesch et al 1963, Cornblath et al 1963) that was distinct from galactosemia but which leads similarly to failure to thrive, vomiting, severe liver, and proximal renal tubular injury. The hypoglycemia was early shown by Dubois and colleagues in elegant metabolic radioactive tracer studies using ^{14}C-labeled glucose, to be caused by a failure of hepatic release of glucose in the interprandial state. The block was unrelated to excessive secretion of insulin and occurred specifically after administration of fructose (Dubois et al 1961).

In the context of a long series of experiments that brilliantly elucidated the specific pathway for fructose metabolism in the liver, kidney and small intestine, Henry-Gery Hers (with Joassin) identified the selective loss of fructose-1-phosphate aldolase-splitting activity in liver biopsies obtained from two patients with fructose intolerance (Hers & Joassin 1961). Penhoet and associates later purified the three tetrameric fructaldolase isozymes, designated A, B, and C, from vertebrate tissues and characterized their structural and catalytic properties (Penhoet et al 1966). It is the isoenzyme aldolase B – formerly known as the liver aldolase – with preferential activity towards fructose-1-phosphate, which is deficient in hereditary fructose intolerance.

Dubois, Froesch, Perheentupa, Levin, Cornblath and many associates reported the key clinical and biochemical features of the disease in adults and children; they demonstrated the acquired defect in glycogen breakdown and gluconeogenesis that accounts for fructose-induced hypoglycemia (Dubois et al 1961, 1965, Froesch et al 1963) and after exposure to fructose, is characteristically resistant to the action of glucagon (Perheentupa et al 1962, Froesch et al 1963, Cornblath et al 1963). The hypoglycemia induced by fructose is relieved by infusion of D-galactose (Cornblath et al 1963). Perheentupa and Levin and associates reported hyperuricemia and hypermagnesemia, respectively, during the acute phase of fructose exposure (Perheentupa and Ravio 1967, Levin et al 1968). Van den Berghe and colleagues later showed that this effect could be explained by the breakdown of purine nucleotides consequent on the sequestration of inorganic phosphate in the liver as the undegraded fructose ester (Van den Berghe et al 1973b) – a metabolic disturbance later confirmed in vivo using magnetic resonance spectroscopy in patients and control subjects after experimental challenge with fructose (Oberhansli et al 1987, Boesiger et al 1994). The hyperuricemic effect of fructose appears to be an exaggeration of the normal response to dietary or infused fructose and its metabolically related sugar alcohol, sorbitol (Perheentupa & Raivio 1967, Mäenpää et al 1968). The effects of uncontrolled ingestion of toxic sugars on the liver structure and function, the natural course of the disease and salutary effects of dietary exclusion therapy have also been documented in 55 patients by Odièvre and colleagues (Odièvre et al 1978).

Renal Manifestations of Fructose Intolerance

After early reports of urinary abnormalities and renal tubulopathy in HFI (Froesch et al 1957, Jeune et al 1961, Dubois et al 1962, Lelong et al 1962, Levin et al 1963, 1968, Ponte et al 1969) as well as the association with frank renal tubular acidosis (Mass et al 1966), the effects of fructose challenge on renal acidification and tubular function were explored in a long series of demanding clinical experiments reported by the distinguished American nephrologist, Curtis Morris, Jr. with the demonstration of fructose-1-phosphate aldolase deficiency in HFI kidney (Morris et al 1967) and the relationship of aldolase expression to the specific pathway of fructose metabolism in the renal cortex (Kranhold et al 1969, Heinz et al 1975), Morris defined the nature and reversibility of the renal acidification defect related to the metabolic disturbance caused by aldolase B deficiency in the proximal tubule (Morris 1968a, 1968b). Morris and colleagues later documented the syndrome of chronic fructose intoxication with growth retardation in children with HFI and defined the maximum content of fructose in the diet that is compatible with satisfactory growth and metabolic control (Mock et al 1983).

Molecular Genetics and Structural Biology of Aldolase B

In the latter era of molecular genetics and after the isolation of the human aldolase B gene and its localization to chromosome 9 (Tolan & Penhoet 1986, Lebo et al 1985), Cross, Tolan and Cox identified the first mutation responsible for hereditary fructose intolerance in 1988; this missense mutation is a prevalent cause of the disease and is globally distributed (Cross et al 1988). Cross and Cox, their associates – and latterly other investigators worldwide – have identified aldolase B mutations of diagnostic significance in HFI and developed molecular analysis of the cognate gene as a facile noninvasive diagnostic tool. The effects of natural and engineered mutations on the integrity of the aldolase B homotetramer and on its catalytic mechanism have been reported by the groups of Cox and Tolan in collaboration with Jurgen Sygusch, who first solved the three-dimensional crystal structure of mammalian class 1 aldolases at atomic resolution (Sygusch et al 1987). Such studies provide a framework for understanding the molecular function of this ancient family of enzymes (Takahashi et al 1989, Kitajima et al 1990, Blom & Sygusch 1997, Rellos et al 2000, Dalby et al 2001, Lafrance-Vanasse & Sygusch 2007).

CLINICAL FEATURES AND PRESENTATION (TABLE 36.1)

In the Infant

While taking breast milk the newborn with aldolase B deficiency does not develop symptoms since although it is relatively sweet, human milk does not contain any fructose. The sweetness of human milk is attributable to its high lactose content (70 g/l, $\sim$0.2 M) – a disaccharide composed of glucose and galactose moieties. Characteristic symptoms of vomiting, nausea, and sweating associated with frank hypoglycemia and metabolic acidosis are induced after transferring infants from breast to sweetened milk formulae or solid foods, particularly those containing added sugars; many natural fruits and vegetables also contain the offending sugars (Froesch et al 1957, 1963, Perheentupa et al 1972, Baerlocher et al 1978).

If large quantities of sugar are consumed, a very severe reaction occurs and the infant or young child becomes lethargic and comatose and may develop seizures (Jeune et al 1961, Perheentupa et al 1962, Sacrez et al 1962, Levin et al 1963, Cornblath et al 1963, Hübschmann & Cobet 1964, Royer et al 1964, Gentil et al 1964). These manifestations, with accompanying biochemical changes including hypoglycemia, are recapitulated in infants with hereditary fructose intolerance after controlled administration of fructose by parenteral infusion (Gitzelmann & Baerlocher 1973).

Continued exposure to the harmful sugars leads to failure to thrive associated with persistent vomiting; it appears

TABLE 36.1 Principal manifestations of hereditary fructose intolerance

Clinical
Vomiting, abdominal pain and diarrhea
Failure to thrive and irritability
Sweating, palpitations, anxiety
Impaired consciousness, coma, seizures
Bruising, jaundice, abdominal swelling
Hepatomegaly, edema
Muscle weakness, acidotic respiration
Rejection of sweet-tasting foods (older patients)
Laboratory findings
Hypoglycemia, hypophosphatemia, hypoalbuminemia
Hypokalemia, hypermagnesemia
Metabolic acidosis, including lactic acidemia
type I pattern of carbohydrate-deficient serum transferrin
Elevated plasma bilirubin, transaminases and aminoacids
Prolonged prothrombin time; reduced coagulation factors
Aminoaciduria, proteinuria, fructosuria, glycosuria

TABLE 36.2 Renal aspects of hereditary fructose intolerance

Bicarbonate wasting and proximal renal tubular acidosis
Complete Debré-Lignac-De Toni-Fanconi syndrome
Proteinuria
Postprandial fructosuria
Nephrocalcinosis
Rickets or osteomalacia in adults

clear that the youngest infants are most susceptible to the toxic effects of dietary sugars (Levin et al 1968). At times the child may be tremulous and irritable as well as pale and sweating due to hypoglycemia and the compensatory adrenergic reaction. The abdomen is often swollen and hepatomegaly may be detected. Jaundice develops and there is a bleeding tendency due to coagulation factor deficiency. Eventually cirrhosis of the liver, with the consequences of hepatic failure and portal hypertension, characterizes this severe metabolic and nutritional illness which can cause death at any age (Baerlocher et al 1978, Odièvre et al 1978).

Irreversible injury to the kidney may also occur with pathological effects on the proximal renal tubule (Lelong et al 1962, Dubois et al 1965, De Vroede et al 1980) (Table 36.2). Frank renal tubular acidosis has been noted (Mass et al 1966, Lamière et al 1978), in the former instance, persistent renal injury due to lifelong indiscriminate exposure to noxious sugars, induced formation of extensive calculi and was associated with chronic renal failure.

The mother or other individuals responsible for care of the infant are critical for the survival and adequate nutrition of individuals affected by hereditary fructose intolerance. By identifying those foods and drinks that cause the disease these carers are best placed to protect the defenseless child by recognizing the relationship of specific dietary articles to episodes of illness or periods of rapid clinical deterioration. In some circumstances, carers other than concerned parents may interfere catastrophically with the nutrition of sick infants with hereditary fructose intolerance by proffering large quantities of honey or sugary 'comfort' drinks to the patient under investigation (Marthaler et al 1967, Köhlin & Melin 1968) – sometimes without the knowledge of nursing staff and parents. On occasion such dietary indiscretions have had disastrous, and even fatal, results.

In Childhood

If the undiagnosed infant with hereditary fructose intolerance survives the challenging initial period of weaning, it may develop an effective self-protective aversion to those foods which engender the most distress. Later, a voluntary exclusion of harmful foods by taste and other association becomes established; over a lifetime, the eating behavior is refined by trial and error (Chambers & Pratt 1956, Froesch et al 1963, Swales & Smith 1966). Voluntary dietary exclusion usually restricts the intake of most sweet-tasting foods and drinks as well as fruits and certain vegetables (Chambers & Pratt 1959, Froesch et al 1963, Cornblath et al 1963, Cox et al 1982).

As the child becomes older they may become difficult to feed with the continuing risk of recurrent ill-health and growth failure (Mock et al 1983). There are several instances where the child, as in the case of the patient first reported by Chambers and Pratt, has been referred for psychiatric advice because of their demanding and fastidious dietary preferences;

the author is aware of several cases where this behavior – or extreme 'fussiness' – has been attributed to rejection of the maternal figure (Figure 36.1) (Cox et al 1983a, Bryan 1988, Cross & Cox 1989, Cox 1990a).

In the Adult

Adults with the disorder who have survived as a result of a self-imposed low-fructose diet often escape diagnosis for many years (Chambers & Pratt 1956, Froesch et al 1963, Swales & Smith 1966, Cox et al 1982, Ali & Cox 1995). Adults with the disease may later come forward in response to articles on the subject in the public domain or as a result of casual observations by healthcare professionals, including dentists (Cox 1990a). The author has encountered several patients in whom HFI has come to light as a result of seizures or episodes of altered conscious provoked by unexpected exposure to fructose or sugar in adult life: these episodes occurred under conditions in which hypoglycemia was strongly implicated (Cox 1990a). Likewise the author has two adult patients with uncontrolled HFI who have had nephrocalcinosis, previously not attributed to the disorder. Many reports in the dental literature showing that the modified dietary practices of many, if not all older children and adults with hereditary fructose intolerance (Newburn et al 1980) are accompanied by reduced or absent dental caries (Marthaler & Froesch 1967, Saxén et al 1989). Even in modern times, frank scurvy, resulting from self-imposed abstention from fruit and vegetables, occurs in adolescents and adults with fructose intolerance (Ali & Cox 1995, Guery et al 2007).

FIGURE 36.1   Eleven-year-old patient with hereditary fructose intolerance 'eating' a pear – a manifestation of unusual feeding habits present since weaning at four months of age accompanied by failure to thrive. Rejection of many foods and drinks had led to accusations of poor parenting and referral to a child psychiatrist. To avoid further disaffection with medical services, the diagnosis of hereditary fructose intolerance was confirmed at the age of four years in the patient and her elder sister by molecular analysis of the human aldolase B gene in DNA obtained from mouthwash samples (see Cross & Cox, 1989). (see also Plate 80)

PERILS OF FRUCTOSE POISONING IN THE CLINICAL ENVIRONMENT

In the past, patients with hereditary fructose intolerance have been exposed to life-threatening infusions of fructose given intravenously and more than 20 patients reported with fatal or near fatal metabolic crises induced by the indiscriminate use of fructose-, invert sugar-, or sorbitol-containing solutions (Cox 1993). However, many fructose-related deaths in hereditary fructose intolerance have probably gone unrecognized (Craig & Crane 1971, Woods & Alberti 1972, Hütteroth et al 1977, Galaske et al 1986). The lethal effects of acute fructose intoxication are characterized by severe (principally lactic) acidosis and hepato-renal failure in the presence of hypoglycemia, electrolyte abnormalities and shock (Lamière et al 1978, Curran & Havill 2002). Examination of necropsy specimens reveals acute (yellow) liver necrosis with gross loss of hepatocytes, proliferation of bile ducts with pigment deposition and activation of Kupffer cells (Müller-Wiefel et al 1983, Ali et al 1993) (see Figure 36.2). In the modern hospital setting, with renal supportive therapy and intensive care facilities available, death results from hepatic failure usually complicated by cerebral edema. Once established, hepato-renal failure is generally irrecoverable but the author is aware of the unexpected survival of a patient with hereditary fructose intolerance who developed severe toxicity as a result of ill-advised postoperative infusions of sorbitol and fructose while in a major German university teaching hospital after undergoing a minor surgical procedure. She was treated for fulminant liver failure by immediate allogeneic hepatic transplantation and survived to be discharged several months later. It is noteworthy in this patient, that she retained her distaste for sweet-tasting foods and drinks; however, although abdominal pain continued to accompany inadvertent dietary indiscretions, she no longer developed hypoglycemic symptoms.

Although the availability of fructose, sorbitol, and solutions of invert sugar for intravenous use was widespread before the late 1970s (Lane & Dodd 1957, Woods & Alberti 1972, Steinmann & Gitzelmann 1976) in many developed countries, most instances of severe intoxication have occurred in German-speaking countries, where the use of these preparations for parenteral administration persisted for many years (Von Heine et al 1969, Danks et al 1972, Steinmann & Gitzelmann 1976, Hüttenroth et al 1977, Schulte & Lenz 1977, Hackl et al 1978, de Vroede et al 1980, Müller-Wiefel et al 1983, Wagner & Wolf 1984, Locher 1987, Rey et al 1988, Steegmanns et al 1990, Ali et al 1993, Cox 1993, 1995, Rosien et al 1993). This historical practice probably owes its origin to the classical experiments of Oscar Minkowski, who showed that in dogs with diabetes induced by pancreatectomy, metabolic assimilation of fructose, unlike glucose, occurred independently of the intact pancreas (Minkowski 1893). By inference, fructose can be taken up independently of insulin and in the past, this has been used to justify its administration to acutely ill patients (Lane & Dodd 1957, Leutenegger et al 1977). Over many years however, the toxicity of fructose has been thoroughly documented in humans without recognized aldolase B deficiency but in whom metabolic (lactic) acidosis and liver injury have followed

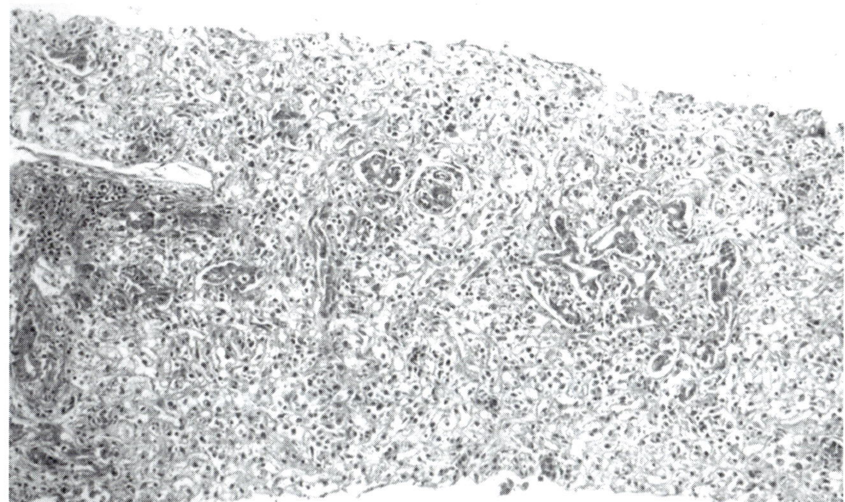

FIGURE 36.2 Photomicrograph of section of needle specimen of liver tissue (stained with Mallory's trichrome reagent) and obtained post-mortem from a 16-year-old female with a dislike of fruit and sugary foods since infancy. 110 g of sorbitol and at least 50 g fructose had been given intravenously in peri-operative fluids at the time of an appendectomy in Germany; despite intensive supportive care, the patient died on the 6th post-operative day as a result of fulminant hepato-renal failure with severe acidosis. Note the complete absence of hepatocytes with bile duct proliferation and abundant activated Kupffer cells. A retrospective diagnosis of hereditary fructose intolerance was made in the patient, and in a surviving brother with similar symptoms, by molecular analysis of the aldolase B gene. The patient and her brother proved to be compound heterozygotes for two inactivating mutations, whose presence was confirmed after death using genomic DNA extracted from minute amounts of this fixed and embedded tissue (see Rosien et al 1993; Ali et al 1993). (see also Plate 81)

high-dose infusions (Craig & Crane 1971, Woods & Alberti 1972, Hütteroth et al 1977, Galaske et al 1986).

To the author's knowledge, the last fatality due to inappropriate infusion of fructose and/or sorbitol in a patient with in hereditary fructose intolerance occurred in Germany in 1993 (Rosien et al 1993) and the availability of these sugars for use in parenteral nutrition has in most countries been greatly restricted (Reimers & Spigset 2003); preparations containing glucose and lipid as a source of energy are now generally preferred. The perils of inadvertent medicinal use of fructose or sorbitol in patients unable to metabolize these ubiquitous compounds nonetheless persist: recently, intravenous infusion of amiodarone caused acute hepatic and renal failure in a patient with hereditary fructose intolerance – a disastrous complication of antiarrhythmic therapy attributed to the polysorbate-80 constituent of the amiodarone preparation (Curran & Havill 2002).

After the identification of molecular defects in the human aldolase B gene responsible for hereditary fructose intolerance by the author's group, facile diagnosis of HFI by molecular analysis of genomic DNA has confirmed the cause of death in several patients with this disorder (Ali et al 1993, Ali & Cox 1995). A retrospective molecular diagnosis by analysis of DNA in the surviving relatives of a 42-year-old man with marked sugar aversion and who, like his younger brother had died after a fatal infusion of fructose- and sorbitol-containing solutions in a German hospital, led to the payment of compensatory damages to the family by the respective hospital authorities (Ali et al 1998). The value of taking a thorough dietary history and heeding patients' expressed food aversions to sweet-tasting articles is emphasized by such tragedies.

THE SYNDROME OF CHRONIC FRUCTOSE INTOXICATION

Not all patients with hereditary fructose intolerance develop an acute metabolic illness as a result of exposure to inappropriate nutrients. Those who continue to consume a low level of fructose in the diet but avoid large excesses of the toxic sugars may develop a chronic toxicity syndrome with delayed puberty or infantilism, growth retardation, low-grade acidosis with aminoaciduria, glycosuria and rickets, or osteomalacia in adults (Mock et al 1983). The metabolic bone disease appears to be due to persistent proximal tubular malfunction with loss of 25-hydroxycholelcalciferol (vitamin D) and phosphate as a result of incomplete reabsorption.

Despite the longstanding nature of these disturbances and the possibility of irreversible injury to the kidney and liver, recognition of HFI remains important since withdrawal of the noxious sugars even in adolescents may lead to improved growth and maturation with salutary resolution of the metabolic abnormalities.

EFFECTS OF HEREDITARY FRUCTOSE INTOLERANCE ON THE KIDNEY

Synopsis

Impaired gluconeogenesis after fructose intake in patients with hereditary fructose intolerance combined with allosteric activation of pyruvate kinase by fructose-1-phosphate leads to the accumulation of the Krebs' cycle precursors, lactate, pyruvate and the amino acid, alanine. It is now recognized that disturbed function of the proximal renal tubule aggravates the metabolic acidosis due to lactate accumulation induced by fructose in HFI and constitutes a key, if not inevitably dominant, component of the biochemical disturbance (Perheentupa et al 1962, Cornblath et al 1963, Lamière et al 1978, Richardson et al 1979).

Renal tubular acidosis is characterized by hyperchloremia in the absence of uremia and the production of alkaline, or minimally acidic urine in the face of metabolic acidosis. It is noteworthy that the acidosis in proximal and distal forms of renal tubular acidosis are normally associated with a normal anion gap (hyperchloremic acidosis); however, in hereditary fructose intolerance, the systemic metabolic disturbance provoked by fructose causes lactic acidosis combined with the proximal renal tubular acidosis to induce an acidosis with an increased anion gap. Where the patient has had severe vomiting with continued exposure to harmful sugars, gastric losses of hydrogen and chloride ions with dehydration and compensatory renal excretion of potassium may further complicate acid–base balance, with the superimposition of a metabolic alkalosis.

Fructose toxicity in the kidney is associated with impaired reabsorption of phosphate, α-amino nitrogen (amino acids), glucose and uric acid – thus indicating a complex set of metabolic disturbances due to generalized malfunction of the proximal nephron. How these effects on the kidney have come to be characterized, is described below.

Nephrocalcinosis and Renal Tubular Acidosis

The association of hereditary fructose intolerance with renal tubular disease was first noted 10 years after the description of the condition. Mass and colleagues reported a 41-year-old woman with a history compatible with fructose-induced hypoglycemia since early childhood (Mass et al 1966). She gave a 1-year history of progressive weakness and was admitted with extreme weakness of all muscle groups. The patient was clinically dehydrated with marked hypokalemia (1.4 mEq/l) and acidosis (serum bicarbonate 8.5 mEq/l, chlorides 105 mEq/l). Although serum calcium concentrations were normal, severe bilateral nephrocalcinosis with functioning kidneys was identified on intravenous urography. The patient had given a history of fructose-induced hypoglycemic symptoms since early childhood and had learned in later life to avoid sweets and certain vegetables. She and her twin sister, who was similarly affected by hereditary fructose intolerance, had nocturnal urinary incontinence until the age of 7 years and had always taken excessive quantities of water; marked nocturia had persisted in adult life. Immediately before admission the proposita had noted the frequent passage of sediment in her urine but had no symptoms of renal colic. The patient was later shown to be unable to acidify her urine in the face of marked acidosis (pH 7.27). By the time of study, creatinine clearance reduced to 71% of predicted. The patient had a positive fructose-loading test which induced hypoglycemia and hypophosphatemia typical of hereditary fructose intolerance – abnormalities which were also confirmed in her twin and another affected sister. The florid renal calculi that occurred in this patient, and their surgical dissolution, were the subject of a contemporaneous report (Higgins & Varney 1966).

Other Reversible Effects on the Kidney

In many of the early reports of hereditary fructose intolerance, especially those occurring in infants and children, aminoaciduria, hypercalciuria, proteinuria, and pyuria were noted; these abnormalities, which indicate disease of the proximal tubule, disappeared promptly after fructose was withdrawn. Several authors had noted the transient induction of these urinary abnormalities after rechallenging HFI patients with the offending sugar but no detailed studies of the renal defect were undertaken. Evidence for transient cellular injury to the kidney as a result of exposure to fructose was provided in later challenge studies by Morris et al (1970) in which acute administration of fructose provoked the appearance of the enzyme activities attributable to lysosomal acid hydrolases (e.g. *N*-acetyl-β-glucosaminidase) in the urine, presumably reflecting cellular injury and fructose-1-phosphate-mediated damage to this organelle (see below).

In the patient with HFI and nephrocalcinosis attributed to renal tubular acidosis first reported by Mass and colleagues (Mass et al 1966), it was noted that progressive acidosis developed during administration of oral potassium chloride but that the patient was unable at the time to excrete an acid urine; the acidosis was later controlled by the daily infusion of a proprietary combination of potassium acetate, bicarbonate, and citrate (total 6 g of each, daily).

Further Characterization of the Renal Defect

The characteristics of the renal component of the metabolic acidosis induced by fructose toxicity in HFI have been defined in a series of fastidious clinical investigations by Morris (see Morris 1968a, 1968b). These studies also confirmed that the effects of fructose on the kidney, at least in most patients with hereditary fructose intolerance, and in the experimental situation, were reversible.

EFFECTS OF FRUCTOSE ON ACID EXCRETION IN FRUCTOSE INTOLERANCE

Morris examined the detailed effects of fructose challenge on the tubular excretion of hydrogen and bicarbonate ions in patients with HFI under controlled acidosis induced by ammonium chloride loading (Morris 1968a). The patients had been maintained on a fructose-free diet and, together with the healthy control subjects, were given 0.08–0.1 g/kg ammonium chloride orally or as a 10% w/v intravenous solution; in all instances arterial pH was shown to enter the acidotic range.

D-fructose (0.25 g/kg) was given as a 25% v/w intravenous solution by bolus infusion over 5 minutes. In some experiments, a bolus of 0.25 g/kg fructose was given as above but this was followed by a slow infusion of a 10% w/v intravenous solution calculated to give fructose at a constant rate of 0.25 g/kg per hour over 2–3 hours. To exclude adverse effects due to hypoglycemia in some of these more prolonged experiments, hypoglycemia was prevented by the infusion of 10% w/v D-glucose starting 10 minutes before the fructose was administered at a rate calculated to deliver 0.10–0.12 g/kg per hour. To obviate the effects of the marked hypophosphatemia normally induced by fructose during sugar challenge in a patient with HFI, buffer phosphate (0.15M Na_2HPO_4/NaH_2PO_4, pH 7.4) was infused intravenously at 0.287–0.75 millimoles/minute) to provoke a phosphate diuresis once the impaired renal acidification was established.

STUDIES IN THE ABSENCE OF FRUCTOSE CHALLENGE

The effects of infusing 0.25 g/kg bodyweight fructose intravenously on the acid excretion by the kidney in otherwise healthy patients with HFI were very rapid. Urinary pH increased from 5.2 or less to greater than 6.2 within 45 minutes of the fructose infusion and remained elevated throughout the period of infusion. One and a half to 3 hours after the experimental infusion of fructose, urinary pH decreased to values of approximately 5.0 and this was accompanied by an increased rate of excretion rate of acid presumably generated in the distal nephron.

EFFECTS OF FRUCTOSE WITHOUT AMMONIUM CHLORIDE LOADING

Administration of fructose to a patient with HFI, with or without simultaneous infusion of glucose to prevent hypoglycemia, caused urinary pH to increase above 7 and the excretion of titratable acid and ammonium ions to decrease to less than 10 mEq/minute. Under these circumstances, calculated net acid excretion became negative but was rapidly restored shortly after termination of the fructose infusion, presumably as the response to the mild fructose-induced metabolic acidosis was restored.

EFFECTS OF FRUCTOSE ON THE RENAL RESPONSE TO ACID LOADING

Compared with healthy subjects, the acidification response to ammonium chloride was rapidly impaired by infusion of fructose in patients with HFI. In the patients who received fructose without glucose, blood glucose concentrations decreased as expected. Within 45 minutes, urinary pH increased from less than 5.2 to greater than 6.2 where it remained during fructose administration; however, within 90–180 minutes after the infusion was stopped, urinary pH returned to about 5.

As the urinary pH increased, the rate of excretion of titratable acid, ammonium ions (and calculated net acid loss) declined and reciprocally increased as the inappropriate urinary pH was corrected. In an HFI patient in whom the fructose infusion was followed by induction of a phosphate diuresis, the excretion rate of titratable acid stimulated by ammonium chloride increased quickly but the urinary pH remained at 6.0 or above.

IMPAIRED REABSORPTION OF BICARBONATE

Having shown that fructose induces a reversible defect of renal acidification in patients with hereditary fructose intolerance, Morris and his assistants conducted further studies to investigate bicarbonate reabsorption in HFI and to compare it with other forms of renal tubular acidosis including cystinosis and classical distal renal tubular acidosis (Morris, 1968b). The patients received intravenous fructose as before and bicarbonate transport was determined over a wide range of plasma bicarbonate concentrations, which were manipulated by intravenous infusion of 3.75% (w/v) sodium bicarbonate in addition to oral bicarbonate loading. The investigations showed clearly that experimental infusion of fructose induced a 20–30% reduction of maximal bicarbonate reabsorption at plasma concentrations of 21–31 mEq/l in patients with HFI. Urinary acidification was impaired but at plasma bicarbonate concentrations less than 14 mEq/l, tubular reabsorption of bicarbonate was complete and the urine was acidified normally to less than pH 5.0.

Marked impairment of bicarbonate reabsorption at plasma bicarbonate concentrations in the normal range strongly implicates the proximal nephron in the renal toxicity of fructose. Even if the activity of the distal nephron were completely abolished, it would not have an effect of this magnitude because 85–90% of the renal reabsorption of filtered bicarbonate occurs in the proximal tubule. Furthermore the disorder resembled the defect in cystinosis, which is accompanied by structural injury to the proximal tubular epithelium and the Fanconi syndrome: reabsorption of phosphate, α-amino nitrogen and urate were all impaired, indicating parallel loss of several functions attributed to the proximal nephron.

The functional defect of acidification due to impaired maximal reabsorption capacity for bicarbonate was restricted to the acute exposure to fructose in patients with HFI and was proportional to the concentration of plasma fructose. In contrast, in patients with classical distal renal tubular acidosis the urinary pH is always inappropriately high (>5.5), and urinary excretion of bicarbonate persists; indeed, the amount of bicarbonate excreted is very small in relation to that filtered and the maximal rate of reabsorption is unimpaired. Thus the tubular defect induced by fructose in HFI and defects observed in other diseases of the proximal nephron such as galactosemia and cystinosis, is distinct from the more common forms of distal renal tubular acidosis, in which there is a primary failure to maintain steep hydrogen ion gradients between the lumen and the pericellular plasma across the tubular epithelium. The disorder in HFI is moreover associated with other proximal tubular transport defects.

Effects of Fructose on Other Aspects of Renal Function

EXCRETION OF PHOSPHATE

In healthy subjects and patients with hereditary fructose intolerance, administration of fructose increased urinary clearance of phosphate but in HFI it is notable that this occurs under circumstances where the plasma concentration of phosphate is markedly decreased; in spite of this, tubular reabsorption of phosphate expressed as the fraction of that filtered is disproportionately reduced in patients with HFI.

ALPHA-AMINO NITROGEN

Levin and colleagues (Levin et al 1963) demonstrated that fructose impairs the tubular reabsorption of α-amino acids in children with HFI; under the conditions utilized to study adult HFI patients by Morris, excretion of α-amino nitrogen was increased about eightfold compared with healthy adult control subjects.

URATE

The heavy deposit of urate crystals in the urine of patients with hereditary fructose intolerance after challenge with fructose, led Perheentupa and Raivio (1967) to discover the hyperuricemic effect of fructose in normal subjects and patients intolerant of fructose. In a single HFI patient in which urinary excretion of uric acid was studied by Morris, fructose provoked a fourfold increase in the excretion of uric acid and a more than twofold enhancement of uric acid clearance (Morris 1968b). Fructose had no measurable effect on these parameters in control subjects.

GLOMERULAR FILTRATION

In the acute experimental situation, the renal glomerular filtration rate, as determined by clearance of inulin, was unaffected. It is noteworthy for future reference that care had to be taken with some of the preparations of inulin. Inulin itself is a natural polymer of fructose and its use during the investigations led to unexpected additional toxicity during the period of study in some HFI patients that was attributed to the release of free fructose monosaccharide, presumably due to spontaneous hydrolysis of the polysaccharide upon storage.

Clinical Implications of Proximal Tubular Disease in HFI

Unless suspected and sought for specifically, the contribution of disordered proximal renal tubular function to the metabolic syndrome of fructose intolerance may easily be overlooked. Since they result from exposure to the noxious sugars, the renal abnormalities are likely to be episodic; moreover on cursory measurement, the urine pH is often unremarkable in patients with unstable acid–base status and the plasma bicarbonate can range between 10 and 20 mM.

On the other hand, metabolic bone disease (rickets or osteomalacia) attributable to renal tubular disease may frequently cripple patients with sufficient exposure to dietary fructose and related sugars to induce a state of mild, but unrecognized, chronic intoxication (Mock et al 1983). However, it is salutary to note that the deleterious effects of fructose on the kidney can be detected within minutes of exposure and that fructose loading may occasionally cause a violent metabolic disturbance with an acute Fanconi syndrome accompanied by hypokalemic paralysis (Steinmann & Gitzelmann 1981).

As in other causes of proximal tubular disease, hypercalciuria and hyperphosphaturia, together with the appearance of other solutes such as glucose, fructose, and amino acids as well as retinol-binding protein and the appearance of lysosomal enzymes in the urine, are likely. Although well documented in the disease, nephrocalcinosis and renal calculi are less common than expected in HFI unless excess vitamin D and calcium supplementation has been given incautiously. The explanation is unknown but increased urinary excretion of citrate (due also to reduced reabsorption in the proximal tubule) may be responsible, since this organic anion inhibits the precipitation of calcium salts and appears to be a limiting factor in the nephrocalcinosis associated with distal renal tubular acidosis. It is likely that increased formation of gluconeogenic precursors in the tricarboxylic acid cycle after exposure to fructose in HFI, may further enhance the urinary excretion of citrate.

Hypokalemia due to the osmotic diuresis consequent upon reduced reabsorption of bicarbonate in the proximal tubule is to be expected when the renal disease is established. Potassium ions are lost principally from the intact distal nephron as a result of the sustained increase in the rate of flow of tubular fluid; for this reason, supplementation with bicarbonate is likely to exacerbate the hypokalemia. Hypokalemia in patients with hereditary fructose intolerance may also be exacerbated by vomiting induced by exposure to fructose with consequential metabolic alkalosis and compensatory loss of potassium from the distal tubule. Hypokalemia

may be severe enough to cause flaccid paralysis and may further compound the misery experienced by patients with fructose intolerance in the absence of a definitive diagnosis and before appropriate dietary and other treatments are introduced (Levin et al 1963, 1968, Mass et al 1966).

INHERITANCE OF HEREDITARY FRUCTOSE INTOLERANCE

In most pedigrees, the pattern of transmission of HFI is compatible with an autosomal recessive trait, and thus with the map position of the aldolase B locus on human chromosome 9q. However, there have been very few instances of parental consanguinity, as would be expected were mutant genes are very rare in the general population; in contrast, there are numerous reports of generation-to-generation inheritance of the disease, e.g. from parent to offspring (Wolf et al 1959, Lindén & Nissell 1964, Barry et al 1968, Rampa & Froesch 1981, Cox et al 1982, Ali & Cox 1995).

In at least two instances, detailed investigations, including molecular analysis of the aldolase B gene, have shown that the disease was inherited in a pseudodominant manner as a result of matings between affected homozygotes and asymptomatic heterozygous partners (Cox et al 1982, Ali & Cox 1995). In one extended but noninbred North American family in which the proband had tragically died due to administration of invert sugar in the perioperative period, investigations conducted by the author's laboratory revealed no less than ten first-degree relatives with symptomatic HFI due to the independent segregation of four mutant aldolase B alleles; parent-to-offspring transmission of the disease occurred twice (Ali & Cox 1995). In the absence of consanguinity, this family indicates the high prevalence of disease-causing alleles, as has been confirmed by studies of random blood samples obtained from the general neonatal population (James et al 1996).

These and other biochemical studies of obligate heterozygotes confirm that HFI is a true recessive disease. In the absence of molecular studies, detection of heterozygous carriers for the condition, for example by enzymatic assay, is difficult (Lindén & Nisell 1964, Raivio et al 1967) moreover, heterozygotes experience no symptoms and only qualitative changes in metabolic parameters on challenge with sugar (Oberhaensli et al 1987).

PATHOLOGICAL INJURY IN HEREDITARY FRUCTOSE INTOLERANCE

In the untreated state, the patient with long-standing hereditary fructose intolerance (usually an infant or young child) has an enlarged liver but is wasted and stunted. In florid cases where the diagnosis has been delayed, signs of dehydration, hepatic decompensation with jaundice, ascites, and generalized edema are present.

There have been few systematic studies of the structural changes in the viscera that bear the principal brunt of the injury in patients with hereditary fructose intolerance. In several instances, the histopathological appearances have been examined at necropsy. However, because such studies are often conducted in the context of legal inquiries into the death of a patient in hospital with an unexplained illness, the findings are usually not placed in the public domain.

Long-standing exposure to harmful sugars principally induces injury in the liver and kidney cortex (Odièvre et al 1978). Limited data are available on the structural changes observed in the renal proximal nephron, where vacuolation and a granular appearance in the cytoplasm of the lining epithelial cells accompanied by slight dilatation of the proximal tubules have been reported (Lelong et al 1962). Examination of the liver shows steatosis and parenchymal loss with focal hepatocyte necrosis; in several instances the diagnosis resembles that of Reye's syndrome, with extensive fatty change in the liver (Yang et al 2000, Thabet et al 2002). Bile duct proliferation with the appearance of epithelial tubule-like structures and intralobular and periportal fibrosis are prominent features and the changes of frank cirrhosis may also be present (Odièvre et al 1978, Müller-Wiefel et al 1983, Rosien et al 1993).

A particular feature of the hepatic changes is the marked increase in the glycogen content of the hepatocytes, reflecting the inhibition of glycogenolysis, which is a central component of the metabolic disturbance induced by exposure to fructose. While the excess of stainable glycogen (>5% by weight) in liver biopsy specimens taken for diagnostic purposes may provide an important clue to hereditary fructose intolerance, its presence might also be misinterpreted and wrongly taken to indicate the presence of a primary glycogen-storage disease (Cain & Ryman 1971). Unlike several, but not all of the glycogen-storage diseases, the hepatic glycogen in HFI has a normal molecular structure and tinctorial properties on histochemical and electron microscopic examination.

Electron microscopy of liver biopsy samples showed parenchymal cells and some Kupffer cells with membrane-bound inclusions; some of these were considered to represent lysosomes. The inclusions contain concentric arrays of dense membraneous material differing markedly in size but with a lucent halo (Rossner & Feist 1971).

The ultrastructural features of fructose-induced injury in the liver and small intestinal mucosa have also been studied in ethically unusual challenge experiments (Phillips et al 1968, 1970). Clear abnormalities were observed on electron microscopy of liver and later small intestinal mucosal tissue taken by biopsy 120 minutes after the *oral* ingestion of a relatively large dose of fructose (50 g) in a patient with HFI. There were abnormalities in the endoplasmic reticulum with irregular membranous arrays; in the cytoplasm there were concentric membranous bodies – some containing electron-dense deposits resembling lysosomal storage material possibly

the product of autophagy, as in galactosemia. The rapid appearance of lysosomal enzymes in the urine after fructose infusion is compatible with the hypothesis that excessive intracellular fructose-1-phosphate may, as with galactose-1-phosphate in galactosemia, stimulate injurious autophagic cellular responses in hereditary fructose intolerance.

In the *acute yellow atrophy of the liver* that follows *massive* fructose intoxication, widespread hepatocyte necrosis is observed; only balloon-like ghosts of the dying or dead cells remain but there is a rapid proliferation of bile ducts with pigment cholestasis and marked activation of Kupffer cells which can be seen scavenging necrotic debris and parenchymal cells (see Figure 36.2 and Müller-Wiefel et al 1983, Ali et al 1993).

METABOLISM OF FRUCTOSE

Transport

Fructose is absorbed from the small intestinal as the free monosaccharide or after release by the luminal hydrolysis of sucrose through the agency of the mucosal disaccharidase, sucrase-isomaltase; the presence of abundant sorbitol dehydrogenase in the intestinal mucosa facilitates the conversion of exogenous sorbitol directly to free fructose. Uptake of fructose from the intestinal lumen involves the GLUT5 transporter isoform and proceeds by facilitated diffusion. Egress of fructose from the intestinal absorptive cell is mediated principally by the GLUT2 sugar transporter molecule. After a sugar-rich meal containing fruit or the consumption of fructose-containing snacks, confectionery or drinks, it is estimated that concentrations of free fructose in the portal blood may exceed 10 mM.

Biochemical Assimilation of Fructose

Utilization of D-fructose occurs independently of insulin but since fructose may lead to an increase in blood glucose concentrations, large fructose loads may ultimately produce a rise in blood glucose concentrations and a transient insulin response. In the liver, kidney and small intestine, three enzymes are present which convert fructose into the metabolic intermediates of glycolysis and gluconeogenesis. These enzymes, ketohexokinase (fructokinase – EC 2.7.13) aldolase B (fructose-1,6-bisphosphate aldolase – EC 4.1.2.13) and triokinase (EC 2.7.1.28) bring about the metabolic assimilation of exogenous fructose in these organs.

Specific Pathway of Fructose Metabolism

Newly absorbed fructose in taken up and rapidly phosphorylated at the carbon-1 position by ketohexokinase (fructokinase) in an energy-dependent reaction requiring ATP in the specific fructose-metabolizing tissues. Aldolase B catalyzes the reversible cleavage of fructose-1-phosphate to form dihydroxyacetone phosphate and D-glyceraldehyde. Triokinase then converts D-glyceraldehyde into D-glyceraldehyde-3-phosphate which serves as the key triose intermediate of glycolysis and gluconeogenesis (Figure 36.3). Thus the intermediary triose products can be incorporated directly into the glycolytic pathway with the eventual formation of lactate and pyruvate – the latter a substrate for the Krebs' cycle. Alternatively, the triose sugars are substrates for gluconeogenesis and serve as a source of new glucose with or without the synthesis of glycogen.

Biochemical and inmmunochemical studies using isozyme-specific antisera confirm that the specialized pathway for fructose metabolism exists in the proximal intestinal tubular epithelial cells, the intestinal absorptive cells of the duodenum and upper small intestine and in the hepatic parenchyma.

In adipose tissue, muscle and other tissues, at high physiological concentrations after meals, fructose may be incorporated into metabolism through the action of hexokinase and hepatic glucokinase (hexokinase D). The sugar is phosphorylated at the 6-carbon position and the fructose 6-phosphate then enters the glycolytic pathways conventionally by the actions of phosphohexokinase (to generate fructose-1,6-bisphosphate) and/or phosphohexoseisomerase (to generate glucose 6-phosphate). There is competition for this pathway by glucose but in patients with HFI lacking the specific pathway mediated by aldolase B, it is quantitatively dominant. It has been estimated that after loading, 80–90% of the administered fructose is ultimately disposed of by this mechanism (Froesch et al 1963). Minor pathways of fructose

disposal include an ill-understood mechanism present in the optic lens and erythrocytes, which leads to the formation of fructose 3-phosphate. Direct conversion of fructose to glucose by way of phosphorylation to fructose 6-phosphate has been proposed, but the presence of this pathway has not been confirmed in animal tissues.

PRIMARY BIOCHEMICAL DEFECT IN FRUCTOSE INTOLERANCE

Hereditary fructose intolerance is caused by deficiency of the specific fructose-1-phosphate cleaving activity of aldolase B, a specialized enzyme of fructose metabolism. Aldolase B is expressed exclusively in the proximal renal tubular epithelial cells, the parenchymal cells of the liver (hepatocytes) and the epithelial cells of the upper small intestine and duodenum. Fructaldolases (EC 4.1.2.13) catalyze the specific reversible cleavage of fructose-1,6-bisphosphate and fructose-1-phosphate into the 3-carbon sugars dihydroxyacetone phosphate, D-glyceraldehyde 3-phosphate, and D-glyceraldehyde. In mammals, there are three genetically distinct but related isozymes of aldolase which are distinguished on the basis of their catalytic and antigenic properties together with their unique tissue distribution. Fructaldolases occur throughout nature; those that are confined to yeasts and prokaryotes contain a bivalent metal cation (usually zinc), that is essential for their activity. The class I aldolases, which are principally of eukaryotic origin, generate a stable

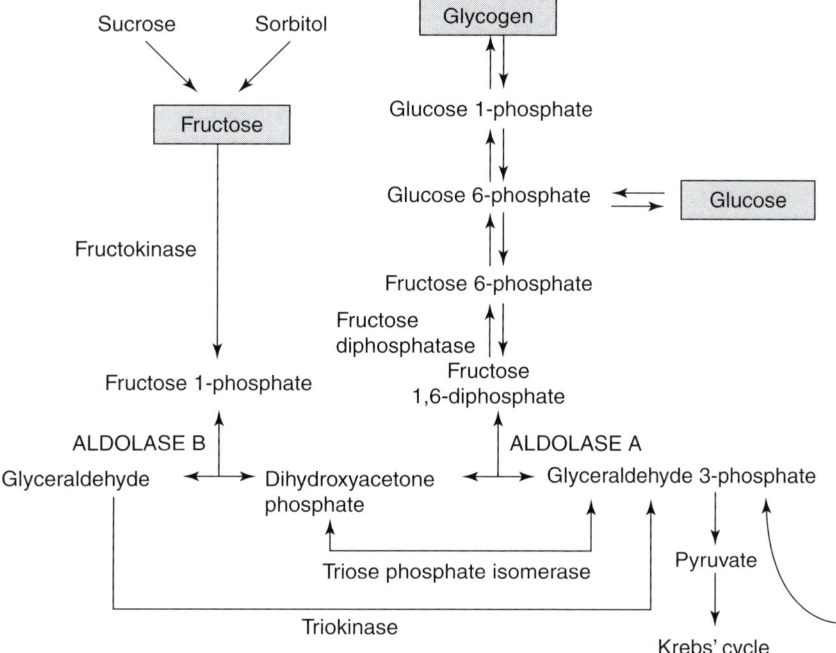

FIGURE 36.3　Overview of the metabolism of dietary fructose, sucrose and sorbitol (from Cox, 2003, Oxford Textbook of Medicine OUP). (see also Plate 82)

carbanion substrate–enzyme complex and transfer a proton from a conserved Schiff base-forming lysine residue which is located at the active site of the enzyme.

In humans, aldolase A (with a principal action on fructose-1,6-bisphosphate, which is cleaved reversibly into dihydroxyacetone phosphate and D-glyceraldehyde 3-phosphate) is found as the housekeeping aldolase in glycolysis. Aldolase A is very active and abundant in skeletal muscle.

Aldolase B, previously known as liver aldolase, accession no. X01098, has approximately equal action against fructose-1-phosphate and fructose-1,6-bisphosphate. Aldolase C is an isozyme expressed principally in the brain and is rich in the cerebellum. As expected for housekeeping enzymes of metabolism the aldolase isozymes A and C are expressed constitutively but the B isoform which is not expressed abundantly at birth is under strict dietary and developmental control. Aldolase C shows approximately tenfold greater activity towards fructose-1,6-bisphosphate and fructose-1-phosphate; aldolase A shows 50 or more greater activity towards the fructose-1,6-bisphosphate. These aldolase isozymes have less affinity for the fructose-1-phosphate substrate with a K_M for aldolase B of approximately 1–5 mmol/l compared with the bisphosphate substrate, for which the K_M is about 50 μmol/l. The aldolase isozymes A, B and C function as homomeric tetramers but it is possible to induce formation chimeric aldolase heterotetramers with intermediate catalytic properties. The structural and catalytic properties of the class 1 mammalian aldolases suggest evolution by a process of gene duplication and divergence; indeed mapping studies of three human aldolase B genes and a pseudo-gene to homeologous chromosomes suggest the emergence of these isozymes occurred during the process of vertebrate tetraploidization.

THE ENZYMATIC DEFECT

Nature of Aldolase B Deficiency

Patients with hereditary fructose intolerance have a selective deficiency of aldolase B which is reduced to less than 10% of mean control values. There is an attendant reduction of fructose-1,6-bisphosphate aldolase activity in the fructose metabolizing tissues as a result of loss of this activity attributable to the aldolase B molecule. However, the presence of the residual aldolase isozymes, A and C, allow gluconeogenesis and glycolysis to continue normally when aldolase B is deficient *provided* that fructose is not present (Dubois et al 1961).

Effects on Intermediary Metabolism

Where aldolase B activity is deficient, newly absorbed fructose is taken up rapidly and phosphorylated by ketohexokinase but cannot be converted to trioses for entry into

glycolysis and gluconeogenesis. As a result, the concentration of fructose-1-phosphate increases rapidly (>15 mM) and inorganic phosphate becomes sequestrated as the sugar ester (Perheentupa et al 1962, Woods et al 1970, Oesige et al 1994). Under these circumstances, the intracellular concentration of free phosphate falls and, combined with the presence of increased concentrations of fructose-1-phosphate, the glycogenolytic enzyme phosphorylase A is inhibited competitively and breakdown of glycogen (glycogenolysis) ceases (Van den Berghe et al 1973b). At the same time, activity of residual aldolase A in the direction of fructose-1,6-bisphosphate formation by condensation of trioses is competitively inhibited so that de novo formation gluconeogenesis is also impaired; a labile and diffusible inhibitor of fructose-1-phosphate aldolase has also been detected in HFI liver tissue (Cox et al 1983c).

Under these circumstances, challenge with fructose in a patient with hereditary fructose intolerance rapidly induces hypoglycemia and hypophosphatemia; the hypoglycemia is refractory to treatment with injected glucagons (Perheentupa et al 1962, Froesch et al 1963, Cornblath et al 1963) (glycogen phosphorylase is inactive) and infusions of dihydroxyacetone phosphate or glycerol (gluconeogenesis is impeded in the path between the reverse action of aldolase B on the condensation of trioses and the activity of glucose 6-phosphatase).

Hypoglycemia induced by fructose challenge in patients with hereditary fructose intolerance may be overcome by the administration of D-galactose (or lactose), indicating that the activities of the galactokinase, phosphoglucomutase and glucose 6-phosphatase pathways are intact (Cornblath et al 1963, Gentil et al 1964); as expected, the hypoglycemia may be easily overcome, if required, by administration of intravenous or oral D-glucose).

In the absence of a diet containing fructose, sucrose or sorbitol, patients with HFI have no defect in the interprandial release of glucose and show a normal hyperglycemic rise in blood glucose after parenteral administration of standard doses (0.2–1.0 mg) of glucagon; moreover, patients with fructose intolerance who have been maintained in good health on an appropriate exclusion diet, withstand starvation normally.

Other Metabolic Disturbances

Exposure of patients with HFI to fructose and related sugars also induces metabolic acidosis, hypophosphatemia, hyperuricemia and hypermagnesemia. Acidosis is caused by a component of lacticacidemia that is due to the accumulation of gluconeogenic precursors of the tricarboxylic acid cycle in the presence of the metabolic block occurring principally in the liver (Cornblath et al 1963). Neither infusions of dihydroxyacetone phosphate nor glycerol can correct the hypoglycemia induced by fructose (Gentil et al 1964), indicating that glucose phosphate isomerase and the triose-condensing reaction of aldolases A and C are inhibited (probably by fructose-1-phosphate). Acute effects of fructose on kidney metabolism also contribute to the acidosis as a result of bicarbonate loss in the proximal nephron (Lamière et al 1978, Richardson et al 1979).

Hypophosphatemia is attributable to the sequestration of inorganic phosphate in the liver and kidney as the undegraded fructose ester; these abnormalities have been demonstrated in living patients with HFI challenged with controlled doses of fructose and examined by ^{31}P – magnetic resonance spectroscopy (Oberhaensli et al 1987, Boesiger et al 1994). In a control patient with hereditary fructosuria, in which exogenous fructose cannot be assimilated as a result of ketohexokinase deficiency and unlike healthy volunteers, the ^{31}P signal attributable to the fructose-1-phosphate ester remained absent after sugar loading (Boesiger et al 1994).

Hyperuricemia, with increased formation of uric acid, results from the degradation of purine nucleotides mediated through activation of the xanthine oxidase pathway – a consequence of intracellular depletion of free phosphate with loss of energy charge. Adenosine monophosphate is degraded to inosine 5′-phosphate and hence to urate, which appears in the plasma and urine (Perheentupa & Raivio 1967, Perheentupa et al 1972). Hypermagnesemia is attributed to the release of chelated pools of intracellular magnesium following the breakdown of the magnesium–adenosine triphosphate (Mg–ATP) complex (Levin et al 1968, Van den Berghe et al 1973a).

High concentrations of unmetabolized fructose 1-phosphate inhibit ketohexokinase action by competitive negative feedback – this prevents further metabolic incorporation of fructose by the aldolase B pathway; as a result transient fructosemia (an outmoded term for hereditary fructose intolerance) and transitory fructosuria are observed. Fructosuria, with the appearance of a nonglucose-reducing sugar in the urine (which is not detected by the specific glucose oxidase methods now in general use), occurs when the concentration of fructose in the peripheral blood exceeds the threshold for renal excretion at approximately 2 mM (Froesch et al 1963). It is notable that although the specific pathway of fructose metabolism is impaired in hereditary fructose intolerance, at least 80% of the administered load of fructose is disposed of by glycolysis, probably mainly in adipose tissue. It appears that fructose can quite readily enter intermediary carbohydrate metabolism since it is phosphorylated at the 6-carbon position by hexokinase and hepatic glucokinase (Froesch et al 1963).

The appearance of reducing sugar in the urine may also indicate the presence of glycosuria. The toxic effects of fructose on the kidney impair reabsorption of filtered glucose in the proximal nephron and glycosuria reflects the generalized Fanconi syndrome resulting from the metabolic disturbance affecting the tubular epithelium.

Latterly, the consequential effects of disturbed carbohydrate metabolism in HFI have been shown to affect other functions, particularly the co-translational glycosylation of proteins by the liver. Thus untreated patients with hereditary fructose intolerance almost invariably show a type I pattern of carbohydrate-deficient serum transferrin on isoelectric focusing, which is corrected within a few weeks of fructose exclusion (Jaeken et al 1996, Pronicka et al 2007). In infants and young children undergoing investigations for ill-defined and undiagnosed severe metabolic illness with multi-system involvement, such abnormalities may easily be identified as a result of investigative screening; in all instances the underlying cause can be established, but the apparent diagnosis of carbohydrate-deficient glycosylation (CDG) syndrome is a misinterpretation until the diagnosis of HFI has been excluded and a appropriate diet instituted. The mechanism of this effect appears again to be related to the intracellular accumulation of fructose-1-phosphate, which is a powerful inhibitor of the phosphomannose isomerase implicated in glycoprotein processing and biosynthesis (Jaeken et al 1996). However, the widespread application of techniques for protein glycosylation analysis, including immuno-isoelectric focusing, in young patients suffering from ostensibly unspecific clinical manifestations, may serve to increase the diagnostic yield for HFI; this is a practical issue that is likely to become ever more important with the uncritical use of multiple diagnostic screening procedures in the modern hospital setting and where formal history-based approaches may be neglected.

THE ENVIRONMENTAL FACTOR: FRUCTOSE, A UBIQUITOUS NUTRIENT

Fructose and sucrose are widely distributed in nature. The free monosaccharide is abundant in honey and fruits, in which it is associated with the disaccharide sucrose (composed of glucose and fructose moieties in β-glycosidic linkage). Fructose and sucrose and, to a lesser extent the sugar alcohol sorbitol, are natural components of many foods, including nuts and vegetables. However, by far the most important sources of these sugars in the modern Western and now also African and Oriental diets, are as commercial additives. Sorbitol is in widespread use as a nonglucose sweetener in diabetic foods and confectionery and is added to many so-called 'sugar-free' sweets and gums; it also used as an excipient in tablets and other medications.

Fructose has the most powerful sweet flavor of the naturally occurring sugars and is 1.4 times sweeter than sucrose, which owes its taste to the fructose moiety. Fructose has an intensity of flavor that is matched by its transitory action; the sugar has other useful properties including reducing power and the propensity to 'brown' when heated. The latter property renders fructose ideally suited to blending in many items of food, including those that are cooked (Cox 1994). Such is the sweetening power of fructose that it is used extensively as corn syrup and high-fructose syrup derived from the commercial hydrolysis of starch in confectionery and popular carbonated drinks. Several ubiquitous well-known brands contain 10–14% w/v sugar.

The human palate appears to have been inured to sweet tastes, since the practice of adding large amounts of sucrose and fructose to foods, drinks and medications is longstanding and widespread. Such is the market for sweet-tasting foods and drinks, that fructose and sucrose are almost universally added in abundant quantities to 'natural' pressed fruit juices and cordials, whose ingestion is typically the major energy-rich component of the first meal of the day for hundreds of millions of persons worldwide.

The addition of sugars on an industrial scale to simple foods is not new: for most of the last century the practice extended to the sweetening of commercial breast milk substitutes for weaning infants (Cox 1994). This practice became ever more evident in the post-war decades and has been an intrinsic part of the burgeoning drive by international food-manufacturing conglomerates to compete in foreign markets for the supply of popular formula feeds to health authorities – as well as individual mothers of young infants. The trend has been associated with changing global practices in breast feeding and ever earlier weaning from sweet human milk: thus infants with HFI are increasingly at risk from exposure to ubiquitous but noxious sugars at times when they are most vulnerable and dependent on their mothers' vigilance. In many countries the individual dietary intake of fructose and its congeners is rising and often exceeds 100 grams daily, even in countries where cane sugar production remains as a shamefully unprofitable relic of former exploitation by colonial European nations (see Table 36.3 and Cox (2002) and accompanying references). These aspects place considerable demands on those that provide dietary advice for patients with hereditary fructose intolerance: considerable vigilance is needed to ensure that up-to-date records of the sugar content of a wide range of food, drinks and medications are maintained. Many manufacturers' labels are misleading and the term 'sugar-free' may fail to report substantial quantities of sorbitol (Ruecker

TABLE 36.3   Annual sugar consumption in 1990s (kg per capita)

Cuba	>80
Brazil	>50
Fiji	>50
Australia	>50
United States	30–40
Western Europe	30–40
China	6.5
Tropical Africa	<10
England 1700	1.7
England 1800	13.9

et al 1981); and when, for example, the item tastes strongly of peppermint, even the most sweet-aversive patient with HFI may be deceived.

MOLECULAR PATHOLOGY OF ALDOLASE B DEFICIENCY

Background

Identification of a molecular defect in the human aldolase B genes obtained from a genomic library from a man with hereditary fructose intolerance was first reported from the author's laboratory (Cross et al 1988). All six genomic aldolase B clones isolated harbored a missense mutation in which a conserved alanine was replaced by an imino acid proline: this is now commonly termed A149P (affecting codon 150 in the revised nomenclature by Antonarakis (1998)), and is due to a G->C transversion in exon 5 of the aldolase B gene. Immunochemical studies using liver and small intestinal mucosa obtained from this patient had detected the expression of aldolase B with reduced specific activity and altered molecular structure; findings that were also confirmed by radioimmunoassay in mucosal tissue obtained from all three of the children of the proband born to an unrelated and apparently healthy woman, who had been shown to be heterozygous for aldolase B deficiency (Cox et al 1982, 1983). Discovery of the mutation confirmed that the preferential loss of fructose-1-phosphate aldolase activity in hereditary fructose intolerance could be ascribed to the aldolase B locus on human chromosome 9q. Moreover, since the patient, his offspring, as well as two other unrelated patients with HFI (in whom segregation as a Mendelian autosomal recessive character was demonstrated) were homozygous for A149P – and the allele was present in another unrelated British HFI patient – it was clear that the mutation was widespread and common in patients with the disease as well as manifestly significant for diagnosis of the disease. The prevalence of the mutation in the normal British population was later confirmed by investigating a random cohort of more than two thousand dried blood spots obtained from newborn infants: the frequency of the A149P allele was 0.00659 ± 0.00248, corresponding to a heterozygote frequency of 1.32% (James et al (1996)). Recent studies in Poland (Gruchota et al 2006)) and central Europe (Santer et al 2005) give comparable frequency estimates in the general population. Molecular analysis of the human aldolase B gene in patients with intolerance of fructose from the United States (Tolan & Brooks 1992), New Zealand (Ali et al 1996), and Anatolia in Turkey (Karabulut et al 2006) demonstrate the widespread distribution of A149P globally, as previously reported from the original study of European patients (Cross et al 1990a, b). Dursun et al (2001) screened 13

Turkish patients with hereditary fructose intolerance for three common mutations. Nine were homozygous for the A149P mutation, predicting a frequency of about 55% in the disease. The A149P mutation creates a novel recognition site for the endonuclease, *Aha* II and its isoschizomers such as *Bsa* HI, and thus its facile detection in exon 5 sequences amplified from genomic DNA as a template in the polymerase chain reaction (PCR) by restriction analysis was introduced rapidly in an early diagnostic application of commercial *Taq* polymerase.

Since the identification of the A149P allele, numerous mutations of the aldolase B gene associated with HFI have been identified. Of these the most predominant appear to be A174D (alanine replaced by an aspartate, Cross et al 1990a), N334K (asparagine replaced by a lysine, Cross et al 1990b), A337V (alanine replaced by a valine, Rellos et al 1999), L256P (leucine replaced by a proline, Ali et al 1994a) and a four base-pair deltion in exon 4, Δ4E4 (Dazzo & Tolan 1990). Widespread mutant alleles of aldolase B that are frequently encountered in patients referred for the molecular identification of HFI and are thus of diagnostic utility are listed in Table 36.4. Other sporadic or apparently private mutations include point mutations and intragenic deletions and initiation codon mutations as well as other null mutations affecting splice sites (Cross & Cox 1990, Brooks et al 1991, Ali et al 1993). Although rare mutations continue to emerge from studies of newly recognized individual patients with HFI all over the world (Cross & Cox 1990, Cross et al 1990a, Ali & Cox 1995, Ali et al 1996, Esposito et al 2004, Chi et al 2007), and while their study by mutagenesis and expression may shed light on the molecular function and structure of the intact aldolase B protein (Rellos et al 1999, 2000), their detection is rarely useful for diagnostic purposes (Ali et al 1998).

Functional Effects of Disease-associated Mutations

Of the missense mutations identified in patients with HFI, many of the mutated residues are invariant in class I aldolases; one exception is A149P, in which the mutated alanine is isozyme B-specific. These findings indicate that the domains in which the residues are located contribute critically to the catalytic activity or structural conformation of the intact wild-type enzyme.

To investigate the effects of the common A149P mutant and other missense variants on enzymatic properties, the cognate cDNA sequences generated from wild type by site-directed mutagenesis were inserted into plasmid vectors that direct heterologous expression of human aldolase B after transformation in *E. coli* (Cox 1994, Brooks & Tolan 1994, Rellos et al 1999, 2000, Malay et al 2005). These studies confirmed as expected the intrinsic loss of function of fructaldolase activity: the expressed A149P protein had markedly reduced activity towards its specific substrate, fructose-1-phosphate (Cox 1994). Thus the A149P mutation has drastic effects on enzyme structure. The protein is thermolabile (Rellos et al 2000, Malay et al 2005) and its altered stability explains the modified electrophoretic mobility, isoelectric focusing properties and reactions of partial identity with isozyme-specific antisera that had been previously reported (Cox et al 1983a, 1983b).

The A149P is the prototype of a structural mutant aldolase; the aldolase B variants A174D, L256P and N334K are other members of this class, since the homotetramers readily dissociate into subunits with greatly impaired enzymatic activity (Rellos et al 2000). The expression of the C134R mutant also retained activity in the bacterial expression system, suggesting that structural instability of the partially active enzyme is responsible for the disease (Brooks & Tolan 1994). Another class of variant aldolases occurring in HFI are best described as catalytic mutants: W147R, R303W and A337V retain the tetrameric structure but possess altered catalytic properties.

The three-dimensional structure of rabbit muscle aldolase A was solved by Sygusch (1987) and further modeling and X-ray analysis has been carried out on the highly homologous human aldolase B and its natural and synthetic variants (Rellos et al 2000, Dalby et al 2001, Malay et al 2005). The mutations C134R, W147R, A149P and A174D on exon 5 are all located in a pocket involved in the binding of substrate (see Figure 36.4; Sygusch et al 1987, Blom & Sygusch 1997) and it seems likely that they would disrupt key residues (K 147 and R 149) implicated in the binding of the C1-phosphate in the active site (Cross et al 1988, Lai et al 1974, Hartmann & Brown 1976). Diffraction studies carried out at atomic resolution in rabbit aldolase A suggest that K147 is sufficiently close to serve as a donor or acceptor of protons during the formation of the carbinolamine intermediate which participates critically in the cleavage or condensation of the C3–C4 carbon–carbon bond in the aldolase reaction (Blom & Sygusch 1997); when this residue is itself mutated, formation of the reaction intermediates is prevented (Morris & Tolan 1994).

The amino acids implicated in the N334K and A337V natural variants of aldolase B arise from mutations in exon 9 of

TABLE 36.4 Widespread aldolase B mutations of diagnostic utility

Common name[1]	Systematic designation	Mutated codon[2]
A149P	c.448G→C	A150
A174D	c.524C→A	A175
N334K	c.1005C→G	N335
R303W	c.910C→T	R304
Δ4E4	c.360–363 del CAAA	N120,K121

[1]Single letter amino acid code
[2]Residue designated according to nomenclature promulgated in S.E. Antonarakis (1998). *Human Mutation* **11**, 1–3.

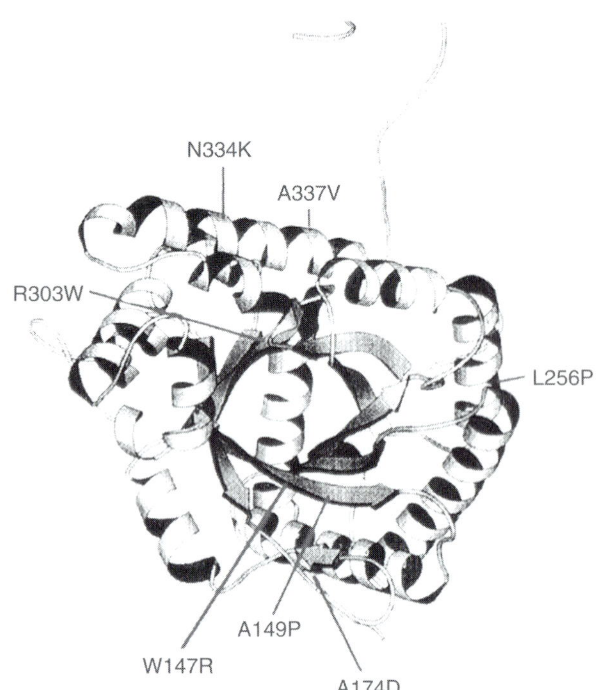

FIGURE 36.4   Model of rabbit aldolase B structure depicting the location of residues mutated in natural missense variants associated with hereditary fructose intolerance (from Rellos et al 2000). (see also Plate 83)

the human gene and are located in a short α-helix where they affect the mobility of the highly flexible carboxyterminal tail which is considered to lie over the active site pocket and promote access of substrate molecules to the enzyme (Sygusch et al 1987, Blom & Sygusch 1997, Lafrance-Vanasse & Sygusch 2007). The residues in this region show considerable differential variation between isozymes A, B, and C and the region is strongly implicated in the functional differentiation of the class I aldolases (Takahashi et al 1989, Kitajima et al 1990). Modeling studies have suggested that the conserved wild type residue R304 encoded on exon 8 takes part in the binding of the phosphate at C6 of the fructose-1,6-bisphosphate substrate (Gamblin et al 1991), but an examination of aldolase A at atomic resolution (1.9 Å) reveals that the arginine interacts with a glutamate at position 355 on the carboxyterminal region to form a hydrogen bond with the C2 carbonyl oxygen of dihydoxyacetone phosphate derived from the C1 moiety of both hexose substrates; the residue appears to stabilize the keto form of the incoming substrate at the active site. This function would be undoubtedly disrupted by replacing the positively charged arginine residue by the large apolar tryptophan as in the HFI-associated mutation, R303W.

Class I aldolase function as tetramers, formation of which involves strong hydrophobic interactions at the subunit interfaces. Expression studies after site-directed mutagenesis, combined with biochemical purification employing molecular exclusion chromatography, have shown that several natural aldolase B variants fall into the category of catalytically active but unstable mutants. The L256P aldolase B variant, in which the drastically mutated hydrophobic residue lies on the surface of the subunit, provides a clear example of disease-associated mutations in this category (Ali et al 1993, Rellos et al 2000); the distorted interface leads to misalignment of overlapping prolines found at the turns of α-helices which form a hydrophobic pocket across adjacent subunits promoting their attachment and stabilization of the tetramer. Intact L256P homotetramers were the most active of all the mutant aldolases but dissociate rapidly into monomers with minimal activity; A149P and L256P showed markedly decreased thermostability. It appears that the formation of tetramers contributes to aldolase B stability as well as catalytic activity by constraining the configuration of each subunit to promote cooperative interactions within the intact quaternary structure (Rellos et al 2000).

Much has been learnt from the mapping of mutations onto existing crystal structures for the class I aldolases but ultimately the complete molecular understanding of the differentiated function of aldolase B will depend on the solution of individual natural variants and engineered synthetic mutants combined with computer modeling studies – and informed by biochemical analysis.

DIAGNOSIS OF HEREDITARY FRUCTOSE INTOLERANCE

Suspicious Clinical Features

The diagnosis of HFI should be suspected in infants after weaning from breast milk with vomiting, failure to thrive, liver disease, renal tubular acidosis with or without Fanconi's syndrome, and rickets; those with hypoglycemia causing post-prandial coma or seizures are also candidates. Pyloric stenosis and other causes of metabolic disturbance may be misdiagnosed and the prominent hepatomegaly and renal tubular disease may lead to a consideration of Reye's syndrome, galactosemia, and, with hypoglycemia, the presence of a glycogen storage disorder. In older children, psychological difficulties with behavioral disturbances and excess parental concern are often considered and the true cause of the symptoms only comes to light as a result of investigations for suspected food allergy. In adults, worsening abdominal pain, faintness after meals or a frank hypoglycemic episode lead to the diagnosis; often the patient gives a vivid account of the specific dietary idiosyncrasies and the diagnosis results from further biochemical testing after the history eventually is seen to be unequivocal. At this stage there may be evidence of stunted development, vitamin deficiencies (e.g. lack of ascorbic or folic acid), hepatic injury and impaired renal function; nephrocalcinosis may be apparent. A rigorous dietary history, exploring as far back as possible, is often decisive in the quest for a definitive diagnosis of HFI; third party evidence from a parent or sibling frequently elucidates the long-standing nature and extent of the dietary intolerance.

Biochemical Testing

Hitherto the unequivocal diagnosis of HFI has depended on the demonstration of deficient (<10%) fructose 2-2 1-phosphate aldolase activity in tissue biopsy specimens. Normally hepatic biopsy has been preferred but biopsy samples of small intestinal mucosa or, more rarely, needle biopsy of the kidney is possible. The enzymatic defect is associated with a modest reduction of fructose-1,6-bisphosphate aldolase activity so that the activity ratio of the measured with the fructose bisphosphate compared with the monophosphate substrates is usually much greater than 2, indicating the residual activity of the aldolase isozymes A and/or C with selective loss of activity attributable to aldolase B.

Controlled Biochemical Challenge with D-fructose

In the absence of enzymatic assays of tissues samples obtained by biopsy, until recently, the intravenous fructose tolerance test has served as an operational but cumbersome method of diagnosing HFI.

The procedure is now difficult to conduct because of the limited commercial availability of suitable fructose solutions for infusion and the elaborate difficulties in ensuring that the challenge can be conducted under conditions of hospital monitoring where, if necessary, procedures for rescue of the adverse metabolic effects, such as hypoglycemia or acute hypokalemia, can be undertaken safely. The lack of availability of D-fructose for intravenous use *rarely, if ever*, justifies use of an oral challenge: oral administration of 0.25 g/kg of this sugar or an equivalent dose of sucrose (0.5 g/kg) to a patient with HFI usually provokes violent reactions with pronounced gastrointestinal effects, a severe metabolic disturbance and, sometimes, a shock-like state. Clearly such reactions are dangerous and are unlikely to promote the professional trust required for effective aftercare and long-term monitoring of patients with this disease.

Administration of 0.25 g/kg body weight as a bolus intravenous infusion of 10–20% w/v D-fructose to a fasting individual over 5–10 minutes is sufficient to induce symptomatic hypoglycemia and measurable hypophosphatemia within 20–60 minutes (Figure 36.5). After establishing

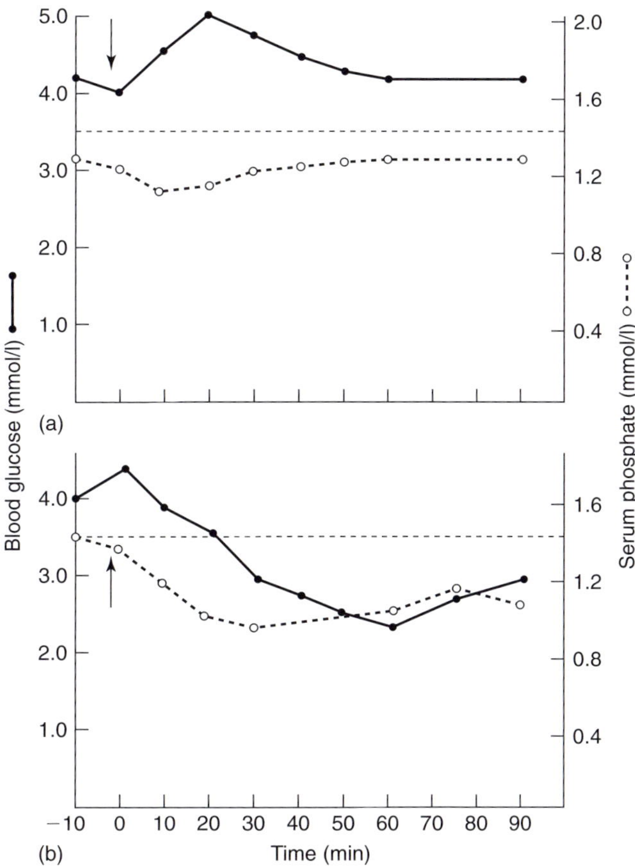

FIGURE 36.5 Intravenous fructose tolerance tests in (a) a 39-year-old woman with alcohol-induced hypoglycemic symptoms given 0.27 g D-fructose/kg body weight as indicated by the arrow and (b) a 39-year-old woman with hereditary fructose intolerance, later confirmed by liver aldolase assay and molecular analysis of the aldolase B gene (0.25 g D-fructose/kg) after Cox 1988.

at least two baseline determinations, blood samples are generally drawn at 10–15-minute intervals after the bolus infusion.

In children and infants the adverse metabolic effects almost invariably occur within 30 minutes but attenuated responses are the general rule in adult patients, who may not develop hypoglycemia until 60 minutes after the infusion; abdominal pain, faintness, and feelings of impending doom are frequently experienced and these are more common and profound in those patients with true aldolase B deficiency, although mild and transient abdominal symptoms are quite often experienced in subjects without hereditary fructose intolerance. Thus the appearance of symptoms after administration of fructose should never be interpreted as a diagnostic indicator; biochemical validation is always necessary.

A positive fructose tolerance test is, for practical purposes, diagnostic of HFI. However the very rare disorder of hepatic gluconeogenesis, fructose-1,6-bisphosphatase deficiency, is associated with episodic severe lactic acidosis and mild-to-moderate hypoglycemia; severe homeostatic failure occurs on starvation (unlike in HFI, where starvation is well tolerated). After challenge with fructose, patients with fructose-1,6-bisphosphatase deficiency also develop hypoglycemia but this itself tends to be moderate to mild. Some patients with this disorder also develop mild aversion to carbohydrate-rich foods but the degree of distaste sugar is not usually comparable to that which characterizes patients with HFI. The primary diagnosis of fructose-1,6-bisphosphatase deficiency usually depends on specific biochemical analysis of tissue biopsy specimens in a laboratory well experienced in micro-analytical enzymatic procedures.

Noninvasive Diagnostic Methods

Magnetic Resonance Spectroscopy

A cumbersome but interesting approach to the in vivo diagnosis of disturbed sugar metabolism in HFI has been the use of magnetic resonance spectroscopy based on the signals obtained from ^{31}P atoms in organic phosphate species present in hepatic tissue. This procedure was used by Oberhaensli and colleagues to demonstrate an increase in the concentration of sugar phosphates, attributable to the presence of fructose 2-2 1-phosphate ester, and a diminished signal from inorganic phosphate in the liver of HFI patients after a fructose load (Oberhaensli et al 1987). Although mass spectroscopy in vivo is an attractive means for investigative research in patients with metabolic disorders such as hereditary fructose intolerance (Boesiger et al 1994) the lack of availability of large spectrometers for human application in all but the most specialized centers means that the technique is unlikely to enter routine practice; moreover the practical difficulties of containing infants and young children in the apparatus without strong sedation also militates against further diagnostic exploration.

Molecular Diagnosis of HFI by Direct Analysis of Genomic DNA

Emergence of the polymerase chain reaction for the facile amplification of DNA sequences has had far-reaching consequences for the direct diagnosis of HFI. Genomic DNA is readily obtained as a template for specific amplification of target sequences and definitive molecular analysis of the genes of diagnostic interest. The template DNA can be extracted from somatic cells in small blood samples, dried blood spots for neonatal screening programs, cultured skin fibroblasts, exfoliated cells present in mouthwashes – or even samples of fixed tissue previous obtained by biopsy or stored in pathological archives. Development of stratagems for the diagnosis of HFI by the molecular analysis of DNA is clearly justified because the other available techniques are inconvenient, cumbersome and expensive or, in the case of visceral biopsy, carry appreciable risks. Biopsy procedures and, when positive, even the formal intravenous fructose tolerance test may also be distressing.

Hierarchical Screening

Although several mutations in the aldolase B gene have been shown to be widespread and are of immediate use in hereditary fructose intolerance (Table 36.4), a comprehensive catalog of the global distribution of these and other mutations in different populations would be advantageous for diagnostic testing. Mutations are assigned in this chapter according to the guidelines for nomenclature promulgated by Antonarakis (1998) (ariel.ucs.unimelb.edu.au:80/-cotton/antonara.htm). Three missense mutations in the aldolase B gene, A150P, A175D and N335K, and one small deletion, Δ4, are prevalent in European related populations; the mutant alleles designated A338V, R304W and R60ter are also widely distributed (Ali et al 1998, Santer et al 2005). Other mutations responsible for the disease, including large intragenic deletions, tend either to be private or restricted to small regions or localities (Cross & Cox 1990, Ali et al 1993, 1994a, 1994b, 1996, Brooks & Tolan 1994, Esposito et al 2004, Chi et al 2007). The pattern of mutation frequencies in HFI patients resembles the familiar prototype of diseases caused by defects other human genes in which a few mutant alleles are prevalent but most are rare. Thus diagnostic methods for HFI by searching for disease-causing alleles depend upon, and are limited by, the confidence with which the distribution of these mutant alleles in a given population can be predicted (Ali et al 1998).

The striking prevalence of the missense mutation A150P in Britain and Northern Europe and its wide global distribution is very remarkable: in several countries, including those with a strong European colonial history, the high proportion of individuals of appropriate ancestry has rendered screening for the presence of this allele useful for the molecular diagnosis of HFI (Cross & Cox 1988, Cross et al 1990c). The A150P variant has so far been detected exclusively on a single informative human aldolase B haplotype, characterized by the presence of linked noncoding polymorphisms in intron 8 of the gene (c->t at nucleotide 84 and a->g at nucleotide 105). These biallelic single base-pair polymorphic variants were found to be in absolute linkage with themselves; 47% of the normal population were heterozygous at these loci but allele specific hybridization studies showed that all of 15 A150P homozygous patients with HFI were also homozygous for the intron 8 84T/105G alleles, indicating absolute linkage with a single haplotype (Brooks & Tolan 1993). Thus the prevalence of the A150P mutation is probably explained by establishment of a founder group derived from a single mutation event in their common ancestor during the early expansion of the European population. As a consequence of latter migration, this allele occurs as the principal disease-associated mutation of aldolase B throughout Europe, the Americas and Australasia. Further combined clinical and genetic studies have shown that hierarchical screening for the other mutations can, in the appropriate population setting, provide a high diagnostic yield for patients in whom hereditary fructose intolerance is suspected.

To determine whether an individual in whom hereditary fructose intolerance is suspected has a molecular lesion in aldolase B ultimately may demand definitive sequencing of the entire locus, but a focused hierarchical scheme, based on the distribution of mutant alleles causing HFI in the relevant population, may expedite practical diagnostic genotyping. Thus for patients with suspected fructose intolerance who are of European origin, we would recommend that initial screening for the previously reported disease-causing mutations (A150P, A175D, N335K and Δ4) should be undertaken by PCR-based analysis of genomic DNA using restriction endonuclease digestion, amplification refractory mutation analysis (ARMS) or mutation-specific oligonucleotide hybridization. Several methods to simplify the strategy have been published (Cross & Cox 1988, Lau & Tolan 1999, Kullberg-Lindh et al 2002), but semiautomated reverse hybridization methods using a representative panel of mutation-specific oligonucleotide probes in a multiplex system have the capacity to expedite molecular diagnosis (Kriegshäuser et al 2007).

Detection of a single copy of a recognized mutant aldolase B allele in an individual patient with symptoms suggestive of fructose intolerance should be taken to indicate the suspected diagnosis of HFI until proved otherwise. Indeed such a finding would mandate an intensified search for the reciprocal (allelic) mutation at the aldolase B locus.

In subjects of European ancestry, a protocol that excluded those who were wild type at these loci could be justified under most circumstances, and if the subjects under investigation were unrelated by blood, it could be argued that further molecular diagnostic procedures should normally await the outcome of an intravenous fructose tolerance test and/or enzymatic assay of fructaldolase activities in tissue biopsy specimens.

Where there is a high degree of consanguinity in particular communities, a more complete ascertainment of the aldolase B gene in patients with symptoms attributable to HFI may be justified: direct sequencing of all the aldolase B exons, including lariat and slice site signal sequences clearly provides a systematic means to identify rare and/or private mutations responsible for disease and inherited by descent from a common ancestor. Evidently, as health services increasingly recognize the value of definitive molecular identification of human diseases, and as sequencing facilities are increasingly dedicated to the study and practice of clinical genetics, automated sequencing of the human aldolase B gene will be an attractive means for pursuing the diagnosis of hereditary fructose intolerance. When two mutant alleles of aldolase B have been detected in a proband, the information may be used to advantage in the screening of siblings for HFI using either cord blood or dried blood spots obtained from Guthrie card as the validated source of template genomic DNA to establish a genetic diagnosis of HFI before exposure to harmful fructose. By these means it should be possible to offer useful advice to the parents of newborn children at risk from symptomatic HFI before they are weaned; we already have preliminary indications that such a strategy is well received by parents of asymptomatic children genetically predisposed to develop fructose intolerance (Ali et al 1998).

TREATMENT AND PROGNOSIS

Once the hereditary fructose intolerance is suspected, a strict diet to exclude fructose, sucrose and sorbitol should be instituted. This is not a bar to the pursuit of the definitive diagnosis by enzymatic assay of fructaldolase activity in fresh or frozen tissue biopsy specimens or, preferably, by molecular analysis of the aldolase B gene. Where doubt exists, for example during the investigation of an infant or child with serious metabolic abnormalities, it is often wise to introduce the exclusion diet presumptively: doing so will not interfere with diagnostic tests and an early response may provide additional evidence that fructose is a key noxious factor in the illness. The regimen extends to all drinks and medicines (Ruecker et al 1981) and detailed professional advice from an experienced metabolic dietitian will be needed from the outset (Table 36.5).

The objective of the treatment is to reduce the intake of fructose and its congeners to below 40 mg/kg body weight daily, since in careful clinical studies, normal growth and metabolism was restored when fructose consumption was reduced to these values, usually amounting to less than 3 grams of monosaccharide per day for a full-grown adult (Mock et al 1983). Many otherwise healthy adult patients with HFI remain completely well while ingesting no more than 1.5 grams of fructose equivalents daily – a rigorous consumption target (Newburn et al 1980).

TABLE 36.5   Foods and drinks containing fructose, sucrose, and sorbitol

Diabetic chocolates and 'sugar-free' candies
Most carbonated fruit-based drinks
Honey, jam, preserves and jellies
Sugar-coated cereals
Fruit, fruit juices and dates
Maple syrup, desserts and puddings
Cakes, pies, cookies
Crackers and biscuits
Ice-cream and sherbet
Sauces
Processed meats
Mayonnaise and salad dressings
Canned vegetables
Stored potatoes
Carrots, onions, leeks
Sweet maize and sweet potatoes
Nuts
Candy and chocolate

Exclusion of fructose in a patient with HFI rapidly corrects the hypoglycemic symptoms; the vomiting and abdominal pain resolve within a matter of days and many of the accompanying metabolic derangements improve. In growing infants and children, the failure to thrive is corrected and catch-up growth occurs briskly, with restoration of normal body mass as the general health improves. Patients in whom a metabolic acidosis due to disease of the proximal renal tubules is present will require additional measures to correct acid–base balance and depletion of electrolytes. Similarly, recovery in patients with impaired function and enlargement of the liver may extend over many weeks or months and clearly, when cirrhosis is established, complete resolution will not occur. However, several reports document salutary responses of severely ill pediatric patients with HFI to withdrawal of the noxious sugars (Perheentupa et al 1972, Odièvre et al 1978).

Patients with acute complications of HFI, such as renal tubular acidosis, need the appropriate specific management of the emergent situation, such as sodium citrate or bicarbonate and parenteral supplementation of potassium to correct hypokalemia. The presence of vitamin D deficiency in these circumstances will require generous supplementation with vitamin D analogs such as calcitriol, given that the toxic effects of fructose on the renal tubule cause loss of the vitamin in the glomerular filtrate due to impaired operation of the specific re-absorption machinery for 25-hydroxycholecalciferol by the megalin pathway. Treatment with vitamin D analogs mandates appropriate blood monitoring of calcium, phosphate, and parathyroid hormone concentrations, as well as alkaline phosphatase activity. Severe and persistent hypophosphatemia might theoretically demand phosphate supplementation but the severe proximal renal tubular acidosis should respond briskly to exclusion of the

noxious sugars in the diet (see above) and the measures set out above are likely to be needed only as temporary interventions while the metabolic disorder is definitively corrected on completion of the withdrawal of fructose.

It is clear that in nearly all patients with the disease, exclusion of fructose brings remarkable benefit and in several the effects have been viewed as near miraculous – particularly in infants and young children with a neglected diagnosis and who would otherwise be beyond hope. With the institution of strict dietary measures, the disorder is compatible with a normal quality and duration of life. The first patient ever to be described with the disorder last reported to the author that she was perfectly fit when she was 67 years old and a grandmother; several individuals, surviving on the basis of self-imposed dietary restrictions and not previously recognized, have been observed in their seventh and eighth decades (Brauman et al 1971). It should be recognized, however, that throughout life, inadvertent exposure to fructose, sucrose, and sorbitol, accompanied by the rapid resurgence of symptoms, will occur on occasion and every effort should be made to reduce the frequency of such episodes to a minimum. For these reasons, scrutiny of all the foods and drinks consumed and the professional advice of a dietitian, including provision of explicit diet sheets, is desirable for every patient in whom the diagnosis is confirmed (Bell & Sherwood 1987).

Many examples of the harmful effects of medication in patients with HFI have been reported and considerable care is required to determine the suitability of all medicines, since sorbitol and sucrose are used to render pediatric suspensions and syrups palatable for non-intolerant individuals; they are also common excipients and coatings for tablets. The fructose-exclusion diet is deficient in the water-soluble vitamins, ascorbic acid (vitamin C) and folic acid; these should be supplemented daily in a suitable formulation (Ketovite™ tablets and liquid supplement are licensed in the United Kingdom).

In the early phases of treatment, it is advisable for the patient to carry warning notices or jewelry for the diagnosis and the appropriate treatment for hypoglycemia (milk or glucose for oral use; glucose by injection, as needed).

PREVENTION OF FRUCTOSE INTOLERANCE BY GENETIC SCREENING

Given the extraordinary abundance of sucrose and fructose in modern foods and drinks, and their resurgence as sweeteners, there is a strong case for identifying individuals in the population who are genetically predisposed to a fatal nutritional disease. Since restriction of exposure to harmful sugars to prevent injury can be undertaken before weaning, neonatal screening would be appropriate. Neonatal screening for hereditary fructose intolerance can be undertaken by molecular analysis of human aldolase B gene sequences present in DNA extracted from cord blood or the dried blood spots available in most countries as a result of mass screening using Guthrie cards for phenylketonuria and other disorders that are preventable in the immediate post-natal period (James et al 1996).

To determine whether screening newborn infants would be practicable, allele-specific dot blot hybridization followed by confirmatory restriction analysis was used in the author's laboratory to investigate the frequency of A150P in a random cohort of neonatal dried blood samples in England. Twenty-seven A150P heterozygotes were identified among 2050 subjects born in the years 1994–95 (1.32 ± 0.49%) which, on the basis of the Hardy-Weinberg equation, predicts a frequency of 1 in 23 000 A159P homozygotes. Given the distribution of mutant aldolase B alleles in the British population, the predicted birth frequency for HFI would be 1 in 18 000 live births and comparable to the incidence of phenylketonuria – another potentially treatable (but perhaps less tractable) nutritional disorder, for which mass neonatal screening has been adopted in all developed countries (James et al 1996).

When combined with preexisting programs that include appropriate counseling and follow-up, mass genetic screening using methods for high-throughput analysis of genomic DNA obtained from the neonatal population for selected HFI alleles is robust and affordable. Since dietary exclusion completely abrogates clinical disease and given the prevalence of A150P and other widespread alleles causing HFI in the populations so far examined (e.g. A175D and N335K in Southern and Eastern Europe, respectively), there is a very strong case for introducing genetic screening at birth to prevent disease (Santer et al 2005; Gruchota et al 2006). Although the ability of nutritional advice to guarantee lifetime health in an individual with HFI who has not been exposed to fructose in infancy and childhood is formally unproven, it should be possible to enhance the probability of success by controlled challenge with noxious sugars at appointed times during development and maturation, thus inducing protective aversion behavior. Without pilot studies in operation, however, the utility of such a screening program must remain theoretical – despite the obvious likelihood that it would reduce suffering and death in numerous infants born with an unrecognized nutritional disease every year.

ACKNOWLEDGMENTS

Much of the research reported from the author's laboratory has been conducted in collaboration with outstanding scientists and physicians. Accordingly, I acknowledge with gratitude the efforts particularly of the late Martin O'Donnell, Drs Michael Camilleri, Manir Ali and Peter Rellos, as well as the pioneer of our molecular genetic studies, Dr (now Professor) Nick Cross. We were supported by grants from the Medical Research Council and the Wellcome Trust.

References

Ali M, Cox TM. Diverse mutations in the aldolase B gene that underlie the prevalence of hereditary fructose intolerance. Independent segregation of four mutant alleles in ten affected members of a large kindred. Am. J. Hum. Genet. 1995; 56: 1002–5.

Ali M, James CL, Cox TM. A newly-identified aldolase B splicing mutation (G->C, 5′ intron 5) in hereditary fructose intolerance from New Zealand. Hum. Mutat. 1996; 7: 155–7.

Ali M, Rellos P, Cox TM. Hereditary fructose intolerance. J. Med. Genet. 1998; 35: 353–65.

Ali M, Rosien U, Cox TM. DNA diagnosis of fatal fructose intolerance from archival tissue. Q. J. Med. 1993; 86: 25–30.

Ali M, Sebastio G, Cox TM. Identification of a novel mutation (Leu 256->Pro) in the human aldolase B gene associated with hereditary fructose intolerance. Hum. Mol. Genet. 1994a; 3: 203–4.

Ali M, Tuncman G, Cross NCP, Vidailhet M, et al. Null alleles of the aldolase B gene in patients with hereditary fructose intolerance. J. Med. Genet. 1994b; 31: 101–5.

Antonarakis SE. Recommendations for a nomenclature system for human gene mutations. Nomenclature Working Group. Hum. Mutat. 1998; 11: 1–3.

Baerlocher K, Gitzelmann R, Steinmann B, Gitzelmann-Cumurasmy N. Hereditary fructose intolerance in early childhood: a major diagnostic challenge. Survey of 20 symptomatic cases. Helv. Paediatr. Acta 1978; 132: 605–8.

Barry RG, St. Columb SR, Magner JW. Hereditary fructose intolerance in parent and child. J. Irish Med. Assoc. 1968; 61: 308–10.

Bell L, Sherwood WG. Current practices and improved recommendations for treating hereditary fructose intolerance. J. Am. Diet. Assoc. 1987; 87: 721–8.

Black JA, Simpson K. Fructose intolerance. BMJ 1967; 4: 138–41.

Blom N, Sygusch J. Product binding and role of the C-terminal region in class I D-fructose-1,6-bisphosphate aldolase. Nat. Struct. Biol. 1997; 4: 36–9.

Bode JC, Zelder O, Rumpelt HJ, Wittkamp U. Depletion of liver adenosine phosphates and metabolic effects of intravenous infusion of fructose or sorbitol in man and in the rat. Eur. J. Clin. Invest. 1973; 3: 436–41.

Boesiger P, Buchli R, Meier D, Steinmann B, Gitzelmann R. Changes of liver metabolite concentrations in adults with disorders of fructose metabolism after intravenous fructose by 31P magnetic resonance spectroscopy. Pediatr. Res. 1994; 36: 436–40.

Brauman J, Kentos P, Frisque P, Gepts W, Verbanck M. Hereditary fructose intolerance in an 83-year-old woman. Acta Clin. Belg. 1971; 26: 65–77.

Brooks CC, Buist N, Tuerck J, Tolan DR. Identification of a splice-site mutation in the aldolase B gene from an individual with hereditary fructose intolerance. Am. J. Hum. Genet. 1991; 49: 1075–81.

Brooks CC, Tolan DR. Association of the widespread A149P hereditary fructose intolerance mutation with newly identified sequence polymorphisms in the aldolase B gene. Am. J. Hum. Genet. 1993; 52: 835–40.

Brooks CC, Tolan DR. A partially active mutant aldolase B from a patient with hereditary fructose intolerance. FASEB J. 1994; 8: 107–13.

Bryan J. When a spoonful of sugar makes the patient feel ill. London: The Independent, December 27th 1988, p. 13.

Cain AR, Ryman BE. High liver glycogen in hereditary fructose intolerance. Gut 1971; 12: 929–32.

Chambers RA, Pratt RTC. Idiosyncrasy to fructose. Lancet 1956; II: 340.

Chi ZN, Hong J, Yang J, et al. Clinical and genetic analysis for a Chinese family with hereditary fructose intolerance. Endocrine 2007; 32: 122–6.

Collins JE, Leonard JV. The dietary management of inborn errors of metabolism. Hum. Nutr. Appl. Nutr. 1985; 39: 255–72.

Cornblath M, Rosenthal IM, Reisner SH, Wybregt SH, Crane RK. Hereditary fructose intolerance. N. Engl. J. Med. 1963; 269: 1271–8.

Cox TM. Hereditary fructose intolerance. Quart. J. Med. 1988; 68: 585–94.

Cox TM. An independent diagnosis: A treatable metabolic disorder is diagnosed by molecular analysis of human genes. BMJ 1990a; 300: 1512–14.

Cox TM. Fructose intolerance: heredity and the environment. In: Randle PJ, Bell J, Scott J, eds. Genetics and Human Nutrition. London: John Libbey and Company, 1990b: pp. 81–92.

Cox TM. Iatrogenic deaths in hereditary fructose intolerance. Arch. Dis. Child. 1993; 69: 413–15.

Cox TM. Aldolase B and fructose intolerance. FASEB J. 1994; 8: 62–71.

Cox TM. Therapeutic use of fructose: professional freedom, pharmacovigilance and Europe. Q. J. Med. 1995; 88: 225–7.

Cox TM. The genetic consequences of our sweet tooth. Nat. Rev. Genet. 2002; 3: 481–7.

Cox TM, Camillieri M, O'Donnell MW, Chadwick VS. Pseudodominant transmission of fructose intolerance in an adult and three offspring: heterozygote detection by intestinal biopsy. N. Engl. J. Med. 1982; 307: 537–40.

Cox TM, O'Donnell MW, Camilleri M, Burghes AH. Isolation and characterization of a mutant liver aldolase in adult hereditary fructose intolerance. Identification of the enzyme variant by radioassay in tissue biopsy specimens. J. Clin. Invest. 1983a; 72: 201–13.

Cox TM, O'Donnell MW, Camilleri M. Allelic heterogeneity in adult hereditary fructose intolerance. Detection of structural mutations in the aldolase B molecule. Mol. Biol. Med. 1983b; 1: 393–400.

Cox TM, O'Donnell MW, Sochor M, Camilleri M, McLean P. Hereditary fructose intolerance: presence of a potent inhibitor of fructose-2-2 1-phosphate aldolase in the liver. Biochem. Soc. Trans. 1983c; 11: 202–3.

Craig GM, Crane CW. Lactic acidosis complicating liver failure after intravenous fructose. BMJ 1971; 4: 211–12.

Cross NCP, Cox TM. Detection of molecular defects in human aldolase B. A study of patients with hereditary fructose intolerance. In: Albertini A, ed. Molecular Probes: Technology and Medical Applications. Fondazione Giovanini Lorenzini. New York: Raven Press Inc., 1988: pp. 157–63.

Cross NCP, Cox TM. Molecular analysis of aldolase B Genes in the diagnosis of hereditary fructose intolerance in the United Kingdom. Q. J. Med. 1989; 73: 1015–20.

Cross NCP, Cox TM. Partial aldolase B gene deletions in hereditary fructose intolerance. Am. J. Hum. Genet. 1990; 47: 101–6.

Cross NCP, de Franchis R, Sebastio G, et al. Molecular analysis of aldolase B genes in hereditary fructose intolerance. Lancet 1990a; 335: 306–9.

Cross NCP, Stojanov LM, Cox TM. A new aldolase B variant, N334K, is a common cause of hereditary fructose intolerance in Yugoslavia. Nuc. Acids Res. 1990b; 18: 1925.

Cross NCP, Tolan DR, Cox TM. Catalytic deficiency of human aldolase B in hereditary fructose intolerance caused by a common missense mutation. Cell 1988; 53: 881–5.

Curran BJ, Havill JH. Hepatic and renal failure associated with amiodarone infusion in a patient with hereditary fructose intolerance. Crit. Care. Resusc. 2002; 4: 112–15.

Dalby AR, Tolan DR, Littlechild JA. The structure of human liver fructose-1,6-bisphosphate aldolase. Acta Crystallogr. D Biol. Crystallogr. 2001; 57: 1526–33.

Danks DM, Connellan JM, Solomon JR. Hereditary fructose intolerance: report of a case and comments on the hazards of fructose infusion. Aust. Paedtr. J. 1972; 8: 282–6.

Dazzo C, Tolan DR. Molecular evidence for compound heterozygosity in hereditary fructose intolerance. Am. J. Hum. Genet. 1990; 46: 1194–9.

De Vroede M, Mozin M-J, Cadranel S, Loeb H, Heinmann R. Découverte d'une fructosémie à l'occasion d'une insouffisance hépatique aigue chez un infant de 16 mois. Pédiatrie 1980; 35: 353–8.

Dubois R, Loeb H, Malaisse-Lagae F, Toppet M. Étude clinique et anatomo-pathologique de deux cas d'intolérance congenitale au fructose. Pédiatrie 1965; 20: 5–14.

Dubois R, Loeb H, Ooms HA, Gillet P, Dartman J, Champenois A. Study of a case of functional hypoglycaemia caused by intolerance to fructose. Helvet. Paediatr. Acta 1961; 16: 90–6.

Dursun A, Kalkanoglu HS, Coskun T, et al. Mutation analysis in Turkish patients with hereditary fructose intolerance. J. Inherit. Metab. Dis. 2001; 24: 523–6.

Esposito G, Santamaria R, Vitagliano L, et al. Six novel alleles identified in Italian hereditary fructose intolerance patients enlarge the mutation spectrum of the aldolase B gene. Hum. Mutat. 2004; 24: 534.

Froesch ER, Prader A, Labhart A, Stuber HW, Wolf HP. Die hereditäre Fructose intoleranz, einer bischer nicht bekannte kongenitale Stoffwechselstörung. Schweiz. Med. Wochenschr. 1957; 87: 1168–71.

Froesch ER, Wolf HP, Baitsch H, Prader A, Labhart A. Hereditary fructose intolerance: an inborn defect of hepatic fructose-1-phosphate splitting aldolase. Am. J. Med. 1963; 34: 151–67.

Galaske RG, Burdelski M, Brodehl J. Primär polyurisches Nierenversagen und acute gelbe Leberdystrophie nach Infusion von Zuckeraustauschstoffen im Kindersalter. Dtsch. Med. Wochenschr. 1986; 111: 978–83.

Gentil C, Colin J, Valette AM, Alagille D, Lelong M. Etude du metabolisme glucidique au cours d'intolérance héréditaire au fructose. Essai d'interprètation de l'hypoglucosemie. Rev. Fr. Etud. Clin. Biol. 1964; 9: 596–607.

Gitzelmann R, Baerlocher K. Vorteile und Nachteile der Fructose in der Nahrung. Pädiatr. Fortbildk. Praxis 1973; 37: 40–55.

Gruchota J, Pronicka E, Korniszewski L, et al. Aldolase B mutations and prevalence of hereditary fructose intolerance in a Polish population. Mol. Genet. Metab. 2006; 87: 376–8.

Guery MJ, Douillard C, Marcelli-Tourvieille S, Dobbelaere D, Wemeau JL, Vantyghem MC. Doctor, my son is so tired…

about a case of hereditary fructose intolerance. Ann. Endocrinol. (Paris) 2007; 68: 456–9.

Hackl JM, Balogh D, Kunz F, Dworzak E, Puschendorf B. Postoperative Fructoseinfusionen bei wahrscheinlich hereditärer Fructosintoleranz. Wein Klin. Wochenschr. 1978; 90: 237–40.

Hartman FC, Brown JB. Affinity labelling on a previously undetected essential lysyl residue in class I fructose bisphosphate aldolase. J. Biol. Chem. 1976; 251: 3057–62.

Heinz F, Schlegel F, Krause PH. Enzymes of fructose metabolism in human kidney. Enzyme 1975; 19: 85–92.

Hers H-G. Le Métabolisme du Fructose. Bruxelles: Editions Arsica, 1957.

Hers H-G, Joassin G. Anomalie de l'aldolase hépatique dans l'intolérance au fructose. Enzymol. Clin. Chem. 1961; 1: 4–14.

Hers H-G, Kusaka T. Le métabolisme du fructose-1-phosphate dans le foie. Biochim. Biophys. Acta 1953; 11: 427–30.

Higgins RB, Varney JK. Dissolution of renal calculi in a case of hereditary fructose intolerance and renal tubular acidosis. J. Urol. 1966; 95: 291–6.

Hübschmann K, Cobet G. Beitrag zur hereditären Fructosintoleranz. Dtsch. Med. Wochenschr. 1964; 89: 938–42.

Hütteroth TH, Wagner R, Knolle J. Toxische leberschädigung nach überdosierung von kohlenhydraten. Med. Klin. 1977; 72: 703–7.

Jaeken J, Pirard M, Adamowicz M, Pronicka E, Van Schaftingen E. Inhibition of phosphomannose isomerase by fructose-1-phosphate: an explanation for defective N-glycosylation in hereditary fructose intolerance. Pediatr. Res. 1996; 40: 764–6.

James CJ, Rellos P, Ali M, Heeley AF, Cox TM. Neonatal screening for hereditary fructose intolerance: frequency of the most common mutant aldolase B allele (A149P) in the British population. J. Med. Genet. 1996; 33: 837–41.

Jeune M, Planson E, Cotte J, Bonnefoy S, Nivelon JL, Skosowsky J. L'intolérance héréditaire au fructose, à propos d'un cas. Pédiatrie. 1961; 16: 605–26.

Karabulut HG, Halsall D, Sayin BD, Tonyukuk V, Cox TM, Bokesoy I. Aldolase B mutations in Turkish families from central Anatolia. Genet. Couns. 2006; 17: 457–60.

Kitajima Y, Takasaki Y, Takahashi I, Hori K. Construction and properties of active chimeric enzymes between human aldolases A and B. Analysis of molecular regions which determine isozyme-specific functions. J. Biol. Chem. 1990; 265: 17493–8.

Köhlin P, Melin K. Hereditary fructose intolerance in four Swedish families. Acta Paediatr. Scand. 1968; 57: 24–32.

Kranhold JF, Loh D, Morris RC Jr. Renal fructose-metabolizing enzymes: significance in hereditary fructose intolerance. Science 1969; 165: 402–3.

Kriegshäuser G, Halsall D, Rauscher B, Oberkanins C. Semi-automated, reverse-hybridization detection of multiple mutations causing hereditary fructose intolerance. Mol. Cell. Probes 2007; 21: 226–8.

Kullberg-Lindh C, Hannoun C, Lindh M. Simple method for detection of mutations causing hereditary fructose intolerance. J. Inherit. Metab. Dis. 2002; 25: 571–5.

Lafrance-Vanasse J, Sygusch J. Carboxy-terminus recruitment induced by substrate binding in eukaryotic fructose bis-phosphate aldolases. Biochemistry 2007; 46: 9533–40.

Lai CY, Nakai N, Chang D. Amino acid sequence of rabbit muscle aldolase and the structure of the active center. Science 1974; 183: 1204–6.

Lamière N, Mussche M, Baele G, Kint G, Ringoir S. Hereditary fructose intolerance: a difficult diagnosis in the adult. Am. J. Med. 1978; 65: 416–23.

Lane HC, Dodd K. Use of glucose, invert sugar and fructose for parenteral feeding of children. Pediatrics 1957; 20: 668–75.

Lau J, Tolan DR. Screening for hereditary fructose intolerance mutations by reverse dot-blot. Mol. Cell. Probes 1999; 13: 35–40.

Lebo RV, Tolan DR, Bruce BD, Cheung MC, Kan YW. Spot-blot anlysis of sorted chromosomes assigns a fructose intolerance disease locus to chromosome 9. Cytometry 1985; 6: 478–83.

Lelong M, Alagille D, Gentil C, Colin J, Tupin J, Bouqier J. Cirrhose hépatique et tubulopathie par absence congenital de l'aldolase hépatique: intolérance au fructose. Bull. Mem. Soc. Méd. Hôp. (Paris) 1962; 113: 58–63.

Leutenegger AF, Göchke H, Stutz K, et al. Comparison between glucose and a combination of glucose, fructose, and xylitol as carbohydrates for total parenteral nutrition of surgical intensive care patients. Am. J. Surg. 1977; 133: 199–25.

Levin B, Oberholzer VG, Snodgrass GJ, Stimmler L, Wilmers MJ. Fructosaemia: an inborn error of fructose metabolism. Arch. Dis. Child. 1963; 38: 220–30.

Levin B, Snodgrass GJ, Oberholzer VG, Burgess EA, Dobbs RH. Fructosaemia: observations on seven cases. Am. J. Med. 1968; 45: 826–38.

Lindén L, Nisell J. Hereditär fructosintolerans. Svenska Läkartidn 1964; 61: 3185–95.

Locher S. Akutes leber- und nierenversagen nach sorbitinfusion bei 28 jährigem patienten mit undiagnozierter fructoseintoleranz. Anästh. Intensivther. Notfallmed. 1987; 22: 194–7.

Mäenpää PH, Raivio KO, Kekomäki MP. Liver adenine nucleotides: fructose-induced depletion and its effect on protein synthesis. Science 1968; 161: 1253–4.

Malay AD, Allen KN, Tolan DR. Structure of the thermolabile mutant aldolase B, A149P: molecular basis of hereditary fructose intolerance. J. Mol. Biol. 2005; 347: 135–44.

Marthaler TM, Froesch ER. Hereditary fructose intolerance. Dental status of eight patients. Br. Dent. J. 1967; 123: 597–9.

Mass RE, Smith WR, Walsh JR. The association of hereditary fructose intolerance and renal tubular acidosis. Am. J. Med. Sci. 1966; 251: 516–23.

Minkowski O. Untersuchungen über den diabetes mellitus nach extirpation des pankreas. Arch. Exp. Pathol. Pharmakol. 1893; 31: 85–189.

Mock DM, Perman JA, Thaler MM, Morris RC Jr. Chronic fructose intoxication after infancy in children with hereditary fructose intolerance: a cause of growth retardation. N. Engl. J. Med. 1983; 309: 764–70.

Morris RC Jr. An experimental renal acidification defect in patients with hereditary fructose intolerance. I. Its resemblance to renal tubular acidosis. J. Clin. Invest. 1968a; 47: 1389–98.

Morris RC Jr. An experimental renal acidification defect in patients with hereditary fructose intolerance. II. Its distinction from classic renal tubular acidosis; its resemblance to the renal acidification defect associated with the Fanconi syndrome of children with cystinosis. J. Clin. Invest. 1968b; 47: 1648–63.

Morris RC Jr., Ueki I, Loh D, Eanes RZ, McLin P. Absence of renal fructose-1-phosphate aldolase activity in hereditary fructose intolerance. Nature 1967; 214: 920–1.

Morris RC, Sandman R, McSherry E, Sebastian A. Urinary lysosomal enzymes in renal tubular disorders. Clin. Res. 1970; 18: 460.

Müller-Wiefel DE, Steinmann B, Holm-Hadula M, Wille L, Schärer K, Gitzelmann R. Infusionsbedingtes nieren- und lebervesagen bei undiagnostizierter hereditäre fructose-intoleranz. Deutsche Med. Wochenschr. 1983; 108: 985–9.

Newburn E, Hoover C, Mettraux G, Graf H. Comparison of the dietary habits and dental health of subjects with hereditary fructose intolerance and control subjects. J. Am. Dent. Assoc. 1980; 101: 619–26.

Nikkilä EA, Somersalso O, Pitkanen E, Perheentupa J. Hereditary fructose intolerance, and inborn deficiency of liver aldolase complex. Metabolism 1962; 11: 727–31.

Oberhaensli RD, Rajagopalan B, Taylor DJ, et al. Study of hereditary fructose intolerance by use of (31)P magnetic resonance spectroscopy. Lancet 1987; II: 931–4.

Odièvre M, Gautier M, Rieu D. Intolérance hereditaire au fructose du nourisson. Evolution des lesions histologiques hépatiques sous traitement diètique prolonge (étude de huit observations). Arch. Fr. Pédiatr. 1969; 26: 433–43.

Odièvre M, Gentil C, Gautier M, Alagille D. Hereditary fructose intolerance in childhood: diagnosis and management and course in 55 patients. Am. J. Dis. Child. 1978; 132: 605–8.

Penhoet E, Rajkumar T, Rutter WJ. Multiple forms of fructose diphosphate aldolase in mammalian tissues. Proc. Nat Acad. Sci. 1966; 56: 1275–82.

Perheentupa J, Pitkänen E, Nikkilä EA, Sommersalo O, Hakosalo J. Hereditary fructose intolerance. A clinical study of four cases. Ann. Paediatr. Fenn. 1962; 8: 221–35.

Perheentupa J, Raivio KO. Fructose-induced hyperuricaemia. Lancet 1967; II: 528–31.

Perheentupa J, Raivio KO, Nikkilä EA. Hereditary fructose intolerance. Acta Med. Scand. Suppl. 1972; 542: 65–75.

Phillips MJ, Hetenyi G Jr., Adachi F. Ultrastructural hepatocellular alterations induced by in vivo fructose infusion. Lab. Invest. 1970; 22: 370–9.

Phillips MJ, Little JA, Ptak TW. Subcellular pathology of hereditary fructose intolerance. Am. J. Med. 1968; 44: 910–24.

Ponté C, Gaudier B, Bonte C, et al. Hereditary fructose intolerance. Difficulties of early detection. Exploration of the tubular function of acid-base regulation (1 case) Pédiatrie 1969; 24: 829–42.

Pronicka E, Adamowicz M, Kowalik A, et al. Elevated carbohydrate-deficient transferrin (CDT) and its normalization on dietary treatment as a useful biochemical test for hereditary fructose intolerance and galactosemia. Pediatr. Res. 2007; 62: 101–5.

Quigley R. Proximal renal tubular acidosis. J. Nephrol. 2006; 19(Suppl. 9): S41–5.

Rampa R, Froesch ER. Eleven cases of hereditary fructose intolerance in one Swiss family with a pair of monozygotic and of dizygotic twins. Helv. Paediatr. Acta 1981; 36: 317–24.

Reimers A, Spigset O. Declaration of fructose and fructose-related adverse effects in commercial drug preparations in European countries. Drug Saf. 2003; 26: 1057–9.

Rellos P, Ali M, Vidailhet M, Sygusch J, Cox TM. Alteration of substrate specificity by a naturally-occurring carboxyterminal mutation (Ala^{337}->Val) in human aldolase B. Biochem. J. 1999; 340: 321–7.

Rellos P, Sygusch J, Cox TM. Expression, purification and characterization of natural mutants of human aldolase B: rôle of quaternary structure in catalysis. Biol. Chem. 2000; 275: 1145–51.

Rey M, Behrens R, Zeilinger G. Fatale folgen einer fructose-infusion bei undiagnostizierter fructose-intoleranz. Dtsch. Med. Wschr. 1988; 113: 945–7.

Richardson RM, Little JA, Patten RL, Goldstein MB, Halperin ML. Pathogenesis of acidosis in hereditary fructose intolerance. Metabolism 1979; 28: 1133–8.

Rosien U, Cox TM, Ali M, et al. Akutes hepatorenales versagen bei hereditärer fructoseintoleranz. Mediz. Klin. 1993; 88: 553–4.

Rossner JA, Feist D. Hereditärer fructoseintoleranz verh. Dtsch. Ges. Pathol. 1971; 55: 376–85.

Rottmann WH, Tolan DR, Penhoet EE. Complete amino acid sequence for human aldolase B derived from cDNA and genomic clones. Proc. Natl Acad. Sci. USA 1984; 81: 2738–42.

Royer P, Lestradet H, Habib R, Lardinois R, Desbuquois B. L'intolérance héréditaire au fructose. Bull. Mem. Soc. Méd. Hôp. Paris 1964; 115: 805–23.

Ruecker AV, Endres W, Shin YS, Butenandt I, Steinmann B, Gitzelmann R. A case of fatal hereditary fructose intolerance: misleading information on formula composition. Hev. Paediatr. Acta 1981; 36: 599–600.

Sacrez R, Juif J-G, Metais P, Sofatzis J, Dourof N. Un cas mortal d'intolérance héréditaire au fructose. Pédiatrie 1962; 17: 875–89.

Sáez DE, Slebe JC. Subcellular localization of aldolase B. J. Cell Biochem. 2000; 78: 62–72.

Santer R, Rischewski J, von Weihe M, et al. The spectrum of aldolase B (ALDOB) mutations and the prevalence of hereditary fructose intolerance in Central Europe. Hum. Mutat. 2005; 25: 594.

Saxén L, Jousimies-Somer H, Kaisla A, Kanervo A, Summanen P, Sippilä I. Subgingival microflora, dental and periodontal conditions in patients with hereditary fructose intolerance. Scand. J. Dent. Res. 1989; 97: 150–8.

Schulte M-J, Lenz W. Fatal sorbitol infusion in a patient with fructose-sorbitol intolerance. Lancet 1977; II: 188.

Sebastio G, de Franchis R, Strisciuglio P, et al. Aldolase B mutations in Italian families affected by hereditary fructose intolerance. J. Med. Genet. 1991; 28: 241–3.

Steegmanns I, Rittmann M, Bayer JR, Gitzelmann R. Erwachsene mit hereditärer fructosintoleranz: Gefhärdung durch fructoseinfusion. Deutsche Med. Wochenschr. 1990; 115: 539–43.

Steinmann B, Baerlocher K, Gitzelmann R. Hereditär storungen des fructosestoffwechsels belastungsproben mit fruktose, sorbitol und dihydroyaceton. Nutr. Metab. 1975; 18(Suppl.1): 115–32.

Steinmann B, Gitzelmann R. Infusions flüssigkeiten sind nicht immer harmlos. Int. J. Vitam. Nutr. Res. 1976 (Suppl. 15): 289–94.

Steinmann B, Gitzelmann R. The diagnosis of hereditary fructose intolerance. Helv. Paediatr. Acta 1981; 36: 297–326.

Swales JD, Smith ADM. Adult fructose intolerance. J. Med. 1966; 35: 455–73.

Sygusch J, Beaudry D, Allaire M. Molecular architecture of rabbit skeletal muscle aldolase at 2.7 Å resolution. Proc. Natl Acad. Sci. USA 1987; 84: 7846–50.

Takahashi I, Takasaki Y, Hori K. Site-directed mutagenesis of human aldolase isozymes: the role of Cys-72 and Cys-338 residues of aldolase A and of the carboxy-terminal Tyr residues of aldolases A and B. J. Biochem. 1989; 105: 281–6.

Thabet F, Durand P, Chevret L, et al. Severe Reye syndrome: report of 14 cases managed in a pediatric intensive care unit over 11 years. Arch. Pediatr. 2002; 9: 581–6.

Tolan DR, Brooks CC. Molecular analysis of common aldolase B alleles for hereditary fructose intolerance in North Americans. Biochem. Med. Metab. Biol. 1992; 48: 19–25.

Tolan DR, Penhoet EE. Characterization of the human aldolase B gene. Mol. Biol. Med. 1986; 3: 245–64.

Van den Berghe G, Bronfman M, Vanneste R, Hers H-G. The mechanism of adenosine triphosphate depletion in the liver after a load of fructose: a kinetic study of liver adenylate deaminase. Biochem. J. 1973a; 162: 601–9.

Van den Berghe G, Hue L, Hers H-G. Effect of the administration of fructose on the glycogenolytic action of Glucagon. An investigation of the pathogeny of hereditary fructose intolerance. Biochem. J. 1973b; 134: 637–45.

Von Heine W, Schill H, Tessmann D, Kupatz H. Letale leberdystrophie bei drei Geschwistern mit hereditärer Fruktosintoleranz nach Dauertropinfusionen mit sorbitolhaltigen Infusionslösungen. Dtsch. Gesundheitsw. 1969; 24:2325–9.

Wagner K, Wolf AS. Todesfall nach fruktose- und sorbitinfusionen. Anaesthesist 1984; 33: 573–8.

Woods HF, Alberti KGMM. Dangers of intravenous fructose. Lancet 1972; II: 1354–7.

Woods HF, Eggleston LV, Krebs H. The cause of hepatic accumulation of fructose-1-phosphate on fructose loading. Biochem. J. 1970; 119: 501–10.

Yang TY, Chen HL, Ni YH, Hwu WL, Chang MH. Hereditary fructose intolerance presenting as Reye's-like syndrome. Acta Paediatr. Taiwan 2000; 41: 218–20.

CHAPTER 37

The Branchio-oto-renal Syndrome

NINE V.A.M. KNOERS AND COR W.J.R. CREMERS

INTRODUCTION

The branchio-oto-renal (BOR) syndrome (OMIM 113650) is a Mendelian developmental disorder with branchial, otic, and renal manifestations. It is also known as the Melnick-Fraser syndrome to reflect the first comprehensive descriptions of the phenotype by these authors in 1975 and 1978, respectively (Melnick et al 1975, Fraser et al 1978). Brief reports of the syndrome, however, date back to the 19th century (Paget 1878). The most characteristic clinical symptoms of BOR syndrome are: (1) hearing loss, (2) branchial arch fistulas or cysts, (3) ear pits, (4) malformations of the external, the middle and/or inner ear, and (5) renal anomalies, of varying severity. Other associated but less common features include facial/palatal abnormalities, lacrimal duct stenosis and external auditory canal stenosis (Melnick et al 1976, Fraser et al 1978, 1980, Chen et al 1995, Stinckens et al 2001).

BOR syndrome is transmitted as an autosomal dominant trait with almost complete penetrance, but with extremely wide intra- and interfamilial variability in clinical expression (Melnick et al 1976, Cremers & Fikkers van Noord 1980, Chen et al 1995). The syndrome has an estimated prevalence of 1 in 40 000 in the general population, and it is reported to occur in 2% of profoundly deaf children (Fraser et al 1980).

Branchio-otic syndrome (BO, OMIM 602588) is characterized by branchial anomalies and hearing impairment, and initially was considered to be distinct from BOR syndrome, because of the lack of renal anomalies and the type of hearing loss (Melnick et al 1978). However, systematic intravenous urography studies performed in individuals considered to have BO syndrome, have revealed renal anomalies in many of them, suggesting that BOR and BO are phenotypic variants of the same disorder (Kalatzis & Petit 1999, Weber & Kousseff 1999). Recent genetic studies have provided ample evidence that this is indeed the case (see below).

The branchio-oto-ureteral (BOU) syndrome termed by Fraser et al to designate a condition in two families with severe bilateral sensorineural hearing loss, pre-auricular pits or tags, and duplication of the collection system, is almost certainly a variable expression of BOR syndrome (Fraser et al 1983). Thus, later reports described BOR families in

which some affected members had a duplicated collecting system while others had other renal anomalies (Heimler & Lieber 1986). The family described by König et al in which individuals had symptoms of BOR, BOU or BO exemplifies best that the three conditions are variable expressions of the same syndrome (König et al 1994).

THE BOR PHENOTYPE

Below we will describe the most characteristic symptoms of the BOR syndrome phenotype in more detail.

Hearing Loss and Functional Defects of the Vestibular System

Hearing loss is the major feature of BOR syndrome and occurs in 93% of patients. The age of onset varies from early childhood to young adulthood. The impairment can be conductive, sensorineural or mixed (Fourman & Fourman 1955, Fraser et al 1978, Chen et al 1995). Interestingly, all three types of hearing loss can be found in individuals within the same family, and the type of hearing loss may even differ between the two ears of one patient. Although the hearing loss was formerly considered to be stable, a few reports have mentioned progressive hearing loss. A recent long-term audiometric follow-up study of suitable patients disclosed that progressive fluctuating hearing loss may be a regular finding in BOR syndrome (Kemperman et al 2001, Stinckens et al 2001). Based on MR imaging and audiometric analysis in several BOR syndrome families, a correlation between progressive fluctuant hearing loss and the presence of a widened vestibular aqueduct has been proposed (Kemperman et al 2001).

Vestibular studies have been rarely reported. In one study reduced or abolished caloric response was established in seven of 11 evaluated patients (Cremers & Fikkers van Noord 1980). More recently, caloric hypofunction was also reported in a small BOR syndrome family (Kemperman et al 2001).

Genetic Diseases of the Kidney
ISBN: 978-0-12-449851-8

Branchial Anomalies

Branchial cysts or fistulas are seen in approximately 65% of patients (Gorlin et al 1995) (Figure 37.1). These hallmark features may be either uni- or bilateral and are usually located on the external lower third of the neck, anterior to the sternocleidomastoid muscle. The fistulas may open into the tonsillar fossa. The openings may be inconspicuous or may drain fluid and become infected.

Ear Pits

Pre-auricular pits are noted as shallow, pinhead sized depressions near the superior attachment of the helix of the ear (Figure 37.1). Pits may also be found in the helix near its upper attachment. Ear pits are found in 70–80% of patients (Fraser et al 1978, Coppage & Smith 1995). Very rarely, they communicate with the tympanic cavity (Cremers 1983). Anterior to the pinna, pre-auricular cartilaginous tags may also be found.

External, Middle and Inner Ear Anomalies

Anomalies of the external ear are present in 30–60% of BOR syndrome patients. These range from severe microtia to minor malformations of the pinnae, resulting in lop-ear deformities or cup-shaped ears. The external canal may be narrow or malformed. Middle ear anomalies are largely responsible for the conductive component of the hearing impairment in BOR syndrome and include malformation, malposition, dislocation or fixation of the ossicles, stapes ankylosis and/or absence of the oval window, and reduction in size or malformation of the middle ear cavity (Cremers et al 1981). Inner ear abnormalities include both cochlear and vestibular compartments, and range from an enlarged vestibular aqueduct, hypodysplastic cochlea, bulbous internal acoustic canals, a deep posterior fossa and acutely-angled promontories to hypoplastic vestibula and/or semicircular canals (Cremers et al 1981, Gimsing & Dymrose 1986, Ostri et al 1991, Dagillas et al 1992, Chitayat et al 1992, Chen et al 1995, Stinckens et al 2001, Kemperman et al 2002).

Renal Anomalies

Renal anomalies are common in BOR syndrome, but the true incidence is difficult to establish because not all affected individuals have undergone image studies. In a systematic study of 21 patients who underwent either abdominal ultrasonography or intravenous pyelography, renal anomalies were found in 67% (Chen et al 1995). In another series of 16 patients structural anomalies were detected in all, and decreased glomerular filtration rate in 25% (Widdershoven et al 1983). The renal component of BOR syndrome can comprise, rather, a range (Melnick et al 1978, Widdershoven et al 1983, Heimler & Lieber 1986, Pierides et al 2002). Most of the anomalies are mild and may be even missed on routine investigations, but severe renal disease is said to be present in approximately 6% of affected individuals (Gorlin et al 1995). The most important abnormalities that lead to end-stage renal disease (ESRD) include renal agenesis with contralateral hypodysplasia or bilateral dysplasia, characterized by decreased renal volume and size, loss of ultrasound cortico-medullary differentiation and hyperechogenicity of the cortex. Bilateral renal agenesis is the extreme, leading to fetal, or immediate neonatal death. Other abnormalities are those of the collecting system and include duplication or absence of the ureter, megaureter, ureteral-pelvic junction obstruction, calyceal cyst or diverticulum and calyectasis, and extra or bifid pelvis. The renal and collecting system anomalies of BOR syndrome can be explained by a disruption of the reciprocal induction of the epithelial ureteric bud and the mesenchymal metanephric blastema in the

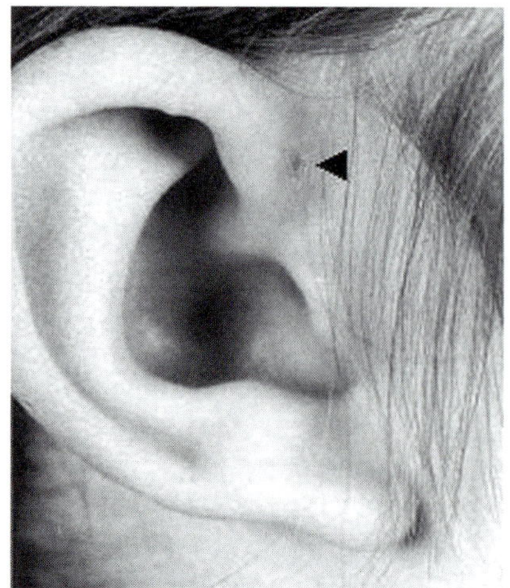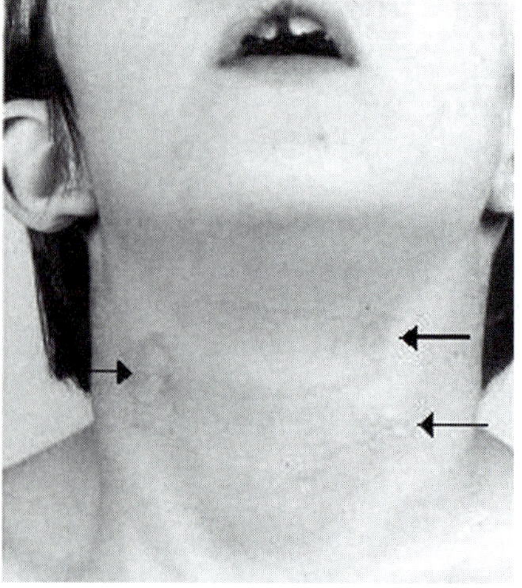

FIGURE 37.1 Left panel: slightly malformed right auricle and pre-auricular sinus (arrowhead) of male BOR syndrome patient at 7 years of age. Right panel: branchial arch fistulas (arrows) of the same patient (adapted from Kemperman et al 2001)

developing embryonic kidney (see below). Neither the presence or absence nor the severity of the renal defect may run true within families.

DIAGNOSIS

The clinical diagnosis of BOR syndrome is based on the presence of two or more of following findings: pre-auricular pits; pinnae deformities; branchial fistulas; hearing loss; and renal anomalies (Chen et al 1995, Yashima et al 2003). If two or more family members are affected, the presence of two of these findings is enough to establish the diagnosis. For an isolated case, however, the presence of three or more findings is necessary for a diagnosis.

Genetic testing can confirm a clinical diagnosis of BOR syndrome and provide genetic recurrence risk information to families. However, in view of the extreme variability of the BOR syndrome phenotype, it is not possible to predict the severity of the features, even when a mutation in one of the known BOR causing genes (see below) is identified in the family. This has of course important implications for prenatal diagnosis; a molecular diagnosis based on the presence of the mutation in the fetus does not say anything about the severity of the syndrome in that particular child. Only antenatal real-time ultrasonography may identify severe renal involvement.

DIFFERENTIAL DIAGNOSIS

BOR syndrome has features overlapping with several other syndromes, which, in some cases, need to be considered in the differential diagnosis. Townes-Brocks syndrome (TBS; OMIM 107480) is an autosomal dominant disorder consisting of imperforate anus, thumb abnormalities, dysplastic ears, hearing loss, and renal hypoplasia/dysplasia and is caused by mutations in the zinc finger transcription factor gene *SALL1* (Kohlhase et al 1998). Patients with a combination of dysplastic ears, hearing loss, and renal anomalies but without anal or thumb anomalies, initially suspected of having BOR syndrome, were shown to harbor *SALL1* mutations, indicating the phenotypic overlap between TBS and BOR syndrome (Engels et al 2000). Branchio-oculo-facial syndrome (BOF; OMIM 113620), an autosomal dominant disorder characterized by branchial sinus defects, dysplastic ears, ear pits, hearing loss, nasolacrimal duct obstruction, microphthalmia, pseudocleft of the upper lip and renal anomalies, may occasionally be confused with BOR syndrome. In most cases, however, the distinct facial phenotype seen in BOF syndrome allows a clear differentiation from BOR syndrome. The results of genetic studies provide ample evidence that BOF and BOR syndromes are distinct entities (Correa-Cerro et al 2000, Lin et al 2000, Trummer et al 2002).

THE GENETICS OF BOR SYNDROME

The First Gene for BOR Syndrome

BOR syndrome is genetically heterogeneous: at least three gene loci have been mapped. In 1992, the first BOR gene locus was localized to 8q12-22 (Kumar et al 1992, Smith et al 1992). In the following years, this locus was further refined by linkage analysis in BOR-affected families and molecular analysis of a deletion carried by a patient, to a 7 cM interval on 8q13.3 (Ni et al 1994, Vincent et al 1994). Subsequently, by the characterization of another deletion associated with an 8q translocation, the candidate region for the BOR gene was narrowed down to ~500 kb (Kalatzis et al 1996). Using a positional cloning strategy, Abdelhak et al identified the BOR gene within this genomic interval and called it *EYA1* (OMIM 601653) on the basis of its strong sequence homology to the *Drosophila* developmental gene *eyes absent* (*eya*), which is required for early *Drosophila* eye morphogenesis (Abdelhak et al 1997a). They identified mutations in the *EYA1* gene in seven people with the BOR syndrome phenotype. *EYA1* belongs to a gene family that includes the *EYA2* and *EYA3* genes, and the gene *EYA4*, which has been shown to cause late-onset deafness at the DFNA10 locus (Duncan et al 1997, Zimmerman et al 1997, Borsani et al 1999, Wayne et al 2001). The *EYA1* gene consists of 16 exons, which encompass 159 kb of genomic DNA, transcribe an mRNA of 3.73 kb and encode a deduced 599 amino-acid protein with a predicted molecular mass of 61.2 kD. Alternative splicing of the *EYA1* gene produces four transcript variants, expressing three different isoforms (isoform a, isoform b and isoform c). These three isoforms share a highly conserved 271 amino-acid carboxy-terminal region, called the *eyes absent* homologous region (eyaHR) or *Eya* domain, encoded by the last eight exons and required for protein–protein interactions. The *EYA1* gene product acts as a transcriptional coactivator; it appears to function through direct interactions with the Six family of homeodomain proteins (Pignoni et al 1997, Xu et al 1997). Mouse *Six* genes are the vertebrate homologs of the *Drosophila sine oculis* (*so*) homeobox gene, and encode transcription factors which act within a genetic network of *EYA* and *PAX* genes to regulate the early development of several organs, among which are the ear and the kidney (see below) (Li et al 2003, Xu et al 2003, Zheng et al 2003, Brodbeck & Englert 2004). It has been shown that the formation of *Eya1–Six1* complexes is necessary for nuclear translocation and transcriptional activation of Six target genes (Ohto et al 1999).

To date, there are at least 61 different mutations in the *EYA1* gene identified in BOR syndrome patients. A full list of mutations is presented on the Pendred/BOR web page (www.medicine.uiowa.edu/prendredandbor). The mutations include large and small deletions and nonsense, missense, frameshift, aberrant splicing and exon skipping mutations (Abdelhak et al 1997a, 1997b, Kumar et al 1997, 1998, Usami et al 1999, Rickard et al 2000, Rodriguez-Soriano et al 2001, Chang et al 2004). Mutations have been found throughout the EYA protein, and the hypothesis by Abdelhak et al that most of the mutations are likely to be situated in the C-terminal eyaHR region, turned out not to be true (Abdelhak et al 1997b). There is no single common mutation, with the majority of mutations being unique to individual families.

In most studies until now, mutations in the *EYA1* gene have been detected in only 20% of patients tested. Based on a large genotype–phenotype analysis of 40 BOR syndrome families, Chang et al have recently proposed to refine the clinical diagnostic criteria for BOR syndrome in order to increase the chance of finding *EYA1* gene mutations (Chang et al 2004). They suggested to include only those persons in the screening of the *EYA1* gene who have: (1) at least three of four major characteristics; (2) two major, and at least two minor characteristics; or (3) at least one major characteristic and a first-degree relative with BOR syndrome (Table 37.1). They identified *EYA1* gene mutations in ~40% of patients meeting these new criteria; of these patients, 80% appeared to have coding sequence variants, whereas 20% had complex rearrangements, such as inversions and large deletions, of the *EYA1* gene. This large proportion of complex genomic rearrangements contributing to BOR syndrome is consistent with an earlier study by Vervoort et al (2002).

Mutations in the *EYA1* gene have also been detected in patients with branchial and otic anomalies, but without the renal anomalies (BO syndrome), thus demonstrating that BOR and BO syndromes are allelic (Vincent et al 1997, Abdelhak et al 1997b, Yashima et al 2003). Azuma et al detected three novel missense mutations in the *EYA1* gene in patients with congenital cataracts and ocular anterior segment anomalies (Azuma et al 2000). One of their patients had clinical features of BOR syndrome as well.

In two patients with complex phenotypes that included features of BOR syndrome, Rickard and coworkers identified de novo microdeletions of the *EYA1* gene and surrounding region (Rickard et al 2001). One of the patients had been diagnosed with oto-facio-cervical (OFC) syndrome, a condition characterized by hearing loss, branchial fistulas, ear pits,

facial abnormalities, hypoplasia of the cervical musculature, variable mental retardation and short stature. The molecular finding in this patient indicates that OFC is allelic with BOR syndrome. The other features observed in this patient are likely to be caused by other genes, in the vicinity of *EYA1*, encompassed by the deletion.

In order to understand how point mutations in the *EYA1* gene cause disease, Buller et al investigated the functional importance of several missense mutations in the conserved EyaHR domain with respect to protein complex formation and cellular localization (Buller et al 2001). They demonstrated that these mutations do not alter subcellular protein localization. The majority of the studied mutations, however, were shown to interrupt the physical interaction between *Eya1* and *Six1* in both yeast and mammalian cells (Buller et al 2001). In view of the importance of the formation of *Eya1–Six1* complexes for nuclear translocation and transcriptional activation of *Six* target genes, it was hypothesized that Eya1 domain mutations cause the disease phenotype by reducing the expression of downstream *Six*-target genes.

A Second Gene for Bor Syndrome

In 2000, an additional locus for BO (*BOS2*) was localized to chromosome 1q31 (OMIM 120502; Kumar et al 2000) and 2 years later a third locus for BOR/BO was mapped to chromosome 14q23.1-q24.3 (*BOS3*, OMIM 608389; Ruf et al 2003). The gene responsible for *BOS2* has not yet been identified. Within the 33-megabase critical genetic interval for *BOS3*, several *SIX* genes (*SIX1*, *SIX4*, and *SIX6*), were located, which were considered excellent candidate genes for *BOS3* in view of their known action within a genetic network of *EYA* and *PAX* genes and the knowledge that the abnormalities seen in *Six1*-deficient mice are very similar to the BOR/BO phenotype in humans (Laclef et al 2003, Xu et al 2003, Zheng et al 2003). By direct sequencing of all exons of all three *SIX* genes in individuals with BOR/BO syndrome from a large number of families, three different *SIX1* mutations were identified in four kindreds, thus identifying *SIX1* (OMIM 601205) as a second gene causing BOR/BO syndrome (Ruf et al 2004). These three *SIX1* mutations, two missense mutations (Y129C and R110W) and one in-frame deletion of three nucleotides (del113E), affect amino acids that are highly conserved among the *Six* gene family, and were absent from at least 90 healthy control individuals. Y129C and del113E are both located in the homeodomain (HD), which is important for DNA binding, and R110W in the Six domain (SD), which is critical for protein–protein interactions of the SIX1 protein. Using a yeast-two hydrid assay, all three mutations were shown to disrupt the physical interaction of *Six1* with *Eya1*. Moreover, in a gel-mobility shift assay the two homeodomain mutations were shown to interfere with *Six1*-DNA binding as well (Ruf et al 2004).

TABLE 37.1   Refined diagnostic criteria for BOR syndrome (adapted from Chang et al 2004)

Major characteristics	Minor characteristics
Branchial anomalies	External ear anomalies
Hearing loss	Middle ear anomalies
Pre-auricular pits	Inner ear anomalies
Renal anomalies	Pre-auricular tags
	Other: facial asymmetry, palatal abnormalities

The Role of EYA1 and SIX1 in Renal Development

The clinical signs of BOR syndrome are indicative of an early developmental defect of the branchial arch apparatus, the otic vesicle and/or surrounding periotic mesenchyme and the mesonephros/metanephros, taking place between the 4th and 10th weeks of human embryonic development. It is therefore very likely that *EYA1* and *SIX1* have critical roles in the development of these tissues. Data from *Eya1* and *Six1* expression studies in mice and the anomalies found in $Eya^{-/-}$ and *Six-1* deficient animals illustrate that this is indeed the case. Here, we will mainly discuss the relevance of these studies for renal development.

The *eya1* gene is expressed very early during mammalian development, between the 4th and 6th weeks in humans (Abdelhak et al 1997a, Kalatzis et al 1998). In the mouse kidney, *Eya1* expression is first detected in the metanephric mesenchyme at E12.5 (Kalatzis et al 1998). *Eya-1* null mice die at birth and show a wide variety of abnormalities, including craniofacial and skeletal defects, and absent ears (Xu et al 1999). Affected mice also lack both kidneys and ureters, because of defective ureteric bud outgrowth from the mesonephric (Wolffian) duct leading to failure of metanephric induction and increased mesenchymal apoptosis (Xu et al 1999). In these $Eya^{-/-}$ mutants, the expression of the gene encoding the glial cell-derived neurotrophic factor (GDNF) was not detected in the metanephric mesenchyme. GDNF, a member of the transforming growth factor β family, is a signaling molecule known to play a pivotal role in the induction and maintenance of ureteric bud branching in the developing metanephros via activation of the RET receptor tyrosine kinase, which is present at the tip of the ureteric bud (Sariola & Saarma 1999). The findings in *Eya1*-null animals indicate that *Eya1* may operate in a genetic regulatory cascade controlling ureteric bud outgrowth, upstream of GDNF.

Six1 is expressed in otic vesicles, nasal epithelia, branchial arches/pouces, somites, a limited set of ganglia, and in the developing kidney (Oliver et al 1995). In the murine embryonic kidney, the gene is expressed in the uninduced metanephric mesenchyme at E10.5 and in the induced mesenchyme around the ureteric bud at E11.5. *Six1* null animals die at birth due to thoracic skeletal defects and severe muscle hypoplasia. All $Six1^{-/-}$ mice have malformations of the auditory system, and bilateral renal agenesis was found in 97.5% (Laclef et al 2003, Ozaki et al 2003, Xu et al 2003, Zheng et al 2003). Similar to what was found for *Eya1* null animals, the renal agenesis in S*ix1*-deficient mice appeared to be due to a failure of ureteric bud invasion into the mesenchyme and subsequent apoptosis of the mesenchyme (Xu et al 2003). Approximately 75% of $Eya1^{+/-}$ $Six1^{+/-}$ double heterozygous mice show renal developmental defects, mostly renal hypoplasia, whereas renal abnormalities occur only at low penetrance and strain-dependent in single heterozygotes for either gene deletion (Xu et al 2003). These data are consistent with the assumption that *Six1* and *Eya1* function together, as components of a complex genetic network, to regulate the development of the mammalian kidney. Based on its expression pattern in wild-type and various knockout animals, including *Eya1* and *Six1* null mice, *Pax2,* another transcription factor with a crucial role in renal development, is assumed to be an additional component of this genetic regulatory network (Xu et al 1999, Laclef et al 2003, Xu et al 2003).

All together, the data of animal and expression studies and the molecular findings in BOR/BO syndrome patients suggest that the BOR-associated mutations in either *EYA1* or *SIX1* may cause developmental defects of the kidney by influencing the formation of transcriptionally active *EYA1-SIX1-PAX2* complexes, which regulate the expression of downstream signaling molecules, such as GDNF, in the metanephric mesenchyme. The resulting reduced expression of these signaling molecules may lead to disturbances of induction of kidney development (Ruf et al 2004). The observation of renal defects at only low penetrance in $Eya1^{+/-}$ and $Six1^{+/-}$ mice is consistent with the incomplete penetrance and wide variability of renal abnormalities in BOR syndrome, even within families. It is assumed that in the case of haploinsufficiency of either *EYA1* or *SIX1* as a result of a BOR-associated mutation, additional modifier loci can influence the activity or function of the gene involved and thereby modulate expression of the mutant phenotype.

There are strong indications that the molecular mechanisms underlying the developmental defects of the auditory and branchial arch systems seen in BOR syndrome show similarities to the mechanism described above for the renal anomalies, with a critical role for the complex genetic network of transcription factors, including *EYA1* and *SIX1* (Ozaki et al 2003, Ruf et al 2004).

References

Abdelhak S, Kalatzis V, Heilig R, et al. A human homologue of the *Drosophila* eyes absent gene underlies branchio-oto-renal (BOR) syndrome and identifies a novel family. Nat. Genet. 1997a; 15: 157–64.

Abdelhak S, Kalatzis V, Helig R, et al. Clustering of mutations responsible for branchio-oto-renal (BOR) syndrome in the *eyes absent* homologous region (eyaHR) of EYA1. Hum. Mol. Genet. 1997b; 6: 2247–55.

Azuma N, Hirakiyama A, Asaka A, Yamada M. Mutations of a human homologue of the *Drosophila* eyes absent gene (EYA1) detected in patients with congenital cataracts and ocular anterior segment anomalies. Hum. Mol. Genet. 2000; 9: 363–6.

Borsani G, DeGrandi A, Ballabio A, et al. EYA4, a novel vertebrate gene related to *Drosophila* eyes absent. Hum. Mol. Genet. 1999; 8: 11–23.

Brodbeck S, Englert C. Genetic determination of nephrogenesis: the Pax/Eya/Six gene network. Pediatr. Nephrol. 2004; 19: 249–55.

Buller C, Xu X, Marquis V, Schwanke R, Xu P-X. Molecular effects of Eya1 domain mutations causing organ defects in BOR syndrome. Hum. Mol. Genet. 2001; 10: 2775–81.

Chang EH, Menezes M, Meyer NC, et al. Branchio-oto-renal syndrome: The mutation spectrum in EYA1 and its phenotypic consequences. Hum. Mutat. 2004; 23: 582–9.

Chen A, Francis M, Ni L, et al. Phenotypic manifestations of the branchio-oto-renal syndrome. Am. J. Med. Genet. 1995; 58: 365–70.

Chitayat D, Hodgkinson KA, Chen M-F, Haber GD, Nakashima S, Sando I. Branchio-oto-renal syndrome: further delineation of an under diagnosed syndrome. Am. J. Med. Genet. 1992; 43: 970–5.

Coppage KB, Smith RJH. Branchio-oto-renal syndrome. J. Am. Acad. Audiol. 1995; 6: 103–10.

Correa-Cerro LS, Kennerknecht I, Just W, Vogel W, Muller D. The gene for branchio-oculo-facial syndrome does not co-localize to EYA1-4 genes. J. Med. Genet. 2000; 37: 620–3.

Cremers CWRJ. Congenital pre-auricular fistula communicating with the tympanic cavity. J. Laryngol. Otol. 1983; 97: 749–53.

Cremers CWRJ, Fikkers van Noord M. The earpits-deafness syndrome. Clinical and genetic aspects. Int. J. Pediatr. Otorhinolaryngol. 1980; 2: 309–22.

Cremers CWJR, Thijssen HO, Fischer AJ, Marres EH. Otological aspects of the earpit-deafness syndrome. ORL Otorhinolaryngol. Relat. Spec. 1981; 43: 223–39.

Dagillas A, Antoniades K, Palasis S. Branchio-oto-renal dysplasia associated with tetralogy of Fallot. Head Neck 1992; 14: 139–42.

Duncan MK, Kos L, Jankins NA, Gilbert DJ, Copeland NG, Tomarev SI. Eyes absent: a gene family found in several metazoan phyla. Mamm. Genom. 1997; 8: 479–85.

Engels S, Kohlhase J, McGaughran J. A SALL1 mutation causes a branchio-oto-renal syndrome like phenotype. J. Med. Genet. 2000; 37: 458–60.

Fourman P, Fourman J. Hereditary deafness in family with earpits (fistula auris congenital). BMJ 1955; 2: 1354–6.

Fraser FC, Ayme S, Halal F, Sproule J. Autosomal dominant duplication of the renal collecting system, hearing loss and external ear anomalies: A new syndrome?. Am. J. Med. Genet. 1983; 14: 473–8.

Fraser FC, Ling D, Clogg D, Nogrady B. Genetic aspects of the BOR syndrome-branchial fistulas, ear pits, hearing loss, and renal anomalies. Am. J. Med. Genet. 1978; 2: 241–52.

Fraser FC, Sproule JR, Halal F. Frequency of the branchio-oto-renal (BOR) syndrome in children with profound hearing loss. Am. J. Med. Genet. 1980; 7: 341–9.

Gimsing S, Dymrose J. Branchio-oto-renal dysplasia in three families. Ann. Otol. Rhinol. Laryngol. 1986; 95: 421–6.

Gorlin RJ, Toriello HV, Cohen MM. Hereditary hearing loss and its syndromes. New York: Oxford University Press, 1995: pp. 234–56.

Heimler A, Lieber E. The branchio-oto-renal syndrome: reduced penetrance and variable expressivity in four generations of a large kindred. Am. J. Med. Genet. 1986; 25: 15–27.

Kalatzis V, Abdelhak S, Compain S, Vincent C, Petit C. Characterization of a translocation-associated deletion defines the candidate region for the gene responsible for branchio-oto-renal syndrome. Genomics 1996; 34: 422–5.

Kalatzis V, Petit C. Branchio-otic syndromes imbroglio. Am. J. Med. Genet. 1999; 82: 440–1.

Kalatzis V, Sahly I, El-Amraoui A, Petit C. Eya1 expression on the developing ear and kidney: towards the understanding of the pathogenesis of Branchio-oto-renal (BOR) syndrome. Dev. Dyn. 1998; 213: 486–99.

Kemperman MH, Koch SMP, Joosten FBM, et al. Inner ear anomalies are frequent but nonobligatory features of the Branchio-oto-renal syndrome. Arch. Otolanryngol. Head Neck Surg. 2002; 128: 1033–8.

Kemperman MH, Stinckens C, Kumar S, Huygen PLM, Joosten FBM, Cremers CWRJ. Progressive, fluctuant hearing loss, enlarged vestibular aqueduct and cochlear hypoplasia in the BOR syndrome. Otol. Neurotol. 2001; 22: 637–43.

Kohlhase J, Wischermann A, Reichenbach H, Froster U, Engel W. Mutations in the SALL1 putative transcription factor gene cause Townes Brocks syndrome. Nat. Genet. 1998; 18: 81–3.

König R, Fuchs S, Dukier C. Branchio-oto-renal (BOR) syndrome: variable expressivity in a five-generation pedigree. Eur. J. Pediatr. 1994; 153: 446–50.

Kumar S, Deffenbacher K, Cremers CW, van Camp G, Kimberling WJ. Branchio-oto-renal syndrome: identifications of novel mutations, molecular characterization, mutation distribution, and prospects for genetic testing. Genet. Test. 1997; 1: 243–51.

Kumar S, Deffenbacher K, Marres HA, Cremers CW, Kimberling WJ. Genomewide search and genetic localization of a second gene associated with autosomal branchio-oto-renal syndrome: clinical and genetic implications. Am. J. Hum. Genet. 2000; 66: 1715–20.

Kumar S, Kimberling WJ, Kenyon JB, Smith RJH, Marres HAM, Cremers CWRJ. Autosomal dominant branchio-oto-renal syndrome-localization of a disease gene to chromosome 8q by linkage in a Dutch family. Hum. Mol. Genet. 1992; 1: 491–5.

Kumar S, Marres HAM, Cremers CWRJ, Kimberling WJ. Identification of three novel mutations in the human EYA1 protein associated with branchio-oto-renal syndrome. Hum. Mutat. 1998; 11: 443–9.

Laclef C, Souil E, Demignon J, Maire P. Thymus, kidney and craniofacial abnormalities in Six1 deficient mice. Mech. Dev. 2003; 120: 669–79.

Li X, Ohgi KA, Zhang J, et al. Eya protein phosphatase activity regulates Six1-Dach-Eya transcriptional effects in mammalian organogenesis. Nature 2003; 426: 247–54.

Lin AE, Semin EV, Daack-Hirsch S, et al. Exclusion of the branchio-oto-renal syndrome locus (EYA1) from patients with branchio-oculo-facial syndrome. Am. J. Med. Genet. 2000; 91: 387–90.

Melnick M, Bixler D, Nance WE, Silk K, Yune H. Familial branchio-oto-renal dysplasia: a new addition to the branchial arch syndromes. Clin. Genet. 1976; 9: 25–34.

Melnick M, Bixler D, Silk K, Yune H, Nance WE. Autosomal dominant branchiootorenal dysplasia. Birth defects original article services XI 1975; 5: 121–8.

Melnick M, Hodes ME, Nance WE, Yune H, Sweeney A. Branchio-oto-renal dysplasia and branchio-otic dysplasia: two distinct autosomal dominant disorders. Clin. Genet. 1978; 13: 425–42.

Ni L, Wagner MJ, Kimberling WJ, et al. Refined localization of the branchiootorenal syndrome gene by linkage and haplotype analysis. Am. J. Med. Genet. 1994; 51: 76–184.

Ohto H, Kamada S, Tago K, et al. Cooperation of six and eya in activation of their target genes through nuclear translocation of Eya. Mol. Cell. Biol. 1999; 19: 6815–24.

Oliver G, Wehr R, Jenkins NA, et al. Homeobox genes and connective tissue patterning. Development 1995; 121: 693–705.

Ostri B, Johnsen T, Bergmann I. Temporal bone findings in a family with branchio-oto-renal syndrome (BOR). Clin. Otolaryngol. 1991; 16: 163–7.

Ozaki H, Nakamura K, Funahashi J-I, et al. Six1 controls patterning of mouse otic vesicle. Development 2003; 131: 551–62.

Paget J. Cases of branchial fistulae in the external ears. Med. Chirurg. Trans. 1978; 61: 50.

Pierides AM, Athanasiou Y, Demetriou K, Koptides M, Deltas CC. A family with the branchio-oto-renal syndrome: clinical and genetic correlations. Nephrol. Dial. Transplant, 2002; 17: 1014–18.

Pignoni F, Hu B, Zavitz KH, Xiao J, Garrity PA, Zipursky SL. The eye-specification proteins So and Eya form a complex and regulate multiple steps in *Drosophila* eye development. Cell 1997; 91: 881–91.

Rickard S, Boxer M, Trompeter ZR, Bitner-Glindzicz M. Importance of clinical evaluation and molecular testing in the branchio-oto-renal (BOR) syndrome and overlapping phenotypes. J. Med. Genet. 2000; 37: 623–727.

Rickard S, Parker M, van't Hoff W, et al. Oto-facio-cervical (OFC) syndrome is a contguous deletion syndrome involving EYA1: molecular analysis confirms allelism with BOR syndrome and further narrows down the Duane syndrome critical region to 1 cM. Hum. Genet. 2001; 108: 398–403.

Rodriguez-Soriano J, Balbao AVJR, Castano L. Branchio-oto-renal syndrome: identification of a novel mutation in the EYA1 gene. Pediatr. Nephrol. 2001; 16: 550–3.

Ruf RG, Berkman J, Wolg MT, et al. A gene locus for branchio-otic syndrome maps to chromosome 14q21.3-q24.3. J. Med. Genet. 2003; 40: 515–19.

Ruf RJ, Xu P-X, Silvius D, et al. SIX1 mutations cause branchio-oto-renal syndrome by disruption of EYA1-SIX1-DNA complexes. Proc. Natl Acad. Sci. USA 2004; 101: 8090–5.

Sariola H, Saarma M. GDNF and its receptors in the regulation of the ureteric branching. Int. J. Dev. Biol. 1999; 43: 413–18.

Smith C, Coppage KB, Ankerstjerne JKB, et al. Localization of the gene for branchiootorenal syndrome to chromosome 8q. Genomics 1992; 14: 841–4.

Stinckens C, Standaert L, Casselman JW, et al. The presence of a widened vestibular aqueduct and progressive sensorineural hearing loss in the branchio-oto-renal syndrome. A family study. Int. J. Pediatr. Otorhinolaryngol. 2001; 59: 163–72.

Trummer T, Muller D, Schilze A, Vogel W, Just W. Branchio-oculo-facial syndrome and branchio-otic/branchio-oto-renal syndromes are distinct entities. J. Med. Genet. 2002; 39: 71–3.

Usami S, Abe S, Shinkawa S, Deffenbacher K, Kumar S, Kimberling WJ. Eya1 nonsense mutation in a Japanese branchio-oto-renal syndrome family. J. Hum. Genet. 1999; 44: 261–5.

Vervoort VS, Smith RJ, O'Brien J, et al. Genomic rearrangements of EYA1 account for a large fraction of families with BOR syndrome. Eur. J. Hum. Genet. 2002; 10: 757–66.

Vincent C, Kalatzis V, Abdelhak S, et al. BOR and BO syndromes are allelic defects of EYA1. Eur. J. Hum. Genet. 1997; 5: 242–6.

Vincent C, Kalatzis V, Compain S, et al. A proposed contiguous gene syndrome on 8q consists of branchio-oto-renal (BOR) syndrome, Duane syndrome, a dominant form of hydrocephalus and trapeze aplasia; implications for the mapping of the BOR gene. Hum. Mol. Genet. 1994; 3: 1859–66.

Wayne S, Robertson NG, DeClau F, et al. Mutations in the transcriptional activator EYA4 cause late-onset deafness at the DFNA10 locus. Hum. Mol. Genet. 2001; 1: 195–200.

Weber KM, Kousseff BG. New manifestations of BOR syndrome. Clin. Genet. 1999; 5: 306–12.

Widdershoven J, Monnens L, Assman K, Cremers C. Renal disorders in the branchio-oto-renal syndrome. Helv. Peadiat. Acta 1983; 38: 513–22.

Xu PX, Adams J, Peters H, Brown MC, Heaney S, Maas R. Eya-1 deficient mice lack ears and kidneys and show abnormal apoptosis of organ primordia. Nat. Genet. 1999; 23: 113–17.

Xu PX, Cheng J, Epstein JA, Maas R. Mouse Eya genes are expressed during limb tendon development and encode a transcriptional activation function. Proc. Natl Acad. Sci. USA 1997; 94: 11974–9.

Xu PX, Zheng W, Huang L, Maire P, Laclef C, Silvius D. Six1 is required for the early organogenesis of mammalian kidney. Development 2003; 130: 3085–94.

Yashima T, Noguchi Y, Ishakawa K, Mizusawa H, Kitamura K. Mutation in the EYA1 gene in patients with Branchio-otic syndrome. Acta Otolaryngol. 2003; 123: 279–82.

Zheng W, Huang L, Wei Z-B, Tang B, Xu P-X. The role of Six1 in mammalian auditory system development. Development 2003; 130: 3989–4000.

Zimmerman JE, Bui QT, Steingrimsson E, et al. Cloning and characterization of two vertebrate homologues of the *Drosophila* eyes absent gene. Genome Res. 1997; 7: 128–41.

CHAPTER 38

Primary Metabolic and Renal Hyperuricemia

KIMIYOSHI ICHIDA, MAKOTO HOSOYAMADA, TATSUO HOSOYA AND HITOSHI ENDOU

INTRODUCTION

Gout is a heterogeneous disease resulting from tissue deposition of monosodium urate or uric acid crystals as a result of supersaturation of extracellular fluids with uric acid. Uric acid is the end product of human purine metabolism and hyperuricemia is the pathogenetic denominator of gout. The clinical manifestations of gout include recurrent attack of a characteristic acute arthritis, urolithiasis, and renal impairment, and accumulation of crystalline aggregates in connective tissue. The effect of uric acid on atherosclerosis is still ambiguous.

Hyperuricemia is classified into overproduction type, underexcretion type, and mixed type having characteristics of overproduction type and underexcretion type (Figure 38.1).

Two-thirds of hyperuricemic patients are diagnosed with underexcretion type hyperuricemia, while 10–15% of the patients have overproduction type and about 25% of patients have mixed type (Wyngaarden & Kelley 1976). The overproduction type hyperuricemia includes single gene disorders in purine metabolic enzymes such as hypoxanthine–guanine phosphoribosyl transferase (HPRT; EC 2.4.2.8) deficiency, (Lesch-Nyhan syndrome, MIM 300322) and phosphoribosyl pyrophosphate synthetase (PRPS; EC 2.7.6.1) overactivity and a complex disorder caused by interaction of polygenic factors and environmental factors such as alcohol and diet intake (Table 38.1). The underexcretion type hyperuricemia includes a single gene disorder, familial juvenile hyperuricemic nephropathy and a complex disorder caused by polygenic and environmental factors. In this chapter, purine metabolism, the

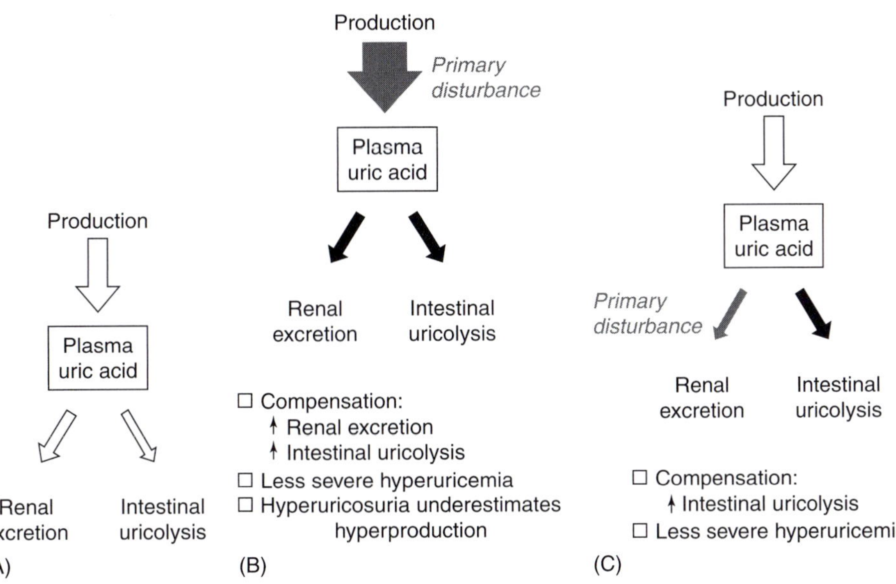

FIGURE 38.1 Schema of uric acid handling in hyperuricemia. (A) Normal uric acid handling in normouricemia. (B) Overproduction type hyperuricemia. (C) Underexcretion type hyperuricemia. (see also Plate 84)

Genetic Diseases of the Kidney
ISBN: 978-0-12-449851-8

Table 38.1   Classification of Hyperuricemia

Condition	Genetic contribution
Increased production	
• *Primary disturbance in purine metabolic enzymes*	
– HGPRT deficiency	HGPRT
Lesch-Nyhan syndrome – complete	
Kelley-Seegmiller syndrome – partial	
– PRPP synthetase overactivity	PRPP synthetase
• *Increased flux through purine metabolic pathway*	
– Increased nucleotide/nucleoside destruction/turnover	
Tissue injury and catabolic states	
Hypoxia, hypoperfusion, burn, rhabdomyolysis	
Malignancies	
Related to malignancy itself	
e.g. myeloprofierative disorders	
Related to therapy	
e.g. tumor lysis	
Psoriasis	
Glycogen storage diseases	
type 1 Von Gierke's	Glucose-6-phosphatase
type V McCardle's	Glycogen phosphorylase
type VI Tarui's	Phosphofructose kinase
Sickle cell disease	β-globin
– Increased purine intake/diet-induced hyperproduction	Alcohol and aldehyde dehydrogenase
– Metabolic syndrome ± uric acid nephrolithiasis	polymorphisms
– Primary gout ± uric acid nephrolithiasis	Complex trait: polygenic + environment
	Candidates?
	5-hydroxytryptamine receptor 2A
	ZNF265 (Sardinian gene)
Decreased excretion	
• *Reduced renal mass*	
All chronic kidney disease	Complex trait: polygenic + environment
• *Increased tubular reabsorption*	
– Primary increase in tubular absorption	URAT1 polymorphism ?
– Secondary to effective volume contraction	Many monogenic salt wasting syndromes
Congenital salt/water wasting syndromes	Congenital diabetes insipidus
Acquired e.g. diuretics, preeclampsia/eclampsia	
– Primary gout ± uric acid nephrolithiasis	Complex trait: polygenic + environment
• *Decreased tubular secretion*	
– Organic anions	
βOH butyrate, acetoacetate, lactate	
– Low dose aspirin, sulfinpyrazone	
– Primary gout ± uric acid nephrolithiasis	Complex trait: polygenic + environment
• *Unknown tubular mechanism*	
– Uromodulin-associated diseases	Uromodulin (Tamm-Horsfall)
Familial juvenile hyperuricemia nephropathy	
Medullary cystic kidney disease type 2	
– Other medullary cystic kidney diseases	
Medullary cystic kidney disease type 1	Locus 1q21
Medullary cystic kidney disease type 3	Locus 1q41
Mixed Overproduction & Underexcretion	
Metabolic syndrome ± uric acid nephrolithiasis	Complex trait: polygenic + environment
Primary gout ± uric acid nephrolithiasis	
Glycogen storage disease type I	Glucose-6-phosphatase

purine metabolic enzyme disorders Lesch-Nyhan syndrome and PRPS (see Figure 38.2) superactivity, and other single gene disorders such as glycogen storage disease types I and VII and the underexcretion type hyperuricemia, familial juvenile hyperuricemic nephropathy, are discussed. Urate handling in human kidney is described in Chapter 9.

CLASSIFICATION OF HYPERURICEMIA

Renal underexcretion is the main mechanism for the development of primary hyperuricemia in most patients, even in the overproduction type patients without genetic disorders (Perez-Ruiz et al 2002). Nevertheless, the type classification is supportive to the selection for antihyperuricemic medicine and the primary disease diagnosis of secondary hyperuricemia. The overproduction type hyperuricemia is judged from the increased uric acid excretion of over 3–4 mg of uric acid per mg creatinine or 800 mg of uric acid per day, while the underexcretion type hyperuricemia is judged from the decreased uric acid clearance or fractional uric acid clearance (Wyngaarden & Kelley 1976, Kaufman et al 1968, Rieselbach et al 1970, Berger & Yu 1975, Simkin 2003) (Figure 38.1).

PURINE METABOLISM

Purine nucleotides are involved in many cellular functions as components of DNA and RNA, as sources of energy, as enzyme cofactors in metabolic pathways, and as components of signal transduction. Net contributions to body pools of purine compounds are provided by dietary purine ingestion and the endogenous pathway of de novo purine nucleotide synthesis in ten enzymatic reactions.

In the de novo purine synthesis pathway, the purine ring is sequentially constructed from small molecule donors on a ribose 5-phosphate backbone provided by 5-phosphoribosyl-1-pyrophosphate (PRPP) to form the first purine product, inosine monophosphate (IMP) (Figure 38.2). Amido phosph-oribosyltransferase (ATase) catalyzes the first step of the de novo purine synthesis pathway, the conversion of PRPP into 5-phosphoribosylamine. ATase is the rate-limiting enzyme in the de novo purine synthesis pathway. ATase is regulated by feedback inhibition by purine nucleotides, IMP, AMP and GMP, and this feedback inhibition is reversed by PRPP, a key regulatory substrate in the de novo purine synthesis pathway (Holmes et al 1973, Smith et al 1994, Yamaoka et al 1997). Increased quantities of intracellular PRPP lead to overproduction of de novo purine synthesis and of uric acid, demonstrated in HPRT deficiency and PRPS overactivity. The preceding enzyme for the de novo purine synthesis pathway, PRPS, catalyzes ribose-5-phosphate and ATP into PRPP. PRPS is also inhibited by purine nucleotides, but less sensitive than ATase (Becker & Kim 1987). PRPP is also used in pyrimidine and pyridine nucleotide synthesis and in salvage of preformed purine bases.

Purines are also obtained through salvage of preformed purine bases, and uptake of cellular degradation products. In the purine salvage, HPRT catalyzes conversion of hypoxanthine and guanine to IMP and GMP respectively

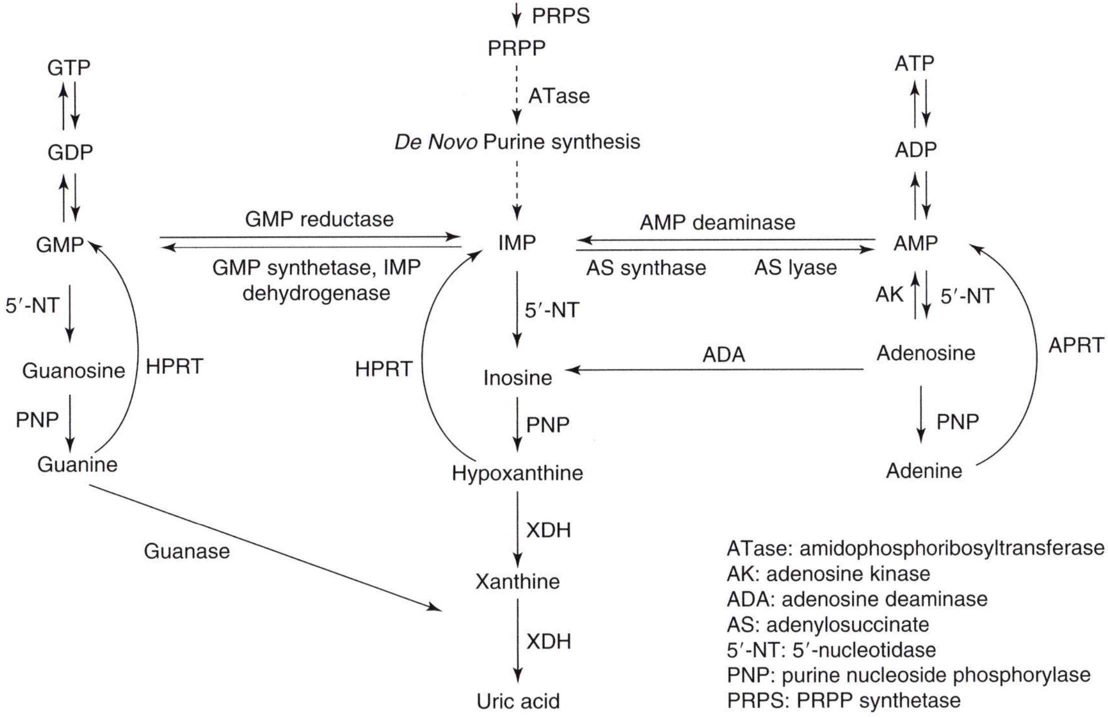

FIGURE 38.2 Purine metabolic pathways and the related enzymes

and adenine phosphoribosyltransferase (APRT) catalyzes conversion of adenine to AMP. The de novo purine synthesis pathway requires several moles of ATP for generation of each mole of purine nucleotide product, while HPRT and APRT require one ATP. Much of the cellular energy is conserved in the purine salvage in comparision with the de novo purine synthesis pathway and 90% of free purines generated during intracellular metabolism are recycled rather than degraded or excreted (Murray 1971).

For balanced production of adenylates and guanylates, interconversion of IMP to adenosine monophosphate (AMP) and guanosine monophosphate (GMP), the precursors of adenosine triphosphate (ATP) and guanosine triphosphate (GTP) are under feedback control. IMP dehydrogenase and adenylosuccinate synthase, catalyzing the first steps leading to the production of guanine nucleotides and adenine nucleotides, are inhibited by GMP and AMP, respectively (Holmes et al 1974, Van der Weyden & Kelly 1974). Furthermore, IMP dehydrogenase and adenylosuccinate synthase are accelerated by ATP and GTP, respectively (Pourquie 1969).

SINGLE GENE DISORDER FOR OVERPRODUCTION TYPE HYPERURICEMIA

Lesch-Nyhan Syndrome

HPRT catalyzes the salvage reactions of hypoxanthine and guanine with PRPP to form IMP and GMP (Figure 38.2). Though HPRT expresses ubiquitously among mammalian tissues, HPRT activity levels markedly change at different developmental stages. HPRT activity is usually higher in rapidly dividing cells. The HPRT gene was cloned and spans approximately 45 kb on Xq26-q27.2 (Shows & Brown 1975, Pai et al 1980, Jolly et al 1982). HPRT gene consists of nine exons (Figure 38.3). A large number of HPRT mutations in HPRT deficiency have been reported since 1983 (Wilson & Kelley 1983, Wilson et al 1983). Though some hot spots for HPRT mutations have been discussed, mutations causing disease appear throughout the HPRT gene (Figure 38.3).

HPRT deficiency, designated Lesch-Nyhan syndrome, is an X-1inked recessive disorder. The clinical features of HPRT deficiency are classified into excessive purine production, and neurological and behavioral manifestations. The approximate clinical phenotype demonstrates a good correlation with the amount of residual enzyme activity. Patients with less 1.5% of residual enzyme activity, designated Lesch-Nyhan disease, have all the above three clinical features. Patients with 1.5–8% of residual enzyme activity, designated Lesch-Nyhan disease variant or neurologic variant, demonstrate uric acid overproduction and neurological abnormalities without the behavioral abnormalities such as self-injurious behaviors. Patients with more than 8% of residual enzyme activity, designated Kelley-Seegmiller syndrome (MIM: 300323) or

HPRT-related hyperuricemia, demonstrate only clinical manifestations of uric acid overproduction. Though the spectra of HPRT deficiency and the manifestation are a continuum and the terms are sometimes confusing, this classification is intelligible and supportive for understanding.

Lesch-Nyhan disease is the most common cause of hyperuricemia in infancy and childhood and the frequency approximates one in 380 000 births (Crawhall et al 1972). The clinical manifestations of Lesch-Nyhan disease include urolithiasis or gout due to uric acid overproduction and overexcretion, mental and growth retardation, choreoathetosis, dystonia, compulsive self-injurious behavior, and sometimes megaloblastic anemia. Infants with Lesch-Nyhan disease appear normal at birth and usually develop normally for the first 3–8 months. The first manifestation is usually the occurrence of large quantities of 'orange sand' in the diaper caused by uric acid crystals and hematuria. The majority of patients with Lesch-Nyhan disease are recognized when they are between 3 and 12 months of age with motor disability or hypotonia. Extrapyramidal signs such as choreoathetosis and dystonia and pyramidal signs such as hyperreflexia and extensor plantar reflex typically begin to develop between 1 and 2 years of age. The patients can eventually neither stand nor sit unassisted because of the motor impairments. Self-injurious behavior is the hallmark of the Lesch-Nyhan disease and occurs in most of the patients with the disease. The average age of the behavior onset was 3 years, with a range of 1 to 8 years (Anderson & Ernst 1994). The characteristic behavior is self-destructive biting of lips, fingers, arms, and tongue. Other self-injurious behaviors include banging or snapping the head, injuring hands or feet with objects. Patients with Lesch-Nyhan disease do not wish to injure themselves.

The absence of HPRT leads to the tendency to accumulate its substrates, hypoxanthine and guanine. Serum concentration and urinary amount of the end product, uric acid, increase in HPRT deficiency. The intermediate product, xanthine, also tends to increase in the blood and urine of the patients with HPRT deficiency. As all patients with Lesch-Nyhan disease overproduce large quantities of uric acid, hyperuricemia – serum uric acid over 10 mg/dl – is present in most patients. Serum uric acid levels, however, occasionally remain in normal range because of efficient renal clearance.

Amount of urinary uric acid excretion and urinary uric acid to creatinine ratio of the patients increase over 50 mg/kg/day and 2, respectively, and are useful for the diagnosis. The measurement of HPRT enzyme activity is the conclusive confirmatory test for HPRT deficiency. The assays can be commonly conducted with cell lysates from any tissues, such as peripheral blood cells, though the lysates rarely provide misleading results. The measurement of HPRT enzyme activity of fibroblasts or lymphocytes in the culture accurately reflects the more actual activity.

Although Lesch-Nyhan disease leads to hyperuricemia resulting in gout, gout is relatively uncommon in children with Lesch-Nyhan disease. However, gout develops in most

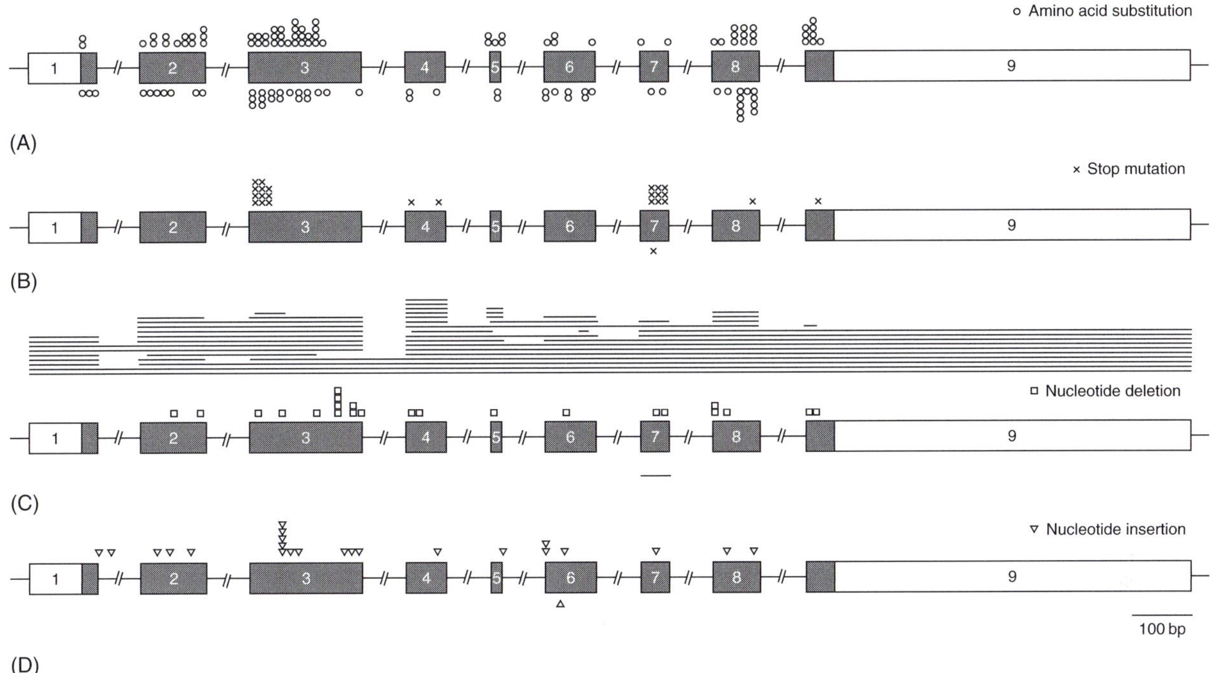

FIGURE 38.3   The nine exons of the gene are shown as individual boxes, with the coding regions in gray and the non-coding regions pale. Those associated with the Lesch-Nyhan disease phenotype are shown above the gene, those associated with HPRT-related hyperuricemia with or without neurogenic dysfunction, below. (A) Single *hprt* base substitutions leading to amino acid substitutions. Individual point mutations are shown as circles. (B) Single *hprt* base alterlations leading to premature stop. Individual point mutations are shown as crosses. (C) Deletion mutations in the *hprt* gene. Short base deletions are shown as squares, and long deletions are shown as a horizontal line spanning the deleted segment. (D) Insertion mutation in the *hprt* gene. Short base insertions are shown as triangles

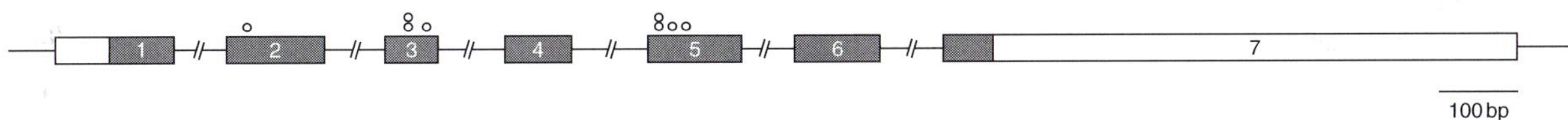

FIGURE 38.4   The seven exons of the gene are shown as individual boxes, with the coding regions in gray and the noncoding regions pale. Single *prps1* base substitutions leading to amino acid substitutions associated with PRPS overactivity. Individual point mutations are shown as circles

patients with Lesch-Nyhan disease without antihyperuricemic treatment. Hyperuricosuria frequently leads to stone formation in the renal medullae or the urological system in untreated patients and even after control of uric acid production with xanthine dehydrogenase inhibitor such as allopurinol.

PRPP Synthetase Overactivity (Superactivity)

PRPS catalyzes the transfer of the pyrophosphate group of ATP to ribose 5-phosphate to form PRPP. PRPP, an important regulator of de novo purine synthesis pathways, is a substrate in the initial step of the de novo pathway and for purine salvage reactions (Figure 38.2). The molecular weight of PRPS, widely expressed, is about 1 000 000 (Becker et al 1977, Taira et al 1989). PRPS is an aggregate complex of two catalytic subunits, PRPS1 and PRPS2, and two associated subunits, PRPP synthetase-associated protein (PRPSAP) 1 and 2 (Becker et al 1990, Ishizuka et al

1996a, Katashima et al 1998). PRPS1 gene and PRPS2 gene are located on the X chromosome at Xq22-24 and Xp22.3-22.2, PRPSAP1 gene and PRPSAP2 gene on the 17q24-q25 and 17p12-p11.2, respectively (Becker et al 1990, Ishizuka et al 1996, Katashima et al 1998). PRPS3 cDNA was also identified and maps to human chromosome 7, but is transcribed only in the testes (Taira et al 1989, 1990).

PRPS overactivity was described at first as a familial disorder characterized by excessive purine production, gout, and accelerated erythrocyte PRPS in the early 1970s (Sperling et al 1972, Becker et al 1973). PRPS overactivity is an X-1inked recessive disorder and about 30 families have been reported. Only a small number of point mutations in PRPS1 has been identified in patients with PRPS overactivity (Figure 38.4). Values for PRPS activity in PRPS overactivity have varied widely and the variant enzymes are insensitive to normal regulation or with catalytic activity 2–4 times greater activity than normal.

The clinical manifestations of PRPS overactivity include urolithiasis or gout due to hyperuricemia and hyperuricosuria similar to HPRT deficiency, and neurological deficits frequently including sensorineural deafness. There is a wide spectrum of the neurological deficits in severity. Patients with the greater severity show symptoms such as sensorineural deafness, cerebellar ataxia, muscular hypotonia, mental and motor retardation since early childhood and signs of uric acid overproduction (Simmonds et al 1982, Christen et al 1992). Two-thirds of the patients with PRPS overactivity, however, have presented with severe gout or kidney stones without neurological deficits in adolescence or early adulthood. The heterozygous females are sometimes hyperuricemic, hyperuricosuric and deaf in the premenopausal period. It should be suspected in any child or young adult of either sex with marked hyperuricemia and hyperuricosuria, but with normal HPRT activity in lysed red cells.

Glycogen Storage Disease

Hyperuricemia and gout frequently occur in glycogen storage disease types I and VII, and less so in types III and V (Mineo et al 1987, Maire et al 1991, Talente et al 1994, Rake et al 2002).

GLYCOGEN STORAGE DISEASE TYPE I

Glycogen storage disease type I (GSD I; Von Gierke disease) is an autosomal recessive inborn error of carbohydrate metabolism caused by defects of the glucose-6-phosphatase (G6Pase) complex. G6Pase catalyzes the hydrolysis of glucose-6-phosphate (G6P) to glucose and phosphate in the terminal steps of gluconeogenesis and glycogenolysis. The genes responsible for GSD I including G6Pase and G6P transporter genes have been identified and various mutations in the genes of GSD I patients have been reported (Lei et al 1993, Lin et al 1998). G6Pase deficiency results in excessive accumulation of glycogen in the liver and kidney, leading to progressive hepatomegaly and renal enlargement. Clinical manifestations of GSD I including coma, seizures, irritability and increased respiratory rate caused by hypoglycemia, lactic acidosis and ketonemia, and hepatomegaly, present in early infancy. The recurrent hypoglycemia leads to the elevation of plasma glucagon levels, activating glycogen phosphorylase. The activation promotes the further elevation of G6P levels, resulting in a decrease of intrahepatic phosphate that inhibits AMP deaminase. This decrease stimulates AMP deaminase, resulting in the degradation of adenine nucleotides and consequent overproduction of uric acid (Greene et al 1978, Cohen et al 1985). Lactic acidosis and ketonemia also decreases renal uric acid excretion by stimulation of uric acid reabsorption via URAT1. Conclusively, both overproduction and underexcretion of uric acid cause hyperuricemia in GSD I.

GLYCOGEN STORAGE DISEASE TYPES III, V, AND VII

Glycogen storage disease types VII (Tarui desease; phosphofructokinase deficiency), V (McArdle disease; glycogen phosphorylase deficiency), and III (glycogen debranching enzyme deficiency) are autosomal recessive disorders. Muscle-related manifestations, muscle cramps with exertion and myoglobinuria with extreme exertion, are typical clinical features of glycogen storage disease types VII and V, while hepatomegaly and hypoglycemia, sometimes muscle weakness and wasting type III, occur. Phosphofructokinase, a rate-limiting enzyme in the glycolysis pathway, and muscle glycogen phosphorylase, catalyze the irreversible conversion of fructose-6-phosphate to fructose-1,6-bisphosphate and the breakdown of glycogen to glucose-1-phosphate, respectively. Glycogen debranching enzyme has the two catalytic activities, amylo-1,6-glucosidase and oligo-1,4-1,4-glucanotransferase, that function independently at separate catalytic sites.

The mechanism of myogenic hyperuricemia in glycogen storage disease types VII, V, and III is related to excessive ATP breakdown in muscle during exercise as a result of impaired ATP synthesis (Mineo et al 1987). Excessive ATP breakdown and an accumulation of ADP or AMP result in the increase of the degradation to purine metabolites leading to overproduction of uric acid, the end metabolite of purines.

SINGLE GENE DISORDER FOR DECREASED EXCRETION TYPE HYPERURICEMIA

Familial Juvenile Hyperuricemic Nephropathy (FJHN, Familial Juvenile Houty Nephropathy: FJGN)

CLINICAL ENTITY

Familial juvenile hyperuricemic nephropathy (MIM 162000) is an autosomal-dominant disease characterized by underexcretion type hyperuricemia irrespective of gender or age, and progressive renal failure. FJHN is a rare disease and dozens of affected families have been reported.

HISTORY AND MOLECULAR BASIS

Since FJHN was first described in 1960 (Duncan & Dixon 1960), FJHN has been reported in many countries. From the facts that FJHN patients had hyperuricemia and a severely reduced FEua disproportionate to renal function, that some FJHN patients presented normal renal function and that the percentage of the FJHN patients with normal renal function in children of FJHN families was higher than that in adults, hyperuricemia due to uric acid underexcretion antedates renal insufficiency (Calabrese et al 1990, Moro et al 1991, McBride et al 1998). The nephropathy was mostly suspected to result from a mechanism other than urate deposition because of few urate crystal deposits and no distinctive feature of gouty nephropathy in renal specimens and no effect of allopurinol for renal deterioration (Puig et al 1993), though some reports described efficacy of allopurinol (Moro et al 1991, Fairbanks et al 2002).

Medullary cystic kidney disease (MCKD) in FJHN-affected families has been reported (Thompson et al 1978, Burke et al 1982) and the possibility of these two diseases being identical was suggested (Burke et al 1982). Autosomal-dominant MCKD is classified into MCKD1 (MIM 174000) mapped to chromosome 1q, with MCKD2 (MIM 603860) mapped to chromosome 16p (Christodoulou et al 1998, Scolari et al 1999). Dahan et al reported the linkage between FJHN and markers within the 16p12 locus in a Belgian family and proposed that FJHN and MCKD2 might be allelic disorders based on the similar location of the gene loci as well as the clinical and pathologic similarity (Dahan et al 2001).

Hart et al identified four novel uromodulin (also known as Tamm-Horsfall protein) gene mutations that segregate with the disease phenotype in three families with FJHN and in one family with MCKD2 and demonstrated that *UMOD* was responsible for FJHN and MCKD2 (Hart et al 2002). After another group independently identified *UMOD* responsible for FJHN (Turner et al 2003), this finding was confirmed by other groups (Bleyer et al 2003, Dahan et al 2003, Rampoldi et al 2003, Wolf et al 2003, Kudo et al 2004, Rezende-Lima et al 2004). Furthermore, *UMOD* was also suggested to be causative of a part of glomerulocystic kidney disease (GCKD) (Rampoldi et al 2003).

On the other hand, a family with FJHN that did not link *UMOD* loci and another family with MCKD and early onset gout linked to neither MCKD1 nor MCKD2 loci have been reported (Auranen et al 2001, Ohno et al 2002). Recently other candidate genes for FJHN, such as the hepatocyte nuclear factor (HNF)-1beta gene, have been reported (Bingham et al 2003). Mutations in the HNF-1beta gene have been associated with renal development disorders including renal cystic disease and early onset diabetes (Bingham et al 2001). Since about 70–80% of families with FJHN are linked with 16p11-p13, *UMOD* is attributed to be responsible for most FJHN and other genes in part (Stacey et al 2003, Kudo et al 2004). However, the criteria and the categorization have been confused because of the diagnosis of these cystic diseases from clinical and pathological features. Thus, responsible gene identification will promote the reclassification of these diseases.

CLINICAL FINDINGS

Renal disease in affected family members is progressive and end-stage renal disease usually occurs in the fourth through seventh decades (Bleyer et al 2003). Women are affected as well as men. The presence of some women and young men with hyperuricemia or gout in a family is highly suggestive of FJHN. Underexcretion type hyperuricemia is found in approximately 50% of FJHN family members

(McBride et al 1998, Hart et al 2002). Hart et al reported that hyperuricemia was found in the vast majority of FJHN family members with *UMOD* deficit, but not all and about 90% of *UMOD* deficit family members had an estimated creatinine clearance of less than 90 ml/min when measured after the age of 18 years (Hart et al 2002). Cyst formation in the kidney of patients with FJHN is about 10%. However, examination for renal cysts in patients with FJHN is not fully performed. Since FJHN and MCKD2 result from a *UMOD* defect, the frequency of renal cysts in patients with FJHN may be higher.

PATHOPHYSIOLOGY OF FJHN

Since most mutations observed in FJHN family were cysteins or calcium-binding sites of uromodulin molecules, it has been hypothesized that these mutations could lead to the misfolding of global protein structure by the loss of disulfide bonds or reducing calcium-binding affinity. This misfolding may delay the maturation of the protein and hence induce intracellular accumulation of mutant protein in the cells of the thick ascending limb (TAL) of Henle in patients (Scolari et al 2004). This hypothesis is supported by the following observations: mutant *UMOD* transfected cells expressed the mutant protein mainly in the endoplasmic reticulum (ER) and reduced dramatically the expression on the plasma membrane observed in wild *UMOD* transfected cells; MCKD2 and GCKD renal biopsy specimens also demonstrated that uromodulin was found in patchy deposits in the cytoplasm along TAL (Rampoldi et al 2003); urinary uromodulin content in patients with FJHN showed a consistent decrease in patients with *UMOD* mutations compared with healthy subjects and patients with chronic renal failure from other causes (Dahan et al 2003). Intracellular accumulation of mutant uromodulin may induce the dysfunction of sodium reabsorption by TAL similarly to the inhibition of the Na-K-2Cl cotransporter (NKCC2) by loop diuretics which induces hyperuricemia as a side effect. Therefore, hyperuricemia observed in an FJHN patient is suggested to be due to the same mechanism of hyperuricemia induced by loop diuretics. Decrease of sodium reabsorption by TAL is also suggested to cause the elevation of NaCl concentration in the luminal fluid delivered to macula densa leading to reduction of GFR by the tubuloglomerular feedback. However, there is a pathophysiological difference between FJHN and loop diuretics treatment. Since loop diuretics inhibit NKCC2, which express on macula densa cells for sensing electrolyte concentration in the tubular fluid (Vallon 2003), the tubuloglomerular feedback function of the macula densa must be attenuated by loop diuretics. On the other hand, the tubuloglomerular feedback function of macula densa cells would be intact in the FJHN patient and be sensitive to the slight change of NaCl concentration in the luminal fluid delivered to macula densa cells, since uromodulin does not express in macula densa cells (Figure 38.5).

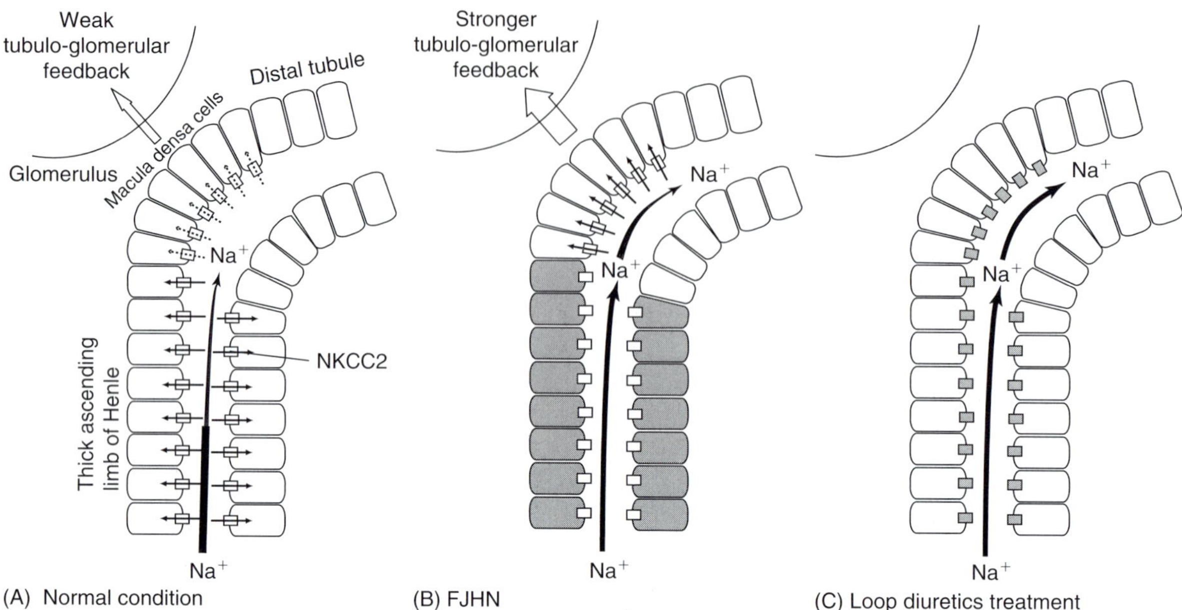

FIGURE 38.5   Hypothesis of pathophysiology of FJHN. (A) Normal NaCl delivery to macula densa through thick ascending limb of Henle (TAL) in a healthy person. Since NaCl in the luminal fluid of TAL is reabsorbed by the Na-K-2Cl co-transporter (NKCC2), low NaCl concentration in the fluid delivered to the macula densa induces weak tubulo-glomerular feedback. (B) Increased sodium delivery to the macula densa through TAL in the FJHN patient. Since the aggregation of uromodulin mutant inhibits NaCl reabsorption through TAL cells (shaded in the figure), higher NaCl concentration in the fluid delivered to intact macula densa induces stronger tubulo-glomerular feedback. (C) Increased NaCl delivery to macula densa through TAL by loop diuretics treatment. Since loop diuretics inhibit sodium reabsorption by NKCC2 (shaded rectangle in the figure) at both TAL and macula densa cells, high NaCl concentration in the fluid delivered to the macula densa does not induce tubulo-glomerular feedback

TREATMENT

Some reports assert that allopurinol might restrain progression to renal failure (Moro et al 1991, McBride et al 1998). Since the pathogenesis of renal failure in FJHN differs from hyperuricemic nephropathy, efficacy of allopurinol is unknown and no candidate medication for FJHN has been reported.

References

Anderson LT, Ernst M. Self-injury in Lesch-Nyhan disease. J. Autism Dev. Disord. 1994; 24(1): 67–81.

Auranen M, Ala-Mello S, et al. Further evidence for linkage of autosomal-dominant medullary cystic kidney disease on chromosome 1q21. Kidney Int. 2001; 6(4): 1225–32.

Becker MA, Heidler SA, et al. Cloning of cDNAs for human phosphoribosylpyrophosphate synthetases 1 and 2 and X chromosome localization of PRPS1 and PRPS2 genes. Genomics 1990; 8(3): 555–61.

Becker MA, Kim M. Regulation of purine synthesis de novo in human fibroblasts by purine nucleotides and phosphoribosylpyrophosphate. J. Biol. Chem. 1987; 262(30): 14531–7.

Becker MA, Meyer LJ, et al. Purine overproduction in man associated with increased phosphoribosylpyrophosphate synthetase activity. Science 1973; 179(78): 1123–6.

Becker MA, Meyer LJ, et al. Human erythrocyte phosphoribosylpyrophosphate synthetase. Subunit analysis and states of subunit association. J. Biol. Chem. 1977; 252(11): 3911–18.

Berger L, Yu TF. Renal function in gout. IV. An analysis of 524 gouty subjects including long-term follow-up studies. Am. J. Med. 1975; 59(5): 605–13.

Bingham C, Bulman MP, et al. Mutations in the hepatocyte nuclear factor-1beta gene are associated with familial hypoplastic glomerulocystic kidney disease. Am. J. Hum. Genet. 2001; 68(1): 219–24.

Bingham C, Ellard S, et al. Atypical familial juvenile hyperuricemic nephropathy associated with a hepatocyte nuclear factor-1beta gene mutation. Kidney Int. 2003; 63(5): 1645–51.

Bleyer AJ, Trachtman H, et al. Renal manifestations of a mutation in the uromodulin (Tamm Horsfall protein) gene. Am. J. Kidney Dis. 2003a; 42(2): E20–6.

Bleyer AJ, Woodard AS, et al. Clinical characterization of a family with a mutation in the uromodulin (Tamm-Horsfall glycoprotein) gene. Kidney Int. 2003b; 64(1): 36–42.

Burke JR, Inglis JA, et al. Juvenile nephronophthisis and medullary cystic disease – the same disease (report of a large family with medullary cystic disease associated with gout and epilepsy). Clin. Nephrol. 1982; 18(1): 1–8.

Calabrese G, Simmonds HA, et al. Precocious familial gout with reduced fractional urate clearance and normal purine enzymes. Q. J. Med. 1990; 75(277): 441–50.

Christen HJ, Hanefeld F, et al. Distinct neurological syndrome in two brothers with hyperuricaemia. Lancet 1992; 340(8828): 1167–8.

Christodoulou K, Tsingis M, et al. Chromosome 1 localization of a gene for autosomal dominant medullary cystic kidney disease. Hum. Mol. Genet. 1998; 7(5): 905–11.

Cohen JL, Vinik A, et al. Hyperuricemia in glycogen storage disease type I. Contributions by hypoglycemia and hyperglucagonemia to increased urate production. J. Clin. Invest. 1985; 75(1): 251–7.

Crawhall JC, Henderson JF, et al. Diagnosis and treatment of the Lesch-Nyhan syndrome. Pediatr. Res. 1972; 6(5): 504–13.

Dahan K, Devuyst O, et al. A cluster of mutations in the UMOD gene causes familial juvenile hyperuricemic nephropathy with abnormal expression of uromodulin. J. Am. Soc. Nephrol. 2003; 14(11): 2883–93.

Dahan K, Fuchshuber A, et al. Familial juvenile hyperuricemic nephropathy and autosomal dominant medullary cystic kidney disease type 2: two facets of the same disease? J Am. Soc. Nephrol. 2001; 12(11): 2348–57.

Duncan H, Dixon AS. Gout, familial hypericaemia, and renal disease. Q. J. Med. 1960; 29: 127–35.

Fairbanks LD, Cameron JS, et al. Early treatment with allopurinol in familial juvenile hyerpuricaemic nephropathy (FJHN) ameliorates the long-term progression of renal disease. Q.J.M. 2002; 95(9): 597–607.

Greene HL, Wilson FA, et al. ATP depletion, a possible role in the pathogenesis of hyperuricemia in glycogen storage disease type I. J. Clin. Invest. 1978; 62(2): 321–8.

Hart TC, Gorry MC, et al. Mutations of the UMOD gene are responsible for medullary cystic kidney disease 2 and familial juvenile hyperuricaemic nephropathy. J. Med. Genet. 2002; 39(12): 882–92.

Holmes EW, McDonald JA, et al. Human glutamine phosphoribosylpyrophosphate amidotransferase. Kinetic and regulatory properties. J. Biol. Chem. 1973; 248(1): 144–50.

Holmes EW, Pehlke DM, et al. Human IMP dehydrogenase. Kinetics and regulatory properties. Biochim. Biophys. Acta 1974; 364(2): 209–17.

Ishizuka T, Ahmad I, et al. The human phosphoribosylpyrophosphate synthetase-associated protein 39 gene (PRPSAP1) is located in the chromosome region 17q24-q25. Genomics 1996a 33(2): 332–4.

Ishizuka T, Kita K, et al. Cloning and sequencing of human complementary DNA for the phosphoribosylpyrophosphate synthetase-associated protein 39. Biochim. Biophys. Acta 1996b 1306(1): 27–30.

Jolly DJ, Esty AC, et al. Isolation of a genomic clone partially encoding human hypoxanthine phosphoribosyltransferase. Proc. Natl Acad. Sci. USA 1982; 79(16): 5038–41.

Katashima R, Iwahana H, et al. Molecular cloning of a human cDNA for the 41-kDa phosphoribosylpyrophosphate synthetase-associated protein. Biochim. Biophys. Acta 1998; 1396(3): 245–50.

Kaufman JM, Greene ML, et al. Urine uric acid to creatinine rtio – a screening test for inherited disorders of purine metabolism. Phosphoribosyltransferase (PRT) deficiency in X-linked cerebral palsy and in a variant of gout. J. Pediatr. 1968; 73(4): 583–92.

Kudo E, Kamatani N, et al. Familial juvenile hyperuricemic nephropathy: detection of mutations in the uromodulin gene in five Japanese families. Kidney Int. 2004; 65(5): 1589–97.

Lei KJ, Shelly LL, et al. Mutations in the glucose-6-phosphatase gene that cause glycogen storage disease type 1a. Science 1993; 262(5133): 580–3.

Lin B, Annabi B, et al. Cloning and characterization of cDNAs encoding a candidate glycogen storage disease type 1b protein in rodents. J. Biol. Chem. 1998; 273(48): 31656–60.

Maire I, Baussan C, et al. Biochemical diagnosis of hepatic glycogen storage diseases: 20 years French experience. Clin. Biochem. 1991; 24(2): 169–78.

McBride MB, Rigden S, et al. Presymptomatic detection of familial juvenile hyperuricaemic nephropathy in children. Pediatr. Nephrol. 1998; 12(5): 357–64.

Mineo I, Kono N, et al. Myogenic hyperuricemia. A common pathophysiologic feature of glycogenosis types III, V, and VII. N. Engl. J. Med. 1987; 317(2): 75–80.

Moro F, Ogg CS, et al. Familial juvenile gouty nephropathy with renal urate hypoexcretion preceding renal disease. Clin. Nephrol. 1991a; 35(6): 263–9.

Moro F, Simmonds HA, et al. Does allopurinol affect the progression of familial juvenile gouty nephropathy?. Adv. Exp. Med. Biol. 1991b; 309A: 199–202.

Murray AW. The biological significance of purine salvage. Annu. Rev. Biochem. 1971; 40: 811–26.

Ohno I, Ichida K, et al. Familial juvenile gouty nephropathy: exclusion of 16p12 from the candidate locus. Nephron 2002; 92(3): 573–5.

Pai GS, Sprenkle JA, et al. Localization of loci for hypoxanthine phosphoribosyltransferase and glucose-6-phosphate dehydrogenase and biochemical evidence of nonrandom X chromosome expression from studies of a human X-autosome translocation. Proc. Natl Acad. Sci. USA 1980; 77(5): 2810–13.

Perez-Ruiz F, Calabozo M, et al. Renal underexcretion of uric acid is present in patients with apparent high urinary uric acid output. Arthritis Rheum. 2002; 47(6): 610–13.

Pourquie J. Regulation of the biosynthesis of purine nucleotides in *Schizosaccharomyces pombe*. II. Kinetic studies of IMP dehydrogenase. Biochim. Biophys. Acta 1969; 185(2): 310–15.

Puig JG, Miranda ME, et al. Hereditary nephropathy associated with hyperuricemia and gout. Arch. Intern. Med. 1993; 153(3): 357–65.

Rake JP, Visser G, et al. Glycogen storage disease type I: diagnosis, management, clinical course and outcome. Results of the European Study on Glycogen Storage Disease Type I (ESGSD I). Eur. J. Pediatr. 2002; 161(Suppl.1): S20–34.

Rampoldi L, Caridi G, et al. Allelism of MCKD, FJHN and GCKD caused by impairment of uromodulin export dynamics. Hum. Mol. Genet. 2003; 12(24): 3369–84.

Rezende-Lima W, Parreira KS, et al. Homozygosity for uromodulin disorders: FJHN and MCKD-type 2. Kidney Int. 2004; 66(2): 558–63.

Rieselbach RE, Sorensen LB, et al. Diminished renal urate secretion per nephron as a basis for primary gout. Ann. Intern. Med. 1970; 73(3): 359–66.

Scolari F, Caridi G, et al. Uromodulin storage diseases: clinical aspects and mechanisms. Am. J. Kidney Dis. 2004; 44(6): 987–99.

Scolari F, Puzzer D, et al. Identification of a new locus for medullary cystic disease, on chromosome 16p12. Am. J. Hum. Genet. 1999; 64(6): 1655–60.

Shows TB, Brown JA. Localization of genes coding for PGK, HPRT, and G6PD on the long arm of the X chromosome in somatic cell hybrids. Cytogenet. Cell Genet. 1975; 14(3–6): 426–9.

Simkin PA. New standards for uric acid excretion and evidence for an inducible transporter. Arthritis Rheum. 2003; 49(5): 735–6, author reply 736–7.

Simmonds HA, Webster DR, et al. An X-linked syndrome characterised by hyperuricaemia, deafness, and neurodevelopmental abnormalities. Lancet 1982; 2(8289): 68–70.

Smith JL, Zaluzec EJ, et al. Structure of the allosteric regulatory enzyme of purine biosynthesis. Science 1994; 264(5164): 1427–33.

Sperling O, Eilam G, et al. Accelerated erythrocyte 5-phosphoribosyl-1-pyrophosphate synthesis. A familial abnormality associated with excessive uric acid production and gout. Biochem. Med. 1972; 6(4): 310–16.

Stacey JM, Turner JJ, et al. Genetic mapping studies of familial juvenile hyperuricemic nephropathy on chromosome 16p11-p13. J. Clin. Endocrinol. Metab. 2003; 88(1): 464–70.

Taira M, Iizasa T, et al. Tissue-differential expression of two distinct genes for phosphoribosyl pyrophosphate synthetase and existence of the testis-specific transcript. Biochim. Biophys. Acta 1989; 1007(2): 203–8.

Taira M, Iizasa T, et al. A human testis-specific mRNA for phosphoribosylpyrophosphate synthetase that initiates from a non-AUG codon. J. Biol. Chem. 1990; 265(27): 16491–7.

Talente GM, Coleman RA, et al. Glycogen storage disease in adults. Ann. Intern. Med. 1994; 120(3): 218–26.

Thompson GR, Weiss JJ, et al. Familial occurrence of hyperuricemia, gout, and medullary cystic disease. Arch. Intern. Med. 1978; 138(11): 1614–17.

Turner JJ, Stacey JM, et al. Uromodulin mutations cause familial juvenile hyperuricemic nephropathy. J. Clin. Endocrinol. Metab. 2003; 88(3): 1398–401.

Vallon V. Tubuloglomerular feedback in the kidney: insights from gene-targeted mice. Pflugers Arch. 2003; 445(4): 470–6.

Van der Weyden MB, Kelly WN. Human adenylosuccinate synthetase. Partial purification, kinetic and regulatory properties of the enzyme from placenta. J. Biol. Chem. 1974; 249(22): 7282–9.

Wilson JM, Kelley WN. Molecular basis of hypoxanthine-guanine phosphoribosyltransferase deficiency in a patient with the Lesch-Nyhan syndrome. J. Clin. Invest. 1983; 71(5): 1331–5.

Wilson JM, Tarr GE, et al. Human hypoxanthine (guanine) phosphoribosyltransferase: an amino acid substitution in a mutant form of the enzyme isolated from a patient with gout. Proc. Natl Acad. Sci. USA 1983; 80(3): 870–3.

Wolf MT, Mucha BE, et al. Mutations of the *Uromodulin* gene in MCKD type 2 patients cluster in exon 4, which encodes three EGF-like domains. Kidney Int. 2003; 64(5): 1580–7.

Wyngaarden JB, Kelley WN. Gout and Hyperuricemia. New York: Grune & Stratton, 1976: pp. 1–1512.

Yamaoka T, Kondo M, et al. Amidophosphoribosyltransferase limits the rate of cell growth-linked de novo purine biosynthesis in the presence of constant capacity of salvage purine biosynthesis. J. Biol. Chem. 1997; 272(28): 17719–25.

CHAPTER 39

Hereditary Cystinosis

VASILIKI KALATZIS AND CORINNE ANTIGNAC

INTRODUCTION

Cystinosis is a lysosomal transport disorder characterized by an intralysosomal accumulation of cystine, the disulfide of the amino acid cysteine. It is the most common inherited cause of the renal Fanconi syndrome. There exist various clinical forms, infantile, juvenile and ocular, based on age of onset and severity of the symptoms. The first clinical description appeared in the early 1900s but it was not until 1998 that the causative gene, *CTNS*, was identified. *CTNS* encodes cystinosin, a novel seven transmembrane domain (TM), lysosomal membrane protein. Cystinosin acts to transport cystine out of the lysosome and its activity is H^+-driven. The functional analysis of *CTNS* mutations associated with all clinical forms, suggests that impaired cystine transport is the cause of pathogenicity and that the level of transport inhibition correlates with the severity of symptoms. A mouse model of cystinosis has been generated and $Ctns^{-/-}$ mice accumulate cystine in various tissues (liver, kidney, lung, spleen, muscle, brain, and heart). However, despite high cystine levels in the kidney, the $Ctns^{-/-}$ mice do not present with proximal tubulopathy or renal dysfunction.

This chapter will review the most recent advances in the molecular biology of cystinosis. There is particular emphasis on the causative *CTNS* mutations identified to date, as well as a correlation between genotype and the differing clinical phenotypes.

CLINICAL COURSE

Cystinosis is a rare autosomal recessive monogenic disorder with an incidence of ~1/200 000 live births, characterized by an intralysosomal accumulation of cystine. It constitutes the most common familial form of the renal Fanconi syndrome. A case of cystinosis was first documented in 1903 (Abderhalden 1903); however, the disease was more clearly defined as a distinct clinical entity in 1949 (Freudenberg 1949). In later years, three clinical forms were delineated based on the age of onset and severity of the renal impairment (for review see Gahl et al 2001). The most severe form, and the one reported in the early 1900s, is now known as infantile cystinosis (MIM 219800). The initial clinical manifestation of infantile cystinosis is proximal tubular dysfunction characterized by fluid and electrolyte loss, aminoaciduria, glycosuria, phosphaturia, hypercalciureia, renal tubular acidosis, carnitine loss, rickets and growth retardation, generally appearing between 6 and 12 months of age. This correlates with severe lesions of proximal tubules, observed even at early stages, with marked heterogeneity in their shape and caliber and focal disappearance of the brush border, leading ultimately to tubular atrophy (Gubler et al 1999). In addition to tubular lesions, peculiar alterations of glomeruli, with multinucleate giant podocytes, can be frequently found. Cystine crystal aggregates, which form when cystine levels reach high concentrations, can be found in interstitial cells as well as, rarely, in podocytes. They are usually absent from tubular cells. Renal glomerular filtration rate begins to deteriorate from 5–6 years and, in the absence of treatment, leads to end-stage renal disease (ERSD) by 10 years of age. The final end-stage kidney is highly atrophic, the glomeruli sclerotic and the tubular structures almost nonexistent (Spear 1974).

Although the kidney is the first organ to appear clinically affected in infantile cystinosis, widespread organ damage, including ocular, endocrinological, hepatic, muscular, and central nervous system (CNS) complications, has been reported. The earliest occurring extrarenal symptom of infantile cystinosis is photophobia, which appears between age 3 and 4 years and worsens with time, leading to varying degrees of discomfort in all patients older than 10 (Dufier et al 1987). This is due to the presence of corneal cystine crystals, detectable by slit lamp examination. These cystine deposits lead to stromal edema and epithelial ulcerations responsible for photophobia, pain, watering, and blepharospasm. Retinopathy can also be seen as early as 3 years,

Genetic Diseases of the Kidney
ISBN: 978-0-12-449851-8

is an almost constant feature by 7 years (Dufier 1992), and results in blindness in 10–15% of cases between the ages of 13 and 40 (Gahl et al 2002). Other clinical features include hypothyroidism, diabetes mellitus, portal hypertension, male hypogonadism, pulmonary dysfunction, myopathy, and CNS calcifications and deterioration (Gahl et al 2001).

The other two well-documented forms of the disease, juvenile (Goldman et al 1971) and ocular (Cogan et al 1957) cystinosis, are less severe and less frequent. Individuals with juvenile cystinosis (MIM 219900) present with a proteinuria between 12 and 15 years of age but do not suffer from severe tubulopathy or growth retardation. Glomerular function deteriorates at a lower rate than for infantile cystinosis and ESRD is reached at variable ages. Photophobia also appears during adolescence in these individuals. Individuals with ocular cystinosis (MIM 219750) do not manifest any renal anomalies. This form is characterized by an adult onset of mild photophobia. Finally, atypical forms, where individuals present with various clinical features of severity between infantile and juvenile cystinosis, have been recently described (Attard et al 1999, Kalatzis et al 2002) and suggest the existence of a larger spectrum of nephropathic disease forms than initially described.

METABOLIC DEFECT

The underlying metabolic defect of cystinosis is an accumulation of cystine in the lysosomes. Cystine, the disulfide of the amino acid cysteine, is a by-product of lysosomal protein hydrolysis, and is reduced to cysteine in the cytoplasm. Lysosome hydrolases, responsible for the digestion of macromolecules, have optimal activity at acid pH and the acidification of the lysosomal lumen is accomplished by the presence of an ATPase H^+-pump in its membrane. The hydrolytic digestion by-products exit the lysosome via specific membrane transporters for metabolic recycling. A defect in one of the lysosomal hydrolases or one of the transporters can result in the accumulation of an undegraded macromolecule or a digestion by-product, respectively. In the case of cystinosis, it is the lysosomal cystine transporter that is defective leading to an abnormal storage of cystine. It is due to its poor solubility that cystine forms crystals as its concentration increases.

The site of cystine storage in the organs of affected children was determined to be the lysosome in the 1960s. This was accomplished by differential centrifugation studies of cystinotic leukocytes (Schneider et al 1967a) and fibroblasts (Schneider et al 1967b), which showed that cystine cosedimented with lysosomal enzymes. As no lysosomal cystine-reducing systems were known to exist, it was suggested that defective lysosomal cystine transport was the likely cause of cystine accumulation. Support for this

hypothesis came in the 1980s from studies performed on whole lysosomes artificially loaded with high levels of cystine. This work showed that cystine was rapidly lost from artificially loaded normal lysosomes, whereas there was no cystine efflux from cystinotic lysosomes (Gahl et al 1982, Steinherz et al 1982). Furthermore, these studies showed that cystine transport must be carrier-mediated because the cystine efflux was saturable (Gahl et al 1982) and cystine countertransport was demonstrated across the lysosomal membrane (Gahl et al 1983) – two hallmarks of carrier-mediated transport. It thus became gradually accepted that lysosomal cystine storage associated with cystinosis was due to a defective transporter of cystine out of the lysosome. This unknown lysosomal cystine transporter was furthermore shown to be indirectly stimulated by Mg-ATP (Gahl et al 1982, Jonas et al 1982, 1983, Greene et al 1987, Smith et al 1987b).

DIAGNOSIS AND TREATMENT

A measure of leukocyte cystine levels using a cystine binding protein assay (Oshima et al 1974, Smith et al 1987a) is generally used to confirm a suspicion of cystinosis. Intraleukocyte cystine concentrations in affected individuals are 50–100-fold higher than in normal individuals; heterozygous carriers have cystine levels fivefold higher than unaffected individuals and are asymptomatic (Schneider et al 1967a). The only existing treatment for cystinosis is the aminothiol cysteamine, a drug that reduces intracellular cystine levels. Cysteamine enters the lysosome by a specific transporter for aminothiols or aminosulfides (Pisoni et al 1995) and reacts with cystine to produce a cysteine–cysteamine mixed disulfide. The mixed disulfide is structurally analogous to lysine and exits via the corresponding carrier (Pisoni et al 1985) to the cytosol where it is reduced to cysteine and cysteamine. This in turn allows the cycling of cysteamine between the cytoplasm and the lysosomes.

Oral cysteamine therapy, if administered early in the course of the disease and in high doses, delays the progression of renal insufficiency (Markello et al 1993). In addition, it prevents the appearance of hypothyroidism (Kimonis et al 1995), and decreases the severity of oral motor and swallowing dysfunction, which are correlated with generalized muscle atrophy (Sonies et al 2005). This indicates the beneficial effects of cysteamine for certain renal and extrarenal complications; however, it has not been in use long enough to know whether it has an effect on later-onset complications, such as CNS anomalies. In contrast, for reasons as yet unknown, it is ineffective against the renal Fanconi syndrome. Moreover, the drug needs to be taken every 6 hours and provokes undesirable side effects – in particular, digestive intolerance and a persistent nauseating odor (Gahl et al 1987). Oral cysteamine is not effective against the ocular anomalies and topical application is required. This is accomplished in the form of cysteamine eye drops, which, similarly to their oral counterparts, need to be administered 10–12 times a day in order to be effective (Gahl et al 2000).

CAUSATIVE GENE AND ENCODED PROTEIN

Our understanding of cystinosis progressed significantly in 1998, with the positional cloning of the causative gene, *CTNS* (Town et al 1998). *CTNS* was first mapped to the short arm of chromosome 17 (17p13) using linkage analysis, by an international consortium in 1995 (the Cystinosis Collaborative Research Group 1995). It is composed of 12 exons, the first two of which are noncoding (Town et al 1998). The remaining ten exons encode a 367 amino acid (aa) protein called cystinosin. Cystinosin is predicted to contain seven transmembrane domains (TM), a 128 aa N-terminal region bearing seven N-glycosylation sites and a noncleavable signal peptide, and a 10 aa cytosolic C-terminal tail. At this time, cystinosin showed no homology to any known proteins and, thus, in the first instance, its presumed lysosomal localization needed to be confirmed.

The subcellular localization of cystinosin was initially determined by fusing an enhanced green fluorescent protein (EGFP) to the C-terminal end and following the fluorescence of the cystinosin–EGFP fusion protein in transfected cells (Figure 39.1) (Cherqui et al 2001). The fluorescence localized to intracellular vesicles, which were identified as lysosomes by colocalization studies with LAMP-2, a lysosomal membrane protein (Figure 39.1A, B). The localization of cystinosin to the lysosomal membrane was later confirmed by the successful generation of anticystinosin antibodies raised in rabbits against the 13 C-terminal amino acids of cystinosin (Haq et al 2002). Furthermore, these antibodies permitted the analysis of the distribution of cystinosin in the human kidney (Figure 39.2). The cytoplasm of the proximal tubular cells contained abundant cystinosin; the labeling was microgranular and colocalized with LAMP-2(Figure 39.2A, B). In contrast, a weaker labeling of the distal tubules and only a faint labeling of the glomeruli could be seen (Figure 39.2C, D). This differential localization pattern suggests that cystinosin is highly required in proximal tubular cells but not in glomeruli, which could explain why the proximal tubulopathy is the first clinical sign of cystinosis, whereas glomerular impairment does not appear until later in the evolution of the disease.

Subsequently, the signals responsible for the targeting of cystinosin to the lysosome were identified. In the

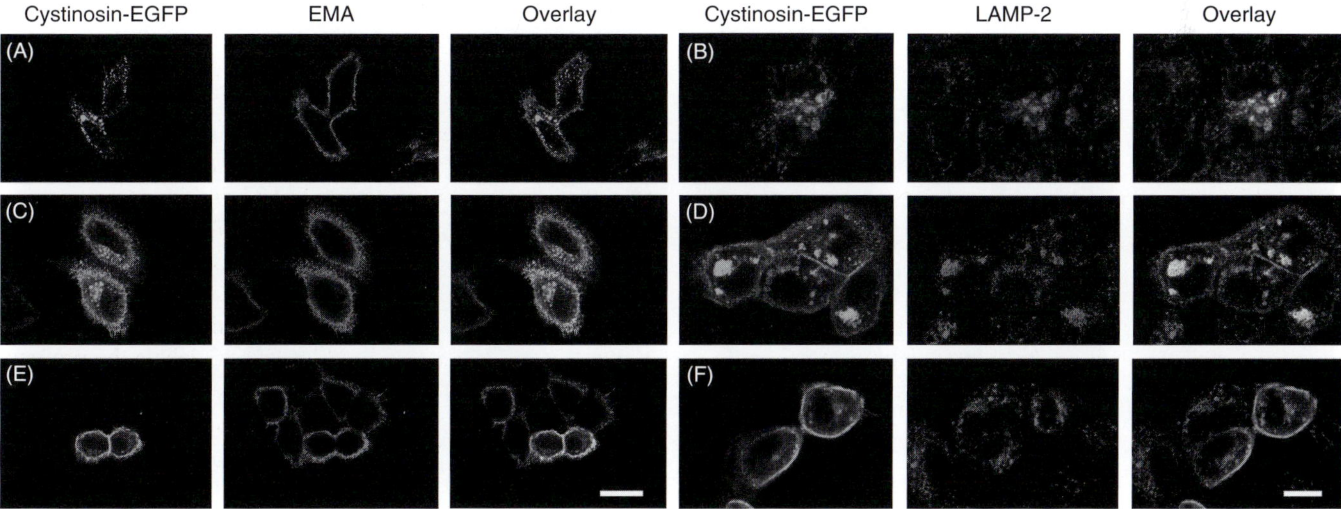

FIGURE 39.1   Subcellular localization of wild-type and mutated cystinosin in HeLa (A, C and E) and MDCK (B, D, and E) cells. For HeLa cells, the cystinosin-EGFP fluorescence (left and right panels) is compared to the immunoreactivity of the endogenous epithelial membrane antigen (EMA) (middle and right panels). Scale bar = 20 μm. For MDCK cells, the cystinosin-EGFP fluorescence (left and right panels) is compared to the immunoreactivity of the endogenous late ensosomal/lysosomal marker LAMP-2 (middle and right panels). Scale bar = 10 μm. (A, B) Wild-type cystinosin-EGFP exhibits an intracellular, punctate distribution corresponding to late endosomes and lysosomes. (C, D) Deletion of the C-terminal GYDQL lysosomal sorting motif (ΔGYDQL) results in the partial localisation of cystinosin-GFP to the plasma membrane with a signal remaining in the late endosomes/lysosomes. (E, F) An example of a complete relocalization of cystinosin to the plasma membrane, as can be seen following deletion of the YFPQA motif (5th inter-TM loop), the Q222R mutation (3rd TM), or the deletions IVFD343–346del and DVVF346–349del (7th TM), coupled to ΔGYDQL. Reproduced from Kalatzis et al 2004 with the permission of Oxford University Press. (see also Plate 85)

C-terminal tail of cystinosin resides a GYDQL motif, which resembles the classic tyrosine-based lysosomal targeting motif (GY-XX-Φ; where X is any aa and Φ is a hydrophobic residue) found in the tail of other lysosomal membrane proteins (Hunziker & Geuze 1996). Mutation or deletion of this motif, in particular of the tyrosine (Y) and lysine (L) residues, results in partial redirection to the plasma membrane (Figure 39.1C, D). A signal remained in the lysosomes suggestive of a second lysosomal targeting motif.

Following site-directed mutagenesis, a second signal was identified and initially mapped to the cytosolic 5th inter-TM loop (Cherqui et al 2001). Deletion of the first half of this loop, in particular a YFPQA pentapeptide, coupled to

the deletion of the GYDQL motif, resulted in a complete relocalization of cystinosin to the plasma membrane (Figure 39.1E, F). In contrast, point mutations of the YFPQA motif did not affect the localization of cystinosin, suggesting that the conformation of this motif is more important than its specific sequence. Therefore, it was considered that the YFPQA motif was likely recognized as part of a secondary structure formed by the 5th inter-TM loop. Support for the existence of this second lysosomal sorting signal came from the observations by Helip-Wooley et al (2002) that a *CTNS* splice site mutation, IVS11 + 2T > C, which results in a truncated protein missing the GYDQL motif but containing an intact YFPQA motif, was still partially localized

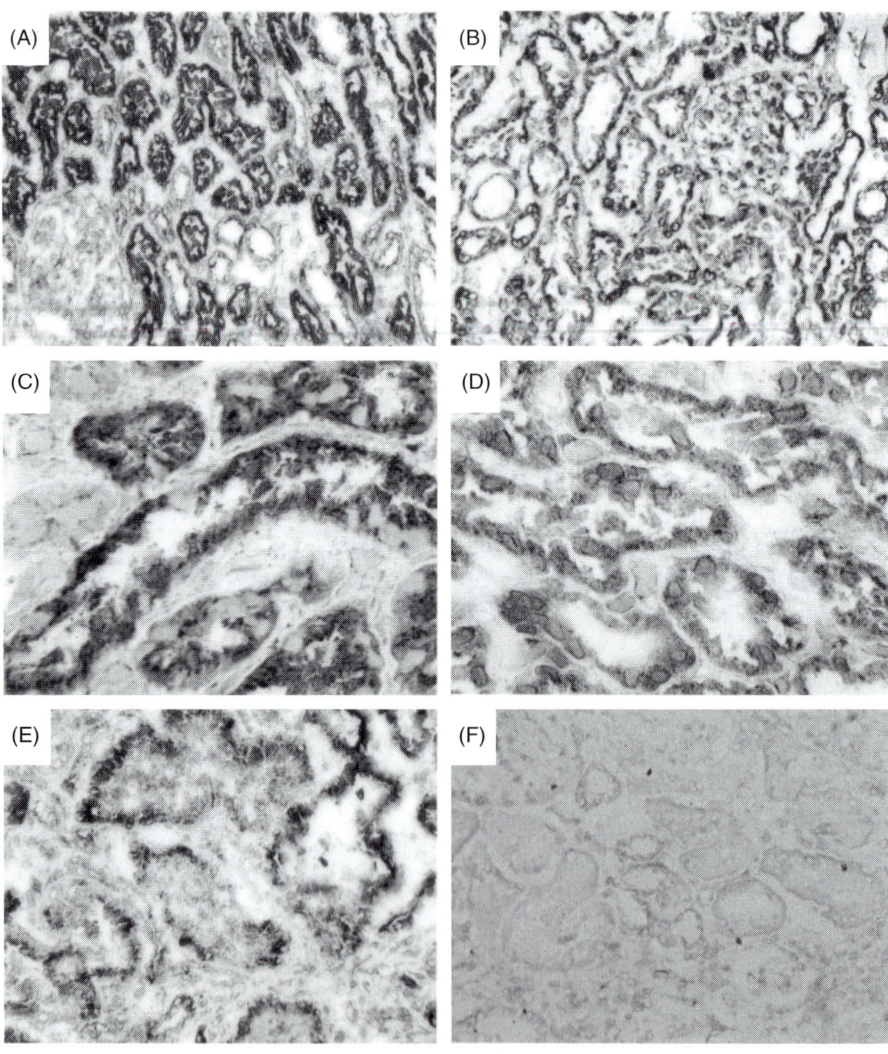

FIGURE 39.2   Distribution of cystinosin in the human kidney. Immunoperoxidase staining of human kidneys with anti-LAMP-2 (A, C, E) or anticystinosin (B, D, F) antibodies. A strong immunostaining of the proximal tubular cells in normal human kidneys can be seen with both anti-LAMP-2 (A) and anti-cystinosin (B) antibodies (magnification = × 120). (C, D) At higher magnification (× 300), this staining can be seen to be granular and diffusely distributed within the cytoplasm for both antibodies. (E) Immunostaining of cystinotic human kidneys using antiLAMP-2 antibodies results in the same labeling of proximal tubule cells (magnification = × 300). (F) In contrast, no significant signal was detected with the anticystinosin antibody in these sections (magnification = × 120). Reprinted from Haq et al 2002 with the permission of the Journal of the American Society of Nephrology. (see also Plate 86)

to lysosomes. Along this line, it has been recently shown that the lysosomal targeting of CLCN3, a membrane protein of unknown function involved in Batten disease, also requires two sorting signals: a conventional dileucine motif (D/E-X_n-LL) located in a cytosolic inter-TM loop and a novel M-X_9-G motif in the C-terminal tail (Kyttala et al 2004, Storch et al 2004).

FUNCTION OF CYSTINOSIN

Once the subcellular localization of cystinosin had been determined, the next question that needed to be addressed was – what role does cystinosin play in the lysosome? Although it has been accepted that the underlying metabolic defect of cystinosis is cystine transport, it was unknown whether this protein was directly or indirectly responsible. It was possible that cystinosin represented the lysosomal cystine transporter itself, however, neither its

sequence nor its predicted topology was similar to known transporters. Due to its similarity with G-protein-coupled receptors, which also have 7 TM, it was suggested that it could be part of a complex that binds, rather than transports, cystine (Attard et al 1999, Mancini et al 2000). Thus its function needed to be determined at the molecular level. As the lysosomal lumen is poorly accessible for transport experiments, cystinosin was redirected to the plasma membrane by deletion of the C-terminal GYDQL motif (cystinosin-ΔGYDQL), creating a cellular model in which the ability of cystinosin to transport cystine could be examined using whole cells (Figure 39.3A, B). Cells expressing cystinosin-ΔGYDQL were equivalent to giant 'inside-out' lysosomes and cystinosin activity could be assayed by its ability to take up ^{35}S-labeled cystine from the extracellular medium (Kalatzis et al 2001a).

Using this model it was shown that when incubated in a neutral extracellular medium, cystinosin-ΔGYDQL expressing cells accumulate ^{35}S-cystine poorly. In contrast, when these cells were placed into an acidic medium,

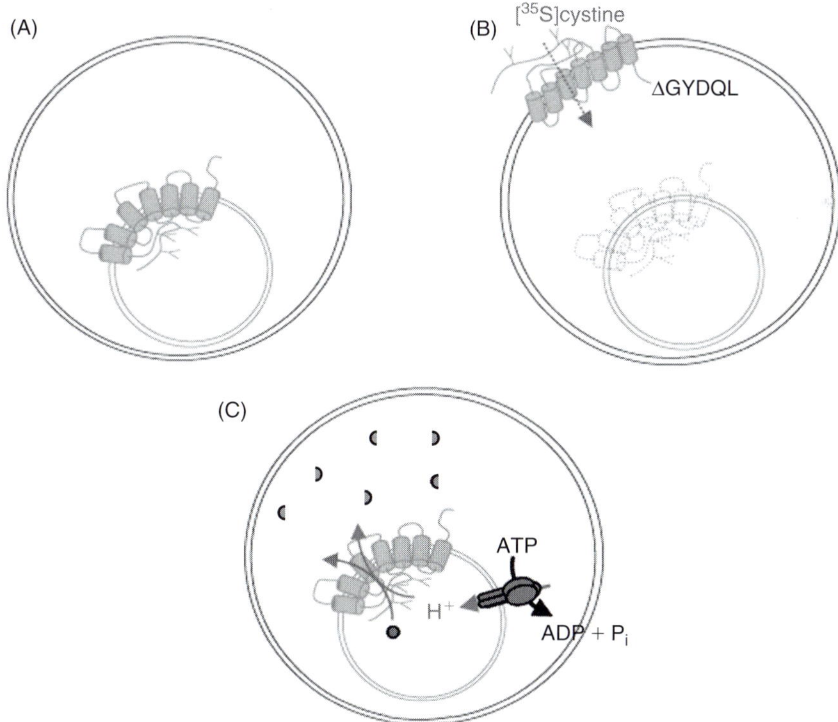

FIGURE 39.3   Function of cystinosin. Schematic representation of the in vitro assay used to determine the function of cystinosin. (A) In vivo, cystinosin (in green) is localized at the lysosomal membrane (smaller gray circle). The N-terminal region is oriented towards the lysosomal lumen and the C-terminal tail towards the cytosol. Cystinosin would transport cystine from the lysosomal lumen to the cytosol. (B) In the in vitro model, cystinosin-ΔGYDQL is relocalized to the plasma membrane (larger gray circle) and, as the extracellular medium is topologically equivalent to the lysosomal lumen, cystinosin-ΔGYDQL is assayed for its ability to transport ^{35}S-cystine from the extracellular medium into the cytosol. (C) Cystinosin is the lysosomal cystine transporter and its activity is H$^+$-driven. The interior of the lysosome is acidified by the H$^+$-ATPase (in red) present in its membrane. Lysosomal cystine (blue circle) efflux by cystinosin is coupled to an efflux of H$^+$. Thus, the influx of H$^+$ by the lysosomal H$^+$-ATPase drives cystinosin-mediated cystine transport toward the cytosol, where it is then reduced to cysteine (blue semi-circles). Reprinted with modifications by permission from The EMBO Journal (Kalatzis et al 2001a), copyright 2001 Macmillan Publishers Ltd. (http://www.nature.com). (see also Plate 87)

mimicking the interior of the lysosome, a dramatic increase in cystine uptake was seen. This observation provided the first direct proof that cystinosin is the lysosomal cystine transporter (Kalatzis et al 2001a). Interestingly, when the transmembrane pH gradient existing between the acid extracellular medium and the neutral cytosol is disrupted, cystine transport is abolished demonstrating that cystinosin cotransports cystine and protons (cystine/H^+-symporter). Thus the H^+-translocating ATPase that acidifies and positively charges the lysosomal lumen actively drives the cystinosin-mediated transport in the efflux direction (Figure 39.3C). The determination that the cystine transport activity of cystinosin is H^+-driven, now explains the initial observations on whole lysosomes that cystine efflux was indirectly stimulated by Mg-ATP (Gahl et al 1982, Greene et al 1987, Jonas et al 1982, 1983, Smith et al 1987b).

Cystinosin-mediated cystine transport is saturable and follows Michaelis-Menten kinetics with a mean K_M of $278 \pm 49\,\mu M$ (Kalatzis et al 2001a), consistent with studies performed on the native cystine transporter from human leukocytes (Gahl et al 1983) and mouse fibroblasts (Greene et al 1990). Furthermore, cystinosin is highly specific for L-cystine (Kalatzis et al 2001a). It does not transport other amino acids, including the monosulfide cysteine, which distinguishes it from the plasma membrane cystine transporters 4F2hc/xCT (Sato et al 1999) and rBAT/$b^{o,+}$AT (International Cystinuria Consortium 1999), and its activity is not stimulated by other ions (Kalatzis et al 2001a). Several distant cystinosin homologs have been identified, all of which share similar sizes and predicted topologies, and which show a weaker homology to the family of bacterial rhodopsins possessing 7 TM and predicted to transport H^+ (Zhai et al 2001). Furthermore, one of these distant homologs, the yeast vacuolar protein ERS1 (Pelham et al 1988, Hardwick & Pelham 1990), was recently shown to be a functional homolog of cystinosin (Gao et al 2005). Indeed, the mutant *ERS1* phenotype can be complemented by the human *CTNS* gene but not by mutant alleles associated with cystinosis. Thus, there is growing evidence that cystinosin defines a novel family of transporters characterized by a 7 TM topology.

The advantage of this in vitro model is that it overcomes the cumbersome task of isolating lysosomes and avoids the need for countertransport experiments, which is not a physiological process, for studying lysosomal transport. Its reliability having been proven, this model now represents a simple functional assay for the analysis of causative *CTNS* point mutations.

CTNS MUTATIONS

To date, approximately 90 different *CTNS* mutations have been detected in all forms of cystinosis (Tables 39.1, 39.2,

and 39.3) (Shotelersuk et al 1998, Town et al 1998, Attard et al 1999, McGowan-Jordan et al 1999, Thoene et al 1999, Anikster et al 2000, Kleta et al 2001, Kalatzis et al 2002, Kiehntopf et al 2002, Mason et al 2003, Pennesi et al 2005). This confirms the allelic status of the different clinical forms as was suggested prior to the cloning of the gene by complementation studies (Pellett et al 1988). In our experience, there is a high mutation detection rate associated with cystinosis, in particular in patients with the infantile form compared to the other clinical forms. We have tested 137 unrelated patients with infantile cystinosis and found mutations in 258 alleles, representing a 98% detection rate. In contrast, we have tested ten patients with juvenile, ocular, or atypical forms and have detected 14 mutant alleles, a mutation detection rate of 70%. This lower detection rate suggests that the mutations segregating in these forms may be situated outside the coding regions, such as in intronic or promoter sequences.

Two large deletions were initially identified as associated with infantile cystinosis, which were instrumental for defining the candidate cystinosis gene interval during the search for the causative gene (Town et al 1998). These are a 57 kb (Touchman et al 2000) and a 13 kb (Forestier et al 1999) deletion that remove the 5′ region of the gene upstream of and including exons 10 and 3, respectively. Interestingly, the 57 kb deletion also removes all of an adjacent gene of unknown function *CARKL* (Touchman et al 2000). However, the removal of this gene does not have an effect on the incidence on disease phenotype as, to date, patients carrying the 57 kb deletion show the same clinical features as those carrying a loss of function point mutation within *CTNS*.

The 13 kb mutation is relatively rare having been described initially in only one French family (Town et al 1998), but more recently in four of 42 patients mainly from Southern Italy (Mason et al 2003). In contrast, the 57 kb deletion is the most common mutation associated with cystinosis. It has been detected in 76% of cystinotic patients of European origin (Table 39.4) and has been shown to be due to a founder effect that probably arose in the middle of the first millennium (Forestier et al 1999), most likely in Germany (Shotelersuk et al 1998). Mason et al (2003) recently reported a lower incidence of this deletion in Northern Italy, as opposed to that reported in French (Forestier et al 1999), Dutch (Heil et al 2001), and Swiss and German (Anikster et al 1999, Kiehntopf et al 2002, Shotelersuk et al 1998) populations (Table 39.4). This could be consistent with a North to South gradient, supporting the hypothesis that the mutation occurred in Northern Europe and subsequently spread throughout Europe (Forestier et al 1999). PCR-based detection assays have been established providing the basis for a rapid molecular diagnostic test for cystinosis in Europe and for patients of European descent in other countries (Anikster et al 1999, Forestier et al 1999, Heil et al 2001). More recently a fluorescent in situ hybridization (FISH) detection assay has been developed, which,

TABLE 39.1   Point mutations associated with infantile cystinosis

Type of mutation	Position	Nucleotide change[a]	Amino acid change	Predicted effect on protein sequence[b]	Transport activity[c]	References
nt substitution	promoter	-295G > C[d]	–	Decreased protein production	nd	Phornphutkul et al 2001
missense	exon 3	340A > C	M1L	removal M start codon	nd	Mason et al 2003 Kalatzis et al 2002
	exon 3	342G > A	M1I	removal M start codon	4 ± 4.8	Kalatzis et al 2002
	exon 7	761C > T	S141F	aa change TM1	2.0 ± 5.3	Kalatzis et al 2004
	exon 8	812T > C	L158P	aa change TM2	1.4 ± 4.2	McGowan-Jordan et al 1999
	exon 8	845G > A	G169D	aa change TM2	-0.82 ± 1.4	Shotelersuk et al 1998
	exon 8	869A > C[e]	N177T	aa change TM2	0.7 ±4.1	Kalatzis et al 2002
	exon 8	883T > C	W182R	aa change TM2	34 ± 5.9	Shotelersuk et al 1998
	exon 9	952G > A	D205N	aa change 1st inter-TM loop	-2.1 ± 2.9	Shotelersuk et al 1998
	exon 9	1004A > G	Q222R	aa change TM3	2.4 ± 1.3	Shotelersuk et al 1998
	exon 11	1200G > A (c.861G > A)	M287I	aa change 5th inter-TM loop	nd	Mason et al 2003
	exon 11	1203C > A	N288K	aa change 5th inter-TM loop	1.6 ± 1.2	Kalatzis et al 2002
	exon 11	1232G > A	S298N	aa change 5th inter-TM loop	77±21	Shotelersuk et al 1998
	exon 11	1252G > T	D305Y	aa change TM6	-4.8 ± 1.7	Attard et al 1999
	exon 11	1253A > G	D305G	aa change TM6	nd	Shotelersuk et al 1998
	exon 11	1261G > A (c.922G > A)	G308R	aa change TM6	-5.3 ± 8.5	Shotelersuk et al 1998 Attard et al 1999 Mason et al 2003
	exon 12	1348G > A (c.1009G > A)	G337R	aa change TM7	nd	Mason et al 2003
	exon 12	1352T > C	L338P	aa change TM7	-8.0 ± 3.9	Attard et al 1999
	exon 12	1354G > A (c.1015GA)	G339R	aa change TM7	-8.0 ± 3.3	Shotelersuk et al 1998 Attard et al 1999 Rupar et al 2001 Kalatzis et al 2002 Kiehntopf et al 2002 Mason et al 2003
nonsense	exon 3	354G > A	W5X	t N-terminal region	nd	Kalatzis et al 2002
	exon 6	622G > T	G95X	t N-terminal region	nd	Town et al 1998
	exon 7	721C > T	Q128X	t N-terminal region	nd	Town et al 1998

(continued)

TABLE 39.1   *(Continued)*

Type of mutation	Position	Nucleotide change[a]	Amino acid change	Predicted effect on protein sequence[b]	Transport activity[c]	References
	exon 7	753G > A	W138X	t TM1	nd	Shotelersuk et al 1998 Attard et al 1999 Town et al 1998 Shotelersuk et al 1998 McGowan-Jordan et al 1999 Thoene et al 1999
	exon 10	1044G > A	W235X	t 3rd inter-TM loop	nd	Pennesi et al 2005
	exon 11	1209C > G	Y290X	t 5th inter-TM loop	nd	Shotelersuk et al 1998 Kleta et al 2001
	exon 11	1229G > A (c.890G > A)	W297X	t 5th inter-TM loop	nd	Mason et al 2003
	exon 12	1317G > A (c.978G > A)	W326X	t 6th inter-TM loop	nd	ud
insertion	exon 6	631-632ins A (c.292-293insA)	T98fsX124	fs N-terminal region	nd	Kiehntopf et al 2002
	exon 7	684-685ins5 (c.345-346ins5)	L116fsX119	fs N-terminal region	nd	Kiehntopf et al 2002
	exon 8	856-857insC (c.516-517insC)	Y173fsX227[f]	fs TM2	nd	Mason et al 2003 ud
	exon 9	985-986insA	T216fsX227	fs TM3	nd	Shotelersuk et al 1998 Attard et al 1999
	exon 10	1035-1036insC (c.696-697insC)	V233fsX295	fs 3rd inter-TM loop	nd	Shotelersuk et al 1998 McGowan-Jordan et al 1999 Attard et al 1999 Mason et al 2003
	exon 10	1036-1037insCG (c.697-698insCG)	V233fsX253	fs 3rd inter-TM loop	nd	Shotelersuk et al 1998 Attard et al 1999 Kiehntopf et al 2002
	exon 11	1261-1262insG (c.926-927insG)	S310fsX364	fs TM6	nd	Shotelersuk et al 1998 Heil et al 2001 Kalatzis et al 2002 Kiehntopf et al 2002
	exon 12	1386-1387ins12	DVEF349–350ins	insertion 4 aa TM7	−1.6 ± 2.4	Kalatzis et al 2002
deletion	exon 3	357-360del4 (c.18-21delGACT)	T7fsX13	fs N-terminal region	nd	Town et al 1998 Shotelersuk et al 1998 Heil et al 2001 Kiehntopf et al 2002 Mason et al 2003
	exon 3	375delT	L14X	fs N-terminal region	nd	Town et al 1998

(continued)

TABLE 39.1   (*Continued*)

Type of mutation	Position	Nucleotide change[a]	Amino acid change	Predicted effect on protein sequence[b]	Transport activity[c]	References
	exon 4	399-400delTG	C20X	fs N-terminal region	nd	Town et al 1998
	exon 5	545-549del5	I69fsX73	fs N-terminal region	nd	Shotelersuk et al 1998 Anikster et al 2000
	exon 6	619delG (c.280delG)	V94fsX118	fs N-terminal region	nd	Kiehntopf et al 2002
	exon 6	651-654del4	H105fsX116	fs N-terminal region	nd	Shotelersuk et al 1998
	exon 7	762delC	F142fsX146	fs TM1	nd	Attard et al 1999
	exon 8	857-858delAC	Y173X	fs TM2	nd	Town et al 1998
	exon 9	908-916del9 (c.569-577del9)	F190X	fs 2nd inter-TM loop	nd	Town et al 1998 Mason et al 2003
	exon 9	931delG	V298X	fs 5th inter-TM loop	nd	ud
	exon 9	953-955del3	D205del	deletion 1 aa 2nd inter-TM loop	1.9 ± 2.4	Shotelersuk et al 1998 Attard et al 1999
	exon 9	998-1004del7 (c.659-665del8)	I220fsX252	fs TM3	nd	Mason et al 2003
	exon 10	1038-1039delGT (c.699-700delGT)	V233fsX295	fs 3rd inter-TM loop	nd	Mason et al 2003
	exon 10	1080delC	F247fsX252	fs TM4	nd	Shotelersuk et al 1998 Attard et al 1999
	exon 10	1109-1131del23	G258fsX288	fs 4th inter-TM loop	nd	Attard et al 1999
	exon 10	1146-1148delCTC	S270del	deletion 1 aa TM5	−1.9 ± 1.0	Attard et al 1999
	exon 12	1331delG	G331fsX366	fs 6th inter-TM loop	nd	Attard et al 1999
	exon 12	1366-1377del12	IVFD343–346del	deletion 4 aa TM7	−0.57 ± 1.1	Attard et al 1999
	exon 12	1375-1386del12	DVVF346–349del	deletion 4 aa TM7	−7.6 ± 5.8	Kalatzis et al 2002
deletion + insertion	exon 12	1367-1374del8insA	I343fsX364	fs TM7	nd	Shotelersuk et al 1998
splice site	IVS3	400–400 + 2delGGT (c.61-62 + 2delGGT)		nd	nd	Kiehntopf et al 2002
	IVS4	479 + 1G > T		t[g]	nd	Town et al 1998
	IVS4	479 + 1G > C (c.140 + 1G > C)		nd	nd	ud
	IVS4	479 + 2T > C (c.140 + 2T > C)		nd	nd	ud
	IVS5	564 + 1G > A		t	nd	Kalatzis et al 2002
	IVS5	564 + 1del4		t	nd	Kalatzis et al 2002
	IVS5	564 + 3A > T (c.225 + 3A > T)		nd	nd	Mason et al 2003
	IVS5	564 + 5GT > CC		t	nd	Kleta et al 2001
	exon 8	898–900 + 24del27		t or altered topology[h]	nd	Attard et al 1999
	IVS8	900 + 1G > C		nd	nd	Town et al 1998

(*continued*)

TABLE 39.1   (*Continued*)

Type of mutation	Position	Nucleotide change[a]	Amino acid change	Predicted effect on protein sequence[b]	Transport activity[c]	References
	IVS8	900 + 1delG		nd	nd	Town et al 1998 Shotelersuk et al 1998
	exon 9	1020G > A		t[g]	nd	Shotelersuk et al 1998
	IVS9	1020 + 1G > A (c.681 + 1G > A)		nd	nd	Attard et al 1999 Mason et al 2003
	IVS10	1192-2A > G		t or insertion 5th inter-TM loop	nd	Kalatzis et al 2002
	IVS11	1310-12G > A		t[g]	nd	Attard et al 1999
polymorphism	IVS6	669 + 22G > C (c.329 + 22G > C)	–	–	nd	Mason et al 2003
	IVS6	668 + 5T > C	–	–	nd	Shotelersuk et al 1998
	exon 7	736A > T[e]	I133P	aa change TM1	92 ± 16	McGowan-Jordan et al 1999
	exon 8	843A > G (c.504A > G)	–	–	nd	Shotelersuk et al 1998 Mason et al 2003
	IVS9	1020 + 9A > G	–	–	nd	Shotelersuk et al 1998
	exon 10	1118T > C[e] (c.779T > C)	I260T	aa change 4th inter-TM loop	74 ± 13	Kalatzis et al 2004 Mason et al 2003
	exon 11	1214A > G	K292R	aa change 5th inter-TM loop	nd	Shotelersuk et al 1998
	exon 11	1224G > A (c.885G > A)	–	–	nd	Mason et al 2003
	exon 11	1299C > T	–	–	nd	Shotelersuk et al 1998

nt – nucleotide; aa-amino acid; ud – unpublished data; nd – not determined; t – truncation; fs – frameshift

[a] The first nt of the cDNA sequence is considered as +1; the ATG start codon is situated at +340. The nucleotide changes indicated here are those proposed in the original publications (except for Kiehntopf et al 2002, Mason et al 2003), even if they do not follow the recommendations that the A of the ATG start codon should be considered as +1 (den Dunnen & Antonarakis 2000, 2001). This differing nomenclature has been used by Kiehntopf et al (2002), by Mason et al (2003) and for our unpublished data, and is indicated in brackets.

[b] According to the original predicted topology reported by Town et al (1998).

[c] Effect on transport activity (expressed as a percentage of wt cystinosin activity ± SEM) reported by Kalatzis et al (2004).

[d] May correspond to a nonpathogenic mutation as recently detected in individuals that carry known pathogenic mutations on both alleles (Mason et al 2003).

[e] Definitively classified following the transport studies reported in Kalatzis et al (2004). N177T was detected in an individual with juvenile cystinosis but the second mutated allele, which was never identified, is likely the mild mutation. I133P was originally erroneously classed as a pathogenic mutation. I260T was identified in the homozygous state in conjunction with S141F in an individual with infantile cystinosis.

[f] Originally reported inaccurately as Y173fsX225 (Mason et al 2003).

[g] Effect on *CTNS* splicing reported by Kalatzis et al (2002).

[h] Effect on *CTNS* splicing reported by Kalatzis et al (2001).

TABLE 39.2 Point mutations associated with juvenile cystinosis or atypical forms

Type of mutation	Position	Nucleotide change[a]	Amino acid change	Predicted effect on protein sequence[b]	Transport activity[c]	References
missense	exon 4	463G > A	V42I[e]	aa change N-terminal tail	97 ± 4.1	Attard et al 1999
	exon 7	755C > T	S139F[e]	aa change TM1	−2.9 ± 1.0	Attard et al 1999
	exon 9	938C > T	P200L	aa change 2nd inter-TM loop	15 ± 4.5	Kalatzis et al 2002
	exon 10	1178A > G	K280R	aa change 5th inter-TM loop	−0.68 ± 0.9	Thoene et al 1999
	exon 11	1261G > T (c.923G > T)	G308V	aa change TM6	nd	Kiehntopf et al 2002
	exon 11	1308C > G	N323K	aa change 6th inter-TM loop	0.14 ± 0.8	Thoene et al 1999
	exon 12	1340C > A	T334N[f]	aa change 6th inter-TM loop	nd	ud
	exon 12	1375G > A	D346N[e]	aa change TM7	61 ± 11	Attard et al 1999
insertion	IVS7	801-10C > G	PCS154–155ins	insertion 3 aa 1st inter-TM loop	9.2 ± 2.1	Attard et al 1999
deletion	exon 5	537-557del21bp[d]	ITILELP67–73del	deletion 7 aa N-terminal tail	19 ± 6.1	Shotelersuk et al 1998 Attard et al 1999 Heil et al 2001 ud
splice site	exon 6	668G > T	G110V[e]	t	120 ± 27	Kalatzis et al 2002
	IVS11	1309 + 2T > C	–	t	nd	Thoene et al 1999

aa – amino acid; nd – not determined; ud – unpublished data; t – truncation; fs – frameshift

[a] The first nt of the cDNA sequence is considered as +1. Differing published nomenclature indicated in brackets (Kiehntopf et al 2002).

[b] According to the original predicted topology reported by Town et al (1998).

[c] Effect on transport activity (expressed as a percentage of wild-type cystinosin activity ± SEM) reported by Kalatzis et al (2004).

[d] Detected in homozygotes and compound heterozygotes with infantile cystinosis (Shotelersuk et al 1998, Heil et al 2001) and compound heterozygotes with juvenile cystinosis (Attard et al 1999, unpublished data); definitively classified following the transport studies reported in Kalatzis et al (2004).

[e] Mutations detected in individuals presenting with atypical features. The G110V missense mutation associated with this patient does not affect cystine transport (Kalatzis et al 2004); it is the splicing defect that gives rise to the associated phenotype (Kalatzis et al 2002).

[f] Also detected in an individual with ocular cystinosis; the 57 kb deletion is the second mutated allele.

TABLE 39.3 Point mutations associated with ocular cystinosis

Type of mutation	Position	Nucleotide change[a]	Amino acid change	Predicted effect on protein sequence[b]	Transport activity[c]	References
nt substitution	promoter	−303G > T[d]	–	decreased protein production	nd	Phornphutkul et al 2001
missense	exon 9	928G > A	G197R	aa change 2nd inter-TM loop	20 ± 7.8	Anikster et al 2000 Kleta et al 2001
	exon 12	1340C > A	T334N[e]	aa change 6th inter-TM loop	nd	ud
insertion	promoter	−303Tins[d]	–	decreased protein production	nd	Phornphutkul et al 2001
splice site	IVS10	1192-3C > G	–	T	nd	Anikster et al 2000

aa – amino acid; t – truncation.

[a] The first nt of the cDNA sequence is considered as +1.

[b] According to the original predicted topology reported by Town et al (1998).

[c] Effect on transport activity (expressed as a percentage of wild-type cystinosin activity ± SEM) reported by Kalatzis et al (2004).

[d] May correspond to severe mutations, as G197R is the mutation on the second allele in these individuals.

[e] Also detected in an individual with juvenile cystinosis; the 57 kb deletion is the second mutated allele.

TABLE 39.4   Frequency of the 57 kb European deletion in various populations

Study	N° unrelated patients tested	N° homozygous deletions	N° heterozygous deletions	% patients with deletions	% alleles with deletions	Ethnicity
Anikster et al 1999	108	48	25	73/108 68%	121/216 56%	European
Forestier et al 1999	147	55	41	96/147 65%	151/294 51%	76% European
McGowan-Jordan et al 1999	20	1	5	6/20 30%	7/40 18%	French-Canadian
Heil et al 2001	9	4	2	6/9 67%	10/18 56%	Dutch
Kiehntopf et al 2002	35	17	13	30/35 86%	47/70 67%	Swiss and German
Mason et al 2003	42	2	6	8/42 19%	10/84 12%	90% Northern Italian

like the PCR-based assays, also allows detection of the less frequent 13 kb deletion (Bendavid et al 2004).

The 57 kb deletion is not the only possible founder mutation associated with cystinosis. A higher incidence of cystinosis has been reported in certain regions, such as the western French province of Brittany (Bois et al 1976, Cochat et al 1999), the United Kingdom (Bickel & Harris 1952), Germany (Manz & Gretz 1985) and (the French-speaking population of) Quebec (DeBraekeleer 1991). In Brittany, the incidence of cystinosis rises to 1 in 26 000 (Bois et al 1976) and a particular mutation, 898–900+24del27 (Attard et al 1999), has been detected in unrelated affected families only from this region. This mutation is a 27 bp deletion that begins 3 bp before the end of exon 8 and continues into IVS8 thus removing the donor splice site (Kalatzis et al 2001b). 898–900 + 24del27 gives rise to aberrant *CTNS* transcripts, which either truncate cystinosin after the 1st or 2nd TM, or alter its topology, thus accounting for the severe phenotype in these families. The identification of the same mutation in seemingly unrelated families from the same closed community is indicative of a common founder effect and would account for the elevated disease incidence observed in these regions. Similarly, in the French-speaking population of Quebec, a common mutation of Irish origin has also been identified, W138X, which truncates cystinosin in the 1st TM (McGowan-Jordan et al 1999). However, W138X has also been detected at a relatively high frequency (~10%) in an American-based population with various ethnic origins (Shotelersuk et al 1998), suggesting that this founder effect arose on a recurrent mutation.

Many of the point mutations associated with the infantile form (Table 39.1) are deletions, insertions, splice site mutations and nonsense mutations that cause premature termination of cystinosin. However, missense mutations and in-frame deletions or insertions have also been detected and these are generally found to cluster in the TM, particularly in the last three (Attard et al 1999). One recurrent missense mutation, G339R, is worth noting. It has been described as

segregating in various ethnic backgrounds (Shotelersuk et al 1998, Attard et al 1999) but most notably in two unrelated families in an Ontario Amish Mennonite population (Rupar et al 2001), in five unrelated families of Jewish-Moroccan origin (Kalatzis et al 2002) and in 11 families of Italian origin (Mason et al 2003). This information could be beneficial for genetic counseling studies as the simple screening for G339R could be carried out on individuals at risk from these regions, providing a rapid identification of heterozygous carriers.

In contrast to the situation for infantile cystinosis, point mutations that do not disrupt the open reading frame of cystinosin are more commonly associated with the juvenile and ocular forms (Tables 39.2 and 39.3, respectively) and are generally found in the inter-TM loops or in the N-terminal region (Attard et al 1999). Affected individuals are either homozygous for such 'mild' mutations or heterozygous for a mild mutation and a 'severe' mutation (i.e. a mutation that segregates in infantile cystinosis). To date, only six patients with ocular cystinosis have been studied at the molecular level (Anikster et al 2000, Kleta et al 2001) and a missense mutation G197R, situated in the 2nd inter-TM loop, segregates in five of these patients. The sixth patient carries a splice site mutation, IVS10-3C > G, which appears to allow sufficient splicing of the normal transcript (Anikster et al 2000) to prevent a severe phenotype. We recently studied a seventh patient that carried a missense mutation T334N in the sixth inter-TM loop (unpublished data). However, we detected this same mutation in a patient with juvenile cystinosis.

Interestingly, three mutations have also been described in the *CTNS* promoter region (Phornphutkul et al 2001). This region extends 316 bp upstream (−316) from the start of exon 1 (+1). One patient with infantile cystinosis carried a G > C transition at position −295(−295G > C) on one allele and the 57 kb deletion on the other. −295G > C results in 19% of wild-type reporter gene activity and decreased CTNS transcription. However, this mutation was recently described

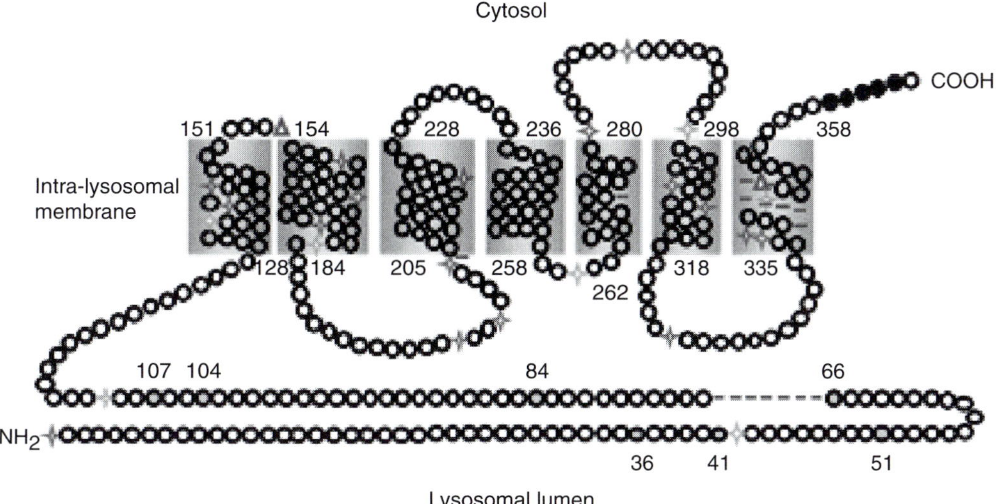

FIGURE 39.4   Position of transport-altering mutations in the topological model of cystinosin. Cystinosin is a 367 aa protein predicted to contain 7 transmembrane domains (TM). The 128 aa N-terminal region with 7 N-glycosylation sites (blue circles) is oriented towards the lysosomal lumen. The 10 aa tail contains a classic GYDQL lysosomal targeting motif (black circles) and is oriented towards the cytosol. Mutations are categorized into three classes according to whether they abolish (red symbols), severely inhibit (<30% of wild-type; blue symbols) or only moderately, or do not, alter (≥30% of wild-type; green symbols) cystine transport. Stars represent point mutations, dashes correspond to deletions of 1 aa, and triangles indicate insertions of 3 (position 154) or 4 (position 349) aa. Numbers correspond to the position of N-glycosylation sites and of the extremities of TMs. Reproduced from Kalatzis et al 2004 with the permission of Oxford University Press. (see also Plate 88)

in five patients of Italian origin, four of which carry known pathogenic mutations on both alleles (Mason et al 2003), questioning the role of –295G > C in disease pathogenesis. The other two mutations, a G > T transition at position −303 (−303G > C) and a T insertion at the same position (−303insT), were detected in individuals with ocular cystinosis carrying G197R on the second allele. These promoter mutations result in 5% and 16% of wild-type reporter gene activity, respectively. It is still unknown whether −303G > C and −303insT correspond to severe or mild mutations, or have no effect on disease pathogenesis.

A number of mutations have been detected that segregate with atypical forms of cystinosis and are worth mentioning (Table 39.2). First, V42I, situated in the N-terminal region, segregates in the homozygous state in an individual who presented with Fanconi syndrome at 3 years and who, in the absence of cysteamine treatment, has normal plasma creatinine concentrations at 18 years (Attard et al 1999). Similarly, S139F, situated in the first TM, segregates in a patient who presented with Fanconi syndrome at 6 years but who, without treatment, presented only a slightly diminished creatinine clearance at 22 years (Attard et al 1999). Thirdly, D346N, a conservative amino acid substitution in the seventh TM, was detected in a patient who developed symptoms at 19 months but developed ESRD later than normal, at age 16 years (Attard et al 1999). Finally, G110V, situated in the N-terminus of cystinosin, was detected in an individual who presented severe features of Fanconi syndrome and developed ESRD at age 13 years, yet the only extrarenal symptoms were mild ocular anomalies

with photophobia occurring at age 23 years (Kalatzis et al 2002). Interestingly, this mutation involves the last nucleotide of exon 6, and it has been shown to affect splicing in leukocyte RNA leading to a truncated protein (Attard et al 1999). It is conceivable that G110V does not lead to the same splicing events in all tissues. Should this be the case, misspliced transcripts in the kidney would account for the severe renal phenotype observed in the carrier, whereas the presence of a majority of correctly spliced forms in other organs would explain the lack of extrarenal disorders.

Functional Assay of *CTNS* Mutations

Thirty-one *CTNS* mutations that did not prematurely terminate protein production (24 missense mutations, seven in-frame deletions or insertions) were recently analyzed for their effect on subcellular localization and function of cystinosin (Tables 39.1, 39.2, and 39.3). As these mutations were associated with all forms of the disease (19 with infantile cystinosis, four with juvenile cystinosis, one with ocular cystinosis, four with atypical forms and three unclassified), this functional analysis provided the first concrete correlation between molecular and clinical phenotype (Kalatzis et al 2004).

First, with regard to subcellular localization, this study provided some insights into the lysosomal sorting signals of cystinosin. Only three mutations of the 31 mutations studied (the point mutation Q222R situated in the third TM, and the two in-frame deletions IVFD343-346del and DVVF346–349del situated in the seventh TM; Figure 39.4) partially relocalized cystinosin to the plasma membrane (Kalatzis et al 2004). These three mutations are associated with infantile cystinosis. It was first considered that the deletions in the seventh TM prevented exposure of the C-terminal GYDQL motif to the cytosolic compartment thus accounting for the partial relocalization. However, deletion of this motif coupled to the deletions IVFD343–346del and DVVF346–349del completely relocalized the mutant protein to the plasma membrane, demonstrating that the activity of the GYDQL motif is unaffected by these deletions. The same observations were made for Q222R, suggesting that these three mutations interfere with the second conformational lysosomal sorting signal.

Secondly, with regard to the functional assay, overall for the three defined forms of cystinosis, the transport data tend to correlate with the clinical phenotype (Kalatzis et al 2004). Globally, 16 of 19 mutations associated with infantile cystinosis abolished transport whereas two of four mutations associated with juvenile, and the one associated with ocular, cystinosis strongly reduced transport to ~9–20% wild-type activity, in agreement with the milder clinical phenotype. These observations validate the generally held hypothesis that infantile cystinosis is caused by two losses of function alleles, whereas patients with milder forms are homozygous or compound heterozygotes for a specific, partially active allele.

Furthermore, the transport data were well correlated with topological and phylogenetic information on cystinosin (Kalatzis et al 2004). All but one (N323K) of the amino acid substitutions that abolished cystine transport affected residues strictly conserved across eukaryotic species. In addition, all but two (N288K, N323K) of these conserved residues are located within or at the border of the TMs, and the remaining two within the fifth inter-TM and sixth inter-TM loop, indicating the importance of these regions (Figure 39.4). Consistent with these results, mutant *CTNS* alleles carrying the transport-abolishing G308R and L338P mutations located in the seventh TM, or lacking the entire sixth and seventh TM, were ineffective or unable to functionally complement the yeast ERS1 mutant phenotype (Gao et al 2005), confirming the functional importance of these regions. Conversely, a mutant *CTNS* allele lacking the N-terminal region could complement in a manner equivalent to wild-type.

The observation that G197R, the point mutation associated with ocular cystinosis, allows 20% of wild-type cystine transport is particularly interesting (Kalatzis et al 2004). First, this level is consistent with measurements made on lysosomes from patients with ocular cystinosis (Gahl & Tietze 1987), which yielded transport levels of 19 ± 10% of wild-type (measurements on infantile cystinosis lysosomes were in the 0–5% range), thus demonstrating the reliability of the in vitro measurements. Second, it defines the threshold for the appearance of clinical signs to between 20–30%, which is consistent with the fact that 50% of wild-type

cystine transport (Gahl et al 1982), which is the level seen in heterozygote carriers, does not give rise to clinical signs.

Of particular interest is the predictive potential of this assay, which was highlighted by the clarification of five unclassified or misclassified alleles (Kalatzis et al 2004). First, the mutation I133F was originally identified in a French-Canadian population as associated with infantile cystinosis (McGowan-Jordan et al 1999). Conversely, its lack of effect on cystine transport suggested that it was a polymorphism (Kalatzis et al 2004). Recent data confirm that this mutation is indeed not causative (McGowan-Jordan et al, personal communication). Similarly, the mutation I260T found in the homozygous state in conjuction with S141F in a patient with infantile cystinosis, which moderately affects transport (Kalatzis et al 2004), was also reported as a polymorphism (http://www.ncbi.nlm.nih.gov/SNP-refSNP ID: rs161400). Thirdly, N177T was identified in an individual with juvenile cystinosis; however, the second mutated allele was unidentified (Kalatzis et al 2002). N177T abolishes cystine transport and, thus, in accordance with the observation that one severe and one mild allele give rise to juvenile cystinosis, N177T can be now classed as a severe mutation. Fourthly, the missense mutation G110V associated with an atypical case of cystinosis (see above), does not affect cystine transport (Kalatzis et al 2004). Thus it is the effect that this mutation has on splicing, and not the amino acid substitution, that gives rise to the associated phenotype.

Finally and most interesting are the conflicting clinical data surrounding the deletion ITILELP67-73del, situated in the N-terminal region (Figure 39.4). This deletion has been described in homozygotes and compound heterozygotes with infantile cystinosis (Shotelersuk et al 1998, Heil et al 2001). However it has also been detected in two cases of juvenile cystinosis and in one of the cases, it is associated with the 57 kb deletion (Attard et al 1999, Kalatzis et al 2004). The functional assay of this mutation showed that it allows ~19% residual transport activity (Kalatzis et al 2004), which, coupled to the fact that no other mutation associated with infantile cystinosis is located in the N-terminal region of cystinosin, reinforces the conclusion that ITILELP67-73del is associated with juvenile cystinosis. The confusion surrounding the clinical status of ITILELP67-73del illustrates the difficulties associated with classifying cystinosis patients into various clinical forms owing to heterogeneous criteria employed by clinicians worldwide.

Overall, this study demonstrated that when the *CTNS* mutation does not prematurely truncate cystinosin, impaired cystine transport is the most frequent cause of pathogenicity. Moreover, the level of transport inhibition often correlates with the severity of symptoms. Therefore, this transport assay has potential prognostic interest when novel mutations are identified. Nonetheless, a minority of mutations showed a discrepancy between molecular and

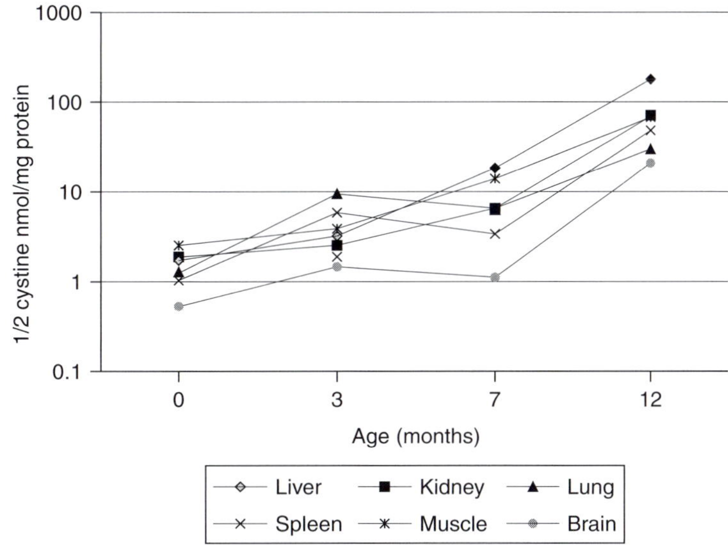

FIGURE 39.5   Intracellular cystine content in various organs of Ctns$^{-/-}$ mice. Graph showing the evolution of cystine levels (expressed in nmol of half cystine/mg protein) in the liver, kidney, lung, spleen, muscle and brain from birth to 12 months. Reproduced from Cherqui et al 2002 with the permission of the American Society for Microbiology

clinical phenotypes. These include the mutations S139F, K280R, and N323K, which abolish cystine transport but are associated with less severe phenotypes, and V42I, W182R, S298N, and D346N, which allow cystine transport within a range not expected to give rise to clinical signs (~40–100%). The further investigation of these discordant mutations might provide deeper insights into the molecular or cellular mechanism that modulate disease severity and could unravel unsuspected aspects of the pathogenic cascade.

CYSTINOSIN-DEFICIENT MOUSE MODEL

The mouse homolog of *CTNS*, *Ctns*, is located on mouse chromosome 11, in a region syntenic to human 17p13, and it consists of ten coding exons (Cherqui et al 2000). *Ctns* encodes a 367 aa sequence that is 92.6% similar to cystinosin and contains the seven N-glycosylation sites, 7 TM, the GYDQL targeting signal and the YFPQA motif. Replacing the last four exons of *Ctns*, which encode the last 5 TM, by an IRES-βgal-neo cassette, generated the first known animal model (*Ctns$^{-/-}$*) for cystinosis (Cherqui et al 2002). This recombinant gene is under control of the endogenous murine *Ctns* promoter and the IRES (internal ribosome entry site) reinitializes translation of the fusion βgal-neo transcript. The truncated cystinosin is not targeted to lysosomes and does not transport cystine.

The mixed background C57Bl6/Sv129 *Ctns$^{-/-}$* mice accumulate cystine at various levels in all tissues tested (liver, kidney, lung, spleen, muscle, brain, and heart) (Cherqui et al 2002) as determined both by the assay of tissue

cystine content (Figure 39.5) and by the detection of cystine crystals from 24 weeks of age (Figure 39.6). Cystine crystals, however, were less abundant than seen in human tissues and were not detected in the lung or the brain. Intertissue variability has also been reported for cystinosis patients (Gahl et al 2001). Cystine accumulation begins from birth and cystine levels increase with age (Cherqui et al 2002). Although the kidney is one of the organs showing the highest levels of cystine, the *Ctns$^{-/-}$* mice, even at 18 months, do not show clinical or biological signs of a proximal tubulopathy. In addition, histological analysis of *Ctns$^{-/-}$* kidney sections (Figure 39.6) did not show the severe and diffuse proximal tubule alterations seen in cystinosis patients (Gubler et al 1999). Instead, focal cystine crystal deposits within proximal tubular cells could be seen (Figure 39.6A) (Cherqui et al 2002). Furthermore, isolated or small groups of proximal tubular cells were vacuolized (Figure 39.6B) and these vacuoles were identified as swollen mitochondria (Figure 39.6C). However, most proximal tubules appeared normal as well as distal tubules, glomeruli, and interstitium (Figure 39.6D).

In contrast to the kidney phenotype, the ocular abnormalities observed in *Ctns$^{-/-}$* mice are similar to those reported in affected children (Dufier et al 1987, Dufier 1992). Corneal cystine crystal deposits (seen from 1 year), as well as impaired and supernormal ERG profiles, have been reported in patients and were detected in 8-month-old mice (Cherqui et al 2002). Associated with the impaired ERG profile in the mice were depigmentation patches in the peripheral retina, indicative of a retinopathy. Retinopathy leading to decreased visual acuity has been reported in cystinosis children from as early as 3 years and was constantly present from 7 years (Dufier 1992). Thus it is possible

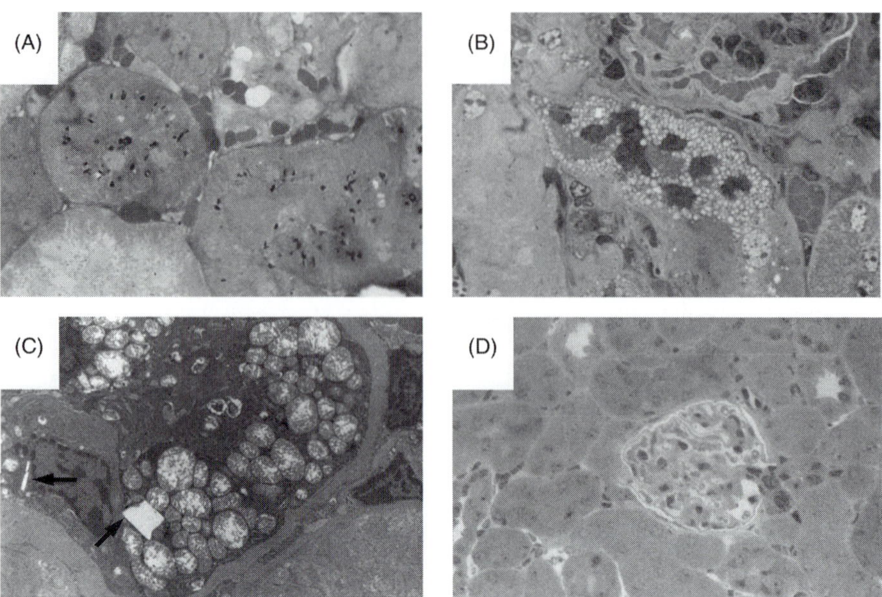

FIGURE 39.6   Histological analysis of renal lesions in $Ctns^{-/-}$ mice. (A) Dark inclusions, corresponding to cystine crystals after osmium fixation, within proximal tubular cells of a 7-month-old mouse (magnification = ×750). (B) Focal microvacuolization of proximal tubular cells of a 7-month-old mouse (magnification = ×750). (C) Swollen mitochondria with clear matrix and disorganized cristae, and rare cystine crystals (arrows) of a 12-month-old mouse (magnification = ×2000). (D) Normal renal parenchyma of a 12-month-old mouse (magnification = ×320). Reproduced from Cherqui et al 2002 with permission of the American Society for Microbiology

that the $Ctns^{-/-}$ mice have defective vision; however, studies are currently underway to more precisely qualify these ocular anomalies.

With regard to the other clinical signs observed in cystinotis patients, the $Ctns^{-/-}$ mice demonstrated normal growth and fertility and no evidence of diabetes or hypothyroidism. A small number of 1-year-old $Ctns^{-/-}$ mice may have muscular abnormalities as indicated by elevated CPK levels and the presence of cystine crystals in interstitial cells that, in one case, were associated with foci of myocyte necrosis. This is consistent with the myopathy and associated histological changes observed in a small proportion of cystinosis patients (Gahl et al 1988). In contrast, the brain of $Ctns^{-/-}$ mice appeared histologically normal; however, neurological anomalies are the last to appear in cystinosis patients (Gahl et al 2001). Comprehensive behavioral studies of $Ctns^{-/-}$ mice are currently underway to assess for the presence of CNS anomalies.

Interestingly, a demineralization and cortical thinning of the vertebrae and long bones was observed in 8-month-old $Ctns^{-/-}$ mice. In cystinotic children, bone abnormalities are classically related to hypophosphatemic rickets and/or hyperparathyroidism secondary to renal failure. As this is not the cause in the $Ctns^{-/-}$ mice, the observed bone defects may be a direct consequence of cystine storage in this tissue. This hypothesis may help explain the growth retardation of cystinotic children, which is more severe than for children with a tubulopathy from other etiologies and which seems to partially improve with cysteamine therapy.

Further Insights into the Renal Pathogenesis of Cystinosis

It cannot be excluded that the C57Bl6/Sv129 background of $Ctns^{-/-}$ mice masks the appearance of renal anomalies, and thus this mouse line needs to be crossed onto other backgrounds and retested. However, the high renal cystine levels and lack of clinical symptoms of these mice, suggest that the proximal tubulopathy associated with cystinosis is not a direct effect of cystine storage but rather a secondary consequence.

It was previously reported that lysosomes of cystine-loaded rabbit proximal tubules display a significant reduction in intracellular ATP concentrations, leading to inhibition of Na/K-ATPase activity (Coor et al 1991). Consequently, the sodium-entry gradient into the tubular cells and the sodium-coupled transport of other solutes, which constitutes the main type of solute transport across the proximal tubular membrane (Cetinkaya et al 2002), is also decreased. These data suggest that the mitochondrial oxidative phosphorylation process responsible for ATP synthesis may be impaired in cystinosis proximal tubular cells. This is consistent with the large mitochondria focally observed in the proximal tubular cells of $Ctns^{-/-}$ mice (Cherqui et al 2002), and the mitochondrial swelling that has been described in kidney biopsies from cystinotic patients (Jackson et al 1962, Spear et al 1971).

Similarly, a recent study shows that the decreased level of cytosolic cysteine, resulting from impaired lysosomal cystine efflux, results in a moderate decrease of glutathione content during exponential growth phase in stably transformed cystinosis cell lines (Chol et al 2004). This in turn results in increased levels of superoxide dismutase, indicative of an augmentation of the oxidative defenses of the cell, although the mitochondria did not show signs of oxidative damage in this experimental model. As the exogenous supply of cysteine precursors was shown to augment glutathione content, and should these observations be confirmed in renal tubular cell lines, the administration of glutathione precursors to cystinosis patients may represent a possible treatment of the Fanconi syndrome associated with infantile cystinosis.

Finally, Park et al (2002) showed that in nephropathic cystinotic fibroblasts, increased lysosomal cystine caused an increased rate of apoptosis. Similarly, when the cystine content of normal fibroblasts and cultured renal proximal tubule epithelial (RPTE) cells was increased by exposure to cystine dimethylester, the apoptotic rate increased to that of cystinotic cells. Moreover, the RPTE cells were more sensitive to increased cystine levels, resulting in higher levels of apoptosis. In contrast, fibroblasts from juvenile or ocular cystinotic fibroblasts did not show increased apoptosis. Taken together, these data suggest that enhanced apoptosis from lysosomal cystine storage may lead to inappropriate cell death and decreased cell numbers, resulting in the nephropathic phenotype. In contrast, the moderate cell loss associated with the variant forms may account for the milder disease phenotype.

CONCLUSIONS

The identification of *CTNS*, and the determination that cystinosin is the lysosomal cystine transporter, represents the culmination of decades of efforts to understand the molecular basis of cystinosis. The catalog of *CTNS* mutations, and, in particular, the identification of founder mutations, have provided the basis for rapid diagnostic screenings of individuals at risks. The development of in vitro functional assay represents a potential prognostic tool as well as a more reliable and uniform method for classifying patients into clinical entities. Finally, the generation of anticystinosin antibodies and of an animal model will provide further insight into the pathogenesis of this disorder. The lack of proximal tubulopathy in *Ctns*$^{-/-}$ mice constitutes a major difference from the infantile form of cystinosis. The comparison of the two species may identify the mechanism of this defect in affected children, the exact origin of which remains elusive. A preliminary therapeutic trial using oral cysteamine has been carried out in *Ctns*$^{-/-}$ mice to examine the efficiency of this treatment to eliminate cystine from different tissues (Cherqui et al 2002). Following a 60-day treatment with 400 mg/kg of cysteamine, a variable decrease was seen in all tissues tested. These encouraging results suggest that this unique *Ctns*$^{-/-}$ animal model will be valuable in testing emerging therapeutics for cystinosis and comparing their efficiency with that of the cysteamine treatment currently in use.

References

Abderhalden E. Familiäre cystindiathese. Z. Physiol. Chem. 1903; 38: 557.

Anikster Y, Lucero C, Guo J, et al. Ocular nonnephropathic cystinosis: clinical, biochemical, and molecular correlations. Pediatr. Res. 2000; 47: 17–23.

Anikster Y, Lucero C, Touchman JW, et al. Identification and detection of the common 65-kb deletion breakpoint in the nephropathic cystinosis gene (CTNS). Mol. Genet. Metab. 1999; 66: 111–16.

Attard M, Jean G, Forestier L, et al. Severity of phenotype in cystinosis varies with mutations in the CTNS gene: predicted effect on the model of cystinosin. Hum. Mol. Genet. 1999; 8: 2507–14.

Bendavid C, Kleta R, Long R, et al. FISH diagnosis of the common 57-kb deletion in CTNS causing cystinosis. Hum. Genet. 2004; 115: 510–14.

Bickel H, Harris H. The genetics of Lignac-Fanconi disease. Acta Paediatr. Scand. 1952; 90(Suppl.): 22–6.

Bois E, Feingold J, Frenay P, Briard M-L. Infantile cystinosis in France: genetics, incidence, geographic distribution. J. Med. Genet. 1976; 13: 434–8.

Cetinkaya I, Schlatter E, Hirsch JR, Herter P, Harms E, Kleta R. Inhibition of Na(+)-dependent transporters in cystine-loaded human renal cells: electrophysiological studies on the Fanconi syndrome of cystinosis. J. Am. Soc. Nephrol. 2002; 13: 2085–93.

Cherqui S, Kalatzis V, Forestier L, Poras I, Antignac C. Identification and characterisation of the murine homologue of the gene responsible for cystinosis, Ctns. BMC Genomics 2000; 1: 2.

Cherqui S, Kalatzis V, Trugnan G, Antignac C. The targeting of cystinosin to the lysosomal membrane requires a tyrosine-based signal and a novel sorting motif. J. Biol. Chem. 2001; 276: 13314–21.

Cherqui S, Sevin C, Hamard G, et al. Intralysosomal cystine accumulation in mice lacking cystinosin, the protein defective in cystinosis. Mol. Cell. Biol. 2002; 22: 7622–32.

Chol M, Nevo N, Cherqui S, Antignac C, Rustin P. Glutathione precursors replenish decreased glutathione pool in cystinotic cell lines. Biochem. Biophys. Res. Commun. 2004; 324: 231–5.

Cochat P, Cordier B, Lacôte C, Saïd M-H. Cystinosis: epidemiology in France. In: Broyer M, ed. Cystinosis. Paris, France: Elsevier, 1999: pp. 28–35.

Cogan DG, Kuwabara T, Kinoshita J, Sheehan L, Merola L. Cystinosis in an adult. J. Am. Med. Assoc. 1957; 164: 394–6.

Coor C, Salmon RF, Quigley R, Marver D, Baum M. Role of adenosine triphosphate (ATP) and NaK ATPase in the inhibition of proximal tubule transport with intracellular cystine loading. J. Clin. Invest. 1991; 87: 955–61.

The Cystinosis Collaborative Research Group. Linkage of the gene for cystinosis to markers on the short arm of chromosome 17. Nat. Genet. 1995; 10: 246–8.

DeBraekeleer M. Hereditary disorders in Saguenay-Lac-St-Jean (Quebec, Canada). Hum. Hered. 1991; 41: 141–6.

den Dunnen JT, Antonarakis SE. Mutation nomenclature extensions and suggestions to describe complex mutations: a discussion. Hum. Mutat. 2000; 15: 7–12.

den Dunnen JT, Antonarakis SE. Nomenclature for the description of human sequence variations. Hum. Genet. 2001; 109: 121–4.

Dufier JL. Ophthalmologic involvement in inherited renal disease. Adv. Nephrol. Necker. Hosp. 1992; 21: 143–56.

Dufier JL, Dhermy P, Gubler MC, Gagnadoux MF, Broyer M. Ocular changes in long-term evolution of infantile cystinosis. Ophthalmic. Paediatr. Genet. 1987; 8: 131–7.

Forestier L, Jean G, Attard M, et al. Molecular characterization of CTNS deletions in nephropathic cystinosis: development of a PCR-based detection assay. Am. J. Hum. Genet. 1999; 65: 353–9.

Freudenberg E. Cystinosis: Cystine disease (Lignac's disease) in children. Adv. Pediatr. 1949; 4: 265.

Gahl WA, Bashan N, Tietze F, Bernardini I, Schulman JD. Cystine transport is defective in isolated leukocyte lysosomes from patients with cystinosis. Science 1982; 217: 1263–5.

Gahl WA, Dalakas MC, Charnas L, et al. Myopathy and cystine storage in muscles in a patient with nephropathic cystinosis. N. Engl. J. Med. 1988; 319: 1461–4.

Gahl WA, Kuehl EM, Iwata F, Lindblad A, Kaiser-Kupfer MI. Corneal crystals in nephropathic cystinosis: natural history and treatment with cysteamine eyedrops [In Process Citation]. Mol. Genet. Metab. 2000; 71: 100–20.

Gahl WA, Reed GF, Thoene JG, et al. Cysteamine therapy for children with nephropathic cystinosis. N. Engl. J. Med. 1987; 316: 971–7.

Gahl WA, Thoene J, Schneider J. Cystinosis: A disorder of lyosomal membrane transport. In: Scriver CJ, Beaudet AL, Sly WS, Valle D, eds. The Metabolic and Molecular Bases of Inherited Disease. New York: McGraw-Hill, 2001: pp. 5085–108.

Gahl WA, Thoene JG, Schneider J. Cystinosis. N. Engl. J. Med. 2002; 347: 111–21.

Gahl WA, Tietze F. Lysosomal cystine transport in cystinosis variants and their parents. Pediatr. Res. 1987; 21: 193–6.

Gahl WA, Tietze F, Bashan N, Bernardini I, Raiford D, Schulman JD. Characteristics of cystine counter-transport in normal and cystinotic lysosome-rich leucocyte granular fractions. Biochem. J. 1983; 216: 393–400.

Gao XD, Wang J, Keppler-Ross S, Dean N. ERS1 encodes a functional homologue of the human lysosomal cystine transporter. FEBS J. 2005; 272: 2497–511.

Goldman H, Scriver CR, Aaron K, Delvin E, Canlas Z. Adolescent cystinosis: comparisons with infantile and adult forms. Pediatrics 1971; 47: 979–88.

Greene AA, Clark KF, Smith ML, Schneider JA. Cystine exodus from normal leucocytes is stimulated by MgATP. Biochem. J. 1987; 246: 547–9.

Greene AA, Marcusson EG, Morell GP, Schneider JA. Characterization of the lysosomal cystine transport system in mouse L-929 fibroblasts. J. Biol. Chem. 1990; 265: 9888–95.

Gubler M-C, Lacoste M, Sich M, Broyer M. The pathology of the kidney in cystinosis. In: Broyer M, ed. Cystinosis. Paris, France: Elsevier, 1999: pp. 42–8.

Haq MR, Kalatzis V, Gubler MC, et al. Immunolocalization of cystinosin, the protein defective in cystinosis. J. Am. Soc. Nephrol. 2002; 13: 2046–51.

Hardwick KG, Pelham HRB. ERS1 a seven transmembrane domain protein from *Saccharomyces cerevisiae*. Nucleic Acids Res. 1990; 18: 2177.

Heil SG, Levtchenko E, Monnens LA, Trijbels FJ, Van der Put NM, Blom HJ. The molecular basis of Dutch infantile nephropathic cystinosis. Nephron 2001; 89: 50–5.

Helip-Wooley A, Park MA, Lemons RM, Thoene JG. Expression of CTNS alleles: subcellular localization and aminoglycoside correction in vitro. Mol. Genet. Metab. 2002; 75: 128–33.

Hunziker W, Geuze HJ. Intracellular trafficking of lysosomal membrane proteins. Bioessays 1996; 18: 379–89.

International Cystinuria Consortium. Non-type I cystinuria caused by mutations in SLC7A9, encoding a subunit (bo, +AT) of rBAT. International Cystinuria Consortium. Nat. Genet. 1999; 23: 52–7.

Jackson JD, Smith FG, Litman NN, Yuile CL, Latta H. The Fanconi syndrome with cystinosis. Electron microsocopy of renal biopsy specimens from five patients. Am. J. Med. 1962; 33: 893–910.

Jonas AJ, Smith ML, Allison WS, Laikind PK, Greene AA, Schneider JA. Proton-translocating ATPase and lysosomal cystine transport. J. Biol. Chem. 1983; 258: 11727–30.

Jonas AJ, Smith ML, Schneider JA. ATP-dependent lysosomal cystine efflux is defective in cystinosis. J. Biol. Chem. 1982; 257: 13185–8.

Kalatzis V, Cherqui S, Antignac C, Gasnier B. Cystinosin, the protein defective in cystinosis, is a H(+)-driven lysosomal cystine transporter. EMBO J. 2001a; 20: 5940–9.

Kalatzis V, Cherqui S, Jean G, et al. Characterization of a putative founder mutation that accounts for the high incidence of cystinosis in Brittany. J. Am. Soc. Nephrol. 2001b; 12: 2170–4.

Kalatzis V, Cohen-Solal L, Cordier B, et al. Identification of 14 novel CTNS mutations and characterization of seven splice site mutations associated with cystinosis. Hum. Mutat. 2002; 20: 439–46.

Kalatzis V, Nevo N, Cherqui S, Gasnier B, Antignac C. Molecular pathogenesis of cystinosis: effect of CTNS mutations on the transport activity and subcellular localization of cystinosin. Hum. Mol. Genet. 2004; 13: 1361–71.

Kiehntopf M, Schickel J, Gonne B, et al. Analysis of the CTNS gene in patients of German and Swiss origin with nephropathic cystinosis. Hum. Mutat. 2002; 20: 237.

Kimonis VE, Troendle J, Rose SR, Yang ML, Markello TC, Gahl WA. Effects of early cysteamine therapy on thyroid function and growth in nephropathic cystinosis. J. Clin. Endocrinol. Metab. 1995; 80: 3257–61.

Kleta R, Anikster Y, Lucero C, et al. CTNS mutations in African American patients with cystinosis. Mol. Genet. Metab. 2001; 74: 332–7.

Kyttala A, Ihrke G, Vesa J, Schell MJ, Luzio JP. Two motifs target Batten disease protein CLN3 to lysosomes in transfected nonneuronal and neuronal cells. Mol. Biol. Cell, 2004; 15: 1313–23.

Mancini GM, Havelaar AC, Verheijen FW. Lysosomal transport disorders. J. Inherit. Metab. Dis. 2000; 23: 278–92.

Manz F, Gretz N. Cystinosis in the Federal Republic of Germany. Coordination and analysis of the data. J. Inher. Metab. Dis. 1985; 8: 2–4.

Markello TC, Bernardini IM, Gahl WA. Improved renal function in children with cystinosis treated with cysteamine. N. Engl. J. Med. 1993; 328: 1157–62.

Mason S, Pepe G, Dall'Amico R, et al. Mutational spectrum of the CTNS gene in Italy. Eur. J. Hum. Genet. 2003; 11: 503–8.

McGowan-Jordan J, Stoddard K, Podolsky L, Orrbine E, et al. Molecular analysis of cystinosis: probable Irish origin of the most common French Canadian mutation. Eur. J. Hum. Genet. 1999; 7: 671–8.

Oshima RG, Willis RC, Furlong CE, Schneider JA. Binding assays for amino acids. The utilization of a cystine binding protein from *Escherichia coli* for the determination of acid-soluble cystine in small physiological samples. J. Biol. Chem. 1974; 249: 6033–9.

Park M, Helip-Wooley A, Thoene J. Lysosomal cystine storage augments apoptosis in cultured human fibroblasts and renal tubular epithelial cells. J. Am. Soc. Nephrol. 2002; 13: 2878–87.

Pelham HR, Hardwick KG, Lewis MJ. Sorting of soluble ER proteins in yeast. EMBO J. 1988; 7: 1757–62.

Pellett OL, Smith ML, Greene AG, Schneider JA. Lack of complementation in somatic cell hybrids between fibroblasts from patients with different forms of cystinosis. Proc. Natl Acad. Sci. USA 1988; 85: 3531–4.

Pennesi M, Marchetti F, Crovella S, et al. A new mutation in two siblings with cystinosis presenting with Bartter syndrome. Pediatr. Nephrol. 2005; 20: 217–19.

Phornphutkul C, Anikster Y, Huizing M, et al. The promoter of a lysosomal membrane transporter gene, CTNS, binds Sp-1, shares sequences with the promoter of an adjacent gene, CARKL, and causes cystinosis if mutated in a critical region. Am. J. Hum. Genet. 2001; 69: 712–21.

Pisoni RL, Park GY, Velilla VQ, Thoene JG. Detection and characterization of a transport system mediating cysteamine entry into human fibroblast lysosomes. Specificity for aminoethylthiol and aminoethylsulfide derivatives. J. Biol. Chem. 1995; 270: 1179–84.

Pisoni RL, Thoene JG, Christensen HN. Detection and characterization of carrier-mediated cationic amino acid transport in lysosomes of normal and cystinotic human fibroblasts. Role in therapeutic cystine removal? J. Biol. Chem. 1985; 260: 4791–8.

Rupar CA, Matsell D, Surry S, Siu V. A G339R mutation in the CTNS gene is a common cause of nephropathic cystinosis in the south western Ontario Amish Mennonite population. J. Med. Genet. 2001; 38: 615–16.

Sato H, Tamba M, Ishii T, Bannai S. Cloning and expression of a plasma membrane cystine/glutamate exchange transporter composed of two distinct proteins. J. Biol. Chem. 1999; 274: 11455–8.

Schneider JA, Bradley K, Seegmiller JE. Increased cystine in leukocytes from individuals homozygous and heterozygous for cystinosis. Science 1967a; 157: 1321–2.

Schneider JA, Bradley K, Seegmiller JE. Increased free-cystine content of fibroblasts cultured from patients with cystinosis. Biochem. Biophys. Res. Commun. 1967b; 29: 527–9.

Shotelersuk V, Larson D, Anikster Y, et al. CTNS mutations in an American-based population of cystinosis patients. Am. J. Hum. Genet. 1998; 63: 1352–62.

Smith M, Furlong CE, Greene AA, Schneider JA. Cystine: binding protein assay. Methods Enzymol. 1987a; 143: 144–8.

Smith ML, Greene AA, Potashnik R, Mendoza SA, Schneider JA. Lysosomal cystine transport. Effect of intralysosomal pH and membrane potential. J. Biol. Chem. 1987b; 262: 1244–53.

Sonies BC, Almajid P, Kleta R, Bernardini I, Gahl WA. Swallowing dysfunction in 101 patients with nephropathic cystinosis: benefit of long-term cysteamine therapy. Medicine (Baltimore) 2005; 84: 137–46.

Spear GS. Pathology of the kidney in cystinosis. Pathol. Annu. 1974; 9: 81–92.

Spear GS, Slusser RJ, Tousimis AJ, Taylor CG, Schulman JD. Cystinosis. An ultrastructural and electron-probe study of the kidney with unusual findings. Arch. Pathol. 1971; 91: 206–21.

Steinherz R, Tietze F, Gahl WA, et al. Cystine accumulation and clearance by normal and cystinotic leukocytes exposed to cystine dimethyl ester. Proc. Natl Acad. Sci. USA 1982; 79: 4446–50.

Storch S, Pohl S, Braulke T. A dileucine motif and a cluster of acidic amino acids in the second cytoplasmic domain of the batten disease-related CLN3 protein are required for efficient lysosomal targeting. J. Biol. Chem. 2004; 279: 53625–34.

Thoene J, Lemons R, Anikster Y, et al. Mutations of CTNS causing intermediate cystinosis. Mol. Genet. Metab. 1999; 67: 283–93.

Touchman JW, Anikster Y, Dietrich NL, et al. The genomic region encompassing the nephropathic cystinosis gene (CTNS): complete sequencing of a 200-kb segment and discovery of a novel gene within the common cystinosis-causing deletion. Genome Res. 2000; 10: 165–73.

Town M, Jean G, Cherqui S, et al. A novel gene encoding an integral membrane protein is mutated in nephropathic cystinosis. Nat. Genet. 1998; 18: 319–24.

Zhai Y, Heijne WH, Smith DW, Saier MH Jr. Homologues of archaeal rhodopsins in plants, animals and fungi: structural and functional predications for a putative fungal chaperone protein. Biochim. Biophys. Acta 2001; 1511: 206–23.

CHAPTER 40

Hepatorenal Tyrosinemia

ROBERT M. TANGUAY, ANNE BERGERON AND ROSSANA JORQUERA

INTRODUCTION

Hereditary tyrosinemia type I (HTI; OMIM 276700), also referred to as hepatorenal tyrosinemia, is the most severe metabolic disorder of the tyrosine degradation pathway (Mitchell et al 2001). This pathway involves five consecutive enzymatic reactions (Figure 40.1). HTI is caused by a deficiency in fumarylacetoacetate hydrolase (FAH), the last enzyme of the pathway, which catalyzes the conversion of fumarylacetoacetate (FAA) in fumarate and acetoacetate (Lindblad et al 1977, Tanguay et al 1990, Phaneuf et al 1992). FAH is mainly expressed in the liver and kidneys,

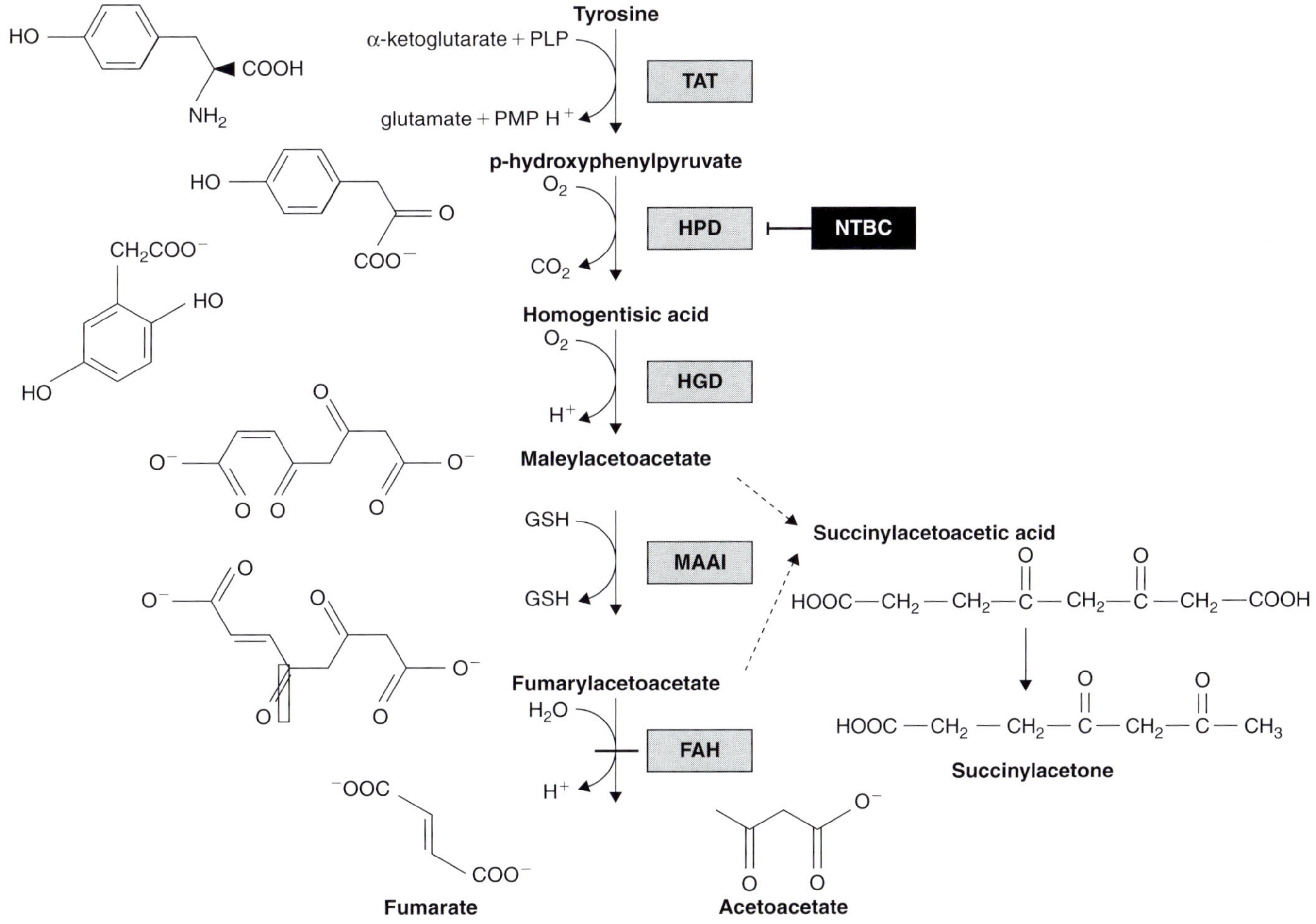

FIGURE 40.1 Pathway of tyrosine degradation. The catabolic breakdown of tyrosine is ensured by the successive action of five enzymes: tyrosine aminotransferase (TAT), 4-hydroxyphenylpyruvate dioxygenase (HPD), homogentisate dioxygenase (HGD), maleylacetoacetate isomerase (MAAI), and fumarylacetoacetate hydrolase (FAH). The therapeutic agent 2-(nitro-4-trifluoromethylbenzoyl)-1,3-cyclohexanedione (NTBC) acts as an inhibitor of HPD and blocks the catabolic pathway

Genetic Diseases of the Kidney
ISBN: 978-0-12-449851-8

but is also found to be expressed in other tissues, such as fibroblasts, amniocytes, chorionic villi, erythrocytes, and oligodendrocytes, albeit at a lower level. FAH is a cytosolic homodimeric enzyme of two 46 kDa subunits whose structure has recently been determined by crystallography (Timm et al 1999, and reviewed in Tanguay 2002). FAH functions as a metalloenzyme and its active site consists of 23 residues, which are rigorously conserved in all known FAHs. HTI shows an autosomal recessive pattern of inheritance and has been reported worldwide with variable frequencies. Globally, the incidence of HTI is around 1/100 000 births but can be much higher in certain regions due to founder effects. Thus in the province of Quebec, Canada, the incidence of HTI in the general population is ~1:16 000 births, but rises to 1:1846 in the region of Saguenay-Lac-St-Jean (SLSJ), where the carrier rate is 1:22 (De Braekeleer & Larochelle 1990, Grompe et al 1994, Poudrier et al 1996). A splice mutation (IVS12 + 5g→a) is found within this population (Grompe et al 1994, Poudrier et al 1996). This mutation is also the predominant one in the cases of HTI reported worldwide (reviewed in St-Louis & Tanguay 1997).

HTI: A SEVERE LIVER AND KIDNEY DISEASE

HTI presents into two main clinical types, acute and chronic (Mitchell et al 2001). The acute form, with early onset at birth or soon after, is mainly characterized by severe liver failure associated with cirrhosis, hepato- and splenomegaly, abnormal blood coagulation and hypoglycemia leading to death in the first months of life. Renal tubular dysfunctions such as Fanconi syndrome and rickets have also been considered hallmarks of HTI (Mitchell et al 2001, Russo et al 2001). In the chronic form, onset is insidious and progress is less aggressive. However, renal manifestations, mainly proximal tubulopathy, are prominent and may even be the presenting problem. Patients show impaired renal tubular reabsorptive function leading to Fanconi syndrome characterized by renal tubular acidosis, generalized aminoaciduria, hypophosphatemic vitamin D-resistant rickets and growth retardation. Histological changes in kidneys include dilatation of proximal tubules, cytoplasmic vacuolization and/or atrophy of tubular cells, nephrocalcinosis, loss of brush border and some degree of glomerulosclerosis and interstitial fibrosis (Paradis et al 1990, Russo & O'Regan 1990, Laine et al 1995). In a retrospective review of 32 children with tyrosinemia, Forget et al (1999) related renal structure obtained by sonography and computed tomography (CT) to glomerular filtration rate (GFR), aminoaciduria, acidosis, and hypercalciuria. Hyperechogenicity of the renal cortex and nephromegaly were noted in 47% of the patients while 17% showed nephrocalcinosis. While the uptake of

the intravenous contrast Omnipaque 300 was normal in all patients surveyed, its excretion was delayed in 64% of patients and was associated with low GFR. The biochemical tests also showed abnormal GFR (48%), aminoaciduria (82%) tubular acidosis (59%), and hypercalciuria (67%) with 13% of children presenting a complete Fanconi syndrome. Seventeen of these patients received a liver transplant and two, a concurrent renal transplant. GFR measured at various times posttransplantation (7–100 months) showed improvement in half of these patients. Tubular acidosis (5/10) and hypercalciuria (5/7) improved within the first 3 months posttransplantation. Renal biopsies performed at transplantation revealed the following anomalies: dilated tubules, mainly proximal (81%), interstitial fibrosis (56%), tubular atrophy (56%), glomerulosclerosis (56%) and vascular sclerosis (19%). In another study of eight Finnish patients, GFR was noted to normalize after orthotopic liver transplantation but some patients still showed signs of tubular dysfunctions 18–36 months posttransplantation (Laine et al 1995). In summary, liver transplantation has been shown to improve renal functions in HTI patients but GFR can remain a problem. The advent of NTBC (2-(2-nitro-4-trifluoro-methylbenzoyl)-1,3 cyclohexanedione) treatment for HTI is advantageous and it seems to prevent dysfunctions in the liver and kidneys. However, it is unsure if this will be sufficient to prevent problems on a long-term basis as evidenced by recent data showing that liver cancer is still a risk even in NTBC-treated patients (van Spronsen et al 2005). Moreover, failure to respond to NTBC has been reported in a child (Mohan et al 1999). Dietary restriction combined with NTBC treatment may also help to prevent renal damage (Lindstedt et al 1992).

Gradual liver alterations also occur along with cirrhosis and hepatocellular carcinoma (HCC) development. In an early study, Weinberg et al (1976) noted HCC in 37% of HTI patients over the age of 2 years. Subsequent studies in Scandinavia (van Spronsen et al 1989) and in Quebec (Russo et al 2001) showed a lower frequency of HCC (~15%) likely due to the advent of transplantation and improved treatment. Dietary restriction, while beneficial at the beginning, does not prevent ulterior liver damage and renal dysfunction. Orthotopic liver transplantation is essentially curative, although renal abnormalities may persist. Only 50% of liver-transplanted patients show partial improvement of renal function and altered kidney size and architecture persist (Paradis et al 1990, Laine et al 1995, Forget et al 1999). Moreover, liver transplantation does not fully correct metabolic perturbations in HTI, as affected kidneys continue to excrete succinylacetoacetic acid (SAA), succinylacetone (SA), and ALA (due to inhibition of δ-ALAD by SA; see Figure 40.1) in urine (Laine et al 1995, Tuchman et al 1987). In patients with persistent or deteriorating renal function despite therapy, combined liver–kidney transplantation may be necessary.

DIAGNOSTIC AND DETECTION OF HEREDITARY TYROSINEMIA

A disruption of the tyrosine catabolic pathway due to a deficiency of fumarylacetoacetate hydrolase causes important metabolic disturbances. These metabolic perturbations give rise to elevated plasma concentrations of amino acids such as tyrosine and methionine as well as excretion of unusual tyrosine metabolites like succinylacetone. Hypertyrosinemia can be caused by numerous other conditions affecting the liver and is also a feature of transient tyrosinemia of the newborn, a condition that resolves spontaneously without significant damage (Mitchell et al 2001, Russo et al 2001). Although elevated levels of tyrosine and plasma α-fetoprotein (AFP) are indicative of hereditary tyrosinemia, the most reliable biochemical diagnostic for HTI consists in the presence of SA in urine, blood and, amniotic fluid (Grenier et al 1976, 1982) and is pathognomonic for this disease. In Quebec, where there is a high incidence of HTI, a neonatal screening program for the disease was established in 1970 and consists of measuring SA levels in dried blood spots. Tandem mass spectrometry (MS/MS) has recently been used as a simple, sensitive, and rapid method for screening HTI by measuring the level of SA (Allard et al 2004). An enzymatic assay based on FAH activity measurements on cultured fibroblasts, blood, or liver specimen has been used (Kvittingen et al 1981, 1983). An appropriate diagnosis can also rely on genetic approaches to confirm the biochemical diagnosis. A genetic diagnostic test is usually performed when the family history or the origin of the patient suggests that one parent may be a carrier of the disorder. The progress achieved in the past decade concerning the molecular study of HTI and the identification of predominant mutations, facilitated the development of appropriate tests. For example, a genetic screening test was designed to detect the most common mutation found in HTI (IVS12 + 5G→a) (Grompe & Al-Dhalimy 1995). This method is based on the amplification of the genomic DNA region containing the mutation by PCR (polymerase chain reaction) followed by enzymatic cleavage of the amplified sequence in order to distinguish the mutated allele from the wild-type. Similar molecular tests have been developed for the Finnish W262X mutation (St-Louis et al 1996) as well as for the G337S (Rootwelt et al 1994), IVS6-1 g→t (Poudrier et al 1999), Q64H (Rootwelt et al 1994), Q279R (Dreumont et al 2001), E357X, E364X, F405H, IVS8-1 g→c, IVS12 + 1g→t, P342L, R341W, R381G and V166G mutations (Tanguay, unpublished). Such mutation analyses can be performed on blood, chorionic villi, or cultured amniocytes. A prenatal biochemical diagnosis can also be done by measuring the level of SA in amniotic fluid sampled between the 14th and 16th weeks of pregnancy. However, some false-positives were reported for this method (Jakobs et al 1990, Grenier et al 1996). Therefore, the genetic approach is a reliable and powerful method for

carrier detection when the mutations are known, while SA measurement remains the ideal diagnostic tool for prenatal diagnosis.

TREATMENT FOR HTI

Dietary treatment alone, while improving early symptoms, is not corrective of the disease as over 90% of the children die before age 12 years (Holme & Lindstedt 1998). Until the introduction of NTBC in 1992, orthotopic liver transplantation was reported to be particularly effective with survival rates exceeding 92%. Moreover, liver transplantation was reported to improve liver function (Tuchman et al 1987, Laine et al 1995). Since 1992, 2-(2-nitro-4-trifluoro-methylbenzoyl)-1,3 cyclohexanedione (NTBC) has been used as HTI therapy. NTBC inhibits the upstream enzyme HPD in the tyrosine degradation pathway impeding the accumulation of the putative toxic metabolites, FAA and SA (see Figure 40.1; Lindstedt et al 1992). NTBC (1–2 mg/kg/day) restores normal hepatic and renal functions, decreases SA excretion and reverses severe neurological crises in days or weeks, although elevated AFP levels may take several months to normalize (Holme & Lindstedt 1998, Paradis 1996). However, some HTI patients do not respond to NTBC and may even develop hepatic dysplasia and HCC (Mohan et al 1999). In an animal model of HTI, the FAH knockout mouse, 100% of NTBC-treated mice develop hepatomas with dysplasia after 7 months (Chen et al 2000) and 13–50% of them develop HCC after 10–24 months, even when using a stringent therapy (i.e. a higher NTBC dose of 4 mg/kg/day plus dietary tyrosine restriction) started prenatally (Grompe et al 1995, Al-Dhalimy et al 2002). Thus, the long-term efficiency of NTBC as a liver cancer-preventing agent in HTI remains uncertain.

MOLECULAR GENETICS OF HTI

The cDNA encoding the FAH protein includes 1259 nucleotides divided into 14 exons spanning over 35 kilobases of DNA in the q23-q25 region of human chromosome 15 (Phaneuf et al 1991, Labelle et al 1993, Awata et al 1994). The *fah* gene is conserved from the fungus *Emericella nidulans* to man whose gene shares high sequence homology with *Mus musculus* and *Rattus norvegicus* (84% and 85% respectively). Recently, two minor alternative transcripts of the *fah* gene have been identified (Dreumont et al 2005). The del100 transcript contains the first seven exons of *fah*, but then exon 8 skipping results in a disturbance of the reading frame and a premature termination codon. A protein resulting from this alternative splicing presents a tissue expression pattern differing from that of FAH. However,

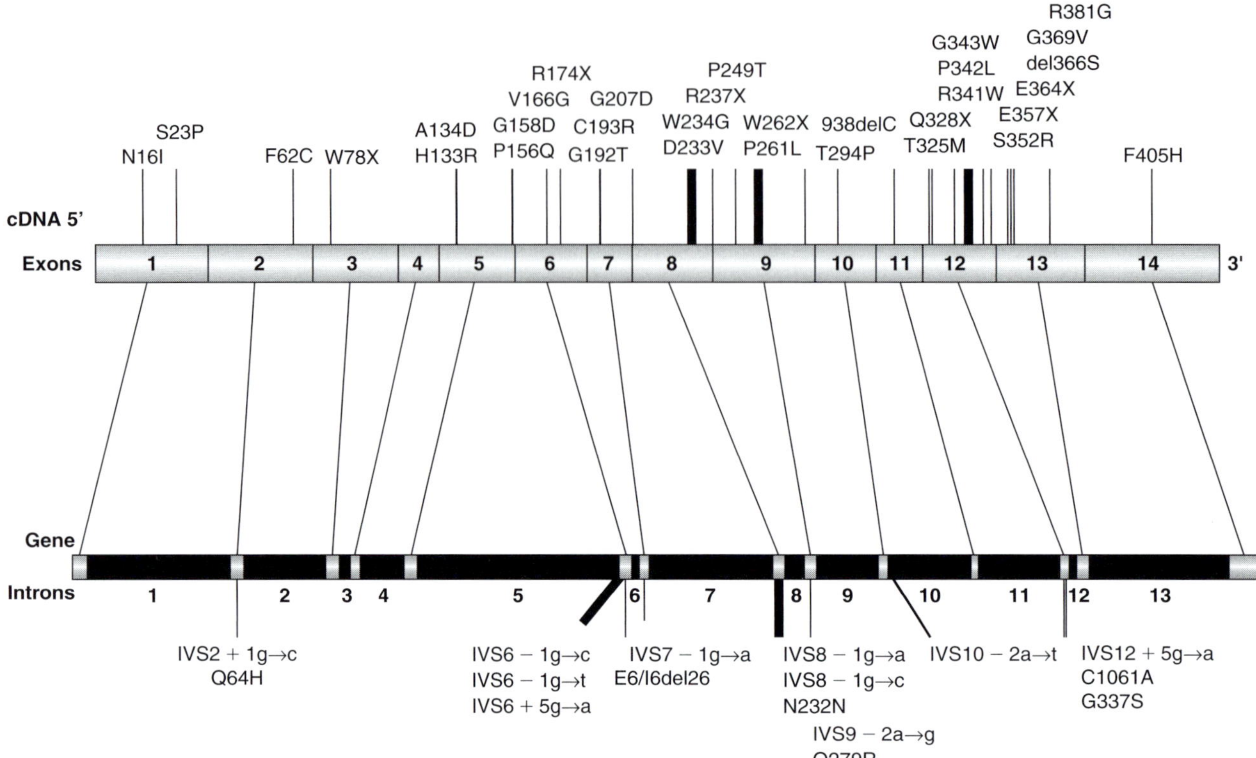

FIGURE 40.2 Mutations within the FAH gene. Among the known HTI-causing mutations, 24 are missense mutations, seven nonsense mutations, 16 splicing mutations, and two are deletions. Splicing mutations are indicated below the FAH gene structure

it remains unclear if this protein is truly functional. No protein related to the second transcript (del231) has been detected yet.

In 1992, Phaneuf and collaborators were the first to demonstrate that a mutation (N16I) in the *fah* gene was responsible for FAH deficiency and the phenotype observed in an HTI patient. They showed, by using an in vitro assay, that introducing the N16I mutation in the FAH cDNA abolished its enzymatic activity. Furthermore, the FAH protein was no longer detectable in the soluble fraction of N16I-FAH transfected cells. Since then, 49 mutations of the *fah* gene have been reported (Bergman et al 1998, Gray et al 2000, Kim et al 2000, Dreumont et al 2001, Arranz et al 2002, Heath et al 2002) (Figure 40.2). These mutations are spread along the gene with no apparent hot spot, although the N-terminal region seems to be less susceptible to mutations. All reported mutations lead to an inactive FAH protein except the R341W mutation, referred to as a pseudodeficiency mutation because patients homozygous for the mutation show no pathological symptoms of the disease (Kvittingen et al 1985). Among the many HTI-causing mutations, the one most frequently encountered worldwide is the IVS12 + 5g→a splicing defect, that is the hallmark mutation in the SLSJ region of Quebec (Grompe et al 1994, Poudrier et al 1996) where 95% of HTI alleles carry the IVS12 + 5g→a mutation as a result of a founder effect. Another frequent mutation found worldwide is the IVS6−1g→t splicing mutation (Ploos van Amstel et al 1996). In addition, there is also a small Finnish population that shows a high prevalence of the disease associated with the W262X nonsense mutation (St-Louis et al 1994). It is noteworthy that no correlation between the genotype and the phenotype of HTI patients has been found in studies of different mutations (Ploos van Amstel et al 1996, Poudrier et al 1998). In fact, in two patients homozygous for the IVS12 + 5g→a mutation, one presented with the chronic form and the other with the acute form (Poudrier et al 1998). Such differences in clinical severity have been suggested to correlate with the pattern of mosaic expression of hepatic FAH expression due to reversion of the causative mutation (Demers et al 2003).

SITE-SPECIFIC REVERSION OF MUTATIONS IN HTI AND RESTORATION OF ENZYME ACTIVITY

In vivo reversion of inherited somatic mutations has been reported in a number of inherited diseases, such as adenosine deaminase deficient severe combined immunodeficiency, Wiskott-Aldrich and Bloom syndromes, epidermolysis

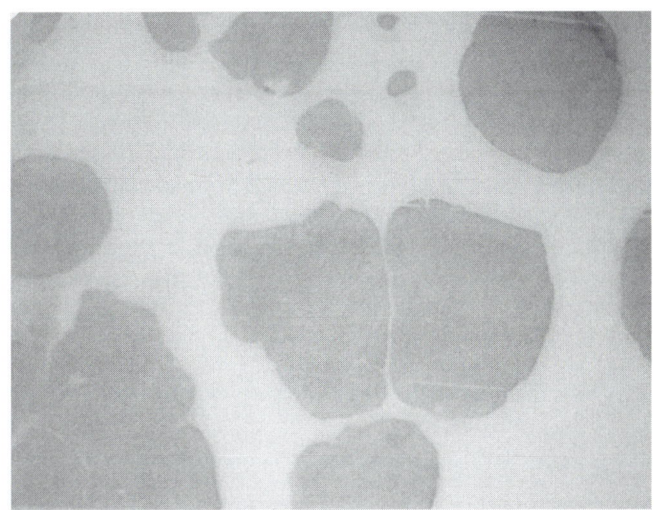

FIGURE 40.3 Mosaic expression of FAH in the liver of an HTI patient. Immunostaining of an liver section shows numerous nodules expressing FAH (dark brown staining). Other regions of the liver are FAH-negative. The section was stained with an antibody specific for FAH. Analysis of DNA obtained from microdissected stained nodules shows that the mutation is corrected in one of the FAH allele. (see also Plate 89)

bullosa, Fanconi anemia and HTI (reviewed in Hirschhorn 2003) (Figure 40.3). It is in HTI, however, that reversion has been shown to occur most frequently. A mosaic pattern of FAH expression was observed in hepatic regenerative nodules of >85% of studied patients (Kvittingen et al 1993, 1994, Demers et al 2003). DNA analysis performed on FAH immunopositive liver nodules showed correction of the HTI-causing mutation in one allele (Kvittingen et al 1994, Demers et al 2003). So far, site-specific reversion has been documented for four different mutations of the *fah* gene (Kvittingen et al 1994, Poudrier et al 1998, Kim et al 2000, Demers et al 2003). Importantly, reversion has never been found in patients' fibroblasts suggesting that a high proliferation capacity coupled to selective growth advantage of FAH$^+$ cells over FAH-deficient cells is required for the generation of reverted nodules. Given the scarcity of kidney samples available for histological analysis, no data are available regarding the occurrence of mutation reversion in kidneys although the limited proliferation capacity of renal cells suggests that it is unlikely that reversion has a significant influence on renal functions. Recently, a study of 26 transplanted French-Canadian HTI patients revealed an inverse correlation between the extent of mutation correction and the clinical severity of the disease (Demers et al 2003).

Mutation reversion is a puzzling feature of HTI and the mechanisms responsible for this event remain unknown. Somatic reversion often results from mitotic recombination but it is unlikely to be a major mechanism of reversion in HTI since the majority of studied patients are homozygous for a given mutation. On the basis of the observation that bilateral cell trafficking can occur between the mother and the fetus (Lo et al 1996, Tanaka et al 1999, Samura et al 2000, Brune et al 2002), it was proposed that implantation of maternal cells in the fetal liver followed by positive growth selection might explain the mosaic expression of FAH in the liver of patients. This hypothesis was recently tested using a genotyping approach on reverted liver sections of HTI patients (Bergeron et al 2004). Studies on four families clearly showed that the cells within the reverted liver regions were not of maternal origin. Thus, mutation reversion in HTI may result from a back-mutation possibly caused by FAA, a metabolite that is known to be mutagenic (Jorquera & Tanguay 1997) or by other toxic metabolites accumulating in HTI livers. The accumulation of such toxic metabolites may generate mutations in vivo and reverted cells would then be favored by a selective growth advantage and allow the formation of regenerating liver nodules. Such positive growth selection of FAH-expressing cells over deficient ones has been shown in a mouse model of HTI where corrected cells were observed to repopulate a liver almost entirely (Overturf et al 1996). Although back-mutations seem to account for the majority of reversion events observed up to now, one exception to the site-specific HTI reversion has recently been reported by Bliksrud et al (2005). In one of the nodules, they found a second-site mutation that may correct the initial splice mutation of the *fah* gene in this patient. This second-site mutation was not observed in the other FAH-positive nodules where a site-specific restoration of the original FAH sequence was observed suggesting that other mechanisms can correct the genomic defect in HTI.

ANIMAL MODELS OF HTI

The lethal albino 14CoS mouse, with its large 3.5-mbp deletion on chromosome 7, has been extensively studied for its phenotype of failure of hormone-dependent genes to respond to glucocorticoids and/or cAMP (the *alf*/hsdr-1 locus) (Gluecksohn-Waelsch 1979). The lethal albino phenotype is complex, affecting gene expression and cell ultrastructure in hepatocytes and proximal tubular renal cells, which show profound alterations of the endoplasmic reticulum (ER) and the Golgi apparatus (Kelsey & Schutz 1993). Mice homozygous for the 14CoS deletion die in the first day postpartum. Vesiculation of the rough ER and Golgi has been observed in the kidneys of newborn 14CoS/14CoS mice (Trigg et al 1973). Gluconeogenesis pathway enzymes (glucose-6-phosphatase, phosphoenolpyruvate carboxykinase and tyrosine amino transferase), urea cycle enzymes (argininosuccinate synthetase and ornithine transcarbamylase), and several liver-enriched transcription factors (C/EBP, HNF-1 and HNF-4) (Tonjes et al 1992) are downregulated in the 14CoS homozygous mouse. In contrast, the levels of several detoxification enzymes such as NAD(P)H:menadione oxidoreductase (NMO-1), UDP-glucuronyltransferase and glutathione-S-transferase-B are increased. Moreover, several growth arrest- or DNA damage-responsive genes like CHOP, GADD45 and c-fos are overexpressed (Fornace et al 1989, Ruppert et al 1992). The finding by positional cloning that the 14CoS deletion disrupted the *fah* gene suggested that this mouse might represent a mouse model for human HTI. Transgenic expression of FAH cDNA in 14CoS homozygotes indeed abolished neonatal lethality and complemented all aspects of the albino phenotype (Kelsey et al 1993). In this study, the degree of phenotypic correction was dependent on levels of transgenic FAH expression. In rescued 14CoS mice, although minimal (0.3–4%) expression of hepatic and renal FAH abolished neonatal lethality, allowed long-term survival, and corrected most of the hepatic and renal alterations, the persistent expression of NMO-I indicates that a low level of FAH is not sufficient to prevent damage from toxic metabolites (Kelsey et al 1993).

Rescue of 14CoS homozygous mice from neonatal lethality is also achieved by null mutation of the HPD gene (the double mutant HPD⁻/FAH⁻ mouse; Endo et al 1997) or by inhibiting HPD with NTBC (Dieter et al 2003, Jorquera et al, unpublished). The former model displays a complete correction of the HTI phenotype, but NTBC-treated 14CoS mice still show liver and kidney alterations and evidence of oxidative stress. In HPD⁻/FAH⁻ mice, retrieval of the tyrosine degradation pathway by HPD gene transfer or acute homogentisic acid (HGA) overloading causes massive apoptosis of hepatocytes and renal tubular cells, acute dysfunction of liver and kidney, and death (Endo et al 1997, Sun et al 2000). High doses of HGA (200–400 mg/kg) interfere with reabsorption of glucose and phosphate and apoptosis is observed in 80% of proximal tubular cells at the high dose. In proximal tubular cells, mitochondrial swelling, vacuolization, and breakage of the brush border are also observed. Interestingly, apoptosis of renal cells, but not renal dysfunction (Fanconi syndrome), is prevented by caspase inhibitors (Sun et al 2000). It was thus suggested that the cell death pathway and the metabolic process leading to Fanconi syndrome are separated in HTI-associated nephropathy. FAA, which is formed and remains intracellular in renal tubules, and SA, which is predominantly derived from glomerular filtration, were suggested to cause apoptotic death and dysfunction of FAH⁻ renal tubular cells, respectively (Sun et al 2000).

An FAH knockout mouse was produced by target inactivation of the *fah* gene by insertion of a *neo* transgene in exon 5 (Grompe et al 1993). This HTI model displays phenotypic alterations similar to those found in 14CoS mice. Neonatal lethality and HTI phenotype can be prevented in FAH knockout mice by blocking the tyrosine catabolic pathway upstream of FAA generation, i.e. by inhibiting HPD with NTBC (Grompe et al 1995) or by mutation inactivation of

the HGO gene (the double mutant HGO⁻/FAH⁻ mouse; Manning et al 1999) (see Figure 40.1). NTBC efficiently prevents acute liver and kidney damage in FAH knockout mice but, as mentioned above, cannot fully prevent long-term liver dysplasia and HCC development (Grompe et al 1995, Chen et al 2000, Al-Dhalimy 2002). Moreover, some molecular markers of the oxidative stress response, such as NMO-1, remain expressed in these mice. NTBC retrieval in the FAH knockout mice causes severe liver and kidney damage/dysfunction and death in 6–8 weeks (Grompe et al 1995). An interesting feature of the double mutant HGO⁻/FAH⁻ model is that liver of mice heterozygous for HGO (HGO⁺/⁻) but homozygous for FAH (FAH⁻/⁻) develop, after NTBC withdrawal, regenerative nodules of healthy hepatocytes (Manning et al 1999). These nodules, probably due to the highly mutagenic environment within hepatocytes, are able to correct the HTI-associated hepatic dysfunction, but do not normalize kidney functions.

Recently, two mice strains containing a point mutation in the *fah* gene were described (Aponte et al 2001). These animals may provide more appropriate models for the study of HTI as opposed to the other models that contain important genomic deletions (14CoS) or transgenes (knockout) which are not found in patients. The *Fah*⁵⁸⁶¹ˢᴮ and *Fah*⁶²⁸⁷ˢᴮ mice strains were generated by N-ethyl-N-nitrosourea treatment, a highly mutagenic compound. The mutation found in the *Fah*⁵⁸⁶¹ˢᴮ strain causes a deletion of exon 7, followed by disruption of the open reading frame and the appearance of a premature termination codon at the 303 position of the protein (Aponte et al 2001). Animals carrying this mutation die within 24 hours after birth. DNA analysis from the *Fah*⁶²⁸⁷ˢᴮ strain revealed the presence of a A→G nucleotide transition causing a missense mutation (E201G) at the 602 position (Aponte et al 2001). The presence of the E201G mutation leads to an inactive FAH (Timm et al 1999). Surprisingly, animals of the *Fah*⁶²⁸⁷ˢᴮ strain were reported to survive 10–20 days without NTBC treatment (Aponte et al 2001).

Thus, evidence arising from all the existing murine HTI models confirms what is observed in HTI patients, namely that an important difference between liver and kidney seems to exist at the level of metabolic processing and toxicity of the tyrosine-derived compounds formed in these tissues due to FAH deficiency. This difference could be mainly determined by involvement of the mercapturate pathway in the detoxification and/or bioactivation of GSH S-conjugates formed from accumulated metabolites in HTI, such as FAA, MAA, and SA.

GENE AND CELLULAR THERAPY AS POTENTIAL TREATMENTS IN HTI

As previously mentioned, the treatment currently used in HTI patients is administration of the drug NTBC combined with a low-protein diet. Despite the observations of strikingly improved liver and kidney functions and general good health of patients under NTBC treatment, a major concern remains as it was reported that long-term use of NTBC did not prevent liver cancer in a murine model of HTI (Grompe et al 1993, 1995). In fact, 17% of mice under a tyrosine and phenylalanine restrictive diet treated with NTBC presented with HCC at 24 months (Grompe et al 1995). Even under NTBC treatment, patients may still have to undergo orthotopic liver transplantation if there is evidence of a tumor from imaging techniques. This surgical procedure has been associated with a 10–15% mortality rate (Whitington & Balistreri 1991), and a risk of organ rejection, and is further complicated by the availability of donor organs and the cost of the procedure. Therefore, considerable efforts were invested into research of alternative therapies such as gene and cellular therapy to treat HTI. Although the adult human liver is generally in a quiescent state, it has an impressive potential for regeneration following injury. The high regenerative capacity of the liver makes chronic liver diseases such as HTI interesting models to consider for gene therapy.

The most widely used animal model for hereditary tyrosinemia is the $Fah^{\Delta exon5}$ mouse strain in which the *fah* gene has been disrupted by insertion of a neomycin gene cassette (Grompe et al 1993). These animals die within hours after birth unless they are rescued by NTBC treatment or by gene therapy. Studies using this murine model showed that FAH-positive cells display an important selective growth advantage and can be selected in vivo (Overturf et al 1996, 1997a). In vivo selection depends on three factors: (1) wild-type FAH-expressing cells must have a growth advantage over mutant cells, (2) tissue regeneration must be occurring and, (3) the defect has to be cell autonomous so the surrounding normal cells are not injured. In vivo selection in FAH-deficient mice was demonstrated when a greater than 50% repopulation of the mutant liver was achieved by transplanting as few as 1000 wild-type hepatocytes in mutant hosts (Overturf et al 1996). An overall hepatic repopulation of 70–95% was obtained in mice injected with 10 000 normal hepatocytes (Overturf et al 1997a). Liver morphology was normal in these rescued animals as well as their hepatic functions following an 8-week selection period off NTBC. The transplanted hepatocytes maintained their regenerative capacity for up to six rounds of serial transplantations, which is equivalent to a minimal number of 69 cell doublings suggesting that they have a regenerative potential similar to that of the hematopoietic stem cells. Importantly, NTBC treatment suppressed repopulation of the host liver by donor hepatocytes (Overturf et al 1996) suggesting that the selection is metabolically related to FAA.

Overturf and colleagues (1996) also attempted a retrovirus-based gene therapy in this HTI murine model. They hypothesized that if FAH-expressing hepatocytes possess a selective growth advantage in the liver of $Fah^{\Delta exon5}$ mice, then a stable expression of FAH in a limited number of hepatocytes selected in vivo could provide subsequent correction in the majority of hepatic cells. They used a retrovirus expressing human FAH. The main advantage of using a retrovirus in gene transfer is its ability to produce long-lasting expression by integration of the transgene in the host genome, although only a small proportion of hepatocytes are transduced in vivo. In such settings, a selective growth advantage would provide a better efficiency in this type of therapy. Mutant mouse livers corrected in situ by multiple retrovirus injections showed a hepatic repopulation of over 90% with FAH corrected cells and restored liver functions. However, multiple retrovirus injections were necessary to obtain efficient correction of the mutant liver and the high level of hepatic correction was attributable to the selection of corrected cells rather than to the high transduction efficiency. Nonetheless, 5–10% of FAH-deficient cells remained in the regenerated liver after a 3-month selection period and no long-term study was done to determine if there might be a tumor risk in treated mice. An ex vivo hepatic correction was also accomplished by transplantation of FAH^- cultured hepatocytes corrected by retroviral transduction (Overturf et al 1998). The efficiency of this method was similar to that of the in vivo method with proportions of corrected FAH^+ liver cells reaching over 85%. Both liver and kidney functions were significantly improved.

Adenoviral vectors also represent an attractive method for gene transfer because they can be used to transduce both dividing and quiescent cells (Li et al 1993). However, as opposed to retroviral vectors, they do not permit stable expression of transgenes on account of their episomal nature. Adenoviral gene transfer was tested in HTI mice (Overturf et al 1997b). Following in vivo transduction with a first-generation adenoviral vector containing a human FAH cDNA, 7/13 animals showed a percentage of FAH-positive hepatocytes of over 50% even after a 9-month observation period. In spite of the liver correction, kidney injuries were observed in these mice and no corrected cells were found in the proximal tubular epithelium of the kidneys. Furthermore, 77% of adenovirus-treated mice presented HCC all in uncorrected regions of the liver at 9 months following viral transduction. This observation again clearly outlines the threat posed by FAH-negative cells in the liver. As compared to the adenoviral approach, the use of retroviruses in the context of HTI proved to be more favorable since it was much more efficient and it also contributed to the prevention of kidney damage.

Another viral-based gene therapy involves the use of adeno-associated viral vectors (AAV). This type of vector has the benefit of conferring stable transgene expression although transduction efficiency is generally low (Miao et al 1998, Xiao et al 1998). Chen and collaborators (2000) used an AAV expressing FAH to transduce $Fah^{\Delta exon5}$ mice. They obtained a hepatic repopulation of FAH^+ cells of 50–90% after 5 months of selection without NTBC. Despite normalized liver functions, they found hepatomas in 5/7 long-term surviving animals.

Finally the use of simple plasmid DNA was also tested as a means of gene transfer in FAH deficient mice (Montini et al 2002, Held et al 2005). Transposases and recombinases allow the integration of transgenes in the host genome and represent easier, more affordable, and less immunogenic alternatives to viral gene transfer therapy. The 'Sleeping Beauty' transposon element was used to correct the liver disease in FAH⁻ mice. However, the effectiveness of gene transfer was low with an integration frequency of ~1%. Only ~60% of animals survived past 3 months and surviving animals displayed 60% of hepatic repopulation with both liver and kidney functions normalized. FAH expression was stable after two rounds of serial transplantation. The site-specific ϕC31 integrase was also tested to correct the mutant liver of Fah$^{\Delta exon5}$ mice (Held et al 2005). This integrase recognizes *attP* sites in the murine and human genomes and since the recombinase is site-specific, it was hypothesized that it would be less likely to cause transformation by random integration in host DNA. Results from this study were comparable to those obtained with the 'Sleeping Beauty' transposon. Although there was a preferred site of integration for the ϕC31 integrase, six other integration sites were found in one mouse. Thus, activation of oncogenes and disruption of tumor suppressor genes in this model remains a risk. Moreover, some recombination between *attP* sites may occur.

In summary, retroviral gene transfer is the most efficient gene therapy in FAH⁻ mice. Nonetheless, random integration in the host genome influences the level of FAH expression and may cause transformation of corrected cells by disrupting genes involved in cell cycle regulation. Adenoviral vectors show high transduction levels but are highly immunogenic. Finally, plasmid DNAs are not immunogenic, but they have low integration rates and may also cause a disruption of important genes in the genome. In addition, gene transfer of plasmid DNA in animals is accomplished by hydrodynamic injection, a method that is considered to be too traumatic to be used in humans.

Given the high efficiency of hepatocyte transplantation in FAH⁻ mice, cellular therapies were also evaluated. Wang et al (2001) showed that the pancreas of normal adult mice contained undifferentiated progenitors of hepatocytes. Whole pancreatic cell transplantation was attempted but showed very poor efficiency with only 10% of animals presenting significant liver repopulation. A more appealing cell type for cellular therapy in HTI is hematopoietic stem cells since these cells can be obtained from living individuals instead of rare organ donors and potential donors are already enlisted in bone-marrow donor registries. In addition, repopulation of the hematopoietic and liver systems at once should reduce the risk of tissue rejection. Results obtained following bone-marrow cell transplantation were disappointing since only 20–30% confluency of FAH-positive hepatocytes was reached after 22 weeks (Wang et al 2002). Previous work had revealed that only highly purified hematopoietic stem cells contributed to liver repopulation

(Lagasse et al 2000). Recent work in the field demonstrated by genetic analyses that bone-marrow-derived hepatocytes in this model originate from cell fusion between bone-marrow donor cells and host hepatocytes rather than from engraftment (Vassilopoulos et al 2003, Wang et al 2003). It was shown that the bone-marrow-derived cells involved in cell fusion were from the myeloid lineage (granulocytes or Kupffer cells) (Camargo et al 2004, Willenbring et al 2004). Therefore, cell fusion between committed myeloid cells and host hepatocytes may be the result of macrophage recruitment to the injured liver followed by fusion via surface receptors on the cells of the immune system.

Generally, following liver injury, regeneration is ensured by existing differentiated hepatocytes that divide and replace lost cells. However, in some instances, oval cells, which are considered as stem cells located near the Hering channel, may participate in hepatic repopulation. These cells are bipotent and can differentiate into bile duct epithelial cells or hepatocytes. Experiments using enriched oval cell populations to rescue FAH⁻ mice were also attempted (Wang et al 2003). Still, the repopulation efficiency was much lower than that observed with an equivalent number of hepatocytes.

In conclusion, although cellular therapy seems to represent a safer alternative to gene therapy, even under the best conditions using normal hepatocytes, liver correction remains incomplete. As previously discussed, the presence of less than 5% of uncorrected cells is sufficient to induce tumor formation. Thus, gene or cellular therapy may not be as promising as initially hoped for the treatment of hereditary tyrosinemia and liver transplantation remains the best solution for cancer prevention until complete liver correction strategies can be realized.

PERSPECTIVES

As noted above, renal functions do not fully improve even after extensive reversion in the livers of HTI patients (Demers et al 2003). The cause of the tubular damage is still unclear. It could be an intracellular process due to the effect of FAH deficiency in proximal tubular cells or an extracellular originating process linked to the concentration of toxic circulating metabolites by glomerular filtration. Thus, additional basic research to gain insights on the pathogenic mechanisms involved is essential. Moreover, the search for alternative or complementary forms of HTI treatments in order to efficiently counteract the toxicity of accumulated metabolites in FAH⁻ hepatocytes and renal cells is of utmost importance for clinical advances regarding HTI.

References

Al-Dhalimy M, Overturf K, Finegold M, Grompe M. Long-term therapy with NTBC and tyrosine-restricted diet in a murine model of hereditary tyrosinemia type I. Mol. Genet. Metab. 2002; 75: 38–45.

Allard P, Grenier A, Korson MS, Zytkovicz TH. Newborn screening for hepatorenal tyrosinemia by tandem mass spectrometry: analysis of succinylacetone extracted from dried blood spots. Clin. Biochem. 2004; 37: 1010–15.

Aponte JL, Sega GA, Hauser LJ, et al. Point mutations in the murine fumarylacetoacetate hydrolase gene: Animal models for the human genetic disorder hereditary tyrosinemia type 1. Proc. Natl Acad. Sci. USA 2001; 98: 641–45.

Arranz JA, Pinol F, Kozak L, et al. Splicing mutations, mainly IVS6-1 (G > T), account for 70% of fumarylacetoacetate hydrolase (FAH) gene alterations, including 7 novel mutations, in a survey of 29 tyrosinemia type I patients. Hum. Mutat. 2002; 20: 180–8.

Awata H, Endo F, Tanoue A, Kitano A, Nakano Y, Matsuda I. Structural organization and analysis of the human fumarylacetoacetate hydrolase gene in tyrosinemia type I. Biochim. Biophys. Acta 2004; 1226: 168–72.

Bergeron A, Russo P, Morissette J, Tanguay RM. No evidence of maternal cell colonization in reverted liver nodules of tyrosinemia type I patients. Gastroenterology 2004; 127: 1381–5.

Bergman AJ, van den Berg IE, Brink W, Poll-The BT, Ploos van Amstel JK, Berger R. Spectrum of mutations in the fumarylacetoacetate hydrolase gene of tyrosinemia type 1 patients in northwestern Europe and Mediterranean countries. Hum. Mutat. 1998; 12: 19–26.

Bliksrud YT, Brodtkorb E, Andresen PA, van den Berg IE, Kvittingen EA. Tyrosinaemia type-I-de novo mutation in liver tissue suppressing an inborn splicing defect. J. Mol. Med. 2005; 83: 406–10.

Brune T, Koch HG, Plumpe U, Coenen-Worch V, Harms E, Louwen F. Effect of pathological perinatal conditions on the maternofetal transfer of mononuclear cells. Fetal Diagn. Ther. 2002; 17: 110–14.

Camargo FD, Finegold M, Goodell MA. Hematopoietic myelomonocytic cells are the major source of hepatocyte fusion partners. J. Clin. Invest. 2004; 113: 1266–71.

Chen SJ, Tazelaar J, Moscioni AD, Wilson JM. In vivo selection of hepatocytes transduced with adeno-associated viral vectors. Mol. Ther. 2000; 1: 414–22.

De Braekeleer M, Larochelle J. Genetic epidemiology of hereditary tyrosinemia in Quebec and in Saguenay-Lac-St-Jean. Am. J. Hum. Genet. 1990; 47: 302–7.

Demers SI, Russo P, Lettre F, Tanguay RM. Frequent mutation reversion inversely correlates with clinical severity in a genetic disease, hereditary tyrosinemia. Hum. Pathol. 2003; 34: 1313–20.

Dieter MZ, Freshwater SL, Miller ML, Shertzer HG, Dalton TP, Nebert DW. Pharmacological rescue of the14CoS/14Cos mouse: hepatocyte apoptosis is likely caused by endogenous oxidative stress. Free Rad. Biol. Med. 2003; 35: 351–67.

Dreumont N, Maresca A, Boisclair-Lachance JF, Bergeron A, Tanguay RM. A minor alternative transcript of the fumarylacetoacetate hydrolase gene produces a protein despite being subjected to nonsense-mediated mRNA Decay. BMC Mol. Biol. 2005; 6: 1.

Dreumont N, Poudrier JA, Bergeron A, Levy HL, Baklouti F, Tanguay RM. A missense mutation (Q279R) in the fumarylacetoacetate hydrolase gene, responsible for hereditary tyrosinemia, acts as a splicing mutation. BMC Genet. 2001; 2: 9.

Endo F, Kubo S, Awata H, et al. Complete rescue of lethal albino c14CoS mice by null mutation of 4-hydroxyphenylpyruvate dioxygenase and induction of apoptosis of hepatocytes in these mice by in vivo retrieval of the tyrosine catabolic pathway. J. Biol. Chem. 1997; 272: 24426–32.

Forget S, Patriquin HB, Dubois J, et al. The kidney in children with tyrosinemia: sonographic, CT and biochemical findings. Pediatr. Radiol. 1999; 29: 104–8.

Fornace AJ Jr., Nebert DW, Hollander MC, et al. Mammalian genes coordinately regulated by growth arrest signals and DNA-damaging agents. Mol. Cell. Biol. 1989; 9: 4196–203.

Gluecksohn-Waelsch S. Genetic control of morphogenetic and biochemical differentiation: lethal albino deletions in the mouse. Cell 1979; 16: 225–37.

Gray RGF, Harper A, Davies P, McKiernan P. Ethnic diversity and mutations in the fumarylacetoacetase gene. J. Inherit. Metab. Dis. 2000; 23: 70.

Grenier A, Bélanger L, Laberge C. α-Fetoprotein measurement in blood spotted on paper: discriminating test for hereditary tyrosinemia in neonatal mass screening. Clin. Chem. 1976; 22: 1001–4.

Grenier A, Cederbaum S, Laberge C, Gagné R, Jakobs C, Tanguay RM. A case of tyrosinaemia type I with normal level of succinylacetone in the amniotic fluid. Prenat. Diagn. 1996; 16: 239–42.

Grenier A, Lescault A, Laberge C, Gagné R, Mamer O. Detection of succinylacetone and the use of its measurement in mass screening for hereditary tyrosinemia. Clin. Chim. Acta 1982; 123: 93–9.

Grompe M, Al-Dhalimy M. Rapid nonradioactive assay for the detection of the common French Canadian tyrosinemia type 1 mutation. Hum. Mutat. 1995; 5: 105.

Grompe M, Al-Dhalimy M, Finegold M, et al. Loss of fumarylacetoacetate hydrolase is responsible for the neonatal hepatic dysfunction phenotype of lethal albino mice. Genes Dev. 1993; 7: 2298–307.

Grompe M, Lindstedt S, Al-Dhalimy M, et al. Pharmacological correction of neonatal lethal hepatic dysfunction in a murine model of hereditary tyrosinaemia type 1. Nat. Genet. 1995; 10: 453–60.

Grompe M, St-Louis M, Demers SI, Al-Dhalimy M, Leclerc B, Tanguay RM. A single mutation of the fumarylacetoacetate hydrolase gene in most French Canadians with hereditary tyrosinemia type I. N. Engl. J. Med. 1994; 331: 353–7.

Heath SK, Gray RG, McKiernan P, Au KM, Walker E, Green A. Mutation screening for tyrosinaemia type 1. J. Inherit. Metab Dis. 2002; 25: 523–4.

Held PK, Olivares EC, Aguilar CP, Finegold M, Calos MP, Grompe M. In vivo correction of murine hereditary tyrosinemia type I by ρC 31 integrase-mediated gene delivery. Mol. Ther. 2005; 11: 399–408.

Hirschhorn R. In vivo reversion to normal of inherited mutations in humans. J. Med. Genet. 2003; 40: 721–8.

Holme E, Lindstedt S. Tyrosinemia type I and NTBC (2-(2-nitro-4trifluoromethylbenzoyl)-1,3-cyclohexanedione). J. Inher. Metab. Dis. 1998; 21: 507–17.

Jakobs C, Stellaard F, Kvittingen EA, Henderson M, Lilford R. First-trimester prenatal diagnosis of tyrosinemia type I by amniotic fluid succinylacetone determination. Prenat. Diagn. 1990; 10: 133–4.

Jorquera R, Tanguay RM. The mutagenicity of the tyrosine metabolite, fumarylacetoacetate, is enhanced by glutathione depletion. Biochem. Biophys. Res. Commun. 1997; 232: 42–8.

Kelsey G, Ruppert S, Beermann F, Grund C, Tanguay RM, Schütz G. Rescue of mice homozygous for lethal albino deletions: Implications for an animal model for the human liver disease tyrosinaemia type 1. Genes Dev. 1993; 7: 2285–97.

Kelsey G, Schutz G. Lessons from lethal albino mice. Curr. Opin. Genet. Dev. 1993; 3: 259–64.

Kim SZ, Kupke KG, Ierardi-Curto L, et al. Hepatocellular carcinoma despite long-term survival in chronic tyrosinaemia I. J. Inherit. Metab. Dis. 2000; 23: 791–804.

Kvittingen EA, Borresen AL, Stokke O, van der Hagen CB, Lie SO. Deficiency of fumarylacetoacetase without hereditary tyrosinemia. Clin. Genet. 1985; 27: 550–4.

Kvittingen EA, Halvorsen S, Jellum E. Deficient fumarylacetoacetate fumarylhydrolase activity in lymphocytes and fibroblasts from patients with hereditary tyrosinemia. Pediatr. Res. 1983; 17: 541–4.

Kvittingen EA, Jellum E, Stokke O. Assay of fumarylacetoacetate fumarylhydrolase in human liver-deficient activity in a case of hereditary tyrosinemia. Clin. Chim. Acta 1981; 115: 311–19.

Kvittingen EA, Rootwelt H, Berger R, Brantzaeg P. Self-induced correction of the genetic defect in tyrosinemia type I. J. Clin. Invest. 1994; 94: 1657–61.

Kvittingen EA, Rootwelt H, Brantzaeg P, Bergan A, Berger R. Hereditary tyrosinemia Self-induced correction of the fumarylacetoacetase deficiency. J. Clin. Invest. 1993; 91: 1816–21.

Labelle Y, Phaneuf D, Leclerc B, Tanguay RM. Characterization of the human fumarylacetoacetate hydrolase gene and identification of a missense mutation abolishing enzymatic activity. Hum. Mol. Genet. 1993; 2: 941–6.

Lagasse E, Connors H, Al-Dhalimy M, et al. Purified hematopoietic stem cells can differentiate into hepatocytes in vivo. Nat. Med. 2000; 6: 1229–34.

Laine J, Salo MK, Krogerus L, Karkkainen J, Wahlroos O, Holmberg C. The neuropathy of type I tyrosinemia after liver transplantation. Pediatr. Res. 1995; 37: 640–5.

Li Q, Kay MA, Finegold M, Stratford-Perricaudet LD, Woo SL. Assessment of recombinant adenoviral vectors for hepatic gene therapy. Hum. Gene. Ther. 1993; 4: 403–9.

Lindblad B, Lindstedt S, Steen G. On the enzymic defects in hereditary tyrosinemia. Proc. Natl Acad. Sci. USA 1977; 74: 4641–5.

Lindstedt S, Holme E, Lock EA, Hjalmarson O, Strandvik B. Treatment of hereditary tyrosinemia type I by inhibition of 4-hydroxyphenylpyruvate dioxygenase. Lancet 1992; 340: 813–17.

Lo YM, Lo ES, Watson N, et al. Two-way cell traffic between mother and fetus: biologic and clinical implications. Blood 1996; 88: 4390–5.

Manning K, al-Dhalimy M, Finegold M, Grompe M. In vivo suppressor mutations correct a murine model of hereditary tyrosinemia type I. Proc. Natl Acad. Sci. USA 1999; 96: 11928–33.

Miao CH, Snyder RO, Schowalter DB, Patijn GA, Donahue B, Kay MA. The kinetics of rAAV integration in the liver. Nat. Genet. 1998; 19: 13–15.

Mitchell GA, Grompe M, Lambert M, Tanguay RM. Hypertyrosinemia. In: Scriver CR, Beaudet AL, Sly WS, Valle D, eds. The Metabolic and Molecular Bases of Inherited Disease, 8th edn, Vol. II. New York: McGraw-Hill, 2001: pp. 1777–806.

Mohan N, Mckiernana P, Preece MA, et al. Indications and outcome of liver transplantation in tyrosinaemia type I. Eur. J. Pediatr. 1999; 158: S49–54.

Montini E, Held PK, Noll M, et al. In vivo correction of murine tyrosinemia type 1 by DNA-mediated transposition. Mol. Ther. 2002; 6: 759–69.

Overturf K, Al-Dhalimy M, Manning K, Ou C-N, Finegold M, Grompe M. Ex vivo hepatic gene therapy of a mouse model of hereditary tyrosinemia type 1. Hum. Gene Ther. 1998; 9: 295–304.

Overturf K, Al-Dhalimy M, Ou CN, Finegold M, Grompe M. Serial transplantation reveals the stem-cell-like regenerative potential of adult mouse hepatocytes. Am. J. Pathol. 1997a; 151: 1273–80.

Overturf K, Al-Dhalimy M, Ou CN, et al. Adenovirus-mediated gene therapy in a mouse model of hereditary tyrosinemia type I. Hum. Gene Ther. 1997b; 8: 513–21.

Overturf K, Al-Dhalimy M, Tanguay RM, et al. Hepatocytes corrected by gene therapy are selected in vivo in a murine model of hereditary tyrosinemia type I. Nat. Genet. 1996; 12: 266–73.

Paradis K. Tyrosinemia: the Quebec experience. Clin. Invest. Med. 1996; 19: 311–16.

Paradis K, Weber A, Seidman EG, et al. Liver transplantation for hereditary tyrosinemia: the Quebec experience. Am. J. Hum. Genet. 1990; 47: 338–42.

Phaneuf D, Labelle Y, Berube D, et al. Cloning and expression of the cDNA encoding human fumarylacetoacetate hydrolase, the enzyme deficient in hereditary tyrosinemia: assignment of the gene to chromosome 15. Am. J. Hum. Genet. 1991; 48: 525–35.

Phaneuf D, Lambert M, Laframboise R, Mitchell G, Lettre F, Tanguay RM. Type 1 hereditary tyrosinemia. Evidence for molecular heterogeneity and identification of a causal mutation in a French Canadian patient. J. Clin. Invest. 1992; 90: 1185–92.

Ploos van Amstel JK, Bergman AJ, van Beurden EA, et al. Hereditary tyrosinemia type 1: novel missense, nonsense and splice consensus mutations in the human fumarylacetoacetate hydrolase gene; variability of the genotype-phenotype relationship. Hum. Genet. 1996; 97: 51–9.

Poudrier J, Lettre F, Scriver CR, Larochelle J, Tanguay RM. Different clinical forms of hereditary tyrosinemia (type I) in patients with identical genotypes. Mol. Genet. Metab. 1998; 64: 119–25.

Poudrier J, Lettre F, St-Louis M, Tanguay RM. Genotyping of a case of tyrosinaemia type 1 with normal level of succinylacetone in amniotic fluid. Prenat. Diagn. 1999; 19: 61–3.

Poudrier J, St-Louis M, Lettre F, Gibson K, Prevost C, Larochelle J, Tanguay RM. Frequency of the IVS12 + 5G→A splice mutation of the fumarylacetoacetate hydrolase gene in carriers of hereditary tyrosinaemia in the French Canadian population of Saguenay-Lac-St-Jean. Prenat. Diagn. 1996; 16: 59–64.

Rootwelt H, Kristensen T, Berger R, Hoie K, Kvittingen EA. Tyrosinemia type 1-complex splicing defects and a missense mutation in the fumarylacetoacetase gene. Hum. Genet. 1994; 94: 235–9.

Ruppert S, Kelsey G, Schedl A, Schmid E, Thies E, Schutz G. Deficiency of an enzyme of tyrosine metabolism underlies altered gene expression in newborn liver of lethal albino mice. Genes Dev. 1992; 6: 1430–43.

Russo PA, Mitchell GA, Tanguay RM. Tyrosinemia: a review. Pediatr. Dev. Pathol. 2001; 4: 212–21.

Russo P, O'Regan S. Visceral pathology of hereditary tyrosinemia type 1. Am. J. Hum. Genet. 1990; 47: 317–24.

Samura O, Pertl B, Sohda S, et al. Female fetal cells in maternal blood: use of DNA polymorphisms to prove origin. Hum. Genet. 2000; 107: 28–32.

St-Louis M, Leclerc B, Laine J, Salo MK, Holmberg C, Tanguay RM. Identification of a stop mutation in five Finnish patients suffering from hereditary tyrosinemia type 1. Hum. Mol. Genet. 1994; 3: 69–72.

St-Louis M, Poudrier J, Tanguay RM. Simple detection of a (Finnish) hereditary tyrosinemia type 1 mutation. Hum. Mutat. 1996; 7: 379–80.

St-Louis M, Tanguay RM. Mutations in the fumarylacetoacetate hydrolase gene causing hereditary tyrosinemia type I: an overview. Hum. Mutat. 1997; 9: 291–9.

Sun MS, Hattori S, Kubo S, Awata H, Matsuda I, Endo F. A mouse model of renal tubular injury of tyrosinemia type I: development of de Toni Fanconi syndrome and apoptosis of renal tubular cells in Fah/Hpd double mutant mice. J. Am. Soc. Nephrol. 2000; 11: 291–300.

Tanaka A, Lindor K, Gish R, et al. Fetal microchimerism alone does not contribute to the induction of primary biliary cirrhosis. Hepatology 1999; 30: 833–8.

Tanguay RM. Fumarylacetoacetate hydrolase, In: Wiley Encyclopedia of Molecular Medicine. London: John Wiley & Sons, 2002: pp. 1338–41.

Tanguay RM, Valet JP, Lescault A, et al. Different molecular basis for fumarylacetoacetate hydrolase deficiency in the two clinical forms of hereditary tyrosinemia (type I). Am. J. Hum. Genet. 1990; 47: 308–16.

Timm DE, Mueller HA, Bhanumoorthy P, Harp JM, Bunick GJ. Crystal structure and mechanism of a carbon-carbon bond hydrolase. Structure Fold Des. 1999; 7: 1023–33.

Tonjes RR, Xanthopoulos KG, Darnell JE Jr., Paul D. Transcriptional control in hepatocytes of normal and c14CoS albino deletion mice. EMBO J. 1992; 11: 127–33.

Trigg MJ, Gluecksohn-Waelsch S. Ultrastructural basis of biochemical effects in a series of lethal alleles in the mouse. J. Cell Biol. 1973; 58: 549–63.

Tuchman M, Freese DK, Sharp HL, Ramnaraine MLR, Asher N, Bloomer JR. Contribution of extrahepatic tissues to biochemical abnormalities in hereditary tyrosinemia type I: study of three patients after liver transplantation. J. Pediatr. 1987; 110: 399–403.

van Spronsen FJ, Berger R, Smit GP, et al. Tyrosinaemia type I: orthotopic liver transplantation as the only definitive answer to a metabolic as well as an oncological problem. J. Inherit. Metab. Dis. 1989; 12: S339–42.

van Spronsen FJ, Bijleveld CM, van Maldegem BT, Wijburg FA. Hepatocellular carcinoma in hereditary tyrosinemia type I despite 2-(2 nitro-4-3 trifluoro-methylbenzoyl)-1,3-cyclohexanedione treatment. J. Pediatr. Gastroenterol. Nutr. 2005; 40: 90–93.

Vassilopoulos G, Wang P-R, Russell DW. Transplanted bone marrow regenerates liver by cell fusion. Nature 2003; 422: 901–4.

Wang X, Al-Dhalimy M, Lagasse E, Finegold M, Grompe M. Liver repopulation and correction of metabolic liver disease by transplanted adult mouse pancreatic cells. Am. J. Pathol. 2001; 158: 571–9.

Wang X, Foster M, Al-Dhalimy M, Lagasse E, Finegold M, Grompe M. The origin and liver repopulating capacity of murine oval cells. Proc. Natl Acad. Sci. USA 2003a; 100: 11881–8.

Wang X, Montini E, Al-Dhalimy M, Lagasse E, Finegold M, Grompe M. Kinetics of liver repopulation after bone marrow transplantation. Am. J. Pathol. 2002; 161: 565–74.

Wang X, Willenbring H, Akkari Y, et al. Cell fusion is the principal source of bone-marrow-derived hepatocytes. Nature 2003b; 422: 897–901.

Weinberg AG, Mize CE, Worthen HG. The occurrence of hepatoma in the chronic form of hereditary tyrosinemia. J. Pediatr. 1976; 88: 434–8.

Whitington PF, Balistreri WF. Liver transplantation in pediatrics: indications, contraindications, and pretransplant management. J. Pediatr. 1991; 118: 169–77.

Willenbring H, Bailey AS, Foster M, et al. Myelomonocytic cells are sufficient for therapeutic cell fusion in liver. Nat. Med. 2004; 10: 744–8.

Xiao W, Berta SC, Lu MM, Moscioni AD, Tazelaar J, Wilson JM. Adeno-associated virus as a vector for liver-directed gene therapy. J. Virol. 1998; 72: 10222–6.

CHAPTER 41

Renal Disease in Type I Glycogen Storage Disease

JANICE Y. CHOU, BRIAN C. MANSFIELD AND DAVID A. WEINSTEIN

HISTORICAL BACKGROUND

GSD-I was described first in 1929 by von Gierke as hepatonephromegalia glycogenia. Later, Cori and Cori (1952) showed the underlying condition of the disease was an absence of G6Pase (also named G6Pase-α or G6PC) activity in hepatic tissues. As increasing numbers of patients with the clinical symptoms of GSD-I were identified, two subgroups emerged. Enzyme assays of frozen liver samples showed that one group of patients had no G6Pase catalytic activity, while the other group retained G6Pase activity. To account for this biochemical heterogeneity, Senior and Loridan (1968) subclassified GSD-I into GSD-Ia, in which patients lack G6Pase activity, and GSD-Ib, in which patients retain G6Pase activity.

GSD-Ia was shown to be due to a loss of activity in the G6Pase catalytic unit but identification of the defect in GSD-Ib was more elusive. From 1972 to 1980, Arion and coworkers (reviewed in Arion et al 1980, Nordlie & Sukalski 1985) performed a series of experiments measuring G6Pase activity in both intact and disrupted liver microsomes. From these studies, they proposed that the hydrolysis of G6P in vivo involves the integration of several membrane proteins, including the G6Pase catalytic unit and its associated transporters. This multicomponent enzyme complex was named the 'G6Pase system' (Figure 41.1). In 1978, Nilsson et al demonstrated that the activity of G6Pase in intact microsomes is resistant to limited proteolysis. This implied that the active site of the enzyme resides within the lumen of the ER and is dependent upon a transmembrane substrate transporter. Refining these studies further, Narisawa et al (1978) and Lange et al (1980) demonstrated that microsomes from GSD-Ib patients only exhibit normal G6Pase activity when disrupted with detergent. They hypothesized that GSD-Ib is caused by a deficiency of a putative G6PT. Molecular genetic studies have since characterized the G6Pase and G6PT genes and confirmed these roles. Inactivating mutations in the

G6Pase gene cause GSD-Ia (Lei et al 1993), and inactivating mutations in the G6PT gene cause GSD-Ib (Hiraiwa et al 1999).

CLINICAL PRESENTATION

While some GSD-I patients present in the neonatal period with hypoglycemia and lactic acidosis, clinical symptoms are not typically detected until the patient is a few months old (Chen 2001, Chou et al 2002). The initial symptoms are usually hepatomegaly and hypoglycemia following a short fast. Physically, affected children have short stature with a significant accumulation of subcutaneous fat and a protuberant abdomen. The kidneys are also symmetrically enlarged but the spleen and heart are of normal size. The plasma may be 'milky' in appearance as a result of a striking elevation of triglycerides.

Easy bruising, epistaxis, and prolonged bleeding times are common as a result of impaired platelet aggregation and adhesion. Skin xanthomas over the upper and lower extremities may be present in infancy and retinal changes characterized by multiple, discrete, yellowish paramacular lesions also occur secondary to overt hyperlipidemia (Chen 2001, Chou et al 2002). The risk for atherosclerosis is low (Lee et al 1994, Ubels et al 2002). Frequent fractures and radiographic evidence of osteopenia are common in adult patients, and radial bone mineral content is significantly reduced in both prepubertal (Lee et al 1995b) and adult (Rake et al 2003) patients. Necrotizing pancreatitis has been observed in several patients with GSD-I, probably because of the massive hypertriglyceridemia (Michels & Beaudet 1980, Kikuchi et al 1991). Hepatocellular adenoma that may undergo malignant transformation (Kudo 2001, Lee 2002, Weinstein & Wolfsdorf 2002) and renal disease (Chen et al 1988, Baker et al 1989, Wolfsdorf et al 1997, Moses 2002, Rake et al 2002) are late complications.

Genetic Diseases of the Kidney
ISBN: 978-0-12-449851-8

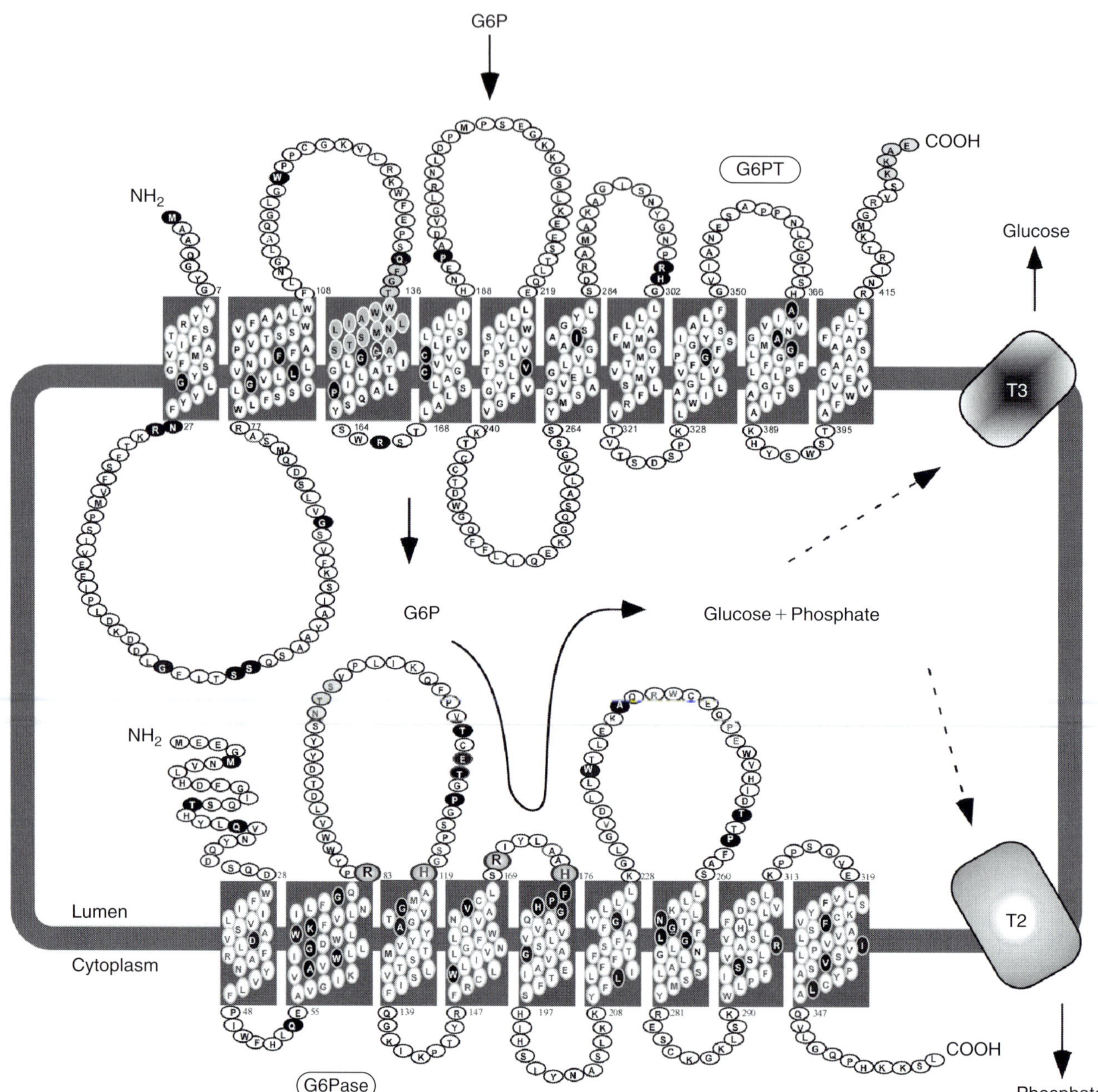

FIGURE 41.1   The G6Pase system, an enzyme complex essential for glucose homeostasis. The components, G6Pase, G6PT, a putative phosphate transporter (T2), and a putative glucose transporter (T3), are shown anchored in the membrane of the ER in contact with both the cytoplasm and ER lumen. The existence of T2 and T3 remains hypothetical. Glucose may be released directly to the blood via the ER/Golgi pathway, rather than into the cytoplasm, as shown. The spatial representation is illustrative only. G6Pase and G6PT are anchored in the ER by 9 and 10 transmembrane helices, respectively (Pan et al 1988a, 1999). Mutations identified in the G6Pase and G6PT genes of GSD-I patients are marked in black. In G6Pase, R83, H119, R170, and H176, which contribute to the active center, are denoted by large shaded circles. In G6PT, amino acid residues comprising the signature motif of transporters of phosphorylated metabolites at amino acid residues 133 to 149 are denoted by shaded circles. Modified from Chou et al (2002) with permission. (see also Plate 90)

Biochemical tests reveal hypoglycemia, hyperlipidemia, hyperuricemia, and lactic acidemia (Chen 2001, Chou et al 2002). Hyperlipidemia in GSD-I is characterized by a combined hypercholesterolemia and hypertriglyceridemia. Increased cholesterol is found in both the very-low-density lipoprotein and the low-density lipoprotein fractions, whereas high-density lipoprotein cholesterol and apolipoprotein A-1 are usually decreased (reviewed in Bandsma et al 2002).

GSD-Ib has a similar clinical course to GSD-Ia, but with the additional findings of neutropenia and myeloid

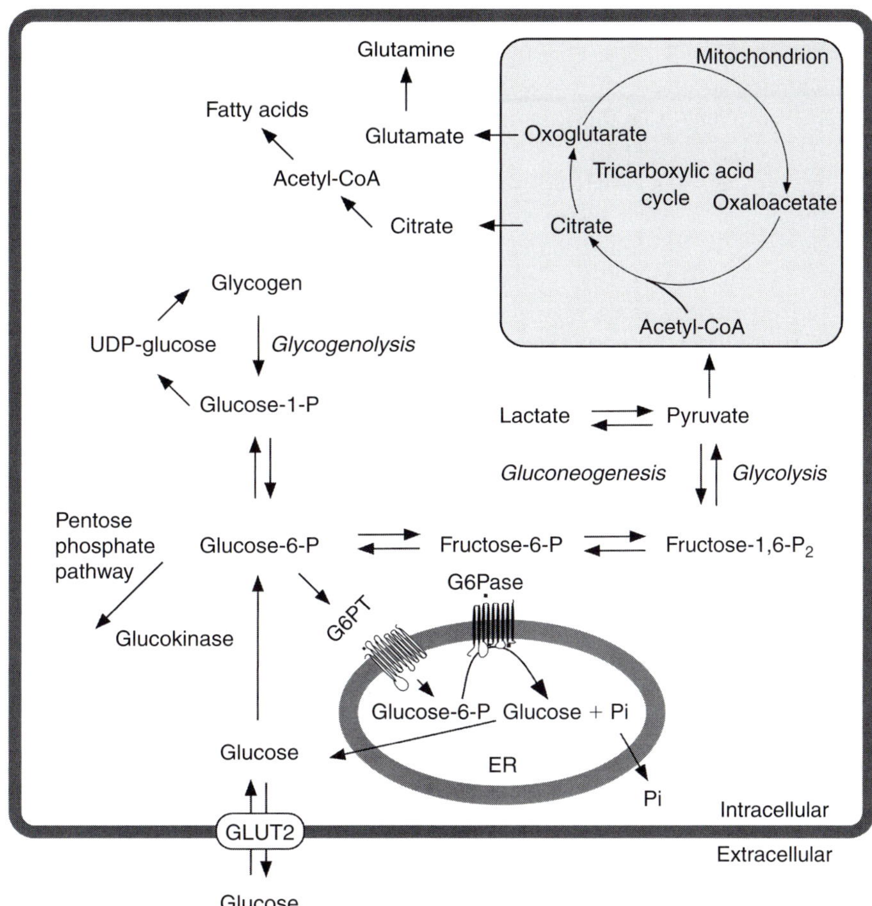

FIGURE 41.2   The primary anabolic and catabolic pathways of G6P in the liver. The G6Pase and G6PT components of the G6Pase system are shown embedded within the membrane of the ER

dysfunctions, which result in recurrent bacterial infections (Beaudet et al 1980, Gitzelmann & Bosshard 1993, Garty et al 1996, Visser et al 2002, Chou & Mansfield 2003). Oral and intestinal mucosal ulcerations are common in GSD-Ib, and most patients also suffer chronic inflammatory bowel disease (Roe et al 1986, Visser et al 2002).

THE G6Pase COMPLEX AND PATHOPHYSIOLOGY OF GSD-I

G6P is the intracellular form of blood glucose. Following tissue uptake of glucose, via facilitated diffusion (Thorens 1993), phosphorylation traps the glucose by placing a negative charge on the molecule, preventing its diffusion back across the cell membrane into the blood. Immediately after eating, excess blood glucose is taken up by the liver, phosphorylated by glucokinase to G6P, and converted to glycogen for storage. Between meals, blood glucose homeostasis is maintained by balancing the tissue uptake of glucose from the blood against the release of G6P-derived glucose to the blood by the liver and kidney. Therefore, the

dephosphorylation of G6P to glucose by the G6Pase system is a key control point for interprandial glucose homeostasis. Consistent with this, the only tissues that express high levels of G6Pase are the liver and kidney. Hepatic release of glucose derived from glycogenolysis (50%) and gluconeogenesis (30%) accounts for the majority of released glucose, with the balance accounted for by renal gluconeogenesis (20%) (reviewed in Gerich et al 2001, Cano 2002).

The key step in the pathways of both glycogenolysis and gluconeogenesis is the terminal one, namely the dephosphorylation of G6P to glucose by the G6Pase complex (Figure 41.2). Both pathways occur predominantly in the cytoplasm, but the active site of G6Pase resides within the lumen of the ER (Shieh et al 2002, Ghosh et al 2002). Therefore, G6P hydrolysis is dependent upon transport of the cytoplasmic G6P across the ER membrane by G6PT (reviewed in Chou et al 2002, Chou & Mansfield 2003). A deficiency in either G6Pase or G6PT will disrupt glucose production and blood glucose homeostasis. The lack of G6P hydrolysis in the liver and kidney leads to the accumulation of excess G6P in the cytoplasm. This in turn stimulates alternative metabolic pathways of G6P leading to the pathophysiological consequences of GSD-I. One obvious consequence is the stimulation of glycogen formation, which gradually becomes excessive and leads to hepatomegaly and nephromegaly.

A second consequence, which leads to a cascade of self-reinforcing events, is the stimulation of glycolysis, which converts G6P to lactate (Figure 41.2). In normal individuals, lactate produces energy in the form of ATP and excess lactate recycles back to glucose via gluconeogenesis. In GSD-I patients, however, this pathway is disrupted, resulting in the accumulation of blood lactate, which leads to lactic acidemia. Lactate competes with uric acid for excretion in the kidney and the elevated lactate in turn promotes hyperuricemia. Moreover, the elevated intracellular G6P concentration depresses intrahepatic phosphate levels, leading to AMP deaminase activation, which exacerbates uric acid production and hyperuricemia (Greene et al 1978, Cohen et al 1985).

Stimulation of glycolysis also leads to increased intracellular accumulation of pyruvate. The resulting elevation in the intracellular concentration of acetyl-CoA stimulates the tricarboxylic acid cycle to drive overproduction of substrates and cofactors (NADH, NADPH) for lipid synthesis (Figure 41.2), which leads to hyperlipidemia. Gradual exhaustion of the liver's capacity to synthesize apolipoproteins and lipoproteins leads to the hallmark GSD-I fat droplets, while a progressive decrease in lipid clearance from the blood contributes further to the hyperlipidemia (reviewed in Bandsma et al 2002).

The mechanisms for immune deficiency in GSD-Ib have not been elucidated conclusively. GSD-Ib patients have unexplained defects in their neutrophil respiratory burst, chemotaxis, and calcium flux, as well as intermittent neutropenia (reviewed in Gitzelmann & Bosshard 1993, Garty et al 1996, Visser et al 2002, Chou & Mansfield 2003). One possible explanation may lie in the observation that, in contrast to the liver and kidney, neutrophils and monocytes from GSD-Ib patients have reduced intracellular concentrations of G6P (Verhoeven et al 1999). Myeloid cells depend upon G6P to stimulate both the glycolytic and pentose phosphate pathways to provide energy for chemotaxis and phagocytosis; reduced G6P levels in GSD-Ib adversely affect these roles. In addition, G6P enhances the ATP-dependent microsomal sequestration of calcium (Chen et al 1998) which is required for second messenger activity in the respiratory burst and chemotaxis (Synderman & Uhing 1988). Normally, sequestration of ATP and calcium feeds back to enhance accumulation of G6P into the ER (Chen et al 1998). In GSD-Ib, however, the low levels of intracellular G6P in neutrophils reduce calcium sequestration. This, in turn, reduces G6P accumulation, yet another example of the self-reinforcing deficiencies which exacerbate the progression of GSD-Ib.

THE MOLECULAR BASIS OF GSD-Ia

G6Pase deficiency (GSD-Ia) is the most prevalent form of GSD-I, representing over 80% of GSD-I cases (Chen 2001, Chou et al 2002), but molecular characterization of G6Pase remained elusive for many years. The breakthrough came with the fortuitous identification of the murine G6Pase cDNA (Shelly et al 1993) and subsequent isolation of the human cDNA and gene. Both human (Lei et al 1993) and mouse (Shelly et al 1993) G6Pase are encoded by single copy genes composed of five exons spanning 10–13 kb of chromosomal DNA. In the human, the gene maps to chromosome 17q21 (Lei et al 1994).

Human G6Pase is a highly hydrophobic, 357 amino acid glycoprotein. It contains nine transmembrane domains (Pan et al 1998a) that anchor it in the ER-membrane (Figure 41.1) and it is expressed primarily in the liver, kidney cortex, and intestine (Nordlie & Sukalski 1985, Pan et al 1998b). The active site of G6Pase encompasses a phosphatase signature motif (Chou & Mansfield 1999, Ghosh et al 2002, Shieh et al 2002, Chou et al 2002), and is localized on the luminal side of the ER membrane (Figure 41.1).

A total of 82 separate G6Pase mutations, including 53 missense, 10 nonsense, 16 insertion/deletion and three splicing mutations (Figure 41.3) have been identified within a group of ~500 GSD-Ia patients (reviewed in Chou et al 2002, Matern et al 2002). Site-directed mutagenesis and transient expression assays of the missense and active site mutants have defined the active site and the role of the transmembrane helices in the activity of G6Pase (Shieh et al 2002). While GSD-Ia is not restricted to any one ethnic group, mutations unique to particular groups have been described. The disease is particularly common in the Ashkenazi Jewish population with a carrier frequency of 1.4% for the R83C mutation (Ekstein et al 2004). No clear genotype–phenotype correlation has been demonstrated in GSD-Ia (Matern et al 2002).

THE MOLECULAR BASIS OF GSD-Ib

The G6PT defects (GSD-Ib) represent approximately 20% of all GSD-I cases (Chen 2001, Chou et al 2002, Rake et al 2002, Chou & Mansfield 2003). Two complementary methods were used to identify the GSD-Ib gene – a computer-based sequence search for liver ESTs with homologies to bacterial phosphate ester transporters (Gerin et al 1997), and a disease linkage analysis study (Annabi et al 1998). The approaches converged when the putative G6PT cDNA identified by homology searches (Gerin et al 1997) was shown to map to the disease locus at chromosome 11q23 (Annabi et al 1998). Subsequent expression-activity studies of the cDNA, combined with site-directed mutagenesis, confirmed the identity of the G6PT activity (Hiraiwa et al 1999).

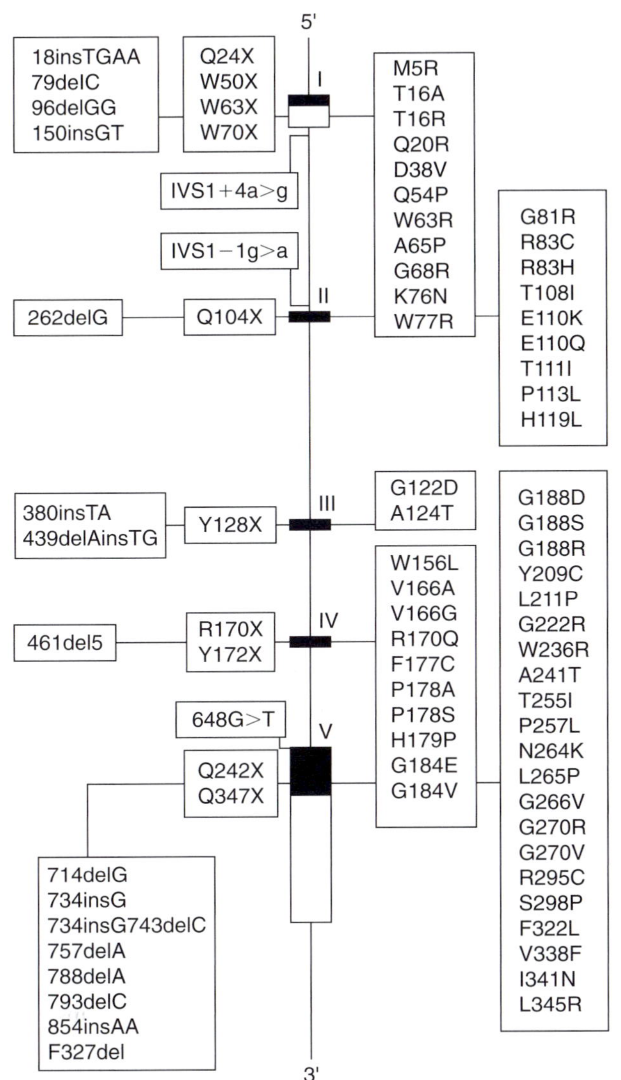

FIGURE 41.3 Mutations identified in the G6Pase gene of GSD-Ia patients. The G6Pase gene is shown as a line diagram with the five exons marked as boxes I to V. Black boxes represent coding regions, white boxes the 5′ and 3′ untranslated regions of the G6Pase transcript. The positions of all known mutations are listed from left to right as insertion/deletion, nonsense, splicing, and missense mutations

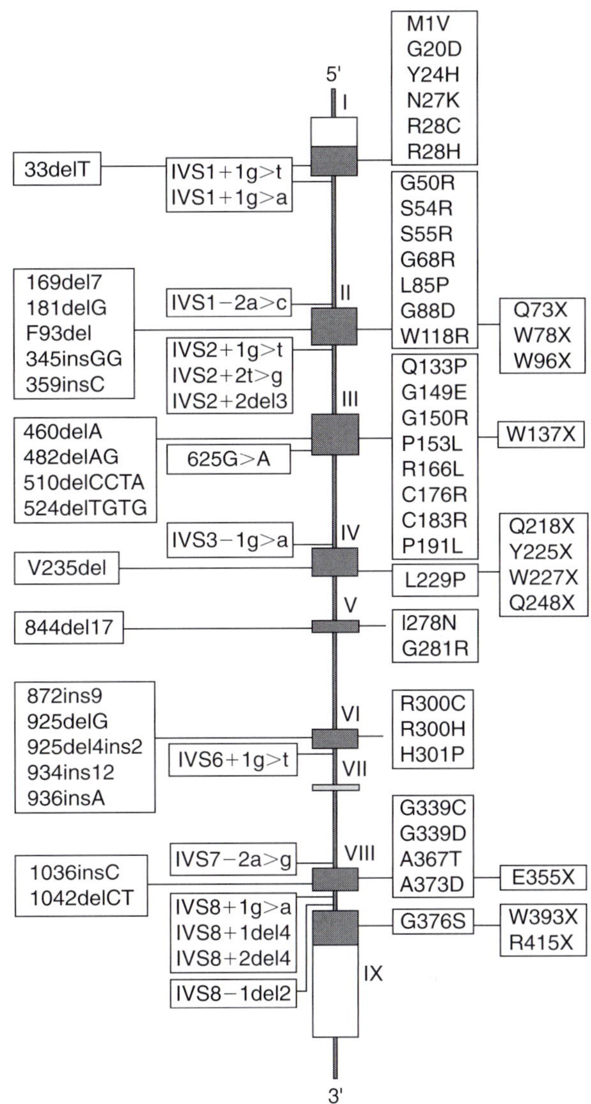

FIGURE 41.4 Mutations identified in the G6PT gene of GSD-Ib patients. The G6PT gene is shown as a line diagram with the nine exons marked as boxes I to IX. Black boxes represent coding regions, white boxes the 5′ and 3′ untranslated regions of the G6PT transcript. The positions of all known mutations are listed from left to right as insertion/deletion, splicing, missense, and nonsense mutations

The human G6PT is encoded by a single copy gene (Marcolongo et al 1998, Hiraiwa et al 1999, Gerin et al 1999) which produces two alternatively spliced transcripts. The primary transcript (G6PT) encodes a 429 amino acid protein, while the variant transcript (vG6PT), which contains an additional 66 bp in exon-7, encodes a 451 amino acid protein. The G6PT transcript is expressed in all tissues examined (Lin et al 1998), but the vG6PT transcript is expressed primarily in the brain, heart, and skeletal muscle (Lin et al 2000). Both G6PT (Figure 41.1) (Pan et al 1999) and vG6PT (Lin et al 2000) are ten transmembrane domain, ER proteins, with a similar topology. Both function as transporters and can couple with G6Pase to form an active G6Pase complex (Hiraiwa et al 1999, Lin et al 2000).

GSD-Ib is not restricted to any one ethnic group, although prevalent mutations have also been described in several groups. A total of 76 separate G6PT mutations, including 32 missense, 11 nonsense, 19 insertion/deletion (including two codon-deletion), and 14 splicing mutations (Figure 41.4) have been identified in more than 150 GSD-Ib patients (reviewed in Chou & Mansfield 2003). Site-directed mutagenesis and expression studies have identified key residues in the signature motif, helical and nonhelical regions of the protein that influence transporter

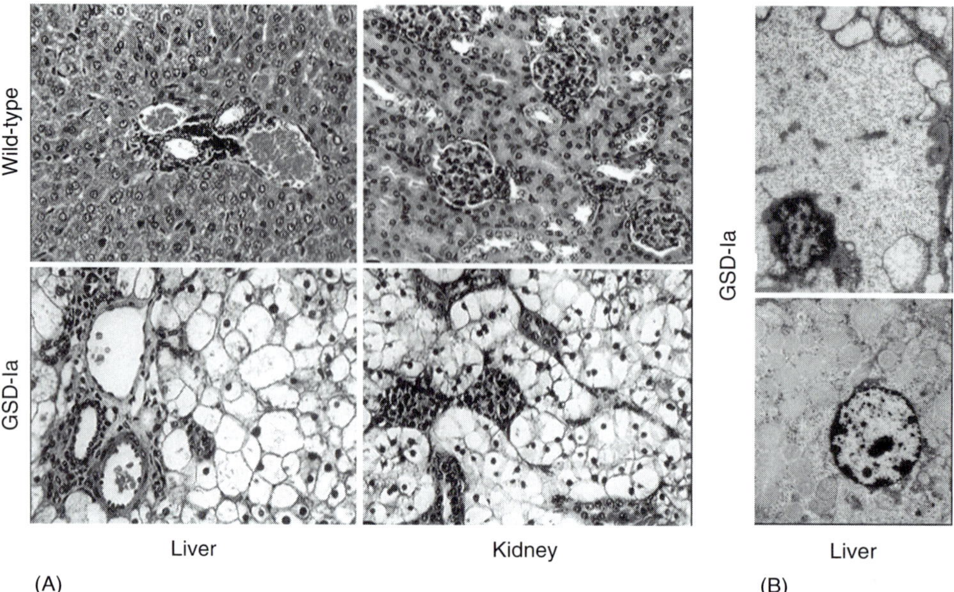

FIGURE 41.5   Histological analyses of liver and kidney from wild-type and GSD-Ia mice. (A) Hematoxylin–eosin-stained liver and kidney from wild-type and GSD-Ia littermates killed at 21 days of age. The liver shows a diffuse mosaic pattern consistent with marked glycogen accumulation in hepatocytes. The enlargement of kidney tubular epithelial cells, indicative of glycogen deposition resulted in compression of the glomeruli. (B) Electron micrographs of liver from a GSD-Ia mouse killed at 26 days of age. Note the abundance of glycogen particles (top) and lipid droplets (bottom) in the cytoplasm

activity either directly, or by reducing the cellular expression of the G6PT protein (Chen et al 2000, 2002, Pan et al 2003). In contrast to GSD-Ia, a phenotype–genotype correlation has gradually emerged in GSD-Ib. Patients with leaky G6PT mutations have metabolic deficiencies, but lack the polymorphonuclear leukocyte dysfunction seen in other patients (reviewed in Chou & Mansfield 2003). A growing database of knowledge about the relative G6P uptake activity of the different G6PT mutants (Chen et al 2000, 2002, Pan et al 2003) should facilitate future genotype–phenotype delineations and help to identify the threshold of G6PT activity required to prevent myeloid dysfunctions.

ANIMAL MODELS OF GSD-I

There are mouse (Lei et al 1996) and dog (Kishnani et al 1997) models of GSD-Ia. Both animals are physiologically similar to humans with respect to G6P metabolism and the animal models are being used to monitor the long-term complications of GSD-I. The G6Pase-deficient (G6Pase$^{-/-}$) mice (Lei et al 1996), generated by gene targeting, manifest all of the symptoms of human GSD-Ia – hypoglycemia, growth retardation, hepatomegaly, nephromegaly, hyperlipidemia, and hyperuricemia – with the exception of hyperlactacidemia. Hematoxylin and eosin staining shows marked glycogen storage in the hepatocytes and renal tubular epithelial cells in G6Pase$^{-/-}$ mice (Figure 41.5A),

similar to those seen in human GSD-Ia patients. Electron microscopic analysis also shows glycogen granules and lipid droplets in hepatocytes (Figure 41.5B). The GSD-Ia dog, a natural breed carrying a null M121I mutation in the G6Pase gene, manifests all the typical symptoms of the human disorder, including lactic acidemia (Kishnani et al 1997).

The GSD-Ib disorder has only been described in humans and there are no naturally occurring animal models for the disease. A G6PT-deficient (G6PT$^{-/-}$) mouse model of GSD-Ib was recently generated by gene targeting (Chen et al 2003). The G6PT$^{-/-}$ mice share the same phenotype as human GSD-Ib patients manifesting phenotypic G6Pase deficiency characterized by hypoglycemia, growth retardation, hepatomegaly, nephromegaly, hyperlipidemia, hyperuricemia, and lactic acidemia, as well as myeloid dysfunctions. A study of the myeloid deficiencies in the G6PT$^{-/-}$ mice has revealed that transient neutropenia, resistance of neutrophils to chemotactic factors, and a reduced production of neutrophil-specific chemokines at the sites of inflammation all contribute to the pathophysiology of GSD-Ib (Chen et al 2003).

KIDNEY DISEASE ASSOCIATED WITH GSD-I

Although renal involvement in GSD-I was noted in von Gierke's pathological description (1929) and kidney

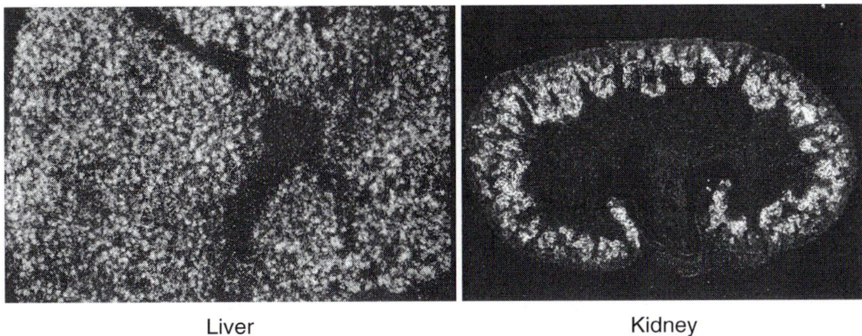

FIGURE 41.6   In situ hybridization analysis of the G6Pase transcript in the mouse liver and kidney. The liver shows a uniform distribution throughout the organ, while the kidney expression is limited to the cortex

enlargement is commonly seen in GSD-I, renal disease was not widely recognized as a major complication until the late 1980s. The renal changes in GSD-I are diverse and include proximal and distal tubular dysfunctions, nephrocalcinosis, nephrolithiasis, glomerular hyperfiltration, and a chronic renal disease leading to renal insufficiency. In the last two and half decades, effective dietary therapies, including nocturnal nasogastric infusion of glucose (Greene et al 1976) and/or oral administration of uncooked cornstarch (Chen et al 1984) have adequately maintained euglycemia in GSD-I and appear to delay significantly the development of chronic renal disease and renal insufficiency. However, glomerular hyperfiltration, hypercalciuria, hypocitraturia that worsens with age, and urinary albumin excretion still occur in metabolically compensated GSD-I patients. Whether dietary therapy will prevent early-stage kidney disease from progressing to chronic renal disease and renal insufficiency, remains to be seen as the first generation of treated GSD-I patients age.

Renal Tubular Acidosis

The renal acid–base homeostasis can be divided broadly into two processes, the reabsorption of filtered bicarbonate occurring fundamentally in the proximal convoluted tubule, and the excretion of ammonium and fixed acids through the titration of urinary buffers in the distal nephron (reviewed in Igarashi et al 2002, Rodriguez Soriano 2002). Impairment in the bicarbonate reabsorption induces increased tubular secretion of hydrogen ions, which causes proximal renal tubular acidosis (type II acidosis). A defect in the excretion of hydrogen ions in the distal tubule causes distal renal tubular acidosis (type I acidosis). While G6Pase is expressed throughout the liver, expression in the kidney is restricted to the cortex (Figure 41.6) (Pan et al 1988b), primarily in the proximal convoluted tubules (Chiquoine 1953, Hume et al 1994). This is consistent with the primary impact of G6Pase deficiency on the proximal tubules, although both type I and type II renal tubular acidosis are described in GSD-I patients.

PROXIMAL RENAL TUBULAR DYSFUNCTION – FANCONI-LIKE SYNDROME

Type II renal tubular acidosis is primarily caused by a loss of bicarbonate in the urine, although a failure to reabsorb phosphate in the proximal tubules can contribute further to the acidosis (reviewed in Igarashi et al 2002, Rodriguez Soriano 2002). In GSD-I, impairment of both bicarbonate and phosphate reabsorption are seen in patients under poor metabolic control. Matsuo et al (1986) first described a GSD-I child suffering from proximal renal tubular acidosis which resulted from a loss of bicarbonate in the urine. Bone radiographs showed no rickets or hyperparathyroidism in this patient, implying a defect of metabolic origin. Matsuo et al (1986) also described four GSD-I children with hyper-phosphaturia who responded to continuous glucose therapy, further supporting the metabolic origin of the proximal tubular dysfunction. In 1990, Chen et al described proximal tubular dysfunctions, including β2-microglobulinuria, generalized amino aciduria and reduced tubular reabsorption of phosphate, in three GSD-I children under 15 years of age who had never received dietary therapy. These defects improved with the initiation of a dietary therapy, solidifying the link between proximal renal tubular defects in GSD-I and poor metabolic control.

The cause of proximal tubular dysfunction in patients under poor metabolic control has not been elucidated, but Lee et al (1995a) have proposed that increased activities of the G6P-mediated metabolic pathways (Figure 41.2) might deplete intracellular high-energy phosphate reserves required for normal tubular functioning. By this reasoning, with improved control, the energy reserves might be preserved, allowing the proximal tubular function to become normal.

DISTAL RENAL TUBULAR ACIDOSIS – NEPHROCALCINOSIS AND NEPHROLITHIASIS

Distal renal tubular acidosis is caused by impaired distal acidification resulting from defects in hydrogen ion secretion. It is characterized by tubular acidosis, growth retardation, polyuria, hypercalciuria, nephrocalcinosis, nephrolithiasis, and K^+ depletion, with a secondary impairment in the secretion of ammonium ions (reviewed in Rodriguez Soriano 2002). While distal renal tubular acidosis is often ascribed to GSD-I, the defects are more truly described as incomplete distal renal tubular acidosis in which nephrocalcinosis and hypocitraturia manifest in the absence of metabolic acidosis (reviewed in Rodriguez Soriano 2002). Only one report (Restaino et al 1993) has addressed this issue of complete versus incomplete renal tubular acidosis.

Renal stones were first described in four GSD-I patients by Chen et al (1988) in a comprehensive study of renal pathogenesis in GSD-I. Renal stones and/or nephrocalcinosis have since been reported in up to 65% of adult GSD-I patients (Talente et al 1994). In 1993, Restaino et al examined renal pathogenesis in 11 GSD-I patients who had received allopurinol to control hyperuricemia. All 11 patients had normal serum levels of calcium, phosphorus, parathyroid hormone, 25-hydroxyvitamin D and urate, but five had renal calculi and/or nephrocalcinosis, five had hypercalciuria, and hypocitraturia was present in all nine subjects who had it measured. Oxalate excretion was normal in all patients, and no patient had evidence of a proximal tubular defect or a metabolic acidosis. However, after acid loading with arginine hydrochloride, ammonia excretion as well as titratable acidity were decreased in eight of nine patients, suggesting that the hypercalciuria and hypocitraturia seen in GSD-I are the consequence of an incomplete distal renal tubular acidosis.

Supporting this, Simoes et al (2001) described a GSD-I patient manifesting hypercalciuria, hyperoxaluria, marked hypocitraturia, recurrent renal stones, and lactic acidosis. The serum levels of creatinine, calcium, potassium, sodium, phosphate, and uric acid were normal. An incomplete distal tubular acidosis was implied for this patient, although the diagnosis was not confirmed by an ammonium loading test. Renal stones were not available for analysis, but urine analysis revealed an abundance of calcium oxalate crystals. Iida et al (2003) further correlated hypocitraturia and incomplete distal tubular acidosis with the recurrence of calcium nephrolithiasis in a 36-year-old GSD-I patient who manifested lactic acidosis but had normal levels of serum calcium, phosphate, and parathyroid hormone. In this case the stones were shown to be composed of calcium oxalate and calcium phosphate. Again, the diagnosis was not confirmed by an ammonium loading test. In the largest reported retrospective survey of a cohort of 231 GSD-Ia and 57 GSD-Ib patients, the European Study on GSD-I (ESGSD-I) reported hypercalciuria in about 33% of the cohort, with two patients exhibiting end-stage renal failure related to renal calcification (Rake et al 2002). However, the existence of proximal and/or distal renal tubule dysfunction was not recorded in this cohort, so the prevalence of these defects is not known.

Neither was it clear which of these patients was receiving dietary therapy control, which could further affect measures of prevalence.

In another study of 15 metabolically compensated GSD-Ia patients, Weinstein et al (2001) investigated the relationship between renal complications of GSD-I, urinary excretion of citrate and calcium, and the systemic acid–base status. They reported that hypocitraturia occurred in all patients over 12 years of age and worsened with age. Hypercalciuria was present in nine of the 15 patients. Thus, unlike proximal tubular dysfunction, which improves with dietary therapies, GSD-I patients under good metabolic control remain at risk for nephrocalcinosis and nephrolithiasis.

Citrate is the most abundant organic anion found in the urine, and is the primary chelator of calcium, preventing nephrolithiasis by inhibiting nucleation, crystal growth, and agglomeration (reviewed in Pajor 1999, Hamm & Hering-Smith 2002). Usually, 65% to 90% of filtered citrate is reabsorbed in the proximal tubules by means of two different sodium-dependent dicarboxylate transporters (Caudarella et al 2003). The activity of these transporters is acutely affected by changes in acid–base status (Aruga et al 2000, Caudarella et al 2003). Acidosis causes increased mitochondrial transport and oxidation of citrate leading to lower cytoplasmic citrate concentrations. This in turn increases tubular reabsorption of citrate leading to hypocitraturia. Normally citrate can be metabolized to bicarbonate to buffer the acidosis. However, in GSD-I, the hypocitraturia caused by enhanced urinary citrate reabsorption may exacerbate the metabolic acidosis.

In GSD-I, renal regulation of calcium homeostasis is also disturbed. Unbound calcium ions are freely filtered through the renal glomeruli and usually reabsorbed throughout the nephron with a 98–99% overall recovery (reviewed in Ba & Friedman 2004). During metabolic acidosis, including renal tubular acidosis and possibly lactic acidemia, the decrease in systemic pH is associated with an increase in urinary calcium excretion (reviewed in Krieger et al 2004). The failure in GSD-I to resolve these conditions leads to a net loss of body calcium and hypercalciuria. In combination with hypocitraturia this can explain, at least in part, the high incidence of nephrolithiasis and nephrocalcinosis in GSD-I patients.

In addition to the distal renal tubular acidosis, the other risk factor for the development of renal calcification and kidney stones in GSD-I is chronic hyperuricemia (reviewed in Reitsma-Bierens 1993). Hyperuricemia in GSD-I probably results from a combination of overproduction of uric acid and underexcretion of urate. Uric acid is the end product of purine metabolism. Its overproduction in GSD-I arises from the excess accumulation of G6P, which enters the pentose phosphate pathway and causes increased production of purine precursors. In the first step of adenine nucleotide degradation, AMP-deaminase catalyzes the deamination of adenine and AMP. This catalysis is stimulated when intrahepatic phosphate levels fall. In GSD-I, the low intrahepatic phosphate concentrations, caused by overaccumulation of G6P, further increase uric acid production (Greene et al 1978, Cohen et al 1985, Mineo & Tarui 1995). Under normal conditions, urate is freely filtered at the renal glomerulus and undergoes reabsorption and secretion in the renal proximal tubules. This process is mediated by members of the organic anion transporter family (reviewed in Rafey et al 2003). Decreased urate secretion is typically seen during lactic acidosis and/or renal insufficiency. GSD-I patients manifest both lactic acidosis and hyperuricemia. The elevated levels of the organic anions, lactate and urate, compete for the organic anion transporters, resulting in a variable concentration of urate in the urine. The primary determinant of uric acid solubility is urinary pH, and a low urinary pH facilitates uric acid stone formation (Polinsky et al 1993, Moran 2003). In GSD-I, acidic urine commonly found in patients increases the risk of nephrolithiasis and nephrocalcinosis.

Chronic Renal Disease

Chronic renal failure usually progresses slowly over a number of years as the internal structures of the kidney become damaged. The early stages are often symptom-free and diagnosis may not occur until kidney function is less than 10% of normal, when accumulation of fluid and waste products in the body causes azotemia and uremia (http://www.nlm.nih.gov/medlineplus/ency/article/000471. htm). Typical symptoms include reduced urinary volume, hyperfiltration, proteinuria, excessive thirst, hematuria, and hypertension.

Chronic renal disease in GSD-I was first reported by Chen et al (1988). In 14 of the 20 patients aged 13 to 47, who had not received adequate dietary therapy within the first decade of life, Chen et al (1988) observed disturbed renal function, characterized by persistent proteinuria occurring at a mean age of 14.8 years. In these patients proteinuria was predominated by albumin and persisted, even after dietary intervention with cornstarch, for a mean duration of 9.2 years. Many also had hypertension, renal stones, hematuria, and altered creatinine clearance. Renal biopsies of the patient group revealed focal segmental glomerulosclerosis, interstitial fibrosis and tubular atrophy with marked glomerular basement-membrane thickening. The glomerular abnormalities consisted of mesangial matrix expansion, tuft collapse, and hyalinosis (Figure 41.7). These are probably not simple mechanical consequences of glycogen accumulation since the glycogen deposit is primarily restricted to the cortical proximal tubular cells. Clinically, the degree of interstitial fibrosis correlated with the degree of renal insufficiency in GSD-I patients. Among the 14 patients manifesting disturbed renal function, six developed progressive renal insufficiency, leading to three deaths from renal failure. In contrast, all 18 patients under 10 years of

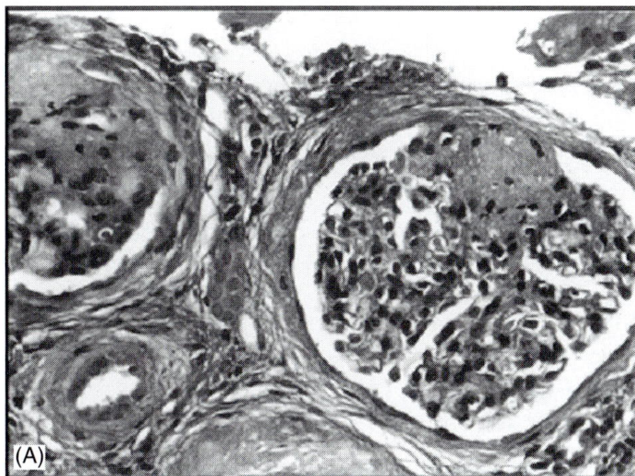
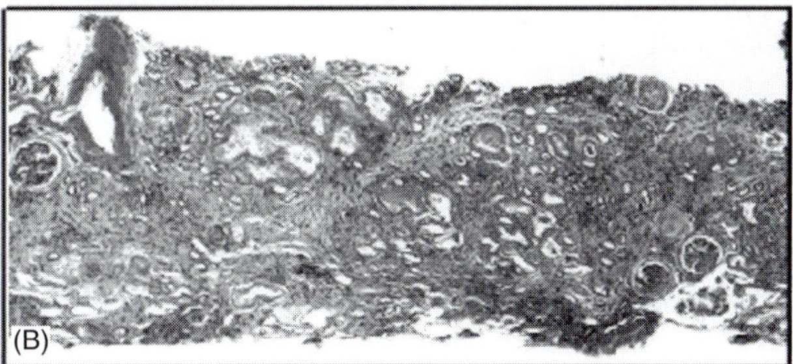

FIGURE 41.7 Histological analysis of the kidneys from GSD-Ia patients. (A) The kidney showed two glomeruli with focal segmental glomerulosclerosis, including segmental tuft collapse and partial adhesion to the capsule (magnification ×200). (B) The kidney showed focal segmental glomerular lesions, interstitial fibrosis, and focal tubular atrophy. Note the uninvolved glomeruli (magnification ×30). Reproduced from Chou et al (2002) with permission

age, who had received adequate dietary therapies, exhibited normal renal function (Chen et al 1988). However, the creatinine clearance, which measures the glomerular filtration rate, was not examined in the young patients with apparently normal kidney function.

Verani and Bernstein (1988) described glomerular basement membrane alterations, representing the early stages of focal segmental glomerulosclerosis in three patients even though one had shown no manifestation of proteinuria. The pathology included thickening and wrinkling of the glomerular basement membrane with focal accumulation of the matrix, widening of the epithelial cell foot processes and partial collapse of the glomerular loops. One patient also manifested clear glomerulosclerosis, interstitial fibrosis, and tubular atrophy, which led to renal failure and necessitated kidney transplantation.

Baker et al (1989) reported albuminuria in three teen-aged patients and frank proteinuria in three patients over 20 years of age. In agreement with the findings of Chen et al (1988) renal biopsy of the proteinuric subjects also revealed focal and global glomerulosclerosis, interstitial fibrosis, mild to moderate diffuse mesangial proliferation, and focal hyalinosis. Obara et al (1993) described focal glomerulosclerosis, interstitial fibrosis, tubular atrophy, marked vacuolization of the tubular epithelial cells, and arteriosclerosis in two GSD-I patients manifesting chronic renal disease with heavy proteinuria. Both also had increased glomerular volume. Electron microscopic analysis of one patient showed thickening of the glomerular basement membrane and lipid droplets within the mesangium.

In summary, the pathology of chronic renal disease in GSD-I includes glomerulosclerosis, interstitial fibrosis, and tubular atrophy with marked glomerular basement-membrane thickening.

The first stage in the development of chronic renal disease is hyperfiltration for which the reported prevalence in GSD-I ranges from 67% to 100% (Baker et al 1989, Reitsma-Bierens et al 1992, Lee et al 1995a, Wolfsdorf & Crigler 1999). The marked increase in the glomerular filtration rate was first documented by Baker et al (1989) in a prospective study of 14 GSD-I patients which included eight young patients who had received adequate dietary therapy since infancy. Despite the intervention, all 14 patients had a marked increase in glomerular filtration rate when estimated by creatinine clearance rates in conjunction with inulin or Tc-DTPA clearance rates. This persistence of glomerular hyperfiltration in metabolically well-controlled GSD-I patients was further documented by Wolfsdorf and Crigler (1999) in a study of 17 GSD-I patients who had received continuous glucose therapy from early infancy. While urinary albumin excretion was increased in only the two older subjects, 16 of the 17 patients examined exhibited glomerular hyperfiltration confirming that hyperfiltration often preceded frank proteinuria, even in young metabolically controlled patients. Chen et al (1988) found that the onset of proteinuria in seven of their older GSD-I patients correlated with an increased renal creatinine clearance, suggesting that glomerular hyperfiltration might lead to glomerulosclerosis. Indeed the association of hyperfiltration and glomerulosclerosis was well established in a number of experimental animals, as well as humans (reviewed in Hostetter 2003).

GSD-I patients have also been reported to have an increased effective renal plasma flow, which is another early sign of renal dysfunction. In one study, 18 of 23 patients showed an elevated flow (Reitsma-Bierens et al 1992) while a second study reported a similar frequency with eight out of nine patients showing flow elevation (Hahn-Ullrich et al 1993).

The etiology of glomerular hyperfiltration and increased effective renal plasma flow in GSD-I is currently unclear although hyperlacticacidemia may be one factor. An increase in blood lactate concentration is known to elevate glomerular filtration rate in normal subjects and in diabetes (Trevisan et al 1987), and lactic acidemia is a common feature of GSD-I, even in patients under good metabolic control.

Hyperlipidemia, which is also universal in GSD-I, may be an independent risk factor for the early development and progression of renal disease. Studies with rats have shown that hypercholesterolemia accelerates the rate of progression of kidney disease (reviewed in Abrass 2004). Both hypercholesterolemia and hypertriglyceridemia are associated with severe podocyte injury, proteinuria, and interstitial injury (Joles et al 2000).

Other factors may be simple mechanical consequences arising from the accumulation of glycogen in the renal tubular epithelial cells which results in their enlargement and compression of the glomeruli, or an apparent reduction in functional renal mass. Another possibility is hyperfunction, a compensatory mechanism seen in renal tissue loss, in which single nephrons grow in size and increase their filtration rate (reviewed in Hostetter 2003). It is also possible that the increased glomerular filtration rate may be a secondary phenomenon in response to tubular epithelial cell energy depletion (Lee et al 1995a). The importance of each of these possibilities remains to be examined.

In conclusion, dietary therapy allows GSD-I patients to maintain euglycemia and removes the early symptomatic signs of the disease. However, patients continue to suffer from hyperlipidemia, hyperuricemia, hypercalciuria, hypocitraturia, and lactic acidemia, which facilitate the slow progression of kidney disease. This is typified by renal stones, nephrocalcinosis, nephrolithiasis, and progressive renal disease that become a frequent and life-threatening complication for older patients.

RENAL DISEASE IN GSD-I AND DIABETES MELLITUS

GSD-I is caused by a deficiency in the G6Pase complex and is characterized by fasting hypoglycemia (Chen 2001, Chou et al 2002). In contrast, diabetes is caused by disturbed insulin functions and is characterized by uncontrolled hyperglycemia (Castano & Eisenbarth 1990, Taylor 2001). Despite the remarkable difference, the renal dysfunction in GSD-I resembles diabetic nephropathy in both its clinical course and pathology (Baker et al 1989, Chen 1991, Obara et al 1993). Diabetic nephropathy follows a well-defined clinical course. It starts with a long period of silent disease during which glomerular hyperfiltration is the only demonstrable renal abnormality. This is followed by the development of microalbuminuria, proteinuria, and ultimately, renal failure (reviewed in Raptis & Viberti 2001, Caramori & Mauer 2003). GSD-I has a similar clinical course. Pathologically, the diseases are also similar. Each exhibits an increased glomerular volume, thickening of the glomerular basement membrane, and expansion of the mesangial matrix, all of which occur as a result of glomerular damage secondary to hyperfiltration, and culminate in focal glomerulosclerosis and interstitial fibrosis. An understanding of the molecular and pathological bases of nephropathy in both disorders may help to unravel the etiology of the kidney dysfunctions common to these conditions.

TREATMENT OF GSD-I

Dietary Treatment

There is no cure for GSD-Ia or GSD-Ib, but many of the disease symptoms can be managed or improved using a dietary therapy. This strategy enables patients to attain near normal growth and pubertal development, with fewer complications as they age. The dietary therapy for afflicted infants typically consists of nocturnal nasogastric infusion of glucose to avoid hypoglycemia (Greene et al 1976). Uncooked cornstarch, which serves as a slow-release carbohydrate to prolong euglycemia between meals, is the mainstay of therapy in children age 6 months or older and adults with GSD-Ia (Chen et al 1984). Cornstarch is rarely tolerated in children age 6 months or younger due to immaturity of the gastrointestinal tract and lack of brush border enzymes (Weinstein & Wolfsdorf 2002). Guidelines for dietary treatment have been issued based on the data of the ESGSD-I study (Rake et al 2002).

Pharmacological Treatment

In GSD-I, pharmacologic therapy is used to decrease the morbidity associated with complications of the disease, especially in patients whose biochemical abnormalities remain despite good metabolic controls. Lipid-lowering agents (niacin, gemfibrozil, or both) are indicated in patients with severe hypertriglyceridemia, and allopurinol is used to control hyperuricemia. Urocit-K (potassium citrate) is the primary pharmacological treatment for patients with severe hypocitraturia and Bicitra (sodium citrate) is used for patients who cannot tolerate Urocit-K. Since renal dysfunction can result in hyperkalemia, assessment of serum potassium concentrations is critical in this therapy. Angiotensin-converting enzyme inhibitors, the mainstay therapy in diabetes and other progressive glomerular diseases (Remuzzi et al 2002), are also beneficial in GSD-I, lowering blood pressure, reducing hyperfiltration, and slowing the progression of renal disease. The consensus guidelines by the ESGSD-I recommend the use of a long-acting angiotensin-converting enzyme inhibitor in GSD-I with hypertension or microalbuminuria (Rake et al 2002).

In GSD-Ib, granulocyte colony stimulating factor therapy is used to improve neutrophil count, recurrent infections, and inflammatory bowel disease (Calderwood et al 2001).

Transplantation

Since the primary organs affected in GSD-I are the liver and kidney, renal, hepatic, and renal–hepatic transplants have been considered as last resort approaches to controlling the disease. By transplanting organs with a functional G6Pase complex, into the patient, the intent is to re-establish normal glucose homeostasis.

Ten renal transplantations in GSD-Ia patients with terminal renal failure have been reported (Emmett & Narins 1978, Chen & Scheinman 1991, Gossmann et al 2001). In all cases, renal transplantation normalized kidney function. However, consistent with the primary role of the liver in glucose homeostasis, the transplants failed to correct the metabolic abnormalities and the patients continued to suffer from hypoglycemia, hyperlipidemia, hyperuricemia, and lactic acidemia.

In patients who fail to respond sufficiently to dietary therapy, or who exhibit multiple liver adenomas, which carry the risk of malignant transformation, orthotopic liver transplantation is advocated (reviewed in Matern et al 1999, Labrune 2002). In the short term, liver transplantation appears to correct metabolic abnormalities, but in the long-term the outcome is mixed. Faivre et al (1999) documented complications in three GSD-Ia patients, 6–8 years after orthotopic liver transplantation, with one patient developing progressive renal insufficiency. In a review of six GSD-Ia patients receiving liver transplantation between ages 12 to 27, Matern et al (1999) reported that all had normal liver function in the absence of renal disease at the time of their writing (0.5–6.8 years posttransplant). However, one patient did reject the transplant 3.5 years posttransplantation. Another patient developed end-stage liver cirrhosis and chronic renal failure 14 years after the first liver transplant and required a second liver and kidney transplantation. Matern et al (1999) also reported that one patient who received a combined liver and kidney transplant suffered chronic rejection. There is insufficient experience yet to conclude whether liver transplantation will be effective in the long term, especially in preventing the development of slowly progressing renal disease. Animal studies suggest it is unlikely. Gene therapy studies using a mouse model of GSD-Ia (Sun et al 2002) have shown that both the liver and kidney must express G6Pase activity to regain normal glucose homeostasis and renal function. However, until better understood, transplantation does remain as an option of last resort for GSD-Ia.

Five GSD-Ib patients received orthotopic liver transplantation and all showed improved metabolic profiles following transplantation (Lachaux et al 1993, Matern et al 1999, Martinez-Olmos et al 2001, Adachi et al 2004) although one patient went on to reject the organ (Matern et al 1999). Renal function was examined in only one patient and remained normal 6.2 years after transplantation (Matern et al 1999). The impact of a hepatic transplantation on myeloid dysfunction is still unclear. In two of the patients, neutropenia appears resolved 3 and 4 years after transplantation (Martinez-Olmos et al 2001, Adachi et al 2004), while in the other three patients, neutropenia persisted following the transplantation. Only more studies will reveal the true potential of transplantation in GSD-Ib.

Gene Therapy

G6Pase and G6PT are highly hydrophobic, multiple-domain transmembrane ER proteins (Pan et al 1998a, 1999), which can not be expressed in soluble forms. To be functional, both proteins must be coexpressed and embedded correctly in the ER membrane (Chou et al 2002). Therefore, protein replacement therapy for either disorder is not an option, although somatic gene therapy, targeting G6Pase or G6PT to the liver and the kidney, is an attractive possibility that has received interest recently.

In 2000, Zingone et al used G6Pase$^{-/-}$ mice to evaluate the feasibility of gene replacement therapy for GSD-Ia using an adenoviral vector carrying murine G6Pase (Ad-mG6Pase) under the control of the tissue-independent RSV promoter. A single intravenous infusion of Ad-mG6Pase into the G6Pase$^{-/-}$ mice at 2 weeks of age restored 19% of normal hepatic G6Pase activity, improved survival and growth of GSD-Ia mice, and transiently corrected the metabolic abnormalities manifested by these mice. However, the benefits were short-lived. To achieve long-term correction of the GSD-Ia disorder, recombinant adeno-associated virus (rAAV)-mediated gene therapies have been investigated in both the murine (Sun et al 2002) and canine (Beaty et al 2002) models.

In the canine study Beaty et al (2002) investigated the effects of a neonatal infusion of a rAAV vector carrying canine G6Pase under the control of a mouse albumin promoter-enhancer. This strategy resulted in a sustained hepatic expression of G6Pase for 11 weeks, an improved liver histology, and an improved biochemical profile compared to controls. Renal histology and function were not evaluated because the recombinant gene was driven by a liver-specific promoter-enhancer.

In the murine study (Sun et al 2002), a two-step regimen was investigated. The first step was a neonatal coinfusion of a rAAV carrying the murine G6Pase (AAV-mG6Pase) with Ad-mG6Pase. This was followed by a second infusion, at 2 weeks of age, with just AAV-mG6Pase. The AAV-mG6Pase gene was expressed under the control of a chicken β-actin promoter driven by a CMV enhancer to give broad tissue specificity. This strategy corrected the murine GSD-Ia disorder for the full 12 months of the study. The G6Pase transgene was expressed in both the liver and kidney, and resulted in the sustained expression of a complete, functional, G6Pase system. Of particular significance was the finding that the Ad/AAV-mG6Pase-infused animals grew normally, with only mild glycogen accumulation in the liver and the kidney.

In the livers of the Ad/AAV-mG6Pase-infused animals there was no histological abnormality throughout the 12-month study (Figure 41.8). Biochemical measurement

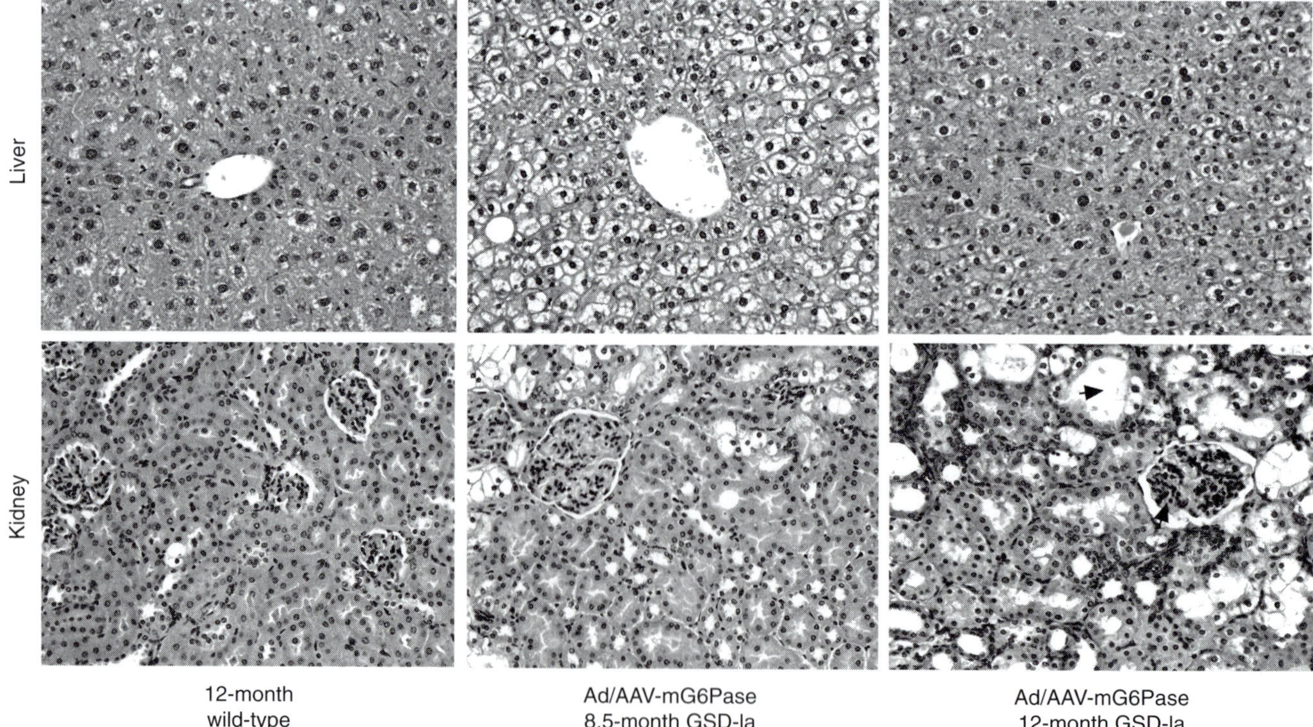

FIGURE 41.8   Histological analyses of the liver and kidney from the Ad/AAV-mG6Pase-infused GSD-Ia mice. Plates show hematoxylin–eosin-stained liver and kidney sections at magnifications of ×200. Note the abnormalities including glomerulosclerosis and tubular dilation (indicated by an arrowhead) in the kidney of GSD-Ia mice at 12 months post-Ad/AAV-mG6Pase-infusion

of microsomes isolated from the treated mice showed sustained restoration of approximately 23% of the normal liver G6Pase activity at 12 months postinfusion. However, all the mice had normal plasma glucose, cholesterol, triglyceride, and uric acid profiles, indicating that glucose homeostasis was being adequately maintained. In the kidneys, biochemical measurements showed that the initial restoration of approximately 34% of the normal kidney G6Pase activity was sustained for 8.5 months but then declined to 14% of normal activity over the subsequent 4 months. Pathology showed that the normal kidney histology was maintained over most of the study, although increased glycogen accumulation and the development of histological abnormalities including glomerulosclerosis and tubular dilatation (Figure 41.8), indicative of kidney disease did start to emerge at 12 months. This would suggest that a restoration of 34% of normal G6Pase activity is protective against kidney disease, but 14% activity is insufficient. However, since the Ad/AAV-mG6Pase-mediated gene transfers direct G6Pase expression to the entire kidney and not specifically to the cortex (Sun et al 2002), care needs to be exercised in interpreting these results. Activities less than 34% may be protective against kidney disease if appropriately localized. Clearly, correction of the initial metabolic symptoms of GSD-Ia can be attained, but liver correction appears to be easier than kidney correction. This may explain why the current symptomatic dietary therapies for human GSD-Ia patients targeting the control of symptomatic hypoglycemia, result in good glucose homeostasis, but fail to prevent longer-term kidney complications.

CONCLUSIONS

The primary complications of GSD-I are a loss of glucose homeostasis due to the loss of functional G6Pase complex in the liver and kidney, followed by secondary renal and liver disease, and, in the case of GSD-Ib, myeloid dysfunctions. While the disease progression can be slowed significantly by the use of dietary therapies, in the long run they are inadequate to completely control the disease. GSD-I patients receiving dietary therapies early in life still manifest hyperlipidemia, hyperuricemia, lactic acidemia, hypercalciuria, and hypocitraturia, all of which facilitate the slow progression of kidney disease. Liver or combined liver–kidney transplantation, to introduce a functional G6Pase complex into the patient, remains a treatment of last resort. It is a highly invasive procedure, requiring matched donors, and carries many potential complications inherent to long-term immunosuppressive therapy, and rejection. The recent development of animal models of both GSD-Ia and GSD-Ib, and the effective use of gene therapy to correct their metabolic imbalances and disease progression, hold promise for the future of gene therapy in humans.

References

Abrass CK. Cellular lipid metabolism and the role of lipids in progressive renal disease. Am. J. Nephrol. 2004; 24: 46–53.

Adachi M, Shinkai M, Ohhama Y, et al. Improved neutrophil function in a glycogen storage disease type 1b patient after liver transplantation. Eur. J. Pediatr. 2004; 163: 202–6.

Annabi B, Hiraiwa H, Mansfield BC, et al. The gene for glycogen storage disease type 1b maps to chromosome 11q23. Am. J. Hum. Genet. 1998; 62: 400–5.

Arion WJ, Lange AJ, Walls HE, Ballas LM. Evidence of the participation of independent translocases for phosphate and glucose-6-phosphate in the microsomal glucose-6-phosphatase system. J. Biol. Chem. 1980; 255: 10396–406.

Aruga S, Wehrli S, Kaissling B, et al. Chronic metabolic acidosis increases NaDC-1 mRNA and protein abundance in rat kidney. Kidney Int. 2000; 58: 206–15.

Ba J, Friedman PA. Calcium-sensing receptor regulation of renal mineral ion transport. Cell Calcium 2004; 35: 229–37.

Baker L, Dahlem S, Goldfarb S, et al. Hyperfiltration and renal disease in glycogen storage disease, type I. Kidney Int. 1989; 35: 1345–50.

Bandsma RHJ, Smit GPA, Kuipers F. Disturbed lipid metabolism in glycogen storage disease type 1. Eur. J. Pediatr. 2002; 161(Suppl. 1): S65–9.

Beaty RM, Jackson M, Peterson D, et al. Delivery of glucose-6-phosphatase in a canine model for glycogen storage disease, type Ia, with adeno-associated virus (AAV) vectors. Gene Ther. 2002; 9: 1015–22.

Beaudet AL, Anderson DC, Michels VV, Arion WJ, Lange AJ. Neutropenia and impaired neutrophil migration in type 1B glycogen storage disease. J. Pediatr. 1980; 97: 906–10.

Calderwood S, Kilpatrick L, Douglas SD, et al. Recombinant human granulocyte colony-stimulating factor therapy for patients with neutropenia and/or neutrophil dysfunction secondary to glycogen storage disease type 1b. Blood 2001; 15: 376–82.

Cano N. Bench-to-bedside review: Glucose production from the kidney. Critical Care 2002; 6: 317–21.

Caramori ML, Mauer M. Diabetes and nephropathy. Curr. Opin. Nephrol. Hypertens. 2003; 12: 273–82.

Castano L, Eisenbarth GS. Type-I diabetes: a chronic autoimmune disease of human, mouse, and rat. Annu. Rev. Immunol. 1990; 8: 647–79.

Caudarella R, Vescini F, Buffa A, Stefoni S. Citrate and mineral metabolism: kidney stones and bone disease. Front. Biosci. 2003; 8: s1084–106.

Chen L-Y, Lin B, Pan C-J, Hiraiwa H, Chou JY. Structural requirements for the stability and microsomal transport activity of the human glucose-6-phosphate transporter. J. Biol. Chem. 2000; 275: 34280–6.

Chen L-Y, Pan C-J, Shieh J-J, Chou JY. Structure-function analysis of the glucose 6-phosphate transporter deficient in glycogen storage disease type Ib. Hum. Mol. Genet. 2002; 11: 3199–207.

Chen L-Y, Shieh J-J, Lin B, et al. Impaired glucose homeostasis, neutrophil trafficking and function in mice lacking the glucose-6-phosphate transporter. Hum. Mol. Genet. 2003; 12: 2547–58.

Chen PY, Csutora P, Veyna-Burke NA, Marchase RB. Glucose-6-phosphate and Ca^{2+} sequestration are mutually enhanced in microsomes from liver, brain, and heart. Diabetes 1998; 47: 874–81.

Chen Y-T. Type I glycogen storage disease: kidney involvement, pathogenesis and it treatment. Pediatr. Nephrol. 1991; 5: 71–6.

Chen Y-T. Glycogen storage diseases. In: Scriver CR, Beaudet AL, Sly WS, et al., eds. The Metabolic and Molecular Bases of Inherited Disease, 8th edn. New York: McGrawHill Inc., 2001: pp. 1521–51.

Chen YT, Coleman RA, Scheinman JI, Kolbeck PC, Sidbury JB. Renal disease in type I glycogen storage disease. N. Engl. J. Med. 1988; 318: 7–11.

Chen YT, Cornblath M, Sidbury JB. Cornstarch therapy in type I glycogen storage disease. N. Engl. J. Med. 1984; 310: 171–5.

Chen YT, Scheinman JI. Hyperglycaemia associated with lactic acidemia in a renal allograft recipient with type I glycogen storage disease. J. Inherit. Metab. Dis. 1991; 14: 80–6.

Chen YT, Scheinman JI, Park HK, Coleman RA, Roe CR. Amelioration of proximal renal tubular dysfunction in type I glycogen storage disease with dietary therapy. N. Engl. J. Med. 1990; 323: 590–3.

Chiquoine AD. The distribution of glucose-6-phosphatase in the liver and kidney of the mouse. J. Histochem. Cytochem. 1953; 1: 429–35.

Chou JY, Mansfield BC. Molecular genetics of type 1 glycogen storage diseases. Trend Endocrinol. Metab. 1999; 10: 104–13.

Chou JY, Mansfield BC. Glucose-6-phosphate transporter: the key to glycogen storage disease type Ib. In: Broer S, Wagner CA, eds. Membrane Transporter Diseases. New York: Kluwer Academic/Plenum Publishers, 2003: pp. 191–205.

Chou JY, Matern D, Mansfield BC, Chen Y-T. Type I glycogen storage diseases: disorders of the glucose-6-phosphatase complex. Curr. Mol. Med. 2002; 2: 121–43.

Cohen JL, Vinik A, Faller J, Fox IH. Hyperuricemia in glycogen storage disease type I. Contributions by hypoglycemia and hyperglucagonemia to increased urate production. J. Clin. Invest. 1985; 75: 251–7.

Cori GT, Cori CF. Glucose-6-phosphatase of the liver in glycogen storage disease. J. Biol. Chem. 1952; 199: 661–7.

Ekstein J, Rubin BY, Anderson SL, et al. Mutation frequencies for glycogen storage disease Ia in the Ashkenazi Jewish population. Am. J. Med. Genet. 2004; 129A: 162–4.

Emmett M, Narins RG. Renal tranplantation in type 1 glycogenosis: failure to improve glucose metabolism. JAMA 1978; 239: 1642–4.

Faivre L, Houssin D, Valayer J, Brouard J, Hadchouel M, Bernard O. Long-term outcome of liver transplantation in patients with glycogen storage disease type Ia. J. Inherit. Metab. Dis. 1999; 22: 723–32.

Garty B, Douglas S, Danon YL. Immune deficiency in glycogen storage disease type 1b. Isr. J. Med. Sci. 1996; 32: 1276–81.

Gerich JE, Meyer C, Woerle HJ, Stumvoll M. Renal gluconeogenesis: its importance in human glucose homeostasis. Diabetes Care 2001; 24: 382–91.

Gerin I, Veiga-da-Cunha M, Achouri Y, Collet J-F, Van Schaftingen E. Sequence of a putative glucose-6-phosphate translocase, mutated in glycogen storage disease type Ib. FEBS Lett. 1997; 419: 235–8.

Gerin I, Veiga-da-Cunha M, Noel G, Van Schaftingen E. Structure of the gene mutated in glycogen storage disease type Ib. Gene 1999; 227: 189–95.

Ghosh A, Shieh J-J, Pan C-J, Sun M-S, Chou JY. The catalytic center of glucose-6-phosphatase: His^{176} is the nucleophile forming the phosphohistidine-enzyme intermediate during catalysis. J. Biol. Chem. 2002; 277: 32837–42.

Gitzelmann R, Bosshard NU. Defective neutrophil and monocyte functions in glycogen storage disease type Ib: a literature review. Eur. J. Pediatr. 1993; 152(Suppl. 1): S33–8.

Gossmann J, Scheuermann EH, Frilling A, Geiger H, Dietrich CF. Multiple adenomas and hepatocellular carcinoma in a renal transplant patient with glycogen storage disease type 1a (von Gierke disease). Transplantation 2001; 72: 343–4.

Greene HL, Slonim AE, O'Neill JA Jr., Burr IM. Continuous nocturnal intragastric feeding for management of type 1 glycogen-storage disease. N. Engl. J. Med. 1976; 294: 423–5.

Greene HL, Wilson FA, Hefferan P, et al. ATP depletion, a possible role in the pathogenesis of hyperuricemia in glycogen storage disease type I. J. Clin. Invest. 1978; 62: 321–8.

Hahn-Ullrich H, Sciuk J, Bartenstein P, Kreysing P, Ullrich K. Effective renal plasma flow in patients with glycogen storage disease Lyne I. Eur. J. Pediatr. 1993; 152: 674–6.

Hamm LL, Hering-Smith KS. Pathophysiology of hypocitraturic nephrolithiasis. Endocrinol. Metab. Clin. North Am. 2002; 31: 885–93.

Hiraiwa H, Pan C-J, Lin B, Moses SW, Chou JY. Inactivation of the glucose-6-phosphate transporter causes glycogen storage disease type 1b. J. Biol. Chem. 1999; 274: 5532–6.

Hostetter TH. Hyperfiltration and glomerulosclerosis. Semin. Nephrol. 2003; 23: 194–9.

Hume R, Bell JE, Hallas A, Burchell A. Immunohistochemical localisation of glucose-6-phosphatase in developing human kidney. Histochemistry 1994; 101: 413–17.

Igarashi T, Sekine T, Inatomi J, Seki G. Unraveling the molecular pathogenesis of isolated proximal renal tubular acidosis. J. Am. Soc. Nephrol. 2002; 13: 2171–7.

Iida S, Matsuoka K, Inoue M, Tomiyasu K, Noda S. Calcium nephrolithiasis and distal tubular acidosis in type 1 glycogen storage disease. Int. J. Urol. 2003; 10: 56–8.

Joles JA, Kunter U, Janssen U, et al. Early mechanisms of renal injury in hypercholesterolemic or hypertriglyceridemic rats. J. Am. Soc. Nephrol. 2000; 11: 669–83.

Kikuchi M, Hasegawa K, Handa I, Watabe M, Narisawa K, Tada K. Chronic pancreatitis in a child with glycogen storage disease type 1. Eur. J. Pediatr. 1991; 150: 852–3.

Kishnani PS, Bao Y, Wu J-Y, Brix AE, Lin J-L, Chen Y-T. Isolation and nucleotide sequence of canine glucose-6-phosphatase mRNA: identification of mutation in puppies with glycogen storage disease type 1a. Biochem. Mol. Med. 1997; 61: 168–77.

Krieger NS, Frick KK, Bushinsky DA. Mechanism of acid-induced bone resorption. Curr. Opin. Nephrol. Hypertens. 2004; 13: 423–36.

Kudo M. Hepatocellular adenoma in type Ia glycogen storage disease. J. Gastroenterol. 2001; 36: 65–6.

Labrune P. Glycogen storage disease type I: indications for liver and/or kidney transplantation. Eur. J. Pediatr. 2002; 161(Suppl. 1): S53–5.

Lachaux A, Boillot O, Stamm D, et al. Treatment with lenograstim (glycosylated recombinant human granulocyte colony-stimulating factor) and orthotopic liver transplantation for glycogen storage disease type Ib. J. Pediatr. 1993; 123: 1005–8.

Lange AJ, Arion WJ, Beaudet AL. Type 1b glycogen storage disease is caused by a defect in the glucose-6-phosphate translocase of the microsomal glucose-6-phosphatase system. J. Biol. Chem. 1980; 255: 8381–4.

Lee PJ. Glycogen storage disease type I: pathophysiology of liver adenomas. Eur. J. Pediatr. 2002; 161(Suppl. 1): S46–9.

Lee PJ, Celermajer DS, Robinson J, McCarthy SN, Betteridge DJ, Leonard JV. Hyperlipidaemia does not impair vascular endothelial function in glycogen storage disease type 1a. Atherosclerosis 1994; 110: 95–100.

Lee PJ, Dalton RN, Shah V, Hindmarsh PC, Leonard JV. Glomerular and tubular function in glycogen storage disease. Pediatr. Nephrol. 1995a; 9: 705–10.

Lee PJ, Patel JS, Fewtrell M, Leonard JV, Bishop NJ. Bone mineralisation in type 1 glycogen storage disease. Eur. J .Pediatr. 1995b; 154: 483–7.

Lei K-J, Chen H, Pan C-J, et al. Glucose-6-phosphatase dependent substrate transport in the glycogen storage disease type 1a mouse. Nat. Genet. 1996; 13: 203–9.

Lei K-J, Pan C-J, Shelly LL, Liu J-L, Chou JY. Identification of mutations in the gene for glucose-6-phosphatase, the enzyme deficient in glycogen storage disease type 1a. J. Clin. Invest. 1994; 93: 1994–9.

Lei K-J, Shelly LL, Pan C-J, Sidbury JB, Chou JY. Mutations in the glucose-6-phosphatase gene that cause glycogen storage disease type 1a. Science 1993; 262: 580–3.

Lin B, Annabi B, Hiraiwa H, Pan C-J, Chou JY. Cloning and characterization of cDNAs encoding a candidate glycogen storage disease type 1b protein in rodents. J. Biol. Chem. 1998; 273: 31656–70.

Lin B, Pan C-J, Chou JY. Human variant glucose-6-phosphate transporter is active in microsomal transport. Hum. Genet. 2000; 107: 526–9.

Marcolongo P, Barone V, Priori G, et al. Structure and mutation analysis of the glycogen storage disease type 1b gene. FEBS Lett. 1998; 436: 247–50.

Martinez-Olmos MA, Lopez-Sanroman A, Martin-Vaquero P, et al. Liver transplantation for type Ib glycogenosis with reversal of cyclic neutropenia. Clin. Nutr. 2001; 20: 375–7.

Matern D, Seydewitz HH, Bali D, Lang C, Chen YT. Glycogen storage disease type I: diagnosis and phenotype/genotype correlation. Eur. J. Pediatr. 2002; 161(Suppl. 1): S10–19.

Matern D, Starzl TE, Arnaout W, et al. Liver transplantation for glycogen storage disease types I, III, and IV. Eur. J. Pediatr. 1999; 158: S43–8.

Matsuo N, Tsuchiya Y, Cho H, Nagai T, Tsuji A. Proximal renal tubular acidosis in a child with type 1 glycogen storage disease. Acta Paediatr. Scand. 1986; 75: 332–5.

Michels VV, Beaudet AL. Hemorrhagic pancreatitis in a patient with glycogen storage disease type I. Clin. Genet. 1980; 17: 220–2.

Mineo I, Tarui S. Myogenic hyperuricemia: what can we learn from metabolic myopathies? Muscle Nerve 1995; 3: S75–81.

Moran ME. Uric acid stone disease. Front. Biosci. 2003; 8: s1339–55.

Moses SW. Historical highlights and unsolved problems in glycogen storage disease type 1. Eur. J. Pediatr. 2002; 161(Suppl. 1): S2–9.

Narisawa K, Igarashi Y, Otomo H, Tada K. A new variant of glycogen storage disease type I probably due to a defect in the glucose-6-phosphate transport system. Biochem. Biophys. Res. Commun. 1978; 83: 1360–4.

Nilsson OS, Arion WJ, Depierre JW, Dallner G, Ernster L. Evidence for the involvement of a glucose-6-phosphate carrier in microsomal glucose-6-phosphatase activity. Eur. J. Biochem. 1978; 82: 627–34.

Nordlie RC, Sukalski KA. Multifunctional glucose-6-phosphatase: a critical review. In: Martonosi AN, ed. The Enzymes of Biological Membranes. New York: Plenum Press, 1985: pp. 349–98.

Obara K, Saito T, Sato H, Ogawa M, Igarashi Y, Yoshinaga K. Renal histology in two adult patients with type I glycogen storage disease. Clin. Nephrol. 1993; 39: 59–64.

Pajor AM. Citrate transport by the kidney and intestine. Semin. Nephrol. 1999; 19: 195–200.

Pan C-J, Chen L-Y, Mansfield BC, Salani B, Varesio L, Chou JY. The signature motif in human glucose-6-phosphate transporter is essential for microsomal transport of glucose-6-phosphate. Hum. Genet. 2003; 112: 430–3.

Pan C-J, Lei K-J, Annabi B, Hemrika W, Chou JY. Transmembrane topology of glucose-6-phosphatase. J. Biol. Chem. 1998a; 273: 6144–8.

Pan C-J, Lei K-J, Chen H, Ward JM, Chou JY. Ontogeny of the murine glucose-6-phosphatase system. Arch. Biochem. Biophys. 1998b; 358: 17–24.

Pan C-J, Lin B, Chou JY. Transmembrane topology of human Glucose-6-phosphate transporter. J. Biol. Chem. 1999; 274: 13865–9.

Polinsky MS, Kaiser BA, Baluarte HJ, Gruskin AB. Renal stones and hypercalciuria. Adv. Pediatr. 1993; 40: 353–84.

Rafey MA, Lipkowitz MS, Leal-Pinto E, Abramson RG. Uric acid transport. Curr. Opin. Nephrol. Hypertens. 2003; 12: 511–16.

Rake JP, Visser G, Huismans D, et al. Bone mineral density in children, adolescents and adults with glycogen storage disease type Ia: a cross-sectional and longitudinal study. J. Inherit. Metab. Dis. 2003; 26: 371–84.

Rake JP, Visser G, Labrune P, Leonard JV, Ullrich K, Smit GP. Glycogen storage disease type I: diagnosis, management, clinical course and outcome. Results of the European Study on Glycogen Storage Disease Type I (ESGSD I). Eur. J. Pediatr. 2002; 161(Suppl. 1): S20–34.

Raptis AE, Viberti G. Pathogenesis of diabetic nephropathy. Exp. Clin. Endocrinol. Diabetes, 2001; 109(Suppl. 2): S424–37.

Reitsma-Bierens WCC. Renal complications in glycogen storage disease type I. Eur. J. Pediatr. 1993; 152(Suppl. 1): S60–2.

Reitsma-Bierens WC, Smit GP, Troelstra JA. Renal function and kidney size in glycogen storage disease type I. Pediatr. Nephrol. 1992; 6: 236–8.

Remuzzi G, Schieppati A, Ruggenenti P. Nephropathy in patients with type 2 diabetes. N. Engl. J. Med. 2002; 346: 1145–51.

Restaino I, Kaplan BS, Stanley C, Baker L. Nephrolithiasis, hypocitraturia, and a distal renal tubular acidification defect in type 1 glycogen storage disease. J. Pediatr. 1993; 122: 392–6.

Rodriguez Soriano J. Renal tubular acidosis: the clinical entity. J. Am. Soc. Nephrol. 2002; 13: 2160–70.

Roe TF, Thomas DW, Gilsanz V, Isaacs H, Atkinson JB. Inflammatory bowel disease in glycogen storage disease type Ib. J. Pediatr. 1986; 109: 55–9.

Senior B, Loridan L. Studies of liver glycogenoses, with particular reference to the metabolism of intravenously administered glycerol. N. Engl. J. Med. 1968; 279: 958–65.

Shelly LL, Lei K-J, Pan C-J, et al. Isolation of the gene for murine glucose-6-phosphatase, the enzyme deficient in glycogen storage disease type 1a. J. Biol. Chem. 1993; 268: 21482–5.

Shieh J-J, Terzioglu M, Hiraiwa H, et al. The molecular basis of glycogen storage disease type 1a: structure and function analysis of mutations in glucose-6-phosphatase. J. Biol. Chem. 2002; 277: 5047–53.

Simoes A, Domingos F, Fortes A, Prata MM. Type 1 glycogen storage disease and recurrent calcium nephrolithiasis. Nephrol. Dial. Transplant. 2001; 16: 1277–9.

Sun M-S, Pan C-J, Shieh J-J, et al. Sustained hepatic and renal glucose-6-phosphatase expression corrects glycogen storage disease type Ia in mice. Hum. Mol. Genet. 2002; 11: 2155–64.

Synderman R, Uhing RJ. Phagocytic cells: stimulus-response coupling mechanisms. In: Gallin GI, Goldstain IM, Snyderman R, eds. Inflammation: Basic Principles and Clinical Correlates. New York: Raven Press, 1988: pp. 309–23.

Talente GM, Coleman RA, Alter C, et al. Glycogen storage disease in adults. Ann. Intern. Med. 1994; 120: 218–26.

Taylor SI. Insulin action, insulin resistance, and type 2 diabetes mellitus. In: Scriver CR, Beaudet AL, Sly WS, et al, eds. The Metabolic and Molecular Bases of Inherited Disease, 8th edn. New York: McGrawHill Inc, 2001: pp. 1433–69.

Thorens B. Facilitated glucose transporters in epithelial cells. Annu. Rev. Physiol. 1993; 55: 591–608.

Trevisan R, Nosadini R, Fioretto P, et al. Metabolic control of kidney hemodynamics in normal and insulin-dependent diabetic subjects. Effects of acetoacetic, lactic, and acetic acids. Diabetes 1987; 36: 1073–81.

Ubels FL, Rake JP, Slaets JP, Smit GP, Smit AJ. Is glycogen storage disease 1a associated with atherosclerosis? Eur. J. Pediatr. 2002; 161(Suppl. 1): S62–4.

Verani R, Bernstein J. Renal glomerular and tubular abnormalities in glycogen storage disease type I. Arch. Pathol. Lab. Med. 1988; 112: 271–4.

Verhoeven AJ, Visser G, van Zwieten R, et al. A convenient diagnostic function test of peripheral blood neutrophils in glycogen storage disease type Ib. Pediatr. Res. 1999; 45: 881–5.

Visser G, Rake JP, Labrune P, et al. Granulocyte colony-stimulating factor in glycogen storage disease type 1b. Results of the European Study on Glycogen Storage Disease Type 1. Eur. J. Pediatr. 2002; 161(Suppl. 1): S83–7.

von Gierke E. Hepato-nephro-megalia glycogenica (Glykogenspeicher-krankheit der Leber und Nieren). Beitr. Pathol. Anat. 1929; 82: 497–513.

Weinstein DA, Somers MJ, Wolfsdorf JI. Decreased urinary citrate excretion in type 1a glycogen storage disease. J. Pediatr. 2001; 138: 378–82.

Weinstein DA, Wolfsdorf JI. Effect of continuous glucose therapy with uncooked cornstarch on the long-term clinical course of type 1a glycogen storage disease. Eur. J. Pediatr. 2002; 161(Suppl. 1): S35–9.

Wolfsdorf JI, Crigler JF Jr. Effect of continuous glucose therapy begun in infancy on the long-term clinical course of patients with type I glycogen storage disease. J. Pediatr. Gastroenterol. Nutr. 1999; 29: 136–43.

Wolfsdorf JI, Laffel LM, Crigler JF Jr. Metabolic control and renal dysfunction in type I glycogen storage disease. J. Inherit. Metab. Dis. 1997; 20: 559–68.

Zingone A, Hiraiwa H, Pan C-J, et al. Correction of glycogen storage disease type 1a in a mouse model by gene therapy. J. Biol. Chem. 2000; 275: 828–32.

CHAPTER 42

Wilson Disease and the Kidney

MICHAEL L. SCHILSKY AND PRAMOD KUMAR MISTRY

INTRODUCTION

Wilson disease is a disorder of copper metabolism that is present in ~1:30000 in almost all populations. Inherited in an autosomal recessive manner, this disorder is due to defective intracellular copper transport in hepatocytes. The mutant gene in Wilson disease encodes a copper-transporting ATPase known as *ATP7B* that is mainly expressed in hepatocytes. Over 300 disease-causing mutations have been described. Mutations that cause reduced function or absence of ATP7B reduce hepatic biliary copper excretion and cause hepatic copper accumulation. When the capacity for storage and cellular protection from excess copper is exceeded, tissue injury results and may be manifest by acute liver injury or chronic injury with development of cirrhosis. In time, copper also accumulates in other sites in the body, most importantly the central nervous system, with resultant neurological or psychiatric signs and symptoms of disease. While this disorder is mainly thought of as one afflicting the liver or the central nervous system, other organs like the kidneys may be affected by the buildup of copper, by the sudden exposure to copper following its release into the circulation during acute tissue injury or by the treatments for Wilson disease itself. The kidneys may suffer direct injury due to copper or copper complexes, or malfunction may occur secondarily due to liver failure in Wilson disease. Timed measurement of urinary copper excretion may also be useful for the diagnosis of Wilson disease. These and other findings will be explored further in light of the newer understanding of the molecular pathophysiology and treatment for Wilson disease.

MOLECULAR PATHOPHYSIOLOGY OF WILSON DISEASE

The molecular defect in Wilson disease is mutation of the gene encoding ATP7B, a copper-transporting ATPase that is mainly but not exclusively expressed in the liver (for review see La Fontaine & Mercer 2007). This protein is highly homologous to another copper-transporting ATPase, ATP7A, which is defective in Menkes disease. In Menkes disease, mutations of ATP7A lead to a maldistribution of copper in many tissues due to the widespread expression of this protein. Furthermore, there is a relative copper deficiency in tissues such as the developing nervous system. By contrast, ATP7B is highly expressed in hepatocytes in the liver (and ATP7A is virtually absent), but some is expressed in other tissues, including the brain and kidneys. For this reason, mutations of ATP7B result in a reduction of biliary copper excretion with resultant copper accumulation in hepatocytes, and later in the circulation where it affects other organ systems.

As well as playing a critical role in cellular copper excretion, the ATP7B and ATP7A proteins transport copper within cells that is utilized for metalloprotein biosynthesis. Examples include ceruloplasmin that is made by the hepatocytes, lysyl oxidase by fibroblasts and dopamine beta hydroxylase in neurons. In the liver, excess copper is transported into bile, either by a lysosomal vesicular pathway or by direct transport from the apical canalicular membrane. In hepatocytes, ATP7B resides mainly in the trans-Golgi, but it relocates in response to copper to vesicles where it transports copper into these vesicles for excretion, or may relocate to the apical or canalicular surface for direct cellular export. This occurs through an extraordinary process of cellular relocalization of the copper-transporting ATPase in response to changes in environmental copper permit microregulation of copper homeostasis. When copper is low, the protein remains mainly at the trans-Golgi where it serves to transport copper used in metalloprotein biosynthesis; when excess copper is present it functions to transport copper for export. In most other cells that express *ATP7A*, copper is transported from intracellular sites to the apical membrane surface where it is excreted via ATP7A.

Genetic Diseases of the Kidney
ISBN: 978-0-12-449851-8

RENAL COPPER EXCRETION AND TRANSPORT

The renal excretion of copper, typically 10–20 µg in a 24-hour period, is a minor component of the overall daily excretion of copper by the gut that includes nonabsorbed copper and copper excreted into bile by the liver (see Figure 42.1; Scheinberg & Sternlieb 1984). When individuals are presented with increased dietary copper, most of the excess copper that is absorbed is excreted in bile, and the urine copper excretion is maintained more or less constant in amount. An exception occurs in untreated Wilson disease and also in acute copper poisoning.

The presence of excessive urinary copper excretion in patients with Wilson disease was first recognized in 1948 by Mandelbrote and his coworkers. In symptomatic patients with Wilson disease, urinary copper excretion typically exceeds 100 µg per day (see Figure 42.2). Since the original recognition of elevated urine copper in patients with Wilson disease, testing for urine copper excretion has become part of the diagnostic testing for this disorder (Roberts & Schilsky 2003). Following treatment with copper chelators such as d-penicillamine or trientine, the urinary excretion of copper is often initially over tenfold the basal values, and decreases with treatment time to about twofold the basal level as tissues stores become partly depleted. This increase in urine copper excretion after chelation treatment is the basis for adjunctive diagnostic testing for Wilson disease, and may help distinguish this disorder from other causes of liver injury (Müller et al 2007). The amount of copper in the urine of patients on chelator is also dose-dependent, with higher dosages yielding higher urine copper excretion. For patients on zinc, the amount of urine copper decreases with time and typically is less than 100 µg per day. Heterozygotes for Wilson disease may have some slight elevations in basal urinary copper excretion but no reported liver or renal disease (Tu et al 1968).

Until the discovery of the ubiquitously expressed copper transporter CTR1 and the highly selective cellular expression of the copper transporting ATPases, ATP7A in Menkes disease and ATP7B in Wilson disease, there was little known about the mechanism of copper transport by the kidneys. Even with the knowledge of the specific transporters, many questions still remain about their expression and integrated function. However, it is possible to speculate how these proteins function in normal states and following exposure to a higher burden of copper as occurs in untreated Wilson disease. A schema depicting how the kidney handles copper in these different conditions is shown in Figure 42.1, and a discussion of the individual proteins and the limited data on their expression follows.

CTR1, the ubiquitous copper transporter typically present on the basolateral surface of cells, is expressed in the kidney (Lee et al 2000). This protein transports circulating copper from albumin, peptides, and other carriers.

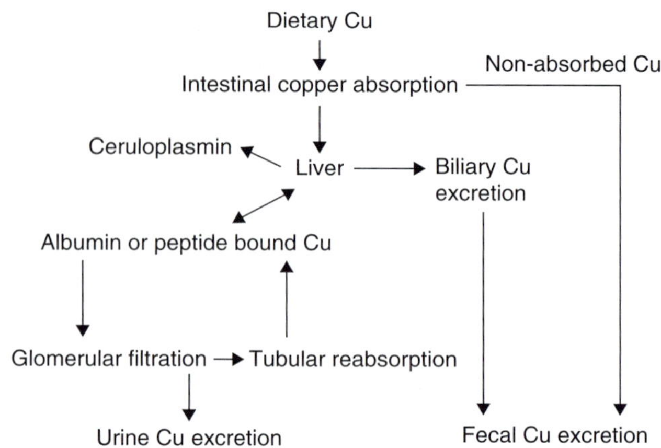

FIGURE 42.1 Copper metabolism. Dietary copper that is absorbed by the gut and transported through the portal circulation is transported mainly in association with albumin. This copper is avidly extracted by liver cells. Some of this copper is incorporated into ceruloplasmin that is secreted by hepatocytes into the bloodstream, some is exchanged back into the bloodstream where it may bind to albumin or other peptides and excess copper is excreted into bile mainly by processes that are dependent on ATP7B function. In Wilson disease, this transport function is diminished or absent and copper increases in the liver cells and then in the bloodstream. Copper in the bloodstream may be extracted by the kidney and excreted in the urine. Uptake of copper by the kidney is thought to be dependent on CTR1 function, while tubular reabsorption of filtered copper may be dependent on ATP7B and ATP7A function

Copper that enters the kidney tubules may be returned to the bloodstream through the efforts of the copper-transporting ATPases, ATP7A and ATP7B. The exact cellular expression of the individual proteins in man is not clearly defined; however, the murine orthologs Atp7a and Atp7b are expressed in glomeruli, and Atp7b is expressed in the kidney medulla (Moore & Cox 2002). It has been suggested that ATP7A and ATP7B may be involved in copper reabsorption in the loops of Henle.

The intracellular trafficking of ATP7B and ATP7A in kidney cells in response to copper is thought to more finely regulate copper homeostasis. There is movement from the trans-Golgi to a site in the cell or on the surface membrane in response to increased environmental copper permits the protein to function in copper export back into the bloodstream or to intracellular storage pools. A study by Linz et al in 2008 demonstrated intracellular trafficking of ATP7B and ATP7A in renal cells as well as in hepatocytes in response to copper. They suggest differences in the localization of the two copper-transporting proteins and their function. ATP7A traffics to the basolateral membrane where it plays a role in copper export by transporting copper for reabsorption into the blood, thereby ridding renal cells of excess copper. While ATP7A moves from the Golgi to the plasma membrane in response to copper, ATP7B may function to redistribute excess copper into vesicles in cells rather than directly export the copper. This could in part account for the increase in

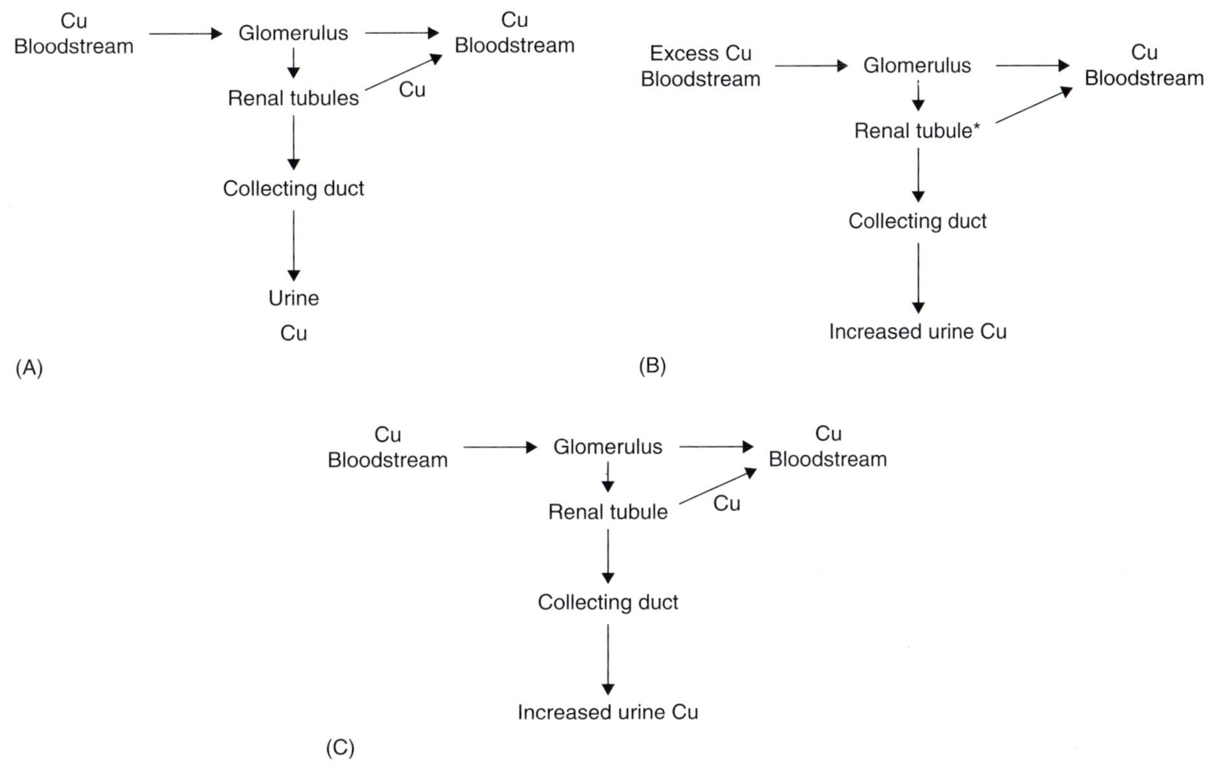

FIGURE 42.2 (A) Normal. In normal individuals, copper is carried in the bloodstream in association with albumin and other peptides. Some of this copper may be transported by CTRI in glomerular cells. Copper is then returned to the bloodstream from the renal tubules by the action of either ATP7A or ATP7B in renal cells. A small amount, ~10–20 μg per day, is excreted in the urine. (B) Untreated Wilson disease. An increased percentage of the total amount of copper in the bloodstream is carried in the bloodstream in association with albumin and other peptides. Some of the circulating copper may be transported by CTRI in glomerular cells. This copper is then returned to the bloodstream by the action of either ATP7A or ATP7B in renal cells. Excess copper, typically bound to amino acids or small peptides, may directly injure renal tubular cells.* In symptomatic Wilson disease urine copper excretion may be >100 μg per day, and in the setting of acute liver failure, this may be increased to greater than 1000 μg per day. (C) Treated Wilson disease. The increased percentage of the total amount of copper in the bloodstream carried in association with albumin and other peptides is reduced to normal levels following treatment. In patients treated with chelation therapy, urine copper excretion may initially be >1000 μmg per day but will reduce over time to ~250 μg to 500 μg per day (the amount being dosage-dependent). Copper-induced renal tubular injury, if present, may improve with therapy

renal tubular copper content in response to excess copper presented to the kidneys in patients with Wilson disease.

RENAL TUBULAR INJURY AND OTHER FINDINGS

There is a paucity of significant histological findings in renal biopsy specimens from patients with Wilson disease (Scheinberg & Sternlieb 1984), but lesions are reported in animal models of Wilson disease. Despite the lack of human evidence for histological findings, functional defects of the renal tubules in Wilson disease due to copper toxicity may be evident. Aminoaciduria was first detected in patients with Wilson disease by Uzman and Denny-Brown in 1948, and since then it is recognized that Fanconi syndrome with aminoaciduria, increased uric acid excretion, phosphate and urate and other small molecules as well as renal tubular acidosis, may also occur. Treatment of patients with reported

tubular defects with d-penicillamine demonstrated disappearance or improvement of the tubular defect, suggesting copper's role in the tubular injury (Leu et al 1970).

Renal copper content has been shown to be increased in Wilson disease. In a report of autopsy specimens from untreated Wilson disease patients, the tissue copper content of kidney was 904 μg/g dry weight (range 100–2045) compared with a normal value of only 15 (range 5–36) (Scheinberg & Sternlieb 1969). Consistent with this finding were reports showing intracytoplasmic granules with positive histochemical staining of copper with rubeanic acid in the tubular epithelium of patients. These findings are consistent with the knowledge that copper induces renal tubular cells to synthesize metallothionein (MT), an endogenous peptide that avidly binds copper and other metals. Copper–MT complexes already formed in the liver are released into the bloodstream following liver injury and are filtered by glomeruli and reabsorbed in the brush border of the proximal convoluted tubule (Kurasaki et al 1998). This peptide–copper complex is toxic due to the release of the copper from the peptide following lysosomal degradation in the proximal convoluted tubules. However, the released metal can also induce MT synthesis in neighboring cells. When these cells are presented with even more copper, then induced MT may serve to protect against further injury. However, large amounts of copper likely overwhelm this protective mechanism, and degradation of the protein–metal complex would lead to an increased amount of intracellular copper that could cause oxidative injury to the cell. Other metals such as zinc may also induce MT in renal cells and help increase safe storage sites for excess copper in renal tubular cells.

Nephrolithiasis may be found in some patients with Wilson disease. This is most likely present in patients with tubular defects in acidification or resultant changes in the urinary excretion of substrates that augment stone formation.

Rhabdomyolysis with resultant renal tubular injury has been reported in a patient with Wilson disease (Propst et al 1995). The rhabdomyolysis was reported to cease with effective treatment with d-penicillamine.

HEPATORENAL SYNDROME

In the setting of either acute or severe chronic liver failure due to Wilson disease, hepatorenal syndrome may develop. This entity is defined by the development of oliguria in association with reduced renal clearance and a low urine sodium excretion of <10 mEq/l that is unresponsive to volume expansion. In patients with liver disease or in the setting of acute liver failure, hepatorenal syndrome typically follows the severe loss of hepatic function and may be exacerbated by concurrent infection or use of diuretics as well as direct renal tubular injury from copper and copper complexes. Treatment options include treatment of any active infection, withdrawal of diuretic therapy if applicable, trials of octreotide and midodrine or terlipressin or if appropriate, liver transplantation that is also curative of Wilson disease.

Patients with acute liver failure due to Wilson disease have a higher incidence of renal dysfunction compared to other etiologies of liver failure as noted above, part of which is due to direct renal tubular injury. However, in some individuals, hepatorenal syndrome may be present and contribute to the renal insufficiency. Hepatorenal syndrome and tubular injury typically reverse following transplantation; however, some patients may require some interval of renal replacement therapy until recovery can occur if there is concurrent renal tubular injury (Emre et al 2001).

INTERVENTIONS FOR WILSON DISEASE

Treatment of Wilson disease with chelating agents (d-penicillamine or trientine) or with zinc salts may stabilize and reverse copper-induced toxicities, including renal tubular injury and Fanconi syndrome as noted above.

During the acute presentation of Wilson disease with liver failure, a large amount of copper is released from the liver and copper and copper–protein complexes cause a nonimmune-mediated hemolytic anemia as well as acute renal tubular injury. Most of these patients will not survive without liver transplantation, and therefore treatments that may allay some of the toxicity and bridge these patients to transplant are helpful. Furthermore, if such treatments can prevent further renal injury, recovery post transplantation is more rapid. Therapies that have been used for these individuals include plasma exchange, plasmapheresis, dialysis (peritoneal, hemodialysis and CVVHD) and albumin dialysis (references in Roberts & Schilsky 2003). Other devices that have been tried include MARS that combines hemofiltration with albumin dialysis and an ion exchange column (Sen et al 2002). Each of these therapies reduces the excess copper released during injury and thereby reduces secondary toxicity to other organs and to the remaining liver. While these treatments have stabilized patients, most still require transplantation due to the severity of the injury to the liver and the presence of cirrhosis or advanced fibrosis on the background of the acute injury that results in a reduced hepatic reserve.

N acetyl cysteine (NAC) is being utilized to prevent oxidative injury to the liver in acetaminophen-induced injury and may be helpful even in nonacetaminophen-induced liver failure as well. Similarly, NAC has been utilized for prevention of renal oxidative injury. In the setting of acute Wilson disease, NAC and other antioxidants that are occasionally tried could also theoretically help with prevention of some of the copper-induced renal injury. Further studies in animal models could perhaps determine whether this treatment could prevent or ameliorate the copper-induced renal tubular damage.

DRUG-INDUCED RENAL INJURY

Renal injury may occur not only from copper, but from some of the treatments for Wilson disease. Most commonly, the copper-chelating drug, d-penicillamine, may cause acute or later-onset renal toxicity. A syndrome with lupus nephritis may occur early on, associated with the appearance of cellular elements in the urinary sediment and renal injury (Scheinberg & Sternlieb 1984). Proteinuria of varying degrees may occur even years following the initiation of therapy with d-penicillamine. Within the first 2 years of therapy, low-molecular-weight proteins are found in the urine, suggesting tubular injury. Later on, the appearance of higher-molecular-weight proteins such as albumin suggests glomerular injury due to d-penicillamine. This may even be as severe as full-blown nephritic syndrome. Vasculitis that has the appearance of lupus nephritis and is associated with anti-DNA antibodies may occur with d-penicillamine treatment. This is often heralded by changes in urine sediment. More recently a description of ANCA-related vasculitis was reported in a patient with Wilson disease treated with d-penicillamine (Bienaime et al 2007). While this has previously been described in patients with rheumatoid arthritis or systemic sclerosis that were treated with d-penicillamine, in some patients with these disorders ANCA can be detected prior to treatment. By contrast this has never been reported for Wilson disease, and therefore the presence of ANCA-vasculitis suggests a direct effect of d-penicillamine treatment.

Trientine may also cause similar lupus-like syndrome as d-penicillamine, however this medication which has an altogether different structure, was used for many years as a replacement therapy for patients that developed intolerance to penicillamine. It is only more recently that this medication has been recommended as first-line therapy for Wilson disease, and therefore the frequency of renal injury from this medication alone is uncertain. In our experience, the de novo appearance of renal injury of patients on trientine therapy for Wilson disease is very rare.

CONCLUSION

Our understanding of renal copper transport has evolved with the new discoveries that led to the understanding of the molecular basis for Wilson and Menkes disease. In Wilson disease, the kidney is susceptible to copper-induced injury, but may also be damaged by medical therapy directed against the copper toxicity or rendered nonfunctional by the hepatorenal syndrome when liver failure is present. Renal tubular injury from copper can improve with medical treatment of Wilson disease. Interventions designed to acutely lower circulating copper or prevent oxidative injury during acute liver failure due to Wilson disease may ameliorate renal toxicity of copper and copper–protein complexes and help support these patients until recovery or transplantation.

References

Azizi E, Eshel G, Aladjem M. Hypercalciuria and nephrolithiasis as a presenting sign in Wilson disease. Eur. J. Pediatr. 1989; 148.

Bienaime F, Clerbaux G, Plaisier E, Mougenot B, Ronco P, Rougier JP. D-penicillamine-induced ANCA-associated crescentic glomerulonephritis in Wilson disease. Am. J. Kidney Dis. 2007; 50: 821–5.

Emre S, Atillasoy EO, Ozdemir S, et al. Orthotopic liver transplantation for Wilson's disease: a single center experience. Transplantation 2001; 72: 1232–6.

Jhang JS, Schilsky ML, Lefkowitch JH, Schwartz J. Therapeutic plasmapheresis as a bridge to liver transplantation in fulminant Wilson disease. J. Clin. Apher 2007; 22: 10–14.

Kurasaki M, Okabe M, Saito S, Suzuki-Kurasaki M. Copper metabolism in the kidney of rats administered copper and copper-metallothionein. Am. J. Physiol. 1998; 274: F783–90.

La Fontaine S, Mercer JF. Trafficking of the copper-ATPases, ATP7A and ATP7B: role in copper homeostasis. Arch. Biochem. Biophys. 2007; 463: 149–67.

Lee J, Prohaska JR, Dagenais SL, Glover TW, Thiele DJ. Isolation of a murine copper transporter gene, tissue specific expression and functional complementation of a yeast copper transport mutant. Gene 2000; 254: 87–96.

Leu ML, Strickland GT, Gutman RA. Renal function in Wilson's disease: response to penicillamine therapy. Am. J. Med. Sci. 1970; 260: 381–98.

Linz R, Barnes NL, Zimnicka AM, Kaplan JH, Eipper B, Lutsenko S. Intracellular targeting of copper-transporting ATPase ATP7A in a normal and Atp7b-/- kidney. Am. J. Physiol. Renal Physiol. 2008; 294: F53–61.

Mandelbrote BM, Stanier MW, Thompson RHS, Thurston MN. Studies on copper metabolism in demyelinating diseases of the central nervous system. Brain 1948; 71: 212–28.

Moore SD, Cox DW. Expression in mouse kidney of membrane copper transporters Atp7a and Atp7b. Nephron 2002; 92(3): 629–34.

Müller T, Koppikar S, Taylor RM, et al. Re-evaluation of the penicillamine challenge test in the diagnosis of Wilson's disease in children. J. Hepatol 2007; 47: 270–6.

Propst A, Propst T, Feichtinger H, Judmaier G, Willeit J, Vogel W. Copper-induced acute rhabdomyolysis in Wilson's disease. Gastroenterology 1995; 108: 885–7.

Roberts E, Schilsky ML. A practice guideline on Wilson disease. Hepatology 2003; 37: 1475–92.

Sozeri E, Feist D, Ruder H, Scharer K. Proteinuria and other renal functions in Wilson's disease. Pediatr. Nephrol. 1997; 11: 307–11.

Scheinberg IH, Sternlieb I. Metabolism of trace metals. In: Bondy B, ed. PK Duncan's Disease of Metabolism, Vol. 2, Endocrinology and Nutrition, 6th edn. Philadelphia, PA: WB Saunders, 1969: pp. 550–64.

Scheinberg IH, Sternlieb I. Wilson's Disease in Major Problems in Internal Medicine, Philadelphia, PA: WB Saunders, 1984: pp. 99–105.

Sen S, Felldin M, Steiner C, et al. Albumin dialysis and Molecular Adsorbents Recirculating System (MARS) for acute Wilson's disease. Liver Transplant. 2002; 8: 962–7.

Tu JB, Blackwell RQ, Fresh JW, Watten RH. In: Birth Defects: Original Article Series, Wilson's Disease, The National Foundation, Vol. IV, No. 2. New York: 1968: pp. 114–21.

Uzman L, Denny-Brown D. Amino-aciduria in hepatolenticular degeneration (Wilson's disease). Am. J. Med. Sci. 1948; 215: 599–611.

Walshe JM. Effect of penicillamine on failure of renal acidification in Wilson's disease. Lancet 1968; 1: 775–9.

CHAPTER 43

Genetic Defects in Renal Phosphate Handling

CARSTEN A. WAGNER, NATI HERNANDO, IAN C. FORSTER, JÜRG BIBER AND HEINI MURER

INTRODUCTION

Body phosphate is distributed between different compartments with about 85% of total phosphate deposited in bone and about 10% in soft tissues. The remaining 2–3% are found in serum, constituting a freely exchangeable and tightly regulated phosphate pool. Serum phosphate levels in adults are maintained in a narrow range of about 0.8–1.45 mmol/l (2.4–4.5 mg/dl) where deviations as found in patients with hypophosphatemia or hyperphosphatemia cause various severe disturbances of cellular and organ function such anemia, ATP depletion, rickets, osteomalacia, nephrolithiasis, or excessive tissue calcifications (Knochel 2000).

Whole bo d y phosphate homeostasis is the product of intestinal uptake of phosphate from diet (usual daily amount in the range of 0.9–1 g) and the rate of reabsorption of phosphate from urine. Both intestinal uptake and renal reabsorption are mediated by a set of specific sodium-dependent phosphate cotransporters (Na^+/Pi-cotransporters) belonging to the SLC34 gene family of solute ca rriers (Murer et al 2004). Their activity and expression is tightly regulated by a large variety of hormones and metabolic factors adjusting uptake and excretion to short- and long-term body requirements. Inborn errors in this delicate balance between phosphate intake and excretion have been identified in the past decade and are caused not only by mutations in the phosphate transporter molecules but also by mutations in several genes that encode factors involved directly and indirectly in their regulation. Identifying these genes and elucidating their functions have greatly contributed to our present understanding of how phosphate homeostasis is achieved. This chapter will first give a brief overview of physiological function and regulation of phosphate transporters. For more extensive reviews the reader is referred to a number of recent reviews and book chapters (Knochel 2000, Silve & Friedlander 2000; Biber et al 2004, Hernando et al 2004, Miyamoto et al 2004, Murer et al 2000, 2003, 2004, Tenenhouse 2005). The second part of this chapter will discuss mutations in several genes that directly or indirectly participate in (renal) control of phosphate balance.

GENERAL ASPECTS OF RENAL PHOSPHATE HANDLING

About 80% of the filtered phosphate is reabsorbed from urine under a normal dietary phosphate intake. Micropuncture and brush border membrane vesicle studies indicated that the bulk of phosphate is reabsorbed in the earlier convoluted parts of the proximal tubule and that the rates of reabsorption may be higher in juxtamedullary nephrons than in superficial nephrons (Baumann et al 1975, Greger et al 1977, Haass et al 1978, Brunette et al 1984, Haramati 1985, Kayne et al 1993) suggesting an axial gradient as well as an internephronal heterogeneity. It has remained controversial if later nephron parts such as the distal tubule also contribute to phosphate reabsorption.

Phosphate absorption in the proximal tubule is mediated by sodium-dependent mechanism(s) (Na/Pi-cotransport) located in the brush border membrane. The stoichiometry of this process was determined to be two or more sodium ions per phosphate ion allowing for the intracellular accumulation of phosphate even against its gradient. Interestingly, NMR studies demonstrated that intracellular phosphate concentrations in the proximal tubular cells may be in the range of 0.7–1.8 mmol/l (Freeman et al 1983).

Genetic Diseases of the Kidney
ISBN: 978-0-12-449851-8

715

The rate of proximal tubular absorption is tightly regulated and adapted to dietary phosphate intake, metabolism, and various conditions. The adaptive response of the proximal tubule is in part mediated by hormones as discussed below.

Dietary Phosphate Intake

Dietary phosphate intake represents a major factor regulating renal phosphate excretion. Higher dietary phosphate intake leads within a few hours to a decrease in renal phosphate reabsorption and hence to phosphaturia (Silve & Friedlander 2000). In rats switched acutely from a low to a high phosphate diet, phosphaturia develops within 4 hours due to a downregulation of renal phosphate cotransporter (Levi et al 1994). Conversely, low-phosphate diet increases renal phosphate reabsorption by upregulating phosphate cotransporters (Levi et al 1994). These adaptive processes can occur even in the absence of measurable changes in serum phosphate levels or altered concentration of phosphaturic hormones (Murer et al 2000, 2001). Along this line, adaption of phosphate transport rates can also be observed in vitro in cell culture models such as opossum kidney (OK) cells that express the opossum ortholog of the human NaPi-IIa cotransporter. Incubation of OK cells in medium containing low or high phosphate adapts Na^+-dependent phosphate transport rates suggesting a cell autonomous regulation (Biber et al 1988, Reshkin et al 1991, Pfister et al 1998).

Hormonal Regulation

The major hormones regulating renal phosphate handling include parathyroid hormone and vitamin D_3. Other hormones affecting renal phosphate excretion are summarized in Table 43.1 and are described in more detail in several reviews (Murer et al 2000, Silve & Friedlander 2000). The role of FGF-23 as a novel hormone is discussed below. Parathyroid hormone (PTH) increases renal phosphate excretion acutely and chronically (Evers et al 1978, Hruska & Hammerman 1981, Brunette et al 1984). Vice versa, parathyroidectomy stimulates renal phosphate reabsorption (Hammerman et al 1980, Brunette et al 1984). As described below, PTH causes a reduction in renal phosphate cotransporter protein expression in the brush border membrane (Murer et al 2000). Similar to the in vivo situation, PTH also suppresses phosphate transport in renal proximal tubular cell culture models such as OK cells (Reshkin et al 1991, Cole 1999, Karim-Jimenez et al 2000). PTH acts also on the expression and activity of the 1α-hydroxylase enzyme catalyzing the hydroxylation of 25-OH-vitamin D_3 to its active form 1,25-(OH)2-vitamin D_3. Several reports indicate that vitamin D_3 directly stimulates renal phosphate reabsorption (Kurnik & Hruska 1984, 1985, Silve &

TABLE 43.1   Factors affecting phosphate reabsorption in the proximal tubule

Factors that decrease Pi reabsorption	Factors that increase Pi reabsorption
Parathyroid hormone/cAMP	Parathyroidectomy
Atrial natriuretic peptide/cGMP	$1\alpha,25(OH)2D_3$
Glucocorticoids	Growth hormone(s)
Dopamine	Insulin-like growth factor
Phosphatonins:	Regulation by dietary intake
Fibroblast growth factor 23	'acute' (minutes to hours)
Fibroblast growth factor 7	'chronic' (hours to days)
Frizzled related protein 4	(involvement of multiple
MEPE	'systemic'
Others	factors)

Friedlander 2000). However, interpretation of data is difficult due to the complex interactions of this hormone with PTH and serum calcium levels that also influence renal phosphate transport. In the intestine a stimulatory effect of vitamin D_3 on phosphate absorption seems to be better established. Stimulatory effects of vitamin D_3 on expression and activity of type II phosphate cotransporter expression and activity has been also reported from cell culture experiments (Allon & Paris 1993, Taketani et al 1998). It is clear from experiments with vitamin D_3 receptor deficient mice and mice lacking 1α-hydroxylase that vitamin D_3 and its receptor are not required for normal adaptation of renal and intestinal phosphate transport activity to changes in dietary intake (Segawa et al 2004, Capuano et al 2005).

Other Factors

A variety of factors have been described that regulate renal phosphate handling directly or indirectly, among them acid-base status, circadian rhythm, age, and gender (for review see Silve & Friedlander 2000). Metabolic acidosis increases phosphaturia, however, the mechanism has not been fully resolved (Ambuhl et al 1998, Stauber et al 2005). Also, respiratory acidosis decreases tubular phosphate absorption, an effect possibly due to the concomitant hyperphosphatemia. In contrast, respiratory alkalosis stimulates proximal tubular phosphate reabsorption through hypophosphatemia because of phosphate shifting from the extracellular space into muscle (Hoppe et al 1982, Haramati & Nienhuis 1984). These changes may be in part explained by alterations in the maximal transport capacity, i.e. the number of active transporters in the brush border membrane. In addition, direct effects of pH on transporter function have been described in brush border membrane vesicles and in heterologously expressed NaPi-IIa transporters in *Xenopus* oocytes demonstrating a direct inhibitory effect of protons on transporter function (Amstutz et al 1985, Busch et al 1995, de la Horra et al 2000, Forster et al 2000).

A MOLECULAR VIEW OF RENAL PHOSPHATE TRANSPORT

Members of three gene families have been identified that mediate Na^+/phosphate transport across plasma membranes. The type I NaPi-cotransporter belongs to the SLC17 family of anion transporters and appears to play no major role in renal phosphate handling (Reimer & Edwards 2004). The SLC34 family members (type II NaPi-cotransporter) mediate phosphate transport across the luminal membrane of many epithelia as found in kidney and small intestine but also in lung, liver, or lactating mammary gland (Murer et al 2004). The type III NaPi-cotransporter belong to the SLC20 family and serves most likely a rather house-keeping role supplying cells with phosphate for metabolism (Collins et al 2004).

The SLC34 Family of Na^+/Phosphate Cotransporters

Three members of the type II Na^+-dependent phosphate cotransporter family have been identified and classified in the SLC34 solute carrier gene family (Murer et al 2004). SLC34A1 (NaPi-IIa) was the first identified Na^+/phosphate cotransporter and is predominantly expressed in kidney where it mediates about 70–80% of overall phosphate reabsorption as evident from a *slc34a1* deficient mouse model (Beck et al 1998). Another member of this family, SLC34A3 (NaPi-IIc), is also almost exclusively found in kidney (Segawa et al 2002). NaPi-IIc may be critical for renal phosphate reabsorption in man as suggested by the recent identification of mutations in this gene in patients with hypercalciuric hypophosphatemic rickets with renal phosphate wasting (Bergwitz et al 2006, Lorenz-Depiereux et al 2006) as discussed below. The third member, SLC34A2 (NaPi-IIb) is not expressed in kidney but in small intestine and in a number of other organs including testis, lung, liver, and lactating mammary glands (Hilfiker et al 1998, Murer et al 2004).

SLC34A1 (NaPi-IIa) and SLC34A3 (NaPi-IIc)

GENE AND PROTEIN STRUCTURE

NaPi-IIa (SLC35A1)

The human SLC34A1 gene lies on chromosome 5q35 (Kos et al 1994), is about 13 kb in length and consists of 13 exons and 12 introns (Hartmann et al 1996). The gene is characterized by having a TATA box at position -51 and a CAAT box at position -143 (Hartmann 1996). For expression in OK cells a 484 bp 5′ fragment was sufficient in a luciferase assay (Hartmann 1996). However, it appears that the promoter is only active in a proximal tubule cell specific environment such as OK cells (Hartmann 1996, Hilfiker et al 1998a, 1998b). Several additional regulatory elements have been identified in the promoter region of type IIa cotransporters. In weaning rats, changes in dietary phosphate intake are associated with altered abundance of NaPi-IIa mRNA and protein. A sequence was mapped about 1 kb upstream of the transcription start site, named phosphate responsive element (PRE) with high similarity to a yeast binding element of the phosphate sensitive transcription factor Pho-4. Binding of the mammalian TFE3 transcription factor to PRE was found. TFE3 expression is transcriptionally regulated by dietary phosphate intake (Kido et al 1999). Furthermore, a vitamin D_3 response element is located ~2 kb upstream from the transcription initiation site of the SLC34A1 gene (Taketani et al 1998). Also, 5′-flanking sequences were described in the OK cell specific NaPi-IIa gene that are sensitive to ambient HCO_3^-/CO_2 tension (Jehle et al 1999). Yamamoto and colleagues described two splice variants (NaPi-IIa-A and NaPi-IIa-B) of the mouse *slc34a3* gene with two alternative first exons. Expression of the NaPi-IIa-B gene in OK cells was regulated by vitamin D_3 requiring the CAAT box (Yamamoto et al 2005).

The SLC34A1 gene codes for a protein of about 635-640 amino acids in mammals (in human 639 amino acids) (Magagnin et al 1993). Based on various biochemical and functional assays it appears that the protein has ten transmembrane helices with two additional re-entry loops and both C- and N-termini located intracellularly (see Figure 43.4) (Radanovic, Gisler, Forster, Biber, Murer, unpublished results).

Alternative splicing of the rat NaPi-IIa gene has been reported yielding three shorter isoforms named NaPi-IIα, NaPi-IIβ, and NaPi-IIγ (Tatsumi et al 1998). The function of these isoforms is unknown but a regulatory role has been proposed inhibiting full NaPi-IIa function (Miyamoto et al 2004). The biological significance of these observations remains to be established.

NaPi-IIc (SLC34A3)

The human SLC34A3 gene spans about 5 kb containing 13 exons and lies on chromosome 9q34 (Bergwitz et al 2006, Lorenz-Depiereux et al 2006). Human NaPi-IIc contains 599 amino acids (Bergwitz et al 2006, Lorenz-Depiereux et al 2006). In the murine slc34a3 gene a transcription initiation site was found at position −170 relative to the 3′-end of exon 1 (Ohkido et al 2003). Furthermore, AP-2 binding motifs and promoter-selective transcription factor-1 (SP-1) binding motifs were described. NaPi-IIc expression in mouse is enhanced in response to a low phosphate diet

(Ohkido et al 2003) and may also be modulated by vitamin D_3 (Capuano, Biber, Murer, Wagner, unpublished results). Dietary Pi-responsive elements (potential TFE-3 binding sites) were located at -1846 and -1822. The same authors also reported that no typical vitamin D_3-responsive element (VDRE) could be detected (Ohkido et al 2003).

TRANSPORT FUNCTION

The transport function of NaPi-IIa and its structure–function relationship have been extensively studied elucidating some basic important features such as cotransport mode, electrogenicity or voltage and pH-dependency. In summary these data demonstrate that NaPi-IIa transports three sodium ions together with one divalent phosphate ion (HPO_4^{2-}) per transport cycle. Two sodium ions bind to the transporter before binding the phosphate and a third sodium ion and translocation to the cytosol occurs. This transport mode is electrogenic (Busch et al 1995, Virkki et al 2005, 2006). The transporter may homodimerize (Gisler et al, unpublished data) but the functional unit is a monomer (Kohler et al 2000). In the third putative transmembrane helix some amino acid residues have been identified that are involved in forming a binding site for one sodium ion and conferring electrogenicity to the transport cycle. These residues are not present in the electroneutral type IIc cotransporter (Bacconi et al 2005, Virkki et al 2005).

The type IIc cotransporter, in contrast, transports only two sodium ions per phosphate and is electroneutral indicating that divalent phosphate is its preferred substrate (Segawa et al 2002, Bacconi et al 2005). Otherwise, NaPi-IIc displays a high degree of homology on protein level with NaPi-IIa.

TISSUE DISTRIBUTION AND LOCALIZATION

Both NaPi-IIa and NaPi-IIc are predominantly expressed in kidney (Magagnin et al 1993, Segawa et al 2002, Ohkido et al 2003) but may also be present in other tissues at much lower abundance. NaPi-IIa-related immunostaining has been reported also from the mouse brain in the region around the third ventricle and the amygdala. There NaPi-IIa was implied in the central control of renal phosphate handling (Woda et al 2004). NaPi-IIc in mouse was also detected by PCR in heart, spleen, and placenta (Ohkido et al 2003).

The localization of NaPi-IIa and IIc has been mainly studied in mouse and rat kidney (Custer et al 1994, Murer et al 2000, Segawa et al 2002, 2005, Ohkido et al 2003, Madjdpour et al 2004) but little is known about their localization and relative abundance in human kidney. In rodents, NaPi-IIa and -IIc localize exclusively to the apical brush border membrane in the proximal tubule. Under a normal phosphate containing diet, NaPi-IIa and -IIc expression is mainly restricted to the early proximal tubule and predominantly to juxtamedullary nephrons (Figures 43.1 and 43.2)

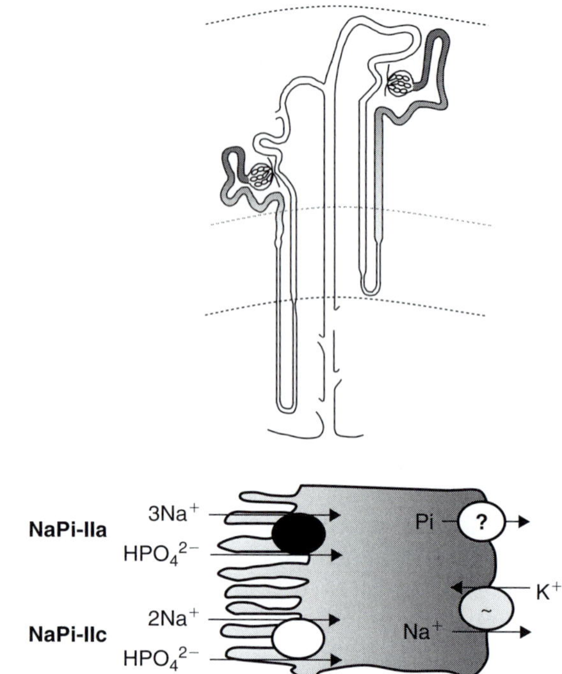

FIGURE 43.1 Scheme of renal phosphate reabsorption in the proximal tubule. The majority of filtered phosphate is reabsorbed in the (early) proximal tubule (shaded region) involving at least two apical sodium-dependent phosphate cotransporters (Na/Pi-cotransporter: NaPi-IIa (SLC34A1) and NaPi-IIc (SLC34A3) expressed in the brush border membrane. Transport is energized by basolateral Na+/K+-ATPases providing a transmembrane electrochemical gradient. The basolateral exit pathway for phosphate has not been identified on a molecular level to date

(Custer et al 1994, Murer et al 2000, Miyamoto et al 2004). During dietary phosphate depletion, NaPi-IIa expression spreads also to the later proximal segments and superficial nephrons (Levi et al 1994, Murer et al 2000, Madjdpour et al 2004). In contrast, NaPi-IIc expression appears to remain confined to the early proximal tubule of both superficial and juxtamedullary nephrons (Segawa et al 2005).

REGULATION OF RENAL PHOSPHATE COTRANSPORTERS

As mentioned above, renal phosphate handling is tightly controlled and adapted to metabolic needs of the whole body as well as to dietary intake. Regulation of renal phosphate reabsorption occurs both acutely as well as chronically. A variety of factors has been identified regulating renal phosphate handling among which PTH, dopamine, serotonin, dexamethasone, metabolic acidosis, and high dietary phosphate intake cause an increase in urinary phosphate excretion (Murer et al 2000, Bacic et al 2005, Stauber et al 2005). A special group of factors also leading to renal phosphate excretion has been identified and termed phosphatonins.

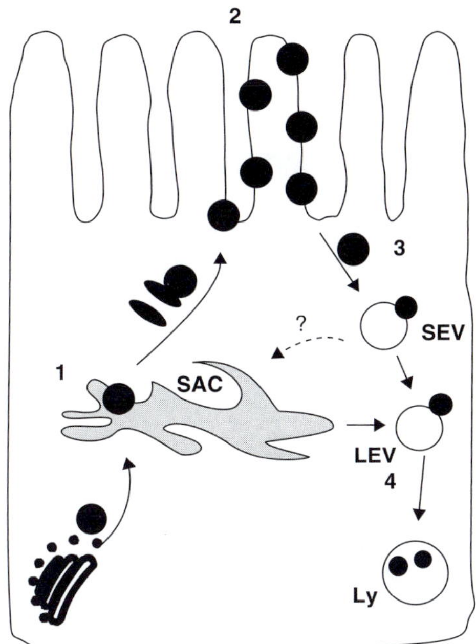

FIGURE 43.2 'Life-cycle' of NaPi-IIa cotransporter protein. After synthesis NaPi-IIa proteins are sorted to the subapical and apical compartment (1; SAC). (2) After insertion in to the apical brush border membrane, NaPi-IIa protein is stabilized as part of a large protein complex (see Figure 43.4). Regulated endocytosis via clathrin-coated vesicles initiates internalization through subapical early endosomes (SEV), late endosomes (LEV) and subsequent sorting to lysosomes (Ly), and degradation (4) completes the 'life-cycle' of NaPi-IIa. No evidence is available for recycling of NaPi-IIa after internalization

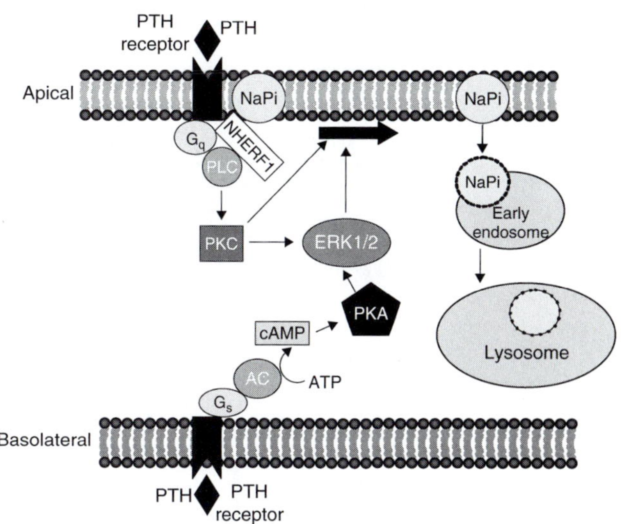

FIGURE 43.3 Regulation of NaPi-IIa cotransporter localization and activity in renal proximal tubule cells by parathyroid hormone (PTH). PTH receptors are expressed on the apical and basolateral membranes of proximal tubular cells. Apical PTH receptors couple to G_q G-proteins and phospholipase C (PLC) which all form part of an apical protein complex including also the Na^+/phosphate cotransporter NaPi-IIa and the PDZ protein NHERF1. Activation of PLC leads to protein kinase C dependent internalization of NaPi-IIa cotransporters from the apical membrane. In contrast, stimulation of basolateral PTH receptors activates protein kinase A via G_s G-proteins, adenylate cyclases (AC) and increased intracellular cAMP levels. PKC and PKA-dependent cascades converge to an ERK1/2 kinase dependent pathway

This group includes FGF-23, FGF-7, sFRP4 (secreted frizzled related protein 4), MEPE, and possibly more unidentified factors (see below) (Kumar 2002, Imel & Econs 2005). In contrast, insulin, low dietary phosphate intake, and 1,25 $(OH)_2$-vitamin D_3 increase renal phosphate reabsorption (Murer et al 2000). For a summary of regulating factors see Table 43.1.

Alterations in the total number of phosphate transporter molecules in the brush border membrane represent the main mechanism leading to changes in urinary phosphate excretion (Murer et al 2000). The cellular and molecular mechanisms by which NaPi-IIa and NaPi-IIc are regulated have been most extensively studied in the case of PTH-induced phosphaturia and during acute and chronic changes in dietary phosphate intake (for review see Miyamoto et al 2004, Murer et al 2000, 2003, 2004).

PTH-induced phosphaturia is a rapid process occurring within minutes due to a decrease in the apical phosphate transport activity. PTH receptors are localized both on the apical and basolateral membrane of proximal tubule cells and signal through different intracellular cascades (Figure 43.3) (Muff et al 1992). Apical PTH receptors are coupled via heterotrimeric G-proteins to phospholipase C (PLC). Activation of PLC (possibly PLCβ1), leads to protein kinase C dependent stimulation of ERK1/2 kinases and subsequent internalization of NaPi-IIa (Murer et al 2000, Traebert et al 2000, Bacic et al 2003). NaPi-IIa, apical PTH receptors, and PLCβ1 are organized together in a macromolecular complex via the PDZ-scaffold protein NHERF1 (NHE regulating factor 1) (Figure 43.4) (Suh et al 2001, Mahon et al 2002, Weinman et al 2005, Capuano et al 2007). Loss of NHERF1 abolishes this complex and impairs apical PTH signaling (Capuano et al 2007). In contrast, basolateral PTH receptors are linked to adenylate cyclase and protein kinase A (PKA). Also, activation of PKA leads to an ERK1/2 dependent internalization of NaPi-IIa (Bacic et al 2003). The actual signal that induces NaPi-IIa internalization has remained elusive. However, all available evidence suggests that NaPi-IIa is not directly phosphorylated but dissociates from NHERF1 which is not internalized upon PTH treatment and appears to be phosphorylated itself (Deliot et al 2005).

An acute reduction in NaPi-IIa transporter molecules is mediated by the internalization of the transporter from the

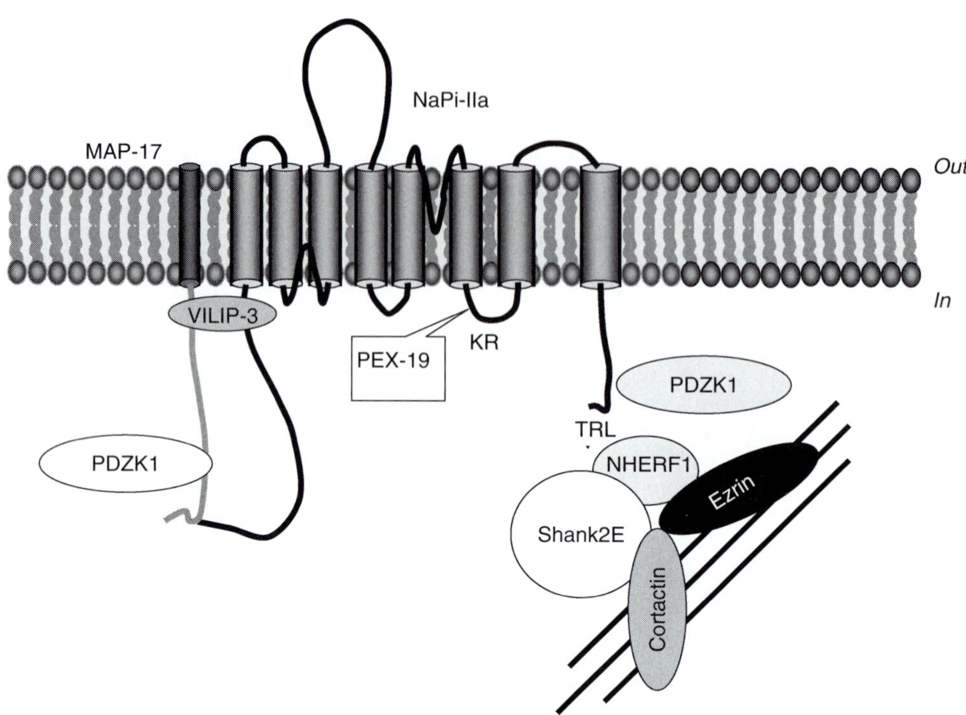

FIGURE 43.4   Protein–protein interactions of the NaPi-IIa cotransporter in renal proximal tubule cells. The NaPi-IIa cotransporter forms part of an apical and subapical protein network in the brush border membrane of proximal tubular cells. NaPi-IIa interacts via a PDZ-binding motif (TRL) at its C-terminus with the PDZ-proteins NHERF1 and PDZK1. NHERF1 may link this complex to the cytoskeleton through binding to ezrin, radixin, and moesin, also involving Shank2E. NaPi-IIa internalization upon stimulation of PTH receptors may involve the interaction of PEX-19, a peroxisomal protein, with a lysine–arginine (KR)-containing motif in the 3rd intracellular loop of NaPi-IIa. At the N-terminus, NaPi-IIa interacts directly with VILIP-3, belonging to a family of neuronal calcium-sensing proteins (Pribanic et al 2003b). The N-terminus is also associated with MAP-17, a protein of unknown function, directly and indirectly via binding of MAP-17 to the fourth domain of PDZK1 (Gisler et al 2003, Pribanic et al 2003a)

apical brush border membrane via the same route as receptor-mediated endocytosis, requiring megalin and involving clathrin-coated vesicles and early endosomes (Bacic et al 2003, 2006, Bachmana et al 2004). Consequently, NaPi-IIa is routed to lysosomes for degradation (Keusch et al 1998, Pfister et al 1998, Traebert et al 2000). The retrieval of NaPi-IIa upon stimulation with PTH depends also on a dibasic amino acid motif containing the KR (lysine-arginine) doublet in the third intracellular loop. This motif is missing in the intestinal NaPi-IIb cotransporter that is not responsive to PTH (Karim-Jimenez et al 2000) and interestingly also in NaPi-IIc. PEX 19, human peroxisomal farnesylated protein, has been suggested to interact with this motif and to participate in NaPi-IIa internalization (Ito et al 2004). Acute or chronic PTH reduces NaPi-IIa mainly on the protein level and does not affect mRNA abundance (Kempson et al 1995). The regulation of NaPi-IIc has not been studied in much detail, although a slow (as compared to NaPi-IIa) downregulation has been suggested (Miyamoto et al 2004). In the NaPi-IIa deficient mouse, NaPi-IIc is upregulated and represents the main phosphate reabsorptive mechanism (Tenenhouse et al 2003). However, NaPi-IIa

null mice are refractory to the phosphaturic action of PTH, suggesting that either NaPi-IIc is less acutely regulated by PTH or that the fact that these animals are hyperphosphaturic and hypophosphatemic induces a state of relative resistance against PTH with respect to regulation of phosphate excretion (Zhao & Tenenhouse 2000). Indeed, acute application of PTH to rats induces internalization of NaPi-IIa whereas NaPi-IIc appears to remain in the brush border membrane (Picard, Capuano, Kaissling, Biber, Murer, Wagner, unpublished results) indicating different modes of regulation for these two phosphate transporters.

Renal phosphate reabsorption matches closely dietary phosphate intake being high during dietary phosphate restriction and decreasing under a high phosphate intake (Trohler et al 1976a). Switching animals acutely from a high phosphate containing diet to a low phosphate containing diet increases NaPi-IIa and NaPi-IIc protein abundance severalfold within 4 hours (Levi et al 1994, Murer et al 2000, Segawa et al 2002, Ohkido et al 2003). In the case of NaPi-IIa, this effect does involve also an increase in NaPi-IIa mRNA abundance which, however, is much less pronounced than the increase in protein. Thus, posttranscriptional changes increasing either

NaPi-IIa translation or stability may contribute to the adaptive response (Levi et al 1994). Also, NaPi-IIc mRNA abundance is increased during dietary phosphate depletion (Segawa et al 2002, Ohkido et al 2003). In animals lacking PTH and/or 1,25 $(OH)_2$-vitamin D_3, the adaptive response of both NaPi-IIa and NaPi-IIc is partially preserved indicating that other factors may mediate this intestine–kidney cross-talk responsible for the regulation of whole body phosphate homeostasis (Trohler et al 1976b, Brautbar et al 1979, Capuano et al 2005). Recent evidence suggests that phosphatonins such as FGF-23 may be involved. Indeed, FGF-23 expression and secretion by bone follows directly dietary intake of phosphate and FGF-23 inhibits renal NaPi-IIa and NaPi-IIc expression allowing adaptation of renal phosphate excretion to dietary intake and needs of bone mineralization (Quarles 2003a, Riminucci et al 2003, Miyamoto et al 2004). Feeding animals with a high phosphate diet induces a rapid downregulation of NaPi-IIa protein expression in the brush border membrane due to internalization and subsequent lysosomal degradation of the transporter (Keusch et al 1998). In contrast, high phosphate diet induces a partial internalization of NaPi-IIc to the subapical compartment where recycling of the transporter back to the membrane may occur (Segawa et al 2005) providing another instance of differential regulation of type IIa and IIc transporters.

Brush border abundance of NaPi-IIa is closely associated with renal phosphate excretion. An increase in NaPi-IIa protein abundance leads to a decrease in phosphate excretion and vice versa (Murer et al 2000). However, a few instances have been described where NaPi-IIa brush border membrane expression and renal phosphate excretion are dissociated. During chronic potassium depletion, renal phosphate reabsorption decreases despite an increase in NaPi-IIa protein expression in the brush border membrane (Zajicek et al 2001, Inoue et al 2004). Similarly, after a PTH bolus, urinary phosphate excretion returns rapidly to low levels whereas NaPi-IIa abundance recovers only slowly (Bielesz, Capuano, Stange, Biber, Murer, Wagner, unpublished results). Regulation of NaPi-IIc may explain partially these apparently paradoxical findings. Alternatively, alterations in membrane lipid composition have been reported to change NaPi-IIa activity (Levi et al 1995, Zajicek et al 2001, Inoue et al 2004, Breusegem et al 2005).

Other Factors Regulating Renal Phosphate Handling

Other factors that increase renal tubular phosphate absorption include growth hormones (Woda et al 2004), insulin-like growth factor 1 (IGF-1) (Jehle et al 1998), insulin (Abraham et al 1990), thyroid hormones (Sorribas et al 1995, Alcalde et al 1999, Schmitt et al 2003). These factors increase NaPi-IIa abundance in the brush border membrane. In contrast, hormones such as dopamine, serotonin, the atrial natriuretic peptide (ANP), and glucocorticoids, induce phosphaturia at least in part through downregulation

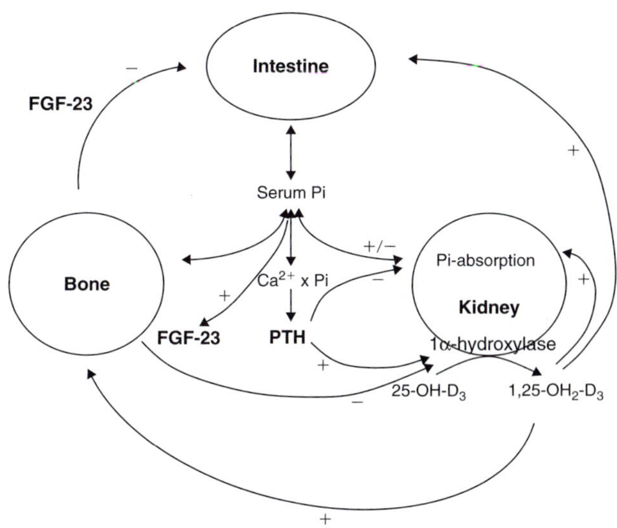

FIGURE 43.5 Hormonal regulation of serum phosphate by parathyroid hormone, vitamin D_3 and FGF-23. Serum phosphate levels are influenced by dietary intake and intestinal absorption, the rate of renal excretion or reabsorption, respectively, and skeletal deposition and release of phosphate from bone. Changes in serum phosphate concentration alter the calcium-phosphate product and a fall in free calcium triggers release of PTH. PTH inhibits renal phosphate reabsorption and at the same time stimulates 1α-hydroxylase expression and activation of 1,25-$(OH)_2$-vitamin D_3 from its inactive precursor 25-OH-vitamin-D_3. The active vitamin D_3 stimulates in turn renal and intestinal phosphate absorption as well as deposition of phosphate in bone. An increase of serum phosphate concentrations also directly increases FGF-23 synthesis and release from bone independent of the PTH axis. FGF-23 lowers phosphate concentrations by inhibiting renal and intestinal phosphate transport as well as preventing activation of 1,25-(OH_2)-vitamin D_3

of NaPi-IIa protein expression in the brush border membrane (Cuche et al 1976, Isaac et al 1992a, 1992b, Glahn et al 1993, Levi et al 1995, LeClaire et al 1998, Berndt et al 2001, Bacic et al 2005).

Role of Interacting Proteins in Regulation of Renal Phosphate Transport

NaPi-IIa has been found to interact with a number of proteins in a conventional yeast-two-hybrid screen using the COOH and NH_2-termini and the third intracellular loop (for review see Biber et al 2004, 2005, Hernando et al 2004a, 2004b, 2005). Some of the C-terminal interactions required a TRL-motif, the last three C-terminal amino acids, representing a PDZ binding motif (Gisler et al 2001). Proteins identified include the Na^+/H^+-exchanger regulatory factors 1 and 2 (NHERF-1, NHERF-2), and PDZK-1 (Gisler et al 2001) (Figures 43.4 and 43.5). A further screen with a split-ubiquitin approach using full-length NaPi-IIa protein confirmed previously identified interaction partners as well as identifying novel partners (Gisler, Fuster, Radanovic,

Hall, Engels, Stagljar, Murer, Biber, Moe, unpublished results).

The role of NHERF1 has gained much attention. Subsequent studies showed that NHERF1 stabilizes apical localization of NaPi-IIa in the renal OK cell line (Hernando et al 2002) and in mouse kidney as evident from a NHERF1 deficient mouse model (Shenolikar et al 2002). Absence of NHERF1 in mouse kidney induces mild phosphaturia (Shenolikar et al 2002). Thus, NHERF-1 may play a role in stabilizing NaPi-IIa in the brush border membrane. In addition, NHERF-1 may also play a role in the spatial organization of signaling molecules that regulate NaPi-IIa localization. The apical PTH receptor as well as phospholipase C interact with NHERF1 (Mahon et al 2002) and the absence of NHERF1 abolishes the PTH induced internalization of NaPi-IIa via apical PTH receptors (Capuano et al 2007).

The role of PDZK1 is less evident; however, a PDZK1 deficient mouse model shows reduced NaPi-IIa expression and a more pronounced phosphaturia under conditions of a high dietary phosphate intake. In addition, NHERF1 expression in the brush border membrane is increased suggesting a partial compensation for PDZK1 loss (Capuano 2005). Clearly, more work is required to clarify the role of interacting proteins in the expression and regulation of NaPi-IIa and possibly also NaPi-IIc.

PRIMARY INHERITED DEFECTS IN RENAL PHOSPHATE HANDING

SLC34A1 (NaPi-IIa)

Prie et al reported two patients presenting with urolithiasis or bone demineralization and persistent idiopathic hypophosphatemia associated with a decrease in maximal renal phosphate reabsorption and described two mutations in the gene encoding SLC34A1 (Prie et al 2002). Mutations were only single nucleotide changes resulting in missense mutations (A48F or V147M) without otherwise altering the protein structure. Surprisingly, mutations were found only on one allele per patient suggesting a dominant effect (Prie et al 2002). Expression of these mutant NaPi-IIa transporters in oocytes was interpreted as showing defective phosphate transport with reduced transport activity, a dominant negative effect on wildtype NaPi-IIa cotransporters, and reduced affinity for phosphate (Prie et al 2008). However, this study received a number of major criticisms. The nature of these putative mutations was not confirmed in other kindreds with similar symptoms, relatives of patients were not genetically investigated, and the expression assay showed only poor expression even in the case of the wild type transporter making a kinetic analysis impossible. Furthermore, mice lacking one allele for NaPi-IIa show mild phosphaturia, but not hypercalciuria or nephrolithiasis (Scheinman & Tenenhouse 2003). In contrast, Virkki et al

expressed the reported mutant NaPi-IIa cotransporters in *Xenopus* oocytes and the renal OK cell line and did not find any evidence for altered subcellular localization, level of expression, or reduced affinity (Virkki et al 2003). The lower transport activity of the mutated phosphate cotransporters was confirmed. Therefore, it was concluded that the initially reported NaPi-IIa mutations could not explain the massive phosphaturia observed in the patients, which most likely represent only polymorphisms. Therefore, the molecular basis for the renal phosphate wasting in the patients reported remains to be uncovered. A similar conclusion was derived from a study screening a large numbers of pedigrees of calcium stone formers with a low renal phosphate reabsorption rate (Lapointe et al 2006). Even though a number of mutations including a N-terminal deletion of 7 amino acids (91del7) in the NaPi-IIa gene was found, no correlation of mutations with the clinical phenotype could be established. Two of the identified mutations decreased phosphate transport activity after expression in *Xenopus* oocytes, which is most likely due to reduced membrane abundance of the mutant transporter. However, the reduction in transport rate was mild and more importantly, only one allele was found to be mutated in stone-forming patients. The interpretation that these mutations may represent rather relatively common polymorphisms was further supported by the fact that screening of a large cohort of stone-forming patients and healthy individuals detected these gene alterations also in several subjects with normal renal phosphate excretion. Thus, polymorphisms in the NaPi-IIa gene may be relatively common; patients with disease-causing mutations have yet to be identified.

SLC34A3 (NaPi-IIc): Hereditary Hypophosphatemic Rickets with Hypercalciuria (HHRH)

Recently, two groups independently identified mutations in the SLC34A3 in patients with hereditary hypophosphatemic rickets with hypercalciuria (HHRH) (OMIM #241530) (Bergwitz et al 2006, Lorenz-Depiereux et al 2006). The disease was first described in a Bedouin family (Tieder et al 1985) and is characterized by the renal wasting of phosphate, causing hypophosphatemia and secondary rickets or osteomalacia, with elevated serum alkaline phosphatase activity. Serum levels of 1,25-(OH_2)-vitamin D_3 are appropriately elevated despite suppressed PTH concentrations which result in increased intestinal calcium absorption and subsequent renal excretion (hypercalciuria), which may promote nephrolithiasis or nephrocalcinosis only in some patients (Tieder et al 1985, 1987, Bergwitz et al 2006, Lorenz-Depiereux et al 2006). In contrast, plasma levels of FGF-23 are normal in HHRH patients (Lorenz-Depiereux et al 2006) (Figure 43.6). Long-term phosphate supplementation appears to reverse the clinical and biochemical

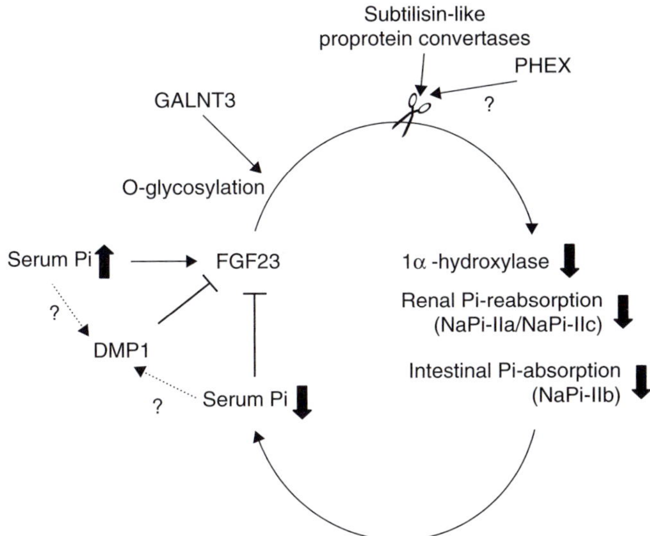

FIGURE 43.6 Regulation and function of FGF-23. FGF-23 is released from bone upon elevated serum phosphate levels. In order to be biologically active FGF-23 requires O-glycosylation by GALNT3 (UDP-N-acetyl-α-D galactosamine:polypeptide N-acetyl-galactosaminyltransferase). FGF-23 lowers serum phosphate levels by decreasing the expression of renal (NaPi-IIa and NaPi-IIc) and intestinal (NaPi-IIb) phosphate transporters thereby reducing phosphate uptake from diet and increasing renal excretion of phosphate. Furthermore, FGF-23 inhibits 1α-hydroxylase expression in kidney preventing the final step in vitamin D_3 activation preventing a compensatory increase in intestinal and renal phosphate transport. High vitamin D_3 levels itself may also directly stimulate FGF-23 synthesis providing for a regulatory circuit and feedback mechanism. FGF-23 is cleaved and inactivated by subtilisin-like proprotein convertases requiring a recognition motif consisting of RXXR at position 176. PHEX is not directly involved in FGF-23 degradation but may indirectly contribute to cleavage

abnormalities in HHRH, with the exception of decreased renal phosphate reabsorption. In contrast, supplementation with vitamin D_3 worsens hypercalciuria and increases the risk for the development of kidney stones.

The SLC34A3 gene is localized on chromosome 9q34. Mutations in the gene were found in several unrelated families including the original Bedouin families in which the disease had been first described (Bergwitz et al 2006, Lorenz-Depiereux et al 2006). Observed changes include frame shifts in the open reading frame with subsequent translation of sequences unrelated to NaPi-IIc and premature termination, missense mutations, intronic deletions, and nucleotide changes in a putative splice site (Bergwitz et al 2006, Lorenz-Depiereux et al 2006).

Inheritance follows an autosomal recessive pattern (Bergwitz et al 2006, Lorenz-Depiereux et al 2006). Some compound heterozygous patients carrying two different missense mutations were also identified (Bergwitz et al 2006). However, patients with only one allele mutated

show normal phosphate balance and excretion but borderline or elevated renal calcium excretion (Bergwitz et al 2006, Lorenz-Depiereux et al 2006).

The fact that mutations in NaPi-IIc are associated with a severe renal phosphate leak suggests that NaPi-IIc may have a more important role in human kidney function than hitherto recognized. Apparently, NaPi-IIc function remains important during adulthood in contrast to what has been described in rodents where it was postulated that NaPi-IIc plays an important role only during growth (Segawa et al 2002).

DEFECTS IN RENAL PHOSPHATE HANDLING SECONDARY TO EXTRARENAL INHERITED DEFECTS

Fibroblast Growth Factor 23 (FGF-23)

Fibroblast growth factor 23 (FGF-23) was first identified by positional cloning in patients with autosomal dominant hypophosphatemic rickets (ADHR) as described below (Consortium 2000). FGF-23 consists of 251 amino acids and represents one candidate for the long-sought phosphatonin(s), hormones regulating systemic phosphate homeostasis (for review see Kumar 2002, Quarles 2003a, 2003b, Schiavi & Kumar 2004, Imel & Econs 2005, Tenenhouse & Sabbagh 2002, 2003, 2005). The highest levels of FGF-23 expression are found in bone cells and it appears that bone represents the major source of circulating FGF-23 levels (Liu et al 2003, Riminucci et al 2003, Mirams et al 2004). However, FGF-23 is also readily detectable in the plasma of healthy persons indicating that it plays a role in physiological processes (Jonsson et al 2003). Synthesis and release of FGF-23 from bone are increased during hyperphosphatemia or high dietary intake of phosphate (Weber et al 2003, Mirams et al 2004, Perwad et al 2005).

An important role of FGF-23 in phosphate homeostasis and bone development and metabolism was demonstrated by Yamashita and colleagues establishing a *fgf-23* deficient mouse model (Shimada et al 2004). *Fgf-23* null mice showed a severe growth defect, high lethality, impaired bone development, hyperphosphatemia, inappropriately high 1,25-$(OH)_2$-vitamin D_3 levels, and suppressed PTH levels. Absence of FGF-23 increased the expression of the 1α-hydroxylase enzyme resulting in hypervitaminosis D which subsequently led to hypercalcemia in *fgf-23* deficient mice. Hyperphosphatemia was at least in part caused by abnormally high expression of NaPi-IIa in kidney. Interestingly, a low phosphate diet corrected hyperphosphatemia and increased life span of mice but failed to affect the severe bone phenotype and hypervitaminosis D, indicating that FGF-23 indeed acts as a negative regulator of vitamin D synthesis and may have direct effects in bone not linked to its role in systemic phosphate homeostasis.

FGF-23 reduced directly the expression of renal (NaPi-IIa and NaPi-IIc) and intestinal (NaPi-IIb) phosphate cotransporters (Bai et al 2003, Saito et al 2003, Miyamoto et al 2004). The time course of this inhibitory action is not fully resolved. Some authors describe an inhibition of phosphate transport in cell culture models within a few hours, whereas other authors detect changes in transport activity and protein expression only after longer time periods (Bowe et al 2001, Yamashita et al 2002). Baum et al., reported that incubation of freshly isolated proximal tubules for three hours with FGF-23R176Q protein, a mutant resistant to proteolytic degradation, decreased tubular phosphate absorption and reduced also NaPi-IIa protein abundance (Baum et al 2005). FGF-23 may act intracellularly via a MAPK kinase dependent pathway (Yamashita et al 2002) similar to previous findings on PTH-dependent inhibition of renal phosphate transport (Bacic et al 2003). It has been also proposed that renal PGE_2 synthesis is enhanced by FGF23 (Syal et al 2006) and treatment of Hyp mice (lacking PHEX) reverses partially renal phosphate wasting (Baum et al 2003).

FGF-23 binds to several receptors of the fibroblast growth factor receptor family and interacts with FGFR1-3 (Yu et al 2005, Kurosu et al 2006). Interestingly, FGF-23 binding to its receptor(s) is enhanced by the aging-suppressor protein klotho thereby indirectly linking regulation of renal phosphate excretion to aging processes (Kuro 2006, Kurosu et al 2006). Klotho is a glycoprotein with a high similarity to β-glucoronidases that is shed of from cells and its important role in the regulation of mineral homeostasis has been suggested by the fact that klotho deficient mice develop hyperphosphatemia and hypercalcemia with vascular calcification (Kuro-o et al 1997, Kuro 2006). Absence of klotho increased serum levels of 1,25-$(OH)_2$ vitamin D3 (Tsujikawa et al 2003). A direct stimulatory effect of klotho on the activity of the renal Ca^{2+}-absorptive TRPV5 channel has been recently demonstrated (Chang et al 2005). FGF-23 often requires sugar-containing molecules as cofactor in in vitro assays such as heparin or glycosaminoglycanes (Yu et al 2005, Kuro 2006, Kurosu et al 2006). Klotho binds directly to FGF23 and thereby converts the unspecific FGF receptor 1 (IIIc) into a specific FGF23 receptor (Urakawa et al 2006).

Degradation of FGF23 is achieved by C-terminal cleavage. It has been initially proposed that PHEX, an endopeptidase, inactivates FGF23 by cleavage. However, more recent in vitro data failed to demonstrate a direct cleavage of FG23 by PHEX (Benet-Pages et al 2004). In contrast, subtilisin-like proprotein convertases appear to be able to cleave FGF23 based on inhibitor and coexpression studies (Benet-Pages et al 2004). FGF23 possesses a recognition site for cleavage by subtilisin-like proprotein convertases consisting of an RXXR motif at position 176. Common mutations in this recognition site such as R176Q, R179W, and R179Q replacing the first or second arginine residue lead to autosomal dominant hypophosphatemic rickets (ADHR) due

to reduced cleavage and subsequently inappropriately high FGF23 levels (see also below) (Bai et al 2003a).

FGF-23 ACTIVATING MUTATIONS: AUTOSOMAL DOMINANT HYPOPHOSPHATEMIC RICKETS (ADHR)

Autosomal dominant hypophosphatemic rickets (OMIM #193100) is a rare but severe inborn disturbance of systemic phosphate homeostasis manifesting itself in childhood as rickets, hypophosphatemia due to renal phosphate wasting, fatigue, bone pain, and lower bone deformities in face of inappropriately low or normal vitamin D_3 levels. In adults, ADHR is characterized by osteomalacia. Positional cloning identified mutations in FGF-23 as the causative factor for ADHR. FGF-23 is located on chromosome 12p13.3 and has three exons. All mutations identified so far destroy the RXXR motif at position 176-179 in the FGF-23 protein which is necessary for cleavage and biological degradation of FGF-23 (Consortium 2000). These mutations include R176Q, R179Q, and R179W (Consortium 2000). Thus, ADHR is caused by a loss of degradation and excessive levels of FGF-23. The fact that mutant FGF-23 is responsible for ADHR was corroborated by studies in which mice were injected with plasmids encoding for normal FGF-23 or FGF-23R179Q. Mutant FGF-23 caused hypophosphatemia, hyperphosphaturia, reduced intestinal phosphate absorption and suppressed serum vitamin D_3 levels. In serum, high levels of mutant FGF-23 were detected whereas normal FGF-23 was degraded (Saito et al 2003). Similarly, implantation of normal FGF-23 and FGF-23R176Q expressing tumor cells into nude mice reproduced all features of ADHR (Bai et al 2003b). A further transgenic mouse model for FGF-23R176Q displayed typical ADHR symptoms and showed signs of secondary hyperparathyroidism (Bai et al 2004). Thus, ADHR is caused by mutations in a motif important for cleavage and degradation of FGF-23 resulting in excessive FGF-23 signaling.

LOSS OF FUNCTION FGF-23 MUTATIONS: FAMILIAL TUMORAL CALCINOSIS

Familial tumoral calcinosis (FTC, OMIM #211900) has been described as the mirror image of phosphate wasting diseases such as ADHR, TIO, and XLH. FTC is characterized by hyperphosphatemia, reduced renal phosphate excretion, ectopic calcifications, and inappropriately normal or elevated 1,25 $(OH)_2$ vitamin D_3 levels. Due to the fact that FTC exhibits features similar to symptoms observed in mice deficient for FGF-23 (*fgf-23* null) (Shimada et al 2004) it had been speculated that loss of FGF-23 may be responsible for FTC. However, *fgf-23* deficient mice show, in addition to the symptoms observed in FTC patients, also a severe growth defect, high lethality, and skeletal abnormalities but lack the ectopic calcifications (Shimada et al 2004). Benet-Pages et al. identified in a subset of FTC patients a homozygous 211A-G transition in the FGF23 gene, resulting in the substitution at an evolutionarily conserved serine residue to glycine (S71G) (Benet-Pages et al 2005). This mutation may cause disease due to loss of function (Benet-Pages et al 2005). Patients have abnormally low FGF-23 plasma concentrations. Mutant FGF-23 protein expressed in vitro, was not secreted and was not retained intracellularly (Benet-Pages et al 2005). As discussed below, mutations in *GALNT3*, an enzyme involved in the O-glycosylation of FGF-23 are also responsible for FTC in other patients.

Fibrous Dysplasia (McCune-Albright Syndrome)

Fibrous dysplasia (FD, OMIM #174800) is caused by non-inherited genetic somatic activating missense mutations in the α-subunit of the stimulatory G protein, G_s, *GNAS1* located on chromosome 20q13.2 (Weinstein et al 1991, Schwindinger et al 1992). FD is characterized by abnormalities in bone (monostotic or polyostotic fibrous dysplasia), in skin (pigmentation), and in the endocrine system (thyrotoxicosis, pituitary gigantism, and Cushing syndrome). The degree of disease and particularly the association with skin and endocrine symptoms show a wide variability. Renal phosphate wasting is detected in about 50% of patients (Ryan et al 1968, Dent & Gertner 1976, Dachille et al 1990).

Riminucci et al investigated the cause of renal wasting in patients with FD and found that serum FGF-23 levels were tightly correlated with disease burden and renal phosphate loss. Elevated FGF-23 levels are thus caused by a large mass of FGF-23 producing cells in the fibrous bone lesions. FGF-23 production in cells isolated from the fibrous lesion was similar to normal bone cells. Hence, abnormal high FGF-23 levels are caused by an increase in FGF-23 producing cells but not by abnormal production of FGF-23 per se (Riminucci et al 2003, Kobayashi et al 2006).

Dentin Matrix Protein 1 (DMP1)

Patients with mutations in DMP1 suffer from hypophosphatemia due to renal phosphate wasting (hyperphosphaturia) with inappropriately normal levels of 1,25-(OH)$_2$ vitamin D$_3$ (Feng et al 2006, Lorenz-Depiereux et al 2006). DMP1 belongs to the large SIBLING family (small integrin-binding ligand, N-linked glycoproteins) of extracellular matrix proteins and is mainly co-expressed with FGF23 in bone. The similar phenotype of patients with X-linked hypophosphatemia (XLH) and autosomal dominant hypophosphatemia with rickets (ADHR) suggested a link to FGF23. Indeed, in patients with loss of intact DMP1, FGF23 levels in serum were high normal or elevated (Feng et al 2006, Lorenz-Depiereux et al 2006). The hypophosphatemia in DMP1 deficient patients causes severe rickets in children or osteomalacia in adults (Feng et al 2006, Lorenz-Depiereux et al 2006) which is characterized by abnormal amounts of osteoid indicating defective mineralization and maturation of

bone (Feng et al 2006). Similarly a mouse model deficient for DMP1 shows hypophosphatemia, hyperphosphaturia, rickets, retarded skeletal growth with abnormal mineralization, disturbed lacunocanicular organization, and defective osteoblast to osteocyte differentiation. This phenotype is only partly rescued by a high phosphate diet suggesting that DMP1 may have an additional role in osteoblast differentiation independent from regulating systemic phosphate homeostasis (Feng et al 2006).

Tumor-Induced Osteomalacia (TIO)

Tumor-induced osteomalacia (TIO) is a rare paraneoplastic syndrome mostly associated with mesenchymal tumors releasing (a) phosphaturic factor(s). Symptoms include renal phosphate wasting causing hypophosphatemia, osteomalacia, and abnormal vitamin D metabolism (Imel & Econs 2005). Surgical removal of the tumor reverses symptoms. In contrast to syndromes of hyperparathyroidism or humoral hypercalcemia of malignancy, the plasma concentrations of calcium, parathyroid hormone, and parathyroid hormone related protein (PTHrp) are in the normal range. Thus, it has been long postulated that one or several phosphaturic factor(s) affecting renal handling of phosphate, also named phosphatonin, are produced and released from TIO-causing tumors. Several proteins, such as FGF-23, sFRP4, FGF-7, and MEPE, have been identified that are produced and secreted from tumors from patients with TIO. Some of these proteins have been shown to regulate phosphate transport in vitro and/or in vivo. However, phosphaturic factors with a physiological role in phosphate homeostasis as demonstrated for FGF-23, should be distinguished from other phosphaturic factors that may be mainly involved in the pathological derangement of renal phosphate handling. The term 'phosphatonin' is more appropriate for the first group of phosphaturic factors and implies a physiological regulator of systemic phosphate homeostasis. However, for factors other than FGF-23 the physiological role of sFRP4, FGF-7, and MEPE, remains to be established.

Secreted Frizzled-Related Protein-4 (sFRP-4)

Secreted frizzled-related protein-4 (sFRP-4) was initially cloned from uterine adenocarcinoma and encodes a secreted 346 amino acids containing protein of approximately 40 kDa (Abu-Jawdeh et al 1999, Kumar 2002). Later, sFRP-4 was identified by SAGE analysis to be highly upregulated in tumor tissue from patients with renal phosphate wasting (Berndt et al 2003). sFRP-4 inhibits phosphate transport in the renal OK cell line as well as in vivo. Infusion of sFRP4 into intact mice or rats as well as into parathyroidectomized rats elicits phosphaturia caused by down-regulation and internalization of the renal Na$^+$/phosphate cotransporter NaPi-IIa (Berndt et al 2003, 2006). Furthermore, long-term infusion of sFRP-4 into normal rats increased renal phosphate excretion and reduced plasma phosphate. Levels of 1,25 (OH)$_2$ vitamin D$_3$ and of enzymes involved in calcitriol production were inappropriately low indicating that sFRP-4 acts independently from PTH and does not alter vitamin D_3 metabolism. Members of the frizzled protein family act on the Wnt-signaling pathway and sFRP4 was shown to induce phosphorylation of β-catenin that may mediate further down-stream events such as internalization of the NaPi-IIa cotransporter (Berndt et al 2003, 2006). Signals regulating sFRP-4 production and release, its physiological function, and its role in tumor induced osteomalacia, need to be further examined.

Fibroblast Growth Factor 7 (FGF-7)

Fibroblast growth factor 7 (FGF-7), also referred to as keratinocyte growth factor (KGF), is a protein of about 28 kDa and was also identified to be overexpressed in mesenchymal tumors from TIO patients (Carpenter et al 2005). FGF-7 inhibits phosphate transport activity in the renal OK cell line; however, the mechanism is not known presently (Carpenter et al 2005).

Matrix Extracellular Phospho-Glycoprotein (MEPE)

MEPE, a glycosylated protein of about 60 kDa, was initially cloned from tumor tissue obtained from a patient with oncogenic hypophosphatemia (OHO) (Rowe et al 2000, Schiavi & Kumar 2004). Bone cells (osteoblasts, osteocytes and odontoblasts) are the major source of MEPE. MEPE, like DMP1, is a member of the extracellular matrix short integrin-binding ligand interacting glycoprotein (SIBLING) family involved in bone regulation. The question whether MEPE represents a phosphaturic factor has not been fully resolved. In one study, MEPE did not inhibit phosphate transport in in-vitro experiments and failed to induce renal phosphate excretion in mice (Bowe et al 2001, Shimada et al 2001). In contrast, a second study reported that injection of MEPE into intact mice results in hypophosphatemia, hyperphosphaturia and mild increases in circulating 1,25 $(OH)_2$ vitamin D_3 levels (Rowe et al 2004). Moreover, implantation of MEPE-producing CHO cells into nude mice caused renal phosphate wasting, whereas MEPE deficient mice have higher bone density (Gowen et al 2003). However, ablation of MEPE in the Hyp mouse lacking the Phex gene did not affect the degree of phosphaturia (Liu et al 2005). The interactions between MEPE and other hormones regulating phosphate homeostasis and handling require more investigation and may explain some of the apparent discrepancies in the reported effects of MEPE. Also, the molecular mechanism by which MEPE causes phosphaturia remains to be elucidated.

PHEX: X-linked Hypophosphatemic Rickets (XLH)

X-linked hypophosphatemic rickets (XLH)(OMIM #307800) is the most common heritable form of rickets with a prevalence of about 1 in 20 000. The disease is characterized by hypophosphatemia due to excessive renal phosphate

wasting leading to rickets, lower extremities deformity, short stature, bone and joint pain, enthesopathy, and dental abscesses. Vitamin D_3 levels are inappropriately normal or even low (Consortium 1995). The disorder is inherited in a dominant manner and the pattern of inheritance and linkage analysis provided evidence for a localization of the underlying gene on chromosome Xp22. Positional cloning identified the PHEX gene (PHosphate regulating gene with homology to Endopeptidases on the X chromosome) in XLH patients (Consortium 1995). Several mouse models of XLH had been available: Hyp, Gy, and Ska1 mice resembling XLH and which were later shown to have mutations in the mouse PHEX homolog (Eicher et al 1976, Strom et al 1997, Carpinelli et al 2002). Early cross-transplant and parabiosis experiments with these mouse models had shown that the defect was extrarenal and due to a circulating factor (Meyer et al 1989a, 1989b, Nesbitt et al 1992). The X-linked Hyp mouse model demonstrated a defect in proximal tubular phosphate absorption, decreased expression of NaPi-IIa protein and mRNA, and also lower NaPi-IIc abundance (Tenenhouse & Sabbagh 2002, Tenenhouse et al 1994, 2003). The adaptive response to dietary phosphate deprivation is conserved (Tenenhouse et al 1995). Serum FGF-23 levels are highly elevated in XLH patients (Jonsson et al 2003) and most likely explain the observed disease. Crossing of hypophosphatemic Hyp mice with hyperphosphatemic *fgf-23* deficient mice produced hyperphosphatemic mice that showed exactly the same phenotype as *fgf-23* null mice, indicating that both mutations affect the same system and that FGF-23 may act downstream of PHEX (Sitara et al 2004, Liu et al 2006). Furthermore, injection of neutralizing anti-FGF-23 antibodies into Hyp may be able to reduce symptoms (Aono et al 2003). However, it is still unclear how mutations in PHEX cause higher FGF-23 levels, as PHEX seems not to mediate direct cleaving of FGF-23 (Liu et al 2003, Benet-Pages et al 2004). Observations in Hyp and *fgf-23* null mice indicate that PHEX and FGF-23 may regulate each other's expression levels and that loss of PHEX may lead to higher expression levels of FGF-23 (Liu et al 2003, 2006, Perwad et al 2005). The biological role of PHEX will require further clarification to understand the pathogenesis of XLH in more detail.

UDP-N-acetyl-α-D Galactosamine:polypeptide N-acetyl-α-D Galactosaminyltransferase (GALNT3): Familial Tumoral Calcinosis

Homozygous loss of function mutations of the UDP-N-acetyl-α-D galactosamine:polypeptide N-acetyl-galactosaminyltransferase (*GALNT3*), a glycosyltransferase involved in mucin-type O-glycosylation, underlie also familial tumoral calcinosis (FTC, OMIM #211900) (Topaz et al 2004, Specktor et al 2006). However, patients carrying only one mutated allele appear to have also mild symptoms, leading to the initial description of FTC as an autosomal dominant disease (Ichikawa et al 2005). Because inactivating mutations in FGF-23 also cause FTC and FGF-23 has some predicted

O-glycosylation sites it has been speculated that GALNT3 is critical for FGF-23 gylcosylation and full function. Recently, Kato et al described that in vitro secretion of FGF-23 from CHO cells deficient in O-glycosylation was impaired and that cotransfection of GALNT3 markedly increased O-glycosylation and secretion of FGF-23. Thus, GALNT3 may play an important role in FGF-23 secretion by mediating its O-glycosylation (Kato et al 2006). This function may also explain how loss of function mutations in either FGF-23 or GALNT3 can produce the same disease, familial tumoral calcinosis.

Fibroblast Growth Factor Receptor 1 (FGFR1): Osteoglophonic Dysplasia

Osteoglophonic dysplasia (OMIM #166250) is caused by mutations in the fibroblast growth factor receptor 1 (FGFR1) that result in an activation of the receptor (White et al 2005). This type of disease is characterized by craniosynostosis, prominent supraorbital ridge, and depressed nasal bridge, as well as the rhizomelic dwarfism and nonossifying bone lesions. White et al identified a number of missense mutations in the FGFR1 gene in four patients. Three patients were hypophosphatemic due to massive renal phosphate wasting (White et al 2005). One patient had inappropriately high FGF23 levels, two patients had high 1,25-$(OH)_2$ vitamin D_3 levels. As two related patients carried the Y372C FGFR1 mutation, the activity of the mutant receptor was tested and found to be highly increased suggesting that overactivity of the FGF1 receptor is responsible for this disease (White et al 2005). The role of the FGF1 receptor in the regulation of renal phosphate handling is not fully understood. However, the recent finding that the aging-related glycoprotein klotho dramatically enhances the affinity and stimulation of the FGF receptors 1, 3, and 4 by FGF23 may suggest that FGF1 represents one of the physiological receptors of FGF23 (Kuro 2006, Kurosu et al 2006).

Hypophosphatemic Bone Disease

Hypophosphatemic bone disease (OMIM 146350) was described in five patients and resembles mild ADHR with hypophosphatemia, modest shortening of stature, bowing of the lower limbs, nonrachitic bone changes, and renal phosphate wasting (Scriver et al 1977). The underlying gene defect has not been identified up to date but mutations in FGF-23 have been excluded (Consortium 2000). Apparently, the disease responds to treatment with vitamin D_3 (Scriver et al 1981).

CONCLUSION AND OUTLOOK

Renal and extrarenal control of systemic phosphate homeostasis requires a complicated network of regulatory factors, phosphate transporters in kidney, intestine, and bone, and intracellular protein–protein interactions. Human and mouse genetics have greatly contributed to the identification of these factors and proteins and elucidating their physiological role. Known mutations cannot be identified in all patients with inherited forms of renal phosphate wasting and more genes may be identified in some of these patients (Table 43.2). In addition, we are only starting to understand the role of FGF-23 and other novel phosphaturic factors and have very little knowledge about the function of PHEX. Also, the role of NaPi-IIc in human kidney function and its overall regulation

TABLE 43.2   Genes/diseases causing altered renal phosphate handling

Gene	Chromosome/ OMIM	Disease name	Symptoms	Inheritance	Tissue distribution of protein	Comments
SLC34A1 (NaPi-IIa)	5q35		Hypophosphatemia, urolithiasis, osteoporosis?	Autosomal dominant or recessive	Kidney (proximal tubule, brain	Relevance not clarified
SLC34A3 (NaPi-IIc)	9q34 OMIM #241530	Hereditary hypophosphatemic rickets with hypercalciuria (HHRH)	Hypophosphatemia, hyperphosphaturia, rickets, hypercalciuria, normal-low FGF23, high 1,25-dihydroxyvitamin D	Autosomal recessive	Kidney (proximal tubule)	
PHEX	Xp22.2-p22.1 OMIM #307800	X-linked hypophosphatemia (XLH)	Hyperphosphaturia Hypophosphatemia Rickets/osteomalacia Normal vit. D, high FGF23	X-linked	Mainly bone	

TABLE 43.2   Continued

Gene	Chromosome/ OMIM	Disease name	Symptoms	Inheritance	Tissue distribution of protein	Comments
FGF23	12p13.3 OMIM #193100 OMIM #211900	– Autosomal dominant hypophosphatemic rickets (ADHR) – Familial tumoral calcinosis (FTC) – tumor induced osteomalacia (TIO)	Hyperphosphaturia Hypophosphatemia Rickets/osteomalacia Normal vit. D_3 High FGF23	Autosomal dominant	Mainly bone	
GALNT3	2q24-q31 OMIM #211900	Familial tumoral calcinosis (FTC)	Hyperphosphatemia, heterotopic calcification	Autosomal recessive	Pancreas, testis	
GNAS1	20q13.2 OMIM #174800	Fibrous dysplasia (McCune-Albright syndrome)	Hyperphosphaturia, high FGF-23 levels, polyostotic fibrous dysplasia, pigmentation, thyrotoxicosis, Cushing	Autosomal dominant, mosaicism	Ubiquitous	
sFRP4	7p14-p13 OMIM 606570	Tumor induced osteomalacia (TIO)	Hypophosphatemia Hyperphosphaturia Osteomalacia	Spontaneous?	Tumor, bone	
FGF7	15q15-q21.1 OMIM 148180	Tumor induced osteomalacia (TIO)	Hypophosphatemia Hyperphosphaturia Osteomalacia	Spontaneous?	Tumor	
MEPE	4q21.1 OMIM 605912	Tumor induced osteomalacia (TIO)	Hypophosphatemia Hyperphosphaturia Osteomalacia	Spontaneous?	Tumor, bone	
FGFR1	8p11.2-p11.1 OMIM #166250	Osteoglophonic dysplasia	Hypophosphatemia Hyperphosphaturia Craniosynostosis	Autosomal dominant?	Ubiquitous, bone, kidney	
DMP1	4q21 OMIM #241520	Autosomal recessive hypophosphatemic rickets (ARHR) Autosomal recessive hypophosphatemia (ARHP)	Hypophosphatemia Hyperphosphaturia Rickets, osteomalacia	Autosomal recessive	Ubiquitous, Bone, heart, kidney	
?	? OMIM 146350	Hypophosphatemic bone disease (HBD)	Hypophosphatemia, modest shortening of stature, bowing of the lower limbs, nonrachitic bone changes, renal phosphate wasting	Autosomal dominant?		

needs further clarification. Another emerging field is the role of dynamic protein–protein interactions in the regulation of renal phosphate cotransporters. Last but not least, identification of protein structures of the transporter molecules and their associated proteins will aid in better understanding the function of these proteins and the impact of mutations.

ACKNOWLEDGMENTS

Work in the laboratories of the authors has been supported by the Swiss National Science Foundation (SNF), the European Union (EuReGene), the Hartmann Müller Foundation, and the Gebert Rüf Foundation.

References

Abraham MI, McAteer JA, Kempson SA. Insulin stimulates phosphate transport in opossum kidney epithelial cells. Am. J. Physiol. 1990; 258: F1592–8.

Abu-Jawdeh G, Comella N, Tomita Y, et al. Differential expression of frpHE: a novel human stromal protein of the secreted frizzled gene family, during the endometrial cycle and malignancy. Lab. Invest. 1999; 79: 439–47.

Alcalde AI, Sarasa M, Raldua D, et al. Role of thyroid hormone in regulation of renal phosphate transport in young and aged rats. Endocrinology 1999; 140: 1544–51.

Allon M, Parris M. Calcitriol stimulates Na(+)-Pi cotransport in a subclone of opossum kidney cells (OK-7A) by a genomic mechanism. Am. J. Physiol. 1993; 264: F404–10.

Ambuhl PM, Zajicek HK, Wang H, Puttaparthi K, Levi M. Regulation of renal phosphate transport by acute and chronic metabolic acidosis in the rat. Kidney Int. 1998; 53: 1288–98.

Amstutz M, Mohrmann M, Gmaj P, Murer H. Effect of pH on phosphate transport in rat renal brush border membrane vesicles. Am. J. Physiol. 1985; 248: F705–10.

Aono Y, Shimada T, Yamazaki Y, et al. The neutralization of FGF-23 ameliorates hypophosphatemia and rickets in Hyp mice (Abstract). J. Bone Miner. Res. 2003; 18: S16.

Bacconi A, Virkki LV, Biber J, Murer H, Forster IC. Renouncing electroneutrality is not free of charge: switching on electrogenicity in a Na$^+$-coupled phosphate cotransporter. Proc. Natl Acad. Sci. USA 2005; 102: 12606–11.

Bachmann S, Schlichting U, Geist B, et al. Kidney-specific inactivation of the megalin gene impairs trafficking of renal inorganic sodium phosphate cotransporter (NaPi-IIa). J. Am. Soc. Nephrol. 2004; 15: 892–900.

Bacic D, Capuano P, Baum M, et al. Activation of dopamine D1-like receptors induces acute internalization of the renal Na$^+$/phosphate cotransporter NaPi-IIa in mouse kidney and OK cells. Am. J. Physiol. Renal Physiol. 2005; 288: F740–7.

Bacic D, Capuano P, Gisler SM, et al. Impaired PTH-induced endocytotic down-regulation of the renal type IIa Na$^+$/P$_i$-cotransporter in RAP deficient mice with reduced megalin expression. Pflügers Arch. 2003a; 446: 475–84.

Bacic D, Lehir M, Biber J, Kaissling B, Murer H, Wagner CA. The renal Na$^+$/phosphate cotransporter NaPi-IIa is internalized via the receptor-mediated endocytic route in response to parathyroid hormone. Kidney Int. 2006; 69: 495–503.

Bacic D, Schulz N, Biber J, Kaissling B, Murer H, Wagner CA. Involvement of the MAPK-kinase pathway in the PTH mediated regulation of the proximal tubule type IIa Na$^+$/P$_i$ cotransporter in mouse kidney. Pflügers Arch. 2003b; 446: 52–60.

Bai X, Miao D, Li J, Goltzman D, Karaplis AC. Transgenic mice overexpressing human fibroblast growth factor 23 (R176Q) delineate a putative role for parathyroid hormone in renal phosphate wasting disorders. Endocrinology 2004; 145: 5269–79.

Bai XY, Miao D, Goltzman D, Karaplis AC. The autosomal dominant hypophosphatemic rickets R176Q mutation in fibroblast growth factor 23 resists proteolytic cleavage and enhances in vivo biological potency. J. Biol. Chem. 2003a; 278: 9843–9.

Bai XY, Miao D, Goltzman D, Karaplis AC. The autosomal dominant hypophosphatemic rickets R176Q mutation in fibroblast growth factor 23 resists proteolytic cleavage and enhances in vivo biological potency. J. Biol. Chem. 2003b; 278: 9843–9.

Baum M, Loleh S, Saini N, Seikaly M, Dwarakanath V, Quigley R. Correction of proximal tubule phosphate transport defect in Hyp mice in vivo and in vitro with indomethacin. Proc. Natl Acad. Sci. USA 2003; 100: 11098–103.

Baum M, Schiavi S, Dwarakanath V, Quigley R. Effect of fibroblast growth factor-23 on phosphate transport in proximal tubules. Kidney Int. 2005; 68: 1148–53.

Baumann K, De Rouffignac C, Roinel N, Rumrich G, Ullrich KJ. Renal phosphate transport: inhomogeneity of local proximal transport rates and sodium dependence. Pflugers Arch. 1975; 356: 287–97.

Beck L, Karaplis AC, Amizuka N, Hewson AS, Ozawa H, Tenenhouse HS. Targeted inactivation of Npt2 in mice leads to severe renal phosphate wasting, hypercalciuria, and skeletal abnormalities. Proc. Natl Acad. Sci. USA 1998; 95: 5372–7.

Benet-Pages A, Lorenz-Depiereux B, Zischka H, White KE, Econs MJ, Strom TM. FGF23 is processed by proprotein convertases but not by PHEX. Bone 2004; 35: 455–62.

Benet-Pages A, Orlik P, Strom TM, Lorenz-Depiereux B. An FGF23 missense mutation causes familial tumoral calcinosis with hyperphosphatemia. Hum. Mol. Genet. 2005; 14: 385–90.

Bergwitz C, Roslin NM, Tieder M, et al. SLC34A3 mutations in patients with hereditary hypophosphatemic rickets with hypercalciuria predict a key role for the sodium-phosphate cotransporter NaP(i)-IIc in maintaining phosphate homeostasis. Am. J. Hum. Genet. 2006; 78: 179–92.

Berndt T, Craig TA, Bowe AE, et al. Secreted frizzled-related protein 4 is a potent tumor-derived phosphaturic agent. J. Clin. Invest. 2003; 112: 785–94.

Berndt TJ, Bielesz B, Craig TA, et al. Secreted frizzled-related protein-4 reduces sodium-phosphate co-transporter abundance and activity in proximal tubule cells. Pflugers Arch. 2006; 451: 579–87.

Berndt TJ, Liang M, Tyce GM, Knox FG. Intrarenal serotonin, dopamine, and phosphate handling in remnant kidneys. Kidney Int. 2001; 59: 625–30.

Biber J, Forgo J, Murer H. Modulation of Na/Pi-cotransport in opossum kidney cells by extracellular phosphate. Am. J. Physiol. Cell Physiol. 1988; 255: C155–61.

Biber J, Gisler SM, Hernando N, Murer H. Protein/protein interactions (PDZ) in proximal tubules. J. Membr. Biol. 2005; 203: 111–18.

Biber J, Gisler SM, Hernando N, Wagner CA, Murer H. PDZ interactions and proximal tubular phosphate reabsorption. Am. J. Physiol. Renal Physiol. 2004; 287: F871–5.

Bowe AE, Finnegan R, Jan de Beur SM, et al. FGF-23 inhibits renal tubular phosphate transport and is a PHEX substrate. Biochem. Biophys. Res. Commun. 2001; 284: 977–81.

Brautbar N, Walling MW, Coburn JW. Interactions between vitamin D deficiency and phosphorus depletion in the rat. J. Clin. Invest. 1979; 63: 335–41.

Breusegem SY, Halaihel N, Inoue M, et al. Acute and chronic changes in cholesterol modulate Na-P$_i$ cotransport activity in OK cells. Am. J. Physiol. Renal Physiol. 2005; 289: F154–65.

Brunette MG, Chan M, Maag U, Beliveau R. Phosphate uptake by superficial and deep nephron brush border membranes. Effect of the dietary phosphate and parathyroid hormone. Pflügers Arch. 1984; 400: 356–62.

Busch AE, Wagner CA, Schuster A, et al. Properties of electrogenic P_i transport by a human renal brush border Na^+/P_i transporter. J. Am. Soc. Nephrol. 1995; 6: 1547–51.

Capuano P, Bacic D, Roos M, et al. Defective coupling of apical PTH – receptors to phospholipase C prevents internalization of the Na^+/phosphate cotransporter NaPi-IIa in NHERF1 deficient mice. Am. J. Physiol. Cell. Physiol. 2007; 292: C927–34.

Capuano P, Bacic D, Stange G, et al. Expression and regulation of the renal Na/phosphate cotransporter NaPi-IIa in a mouse model deficient for the PDZ protein PDZK1. Pflugers Arch. 2005a; 449: 392–402.

Capuano P, Radanovic T, Wagner CA, et al. Intestinal and renal adaptation to a low-Pi diet of type II NaPi cotransporters in vitamin D receptor- and 1alphaOHase-deficient mice. Am. J. Physiol. Cell Physiol. 2005b; 288: C429–34.

Carpenter TO, Ellis BK, Insogna KL, Philbrick WM, Sterpka J, Shimkets R. Fibroblast growth factor 7: an inhibitor of phosphate transport derived from oncogenic osteomalacia-causing tumors. J. Clin. Endocrinol. Metab. 2005; 90: 1012–20.

Carpinelli MR, Wicks IP, Sims NA, et al. An ethyl-nitrosourea-induced point mutation in phex causes exon skipping, x-linked hypophosphatemia, and rickets. Am. J. Pathol. 2002; 161: 1925–33.

Chang Q, Hoefs S, van der Kemp AW, Topala CN, Bindels RJ, Hoenderop JG. The beta-glucuronidase klotho hydrolyzes and activates the TRPV5 channel. Science 2005; 310: 490–3.

Cole JA. Parathyroid hormone activates mitogen-activated protein kinase in opossum kidney cells. Endocrinology 1999; 140: 5771–9.

Collins JF, Bai L, Ghishan FK. The SLC20 family of proteins: dual functions as sodium-phosphate cotransporters and viral receptors. Pflugers Arch. 2004; 447: 647–52.

Consortium TA. Autosomal dominant hypophosphataemic rickets is associated with mutations in FGF23. Nat. Genet. 2000; 26: 345–8.

Consortium TH. A gene (PEX) with homologies to endopeptidases is mutated in patients with X-linked hypophosphatemic rickets. Nat. Genet. 1995; 11: 130–6.

Cuche JL, Marchand GR, Greger RF, Lang RC, Knox FG. Phosphaturic effect of dopamine in dogs. Possible role of intrarenally produced dopamine in phosphate regulation. J. Clin. Invest. 1976; 58: 71–6.

Custer M, Lötscher M, Biber J, Murer H, Kaissling B. Expression of $Na-P_i$ cotransport in rat kidney: localization by RT-PCR and immunohistochemistry. Am. J. Physiol. 1994; 266: F767–74.

Dachille RD, Goldberg JS, Wexler ID, Shons AR. Fibrous dysplasia-induced hypocalcemia/rickets. J. Oral Maxillofac. Surg. 1990; 48: 1319–22.

de la Horra C, Hernando N, Lambert G, Forster I, Biber J, Murer H. Molecular determinants of pH sensitivity of the type IIa Na/P_i cotransporter. J. Biol. Chem. 2000; 275: 6284–7.

Deliot N, Hernando N, Horst-Liu Z, et al. PTH treatment induces dissociation of NaPi-IIa/NHERF1 complexes. Am. J. Physiol. Cell Physiol. 2005; 289: C159–67.

Dent CE, Gertner JM. Hypophosphataemic osteomalacia in fibrous dysplasia. Quart. J. Med. 1976; 45: 411–20.

Eicher EM, Southard JL, Scriver CR, Glorieux FH. Hypophosphatemia: mouse model for human familial hypophosphatemic (vitamin D-resistant) rickets. Proc. Natl Acad. Sci. USA 1976; 73: 4667–71.

Evers C, Murer H, Kinne R. Effect of parathyrin on the transport properties of isolated renal brush-border vesicles. Biochem. J. 1978; 172: 49–56.

Feng JQ, Ward LM, Liu S, et al. Loss of DMP1 causes rickets and osteomalacia and identifies a role for osteocytes in mineral metabolism. Nat. Genet. 2006; 38: 1310–15.

Forster IC, Biber J, Murer H. Proton-sensitive transitions of renal type II Na^+-coupled phosphate cotransporter kinetics. Biophys. J. 2000; 79: 215–30.

Freeman D, Bartlett S, Radda G, Ross B. Energetics of sodium transport in the kidney: saturation transfer ^{31}P-NMR. Biochim. Biophys. Acta 1983; 762: 325–36.

Gisler SM, Pribanic S, Bacic D, et al. PDZK1: I. a major scaffolder in brush borders of proximal tubular cells. Kidney Int. 2003; 64: 1733–45.

Gisler SM, Stagljar I, Traebert M, Bacic D, Biber J, Murer H. Interaction of the type IIa Na/P_i cotransporter with PDZ proteins. J. Biol. Chem. 2001; 276: 9206–13.

Glahn RP, Onsgard MJ, Tyce GM, Chinnow SL, Knox FG, Dousa TP. Autocrine/paracrine regulation of renal Na^+-phosphate cotransport by dopamine. Am. J. Physiol. 1993; 264: F618–22.

Gowen LC, Petersen DN, Mansolf AL, et al. Targeted disruption of the osteoblast/osteocyte factor 45 gene (OF45) results in increased bone formation and bone mass. J. Biol. Chem. 2003; 278: 1998–2007.

Greger R, Lang F, Marchand G, Knox FG. Site of renal phosphate reabsorption. Micropuncture and microinfusion study. Pflugers Arch. 1977; 369: 111–18.

Haas JA, Bernt T, Knox FG. Nephron heterogeneity of phosphate reabsorption. Am. J. Physiol. Renal Physiol. 1978; 234: F287–90.

Hammerman MR, Karl IE, Hruska KA. Regulation of canine renal vesicle Pi transport by growth hormone and parathyroid hormone. Biochim. Biophys. Acta 1980; 603: 322–35.

Haramati A. Tubular capacity for phosphate reabsorption in superficial and deep nephrons. Am. J. Physiol. Renal Physiol. 1985; 248: F729–33.

Haramati A, Nienhuis D. Renal handling of phosphate during acute respiratory acidosis and alkalosis in rat. Am. J. Physiol. Renal Physiol. 1984; 247: F596–601.

Hartmann CM, Hewson AS, Kos CH, et al. Structure of murine and human renal type II $Na+$-phosphate cotransporter genes (Npt2 and NPT2). Proc. Natl Acad. Sci. USA 1996; 93: 7409–14.

Hernando H, Wagner CA, Gisler SM, Biber J, Murer H. PDZ proteins and proximal ion transport. Curr. Opin. Nephrol. Hypertens. 2004a; 13: 569–54.

Hernando N, Deliot N, Gisler SM, et al. PDZ-domain interactions and apical expression of type IIa Na/Pi cotransporters. Proc. Natl Acad. Sci. USA 2002; 99: 11957–62.

Hernando N, Gisler SM, Pribanic S, et al. NaPi-IIa and interacting partners. J. Physiol. 2005; 567: 21–6.

Hernando N, Wagner CA, Gisler SM, Biber J, Murer H. PDZ proteins and proximal ion transport. Curr. Opin. Nephrol. Hypertens. 2004b; 13: 569–74.

Hilfiker H, Hartmann C, Stange G, Murer H. Characterization of the 5'-flanking region of the OK-cell type II Na/P_i cotransport gene. Am. J. Physiol. Renal Physiol. 1998a; 274: F197–204.

Hilfiker H, Hattenhauer O, Traebert M, Forster I, Murer H, Biber J. Characterization of a murine type II sodium-phosphate cotransporter expressed in mammalian small intestine. Proc. Natl Acad. Sci. USA. 1998c; 95: 14564–9.

Hilfiker H, Kvietikova I, Hartmann CM, Stange G, Murer H. Characterization of the human type II Na/P$_i$ cotransporter promotor. Pflugers Arch. 1998b; 436: 591–8.

Hoppe A, Metler M, Berndt TJ. Efefct of respiratory alkalosis on renal phosphate excretion. Am. J. Physiol. Renal Physiol. 1982; 243: F471–5.

Hruska KA, Hammerman MR. Parathyroid hormone inhibition of phosphate transport in renal brush border vesicles from phosphate-depleted dogs. Biochim. Biophys. Acta 1981; 645: 351–6.

Ichikawa S, Lyles KW, Econs MJ. A novel GALNT3 mutation in a pseudo-autosomal dominant form of tumoral calcinosis: evidence that the disorder is autosomal recessive. J. Clin. Endocrinol. Metab. 2005; 90: 2420–3.

Imel EA, Econs MJ. Fibroblast growth factor 23: roles in health and disease. J. Am. Soc. Nephrol. 2005; 16: 2565–75.

Inoue M, Digman MA, Cheng M, et al. Partitioning of NaPi cotransporter in cholesterol, sphingomyelin and glycosphingolipid enriched membrane domains modulates NaPi protein diffusion, clustering and activity. J. Biol. Chem. 2004 (in press).

Isaac J, Berndt TJ, Chinnow SL, Tyce GM, Dousa TP, Knox FG. Dopamine enhances the phosphaturic response to parathyroid hormone in phosphate-deprived rats. J. Am. Soc. Nephrol. 1992a; 2: 1423–9.

Isaac J, Glahn RP, Appel MM, Onsgard M, Dousa TP, Knox FG. Mechanism of dopamine inhibition of renal phosphate transport. J. Am. Soc. Nephrol. 1992b; 2: 1601–7.

Ito M, Iidawa S, Izuka M, et al. Interaction of a farnesylated protein with renal type IIa Na/P$_i$ cotransporter in response to parathyroid hormone and dietary phosphate. Biochem. J. 2004; 377: 607–16.

Jehle AW, Forgo J, Biber J, Lederer E, Krapf R, Murer H. IGF-I and vanadate stimulate Na/P$_i$-cotransport in OK cells by increasing type II Na/P$_i$-cotransporter protein stability. Pflügers Arch. 1998; 437: 149–54.

Jehle AW, Hilfiker H, Pfister M, et al. Type II Na$^+$/Pi cotransport is regulated at the transcriptional level by ambient bicarbonate/carbon dioxide tension in Opossum kidney cells. Am. J. Physiol. Renal Physiol. 1999; 276: F46–53.

Jonsson KB, Zahradnik R, Larsson T, et al. Fibroblast growth factor 23 in oncogenic osteomalacia and X-linked hypophosphatemia. N. Engl. J. Med. 2003; 348: 1656–63.

Karim-Jimenez Z, Hernando N, Biber J, Murer H. A dibasic motif involved in parathyroid hormone-induced down-regulation of the type IIa NaPi cotransporter. Proc. Natl Acad. Sci. USA 2000; 97: 12896–901.

Kato K, Jeanneau C, Tarp MA, et al. Polypeptide GalNAc-transferase T3 and familial tumoral calcinosis: secretion of FGF23 requires O-glycosylation. J. Biol. Chem. 2006; 281: 18370–7.

Kayne LH, D'Argenio DZ, Meyer JH, Hu MS, Jamgotchian N, Lee DBN. Analysis of segmental phosphate absorption in intact rats. A compartmental analysis approach. J. Clin. Invest. 1993; 91: 915–22.

Kempson SA, Lötscher M, Kaissling B, Biber J, Murer H, Levi M. Parathyroid hormone action on phosphate transporter mRNA and protein in rat renal proximal tubules. Am. J. Physiol. 1995; 268: F784–91.

Keusch I, Traebert M, Lötscher M, Kaissling B, Murer H, Biber J. Parathyroid hormone and dietary phosphate provoke a

lysosomal routing of the proximal tubular Na/Pi-cotransporter type II. Kidney Int. 1998; 54: 1224–32.

Kido S, Miyamoto K, Mizobuchi H, et al. Identificaio of regulatory sequences and binding prteins in the type II sodium/phosphate cotransporter NPT2 gene responsive to dietary phosphate. J. Biol. Chem. 1999; 274: 28256–63.

Knochel JP. Clinical and physiologic phosphate disturbances. In: Seldin DW, Giebisch GH, eds. The Kidney, 3rd edn. Philadelphia, PA: Lippincott Williams & Wilkins, 2000: pp. 1905–34.

Kobayashi K, Imanishi Y, Koshiyama H, et al. Expression of FGF23 is correlated with serum phosphate levels in isolated fibrous dysplasia. Life Sci. 2006; 78: 2295–301.

Köhler K, Forster IC, Lambert G, Biber J, Murer H. The functional unit of the renal type IIa Na$^+$/P$_i$ cotransporter is a monomer. J. Biol. Chem. 2000; 275: 26113–20.

Kos CH, Tihy F, Econs MJ, Murer H, Lemieux N, Tenenhouse HS. Localization of a renal sodium-phosphate cotransporter gene to human chromosome 5q35. Genomics 1994; 19: 176–7.

Kumar R. New insights into phosphate homeostasis: fibroblast growth factor 23 and frizzled-related protein-4 are phosphaturic factors derived from tumors associated with osteomalacia. Curr. Opin. Nephrol. Hypertens. 2002; 11: 547–53.

Kurnik BR, Hruska KA. Effects of 1,25-dihydroxycholecalciferol on phosphate transport in vitamin D-deprived rats. Am. J. Physiol. 1984; 247: F177–84.

Kurnik BR, Hruska KA. Mechanism of stimulation of renal phosphate transport by 1,25-dihydroxycholecalciferol. Biochim. Biophys. Acta 1985; 817: 42–50.

Kuro OM. Klotho as a regulator of fibroblast growth factor signaling and phosphate/calcium metabolism. Curr. Opin. Nephrol. Hypertens. 2006; 15: 437–41.

Kuro-o M, Matsumura Y, Aizawa H, et al. Mutation of the mouse klotho gene leads to a syndrome resembling ageing. Nature 1997; 390: 45–51.

Kurosu H, Ogawa Y, Miyoshi M, et al. Regulation of fibroblast growth factor-23 by klotho. J. Biol. Chem. 2006; 281: 6120–3.

Lapointe JY, Tessier J, Paquette Y, et al. NPT2a gene variation in calcium nephrolithiasis with renal phosphate leak. Kidney Int. 2006; 69: 2261–7.

LeClaire MM, Berndt TJ, Knox FG. Effect of renal interstitial infusion of L-dopa on sodium and phosphate excretions. J. Lab. Clin. Med. 1998; 132: 308–12.

Levi M, Lötscher M, Sorribas V, et al. Cellular mechanisms of acute and chronic adaptation of rat renal P$_i$ transporter to alterations in dietary P$_i$. Am. J. Physiol. 1994; 267: F900–8.

Levi M, Shayman JA, Abe A, et al. Dexamethasone modulates rat renal brush border membrane phosphate transporter mRNA and protein abundance and glycosphingolipid composition. J. Clin. Invest. 1995; 96: 207–16.

Liu S, Brown TA, Zhou J, et al. Role of matrix extracellular phosphoglycoprotein in the pathogenesis of X-linked hypophosphatemia. J. Am. Soc. Nephrol. 2005; 16: 1645–53.

Liu S, Guo R, Simpson LG, Xiao ZS, Burnham CE, Quarles LD. Regulation of fibroblastic growth factor 23 expression but not degradation by PHEX. J. Biol. Chem. 2003; 278: 37419–26.

Liu S, Zhou J, Tang W, Jiang X, Rowe DW, Quarles LD. Pathogenic role of Fgf23 in Hyp mice. Am. J. Physiol. Endocrinol. Metab. 2006; 291: E38–49.

Lorenz-Depiereux B, Bastepe M, Benet-Pages A, et al. DMP1 mutations in autosomal recessive hypophosphatemia implicate a bone matrix protein in the regulation of phosphate homeostasis. Nat. Genet. 2006a; 38: 1248–50.

Lorenz-Depiereux B, Benet-Pages A, Eckstein G, et al. Hereditary hypophosphatemic rickets with hypercalciuria is caused by mutations in the sodium-phosphate cotransporter gene SLC34A3. Am. J. Hum. Genet. 2006b; 78: 193–201.

Madjdpour C, Bacic D, Kaissling B, Murer H, Biber J. Segment-specific expression of sodium-phosphate cotransporters NaPi-IIa and -IIc and interacting proteins in mouse renal proximal tubules. Pflugers Arch. 2004; 448: 402–10.

Magagnin S, Werner A, Markovich D, et al. Expression cloning of human and rat renal cortex Na/P$_i$ cotransport. Proc. Natl Acad. Sci. USA 1993; 90: 5979–83.

Mahon MJ, Donowitz M, Yun CC, Segre GV. Na$^+$/H$^+$ exchanger regulatory factor 2 directs parathyroid hormone 1 receptor signalling. Nature 2002; 417: 858–61.

Meyer RA Jr. Meyer MH, Gray RW. Parabiosis suggests a humoral factor is involved in X-linked hypophosphatemia in mice. J. Bone Miner. Res. 1989a; 4: 493–500.

Meyer RA Jr. Tenenhouse HS, Meyer MH, Klugerman AH. The renal phosphate transport defect in normal mice parabiosed to X-linked hypophosphatemic mice persists after parathyroidectomy. J. Bone Miner. Res. 1989b; 4: 523–32.

Mirams M, Robinson BG, Mason RS, Nelson AE. Bone as a source of FGF23: regulation by phosphate? Bone 2004; 35: 1192–9.

Miyamoto KI, Segawa H, Ito M, Kuwahata M. Physiological regulation of renal sodium-dependent phosphate cotransporters. Japan J. Physiol. 2004; 54: 93–102.

Muff R, Fischer JA, Biber J, Murer H. Parathyroid hormone receptors in control of proximal tubule function. Annu. Rev. Physiol. 1992; 54: 67–79.

Murer H, Forster I, Biber J. The sodium phosphate cotransporter family SLC34. Pflugers Arch. 2004; 447: 763–7.

Murer H, Hernando N, Forster I, Biber J. Proximal tubular phosphate reabsorption: molecular mechanisms. Physiol. Rev. 2000; 80: 1373–409.

Murer H, Hernando N, Forster I, Biber J. Molecular mechanisms in proximal tubular and small intestinal phosphate reabsorption (plenary lecture). Mol. Membr. Biol. 2001; 18: 3–11.

Murer H, Hernando N, Forster I, Biber J. Regulation of Na/Pi transporter in the proximal tubule. Annu. Rev. Physiol. 2003; 65: 531–42.

Nesbitt T, Coffman TM, Griffiths R, Drezner MK. Crosstransplantation of kidneys in normal and Hyp mice. Evidence that the Hyp mouse phenotype is unrelated to an intrinsic renal defect. J. Clin. Invest. 1992; 89: 1453–9.

Ohkido I, Segawa H, Yanagida R, Nakamura M, Miyamoto K. Cloning, gene structure and dietary regulation of the type-IIc Na/Pi cotransporter in the mouse kidney. Pflugers Arch. 2003; 446: 106–15.

Perwad F, Azam N, Zhang MY, Yamashita T, Tenenhouse HS, Portale AA. Dietary and serum phosphorus regulate fibroblast growth factor 23 expression and 1,25-dihydroxyvitamin D metabolism in mice. Endocrinology 2005; 146: 5358–64.

Pfister MF, Hilfiker H, Forgo J, Lederer E, Biber J, Murer H. Celular mechanisms involved in the acute adaptation of Na/Pi cotransport to high or low-P$_i$ medium in OK cells. Pflugers Arch. 1998a; 435: 713–19.

Pfister MF, Ruf I, Stange G, Ziegler U, Lederer E, Biber J, Murer H. Parathyroid hormone leads to the lysosomal degradation of the renal type II Na/Pi cotransporter. Proc. Natl Acad. Sci. USA 1998b; 95: 1909–14.

Pribanic S, Gisler SM, Bacic D, et al. Interactions of MAP17 with the NaPi-IIa/PDZK1 protein complex in renal proximal tubular cells. Am. J. Physiol. Renal Physiol. 2003a; 285: F784–91.

Pribanic S, Loffing J, Madjdpour C, et al. Expression of visinin-like protein-3 in mouse kidney. Nephron Physiol. 2003b; 95: 76–82.

Prie D, Huart V, Bakouh N, et al. Nephrolithiasis and osteoporosis associated with hypophosphatemia caused by mutations in the type 2a sodium-phosphate cotransporter. N. Engl. J. Med. 2002; 347: 983–91.

Quarles LD. Evidence for a bone–kidney axis regulating phosphate homeostasis. J. Clin. Invest. 2003a; 112: 642–6.

Quarles LD. FGF23, PHEX, and MEPE regulation of phosphate homeostasis and skeletal mineralization. Am. J. Physiol. Endocrinol. Metab. 2003b; 285: E1–9.

Reimer RJ, Edwards RH. Organic anion transport is the primary function of the SLC17/type I phosphate transporter family. Pflugers Arch. 2004; 447: 629–35.

Reshkin SJ, Forgo J, Biber J, Murer H. Functional assymetry of phosphate transport and its regulation in OK cells: phosphate adaptation. Pflugers Arch. 1991a; 419: 256–62.

Reshkin SJ, Forgo J, Murer H. Apical and basolateral effects of PTH in OK cells: transport inhibition, messenger production, effects of pertussis toxin, and interaction with a PTH analog. J. Membr. Biol. 1991b; 124: 227–37.

Riminucci M, Collins MT, Fedarko NS, et al. FGF-23 in fibrous dysplasia of bone and its relationship to renal phosphate wasting. J. Clin. Invest. 2003; 112: 683–92.

Rowe PS, de Zoysa PA, Dong R, et al. MEPE, a new gene expressed in bone marrow and tumors causing osteomalacia. Genomics 2000; 67: 54–68.

Rowe PSN, Kumagai Y, Gutierrez G, et al. MEPE has properties of an osteoblastic phosphatonin and minhibin. Bone 2004; 34: 303–19.

Ryan WG, Nibbe AF, Schwartz TB, Ray RD. Fibrous dysplasia of bone with vitamin D resistant rickets: a case study. Metabolism 1968; 17: 988–98.

Saito H, Kusano K, Kinosaki M, et al. Human fibroblast growth factor-23 mutants suppress Na$^+$-dependent phosphate co-transport activity and 1alpha,25-dihydroxyvitamin D$_3$ production. J. Biol. Chem. 2003; 278: 2206–11.

Scheinman SJ, Tenenhouse HS. Nephrolithiasis, osteoporosis, and mutations in the type 2a sodium-phosphate cotransporter. N. Engl. J. Med. 2003; 348: 264–5; author reply 265

Schiavi SC, Kumar R. The phosphatonin pathway: new insights in phosphate homeostasis. Kidney Int. 2004; 65: 1–14.

Schmitt R, Klussmann E, Kahl T, Ellison DH, Bachmann S. Renal expression of sodium transporters and aquaporin-2 in hypothyroid rats. Am. J. Physiol. Renal. Physiol. 2003; 284: F1097–104.

Schwindinger WF, Francomano CA, Levine MA. Identification of a mutation in the gene encoding the alpha subunit of the stimulatory G-protein of adenylyl cyclase in McCune-Albright syndrome. Proc. Natl Acad. Sci. 1992; 89: 5152–6.

Scriver CR, MacDonald W, Reade T, Glorieux RH, Nogrady B. Hypophosphatemic nonrachitic bone disease: an entity distinct from X-linked hypophosphatemia in the renal defect, bone involvement, and inheritance. Am. J. Med. Genet. 1977; 1: 101–17.

Scriver CR, Reade T, Halal F, Costa T, Cole DE. Autosomal hypophosphataemic bone disease responds to 1,25-(OH)$_2$D$_3$. Arch. Dis. Child, 1981; 56: 203–7.

Segawa H, Kaneko I, Yamanaka S, et al. Intestinal Na-P(i) cotransporter adaptation to dietary P(i) content in vitamin D receptor null mice. Am. J. Physiol. Renal Physiol. 2004; 287: F39–F7.

Segawa H, Kaneko I, Takahashi A, et al. Growth-related renal type II Na/Pi cotransporter. J. Biol. Chem. 2002; 277: 19665–72.

Segawa H, Yamanaka S, Ito M, et al. Internalization of renal type IIc Na-Pi cotransporter in response to a high-phosphate diet. Am. J. Physiol. Renal Physiol. 2005; 288: F587–96.

Shenolikar S, Voltz JW, Minkoff CM, Wade JB, Weinman EJ. Targeted disruption of the mouse NHERF-1 gene promotes internalization of proximal tubule sodium-phosphate cotransporter type IIa and renal phosphate wasting. Proc. Natl Acad. Sci. USA 2002; 99: 11470–5.

Shimada T, Kakitani M, Yamazaki Y, et al. Targeted ablation of Fgf23 demonstrates an essential physiological role of FGF23 in phosphate and vitamin D metabolism. J. Clin. Invest. 2004; 113: 561–8.

Shimada T, Mizutani S, Muto T, et al. Cloning and characterization of FGF23 as a causative factor of tumor-induced osteomalacia. Proc. Natl Acad. Sci. USA 2001; 98: 6500–5.

Silve C, Friedlander G. Renal regulation of phosphate excretion. In: Seldin DW, Giebisch GH, eds. The Kidney, 3rd edn. Philadelphia, PA: Lippincott Williams & Wilkins, 2000: pp. 1885–903.

Sitara D, Razzaque MS, Hesse M, et al. Homozygous ablation of fibroblast growth factor-23 results in hyperphosphatemia and impaired skeletogenesis, and reverses hypophosphatemia in Phex-deficient mice. Matrix Biol. 2004; 23: 421–32.

Sorribas V, Markovich D, Verri T, Biber J, Murer H. Thyroid hormone stimulation of Na/Pi-cotransport in opossum kidney cells. Pflügers Arch. 1995; 431: 266–71.

Specktor P, Cooper JG, Indelman M, Sprecher E. Hyperphosphatemic familial tumoral calcinosis caused by a mutation in GALNT3 in a European kindred. J. Hum. Genet. 2006; 51: 487–90.

Stauber A, Radanovic T, Stange G, Murer H, Wagner CA, Biber J. Regulation of intestinal phosphate transport. II. Metabolic acidosis stimulates Na$^+$-dependent phosphate absorption and expression of the Na$^+$-P$_i$ cotransporter NaPi-IIb in small intestine. Am. J. Physiol. Gastrointest. Liver Physiol. 2005; 288: G501–6.

Strom TM, Francis F, Lorenz B, et al. Pex gene deletions in Gy and Hyp mice provide mouse models for X-linked hypophosphatemia. Hum. Mol. Genet. 1997; 6: 165–71.

Suh PG, Hwang JI, Ryu SH, Donowitz M, Kim JH. The roles of PDZ-containing proteins in PLC-beta-mediated signaling. Biochem. Biophys. Res. Commun. 2001; 288: 1–7.

Syal A, Schiavi S, Chakravarty S, Dwarakanath V, Quigley R, Baum M. Fibroblast growth factor-23 increases mouse PGE2 production in vivo and in vitro. Am. J. Physiol. Renal Physiol. 2006; 290: F450–5.

Taketani Y, Segawa H, Chikamori M, et al. Regulation of type II renal Na$^+$-dependent inorganic phosphate transporters by 1,25-dihydroxyvitamin D$_3$: identification of a vitamin D-responsive element in the human NAPI-3 gene. J. Biol. Chem. 1998; 273: 14575–81.

Tatsumi S, Miyamoto KI, Kouda T, et al. Identification of three isoforms for the Na$^+$-dependent phosphate cotransporter (NaPi-2) in rat kidney. J. Biol. Chem. 1998; 273: 28568–75.

Tenenhouse HS. Regulation of phosphorus homeostasis by the type IIa Na/phosphate cotransporter. Annu. Rev. Nutr. 2005; 25: 197–214.

Tenenhouse HS, Martel J, Biber J, Murer H. Effect of P$_i$ restriction on renal Na$^+$-P$_i$ cotransporter mRNA and immunoreactive protein in X-linked Hyp mice. Am. J. Physiol. 1995; 268: F1062–9.

Tenenhouse HS, Martel J, Gauthier C, Segawa H, Miyamoto K. Differential effects of Npt2a gene ablation and X-linked Hyp mutation on renal expression of Npt2c. Am. J. Physiol. Renal Physiol. 2003; 285: F1271–8.

Tenenhouse HS, Murer H. Disorders of renal tubular phosphate transport. J. Am. Soc. Nephrol. 2003; 14: 240–8.

Tenenhouse HS, Sabbagh Y. Novel phosphate-regulating genes in the pathogenesis of renal phosphate wasting disorders. Pflügers Arch. 2002; 444: 317–26.

Tenenhouse HS, Werner A, Biber J, et al. Renal Na$^+$-phosphate cotransport in murine X-linked hypophosphatemic rickets. Molecular characterization. J. Clin. Invest. 1994; 93: 671–6.

Tieder M, Modai D, Samuel R, et al. Hereditary hypophosphatemic rickets with hypercalciuria. N. Eng. J. Med. 1985; 312: 611–17.

Tieder M, Modai D, Shaked U, et al. 'Idiopathic' hypercalciuria and hereditary hypophosphatemic rickets: two phenotypical expressions of a common genetic defect. N. Eng. J. Med. 1987; 316: 125–9.

Topaz O, Shurman DL, Bergman R, et al. Mutations in GALNT3, encoding a protein involved in O-linked glycosylation, cause familial tumoral calcinosis. Nat. Genet. 2004; 36: 579–81.

Traebert M, Roth J, Biber J, Murer H, Kaissling B. Internalization of proximal tubular type II Na-P$_i$ cotransporter by PTH: immunogold electron microscopy. Am. J. Physiol. Renal Physiol. 2000a; 278: F148–54.

Traebert M, Völkl H, Biber J, Murer H, Kaissling B. Luminal and contraluminal action of 1-34 and 3-34 PTH peptides on renal type IIa Na-P(i) cotransporter. Am. J. Physiol. Renal Physiol. 2000b; 278: F792–8.

Trohler U, Bonjour JP, Fleisch H. Inorganic phosphate homeostasis. Renal adaptation to the dietary intake in intact and thyroparathyroidectomized rats. J. Clin. Invest. 1976a; 57: 264–73.

Trohler U, Bonjour JP, Fleisch H. Renal tubular adaptation to dietary phosphorus. Nature 1976b; 261: 145–6.

Tsujikawa H, Kurotaki Y, Fujimori T, Fukuda K, Nabeshima Y. Klotho, a gene related to a syndrome resembling human premature aging, functions in a negative regulatory circuit of vitamin D endocrine system. Mol. Endocrinol. 2003; 17: 2393–403.

Urakawa I, Yamazaki Y, Shimada T, et al. Klotho converts canonical FGF receptor into a specific receptor for FGF23. Nature 2006; 444: 770–4.

Virkki LV, Forster IC, Bacconi A, Biber J, Murer H. Functionally important residues in the predicted 3(rd) transmembrane domain of the type IIa sodium-phosphate co-transporter (NaPi-IIa). J. Membr. Biol. 2005a; 206: 227–38.

Virkki LV, Forster IC, Biber J, Murer H. Substrate interactions in the human type IIa sodium-phosphate cotransporter (NaPi-IIa). Am. J. Physiol. Renal Physiol. 2005b; 288: F969–81.

Virkki LV, Forster IC, Hernando N, Biber J, Murer H. Functional characterization of two naturally occurring mutations in the human sodium-phosphate cotransporter type IIa. J. Bone Miner. Res. 2003; 18: 2135–41.

Virkki LV, Murer H, Forster IC. Voltage clamp fluorometric measurements on a type II Na$^+$-coupled Pi cotransporter: shedding light on substrate binding order. J. Gen. Physiol. 2006; 127: 539–55.

Weber TJ, Liu S, Indridason OS, Quarles LD. Serum FGF23 levels in normal and disordered phosphorus homeostasis. J. Bone Miner. Res. 2003; 18: 1227–34.

Weinman EJ, Cunningham R, Wade JB, Shenolikar S. The role of NHERF-1 in the regulation of renal proximal tubule sodium-hydrogen exchanger 3 and sodium-dependent phosphate cotransporter 2a. J. Physiol. 2005; 567: 27–32.

Weinstein LS, Shenker A, Gejman PV, Merino MJ, Friedman E, Spiegel AM. Activating mutations of the stimulatory G protein in the McCune-Albright syndrome. N. Engl. J. Med. 1991; 325: 1688–95.

White KE, Cabral JM, Davis SI, et al. Mutations that cause osteoglophonic dysplasia define novel roles for FGFR1 in bone elongation. Am. J. Hum. Genet. 2005; 76: 361–7.

Woda CB, Halaihel N, Wilson PV, Haramati A, Levi M, Mulroney SE. Regulation of renal NaPi-2 expression and tubular phosphate reabsorption by growth hormone in the juvenile rat. Am. J. Physiol. Renal Physiol. 2004; 287: F117–23.

Yamamoto H, Tani Y, Kobayashi K, et al. Alternative promoters and renal cell-specific regulation of the mouse type IIa sodium-dependent phosphate cotransporter gene. Biochim. Biophys. Acta 2005; 1732: 43–52.

Yamashita T, Konishi M, Miyake A, Inui K, Itoh N. Fibroblast growth factor (FGF)-23 inhibits renal phosphate reabsorption by activation of the mitogen-activated protein kinase pathway. J. Biol. Chem. 2002; 277: 28265–70.

Yu X, Ibrahimi OA, Goetz R, et al. Analysis of the biochemical mechanisms for the endocrine actions of fibroblast growth factor-23. Endocrinology 2005; 146: 4647–56.

Zajicek HK, Wang H, Puttaparthi K, et al. Glycosphingolipids modulate renal phosphate transport in potassium deficiency. Kidney Int. 2001; 60: 694–704.

Zhao N, Tenenhouse HS. Npt2 gene disruption confers resistance to the inhibitory action of parathyroid hormone on renal sodium-phosphate cotransport. Endocrinology 2000; 141: 2159–65.

Systemic Hereditary Diseases with Renal Involvement: Multifactorial Diseases

Genetic Susceptibility to Kidney Disease as a Consequence of Systemic Autoimmunity

ANDREW WANG, CHANDRA MOHAN AND EDWARD K. WAKELAND

INTRODUCTION

Systemic lupus erythematosus (SLE) is a chronic autoimmune disease that is classically associated with the production of pathogenic autoantibodies to a spectrum of nuclear antigens (Theofilopoulos & Kono 2001, Banchereau & Pascual 2006, Fairhurst et al 2006). SLE presents with a diverse array of clinical symptoms, which often reflect the consequences of injury to multiple organ systems. This clinical heterogeneity results from tissue damage targeted by autoantibody and inflammatory processes initiated as a consequence of deposition of complement-fixing immune complexes. Severe complications, which ultimately develop in about 50% of lupus patients, can manifest as a variety of clinical problems, including nephritis, central nervous system vasculitis, pulmonary hypertension, interstitial lung disease, and stroke. Current treatments for SLE involve a variety of immunosuppressive drug therapies, including hydroxychloroquine, steroids, and cytotoxic drugs. Although these therapies allow management of disease severity for many patients, a variety of deleterious drug side effects and therapy-resistant disease symptoms significantly diminish the quality of life for many SLE patients.

The renal manifestations of SLE, termed lupus nephritis, occur in approximately 50% of patients (Cameron 1999), and pose the greatest risk to the patient as it is the leading cause of morbidity and mortality (Davis et al 1996). The pathogenesis of lupus nephritis is poorly understood given the complex nature of the disease and its highly heterogeneous presentation. It is thought that autoantibodies are critical in the initiation of lupus nephritis. These autoantibodies include antinuclear antibodies (ANA), a diagnostic hallmark of SLE present in over 98% of patients, and nonantinuclear antibodies such as antiglomerular basement membrane autoantibodies. It has been postulated that the deposition of these antibodies as well as their immune complexes in the glomeruli are required for the development of glomerulonephritis (Deshmukh et al 2006). Recent work has also implicated a role for various innate stimuli and other antibody-independent factors. Moreover, genetic studies in both humans and mouse models of SLE have revealed kidney-intrinsic factors that may also play an important role in lupus nephritis.

Indeed, a large body of literature has sought to understand the genetic basis and immunological mechanisms underlying lupus nephritis. The focus of this review is the genetic basis of nephritis, utilizing findings from both human studies and animal models. We will discuss the overall features of the disease, the current understanding of the role of genetics in susceptibility, and the role of genetic epistasis in the development of severe kidney pathology.

CLINICAL OVERVIEW OF LUPUS NEPHRITIS

The renal manifestations of SLE are highly heterogeneous. The disease may affect glomeruli, tubules, interstitium, and blood vessels with varying degrees of chronicity, severity, and kinetics, and has the ability to transform from one morphological form to another in a spontaneous or treatment-induced fashion. Subsequently, clinical presentation of lupus nephritis is also highly varied, ranging from asymptomatic hematuria and proteinuria to nephrotic syndrome and renal failure. Given the wide spectrum of disease, the World Health Organization (WHO) classification of lupus nephritis was devised to help standardize interpretation of renal biopsies (Churg & Sobin 1982, D'Agati 1997). It is the most widely used and accepted classification of disease status by clinicians and pathologists.

Genetic Diseases of the Kidney
ISBN: 978-0-12-449851-8

Benign autoimmunity

	ANA	Mesangial alterations	Subendothelial deposits	Subepithelial deposits	Sclerosing lesions	Renal failure
Class I	−/+	−	−	−	−	−
Class II	+	−/+	−	−	−	−

Genes and environment

Pathogenic autoimmunity

	ANA	Mesangial alterations	Subendothelial deposits	Subepithelial deposits	Sclerosing lesions	Renal failure
Class III	+	+	+	−/+	+	−/+
Class V	−/+	+	−/+	+	−/+	−/+
Class IV	+	+	+	−/+	+	+
Class VI	+	+	+	−/+	+	+

Disease course

FIGURE 44.1 Diagram illustrating the WHO classification criteria in relation to disease course

There are six classes of lupus nephritis defined by WHO guidelines that essentially describe a continuum of disease. Figure 44.1 presents the most salient features of the different WHO classes of lupus nephritis. Importantly, Figure 44.1 highlights two distinct stages of lupus of relevance to kidney function – a 'benign' stage in which patients are seropositive for autoantibodies but lack renal manifestations, and a 'pathogenic' stage where renal disease is active with or without seropositivity (Nguyen et al 2002). Class I and II nephritis belong to the 'benign' category. Class I lupus nephritis is limited to the early course of SLE and is characterized by an absence of glomerular abnormalities and the lack of detectable renal manifestations. Class II is defined as glomerular disease limited to the mesangium, and is further subdivided based on the presence or absence of mesangial hypercellularity. These patients have detectable ANAs and hypocomplementemia, indicative of active disease, but typically have no renal manifestations. Indeed, these patients typically have no clinical symptoms of renal disease and exhibit normal renal function by all conventional measures, although some may exhibit mild hematuria, proteinuria, or leukocyturia.

Class I and II patients highlight the fact that a dysregulation of the immune system, and specifically the formation of ANAs, typically precedes renal disease but may not precipitate it (Arbuckle et al 2003). This observation is consistent with the fact that about 5% of the normal population exhibit positive ANAs titers without ever developing lupus (Shmerling 2003, Wandstrat et al 2006). There are two important implications from these observations. First, it indicates that the progression of lupus from benign to severe pathology involves a transitional threshold. That is, an accumulation of molecular events, dictated by a combination of genetic predisposition and environmental stimuli, is needed to transition to severe pathology. Second, ANAs may not be in themselves pathogenic. Thus, there appears to be a qualitative difference between antibodies which can and cannot precipitate renal disease. Class V lupus patients exemplify this interpretation.

Class V lupus nephritis is defined as membranous proliferative glomerulonephritis. Interestingly, these patients are typically negative for or have low titers of ANA, and often present with renal manifestations well before the development of other SLE clinical features. Class V disease is characterized by widespread epimembranous immune deposits in the absence of endocapillary proliferation and a thickening of the glomerular basement membrane. This subset of patients provides anecdotal demonstration that other autoantibodies, such as antiglomerular antibodies or other kidney-specific antibodies, and not ANAs, may be the key players in disease pathogenesis. Rephrased, ANAs, or at least those screened for by conventional methods, may be dispensable in the development of severe kidney disease. Class V patients provide a case wherein genetic and environmental factors can 'short-circuit' the 'benign' phase of the disease, or at least have benign autoantibodies which are unconventional and therefore missed by routine diagnostics. A more thorough discussion on the pathogenicity of autoantibodies is found later in the text.

Class III and IV lupus have transitioned to pathogenic autoimmunity. Class III and IV nephritis are both defined as endocapillary proliferative glomerulonephritis, and are distinguished based on the distribution of endocapillary proliferation. The glomerular capillary lumina are narrowed or obstructed by hyperproliferation of endothelial and mesangial cells and infiltrated by mononuclear and polymorphonuclear cells. There is a varying degree of tubular atrophy, interstitial fibrosis, and inflammation. Up to two-thirds of class III and IV patients are positive for SLE serologies, and typically have hematuria, leukocyturia, cellular casts, and proteinuria, as well as renal hypertension. At this stage of disease, patients typically have impaired renal function as assessed by serum creatinine and measures of glomerular filtration rates. Finally, the WHO classifies patients with advanced, chronic disease, extensive glomerular scarring, and renal insufficiency as having class VI lupus nephritis.

GENETIC PREDISPOSITION TO SLE

Since renal manifestations are a sequela of SLE, genetic predisposition to lupus nephritis is intertwined with the genetics of susceptibility to SLE. It is clear from studies of both man and mouse that genetic factors play a dominant role in dictating predisposition to SLE. The high concordance rate of SLE in monozygotic twins (30–57%) and increased familial incidence provide strong evidence for genetic predisposition as a major factor in disease susceptibility (Deapen et al 1992, Wandstrat & Wakeland 2001). Moreover, the occurrence of spontaneous lupus in specific inbred mouse strains, such as the F1 hybrid of the New Zealand Black (NZB) and New Zealand White (NZW) strains, the MRL/*lpr/lpr* strain, and BXSB/*yaa* mice, further point to the importance of predisposing disease alleles.

Genetic analysis of SLE susceptibility in humans has progressed significantly over the past few years and several disease alleles have been identified. The most robust associations detected in genetic analyses have been with deficiencies in the C2, C4, and C1q components of the complement system (Manderson et al 2004). However, the disease-associated alleles of these complement components are relatively rare in human populations and, thus, only account for a small portion of disease incidence. Nonetheless, genetic studies indicate that 45% of all white SLE patients are homozygous or heterozygous for defective alleles of complement component C4a, indicating that even small deficiencies in this component may be an important component of disease progression. Furthermore, >90% of individuals deficient for the expression of C1q develop severe lupus-like disease (Agnello 1978, Morgan & Walport 1991). Although this genetic deficiency mediates SLE susceptibility in a highly penetrant fashion, the C1q-deficient allele is extremely rare in the human population and thus this genetic system only accounts for a very small proportion of affected individuals. The precise roles that any of these complement deficiencies play in the development of SLE remains unclear, although they are thought to cause impaired clearance of immune complexes and apoptotic blebs, which may lead to a breach in immune tolerance due to the accumulation of an excess of these self-antigens in regions of immune activation (Nelson 1953, Casciola-Rosen et al 1994).

Linkage and/or association analyses in humans also have associated alleles of HLA-DR (DR2 and DR3) (Schur et al 1982), PDCA (Prokunina et al 2002), PTPN22 (Kyogoku et al 2004), Fc receptors (Salmon et al 1996, Edberg et al 2002) and IRF5 (Graham et al 2006) with susceptibility to SLE. In this instance, the disease alleles associated with SLE are often quite common in human populations, consistent with the multifactorial nature of genetic predisposition in this disease. Thus, although these disease alleles are relatively common in SLE patients, they are also quite frequent in unaffected individuals as well. This is consistent with the possibility that complex epistatic interactions between these disease alleles and a multitude of other susceptibility and/or suppressive allelic modifiers and environmental factors are essential for the development of severe autoimmunity. As will be discussed below, these types of interaction have been clearly delineated in animal models of systemic autoimmunity (Wakeland et al 2001). Finally, the roles that these disease alleles play in SLE pathogenesis are either unknown or very poorly characterized, which represents a major deficiency in our understanding of genetic predisposition to SLE in humans.

Genetic linkage studies in murine models of lupus, on the other hand, have been particularly fruitful in defining the roles that individual susceptibility alleles play in disease progression. Most notably, congenic dissection, in which genomic segments containing known susceptibility/resistant loci are introgressed between defined genetic backgrounds, has been a valuable tool. This genetic strategy, which was pioneered by George Snell more than 50 years ago for the analysis of histocompatibility alleles, involves the production of a series of strains with identical genomes that only differ by the incorporation of a small genomic segment known to contain a specific susceptibility locus. These strains can be used to characterize the phenotypes mediated by individual susceptibility loci, even before the causative genetic variation has been identified (Snell 1948). Studies employing these strategies to analyze NZB/NZW-derived MRL/*lpr/lpr* and BXSB/yaa strains have been fruitful in identifying candidate genes and understanding disease mechanisms. The inhibitory Fc receptor gamma IIb (Bolland & Ravetch 2000), complement components (Botto et al 1998), the pro-apoptotic Fas receptor and Fas ligand (Cohen & Eisenberg 1991), members of the SLAM family of receptors (Wandstrat et al 2004), and Toll-like receptors (TLR) 7 (Pisitkun et al 2006, Subramanian et al 2006) and 9 (Christensen et al 2005) are but a handful of genes that have been implicated from studies in the mouse. Importantly, this strategy has been invaluable in delineating the roles that individual disease loci play in the development of nephritis in lupus-prone mice.

We have utilized congenic dissection to define the roles that individual susceptibility loci play in disease progression in the NZB/NZW-derived NZM2410 strain (Wakeland et al 2001). An initial linkage analysis in crosses with the autoimmune-resistant C57/BL6 (B6) strain identified three susceptibility loci, termed *Sle1*, *Sle2*, and *Sle3*, that contributed to disease in the NZM2410 lupus-prone mouse strain. These loci were introgressed individually onto the B6 genetic background, which produced a series of B6-congenic strains, each carrying a single susceptibility locus in isolation on the B6 background. Over the past 10 years, we have performed detailed characterizations of the autoimmune phenotypes expressed by these congenic intervals, both individually and in various combinations (Mohan et al 1997, 1998, 1999a, 1999b, 2006, Morel et al 1997, 1999, 2000, 2001, Wandstrat et al 2004, Kumar et al 2006, Subramanian et al 2006).

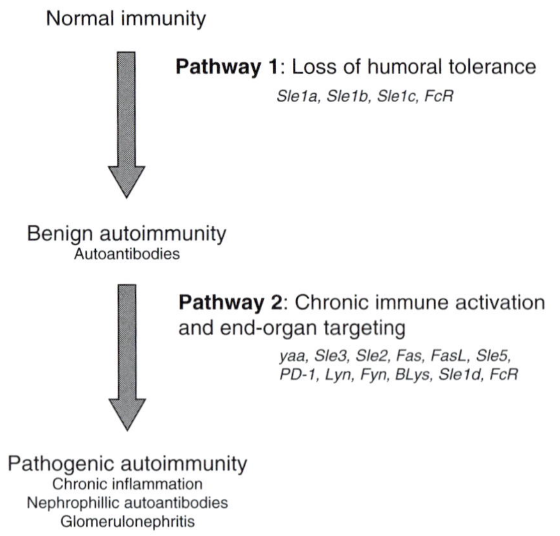

Normal immunity

Pathway 1: Loss of humoral tolerance
Sle1a, Sle1b, Sle1c, FcR

Benign autoimmunity
Autoantibodies

Pathway 2: Chronic immune activation
and end-organ targeting
*yaa, Sle3, Sle2, Fas, FasL, Sle5,
PD-1, Lyn, Fyn, BLys, Sle1d, FcR*

Pathogenic autoimmunity
Chronic inflammation
Nephrophillic autoantibodies
Glomerulonephritis

FIGURE 44.2 Hypothetical model defining two interactive genetic pathways that mediate the development of autoimmunity and transition to severe pathology

These studies have led to a model of lupus pathogenesis, which is presented in Figure 44.2 (adapted from Wakeland et al (2001) and Arbuckle et al (2003)). In this model, lupus susceptibility genes are postulated to modulate two separate immunologic pathways, the first mediating a breach in immune tolerance leading to benign autoimmunity (pathway 1) and the second driving the transition from benign to pathogenic autoimmunity (pathway 2). Genes that confer susceptibility to nephritis, some of which may be kidney-intrinsic, belong to the second category.

The experimental observations that have led to this model can be illustrated by the genetic interactions that lead to fatal disease among some of the B6-congenic strains that we have developed. The most potent susceptibility locus identified in our original linkage analysis was *Sle1*, which is the best developed example of a gene in the first pathway. B6.*Sle1* mice develop high-titer IgG ANAs in the absence of significant renal disease and thus are reminiscent of the 'benign' autoimmunity described above. This mouse illustrates the concept that genes predisposing to ANAs may not in themselves be capable of mediating a transition to lupus nephritis. Our ongoing analysis of this locus has determined that it contains a linked cluster of four loci that impact susceptibility to autoimmunity and that polymorphisms in a cluster of seven SLAM/CD2 genes play a predominant role in the development of ANAs. To date, *Ly108* within the SLAM/CD2 gene cluster has been implicated as the strongest candidate gene in this interval (Wandstrat et al 2004, Kumar et al 2006). The disease allele of *Ly108* has been demonstrated to impair the induction of immune tolerance in the immature B cell compartment, leading to an increased frequency of autoreactive B cells in the spleen and lymph nodes. As a result of this

lesion in an early checkpoint in B cell tolerance, these mice are highly predisposed to spontaneously develop ANAs by about 7 months of age (penetrance >90%). As discussed above, a similar phenotype develops in roughly 5% of the aging human population, suggesting that a lesion in B cell tolerance may be a relatively common occurrence.

As shown in Figure 44.2, the second pathway contains a series of disease alleles or susceptibility loci that drive chronic immune activation. The phenotypes mediated by *Sle3* and *yaa* have been thoroughly investigated and their properties provide important insights into the types of genetic interaction that can lead to severe autoimmunity. Both B6.*Sle3* and B6.*yaa* mice have mild autoimmune phenotypes that lead to little or no autoantibody production. This characteristic is shared with the other disease alleles and susceptibility loci included in this pathway, all of which exhibit variable but generally minimal propensity to cause autoimmunity when isolated on an otherwise normal genome. Interestingly, B6.*Sle3* mice may develop low levels of GN, despite the absence of significant levels of IgG ANAs (Mohan et al 1999b). The lack of high-titered ANAs but presence of GN seen in the B6.*Sle3* mouse may be similar to class V GN observed in the class V subset of SLE patients and points to the importance of kidney-intrinsic genes and/or genes that predispose to the production of nonantinuclear autoantibodies or other nephrotoxic factors. B6.*yaa* mice have virtually no autoimmune phenotypes, although a careful analysis of their serum IgM reveals the presence of autoantibodies preferentially recognizing RNA containing antigens (Subramanian et al 2006).

The combination of either of these pathway 2 loci with *Sle1* in the bicongenic B6.*Sle1Sle3* and B6.*Sle1yaa* strains results in the development of a potent systemic autoimmunity, leading to severe levels of GN and highly penetrant fatal disease by 9 months of age (Mohan et al 1999a, Subramanian et al 2006). Thus, although none of these disease alleles is capable of driving severe pathology individually, a combination of *Sle1* with either *Sle3* or *yaa* results in the transition of the benign autoimmunity mediated by *Sle1* into a potent systemic autoimmunity culminating in lupus nephritis and kidney failure. These analyses clearly illustrate the dramatic impact of epistatic interactions between genes in pathways 1 and 2 on the development of fatal disease in this mouse model of lupus nephritis.

Recent work on the role of the innate immune compartment in SLE has provided new insights into at least one type of disease allele within pathway 2. Genes in the toll-like receptor (TLR) family have been recently implicated in playing a critical role in SLE. TLRs normally function within the innate immune system to recognize pathogen-derived molecules, serving to initiate immediate host responses to infection and to drive the activation of the adaptive immune system against invading pathogens (Hargreaves & Medzhitov 2005, Pasare & Medzhitov 2005). The TLR9, TLR7, and TLR3 molecules bind specifically with dsDNA, ssRNA, and dsRNA, respectively, but have also been shown to recognize self-derived nuclear antigens (Marshak-Rothstein 2006). A clear role for TLR7 and TLR9 in dictating humoral response to RNA and DNA-containing antigens, respectively, has been demonstrated by a number of investigators (Christensen et al 2005, Pisitkun et al 2006, Subramanian et al 2006). We and others have recently demonstrated that a twofold increase in the level of expression of TLR7 is the genetic lesion underlying the potent autoimmune accelerating phenotype of *yaa* (Pisitkun et al 2006, Subramanian et al 2006). These studies demonstrate that an incremental increase in the signaling of the TLR7 pathway is sufficient to drive the development of fatal lupus nephritis, when coupled with breach in B cell tolerance mediated by *Sle1*. In addition, we have shown that *Sle3* mediates increased responsiveness of myeloid cells to TLR4 ligands, suggesting that the synergistic interactions of *Sle3* with *Sle1* may also involve increased signaling by the Toll receptor pathway (Zhu et al 2005, Subramanian et al 2006). These and other studies demonstrate that dysregulation of the innate immune system is a potent driver of the transition from benign to pathogenic autoimmunity, thus suggesting that simultaneous dysregulation of the adaptive and innate immune systems may be a common feature in autoimmune-prone genomes. Since TLRs have an impact in dictating autoantibody repertoire as well as in activating effector cells such as macrophages, dendritic cells, and granulocytes, they thus straddle the two categories of genes. As will be discussed later, TLRs have also been shown to be expressed on renal cells, and it remains to be seen what effect TLR dysregulation has on the kidney intrinsically. In a similar fashion, defects in complement components and apoptosis affect both the development of benign autoimmunity and also the transition and perpetuation of pathogenic autoimmunity. Complement is thought to play two important roles in lupus pathogenesis. First, complement modulates adaptive immunity. C3b can bind foreign or self-antigen and signal via the complement receptor complex CD21/35 on B-cells (Molina et al 1994). The simultaneous coligation of the BCR and CD21/35 with antigen complexed with complement leads to enhanced signaling in B cells (Carter & Fearon 1992). It has also been shown that complement-fixation of antigen plays a major role in localizing antigen to the lymphoid follicles via engagement of complement receptors on follicular dendritic cells, which further serve to skew B-cell function (Fang et al 1998). Secondly, complement plays an important role in immune clearance. This function of complement plays a critical role in SLE, where ICs are inappropriately cleared and deposit in tissues where they cause pathologies. Similarly, apoptosis defects likely play a role in both pathway 1 and pathway 2. Faulty negative selection due to defects in apoptosis, as exemplified in the *fas* and *gld*-defective murine models, leads to inappropriate release of autoreactive immune cells from the bone marrow and thymus. This same inability to apoptose perpetuates chronically activated effector cells.

The most critical finding from these studies is that disease alleles from both pathways must be present in the genome to cause fatal disease and that these pathways interact in a non-additive fashion to produce pathogenic autoimmunity. Thus, understanding the epistatic interactions between lupus genes represents one of the greatest challenges facing immunogeneticists today.

PATHOGENESIS OF LUPUS NEPHRITIS

Lupus nephritis is thought to be initiated by deposition of preformed circulating immune complexes (IC) in the glomerular capillary wall. Since ICs vary widely in both their stereochemical properties and serum concentrations, ICs may preferentially deposit in mesangial, subendothelial, or subepithelial sites. Clearly, not all ICs are pathogenic. The 5% of the population bearing ANAs, and specifically anti-chromatin and anti-dsDNA ANAs, never develop nephritis. As described above, *Sle1* congenic mice also do not develop nephritis despite the persistence of high-titer IgG ANA. Thus, understanding the pathogenicity of autoantibodies has been an active field of research and remains one of the most controversial topics in the field.

Work in the late 1980s by Weber and colleagues showed that histones, which are highly cationic, had strong affinity for the negatively charged sites in the glomerular capillary wall (Schmiedeke et al 1989). Work in the late 1970s by Koffler and colleagues demonstrated that antiDNA antibodies may also cross-react with intrinsic glomerular basement membrane components such as heparin sulfate proteoglycans (Winfield et al 1977). These findings have been supported by recent work by Rekvig and colleagues, which demonstrated that anti-dsDNA and antihistone H1 antibodies and not a variety of other autoantibodies could cross-react with kidney components (Kalaaji et al 2006).

Antibodies formed against complement itself have also been implicated in the pathogenesis of nephritis. Studies have demonstrated the strong correlation between anti-C1q autoantibody positivity and renal involvement (Siegert et al 1991), the predictive value of anti-C1q autoantibody titers for flares of nephritis (Moroni et al 2001), and the accumulation of anti-C1q antibodies in the kidneys of both men and mice with lupus (Mannik & Wener 1997, Mannik et al 2003). Recent work by Daha and colleagues has provided evidence that anti-C1q antibodies are pathogenic only in concert with other IC deposits (Trouw et al 2004).

Direct engagement of kidney components has also been shown to play an important role in GN, especially in the membranous form. Nephrophilic antibodies, such as those formed against the glomerular basement membrane and intact glomeruli, have been demonstrated to arise later in the disease course, and may indicate the immune system's response to previously inaccessible antigens released during initial kidney damage. Our recent work has begun to shed light on the identity of nephrophilic antibodies. Using a proteomic array-based approach showed that specific clusters of serum nephrophilic antibodies, but not all kidney-recognizing antibodies, correlated with lupus severity (Li et al 2007). It was demonstrated that IgG autoantibodies to glomeruli and laminin, myosin, matrigel, vimentin, and heparin sulfate correlated with disease activity while other antibodies such as vitronectin and entactin did not. These studies pave the way for future work needed to unravel the pathogenic potential of lupus autoantibodies, which may in turn lead to a better understanding of disease mechanism.

After IC deposition, it is thought that the complement cascade plays an important role, although whether this event occurs in the circulation or in the kidney itself is yet unknown. Complement fixation results in the release of chemotactins C3a and C5a, which may promote recruitment of immune cells which then insult the kidney via the release of reactive oxygen species and inflammatory cytokines. Consistent with this hypothesis, Cook and colleagues have recently demonstrated in vivo that ablation of the classical pathway component C3 as well as the alternative pathway component Factor D rescued nephritis (Turnberg et al 2006). These sets of studies indicate the important role of complement in the pathogenesis of lupus nephritis.

Recently, with the recognition of the importance of TLRs in SLE pathogenesis, researchers have identified a renal-intrinsic role for TLRs. Tubular epithelial cells have been shown to express TLR4 and can respond to LPS stimulation by releasing chemokines and cytokines (Tsuboi et al 2002, Chowdhury et al 2006). Glomerular mesangial cells have been shown to express TLR3, which recognizes double-stranded RNA, and Anders and colleagues demonstrated that stimulation of these cells in vitro with synthetic TLR3 ligand induced secretion of IL6 and other chemokines (Patole et al 2005). Importantly, they found that by blocking NFkB – the most distal signaling event following TLR3 ligation – they were able to ablate these responses. Given the large body of work that has focused on the immune consequences of TLR and TLR-pathway dysregulation, it is worth closely examining the effect of inappropriate TLR signaling in the kidney as well. Indeed, genetic lesions in innate components may be so potent in aggravating lupus kinetics precisely because they affect both the immune and kidney compartments simultaneously.

What should be obvious from the present discussion is that it is extremely difficult to dissect renal-intrinsic mechanisms from immunological mechanisms when studying lupus nephritis pathogenesis. In almost every study, the events of the kidney are vitally linked to the events in the immune system. Subsequently, studies aimed at identifying renal-intrinsic factors leading to increased susceptibility are largely obscured by the inability to dissociate inflammatory events mediated by the immune system from those mediated by the kidney.

However, there is good evidence to date that suggests a crucial role for renal endogenous mechanisms. These studies, done primarily in the mouse, exploit bone marrow chimera technology and, more recently, kidney transplantation to assess if the contributions of certain genes to kidney pathology are intrinsic to the kidney itself. Tipping and colleagues have used bone marrow chimera experiments to show a clear role for kidney-endogenous sources of IL1-beta, IL-12, IFN-gamma, TNF-alpha, and CD40 in mediating GN (Li et al 1998, Timoshanko et al 2001, 2002, 2003, 2004). The role of FcR-gamma was also clarified recently using this approach by Clynes and colleagues, who showed that Fcr-gamma expression was needed only in hematopoetic cells and not renal cells, minimizing the purported role of mesangial cell activation via FcR cross-linking (Bergtold et al 2006). Given these observations, it follows logically that any alteration in kidney responses to pathogenic insult may have a profound impact on the kinetics and severity of GN. Indeed, this experimental strategy will be needed to establish whether candidate genes identified in murine lupus are bona fide nephritis susceptibility genes and to gain insight into how these genes contribute to kidney pathology.

NEPHRITIC SUSCEPTIBILITY LOCI MAPPED IN MURINE LUPUS MODELS

Table 44.1 summarizes the statistically significant loci for nephritis that have been uncovered using murine mapping studies (Watson et al 1993, Drake et al 1994, 1995, Kono et al 1994, Morel et al 1994, Rigby et al 2004, Rozzo et al 1996, 2000, Vyse & Kotzin 1996, Vyse et al 1996a, 1996b, 1997, Wang et al 1997, Gu et al 1998, Hogarth et al 1998, Santiago et al 1998, Vidal et al 1998, Morel et al 1994, 1999, Nakatsuru et al 1999, Haywood et al 2000, 2001, Tucker et al 2000, Waters et al 2001, Xie et al 2001, 2005, Rahman et al 2002, Kono et al 2003). The table summarizes the results of mapping studies done in the MRL/lpr/lpr, BXSB.yaa, and the NZB/NZW-derived strains. As can be noted, all mouse strains that develop lupus spontaneously harbor nephritis susceptibility loci. Whereas some of the mapped loci confer susceptibility to ANAs as well as nephritis, others confer susceptibility to nephritis but not ANAs. Chromosomes 1, 4, 7, and 17 appear to be the most commonly implicated chromosomes harboring nephritis susceptibility loci, with these loci originating from several different strain backgrounds. These include repeatedly mapped loci on the distal chromosome 1 (88–101cM), mid-chromosome 4 (31–48cM), proximal chromosome 7 (16–31cM), and centromeric chromosome 17 (around H2 and complement).

Sle1, located on a segment of mouse chromosome 1 syntenic with human chromosome 1, has been further resolved to four sub-loci, termed *Sle1a, Sle1b, Sle1c,* and *Sle1d,*

TABLE 44.1 Genetic loci associated with lupus nephritis in murine studies

Name of locus	Chr[1]	cM[2]	Disease strain	Mapped phenotype[3]	Ref
Bxs4/Sle10	1	11	BXSB	nephritis	71
Bxs1/Yaa2	1	32.8	BXSB	nephritis, ANA, etc.	72
Bxs2/Yaa3	1	63.1	BXSB	nephritis, anti-dsDNA, etc.	71, 72
Bxs3	1	71	BXSB	nephritis, anti-dsDNA, etc.	71, 72
Sle1	1	88	NZM2410	nephritis	75
Lbw7	1	92	NZB	anti-dsDNA	76
Sbw1	1	92	NZB	splenomegaly	76
Cgnz1	1	92.3	NZM2328	chronic nephritis, proteinuria	86
Nba2	1	94.2	NZB	nephritis, anti-dsDNA, gp70IC	81, 87–89
Agnz1	1	101	NZM2328	acute nephritis	86
Wbw1	2	86	NZW	proteinuria	77
Mrl	2	50	MRL-lpr	anti-dsDNA	78
Sles3	2	16	NZW	suppressive (nephritis)	41
no name	3	32.8	BXSB	nephritis, ANA	71
Bxs5	3	63	BXSB	nephritis, ANA	71
Lprm2	3	66	MRL-lpr	vasculitis	80
Lprm1	4	35.5	MRL-lpr	vasculitis	80
Nbwa2/Sle15	4	31.2	NZB	nephritis	90
Lbw2	4	42.6	NZB	nephritis, mortality	76
Sle2	4	44.5	NZM2410	nephritis	75
no name	4	48.5	NZB	nephritis	88
Sbw2	4	53	NZB	splenomegaly	76
Lmb1	4	54	C57BL/6	splenomegaly, ANA	80
Sles2	4	57.6	C57BL/6	suppressive (nephritis, ANA)	41
Nba1	4	70	NZB	nephritis	81
Nba4	5	15	NZB	nephritis	91
Sle6	5	20	NZW	nephritis	41
Lmb2	5	27	MRL-lpr	splenomegaly, ANA	80
Lprm4	5	54	MRL-lpr		79
Lbw3	5	88	NZW	disease accelerator	76
Lxw2	6	25.5	NZW	nephritis	92
no name	6	35	C3H	nephritis	93
Lbw4	6	60	NZB	disease accelerator	76
Sle5	7	4	NZM2410	nephritis, anti-dsDNA	82
Lrdm1	7	6	MRL-lpr	nephritis, anti-DNA, etc.	83
no name	7	16	NZM2410	nephritis, anti-dsDNA	82
Lbw5	7	22	NZW	disease accelerator	76
no name	7	25	NZW	nephritis, anti-DNA, gp70 IC	84
Lmb3	7	26	MRL-lpr	splenomegaly, ANA	81
Sle3	7	28	NZM2410	nephritis, autoantibodies	80
Nba3	7	31	NZB	nephritis	81
Sles4	9	2	NZW	suppressive (nephritis)	41
Sle12	10	69	NZM2410	nephritis	82
Lmb4	10	50	MRL-lpr	splenomegaly	80
Sle13	11	20	NZM2410	acute nephritis, anti-dsDNA	82
Nba	11	17	NZB	nephritis	95
Lbw8	11	37	NZB	ANA	76
Mrl	11	54	MRL-lpr	ANA, vasculitis	78
Nbwa1/Sle14	12	3.5	NZB	nephritis, ANA	90
Lrdm2	12	61.8	MRL-lpr	nephritis, anti-DNA, etc.	83
Bxs6	13	24	BXSB	nephritis	94
Nwa	14	19.5	NZW	ANA	95
Swrl2	14	27.5	SWR	nephritis, IgG ANA	74

(Continued)

TABLE 44.1   (*Continued*)

Name of locus	Chr[1]	cM[2]	Disease strain	Mapped phenotype[3]	Ref
Nba	14	40	NZB	nephritis	95
Lprm3	14	44.3	C3H	nephritis	79
Lprm5	16	33.5	MRL-lpr	ANA	79
Nwa1	16	38	NZW	nephritis, anti-dsDNA, etc.	95
H2	17	19	several[4]	nephritis, mortality, ANA, etc.	several[4]
Sles1	17	19	several	fully suppressive	41, 74–75
Lbw6	18	47	NZW	nephritis, mortality	76

[1]Indicated is the chromosomal location of the loci, ordered according to the chromosome numbers
[2]Indicated is the position of the locus on the chromosome in centimorgans (cM)
[3]Where multiple phenotypes were mapped, only 'nephritis' and a couple of the other phenotypes have been listed
[4]'Several' nephritis loci have been mapped to 'H2'

TABLE 44.2   Monocongenic strains in which lupus-associated nephritis is aggravated or ameliorated

Congenic strain[1]	Disease strain	Control strain	Locus	Chromosomal position[2]	Phenotypes noted in congenics[3]: proteinuria	nephritis	deposits	auto-Ab	Ref
Introgression of disease interval onto 'normal' strain background									
B6.NZBc1 (35–106)	NZB	C57BL/6	Nba2, others	1 (35–106)	Yes (low)	Yes	Yes	Yes	87
B6.NZBc1 (85–106)	NZB	C57BL/6	Nba2	1 (85–106)	no	Yes	Yes	Yes	87
B6.Sle1	NZM2410	C57BL/6	Slc1	1 (85–122)	no	Mild	no	Yes	40
B6.Sle3	NZM2410	C57BL/6	Sle3	1 (15–45)	no	Modest	Yes	Yes	40, 43
B6.MRLc7	MRL	C57BL/6	Lmb3	7 (1–28)	na[4]	Yes	Yes	Yes	98
Introgression of 'normal' interval onto disease-strain background									
NZM2328. Cgnz1	NZM2328	C57L/J	Cgnz1	1 (88–112)	decreased	decreased	decreased	decreased	99
NZM2328. Adnz1	NZM2328	C57L/J	Adnz1	4 (16–60)	no change	no change	no change	decreased	99

[1]Indicated are congenic mouse strains in which a disease susceptibility interval has been backcrossed onto a control strain background (i.e. the first 5 strains) or in which a disease-resistant interval has been backcrossed onto a disease susceptible strain background (i.e. the last 2 strains)
[2]Indicated are the chromosomal locations of the introgressed genetic interval in centimorgans (cM)
[3]Listed are the salient phenotypes in the congenic strains compared to the phenotype of the background strain
[4]na = not available

of which *Sle1b* seems to be the strongest loci that predisposes a loss of tolerance to chromatin (Morel et al 2001). Of interest, *Sle1d* has been purported to play a role in the development of GN in the absence of predisposing ANAs.

What is obvious from these studies is that conventional mapping studies are inconclusive in differentiating nephritis-mediating genes from other lupus genes. To overcome the limitations, congenic analyses as an alternative strategy have been used to study purported nephritis-loci *in vacuo* (Morel et al 1997, Mohan et al 1999, Wither et al 2003, Kong et al 2004, Waters et al 2004). The results of these studies are presented in Table 44.2. However, due to the expression of many of these susceptibility genes in both blood and renal cells, congenic analyses has also failed to confidently distinguish renal-intrinsic factors from hematopoietic factors. As described above, bone marrow chimera

TABLE 44.3   Nephritis-susceptibility loci mapped in human lupus nephritis

Locus	Ethnic group	LOD score of P-value
2q34-35 (SLEN2)	African-American	P = 0.000001
3q23	African-American	P = 0.00007
4q13.1	European-American	P = 0.00003
10q22.3 (SLEN1)	European-American	LOD = 3.16
11p13	African-American	P = 0.00003
11p15.6 (SLEN3)	African-American	LOD = 3.34
16q12	Multiple	P = 0.006

and renal transplantation experiments are needed to adequately assess the contributions of nephritis genes in kidney-intrinsic pathology.

TABLE 44.4   Candidate genes associated with nephritis in human SLE

Gene	Chrom Position	Ethnic Group	Nature of association with nephritis in SLE patients
FcγRIIa	1q23	Dutch Caucasians	R131 allele, P = 0.03
FcγRIIa	1q23	African-Americans	R131 allele, X^2 = 11.3
FcγRIIa	1q23	Hispanics	R131 allele, P < 0.002
FcγRIIa	1q23	Koreans	R131 allelic, X^2 = 9.29, P = 0.00959
FcγRIIa	1q23	Brazilians	R131 allele, P < 0.02
FcγRIIIA	1q23	Koreans	F176, P < 0.022
FcγRIIIA	1q23	Multiple	F158, P = 0.003
PARP	1q41-42	Koreans	SNP-1963:A > G, P = 0.03, + 28077G > A P = 0.0008
PDCD1	2q37	Swedish	PD-1.3A, X^2 = 10.2, P = 0.002
TNFB	6p21	Koreans	TNFB* 2, P < 0.0001
ACE	17q23	Hispanic, Asian	Position 23949 $(CT)_{2/3}$, P = 0.014
Dnase II	19p13.2-q13.4	Koreans	SNP-1066:G > C, P = 0.04
			SNP + 2630:T > C, P = 0.04
			SNP + 6235:G > C, P = 0.05

LUPUS NEPHRITIS LOCI IDENTIFIED IN HUMAN LUPUS

As executed in murine lupus, a limited number of mapping studies have also been executed in human lupus nephritis (Quintero-del-Rio et al 2002, 2004, Behrens et al 2005, Gillett et al 2005). As detailed in Table 44.3, these studies point to the existence of nephritis susceptibility loci on human chromosomes 2 (SLEN2), 3, 4, 10 (SLEN1), 11 (SLEN3), and 16. Since most of these loci are not associated with high serum ANA, a subset of these may harbor genetic elements that promote renal disease in a kidney-intrinsic fashion.

Finally, several research groups have identified allelic polymorphisms in specific genes (e.g. *FcRII, FcRIII, PARP, PDCD1*) that are highly associated with nephritis in SLE patients (Duits et al 1995, Kim et al 1996, Song et al 1998, Salmon et al 1999, Zuniga et al 2001, Parsa et al 2002, Karassa et al 2003, Bazilio et al 2004, Prokunina et al 2004, Shin et al 2005, Hur et al 2006). These findings are summarized in Table 44.4. One important point on these data is that it still remains likely that the reported genes may not actually be the bona fide lupus genes, and that they simply are in linkage disequilibrium with the actual causative genes.

genomic intervals identified in both murine and human studies which appear to potentiate nephritis. The extent to which these loci contribute to nephritis specifically remains the most pressing work to be done in this field.

It also clear that IC deposition plays a crucial role in disease initiation. The body of work to date has shown that there are clear difference between autoantibodies and their pathogenic potentials. Studies have uncovered loci that solely predispose to a loss of tolerance to nuclear antigen. Mice and humans with this form of benign autoimmunity are seropositive and yet never develop nephritis. However, it is clear from work done in murines that the simultaneous presence of other susceptibility loci can then drive the transition of benign autoimmunity to pathogenic autoimmunity. In particular, this transition is characterized by the production of nephrophillic antibodies. Understanding the genetic components and immunologic mechanisms mediating this transition is another important goal. The process of lupus nephritis is a dynamic one. Immunologic events are occurring simultaneously with renal events which, in turn, affect immunologic events ad infinitum. Thus, to understand a constantly evolving – and therefore unobservable – process, it is necessary to identify the genetic seeds from which this process is derived.

CONCLUSION

Much work remains to be done to unravel the pathogenesis of lupus nephritis. What is clear from the present body of work is that clearly there are pathways endogenous to the kidney which can affect disease severity. Given this, it is likely that certain alterations in kidney-intrinsic genes play a role in potentiating disease. To date, there have been several

References

Agnello V. Association of systemic lupus erythematosus and SLE-like syndromes with hereditary and acquired complement deficiency states. Arthritis Rheum. 1978; 21: S146–52.

Arbuckle MR, et al. Development of autoantibodies before the clinical onset of systemic lupus erythematosus. N. Engl. J. Med. 2003; 349: 1526–33.

Banchereau J, Pascual V. Type I interferon in systemic lupus erythematosus and other autoimmune diseases. Immunity 2006; 25: 383–92.

Bazilio AP, et al. Fc gamma RIIa polymorphism: a susceptibility factor for immune complex-mediated lupus nephritis in Brazilian patients. Nephrol. Dial. Transplant. 2004; 19: 1427–31.

Behrens TW, et al. Progress towards understanding the genetic pathogenesis of systemic lupus erythematosus. Novartis. Found. Symp. 2005; 267: 145–60.

Bergtold A, Gavhane A, D'Agati V, Madaio M, Clynes R. FcR-bearing myeloid cells are responsible for triggering murine lupus nephritis. J. Immunol. 2006; 177: 7287–95.

Bolland S, Ravetch JV. Spontaneous autoimmune disease in Fc(gamma)RIIB-deficient mice results from strain-specific epistasis. Immunity 2000; 13: 277–85.

Botto M, et al. Homozygous C1q deficiency causes glomerulonephritis associated with multiple apoptotic bodies [see comments]. Nat. Genet. 1998; 19: 56–9.

Cameron JS. Lupus nephritis. J. Am. Soc. Nephrol. 1999; 10: 413–24.

Carter RH, Fearon DT. CD19: lowering the threshold for antigen receptor stimulation of B lymphocytes. Science 1992; 256: 105–7.

Casciola-Rosen LA, Anhalt G, Rosen A. Autoantigens targeted in systemic lupus erythematosus are clustered in two populations of surface structures on apoptotic keratinocytes. J. Exp. Med. 1994; 179: 1317–30.

Chowdhury P, Sacks SH, Sheerin NS. Toll-like receptors TLR2 and TLR4 initiate the innate immune response of the renal tubular epithelium to bacterial products. Clin. Exp. Immunol. 2006; 145: 346–56.

Christensen SR, et al. Toll-like receptor 9 controls anti-DNA autoantibody production in murine lupus. J. Exp. Med. 2005; 202: 321–31.

Churg J, Sobin LH. Renal Disease: Classification and Atlas of Glomerular Disease. W.H. Organization. Tokyo: Igaku-Shoin, 1982.

Cohen PL, Eisenberg RA. Lpr and gld: Single gene models of systemic autoimmunity and lymphoproliferative disease. Annu. Rev. Immunol. 1991; 9: 243–69.

D'Agati VD. Systemic lupus erythematosus. In: Silva FG, Nadasdy T, eds. Renal Biopsy Interpretation. New York: Churchill Livingston, Inc, 1997: pp. 181–220.

Davis JC, Tassiulas IO, Boumpas DT. Lupus nephritis. Curr. Opin. Rheumatol. 1996; 8: 415–23.

Deapen D, et al. A revised estimate of twin concordance in systemic lupus erythematosus. Arthritis Rheum. 1992; 35: 311–18.

Deshmukh US, Bagavant H, Fu SM. Role of anti-DNA antibodies in the pathogenesis of lupus nephritis. Autoimmun. Rev. 2006; 5: 414–18.

Drake CG, Babcok SK, Palmer E, Kotzin BL. Genetic analysis of the NZB contribution to lupus-like autoimmune disease. Proc. Natl Acad. Sci. USA 1994; 91: 4062–5.

Drake CG, et al. Analysis of the New Zealand Black contribution to lupus-like renal disease. Multiple genes that operate in a threshold manner. J. Immunol. 1995; 154: 2441–7.

Duits AJ, et al. Skewed distribution of IgG Fc receptor IIa (CD32) polymorphism is associated with renal disease in systemic lupus erythematosus patients. Arthritis Rheum. 1995; 38: 1832–6.

Edberg JC, et al. Genetic linkage and association of Fcgamma receptor IIIA (CD16A) on chromosome 1q23 with human systemic lupus erythematosus. Arthritis Rheum. 2002; 46: 2132–40.

Fairhurst AM, Wandstrat AE, Wakeland EK. Systemic lupus erythematosus: multiple immunological phenotypes in a complex genetic disease. Adv. Immunol. 2006; 92: 1–69.

Fang Y, Xu C, Fu YX, Holers VM, Molina H. Expression of complement receptors 1 and 2 on follicular dendritic cells is necessary for the generation of a strong antigen-specific IgG response. J. Immunol. 1998; 160: 5273–9.

Gillett CD, et al. Fine mapping chromosome 16q12 in a collection of 231 systemic lupus erythematosus sibpair and multiplex families. Genes Immun. 2005; 6: 19–23.

Graham RR, et al. A common haplotype of interferon regulatory factor 5 (IRF5) regulates splicing and expression and is associated with increased risk of systemic lupus erythematosus. Nat. Genet. 2006; 38: 550–5.

Gu L, et al. Genetic determinants of autoimmune disease and coronary vasculitis in the MRL-lpr/lpr mouse model of systemic lupus erythematosus. J. Immunol. 1998; 161: 6999–7006.

Hargreaves DC, Medzhitov R. Innate sensors of microbial infection. J. Clin. Immunol. 2005; 25: 503–10.

Haywood ME, et al. Identification of intervals on chromosomes 1, 3, and 13 linked to the development of lupus in BXSB mice. Arthritis Rheum. 2000; 43: 349–55.

Haywood ME, et al. Autoantigen glycoprotein 70 expression is regulated by a single locus, which acts as a checkpoint for pathogenic anti-glycoprotein 70 autoantibody production and hence for the corresponding development of severe nephritis, in lupus-prone PXSB mice. J. Immunol. 2001; 167: 1728–33.

Hogarth MB, et al. Multiple lupus susceptibility loci map to chromosome 1 in BXSB mice. J. Immunol. 1998; 161: 2753–61.

Hur JW, et al. Poly(ADP-ribose) polymerase (PARP) polymorphisms associated with nephritis and arthritis in systemic lupus erythematosus. Rheumatology (Oxford) 2006; 45: 711–17.

Kalaaji M, Sturfelt G, Mjelle JE, Nossent H, Rekvig OP. Critical comparative analyses of anti-alpha-actinin and glomerulus-bound antibodies in human and murine lupus nephritis. Arthritis Rheum. 2006; 54: 914–26.

Karassa FB, Trikalinos TA, Ioannidis JP. The Fc gamma RIIIA-F158 allele is a risk factor for the development of lupus nephritis: a meta-analysis. Kidney Int. 2003; 63: 1475–82.

Kim TG, et al. Systemic lupus erythematosus with nephritis is strongly associated with the TNFB*2 homozygote in the Korean population. Hum. Immunol. 1996; 46: 10–17.

Kong PL, Morel L, Croker BP, Craft J. The centromeric region of chromosome 7 from MRL mice (Lmb3) is an epistatic modifier of Fas for autoimmune disease expression. J. Immunol. 2004; 172: 2785–94.

Kono DH, et al. Lupus susceptibility loci in New Zealand mice. Proc. Natl Acad. Sci. USA 1994; 91: 10168–72.

Kono DH, Park MS, Theofilopoulos AN. Genetic complementation in female (BXSB x NZW)F2 mice. J. Immunol. 2003; 171: 6442–7.

Kumar KR, et al. Regulation of B cell tolerance by the lupus susceptibility gene Ly108. Science 2006; 312: 1665–9.

Kyogoku C, et al. Genetic association of the R620W polymorphism of protein tyrosine phosphatase PTPN22 with human SLE. Am. J. Hum. Genet. 2004; 75: 504–7.

Li QZ, et al. Protein array autoantibody profiles for insights into systemic lupus erythematosus and incomplete lupus syndromes. Clin. Exp. Immunol. 2007; 147: 60–70.

Li S, Kurts C, Kontgen F, Holdsworth SR, Tipping PG. Major histocompatibility complex class II expression by intrinsic renal cells is required for crescentic glomerulonephritis. J. Exp. Med. 1998; 188: 597–602.

Manderson AP, Botto M, Walport MJ. The role of complement in the development of systemic lupus erythematosus. Annu. Rev. Immunol. 2004; 22: 431–56.

Mannik M, Merrill CE, Stamps LD, Wener MH. Multiple autoantibodies form the glomerular immune deposits in patients with systemic lupus erythematosus. J. Rheumatol. 2003; 30: 1495–504.

Mannik M, Wener MH. Deposition of antibodies to the collagen-like region of C1q in renal glomeruli of patients with proliferative lupus glomerulonephritis. Arthritis Rheum. 1997; 40: 1504–11.

Marshak-Rothstein A. Toll-like receptors in systemic autoimmune disease. Nat. Rev. Immunol. 2006; 6: 823–35.

Mohan C, Alas E, Morel L, Yang P, Wakeland EK. Genetic dissection of SLE pathogenesis: *Sle1* on murine chromosome 1 leads to a selective loss of tolerance to H2A/H2B/DNA subnucelosomes. J. Clin. Invest. 1998; 101: 1362–72.

Mohan C, et al. Genetic dissection of lupus pathogenesis: A recipe for nephrophilic autoantibodies. J. Clin. Invest. 1999a; 103: 1685–95.

Mohan C, Morel L, Yang P, Wakeland EK. Genetic dissection of SLE pathogenesis: *Sle2* on murine chromosome 4 leads to B-cell hyperactivity. J. Immunol. 1997; 159: 454–65.

Mohan C, Yu Y, Morel L, Yang P, Wakeland EK. Genetic dissection of SLE pathogenicity: *Sle3* on murine chromosome 7 impacts T cell activation, differentiation, and cell death. J. Immunol. 1999b; 162: 6492–502.

Molina H, Kinoshita T, Webster CB, Holers VM. Analysis of C3b/C3d binding sites and factor I cofactor regions within mouse complement receptors 1 and 2. J. Immunol. 1994; 153: 789–95.

Morel L, Blenman KR, Croker BP, Wakeland EK. The major murine systemic lupus erythematosus susceptibility locus, Sle1, is a cluster of functionally related genes. Proc. Natl Acad. Sci. USA 2001; 98: 1787–92.

Morel L, et al. Functional dissection of systemic lupus erythematosus using congenic mouse strains. J. Immunol. 1997; 158: 6019–28.

Morel L, et al. Multiplex inheritance of component phenotypes in a murine model of lupus. Mamm. Genome 1999a; 10: 176–81.

Morel L, et al. Genetic reconstitution of systemic lupus erythematosus immunopathology with polycongenic murine strains. Proc. Natl Acad. Sci. USA 2000; 97: 6670–5.

Morel L, Rudofsky UH, Longmate JA, Schiffenbauer J, Wakeland EK. Polygenic control of susceptibility to murine systemic lupus erythematosus. Immunity 1994; 1: 219–29.

Morel L, Tian XH, Croker BP, Wakeland EK. Epistatic modifiers of autoimmunity in a murine model of lupus nephritis. Immunity 1999b; 11: 131–9.

Morgan BP, Walport MJ. Complement deficiency and disease. Immunol. Today 1991; 12: 301–6.

Moroni G, et al. Anti-C1q antibodies may help in diagnosing a renal flare in lupus nephritis. Am. J. Kidney Dis. 2001; 37: 490–8.

Nakatsuru S, et al. Genetic dissection of the complex pathological manifestations of collagen disease in MRL/lpr mice. Pathol. Int. 1999; 49: 974–82.

Nelson RA Jr.. The immune-adherence phenomenon; an immunologically specific reaction between microorganisms and erythrocytes leading to enhanced phagocytosis. Science 1953; 118: 733–7.

Nguyen C, Limaye N, Wakeland EK. Susceptibility genes in the pathogenesis of murine lupus. Arthritis Res. 2002; 4(Suppl. 3): S255–63.

Parsa A, et al. Association of angiotensin-converting enzyme polymorphisms with systemic lupus erythematosus and nephritis: analysis of 644 SLE families. Genes Immun. 2002; 3(Suppl. 1): S42–6.

Pasare C, Medzhitov R. Toll-like receptors: linking innate and adaptive immunity. Adv. Exp. Med. Biol. 2005; 560: 11–18.

Patole PS, et al. Toll-like receptor-4: renal cells and bone marrow cells signal for neutrophil recruitment during pyelonephritis. Kidney Int. 2005; 68: 2582–7.

Pisitkun P, et al. Autoreactive B cell responses to RNA-related antigens due to TLR7 gene duplication. Science 2006; 312: 1669–72.

Prokunina L, et al. A regulatory polymorphism in PDCD1 is associated with susceptibility to systemic lupus erythematosus in humans. Nat. Genet. 2002; 32: 666–9.

Prokunina L, et al. The systemic lupus erythematosus-associated PDCD1 polymorphism PD1.3A in lupus nephritis. Arthritis Rheum. 2004; 50: 327–8.

Quintero-del-Rio AI, et al. SLEN2 (2q34-35) and SLEN1 (10q22.3) replication in systemic lupus erythematosus stratified by nephritis. Am. J. Hum. Genet. 2004; 75: 346–8.

Quintero-del-Rio AI, Kelly JA, Kilpatrick J, James JA, Harley JB. The genetics of systemic lupus erythematosus stratified by renal disease: linkage at 10q22.3 (SLEN1), 2q34-35 (SLEN2), and 11p15.6 (SLEN3). Genes Immun. 2002; 3(Suppl. 1): S57–62.

Rahman ZS, et al. A novel susceptibility locus on chromosome 2 in the (New Zealand Black x New Zealand White)F1 hybrid mouse model of systemic lupus erythematosus. J. Immunol. 2002; 168: 3042–9.

Rigby RJ, et al. New loci from New Zealand Black and New Zealand White mice on chromosomes 4 and 12 contribute to lupus-like disease in the context of BALB/c. J. Immunol. 2004; 172: 4609–17.

Rozzo SJ, Vyse TJ, Drake CG, Kotzin BL. Effect of genetic background on the contribution of New Zealand black loci to autoimmune lupus nephritis. Proc. Natl Acad. Sci. USA 1996; 93: 15164–8.

Rozzo SJ, Vyse TJ, Menze K, Izui S, Kotzin BL. Enhanced susceptibility to lupus contributed from the nonautoimmune C57BL/10, but not C57BL/6, genome. J. Immunol. 2000; 164: 5515–21.

Salmon JE, et al. Altered distribution of Fcgamma receptor IIIA alleles in a cohort of Korean patients with lupus nephritis. Arthritis Rheum. 1999; 42: 818–19.

Salmon JE, et al. Fc gamma RIIA alleles are heritable risk factors for lupus nephritis in African Americans. J. Clin. Invest 1996; 97: 1348–54.

Santiago ML, et al. Linkage of a major quantitative trait locus to Yaa gene-induced lupus-like nephritis in (NZW x C57BL/6) F1 mice. Eur. J. Immunol. 1998; 28: 4257–67.

Schmiedeke TM, Stockl FW, Weber R, Sugisaki Y, Batsford SR, Vogt A. Histones have high affinity for the glomerular basement membrane. Relevance for immune complex formation in lupus nephritis. J. Exp. Med. 1989; 169: 1879–94.

Schur PH, Meyer I, Garovoy M, Carpenter CB. Associations between systemic lupus erythematosus and the major histocompatibility complex: clinical and immunological considerations. Clin. Immunol. Immunopathol. 1982; 24: 263–75.

Shin HD, et al. DNase II polymorphisms associated with risk of renal disorder among systemic lupus erythematosus patients. J. Hum. Genet. 2005; 50: 107–11.

Shmerling RH. Autoantibodies in systemic lupus erythematosus–there before you know it. N. Engl. J. Med. 2003; 349: 1499–500.

Siegert C, Daha M, Westedt ML, d.van V, Breedveld F. IgG autoantibodies against C1q are correlated with nephritis, hypocomplementemia, and dsDNA antibodies in systemic lupus erythematosus. J. Rheumatol. 1991; 18: 230–4.

Snell GD. Methods for the study of histocompatibility genes. J. Genetics 1948; 49: 87.

Song YW, et al. Abnormal distribution of Fc gamma receptor type IIa polymorphisms in Korean patients with systemic lupus erythematosus. Arthritis Rheum. 1998; 41: 421–6.

Subramanian S, et al. From the Cover: A *Tlr7* translocation accelerates systemic autoimmunity in murine lupus. Proc. Natl Acad. Sci. USA 2006; 103: 9970–75.

Theofilopoulos AN, Kono DH. Genetics of systemic autoimmunity and glomerulonephritis in mouse models of lupus. Nephrol. Dial. Transplant. 2001; 16(Suppl. 6): 65–7.

Timoshanko JR, Holdsworth SR, Kitching AR, Tipping PG. IFN-gamma production by intrinsic renal cells and bone marrow-derived cells is required for full expression of crescentic glomerulonephritis in mice. J. Immunol. 2002; 168: 4135–41.

Timoshanko JR, Kitching AR, Holdsworth SR, Tipping PG. Interleukin-12 from intrinsic cells is an effector of renal injury in crescentic glomerulonephritis. J. Am. Soc. Nephrol. 2001; 12: 464–71.

Timoshanko JR, Kitching AR, Iwakura Y, Holdsworth SR, Tipping PG. Leukocyte-derived interleukin-1beta interacts with renal interleukin-1 receptor I to promote renal tumor necrosis factor and glomerular injury in murine crescentic glomerulonephritis. Am. J. Pathol. 2004; 164: 1967–77.

Timoshanko JR, Sedgwick JD, Holdsworth SR, Tipping PG. Intrinsic renal cells are the major source of tumor necrosis factor contributing to renal injury in murine crescentic glomerulonephritis. J. Am. Soc. Nephrol. 2003; 14: 1785–93.

Trouw LA, et al. Anti-C1q autoantibodies deposit in glomeruli but are only pathogenic in combination with glomerular C1q-containing immune complexes. J. Clin. Invest. 2004; 114: 679–88.

Tsuboi N, et al. Roles of toll-like receptors in C-C chemokine production by renal tubular epithelial cells. J. Immunol. 2002; 169: 2026–33.

Tucker RM, et al. Genetic control of glycoprotein 70 autoantigen production and its influence on immune complex levels and nephritis in murine lupus. J. Immunol. 2000; 165: 1665–72.

Turnberg D, et al. Complement activation contributes to both glomerular and tubulointerstitial damage in adriamycin nephropathy in mice. J. Immunol. 2006; 177: 4094–102.

Vidal S, Kono DH, Theofilopoulos AN. Loci predisposing to autoimmunity in MRL-Fas lpr and C57BL/6-Faslpr mice. J. Clin. Invest. 1998; 101: 696–702.

Vyse TJ, et al. Genetic linkage of IgG autoantibody production in relation to lupus nephritis in New Zealand hybrid mice. J. Clin. Invest 1996a; 98: 1762–72.

Vyse TJ, Kotzin BL. Genetic basis of systemic lupus erythematosus. Curr. Opin. Immunol. 1996; 8: 843–51.

Vyse TJ, Morel L, Tanner FJ, Wakeland EK, Kotzin BL. Backcross analysis of genes linked to autoantibody production in New Zealand white mice. J. Immunol. 1996b; 157: 2719–27.

Vyse TJ, Rozzo SJ, Drake CG, Izui S, Kotzin BL. Control of multiple autoantibodies linked with a lupus nephritis susceptibility locus in New Zealand black mice. J. Immunol. 1997; 158: 5566–74.

Wakeland EK, Liu K, Graham RR, Behrens TW. Delineating the genetic basis of systemic lupus erythematosus. Immunity 2001; 15: 397–408.

Wandstrat A, Wakeland E. The genetics of complex autoimmune diseases: non-MHC susceptibility genes. Nat. Immunol. 2001; 2: 802–9.

Wandstrat AE, et al. Association of extensive polymorphisms in the SLAM/CD2 gene cluster with murine lupus. Immunity 2004; 21: 769–80.

Wandstrat AE, et al. Autoantibody profiling to identify individuals at risk for systemic lupus erythematosus. J. Autoimmun. 2006; 27: 153–60.

Wang Y, Nose M, Kamoto T, Nishimura M, Hiai H. Host modifier genes affect mouse autoimmunity induced by the lpr gene. Am. J. Pathol. 1997; 151: 1791–8.

Waters ST, et al. Breaking tolerance to double stranded DNA, nucleosome, and other nuclear antigens is not required for the pathogenesis of lupus glomerulonephritis. J. Exp. Med. 2004; 199: 255–64.

Waters ST, et al. NZM2328: a new mouse model of systemic lupus erythematosus with unique genetic susceptibility loci. Clin. Immunol. 2001; 100: 372–83.

Watson ML, et al. Genetic analysis of MRL-lpr mice: Relationship of the Fas apoptosis gene to disease manifestations and renal disease- modifying loci. J. Exp. Med. 1993; 176: 1645–56.

Winfield JB, Faiferman I, Koffler D. Avidity of anti-DNA antibodies in serum and IgG glomerular eluates from patients with systemic lupus erythematosus. Association of high avidity anti-native DNA antibody with glomerulonephritis. J. Clin. Invest. 1977; 59: 90–6.

Wither JE, et al. Functional dissection of lupus susceptibility loci on the New Zealand black mouse chromosome 1: evidence for independent genetic loci affecting T and B cell activation. J. Immunol. 2003; 171: 1697–706.

Xie S, et al. Genetic contributions of nonautoimmune SWR mice toward lupus nephritis. J. Immunol. 2001; 167: 7141–9.

Xie S, et al. Genetic origin of lupus in NZB/SWR hybrids: lessons from an intercross study. Arthritis Rheum. 2005; 52: 659–67.

Zhu J, et al. T cell hyperactivity in lupus as a consequence of hyperstimulatory antigen-presenting cells. J. Clin. Invest. 2005; 115: 1869–78.

Zuniga R, et al. Low-binding alleles of Fcgamma receptor types IIA and IIIA are inherited independently and are associated with systemic lupus erythematosus in Hispanic patients. Arthritis Rheum. 2001; 44: 361–7.

CHAPTER 45

IgA Nephropathy

ISABEL BEERMAN, FRANCESCO SCOLARI AND ALI GHARAVI

DEFINITION AND CLINICAL FEATURES

IgA Nephropathy (OMIM #161950) was first described in 1968 by Berger and Hinglais, who utilized the novel immun-ofluorescence technique to describe a form of mesangial proliferative glomerulonephritis characterized by diffuse mesangial deposits of immunoglobulin A (IgA) (Berger & Hinglais 1968). These observations have provided the criteria for diagnosis of IgA nephropathy (IgAN) ever since (Jennette 1988). This disorder was soon recognized as the most common primary glomerulonephritis in the world (D'Amico 1987). IgAN has an estimated population prevalence of 5–50/100 000 with great geographical variations (Pettersson et al 1984, Schena 1990): it accounts for about 5% of cases of primary glomerulonephritis in the USA but over 40% of cases in certain parts of Europe and Asia (Glassock et al 1985, Power et al 1985, D'Amico 1987, Taguchi et al 1987, Levy & Berger 1988, Abu-Romeh et al 1989, Schena 1990, Cheong et al 1991, Research Group on PCRD 1999, Briganti et al 2001, Yamagata et al 2002, Gesualdo et al 2004). Due to clinical variability and requirement for renal biopsy for diagnosis, the true incidence of this disorder is not known (Stratta et al 1996); however, autopsy studies and transplant biopsy series suggest that it may have a prevalence of up to 1.3% (Rosenberg et al 1990, Cosyns et al 1998, Suzuki et al 2003). The diagnosis is made by renal biopsy, demonstrating mesangial proliferation with glomerular immune complexes that include IgA, as the sole of predominant immunoglobulin deposited; C3 deposition is almost always present and IgG and IgM are also often present. There is no reduction in serum complement levels and no detectable autoantibodies. Immune complexes composed of IgA, complement, fibronectin and collagen are detected in circulation (van Es et al 1988, Baldree et al 1993).

IgAN can occur as a primary (idiopathic) disorder or as a secondary trait due to infections by organisms such as HIV or *Schistosoma*, or in association with systemic diseases such as Crohn's disease, celiac sprue, ankylosing spondylitis, or liver cirrhosis (Wall et al 1984, Sobh et al 1988, Chin et al 1992, Hirsch et al 1992, Nochy et al 1993, Beaufils et al 1995, Barsoum et al 1996, Dash et al 1997, Praditpornsilpa et al 1999, Takemura et al 2002). Among patients with primary IgAN, interindividual variation in clinical presentation, progression and therapeutic response has suggested that the diagnosis encompasses heterogeneous disorders that cannot be distinguished on the basis of renal histology alone. However, there are no universally accepted clinical or pathologic findings that can clearly subclassify the disease or predict outcome (Haas 1997, Bartosik et al 2001, Lai et al 2002).

Manifestations of disease include persistent or episodic microscopic hematuria (60–70%), episodic macroscopic hematuria (15–45%), nephrotic syndrome (5%) and, rarely, acute reversible renal failure (3%) (Galla 1995). Disease onset commonly occurs in the second or third decade of life and most studies report a 2:1 male predominance (Galla 1995). Mucosal infections are a precipitating or exacerbating factor in the development of disease. Long-term studies indicate that 15–25% of patients with biopsy-proven IgAN eventually develop ESRD (Nicholls et al 1984, Clarkson et al 1987, D'Amico et al 1987, Propper et al 1987, Tojo et al 1987, Velo et al 1987, Johnston et al 1992, Research Group on PCRD 1999, Geddes et al 2003, D'Amico 2004). In addition, IgAN may recur after renal transplantation in 50–60% of cases, and can lead to graft loss (Mathew 1988, O'Meara et al 1989, Lim & Terasaki 1991, Kessler et al 1996, Cosyns et al 1998).

Previous studies have demonstrated substantial overlap between IgAN and Henoch-Schönlein purpura (HSP) (Waldo 1988, Galla 1995, Fervenza 2003). The renal pathology is identical between the two disorders. HSP occurs predominantly in children and is characterized by the tetrad of palpable purpura, arthritis, abdominal pain and, sometimes, renal disease. HSP nephritis mostly manifests as hematuria and rarely progresses to renal failure in children (Tancrede-Bohin et al 1997, Pillebout et al 2002). Skin biopsy in HSP patients demonstrates deposition of IgA in dermal capillaries and, in the presence of compatible clinical feature, is diagnostic of the disease. Similarly, 30–50% of patients with IgAN demonstrate IgA deposition in dermal capillaries, although the sensitivity and specificity of this finding is too low to be diagnostic of the disease.

Genetic Diseases of the Kidney
ISBN: 978-0-12-449851-8

Finally, IgAN and HSP nephritis can occur within the same family and in identical twins, providing further evidence of a common pathogenic basis (Meadow & Scott 1985, Levy 1989, 1993, 2001, Miyagawa et al 1989). Hence, many investigators believe that HSP is a systemic form of IgAN (Waldo 1988). Since HSP usually occurs in children and IgAN in adults, differences in clinical manifestations may be related to different antigenic exposure at various stages of life (Hoffjan et al 2004, 2005).

PATHOGENESIS

The IgA System

Humans have two IgA subtypes, IgA1 and IgA2 (Kerr 1990). IgA1 contains an additional 13 amino acids in the hinge region that can be targeted by bacterial proteases. This hinge region contains O-linked glycans, a feature unique to IgA and IgD (Kerr 1990). Serum IgA is mostly composed of IgA1 subclass which circulates in monomeric form and is produced by the bone marrow and some peripheral lymphoid organs. Mucosal IgA is the most abundant immunoglulin in secretions and is produced by lymphocytes and plasma cells in lymphoid tissue at mucosal surfaces, such as Peyer's patches (Kerr 1990, van Egmond et al 2001, Monteiro & Van De Winkel 2003). In contrast to circulating IgA, mucosal IgA is composed of both IgA1 and IgA2 subclasses and is produced as a dimer joined by a joining (J) chain. The IgA dimer + J chain interact with the polymeric immunoglobulin receptor at the basolateral surface the mucosal epithelial cells (Kerr 1990, Johansen et al 2001, van Egmond et al 2001, Monteiro & Van De Winkel 2003). As this complex is translocated across the epithelial cell, the transmembrane domain of the polymeric immunoglobulin receptor is cleaved off, leaving the remaining portion (called secretory component) covalently bound to IgA. This IgA-secretory component complex (now called secretory IgA) forms the first line of defense against microorganisms at mucosal surfaces, preventing both bacterial and viral infections (Norderhaug et al 1999). In addition to the traditional Fab binding regions, IgA1 bound to secretory component possesses adhesin binding glycan epitopes that provide supplementary bacterial binding sites (Royle et al 2003). The binding of secretory IgA to antigen does not trigger an immune response, consistent with the need to prevent chronic inflammation at mucosal surfaces. The role of circulating monomeric IgA has not been well defined, but it was postulated to enable clearance of antigens via the phagocytic system (Hammarstrom et al 2000, Cunningham-Rundles 2001). Serum IgA may also help in removal of antigens that have invaded the circulation in pathologic conditions and enables clearance of partially degraded self-antigens (Weisbart et al 1988). By eliminating antigen entry at the mucosal barrier, IgA simultaneously plays an 'immune exclusion role' that prevents immune

activation and immune complex formation (Mestecky et al 1999, Favre et al 2005). In support of this hypothesis, in individuals with selective IgA deficiency allergic or autoimmune disorders predominate compared to infectious complications (Hammarstrom et al 2000, Cunningham-Rundles 2001).

While monomeric IgA or secretory IgA generally do not cause immune activation, polymeric IgA can efficiently trigger the immune system via interaction with IgA receptors. There are five types of IgA receptors, of which three are Fc receptors: the polymeric immunoglobulin receptor (pIgR), the Fcα/μR and the FcαRI (also called CD89); two other alternative receptors are the asialoglycoprotein and the transferrin receptors (van Egmond et al 2001, Monteiro & Van De Winkel 2003). The pIgR, whose cleavage yields the secretory component, transports IgM and IgA across epithelial barriers. CD89 is a transmembrane receptor of the immunoglobulin receptor gene superfamily. It is specific for IgA and can bind both IgA1 and IgA2 subclasses. Its expression is restricted to myeloid lineage with signaling mediated by association with the FcRγ chain. The recently described Fcα/μR binds both IgA and IgM; it is constitutively expressed on B lymphocytes and macrophages, but also in secondary lymphoid organs, such as lymph node and appendix, kidney and intestine. The asialoglycoprotein receptor is expressed in the liver, and mediates clearance of serum glycoproteins such as IgA via recognition of terminal galactose residues. The transferrin receptor selectively binds IgA1. Because the transferrin receptor and Fcα/μR are expressed in glomerular mesangial cells, they may mediate interaction with IgA1 complexes and trigger glomerular inflammation (McDonald et al 2002, Moura et al 2004, 2005).

Insight From Human Studies

The pathogenesis of IgAN and HSP remains largely unknown. Clinical features such as the presence of IgA deposits in dermal tissue (O'Neill et al 1982), the polyclonal origin of mesangial IgA deposits (Sakai 1988) and the disappearance of disease in IgAN kidneys inadvertently transplanted in patients with other forms of renal disease (Cuevas et al 1987, Koselj et al 1997), all suggest that a systemic process contributes to the renal disease (Galla 1995, Julian et al 1999, Barratt & Feehally 2005). Over the years, numerous pathogenic mechanisms have been proposed for this disease, suggesting either a primary immune defect (e.g. autoimmunity) that results in increased IgA immune complex formation, deficiency of a circulating factor which normally prevents IgA aggregation, or a defect in IgA structure leading to reduced clearance and increased tissue deposition.

IgA Levels

Data indicate that ~50% of IgAN patients have increased circulating IgA levels, with a higher proportion of the λ light chain compared to normal serum (Lai et al 1986, 1988, Galla 1995, Julian et al 1999, Barratt & Feehally 2005). The circulating IgA is in polymeric form which favors immune complex formation (Jones et al 1990). Overproduction by the bone marrow has been suspected because circulating IgA immune complexes are almost exclusively composed of the IgA1 subclass (Czerkinsky et al 1986, van den Wall Bake et al 1988). Moreover, mucosal IgA levels are not usually elevated and mucosal plasma cell numbers are not increased (van den Wall Bake et al 1988, Layward et al 1993). Numerous studies have shown that circulating IgA levels do not correlate with the clinical manifestations of disease, such as development of macroscopic hematuria, or proteinuria or progression to renal failure (Galla 1995, Julian et al 1999, Barratt & Feehally 2005). While some studies have shown an exaggerated serum IgA response to antigen challenge in IgAN patients, correction for higher preimmunization levels actually revealed reduced response to antigen (Jackson et al 1992). This suggested that repeated suboptimal response to antigenic challenge may lead to persistent stimulation of bone marrow and memory cells leading to a polyclonal IgA response. However, this hypothesis does not explain the mechanism of IgA immune complex formation and deposition in the renal mesangium, because conditions associated with high circulating IgA levels, such as HIV-1 infection or liver disease, rarely lead to renal disease. Hence additional defects that promote IgA aggregation and deposition must be present.

Immune Defects

Because IgAN is characterized by IgA deposition in tissues, autoimmunity has been evoked as a possible etiology. However, circulating IgA1 antibodies recognize a variety of antigens suggesting that immune complex formation is not elicited by specific mesangial antigen (Galla 1995, Julian et al 1999, Barratt & Feehally 2005). Examination of the relationship between the histocompatibility antigens and IgAN has also yielded conflicting results (Luger et al 1994, Schena 1995). Although associations with HLA class I and II antigens have been reported, no consistent relationship has been detected and most findings have not been replicated, sometimes within the same population (Luger et al 1994, Schena 1995). For example, Fennessy et al studied IgAN patients from three European populations but found no consistent association of the HLA genotypes with disease (Fennessy et al 1996). Thus, the role of the HLA locus in IgAN is unclear. Finally, deficiencies in several complement proteins have also been associated with IgAN or HSP. These include complete C3 or C1q deficiency, or partial deficiencies of H, P, I, C2 and C4BP (Imai et al 1991, Jin et al 1992, Yoshioka et al 1992, Wyatt 1993, Galla 1995, Topaloglu et al 1996, Kanda et al 2001). Because these complement defects mostly result in membranoproliferative glomerulonephritis or systemic lupus erythematosus, these

rare associations suggest that the complement pathway normally plays a protective role in preventing the development of IgA nephropathy.

Glycosylation Defects (Figure 45.1)

Recurrence of IgAN after transplantation implicates extrarenal factors in the development of disease (Mathew 1988, O'Meara et al 1989, Lim & Terasaki 1991, Kessler et al 1996, Cosyns et al 1998). Numerous studies have reported significant abnormalities in the carbohydrate moieties in the hinge regions of IgA1 in patients with IgAN (Tomana et al 1997, 1999). The hinge region of IgA1, located between the first and second heavy chain domain, contains three to five O-glycosylation sites, a structural characteristic unique to higher primates (Kerr 1990, Mattu et al 1998). These sites contain N-acetyl-galatosamine (GalNAc) with ß1,3-linked galactose which is attached by core 1-galactosyltransferase (ß1,3GalT) enzyme; these GalNAc moieties may be further modified by another enzyme, α2,6-sialyltransferase (Field et al 1989, Julian et al 1999) (Figure 45.1). The galactose residues on the O-linked glycan chains are thought to mediate interaction with the asialoglycoprotein receptor, permitting clearance of IgA1 by the liver (Julian et al 1999). Multiple studies, using independent methodologies, have shown that circulating and mesangial IgA1 in IgAN patients demonstrate incompletely galactosylated O-glycans in the hinge regions (Allen et al 1995, 2001, Tomana et al 1997, 1999, Hiki et al 2001). These glycosylation defects are absent in other O-glycosylated proteins, such as C1 inhibitors, suggesting a specific defect. These galactose-deficient IgA1 molecules are more likely to self-aggregate and are recognized by naturally occurring antibodies to reactive undergalactosylated GalNAc, resulting in formation of immune complexes (Tomana et al 1999). Immune complexes containing these Gal deficient IgA1 are cleared less effectively clearance receptors. In addition, they efficiently bind to mesangial cells, resulting in glomerular deposition (Moura et al 2004, 2005) and activation of inflammatory cytokines that can initiate or perpetuate the course of glomerulonephritis perhaps via interaction via the transferrin receptor (Iwase et al 1999, Sano et al 2002, Moura et al 2004, 2005, Novak et al 2005). Interestingly, other rheumatologic diseases, such as rheumatoid arthritis or systemic lupus erythematosus, also demonstrate alterations in immunoglobulin glycosylation, such as abnormal galactosylation of serum IgG (Axford 1999). These findings implicate glycosylation defects in the development and propagation of glomerulonephritis. Consistent with these data, mice with defective O-glycan synthesis have been shown to have autoimmunity and glomerulonephritis (Lowe 2001). To date, investigations into the causes of aberrant galactosylation have not uncovered variations in the coding sequence of the IgA1 hinge region (Greer et al 1998). However, studies have shown that the ß1,3GalTransferase

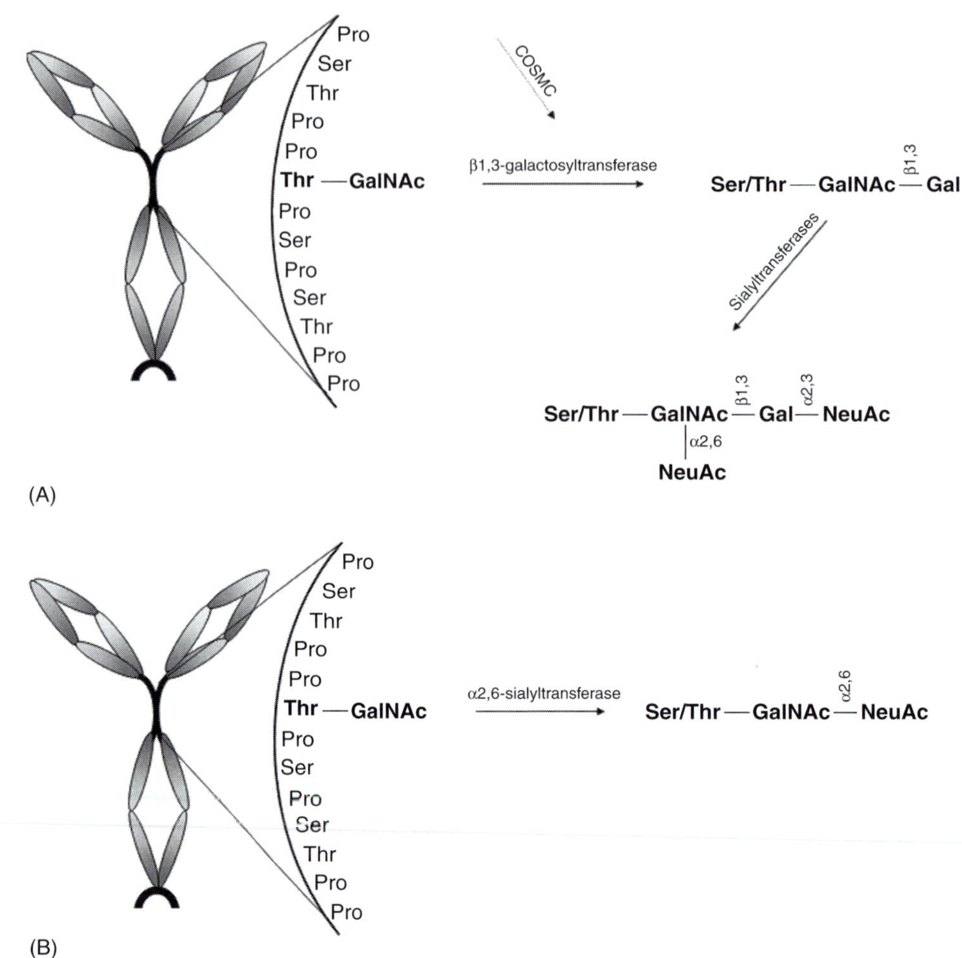

FIGURE 45.1   The structure of IgA1. The heavy chain is in blue, the light chain in yellow. Compared to IgA2, the hinge region of IgA1 contains an additional 13 amino acids; the serines and threonines in this region serve as O-glycosylation sites. Three to five sites are occupied by O-glycans in each IgA1 molecule. (A) N-acetylgalactosamine (GalNAc) is attached to Ser/Thr via N-acetylgalactosaminyltransferase 2. GalNAc can be modified by addition of galactose (Gal) by β1,3-galactosyltransferase and further expanded by sialyltransferase mediated attachment of sialyic acid (NeuAc) to GalNAc and Gal by α2,6 or α2,3 linkage, respectively. COSMC is require for proper folding of β1,3-galactosyltransferase. (B) Alternatively, α2,6-sialyltransferase can directly attach NeuAc to GalNAc; this arrangement prevents further modification. Normal serum IgA1 mostly comprises Gal-GalNAc disaccharides and its mono- and di-sialylated forms. Variants with terminal GalNAc or sialylated GalNAc are more common in IgAN. (see also Plate 91)

activity is reduced in IgAN and in HSP patients with nephritis (but not HSP without nephritis) suggesting that acquired defects in this enzyme or impairment of regulatory pathways controlling enzyme activity may cause the under-galactosylation phenotype (Allen et al 1997, 1998). There are several ß1,3GalTransferase with specificities for different carbohydrate substrates (Iwasaki et al 2003); the main candidate for the core ß1,3GalT that adds galactose to the O-linked carbohydrate moieties of IgA1 is the C1GALT1 gene located on 7p14-p13 (Ju et al 2002). This gene does not show linkage to familial IgAN, but pathways regulating core ß1,3GalT activity have now been implicated in human disease. Recently, it was shown that proper activity of C1GALT1 requires expression of a molecular chaperone

designated COSMC, which enables proper protein folding (Ju & Cummings 2002). Without COSMC, the C1GALT1 is targeted to proteasomes, resulting in the incomplete glycosylation of membrane proteins (Ju & Cummings 2002). Somatic mutations in COSMC in blood cells are associated with the Tn syndrome, an acquired defect resulting in anemia, leukopenia and thrombocytopenia (Ju & Cummings 2005). So far, one study has found reduced expression of COSMC in patients with IgA nephropathy (Qin et al 2005). Taken together, these data suggest that inherited or acquired defects in glycosylation may be the critical factor linking increased IgA production to immune complex formation, mesangial deposition and glomerular inflammation.

Insight From Animal Models

Animal models have traditionally provided powerful tools for dissecting the pathogenesis of complex traits. Several models of IgA nephropathy have been produced in the mouse, each recapitulating some of the features of human disease and highlighting potential pathogenic mechanisms. However, several fundamental differences in the murine IgA system must be taken into account (Kerr 1990, van Egmond et al 2001, Monteiro & Van De Winkel 2003). Mice only produce a single IgA isotype. Whereas circulating IgA in humans is mostly in monomeric form, circulating IgA in mice is mostly in polymeric form. Mice do not have a homolog of the human IgA receptor FcαRI (CD89), perhaps because interaction with polymeric IgA would cause immune activation. Finally, murine IgA also lacks the hinge region in the immunoglobulin heavy chain, preventing study of glycosylation defects observed in humans. Despite these caveats, some useful insight has emerged from animal studies.

SPONTANEOUS IgAN IN THE MOUSE

The DDY mouse isg one of the best studied models of IgAN. This strain originated from an outbred stock which was noted to have high circulating IgA levels (Imai et al 1985). With aging, these mice develop mesangial proliferative glomerulonephritis with spontaneous IgA deposition (Imai et al 1985). Selective mating of mice with high IgA levels resulted in a substrain (named HIGA) with high serum polymeric IgA and the development of IgA deposition (Muso et al 1996, Miyawaki et al 1997). The high IgA levels may be due to dysregulation of mucosal immunity: these mice display high IgA-producing plasma cells in the intestine. Several genome scans have now been reported in the original DDY stock as well as the HIGA substrain (Nogaki et al 2005, Suzuki et al 2005). Interestingly, the genome scan in the outbred stock has yielded linkage peaks in regions that correspond to the human *IGAN1* locus and the selectin loci, suggesting that the model may be useful for elucidation of human disease (Suzuki et al 2005). Genome scans in the HIGA mouse have also identified loci that control serum IgA levels, which are distinct from another locus determining glomerular IgA deposition (Nogaki et al 2005). This model may therefore prove useful for dissecting mechanisms of IgA overproduction from those governing deposition. Although HIGA strain was derived from the DDY stock, it is worth noting that there is little overlap between linkage peaks identified in the two models, suggesting significant multigenic determination of disease.

UTEROGLOBIN AND IgAN

A number of genetically engineered mice have also recapitulated features of IgA nephropathy. A potential candidate molecule was suggested by targeted deletion of the uteroglobin gene, which causes severe renal disease with fibronectin (Fn) deposition in mice (Zhang et al 1997); this was initially thought to constitute a model of fibronectin deposition disease, a variant of membranoproliferative glomerulonephritis. Subsequent immunofluorescence analysis of these uteroglobin null mice revealed mesangial IgA and C3 deposition, a finding confirmed in a second mouse expressing an antisense transgene to uteroglobin (Zheng et al 1999). Uteroglobin is an anti-inflammatory secreted protein which is expressed by mucosal epithelial cells of most organs and is detectable in serum and urine (Shijubo et al 2003). Its expression is upregulated by mucosal inflammation. In airway epithelial cells, uteroglobin represses allergen-induced immune signaling, suggesting a homeostatic role in repressing inflammatory response (Mandal et al 2004, 2005, Ray et al 2005). Additional studies also demonstrated that circulating uteroglobin normally prevents formation of IgA-fibronectin complexes in serum (Zheng et al 1999). Inactivation of the gene therefore leads to increased IgA-fibronectin aggregation, immune complex formation, and subsequent glomerular deposition resulting in inflammation. These data identified a major circulating protein involved in immune complex formation. Humans with IgAN, however, have elevated uteroglobin levels, suggesting a compensatory increase in response to high circulating IgA–fibronectin complexes (Coppo et al 2002). Similarly, association studies of IgA nephropathy with common variants in uteroglobin gene have yielded conflicting results (Szelestei et al 2000, Matsunaga et al 2002, Menegatti et al 2002, Narita et al 2002).

CD89 AND IgAN

Other studies have implicated the FcαRI receptor (CD89) in the pathogenesis of IgAN. CD89 is specific for IgA and binds both IgA1 and IgA2 subclasses, with expression restricted in cells of the myeloid lineage (van Egmond et al 2001, Monteiro & Van De Winkel, 2003). Monomeric IgA is a weak activator of CD89, whereas polymeric IgA binds CD89 with avidity. In patients with IgAN, CD89 expression is reduced in monocytes, but CD89 occupancy is increased and there is increased expression of soluble CD89 complexed with IgA (Grossetete et al 1998, Launay et al 2000). Hence, CD89 is one of the components of IgA immune complexes. This suggested that IgA1–CD89 interaction leads to cleavage of the CD89 extracellular domain and shedding of complexes into the circulation. Consistent with these data, overexpression of the human CD89 receptor in the macrophage/monocyte lineage results in circulating CD89–IgA complexes, which deposit in the renal mesangium, recapitulating histologic and clinical features of IgAN (Launay et al 2000). The mechanism for spontaneous disease in this model can be explained by the absence of the CD89 homolog in mice and the fact that the circulating mouse IgA is predominantly in polymeric form. In immunodeficient (SCID) CD89 transgenic mice which are immunoglobulin deficient, disease could be induced only by transfer of IgA from IgAN patients but not IgA from normal controls, suggesting an additional requirement for abnormal IgA structure in immune complex formation (Launay et al 2000).

INTESTINAL INFLAMMATION AND SECONDARY FORMS OF IgAN

Inflammatory bowel disease is associated with increased incidence of IgAN. Recent data from LIGHT transgenic mice have provided insight into such secondary forms of IgAN associated with intestinal disease. LIGHT is a ligand for the lymphotoxin ß receptor (LTßR) and is expressed on activated T cells and immature dendritic cells. LTßR expression in the intestine is required for the development of mucosal lymphoid organs, and consequently for the production of IgA and recruitment of IgA precursors; LTßR null mice display low levels of baseline serum and fecal IgA (Kang et al 2002). On the other hand, mice overexpressing human LIGHT in the T cell lineage spontaneously develop severe intestinal inflammation and autoimmune disease manifested by splenomegaly, lymphadenopathy and glomerulonephritis (Wang et al 2001). The glomerulonephritis is associated with mesangial IgA deposition and 30–40-fold elevation of serum polymeric IgA levels (Wang et al 2004). Increased IgA precedes overt intestinal inflammation, suggesting that increased LIGHT mediated LTßR signaling causes IgA overproduction. Intestinal inflammation further amplified IgA overproduction: these mice showed increased numbers of IgA-positive Peyer's patches and reduced IgA secretion into the gut lumen, suggesting that impaired transport of IgA due to epithelial inflammation results in inappropriate release of polymeric IgA into the serum (Wang et al 2004). Consistent with this hypothesis, two other models of intestinal inflammation resulted in increased serum polymeric IgA and glomerular IgA deposition; the IgA overproduction could be prevented by administration of anti-LIGHT blocking antibodies (Wang et al 2004). Finally, examination of patients with Crohn's disease revealed elevated serum IgA in the majority of these patients, with high IgA levels correlating with increased incidence of hematuria (Wang et al 2004). These data provide a mechanism explaining clinical association of inflammatory bowel disorders with IgA nephropathy and implicate dysregulation of intestinal mucosal epithelium with consequent release of IgA complexes into the circulation as the cause of some secondary forms of IgAN.

GENETIC EPIDEMIOLOGY

The striking variation in ethnic prevalence of IgAN provided initial evidence for a genetic effect (Galla et al 1984).

IgAN is the most common form of glomerulonephritis in Asia, where it accounts for up to 40% of cases (Glassock et al 1985). It is also highly prevalent in the Zuni and Manitoba Native Americans and the Australian Aborigines, where familial cases have also been reported (Smith & Tung 1985, O'Connell et al 1987, Casiro et al 1988, Hoy & Megill 1989, Hughson et al 1989, Smith & Hoy 1989, Hoy et al 1993). On the other hand, IgAN was reported to be uncommon in populations originating from Africa and the Indian subcontinent (Galla et al 1984, Jennette et al 1985, Date et al 1987, Srivastava 1987, Seedat et al 1988, Swanepoel et al 1989). More recent studies have shown that in Kentucky and Tennessee the incidence rates of newly diagnosed IgAN are similar between African-Americans and Caucasians (Sehic et al 1997, Wyatt et al 1998), suggesting that variable incidence in the African-American population may be due to variation in biopsy policy. However, the higher incidence rates of IgAN both in Caucasian and African-Americans from Kentucky and Tennessee may also be compatible with the possibility of a founder mutation among both groups.

Multiplex Kindreds

Familial aggregation has been reported from most parts of the world, including description of sib-pairs, mutiplex kindreds as well as more extended kindreds originating from isolated populations (Figures 45.2, 45.3). Because of the requirement for renal biopsy for diagnosis and the intermittence of urinary abnormalities, there are no systematic studies reporting the prevalence of familial IgAN or sibling recurrence risk to formally quantify genetic contribution to disease.

Initial reports documented a higher prevalence of urinary abnormalities or history of renal disease among relatives of IgAN patients. In 1973, a series of 96 IgAN patients described a patient whose maternal grandmother, mother, two brothers and a nephew had proteinuria and hematuria (de Werra et al 1973). In addition, in a second family, the brother of the patient had hematuria. Six other patients in that series had at least one relative with renal disease. A 1975 report observed that 5 of 20 IgAN patients had a first degree relative with urinary abnormalities (Sissons et al 1975). In 1984, it was reported that 3.8% of 290 patients with IgAN compiled by the UK glomerulonephritis registry had a family history of renal disease (Johnston et al 1992). In 1989, Rambausek found that 9.6% of IgAN patients followed in Heidelberg, Germany had one or more siblings with glomerulonephritis (Rambausek et al 1987). In 1995, Schena reported urinary abnormalities in 61 out 269 asymptomatic first-degree relatives from 48 families of IgAN patients (Schena 1995).

Familial aggregation of biopsy-confirmed IgAN was independently reported in 1978 in two families with affected HLA-identical brothers (Tolkoff-Rubin et al 1978, Sabatier et al 1979). Since then, familial IgAN has been increasingly

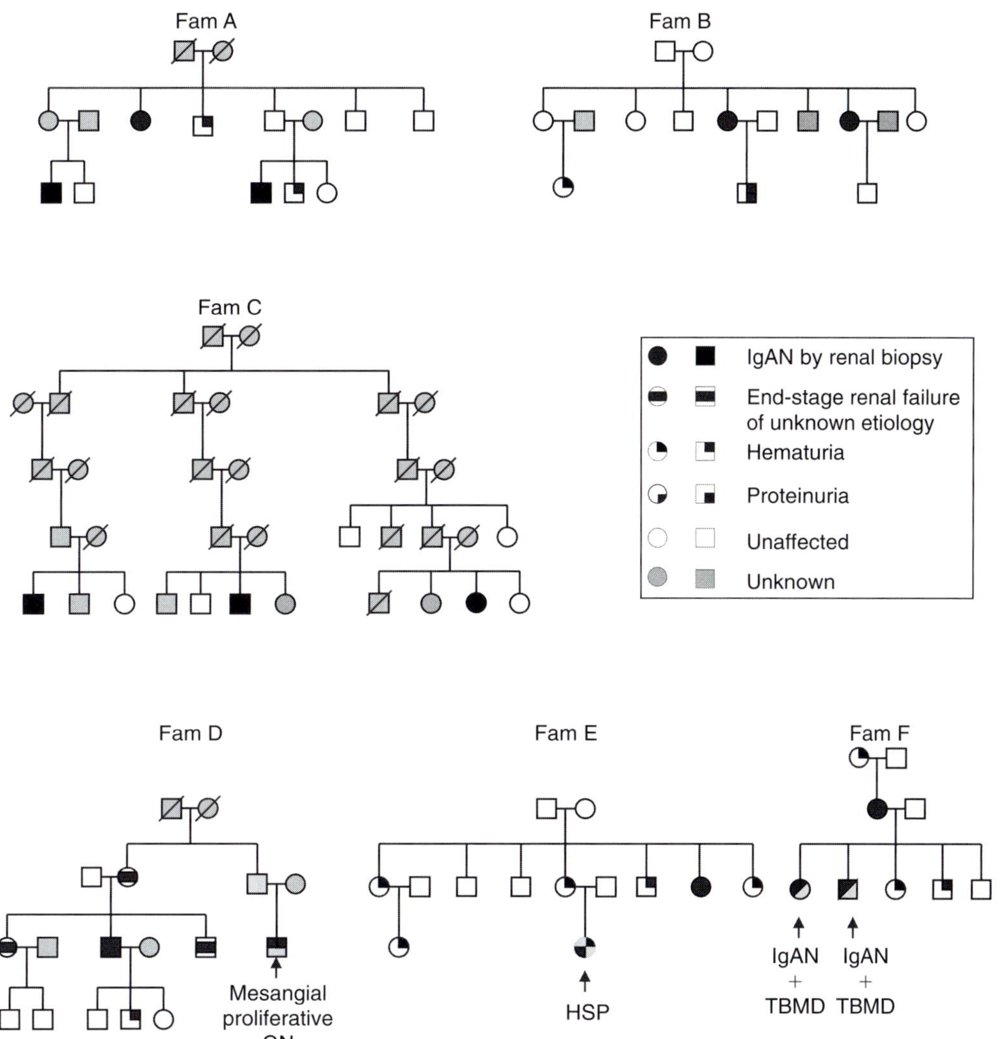

FIGURE 45.2 Multiplex families with IgAN. Families A and B are representative of multiplex families reported: the index cases present with biopsy documented disease, and screening of first and second degree relatives reveals family members with hematuria and/or proteinuria on repeated exams. Family C was ascertained by reviewing biopsy records in Brescia, Italy: it was discovered that three IgAN patients thought to have sporadic disease are in fact 3rd degree cousins. Families C–F represent potential subtypes of familial IgAN in which there is aggregation with another glomerular disease: mesangial proliferative IgA with no IgA deposits (Fam D), Henoch-Schönlein purpura (Fam E) and thin basement membrane disease (Fam F).

described. From 1978 to 1992, 35 families with at least two members with biopsy-proven IgAN were reported from Europe, USA and Asia (Levy 1993). In a systematic survey of renal biopsies from Brescia, a district of Northern Italy, IgAN was diagnosed in 185 patients between 1972 and 1997 (Scolari et al 1999). Twenty-six of these patients (14%) fit into ten pedigrees; screening of family members resulted in diagnosis of many individuals with clinical glomerulonephritis (Scolari et al 1999). Similarly, nuclear families with multiple cases of IgAN have been reported from China, Japan, or Turkey (Rambausek et al 1985, Masuda et al 1996, Kabasakal et al 1997). Levy also reported the final results of a French cooperative study that discovered 40 families with two or three members who had biopsy-documented IgAN. The majority of families were of European origin, but there were also families from Africa and Central America. Relationships among patients with IgAN varied greatly, including parent–child pairs, sibling pairs and more distant relationships (Levy 1993). Interestingly, in five of these 40 families, one or more additional relatives had a diagnosis of HSP, strongly suggesting a biological link between the two diseases (Levy 1989). In one hundred additional families, one patient had IgAN while a relative had a nonspecified glomerular disease (Levy 1989, 1993, 2001). These data indicated that familial IgAN is quite common but is probably underdiagnosed.

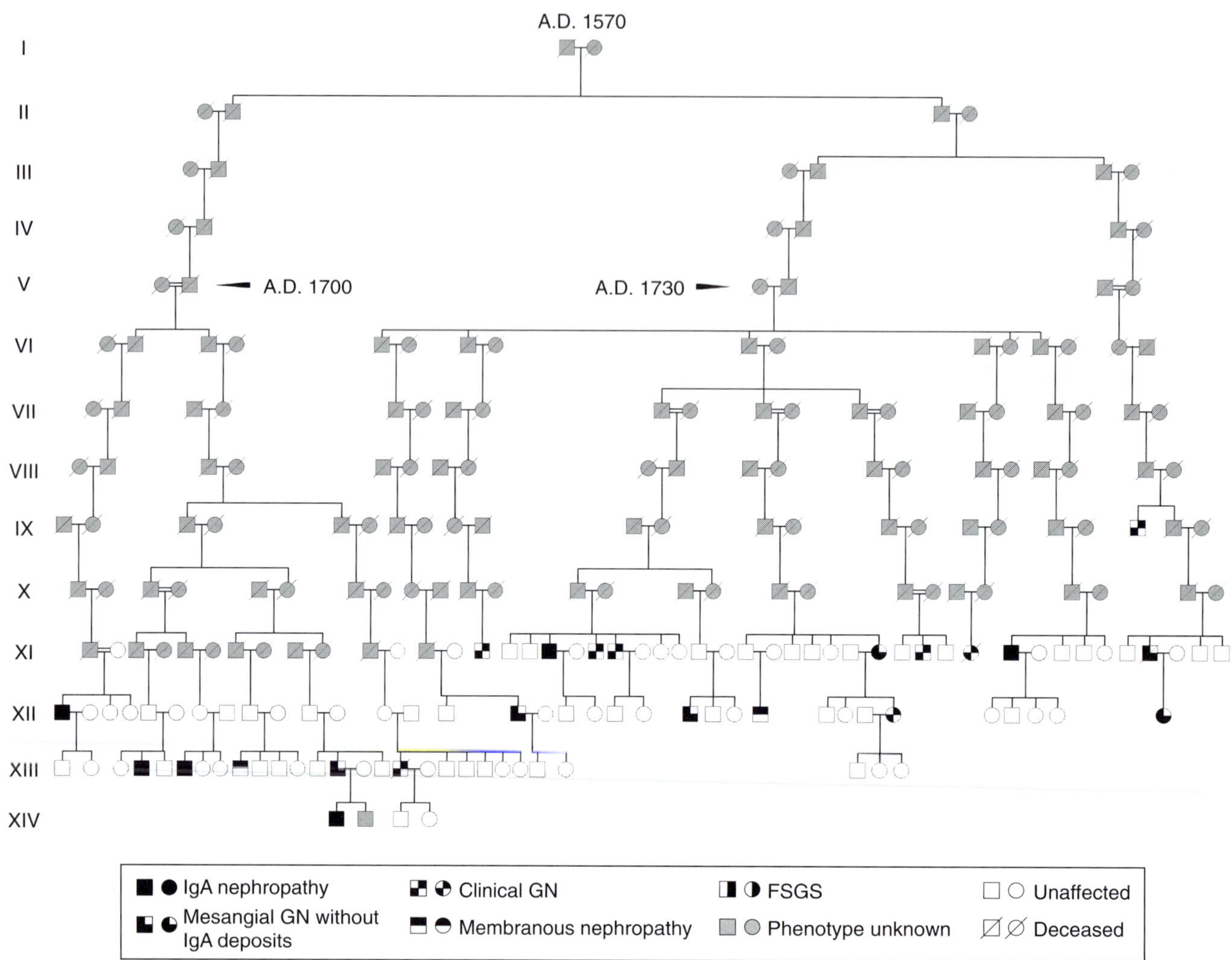

FIGURE 45.3   Extended kindred identified from a village in the Valtrompia valley in northern Italy. Genealogical investigations demonstrated that patients with IgAN and various forms of glomerulonephritis could be traced back to a common ancestral pair born in the same village in 1570 (adapted from Izzi et al 2006)

Extended Kindreds with IgAN

In 1985, Julian et al described a large pedigree from Kentucky with six patients with IgAN descending from a common ancestor; eight other patients with IgAN fit into pedigrees that were potentially related to one another (Julian et al 1985). Seventeen other pedigree members had clinical glomerulonephritis and six had a diagnosis of chronic nephritis on their death certificates. Male-to-male transmission and equal numbers of affected males and females was consistent with a simple autosomal dominant model (Julian et al 1985). Studies of HLA haplotype were inconclusive: increased frequencies of some antigens in both affected and unaffected family members compared to healthy controls could be ascribed to distortions produced by the close family relationships (Julian et al 1985). Further investigations did not reveal cosegregation of IgA levels with kidney disease (Julian et al 1988). Epidemiological investigation did not reveal a common environmental factor. A detailed genealogical investigation extended these observations, showing that of 96 IgAN patients with ancestry in eastern or central Kentucky, 53 (55%) had at least one relative with IgAN (Wyatt et al 1987); the birthplaces of 37 patients in a complex pedigree clustered in the extreme eastern portion of Kentucky. This familial and geographic clustering could be explained by a founder mutation: the eastern Kentucky population experienced little admixture since the original settlers arrived in the late 18th century, suggesting that the patients with IgAN share a common susceptibility allele that was transmitted from one or several common ancestors.

Other extended kindreds originating in isolated ethnic groups have been reported. In 1987, a four-generation Australian Aboriginal family with a high incidence of renal disease was described (O'Connell et al 1987). Twenty-eight of 114 family members screened had urinary abnormalities and IgAN was diagnosed in five of eight who underwent renal biopsy: three brothers, the father and a fourth-degree female cousin. The pattern of inheritance did not fit any Mendelian model, and renal disease did not correlate with diabetes.

Hoy et al. also observed high rates of mesangial proliferative glomerulonephritis in the Zuni Native-American tribes, an isolated community from Western New Mexico in the United States derived from a small founding population (Hoy & Megill 1989, Hughson et al 1989, Hoy et al 1993). Contrary to the general population where diabetic nephropathy predominates, primary glomerular disorders accounted for the majority of cases of renal failure in this community. Renal biopsy findings on light microscopy ranged from mild mesangial expansion to segmental and/or global glomerular sclerosis, crescents, and membrano-proliferative change. Mesangial IgA was present in most but not all biopsies. Three families were identified with multiple members with IgAN. In one family, IgA mesangial deposits were found in a father and one son; in a second family, a brother, sister and a distant female cousin were affected. In a third large family, the mother, a son and a maternal aunt had IgA deposits together with an aunt and nephew in another part of the pedigree. Recurrence after transplantation may also be more frequent in this population compared to regional controls (Smith & Harford 1995). The high incidence of IgA nephropathy in a community that experienced a population bottleneck, together with familial aggregation of disease, strongly suggests genetic predisposition due to a founder mutation.

In 1992, three extended pedigrees from the Valsaviore valley of the district of Brescia (northern Italy) were reported (Scolari et al 1992). The Valsaviore population possesses certain demographic characteristics of a genetic isolate, such as a partial geographical isolation, low immigration rate, and a high degree of inbreeding. These kindreds contained a variety of patients with primary glomerulonephritis, with IgAN being the most frequent form. The affected individuals were related by 1st or 2nd degree relationships or displayed a more distant connection, sharing ancestors 7–10 generations back. Since the initial description of the 12 IgAN patients from these three pedigrees, two other pedigree members have been diagnosed with IgAN (Scolari et al 1999). In addition to the high number of relatives with clinical glomerulonephritis, renal biopsy in some family members showed various types of primary glomerulonephritis, including IgM nephropathy, mesangial glomerulonephritis with isolated C3 deposits, membrano-proliferative glomerulonephritis and focal glomerular sclerosis. The majority of IgAN families from the Valsaviore valley exhibited a male predominance among affected members, consistent with the reported gender distribution of sporadic disease. No obvious environmental factors could be implicated in pathogenesis of

glomerulonephritis, and it is noteworthy that five affected relatives were born outside Valsaviore. The genealogies of patients indicated a limited number of ancestors; in addition, their birthplaces and those of their ancestors clustered in Valsaviore. Examination of serum IgA levels in the first Valsaviore pedigree revealed higher correlations among sibling pairs than among unrelated individuals living in the same household. However, IgA serum levels were elevated in both the affected and unaffected family members compared to local controls. Thus familial aggregation of high IgA serum levels in these families suggested a genetic control of serum IgA concentration, but this could not be implicated in the pathogenesis of IgAN.

A systematic epidemiologic investigation of the prevalence of IgA nephropathy and other forms of glomerulonephritis was recently reported (Izzi 2006). These investigators screened 5642 residents of the most northern portion of the Valtrompia valley (in the district of Brescia, northern Italy), focusing on five villages where geographical and cultural isolation have generated a great deal of homogeneity in lifestyle and eating habits. A community screening program with dipstick urinalysis was performed in these five villages. The screening program, together with investigation of medical records identified 47 patients with primary glomerulonephritis. Strikingly, 42 patients originated from three mountain villages (Collio, San Colombano, and Bovegno), including 24 with biopsy-documented disease (10 IgA nephropathy; 8 mesangial proliferative glomerulonephritis without IgA deposits; 4 focal segmental glomerular sclerosis; 2 membranous nephropathy) and 18 individuals diagnosed with clinical glomerulonephritis based on the presence of abnormal urinalyses on repeated examinations, (Figure 45.3). In contrast, there were only five cases of primary glomerulonephritis in the other two nearby villages (Pezzaze and Tavernole) which shared population histories and lifestyles. Detailed genealogical investigations were performed, enabling the construction of three large pedigrees that connected the affected individuals from each high incidence village to a common ancestral pair born in the XVI–XVII centuries in the same village from which patients were ascertained. Consistent with these data, genetic analysis with unlinked autosomal markers showed lower marker heterozygosity and Fis values in Valtrompia villages compared to the nearby urban center (Brescia), demonstrating that the Valtrompia population is derived from a more limited set of founders resulting in lower genetic diversity and increased inbreeding. These data strongly suggested the existence of a founder effect and transmission of common susceptibility allele(s) from one or several common ancestors to patients with primary glomerulonephritis. Moreover, familial aggregation of different forms of glomerulonephritis suggested that some of these disorders may have a common pathogenic basis.

In studies of familial disease, the clustering of IgAN could not be attributed to any obvious common environmental triggers, such as a viral epidemic. Although in some reports, alcohol use and liver disease were suspected as predisposing elements (Smith & Hoy 1989), these factors could not be implicated as mesangial glomerulonephritis occurred in infants, children, and abstainers. The temporal onset of disease varied widely within families, making exposure to a single environmental factor at a common time point unlikely. However, one cannot exclude a yet-unidentified environmental element that may still account for the broad geographic variation in the prevalence of IgAN.

In aggregate, studies document significant familial clustering of IgAN, and this clustering does not appear to be simply explained by a common-source environmental exposure, therefore suggesting a genetic component. Furthermore, the genealogical investigations of IgAN in five relatively isolated, but ethnically distinct populations suggested that regional and racial clustering of sporadic IgAN might be due to a founder effect. However, these latter studies could not formally prove the existence of a genetic mechanism as the degree of kinship among the IgAN patients could also be explained by the inherent structure of population isolates, i.e. most individuals from such populations are related to one another regardless of phenotype.

Phenotypic Features of Familial Disease and Potential Subtypes

Data suggest that familial IgAN is much more frequent than originally estimated but may be under-recognized due to subclinical manifestation of disease in affected relatives (Julian et al 1988, Levy 2001). In general, familial IgAN is identified more frequently at centers where nephrologists are attuned to the notion that IgAN may be an inherited disease. If hematuria is considered diagnostic of IgAN in family members of biopsy-documented IgAN cases, familial disease may constitute 10–15% of all cases (Schena 1995, Levy 2001). However, one cannot differentiate between familial and nonfamilial IgAN with respect to the presenting features such as the mean age at apparent clinical onset of the disease, age at the time of renal biopsy, degree of microscopic hematuria, history of macroscopic hematuria, magnitude of proteinuria. In addition, the histologic findings and the frequency of the immunoglobulin isotype and C3 in the renal biopsy specimens do not differ. One study, however, has reported that renal prognosis is significantly worse for familial IgAN, resulting in 41% renal survival rate, 20 years after the apparent onset of the disease, compared with 94% in patients with sporadic IgAN (Schena et al 2002). Other studies have reported that a family history of vascular disease or intrauterine growth retardation may predict worse outcome (Geddes et al 1997, Zidar et al 1998). If confirmed, these data would provide a strong argument for systematic identification of familial cases of IgAN, because such patients should be targeted for more aggressive

follow-up and therapy. Moreover, recognition of familial disease has clinical implications for selecting donors for transplantation, motivating repeated screening of family members to avoid donation from individuals with mild, intermittent symptoms. However, at present, most nephrologists do not routinely obtain detailed family histories in patients with IgAN or perform urinalysis of healthy family members. Even if family members are screened, hematuria may be intermittent, necessitating repeated screening for detection of affected individuals. Hence, routine diagnosis of familial IgAN may necessitate the development of better diagnostic tools to identify sub-clinical cases.

Most large series also report families in which IgAN aggregates with other forms of glomerular disease, particularly with mesangial proliferative glomerulonephritis (Scolari et al 1992, Levy 1993, Scolari 1999, Frasca et al 2000) (Figure 45.2, Fam D). For example, the Valtrompia study cited above identified many patients with mesangial proliferative glomerulonephritis (Izzi et al 2006). Among these, some were IgA positive and some IgA negative on immunofluorescence studies but could not be distinguished based on clinical phenotypes, which were characterized by hematuria and/or non-nephrotic proteinuria (Izzi 2006). Moreover, these phenotypes segregate both within the extended pedigrees and within two nuclear families. Finally, in one patient who had serial renal biopsies mesangial proliferative glomerulonephritis was initially diagnosed but IgA deposits were subsequently detected on the follow-up biopsy. These data are consistent with other data reporting clinical onset of IgAN may precede the detection of IgA deposits, suggesting that at least in a subset of patient, mesangial proliferative glomerulonephritis may be independent of IgA deposition.

Familial aggregation of IgAN with HSP has been long recognized, suggesting another potential subtype (Levy 1989, 1993, Miyagawa et al 1989, Lasseur et al 1997) (Figure 45.2, Fam E). Parent–offspring pairs as well as monozygotic twins with these phenotypes have been reported (Levy 1989, Miyagawa et al 1989, Levy 1993, Lasseur et al 1997). While renal pathologies in these disorders are indistinguishable, HSP is essentially a pediatric disease with additional clinical manifestations which may occur independent of nephritis. The presence of these additional clinical manifestations remains unexplained but might be attributed to variation in background genes or more likely age specific susceptibility to environmental stimuli. For example, studies have reported association of HSP with seasonal variation, vaccinations, bacterial infections, or drug exposures (e.g. acetylsalicylic acid) suggesting that a specific environmental exposure in early stages of life may trigger the disease (Nielsen 1988, Patel et al 1988, Sola Alberich et al 1997, Coppo et al 1999, Fervenza 2003, Yang et al 2005). In another interesting example, HSP developed in 2 brothers shortly after they were incarcerated in the same jail, further highlighting the possibility of gene–environment interaction in the development of this trait (Cakir et al 2004).

Finally, several reports have noted abnormalities in the glomerular basement membrane in a subset of patients with familial or sporadic IgAN (Cosio et al 1994, Berthoux et al 1995, Yoshida et al 1998) (Figure 45.2, Fam F). Examination of 1078 renal biopsies found 54 (5%) which had segmental or diffuse thinning of the glomerular basement membrane (Cosio et al 1994). Most of these patients fulfilled diagnostic criteria for IgA nephropathy (18%), mesangial proliferative glomerulonephritis (22%) or focal segmental glomerulosclerosis (18%). Furthermore, in 92% of these patients, systematic screening programs revealed urinary abnormalities (mostly hematuria) in family members. This suggests overlap between various forms of glomerulonephritis and thin basement membrane disease (TBMD), a common autosomal dominant trait characterized by hematuria due to structural abnormalities of the glomerular basement membrane (Savige et al 2003). Among multiplex families with IgA nephropathy, a subset also demonstrates diffuse glomerular basement membrane thinning characteristic of TBMD (Frasca et al 2004). These families do not have mutations in the collagen genes responsible for thin basement membrane disease nor do they display linkage to the IGAN1 locus on chromosome 6 (Frasca et al 2004). The presence of thin glomerular basement membrane did not alter clinical presentation or affect renal prognosis. These observations can be attributed to random coaggregation of two fairly common nephropathies, a hypothesis supported by bilineal inheritance in one family. Alternatively, these families may represent a distinct subset of familial IgAN, resulting from primary structural abnormalities of the glomerular basement membrane which promote 'trapping' and deposition of circulating IgA complexes. Consistent with this hypothesis, one study has reported that IgA1 glycosylation defects were identified solely in IgAN patients with normal basement membrane thickness, while those with thin basement membranes displayed no abnormalities in glycosylation, suggesting interindividual variation in mechanisms of IgA1 deposition (Linossier et al 2003). Whether a chance association or a distinct pathologic subset, these data suggest that IgAN with abnormal basement membranes should be recognized in the design and execution of clinical and genetic studies.

Potential Inheritance Models

Consideration of inheritance models in familial disease suggests that autosomal recessive transmission is most unlikely (Figure 45.2). Many families have affected members in different branches or generations of the pedigree, arguing against autosomal recessive transmission; moreover, there are no reports of IgAN occurring more frequently among offspring of consanguineous union, as would be expected

of a trait with a significant autosomal recessive component. X-linked transmission is a consideration as the IgAN is more frequent in males; however, this can be excluded in the majority of pedigrees because of male to male transmission of disease. Autosomal dominant transmission with complete penetrance seems unlikely, as described pedigrees have many instances of what would be unaffected obligate carriers. Autosomal dominant transmission with incomplete penetrance seems a likely possible genetic model that would fit most pedigrees. Such a model could explain the presence of cases in many branches of kindreds and the occurrence of disease in sequential generations. The incomplete penetrance could be accounted for by the necessity for another genetic factor or an environmental exposure. The other major model most consistent with the observed familial aggregation is multifactorial determination, in which the combined effects of many genes and/or environmental factors are required to produce a disease phenotype in a single individual. Unfortunately, as for other common diseases, it is impossible to distinguish between dominant transmission with incomplete penetrance and multifactorial determination from the pattern of aggregation. The development of the human genetic map has had a major impact on the ability to distinguish among these possibilities. Under an autosomal dominant model of transmission, one would expect particular chromosome segments to segregate with the disease significantly more often than expected by chance. Parametric models of analysis can accommodate incomplete penetrance as well as locus heterogeneity, in which some, but not all families have disease attributable to a particular chromosome segment. Such analysis can also help validate clinical observations by determining whether potential familial subtypes demonstrate linkage to the same genetic interval.

MOLECULAR GENETICS

In the past decade, studies of linkage have been successfully applied to a large number of Mendelian traits, resulting in identification of genes underlying nearly two thousand inherited syndromes (Online Mendelian Inheritance in Man; Peltonen & McKusick 2001). These studies have the advantage of identifying the chromosomal positions of disease loci in the absence of knowledge of disease pathogenesis. More recently, the publication of the sequence of the human genome (Lander et al 2001) and characterization of haplotype maps (Altshuler et al 2005) of the genome have enabled the application of association studies, which can localize disease contributing alleles in a population with non-familial disease. Results from such studies are emerging for human IgA nephropathy and related phenotypes, providing the initial steps to dissect the genetic components of the trait.

Molecular Genetics Studies for Traits Relevant to IgAN

Epidemiologic data have demonstrated that serum immunoglobulin levels are heritable and segregate as a multifactorial trait (Grundbacher 1974, Asamoah et al 1987, Borecki et al 1994, Kacprazak-Bergman 1994, Hatagima et al 1999). Most genome-scans have focused on IgE levels because this is a marker for various atopic traits such as dermatitis or asthma (Palmer et al 2000). One genome scan performed in a normal population has identified suggestive linkage of IgA levels to chromosomes 10 and 13 (Wiltshire et al 1998). The chromosome 13 locus coincided with a locus for IgE levels, suggesting a master gene regulating immunoglobulin levels. Since immunoglobulin levels have polygenic determination, gene identification will be complicated.

On the other hand, investigations of Mendelian traits at the extreme of the distribution have been more successful. Selective IgA deficiency is the most common primary immunodeficiency, with a prevalence of 1/600 in Whites (Hammarstrom et al 2000, Cunningham-Rundles 2001). This increases susceptibility to mucosal infections but it is more often associated with autoimmunity (including glomerulonephritis), suggesting an antiinflammatory role for IgA. The mode of inheritance is most likely dominant, with variable penetrance and expressivity; there is also evidence for locus heterogeneity. Contrary to IgA nephropathy, dominant IgA deficiency is strongly linked to the HLA locus on chr 6p21 but the identity of the gene is not yet known (Vorechovsky et al 1999, 2000). IgA deficiency also overlaps with common variable immunodeficiency (CVID), a more severe immune defect characterized by deficiency of all immunoglobulin classes. It was recently shown that some cases of CVID and selective IgA deficiency are caused by inactivating mutation in *TNFRSF13B*, a member of the tumor necrosis factor receptor family (Castigli et al 2005, Salzer et al 2005). In naïve B lymphocytes TNFRSF13B signaling is required for class switch recombination to generate mature IgG, IgA and IgE producing B-cells (Castigli et al 2004). TNFRSF13B deficient B lymphocytes could produce IgG, but not IgA, suggesting alternative mechanisms for IgG production. Hence, these studies have identified a molecular pathway that regulate IgA production. This may be relevant to IgAN because defects in immunoglobulin isotype switching may participate in the pathogenesis of the trait (Moore et al 1990). However, abnormalities in class switching have not been conclusively documented in lymphocytes of patients with IgA nephropathy.

Wiskott Aldrich syndrome shows association with IgA nephropathy and HSP, with intra-familial concordance for the IgA nephropathy phenotype (DeSanto et al 1988, Lasseur et al 1997, Duzova et al 2001). Wiskott Aldrich syndrome is a rare X-linked immune disorder primarily characterized by eczema, thrombocytopenia, susceptibility to infection, and bloody diarrhea; these manifestations can occur in isolation or as a complex (Nonoyama & Ochs 2001). Autoimmune disorders also occur in 40% of patients, predicting a poorer prognosis due to increased incidence of malignancy. Serum IgG and IgM levels are reduced, but IgA and IgE levels are increased. The underlying gene (*WASP*) was identified in 1994 and encodes a protein whose expression is restricted to the lymphocytic and megakaryocytic cell lineages (Derry et al 1994). The WASP protein coordinates signal transduction by enabling with actin polymerization and cytoskeletal remodeling downstream of receptor engagement, with mutations resulting in defective T-cell and B-cell proliferation (Nonoyama & Ochs 2001). This may explain the autoimmune disease and increased frequency of malignancies. The mechanisms of dysregulated immunoglobulin production are still unclear, and have been attributed to impaired B-cell/T-cell interaction. The variability in the clinical presentations of Wiskott Aldrich syndrome, including the renal complications, also remains largely unexplained. Although circulating IgA levels are increased, glomerulonephritis sometimes, but not always, presents as IgA nephropathy (Nonoyama & Ochs 2001). This suggests that WASP mutations may be permissive but additional factors are required to determine the renal phenotype. Consistent with this notion, abnormalities of IgA glycosylation have been reported in patients with WASP and IgA nephropathy/HSP, again implicating alterations in IgA structure as a requirement for manifestation of the trait (Piller et al 1991, Lasseur et al 1997).

Molecular Genetic Studies of Primary IgAN

ASSOCIATION STUDIES

In most studies the role of genetic factors in the pathogenesis of IgAN was examined using studies of association at candidate loci. Association studies typically compare allele frequencies at candidate loci between cases and controls. Such studies are predicated on the presence of linkage disequilibrium between the locus under investigation and the trait, and the assumption of few disease-contributing alleles in the population studied (Risch & Merikangas 1996, Risch 2000). Association studies usually require prior insight into disease pathogenesis as candidate loci are selected based on information provided by traditional biologic investigations. In association studies of IgAN, patients with sporadic (non-familial) disease were examined and candidate loci were selected from the broad pathogenic pathways proposed to account for the development of disease, including the HLA (Luger et al 1994, Schena 1995), imunoglobulin switch regions (Moore et al 1990), uteroglobin (Narita et al 2002), T-cell receptor (Deenitchina et al 1999), Megsin (Li et al 2004), tumor necrosis factor alpha (Tuglular et al 2003), and renin angiotensin system (Pei et al 1997, Syrjanen et al 2000, Schena et al 2001, Goto et al 2002, Narita et al 2003). Genome-wide studies which do not require a priori knowledge of pathogenesis are now emerging (Ohtsubo et al 2005).

As exemplified by studies of the histocompatibility (HLA) locus in IgAN, association studies of candidate loci have yielded conflicting results (Galla 2001). Although associations with class I and II antigens have been reported, no consistent relationship has been detected and most findings have not been replicated, even within the same population (Fennessy et al 1996). Thus, the first two instances of familial IgAN describing HLA-identical siblings were interpreted as suggesting a genetic linkage between HLA and disease (Tolkoff-Rubin et al 1978, Sabatier et al 1979). Subsequently however, most sib pairs reported were not HLA identical, and no single HLA allele was found to be associated with familial IgAN. In sporadic IgAN patients from three European populations, Fennessy et al also found no consistent relationship between the HLA genotypes and sporadic disease (Fennessy et al 1996). More recently, association studies of the uteroglobin gene and the pathogenesis of IgAN have yielded similarly disparate results, with the same allele conferring both protection and risk depending on the population studied (Szelestei et al 2000, Matsunaga et al 2002, Menegatti et al 2002, Narita et al 2002). Finally, the oft-studied polymorphism in the renin-angiotensin systems has been associated with worse prognosis in IgA nephropathy (Pei et al 1997, Syrjanen et al 2000, Schena et al 2001, Goto et al 2002, Narita et al 2003).

In the most robust association study reported to date, Takei et al conducted a comprehensive study of the selectin genes in a Japanese population (Takei et al 2002). Selectins compose a family of adhesion molecules expressed in leukocytes (L-selectin), endothelial cells (E-selectin and P-selectin) and platelets (P-selectin); these three genes form a cluster on chromosome 1q24-25 and have been implicated in the pathogenesis of inflammation, immunity and atherosclerosis. Selectin expression is increased in renal biopsies with leukocytic interstitial infiltrates, suggesting a role in the inflammatory process which can exacerbate disease. Comparison of 346 IgAN patients with 408 controls with 34SNPs spanning the selectin gene cluster revealed significant association with eight single nucleotide polymorphisms (SNPs) located in the E-selectin and L-selectin genes (Takei et al 2002). The SNPs were in linkage disequilibrium, and included two missense variants and a promoter variant with potential functional significance. These latter three SNPs formed a haplotype which significantly increased risk of disease ($P = 0.00003$). It will be interesting to examine whether variation in selectin genes similarly contribute to IgAN in a population outside Japan.

A single low-density genome-wide association study of IgA nephropathy with 88 148 SNPs has been reported to date (Ohtsubo et al 2005). In this study the genome was analyzed in three successive stages: 94 IgAN patients were initially surveyed at the genome-wide level, and all SNPs yielding $P < 0.01$ were typed in an additional 94 cases;

the 581 SNPs that continued to be significant at $P < 0.01$ were then evaluated in a full cohort of 465 cases and 634 controls. This analysis yielded significant association with a missense variant E928K in the immunoglobulin µ-binding protein 2 (*IGHMBP2*) on chr 11q13.2-q13.4 ($P = 0.00003$), a finding which was robust to correction for multiple testing. *IGHMBP2* encodes a mammalian transcription factor with helicase-like motifs which may have a role in immunoglobulin class switching. However, the link between this gene and IgA nephropathy is not yet clear because inactivating mutations in *IGHMBP2* were shown to cause spinal muscle atrophy with respiratory distress type 1 in humans (Grohmann et al 2001).

In general, association studies suffer from false positive results owing to unrecognized stratification (situations in which cases and controls come from either overtly or subtly different genetic backgrounds) and subgroup analyses that test multiple hypotheses (Risch 2000, Freedman et al 2004). This situation is not unique to the IgAN literature but underscores the need for studies of large, well-phenotyped cohorts of patients and appropriately matched controls, the necessity for additional cohorts for replication of initial findings or the utilization of the more robust design of the transmission disequilibrium test to examine the contribution of candidate loci to disease pathogenesis (Risch & Merikangas 1996, Risch 2000). With the advent of high density genotyping techniques, haplotype maps of the human genome, and the novel analytic techniques to handle large amounts of data, it is likely that genome-wide association studies will be more successful.

Linkage Studies

Analysis of linkage in pedigrees segregating IgAN provides an alternative approach. This is ideally suited for studies of familial IgAN, because selection of the subgroup with the strongest genetic predisposition would maximize the chances of detecting a single gene with large effect. Moreover, until genome-wide studies were performed, one could not know whether there might be a gene with sufficient effect on a trait to be detected by linkage. In such situations, similarly to investigations of families with breast cancer (Hall et al 1990), analysis of genetic linkage has the capacity to clarify the mode of transmission and also identify loci imparting large effects on disease risk. A single study has been reported to date, examining families with two or more members with IgAN (Gharavi et al 2000). All families were ascertained through biopsy-documented IgAN cases; families with at least one additional affected individual were selected for study. Relatives provided medical records and were screened for hematuria, proteinuria and renal function. Relatives were classified as affected if they had IgAN on renal biopsy, or alternatively hematuria (≥ 5 red blood cells per high-power field) or proteinuria (≥ 3 + proteinuria on urine dipstick) on at least three occasions, or ESRD without other identifiable causes. Because the peak risk of disease is in the second and third decade of life, relatives ≥ 40 years of age with normal urinalysis and no history of renal disease were classified as unaffected; unaffected relatives ≥ 40 years of age were classified as phenotype unknown.

Thirty multiplex kindred (24 from Italy and 6 from the United States) were studied, including 94 affected members, 48 unaffected members and 21 with phenotype unknown owing to young age. Among affected members, 60 had biopsy-documented IgAN (12 with ESRD), 29 had persistent hematuria/proteinuria (16 with episodic gross hematuria) and 5 had idiopathic ESRD. The mean age of diagnosis was 33 years and the male:female ratio was 1.5:1. All phenotypes were assigned prospectively before initiation of genotyping. Consistent with the reported literature, the pattern of disease segregations was consistent with either multifactorial inheritance or autosomal dominant transmission with reduced penetrance (Figure 45.1).

There was no evidence of linkage at candidate loci such as the HLA regions, the uteroglobin gene, the galactosyl transferase genes, and the immunoglobulin loci, demonstrating that these loci do not constitute a major inherited risk factor for the development of IgAN in our population (Gharavi et al 2000). A 10 cM genome-wide screen was next performed to search for novel loci imparting large effects on the risk of development of disease. There was no evidence of linkage under models of genetic homogeneity (i.e. a single locus influencing the trait in all families). However, five chromosomal regions yielded LOD scores (the decimal logarithm of the likelihood ratio) >1 under models of genetic heterogeneity; these were further analyzed by genotyping additional family members and adding more markers to increase informativeness. With addition of more markers, one chromosome segment on chromosome 6q22-23 yielded significant evidence of linkage under a model of genetic heterogeneity. After the addition of 20 polymorphic markers in this region, multipoint linkage analysis yielded a peak LOD score of 5.6 (odds of 398 107:1 in favor of linkage), under a parametric model in which 60% of families linked to this locus under a model of autosomal dominant inheritance with reduced penetrance. This locus on chromosome six was named *IGAN1*. To confirm these data, the analysis was repeated by using more stringent phenotypic criteria for affection status, and nonparametric, model-free analysis. There was robust evidence of linkage at 6q22-23, far exceeding all thresholds for significant linkage under all models examined (Gharavi et al 2000). This study thus simultaneously established the presence of a single gene with large effect on the risk of development of IgAN and identified the chromosomal position of this gene.

There was no effect of ethnicity on evidence of linkage to 6q22-23, as both Italian and US kindreds contributed to linkage at *IGAN1*. Furthermore, there were no differences in the clinical manifestation of disease (such as penetrance, age of onset, development of end-stage renal failure), based

on linkage status to *IGAN1*. Interestingly, isolated regions such as eastern Kentucky or Valsaviore comprised families with both positive and negative LOD scores at *IGAN1*. This suggested that mutations at both *IGAN1* and loci other than *IGAN1* account for disease in these populations; alternatively, because parametric linkage analysis depends on correct specification of family structure and inheritance models, the presence of complex population structure (such as bilineal inheritance) could have confounded the analysis in kindreds with negative LOD scores. These issues will be clarified once the identity of *IGAN1* is determined. The analysis of linkage confined the location of *IGAN1* to a 6.5 cM region, a relatively large genomic segment that contains hundreds of genes. Examination of known genes in the interval revealed no compelling candidates, suggesting that the gene underlying this linkage will likely identify a novel pathway for the development of IgAN.

The finding of genetic heterogeneity showed that mutation in additional loci can be responsible for familial IgAN. One other genomic region, on chromosome 3p23-24, yielded LOD scores suggestive of linkage, but these results did not reach genome-wide significance level (Gharavi et al 2000). Results of additional linkage studies in independent cohorts of familial IgAN are now emerging (Schena 2003, Pei 2005) resulting in identification of additional loci. Studying 26 multiplex families with IgAN, Schena reported suggestive linkage to chromosome 4q22-31 (Schena 2003). Studying a single large Canadian kindred with IgAN, Pei et al have also reported linkage to chr 2q36 with LOD score >3 (Pei 2005). Once these genes underlying familial IgA nephropathy are identified they may delineate a biological pathway that can provide insight into the pathogenesis of sporadic disease. Moreover, the finding of genetic heterogeneity in familial IgAN is not surprising given the observed variability in the clinical manifestation and outcome of disease. Studies are under way to determine whether families that link to different chromosomal regions manifest different clinical variables.

Approaches to identifying the genes underlying IgAN will encompass traditional meiotic mapping to refine the disease interval to permit systematic sequencing of positional candidates. In addition, the incomplete penetrance of disease in the kindreds examined suggests mutations in genes underlying familial IgAN will likely explain a fraction of sporadic disease. This may be particularly true in isolated populations with a high likelihood of a founder mutation such as those described in northern Italy or eastern Kentucky. In such populations one can study sporadic cases of IgAN to search for linkage disequilibrium (LD). LD mapping takes advantage of the possibility that common alleles transmitted from a remote common ancestor account for disease susceptibility in a significant fraction of the disease population (Risch 2000). In this setting, affected individuals will have inherited the same chromosome segment harboring the disease-causing mutation, enabling localization of the gene by haplotype analysis. The success of these studies depends on the age of the population under study, the number of founding alleles, and the risk imparted by such alleles (Risch 2000). These parameters are difficult to predict a priori, but the feasibility of LD mapping for identification of genes underlying human complex traits is demonstrated by successful identification of genes for Crohn's disease or macular degeneration (Ogura et al 2001, Rioux et al 2001, Edwards et al 2005, Haines et al 2005, Klein et al 2005). This task has also been greatly facilitated by the availability of the haplotype map of the genome, which enables rapid selection of 'tag' SNPs to capture most of the genetic variation in genetic intervals (Altshuler et al 2005). Hence these methods may facilitate identification of IgAN intervals either by refinement of the genetic interval mapped in familial cases, or by de novo localization of disease loci in genome-wide association studies.

CONCLUSIONS

The literature suggests that significant genetic predisposition to IgAN with familial disease represents up to 10–15% of cases of primary disease in some continents. The description of this subgroup of patients has provided an important clinical observation, but has also enabled genetic studies that may eventually help elucidate the pathogenesis of the disease. In 50–60% of families with more than one case of IgAN, the disease may be due to mutation in a gene on chromosome 6q22-23. Additional studies taking the same approach are identifying additional loci underlying familial IgAN. The identification of these genes will constitute a first step in understanding the molecular basis of IgAN. Moreover, the possibility to carry out genome-wide association studies will now enable analysis of the genetic contribution to non-familial disease and help distinguish the role of genetic and environmental modifiers. These studies offer the prospect of better diagnostic tools and improved clinical classification of disease; in the long term, they may also contribute to the development of more effective therapies for this common disorder.

References

Abu-Romeh SH, van der Meulen J, Cozma MC, et al. Renal diseases in Kuwait. Experience with 244 renal biopsies. Int. Urol. Nephrol. 1989; 21: 25–9.

Allen AC, Bailey EM, Brenchley PE, Buck KS, Barratt J, Feehally J. Mesangial IgA1 in IgA nephropathy exhibits aberrant O-glycosylation: observations in three patients. Kidney Int. 2001; 60: 969–73.

Allen AC, Harper SJ, Feehally J. Galactosylation of N- and O-linked carbohydrate moieties of IgA1 and IgG in IgA nephropathy. Clin. Exp. Immunol. 1995; 100: 470–4.

Allen AC, Topham PS, Harper SJ, Feehally J. Leucocyte beta 1,3 galactosyltransferase activity in IgA nephropathy. Nephrol. Dial. Transplant. 1997; 12: 701–6.

Allen AC, Willis FR, Beattie TJ, Feehally J. Abnormal IgA glycosylation in Henoch-Schonlein purpura restricted to patients with clinical nephritis. Nephrol. Dial. Transplant. 1998; 13: 930–4.

Altshuler D, Brooks LD, Chakravarti A, Collins FS, Daly MJ, Donnelly P. A haplotype map of the human genome. Nature 2005; 437: 1299–320.

Asamoah A, Wyatt RJ, Julian BA, Quiggins PA, Wilson AF, Elston RC. A major gene model for the familial aggregation of plasma IgA concentration. Am. J. Med. Genet. 1987; 27: 857–66.

Axford JS. Glycosylation and rheumatic disease. Biochim. Biophys. Acta 1999; 1455: 219–29.

Baldree LA, Wyatt RJ, Julian BA, Falk RJ, Jennette JC. Immunoglobulin A-fibronectin aggregate levels in children and adults with immunoglobulin A nephropathy. Am. J. Kidney Dis. 1993; 22: 1–4.

Barratt J, Feehally J. IgA nephropathy. J. Am. Soc. Nephrol. 2005; 16: 2088–97.

Barsoum R, Nabil M, Saady G, et al. Immunoglobulin-A and the pathogenesis of schistosomal glomerulopathy. Kidney Int. 1996; 50: 920–8.

Bartosik LP, Lajoie G, Sugar L, Cattran DC. Predicting progression in IgA nephropathy. Am. J. Kidney Dis. 2001; 38: 728–35.

Beaufils H, Jouanneau C, Katlama C, Sazdovitch V, Hauw JJ. HIV-associated IgA nephropathy – a post-mortem study. Nephrol. Dial. Transplant. 1995; 10: 35–8.

Berger J, Hinglais N. [Intercapillary deposits of IgA-IgG]. J. Urol. Nephrol. (Paris) 1968; 74: 694–5.

Berthoux FC, Laurent B, Koller JM, et al. Primary IgA glomerulonephritis with thin glomerular basement membrane: a peculiar pathological marker versus thin membrane nephropathy association. Contrib. Nephrol. 1995; 111: 1–6; discussion 6–7.

Borecki IB, McGue M, Gerrard JW, Lebowitz MD, Rao DC. Familial resemblance for immunoglobulin levels. Hum. Genet. 1994; 94: 179–5.

Briganti EM, Dowling J, Finlay M, et al. The incidence of biopsy-proven glomerulonephritis in Australia. Nephrol. Dial. Transplant. 2001; 16: 1364–7.

Cakir N, Pamuk ON, Donmez S. Henoch-Schonlein purpura in two brothers imprisoned in the same jail: presentation two months apart. Clin. Exp. Rheumatol. 2004; 22: 235–7.

Casiro OG, Stanwick RS, Walker RD. The prevalence of IgA nephropathy in Manitoba Native Indian children. Can. J. Public Health 1988; 79: 308–10.

Castigli E, Scott S, Dedeoglu F, et al. Impaired IgA class switching in APRIL-deficient mice. Proc. Natl Acad. Sci. USA 2004; 101: 3903–8.

Castigli E, Wilson SA, Garibyan L, et al. TACI is mutant in common variable immunodeficiency and IgA deficiency. Nat. Genet. 2005; 37: 829–34.

Cheong I, Kong N, Segasothy M, Moras Z, Menon P, Suleiman AB. IgA nephropathy in Malaysia. Southeast Asian J. Trop. Med. Public Health 1991; 22: 120–2.

Chin SE, Axelsen RA, Crawford DH, et al. Glomerular abnormalities in children undergoing orthotopic liver transplantation. Pediatr. Nephrol. 1992; 6: 407–11.

Clarkson AR, Woodroffe AJ, Aarons I. IgA nephropathy in patients followed-up for at least ten years. Semin. Nephrol. 1987; 7: 377–8.

Coppo R, Amore A, Gianoglio B. Clinical features of Henoch-Schonlein purpura. Italian Group of Renal Immunopathology. Ann. Med. Interne. (Paris) 1999; 150: 143–50.

Coppo R, Chiesa M, Cirina P, Peruzzi L, Amore A. In human IgA nephropathy uteroglobin does not play the role inferred from transgenic mice. Am. J. Kidney Dis. 2002; 40: 495–503.

Cosio FG, Falkenhain ME, Sedmak DD. Association of thin glomerular basement membrane with other glomerulopathies. Kidney Int. 1994; 46: 471–4.

Cosyns JP, Malaise J, Hanique G, et al. Lesions in donor kidneys: nature, incidence, and influence on graft function. Transpl. Int. 1998; 11: 22–7.

Cuevas X, Lloveras J, Mir M, Aubia J, Masramon J. Disappearance of mesangial IgA deposits from the kidneys of two donors after transplantation. Transplant Proc. 1987; 19: 2208–9.

Cunningham-Rundles C. Physiology of IgA and IgA deficiency. J. Clin. Immunol. 2001; 21: 303–9.

Czerkinsky C, Koopman WJ, Jackson S, et al. Circulating immune complexes and immunoglobulin A rheumatoid factor in patients with mesangial immunoglobulin A nephropathies. J. Clin. Invest. 1986; 77: 1931–8.

D'Amico G. The commonest glomerulonephritis in the world: IgA nephropathy. Q. J. Med. 1987; 64: 709–27.

D'Amico G. Natural history of idiopathic IgA nephropathy and factors predictive of disease outcome. Semin. Nephrol. 2004; 24: 179–96.

D'Amico G, Colasanti G, Barbiano di Belgioioso G, et al. Long-term follow-up of IgA mesangial nephropathy: clinico-histological study in 374 patients. Semin. Nephrol. 1987; 7: 355–8.

Dash SC, Bhuyan UN, Dinda AK, et al. Increased incidence of glomerulonephritis following spleno-renal shunt surgery in non-cirrhotic portal fibrosis. Kidney Int. 1997; 52: 482–5.

Date A, Raghavan R, John TJ, Richard J, Kirubakaran MG, Shastry JC. Renal disease in adult Indians: a clinicopathological study of 2,827 patients. Q. J. Med. 1987; 64: 729–37.

de Werra P, Morel-Maroger L, Leroux-Robert C, Richet G. [Glomerulonephritis with diffuse IgA deposits in the mesangium. Study of 96 adult cases]. Schweiz. Med. Wochenschr. 1973; 103: 761–8.

Deenitchina SS, Shinozaki M, Hirano T, et al. Association of a T-cell receptor constant alpha chain gene polymorphism with progression of IgA nephropathy in Japanese patients. Am. J. Kidney Dis. 1999; 34: 279–88.

Derry JM, Ochs HD, Francke U. Isolation of a novel gene mutated in Wiskott-Aldrich syndrome. Cell 1994; 79: 922.

DeSanto NG, Sessa A, Capodicasa G, et al. IgA glomerulonephritis in Wiskott-Aldrich syndrome. Child Nephrol. Urol. 1988; 9: 118–20.

Duzova A, Topaloglu R, Sanal O, et al. Henoch-Schonlein purpura in Wiskott-Aldrich syndrome. Pediatr. Nephrol. 2001; 16: 500–2.

Edwards AO, Ritter R 3rd, Abel KJ, Manning A, Panhuysen C, Farrer LA. Complement factor H polymorphism and age-related macular degeneration. Science 2005; 308: 421–4.

Favre L, Spertini F, Corthesy B. Secretory IgA possesses intrinsic modulatory properties stimulating mucosal and systemic immune responses. J. Immunol. 2005; 175: 2793–800.

Fennessy M, Hitman GA, Moore RH, et al. HLA-DQ gene polymorphism in primary IgA nephropathy in three European populations. Kidney Int. 1996; 49: 477–80.

Fervenza FC. Henoch-Schonlein purpura nephritis. Int. J. Dermatol. 2003; 42: 170–7.

Field MC, Dwek RA, Edge CJ, Rademacher TW. O-linked oligosaccharides from human serum immunoglobulin A1. Biochem. Soc. Trans. 1989; 17: 1034–5.

Frasca GM, Onetti-Muda A, Renieri A. Thin glomerular basement membrane disease. J. Nephrol. 2000; 13: 15–19.

Frasca GM, Soverini L, Gharavi AG, et al. Thin basement membrane disease in patients with familial IgA nephropathy. J. Nephrol. 2004; 17: 778–85.

Frasca GM, Soverini L, Preda P, et al. Two different glomerular diseases in the same patient at an interval of 7 years. Nephrol. Dial. Transplant. 2002; 17: 2014–16.

Freedman ML, Reich D, Penney KL, et al. Assessing the impact of population stratification on genetic association studies. Nat. Genet. 2004; 36: 388–93.

Galla JH. IgA nephropathy. Kidney Int. 1995; 47: 377–87.

Galla JH. Molecular genetics in IgA nephropathy. Nephron 2001; 88: 107–12.

Galla JH, Kohaut EC, Alexander R, Mestecky J. Racial difference in the prevalence of IgA-associated nephropathies. Lancet 1984; 2: 522.

Geddes CC, Rauta V, Gronhagen-Riska C, et al. A tricontinental view of IgA nephropathy. Nephrol. Dial. Transplant. 2003; 18: 1541–8.

Geddes CC, Warwick GL, Tulloch I, Boulton-Jones JM. The influence of familial factors on the progression of IgA nephropathy. Nephrol. Dial. Transplant. 1997; 12: 1963–7.

Gesualdo L, Di Palma AM, Morrone LF, Strippoli GF, Schena FP. The Italian experience of the national registry of renal biopsies. Kidney Int. 2004; 66: 890–4.

Gharavi AG, Yan Y, Scolari F, et al. IgA nephropathy, the most common cause of glomerulonephritis, is linked to 6q22-23. Nat. Genet. 2000; 26: 354–7.

Glassock RJ, Kurokawa K, Yoshida M, et al. IgA nephropathy in Japan. Am. J. Nephrol. 1985; 5: 127–37.

Goto S, Narita I, Saito N, et al. A(-20)C polymorphism of the angiotensinogen gene and progression of IgA nephropathy. Kidney Int. 2002; 62: 980–5.

Greer MR, Barratt J, Harper SJ, Allen AC, Feehally J. The nucleotide sequence of the IgA1 hinge region in IgA nephropathy. Nephrol. Dial. Transplant. 1998; 13: 1980–3.

Grohmann K, Schuelke M, Diers A, et al. Mutations in the gene encoding immunoglobulin mu-binding protein 2 cause spinal muscular atrophy with respiratory distress type 1. Nat. Genet. 2001; 29: 75–7.

Grossetete B, Launay P, Lehuen A, Jungers P, Bach JF, Monteiro RC. Down-regulation of Fc alpha receptors on blood cells of IgA nephropathy patients: evidence for a negative regulatory role of serum IgA. Kidney Int. 1998; 53: 1321–35.

Grundbacher FJ. Heritability estimates and genetic and environmental correlations for the human immunoglobulins G, M, and A. Am. J. Hum. Genet. 1974; 26: 1–12.

Haas M. Histologic subclassification of IgA nephropathy: a clinicopathologic study of 244 cases. 1997.

Haines JL, Hauser MA, Schmidt S, et al. Complement factor H variant increases the risk of age-related macular degeneration. Science 2005; 308: 419–21.

Hall JM, Lee MK, Newman B, et al. Linkage of early-onset familial breast cancer to chromosome 17q21. Science 1990; 250: 1684–9.

Hammarstrom L, Vorechovsky I, Webster D. Selective IgA deficiency (SIgAD) and common variable immunodeficiency (CVID). Clin. Exp. Immunol. 2000; 120: 225–31.

Hatagima A, Cabello PH, Krieger H. Causal analysis of the variability of IgA, IgG, and IgM immunoglobulin levels. Hum. Biol. 1999; 71: 219–29.

Hiki Y, Odani H, Takahashi M, et al. Mass spectrometry proves under-O-glycosylation of glomerular IgA1 in IgA nephropathy. Kidney Int. 2001; 59: 1077–85.

Hirsch DJ, Jindal KK, Trillo A, Cohen AD. Acute renal failure in Crohn's disease due to IgA nephropathy. Am. J. Kidney Dis. 1992; 20: 189–90.

Hoffjan S, Nicolae D, Ostrovnaya I, et al. Gene–environment interaction effects on the development of immune responses in the 1st year of life. Am. J. Hum. Genet. 2005; 76: 696–704.

Hoffjan S, Ostrovnaja I, Nicolae D, et al. Genetic variation in immunoregulatory pathways and atopic phenotypes in infancy. J. Allergy Clin. Immunol. 2004; 113: 511–18.

Hoy WE, Hughson MD, Smith SM, Megill DM. Mesangial proliferative glomerulonephritis in southwestern American Indians. Am. J. Kidney Dis. 1993; 21: 486–96.

Hoy WE, Megill DM. End-stage renal disease in southwestern Native Americans, with special focus on the Zuni and Navajo Indians. Transplant Proc. 1989; 21: 3906–8.

Hughson MD, Megill DM, Smith SM, Tung KS, Miller G, Hoy WE. Mesangiopathic glomerulonephritis in Zuni (New Mexico) Indians. Arch. Pathol. Lab. Med. 1989; 113: 148–57.

Imai H, Nakamoto Y, Asakura K, Miki K, Yasuda T, Miura AB. Spontaneous glomerular IgA deposition in ddY mice: an animal model of IgA nephritis. Kidney Int. 1985; 27: 756–61.

Imai K, Nakajima K, Eguchi K, et al. Homozygous C3 deficiency associated with IgA nephropathy. Nephron 1991; 59: 148–52.

Iwasaki H, Zhang Y, Tachibana K, et al. Initiation of O-glycan synthesis in IgA1 hinge region is determined by a single enzyme, UDP-N-acetyl-alpha-D-galactosamine:polypeptide N-acetylgalactosaminyltransferase 2. J. Biol. Chem. 2003; 278: 5613–21.

Iwase H, Ohkawa S, Ishii-Karakasa I, et al. Study of the relationship between sticky human serum IgA1 and its O-glycan glycoform. Biochem. Biophys. Res. Commun. 1999; 261: 472–7.

Izzi C, Sanna-Cherchi S, Prati E, et al. Familial aggregation of primary glomerulonephritis in an Italian population isolate: the Valtrompia Study. Kidney Int. 2006.

Jackson S, Moldoveanu Z, Kirk KA, et al. IgA-containing immune complexes after challenge with food antigens in patients with IgA nephropathy. Clin. Exp. Immunol. 1992; 89: 315–20.

Jennette JC. The immunohistology of IgA nephropathy. Am. J. Kidney Dis. 1988; 12: 348–52.

Jennette JC, Wall SD, Wilkman AS. Low incidence of IgA nephropathy in blacks. Kidney Int. 1985; 28: 944–50.

Jin DK, Kohsaka T, Jun A, Kobayashi N. Complement 4 gene deletion in patients with IgA nephropathy and Henoch-Schonlein nephritis. Child Nephrol. Urol. 1992; 12: 208–11.

Johansen FE, Braathen R, Brandtzaeg P. The J chain is essential for polymeric Ig receptor-mediated epithelial transport of IgA. J. Immunol. 2001; 167: 5185–92.

Johnston PA, Brown JS, Braumholtz DA, Davison AM. Clinicopathological correlations and long-term follow-up of 253 United Kingdom patients with IgA nephropathy. A report from the MRC Glomerulonephritis Registry. Q. J. Med. 1992; 84: 619–27.

Jones CL, Powell HR, Kincaid-Smith P, Roberton DM. Polymeric IgA and immune complex concentrations in IgA-related renal disease. Kidney Int. 1990; 38: 323–31.

Ju T, Brewer K, D'Souza A, Cummings RD, Canfield WM. Cloning and expression of human core 1 beta1,3-galactosyltransferase. J. Biol. Chem. 2002a; 277: 178–86.

Ju T, Cummings RD. A unique molecular chaperone Cosmc required for activity of the mammalian core 1 beta 3-galactosyltransferase. Proc. Natl Acad. Sci. USA 2002; 99: 16613–18.

Ju T, Cummings RD. Protein glycosylation: chaperone mutation in Tn syndrome. Nature 2005; 437: 1252.

Ju T, Cummings RD, Canfield WM. Purification, characterization, and subunit structure of rat core 1 Beta1,3-galactosyltransferase. J. Biol. Chem. 2002b; 277: 169–77.

Julian BA, Quiggins PA, Thompson JS, et al. Familial IgA nephropathy. Evidence of an inherited mechanism of disease. N. Engl. J. Med. 1985; 312: 202–8.

Julian BA, Tomana M, Novak J, Mestecky J. Progress in the pathogenesis of IgA nephropathy. Adv. Nephrol. Necker. Hosp. 1999; 29: 53–72.

Julian BA, Woodford SY, Baehler RW, McMorrow RG, Wyatt RJ. Familial clustering and immunogenetic aspects of IgA nephropathy. Am. J. Kidney Dis. 1988; 12: 366–70.

Kabasakal C, Keskinoglu A, Mir S, Basdemir G. IgA nephropathy occurring in two siblings of three families. Turk. J. Pediatr. 1997; 39: 395–401.

Kacprazak-Bergman I. Sexual dimorphism of heritability of immunoglobulin levels. Ann. Hum. Biol. 1994; 21: 563–9.

Kanda E, Shimamura H, Tamura H, et al. IgA nephropathy with complement deficiency. Intern. Med. 2001; 40: 52–55.

Kang HS, Chin RK, Wang Y, et al. Signaling via LTbetaR on the lamina propria stromal cells of the gut is required for IgA production. Nat. Immunol. 2002; 3: 576–82.

Kerr MA. The structure and function of human IgA. Biochem. J. 1990; 271: 285–96.

Kessler M, Hiesse C, Hestin D, Mayeux D, Boubenider K, Charpentier B. Recurrence of immunoglobulin A nephropathy after renal transplantation in the cyclosporine era. Am. J. Kidney Dis. 1996; 28: 99–104.

Klein RJ, Zeiss C, Chew EY, et al. Complement factor H polymorphism in age-related macular degeneration. Science 2005; 308: 385–9.

Koselj M, Rott T, Kandus A, Vizjak A, Malovrh M. Donor-transmitted IgA nephropathy: long-term follow-up of kidney donors and recipients. Transplant Proc. 1997; 29: 3406–7.

Lai FM, Szeto CC, Choi PC, et al. Primary IgA nephropathy with low histologic grade and disease progression: is there a "point of no return"? Am. J. Kidney Dis. 2002; 39: 401–6.

Lai KN, Chan KW, Mac-Moune F, et al. The immunochemical characterization of the light chains in the mesangial IgA deposits in IgA nephropathy. Am. J. Clin. Pathol. 1986; 85: 548–51.

Lai KN, Chui SH, Lai FM, Lam CW. Predominant synthesis of IgA with lambda light chain in IgA nephropathy. Kidney Int. 1988; 33: 584–9.

Lander ES, Linton LM, Birren B, et al. Initial sequencing and analysis of the human genome. Nature 2001; 409: 860–921.

Lasseur C, Allen AC, Deminiere C, Aparicio M, Feehally J, Combe C. Henoch-Schonlein purpura with immunoglobulin A nephropathy and abnormalities of immunoglobulin A in a Wiskott-Aldrich syndrome carrier. Am. J. Kidney Dis. 1997; 29: 285–7.

Launay P, Grossetete B, Arcos-Fajardo M, et al. Fcalpha receptor (CD89) mediates the development of immunoglobulin A (IgA) nephropathy (Berger's disease). Evidence for pathogenic soluble receptor-Iga complexes in patients and CD89 transgenic mice. J. Exp. Med. 2000; 191: 1999–2009.

Layward L, Allen AC, Hattersley JM, Harper SJ, Feehally J. Elevation of IgA in IgA nephropathy is localized in the serum and not saliva and is restricted to the IgA1 subclass. Nephrol. Dial. Transplant. 1993; 8: 25–8.

Levy M. Familial cases of Berger's disease and anaphylactoid purpura: more frequent than previously thought. Am. J. Med. 1989; 87: 246–8.

Levy M. Multiplex families in IgA nephropathy. Contrib. Nephrol. 1993; 104: 46–53.

Levy M. Familial cases of Berger's disease and anaphylactoid purpura. Kidney Int. 2001; 60: 1611–12.

Levy M, Berger J. Worldwide perspective of IgA nephropathy. Am. J. Kidney Dis. 1988; 12: 340–7.

Li YJ, Du Y, Li CX, et al. Family-based association study showing that immunoglobulin A nephropathy is associated with the polymorphisms 2093C and 2180T in the 3' untranslated region of the Megsin gene. J. Am. Soc. Nephrol. 2004; 15: 1739–43.

Lim EC, Terasaki PI. Outcome of renal transplantation in different primary diseases. Clin. Transpl. 1991: 293–303.

Linossier MT, Palle S, Berthoux F. Different glycosylation profile of serum IgA1 in IgA nephropathy according to the glomerular basement membrane thickness: normal versus thin. Am. J. Kidney Dis. 2003; 41: 558–64.

Lowe JB. Glycosylation, immunity, and autoimmunity. Cell 2001; 104: 809–12.

Luger AM, Komathireddy G, Walker RE, Pandey JP, Hoffman RW. Molecular and serologic analysis of HLA genes and immunoglobulin allotypes in IgA nephropathy. Autoimmunity 1994; 19: 1–5.

Mandal AK, Ray R, Zhang Z, Chowdhury B, Pattabiraman N, Mukherjee AB. Uteroglobin inhibits prostaglandin F2alpha receptor-mediated expression of genes critical for the production of pro-inflammatory lipid mediators. J. Biol. Chem. 2005; 280: 32897–904.

Mandal AK, Zhang Z, Ray R, et al. Uteroglobin represses allergen-induced inflammatory response by blocking PGD2 receptor-mediated functions. J. Exp. Med. 2004; 199: 1317–30.

Masuda J, Shiiki H, Fujii Y, Dohi K, Harada A. Identical twin sisters with IgA nephropathy. Nippon Jinzo Gakkai Shi 1996; 38: 52–6.

Mathew TH. Recurrence of disease following renal transplantation. Am. J. Kidney Dis. 1988; 12: 85–96.

Matsunaga A, Numakura C, Kawakami T, et al. Association of the uteroglobin gene polymorphism with IgA nephropathy. Am. J. Kidney Dis. 2002; 39: 36–41.

Mattu TS, Pleass RJ, Willis AC, et al. The glycosylation and structure of human serum IgA1, Fab, and Fc regions and the role of N-glycosylation on Fc alpha receptor interactions. J. Biol. Chem. 1998; 273: 2260–72.

McDonald KJ, Cameron AJ, Allen JM, Jardine AG. Expression of Fc alpha/mu receptor by human mesangial cells: a candidate receptor for immune complex deposition in IgA nephropathy. Biochem. Biophys. Res. Commun. 2002; 290: 438–42.

Meadow SR, Scott DG. Berger disease: Henoch-Schonlein syndrome without the rash. J. Pediatr. 1985; 106: 27–32.

Menegatti E, Nardacchione A, Alpa M, et al. Polymorphism of the uteroglobin gene in systemic lupus erythematosus and IgA nephropathy. Lab Invest 2002; 82: 543–6.

Mestecky J, Russell MW, Elson CO. Intestinal IgA: novel views on its function in the defence of the largest mucosal surface. Gut 1999; 44: 2–5.

Miyagawa S, Dohi K, Hanatani M, et al. Anaphylactoid purpura and familial IgA nephropathy. Am. J. Med. 1989; 86: 340–2.

Miyawaki S, Muso E, Takeuchi E, et al. Selective breeding for high serum IgA levels from noninbred ddY mice: isolation of a strain with an early onset of glomerular IgA deposition. Nephron 1997; 76: 201–7.

Monteiro RC, Van De Winkel JG. IgA Fc receptors. Annu. Rev. Immunol. 2003; 21: 177–204.

Moore RH, Hitman GA, Sinico RA, et al. Immunoglobulin heavy chain switch region gene polymorphisms in glomerulonephritis. Kidney Int. 1990; 38: 332–6.

Moura IC, Arcos-Fajardo M, Gdoura A, et al. Engagement of transferrin receptor by polymeric IgA1: evidence for a positive feedback loop involving increased receptor expression and mesangial cell proliferation in IgA nephropathy. J. Am. Soc. Nephrol. 2005; 16: 2667–76.

Moura IC, Arcos-Fajardo M, Sadaka C, et al. Glycosylation and size of IgA1 are essential for interaction with mesangial transferrin receptor in IgA nephropathy. J. Am. Soc. Nephrol. 2004; 15: 622–34.

Muso E, Yoshida H, Takeuchi E, et al. Enhanced production of glomerular extracellular matrix in a new mouse strain of high serum IgA ddY mice. Kidney Int. 1996; 50: 1946–57.

Narita I, Goto S, Saito N, et al. Interaction between ACE and ADD1 gene polymorphisms in the progression of IgA nephropathy in Japanese patients. Hypertension 2003; 42: 304–9.

Narita I, Saito N, Goto S, et al. Role of uteroglobin G38A polymorphism in the progression of IgA nephropathy in Japanese patients. Kidney Int. 2002; 61: 1853–8.

Nicholls KM, Fairley KF, Dowling JP, Kincaid-Smith P. The clinical course of mesangial IgA associated nephropathy in adults. Q. J. Med. 1984; 53: 227–50.

Nielsen HE. Epidemiology of Schonlein-Henoch purpura. Acta Paediatr. Scand. 1988; 77: 125–31.

Nochy D, Glotz D, Dosquet P, et al. Renal disease associated with HIV infection: a multicentric study of 60 patients from Paris hospitals. Nephrol. Dial. Transplant. 1993; 8: 11–19.

Nogaki F, Oida E, Kamata T, et al. Chromosomal mapping of hyperserum IgA and glomerular IgA deposition in a high IgA (HIGA) strain of DdY mice. Kidney Int. 2005; 68: 2517–25.

Nonoyama S, Ochs HD. Wiskott-Aldrich syndrome. Curr. Allergy Asthma Rep. 2001; 1: 430–7.

Norderhaug IN, Johansen FE, Schjerven H, Brandtzaeg P. Regulation of the formation and external transport of secretory immunoglobulins. Crit. Rev. Immunol. 1999; 19: 481–508.

Novak J, Tomana M, Matousovic K, et al. IgA1-containing immune complexes in IgA nephropathy differentially affect proliferation of mesangial cells. Kidney Int. 2005; 67: 504–13.

O'Connell PJ, Ibels LS, Thomas MA, Harris M, Eckstein RP. Familial IgA nephropathy: a study of renal disease in an Australian aboriginal family. Aust. NZ J. Med. 1987; 17: 27–33.

O'Meara Y, Green A, Carmody M, et al. Recurrent glomeru-lonephritis in renal transplants: fourteen years' experience. Nephrol. Dial. Transplant. 1989; 4: 730–4.

O'Neill S, Walker F, Tanner A, Browne O, O'Dwyer WF. Dermal IgA deposits in IgA nephropathy. Ir. Med. J. 1982; 75: 327.

Ogura Y, Bonen DK, Inohara N, et al. A frameshift mutation in NOD2 associated with susceptibility to Crohn's disease. Nature 2001; 411: 603–6.

Ohtsubo S, Iida A, Nitta K, et al. Association of a single-nucle-otide polymorphism in the immunoglobulin mu-binding pro-tein 2 gene with immunoglobulin A nephropathy. J. Hum. Genet. 2005; 50: 30–5.

Online Mendelian Inheritance in Man, O. T. McKusick-Nathans Institute for Genetic Medicine, Johns Hopkins University (Baltimore, MD) and National Center for Biotechnology Information, National Library of Medicine (Bethesda, MD), 2000. World Wide Web URL: http://www.ncbi.nlm.nih.gov/omim/.

Palmer LJ, Burton PR, James AL, Musk AW, Cookson WO. Familial aggregation and heritability of asthma-associated quantitative traits in a population-based sample of nuclear fam-ilies. Eur. J. Hum. Genet. 2000; 8: 853–60.

Patel U, Bradley JR, Hamilton DV. Henoch-Schonlein purpura after influenza vaccination. Br. Med. J. (Clin. Res. Ed.) 1988; 296: 1800.

Pei Y, Magistroni R, Klassen J, Kappel J, Cattran D, St.George P. Localization of a novel disease susceptibility locus to chromo-some 2q36 by genome scan of a large extended Canadian fam-ily with IgA nephropathy (IgAN). J. Am. Soc. Nephrol. 2005; 16(Suppl. S).

Pei Y, Scholey J, Thai K, Suzuki M, Cattran D. Association of angiotensinogen gene T235 variant with progression of immu-noglobin A nephropathy in Caucasian patients. J. Clin. Invest. 1997; 100: 814–20.

Peltonen L, McKusick VA. Genomics and medicine. Dissecting human disease in the postgenomic era. Science 2001; 291: 1224–9.

Pettersson E, von Bonsdorff M, Tornroth T, Lindholm H. Nephritis among young Finnish men. Clin. Nephrol. 1984; 22: 217–22.

Pillebout E, Thervet E, Hill G, Alberti C, Vanhille P, Nochy D. Henoch-Schonlein purpura in adults: outcome and prognostic factors. J. Am. Soc. Nephrol. 2002; 13: 1271–8.

Piller F, Le Deist F, Weinberg KI, Parkman R, Fukuda M. Altered O-glycan synthesis in lymphocytes from patients with Wiskott-Aldrich syndrome. J. Exp. Med. 1991; 173: 1501–10.

Power DA, Muirhead N, Simpson JG, et al. IgA nephropathy is not a rare disease in the United Kingdom. Nephron 1985; 40: 180–4.

Praditpornsilpa K, Napathorn S, Yenrudi S, Wankrairot P, Tungsaga K, Sitprija V. Renal pathology and HIV infection in Thailand. Am. J. Kidney Dis. 1999; 33: 282–6.

Propper DJ, Power DA, Simpson JG, Edward N, Catto GR. The incidence, mode of presentation, and prognosis of IgA neph-ropathy in northeast Scotland. Semin. Nephrol. 1987; 7: 363–6.

Qin W, Zhou Q, Yang LC, et al. Peripheral B lymphocyte beta1,3-galactosyltransferase and chaperone expression in immu-noglobulin A nephropathy. J. Intern. Med. 2005; 258: 467–77.

Rambausek M, Hartz G, Waldherr R, Andrassy K, Ritz E. Familial glomerulonephritis. Pediatr. Nephrol. 1987; 1: 416–18.

Rambausek M, Waldherr R, Andrassy K, Ritz E. Hypertension in mesangial IgA glomerulonephritis. Proc. Eur. Dial. Transplant Assoc. Eur. Ren. Assoc. 1985; 21: 693–7.

Ray R, Choi M, Zhang Z, Silverman GA, Askew D, Mukherjee AB. Uteroglobin suppresses SCCA gene expression associated with allergic asthma. J. Biol. Chem. 2005; 280: 9761–4.

Research Group on Progressive Chronic Renal Disease. Nationwide and long-term survey of primary glomerulone-phritis in Japan as observed in 1,850 biopsied cases. Nephron 1999; 82: 205–13.

Rioux JD, Daly MJ, Silverberg MS, et al. Genetic variation in the 5q31 cytokine gene cluster confers susceptibility to Crohn dis-ease. Nat. Genet. 2001; 29: 223–8.

Risch N, Merikangas K. The future of genetic studies of complex human diseases. Science 1996; 273: 1516–17.

Risch NJ. Searching for genetic determinants in the new millen-nium. Nature 2000; 405: 847–56.

Rosenberg HG, Martinez PS, Vaccarezza AS, Martinez LV. Morphological findings in 70 kidneys of living donors for renal transplant. Pathol. Res. Pract. 1990; 186: 619–24.

Royle L, Roos A, Harvey DJ, et al. Secretory IgA N- and O-gly-cans provide a link between the innate and adaptive immune systems. J. Biol. Chem. 2003; 278: 20140–53.

Sabatier JC, Genin C, Assenat H, Colon S, Ducret F, Berthoux FC. Mesangial IgA glomerulonephritis in HLA-identical brothers. Clin. Nephrol. 1979; 11: 35–8.

Sakai H. Cellular immunoregulatory aspects of IgA nephropathy. Am. J. Kidney Dis. 1988; 12: 430–2.

Salzer U, Chapel HM, Webster AD, et al. Mutations in TNFRSF13B encoding TACI are associated with common variable immunodeficiency in humans. Nat. Genet. 2005; 37: 820–8.

Sano T, Hiki Y, Kokubo T, Iwase H, Shigematsu H, Kobayashi Y. Enzymatically deglycosylated human IgA1 molecules accu-mulate and induce inflammatory cell reaction in rat glomeruli. Nephrol. Dial. Transplant. 2002; 17: 50–6.

Savige J, Rana K, Tonna S, Buzza M, Dagher H, Wang YY. Thin basement membrane nephropathy. Kidney Int. 2003; 64: 1169–78.

Schena FP. A retrospective analysis of the natural history of pri-mary IgA nephropathy worldwide. Am. J. Med. 1990; 89: 209–15.

Schena FP. Immunogenetic aspects of primary IgA nephropathy. Kidney Int. 1995; 48: 1998–2013.

Schena FP, Di Perna M, Mazzucco E, et al. Two novel loci effect-ing on the development of IgA nephropathy (IgAN): 4q22.1–31.21 and 7q33–36.3. J. Am. Soc. Nephrol. 2003; 14(Suppl. S): 25–26A.

Schena FP, Cerullo G, Rossini M, Lanzilotta SG, D'Altri C, Manno C. Increased risk of end-stage renal disease in familial IgA nephropathy. J. Am. Soc. Nephrol. 2002; 13: 453–60.

Schena FP, D'Altri C, Cerullo G, Manno C, Gesualdo L. ACE gene polymorphism and IgA nephropathy: an ethnically homo-geneous study and a meta-analysis. Kidney Int. 2001; 60: 732–40.

Scolari F. Familial IgA nephropathy. J. Nephrol. 1999; 12: 213–19.

Scolari F, Amoroso A, Savoldi S, et al. Familial clustering of IgA nephropathy: further evidence in an Italian population. Am. J. Kidney Dis. 1999; 33: 857–65.

Scolari F, Amoroso A, Savoldi S, et al. Familial occurrence of primary glomerulonephritis: evidence for a role of genetic factors. Nephrol. Dial. Transplant. 1992; 7: 587–96.

Seedat YK, Nathoo BC, Parag KB, Naiker IP, Ramsaroop R. IgA nephropathy in blacks and Indians of Natal. Nephron 1988; 50: 137–41.

Sehic AM, Gaber LW, Roy S 3rd, et al. Increased recognition of IgA nephropathy in African-American children. Pediatr. Nephrol. 1997; 11: 435–7.

Shijubo N, Kawabata I, Sato N, Itoh Y. Clinical aspects of Clara cell 10-kDa protein/uteroglobin (secretoglobin 1A1). Curr. Pharm. Des. 2003; 9: 1139–49.

Sissons JG, Woodrow DF, Curtis JR, et al. Isolated glomerulonephritis with mesangial IgA deposits. BMJ 1975; 3: 611–14.

Smith SM, Harford AM. IgA nephropathy in renal allografts: increased frequency in Native American patients. Ren. Fail. 1995; 17: 449–56.

Smith SM, Hoy WE. Frequent association of mesangial glomerulonephritis and alcohol abuse: a study of 3 ethnic groups. Mod. Pathol. 1989; 2: 138–43.

Smith SM, Tung KS. Incidence of IgA-related nephrities in American Indians in New Mexico. Hum. Pathol. 1985; 16: 181–4.

Sobh MA, Moustafa FE, Sally SM, Deelder AM, Ghoniem MA. Characterisation of kidney lesions in early schistosomal-specific nephropathy. Nephrol. Dial. Transplant. 1988; 3: 392–8.

Sola Alberich R, Jammoul A, Masana L. Henoch-Schonlein purpura associated with acetylsalicylic acid. Ann. Intern. Med. 1997; 126: 665.

Srivastava RN. Pediatric renal problems in India. Pediatr. Nephrol. 1987; 1: 238–44.

Stratta P, Segoloni GP, Canavese C, et al. Incidence of biopsy-proven primary glomerulonephritis in an Italian province. Am. J. Kidney Dis. 1996; 27: 631–9.

Suzuki H, Suzuki Y, Yamanaka T, et al. Genome-wide scan in a novel IgA nephropathy model identifies a susceptibility locus on murine chromosome 10, in a region syntenic to human IGAN1 on chromosome 6q22–23. J. Am. Soc. Nephrol. 2005; 16: 1289–99.

Suzuki K, Honda K, Tanabe K, Toma H, Nihei H, Yamaguchi Y. Incidence of latent mesangial IgA deposition in renal allograft donors in Japan. Kidney Int. 2003; 63: 2286–94.

Swanepoel CR, Madaus S, Cassidy MJ, et al. IgA nephropathy – Groote Schuur Hospital experience. Nephron 1989; 53: 61–4.

Syrjanen J, Huang XH, Mustonen J, Koivula T, Lehtimaki T, Pasternack A. Angiotensin-converting enzyme insertion/deletion polymorphism and prognosis of IgA nephropathy. Nephron 2000; 86: 115–21.

Szelestei T, Bahring S, Kovacs T, et al. Association of a uteroglobin polymorphism with rate of progression in patients with IgA nephropathy. Am. J. Kidney Dis. 2000; 36: 468–73.

Taguchi T, von Bassewitz DB, Takebayashi S, Harada T. A comparative study of IgA nephritis in Japan and Germany. An approach to its geopathology. Pathol. Res. Pract. 1987; 182: 358–67.

Takei T, Iida A, Nitta K, et al. Association between single-nucleotide polymorphisms in selectin genes and immunoglobulin A nephropathy. Am. J. Hum. Genet. 2002; 70: 781–6.

Takemura T, Okada M, Yagi K, Kuwajima H, Yanagida H. An adolescent with IgA nephropathy and Crohn disease: pathogenetic implications. Pediatr. Nephrol. 2002; 17: 863–6.

Tancrede-Bohin E, Ochonisky S, Vignon-Pennamen MD, Flageul B, Morel P, Rybojad M. Schonlein-Henoch purpura in adult patients. Predictive factors for IgA glomerulonephritis in a retrospective study of 57 cases. Arch. Dermatol. 1997; 133: 438–42.

Tojo S, Hatano M, Honda N, et al. Natural history of IgA nephropathy in Japan. Semin. Nephrol. 1987; 7: 386–8.

Tolkoff-Rubin NE, Cosimi AB, Fuller T, Rublin RH, Colvin RB. IGA nephropathy in HLA-identical siblings. Transplantation 1978; 26: 430–3.

Tomana M, Matousovic K, Julian BA, Radl J, Konecny K, Mestecky J. Galactose-deficient IgA1 in sera of IgA nephropathy patients is present in complexes with IgG. Kidney Int. 1997; 52: 509–16.

Tomana M, Novak J, Julian BA, Matousovic K, Konecny K, Mestecky J. Circulating immune complexes in IgA nephropathy consist of IgA1 with galactose-deficient hinge region and antiglycan antibodies. J. Clin. Invest. 1999; 104: 73–81.

Topaloglu R, Bakkaloglu A, Slingsby JH, et al. Molecular basis of hereditary C1q deficiency associated with SLE and IgA nephropathy in a Turkish family. Kidney Int. 1996; 50: 635–42.

Tuglular S, Berthoux P, Berthoux F. Polymorphisms of the tumour necrosis factor alpha gene at position -308 and TNFd microsatellite in primary IgA nephropathy. Nephrol. Dial. Transplant. 2003; 18: 724–31.

van den Wall Bake AW, Daha MR, Evers-Schouten J, van Es LA. Serum IgA and the production of IgA by peripheral blood and bone marrow lymphocytes in patients with primary IgA nephropathy: evidence for the bone marrow as the source of mesangial IgA. Am. J. Kidney Dis. 1988a; 12: 410–14.

van den Wall Bake AW, Daha MR, van der Ark A, Hiemstra PS, Radl J, van Es LA. Serum levels and in vitro production of IgA subclasses in patients with primary IgA nephropathy. Clin. Exp. Immunol. 1988b; 74: 115–20.

van Egmond M, Damen CA, van Spriel AB, Vidarsson G, van Garderen E, van de Winkel JG. IgA and the IgA Fc receptor. Trends Immunol. 2001; 22: 205–11.

van Es LA, van den Wall Bake AW, Valentijn RM, Daha MR. Composition of IgA-containing circulating immune complexes in IgA nephropathy. Am. J. Kidney Dis. 1988; 12: 397–401.

Velo M, Lozano L, Egido J, Gutierrez-Millet V, Hernando L. Natural history of IgA nephropathy in patients followed-up for more than ten years in Spain. Semin. Nephrol. 1987; 7: 346–50.

Vorechovsky I, Cullen M, Carrington M, Hammarstrom L, Webster AD. Fine mapping of IGAD1 in IgA deficiency and common variable immunodeficiency: identification and characterization of haplotypes shared by affected members of 101 multiple-case families. J. Immunol. 2000; 164: 4408–16.

Vorechovsky I, Webster AD, Plebani A, Hammarstrom L. Genetic linkage of IgA deficiency to the major histocompatibility complex: evidence for allele segregation distortion, parent-of-origin penetrance differences, and the role of anti-IgA antibodies in disease predisposition. Am. J. Hum. Genet. 1999; 64: 1096–109.

Waldo FB. Is Henoch-Schonlein purpura the systemic form of IgA nephropathy? Am. J. Kidney Dis. 1988; 12: 373–7.

Wall BA, Agudelo CA, Pisko EJ. Increased incidence of recurrent hematuria in ankylosing spondylitis: a possible association with IgA nephropathy. Rheumatol. Int. 1984; 4: 27–9.

Wang J, Anders RA, Wu Q, et al. Dysregulated LIGHT expression on T cells mediates intestinal inflammation and contributes to IgA nephropathy. J. Clin. Invest. 2004; 113: 826–35.

Wang J, Lo JC, Foster A, et al. The regulation of T cell homeostasis and autoimmunity by T cell-derived LIGHT. J. Clin. Invest. 2001; 108: 1771–80.

Weisbart RH, Kacena A, Schuh A, Golde DW. GM-CSF induces human neutrophil IgA-mediated phagocytosis by an IgA Fc receptor activation mechanism. Nature 1988; 332: 647–8.

Wiltshire S, Bhattacharyya S, Faux JA, et al. A genome scan for loci influencing total serum immunoglobulin levels: possible linkage of IgA to the chromosome 13 atopy locus. Hum. Mol. Genet. 1998; 7: 27–31.

Wyatt RJ. The complement system in IgA nephropathy and Henoch-Schonlein purpura: functional and genetic aspects. Contrib. Nephrol. 1993; 104: 82–91.

Wyatt RJ, Julian BA, Baehler RW, et al. Epidemiology of IgA nephropathy in central and eastern Kentucky for the period 1975 through 1994. Central Kentucky Region of the Southeastern United States IgA Nephropathy DATABANK Project. J. Am. Soc. Nephrol. 1998; 9: 853–8.

Wyatt RJ, Rivas ML, Julian BA, et al. Regionalization in hereditary IgA nephropathy. Am. J. Hum. Genet. 1987; 41: 36–50.

Yamagata K, Takahashi H, Tomida C, Yamagata Y, Koyama A. Prognosis of asymptomatic hematuria and/or proteinuria in men. High prevalence of IgA nephropathy among proteinuric patients found in mass screening. Nephron 2002; 91: 34–42.

Yang YH, Hung CF, Hsu CR, et al. A nationwide survey on epidemiological characteristics of childhood Henoch-Schonlein purpura in Taiwan. Rheumatology (Oxford) 2005; 44: 618–22.

Yoshida K, Suzuki J, Suzuki S, Kume K, Suzuki H, Hujiki T. A case of IgA nephropathy in three sisters with thin basement membrane disease. Am. J. Nephrol. 1998; 18: 422–4.

Yoshioka K, Takemura T, Akano N, et al. IgA nephropathy in patients with congenital C9 deficiency. Kidney Int. 1992; 42: 1253–8.

Zhang Z, Kundu GC, Yuan CJ, et al. Severe fibronectin-deposit renal glomerular disease in mice lacking uteroglobin. Science 1997; 276: 1408–12.

Zheng F, Kundu GC, Zhang Z, Ward J, DeMayo F, Mukherjee AB. Uteroglobin is essential in preventing immunoglobulin A nephropathy in mice. Nat. Med. 1999; 5: 1018–25.

Zidar N, Cavic MA, Kenda RB, Koselj M, Ferluga D. Effect of intrauterine growth retardation on the clinical course and prognosis of IgA glomerulonephritis in children. Nephron 1998; 79: 28–32.

CHAPTER 46

Susceptibility to Diabetic Nephropathy

BARRY I. FREEDMAN, DONALD W. BOWDEN AND STEPHEN S. RICH

OVERVIEW

Epidemiology of Diabetic Complications

Diabetic nephropathy (DN) remains the most common and rapidly increasing etiology of end-stage renal disease (ESRD) in developed nations. Approximately 35% of European-Americans with diabetes mellitus (DM) are at risk for the development of overt proteinuria, chronic renal insufficiency and ESRD, with higher proportions observed among diabetic African-Americans, Hispanic-Americans and Native-Americans (USRDS 2003). Diabetic patients on dialysis suffer markedly higher morbidity and mortality rates, compared to age-, gender-, and race-matched patients with nondiabetic etiologies of ESRD (USRDS 2003). The relatively recent epidemic of DM and nephropathy has adversely impacted healthcare delivery systems worldwide and caused severe personal suffering for affected individuals and their families. Mounting evidence suggests a role for inherited, as well as environmental, factors in the development of DN.

Given recent trends, the lifetime risk for developing DM in those born in the year 2000 is projected to be 30% for European-Americans and 40–49% for African-Americans (Narayan et al 2003). Annual DM-related healthcare expenditures in the United States currently exceed $130 billion (Hogan et al 2003), with yearly costs for renal replacement therapy exceeding $23 billion (USRDS 2003).

Despite improvements in medical therapy for hyperglycemia, hypertension and elevated serum lipid levels, the incidence rate of diabetic ESRD in the USA has doubled to 148 cases per million Americans in the last decade and more than 44% of incident dialysis patients now have diabetic ESRD (USRDS 2003). This rate is expected to continue to climb. The incidence rates of type 2 diabetic ESRD are fourfold higher in African-Americans and Native-Americans, compared to European-Americans (USRDS 2003). Racial differences in the prevalence and control of DM and hypertension, socioeconomic status and access to health care are clearly present. However, these factors do not fully account for the excess incidence

rate of DN in minority populations (Brancati et al 1992). In striking contrast, diabetic macrovascular disease, especially coronary artery disease, appears to be less frequent in African-Americans and Native-Americans relative to European-Americans, given equal access to medical care (Karter et al 2002, Young et al 2003).

Pathology of Advanced Diabetic Nephropathy

The renal histologic changes that are observed in type 1 and type 2 diabetic patients with DN are identical and include diffuse and/or nodular increases in mesangial matrix, thickening of glomerular and tubular basement membranes, and hyalinosis of afferent and efferent arterioles. These changes are observed in the absence of mesangial, glomerular basement membrane (GBM) or subendothelial immune complex deposition on immunofluorescence and electron microscopy, and in the absence of other systemic diseases causing nephropathy (Heptinstall 1983).

The role of hyperglycemia in the causation of these histologic changes is likely multifactorial (Caramori et al 2003). Elevated plasma glucose concentrations activate protein kinase C (PKC) in cultured vascular smooth muscle, epithelial and mesangial cells (Natarajan et al 1997, Hoshi et al 2002, Cha et al 2000), increase the activity of growth factors and cytokines (Ding et al 2003, Ziyadeh 2004), lead to formation of reactive oxygen species (Kiritoshi et al 2003) and advanced glycation end-products (Brownlee et al 1984), and stimulate the aldose reductase pathway (Dunlop 2000, Hodgkinson et al 2001). It is likely that interactions exist between these pathways and genetic susceptibility to DN (Caramori et al 2003).

There is general agreement that DM is the cause of nephropathy in subjects having a long duration of diabetes (peak incidence at 15–20 years DM duration in type 1 disease), persistent albuminuria greater than 300 mg/24 hours (often >1.5 g/day), and with proliferative diabetic retinopathy in the absence of other risk factors for kidney disease. Most clinicians do not advocate renal biopsy in type 2 diabetic patients with classic clinical presentations (Parving et al

Genetic Diseases of the Kidney
ISBN: 978-0-12-449851-8

1992), although some have argued for more liberal biopsy criteria to minimize misclassification of patients in research studies and to improve treatment (Mazzucco et al 2002). Type 2 diabetic subjects more often present with albuminuria at initial diagnosis than type 1 diabetics, likely reflecting renal injury from prolonged phases of insulin resistance or an incorrect date of DM onset (UK Prospective Diabetes Group 1998).

Phenotypes in Diabetic Kidney Disease

Care in defining the phenotype is critically important when undertaking a search for the genetic basis of susceptibility to DN. An individual with long-standing DM accompanied by proliferative diabetic retinopathy and nephrotic range proteinuria clearly has DN. However, many diabetic individuals present with lesser degrees of proteinuria, lacking diabetic retinopathy or having shorter DM duration when proteinuria is initially detected. This is particularly common in type 2 diabetic individuals where the onset of DM may be difficult to determine and the renal histologic lesions are more variable than in type 1 DM (Parving et al 1992, Mazzucco et al 2002). Genetic analyses that classify subjects with typical and atypical lesions as 'affected' would likely introduce clinical (and genetic) heterogeneity and make gene identification difficult. The Family Investigation in Nephropathy and Diabetes (FIND) and Genetics of Kidneys in Diabetes (GoKinD) studies utilized strict entry criteria in order to maximize enrollment of patients expected to have the typical renal histologic lesions of DN (The Family Investigation of Nephropathy and Diabetes Research Group 2003). Lesions such as mesangial matrix expansion or vascular hyalinosis could be under separate genetic control; however, it is not feasible to perform large-scale renal biopsy trials in order to examine the genes impacting distinct renal pathologic lesions. Other useful surrogates for the renal histologic lesions of DN include measures of renal function and albuminuria. Elevated urinary albumin excretion (UAE) and diminished glomerular filtration rate (GFR) are recognized markers of risk for progressive kidney failure (Kussman et al 1976). Additionally, these measures are also major risk factors for the competing complications of premature cardiovascular morbidity and mortality (Rossing et al 1996, Gerstein et al 2001, Adler et al 2003, Weiner et al 2004). Microalbuminuria (UAE of 30–299 mg/day) appears to be a marker of generalized systemic endothelial dysfunction, as well as for progressive renal disease.

The natural history of DN has been altered by the use of medications that improve blood pressure (Christensen et al 1987) and glycemic control (The Diabetes Control and Complications Trial/Epidemiology of Diabetes Intervention and Complications Research Group 2000, Stratton et al 2000, Writing Team for the Diabetes Control and Complications Trial/Epidemiology of Diabetes Interventions and Complications Research Group 2003), angiotensin converting enzyme inhibitors (Lewis et al 1993) and angiotensin receptor blockers (Lewis et al 2001, Brenner et al 2001), lipid-lowering agents (particularly statins) (Bianchi et al 2003) and smoking cessation (Bleyer et al 2000). These medications complicate the study of DN susceptibility, since they reduce albuminuria, increase the duration of diabetes prior to clinical symptoms, and stabilize renal function. Therefore, diabetics who are genetically susceptible to kidney disease may significantly delay the manifestation of the nephropathy phenotype in an environment of normoglycemia, normotension, and controlled serum lipid levels.

Albumin excretion rates vary widely, up to 40% on repeated measures (Mogensen et al 1995) and a significant proportion of type 1 diabetics may have spontaneous resolution of microalbuminuria (Perkins et al 2003). Diabetes duration at the onset of microalbuminuria is also a determinant for risk of progressive kidney failure. Type 1 diabetic subjects who develop microalbuminuria after 25 years of postpubertal DM appear to be at markedly lower risk for development of progressive renal disease (Warram et al 1996).

Quantification of renal function in patients with DN may also be problematic. Serum creatinine concentration is a poor measure of kidney function, reflecting underlying muscle mass and impacted by gender, age, nutritional status, and race. Twenty-four-hour urine studies for urea and creatinine clearance are prone to collection error. Therefore, a series of equations utilizing age, race, nutritional status, and serum creatinine concentration were devised to more accurately reflect GFR and have been validated for use in diabetic patients (Levey et al 1999).

ANIMAL MODELS OF DIABETIC NEPHROPATHY

Several diabetic rodents develop proteinuria and/or renal disease. These animal models have been studied in the hope that they would provide insights into human DN. Until recently, these models failed to exactly reproduce the human renal histologic changes and the timing of their development.

The obese Zucker rat, Goto-Kakizaki (GK) rat, and db/db mouse, spontaneous models of DM, develop glomerular hypertrophy, GBM thickening, and mesangial matrix expansion (Kasiske et al 1985, Yagihashi et al 1978, Bower et al 1980). However, these inbred strains fail to develop diffuse glomerulosclerosis, nodule formation (Kimmelstiel Wilson disease) and progressive elevations in blood urea nitrogen and serum creatinine concentration (Janssen et al 1999). An autosomal recessive mutation in the *db* gene encoding the leptin receptor causes obesity and type 2 DM in the db/db mouse (Chen et al 1996). An autosomal recessive mutation of the *fa* gene, also encoding the leptin receptor, causes hyperphagia, obesity and hyperglycemia in the Zucker rat (Iida et al 1996). Zucker rats develop proteinuria, focal and segmental glomerulosclerosis (FSGS) and progressive renal insufficiency with aging (FSGS is atypical in human DN) (Kasiske et al 1985). Lipid lowering in Zucker rats ameliorates glomerular injury and reduces proteinuria. A recent review discusses additional rodent models of type 2 DN (Janssen et al 1999).

Recently, a novel diabetic rat, T2DN/Mcwi, has been developed that closely mimics the course of human DN (Nobrega et al 2004). The strain was created by combining the genomes of the fawn hooded rat (FHH), a model of hypertension-associated renal failure/focal and segmental glomerulosclerosis, and the diabetic GK rat. Mitochondrial DNA and a small number of passenger loci from the FHH were introgressed into the GK strain. The resulting offspring demonstrated early onset DM with overt proteinuria (worsening with age), glomerular hypertrophy, thickened glomerular and tubular basement membranes, mesangial matrix expansion and focal glomerulosclerosis progressing to global glomerulosclerosis with nodule formation. In contrast, diabetic GK rats fail to develop progressive renal failure and only display glomerular hypertrophy and GBM thickening. Transmission of the known renal failure loci from the FHH (i.e. *Rf*-1, *Rf*-2) did not occur in the T2DN/Mcwi strain (Brown et al 1996). A 57 centiMorgan (cM) segment on rodent chromosome 1 containing the major QTL for hyperglycemia (*Niddm1*) differed between GK and T2DN rats. *Niddm1*, and other loci that differ between GK and T2DN strains, provide an opportunity to identify the inherited factors beyond hyperglycemia that cause DN in these animals. Homologous regions of the human genome can then be evaluated to determine whether they play a role in the pathogenesis of the human disease.

FAMILIAL FACTORS IN DIABETIC NEPHROPATHY

The natural history of microvascular disease in DM differs based on organ system. Given prolonged hyperglycemia, the vast majority of diabetic individuals will develop background diabetic retinopathy; however, only one third of type 1 or type 2 diabetic individuals will develop overt kidney disease and progress to dialysis (EURODIAB IDDM Complications Study 1994). The factors that *initiate* overt nephropathy in this subset of patients are unclear, but consistent familial aggregation of kidney disease (albuminuria, creatinine clearance, ESRD) and renal histologic changes have been observed. Hypertension in the parents of diabetic subjects predisposes diabetic children to nephropathy (Krolewski et al 1988). Hyperglycemia, hyperlipidemia, hypertension, and tobacco abuse are all well accepted risk factors causing the *progression* of DN (Christensen et al

1987, The Diabetes Control and Complications Trial/ Epidemiology of Diabetes Intervention and Complications Research Group 2000, Stratton et al 2000, Writing Team for the Diabetes Control and Complications Trial/Epidemiology of Diabetes Interventions and Complications Research Group 2003, Lewis et al 1993, 2001, Brenner et al 2001, Bianchi et al 2003, Bleyer et al 2000).

Diabetic family members, even in the absence of clinical nephropathy, demonstrate similar patterns of glomerular involvement. Fioretto et al evaluated renal biopsy material from 21 type 1 diabetic siblings (Fioretto et al 1999). They found familial concordance in the percentage of the glomerulus occupied by mesangial matrix and the patterns of injury (presence of thickened GBM). Similar histologic patterns were observed within families despite differing DM duration and glycemic control.

The heritability of UAE (often measured as a spot urine albumin:creatinine ratio [ACR]) has been evaluated in type 2 diabetic Pima Indian and European-American families. There are limitations to assessing the heritability of albuminuria, including the variation on repeat measurement and spontaneous resolution, medication effects (angiotensin converting enzyme inhibitors and angiotensin receptor blockers), and the fact that elevated urinary albumin predicts cardiovascular death more strongly than renal disease in hypertensive and diabetic individuals. Nonetheless, albuminuria is clearly a marker of diabetic renal disease and a large proportion of albuminuric patients will progress to require renal replacement therapy (Kussman et al 1976, Adler et al 2003). Strong and consistent evidence across multiple studies supports familial aggregation of albuminuria.

A segregation analysis for urine ACR was performed in 630 European-American type 2 diabetic subjects from 96 large families treated at the Joslin Diabetes Center (1269 total relatives evaluated) (Fogarty et al 2000). Evidence supporting a major gene effect on the regulation of ACR was identified. A single major locus with multifactorial inheritance was most strongly supported in diabetic family members, although evidence for Mendelian transmission was also observed. The overall heritability of urine ACR was 0.27 in all families, increasing to 0.31 when the analyses were restricted to diabetic subjects. Similar results were reported in diabetic Pima Indians (Imperatore et al 2000). In a segregation analysis of urine ACR in 715 Pima families, support was found for the existence of a single major (Mendelian) gene contributing to variation in ACR levels. The heritability of ACR in Pima families was 0.21.

Familial aggregation of urine ACR was examined in European-American siblings concordant for type 2 DM in the Diabetes Heart Study (DHS) (Langefeld et al 2004) and in a biracial sample of multiply affected hypertensive families from the Hypertension Genetic Epidemiology Network (HyperGEN) (Freedman et al 2003). In 310 DHS families (662 European American diabetic individuals and 422

DM-concordant sibling pairs) the heritability of urine ACR was 0.46 (Langefeld et al 2004). Age, gender, use of medicines affecting ACR and mean blood pressure contributed to only 9% of the variation in ACR level. Similar heritability estimates for urine ACR were observed in 1164 non-diabetic, hypertensive sibling pairs from HyperGEN families. The heritability of urine ACR was 0.49, after controlling for the significant effects of covariates (age, gender, medications, blood pressure) and race (Freedman et al 2003). Together, these reports strongly support the presence of genes influencing urine ACR in diabetic and hypertensive families. Whether these genes underlie generalized endothelial dysfunction with microalbuminuria and resultant risk for macrovascular disease, risk for progressive renal disease, or both, is less clear.

The heritability of estimated GFR or creatinine clearance has been evaluated in four reports, only one containing diabetic families. In 539 monozygotic and 1208 dizygotic twins, the heritability of calculated creatinine clearance using the Cockcroft Gault equation was 0.63 (Hunter et al 2002). In HyperGEN and other European American families enriched for members at increased risk for premature cardiovascular disease, heritability estimates for measured creatinine clearance (using 24-hour urine collections) ranged from 0.17 to 0.53 (DeWan et al 2000, Hunt et al 2002). In DHS families, the heritability of GFR calculated using the Modification of Diet in Renal Disease (MDRD) equation in 422 European-American type 2 diabetic sibling pairs was 0.75 (Langefeld et al 2004). The proportion of the variance in GFR attributed to mean arterial blood pressure, medications and hemoglobin A_{1c} in the DHS was only 2%. These reports support the concept that genetic factors influence renal function in diabetic and nondiabetic families. It is likely that the familial aggregation of GFR may be less subject to variation than urine ACR. While reduced kidney function is also a risk factor for cardiovascular disease, it may prove to be a better predictor of risk for progressive renal disease than microalbuminuria.

The familial clustering of overt DN and diabetic ESRD has been widely observed in multiple racial and ethnic groups. The earliest reports detecting a familial aggregation of diabetic kidney disease were in type 1 diabetic subjects. Seaquist et al demonstrated that 83% of the diabetic siblings of European American probands with DN had overt nephropathy, in contrast to only 17% of the diabetic siblings of probands without nephropathy but with longstanding DM (Seaquist et al 1989). Quinn et al reported that the lifetime risk for developing nephropathy (after 30 years of DM) was 35% among European-American type 1 diabetics (Quinn et al 1996). The risk of nephropathy in diabetic siblings was strongly influenced by the presence of proteinuria in their diabetic index case, with a 72% cumulative risk of nephropathy in those whose sibling had proteinuria compared with a 25% cumulative risk in those having a normoalbuminuric sibling. This nearly 50% difference in nephropathy

risk based on familial factors was larger than that which could plausibly be attributed solely to the familial clustering of environmental risk factors. The familial aggregation of DN in type 1 DM and type 2 DM has also been reported in European (Borch-Johnsen et al 1992, Strojek et al 1997), Pima Indian (Pettitt et al 1990), African-American (Satko et al 2002), South American (Canani et al 1999), South East Asian (Ramirez et al 2002), and Indian populations (Vijay et al 1999).

African-American families with type 2 diabetic ESRD members from Los Angeles (Ferguson et al 1988), Winston-Salem (Freedman et al 1995), Cleveland (Covic et al 2001), the Southeastern US (Freedman et al 1997b), and Baltimore (Lei et al 1998) exhibit familial clustering of ESRD despite differing environments. The Baltimore report was population-based, and controlled for the prevalence of DM and hypertension within families (Lei et al 1998). The observed familial clustering of kidney disease in African Americans families was in marked excess of that which could simply be due to the combined high prevalence of DM and hypertension within families. African-Americans with first degree relatives having ESRD are at nine-fold increased risk for developing ESRD compared to African-Americans without a family history (Freedman et al 1993). This is markedly higher than the threefold increase in risk that was observed among European-Americans having first-degree ESRD relatives (Spray et al 1995). Another potentially important clinical observation is that African American multiplex ESRD families often contain members with type 2 diabetic ESRD and nondiabetic etiologies of ESRD (Freedman et al 1993, 1999, Bergman et al 1997). This suggests that 'renal failure genes,' independent from type 2 DM, may exist (Freedman 2002). If genetically susceptible family members develop hyperglycemia, hypertension or other 'triggers' that can initiate renal disease (such as antinuclear antibodies or human immunodeficiency virus infection), they are far more likely to develop clinical nephropathy than are others with the same degree of hypertension or hyperglycemia.

CANDIDATE PATHWAYS OF DIABETIC NEPHROPATHY

Multiple strategies exist for interrogating the human genome for evidence of genetic factors that contribute to risk for DN. There are two typical approaches of investigation – a candidate gene evaluation and the genome scan. The evaluation of candidate genes has evolved into two (often overlapping) strategies. The first strategy involves characterization of individual genes that should (based upon the underlying biological knowledge) have an effect on risk. The second strategy is to characterize the genes that may be involved in a pathophysiological pathway, recognizing that multiple genes in the pathway may have recognized effects with each other and unrecognized effects on genes in the pathway and other genes (Hill 1965, Clayton & McKeigue 2001, Chakravarti 2003).

Candidate gene association studies and genome scan linkage studies have different strengths and limitations. It is widely accepted that association studies are very sensitive and have the ability to detect what could be considered minor (e.g. approximately 5% or less of the risk) genetic contributions (Risch & Merikangas 1996). Results from association studies have been difficult to replicate. If such studies reflect true associations, one can envision that minor genetic components (multiple genes with individually small effects) may differ significantly in their frequency from one population to another, even within a population. In addition, populations in different studies are frequently ascertained using different clinical criteria or sampling designs. Further, the results of genotyping (often with different markers) are analyzed using different statistical genetic methods. These differences, in combination with the fact that many published candidate gene analyses are likely too small for adequate statistical power to detect small genetic effects, increases the likelihood that genetic associations will be difficult to replicate. At this point in time, association studies have been limited to evaluations of candidate genes, rather than a genome-wide perspective. This is due to the fact that genetic association is dependent upon linkage disequilibrium (LD). Recent studies have suggested that high LD is confined to blocks of genomic sequence averaging about 20–50 kb (Daly et al 2001, Gabriel et al 2002) (depending upon population) and consequently association studies are very 'short-sighted.' On the positive side, this short-sightedness is an advantage if true association is observed since it suggests that the susceptibility gene is close at hand.

Linkage analysis is capable of detecting genes of relatively large effect (e.g. loci contributing at least 25% of the risk). Linkage analysis in family sets is 'far-sighted,' e.g. single marker loci can detect linkage to a susceptibility locus on the order of 5 centiMorgans (cM; approximately 5 megabases; Mb) distant, but has little or no ability to precisely locate the susceptibility locus within that region. Nonparametric linkage peaks observed in most studies of complex disorders extend over regions of 5–20 Mb. As a consequence, linkage analysis of complex disorders provides little help in precisely defining the location of the hypothesized susceptibility gene. From these observations on association and linkage analysis it can be concluded that molecular genetic analysis combining association and linkage is the most powerful approach for assessing genetic contributions to diabetic complications.

The most frequent studies of candidate genes involve functional polymorphisms that affect levels (or activity) in many pathways, including the renin–angiotensin, nitric oxide, and bradykinin systems. Other genes have been tested that are associated with pathways involved with insulin resistance, lipid production and metabolism and glucose homeostasis, including aldose reductase (AR), transforming growth factor-beta (TGFβ), and glucose transporter-1 (GLUT-1).

Currently, there has been no consistent and replicated evidence for the contributions by a set of candidate gene(s) or surrogate markers for DN risk (or protection). As discussed, failure to identify risk genes could be due to small sample size, incomplete genetic dissection of the polymorphisms in the candidate gene, or extensive genetic and phenotypic heterogeneity. Ongoing initiatives, such as FIND and GoKinD, will provide DNA resources for enhancing sample size. Continued sequencing and analysis of the human genome by the Human Genome Project (Collins et al 2003a, 2003b) and the HapMap project (The International HapMap Consortium 2003) will allow better exploitation of genetic variation.

Candidate Genes for Diabetic Nephropathy

The genetic hypothesis that is often tested is whether there are polymorphisms in candidate genes that exhibit differences in frequency in cases (DN) and controls (DM without evidence of nephropathy). The selection of cases and controls is critical in the interpretation of the genetic analyses. Although not discussed in this chapter, the selection of controls may be more difficult than that of cases, as defining an 'absence' of nephropathy in diabetics may be a function of duration of hyperglycemia, amount or variation in glycemic control, or underlying variation in renal physiology (Caramori et al 2000, 2002). Rather than focus on nephropathy occurrence (initiation), other designs attempt to identify genes that influence rate of decay of kidney function (progression). Although the number of candidate genes is nearly limitless, the following sections illustrate the rationale and findings related to selective candidates and risk of DN.

NHE-1

Na^+/H^+ antiport-1 (NHE-1) activity is increased in skin fibroblasts of patients with type 1 DM with nephropathy, compared to those without evidence of nephropathy (Trevisan et al 1992, Lurbe et al 1996). Skin fibroblast NHE-1 activity in sib pairs concordant for type 1 DM is highly correlated within the pairs, independent of age, duration of DM, gender and glycemic control (Trevisan et al 1999). The NHE-1 system parallels the red blood cell (RBC) sodium/lithium (Na^+/Li^+) counter transport system (Mahnensmith & Aronson 1985), which is associated with hypertension and cardiovascular disease in nondiabetic subjects (Adragna et al 1982). Na^+/Li^+ activity is also increased in activity in the RBCs of subjects with DN and their nondiabetic parents (Walker et al 1990). Many studies have focused on the NHE family of genes as candidates for essential hypertension and renal disease, based upon its role in pH homeostasis and regulation of cellular volume. Support for an association of *NHE5* with ESRD has been demonstrated (Yu et al 2000b), although evidence of association with other genes in the NHE pathway (*NHE1*, *NHE3*) have been equivocal.

TGFβ

TGF-β (TGF-β1, TGF-β2, and TGF-β3) mRNA and proteins, as well as TGF-β-receptor mRNA, have been identified in rodent glomerular cells (Nakamura et al 1993, Hilll et al 2000). Glomerular mesangial and epithelial cells exposed in vitro to TGF-β demonstrate increased extracellular matrix (ECM) protein synthesis (type IV collagen, fibronectin, laminin, and proteoglycans), decreased synthesis of matrix metalloproteinases (MMPs), and increased production of tissue inhibitor of metalloproteinases (TIMPs) (Nakamura et al 1992, Roberts et al 1992). TGF-β1 also stimulates glucose uptake by enhancing GLUT1 in mesangial cells (Inoki et al 1999), which could further disrupt metabolic homeostasis in the diabetic environment. Several biological factors may be involved in elevating TGF-β expression in DM, including hyperglycemia, increased intraglomerular pressure, presence of glycated proteins, and activation of PKC (Chiarelli et al 2000). TGF-β1 gene over-expression has been described in rodent (Nakamura et al 1993, Shankland et al 1994, Koya et al 2000) and human diabetic glomeruli (Murphy et al 1999, Iwano et al 1996). These data suggest that polymorphisms in the TGF-β gene family may account for a portion of the observed risk of nephropathy in DM.

Studies in African-American sibling pairs concordant for ESRD exhibited marginal evidence supporting linkage of *TGF-β2* polymorphisms with chronic glomerular disease, but not DN (Freedman et al 1997a). In type 1 diabetic subjects from Denmark, a weak association was observed in TGF-β1 with DN (Pociot et al 1998). In a Caucasian sample with advanced DN compared to type 1 diabetic subjects with at least 15 years DM duration and normal albumin excretion, no evidence of association of DN with polymorphisms in TGF-β1 was observed (Ng et al 2003). Despite the weak genetic evidence for association and linkage of polymorphisms in TGF-β genes with nephropathy, animal models of DM consistently show increased TGF-β1 and TGF-β2 mRNA levels and TGF-β2 and TGF-β type II receptor proteins (Hill et al 2000). It is likely that other cellular and genetic factors modulate the local activity of TGF-β (Huang et al 2002). Evidence of interactive effects has recently been shown between TGF-β1 and ACE expression in renal arterioles in the rat (Pueyo et al 2004). These results indicate that the complex interactions between TGF-β genes and other genetic and/or environmental factors may play an important role in regulation of cell growth and response to high glucose, so that the individual effects of genes may be diminished and difficult to identify and replicate in humans.

GROWTH HORMONE (GH) AND INSULIN-LIKE GROWTH FACTOR-1 (IGF-1)

The somatomedins, or insulin-like growth factors (IGFs), comprise a family of peptides that play important roles in mammalian growth and development. IGF-1 mediates many of the growth-promoting effects of growth hormone.

Diabetes leads to decreased hepatic production of IGF-1 and the subsequent decrease in serum IGF-1 leads to an excess in growth hormone secretion (Janssen et al 2000). The increase in growth hormone can, in turn, stimulate local IGF1 pathways in other tissues (i.e. the kidney). IGF-1 in vitro has been shown to increase mesangial cell proliferation (Conti et al 1988). Mesangial cells from nonobese diabetic (NOD) mice secrete increased IGF-1 (Elliott et al 1993), and the consequent reduction in MMP-2 activity (Lupia et al 1999) is thought to lead to glomerular ECM accumulation. Increased renal IGF-1 could be caused by changes in renal IGF-1 receptors and IGF-1 binding proteins, rather than by increased local kidney IGF-1 production (Flyvbjerg et al 1994, Gronbaek et al 1996). *IGF1* and *IGF1R* genes have been utilized as candidates for low birth weight, insulin resistance, and type 2 DM (Rasmussen et al 2000). In 82 Danish families with type 2 DM, no mutations were identified in diabetic subjects that had changes in the amino acid sequence in the *IGF1* or *IGF1R* genes. This failure to identify functional variants segregating with disease suggested that variability in the coding regions of *IGF1* and *IGF1R* do not associate with reduced birth weight, insulin sensitivity index, or type 2 DM. There is little evidence suggesting that polymorphisms in the *IGF1* locus contribute to risk of DN, despite compelling roles for IGF-1 in pathogenesis and treatment.

VASCULAR ENDOTHELIAL GROWTH FACTOR (VEGF)

In the kidney, vascular endothelial growth factor (VEGF) is almost exclusively expressed in glomerular and tubular epithelial cells (Simon et al 1998, Cooper et al 1999). Glucose stimulates VEGF expression in vitro in mesangial cells (Cha et al 2000, Kim et al 2000), probably through the PKC pathway (Williams et al 1997, Cha et al 2000, Kim et al 2000). TGF-β, PKC, and nitric oxide (NO) enhance VEGF expression in cultured mesangium (Gruden et al 1999, Flyvbjerg et al 2000). VEGF expression is increased in glomeruli of diabetic animals (Cooper et al 1999) and diabetic rats treated with anti-VEGF antibody do not exhibit hyperfiltration and have a lesser increase in albumin excretion rate. Thus, VEGF is a strong candidate gene for DN risk.

The role of VEGF in the pathogenesis of DN remains unresolved, and it is uncertain whether VEGF expression is the cause of pathologic changes or represents a reparative response. Relatively few genetic studies have been undertaken, in part due to the incomplete characterization of the *VEGF* gene and absence of common polymorphic sites. This situation is only now being rectified by the sequencing of the gene (http://pga.gs.washington.edu/) and subsequent identification of both microsatellite and single nucleotide polymorphisms (SNPs). In a series of 232 patients with type 1 DM and 141 normal controls, the 18 base pair (bp) promoter insertion/deletion (I/D) polymorphism was tested for association with disease state and presence of nephropathy (Yang 2003). The D/D genotype was significantly increased in those subjects with DN compared to those diabetic subjects without evidence of renal disease after 20 years of DM duration. A further incremental increase in risk was observed when the *VEGF* genotype was combined with an aldose reductase (*ALR2*) polymorphism. In contrast, two *VEGF* promoter SNPs exhibited no association with proteinuria (DN); however, one VEGF polymorphism (-460 bp) was associated with proliferative (versus non-proliferative) retinopathy (Ray et al 2004).

RENIN-ANGIOTENSIN-ALDOSTERONE SYSTEM

Genes in the renin-angiotensin-aldosterone (RAA) system have long been hypothesized to play an important role in risk of DN. The RAA system is involved in regulation of blood pressure and sodium homeostasis, features thought to be crucial to the progression of nephropathy. The RAA system hormone, angiotensin II, also plays a critical role in hypertension, congestive heart failure, and coronary disease. In addition, two subtypes of the angiotensin II receptor, type 1 (AT1) and type 2 (AT2), have been shown to be part of the G-protein-coupled receptor superfamily, and is rapidly internalized and desensitized upon agonist stimulation (Miura et al 2003). Glucose exposure has been shown to increase mesangial angiotensin II production (Zhang et al 1999), and angiotensin II stimulates cellular glucose uptake and transcription of GLUT-1 (Tang et al 1995), leading to high intracellular glucose concentrations. It has been hypothesized that angiotensin II can stimulate in vitro ECM synthesis through TGF-β activity in mesangial and tubular cells (Wolf et al 1993, Kagami et al 1994) and can inhibit mesangial cell collagenase activity (Singh et al 1999), reducing ECM turnover. Thus, the angiotensin II receptor and its subtypes, AT1 and AT2, are viable candidate genes for DN.

Much of the genetic research on genes in the RAA system has focused on associations of DN with the *ACE* insertion/deletion (I/D) polymorphism, the angiotensinogen (*AGT*) M235T polymorphism and, to a lesser extent, variants in the angiotensin II type 1 (*AT1*) and type 2 (*AT2*) receptor genes. The impact of *ACE* (I/D) on risk of DN has been equivocal. Early studies (Doria et al 1994, Marre et al 1994) reported that the ACE I/D genotype was associated with DN, but this association was disputed by others (Schmidt et al 1995, Tarnow et al 1995). In a large-scale, multicenter study on subjects with type 1 DM with long-term hyperglycemia who are at risk for renal disease (GENEDIAB), the *ACE* I/D genotype did appear to be associated with susceptibility to DN and its progression toward renal failure (Marre et al 1997). Polymorphisms in *AGT* and *AT1* were not associated individually with DN; however, an interaction between the *ACE* I/D and *AGT* M235T genotype was observed that could account for the degree of renal involvement. Other studies of linkage and association of DN with *AGT* have been similarly equivocal (Fogarty et al 1996, Ringel et al 1997, Rogus et al 1998, Lovati et al 2001, van Ittersum et al 2000).

Pathways

Current knowledge of the pathophysiology of DN suggests a complex interaction of multiple pathways associated with alterations in the balance between ECM production and removal, with interactions between genetic and environmental factors. These pathophysiological concepts have been previously reviewed (Baynes & Thorpe 1999, Flyvbjerg 2000, Brownlee & Livingston 2003, Caramori et al 2003a) and are summarized below.

Renal ECM components do not change in parallel with the development of DN. The α3(IV) and α4(IV) collagen chains persist or increase in density in the GBM of patients with rapidly developing DN lesions, while the α1(IV) and α2(IV) chains decrease in density in the peripheral capillary wall and mesangial matrix (Kim et al 1991, Zhu et al 1994), The ECM changes in DM are site-specific, differing in direction in the GBM compared to mesangial matrix. The observed variation in structural features such as the thickness in the capillary wall, mesangial matrix volume, and width of the GBM has been shown to be partially determined by genetic factors (Fioretto et al 1999). Thus, the hyperglycemic environment of the diabetic state can modify genetically-determined morphologic structures in the kidney.

The major hypotheses as to how hyperglycemia causes DN include: (1) increased activity of a variety of growth factors, including TGF-β, GH, IGF1, VEGF, and epidermal growth factor (EGF); (2) activation of PKC isoforms; (3) increased release of hormones, including renin, angiotensin, endothelin, and bradykinin; (4) formation of reactive oxygen species (ROS); (5) increased formation of advanced glycation end-products (AGE); (6) increased activity of the aldose reductase pathway; and (7) abnormalities in glucose transport mechanisms. The various hypotheses are not independent and key components of each pathway are determined, in part, by genes. Recently, a unifying hypothesis has been developed which links several important pathways in the pathogenesis of DN. This concept theorizes that hyperglycemia-induced mitochondrial superoxide overproduction results in (1) increased activation of PKC isoforms, (2) increased formation of AGEs, (3) acceleration of glucose flux through the aldose reductase pathway and (4) increased glucose flux into the hexosamine pathway. It is theorized that these perturbations stimulate growth factors that ultimately result in ECM accumulation.

THE HEXOSAMINE PATHWAY

Shunting of extracellular glucose into the hexosamine pathway could contribute to diabetic complications through a variety of actions. Fructose-6-phosphate is converted to glucosamine-6-phosphate by glutamine:fructose-6-phosphate aminotransferase (GFAT) and provides substrates for proteoglycan and O-linked glycoprotein synthesis. Inhibition of the rate limiting enzyme, GFAT, prevents high glucose enhanced transcription of TGF-β, TGF-β1 and PAI-1, the latter two closely linked to the pathogenesis of DN

(Sayeski & Kudlow 1996, Chen et al 1998, Kolm-Litty et al 1998). Hyperglycemia-induced mitochondrial superoxide dismutase (SOD) production in bovine aortic endothelial cells was accompanied by an increase in activity of the hexosamine pathway and this was prevented by various inhibitors of mitochondrial electron transport (Du et al 2000). These inhibitors also blocked hyperglycemia-induced TGF-β1 promoter and plasminogen activator inhibitor (PAI-1) promoter gene expression. Mitochondrial SOD stimulation of the hexosamine pathway appeared to be mediated by modification of the transcription factor Sp1, which in turn regulates *TGF-β1*, *PAI-1* and other glucose-responsive genes (Du et al 2000).

GLYCOSYLATION PRODUCTS

A major effect of DM is the nonenzymatic reaction between reducing sugars, such as glucose, and free amino groups, lipids, or nucleic acids (Day et al 1979). This early glycation process proceeds through labile Schiff base adduct formation and intramolecular Amadori rearrangement to form a stable glucose-modified protein. Amadori products comprise the majority of plasmatic glucose-modified proteins, and these are receptors for some of those modified proteins (Monnier et al 1986, Brownlee et al 1988). Further modifications of Maillard reaction products lead to inter- and intramolecular cross-linkage and AGE formation. The accumulation of glycation products may be a major contributor to diabetic complications (Monnier et al 1986, Brownlee et al 1988, Soulis-Liparota et al 1991). Increased AGE formation can stimulate the synthesis of growth factors, including IGF-1 and TGF-β. Binding of AGE proteins by cell surface AGE receptors (the *RAGE* gene is located on human chromosome 6p) induces intracellular oxidative responses (Yan et al 1994).

Serum AGE levels have been shown to be independent predictors of progression of early glomerular lesions in Mexican-Americans with type 1 DM (Berg et al 1997). Further, renal cells grown in high glucose have been shown to upregulate TGF-β and PKC (Ayo et al 1991, Fumo et al 1994) and exhibit ECM overproduction (Ziyadeh et al 1993, 1994). Amadori-glycated proteins, which are earlier products of the glycation process than AGEs, have similar cellular effects even without high-glucose media. Glycated albumin and high-glucose media together cause an even greater in vitro increase in TGF-β1 and type II TGF-β-receptor mRNAs than either alone (Singh et al 1998).

ALDOSE REDUCTASE (AR)

Aldose reductase catalyzes the reduction of glucose to sorbitol in the polyol pathway and has been associated with development of DN (Caramori et al 2003). Aldose reductase is present in all target tissues that develop diabetic complications. Increased aldose reductase activity leads to increased concentrations of diacylglycerol, a PKC activator that may regulate ECM synthesis and removal. Increased

aldose reductase activity also consumes NADPH, leading to decreased glutathione, an antioxidant coenzyme. PKC controls aldose reductase (*AR*) gene transcription (Fazzio et al 1999). An aldose reductase inhibitor prevents type IV collagen and fibronectin accumulation in human proximal tubular epithelial cells (PTEC) exposed to high glucose (Phillips 1997). Interestingly, aldose reductase inhibitors have been shown to have only partial (Passariello et al 1993) or no (Ranganathan et al 1993) effect in reducing the risk of human DN. Multiple studies have evaluated polymorphisms in *AR* for association with DN. In a large series of cases and controls, two variants (5′*ALR2* and -C106T) were shown to be associated with increased risk of DN (Moczulski et al 2000). A meta-analysis of recent studies (Neamat-Allah et al 2001) further supported the association of the -C106T polymorphism with DN risk. Thus, there appears to be compelling evidence that high glucose flux through aldose reductase is modulated by variation in the *AR* gene, providing support for the role of the polyol pathway in inhibition of expression of antioxidant enzymes.

GLUCOSE TRANSPORTERS (GLUT)

Mammalian cells require specific carriers in the plasma membrane to allow glucose entry into cells. Two types have been identified: facilitated glucose transporters (GLUTs) and Na⁺-glucose cotransporters (SGLTs). GLUTs are found in all mammalian cells and allow the cellular glucose entry across the plasma membrane along its concentration gradient. At least six isoforms of GLUTs have been identified: GLUT1–GLUT7 (GLUT6 is a pseudogene). GLUT1 is localized mostly at the plasma membrane in the basal state in contrast to GLUT4, which is intracellular in the basal state and is translocated to the plasma membrane by insulin stimulation. GLUT1 is ubiquitous and serves as a main glucose transporter in cultured mesangial cells and skin fibroblasts (Longo et al 1998). Upregulation of GLUT1 expression may be important in the genesis of DN. Both high glucose and increased levels of TGB-β stimulate cellular glucose uptake by enhancing mesangial cell GLUT1 expression (Heilig et al 1997, Inoki et al 1999). Conversely, suppression of GLUT1 expression by anti-sense GLUT1 protects mesangial cells from glucose induction of GLUT1 and increased ECM synthesis (Heilig et al 2001). Studies of *Xba*I, a *GLUT1* restriction fragment length polymorphism (RFLP), have demonstrated an association with susceptibility to DN in type 1 DM patients (Liu et al 1999, Hodgkinson et al 2001, Ng et al 2002).

Oxidative Stress

Oxidative stress, leading to production of reactive oxygen species (ROS), may play an important role in the pathogenesis of DN. Variation in the critical morphologic and biological processes important in DN appears to be influenced by genetic factors, independent of duration of DM and glycemic control.

Increased net ROS production in DM may involve several mechanisms. High glucose results in overproduction of electron donors in the tricarboxylic acid (TCA) cycle which, in turn, generates a high mitochondrial membrane potential. High membrane potential inhibits electron transfer at mitochondrial respiratory chain complex III and increases the half life of superoxide-generating intermediates such as ubisemiquinone (Brownlee 2001, Du et al 2001). Genes regulating mitochondrial biogenesis and oxidative phosphorylation (OxPhos) are encoded both in mitochondrial and nuclear DNA. Mutations of mitochondrial and nuclear DNA due to either oxidative damage or genetic abnormalities result in OxPhos pathway dysfunction and increased ROS production. Recently, DNA microarray studies of muscle tissue from type 2 DM patients identified a coordinated downregulation of peroxisome proliferator-activated receptor-γ coactivator-1α (PGC-1α) and its responsive genes involved in mitochondrial biogenesis and OxPhose (Mootha et al 2003). It remains unclear whether downregulation of PGC-1α in DM occurs in other cell types, specifically glomerular or other renal cells that may be important in the pathogenesis of DN.

Genetic variants of the antioxidant regulatory system may also play a role in the pathogenesis of diabetic complications. The response of antioxidant genes to high glucose was shown to be abnormal in type 1 DM patients with DN. Glucose incubation of human PBMC increased the mRNA and enzyme activity levels of catalase, Cu/Zn SOD and GPX in samples from normal controls and type 1 DM patients without nephropathy; conversely, incubation decreased these levels in samples from patients with DN (Hodgkinson et al 2003). In the GK rat model of type 2 DM, there is a significant decrease in the heart mitochondria contents of CoQ and glutathione (Santos et al 2003). Intracellular ROS are overproduced by the mitochondrial electron transport chain in response to hyperglycemia. ROS not only appear necessary for the damage caused by hyperglycemia but also are common to several pathophysiologic hypotheses in DN. There is increased oxidative stress in families of type 1 DM patients (Matteucci & Giampietro 2001) suggesting that an abnormal redox state could precede the onset of DM. Genetic analysis of the OxPhos pathway has yet to be extensively explored; thus, dissection of candidate genes in this complex may represent an important component of integrating genetic control of complex physiologic pathways.

THE SEARCH FOR GENETIC FACTORS IN TYPE 1 DIABETIC NEPHROPATHY

The search for genes that influence risk for DN can take place at many levels. The two classical study designs used for this search – case-control and family-based analyses – are discussed above. Within the case-control design, the vast majority of studies have used a candidate gene approach (discussed above), whether interrogating individual candidate genes or genes related to specific biological pathways. The case-control design determines the evidence of association between a genetic variant and the outcome of DN (or an intermediate phenotype). Within the family-based approach, the unit of analysis is typically a pair of siblings, as parents are often unavailable for study (except in the case of type 1 DN). Several general analyses of family data have been employed – the genome scan for linkage of DN, the candidate gene evaluation of linkage, and the family-based association approach. To date, there has been no large scale, adequately powered study of linkage or association at the genomic level for type 1 or type 2 DN.

Diabetic Nephropathy and Intermediate Phenotypes

When considering genetic determinants of risk for DN, there are a number of advantages to studying intermediate phenotypes as opposed to clinical outcome. Renal lesions are the necessary and specific prerequisite for most clinical events (i.e. microalbuminuria or diminished GFR). In contrast, the expression of kidney disease also involves a variety of additional factors that may only be relevant in the setting of pre-existing subclinical disease. For this reason, DN is a more complex, multifactorial and imprecise phenotype than anatomically defined (or subclinically defined) intermediate phenotypes (such as renal morphology or functional parameters) and is, therefore, a less efficient phenotype for analysis in genetic association studies. Another advantage of intermediate renal phenotypes is that they can be defined as a quantitative trait for genetic analysis, providing greater power for association studies compared to dichotomous clinical events. Finally, many of the intermediate phenotypes, similar to DN, are heritable traits (perhaps more heritable than DN per se). The utility of intermediate phenotypes (quantitative traits) for DN is presented in greater detail below.

Genome-Wide Association Studies in Diabetic Nephropathy

As discussed above, there have been many case-control studies of selected candidate genes in nephropathy, primarily resulting from type 1 DM. Genetic variation can modify disease susceptibility or interact with environmental or other genetic risk factors. The general association studies of candidate genes with DN have been less successful for identification of genetic determinants of common complex disorders compared to identifying variants predisposing to Mendelian (single gene) disorders, particularly where many gene variants may each contribute a relatively small amount to overall risk (Risch 1990, Altmuller et al 2001). The genetic variants with small individual effect sizes remain important because of the large attributable risk they confer in common disorders. Thus, association studies should be more efficient for detection of variants with modest or small effects, and can be expanded from a candidate gene to the human genome (Risch & Merikangas 1996). As large-scale genotyping capacity intersects with decreased cost of genotyping, the ability to perform a robustly powered, cost-efficient whole genome association study of DN (and its intermediate phenotypes) becomes closer to reality.

Genome-Wide Linkage Studies of Diabetic Nephropathy

In a genome screen (or scan), no a priori hypothesis is made as to the nature or location of the specific susceptibility genes for type 1 DN or its intermediate phenotypes. Instead, the genetic determinants of these phenotypes are tested on a large number of highly polymorphic, anonymous DNA markers scattered at roughly equal intervals throughout the human genome. This approach has been successfully applied in a number of Mendelian disorders, with investigations proceeding from positive linkage results (at the level of genetic markers every 10 Mb) to linkage disequilibrium mapping (markers every Mb or less) to gene identification. It is, however, a much greater challenge to apply this approach to genetically complex disorders. For most complex diseases, it is likely that multiple genes influence disease expression, making it difficult to isolate and characterize the effects of each disease-determining locus. Genes may interact (epistasis) or induce susceptibility independently (heterogeneity). In addition, complex diseases such as DN are influenced by environmental factors. In spite of these complicating factors, the genome screen strategy has been successful for type 2 DM (Horikawa et al 2000), Crohn's disease (Hugot et al 2001), asthma (Van Eerdewegh et al 2002), and schizophrenia (Straub et al 2002). The likelihood of success has increased with advances in high-throughput genotyping, development of semiparametric genetic analysis methods, the potential to accumulate sufficient patient and family materials, and more refined definition and measurement of phenotypes.

Genomic Strategies for Diabetic Nephropathy

The advent of DNA microarray technology has revolutionized detection of common variants in studies of disease susceptibility, progression and prediction. DNA array-based methods for gene profiling in DN are now becoming a reality. Recent work has focused on mouse models of DN (Susztak et al 2003, 2004), a system that allows use of genetically defined strains, carefully characterized tissues and controlled environments. The utility of the DNA array approach is that thousands of candidate genes and pathways can be targeted as well as random sets of expressed sequence tags (ESTs) and anonymous sequences. The translation to human DN becomes more difficult, as the tissue procurement for use in the experiments is limited and the samples over the time course for tracking disease progression are often not available. Nonetheless, the application of this new technology holds much promise for detection of genes that influence initiation and progression of DN.

THE SEARCH FOR TYPE 2 DIABETIC NEPHROPATHY GENES

With compelling evidence that there is a genetic contribution to susceptibility to type 2 DN, multiple research groups have embarked on efforts to locate and identify susceptibility genes. As previously discussed, the genes that contribute to DN risk are likely to be heterogeneous in nature and have patterns of inheritance that are more complex than simple monogenic (Mendelian) disorders. Approaches that have been taken to identify these genes have ranged from limited evaluations of single genetic polymorphisms in small case-control studies to systematic evaluations of the human genome using genome scans in large collections of families.

Candidate Gene Approaches in Type 2 Diabetic Nephropathy

Candidate gene analyses have been used extensively to assess whether specific genes contribute to the complications of type 2 DM. The rationale for these studies is that biochemical and physiological analysis has identified and characterized the involvement of many proteins in the pathogenesis of DM complications. Table 46.1 summarizes a number of classes of genes that have been the focus of analysis. The actual number of genes that have been tested at some level is in excess of fifty and continues to grow. A logical next step is to ask whether a DNA sequence difference, i.e. allele or polymorphism, in a specific gene is the source of effect risk for DN. One can envision this effect if a gene sequence difference leads to altered protein function or increased/decreased expression of the protein coded by the gene. For example a -C863A single nucleotide polymorphism (SNP) in the tumor necrosis α (*TNFA*) gene promoter has been reported to be associated with lower levels of circulating TNFα protein (Skoog et al 1999). Several recent studies have characterized a SSTR (short sequence tandem repeat) in the promoter region of a matrix metalloproteinase gene, *MMP9*, in which alleles with CA$_n$-repeat less than 21 have reduced expression of the MMP9 protein (Shimajiri et al 1999, Maeda et al 2001).

The great majority of polymorphisms analyzed in candidate gene studies have no proven biological difference between different alleles. It should be noted that much of the biological literature that presents evidence of 'functional' polymorphisms also has conflicting results. In addition the evidence for 'functional' polymorphisms is frequently derived from in vitro studies (e.g. tissue culture

TABLE 46.1 Candidate genes for contributions to diabetic nephropathy

Gene class	Representative genes	Loci	Representative references
Oxidases/esterases/ reductases/ion exchangers			
	Paraoxonase	PON	(Araki 2000a)
	Apolipoprotein (apo) E	APOE	(Araki 2000b)
	Methylenetetrahydrofolate reductase	MTHFR	(Moczulski 2003)(Makita 2003)
	Na^+/H^+ exchanger genes	NHE1-NHE5	(Yu 2000b)
Renin-angiotensin system components			
	Angiotensin converting enzyme	ACE	(Staessen 1997)(Kunz 1998)
	Angiotensinogen	AGT	(Lovati 2001)
	Angiotensin receptor	AGTR1	(Jacobsen 2003)
	Plasma kallikrein	KLKB1	(Yu 2000)
	Tissue kallikrein	KLK1	(Yu 2002)
Cytokines and growth factors			
	Interleukins	IL1, IL6	(Bensen 2003a, 2003b)
	Interleukin receptors	ILR	
	Interleukin receptor antagonists	IL1RN	(Blakemore 1996)
	Transforming growth factor beta	TGFB1, TGFB2, TGFB3	(Freedman 1997b)(Pociot 1998) (Ng 2003)
	Platelet-derived growth factor	PDGF	(Freedman 1997b)
	Tumor necrosis factors	TNFA, TNFB	(Freedman 1997b)
	Inflammatory (non-cytokine) and vascular factors		
	Plasminogen activator inhibitor	PAI-1	(Wong 2000)
	Matrix metalloproteinase-9	MMP9	(Maeda 2001)(Hirakawa 2003)
	Atrial natriuretic factor	ANT	(Roussel 2004)
	Endothelin	EDN1	(Freedman 2000)
	Nitric oxide synthases	NOS1, NOS2, NOS3	(Rippin 2003)(Degen 2001)
Advanced glycation endproducts (AGE), glucose metabolism, metabolic regulation			
	Aldose reductase	ADR	(Lajer 2004)(Neamat-Allah 2001) (Matsunaga-Irie 2004) (Pettersson-Fernholm 2003)
	AGE receptor	RAGE	(Poirier 2002)
	Glucose transporters	GLUT-1	(Hodgkinson 2001a)(Tarnow 2001)
	Beta3-adrenergic receptor	ADRB3	(Nakajima et al 2000)(Tarnow 1999)

or test tube experiments) that may have limited relevance to intact living systems. It is important to note that SNPs or SSTRs that are located within or near a gene of interest can still be used to test for linkage or association of a candidate gene. In this case it is assumed that the nearby SNP or SSTR can act as a surrogate marker for a susceptibility allele of the candidate gene that has not been discovered or is poorly characterized.

The number of candidate gene analyses that have been performed with DN has been large, particularly in samples of cases and controls using a relatively small number of gene classes (summarized in Table 46.1). These genetic studies are the easiest to perform and analyze, which is reflected in well over 125 manuscripts testing association of candidate genes with DN published since the beginning of 2002. The majority of the gene classes and their products have been implicated in inflammatory processes and reflect the general impression that inflammation plays a key role in the pathogenic processes leading to the complications of DM.

The evaluation of the angiotensin converting enzyme insertion/deletion polymorphism in intron 16 of the gene (*ACE* I/D) serves as a case in point (as noted above). More than 30 studies have assessed association between this polymorphism and diabetic complications. The D allele of the gene is associated with higher levels of angiotensin converting enzyme (*ACE*) that could be pro-inflammatory. So extensively has *ACE* I/D been evaluated that meta-analyses are now being performed. Staessen et al (1997) evaluated studies including 49 959 subjects (both nondiabetic and diabetic) and concluded that the D allele is associated with diabetes-associated renal complications and increased risk of cardiovascular disease. The excess risk for the D/D genotype was 56% for DN (11 studies) and for the D/I genotype was 40% (a significant difference). Fujisawa et al (1998) evaluated 18 studies looking for association between *ACE* I/D and nephropathy and retinopathy. They found significant evidence for association of the D allele with nephropathy in both type 1 DM and type 2 DM patients but found no evidence for association with retinopathy. In contrast, Kunz et al (1998) evaluated 19 studies and came to the opposite conclusion – no evidence of association between the *ACE* I/D allele frequencies and DN was observed.

Unfortunately the conflicting results of the ACE gene meta-analyses are a consistent summary of candidate gene studies. It has proven difficult to replicate the results of such studies. Some potential reasons for this were described above – poorly powered studies, different patient ascertainment schemes, and different analytical methods. In addition, it is often thought that positive associations are more likely to be published than negative studies; thus, it becomes difficult to detect any clear pattern of candidate gene association with DM complications to date. There is an emerging consensus that relatively large populations (greater than 1000 cases and 1000 controls) will be necessary to confirm evidence of association when the true odds ratios for susceptibility frequently reside in the 1.3–1.5 range.

Most investigators acknowledge the limitations of candidate gene studies. Several groups have tried to address these limitations by carrying out their analyses of candidate genes in linkage studies in family collections, as well as in (or in parallel with) case-control populations. In addition, classes of genes (common pathways) can be studied as a group (Yu et al 1996, 2000b, Freedman et al 1997a, 2000) in a manner similar to that outlined in Table 46.1. Evaluating multiple genes simultaneously in both families and case-control designs provides an opportunity to evaluate whether the genes are linked to inheritance of DN in families in addition to association in populations. Simultaneous evaluation of multiple genes provides an internal metric for assessing the relative impact of a positive result. Consistent with the assertion that most candidate genes tested to date contribute a relatively small part of the genetic risk, there are no apparent instances in which linkage analysis in families suggests co-inheritance with the trait under study, ESRD. In contrast, several genes, insulin 1 receptor antagonist (*IL1RN*), sodium hydrogen exchanger 5 (NHE5), nitric oxide synthase (*NOS1*), and plasma kallikrein (*KLKB1*) (Freedman et al 1997a, 2000, Yu et al 1998, 2000a) have exhibited statistical evidence of association, and, as outlined above, may be minor contributors to genetic risk. A detailed evaluation of the plasma kallikrein gene (KLKB1) identified several uncommon amino acid coding changes that are observed only or primarily in multiplex ESRD families (Yu et al 2000a). An extensive follow-up analysis of *KLKB1* now suggests that there is little compelling evidence for association with DN (unpublished data).

With the recent realization that much of the human genome is comprised of blocks of linkage disequilibrium (LD), the results of previous candidate gene analyses in DN should now be reassessed. Even more important is the emerging realization that compelling evidence of genetic association with complex diseases is frequently observed with noncoding polymorphisms, i.e. polymorphisms outside the coding region of the gene and without obvious functional impact on the primary protein sequence or expression of the protein. This finding suggests that many of the previous candidate genes studies were not adequately comprehensive since they typically focused on a small number of SNPs or relatively uncommon polymorphisms with some literature evidence for functionality. A thorough evaluation of a candidate gene should now encompass genotyping of multiple SNPs across the genomic region encompassing the coding and 5′- and 3′- neighboring regions of a gene. If performed appropriately, this approach should reveal any existing LD across the gene and through both single SNP and haplotype analysis will enable a comprehensive assessment of whether a gene is truly associated with DN.

Genome Scans and Family Studies in Type 2 Diabetic Nephropathy

Though the most complex to execute, genome scans of appropriate family collections with significant incidence of DN offer the greatest hope for identifying the principal genetic components of DN. For complex genetic entities such as DN, analysis of 1000 DNA samples (or more) from many different families are necessary. This corresponds to more than 400 000 genotypes using SSTR markers which are not by intent specifically associated with genes. When the genome scan is completed, the result is a dataset that allows simultaneous tracking of every part of every chromosome in each individual with disease status. The chromosomal regions with the strongest evidence of linkage would be assumed to carry the most important susceptibility genes. The advantage of the genome screen strategy is the comprehensive evaluation of the entire genome. Compared to the effort that has gone into genome scans of type 1 DM and type 2 DM, the efforts in DN have thus far been modest. A major reason for this is the rarity of families with multiple living DN-affected family members that meet appropriate criteria – ESRD or overt nephropathy with macroalbuminuria, making accumulation of useful family collections a significant recruiting challenge.

Only a few descriptions of genome scans for type 2 DN have been published. The first genome scan searching for nephropathy and retinopathy loci, i.e. microvascular complications, was performed in Pima Indian families (Imperatore et al 1998). Genome scan data from 98 sibling pairs in which each sibling had type 2 DM and renal failure were analyzed. The strongest evidence for linkage was observed on chromosome 7q (maximum LOD score of 2.7 at the genetic marker D7S500). Additional evidence for linkage was found on chromosomes 3 and 9 (also linked to retinopathy), and chromosome 20.

Although not a complete genome scan, a similar approach in searching for nephropathy linkage in type 1 DM families (Moczulski et al 1998) was employed when evidence of linkage was found in the region containing the angiotensin II type 1 receptor gene (*AGTR1*) on chromosome 3. More extensive analysis of the chromosome 3 region suggested strong evidence of a nephropathy gene, but a detailed evaluation of the *AGTR1* gene suggested that alleles of *AGTR1* were not the primary source of the linkage evidence.

Other genome scans have recently been completed in different populations. The results of a genome scan of 18 Turkish kindreds containing multiple individuals affected with type 2 DN have been reported (Vardarli et al 2002). These unusual families had multiple cousin–cousin marriages and consequently, a significant inbreeding component. Using an autosomal dominant model for inheritance of DN, a strong linkage peak was observed on chromosome 18 (LOD score of 6.6 between D18S43 and D18S50). Evaluation of these loci in the Pima Indian families also showed evidence of linkage (confirmation), although this region was not apparent in the original Pima genome scan.

The first genome scan of DN in African-American families has only recently been completed (Bowden et al 2004). DNA samples were collected from 206 type 2 DM sibling pairs that were concordant for ESRD or ESRD and overt nephropathy from 166 African-American families (355 total affected individuals). In the initial analyses no LOD scores greater than 2.0 were observed. Four loci had LOD scores greater than 1.0, with a region on chromosome 7p (LOD score of 1.43 at 29 cM) being the highest. Multilocus interaction analysis detected six additional loci (on chromosomes 7p, 12p, 14q, 16p, 18q, and 21q) with LOD scores between 1.15–1.63. Additional analyses evaluating phenotypic interactions revealed multiple locations in the genome with evidence for interactions with age-at-onset of ESRD (9 loci), duration of DM before onset of ESRD (19 loci), and age-at-onset of DM (14 loci). Further analyses revealed evidence for a DN susceptibility locus on chromosome 3q in ~29% of the families (LOD score of 4.55 at 135 cM in the optimal subset of families). Additional linkage evidence of linkage was observed on chromosome 7p (LOD score of 3.59) in 37%

of the families with longer duration of DM prior to diagnosis of ESRD, and on chromosome 18q (LOD score of 3.72) in ~64% of the families with early age of DM diagnosis. Additional evidence for linkage was observed on chromosome 10q (LOD score of 2.65) in ~54% of the families with a later age of DM diagnosis). This is noteworthy since this region is the location for the human homolog for the rat *Rf1* (renal failure) gene (Brown et al 1996). The 7p linkage has been observed in a recent genome scan of African-American type 2 DM (Sale et al 2004) and, since the affected subjects also have type 2 DM, it is unclear at this time whether this latter locus represents a DM or nephropathy locus. From this study it was concluded that there is good evidence for DN susceptibility loci on chromosomes 3q, 7p, and 18q, although the 7p locus may represent a type 2 DM susceptibility locus.

Although these genome scan studies have been performed in diverse populations with different study designs and analytical methods, investigators can take some encouragement from the results. Several of these locations are consistent between studies. The linkage regions that were observed on chromosomes 3q, 10q, and 18q had evidence for overlap of the peaks from different genome scans, candidate linkage analyses, and association analyses. The chromosome 3q peak has been linked to type 1 DN (Moczulski et al 1998), and type 2 DN in Pima Indians (Imperatore et al 1998) and African-Americans (Bowden et al 2004). Interestingly, creatinine clearance (an associated quantitative trait for renal function) has also been mapped to chromosome 3q in the Framingham study (Fox et al 2004). In addition to the region on chromosome 10q being the location of the human homolog of the rodent *Rf1* gene (Brown et al 1996), there is evidence for DN linkage in an African-American family collection (Bowden et al 2004) and in an independent DN family collection (Iyengar et al 2003); further, the 10q region is also linked to creatinine clearance in European-American families from Utah (Hunt et al 2002). Finally, the 18q linkage to DN has been observed in Turkish (Vardarli et al 2002) and African-American families using both linkage analysis and association analysis with DN in Pima Indians. These complementary data increase the likelihood that several of chromosomal locations mapped by linkage analysis reflect the presence of true DN genes.

It is clear from linkage studies of other diseases, such as type 1 DM and type 2 DM, that multiple genome scans in different populations are of great value in trying to assess the validity of individual linkage analysis results. In essence, evidence of confirmation from multiple studies is the most widely used metric for accepting linkage analysis results. A major ongoing NIH-sponsored multicenter study, the Family Investigation of Nephropathy and Diabetes (FIND) should contribute to expanding the number and size of genome scan results. The FIND will contain immortalized cell lines from approximately 6000 individuals in multiply affected DM families that are enriched for the presence of overt DN (families are recruited based upon the presence of a case with advanced DN). African-American, European-American, Hispanic-American and American-Indian families are being recruited in nearly equal numbers. The final genome scan (expected to be complete in late 2005) will allow for the replication of results from prior small genome scans and determination of whether DN susceptibility alleles are consistent between ethnic groups. FIND genome scans will be performed using associated quantitative traits (albuminuria, proteinuria, and GFR), as well as for overt DN.

The technology for carrying out large scale molecular genetic studies is rapidly improving. In the near future it will be possible to combine the strength of case-control studies, which is sensitivity, with comprehensiveness, heretofore the major strength of genome scan linkage analyses, in whole genome association analyses. One intriguing study utilizing elements of this approach has been reported (Tanaka et al 2003) in which greater than 55 000 gene-based SNP loci were genotyped on 94 DN cases and 94 control DNAs. A single gene encoding solute carrier family 12 member 3 (*SLC12A3*), the gene associated with Gitelman's syndrome, showed consistent evidence for association with DN susceptibility with a single SNP in intron 24 showing significant association. Analysis of additional SNPs in the gene revealed several additional SNPs that were also significantly associated with nephropathy, especially one in exon 23 (+78 G to A leading to Arg913Gln), but haplotype analysis suggested only modest evidence of association with a single uncommon (4% in controls) haplotype. The results suggested that the Arg913Gln substitution in the *SLC12A3* gene provided protection from development of DN. These types of studies present many challenges both technically and analytically (e.g. how to assess levels of significance in association studies with 55 000 SNPs), but will likely become an important component of future searches for DN genes.

The identification of genes underlying susceptibility to DN has, to date, been a difficult task. However, with evolving clinical data sources containing immortalized cell lines from multiply affected families with members having diabetic kidney disease, coupled with improvements in phenotyping, high throughput genotyping, increased genetic information from the HapMap project, and novel analytic methods, it is becoming feasible. Replication of linkage peaks from genome scans in families from different racial and ethnic groups provides reassurance that regions of the genome that contain putative diabetic nephropathy susceptibility loci can be identified, and will ultimately lead to gene identification.

References

Adler AI, Stevens RJ, Manley SE, et al. Development and progression of nephropathy in type 2 diabetes: The United Kingdom Prospective Diabetes Study (UKPDS 64). Kidney Int. 2003; 63: 225–32.

Adragna NC, Canessa ML, Solomon H, Slater E, Tosteson DC. Red cell lithium-sodium countertransport and sodium-potassium cotransport in patients with essential hypertension. Hypertension 1982; 4: 795–804.

Altmuller J, Palmer LJ, Fischer G, Scherb H, West M. Genomewide scans of complex human diseases. True linkage is hard to find. Am. J. Hum. Genet. 2001; 69: 936–50.

Araki S, Makita Y, Canani L, Ng D, Warram JH, Krolewski AS. Polymorphisms of human paraoxonase 1 gene (PON1) and susceptibility to diabetic nephropathy in type 1 diabetes mellitus. Diabetologia 2000a; 43: 1540–3.

Araki S, Moczulski DK, Hanna L, Scott LJ, Warram JH, Krolewski AS. APOE polymorphisms and the development of diabetic nephropathy in type 1 diabetes: Results of case-control and family-based studies. Diabetes 2000b; 49: 2190–5.

Ayo SH, Radnik R, Garoni JA. High glucose increases diacylglycerol mass and activates protein kinase C in mesangial cell cultures. Am. J. Physiol. 1991; 261: F571–7.

Baynes JW, Thorpe SR. Role of oxidative stress in diabetic complications: A new perspective on an old paradigm. Diabetes 1999; 48: 1–9.

Bensen JT, Langefeld CD, Hawkins GA, et al. Nucleotide variation, haplotype structure, and association with end-stage renal disease of the human intereukin-1 gene cluster. Genomics 2003a; 82: 194–217.

Bensen JT, Langefeld CD, Li L, et al. Association of an IL-1A 3' UTR polymorphism with end-stage renal disease and IL-1 alpha expression. Kidney Int. 2003b; 63: 1211–19.

Berg TJ, Bangstad HJ, Torjesen PA, Osterby R, Bucala R, Hanssen KF. Advanced glycation end products in serum predict changes in the kidney morphology of patients with insulin-dependent diabetes mellitus. Metabolism 1997; 46: 661–5.

Bergman R, Key BO, Kirk KA, Warnock DG, Rostand SG. Kidney disease in the first-degree relatives of African-Americans with hypertensive end-stage renal disease. Am. J. Kidney Dis. 1997; 27: 341–6.

Bianchi A, Bigazzi R, Caiazza A, Campese VM. A controlled, prospective study of the effects of atorvastatin on proteinuria and progression of kidney disease. Am. J. Kidney Dis. 2003; 41: 565–70.

Blakemore AI, Cox A, Gonzalez AM, et al. Interleukin-1 receptor antagonist allele (IL1RN*2) associated with nephropathy in diabetes mellitus. Hum. Genet. 1996; 97: 369–74.

Bleyer AJ, Shemanski LR, Burke GL, Hansen KJ, Appel RG. Tobacco, hypertension and vascular disease: Risk factors for renal functional decline in an older population. Kidney Int. 2000; 57: 2072–9.

Borch-Johnsen K, Norgaard K, Hommel E, et al. Is diabetic nephropathy an inherited complication?. Kidney Int. 1992; 41: 719–22.

Bowden DW, Colicigno CJ, Langefeld CD, et al. A genome scan for diabetes associated end-stage renal disease in African Americans. Kidney Int. 2004; 66: 1517–26.

Bower G, Brown DM, Steffes MW, Vernier RL, Mauer SM. Studies of the glomerular mesangium and the juxtaglomerular apparatus in the genetically diabetic mouse. Lab. Invest. 1980; 43: 333–41.

Brancati FL, Whittle JC, Whelton PK, Seidler AJ, Klag MJ. The excess incidence of diabetic end-stage renal disease among blacks. A population-based study of potential explanatory factors. JAMA 1992; 268: 3079–84.

Brenner BM, Cooper ME, de Zeeuw D, et al. Effects of losartan on renal and cardiovascular outcomes in patients with type 2 diabetes and nephropathy. N. Engl. J. Med. 2001; 345: 861–9.

Brown DM, Provoost P, Daly MJ, Lander E, Jacob HJ. Renal disease susceptibility and hypertension are under independent genetic control in the fawn-hooded rat. Nat. Genet. 1996; 12: 44–51.

Brownlee M. Biochemistry and molecular cell biology of diabetic complications. Nature 2001; 414: 813–20.

Brownlee M, Cerami A, Vlassara H. Advanced glycosylation end products in tissue and the biochemical basis of diabetic complications. N. Engl. J. Med. 1988; 318: 1315–21.

Brownlee M, Livingston JN. Biochemical mechanisms of microvascular disease. In: Porte D Jr., Sherwin RS, Baron A, eds. Ellenberg & Rifkin's Diabetes Mellitus. New York: McGraw Hill, 2003: pp. 181–96.

Brownlee M, Vlassara H, Cerami A. Nonenzymatic glycosylation and the pathogenesis of diabetic complications. Ann. Intern. Med. 1984; 101: 527–37.

Canani LH, Gerchman F, Gross JL. Familial clustering of diabetic nephropathy in Brazilian type 2 diabetic patients. Diabetes 1999; 48: 909–13.

Caramori ML, Fioretto P, Mauer M. The need for early predictors of diabetic nephropathy risk: Is albumin excretion rate sufficient? Diabetes 2000; 49: 1399–408.

Caramori ML, Kim Y, Huang C, et al. Cellular basis of diabetic nephropathy: Study design and renal structural-functional relationships in patients with long-standing diabetes. Diabetes 2002; 51: 506–13.

Caramori ML, Mauer M. Diabetes and nephropathy. Curr. Opin. Nephrol. Hypertens. 2003a; 12: 273–82.

Caramori ML, Mauer M. Pathophysiology of renal complications. In: Porte D Jr., Sherwin RS, Baron A, eds. Ellenberg and Rifkin's Diabetes Mellitus. New York: McGraw-Hill, 2003b: pp. 697–722.

Cha DR, Kim NH, Yoon JW, et al. Role of vascular endothelial growth factor in diabetic nephropathy. Kidney Int. 2000; 58(Suppl. 77): S104–12.

Chakravarti A. Nature, nurture and human disease. Nature 2003; 421: 413–14.

Chen H, Charlat O, Tartaglia LA, et al. Evidence that the diabetes gene encodes the leptin receptor: Identification of a mutation in the leptin receptor gene in db/db mice. Cell 1996; 84: 491–5.

Chen YQ, Su M, Walia RR, Hao Q, Covington JW, Vaughan DE. Sp1 sites mediate activation of the plasminogen activator inhibitor-1 promoter by glucose in vascular smooth muscle cells. J. Biol. Chem. 1998; 273: 8225–31.

Chiarelli F, Santilli F, Mohn A. Role of growth factors in the development of diabetic complications. Horm. Res. 2000; 53: 53.

Christensen CK, Mogensen CE. Antihypertensive treatment: Long-term reversal of progression of albuminuria in incipient diabetic nephropathy. A longitudinal study of renal function. J. Diabetes Complications 1987; 1: 45–52.

Clayton D, McKeigue PM. Epidemiological methods for studying genes and environmental factors in complex diseases. Lancet 2001; 358: 1356–60.

Collins FS, Green ED, Guttmacher AE, Guyer MS. A vision for the future of genomics research. Nature 2003a; 442: 835–47.

Collins FS, Morgan M, Patrinos A. The Human Genome Project: Lessons from large-scale biology. Science 2003b; 300: 286–90.

Conti FG, Striker LJ, Lesniak MA, Mackay K, Roth J, Striker GE. Studies on binding and mitogenic effect of insulin and insulin-like growth factor I in glomerular mesangial cells. Endocrinology 1988; 122: 2788–95.

Cooper ME, Vranes D, Youssef S, et al. Increased renal expression of vascular endothelial growth factor (VEGF) and its receptor VEGFR-2 in experimental diabetes. Diabetes 1999; 48: 2229–39.

Covic AM, Iyengar SK, Olson JM, et al. A family-based strategy to identify genes for diabetic nephropathy. Am. J. Kidney Dis. 2001; 37: 637–47.

Daly MJ, Rioux JD, Schaffner SF, Hudson TJ, Lander ES. High-resolution haplotype structure in the human genome. Nat. Genet. 2001: 29229–32.

Day JF, Thorpe SR, Baynes JW. Nonenzymatically glycosylated albumin. In vitro preparation and isolation from normal human serum. J. Biol. Chem. 1979; 254: 595–7.

Degen B, Schmidt S, Ritz E. A polymorphism in the gene for the endothelial nitric oxide synthase and diabetic nephropathy. Nephrol. Dial. Transplant. 2001; 16: 185.

DeWan AT, Arnett DK, Atwood LD, et al. A genome scan for renal function among hypertensives: The HyperGEN study. Am. J. Hum. Genet. 2000; 68: 136–44.

The Diabetes Control and Complications Trial/Epidemiology of Diabetes Intervention and Complications Research Group. Retinopathy and nephropathy in patients with type 1 diabetes four years after a trial of intensive therapy. N. Engl. J. Med. 2000; 342: 381–9.

Ding SS, Qiu C, Hess P, Xi JF, Zheng N, Clozel M. Chronic endothelin receptor blockade prevents both early hyperfiltration and late overt diabetic nephropathy in the rat. J. Cardiovasc. Pharmacol. 2003; 42: 48–54.

Doria A, Warram JH, Krolewski AS. Genetic predisposition to diabetic nephropathy: Evidence for a role of the angiotensin I-converting enzyme gene. Diabetes 1994; 43: 690–5.

Dunlop M. Aldose reductase and the role of the polyol pathway in diabetic nephropathy. Kidney Int. 2000; 77(Suppl.): S3–12.

Du XL, Edelstein D, Dimmeler S, Ju Q, Brownlee M. Hyperglycemia inhibits endothelial nitric oxide synthase activity by posttranslational modification at the Akt site. J. Clin. Invest. 2001; 108: 1341–8.

Du XL, Edelstein D, Rossetti L, et al. Hyperglycemia-induced mitochondrial superoxide overproduction activates the hexosamine pathway and induces plasminogen activator inhibitor-1 expression by increasing Sp1 glycosylation. PNAS 2000; 97: 12222–6.

Elliott SJ, Striker LJ, Hattori M, et al. Mesangial cells from diabetic NOD mice constitutively secrete increased amounts of insulin-like growth factor-I. Endocrinology 1993; 133: 1783–8.

The EURODIAB IDDM Complications Study. Microvascular and acute complication in IDDM patients. Diabetologia 1994; 37: 278–85.

The Family Investigation of Nephropathy and Diabetes Research Group. Genetic determinants of diabetic nephropathy: The family investigators of nephropathy and diabetes (FIND). J. Am. Soc. Nephrol. 2003; 14(Suppl. 2): S202–4.

Fazzio A, Spycher SE, Azzi A. Signal transduction in rat vascular smooth muscle cells: Control of osmotically induced aldose reductase expression by cell kinases and phosphatases. Biochem. Biophys. Res. Commun. 1999; 255: 12–16.

Ferguson RM, Grim CE, Opgenorth TJ. A familial risk of chronic renal failure among blacks on dialysis? J. Clin. Epidemiol. 1988; 41: 1189–96.

Fioretto P, Steffes MW, Barboso J, Rich SS, Miller ME, Mauer M. Is diabetic nephropathy inherited? Studies of glomerular structure in type 1 diabetic sibling pairs. Diabetes 1999; 48: 865–9.

Flyvbjerg A. Putative pathophysiological role of growth factors and cytokines in experimental diabetic kidney disease. Diabetologia 2000; 43: 1205–23.

Flyvbjerg A, Kessler U, Kiess W. Increased kidney and liver insulin-like growth factor II/mannose-6-phosphate receptor concentration in experimental diabetes in rats. Growth Regul. 1994; 4: 188–93.

Fogarty D, Hanna LS, Wantman M, Warram JH, Krolewski AS, Rich SS. Segregation analysis of urinary albumin excretion in families with type 2 diabetes. Diabetes 2000; 49: 1057–63.

Fogarty DG, Harron JC, Hughes AE, Nevin NC, Doherty CC, Maxwell AP. A molecular variant of angiotensinogen is associated with diabetic nephropathy in IDDM. Diabetes 1996; 45: 1204–8.

Fox CS, Yang Q, Cupples LA, et al. Genomewide linkage analysis to serum creatinine, GFR, and creatinine clearance in a community based population: The Framingham Heart Study. J. Am. Soc. Nephrol. 2004; 15: 2457–61.

Freedman BI. End-stage renal failure in African Americans: Insights in kidney disease susceptibility. Nephrol. Dial. Transplant. 2002; 17: 198–200.

Freedman BI, Beck SR, Rich SS, et al. A genome-wide scan for urinary albumin excretion in hypertensive families. Hypertension 2003; 42: 291–6.

Freedman BI, Soucie JM, McClellan W. Family history of end-stage renal disease among incident dialysis patients. J. Am. Soc. Nephol. 1997a; 8: 1942–5.

Freedman BI, Soucie JM, Stone SM, Pegram S. Familial clustering of end-stage renal disease in blacks with HIV-associated nephropathy. Am. J. Kidney Dis. 1999; 34: 254–8.

Freedman BI, Spray BJ, Tuttle AB, Buckalew VM. The familial risk of end-stage renal disease in African Americans. Am. J. Kidney Dis. 1993; 21: 387–93.

Freedman BI, Tuttle AB, Spray BJ. Familial predisposition to nephropathy in African-Americans with non-insulin-dependent diabetes mellitus. Am. J. Kidney Dis. 1995; 25: 710–13.

Freedman BI, Yu H, Anderson PJ, Roh BH, Rich SS, Bowden DW. Genetic analysis of nitric oxide and endothelin in end-stage renal disease. Nephrol. Dial. Transplant. 2000; 15: 1794–1800.

Freedman BI, Yu H, Rothschild CB, Spray BJ, Rich SS, Bowden DW. Genetic linkage analysis of growth factor loci and end stage renal disease in African Americans. Kidney Int. 1997b; 51: 819–25.

Fujisawa T, Ikegami H, Kawaguchi Y, et al. Meta-analysis of association of insertion/deletion polymorphism of angiotensin I-converting enzyme gene with diabetic nephropathy and retinopathy. Diabetologia 1998; 41: 47–53.

Fumo P, Kuncio GS, Ziyadeh FN. PKC and high glucose stimulate collagen alpha 1(IV) transcriptional activity in a reporter mesangial cell line. Am. J. Physiol. 1994; 267: F632–8.

Gabriel SB, Schaffner SF, Nguyen H, et al. The structure of haplotype blocks in the human genome. Science 2002; 296: 2225–9.

Gerstein HC, Mann JF, Yi Q, et al. Albuminuria and risk of cardiovascular events, death, and heart failure in diabetic and nondiabetic individuals. JAMA 2001; 286: 421–6.

Gronbaek H, Nielsen B, Frystyk J, Flyvbjerg A, Orskov H. Effect of lanretoide on local kidney IGF-I and renal growth in experimental diabetes in the rat. Exp. Nephrol. 1996; 4: 295–303.

Gruden G, Thomas S, Burt D, et al. Interaction of angiotensin II and mechanical stretch on vascular endothelial growth factor production by human mesangial cells. J. Am. Soc. Nephrol. 1999; 10: 730–7.

Han DC, Isono M, Hoffman BB, Ziyadeh FN. High glucose stimulates proliferation and collagen type I synthesis in renal cortical fibroblasts: Mediation by autocrine activation of TGF-beta. J. Am. Soc. Nephrol. 1999; 10: 1891–9.

Heilig CW, Concepcion LA, Riser BL, et al. D-glucose stimulates mesangial cell GLUT1 expression and basal and IGF-I sensitive glucose uptake in rat mesangial cells: Implications for diabetic nephropathy. Diabetes 1997; 46: 1030–9.

Heilig CW, Kresiberg JI, Freytag S, et al. Antisense GLUT-1 protects mesangial cells from glucose induction of GLUT-1 and fibronectin expression. Am. J. Physiol. Renal Physiol. 2001; 280: F657–66.

Heptinstall RH. Pathology of One Kidney. Boston, MA: Little and Brown, 1983.

Hill ΛB. The environment and disease: Association or causation. Proc. R. Soc. Med. 1965; 58: 295–300.

Hill C, Flyvbjerg A, Gronbaek H, et al. The renal expression of transforming growth factor-beta isoforms and their receptors in acute and chronic experimental diabetes in rats. Endocrinology 2000; 141: 1196–208.

Hirakawa S, Lange EM, Colicigno CJ, Freedman BI, Rich SS, Bowden DW. Evaluation of genetic variation and association in the matrix metalloproteinase 9 (MMP9) gene in ESRD patients. Am. J. Kidney Dis. 2003; 42: 133–42.

Hodgkinson AD, Bartlett T, Oates PJ, Millward BA, Demaine AG. The response of antioxidant genes to hyperglycemia is abnormal in patients with type 1 diabetes and diabetic nephropathy. Diabetes 2003; 52: 846–51.

Hodgkinson AD, Millward BA, Demaine AG. Polymorphisms of the glucose transporter (GLUT1) gene are associated with diabetic nephropathy. Kidney Int. 2001a; 59: 985–9.

Hodgkinson AD, Sondergaard KL, Yang B, Cross DF, Millward BA, Demaine AG. Aldose reductase expression is induced by hyperglycemia in diabetic nephropathy. Kidney Int. 2001b; 60: 211–18.

Hogan PJ, Dall T, Nikolov P. American Diabetes Association. Economic costs of diabetes in the US in 2002. Diabetes Care 2003; 26: 917–32.

Horikawa Y, Oda N, Cox NJ, et al. Genetic variation in the gene encoding calpain-10 is associated with type 2 diabetes mellitus. Nat. Genet. 2000; 26: 163–75.

Hoshi S, Nomoto K, Kuromitsu J, Tomari S, Nagata M. High glucose induced VEGF expression via PKC and ERK in glomerular podocytes. Biochem. Biophys. Res. Commun. 2002; 290: 177–84.

Huang C, Kim Y, Caramori MLA, et al. Cellular basis of diabetic nephropathy II: The TGF-b system and diabetic nephropathy lesions in type 1 diabetes. Diabetes 2002; 51: 3577–81.

Hugot J, Chamaillard M, Zouali H, et al. Association of Nod2 leucine-rich repeat variants with susceptibility to Crohn's disease. Nature 2001; 411: 599–602.

Hunt SC, Hasstedt SJ, Coon H, et al. Linkage of creatinine clearance to chromosome 10 in Utah pedigrees replicates a locus for end-stage renal disease in humans and renal failure in the fawn-hooded rat. Kidney Int. 2002; 62: 1143–8.

Hunter DJ, Lange M, Snieder H, et al. Genetic contribution to renal function and electrolyte balance: A twin study. Clin. Sci. 2002; 103: 259–65.

Iida M, Murakami T, Ishida K, Mizuno A, Kuwajima M, Shima K. Phenotype-linked amino acid alteration in leptin receptor cDNA from Zucker fatty (fa/fa) rat. Biochem. Biophys. Res. Commun. 1996; 222: 19–26.

Imperatore G, Hanson RL, Pettitt DJ, Kobes S, Bennett PH, Knowler WC. Sib-pair linkage analysis for susceptibility genes for microvascular complications among Pima Indians with type 2 diabetes. Pima Diabetes Genes Group. Diabetes 1998; 47: 821–30.

Imperatore G, Knowler WC, Pettitt DJ, Kobes S, Bennett PH, Hanson RL. Segregation analysis of diabetic nephropathy in Pima Indians. Diabetes 2000; 49: 1049–56.

Inoki K, Haneda M, Maeda S, Koya D, Kikkawa R. TGF-beta 1 stimulates glucose uptake by enhancing GLUT1 expression in mesangial cells. Kidney Int. 1999; 55: 1704–12.

The International HapMap Consortium. The International HapMap Project. Nature 2003; 426: 789–96.

Iwano M, Kubo A, Nishino T, et al. Quantification of glomerular TGF-beta 1 mRNA in patients with diabetes mellitus. Kidney Int. 1996; 49: 1120–6.

Iyengar SK, Fox KA, Schacher M, et al. Linkage analysis of candidate loci for end-stage renal disease due to diabetic nephropathy. J. Am. Soc. Nephrol. 2003; 14(Suppl. 2): S195–201.

Jacobsen P, Tarnow L, Carstensen B, Hovind P, Poirier O, Parving HH. Genetic variation in the renin-angiotensin system and progression of diabetic nephropathy. J. Am. Soc. Nephrol. 2003; 14: 2843–50.

Janssen JA, Lamberts SW. Circulating IGF-I and its protective role in the pathogenesis of diabetic angiopathy. Clin. Endocrinol. 2000; 52: 1–9.

Janssen U, Phillips AO, Floege J. Rodent models of nephropathy associated with type II diabetes. J. Nephrol. 1999; 12: 159–72.

Kagami S, Border WA, Miller DE, Noble NA. Angiotensin II stimulates extracellular matrix protein synthesis through induction of transforming growth factor-beta expression in rat glomerular mesangial cells. J. Clin. Invest. 1994; 93: 2431–7.

Karter AJ, Ferrara A, Liu JY, Moffet HH, Ackerson LM, Selby JV. Ethnic disparities in diabetic complications in an insured population. JAMA 2002; 287: 2519–27.

Kasiske BL, Cleary MP, O'Donnell MP, Keane WF. Effects of genetic obesity on renal structure and function in the Zucker rat. J. Lab. Clin. Med. 1985; 106: 598–604.

Kim NH, Jung HH, Cha DR, Choi DS. Expression of vascular endothelial growth factor in response to high glucose in rat mesangial cells. J. Endocrinol. 2000; 165: 617–24.

Kim Y, Kleppel MM, Butkowski R, Mauer SM, Wieslander J, Michael AF. Differential expression of basement membrane collagen chains in diabetic nephropathy. Am. J. Pathol. 1991; 138: 413–20.

Kiritoshi S, Nishikawa T, Sonoda K, et al. Reactive oxygen species from mitochondria induce cyclosygenase-2 gene expression in human mesangial cells: Potential role in diabetic nephropathy. Diabetes 2003; 52: 2570–7.

Kolm-Litty V, Sauer U, Nerlich A, Lehmann R, Schleicher ED. High glucose-induced transforming growth factor β1 production is mediated by the hexosamine pathway in porcine glomerular mesangial cells. J. Clin. Invest. 1998; 101: 160–9.

Koya D, Haneda M, Nakagawa H, et al. Amelioration of accelerated diabetic mesangial expansion by treatment with a PKC beta inhibitor in diabetic db/db mice, a rodent model for type 2 diabetes. FASEB J 2000; 14: 439–47.

Krolewski AS, Canessa M, Warram JH, et al. Predisposition to hypertension and susceptibility to renal disease in insulin-dependent diabetes mellitus. N. Engl. J. Med. 1988; 318: 140–5.

Kunz R, Bork JP, Fritsche L, Ringel J, Sharma AM. Association between the angiotensin-converting enzyme-insertion/deletion polymorphism and diabetic nephropathy: A methodologic appraisal and systematic review. J. Am. Soc. Nephrol. 1998; 9: 1653–63.

Lajer M, Tarnow L, Fleckner J, et al. Association of aldose reductase gene Z + 2 polymorphism with reduced susceptibility to diabetic nephropathy in Caucasian Type 1 diabetic patients. Diabet. Med. 2004; 21: 867–73.

Langefeld CD, Beck SR, Bowden DW, Rich SS, Wagenknecht LE, Freedman BI. Heritability of GFR and albuminuria in Caucasians with type 2 diabetes mellitus. Am. J. Kidney Dis. 2004; 5: 796–800.

Lei HH, Perneger TV, Klag MJ, Whelton P, Coresh J. Familial aggregation of renal disease in a population-based case-control study. J. Am. Soc. Nephrol. 1998; 9: 1270–6.

Levey AS, Bosch JP, Lewis JB, Greene T, Rogers N, Roth D. A more accurate method to estimate glomerular filtration rate from serum creatinine: A new prediction equation. Ann. Intern. Med. 1999; 130: 461–70.

Lewis EJ, Hunsicker LG, Bain RP, Rohde RD. The effect of angiotensin-converting enzyme inhibition on diabetic nephropathy. N. Engl. J. Med. 1993; 329: 1456–62.

Lewis EJ, Hunsicker LG, Clarke WR. Collaborative Study Group. Renoprotective effect of the angiotensin-receptor antagonist irbesartan in patients with nephropathy due to type 2 diabetes. N. Engl. J. Med. 2001; 345: 851–60.

Liu ZH, Guan TJ, Chen ZH, Li LS. Glucose transporter (GLUT1) allele (Xbal)-associated with nephropathy in non-insulin-dependent diabetes mellitus. Kidney Int. 1999; 55: 1843–8.

Longo N, Griggin LD, Schuster RC, Langley S, Elsas LJ. Increased glucose transport by human fibroblasts with a heritable defect in insulin binding. Metabolism 1998; 38: 690–7.

Lovati E, Richard A, Frey BM, Frey FJ, Ferrari P. Genetic polymorphisms of the renin-angiotensin-aldosterone system in end-stage renal disease. Kidney Int. 2001; 60: 46–54.

Lupia E, Ellit SJ, Lenz O, et al. IGF-1 decrease collagen degradation in diabetic NOD mesangial cells: Implications for diabetic nephropathy. Diabetes 1999; 48: 1638–44.

Lurbe A, Fioretto P, Mauer M, LaPointe MS, Batlle D. Growth phenotype of cultured skin fibroblasts from IDDM patients with and without nephropathy and overactivity of the Na^+/H^+ antiporter. Kidney Int. 1996; 50: 1684–93.

Maeda S, Haneda M, Guo B, et al. Dinucleotide repeat polymorphism of matrix metalloproteinase-9 gene is associated with diabetic nephropathy. Kidney Int. 2001; 60: 1428–34.

Mahnensmith RL, Aronson PS. The plasma membrane sodium-hydrogen exchanger and its role in physiological and pathophysiological processes. Circ. Res. 1985; 56: 773–88.

Makita Y, Moczulski DK, Bochenski J, Smiles AM, Warram JH, Krolewski AS. Methylenetetrahydrofolate reductase gene polymorphism and susceptibility to diabetic nephropathy in type 1 diabetes. Am. J. Kidney Dis. 2003; 41: 1189–94.

Marre M, Jeunemaitre X, Gallois Y, et al. Contribution of genetic polymorphism in the renin-angiotensin system to the development of renal complications in insulin-dependent diabetes. J. Clin. Invest. 1997; 99: 1585–95.

Marre M, Bernadet P, Gallois Y, et al. Relationship between angiotensin I converting enzyme gene polymorphism, plasma levels and diabetic retinal and renal complications. Diabetes 1994; 43: 384–8.

Matsunaga-Irie S, Maruyama T, Yamamoto Y, et al. Relation between development of nephropathy and the p22phox c242T and receptor for advanced glycation end product G1704T gene polymorphisms in type 2 diabetic patients. Diabetes Care 2004; 27: 303–7.

Matteucci E, Giampietro O. Oxidative stress in families of type 1 diabetic patients. Diabetes Care 2001; 24: 167–8.

Mazzucco G, Bertani T, Fortunato M, et al. Different patterns of renal damage in type 2 diabetes mellitus: A multicentric study on 393 biopsies. Am. J. Kidney Dis. 2002; 39: 713–20.

Miura S, Saku K, Karnik S. Molecular analysis of the structure and function of the angiotensin II type 1 receptor. Hypertens. Res. 2003; 26: 937–43.

Moczulski DK, Fojcik H, Zubowska-Szczechoska E, Szydlowska I, Grzeszczak W. Effects of the C677T and A1298C polymorphisms of the MTHFR gene on the genetic predisposition for diabetic nephropathy. Nephrol. Dial. Transplant. 2003; 18: 1535–40.

Moczulski DK, Rogus JJ, Antonellis A, Warram JH. Major susceptibility locus for nephropathy in type 1 diabetes on chromosome 3q: results of novel discordant sib-pair analysis. Diabetes 1998; 47: 1164–9.

Moczulski DK, Scott L, Antonellis A, et al. Aldose reductase gene polymorphisms and susceptibility to diabetic nephropathy in type 1 diabetes mellitus. Diabetic Med. 2000; 17: 111–18.

Mogensen CE, Vestbo E, Poulsen PL, et al. Microalbuminuria and potential confounders. A review and some observations on variability of urinary albumin excretion. Diabetes Care 1995; 18: 572–81.

Monnier VM, Vishwanath V, Frank KE, Elmets CA, Dauchot P, Kohn RR. Relation between complications of type 1 diabetes mellitus and collagen-linked fluorescence. N. Engl. J. Med. 1986; 314: 403–8.

Mootha VK, Lindgren CM, Eriksson KF, et al. PGC-1 alpha-responsive genes involved in oxidative phosphorylation are coordinately downregulated in human diabetes. Nat. Genet. 2003; 34: 267–73.

Murphy M, Godson C, Cannon S, et al. Suppression subtractive hybridization identifies high glucose levels as a stimulus for expression of connective tissue growth factor and other genes in human mesangial cells. J. Biol. Chem. 1999; 274: 5830–4.

Nakajima S, Baba T. Trp64Arg polymorphism of the beta3-adrenegic receptor is not associated with diabetic nephropathy in Japanese patients with type 2 diabetes. Diabetes Care 2000; 23: 862–3.

Nakamura T, Fukui M, Ebihara I, et al. mRNA expression of growth factors in glomeruli from diabetic rats. Diabetes 1993; 42: 450–6.

Nakamura T, Miller D, Ruoslahti E, Border WA. Production of extracellular matrix by glomerular epithelial cells is regulated by transforming growth factor-beta 1. Kidney Int. 1992; 41: 1213–21.

Narayan KM, Boyle JP, Thompson TJ, Sorensen SW, Williamson DF. Lifetime risk for diabetes mellitus in the United States. JAMA 2003; 290: 1884–90.

Natarajan R, Bai W, Lanting L, Gonzales N, Nadler J. Effects of high glucose on vascular endothelial growth factor expression in vascular smooth muscle cells. Am. J. Physiol. 1997; 273: H2224–H31.

Neamat-Allah M, Feeney SA, Savage DA, et al. Analysis of the association between diabetic nephropathy and polymorphisms in the aldose reductase gene in type 1 and type 2 diabetes mellitus. Diabet. Med. 2001; 18: 906–14.

Ng DP, Canani L, Araki S, et al. Minor effect of GLUT1 polymorphisms on susceptibility to diabetic nephropathy in type 1 diabetes. Diabetes 2002; 51: 2264–9.

Ng DP, Warram JH, Krolewski AS. TGF-beta 1 as a genetic susceptibility locus for advanced diabetic nephropathy in type 1 diabetes mellitus: An investigation of multiple known DNA sequence variants. Am. J. Kidney Dis. 2003; 41: 22–8.

Nobrega MA, Fleming S, Roman RJ, et al. Initial characterization of a rat model of diabetic nephropathy. Diabetes 2004; 53: 735–42.

Parving HH, Gall MA, Skott P, et al. Prevalence and causes of albuminuria in non-insulin-dependent diabetic patients. Kidney Int. 1992; 41: 758–62.

Passariello N, Sepe J, Marrazzo G, et al. Effect of aldose reductase inhibitor (tolrestat) on urinary albumin excretion rate and glomerular filtration rate in IDDM subjects with nephropathy. Diabetes Care 1993; 16: 789–95.

Perkins BA, Ficociello LH, Silva KH, et al. Regression of microalbuminuria in type 1 diabetes. N. Engl. J. Med. 2003; 348: 2285–93.

Pettersson-Fernholm K, Forsblom C, Hudson BI, et al. Finn-Diane Study Group. The functional -374 T/A RAGE gene polymorphism is associated with proteinuria and cardiovascular disease in type 1 diabetic patients. Diabetes 2003; 52: 891–4.

Pettitt DJ, Saad MF, Bennett PH, Nelson RG, Knowler WC. Familial predisposition to renal disease in two generations of Pima Indians with type 2 (non-insulin-dependent) diabetes mellitus. Diabetologia 1990; 33: 438–43.

Phillips AO, Steadman R, Morrisey K, Martin J, Eynstone L, Williams JD. Exposure of human renal proximal tubular cells to glucose leads to accumulation of type IV collagen and fibronectin by decreased degradation. Kidney Int 1997; 52: 973–84.

Pociot F, Hansen PM, Karlsen AE, Langdahl BL, Johannesen J, Nerup J. TGF-beta 1 gene mutations in insulin-dependent diabetes mellitus and diabetic nephropathy. J. Am. Soc. Nephrol. 1998; 9: 2302–7.

Poirier O, Nicaud V, Vionnet N, et al. Polymorphism screening of four genes encoding advanced glycation end-product putative receptors. Association study with nephropathy in type 1 diabetic patients. Diabetes 2002; 51: 256.

Pueyo ME, Challah M, Gauguier D, et al. Transforming growth factor-beta 1 production is correlated with genetically determined ACE expression in congenic rats: A possible link between ACE genotype and diabetic nephropathy. Diabetes 2004; 53: 1111–18.

Quinn M, Angelico MC, Warram JH, Krolewski AS. Familial factors determine the development of diabetic nephropathy in patients with IDDM. Diabetologia 1996; 39: 940–5.

Ramirez SPB, McClellan W, Port FK, Hsu SIH. Risk factors for proteinuria in a large, multiracial, Southeast Asian population. J. Am. Soc. Nephrol. 2002; 13: 1907–17.

Ranganathan S, Kremph M, Feraille E, Charbonnel B. Short term effect of an aldose reductase inhibitor on urinary albumin excretion rate (UAER) and glomerular filtration rate (GFR) in type 1 diabetic patients with incipient nephropathy. Diab. Metab. 1993; 19: 257–61.

Rasmussen SK, Lautier C, Hansen L, et al. Studies of the variability of the genes encoding the insulin-like growth factor I receptor and its ligand in relation to type 2 diabetes mellitus. J. Clin. Endocrinol. Metab. 2000; 85: 1606–10.

Ray D, Mishra M, Ralph S, Read I, Davies R, Brenchley P. Association of the VEGF gene with proliferative diabetic retinopathy but not proteinuria in diabetes. Diabetes 2004; 53: 861–4.

Ringel J, Beige J, Kunz R, Distler A, Sharma AM. Genetic variants of the renin-angiotensin system, diabetic nephropathy and hypertension. Diabetologia 1997; 40: 193–9.

Rippin JD, Patel A, Belyaev ND, Gill GV, Barnett AH, Bain SC. Nitric oxide synthase gene polymorphisms and diabetic nephropathy. Diabetologia 2003; 46: 426–8.

Risch N. Linkage strategies for genetically complex traits. Multilocus models. Am. J. Hum. Genet. 1990; 46: 222–8.

Risch N, Merikangas K. The future of genetic studies of complex human diseases. Science 1996; 273: 1516–17.

Roberts AB, McCune BK, Sporn MB. TGF-beta: Regulation of extracellular matrix. Kidney Int. 1992; 41: 557–9.

Rocco MV, Chen Y, Goldfarb S, Ziyadeh FN. Elevated glucose stimulates TGF-beta gene expression and bioactivity in proximal tubule. Kidney Int. 1992; 41: 107–14.

Rogus JJ, Moczulski D, Freire MBS, Yang Y, Warram JH, Krolewski AS. Diabetic nephropathy is associated with AGT polymorphism T235. Results of a family-based study. Hypertension 1998; 31: 627–31.

Rossing P, Hougaard P, Borch-Johnsen K, Parving HH. Predictors of mortality in insulin dependent diabetes: 10 year observational follow up study. BMJ 1996; 313: 779–84.

Roussel R, Tregouet DA, Hadjadj S, Jeunemaitre X, Marre M. Investigation of the human ANP gene in type 1 diabetic nephropathy: Case-control and follow-up studies. Diabetes 2004; 53: 1394–8.

Sale MM, Freedman BI, Langefeld CD, et al. A genome-wide scan for type 2 diabetes in African Americans families reveals evidence for a locus on chromosome 6q. Diabetes 2004; 53: 830–7.

Santos DL, Palmeira CM, Seica R, et al. Diabetes and mitochondrial oxidative stress: A study using heart mitochondria from the diabetic Goto-Kakizaki rat. Mol. Cell Biochem. 2003; 246: 163–70.

Satko SG, Langefeld CD, Daeihagh PD, Bowden DW, Rich SS, Freedman BI. Nephropathy in siblings of African-Americans with overt type 2 diabetic nephropathy. Am. J. Kidney Dis. 2002; 40: 489–94.

Sayeski PP, Kudlow JE. Glucose metabolism to glucosamine is necessary for glucose stimulation of transforming growth factor-alpha gene transcription. J. Biol. Chem. 1996; 271: 15237–43.

Schmidt S, Schone N, Ritz E, et al. Association of ACE gene polymorphism and diabetic nephropathy? Kidney Int 1995; 47: 1176–81.

Seaquist ER, Goetz FC, Rich SS, Barboso J. Familial clustering of diabetic kidney disease: Evidence for genetic susceptibility to diabetic nephropathy. N. Engl. J. Med. 1989; 320: 1161–5.

Shankland SJ, Scholey JW, Ly H, Thai K. Expression of transforming growth factor-beta 1 during diabetic renal hypertrophy. Kidney Int. 1994; 46: 430–42.

Shimajiri S, Arima N, Tanimoto A, et al. Shortened microsatellite d(CA)21 sequence down-regulates promoter activity of matrix metalloproteinase. FEBS Lett. 1999; 455: 70–4.

Simon M, Rockl W, Hornig C, et al. Receptors of vascular endothelial growth factor/vascular permeability factor (VEGF/VPF) in fetal and adult human kidney: Localization and [125I]VEGF binding sites. J. Am. Soc. Nephrol. 1998; 9: 1032–44.

Singh AK, Mo W, Dunea G, Arruda JA. Effect of glycated proteins on the matrix of glomerular epithelial cells. J. Am. Soc. Nephrol. 1998; 9: 802–10.

Singh R, Alavi N, Singh AK, Leehey DJ. Role of angiotensin II in glucose-induced inhibition of mesangial matrix degradation. Diabetes 1999; 48: 2066–73.

Skoog T, van't Hooft FM, Kallin B, et al. A common functional polymorphism (C–A substitution at position-863) in the promoter region of the tumour necrosis factor-alpha (TNF-alpha) gene associated with reduced circulating levels of TNF-alpha. Hum. Mol. Genet. 1999; 8: 1443–9.

Soulis-Liparota T, Cooper M, Papazoglou D, Clarke B, Jerums G. Retardation by aminoguanidine of development of albuminuria, mesangial expansion, and tissue fluorescence in streptozocin-induced diabetic rat. Diabetes 1991; 40: 1328–34.

Spray BJ, Atassi NG, Tuttle AB, Freedman BI. Familial risk, age at onset, and cause of end-stage renal disease in white Americans. J. Am. Soc. Nephrol. 1995; 5: 1806–10.

Staessen JA, Wang JG, Ginocchio G, et al. The deletion/insertion polymorphism of the angiotensin converting enzyme gene and cardiovascular-renal risk. J. Hypertens. 1997; 15: 1579–92.

Stratton IM, Adler AI, Neil HA, et al. Association of glycaemia with macrovascular and microvascular complications of type 2 diabetes (UKPDS 35): Prospective observational study. BMJ 2000; 321: 405–12.

Straub RE, Jiang Y, MacLean CJ, et al. Genetic variation in the 6p22.3 gene DTNBP1, the human ortholog of the mouse dysbindin gene, is associated with schizophrenia. Am. J. Hum. Genet. 2002; 71: 337–48.

Strojek K, Grzeszczak W, Morawin E, et al. Nephropathy of type II diabetes: Evidence for hereditary factors? Kidney Int. 1997; 51: 1602–7.

Susztak K, Bottinger EP, Novetsky A, et al. Molecular profiling of diabetic mouse kidney reveals novel genes linked to glomerular disease. Diabetes 2004; 53: 784–94.

Susztak K, Sharma K, Schiffer M, McCue P, Ciccone E, Bottinger EP. Genomic strategies for diabetic nephropathy. J. Am. Soc. Nephrol. 2003; 14: S271–8.

Tanaka N, Babazono T, Saito S, et al. Association of solute carrier family 12 (sodium/chloride) member 3 with diabetic nephropathy, identified by genome-wide analyses of single nucleotide polymorphisms. Diabetes 2003; 52: 2848–53.

Tang SS, Diamant D, Rhoads DB, Ingelgfinger JR. Angiotensin II regulates glucose uptake in immortalized rat proximal tubular cells. J. Am. Soc. Nephrol. 1995; 6: 748A.

Tarnow L, Cambien F, Rossing P, et al. Lack of relationship between an insertion/deletion polymorphism in the angiotensin I-converting enzyme gene and diabetic nephropathy and proliferative retinopathy in IDDM patients. Diabetes 1995; 44: 489–94.

Tarnow L, Grarup N, Hansen T, Parving HH, Pedersen O. Diabetic microvascular complications are not associated with two polymorphisms in the GLUT-1 and PC-1 genes regulating glucose metabolism I Caucasian type 1 diabetic patients. Nephrol. Dial. Transplant. 2001; 16: 1653–6.

Tarnow L, Urhammer SA, Mottalu B, Hansen BV, Pedersen O, Parving HH. The Trp64Arg amino acid polymorphism of the beta3-adrenergic receptor gene does not contribute to the genetic susceptibility of diabetic microvascular complications in Caucasian type 1 diabetic patients. Nephrol. Dial. Transplant. 1999; 14: 895–7.

Trevisan R, Fioretto P, Barboso J, Maucr M. Insulin dependent diabetic sibling pairs are concordant for sodium-hydrogen antiport activity. Kidney Int. 1999; 55: 2383–9.

Trevisan R, Li LK, Messent J, et al. Na^+/H^+ antiport activity and cell growth in cultured skin fibroblasts of IDDM patients with nephropathy. Diabetes 1992; 41: 1239–46.

UK Prospective Diabetes Study Group. Tight blood pressure control and risk of macrovascular and microvascular complications in type 2 diabetes: UKPDS 38. BMJ 1998; 31: 703–13.

US Renal Data System, USRDS 2003 Annual Report: Atlas of End-Stage Renal Disease in the United States, National Institutes of Health, National Institutes of Diabetes and Digestive and Kidney Diseases. Bethesda, MD: 2003.

Van Eerdewegh P, Little RD, Dupuis J, et al. Association of the ADAM33 gene with asthma and bronchial hyperresponsiveness. Nature 2002; 418: 426–30.

van Ittersum FJ, de Man AM, Thijssen S, et al. Genetic polymorphisms of the renin-angiotensin system and complications of insulin-dependent diabetes mellitus. Nephrol. Dial. Transplant. 2000; 15: 1000–7.

Vardarli I, Baier L, Hanson JA, et al. Gene for susceptibility to diabetic nephropathy in type 2 diabetes maps to 18q22.3-23. Kidney Int. 2002; 62: 2176–83.

Vijay V, Snehalatha C, Shina K, Lalitha S, Ramachandran A. Familial aggregation of diabetic kidney disease in Type 2 diabetes in south India. Diabetes Res. Clin. Pract. 1999; 43: 167–71, www.gokind.org/about.html, 2004.

Walker JD, Tariq T, Viberti G. Sodium-lithium countertransport activity in red cells of patients with insulin dependent diabetes and nephropathy and their parents. BM J 1990; 301: 635–8.

Warram JH, Gearin G, Laffel L, Krolewski AS. Effect of duration of type 1 diabetes on the prevalence of stages of diabetic nephropathy defined by urinary albumin/creatinine ratio. J. Am. Soc. Nephrol. 1996; 7: 930–7.

Weiner DE, Tighiouart H, Amin MG, et al. Chronic kidney disease as a risk factor for cardiovascular disease and all-cause mortality: A pooled analysis of community-based studies. J. Am. Soc. Nephrol. 2004; 15: 1307–15.

Williams B, Gallacher B, Patel H, Orme C. Glucose-induced protein kinase C activation regulates vascular permeability factor mRNA expression and peptide production by human vascular smooth muscle cells in vitro. Diabetes 1997; 46: 1497–503.

Wolf H, Mueller E, Stahl RA, Ziyadeh FN. Angiotensin II-induced hypertrophy of cultured murine proximal tubular cells is mediated by endogenous transforming growth factor-beta. J. Clin. Invest. 1993; 92: 1366–72.

Wong TY, Poon P, Szeto CC, Chan JC, Li PK. Association of plasminogen activator inhibitor-1 4G/4G genotype and type 2 diabetic nephropathy in Chinese patients. Kidney Int. 2000; 57: 632–8.

Writing Team for the Diabetes Control and Complications Trial/Epidemiology of Diabetes Interventions and Complications Research Group. Sustained effect of intensive treatment of type 1 diabetes mellitus on development and progression of diabetic nephropathy: The Epidemiology of Diabetes Interventions and Complications (EDIC) study. JAMA 2003; 290: 2159–67.

Yagihashi S, Goto Y, Kakizaki M, Kaseda N. Thickening of glomerular basement membrane in spontaneously diabetic rats. Diabetologia 1978; 15: 309–12.

Yan SD, Schmidt AM, Anderson GM, et al. Enhanced cellular oxidant stress by the interaction of advanced glycation end products with their receptors/binding proteins. J. Biol. Chem. 1994; 269: 9889–97.

Yang B, Cross DF, Ollerenshaw M, Millward BA, Demaine AG. Polymorphisms of the vascular endothelial growth factor and susceptibility to diabetic microvascular complications in patients with type 1 diabetes mellitus. J. Diabetes Complications 2003; 17: 1–6.

Young BA, Maynard C, Boyko EJ. Racial differences in diabetic nephropathy, cardiovascular disease, and mortality in a national population of veterans. Diabetes Care 2003; 26: 2392–9.

Yu H, Anderson PJ, Freedman BI, Rich SS, Bowden DW. Genomic structure of the human plasma prekallikrein gene, identification of allelic variants, and analysis in end-stage renal disease. Genomics 2000a; 69: 225–34.

Yu H, Bowden DW, Spray BJ, Rich SS, Freedman BI. Linkage analysis between loci in the renin-angiotensin axis and end-stage renal disease in African Americans. J. Am. Soc. Nephrol. 1996; 7: 2559–64.

Yu H, Bowden DW, Spray BJ, Rich SS, Freedman BI. Identification of human plasma kallikrein gene polymorphisms and evaluation of their role in end-stage renal disease. Hypertension 1998; 31: 906–11.

Yu H, Freedman BI, Rich SS, Bowden DW. Human Na^+/H^+ exchanger genes: Identification of polymorphisms by radiation hybrid mapping and analysis of linkage in end-stage renal disease. Hypertension 2000b; 35: 135–43.

Yu H, Song Q, Freedman BI, Chao J, Chao L, Bowden DW. Association of the tissue kallikrein gene promoter with ESRD and hypertension. Kidney Int. 2002; 61: 1030–9.

Zhang SL, Filep JG, Hohman TC, Tang SS, Ingelfinger JR, Chan JS. Molecular mechanisms of glucose action on angiotensinogen gene expression in rat proximal tubular cells. Kidney Int. 1999; 55: 454–64.

Zhu D, Kim Y, Steffes MW, Groppoli TJ, Butkowski R, Mauer SM. Glomerular distribution of type IV collagen in diabetes by high resolution quantitative immunochemistry. Kidney Int. 1994; 45: 425–33.

Ziyadeh FN. Mediators of diabetic renal disease: The case for tgf-Beta as the major mediator. J. Am. Soc. Nephrol. 2004; 15(Suppl. 1): S55–7.

Ziyadeh FN, Mo W, Dunea G, Arruda JA. The extracellular matrix in diabetic nephropathy. Am. J. Kidney Dis. 1993; 22: 736–44.

Ziyadeh FN, Sharma K, Ericksen M, Wolf G. Stimulation of collagen gene expression and protein synthesis in murine mesangial cells by high glucose is mediated by autocrine activation of transforming growth factor-beta. J. Clin. Invest. 1994; 93: 536–42.

CHAPTER 47

HIV-associated Nephropathy

CHRISTINA M. WYATT AND PAUL E. KLOTMAN

INTRODUCTION

HIV-associated nephropathy, or HIVAN, has emerged as an important cause of chronic kidney disease and end-stage renal disease (ESRD) since it was first described in 1984 (Gardenschwartz et al 1984, Pardo et al 1984, Rao et al 1984). The classic histologic pattern of HIVAN, a collapsing form of focal segmental glomerulosclerosis (FSGS), shows a striking racial predilection for individuals of African descent. While this suggests that host genetic factors are important in the development of disease, it is likely that HIVAN results from complex interactions between host and viral genetic factors.

HISTOLOGY

HIV infection has been associated with a number of renal lesions. In this chapter, we reserve the term HIVAN to refer to the classic pattern of collapsing FSGS with associated tubular microcysts and interstitial fibrosis (Cohen & Nast 1988). While the majority of patients with HIV-associated proteinuric renal disease have classic HIVAN, nearly 40% will have an alternative diagnosis on biopsy. Isolated FSGS and membranoproliferative glomerulonephritis account for an additional 20–25% of HIV-associated glomerular disease (D'Agati & Appel 1998).

In classic HIVAN, the kidneys are enlarged or normal in size on gross examination, consistent with the histologic findings of glomerular and tubular epithelial proliferation.

The glomerular lesion of HIVAN is a collapsing form of FSGS, characterized by podocyte proliferation and dedifferentiation (Figure 47.1). Renal tissue from patients with idiopathic or HIV-associated collapsing FSGS demonstrates loss of podocyte maturity markers, including synaptopodin, podocalyxin, and the Wilms' tumor antigen WT-1. Affected podocytes express the nuclear cell-cycle protein Ki-67, a marker of cell proliferation not normally expressed in the adult kidney. The absence of podocyte dedifferentiation and

proliferation in tissue from patients with minimal change disease and membranous nephropathy suggests that podocyte dysregulation may be of unique pathogenic importance in collapsing FSGS (Barisoni et al 1999).

The collapsing FSGS of HIVAN is classically associated with microcystic tubular changes and interstitial inflammation, with varying degrees of interstitial fibrosis (Laurinavicius et al 1999). Microcyst formation has been demonstrated in multiple nephron segments by immunostaining for segment-specific tubular antigens. While microcysts derived from proximal tubule and collecting duct epithelium demonstrate normal expression of aquaporin-1 or aquaporin-2, respectively, microcysts derived from thick ascending limb epithelium demonstrate heterogeneous expression of Tamm-Horsfall protein. This heterogeneity of expression may suggest a pathogenic role for virus-induced effects on tubular epithelial cell differentiation (Ross et al 2001).

CLINICAL PRESENTATION AND EPIDEMIOLOGY

HIVAN was originally described as a disease of patients with advanced AIDS (Rao et al 1984, Winston et al 1999). Following the introduction of highly active antiretroviral therapy (HAART) in 1996, HIVAN has been described in earlier stages of HIV infection (Levin et al 2001, Winston et al 2001). Patients typically present with moderate proteinuria and progressive renal insufficiency, although there is increasing evidence that treatment of HIV infection may delay the progression of HIVAN to ESRD. Unlike other renal diseases associated with proteinuria and rapid progression, HIVAN does not commonly present with severe hypertension or edema (Laurinavicius et al 1999, Abbott et al 2001). In the post-HAART era, the diagnosis of HIVAN should also be considered in the setting of proteinuria and slowly progressive renal disease.

Genetic Diseases of the Kidney
ISBN: 978-0-12-449851-8

793

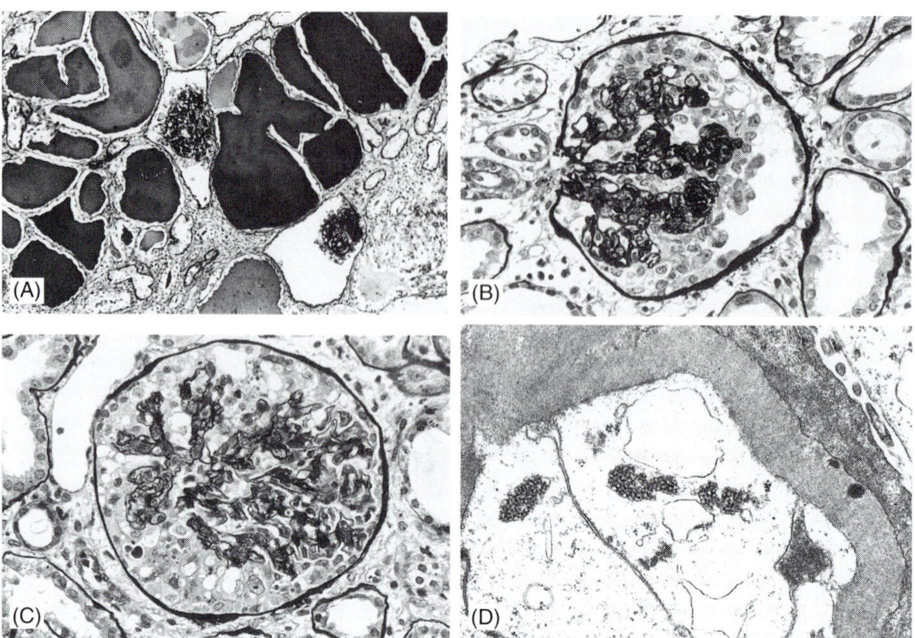

FIGURE 47.1   Histopathology in HIV-associated renal disease. (A) Low-power view shows global collapsing sclerosis with dilatation of Bowman's space. There are multiple dilated tubules containing proteinaceous casts, forming tubular microcysts. Periodic acid–Schiff; magnification × 125. (B) The glomerular capillary lumina are obliterated by global collapse of glomerular basement membranes, with hypertrophy and hyperplasia of the overlying podocytes. Jones, methenamine silver stain; magnification × 325. (C) This example of collapsing sclerosis has marked podocyte hyperplasia, forming a pseudocrescent that obliterates the urinary space. Many of the podocytes contain intracytoplasmic protein resorption droplets. Jones, methenamine silver stain; magnification × 325. (D) High power view of a glomerular endothelial cell by electron microscopy shows multiple intracytoplasmic tubuloreticular inclusions ('interferon footprints'). Photomicrographs graciously provided by Vivette D'Agati

Prevalence of HIV and AIDS

As the incidence of HIV continues to increase and mortality from AIDS declines due to HAART, the number of people living with HIV and AIDS continues to rise (Figure 47.2). In the United States, nearly 385 000 people were living with AIDS at the end of 2002, the last year for which data are available (CDC 2003). Worldwide, an estimated 40 million people are living with HIV, with 5 million new cases and 3 million deaths due to AIDS in 2003. The HIV/AIDS epidemic is most pronounced in sub-Saharan Africa, where at least 25 million people were estimated to be living with HIV infection or AIDS at the end of 2003. The prevalence of HIV/AIDS has not risen as sharply in Africa, due to limited access to HAART and subsequent high mortality rates among those infected with HIV (UNAIDS 2003).

Prevalence and Incidence of HIV-associated Renal Disease

Despite increases in the prevalence of HIV and AIDS, the incidence of ESRD attributed to HIVAN has remained relatively stable in the United States since the introduction of HAART in 1996 (Eggers & Kimmel 2004). In 2002, the last year for which data are available, 836 patients were started on renal replacement therapy with a diagnosis of AIDS nephropathy (USRDS 2004). The number of patients with HIVAN who have not reached ESRD is more difficult to estimate. A retrospective cohort study from Boston documented progression to ESRD in 14/18 patients with biopsy-proven HIVAN; however, this cohort was diagnosed prior to the introduction of HAART, and less than half received any antiretroviral therapy. Of note, 94% of patients with HIVAN in this cohort were African-American, compared to only 57% of patients with idiopathic collapsing FSGS (Laurinavicius et al 1999). In a larger retrospective cohort of patients with newly diagnosed HIV infection, approximately 2% of those with initially normal renal function developed some degree of renal insufficiency on follow-up. This cohort was assembled between 1990 and 1998, including patients diagnosed before and after the introduction of HAART (Gupta et al 2004).

Many studies of HIV-associated renal disease have been limited by selection bias because they included only subjects with known renal disease. By contrast, the Women's Interagency HIV Study enrolled 2059 seropositive women from urban centers in 1994–1995, with recruitment independent of the presence of renal disease. More than half of the cohort subjects were African-American. While subjects were recruited prior to the introduction of HAART, 93% of the cohort was on HAART at the 5-year follow-up. Initial cross-sectional data demonstrated persistent dipstick proteinuria

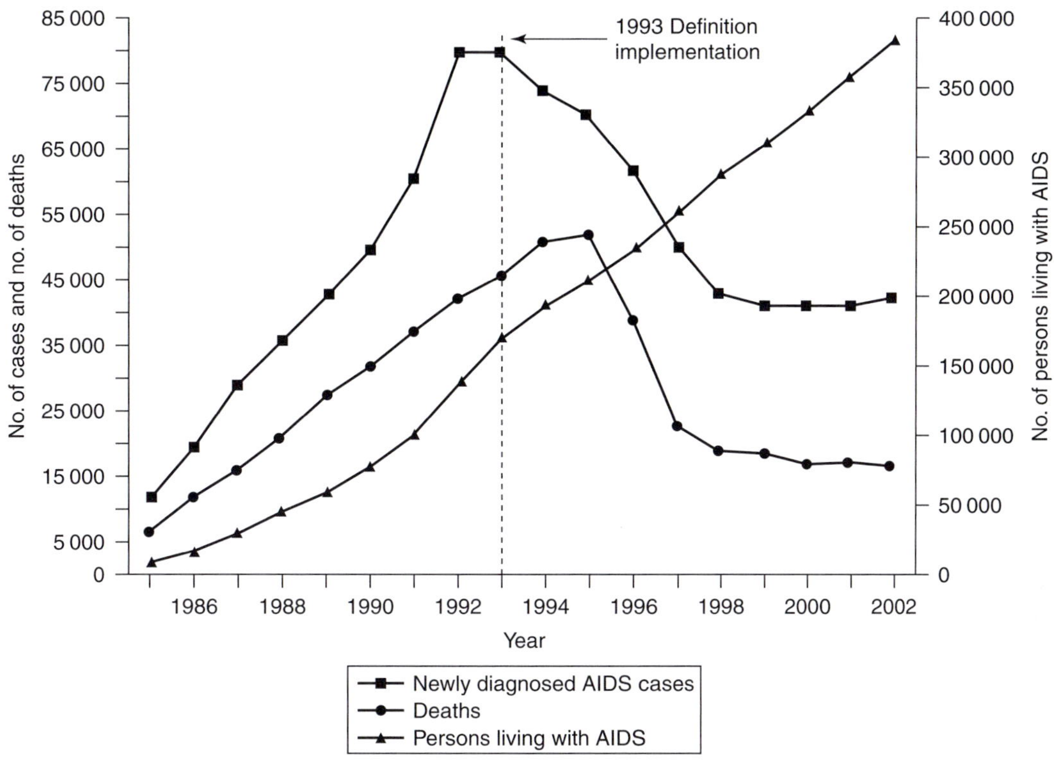

FIGURE 47.2 AIDS cases, deaths, and persons living with AIDS in the United States. Reproduced from the 2003 CDC HIV/AIDS Surveillance Report

in nearly one-third of seropositive women. In multivariate regression analysis, proteinuria was associated with CD4 T-lymphocyte depletion, hepatitis C virus exposure, and black race. As measured by doubling of serum creatinine, renal failure occurred in 2% of the overall cohort and in 5% of women with proteinuria. In the subgroup with proteinuria, renal failure was associated with markers of more advanced HIV disease, including CD4 lymphocyte count <200 cells/mm^3 and detectable HIV viral load (Szczech et al 2002).

Another large urban cohort demonstrated a similar incidence of HIVAN among a largely African-American population (76%), with approximately 2% of the 3976 subjects developing clinical or biopsy-proven HIVAN during the study period. The overall incidence of HIVAN during the 12-year study was 8 per 1000 person years, but was significantly lower in the last 3 years of the study. HIVAN incidence was highest among subjects with AIDS who received no antiretroviral therapy, and lowest among those treated with HAART prior to the development of severe immunosuppression. As in previous cohort studies, HIVAN was associated with black race and advanced HIV disease. Of note, more than three-quarters of the cohort subjects were African-American (Lucas et al 2004).

The majority of studies exploring the prevalence and natural history of HIV-related renal disease in the United States have not uniformly reported histology, making it difficult to distinguish HIVAN from other HIV-associated renal lesions. A retrospective autopsy series from Texas identified HIVAN in nearly 7% of patients dying with AIDS prior to the introduction of HAART. Of note, 93% of cases were in African-Americans, with a prevalence of 12% in that subgroup (Shahinian et al 2000). Autopsy series are limited by inherent selection bias, so these data likely overestimate the prevalence of HIVAN in the pre-HAART era.

While most of the large cohort studies of HIVAN spanned the years before and after the introduction of HAART, smaller cohort studies have explored the prevalence of HIV-associated renal disease and HIVAN in the post-HAART era. Screening urinalyses detected dipstick proteinuria in 7% of 557 HIV-infected patients followed at a single center in 1998. Renal biopsy was performed in 14/15 patients with 24-hour urine protein excretion of >1.5 g/day, with a diagnosis of HIVAN in 9 cases (1.6%). All patients with a confirmed diagnosis and one patient with a clinical diagnosis of HIVAN were African-American (Ahuja et al 1999).

Racial Predilection of HIVAN

As suggested by these studies, HIVAN shows a marked predilection for seropositive patients of African descent; only sickle cell disease is more strongly associated with African-American race as a cause of ESRD (Abbott et al 2001). In 1999, HIVAN became the third leading cause of ESRD among African-Americans between the ages of 20–64 years

(Monahan et al 2001). While African-Americans accounted for a disproportionate 42% of prevalent AIDS cases in the United States in 2002 (CDC 2003), they accounted for more than 90% of the ESRD due to HIVAN. A similar racial predilection has been documented in biopsy series from other countries with significant black minorities, including Brazil and France (Lopes et al 1992, Laradi et al 1998). Studies in predominantly Asian or Caucasian populations have demonstrated much lower rates of HIVAN, even in the absence of effective antiretroviral therapy. In a small biopsy series from Thailand, there were no cases of HIVAN identified among 26 seropositive patients with proteinuria, consistent with the lack of reported HIVAN in that region (Praditpornsilpa et al 1999). A prospective autopsy series from Switzerland demonstrated a high prevalence of renal pathology among 239 patients dying of AIDS, but documented only one case of classic HIVAN in a patient of African descent (Hailemariam et al 2001).

The factors contributing to this racial disparity are not well understood, but there is evidence that a family history of kidney disease may be associated with an increased propensity to develop HIV-associated renal disease. A historical case-control study found that African-Americans with ESRD attributed to HIV were significantly more likely to have a first- or second-degree relative with ESRD when compared to seropositive patients without renal disease. After adjusting for age, gender, and family size, African-Americans with HIV-associated renal disease were more than five times as likely to have a close relative with ESRD from any cause (Freedman et al 1999).

While it is clear that black race is a strong risk factor for the development of ESRD due to HIVAN, it is possible that this discrepancy reflects more aggressive disease, rather than increased prevalence of disease in patients of African descent. Retrospective analysis of a large urban AIDS population suggested that the occurrence of HIV-associated acute renal failure was independent of race, while chronic renal failure occurred almost exclusively in African-Americans (Cantor et al 1991). Since renal histopathology was not available, it is unclear what proportion of cases represented HIVAN. More recently, analysis of the prospective Women's Interagency HIV Study cohort found that black race was associated with an odds ratio of 2.0 for the presence of proteinuria in seropositive women (Szczech et al 2002). When compared to the 18-fold increase in relative risk of ESRD due to HIVAN (Kopp & Winkler 2003), these data may suggest that black race is linked to both development and progression of HIV-associated renal disease.

PATHOGENESIS

The striking predilection of HIVAN for seropositive patients of African descent suggests that host genetic factors play an important role in the pathogenesis of HIVAN. In addition, a growing body of literature supports a direct role for HIV-1 infection in the development of HIVAN. Together, these data suggest that HIVAN results from complex interactions between host and viral factors.

HIV-1 Infection of Renal Epithelial Cells

Early reports of viral DNA detection in human renal tissue were controversial (Cohen et al 1989, Kimmel et al 1993), but more sensitive techniques have supported the susceptibility of intrinsic renal cells to HIV infection. Productive HIV-1 infection of cultured tubular epithelial cells was originally hypothesized to occur via CD4-independent mechanisms, as these cells were not thought to express the CD4 receptor (Ray et al 1998). More recently, debate has focused on renal epithelial cell expression of coreceptors for viral entry, in particular the chemokine receptors CXCR4, CCR3, and CCR5. Analysis of renal tissue from patients with HIVAN failed to demonstrate the presence of CXCR4, CCR5, or HIV-1 RNA in intrinsic renal cells by in situ hybridization (Eitner et al 2000); however, in vitro studies have demonstrated the expression of CD4, CXCR4, CCR3, and CCR5 in cultured human epithelial and mesangial cells by indirect immunofluorescence. Both cell types also appear to support active viral replication in vitro, as demonstrated by PCR (Conaldi et al 1998). Studies employing these more sensitive techniques in vivo are needed to clarify the role of CD4 and chemokine receptors in the susceptibility of renal epithelial cells to HIV-1 infection.

Analysis of human HIVAN biopsy samples by PCR and in situ hybridization has confirmed the ability of HIV-1 to infect renal epithelial cells (Bruggeman et al 2000). Detection of circular unintegrated HIV-1 DNA in human renal tissue was felt to indicate active viral replication, but did not differentiate between infection of intrinsic renal cells and infection of infiltrating leukocytes. In situ hybridization localized the expression of viral mRNA to tubular and glomerular epithelial cells (Figure 47.3). The detection of HIV-1 mRNA in renal epithelial cells of patients with undetectable viral loads also suggested that the kidney may serve as a reservoir for HIV-1 infection.

Expression of HIV-1 mRNA in tubular epithelial cells can be detected throughout the renal tubule in biopsies from patients with HIVAN. The pattern of viral mRNA expression is similar to the development of tubular microcysts, consistent with a pathogenic role for viral infection. HIV-1 mRNA is detected in normal-appearing tubular epithelium and in mildly dilated tubules, but not in areas of severe microcystic dilatation (Ross et al 2001). The absence of viral mRNA expression in the most advanced lesions could help to explain the failure of earlier studies to detect viral genomic material in human HIVAN samples, and may also explain why tubular proliferation is limited to microcyst formation.

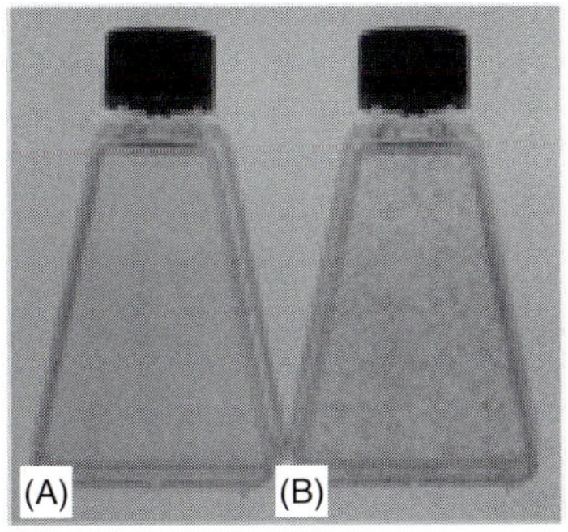

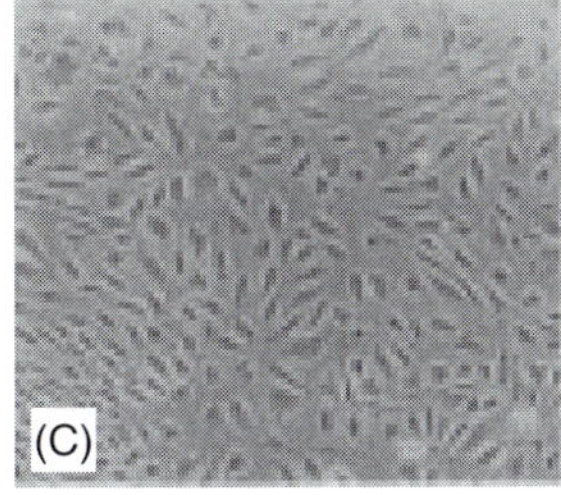

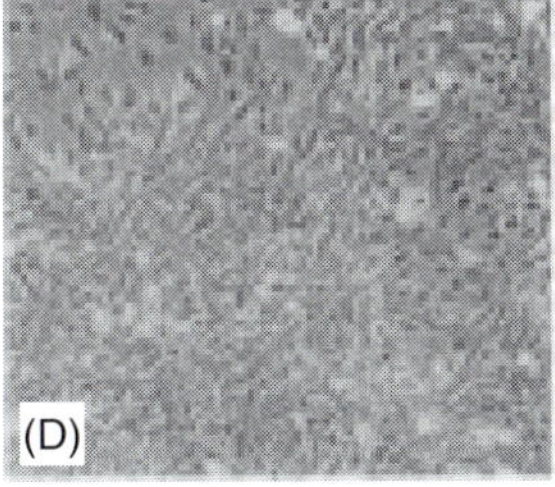

FIGURE 47.10 Podocytes infected with Babe-Puro vector (A and C) or Babe-Puro/Nef (B and D). Upper panel (A and B), the cells stained with Wright Giemsa stain after 10 days of incubation at permissive temperature. Lower panel (C and D), the same cells observed under light microscope with × 10 objective. The Nef-expressing podocytes form foci, whereas podocytes with vector alone show a monolayer with distinct boundaries. Reproduced with permission from Hussain et al (2002). HIV-1 Nef induces proliferation and anchorage-independent growth in podocytes. J. Am. Soc. Nephrol. 13: 1806–15

mutation of *Vpr* (ΔVpr) or inactivating mutations of both *Vpr* and *Nef* (ΔVprΔNef). These Vpr-deficient mice do not develop renal disease. A third Tg26-derived mouse line carries a sequentially deleted transgene encoding only *Tat* and *Vpr*. While homozygous Tat/Vpr mice develop lethal proteinuric renal disease by three months of age, heterozygotes only develop significant renal disease postpartum. The phenotype of mice expressing Vpr in the absence of Nef is similar in Tat/Vpr heterozygotes and dual transgenic mice bearing both ΔVprΔNef and Tat/Vpr. Dual transgenic mice carrying ΔVpr and Tat/Vpr express both Nef and Vpr, and these mice develop more severe renal disease (Figure 47.11). Although these models suggest that renal expression of *Vpr* is sufficient for the development of mild FSGS, the difference in disease severity further supports a synergistic role for *Nef*.

To evaluate the potential role of *Vpr* expression in infiltrating macrophages, the same investigators developed a fourth transgenic mouse line expressing *Vpr* under the control of the macrophage-specific promoter c-*fms*. While these mice develop only transient proteinuria, histological examination reveals FSGS without tubulointerstitial disease. The failure of any of these models to develop the full spectrum of renal lesions observed in HIVAN further supports the proposed synergism between multiple viral genes expressed in both renal and non-renal tissues.

Tat

Data suggesting a pathogenic role for a third viral gene, the regulatory gene *Tat*, were derived from in vitro studies in cultured human mesangial cells and podocytes (Conaldi et al 2002). Exposure to extracellular Tat stimulates proliferation of cultured podocytes, and to a lesser extent, mesangial cells. As in other models, proliferating podocytes demonstrate increased expression of the cell-cycle protein Ki-67 and decreased expression of synaptopodin. Subsequent cytoskeletal rearrangements, including depolymerization of actin fibers, are associated with increased permeability of podocyte monolayers. This disruption of the actin cytoskeleton may be responsible for the increased glomerular permeability that characterizes HIVAN. Basic fibroblast growth factor has been hypothesized to mediate Tat effects, although this remains unclear (Ray et al 1994). In vivo studies are needed to clarify the role of Tat in the pathogenesis of HIVAN.

Host Genetic Factors

Although the striking racial predilection of HIVAN suggests an important genetic component, less is known about the host genetic factors involved.

Genetic Factors Associated with Black Race

It is appealing to postulate an association with genetic polymorphisms that have resulted from unique selective pressures, such as the hemoglobinopathies and the Duffy antigen polymorphism that arose in regions where malaria is endemic. There is no reported association between the hemoglobinopathies and HIVAN, and a genotyping study of 20 patients with HIVAN failed to find disproportionate expression of the Duffy antigen polymorphism (Woolley et al 2001). While one cohort study has suggested an association between HLA type and collapsing FSGS in whites, there were limited data in African-Americans with idiopathic or HIV-associated disease (Laurinavicius 1999). Linkage disequilibrium mapping has been suggested as a method to study the genetic factors predisposing individuals of African descent to HIVAN (Kopp & Winkler 2003), but such studies have yet to be reported.

Genetic Factors Involved in Viral Cell Entry

Host susceptibility to viral cell entry has been proposed as an alternative explanation for the disproportionate burden of HIVAN in Blacks, although there is little evidence to support this hypothesis. Interest has focused on coreceptors for viral cell entry, in particular the chemokine receptors. A CCR5 polymorphism has been associated with variable resistance to systemic HIV infection in Caucasians, but does not appear to play a significant role in populations of African or Asian descent (Liu et al 1996, Samson et al 1996). More recently, genetic variability in the CCR5 ligand RANTES has been associated with differences in disease susceptibility and severity in some populations

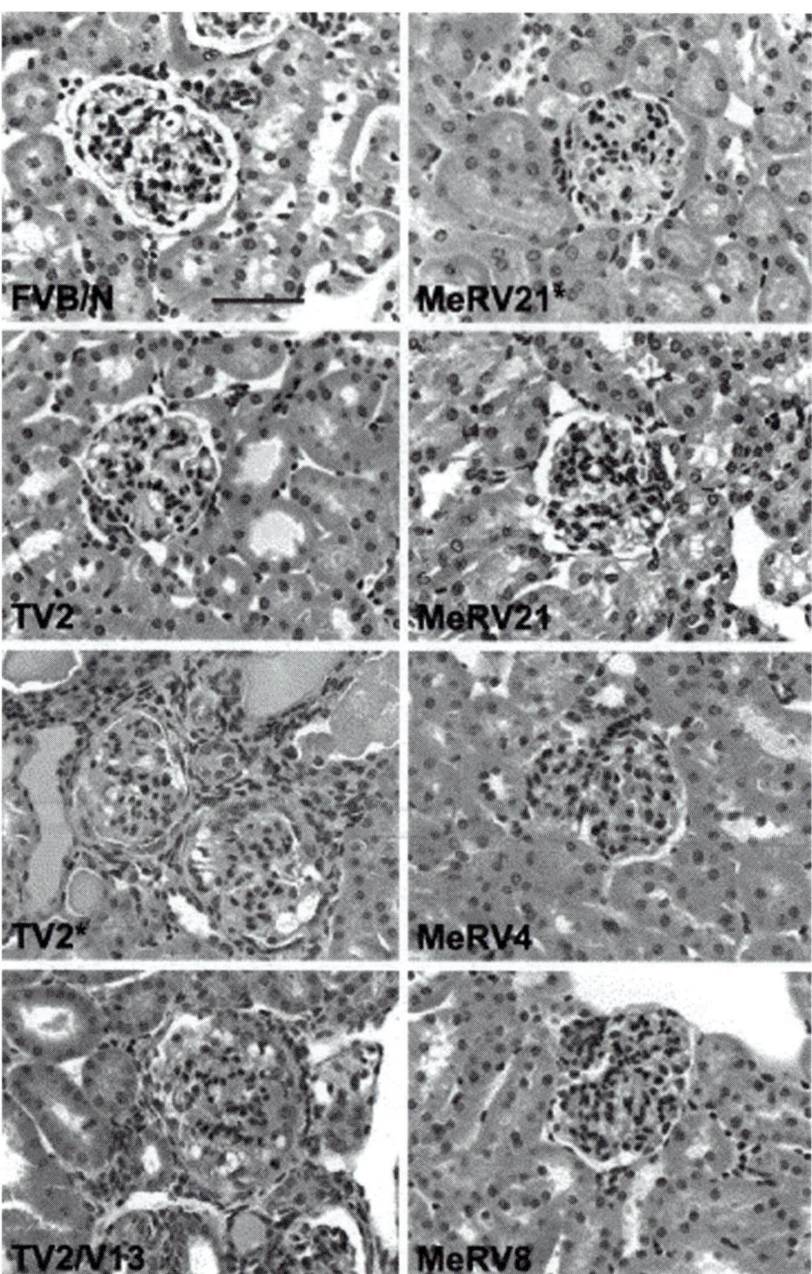

FIGURE 47.11   Renal histopathology in nontransgenic mice (FVB/N), in transgenic mice expressing only HIV-1 *Vpr* and *Tat* (TV2), and in transgenic mice expressing HIV-1 *Vpr* under the control of *c-fms* (MeRV). Kidneys were fixed in 10% buffered formalin, sectioned and stained with hematoxylin and eosin. A hemizygous TV2 female and a nursing TV2* female are compared to a nontransgenic female (FVB/N) to demonstrate varying degrees of histopathology. Glomerular condensation (obscuring capillary lumen) is visible in both TV2 mice but tubular anomalies (acid-staining protein casts and fibrotic lesions) are present predominantly in the nursing mouse (TV2*). Dual transgenic TV2/V13 mice displayed focal glomerulosclerosis (FGS) and tubular atrophy more severe than in TV2 mice. MeRV (*c-fms/vpr*) mice from three independent lines (4, 8, and 21) displayed FGS absent signs of tubular histopathology. Mice were 5 to 10 weeks old. Original magnification was 400 × . The magnification scale bar in top left panel is equivalent to 30 μm. Adapted with permission from Dickie et al (2004). Focal glomerulosclerosis in proviral and *c-fms* transgenic mice links Vpr expression to HIV-associated nephropathy. Virology 322: 69–81. (see also Plate 95)

(Gonzalez et al 2001). As mentioned previously, renal epithelial cell expression of the chemokine receptors has been debated, and published studies have not explored the role of chemokine or chemokine receptor polymorphisms in HIVAN (Conaldi et al 1998, Eitner et al 2000).

CHEMOKINES AND CYTOKINES

In addition to a putative role as cofactors in renal epithelial cell infection, chemokine receptors and their ligands have also been implicated in the downstream pathogenic effects of renal HIV infection. Analysis of microdissected renal tissue by chemiluminescent-enhanced ELISA demonstrated higher levels of the chemokines MCP-1, interleukin-8, and RANTES in tissue from patients with HIVAN or HIV infection, compared to uninfected patients with idiopathic FSGS or without renal disease (Kimmel et al 2003). Increased expression of chemokines in pathologically normal renal tissue from HIV-infected patients may suggest that chemokines alone are not sufficient to cause HIVAN.

In contrast, interferon-α and the interferon-γ receptor were markedly increased only in tissue from patients with HIVAN, particularly in the interstitium. This increase in cytokines and cytokine receptors was accompanied by upregulation of MHC class II, primarily in glomerular tissue, suggesting a role for T-lymphocyte recruitment and activation in the pathogenesis of HIVAN (Kimmel et al 2003). Cell-mediated cytotoxicity was previously reported to be non-essential but synergistic in the development of HIVAN, based on studies in athymic Tg26 mice (Shrivastav et al 2000).

The prominent fibrosis of HIVAN has also been associated with increased expression of transforming growth factor-β (TGF-β). Enhanced expression of TGF-β was originally noted in the kidneys of HIV-1 transgenic mice (Kopp et al 1992). Analysis of human biopsy samples also demonstrated increased expression of TGF-β in patients with HIVAN, compared to patients with other glomerular diseases (Yamamoto et al 1999). Expression of TGF-β was similar in normal renal tissue from HIV-infected and uninfected patients. Overexpression of TGF-β in HIVAN was accompanied by increased levels of the viral protein Tat. With effects on both viral and host gene transcription, Tat may also induce the expression of other cytokines, including basic fibroblast growth factor (Conaldi et al 2002).

Novel Host Genes Involved in HIVAN

Comparison of gene expression in HIV-1 transgenic and control podocytes has identified several novel host genes which are differentially expressed and which may play a role in the development of the glomerular lesions of HIVAN. A small leucine-rich repeat protein, podocan, is overexpressed in HIV-1 transgenic podocytes as determined by representational difference analysis of cDNA (Ross et al 2003). This novel gene product localizes to the glomerular basement membrane in normal kidneys, and is increased in sclerotic glomeruli in an experimental murine model of HIVAN (Figure 47.12).

A second candidate gene, sidekick-1, is not normally expressed in adult podocytes but is highly expressed in HIV-1 transgenic podocytes. Originally described in *Drosophila*, sidekick genes appear to encode type I transmembrane proteins. A highly conserved sequence at the carboxy terminus is thought to represent a PDZ binding domain, but the function of these proteins in renal tissue has not been well described. Immunohistochemical staining of mouse and human fetal tissue demonstrates abundant expression of sidekick genes in ureteric bud-derived tissues, with no glomerular expression in the developing kidney. In the normal adult kidney, sidekick expression appears to be limited to mesangial cells. In HIV-1 transgenic mouse kidneys and in human HIVAN, sidekick-1 is overexpressed in diseased glomeruli, with ectopic expression in podocytes and parietal epithelial cells as well as increased expression in mesangial cells (Kaufman et al 2004).

Identification of a Genetic Locus for Susceptibility to HIVAN

While the identification of differentially expressed host genes has contributed to our understanding of the pathogenesis of HIVAN, these studies do not explain the marked racial variation in susceptibility to HIVAN. In an effort to identify host genetic factors associated with disease susceptibility and severity, HIV-1 transgenic mice were bred onto different genetic backgrounds by crossing heterozygous Tg26 HIV-1 transgenic mice with five other inbred strains (Gharavi et al 2004). The renal phenotype varied significantly among the resulting transgenic hybrids, with no renal disease in transgenic mice bred onto the CAST/Ei and BALB/c backgrounds, variable disease in the three remaining hybrid strains, and severe disease in Tg26 mice bred onto the parent background. Disease severity did not correlate with the level of transgene expression, suggesting that genetic background rather than differences in transgene expression accounted for the variability in phenotype. Backcross of the CAST/Ei hybrid with Tg26 produced a range of renal phenotypes, suggesting segregation of genetic loci associated with disease susceptibility and severity. Linkage analysis in transgenic backcross mice identified a locus on chromosome 3, *HIVAN1*, which accounted for 15% of the variability in renal phenotype. Additional loci on chromosomes 11, 14, and 16 were suggestive of linkage to renal disease, and a locus on chromosome 9 was linked to serum cholesterol. The *HIVAN1*- containing interval of mouse chromosome 3 corresponds to human chromosome 3q25–27, a region with suggestive linkage to other human renal diseases. This region contains a number of potential candidate genes, and further studies may identify a common gene involved in the development or progression of renal disease due to diverse etiologies.

In addition to identifying a genetic locus for susceptibility to HIVAN, the studies in transgenic backcross mice also highlight the complexity of genetic predisposition to disease. The observed phenotypic variability suggests that multiple genes interact to determine disease development and progression, and that those diverse genes do not co-segregate. Of particular interest, inheritance of the CAST/Ei allele of *HIVAN1* was associated with more severe renal disease in backcross mice, despite the absence of renal pathology in CAST/Ei transgenic hybrids (Gharavi et al 2004). Further analysis of the CAST/Ei genome may identify protective

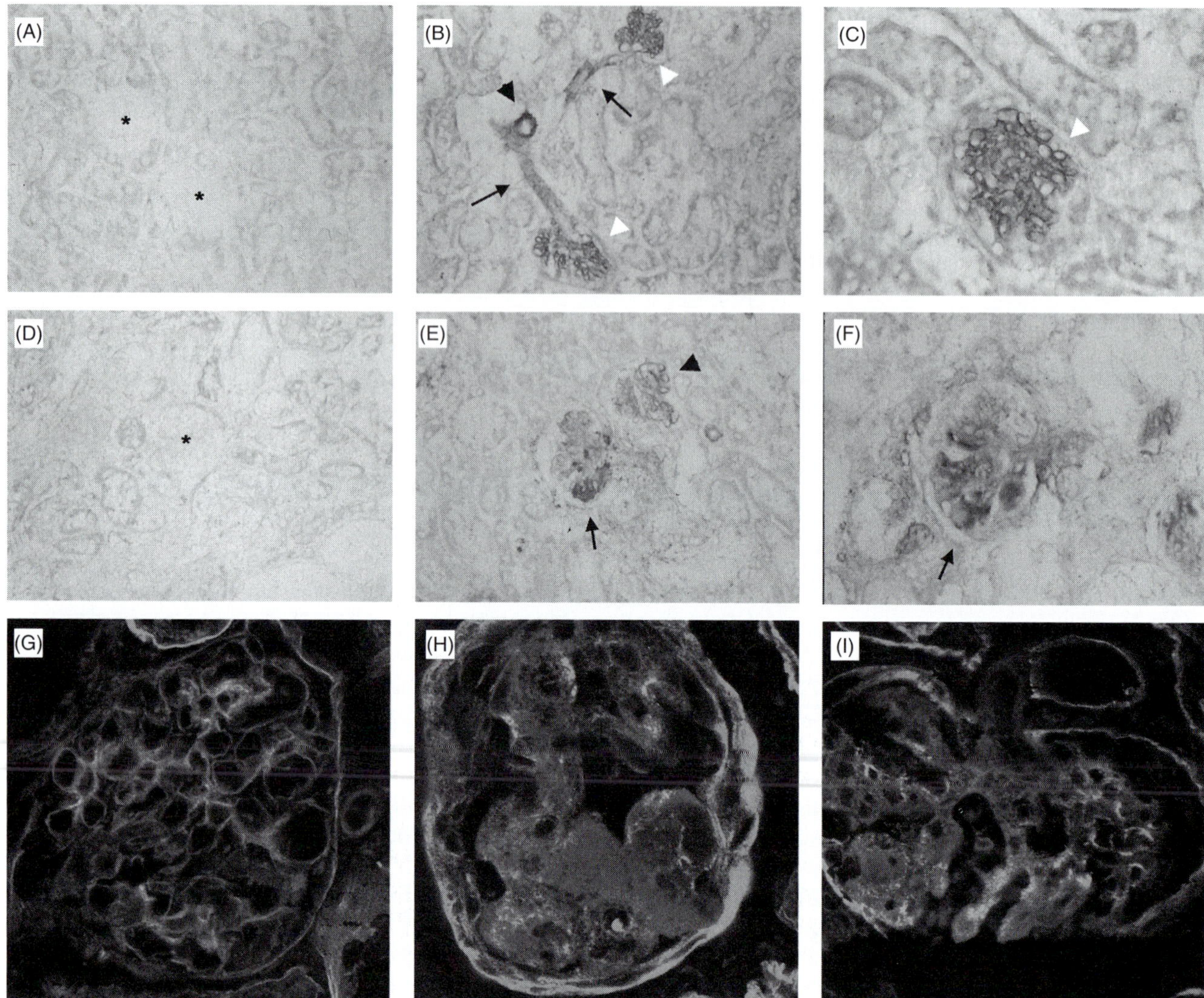

FIGURE 47.12   Immunolocalization of podocan protein in normal mouse kidney and in diseased kidney from the HIV-1 transgenic HIVAN model. In the normal mouse kidney (A–C), podocan staining was observed surrounding capillary loops in the glomerulus in a pattern consistent with that of the GBM (B and C, white arrowheads). Staining of afferent arterioles (black arrows) and interlobular arteries (black arrowhead) was also seen (B). In the HIV-1 transgenic kidney (D–F), the distribution of podocan in nondiseased glomeruli was comparable with that in the nontransgenic kidney (E, arrowhead). In affected glomeruli, a marked increase in podocan deposition was clearly evident in the sclerotic glomerular lesions (E and F, arrows). To illustrate the increased level of podocan deposition in the FSGS lesions of HIV-1 transgenic kidneys, confocal images of a nondiseased glomerulus (G) and two sclerotic glomeruli (H and I) are shown. To allow a direct comparison, all three images were taken from the same section using identical settings on the confocal microscope. The sections were co-labeled for laminin A chain to delineate Bowman's capsule. No staining was detected in glomeruli (*), afferent arterioles, or interlobular arteries with the preimmune serum from the animal in which the anti-podocan antibodies were raised (A and D). Reproduced with permission from Ross et al (2003). Podocan, a novel small leucine-rich repeat protein expressed in the sclerotic glomerular lesion of experimental HIV-associated nephropathy. J. Biol. Chem. 278: 33248–55. Copyright © 2003 American Society for Biochemistry and Molecular Biology. (see also Plate 96)

genes that outweigh the permissive effect of *HIVAN1*. Genetic predisposition to HIVAN likely involves an intricate balance between numerous protective and permissive factors, with the balance favoring disease development and progression in individuals of African descent. Currently recognized host and viral factors involved in the pathogenesis of HIVAN are summarized schematically in Figure 47.13.

TREATMENT

There are currently no randomized clinical trial data to guide the treatment of HIVAN. Limited observational data have suggested a potential role for corticosteroids and inhibitors of the renin-angiotensin-aldosterone system. The introduction of HAART in 1996 was associated with a

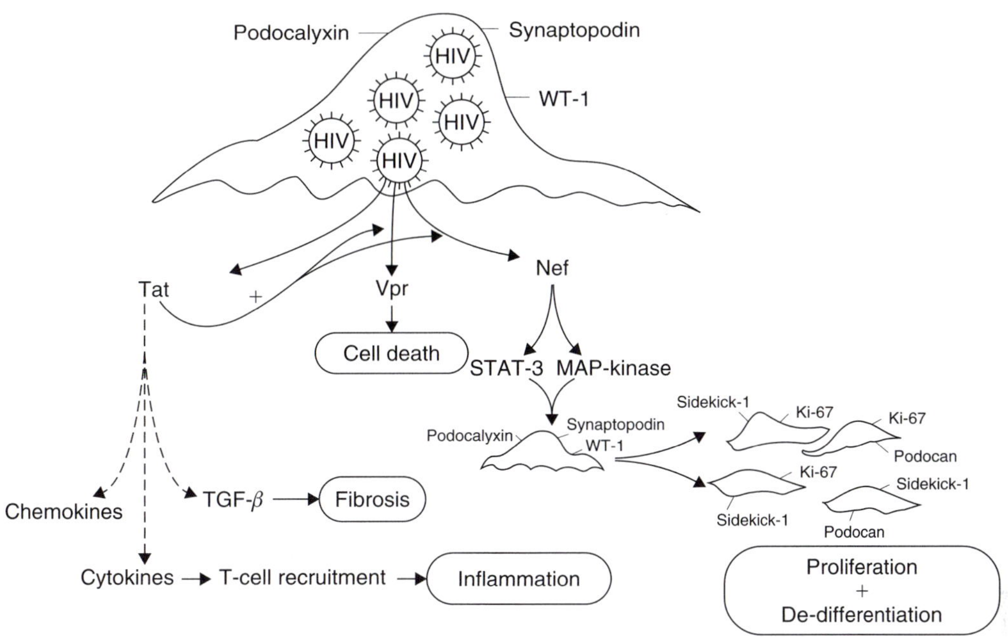

FIGURE 47.13   Current concepts in the pathogenesis of HIVAN. The development of HIVAN involves direct HIV infection, replication and gene expression in renal epithelial cells. The viral protein Tat may increase production of host chemokines and cytokines through nuclear mechanisms, resulting in interstitial inflammation and fibrosis. Tat also upregulates the expression of other viral genes such as *Vpr* and *Nef*. Expression of viral *Vpr* in renal epithelium or in immune cells may be involved in apoptosis. HIV-1 *Nef* expression in podocytes induces proliferation and de-differentiation via the MAP-kinase and STAT-3 pathways, with loss of maturity markers and expression of the cell cycle protein Ki-67. In addition, the proliferating cells overexpress the novel host genes *sidekick-1* and *podocan*

decline in incidence of ESRD attributed to HIV, suggesting a role for HAART in the treatment of HIVAN.

Corticosteroids

The rationale for use of corticosteroids in HIVAN is extrapolated from limited data in other renal diseases with overlapping clinical or histologic features, including hepatitis C virus-related membranoproliferative glomerulonephritis and idiopathic FSGS. In addition, corticosteroids have been hypothesized to ameliorate the inflammatory interstitial component of HIVAN. Prior to the introduction of HAART, several case reports and small cohort studies suggested stabilization or improvement in HIVAN with corticosteroids; however, these studies were limited by small sample size and lack of comparable controls (Smith et al 1996, Watterson et al 1997, Eustace et al 2000). One retrospective cohort study has suggested that corticosteroid therapy remains associated with improved renal outcomes in the post-HAART era, although the cohort included a heterogeneous mix of HIV-related renal lesions (Szczech et al 2002).

The rate of observed infectious complications has varied widely among reports, and the potential risks of further immunosuppression have limited the widespread use of corticosteroids in clinical practice. In addition, growing insight into the pathogenesis of HIVAN has raised questions about the rationale for corticosteroid therapy. Unlike hepatitis C virus-related glomerulonephritis, HIVAN is not an immune complex-mediated disease. Indeed, immunosuppressive therapy may be counterproductive in light of growing evidence for the direct pathogenicity of HIV infection. Finally, the utility of corticosteroids in the idiopathic form of collapsing FSGS has been questioned (Franceschini et al 2003). With the availability of specific antiviral therapy, the potential risk of immunosuppression, and the lack of randomized controlled trials, the therapeutic role of corticosteroids in the post-HAART era is unclear.

Inhibitors of the Renin-Angiotensin-Aldosterone System

The established benefit of ACE inhibitors in other proteinuric renal diseases formed the basis for their initial use in HIVAN. An early report of decreased proteinuria with fosinopril suggested the need for further study; over a decade later, there are still no randomized trial data on the use of ACE inhibitors in HIVAN (Burns et al 1994). Prior to the introduction of HAART, a small retrospective cohort study demonstrated prolonged renal survival in captopril-treated patients with biopsy-proven HIVAN, compared to controls matched for age, gender, race, and baseline creatinine (Kimmel et al 1996).

A larger prospective cohort study enrolled 44 patients with biopsy-proven HIVAN and early renal insufficiency. Treatment assignment was based on patient consent, but the treatment and control groups were comparable with respect to age, severity of immunosuppression, and baseline creatinine and proteinuria. The analyses were adjusted for a nonsignificant increase in exposure to antiretroviral therapy among treated patients. While all 16 untreated patients developed ESRD during the 5-year study period, only one of 28 fosinopril-treated patients required renal replacement therapy. There was also a significant increase in mortality in the untreated group, which could be attributed to complications of HIV infection or ESRD. Alternatively, it is possible that there were significant unmeasured differences between the two groups at baseline, particularly with the use of self-selected controls (Wei et al 2003).

Despite the potential for bias in currently available data, ACE inhibitor therapy is generally accepted as a reasonable therapeutic option in patients with HIVAN. In carefully selected patients without hypotension or hyperkalemia, ACE inhibitors and angiotensin receptor blockers likely have more favorable risk–benefit profiles than other nonspecific therapies, corticosteroids in particular. Nonetheless, definitive recommendations must await prospective randomized trials of ACE inhibitors and angiotensin receptor blockers in the treatment of HIVAN.

Antiretroviral Therapy

While there have been no prospective randomized trials of specific antiviral therapy in the treatment of HIVAN, there is powerful observational evidence that effective antiretroviral therapy may prevent the development or progression of HIV-associated renal disease. In addition, there is some evidence that antiretroviral therapy might actually induce regression of established renal disease. While fibrosis has traditionally been considered irreversible, there is a precedent for reversal of fibrosis with antiviral therapy; treatment of hepatitis B cirrhosis with adefovir causes regression of fibrosis in over 50% of patients (Hadziyannis et al 2003, Marcellin et al 2003).

Prior to the introduction of protease inhibitors and the use of combination therapy, a small retrospective cohort study suggested that zidovudine treatment might slow the progression of HIVAN when initiated early in the course of the disease (Michel et al 1992). A subsequent prospective cohort study supported the potential benefit of zidovudine therapy (Ifudu et al 1995); however, both studies were limited by small sample size, non-random treatment assignment, and heterogeneous populations.

Since the introduction of HAART in 1996, two groups have reported recovery of renal function and reversal of histological changes of HIVAN after the initiation of HAART (Wali et al 1998, Winston et al 2001). Both cases were diagnosed early in the course of rapidly progressive, dialysis-dependent renal failure, and both patients had classic histological changes of HIVAN on initial biopsy. Following treatment with HAART, repeat biopsies demonstrated

reversal of collapsing glomerulopathy, decrease or disappearance of tubular microcysts, and marked reduction in interstitial fibrosis. In addition, reversal of podocyte dedifferentiation and proliferation was confirmed by the absence of nuclear staining for the proliferation marker Ki-67 and restoration of staining for synaptopodin (Figure 47.14).

While these and other small studies have suggested a role for antiretroviral therapy in the treatment of HIVAN (Szczech et al 2002), the most powerful evidence comes from the observed decline in incidence of ESRD attributed to HIV. In the 5 years prior to the introduction of HAART, the incidence of ESRD due to HIV/AIDS increased more rapidly than any other etiology of chronic kidney disease. The annual incidence of ESRD due to HIV/AIDS decreased for the first time in 1996, coinciding with the introduction of HAART. Despite the rising prevalence of HIV infection in the United States, the incidence of ESRD due to HIV/AIDS has remained relatively stable since that time (USRDS 2003). A recent analysis of USRDS data from the period 1995–2000 noted only a 10% decrease in the incidence of ESRD attributed to HIV/AIDS, despite marked improvements in mortality and other HIV-associated complications. The 159% increase in the prevalence of ESRD due to HIV is largely attributable to improved survival of patients with HIV and ESRD (Eggers & Kimmel 2004).

While compelling, national survey data from the USRDS are limited to patients reaching end-stage renal disease. In the absence of standardized diagnostic criteria or biopsy confirmation, reported cases may represent a heterogeneous mix of HIVAN and other HIV-associated renal lesions. A recently published study attempted to address these concerns in a large urban cohort followed between 1989 and 2001 (Lucas et al 2004). Among 3976 HIV-infected individuals, the investigators identified 135 cases of HIVAN, defined by renal biopsy (30 cases) or uniformly applied clinical criteria (105 cases). Clinical diagnosis of HIVAN required nephrotic range proteinuria and subacute renal insufficiency, in the absence of an alternative etiology such as long-standing diabetes or cryoglobulinemia. Excluding patients with evidence of HIVAN at enrollment, 94 patients developed HIVAN during the study period. After adjusting for AIDS status and other clinical and demographic characteristics, the incidence of HIVAN declined in the period 1998–2001 compared to 1995–1997. Further analysis demonstrated that treatment with HAART was associated with a 60% reduction in the risk of HIVAN in this cohort.

Renal Replacement Therapy

Despite the promise of improved outcomes with antiretroviral therapy, more than 800 new cases of HIV-related ESRD were reported in the United States in 2002, the last year for which data are available (USRDS 2004). While the majority of HIV-infected patients with ESRD are treated with hemodialysis, outcomes are similar among those treated

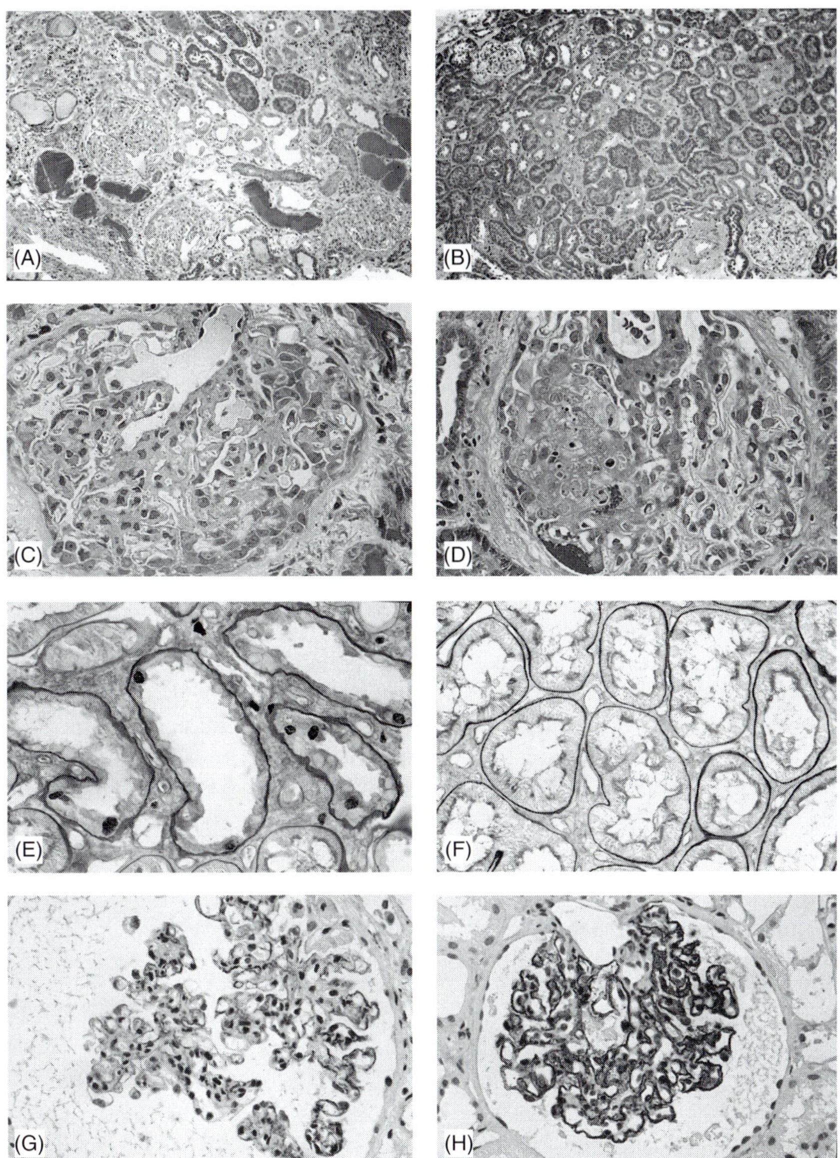

FIGURE 47.14 Kidney biopsy specimens obtained before and after the initiation of highly active antiretroviral therapy. Panels A and C are low-power and high-power views, respectively, of the pretreatment biopsy specimen. Panel A shows one of three glomeruli with collapsing sclerosis and marked hyperplasia of podocytes (trichrome stain, $\times 125$). The tubules are separated by edema, mild fibrosis, and patchy interstitial inflammatory infiltrates. Many proximal tubules show degenerative changes, and there are focal tubular microcysts containing large casts. Panel C shows a glomerulus with segmental collapse of the glomerular tuft and hyperplasia of the overlying podocytes (trichrome stain, $\times 400$). Panels B and D are low-power and high-power views, respectively, of the biopsy specimen obtained three months after the initiation of highly active antiretroviral therapy. Panel B shows normal glomeruli and mild focal interstitial fibrosis, with restoration of normal tubular architecture (trichrome stain, $\times 125$). No tubular microcysts or interstitial inflammation is apparent. In panel D, a glomerulus contains a discrete segmental scar (trichrome stain, $\times 400$). Some of the overlying podocytes contain protein-resorption droplets, but without the hyperplasia that was prominent before the initiation of treatment. Panels E, F, G, and H show the results of immunohistochemical staining of kidney-biopsy specimens obtained before and after the initiation of antiretroviral therapy. In the pretreatment biopsy specimen (panel E), many nuclei in the renal tubular epithelial cells stain for Ki-67. There is diffuse loss from the proximal tubules of brush border staining for periodic acid–Schiff (periodic acid–Schiff counterstain, $\times 400$). In the biopsy specimen obtained after the initiation of antiretroviral therapy (panel F), a representative field shows no staining for Ki-67. The proximal tubular brush border has been restored (periodic acid–Schiff counterstain, $\times 400$). Immunostaining for synaptopodin in the pretreatment biopsy specimen shows weak staining or no staining in the podocytes of a collapsed glomerulus (Panel G, $\times 400$). There is strong, global positivity for synaptopodin in the podocytes of a representative glomerulus from the biopsy specimen obtained after three months of highly active antiretroviral therapy (panel H, $\times 400$). Reproduced with permission from Winston et al (2001) Nephropathy and establishment of a renal reservoir of HIV type 1 during primary infection. N. Engl. J. Med. 344: 1979–84. Copyright © 2001 Massachusetts Medical Society. (see also Plate 97)

with peritoneal dialysis (Ahuja et al 2003). After adjusting for demographics and year of initiation of dialysis, one-year survival of HIV-infected patients on dialysis improved from 56% to 74% between 1990 and 1999 (Ahuja et al 2002).

Although HIV infection has traditionally been considered a contraindication to kidney transplantation, improved clinical outcomes with the introduction of HAART have prompted interest in this modality of renal replacement therapy. Analysis of the USRDS cohort suggests that transplantation may be a reasonable alternative in selected patients in the post-HAART era (Abbott et al 2004). A small pilot study also supported the safety and feasibility of kidney transplantation in HIV-infected patients with undetectable viral loads, preserved immune function, and no prior opportunistic infection (Stock et al 2003). Additional interest in transplantation stems from suggestive evidence that commonly used immunosuppressive agents may have a prohibitive effect on HIV replication. Both calcineurin inhibitors and mycophenolate mofetil have been shown to inhibit HIV replication in vitro (Karpas et al 1992, Margolis et al 1999), and a small prospective cohort study suggested more rapid immune reconstitution in patients treated with both HAART and cyclosporin A (Rizzardi et al 2002). The clinical utility of this approach remains unclear, and prospective multicenter trials are currently exploring the role of renal transplantation in the treatment of patients with HIV infection and ESRD.

Future Directions in the Treatment of HIVAN

With increasing insight into the complex pathogenesis of HIVAN, it is clear that both host and viral pathways are potential therapeutic targets. Research and development of improved antiretroviral therapy could have implications for both systemic HIV infection and HIV-related renal disease. In addition, specific viral proteins implicated in the development of HIVAN could serve as potential targets for the prevention or treatment of HIVAN. Host pathways involved in HIV gene expression and in the host immune response are also potential therapeutic targets, and have been explored in vitro and in animal models of HIVAN.

Increased chemokine expression has been described in human HIVAN and in transgenic mice expressing the HIV-1 genome under the control of the CD4 promoter (Hanna et al 1998). This mouse model allows preliminary study of the chemokine pathway as a potential therapeutic target in HIVAN. The transcription factor NF-kappa B and its regulatory protein kinase I kappa B are involved in immune activation and inflammation (Karin & Delhase 2000). Inhibition of this pathway has been shown to decrease RANTES expression, ameliorate renal disease and cachexia, and improve survival in the CD4-HIV transgenic mouse model (Heckmann et al 2004).

While interference with the NF-kappa B pathway may act through decreased chemokine expression, there is also evidence that NF-kappa B regulates HIV-1 gene transcription in the kidney (Bruggeman et al 2001). Viral transcription requires a number of other cellular cofactors, including cyclin-dependent kinases, which have been proposed as potential therapeutic targets. Inhibitors of cyclin-dependent kinase-9 have been shown to suppress HIV gene expression in podocytes in vitro, with subsequent reversal of the podocyte proliferation and dedifferentiation characteristic of HIVAN (Nelson et al 2001). While the cyclin-dependent kinase inhibitor flavopiridol has a reasonable therapeutic window, interference with host processes may have unrecognized implications for the clinical safety and utility of such therapies.

CONCLUSION

HIV-associated nephropathy is an important complication of HIV infection, particularly among blacks. Characterized by collapsing FSGS, tubular microcysts, and interstitial inflammation, HIVAN classically presents with proteinuria and rapid progression to ESRD, although an indolent course is more characteristic in the post-HAART era. Despite increases in the prevalence of HIV infection, the incidence of ESRD due to HIVAN has remained relatively stable in the United States since the introduction of HAART in 1996. Increased awareness of renal disease among physicians specializing in HIV/AIDS, and increased penetrance of HAART into high-risk patient populations are likely responsible for the stabilization in incidence of HIV-related ESRD.

The pathogenesis of HIVAN involves direct HIV infection of intrinsic renal cells, expression of viral genes, and subsequent interactions between host and viral factors.

Studies in human renal tissue and transgenic animal models, and in vitro, have implicated several viral genes in the pathologic changes of HIVAN, including *Nef*, *Vpr*, and *Tat*. In addition, the pathogenesis of HIVAN involves well-described host processes required for viral gene expression and host immune response, as well as the expression of several novel gene products. Recently, a genetic locus for susceptibility to HIVAN has been identified in a chromosome region previously associated with other forms of renal disease. This evidence of a genetic basis for HIVAN may help to explain the marked racial predilection, and may also provide insight into other renal diseases.

The treatment of HIVAN is based on observational data, and includes nonspecific therapies such as glucocorticoids and ACE inhibitors, as well as specific antiretroviral therapy. Outcomes have improved dramatically since the introduction of HAART, but prospective randomized trials are currently ongoing to determine the optimal treatment of HIVAN. In the meantime, a diagnosis of HIVAN should prompt initiation of HAART, independent of viral load or CD4 lymphocyte count.

References

Abbott KC, Hypolite I, Welch PG, Agodoa LY. Human immunodeficiency virus/acquired immunodeficiency syndrome-associated nephropathy at end-stage renal disease in the United States: patient characteristics and survival in the pre highly active antiretroviral therapy era. J. Nephrol. 2001; 14: 377–383.

Abbott KC, Swanson SJ, Agodoa LY, Kimmel PL. Human immunodeficiency virus infection and kidney transplantation in the era of highly active antiretroviral therapy and modern immunosuppression. J. Am. Soc. Nephrol. 2004; 15: 1633–9.

Ahuja TS, Borucki M, Funtanilla M, Shahinian V, Hollander M, Rajaraman S. Is the prevalence of HIV-associated nephropathy decreasing? Am. J. Nephrol. 1999; 19: 655–9.

Ahuja TS, Collinge N, Grady J, Khan S. Is dialysis modality a factor in survival of patients with ESRD and HIV-associated nephropathy? Am. J. Kidney Dis. 2003; 41: 1060–4.

Ahuja TS, Grady J, Khan S. Changing trends in the survival of dialysis patients with human immunodeficiency virus in the United States. Am. J. Kidney Dis. 2002; 13: 1889–93.

Barisoni L, Kriz W, Mundel P, D'Agati V. The dysregulated podocyte phenotype: A novel concept in the pathogenesis of collapsing focal segmental glomerulosclerosis and HIV-associated nephropathy. J. Am. Soc. Nephrol. 1999; 10: 51–61.

Bruggeman LA, Adler SH, Klotman PE. Nuclear factor-κB binding to the HIV-1 LTR in kidney: implications for HIV-associated nephropathy. Kidney Int. 2001; 59: 2174–81.

Bruggeman LA, Dikman S, Meng C, Quaggin SE, Coffman TM, Klotman PE. Nephropathy in human immunodeficiency virus-1 transgenic mice is due to renal transgene expression. J. Clin. Invest. 1997; 100: 84–92.

Bruggeman LA, Ross MD, Tanji N, et al. Renal epithelium is a previously unrecognized site of HIV-1 infection. J. Am. Soc. Nephrol. 2000; 11: 2079–87.

Burns MC, Matute R, Onyema D, Davis I, Toth I. Response to inhibition of angiotensin-converting enzyme in human immunodeficiency virus-associated nephropathy: a case report. Am. J. Kidney Dis. 1994; 23: 441–3.

Cantor ES, Kimmel PL, Bosch JP. Effect of race on expression of acquired immunodeficiency syndrome-associated nephropathy. Arch. Intern. Med. 1991; 151: 125–8.

Centers for Disease Control and Prevention (CDC) HIV/AIDS Surveillance Report. Atlanta, GA: 2003.

Cohen AH, Nast CC. HIV-associated nephropathy. A unique combined glomerular, tubular, and interstitial lesion. Mod. Pathol. 1988; 1: 87–97.

Cohen AH, Sun NC, Shapsak P, Imagawa DT. Demonstration of human immunodeficiency virus in renal epithelium in HIV-associated nephropathy. Mod. Pathol. 1989; 2: 125–8.

Conaldi PG, Biancone L, Bottelli A, et al. HIV-1 kills renal tubular epithelial cells in vitro by triggering an apoptotic pathway involving caspase activation and Fas upregulation. J. Clin. Invest. 1998; 102: 2041–9.

Conaldi PG, Bottelli A, Baj A, et al. Human immunodeficiency virus-1 Tat induces hyperproliferation and dysregulation of renal glomerular epithelial cells. Am. J. Pathol. 2002; 161: 53–61.

D'Agati V, Appel GB. Renal pathology of human immunodeficiency virus infection. Semin. Nephrol. 1998; 18: 406–21.

Dickie P, Felser J, Eckhaus M, et al. HIV-associated nephropathy in transgenic mice expressing HIV-1 genes. Virology 1991; 185: 109–19.

Dickie P, Roberts A, Uwiera R, Witmer J, Sharma K, Kopp JB. Focal glomerulosclerosis in proviral and c-*fms* transgenic mice links Vpr expression to HIV-associated nephropathy. Virology 2004; 322: 69–81.

Eggers PW, Kimmel PL. Is there an epidemic of HIV infection in the US ESRD program? J. Am. Soc. Nephrol. 2004; 15: 2477–85.

Eitner F, Cui Y, Hudkins KL, et al. Thrombotic microangiopathy in the HIV-2-infected macaque. Am. J. Pathol. 1999; 155: 649–61.

Eitner F, Cui Y, Hudkins KL, et al. Chemokine receptor CCR5 and CXCR4 expression in HIV-associated kidney disease. Am. Soc. Nephrol. 2000; 11: 856–67.

Eustace JA, Nuermberger E, Choi M, Scheel PJ, Moore R, Briggs WA. Cohort study of the treatment of severe HIV-associated nephropathy with corticosteroids. Kidney Int. 2000; 58: 1253–60.

Franceschini N, Hogan SL, Falk RJ. Primum non nocere: Should adults with idiopathic FSGS receive steroids?. Semin. Nephrol. 2003; 23: 229–33.

Freedman BI, Soucie JM, Stone SM, Pegram S. Familial clustering of end-stage renal disease in blacks with HIV-associated nephropathy. Am. J. Kidney Dis. 1999; 34: 254–8.

Gardenschwartz MH, Lerner CWK, Seligson GR, et al. Renal disease in patients with AIDS: a clinicopathologic study. Clin. Nephrol. 1984; 21: 197–204.

Gharavi AG, Ahmad T, Wong RD, et al. Mapping a locus for susceptibility to HIV-1-associated nephropathy to mouse chromosome 3. Proc. Natl Acad. Sci. USA 2004; 101: 2488–93.

Gonzalez E, Dhanda R, Bamshad M, et al. Global survey of genetic variation in CCR5, RANTES, and MIP-1alpha: impact on the epidemiology of the HIV-1 pandemic. Proc. Natl Acad. Sci. USA 2001; 98: 5199–204.

Gupta SK, Mamlin BW, Johnson CS, Dollins MD, Topf JM, Dube MP. Prevalence of proteinuria and the development of chronic kidney disease in HIV-infected patients. Clin. Nephrol. 2004; 61: 1–6.

Hadziyannis SJ, Tassopoulos NC, Heathcote EJ et al. Adefovir Dipivoxil 438 Study Group. Adefovir dipivoxil for the treatment of hepatitis B e antigen-negative chronic hepatitis B. N. Engl. J. Med. 2003; 348: 800–7.

Hailemariam S, Walder M, Burger HR, et al. Renal pathology and premortem clinical presentation of Caucasian patients with AIDS: an autopsy study from the era prior to antiretroviral therapy. Swiss Med. Wkly. 2001; 131: 412–17.

Hanna Z, Kay DG, Cool M, Jothy S, Rebai N, Jolicoeur P. Transgenic mice expressing human immunodeficiency virus type 1 in immune cells develop a severe AIDS-like syndrome. J. Virology 1998a; 72: 121–32.

Hanna Z, Kay DG, Rebai N, Guimond A, Jothy S, Jolicoeur P. Nef harbors a major determinant of pathogenicity for an AIDS-like disease induced by HIV-1 in transgenic mice. Cell 1998b; 95: 163–75.

He JC, Husain M, Sunamoto M, et al. Nef stimulates proliferation of glomerular podocytes through activation of Src-dependent Stat3 and MAPK1,2 pathways. J. Clin. Invest. 2004; 114: 643–51.

Heckmann A, Waltzinger C, Jolicoeur P, Dreano M, Kosco-Vilbois MH, Sagot Y. IKK2 inhibitor alleviates kidney and wasting diseases in a murine model of human AIDS. Am. J. Pathol. 2004; 164: 1253–62.

Husain M, Gusella GL, Klotman ME, et al. HIV-1 Nef induces proliferation and anchorage-independent growth in podocytes. J. Am. Soc. Nephrol. 2002; 13: 1806–15.

Ifudu O, Rao TK, Tan CC, Fleischman H, Chirgwin K, Friedman EA. Zidovudine is beneficial in human immunodeficiency virus associated nephropathy. Am. J. Nephrol. 1995; 15: 217–21.

Kajiyama W, Kopp JB, Marinos NJ, Klotman PE, Dickie P. HIV-transgenic mice lacking gag-pol-nef develop glomerulosclerosis and express viral RNA and protein in glomerular epithelial and tubular cells. Kidney Int. 2000; 58: 1148–59.

Karin M, Delhase M. The I kappa B kinase (IKK) and NF-kappa B: key elements of pro-inflammatory signaling. Semin. Immunol. 2000; 12: 85–98.

Karpas A, Lowdell M, Jacobson SK, Hill F. Inhibition of human immunodeficiency virus and growth of infected T cells by the immunosuppressive drugs cyclosporin A and FK506. Proc. Natl Acad. Sci. USA 1992; 89: 8351–5.

Kaufman L, Hayashi K, Ross MJ, Ross MD, Klotman PE. Sidekick-1 is upregulated in glomeruli in HIV-associated nephropathy. J. Am. Soc. Nephrol. 2004; 15: 1721–30.

Kimmel PL, Cohen DJ, Abraham AA, Bodi I, Schwartz AM, Phillips TM. Upregulation of MHC class II, interferon-α and interferon-γ receptor protein expression in HIV-associated nephropathy. Nephrol. Dial. Transplant. 2003; 18: 285–92.

Kimmel PL, Ferreira-Centeno A, Farkas-Szallasi T, Abraham AA, Garrett CT. Viral DNA in microdissected renal biopsy tissue from HIV infected patients with nephrotic syndrome. Kidney Int. 1993; 43: 1347–52.

Kimmel PL, Mishkin GJ, Umana WO. Captopril and renal survival in patients with human immunodeficiency virus nephropathy. Am. J. Kidney Dis. 1996; 28: 202–8.

Kopp JB, Klotman ME, Adler SH, et al. Progressive glomerulosclerosis and enhanced renal accumulation of basement membrane components in mice transgenic for human immunodeficiency virus type 1 genes. Proc. Natl Acad. Sci. USA 1992; 89: 1577–81.

Kopp JB, Winkler C. HIV-associated nephropathy in African Americans. Kidney Int. 2003; 63: 43–9.

Laradi A, Mallet A, Beaufils H, Allouache M, Martinez F. HIV-associated nephropathy: outcome and prognosis factors. Groupe d'Etudes Nephrologiques d'Ile de France. J. Am. Soc. Nephrol. 1998; 9: 2327–35.

Laurinavicius A, Hurwitz S, Rennke HG. Collapsing glomerulopathy in HIV and non-HIV patients: A clinicopathological and follow-up study. Kidney Int. 1999; 56: 2203–13.

Levin ML, Palella F, Shah S, Lerma E, Butter J, Kanwar YS. HIV-associated nephropathy occurring before HIV antibody seroconversion. Am. J. Kidney Dis. 2001; 31: E59.

Lopes GS, Marques LP, Rioja LS, Basilio-de-Oliveira CA, Nery AC, Santos Oda R. Glomerular disease and human immunodeficiency virus infection in Brazil. Am. J. Nephrol. 1992; 12: 281–7.

Lucas GM, Eustace JA, Sozio S, Mentari EK, Appiah KA, Moore RD. Highly active antiretroviral therapy and the incidence of HIV-1-associated nephropathy: a 12-year cohort study. AIDS 2004; 20: 516–41.

Mandell GI, Douglas RG, Bennett JE. Principles and Practice of Infectious Diseases, 5th edn. New York: Churchill Livingstone, 2000.

Marcellin P, Chang TT, Lim SG et al. Adefovir Dipivoxil 437 Study Group. Adefovir dipivoxil for the treatment of hepatitis B e antigen-positive chronic hepatitis B. N. Engl. J. Med. 2003; 348: 808–16.

Margolis D, Heredia A, Gaywee J, Drusano G, Oldach D, Redfield RR. Abacavir and mycophenolic acid, an inhibitor of inosine monophosphate dehydrogenase, have profound and synergistic anti-HIV activity. J. Acquir. Immune Defic. Syndr. 1999; 21: 362–70.

Marras D, Bruggeman LA, Gao F, et al. Replication and compartmentalization of HIV-1 in kidney epithelium of patients with HIV-associated nephropathy. Nat. Med. 2002; 8: 522–6.

Michel C, Dosquet P, Ronco P, Mougenot B, Viron B, Mignon F. Nephropathy associated with infection by human immunodeficiency virus: a report on 11 cases including 6 treated with zidovudine. Nephron 1992; 62: 434–40.

Monohan M, Tanji N, Klotman PE. HIV-associated nephropathy: an urban epidemic. Sem. Nephrol. 2001; 21: 394–402.

Nelson PJ, Gelman IH, Klotman PE. Suppression of HIV-1 expression by inhibitors of cyclin-dependent kinases promotes differentiation of infected podocytes. J. Am. Soc. Nephrol. 2001; 12: 2827–31.

Pardo V, Aldana M, Colton RM, et al. Glomerular lesions in the acquired immunodeficiency syndrome. Ann. Intern. Med. 1984; 101: 429–34.

Praditpornsilpa K, Napathorn S, Yenrudi S, Wankrairot P, Tungsaga K, Sitprija V. Renal pathology and HIV infection in Thailand. Am. J. Kidney Dis. 1999; 33: 282–6.

Rao TK, Filippone EJ, Nicastri AD, et al. Associated focal and segmental glomerulosclerosis in the acquired immunodeficiency syndrome. N. Engl. J. Med. 1984; 310: 669–73.

Ray PE, Bruggeman LA, Weeks BS, et al. BFGF and its low affinity receptors in the pathogenesis of HIV-associated nephropathy in transgenic mice. Kidney Int. 1994; 46: 759–72.

Ray PE, Liu X, Robinson LR, et al. A novel HIV-1 transgenic rat model of childhood HIV-1 associated nephropathy. Kidney Int. 2003; 6: 2242–51.

Reid W, Sadowska M, Denaro F, et al. An HIV-1 transgenic rat that develops HIV-related pathology and immunologic dysfunction. Proc. Natl Acad. Sci. USA 2001; 98: 9271–6.

Rizzardi GP, Harari A, Capiluppi B, et al. Treatment of primary HIV-1 infection with cyclosporin A coupled with highly active antiretroviral therapy. J. Clin. Invest. 2002; 109: 681–8.

Ross MD, Bruggeman LA, Hanss B, et al. Podocan, a novel small leucine-rich repeat protein expressed in the sclerotic glomerular lesion of experimental HIV-associated nephropathy. J. Biol. Chem. 2003; 278: 33248–55.

Ross MJ, Bruggeman LA, Wilson PA, Klotman PE. Microcyst formation and HIV-1 gene expression occur in multiple nephron segments in HIV-associated nephropathy. J. Am. Soc. Nephrol. 2001; 12: 2645–51.

Samson M, Libert F, Doranz BJ, et al. Resistance to HIV-1 infection in caucasian individuals bearing mutant alleles of the CCR-5 chemokine receptor gene. Nature 1996; 382: 722–5.

Schwartz EJ, Cara A, Snoeck H, et al. Human Immunodeficiency Virus-1 induces loss of contact inhibition in podocytes. J. Am. Soc. Nephrol. 2001; 12: 1677–84.

Segerer S, Eitner F, Cui Y, Hudkins KL, Alpers CE. Cellular injury associated with renal thrombotic microangiopathy in human immunodeficiency virus-infected macaques. J. Am. Soc. Nephrol. 2002; 13: 370–8.

Shahinian V, Rajaraman S, Borucki M, Grady J, Hollander WM, Ahuja TS. Prevalence of HIV-associated nephropathy in autopsies of HIV-infected patients. Am. J. Kidney Dis. 2000; 35: 884–8.

Shrivastav S, Cusumano A, Kanno Y, Chen G, Bryant JL, Kopp JB. Role of T lymphocytes in renal tissue in HIV-transgenic mice. Am. J. Kidney Dis. 2000; 35: 408–17.

Smith MC, Austen JL, Carey JT, et al. Prednisone improves renal function and proteinuria in human immunodeficiency virus-associated nephropathy. Am. J. Med. 1996; 101: 41–8.

Stock PG, Roland ME, Carlson L, et al. Kidney and liver transplantation in human immunodeficiency virus-infected patients: a pilot safety and efficacy study. Transplant 2003; 76: 370–5.

Szczech LA, Edwards LJ, Sanders LL, et al. Protease inhibitors are associated with a slowed progression of HIV-related renal diseases. Clin. Nephrol. 2002a; 57: 336–41.

Szczech LA, Gange SJ, van der Horst C, et al. Predictors of proteinuria and renal failure among women with HIV infection. Kidney Int. 2002b; 61: 195–202.

UNAIDS: AIDS epidemic update 2003. Joint United Nations Programme on HIV/AIDS (UNAIDS), 2003.

US Renal Data System (USRDS): USRDS 2004 Annual Data Report: Atlas of End-stage Renal Disease in the United States. The National Institutes of Health, National Institute of Diabetes and Digestive and Kidney Diseases, Bethesda, MA: 2004.

Wali RK, Drachenberg CI, Papadimitriou JC, Keay S, Ramos E. HIV-1-associated nephropathy and response to highly-active antiretroviral therapy. Lancet 1998; 352: 783–4.

Watterson MK, Detwiler RK, Bolin P. Clinical response to prolonged corticosteroids in a patient with human immunodeficiency virus-associated nephropathy. Am. J. Kidney Dis. 1997; 29: 624–6.

Wei A, Burns GC, Williams BA, Mohammed NB, Visintainer P, Sivak SL. Long-term renal survival in HIV-associated nephropathy with angiotensin-converting enzyme inhibition. Kidney Int. 2003; 64: 1462–71.

Winston JA, Bruggeman LA, Ross MD, et al. Nephropathy and establishment of a renal reservoir of HIV type 1 during primary infection. N. Engl. J. Med. 2001; 344: 1979–84.

Winston JA, Klotman ME, Klotman PE. HIV-associated nephropathy is a late, not early, manifestation of HIV-1 infection. Kidney Int. 1999; 55: 1036–40.

Woolley IJ, Kalayjian R, Valdez H, et al. HIV nephropathy and the Duffy antigen/receptor for chemokines in African-Americans. J. Nephrol. 2001; 14: 384–7.

Yamamoto T, Noble NA, Miller DE, et al. Increased levels of transforming growth factor-β in HIV-associated nephropathy. Kidney Int. 1999; 55: 579–92.

Index

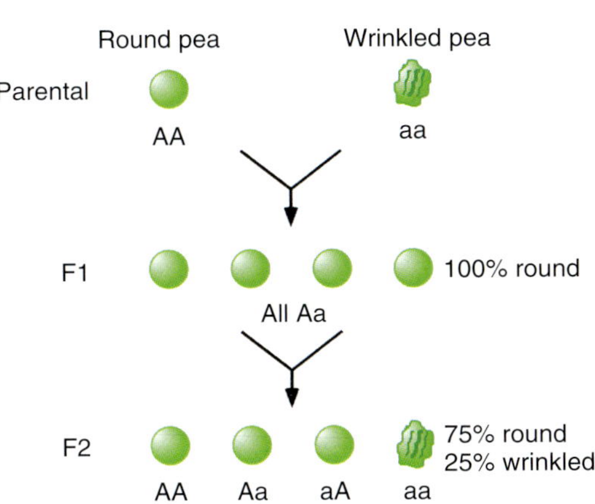

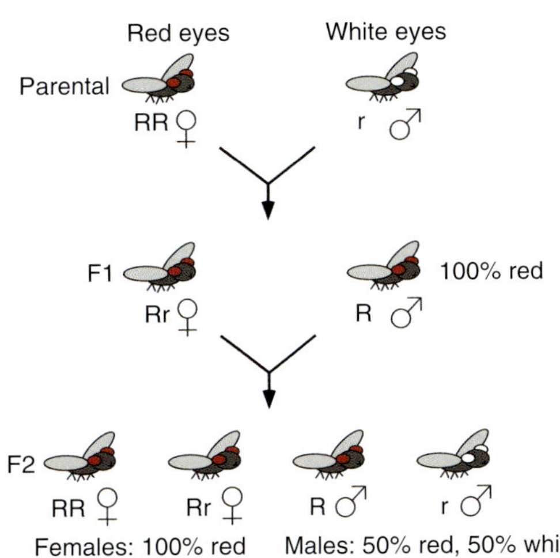

PLATE 1 Mendelian segregation. An example of one of Mendel's pea plant crosses is shown, in which plants that produced round peas were crossed with plants that produced wrinkled peas, resulting in a first generation of progeny (F_1) that all produced round peas. Intercrossing of these F_1 plants resulted in an F_2 generation in which 75% of plants produced round peas and 25% produced wrinkled peas. From such experiments Mendel concluded that traits like pea shape were determined by pairs of factors (now called genes) that occur in different forms (or alleles – 'A' and 'a' in the PLATE), that one allele was transmitted from each parent to its offspring, and that the different combinations of alleles (now called genotypes) determined each trait. In this example, genotypes 'AA', 'Aa', and 'aA' result in round peas, and genotype 'aa' results in wrinkled peas. (see also Figure 1.1)

PLATE 2 Morgan's demonstration of the chromosomal basis of inheritance. An example of one of Morgan's fruit fly crosses is shown, in which a cross of red-eyed females with white-eyed males resulted in F_1 offspring that all had red eyes. Intercrossing of the F_1 flies resulted in an F_2 generation in which all of the females and 50% of the males had red eyes, but the other 50% of males had white eyes. From these experiments, and the observation that male and female fruit flies have different numbers of chromosomes, Morgan concluded that genes are carried on chromosomes, and that genes for eye color in the fruit fly are on the sex-linked chromosome. In the PLATE, females inherit two alleles, one from each parent, and their genotypes 'RR' and 'Rr' result in red eyes. Males inherit only one allele, from their mother; genotype 'R' results in red eyes, while genotype 'r' results in white eyes. (see also Figure 1.2)

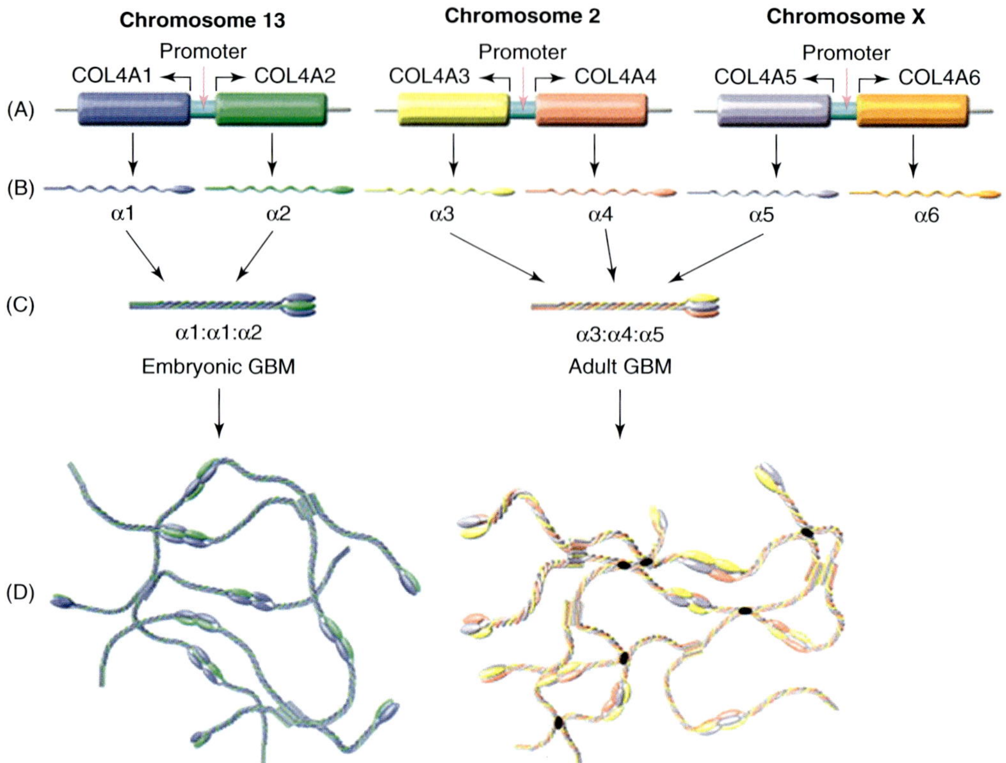

PLATE 3 Type IV collagen genes, α chains and GBM specific isoforms. (A) The six collagen IV genes (*COL4A1–COL4A6*) located pairwise in a head-to-head fashion on three different chromosomes generate six different α chains that have a globular noncollagenous domain at their C-terminus (B). Three chains form triple-helical molecules that can have different combinations (C). (D) Extracellularly, the triple-helical type IV collagen molecules form a network by associating with each other at their ends so that two molecules are cross-linked through their C-terminal globular domain (NC1) and for trimers, associate with each other at the N-termini. In the embryonic GBM, the ubiquitous α1:α1:α2 trimer is the only isoform. After birth, this isoform is gradually replaced by an α3:α4:α5 isoform, which is more cross-linked through disulfide bonds within the collagenous regions (illustrated by black spots) and more resistant to extracellular proteolysis. Defects in the α3:α4:α5 trimers lead to AD or TBMN, depending on the extensiveness of alleles involved. Reproduced from Tryggvason & Patrakka (2006). (see also Figure 4.2)

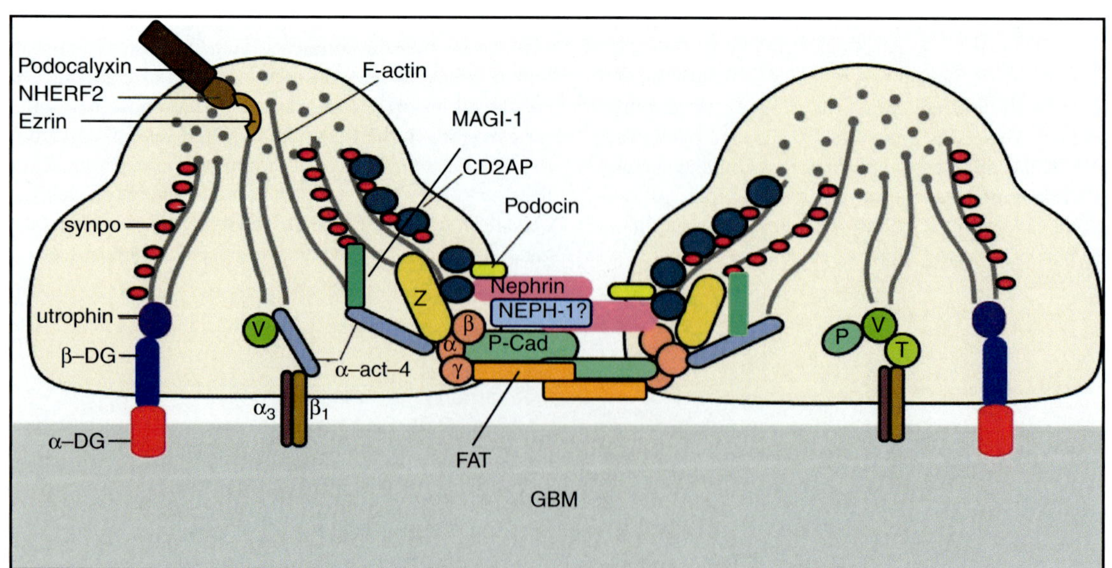

PLATE 4 Electromicrographs of renal biopsy in a patient with ACTN4 mutation. A portion of the glomerular capillary loop is depicted. There is extensive foot process effacement with areas of preservation. Significant vacuolization is also present indicating podocyte injury. Courtesy of Joel Henderson, Department of Pathology, Brigham and Women's Hospital. (see also Figure 6.1)

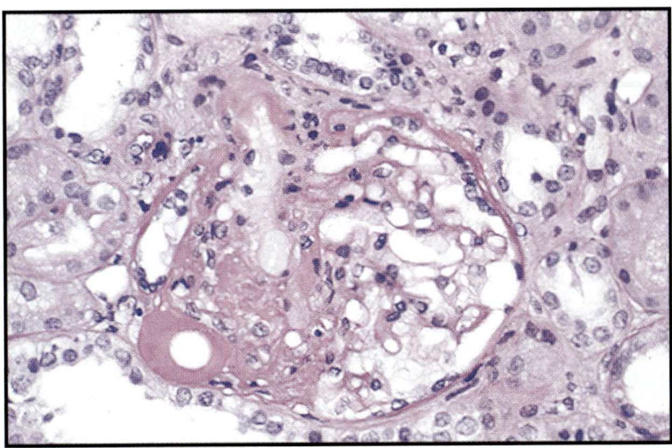

PLATE 5 Podocyte proteins. Schematic diagram showing proteins of demonstrated importance in podocyte and slit-diaphragm function. Courtesy of Peter Mundel, Mt Sinai School of Medicine. (see also Figure 6.3)

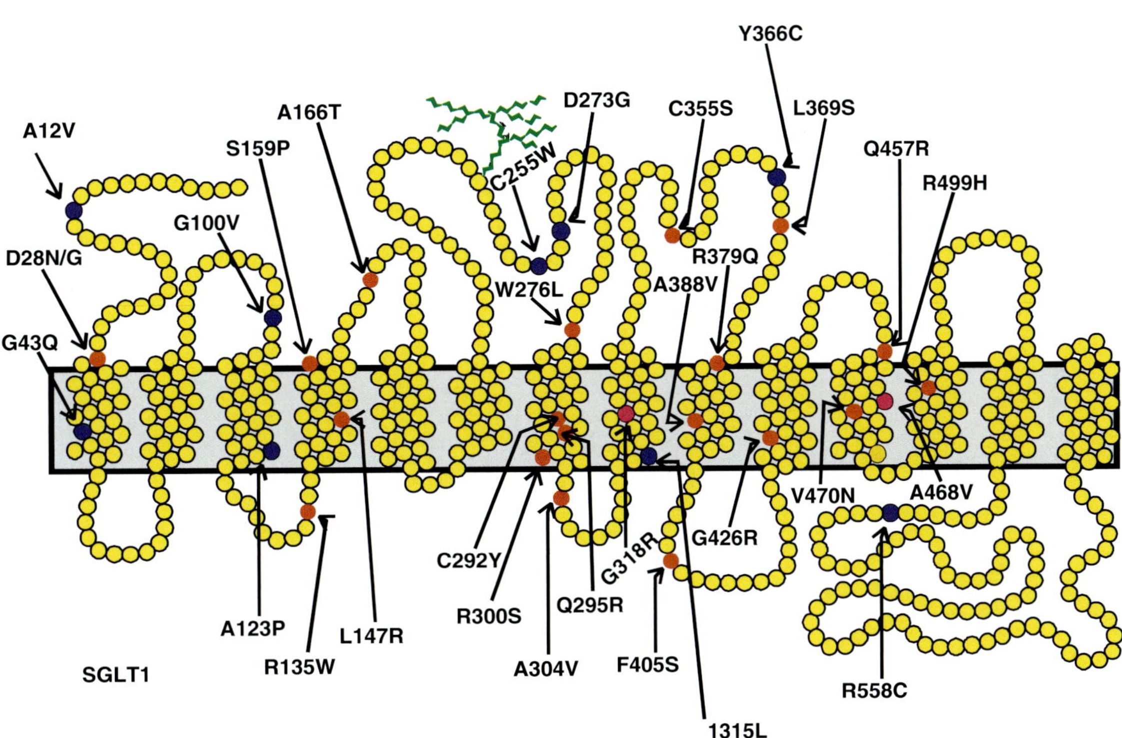

PLATE 6 A secondary structure model of SGLT1 showing the location of the missense mutations identified in patients with glucose-galactose malabsorption (GGM). The 664 amino acid protein is predicted to contain 14 transmembrane helices and one external N-linked glycosylation site. Both the N- and C-termini are shown on the extracellular side of the membrane. (see also Figure 7.3)

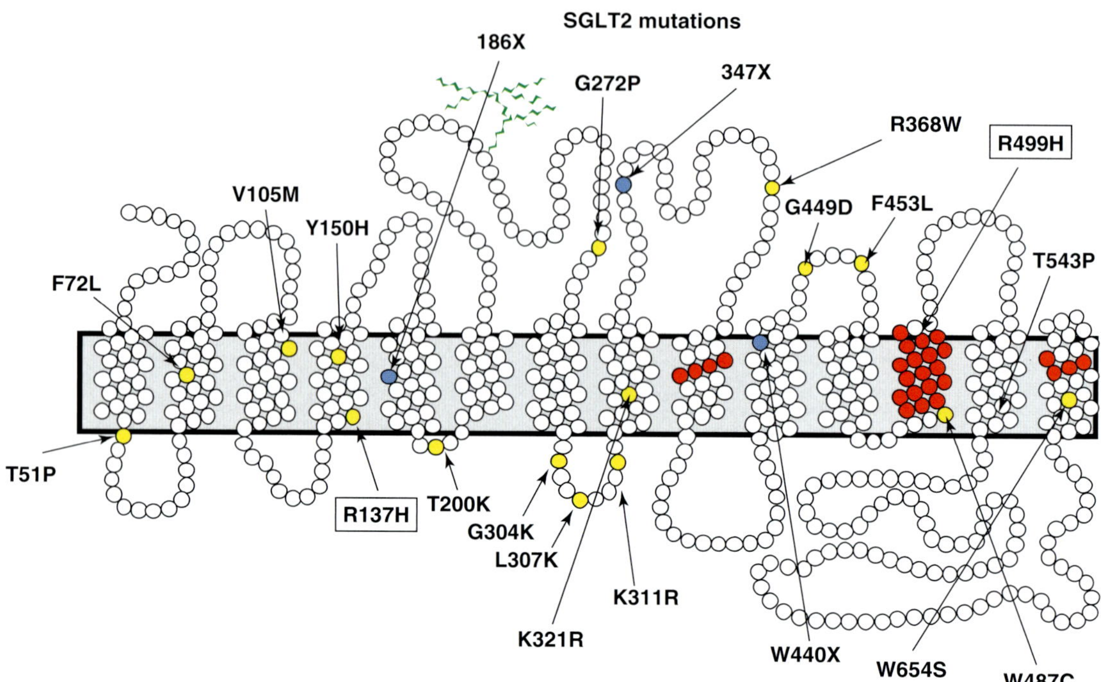

PLATE 7 A secondary structure model of SGLT2 showing the location of the missense, nonsense, and deletion mutations identified in patients with familial renal glucosuria (FRG). The 672 membrane protein is assumed to have the same secondary structure as SGLT1 (PLATE 6). It should be noted that there are no functional data available for the SGLT2 mutants (see Table 7.1) as it has been difficult to express in heterologous expression systems. The boxed mutation R137W is the only residue that is found mutated in GGM patients (R135H). (see also Figure 7.4)

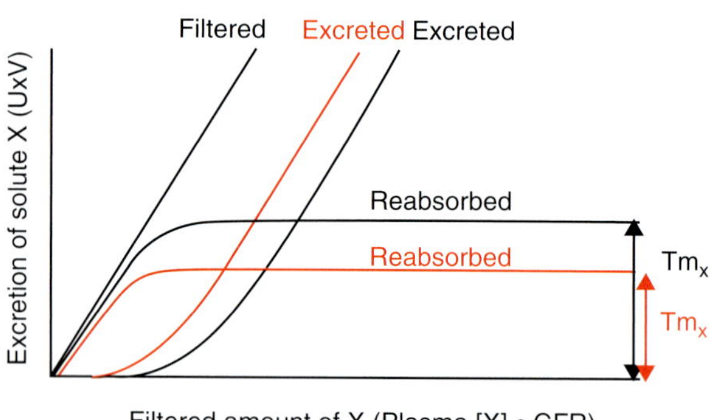

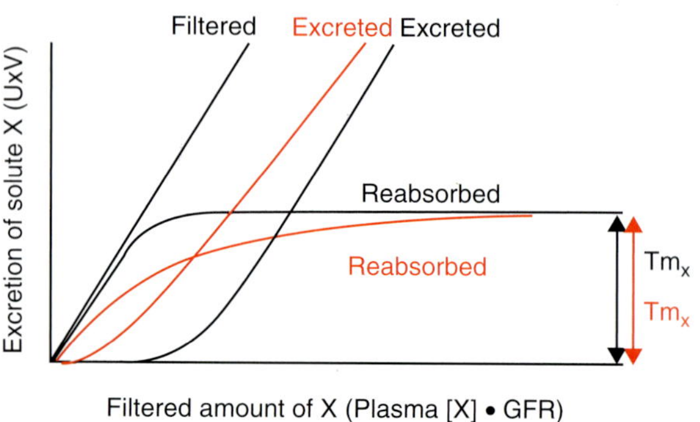

PLATE 8 Threshold of renal reabsorption of solute X. The black lines depict normal filtration, reabsorption, and excretion of solutes as functions of the filtered load (product of the glomerular filtration rate and the solute concentration in the filtrate). Reabsorption of solutes increases as more solute is delivered to the nephron until the tubular reabsorptive capacity (Tm_x) is reached so further increases in filtered load results in excretion of X in the urine. The red lines represent the possible defects in the proximal tubule. The upper panel shows a reduction in transport capacity while the lower panel shows a reduction in affinity for substrate. In either case, increased excretion of X is observed for a given filtered load. The two types of theoretical defects are not mutually exclusive. (see also Figure 10.3)

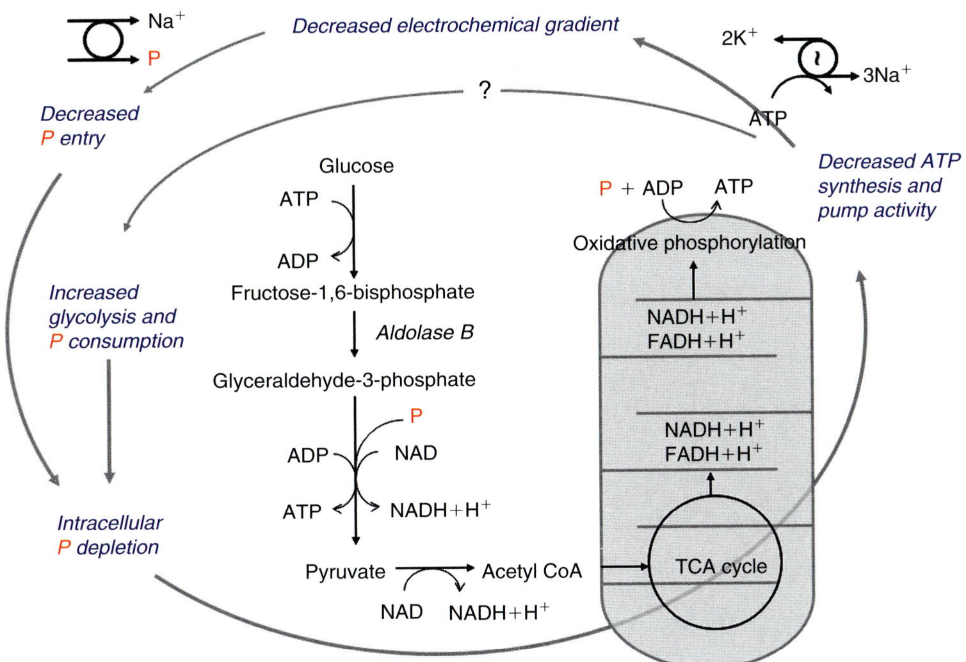

PLATE 9   Potential vicious cycle of phosphate depletion, increased glycolysis and decreased ATP generation in the proximal tubule. Depicted is a 'circular model' where intracellular phosphate depletion can amplify the effects of impaired ATP generation and/or increased glycolysis. Regardless of where the primary defect is situated, this vicious cycle amplifies phosphate trapping and ATP depletion. Glycolysis is divided into two stages separated by the aldolase step. The initial part of glycolysis consumes ATP with no net exchange of phosphate. The second part consumes phosphate, and generates ATP and reduced equivalents. ATP generation by oxidative phosphorylation requires adequate free intracellular phosphate. (see also Figure 10.7)

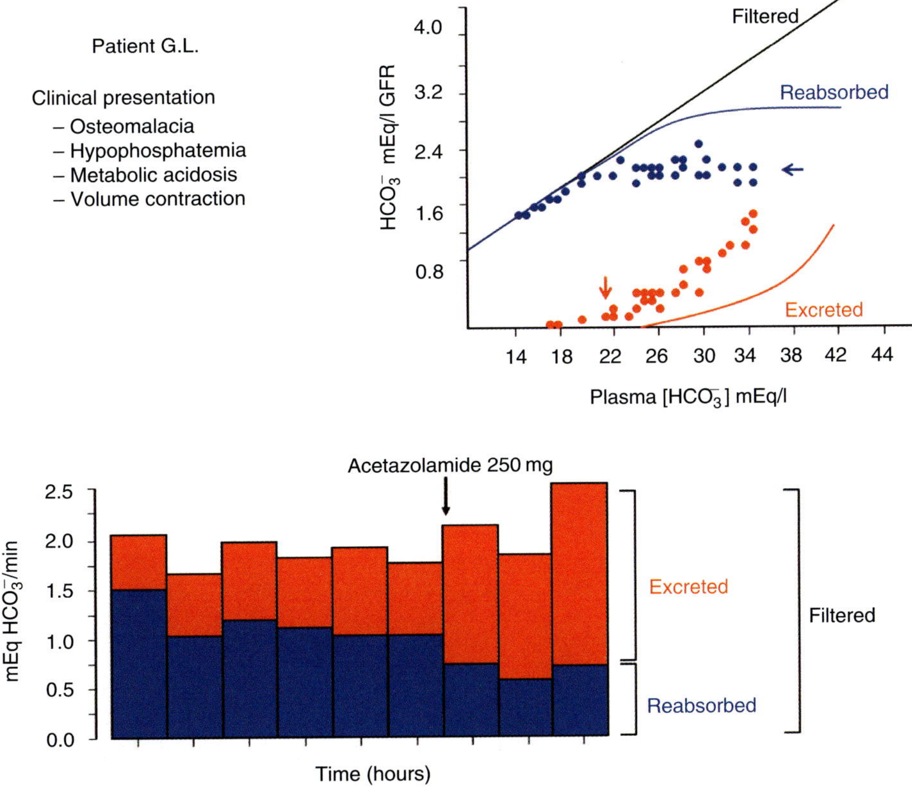

PLATE 10   Clinical investigation of a patient with idiopathic Fanconi syndrome. Presenting features are shown in the upper left. The upper right panel depicts a titration study where excreted and reabsorbed HCO_3^- was measured and calculated over a range of varying filtered loads. The expected normal relationship is shown as solid lines. The actual data points are shown as dots. Both proximal capacity (blue arrow) and affinity defects (red arrow) were present. In the bottom panel, the patient's serum HCO_3^- was maintained between 22–24 mM by continuous HCO_3^- infusion and glomerular filtration was measured by inulin clearance. HCO_3^- excretion and reabsorption rates were measured and calculated respectively. A carbonic anhydrase inhibitor was given at the time indicated which resulted in further HCO_3^- wasting (unpublished studies: Seldin DW and Rector FC Jr.). (see also Figure 10.8)

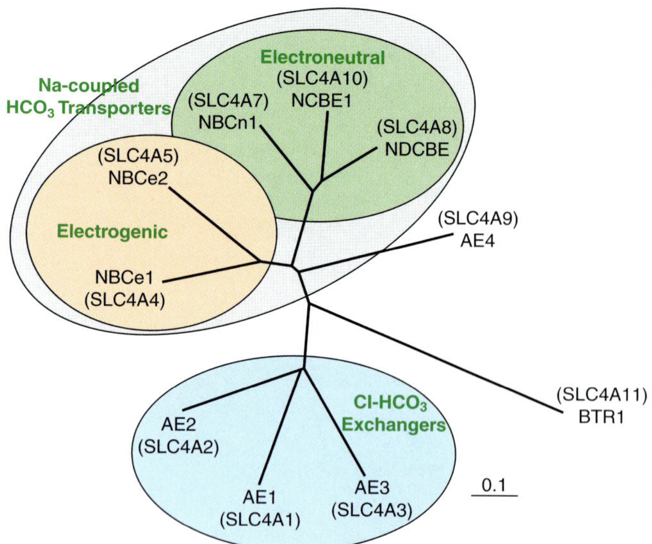

PLATE 11 SLC4 family of bicarbonate transporters (from Boron 2006). (see also Figure 11.2)

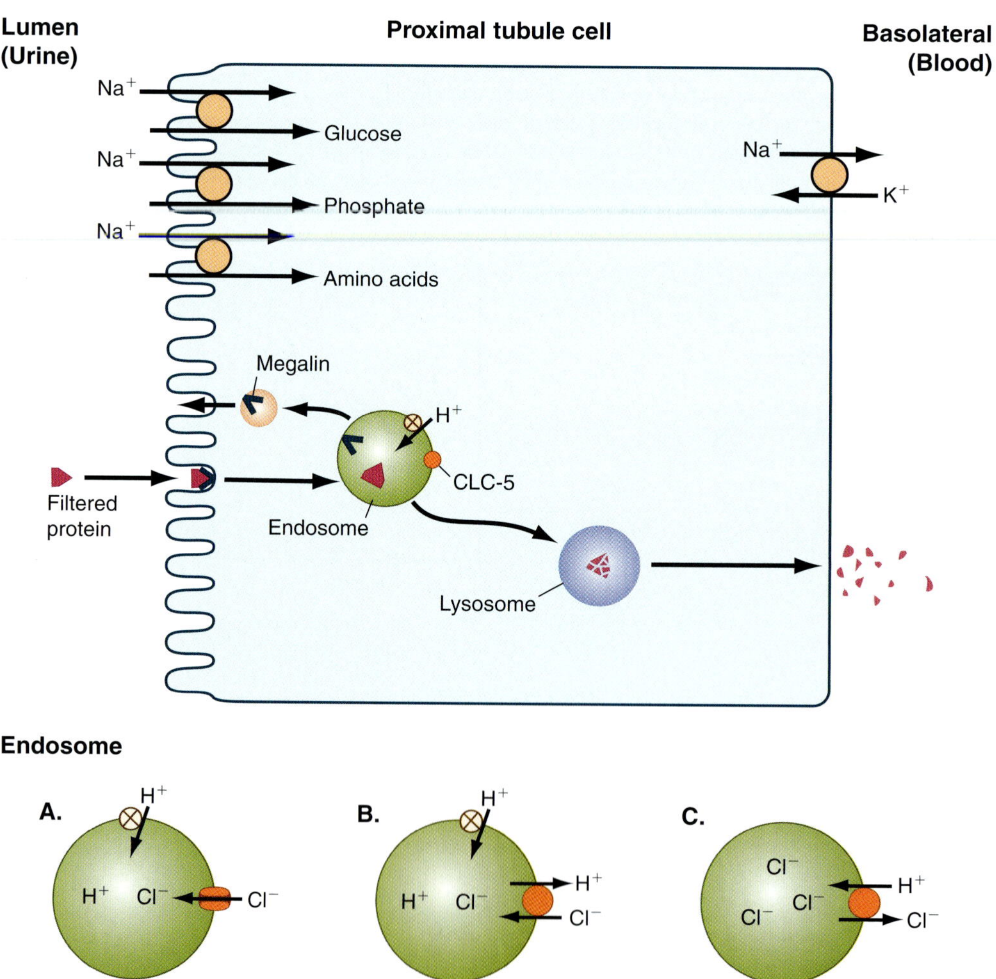

PLATE 12 Possible roles of chloride–proton antiporter function in endosomal acidification. In proximal tubular epithelial cells, CLC-5 is expressed in subapical endosomes in which the V-type H^+-ATPase is responsible for electrogenic transport of protons into the endosomal lumen. CLC-5 facilitates this acidification. (A) Until recently CLC-5 was thought to function as a chloride channel, allowing entry of chloride which would shunt the positive charge. (B and C) CLC-5 is now known to act as a Cl^-/H^+ exchanger. (B) In the presence of an active H^+-ATPase, the proton gradient would drive exit of H^+ coupled with entry of Cl^-, and this would effectively dissipate the interior positive charge. (C) In the nascent endosome, an initially high luminal Cl^- concentration would drive exit of Cl^- coupled with entry of H^+ and thereby achieve acidification in the absence of a proton pump, if there were a mechanism to shunt the inside-positive charge generated by the antiporter (Brandt & Jentsch 1995, Picollo & Pusch 2005, Scheel et al 2005). (see also Figure 12.2)

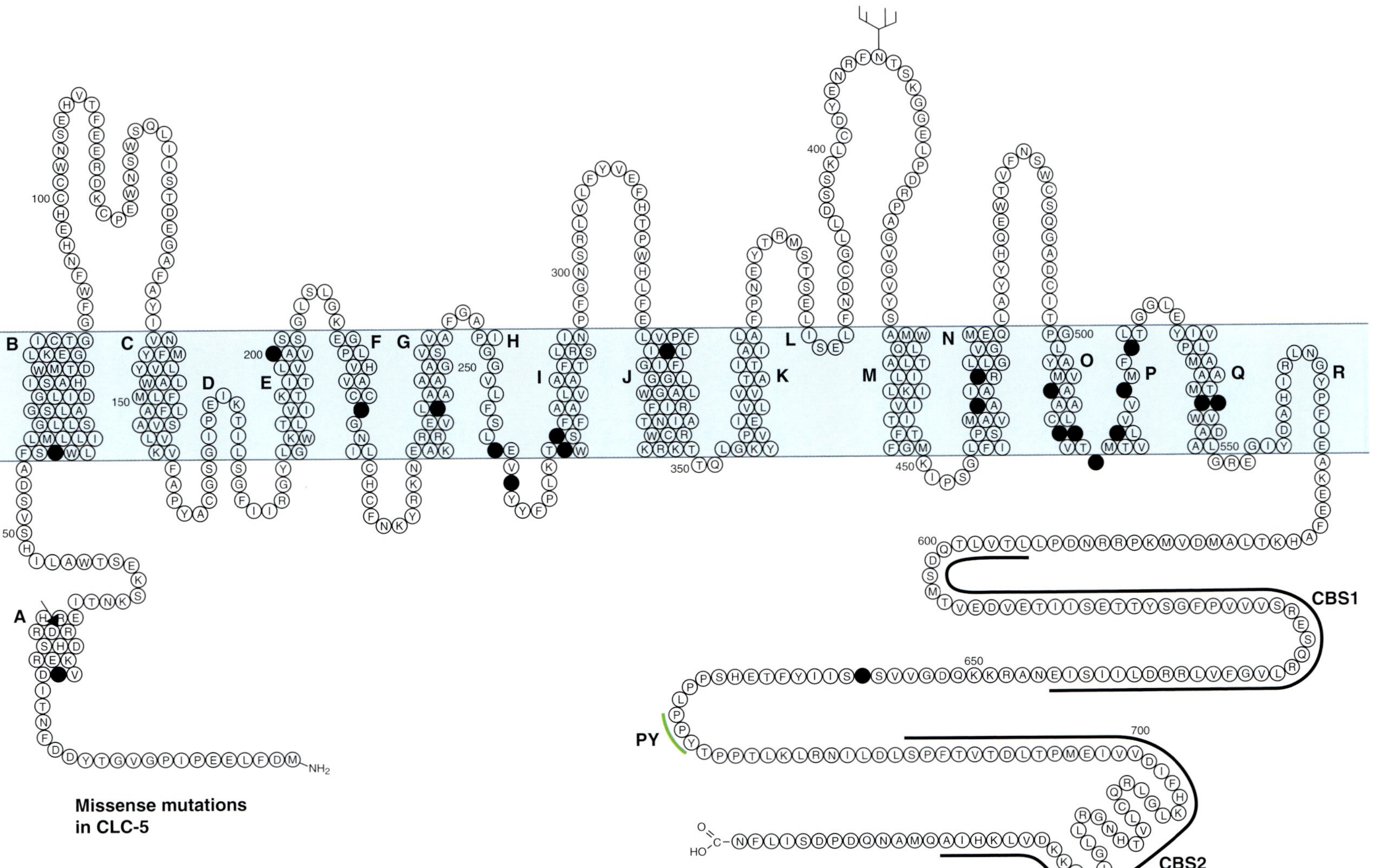

PLATE 13   Missense mutations in *CLCN5*. The CLC-5 protein is shown schematically with 18 α-helices as predicted by Dutzler et al (2002) and Wu et al (2003). Helices are labeled A through R, and represented relative to the membrane (blue shading); intracellular domains are below and extracellular above the membrane. Helices B, G, H, I, P, and Q form the interface between subunits. Other functionally significant domains include a signal sorting sequence at helix A in the intracellular N-terminal region, two CBS domains in the C-terminal chain, and a PY motif between them. Twenty-six missense mutations, in which a single base change results in an amino acid substitution, have been reported at 22 codons, indicated in filled circles; an additional in-frame insertion of histidine at codon 30 is represented by a triangle. (see also Figure 12.3)

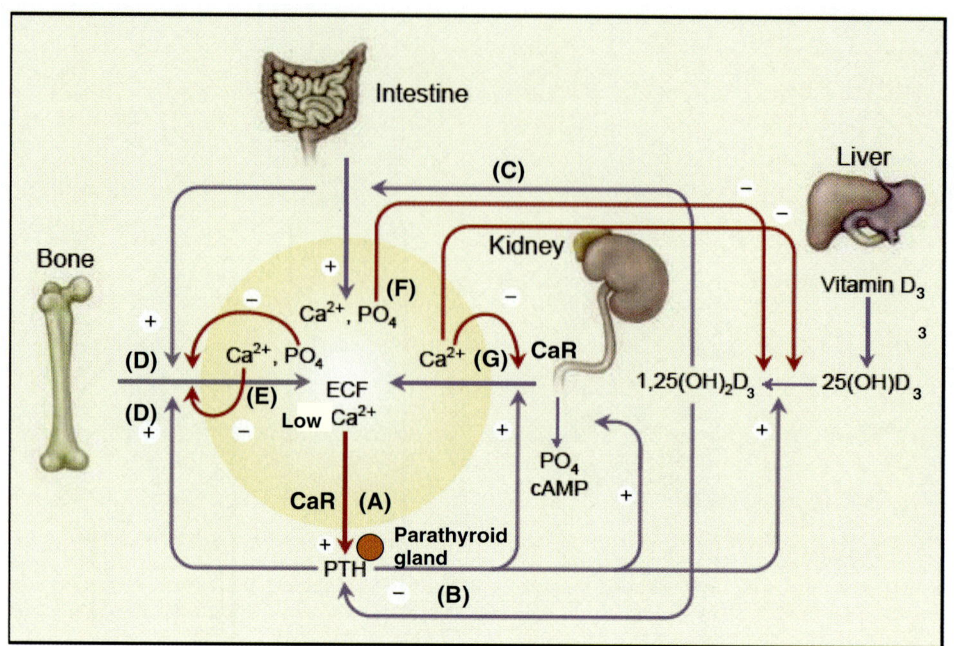

PLATE 14 Schematic diagram showing the principal components involved in extracellular calcium homeostasis. Actions that elevate extracellular calcium levels are illustrated by purple arrows; actions that reduce those levels are indicated by red arrows. In this example, (A) a reduction in blood calcium concentration promotes parathyroid hormone secretion, which (B) stimulates the kidney to increase tubular calcium reabsorption in the distal tubule, enhance phosphaturia and promote synthesis of 1,25-dihydroxyvitamin D$_3$. The 1,25-dihydroxyvitamin D$_3$ can (C) stimulate intestinal calcium and phosphate absorption and (D) act with PTH to enhance skeletal calcium and phosphate release. In addition to being controlled by PTH and 1,25-dihydroxyvitamin D$_3$, extracellular calcium and phosphate themselves can act as extracellular or 'first' messengers, regulating many of the same functions of kidney and parathyroid that PTH and 1,25-dihydroxyvitamin D$_3$ act on. Increases in extracellular calcium and phosphate levels both (E) decrease bone resorption and enhance bone formation and also (F) inhibit the 1-hydroxylation of 25-hydroxyvitamin D$_3$. An additional effect of extracellular calcium on the kidney is (G) to inhibit its own reabsorption in the distal tubule. Among these first messenger functions of extracellular calcium, the regulation of PTH release and renal tubular calcium reabsorption are known to be mediated by the CaR in the PLATE. The mechanisms underlying the other effects of extracellular calcium and phosphate remain to be elucidated. Modified from Brown EM et al. (1995) Calcium-ion-sensing cell-surface receptors. N. Engl. J. Med. 333: 234–240. © 1995 Massachusetts Medical Society with permission. cAMP, cyclic AMP; ECF, extracellular fluid; PTH, parathyroid hormone. (see also Figure 15.1)

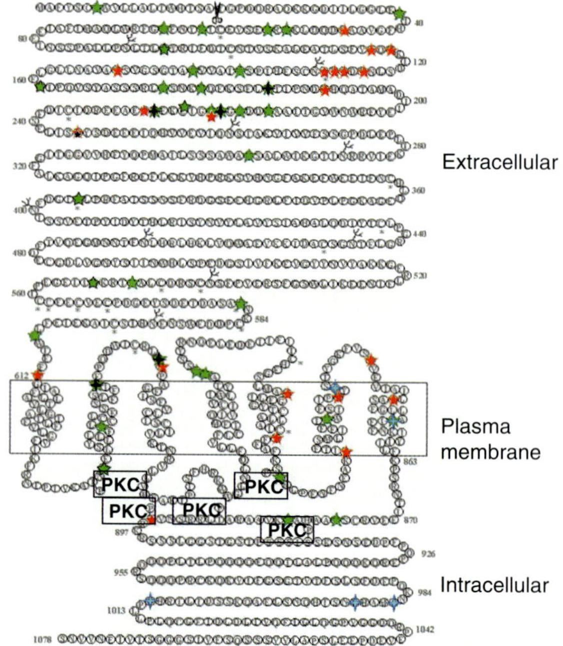

PLATE 15 Schematic diagram of the CaR, delineating the locations of activating and inactivating mutations and polymorphisms. the circles indicate individual amino acids; blue symbols are polymorphisms; red symbols are activating mutations; black symbols over green are two inactivating mutations at the same codon; asterisks show conserved cysteines; the 'P' symbols shown in circles that are attached to the main amino-acid chain are protein kinase C phosphorylation sites and are highlighted by 'PKC' within boxes; branched symbols are glycosylation sites; scissors denote the cleavage site of the signal peptide. Reproduced with permission from Pidasheva S et al (2004) CASRdb: calcium-sensing receptor locus-specific database for mutations causing familial (benign) hypocalciuric hypercalcemia, neonatal severe hyperparathyroidism, and autosomal dominant hypocalcemia. Hum Mutat 24: 107–111. Copyright © 2004 Wiley-Liss, Inc. (see also Figure 15.2)

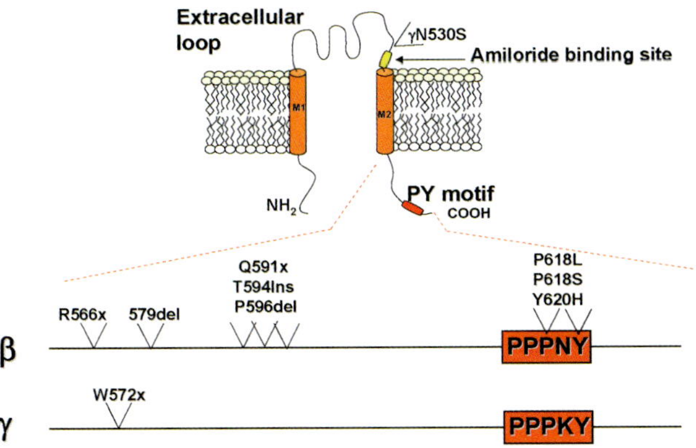

PLATE 16   Membrane topology of an ENaC subunit and locations of the mutations causing Liddle syndrome. Mutations are found in the intracellular carboxy terminus of the β or γ ENaC subunit, deleting or mutating a conserved proline-rich PY motif. One mutation (γN530S) has been described in the extracellular loop of γENaC, close to the amiloride binding site. (see also Figure 16.1)

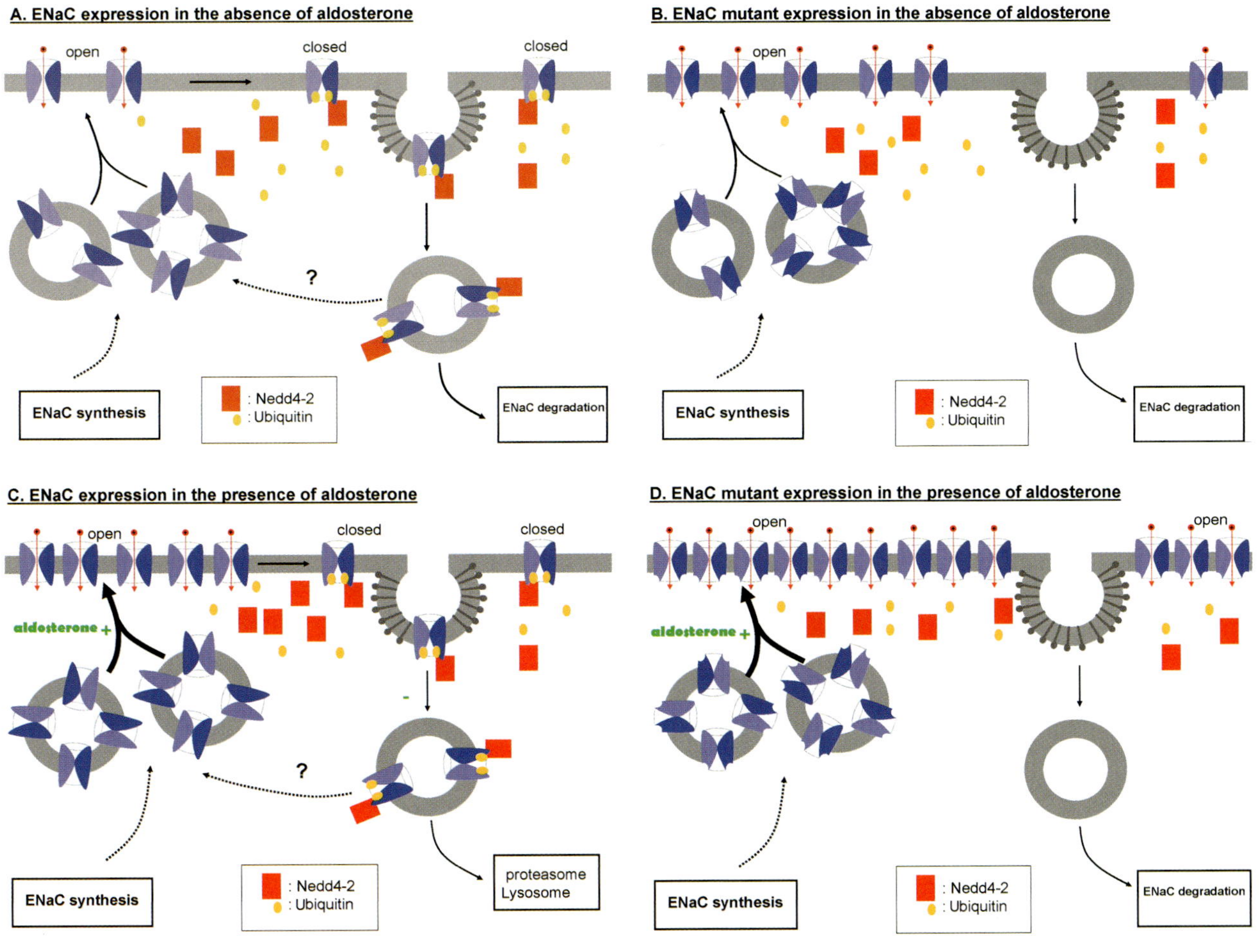

PLATE 17   This PLATE illustrates, in principal cells of the ASDN, the apical and subapical pool of ENaC wt and ENaC with mutations causing Liddle's syndrome. (A, B): principal cells are not stimulated by aldosterone and a few ENaC wt (A) are active at the apical membrane with a 1/2 life of approximately 0.5 to 1 h; upon binding Nedd4-2 the ubiquitylated channels close, are internalized in clathrin-coated pits, and may either be degraded or recycled back to the plasma membrane. The ENaC mutants (B) lacking the PY motif have lost their ability to interact with Nedd4-2, remain open, and are less efficiently internalized in clathrin-coated pits, leading to the retention of active ENaC at the apical membrane. (C, D) In the presence of aldosterone the subapical pool of ENaC is translocated into the apical membrane, resulting in a higher number of active channels at the cell surface. Aldosterone may potentially also control ENaC removal from the cell surface by decreasing ENaC retrieval (C). ENaC lacking the PY motif are less efficiently removed from the apical membrane and tend to accumulate, leading to an increased and prolonged (>1/2 life of ENaC wt) Na$^+$ entry into the cell. (see also Figure 16.2)

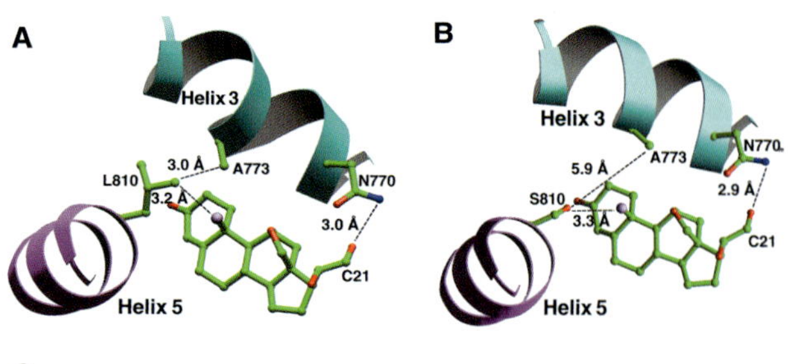

C

Helix 5		Helix 3		Aldosterone	19-NP
S	$-CH_2-OH$	CH_3-	A	100.0 ± 5.8	6.1 ± 1.2
M	$-CH_2-CH-S-CH_3$	CH_3-	A	120.3 ± 9.8	111.0 ± 5.5
L	$-CH_2-CH-CH_3$ (CH₃)	CH_3-	A	123.1 ± 15.7	109.2 ± 12.2
V	$-CH-CH_3$ (CH₃)	CH_3-	A	116.8 ± 30.4	54.4 ± 4.9
A	$-CH_3$	CH_3-	A	40.0 ± 7.2	3.2 ± 0.2
V	$-CH-CH_3$ (CH₃)	$H-$	G	138.2 ± 38.9	5.7 ± 2.0
L	$-CH_2-CH-CH_3$ (CH₃)	$H-$	G	108.1 ± 8.4	96.8 ± 6.9

PLATE 18 Helix 3–Helix 5 interaction in progesterone-mediated activation of MR_{L810}. (A) Structural model of a portion of the hormone-binding domain (HBD) of MR_{L810} bound to aldosterone. Using the crystal structure of the progesterone receptor (PR), a model of the MR LBD was created by substituting MR-specific residues in the ligand-binding cavity for their corresponding residues in PR. The side chain of L810 lies in sufficiently close proximity to A773 and the C-19 methyl group of the steroid to form van der Waals interactions. (B) Model of MR_{WT}. The side chain of S810 does not interact with either A773 or the steroid. (C) Activity of MRs with various amino acid substitutions at residues 810 and 773. Mutant receptors containing the indicated substitutions at positions 810 in helix 5 and 773 in helix 3 were tested for their ability to induce luciferase activity in the presence of 1 nM aldosterone or 19-norprogesterone. The length of each side chain is approximated. Each data point represents the mean of 9 independent transfections and is expressed as mean ± SEM of the percentage of luciferase induction by MRWT in response to aldosterone. (see also Figure 18.3)

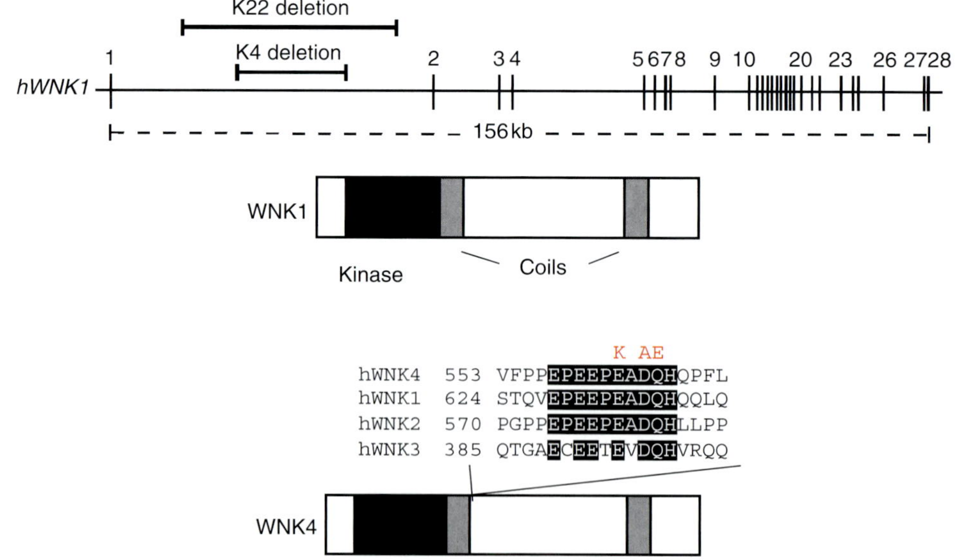

PLATE 19 Mutations in WNK1 and WNK4 cause pseudohypoaldosteronism type II. The general protein structures of WNK1 (encoded on human chromosome 12) and WNK4 (encoded on human chromosome 17) are shown; they have highly homologous N-terminal kinase domains (shown in black), and two coil-coil domains (shown in gray). At the top of the PLATE, the genomic organization of WNK1 is shown, with its 28 exons spanning 156 kb in genomic DNA. The positions of deletions in the first intron of WNK1 found in two families (K22 and K4) that show genetic linkage to 12p are shown. Above the diagram of WNK4, the amino acid sequence of a short acidic segment of human WNK4 (hWNK4) is shown and compared to the other human WNKs. Missense mutations found in three different PHAII kindreds linked to chromsome 17 are indicated above the sequence – all occur at completely conserved positions and change the charge of the encoded amino acid. (see also Figure 19.1)

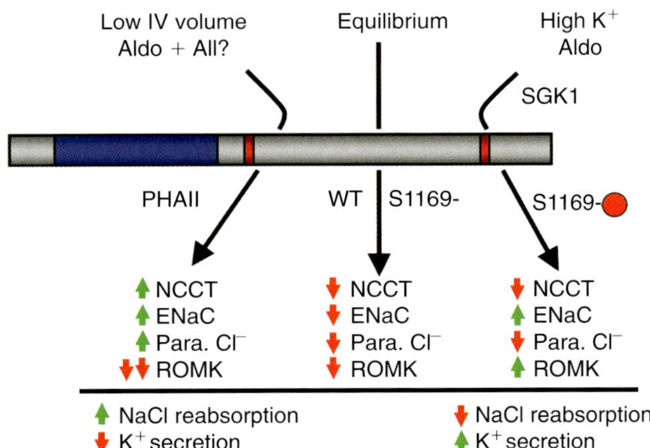

PLATE 20 WNK4 is a switch that regulates the balance between renal salt reabsorption and K^+ secretion. WNK4 is shown to exist in three distinct states that enable it to direct the kidney to modulate the balance between renal Na–Cl reabsorption and K^+ secretion. In a ground or equilibrium state, WNK4 inhibits mediators of both Na–Cl reabsorption and K^+ secretion. In a second state, exemplified by WNK4 harboring PHAII mutations, WNK4 loses inhibition of Na–Cl reabsorbing pathways, including NCCT, ENaC and paracellular Cl^- flux, but augments inhibition of K^+ secretion via ROMK, resulting in increased salt reabsorption and decreased K^+ secretion. This state is the appropriate response to intravascular volume depletion, a condition accompanied by high levels of aldosterone and angiotensin II. A third state, characterized by WNK4 with phosphorylation at S1169 via the kinase SGK1, results in inhibition of the Na–Cl cotransporter NCCT, but increased activity of the epithelial sodium channel (ENaC) and increased activity of the K^+ channel ROMK, resulting in increased K^+ secretion without increased Na^+ reabsorption. (Reprinted with permission from Annual Reviews of Physiology, 2007 15:48). (see also Figure 19.2)

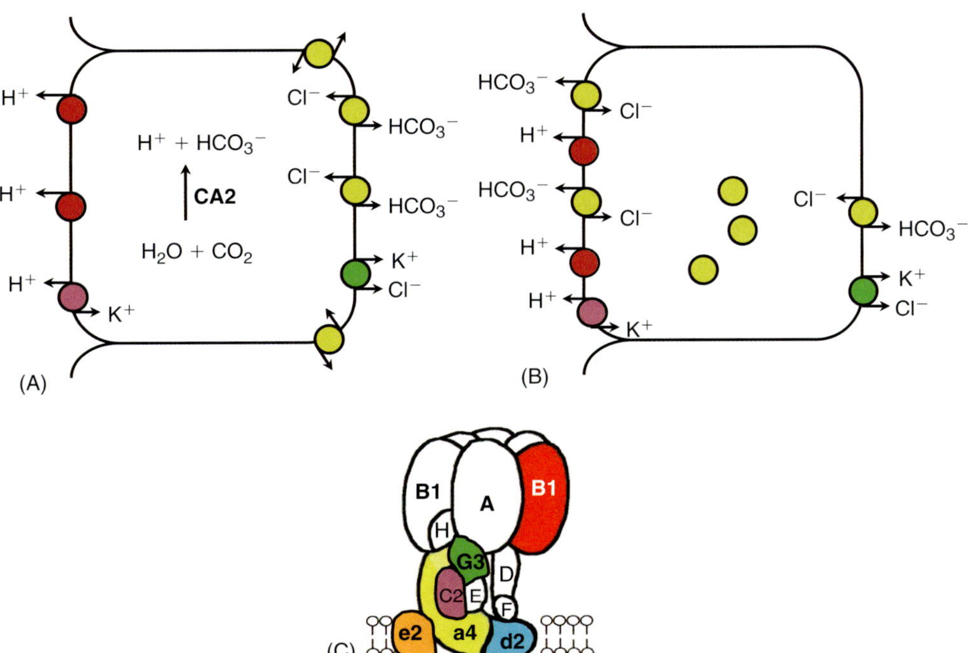

PLATE 21 Polarized function of α-IC and molecular contributors to dRTA. (A) In cortical collecting duct α-IC, acid–base balance is tightly regulated by proton pumps on the apical surface of the cell that actively secrete acid into the duct. This process is coupled to the reclamation of bicarbonate ions into the interstitium, via the chloride–bicarbonate exchanger AE1, whose expression in the kidney is confined to the basolateral surface of these cells. Other molecular residents are also illustrated, with the exception of the ubiquitously expressed Na/K-ATPase. The KCC4 potassium-chloride co-transporter has to date only been localized in rodents. (B) In at least two forms of autosomal dominant distal renal tubular acidosis, loss of a C-terminal basolateral targeting sequence results in a non-polarized distribution of AE1, such that a proportion of the protein reaches the apical surface, thus probably disrupting the electrochemical balance of the cell in other cases. (C) The apical proton pump is a multimeric complex of at least 13 subunits. At least six of these are the product of alternate genes (shown in color) that are either solely or mainly expressed in the kidney. (see also Figure 20.1)

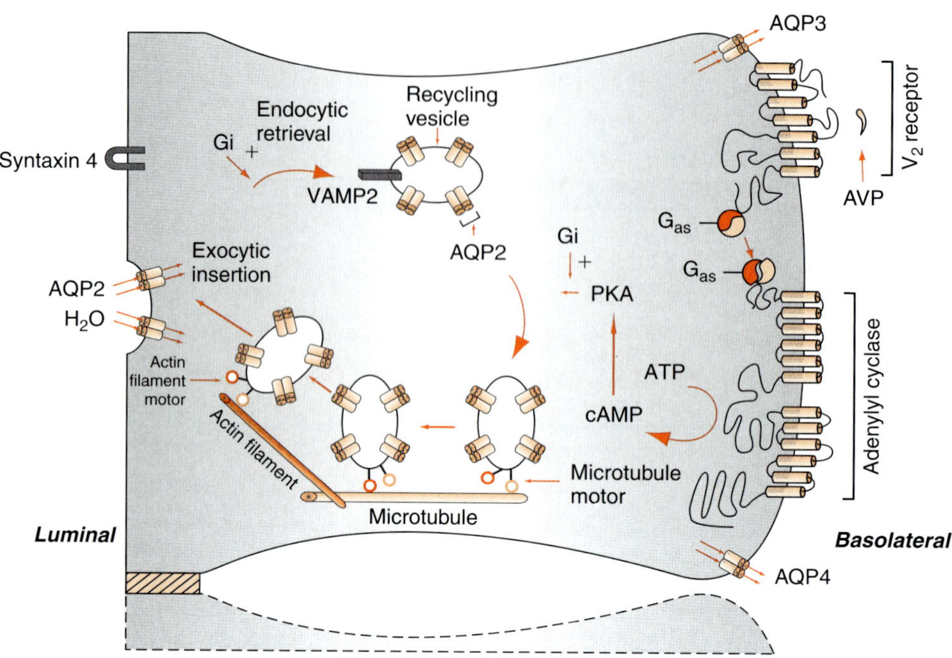

PLATE 22 Schematic representation of the effect of arginine vasopressin (AVP) to increase water permeability in the principal cells of the collecting duct. AVP is bound to the V_2 receptor (a G-protein-linked receptor) on the basolateral membrane. The basic process of G-protein-coupled receptor signaling consists of three steps: a hepta-helical receptor that detects a ligand (in this case, AVP) in the extracellular milieu, a G-protein that dissociates into α-subunits bound to GTP and $\beta\gamma$-subunits after interaction with the ligand-bound receptor, and an effector (in this case, adenylyl cyclase) that interacts with dissociated G-protein subunits to generate small-molecule second messengers. AVP activates adenylyl cyclase increasing the intracellular concentration of cyclic adenosine monophosphate (cAMP). The topology of adenylyl cyclase is characterized by two tandem repeats of six hydrophobic transmembrane domains separated by a large cytoplasmic loop and terminates in a large intracellular tail. Generation of cAMP follows receptor-linked activation of the heteromeric G-protein (G_s) and interaction of the free $G_{\alpha s}$-chain with the adenylyl cyclase catalyst. Protein kinase A (PKA) is the target of the generated cAMP. Cytoplasmic vesicles carrying the water channel proteins (represented as homotetrameric complexes) are fused to the luminal membrane in response to AVP, thereby increasing the water permeability of this membrane. Microtubules and actin filaments are necessary for vesicle movement toward the membrane. The mechanisms underlying docking and fusion of aquaporin-2 (AQP2)-bearing vesicles are not known. The detection of the small GTP binding protein Rab3a, synaptobrevin 2, and syntaxin 4 in principal cells suggests that these proteins are involved in AQP2 trafficking (Valenti et al 1998). When AVP is not available, water channels are retrieved by an endocytic process, and water permeability returns to its original low rate. AQP3 and AQP4 water channels are expressed on the basolateral membrane. (see also Figure 21.1)

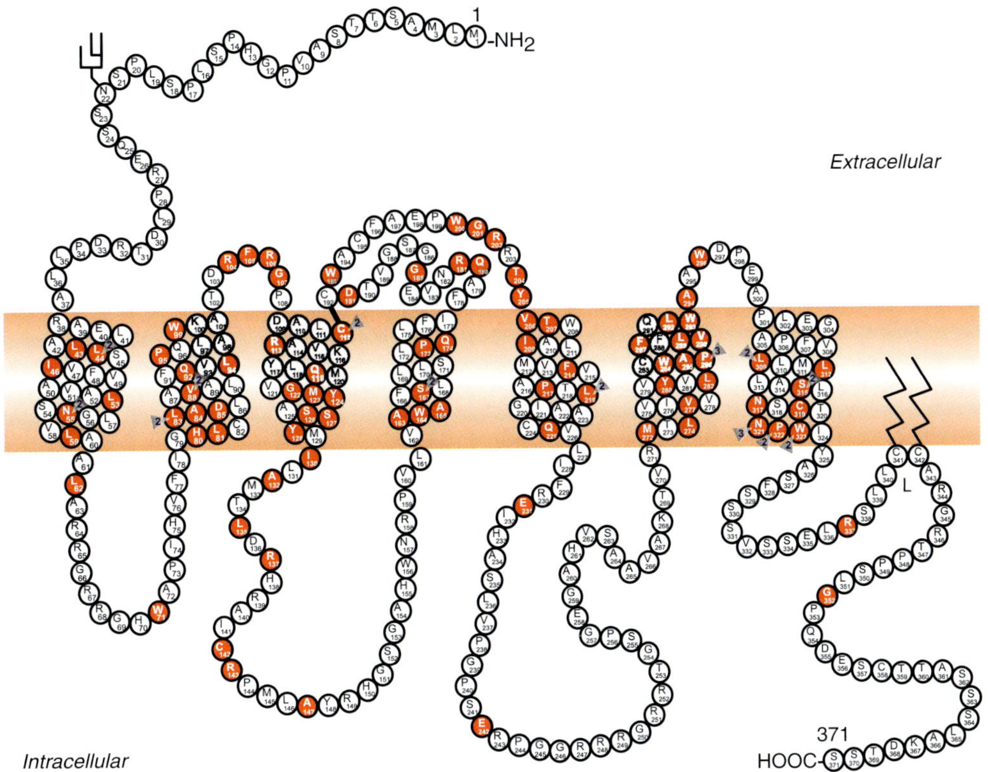

PLATE 23 Schematic representation of the V_2 receptor and identification of 183 putative disease-causing *AVPR2* mutations. Predicted amino acids are given as the one-letter amino acid code. Solid symbols indicate missense or nonsense mutations; a number indicates more than one mutation in the same codon; other types of mutations are not indicated on the PLATE. The extracellular, transmembrane, and cytoplasmic domains are defined according to Mouillac et al (1995). There are 89 missense, 18 nonsense, 45 frameshift deletion or insertion, seven inframe deletion or insertion, four splice-site, and 19 large deletion mutations, and one complex mutation. See http://www.medicine.mcgill.ca/nephros for a list of mutations. (see also Figure 21.3)

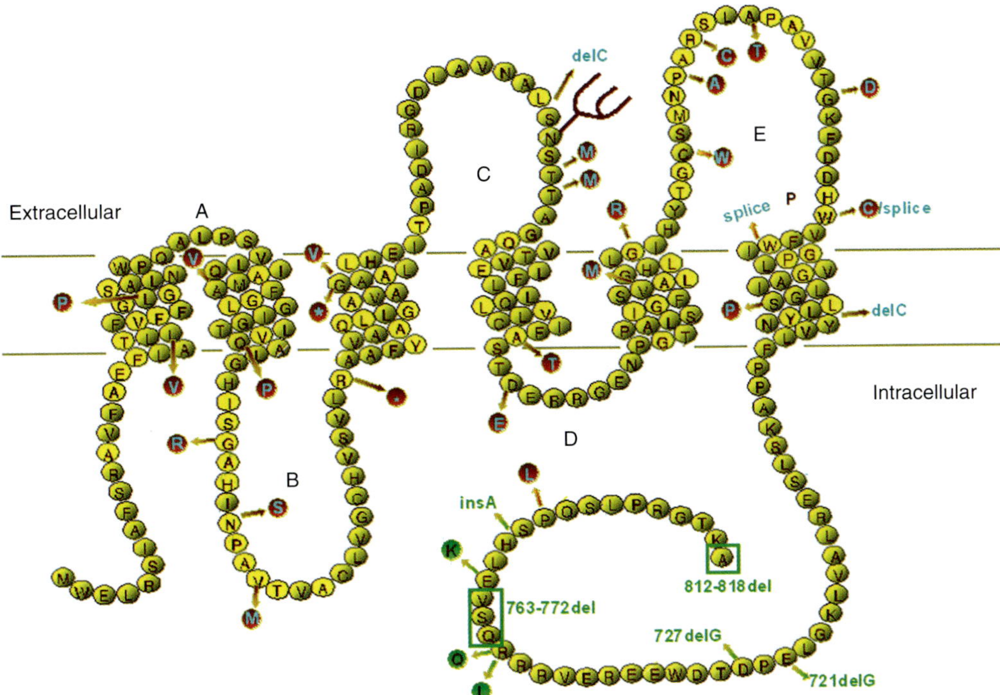

PLATE 24 Mutations in Aquaporin-2 in nephrogenic diabetes insipidus. An AQP2 monomer consists of six transmembrane domains connected by five loops (A through E) and with its N- and C-termini located intracellularly. Loops B and E meet each other with their characteristic NPA boxes in the membrane and form the water selective pore. The N-glycosylation site (tree at N123) is indicated. Bright amino acids are conserved between different AQPs (Heymann & Engel 2000). Red amino acids and blue words or green amino acids and words outside the topology model represent mutations found in patients with recessive NDI or dominant NDI, respectively. Missense (amino acid), stop codon (*), nucleotide deletions (del) or insertions (ins), or splice site (splice) mutations are indicated. (see also Figure 22.1)

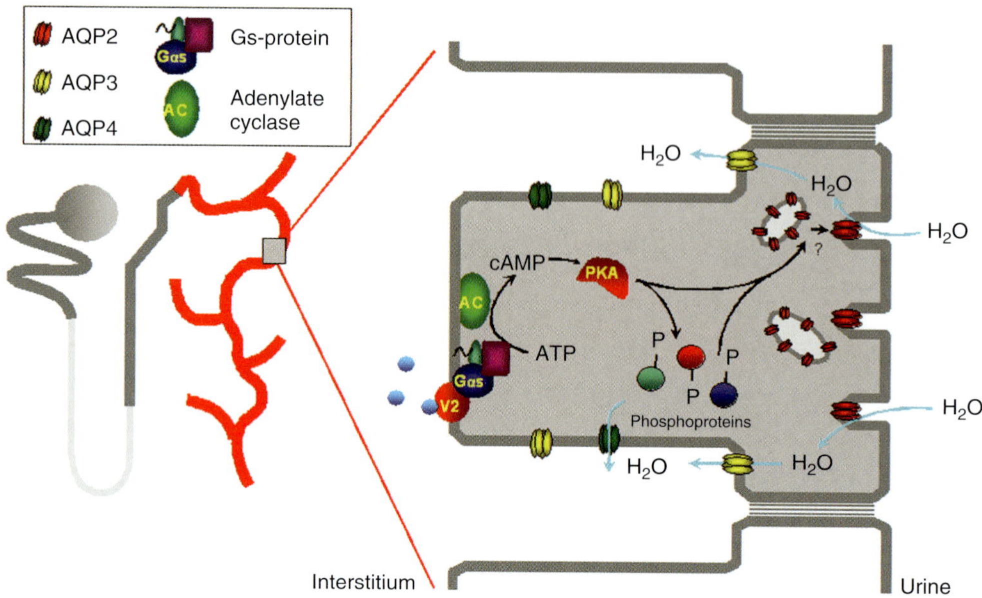

PLATE 25 Regulation of the Aquaporin-2 mediated water transport by vasopressin. Vasopressin, vasopressin V2 receptor (V2), stimulatory GTP binding protein (Gs-protein), adenylate cyclase (AC), adenosine triphosphate (ATP), cyclic adenosine monophosphate (cAMP), protein kinase A (PKA) and phosphorylated proteins (O-P) are indicated. For details see text. (see also Figure 22.2)

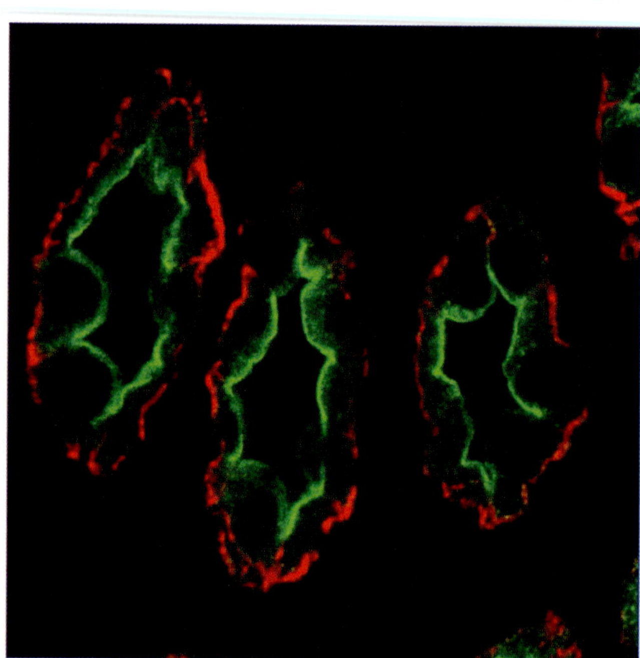

PLATE 26 Expression of AQP2 and AQP3 in renal collecting ducts. Rabbit anti-AQP3 (red) and guinea pig anti-AQP2 (green) antibodies were used to specifically stain the kidney of rats dehydrated for 24 hours for AQP2 and AQP3. Note the apical membrane localization of AQP2 and the basolateral membrane localization of AQP3 in the collecting duct principal cells. (see also Figure 22.3)

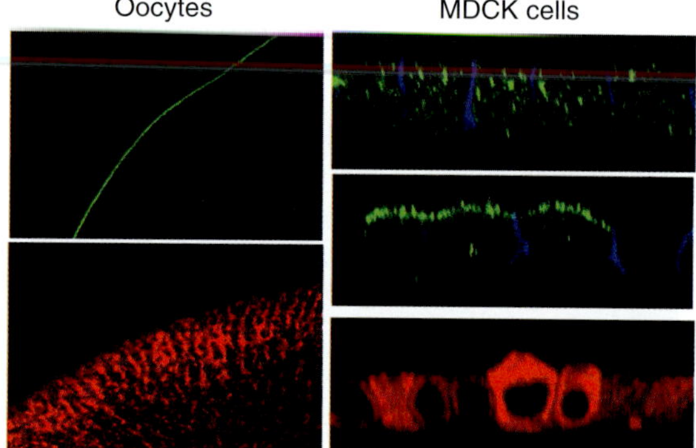

PLATE 27 Localization of wild-type AQP2 and an AQP2 mutant in recessive NDI. Following injection of their cRNAs, wild-type AQP2 (upper left) or an AQP2 mutant in recessive NDI (lower left) were expressed in oocytes. Also, MDCK cells were stably transfected with an AQP2 mutant in recessive NDI or with wt-AQP2, and grown to confluence. All cells were subjected to immunocytochemistry and analyzed by confocal laser scanning microscopy. In oocytes, wt-AQP2 is located in the plasma membrane, whereas the mutant in recessive NDI shows the reticular pattern of the ER. In MDCK cells a similar reticular pattern (only the nucleus and plasma membranes are unstained) is observed for the AQP2 mutant in NDI (lower right). In MDCK cells, wild-type AQP2 resides in intracellular vesicles without stimulation (upper right), but is redistributed to the apical membrane with forskolin (middle right). Wild-type-AQP2 (green), the mutant in recessive NDI (red) and the basolateral marker E-cadherin (blue) are indicated. (see also Figure 22.4)

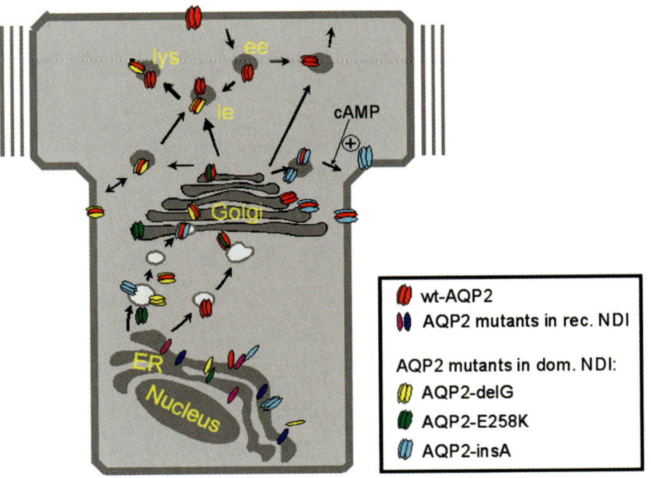

PLATE 28 Cellular mechanism underlying recessive and dominant NDI. In healthy individuals, wild-type AQP2 is assembled to homotetramers in the ER, and traverses through the Golgi complex to intracellular vesicles, from where it will be re-distributed to the apical membrane upon activation of the cAMP signaling cascade by AVP. Upon removal of AVP, AQP2 recycles via early endosomes (ee) to its storage compartment. AQP2 mutants in recessive NDI are misfolded and retained in the endoplasmic reticulum (ER), from where they are targeted for degradation by proteasomes in the cytosol, and are not able to heteroligomerize with wt-AQP2. In contrast, AQP2 mutants in dominant NDI (here represented by AQP2-delG, AQP2-E258K and AQP2-insA), heteroligomerize with wt-AQP2 in the ER and, following trafficking through the Golgi complex, missort the formed wt-AQP2/mutant complex to the late endosomes (LE)/lysosomes (Lys) (AQP2-delG and AQP2-E258K) or to vesicles that translocate the AQP2 complexes to the basolateral membrane upon stimulation with AVP (AQP2-insA). (see also Figure 22.5)

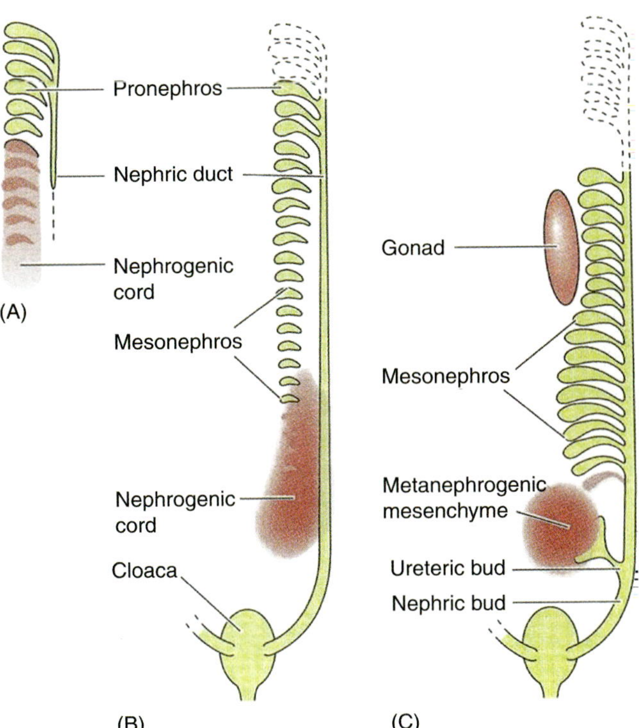

PLATE 29 Development of the mammalian excretory system progresses in a cranial-caudal fashion with the formation of the nephric duct and associated pronephros (A), mesonephros (B), and finally the metanephros (C). Adapted from Saxen (1987). (see also Figure 23.1)

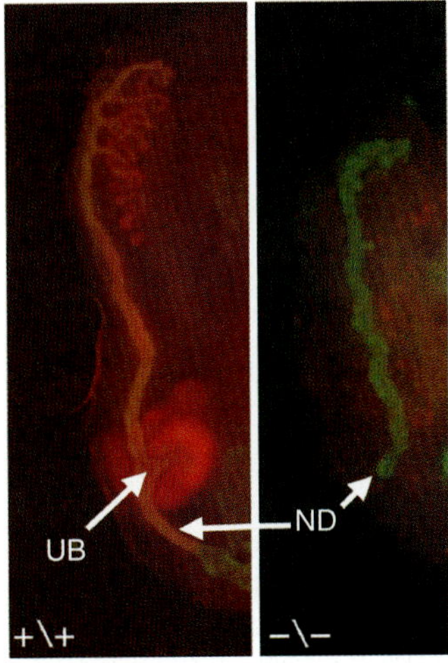

PLATE 30 Examination of mesonephric and metanephric kidney development in wild-type (+/+) and *Pax2* deficient (-/-) mice. Longitudinal sections from E11 mice were immunostained with anti-pancytokeratin (green) and anti-Pax2 (red). Mice lacking Pax2 expression fail to develop both mesonephric tubules and the metanephric kidney. UB, ureteric bud; ND, nephric duct. Image courtesy of Dr Greg Dressler. (see also Figure 23.3)

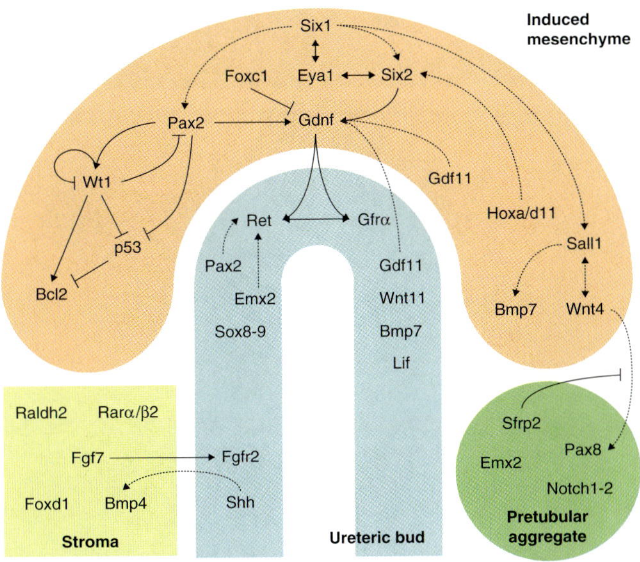

PLATE 31 A schematic depiction of representative transcription factors and signaling pathways involved in the molecular crosstalk between the four cellular compartments in the early metanephros. (see also Figure 23.4)

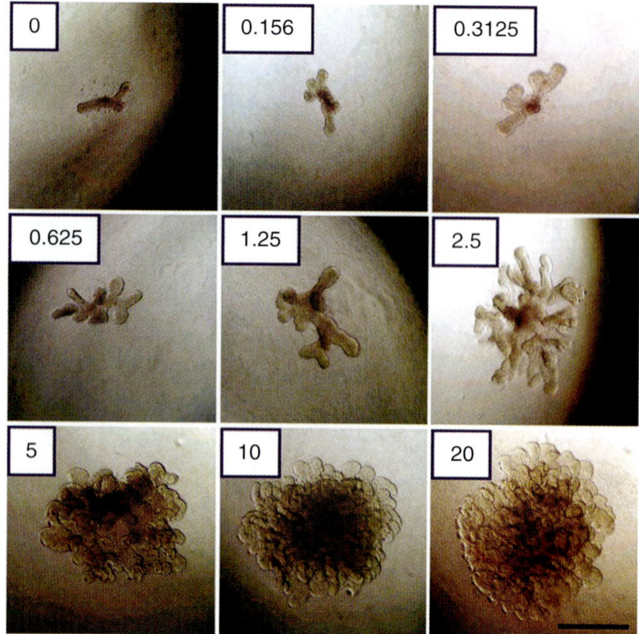

PLATE 32 Culture of the isolated ureteric bud in vitro in the presence of Gdnf, Fgf1, and increasing doses of pleiotrophin (in μg/ml) results in extensive branching and bud elongation. Used with permission from H. Sakurai (Sakurai et al 2001). (see also Figure 23.5)

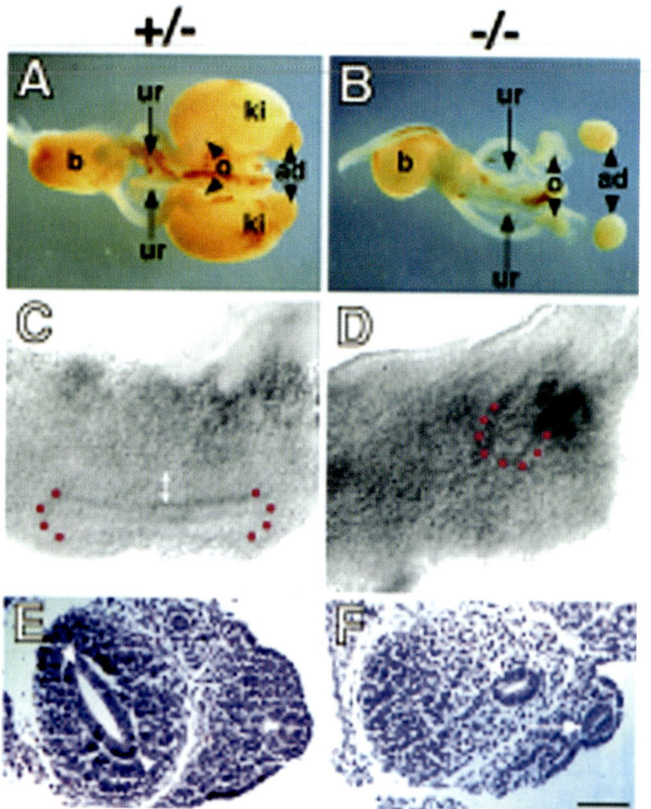

PLATE 33 Dissected urogenital systems of normal (heterozygous) (A) and heparan sulfate 2-O-sulfotransferase homozygous mutant (B) newborn mice reveals bilateral renal agenesis in the absence of *Hs2st* expression. Homozygotes display bilateral renal agenesis and possess a blind-ended ureter (ur). The remainder of the mutant urogenital system is normal. (ad) Adrenal gland; (b) bladder; (ki) kidney; (o) ovary. (C–F) Morphology of representative E11.5 heterozygous (C, E) and mutant (D, F) embryonic kidney rudiments viewed in whole-mount (lateral view) (C, D) or as transverse sections stained with hematoxylin and eosin (E, F). (C) In heterozygotes, the ureteric bud has invaded the mesenchyme and bifurcated once to form a T-shape surrounded by a mantle of condensed mesenchyme (double-headed arrow). (D) In mutants, the ureteric bud has emerged from the Wolffian duct and invaded the mesenchymal blastema but failed to bifurcate and there are no overt signs of mesenchymal condensation. In C and D the position of the ureteric bud tip(s) is highlighted by red dots. (E) A dilated ureteric bud (arrowheads) is present in the heterozygotes surrounded by condensed mesenchyme. (F) In mutants, the ureteric bud (arrowhead) has not dilated, and there are no overt signs of condensation in the mesenchyme. (Arrow in E and F) Wolffian duct. Scale bar: 1300 μm (A and B); 30 μm (C–F). Used with permission from Bullock et al 1998. (see also Figure 23.6)

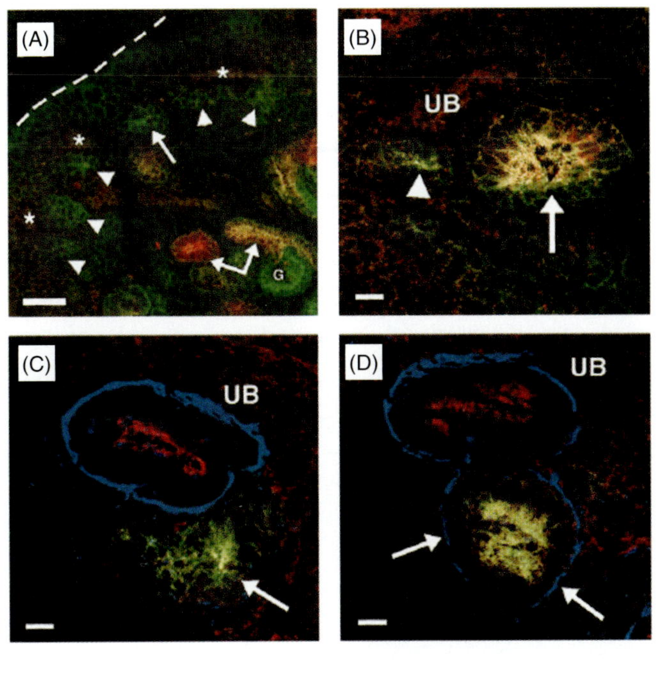

PLATE 34 Mesenchymal–epithelial transformation of the renal vesicle is accompanied by upregulation of cadherin-6 and expression of laminin in the basement membrane. (A, B) Immunostaining for R-cadherin (green) and cadherin-6 (red) demonstrates primarily R-cadherin expression in the early mesenchymal aggregates (arrowheads) adjacent to the ureteric bud tips (asterisks in A, UB in B), whereas regions that have undergone MET to form tubules express increasing amounts of cadherin-6 (double-headed arrow in A, arrow in B; yellow indicates dual expression, red indicates primarily cadherin-6). G, glomerulus; dashed line, edge of the kidney. (C, D) Immunostaining the region of the ureteric bud tips for R-cadherin (green), Zo-1 (red), and laminin (blue) demonstrates that the ureteric bud (UB) is fully epithelialized with a laminin-containing basement membrane (blue ring in C and D) and Zo-1 expression at the apical cell junctions, while a newly formed mesenchymal aggregate (arrow in C) has no distinct basement membrane, and expresses primarily R-cadherin (green cells in C) with some areas of co-expression of R-cadherin and Zo-1 (yellow cells in C). In a vesicle undergoing mesenchymal–epithelial transformation (arrows in D), the laminin-containing basement membrane is detectable and Zo-1 expression is upregulated resulting in most of the cells appearing yellow. Used with permission from Mah et al 2000. (see also Figure 23.7)

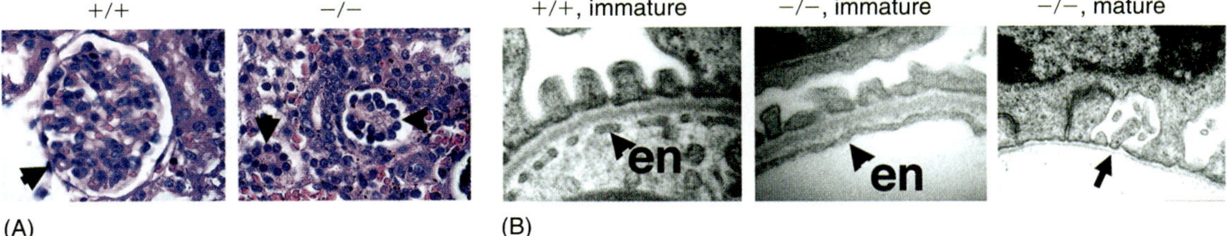

PLATE 35 Endothelial cell development within the glomerulus requires Vegf-A expression by the podocyte. (A) Glomeruli from wild-type mice (+/+) demonstrate normal capillary loops (arrow) while glomeruli from mice lacking Vegf-A expression by the podocyte (−/−) reveal failure of endothelial cell expansion, with a simple layer of immature podocytes surrounding a rudimentary mesangium (arrowhead). (B) Transmission EM reveals a normal basement membrane and capillary endothelial fenestrations in the wild-type glomerulus (+/+, en), whereas endothelial fenestrations are absent in the immature Vegf-A deficient glomerulus (−/−, arrowhead) and the entire endothelial layer is lost in the mature glomerulus (−/−, arrow). Used with permission from Eremina et al (2003). (see also Figure 23.10)

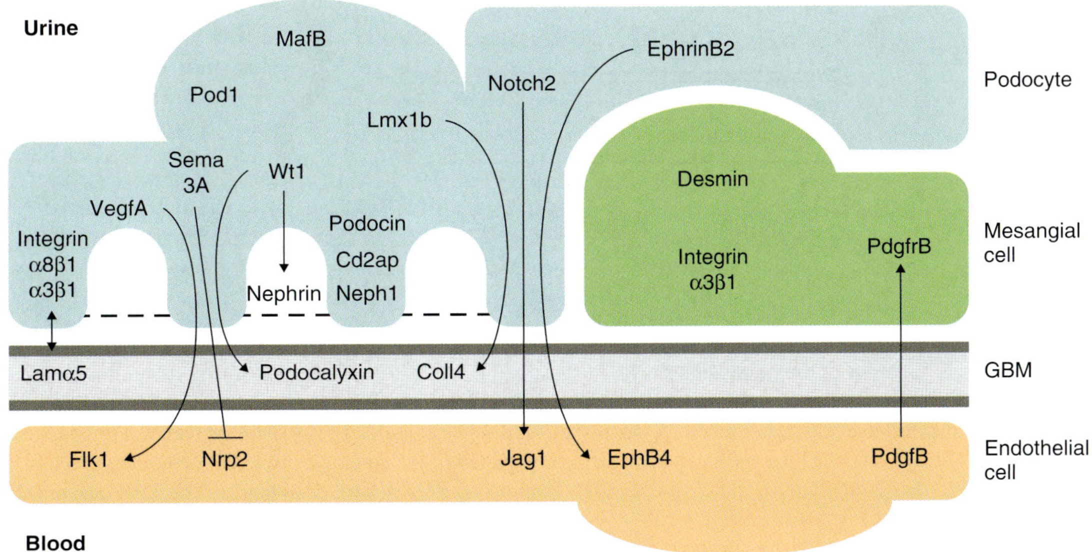

PLATE 36 A schematic depiction of representative pathways involved in the molecular control of podocyte, endothelial and mesangial development in the glomerulus. (see also Figure 23.11)

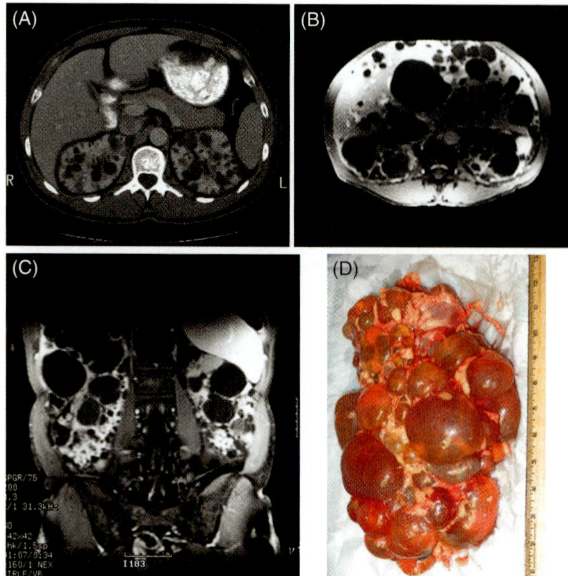

PLATE 37 Autosomal dominant polycystic kidney disease. Axial CT (A) and MRI (B) and coronal MRI (C) images demonstrating moderate polycystic kidney disease and absence of hepatic cysts (A) and more advanced polycystic kidney and liver disease with marked enlargement of both organs (B, C). In both the CT and MR images, cysts appear as hypodense areas within the organ parenchyma. (D) Nephrectomy specimen from a patient with ADPKD and end stage renal disease showing profound cystic deformation of the kidney. (Modified from Somlo S, Torres VE and Caplan MJ. (2007). In: Seldin and Giebisch's The Kidney: Physiology and Pathophysiology, Alpern, RJ and Hebert SC. Academic Press, pp. 2283–2314). (see also Figure 24.1)

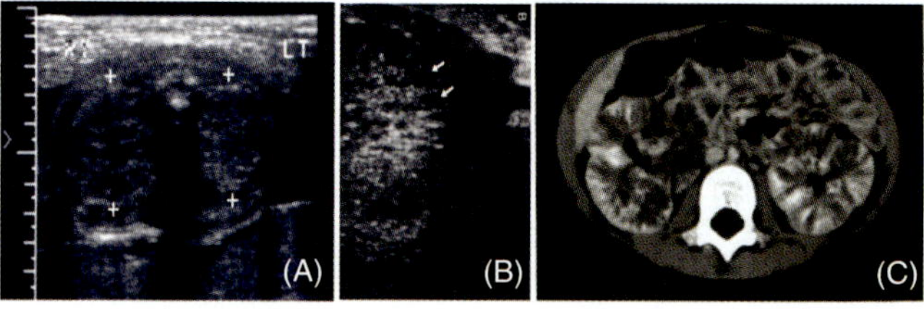

PLATE 38 Autosomal recessive polycystic kidney disease. In utero sonogram (A) of an ARPKD fetus at 26 weeks gestational age. Transverse view reveals enlarged, diffusely echogenic kidneys. ARPKD in a neonate (B). High resolution sonography reveals radially arrayed dilated collecting ducts (arrowheads). ARPKD in a symptomatic four year old girl (C). Contrast-enhanced CT shows a striated nephrogram and prolonged corticomedullary differentiation. (Modified from Guay-Woodford LM, Jafri Z and Bernstein J. Other cystic diseases. In: Comprehensive Clinical Nephrology. Johnson R and Feehally J (eds), 3rd edn.). (see also Figure 24.2)

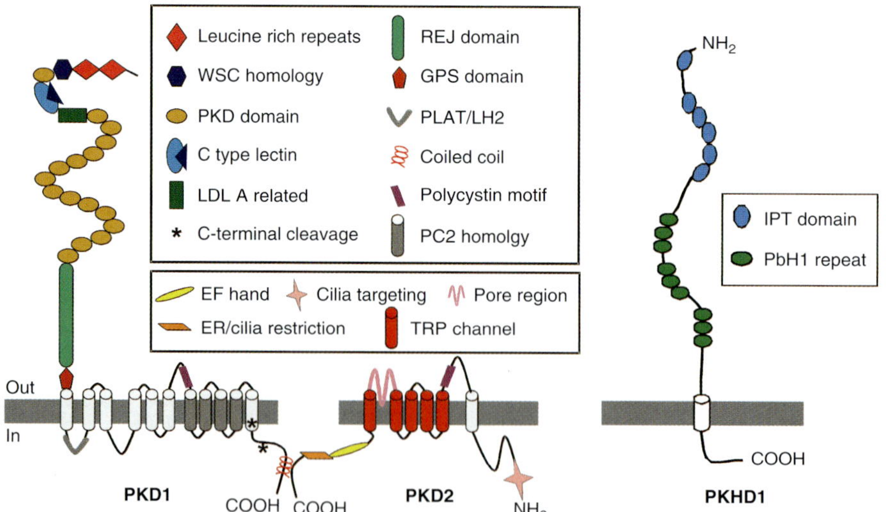

PLATE 39 Schematic representation of the predicted structural domains of *PKD1, PKD2* and *PKHD1* gene products. There are a combination of structurally predicted domains and experimentally identified or verified functional domains shown. In PC1, the immunoglobulin-like structure of the PKD domains have been determined by structural studies. Cleavage at the GPS site, as well as at two C-terminal cleavage locations, have been demonstrated. The REJ has been experimentally implicated in the autoproteolytic GPS cleavage process. The interaction of the COOH-terminal coiled coil domain of PC1 with the COOH-terminus of PC2 is well established. The TRP channel function of PC2 has been shown in several systems to be a non-selective cation channel permeable to Ca^{2+}. The EF-hand can bind Ca^{2+} in vitro (S. Somlo, unpublished observations). The functions of the cilia targeting and the ER/cilia restriction domains in PC2 have been defined. Polyductin/fibrocystin has a short cytoplasmic carboxyl terminus, single transmembrane (TM) domain, and a long extracellular domain with tandemly repeated immunoglobulin-like-plexin transcription factor (IPT) domains and a minimum of 9 parallel β-helix 1 (PbHl) repeats clustered within three groups between the last IPT domain and the TM domain. The carboxy terminus includes putative cAMP/cGMP-dependent protein kinase phosphorylation sites. (Modified from Somlo S, Torres VE and Caplan MJ. (2007). In: Seldin and Giebisch's The Kidney: Physiology & Pathophysiology, Alpern RJ and Hebert SC. Academic Press, pp. 2283–2314). (see also Figure 24.3)

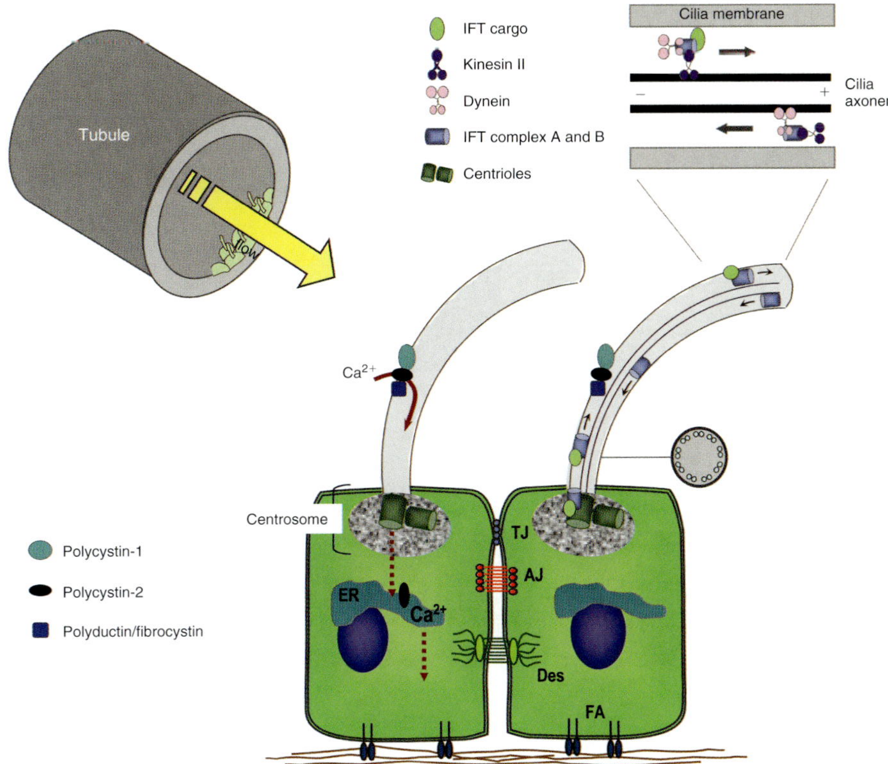

PLATE 40 Structure and function of cilia. Protein synthetic machinery is not present in cilia so proteins must enter cilia by intra-flagellar transport (IFT) from the cytosol. IFT involves the movement of large multi-protein complexes along the outer doublet microtubules beneath the ciliary membrane. Proteins destined for the ciliary membrane (e.g. the polycystins and polyductin) are synthesized in the endoplasmic reticulum, processed through the Golgi stack and traf-ficked into vesicles. These vesicles are transported to the base of the cilium comprised of the centrosome/basal body complex, where they interact with the IFT-particle proteins concentrated in a cytoplasmic pool. The vesicles fuse with the plasma membrane and the particles, and associated membrane are transported to the fiagellar tip by the plus strand hetertrimeric kinesin-II motor. Once at the tip, the kinesin-II is replaced by cytoplasmic dynein lb/2 which acts as the minus strand motor to carry the particles back to the cell body. (see also Figure 24.4)

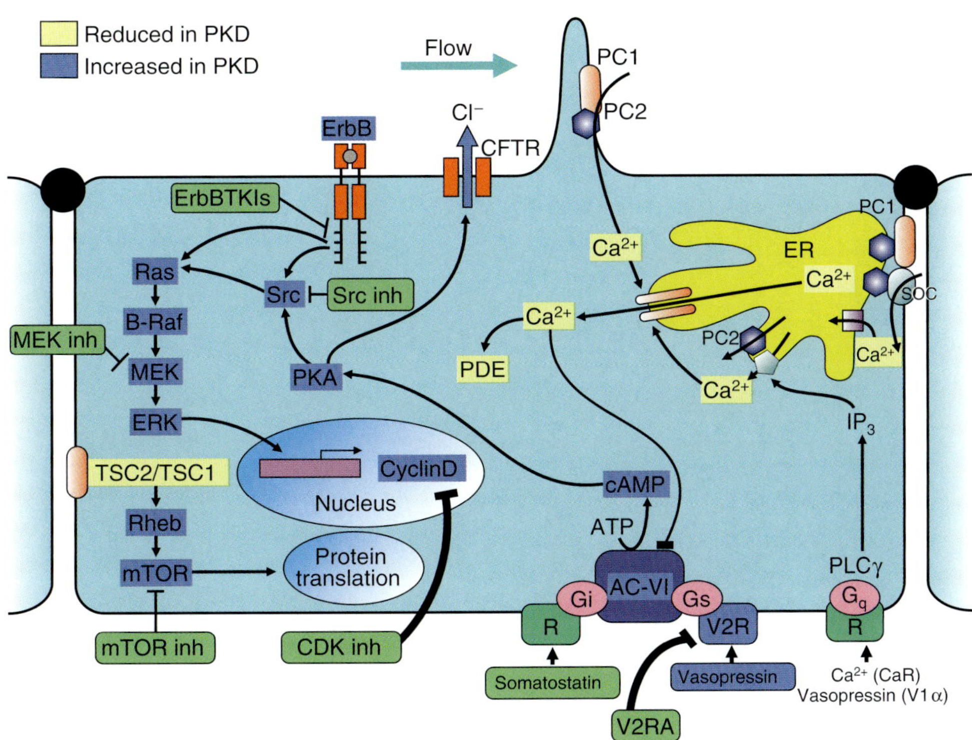

PLATE 41 Polycystin signaling pathways and PKD. Schematic representation of cellular signaling pathways that involve polycystins and related potential targets for therapy. The polycystin complex appears in the cilium where it may participate in flow stimulated Ca²⁺ entry; PC2 is abundant in the ER and intracellular endoplasmic reticulum (ER) Ca²⁺ stores are likely subject to regulation by the polycystins as well. There are direct or indirect interactions between the polycystin proteins and components of the MAPK/ERK, cAMP and mTOR pathways. These response pathways have come to be seen as potentially valuable targets in the search for PKD therapeutics. AC–VI = adenylyl cyclase 6. CDK = cyclin-dependent kinase. ER = endoplasmic reticulum. MAPK = mitogen-activated protein kinase. mTOR = mammalian target of rapamycin. PC1 = polycystin-1. PC2 = polycystin-2. PDE = phosphodiesterase. PKA = protein kinase A. R = receptor. TSC = tuberous sclerosis proteins tuberin (TSC2) and hamartin (TSCl). V2R = vasopressin V2 receptor. V2RA = vasopressin V2 receptor antagonists. (Modified from Torres VE, Harris PC, Pirson Y. Autosomal dominant polycystic kidney disease. Lancet 2007; 369: 1287–1301). (see also Figure 24.5)

PLATE 42 Genetics and pathophysiology of nephronophthisis and other cystic diseases of the kidney (CDK): Disease groups are color coded as they relate to Figure 25.2. Columns represent disease names, genes, encoded proteins, cell biological data, clinical phenotypes, and evolutionary conservation of genes in mouse, zebrafish, and *Cenorrhabditis elegans*. Human genes mutated in CDK are marked in different colors for PKD (green), MCKD and GCKD (purple), NPHP, JBTS, ALMS (blue), BBS, and OFD (orange). Colors are repeated in animal model columns if mutations in the respective gene also lead to CDK in mouse or zebrafish, or if the gene is expressed in head and tail of *C. elegans* ciliated neurons. Genes known so far to be mutated in CDK of animal models only (and not in human disease), are marked for mouse models (yellow) and zebrafish models (dark blue). Citations of the literature are shown in italic. (see also Table 25.1)

[a]Human cystic kidney disease	Human *gene* (inheritance)	Protein (domains)	Protein interaction partner	Subcellular localization	Human Renal disease (median ESRF)	Human extrarenal disease	Mouse CDK model	Mouse gene	Zebrafish CDK *gene* (phenotype)	*C. elegans* orthologous gene[b]
ADPKD1	*PKD1* (*Hughes et al 1995*) (AD)	Polycystin-1 (CC, multiple others)	[b]PKD2, [b]ATP2 (*Hu and Barr, 2005*), PACS1 and 2 (Kottgen et al 2005), [b]Siah-1 (*Kim et al 2004*)	Cilia, Golgi, FA, desmosomes	Cysts (≈55 yr)	Liver cysts, pancreatic cysts, cerebral aneurysms	*Pkd1−/−* (*Lu et al 1997*)	*Pkd1*	*pkd1* (*Kim et al 1999*)	*lov-1* (*Barr et al 2001*)
ADPKD2	*PKD2* (*Mochizuki et al 1996*) (AD)	Polycystin-2 (CC, multiple others)	[b]PKD1, [b]Hax-1 (*Gallagher et al 2000*), [b]PIGEA-14 (*Hidaka et al 2004*)	Cilia, Golgi, FA	Cysts (≈70 yr)	Liver cysts, pancreatic cysts	*Pkd2−/−* (*Wu et al 1998*) *Pkd1+/− Pkd2+/−* (*Wu et al 2002*) (rat model: pck)	*Pkd2*	*pkd2* (*Sun et al 2004*) (body curvature only)	*pkd-2* (*Barr et al 2001*)
ARPKD	*PKHD1* (*Ward et al 2002*) (AR)	Fibrocystin/ polyductin (multiple)	?	Cilia, cytoplasm (*Menezes et al 2004*)	Cysts, (42% at 20 yrs) (*Bergmann et al 2005*)	Liver fibrosis (>44%)		*Pkhd1*	?	?
GCDK	*HNF1β/ TCF2* (AD)	HNF1β	?			MODY diabetes type 5	*Hnf-1β* (*Hiesberger et al 2004*)	*Hnf-1β*	*vhnf1* (*Sun et al 2004*)	?
MCDK1	*UMOD* (*Hart et al 2002*) (AD/AR)	Uromodulin/ THP	?	Apical membrane	CM Cysts	Gout	*Umod−/−* (*Bachmann et al 2005*)	*Umod*	? (*crumbs* in fruit fly)	?
NPHP1	*NPHP1* (*Hildebrandt et al 1997a Saunier et al 1997a*) (AR)	NPHP1 (CC, SH3, E-rich, NPHP homology domain)	NPHP2/inversin, NPHP3, NPHP4, β-tubulin, [b]p130Cas, Pyk2, tensin, filamin A, filamin B	Cilia, BB, FA, AJ	NPHP (13 yr)	RP (10%), OMA (≈2%), CVA (1/157) (*Parisi et al 2004*)	*Nphp1−/−* (*Otto, unpubl.*)	*Nphp1*	*nphp1 mating phenotype*)	*nph-1*, M28.7
NPHP2	*INVS/NPHP2* (*Otto et al 2003b*) (AR)	Inversin/ NPHP2 (ANK, CC, NLS, SP)	NPHP1, [b]Apc2 [b]CALM, N-cadherin, catenins	Cilia, BB, CEN	CM cysts (<1 yr)	RP (10%), LF	*inv/inv* (*Mochizuki et al 1998, Morgan et al 1998*)	*Invs* (*Mochizuki et al 1998, Morgan et al 1998*)	*invs* (*Otto et al 2003b*) (CDK, laterality defect)	ZC15.7
NPHP3	*NPHP3*	NPHP3 (CC,	NPHP1	Cilia, retinal	CM NPHP	RP (10%), LF?	*pcy* (*Olbrich*	*Nphp3*	?	

NPHP4	*NPHP4* (AR)	NPHP4	NPHP1	Cilia, BB, AJ	CM NPHP (21 yr)	RP (10%), OMA, LF	–	*Nphp4*	?	*nph-4,* R13H4.1
NPHP5	*NPHP5* (AR)	NPHP5 (IQ, CC)	[b]CALM, RPGR	Cilia, retinal cilia	CM NPHP (13 yr)	RP with early onset (100%)	–	*Nphp5 (Otto, 2005)*	*zgc:101792*	?
JBTS3	*AHI1(Dixon-Salazar et al 2004 Ferland et al 2004)* (AR)	AHI protein (SH3, WD40, SH3-binding, CC)	?	?	CM NPHP	CA	–	*Ahi1*	?	*C14B1.4?*
ALMS1	*ALMS1 (Collin et al 2002 Hearn et al 2002)* (AR)	ALMS1 (CC)	?	CEN	NPHP, UTM	MR, RP, HG/IF, OB, PDM				
BBS1	*BBS1 (Mykytyn et al 2002)* (AR)	?	?	Olfactory cilia	NPHP, UTM	MR, RP, PD, HG/IF, OB, DM, CHD		*Bbs1*	?	*bbs-1 (Ansley et al 2003)*
BBS2	*BBS2 (Nishimura et al 2004)* (AR)	BBS2 (CC)	?	Olfactory and retinal cilia *(Nishimura et al 2004)*	NPHP, UTM	MR, RP, PD, HG/IF, OB, DM, CHD	*Bbs2−/− (Nishimura et al 2004)*	*Bbs2*	*bbs2*	*bbs-2 (Ansley et al 2003)*
BBS3	*BBS3 (Fan et al 2004),* ARL6 (AR)	ADP-ribosylation factor-like 6, ARF	?	Ciliated neurons	NPHP, UTM	MR, RP, PD, HG/IF, OB, DM, CHD	–	*Bbs3*	*zgc:101762*	C38D4.8
BBS4	*BBS4 (Mykytyn et al 2001)* (AR)	BBS4 (TPR)	?	Olfactory cilia, BB, CEN	NPHP, UTM	MR, RP, PD, HG/IF, OB, DM, CHD	*Bbs4−/− (Mykytyn et al 2004)*	*Bbs4*	*ttc8*	K04G7.3
BBS5	*BBS5 (Li et al 2004)* (AR)	BBS5	?	BB	NPHP, UTM	MR, RP, PD, HG/IF, OB, DM, CHD	–	*Bbs5*	*zgc:56578*	*bbs-5,* R01H10.6
BBS6, MKKS	*BBS6 (Katsanis et al 2000)* (AR)	BBS6, MKKS (chaperonin)	?	? BB	NPHP, UTM	MR, RP, PD, HG/IF, OB, DM, CHD	–	*Bbs6*	*zgc:55608*	?
BBS7	*BBS7 (Badano et al 2003)* (AR)	BBS7 (CC)	?	Ciliated neurons *(Blacque et al 2004)*	NPHP, UTM	MR, RP, PD, HG/IF, OB, DM, CHD	–	*Bbs7*	?	*bbs-7/osm-12 (Ansley et al 2003)*

(Continued)

[a]Human cystic kidney disease	Human *gene* (inheritance)	Protein (domains)	Protein interaction partner	Subcellular localization	Human Renal disease (median ESRF)	Human extrarenal disease	Mouse CDK model	Mouse *gene*	Zebrafish CDK *gene* (phenotype)	*C. elegans* orthologous *gene*[b]
BBS8	*BBS8 (Ansley et al 2003)* (AR)	BBS8, TTC8 (TPR)	?	Ciliated neurons *(Blacque et al 2004)*	NPHP, UTM	MR, RP, PD, HG/IF, OB, DM, CHD	–	*Bbs8*	?	*bbs-8 (Ansley et al 2003)*
OFD1	*OFD1 (Romio et al 2003)* (XLD)	OFD1 (CC)	?	Cilia, CEN *(Romio et al 2004)*	PKD	MR, PD, deafness, dental, hydroceph.	–	*Ofd1*	*ofd1*	?
?	*PKDR1*	Sam-Cystin					*cy/+ (rat)*	*Pkdr1*		?
?	?	?					*wpk (rat)*	*Wpk*		?
?	*NEK8*	NEK8	?	?	?	?	*Jck (Atala et al 1993)*	*Nek8*	*jck (Atala et al 1993) (CDK) LOC402785*	*ZC581.1R*
?	*BICC1*	bicaudal C homolog 1					*Bpk, Jcpk (Cogswell et al 2003)*	*Bicc1*		?
?	*NEK1*	NEK1	?	?	?	?	*Kat (Upadhya et al 2000 Upadhya et al 1999)*	*Nek1*	?	*Y39G10AR.3*
?		polaris	?	?	?	?	*Orpk (Moyer et al 1994), fxo (Huangfu et al 2003)*	*Tg737*	*ttc10*	*osm-5*
?	*KIF3A*	kinesin family member 3A	?	?	?	?	*Kif3A−/− (Lin et al 2003)*	*Kif3A*	*ENSDARG 00000021329*	*Klp-20, Y50D7A.6*
?	*KIF3B*	kinesin family member 3B	?	?	?	?	*Kif3B⁻/⁻*	*Kif3B*	*ENSDARG 00000012081*	*Klp-11, F20C5.2*
?	*TNS*	tensin	?	?	?	?	*Tns−/− (Lo et al 1997)*	*Tns*	*ENSDARG 00000024585*	*F46F11.3*
?	*ACE*	ACE	?	?	?	?	*Ace−/− (Carpenter et al 1996)*	*ACE*	*ENSDARG 00000002836*	*C42D8.5*
?	*EGFR*	EGFR	?	?	?	?	*Egfr −/−*	*Egfr*	?	?
?	*CYS*	cystin	?	?	?	?	*cpk*	*Cys1*	?	?
?	*CTNNB1*	β-catenin	?	?	?	?	*β-catenin over-expression*	*β-catenin*	*β-catenin*	?
?	*BCL2*	BCL-2	?	?	?	?	*Bcl-2 −/−*	*Bcl-2*	*bcl2l*	?

?	PLMP	prenyl transferase-like mitochondrial protein	?	?	?	?	kd	Plmp	?	?
?	CDX1	homoebox protein CDX-1	?	?	?	?	–	Cdx1	Cad1/caudal (Sun et al 2004)	?
?	ESRRBL1/ HIPPI	estrogen receptor β-like 1	?	?	?	?	–	Hippi	hi3417/IFT57 (Sun et al 2004)	IFT57 (Sun et al., 2004)
?	CDV1	CDV-1	?	?	?	?	–	Cdv1	hi409/IFT81 (Sun et al 2004)	IFT81 (Sun et al., 2004)
?	SLB	selective LIM binding factor	?	?	?	?	wim (Huangfu et al 2003)	Slb	Hi2211/IFT172 (Sun et al 2004)	IFT172 (Sun et al., 2004)
?	ARL2L1	ADP-ribosylation factor-like 2-like 1	?	?	?	?	–	Arl2l1	hi459/scorpion (Sun et al 2004)	?
?	RUVBL1	RuvB-like 1	?	?	?	?	–	?	hi1055b/pontin (Sun et al 2004)	?
?	RUVBL2	RuvB-like 2	?	?	?	?	–	?	reptin (Sun et al 2004)	?
?	LRRC6L		?	?	?	?	–	Lrrc6l	hi3308/seahorse (Sun et al 2004)	?
?	CLUAP1	Clusterin associated protein 1	BBS-7 (zebrafish)	?	?	?	–	Cluap1	hi3959/qilin (Sun et al 2004)	C04C3.5
?	?		?	?	?	?	–	?	hi1392/twister (Sun et al 2004)	?
?	PDLIM7	enigma protein	?	?	?	?	–	Pdlim7	hi2005/enigma (Sun et al 2004)	?

[a]ADPKD, autosomal dominant polycystic kidney disease; ARPKD, autosomal recessive polycystic kidney disease; NPHP, nephronophthisis; JBTS, Joubert syndrome; ALMS, Alstrom syndrome; BBS, Bardet-Biedl syndrome; MKKS, McKusick-Kaufman syndrome; OFD, oral-facial-digital syndrome
[b]Direct interaction.

AD, autosomal dominant; AJ, adherens junction; AR, autosomal recessive; BB, basal bodies; CC, coiled-coil; CEN, centrosomes; CHD (congenital heart defects); CVA, cerebellar vermis aplasia/hypoplasia; DM, diabetes mellitus; ESRF, end-stage renal failure; FA, focal adhesion; HG hypogenitalism; IF, infertility; MKKS, McKusick-Kaufman (hydrometrocolpos, postaxial polydactyly and congenital heart defects); MR, mental retardation; MTOC, microtubule organization center; NLS, nuclear localization signal; NPHP, nephronophthisis (tubular basement disruption, interstitial nephritis and fibrosis, cortico-medullary cysts); OMA, oculomotor apraxia type Cogan; PKD, polycystic kidneys; PD, polydactyly; RM, renal malformation; RP retinitis pigmentosa; SI, situs inversus; SP, signal peptide; TPR, tetratricopeptide repeat; TL, tyrosine ligase; UTM, malformations of the kidney and urinary tract; XLD, X-linked dominant

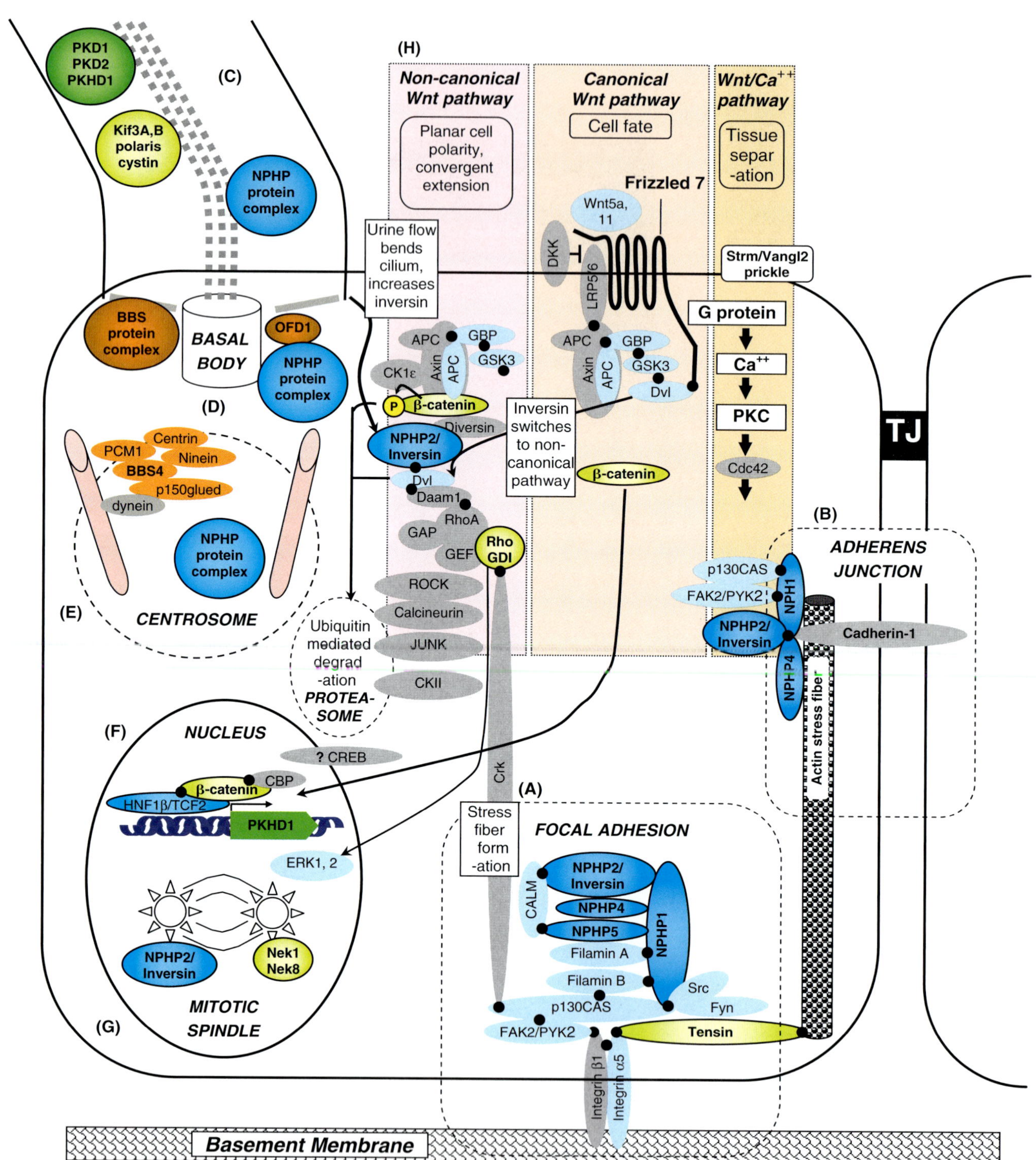

PKD1
PKD2
PKHD1
Kif3A,B
polaris
cystin
NPHP
protein
complex
(C)
BBS
protein
complex
BASAL
BODY
OFD1
NPHP
protein
complex
(D)
PCM1
Centrin
Ninein
BBS4
p150glued
dynein
NPHP
protein
complex
CENTROSOME
(E)
NUCLEUS
(F)
? CREB
β-catenin
CBP
HNF1β/TCF2
PKHD1
ERK1, 2
NPHP2/
Inversin
Nek1
Nek8
MITOTIC
SPINDLE
(G)
(H)
Non-canonical
Wnt pathway
Planar cell
polarity,
convergent
extension
Canonical
Wnt pathway
Cell fate
Wnt/Ca++
pathway
Tissue
separ
-ation
Frizzled 7
Wnt5a,
11
Urine flow
bends
cilium,
increases
inversin
DKK
LRP5/6
Strm/Vangl2
prickle
G protein
Ca++
PKC
Cdc42
APC
GBP
Axin
APC
CK1ε
GSK3
APC
GBP
Axin
APC
GSK3
Dvl
P
β-catenin
Diversin
Inversin
switches
to non-
canonical
pathway
β-catenin
NPHP2/
Inversin
Dvl
Daam1
RhoA
GAP
GEF
Rho
GDI
ROCK
Calcineurin
JUNK
CKII
Ubiquitin
mediated
degrad
-ation
PROTEA-
SOME
Crk
(A)
Stress
fiber
form
-ation
FOCAL ADHESION
NPHP2/
Inversin
CALM
NPHP4
NPHP5
Filamin A
Filamin B
p130CAS
FAK2/PYK2
NPHP1
Src
Fyn
Tensin
Integrin β1
Integrin α5
(B)
ADHERENS
JUNCTION
p130CAS
FAK2/PYK2
NPHP1
NPHP2/
Inversin
NPHP4
Cadherin-1
Actin stress fiber
TJ
Basement Membrane

PLATE 43   Localization of cystoproteins to subcellular locations that play a role in planar cell polarity (PCP) pathways (focal adhesion, adherens junction, cilia, basal body, centrosome, nucleus, mitotic spindle, and the Wnt pathways). Cystoproteins (proteins that are mutated in cystic kidney diseases) are shown on colored background and in bold type. Colors for groups of diseases or mouse models (yellow) match colors from Table 25.1. Proteins appear in non-bold if they have been described in the context of CDK pathogenesis (e.g. as a binding partner to a bona fide cystoprotein), but they have not been shown to be mutated in CDK. Associated proteins with no known role in CDK are shown on gray background. For references on protein function in the literature see Table 25.1. Black dots connect proteins that are direct interaction partners. (A) *Focal adhesions.* (i) Nephrocystin-1 directly interacts with the focal adhesion adapter protein p130Cas ('crk-associated sub-strate') (Donaldson et al 2000, 2002, Hildebrandt & Otto 2000), which is a major mediator of focal adhesion assembly, binds to focal adhesion kinase, and mediates stress fiber formation (Brugge 1998). Nephrocystin-1 competes for binding to p130Cas with the proto-oncogene products Src and Fyn (Donaldson et al 2000). (ii) Nephrocystin-1 is in a protein complex with the focal adhesion proteins Pyk2/Fak2 (focal adhesion kinase 2), tensin (Benzing et al 2001), and filamin A and B. Its overexpression leads to activation of extracellular regulated kinases 1 and 2 (ERK1 and ERK2) (Benzing et al 2001). (iii) In children with NPHP overexpression of $\alpha5\beta1$ integrin was described in proximal tubules, which most likely results from defective $\alpha6$ integrin expression (Rahilly & Fleming 1995). (iv) The knockout mouse models for tensin (Lo et al 1997) and for the Rho GDIα gene (Togawa et al 1999) both exhibit an NPHP-like phenotype, thereby implicating further proteins of the focal adhesion signaling cascade in the pathogenesis of NPHP-like diseases. (B) *Adherens junctions.* (i) Nephrocystin-1 co-localizes with E-cadherin and p130Cas to adherens junctions. (ii) The C-terminal half of nephrocystin, the 'nephrocystin homology domain,' is able to promote nephrocystin self-association and epithelial adherens junctional targeting (Donaldson et al 2002). (iii) Disruption of this targeting leads to reduced transepithelial resistance (Donaldson et al 2002). (iv) Nephrocystin-4 is in a protein complex with nephrocystin-1, NPHP2/inversin, p130Cas, and Pyk2/Fak2, and has been localized to adherens junctions in confluent MDCK cells (Mollet et al 2005). (C) *Primary cilia.* Recently, the development of a unifying hypothesis of renal cystogenesis has been established (Watnick & Germino 2003). This hypothesis states that proteins that, if mutated, cause renal cystic disease in humans, in mice or in zebrafish, are part of a functional module, as defined by their subcellular localization to primary cilia, basal bodies or centrioles. This applies to polycystin-1 and -2, fibrocystin/ polyductin, nephrocystin-1, -2 (inversin), -3, -4, -5, Bardet-Biedl syndrome (BBS)-associated proteins, cystin, polaris, ALMS1, OFD1, and many others (Table 25.1). Since nephrocystins interact, they are represented here as the 'nephrocystin complex.' Based on positional clon-ing, mutations in the inversin gene were identified as causing infantile NPHP (type 2) (Otto et al 2003b). This established a link between the pathogenesis of NPHP and disease mechanisms of polycystic kidney disease (Otto et al 2003a). Within these studies it was found that nephrocystin-1 interacts with both inversin and with β-tubulin. Since β-tubulin is a major component of primary cilia, this led to demonstra-tion of colocalization of all three proteins in the primary renal cilia of epithelial cells (Otto et al 2003a). The complex also contains NPHP3, which is mutated in adolescent NPHP (type 3) and in the renal cystic mouse model pcy. The ciliary hypothesis of CDL was confirmed, by revealing that Nphp3 mRNA was expressed in kidney, retinal connecting cilia, ciliated bile ducts, and the node, which regulates left–right body axis in mice (Olbrich et al 2003), and by identifying mutations in nephrocystin-5 (IQCB1), which colocalizes with calmodulin to primary cilia of renal epithelial cells and retinal connecting cilia (Otto et al 2005). (D) *Basal bodies.* Basal bodies are short cylindrical arrays of micro-tubules and other proteins, found at the base of a cilia, that organize the assembly of the ciliary axoneme. They are analogous to centrosomes. Proteins mutated in Bardet-Biedl syndrome (BBS) are components of the basal body transitional zone and are highly conserved in evolution. (E) *Centrosomes.* For a protein mutated in the related disease Bardet-Biedl syndrome type 4 (BBS4) it was shown that BBS4 is instrumental in recruiting proteins to the pericentrosomal matrix, confirming the role of centrosomal function in the pathogenesis of BBS (Ansley et al 2003). (F) *Transcriptional programs.* The transcription factor HNF1β regulates transcription of multiple CDK genes (Hiesberger et al 2005). (G) *Cell cycle regulation.* The hypotheses of functional involvement of the nephrocystin complex at focal adhesions and adherens junction on one hand and the ciliary/centrosomal hypotheses on the other hand may be integrated by demonstration that different locations of the com-plex may predominate in a cell cycle-dependent manner. This is evidenced by the findings that inversin/nephrocystin-2 and nephrocystin-4 expression occurs in a cell-cycle-dependent manner. Inversin exhibits a dynamic expression pattern in MDCK cells showing expression at centrosomes in early prophase, at spindle poles in metaphase and anaphase, and at the midbody in cells undergoing cytokinesis (Morgan et al 2002a, Nurnberger et al 2002). Nephrocystin-4 was detected in MDCK cells at centrosomes of dividing cells, and in polarized cells close to the cytoskeleton and in the vicinity of the cortical actin cytoskteton, with colocalization of p130Cas, Pyk2, and β-catenin (Mollet et al 2005). (H) *Wnt pathways.* Recent data demonstrated that in renal tubules inversin may induce switching from the canonical to the non-canonical Wnt signaling pathway in response to flow sensing by primary cilia of renal tubular cells (Simons et al 2005). It is thought that this function of inversin is important to maintain renal epithelial cell polarity (Germino 2005). The hypothesis that the renal cystic disease phenotype is due to defects in the maintenance of planar cell polarity seems plausible at present for multiple reasons: (i) it would reconcile previous func-tional hypotheses, since focal adhesions, adherens junctions, cilia, centrosomes, and basal bodies, as well as regulation of the cell cycle, all play a pivotal role in the regulation and maintenance of planar cell polarity (Keller 2002); (ii) since planar cell polarity plays an important role in developmental morphogenesis and also in the regeneration of differentiated tissue, a defect in planar cell polarity may explain both occurrence of cysts during organogenesis and degenerative cystogenesis as it occurs in nephronophthisis; (iii) the mechanism of convergent extension, which may be central to renal tubular morphology, was shown to be disturbed in many renal cystic diseases (Ross et al 2005). APC, adenomatous polyposis coli protein; APC2, anaphase promoting complex subunit 2; CBP, CREB binding protein; CREB, cAMP response element binding protein; CK1ε, casein kinase epsilon; DKK, Dickkopf-related protein; CK2, casein kinase II; GAP, RhoGTPase activating protein; GEF, guanine nucleotide exchange factor; GSK3, glycogen synthase kinase 3; JUNK, JUN kinase; PCM1, pericentriolar material 1; p150, p150glued/dynactin-1; PKC, protein kinase C; ROCK, Rho-associated protein kinase. (see also Figure 25.2)

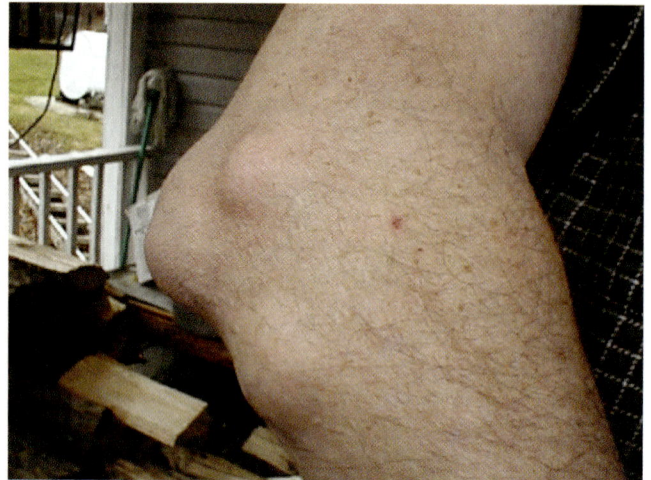

PLATE 44   (see also Figure 26.1)

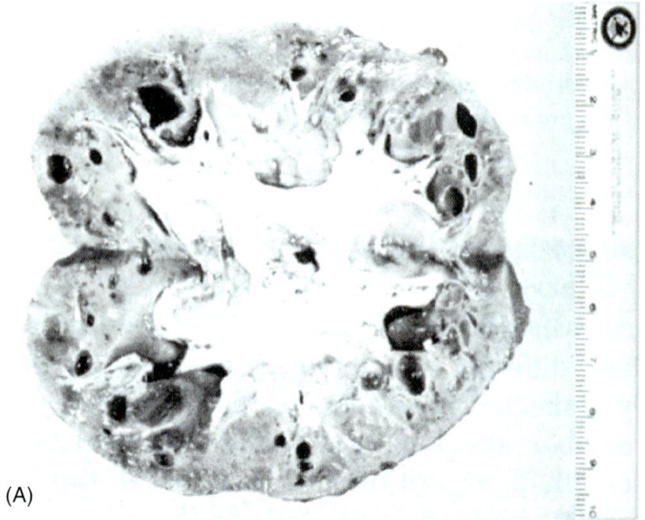

(A)

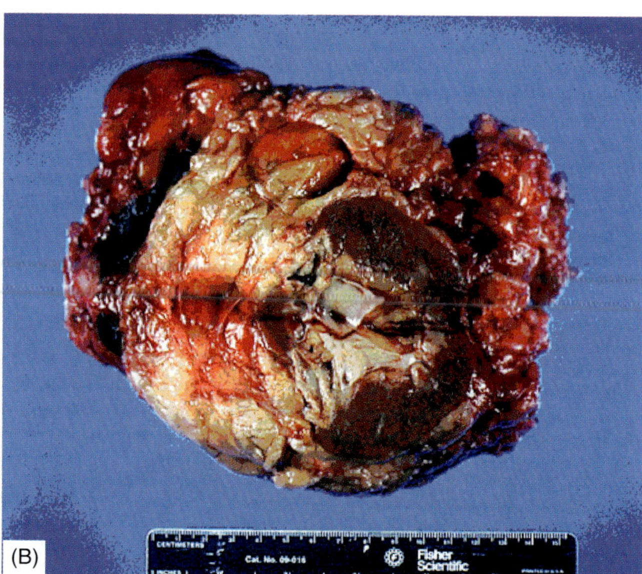

(B)

PLATE 45   (A) Gross findings in a patient with medullary cystic kidney disease. Note cysts at cortico-medullary junction. Cysts are actually an inconsistent finding in medullary cystic kidney disease. Adapted from Goldman SH, Walker SR, Merigan TC, Gardner KD, Bull JMC. Hereditary occurrence of cystic disease of the renal medulla. N. Engl. J. Med. 1966; 274: 984–92. (B) Gross kidney specimen from patient with MCKD2. (see also Figure 26.3)

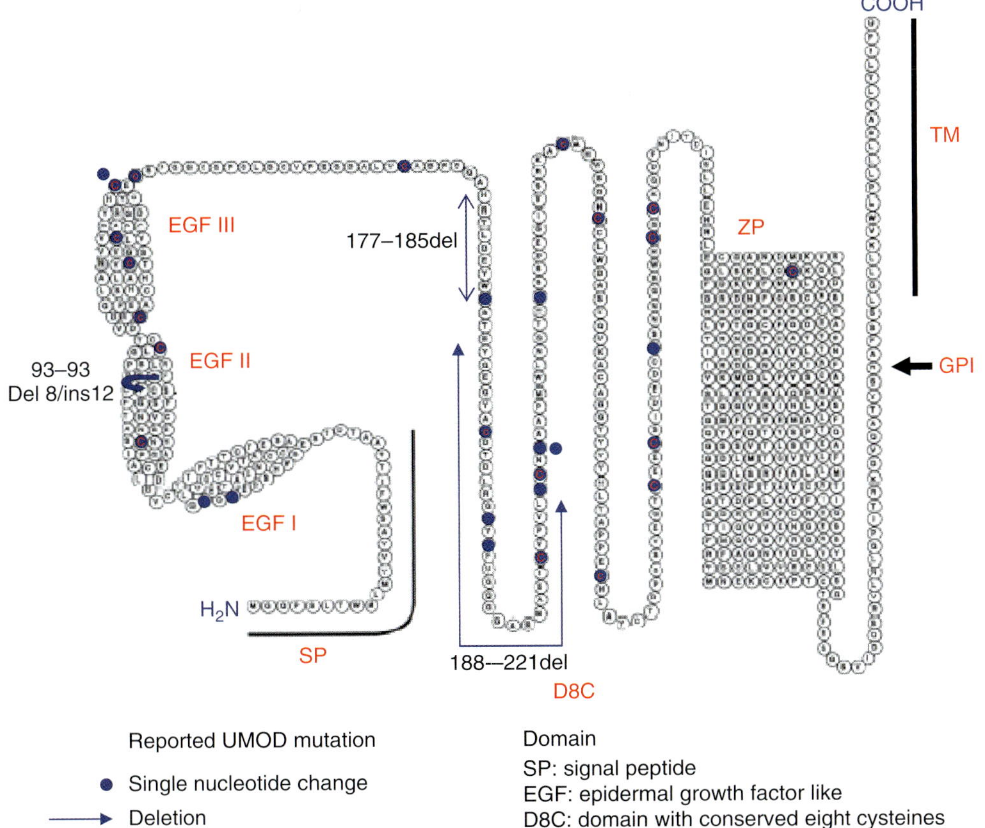

PLATE 46   Reported amino acid changes caused by UMOD mutations showing affected domains. (see also Figure 26.4)

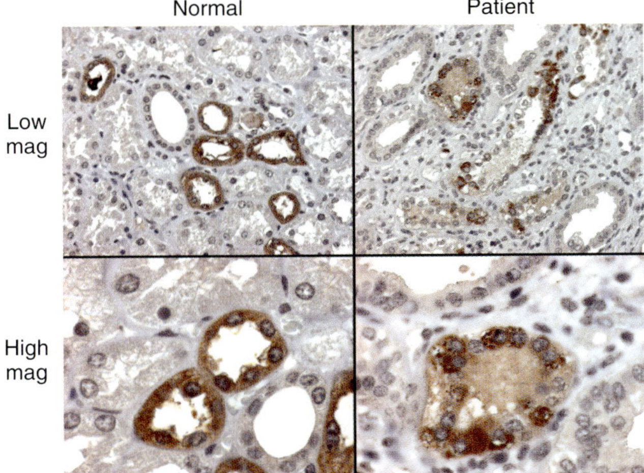

PLATE 47   Uromodulin deposits seen in the thick ascending limb. Immunohistochemical staining with antibody to Tamm–Horsfall protein (uromodulin). In the normal section, antibody staining is diffuse throughout all tubular cells. In contrast, in the patient with a uromodulin mutation, tubular atrophy is present (see low magnification). Under high magnification, antibody staining reveals irregular, dense deposits in tubular cells. (Top panels magnified $\times 100$, bottom panels magnified $\times 400$). (see also Figure 26.5)

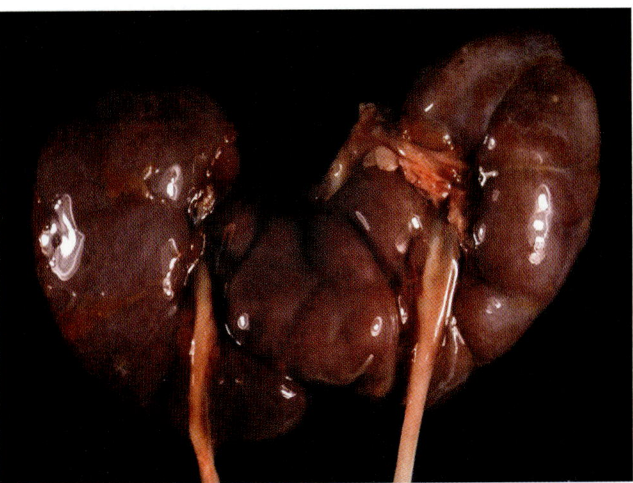

PLATE 48   Horseshoe kidney. Horseshoe kidney formed by fusion of two kidney moieties, each drained by a separate ureter. Photo courtesy of Arthur Weinberg. (see also Figure 27.1)

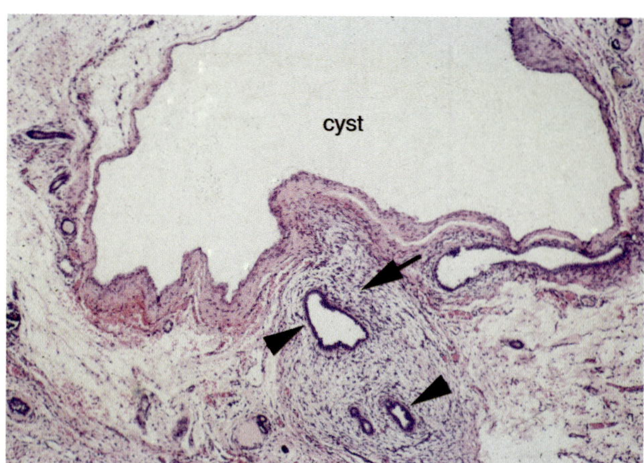

PLATE 49 Renal dysplasia. Histology of a dysplastic kidney showing typical features including large cysts lined by flattened epithelium and primitive ducts (arrowheads) surrounded by fibromuscular collars (arrow). Photo courtesy of Arthur Weinberg. (see also Figure 27.2)

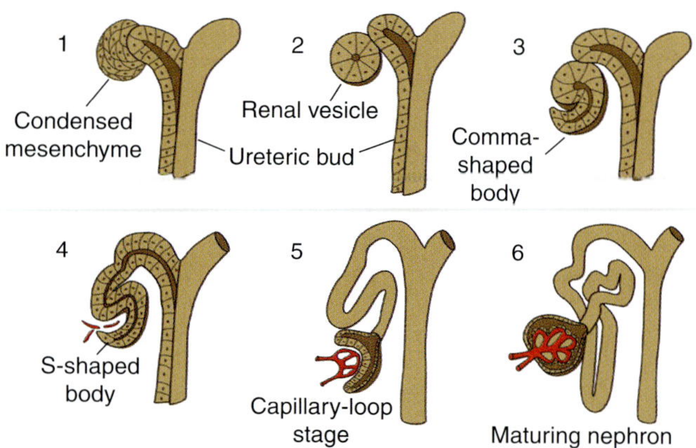

PLATE 51 Stages of nephrogenesis. 1) Condensed mesenchyme. 2) Renal vesicle. 3) Comma-shaped body. 4) S-shaped body. 5) Capillary-loop stage. 6) Maturing nephron. (see also Figure 27.4)

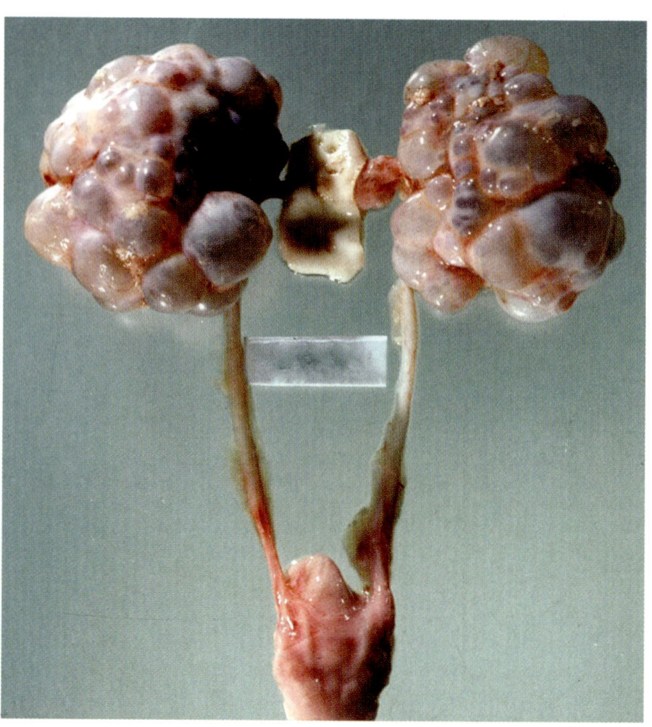

PLATE 50 Bilateral multicystic dysplastic kidneys. Gross appearance of multicystic dysplastic kidneys showing numerous fluid-filled cysts. Photo courtesy of Arthur Weinberg. (see also Figure 27.3)

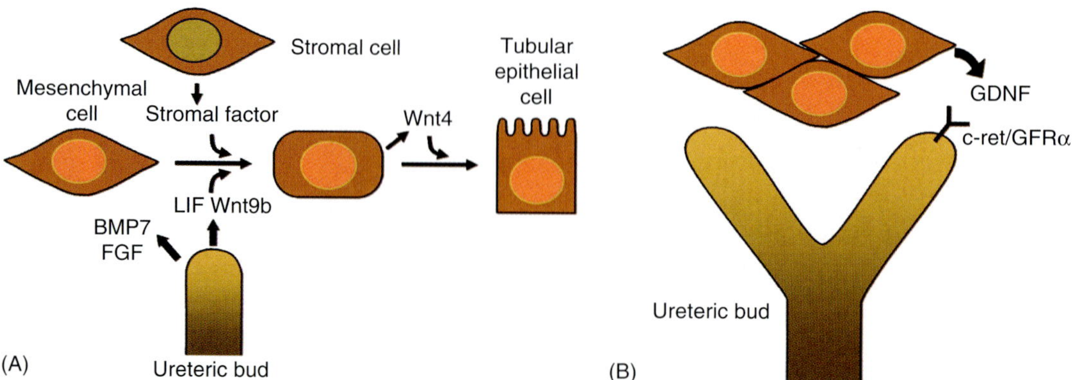

PLATE 52 Molecular regulation of kidney development. (A) Nephron induction. The ureteric bud secretes leukemia inhibitory factor (LIF) and Wnt9b, which induce metanephric mesenchymal cells to differentiate into renal epithelial cells. Wnt4 and stromal factors are also required. BMP7 and FGF are required for cell survival and proliferation. (B) Branching morphogenesis of the ureteric bud. Metanephric mesenchymal cells secrete glial cell-derived neurotrophic factor (GDNF), which binds to its receptor, c-ret/GFRα, located on the ureteric bud. Deletion of either GDNF or c-ret prevents ureteric branching. (see also Figure 27.5)

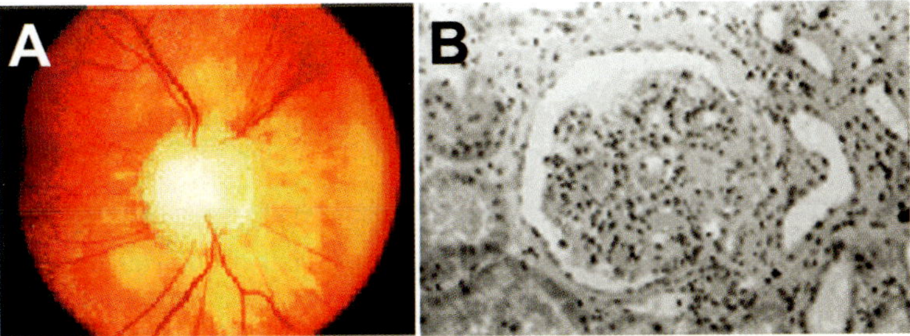

PLATE 53 Optic nerve coloboma (A) and renal histology (B) of a patient with renal-coloboma syndrome. The optic nerve head contains a large excavation, and retinal vessels emerge from the periphery. The glomerulus shows mesangial sclerosis. Reproduced with permission from Schimmenti et al (1995). (see also Figure 27.6)

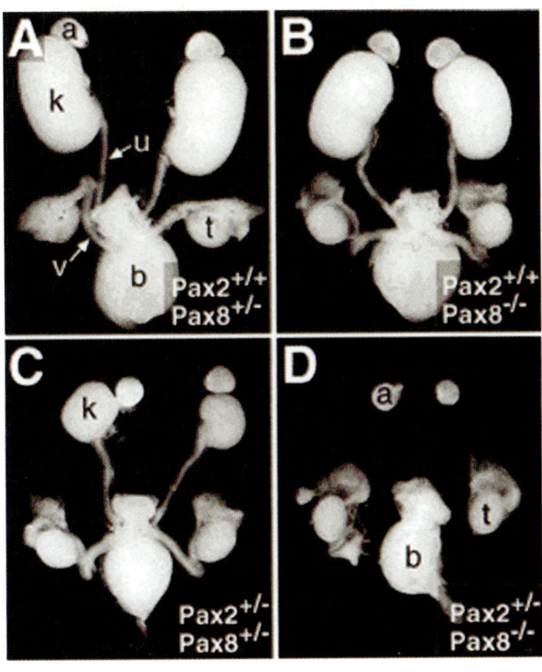

PLATE 54 Renal hypoplasia in *Pax2/Pax8* mutant mice. The kidneys in $Pax2^{+/-}Pax8^{+/-}$ embryos (C) are smaller than control kidneys (A, B). $Pax2^{+/-}Pax8^{-/-}$ embryos (D) fail to develop a kidney, ureter, and genital tract, whereas the adrenal gland, testis, and bladder form normally. a, adrenal gland; b, bladder; k, kidney; t, testis; u, ureter; v, vas deferens. Reproduced with permission from Bouchard et al (2002). (see also Figure 27.7)

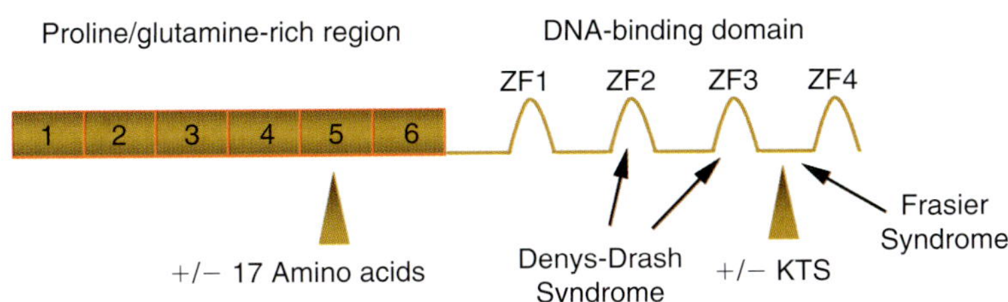

PLATE 55 Mutations of WT1 in Denys-Drash and Frasier syndromes. Structure of WT1 is depicted showing the amino-terminal proline-glutamine-rich region and four zinc finger (ZF) DNA-binding domains. Locations of alternative splicing are indicated by the arrowheads. Arrows indicate the locations of mutations causing Denys-Drash syndrome and Frasier syndrome. (see also Figure 27.8)

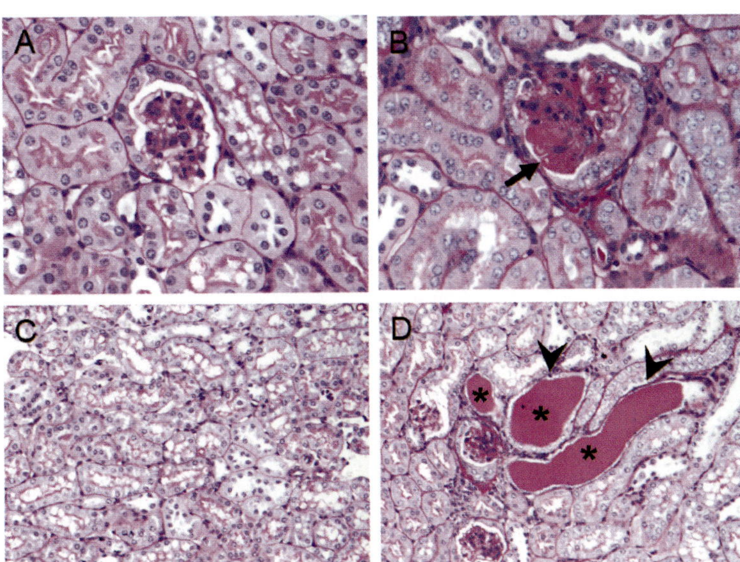

PLATE 56 Mouse model of Denys-Drash syndrome. Normal glomerulus in wild-type mice (A) and glomerulosclerosis (arrow) in $Wt1^{+/R394W}$ mice (B). Renal tubules are dilated (arrowheads) and contain proteinaceous casts (asterisks) in $Wt1^{+/R394W}$ mice (D) compared to wild-type mice (C). Kidneys are stained periodic acid and Schiff's (PAS). Photo courtesy of Vicki Huff and Hao Zhang. (see also Figure 27.9)

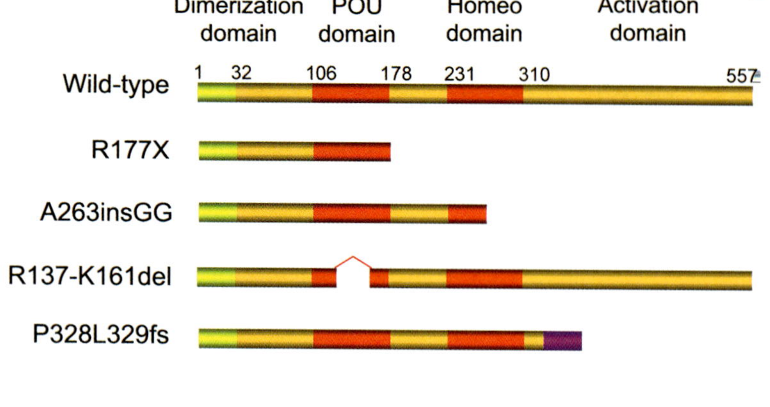

PLATE 57 Mutations of TCF2 in MODY5. Upper panel shows wild-type HNF-1β containing an amino-terminal dimerization domain, DNA-binding POU/homeodomain, and carboxy-terminal activation domain. Four representative mutations of HNF-1β in families with MODY5 are shown. Most mutations affect the POU-homeodomain and prevent DNA binding. The P328L329fsdelCCTCT mutant (bottom) contains an intact DNA-binding domain but lacks the activation domain. Pink box indicates ectopic amino acids introduced by the frameshift mutation. (see also Figure 27.11)

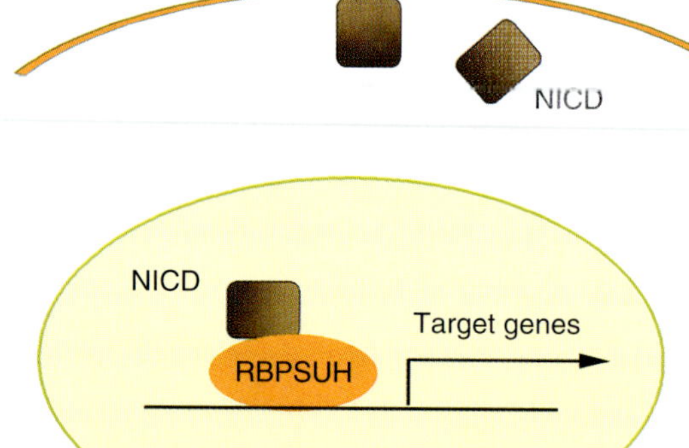

PLATE 58 Notch signaling pathway. Notch family receptors (NOTCH1–NOTCH4) are expressed on the cell surface and exist as a proteolytically cleaved extracellular domain tethered to a membrane domain. Binding of ligands of the Jagged (JAG1 and JAG2) and Delta-like (DLL1, DLL3, DLL4) families induces two additional proteolytic cleavages that release the Notch intracellular domain (NICD). The Notch intracellular domain translocates to the nucleus where it forms a complex with RBPSUH, displacing a histone deacetylase/corepressor complex (not shown), leading to the transcriptional activation of Notch target genes. (see also Figure 27.13)

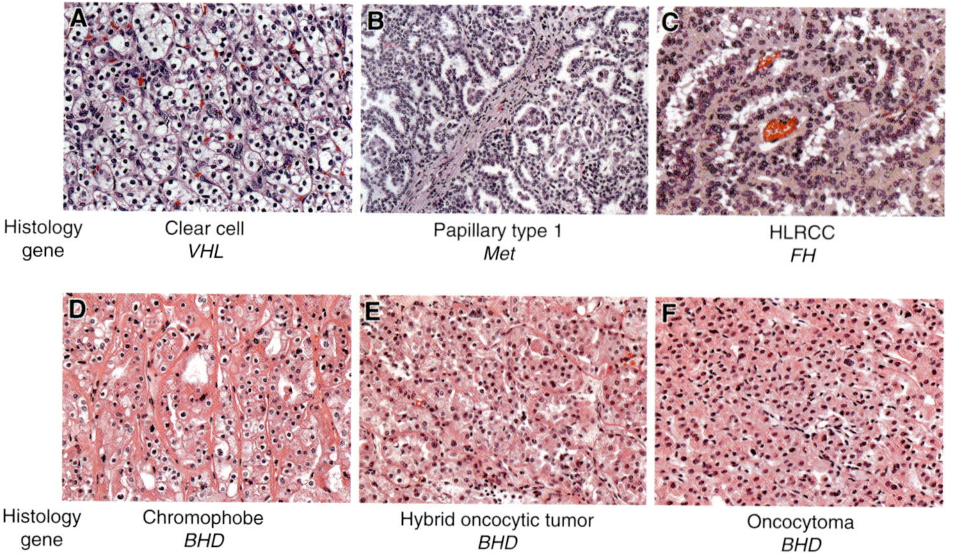

PLATE 59 Hereditary cancer syndromes display distinct histology. Von Hippel-Lindau (VHL) is associated with clear cell RCC (A). Hereditary papillary renal cancer (HPRC) patients develop papillary type I RCC (B), while patients with hereditary leiomyomatosis and renal cell cancer (HLRCC) develop characteristic tubulopapillary tumors suggestive of type II papillary RCC (C). Birt-Hogg-Dubé (BHD)-associated tumors may have multiple histologies, including chromophobe (D), hybrid oncocytic tumors (E), and oncocytoma (F). (see also Figure 28.1)

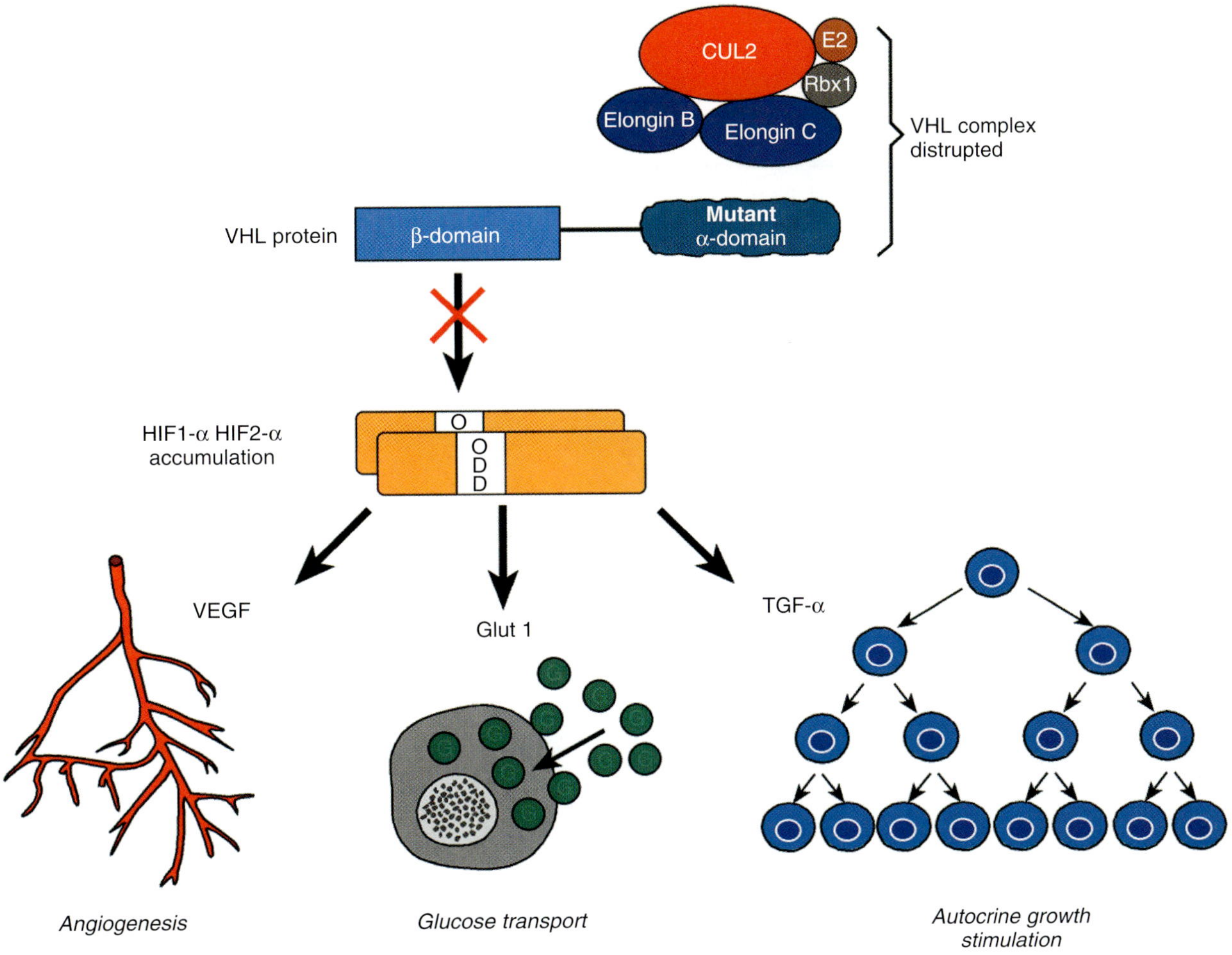

PLATE 60   Molecular abnormalities of VHL. Absence of functional *VHL* gene product leads to decreased degradation of HIF-1α and subsequent transcription of downstream hypoxia-inducible genes such as VEGF, EGFR, Glut-1 and TGF-α. Adapted from Linehan et al (2004). (see also Figure 28.3)

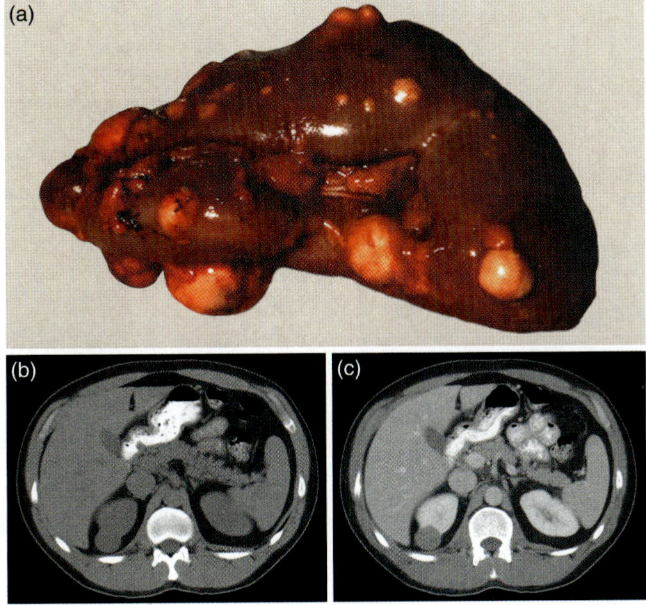

PLATE 61   Hereditary papillary renal cancer. Patients with HPRC may form multiple bilateral tumors (A) (adapted from Linehan et al 2003). Papillary renal cancers may appear as hypovascular lesions on CT imaging (B). Adapted from Grubb et al (2005). (see also Figure 28.4)

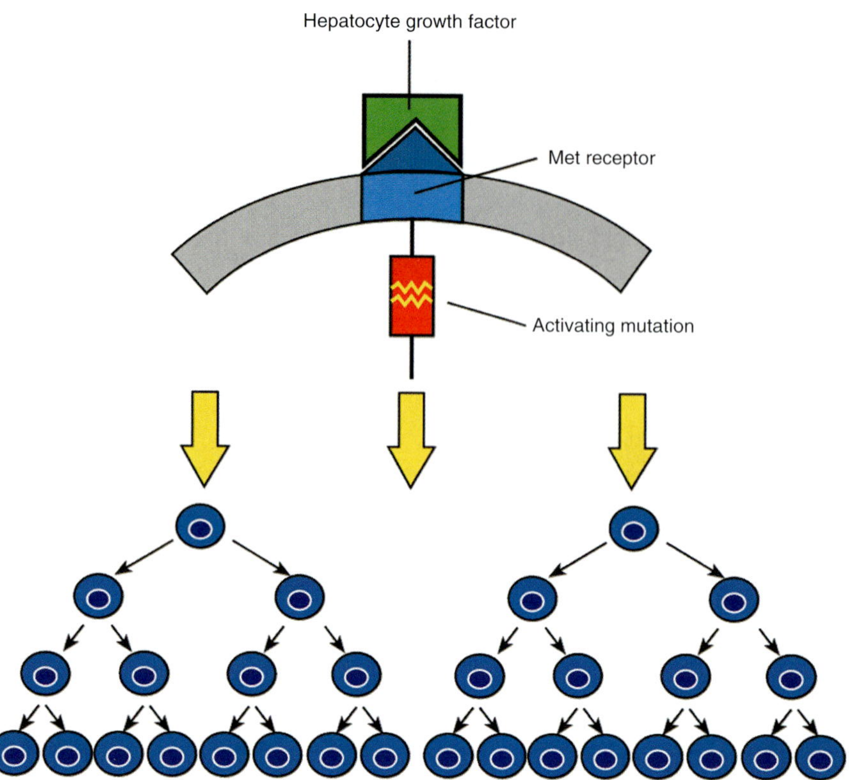

PLATE 62 HPRC tumorigenesis results from activating mutations in *MET*. Adapted from Linehan et al (2003). (see also Figure 28.5)

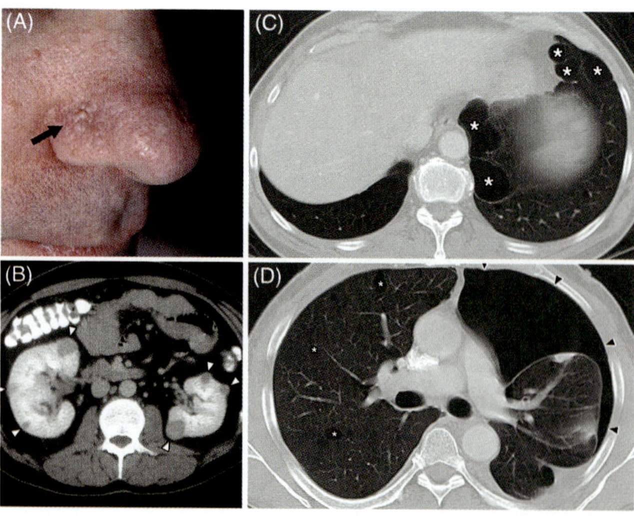

PLATE 63 BHD manifestations. The characteristic lesion of Birt-Hogg-Dubé (BHD) is the cutaneous fibrofolliculoma, a benign hair follicle tumor of the face and upper trunk (A, arrow). Patients are also at risk for forming multifocal and bilateral renal tumors (B, arrowheads). Patients are also susceptible to form pulmonary cysts (C, D, asterisks) and spontaneous pneumothoraces (D, arrowheads). (see also Figure 28.6)

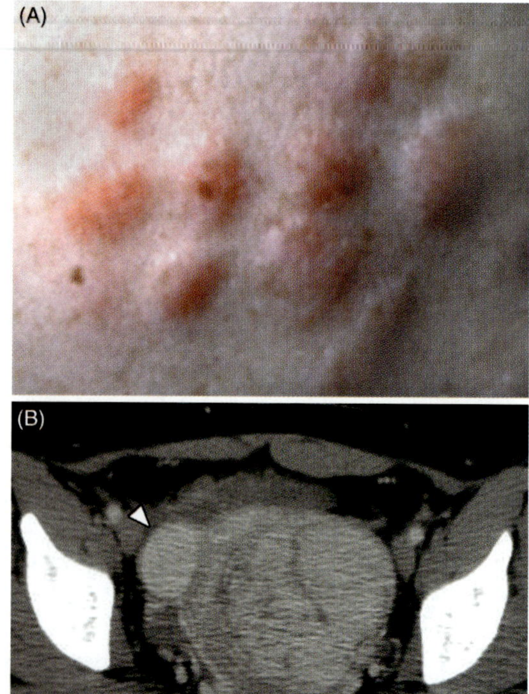

PLATE 64 Cutaneous and uterine leiomyomas in HLRCC. The characteristic cutaneous lesion of HLRCC is the leiomyoma (A) (from Linehan et al 2003). Uterine leiomyomas are found in up to 98% of female patients affected with HLRCC (B, arrowhead). (see also Figure 28.7)

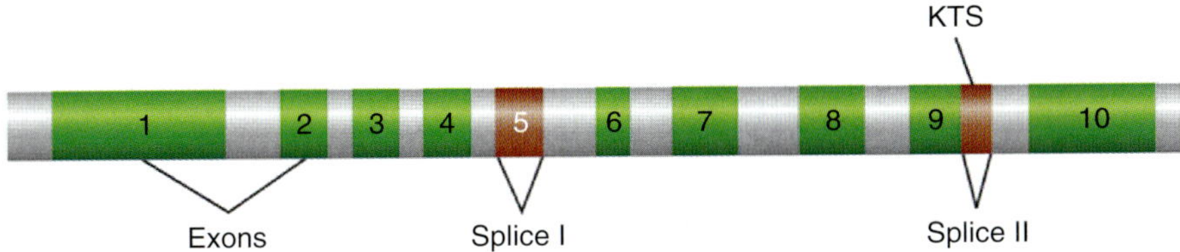

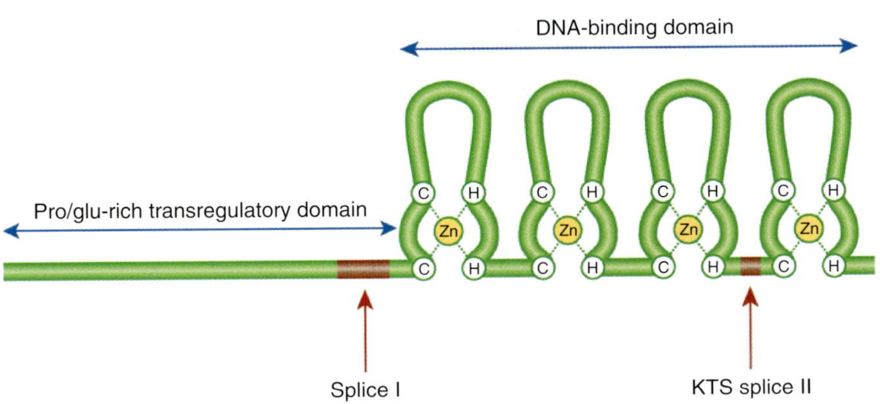

PLATE 65 Structure of *WT1* mRNA and protein. Red boxes represent nucleotides or amino acids included as the result of alternative mRNA splicing. (see also Figure 29.1)

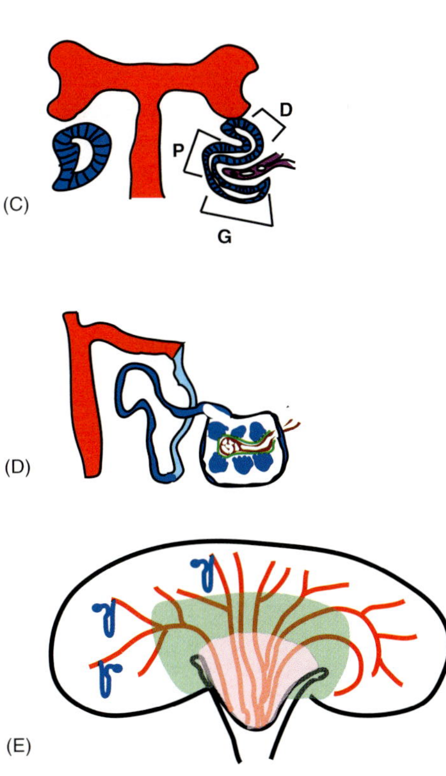

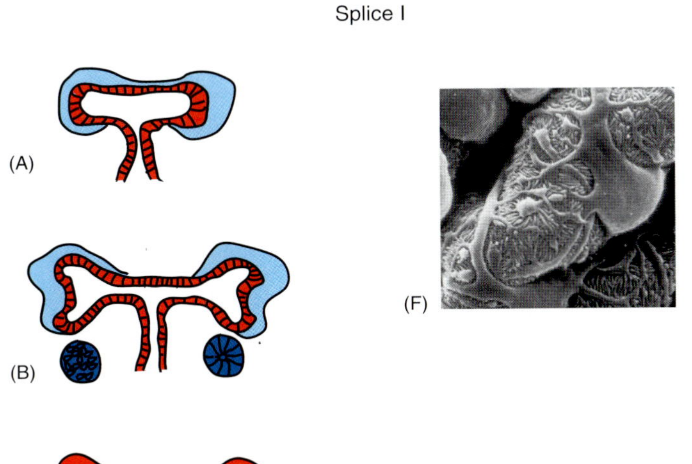

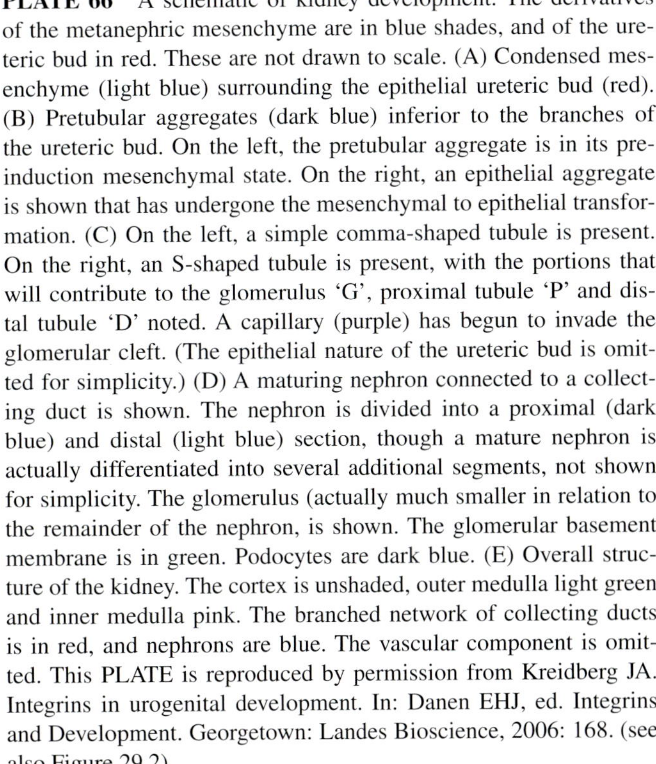

PLATE 66 A schematic of kidney development. The derivatives of the metanephric mesenchyme are in blue shades, and of the ureteric bud in red. These are not drawn to scale. (A) Condensed mesenchyme (light blue) surrounding the epithelial ureteric bud (red). (B) Pretubular aggregates (dark blue) inferior to the branches of the ureteric bud. On the left, the pretubular aggregate is in its pre-induction mesenchymal state. On the right, an epithelial aggregate is shown that has undergone the mesenchymal to epithelial transformation. (C) On the left, a simple comma-shaped tubule is present. On the right, an S-shaped tubule is present, with the portions that will contribute to the glomerulus 'G', proximal tubule 'P' and distal tubule 'D' noted. A capillary (purple) has begun to invade the glomerular cleft. (The epithelial nature of the ureteric bud is omitted for simplicity.) (D) A maturing nephron connected to a collecting duct is shown. The nephron is divided into a proximal (dark blue) and distal (light blue) section, though a mature nephron is actually differentiated into several additional segments, not shown for simplicity. The glomerulus (actually much smaller in relation to the remainder of the nephron, is shown. The glomerular basement membrane is in green. Podocytes are dark blue. (E) Overall structure of the kidney. The cortex is unshaded, outer medulla light green and inner medulla pink. The branched network of collecting ducts is in red, and nephrons are blue. The vascular component is omitted. This PLATE is reproduced by permission from Kreidberg JA. Integrins in urogenital development. In: Danen EHJ, ed. Integrins and Development. Georgetown: Landes Bioscience, 2006: 168. (see also Figure 29.2)

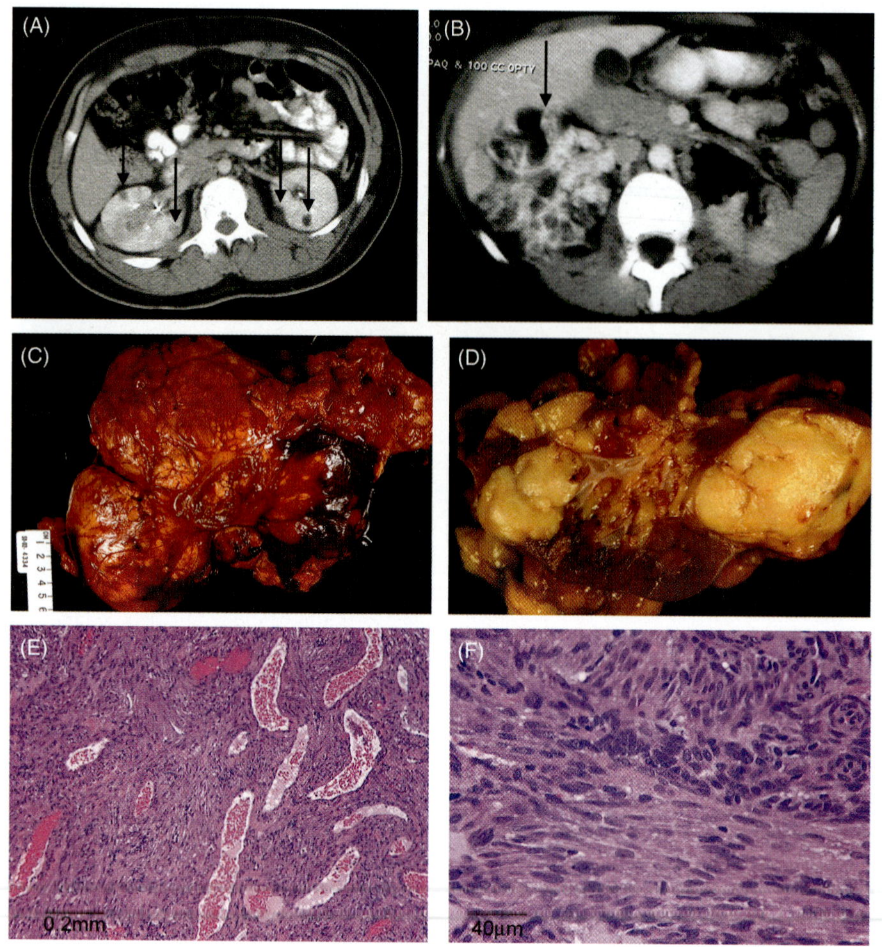

PLATE 67 Renal angiomyolipoma. A, B. CT scans following contrast administration. A. Four small AML with variable fat and vascular content (arrows). B. Extensive AML involving the right kidney. Note areas of both high enhancement reflecting high vascularity and regions of low enhancement reflecting fat. Arrow points to kidney. C, D. Gross pictures of a massively enlarged and distorted kidney due to extensive involvement by AML. Yellowish material is fatty AML. B, external view; C, after sectioning. E, F. Histologic sections of AML. Note numerous vascular channels in E; and proliferating spindle-shaped SMCs in F. Images courtesy of John Bissler, Frank McCormack, Elizabeth P. Henske, Nisreen El-Hashemite. (see also Figure 30.1)

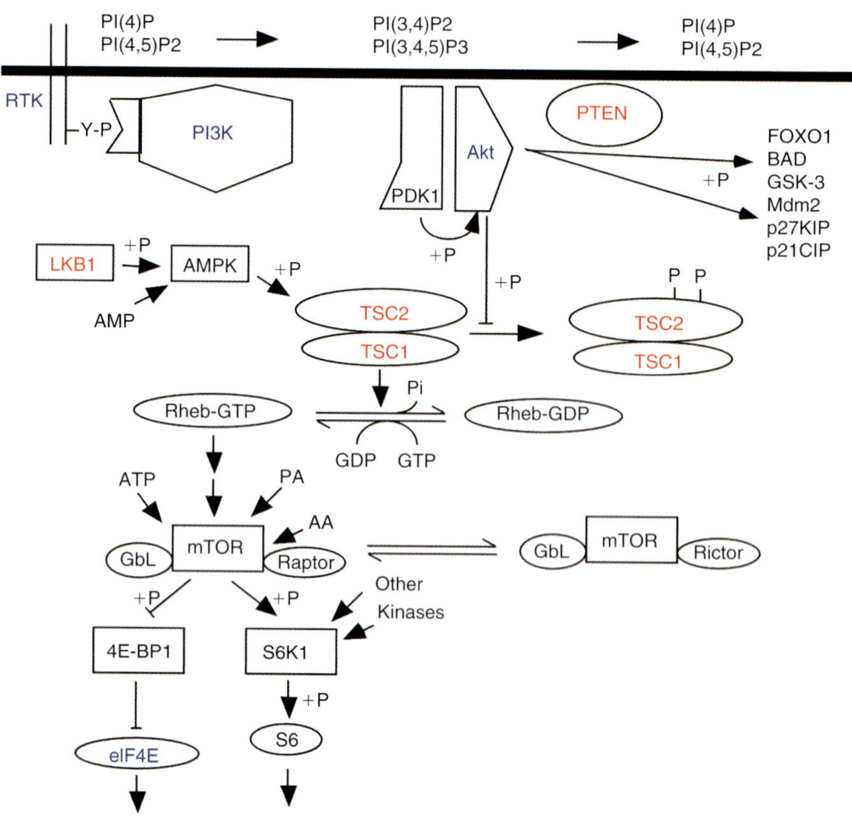

PLATE 68 The PI3K-Akt-TSC1/TSC2-mTOR signaling pathway. This PLATE portrays a portion of a more complex pathway. Receptors illustrated by the parallel lines at left undergo tyrosine phosphorylation (Y-P), which leads to recruitment of PI3kinase, which phosphorylates phosphoinositides at the 3' position to generate $PI(3,4)P_2$ and $PI(3,4,5)P_3$. These molecules lead to recruitment of Akt to the membrane which is then activated by phosphorylation by PDK1, and phosphorylates many downstream targets including TSC2. Phosphorylation of TSC2 reduces its GAP activity, leading to high levels of Rheb-GTP, which through unknown mechanisms positively activates mTOR. mTOR then phosphorylates the S6Ks and 4E-BP1, leading to enhanced protein translation and cell growth. LKB1 connects with this pathway by phosphorylating AMPK in the presence of high AMP/ATP levels, and pAMPK then phosphorylates TSC2 and activates its GAP activity. See text for additional details. (see also Figure 30.3)

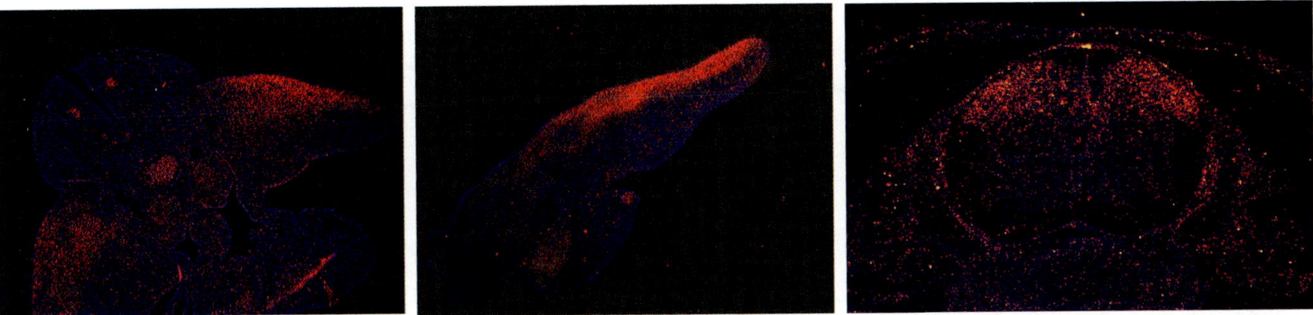

PLATE 69   Temporal-spatial of *Lmx1b* expression during mouse development. RNA in situ hybridization of *Lmx1b* in transverse sections at embryonic day 11.5 (left panel) and dorso-ventral sections of the forelimb at embryonic day 12.5 (middle panel), both showing strong expression in dorsal mesenchyme. *Lmx1b* expression in transverse section of the spinal cord at embryonic day 15.5 showing dorsal expression (right panel). (see also Figure 31.2)

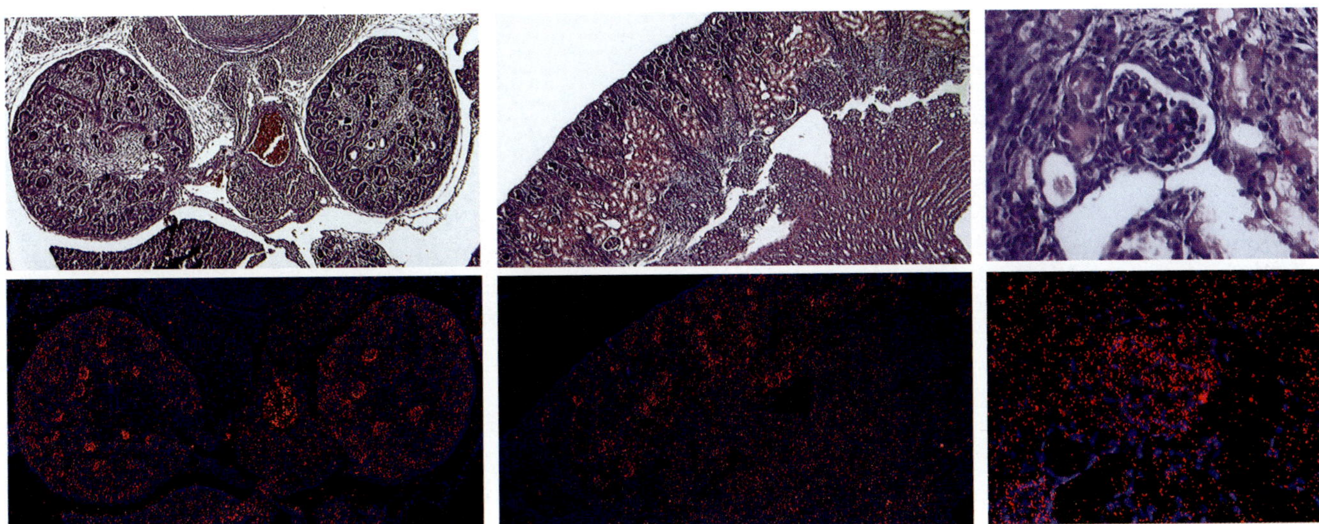

PLATE 70   Expression of *Lmx1b* during mouse kidney development. Top panels are hematoxylin & eosin sections. Bottom panels show sequential sections of RNA in situ hybridization to *Lmx1b* in metanephric kidney at embryonic day 14.5 (left panel), postnatal day 3 (middle panel), and higher magnification of the postnatal glomerulus (right panel). (see also Figure 31.3)

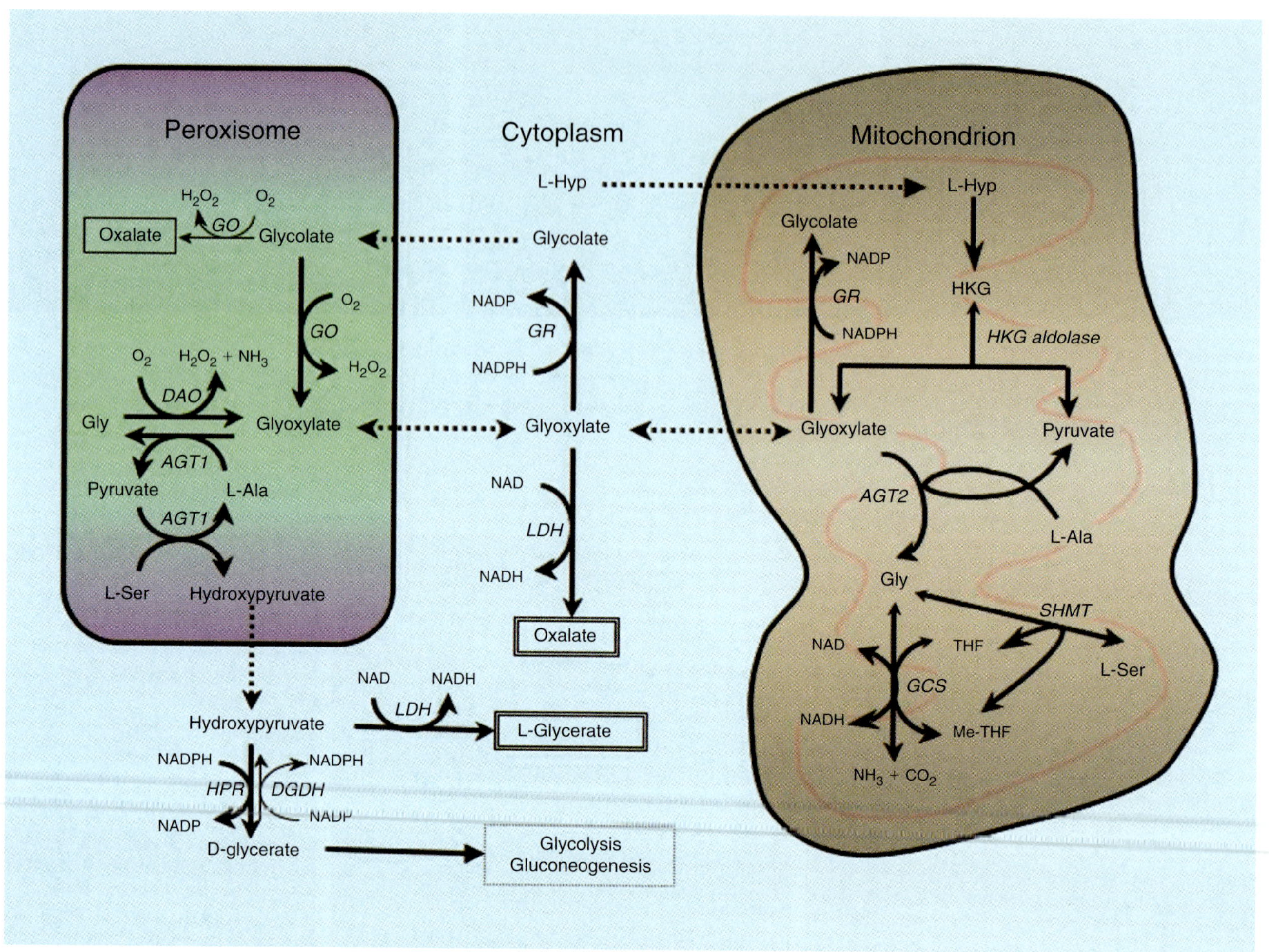

PLATE 71   Pathways associated with oxalate synthesis. Transport processes are shown as the thin, dark arrows and enzymatic reactions as the thicker, lighter arrows. DAO, D-amino acid oxidase; GCS, glycine cleavage system; GO, glycolate oxidase; GOA, glutamate:oxaloacetate aminotransferase; GR, glyoxylate reductase; HKGA, 4-hydroxy-2-ketoglutarate lyase; LDH, lactate dehydrogenase; SHMT, serine hydroxymethyltransferase; THF, tetrahydrofolate. Modified from Baker et al (2004). (see also Figure 33.1)

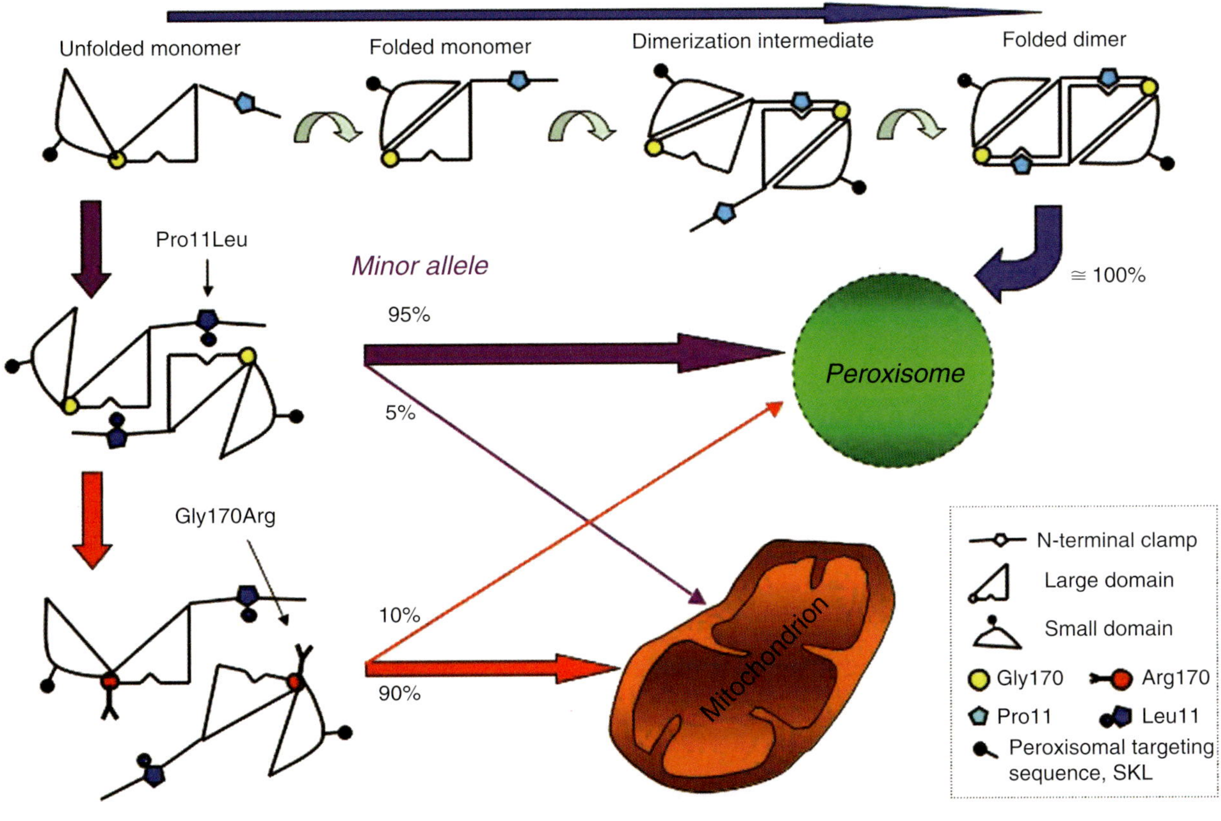

PLATE 72   Schematic of human AGT subcellular distribution as a function of phenotype. The minor allele facilitates mistargeting of AGT with the Gly170Arg mutation to the mitochondrion. See text for discussion. (see also Figure 33.4)

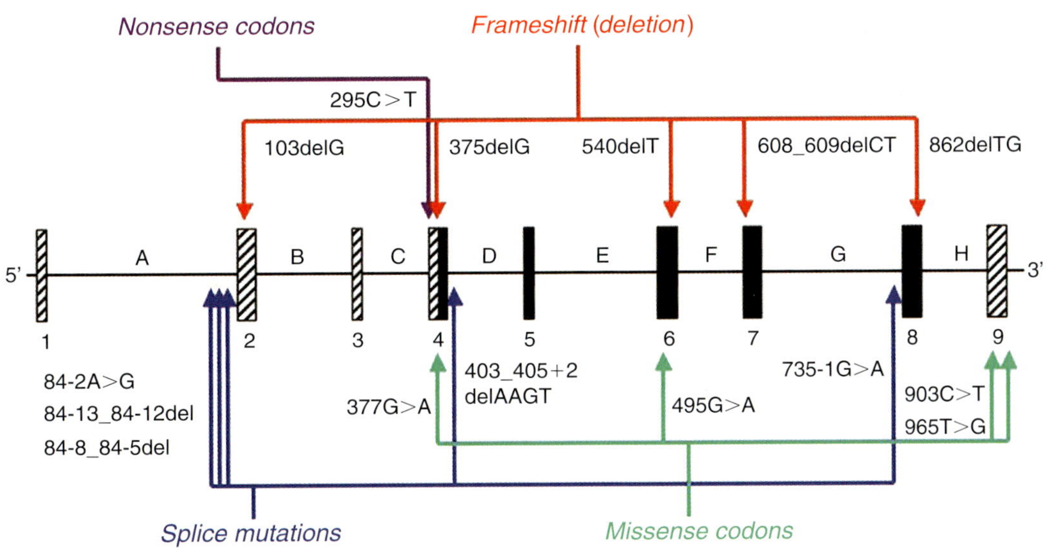

PLATE 73   Summary of described GRHPR mutations. To date, 14 mutations have been identified in PH2 patients. A novel 377G→A missense mutation in exon 4 of a Japanese patient is also included (Takayama & Cramer, unpublished data). This patient was a compound heterozygote with the 495G→A missense mutation. All mutations identified thus far lead to complete loss of GRHPR enzyme activity. (see also Figure 33.5)

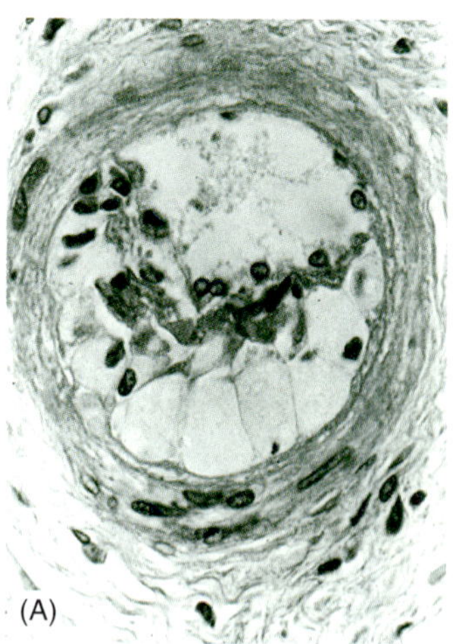

PLATE 74 (A) Photomicrograph of a prostatic arteriole from a classically affected male, showing the hypertrophied, lipid-laden endothelial cells encroaching on the vascular lumen. (B) Electromicrograph showing an occluded vessesl due to the progressive lysosomal deposition of the glycosphingolipid substrate which leads to the narrowing and eventual occlusion of the vascular lumen. (see also Figure 35.2(A))

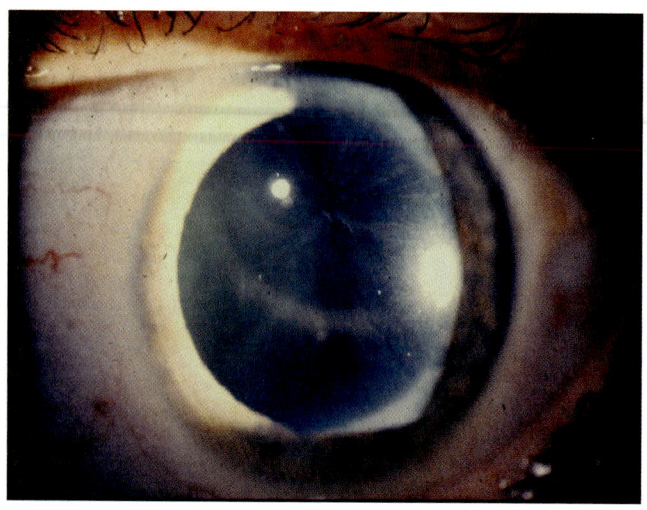

PLATE 76 Corneal opacity in a heterozygous female for the classic phenotype observed by slit-lamp microscopy. (see also Figure 35.4)

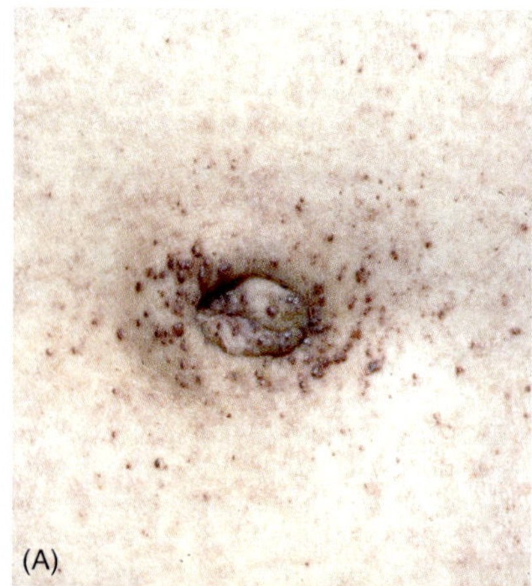

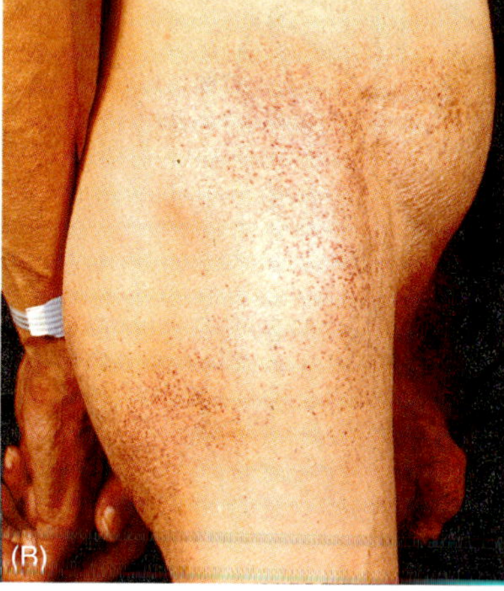

PLATE 75 (A) Clusters of dark-red to blue angiokeratomas (telangiectases) in the umbilical area. (B) A lateral view showing the typical distribution of the skin lesions in the 'bathing trunk' area of a classically affected male with Fabry disease. (see also Figure 35.3)

PLATE 77 (A) Photomicrograph of a glomerulus from a 35-year-old classically affected male. (B) Photomicrograph of the end-stage renal disease in a 42-year-old classically affected male. Zenker's fixation, paraffin embedding, and H & E staining. (see also Figure 35.5)

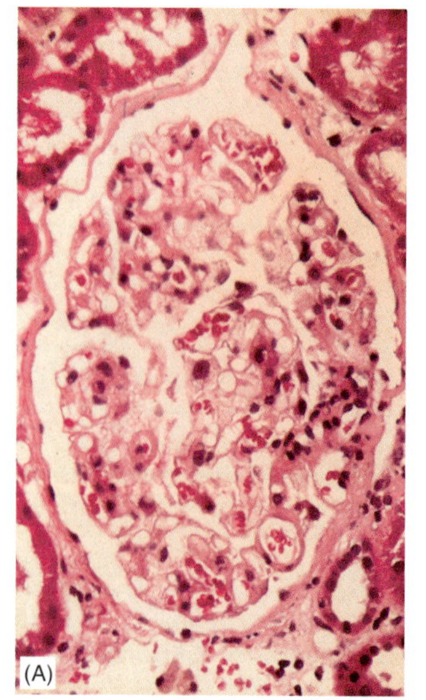

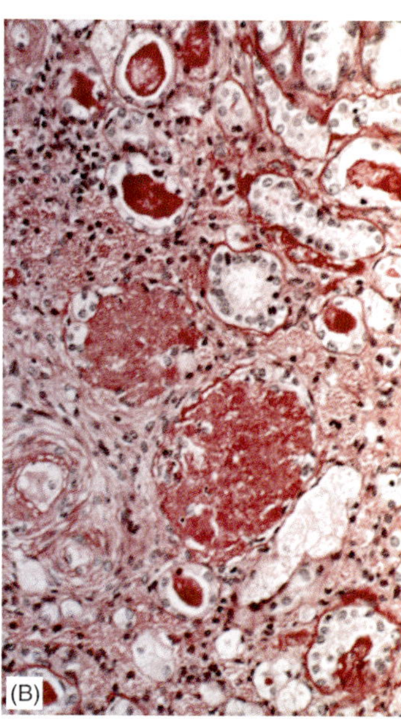

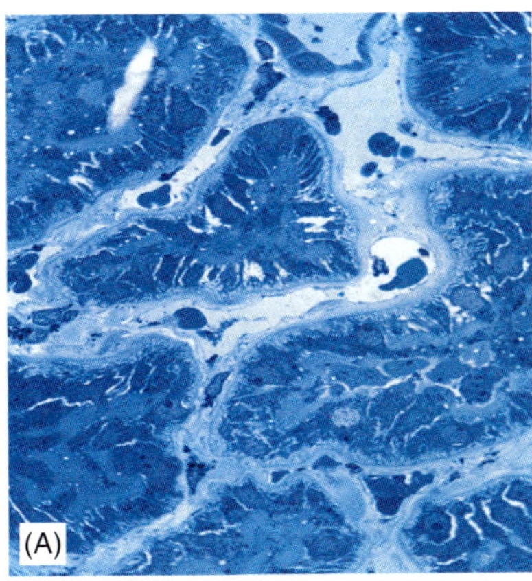
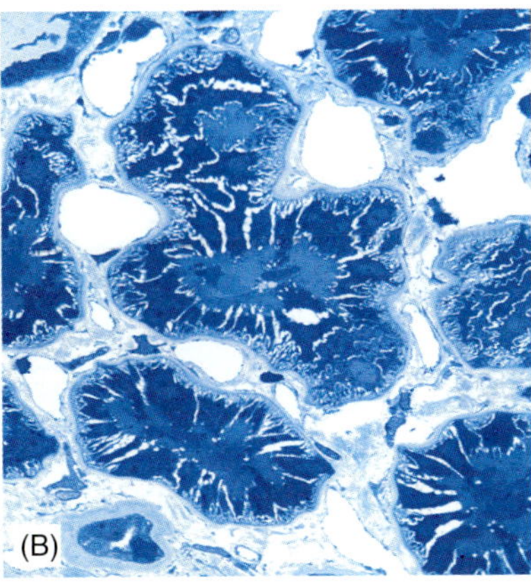

PLATE 78 Paired photomicrographs of the renal cortex pre-treatment (A) and post-treatment (B) from a patient who received 11 treatments of agalsidase beta in the double-blind study (Eng et al 2001b). Pre-treatment, the intertubular capillaries demonstrate numerous GL-3 inclusions (arrows). Post-treatment the capillary endothelium has been cleared of GL-3 inclusions (arrows), resulting in a '0' score. (see also Figure 35.8)

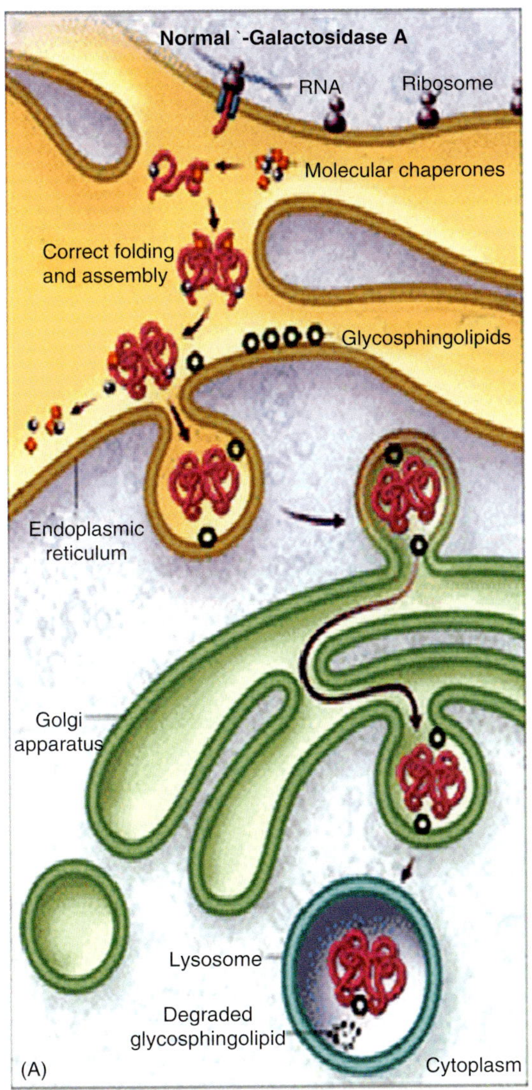

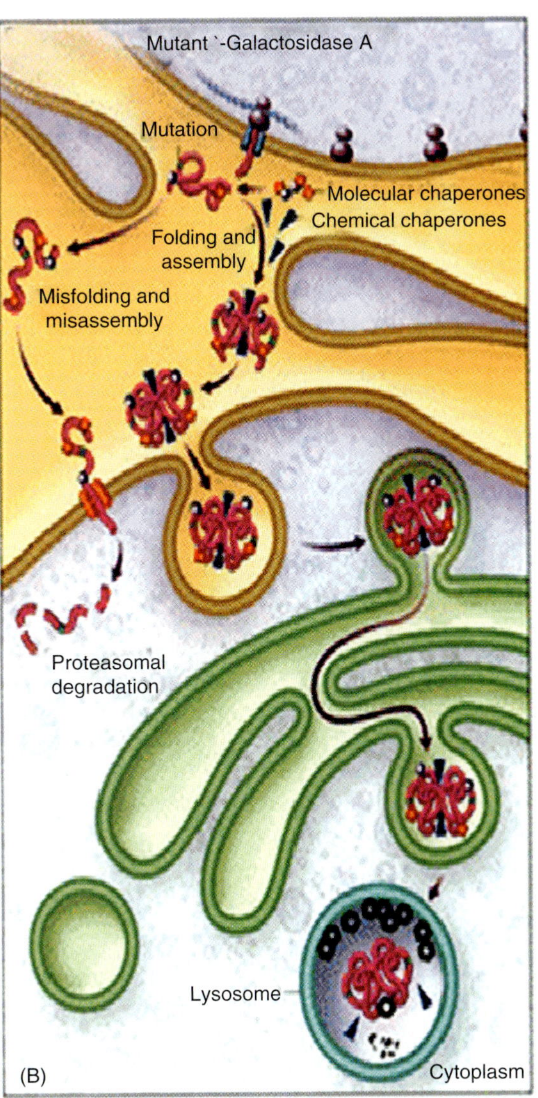

PLATE 79 Proposed mechanism of enzyme stabilization by pharmacologic chaperones. (A) shows the processing of normal α-Gal A. Newly synthesized α-Gal A is translocated into the endoplasmic reticulum, where molecular chaperones facilitate its proper folding and dimerization by specialized processing enzymes. The molecular chaperones then dissociate from the folded, dimerized enzyme, which moves to the Golgi apparatus and then to lysosomes, where the enzyme is stable and active in the acidic environment of these organelles. (B) shows the processing of mutant α-Gal A. Most mutations in the α-Gal A gene encode enzyme molecules that are misfolded, misassembled, or aggregated in the endoplasmic reticulum, where they are degraded, presumably by the ubiquitin–proteasome pathway. However, certain missense mutations decrease the stability of the enzyme, but the conformation of the active site is retained. Although most of these mutant enzymes are degraded in the endoplasmic reticulum, they may be stabilized by site-specific pharmacologic chaperones, such as galactose or deoxygalactonorjirimycin (Fan et al 1999), that bind to the active site of the enzyme, promote folding, and stabilize the mutant enzyme. Some of the enzyme then reaches the lysosomes, where it retains low levels of activity. In the lysosomes, the accumulated glycosphingolipid substrates displace the pharmacologic chaperones and are hydrolyzed by the enzyme. From Frustaci et al 2001. (see also Figure 35.9)

PLATE 80   Eleven-year-old patient with hereditary fructose intolerance 'eating' a pear – a manifestation of unusual feeding habits present since weaning at four months of age accompanied by failure to thrive. Rejection of many foods and drinks had led to accusations of poor parenting and referral to a child psychiatrist. To avoid further disaffection with medical services, the diagnosis of hereditary fructose intolerance was confirmed at the age of four years in the patient and her elder sister by molecular analysis of the human aldolase B gene in DNA obtained from mouthwash samples (see Cross & Cox, 1989). (see also Figure 36.1)

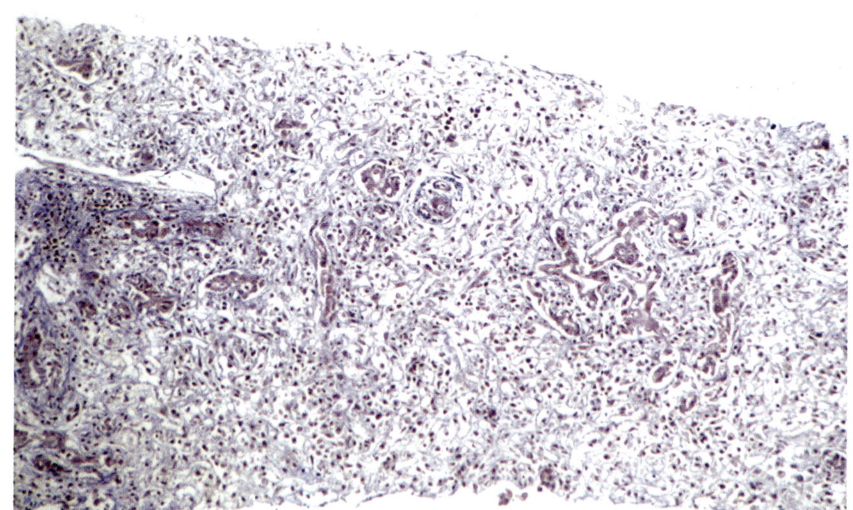

PLATE 81   Photomicrograph of section of needle specimen of liver tissue (stained with Mallory's trichrome reagent) and obtained post-mortem from a 16-year-old female with a dislike of fruit and sugary foods since infancy. 110 g of sorbitol and at least 50 g fructose had been given intravenously in peri-operative fluids at the time of an appendectomy in Germany; despite intensive supportive care, the patient died on the 6th post-operative day as a result of fulminant hepato-renal failure with severe acidosis. Note the complete absence of hepatocytes with bile duct proliferation and abundant activated Kupffer cells. A retrospective diagnosis of hereditary fructose intolerance was made in the patient, and in a surviving brother with similar symptoms, by molecular analysis of the aldolase B gene. The patient and her brother proved to be compound heterozygotes for two inactivating mutations, whose presence was confirmed after death using genomic DNA extracted from minute amounts of this fixed and embedded tissue (see Ali et al 1993). (see also Figure 36.2)

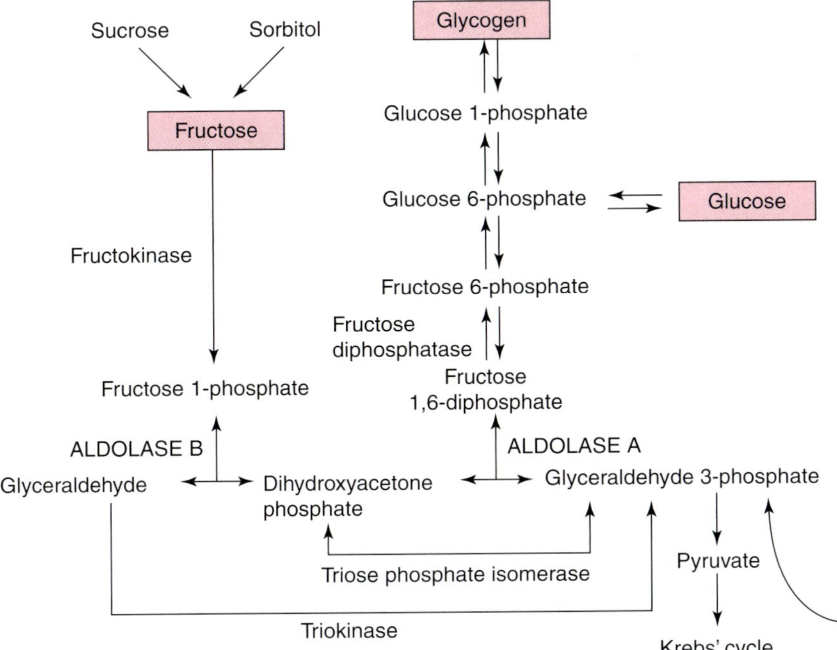

PLATE 82 Overview of the metabolism of dietary fructose, sucrose and sorbitol. (see also Figure 36.3)

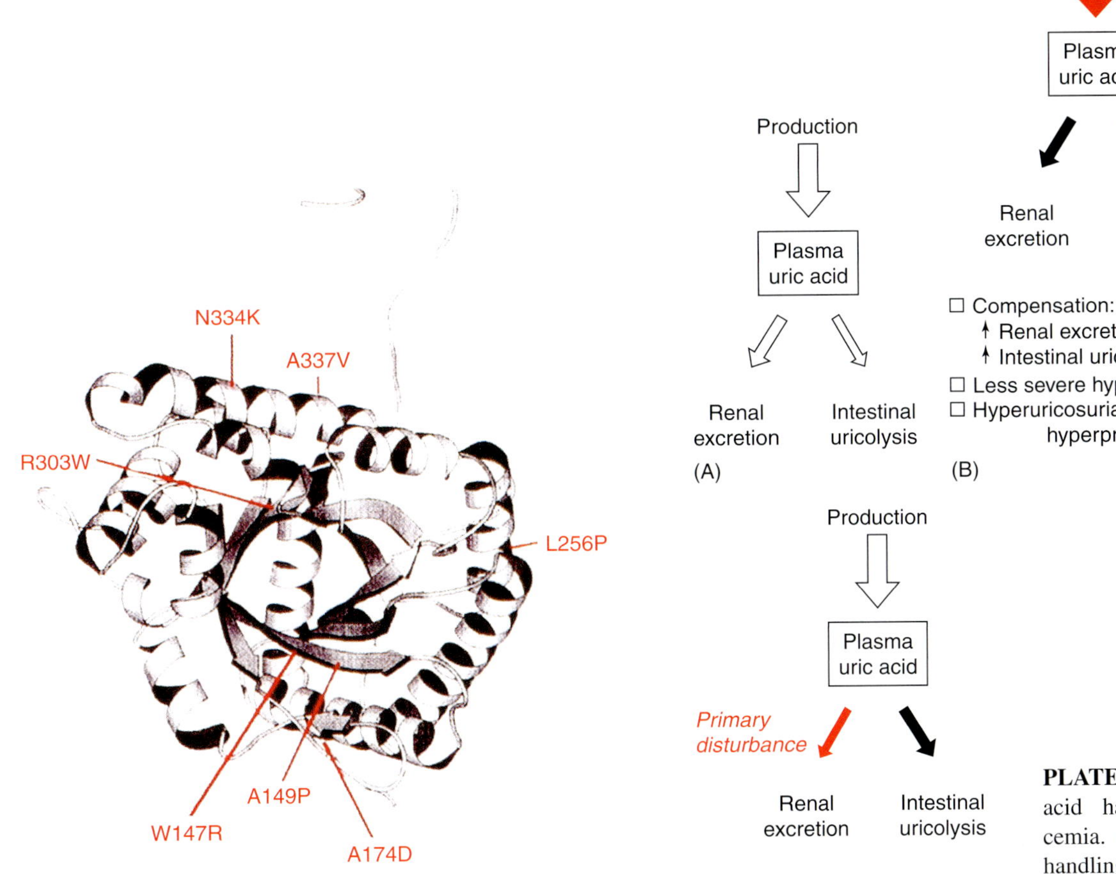

PLATE 83 Model of rabbit aldolase B structure depicting the location of residues mutated in natural missense variants associated with hereditary fructose intolerance (from Rellos et al 2000). (see also Figure 36.4)

PLATE 84 Schema of uric acid handling in hyperuricemia. (A) Normal uric acid handling in normouricemia. (B) Overproduction type hyperuricemia. (C) Underexcretion type hyperuricemia. (see also Figure 38.1)

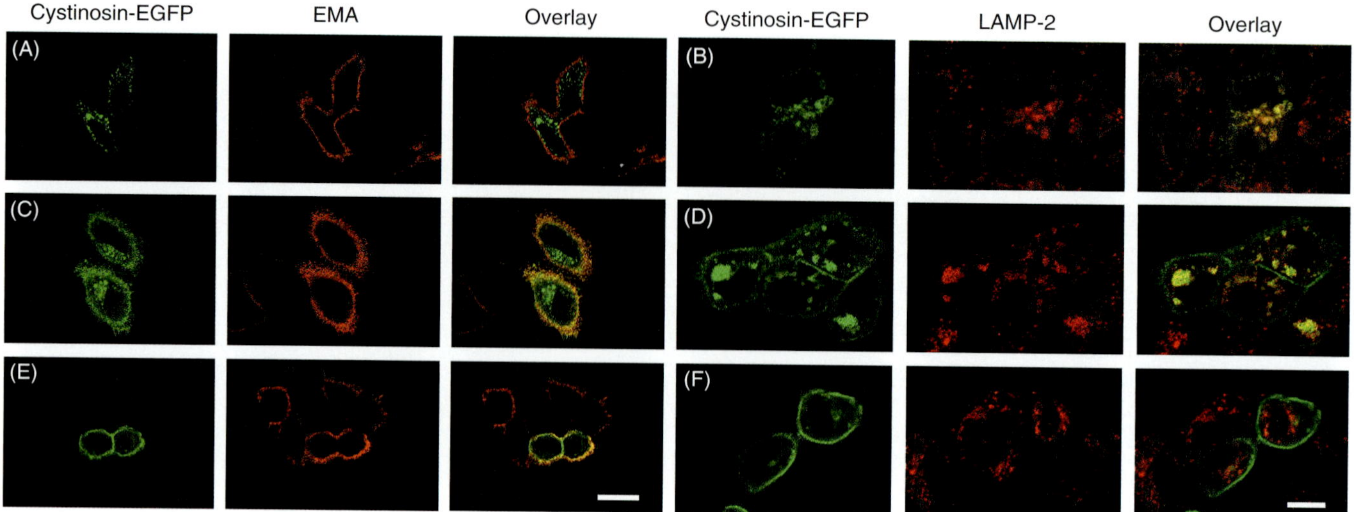

PLATE 85 Subcellular localization of wild-type and mutated cystinosin in HeLa (A, C and E) and MDCK (B, D, and E) cells. For HeLa cells, the cystinosin-EGFP fluorescence (left and right panels) is compared to the immunoreactivity of the endogenous epithelial membrane antigen (EMA) (middle and right panels). Scale bar = 20 μm. For MDCK cells, the cystinosin-EGFP fluorescence (left and right panels) is compared to the immunoreactivity of the endogenous late ensosomal/lysosomal marker LAMP-2 (middle and right panels). Scale bar = 10 μm. (A, B) Wild-type cystinosin-EGFP exhibits an intracellular, punctate distribution corresponding to late endosomes and lysosomes. (C, D) Deletion of the C-terminal GYDQL lysosomal sorting motif (ΔGYDQL) results in the partial localisation of cystinosin-GFP to the plasma membrane with a signal remaining in the late endosomes/lysosomes. (E, F) An example of a complete relocalization of cystinosin to the plasma membrane, as can be seen following deletion of the YFPQA motif (5th inter-TM loop), the Q222R mutation (3rd TM), or the deletions IVFD343–346del and DVVF346–349del (7th TM), coupled to ΔGYDQL. Reproduced from Kalatzis et al 2004 with the permission of Oxford University Press. (see also Figure 39.1)

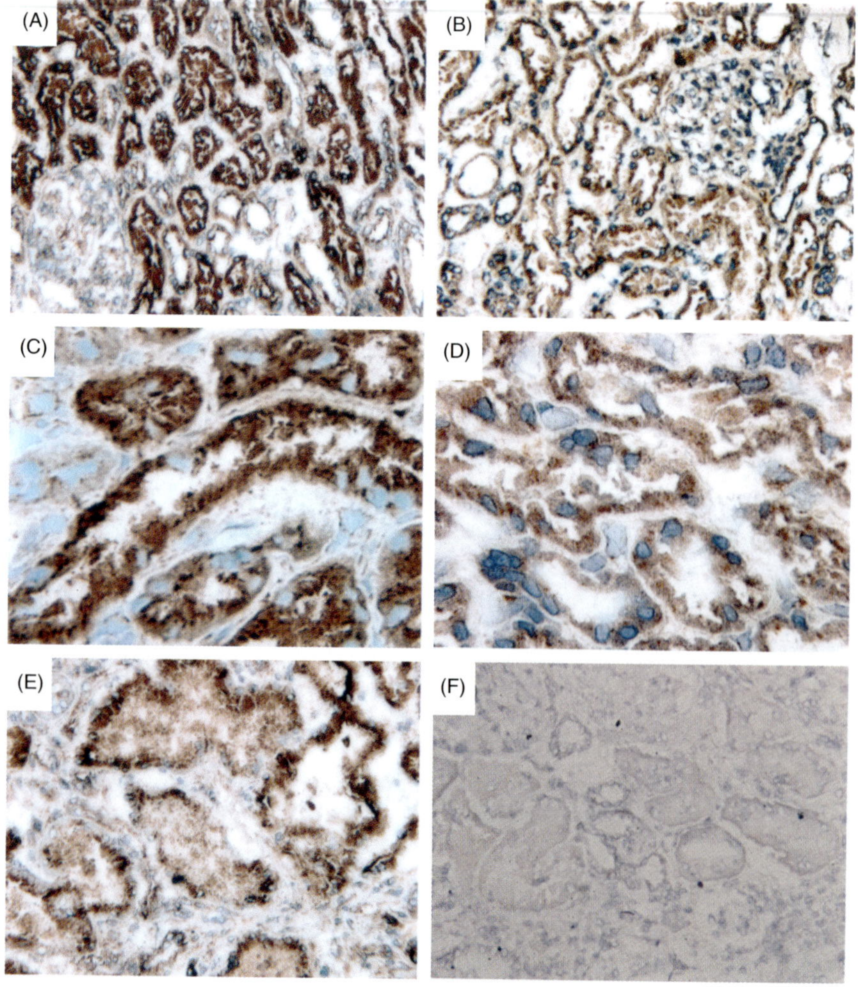

PLATE 86 Distribution of cystinosin in the human kidney. Immunoperoxidase staining of human kidneys with anti-LAMP-2 (A, C, E) or anticystinosin (B, D, F) antibodies. A strong immunostaining of the proximal tubular cells in normal human kidneys can be seen with both anti-LAMP-2 (A) and anti-cystinosin (B) antibodies (magnification = × 120). (C, D) At higher magnification (× 300), this staining can be seen to be granular and diffusely distributed within the cytoplasm for both antibodies. (E) Immunostaining of cystinotic human kidneys using antiLAMP-2 antibodies results in the same labeling of proximal tubule cells (magnification = × 300). (F) In contrast, no significant signal was detected with the anticystinosin antibody in these sections (magnification = × 120). Reprinted from Haq et al 2002 with the permission of the Journal of the American Society of Nephrology. (see also Figure 39.2)

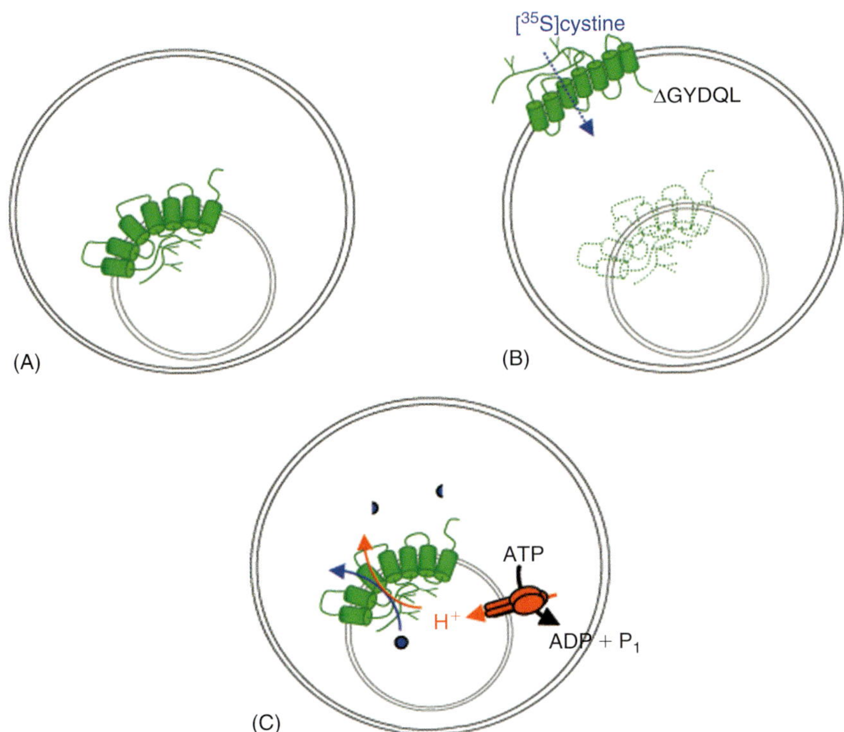

PLATE 87 Function of cystinosin. Schematic representation of the in vitro assay used to determine the function of cystinosin. (A) In vivo, cystinosin (in green) is localized at the lysosomal membrane (smaller gray circle). The N-terminal region is oriented towards the lysosomal lumen and the C-terminal tail towards the cytosol. Cystinosin would transport cystine from the lysosomal lumen to the cytosol. (B) In the in vitro model, cystinosin-ΔGYDQL is relocalized to the plasma membrane (larger gray circle) and, as the extracellular medium is topologically equivalent to the lysosomal lumen, cystinosin-ΔGYDQL is assayed for its ability to transport ^{35}S-cystine from the extracellular medium into the cytosol. (C) Cystinosin is the lysosomal cystine transporter and its activity is H$^+$-driven. The interior of the lysosome is acidified by the H$^+$-ATPase (in red) present in its membrane. Lysosomal cystine (blue circle) efflux by cystinosin is coupled to an efflux of H$^+$. Thus, the influx of H$^+$ by the lysosomal H$^+$-ATPase drives cystinosin-mediated cystine transport toward the cytosol, where it is then reduced to cysteine (blue semi-circles). Reprinted with modifications by permission from The EMBO Journal (Kalatzis et al 2001a), copyright 2001 Macmillan Publishers Ltd. (http://www.nature.com). (see also Figure 39.3)

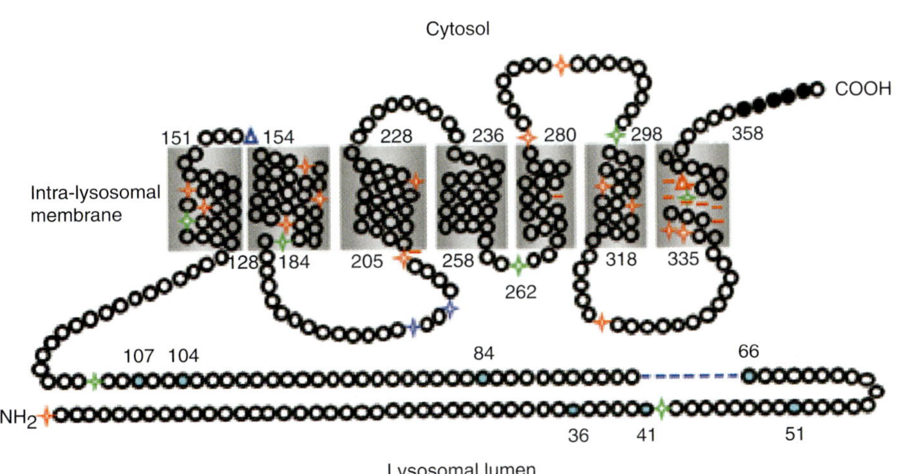

PLATE 88 Position of transport-altering mutations in the topological model of cystinosin. Cystinosin is a 367 aa protein predicted to contain 7 transmembrane domains (TM). The 128 aa N-terminal region with 7 N-glycosylation sites (blue circles) is oriented towards the lysosomal lumen. The 10 aa tail contains a classic GYDQL lysosomal targeting motif (black circles) and is oriented towards the cytosol. Mutations are categorized into three classes according to whether they abolish (red symbols), severely inhibit (<30% of wild-type; blue symbols) or only moderately, or do not, alter ($\geq$30% of wild-type; green symbols) cystine transport. Stars represent point mutations, dashes correspond to deletions of 1 aa, and triangles indicate insertions of 3 (position 154) or 4 (position 349) aa. Numbers correspond to the position of N-glycosylation sites and of the extremities of TMs. Reproduced from Kalatzis et al 2004 with the permission of Oxford University Press. (see also Figure 39.4)

PLATE 89 Mosaic expression of FAH in the liver of an HTI patient. Immunostaining of a liver section shows numerous nodules expressing FAH (dark brown staining). Other regions of the liver are FAH-negative. The section was stained with an antibody specific for FAH. Analysis of DNA obtained from microdissected stained nodules shows that the mutation is corrected in one of the FAH allele. (see also Figure 40.3)

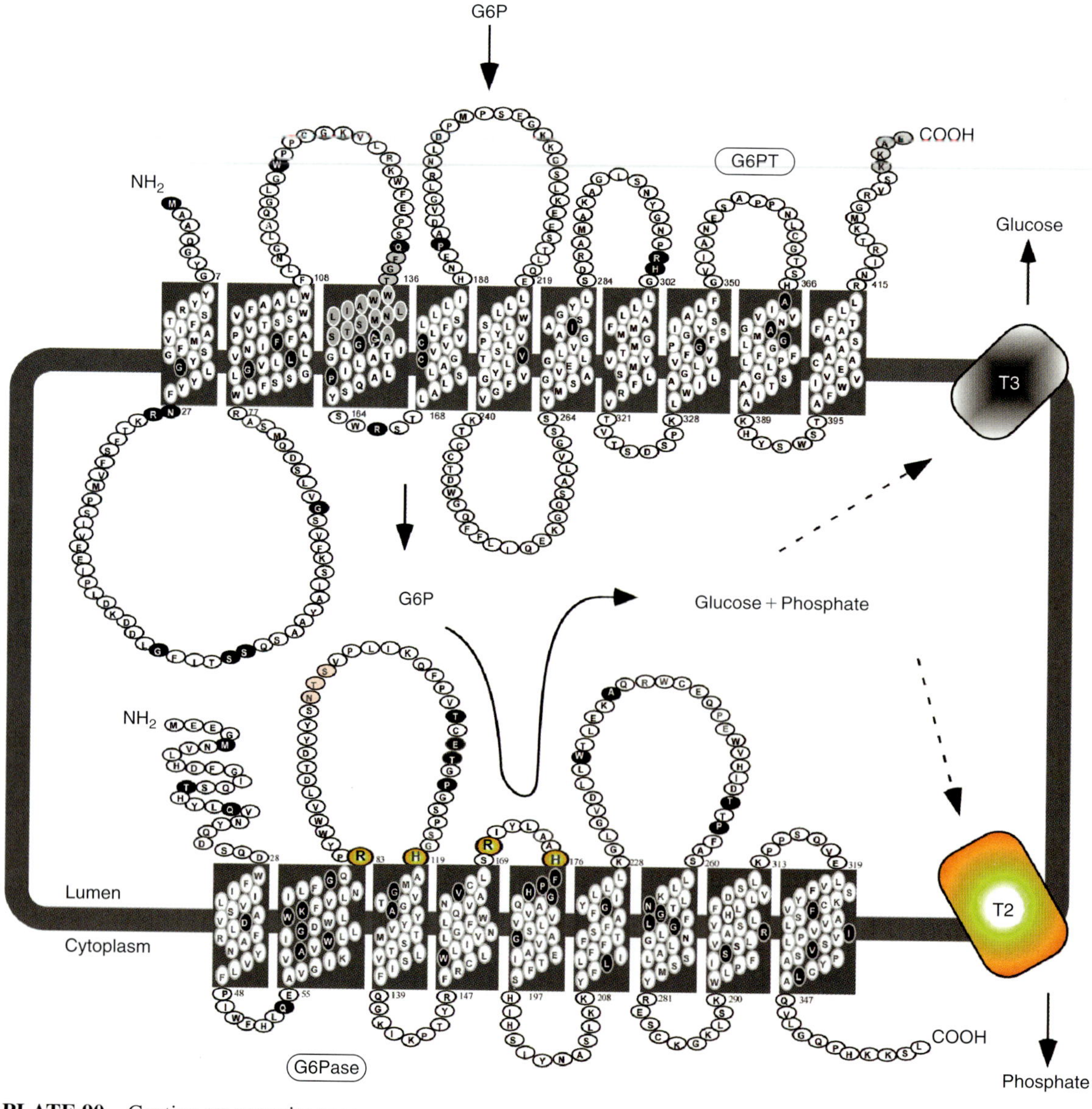

PLATE 90 Caption on opposite page

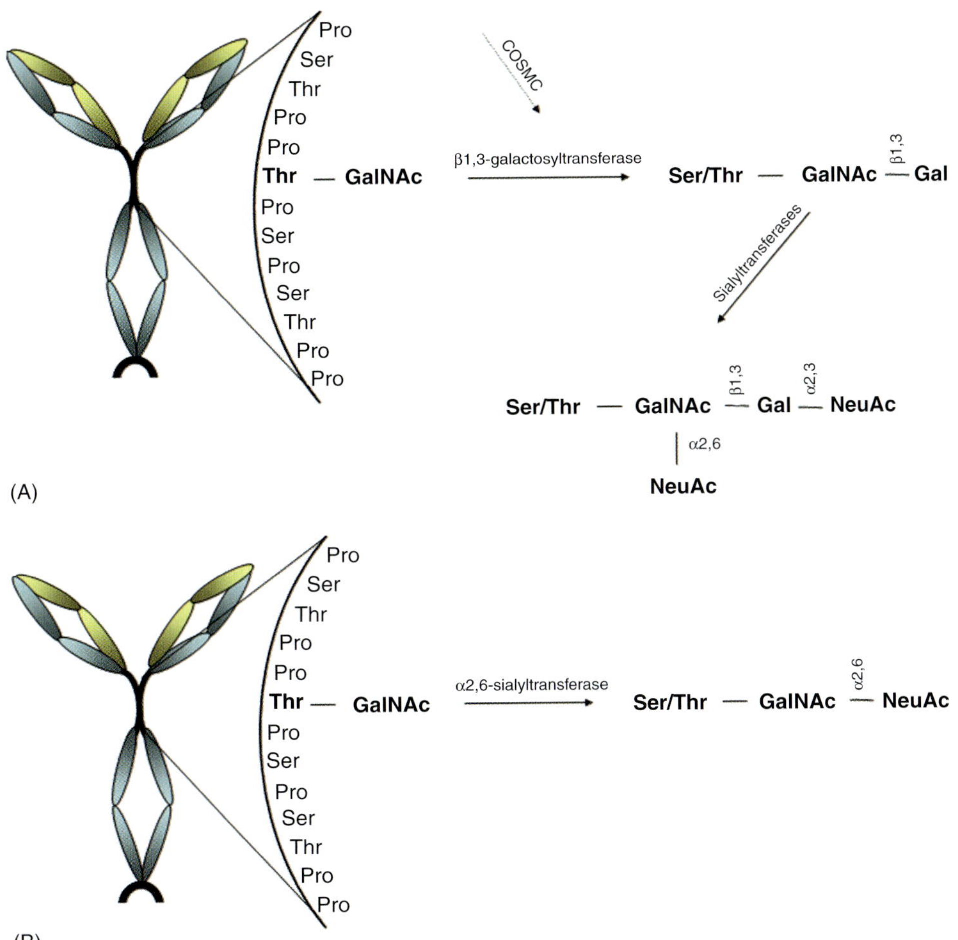

PLATE 91 The structure of IgA1. The heavy chain is in blue, the light chain in yellow. Compared to IgA2, the hinge region of IgA1 contains an additional 13 amino acids; the serines and threonines in this region serve as O-glycosylation sites. Three to five sites are occupied by O-glycans in each IgA1 molecule. (A) N-acetylgalactosamine (GalNAc) is attached to Ser/Thr via N-acetylgalactosaminyltransferase 2. GalNAc can be modified by addition of galactose (Gal) by β1,3-galactosyltransferase and further expanded by sialyltransferase mediated attachment of sialyic acid (NeuAc) to GalNAc and Gal by α2,6 or α2,3 linkage, respectively. COSMC is require for proper folding of β1,3-galactosyltransferase. (B) Alternatively, α2,6-sialyltransferase can directly attach NeuAc to GalNAc; this arrangement prevents further modification. Normal serum IgA1 mostly comprises Gal-GalNAc disaccharides and its mono- and di-sialylated forms. Variants with terminal GalNAc or sialylated GalNAc are more common in IgAN. (see also Figure 45.1)

PLATE 90 The G6Pase system, an enzyme complex essential for glucose homeostasis. The components, G6Pase, G6PT, a putative phosphate transporter (T2), and a putative glucose transporter (T3), are shown anchored in the membrane of the ER in contact with both the cytoplasm and ER lumen. The existence of T2 and T3 remains hypothetical. Glucose may be released directly to the blood via the ER/Golgi pathway, rather than into the cytoplasm, as shown. The spatial representation is illustrative only. G6Pase and G6PT are anchored in the ER by 9 and 10 transmembrane helices, respectively (Pan et al 1988a, 1999). Mutations identified in the G6Pase and G6PT genes of GSD-I patients are marked in black. In G6Pase, R83, H119, R170, and H176, which contribute to the active center, are denoted by large shaded circles. In G6PT, amino acid residues comprising the signature motif of transporters of phosphorylated metabolites at amino acid residues 133 to 149 are denoted by shaded circles. Modified from Chou et al (2002) with permission. (see also Figure 41.1)

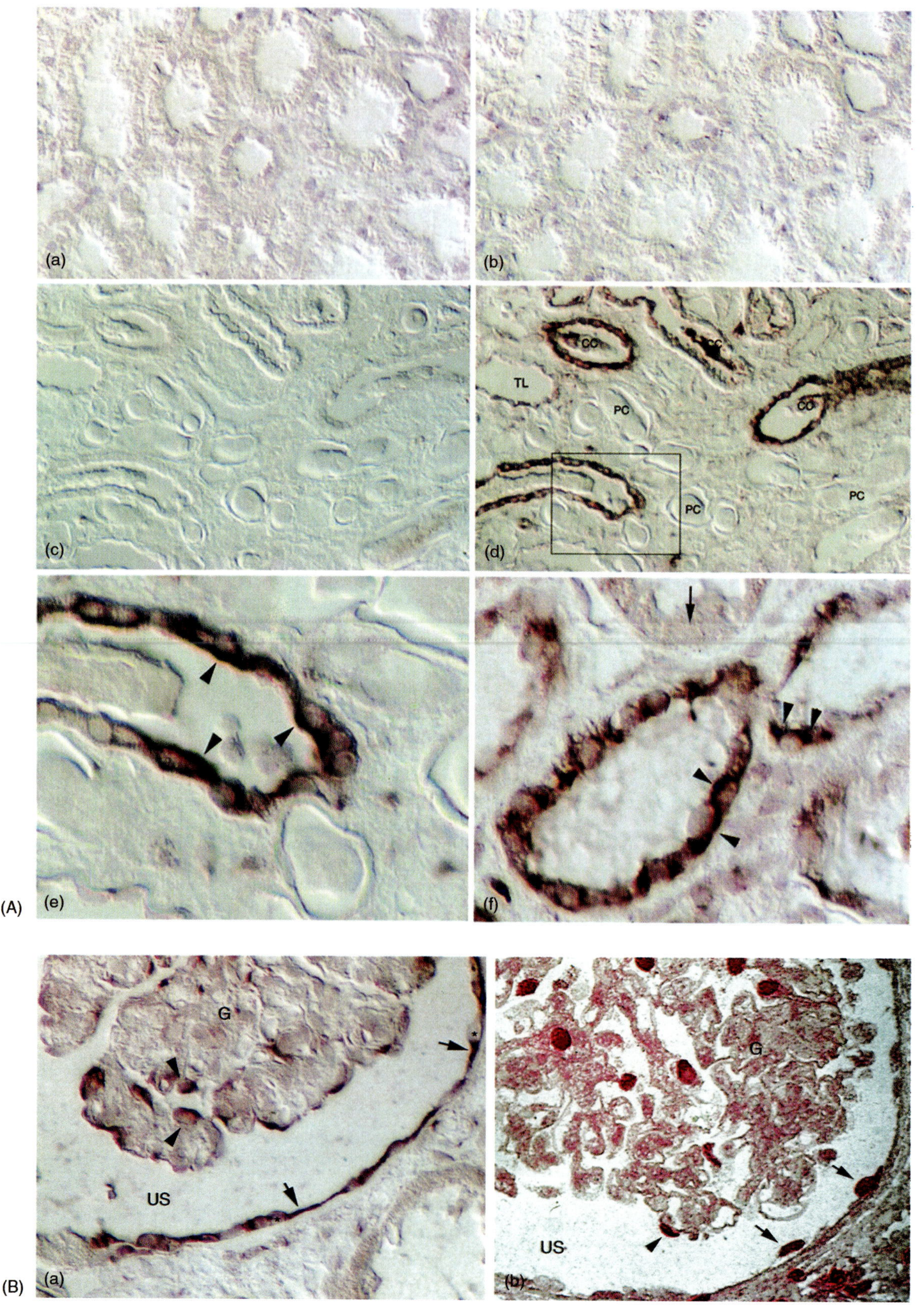

PLATE 92 Detection of HIV-1 mRNA in renal biopsies by in situ hybridization. Viral mRNA is detected in tubular epithelial cells (A) and podocytes (B). G, glomerulus; US, urinary space. Original magnification ×200. Adapted with permission from Bruggeman et al (2000). Renal epithelium is a previously unrecognized site of HIV-1 infection. J. Am. Soc. Nephrol. 11: 2079–87. (see also Figure 47.3)

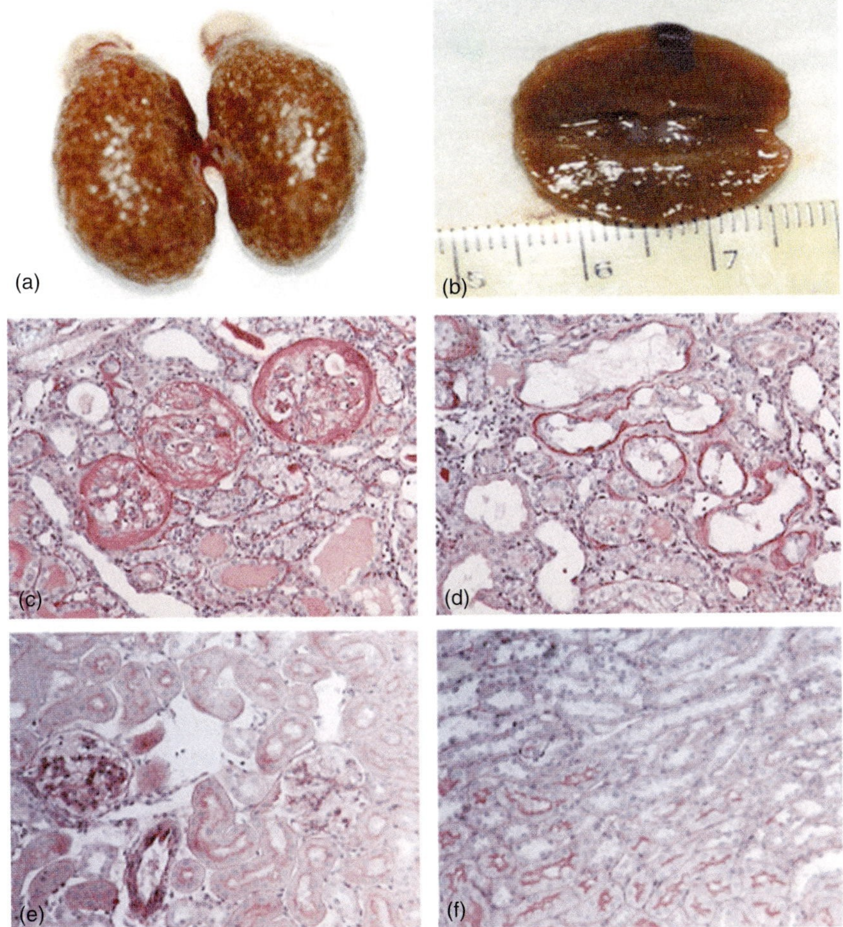

PLATE 93 Late renal disease in human immunodeficiency virus transgenic (HIV-Tg) rats. (A and B) Macroscopic pictures of the kidneys of HIV-Tg rat with late renal disease. Note the pitted external and internal renal surfaces. A representative periodic acid–Schiff (PAS) staining of the renal cortex (C) and medulla (D) in sections of HIV-Tg rats with late renal disease. A representative PAS staining of the renal cortex (E) and medulla (F) in sections of HIV-Tg rats without renal disease. Increased PAS+ staining is detected in sclerotic glomeruli (C) and dilated tubules (D) of HIV-Tg rats with late renal disease. Original magnification ×100. Reproduced with permission from Ray et al (2003) A novel HIV-1 transgenic rat model of childhood HIV-1 associated nephropathy. Kidney Int. 6: 2242–51. Copyright © 2003 Blackwell Publishing. (see also Figure 47.7)

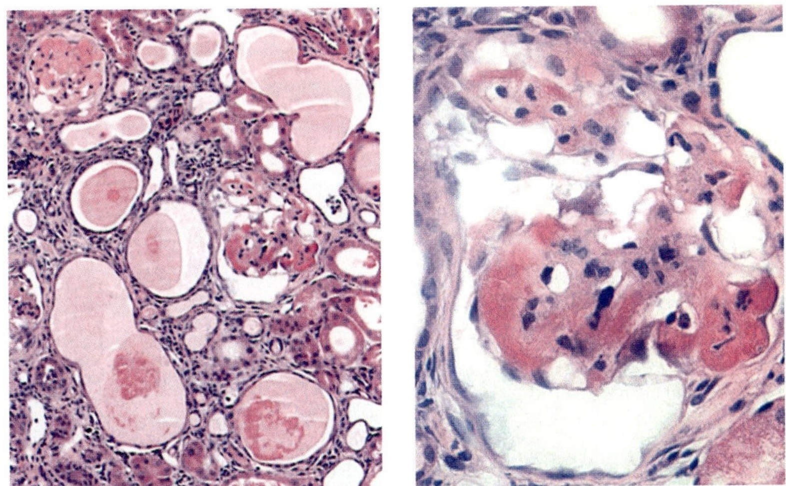

PLATE 94 Histopathology in kidneys of CD4C/HIV^{WT} Tg mice. Transgenic mice expressing the HIV genome under the control of the CD4 promoter develop severe tubulointerstitial disease and microcystic dilatation, in addition to mild focal glomerulosclerosis. Images graciously provided by Chunyan Hu, Zaher Hanna, and Paul Jolicoeur. (see also Figure 47.8)

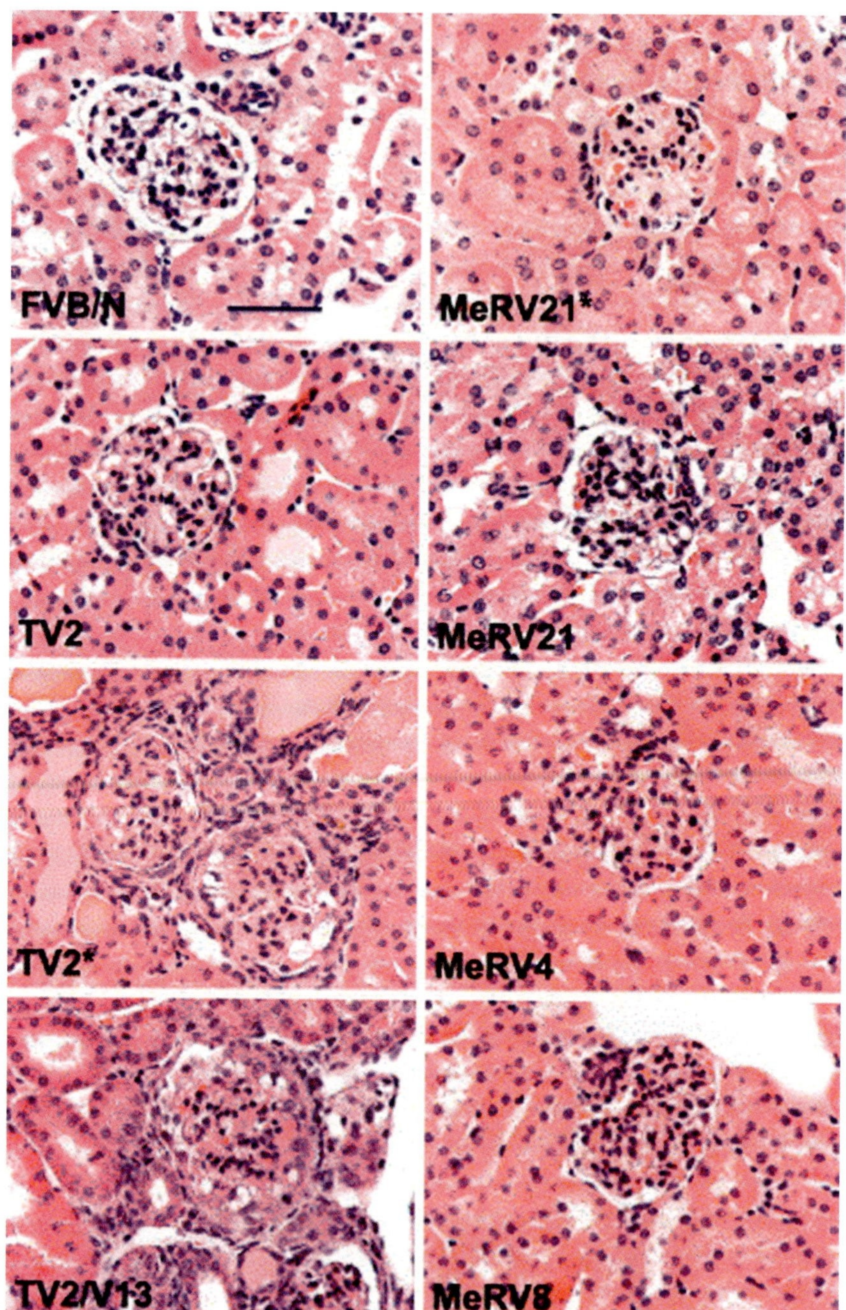

PLATE 95 Renal histopathology in nontransgenic mice (FVB/N), in transgenic mice expressing only HIV-1 *Vpr* and *Tat* (TV2), and in transgenic mice expressing HIV-1 *Vpr* under the control of *c-fms* (MeRV). Kidneys were fixed in 10% buffered formalin, sectioned and stained with hematoxylin and eosin. A hemizygous TV2 female and a nursing TV2* female are compared to a nontransgenic female (FVB/N) to demonstrate varying degrees of histopathology. Glomerular condensation (obscuring capillary lumen) is visible in both TV2 mice but tubular anomalies (acid-staining protein casts and fibrotic lesions) are present predominantly in the nursing mouse (TV2*). Dual transgenic TV2/V13 mice displayed focal glomerulosclerosis (FGS) and tubular atrophy more severe than in TV2 mice. MeRV (*c-fms/vpr*) mice from three independent lines (4, 8, and 21) displayed FGS absent signs of tubular histopathology. Mice were 5 to 10 weeks old. Original magnification was ×400. The magnification scale bar in top left panel is equivalent to 30 μm. Adapted with permission from Dickie et al (2004) Focal glomerulosclerosis in proviral and c-*fms* transgenic mice links Vpr expression to HIV-associated nephropathy. Virology 322: 69–81. (see also Figure 47.11)

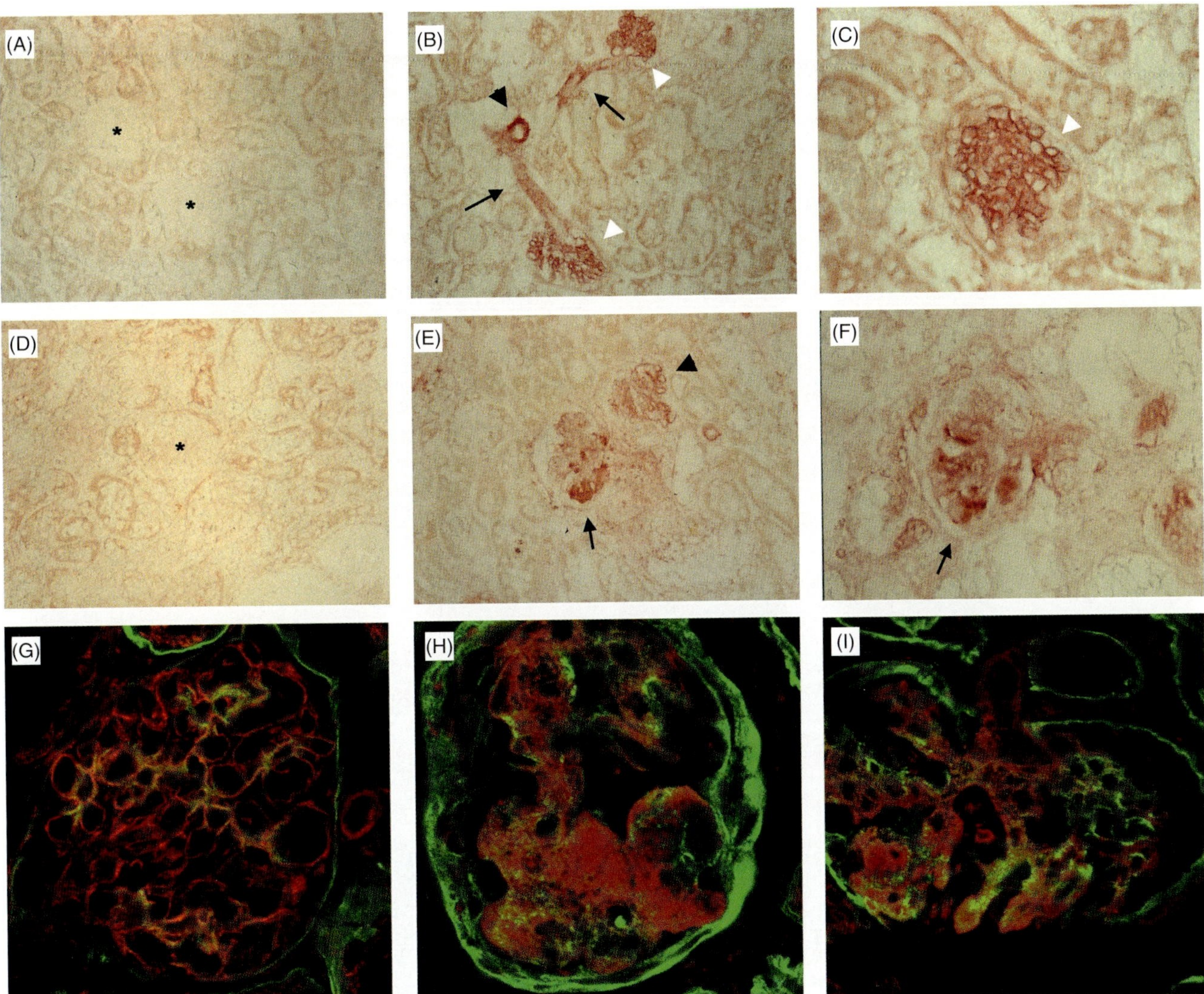

PLATE 96 Immunolocalization of podocan protein in normal mouse kidney and in diseased kidney from the HIV-1 transgenic HIVAN model. In the normal mouse kidney (A–C), podocan staining was observed surrounding capillary loops in the glomerulus in a pattern consistent with that of the GBM (B and C, white arrowheads). Staining of afferent arterioles (black arrows) and interlobular arteries (black arrowhead) was also seen (B). In the HIV-1 transgenic kidney (D–F), the distribution of podocan in nondiseased glomeruli was comparable with that in the nontransgenic kidney (E, arrowhead). In affected glomeruli, a marked increase in podocan deposition was clearly evident in the sclerotic glomerular lesions (E and F, arrows). To illustrate the increased level of podocan deposition in the FSGS lesions of HIV-1 transgenic kidneys, confocal images of a nondiseased glomerulus (G) and two sclerotic glomeruli (H and I) are shown. To allow a direct comparison, all three images were taken from the same section using identical settings on the confocal microscope. The sections were co-labeled for laminin A chain to delineate Bowman's capsule. No staining was detected in glomeruli (asterisks), afferent arterioles, or interlobular arteries with the preimmune serum from the animal in which the anti-podocan antibodies were raised (A and D). Reproduced with permission from Ross et al (2003) Podocan, a novel small leucine-rich repeat protein expressed in the sclerotic glomerular lesion of experimental HIV-associated nephropathy. J. Biol. Chem. 278: 33248–55. Copyright © 2003 American Society for Biochemistry and Molecular Biology. (see also Figure 47.12)

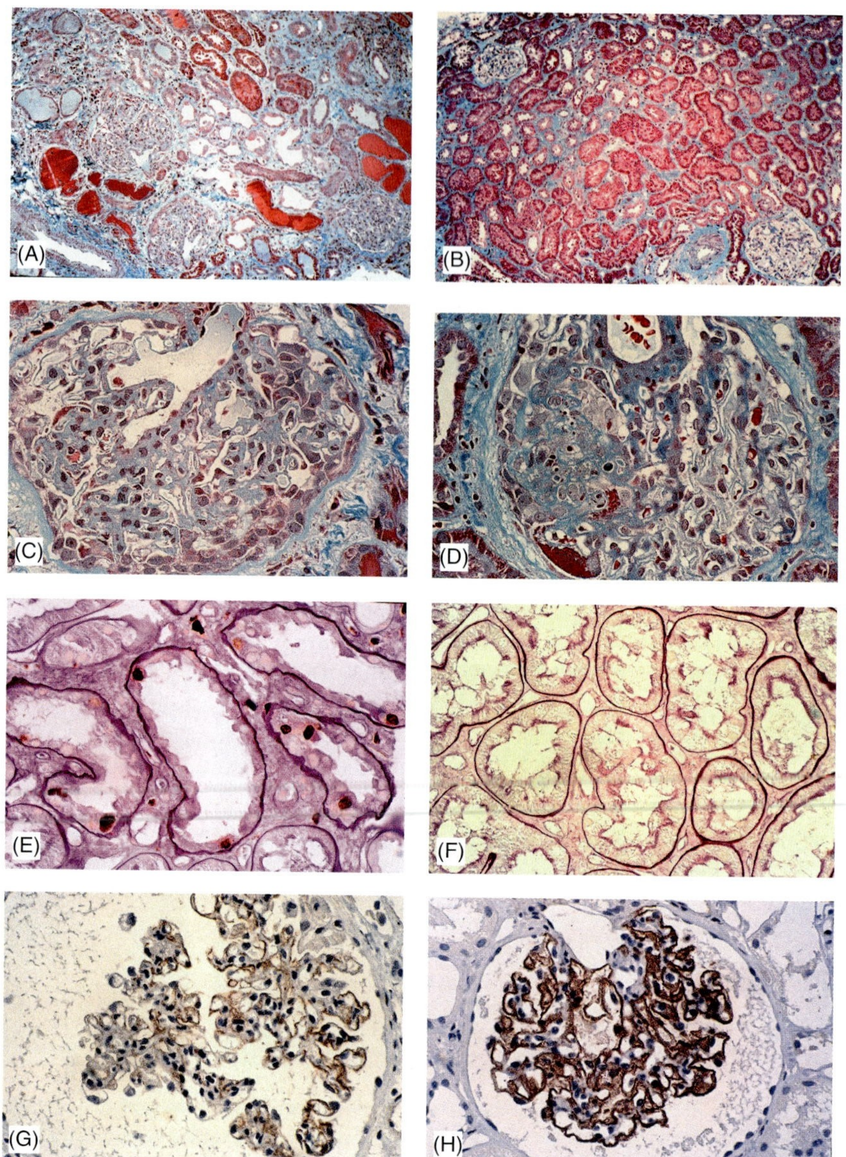

PLATE 97 Kidney biopsy specimens obtained before and after the initiation of highly active antiretroviral therapy. Panels A and C are low-power and high-power views, respectively, of the pretreatment biopsy specimen. Panel A shows one of three glomeruli with collapsing sclerosis and marked hyperplasia of podocytes (trichrome stain, ×125). The tubules are separated by edema, mild fibrosis, and patchy interstitial inflammatory infiltrates. Many proximal tubules show degenerative changes, and there are focal tubular microcysts containing large casts. Panel C shows a glomerulus with segmental collapse of the glomerular tuft and hyperplasia of the overlying podocytes (trichrome stain, ×400). Panels B and D are low-power and high-power views, respectively, of the biopsy specimen obtained three months after the initiation of highly active antiretroviral therapy. Panel B shows normal glomeruli and mild focal interstitial fibrosis, with restoration of normal tubular architecture (trichrome stain, ×125). No tubular microcysts or interstitial inflammation is apparent. In panel D, a glomerulus contains a discrete segmental scar (trichrome stain, ×400). Some of the overlying podocytes contain protein-resorption droplets, but without the hyperplasia that was prominent before the initiation of treatment. Panels E, F, G, and H show the results of immunohistochemical staining of kidney-biopsy specimens obtained before and after the initiation of antiretroviral therapy. In the pretreatment biopsy specimen (panel E), many nuclei in the renal tubular epithelial cells stain for Ki-67. There is diffuse loss from the proximal tubules of brush border staining for periodic acid–Schiff (periodic acid–Schiff counterstain, ×400). In the biopsy specimen obtained after the initiation of antiretroviral therapy (panel F), a representative field shows no staining for Ki-67. The proximal tubular brush border has been restored (periodic acid–Schiff counterstain, ×400). Immunostaining for synaptopodin in the pretreatment biopsy specimen shows weak staining or no staining in the podocytes of a collapsed glomerulus (panel G, ×400). There is strong, global positivity for synaptopodin in the podocytes of a representative glomerulus from the biopsy specimen obtained after three months of highly active antiretroviral therapy (panel H, ×400). Reproduced with permission from Winston et al Nephropathy and establishment of a renal reservoir of HIV type 1 during primary infection. N. Engl. J. Med. (2001) 344: 1979–84. Copyright © 2001 Massachusetts Medical Society. (see also Figure 47.14)